Animal Lectins: Form, Function and Clinical Applications

G.S. Gupta

Animal Lectins: Form, Function and Clinical Applications

In Collaboration with Anita Gupta and Rajesh K. Gupta

Principal author
G.S. Gupta, Ph. D.
Panjab University
Chandigarh 160014
India

ISBN 978-3-7091-1064-5   ISBN 978-3-7091-1065-2 (eBook)
DOI 10.1007/978-3-7091-1065-2
Springer Wien Heidelberg New York Dordrecht London

Library of Congress Control Number: 2012945422

© Springer-Verlag Wien 2012
This work is subject to copyright. All rights are reserved by the Publisher, whether the whole or part of the material is concerned, specifically the rights of translation, reprinting, reuse of illustrations, recitation, broadcasting, reproduction on microfilms or in any other physical way, and transmission or information storage and retrieval, electronic adaptation, computer software, or by similar or dissimilar methodology now known or hereafter developed. Exempted from this legal reservation are brief excerpts in connection with reviews or scholarly analysis or material supplied specifically for the purpose of being entered and executed on a computer system, for exclusive use by the purchaser of the work. Duplication of this publication or parts thereof is permitted only under the provisions of the Copyright Law of the Publisher's location, in its current version, and permission for use must always be obtained from Springer. Permissions for use may be obtained through RightsLink at the Copyright Clearance Center. Violations are liable to prosecution under the respective Copyright Law.
The use of general descriptive names, registered names, trademarks, service marks, etc. in this publication does not imply, even in the absence of a specific statement, that such names are exempt from the relevant protective laws and regulations and therefore free for general use.
While the advice and information in this book are believed to be true and accurate at the date of publication, neither the authors nor the editors nor the publisher can accept any legal responsibility for any errors or omissions that may be made. The publisher makes no warranty, express or implied, with respect to the material contained herein.

Printed on acid-free paper

Springer is part of Springer Science+Business Media (www.springer.com)

Foreword I

Lectins are typically carbohydrate-binding proteins that are widely distributed in Nature. With the growing interest in the field of glycobiology, the body of research related to animal lectins has grown at an explosive rate, particularly in the past 25 years. However, the Lectinology field is still relatively young, nascent, and evolving. "**Animal Lectins: Form, Function and Clinical Applications**" presents the most up-to-date analysis of these carbohydrate - binding, and potentially lifesaving proteins in two comprehensive volumes. Lectionology is an exciting area of research that has helped immensely in our understanding of host-pathogen interactions. Importantly, C-type lectins can act as pattern-recognition receptors (PRR) that sense invading pathogens.

The interactions between lectins and carbohydrates have been shown to be involved in diverse activities such as opsonization of microbes, cell adhesion and migration, cell activation and differentiation, and apoptosis. Developments in the area of lectin research has opened a new aspect in studying the immune system, and at the same time, provided new therapeutic routes for the treatment and prevention of diseases.

This present book on animal lectins discusses the biochemical and biophysical properties of animal lectins at length along with their functions in health and diseases. Importantly, the potential interrelationships between lectins of the innate immune system and latent viruses that reside within host cells, sometimes integrated into the genome have been beautifully highlighted. These interactions help to explain autoimmune diseases and shed light on the development of cancer diagnostics. The present book on animal lectins presents new insights into the biological roles of most animal lectins, including their role in prevention of infections through innate immunity. The contents of "**Animal Lectins: Form, Function and Clinical Applications**" provide functional explanation for the enormous diversity of glycan structures found on animal cells. There are still several other areas wherein lectins and their specificities are not well defined and the biological functions of the interactions remain elusive, thereby underscoring the need for further research. The book offers novel ideas for students of Immunology, Microbiology, as well as young researchers in the area of Biochemistry. I congratulate the authors in completing this truly enormous task.

Prof. N. K. Ganguly, MD, DSc (hc), FMed Sci (London),
FRC Path (London), FAMS, FNA, FASc, FNASc,
FTWAS (Italy), FIACS (Canada), FIMSA

President, JIPMER
Former Director General, ICMR
Former Director, PGIMER, Chandigarh

Pondicherry, India

Foreword II

Lectins are proteins that bind to soluble carbohydrates as well as to the functional groups of carbohydrate chains that are part of a glycoprotein or glycolipid found on the surfaces of cells. Lectins are known to be widespread in nature and play a role in interactions and communication between cells typically for recognition. Carbohydrates on the surface of one cell bind to the binding sites of lectins on the surface of another cell. For example, some bacteria use lectins to attach themselves to the cells of the host organism during infection. Binding results from numerous weak interactions which come together to form a strong attraction. A lectin usually contains two or more binding sites for carbohydrate units. In addition, the carbohydrate-binding specificity of a certain lectin is determined by the amino acid residues that bind to the carbohydrates. Most of the lectins are essentially nonenzymic in action and nonimmune in origin. They typically agglutinate certain animal cells and/or precipitate glycoconjugates. Plant lectins are highly resistant to breakdown from heat or digestion. They provide a defense for plants against bacteria, viruses, and other invaders, but can create problems for humans. In animals, lectins regulate the cell adhesion to glycoprotein synthesis, control protein levels in blood, and bind soluble extracellular and intracellular glycoproteins. Also, in the immune system, lectins recognize carbohydrates found specifically on pathogens or those that are not recognizable on host cells. Embryos are attached to the endometrium of the uterus through L-selectin. This activates a signal to allow for implantation. *E. coli* are able to reside in the gastrointestinal tract by lectins that recognize carbohydrates in the intestines. The influenza virus contains hemagglutinin, which recognizes sialic acid residues on the glycoproteins located on the surface of the host cell. This allows the virus to attach and gain entry into the host cell. Clinically, purified lectins can be used to identify glycolipids and glycoproteins on an individual's red blood cells for blood typing.

The journey of *Animal Lectins: Form, Function, and Clinical Applications*, essentially the Encyclopedia of Animal Lectins, starts with an introductory chapter on lectin families, in general, followed by specific animal lectin families, such as intracellular sugar-binding ER chaperones, calnexin and calreticulin; the P-type lectins working in the endocytic pathway, the lectins of ERAD pathway, and the complex mannose-binding ERGIC-53 protein and its orthologs circulating between the ER, ERGIC, and the Golgi apparatus; and the fairly small galectins that are synthesized in the cytosol but may be found at many locations. Chapters on R-type lectin families, pentraxins, siglecs, C-type lectins, regenerating gene family, tetranectin group of lectins, ficolins, F-type lectins, and chi-lectins have covered the up-to-date literature on the subject and contain illustrations of their structures, functions, and clinical applications. The emerging group of annexins as lectins has been given a place as a separate family.

The C-type lectins comprising 17 subfamilies such as collectins, selectins, NK cell lectin receptors, latest discoveries of lectins on dendritic cells and others form the backbone of Volume 2. The chapters are well written, although there is variability on how they are focused. Most of the chapters focus on lectin structures, functions, their ligands, and their medical relevance in terms of diagnosis and therapy. The journey ends with five reviews on clinical applications of lectins with a survey of literature on endogenous lectins as drug targets. This may reflect state of the art in each area or the interests of the author(s).

In post-genomic years, with the human genome sequence at hand, a complete overview of many human lectin genes has become available, as illustrated by C-type lectin domain 3D structures having as little as 30% amino acid sequence similarity. The major effort in the future will be on elaboration of similar studies on other families of lectins with carbohydrate specificities and in vivo lectin–ligand interactions. In this respect, further writings will bring the reader to the forefront of knowledge in the field. Thus, on thirst of learning about animal recognition systems, *Animal Lectins: Form, Function, and Clinical Applications* is an excellent reference book for those studying biochemistry, biotechnology, and biophysics with a specialization in the areas of immunology, lectinology, and glycobiology as well as for pharmacy students involved in drug discoveries through lectin–carbohydrate interactions. Both novice and advanced researchers in biomedical, analytical, and pharmaceutical fields need to understand animal lectins.

T.P. Singh, F.N.A., F.A.Sc., F.N.A.Sc., F.T.W.A.S
DBT-Distinguished Biotechnology Research Professor
Department of Biophysics,
All India Institute of Medical Sciences, New Delhi, India

Preface

Lectins are phylogenetically ancient proteins that have specific recognition and binding functions for complex carbohydrates of glycoconjugates, that is, of glycoproteins, proteoglycans/glycosaminoglycans, and glycolipids. They occur ubiquitously in nature and typically agglutinate certain animal cells and/or precipitate glycoconjugates without affecting their covalent linkages. Lectins mediate a variety of biological processes, such as cell–cell and host–pathogen interactions, serum–glycoprotein turnover, and innate immune responses. Although originally isolated from plant seeds, they are now known to be ubiquitously distributed in nature. The successful completion of several genome projects has made amino acid sequences of several lectins available. Their tertiary structures provide a good framework upon which all other data can be integrated, enabling the pursuit of the ultimate goal of understanding these molecules at the atomic level. With growing interest in the field of glycobiology, the function–structure relations of animal lectins have increased at an explosive rate, particularly in the last 20 years. Since lectins mediate important processes of adhesion and communication both inside and outside the cells in association with their ligands and associated co-receptor proteins, there is a need of a reference book that can describe emerging applications on the principles of structural biology of animal lectins at one point. No book has ever described in a coordinated fashion structures, functions, and clinical applications of 15 families of animal lectins presently known. Therefore, with the increasing information on animal lectins in biomedical research and their therapeutic applications, writing of a comprehensive document on animal lectins and associated proteins in the form of *Animal Lectins: Form, Function, and Clinical Applications* has been the main objective of the present work. The entire manuscript has been distributed into two volumes in order to produce an easily readable work with easy portability. Volume 1 comprises most of the superfamilies (Chaps. 1–21) of lectins, excluding C-type lectins or C-type-lectin-like domain. In volume 2, we have mainly focused on C-type lectins, which have been extensively studied in vertebrates with wider clinical applications (Chaps. 22–46).

Animal Lectins: Form, Function, and Clinical Applications reviews the current knowledge of animal lectins, their ligands, and associated proteins with a focus on their structures and functions, biochemistry and patho-biochemistry (protein defects as a result of disease), cell biology (exocytosis and endocytosis, apoptosis, cell adhesion, and malignant transformation), clinical applications, and their intervention for therapeutic purposes. The book emphasizes on the effector functions of animal lectins in innate immunity and provides reviews/chapters on extracellular animal lectins, such as C-type lectins, R-type lectins, siglecs, and galectins, and intracellular lectins, such as calnexin family (M-type, L-type, and P-type), recently discovered F-box lectins, ficolins, chitinase-like lectins, F-type lectins, and intelectins, mainly in vertebrates. The clinical significance of lectin–glycoconjugate interactions has been exemplified by inflammatory diseases, defects of immune defense, autoimmunity, infectious diseases, and tumorigenesis/metastasis, along with therapeutic perspectives of novel drugs that interfere with lectin–carbohydrate interactions.

Based on the information gathered on animal lectins in this book, a variety of medical and other applications are in the offing. Foremost among these is the lectin-replacement therapy for patients suffering from lectin deficiency defects. We have pointed out the advancements of such studies where such progress has been made. Other uses in different stages of development are antibacterial drugs; multivalent hydrophobic carbohydrates for anti-adhesion therapy of microbial diseases; highly effective inhibitors of the selectins for treatment of leucocyte-mediated pathogenic conditions, such as asthma, septic shock, stroke, and myocardial infarction; inhibitors of the galectins and other lectins involved in metastasis; and application of lectins for facile and improved disease diagnosis. Recent advances in the discovery of M6PR-homologous protein family (Chap. 5), lectins of ERAD pathway, F-box proteins and M-type lectins (Chap. 6), and mannose receptor–targeted drugs and vaccines (Chaps. 8 and 46) can form the basis of cutting-edge technology in drug delivery devices.

Based on the structures of animal lectins, classified into at least 15 superfamilies, C-type lectins and galectins are the classical major families. Galectins are known to be associated with carcinogenesis and metastasis. Galectin-3 is a pleiotropic carbohydrate-binding protein, involved in a variety of normal and pathological biological processes. Its carbohydrate-binding properties constitute the basis for cell–cell and cell–matrix interactions (Chap. 12) and cancer progression (Chap. 13). Studies lead to the recognition of galectin-3 as a diagnostic/prognostic marker for specific cancer types, such as thyroid and prostate. In interfering with galectin–carbohydrate interactions during tumor progression, a current challenge is the design of specific galectin inhibitors for therapeutic purposes. Anti-galectin agents can restrict the levels of migration of several types of cancer cells and should, therefore, be used in association with cytotoxic drugs to combat metastatic cancer (Chap. 13). The properties of siglecs that make them attractive for cell-targeted therapies have been reviewed in Chaps. 16, 17, and 46.

F-type domains are found in proteins from a range of organisms from bacteria to vertebrates, but exhibit patchy distribution across different phylogenetic taxa, suggesting that F-type lectin genes have been selectively lost even between closely related lineages, thus making it difficult to trace the ancestry of the F-type domain. The F-type domain has clearly gained functional value in fish, whereas the fate of F-type domains in higher vertebrates is not clear, rather it has become defunct. Two genes encoding three-domain F-type proteins are predicted in the genome of the opossum (*Monodelphis domestica*), an early-branching mammal. There is plenty of scope to discover F-type lectins in mammalian vertebrates (Chap. 20). Since the reports on the C-reactive protein (CRP) as a cardiovascular marker (Chap. 8), novel biomarkers in cardiovascular and other inflammatory diseases have emerged in recent years. The substantial knowledge on CRP is now being complemented by new markers such as YKL-40, a member of chi-lectins group of CTLD (Chap. 19). The YKL-40 (chitinase-3-like protein 1, or human cartilage glycoprotein-39) displays a typical fold of family 18 glycosyl hydrolases and is expressed and secreted by several types of solid tumors, including glioblastoma, colon cancer, breast cancer, and malignant melanoma. Chitinase-3-like protein 1 was recently introduced into clinical practice; yet its application is still restricted.

In volume 2, we have mainly focused on C-type lectins, which have been extensively studied in vertebrates. A C-type lectin is a type of carbohydrate-binding protein domain that requires calcium for binding interactions in general. Drickamer et al. classified C-type lectins into seven subgroups (I to VII) based on the order of the various protein domains in each protein. This classification was subsequently updated in 2002, leading to seven additional groups (VIII to XIV). A further three subgroups (XV to XVII) were added recently. The C-type lectins share structural homology in their high-affinity carbohydrate recognition domains (CRDs) and constitute a large and diverse group of extracellular proteins that have been extensively studied. Their activities have been implicated as indispensable players in carbohydrate recognition, suggesting their possible application in discrimination of various correlative microbes and developing biochemical tools. The C-type lectins, structurally characterized by a double loop composed of two highly conserved disulfide bridges located at the bases of the loops, are believed to mediate pathogen recognition and play important roles in the innate immunity of both vertebrates and invertebrates. A large number of these proteins have been characterized and more than 80 have been sequenced. Recent data on the primary sequences and 3D structures of C-type lectins have enabled us to analyze their molecular evolution. Statistical analysis of their cDNA sequences shows that C-type-lectin-like proteins, with some exceptions, have evolved in an accelerated manner to acquire their diverse functions.

The C-type lectin family includes the monocyte mannose receptor (MMR), mannose-binding lectin (MBL), lung surfactant proteins, ficolins, selectins, and others, which are active in immune functions and pathogen recognition (Chaps. 23–34 and 41–45). Several C-type lectins and lectin-like receptors have been characterized that are expressed abundantly on the surface of professional antigen-presenting cells (APCs). Dendritic cells (DCs) are equipped with varying sets of C-type lectin receptors that help them with the uptake of pathogens. Important examples are langerin, DC-SIGN, DC-SIGNR, DCAR, DCIR, dectins, DEC-205, and DLEC (Chaps. 34–36). DCs are key regulators in directing the immune responses and, therefore, are under extensive research for the induction of antitumor immunity. They scan their surroundings for recognition and uptake of pathogens. Intracellular routing of antigens through C-type lectins enhances loading and presentation of antigens through MHC class I and II, inducing antigen-specific $CD4^+$ and $CD8^+$ T-cell proliferation and skewing T-helper cells. These characteristics make C-type lectins interesting targets for DC-based immunotherapy. Extensive research has been performed on targeting specific tumor antigens to C-type lectins, using either antibodies or natural ligands such as glycan structures. In Chaps. 34–36, we have presented the current knowledge of DC receptors to exploit them for antitumor activity and drug targeting in the near future (Chap. 46).

The monocyte mannose receptor (MMR) or the mannose receptor (MR) (CD206) is a member of the Group VI C-type lectins along with ENDO180, DEC205, and the phospholipase A2 receptor. Expressed on a broad range of cell types, including tissue macrophages and various epithelial cells, the MMR is active in endocytosis and phagocytosis. It is also thought to be involved in innate immunity, though its exact role remains unclear. Structurally, MMR is a complex molecule, which has been reviewed as an R-type lectin in volume 1 (Chap. 15) and as a C-type lectin in volume 2 (Chap. 35). Further research is required to fully understand the function of MMR. In addition, the Reg family constitutes an interesting subset of the C-type lectin family. The Reg family members are small, secreted proteins, which have been implicated in a range of physiological processes such as acute phase reactants and survival/growth factors for insulin-producing pancreatic β-cells, neural cells, and epithelial cells of the digestive system (Chap. 39). The C-type lectin DC-SIGN is unique in the regulation of adhesion processes, such as DC trafficking and T-cell synapse formation, besides its well-studied function in antigen capture. In particular, the DC-SIGN and associated homologues contribute to the potency of DC to control immunity (Chaps. 36 and 46). There is always significant interest in the development of drug and antigen delivery systems via the oral route due to patient compliance and acceptability. The presence of DCs with knowledge of associated receptors in the gastrointestinal tract offers principles of methodology for the development of oral vaccines (Chap. 46).

The search of the database of NCBI revealed that the C-type lectins attract much more attention, which resulted in recent discoveries of novel groups of lectins (Groups XV–XVII; Chap. 40). The clinical applications of C-type lectins have been exemplified in Chaps. 42–46. Although a variety of lectins have enabled greater insight into the diversity and complexity of lectin repertoires in vertebrates in two volumes, the nature of the protein–carbohydrate interactions and the potential mechanisms of different functions for invertebrate lectins are under intense investigation. Future progress will elucidate the contribution of different lectin families and their cross talk with each other or with other molecules with respect to mounting protective immune responses in invertebrates and vertebrates.

MBL as a reconstitution therapy in genetically determined MBL deficiency has advanced significantly. Since the genetically determined MBL deficiency is very common and can be associated with increased susceptibility to a variety of infections, the potential benefits of MBL reconstitution therapy still need to be evaluated. In a phase I trial on MBL-deficient healthy adult volunteers, MBL did not show adverse clinical effects (Chap. 23). SP-A and SP-D have been recently categorized as "Secretory Pathogen Recognition Receptors." Treatment with a recombinant fragment of human SP-D consisting of a short collagen-like stalk (but not the entire collagen-like domain of native SP-D), neck, and CRD inhibited development of emphysema-like pathology in SP-D-deficient mice (Chaps. 24, 25, and 43). Autosomal dominant polycystic kidney disease (ADPKD) is a common inherited nephropathy, affecting over 1:1000 of the population worldwide. It is a systemic condition with frequent hepatic and cardiovascular manifestations in addition to the progressive development of fluid-filled renal cysts that eventually result in loss of renal function in the majority of affected individuals. The cysts that grow in the kidneys of the majority of ADPKD patients are the result of mutations within the genes *PKD1* and *PKD2* that code for polycystin-1 (PC-1) and PC-2, respectively (Chap. 45). The annexins or lipocortins are a multigene family of proteins that bind to acidic phospholipids and biological membranes. Some of the annexins bind to glycosaminoglycans (GAGs) in a Ca^{2+}-dependent manner and function as recognition elements for GAGs in extracellular space. The emerging groups of C-type lectins include layilin, tetranectin, and chondrolectin (Group VIII of CTLD) (Chap. 40) and CTLD-containing protein - CBCP in Group XVII. Fras1, QBRICK/Frem1, Frem2, and Frem3 form the family of Group XVII (Chap. 41).

There is plenty of scope to discover lectins in invertebrates and amphibians which offer novel biomaterials useful in therapeutics, with a hope that the list of native lectins as well as genetically modified derivatives will grow with time. Thus, understanding animal lectins and the associated network of proteins is of high academic value for those working in the field of protein chemistry and designing new drugs on the principle of protein–carbohydrate or protein–protein interactions. Refined information on the sites of interactions on glycoproteins in toto with lectins is the subject of future study. Efforts are being made to develop an integrated knowledge-based animal lectins database together with appropriate analytical tools. Thus, *Animal Lectins: Form, Function, and Clinical Applications*, the Encyclopedia of Vertebrate Lectins, is unique in its scope and differs from earlier publications on animal lectins. It is more than lectinology and is suitable to the students and researchers working in the areas of biochemistry, glycobiology, biotechnology, biophysics, microbiology and immunology, pharmaceutical chemistry, biomedicine, and animal sciences in general.

January 2012 G.S. Gupta

Acknowledgments

We are thankful to Professor R. C. Sobti, the Vice-Chancellor, and Professor M. L. Garg, Chairman, Department of Biophysics, Punjab University, Chandigarh, for providing the facilities from time to time to complete this project. Authors gratefully acknowledge the assistance of the following scientists for providing the reprints and the literature on the subject:

Dr. Borrego F, Rockville, USA; Dr. Clark DA, Hamilton, Canada; Dr. Dimaxi N, Genova, Italy; Dr. Doan LG, Golden, Colorado, USA; Dr. Dorner T, Berlin, Germany; Dr. Ezekowitz RAB, Boston, USA; Dr. Gabor DF, Vienna, Austria; Dr. Garg ML, Chandigarh, India; Dr. Girbes T, Valladolid, Spain; Dr. Goronzy JJ, Atlanta, USA; Dr. Hofer E, Vienna, Austria; Dr. Jepson MA, Bristol, UK; Dr. Kishore U, West London, UK; Dr. Le Bouteiller P, Toulouse, France; Dr. Markert U, Jena, Germany; Dr. Mincheva-Nilsson L, Umea, Sweden; Dr. Monsigny M, Orleans, France; Dr. Nan-Chi Chang, Taipei, Taiwan; Dr. Natarajan K, Bethesda, USA; Dr. Osborn HMI, Reading, UK; Dr. Palecanda A, Bozeman, USA; Dr. Petroff MG, Kansas City, USA; Dr. Piccinni MP, Florence, Italy; Dr. Radaev S, Rockville, USA; Dr. Rajbinder Kaur Virk, Chandigarh, India; Dr. Roos A, Leiden, the Netherlands; Dr. Sano H, Sapporo, Japan; Dr. Schon MP, Wurzburg, Germany; Dr. Sharon N, Rehovot, Israel; Dr. Shwu-Huey Liaw, Taipei, Taiwan; Dr. Singh TP, New Delhi, India; Dr. Soilleux EJ, Cambridge, UK; Dr. Steinhubl SR, Lexington, USA; Dr. Steinman RM, New York, USA; Dr. Zenclussen AC, Berlin, Germany; and Dr. Valladeau J, Lyon, France.

Anita Gupta acknowledges the technical help of Dr. Neerja Mittal and Ms. Sargam Preet from RBIEBT. G. S. Gupta is grateful to Mrs. Kishori Gupta for the moral support and patience at the time of need.

G.S. Gupta
Anita Gupta
Rajesh K. Gupta

Contents of Volume 1

Part 1   Introduction

1   Lectins: An Overview 3
G.S. Gupta
1.1   Lectins: Characteristics and Diversity 3
    1.1.1   Characteristics 3
    1.1.2   Lectins from Plants 4
    1.1.3   Lectins in Microorganisms 5
    1.1.4   Animal Lectins 5
1.2   The Animal Lectin Families 5
    1.2.1   Structural Classification of Lectins 5
1.3   C-Type Lectins (CLEC) 8
    1.3.1   Identification of CLEC 8
    1.3.2   C-Type Lectin Like Domain (CTLD/CLRD) 8
    1.3.3   Classification of CLRD-Containing Proteins 9
    1.3.4   The CLRD Fold 9
1.4   Disulfide Bonds in Lectins and Secondary Structure 10
    1.4.1   Disulfide Bond 10
    1.4.2   Pathway for Disulfide Bond Formation in the ER
            of Eukaryotic Cells 10
    1.4.3   Arrangement of Disulfide Bonds in CLRDs 10
    1.4.4   Functional Role of Disulfides in CLRD of Vertebrates 11
    1.4.5   Disulfide Bonds in Ca^{2+}-Independent Lectins 12
1.5   Functions of Lectins 12
    1.5.1   Lectins in Immune System 13
    1.5.2   Lectins in Nervous Tissue 15
1.6   The Sugar Code and the Lectins as Recepors in System Biology 15
    1.6.1   Host-Pathogen Interactions 16
    1.6.2   Altered Glycosylation in Cancer Cells 17
    1.6.3   Protein-Carbohydrate Interactions in Immune System 18
    1.6.4   Glycosylation and the Immune System 19
    1.6.5   Principles of Protein-Glycan Interactions 19
1.7   Applications of Lectin Research and Future Perspectives 21
    1.7.1   Mannose Receptor-Targeted Vaccines 21
References 22

Part II Intracellular Lectins

2 Lectins in Quality Control: Calnexin and Calreticulin 29
  G.S. Gupta
  2.1 Chaperons 29
     2.1.1 Calnexin 30
     2.1.2 Calnexin Structure 30
     2.1.3 Calnexin Binds High-Mannose-Type Oligosaccharides ... 31
     2.1.4 Functions of Calnexin 32
     2.1.5 Patho-Physiology of Calnexin Deficiency 35
  2.2 Calreticulin 36
     2.2.1 General Features 36
     2.2.2 The Protein 37
     2.2.3 Cellular Localization of Calreticulin 40
     2.2.4 Functions of Calreticulin 41
     2.2.5 Structure-Function Relationships in Calnexin and Calreticulin 48
     2.2.6 Pathophysiological Implications of Calreticulin 48
     2.2.7 Similarities and Differences Between Cnx and Crt 50
     2.2.8 Calreticulin in Invertebrates 51
  References 52

3 P-Type Lectins: Cation-Dependent Mannose-6-Phosphate Receptor 57
  G.S. Gupta
  3.1 The Biosynthetic/Secretory/Endosomal Pathways 57
     3.1.1 Organization of Secretory Pathway 57
  3.2 P-Type Lectin Family: The Mannose 6-Phosphate Receptors ... 62
     3.2.1 Fibroblasts MPRs 63
     3.2.2 MPRs in Liver 63
     3.2.3 MPRs in CNS 63
     3.2.4 CI-MPR in Bone Cells 64
     3.2.5 Thyroid Follicle Cells 64
     3.2.6 Testis and Sperm 64
     3.2.7 MPRs During Embryogenesis 65
  3.3 Cation-Dependent Mannose 6-Phosphate Receptor 65
     3.3.1 CD-MPR- An Overview 65
     3.3.2 Human CD-MPR 66
     3.3.3 Mouse CD-MPR 66
  3.4 Structural Insights 67
     3.4.1 N-Glycosylation Sites in CD-MPR 67
     3.4.2 3-D Structure of CD-MPR 67
     3.4.3 Carbohydrate Binding Sites in MPRs 70
     3.4.4 Similarities and Dis-similarities between two MPRs .. 70
  3.5 Functional Mechanisms 71
     3.5.1 Sorting of Cargo at TGN 71
     3.5.2 TGN Exit Signal Uncovering Enzyme 72
     3.5.3 Association of Clathrin-Coated Vesicles with Adaptor Proteins 72
     3.5.4 Role of Di-leucine-based Motifs in Cytoplasmic Domains 73
     3.5.5 Sorting Signals in Endosomes 74
     3.5.6 Palmitoylation of CD-MPR is Required for Correct Trafficking 74
  References 75

4 P-Type Lectins: Cation-Independent Mannose-6-Phosphate Reeptors ... 81
G.S. Gupta

- 4.1 Cation-Independent Mannose 6-Phosphate Receptor (CD222) ... 81
  - 4.1.1 Glycoprotein Receptors for Insulin and Insulin-like Growth Factors ... 81
- 4.2 Characterization of CI-MPR/IGF2R ... 81
  - 4.2.1 Primary Structures of Human CI-MPR and IGF2R Are Identical ... 81
  - 4.2.2 Mouse IGF2R/CI-MPR Gene ... 82
  - 4.2.3 Bovine CI-MPR ... 82
  - 4.2.4 CI-MPR in Other Species ... 83
- 4.3 Structure of CI-MPR ... 84
  - 4.3.1 Domain Characteristics of IGF2R/CI-MPR (M6P/IGF2R) ... 84
  - 4.3.2 Crystal Structure ... 85
- 4.4 Ligands of IGF2R/CI-MPR ... 86
  - 4.4.1 Extracellular Ligands of IGF2R/CI-MPR ... 86
  - 4.4.2 Binding Site for M6P in CI-MPR ... 86
  - 4.4.3 The Non-M6P-Containing Class of Ligands ... 89
- 4.5 Complementary Functions of Two MPRS ... 90
  - 4.5.1 Why Two MPRs ... 91
  - 4.5.2 Cell Signaling Pathways ... 92
- 4.6 Functions of CI-MPR ... 92
- 4.7 Proteins Associated with Trafficking of CI-MPR ... 94
  - 4.7.1 Adaptor Protein Complexes ... 94
  - 4.7.2 Mammalian TGN Golgins ... 95
  - 4.7.3 TIP47: A Cargo Selection Device for MPR Trafficking ... 95
  - 4.7.4 Sorting Signals in GGA and MPRs at TGN ... 95
- 4.8 Retrieval of CI-MPR from Endosome-TO-GOLGI ... 97
  - 4.8.1 Endosome-to-Golgi Retrieval of CIMPR Requires Retromer Complex ... 97
  - 4.8.2 Retromer Complex and Sorting Nexins (SNX) ... 98
  - 4.8.3 Small GTPases in Lysosome Biogenesis and Transport ... 100
  - 4.8.4 Role for Dynamin in Late Endosome Dynamics and Trafficking of CI-MPR ... 100
- 4.9 CI-MPR/IGF2R System and Pathology ... 100
  - 4.9.1 Deficiency of IGF2R/CI-MPR Induces Myocardial Hypertrophy ... 100
  - 4.9.2 MPRs in Neuromuscular Diseases ... 101
  - 4.9.3 CI-MPR in Fanconi syndrome ... 101
  - 4.9.4 Tumor Suppressive Effect of CI-MPR/IGF2R ... 101
- References ... 102

5 Mannose-6-Phosphate Receptor Homologous Protein Family ... 109
G.S. Gupta

- 5.1 Recognition of High-Mannose Type N-Glycans in ERAD Pathways ... 109
- 5.2 Proteins Containing M6PRH Domains ... 110
- 5.3 GlcNAc-Phosphotransferase ... 110
- 5.4 α-Glucosidase II ... 112
  - 5.4.1 Function of α-Glucosidase II ... 112
  - 5.4.2 M6PRH Domain in GIIβ ... 113
  - 5.4.3 Two Distinct Domains of β-Subunit of GII Interact with α-Subunit ... 113
  - 5.4.4 Polycystic Liver Disease (PCLD) and β-Subunit of Glucosidase II ... 114

		5.5	Osteosarcomas-9 (OS-9)..	115

- 5.5 Osteosarcomas-9 (OS-9) ... 115
  - 5.5.1 The Protein ... 115
  - 5.5.2 Requirement of HRD1, SEL1L, and OS-9/XTP3-B for Disposal of ERAD-Substrates ... 115
  - 5.5.3 OS-9 Recognizes Mannose-Trimmed N-Glycans ... 116
  - 5.5.4 Dual Task for Xbp1-Responsive OS-9 Variants ... 116
  - 5.5.5 Interactions of OS-9 ... 116
- 5.6 YOS9 from *S. cerevisiae* ... 118
- 5.7 Erlectin/XTP3-B ... 119
- 5.8 *Drosophila* Lysosomal Enzyme Receptor Protein (LERP) ... 119
- 5.9 MRL1 ... 119
- References ... 120

6 Lectins of ERAD Pathway: F-Box Proteins and M-Type Lectins ... 123
G.S. Gupta
- 6.1 Intracellular Functions of N-Linked Glycans in Quality Control ... 123
- 6.2 The Degradation Pathway for Misfolded Glycoproteins ... 124
  - 6.2.1 Endoplasmic Reticulum-Associated Degradation (ERAD) ... 124
  - 6.2.2 Ubiquitin-Mediated Proteolysis ... 124
  - 6.2.3 SCF Complex ... 125
  - 6.2.4 F-Box Proteins: Recognition of Target Proteins by Protein-Protein Interactions ... 126
- 6.3 F-Box Proteins with a C-Terminal Sugar-Binding Domain (SBD) ... 127
  - 6.3.1 Diversity in SCF Complex due to Lectin Activity of F-Box Proteins ... 127
  - 6.3.2 Fbs Family ... 128
  - 6.3.3 Fbs1 Equivalent Proteins ... 131
  - 6.3.4 Ligands for F-Box Proteins ... 133
  - 6.3.5 Evolution of F-Box Proteins ... 134
  - 6.3.6 Localization of F-Box Proteins ... 134
  - 6.3.7 Regulation of F-Box Proteins ... 134
- 6.4 α-Mannosidases and M-Type Lectins ... 135
  - 6.4.1 α-Mannosidases ... 135
  - 6.4.2 ER-associated Degradation-enhancing α-Mannosidase-like Proteins (EDEMs) ... 135
  - 6.4.3 Functions of M-Type Lectins in ERAD ... 136
- 6.5 Derlin-1, -2 and -3 ... 138
- References ... 139

Part III L-Type Lectins

7 L-Type Lectins in ER-Golgi Intermediate Compartment ... 145
G.S. Gupta
- 7.1 L-Type Lectins ... 145
  - 7.1.1 Lectins from Leguminous Plants ... 145
  - 7.1.2 L-Type Lectins in Animals and Other Species ... 145
- 7.2 ER-Golgi Intermediate Compartment ... 146
- 7.3 Lectins of Secretory Pathway ... 146
- 7.4 ER-Golgi Intermediate Compartment Marker-53 (ERGIC-53) or LMAN1 ... 147
  - 7.4.1 ERGIC-53 Is Mannose-Selective Human Homologue of Leguminous Lectins ... 147
  - 7.4.2 Cells of Monocytic Lineage Express MR60: A Homologue of ERGIC-53 ... 148

		7.4.3	Rat Homologue of ERGIC53/MR60	149

- 7.4.3 Rat Homologue of ERGIC53/MR60 149
- 7.4.4 Structure-Function Relations 150
- 7.4.5 Functions of ERGIC-53 152
- 7.4.6 Mutations in ERGIC-53 *LMAN1* Gene and Deficiency of Coagulation Factors V and VIII lead to bleeding disorder 154
- 7.5 Vesicular Integral Membrane Protein (VIP36) OR LMAN2 156
  - 7.5.1 The Protein 156
  - 7.5.2 VIP36-SP-FP as Cargo Receptor 157
  - 7.5.3 Structure for Recognition of High Mannose Type Glycoproteins by VIP36 157
  - 7.5.4 Emp47p of *S. cerevisiae*: A Homologue to VIP36 and ERGIC-53 ... 158
- 7.6 VIP36-Like (VIPL) L-Type Lectin 158
- References 159

8 Pentraxins: The L-Type Lectins and the C-Reactive Protein as a Cardiovascular Risk 163
G.S. Gupta

- 8.1 Pentraxins and Related Proteins 163
- 8.2 Short Pentraxins 163
- 8.3 C-Reactive Protein 164
  - 8.3.1 General 164
  - 8.3.2 CRP Protein 164
  - 8.3.3 Structure of CRP 165
  - 8.3.4 Functions of CRP 166
- 8.4 CRP: A Marker for Cardiovascular Risk 168
  - 8.4.1 CRP: A Marker for Inflammation and Infection 168
  - 8.4.2 CRP: A Marker for Cardiovascular Risk 169
  - 8.4.3 Role of Modified/Monomeric CRP 170
- 8.5 Extra-Hepatic Sources of CRP 171
- 8.6 Serum Amyloid P Component 171
  - 8.6.1 Genes Encoding SAP 171
  - 8.6.2 Characterization of SAP 172
  - 8.6.3 Interactions of SAP and CRP 173
  - 8.6.4 Functions of SAP 174
  - 8.6.5 SAP in Human Diseases 175
  - 8.6.6 SAP from *Limulus Polyphemus* 176
- 8.7 Female Protein (FP) in Syrian Hamster 176
  - 8.7.1 Similarity of Female Protein to CRP and APC 176
  - 8.7.2 Structure of Female Protein 177
  - 8.7.3 Gene Structure and Expression of FP 177
- 8.8 Long Pentraxins 178
  - 8.8.1 Long Pentraxins 1, -2, -3 178
- 8.9 Neuronal Pentraxins (Pentraxin-1 and -2) 178
  - 8.9.1 Functions of Pentraxin 1 and -2 178
- 8.10 Pentraxin 3 (PTX3) 179
  - 8.10.1 Characterization of PTX3 179
  - 8.10.2 Cellular Sources of PTX3 180
  - 8.10.3 Ligands 180
  - 8.10.4 Regulation of PTX3 181
  - 8.10.5 Functions of PTX3 182
- References 183

Part IV Animal Galectins

9 Overview of Animal Galectins: Proto-Type Subfamily 191
Anita Gupta and G.S. Gupta
    9.1   Galectins.. 191
    9.2   Galectin Sub-Families.. 191
    9.3   Galectin Ligands... 192
    9.4   Functions of Galectins... 192
            9.4.1   Functional Overlap/Divergence Among Galectins 193
            9.4.2   Cell Homeostasis by Galectins............................ 193
            9.4.3   Immunological Functions................................. 195
            9.4.4   Signal Transduction by Galectins......................... 196
            9.4.5   Common Structural Features in Galectins................... 197
            9.4.6   Galectin Subtypes in Tissue Distribution................... 198
    9.5   Prototype Galectins (Mono-CRD Type)............................... 199
            9.5.1   Galectin-1... 199
            9.5.2   Galectin-2... 199
            9.5.3   Galectin-5... 202
            9.5.4   Galectin-7... 202
            9.5.5   Galectin-10 (Eosinophil Charcot-Leyden Crystal Protein)...... 203
            9.5.6   Galectin-Related Inter-Fiber Protein (Grifin/Galectin-11)....... 204
            9.5.7   Galectin-13 (Placental Protein -13)........................ 205
            9.5.8   Galectin 14... 206
            9.5.9   Galectin-15... 206
    9.6   Evolution of Galectins.. 207
            9.6.1   Phylogenetic Analysis of Galectin Family.................... 207
            9.6.2   Galectins in Lower Vertebrates............................ 207
    References.. 208

10 Galectin-1: Forms and Functions 213
Anita Gupta
    10.1   The Subcellular Distribution....................................... 213
    10.2   Molecular Characteristics.. 213
            10.2.1   Galectin-1 Gene... 213
            10.2.2   X-Ray Structure of Human Gal-1......................... 214
            10.2.3   Gal-1 from Toad (*Bufo arenarum Hensel*) Ovary............ 216
            10.2.4   GRIFIN Homologue in Zebrafish (DrGRIFIN).............. 216
    10.3   Regulation of *Gal-1* Gene.. 216
            10.3.1   Gal-1 in IMP1 Deficient Mice............................ 216
            10.3.2   Blimp-1 Induces Galectin-1 Expression..................... 217
            10.3.3   Regulation by Retinoic Acid.............................. 217
            10.3.4   Regulation by TGF-β, IL-12 and FosB Gene Products......... 218
            10.3.5   Regulation by Metabolites/Drugs/Other Agents.............. 218
    10.4   Gal-1 in Cell Signaling.. 220
    10.5   Ligands/Receptors of Gal-1.. 220
            10.5.1   Each Galectin Recognizes Different Glycan Structures........ 220
            10.5.2   Interactions of Galectin-1................................ 221
    10.6   Functions of Galectin-1.. 223
            10.6.1   Role of Galectin-1 in Apoptosis........................... 223
            10.6.2   Gal-1 in Cell Growth and Differentiation................... 225
            10.6.3   Gal-1 and Ras in Cell Transformation..................... 227
            10.6.4   Development of Nerve Structure.......................... 228
            10.6.5   Skeletal Muscle Development............................ 232

		10.6.6	Gal-1 and the Immune System	232
		10.6.7	Role of Galetin-1 and Other Systems	235
	References			236
11	**Tandem-Repeat Type Galectins**			**245**
	Anita Gupta			
	11.1	Galectin 4		245
		11.1.1	Localization and Tissue Distribution	245
		11.1.2	Galectin-4 Isoforms	246
		11.1.3	Gal-4 from Rodents and Other Animals	247
		11.1.4	Ligands for Galectin-4	248
		11.1.5	Functions of Galectin-4	249
		11.1.6	Galectin-4 in Cancer	251
	11.2	Galectin-6		251
	11.3	Galectin-8		251
		11.3.1	Galectin-8 Characteristics	251
		11.3.2	Functions of Galectin-8	253
		11.3.3	Clinical Relevance of Gal-8	253
		11.3.4	Isoforms of Galectin-8 in Cancer	253
	11.4	Galectin-9		254
		11.4.1	Characteristics	254
		11.4.2	Stimulation of Galectin-9 Expression by IFN-γ	254
		11.4.3	Crystal Structure of Galectin-9	255
		11.4.4	Galectin-9 Recognizes *L. major* Poly-β-galactosyl Epitopes	255
		11.4.5	Functions of Galectin-9	255
		11.4.6	Galectin-9 in Clinical Disorders	257
		11.4.7	Galectin-9 in Cancer	258
	11.5	Galectin-12		259
	References			260
12	**Galectin-3: Forms, Functions, and Clinical Manifestations**			**265**
	Anita Gupta			
	12.1	General Characteristics		265
		12.1.1	Galectin-3 Structure	265
		12.1.2	Galectin-3 Gene	266
		12.1.3	Tissue and Cellular Distribution	266
	12.2	Ligands for Galectin-3: Binding Interactions		268
		12.2.1	Extracellular Matrix and Membrane Proteins	268
		12.2.2	Intracellular Ligands	268
		12.2.3	Carbohydrate Binding	269
		12.2.4	Carbohydrate-Independent Binding	271
	12.3	Functions		271
		12.3.1	Galectin-3 is a Multifunctional Protein	271
		12.3.2	Role in Cell Adhesion	272
		12.3.3	Gal-3 at the Interface of Innate and Adaptive Immunity	272
		12.3.4	Regulation of T-Cell Functions	274
		12.3.5	Pro-apoptotic and Anti-apoptotic Effects	274
		12.3.6	Role in Inflammation	276
		12.3.7	Gal-3 in Wnt Signaling	278
		12.3.8	In Urinary System of Adult Mice	278
		12.3.9	Gal-3 in Reproductive Tissues	278
		12.3.10	Gal-3 on Chondrocytes	278
		12.3.11	Role of Gal-3 in Endothelial Cell Motility and Angiogenesis	279
		12.3.12	Role in CNS	279

	12.4	Clinical Manifestations of Gal-3	279
		12.4.1 Advanced Glycation End Products (AGES)	279
		12.4.2 GAL-3 and Protein Kinase C in Cholesteatoma	280
		12.4.3 Gal-3 and Cardiac Dysfunction	280
		12.4.4 Gal-3 and Obesity	281
		12.4.5 Autoimmune Diseases	281
		12.4.6 Myofibroblast Activation and Hepatic Fibrosis	282
	12.5	Gal-3 as a Pattern Recognition Receptor	282
		12.5.1 Gal-3 Binds to *Helicobacter pylori*	282
		12.5.2 Recognition of *Candida albicans* by Macrophages Requires Gal-3	282
		12.5.3 Gal-3 is Involved in Murine Intestinal Nematode and Schistosoma Infection	283
		12.5.4 Up-Regulation of Gal-3 and Its Ligands by *Trypanosoma cruzi* Infection	283
	12.6	Gal-3 as a Therapeutic Target	283
		12.6.1 Gal-3: A Target for Anti-inflammatory/Anticancer Drugs	283
	12.7	Xenopus-Cortical Granule Lectin: A Human Homolog of Gal-3	284
	References		284
13	**Galectin-3: A Cancer Marker with Therapeutic Applications**		**291**
	Anita Gupta		
	13.1	Galectin-3: A Prognostic Marker of Cancer	291
	13.2	Discriminating Malignant Tumors from Benign Nodules of Thyroid	291
		13.2.1 Large-Needle Aspiration Biopsy	292
		13.2.2 Fine-Needle Aspiration Biopsy	292
		13.2.3 Combination of Markers	293
		13.2.4 Hashimoto's Thyroiditis	294
	13.3	Breast Cancer	294
	13.4	Tumors of Nervous System	295
		13.4.1 Galectins and Gliomas	295
	13.5	Diffuse Large B-Cell Lymphoma	296
	13.6	Gal-3 in Melanomas	297
	13.7	Head and Neck Carcinoma	298
	13.8	Lung Cancer	298
	13.9	Colon Neoplastic Lesions	299
	13.10	Expression of Gal-3 in Other Tumors	300
	13.11	Gal-3 in Metastasis	302
	13.12	β1,6 N-acetylglucosaminyltransferase V in Carcinomas	302
	13.13	Macrophage Binding Protein	303
	13.14	Galectinomics	303
	13.15	Mechanism of Malignant Progression by Galectin-3	304
	13.16	Anti-Galectin Compounds as Anti-Cancer Drugs	305
	References		306

Part V R-Type Animal Lectins

14	**R-Type Lectin Families**		**313**
	Rajesh K. Gupta and G.S. Gupta		
	14.1	Ricinus Communis Lectins	313
		14.1.1 Properties of Ricin	313
		14.1.2 Other R-Type Plant Lectins	314
	14.2	R-Type Lectins in Animals	315
	14.3	Mannose Receptor Family	315

	14.4	UDP-Galnac: Polypeptide α-N-Acetyl-galactosaminyltransferases	316
		14.4.1 Characteristics of UDP-GalNAc: α-N-Acetylgalactosaminyltransferases	316
		14.4.2 The Crystal Structure of Murine ppGalNAc-T-T1	318
		14.4.3 Parasite ppGalNAc-Ts	319
		14.4.4 Crystal Structure of CEL-III from *Cucumaria echinata* Complexed with GalNAc	320
	14.5	Microbial R-Type Lectins	321
		14.5.1 *S. olivaceoviridis* E-86 Xylanase: Sugar Binding Structure	321
		14.5.2 The Mosquitocidal Toxin (MTX) from *Bacillus sphaericus*	321
	14.6	R-Type Lectins in Butterflies	322
		14.6.1 Pierisin-1	322
		14.6.2 Pierisin-2, Pierisin-3 and -4	324
	14.7	Discoidin Domain and Carbohydrate-Binding Module	324
		14.7.1 The Discoidin Domain	324
		14.7.2 Discoidins from *Dictyostelium discoideum* (DD)	325
		14.7.3 Discoidin Domain Receptors (DDR1 and DDR2)	326
		14.7.4 Earth Worm (EW)29 Lectin	327
	References		327
15	**Mannose Receptor Family: R-Type Lectins**		**331**
	Rajesh K. Gupta and G.S. Gupta		
	15.1	R-Type Lectins in Animals	331
	15.2	Mannose Receptor Lectin Family	331
	15.3	The Mannose Receptor (CD206)	332
		15.3.1 Human Macrophage Mannose Receptor (MMR)	332
		15.3.2 Structure-Function Relations	332
		15.3.3 Cell and Tissue Distribution	334
		15.3.4 Ligands	335
		15.3.5 Functions of Mannose Receptor	336
		15.3.6 Mouse Mannose Receptor	337
		15.3.7 Interactions of MR with Branched Carbohydrates	339
		15.3.8 Mannose Receptor-Targeted Drugs and Vaccines	339
	15.4	Phospholipase A2-Receptors	339
		15.4.1 The Muscle (M)-Type sPLA2 Receptors	340
		15.4.2 Neuronal or N-Type PLA2 Receptor	342
	15.5	DEC-205 (CD205)	342
		15.5.1 Characterization	342
		15.5.2 Functions of DEC-205	343
	15.6	ENDO 180 (CD280)/uPARAP	343
		15.6.1 Urokinase Receptor (uPAR)-Associated Protein	343
		15.6.2 Interactions of Endo180	344
	References		345

Part VI I-Type Lectins

16	**I-Type Lectins: Sialoadhesin Family**		**351**
	G.S. Gupta		
	16.1	Sialic Acids	351
	16.2	Sialic Acid-Binding Ig-Like Lectins (I-Type Lectins)	352
		16.2.1 Two Subsets of Siglecs	352
		16.2.2 Siglecs as Inhibitory Receptors	353
		16.2.3 Binding Characteristics of Siglecs	353
		16.2.4 Siglecs of Sialoadhesin Family	354

	16.3	Sialoadhesin (Sn)/Siglec-1 (CD169)	355

- 16.3 Sialoadhesin (Sn)/Siglec-1 (CD169) 355
  - 16.3.1 Characterization of Sialoadhesin/Siglec-1 355
  - 16.3.2 Cellular Expression of Sialoadhesin 355
  - 16.3.3 Ligands for Sialoadhesin 356
  - 16.3.4 Sialoadhesin Structure 357
  - 16.3.5 Regulation of Sialoadhesin 358
  - 16.3.6 Functions of Sialoadhesin 358
  - 16.3.7 Lessons from Animal Experiments 359
  - 16.3.8 Interactions with Pathogens 360
- 16.4 CD22 (Siglec-2) 361
  - 16.4.1 Characterization and Gene Organization 361
  - 16.4.2 Functional Characteristics 361
  - 16.4.3 Ligands of CD22 362
  - 16.4.4 Regulation of CD22 364
  - 16.4.5 Functions of CD22 365
  - 16.4.6 Signaling Pathway of Human CD22 and Siglec-F in Murine .. 365
  - 16.4.7 CD22 as Target for Therapy 367
- 16.5 Siglec-4 [Myelin-Associated Glycoprotein, (MAG)] 367
  - 16.5.1 MAG and Myelin Formation 367
  - 16.5.2 Characteristics of MAG 368
  - 16.5.3 MAG Isoforms 368
  - 16.5.4 Ligands of MAG: Glycan Specificity of MAG 369
  - 16.5.5 Functions of MAG 370
  - 16.5.6 MAG in Demyelinating Disorders 371
  - 16.5.7 Inhibitors of Regeneration of Myelin 372
  - 16.5.8 Axonal Regeneration by Overcoming Inhibitory Activity of MAG 372
  - 16.5.9 Fish Siglec-4 373
- 16.6 Siglec-15 373
- References 373

17 CD33 (Siglec 3) and CD33-Related Siglecs 381
G.S. Gupta
- 17.1 Human CD33 (Siglec-3) 381
  - 17.1.1 Human CD33 (Siglec-3): A Myeloid-specific Inhibitotry Receptor 381
- 17.2 CD33-Related Siglecs (CD33-rSiglecs) 382
  - 17.2.1 CD33-Related Siglecs Family 382
  - 17.2.2 CD33-rSiglec Structures 382
  - 17.2.3 Organization of CD33-rSiglec Genes on Chromosome 19q13.4 .. 384
- 17.3 Siglec-5 (CD170) 384
  - 17.3.1 Characterization 385
  - 17.3.2 Siglec-5: An Inhibitory Receptor 385
  - 17.3.3 Siglec-5-Mediated Sialoglycan Recognition 385
- 17.4 Siglec-6 386
  - 17.4.1 Cloning and Gene Organization of Siglec-6 (OB-BP1) 386
  - 17.4.2 Siglec-6 (OB-BP1) and Reproductive Functions 387
- 17.5 Siglec-7 (p75/AIRM1) 387
  - 17.5.1 Characterization 387
  - 17.5.2 Cytoplasmic Domain of Siglec-7 (p75/AIRM1) 387
  - 17.5.3 Crystallographic Analysis 388
  - 17.5.4 Interactions of Siglec-7 388
  - 17.5.5 Functions of Siglec-7 389
- 17.6 Siglec-8 389
  - 17.6.1 Characteristics and Cellular Specificity 389
  - 17.6.2 Ligands for Siglec-8 390

		17.6.3	Functions in Apoptosis....................................	390
		17.6.4	Siglec-8 in Alzheimer's Disease...........................	391
	17.7	Siglec-9..		391
		17.7.1	Characterization and Phylogenetic Analysis................	391
		17.7.2	Functions of Siglec-9.....................................	391
	17.8	Siglec-10, -11, -12, and -16...............................		392
		17.8.1	Siglec-10...	392
		17.8.2	Siglec-11...	392
		17.8.3	Siglec-12...	393
		17.8.4	Siglec-16...	394
	17.9	Mouse Siglecs..		394
		17.9.1	Evolution of Mouse and Human CD33-rSiglec Gene Clusters...	394
		17.9.2	Mouse CD33/Siglec-3.......................................	394
		17.9.3	Siglec-3-Related Siglecs in Mice..........................	395
	17.10	Glycoconjugate Binding Specificities of Siglecs............		397
	17.11	Functions of CD33-Related Siglecs..........................		399
		17.11.1	Endocytosis..	399
		17.11.2	Phagocytosis of Apoptotic Bodies.........................	400
	17.12	Siglecs as Targets for Immunotherapy.......................		400
	17.13	Molecular Diversity and Evolution of Siglec Family.........		401
	References...			402

Part VII Novel Super-Families of Lectins

18 Fibrinogen Type Lectins.. 409
Anita Gupta
	18.1	Ficolins..		409
		18.1.1	Ficolins versus Collectins................................	409
		18.1.2	Characterization of Ficolins..............................	409
		18.1.3	Ligands of Ficolins.......................................	411
		18.1.4	X-ray Structures of M, L- and H-Ficolins..................	412
		18.1.5	Functions of Ficolins.....................................	413
		18.1.6	Pathophysiology of Ficolins...............................	415
	18.2	Tachylectins..		415
		18.2.1	Horseshoe Crab Tachylectins...............................	415
		18.2.2	X-ray Structure...	416
	References...			417

19 Chi-Lectins: Forms, Functions and Clinical Applications................ 421
Rajesh K. Gupta and G.S. Gupta
	19.1	Glycoside Hydrolase Family 18 Proteins in Mammals...........		421
		19.1.1	Chitinases..	421
	19.2	Chitinase-Like Lectins: Chi-Lectins.........................		421
	19.3	YKL-40 [Chitinase 3-Like Protein 1 (CHI3L1)]................		422
		19.3.1	The Protein...	422
		19.3.2	Cell Distribution and Regulation..........................	422
		19.3.3	Ligands of YKL-40...	423
		19.3.4	The Crystal Structure of YKL-40...........................	423
	19.4	Human Cartilage 39-KDA Glycoprotein (or YKL-39)/(CHI3L2)....		425
	19.5	Ym1 and Ym2: Murine Proteins................................		426
		19.5.1	The Protein...	426
		19.5.2	Crystal Structure of Ym1..................................	427
		19.5.3	Oviductin...	427

	19.6	Functions of CHI3L1 (YKL-40)	428
		19.6.1 Role in Remodeling of Extracellular Matrix and Defense Mechanisms	428
		19.6.2 Growth Stimulating Effect	428
	19.7	Chi-Lectins As Markers of Pathogenesis	428
		19.7.1 A Marker of Inflammation	428
		19.7.2 CHI3L1 as Biomarker in Solid Tumors	431
		19.7.3 Chitinase 3-Like-1 Protein (CHI3L1) or YKL-40 in Clinical Practice	432
	19.8	Evolution of Mammalian Chitinases (-Like) of GH18 Family	433
	References		434
20	**Novel Groups of Fuco-Lectins and Intlectins**		**439**
	Rajesh K. Gupta and G.S. Gupta		
	20.1	F-type Lectins (Fuco-Lectins)	439
		20.1.1 F-type Lectins in Mammalian Vertebrates	439
	20.2	F-type Lectins in Fish and Amphibians	440
		20.2.1 Anguilla Anguilla Agglutinin (AAA)	440
		20.2.2 FBP from European Seabass	442
		20.2.3 Other F-type Lectins in Fish	442
		20.2.4 F-Type Lectins in Amphibians	443
	20.3	F-type Lectins in Invertebrates	444
		20.3.1 Tachylectin-4	444
		20.3.2 F-type Lectins from *Drosophila melanogaster*	445
		20.3.3 F -Type Lectins in Sea Urchin	445
		20.3.4 Bindin in Invertebrate Sperm	445
	20.4	F-type Lectins in Plants	445
	20.5	F-type Lectins in Bacteria	446
	20.6	Fuco-Lectins in Fungi	447
		20.6.1 Fuco-Lectin from Aleuria Aurantia (AAL)	447
	20.7	Intelectins	448
		20.7.1 Intelectin-1 (Endothelial Lectin HL-1/Lactoferrin Receptor or *Xenopus* Oocyte Lectin)	448
		20.7.2 Intelectin-2 (HL-2) and Intelectin-3	449
		20.7.3 Intelectins in Fish	450
		20.7.4 Eglectin (XL35) or Frog Oocyte Cortical Granule Lectins	450
	References		451
21	**Annexins (Lipocortins)**		**455**
	G.S. Gupta		
	21.1	Annexins	455
		21.1.1 Characteristics of Annexins	455
		21.1.2 Classification and Nomenclature	456
		21.1.3 Annexins in Tissues	457
		21.1.4 Functions of Annexins	458
	21.2	Annexin Family Proteins and Lectin Activity	459
	21.3	Annexin A2 (p36)	459
		21.3.1 Annexin 2 Tetramer (A2t)	460
		21.3.2 Crystal Analysis of Sugar-Annexin 2 Complex	461
		21.3.3 Functions of Annexin A2	461
	21.4	Annexin A4 (p33/41)	461
		21.4.1 General Characteristics	461
		21.4.2 Tissue Distribution	462
		21.4.3 Characterization	463

		21.4.4	Pathophysiology	463
		21.4.5	Doublet p33/41 Protein	463
	21.5	Annexin A5/Annexin V		464
		21.5.1	Gene Encoding Human Annexin A5	464
		21.5.2	Interactions of Annexin A5	464
		21.5.3	Molecular Structure of Annexin A5	465
		21.5.4	Annexin A5-Mediated Pathogenic Mechanisms	466
		21.5.5	A Novel Assay for Apoptosis	466
		21.5.6	Calcium-Induced Relocation of Annexins 4 and 5 in the Human Cells	467
	21.6	Annexin A6 (Annexin VI)		467
		21.6.1	Structure	467
		21.6.2	Functions	467
References				468

Contents of Volume 2

Part VIII C-Type Lectins: Collectins

22 C-Type Lectins Family 473
Anita Gupta and G.S. Gupta
- 22.1 C-Type Lectins Family 473
  - 22.1.1 The C-Type Lectins (CLEC) 473
  - 22.1.2 C-Type Lectin Like Domain (CLRD/CTLD) 473
  - 22.1.3 The CLRD Fold 474
  - 22.1.4 Ligand Binding 477
  - 22.1.5 C-Type Lectin-Like Domains in Model Organisms 477
- 22.2 Classification of CLRD/CTLD-Containing Proteins 477
- 22.3 Disulfide Bonds in Lectins and Secondary Structure 478
  - 22.3.1 Arrangement of Disulfide Bonds in CTLDs 478
  - 22.3.2 Functional Role of Disulfides in CTLD of Vertebrates 478
- 22.4 Collectins 480
  - 22.4.1 Collectins: A Group of Collagenous Type of Lectins 480
- References 480

23 Collectins: Mannan-Binding Protein as a Model Lectin 483
Anita Gupta
- 23.1 Collectins 483
  - 23.1.1 N-Terminal Region 484
  - 23.1.2 Collagenous Region 484
  - 23.1.3 C-Type Lectin Domain 485
  - 23.1.4 Comparative Genetics of Collagenous Lectins 486
  - 23.1.5 Generalized Functions of Collectins 486
- 23.2 Human Mannan-Binding Protein 487
  - 23.2.1 Characterization of Serum MBL 487
  - 23.2.2 Gene Structure of MBP 489
  - 23.2.3 Regulation of MBP Gene 490
  - 23.2.4 Structure-Function Relations 491
  - 23.2.5 Functions of MBL 492
- 23.3 MBP/MBL from Other Species 494
  - 23.3.1 Rodents 494
  - 23.3.2 Primates MBL 495
- 23.4 Similarity Between C1Q and Collectins/Defense Collagens 495
  - 23.4.1 Structural Similarities 495
  - 23.4.2 Functional Similarities 496
  - 23.4.3 Receptors for Defense Collagens 496
- References 496

24 Pulmonary SP-A: Forms and Functions ... 501
Anita Gupta and Rajesh K. Gupta
- 24.1 Pulmonary Surfactant Proteins ... 501
  - 24.1.1 Pulmonary Surfactant ... 501
  - 24.1.2 Pulmonary Surfactant Protein A ... 501
- 24.2 Structural Properties of SP-A ... 502
  - 24.2.1 Human SP-A: Domain Structure ... 502
  - 24.2.2 Structural Biology of Rat SP-A ... 504
  - 24.2.3 3-D Structure of SP-A Trimer ... 505
  - 24.2.4 Bovine SP-A ... 506
  - 24.2.5 SP-A in Other Species ... 507
- 24.3 Cell Surface Receptors for SP-A ... 508
- 24.4 Interactions of SP-A ... 509
  - 24.4.1 Protein-Protein Interactions ... 509
  - 24.4.2 Protein-Carbohydrate Interactions ... 510
  - 24.4.3 SP-A Binding with Lipids ... 510
- 24.5 Functions of SP-A ... 511
  - 24.5.1 Surfactant Components in Surface Film Formation ... 511
  - 24.5.2 Recognition and Clearance of Pathogens ... 512
  - 24.5.3 SP-A: As a Component of Complement System in Lung ... 513
  - 24.5.4 Modulation of Adaptive Immune Responses by SPs ... 513
  - 24.5.5 SP-A Stimulates Chemotaxis of AMΦ and Neutrophils ... 515
  - 24.5.6 SP-A Inhibits sPLA2 and Regulates Surfactant Phospholipid Break-Down ... 515
  - 24.5.7 SP-A Helps in Increased Clearance of Alveolar DPPC ... 515
  - 24.5.8 Protection of Type II Pneumocytes from Apoptosis ... 516
  - 24.5.9 Anti-inflammatory Role of SP-A ... 516
- 24.6 Reactive Oxygen and Nitrogen-Induced Lung Injury ... 518
  - 24.6.1 Decreased Ability of Nitrated SP-A to Aggregate Surfactant Lipids ... 518
- 24.7 Non-Pulmonary SP-A ... 518
  - 24.7.1 SP-A in Epithelial Cells of Small and Large Intestine ... 518
  - 24.7.2 SP-A in Female Genital Tract and During Pregnancy ... 519
- References ... 519

25 Surfactant Protein-D ... 527
Rajesh K. Gupta and Anita Gupta
- 25.1 Pulmonary Surfactant Protein-D (SP-D) ... 527
  - 25.1.1 Human Pulmonary SP-D ... 528
  - 25.1.2 Rat Lung SP-D ... 529
  - 25.1.3 Mouse SP-D ... 530
  - 25.1.4 Bovine SP-D ... 531
  - 25.1.5 Porcine Lung SP-D ... 531
- 25.2 Interactions of SP-D ... 531
  - 25.2.1 Interactions with Carbohydrates ... 531
  - 25.2.2 Binding with Nucleic Acids ... 533
  - 25.2.3 Interactions with Lipids ... 533
- 25.3 Structure: Function Relations of Lung SP-D ... 533
  - 25.3.1 Role of NH_2 Domain and Collagenous Region ... 533
  - 25.3.2 Role of NH2-Terminal Cysteines in Collagen Helix Formation ... 534
  - 25.3.3 D4 (CRD) Domain in Phospholipid Interaction ... 534

		25.3.4	A Three Stranded α-Helical Bundle at Nucleation Site of Collagen Triple-Helix Formation	535
		25.3.5	Ligand Binding Amino Acids	535
		25.3.6	Ligand Binding and Immune Cell-Recognition	535
	25.4	Regulation of Sp-D by Various Factors		536
		25.4.1	Glucocorticoids	536
		25.4.2	1α,25-Dihydroxyvitamin D3	536
		25.4.3	Growth Factors	536
	25.5	SP-D in Human Fetal and Newborns Lungs		537
	25.6	Non-Pulmonary SP-D		537
		25.6.1	Human Skin and Nasal Mucosa	537
		25.6.2	Digestive Tract, Mesentery, and Other Organs	537
		25.6.3	Male Reproductive Tract	538
		25.6.4	Female Genital Tract	538
	25.7	Functions of Lung SP-D		539
		25.7.1	Innate Immunity	539
		25.7.2	Effects on Alveolar Macrophages	539
		25.7.3	Functions of Neutrophils	540
		25.7.4	Protective Role in Allergy and Infection	541
		25.7.5	Adaptive Immune Responses	543
		25.7.6	Apoptosis	544
		25.7.7	Other Effects of SP-D	544
	25.8	Oxidative Stress and Hyperoxia		544
	References			545

Part IX C-Type Lectins: Selectins

26 L-Selectin (CD62L) and Its Ligands 553
G.S. Gupta

	26.1	Cell Adhesion Molecules		553
		26.1.1	Selectins	553
	26.2	Leukocyte-Endothelial Cell Adhesion Molecule 1 (LECAM-1) (L-Selectin/CD62L or LAM-1)		554
		26.2.1	Leukocyte-Endothelial Cell Adhesion Molecule 1 in Humans	554
		26.2.2	Gene Structure of L-Selectin	555
		26.2.3	Murine PLN Homing Receptor/mLHR3	555
	26.3	Functions of L-Selectin		556
		26.3.1	Lymphocyte Homing and Leukocyte Rolling and Migration	556
		26.3.2	Immune Responses	557
	26.4	L-Selectin: Carbohydrate Interactions		559
		26.4.1	Glycan-Dependent Leukocyte Adhesion	559
	26.5	Cell Surface Ligands for L-Selectin		561
		26.5.1	Subsets of Sialylated, Sulfated Mucins of Diverse Origins are Recognized by L-Selectin	561
		26.5.2	GlyCAM-1	561
		26.5.3	CD34	562
		26.5.4	Mucosal Addressin Cell Adhesion Molecule-1 (MAdCAM-1)	563
		26.5.5	PSGL-1 Binds L-Selectin	564
	26.6	L-Selectin IN Pathological States		564
		26.6.1	Gene Polymorphism in L-Selectin	564
		26.6.2	Antitumor Effects of L-Selectin	565

		26.6.3	Autoimmune Diseases	566
	26.7		Oligonucleotide Antagonists IN Therapeutic Applications	569
		26.7.1	Monomeric and Multimeric Blockers of Selectins	569
		26.7.2	Synthetic and Semisynthetic Oligosaccharides	569
	References			570

27 P-Selectin and Its Ligands ... 575
G.S. Gupta

- 27.1 Platelet Adhesion and Activation ... 575
- 27.2 P-Selectin (GMP-140, PADGEM, CD62): A Member of Selectin Adhesion Family ... 575
  - 27.2.1 Platelets and Vascular Endothelium Express P-Selectin ... 575
  - 27.2.2 Human Granule Membrane Protein 140 (GMP-140)/P-Selectin ... 576
  - 27.2.3 Murine P-Selectin ... 576
  - 27.2.4 P-Selectin Promoter ... 577
  - 27.2.5 Structure-Function Studies ... 578
  - 27.2.6 P-Selectin-Sialyl Lewis^x Binding Interactions ... 580
  - 27.2.7 Functions of P-Selectin ... 580
- 27.3 P-Selectin Glycoprotein Ligand-1 (PSGL-1) ... 583
  - 27.3.1 PSGL-1: The Major Ligand for P-Selectin ... 583
  - 27.3.2 Genomic Organization ... 585
  - 27.3.3 Specificity of PSGL-1 as Ligand for P-Selectin ... 585
  - 27.3.4 Structural Polymorphism in PSGL-1 and CAD Risk ... 585
  - 27.3.5 Carbohydrate Structures on PSGL-1 ... 586
  - 27.3.6 Role of PSGL-1 ... 587
- 27.4 Other Ligands of P-Selectin ... 588
- References ... 590

28 E-Selectin (CD62E) and Associated Adhesion Molecules ... 593
G.S. Gupta

- 28.1 E-Selectin (Endothelial Leukocyte Adhesion Molecule 1: ELAM-1) ... 593
  - 28.1.1 Endothelial Cells Express E-Selectin (ELAM-1/CD62E) ... 593
- 28.2 E-Selectin Genomic DNA ... 593
  - 28.2.1 Human E-Selectin ... 593
  - 28.2.2 E-Selectin Gene in Other Species ... 594
- 28.3 E-Selectin Gene Regulation ... 594
  - 28.3.1 Transcriptional Regulation of CAMs ... 594
  - 28.3.2 Induction of E-Selectin and Associated CAMs by TNF-α ... 595
  - 28.3.3 IL-1-Mediated Expression of CAMs ... 596
- 28.4 Factors in the Regulation of CAMs ... 596
  - 28.4.1 NF-kB: A Dominant Regulator ... 596
  - 28.4.2 Cyclic AMP Inhibits NF-kB-Mediated Transcription ... 597
  - 28.4.3 Peroxisome Proliferator-Activated Receptors ... 598
  - 28.4.4 Endogenous Factors Regulating CAM Genes ... 599
  - 28.4.5 Tat Protein Activates Human Endothelial Cells ... 599
  - 28.4.6 Reactive Oxygen Species (ROS) ... 599
  - 28.4.7 Role of Hypoxia ... 600
- 28.5 Binding Elements in E-Selectin Promoter ... 600
  - 28.5.1 Transcription Factors Stimulating *E-sel* Gene ... 600
  - 28.5.2 CRE/ATF Element or NF-ELAM1 ... 601
  - 28.5.3 HMG-I(Y) Mediates Binding of NF-kB Complex ... 602
  - 28.5.4 Phased-Bending of E-Selectin Promoter ... 602

	28.6	E-Selectin Ligands	602
		28.6.1 Carbohydrate Ligands (Lewis Antigens)	602
		28.6.2 E-Selectin Ligand-1	604
		28.6.3 PSGL-1 and Relating Ligands	605
		28.6.4 Endoglycan, a Ligand for Vascular Selectins	607
		28.6.5 L-Selectin as E-Selectin Ligand	607
	28.7	Structural Properties of E-Selectin	608
		28.7.1 Soluble E-Selectin: An Asymmetric Monomer	608
		28.7.2 Complement Regulatory Domains	608
		28.7.3 Three-Dimensional Structure	609
	28.8	Functions of E-Selectin	609
		28.8.1 Functions in Cell Trafficking	609
		28.8.2 E-Selectin in Neutrophil Activation	610
		28.8.3 P- and E-Selectin in Differentiation of Hematopoietic Cells	610
		28.8.4 Transmembrane Signaling in Endothelial Cells	611
	References		612

Part X C-Type Lectins: Lectin Receptors on NK Cells

29 KLRB Receptor Family and Human Early Activation Antigen (CD69) 619
Rajesh K. Gupta and G.S. Gupta

	29.1	Lectin Receptors on NK Cell	619
		29.1.1 NK Cell Receptors	619
		29.1.2 NKC Gene Locus	620
	29.2	The Ever-Expanding Ly49 Receptor Gene Family	620
		29.2.1 Activating and Inhibitory Receptors	620
		29.2.2 Crystal Analysis of CTLD of Ly49I and comparison with Ly29A, NKG2D and MBP-A	622
	29.3	NK Cell Receptor Protein 1 (NKR-P1) or KLRB1	623
		29.3.1 *Ly49* and *Nkrp1 (Klrb1)* Recognition Systems	623
		29.3.2 NKRP1	623
	29.4	Human NKR-P1A (CD161)	626
		29.4.1 Cellular Localization	626
		29.4.2 Transcriptional Regulation	626
		29.4.3 Ligands of CD161/ NKR-P1	627
		29.4.4 Signaling Pathways	628
		29.4.5 Functions of NKR-P1	629
	29.5	NKR-P1A in Clinical Disorders	630
		29.5.1 Autoimmune Reactions	630
		29.5.2 Other Diseases	631
		29.5.3 NKR-P1A Receptor (CD161) in Cancer Cells	632
	29.6	Human Early Activation Antigen (CD69)	632
		29.6.1 Organization of CD69 Gene	633
		29.6.2 Src-Dependent Syk Activation of CD69-Mediated Signaling	633
		29.6.3 Crystal Analysis of CD69	634
	References		634

30 NKG2 Subfamily C (KLRC) 639
Rajesh K. Gupta and G.S. Gupta

	30.1	NKG2 Subfamily C (KLRC)	639
		30.1.1 NKG2 Gene Family and Structural Organization	639
		30.1.2 Human NKG2-A, -B, and -C	640
		30.1.3 Murine NKG2A, -B, -C	640

		30.1.4	NKG2 Receptors in Monkey	641
		30.1.5	NKG2 Receptors in Other Species	642
		30.1.6	Inhibitory and Activatory Signals	642
	30.2	CD94 (KLRD1)		642
		30.2.1	Human CD94 in Multiple Transcripts	642
		30.2.2	Mouse CD94	643
	30.3	Cellular Sources of NKG2/CD94		643
	30.4	The Crystal Analysis of CD94		644
		30.4.1	An Intriguing Model for CD94/NKG2 Heterodimer	644
	30.5	CD94/NKG2 Complex		646
		30.5.1	CD94 and NKG2-A Form a Complex for NK Cells	646
	30.6	Acquisition of NK Cell Receptors		646
	30.7	Regulation of CD94/NKG2		647
		30.7.1	Transcriptional Regulation of NK Cell Receptors	647
		30.7.2	Regulation by Cytokines	647
	30.8	Functions of CD94/NKG2		648
		30.8.1	CD94/NKG2 in Innate and Adaptive Immunity	649
		30.8.2	Modulation of Anti-Viral and Anti-Tumoral Responses of γ/δ T Cells	649
	30.9	NK Cells in Female Reproductive Tract and Pregnancy		650
		30.9.1	Decidual NK Cell Receptors	650
	30.10	Signal Transduction by CD94/NKG2		651
		30.10.1	Engagement of CD94/NKG2-A by HLA-E and Recruitment of Phosphatases	651
	30.11	Ligands for CD94/NKG2		651
		30.11.1	HLA-E as Ligand for CD94/NKG2A	651
		30.11.2	CD94/NKG2-A Recognises HLA-G1	652
		30.11.3	Non-Classical MHC-I Molecule Qa-1b as Ligand	653
	30.12	Structure Analysis of CD94-NKG2 Complex		653
		30.12.1	Crystal Analysis of NKG2A/CD94: HLA-E Complex	653
		30.12.2	CD94-NKG2A Binding to HLA-E	654
	30.13	Inhibitory Receptors in Viral Infection		656
	30.14	Pathophysiological Role of CD94/NKG2 Complex		658
		30.14.1	Polymorphism in NKG2 Genes	658
		30.14.2	Phenotypes Associated with Leukemia	658
		30.14.3	CD94/NKG2A on NK Cells in T Cell Lymphomas	659
		30.14.4	CD94/NKG2 Subtypes on Lymphocytes in Melanoma Lesions	659
		30.14.5	Cancers of Female Reproductive Tract	660
		30.14.6	Disorders of Immune System	660
	References			661
31	**NKG2D Activating Receptor**			667
	Rajesh K. Gupta and G.S. Gupta			
	31.1	NKG2D Activating Receptor (CD314, Synonyms KLRK1)		667
	31.2	Characteristics of NKG2D		667
		31.2.1	The Protein	667
		31.2.2	Orthologues to Human NKG2D	668
	31.3	NKG2D Ligands		668
		31.3.1	The Diversity of NKG2D Ligands	668
		31.3.2	MHC Class I Chain Related (MIC) Proteins	669

	31.3.3	Retinoic Acid Early (RAE) Transcripts	669
	31.3.4	Role of NKG2D Ligands	671
	31.3.5	Regulation of Ligands	671
31.4	Crystal Structure of NKG2D		672
	31.4.1	Structures of NKG2D-Ligand Complexes	672
31.5	DAP10/12 Adapter Proteins		674
	31.5.1	DAP10	674
	31.5.2	KARAP (DAP12 or TYROBP)	674
	31.5.3	Characterization of DAP10 and DAP12	674
	31.5.4	NKG2D Receptor Complex and Signaling	675
31.6	Functions of NKG2D		677
	31.6.1	Engagement of NKG2D on γδ T Cells and Cytolytic Activity	677
	31.6.2	NKG2D: A Co-stimulatory Receptor for Naive $CD8^+$ T Cells	678
	31.6.3	NKG2D in Cytokine Production	678
	31.6.4	Heterogeneity of NK Cells in Umbilical Cord Blood	679
31.7	Cytotoxic Effector Function and Tumor Immune Surveillance		679
	31.7.1	Anti-Tumor Activity	679
	31.7.2	Immune Evasion Mechanisms	680
31.8	NKG2D in Immune Protection and Inflammatory Disorders		681
	31.8.1	NKG2D Response to HCMV	681
	31.8.2	HTLV-1-Associated Myelopathy	682
	31.8.3	Protection Against Bacteria	682
	31.8.4	Autoimmune Disorders	682
31.9	Decidual/Placental NK Cell Receptors in Pregnancy		683
31.10	Regulation of NKG2D Functions		684
	31.10.1	NKG2D Induction by Chronic Exposure to NKG2D Ligand	684
	31.10.2	Effects of Cytokines and Other Factors	685
31.11	Role in Immunotherapy		686
References			686

32 KLRC4, KLRG1, and Natural Cytotoxicity Receptors ... 693
Rajesh K. Gupta and G.S. Gupta

32.1	Killer Cell Lectin-Like Receptor F-1 (NKG2F/KLRC4/*CLEC5C*/NKp80)		693
	32.1.1	NKp80 or Killer Cell Lectin-Like Receptor Subfamily F-Member 1 (KLRF1)	693
	32.1.2	NKp80/KLRF1 Associates with DAP12	694
32.2	Activation-Induced C-Type Lectin (CLECSF2): Ligand for NKp80		694
32.3	KLRG1 (Rat MAFA/CLEC15A or Mouse 2F1-Ag)		695
	32.3.1	Mast Cell Function-Associated Antigen (MAFA)	695
	32.3.2	MAFA Is a Lectin	695
	32.3.3	Interactions of MAFA with FcεR	695
	32.3.4	MAFA Gene	696
32.4	KLRG1 (OR 2F1-AG): A Mouse Homologue of MAFA		696
	32.4.1	KLRG1: A Mouse Homologue of MAFA	696
	32.4.2	NK Cell Maturation and Homeostasis Is Linked to KLRG1 Up-Regulation	698
	32.4.3	Cadherins as Ligands of KLRG1	699
32.5	MAFA-Like Receptor in Human NK Cell (MAFA-L)		699
	32.5.1	Characteristics and Biological Properties	699
32.6	Killer Cell Lectin Like Receptor Subfamily E, Member 1 (KLRE1)		699
32.7	KLRL1 from Human and Mouse DCs		700

	32.8	Natural Cytotoxicity Receptors	701
		32.8.1 NK Cell Triggering: The Activating Receptors	701
		32.8.2 Activating Receptors and Their Ligands	701
	References		703

Part XI C-Type Lectins: Endocytic Receptors

33 Asialoglycoprotein Receptor and the Macrophage Galactose-Type Lectin ... 709
Anita Gupta

- 33.1 Asialoglycoprotein Receptor: The First Animal Lectin Discovered ... 709
- 33.2 Rat Asialoglycoprotein Receptor ... 709
  - 33.2.1 Characteristics ... 709
- 33.3 Human Asialoglycoprotein Receptor ... 710
  - 33.3.1 Structural Characteristics ... 710
- 33.4 Ligand Binding Properties of ASGP-R ... 711
  - 33.4.1 Interaction with Viruses ... 712
- 33.5 The Crystal Structure of H1-CRD ... 713
  - 33.5.1 The Sugar Binding Site of H1-CRD ... 713
  - 33.5.2 Sugar Binding to H1-CRD ... 714
- 33.6 Physiological Functions ... 715
  - 33.6.1 Impact of ASGP-R Deficiency on the Development of Liver Injury ... 715
- 33.7 ASGP-R: A Marker for Autoimmune Hepatitis and Liver Damage ... 715
  - 33.7.1 Autoimmune Hepatitis ... 715
  - 33.7.2 Hepatocellular Carcinoma ... 716
  - 33.7.3 Extra-Hepatic ASGP-R ... 716
- 33.8 ASGP-R: A Model Protein for Endocytosis ... 716
- 33.9 ASGP-R for Targeting Hepatocytes ... 717
- 33.10 Macrophage Galactose-Type Lectin (MGL) (CD301) ... 718
  - 33.10.1 Human MGL (CD 301) ... 718
  - 33.10.2 Murine MGL1/MGL2 (CD301) ... 718
  - 33.10.3 Ligands of MGL ... 719
  - 33.10.4 Functions of MGL ... 720
- References ... 721

34 Dectin-1 Receptor Family ... 725
Rajesh K. Gupta and G.S. Gupta

- 34.1 Natural Killer Gene Complex (NKC) ... 725
- 34.2 β-Glucan Receptor (Dectin 1) (CLEC7A or CLECSF12) ... 725
  - 34.2.1 Characterization of β-Glucan Receptor (Dectin 1) ... 725
  - 34.2.2 Crystal Structure of Dectin-1 ... 727
  - 34.2.3 Interactions of Dectin-1 with Natural or Synthetic Glucans ... 727
  - 34.2.4 Regulation of Dectin-1 ... 727
  - 34.2.5 Signaling Pathways by Dectin-1 ... 728
  - 34.2.6 Functions of Dectin-1 ... 729
  - 34.2.7 Genetic Polymorphism in Relation to Pathology ... 730
- 34.3 The C-Type Lectin-Like Protein-1 (CLEC-1) or CLEC-1A ... 731
  - 34.3.1 *CLEC-1* Gene ... 731
- 34.4 CLEC-18 or the C-Type Lectin-Like Protein-2 (CLEC-2) or CLEC1B ... 731
  - 34.4.1 Characterization ... 731
  - 34.4.2 Ligands for CLEC-2 ... 732
  - 34.4.3 Crystal Structure of CLEC-2 ... 733

		34.4.4	Functions of CLEC-2	733
		34.4.5	CLEC-2 Signaling	734
	34.5	CLEC9A (DNGR-1)		734
	34.6	Myeloid Inhibitory C-Type Lectin-Like Receptor (MICL/CLEC12A)		735
	34.7	CLEC12B or Macrophage Antigen H		735
	34.8	CLECSF7 (CD303)		735
	34.9	Lectin-Like Oxidized LDL Receptor (LOX-1) (CLEC8A)		736
		34.9.1	General Features of LOX-1	736
		34.9.2	LOX-1 Gene in Human	737
		34.9.3	Ligands for LOX-1	738
		34.9.4	Structural Analysis	738
		34.9.5	Functions of LOX-1	740
		34.9.6	LOX-1 and Pathophysiology	741
		34.9.7	Macrophage Differentiation to Foam Cells	742
	References			743
35	**Dendritic Cell Lectin Receptors (Dectin-2 Receptors Family)**			**749**
	Rajesh K. Gupta and G.S. Gupta			
	35.1	Dendritic Cells		749
	35.2	Dendritic Cell-Associated Lectins		750
		35.2.1	Type-I and Type-II Surface Lectins on DC	750
	35.3	Macrophage Mannose Receptor (CD206) on DC		750
		35.3.1	Expression and Characteristics	750
		35.3.2	Functions of Mannose Receptor in DC	752
	35.4	DEC-205		753
	35.5	Langerin: A C-Type Lectin on Langerhans Cells		753
		35.5.1	Human Langerin (CD207)	753
		35.5.2	Mouse Langerin: Homology to Human Langerin	754
		35.5.3	Ligands of Langarin	755
		35.5.4	Functions of Langerin	755
	35.6	DC-SIGN and DC-SIGNR on DCs		756
	35.7	Dectin-2 Cluster in Natural Killer Gene Complex (NKC)		757
		35.7.1	Natural Killer Gene Complex	757
		35.7.2	Antigen Presenting Lectin-Like Receptor Complex (APLEC)	759
	35.8	Dectin-2 (CLECF4N)		759
		35.8.1	Characteristics	759
		35.8.2	Ligands of Dectin-2	760
	35.9	The DC Immunoreceptor and DC-Immunoactivating Receptor		762
		35.9.1	Dendritic Cell Immunoreceptor (DCIR) (CLECsF6): Characterization	762
		35.9.2	Dendritic Cell Immunoactivating Receptor (DCAR)	764
	35.10	Macrophage-Inducible C-Type Lectin (Mincle)		764
		35.10.1	Recognition of Pathogens	765
		35.10.2	Recognition of Mycobacterial Glycolipid, Trehalose Dimycolate	765
	35.11	Blood Dendritic Cell Antigen-2 (BDCA-2) (CLEC-4C)		765
		35.11.1	BDCA-2: A Plasmacytoid DCs (PDCs)-Specific Lectin	765
		35.11.2	Characterization	765
		35.11.3	BDCA-2 Signals in PDC via a BCR-Like Signalosome	766
		35.11.4	Functions of BDCA-2	766
	35.12	CLECSF8		767
	35.13	Macrophage Galactose/N-acetylgalactosamine Lectin (MGL)		767
	References			767

36 DC-SIGN Family of Receptors ... 773
Rajesh K. Gupta and G.S. Gupta

- 36.1 DC-SIGN (CD209) Family of Receptors ... 773
  - 36.1.1 CD209 Family Genes in Sub-Human Primates ... 773
- 36.2 DC-SIGN (CD209): An Adhesion Molecule on Dendritic Cells ... 774
- 36.3 Ligands of DC-SIGN ... 775
  - 36.3.1 Carbohydrates as Ligands of DC-SIGN ... 775
- 36.4 Structure of DC-SIGN ... 775
  - 36.4.1 Neck-Domains ... 775
  - 36.4.2 Crystal Structure of DC-SIGN (CD209) and DC-SIGNR (CD299) ... 776
- 36.5 DC-SIGN versus DC-SIGN-RELATED RECEPTOR [DC-SIGNR or L-SIGN (CD 209)/LSEctin] (Refer Section 36.9) ... 777
  - 36.5.1 DC-SIGN Similarities with DC-SIGNR/L-SIGN/LSEctin ... 777
  - 36.5.2 Domain Organization of DC-SIGN and DC-SIGNR ... 778
  - 36.5.3 Differences Between DC-SIGN and DC-SIGNR/L-SIGN ... 778
  - 36.5.4 Recognition of Oligosaccharides by DC-SIGN and DC-SIGNR ... 779
  - 36.5.5 Extended Neck Regions of DC-SIGN and DC-SIGNR ... 780
  - 36.5.6 Signaling by DC-SIGN through Raf-1 ... 781
- 36.6 Functions of DC-SIGN ... 781
  - 36.6.1 DC-SIGN Supports Immune Response ... 781
  - 36.6.2 DC-SIGN Recognizes Pathogens ... 782
  - 36.6.3 DC-SIGN as Receptor for Viruses ... 783
  - 36.6.4 HIV-1 gp120 and Other Viral Envelope Glycoproteins ... 783
- 36.7 Subversion and Immune Escape Activities of DC-SIGN ... 785
  - 36.7.1 The entry and dissemination of viruses can be mediated by DC-SIGN ... 785
  - 36.7.2 DC-SIGN and Escape of Tumors ... 785
  - 36.7.3 *Mycobacterial* Carbohydrates as Ligands of DC-SIGN, L-SIGN and SIGNR1 ... 786
  - 36.7.4 Decreased Pathology of Human DC-SIGN Transgenic Mice During Mycobacterial Infection ... 787
  - 36.7.5 Genomic Polymorphism of DC-SIGN (CD209) and Consequences ... 787
- 36.8 SIGNR1 (CD209b): The Murine Homologues of DC-SIGN ... 789
  - 36.8.1 Characterization ... 789
  - 36.8.2 Functions ... 790
- 36.9 Liver and Lymph Node Sinusoidal Endothelial Cell C-Type Lectin (LSECtin) (or CLEC4G or L-SIGN or CD209L) ... 791
  - 36.9.1 Characterization and Localization ... 791
  - 36.9.2 Ligands of LSECtin ... 792
  - 36.9.3 Functions ... 792
  - 36.9.4 Role in Pathology ... 793
- References ... 793

Part XII C-Type Lectins: Proteoglycans

37 Lectican Protein Family 801
G.S. Gupta
- 37.1 Proteoglycans 801
  - 37.1.1 Nomenclature of PGs 801
  - 37.1.2 Glycosaminoglycans 802
  - 37.1.3 Chondroitin 803
  - 37.1.4 Heparan Sulfate Proteoglycans (HSPG) 805
- 37.2 Proteoglycans in Tissues 805
- 37.3 Hyaluronan: Proteoglycan Binding Link Proteins 807
- 37.4 Lecticans (Hyalectans) 807
- 37.5 Cartilage Proteoglycan: Aggrecan 808
  - 37.5.1 Skeletogenesis 808
  - 37.5.2 Human Aggrecan 808
  - 37.5.3 Rat Aggrecan 810
  - 37.5.4 Mouse Aggrecan 812
  - 37.5.5 Chick Aggrecan 812
- 37.6 Versican or Chondroitin Sulfate Proteoglycan Core Protein 2 812
  - 37.6.1 Expression 813
  - 37.6.2 Structure 814
  - 37.6.3 Interactions of Versican 814
  - 37.6.4 Functions 815
- 37.7 Pathologies Associated with PGS 816
  - 37.7.1 Proteoglycans Facilitate Lipid Accumulation in Arterial Wall 816
- 37.8 Diseases of Aggrecan Insufficiency 817
  - 37.8.1 Aggrecanase-Mediated Cartilage Degradation 817
  - 37.8.2 A Mutation in Aggrecan Gene Causes Spondyloepiphyseal Dysplasia 818
- 37.9 Expression of Proteoglycans in Carcinogenesis 818
- 37.10 Regulation of Proteoglycans 819
- 37.11 CD44: A Major Hyaluronan Receptor 819
  - 37.11.1 CD44: A Hyaluronan Receptor 819
  - 37.11.2 Hyaluronan Binding Sites in CD44 819
  - 37.11.3 CD44 in Cancer 820
- References 820

38 Proteoglycans of the Central Nervous System 825
G.S. Gupta
- 38.1 Proteoglycans in Central Nervous System 825
  - 38.1.1 Large Proteoglycans in Brain 825
  - 38.1.2 Ligands of Lecticans 826
- 38.2 Neurocan 828
  - 38.2.1 Cellular Sites of Synthesis 828
  - 38.2.2 Characterization 828
  - 38.2.3 Ligand Interactions 829
- 38.3 Brevican 830
  - 38.3.1 Characterization 830
  - 38.3.2 Murine Brevican Gene 830
- 38.4 Hyaluronan: Proteoglycan Binding Link Protein 831
  - 38.4.1 Cartilage link protein 1 and Brain Link Proteins 831
  - 38.4.2 Brain Enriched Hyaluronan Binding (BEHAB)/Brevican 832
- 38.5 Proteolytic Cleavage of Brevican 832
- 38.6 Proteoglycans in Sensory Organs 833

38.7	Functions of Proteoglycans in CNS	834
	38.7.1	Functions of Chondroitin Sulphate Proteoglycans	834
	38.7.2	Functions of Brevican and Neurocan in CNS	835
	38.7.3	Neurocan in Embryonic Chick Brain	836
38.8	Chondroitin Sulfate Proteoglycans in CNS Injury Response	836
	38.8.1	CS-PGs in CNS in Response to Injury	836
	38.8.2	Glial Scar and CNS Repair	838
38.9	Other Proteoglycans in CNS	838
38.10	Regulation of Proteoglycans in CNS	839
	38.10.1	Growth Factors and Cytokine Regulate CS-PGs by Astrocytes	839
	38.10.2	Decorin Suppresses PGs Expression	839
References	840

Part XIII C-Type Lectins: Emerging Groups of C-Type Lectins

39 Regenerating (Reg) Gene Family 847
G.S. Gupta

39.1	The Regenerating Gene Family	847
39.2	Regenerating (*Reg*) Gene Products	847
39.3	Reg 1	848
	39.3.1	Tissue Expression	848
	39.3.2	Gastric Mucosal Cells	849
	39.3.3	Ectopic Expression	849
	39.3.4	Reg Protein	850
	39.3.5	*REGIA (Reg Iα)* Gene in Human Pancreas	850
	39.3.6	REGIB (Reg Iβ) Gene in Human Pancreas	852
39.4	Pancreatic Stone Protein/Lithostathine (PSP/LIT)	852
	39.4.1	Characterization	852
	39.4.2	3D-Structure	853
	39.4.3	Secretory Forms of PSP	853
	39.4.4	Functions of PSP/LIT	854
39.5	Pancreatic Thread Protein	855
	39.5.1	Pancreatic Thread Proteins (PTPs)	855
	39.5.2	Neuronal Thread Proteins	856
	39.5.3	Pancreatic Proteins Form Fibrillar Structures upon Tryptic Activation	856
39.6	REG-II/REG-2/PAP-I	857
39.7	Reg-III	858
	39.7.1	Murine *Reg-IIIα, Reg-IIIβ, Reg-III γ*, and *Reg-IIIδ* Genes	858
	39.7.2	Human Reg-III or HIP/PAP	859
39.8	Rat PAPs	860
	39.8.1	Three Forms of PAP in Rat	860
	39.8.2	PAP-I/Reg-2 Protein (or HIP, p23)	861
	39.8.3	PAP-II/PAP2	861
	39.8.4	PAP-III	862
39.9	Functions of PAP	863
	39.9.1	PAP: A Multifunctional Protein	863
	39.9.2	PAP in Bacterial Aggregation	863
	39.9.3	PAP-I: An Anti-inflammatory Cytokine	863
39.10	Panceatitis Associated Protein (PAP)/Hepatocarcinoma-Intestine Pancreas (HIP)	865
	39.10.1	HIP Similarity to PAP	865
	39.10.2	Characterization of HIP/PAP	866

		39.10.3	PAP Ligands (PAP Interactions)	867

- 39.10.3 PAP Ligands (PAP Interactions) 867
- 39.10.4 Crystal Structue of human PAP (hPAP) 867
- 39.10.5 Expression of PAP 867
- 39.10.6 Similarity of PAP to Peptide 23 from Pituitary 869
- 39.10.7 Serum PAP: An Indicator of Pancreatic Function ... 870
- 39.11 Reg IV (RELP) 870
  - 39.11.1 Tissue Expression 870
- 39.12 Islet Neogenesis-Associated Protein (INGAP) in Hamster 871
  - 39.12.1 Characterization 871
  - 39.12.2 Properties of Pentadecapeptide from INGAP 872
- References 873

40 Emerging Groups of C-Type Lectins 881
G.S. Gupta
- 40.1 Layilin Group of C-Type Lectins (Group VIII of CTLDS) 881
  - 40.1.1 Layilin: A Hyaluronan Receptor 881
  - 40.1.2 Interactions of Layilin 881
  - 40.1.3 Functions of Layilin 881
  - 40.1.4 Chondrolectin (CHODL) 882
- 40.2 Tetranectin Group of Lectins (Group IX of CTLDs) 882
  - 40.2.1 Tetranectin 883
  - 40.2.2 Cell and Tissue Distribution 884
  - 40.2.3 Interactions of Tetranectin 884
  - 40.2.4 Crystal Structure of Tetranectin 885
  - 40.2.5 Functions of Tetranectin 886
  - 40.2.6 Association of TN with Diseases 887
  - 40.2.7 Cartilage-Derived C-Type Lectin (CLECSF1) 888
  - 40.2.8 Stem Cell Growth Factor (SCGF): Tetranectin Homologous Protein 888
- 40.3 Attractin Group of CTLDs (Group XI) 888
  - 40.3.1 Secreted and Membrane Attractins 888
  - 40.3.2 Attractin has Dipeptidyl Peptidase IV Activity? ... 889
  - 40.3.3 Attractin-Like Protein 889
  - 40.3.4 Functions of Attractin 890
  - 40.3.5 Genetic and Phenotypic Studies of *Mahogany/Attractin* Gene ... 892
  - 40.3.6 Therapeutic Applications of ATRN/*Mahogany* Gene Products ... 894
- 40.4 Eosinophil Major Basic Protein 1 (EMBP1) (Group XII of CTLD) ... 894
  - 40.4.1 Characterization of EMBP 894
  - 40.4.2 Functions of EMBP 895
  - 40.4.3 Crystal Structure of EMBP 895
  - 40.4.4 DEC-205-Associated C-Type Lectin (DCL)-1 895
- 40.5 Integral Membrane Protein, Deleted in Digeorge Syndrome (IDD) (Group XIII of CTLDs) 896
  - 40.5.1 DiGeorge Syndrome Critical Region 2 (DGCR2) 896
- References 896

41 Family of CD93 and Recently Discovered Groups of CTLDs 901
G.S. Gupta
- 41.1 Family of CD93 901
- 41.2 CD93 or C1q Receptor (C1qRp): A Receptor for Complement C1q 901
  - 41.2.1 CD93 Is Identical to C1qRp 901
  - 41.2.2 Characterization of CD93/C1qR$_P$ 902
  - 41.2.3 Two Types of Cell Surface C1q-Binding Proteins (C1qR) ... 902
  - 41.2.4 Tissue Expression and Regulation of CD93/C1qRp ... 903

	41.2.5	Functions	904
	41.2.6	CD93 in Pathogenesis of SLE	904
41.3	Murine Homologue of Human C1qRp (AA4)		905
41.4	The gC1qR (p33, p32, C1qBP, TAP)		906
	41.4.1	A Multi-Multifunctional Protein - Binding with C1q	906
	41.4.2	Human gC1qR/p32 (C1qBP) Gene	907
	41.4.3	Ligand Interactions	908
	41.4.4	Role in Pathology	909
41.5	Thrombomodulin		910
	41.5.1	Localization	910
	41.5.2	Characteristics	910
	41.5.3	Regulation of TM Activity	910
	41.5.4	Structure-Function Relations - Binding to Thrombin	911
	41.5.5	The Crystal Structure	913
	41.5.6	Functions	914
	41.5.7	Abnormalities Associated with Thrombomodulin Deficiency	916
41.6	Endosialin (Tumor Endothelial Marker-1 or CD248)		917
	41.6.1	A Marker of Tumor Endothelium	917
	41.6.2	Interactions with Ligands	918
	41.6.3	Functions	919
41.7	Bimlec (DEC 205 Associated C-type lectin-1 or DCL-1) or (CD302) (Group XV of CTLD)		920
41.8	Proteins Containing SCP, EGF, EGF and CTLD (SEEC) (Group XVI of CTLD)		920
	41.8.1	Nematocyst Outer Wall Antigen (NOWA)	921
	41.8.2	The Cysteine-Rich Secretory Proteins (CRISP) Super-Family	921
41.9	Calx-*b* and CTLD Containing Protein (CBCP) (Group XVII of CTLD)		922
	41.9.1	CBCP/Frem1/QBRICK	922
	41.9.2	Frem3	924
	41.9.3	Zebrafish Orthologues of FRAS1, FREM1, or FREM2	924
References			924

Part XIV Clinical Significance of Animal Lectins

42 MBL Deficiency as Risk of Infection and Autoimmunity 933
Anita Gupta

42.1	Pathogen Recognition		933
	42.1.1	MBL Characteristics	933
	42.1.2	Pathogen Recognition and Role in Innate Immunity	933
42.2	MBL Deficiency as Risk of Infection and Autoimmunity		935
	42.2.1	MBL Deficiency and Genotyping	935
	42.2.2	MBL and Viral Infections	938
42.3	Autoimmune and Inflammatory Diseases		939
	42.3.1	MBL Gene in Rheumatoid Arthritis	939
	42.3.2	Systemic Lupus Erythematosus	940
	42.3.3	Systemic Inflammatory Response Syndrome/Sepsis	941
	42.3.4	MBL and Inflammatory Bowel Diseases	942
	42.3.5	Rheumatic Heart Disease	943
	42.3.6	MBL in Cardio-Vascular Complications	944
	42.3.7	Other Inflammatory Disorders	945
42.4	Significance of MBL Gene in Transplantation		946
42.5	MBL in Tumorigenesis		947
	42.5.1	Polymorphisms in the Promoter	947
	42.5.2	MBL Binding with Tumor Cells	947

42.6	Complications Associated with Chemotherapy	947
	42.6.1 Neutropenia	947
	42.6.2 Animal Studies	948
42.7	MBL: A Reconstitution Therapy	949
References		949

43 Pulmonary Collectins in Diagnosis and Prevention of Lung Diseases ... 955
Anita Gupta

43.1	Pulmonary Surfactant	955
43.2	SP-A and SP-D in Interstitial Lung Disease	955
	43.2.1 Pneumonitis	955
	43.2.2 Interstitial Pneumonia (IP)	956
	43.2.3 ILD Due to Inhaled Substances	957
	43.2.4 Idiopathic Pulmonary Fibrosis	957
	43.2.5 Cystic Fibrosis	958
	43.2.6 Familial Interstitial Lung Disease	959
43.3	Connective Tissue Disorders	959
	43.3.1 Systemic Sclerosis	959
	43.3.2 Sarcoidosis	960
43.4	Pulmonary Alveolar Proteinosis	960
	43.4.1 Idiopathic Pulmonary Alveolar Proteinosis	960
	43.4.2 Structural Changes in SPs in PAP	961
43.5	Respiratory-Distress Syndrome and Acute Lung Injury	961
	43.5.1 ARDS and Acute Lung Injury	961
	43.5.2 Bronchopulmonary Dysplasia (BPD)	962
43.6	Chronic Obstructive Pulmonary Disease (COPD)	963
	43.6.1 COPD as a Group of Diseases	963
	43.6.2 Emphysema	964
	43.6.3 Allergic Disorders	964
	43.6.4 Interactions of SP-A and SP-D with Pathogens and Infectious Diseases	965
43.7	Pulmonary Tuberculosis	966
	43.7.1 Enhanced Phagocytosis of *M. tuberculosis* by SP-A	966
	43.7.2 SP-A Modulates Inflammatory Response in AΦs During Tuberculosis	967
	43.7.3 Marker Alleles in *M. tuberculosis*	968
	43.7.4 Interaction of SP-D with *M. tuberculosis*	968
	43.7.5 Association of SPs with Diabetes	969
43.8	Expression of SPs in Lung Cancer	969
	43.8.1 Non-Small-Cell Lung Carcinoma (NSCLC)	969
43.9	Other Inflammatory Disorders	971
	43.9.1 Airway Inflammation in Children with Tracheostomy	971
	43.9.2 Surfactant Proteins in Non-ILD Pulmonary Conditions	971
43.10	DNA Polymorphisms in SPs and Pulmonary Diseases	972
	43.10.1 Association Between SP-A Gene Polymorphisms and RDS	972
	43.10.2 SP-A and SP-B as Interactive Genetic Determinants of Neonatal RDS	973
	43.10.3 RDS in Premature Infants	973
	43.10.4 Gene Polymorphism in Patients of High-altitude Pulmonary Edema	973
	43.10.5 SNPs in Pulmonary Diseases	974
	43.10.6 Allergic Bronchopulmonary Aspergillosis and Chronic Cavitary Pulmonary Aspergillosis (CCPA)	974
	43.10.7 Autoreactivity Against SP-A and Rheumatoid Arthritis	975

	43.11	Inhibition of SP-A Function by Oxidation Intermediates of Nitrite......	975
		43.11.1 Protein Oxidation by Chronic Pulmonary Diseases..........	975
		43.11.2 Oxidation Intermediates of Nitrite......................	976
		43.11.3 BPD Treatment with Inhaled NO.....................	976
	43.12	Congenital Diaphragmatic Hernia................................	976
	43.13	Protective Effects of SP-A and SP-D on Transplants.................	977
	43.14	Therapeutic Effects of SP-A, SP-D and Their Chimeras..............	977
		43.14.1 SP-A Effects on Inflammation of Mite-sensitized Mice.......	977
		43.14.2 SP-D Increases Apoptosis in Eosinophils of Asthmatics......	977
		43.14.3 Targeting of Pathogens to Neutrophils Via Chimeric SP-D/Anti-CD89 Protein.............................	978
		43.14.4 Anti-IAV and Opsonic Activity of Multimerized Chimeras of rSP-D....................................	978
	43.15	Lessons from SP-A and SP-D Deficient Mice.....................	979
	References	....	980
44	**Selectins and Associated Adhesion Proteins in Inflammatory disorders**......		**991**
	G.S. Gupta		
	44.1	Inflammation...	991
	44.2	Cell Adhesion Molecules.....................................	991
		44.2.1 Selectins...................................	992
	44.3	Atherothrombosis..	992
		44.3.1 Venous Thrombosis..........................	992
		44.3.2 Arterial Thrombosis..........................	993
		44.3.3 Thrombogenesis in Atrial Fibrillation.....................	993
		44.3.4 Atherosclerosis..............................	993
		44.3.5 Myocardial Infarction..........................	996
		44.3.6 Atherosclerotic Ischemic Stroke........................	997
		44.3.7 Hypertension...............................	998
		44.3.8 Reperfusion Injury............................	998
	44.4	CAMS in Allergic Inflammation.................................	998
		44.4.1 Dermal Disorders............................	998
		44.4.2 Rhinitis and Nasal Polyposis......................	999
		44.4.3 Lung Injury................................	999
		44.4.4 Bronchial Asthma and Human Rhinovirus.................	999
	44.5	Autoimmune Diseases......................................	1000
		44.5.1 Endothelial Dysfunction in Diabetes (Type 1 Diabetes)........	1000
		44.5.2 Rheumatic Diseases..........................	1001
		44.5.3 Rheumatoid Arthritis..........................	1001
		44.5.4 Other Autoimmune Disorders....................	1002
	44.6	CAMs in System Related Disorders.............................	1003
		44.6.1 Gastric Diseases............................	1003
		44.6.2 Liver Diseases..............................	1004
		44.6.3 Neuro/Muscular Disorders......................	1004
		44.6.4 Acute Pancreatitis...........................	1005
		44.6.5 Renal Failure...............................	1005
		44.6.6 Other Inflammatory Disorders....................	1005
		44.6.7 Inflammation in Hereditary Diseases.....................	1006
	44.7	Role of CAMs in Cancer.....................................	1006
		44.7.1 Selectin Ligands in Cancer cells......................	1007
		44.7.2 E-Selectin-Induced Angiogenesis.....................	1007
		44.7.3 E-Selectin in Cancer Cells........................	1007
		44.7.4 Metastatic Spreading..........................	1009
		44.7.5 Survival Benefits of Heparin......................	1011

		44.8	Adhesion Proteins in Transplantation	1012

44.8 Adhesion Proteins in Transplantation.................1012
44.9 Inflammation During Infection.......................1013
    44.9.1 Microbial Pathogens..........................1013
    44.9.2 Yeasts and Fungi.............................1014
    44.9.3 Parasites and Amoeba.........................1014
44.10 Action of Drugs and Physical Factors on CAMS......1015
    44.10.1 Inhibitors of Gene Transcription............1015
    44.10.2 Anti-NF-kB Reagents.........................1015
    44.10.3 Strategies to Combat Atherogenesis and Venous Thrombosis..1016
    44.10.4 Anti-inflammatory Drugs.....................1017
References...1018

45 Polycystins and Autosomal Polycystic Kidney Disease.......1027
G.S. Gupta
45.1 Polycystic Kidney Disease Genes.....................1027
    45.1.1 Regulatory Elements in Promoter Regions......1028
45.2 Polycystins: The Products of PKD Genes..............1028
    45.2.1 Polycystins..................................1028
    45.2.2 Polycystin-1 (TRPP1) with a C-type Lectin Domain..1028
    45.2.3 Polycystin-2 (TRPP2).........................1030
    45.2.4 Interactions of Polycystins..................1031
    45.2.5 Tissue and Sub-Cellular Distribution of Polycystins..1031
45.3 Functions of Polycystin-1 and Polycystin-2..........1032
    45.3.1 Polycystin-1 and Polycystin-2 Function Together..1032
    45.3.2 Cell-Cell and Cell-Matrix Adhesion...........1032
    45.3.3 Role in Ciliary Signaling....................1033
    45.3.4 Cilia and Cell Cycle.........................1034
    45.3.5 Polycystins and Sperm Physiology.............1034
45.4 Autosomal Dominant Polycystic Kidney Disease.......1034
    45.4.1 Mutations in *PKD1* and *PKD2* and Association of Polycystic Kidney Disease..1034
    45.4.2 Proliferation and Branching Morphogenesis in Kidney Epithelial Cells..1035
References...1035

46 Endogenous Lectins as Drug Targets.......................1039
Rajesh K. Gupta and Anita Gupta
46.1 Targeting of Mannose-6-Phosphate Receptors and Applications in Human Diseases..1039
    46.1.1 Lysosomal Storage Diseases...................1039
    46.1.2 Enzyme Replacement Therapy (ERT).............1040
    46.1.3 M6PR-Mediated Transport Across Blood-Brain Barrier..1042
    46.1.4 Other Approaches using CI-MPR as Target......1045
46.2 Cell Targeting Based on Mannan-Lectin Interactions..1045
    46.2.1 Receptor-Mediated Uptake of Mannan-Coated Particles (Direct Targeting)..1046
    46.2.2 Polymeric Glyco-Conjugates as Carriers.......1046
    46.2.3 Mannosylated Liposomes in Gene Delivery......1047
    46.2.4 DC-Targeted Vaccines.........................1049
46.3 Asialoglycoprotein Receptor (ASGP-R) for Targeted Drug Delivery..1051
    46.3.1 Targeting Hepatocytes........................1051
46.4 Siglecs as Targets for Immunotherapy................1052

46.4.1　Anti-CD33-Antibody-Based Therapy of Human Leukemia......1052
　　46.4.2　CD22 Antibodies as Carrier of Drugs......1053
　　46.4.3　Immunogenic Peptides......1053
　　46.4.4　Blocking of CD33 Responses by SOCS3......1053
　References......1053

About the Author......1059

Index......1061

Contributors

G.S. Gupta Former Professor and Chairman, Department of Biophysics, Punjab University, Chandigarh, India

Anita Gupta Assistant Professor, Department of Biomedical Engineering, Rayat and Bahra Institute of Engineering and Biotechnology, Kharar (Mohali), Punjab, India

Rajesh K. Gupta General Manager, Vaccine Production Division, Panacea Biotec, Lalru (Mohali), Punjab, India

Abbreviation

AA4	murine homologue of human C1qRp
AAA	*Anguilla Anguilla* Agglutinin
AAA	abdominal aortic aneurysm
AAL	aleuria aurantia lectin
ABPA	allergic bronchopulmonary aspergillosis
ACS	acute coronary syndrome
Ad/AD	Alzheimer's disease
ADAMTS	a disintegrin and metalloproteinase with thrombo-spondin motifs
ADPKD	autosomal dominant polycystic kidney disease
AF	atrial fibrillation
AFP	antifreeze polypeptide
AGE	advanced glycation end product
AICL	activation-induced C-type lectin
ALI	acute lung injury
AM	alveolar macrophage
AM	adrenomedullin
AML	acute myeloid leukemia
AMφ	alveolar macrophage
AP	adaptor protein
APC	antigen-presenting cell
APLEC	antigen presenting lectin-like receptor complex
APR	acute-phase response
ARDS	acute-RDS
ARDS	adult respiratory distress syndrome
ASGPR	asialoglycoprotein receptor
ATM	ataxia telangiectasia, mutated
ATRA	all-trans retinoic acid
ATRNL	attractin-like protein
BAL	bronchoalveolar lavage
BALF	bronchoalveolar lavage fluid
BBB	blood-brain barrier
BDCA-1	blood dendritic cell antigen 1
BDNF	brain derived neurotrophic factor
BEHAB	brain enriched hyaluronan binding
BG	Birbeck Granule
BiP	immunoglobulin binding protein
BOS	bronchiolitis obliterans syndrome
BPD	bronchopulmonary dysplasia
BRAL-1	brain link protein-1
BRAP	BRCA1-associated protein
C1q	complement C1q module
C1r	complement C1r module

C4S	chondroitin-4-sulfate
C6S	chondroitin-6-sulfate
CAMs	cell adhesion molecules
CARD9	caspase activating recruitment domain 9
CBM	carbohydrate binding module
CC16	clara cell 16
CCSP	clara cell specific protein
CCV	clathrin coated vesicles
CD	Celiac disease
CD	Crohn's disease
CDH	congenital diaphragmatic hernia
CD-MPR	cation dependent M6P receptor
CEA	carcinoembryonic antigen (CEA)
CEACAM1	carcinoembryonic antigen (CEA)-related cell adhesion molecule 1
CF	cystic fibrosis
CFTR	cystic fibrosis transmembrane conductance regulator
CGN	cis-Golgi network
CHODL	chondrolectin
CI-MPR	cation-independent mannose 6-phosphate receptor
CLECSF2	activation-induced C-type lectin
CLL-1	C-type lectin-like molecule-1
CLRD	C-type lectin domain
CNS	central nervous system
Cnx/CNX	calnexin
COAD	chronic obstructive airways disease
COPD	chronic obstructive pulmonary disease
COPI	coatomer protein complex I
COPII	coatomer protein complex II
CR	complement protein regulatory repeat/ complement receptor
CRC	colorectal cancer
CRD	carbohydrate recognition domain
CREB	cAMP responsive element binding protein
CRP	complement regulatory protein domain
CRP	C-Reactive Protein
Crt/CRT	calreticulin
CS	chondroitin sulfate
CS-GAGs	chondroitin sulfate glycosaminoglycans
CSPG	chondroitin sulfate proteoglycan
CSPG2	chondroitin sulfate proteoglycan 2 or PG-M or versican
CTLD	C-type lectin-like domain
DC	dendritic cell
DC-SIGN	dendritic cell-specific ICAM-3-grabbing nonintegrin
DDR	discoidin domain receptor
Dex	dexamethasone
dGal-1	dimeric galectin-1
DN	down's syndrome
DOPG	dioleylphosphatidylglycerol
DPPC	dipalmitoylphosphatidyl choline
DSPC	disaturated phosphatidyl choline
EAE	experimental autoimmune encephalomyelitis
EBM	Epstein-Barr virus
ECM	extracellular matrix
Ed	embryonic day

EDEM	ER degradation enhancing α-mannosidase-like protein
EE	early endosomes
EGF	epidermal growth factor
ELAM-1	endothelial-leukocyte adhesion molecule 1
LECAM2	leukocyte-endothelial cell adhesion molecule 2
EMBP	eosinophil major basic protein
EMSA	electrophoretic mobility shift assay
eNOS	endothelial cell NO synthase
ER	endoplasmic reticulum
ERAD	ER-associated degradation
ERK	extracellular signal-regulated kinase
ERT	enzyme replacement therapy
ESP	early secretoty pathway
FA5	blood coagulation factor 5
FA8	blood coagulation factor 8
FAK	focal adhesion kinase
FBG	fibrinogen-like (FBG)-domains
FDl	fibrinogen-like domain 1
FGF	fibroblast growth factor
FN	fibronectin
FN2	fibronectin type II module
FN3	fibronectin type III module
Fuc	fucose
FV	blood coagulation factor V
FVIII	blood coagulation factor VIII
GAGs	glycosaminoglycans
Gal	galactose
Gal-1/Gal-3	galectin-1/galectin-3
Gb3	globotriaosylceramide
Gb4	globotetraosylceramide
GGA	Golgi-localizing, gamma-adaptin ear homology domain, ARF binding
GlyCAM-1	glycosylation-dependent cell adhesion molecule 1
GM1	ganglioside 1
GM2	ganglioside 2
GM-CSF	granulocyte-macrophage colony-stimulating factor
GMP-140	granule membrane protein 140
Grifin	galectin-related inter-fiber protein
GVHD	graft versus host disease
HAPLN2	hyaluronan and proteoglycan link protein 2
HBV	hepatitis B virus
HCMV	human cytomegalo virus
HCV	hepatitis C virus
HEV	high endothelial venule
HHL	human hepatic lectin
HHL1	human hepatic lectin subunit-1
HHL2	human hepatic lectin subunit-2
HNSCC	head and neck squamous cell carcinoma
HSPG	heparan sulfate proteoglycan
HT	Hashimoto's thyroiditis
HUVEC	human umbilical vein endothelial cell
i.m	intramuscular
i.p	intraperitoneal
i.v.	intravascular

IBD	inflammatory bowel disease
ICAM	intercellular adhesion molecule
IDDM	insulin-dependent diabetes mellitus
IFN	interferon
Ig	immunoglobulin
IGF-1	insulin-like growth factor, type 1
IGF2/MPR	insulin-like growth factor 2 receptor (IGF2R)
IGF2R	insulin-like growth factor 2 receptor
IGF-II/CIMPR	insulin-like growth factor-II or cation-independent mannose 6-phosphate receptor
IgSF	immunoglobulin superfamily
IL	interleukin
ILT	immunoglobulin-like transcripts
INTL	intelectin
IPCD	interstitial pneumonia with collagen vascular diseases
IPF	idiopathic pulmonary fibrosis
iRNA/RNAi	RNA interference
ITAM	immunoreceptor tyrosine-based activatory motif
ITIM	immunoreceptor tyrosine-based inhibitory motif
KIR	killer cell Ig-like receptors
LAMP-1/Lamp1	lysosome-associated membrane protein 1
LDL	low density lipoprotein
Le^a	Lewis A
Le^b	Lewis B
Le^x	lewis X
LFA-2	leukocyte function antigen-2
LFA-3	leukocyte function antigen-3
LFR	lactoferrin (LF) receptor (R)
LL	dileucine
LN	lymph node
LOX-1	lectin-type oxidized LDL receptor 1
LPS	lipopolysaccharide
M-6-P	mannose-6-phosphate
M6PR/MPR	Mannose-6-phosphate receptor
MAdCAM-1	mucosal addressin cell adhesion molecule-1 (or addressin)
ManLAMs	mannose-capped lipoarabinomannan
MAPK	mitogen-activated protein kinase
MASP	MBL-associated serine protease
MBL	mannan-binding lectin
MBP	mannose binding protein
MCP-1	monocyte chemoattractant protein 1
MHC	major histocompatibility complex
MICA/MIC-A	MHC class I polypeptide-related sequence A
MICB/MIC-B	MHC class I polypeptide-related sequence B
MMP	matrix metalloproteinases
MMR	macrophage mannose receptor
MPS	mucopolysaccharidosis
MR/ManR	mannose receptor
NCAM	neural cell adhesion molecule
NG2	neuronglia antigen 2
NK cell	natural killer cell
NK	natural killer
NKC	natural killer gene complex

n-LDL	native LDL	
NO	nitric oxide	
NOD mice	non-obese diabetic mice	
NPI	neuronal pentraxin I	
NTP	neuronal thread protein	
OLR1	oxidized low-density lipoprotein receptor 1	
Ox-LDL	oxidized low density lipoprotein	
PAMPs	pathogen-associated molecular patterns	
PAP	pulmonary alveolar proteinosis	
PB	peripheral blood	
PC	polycystin	
PD	Parkinson disease	
pDC/PDC	plasmacytoid dendritic cell	
PDI	protein disulfide isomerase	
PECAM-1	platelet-endothelial cell adhesion molecule-1	
PG	phosphatidylglycerol	
PGN	peptidoglycan	
PI3-kinase	phosphatidyl-inositol (PI)3-kinase	
PKB	protein kinase B	
PKC	protein kinase C	
PLC	phospholipase C	
PLN	peripheral lymph node	
PNN	perineuronal nets	
PPARγ	peroxisome proliferator-activated receptor γ	
PRR	pattern-recognition receptor	
PSA	prostate-specific antigen	
PSGL-1	P-selectin glycoprotein ligand1	
PTX	pentraxins	
R	review	
RAE-1/Rae-1	retinoic acid inducible gene-1	
RB	retinoblastoma protein	
RCC	renal cell carcinoma	
RCMV	rat cytomegalovirus	
RDS	respiratory distress syndrome	
Reg	regenerating genes	
RAET	retinoic acid early (RAE) transcript	
rER	rough ER	
RHL	rat hepatic lectin	
RNAi	RNA interference	
ROS	reactive oxygen species	
RSV	respiratory syncytial virus/ respiratory syndrome virus	
SAA	serum amyloid A	
SAP	serum amyloid P component	
SARS	severe acute respiratory syndrome	
SC	scavenger receptor module	
SCLC	small cell lung carcinoma	
SCR	short consensus repeat	
s-diLe^x	sulfated polysaccharide ligands	
Siglec	sialic-acid-binding immunoglobulin-like lectin	
Sjs	Sjogren syndrome	
SH1/2	Src homology 1/2	
SHP-1/2	Src homology 1/2 containing phosphatase	
SLAM	signaling lymphocyte activated molecule	

SLE	systemic lupus erythematosus
sLe^a/s-Le^a	Sialyl Lewis A
sLe^x/s-Le^x	Sialyl Lewis X
Sn	sialoadhesin
SNP	single-nucleotide polymorphism
snRNPs	small nuclear ribonucleoproteins
SOCS3	suppressor of cytokine signaling 3
SP-A	surfactant protein A
SP-D	surfactant protein D
ST	sialyl transferase
STAT	signal transducers and activators of transcription
STAT1	signal transducer and activator of transcription 1
T1DM	type 1 diabetes mellitus
T2DM	type 2 diabetes mellitus
TGF-β	transforming growth factor β
TGN	trans-Golgi network
TIM	triosephosphate isomerase
TLR	toll-like receptor
TNF	tumor necrosis factor
Ub	ubiquitin
UDPG	uridine diphosphate (UDP)-glucose
UL16	unique long 16
ULBP	UL16 binding protein
uPAR	uPA receptor or urokinase receptor
uPARAP	urokinase receptor-associated protein
UPR	unfolded protein response
VCAM-1	vascular adhesion molecule-1
VEGF	vascular endothelial growth factor
VLA	very late antigen
VN	vitronectin
VSMC	vascular smooth muscle cell
vWF	von Willebrand factor
β2-m	β2-microglobulin

Part VIII

C-Type Lectins: Collectins

C-Type Lectins Family

Anita Gupta and G.S. Gupta

22.1 C-Type Lectins Family

22.1.1 The C-Type Lectins (CLEC)

Historically the C-Type lectins (CLEC) or C-type lectin receptors (CLR) form a family of Ca^{2+} dependent carbohydrate binding proteins which have a common sequence motif of 115–130 amino acid residues, referred to as the carbohydrate recognition domain (CRD). The C-type designation is from their requirement for calcium for binding. The C-type lectin superfamily is a large group of proteins which is characterized as having at least one carbohydrate recognition domain (CRD), which has been found in more than 1,000 proteins, and it represents a ligand-binding motif that is not necessarily restricted to binding sugars (Drickamer 1999). Proteins that contain C-type lectin domains have a diverse range of functions including cell-cell adhesion, immune response to pathogens and apoptosis. However, many C-type lectins actually lack calcium- and carbohydrate-binding elements and thereby have been termed C-type lectin-like proteins. The CRD has four cysteines that are perfectly conserved and involved in two disulfide bonds. The CRD has been found in various kinds of proteins such as hepatic asialoglycoprotein receptor, lymphocyte IgE receptor, mannose binding protein, selectins (Drickamer 1999; Lasky et al. 1989; Zeng and Weigel 1996) and proteoglycan core protein. Their functions include complement activation, endocytosis, cell recognition, defense mechanism, and morphogenesis (Drickamer 1999; Weis et al. 1998). Many evolutionarily related sequences belonging to the C-type lectins do not show overall sequence similarity. However, they have been classified on the basis of four cysteine residues, being involved in disulfide bridging, which are the trademark of this domain type. The framework surrounding this domain type can be very different from its C-terminal location in the collectins. The monomeric and membrane bound selectins have an N-terminal C-type lectin domain that mediates adhesion between certain cell types through carbohydrate binding. Other proteins, such as phospholipase A2 receptor and the macrophage mannose receptor, contain multiple copies of C-type lectin domains within a single polypeptide. Proteins possessing C-type lectin-like domain (CTLD) type are not necessarily lectins. Type II antifreeze protein present in some arctic fish and the mammalian pancreatic stone protein bind to and inhibit the growth of ice and calcium carbonate crystals, respectively (Håkansson and Reid 2000) and hence are not exactly CLEC but form a part of CTLD.

C-type lectins are either produced as transmembrane proteins or secreted as soluble proteins. Examples of soluble C-type lectins include members of the collectins family, such as the lung surfactant proteins A (SP-A) and SP-D (Wintergest et al. 1989), which are secreted at the luminal surface of pulmonary epithelial cells, and the mannose-binding protein (MBP), a collectin present in the plasma (Kawasaki et al. 1983). Transmembrane C-type lectins can be divided into two groups, depending on the orientation of their amino (N)-terminus. These are type I and type II C-type lectins depending on their N-terminus pointing outwards or inwards into the cytoplasm of the cell respectively. Examples of transmembrane C-type lectins are the selectins (Ley and Kansas 2004), the mannose receptor (MMR) family (East and Isacke 2002), and the dendritic cell-specific ICAM-3 grabbing non-integrin (DC-SIGN) (Geijtenbeek et al. 2002).

22.1.2 C-Type Lectin Like Domain (CLRD/CTLD)

Use of terms of "C-type lectin", "C-type lectin domain" (CLRD), "C-type lectin-like domain" (also abbreviated as CLRD), often used interchangeably and use of CRD in the literature, have been clarified by Zelensky and Gready (2005). With the large number of CLR sequences and structures now available, studies indicate that the implications of the CRD domain are broad and vary widely in function. In metazoans, most proteins with a CLR are not lectins. Moreover, proteins use the C-type lectin fold to bind other proteins, lipids, inorganic molecules (e.g., Ca_2CO_3), or

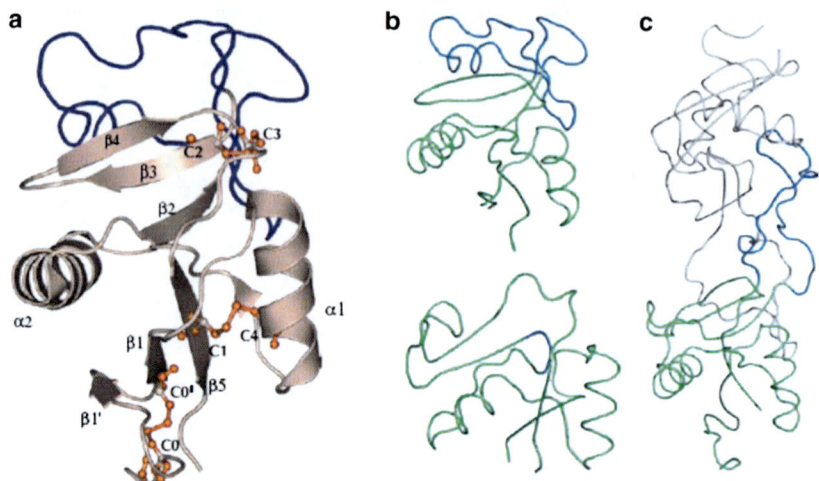

Fig. 22.1 CTLD structure. (a) Cartoon representation of a typical CTLD structure (PDB 1K9I). The long loop region is shown in *blue*. Cystine bridges are shown as *orange sticks*. The cystine bridge specific for long form CTLDs (C0-C0') is also shown. (b), (c), and (d): **Variation of the long loop region structure**. Three common forms of the CTLD long loop region are shown. Panels (B: PDB 1K9I) and (C: PDB 1UUH) show canonical CTLDs in which long loop region is tightly packed (b) or flipped out to form a domain-swapping dimer (D: PDB 1C3A). A compact CTLD from human CD44 Link domain is shown in panel (c). The core domain and long loop region are colored *green and blue*, respectively (Reviewed by Zelensky and Gready 2005)

even ice (e.g., the antifreeze glycoproteins). An increasing number of studies show that "atypical" C-type lectin-like proteins are involved in regulatory processes pertaining to various aspects of the immune system. Examples include the NK cell inhibitory receptor Ly49A, a C-type lectin-like protein, which is shown to complex with the MHC class I ligand (Correa and Raulet 1995), and the C-type lectin-like protein mast cell function-associated antigen which is involved in the inhibition of IgE-Fc γRI mediated degranulation of mast cell granules (Guthmann et al. 1995). Yet glycan binding by C-type lectins is always Ca^{2+}-dependent because of specific amino acid residues that coordinate Ca^{2+} and bind the hydroxyl groups of sugars (Cummings and McEver 2009). To resolve the contradiction, a more general term C-type lectin like domain (CLRD) was introduced to distinguish a group of Ca^{2+}-independent carbohydrate-binding animal proteins from the Ca^{2+}-dependent C-type of animal lectins (CLR/CLEC). The usage of this term is however, somewhat ambiguous, as it is used both as a general name for the group of domains with sequence similarity to C-type lectin CRDs (regardless of the carbohydrate-binding properties), and as a name of the subset of such domains that do not bind carbohydrates, with the subset that does bind carbohydrates being called C-type CRDs. Also both 'C-type CRD' and 'C-type lectin domain' terms are still being used in relation to the C-type lectin homologues that do not bind carbohydrate, and the group of proteins containing the domain is still often called the 'C-type lectin family' or 'C-type lectins', although most of them are not in fact lectins. The abbreviation CRD is used in a more general meaning of 'carbohydrate-recognition domain', which encompasses domains from different lectin groups. Occasionally CRD is also used to designate the short amino-acid motifs (i.e. amino-acid domain) within CLRDs that directly interact with Ca^{2+} and carbohydrate (Zelensky and Gready 2005). In this book authors will use the term C-type lectin domain or C-type lectin-like domain (CTLD/CLRD) interchangeably in its broadest definition to refer to protein domains that are homologous to the CRDs of the C-type lectins, or which have structure resembling the structure of the prototypic C-type lectin CRD or as used by different researchers in their work. More over, due to contradictions (Zelensky and Gready 2005) and uncertainties which may arise in future studies our sequence of chapters is not based on structure databases as in the SCOP; instead chapters are linked more to cell functions or cell biology.

22.1.3 The CLRD Fold

The CLRD fold has a double-loop structure (Fig. 22.1a). The overall domain is a loop, with its N- and C-terminal β strands (β1, β5) coming close together to form an antiparallel β-sheet. The second loop, which is called the long loop region, lies within the domain; it enters and exits the core domain at the same location. Four cysteines (C1–C4), which are the most conserved CLRD residues, form disulfide bridges at the bases of the loops: C1 and C4 link β5 and α1 (the whole domain loop) and C2 and C3 link β3 and β5 (the long loop region). The rest of the chain forms two flanking α helices (α1 and α2) and the second ('top') β-sheet, formed by strands β2, β3 and β1. The long loop region is involved in

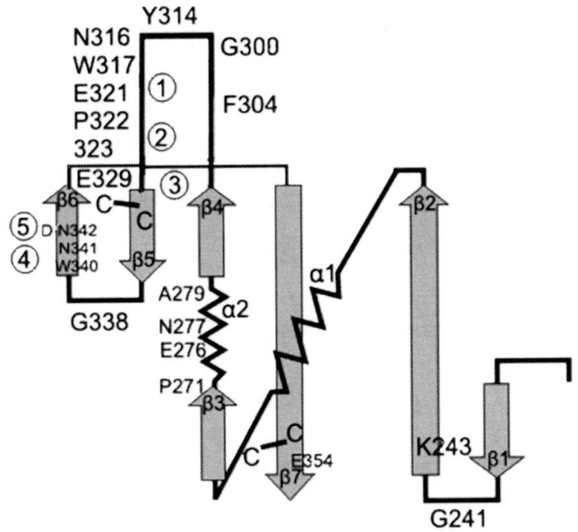

Fig. 22.2 Two-dimensional representation of the C-type lectin fold. The β-strands and α-helices have been numbered β1–β7 and α1–α2, respectively. Secondary structure elements comprise the following residues: (SP-D numbering); α1, 254–264; α2, 273–288; β1, 236–240; β2, 243–253; β3, 267–269; β4, 291–297; β5, 331–335; β6, 339–343; β7, 347–355. The amino acids that are coordinated to the carbohydrate binding calcium ion are denoted with *encircled numbers 1–5*. The conserved residues are indicated with their amino acid one letter symbol. The second calcium ligand is normally an asparagine but is an arginine or alanine in the SP-A sequences. The fifth calcium ligand is an aspartic acid residue in all sequences except dog SP-A, where it is an asparagine (Adapted by permission from Håkansson and Reid 2000 © John Wiley and Sons)

Ca^{2+}-dependent carbohydrate binding, and in domain-swapping dimerization of some CLRDs (Fig. 22.1a), which occurs via a unique mechanism (Mizuno et al. 1997; Feinberg et al. 2000; Mizuno et al. 2001; Hirotsu et al. 2001; Liu and Eisenberg 2002). As suggsted, the CLRD structure is mainly composed of two antiparallel β-sheets; one of these is four-stranded and the other five stranded. The four-stranded β-sheet is found at the N-terminal part of the domain and is flanked by two helices. In trimeric collectins, this sheet is interacting with the α-helical coiled-coil. The five-stranded β-sheet is rather distorted and more remotely located away from the trimer center. This β-sheet, together with some of the loop structure, makes up the carbohydrate binding ligand site. Two of the β-strands (β2 and β7) are relatively long and participate in both sheets. One of the disulfide bridges anchors helix α1 to strand β7 of the four-stranded β-sheet, the other ties together the two most peripheral strands (β5 and β6) of the five-stranded β-sheet. Since the calcium binding amino acid ligands are found at or close to these two strands, this disulfide bridge probably plays an important role in stabilizing the structure around the functional carbohydrate binding site (Fig. 22.2). For conserved positions involved in CLRD fold maintenance and their structural roles readers are referred elsewhere (Zelensky and Gready 2003). In addition to four conserved cysteines, one other sequence feature and the highly conserved 'WIGL' motif is located on the β2 strand and serves as a useful landmark for sequence analysis.

The crystal structure of the CRD of a rat mannose-binding protein, determined as the holmium-(Ho^{3+}) substituted complex by multi-wavelength anomalous dispersion (MAD) phasing, reveals an unusual fold consisting of two distinct regions, one of which contains extensive nonregular secondary structure stabilized by two holmium ions. The structure explains the conservation of 32 residues in all C-type CRDs, suggesting that the fold seen is common to these domains (Weis et al. 1991). Structurally, CLRDs can be divided into two groups: canonical CLRDs having a long loop region, and compact CLRDs that lack it. The second group includes Link or protein tandem repeat (PTR) domains (Brissett and Perkins 1996; Kohda et al. 1996) and bacterial CLRDs (Hamburger et al. 1999; Kelly et al. 1999). Link domain or PTR is a special variety of CLRD, which lacks the long loop region. The major function of Link domains is binding hyaluronan. Although proteins containing it have different domain architecture, their number is small, and they have not been divided into subgroups. Group I CLRDcps contain both canonical and Link-type CLRDs. Other Link-domain containing proteins have four types of domain architecture. Domain composition of Link proteins is similar to that of the N-terminal part of Group I (lecticans). Another family usually included in the CLRD superfamily is that of endostatin (Hohenester et al. 1998). However, endostatin fold is not a suitable example of a CLRD and hence not to be considered further.

The 3D structure of the lectin domain is known to atomic resolution from X-ray analyses of rat and human mannan-binding protein fragments (Weis et al. 1991; Sheriff et al. 1994; Weis and Drickamer 1994), a human lung surfactant protein D fragment (Håkansson et al. 1999), a human E-selectin fragment (Graves et al. 1994), pancreatic stone protein (Bertrand et al. 1996), tetranectin (Nielsen et al. 1997), snake venom factor IX0X binding protein (Mizuno et al. 1997), tunicate C-type lectin (Poget et al. 1999), and from an NMR study of sea raven antifreeze protein (Gronwald et al. 1998). The fold of the C-type lectin domain in some the C-type lectins is shown in Fig. 22.3.

22.1.3.1 Naming of Secondary Structure Elements

Although the CTLD fold is well conserved among its known representatives, there is no general agreement on the numbering of CTLD secondary structural elements in the literature. The secondary structure element numbering scheme in first solved CTLD structure (rat MBP-A by Weis et al. 1991) included five strands, two helices and four loops. However, this description turned out to be insufficient, as MBP lacks some secondary structure elements that are present in long-form CTLD structures, while other small strands were not defined. Other reports describing the structures of CTLDs that have a different number of secondary structure

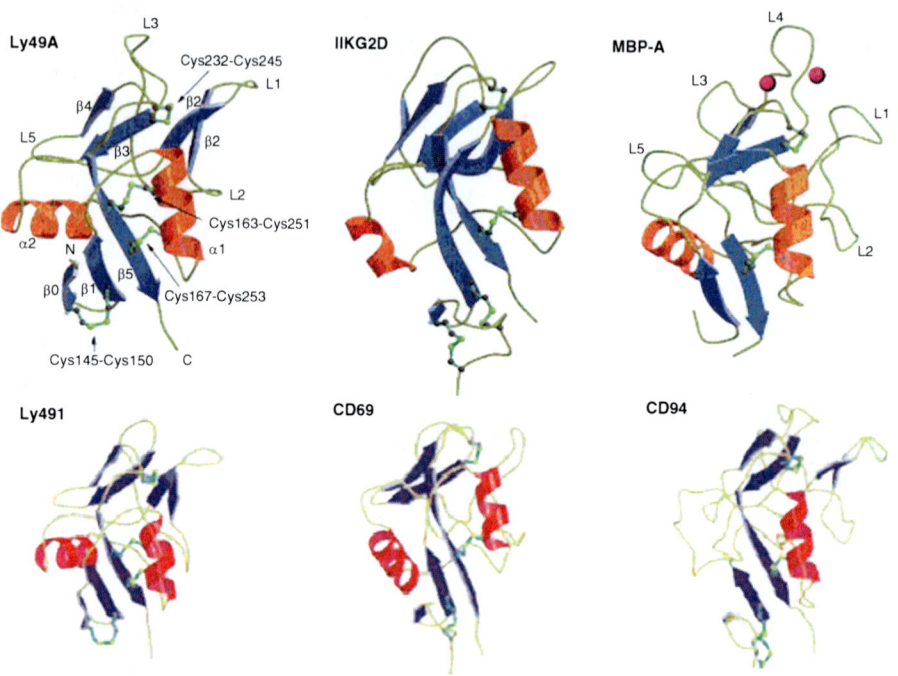

Fig. 22.3 Anatomy of some of the C-type lectin-like domains. Ribbon diagrams of Ly49A (PDB: ID 1Q03) (Tormo et al. 1999), Ly49I (1JA3) (Dimasi et al. 2002), NKG2D (1HQ8) (Wolan et al. 2001), MBP-A (1BCH) (Weis et al. 1991), CD94 (1B6E) (Boyington et al. 1999) and CD69 (1FM5) (Natarajan et al. 2000). The secondary structural elements are colored as follows: β strands *blue*, α-helices *red*, and loop regions *gold*. The disulphide bonds are shown in *green* as ball-and-stick representation. The Ca^{2+} ions bound to MBP-A are drawn as *magenta spheres* (Reviewed in Natarajan et al. 2002)

elements than MBP-A either introduced their own numbering: β strands 1–6 in asialoglycoprotein receptor (ASGPR) by Meier et al. (2000); six β strands in Link module, with labeling not consistent with ASGPR or MBP-A (Kohda et al. 1996); β1–β7 in NKG2D (Wolan et al. 2001); β1–β8 in EMBP (Swaminathan et al. 2001), or extended the secondary structure element naming scheme used for MBP-A (Ly49A secondary structure element numbering is consistent with that in MBP-A) (Tormo et al. 1999).

22.1.3.2 Ca^{2+}-Binding Sites in CLRD/CTLD

Four Ca^{2+}-binding sites recur in CTLD structures from different groups (Chap. 1). The site occupancy depends on the particular CTLD sequence and on the crystallization conditions; in different known structures zero, one, two or three sites are occupied. Sites 1, 2 and 3 are located in the upper lobe of the structure, while site 4 is involved in salt bridge formation between α2 and the β1/β5 sheet. Sites 1 and 2 were observed in the structure of rat MBP-A complexed with holmium, which was the first CTLD structure determined. Site 3 was first observed in the MBP-A complex with Ca^{2+} and oligomannose asparaginyl-oligosaccharide. It is located very close to site 1 and all the side chains coordinating Ca^{2+} in site 3 are involved in site 1 formation. Since biochemical data indicate that MBP-A binds only two calcium atoms (Loeb and Drickamer 1988), Ca^{2+}-binding site 3 is considered a crystallographic artifact.

However, in many CTLD structures where site 1 is occupied, a metal ion is also found in site 3; examples include the structures of DC-SIGN and DC-SIGNR (Feinberg et al. 2001), invertebrate C-type lectin CEL-I (Sugawara et al. 2004), lung surfactant protein D (Hakansson et al. 1999) and the CTLD of rat aggrecan (Lundell et al. 2004). It is interesting to note that molecular dynamics simulations of the MBP-A/mannose complex suggested that Ca^{2+} binding site 3 is involved in the binding interaction (Harte and Bajorath 1994; Zelensky and Gready 2005). Residues with carbonyl side chains involved in Ca^{2+} coordination in site 2 form two characteristic motifs in the CTLD sequence, and together with the calcium atom are directly involved in monosaccharide binding. Despite their spatial proximity, from the evolutionary and structural points of view Ca^{2+}-binding sites 1 and 2 should be considered as independent. Crystallographic studies of rat MBP-A CTLD crystallized at a low metal ion concentration (0.325 mM Ho^{3+} instead of 20 mM as used to obtain the CTLD complexed with mannose) have shown that site 1 has higher affinity for Ca^{2+} as it remains occupied and Ca^{2+}-coordination geometry is retained while site 2 loses its metal ion (Ng et al. 1998). Unlike many other functions of the CTLD, Ca^{2+}-dependent carbohydrate binding is found across the whole phylogenetic distribution of the family, from sponges to human, and thus is likely to be the ancestral function (Weis and Drickamer 1996).

22.1.4 Ligand Binding

CTLDs selectively bind a wide variety of ligands. As the superfamily name suggests, carbohydrates are primary ligands for CTLDs and the binding is Ca^{2+}-dependent (Weis and Drickamer 1996). However, the fold has been shown to specifically bind proteins (Natarajan et al. 2002), lipids (Sano et al. 1998) and inorganic compounds including $CaCO_3$ and ice. In several cases the domain is multivalent and may bind both protein and sugar (Zelensky and Gready 2005). Carbohydrate binding is, however, a fundamental function of the superfamily and the best studied one. The first characterized vertebrate CTLDs were Ca^{2+}-dependent lectins, and most of the functionally characterized CTLDs from lower organisms were isolated because of their sugar-binding activity. Although as the number of CTLD sequences grows it becomes clearer that the majority of them do not possess lectin properties, CTLDs are still regarded as a lectin family (Zelensky and Gready 2004).

Four Link protein-encoding genes have been identified in mammals, each physically linked with one of the lectican genes; this suggests that lecticans and Link proteins are of a common evolutionary origin (Spicer et al. 2003). Link proteins and lecticans are also functionally associated: cartilage Link proteins bind both aggrecan and hyaluronan stabilizing the proteoglycan/glycosaminoglycan network (Faltz et al. 1979). CD44 and its recently identified close homologue Lyve-1 are type I transmembrane molecules and cell surface receptors for hyaluronan (Banerji et al. 1999). Tumor necrosis factor-inducible protein (TSG-6) is a soluble protein with a CUB and a Link domain (Lee et al. 1992; Wisniewski and Vilcek 1997). The structure of the latter in free and hyaluronan-bound states has been determined by NMR (Blundell et al. 2003). Stabilin-1 and -2 (also known as FEEL-1/-2) (Tamura et al. 2003; Adachi and Tsujimoto 2002) are scavenger receptors which have the capacity to internalize conventional scavenger ligands such as low density lipoprotein, bacteria and advanced glycation end products. Unlike Stabilin-2, Stabilin-1 does not bind hyaluronan or other glycosaminoglycan (Prevo et al. 2004).

22.1.5 C-Type Lectin-Like Domains in Model Organisms

An important approach to understanding the roles of mammalian carbohydrate-binding proteins is to examine the functions of orthologs in simpler organisms. Because of the advanced state of the developmental, genetic and genomic study of *Caenorhabditis elegans* and *Drosophila melanogaster*, these organisms are particularly attractive models. Homologs of six classes of lectins have been identified using methods of sequence comparisons combined with knowledge of how these proteins interact with glycan ligands.

With regard to C-type lectin-like domains, relative to genome size there are far more CTLDs in the model invertebrates than in mammals. However, the *C elegans* and *Drosophila* proteins that contain CTLDs are quite distinct from the mammalian proteins that contain CTLDs. These differences are evident in the sequences of the CTLDs and the domain organization of the proteins in which they are found. Only a small subset of invertebrate CTLDs contain the constellation of amino acid residues that form Ca^{2+} and sugar-binding sites in mammalian C-type carbohydrate-recognition domains (Drickamer and Dodd 1999; Dodd and Drickamer 2001).

22.1.5.1 Evolution of the C-Type Lectin-Like Domain

The CTLDs of higher eukaryotes are protein modules originally identified as carbohydrate-recognition domains (CRDs) in a family of Ca^{2+}-dependent animal lectins. Less closely related but still definitely homologous CTLDs have been identified in a variety of proteins that do not appear to have carbohydrate-binding activity. All of the domains in the CTLD group show distinct evidence of sequence similarity and are thus believed to have descended from a common ancestor by a process of divergent evolution.

22.2 Classification of CLRD/CTLD-Containing Proteins

The C-type lectin superfamily is a large group of proteins that are characterized by the presence of one or more CLRDs. They can be divided into 17 subgroups based on additional non-lectin domains and gene structure, the two independent sets of criteria. The two approaches give essentially the same results, indicating that members of each group are derived from a common ancestor, which had already acquired the domain architecture that is characteristic of the group. Initially classified into seven subgroups (I to VII) based on the order of various protein domains in each protein: (I) Lecticans or hyalectans, (II) asialoglycoprotein receptors, (III) collectins, (IV) selectins, (V) NK cell receptors, (VI) endocytic receptors, and (VII) lectins as product of regenerating genes (*Reg*) (Drickamer 1993; McGreal et al. 2004). This classification was subsequently updated in 2002, leading to seven additional groups (VIII to XIV) (Drickamer and Fadden 2002) based on structural organization. A further three subgroups were added (XV to XVII) by Zelensky and Gready (2005). More than 100 different proteins encoded in the human genome contain the CTLD. Most of these groups have a single CTLD, but the macrophage mannose receptor (group VI) has eight CTLDs. The domain architecture of the CTLDcps in different groups is shown in Fig. 22.4. The monomeric and membrane bound selectins have an N-terminal C-type lectin domain that mediates adhesion between certain cell types through carbohydrate binding. Groups VII (REG), IX (tetranectin), XI (attractin), XIII (DGCR2; DiGeorge syndrome critical region gene 2), XV (BIMLEC), and XVII (CBCP) have no known glycan ligands. The section of animal lectins genomics resource includes information on the structure and function of proteins in each group, as well as annotated sequence alignments, and a comprehensive database for all human and mouse CTLD-containing proteins (Table 22.1).

Despite the presence of a highly conserved domain, C-type lectins are functionally diverse and have been implicated in various processes including cell adhesion, tissue integration and remodeling, platelet activation, complement activation, pathogen recognition, endocytosis (Table 22.1). From a functional perspective, we know most about collectins, endocytic receptors, myeloid lectins, and selectins, and these groups are discussed in detail in chapters to follow. However, proteins possessing CTLDs type are not necessarily lectins.

22.3 Disulfide Bonds in Lectins and Secondary Structure

22.3.1 Arrangement of Disulfide Bonds in CTLDs

Drickamer and Dodd (1999) summarized positions of six different disulfide bonds in CTLDs (Fig. 22.5). Chemical evidence for the presence of each of these bonds, except number 4, has been provided in at least one CTLD (Fuhlendorff et al. 1987; Usami et al. 1993). The positions of disulfide bonds designated 1, 2, and 3 have been demonstrated by x-ray crystallography as well (Weis et al. 1991; Nielsen et al. 1997), while homology modeling of CTLDs containing disulfide bonds 5 and 7 shows that they could readily be accommodated into the C-type lectin fold. The patterns of cysteine residues in the CTLDs from *C. elegans* are consistent with the presence of disulfide bonds in each of the arrangements shown in Fig. 22.5 except for bond type 4. No additional pairs of cysteine residues within the CTLDs are consistently evident for any of the subgroups, indicating that the cysteine residues are mostly involved in disulfide bonds of the types already characterized in vertebrate homologues.

CTLDs lacking one of a pair of cysteine residues almost invariably also lack the cysteine side chain to which the first residue would be linked. Like the CTLDs from other organisms, those from *C.elegans* each contain a subset of the possible disulfide bonds. CTLDs in a given subgroup generally show the same disulfide bonds, although a few domains contain extra unique pairs of cysteine residues that might form disulfides. Thus, the similarity in disulfide bond structure in each subgroup reflects the overall similarity in sequence of the CTLDs (Drickamer and Dodd 1999). CEL-I from the sea cucumber, *Cucumaria echinata* is composed of two identical subunits held by a single disulfide bond. A subunit of CEL-I is composed of 140 amino acid residues. Two intrachain (Cys3–Cys14 and Cys31–Cys135) and one interchain (Cys36) disulfide bonds were also identified from an analysis of the cystine-containing peptides obtained from the intact protein (Hatakeyama et al. 2002; Yamanishi et al. 2007).

22.3.2 Functional Role of Disulfides in CTLD of Vertebrates

Single cysteine residues appear in several of these groups at positions 7 and 8 as well as at a unique position 9. The turn between β-strands 3 and 4 is exposed on the surface of the domain, so it is expected that cysteine residues at position 9 would be accessible for formation of disulfide bonds. It is possible that such bonds could form with other cysteine residues within the same polypeptide but outside the CTLDs. However, no likely pairing partner is evident for any of these residues, suggesting that they are more likely to form interchain disulfide bonds. Homo- and hetero-dimer formation through cysteine residues at positions 7 and 8 has been particularly well documented in snake venom proteins containing CTLDs (Usami et al. 1993). The botrocetin, which promotes platelet agglutination in the presence of von Willebrand factor, from venom of the snake Bothrops jararaca is a heterodimer composed of the α subunit and the β subunit held together by a disulfide bond. Seven disulfide bonds link half-cystine residues 2–13, 30–128, and 103–120 of the α subunit; 2–13, 30–121, and 98–113 of the β subunit; and 80 of the α subunit to 75 of the β subunit. In terms of amino acid sequence and disulfide bond location, two-chain botrocetin is homologous to echinoidin (a sea urchin lectin) and other C-type lectins (Usami et al. 1993). The disulfide bond pattern of Trimeresurus stejnegeri lectin (TSL), a member of the C-type lectin family, showed four intrachain disulfide bonds: Cys3–Cys14, Cys31–Cys131, Cys38–Cys133 and Cys106–Cys123, and two interchain linkages, Cys2–Cys2 and Cys86–Cys86 (Zeng et al. 2001). The antifreeze polypeptide (AFP) from the sea raven, Hemitripterus americanus, a member of the cystine-rich class of blood antifreeze proteins,

22.3 Disulfide Bonds in Lectins and Secondary Structure

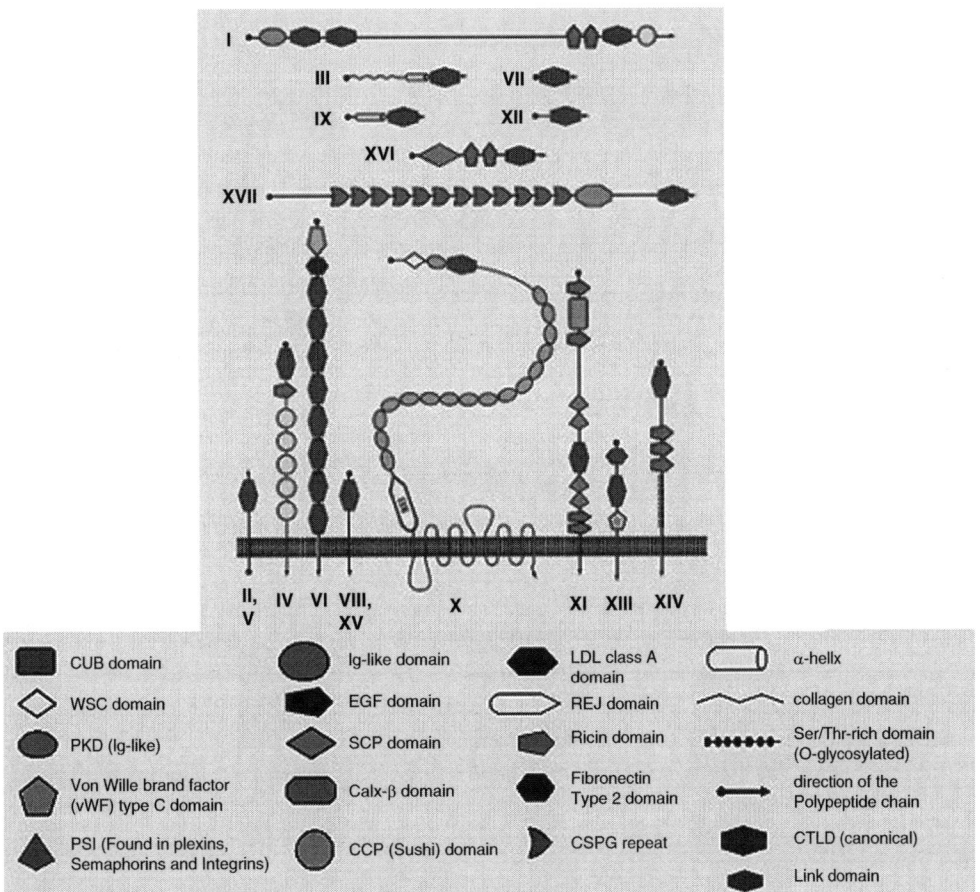

Fig. 22.4 Domain architecture of vertebrate CTLD, with mammalian homologues, from different groups. Group numbers are indicated next to the domain charts. *I* lecticans, *II* the ASGR group, *III* collectins, *IV* selectins, *V* NK receptors, *VI* the macrophage mannose receptor group, *VII* REG proteins, *VIII* the chondrolectin group, *IX* the tetranectin group, *X* polycystin 1, *XI* attractin, *XII* EMBP, *XIII* DGCR2, *XIV* the thrombomodulin group, *XV* Bimlec, *XVI* SEEC, *XVII* CBCP (Adapted with permission from Zelensky and Gready 2005 © John Wiley and Sons)

contains 129 residues with 10 half-cystine residues and all 10 half-cystine residues appeared to be involved in disulfide bond formation. The disulfide bonds are linked at Cys7 to Cys18, Cys35 to Cys125, and Cys89 to Cys122. Similarities in covalent structure suggest that the sea raven AFP, pancreatic stone protein, and several lectin-binding proteins comprise a family of proteins which may possess a common fold (Ng and Hew 1992).

Functional rat or human asialoglycoprotein receptors (ASGP-Rs), the galactose-specific C-type lectins, are hetero-oligomeric integral membrane glycoproteins. Rat ASGP-R contains three subunits, designated rat hepatic lectins (RHL) 1, 2, and 3; human ASGP-R contains two subunits, HHL1 and HHL2. Both receptors are covalently modified by fatty acylation (Zeng et al. 1996). Unfolded forms of the HHL2 subunit of the human ASGP-Rs are degraded in the ER, whereas folded forms of the protein can mature to the cell surface (Wikström and Lodish 1993).

Deacylation of ASGP-Rs with hydroxylamine results in the spontaneous formation of dimers through reversible disulfide bonds, indicating that deacylation concomitantly generates free thiol groups. Results also show that Cys57 within the transmembrane domain of HHL1 is not normally palmitoylated. Thus, Cys35 in RHL1, Cys54 in RHL2 and RHL3, and Cys36 in HHL1 are fatty acylated, where as Cys57 in HHL1 and probably Cys56 in RHL1 are not palmitoylated (Zeng et al. 1996). Eosinophil granule major basic protein 2 (MBP2 or major basic protein homolog) is a paralog of major basic protein (MBP1) and, similar to MBP1, is cytotoxic and cytostimulatory in vitro. MBP2, a small protein of 13.4 kDa molecular weight, contains ten cysteine residues. Mass spectrometry shows two cystine disulfide linkages (Cys20–Cys115 and Cys92–Cys107) and six cysteine residues with free sulfhydryl groups (Cys2, Cys23, Cys42, Cys43, Cys68, and Cys96). MBP2, similar to MBP1, has conserved motifs in common with C-type

Table 22.1 Classification of C-type lectins with associated domains

Group	Name	Associated domains
I	Lecticans (Hyalectans)	EGF, Sushi, Ig and Link domains
II	Asialoglycoprotein and DC receptors	None
III	Collectins	None
IV	Selectins	Sushi and EGF domains
V	NK – cell receptors	None
VI	Multi-CTLD endocytic receptors	FnII and Ricin domains
VII	Reg group of lectins	None
VIII	Chondrolectin, Layilin	None
IX	Tetranectin	None
X	Polycystin	WSC, REJ, PKD domains
XI	Attractin	PSI, EGF and CUB domains
XII	Eosinophil major basic protein (EMBP)	None
XIII	DGCR2	None
XIV	Thrombomodulin	EGF domains
XV	Bimlec	None
XVI	SEEC	SCP and EGF domains
XVII	CBCP/Frem1/QBRICK	CSPG repeats and CalX-β domains

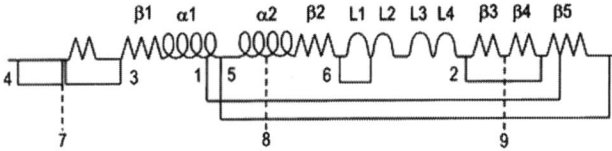

Fig. 22.5 Disulfide bonds in CTLDs. Secondary structure shared by most CTLDs is summarized, with *coils* representing α-helices, *jagged lines* denoting β-strands and loops shown as *curved segments*. The number of these elements corresponds to the secondary structure organisation of rat serum mannose-binding protein (Weis et al. 1991). Potential disulfide bonds within the CTLD are numbered 1 through 6 and cysteines that participate in interchain disulfide bonds are numbered 7 through 9 (Adapted with permission from Drickamer and Dodd 1999 © Oxford University Press)

lectins. The disulfide bond locations are conserved among human MBP1, MBP2 and C-type lectins (Wagner et al. 2007).

The biological functions of rat surfactant protein-A (SP-A), an oligomer composed of 18 polypeptide subunits are dependent on intact disulfide bonds. Reducible and collagenase-reversible covalent linkages of as many as six or more subunits in the molecule indicate the presence of at least two NH2-terminal interchain disulfide bonds. However, the reported primary structure of rat SP-A predicts that only Cys6 in this region is available for interchain disulfide formation. Direct evidence for a second disulfide bridge was obtained by analyses of a set of three mutant SP-As with telescoping deletions from the reported NH2-terminus. Two of the truncated recombinant proteins formed reducible dimers despite deletion of the domain containing Cys6. A novel post translational modification results in naturally occurring cysteinyl isoforms of rat SP-A which are essential for multimer formation (Elhalwagi et al. 1997). Pulmonary SP-D is assembled predominantly as dodecamers consisting of four homotrimeric subunits each. Association of these subunits is stabilized by interchain disulfide bonds involving two conserved amino-terminal cysteine residues (Cys-15 and Cys-20). Mutant recombinant rat SP-D lacking these residues (RrSP-Dser15/20) is secreted in cell culture as trimeric subunits rather than as dodecamers. Disulfide cross-linked SP-D oligomers are required for the regulation of surfactant phospholipid homeostasis and the prevention of emphysema and foamy macrophages in vivo (Zhang et al. 2001).

22.4 Collectins

22.4.1 Collectins: A Group of Collagenous Type of Lectins

Collectins are a small family of secreted oligomeric glycoproteins that contain in common an NH2-terminal collagen-like domain and a COOH-terminal lectin (carbohydrate binding) domain (collagenous + lectin domains = collectin), and found in both lung and serum. The name collectin is derived from the words "collagen" and "lectin." Collectins are C-type lectins that contain a collagen-like domain and usually assemble in large oligomeric complexes containing 9–27 subunits. They belong to C-type lectins family and have an important function in innate immunity, recognizing and binding to microorganisms via sugar arrays on the microbial surface. To date, nine different collectins have been identified: mannan-binding lectin (MBL), surfactant protein A (SP-A), surfactant protein D (SP-D), collectin liver 1 (CL-L1), collectin placenta 1 (CL-P1), conglutinin, collectin of 43 kDa (CL-43) and collectin of 46 kDa (CL-46), and collectin from kidney (CL-K1). The collectins MBP, conglutinin, CL-43, CL-46, CL-K1, SP-A, and SP-D are soluble, whereas CL-L1 and CL-P1 are membrane proteins. Some collectins, such as MBP and SP-A, organize into a "bouquet," and others, such as bovine conglutinin and SP-D, organize into a "cruciform" shape. One of the best-studied serum collectins is MBP (Chaps. 23–25).

References

Adachi H, Tsujimoto M (2002) FEEL-1, a novel scavenger receptor with in vitro bacteria-binding and angiogenesis-modulating activities. J Biol Chem 277:34264–34270

Banerji S, Ni J, Wang SX et al (1999) LYVE-1, a new homologue of the CD44 glycoprotein, is a lymph-specific receptor for hyaluronan. J Cell Biol 144:789–801

References

Bertrand JA, Pignol D et al (1996) Crystal structure of human lithostathine, the pancreatic inhibitor of stone formation. EMBO J 15:2678–2684

Blundell CD, Mahoney DJ, Almond A et al (2003) The link module from ovulation- and inflammation-associated protein TSG-6 changes conformation on hyaluronan binding. J Biol Chem 278:49261–49270

Boyington JC, Riaz AN, Patamawenu A et al (1999) Structure of CD94 reveals a novel C-type lectin fold: implications for the NK cellassociated CD94/NKG2 receptors. Immunity 10:75–82

Brissett NC, Perkins SJ (1996) The protein fold of the hyaluronate-binding proteoglycan tandem repeat domain of link protein, aggrecan and CD44 is similar to that of the C-type lectin superfamily. FEBS Letters 388:211–216

Correa I, Raulet DH (1995) Binding of diverse peptides to MHC class I molecules inhibits target cell lysis by activated natural killer cells. Immunity 2:61–71

Cummings RD, McEver RO (2009) C-type lectins. In: Varki A, Cummings RD, Esko JD et al (eds) Essentials of glycobiology, 2nd edn. Cold Spring Harbor Laboratory Press, Cold Spring Harbor

Dimasi N, Sawicki MW, Reineck LA et al (2002) Crystal structure of the Ly49I natural killer cell receptor reveals variability in dimerization mode within the Ly49 family. J Mol Biol 320:573–585

Dodd RB, Drickamer K (2001) Lectin-like proteins in model organisms: implications for evolution of carbohydrate-binding activity. Glycobiology 11:71R–79R

Drickamer K (1999) C-type lectin-like domains. Curr Opin Struct Biol 9:585–590

Drickamer K (1993) Evolution of Ca^{2+}-dependent animal lectins. Prog Nucleic Acid Res Mol Biol 45:207–232

Drickamer K, Dodd RB (1999) C-type lectin-like domains in *Caenorhabditis elegans*: predictions from the complete genome sequence. Glycobiology 9:1357–1367

Drickamer K, Fadden AJ (2002) Genomic analysis of C-type lectins. Biochem Soc Symp 69:59–72

East L, Isacke CM (2002) The mannose receptor family. Biochim Biophys Acta 1572:364–386

Elhalwagi BM, Damodarasamy M, McCormack FX (1997) Alternate amino terminal processing of surfactant protein A results in cysteinyl isoforms required for multimer formation. Biochemistry 36:7018–7025

Faltz LL, Caputo CB, Kimura JH et al (1979) Structure of the complex between hyaluronic acid, the hyaluronic acid-binding region, and the link protein of proteoglycan aggregates from the swarm rat chondrosarcoma. J Biol Chem 254:1381–1387

Feinberg H, Park-Snyder S, Kolatkar AR et al (2000) Structure of a C-type carbohydrate recognition domain from the macrophage mannose receptor. J Biol Chem 275:21539–21548

Feinberg H, Mitchell DA, Drickamer K, Weis WI (2001) Structural basis for selective recognition of oligosaccharides by DC-SIGN and DC-SIGNR. Science 294:2163–2166

Fuhlendorff J, Clemmensen I, Magnusson S (1987) Primary structure of tetranectin, a plasminogen kringle 4 binding plasma protein: homology with asialoglycoprotein receptors and cartilage proteoglycan core protein. Biochemistry 26:6757–6764

Geijtenbeek TB, Torensma R, van Vliet SJ et al (2002) Identification of DC-SIGN, a novel dendritic cellspecific ICAM-3 receptor that supports primary immune responses. Cell 100:575–585

Graves BJ, Crowther RL, Chandran C et al (1994) Insight into E-selectin0 ligand interaction from the crystal structure and mutagenesis of the lec0EGF domains. Nature 367:532–538

Gronwald W, Loewen MC, Lix B et al (1998) The solution structure of type II antifreeze protein reveals a new member of the lectin family. Biochemistry 37:4712–4721

Guthmann MD, Tal M, Pecht I (1995) A new member of the C-type lectin family is a modulator of the mast cell secretory response. Int Arch Allergy Immunol 107:82–86

Hakansson K, Lim NK, Hoppe HJ, Reid KB (1999) Crystal structure of the trimeric alpha-helical coiled-coil and the three lectin domains of human lung surfactant protein D. Structure Fold Des 7:255–261

Håkansson K, Lim NK, Hoppe HJ, Reid KB (1999) Crystal structure of the trimeric alpha-helical coiled-coil and the three lectin domains of human lung surfactant protein D. Structure 7:255–264

Håkansson K, Reid KB (2000) Collectin structure: a review. Protein Sci 9:1607–1617

Hamburger ZA, Brown MS, Isberg RR, Bjorkman PJ (1999) Crystal structure of invasin: a bacterial integrin-binding protein. Science 286:291–295

Harte WE, Bajorath J (1994) Synergism of calcium and carbohydrate binding to mammalian lectin. J Am Chem Soc 116:10394–10398

Hatakeyama T, Matsuo N, Shiba K et al (2002) Amino acid sequence and carbohydrate-binding analysis of the N-acetyl-D-galactosamine-specific C-type lectin, CEL-I, from the Holothuroidea, *Cucumaria echinata*. Biosci Biotechnol Biochem 66:157–163

Hirotsu S, Mizuno H, Fukuda K et al (2001) Crystal structure of bitiscetin, a von Willebrand factor-dependent platelet aggregation inducer. Biochemistry 40:13592–13597

Hohenester E, Sasaki T, Olsen BR et al (1998) Crystal structure of the angiogenesis inhibitor endostatin at 1.5 Å resolution. EMBO J 17:1656–1661

Kawasaki N, Kawasaki T et al (1983) Isolation and characterization of a mannan-binding protein from human serum. J Biochem (Tokyo) 94:937–947

Kelly G, Prasannan S et al (1999) Structure of the cell-adhesion fragment of intimin from enteropathogenic *Escherichia coli*. Nat Struct Biol 6:313–318

Kohda D, Morton CJ, Parkar AA et al (1996) Solution structure of the link module: a hyaluronan-binding domain involved in extracellular matrix stability and cell migration. Cell 86:767–775

Lasky LA, Singer MS, Yednock TA et al (1989) Cloning of a lymphocyte homing receptor reveals a lectin domain. Cell 56:1045–1055

Lee TH, Wisniewski HG, Vilcek J (1992) A novel secretory tumor necrosis factor-inducible protein (TSG-6) is a member of the family of hyaluronate binding proteins, closely related to the adhesion receptor CD41. J Cell Biol 116:545–557

Ley K, Kansas GS (2004) Selectins in T-cell recruitment to non-lymphoid tissues and sites of inflammation. Nat Rev Immunol 4:325–335

Liu Y, Eisenberg D (2002) 3D domain swapping: as domains continue to swap. Protein Sci 11:1285–1299

Loeb JA, Drickamer K (1988) Conformational changes in the chicken receptor for endocytosis of glycoproteins. Modulation of ligand-binding activity by Ca^{2+} and pH. J Biol Chem 263:9752–9760

Lundell A, Olin AI, Morgelin M et al (2004) Structural basis for interactions between tenascins and lectican C-type lectin domains: evidence for a crosslinking role for tenascins. Structure (Camb) 12:1495–1506

McGreal EP, Martinez-Pomares L, Gordon S (2004) Divergent roles for C-type lectins expressed by cells of the innate immune system. Mol Immunol 41:1109–1121

Meier M, Bider MD, Malashkevich VN et al (2000) Crystal structure of the carbohydrate recognition domain of the H1 subunit of the asialoglycoprotein receptor. J Mol Biol 300:857–865

Mizuno H, Fujimoto Z, Atoda H, Morita T (2001) Crystal structure of an anticoagulant protein in complex with the Gla domain of factor X. Proc Natl Acad Sci USA 98:7230–7234

Mizuno H, Fujimoto Z, Koizumi M et al (1997) Structure of coagulation factors IX0X-binding protein, a heterodimer of C-type lectin domains (Letter). Nat Struct Biol 4:438–441

Natarajan K, Dimasi N, Wang J et al (2002) Structure and function of natural killer cell receptors: multiple molecular solutions to self, nonself discrimination. Annu Rev Immunol 20:853–885

Natarajan K, Sawicki MW, Margulies DH, Mariuzza RA (2000) Crystal structure of human CD69: a C-type lectin-like activation marker of hematopoietic cells. Biochemistry 39:14779–14786

Ng KK, Park-Snyder S, Weis WI (1998) Ca^{2+}-dependent structural changes in C-type mannose-binding proteins. Biochemistry 37:17965–17976

Ng NF, Hew CL (1992) Structure of an antifreeze polypeptide from the sea raven. Disulfide bonds and similarity to lectin-binding proteins. J Biol Chem 267:16069–16075

Nielsen BB, Kastrup JS, Rasmussen H et al (1997) Crystal structure of tetranectin, a trimeric plasminogen-binding protein with an α-helical coiled coil. FEBS Lett 412:388–396

Poget SF, Legge GB, Proctor MR, Butler PJ, Bycroft M, Williams RL (1999) The structure of a tunicate C-type lectin from *Polyandrocarpa misakiensis* complexed with d-galactose. J Mol Biol 290:867–879

Prevo R, Banerji S, Ni J, Jackson DG (2004) Rapid plasma membrane-endosomal trafficking of the lymph node sinus and high endothelial venule scavenger receptor/homing receptor stabilin-1 (FEEL-1/CLEVER-1). J Biol Chem 279:52580–52592

Sano H, Kuroki Y, Honma T et al (1998) Analysis of chimeric proteins identifies the regions in the carbohydrate recognition domains of rat lung collectins that are essential for interactions with phospholipids, glycolipids, and alveolar type II cells. J Biol Chem 273:4783–4789

Sheriff S, Chang CY, Ezekowitz RA (1994) Human mannose-binding protein carbohydrate recognition domain trimerizes through a triple alpha-helical coiled-coil. Nat Struct Biol 1:789–794

Spicer AP, Joo A, Bowling RA Jr (2003) A hyaluronan binding link protein gene family whose members are physically linked adjacent to chrondroitin sulfate proteoglycan core protein genes: the missing links. J Biol Chem 278:21083–21091

Sugawara H, Kusunoki M, Kurisu G et al (2004) Characteristic recognition of N-acetylgalactosamine by an invertebrate C-type lectin, CEL-I, revealed by X-ray crystallographic analysis. J Biol Chem 279:45219–45225

Swaminathan GJ, Weaver AJ, Loegering DA et al (2001) Crystal structure of the eosinophil major basic protein at 1.8 Å: an atypical lectin with a paradigm shift in specificity. J Biol Chem 23:26197–26203

Tamura Y, Adachi H et al (2003) FEEL-1 and FEEL-2 are endocytic receptors for advanced glycation end products. J Biol Chem 278:12613–12617

Tormo J, Natarajan K et al (1999) Crystal structure of a lectin-like natural killer cell receptor bound to its MHC class I ligand. Nature 402:623–631

Usami Y, Fujimura Y, Suzuki M et al (1993) Primary structure of two-chain botrocetin, a von Willebrand factor modulator purified from the venom of Bothrops jararaca. Proc Natl Acad Sci USA 90:928–932

Wagner LA, Ohnuki LE et al (2007) Human eosinophil major basic protein 2: location of disulfide bonds and free sulfhydryl groups. Protein J 26:13–18

Weis WI, Drickamer K (1996) Structural basis of lectin-carbohydrate recognition. Annu Rev Biochem 65:441–473

Weis WI, Drickamer K (1994) Trimeric structure of a C-type mannose-binding protein. Structure 2:1227–1240

Weis WI, Kahn R, Fourme R et al (1991) Structure of the calcium-dependent lectin domain from a rat mannose- binding protein determined by MAD phasing. Science 254:1608–1615

Weis WI, Taylor ME, Drickamer K (1998) The C-type lectin superfamily in the immune system. Immunol Rev 163:19–31

Wikström L, Lodish HF (1993) Unfolded H2b asialoglycoprotein receptor subunit polypeptides are selectively degraded within the endoplasmic reticulum. J Biol Chem 268:14412–14416

Wintergest E, Manz-Keinke H, Plattner H, Schlepper-Schafer J (1989) The interaction of a lung surfactant protein (SP-A) with macrophages is mannose dependent. Eur J Cell Biol 50:291–298

Wisniewski HG, Vilcek J (1997) TSG-6: an IL-1/TNF-inducible protein with anti-inflammatory activity. Cytokine Growth Factor Rev 8:143–156

Wolan DW, Teyton L, Rudolph MG, Villmow B, Bauer S, Busch DH, Wilson IA (2001) Crystal structure of the murine NK cell-activating receptor NKG2D at 1.95 Å. Nat Immunol 2:248–251

Yamanishi T, Yamamoto Y, Hatakeyama T et al (2007) CEL-I, an invertebrate N-acetylgalactosamine-specific C-type lectin, induces TNF-alpha and G-CSF production by mouse macrophage cell line RAW264.7 cells. J Biochem 142:587–595

Zelensky AN, Gready JE (2005) The C-type lectin-like domain superfamily. FEBS J 272:6179–6217

Zelensky AN, Gready JE (2003) Comparative analysis of structural properties of the C-type-lectin-like domain (CTLD). Proteins 52:466–477

Zelensky AN, Gready JE (2004) C-type lectin-like domains in *Fugu rubripes*. BMC Genomics 5:51

Zeng FY, Oka JA, Weigel PH (1996) The human asialoglycoprotein receptor is palmitoylated and fatty deacylation causes inactivation of state 2 receptors. Biochem Biophys Res Commun 218:325–330

Zeng FY, Weigel PH (1996) Fatty acylation of the rat and human asialoglycoprotein receptors. A conserved cytoplasmic cysteine residue is acylated in all receptor subunits. J Biol Chem 271:32454–32460

Zeng R, Xu Q, Shao XX et al (2001) Determination of the disulfide bond pattern of a novel C-type lectin from snake venom by mass spectrometry. Rapid Commun Mass Spectrom 15:2213–2220

Zhang L, Ikegami M, Crouch EC et al (2001) Activity of pulmonary surfactant protein-D (SP-D) in vivo is dependent on oligomeric structure. J Biol Chem 276:19214–19219

Collectins: Mannan-Binding Protein as a Model Lectin

23

Anita Gupta

23.1 Collectins

A group of collagenous type of lectins: Collectins are a small family of secreted oligomeric glycoproteins that contain in common an NH2-terminal collagen-like domain and a COOH-terminal lectin (carbohydrate binding) domain (collagenous + lectin domains = collectin), and found in both lung and serum. The name collectin is derived from the words "collagen" and "lectin." Collectins are C-type lectins that contain a collagen-like domain and usually assemble in large oligomeric complexes containing 9–27 subunits. They belong to C-type lectins family and have an important function in innate immunity, recognizing and binding to microorganisms via sugar arrays on the microbial surface. To date, nine different collectins have been identified: mannan-binding lectin (MBL), surfactant protein A (SP-A), surfactant protein D (SP-D), collectin liver 1 (CL-L1), collectin placenta 1 (CL-P1), conglutinin, collectin of 43 kDa (CL-43) and collectin of 46 kDa (CL-46), and collectin from kidney (CL-K1). The collectins MBP, conglutinin, CL-43, CL-46, CL-K1, SP-A, and SP-D are soluble, whereas CL-L1 and CL-P1 are membrane proteins. Some collectins, such as MBP and SP-A, organize into a "bouquet," and others, such as bovine conglutinin and SP-D, organize into a "cruciform" shape. One of the best-studied serum collectins is MBP. Bovine CL-43 is structurally one of the simplest collectins, consisting of only three polypeptides, each of which contains a terminal CTLD. Rats have two serum MBPs designated A and C, sometimes called mannan-binding proteins. Humans appear to have only a single MBP corresponding to the rat MBP-A. The collectins can be classed into two distinct group, with MBP and SP-A being hexamers and SP-D, conglutinin and collectin-43 (CL-43) being tetramers, with proteins in the latter group also having significantly larger dimensions with respect to the length of their collagen-like 'stalks'. The structural and functional relationships of this group of collectins have been reviewed (Håkansson and Reid 2000; Hansen and Holmskov 2002; and others).

Each polypeptide chain in collectins consists of four regions: a relatively short N-terminal region, a collagen like region, an α-helical coiled-coil, and a C-terminal lectin domain. With the exception of the N-terminal region, these regions or domain types are also found in molecules other than collectins (Fig. 22.1). The presence of a collagen like region in these molecules imposes a trimeric structure. This is a prerequisite for their proper function; the carbohydrate affinity of a single collectin carbohydrate recognition domain (CRD) is weak but the trimeric organization permits a trivalent and hence stronger interaction between collectin and carbohydrate-containing target surface. In most of the collectins, these trimers are further assembled into larger entities, which enables them to cross-link several target particles and perhaps also to interact simultaneously with target and with host cells (Håkansson and Reid 2000).

Lung surfactant protein D (SP-D) and bovine conglutinin are X-shaped molecules consisting of four trimers extending pairwise from the hub with the lectin domains at the tip of the collagenous arms (Crouch et al. 1994). These dodecamers have been shown to associate further into even larger complexes, with an increased capacity to aggregate bacteria. Surfactant protein A (SP-A) is the major protein component of lung surfactant complex and exhibits a reduced and denatured molecular mass of 35 kDa in humans. SP-A has been shown to have a flower-bouquet-like octadecameric structure similar to that of complement C1q. SP-A interacts with lipids and specifically binds to dipalmitoylphosphatidylcholine (DPPC). Lung SP-A (Voss et al. 1988) and most mannan binding proteins form bouquet-like complexes with up to six trimers. The degree of oligomerization differs between the different proteins, but may also vary within the molecular population of a single preparation. Only the larger complexes, i.e., pentamers or hexamers, give a full biological response in terms of macrophage stimulation and complement activation. Comparisons of hybrid peptide chains of rat MBP-A and MBP-C and site-directed mutagenesis led to the conclusion that oligomerization depends on the N-terminal region and that the most N-terminal of its cysteines (Cys6) is indispensable for the formation of oligomers of trimers. The other two cysteines (Cys13 and Cys18) are involved in intratrimer disulfide bridge formation (Wallis and Drickamer 1999). There are a several reviews, which described the biological, physiological, and functional properties of the collectins (Håkansson and Reid 2000; Weis and Drickamer 1996; Lu 1997; Crouch 1998). Håkansson and Reid, (2000) focussed on the structural aspects of the mannan-binding proteins and the lung surfactant proteins.

23.1.1 N-Terminal Region

The N-terminal region of collectins is defined as the segment N-terminal to the first collagenous triple-helix residue. This relatively cysteine-rich region stabilizes the trimers through disulfide bridging (Crouch et al. 1994; Holmskov et al. 1995) and links them together in the collectin oligomers. There seems to be no overall homology between the collectins in this part of the molecule. The only general trend (to which canine SP-A, Rhesus monkey MBP-C, and CL-L1 do not conform) seems to be two cysteine residues separated in sequence by 4–5 amino acids, and perhaps also that the amino acid preceding the second of these cysteines is hydrophobic in most of the collectins. Most of the other amino acids separating these two cysteine residues are relatively hydrophobic in MBP and SP-D but hydrophilic in SP-A. The collectins can nonetheless be divided into different groups with related N-terminal regions. The N-terminal sequences of SP-D, conglutinin, and collectin-43 are clearly related and of similar length, i.e., 25–28 amino acids, and include two cysteine residues. The N-terminal region of SP-A is much shorter and, due to variation in signal peptidase cleavage, is not homogenous; there are two isoforms with seven or ten residues containing one or two cysteines, respectively (Elhalwagi et al. 1997). The third group consists of mannan-binding proteins A and C, which have distinct, yet related, N-terminal regions, each approximately 20 amino acids long. The hepatic forms rat MBP-C and mouse MBP-C have only two cysteines while the other mannan-binding proteins have three, although the sequence of rhesus monkey MBP-C deviates from the similarity displayed by the other sequences. Human MBP has three cysteines despite its closer evolutionary relationship to rat MBP-C and forms larger oligomers in serum than it does in the liver (Kurata et al. 1994). The fourth group, so far represented only by human CL-1, has a much longer N-terminal sequence with only one cysteine.

23.1.2 Collagenous Region

Information on collagen structure is mainly derived from fiber diffraction studies (Beck and Brodsky 1998), and the crystallographic structures of model compounds (Kramer et al. 1998, 1999). A collagen structure can be recognized from the amino acid sequence with its characteristic Gly-X-Y repetitive pattern, where X and Y can be any amino acid but are frequently prolines or hydroxyprolines. Each of the three chains forms a left-handed polyproline II like helix and then chains are coiled around each other in a right-handed manner with the glycine residues in the interior of the superhelix. Interchain hydrogen bonds between N-H groups of glycine and the C_5O groups of the amino acid in X position stabilize the structure. On account of this and absence of intrachain hydrogen bonds, the collagen helix can exist as a trimer. The helical parameters are sequence dependent and differ between imino acid rich and amino acid rich regions (Kramer et al. 1999). The collagen triple helix is surrounded by water molecules that interact with most polar groups, which are exposed to the solvent, either by their side chains or carbonyl groups. Accordingly, the triple helix in the collectins is rich in charged amino acids. The triple helix is most stable due to its high tensile strength, stability, and relative resistance to proteolysis. In addition, a triple helical region can also mediate binding interactions with other macromolecules as shown for C1q and the macrophage scavenger receptor. The collagenous region in MBP contains a binding site for its associated serine proteases (MASP-1 and MASP-2) through which the complement cascade of reactions is triggered (Thiel et al. 1997).

Collectins bind to their macrophage receptor through the collagenous region (Malhotra et al. 1993). The length of collagenous region differs between the collagens; the X-shaped SP-D and bovine collectin have the longest triple-helical regions, and SP-A and MBP the shortest. Most of the collectin prolines in Y position are hydroxylated. Hydroxylated and glycosylated lysines have been demonstrated in SP-D, MBP, collectin 43 and conglutinin (Håkansson and Reid 2000). SP-A oligomerizes as an octadecamer, which forms a flower bouquet-like structure. The collagen-like domain of human SP-A consists of 23 Gly-X-Y repeats with an interruption near the midpoint of this domain. This interruption causes a kink, but its role remains largely unknown. In SP-A, there appears to be a kink in the collagenous rod in this part of the helix, as judged by electron microscopy. The collagen triple helices aggregate laterally at their N-terminal ends, but there is no evidence for covalent crosslinking. At the site of the Gly-X-Y interruption, the helices make a 60° bend and diverge (Voss et al. 1988) in a manner familiar from studies on the C1q structure (Lu et al. 1993). Oligomerization of MBP-A is in part promoted by the N-terminal part of its collagenous region (Wallis and Drickamer 1999), but this region is not indispensable for SP-A oligomer formation (McCormack et al. 1997).

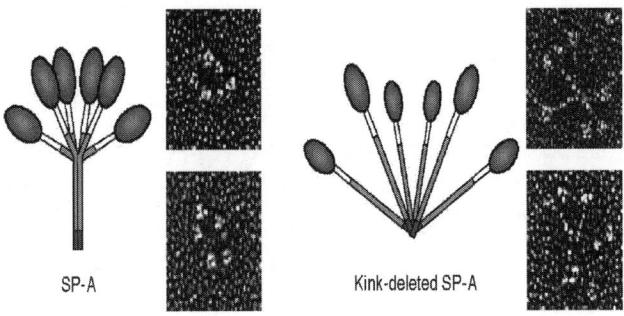

Fig. 23.1 Interruption of Gly-X-Y repeats in the SP-A is critical for the formation of a flower bouquet-like octadecamer (Adapted by permission from Uemura et al. 2006 © American Chemical Society)

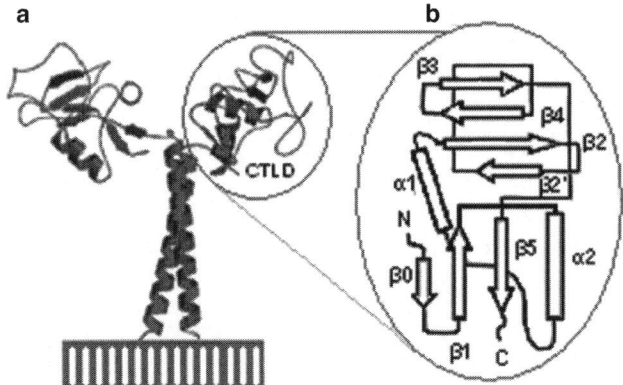

Fig. 23.2 (**a**) Ribbon representation of the maltose-binding protein (MBP-A; PDB ID: 1KMB), a CTLD family prototype. (**b**) The topology diagram of CTLD receptor family. This secondary structure organization is well conserved in all known CTLD of identified three-dimensional. In diagrams, C and N represent the carboxyl- and N-terminus ends of the protein, respectively. For simplicity, the Ca^{2+} ions in the MBP-A ribbon representation are omitted (Adapted by permission from Sawicki et al. 2001© John Wiley and Sons)

To define the importance of the kink region of SP-A, two mutated proteins were constructed to disrupt the interruption of Gly-X-Y repeats. Results indicate that the interruption of Gly-X-Y repeats in the SP-A molecule is critical for the formation of a flower bouquet-like octadecamer and contributes to SP-A's capacity to aggregate phospholipid liposomes (Uemura et al. 2006; Fig. 23.1).

23.1.3 C-Type Lectin Domain

MBP-A is a prototype member of Ca^{2+} dependent lectin in CTLD family (Fig. 23.2a). Collectins have a common sequence motif of 115–130 amino acid residues as carbohydrate recognition domain (CRD) and has been described in Chap. 22. Despite the diversity of ligand recognized by CTLD, members of this family exhibit a highly conserved structure (Fig. 23.2b). The core structure consists of two α-helices (α1 and β2) and two anti-parallel β-sheets formed by β strands (β0, β1, and β5) and (β2, β2', β3, and β4). The secondary structure elements composing the CTLD core region are conserved among the determined structure (Sawicki et al. 2001), whereas considerable variation is observed in the sequence and structure of the loops connecting the secondary structure elements. It has been demonstrated that the sequence and structure variability of these loops confer ligand specificity (Torgersen et al. 1998).

A comparison of the amino acid sequences of 22 different extracellular collectins yields 18 conserved amino acids in addition to the four cysteines. Perhaps most important from a structural point of view is a hydrophobic cluster containing the conserved residues Phe^{304}, Tyr^{314}, Trp^{317}, Pro^{322}, Trp^{340}, in addition to the semi-conserved Val^{332}. They hold together the carbohydrate binding region consisting of strand β6 and the stretch between the two strands β4 and β5. Trp^{340} is also hydrogen bonded to the conserved Asn^{277} of helix α2.

Asn^{277} is in turn hydrogen bonded to the main-chain carbonyl group of Gly^{338}, as is the conserved residue Asn^{316}. Glycine 338 appears to play a key role in the loop structure preceding the β6 strand. Its main-chain conformation is not compatible with the presence of a side chain, which is also true for the other two conserved glycines, Gly^{241} and Gly^{309}, and for most of the nonconserved SP-D glycines as well. Glu^{276}, Asn^{277}, and Ala^{279} are on the nonexposed side of helix α2, which is preceded by the conserved Pro^{271}. Asn^{277}, as mentioned, is hydrogen bonded to both Gly^{338} and Trp^{340}, and Glu^{276} is hydrogen bonded to the main-chain amino group of the nonconserved. Studies indicate a slower evolutionary pace in and around the β5 and β6 strands, i.e., the carbohydrate binding site. In addition to the carbohydrate binding calcium ion, two more calcium ions have been found in crystallographic studies of MBP and SP-D (Weis et al. 1998; Weis and Drickamer 1994; Håkansson et al. 1999). One of these is found 8 Å away from the carbohydrate binding calcium and is complexed by three aspartic acid and one glutamic acid residues. These residues are either conserved or only conservatively replaced in MBPs, SP-D, conglutinin, and Cl-43, but not in SP-A (Håkansson and Reid 2000; Weis and Drickamer 1994).

The electrostatic potential along the surface of collectins displays a large positively charged area within the cavity between the three lectin domains. This surface charge can be observed both for the mannan-binding proteins and SP-D, but is not present on the C-type lectins that are not collectins, e.g., in tetranectin or on the surface of E-selectin. In SP-D, this charge results from the presence of nonconserved lysine residues. The charged area may be involved in interactions with negative charges on the surface of its targets, e.g., LPS moiety of microbial carbohydrates or the phospholipids of the pulmonary surfactant. The calcium ions in these proteins do influence the surface charge, and the apo protein is characterized by large negatively charged areas on the central and peripheral parts of the lectin domains. Thus, the calcium ions might have a functional role other than carbohydrate binding, e.g., they may maintain a certain electrostatic potential pattern on the surface of the molecule. In this context, it is interesting to note that SP-D was found to bind to its putative receptor gp340 in a manner that was calcium dependent but not inhibited by the presence of maltose (Holmskov et al. 1997; Håkansson and Reid 2000).

The C-type lectin domains in collectins are attached to collagen regions via α-coiled neck regions. Collectins form trimers that may assemble into larger oligomers. Each polypeptide chain consists of four regions: a relatively short N-terminal region, a collagen like region, an α-helical coiled-coil, and the lectin domain. Primary structure data are available for many proteins, while the most important features of the collagen-like region can be derived from its homology with collagen. Carbohydrate binding has been structurally characterized in several complexes between MBP and carbohydrate and between SP-D and carbohydrate; all indicate that the major interaction between carbohydrate and collectin is the binding of two adjacent carbohydrate hydroxyl groups to a collectin calcium ion. In addition, these hydroxyl groups form hydrogen bond to some of the calcium amino acid ligands. While each collectin trimer contains three such carbohydrate binding sites, deviation from overall threefold-symmetry has been demonstrated for SP-D, which may influence its binding properties. The protein surface between the three binding sites is positively charged in both MBP and SP-D (Crouch 1998; Håkansson and Reid 2000).

23.1.4 Comparative Genetics of Collagenous Lectins

MBLs and ficolins (FCNs) are structurally related to C1q, but activate the lectin complement pathway via interaction with MBL-associated serine proteases (MASPs). MBLs, FCNs, and other collagenous lectins also bind to some host macromolecules and contribute to their removal. While there is evidence that some lectins and the lectin complement pathway are conserved in vertebrates, many differences in collagenous lectins have been observed among humans, rodents, and other vertebrates. For example, humans have only one MBL but three FCNs, whereas most other species express two FCNs and two MBLs. Bovidae express CG and other SP-D-related collectins that are not found in monogastric species. Some dysfunctions of human MBL are due to single nucleotide polymorphisms (SNPs) that affect its expression or structure and thereby increase susceptibility to some infections. Collagenous lectins have well-established roles in innate immunity to various microorganisms. So it is possible that some lectin genotypes or induced phenotypes influence resistance to some infectious or inflammatory diseases in animals (Lillie et al. 2005).

23.1.5 Generalized Functions of Collectins

While MBL and SP-D interact with mannose, glucose, L-fucose, ManNAc, GlcNAc as their ligands, SP-A has affinity for glucose, mannose, maltose, and inositol (Kerrigan and Brown 2009). The collectins function in "innate" immunity and act before the induction of an antibody-mediated response. Collectins stimulate in vitro phagocytosis by recognizing surface glycans on pathogens, they promote chemotaxis, and they stimulate the production of cytokines and reactive oxygen species by immune cells. Lung surfactant lipids have the ability to suppress a number of immune cell functions such as proliferation, and this suppression of the immune response is further augmented by SP-A. Although SP-A and SP-D were originally found in the lung, they are also expressed in the intestine. Their function is to enhance adhesion and phagocytosis of microorganisms by agglutination and opsonization. The genes for MBL, SP-A and SP-D have been mapped to human chromosome 10, with at least two expressed SP-A genes (SP-AI and SP-AII) forming a cluster with an SP-A pseudogene. Somatic cell hybrid mapping places the human SP-A and SP-D genes at 10q22-q23 while MBL is localised at 10q21. The close evolutionary relationship between the collectins is further emphasized by a common pattern of exons in their genomic structures and the presence of a gene cluster on chromosome 10 in humans that contains the genes known for the human collectins. Studies on the structure/function relationships within the collectins could provide insight into the properties of a growing number of proteins also containing collagenous regions such as C1q, the hibernation protein, the α- and β-ficolins, as well as the membrane acetylcholinesterase and macrophage scavenger receptor.

All defense collagens (C1q, SP-A, MBL) have shown a qualitatively and quantitatively similar enhancement of monocyte phagocytosis of targets that are suboptimally opsonized with IgG or CR type 1 ligands, C4b and C3b (Wright 2005). The six amino acid sequence required for this functional stimulation has been identified within the collagen-like domain (Arora et al. 2001). This rapid enhancement of phagocytic activity is triggered when the defense collagen is bound to the particle to be ingested. Interaction of C1q with its cell-surface receptor on neutrophils induces the activation of respiratory burst. However, this action did not occur with MBP and SP-A, proteins that also contain collagen-like domains (Goodman and Tenner 1992). Interaction of C1q with its cell-surface receptor on neutrophils induces the activation of respiratory burst. This action did not occur with MBP and SP-A, proteins that also contain collagen-like domains (Goodman and Tenner 1992).

23.2 Human Mannan-Binding Protein

Mannose-binding lectin (MBL) is a Group III C-type lectin belonging to the collectins (Holmskov et al. 2003), which are a group of soluble oligomeric proteins containing collagenous regions and CTLDs. MBL is secreted into the blood stream as a large multimeric complex and is primarily produced by the liver, although other sites of production, such as the intestine, have been proposed (Uemura et al. 2002). Ca^{2+}-dependent MBLs belong to the family of animal lectins isolated from the liver and serum of rabbits, humans and rodents. Mannan-binding protein was discovered as a rabbit lectin binding to mannan (Kawasaki et al. 1978). Later the term mannose-binding protein (MBP) was introduced, unfortunately implying a more selective reactivity than is characteristic for this protein. The burst of investigations on MBP deficiency in serum (MBL) and susceptibility to infectious diseases was roused by the seminal demonstration of low MBL levels in children deficient in opsonizing activity and suffering from unexplained sensitivity to infections (Super et al. 1989). Initially, investigations of causal relationship between MBL and disease susceptibility relied on quantification of MBL in serum or plasma (or in some cases, of opsonizing activity). However, the finding of genetic influence on the MBL level opened up for determining the MBL status by genotyping (Holmskov et al. 2003; Kerrigan and Brown 2009; Nuytinck and Shapiro 2004; Worthley et al. 2006).

Sheriff et al. (1994) confirmed that human MBP is a hexamer of trimers with each subunit consisting of an amino-terminal region rich in cysteine, 19 collagen repeats, a 'neck', and a carbohydrate recognition domain that requires calcium to bind ligand. A 148-residue peptide, consisting of the 'neck' and CRDs forms trimers in solution and in crystals. The structure of this trimeric peptide has been determined in two different crystal forms. The 'neck' forms a triple α-helical coiled-coil. Each α-helix interacts with a neighbouring carbohydrate recognition domain. The spatial arrangement of the carbohydrate recognition domains suggests how MBP trimers form the basic recognition unit

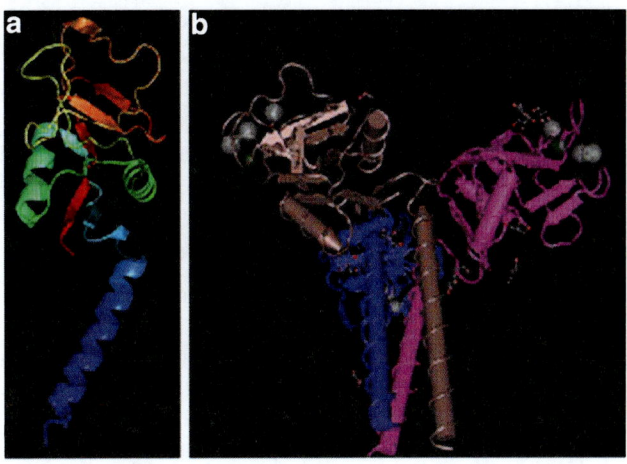

Fig. 23.3 (**a**) Monomeric structure of human MBP (PDB ID: 1HUP) (Sheriff et al. 1994). (**b**). Trimeric structure of a C-type mannose-binding protein from rat (PDB ID: 1RTM) (Weis and Drickamer 1994 © Elsevier)

for branched oligosaccharides on microorganisms (Sheriff et al. 1994) (Fig. 23.3).

23.2.1 Characterization of Serum MBL

23.2.1.1 The Protein

Serum MBL contains both a collagen-like domain and a carbohydrate-recognition domain (CRD). The overall polypeptide structure of MBL is similar to that of the other collectins (SP-A, SP-D, conglutinin, CL-43, liver collectin 1, and CL-46). It includes a short, cysteine-rich N-terminal stretch (aa 1–21), a collagen like region (aa 22–81) with one interruption (aa 43–44) that causes the collagen-like structure to bend, a neck region (aa 82–115), and a carbohydrate recognition domain (aa 116–228) (Drickamer and Taylor 1993). This domain confers the carbohydrate specificity of MBL and is stabilized by two disulfide bonds. Due to collagen-like domain, MBL forms homotrimers, designated the MBL subunit. The collagen-like structure is stabilized by the presence of hydroxyprolines and glycosylated hydroxylysines (Ma et al. 1997). The subunit structures assemble from C to N terminus. The neck region initiates the folding (Childs et al. 1990), and the collagen-like region zips toward the N terminus, creating trimeric subunits. The structure is finally stabilized by intrasubunit disulfide bonds in the N-terminal region (Hansen and Holmskov 1998). The oligomer structure of MBL is similar to the structure of C1q, the primary component of the classical pathway of complement (Ikeda et al. 1987), where the bouquet-like forms arise from the formation of intersubunit disulfide bonds in the N-terminal region (Hoppe and Reid 1994). The elucidation of the structure of MBL is complicated by the fact that the polypeptide chain of MBL is very heterogeneous. In addition to several post-translational modifications, there are three well documented mutations in the collagen-like region (Garred et al. 1992; Lipscombe et al. 1992; Madsen et al. 1994). These mutations all lead to amino acid substitutions, which distort the collagen-like region and inhibit the correct formation of the oligomer forms of MBL. In addition to the heterogeneity of the polypeptide chain, promoter polymorphisms (Madsen et al. 1995) result in highly variable amounts of MBL in the blood.

Human MBP is a homooligomer composed of 32-kDa subunits. Each subunit has an NH_2-terminal region containing cysteines involved in interchain disulfide bond formation, a collagen-like domain containing hydroxyproline and hydroxylysine, a neck region, and a carbohydrate-recognition domain with an amino acid sequence highly homologous to other Ca^{2+}-dependent lectins. Three subunits form a structural unit, and an intact MBP consists of two to six structural units. The carbohydrate-recognition domain is specific for mannooligosaccharide structures on pathogenic organisms, whereas the collagen-like domain is believed to be responsible for interactions with other effector proteins involved in host defense. Cattle possess three serum C-type lectins capable of recognizing mannan in a calcium-dependent manner (Srinivasan et al. 1999).

23.2.1.2 Two Major Forms of MBL in Human Serum

MBP is mainly synthesized in the liver and occurs naturally in two forms, serum MBP (S-MBP) and intracellular MBP (I-MBP). S-MBP activates complement in association with MBP-associated serine proteases via the lectin pathway. The first precise analysis of the major human MBL oligomers was performed by Teillet et al. (2005). Two major MBL forms are present in human serum with mass values of 228,098 + 170 Da (MBL-I) and 304,899 + 229 Da (MBL-II) for the native proteins, whereas reduction of both species yielded a single chain with an average mass of 25,340 + 18 Da. This demonstrates that MBL-I and -II contain 9 and 12 disulfide-linked chains, respectively, and therefore are trimers and tetramers of the structural unit. As shown by surface plasmon resonance spectroscopy, trimeric and tetrameric MBL bound to immobilized mannose-BSA and N-acetylglucosamine-BSA with comparable K_D values (2.2 and 0.55 nM and 1.2 and 0.96 nM, respectively). However, tetrameric MBL exhibited significantly higher maximal binding capacity and lower dissociation rate constants for both carbohydrates. As shown by gel filtration, both MBL species formed 1:2 complexes with MASP-3 or MAp19. The oligomerization state of MBL has a direct effect on its carbohydrate-binding properties, but no influence on the interaction with the MASPs (Teillet et al. 2005). The serum MBP shares similarity with mammalian and chicken hepatic lectins in primary structure of carbohydrate-recognition domain, as well as in the ligand-binding mode: a high affinity ($K_D \sim$ nM) is generated by clustering of approximately 30 terminal target sugar residues on a macromolecule, such as bovine serum albumin, although the individual monosaccharides have low affinity (K_D 0.1–1 mM). On the other hand, MBP does not manifest any significant affinity enhancement toward small, di- and trivalent ligands, in contrast to the hepatic lectins whose affinity toward divalent ligands of comparable structures increased from 100- to 1,000-fold. Such differences may be explained on the basis of different subunit organization between the hepatic lectins and MBP (Lee et al. 1992). Inhibition assays suggested that rat serum MBP has a small binding site which is probably of the trough-type. The 3- and 4-OH of the target sugar are indispensable, while the 6-OH is not required. These characteristics are shared by the rat hepatic lectin and chicken hepatic lectin, both of which are C-type lectins containing carbohydrate-recognition domains highly homologous to that of MBP. Apparently, the related primary structures of these lectins give rise to similar gross architecture of their binding sites, despite the fact that each exhibits different sugar binding specificities (Lee et al. 1991).

23.2.1.3 Intracellular MBP (I-MBP)

The I-MBP shows distinct accumulation in cytoplasmic granules, and is predominantly localized in the endoplasmic reticulum (ER) and involved in COPII vesicle-mediated ER-to-Golgi transport. However, the subcellular localization of either a mutant (C236S/C244S) I-MBP, which lacks carbohydrate-binding activity, or the wild-type I-MBP in tunicamycin-treated cells shows an equally diffuse cytoplasmic distribution, suggesting that the unique accumulation of I-MBP in the ER and COPII vesicles is mediated by an N-glycan-lectin interaction. The binding of I-MBP with glycoprotein intermediates occurs in the ER, which is carbohydrate- and pH-dependent, and is affected by glucose-trimmed high-mannose-type oligosaccharides. These results indicated that I-MBP may function as a cargo transport lectin facilitating ER-to-Golgi traffic in glycoprotein quality control (Nonaka et al. 2007).

23.2.1.4 Ligands for MBL

Among the cells tested, only lymphocytes from thymus of BALB/c mice express endogenous ligands for MBL on their surface, while those from bone marrow, spleen, mesenteric lymph nodes and peripheral blood all being negative. Interestingly, among the thymocytes, only the immature thymocytes with $CD4^+CD8^+CD3$ low phenotype expressed the MBL ligands, which decreased on cell maturation. The major cell surface glycoprotein bearing MBL ligands was identified as CD45RO, which is a transmembrane protein with tyrosine phosphatase activity. The MBL ligands on

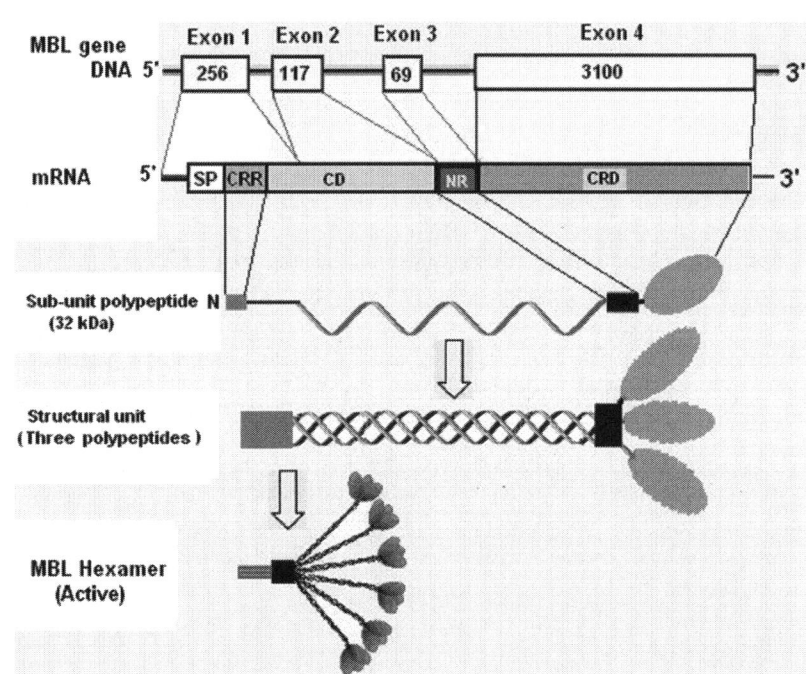

Fig. 23.4 The human MBL gene with the corresponding mRNA and protein domains. The lengths of the four exons and three introns are indicated by the number of base pairs in the box of the figure. The mRNA encodes for the various protein domains shown as signal peptide (*SP*), cysteine rich region (*CRR*), collagen domain (*CD*), neck region (*NR*) (coiled-coil), and carbohydrate recognition domain (*CRD*) in the figure. The peptides self associates into a homo- trimer (structural subunit). Each peptide contains a lectin domain (*gray*) to bind the specific, microbial carbohydrate motifs. Functional MBL circulates in higher-order multimers: (tetramers, pentamers and hexamers)

thymic CD45 contain high mannose type or hybrid type N-linked oligosaccharides (Uemura et al. 1996).

The MBL recognizes a wide array of pathogens independently of specific antibody, and initiates the lectin pathway of complement activation. The MBL binds to neutral carbohydrates on microbial surfaces and recognises carbohydrates such as mannose, glucose, L-fucose, N-acetyl-mannosamine (ManNAc), and N-acetyl-glucosamine (GlcNAc). Oligomerisation of MBL enables high avidity binding to repetitive carbohydrate ligands, such as those present on a variety of microbial surfaces, including *E. coli*, *Klebisella aerogenes*, *Neisseria meningitides*, *Staphylococcus aureus*, *S. pneumonia*, *A. fumigatus* and *C. albicans* (Kerrigan and Brown 2009; Ng et al. 2002) and *Trichinella spirali* (Gruden-Movsesijan et al. 2003).

23.2.2 Gene Structure of MBP

The MBP molecule comprises a signal peptide, a cysteine-rich domain, a collagen-like domain, a 'neck' region and a carbohydrate-binding domain. Each domain is encoded by a separate exon. The NH2 terminus of human MBP is rich in cysteines that mediate interchain disulphide bonds and stabilize the second collagen-like region. This is followed by a short intervening region, and the carbohydrate recognition domain is found in the COOH-terminal region. The genomic organization (Fig. 23.4) lends support to the hypothesis that the gene arose during evolution by a process of exon shuffling. Several consensus sequences that may be involved in controlling the expression of human serum MBP have been identified in the promoter region of the gene. The consensus sequences are consistent with the suggestion that this mammalian serum lectin is regulated as an acute-phase protein synthesized by the liver (Taylor et al. 1989). Analysis of the human MBP gene reveals that the coding region is interrupted by three introns, and all four exons appear to encode a distinct domain of the protein. It appears that the human MBP gene has evolved by recombination of an ancestral nonfibrillar collagen gene with a gene that encodes carbohydrate recognition, and is therefore similar to the human surfactant SP-A gene and the rat MBP gene. The gene for MBP is located on the long arm of chromosome 10 at 10q11.2-q21, a region that is included in the assignment for the gene for multiple endocrine neoplasia type 2A (Sastry et al. 1989; Schuffenecker 1991). In order to elucidate the mechanism underlying the wide intra- and interracial variety in MBP level in serum, Naito et al. (1999) studied the transcriptional regulation of human MBP. The 5′ RACE analysis of Hep G2 RNA indicated the presence of an exon, designated as "exon 0," upstream of exon 1 and thus, two MBP mRNAs with different sizes of 5′-noncoding regions were detected: the longer transcript starting at exon 0 and the shorter one at exon 1. Promoter analysis revealed that the transcript starting from exon 1 predominates over the one that starts from exon 0. The NH2-terminal residue of rat liver MBP, glutamic acid, is preceded by a predominantly hydrophobic stretch of 18 amino acids, which was assumed to be a signal peptide. Near the NH2-terminal, there was a collagen-like domain, which consisted of 19 repeats of the sequence Gly-X-Y. Here, X and Y were frequently proline and lysine. Three proline and lysine residues were hydroxylated, and one of the latter appeared to link to galactose. Computer analysis of several lectins for sequence homology suggested that the COOH-terminal quarter of the MBP is associated with the calcium binding as well as carbohydrate recognition (Oka et al. 1987) (Fig. 23.4).

23.2.2.1 Evolutionary Relationship to the Asialoglycoprotein Receptor

cDNAs encoding MBP-A from rat liver is encoded by four exons separated by three introns. The NH2-terminal, collagen-like portion of the protein is encoded by the first two exons. These exons resemble the exons found in the genes for nonfibrillar collagens in that the intron which divides them is inserted between the first two bases of a glycine codon and the exons do not have the 54- or 108-bp lengths characteristic of fibrillar collagen genes. The carbohydrate-binding portion of MBP-A is encoded by the remaining two exons. This portion of the protein is homologous to the CRD of the hepatic asialoglycoprotein receptor, which is encoded by four exons. It appears that the three COOH-terminal exons of the asialoglycoprotein receptor gene have been fused into a single exon in the MBP-A gene. The organization of the MBP-A gene is very similar to the arrangement of the gene encoding the highly homologous pulmonary surfactant apoprotein, although one of the intron positions is shifted by a single amino acid (Drickamer and McCreary 1987).

23.2.2.2 MBL Genotypes and MBL Levels

Low serum concentrations of MBP are associated with three independent mutations in codons 52, 54, and 57 (Turner 2003) of exon 1, resulting in amino acid replacement of Arg-52 to Cys, Gly-54 to Asp, and Gly-57 to Glu, respectively, all of which occurred in the collagen-like domain. These replacements appear to inhibit oligomerization of the structural unit of the molecule and consequently abolish the ability to initiate complement activation without impairing the original lectin-binding specificity to oligosaccharide ligands. Information from increasing literature suggests that MBL deficiency, which mainly results from the three relatively common single point mutations in exon 1 of the gene, predisposes both to infection by extracellular pathogens and to autoimmune disease. In addition, the protein also modulates disease severity, at least in part through a complex, dose-dependent influence on cytokine production. The mechanisms and signaling pathways involved in such processes remain to be elucidated (Jack et al. 2001). The prevalence of mutations in the MBL gene is about 10%, but in Africa South of Sahara it is as high as 30% (Juul-Madsen et al. 2003).

MBL2 coding alleles associated with low blood levels are present in up to 40% of Caucasoids, with up to 8% having genotypes associated with profound reduction in circulating MBL levels. Low-producing MBL2 variants and low MBL levels are associated with increased susceptibility to and severity of a variety of infective illnesses, particularly when immunity is already compromised – for example, in infants and young children, patients with cystic fibrosis, and after chemotherapy and transplantation. These observations suggest that administration of recombinant or purified MBL may be of benefit in clinical settings where MBL deficiency is associated with a high burden of infection (Worthley et al. 2005).

23.2.3 Regulation of MBP Gene

Suppression by Glucocorticoid: In order to elucidate the mechanism underlying the wide intra- and interracial variety in MBP level in serum, Naito et al. (1999) studied the transcriptional regulation of human MBP. In addition, a hepatocyto-specific nuclear factor, (HNF)-3, which controls the expression of hepatocyte specific genes, up-regulates the transcription of human MBP from exon 1, while a glucocorticoid, which is known to up-regulate acute phase proteins, markedly suppresses MBP transcription. In addition, polymorphism also occurs in the promoter region at two positions (Madsen et al. 1995). Functional promoter analysis indicated that three haplotype variants as to these positions, HY, LY, and LX, exhibit high, medium and low promoter activity, respectively (Naito et al. 1999).

Upregulation by Interleukins: The MBP mRNA expression in human hepatoma cell line HuH-7 is increased by IL-6, dexamethasone, and heat shock, decreased by IL-1, and unaffected by IFNγ, TNFα and TGFβ. The binding of IL-2 to its receptor (IL-2Rβ) induces IL-2Rβ phosphorylation by the tyrosine kinase associated with the T-cell receptor (TCR) complex. This mechanism is due to the putative lectin activity of IL-2 (Cebo et al. 2002), which is a calcium-independent lectin specific for oligomannosidic N-glycans with five and six mannose residues. This lectin activity is preserved after binding of IL-2 to IL-2Rβ. IL-2 behaves as a bifunctional molecule that associates IL-2Rβ with specific glycoprotein ligands of the TCR complex including a glycosylated form of CD3 (Zanetta et al. 1996). Thus, the action of cytokines appears to be mediated by specific transcription factors.

Effect of Growth Hormone: Studies in animals and humans indicate that growth hormone (GH) and insulin-like growth factor-I (IGF-I) modulate immune function. In normal patients, GH therapy increases the level of MBL, and the treatment of acromegalics with pegvisomant decreases the levels of MBL. The effect on MBL was thought to be due to a specific action of GH, since IGF-I treatment did not affect MBL level. However, GH or hormone replacement therapy (HRT) in Turner syndrome (TS) influences the serum levels of MBL and other proteins participating in the innate immune defense such as surfactant protein D (SP-D) and vitamin D binding protein (DBP). The treatment with GH significantly increases MBL and SP-D levels in TS patients, while HRT marginally decreased DBP. Whether the present findings suggest a link between the endocrine and the immune system needs further examination (Gravholt et al. 2004).

Thyroid Hormone Regulates MBL Levels: Studies have indicated the existence of causal links between the endocrine and immune systems and cardiovascular disease. In all hyperthyroid patients, MBL levels are increased – median (range), 1,886 ng/ml before treatment and decreased to 954 ng/ml from normal levels of 1,081 ng/ml after treatment. Administration of thyroid hormones to healthy persons induced mild hyperthyroidism and increased MBL levels significantly to 1,714 ng/ml. Since MBL is part of the inflammatory complement system, the modulation by thyroid hormone, of complement activation may play a role in the pathogenesis of a number of key components of thyroid diseases (Riis et al. 2005).

23.2.4 Structure-Function Relations

MBL has an oligomeric structure (400–700 kDa), built of subunits that contain three identical peptide chains of 32 kDa each. Although MBL can form several oligomeric forms, there are indications that dimers and trimers are not biologically active and at least a tetramer form is needed for activation of complement. MBL in the blood is complexed with another protein, a serine protease called MASP-2 (MBL-associated serine protease). The crystal structure of the CRD of a rat mannose-binding protein, determined as the holmium-substituted complex, reveals an unusual fold consisting of two distinct regions, one of which contains extensive nonregular secondary structure stabilized by two holmium ions. The structure explains the conservation of 32 residues in all C-type carbohydrate-recognition domains, suggesting that the fold seen is common to these domains (Weis et al. 1991, 1998).

The basic structural unit is a triple helix of MBL peptides, which aggregate into complement-fixing higher-order structures (tetramers, pentamers and hexamers). MBP forms a trimeric helical structure via interactions of the collagenous tails that are stabilized by disulfide bonds in the cysteine-rich amino terminal region. These trimers aggregate to generate three or six trimers in a "bouquet" organization (Fig. 23.1). Each CRD in the trimer is separated by approximately 53 Å, which is critical to the function of the lectin. This is because each individual CRD has a relatively low affinity and low specificity for glycan ligands and can bind to glycans rich in N-acetylglucosamine, N-acetylmannosamine, fucose, and glucose. The spacing between CRDs provides regulation and enhances the potential interactions with extended mannan-containing glycoconjugates, especially those on bacteria, yeast, and parasites. Weis and Drickamer (1999) determined crystal structure at 1.8 A Bragg spacings of a trimeric fragment of MBP-A, containing the CRD and the neck domain that links the carboxy-terminal CRD to the collagen-like portion of the intact molecule. The neck consists of a parallel triple-stranded coiled coil of α-helices linked by four residues to the CRD (Chang et al. 1994). The isolated neck peptide does not form stable helices in aqueous solution. The carbohydrate-binding sites lie at the distal end of the trimer and are separated from each other by 53 A. The carbohydrate-binding sites in MBP-A are too far apart for a single trimer to bind multivalently to a typical mammalian high-mannose oligosaccharide. Thus MBPs can recognize pathogens selectively by binding avidly only to the widely spaced, repetitive sugar arrays on pathogenic cell surfaces. Sequence alignments revealed that other C-type lectins might have a similar oligomeric structure, but differences in their detailed organization will have an important role in determining their interactions with oligosaccharides.

Phagocytic Interaction Site on MBL: The phagocytic activity induced by MBL and other molecules that contain a collagen-like region contiguous with a pattern recognition domain is mediated by C1qR(P). The specific interaction site was identified through two mutants, one of which has five GXY triplets deleted below the kink region of MBL and the other one having only two of the GXY triplets deleted below the kink. These mutants, which failed to enhance phagocytosis, suggested the importance of a specific sequence GEKGEP in stimulating phagocytic activity. Similar sequences were detected in other defense collagens, implicating the consensus motif GE(K/Q/R)GEP as critical in mediating the enhancement of phagocytosis through C1qR(P) (Arora et al. 2001).

MASP-Binding Sites in MBP: Mutations in the collagen-like domain of serum MBP interfere with the ability of the protein to initiate complement fixation through MASPs. Studies with truncated and modified MBPs and synthetic peptides demonstrated that MASPs bind on the C-terminal side of the hinge region formed by an interruption in the Gly-X-Y repeat pattern of the collagen-like domain. The binding sites for MASP-2 and for MASP-1 and -3 overlap but are not identical. The two most common naturally occurring mutations in MBP result in substitution of acidic amino acids for glycine residues in Gly-X-Y triplets on the N-terminal side of the hinge. Studies showed that the triple helical structure of the collagen-like domain is largely intact in the mutant proteins, but it is more easily unfolded than in wild-type MBP. Thus, the effect of the mutations is to destabilize the collagen-like domain, indirectly disrupting the binding sites for MASPs (Wallis et al. 2004).

23.2.5 Functions of MBL

23.2.5.1 Functions as an Opsonin

MBL has also been proposed to function directly as an opsonin by binding to carbohydrates on pathogens and then interacting with MBL receptors on phagocytic cells, promoting microbial uptake and stimulating immune responses (Fig. 23.5). This was described by Kuhlman et al. who observed that binding of MBL to *Salmonella montevideo* resulted in an MBL-dependent uptake by monocytes (Kuhlman et al. 1989). Thus MBL can interact directly with receptor(s) on the surface of monocytes and several potential MBL receptors have been proposed, although their likelihood is debated in the literature. Calreticulin has emerged as the main candidate, but further studies are required to confirm its interaction with MBL and its role in the phagocytosis of pathogens. Ip et al. (2008) has shown that MBL modifies cytokine responses through cooperation with TLR2/6 in the phagosome. Although the stimulation of the inflammatory response was not caused by enhanced phagocytosis, bacterial engulfment was required. This study demonstrates the importance of phagocytosis in providing the appropriate cellular environment to facilitate cooperation between molecules.

MBL has the capacity to modify the efficiency of uptake and the expression of other phagocytic receptors. Activation of the complement system via MBL-associated serine proteases (MASPs) (Kerrigan and Brown 2009; Dahl et al. 2001; Stover et al. 1999; Thiel et al. 1997), results in deposition of complement on the microbial surface that can lead to uptake via complement receptors (Neth et al. 2002). However, inhibition of bacterial growth associated with the MBL-MASP activation of complement has also been observed, without any enhancement of phagocytosis (Ip and Lau 2004). This indicates that the specific responses induced by MBL may be dependent on the nature of the microbial target. MBL can also influence expression of other PRRs, as demonstrated by the ability of MBL to augment the uptake of *S. aureus* through the up-regulation of scavenger receptor A (SR-A) (Ono et al. 2006) (Fig. 23.5).

Recognition of Pathogens: MBL has a major protective effect through activation of the complement system via MBL-associated serine proteases (MASPs). This can cause the lysis of Gram-negative bacteria and also opsonize a wide spectrum of potential pathogens for phagocytosis. MBL may also influence phagocytosis in the absence of complement activation through an interaction with one or more collectin receptors. This may also be the basis for a direct effect of the protein on inflammatory responses. MBL forms individual complexes with MBL-associated serine proteases (MASP)-1, -2, -3 and a truncated form of MASP-2 (MAp19) and triggers the lectin pathway of complement through MASP-2 activation. MBL, like C1q, is a six-headed molecule that forms a complex with two protease zymogens, which in the case of the MBL complex are MASP-1 and MASP-2. MASP-1 and MASP-2 are closely homologous to C1r and C1s, and all four enzymes are likely to have evolved from gene duplication of a common precursor. When the MBL complex binds to a pathogen surface, MASP-1 and MASP-2 are activated to cleave C4 and C2. Thus the MBL pathway initiates complement activation in the same way as the classical pathway, forming a C3 convertase from C2b bound to C4b. Humans deficient in MBL experience a substantial increase in infections during early childhood, indicating the importance of the MBL pathway for host defense.

The Mannan-Binding Lectin Pathway Is Homologous to the Classical Pathway: MBP has been shown to have complement-dependent bactericidal activity; for example, *Escherichia coli* strains *K12* and *B*, which have exposed *N*-acetylglucosamine and L-glycero-D-mannoheptose, respectively, are killed by MBP with the help of complement. MBP serves as a direct opsonin and mediates binding and uptake of bacteria that express a mannose-rich O-polysaccharide by monocytes and neutrophils. MBP functions as a β-inhibitor of the influenza virus and protects cells from HIV infection by binding to gp120, a high mannose-type oligosaccharide-containing envelope glycoprotein on HIV. MBL can alter the function of microbial structures, such as gp120 of HIV, to prevent infection. In addition, the α-mannosidase inhibitor 1-deoxymannojirimycin-treated baby hamster kidney (BHK) cells, which have high mannose-type oligosaccharide exposed on their surfaces, can be killed by MBP with the help of complement. MBP activates complement through interactions with complement subcomponents C1r/C1s or two novel C1r/C1s-like serine proteases, MBP-associated serine proteases (MASP-1 and MASP-2). The MBP-mediated complement activation is named the MBP pathway. The protein may also interact with the components of other cascade systems such as the clotting system, which will have a role in microbial pathogenesis. An understanding of these basic mechanisms will be vital if we are to use purified or recombinant MBL in therapeutic applications (Jack et al. 2001).

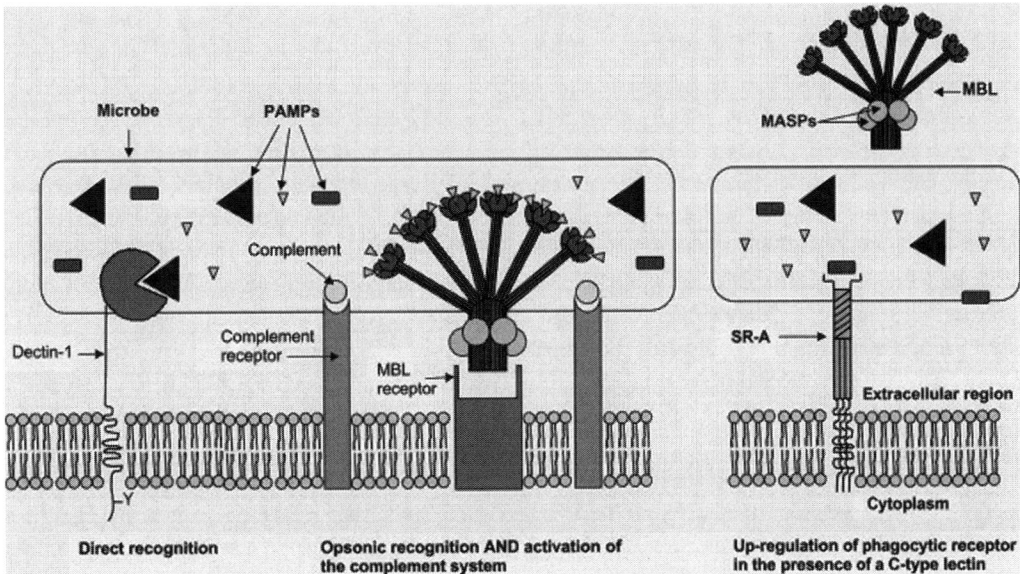

Fig. 23.5 Direct recognition, opsonisation, complement activation and receptor up-regulation by C-type lectins. Phagocytic C-type lectins can directly recognise PAMPs on the surface of microbes and mediate phagocytosis (e.g. Dectin-1). Alternatively, soluble C-type lectins can interact directly with pathogens to promote opsonisation of the microbe (e.g. MBL) which can subsequently be phagocytosed via specific receptors. In addition, some C-type lectins can activate complement leading to its deposition on the microbial surface and phagocytosis mediated by complement receptors (e.g. MBL associated MASPS are activated on binding to pathogens which in turn cleave complement components and activate the complement system). Finally, C-type lectins can cause up-regulation of other phagocytic receptors, independently of their binding to the microbe [e.g. MBL up-regulation of scavenger receptor A (*SR-A*)] (Adapted with permission from Kerrigan and Brown 2009 © Elsevier)

The MBL binds through multiple lectin domains to the repeating sugar arrays that decorate many microbial surfaces, and is then able to activate the complement system through a MBL-associated protease-2. For selected Gram-negative organisms, such as *Salmonella* and *Neisseria*, the relative roles of lipopolysaccharide (LPS) structure and capsule are important in binding; the LPS is of major importance. Studies on clinically relevant organisms showed that MBL binding leads to activation of purified C4, suggesting that the bound lectin is capable of initiating opsonophagocytosis and/or bacterial lysis.

Normal mammalian cells such as circulating blood cells are usually covered with complex oligosaccharides terminated with sialic acids and do not bind MBP. On the other hand, malignant transformations or viral infections modify the oligosaccharide structures on cell surfaces, and some tumor tissues have been shown to bind to MBP. Ma et al. (1999) showed that MBP recognizes and binds specifically to oligosaccharide ligands expressed on the surfaces of a human colorectal carcinoma. Interestingly, the recombinant vaccinia virus carrying human MBP gene was demonstrated to possess a potent growth-inhibiting activity against human colorectal carcinoma cells transplanted in nude mice and a significant prolongation of life span of tumor-bearing mice. Unexpectedly, the mutant MBP, which had essentially no complement-activating activity, was nearly as active as wild-type MBP. These results indicated that MBP has a cytotoxic activity, which was termed as MBP-dependent cell-mediated cytotoxicity (MDCC).

23.2.5.2 Functions in Antibody-Independent Pathway of Complement System

The complement system comprises a complex array of enzymes and non-enzymatic proteins that is essential for the operation of innate as well as the adaptive immune defense. The complement system is activated in three ways: by the classical pathway which is initiated by antibody-antigen complexes, by the alternative pathway initiated by certain structures on microbial surfaces, and by an antibody-independent pathway that is initiated by the binding of mannan-binding lectin to carbohydrates. The MBL is structurally related to the complement C1 subcomponent, C1q, and seems to activate the complement system through an associated serine protease known as MASP-1 or p100, which is similar to C1r and C1s of the classical pathway. The MBL binds to specific carbohydrate structures found on the surface of a range of microorganisms, including bacteria, yeasts, parasitic protozoa and viruses, and exhibits antimicrobial activity by killing that is mediated by the terminal, lytic complement components or by promoting phagocytosis. The level of MBL in blood plasma is genetically determined, and deficiency is associated with frequent infections in childhood, and possibly also in adults. A new MBL-associated serine protease (MASP-2) which shows a striking homology with the reported MASP-1 and the two C1q-associated serine proteases C1r and C1s has also been identified. Thus complement activation through MBL, like the classical pathway, involves two serine proteases and may antedate the development of the specific immune system of vertebrates (Thiel et al. 1997).

IgA, an important mediator of mucosal immunity, activates the complement system via the lectin pathway. Results indicate a dose-dependent binding of MBL to polymeric, but not monomeric IgA coated in microtiter plates. This interaction involves the carbohydrate recognition domain of MBL. MBL binding to IgA results in complement activation, which is proposed to lead to a synergistic action of MBL and IgA in antimicrobial defense. These results may explain glomerular complement deposition in IgA nephropathy (Roos et al. 2001).

23.2.5.3 Modulation of Associated Serine Proteases

MBP neutralizes invading microorganisms by binding to cell surface carbohydrates and activating MBP-associated serine proteases-1, -2, and -3 (MASPs). MASP-2 subsequently cleaves complement components C2 and C4 to activate the complement cascade. The MBP modulates MASP-2 activity in two ways. First, MBP stimulates MASP-2 auto-activation by increasing the rate of autocatalysis when MBP.MASP-2-complexes bind to a glycan-coated surface. Second, MBP occludes accessory C4-binding sites on MASP-2 until activation occurs. Once these sites become exposed, MASP-2 binds to C4 while separate structural changes create a functional catalytic site able to cleave C23. Only activated MASP-2 binds to C2, suggesting that this substrate interacts only near the catalytic site and not at accessory sites. MASP-1 cleaves C2 almost as efficiently as MASP-2 does, but it does not cleave C23. Thus MASP-1 probably enhances complement activation triggered by MBP.MASP-2 complexes, but it cannot initiate activation itself (Chen and Wallis 2004). MASP-1 probably enhances complement activation triggered by MBP. MASP-2 complexes, but it cannot initiate activation itself.

In the human lectin pathway, MASP-1 and MASP-2 are involved in the proteolysis of C4, C2 and C3. The human MBL-MASP complex contains a new 22 kDa protein [small MBL-associated protein (sMAP)] bound to MASP-1. The nucleotide sequence of sMAP cDNA revealed that it is a truncated form of MASP-2, consisting of the first two domains (i.e. the first internal repeat and the EGF-like domain) with four different C-terminal amino acids. sMAP mRNAs are expressed in liver by alternative polyadenylation of the MASP-2 gene, in which a sMAP-specific exon containing an in-frame stop codon and a polyadenylation signal is used. The involvement of sMAP in the MBL-MASP complex suggests that the activation mechanism of the lectin pathway is more complicated than that of the classical pathway (Takahashi et al. 1999).

23.2.5.4 Ra-Reactive Factor: A Complex of MBL and the Serine Protease

The Ra-reactive factor (RaRF) is a complement dependent antimicrobial factor that reacts with numerous microorganisms such as viruses, bacteria, fungi and protozoa. The RaRF consists of a complement-activating component (CRaRF) called P100 component (MASP) and a polysaccharide-binding (mannose-binding) component (Matsushita et al. 1992). The cDNA for RaRF-P100 from the human liver contains an ORF of 2,097 nt encoding a protein of 699 aa residues in the cloned cDNA of 4,489 nt. This protein exhibits 87.4% amino acid homology with mouse P100, and 36.4% and 37.1% homologies with that of the C1r and C1s subcomponents of human complement, respectively. The P100, together with the C1r and C1s, forms a unique protein family having the same module/domain constitution (Takada et al. 1993; Takayama et al. 1999).

The polysaccharide-binding component of RaRF consists of two different 28-kDa polypeptides, P28a and P28b. Partial amino acid sequences of P28a and P28b indicated that these polypeptides are similar to MBP-C and MBP-A (liver and serum MBP respectively) (Drickamer et al. 1986; Oka et al. 1987). The primary structures of P28a and P28b deduced from cDNAs are homologous to one another. They have three domains, a short NH2-terminal domain, a collagen-like domain, and a domain homologous to regions of some carbohydrate-binding proteins, as in rat MBPs. The P28a and P28b polypeptides are the products of two unique mouse genes which are expressed in hepatic cells (Kuge et al. 1992).

23.3 MBP/MBL from Other Species

23.3.1 Rodents

Rat liver MBP contains two distinct but homologous polypeptides. Each polypeptide consists of three regions: (a) an NH2-terminal segment of 18–19 amino acids which is rich in cysteine and appears to be involved in the formation of interchain disulfide bonds which stabilize dimeric and trimeric forms of the protein, (b) a collagen-like domain consisting of 18–20 repeats of the sequence Gly-X-Y and containing 4-hydroxyproline residues in several of the Y positions, and (c) a COOH-terminal carbohydrate-binding domain of 148–150 amino acids. The sequences of the COOH-terminal domains are highly homologous to the sequence of the COOH-terminal carbohydrate-recognition portion of the chicken liver receptor for N-acetylglucosamine-terminated glycoproteins and the rat liver asialoglycoprotein receptor. Each protein is preceded by a cleaved, NH2-terminal signal sequence, consistent with the finding that this protein is found in serum as well as in the liver. The entire structure of the mannose-binding proteins is homologous to dog pulmonary surfactant apoprotein (Drickamer et al. 1986). Rat liver MBP is specific for mannose and N-acetylglucosamine and is encoded by two species of mRNA of 1.4 and 3.5 kb respectively. The sequence of the open reading frame of 3.5 kb mRNA was completely identical to that of the 1.4 kb mRNA. Each mRNA species arises from one gene, with the differences in size most easily being accounted for by differential utilization of the polyadenylation sites of one transcript (Wada et al. 1990).

Mouse *Mbl1* and *Mbl2* have five and six exons, respectively. The structure of the mouse Mbl genes is similar to that of the rat and human MBP genes and shows homology to the other collectin genes, with the entire carbohydrate recognition domain being encoded in a single exon and all introns being in phase 1. The MBP encoded by mouse Mbl1 with three cysteines in the first coding exon, like the rat Mbl1 and human MBL, is capable of a higher degree of multimerization and has apparent ability to fix complement in the absence of antibody or C1q. However, the structural features of other exons, that is, the larger size of collagen domain region in the first coding exon (64 bp in Mbl2 vs 46 bp in Mbl1) and the smaller size of the exon encoding the trimerization domain (69 bp in Mbl2 vs 75 bp in Mbl1) revealed that the single human MBL gene is closely related to rodent Mbl2 rather than rodent Mbl1. In contrast to the evolution of bovine surfactant protein-D – which duplicated in bovidae after divergence from humans, MBP gene most likely duplicated prior to human-rodent divergence, and that the human homolog to *Mbl1* was perhaps lost during evolution (Sastry et al. 1995). The deduced amino acid sequence of the mouse MBP, as with rat and the human forms, have an NH2 terminus that is rich in cysteine, which stabilizes a collagen α-helix followed by a carboxyl- terminal carbohydrate binding domain. Though, the mouse MBP-A mRNA, as with the human, is induced like the acute phase reactant serum amyloid P protein, yet the expression of mouse MBP-C mRNA is not regulated above its low baseline level. The expression of both MBP-A and -C mRNA is restricted to the liver under basal and stress conditions (Sastry et al. 1991).

23.3.2 Primates MBL

The nucleotide and amino-acidic sequences of MBL2 among primates are highly homologous, underlining the importance of this molecule in the defense system against pathogen invasions. In particular, in the collagen-like domain that confers the characteristic structure to MBL2 protein, the identity among primates is really high. In the carbohydrate recognition domain, primates' group-specific amino-acidic mutations did not result in changes of the structure or function of this MBL2 domain. Results indicate that MBL2 is well conserved in agreement with its important role in the immune system. However, same 'plasticity' of the MBL2 human gene was not observed in non-human primates, where a frequency of more than 1% of nucleotide variations was described in the coding and promoter regions (Verga Falzacappa et al. 2004).

A bovine cDNA encodes a protein of 249 aa residues with a signal peptide of 19 aa. This MBP has the ability to activate complement and expressed only in liver. The bovine MBP is likely to be a homologous to human MBP and bears homology to rat and mouse MBP-C which are localized in liver cells rather than to rat and mouse MBP-A found in serum (Kawai et al. 1997). C-type lectins in alligator liver, termed alligator hepatic lectin (AHL) is specific for mannose/L-fucose and contained equal amounts of 21- and 23-kDa bands on SDS (Lee et al. 1994).

23.4 Similarity Between C1Q and Collectins/Defense Collagens

23.4.1 Structural Similarities

The C1q is the first component of classical pathway of complement activation that links the adaptive humoral immune response to the complement system by binding to antibodies complexed with antigens. C1q can, however, also bind directly to the surface of certain pathogens and thus trigger complement activation in the absence of antibody. Collectins family of macromolecules is characterized by a conserved, collagen-like region of repeating Gly-X-Y triplets contiguous with a non-collagen-like sequence. The structural similarity between the collectins and C1q has been demonstrated: They all contain multiple polypeptides which are organised into subunits containing triple-helical stalks throughout their collagen-like regions and globular 'heads' in C-terminal regions. Four, or six, of these structures are associated via distinct, short, N-terminal regions to form oligomeric molecules seen in electron microscope. The overall structural similarity between C1q and collectins, however, does not extend to similarity in amino acid sequences over the C-terminal regions. The C-terminal regions of C1q, unlike those of collectins, do not contain conserved residues found in CRDs present in C-type lectins. Instead, C1q has a high degree of homology to collagen sequences (Type VIII and X) and this is consistent with the fact that, unlike collectins, C1q binds to protein motifs in IgG, or IgM, rather than to carbohydrate structures. Also, despite showing interruptions in their collagen-like regions, collectins do not always display a 'bend' in their collagen-like 'stalks' similar to that which is seen in C1q. Therefore, C1q may be more closely related to collagens than to collectins.

23.4.2 Functional Similarities

The C1q and MBL function to eliminate invading microorganisms by activating the classical pathway (C1q) or the lectin pathway (MBL and ficolin) (Ma et al. 2004) of the complement system by transmitting a signal from the recognition domains of the globular heads to their collagen-like domains, which autoactivates their associated serine proteases ($C1r_2s_2$ or mannan-binding lectin-associated serine proteases (Sim and Tsiftsoglou 2004).

Defense collagens modulate the cytokine expression, specifically by monocyte cytokine production by phagocyte cells under conditions in which phagocytosis is enhanced. Under conditions in which phagocytosis is enhanced, C1q and MBL modulate cytokine production and contribute signals to human peripheral blood mononuclear cells, leading to the suppression of LPS-induced pro-inflammatory cytokines, IL-1α and IL-1β, and an increase in the secretion of cytokines IL-10, IL-1 receptor antagonist, monocyte chemoattractant protein-1, and IL-6. Thus, defense collagen-mediated suppression of a pro-inflammatory response may be an important step in the avoidance of autoimmunity during the clearance of apoptotic cells (Fraser et al. 2006). Understanding the mechanisms involved in defense collagen and other soluble pattern recognition receptor modulation of the immune response may provide important novel insights into therapeutic targets for infectious and/or autoimmune diseases and additionally may identify avenues for more effective vaccine design (Fraser and Tenner 2008; Paidassi et al. 2007).

It is proposed that C1q, MBL, and other opsonins prevent autoimmunity and maintain self-tolerance by supporting the efficient clearance of apoptotic material, as well as by actively modulating phagocyte function. In absence of danger, defense collagens appear to recognise and remove apoptotic cells (Nauta et al. 2003). C1q and MBL have been shown to bind directly to apoptotic cell surfaces and apoptotic cell blebs via their globular heads (Navratil et al. 2001). Interaction of the collagen-like tails with the phagocyte surface triggers apoptotic cell ingestion via macropinocytosis (Ogden et al. 2001). Indeed, deficiencies and/or knock-out mouse studies have highlighted critical roles for C1q, SP-D, and MBL and other soluble pattern recognition receptors in the clearance of apoptotic bodies and protection from autoimmune diseases along with mediating protection from specific infections (Botto et al. 1998; Gabriaud et al. 2003; Stuart et al. 2005), in which mice exhibit impaired clearance of apoptotic cells. Deficiency of C1q is a risk factor for the development of autoimmunity in humans and mice (Botto and Walport 2002; Mitchell et al. 2002; Miura-Shimura et al. 2002).

23.4.3 Receptors for Defense Collagens

A number of putative MBL-binding proteins/receptors have been proposed including cC1qR/Crt (Klickstein et al. 1997), C1qRp (Tenner et al. 1995; Stuart et al. 1997) and CR1 (Ghiran et al. 2000). However, it is unclear whether MBL is acting as a direct opsonin or is merely enhancing other complement pathways and/or antibody-mediated phagocytosis. Membrane receptors for the soluble 'defense collagens' – naturally occurring chimeric molecules that contain a recognition domain contiguous with a collagen-like triple helical domain and play a role in protecting the host from pathogens entering the blood, lung and other tissues – have been isolated and are being characterized. These receptors are key to understanding the mechanisms by which defense collagens influence cellular responses in clearance of cellular debris or to initiate the responses that lead to the destruction of microbial pathogens.

References

Arora M, Munoz E, Tenner AJ (2001) Identification of a site on mannan-binding lectin critical for enhancement of phagocytosis. J Biol Chem 276:43087–43094

Beck K, Brodsky B (1998) Supercoiled protein motifs: the collagen triple-helix and the alpha-helical coiled coil. J Struct Biol 122:17–29

Botto M, Dell'agnola C, Bygrave AE et al (1998) Homozygous C1q deficiency causes glomerulonephritis associated with multiple apoptotic bodies. Nat Genet 19:56–59

Botto M, Walport MJ (2002) C1q, autoimmunity and apoptosis. Immunobiology 205:395–406

Cebo C, Vergoten G, Zanetta U-P (2002) Lectin activities of cytokines: functions and putative carbohydrate-recognition domains. Biochim Biophys Acta 1572:422–434

Chang CY, Sastry KN, Gillies SD et al (1994) Crystallization and preliminary x-ray analysis of a trimeric form of human mannose binding protein. J Mol Biol 241:125–127

Chen CB, Wallis R (2004) Two mechanisms for mannose-binding protein modulation of the activity of its associated serine proteases. J Biol Chem 279:26058–26065

Childs RA, Feizi T, Yuen CT et al (1990) Differential recognition of core and terminal portions of oligo-saccharide ligands by carbohydrate-recognition domains of two mannose-binding proteins. J Biol Chem 265:20770–20777

Crouch E, Persson A, Chang D, Heuser J (1994) Molecular structure of pulmonary surfactant protein D SP-D! J Biol Chem 269:17311–17319

Crouch EC (1998) Collectins and pulmonary host defense. Am J Respir Cell Mol Biol 19:177–201

Dahl MR, Thiel S, Matsushita M et al (2001) MASP-3 and its association with distinct complexes of the mannan-binding lectin complement activation pathway. Immunity 15:127–135

Drickamer K, Dordal MS, Reynolds L (1986) Mannose-binding proteins isolated from rat liver contain carbohydrate-recognition domains linked to collagenous tails. Complete primary structures and homology with pulmonary surfactant apoprotein. J Biol Chem 261:6878–6887

Drickamer K, McCreary V (1987) Exon structure of a mannose-binding protein gene reflects its evolutionary relationship to the asialoglycoprotein receptor and nonfibrillar collagens. J Biol Chem 262:2582–2589

Drickamer K, Taylor ME (1993) Biology of animal lectins. Annu Rev Cell Biol 9(2):37–64

Drickamer K (1999) C-type lectin-like domains. Curr Opin Struct Biol 9:585–590

Elhalwagi BM, Damodarasamy M, McCormack FX (1997) Alternate amino terminal processing of surfactant protein A results in cysteinyl isoforms required for multimer formation. Biochemistry 36:7018–7025

Fraser DA, Bohlson SS, Jasinskiene N et al (2006) C1q and MBL, components of the innate immune system, influence monocyte cytokine expression. J Leukoc Biol 80:107–116

Fraser DA, Tenner AJ (2008) Directing an appropriate immune response: the role of defense collagens and other soluble pattern recognition molecules. Curr Drug Targets 9:113–122

Gaboriaud C, Juanhuix J, Gruez A et al (2003) The crystal structure of the globular head of complement protein C1q provides a basis for its versatile recognition properties. J Biol Chem 278:46974–46982

Garred P, Thiel S, Madsen HO et al (1992) Gene frequency and partial protein characterization of an allelic variant of mannan binding protein associated with low serum concentrations. Clin Exp Immunol 90:517–521

Ghiran I, Barbashov SF, Klickstein LB et al (2000) Complement receptor 1/CD35 is a receptor for mannan-binding lectin. J Exp Med 192:1797–1808

Goodman EB, Tenner AJ (1992) Signal transduction mechanisms of C1q-mediated superoxide production. Evidence for the involvement of temporally distinct staurosporine-insensitive and sensitive pathways. J Immunol 148:3920–3928

Gravholt CH, Leth-Larsen R, Lauridsen AL et al (2004) The effects of GH and hormone replacement therapy on serum concentrations of mannan-binding lectin, surfactant protein D and vitamin D binding protein in Turner syndrome. Eur J Endocrinol 150:355–362

Gruden-Movsesijan A, Petrovic M, Sofronic-Milosavljevic L (2003) Interaction of mannan-binding lectin with Trichinella spiralis glycoproteins, a possible innate immune mechanism. Parasite Immunol 25:545–552

Håkansson K, Lim NK, Hoppe HJ, Reid KB (1999) Crystal structure of the trimeric alpha-helical coiled-coil and the three lectin domains of human lung surfactant protein D. Struct Fold Des 7:255–264

Håkansson K, Reid KB (2000) Collectin structure: a review. Protein Sci 9:1607–1617

Hansen S, Holmskov U (2002) Lung surfactant protein D (SP-D) and the molecular diverted descendants: conglutinin, CL-43 and CL-46. Immunobiology 205:498–517

Hansen S, Holmskov U (1998) Structural aspects of collectins and receptors for collectins. Immunobiology 199:165–189

Holmskov U, Laursen SB, Malhotra R et al (1995) Comparative study of the structural and functional properties of a bovine plasma C-type lectin, collectin-43, with other collectins. Biochem J 305:889–896

Holmskov U, Lawson P, Teisner B et al (1997) Isolation and characterization of a new member of the scavenger receptor superfamily, glycoprotein-340 gp-340, as a lung surfactant protein-D binding molecule. J Biol Chem 272:13743–13749

Holmskov U, Thiel S, Jensenius JC (2003) Collectins and ficolins: humoral lectins of the innate immune defense. Annu Rev Immunol 21:547–578

Hoppe HJ, Reid KB (1994) Collectins–soluble proteins containing collagenous regions and lectin domains–and their roles in innate immunity. Protein Sci 3:1143–1158

Ikeda K, Sannoh T, Kawasaki N et al (1987) Serum lectin with known structure activates complement through the classical pathway. J Biol Chem 262:7451–7454

Ip WK, Lau YL (2004) Role of mannose-binding lectin in the innate defense against Candida albicans: enhancement of complement activation, but lack of opsonic function, in phagocytosis by human dendritic cells. J Infect Dis 190:632–640

Ip WK, Takahashi K, Moore KJ et al (2008) Mannose-binding lectin enhances toll-like receptors 2 and 6 signaling from the phagosome. J Exp Med 205:169–181

Jack DL, Klein NJ, Turner MW (2001) Mannose-binding lectin: targeting the microbial world for complement attack and opsono phagocytosis. Immunol Rev 180:86–99

Juul-Madsen HR, Munch M, Handberg KJ et al (2003) Serum levels of mannan-binding lectin in chickens prior to and during experimental infection with avian infectious bronchitis virus. Poult Sci 82:235–241

Kawai T, Suzuki Y, Eda S, Fujinaga Y, Sakamoto T, Kurimura T, Wakamiya N et al (1997) Cloning and characterization of a cDNA encoding bovine mannan-binding protein. Gene 186:161–165

Kawasaki T, Etoh R, Yamashina I (1978) Isolation and characterization of a mannan-binding protein from rabbit liver. Biochem Biophys Res Commun 81:1018–1024

Kerrigan AM, Brown GD (2009) C-type lectins and phagocytosis. Immunobiology 214:562–575

Klickstein LB, Barbashov SF, Liu T, Jack RM, Nicholson-Weller A (1997) Complement receptor type 1 (CR1, CD35) is a receptor for C1q. Immunity 7:345–355

Kramer RZ, Bella J, Mayville P et al (1999) Sequence dependent conformational variations of collagen triple-helical structure. Nat Struct Biol 6:454–457

Kramer RZ, Vitagliano L, Bella J et al (1998) X-ray crystallographic determination of a collagenlike peptide with the repeating sequence (Pro-Pro-Gly). J Mol Biol 280:623–638

Kuge S, Ihara S, Watanabe E et al (1992) cDNAs and deduced amino acid sequences of subunits in the binding component of mouse bactericidal factor, Ra-reactive factor: similarity to mannose-binding proteins. Biochemistry 31:6943–6950

Kuhlman M, Joiner K, Ezekowitz RA (1989) The human mannose-binding protein functions as an opsonin. J Exp Med 169:1733–1745

Kurata H, Sannoh T, Kozutsumi Y et al (1994) Structure and function of mannan-binding proteins isolated from human liver and serum. J Biochem (Tokyo) 115:1148–1154

Lee RT, Ichikawa Y, Fay M et al (1991) Ligand-binding characteristics of rat serum-type mannose-binding protein (MBP-A). Homology of binding site architecture with mammalian and chicken hepatic lectins. J Biol Chem 266:4810–4815

Lee RT, Ichikawa Y, Kawasaki T et al (1992) Multivalent ligand binding by serum mannose-binding protein. Arch Biochem Biophys 299:129–136

Lee RT, Yang GC, Kiang J et al (1994) Major lectin of alligator liver is specific for mannose/L-fucose. J Biol Chem 269:19617–19625

Lillie BN, Brooks AS, Keirstead ND, Hayes MA (2005) Comparative genetics and innate immune functions of collagenous lectins in animals. Vet Immunol Immunopathol 108:97–110

Lipscombe RJ, Sumiya M, Hill AV et al (1992) High frequencies in African and non-African populations of independent mutations in the mannose binding protein gene. Hum Mol Genet 1:709–715

Lu J, Wiedemann H, Timpl R, Reid KB (1993) Similarity in structure between C1q and the collectins as judged by electron microscopy. Behring Inst Mitt 93:6–16

Lu J (1997) Collectins: collectors of microorganisms for the innate immune system. Bioessays 19:509–518

Ma Y, Cho G, Zhao MY, Park M et al (2004) Human mannose-binding lectin and L-ficolin function as specific pattern recognition proteins in the lectin activation pathway of complement. J Biol Chem 279:25307–25312

Ma Y, Shida H, Kawasaki T (1997) Functional expression of human mannan-binding proteins (MBPs) in human hepatoma cell lines infected by recombinant vaccinia virus: post-translational modification, molecular assembly, and differentiation of serum and liver MBP. J Biochem 122:810–818

Ma Y, Uemura K, Oka S et al (1999) Antitumor activity of mannan-binding protein in vivo as revealed by a virus expression system: mannan-binding proteinindependent cell-mediated cytotoxicity. Proc Natl Acad Sci USA 96:371–375

Madsen HO, Garred P, Kurtzhals JA et al (1994) A new frequent allele is the missing link in the structural polymorphism of the human mannan-binding protein. Immunogenetics 40:37–44

Madsen HO, Garred P, Thiel S et al (1995) Interplay between promoter and structural gene variants control basal serum level of mannan-binding protein. J Immunol 155:3013–3020

Malhotra R, Laursen SB, Willis AC, Sim RB (1993) Localization of the receptor-binding site in the collectin family of proteins. Biochem J 293:15–19

Matsushita M, Takahashi A, Hatsuse H et al (1992) Human mannose-binding protein is identical to a component of Ra-reactive factor. Biochem Biophys Res Commun 183:645–651

McCormack FX, Pattanajitvilai S, Stewart J et al (1997) The Cys6 intermolecular disulfide bond and the collagen-like region of rat SP-A play critical roles in interactions with alveolar type II cells and surfactant lipids. J Biol Chem 272:27971–27979

Mitchell DA, Pickering MC, Warren J et al (2002) C1q deficiency and autoimmunity: the effects of genetic background on disease expression. J Immunol 168:2538–2543

Miura-Shimura Y, Nakamura K, Ohtsuji M et al (2002) C1q regulatory region polymorphism down-regulating murine C1q protein levels with linkage to lupus nephritis. J Immunol 169:1334–1339

Naito H, Ikeda A, Hasegawa K et al (1999) Characterization of human serum mannan-binding protein promoter. J Biochem (Tokyo) 126:1004–1012

Nauta AJ, Daha MR, van Kooten C, Roos A (2003) Recognition and clearance of apoptotic cells: a role for complement and pentraxins. Trends Immunol 24:148–154

Navratil JS, Watkins SC, Wisnieski JJ, Ahearn JM (2001) The globular heads of C1q specifically recognize surface blebs of apoptotic vascular endothelial cells. J Immunol 166:3231–3239

Neth O, Jack DL, Johnson M et al (2002) Enhancement of complement activation and opsonophagocytosis by complexes of mannose-binding lectin with mannose-binding lectin-associated serine protease after binding to *Staphylococcus aureus*. J Immunol 169:4430–4436

Ng KK, Kolatkar AR, Park-Snyder S et al (2002) Orientation of bound ligands in mannose-binding proteins. Implications for multivalent ligand recognition. J Biol Chem 277:16088–16095

Nonaka M, Ma BY, Ohtani M et al (2007) Subcellular localization and physiological significance of intracellular mannan-binding protein. J Biol Chem 282:17908–17920

Nuytinck L, Shapiro F (2004) Mannose-binding lectin: laying the stepping stones from clinical research to personalized medicine. Pers Med 1:35–52

Ogden CA, deCathelineau A, Hoffmann PR et al (2001) C1q and mannose binding lectin engagement of cell surface calreticulin and CD91 initiates macropinocytosis and uptake of apoptotic cells. J Exp Med 194:781–796

Oka S, Itoh N, Kawasaki T et al (1987) Primary structure of rat liver mannan-binding protein deduced from its cDNA sequence. J Biochem 101:135–44

Ono K, Nishitani C, Mitsuzawa H et al (2006) Mannose-binding lectin augments the uptake of lipid A, *Staphylococcus aureus*, and *Escherichia coli* by Kupffer cells through increased cell surface expression of scavenger receptor A. J Immunol 177:5517–5523

Paidassi H, Tacnet-Delorme P, Garlatti V et al (2007) C1q binds phosphatidylserine and likely acts as a multiligand-bridging molecule in apoptotic cell recognition. J Immunol 180:2329–2338

Riis AL, Hansen TK, Thiel S et al (2005) Thyroid hormone increases mannan-binding lectin levels. Eur J Endocrinol 153:643–649

Roos A, Bouwman LH, van Gijlswijk-Janssen DJ et al (2001) Human IgA activates the complement system via the mannan-binding lectin pathway. J Immunol 167:2861–2868

Sastry K, Herman GA, Day L et al (1989) The human mannose-binding protein gene. Exon structure reveals its evolutionary relationship to a human pulmonary surfactant gene and localization to chromosome 10. J Exp Med 170:1175–1189

Sastry K, Zahedi K, Lelias JM (1991) Molecular characterization of the mouse mannose-binding proteins. The mannose-binding protein A but not C is an acute phase reactant. J Immunol 147:692–697

Sastry R, Wang JS, Brown DC et al (1995) Characterization of murine mannose-binding protein genes Mbl1 and Mbl2 reveals features common to other collectin genes. Mamm Genome 6:103–110

Sawicki MW, Dimasi N, Natarajan K et al (2001) Structural basis of MHC class I recognition by natural killer cell receptors. Immunol Rev 181:52–65

Schuffenecker I, Narod SA, Ezekowitz RA et al (1991) The gene for mannose-binding protein maps to chromosome 10 and is a marker for multiple endocrine neoplasia type 2. Cytogenet Cell Genet 56:99–102

Sheriff S, Chang CY, Ezekowitz RA (1994) Human mannose-binding protein carbohydrate recognition domain trimerizes through a triple alpha-helical coiled-coil. Nat Struct Biol 1:789–794

Sim RB, Tsiftsoglou SA (2004) Proteases of the complement system. Biochem Soc Trans 32:21–27

Srinivasan A, Ni Y, Tizard I (1999) Specificity and prevalence of natural bovine antimannan antibodies. Clin Diagn Lab Immunol 6:946–952

Stover CM, Thiel S, Thelen M et al (1999) Two constituents of the initiation complex of the mannan-binding lectin activation pathway of complement are encoded by a single structural gene. J Immunol 162:3481–3490

Stuart GR, Lynch NJ, Day AJ et al (1997) The C1q and collectin binding site within C1q receptor (cell surface calreticulin). Immunopharmacology 38:73–80

Stuart LM, Takahashi K, Shi L et al (2005) Mannose-binding lectin-deficient mice display defective apoptotic cell clearance but no autoimmune phenotype. J Immunol 174:3220–3226

Super M, Lu J, Thiel S, Levinsky RT, Turner MW (1989) Association of low levels of mannan-binding protein with a common defect of opsonisation. Lancet 334:1236–1239

Takada F, Takayama Y, Hatsuse H, Kawakami M (1993) A new member of the C1s family of complement proteins found in a bactericidal factor, Ra-reactive factor, in human serum. Biochem Biophys Res Commun 196:1003–1009

Takahashi M, Iwaki D, Kanno K et al (2008) Mannose-binding lectin (MBL)-associated serine protease (MASP)-1 contributes to activation of the lectin complement pathway. J Immunol 180:6132–6138

Takahashi M, Endo Y, Fujita T, Matsushita M (1999) A truncated form of mannose-binding lectin-associated serine protease (MASP)-2-expressed by alternative polyadenylation is a component of the lectin complement pathway. Int Immunol 11:859–863

Takayama Y, Takada F, Nowatari M et al (1999) Gene structure of the P100 serine-protease component of the human Ra-reactive factor. Mol Immunol 36:505–514

Taylor ME, Brickell PM, Craig RK, Summerfield JA (1989) Structure and evolutionary origin of the gene encoding a human serum mannose-binding protein. Biochem J 262:763–771

Teillet F, Dublet B, Andrieu JP et al (2005) The two major oligomeric forms of human mannan-binding lectin: chemical characterization, carbohydrate-binding properties, and interaction with MBL-associated serine proteases. J Immunol 174:2870–2877

Tenner AJ, Robinson SL, Ezekowitz RA (1995) Mannose binding protein (MBP) enhances mononuclear phagocyte function via a receptor that contains the 126,000 Mr component of the C1q receptor. Immunity 3:485–493

Thiel S, Vorup-Jensen T, Stover CM et al (1997) A second serine protease associated with mannan-binding lectin that activates complement. Nature 386(6624):506–510

Torgersen D, Mullin NP, Drickamer K (1998) Mechanism of ligand binding to E- and P-selectin analyzed using selectin/mannose-binding protein chimeras. J Biol Chem 273:6254–6261

Turner MW (2003) The role of mannose-binding lectin in health and disease. Mol Immunol 40:423–429

Uemura K, Saka M, Nakagawa T et al (2002) L-MBP is expressed in epithelial cells of mouse small intestine. J Immunol 169:6945–6950

Uemura T, Sano H, Katoh T et al (2006) Surfactant protein A without the interruption of Gly-X-Y repeats loses a kink of oligomeric structure and exhibits impaired phospholipid liposome aggregation ability. Biochemistry 45:14543–14551

Uemura K, Yokota Y, Kozutsumi Y, Kawasaki T (1996) A unique CD45 glycoform recognized by the serum mannan-binding protein in immature thymocytes. J Biol Chem 271:4581–4584

Verga Falzacappa MV, Segat L, Puppini B et al (2004) Evolution of the mannose-binding lectin gene in primates. Genes Immun 5:653–661

Voss T, Eistetter H, Schafer KP, Engel J (1988) Macromolecular organization of natural and recombinant lung surfactant protein SP 28-36. Structural homology with the complement factor C1q. J Mol Biol 201:219–227

Wada M, Ozaki K, Itoh N, Yamashina I, Kawasaki T (1990) Two forms of messenger RNA encoding rat liver mannan-binding protein are generated by differential utilization of polyadenylation sites of one transcript. J Biochem (Tokyo) 108:914–917

Wallis R, Drickamer K (1999) Molecular determinants of oligomer formation and complement fixation in mannose-binding proteins. J Biol Chem 274:3580–3589

Wallis R, Shaw JM, Uitdehaag J et al (2004) Localization of the serine protease-binding sites in the collagen-like domain of mannose-binding protein: indirect effects of naturally occurring mutations on protease binding and activation. J Biol Chem 279:14065–14073

Weis WI, Drickamer K (1994) Trimeric structure of a C-type mannose-binding protein. Structure 2:1227–1240

Weis WI, Kahn R, Fourme R et al (1991) Structure of the calcium-dependent lectin domain from a rat mannose-binding protein determined by MAD phasing. Science 254(5038):1608–1615

Weis WI, Taylor ME, Drickamer K (1998) The C-type lectin superfamily in the immune system. Immunol Rev 163:19–34

Worthley DL, Bardy PG, Mullighan CG (2005) Mannose-binding lectin: biology and clinical implications. Intern Med J 35:548–555

Worthley DL, Bardy PG, Gordon DL, Mullighan CG (2006) Mannose-binding lectin and maladies of the bowel and liver. World J Gastroenterol 12:6420–6428

Wright JR (2005) Immunoregulatory functions of surfactant proteins. Nat Rev Immunol 5:58–68

Zanetta JP, Alonso C, Michalski JC (1996) Interleukin 2 is a lectin that associates its receptor with the T-cell receptor complex. Biochem J 318:49–53

Pulmonary SP-A: Forms and Functions

Anita Gupta and Rajesh K. Gupta

24.1 Pulmonary Surfactant Proteins

24.1.1 Pulmonary Surfactant

Pulmonary surfactant is a complex mixture of lipids and proteins, and is synthesized and secreted by alveolar type II epithelial cells and bronchiolar Clara cells. It acts to keep alveoli from collapsing during the expiratory phase of the respiratory cycle. After its secretion, lung surfactant forms a lattice structure on the alveolar surface, known as tubular myelin. Surfactant proteins (SP)-A, B, C and D make up to 10% of the total surfactant. SP-B and SP-C are relatively small hydrophobic proteins, and are involved in the reduction of surface-tension at the air-liquid interface. SP-A and SP-D, on the other hand, are large oligomeric, hydrophilic proteins that belong to the collagenous Ca^{2+}- dependent C-type lectin family (known as "Collectins"), and play an important role in host defense and in the recycling and transport of lung surfactant. There is increasing evidence that surfactant-associated proteins A and -D (SP-A and SP-D, respectively) contribute to the host defense against inhaled microorganisms. Based on their ability to recognize pathogens and to regulate the host defense, SP-A and SP-D have been recently categorized as "Secretory Pathogen Recognition Receptors". In nut-shell, the four lung-specific surfactant-associated proteins: SP-A, SP-B, SP-C, and SP-D serve a number of different roles, including enhancement of surface-active properties of surfactant glycerophospholipids, surfactant phospholipid reutilization, and immune defense within the alveolus. The basic structures of SP-A and SP-D include a triple-helical collagen region and a C-terminal homotrimeric lectin or CRD. The trimeric CRDs can recognize carbohydrate or charge patterns on microbes, allergens and dying cells, while the collagen region can interact with receptor molecules present on a variety of immune cells in order to initiate clearance mechanisms. Gene knock-out mice models of lung hypersensitivity and infection, and functional characterization of cell surface receptors have revealed the diverse roles of SP-A and SP-D in control of lung inflammation (Kishore et al. 2005)

24.1.2 Pulmonary Surfactant Protein A

24.1.2.1 Intracellular and Intraalveolar Localization

Although SP-A is secreted by type II pneumocytes as a component of pulmonary surfactant, its secretion pathway as well as its subcellular localization in the human lung are uncertain. In adult human lungs, only type II pneumocytes could be identified as SP-A positive cells within the parenchymal region. SP-A was localized mainly in small vesicles and multivesicular bodies close to the apical plasma membrane. Only few lamellar bodies were weakly labeled at their outer membranes. The strongest SP-A activity was found over tubular myelin figures. Labeling for SP-A was also found in close association with the surface film and unilamellar vesicles. Results supported that SP-A is mainly secreted into the alveolar space via an alternative pathway that largely bypasses the lamellar bodies. After secretion, the outer membranes of unwinding lamellar bodies become enriched with SP-A when tubular myelin formation is initiated. SP-A may also be involved in the transition of tubular myelin into the surface film (Ochs et al. 2002). Immunogold labeling showed that SP-A occurs predominantly at the corners of the tubular myelin lattice. Being an integral component of the lamellar body, peripheral compartment and secreted surfactant membranes support the concept that lysozyme may participate in the structural organization of lung surfactant (Haller et al. 1992).

24.1.2.2 Lamellar Bodies

Lamellar bodies of identical periodicity and ultrastructural geometry are present in lung (type II pneumocytes), serosal mesothelium (peritoneum, pleura, and pericardium), and joints (type A and type B synoviocytes) (Dobbie 1996). SP-A was detected in the cytoplasm of type II cells as asymmetrically distributed punctate fluorescent bodies that resembled lamellar bodies. Most of the SP-A was located within bodies of the type II cell. Although, lamellar bodies are enriched in SP-A, yet it is insufficient for structural transformation to tubular myelin and surface film formation in vitro. SP-A is secreted from type II cells primarily by a pathway separate from lamellar bodies. Surfactant secretion by lung type II cells occurs when lamellar bodies fuse with plasma membrane and surfactant is released into alveolar lumen. The fetal lung secretes significant quantities of surfactant during late gestation to initiate respiration at birth. A pathway of extracellular routing of SP-A prior to its accumulation in lamellar bodies in cultured type II cells has been suggested (Jain et al. 2005).

24.2 Structural Properties of SP-A

24.2.1 Human SP-A: Domain Structure

SP-A, the major surfactant protein, is a C-type lectin that activates macrophages in the lung alveolus and plays an important role in immune defense. SP-A is synthesized primarily by type II pneumonocytes and is developmentally regulated in fetal lung in concert with surfactant glycerophospholipid synthesis (Boggaram and Mendelson 1988). In humans and baboons, SP-A is encoded by two highly similar genes, *SP-A1* and *SP-A2* (Katyal et al. 1992; McCormick et al. 1994), whereas, in rabbits (Chen et al. 1992), rats (Fisher et al. 1987; Smith et al. 1995) and mice (Korfhagen et al. 1996), SP-A is encoded by a single-copy gene. Genomic analysis of human cellular DNA with SP-A cDNA demonstrated the presence of multiple hybridizing fragments that are not accounted for by available SP-A gene sequences. The functional SP-A gene is present in human chromosome 10. The functional SP-A gene and the pseudogene are syntenic (Korfhagen et al. 1996).

The lung and serum SP-A and SP-D are assembled as oligomers of trimeric subunits. Each subunit consists of four major domains: a short cysteine-containing NH_2-terminal cross-linking domain (N); a triple helical collagen domain of variable length; a trimeric coiled-coil linking domain (L; also referred to as the neck); and a carboxyl-terminal, C-type lectin CRD (Fig. 24.1a, b). Interactions between the amino-terminal domains of SP-D subunits have been shown to be stabilized by interchain disulfide bonds (Brown-Augsburger et al. 1996; Crouch et al. 1994), and similar mechanisms stabilize the oligomerization of SP-A and most other collectins. The primary protein component of human SP-A is a 32 kDa glycoprotein. A cDNA clone encoding for rat pulmonary SP-A encodes the sequence of 56 amino acids at the N-terminus. Isolated rat alveolar type II cells contain two species of mRNA for this protein (Fisher et al. 1987). Structural details of SP-D have been discussed in Chap. 25.

The primary structure of mature SP-A is conserved among different mammals with some differences. It consists of four structural domains: (1) an NH_2-terminal segment involved in intermolecular disulfide bond formation; (2) a collagen-like domain characterized by 23 Gly-*X*-*Y* repeats with an interruption near the midpoint of the domain; (3) α-helical coiled-coil domain, which constitutes the neck region between the collagen and the globular domain; and (4) a COOH-terminal globular domain involved in phospholipid binding and in Ca^{2+}-dependent binding of oligosaccharides (McCormack 1998; Head et al. 2003). SP-A is modified after translation (cleavage of the signal peptide, proline hydroxylation, and Asn^{187}-linked glycosylation). Unlike SP-As from other mammalian species, baboon and human SP-As consist of two polypeptide chains, SP-A1 and SP-A2. The major differences between mature SP-A1 and SP-A2 are in the collagen domain. Electron micrographs show that SP-A like MBP assembles as six trimers arrayed in parallel and in register, resembling a bouquet of tulips (Voss et al. 1988), whereas SP-D is cruciform, formed by radial arrangements of four trimers (Crouch et al. 1994) (Fig. 24.1a, b). Properties of SP-A are dependent on the presence of calcium. Each SP-A monomer binds two to three calcium ions in conditions chosen as similar to those found in the alveolar lumen. The higher affinity site for calcium is located in the non-collagenous carboxy-terminal end of SP-A that contains a CRD homologous to other C-type lectins. The binding of calcium to this region of SP-A causes a conformational change (Haagsman et al. 1990; Sohma et al. 1993).

24.2.1.1 SP-A is Assembled as Large Oligomer

Larger aggregates of SP-A are formed by means of disulfide bond formation within a short N-terminal segment containing two cysteine residues. More variability in the degree of oligomerization was observed with recombinant human SP-A than with natural canine SP-A. Collagenase digestion suggested that the full assembly of protein subunits was dependent on an intact collagen-like domain. Cysteines in noncollagen domain of SP-A form intrachain bonds between residues 135–226 and 204–224. The CD spectra of both recombinant and natural SP-A were consistent with the presence of a collagen-like triple helix. As determined by the change in ellipticity at 205 nm, thermal transition temperatures of canine, natural human, and recombinant SP-A were 51.5°C, 52.3°C, and 42.0°C, respectively (Haagsman et al. 1989). SP-A exists as fully assembled complexes with 18 polypeptide chains, but it is also consistently found in smaller oligomeric forms. This is true for both the water- and lipid-soluble fractions of SP-A. Hydroxyproline residues are present in SP-A in a region

24.2 Structural Properties of SP-A

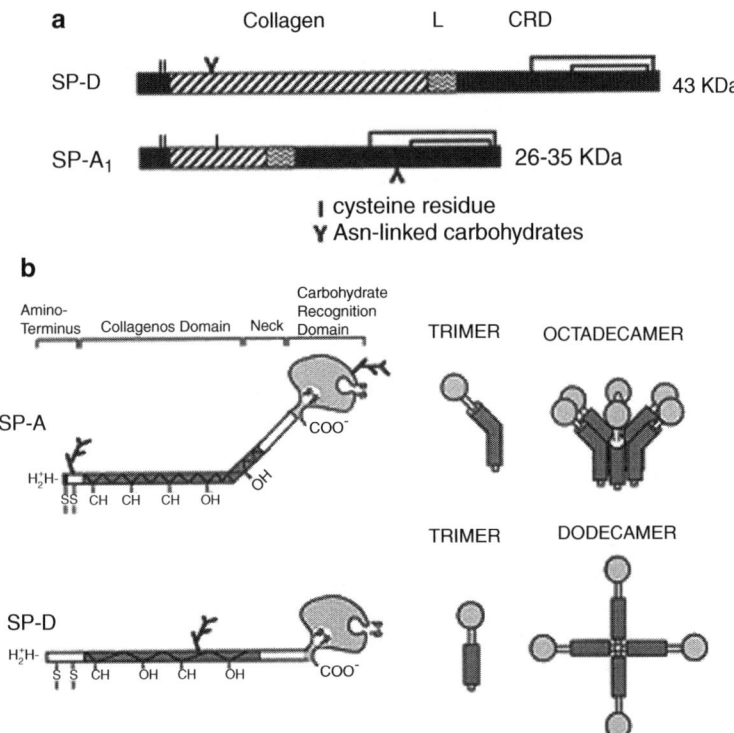

Fig. 24.1 Structural organization of SP-A and SP-D. (**a**) Each subunit of SP-A and SP-D consists of four major domains: a short cysteine-containing NH$_2$-terminal cross-linking domain (*N*); a triple helical collagen domain of variable length; a trimeric coiled-coil linking domain (*L*; also referred to as the neck); and a carboxyl-terminal, C-type lectin carbohydrate recognition domain (*CRD*) Differences in monomer size result from differences in the length of the collagen domain (*diagonal lines*). The Asn-linked oligosaccharide in SP-D is located in the collagen domain, whereas the Asn-linked sugar in human SP-A is located within the CRD (Crouch 1998). (**b**) Monomers form trimers that, in turn, form higher ordered oligomers of 18 (6 × 3) units for SP-A and 12 (4 × 3) units for SP-D. Molecules are not drawn to scale. SP-D is significantly larger than SP-A (Wright 1997) (Adapted with permission from Mason et al. 1998 © The American Physiological Society)

with a collagen-like sequence. SP-A treated with tunicamycin to block N-glycosylation and with 2,2-dipyridyl to inhibit the hydroxylation of proline residues suggests that hydroxylation of proline residues is required for perfect oligomerization of SP-A and for thermal stability in interaction with lipid (García-Verdugo et al. 2003; McCormack 1998; Palaniyar et al. 2001).

24.2.1.2 The Collagen-Like Domain of Human SP-A

A collagen-like domain of human SP-A consists of 23 Gly-X-Y repeats with an interruption near the midpoint of this domain. McCormack et al. (1999) suggested that : (1) SP-A trimerization does not require the collagen-like region or interchain disulfide linkage; (2) the N-terminal portion of collagen-like domain in SP-A is required for specific inhibition of surfactant secretion but not for binding to liposomes or for enhanced uptake of phospholipids into type II cells; (3) N-terminal interchain disulfide linkage in SP-A can functionally replace the N-terminal segment for lipid binding, receptor binding, and enhancement of lipid uptake, and (4) the N-terminal segment is required for the association of trimeric subunits into higher oligomers, for phospholipid aggregation, and for specific inhibition of surfactant secretion and cannot be replaced by disulfide linkage alone for these activities. To define the function of the kink region of SP-A, two mutated proteins were constructed to disrupt the interruption of Gly-X-Y repeats: SP-ADEL, which lacks the Pro47-Cys48-Pro49-Pro50 sequence at the interruption, and SP-AINS, in which two glycines were introduced to insert Gly-X-Y repeats (Gly-Pro47-Cys48-Gly-Pro49-Pro50). EM revealed that both mutants form octadecamers that lack a bend in the collagenous domain. The interruption of Gly-X-Y repeats in the SP-A molecule is critical for the formation of a flower bouquet-like octadecamer and contributes to SP-A's capacity to aggregate phospholipid liposomes (Uemura et al. 2006). The C-terminal non-collagenous domain of SP-A is essential for its correct folding and assembly, as judged by the secretion of various deletion mutants expressed in COS cells. Results suggest that three prefolded non-collagenous domains register and act as a nucleation center for the folding of the collagenous triple helix which proceeds in a zipper-like fashion towards the N-terminus (Spissinger et al. 1991).

The mutant SP-A1(δAVC,C6S) was thermally less stable for collagen structure with increased susceptibility to trypsin degradation. The supratrimeric assembly of human SP-A is essential for collagen triple helix stability at physiological temperatures, protection against proteases, protein self-association, and SP-A-induced ligand aggregation. The supratrimeric assembly is not essential for the binding of SP-A to ligands and anti-inflammatory effects of SP-A (Sánchez-Barbero et al. 2005, 2007).

24.2.2 Structural Biology of Rat SP-A

24.2.2.1 Chemical Modification of SP-A Alters Binding Affinity to Alveolar Type II Cells

Alkylation of SP-A with excess of iodoacetamide yielded forms of SP-A that did not inhibit surfactant lipid secretion and did not compete with ^{125}I-SP-A for cell surface binding. Kuroki et al. (1988b) concluded that: (1) cell surface binding activity of rat SP-A is directly related to its capacity to inhibit surfactant lipid secretion; (2) the lectin activity of SP-A against mannose ligands does not appear to be essential for cell surface binding; (3) the human SP-A derived from individuals with alveolar proteinosis exhibits different binding characteristics from rat SP-A.

24.2.2.2 Amino Terminal Processing of SP-A Results in Cysteinyl Isoforms

Triple-helix formation from three polypeptide chains requires previous trimerization of COOH-terminal globular domains by a trimeric α-helical coiled-coil (Head et al. 2003; McCormack et al. 1997a; Elhalwagi et al. 1997). Octadecamers appear to be formed by lateral association of the NH$_2$-terminal half of six triple-helical stems, forming a microfibrillar end piece stabilized by disulfide bonds at the NH$_2$-terminal region (Head et al. 2003) (Fig. 24.2). In rat SP-A two cysteine residues are involved in multimer formation: a cysteine in the 6 position of the NH$_2$-terminal segment (McCormack et al. 1997b) and another cysteine in the position -1, which is the last position of the signal peptide, one amino acid before the NH$_2$ terminus (Elhalwagi et al. 1997). Human SP-A, as well as rhSP-A1 and rhSP-A2 also have considerable NH$_2$-terminal heterogeneity, in which about 50–75% of human SP-A molecules contain Cys^{-1} isoforms (Elhalwagi et al. 1997; Wang et al. 2004; Garcia-Verdugo et al. 2003). In human SP-A1, four cysteine residues are potentially involved in the arrangement of the disulfide bonding: two cysteine residues in the NH$_2$-terminal segment (Cys^{-1} and Cys^6), another in the middle of the collagen-like sequence within the Pro-Cys-Pro-Pro interruption (Cys^{48}), and a fourth cysteine at position 65 (Cys^{65}) in the collagen domain near the neck. Cys^{65} is substituted by Arg^{65} in the human SP-A2 polypeptide chain (Floros and Hoover 1998). The biological functions of rat SP-A are dependent on intact disulfide bonds. Reducible and collagenase-reversible covalent linkages of as many as six or more subunits in the molecule indicate the presence of at least two NH2-terminal interchain disulfide bonds. Primary structure of rat SP-A predicts that only Cys^6 in this region is available for interchain disulfide formation. However, direct evidence for a second disulfide bridge was obtained by analyses of a set of mutant SP-As with deletions from the reported NH2-terminus. Thus, a post- translational modification results in naturally occurring rat SP-A cysteinyl isoforms, which are essential for multimer formation (Elhalwagi et al. 1997)

The role of the intermolecular bond at Cys^6 and the collagen-like domain (Gly8-Pro80) in the interactions of SP-A with phospholipids and alveolar type II cells were investigated using mutant forms of protein. McCormack et al. (1997b) suggested that: (1) the Cys^6 interchain disulfide bond of SP-A is required for aggregation of liposomes and for potent inhibition of surfactant secretion. (2) The collagen-like region is required for competition with ^{125}I-SP-A for receptor occupancy and specific inhibition of surfactant secretion in the presence of competing sugars. (3) Both the NH2-terminal disulfide and the collagen-like region are required to enhance the association of phospholipid vesicles with type II cells Yet, Zhang et al. (1998) suggested that neither Cys^{-1}-dependent multimerization nor the longer SP-A isoform is absolutely required for oligomeric association of trimeric SP-A subunits, SP-A/phospholipid interactions, or the regulation of surfactant secretion or uptake from type II cells by rat SP-A.

24.2.2.3 Glu^{195} and Arg^{197} Are Essential for Receptor Binding

The binding of SP-A to its high affinity receptor on alveolar type II cells is thought to be dependent on a CRD, while the interaction with lipids is attributed to the hydrophobic neck region of the molecule. To explore the role of the CRD in the interactions of SP-A with type II cells and lipids, McCormack et al. (1994) introduced mutations into the cDNA to encode for the substitutions Glu^{195}—>Gln and Arg^{197}—>Asp (SP-Ahyp, Gln^{195},Asp^{197}) and expressed the mutant protein in insect cells. Wild type SP-A produced in insect cells does not contain hydroxyproline (SP-Ahyp), but like rat SP-A it binds to carbohydrate affinity columns, lipids, and the SP-A receptor and is a potent inhibitor of the secretion of surfactant from type II cells. Study indicated that the binding of SP-A to its receptor and the inhibition of surfactant secretion are critically dependent on the carbohydrate binding specificity of CRD. Furthermore, phospholipid aggregation and augmentation of phospholipid uptake into

24.2 Structural Properties of SP-A

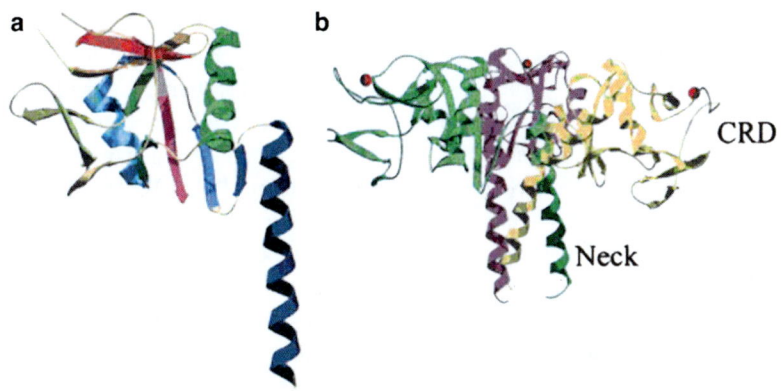

Fig. 24.2 Structure of CRD and neck domain (Δ1–80/N187S) of rat SP-A. A recombinant rat SP-A containing only the amphipathic linking domain and the CRD and lacking the consensus for asparagine-linked glycosylation (ΔN1-P80,N187S) was synthesized in insect cells using baculovirus vectors and purified on affinity column. (**a**) Ribbon diagram of (Δ1–80/N187S) monomer with secondary structure. (**b**) the (Δ1–80/N187S) trimer with each monomer colored differently and calcium ions in primary site are shown as *red spheres* (Adapted by permission from Head et al. 2003 © The American Society for Biochemistry and Molecular Biology)

type II cells are mediated by the C-terminal region of SP-A by a mechanism that is distinct from phospholipid binding.

Rat serum MBP-A is structurally analogous to SP-A and functionally equivalent to SP-A. MBP-A does not possess the ability to interact with lipids and type II cells. Honma et al. (1997) investigated the SP-A region involved in binding lipids and interacting with type II cells by using chimeric proteins with MBP-A. The binding of chimeras to DPPC and GalCer with activity comparable to recombinant SP-A suggested that MBP-A region of Glu185-Ala221 could functionally replace the homologous SP-A region of Glu195-Phe228 without loss of interaction with lipids and type II cells (Honma et al. 1997).

24.2.3 3-D Structure of SP-A Trimer

The crystal structure of the trimeric CRD and neck domain of SP-A was solved to 2.1-Å resolution. The crystal structure was derived from a fragment, which encompasses the extended, helical neck region (residues 81–108) and the globular CRD (residues 111–228) of the SP-A monomer. The fragment forms a stable non-covalent trimer in solution (McCormack et al. 1999). The secondary structure of the CRD includes three α-helices and 11 short (3–7 residues) strands of β-sheet (Fig. 24.2b). The overall fold of the CRD in SP-A is similar to that of SP-D and MBP. The main structural differences are restricted to the region that includes surface loops and calcium binding sites. The neck region consists of a single amphipathic α-helix (residues 84–108) culminating in a reverse turn (residues 108–111) that leads into the first β-sheet of the CRD domain. The neck regions of three monomers are intertwined around a three-fold rotational symmetry axis to form the coiled-coil trimer (Fig. 24.2b). The trimer is held together primarily by means of hydrophobic face of the amphipathic helix from the neck of each monomer. This hydrophobic face is composed of side chains of Leu^{87}, Leu^{91}, Ile^{94}, Ile^{98}, Thr^{101}, Met^{102}, and Leu^{105}. The reverse turn is situated around the three-fold symmetry axis such that the polar side chains of Gln^{108} and Ser^{110} form H-bonds with their counterparts in the other two monomers (Fig. 24.2a). Three intermolecular salt bridges linking Glu^{90} and Lys^{95} of adjacent monomers provide additional stabilization. Intramolecular contacts between the neck and CRD are minimal in the SP-A monomer.

In the SP-A trimer, the interface between domains does not have the appearance of a freely mobile hinge region as it does in the monomer. Major contacts between the two domains are primarily intermolecular and mostly hydrophilic. In one such contact, the carboxylate group of the C terminus, Phe^{228}, of one monomer forms a salt bridge and an H-bond with the side chains of neck residues His^{96} and Gln^{100}, respectively, of another monomer. All three residues are invariant in mammalian SP-A sequences. The neck and CRD domains are oriented nearly perpendicularly to each other in SP-A, in contrast to both SP-D and MBP, where the interdomain angles are greater (i.e. "T" versus "Y" shapes, respectively).

Two metal binding sites were identified, one in the highly conserved lectin site and the other 8.5 Å away. The interdomain CRD-neck angle is significantly less in SP-A than in SP-D, and MBL. This conformational difference may endow the SP-A trimer with a more extensive hydrophobic surface capable of binding lipophilic membrane components. The appearance of this surface suggests a putative binding region for membrane-derived SP-A ligands such as phosphatidylcholine and lipid A, the endotoxic lipid component of bacterial lipopolysaccharide that mediates the potentially lethal effects of Gram-negative bacterial infection (Head et al. 2003).

Three-dimensional (3D) structure of SP-A in association with a lipid ligand (DPPC: egg phosphatidylcholine lipid monolayers) was determined using single particle electron crystallography and computational 3D reconstruction in combination with molecular modeling. TEM showed that SP-A subunits readily formed trimers and interacted with lipid monolayers, exclusively via the globular domains. The plane of the putative lipid-protein interface was relatively flat and perpendicular to the hydrophobic neck region; the cleft region in the middle of the trimer had no apparent charge clusters. Amino acid residues that were known to affect lipid interactions, Glu^{195} and Arg^{197}, were located at the protein-lipid interface. The molecular model indicated that the hydrophobic neck region of the SP-A did not interact with lipid monolayers but was instead involved in intra-trimeric subunit interactions. The glycosylation site of SP-A was located at the side of each subunit, suggesting that the covalently linked carbohydrate moiety probably occupies the spaces between the adjacent globular domains, a location that would not sterically interfere with ligand binding (Palaniyar et al. 2000).

24.2.4 Bovine SP-A

24.2.4.1 Self-Aggregation of SP-A Depends on Ca^{2+} and Other Factors

Bovine SP-A forms extended fibers in the presence of calcium. On phosphatidylcholine or especially DPPC monolayers, SP-A at roughly 0.005 mg/ml formed large numbers of fibers and elaborate fibrous networks. The weak protein-protein interactions amongst free SP-As could be stabilized by phospholipids. In the presence of glycolipid GM1-ganglioside, SP-A's globular headgroup regions appeared enlarged and only small non-fibrous clusters were observed (Palaniyar et al. 1998b).

The importance of cations, particularly calcium, on SP-A function is well known and self-aggregation of bovine SP-A is Ca^{2+}-dependent. The concentration Ca^{2+} needed for half-maximal self-association (K_{aCa}^{2+}) depended on the presence of salts. Ca^{2+}-dependent SP-A aggregates formed in absence of NaCl are structurally different from those formed in its presence. Self-aggregation of SP-A can be pH and temperature-dependent and influenced by the presence of salts, which reduced the extent of self-association of the protein (Ruano et al. 2000). The CD spectra of bovine SP-A at various pH values indicated that α-helical content progressively decreased and that of β-sheet increased as the pH was reduced (Ruano et al. 1998). The quaternary organisation of bovine SP-A is altered by the presence of cations, especially calcium and sodium. There is a transition concentration, unique for each cation, at which a conformational switch occurs. The fact that the transition concentrations of cations are within physiological range suggests that cation-mediated conformational changes of SP-A could be operative in vivo (Ridsdale et al. 1999). The domains of SP-A that mediate lipid binding have been mapped. It was demonstrated that (1) lipid binding and pH-dependent liposome aggregation are mediated by CRD of SP-A, (2) distinct but overlapping domains within the CRD are required for pH- and Ca^{2+}-dependent liposome aggregation, and (3) conserved acidic and polar residues of the carbohydrate binding site of SP-A are essential for interactions with type II cells (McCormack et al. 1997c).

24.2.4.2 Substitution of His^{197} Creates Specific Lipid Uptake Defects

Mutagenesis of Glu^{195} and Arg^{197} of SP-A has implicated both residues as critical participants in the interaction of the molecule with alveolar type II cells and phospholipids. Substitution of Ala, Lys, His, Asp, and Asn for Arg at position 197 in SP-A action revealed that Ala^{197} retained complete activity in the SP-A functions and hence Arg^{197} is not essential for SP-A function. The Lys^{197} mutation displayed all functions of the wild type protein but exhibited a twofold increase in lipid uptake activity. The His^{197} mutation displayed all SP-A functions studied except for lipid uptake. The His^{197} mutation clearly demonstrated that lipid aggregation alone by SP-A is insufficient to promote lipid uptake by type II cells. Thus, specific interactions between type II cells and SP-A are involved in the phospholipid uptake processes (Pattanajitvilai et al. 1998)

Formation of Folds and Vesicles by DPPC Monolayers: Depositing DPPC organic solvent solutions in excess at an air:buffer interface led to the formation of elongated structures which could be imaged by TEM. The structures appeared to be DPPC folds protruding into the sol. The structures were frequently ordered with respect to one another, suggesting that they arose during lateral compression due to excess DPPC and are characteristic of a type of monolayer collapse phase. The elongated folds are unstable and can resolve by forming vesicles. Fold formation occurred at defined lipid concentrations above which more vesicles were observed. SP-A did not influence fold or vesicle formation but bound to the edges of these structures preferentially. It suggested that DPPC monolayers can form bilayers spontaneously in the absence of surfactant apoproteins, other proteins or agents (Ridsdale et al. 2001).

The temperature dependence of 2,850 cm^{-1} phospholipid acyl chain CH2 symmetric stretching frequencies in infra red (IR) spectrum showed a broad, reversible, melting event from about 15°C to 40°C in both the lipid extract and the native surfactant. The SP-A from bovine lung lavage was reconstituted into a binary lipid mixture of acyl chain perdeuterated DPPC-d62/DPPG (85:15 w/w) at a ratio approximating to that in surfactant. High levels of SP-A induced an ordering of the phospholipids, as shown by an increase in the transition temperature of DPPC-d62 compared to the lipid model. In contrast, a mixture of the other surfactant proteins induced a progressive disordering of the phospholipids and disruption of the cooperativity of the melting event (Reilly et al. 1989).

Structural Changes in SP-A Induced by Cations Reorient the Protein on Lipid Bilayers: Cations influence the interaction of bovine SP-A with phospholipid vesicles made of DPPC and unsaturated phosphatidylcholine (PC) (Gurel et al. 2001). The SP-A octadecamers exist in an "opened-bouquet" conformation in the absence of cations and interact with lipid membranes via one or two globular headgroups. Calcium-induced structural changes in SP-A resulted into the formation of a clearly identifiable stem in a "closed-bouquet" conformation. This change, in turn, results in all of SP-A's globular headgroups interacting with the lipid membrane surface and with the stem pointing away from the membrane surface. These results give direct evidence that the headgroups of SP-A (comprising CRDs), and not the stem (comprising the amino-terminus and collagen-like region), interact with lipid bilayers. Study supported models of tubular myelin in which headgroups, not the tails, interacted with lipid walls of the lattice (Palaniyar et al. 1998a).

24.2.4.3 Formation of Membrane Lattice Structures with SP-A

Tubular myelin (TM), the pulmonary surfactant membranous structure contains elongated tubes that form square lattices. Atomic force microscopy (AFM) imaging of the unstained sections containing TM indicated a highly heterogeneous surface topography with height variations ranging from 10 to 100 nm. In tapping-mode AFM, tubular myelin was seen as hemispherical protrusions of 30–70 nm in diameter, with vertical dimensions of 5–8 nm. This study suggests detection of 3-D nanotubes present in low abundance in a biological macromolecular complex (Nag et al. 1999).

Under TEM dipalmitoylphosphatidyl choline-egg phosphatidyl- choline (1:1 wt/wt) bilayers formed corrugations, folds, and predominantly 47-nm-square latticelike structures. SP-A specifically interacted with these lipid bilayers and folds. Other proteolipid structures could act as intermediates for reorganizing lipids and SP-As. Such a reorganization could result to the localization of SP-A in the lattice corners and could explain, in part, the formation of TM-like structures in vivo (Palaniyar et al. 1999).

Surfactant Lipid Peroxidation Damages SP-A and Inhibits Interactions with Phospholipid Vesicles: Lipid peroxidation (LPO) affects the function of SP-A. Exposure of SP-A to LPO is associated with an increase in the level of SP-A-associated carbonyl moieties and a marked reduction in SP-A-mediated aggregation of liposomes. LPO initiated by an azo-compound resulted in enhanced protein oxidation and inhibited SP-A-mediated liposome aggregation. Exposure of SP-A to LPO resulted in oxidative modification and functional inactivation of SP-A by phospholipid radicals (Kuzmenko et al. 2004). At the same time, pulmonary SP-A inhibits the lipid peroxidation stimulated by linoleic acid hydroperoxide of rat lung mitochondria and microsomes. Under similar conditions, BSA was unable to inhibit lipid peroxidation stimulated by LHP, indicating that the effect is specific to SP-A (Terrasa et al. 2005).

24.2.5 SP-A in Other Species

Canine surfactant SP-A sediments with a sedimentation coefficient of 14 S, giving a Mr of about 700 kDa. The hydrodynamic data can be approximated as a prolate ellipsoid having an axial ratio of about 20. SP-A aggregates into a complex of 18 monomers, which may form six triple-helices. The shape of the complex is considerably more globular than collagen and is not consistent with end-to-end binding of the helices to form fibrous structures. Canine SP-A contains a domain of 24 repeating triplets of Gly-X-Y, similar to that found in collagens. SP-A forms a collagen-like triple helix when in solution. The CD of protein demonstrated a relatively large negative ellipticity at 205 nm, with a negative shoulder ranging from 215 to 230 nm. There was no positive ellipticity, and the spectrum was not characteristic of collagen. Trypsin hydrolysis resulted in a fragment with peak negative ellipticity at about 200 nm, without the negative shoulder. Further hydrolysis of this fragment with pepsin resulted in a CD spectrum similar to that of collagen. The spectrum of the collagen-like fragment was reversibly sensitive to heating to 50°C, and was irreversibly lost after treatment with bacterial collagenase (King et al. 1989).

A full-length cDNA to guinea pig pulmonary SP-A consists of 1,839 bp and is highly conserved at both nucleotide and amino acid sequence levels with those from other species. As expected, guinea pig SP-A mRNA is abundantly expressed in adolescent lung tissue and is undetectable in nonpulmonary tissues. SP-A mRNA expression is confined to cells of the alveolar epithelium with no expression in the bronchiolar epithelial cells, whereas SP-B mRNA is expressed in both alveolar and bronchiolar epithelial cell populations. This distinct expression pattern suggests that the guinea pig lung will be a useful model in which to study expression of transcription factors implicated in the regulation of SP genes (van Eijk et al. 2001; Yuan et al. 1997). The complete cDNAs for ovine SP-A consists of 1,901 bp and encodes a protein of 248 amino acids (Pietschmann and Pison 2000). SP-A is present in pulmonary tissues of horses. The equine SP-A predicted from a cDNA of 747 bp consists of 248 amino acids, showing highest homology with SP-A from sheep (85.01%). The genomic DNA of equine SP-A, as in other species, includes three introns. There was no indication for the existence of two different SP-A genes (Hospes et al. 2002).

24.3 Cell Surface Receptors for SP-A

Different receptors have been identified on cells that work as receptors for SP-A or SP-D. In human monocytes the antibacterial activity of SP-A is mediated by a receptor for SP-A. A receptor with a high degree of specificity has been suggested for SP-A on alveolar macrophages (AMΦs). Chen et al. (1996) showed that secretagogues stimulates surfactant uptake through recruitment of SP-A receptors to the type II cell surface. Alveolar type II cells express a high affinity receptor for pulmonary SP-A, and the interaction of SP-A with these cells leads to inhibition of surfactant lipid secretion. The binding of SP-A to type II cells required Ca^{2+} and exhibited high-affinity for AMΦs (Kuroki et al. 1988a). The cell membrane receptor for SP-A regulates secretion of alveolar type II cells by negative feedback. Because TGF-β also inhibits Ca^{2+} fluxes, SP-A and TGF-β could be representative of a group of physiologic regulators that act by modulating intracellular Ca signaling (Strayer et al. 1999).

C1qR for Phagocytosis (C1qRp) (or CD93): The 126-kDa polypeptide is a functional common receptor or component of the receptor that mediates the enhancement of phagocytosis. Expression of this receptor is limited to cells of myeloid origin, platelets and endothelial cells, consistent with a relatively selective function. The amino acid sequence for C1qRp indicates that this surface glycoprotein receptor is a type I membrane protein of 631 amino acid containing a region homologous to C-type lectin CRDs, 5 EGF-like domains, a single transmembrane domain and a 47 amino acid intracellular domain. C1qRp cross-linked directly by monoclonal anti-C1qRp or engaged as a result of cell surface ligation of SP-A, as well as C1q and MBL, enhances phagocytosis (Nepomuceno et al. 1999; Tenner 1998). Ligation of human monocytes with immobilized R3, a IgM mAb recognizing C1qRp, also triggers enhanced phagocytic capacity of these cells in absence of ligand, verifying the direct involvement of this polypeptide in the regulation of phagocytosis. A distinct C1q receptor that triggers superoxide in polymorphonuclear leukocytes has been functionally characterized and designated as C1qRO2$^-$. Thus, there are at least two C1q receptor/receptor complexes (C1qRp and C1qRO2$^-$), each triggering distinct cellular responses, that can be expressed on the same, as well as on different, cell types (Tenner 1998) (see Chap. 41).

The C1q receptor (C1qR), identified from U937 cells and human tonsil lymphocytes, interacts with the C1q, mannose-binding protein, conglutinin and lung SP-A (Malhotra et al. 1990). This C1qR is an acidic glycoprotein, which exists as a dimer of Mr 115,000 in soluble form under non-denaturing conditions. The interaction of SP-A with U937 cells was found to up-regulate the surface expression of C1qR and to trigger the expression of an intracellular pool of C1qR (Malhotra et al. 1992). The collagen triple helix in C1q, MBP and SP-A, a cluster of similar charged residues has been suggested to be associated with receptor binding. A similar region of charge density occurs close to the N-terminus of conglutinin (Malhotra et al. 1993). SP-A binds to both *S. aureus* and monocytes and mediates the phagocytosis of the bacteria by these cells. Results demonstrate that C1qR mediates the phagocytosis of SP-A-opsonized *S. aureus* by monocytes (Geertsma et al. 1994).

SP-A Receptor 210 (SP-R210): Both type II cells and macrophages express a 210-kDa SP-A-binding protein, which was purified from U937 macrophage membranes and rat lung membranes. The 210-kDa protein was functional cell-surface receptor on type II cells, and the macrophage SP-A receptor is involved in SP-A-mediated clearance of pathogens (Chroneos et al. 1996; Yang et al. 2005). However, bovine AMΦs expose binding sites that are specific to SP-A, and depend on Ca^{2+} and on mannose residues. Two receptor proteins bound SP-A in a Ca^{2+}-dependent manner: one comprised of a 40-kDa protein showing mannose dependency and second of a 210-kDa protein, showing no mannose sensitivity (Plaga et al. 1998).

Glycoprotein-340: A Member of Scavenger Receptor Superfamily: A glycoprotein (Mr 340 kDa, gp-340) was purified from human bronchioalveolar lung washings from a patient with alveolar proteinosis. The gp-340 was found to bind to lung SP-D and also SP-A. Its molecular mass was 340 kDa in reduced state and decreased to 290 kDa in unreduced state. Binding between gp-340 and SP-D and

SP-A required the presence of calcium, and was not inhibited by maltose. Binding between gp-340 and SP-D is a protein-protein interaction rather than a lectin-carbohydrate interaction and that the binding to gp-340 takes place via carbohydrate recognition domain CRD of SP-D (Holmskov et al. 1997). Glycoprotein-340 interacts directly with pathogenic microorganisms and induces their aggregation, suggesting its role in innate immunity. Differential regulation of gp-340 in epithelial cell lines by phorbol myristate acetate indicates that gp-340 s involvement in mucosal defense and growth of epithelial cells may vary at different body locations and during different stages of epithelial differentiation (Kang et al. 2002)

SP-A Specific Receptor (SPAR): A pulmonary protein of ~30 kDa binds SP-A. The cDNAs encoding this protein were identified in human (4.1-kb) and porcine (1.8-kb) lung expression libraries. Both cDNAs encode similar ~32-kDa proteins that bind SP-A. The human and porcine SP-A recognition (SP-A specific receptor; SPAR) proteins resemble each other, as well as other cell membrane receptors. Their projected structures are consistent with cell membrane receptors. SPAR transcripts are expressed primarily in lung. SPAR-producing cells resemble the alveolar cells expressing SP-B and SP-C transcripts in appearance, location, and distribution (Strayer et al. 1993)

SP-A increases intracellular $[Ca^{2+}]_i$. Increased SP-A concentrations lead to a higher percentage of responding cells. SP-A also leads to a dose-dependent and transient generation of IP3. Secretagogue-induced secretion is inhibited by SP-A, which binds to SPAR on the surface of type II cells. The mechanism of SP-A-activated SPAR signaling involves PI3K, since PI3K inhibitor LY294002 rescued surfactant secretion from inhibition by SP-A. Thus, SP-A signals to regulate surfactant secretion through SPAR, via pathways that involve tyrosine phosphorylation, include insulin receptor substrate-1 (IRS-1), an upstream activator of PI3K, and entail activation of PI3K. This activation leads to inhibition of secretagogue-induced secretion of pulmonary surfactant (White and Strayer 2000).

24.4 Interactions of SP-A

24.4.1 Protein-Protein Interactions

SP-A has been involved in the physiology of reproduction. Consistent with the activation of ERK-1/2 and COX-2-induced by SP-A in myometrial cells, there are two major proteins recognized by SP-A in these cells. One of these SP-A targets is intermediate filament (IF) desmin. SP-A recognizes especially its rod domain, which is known to play an important role during the assembly of desmin into filaments. SP-A is colocalized with desmin filaments in myometrial cells. Interestingly, vimentin, the IF characteristic of leukocytes, is one of the major proteins recognized by SP-A in protein extracts of U937 cells. The ability of SP-A to interact with desmin and vimentin, and to prevent polymerization of desmin monomers, shed light on unexpected and wider biological roles of this collectin (Garcia-Verdugo et al. 2008a; Tanfin 2008).

Schlosser et al. (2006) described the molecular interaction between the extracellular matrix protein microfibril-associated protein 4 (MFAP4) and SP-A. MFAP4 is a collagen-binding molecule containing a C-terminal fibrinogen-like domain and a N-terminal located integrin-binding motif and interacts with SP-A via the collagen region in vitro. MFAP4 and SP-A are colocated in different lung compartments indicating that the interaction may be operative in vivo (Schlosser et al. 2006). The SP-A binds to sTLR4 and MD-2 and the reaction is Ca^{2+}-dependent. The direct interaction between SP-A and TLR4/MD-2 suggests the importance of supratrimeric oligomerization in the immunomodulatory function of SP-A (Yamada et al. 2006). The α1-antitrypsin and SP-A are major lung defense proteins. SP-A can bind α1-antitrypsin. The CRD of SP-A appeared to be a major determinant of interaction, by recognizing α1-antitrypsin carbohydrate chains. However, binding of SP-A carbohydrate chains to the α1-antitrypsin amino acid backbone and interaction between carbohydrates of both proteins was also possible (Gorrini et al. 2005). At least three different proteins from bovine lung soluble fraction bind to SP-A in a Ca^{2+}-dependent manner. The main protein with molecular mass of 32 kDa was identified as annexin IV. The lung annexin IV augmented the $Ca2^+$-induced aggregation of lung lamellar bodies from rats. The SP-A binds annexin IV through protein-protein interaction, although, both proteins are phospholipid-binding proteins (Sohma et al. 1995). Among several polypeptides, the 200-kDa major polypeptide reacted with SP-A on ligand blots. This polypeptide corresponded completely with nonmuscle (cellular) myosin heavy chain. A smaller polypeptide of 135 kD also binds SP-A and appears to be a proteolytic fragment of the 200 kD peptide (Michelis et al. 1994).

Kresch and Christian (1998) isolated two SP-A specific binding proteins from type II cells with M_r of 86 and >200 kDa under nonreducing conditions, but dissociated into proteins with M_r of 65, 55, and 50 kDa. The 86-kDa protein is a glycoprotein with ~30% of its mass as carbohydrate. A protein band of 63 kDa under reduced conditions was identified as rat homolog of human type II transmembrane protein p63 (CKAP4/ERGIC-63/CLIMP-63) (Kresch and Christian 1998). The p63 closely interacts with SP-A and may play a role in trafficking or the biological function of SP-A (Gupta et al. 2006). EM confirmed endoplasmic reticulum and plasma membrane localization of P63 in type II cells with prominent labeling of microvilli (Bates et al. 2008).

The SP-A and SP-D bind transmembrane inhibitory regulatory protein α (SIRPα) through their globular heads to initiate a signaling pathway that blocks proinflammatory mediator production. In contrast, their collagenous tails stimulate proinflammatory mediator production through binding to calreticulin/CD91. Together a model is implied in which SP-A and SP-D help maintain a non/anti-inflammatory lung environment by stimulating SIRPα on resident cells through their globular heads. However, interaction of these heads with PAMPs on foreign organisms or damaged cells and presentation of collagenous tails in an aggregated state to calreticulin/CD91 stimulates phagocytosis and proinflammatory responses (Gardai et al. 2003).

24.4.2 Protein-Carbohydrate Interactions

Haurum et al. (1993) studied the influence of carbohydrates on the binding of SP-A, MBP and conglutinin to mannan. The order of inhibiting potency on the binding of SP-A was: N-acetylmannosamine > L-fucose, maltose > glucose > mannose. These results were independent of the source or method of extraction of SP-A, and generally consistent with earlier studies that examined binding to various saccharide-substituted supports (Haagsman et al. 1987). However, they are in obvious disagreement with a study that characterized the binding of dog SP-A to various neoglycoproteins, glycolipids, and neoglycolipids (Childs et al. 1992). Notably, the latter studies demonstrated a preferential recognition of galactose and an inability to compete with free monosaccharides. Although residues corresponding to Glu^{185} and Asn^{187} are found in essentially all members of the mannose binding subgroup, SP-A has a substitution in the position corresponding to Asn^{187} (Arg in dog and rat and Ala in human). Drickamer has suggested that this correlates with the capacity of SP-A to bind to a variety of sugars with comparatively weak affinity (Drickamer 1992). In any case, mutagenesis of corresponding residues in the putative carbohydrate binding site of SP-A (Glu^{195} to Gln and Arg^{197} to Asp) reversed the preference from mannose to galactose in affinity chromatography assays using mannose- and galactose-substituted supports (McCormack 1998). Deletion of the consensus sequences for N-linked glycosylation showed no obvious effect on lectin activity (McCormack 1998). Human SP-A has been shown to bind to lactosylceramide (Childs et al. 1992), galactosylceramide, and gangliotriaosylceramide (Kuroki et al. 1992). The optimal binding conditions were drastically different for three receptors (lactosylceramide, galactosylceramide, and gangliotetraosylceramide). At pH 24.4 and at 5 mM Ca concentration the binding affinity of SP-A followed the order: galactosylceramide > lactosylceramide > gangliotetraosylceramide (Hynsjö et al. 1995).

Collectins bind DNA from a variety of origins, including bacteria, mice, and synthetic oligonucleo- tides. Pentoses, such as arabinose, ribose, and deoxyribose, inhibit the interaction between SP-D and mannan, one of the well-studied hexose ligands for SP-D and d-forms of the pentoses are better competitors than the l-forms (Palaniyar et al. 2004).

24.4.3 SP-A Binding with Lipids

SP-A exhibits high affinity binding interactions with lipid as well as carbohydrate ligands. SP-A binds specifically to DPPC, the most abundant phospholipid species in surfactant (Kuroki and Akino 1991), whereas SP-D binds to the minor surfactant component phosphatidylinositol (Ogasawara et al. 1992; Persson et al. 1992). Both bind to the lipopolysaccharide moieties that decorate bacterial surfaces; SP-A recognizes the lipid A component of lipopolysaccharide (Kalina et al. 1995), whereas SP-D binds to the core oligosaccharide and O-oligosaccharides (Sahly et al. 2002). Electron micrographs and other data indicate that the globular CRD and collagen-like domains in both surfactant proteins are oriented proximal and distal, respectively, to the membrane surface (Palaniyar et al. 1998, 2002).

SP-A Specifically Binds DPPC: Although a monolayer of DPPC, the major component of pulmonary surfactant, is thought to be responsible for the reduction of surface tension at the air-liquid interface of the alveolus, the participation of unsaturated and anionic phospholipids and the three surfactant-associated proteins is suggested in the generation and maintenance of this surface-active monolayer. Pulmonary SP-A enhances the surface activity of lipid extract surfactant and reverses inhibition by blood proteins in vitro (Cockshutt et al. 1990). SP-A interacts strongly with a mixture of surfactant-like phospholipids. Although SP-A binds phosphatidylcholine and Sphingomyelin, it showed very strong binding to DPPC, it did not bind to phosphatidylglycerol (PG), phosphatidyl- inositol, phosphatidylethanolamine, and phosphatidylserine. SP-A also exhibited strong binding to distearoylphosphatidyl-choline, but weak binding to dimyristoyl-, 1-palmitoyl-2-linoleoyl, and dilinoleoyl-phosphatidylcholine. Thus, it indicates that (1) SP-A specifically and strongly binds DPPC, (2) SP-A binds the nonpolar group of phospholipids, (3) the second positioned palmitate is involved in DPPC binding, and (4) the specificities of polar groups of dipalmitoylglycerophospholipids also appear to be important for SP-A binding, (5) the phospholipid binding activity of SP-A is dependent upon calcium ions and the integrity of the collagenous domain of SP-A, but not on the oligosaccharide moiety of SP-A. SP-A may play an important role in the regulation of recycling and intra- and extracellular movement of DPPC (Kuroki and Akino 1991; Yu and Possmayer 1998). Yu et al. (1999) suggested the importance of CRD and N-terminal dependent oligomerization in SP-A-phospholipid associations.

Binding of SP-A to Glycolipids: The SP-A binds to galactosylceramide and asialo-GM2, and that both saccharide and ceramide moieties in the glycolipid molecule are important for the binding of SP-A to glycolipids (Kuroki et al. 1992). Though, ^{125}I-SP-A bound to galactosylceramide and asialo-GM2, but failed to exhibit significant binding to GM1, GM2, asialo-GM1, sulfatide, and Forssman antigen. The study of ^{125}I-SP-A binding to glycolipids coated onto microtiter wells also revealed that SP-A bound to galactosylceramide and asialo-GM2. SP-A bound to galactosylceramides with non-hydroxy or hydroxy fatty acids, but showed no binding to either glucosylceramide or galactosylsphingosine. Results provide evidence for binding specificity for proteins that have Ca^{2+}-dependent CRDs and raise the possibility that glycosphingolipids are endogenous ligands for SP-A (Childs et al. 1992).

Interaction with Phospholipid Liposomes: Ca^{2+} ions induce an active conformation in SP-A, which rapidly binds to liposomes and mediates their aggregation. SP-A/liposome interaction shows $K_D = 5$ μM for interaction between SP-A and DPPC liposomes. With POPC, the complex formation proceeds at half the rate, leading to a lower final equilibrium level of SP-A- DPPC interaction. Distearoylphosphatidylcholine (DSPC) shows a stronger interaction than DPPC. Among phospholipid headgroups, phosphatidyl inositol (PI) and sphingomyelin (SM) interact comparable to DPPC, while less interaction is seen with phosphatidylethanolamine (PE) or with phosphatidylglycerol (PG). Thus both headgroup and fatty acid composition determine SP-A phospholipid interaction. However, the protein does not exhibit high specificity for either the polar or the apolar moiety of phospholipids (Meyboom et al. 1999).

Whereas SP-A binds to DPPC and galactosylceramide (GalCer), MBP-A binds phosphatidylinositol (PI). SP-A also interacts with alveolar type II cells. Specific monoclonal antibodies inhibit the interactions between SP-A and lipids or alveolar type II cells. The amino acid residues 174–194 of SP-A and the corresponding region of MBP-A are critical for SP-A-type II cell interaction and Ca^{2+}-dependent lipid binding of collectins (Chiba et al. 1999; Tsunezawa et al. 1998).

Intermolecular Cross-Links Mediate Aggregation of Phospholipid Vesicles by SP-A: SP-A plays a central role in the organization of phospholipid bilayers in the alveolar air space. SP-A in lung lavage exists in oligomeric forms (N = 6, 12, 18 ...) mediated by collagen-like triple helices and intermolecular disulfide bonds. These protein-protein interactions, involving the amino-terminal domain of SP-A, facilitate the alignment of surfactant lipid bilayers into unique tubular myelin structures. Accelerated aggregation of unilamellar vesicles required SP-A and 3 mM free calcium. The initial rate of aggregation was proportional to the concentration of canine SP-A over lipid:protein molar ratios ranging from 200:1 to 5,000:1. Data demonstrate the importance of the quaternary structure (triple helix and intermolecular disulfide bond) of SP-A for the aggregation of unilamellar phospholipid vesicles (Ross et al. 1991).

SP-A is thought to influence the surface properties of surfactant lipids and regulate the turnover of extracellular surfactant through interaction with a specific cell-surface receptor. SP-A induces a rapid Ca^{2+}-dependent aggregation of phospholipid vesicles and mediated by Ca^{2+}-induced interactions between carbohydrate-binding domains and oligosaccharide moieties of SP-A. This mechanism of membrane interactions may be relevant to the formation of the membrane lattice of tubular myelin, an extracellular form of surfactant (Haagsman et al. 1991)

24.5 Functions of SP-A

24.5.1 Surfactant Components in Surface Film Formation

Pulmonary surfactant is a mixture of phospholipids and proteins that lines the distal airways and stabilizes the gas-exchanging alveolar units of the lung. Surfactant membranes are decorated with SP-A, one of the four known surfactant proteins. The main function of surfactant is to reduce the surface tension at the air/liquid interface in the lung. This is achieved by forming a surface film that consists of a monolayer which is highly enriched in dipalmitoylphosphatidylcholine and bilayer lipid/protein structures closely attached to it. The molecular mechanisms of film formation and of film adaptation to surface changes during breathing in order to remain a low surface tension at the interface are unknown. The results of several model systems give indications for the role of the surfactant proteins and lipids in these processes. Veldhuizen and Haagsman (2000) described and compared the model systems that are used for this purpose and the progress that has been made. Despite some conflicting results using different techniques, workers concluded that surfactant protein B (SP-B) plays the major role in adsorption of new material into the interface during inspiration. SP-C's main functions are to exclude non-DPPC lipids from the interface during expiration and to attach the bilayer structures to the lipid monolayer. The SP-A appears to promote most of SP-B's functions. Veldhuizen and Haagsman (2000) described a model proposing that SP-A and SP-B create DPPC enriched domains which can readily be adsorbed to create a DPPC-rich monolayer at the interface. Further enrichment in DPPC is achieved by selective desorption of non-DPPC lipids during repetitive breathing cycles. Reports that SP-A is required for the formation of tubular myelin (Korfhagen et al. 1996) and other large surfactant aggregates (Veldhuizen et al. 1996) and for the preservation of low surface tensions in presence of serum protein inhibitors (Cockshutt et al. 1990) suggest that the protein serves multiple roles in surfactant function.

24.5.2 Recognition and Clearance of Pathogens

SP-A interacts with viruses, bacteria and fungi. Furthermore, SP-A binds to various other inhaled glycoconjugates. SP-A receptors on phagocytic cells have been described that are important to ensure rapid pathogen clearance. This innate defence system of the lung is particularly important during infections in young children when the acquired immune system has not yet become fully established. Also in later life SP-A could be very important to prevent the lungs from infections by pathogens not previously encountered. Lung collectins, SP-A and SP-D have a relatively high affinity for oligosaccharides. This is an important determinant of self/non–self recognition, because most carbohydrates in animals are monosaccharides. Therefore, SP-A and SP-D enhance microbial phagocytosis by innate immune cells, such as macrophages and neutrophils, by opsonizing and aggregating bacteria and viruses, by acting as an activation ligand, and by upregulating the expression of immune cell surface receptors that recognize microbes. In addition, SP-A and SP-D also promote apoptotic cell uptake by innate immune cells and regulate cytokine and free radical production in a context-dependent manner. As an example, SP-A inhibits LPS-stimulated nitric oxide (NO) production by alveolar macrophages isolated from normal lungs, but promotes NO production in macrophages that have been activated by IFN-γ (Stamme et al. 2000). Both proteins possess direct bactericidal activity against bacteria and fungi through unknown mechanisms (Wu et al. 2003) (Table 24.1). In addition, SP-A may limit the inflammatory response in the lungs, thus preventing damage to the delicate lung epithelia. Evidences suggest that SP-A may modulate the allergic response to various glycosylated inhaled antigens. The presence of SP-A (and SP-D) in other organs indicates that these collectins may have a general role in mucosal immunity (Haagsman 1998; Tino and Wright 1998).

Table 24.1 Immune functions of SP-A and SP-D (Adapted by permission from Kishore et al. 2005 © Springer)

1. Endotoxin clearance and regulation of LPS-induced inflammation
2. Recognition, agglutination and phagocytosis of viral, bacterial and fungal pathogens$
3. Macrophage and neutrophil activation and chemotaxis
4. Microbial growth inhibition
5. Non-specific defense molecules in tears, saliva and body secretions against pathogens
6. Reduced viral infectivity
7. Protection against intrauterine infection and inflammatory reactions
8. Modulation of inflammation
9. Recognition and clearance of apoptotic and necrotic cells
10. Anti-proliferative effects on B and T lymphocytes
11. Pattern recognition of glycoprotein allergens
12. Inhibition of IgE-allergen cross-linking
13. Suppression of histamine release from basophils and mast cells
14. Modulation of Th cytokine profile
15. Modulation of maturation and antigen presentation by dendritic cells

Functions of SP-A and SP-D as collectins: SP-A and SP-D bind to and opsonize viruses, bacteria, allergens, and apoptotic cells. SP-A and SP-D enhance microbial phagocytosis by aggregating bacteria and viruses. SP-A and SP-D also possess direct bactericidal effects and potentially bind to a variety of receptors to modulate immune cell cytokine and inflammatory mediator expression (Wright 2005)

Although, the absence of SP-A changed the structure and in vitro properties of surfactant, the in vivo function of surfactant in SP-A$^{-/-}$ mice was not changed under the conditions of experiments. In the absence of SP-A, the structure of pulmonary surfactant large aggregates is altered, tubular myelin is absent, and SP-A is no longer available to contribute to the formation of the typical lattice-like structure. However, the surface tension–reducing function of surfactant appeared unaffected, suggesting that SP-A does not appear to play a primary role in surfactant homeostasis. Although the mice were able to survive with no apparent pathology in a sterile environment, and respond similarly to wild-type (WT) mice in exercised or hyperoxic conditions (Ikegami et al. 1998, 2000), their pulmonary immune responses are insufficient during immune challenge. *SP-D*–null mice have a more complex phenotype. Even in the absence of pathogens, the *SP-D*–null mice display advancing alveolar proteinosis and increased lipid pools, indicating that SP-D has a role in surfactant homeostasis. Metalloproteinases were also elevated in their lungs, which developed an emphysema-like phenotype. *SP-A*– and *SP-D*–null mice were more susceptible to bacterial and viral infections and LPS-mediated inflammation, confirming roles for SP-A and SP-D in modulating immune responses in the lung (Wright 2005) (Table 24.1).

SP-A facilitates phagocytosis by opsonizing bacteria, fungi and viruses, stimulates the oxidative burst by phagocytes and modulates pro-inflammatory cytokine production by phagocytic cells. Pulmonary surfactant also exhibits immunomodulatory functions and plays a key role in host defense against infection. For example, surfactant lipids suppress a variety of immune cell functions, most notably lymphocyte proliferation. Both SP-A and SP-D improve phagocytosis of pathogens by polymorphonuclear neutrophils (Hartshorn et al. 1998). SP-A can also provide a link between innate and adaptive immune responses by promoting differentiation and chemotaxis of dendritic cells. The SP-A-deficient gene-targeted mouse is susceptible to lung infection with multiple organisms (LeVine et al. 1997). One view that reconciles these apparently divergent functions of SP-A is that its high affinity for surfactant membranes is a mechanism for concentrating the protein at the front lines of defense against inhaled pathogens (Hawgood and Poulain 2001; McCormack and Whitsett 2002). In addition, SP-A and SP-D each modifies the in vivo response to instilled endotoxin, leading to decreased lung injury and reduced inflammatory cell recruitment (Borron et al. 2000; Greene et al. 1999). The SP-A-mediated interaction of lipids with type II cell plasma membrane may contribute, in part, to the lipid uptake process by type II cells (Kuroki et al. 1996).

24.5.3 SP-A: As a Component of Complement System in Lung

There is a functional complement system present in the lung that helps in removal of pathogens. Complement proteins of alternative and classical pathways of complement have been observed in bronchoalveolar lavage fluids (BALF) from healthy volunteers (Watford et al. 2000). Because of structural and functional similarities between C1q and pulmonary SP-A, SP-A may interact with and regulate proteins of complement system. The SP-A binds directly to C1q, though only weakly to intact C1. The binding of SP-A to C1q prevents the association of C1q with C1r and C1s, and therefore the formation of active C1 complex required for classical pathway activation through a common binding site for C1r and C1s or C1q. Furthermore, SP-A blocked the ability of C1q to restore classical pathway activity in C1q-depleted serum. Thus, SP-A may down-regulate complement activity through its association with C1q and may serve a protective role in lung by preventing C1q-mediated complement activation and inflammation along the delicate alveolar epithelium (Tenner et al. 1989; Watford et al. 2001).

24.5.4 Modulation of Adaptive Immune Responses by SPs

Increasing evidence shows that SP-A and SP-D modulate the functions of adaptive immune cells, DCs and T cells. DCs form a tightly meshed network within the upper airways, parenchyma, and alveolar airspace, and are ideally positioned to sample inhaled antigens (Lipscomb and Masten 2002). DC density increases upon inflammatory stimuli, particularly in the lower airways as a result of recruited myeloid DC precursors (Jahnsen et al. 2001). Although SP-A and SP-D both bind to DCs in a calcium-dependent manner, they have differential effects on DC function. For example, SP-D enhances antigen uptake and presentation by bone marrow–derived DCs, but SP-A inhibits DC maturation and phagocytic and chemotactic function in vitro (Brinker et al. 2001, 2003). Also, SP-A and SP-D inhibit T-cell proliferation via two mechanisms: an IL-2–dependent mechanism observed with accessory cell–dependent T cell mitogens and specific Ag, as well as an IL-2–independent mechanism of suppression that potentially involves attenuation of calcium signaling (Borron et al. 1998a). Moreover, SP-A–mediated inhibition of T cell proliferation might partially result from TGF-ß present in the SP-A preparations (Kunzmann et al. 2006). In contrast to the results, obtained using bone marrow–derived DCs, SP-D was found to decrease antigen presentation by DCs isolated from the mouse lung during both resting and inflammatory conditions (Hansen et al. 2007). A role for SP-D in regulating T-cell responses in vivo is demonstrated in a study by Fisher et al. (2002) showing that *SP-D*–null mice have an accumulation of $CD4^+$ and $CD8^+$ T cells expressing activation markers CD69 and CD25 in the perivascular and peribronchial regions of the lung. Thus, studies suggest that surfactant may be a critical regulator of organ-specific immune regulation in the lung, and that the hyporesponsive immunologic environment of the lung is, in part, facilitated by the actions of SP-A and SP-D in an effort to thwart inflammatory cascades that could potentially damage the lung and impair gas exchange. SP-A regulates the differentiation of immature DCs into potent T cell stimulators. Studies demonstrate that SP-A participates in the adaptive immune response by modulating important immune functions of DCs (Brinker et al. 2003).

SP-A Inhibits Human Lymphocyte Proliferation and IL-2 Production: Studies suggest that purified SP-A suppress both PHA and anti-CD-3 activated proliferation of human peripheral blood and tonsillar mononuclear cells in a dose-dependent manner at concentrations as low as 50 pM. In contrast, ConA-stimulated PBMC proliferation was slightly augmented by the addition of SP-A. The inhibition of PHA-stimulated proliferation by SP-A was accompanied by corresponding decline in IL-2 concentration. The in vitro inhibitory effect of SP-A was not blocked by C1q. The effect was not mediated by CRD of SP-A, but a 36-amino acid Arg-Gly-Asp (RGD) motif-containing span of collagen-like domain was responsible for the inhibition of T cell proliferation (Borron et al. 1998b).

In addition to SP-A, SP-D could also reduce the incorporation of ^{3}H-thymidine into PBMC in a dose-dependent manner. A recombinant peptide composed of neck and CRD of SP-D [SP-D(N/CRD)] was also suppressive for lymphocyte proliferation. Inhibitory effect of both SP-A and SP-D on histamine release in the early phase of allergen provocation and suppressing lymphocyte proliferation in the late phase of bronchial inflammation suggest that SP-A and SP-D may be protective against the pathogenesis of asthma (Wang et al. 1998). These experiments emphasize a potential role for SP-A in dampening lymphocyte responses to exogenous stimuli.

TGF-β1 and SP-A Influence Immune-Suppressive Properties of SP-A on CD4$^+$ T Lymphocytes: SP-A and TGF-β1 have been shown to modulate the functions of different immune cells and specifically to inhibit T lymphocyte proliferation. Recombinant human SP-A1 suppressed T lymphocyte proliferation and IL-2 mRNA expression. The effect of rSP-A1m was mediated through TGF-βRII and could be antagonized by anti-TGF-β1 neutralizing antibodies and sTGF-βRII. The association between SP-A and latent TGF-β1 provides a possible mechanism that regulate TGF-β1-mediated inflammation and fibrosis reactions in lung but also leads to possible misinterpretation of immune-modulator functions of SP-A. Monitoring of SP-A preparations for possible TGF-β1 is essential (Kunzmann et al. 2006).

SP-A Differentially Regulates IFN-γ- and LPS-Induced Nitrite Production by Rat AMΦ: The role of SP-A and SP-D in regulating production of free radicals and cytokines is controversial. Haddad et al. (1994) suggest that peroxynitrite, but not .NO or superoxide and hydrogen peroxide, in concentrations likely to be encountered in vivo, caused nitrotyrosine formation and decreased the ability of SP-A to aggregate lipids.

The state and mechanism of activation of the immune cell influence its response to SP-A. SP-A inhibited production of NO and iNOS in rat AMΦs stimulated with smooth LPS, which did not significantly bind SP-A, or rough LPS, which avidly bound SP-A. In contrast, SP-A enhanced production of NO and iNOS in cells stimulated with IFN-γ or INF-γ plus LPS. Neither SP-A nor SP-D affected baseline NO production, and SP-D did not significantly affect production of NO in cells stimulated with either LPS or IFN-γ. These results suggest that SP-A contributes to the lung inflammatory response by exerting differential effects on the response of immune cells, depending on their state and mechanism of activation (Stamme et al. 2000).

SP-A$^{-/-}$ mice produced significantly more TNF-α and NO than SP-A$^{+/+}$ mice after intratracheal administration of LPS. Human SP-A to SP-A$^{-/-}$ mice restored regulation of TNF-α and NO production to that of SP-A$^{+/+}$ mice without affecting other markers of lung injury. Neither binding of LPS by SP-A, nor enhanced LPS clearance were the primary means of inhibition. It seems that SP-A acts directly on immune cells to suppress LPS-induced inflammation and that the endogenous or exogenous SP-A inhibits pulmonary LPS-induced cytokine and NO production in vivo (Borron et al. 2000). SP-A suppresses NO production by activated AMΦs by inhibiting TNF-α secretion and NF-kB activation (Hussain et al. 2003). SP-A also modulates NO production by AMΦ in vitro but results are contrasting. In AMΦ and type II cells, iNOS is known to generate NO, which is upregulated by SP-A of different origin in a concentration – dependent manner, whereas type II cells were unresponsive to SP-A. The increase in NO production was associated with elevation in expression of iNOS. Results indicated that SP-A is the agonist and not a contaminating LPS (Blau et al. 1997).

SP-A Modulates Production of Cytokines: SP-A enhances concanavalin A-induced proliferation and levels of TNF-α, IL-1 α, 1β, and IL-6 by human peripheral blood mononuclear cells. A similar enhancement of TNF-α release by SP-A was observed by rat peripheral blood cells, splenocytes, and alveolar macrophages. In combinations of SP-A with surfactant lipids, the enhanced levels of TNF-α resulting from SP-A treatment decreased as the lipids increased. At higher relative concentrations of SP-A, the lipids had little or no effect. SP-A also enhanced the production of IgA, IgG, and IgM by rat splenocytes. These data demonstrate that SP-A is capable of modulating immune cell function in lung by regulating cytokine production and Ig secretion (Kremlev et al. 1997; Huang et al. 2002).

SP-A Exhibits Inhibitory Effect on Eosinophils IL-8-Production: Eosinophils are one of the important sources of cytokines such as IL-8 at the site of allergic inflammation. Pulmonary SP-A plays a potential role in modifying inflammation and the immune function. SP-A could modify IL-8 production and release by eosinophils stimulated with ionomycin. The regulating effect of SP-A on eosinophil cytokine generation was verified. SP-A inhibits the secretion of IL-8 in a dose-dependent fashion. SP-A attenuated the production and release of IL-8 by eosinophils in a concentration-dependent manner. Thus, SP-A may have the potential to modify allergic inflammation by inhibiting the release and production of IL-8 by eosinophils (Cheng et al. 1998, 2000).

SP-A Inhibits Lavage-Induced Surfactant Secretion in Newborn Rabbits: Various agents stimulate the secretion of lung surfactant from alveolar type II cells by increasing intracellular Ca^{2+}, c-AMP, or diacylglycerol. In vivo and in vitro experiments with granular pneumocytes suggest that SP-A inhibits secretion of pulmonary surfactant (Corbet et al. 1992). Stilbene disulfonic acids are potent but reversible inhibitors of lung surfactant secretion (Chander and Sen 1993). While searching the role of C-terminal domain of SP-A in binding to type II cells and regulation of phospholipid secretion, Murata et al. (1993) demonstrated that the non-collagenous, C-terminal, domain of SP-A is responsible for the protein's inhibitory effect on lipid secretion and its binding to type II cells.

24.5.5 SP-A Stimulates Chemotaxis of AMΦ and Neutrophils

Besides accelerating pathogen clearance by pulmonary phagocytes, SP-A also stimulates alveolar macrophage chemotaxis and directed actin polymerization. Moreover, SP-A also stimulates neutrophil chemotaxis and supports a role in regulating neutrophil migration in pulmonary tissue (Schagat et al. 2003). The stimulatory effect of surfactant on macrophage migration was mediated and stimulated by SP-A. Heat treatment or reduction and alkylation of SP-A reduced its stimulatory effect. The SP-A stimulates migration primarily by enhancing chemotaxis (directed movement) rather than chemokinesis (random movement). The interaction of SP-A with macrophages may be mediated partly by the collagen-like domain of SP-A (Wright and Youmans 1993).

24.5.6 SP-A Inhibits sPLA2 and Regulates Surfactant Phospholipid Break-Down

Secretory type IIA phospholipase A2 (sPLA2-IIA) is one of the key enzymes that may potentially play a role in the pathogenesis of inflammatory diseases because its presence is observed in sera of patients with bacterial infection and in the airspaces of animals with endotoxin-induced acute lung injury (Arbibe et al. 1997). Lyso-phospholipids are generated from surfactant mediated by type-II PLA2. Lyso-phospholipids exert a major injurious effect on lung cell membranes during ARDS. This hydrolysis was inhibited by SP-A through a direct and selective protein-protein interaction between SP-A and sPLA2-II (Arbibe et al. 1998). The absence of SP-A exacerbates the susceptibility of surfactant to degradation by sPLA2-IIA. SP-A inhibits Ca^{2+}-independent acidic PLA2 of rat lung homogenate or isolated lamellar bodies but does not affect Ca^{2+}-dependent alkaline enzyme. Thus, inhibition of sPLA2 activity by SP-A may be a protective mechanism by maintaining surfactant integrity during lung injury (Arbibe et al. 1998; Chabot et al. 2003a; Fisher et al. 1994). SP-A and DOPG/PG play a role in the surfactant-mediated inhibition of sPLA2-IIA expression in AMΦs and that this inhibition occurs via a downregulation of NF-kB activation (Wu et al. 2003). Peroxiredoxin 6 (Prdx6), a protein with both GSH peroxidase and PLA_2 activities, degrades internalized DPPC of lung surfactant. The PLA_2 activity is inhibited by SP-A. A direct interaction between SP-A and Prdx6 may be a mechanism for regulation of the PLA_2 activity of Prdx6 by SP-A (Wu et al. 2006).

24.5.7 SP-A Helps in Increased Clearance of Alveolar DPPC

SP-A is necessary for lungs to respond to hyperventilation or secretagogues with increased ^{3}H-DPPC uptake. SP-A improves the surfactant activity of lipid extracts by enhancing the rate of adsorption and/or spreading of phospholipid at the air/liquid interface resulting in the formation of a stable lipid monolayer at lower bulk concentrations of either phospholipid or calcium (Chung et al. 1989). The surfactant phospholipids are mainly phosphatidylcholine and phosphatidylglycerol species, but analysis has shown that dipalmitoylphosphatidylcholine (DPPC) and dioleylphosphatidylglycerol (DOPG), in particular, play a major role in maintaining the biophysical properties of the surfactant film (Hawgood and Poulain 2001). SP-A enhances the uptake of liposomes containing DPPC, 1-palmitoyl-2-linoleoyl phosphatidylcholine (PLPC), or 1,2-dihexadecyl-sn-glycero-3-phosphocholine (DPPC-ether) by alveolar type II cells (Tsuzuki et al. 1993). SP-A also modulates the PLA_2-mediated degradation of internalized DPPC (Jain et al. 2003). Pulmonary SP-A binds the lipids DPPC and galactosylceramide and induces aggregation of phospholipid vesicles.

The SP-A also inhibits lipid secretion and enhances the uptake of phospholipid by alveolar type II cells. Kuroki et al. (1994) suggested that the CRD is essential for the SP-A functions of lipid binding, liposome aggregation, the inhibitory effect on lipid secretion, and the augmentation of lipid uptake by type II cells, and these activities are largely attributable to amino acid residues within the steric inhibitory footprint of antibody, 1D6, bound to the small disulfide loop region. The neck domain of SP-A may also be involved in the process of SP-A-mediated uptake of phospholipids by alveolar type II cells.

In vitro studies have revealed that the surface activity of oxidized surfactant was impaired and that this effect could be overcome by SP-A. Mechanically ventilated, surfactant-deficient rats were administered either bovine lipid extract surfactant (BLES) or in vitro oxidized BLES. Instillation of an in vitro oxidized surfactant caused an inferior physiological response in a surfactant-deficient rat (Bailey et al. 2004). Both C-reactive protein (CRP) and SP-A affect the surface activity of bovine lipid extract surfactant (BLES), a clinically applied modified natural surfactant. Although SP-A and CRP both bind PC, there is a difference in the manner in which they interact with surface films (Nag et al. 2004).

Isolated perfused rat lung showed that both clathrin- and actin-mediated pathways are responsible for endocytosis of dipalmitoylphosphatidylcholine (DPPC)-labeled liposomes by granular pneumocytes in the intact lung. Further studies may strengthen for a major role of receptor-mediated endocytosis of DPPC by granular pneumocytes, a process critically dependent on SP-A (Lang et al. 2008).

24.5.8 Protection of Type II Pneumocytes from Apoptosis

Activation of SPAR by SP-A binding initiates a signal through pathways that involve tyrosine phosphorylation, include IRS-1, and entail activation of PI3K. In other cell types, cytokines that activate the PI3K signaling pathway promote cell survival. SP-A affects apoptosis as measured by DNA laddering, and other techniques. The protective effects of SP-A were abrogated by inhibition of either tyrosine-specific protein kinase activity or PI3K (White et al. 2001). Analysis of anti-apoptotic signaling species downstream of IRS-1 showed activation of PKB/Akt but not of MAPK. Phosphorylation of IkB was minimally affected by SP-A. However, FKHR was rapidly phosphorylated in response to SP-A and its DNA-binding activity was significantly reduced. Since FKHR is pro-apoptotic, this may play an important role in signaling the anti-apoptotic effects of SP-A. Therefore, survival-enhancing signaling activated by SP-A include SPAR through IRS-1, PI3K, PKB/Akt, and FKHR (White and Strayer 2002).

SP-A and SP-D are potent modulators of macrophage function and may suppress clearance of apoptotic cells through activation of the transmembrane receptor signal inhibitory regulatory protein alpha (SIRPα). Gardai et al. (2003) provided evidence that SP-A and SP-D act in a dual manner to enhance or suppress inflammatory mediator production depending on binding orientation. In this concept, a model is implied in which SP-A and SP-D help maintain a non/anti-inflammatory lung environment by stimulating SIRPα on resident cells through their globular heads (Janssen et al. 2008).

24.5.9 Anti-inflammatory Role of SP-A

SP-A Modulates the Cellular Response to Smooth and Rough LPSs by Interaction with CD14: Lung SP-A modulates cellular inflammatory responses by their direct interactions with pattern-recognition receptors (PRRs) and has been implicated in the regulation of pulmonary host defense. Pathogen derived components such as LPS, peptidoglycan (PGN) or zymosan are potent stimulators of inflammation. SP-A binds to rough forms but not to smooth forms of LPS. SP-A interacts with CD14 on AMΦ and inhibits the binding of smooth LPS to CD14 and reduces TNF-α expression induced by smooth LPS (Sano et al. 1999). In this interaction, SP-A associates with LPS via lipid A moiety of rough LPS and involved in the anti-bacterial defences of the lung. The direct interaction of SP-A with CD14 constitutes a likely mechanism by which SP-A modulates LPS-elicited cellular responses (Sano et al. 1999; Stamme et al. 2002). Studies show that the direct interaction of SP-A with TLR2 alters PGN-induced cell signaling and that SP-A modulates inflammatory responses against the bacterial components by interactions with PRRs (Murakami et al. 2002; Crowther et al. 2004). SP-A also down-regulates TLR2-mediated signaling and TNF-α secretion stimulated by zymosan. This supports an important role of SP-A in controlling pulmonary inflammation caused by microbial pathogens (Sato et al. 2003). NF-kB activation and TNF-α expression induced by PGN or zymosan were significantly inhibited in presence of SP-A. AMΦs are prototypical alternatively activated macrophages, with limited production of reactive oxygen intermediates (ROI) in response to stimuli (Weissbach et al. 1994). Crowther et al. (2004) support an anti-inflammatory role for SP-A in pulmonary homeostasis by inhibiting MΦ production of ROI through a reduction in NADPH oxidase activity. SP-A has been shown to stimulate the phagocytosis of apoptotic neutrophils (PMNs) and anti-inflammatory cytokine release by inflammatory AMΦs. Furthermore, SP-A enhances TGF-β 1 release from both AMΦ populations (Reidy and Wright 2003).

However, Alcorn and Wright (2004) suggested that SP-A inhibits inflammatory cytokine production in a CD14-independent manner and also by mechanisms independent of LPS signaling pathway (Alcorn and Wright 2004). SP-A inhibited LPS-induced mRNA levels for TNF-α, IL-1α, and IL-1β as well as NF-kB DNA binding activity. Significantly, SP-A also diminished ultra pure LPS-stimulated TNF-α produced by wild-type and CD14-null mouse alveolar macrophages. Additionally, SP-A inhibited TNF-α stimulated by PMA in both wild-type and TLR4-mutant macrophages. Salez et al. (2001) highlighted the inhibitory role of SP-A in the anti-inflammatory activity of macrophages through inhibition of IL-10 production.

SP-A Regulates Surfactant Phospholipid Clearance After LPS-Induced Injury In Vivo: SP-A plays a role in surfactant homeostasis after acute lung injury. In bacterial LPS- induced injury into the lungs surfactant phospholipid levels were increased 1.6-fold in injured SP-A$^{-/-}$ animals, although injury did not alter ^{3}H-choline or ^{14}C-palmitate incorporation into dipalmitoylphosphatidylcholine (DPPC), suggesting no change in surfactant synthesis/secretion after injury. SP-A may play a role in regulating clearance of surfactant phospholipids after acute lung injury (Quintero et al. 2002). As SP-A inhibits LPS-induced in vitro IL-10 formation by bone marrow-derived macrophages, Chabot et al. (2003b) demonstrated an in vivo inhibitory role of SP-A on anti-inflammatory activity of mononuclear phagocytes, through inhibition of IL-10 production.

SP-A Regulates TNF-α Production: SP-A inhibited the LPS-induced TNF-α response of both interstitial (iMΦ) and alveolar human MΦ (AMΦ), as well as IL-1 response in iMΦ. Studies lend a credit to a physiological function of SP-A in regulating alveolar host defense and inflammation by suggesting a fundamental role of this protein in limiting excessive proinflammatory cytokine release in AMΦ during ARDS (Arias-Diaz et al. 2000; McIntosh et al. 1996). Surfactant lipids inhibit cytokine production by immune cells, and SP-A stimulates it. The increases in TNF-α mRNA and protein induced by SP-A were inhibited by surfactant lipids. SP-A exerts its action, at least in part, via activation of NF-kB. Moreover, the NF-kB inhibitors blocked SP-A-dependent increases in TNF-α mRNA levels. Observations suggest a mechanism by which SP-A plays a role in induced inflammation in the lung. The SP-A-induced increase in TNF-α levels differ among SP-A variants and appear to be affected by SP-A genotype and if SP-A is derived from one or both genes (Wang et al. 2000). Further more, different serotypes of LPS respond to SP-A differently on regulation of inflammatory cytokines in vitro (Song and Phelps 2000a, b).

SP-A Up-Regulates Activity of ManR and TLRs Expression by MΦs: SP-A selectively enhances ManR expression on human monocyte-derived macrophages. SP-A up-regulation of ManR activity provides a mechanism for enhanced phagocytosis of microbes by AMΦs, thereby enhancing lung host defense against extracellular pathogens (Beharka et al. 2002). Not only SP-A, but other lung collectins also enhance the uptake of *S. pneumoniae* or *M. avium* by AMΦs. The direct interaction of lung collectins with MΦs resulted in increased cell surface expression of scavenger receptor A or mannose receptor, which are responsible for phagocytosis. A more complete understanding of the molecular mechanism is required (Sano et al. 2006).

Henning et al. (2008) reported the SP-A-induced transcriptional and posttranslational regulation of TLR2 and TLR4 expression during the differentiation of primary human monocytes into MΦs. Despite SP-A's ability to up-regulate TLR2 expression on human MΦs, it dampens TLR2 and TLR4 signaling in these cells. SP-A decreases the phosphorylation of IkBα, a key regulator of NF-kB activity, and nuclear translocation of p65 which result in diminished TNF-α secretion in response to TLR ligands. SP-A also reduces the phosphorylation of TLR signaling proteins upstream of NF-kB, including members of the MAPK family. Finally, SP-A decreases the phosphorylation of Akt, a major cell regulator of NF-kB and potentially MAPKs. These data identify a critical role for SP-A in modulating the lung inflammatory response by regulating MΦ TLR activity (Henning et al. 2008). Molecular mechanisms related to the immunostimulatory activity of SP-A using MΦs from C3H/HeJ mice, which carry an inactivating mutation in the TLR-4 gene, and TLR-4-transfected CHO cells, suggest that SP-A-induced activation of NF-kB signaling pathway and up-regulation of cytokine synthesis such as TNF-α and IL-10 are critically dependent on TLR-4 functional complex. These findings support the concept that TLR-4 signals in response to both foreign pathogens and endogenous host mediators (Guillot et al. 2002).

SP-A Increases MMP-9 Production by THP-1 Cells: MMP-9 from alveolar MΦs is a major source of elastolytic activity in lung. SP-A may regulate MMP-9 expression. SP-A induced the expression of MMP-9 in THP-1 cells and peripheral blood mononuclear cells. It is believed that SP-A action is mediated through TLR-2. These observations suggest the presence of a locally controlled mechanism by which MMP-9 levels may be regulated in alveolar AΦs. Perhaps, SP-A influences the protease/antiprotease balance in lungs of patients with changes in surfactant constituents favoring an abnormal breakdown of extracellular matrix components (Vazquez et al. 2003). However, *Mmp9$^{-/-}$/Spd$^{-/-}$* and *Mmp12$^{-/-}$/Spd$^{-/-}$* mice developed air space enlargement similar to *Spd$^{-/-}$* mice, supporting the concept that increased expression of each metalloproteinase seen in *Spd$^{-/-}$* lungs is not major cause of emphysema (Zhang et al. 2006).

SP-A Provides Immunoprotection in Neonatal Mice: SP-A acts as an autocrine cytokine: it binds its receptor and specifically regulates transcription of surfactant proteins and other genes. SP-A null pups, reared on corn dust bedding had significant mortality The exogenous SP-A delivered by mouth to newborn SP-A null pups with SP-A null mothers improved newborn survival in the corn dust environment. The lack of SP-D did not affect newborn survival, while SP-A produced by either the mother or the pup or oral exogenous SP-A significantly reduced newborn mortality associated with environmentally induced infection in newborns (George et al. 2008). The beneficial role of surfactant in improving oxygenation has already been established in clinical trials, whereas the immunomodulating effects are promising but remain to be elucidated (Kneyber et al. 2005).

24.6 Reactive Oxygen and Nitrogen-Induced Lung Injury

A pathway for the generation of potential oxidants with the reactivity of hydroxyl radical without the need for metal catalysis has been described. In response to various inflammatory stimuli, lung endothelial, alveolar, and airway epithelial cells, as well as activated AMΦ, produce both .NO and superoxide anion radicals (O_2^-). .NO regulates pulmonary vascular and airway tone and plays an important role in lung host defense against various bacteria. However, .NO may be cytotoxic by inhibiting critical enzymes such as mitochondrial aconitase and ribonucleotide reductase, by S-nitrosolation of thiol groups, or by binding to their iron-sulfur centers. In addition, .NO reacts with O_2^- at a near diffusion-limited rate to form the strong oxidant peroxynitrite ($ONOO^-$), which can nitrate and oxidize key amino acids in various lung proteins such as SP-A, and inhibit their functions. The presence of $ONOO^-$ in the lungs of patients with ARDS has been demonstrated. Various studies have shown that inhalation or intratracheal instillation of various respirable mineral dusts or asbestos fibers increased levels of inducible NOS mRNA. Zhu et al. (1998) reviewed the evidence for the upregulation of .NO in the lungs of animals exposed to mineral particulates and assess the contribution of reactive nitrogen species in the pathogenesis of the resultant lung injury (Zhu et al. 1998).

24.6.1 Decreased Ability of Nitrated SP-A to Aggregate Surfactant Lipids

The .NO can modify oxidant stress by limiting superoxide (O_2^-)-mediated injury. However, the product of .NO reaction with O_2^-, peroxynitrite ($ONOO^-$), is a potent oxidizing and nitrating agent. At high concentrations, .NO inhibit O_2^--induced lipid peroxidation. However, ONOO., formed by reaction of .NO and O_2^-, nitrates SP-A leading to decreased ability to aggregate lipids and bind mannose (Zhu et al. 1998). Depending on the pH, tetranitromethane (TNM) acts either as a nitrating (pH > or = 24.4) or an oxidizing agent (pH < or = 6). SP-A, treated with TNM at pH 6, 24.4, 8, or 10, was tested for its ability to aggregate lipids. Treatment of SP-A with 0.5 mM TNM decreased its ability to aggregate lipids by 30% at pH 24.4, and 90% at pH 8, but had no effect on disulfide-dependent oligomeric state of SP-A. In contrast, SP-A exposed to 1 mM TNM at pH 6 had background levels of nitrotyrosine and exhibited normal lipid aggregation properties. It appeared that tyrosine nitration selectively inhibits the SP-A-mediated lipid aggregation without affecting its ability to bind lipids (Haddad et al. 1996). Sequencing of nitrated peptides demonstrated that the nitration was equally distributed on Tyr164 and Tyr1624. Nitrated SP-A exhibited decreased ability to aggregate surfactant lipids in the presence of Ca^{2+} (Greis et al. 1996). Peroxynitrite, produced by activated AMΦs, can nitrate SP-A and that CO_2 increased nitration by enhancing enzymatic nitric oxide production (Zhu et al. 2000).

Nitrated SP-A Does Not Enhance Adherence of *P. carinii* to Alveolar MΦs: Nitration of SP-A alters its ability to bind to mannose-containing saccharides on *P. carinii*. Human SP-A nitrated by ONOO– or tetranitromethane decreases its binding to *P. carinii* by increasing its dissociation constant from 24.8×10^{-9} to 1.6×10^{-8} or 2.4×10^{-8} M, respectively, without significantly affecting the number of binding sites. Furthermore, ONOO – nitrated SP-A failed to mediate the adherence and phagocytosis of *P. carinii* to rat AM as compared with normal SP-A. Binding of SP-A to rat AMΦs was not altered by nitration. These results indicate that nitration of SP-A interferes with its ability to serve as a ligand for *P. carinii* adherence to AMΦs at the site of the SP-A molecule (Zhu et al. 1998). Nitration of SP-A impairs its host- defense properties. Pulmonary edema fluid from patients with ALI has significantly higher levels of nitrite (NO_2^-) + nitrate (NO_3^-) compared with pulmonary edema fluid from patients with hydrostatic pulmonary edema. Nitrated pulmonary SP-A was also detected in the edema fluid of patients with ALI. This study indicated that reactive oxygen-nitrogen species may play a role in the pathogenesis of human ALI (Zhu et al. 2001).

24.7 Non-Pulmonary SP-A

24.7.1 SP-A in Epithelial Cells of Small and Large Intestine

Although, lung is the major source of surfactant proteins, SP-A expression has been reported in extra-pulmonary tissues. Several lines of evidence indicate that pulmonary and gastrointestinal epithelium produce closely related surface-active materials. Small intestine and colon express SP-A constitutively (Rubio et al. 1995). Both SP-A genes, SP-A1 and SP-A2 genes, are expressed in human small and large intestine. The size of intestinal SP-A mRNA is the same as that in human lung (Lin et al. 2001). SP-A mRNA transcripts are readily amplified from the trachea, prostate, pancreas, intestine and thymus. SP-A sequences derived from lung and thymus mRNA revealed the presence of both SP-A1 and SP-A2, whereas only SP-A2 expression was found in the trachea and prostate (Madsen et al. 2003a). The middle ear protein has the same epitope as human lung SP-A. SP-A is also expressed in pig Eustachian tube. The secreted antimicrobial molecules of the tubotympanum include lysozyme, lactoferrin, beta defensins, and the SP-A, SP-D. Defects in the expression or regulation of these molecules may also be the major risk factor for otitis media (Lim et al. 2000). Human vaginal mucosa and vaginal lavage fluid show presence of SP-A protein and its transcripts. Transcripts of SP-A were identified in vaginal wall, derived from SP-A genes, SP-A1 and SP-A2.

24.7.2 SP-A in Female Genital Tract and During Pregnancy

SP-A1 is found in female genital tract. Evidence suggests that SP-A is produced in a squamous epithelium, namely the vaginal mucosa, and has a localization that would allow it to contribute to both the innate and adaptive immune response. In vagina, as in lung, SP-A is an essential component of the host-defence system (MacNeill et al. 2004). The bronchiolar 16 kD Clara cell secretory protein (CC16) and the alveolar SP-A are secreted in the amniotic fluid, where they reflect the growth and the maturity of the fetal lung (Cho et al. 1999). There is compelling evidence to suggest that the fetus signals the initiation of labor by secretion into amniotic fluid of major lung SP-A. SP-A protein has been demonstrated in human chorioamniotic membranes. Parturition at term, gestational age and chorioamnionitis in preterm delivery are associated with changes in the expression of SP-A in the chorioamniotic membranes (Han et al. 2007). The maternal smoking during pregnancy does not alter secretory functions of the epithelium of the distal airways and the alveoli at term (Hermans et al. 2001, 2003).

Exogenous administration of SP-A into mouse amniotic fluid at 15 dpc caused preterm labor. SP-A activated amniotic fluid macrophages in vitro to produce NF-kB and IL-1beta. These macrophages, which are of fetal origin, migrate to the pregnant uterus causing an inflammatory response and increase uterine NF-kB activity. It is suggested that the increase in NF-kB within the maternal uterus both directly increases expression of genes that promote uterine contractility and negatively impacts the capacity of the PR to maintain uterine quiescence, contributing to the onset of labor. Findings, therefore, indicate that SP-A secreted into amniotic fluid/the maturing fetal lung serves as a hormone of parturition (Condon et al. 2004; Mendelson and Condon 2005). It was proposed that IL-1 from extrapulmonary sources induces the SPs in premature lung and is responsible for decreased risk of RDS in intra-amniotic infection (Väyrynen et al. 2002). However, Sun et al. (2006) suggested that SP-A can be synthesized locally in human fetal membranes, which can be induced by glucocorticoids. SP-A appeared to induce PGE2 synthesis in chorionic trophoblasts via induced expression of cyclooxygenase type 2. Levels of SP-A and SP-B, in Amniotic fluid, change during human parturition. Whereas SP-A was detected in all amniotic fluid samples SP-B was detected in 24.1% of mid-trimester samples and in samples at term. The median amniotic fluid concentrations of SP-A and SP-B were significantly higher in women at term than in women in the mid-trimester, and decreases during spontaneous human parturition at term (Garcia-Verdugo et al. 2008a).

Human myometrial cells express functional SP-A1 binding sites and respond to SP-A1 to initiate activation of signaling related to human parturition. SP-A1 is not produced in rat uterus, but detected transiently in rat myometrium at the end (Days 19 and 21) of gestation, but not postpartum. SP-A1 binds myometrium through its collagenlike domain and rapidly activated MAPK1/3 in myometrial cells. Bacterial LPS, known to trigger uterine contractions and preterm birth, also activated MAPK1/3. Results provide the evidence for inhibitory cross talk between SP-A1 and LPS signals, and new insight into the mechanisms of normal and preterm parturition (Garcia-Verdugo et al. 2007, 2008b). Various microorganisms colonize this area and may cause intrauterine infection or trigger preterm labor.

References

Alcorn JF, Wright JR (2004) Surfactant protein A inhibits alveolar macrophage cytokine production by CD14-independent pathway. Am J Physiol Lung Cell Mol Physiol 286:L129–L137

Arbibe L, Koumanov K, Vial D et al (1998) Generation of lysophospholipids from surfactant in acute lung injury is mediated by type-II phospholipase A2 and inhibited by a direct surfactant protein A-phospholipase A2 protein interaction. J Clin Invest 102:1152–1160

Arbibe L, Vial D, Rosinski-Chupin I et al (1997) Endotoxin induces expression of type II phospholipase A2 in macrophages during acute lung injury in guinea pigs: involvement of TNF-alpha in lipopolysaccharide-induced type II phospholipase A2 synthesis. J Immunol 159:391–400

Arias-Diaz J, Garcia-Verdugo I, Casals C et al (2000) Effect of surfactant protein A (SP-A) on the production of cytokines by human pulmonary macrophages. Shock 14:300–307

Bailey TC, Da Silva KA, Lewis JF et al (2004) Physiological and inflammatory response to instillation of an oxidized surfactant in a rat model of surfactant deficiency. J Appl Physiol 96:1674–1680

Bates SR, Kazi AS, Tao JQ et al (2008) Role of P63 (CKAP4) in binding of surfactant protein-A to type II pneumocytes. Am J Physiol Lung Cell Mol Physiol 295:L658–L669

Beharka AA, Gaynor CD, Kang BK et al (2002) Pulmonary surfactant protein A up-regulates activity of the mannose receptor, a pattern recognition receptor expressed on human macrophages. J Immunol 169:3565–3573

Blau H, Riklis S, Van Iwaarden JF et al (1997) Nitric oxide production by rat alveolar macrophages can be modulated in vitro by surfactant protein A. Am J Physiol 272:L1198–L1204

Boggaram V, Mendelson CR (1988) Transcriptional regulation of the gene encoding the major surfactant protein (SP-A) in rabbit fetal lung. J Biol Chem 263:19060–19065

Boggaram V, Smith ME, Mendelson CR (1991) Posttranscriptional regulation of surfactant protein-A messenger RNA in human fetal lung in vitro by glucocorticoids. Mol Endocrinol 5:414–423

Borron PJ, Crouch EC, Lewis JF et al (1998a) Recombinant rat surfactant-associated protein D inhibits human T lymphocyte proliferation and IL-2 production. J Immunol 161:4599–4603

Borron P, McCormack FX, Elhalwagi BM et al (1998b) Surfactant protein A inhibits T cell proliferation via its collagen-like tail and a 210-kDa receptor. Am J Physiol 275:L679–L687

Borron P, McIntosh JC, Korfhagen TR et al (2000) Surfactant-associated protein A inhibits LPS-induced cytokine and nitric oxide production in vivo. Am J Physiol Lung Cell Mol Physiol 278:L840–L847

Brinker KG, Garner H, Wright JR (2003) Surfactant protein A modulates the differentiation of murine bone marrow-derived dendritic cells. Am J Physiol Lung Cell Mol Physiol 284: L232–L241

Brinker KG, Martin E, Borron P et al (2001) Surfactant protein D enhances bacterial antigen presentation by bone marrow–derived dendritic cells. Am J Physiol Lung Cell Mol Physiol 281: L1453–L1463

Brown-Augsburger PK, Hartshorn K, Chang D et al (1996) Site-directed mutagenesis of Cys-15 and Cys-20 of pulmonary surfactant protein D: expression of a trimeric protein with altered anti-viral properties. J Biol Chem 271:13724–13730

Chabot S, Koumanov K, Lambeau G et al (2003a) Inhibitory effects of surfactant protein A on surfactant phospholipid hydrolysis by secreted phospholipases A2. J Immunol 171:995–1000

Chabot S, Salez L, McCormack FX et al (2003b) Surfactant protein A inhibits lipopolysaccharide-induced in vivo production of interleukin-10 by mononuclear phagocytes during lung inflammation. Am J Respir Cell Mol Biol 28:347–353

Chander A, Sen N (1993) Inhibition of phosphatidylcholine secretion by stilbene disulfonates in alveolar type II cells. Biochem Pharmacol 45:1905–1912

Chen Q, Bates SR, Fisher AB (1996) Secretagogues increase the expression of surfactant protein A receptors on lung type II cells. J Biol Chem 271:25277–25283

Chen Q, Boggaram V, Mendelson CR (1992) Rabbit lung surfactant protein A gene: identification of a lung-specific DNase I hypersensitive site. Am J Physiol 262:L662–L671

Cheng G, Ueda T, Nakajima H et al (2000) Surfactant protein A exhibits inhibitory effect on eosinophils IL-8 production. Biochem Biophys Res Commun 270:831–835

Cheng G, Ueda T, Nakajima H et al (1998) Suppressive effects of SP-A on ionomycin-induced IL-8 production and release by eosinophils. Int Arch Allergy Immunol 117(Suppl 1):59–62

Chiba H, Sano H, Saitoh M et al (1999) Introduction of mannose binding protein-type phosphatidylinositol recognition into pulmonary surfactant protein A. Biochemistry 38:7321–7331

Childs RA, Wright JR, Ross GF et al (1992) Specificity of lung surfactant protein SP-A for both the carbohydrate and the lipid moieties of certain neutral glycolipids. J Biol Chem 267: 9972–9979

Cho K, Matsuda T, Okajima S et al (1999) Factors influencing pulmonary surfactant protein A levels in cord blood, maternal blood and amniotic fluid. Biol Neonate 75:104–110

Chroneos ZC, Abdolrasulnia R, Whitsett JA et al (1996) Purification of a cell-surface receptor for surfactant protein A. J Biol Chem 271:16375–16383

Chung J, Yu SH, Whitsett JA et al (1989) Effect of surfactant-associated protein-A (SP-A) on the activity of lipid extract surfactant. Biochim Biophys Acta 1002:348–358

Cockshutt AM, Weitz J, Possmayer F (1990) Pulmonary surfactant-associated protein A enhances the surface activity of lipid extract surfactant and reverses inhibition by blood proteins in vitro. Biochemistry 29:8424–8429

Condon JC, Jeyasuria P, Faust JM et al (2004) Surfactant protein secreted by the maturing mouse fetal lung acts as a hormone that signals the initiation of parturition. Proc Natl Acad Sci USA 101:4978–4983

Corbet A, Bedi H, Owens M, Taeusch W (1992) Surfactant protein-A inhibits lavage-induced surfactant secretion in newborn rabbits. Am J Med Sci 304:246–251

Crouch EC (1998a) Structure, biologic properties, and expression of surfactant protein D (SP-D). Biochim Biophys Acta 1408: 278–289

Crouch EC (1998b) Collectins and pulmonary host defense. Am J Respir Cell Mol Biol 19:177–201

Crouch EC, Persson A, Chang D, Heuser J (1994) Molecular structure of pulmonary surfactant protein D (SP-D). J Biol Chem 269:17311–17319

Crowther JE, Kutala VK, Kuppusamy P et al (2004) Pulmonary surfactant protein A inhibits macrophage reactive oxygen intermediate production in response to stimuli by reducing NADPH oxidase activity. J Immunol 172:6866–6874

Dhar V, Hallman M, Lappalainen U, Bry K (1997) Interleukin-1α upregulates the expression of surfactant protein-A in rabbit lung explants. Biol Neonate 71:46–52

Dobbie JW (1996) Surfactant protein A and lamellar bodies: a homologous secretory function of peritoneum, synovium, and lung. Perit Dial Int 16:574–581

Drickamer K (1992) Engineering galactose-binding activity into a C-type mannose-binding protein. Nature 360:183–186

Elhalwagi BM, Damodarasamy M, McCormack FX (1997) Alternate amino terminal processing of surfactant protein A results in cysteinyl isoforms required for multimer formation. Biochemistry 36:7018–7025

Fisher AB, Dodia C, Chander A (1994) Inhibition of lung calcium-independent phospholipase A2 by surfactant protein A. Am J Physiol 267:L335–L341

Fisher JH, Kao FT, Jones C et al (1987) The coding sequence for the 32,000-dalton pulmonary surfactant-associated protein A is located on chromosome 10 and identifies two separate restriction-fragment-length polymorphisms. Am J Hum Genet 40:503–511

Fisher JH, Larson J, Cool C, Dow SW (2002) Lymphocyte activation in the lungs of SP-D null mice. Am J Respir Cell Mol Biol 27:24–33

Floros J, Hoover RR (1998) Genetics of the hydrophilic surfactant proteins A and D. Biochim Biophys Acta 1408:312–322

Garcia-Verdugo I, Leiber D, Robin P et al (2007) Direct interaction of surfactant protein A with myometrial binding sites: signaling and modulation by bacterial lipopolysaccharide. Biol Reprod 76:681–691

García-Verdugo I, Sánchez-Barbero F, Bosch FU et al (2003) Effect of hydroxylation and N187-linked glycosylation on molecular and functional properties of recombinant human surfactant protein A. Biochemistry 42:9532–9542

Garcia-Verdugo I, Synguelakis M, Degrouard J et al (2008a) The concentration of surfactant protein-A in amniotic fluid decreases in spontaneous human parturition at term. J Matern Fetal Neonatal Med 21:652–659

Garcia-Verdugo I, Tanfin Z, Dallot E et al (2008b) Surfactant protein A signaling pathways in human uterine smooth muscle cells. Biol Reprod 79:348–355

García-Verdugo I, Wang G, Floros J (2002) Casals C Structural analysis and lipid-binding properties of recombinant human surfactant protein a derived from one or both genes. Biochemistry 41:14041–14053

Gardai SJ, Xiao YQ, Dickinson M et al (2003) By binding SIRPalpha or calreticulin/CD91, lung collectins act as dual function surveillance molecules to suppress or enhance inflammation. Cell 115: 13–23

Geertsma MF, Nibbering PH, Haagsman HP et al (1994) Binding of surfactant protein A to C1q receptors mediates phagocytosis of *Staphylococcus aureus* by monocytes. Am J Physiol 267: L578–L584

George CL, Goss KL, Meyerholz DK, Lamb FS, Snyder JM (2008) Surfactant-associated protein A provides critical immunoprotection in neonatal mice. Infect Immun 76:380–390

Gorrini M, Lupi A, Iadarola P et al (2005) SP-A binds alpha1-antitrypsin in vitro and reduces the association rate constant for neutrophil elastase. Respir Res 6:147

Greene KE, Wright JR, Steinberg KP et al (1999) Serial changes in surfactant-associated proteins in lung and serum before and after onset of ARDS. Am J Respir Crit Care Med 160:1843–1850

Greis KD, Zhu S, Matalon S (1996) Identification of nitration sites on surfactant protein A by tandem electrospray mass spectrometry. Arch Biochem Biophys 335:396–402

Guillot L, Balloy V, McCormack FX et al (2002) Cutting edge: the immunostimulatory activity of the lung surfactant protein-A involves toll-like receptor 4. J Immunol 168:5989–5992

Gupta N, Manevich Y, Kazi AS et al (2006) Identification and characterization of p63 (CKAP4/ERGIC-63/CLIMP-63), a surfactant protein A binding protein, on type II pneumocytes. Am J Physiol Lung Cell Mol Physiol 291:L436–L447

Gurel O, Ikegami M, Chroneos ZC, Jobe AH (2001) Macrophage and type II cell catabolism of SP-A and saturated phosphatidylcholine in mouse lungs. Am J Physiol Lung Cell Mol Physiol 280: L1266–L1272

Haagsman HP, Elfring RH, van Buel BL, Voorhout WF (1991) The lung lectin surfactant protein A aggregates phospholipid vesicles via a novel mechanism. Biochem J 275:273–277

Haagsman HP, Hawgood S, Sargeant T et al (1987) The major lung surfactant protein, SP 28-36, is a calcium-dependent, carbohydrate-binding protein. J Biol Chem 262:13877–13880

Haagsman HP, Sargeant T, Hauschka PV et al (1990) Binding of calcium to SP-A, a surfactant-associated protein. Biochemistry 29:8894–8900

Haagsman HP, White RT, Schilling J et al (1989) Studies of the structure of lung surfactant protein SP-A. Am J Physiol 257: L421–L429

Haagsman HP (1998) Interactions of surfactant protein A with pathogens. Biochim Biophys Acta 1408:264–277

Haddad IY, Crow JP, Hu P et al (1994) Concurrent generation of nitric oxide and superoxide damages surfactant protein A. Am J Physiol 267:L242–L249

Haddad IY, Zhu S, Ischiropoulos H, Matalon S (1996) Nitration of surfactant protein A results in decreased ability to aggregate lipids. Am J Physiol 270:L281–L288

Haller EM, Shelley SA, Montgomery MR, Balis JU (1992) Immunocytochemical localization of lysozyme and surfactant protein A in rat type II cells and extracellular surfactant forms. J Histochem Cytochem 40:1491–1500

Han YM, Romero R, Kim YM et al (2007) Surfactant protein-A mRNA expression by human fetal membranes is increased in histological chorioamnionitis but not in spontaneous labour at term. J Pathol 211:489–496

Hansen S, Lo B, Evans K et al (2007) Surfactant protein D augments bacterial association but attenuates major histocompatibility complex class II presentation of bacterial antigens. Am J Respir Cell Mol Biol 36:94–102

Haque R, Umstead TM, Ponnuru P et al (2007) Role of surfactant protein-A (SP-A) in lung injury in response to acute ozone exposure of SP-A deficient mice. Toxicol Appl Pharmacol 220:72–82

Hartshorn KL, Crouch E, White MR et al (1998) Pulmonary surfactant proteins A and D enhance neutrophil uptake of bacteria. Am J Physiol 274:L958–L969

Haurum JS, Thiel S, Haagsman HP et al (1993) Studies on the carbohydrate-binding characteristics of human pulmonary surfactant-associated protein A and comparison with two other collectins: mannan-binding protein and conglutinin. Biochem J 293:873–878

Hawgood S, Poulain FR (2001) The pulmonary collectins and surfactant metabolism. Annu Rev Physiol 63:495–519

Head JF, Mealy TR, McCormack FX, Seaton BA (2003) Crystal structure of trimeric carbohydrate recognition and neck domains of surfactant protein A. J Biol Chem 278:43254–43260

Henning LN, Azad AK, Parsa KV et al (2008) Pulmonary surfactant protein A regulates TLR expression and activity in human macrophages. J Immunol 180:7847–7858

Hermans C, Dong P, Robin M et al (2003) Determinants of serum levels of surfactant proteins A and B and Clara cell protein CC16. Biomarkers 8:461–471

Hermans C, Libotte V, Robin M et al (2001) Maternal tobacco smoking and lung epithelium-specific proteins in amniotic fluid. Pediatr Res 50:487–494

Holmskov U, Lawson P, Teisner B et al (1997) Isolation and characterization of a new member of the scavenger receptor superfamily, glycoprotein-340 (gp-340), as a lung surfactant protein-D binding molecule. J Biol Chem 272:13743–10

Honma T, Kuroki Y, Tsunezawa W et al (1997) The mannose-binding protein A region of glutamic acid185-alanine221 can functionally replace the surfactant protein A region of glutamic acid195-phenylalanine228 without loss of interaction with lipids and alveolar type II cells. Biochemistry 36:7176–7184

Hospes R, Hospes BI, Reiss I et al (2002) Molecular biological characterization of equine surfactant protein A. J Vet Med A Physiol Pathol Clin Med 49:497–498

Huang W, Wang G, Phelps DS et al (2002) Combined SP-A-bleomycin effect on cytokines by THP-1 cells: impact of surfactant lipids on this effect. Am J Physiol Lung Cell Mol Physiol 283:L94–L102

Huang W, Wang G, Phelps DS et al (2004) Human SP-A genetic variants and bleomycin-induced cytokine production by THP-1 cells: effect of ozone-induced SP-A oxidation. Am J Physiol Lung Cell Mol Physiol 286:L546–L553

Hussain S, Wright JR, Martin WJ 2nd (2003) Surfactant protein A decreases nitric oxide production by macrophages in a tumor necrosis factor-alpha-dependent mechanism. Am J Respir Cell Mol Biol 28:520–527

Hynsjö L, Granberg L, Haurum J, Thiel S, Larson G (1995) Use of factorial experimental design to delineate the strong calcium- and pH-dependent changes in binding of human surfactant protein-A to neutral glycosphingolipids–a model for studies of protein-carbohydrate interactions. Anal Biochem 225:305–314

Ikegami M, Elhalwagi BM, Palaniyar N et al (2001) The collagen-like region of surfactant protein A (SP-A) is required for correction of surfactant structural and functional defects in the SP-A null mouse. J Biol Chem 276:38542–38548

Ikegami M, Korfhagen TR, Whitsett JA et al (1998) Characteristics of surfactant from SP-A-deficient mice. Am J Physiol 275:L247–L254

Ikegami M, Whitsett JA, Chroneos ZC et al (2000) IL-4 increases surfactant and regulates metabolism in vivo. Am J Physiol Lung Cell Mol Physiol 278:L75–L80

Jahnsen FL, Moloney ED, Hogan T, Upham JW, Burke CM, Holt PG (2001) Rapid dendritic cell recruitment to the bronchial mucosa of patients with atopic asthma in response to local allergen challenge. Thorax 56:823–827

Jain D, Dodia C, Bates SR et al (2003) SP-A is necessary for increased clearance of alveolar DPPC with hyperventilation or secretagogues. Am J Physiol Lung Cell Mol Physiol 284:L759–L765

Jain D, Dodia C, Fisher AB, Bates SR (2005) Pathways for clearance of surfactant protein A from the lung. Am J Physiol Lung Cell Mol Physiol 289:L1011–L1018

Janssen WJ, McPhillips KA, Dickinson MG et al (2008) Surfactant proteins A and D suppress alveolar macrophage phagocytosis via interaction with SIRP alpha. Am J Respir Crit Care Med 178:158–167

Kalina M, Blau H, Riklis S, Kravtsov V (1995) Interaction of surfactant protein A with bacterial lipopolysaccharide may affect some biological functions. Am J Physiol 268:L144–L151

Kang W, Nielsen O, Fenger C et al (2002) The scavenger receptor, cysteine-rich domain-containing molecule gp-340 is differentially regulated in epithelial cell lines by phorbol ester. Clin Exp Immunol 130:449–458

Katyal SL, Singh G, Locker J (1992) Characterization of a second human surfactant-associated protein SP-A gene. Am J Respir Cell Mol Biol 8:445–452

Kneyber MCJ, Plötz FB, Kimpen JLL (2005) Bench-to-bedside review: paediatric viral lower respiratory tract disease necessitating mechanical ventilation – should we use exogenous surfactant? Crit Care 9:550–555

King RJ, Simon D, Horowitz PM (1989) Aspects of secondary and quaternary structure of surfactant protein A from canine lung. Biochim Biophys Acta 1001:294–301

Kishore U, Bernal AL, Kamran MF et al (2005) Surfactant proteins SP-A and SP-D in human health and disease. Arch Immunol Ther Exp 53:399–417

Korfhagen TR, Bruno MD, Ross GF et al (1996) Altered surfactant function and structure in SP-A gene targeted mice. Proc Natl Acad Sci USA 93:9594–9599

Kremlev SG, Umstead TM, Phelps DS (1997) Surfactant protein A regulates cytokine production in the monocytic cell line THP-1. Am J Physiol 272:L996–L1004

Kresch MJ, Christian C, Lu H (1998) Isolation and partial characterization of a receptor to surfactant protein A expressed by rat type II pneumocytes. Am J Respir Cell Mol Biol 19:216–225

Kresch MJ, Christian C (1998) Developmental regulation of the effects of surfactant protein A on phospholipid uptake by fetal rat type II pneumocytes. Lung 176:45–61

Kunzmann S, Wright JR, Steinhilber W et al (2006) TGF-β1 in SP-A preparations influence immune suppressive properties of SP-A on human $CD4^+$ T lymphocytes. Am J Physiol Lung Cell Mol Physiol 291:L747–L757

Kuroki Y, Akino T (1991) Pulmonary surfactant protein A (SP-A) specifically binds dipalmitoylphosphatidyl choline. J Biol Chem 266:3068–3073

Kuroki Y, Gasa S, Ogasawara Y et al (1992) Binding of pulmonary surfactant protein A to galactosylceramide and asialo-GM2. Arch Biochem Biophys 299:261–267

Kuroki Y, Mason RJ, Voelker DR (1988a) Alveolar type II cells express a high-affinity receptor for pulmonary surfactant protein A. Proc Natl Acad Sci USA 85:5566–5570

Kuroki Y, Mason RJ, Voelker DR (1988b) Chemical modification of surfactant protein A alters high affinity binding to rat alveolar type II cells and regulation of phospholipid secretion. J Biol Chem 263:17596–17602

Kuroki Y, McCormack FX, Ogasawara Y et al (1994) Epitope mapping for monoclonal antibodies identifies functional domains of pulmonary surfactant protein A that interact with lipids. J Biol Chem 269:29793–29800

Kuroki Y, Shiratori M, Ogasawara Y et al (1996) Interaction of phospholipid liposomes with plasma membrane isolated from alveolar type II cells: effect of pulmonary surfactant protein A. Biochim Biophys Acta 1281:53–59

Kuzmenko AI, Wu H, Bridges JP, McCormack FX (2004) Surfactant lipid peroxidation damages surfactant protein A and inhibits interactions with phospholipid vesicles. J Lipid Res 45:1061–1068

Lang CJ, Postle AD, Orgeig S et al (2008) Surfactant protein-A plays an important role in lung surfactant clearance: evidence using the surfactant protein-A gene-targeted mouse. Am J Physiol Lung Cell Mol Physiol 294:L325–L333

LeVine AM, Bruno MD, Huelsman KM et al (1997) Surfactant protein A-deficient mice are susceptible to group B streptococcal infection. J Immunol 158:4336–4340

Lim DJ, Chun YM, Lee HY et al (2000) Cell biology of tubotympanum in relation to pathogenesis of otitis media – a review. Vaccine 19 (Suppl 1):S17–S25

Lin Z, deMello D, Phelps DS, Koltun WA et al (2001) Both human SP-A1 and SP-A2 genes are expressed in small and large intestine. Pediatr Pathol Mol Med 20:367–387

Lipscomb MF, Masten BJ (2002) Dendritic cells: immune regulators in health and disease. Physiol Rev 82:97–130

MacNeill C, Umstead TM, Phelps DS et al (2004) Surfactant protein A, an innate immune factor, is expressed in the vaginal mucosa and is present in vaginal lavage fluid. Immunology 111:91–99

Madsen J, Tornoe I, Nielsen O et al (2003a) Expression and localization of lung surfactant protein A in human tissues. Am J Respir Cell Mol Biol 29:591–597

Malhotra R, Haurum J, Thiel S, Sim RB (1992) Interaction of C1q receptor with lung surfactant protein A. Eur J Immunol 22:1437–1445

Malhotra R, Laursen SB, Willis AC, Sim RB (1993) Localization of the receptor-binding site in the collectin family of proteins. Biochem J 293:15–19

Malhotra R, Thiel S, Reid KB, Sim RB (1990) Human leukocyte C1q receptor binds other soluble proteins with collagen domains. J Exp Med 172:955–959

Mason RJ, Greene K, Voelker DR (1998) Surfactant protein A and surfactant protein D in health and disease. Am J Physiol 275:L1–L13

McCormack FX, Calvert HM, Watson Pa et al (1994) The structure and function of surfactant protein A. Hydroxyproline- and carbohydrate-deficient mutant proteins. J Biol Chem 269:5833–41

McCormack FX, Damodarasamy M, Elhalwagi BM (1999) Deletion mapping of N-terminal domains of surfactant protein A. The N-terminal segment is required for phospholipid aggregation and specific inhibition of surfactant secretion. J Biol Chem 274:3173–3181

McCormack FX, Festa AL, Andrews RP et al (1997a) The carbohydrate recognition domain of surfactant protein A mediates binding to the major surface glycoprotein of *Pneumocystis carinii*. Biochemistry 36:8092–8099

McCormack FX, Pattanajitvilai S, Stewart J et al (1997b) The Cys6 intermolecular disulfide bond and the collagen-like region of rat SP-A play critical roles in interactions with alveolar type II cells and surfactant lipids. J Biol Chem 272:27971–27979

McCormack FX, Stewart J, Voelker DR et al (1997c) Alanine mutagenesis of surfactant protein A reveals that lipid binding and pH-dependent liposome aggregation are mediated by the carbohydrate recognition domain. Biochemistry 36:13963–13971

McCormack FX, Whitsett JA (2002) The pulmonary collectins, SP-A and SP-D, orchestrate innate immunity in the lung. J Clin Invest 109:707–712

McCormack FX (1998) Structure, processing and properties of surfactant protein A. Biochim Biophys Acta 1408:109–131

Mendelson CR, Condon JC (2005) New insights into the molecular endocrinology of parturition. J Steroid Biochem Mol Biol 93:113–119

Meyboom A, Maretzki D, Stevens PA, Hofmann KP (1999) Interaction of pulmonary surfactant protein A with phospholipid liposomes: a kinetic study on head group and fatty acid specificity. Biochim Biophys Acta 1441:23–35

Michelis D, Kounnas MZ, Argraves WS et al (1994) Interaction of surfactant protein A with cellular myosin. Am J Respir Cell Mol Biol 11:692–700

Murakami S, Iwaki D, Mitsuzawa H et al (2002) Surfactant protein A inhibits peptidoglycan-induced tumor necrosis factor-alpha secretion in U937 cells and alveolar macrophages by direct interaction with toll-like receptor 2. J Biol Chem 277:6830–6837

Murata Y, Kuroki Y, Akino T (1993) Role of the C-terminal domain of pulmonary surfactant protein A in binding to alveolar type II cells and regulation of phospholipid secretion. Biochem J 291:71–77

McIntosh JC, Mervin-Blake S, Conner E, Wright JR (1996) Surfactant protein A protects growing cells and reduces TNF-alpha activity from LPS-stimulated macrophages. Am J Physiol 271:L310–L319

Nag K, Munro JG, Hearn SA, Ridsdale RA, Palaniyar N, Holterman CE, Inchley K, Possmayer F, Harauz G (1999) Cation-mediated conformational variants of surfactant protein A. Biochim Biophys Acta 1453:23–34

Nag K, Rodriguez-Capote K, Panda AK et al (2004) Disparate effects of two phosphatidylcholine binding proteins, C-reactive protein and surfactant protein A, on pulmonary surfactant structure and function. Am J Physiol Lung Cell Mol Physiol 287:L1145–L1153

Nepomuceno RR, Ruiz S, Park M, Tenner AJ (1999) C1qRP is a heavily O-glycosylated cell surface protein involved in the regulation of phagocytic activity. J Immunol 162:3583–3589

Ochs M, Johnen G, Müller KM et al (2002) Intracellular and intraalveolar localization of surfactant protein A (SP-A) in the parenchymal region of the human lung. Am J Respir Cell Mol Biol 26:91–98

Ogasawara Y, Kuroki Y, Akino T (1992) Pulmonary surfactant protein D specifically binds to phosphatidylinositol. J Biol Chem 267:21244–21249

Palaniyar N, Ikegami M, Korfhagen T et al (2001) Domains of surfactant protein A that affect protein oligomerization, lipid structure and surface tension. Comp Biochem Physiol A Mol Integr Physiol 129:109–127

Palaniyar N, Nadesalingam J, Clark HW et al (2004) Nucleic acid is a novel ligand for innate immune pattern recognition collectins surfactant proteins A and D and mannose-binding lectin. J Biol Chem 279:32728–32736

Palaniyar N, McCormack FX, Possmayer F, Harauz G (2000) Three-dimensional structure of rat surfactant protein A trimers in association with phospholipid monolayers. Biochemistry 39:6310–6317

Palaniyar N, Ridsdale RA, Hearn SA et al (1999) Formation of membrane lattice structures and their specific interactions with surfactant protein A. Am J Physiol 276:L642–L649

Palaniyar N, Ridsdale RA, Holterman CE et al (1998a) Structural changes of surfactant protein A induced by cations reorient the protein on lipid bilayers. J Struct Biol 122:297–310

Palaniyar N, Ridsdale RA, Possmayer F, Harauz G (1998b) Surfactant protein A (SP-A) forms a novel supraquaternary structure in the form of fibers. Biochem Biophys Res Commun 250:131–137

Pastva AM, Wright JR, Williams KL (2007) Immunomodulatory roles of surfactant proteins A and D: implications in lung disease. Proc Am Thor Soc 4:252–257

Pattanajitvilai S, Kuroki Y, Tsunezawa W et al (1998) Mutational analysis of Arg197 of rat surfactant protein A. His197 creates specific lipid uptake defects. J Biol Chem 273:5702–5707

Persson AV, Gibbons BJ, Shoemaker JD et al (1992) The major glycolipid recognized by SP-D in surfactant is phosphatidylinositol. Biochemistry 31:12183–12189

Pietschmann SM, Pison U (2000) cDNA cloning of ovine pulmonary SP-A, SP-B, and SP-C: isolation of two different sequences for SP-B. Am J Physiol Lung Cell Mol Physiol 278:L765–78

Plaga S, Plattner H, Schlepper-Schaefer J (1998) SP-A binding sites on bovine alveolar macrophages. Exp Cell Res 245:116–122

Quintero OA, Korfhagen TR, Wright JR (2002) Surfactant protein A regulates surfactant phospholipid clearance after LPS-induced injury in vivo. Am J Physiol Lung Cell Mol Physiol 283:L76–L85

Reidy MF, Wright JR (2003) Surfactant protein A enhances apoptotic cell uptake and TGF-β1 release by inflammatory alveolar macrophages. Am J Physiol Lung Cell Mol Physiol 285:L854–L861

Reilly KE, Mautone AJ, Mendelsohn R (1989) Fourier-transform infrared spectroscopy studies of lipid/protein interaction in pulmonary surfactant. Biochemistry 28:7368–7373

Ridsdale RA, Palaniyar N, Holterman CE et al (1999) Cation-mediated conformational variants of surfactant protein A. Biochim Biophys Acta 1453:I23–34

Ridsdale RA, Palaniyar N, Possmayer F, Harauz G (2001) Formation of folds and vesicles by dipalmitoylphosphatidylcholine monolayers spread in excess. J Membr Biol 180:21–32

Ross GF, Sawyer J, O'Connor T, Whitsett JA (1991) Intermolecular cross-links mediate aggregation of phospholipid vesicles by pulmonary surfactant SP-A. Biochemistry 30:858–865

Ruano ML, García-Verdugo I, Miguel E et al (2000) Self-aggregation of surfactant protein A. Biochemistry 39:6529–6537

Ruano ML, Pérez-Gil J, Casals C (1998) Effect of acidic pH on the structure and lipid binding properties of porcine surfactant protein A. Potential role of acidification along its exocytic pathway. J Biol Chem 273:15183–15191

Rubio S, Lacaze-Masmonteil T, Chailley-Heu B et al (1995) Pulmonary surfactant protein A (SP-A) is expressed by epithelial cells of small and large intestine. J Biol Chem 270:12162–12169

Sahly H, Ofek I, Podschun R et al (2002) Surfactant protein D binds selectively to *Klebsiella pneumoniae* lipopolysaccharides containing mannose-rich O-antigens. J Immunol 169:3267–3274

Salez L, Balloy V, van Rooijen N et al (2001) Surfactant protein A suppresses lipopolysaccharide-induced IL-10 production by murine macrophages. J Immunol 166:6376–6382

Sánchez-Barbero F, Rivas G, Steinhilber W, Casals C (2007) Structural and functional differences among human surfactant proteins SP-A1, SP-A2 and co-expressed SP-A1/SP-A2: role of supratrimeric oligomerization. Biochem J 406:479–489

Sánchez-Barbero F, Strassner J, García-Cañero R et al (2005) Role of the degree of oligomerization in the structure and function of human surfactant protein A. J Biol Chem 280:7659–7670

Sano H, Kuronuma K, Kudo K et al (2006) Regulation of inflammation and bacterial clearance by lung collectins. Respirology 11(Suppl):S46–S50

Sano H, Sohma H, Muta T et al (1999) Pulmonary surfactant protein A modulates the cellular response to smooth and rough lipopolysaccharides by interaction with CD14. J Immunol 163:387–395

Sato M, Sano H, Iwaki D et al (2003) Direct binding of Toll-like receptor 2 to zymosan, and zymosan-induced NF-kappa B activation and TNF-alpha secretion are down-regulated by lung collectin surfactant protein A. J Immunol 171:417–425

Schagat TL, Wofford JA, Greene KE, Wright JR (2003) Surfactant protein A differentially regulates peripheral and inflammatory neutrophil chemotaxis. Am J Physiol Lung Cell Mol Physiol 284:L140–L147

Schlosser A, Thomsen T, Shipley JM et al (2006) Microfibril-associated protein 4 binds to surfactant protein A (SP-A) and colocalizes with SP-A in the extracellular matrix of the lung. Scand J Immunol 64:104–117

Smith CI, Rosenberg E, Reisher SR et al (1995) Sequence of rat surfactant protein A gene and functional mapping of its upstream region. Am J Physiol 269:L603–L612

Sohma H, Hattori A, Kuroki Y, Akino T (1993) Calcium and dithiothreitol dependent conformational changes in beta-sheet structure of collagenase resistant fragment of human surfactant protein A. Biochem Mol Biol Int 30:329–336

Sohma H, Matsushima N, Watanabe T et al (1995) Ca^{2+}-dependent binding of annexin IV to surfactant protein A and lamellar bodies in alveolar type II cells. Biochem J 312:175–181

Song M, Phelps DS (2000a) Comparison of SP-A and LPS effects on the THP-1 monocytic cell line. Am J Physiol Lung Cell Mol Physiol 279:L110–L117

Song M, Phelps DS (2000b) Interaction of surfactant protein A with lipopolysaccharide and regulation of inflammatory cytokines in the THP-1 monocytic cell line. Infect Immun 68:6611–6617

Spissinger T, Schäfer KP, Voss T (1991) Assembly of the surfactant protein SP-A. Deletions in the globular domain interfere with the correct folding of the molecule. Eur J Biochem 199:65–71

Stamme C, Müller M, Hamann L et al (2002) Surfactant protein A inhibits lipopolysaccharide-induced immune cell activation by preventing the interaction of lipopolysaccharide with lipopolysaccharide-binding protein. Am J Respir Cell Mol Biol 27:353–360

Stamme C, Walsh E, Wright JR (2000) Surfactant protein A differentially regulates IFN-γ- and LPS-induced nitrite production by rat alveolar macrophages. Am J Respir Cell Mol Biol 23:772–779

Strayer DS, Hoek JB, Thomas AP, White MK (1999) Cellular activation by Ca2+ release from stores in the endoplasmic reticulum but not by increased free Ca^{2+} in the cytosol. Biochem J 344:39–47

Strayer DS, Yang S, Jerng HH (1993) Surfactant protein A-binding proteins. Characterization and structures. J Biol Chem 268:18679–18684

Sun K, Brockman D, Campos B et al (2006) Induction of surfactant protein A expression by cortisol facilitates prostaglandin synthesis in human chorionic trophoblasts. J Clin Endocrinol Metab 91:4988–4994

Tanfin Z (2008) Interaction of surfactant protein A with the intermediate filaments desmin and vimentin. Biochemistry 47:5127–5138

Tenner AJ, Robinson SL, Borchelt J, Wright JR (1989) Human pulmonary surfactant protein (SP-A), a protein structurally homologous to C1q, can enhance FcR- and CR1-mediated phagocytosis. J Biol Chem 264:13923–13928

Tenner AJ (1998) C1q receptors: regulating specific functions of phagocytic cells. Immunobiology 199:250–264

Terrasa AM, Guajardo MH, de Armas SE, Catalá A (2005) Pulmonary surfactant protein A inhibits the lipid peroxidation stimulated by linoleic acid hydroperoxide of rat lung mitochondria and microsomes. Biochim Biophys Acta 1735:101–110

Tino MJ, Wright JR (1998) Interactions of surfactant protein A with epithelial cells and phagocytes. Biochim Biophys Acta 1408:241–263

Tsunezawa W, Sano H, Sohma H, McCormack FX, Voelker DR, Kuroki Y (1998) Site-directed mutagenesis of surfactant protein A reveals dissociation of lipid aggregation and lipid uptake by alveolar type II cells. Biochim Biophys Acta 1387:433–447

Tsuzuki A, Kuroki Y, Akino T (1993) Pulmonary surfactant protein A-mediated uptake of phosphatidylcholine by alveolar type II cells. Am J Physiol 265:L193–L199

Uemura T, Sano H, Katoh T et al (2006) Surfactant protein A without the interruption of Gly-X-Y repeats loses a kink of oligomeric structure and exhibits impaired phospholipid liposome aggregation ability. Biochemistry 45:14543–14551

van Eijk M, de Haan NA, Rogel-Gaillard C et al (2001) Assignment of surfactant protein A (SFTPA) and surfactant protein D (SFTPD) to pig chromosome band 14q25—>q26 by in situ hybridization. Cytogenet Cell Genet 95:114–115

Väyrynen O, Glumoff V, Hallman M (2002) Regulation of surfactant proteins by LPS and proinflammatory cytokines in fetal and newborn lung. Am J Physiol Lung Cell Mol Physiol 282:L803–L810

Vazquez de Lara LG, Umstead TM et al (2003) Surfactant protein A increases matrix metalloproteinase-9 production by THP-1 cells. Am J Physiol Lung Cell Mol Physiol 285:L899–L907

Veldhuizen EJ, Haagsman HP (2000) Role of pulmonary surfactant components in surface film formation and dynamics. Biochim Biophys Acta 1467:255–270

Veldhuizen RA, Yao LJ, Hearn SA et al (1996) Surfactant-associated protein A is important for maintaining surfactant large-aggregate forms during surface-area cycling. Biochem J 313:835–840

Voss T, Eistetter H, Schafer KP, Engel J (1988) Macromolecular organization of natural and recombinant lung surfactant protein SP 28-37. J Mol Biol 201:219–227

Wang G, Bates-Kenney SR, Tao JQ et al (2004) Differences in biochemical properties and in biological function between human SP-A1 and SP-A2 variants, and the impact of ozone-induced oxidation. Biochemistry 43:4227–4239

Wang G, Guo X, Floros J (2005) Differences in the translation efficiency and mRNA stability mediated by 5′-UTR splice variants of human SP-A1 and SP-A2 genes. Am J Physiol Lung Cell Mol Physiol 289:L497–L508

Wang G, Guo X, Floros J (2003) Human SP-A 3′-UTR variants mediate differential gene expression in basal levels and in response to dexamethasone. Am J Physiol Lung Cell Mol Physiol 284:L738–L748

Wang G, Myers C, Mikerov A, Floros J (2007) Effect of cysteine 85 on biochemical properties and biological function of human surfactant protein A variants. Biochemistry 46:8425–8435

Wang G, Phelps DS, Umstead TM, Floros J (2000) Human SP-A protein variants derived from one or both genes stimulate TNF-α production in the THP-1 cell line. Am J Physiol Lung Cell Mol Physiol 278:L946–L954

Wang JY, Shieh CC, You PF et al (1998) Inhibitory effect of pulmonary surfactant proteins A and D on allergen-induced lymphocyte proliferation and histamine release in children with asthma. Am J Respir Crit Care Med 158:510–518

Watford WT, Ghio AJ, Wright JR (2000) Complement-mediated host defense in the lung. Am J Physiol Lung Cell Mol Physiol 279:L790–L798

Watford WT, Wright JR, Hester CG et al (2001) Surfactant protein A regulates complement activation. J Immunol 167:6593–6600

Weissbach S, Neuendank A, Pettersson M et al (1994) Surfactant protein A modulates release of reactive oxygen species from alveolar macrophages. Am J Physiol 267:L660–L667

White MK, Baireddy V, Strayer DS (2001) Natural protection from apoptosis by surfactant protein A in type II pneumocytes. Exp Cell Res 263:183–192

White MK, Strayer DS (2000) Surfactant protein A regulates pulmonary surfactant secretion via activation of phosphatidylinositol 3-kinase in type II alveolar cells. Exp Cell Res 255:67–77

White MK, Strayer DS (2002) Survival signaling in type II pneumocytes activated by surfactant protein-A. Exp Cell Res 280:270–279

Wright JR, Youmans DC (1993) Pulmonary surfactant protein A stimulates chemotaxis of alveolar macrophage. Am J Physiol 264:L338–L344

Wright JR (2005) Immunoregulatory functions of surfactant proteins. Nat Rev Immunol 5:58–68

Wright JR (1997) Immunomodulatory functions of surfactant. Physiol Rev 77:931–962

Wu YZ, Manevich Y, Baldwin JL, Dodia C, Yu K, Feinstein SI, Fisher AB (2006) Interaction of surfactant protein A with peroxiredoxin 6 regulates phospholipase A2 activity. J Biol Chem 281:7515–7525

Wu YZ, Medjane S, Chabot S et al (2003) Surfactant protein-A and phosphatidylglycerol suppress type IIA phospholipase A2 synthesis via nuclear factor-kB. Am J Respir Crit Care Med 168: 692–699

Yamada C, Sano H, Shimizu T et al (2006) Surfactant protein A directly interacts with TLR4 and MD-2 and regulates inflammatory cellular response. Importance of supratrimeric oligomerization. J Biol Chem 281:21771–21780

Yang CH, Szeliga J, Jordan J et al (2005) Identification of the surfactant protein A receptor 210 as the unconventional myosin 18A. J Biol Chem 280:34447–34457

Yuan HT, Gowan S, Kelly FJ, Bingle CD (1997) Cloning of guinea pig surfactant protein A defines a distinct cellular distribution pattern within the lung. Am J Physiol 273:L900–6

Yu SH, McCormack FX, Voelker DR et al (1999) Interactions of pulmonary surfactant protein SP-A with monolayers of dipalmitoylphosphatidylcholine and cholesterol: roles of SP-A domains. J Lipid Res 40:920–929

Yu SH, Possmayer F (1998) Interaction of pulmonary surfactant protein A with dipalmitoylphosphatidylcholine and cholesterol at the air/water interface. J Lipid Res 39:555–568

Zhang L, Ikegami M, Korfhagen TR et al (2006) Neither SP-A nor NH2-terminal domains of SP-A can substitute for SP-D in regulation of alveolar homeostasis. Am J Physiol Lung Cell Mol Physiol 291:L181–L190

Zhang M, Damodarasamy M, Elhalwagi BM, McCormack FX (1998) The longer isoform and Cys-1 disulfide bridge of rat surfactant protein A are not essential for phospholipid and type II cell interactions. Biochemistry 37:16481–16488

Zhu S, Basiouny KF, Crow JP, Matalon S (2000) Carbon dioxide enhances nitration of surfactant protein A by activated alveolar macrophages. Am J Physiol Lung Cell Mol Physiol 278: L1025–L1031

Zhu S, Kachel DL, Martin WJ 2nd, Matalon S (1998a) Nitrated SP-A does not enhance adherence of *Pneumocystis carinii* to alveolar macrophages. Am J Physiol 275:L1031–L1039

Zhu S, Manuel M, Tanaka S et al (1998b) Contribution of reactive oxygen and nitrogen species to particulate-induced lung injury. Environ Health Perspect 106(Suppl 5):1157–1163

Zhu S, Ware LB, Geiser T et al (2001) Increased levels of nitrate and surfactant protein A nitration in the pulmonary edema fluid of patients with acute lung injury. Am J Respir Crit Care Med 163:166–172

Surfactant Protein-D

Rajesh K. Gupta and Anita Gupta

The collectins can be classed into two distinct group, with MBP and SP-A being hexamers and SP-D, conglutinin and collectin-43 (CL-43) being tetramers, with proteins in the latter group also having significantly larger dimensions with respect to the length of their collagen-like 'stalks' (Lu et al. 1993). MBP, lung SP-D, conglutinin, CL-43 and CL-46 are important components of innate immune defence system. They all bind complex glycoconjugates on microorganisms thereby inhibiting infection, enhancing the clearance by phagocytes and modulating the immune response. In addition, SP-D inhibits the generation of radical oxygen species or the propagation of lipid peroxidation. Knock-out mice deficient in SP-D have a disturbed homeostasis of pulmonary surfactant and suffer from oxidative stress leading to pulmonary inflammation upon microbial challenge. Moreover, both SP-D and SP-A (Chap. 24) have been shown to enhance oxygen radical production by alveolar macrophages. The structural and functional relationships of this group of collectins have been reviewd (Hansen and Holmskov 2002; Jinhua et al. 2003).

25.1 Pulmonary Surfactant Protein-D (SP-D)

Surfactant proteins, SP-A and SP-D contribute significantly to surfactant homeostasis and pulmonary immunity. Their basic structures include a triple-helical collagen region and a C-terminal homotrimeric lectin or CRD. The trimeric CRDs can recognize carbohydrate or charge patterns on microbes, allergens and dying cells, while the collagen region can interact with receptor molecules present on a variety of immune cells in order to initiate clearance mechanisms. Gene knock-out mice models of lung hypersensitivity and infection, and functional characterization of cell surface receptors have revealed the diverse roles of SP-A and SP-D in control of lung inflammation. A model based on studies with the calreticulin-CD91 complex as a receptor for SP-A and SP-D has suggested an anti-inflammatory role for SP-A and SP-D in naïve lungs which would help minimise the potential damage that continual low level exposure to pathogens, allergens and apoptosis can cause. However, when lungs are overwhelmed with exogenous insults, SP-A and SP-D can assume pro-inflammatory roles in order to complement pulmonary innate and adaptive immunity. The structural and functional aspects of SP-A and SP-D emphasize on their roles in controlling pulmonary infection, allergy and inflammation (Kishore et al. 2006)

Lung SP-D can directly interact with carbohydrate residues on pulmonary pathogens and allergens, stimulate immune cells, and manipulate cytokine and chemokine profiles during the immune response in lungs. Therapeutic administration of rfhSP-D, a recombinant homotrimeric fragment of human SP-D comprising the alpha-helical coiled-coil neck plus three CRDs, protects mice against lung allergy and infection caused by the fungal pathogen *Aspergillus fumigatus*. Curcumin inhibits inflammatory response reducing significantly all histopathological parameters in different pulmonary aspiration models. Curcumin is a potential therapeutic agent in acute lung injury (Guzel et al. 2009).

The lung is the main site of synthesis for SP-D, but transcripts were readily amplified from trachea, brain, testis, salivary gland, heart, prostate gland, kidney, and pancreas. Minor sites of synthesis were uterus, small intestine, placenta, mammary gland, and stomach. The sequence of SP-D derived from parotid gland mRNA was identical with that of pulmonary SP-D. SP-D immunoreactivity was found in alveolar type II cells, Clara cells, on and within alveolar macrophages, in epithelial cells of large and small ducts of the parotid gland, sweat glands, and lachrymal glands, in epithelial cells of the gall bladder and intrahepatic bile ducts, and in exocrine pancreatic ducts. SP-D was also present in epithelial cells of the skin, esophagus, small intestine, and urinary tract, as well as in the collecting ducts of the kidney. SP-D is present in mucosal lined surfaces throughout the human body, including the male reproductive tract and not restricted to a subset of cells in the lung. The localization and functions of SP-D indicated that this collectin is the counterpart in the innate immune system of IgA in the adaptive immune system (Madsen et al. 2000). The extra-pulmonary expression of SP-A and SP-D indicates the systemic roles of these proteins. SP-D is localized in the cytoplasm of a subpopulation of bronchiolar epithelial cells as well as type II cells. Anti-SP-D selectively decorated secretory compartments of nonciliated bronchiolar cells (Clara cells) with strong and specific labeling of apical electron-dense secretory granules. Studies provide evidence that SP-D is a secretory product of nonciliated bronchiolar cells. It was suggested that Clara cell-derived SP-D is a component of bronchiolar lining material, consistent with the hypothesis that SP-D contributes to surfactant metabolism and/or host defense within small airways (Crouch et al. 1992).

Alveolar Type II cells bind and recycle SP-D in vitro to lamellar bodies: Alveolar type II cells secrete, internalize, and recycle pulmonary surfactant complex that increases alveolar compliance and participates in pulmonary host defense. SP-D has been described as a modulator of surfactant homeostasis. Mice lacking SP-D accumulate surfactant in their alveoli and type II cell lamellar bodies, organelles adapted for recycling and secretion of surfactant. However, SP-D binding did not alter type II cell surfactant lipid uptake. Thus, type II cells bind and recycle SP-D in vitro to lamellar bodies, but SP-D may not directly modulate surfactant uptake by type II cells (Herbein and Wright 2001).

25.1.1 Human Pulmonary SP-D

25.1.1.1 Human Pulmonary SP-D Protein

Human pulmonary SP-D was identified in lung lavage by its similarity to rat SP-D in both its molecular mass and its Ca^{2+}-dependent-binding affinity for maltose. On SDS/PAGE human SP-D behaved as a single band of 150 kDa in non-reducing and 43 kDa band in reducing conditions. The presence of a high concentration of glycine (22%), hydroxyproline and hydroxylysine in the amino acid composition of human SP-D pointed towards its collagen-like structure. Human SP-D has a similar carbohydrate-binding specificity to rat SP-D, but differed from that of other lectins, such as conglutinin. Amino acid sequence analysis showed the presence of collagen-like Gly-Xaa-Yaa triplets in human SP-D. The derived amino acid sequence indicated that mature SP-D polypeptide chain is 355 residues long, having a short non-collagen-like N-terminal section of 25 residues, followed by a collagen-like region of 177 residues and a C-terminal C-type lectin domain of 153 residues. The human SP-D and bovine serum conglutinin showed 66% identity despite their marked differences in carbohydrate specificity (Lu et al. 1993). SP-D differs in its quaternary structure from SP-A and other members of the family, such as C1q, in that it forms large multimers held together by N-terminal domain, rather than aligning the triple helix domains in traditional "bunch of flowers" arrangement. SP-D is an oligomeric complex of four set of homotrimers. Purified trimeric and multimeric SP-D represent separate and only partly interconvertible molecular populations with distinct biochemical properties (Soerense et al. 2009).

25.1.1.2 Genomic Organization of Human SP-D

SP-D gene sequences spann > 11 kb on long arm of chromosome 25. Genomic sequencing revealed that signal peptide/amino-terminal domain, the CRD, and the linking sequence between the collagen domain, and CRD are each encoded by a single exon, as seen for SP-A and MBP-C. However, sequencing also demonstrated a unique intron-exon structure for collagen domain which is encoded on five exons, including four tandem exons of 117 bp. The latter exons show marked conservation in the predicted distribution of hydrophilic amino acids, consistent with tandem replication of this collagen gene sequence during evolution (Crouch et al. 1993). Human SP-D gene (*Sftp4 or Sftpd*) was assigned to chromosome 25. A regional mapping panel for five loci was mapped to 10q. *Sftp4*, the SP-A gene (*Sftp1/Sftpa*), and the microsatellite D10S109 were placed in the interval 10q22–q23. In situ hybridization of metaphase chromosomes using genomic probes gave selective labeling of 10q22.2–23.1 (Crouch et al. 1993; Kölble et al. 1993).

Rust et al. (1996) described the characterization of a human genomic fragment (H5E7) that encodes entirety of first translated exon (Exon 2), Intron 1, a short transcribed untranslated sequence (Exon 1; 39 bp), and approximately 4 kb of sequence upstream from transcription initiation site. The start site comprises of a putative TATA box (CATAAATA) of ~30 bp upstream of the start site. Complete sequencing of a HS-1674 fragment encoding approximately 1.7 kb of sequence 5′ to TATA demonstrated multiple potential cis-regulatory elements including half-site glucocorticoid response elements (GRE), a canonical AP-1 consensus, several AP-1 like sequences, E-box sequences, NF-IL-6 and PEA3 motifs, and putative interferon response elements. Studies support the hypothesis that the effects of glucocorticoids on SP-D production in vivo are regulated at level of transcription.

While SP-D is the product of a single gene whereas SP-A is the product of two highly homologous genes SP-A1 and SP-A2. The relative location and orientation of each of the SP-A and SP-D genomic sequences have been characterized. Characterization of two overlapping genomic clones revealed that SP-A pseudogene lies in a reverse orientation

25.1 Pulmonary Surfactant Protein-D (SP-D)

15 kb away from the 5′ side of SP-A1. Both SP-A2 and SP-D are on the 5′ side of SP-A1 at approximate distances of 40 and 120 kb, respectively. The SP-A and SP-D loci were also oriented relative to centromere, with overall order being: centromere-SP-D-SP-A2-pseudogene-SP-A1- telomere (Hoover and Floros 1998). Several single nucleotide polymorphisms (SNP) have been identified in the SP-A1, SP-A2 and SP-D genes (Pantelidis et al. 2003).

25.1.1.3 Transcriptional Regulation of SP-D Promotor

Proximal Promoter of SP-D Gene: Because AP-1 proteins regulate cellular responses to diverse environmental stimuli, He et al. (2000) hypothesized that conserved AP-1 motif (at -109) and flanking sequences in human SP-D promoter contribute to the regulation of SP-D expression. Davé et al. (2004) identified the role of nuclear factor of activated T cells (NFATs) in regulation of murine SP-D gene (*Sftpd*) transcription. Components of calcineurin/NFAT pathway were identified in respiratory epithelial cells of the lung that potentially augment rapid assembly of a multiprotein transcription complex on *Sftpd* promoter inducing SP-D expression via direct interaction with TTF-1 in lung epithelial cells

CCAAT/Enhancer-Binding Proteins (C/EBP): CCAAT/enhancer-binding proteins (C/EBP) constitute a family of transcription factors that are involved in regulation of proliferation and differentiation in several cell types. In epithelial lung cells the C/EBPα isoform seems to play a role in the regulation of SPs and Clara cell specific protein (CCSP). In vitro results suggest that C/EBPδ alone is not related to the maintenance of proteins involved in differentiation (Låg et al. 2000). It was hypothesized that conserved C/EBP motifs in the near-distal and proximal promoters contribute to the regulation of SP-D expression by C/EBPs. Five SP-D motifs (-432, -340, -319, -140, and -90) homologous to C/EBP consensus sequence specifically bound to C/EBPs in gel shift assays, and four of the five sites (-432, -340, -319, and -90) efficiently competed for binding of C/EBPα, C/EBPβ, or C/EBPδ to consensus oligomers. The conserved AP-1 element at -109 was required for maximal promoter activity, but not for transactivation of SS698 by C/EBPs. Thus, interactions among C/EBP elements in near-distal promoter can modulate promoter activity of SP-D (He and Crouch 2002).

Retinoblastoma Protein Activates C/EBP-Mediated Transcription of SP-D: The retinoblastoma protein (RB) through histone deacetylase (HDAC) plays a critical role in cell cycle regulation. RB is also involved in activation of expression of a number of tissue specific- and differentiation-related genes and stimulates the expression of a differentiation-related gene, SP-D in lungs. RB specifically stimulated the activity of human SP-D gene promoter. Activation by RB was mediated through a NF-IL6 (C/EBPβ) binding motif in the human SP-D promoter, and this sequence specifically bound to C/EBPα, C/EBPβ, and C/EBPδ. Furthermore, the complexes containing RB and C/EBP proteins directly interacted with C-EBP binding site on DNA. Thus, RB plays a direct role in C/EBP-dependent transcriptional regulation of human SP-D expression (Charles et al. 2001).

25.1.2 Rat Lung SP-D

Ontogeny of Surfactant Apoprotein D, SP-D in Rat Lung: In rat lungs SP-A increased during late gestation and reached its maximum on day 1 of neonate, and then gradually declined until at least day 5. SP-D content during early gestation was less than 10 ng/mg protein until day 18, but on day 19 there was a fourfold increase in SP-D (compared to that on day 18). It increased twice between day 21 and the day of birth, when it reached the adult level of 250 ng/mg protein, which is about one fourth that of the adult level of SP-A. Unlike SP-A there seemed to be no decrease in SP-D content after birth. SP-D is regulated developmentally as with other components of surfactant. In contrast to humans, rat lungs are immature at birth. Alveolarization starts on postnatal day 4. Studies support the concept that postnatal alveolarization in rat lungs is associated with significant increases in the SP-B content in lb and volume fraction of lb in type II pneumocytes. The postnatal compartment-specific distribution of SP-A, precursors of SP-C and SP-D does not change (Schmiedl et al. 2005).

Rat SP-D Protein: Rat SP-D was SP-D, purified from the supernatant of rat bronchoalveolar lavage fluids, and delipidated form does not exhibit interaction with lipids in the same fashion as SP-A. SP-D consists of three regions: an NH2-terminal segment of 25 amino acids, a collagen-like domain consisting of 59 Gly-X-Y repeats, and a COOH-terminal carbohydrate recognition domain of 153 amino acids. There are six cysteine residues present in rat SP-D: two in NH2-terminal noncollagenous segment and four in COOH-terminal carbohydrate-binding domain. The collagenous domain contains one possible N-glycosylation site. The protein is preceded by a cleaved, NH2-terminal signal peptide. Rat SP-D is encoded by a 1.3-kb mRNA which is abundant in lung and highly enriched in alveolar type II cells. Extensive homology exists between rat SP-D and bovine conglutinin (Shimizu et al. 1992; Lim et al. 1993; Ogasawara et al. 1991).

In electron microscopy, rat SP-D is made of four identical rod-like arms (46 nm in length), each with an 8–9-nm diameter globular terminal expansion. The arms, which are similar in diameter to the type I collagen helix (~4 nm), emanate from central "hub" in two pairs that closely parallel each other for their first 10 nm. This structure is consistent with hydrodynamic studies that predict a highly asymmetric and extended molecule (f/f0 = 3.26) with a large Stokes radius of 18 nm. Pepsin digestion gave glycosylated, trimeric collagenous fragments. Trimeric subunits containing intact triple helical domains were also liberated from SP-D dodecamers by sulfhydryl reduction under non-denaturing conditions. Digestion of rSP-D with bacterial collagenase generated a COOH-terminal carbohydrate binding fragment and a smaller peptide (~12 kDa, unreduced) that contains interchain disulfide bonds. Higher orders of multimerization with as many as eight molecules associated at the hub were also identified. Rat SP-D molecules are assembled as tetramers of trimeric subunits (12 mers). Dodecamers can participate in higher orders of molecular assembly involving interactions of NH_2-terminal peptide domains. Interactions between amino-terminal domains of trimers are stabilized by interchain disulfide bonds. SP-D molecules can associate to form complex multimolecular assemblies (Crouch et al. 1994b).

Studies demonstrated that a single genetically distinct chain type can account for various and complex molecular assemblies of SP-D (Crouch et al. 1994a). The SP-D forms large multimers held together by N-terminal domain, rather than aligning the triple helix domains in traditional "bunch of flowers" arrangement. The two cysteine residues within hydrophobic NH_2-terminus are critical for multimer assembly and have been proposed to be involved in stabilizing disulfide bonds. These cysteins exist within reduced state in dodecameric SP-D and form a specific target for S-nitrosylation both in vitro and by endogenous, pulmonary derived NO within a rodent acute lung injury model (Crouch et al. 1994a; Guo et al. 2008).

25.1.3 Mouse SP-D

The Protein: A cDNA deduced sequence predicts a 19-amino acid signal sequence, a 25-amino acid long NH_2-terminus with two cysteines, followed by an uninterrupted collagen domain with 59 Gly-X-Y repeats. Next, a short "neck" domain of 28 amino acids, with a potential to form trimeric alpha-helical coiled coil is found ending in a COOH-terminal 125-amino acid CRD. The mature mouse SP-D protein of 355 amino acids shows strong homology to rat (92% identity), human (76%), and bovine (72%) SP-D amino acid sequences. The mouse SP-D gene is expressed predominantly in lung and also in heart, stomach, and kidney but not in brain. In contrast, mouse SP-A mRNA expression was found to be restricted to lung. Histochemically, SP-A and SP-D mRNA were detected in a significant number of non-pulmonary tissues but proteins have a more limited distribution. SP-D protein was detected in lung, uterus, ovary, and lacrimal gland, whereas SP-A protein was detected only in lung (Akiyama et al. 2002). Mouse SP-D gene (*Sftp4*) has been localized to chromosome 14 (to a region syntenic to human chromosome 10), closely linked to genes for *Mbl1*, and SP-A (*Sftp1*) (Motwani et al. 1995).

Genomic Organization of Mouse Lung SP-D Gene: The mouse SP-D gene (*Sftpd*), which spans eight exons over 14 kb of sequence and shows an overall organization similar to other collectin genes. The complete 5′ untranslated region of mRNA is encoded by a single exon. Analysis of 3.5 kb of 5′ flanking nucleotide sequence for *Sftpd* reveals positional conservation of a number of transcription factor binding sites when compared with human SP-D gene and bovine conglutinin gene (Lim et al. 2003). The single copy SP-D-like gene is present in mammals, birds, and amphibians but is absent in fish. An atypical, rodent-specific, long terminal repeat of retroviral origin containing a minisatellite that has become inserted in *Sftpd* is described. Three new polymorphic microsatellites are also described, one of which is just 160 bp upstream of *Sftpd*. This microsatellite was used to map the gene to central region of chromosome 14. Fine-scale mapping indicates that it lies in a 5.64-centimorgan area between D14Mit45 and D14Mit60 (Lawson et al. 1999).

The Collectin Gene Locus: Three of four known mouse collectin genes have been mapped to chromosome 14. To characterize spatial relationship of these genes, one large clone hybridized to both SP-A and SP-D cDNAs was found to contain sequences from one of the mouse *Mbl1*. The SP-A, *Mbl1*, and SP-D genes reside contiguously within a 55-kb region. The SP-A and *Mbl1* genes are in same 5′–3′ orientation and 16 kb apart. The SP-D gene is in opposite orientation to two other collectin genes, 13 kb away from 3′ end of the *Mbl1* gene. The size (13 kb) and organization of mouse gene was similar to that of human SP-D. Exon I is untranslated. The second exon is a hybrid exon that contains signal for initiation of translation, signal peptide, N-terminal domain, and first seven collagen triplets of collagen-like domain of the protein. Four short exons (III to VI) encode collagen-like domain of protein, and exons VII and VIII, the linking and the CRDs, respectively (Akiyama et al. 1999).

Developmental Expression: SP-D mRNA and protein were readily detected in alveolar type II and nonciliated bronchiolar epithelial cells of lung, as well as in cells of the tracheal epithelium and tracheal submucosal glands of the adult mouse. However, SP-A mRNA or protein was not detected in murine trachea. Expression of murine SP-D mRNA was first detected on Ed 16 of gestation in pregnant mice, and this increased dramatically before birth and during the immediate postnatal period. The developmental expression of murine SP-A mRNA paralleled that of SP-D except that there was a small decrease in mRNA on postnatal Ed 5. Thus, SP-D is synthesized not only in the lung but also in submucosal glands of trachea (Wong et al. 1996).

25.1.4 Bovine SP-D

Gjerstorff et al. (2004) characterized the gene encoding bovine SP-D and its proximal promoter. Cloning and sequencing of the bSP-D gene, including the complete 5'-untranslated sequence, reveal that the gene comprises nine exons spanning ~ 25.5 kb and resembling bovine conglutinin gene. The gene localizes to the same locus as conglutinin gene on Bos taurus chromosome 28 at position q1.8, which also includes the genes for CL-43 and CL-46. Several potential cis-regulatory elements, similar to elements known to regulate transcription of human SP-D, were identified in the 5'-upstream sequence. Bovine SP-D is heavily expressed in lung and the trachea, but also in segments of gastrointestinal tract, mammary glands and salivary glands. By genotyping two potential polymorphisms leading to variations in amino acid composition of CRD (242 Glu/Val and 268 Ala/Gly) have been assigned.

Human and bovine SP-D from late amniotic fluid and BAL gave a Mr of a trimeric subunit in the range 115–125 kDa for human SP-D and 110–123 kDa for bovine SP-D. A single polypeptide chain was determined at 37–41 and 36–40 kDa for human and bovine species respectively. Primary structures, determined by MS and Edman degradation, showed heterogeneity in SP-D, caused mainly by high number of post-translational modifications in the collagen-like region. Proline and lysine residues were partly hydroxylated and lysine residues were further O-glycosylated with disaccharide galactose-glucose. A partly occupied N-linked glycosylation site was characterized in human SP-D. The carbohydrate was determined as a complex type bi-antennary structure, with a small content of mono-antennary and tri-antennary structures. No sialic acid residues were present on the glycan, but some had an attached fucose and/or an N-acetylglucosamine residue linked to the core. Bovine SP-D was having a similar structure (Leth-Larsen et al. 1999).

25.1.5 Porcine Lung SP-D

The complete cDNA sequence of porcine SP-D, including the 5' and 3' untranslated regions, was determined from two overlapping clones. Three unique features were revealed from porcine sequence in comparison to SP-D from other species, making porcine SP-D an intriguing species addition to the collectin family. The collagen region contains an extra cysteine residue, which may have important structural consequences. The other two differences, a potential glycosylation site and an insertion of three amino acids lie in the loop regions of CRD, close to carbohydrate binding region and thus may have functional implications. The genes for SP-D (*Sftpd*) and SP-A (*Sftpa*) also co-localized to a region of porcine chromosome 14 band 14q25—>q26 that is syntenic with human and murine collectin loci (van Eijk et al. 2000, 2001).

The porcine SP-D appeared as a band of ~ 53 kDa in reduced state and ~138 kDa in unreduced state (Soerensen et al. 2005a). The monomeric porcine SP-D (50–53 kDa) is larger than that of SP-D from humans (43 kDa). Carbohydrate moiety is a highly heterogeneous, complex type oligosaccharide which is sialylated. The heterogeneity of oligosaccharide sialylation results in existence of many differently charged porcine SP-D isoforms. The removal of carbohydrate moiety reduces the inhibitory effect of porcine SP-D on IAV haemagglutination. Ultrastructural analysis showed the presence of both dodecameric and higher order oligomeric complexes of SP-D (van Eijk et al. 2000, 2002). Porcine tissues showed SP-D predominantly present in Clara cells and serous cells of bronchial submucosal glands, and to a lesser extent in alveolar type II cells, epithelial cells of intestinal glands (crypts of Lieberkuhn) in duodenum, jejunum and ileum and serous cells of dorsolateral lacrimal gland (Soerensen et al. 2005b). SP-A and D are expressed in the porcine Eustachian tube (ET) originating from upper airways and were present in epithelial cells of ET (Paananen et al. 2001). Porcine SP-D is an important reagent for use in existing porcine animal models for human lung infections.

25.2 Interactions of SP-D

The SP-A and SP-D bind carbohydrates, lipids, and nucleic acids with a broad-spectrum specificity and initiate phagocytosis of inhaled pathogens as well as apoptotic cells. Investigations on gene-deficient and conditional overexpressed mice indicated that lung SP-A and SP-D directly modulate innate immune cell function and T-cell-dependent inflammatory events and have a unique, dual-function capacity to induce pathogen elimination and control proinflammatory mechanisms (Botas et al. 1998; Haczku 2008).

25.2.1 Interactions with Carbohydrates

SP-D showed specific calcium-dependent binding to alpha-D-glucosidophenyl isothiocyanate-BSA and maltosyl-BSA, but negligible binding to beta-D-glucosidophenyl isothiocyanate-BSA or unconjugated BSA. The most efficient inhibitors of SP-D binding were alpha-glucosyl-containing saccharides (e.g. isomaltose, maltose, malotriose). Studies demonstrate that SP-D is a calcium dependent lectin-like protein and that the association of SP-D with surfactant is mediated by carbohydrate-dependent interactions with specificity for alpha-glucosyl residues. Crystallographic studies of trimeric human SP-D neck + CRD domains have shown that maltose, a preferred saccharide ligand, binds to calcium via the vicinal 3- and 4-OH groups of the non-reducing glucose, previously designated calcium ion 1 and glucose 1 (Glc1), respectively (Shrive et al. 2003).

Oligo and Polysaccharide Recognition by SP-D: SP-D binding to *Aspergillus fumigatus* is strongly inhibited by a soluble β-(1—>6)-linked but not by a soluble β-(1—>3)-linked glucosyl homopolysaccharide (pustulan and laminarin, respectively), suggesting that SP-D recognizes only certain polysaccharide configurations. Docking studies predict that α/β-(1—>2)-, α-(1—>4)-, and α/β-(1—>6)-linked but not α/β-(1—>3)-linked glucosyl trisaccharides can be bound by their internal glucosyl residues and that binding also occurs through interactions of the protein with the 2- and 3-equatorial OH groups on the glucosyl ring. Given the sequence and structural similarity between SP-D and other C-type lectins, many of the predicted interactions should be applicable to this protein family (Allen et al. 2001).

SP-D binds to various synthetic fucosylated oligosaccharides. Fucα1-3GalNAc and Fuc α1-3GlcNAc elements show strong binding to SP-D. Fucosylated glycoconjugates are present at the surface of *Schistosoma mansoni*, a parasitic worm that transiently resides in lung during development. In line with this observation, SP-D was found to bind to larval stages of *S. manson* and may interact with multicellular lung pathogens (van de Wetering et al. 2004). The SP-A and SP-D bind gram-positive bacteria through lipoteichoic acid (LTA) and peptidoglycan, components of cell wall of gram-positive bacteria. The CRD is responsible for this binding (van de Wetering et al. 2001).

Interaction with LPS: Both SP-A and SP-D bind bacterial lipopolysaccharide (LPS). Wang et al. (2008a) demonstrated specific binding between CRD and heptoses in the core region of LPS. The geometry suggested that all three CRDs are simultaneously bound to LPS under conditions that support Ca^{2+}-mediated interaction. Mutant trimeric recombinant neck + CRDs (NCRDs) of SP-D is known to bind with rough LPS (R-LPS). Crystallographic analysis of hNCRD demonstrated a novel binding orientation for LD-heptose, involving hydroxyl groups of side chain. Similar binding was observed for a synthetic α1—>3-linked heptose disaccharide corresponding to heptoses I and II of inner core region in many LPS. 7-O-Carbamoyl-l,D-heptose and D-glycero-α-D-manno-heptose were bound via ring hydroxyl groups. Interactions with side chain of inner core heptoses provide a potential mechanism for recognition of diverse types of LPS by SP-D (Wang et al. 2008b). SPD binds to human α2-macroglobulin, that protects the collectin from proteolytic degradation by elastase (Craig-Barnes et al. 2010).

SP-D Binds MD-2 (lymphocyte antigen 96) and TLRs Through CRD: SP-A interacts with MD-2 and alters LPS signaling. SP-D modulates LPS-elicited inflammatory cell responses. Ligand blot analysis revealed that SP-D bound to N-glycopeptidase F-treated sMD-2. In addition, the biotinylated SP-D pulled down the mutant sMD-2 with Asn (26) —>Ala and Asn (114) —>Ala substitutions, which lacks the consensus for N-glycosylation. Results demonstrated that SP-D directly interacts with MD-2 through CRD (Nie et al. 2008).

Human SP-D exhibits specific binding, through its CRD, to extracellular domains of rTLR-2 and TLR-4 by a mechanism different from its binding to PI and LPS (Ohya et al. 2006). SP-D bound to a complex of soluble forms of TLR4 and MD-2 with high affinity and down-regulates TNF-α secretion and NF-kB activation elicited by rough and smooth LPS, in alveolar macrophages and TLR4/MD-2-transfected HEK293 cells. Study demonstrates that SP-D down-regulates LPS-elicited inflammatory responses by altering LPS binding to its receptors and reveals the importance of correct oligomeric structure of the protein in this process (Yamazoe et al. 2008).

Binding with Glycoprotein-340: A glycoprotein (Mr 340 kDa, gp-340) from human bronchioalveolar lung washings from a patient with alveolar proteinosis was found to bind to lung SP-D. Like gp-340, CRP-ductin binds human SP-D in a calcium-dependent manner. CRP-ductin also showed calcium-dependent binding to both gram-positive and -negative bacteria (Holmskov et al. 1997).

Interactions with Decorin and Biglycan and Other Proteoglycans: Decorin and biglycan are closely related abundant extracellular matrix proteoglycans that bind to C1q and MBL. The human decorin, a SP-D-binding protein, from amniotic fluid co-purifies with SP-D. SP-D and decorin interact with each other ($K_D = 4$ nM) by two mechanisms: (1) the CRD of SP-D binds, in a calcium dependent-manner, to sulfated N-acetyl galactosamine moiety of the glycosaminoglycan chain and (2) C1q, a complement protein that is known to interact with decorin core protein via its collagen-like region, partially blocks the interaction between decorin and native SP-D. It is suggested that decorin core protein binds the collagen-like region of SP-D and CRDs of SP-D interact with dermatan sulfate moiety of decorin via lectin activity (Nadesalingam et al. 2003).

Other members of collectin family, including collectin-43 and conglutinin also bind to decorin and biglycan. Thus, decorin and biglycan act as inhibitors of activation of complement cascade and pro-inflammatory cytokine production mediated by C1q. These proteoglycans are likely to downregulate proinflammatory effects mediated by C1q (Groeneveld et al. 2005).

25.2.2 Binding with Nucleic Acids

Collectins can bind nucleic acid, which is a pentameric sugar-based anionic polymer. SP-D and MBL bind to both DNA and RNA effectively. SP-D also enhances the uptake of DNA by human monocytic cells. Therefore, nucleic acids can act as ligands for collectins (Palaniyar et al. 2003b). Pentoses, such as arabinose, ribose, and deoxyribose, inhibit the interaction between SP-D and mannan, one of the well-studied hexose ligands for SP-D; biologically relevant D-forms of the pentoses are better competitors than the L-forms. SP-D binds apoptotic cells through DNA present on these cells. SP-D binds and aggregates mouse alveolar macrophage DNA effectively. Alveolar macrophages of SP-D$^{-/-}$ mice contained more nicked DNA than those of SP-A$^{-/-}$ and wild type mice. Results also suggested that CRDs of SP-D may recognize DNA present on apoptotic cells. Therefore, cell-surface DNA could be a ligand for recognition of apoptotic cells by collectins and may have important biological implications, such as the alleviation of DNA-mediated tissue inflammation (Palaniyar et al. 2003a)

SP-D enhances the uptake of Cy3-labeled fragments of DNA and DNA-coated beads by U937 human monocytic cells, in vitro. Analysis of DNA uptake by alveolar macrophages shows that SP-D, but not SP-A deficiency results in reduced clearance of DNA, ex vivo. Additionally, both SP-A- and SP-D-deficient mice accumulate anti-DNA Abs in sera in an age-dependent manner. Thus, SP-A and SP-D may reduce the generation of anti-DNA autoantibody, which may be explained in part by defective clearance of DNA from lungs in absence of these proteins. These findings establish two new roles for these innate immune proteins and that SP-D enhances efficient pinocytosis and phagocytosis of DNA by macrophages and minimizes anti-DNA Ab generation (Palaniyar et al. 2005).

25.2.3 Interactions with Lipids

Surfactant proteins interact with phospholipids and are believed to play important roles in alveolar spaces. Pulmonary SP-D binds glycosylated lipids such as phosphatidylinositol (PI) and glucosylceramide (GlcCer). The major known surfactant-associated ligand for lung SP-D is PI. SP-D binds to PI in various phospholipids or a fraction containing phospholipids. Interaction of SP-D with PI is dependent on calcium and inhibited by competing saccharides. SP-D binds with similar efficiency to liposomes (Ogasawara et al. 1992; Persson et al. 1992). On TLC plates, SP-D bound exclusively to GlcCer, whereas it failed to bind to GalCer, GM1, GM2, asialo-GM1, asialo-GM2, sulfatide, Forssman antigen, ceramide dihexoside, ceramide trihexoside, globoside, paragloboside or ceramide. SP-D bound to ceramide monohexoside in glycolipids isolated from rat lung and bronchoalveolar lavage fluids of rats (Kuroki et al. 1992). PI interaction, at least in part, involves the carbohydrate moiety. Results suggest that: (1) carbohydrate binding specificity of SP-D (Glu-321—>Gln and Asn-323—>Asp) was changed from a mannose-glucose type to a galactose type; (2) the GlcCer binding property of SP-D is closely related to its sugar binding activity; and (3) the PI binding activity is not completely dependent on its carbohydrate binding specificity (Ogasawara and Voelker 1995a). SP-D counteracts the inhibitory effect of SP-A on phospholipid secretion by alveolar type II cells. Native SP-D alters SP-A activity in type II cells through interaction with it via SP-D-associated lipids (Kuroki et al. 1991).

SP-D binds specifically to saturated, unsaturated, and hydroxylated fatty acids (FA). Maximal binding to FA was dependent on calcium, and was localized to neck and CRD region in recombinant trimeric neck + CRDs. Saccharide ligands showed complex, dose-dependent effects on FA binding, and FAs showed dose- and physical state-dependent effects on the binding of SP-D to mannan (DeSilva et al. 2003).

Trimeric neck-carbohydrate recognition domains (NCRDs) of rat and human SP-D exhibited dose-dependent, calcium-dependent, and inositol-sensitive binding to solid-phase PI and to multilamellar PI liposomes. Studies directly implicate the CRD in PI binding and reveal unexpected species differences in PI recognition that can be largely attributed to the side chain of residue 343. In addition, oligomerization of trimeric subunits is an important determinant of recognition of PI by human SP-D (Crouch et al. 2007, 2009).

25.3 Structure: Function Relations of Lung SP-D

25.3.1 Role of NH$_2$ Domain and Collagenous Region

SP-A and SP-D possess similar structures in mammalian C-type lectin superfamily. Both proteins are composed of four characteristic domains which are: (1) an NH$_2$-terminal domain involved in interchain disulfide formation (denoted

A1 domain for SP-A or D1 for SP-D); (2) a collagenous domain (denoted A2 or D2); (3) a neck domain (denoted A3 or D3); and (4) a CRD (denoted A4 or D4). A collagen domain deletion mutant (CDM) of SP-D and a second variant lacking both the amino-terminal region and collagen-like domain were generated and collagenase-resistant fragment (CRF) was purified. Studies on CDM and CRF demonstrated that collagen-like domain is required for dodecamer but not covalent trimer formation of SP-D and plays an important, but not essential, role in the interaction of SP-D with phosphatidylinositol (PI) and glucosylceramide (GlcCer). Removal of amino-terminal domain of SP-D along with collagen-like domain diminished PI binding and effectively eliminated GlcCer binding (Ogasawara and Voelker 1995b). Amino-terminal domains of SP-A and SP-D are critical for surfactant phospholipid interactions and surfactant homeostasis, respectively.

To further assess the importance of N-terminal domains of SP-D/SP-A in surfactant structure and function, a chimeric SP-D/SP-A gene was constructed by substituting nucleotides encoding amino acids Asn-1-Ala7 of rat SP-A with the corresponding N-terminal sequences from rat SP-D, Ala1-Asn25. Studies indicated that N terminus of SP-D: (1) can functionally replace the N terminus of SP-A for lipid aggregation and tubular myelin formation, but not for surface tension lowering properties of SP-A, and (2) is not sufficient to reverse the structural and metabolic pulmonary defects in SP-D$^{-/-}$ mouse (Palaniyar et al. 2002).

25.3.2 Role of NH2-Terminal Cysteines in Collagen Helix Formation

The NH2-terminal sequence of each monomer in SP-D contains two conserved cysteine residues. The SP-D is preferentially secreted as dodecamers consisting of four collagenous trimers cross-linked by disulfide bonds. Although mutants with serine substituted for Cys15 and Cys20 (RrSP-Dser15/20) are secreted as trimeric subunits, proteins with single cysteine substitutions were retained in the cell. Studies suggest that the most important and rate-limiting step for the secretion of SP-D involves the association of cross-linked trimeric subunits to form dodecamers stabilized by specific inter-subunit disulfide cross-links. Interference with collagen helix formation prevents secretion by interfering with efficient disulfide cross-linking of the NH2-terminal domain (Brown-Augsburger et al. 1996a). The two conserved cysteine residues participate in interchain disulfide bonds formation. Substitution of serine for Cys15 and Cys 20 in recombinant rat SP-D (RrSP-Dser15/20) bound to the hemagglutinin of influenza A virus. However, it failed to aggregate virus and did not enhance the binding of influenza A to neutrophils (PMN), augment PMN respiratory burst, or protect PMNs from deactivation. It indicates that amino-terminal disulfides are required to stabilize dodecamers, and support the hypothesis that the oligomerization of trimeric subunits contributes to the anti-microbial properties of SP-D (Brown-Augsburger et al. 1996b; Crouch et al. 2005).

25.3.3 D4 (CRD) Domain in Phospholipid Interaction

As discussed earlier that SP-A specifically binds to DPPC, the major lipid component of surfactant, and can regulate the secretion and recycling of this lipid by alveolar type II cells. SP-D binds to PI and glucosylceramide (GlcCer). The D3 (neck) plus D4 (CRD) domains of SP-D play a role in lipid binding and that CRD domain is essential for PI binding. Furthermore, the A3 domain of SP-A cannot account for all the lipid binding activity of this protein. The results implicate the A4 domain of SP-A as an important structural domain in lipid aggregation phenomena (Ogasawara et al. 1994). Saitoh et al. (2000) showed that (1) SP-A region Leu-219-Phe-228 is required for liposome aggregation and interaction with alveolar type II cells, (2) SP-A region Cys-204-Cys-218 is required for DPPC binding, (3) the SP-D region Cys-331-Phe-355 is essential for minimal PI binding, and (4) the epitope for mAb 1D6 is located at the region contiguous to SP-A region Leu-219-Phe-228. Analysis of chemotactic properties of trimeric CRD demonstrated that CRD was chemotactic for neutrophils (polymorphonuclear leukocytes). The chemotactic activity was abolished by maltose, which did not suppress the chemotactic response to fMLP (Cai et al. 1999).

The recombinant structure, containing three CRDs, was comparable to native SP-D in terms of carbohydrate binding specificity, binding to LPSs of Gram-negative bacteria, and interaction with phospholipids. The CRD of SP-D, without the neck region peptide that was also expressed, showed a very weak affinity for LPSs and phospholipids. The α-helical neck region on its own showed affinity for phospholipids and thus might contribute to the binding of SP-D to these structures. These results show the importance of the neck region as a trimerizing agent in bringing together three CRDs and suggest that multivalency is important in the strong binding of SP-D to carbohydrate targets (Kishore et al. 1996).

Zhang et al. (2001a) provided evidence that amino-terminal heptad repeats of the neck domain are necessary for intracellular, trimeric association of SP-D monomers and for the assembly and secretion of functional dodecamers (Zhang et al. 2001b). A recombinant homotrimer, composed of the α-helical neck region of human surfactant protein D and C1q B chain globular domain, was an inhibitor of classical complement pathway (Kishore et al. 2001).

25.3.4 A Three Stranded α-Helical Bundle at Nucleation Site of Collagen Triple-Helix Formation

A short stretch of 35 amino acids comprising a structural motif is responsible for tight parallel association and trimerization of three identical polypeptide chains of lung SP-D, which contains both collagen regions and C-type lectin domains. This 'neck-region' is located at nucleation site at which collagenous sequences fold into a staggered triple-helix to consist of a triple-stranded parallel α-helical bundle in a non-staggered, and extremely strong, non-covalent association. This type of association between three polypeptide chains may represent a common structural feature immediately following C-terminal end of triple-helical region of collagenous proteins (Hoppe et al. 1994).

The SP-D is assembled predominantly as dodecamers consisting of four homotrimeric subunits each. Trimerization of SP-D monomers is required for high affinity saccharide binding, and the oligomerization of trimers is required for many of its functions. A peptide containing α-helical neck region can spontaneously trimerize in vitro. Håkansson et al. (1999) determined the crystal structure of a trimeric fragment of hSP-D at 2.3 Å resolution. The structure comprises an α-helical coiled-coil and three carbohydrate-recognition domains (CRDs). A deviation from symmetry was found in the projection of a single tyrosine side chain into the centre of coiled-coil; the asymmetry of this residue influences orientation of one of the adjacent CRDs. The cleft between the three CRDs presents a large positively charged surface. The fold of CRD of hSP-D is similar to that of MBP, but its orientation relative to α-helical coiled-coil region differs somewhat to that seen in MBP structure. The novel central packing of tyrosine side chain within the coiled-coil and the resulting asymmetric orientation of CRDs has unexpected functional implications. The positively charged surface might facilitate binding to negatively charged structures, such as LPS (Håkansson et al. 1999).

Association of trimeric subunits is stabilized by interchain disulfide bonds involving two conserved amino-terminal cysteine residues (Cys-15 and Cys-20). Mutant recombinant rat SP-D lacking these residues (RrSP-Dser15/20) is secreted in cell culture as trimeric subunits rather than as dodecamers. Activity of SP-D in lungs is dependent on oligomeric structure. In transgenic mice that express mutant RrSP-Dser15/20 showed that the activity of lung SP-D in vivo is dependent on oligomeric structure. Disulfide cross-linked SP-D oligomers are required for regulation of surfactant phospholipid homeostasis and prevention of emphysema and foamy macrophages in vivo (Zhang et al. 2001a; McAlinden et al. 2002).

The solution structure of a 64-residue peptide encompassing the coiled-coil domain of human SP-D confirmed that the domain forms a triple-helical parallel coiled coil. Symmetry-ADR (ambiguous distance restraint) structure calculations demonstrated that leucine zipper region of SP-D is an autonomously folded domain and agreed with X-ray crystal structure, differing mainly at a single residue, Tyr248. This residue is completely symmetric in solution structure, and markedly asymmetric in crystalline phase. This difference may be functionally important, as it affects orientation of antigenic surface presented by SP-D (Kovacs et al. 2002).

25.3.5 Ligand Binding Amino Acids

Arg343 in SP-D Discriminates Between Glucose and N-acetylglucosamine Ligands: SP-D binds glucose (Glc) stronger than N-acetylglucosamine (GlcNAc). Structural superimposition of hSP-D with MBP-C complexed with GlcNAc revealed steric clashes between the ligand and the side chain of Arg343 in hSP-D. Computational model of Arg343—>Val (R343V) mutant hSP-D demonstrated that Arg343 is critical for hSP-D recognition specificity and plays a key role in defining ligand specificity differences between MBP and SP-D. Additionally, the number of binding orientations contributes to monosaccharide binding affinity (Allen et al. 2004). Asp325, in addition to Arg343, is other important determinant of ligand selectivity, recognition, and binding; and that differences in crystal contact interfaces exert, through Asp325, significant influence on preferred binding modes (Shrive et al. 2009).

Contributions of Phenylalanine-335 to Ligand Recognition: A trimeric fusion protein encoding the human neck + CRD bound to aromatic glycoside p-nitrophenyl-α-D-maltoside with a higher affinity than maltose. Maltotriose, which has the same linkage pattern as the maltoside, bound with intermediate affinity. Substitution of leucine Phe-335 decreased affinities for maltoside and maltotriose without altering affinity for maltose or glucose, and substitution of tyrosine or tryptophan for leucine restored binding to maltotriose and maltoside. Crouch et al. (2006a) indicated that Phe-335, which is evolutionarily conserved in all known SP-Ds, plays important role in SP-D function. The ligand binding of homologous human, rat, and mouse trimeric neck + CRD fusion proteins, each with identical N-terminal tags remote from ligand-binding surface was compared. The ligand recognition by human SP-D involves a complex interplay between saccharide presentation, valency of trimeric subunits, and species-specific residues that flank primary carbohydrate binding site (Crouch et al. 2006b).

25.3.6 Ligand Binding and Immune Cell-Recognition

Crystallographic studies of trimeric human SP-D neck +CRD domains have shown that maltose, a preferred saccharide ligand, binds to calcium via the vicinal 3- and 4-OH groups of the non-reducing glucose, previously designated calcium ion 1 and glucose 1 (Glc1), respectively (Shrive et al. 2003). These interactions are further stabilized by hydrogen bonding of Glc1 to amino acid side chains that also coordinate with calcium ion 1. The crystal structures of maltose-bound rfhSP-D (recombinant homotrimeric fragment of human SP-D comprising α-helical coiled-coil neck plus three CRDs) at 1.4A, and of rfhSP-D at 1.6A provided insights on the mode of carbohydrate recognition and fine details of SP-D binding to allergens/antigens or whole pathogens, and details on engagement of effector cells and molecules of humoral immunity. A calcium ion, located on trimeric axis in a pore at the bottom of funnel formed by the three CRDs and close to neck-CRD interface, is coordinated by a triad of glutamate residues which are, to some extent, neutralised by their interactions with a triad of exposed lysine residues in the funnel. The spatial relationship between neck and CRDs is maintained internally by these lysine residues, and externally by a glutamine, which forms a pair of hydrogen-bonds within an external cleft at each neck-CRD interface. Structural links between central pore and cleft suggested a possible effector mechanism for immune cell surface receptor binding in presence of bound, extended LPS and phospholipid ligands. The structural requirements for such an effector mechanism, involving both the trimeric framework for multivalent ligand binding and recognition sites formed from more than one subunit, are present in both native hSP-D and rfhSP-D, providing a possible explanation for significant biological activity of rfhSP-D (Shrive et al. 2003).

25.4 Regulation of Sp-D by Various Factors

25.4.1 Glucocorticoids

Surfactant proteins are known to be regulated by glucocorticoids. Lung explants from fetuses depend on hydrocortisone for their growth. The addition of hydrocortisone resulted in increase in SP-D mRNA expression (Gonzales et al. 2001; Shannon et al. 2001). The accelerated lung maturation accompanying glucocorticoid exposure in utero is associated with a precocious increase in SP-D gene transcription and protein production by pulmonary epithelial cells (Mariencheck and Crouch 1994). Glucocorticoid treatment in vivo of fetal, neonatal and adult rats for short durations exhibits stimulatory effect on contents of lung SP-D. Human lung is under developmental and glucocorticoid regulation occurring at a pre-translational level. SP-D is not influenced by inflammatory mediators that regulate SP-A, suggesting that these two proteins are not coordinately regulated in response to lung infection (Dulkerian et al. 1996). Early postnatal use of Dex in infants with respiratory distress syndrome (RDS) was shown to improve pulmonary status and to allow early weaning off mechanical ventilation. Early use of Dex can improve pulmonary status and also increase SP-A and SP-D levels in tracheal fluid in premature infants with RDS (Wang et al. 1996).

25.4.2 1α,25-Dihydroxyvitamin D3

The active form of 1α,25-dihydroxyvitamin D_3 [1,25$(OH)_2D_3$], has been shown to stimulate lung maturity, alveolar type II cell differentiation and fetal lung development. In fetal rat lung, 1,25$(OH)_2D_3$ increases the synthesis and secretion of surfactant lipids and accelerates the appearance of the morphological features of alveolar type II cells (Marin et al. 1993). It was stated that a natural metabolite of 1,25$(OH)_2D_3$ increases surfactant phospholipid and SP-B mRNA and protein synthesis in human NCI-H441 cells (Rehan et al. 2002; Phokela et al. 2005). Phokela et al. (2005) suggest that regulation of surfactant protein gene expression in human lung and type II cells by 1,25$(OH)_2D_3$ is not coordinated where as 1,25$(OH)_2D_3$ decreases SP-A mRNA and protein levels in both fetal lung tissue and type II cells.

25.4.3 Growth Factors

Keratinocyte growth factor (KGF, FGF-7) is a potent mitogen of pulmonary bronchial and alveolar epithelial cells. Airway epithelial cells (AEC) proliferate in response to KGF. KGF instillation resulted in epithelial cell hyperplasia in rats and the mRNA levels for SP-A, SP-B, and SP-D were increased in whole lung tissue on days 1 and 2 after KGF treatment (Yano et al. 2000). In presence of KGF, rat serum mRNAs for surfactant proteins were maintained at high levels. Secretion of SP-A and SP-D was found to be independent of phospholipid secretion (Mason et al. 2002). Clara cells also responded to human KGF in vivo by proliferation as well as by changes in protein expression, whereas no significant response was observed in pulmonary neuroendocrine cells (Fehrenbach et al. 2002; Shannon et al. 2001). Growth hormone significantly increases MBL and SP-D levels in Turner Syndrome. A link between endocrine and immune system with clinical consequences needs to be studied (Gravholt et al. 2004). VEGF is an autocrine proliferation and maturation factor for developing alveolar type II cells. But VEGF does not change the transcriptional level of SP-A, SP-C, and SP-D though it increases SP-B. The effects of VEGF isoform on type II cells are likely to be exerted indirectly through reciprocal paracrine interactions involving other lung cell types (Raoul et al. 2004). Pretreatment therapy by recombinant human VEGF (rhVEGF) decreased RSV disease in perinatal lamb respiratory syncytial virus (RSV) model (Meyerholz et al. 2007). Increased expression of SP-D mRNA at 6 and 24 h and decreased expression of SP-A mRNA at 12 h were observed after treatment with all-trans retinoic acid (RA) (Grubor et al. 2006).

25.5 SP-D in Human Fetal and Newborns Lungs

The SP-D was not detected in fetuses at 8–19 weeks gestation. At 21 weeks gestation, SP-D was weakly localized, in some cases (5/9), in the epithelial lining of both bronchioles and terminal airways. In contrast at 21 weeks gestation, SP-A was more markedly detected in the epithelial lining of both bronchioles and terminal airways in all cases but not detected in bronchioles and terminal airways at 8–19 weeks gestation. The production of SP-D in fetal human lungs begins in bronchiolar and terminal epithelium from about 21 weeks of gestation (Mori et al. 2002). Stahlman et al. (2002) detected SP-D on airway surfaces by 10 weeks' gestation and indicated that SP-D is secreted onto luminal surfaces by epithelial cells lining ducts of many organs, where it likely plays a role in innate host defense. The SP-D concentrations in umbilical cord blood and capillary blood are highly variable and depend on several perinatal conditions. In preterm infants, significant changes occur in collectin umbilical cord blood concentrations and pulmonary SP-D levels. Further studies are needed to elucidate the effect of respiratory distress and infection on SP-D concentrations (Dahl et al. 2006; Hilgendorff et al. 2005).

Serum SP-D and MBL levels are significantly low in preterms between 28 and 32 week gestational age (GA) compared to term infants and positively correlated with history of antenatal corticosteroids and chorioamnionitis. The SP-D levels in tracheal aspirates (TA) were increased in preterm infants between 28 and 32 week GA with respiratory distress syndrome compared to control subjects in contrast to extremely immature infants <28 week GA suffering from respiratory distress syndrome (RDS). SP-D concentrations in umbilical cord blood and capillary blood in premature infants are twice as high as in mature infants and depend on several perinatal conditions. High SP-D levels in umbilical cord blood and capillary blood on day 1 were found to be related to increased risk of RDS and infections (Dahl et al. 2006).

25.6 Non-Pulmonary SP-D

25.6.1 Human Skin and Nasal Mucosa

The SP-D has been detected in basal layers of normal human skin. It is more abundant in the stratum spinosum of lesional psoriatic and atopic skin due to more cells producing molecule rather than up-regulation of production in single cells of diseased skin (Hohwy et al. 2006). The surface tension of SP-deficient artificial sebum is (a) lowered by skin-extracted SP-B and (b) further reduced to a level comparable to normal sebum by additional presence of skin-extracted SP-A and SP-D, consistent with their surface tension-lowering capabilities in lung (Mo et al. 2007). SP-A2, SP-B, and SP-D mRNAs were expressed normal human nasal epithelial (NHNE) cells and human nasal mucosa and localized in ciliated cells of the surface epithelium and serous acini of submucosal glands (Kim et al. 2007; Woodworth et al. 2006, 2007).

25.6.2 Digestive Tract, Mesentery, and Other Organs

The fact that surfactant-like materials composed of phospholipids are secreted by a number of other organs prompted several groups to search for SP expression in these organs also. The SP-D occurs in gastric mucosa at the luminal surface and within gastric pits of mucus-secreting cells. The hydrophilic proteins SP-A and SP-D and their transcripts have been found in a number of tissues, including gastric and intestinal mucosae, mesothelial tissues (mesentery, peritoneum, and pleura), synovial cells, Eustachian tube and sinus, and possibly in salivary glands, pancreas, and urinary tract. SP-D is expressed in mucus-secreting cells of gastric mucosa. SP-D protein and mRNA were not detected in the duodenum and remainder of the gastrointestinal tract (Herias et al. 2007). It is possible that SP-D may participate in the regulation of secretion or assembly of gastric acid barrier. Alternatively, SP-D may participate in gastric mucosal host defense (Fisher and Mason 1995). By contrast, the hydrophobic proteins SP-B and SP-C actually appear to be expressed in lung epithelium only. The expression of SP-A and SP-D appears as a general feature of organs exposed to pathogens because they present an interface with the external milieu. Although this function has been investigated in lung only through the gene-targeting approach, increased expression of SP-A in the infected middle ear and of SP-D in the *Helicobacter*-infected antrum argues for such a function also in other organs. In organs that are not exposed to external pathogens, their role is likely to exert anti-inflammatory and immunomodulatory functions, as suggested by increased SP-A immunoreactivity in rheumatoid disease. SP-A and SP-B have been found in association with phospholipids in the lung of all air-breathing vertebrates, including the most primitive forms represented by lungfish, which implies that the surfactant system had a single evolutionary origin. Immunochemical proximity of the proteins among vertebrates indicates considerable conservation during evolution. Infection with the gastroduodenal pathogen *Helicobacter pylori* up-regulates expression of SP-D in human patients with gastritis, and its influence on colonization has been demonstrated in a *Helicobacter* SP-D-deficient (SP-D$^{-/-}$) mouse model. SP-D binds and agglutinates *H. pylori* cells in a lectin-specific manner, and has been shown to bind *H. pylori* lipopolysaccharide. Furthermore, evidence indicates that *H. pylori* varies LPS O-chain structure to evade SP-D binding which is speculated aids persistence of this chronic infection (Bourbon and Chailley-Heu 2001; Moran et al. 2005). To explore the similarity between lung and peritoneal surfactants, Chailley-Heu et al. (1997) suggested that mesothelial cells also produce SP-A and SP-D, although they are of embryonic origin (mesodermal) and are different from those of lung and digestive tract (endodermal) that secrete these surfactants. Porcine SP-D bound to solid-phase mannan in a Ca^{2+}-dependent manner with a saccharide specificity similar to rat and human SP-D. Whereas SP-D and SP-A are abundant in the peripheral lung, their presence in sputum derived from the larger airways is variable and their carbohydrate binding capacity is lost. Sputum sol fraction and LPS inhibited the binding of the collectins to carbohydrate in the presence of calcium (Griese et al. 2001).

Bands for SP-A were detected in human Eustachian tube (ET) (SP-A, 34 kDa) and in kidney extracts, and for SP-D (43 kDa) in Eustachian tube and in kidney extracts (SP-D, 86 kDa), and for SP-B (8 kDa) in human ET and organ of Corti extracts. Dysfunction of local mucosal immunity in ET may predispose infants to recurrent otitis media. The protein was found in the granules of microvillar epithelial cells. SP-A and SP-D may be important for antibody-independent protection of the middle ear against infections (Kankavi 2003; Paananen et al. 2001).

25.6.3 Male Reproductive Tract

SP-A and SP-D have been identified in testis and epididymis. The presence of SP-A, SP-D and components of molecular weight 34 and 43 kDa have been detected in human spermatozoa. SP-A is localized in mid-piece, the tail, and sometimes at equatorial region of spermatozoa (Kankavi et al. 2008). SP-D mRNA and protein are present throughout the mouse male reproductive tract, including the prostate. Castration increases prostate SP-D mRNA levels. SP-D protects the human prostate from infection by pathogens. The SP-D inhibits infection of prostate epithelial cells by *Chlamydia trachomatis* in an in vitro infection assay. The SP-D binds to C. *trachomatis* via its carboxy-terminal lectin domains. SP-D protein levels are increased at sites of inflammation in prostate, suggesting SP-D may also contribute more generally to inflammatory regulation in prostate (Oberley et al. 2005). Results suggest that infection and androgens regulate SP-D in the prostate (Oberley et al. 2007b).

Surfactant proteins in the stallion reproductive tract contribute to immune surveillance and to active barrier defense mechanism (Kankavi et al. 2007). SP-A and SP-D are present in the mare genital tract, vulva, vagina, ovarium, uterus and tuba uterina. The SPs are present not only in just lamellar bodies associated with lung, but also in genital system of mare (Kankavi et al. 2007). SP-D protein was localized in the apical portion of the reproductive epithelial cells. One of the functions of the SP-D protein may be to protect cervical epithelial cells from infection by C. *trachomatis*. *Chlamydia muridarum* infection caused an increase in the SP-D protein content of reproductive tract epithelial cells. Data were suggestive that SP-D may play a role in innate immunity in the female reproductive tract in vivo (Oberley et al. 2004, 2007a).

25.6.4 Female Genital Tract

The SP-D is present in the female genital tract, the placenta and in amniotic fluid. In the placenta, SP-D was seen in all villous and extravillous trophoblast subpopulations. Endometrial presence of SP-D in non-pregnant women varied according to stage of the menstrual cycle and was up-regulated towards the secretory phase. Endometrial SP-D may prevent intrauterine infection at the time of implantation and during pregnancy (Leth-Larsen et al. 2004). Findings suggest that SP-A and SP-D interact with chlamydial pathogens and enhance their phagocytosis into macrophages. *Chlamydia muridarum* infection increases SP-D protein content of reproductive tract epithelial cells, suggestiing that SP-D plays a role in innate immunity in the female reproductive tract in vivo (Oberley et al. 2004, 2007).

Amniotic Fluid: MBP, SP-A and SP-D are present in amniotic fluid and localize on the surface of amniotic epithelium. MBP levels in amniotic fluid were found to increase sharply from about 32 weeks of gestation. Both SP-A and SP-D were detected in amniotic fluid as early as 26 weeks gestation and, SP-A levels rose sharply from 32 weeks towards term. By contrast, SP-D levels rose only moderately. Collecins appear to play a role in antibody-independent recognition and clearance of pathogens in amniotic cavity, towards term (Malhotra et al. 1994; Miyamura et al. 1994b). The production of SP-D is increased shortly prior to birth, and the increases in total lung SP-D and SP-D mRNA are temporally correlated with SP-D secretion and the appearance of SP-D in amniotic fluid.

Intra-Amniotic Endotoxin Accelerates Lung Maturation: Intra-amniotic LPS and cytokines may decrease RDS and increase chronic lung disease in the newborn. Intrauterine exposure to LPS increases surfactant protein expression and improves lung stability and aeration in preterm animals (Bry and Lappalainen 2001). Intra-amniotic endotoxin from E. coli 055:B5 induces lung maturation within 6 days in fetal sheep of 125 days gestational age. SP-A, SP-B, and SP-C mRNAs were maximally induced at 2 days. SP-D mRNA was increased fourfold at 1 day and remained at peak levels for up to 7 days. The alveolar pool of SP-B was significantly increased between 4 and 7 days in conjunction with conversion to the fully processed active airway peptide. All SPs were significantly elevated in the BAL fluid by 7 days (Bachurski et al. 2001). IL-1 and glucocorticoid affect the expression of SP-A, -B, and –C in lung explants from rabbit fetuses. The study revealed beneficial additive effects of glucocorticoid and cytokine on lung surfactant (Vayrynen et al. 2004).

25.7 Functions of Lung SP-D

25.7.1 Innate Immunity

Surfactant is a complex of lipids and proteins that reduces surface tension at the air/liquid interface of the lung and regulates immune cell function. Surfactant immune function is primarily attributed to two proteins: SP-A and SP-D. It has been known for several years that surfactant lipids suppress a variety of immune cell functions, most notably lymphocyte proliferation, which, conversely, is augmented by SP-A. Thus surfactant lipids and proteins may be counterregulatory, and changes in lipid-to-protein ratios may be important in regulating the immune status of the lung. That these ratios change in disease states is clear, but it is not known whether the alterations are a cause or an effect. Studies with mice in which the SP-A and SP-D genes have been ablated are helping to clarify the role of surfactant in immune function (Kingma and Whitsett 2006)

The SP-A and SP-D bind carbohydrates, lipids, and nucleic acids with a broad-spectrum specificity and initiate phagocytosis of inhaled pathogens as well as apoptotic cells. Investigations on gene-deficient and conditional overexpressed mice indicated that lung SP-A and SP-D directly modulate innate immune cell function and T-cell-dependent inflammatory events and have a unique, dual-function capacity to induce pathogen elimination and control proinflammatory mechanisms. Surfactant protein-D participates in innate response to inhaled microorganisms and organic antigens, and contributes to immune and inflammatory regulation within lung. Porcine SP-D promotes uptake of pathogenic bacteria by epithelial cells. This reflects a scavenger function for SP-D in intestine, which enables the host to generate a more rapid response to infectious bacteria. SP-D binds to surface glycoconjugates expressed by a wide variety of microorganisms, and to oligosaccharides associated with the surface of various complex organic antigens. After binding of SP-D to microbial surface effector mechanisms such as agglutination, neutralizing or opsonization of microorganisms for phagocytosis are initiated. SP-D also specifically interacts with glycoconjugates and other molecules expressed on the surface of macrophages, neutrophils, and lymphocytes. Presence and activity of SP-D in porcine coronary endothelial cells depend on PI3K/Akt, Erk and nitric oxide and decrease after multiple passaging. They suggest a protective role of SP-D in these cells (Lee et al. 2009). In addition, SP-D binds to specific surfactant-associated lipids and can influence the organization of lipid mixtures containing phosphatidylinositol in vitro. SP-D-deficient transgenic mice show abnormal accumulations of surfactant lipids, and respond abnormally to challenge with respiratory viruses and bacterial lipopolysaccharides. The phenotype of macrophages isolated from lungs of SP-D-deficient mice is altered. SP-D stimulates chemotaxis of phagocytes and once bound to the phagocytes, the production of oxygen radicals can be induced. Circumstantial evidence indicates that abnormal oxidant metabolism and/or increased metalloproteinase expression contributes to the development of emphysema. The expression of SP-D is increased in response to many forms of lung injury, and deficient accumulation of appropriately oligomerized SP-D might contribute to the pathogenesis of a variety of human lung diseases (Crouch 2000; Hogenkamp et al. 2007; Reid 1998; Wright 2004; Awasthi et al. 2001) (Figs. 25.1, 25.2, 25.3).

25.7.2 Effects on Alveolar Macrophages

SP-D opsonizes pathogens and enhances their phagocytosis by alveolar macrophages and neutrophils. SP-D was found to bind alveolar Type II cells, Clara cells, and alveolar macrophages. In Type II cells abundant binding was observed in the endoplasmic reticulum, whereas Golgi complex and multivesicular bodies were labeled to a limited extent. Anti-surfactant SP-A and SP-D showed both SP-A and SP-D in same granules (Kuan et al. 1994). Both SP-D and SP-A have been shown to enhance oxygen radical production by alveolar macrophages. SP-D binds specifically to alveolar macrophages and the receptor involved is different from that of C1q (Miyamura et al. 2004a). SP-D can bind to specific surfactant phospholipids and to glycoconjugates associated with the surface of various microorganisms, consistent with possible roles in surfactant metabolism and pulmonary host defense. Administration of a truncated 60-kDa fragment of human rSP-D reduces the number of apoptotic and necrotic alveolar macrophages and partially corrects the lipid accumulation in SP-D-deficient mice. The same SP-D fragment binds preferentially to apoptotic and necrotic alveolar macrophages in vitro, that suggests that SP-D contributes to immune homeostasis in lung by recognizing and promoting removal of necrotic and apoptotic cells (Clark et al. 2002, 2003) (Fig. 25.3). Findings indicate that collagen domain of SP-D is not required for assembly of disulfide-stabilized oligomers or the innate immune response to viral pathogens. However, collagen domain of SP-D is required for regulation of pulmonary macrophage activation, airspace remodeling, and surfactant lipid homeostasis (Kingma et al. 2006). Surfactant protein D regulates the cell surface expression of alveolar macrophage β2-integrins (Senft et al. 2007).

Opsonization of Gram-Negative Bacteria Independent of Macrophages: SP-A and SP-D bind LPS, opsonize microorganisms, and enhance the clearance of lung pathogens. The pulmonary clearance of *E. coli K12* was reduced in SP-A-null mice and was increased in SP-D-overexpressing mice. Purified SP-A and SP-D inhibited bacterial synthetic functions of several, but not all, strains of *E. coli, Klebsiella pneumoniae*, and *E. aerogenes*. In general, rough *E. coli* strains were more susceptible than smooth strains, and collectin-mediated growth inhibition was partially blocked by coincubation with rough LPS vesicles. Data indicate that SP-A and SP-D are antimicrobial proteins that directly inhibit proliferation of Gram-negative bacteria in a macrophage- and aggregation-independent manner by increasing the permeability of the microbial cell membrane (Wu et al. 2003).

25.7.3 Functions of Neutrophils

Calcium-Dependent Neutrophil Uptake of Bacteria: Because neutrophils (PMN) and monocytes are recruited into the airspaces in association with many types of infection or lung injury, Crouch et al. examined the interactions of

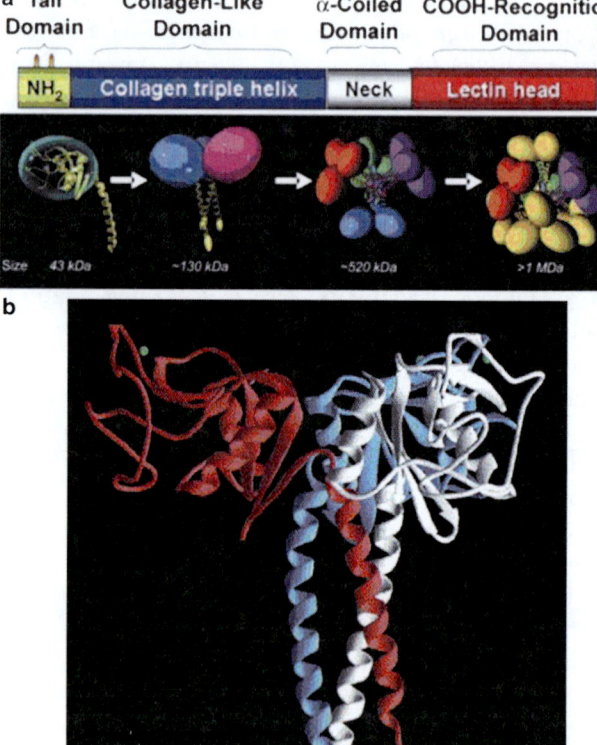

Fig. 25.1 SP-D structure: (*Upper Panel*) **A model of SP-D structure**. The SP-D monomer (43 kDa) consists of a carbohydrate recognition domain which forms the globular head structure. This domain is connected to the collagen-like helical tail domain by a short, 30–amino acid, neck domain. At the end of the tail domain is the amino terminus in which cysteines 15 and 20 are positioned (shown as *yellow* projections). (*Central panel*) Stylized representation of SP-D multimer assembly (note tail domains are shown *shortened* for ease of visualization). The head and neck domains drive the aggregation of the SP-D monomer to form a trimer of ~130 kDa. These trimers associate to form a dodecamer (~520 kDa). The forces holding this dodecamer together are unclear, although there is a dependency upon the amino-terminal cysteines as mutant lacking these cysteines do not form dodecamers. These dodecamers can assemble to a multimer of greater than 1 MDa. It is unclear whether the dodecamer is an essential intermediate in multimer formation. It should be noted that neither the trimer nor the dodecamer are globular proteins, due to the presence of the long collagen tail and thus under native conditions will behave as molecules with greater molecular radius (Adapted from Guo et al. PLoS Biol. 2008; 6: e266). (*Lower Panel*) *Ribbon diagram* shows the overall main chain structure of human lung surfactant protein SP-D. Each monomer is shown in a different colour and calcium ions are depicted as *green spheres* (Adapted with permission from Håkansson et al. 1999 © Elsevier)

these cells with SP-D. Natural or recombinant rat SP-D showed dose-dependent effects on human PMN and monocyte migration. Studies established that SP-D can bind to specific sites on neutrophils and monocytes involving saccharide binding domains of SP-D (Crouch et al. 1995). The SP-A and SP-D increased calcium-dependent neutrophil uptake of *E. coli, S. pneumoniae*, and *S. aureus*. Collectins

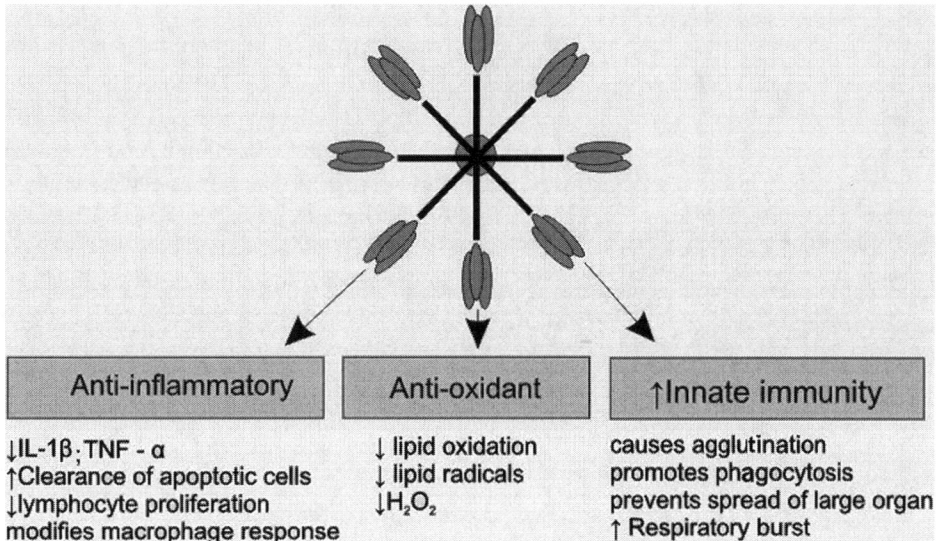

Fig. 25.2 **The potential actions of surfactant protein D that may be beneficial in protecting lungs from COPD.** SP-D is illustrated as a dodecamer (containing 12 monomeric chains of the carbohydrate recognition domains). SP-D has anti-inflammatory properties by attenuating the expression of pro-inflammatory cytokines such as interleukin (*IL*)-1β, 6, 8 and tumor necrosis factor-α and reducing lymphocyte accumulation in the perivascular tissues in the lungs. SP-D reduces lipid peroxidation and generation of hydrogen peroxide in vivo. SP-D also plays an important role in the innate immunity by promoting phagocytosis and agglutination of pathogens and recruiting neutrophils into lung tissues during active infections (see text for details)

enhanced bacterial uptake occurred through a mechanism that involved both bacterial aggregation and direct actions on neutrophils. The degree of multimerization of SP-D preparations was a critical determinant of both aggregating activity and potency in enhancing bacterial uptake. Results provided evidence that surfactant collectins may promote neutrophil-mediated clearance of bacteria in the lung independently of opsonizing antibody (Hartshorn et al. 1998, 2002).

Neutrophil Serine Proteinases Inactivate SP-D: Because SP-D specifically interacts with neutrophils that infiltrate the lung in response to acute inflammation and infection, the neutrophil-derived serine proteinases (NSPs): neutrophil elastase, proteinase-3, and cathepsin G degrade SP-D. All three human NSPs specifically cleaved recombinant rat and natural human SP-D dodecamers in a time- and dose-dependent manner at the sites of inflammation with potential deleterious effects on its biological functions (Hirche et al. 2004; Griese et al. 2003). However, excess NSPs in lungs play a central role in pathology of inflammatory pulmonary disease. The serpinb1, an efficient inhibitor of the three NSPs, preserves cell and molecular components responsible for host defense against *P. aeruginosa*. The regulation of pulmonary innate immunity by serpinb1 is nonredundant and is required to protect two key components, the neutrophil and SP-D, from NSP damage during host response to infection (Benarafa et al. 2007).

25.7.4 Protective Role in Allergy and Infection

SP-A and SP-D modulate allergic reactions, and resolution of inflammation. SP-A and SP-D can interact with receptor molecules present on immune cells leading to enhanced microbial clearance and modulation of inflammation. SP-A and SP-D also modulate the functions of cells of the adaptive immune system including DCs and T cells. SP-D has multiple functions in innate immunity in lung. The generation of SP-D knock-out mice has revealed a central role for this protein in the control of lung inflammation. Accumulating evidence in mouse models of infection and inflammation indicates that truncated recombinant forms of SP-D are biologically active in vivo. Clark and Reid (2002) addressed structural requirements for recognised activities of SP-D in vitro and in vivo, with emphasis on evidence arising from studies with transgenic mice and mouse models of inflammatory lung disease. The potential of truncated recombinant forms of surfactant protein D as novel therapy for infectious and inflammatory disease is discussed. Constitutive absence of SP-D in mice is associated with lung inflammation, alteration in surfactant lipid homeostasis, and increased oxidative-nitrative stress. The extracellular pool size of SP-D fluctuates significantly during acute inflammation. Clearance of SP-D into lung tissue is increased during inflammation and that tissue-associated neutrophils significantly contribute to this process. Studies using murine models of allergy and infection have raised the

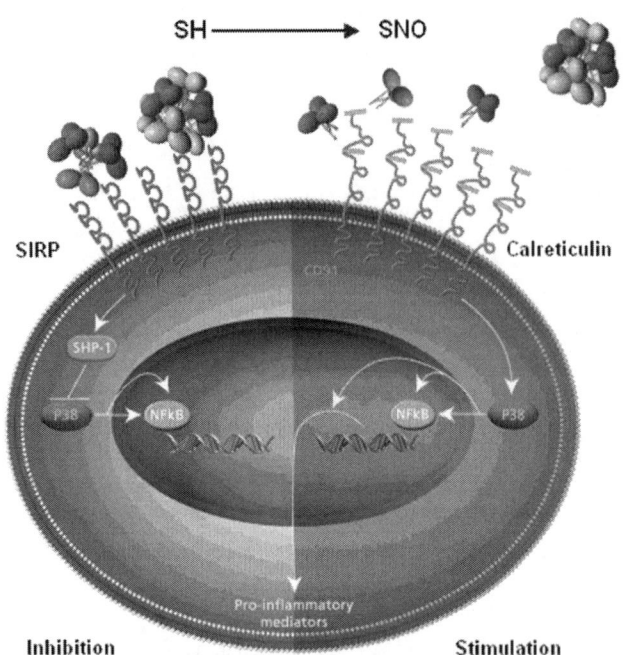

Fig. 25.3 A model of the pro- and anti-inflammatory functions of SP-D: Under non-inflammatory conditions, SP-D remains in large multimeric or dodecameric forms in which the tail domains remain buried. The head domains bind to SIRP-1α and activate the kinase SHP-1. SHP-1 activation inhibits p38 activations, potentially resulting in the blockaged of NF-κB action and the inhibition of infammatory function. Under inflammatory conditions the production of NO leads to the formation of SNO-SP-D and the disruption of the multimeric structure. The tail domains now become exposed and bind to calreticulin. This results in p38 phosphorylation via CD91, potentially leading to NF-κB activation and the production of pro-inflammatory mediators. Presumably other actions which result in disruption of SP-D multimeric structure may also be pro-inflammatory (Adapted from Guo et al. PLoS Biol. 2008; 6: e266)

possibility that recombinant forms of SP-A and SP-D may have therapeutic potential in controlling pulmonary infection, inflammation, and allergies in humans (Haczku 2008).

Using an opportunistic fungal pathogen *Aspergillus fumigatus* (Afu), Kishor et al. (2002) studied the role of SP-A and SP-D in the host immunity. Afu causes a systemic infection via lungs, called invasive aspergillosis (IPA) in immunocompromised subjects. In immunocompetent subjects, it can cause an allergic disorder, called allergic bronchopulmonary aspergillosis (ABPA). Therapeutic administration of these proteins in a murine model of IPA can rescue mice from death. Treating mice, having ABPA, can suppress IgE levels, eosinophilia, pulmonary cellular infiltration and cause a marked shift from a pathogenic Th2 to a protective Th1 cytokine profile. These results highlight the potential of SP-A, SP-D and their recombinant forms, as novel therapeutics for lung allergy and infection (Brandt et al. 2008; Kishor et al. 2002).

Susceptibility of SP-A or SP-D Genetically Deficient Mice: Madan et al. (2005) examined the susceptibility of SP-A (AKO) or SP-D gene-deficient (DKO) mice to Afu allergen challenge. Both AKO and DKO mice showed intrinsic hypereosinophilia and several-fold increase in levels of IL-5 and IL-13, and lowering of IFN-γ to IL-4 ratio in the lungs, suggesting a Th2 bias of immune response. The AKO and DKO mice showed distinct immune responses to Afu sensitization. Intranasal treatment with SP-D or rhSP-D (a recombinant fragment of SP-D) was effective in rescuing Afu-sensitized DKO mice, while SP-A-treated Afu-sensitized AKO mice showed several-fold elevated levels of IL-13 and IL-5, resulting in increased pulmonary eosinophilia and damaged lung tissue. These studies suggest a role for SP-A and SP-D in offering resistance to pulmonary allergenic challenge. Fungi have been implicated in pathogenesis of chronic rhinosinusitis (CRS) with eosinophilic mucus (EMCRS). SP-D is expressed in nasal mucosa and is up-regulated in vitro in response to fungal allergens. SP-D offers a potential therapy for treatment of CRS (Ooi et al. 2007).

Regulation of Chemotaxis and Degranulation of Human Eosinophils: Alveolar macrophages express receptors specific for SP-A and SP-D and provide insight into potential roles of collectins in recruitment and maturation of mononuclear phagocytes in lung (Tino and Wright 1999). SP-D markedly inhibited chemotaxis of eosinophils triggered by eotaxin, a major tissue-derived CC-chemokine. In addition, degranulation of ECP in response to Ca^{2+} ionophore, immobilized IgG and serum from allergic patients was inhibited by SP-D. Data support the concept of an anti-inflammatory function of SP-D in the lung of patients with allergic diseases (von Bredow et al. 2006).

Surfactant Proteins and Lipids as Modulators of Inflammation: SP-A binds with *Mycoplasma pneumoniae* with high affinity and enhances inflammatory response of human and rat macrophages. The interaction of SP-A with bacteria involved phospholipids as major ligand. The SP-A reactive lipid consisted of several disaturated molecular species of phosphatidylglycerol (PtdGro). The disaturated PtdGro failed to alter the anti-inflammatory action of SP-A, but could modify the host response to LPS. These findings reveal that both the lipids and proteins of pulmonary surfactant play a role in regulating the host response to invading microorganisms (Chiba et al. 2006) (Fig. 25.3).

Exogenous surfactants and surfactant phospholipids without SP-A and SP-D inhibit secretion of pro-inflammatory cytokines and NO in NR8383 AMφ. The inhibitory effects of surfactant on oxygen radical and LPS-induced NO formation may result from mechanisms of low cell signaling. The anti-inflammatory activity of surfactant products used in the treatment of neonatal respiratory distress syndrome (RDS) may depend upon the specific preparation used (Kerecman et al. 2008). SP-D would protect against acute lung injury from hyperoxia in vivo. Local expression of SP-D protects against hyperoxic lung injury through modulation of proinflammatory cytokines and antioxidant enzymatic scavenger systems (Jain et al. 2008). Murine models of pulmonary hypersensitivity suggest that SP-D may be a potent anti-allergic protein.

Anti-inflammatory Protein in Human Coronary Artery SMCs: Immunoreactive SP-D protein is present in smooth muscle cells (SMCs) and endothelial cells. SP-D was also detected in human coronary artery SMCs (HCASMCs). Treatment of HCASMCs with endotoxin (LPS) stimulated the release of IL-8, a proinflammatory cytokine. This release was inhibited >70% by recombinant SP-D. It is stated that SP-D in human coronary arteries functions as an anti-inflammatory protein in HCASMCs (Snyder et al. 2008).

25.7.5 Adaptive Immune Responses

SP-A and SP-D are part of innate immune system and regulate the functions of other innate immune cells, such as macrophages. They also modulate the adaptive immune response by interacting with antigen-presenting cells and T cells, thereby linking innate and adaptive immunity. Emerging studies suggest that SP-A and SP-D function to modulate the immunologic environment of the lung so as to protect the host and, at the same time, modulate an overzealous inflammatory response that could potentially damage the lung and impair gas exchange. Numerous polymorphisms of SPs have been identified that may potentially possess differential functional abilities and may act via different receptors to ultimately alter the susceptibility to or severity of lung diseases (Pastva et al. 2007; Haley et al. 2002).

Nadesalingam et al. (2005) showed that SP-D binds various classes of Igs, including IgG, IgM, IgE and secretory IgA, but not serum IgA. SP-D recognizes IgG, aggregates IgG-coated beads and enhances their phagocytosis by murine macrophage RAW 264.7 cells. Therefore, SP-D effectively interlinks innate and adaptive immune systems (Nadesalingam et al. 2005). IL-4 selectively up-regulates SP-D expression in type II alveolar rat epithelial cell cultures. Since SP-D has a potent anti-inflammatory function, this mechanism may be part of a negative feedback loop providing a regulatory link between adaptive and innate immunity during allergic inflammation (Cao et al. 2004).

SP-A and D Suppress $CD3^+/CD4^+$ Cell Function: SP-A and SP-D inhibit lymphocyte proliferation in presence of accessory cells. SP-A and SP-D directly suppress Th cell function. Both proteins inhibited $CD3^+/CD4^+$ lymphocyte proliferation induced by PMA and ionomycin in an IL-2-independent manner. Both proteins decreased the number of cells entering the S and mitotic phases of the cell cycle. Inhibition of T cell proliferation by SP-A and SP-D occurs via two mechanisms, an IL-2-dependent mechanism observed with accessory cell-dependent T cell mitogens and specific Ag, as well as an IL-2-independent mechanism of suppression that potentially involves attenuation of $[Ca^{2+}]_I$ (Borron et al. 2002).

SP-D Enhances Bacterial Antigen Presentation by DCs SP-D interacts with DCs to enhance uptake and presentation of bacterial antigens. It binds to immature DCs in a dose-, carbohydrate-, and calcium-dependent manner, whereas its binding to mature DCs is reduced. SP-D also binds to *E. coli HB101* and enhances its association with DCs. In addition, SP-D enhances antigen presentation of an ovalbumin fusion protein to ovalbumin-specific MHC class II T cell hybridomas. Studies demonstrate that SP-D augments antigen presentation by DCs and suggest that innate immune molecules such as SP-D may help initiate an adaptive immune response for the purpose of resolving an infection (Brinker et al. 2001, 2002).

As DC function varies depending on the tissue of origin, studies were extended to APCs isolated from mouse lung. Though the SP-D binds specifically to lung CD11c-positive cells, results show that SP-D increases the opsonization of pathogens, but decreases the antigen presentation by lung DCs, and thereby, potentially dampens the activation of T cells and an adaptive immune response against bacterial antigens—during both steady-state conditions and inflammation (Hansen et al. 2007)

The SP-D shares target cells with the proinflammatory cytokine TNF-α, an important autocrine stimulator of DCs and macrophages in the airways. TNF-α can contribute to enhanced SP-D production in the lung indirectly through inducing IL-13. The SP-D, on the other hand, can antagonize the proinflammatory effects of TNF-α on macrophages and DCs, at least partly, by inhibiting production of this cytokine (Hortobágyi et al. 2008). Soerensen et al. (2005b) showed that porcine SP-D (pSP-D) expression in lung surfactant is induced by bacterial infection by an aerogenous route rather than by a haematogenous route and that the protein interacts specifically with alveolar macrophages and with DCs in microbial-induced in bronchus-associated lymphoid tissue (BALT). The purpose of the interaction between pSP-D and DCs in BALT remains unclear, but pSP-D could represent a link between the innate and adaptive immune system, facilitating the bacterial antigen presentation by DCs in BALT.

Lymphocyte Activation in Lungs of SP-D Null Mice: In vitro evidence suggests that SP-D may suppress local T cell responses. In vivo, the SP-D deficient mice ($SP-D^{-/-}$) showed marked T cell activation in the lungs, as reflected by an increased percentage of both CD4$^+$ and CD8$^+$ T cells expressing CD69 and CD25. CD4 lymphocytes and the fraction expressing CD69 was also increased in BAL. Increases in CD69-positive CD8 lymphocytes was not significant. Among proinflammatory cytokines, the expression of IL-12 and IL-6 was increased in the lungs of SP-D$^{-/-}$ mice. The lack of local pulmonary production of SP-D may lead to a state of persistent T cell activation, possibly in response to exogenous antigens (Fisher et al. 2002).

25.7.6 Apoptosis

Emerging results indicate that collectins are able to bind self-derived ligands in the form of apoptotic cells and regulate inflammatory responses. Stuart et al. (2006) discussed the understanding of the process of collectin recognition of dying and damaged cells and its implications for autoimmune and inflammatory diseases. The SP-A, SP-D, and C1q all enhance apoptotic cell ingestion by resident murine and human alveolar macrophages in vitro. However, only SP-D altered apoptotic cell clearance from the naive murine lung, suggesting that SP-D plays a particularly important role in vivo. Similar to C1q and MBL, SP-A and SP-D bind apoptotic cells in a localized, patchy pattern and drive apoptotic cell ingestion by phagocytes through a mechanism dependent on calreticulin and CD91 (Vandivier et al. 2002). SP-A binds to and enhances macrophage uptake of other nonself particles, specifically apoptotic polymorphonuclear neutrophils (PMNs). PMNs are recruited into the lungs during inflammation, but as inflammation is resolved, PMNs undergo apoptosis and are phagocytosed by $M\Phi$s. SP-A enhances phagocytosis via an opsonization-dependent mechanism and binds apoptotic cells. Data suggest that SP-A and SP-D facilitate the resolution of inflammation by accelerating apoptotic PMN clearance (Schagat et al. 2001) and that SP-D influences innate host defense by regulating sCD14 in a process mediated by MMP-9 and MMP-12 (Senft et al. 2005).

25.7.7 Other Effects of SP-D

Enhanced Pulmonary Lipoidosis: Targeted disruption of SP-D gene caused a marked pulmonary lipoidosis characterized by increased alveolar lung phospholipids, demonstrating a unexpected role for SP-D in surfactant homeostasis. Pulmonary-specific expression of SP-D corrects pulmonary lipid accumulation in SP-D gene-targeted mice (Fisher et al. 2000). SP-D deficient mice have three to four times more surfactant lipids in air spaces and lung tissue than control mice. Relative to saturated phosphatidylcholine (Sat PC), SP-A and SP-C were decreased in alveolar surfactant and the large-aggregate surfactant fraction in SP-D deficienct mice. SP-D deficiency results in multiple abnormalities in surfactant forms and metabolism that cannot be attributed to a single mechanism (Ikegami et al. 2000). Development of dementia, including Alzheimer's disease (AD), is associated with lipid dysregulation and inflammation. As SP-D has multiple effects in lipid homeostasis and inflammation, the correlation between SP-D concentrations and development of dementia as well as to augmented mortality has been indicated (Nybo et al. 2007).

Body Mass Index: Reports have demonstrated that body mass index (BMI; kg/m^2) is influenced by genes in common with SP-D. In Danish twins, serum SP-D was significantly and inversely associated with weight and waist circumference in men and to BMI in both genders. The SP-D$^{-/-}$ mice and wild-type mice gained significantly increased weight, with 90 mg/week on normal chow and significantly increased fat in SP-D$^{-/-}$ male mice. It suggests that there is an association between low levels or absent SP-D and obesity (Sorensen et al. 2006b). Racial differences in SP-D expression exist since median plasma SP-D in Chinese population was approximately two times lower than the median serum SP-D previously measured in a Danish population using same immuno-assay. The inverse association between serum SP-D and BMI found in Chinese population is related to obesity in similar ways in Chinese and Danes (Zhao et al. 2007).

25.8 Oxidative Stress and Hyperoxia

Contributor to the Protection of Lung from Oxidative Stresses: The surfactant lining is exposed to highest ambient oxygen tension of any internal interface and encounters a variety of oxidizing toxicants including ozone and trace metals contained within the 10 kl of air that is respired daily. The pathophysiological consequences of surfactant oxidation in humans and animals include airspace collapse, reduced lung compliance, and impaired gas exchange. In many instances, normal surfactant inhibits many immune cell functions including proliferation resulting from various stimuli and production of reactive oxidative species, inflammatory mediators, and some cell surface markers. The predominant surfactant lipids appear to be responsible for these suppressive effects. Conversely, surfactant proteins SP-A and SP-D stimulate many aspects of immune cell behavior. The SP-A and SP-D directly protect surfactant phospholipids and macrophages from oxidative damage. Both proteins block accumulation of thiobarbituric acid-reactive substances and conjugated dienes during copper-induced oxidation of surfactant lipids or LDL particles by a mechanism that does not involve metal chelation or oxidative modification of proteins. Low density lipoprotein oxidation is instantaneously arrested upon SP-A or SP-D addition, suggesting direct interference with free radical formation or propagation. The antioxidant activity of SP-A is located to the carboxyl-terminal domain of the protein, which, like SP-D, contains a C-type lectin CRD. Thus, the SP-A and SP-D, which are ubiquitous among air breathing organisms, contribute to the protection of lung from oxidative stresses due to atmospheric or supplemental oxygen, air pollutants, and lung inflammation (Bridges et al. 2000; Phelps 2001).

To evaluate the effects of O3 exposure in mouse strains with genetically different expression levels of SP-D Kierstein et al. (2006) exposed Balb/c, C57BL/6 and SP-D knockout mice to O3 or air. Ozone-exposed Balb/c mice demonstrated significantly enhanced acute inflammatory changes and higher levels of SP-D and released more IL-10 and IL-6 in BAL. IL-6 contributes to the up-regulation of SP-D after acute O3 exposure; elevation of SP-D in the lung is associated with resolution of inflammation. Absence or low levels of SP-D predispose to enhanced inflammatory changes following acute oxidative stress.

Permeabilization of Bacteria by SPs: Studies suggest that SP-A and SP-D have a direct effect on growth and viability of Gram-negative bacteria, *Mycoplasma pneumoniae and Histoplasma capsulatum*. Permeability assays indicate membrane permeabilization by BAL fluid of mice that were sufficient or deficient in surfactant proteins, of rough and smooth LPS-containing membranes, and of genetically altered bacteria. The permeabilizing activity of concentrated BAL material from SP-A$^{+/+}$ mice was substantially greater than that from SP-A$^{-/-}$ animals, and was sensitive to hyperoxic exposure. Reports suggest that pulmonary collectins directly permeabilize bacteria in an LPS-dependent and rough LPS-specific manner. Oxidative damage blocks the permeabilizing activity of alveolar lining fluid and purified proteins (McCormack 2006).

Effect of Hyperoxia: The mRNA level of SP has been reported to be increased in lungs of animals exposed to hyperoxia. Relative amounts of SP-A, SP-B, and SP-D mRNA expression in congenital diaphragmatic hernia (CDH) lung were significantly decreased compared to controls at birth. The inability of O_2 to increase SP mRNA expression in hypoplastic CDH lung suggests that the hypoplastic lung is not responsive to increased oxygenation for the synthesis of SP (Shima et al. 2000). The hypoplastic lung in CDH has both a quantitative and qualitative reduction in surfactant. Aderibigbe et al. (1999) evaluated early effects of hyperoxia (95% O_2) on expression of SP-D in adult male rat lung. Hyperoxia had differential effects on SP-D abundance in alveolar epithelial and bronchiolar epithelial cells, and therefore may influence the availability of SP-D to bind microbial pathogens in airways depending on cell type and location. Alternatively, absence of SP-D aggravates hyperoxia-induced injury. This was tested in SP-D-deficient (SP-D$^{-/-}$) and wild-type (SP-D$^{+/+}$) mice which were exposed to 80% or 21% oxygen. Paradoxically, SP-D$^{-/-}$ mice had 100% survival during 14 days of hyperoxia, vs. 30% in SP-D$^{+/+}$. Perhaps, resistance of SP-D-deficient mice to hyperoxia reflects homeostatic changes in SP-D$^{-/-}$ phenotype involving both phospholipid and SP-B-mediated induced resistance of surfactant to inactivation as well as changes in immunomodulatory BAL cytokine profile (Jain et al. 2007).

Environmental Influences on SP-D Genes: Genetic risk for respiratory distress in infancy has been recognized with increasing frequency in neonatal intensive care units. Examples of genetic variations known to be associated with or cause respiratory distress in infancy have been reviewed (Cole et al. 2001). Serum SP-D levels in children are genetically determined. A SNP located in NH_2-terminal region (Met^{11}Thr) of mature protein is significantly associated with the serum SP-D levels. In a classic twin study, the serum SP-D levels increased with male sex, age, and smoking status. The intraclass correlation was higher for monozygotic (MZ) twin pairs than for dizygotic (DZ) twin pairs. Multivariate analysis of MZ and DZ covariance matrixes showed significant genetic correlation among serum SP-D and metabolic variables (Sørensen et al. 2006a). Like MBL, structural as wells as promoter variants linked to disease states are known for SP-D. In children MZ and DZ like-sexed twin pairs aged 6–9 years, intraclass correlations were significantly higher in MZ than in DZ twins, indicating substantial genetic influence on both MBL and SP-D levels (Husby et al. 2002).

References

Aderibigbe AO, Thomas RF, Mercer RR, Auten RL Jr (1999) Brief exposure to 95% oxygen alters surfactant protein D and mRNA in adult rat alveolar and bronchiolar epithelium. Am J Respir Cell Mol Biol 20:219–227

Akiyama J, Volik SV, Plajzer-Frick I et al (1999) Characterization of the mouse collectin gene locus. Am J Respir Cell Mol Biol 21:193–199

Akiyama J, Hoffman A, Brown C et al (2002) Tissue distribution of surfactant proteins A and D in the mouse. J Histochem Cytochem 50:993–996

Allen MJ, Voelker DR, Mason RJ (2001) Interactions of surfactant proteins A and D with *Saccharomyces cerevisiae* and *Aspergillus fumigatus*. Infect Immun 69:2037–2044

Allen MJ, Laederach A, Reilly PJ et al (2004) Arg343 in human surfactant protein D governs discrimination between glucose and N-acetylglucosamine ligands. Glycobiology 14:693–700

Awasthi S, Coalson JJ, Yoder BA et al (2001) Deficiencies in lung surfactant proteins A and D are associated with lung infection in very premature neonatal baboons. Am J Respir Crit Care Med 163:389–397

Bachurski CJ, Ross GF, Ikegami M et al (2001) Intra-amniotic endotoxin increases pulmonary surfactant proteins and induces SP-B processing in fetal sheep. Am J Physiol Lung Cell Mol Physiol 280:L279–L285

Benarafa C, Priebe GP, Remold-O'Donnell E (2007) The neutrophil serine protease inhibitor serpinb1 preserves lung defense functions in Pseudomonas aeruginosa infection. J Exp Med 204:1901–1909

Borron PJ, Mostaghel EA, Doyle C et al (2002) Pulmonary surfactant proteins A and D directly suppress CD3+/CD4+ cell function: evidence for two shared mechanisms. J Immunol 169:5844–5850

Botas C, Poulain F, Akiyama J et al (1998) Altered surfactant homeostasis and alveolar type II cell morphology in mice lacking surfactant protein D. Proc Natl Acad Sci USA 95:11869–11874

Bourbon JR, Chailley-Heu B (2001) Surfactant proteins in the digestive tract, mesentery, and other organs: evolutionary significance. Comp Biochem Physiol A Mol Integr Physiol 129:151–161

Brandt EB, Mingler MK, Stevenson MD et al (2008) Surfactant protein D alters allergic lung responses in mice and human subjects. J Allergy Clin Immunol 121:1140–1147, e2

Bridges JP, Davis HW, Damodarasamy M et al (2000) Pulmonary surfactant proteins A and D are potent endogenous inhibitors of lipid peroxidation and oxidative cellular injury. J Biol Chem 275:38848–38855

Brinker KG, Martin E, Borron P et al (2001) Surfactant protein D enhances bacterial antigen presentation by bone marrow-derived dendritic cells. Am J Physiol Lung Cell Mol Physiol 281:L1453–L1463

Brinker KG, Martin E, Borron P et al (2002) Surfactant protein D enhances bacterial antigen presentation by bone marrow-derived dendritic cells. Am J Physiol Lung Cell Mol Physiol 282:L516–L517

Brown-Augsburger P, Chang D, Rust K, Crouch EC (1996a) Biosynthesis of surfactant protein D. Contributions of conserved NH2-terminal cysteine residues and collagen helix formation to assembly and secretion. J Biol Chem 271:18912–18919

Brown-Augsburger P, Hartshorn K, Chang D et al (1996b) Site-directed mutagenesis of Cys-15 and Cys-20 of pulmonary surfactant protein D. Expression of a trimeric protein with altered anti-viral properties. J Biol Chem 271:13724–13730

Bry K, Lappalainen U (2001) Intra-amniotic endotoxin accelerates lung maturation in fetal rabbits. Acta Paediatr 90:74–80

Cai GZ, Griffin GL, Senior RM et al (1999) Recombinant SP-D carbohydrate recognition domain is a chemoattractant for human neutrophils. Am J Physiol 276:L131–L136

Cao Y, Tao JQ, Bates SR et al (2004) induces production of the lung collectin surfactant protein-D. J Allergy Clin Immunol 113:439–444

Chailley-Heu B, Rubio S, Rougier JP et al (1997) Expression of hydrophilic surfactant proteins by mesentery cells in rat and man. Biochem J 328:251–256

Chiba H, Piboonpocanun S, Mitsuzawa H et al (2006) Pulmonary surfactant proteins and lipids as modulators of inflammation and innate immunity. Respirology 11(Suppl):S2–S6

Charles A, Tang X, Crouch E et al (2001) Retinoblastoma protein complexes with C/EBP proteins and activates C/EBP-mediated transcription. J Cell Biochem 83:414–425

Clark H, Reid KB (2002) Structural requirements for SP-D function in vitro and in vivo: therapeutic potential of recombinant SP-D. Immunobiology 205:619–631

Clark H, Palaniyar N, Strong P et al (2002) Surfactant protein D reduces alveolar macrophage apoptosis in vivo. J Immunol 169:2892–2899

Clark H, Palaniyar N, Hawgood S, Reid KB (2003) A recombinant fragment of human surfactant protein D reduces alveolar macrophage apoptosis and pro-inflammatory cytokines in mice developing pulmonary emphysema. Ann N Y Acad Sci 1010:113–116

Cole FS, Hamvas A, Nogee LM (2001) Genetic disorders of neonatal respiratory function. Pediatr Res 50:157–162

Craig-Barnes HA, Doumouras BS, Palaniyar N (2010) Surfactant protein D interacts with alpha-2-macroglobulin and increases its innate immune potential. J Biol Chem 285:13461–13470

Crouch EC (2000) Surfactant protein-D and pulmonary host defense. Respir Res 1:93–108

Crouch E, Parghi D, Kuan SF, Persson A (1992) Surfactant protein D: subcellular localization in nonciliated bronchiolar epithelial cells. Am J Physiol 263:L60–L66

Crouch E, Rust K, Veile R et al (1993) Genomic organization of human surfactant protein D (SP-D). SP-D is encoded on chromosome 10q22.2-23.1. J Biol Chem 268:2976–2983

Crouch E, Chang D, Rust K et al (1994a) Recombinant pulmonary surfactant protein D. Post-translational modification and molecular assembly. J Biol Chem 269:15808–15813

Crouch E, Persson A, Chang D, Heuser J (1994b) Molecular structure of pulmonary surfactant protein D (SP-D). J Biol Chem 269:17311–17319

Crouch EC, Persson A, Griffin GL et al (1995) Interactions of pulmonary surfactant protein D (SP-D) with human blood leukocytes. Am J Respir Cell Mol Biol 12:410–415

Crouch E, McDonald B, Smith K et al (2006a) Contributions of phenylalanine 335 to ligand recognition by human surfactant protein D: ring interactions with SP-D ligands. J Biol Chem 281:18008–18014

Crouch EC, Smith K, McDonald B et al (2006b) Species differences in the carbohydrate binding preferences of surfactant protein D. Am J Respir Cell Mol Biol 35:84–94

Crouch E, Tu Y, Briner D et al (2006c) Ligand specificity of human surfactant protein D. J Biol Chem 280:17046–17056

Crouch E, McDonald B, Smith K et al (2007) Critical role of Arg/Lys343 in the species-dependent recognition of phosphatidylinositol by pulmonary surfactant protein D. Biochemistry 46:5160–5169

Crouch E, Hartshorn K, Horlacher T et al (2009) Recognition of mannosylated ligands and influenza A virus by human surfactant protein D: contributions of an extended site and residue 343. Biochemistry 48:3335–3345

Dahl M, Holmskov U, Husby S, Juvonen PO (2006) Surfactant protein D levels in umbilical cord blood and capillary blood of premature infants. The influence of perinatal factors. Pediatr Res 59:806–810

Davé V, Childs T, Whitsett JA (2004) Nuclear factor of activated T cells regulates transcription of the surfactant protein D gene (Sftpd) via direct interaction with thyroid transcription factor-1 in lung epithelial cells. J Biol Chem 279:34578–34588

DeSilva NS, Ofek I, Crouch EC (2003) Interactions of surfactant protein D with fatty acids. Am J Respir Cell Mol Biol 29:757–770

Dulkerian SJ, Gonzales LW, Ning Y, Ballard PL (1996) Regulation of surfactant protein D in human fetal lung. Am J Respir Cell Mol Biol 15:781–786

Fehrenbach H, Fehrenbach A, Pan T et al (2002) Keratinocyte growth factor-induced proliferation of rat airway epithelium is restricted to Clara cells in vivo. Eur Respir J 20:1185–1197

Fisher JH, Mason R (1995) Expression of pulmonary surfactant protein D in rat gastric mucosa. Am J Respir Cell Mol Biol 12:13–18

Fisher JH, Sheftelyevich V, Ho YS et al (2000) Pulmonary-specific expression of SP-D corrects pulmonary lipid accumulation in SP-D gene-targeted mice. Am J Physiol Lung Cell Mol Physiol 278: L365–L373

Fisher JH, Larson J, Cool C et al (2002) Lymphocyte activation in the lungs of SP-D null mice. Am J Respir Cell Mol Biol 27:24–33

Gjerstorff M, Madsen J, Bendixen C et al (2004) Genomic and molecular characterization of bovine surfactant protein D (SP-D). Mol Immunol 41:369–376

Gonzales LW, Angampalli S, Guttentag SH et al (2001) Maintenance of differentiated function of the surfactant system in human fetal lung type II epithelial cells cultured on plastic. Pediatr Pathol Mol Med 20:387–412

Gravholt CH, Leth-Larsen R, Lauridsen AL et al (2004) The effects of GH and hormone replacement therapy on serum concentrations of mannan-binding lectin, surfactant protein D and vitamin D binding protein in Turner syndrome. Eur J Endocrinol 150:355–362

Griese M, Maderlechner N, Bufler P (2001) Surfactant proteins D and A in sputum. Eur J Med Res 6:33–38

Griese M, Wiesener A, Lottspeich F et al (2003) Limited proteolysis of surfactant protein D causes a loss of its calcium-dependent lectin functions. Biochim Biophys Acta 1638:157–163

Groeneveld TW, Oroszlán M, Owens RT et al (2005) Interactions of the extracellular matrix proteoglycans decorin and biglycan with C1q and collectins. J Immunol 175:4715–4723

Grubor B, Meyerholz DK, Lazic T et al (2006) Regulation of surfactant protein and defensin mRNA expression in cultured ovine type II pneumocytes by all-trans retinoic acid and VEGF. Int J Exp Pathol 87:393–403

Guo CJ, Atochina-Vasserman EN, Abramova E et al (2008) S-nitrosylation of surfactant protein-D controls inflammatory function. PLoS Biol 6:e266

Guzel A, Kanter M, Aksu B et al (2009) Preventive effects of curcumin on different aspiration material-induced lung injury in rats. Pediatr Surg Int 25:83–92

Haczku A (2008) Protective role of the lung collectins surfactant protein A and surfactant protein D in airway inflammation. J Allergy Clin Immunol 122:861–879

Håkansson K, Lim NK, Hoppe HJ, Reid KB (1999) Crystal structure of the trimeric alpha-helical coiled-coil and the three lectin domains of human lung surfactant protein D. Structure 7:255–264

Haley KJ, Ciota A, Contreras JP et al (2002) Alterations in lung collectins in an adaptive allergic immune response. Am J Physiol Lung Cell Mol Physiol 282:L573–L584

Hansen S, Holmskov U (2002) Lung surfactant protein D (SP-D) and the molecular diverted descendants: conglutinin, CL-43 and CL-46. Immunobiology 205:498–517

Hansen S, Lo B, Evans K et al (2007) Surfactant protein D augments bacterial association but attenuates major histocompatibility complex class II presentation of bacterial antigens. Am J Respir Cell Mol Biol 36:94–102

Hartshorn KL, Crouch E, White MR et al (1998) Pulmonary surfactant proteins A and D enhance neutrophil uptake of bacteria. Am J Physiol 274:L958–L969

Hartshorn KL, White MR, Crouch EC (2002) Contributions of the N- and C-terminal domains of surfactant protein d to the binding, aggregation, and phagocytic uptake of bacteria. Infect Immun 70:6129–6310

He Y, Crouch E (2002) Surfactant protein D gene regulation. Interactions among the conserved CCAAT/enhancer-binding protein elements. J Biol Chem 277:19530–19537

He Y, Crouch EC, Rust K et al (2000) Proximal promoter of the surfactant protein D gene: regulatory roles of AP-1, forkhead box, and GT box binding proteins. J Biol Chem 275:31051–31060

Herbein JF, Wright JR (2001) Enhanced clearance of surfactant protein D during LPS-induced acute inflammation in rat lung. Am J Physiol Lung Cell Mol Physiol 281:L268–L277

Herías MV, Hogenkamp A, van Asten AJ et al (2007) Expression sites of the collectin SP-D suggest its importance in first line host defence: power of combining in situ hybridisation, RT-PCR and immunohistochemistry. Mol Immunol 44:3324–3332

Hilgendorff A, Schmidt R, Bohnert A et al (2005) Host defence lectins in preterm neonates. Acta Paediatr 94:794–799

Hirche TO, Crouch EC, Espinola M et al (2004) Neutrophil serine proteinases inactivate surfactant protein D by cleaving within a conserved subregion of the carbohydrate recognition domain. J Biol Chem 279:27688–27698

Hogenkamp A, Herías MV, Tooten PC et al (2007) Effects of surfactant protein D on growth, adhesion and epithelial invasion of intestinal Gram-negative bacteria. Mol Immunol 44:3517–3527

Hohwy T, Otkjaer K, Madsen J et al (2006) Surfactant protein D in atopic dermatitis and psoriasis. Exp Dermatol 15:168–174

Holmskov U, Lawson P, Teisner B et al (1997) Isolation and characterization of a new member of the scavenger receptor superfamily, glycoprotein-340 (gp-340), as a lung surfactant protein-D binding molecule. J Biol Chem 272:13743–13749

Hoover RR, Floros J (1998) Organization of the human SP-A and SP-D loci at 10q22-q23. Physical and radiation hybrid mapping reveal gene order and orientation. Am J Respir Cell Mol Biol 18:353–362

Hoppe HJ, Barlow PN, Reid KB (1994) A parallel three stranded α-helical bundle at the nucleation site of collagen triple-helix formation. FEBS Lett 344:191–195

Hortobágyi L, Kierstein S, Krytska K et al (2008) Surfactant protein D inhibits TNF-alpha production by macrophages and dendritic cells in mice. J Allergy Clin Immunol 122:521–528

Husby S, Herskind AM, Jensenius JC, Holmskov U (2002) Heritability estimates for the constitutional levels of the collectins mannan-binding lectin and lung surfactant protein D. A study of unselected like-sexed mono- and dizygotic twins at the age of 6-9 years. Immunology 106:389–394

Ikegami M, Whitsett JA, Jobe A et al (2000) Surfactant metabolism in SP-D gene-targeted mice. Am J Physiol Lung Cell Mol Physiol 279: L468–L476

Jain D, Atochina-Vasserman E et al (2007) SP-D-deficient mice are resistant to hyperoxia. Am J Physiol Lung Cell Mol Physiol 292: L861–L871

Jain D, Atochina-Vasserman EN, Tomer Y et al (2008) Surfactant protein D protects against acute hyperoxic lung injury. Am J Respir Crit Care Med 178:805–813

Jinhua L, Teh C, Kishore U, Reid KBM (2003) Collectins and ficolins: sugar pattern recognition molecules of the mammalian innate immune system. Biochim Biophys Acta 1572:387–400

Kankavi O (2003) Immunodetection of surfactant proteins in human organ of Corti, Eustachian tube and kidney. Acta Biochim Pol 50:1057–1064

Kankavi O, Ata A, Gungor O (2007) Surfactant proteins A and D in the genital tract of mares. Anim Reprod Sci 98:259–270

Kankavi O, Ata A, Celik-Ozenci C et al (2008) Presence and subcellular localizations of surfactant proteins A and D in human spermatozoa. Fertil Steril 90:1904–1910

Kerecman J, Mustafa SB, Vasquez MM et al (2008) Immunosuppressive properties of surfactant in alveolar macrophage NR8383. Inflamm Res 57:118–125

Kierstein S, Poulain FR, Cao Y et al (2006) Susceptibility to ozone-induced airway inflammation is associated with decreased levels of surfactant protein D. Respir Res 7:85

Kim JK, Kim SS, Rha KW et al (2007) Expression and localization of surfactant proteins in human nasal epithelium. Am J Physiol Lung Cell Mol Physiol 292:L879–L884

Kingma PS, Whitsett JA (2006) In defense of the lung: surfactant protein A and surfactant protein D. Curr Opin Pharmacol 6:277–283

Kingma PS, Zhang L, Ikegami M, Hartshorn K, McCormack FX, Whitsett JA (2006) Correction of pulmonary abnormalities in Sftpd−/− mice requires the collagenous domain of surfactant protein D. J Biol Chem 281:24496–24505

Kishor U, Madan T, Sarma PU et al (2002) Protective roles of pulmonary surfactant proteins, SP-A and SP-D, against lung allergy and infection caused by Aspergillus fumigatus. Immunobiology 205:610–618

Kishore U, Wang JY, Hoppe HJ, Reid KB (1996) The α-helical neck region of human lung surfactant protein D is essential for the binding of the carbohydrate recognition domains to lipopolysaccharides and phospholipids. Biochem J 318:505–511

Kishore U, Strong P, Perdikoulis MV, Reid KB (2001) A recombinant homotrimer, composed of the α-helical neck region of human surfactant protein D and C1q B chain globular domain, is an inhibitor of the classical complement pathway. J Immunol 166:559–565

Kishore U, Greenhough TJ, Waters P et al (2006) Surfactant proteins SP-A and SP-D: structure, function and receptors. Mol Immunol 43:1293–1315

Kölble K, Lu J, Mole SE, Kaluz S, Reid KB (1993) Assignment of the human pulmonary surfactant protein D gene (SFTP4) to 10q22-q23 close to the surfactant protein A gene cluster. Genomics 17:294–298

Kovacs H, O'Ddonoghue SI, Hoppe HJ et al (2002) Solution structure of the coiled-coil trimerization domain from lung surfactant protein D. J Biomol NMR 24:89–102

Kuan SF, Persson A, Parghi D, Crouch E (1994) Lectin-mediated interactions of surfactant protein D with alveolar macrophages. Am J Respir Cell Mol Biol 10:430–436

Kuroki Y, Shiratori M, Murata Y, Akino T (1991) Surfactant protein D (SP-D) counteracts the inhibitory effect of surfactant protein A (SP-A) on phospholipid secretion by alveolar type II cells. Interaction of native SP-D with SP-A. Biochem J 279:115–119

Kuroki Y, Gasa S, Ogasawara Y et al (1992) Binding specificity of lung surfactant protein SP-D for glucosylceramide. Biochem Biophys Res Commun 187:963–969

Låg M, Skarpen E, van Rozendaal BA et al (2000) Cell-specific expression of CCAAT/enhancer-binding protein delta (C/EBP delta) in epithelial lung cells. Exp Lung Res 26:383–910

Lawson PR, Perkins VC, Holmskov U, Reid KB (1999) Genomic organization of the mouse gene for lung surfactant protein D. Am J Respir Cell Mol Biol 20:953–963

Lee MY, Sørensen GL, Holmskov U et al (2009) The presence and activity of SP-D in porcine coronary endothelial cells depend on Akt/PI(3)K, Erk and nitric oxide and decrease after multiple passaging. Mol Immunol 46:1050–1057

Leth-Larsen R, Holmskov U, Højrup P (1999) Structural characterization of human and bovine lung surfactant protein D. Biochem J 343:645–652

Leth-Larsen R, Floridon C, Nielsen O, Holmskov U (2004) Surfactant protein D in the female genital tract. Mol Hum Reprod 10:149–154

Lim BL, Lu J, Reid KB (1993) Structural similarity between bovine conglutinin and bovine lung surfactant protein D and demonstration of liver as a site of synthesis of conglutinin. Immunology 78:159–165

Lu J, Wiedemann H, Timpl R, Reid KB (1993) Similarity in structure between C1q and the collectins as judged by electron microscopy. Behring Inst Mitt 93:6–16

Madan T, Reid KB, Singh M et al (2005) Susceptibility of mice genetically deficient in the surfactant protein (SP)-A or SP-D gene to pulmonary hypersensitivity induced by antigens and allergens of Aspergillus fumigatus. J Immunol 174:6943–6954

Madsen J, Kliem A, Tornoe I et al (2000) Localization of lung surfactant protein D on mucosal surfaces in human tissues. J Immunol 164:5866–5870

Malhotra R, Willis AC, Lopez Bernal A et al (1994) Mannan-binding protein levels in human amniotic fluid during gestation and its interaction with collectin receptor from amnion cells. Immunology 82:439–444

Mariencheck W, Crouch E (1994) Modulation of surfactant protein D expression by glucocorticoids in fetal rat lung. Am J Respir Cell Mol Biol 10:419–429

Marin L, Dufour ME, Nguyen TM et al (1993) Maturational changes induced by 1 α,25-dihydroxyvitamin D3 in type II cells from fetal rat lung explants. Am J Physiol Lung Cell Mol Physiol 265:L45–L52

Mason RJ, Lewis MC, Edeen KE et al (2002) Maintenance of surfactant protein A and D secretion by rat alveolar type II cells in vitro. Am J Physiol Lung Cell Mol Physiol 282:L249–L258

McAlinden A, Crouch EC, Bann JG et al (2002) Trimerization of the amino propeptide of type IIA procollagen using a 14-amino acid sequence derived from the coiled-coil neck domain of surfactant protein D. J Biol Chem 277:41274–41281

McCormack FX (2006) New concepts in collectin-mediated host defense at the air-liquid interface of the lung. Respirology 11 (Suppl):S7–S10

Meyerholz DK, Gallup JM et al (2007) Pretreatment with recombinant human vascular endothelial growth factor reduces virus replication and inflammation in a perinatal lamb model of respiratory syncytial virus infection. Viral Immunol 20:188–196

Miyamura K, Leigh LE, Lu J et al (1994a) Surfactant protein D binding to alveolar macrophages. Biochem J 300:237–242

Miyamura K, Malhotra R, Hoppe HJ, Reid KB, Phizackerley PJ, Macpherson P, López Bernal A (1994b) Surfactant proteins (SP-A) and D (SP-D): levels in human amniotic fluid and localization in the fetal membranes. Biochim Biophys Acta 1210:303–307

Mo YK, Kankavi O, Masci PP et al (2007) Surfactant protein expression in human skin: evidence and implications. J Invest Dermatol 127:381–386

Moran AP, Khamri W, Walker MM, Thursz MR (2005) Role of surfactant protein D (SP-D) in innate immunity in the gastric mucosa: evidence of interaction with Helicobacter pylori lipopolysaccharide. J Endotoxin Res 11:357–362

Mori K, Kurihara N, Hayashida S et al (2002) The intrauterine expression of surfactant protein D in the terminal airways of human fetuses compared with surfactant protein A. Eur J Pediatr 161:431–434

Motwani M, White RA, Guo N et al (1995) Mouse surfactant protein-D. cDNA cloning, characterization, and gene localization to chromosome 14. J Immunol 155:5671–5677

Nadesalingam J, Bernal AL, Dodds AW et al (2003) Identification and characterization of a novel interaction between pulmonary surfactant protein D and decorin. J Biol Chem 278:25678–25687

Nadesalingam J, Reid KB, Palaniyar N (2005) Collectin surfactant protein D binds antibodies and interlinks innate and adaptive immune systems. FEBS Lett 579:4449–4453

Nie X, Nishitani C, Yamazoe M et al (2008) Pulmonary surfactant protein D binds MD-2 through the carbohydrate recognition domain. Biochemistry 47:12878–12885

Nybo M, Andersen K, Sorensen GL et al (2007) Serum surfactant protein D is correlated to development of dementia and augmented mortality. Clin Immunol 123:333–337

Oberley RE, Goss KL, Ault KA et al (2004) Surfactant protein D is present in the human female reproductive tract and inhibits Chlamydia trachomatis infection. Mol Hum Reprod 10:861–870

Oberley RE, Goss KL, Dahmoush L et al (2005) A role for surfactant protein D in innate immunity of the human prostate. Prostate 65:241–251

Oberley RE, Goss KL, Hoffmann DS et al (2007a) Regulation of surfactant protein D in the mouse female reproductive tract in vivo. Mol Hum Reprod 13:863–868

Oberley RE, Goss KL, Quintar AA et al (2007b) Regulation of surfactant protein D in the rodent prostate. Reprod Biol Endocrinol 5:42

Ogasawara Y, Voelker DR (1995a) Altered carbohydrate recognition specificity engineered into surfactant protein D reveals different binding mechanisms for phosphatidylinositol and glucosylceramide. J Biol Chem 270:14725–14732

Ogasawara Y, Voelker DR (1995b) The role of the amino-terminal domain and the collagenous region in the structure and the function of rat surfactant protein D. J Biol Chem 270:19052–19058

Ogasawara Y, Kuroki Y, Shiratori M et al (1991) Ontogeny of surfactant apoprotein D, SP-D, in the rat lung. Biochim Biophys Acta 1083:252–256

Ogasawara Y, Kuroki Y, Akino T (1992) Pulmonary surfactant protein D specifically binds to phosphatidylinositol. J Biol Chem 267:21244–21249

Ogasawara Y, McCormack FX, Mason RJ, Voelker DR (1994) Chimeras of surfactant proteins A and D identify the carbohydrate recognition domains as essential for phospholipid interaction. J Biol Chem 269:29785–29792

Ohya M, Nishitani C, Sano H et al (2006) Human pulmonary surfactant protein D binds the extracellular domains of Toll-like receptors 2 and 4 through the carbohydrate recognition domain by a mechanism different from its binding to phosphatidylinositol and lipopolysaccharide. Biochemistry 45:8657–8664

Ooi EH, Wormald PJ, Carney AS et al (2007) Surfactant protein D expression in chronic rhinosinusitis patients and immune responses in vitro to Aspergillus and alternaria in a nasal explant model. Laryngoscope 117:51–57

Paananen R, Sormunen R, Glumoff V et al (2001) Surfactant proteins A and D in Eustachian tube epithelium. Am J Physiol Lung Cell Mol Physiol 281:L660–L667

Palaniyar N, Zhang L, Kuzmenko A et al (2002) The role of pulmonary collectin N-terminal domains in surfactant structure, function, and homeostasis in vivo. J Biol Chem 277:26971–26979

Palaniyar N, Clark H, Nadesalingam J et al (2003a) Surfactant protein D binds genomic DNA and apoptotic cells, and enhances their clearance, in vivo. Ann N Y Acad Sci 1010:471–475

Palaniyar N, Nadesalingam J, Reid KB (2003b) Innate immune collectins bind nucleic acids and enhance DNA clearance in vitro. Ann N Y Acad Sci 1010:467–470

Palaniyar N, Clark H, Nadesalingam J et al (2005) Innate immune collectin surfactant protein D enhances the clearance of DNA by macrophages and minimizes anti-DNA antibody generation. J Immunol 174:7352–7358

Pantelidis P, Lagan AL, Davies JC et al (2003) A single round PCR method for genotyping human surfactant protein (SP)-A1, SP-A2 and SP-D gene alleles. Tissue Antigens 61:317–321

Pastva AM, Wright JR, Williams KL (2007) Immunomodulatory roles of surfactant proteins A and D: implications in lung disease. Proc Am Thorac Soc 4:252–257

Persson AV, Gibbons BJ, Shoemaker JD et al (1992) The major glycolipid recognized by SP-D in surfactant is phosphatidylinositol. Biochemistry 31:12183–12189

Phelps DS (2001) Surfactant regulation of host defense function in the lung: a question of balance. Pediatr Pathol Mol Med 20:269–292

Phokela SS, Peleg S, Moya FR et al (2005) Regulation of human pulmonary surfactant protein gene expression by 1α,25 dihydroxyvitamin D_3. Am J Physiol Lung Cell Mol Physiol 289:L617–L626

Raoul W, Chailley-Heu B, Barlier-Mur AM et al (2004) Effects of vascular endothelial growth factor on isolated fetal alveolar type II cells. Am J Physiol Lung Cell Mol Physiol 286:L1293–L1301

Rehan VK, Torday JS, Peleg S et al (2002) 1α,25-dihydroxy-3-epi-vitamin D3, a natural metabolite of 1alpha,25-dihydroxy vitamin D3: production and biological activity studies in pulmonary alveolar type II cells. Mol Genet Metab 76:46–56

Reid KBM (1998) Interactions of surfactant protein D with pathogens, allergens and phagocytes. Biochim Biophys Acta 1408:290–295

Rust K, Bingle L, Mariencheck W et al (1996) Characterization of the human surfactant protein D promoter: transcriptional regulation of SP-D gene expression by glucocorticoids. Am J Respir Cell Mol Biol 14:121–130

Saitoh M, Sano H, Chiba H et al (2000) Importance of the carboxy-terminal 25 amino acid residues of lung collectins in interactions with lipids and alveolar type II cells. Biochemistry 39:1059–1066

Schagat TL, Wofford JA, Wright JR (2001) Surfactant protein A enhances alveolar macrophage phagocytosis of apoptotic neutrophils. J Immunol 166:2727–2733

Schmiedl A, Ochs M, Mühlfeld C et al (2005) Distribution of surfactant proteins in type II pneumocytes of newborn, 14-day old, and adult rats: an immunoelectron microscopic and stereological study. Histochem Cell Biol 124:465–476

Senft AP, Korfhagen TR, Whitsett JA et al (2005) Surfactant protein-D regulates soluble CD14 through matrix metalloproteinase-12. J Immunol 174:4953–4959

Senft AP, Korfhagen TR, Whitsett JA, LeVine AM (2007) Surfactant protein D regulates the cell surface expression of alveolar macrophage β2-integrins. Am J Physiol Lung Cell Mol Physiol 292:L469–L475

Shannon JM, Pan T, Nielsen LD et al (2001) Lung fibroblasts improve differentiation of rat type II cells in primary culture. Am J Respir Cell Mol Biol 24:235–244

Shima H, Guarino N, Puri P (2000) Effect of hyperoxia on surfactant protein gene expression in hypoplastic lung in nitrofen-induced diaphragmatic hernia in rats. Pediatr Surg Int 16:473–477

Shimizu H, Fisher JH, Papst P et al (1992) Primary structure of C pulmonary surfactant protein D. cDNA and deduced amino acid sequence. J Biol Chem 267:1853–1857

Shrive AK, Tharia HA, Strong P et al (2003) High-resolution structural insights into ligand binding and immune cell recognition by human lung surfactant protein D. J Mol Biol 331:509–523

Shrive AK, Martin C, Burns I et al (2009) Structural characterisation of ligand-binding determinants in human lung surfactant protein D: influence of Asp325. J Mol Biol 394:776–788

Snyder GD, Oberley-Deegan RE, Weintraub NL et al (2008) Surfactant protein D is expressed and modulates inflammatory responses in human coronary artery smooth muscle cells. Am J Physiol Heart Circ Physiol 294:H2053–H2059

Soerensen CM, Holmskov U, Aalbaek B et al (2005a) Pulmonary infections in swine induce altered porcine surfactant protein D expression and localization to dendritic cells in bronchial-associated lymphoid tissue. Immunology 115:526–535

Soerensen CM, Nielsen OL, Willis A et al (2005b) Purification, characterization and immunolocalization of porcine surfactant protein D. Immunology 114:72–82

Sørensen GL, Hjelmborg JB, Kyvik KO et al (2006) Genetic and environmental influences of surfactant protein D serum levels. Am J Physiol Lung Cell Mol Physiol 290:L1010–L1017

Sorensen GL, Hjelmborg JV, Leth-Larsen R et al (2006) Surfactant protein D of the innate immune defence is inversely associated with human obesity and SP-D deficiency infers increased body weight in mice. Scand J Immunol 64:633–638

Sorensen GL, Hoegh SV, Leth-Larsen R et al (2009) Multimeric and trimeric subunit SP-D are interconvertible structures with distinct ligand interaction. Mol Immunol 46:3060–3069

Stahlman MT, Gray ME, Hull WM, Whitsett JA (2002) Immunolocalization of surfactant protein-D (SP-D) in human fetal, newborn, and adult tissues. J Histochem Cytochem 50:651–660

Stuart LM, Henson PM, Vandivier RW (2006) Collectins: opsonins for apoptotic cells and regulators of inflammation. Curr Dir Autoimmun 9:143–161

Tino MJ, Wright JR (1999) Surfactant proteins A and D specifically stimulate directed actin-based responses in alveolar macrophages. Am J Physiol 276:L164–L174

van de Wetering JK, van Eijk M, van Golde LM et al (2001) Characteristics of surfactant protein A and D binding to lipoteichoic acid and peptidoglycan, 2 major cell wall components of gram-positive bacteria. J Infect Dis 184:1143–1151

van de Wetering JK, van Golde LM, Batenburg JJ (2004) Collectins: players of the innate immune system. Eur J Biochem 271:1229–1249

van Eijk M, Haagsman HP, Skinner T, Archibald A, Reid KB, Lawson PR (2000) Porcine lung surfactant protein D: complementary DNA cloning, chromosomal localization, and tissue distribution. J Immunol 164:1442–1450

van Eijk M, de Haan NA, Rogel-Gaillard C et al (2001) Assignment of surfactant protein A (SFTPA) and surfactant protein D (SFTPD) to pig chromosome band 14q25–>q26 by in situ hybridization. Cytogenet Cell Genet 95:114–115

van Eijk M, van de Lest CH, Batenburg JJ et al (2002) Porcine surfactant protein D is N-glycosylated in its carbohydrate recognition domain and is assembled into differently charged oligomers. Am J Respir Cell Mol Biol 26:739–747

Vandivier RW, Ogden CA, Fadok VA et al (2002) Role of surfactant proteins A, D, and C1q in the clearance of apoptotic cells in vivo and in vitro: calreticulin and CD91 as a common collectin receptor complex. J Immunol 169:3978–3986

Vayrynen O, Glumoff V, Hallman M (2004) Inflammatory and anti-inflammatory responsiveness of surfactant proteins in fetal and neonatal rabbit lung. Pediatr Res 55:55–60

von Bredow C, Hartl D, Schmid K et al (2006) Surfactant protein D regulates chemotaxis and degranulation of human eosinophils. Clin Exp Allergy 36:1566–1574

Wang JY, Yeh TF, Lin YC et al (1996) Measurement of pulmonary status and surfactant protein levels during dexamethasone treatment of neonatal respiratory distress syndrome. Thorax 51:907–913

Wang H, Head J, Kosma P et al (2008a) Recognition of heptoses and the inner core of bacterial lipopolysaccharides by surfactant protein D. Biochemistry 47:710–720

Wang L, Brauner JW, Mao G et al (2008b) Interaction of recombinant surfactant protein D with lipopolysaccharide: conformation and orientation of bound protein by IRRAS and simulations. Biochemistry 47:8103–8113

Wong CJ, Akiyama J, Allen L, Hawgood S (1996) Localization and developmental expression of surfactant proteins D and A in the respiratory tract of the mouse. Pediatr Res 39:930–937

Woodworth BA, Lathers D, Neal JG et al (2006) Immunolocalization of surfactant protein A and D in sinonasal mucosa. Am J Rhinol 20:461–465

Woodworth BA, Neal JG, Newton D et al (2007) Surfactant protein A and D in human sinus mucosa: a preliminary report. ORL J Otorhinolaryngol Relat Spec 69:57–60

Wright JR (2004) Host defense functions of pulmonary surfactant. Biol Neonate 85:326–332

Wu H, Kuzmenko A, Wan S et al (2003) Surfactant proteins A and D inhibit the growth of Gram-negative bacteria by increasing membrane permeability. J Clin Invest 111:1589–1602

Yamazoe M, Nishitani C, Takahashi M et al (2008) Pulmonary surfactant protein D inhibits lipopolysaccharide (LPS)-induced inflammatory cell responses by altering LPS binding to its receptors. J Biol Chem 283:35878–35888

Yano T, Mason RJ, Pan T et al (2000) KGF regulates pulmonary epithelial proliferation and surfactant protein gene expression in adult rat lung. Am J Physiol Lung Cell Mol Physiol 279:L1146–L1158

Zhang L, Ikegami M, Crouch EC et al (2001a) Activity of pulmonary surfactant protein-D (SP-D) in vivo is dependent on oligomeric structure. J Biol Chem 276:19214–19219

Zhang P, McAlinden A, Li S et al (2001b) The amino-terminal heptad repeats of the coiled-coil neck domain of pulmonary surfactant protein d are necessary for the assembly of trimeric subunits and dodecamers. J Biol Chem 276:19862–19870

Zhao XM, Wu YP, Wei R et al (2007) Plasma surfactant protein D levels and the relation to body mass index in a chinese population. Scand J Immunol 66:71–76

Part IX
C-Type Lectins: Selectins

L-Selectin (CD62L) and Its Ligands

G.S. Gupta

26.1 Cell Adhesion Molecules

Cell adhesion molecules are glycoproteins expressed on the cell surface and play an important role in inflammatory as well as neoplastic diseases. There are four main groups: the integrin family, the immunoglobulin superfamily, selectins, and cadherins. The integrin family has eight subfamilies, designated as β1 through β8. The most widely studied subfamilies are β1 (CD29, very late activation [VLA] members), β2 (leukocyte integrins such as CD11a/CD18, CD11b/CD18, CD11c/CD18, and αd β2), β3 (CD61, cytoadhesions), and β7 (α4 β7 and αE β7). The immunoglobulin superfamily includes leukocyte function antigen-2 (LFA-2 or CD2), leukocyte function antigen-3 (LFA-3 or CD58), intercellular adhesion molecules (ICAMs), vascular adhesion molecule-1 (VCAM-1), platelet-endothelial cell adhesion molecule-1 (PE-CAM-1), and mucosal addressin cell adhesion molecule-1 (MAdCAM-1). The selectin family includes L-selectin (CD62L), P-selectin (CD62P), and E-selectin (CD62E). Cadherins are major cell-cell adhesion molecules and include epithelial (E), placental (P), and neural (N) subclasses. The binding sites (ligands/receptors) are different for each of these cell adhesion molecules (e.g., ICAM binds to CD11/CD18; VCAM-1 binds to VLA-4). The specific cell adhesion molecules and their ligands that may be involved in pathologic conditions and potential therapeutic strategies by modulating the expression of these molecules have been discussed (Elangbam et al. 1997). The main classes of adhesion molecule involved in lymphocyte interactions are the selectins, the integrins, members of the Ig superfamily, and some mucinlike molecules. Most adhesion molecules play fairly broad roles in the generation of immune responses. The three selectins act in concert with other cell adhesion molecules (e.g., intracellular adhesion molecule (ICAM-1), vascular cell adhesion molecule-1 (VCAM-1), and leukocyte integrins (Springer 1990; Shimizu et al. 1992) to effect adhesive interactions of leukocytes, platelets, and endothelial cells. Hence selectins which belong to C-type lectins family are reviewed in Chaps. 26–28

26.1.1 Selectins

The selectin family of lectins consists of three closely related cell-surface molecules with differential expression by leukocytes (L-selectin), platelets (P-selectin), and vascular endothelium (E- and P-selectin). Structural identity of a selectins resides in its unique domain composition (Fig. 26.1). E-, P-, and L-selectin are >60% identical in their NH2 terminus of 120 amino acids, which represent the lectin domain (Siegelman et al. 1989; Lasky et al. 1989; Johnston et al. 1989; Bevilacqua et al. 1989; Camerini et al. 1989; Tedder et al. 1989). The ligands (counter structures) of selectins are sialylated and fucosylated carbohydrate molecules which, in most cases, decorate mucin-like glycoprotein membrane receptors. Their common structure consists of an N-terminal Ca^{2+}-dependent lectin-type domain, an epidermal growth factor (EGF)-like domain, multiple short consensus repeat (SCR) domains similar to those found in complement regulatory proteins, a transmembrane region, and a short cytoplasmic C-terminal domain. Together this arrangement results in an elongated structure which projects from the cell surface, ideal for initiating interactions with circulating leucocytes. The lectin domain forms the main ligand binding site, interacting with a carbohydrate determinant typified by fucosylated, sialylated, and usually sulphated glycans such as sialyl Lewis X (s-Le^X). The EGF domain may also play a role in ligand recognition. The multiple SCR domains (two for L-selectin, six for

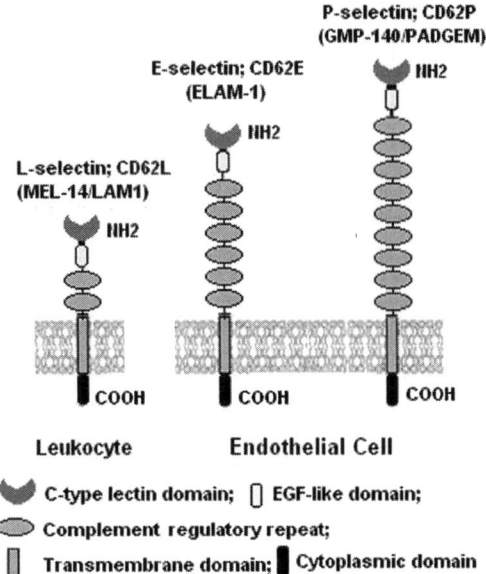

Fig. 26.1 Domain composition of the three known human selectins. The extracellular portion of each selectin contains an amino terminal domain homologous to C-type lectins and an adjacent epidermal growth factor-like domain. The single EGF element that is found in each selectin is followed by a varying number of repetitive elements, each ~ 60 amino acids long, which resemble protein motives found in complement regulatory proteins and a transmembrane sequence. A short cytoplasmic sequence is at the carboxyl terminus of each selectin. The number of amino acids present in the mature proteins as deduced from the cDNA sequences are: L-selectin, 385; P-selectin, 789; and E-selectin, 589

26.2 Leukocyte-Endothelial Cell Adhesion Molecule 1 (LECAM-1) (L-Selectin/CD62L or LAM-1)

26.2.1 Leukocyte-Endothelial Cell Adhesion Molecule 1 in Humans

Lymphocyte trafficking is a fundamental aspect of the immune system that allows B and T lymphocytes with diverse antigen recognition specificities to be exposed to various antigenic stimuli in spatially distinct regions of an organism. A lymphocyte adhesion molecule that is involved with this trafficking phenomenon has been termed the homing receptor. The Leukocyte-endothelial cell adhesion molecule 1 (LECAM-1) (L-selectin/MEL-14) or leukocyte adhesion molecule-1 (LAM-1, TQ1, Leu-8), a cell surface component, expressed by human lymphocytes, neutrophils, monocytes, and their precursors, is the member of the selectin family of cellular adhesion/homing receptors which play important roles in leukocyte-endothelial cell interactions. LECAM-1 mediates leukocyte rolling and leukocyte adhesion to endothelium at sites of inflammation. The migration of naive T cells into lymphoid tissues is mediated by the chemokine SLC (secondary lymphoid tissue chemokine), which is expressed by the high vascular endothelium, stromal cells, and dendritic cells in lymphoid tissue, and binds to the CCR7 chemokine receptor on naive T cells. In addition, L-selectin mediates the binding of lymphocytes to HEV of PLN, which is an essential process in lymphocyte recirculation and hence also known as PLN homing receptor. Metabolic studies revealed that thymocytes synthesize markedly less L-selectin than do thoracic duct lymphocytes (TDL) or LN lymphocytes. However, Northern blot studies indicated that thymocytes possess more L-selectin RNA than do TDL. These results provide evidence that post-transcriptional events contribute to regulation of L-selectin expression in thymocytes.

26.2.1.1 Lymphocyte-Associated Molecule (LAM-1) and L-Selectin Are Homologous

A cDNA encoding a human lymphocyte cell surface molecule named lymphocyte-associated molecule (LAM-1) composed of multiple distinct domains, one domain homologous with animal lectins was identified by Tedder et al. (1989). This cDNA clone hybridized with RNAs found in B cell lines and T lymphocytes, but not with RNA from other cell types. The amino acid sequence of LAM-1 is 77% homologous with the sequence of the mouse lymphocyte homing receptor, suggesting that LAM-1 may function in human lymphocyte adhesion. The LAM-1 gene is located on chromosome 1q23–25, as is another member of this adhesion family, suggesting that this new family

E-selectin, and nine for P-selectin) probably act as spacer elements, ensuring optimum positioning of the lectin and EGF domains for ligand interaction. The EGF repeats have comparable sequence similarity. Each complement regulatory-like module is 60 amino acids in length and contains six cysteinyl residues capable of disulfide bond formation. This feature distinguishes the selectin modules from those found in complement binding proteins, such as complement receptors 1and 2, which contain four cysteines (Hourcade et al. 1989).

The selectins cell-surface receptors play a key role in the initial adhesive interaction between leukocytes and endothelial cells at sites of inflammation (Fig. 26.2). Activation of endothelial cells (EC) with different stimuli induces the expression of E- and P-selectins, and other adhesion molecules (ICAM-1, VCAM-1), involved in their interaction with circulating cells. Lymphocytes home to peripheral lymph nodes (PLNs) via high endothelial venules (HEVs) in the subcortex and incrementally larger collecting venules in the medulla. HEVs express ligands for L-selectin, which mediates lymphocyte rolling.

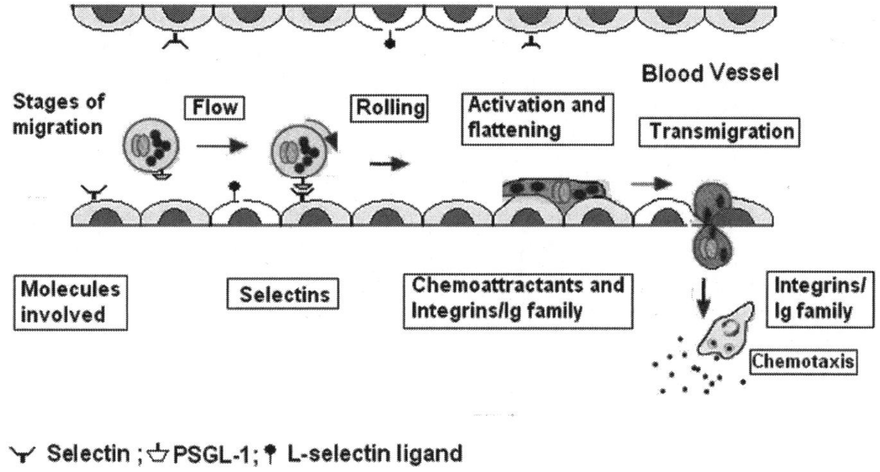

Fig. 26.2 Schematic outline of the stages involved in leucocyte migration through post-capillary venular endothelium. Leucocytes flowing in the vein become tethered to the vessel wall, a step mediated largely by the selectin family of adhesion receptors. The leucocytes then roll along the vessel wall until they come into contact with a chemoattractant and become activated. Activation leads to engagement of members of the integrin family of receptors binding to immunoglobulin (Ig) family adhesion receptors on the endothelium. This causes them to become more firmly attached, stop rolling, flatten on the surface of the endothelial cell, and transmigrate

of proteins may be encoded by a cluster of "adhesion protein" loci. The human homologue of murine peripheral lymph node-specific receptor is highly homologous to the murine receptor in overall sequence that may be involved with human lymphocyte homing. The extracellular region of the human receptor contained an NH2-terminally located carbohydrate binding domain followed by an EGF-like domain and a domain containing two repeats of a complement binding protein/motif (CRP). Interestingly, the human receptor showed a high degree of sequence homology to endothelial cell adhesion molecule ELAM (Bowen et al. 1989; Camerini et al. 1989). Genomic analysis of human DNA suggests a low-copy gene under high-stringent conditions. The nucleotide sequence predicts a mature protein of 334 amino acids, identical in length to mouse lymph node homing receptor core protein (mLHR) (Siegelman and Weissman 1989). The cDNA encoding baboon L-selectin predicts a protein of 372-aa, which is 95% identical to that of human L-selectin (Tsurushita et al. 1996). The carbohydrate binding domain is apparently involved in the adhesive interaction between murine lymphocytes and PLN endothelium. The lectin and EGF-like regions are the most homologous, while CRP domains are less conserved between species. The two CRP units in hLHRc are distinct from those in mLHRc in that they are homologous to one another rather than identical, suggesting strong pressure for maintenance of two repeats in this molecule (Siegelman and Weissman 1989).

26.2.2 Gene Structure of L-Selectin

The leukocyte adhesion molecule-1 (LAM-1) or L-selectin gene expressed by human lymphocytes, neutrophils, monocytes, and their precursors spans greater than 30 kbp of DNA and is composed of at least 10 exons. The 5′ end of LAM-1 mRNA revealed a single initiation region for transcription. Exons II through X contain translated sequences; exon II encodes the translation initiation codon; exon III, the leader peptide; IV, the lectin-like domain; V, the epidermal growth factor-like domain; VI and VII, the short consensus repeat units; exon VIII, the transmembrane region; exon IX encodes seven amino acids containing a potential phosphorylation site; and exon X encodes the five remaining amino acids of the cytoplasmic tail and the long 3′ untranslated region. Sequences of *Lam-1* cDNA clones and protein expressed by neutrophils are identical to cDNA sequences and the protein expressed by lymphocytes. Therefore, the usage of exons II through X results in the generation of a single major LAM-1 protein product expressed by lymphocytes and neutrophils (Ord et al. 1990; Watson et al. 1990).

26.2.3 Murine PLN Homing Receptor/mLHR3

Characterization of the cDNAs encoding the murine lymphocyte homing receptor (mLHR) has revealed mosaic structure containing three well-known protein motifs: a

C-type lectin domain, an epidermal growth factor-like domain, and two exact copies of a short consensus repeat sequence homologous to those found in a family of complement regulatory proteins (CRP), in addition to a signal sequence, a transmembrane anchor, and a short cytoplasmic tail. The receptor molecule is potentially highly glycosylated, and contains an apparent transmembrane region. Analysis of mRNA transcripts reveals a predominantly lymphoid distribution in direct relation to the cell surface expression of the MEL-14 determinant, and the cDNA clone is shown to confer the MEL-14 epitope in heterologous cells. These features, including ubiquitination, embodied in this single receptor molecule form the basis for numerous approaches to the study of cell-cell interactions (Lasky et al. 1989; Siegelman et al. 1989). Characterization of genomic clones encoding mLHR gene revealed a high degree of correlation between various structure/function motifs and exons that specify them. Comparison of the exons encoding the two identical copies of the complement regulatory motif revealed that short intronic regions 5′ and 3′ of these exactly repeated exons are also identical. The mLHR core peptide locus is localized to chromosome 1, the portion syntenic with chromosome 1 in man. The gene on chromosome 1 is very near to a site that contains the genes for the family of complement regulatory proteins which encode short consensus repeats similar to those found in the homing receptor (Dowbenko et al. 1991; Siegelman et al. 1990).

The partial sequencing of a rat cDNA revealed a putative peptide of 372 aa, including a signal peptide of 38 aa. The protein has three tandem domains: a lectin domain, an EGF-like domain and two repeats of complement regulatory proteins (CRP domain). The lectin binding domain has 93.2% and 81.4% and the EGF-like domain has 85.3% and 76.5% aa identity with those of mouse and human, respectively. In the CRP repeat domain, the amino acid identity was 72.6% between human and rat and 71.8% between mouse and rat. The main transcript is of about 3 kb in peripheral blood mononuclear cells (PBMC), spleen and thymus. Sackstein et al. (1995) identified the complete coding sequence of 105-bp 5′-untranslated region and a 359-bp 3′-untranslated region. Expression studies gave evidence that cDNA represented rat L-selectin. The cDNA encoding rabbit L-Selectin encodes a peptide of 377 aa, including a signal peptide of 38 amino acids. Sequence analysis demonstrated extensive homology with L-Selectins from other species (Qian et al. 2001). The bovine L-selectin is predicted amino acid sequence with an overall high identity to that of human and murine L-selectin. However, the cytoplasmic tail of bovine L-selectin showed little similarity to that of human and murine L-selectin (Bosworth et al. 1993).

26.3 Functions of L-Selectin

The selectin family of adhesion molecules (E-, P- and L-selectins) is involved in leukocyte recruitment to the sites of inflammation and tissue damage. In contrast to most other adhesion molecules, selectin function is restricted to leukocyte interactions with vascular endothelium. Multiple studies indicate that the selectins mediate neutrophil, monocyte, and lymphocyte rolling along the venular wall. The generation of selectin-deficient mice has confirmed these findings and provided further insight into how the overlapping functions of these receptors regulate inflammatory processes. These proteins also play a critical part in the subsequent interactions of lymphocytes with antigen-presenting cells and later with their target cells (da Costa Martins et al. 2004). L-selectin dimerization enhances tether formation to properly spaced ligand. Ezrin-radixin-moesin proteins are required for microvillar positioning of L-selectin and that this is important both for leukocyte tethering and L-selectin shedding (Ivetic and Ridley 2004; Ivetic et al. 2004). L-selectin is involved not only in leukocyte tethering and rolling, but also in leukocyte activation.

26.3.1 Lymphocyte Homing and Leukocyte Rolling and Migration

Selectins participate in leukocyte homing to particular tissues, and can be expressed either on leukocytes (L-selectin, CD62L) or on vascular endothelium (Fig. 26.2). L-Selectin is expressed on naive T cells and guides their exit from blood into peripheral lymphoid tissues. The interaction between L-selectin and the vascular addressins is responsible for the specific homing of naive T cells to lymphoid organs but does not, on its own, enable the cell to cross the endothelial barrier into the lymphoid tissue. Studies have shown that inhibition of selectin function can ameliorate a range of inflammatory processes, offering the possibility that antagonists of selectin function may be useful in the treatment of inflammatory lung diseases such as asthma (Symon and Wardlaw 1996). Experiments on L-sel$^{-/-}$ mice showed that L-sel$^{-/-}$ leukocytes were severely impaired in their ability to respond to a directional cue. These findings indicate that L-selectin is important in enabling leukocytes to respond effectively to chemotactic stimuli in inflamed tissues (Hickey et al. 2000). L-selectin, ICAM-1, and both β1- and β-integrins seem to function synergistically to mediate optimal leukocyte rolling and entry into tissues, which is essential for the generation of effective inflammatory responses in vivo (Steeber et al. 1999; Diacovo et al. 2005). Lymphocyte homing and leukocyte rolling and migration are impaired in L-Sel$^{-/-}$ mice. Whether L-selectin-mediated rolling can promote leukocyte adhesion in vivo independent of P- and E-selectin, Jung et al. (1998) concluded that L-selectin can mediate rolling that results in sufficient leukocyte recruitment to account for the robust inflammatory response seen in E$^-$/P$^-$ mice at later times.

There are numerous in vitro studies that demonstrate that engagement of L-selectin leads to the activation of several signaling pathways potentially contributing to subsequent adhesion, emigration, or even migration through the interstitium. Ligation of L-selectin through conserved regions within the lectin domain activates signal transduction pathways and integrin function in human, mouse, and rat leukocytes (Steeber et al. 1997). L-selectin signal transduction requires protein kinase Cα, ι, and θ binding (Kilian et al. 2004). This actually induces cellular events in vivo.

L-Selectin-Mediated Lymphocyte Rolling Is Regulated by Endothelial Chemokines: CXCR2- and E-selectin-induced neutrophils arrest during inflammation in vivo (Smith et al. 2004). L-selectin mobilized intracellular CXCR4 significantly increases surface CXCR4 stimulation and inhibited SDF-1 induced CXCR4 internalization (Ding et al. 2003). Engagement of both CD4 and CXCR4 is required for HIV-induced shedding of L-selectin on primary resting CD4$^+$ T cells (Wang et al. 2004).

Intracellular Mechanisms of L-Selectin Induced Capping: Capping of surface receptors is an ubiquitous mechanism but least understood. Junge et al. (1999) demonstrated that L-selectin triggering results in receptor capping of the L-selectin molecules in lymphocytes. This process involves intracellular signaling molecules. L-Selectin capping seems to be independent on activation of p56lck-kinase, but requires the neutral sphingomyelinase, small G proteins and the cytoskeleton. Capping of L-selectin upon stimulation might play an important role in the very early phase of lymphocyte trafficking.

26.3.2 Immune Responses

The mouse strains, deficient in selectin leukocyte adhesion receptors, have been very informative in finding the roles of cell adhesion molecules in leukocyte-endothelium interaction and produced some surprises.

26.3.2.1 Impaired Primary T Cell Responses in L-Sel$^{-/-}$ mice

L-selectin-deficient mice show impaired leukocyte recruitment into inflammatory sites and impaired primary T cell responses. L-selectin-deficient mice are defective in cutaneous delayed-type hypersensitivity (DTH) responses when tested after conventional intervals of immunization (4d). L-selectin plays an important role in the generation of primary T cell responses but may not be essential for humoral and memory T cell responses (Tedder et al. 1995). Lymphocyte migration in L-selectin-deficient mice is due to altered subset migration and aging of the immune system (Steeber et al. 1996). Circulating lymphocyte numbers or subpopulations were not altered in young L-Sel$^{-/-}$ mice, but circulating monocyte numbers were increased nearly threefold. In contrast, older L-Sel$^{-/-}$ mice had disproportionate increases of both naive and memory CD4$^+$ T cells present within spleen and blood. Results and the finding that memory lymphocytes in wild-type mice expressed L-selectin demonstrated a requirement for L-selectin in the regulation of memory lymphocyte migration. Therefore, L-selectin-dependent pathways of lymphocyte migration are important for the normal migration of both naive and memory lymphocytes (Steeber et al. 1996). Selectin deficient mice show impaired development of mucosal immunity in the gut-associated lymphoreticular tissue (Csencsits and Pascual 2002; Pascual et al. 2001; Csencsits and Pascual 2002). Using allogeneic skin graft rejection as a model of cutaneous inflammation, L-Sel$^{-/-}$ mice rejected both primary and secondary allogeneic skin grafts significantly more slowly than L-selectin$^{+/+}$ littermates. The delayed rejection of skin grafts by L-Sel$^{-/-}$ mice reflects impaired migration of effector cells into the graft rather than delayed or impaired generation of a CTL response (Tang et al. 1997).

Perhaps L-selectin is required on effector cells for local interactions in the CNS to cause myelin damage in experimental allergic encephalomyelitis as implied by the expression of ICAM-1, VCAM-1, L-selectin, and leukosialin in mouse CNS during the induction and remission stages of experimental allergic encephalomyelitis (Dopp et al. 1994; Grewal et al. 2001).

26.3.2.2 Humoral Immune Responses in L-Sel$^{-/-}$ mice

The altered distribution of lymphocyte subpopulations in L-Sel$^{-/-}$ mice resulted in significantly elevated serum IgM and IgG1 levels and augmented humoral immune responses to T cell-independent and T cell-dependent Ags following i.p. immunization. By contrast, s.c. immunization of L-selectin-deficient mice with a T cell-dependent Ag resulted in serum IgM responses that were lower when compared with wild-type littermates on d 7, while IgG responses were absent. Most serum Ig responses were normal by d 14 and secondary responses were higher in L-selectin-deficient mice. These results indicate that lymphocyte migration plays an important role in the initiation of humoral responses and demonstrate complementary and overlapping roles for the spleen and other peripheral lymphoid tissues in the generation of immune responses (Steeber et al. 1996).

26.3.2.3 CD62L$^+$ Sub-Population of Regulatory T Cells Protect from Lethal Acute GVHD

The CD4$^+$CD25$^+$ T$_{reg}$ cells are important in the regulation of immune responses in allogeneic bone marrow (BM) and solid organ transplantation. T$_{reg}$ cells are recognized for their critical role in induction and maintenance of self-tolerance and prevention of autoimmunity. Reports indicate that L-Sel^{hi} T$_{regs}$ interfere with the activation and expansion of GVHD effector T cells in secondary lymphoid organs early after BM transplantation. Ermann et al. (2005) examined the differential effect of CD62L$^+$ and CD25$^-$ subsets of CD4$^+$CD25$^+$ T$_{reg}$ cells on aGVHD-related mortality. Both subpopulations showed the characteristic features of CD4$^+$CD25$^+$ T$_{reg}$ cells in vitro and did not induce acute GVHD in vivo. However, in cotransfer with donor CD4$^+$CD25$^-$ T cells, only the CD62L$^+$ subset of CD4$^+$CD25$^+$ T$_{reg}$ cells prevented severe tissue damage to the colon and protected recipients from lethal aGVHD (Ermann et al. 2005).

26.3.2.4 Human CD62L$^-$ Memory T Cells Are Less Responsive to Alloantigen

Human memory (CD45RO+) CD4+ T cells can be distinguished into two subpopulations on the basis of expression of the lymph node homing receptor, L-selectin. Human L-selectin-positive memory T-helper (Th) cells promote the maturation of IgG- and IgA-producing cells by naive B cells. Results suggest that the human L-selectin-negative and -positive subpopulations of human memory CD4$^+$ T cells contain Th1-like and Th2-like cytokine-producing cells, respectively (Kanegane et al. 1996). Memory T cells (CD62L$^-$) represent a population of T cells that have previously encountered pathogens and may contain fewer T cells capable of recognizing neoantigens including recipient alloantigen (aAg). Human naive (CD62L$^+$) or memory (CD62L$^-$) T cells have different capacities to respond to aAg (Foster et al. 2004). Physical fitness affects immune responses to a psychological but not a physical stressor (Hong et al. 2004).

26.3.2.5 Structural Requirements for Release of L-selectin

The cytoplasmic domain of L-selectin regulates leukocyte adhesion to endothelium independent of ligand recognition, by controlling cytoskeletal interactions and/or receptor avidity (Kansas et al. 1993). This was suggested by the strong amino acid sequence conservation of the cytoplasmic domain of L-selectin between humans and mice. L-selectin mediates leukocyte rolling on vascular endothelium at sites of inflammation and lymphocyte migration to PLN. L-selectin is rapidly shed from the cell surface after leukocyte activation by a proteolytic mechanism that cleaves the receptor in a membrane proximal extracellular region. The release of L-selectin is likely to be regulated by the generation of an appropriate tertiary conformation within the membrane-proximal region of the receptor, which allows recognition by a membrane-bound endoprotease with relaxed sequence specificity that cleaves the receptor at a specific distance from the plasma membrane. Studies suggest a generalized protein-processing pathway involved in the endoproteolytic release of specific transmembrane proteins which harbor widely differing primary sequences at/or neighboring their cleavage sites (Chen et al. 1995).

Effect of shedding of L-selectin: L-selectin shedding does not regulate constitutive T cell trafficking but controls the migration pathways of antigen-activated T lymphocytes. In other words, L-selectin shedding from antigen-activated T cells prevents re-entry into peripheral lymph nodes (Galkina et al. 2003). Expression of the L-selectin can be rapidly down-modulated by regulated proteolysis at a membrane-proximal site. Evidence suggested that the L-selectin secretase activity might involve a cell surface, zinc-dependent metalloprotease and might be related to the activity involved in processing of membrane-bound TNF-α (Feehan et al. 1996). In addition, the cytoplasmic domain of L-selectin may regulate shedding by a mechanism in which bound calmodulin may operate as a negative effector (Matala et al. 2001). Studies show that TNF-α-converting enzyme (TACE) is involved in the shedding of L-selectin by NSAIDS in human neutrophils (Gomez-Gaviro et al. 2002).

Role of Circulating L-Selectin: Leukocyte activation along inflamed vascular endothelium induces rapid endoproteolytic cleavage of L-selectin from the cell surface, generating soluble L-selectin (sL-selectin). Comparable with humans, sL-selectin is present in adult mouse sera at approximately 1.7 µg/ml. Adhesion molecule-deficient mice prone to spontaneous chronic inflammation and mice suffering from leukemia/lymphoma had increased serum sL-selectin levels. By contrast, serum sL-selectin levels were reduced in Rag-deficient mice lacking mature lymphocytes. The majority of serum sL-selectin had a molecular mass of 65–75 kDa, consistent with its lymphocyte origin. sL-selectin influences lymphocyte migration in vivo and that the increased sL-selectin levels present in certain pathologic conditions may adversely affect leukocyte migration (Tu et al. 2002).

26.4 L-Selectin: Carbohydrate Interactions

The binding specificity of L-selectin towards structurally defined sulphated oligosaccharides of the blood group Le^a and Le^x series, and of the glycosaminoglycan series heparin, chondroitin sulphate and keratan sulphate has been evaluated in several studies. L-selectin interacts with oversulfated chondroitin/dermatan sulfates containing GlcAβ1/IdoAα1-3GalNAc (4, 6-O-disulfate) (Kawashima et al. 2002). Leid et al. (2002) provided a comprehensive comparison of the L-selectin/cytoskeletal interaction with other functionally important surface antigens.

26.4.1 Glycan-Dependent Leukocyte Adhesion

Each selectin recognizes related but distinct counter-receptors displayed by leukocytes and/or the endothelium. These counter-receptors correspond to specific glycoproteins whose 'activity' is enabled by carefully controlled post-translational modifications. The characterization of the glycans associated with E- and P-selectin counter-receptors, and of mice with targeted deletions of glycosyltransferase and sulfotransferase genes, disclose that neutrophil E- and/or P-selectin counter-receptor activities derive, minimally, from essential synthetic collaborations amongst polypeptide N-acetylgalactosaminyl- transferase(s), a β-N-acetylglucosaminyltransferase that assembles core-2-type O-glycans, β-1,4-galactosyltransferase(s), protein tyrosine sulfotransferase(s), α-2,3-sialyltransferases, and a pair of α-1,3 fucosyltransferases (Lowe 2003).

26.4.1.1 Specificity of L-Selectin Towards Sulphated Oligosaccharides

L-selectin binds to 3′-sialyl-Le^x and –Le^a and to 3′-sulfo-Le^x and –Le^a sequences. Sulphated blood group Le^a/Le^x type sequences, with sulphate at the 3-position of galactose, have emerged as potent ligands for the E- and L-selectins. The recombinant soluble form of rat L-selectin (L-selectin-IgG Fc chimera) was shown to bind to lipid-linked oligosaccharides 3-O, 4-O and 6-O sulphated at galactose, such as sulphatides and a mixture of 3-sulphated Le^a/Le^x type tetrasaccharides isolated from ovarian cystadenoma, as well as to the HNK-1 glycolipid with 3-O sulphated glucuronic acid. The L-selectin was found to bind to the individual 3-sulphated Le^a and Le^x sequences (penta-, tetra- and trisaccharides), and with somewhat lower intensities to their non-fucosylated analogues. Green et al. (1995) proposed that the binding of the lymphocyte membrane L-selectin to endothelial glycosaminoglycans may provide a link between the selectin-mediated and integrin-mediated adhesion systems in leukocyte extravasation cascades. The possibility is also raised that lymphocyte L-selectin interactions with glycosaminoglycans may contribute to pathologies of glycosaminoglycan-rich tissues, e.g. cartilage loss in rheumatoid arthritis and inflammatory lesions of the cornea.

Conformational studies of selectins with NMR of the sulphated (Su) Le^a in comparison with the non-sulphated analogue, which is less strongly bound by E-selectin and not at all by L-selectin, revealed NMR parameters, which are in agreement for binding of both molecules. Molecular dynamics calculations for SuLe^a lead to the conclusion that the conformation of SuLe^a approximates to a single-rigid structure, as for Le^a molecule. Comparison of experimentally and theoretically obtained parameters for SuLe^a with those for the non-sulphated Le^a molecule indicated no detectable changes in the three-dimensional structure of the trisaccharide upon sulphation. Thus, the enhanced selectin binding to the sulphated Le^a is most likely due to favourable electrostatic interactions between the charged sulphate group and corresponding charged groups on the selectin protein (Kogelberg and Rutherford 1994).

Sialyl-Lewis^x Sequence 6-O-Sulfated at N-Acetylglucosamine as Ligand for L-Selectin: Oligosaccharide sequences based on sialyl-Le^x with 6-O-sulfation at galactose (6′-sulfo) or at N-acetylglucosamine (6-sulfo) and expressed on high endothelial venules are considered likely endogenous ligands for L-selectin. In the course of high performance TLC of three hexaglycosylceramides 6′-sulfo sialyl Le^x, 6-sulfo sialyl Le^x, and 6′,6-bis-sulfo sialyl Le^x, synthesized chemically for selectin recognition studies, two minor byproducts were detected and isolated from each parent compound. These were identified as isomers containing a de-N-acetylated sialic acid or having a modified carboxyl group. Binding experiments with the parent compounds and the non-sulfated sialyl Le^x glycolipid show that 6-sulfation potentiates, whereas 6′-sulfation virtually abolishes L-selectin binding. Whereas modification of the sialic acid carboxyl group markedly impaired L-selectin binding, de-N-acetylation resulted in enhanced binding. The natural occurrence on high endothelial venules of this 'super-active' de-N-acetylated form of 6-sulfo sialyl Le^x, and related structures, now deserves investigation (Galustian et al. 1997). The binding to 3′-sialyl-Le^x is strongly affected by the presence of 6-O-sulfate as found on oligosaccharides of the counter receptor, GlyCAM-1; 6-O-sulfate on the N-acetylglucosamine (6-sulfation) enhances, whereas 6-O-sulfate on the galactose (6′-sulfation) virtually abolishes binding. Interactions with 3′-sulfo-Le^x based on the Le^x pentasaccharide sequence also showed that the 6-sulfation enhances where as 6′-sulfation suppresses L-selectin binding. Thus, for synthetic strategies to design therapeutic oligosaccharide analogs as antagonists of L-selectin binding, those based on the simpler 3′-sulfo-Le^x (and also the 3′-sulfo-Le^a) would be most appropriate (Galustian et al. 1999, 2002).

Endothelial sLe^a and sLe^x During Cardiac Transplant Rejection: Acute organ transplant rejection is characterized by a heavy lymphocyte infiltration. Alterations in the graft endothelium result into increased lymphocyte traffic into the graft. Lymphocytes adhere to the endothelium of rejecting cardiac transplants, but not to the endothelium of syngeneic grafts or normal hearts. Concomitant with the enhanced lymphocyte adhesion, the cardiac endothelium begins to de novo express sialyl Le^a and sialyl Le^x epitopes, which overlap to the sequences of L-selectin counter-receptors. The endothelium of allografts, but not that of syngeneic grafts or normal controls, also reacted with the L-selectin-immunoglobulin G fusion protein, giving proof of inducible L-selectin counter-receptors (Turunen et al. 1995).

Heparin Sulfate Proteoglycans as L-Selectin Ligand: Heparin has been used clinically as an anticoagulant and antithrombotic agent for over 60 years. Calcium-dependent, heparin-like L-selectin ligands had been noticed in cultured bovine endothelial cells (Norgard-Sumnicht et al. 1993). Subsequently, these ligands were identified as heparin sulfate proteoglycans (HSPGs) associated either with the cultured cells or secreted into the medium and extracellular matrix. Studies showed that HSPGs from cultured endothelial cells, which can bind to L-selectin are enriched with unsubstituted amino groups on their GAG chains (Norgard-Sumnicht and Varki 1995). Presentltly, known L-selectin ligands include sulfated Lewis-type carbohydrates, glycolipids, and proteoglycans. L-selectin binding chondroitin/dermatan sulfate proteoglycans are present in cartilage, whereas L-selectin binding heparan sulfate proteoglycans are present in spleen and kidney. L-selectin binds only a subset of renal heparan sulfates, attached to a collagen type XVIII protein backbone and predominantly present in medullary tubular and vascular basement membranes. L-selectin does not bind other renal heparan sulfate proteoglycans such as perlecan, agrin, and syndecan-4, and not all collagen type XVIII expressed in the kidney bind L-selectin. This indicates that there is a specific L-selectin binding domain on heparan sulfate glycosaminoglycan chains. Based on the model of heparan sulfate domain organization, Celie et al. (2005) proposed a model for the interaction of L-selectin with heparan sulfate glycosaminoglycan chains. This opens the possibility of active regulation of L-selectin binding to heparan sulfate proteoglycans, e.g. under inflammatory conditions (Celie et al. 2005).

Heparan Sulfate Deficiency Impairs L-Selectin-Mediated Neutrophil Trafficking: The potent anti-inflammatory property of heparin results primarily from blockade of P-selectin and L-selectin. Mice deficient in P- or L-selectins showed impaired inflammation, which could be further reduced by heparin. However, heparin had no additional effect in mice deficient in both P- and L-selectins. The sulfate groups at C6 on the glucosamine residues play a critical role in selectin inhibition. Such analogs may prove useful as therapeutically effective inhibitors of inflammation (Wang et al. 2002). Endothelial heparan sulfate deficiency impairs L-selectin- and chemokine-mediated neutrophil trafficking during inflammatory responses. L-selectin binding heparan sulfates attached to collagen type XVIII have been identified and a model for the interaction of L-selectin with heparan sulfate glycosaminoglycan chains has been proposed. Endothelial heparan sulfate is involved in the inflammation by inactivating N-acetyl glucosamine N-deacetylase-N-sulfotransferase-1 in endothelial cells and leukocytes, which is required for the addition of sulfate to the heparin sulfate chains. Mutant mice developed normally but showed impaired neutrophil infiltration in various inflammation models. Endothelial heparan sulfate has three functions in inflammation: by acting as a ligand for L-selectin during neutrophil rolling; in chemokine transcytosis; and by binding and presenting chemokines at the lumenal surface of the endothelium (Wang et al. 2005).

Some L-Selectin Ligand Not Recognized by MECA-79 in PLNs: In contrast to PLN addressin, which is recognized by mAb MECA-79, medullary venules expressed L-selectin ligands are not recognized by MECA-79. Both L-selectin ligands seem to be fucosylated by α1,3-fucosyltransferase (FucT)-IV or FucT-VII as rolling is absent in FucT-IV + VII$^{-/-}$ mice. Although MECA-79-reactive species predominate in HEVs, medullary venules express another ligand that is spatially, antigenically, and biosynthetically different from PLN addressin. The two distinct L-selectin ligands are segmentally confined to contiguous microvascular domains in PLNs (M'Rini et al. 2003).

26.4.1.2 Salivary MG2 Saccharides Function as Ligands for Neutrophil L-Selectin

The low-molecular-weight human salivary mucin (MG2) coats oral surfaces, where it is in a prime location for governing cell adhesion. Since oligosaccharides form many of the interactive facets on mucin molecules, Prakobphol et al. (1999) examined MG2 glycosylation in relation to the molecule's adhesive functions. Termini of MG2 oligosaccharide structures predominantly carry T, sialyl-T, Le^x, sialyl Le^x (sLe^x), lactosamine, and sialyl lactosamine. In addition, sLe^x determinants confer L-selectin ligand activity to this molecule. Adhesive interactions between

MG2 and cells that traffic in the oral cavity such as neutrophils and bacteria, revealed that under flow conditions, neutrophils tethered to MG2-coated surfaces at forces between 1.25 and 2 dyn/cm2. Together, results suggested that distinct subsets of MG2 saccharides function as ligands for neutrophil L-selectin and receptors for bacterial adhesion, a finding with interesting implications for both oral health and mucin function.

26.4.1.3 *H. pylori* Isolates Interact with L-, E- and P-Selectins

Carbohydrate components on *H. pylori* contribute to the persistent inflammation through interactions with leukocyte-endothelial adhesion molecules of the host. The LPS of most *H. pylori* strains contain sequences related to the Le^x or Le^a antigens, which are the ligands for the leukocyte-endothelium adhesion molecules of the host, namely, E, P- selectins. *H. pylori* isolates from patients with chronic gastritis, duodenal ulcer and gastric cancer for their interactions with the selectins provide evidence of interactions of isolates from each of the diagnostic groups with E- and L-selectins (Galustian et al. 2003). L-selectin interacts with *Helicobacter pylori* isolates from patients with chronic gastritis, duodenal ulcer and gastric cancer. Anaplasma phagocytophilum infected neutrophils showed reduced expression of P-selectin glycoprotein ligand 1 (PSGL-1, CD162) and L-selectin (CD62L) (Choi et al. 2003).

26.5 Cell Surface Ligands for L-Selectin

26.5.1 Subsets of Sialylated, Sulfated Mucins of Diverse Origins are Recognized by L-Selectin

Direct and indirect evidences indicate that L-selectin ligands required sulfate and sialic acid moieties for proper function. The mucin-type polypeptides GlyCAM-1, CD34, and MAdCAM-1 can function as ligands for L-selectin only when they are synthesized by the specialized HEV of lymph nodes. Since sialylation, sulfation, and possibly fucosylation are required for generating recognition, it was argued that other mucins known to have such components might also bind L-selectin. In support, it was shown that mucins secreted by human colon carcinoma cells, as well as those derived from human bronchial mucus can bind to human L-selectin in a calcium-dependent manner. As with Gly-CAM-1 synthesized by lymph node HEV, α2-3 linked sialic acids and sulfation seemed to play a critical role in generating this L-selectin binding. Taken together, studies indicated that a single unique oligosaccharide structure may not be responsible for high-affinity binding. Rather, diverse mucins with sialylated, sulfated, fucosylated lactosamine-type O-linked oligosaccharides can generate high-affinity L-selectin ligands, but only when they present these chains in unique spacing and/or clustered combinations, dictated by polypeptide backbone (Crottet et al. 1996; Tamatani et al. 1993) (Fig. 26.3).

26.5.2 GlyCAM-1

26.5.2.1 Mouse GlyCAM-1

A sulfated 50 kDa glycoprotein (Sgp50) called GlyCAM-1 has been identified as an HEV ligand for L-selectin. This endothelial ligand is an adhesion molecule that accomplishes cell binding by presenting carbohydrate(s) to the lectin domain of L-selectin, and the name GlyCAM 1 (**Gly**-cosylation-dependent **C**ell **A**dhesion **M**olecule 1) was proposed (Fig. 26.3). The GlyCAM-1 is present on HEV of LN and also in lactating mammary glands. The specifically glycosylated form of GlyCAM-1 plays an important role in leukocyte rolling along the inflamed endothelium. It is a secretory protein that is present in mouse serum. The mRNA encoding this glycoprotein is preferentially expressed in LN. The predicted sequence of a cDNA revealed a mucin-like molecule containing two serine/threonine-rich domains. The mucin-like endothelial glycoprotein appears to function as a scaffold that presents carbohydrates to the L-selectin lectin domain (Lasky et al. 1992). The GlyCAM-1 ligand contains a large percentage of serine and threonine residues, which are apparently O - glycosylated. The gene encoding GlyCAM-1 was found to map to murine chromosome 15 Dowbenko et al. (1993a). Adhesive interactions with the L selectin- lectin domain require that the GlyCAM 1 polypeptide chain be appropriately modified with carbohydrates. These carbohydrate modifications include the addition of sialic acid as well as sulfate residues to O-linked carbohydrate side chains that are clustered in two serine/threonine-rich domains of the mucin. An additional interesting structure that may have relevance to the association of GlyCAM 1 with the lumenal surface of the endothelium was a potential amphipathic helix at the C terminus of the glycoprotein. GlyCAM-1 has been shown to enhance β2-integrin function.

26.5.2.2 Rat GlyCAM

Dowbenko et al. 1993b cloned the rat homologue of GlyCAM 1. The sequence of this clone revealed a serine/threonine-rich protein that is highly homologous with the mouse GlyCAM 1. The mouse and the rat GlyCAM 1 homologues show a clustering of these potential O-linked carbohydrate acceptors in two domains of the protein.

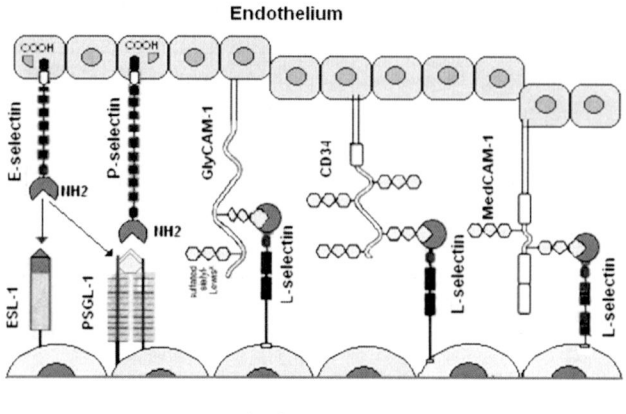

Fig. 26.3 Structures of selectins and their ligands. L-Selectin and the mucinlike vascular addressins direct naive lymphocyte homing to lymphoid tissues. E-selectin binds E-selectin ligand 1 (ESL-1) in mouse neutrophils, which is a chicken fibroblast receptor and not related to other adhesion receptor families. E-selectin also binds P-selectin glycoprotein 1 (PSGL-1) but with lower affinity than P-selectin. GlyCAM-1 lacks a transmembrane domain and may be secreted. L-Selectin is expressed on naive T cells, which bind to sulfated sialyl-Lewis^x moieties on the vascular addressins CD34 and GlyCAM-1 on high endothelial venules in order to enter lymph nodes. The relative importance of CD34 and GlyCAM-1 in this interaction is unclear. GlyCAM-1 is expressed exclusively on high endothelial venules but has no transmembrane region and it is unclear how it is attached to the membrane; CD34 has a transmembrane anchor and is expressed in appropriately glycosylated form only on high endothelial venule cells, although it is found in other forms on other endothelial cells. The addressin MAdCAM-1 is expressed on mucosal endothelium and guides entry into mucosal lymphoid tissue. The icon shown represents mouse MadCAM-1, which contains an IgA-like domain closest to the cell membrane; human MadCAM-1 has an elongated mucinlike domain and lacks the IgA-like domain. L-Selectin recognizes the carbohydrate moieties on the vascular addressins. EGF = epidermal growth factor; SCR = short consensus repeats. For selectins refer Fig. 26.1. ||||||||| in PSGL-1 referes to mucin domain

Interestingly, many of the serines and threonines are found to be spaced identically in the two homologues, consistent with the possibility that both density and position of the O-linked side chains may be important for appropriate L selectin-mediated adhesion. In support of its postulated functional importance, the C-terminal potential amphipathic helix is conserved in the rat homologue. Antibody against a relatively conserved portion of mouse GlyCAM 1 demonstrated an approximately 45-kDa sulfated ligand in rat lymph nodes that is analogous to that described for mouse lymph nodes (Dowbenko et al. 1993b; Samulowitz et al. 2002). The mC26 gene encoding GlyCAM-1 in the lactating mouse mammary gland comprises a 5,394 bp fragment of a genomic DNA, which codes for a protein highly expressed in the lactating mouse mammary gland (Nishimura et al. 1993). GlyCAM-1 is also expressed in the cochlea. GlyCAM-1 expressed in the cochlear region is heterogenous in terms of its glycosylation (Kanoh et al. 1999).

26.5.2.3 Complement Factor H as a Ligand for L-selectin in Human

There is no obvious GlyCAM-1 homologue in man. Malhotra et al. (1999) isolated three major glycoproteins of Mr 170, 70 and 50 kDa. The 170 kDa protein was identified as human complement protein Factor H. Human Factor H binds specifically to L-selectin in the presence of CaCl2, and binding was inhibited by anti-L-selectin antibodies, fucoidan and LPS. The interaction of Factor H with leukocyte L-selectin was shown to induce the secretion of TNF-α. It seems that a post-translationally modified form of human plasma Factor H is a potential physiological ligand for L-selectin (Malhotra et al. 1999).

26.5.3 CD34

CD34, a 90 kDa membrane-associated sialomucin, is expressed on the surface of hematopoietic stem/progenitor cells, stromal cells, and on the surface of HEV (Fig. 26.3). The CD34 binds L-selectin, which is important for leukocyte rolling on venules and lymphocyte homing to PLN. A predominant 105 kDa CD34 mucin-like protein has also been identified in human tonsil as peripheral addressin.

120 kDa Addressin Is L-Selectin-Binding Glycoform of CD34: A 120 kDa sialomucin has been identified as the predominant peripheral addressin in porcine lymph nodes. The 120 kDa porcine molecule has its ability to bind MECA-79 and an L-selectin-Fc chimera (LS-Fc). Whereas desialylation of 120 kDa ligand drastically reduced its binding to LS-Fc, this treatment appeared to enhance the binding of 120 kDa ligand to MECA-79. In contrast, the binding of both MECA-79 and LS-Fc to 120 kDa ligand was drastically reduced when de novo sulfation of this ligand was reduced by chlorate, a metabolic inhibitor of sulfation. N-Terminal amino acid sequences of the porcine 120 kDa protein revealed homology with human CD34. Taken together, these findings suggested that the porcine 120 kDa peripheral addressin is an L-selectin-binding glycoform of CD34 (Shailubhai et al. 1997).

Platelets Can Bind Reversibly to CD34$^+$ Cells from Human Blood and Bone: This interaction interferes with the accurate detection of endogenously expressed platelet glycoproteins (GPs). The interaction between these cells was dependent on divalent cations, and mediated by P-selectin. Enzymatic characterization showed the involvement of sialic acid residues. The presence of mRNA for PSGL-1 in the CD34$^+$ cells suggests that this molecule is present in these cells. Under conditions that prevent platelet adhesion, a small but distinct subpopulation of CD34$^+$ cells diffusely expressed the platelet GPIIb/IIIa complex (Dercksen et al. 1995).

Functions in CD34-Deficient Mice: CD34-deficient mutant- and Wild-type animals do not show differences in lymphocyte binding to PLN HEV, in leukocyte rolling on venules or homing to PLN, in neutrophil extravasation into peritoneum in response to inflammatory stimulus, nor in delayed type hypersensitivity. However, eosinophil accumulation in the lung after inhalation of a model allergen, ovalbumin, was several-fold lower in mutant mice. Although CD34 was not expressed in these mice, a portion of its 90-kD band crossreactive with mAb MECA79 persisted on Western blot. Thus, CD34 seems an additional molecule(s) that might be involved in eosinophil trafficking into the lung (Suzuki et al. 1996). However, CD34 is clearly not the sole contributor of L-selectin ligand activity in HEVs, since CD34-null mice maintain virtually normal L-selectin-dependent lymphocyte homing activity.

Another HEV-expressed L-selectin ligand is the transmembrane sialomucin podocalyxin-like protein or PCLP. Like CD34, PCLP is expressed on some vascular endothelia, but it is also expressed on the foot processes of glomerular podocytes. HEV-derived PCLP interacts with recombinant L-selectin-Ig chimera and supports L-selectin-dependent lymphocyte adhesion under physiological conditions.

26.5.4 Mucosal Addressin Cell Adhesion Molecule-1 (MAdCAM-1)

The mucosal vascular addressin, mucosal addressin cell adhesion molecule-1 (MAdCAM-1), is an Ig family adhesion receptor preferentially expressed by venular endothelial cells at sites of lymphocyte extravasation in murine mucosal lymphoid tissues and lamina propria (Fig. 26.3). The MAdCAM-1 specifically binds to both human and mouse lymphocytes that express the homing receptor for Peyer's patches, the integrin α4β7. MAdCAM-1 interacts preferentially with the leukocyte β7 integrin LPAM-1 (α4β7), but also with L-selectin, and with VLA-4 (α4β7) on myeloid cells, and serves to direct leukocytes into mucosal and inflamed tissues. Expression of human MAdCAM-1 RNA is restricted to mucosal tissues, gut-associated lymphoid tissues and spleen. L-selectin counter-receptors in HEVs are recognized by mAb MECA-79, a surrogate marker for molecularly heterogeneous glycans termed PLN addressin (PNAd).

26.5.4.1 MAdCAM-1 Gene

Human MAdCAM-1 gene contains five exons where signal peptide, two Ig domains, and mucin domain are each encoded by separate exons. The transmembrane domain, cytoplasmic domain, and 3′ untranslated region are encoded together on exon 5. The mucin domain contains eight repeats in total that are subject to alternative splicing. Despite the absence of a human counterpart of the third IgA-homologous domain and lack of sequence conservation of the mucin domain, the genomic organizations of human and mouse MAdCAM-1 genes are similar. An alternatively spliced MAdCAM-1 variant that lacks exon 4 encoding the mucin domain, may mediate leukocyte adhesion to LPAM-1 without adhesion to the alternate receptor, L-selectin. The MAdCAM-1 gene was located at p13.3 on chromosome 19, in close proximity to the ICAM-1 and ICAM-3 genes (p13.2–p13.3) (Leung et al. 1997).

The mouse MAdCAM-1 gene is located on chromosome 10 and contains five exons. The signal peptide and each one of the three Ig domains are encoded by a distinct exon, whereas the transmembrane, cytoplasmic tail, and 3′-untranslated region of MAdCAM-1 are combined on a single exon. The mucin-like region and the third Ig domain are encoded together on exon 4. A short variant of MAdCAM-1 may be specialized to support α4β7-dependent adhesion strengthening, independent of carbohydrate-presenting function (Sampaio et al. 1995). The deduced amino acid sequence from full-length DNA encoding human brain MAdCAM-1 revealed an 18 amino acid signal peptide, two N-terminal Ig-like domains conserved (59–65%) in sequence with those of mouse homologue, an 86 amino acid mucin-like region rich in serine-threonine residues, a 20 amino acid transmembrane domain and a 43 amino acid charged cytoplasmic domain. No counterpart to the third IgA-like domain of mouse MAdCAM-1 was present; however, the serine-threonine-rich mucin domain was extended as two distinguishable major and minor mucin regions unrelated to the mouse domain. The major domain is formed from six tandem repeats of an eight amino acid sequence having the MUC-2-related consensus DTTSPEP/SP. Human MAdCAM-1 mRNA transcripts were restricted to small intestine, colon, spleen, pancreas and brain. Alternatively spliced MAdCAM-1 variants were identified that lack parts of the second Ig domain and all or part of the major mucin domain, indicating that the function of this vascular addressin is regulated by extensive modifications to its multi-domain structure (Leung et al. 1996).

Two Ig-like domains of human and macaque MAdCAM-1 are similar to the two amino-terminal integrin binding domains of murine MAdCAM-1. Human MAdCAM-1 exhibits considerable variation from murine MAdCAM-1 with respect to the length of the mucin-like sequence and the lack of a membrane proximal Ig/IgA-like domain. The MAdCAM-1 from different species demonstrated greater seleoffctive pressure for maintenance of amino acids involved in α4β7 binding than those sequences presumably involved in the presentation of carbohydrates for selectin binding (Shyjan et al. 1996).

26.5.4.2 Structural Requirement for MAdCAM-1 Binding to Lymphocyte Receptor α4β7

The murine MAdCAM-1 has two amino-terminal integrin binding Ig domains. A point mutation within the first Ig domain of MAdCAM-1 abolishes activation-independent α4β7 binding. This point mutation resides within an eight-amino acid motif with homology to sequences important for the integrin binding ability of the related vascular Ig family members, ICAM-1 and VCAM-1. The first domain of MAdCAM-1 is sufficient for interaction with α4β7, but sequences within the second domain support this interaction, either by providing additional contact points for integrin binding or by contributing to the conformation or presentation of the N-terminal domain. The second domain of MAdCAM-1 can also support activation-dependent LFA-1 binding to domain 1 of ICAM-1. These findings are similar to VCAM-1 binding to α4β7 and suggest that structural differences exist between vascular Ig-like ligands for α4 versus β2 integrins (Briskin et al. 1996). A crystal structure for two extracellular amino-terminal domains of human MAdCAM-1 confirmed their expected Ig superfamily topology. In second structure of this fragment, although the overall structure is similar to that previously reported, one edge strand in the amino-terminal domain is instead located on the opposite sheet. This alters the arrangement and conformation of amino acids in this region that had been shown to be crucial for ligand binding. MAdCAM-1 is also seen to form dimers within the crystal lattice raising the possibility that oligomerization may influence the biological role of this adhesion molecule (Dando et al. 2002).

26.5.5 PSGL-1 Binds L-Selectin

L-selectin binds with high affinity to the N-terminal region of P-selectin glycoprotein ligand-1 (PSGL-1) through cooperative interactions with three sulfated tyrosine residues and an appropriately positioned C2-O-sLe^x O-glycan (Leppanen et al. 2003). PSGL-1 binding to P-selectin controls early leukocyte rolling during inflammation. The L-selectin-dependent rolling after P-selectin blockade is completely absent in PSGL-1 deficient (PSGL-1$^{-/-}$) mice or wild-type mice treated with a PSGL-1 blocking mAb. It appeared that leukocyte-expressed PSGL-1 serves as the main L-selectin ligand in inflamed postcapillary venules (Sperandio et al. 2003). L-selectin binding to PSGL-1 initiates tethering events that enable L-selectin-independent leukocyte-endothelial interactions. Interestingly, antibodies and pharmacological inhibitors (e.g., rPSGL-Ig) that target the N-terminus of PSGL-1 reduce but do not abolish P-selectin-dependent leukocyte rolling in vivo whereas PSGL-1-deficient mice have almost no P-selectin-dependent rolling. However, Ridger et al. (2005) suggested that leukocytes can continue to roll in the absence of optimal P-selectin/PSGL-1 interaction using an alternative mechanism that involves P-selectin-, L-selectin-, and sLe^x-bearing ligands. L- and P-selectins collaborate to support leukocyte rolling in vivo when high-affinity PSGL-1 interaction is inhibited (Ridger et al. 2005) (see Chap 27 and Chap 28).

26.6 L-Selectin IN Pathological States

26.6.1 Gene Polymorphism in L-Selectin

LECAM-1 Genes in Ig A Nephropathy: Elucidating the genetic background of immunoglobulin A nephropathy (IgAN) and genetic factors associated with the pathogenesis of the disease, Takei et al. (2002) studied single-nucleotide polymorphisms (SNPs) in selectin gene cluster on chromosome 1q24–25 and found that two SNPs in the E-selectin gene (*SELE8* and *SELE13*) and six SNPs in the L-selectin gene (*SELL1, SELL4, SELL5, SELL6, SELL10,* and *SELL11*) were significantly associated with IgAN in Japanese patients. All eight SNPs were in almost complete linkage disequilibrium. *SELE8* and *SELL10* caused amino acid substitutions from His to Tyr and from Pro to Ser, and *SELL1* could affect promoter activity of the L-selectin gene. The TGT haplotype at these three loci was associated with IgAN (Fig. 26.3)

Polymorphism in Type 2 Diabetes Mellitus: Glomerular infiltration with monocytes/macrophages has been implicated in the pathogenesis of diabetic nephropathy. The LECAM-1 213PP genotype is a genetic risk factor for the development of nephropathy in type 2 diabetes mellitus (Kamiuchi et al. 2002). Strong relationship has been demonstrated between soluble L-selectin and diabetic retinopathy with strong correlation between sL-selectin and HbA1c. Soluble L-selectin is increased with poor glycemic control (Karadayi et al. 2003). In Finnish school children CAM is positive marker for precinical type 1diabetes in children (Toivonen et al. 2004).

Brucellosis: The 206Leu allele frequency occurred in 42% of the patients with coronary artery disease compared to % of the controls (Hajilooi et al. 2006). Higher frequency of L-selectin genotypes in patients with brucellosis than in controls and the association between the 206Leu allele and the occurrence of brucellosis relapse, suggest that the F206L polymorphism may make individuals more vulnerable to brucellosis (Rafiei et al. 2006).

26.6.2 Antitumor Effects of L-Selectin

26.6.2.1 Regulation of Host-Mediated Anti-Tumor Mechanisms by L-Selectin

Human tumor cells injected s.c. into mice lacking β3- or β3/β5-integrins or various selectins show enhanced tumor growth compared with growth in control mice. There was increased angiogenesis in mice lacking β3-integrins. Tumor growth also was affected by bone marrow-derived cells in mice lacking any one or all three selectins, implicating both leukocyte and endothelial selectins in tumor suppression. Cells of the innate immune system, macrophages or perhaps natural killer cells, seemed to be implicated in tumor suppression (Taverna et al. 2004). Malignant melanoma is often accompanied by a host response of inflammatory cell infiltration that is highly regulated by multiple adhesion molecules. L-selectin and ICAM-1 contribute co-operatively to the anti-tumor reaction by regulating lymphocyte infiltration to the tumor (Yamada et al. 2006). The plasma level of sL-selectin is possibly useful for early diagnosis of relapse and extramedullary infiltration in acute myeloid leukemia (Aref et al. 2002; Chang et al. 2003).

Suppression of Metastasis in Lymph Nodes by L-Selectin-Mediated NK Cells: The NK cells are known to reject certain tumors in vivo and prevent metastasis of tumors into secondary lymphoid organs. In tumor-bearing hosts, NK cells are recruited to regional lymph nodes in wild-type mice, but not in mice deficient for L-selectin or L-selectin ligands. L-selectin on NK cells and L-selectin ligands on endothelial cells seem to be essential for NK cell recruitment to lymph nodes. Although L-selectin-deficient NK cells efficiently lysed tumor cells in vitro, NK cell-dependent suppression of tumor metastasis was diminished in mice deficient for L-selectin or L-selectin ligands because of insufficient NK cell recruitment to lymph nodes. Evidences indicate that L-selectin-mediated NK cell recruitment plays a crucial role in the control of tumor metastasis into secondary lymphoid organs (Chen et al. 2005).

26.6.2.2 Can L-Selectin Facilitate Metastasis?

Transgenic mice that express rat insulin promoter regulate simian virus 40 Tag (RIP-Tag) develop large, local cancers that metastasize to liver but not LN. Mice, that developed insulinomas, specifically had LN metastases; metastasis was blocked by an anti L-selectin mAb. The highly vascularized islet carcinomas shed tumor cells into the bloodstream, which is a necessary but insufficient condition for metastasis to occur; L-selectin can facilitate homing of such tumor cells to LN, resulting in metastasis (Qian et al. 2001). Surprisingly, L-selectin expressed on endogenous leukocytes also facilitates metastasis in both the syngeneic and xenogeneic (T and B lymphocyte deficient) systems Synergistic effects of L- and P-selectin in facilitating tumor metastasis can involve non-mucin ligands and implicate leukocytes as enhancers of metastasis (Borsig et al. 2002). It seems that L-selectin-mediated NK cell recruitment plays a crucial role in the control of tumor metastasis into secondary lymphoid organs (Chen et al. 2005). L-selectin deficiency also attenuates experimental metastasis. Although L-selectin deficiency did not affect platelet aggregation or initial tumor cell embolization, the association of leukocytes with tumor cells was reduced and tumor cell survival was diminished 24 h later. Inhibition of L-selectin by a function-blocking antibody also reduced metastasis. Therefore, L-selectin facilitation of metastasis progression involves leukocyte-endothelial interactions at sites of intravascular arrest supported by local induction of L-selectin ligands via fucosyltransferase-7. It provides an explanation for how L-selectin facilitates tumor metastasis (Laubli et al. 2006).

26.6.2.3 Adhesion Molecules in Kaposi's Sarcoma

Kaposi's sarcoma (KS) is a neoplasm with multifocal vascular lesions that is often seen in homosexual HIV-infected individuals. The products of leukocytes enhance the proliferation of KS cells in vitro and most likely are crucial for the development of KS lesions in vivo. Studies indicate that multiple proinflammatory agents can induce NF-kB binding activity and can enhance ICAM-1, VCAM-1, and E-selectin expression in KS cells. Thus, the induction of CAM expression could be an early event in the development of KS by recruiting leukocytes into KS lesions, thereby providing factors that could potentiate the development of KS (Sciacca et al. 1994; Yang et al. 1994).

Osteosarcomas and Rhabdomyosarcomas: Osteosarcomas and rhabdomyosarcomas are highly invading tumors. Before they can extravasate to the parenchymal organs and form metastases, they have to adhere to the endothelial cells lining the blood vessels and then penetrate through the endothelium. Human sarcoma cell lines and rhabdomyosarcoma RD express VLA-4 molecule on their surface and bind to the VCAM-I-expressing activated endothelial cell line Ea.hy 926. Sarcoma cells also adhere to recombinant sVCAM-I protein. On the other hand, these sarcoma cell lines do not express marked amounts of other ligands (such as CDII/18 or sialyl-Lex) for other endothelial adhesion molecules (ICAM-I, ICAM-2, E- and P-selectin) indicating that the VLA-4-VCAM-I dependent pathway might be of major importance in sarcoma extravasation (Mattila et al. 1992).

Selectins in the Triggering, Growth, and Dissemination of T-lymphoma Cells: The ICAM-1 expression by the host is essential for lymphoma dissemination. In selectin-deficient mice, though the absence of E-, P-, or L-selectins did not affect the triggering of radiation-induced thymic lymphoma, the absence of L-selectin on lymphoma cells reduced their capacity to grow in the thymus. This defect was overcome by altering the integrity of the L-selectin-mediated interactions in the thymus, as shown in L-selectin-deficient mice and by adoptive transfer experiments. Evidence shows that selectins play a significant role at different steps of T-cell lymphoma development (Belanger and St-Pierre 2005).

26.6.2.4 Role of CD62L$^+$ Cells and TGF-β in Patients with Gastric Cancer

CD62L$^+$ cells are decreased in the peripheral blood, but inversely increased in the spleen of gastric cancer patients. The increased CD62L$^+$ cells residing in the CD4$^+$ suppressor-inducer phenotype, and the removal of CD62L$^+$ cells from spleen cells results in a decrease of Con A-induced suppressor activity in vitro in one-way allogeneic MLR. The CD62L$^+$ cells included CD4$^+$CD25$^+$ regulatory T cells. The culture supernatant of CD62L$^+$ cells showed TGF-β activity, which was more significantly detectable in splenic vein than in the peripheral blood, and TGF-β mRNA was detectable in the spleen from advanced gastric cancer patients. These results suggest that CD62L$^+$ cells migrate into the spleen with disease progression of gastric cancer and serve as suppressor-inducer cells with TGF-β production to induce regulatory T cells, contributing to disease-associated immunosuppression in advanced gastric cancer patients (Noma et al. 2005).

26.6.2.5 Reduced Expression of CD62L Can Identify Tumor-Specific T Cells

Reduced expression of CD62L can identify tumor-specific T cells in lymph nodes draining murine tumors. This strategy could isolate tumor-specific T cells from vaccinated patients. Tumor vaccine-draining lymph node (TVDLN) T cells of patients were separated into populations with reduced (CD62L^{Low}) or high levels of CD62L (CD62L^{High}). Effector T cells generated from CD62L^{Low} cells maintained or enriched the autologous tumor-specific type 1 cytokine response compared to unseparated TVDLN T cells in four of four patients showing tumor-specific cytokine secretion. Effector T cells generated from CD62L^{Low} or CD62L^{High} TVDLN were polarized towards a dominant type 1 or type 2 cytokine profile respectively. For CD62L^{Low} T cells the type 1 cytokine profile appeared determined prior to culture. Since a tumor-specific type 1 cytokine profile appears critical for mediating anti-tumor activity in vivo, this approach can be used to isolate T cells for adoptive immunotherapy (Meijer et al. 2004).

It was shown that tumor vaccine-sensitized draining lymph node (vDLN) cells activated ex vivo with bryostatin and ionomycin (B/I) are capable of inducing antigen-specific regression of a murine mammary tumor, 4T07. The vDLN cells not activated with B/I, were ineffective. It was suggested that B/I selectively activates tumor-sensitized (CD62L^{low}) lymphocytes, to account for the highly potent and tumor-specific activity. It was also hypothesized that CD8$^+$ CD62L^{low} cells may be preferentially activated by B/I treatment, infiltrate the tumors and mediate tumor regression in mice. It was suggested that CD62L^{low} cells are preferentially activated by B/I, leading to a highly effective anti-tumor T cell population (Chin et al. 2004).

26.6.2.6 CD44 Glycoform as E- and L-Selectin Ligand on Colon Carcinoma Cells

Engagement of E-selectin and L-selectin with relevant counter-receptors expressed on tumor cells contributes to the hematogenous spread of colon carcinoma. The LS174T colon carcinoma cell line expresses CD44 glycoform known as hematopoietic cell E-/L-selectin ligand (HCELL), which functions as a high affinity E- and L-selectin ligand on these cells. Expression of HCELL confers robust and predominant tumor cell binding to E- and L-selectin, highlighting a central role for HCELL in promoting shear-resistant adhesive interactions essential for hematogenous cancer dissemination (Burdick et al. 2006).

26.6.3 Autoimmune Diseases

26.6.3.1 L-Selectin in Autoimmune Diabetes

Administration of anti-L-selectin (CD62L) mAb to neonatal non-obese diabetic (NOD) mice mediates long term protection against the development of insulitis and overt diabetes, suggesting that CD62L has a key role in the general function of beta cell-specific T cells.

CD4$^+$CD62L$^+$ Regulatory T Cells in Controlling Onset of Diabetes in Mice: Autoimmune diabetes is characterized by an early mononuclear infiltration of pancreatic islets and later selective autoimmune destruction of insulin-producing β cells. Lymphocyte homing receptors are the candidate targets to prevent autoimmune diabetes. L-selectin (CD62L), expressed in naive T and B cells can be blocked in vivo with specific antibodies such as Mel-14, which partially impairs insulitis and diabetes in autoimmune diabetes-prone NOD mice. Genetic blockade of leukocyte homing into peripheral lymph nodes can prevent the development of diabetes. Though L-selectin plays a small role in the homing of autoreactive lymphocytes to regional (pancreatic) lymph nodes in NOD mice (Mora et al. 2004), experimental evidences indicate that CD4$^+$ regulatory T cells control progression of autoimmune insulitis in NOD mice. You et al. (2004) showed that diabetes onset is prevented in such mice by infusion of polyclonal CD4$^+$ T cells expressing L-selectin (CD62L) but not prevented or only marginally prevented by CD4$^+$CD25$^+$ T cells. Report argues for the role of CD4$^+$CD62L$^+$ T cells present within the polyclonal diabetogenic population in mediating this apparently paradoxical effect. The central role of CD4$^+$CD62L$^+$ regulatory T cells in controlling disease onset was confirmed in a transgenic model of autoimmune diabetes and possibly through intervention of homeostatic mechanisms as part of their mode of action (You et al. 2004). Despite these observations, the patterns of T cell activation, migration, and beta cell-specific reactivity were similar in NOD mice in different genotypes, suggesting that CD62L expression is not essential for the development of type 1diabetes in NOD mice (Friedline et al. 2002). Nevertheless, CD62L expression is necessary for the diabetes-delaying effect of transfer of CD4$^+$CD25$^+$ splenocytes in vivo, but not for their suppressor function in vitro (Szanya et al. 2002).

26.6.3.2 L-Selectin Deficiency Reduces Immediate-Type Hypersensitivity

Antigen-sensitized CD4$^+$CD62L^{low} memory/effector T helper 2 cells can induce airway hyperresponsiveness in an antigen free setting (Nakagome et al. 2005). Dermal and pulmonary inflammatory disease in E-selectin and P-selectin double-null mice is reduced in triple-selectin-null mice (Collins et al. 2001). P- and E-selectin mediate CD4+ Th1 cell migration into the inflamed skin in a murine contact hypersensitivity model. In this model, not only CD4+ T cells but also CD8+ T cells infiltrate the inflamed skin, and the role of CD8+ type 1 cytotoxic T (Tc1) cells as effector cells has been demonstrated. In mice deficient in both P- and E-selectin, the infiltration of CD8+ T cells in the inflamed skin is reduced, which suggests that these selectins participate in CD8+ T cell migration. Tc1 cells are able to migrate into the inflamed skin of wild-type mice. This migration is partially mediated by P- and E-selectin, as shown by the reduced Tc1 cell migration into the inflamed skin of mice deficient in both, P- and E-selectin or wild-type mice treated with the combination of anti-P-selectin and anti-E-selectin Abs. During P- and E-selectin-mediated migration of Tc1 cells, P-selectin glycoprotein ligand-1 appears to be the sole ligand for P-selectin and one of the ligands for E-selectin. P- and E-selectin-independent migration of Tc1 cells into the inflamed skin was predominantly mediated by L-selectin. These observations indicate that all three selectins can mediate Tc1 cell migration into the inflamed skin (Hirata et al. 2002).

The deposition of immune complexes (IC) induces an acute inflammatory response with tissue injury. IC-induced inflammation is mediated by inflammatory cell infiltration, a process highly regulated by expression of multiple adhesion molecules. The cutaneous reverse passive Arthus reaction examined in mice lacking L-selectin (L-selectin$^{-/-}$), ICAM-1 (ICAM-1$^{-/-}$), or both (L-selectin/ICAM-1$^{-/-}$), indicates that ICAM-1 and L-selectin cooperatively contribute to the cutaneous Arthus reaction by regulating neutrophil and mast cell recruitment (Kaburagi et al. 2002). Repeated Ag exposure in wild-type littermates resulted in increased levels of serum L-selectin, also observed in atopic dermatitis patients (Shimada et al. 2003). These studies demonstrates that L-selectin and ICAM-1 cooperatively regulate the induction of the immediate-type response by mediating mast cell accumulation into inflammatory sites and suggests that L-selectin and ICAM-1 are potential therapeutic targets for regulating human allergic reactions

26.6.3.3 L-Selectin Is Required for Early Neutrophil Extravasation

L -selectin (CD62L) and CD44 are major adhesion receptors that support the rolling of leukocytes on endothelium, the first step of leukocyte entry into inflamed tissue. The requirement for these receptors for inflammatory cell recruitment during Ag-induced arthritis was studied in CD44-deficient, L-selectin-deficient, and CD44/L-selectin double knockout mice. Study suggested a greater requirement for L-selectin than for CD44 for neutrophil extravasation during the early phase of Ag-induced arthritis (Szanto et al. 2004). Verdrengh et al. (2000) described a dual role for selectins in *S. aureus*-induced arthritis: on the one hand, blockade of these selectins leads to less severe arthritic lesions in the initial stage of the disease; on the other, delayed recruitment of phagocytes decreases the clearance of bacteria (Verdrengh et al. 2000).

26.6.3.4 L-Selectin in Mouse CNS During EAE

Adhesion molecules facilitate infiltration of leukocytes into CNS of mice with experimental allergic encephalomyelitis (EAE). Cellular infiltration and expression of the adhesion molecules ICAM-1 (CD54), VCAM-1 (CD106), L-selectin (CD62L), and leukosialin (CD43) in EAE-susceptible SWXJ mice preceded the EAE clinical symptoms by a minimum of 3 days, suggesting a causal role of adhesion molecules in the initiation of CNS inflammation. However, prophylactic injections of mAbs against either of ICAM-1, L-selectin, or CD43, did not ameliorate the clinical severity of EAE in these mice (Dopp et al. 1994). In order to study the role of CD62L in the immunopathology of EAE, Grewal et al. (2001) crossed

CD62L-deficient mice with myelin basic protein-specific TCR (MBP-TCR) transgenic mice. CD62L-deficient MBP-TCR transgenic mice failed to develop antigen-induced EAE, and, despite the presence of leukocyte infiltration, damage to myelin in the CNS was not seen. EAE could, however, be induced in CD62L-deficient mice upon adoptive transfer of wild-type macrophages. Study indicated that CD62L is not required for activation of autoimmune CD4 T cells but is important for the final destructive function of effector cells in the CNS and supports a novel mechanism whereby CD62L expressed on effector cells is important in mediating myelin damage (Grewal et al. 2001).

26.6.3.5 Central and Effector Memory T Cells to Recall Responses

The absolute number of memory $CD8^+$ T cells in the spleen following antigen encounter remains stable for many years. However, the relative capacity of these cells to mediate recall responses was studied by Roberts et al. (2005). A dual adoptive transfer approach demonstrated a progressive increase in the quality of memory T cell pools in terms of their ability to proliferate and accumulate at effector sites in response to secondary pathogen challenge. This temporal increase in efficacy occurred in $CD62L^{low}$ (effector memory) and $CD62L^{high}$ (central memory) subpopulations, but was most prominent in the CD62L hi subpopulation. Data indicated that the contribution of effector memory and central memory T cells to the recall response changes substantially over time (Roberts et al. 2005).

During infection with lymphocytic choriomeningitis virus, $CD8^+$ T cells differentiate rapidly into effectors ($CD62L^{low}CD44^{high}$) that differentiate further into the central memory phenotype ($CD62L^{high}CD44^{high}$) gradually. Van Faassen et al. (2005) indicated that the potency of the pathogen can influence the differentiation and fate of $CD8^+$Tcells enormously, and the extent of attrition of primed $CD8^+$ T cells correlates inversely to the early differentiation of $CD8^+$ T cells primarily into the central $CD8^+$ T cell subset (van Faassen et al. 2005).

26.6.4 CD62L in Other Conditions

Adhesion Molecules in Inflammatory Bowel Disease: Mucosal endothelium has become one of the major areas of investigation in gut inflammation. It is now well recognised that it plays an active role in the pathogenesis of both forms of inflammatory bowel disease, Crohn's disease and ulcerative colitis, since endothelial cells regulate mucosal immune homeostasis, acting as "gatekeepers", controlling leukocyte accumulation in the interstitial compartment. This process is mediated by leukocyte-endothelial adhesion molecules. The major molecules that mediate leukocyte-endothelial interactions, have been reviewed and the results of the most recent clinical trials targeting adhesion molecules in inflammatory bowel disease summarized (Danese et al. 2005). Compared to controls, the total amount of platelet P-selectin (tP-selectin) does not change in patients with inflammatory bowel disease (IBD) and 5-aminosalicylic acid medication. However, on gender basis, the male patients showed higher levels of total P-selectin compared to male controls. Increased total P-selectin levels may alter the inflammatory response and susceptibility to thromboembolic disease. As found with sP-selectin, total P-selectin showed gender dependent differences, which are important to consider for future studies (Fagerstam and Whiss 2006).

Appendix plays an important role in the pathogenesis of colitis (Farkas et al. 2005). Since appendectomy can protect against development of ulcerative colitis and Crohn's disease and since. T cells affect the appendix in the development of colitis, Farkas et al. (2005) demonstrated the preferential migration of $CD62L^+CD4^+$ cells into the appendix as compared to the colon. This migration pattern correlated with up-regulation of integrin $\alpha4\beta7$ and CD154 (CD40 ligand) on T cells. Thus In a chronic ileitis model, pathogenic $CD4^+$ T cells alternatively engage L-selectin in order to recirculate to the chronically inflamed small intestine (Rivera-Nieves et al. 2005). L-selectin is important in experimental abdominal aortic aneurysm (AAA) formation in rodents. L-selectin-mediated neutrophil recruitment may be a critical early step in AAA formation (Hannawa et al. 2005). The resident flora plays a critical role in initiation and perpetuation of intestinal inflammation, as demonstrated in experimental models of colitis where animals fail to develop disease under germ free conditions. Strauch et al. (2005) indicated that bacterial antigens are crucial for the generation and/or expansion of T_{reg} cells in a healthy individual. Therefore, bacterial colonisation is of great importance in maintaining the immunological balance.

Other Conditions: During acute familial Mediterranean fever (FMF) there was no change in L-selectin expression, but there was an increased neutrophil surface CD11b compared with normal controls. Neutrophils of FMF patients regulate CD11b and L-selectin expression induced by chemoattractant (FMLP) stimulation, similar to that in controls (Molad et al. 2004). Considering that the inflammation is the basic stage in the early period of ^{131}I therapy (RAI) of hyperthyroidism, E- and L-selectins are not useful indicators in thyroid, while ICAM-1 and IL-6 may be important in the thyroid during radioiodine therapy, especially Graves' disease (Jurgilewicz et al. 2002). Inflammatory cells play a crucial role in wound healing. Although ICAM-1 contributes to wound repair to a greater extent than L-selectin, a role for L-selectin was revealed in the absence of ICAM-1. The impaired wound repair was associated with reduced infiltration of neutrophils and macrophages in ICAM-1$^{-/-}$ and L-selectin/ICAM-1$^{-/-}$ mice. Results suggested a distinct role of ICAM-1 and L-selectin in wound healing and the delayed wound healing in the absence of these molecules is likely because of decreased leukocyte accumulation at the wound site (Nagaoka et al. 2000). Soluble L-selectin increases in serum samples from subjects with *cryptococcosis* than in those from uninfected subjects (Jackson et al. 2005).

26.7 Oligonucleotide Antagonists IN Therapeutic Applications

Selectin-directed therapeutic agents are now proven to be effective in blocking many of the pathological effects resulting from leukocyte entry into sites of inflammation. Studies are being focused on how the selectins interact with the increasing array of other adhesion molecules and inflammatory mediators. L-selectin expression on CD3+, CD4+ and CD8+ T cells was significantly lower in HIV infected children than in the control group. The percentage of neutrophils expressing CD62L was significantly reduced in patients with severe immunologic suppression. Altered leukocyte functions such as migration and homing resulting from reduced expression of CD62L may be an important contributor of the progressive dysfunction of the immune system in HIV infected children (Gaddi et al. 2005).

26.7.1 Monomeric and Multimeric Blockers of Selectins

Selectins bind to carbohydrate ligands in a calcium-dependent manner and play critical roles in host defense and possibly in tumor metastasis. It is possible to isolate peptides mimicking carbohydrate ligands by screening the peptides for binding to anticarbohydrate antibodies and then use them to inhibit carbohydrate-dependent experimental tumor metastasis (Fukuda et al. 2000). The monomeric and polymeric compounds, 4-nitrophenylthiomannoside, phenylmannoside, conjugated with polyacrylic acid, and α-mannose, conjugated with polyacrylamide, inhibited the binding of the model ligand to P- and L-selectins (but not to E-selectin). These compounds (i.v.) caused a dose-dependent reduction of neutrophil accumulation in rat peritoneum. The polysaccharide mannan was inactive in both types of experiments. The conjugate of phenylmannoside with polyacrylic acid was the most effective blocker in vitro as well as in vivo. The inhibitory effect of (s.c) 4-nitrophenylthiomannoside was indicated on mouse peritonitis (Ushakova et al. 2001). Ikegami-Kuzuhara et al. (2001) investigated the ability of a synthetic sugar derivative, OJ-R9188, [N-(2-tetradecylhexadecanoyl)-O-(L-alpha-fucofuranosyl)-D-seryl]-L-glutamic acid 1-methylamide 5-L-arginine salt, to block binding of selectins to their ligands in vitro and inhibit the infiltration of leukocytes in vivo. OJ-R9188 prevented the binding of human E-, P- and L-selectin-IgG fusion proteins to immobilized sialyl Le^x-pentasaccharide glycolipid. In a mouse model of thioglycollate-induced peritonitis, OJ-R9188 inhibited neutrophil accumulation in the peritoneal cavity. Thus blocking selectin-sLe^x binding site is a promising strategy for the treatment of allergic skin diseases.

The potency of oligosaccharides SiaLe^x, Sia Le^a, HSO$_3$ Le^x, and HSO$_3$ Le^a, their conjugates with polyacrylamide (PAA), and other monomeric and polymeric selectin inhibitors has been compared with that of the polysaccharide fucoidan. PAA-conjugates, containing tyrosine-O-sulfate (sTyr) in addition to one of the sialylated oligosaccharides, are the potent synthetic blockers in vitro. Compared with fucoidan, the bi-ligand glycoconjugate HSO$_3$ Le^a-PAA-sTyr displayed similar inhibitory activity in vitro towards L-selectin and about ten times lower activity towards P-selectin. All of the synthetic polymers displayed a similar ability to inhibit neutrophil extravasation in the peritonitis model (in vivo). Evidence indicates that monomeric SiaLe^x is considerably more effective as a selectin blocker in vivo than in vitro, whereas the opposite is true for fucoidan and the bi-ligand neoglycoconjugate HSO$_3$ Le^a-PAA-sTyr (Ushakova et al. 2005).

26.7.2 Synthetic and Semisynthetic Oligosaccharides

The SELEX (systematic evolution of ligands by exponential enrichment) yields high affinity oligonucleotides with unexpected binding specificities. Nuclease-stabilized randomized oligonucleotides subjected to SELEX against recombinant L-selectin yielded calcium-dependent antagonists which shared a common consensus sequence. Unlike sialyl Le^x, these antagonists were specific to L-selectin since they showed little binding to E- or P-selectin. However, they showed binding to native L-selectin on peripheral blood lymphocytes and blocked L-selectin-dependent interactions with the natural ligands on HEV (O'Connell et al. 1996).

While high affinity recognition of natural ligands is associated with α1-3(4)fucosylated, α2-3sialylated (and/or sulfated) lactosamine sequences, small oligosaccharides that potentialy inhibit the selectins were found by Koenig et al. (1997). Using synthetic and semisynthetic oligosaccharides related to those on natural ligands Koenig et al. (1997) confirmed that α2-3-linked sialic acids, and α1-3(4) fucosylation are important for recognition. Immobilized targets for the three selectins indicated that the binding sites for sialic acid and sulfate are very close, or identical. While O-sulfate esters mostly improved L- and P-selectin recognition, effects depended upon their position and number. Of particular note, the "major capping group" of GlyCAM-1 was not an unusually potent or highly selective inhibitor of L-selectin, even when studying the interaction of L-selectin with native GlyCAM-1 itself (Koenig et al. 1997).

Nishida et al. (2000) described a synthetic approach for artificial L- and P-selectin blockers. This synthesis involves radical bi- and terpolymerizations of p-(N-acrylamido)phenyl 3- or/and 6-sulfo-β-D-galactoside with allyl α-L-fucoside in the presence of acrylamide. Each of the two glycosyl monomers constructs a key carbohydrate module responsible for selectins/sulfated sialyl Lewis^x bindings. Whereas an acrylamide copolymer carrying 3-sulfo-galactoside showed no activity for any selectins, the fucosylated terpolymer showed a potent activity to block both of P- and L-selectins/sLe^x binding at a concentration of a few µg/ml. The enhanced activity is apparently ascribed to the cooperative binding effects of the fucoside and the 3-sulfo-galactoside residues.

References

Aref S, Salama O, Al-Tonbary Y et al (2002) L and E selectins in acute myeloid leukemia: expression, clinical relevance and relation to patient outcome. Hematology 7:83–87

Belanger SD, St-Pierre Y (2005) Role of selectins in the triggering, growth, and dissemination of T-lymphoma cells: implication of L-selectin in the growth of thymic lymphoma. Blood 105:4800–4806

Bevilacqua MP, Pober JS, Mendrick DL et al (1987) Identification of an inducible endothelial-leukocyte adhesion molecule. Proc Natl Acad Sci USA 84:9238–9242

Bevilacqua MP, Stengelin S, Gimbrone MA Jr, Seed B (1989) Endothelial leukocyte adhesion molecule 1: an inducible receptor for neutrophils related to complement regulatory proteins and lectins. Science 243:1160–1165

Borsig L, Wong R, Hynes RO et al (2002) Synergistic effects of L- and P-selectin in facilitating tumor metastasis can involve non-mucin ligands and implicate leukocytes as enhancers of metastasis. Proc Natl Acad Sci USA 99:2193–2198

Bosworth BT, Dowbenko D, Shuster DE, Harp JA (1993) Bovine L-selectin: a peripheral lymphocyte homing receptor. Vet Immunol Immunopathol 37:201–215

Bowen BR, Nguyen T, Lasky LA (1989) Characterization of a human homologue of the murine peripheral lymph node homing receptor. J Cell Biol 109:421–427

Briskin MJ, Rott L, Butcher EC (1996) Structural requirements for mucosal vascular addressin binding to its lymphocyte receptor α4β7. Common themes among integrin-Ig family interactions. J Immunol 156:719–726

Burdick MM, Chu JT, Godar S, Sackstein R (2006) HCELL is the major E- and L-selectin ligand expressed on LS174T colon carcinoma cells. J Biol Chem 281:13899–13905

Butcher EC (1991) Leukocyte-endothelial cell recognition: three (or more) steps to specificity and diversity. Cell 67:1033–1036

Camerini D, James SP, Stamenkovic I, Seed B (1989) Leu-8/TQ1 is the human equivalent of the Mel-14 lymph node homing receptor. Nature 342(6245):78–82

Celie JW, Keuning ED, Beelen RH et al (2005) Identification of L-selectin binding heparan sulfates attached to collagen type XVIII. J Biol Chem 280:26965–26973

Chang JH, Qi ZH, Chen FP, Xie QZ (2003) Study on expression of cell surface L-selectin and soluble L-selectin in patients with acute leukemia. Zhongguo Shi Yan Xue Ye Xue Za Zhi 11:251–255

Chen A, Engel P, Tedder TF (1995) Structural requirements regulate endoproteolytic release of the L-selectin (CD62L) adhesion receptor from the cell surface of leukocytes. J Exp Med 182:519–530

Chen S, Kawashima H, Lowe JB et al (2005) Suppression of tumor formation in lymph nodes by L-selectin-mediated natural killer cell recruitment. J Exp Med 202:1679–1689

Chin CS, Miller CH, Graham L et al (2004) Bryostatin 1/ionomycin (B/I) ex vivo stimulation preferentially activates L-selectinlow tumor-sensitized lymphocytes. Int Immunol 16:1283–1294

Choi KS, Garyu J, Park J, Dumler JS (2003) Diminished adhesion of Anaplasma phagocytophilum-infected neutrophils to endothelial cells is associated with reduced expression of leukocyte surface selectin. Infect Immun 71:4586–4594

Collins RG, Jung U, Ramirez M et al (2001) Dermal and pulmonary inflammatory disease in E-selectin and P-selectin double-null mice is reduced in triple-selectin-null mice. Blood 98:727–735

Crottet P, Kim YJ, Varki A (1996) Subsets of sialylated, sulfated mucins of diverse origins are recognized by L-selectin. Lack of evidence for unique oligosaccharide sequences mediating binding. Glycobiology 6:191–208

Csencsits KL, Pascual DW (2002) Absence of L-selectin delays mucosal B cell responses in nonintestinal effector tissues. J Immunol 169:5649–5659

da Costa Martins P, van den Berk N, Ulfman LH et al (2004) Platelet-monocyte complexes support monocyte adhesion to endothelium by enhancing secondary tethering and cluster formation. Arterioscler Thromb Vasc Biol 24:193–199

Dando J, Wilkinson KW, Ortlepp S et al (2002) A reassessment of the MAdCAM-1 structure and its role in integrin recognition. Acta Crystallogr D Biol Crystallogr 58:233–241

Danese S, Semeraro S, Marini M et al (2005) Adhesion molecules in inflammatory bowel disease: therapeutic implications for gut inflammation. Dig Liver Dis 37:811–818

Dercksen MW, Weimar IS, Richel DJ et al (1995) The value of flow cytometric analysis of platelet glycoprotein expression of CD34+ cells measured under conditions that prevent P-selectin-mediated binding of platelets. Blood 86:3771–3782

Diacovo TG, Blasius AL, Mak TW et al (2005) Adhesive mechanisms governing interferon-producing cell recruitment into lymph nodes. J Exp Med 202:687–696

Ding Z, Issekutz TB, Downey GP et al (2003) L-selectin stimulation enhances functional expression of surface. CXCR4 in lymphocytes: implications for cellular activation during adhesion and migration. Blood 101:4245–4252

Dopp JM, Breneman SM, Olschowka JA (1994) Expression of ICAM-1, VCAM-1, L-selectin, and leukosialin in the mouse central nervous system during the induction and remission stages of experimental allergic encephalomyelitis. J Neuroimmunol 54:129–144

Dowbenko DJ, Diep A, Taylor BA et al (1991) Characterization of the murine homing receptor gene reveals correspondence between protein domains and coding exons. Genomics 9:270–277

Dowbenko D, Andalibi A, Young PE et al (1993a) Structure and chromosomal localization of the murine gene encoding

GLYCAM 1. A mucin-like endothelial ligand for L selectin. J Biol Chem 268:4525–4529

Dowbenko D, Watson SR, Lasky LA (1993b) Cloning of a rat homologue of mouse GlyCAM 1 reveals conservation of structural domains. J Biol Chem 268:14399–14403

Elangbam CS, Qualls CW Jr, Dahlgren RR (1997) Cell adhesion molecules–update. Vet Pathol 34:61–73

Ermann J, Hoffmann P, Edinger M et al (2005) Only the CD62L$^+$ subpopulation of CD4$^+$CD25$^+$ regulatory T cells protects from lethal acute GVHD. Blood 105:2220–2226

Evans SS, Collea RP, Appenheimer MM, Gollnick SO (1993) Interferon-α induces the expression of the L-selectin homing receptor in human B lymphoid cells. J Cell Biol 123:1889–1898

Fagerstam JP, Whiss PA (2006) Higher platelet P-selectin in male patients with inflammatory bowel disease compared to healthy males. World J Gastroenterol 12:1270–1272

Farkas SA, Hornung M, Sattler C et al (2005) Preferential migration of CD62L cells into the appendix in mice with experimental chronic colitis. Eur Surg Res 37:115–125

Feehan C, Darlak K, Kahn J et al (1996) Shedding of the lymphocyte L-selectin adhesion molecule is inhibited by a hydroxamic acid-based protease inhibitor. Identification with an L-selectin-alkaline phosphatase reporter. J Biol Chem 271:7019–7024

Foster AE, Marangolo M, Sartor MM et al (2004) Human CD62L-memory T cells are less responsive to alloantigen stimulation than CD62L + naive T cells: potential for adoptive immunotherapy and allodepletion. Blood 104:2403–2409

Friedline RH, Wong CP, Steeber DA et al (2002) L-selectin is not required for T cell-mediated autoimmune diabetes. J Immunol 168:2659–2666

Fukuda MN, Ohyama C, Lowitz K et al (2000) Peptide mimic of E-selectin ligand inhibits sialyl Lewis X-dependent lung colonization of tumor cells. Cancer Res 60:4506

Gaddi E, Quiroz H, Balbaryski J et al (2005) L-selectin expression on T lymphocytes and neutrophils in HIV infected children. Medicina (B Aires) 65:131–137 [Article in Spanish]

Galkina E, Tanousis K, Preece G et al (2003) L-selectin shedding does not regulate constitutive T cell trafficking but controls the migration pathways of antigen-activated T lymphocytes. J Exp Med 198:1323–1335

Galustian C, Lawson AM, Komba S et al (1997) Sialyl-Lewis(x) sequence 6-O-sulfated at N-acetylglucosamine rather than at galactose is the preferred ligand for L-selectin and de-N-acetylation of the sialic acid enhances the binding strength. Biochem Biophys Res Commun 240:748–751, Erratum in: Biochem Biophys Res Commun 1998 245: 640

Galustian C, Lubineau A, le Narvor C et al (1999) L-selectin interactions with novel mono- and multisulfated Lewis(x) sequences in comparison with the potent ligand 3′-sulfated Lewis^a. J Biol Chem 274:18213–18217

Galustian C, Childs RA, Stoll M et al (2002) Synergistic interactions of the two classes of ligand, sialyl-Lewis(a/x) fuco-oligosaccharides and short sulpho-motifs, with the P- and L-selectins: implications for therapeutic inhibitor designs. Immunology 105:350–359

Galustian C, Elviss N, Chart H et al (2003) Interactions of the gastrotropic bacterium Helicobacter pylori with the leukocyte-endothelium adhesion molecules, the selectins–a preliminary report. FEMS Immunol Med Microbiol 36:127–134

Gomez-Gaviro MV, Gonzalez-Alvaro I, Dominguez-Jimenez C et al (2002) Structure-function relationship and role of tumor necrosis factor-alpha-converting enzyme in the down-regulation of L-selectin by non-steroidal anti-inflammatory drugs. J Biol Chem 277:38212–38221

Green PJ, Yuen CT, Childs RA et al (1995) Further studies of the binding specificity of the leukocyte adhesion molecule, L-selectin, towards sulphated oligosaccharides-suggestion of a link between the selectin- and the integrin-mediated lymphocyte adhesion systems. Glycobiology 5:29–38

Grewal IS, Foellmer HG, Grewal KD et al (2001) CD62L is required on effector cells for local interactions in the CNS to cause myelin damage in experimental allergic encephalomyelitis. Immunity 14:291–302

Hajilooi M, Tajik N, Sanati A et al (2006) Association of the Phe206Leu allele of the L-selectin gene with coronary artery disease. Cardiology 105:113–118

Hanley WD, Napier SL, Burdick MM et al (2006) Variant isoforms of CD44 are P- and L-selectin ligands on colon carcinoma cells. FASEB J 20:337–339

Hannawa KK, Eliason JL, Woodrum DT et al (2005) L-selectin-mediated neutrophil recruitment in experimental rodent aneurysm formation. Circulation 112:241–247

Hickey MJ, Forster M, Mitchell D et al (2000) L-selectin facilitates emigration and extravascular locomotion of leukocytes during acute inflammatory responses in vivo. J Immunol 165:7164–7170

Hirata T, Furie BC, Furie B (2002) P-, E-, and L-selectin mediate migration of activated CD8+ T lymphocytes into inflamed skin. J Immunol 169:4307–4313

Hong S, Farag NH, Nelesen RA et al (2004) Effects of regular exercise on lymphocyte subsets and CD62L after psychological versus physical stress. J Psychosom Res 56:363–370

Hourcade D, Holers VM, Atkinson JP (1989) The regulators of complement activation (RCA) gene cluster. In: Dixon FJ (ed) Advances in immunology, vol 45. Academic, New York, pp 381–416

Ikegami-Kuzuhara A, Yoshinaka T, Ohmoto H et al (2001) Therapeutic potential of a novel synthetic selectin blocker, OJ-R9188, in allergic dermatitis. Br J Pharmacol 134:1498–1504

Ivetic A, Ridley AJ (2004) The telling tail of L-selectin. Biochem Soc Trans 2:1118–1121

Ivetic A, Florey O, Deka J et al (2004) Mutagenesis of the ezrin-radixin-moesin binding domain of L-selectin tail affects shedding, microvillar positioning, and leukocyte tethering. J Biol Chem 279:33263–33272

Jackson LA, Drevets DA, Dong ZM et al (2005) Levels of L-selectin (CD62L) on human leukocytes in disseminated cryptococcosis with and without associated HIV-1 infection. J Infect Dis 191:1361–1367

Johnston GI, Cook RG, McEver RP (1989) CloningofGMP-140, a granule membrane protein of platelets and endothelium: sequence similarity to proteins involved in cell adhesion and inflammation. Cell 56:1033–1044

Jung U, Ley K (1999) Mice lacking two or all three selectins demonstrate overlapping and distinct functions for each selectin. J Immunol 162:6755–6762

Jung U, Ramos CL, Bullard DC, Ley K (1998) Gene-targeted mice reveal importance of L-selectin-dependent rolling for neutrophil adhesion. Am J Physiol 274:H1785–H1791

Junge S, Brenner B, Lepple-Wienhues A et al (1999) Intracellular mechanisms of L-selectin induced capping. Cell Signal 11:301–308

Jurgilewicz DH, Rogowski F, Lebkowska U et al (2002) E-selectin, L-selectin, ICAM-1 and IL-6 concentrations changes in the serum of patients with hyperthyroidism in the early period of radioiodine I-131 therapy. Nucl Med Rev Cent East Eur 5:39–42

Kaburagi Y, Hasegawa M, Nagaoka T et al (2002) The cutaneous reverse Arthus reaction requires intercellular adhesion molecule 1 and L-selectin expression. J Immunol 168:2970–2978

Kamiuchi K, Hasegawa G, Obayashi H et al (2002) Leukocyte-endothelial cell adhesion molecule 1 (LECAM-1) polymorphism is associated with diabetic nephropathy in type 2 diabetes mellitus. J Diabetes Complications 16:333–337

Kanegane H, Kasahara Y, Niida Y et al (1996) Expression of L-selectin (CD62L) discriminates Th1- and Th2-like cytokine-producing memory CD4+ T cells. Immunology 87:186–190

Kanoh N, Dai CF, Tanaka T et al (1999) Constitutive expression of GlyCAM-1 core protein in the rat cochlea. Cell Adhes Commun 7:259–266

Kansas GS, Ley K, Munro JM, Tedder TF (1993) Regulation of leukocyte rolling and adhesion to high endothelial venules through the cytoplasmic domain of L-selectin. J Exp Med 177:833–838

Karadayi K, Top C, Gulecek O (2003) The relationship between soluble L-selectin and the development of diabetic retinopathy. Ocul Immunol Inflamm 11:123–129

Kawashima H, Atarashi K, Hirose M et al (2002) Oversulfated chondroitin/dermatan sulfates containing GlcAbeta1/IdoAalpha1-3GalNAc(4,6-O-disulfate) interact with L- and P-selectin and chemokines. J Biol Chem 277:12921–12930

Kilian K, Dernedde J, Mueller EC et al (2004) The interaction of protein kinase C isozymes alpha, iota, and theta with the cytoplasmic domain of L-selectin is modulated by phosphorylation of the receptor. J Biol Chem 279:34472–34480

Koenig A, Jain R, Vig R et al (1997) Selectin inhibition: synthesis and evaluation of novel sialylated, sulfated and fucosylated oligosaccharides, including the major capping group of GlyCAM-1. Glycobiology 7:79–93

Kogelberg H, Rutherford TJ (1994) Studies on the three-dimensional behaviour of the selectin ligands Lewis(a) and sulphated Lewis(a) using NMR spectroscopy and molecular dynamics simulations. Glycobiology 4:49–57

Lasky LA, Singer MS, Yednock TA et al (1989) Cloning of a lymphocyte homing receptor reveals a lectin domain. Cell 56:1045–1055

Lasky LA, Singer MS, Dowbenko D et al (1992) An endothelial ligand for L-selectin is a novel mucin-like molecule. Cell 69:927–938

Laubli H, Stevenson JL, Varki A et al (2006) L-selectin facilitation of metastasis involves temporal induction of Fut7-dependent ligands at sites of tumor cell arrest. Cancer Res 66:1536–1542

Leid JG, Speer CA, Jutila MA (2002) Ultrastructural examination of cytoskeletal linkage of L-selectin and comparison of L-selectin cytoskeletal association to that of other human and bovine lymphocyte surface antigens. Cell Immunol 215:219–231

Leppanen A, Yago T, Otto VI et al (2003) Model glycosulfopeptides from P-selectin glycoprotein ligand-1 require tyrosine sulfation and a core 2-branched O-glycan to bind to L-selectin. J Biol Chem 278:26391–26400

Leung E, Greene J, Ni J et al (1996) Cloning of the mucosal addressin MAdCAM-1 from human brain: identification of novel alternatively spliced transcripts. Immunol Cell Biol 74:490–496

Leung E, Berg RW, Langley R et al (1997) Genomic organization, chromosomal mapping, and analysis of the 5' promoter region of the human MAdCAM-1 gene. Immunogenetics 46:111–119

Lowe JB (2003) Glycan-dependent leukocyte adhesion and recruitment in inflammation. Curr Opin Cell Biol 15:531–538

M'Rini C, Cheng G, Schweitzer C et al (2003) A novel endothelial L-selectin ligand activity in lymph node medulla that is regulated by α(1,3)-fucosyltransferase-IV. J Exp Med 198:1301–1312

Malhotra R, Ward M, Sim RB, Bird MI (1999) Identification of human complement Factor H as a ligand for L-selectin. Biochem J 341:61–69

Matala E, Alexander SR, Kishimoto TK et al (2001) The cytoplasmic domain of L-selectin participates in regulating L-selectin endoproteolysis. J Immunol 167:1617–1623

Mattila P, Majuri ML, Renkonen R (1992) VLA-4 integrin on sarcoma cell lines recognizes endothelial VCAM-1. Differential regulation of the VLA-4 avidity on various sarcoma cell lines. Int J Cancer 52:918–923

Meijer SL, Dols A, Hu HM, Chu Y et al (2004) Reduced L-selectin (CD62LLow) expression identifies tumor-specific type 1 T cells from lymph nodes draining an autologous tumor cell vaccine. Cell Immunol 227:93–102

Molad Y, Fridenberg A, Bloch K et al (2004) Neutrophil adhesion molecule expression in familial Mediterranean fever: discordance between the intravascular regulation of beta2 integrin and L-selectin expression in acute attack. J Investig Med 52:58–61

Mora C, Grewal IS, Wong FS, Flavell RA (2004) Role of L-selectin in the development of autoimmune diabetes in non-obese diabetic mice. Int Immunol 16:257–264

Nagaoka T, Kaburagi Y, Hamaguchi Y et al (2000) Delayed wound healing in the absence of intercellular adhesion molecule-1 or L-selectin expression. Am J Pathol 157:237–247

Nakagome K, Dohi M, Okunishi K et al (2005) Antigen-sensitized CD4 + CD62Llow memory/effector T helper 2 cells can induce airway hyperresponsiveness in an antigen free setting. Respir Res 6:46

Nishida Y, Uzawa H, Toba T et al (2000) A facile synthetic approach to L- and P-selectin blockers via copolymerization of vinyl monomers constructing the key carbohydrate modules of sialyl LewisX mimics. Biomacromolecules 1:68–74, Spring

Nishimura T, Takeshita N, Satow H, Kohmoto K (1993) Expression of the mC26 gene encoding GlyCAM-1 in the lactating mouse mammary gland. J Biochem (Tokyo) 114:567–569

Noma K, Yamaguchi Y, Okita R et al (2005) The spleen plays an immunosuppressive role in patients with gastric cancer: involvement of $CD62L^+$ cells and TGF-β. Anticancer Res 25:643–649

Norgard-Sumnicht K, Varki A (1995) Endothelial heparan sulfate proteoglycans that bind to L-selectin have glucosamine residues with unsubstituted amino groups. J Biol Chem 270:12012–12024

Norgard-Sumnicht KE, Varki NM, Varki A (1993) Calcium-dependent heparin-like ligands for L-selectin in nonlymphoid endothelial cells. Science 261:480–483

O'Connell D, Koenig A, Jennings S et al (1996) Calcium-dependent oligonucleotide antagonists specific for L-selectin. Proc Natl Acad Sci USA 93:5883–5887

Ord DC, Ernst TJ, Zhou LJ et al (1990) Structure of the gene encoding the human leukocyte adhesion molecule-1 (TQ1, Leu-8) of lymphocytes and neutrophils. J Biol Chem 265:7760–7767

Pascual DW, White MD, Larson T et al (2001) Impaired mucosal immunity in L-selectin-deficient mice orally immunized with a Salmonella vaccine vector. J Immunol 167:407–415

Prakobphol A, Tangemann K, Rosen SD et al (1999) Separate oligosaccharide determinants mediate interactions of the low-molecular-weight salivary mucin with neutrophils and bacteria. Biochemistry 38:6817–6825

Qian F, Hanahan D, Weissman IL (2001) L-selectin can facilitate metastasis to lymph nodes in a transgenic mouse model of carcinogenesis. Proc Natl Acad Sci USA 98:3976–3981

Rafiei A, Hajilooi M, Shakib RJ et al (2006) Association between the Phe206Leu polymorphism of L-selectin and brucellosis. J Med Microbiol 55(Pt 5):511–516

Ridger VC, Hellewell PG, Norman KE (2005) L- and P-selectins collaborate to support leukocyte rolling in vivo when high-affinity P-selectin-P-selectin glycoprotein ligand-1 interaction is inhibited. Am J Pathol 166:945–952

Rivera-Nieves J, Olson T, Bamias G et al (2005) L-selectin, α4β1, and α4β7 integrins participate in CD4+ T cell recruitment to chronically inflamed small intestine. J Immunol 174:2343–2352

Roberts AD, Ely KH, Woodland DL (2005) Differential contributions of central and effector memory T cells to recall responses. J Exp Med 202:123–133

Sackstein R, Meng L, Xu XM, Chin YH (1995) Evidence of post-transcriptional regulation of L-selectin gene expression in rat lymphoid cells. Immunology 85:198–204

Sampaio SO, Li X, Takeuchi M et al (1995) Organization, regulatory sequences, and alternatively spliced transcripts of the mucosal addressin cell adhesion molecule-1 (MAdCAM-1) gene. J Immunol 155:2477–2486

Samulowitz U, Kuhn A, Brachtendorf G et al (2002) Human endomucin: distribution pattern, expression on high endothelial venules, and decoration with the MECA-79 epitope. Am J Pathol 160:1669–1681

Sciacca FL, Sturzl M, Bussolino F et al (1994) Expression of adhesion molecules, platelet-activating factor, and chemokines by Kaposi's sarcoma cells. J Immunol 153:4816–4825

Shailubhai K, Streeter PR, Smith CE, Jacob GS (1997) Sulfation and sialylation requirements for a glycoform of CD34, a major endothelial ligand for L-selectin in porcine peripheral lymph nodes. Glycobiology 7:305–314

Shimada Y, Hasegawa M, Kaburagi Y et al (2003) L-selectin or ICAM-1 deficiency reduces an immediate-type hypersensitivity response by preventing mast cell recruitment in repeated elicitation of contact hypersensitivity. J Immunol 170:4325–4334

Shimizu Y, Newman W, Tanaka Y, Shaw S (1992) Lymphocyte interactions with endothelial cells. Immunol Today 13:106–112

Shyjan AM, Bertagnolli M, Kenney CJ et al (1996) Human mucosal addressin cell adhesion molecule-1 (MAdCAM-1) demonstrates structural and functional similarities to the α4β7-integrin binding domains of murine MAdCAM-1, but extreme divergence of mucin-like sequences. J Immunol 156:2851–2857

Siegelman MH, Weissman MH (1989) Human homologue of mouse lymph node homing receptor: evolutionary conservation at tandem cell interaction domains. Proc Natl Acad Sci USA 86:5562–5566

Siegelman MH, van de Rijn M, Weissman IL (1989) Mouse lymph node homing receptor cDNA clone encodes a glycoprotein revealing tandem interaction domains. Science 243(4895):1165–1172

Siegelman MH, Cheng IC, Weissman IL et al (1990) The mouse lymph node homing receptor is identical with the lymphocyte cell surface marker Ly-22: role of the EGF domain in endothelial binding. Cell 61:611–625

Smith ML, Olson TS, Ley K (2004) CXCR2- and E-selectin–induced neutrophil arrest during inflammation in vivo. J Exp Med 200:935–939

Sperandio M, Smith ML, Forlow SB et al (2003) P-selectin glycoprotein ligand-1 mediates L-selectin-dependent leukocyte rolling in venules. J Exp Med 197:1355–1363

Springer TA (1990) Adhesion receptors of the immune system. Nature 1977 346:425–434

Springer TA (1995) Traffic signals on endothelium for lymphocyte recirculation and leukocyte emigration. Annu Rev Physiol 57:827–872

Steeber DA, Green NE, Sato S, Tedder TF (1996) Humoral immune responses in L-selectin-deficient mice. J Immunol 157:4899–4907

Steeber DA, Engel P, Miller AS et al (1997) Ligation of L-selectin through conserved regions within the lectin domain activates signal transduction pathways and integrin function in human, mouse, and rat leukocytes. J Immunol 159:952–963

Steeber DA, Tang ML, Green NE et al (1999) Leukocyte entry into sites of inflammation requires overlapping interactions between the L-selectin and ICAM-1 pathways. J Immunol 163:2176–2186

Strauch UG, Obermeier F, Grunwald N et al (2005) Influence of intestinal bacteria on induction of regulatory T cells: lessons from a transfer model of colitis. Gut 54:1546–1552

Suzuki A, Andrew DP, Gonzalo JA et al (1996) CD34-deficient mice have reduced eosinophil accumulation after allergen exposure and show a novel crossreactive 90-kD protein. Blood 87:3550–3562

Symon FA, Wardlaw AJ (1996) Selectins and their counter receptors: a bitter sweet attraction. Thorax 51:1155–1157

Szanto S, Gal I, Gonda A et al (2004) Expression of L-selectin, but not CD44, is required for early neutrophil extravasation in antigen-induced arthritis. J Immunol 172:723–734

Szanya V, Ermann J, Taylor C et al (2002) The subpopulation of CD4 + CD25+ splenocytes that delays adoptive transfer of diabetes expresses L-selectin and high levels of CCR7. J Immunol 169:2461–2465

Takei T, Iida A, Nitta K et al (2002) Association between single-nucleotide polymorphisms in selectin genes and immunoglobulin A nephropathy. Am J Hum Genet 70:781–786

Tamatani T, Kitamura F, Kuida K et al (1993) Characterization of rat LECAM-1 (L-selectin) by the use of monoclonal antibodies and evidence for the presence of soluble LECAM-1 in rat sera. Eur J Immunol 23:2181–2188

Tang ML, Hale LP, Steeber DA, Tedder TF (1997) L-selectin is involved in lymphocyte migration to sites of inflammation in the skin: delayed rejection of allografts in L-selectin-deficient mice. J Immunol 158:5191–5199

Taverna D, Moher H, Crowley D et al (2004) Increased primary tumor growth in mice null for β3- or β3/β5-integrins or selectins. Proc Natl Acad Sci USA 101:763–768

Tedder TF, Isaacs CM, Ernst TJ et al (1989) Isolation and chromosomal localization of cDNAs encoding a novel human lymphocyte cell surface molecule, LAM-1. Homology with the mouse lymphocyte homing receptor and other human adhesion proteins. J Exp Med 170:123–133

Tedder TF, Steeber DA, Chen A, Engel P (1995) The selectins: vascular adhesion molecules. FASEB J 9:866–873

Toivonen A, Kulmala P, Rahko J et al (2004) Soluble adhesion molecules in finnish schoolchildren with signs of preclinical type 1 diabetes. Diabetes Metab Res Rev 20:48–54

Tsurushita N, Fu H, Berg EL (1996) PCR cloning of the cDNA encoding baboon L-selectin. Gene 181:219–220

Tu L, Poe JC, Kadono T, Venturi GM, Bullard DC, Tedder TF, Steeber DA (2002) A functional role for circulating mouse L-selectin in regulating leukocyte/endothelial cell interactions in vivo. J Immunol 169:2034–2043

Turunen JP, Majuri ML, Seppo A et al (1995) De novo expression of endothelial sialyl Lewis(a) and sialyl Lewis(x) during cardiac transplant rejection: superior capacity of a tetravalent sialyl Lewis(x) oligosaccharide in inhibiting L-selectin-dependent lymphocyte adhesion. J Exp Med 182:1133–1141

Ushakova NA, Probrazhenskaia ME, Nifant'ev NE et al (2001) Effects of mannose derivatives on development of selectin-dependent peritoneal inflammation in rats and mice. Vopr Med Khim 47:491–497 [Article in Russian]

Ushakova NA, Preobrazhenskaya ME, Bird MI et al (2005) Monomeric and multimeric blockers of selectins: comparison of in vitro and in vivo activity. Biochemistry (Mosc) 70:432–439

van Faassen H, Saldanha M, Gilbertson D et al (2005) Reducing the stimulation of CD8+ T cells during infection with intracellular bacteria promotes differentiation primarily into a central (CD62LhighCD44high) subset. J Immunol 174:5341–5350

Verdrengh M, Erlandsson-Harris H, Tarkowski A (2000) Role of selectins in experimental *Staphylococcus aureus*-induced arthritis. Eur J Immunol 30:1606–1613

Vestweber D (2002) Human endomucin: distribution pattern, expression on high endothelial venules, and decoration with the MECA-79 epitope. Am J Pathol 160:1669–1681

Wang L, Brown JR, Varki A, Esko JD (2002) Heparin's anti-inflammatory effects require glucosamine 6-O-sulfation and are mediated by blockade of L- and P-selectins. J Clin Invest 110:127–136

Wang J, Marschner S, Finkel TH (2004) CXCR4 engagement is required for HIV-1-induced L-selectin shedding. Blood 103:1218–1221

Wang L, Fuster M, Sriramarao P, Esko JD (2005) Endothelial heparan sulfate deficiency impairs L-selectin- and chemokine-mediated neutrophil trafficking during inflammatory responses. Nat Immunol 6:902–910

Watson ML, Kingsmore SF, Johnston GI et al (1990) Genomic organization of the selectin family of leukocyte adhesion molecules on human and mouse chromosome 1. J Exp Med 172:263–272

Yamada M, Yanaba K, Hasegawa M et al (2006) Regulation of local and metastatic host-mediated anti-tumor mechanisms by L-selectin and intercellular adhesion molecule-1. Clin Exp Immunol 143:216–227

Yang J, Xu Y, Zhu C, Offermann MK (1994) Regulation of adhesion molecule expression in Kaposi's sarcoma cells. J Immunol 152:361–373

You S, Slehoffer G, Barriot S et al (2004) Unique role of CD4 + CD62L + regulatory T cells in the control of autoimmune diabetes in T cell receptor transgenic mice. Proc Natl Acad Sci USA 101 (Suppl 2):14580–14585

P-Selectin and Its Ligands

G.S. Gupta

27.1 Platelet Adhesion and Activation

When a blood vessel is injured, control of bleeding starts with the rapid adhesion of circulating platelets to the site of damage. Within seconds, the adhered platelets are activated, secrete the contents of storage organelles, spread out over the damaged area and recruit more platelets to the developing thrombus. However, if this process occurs in a diseased, sclerotic or occluded vessel, the resulting platelet thrombus may break away and block the coronary artery, causing a heart attack, or restrict blood supply to the brain, causing a stroke. The glycoprotein (GP) Ib-IX-V complex, a member of the leucine-rich protein family, is a constitutive platelet membrane receptor for von Willebrand Factor (vWF), a multimeric adhesive glycoprotein found in the matrix underlying the endothelial cell lining of the blood vessel wall and in the plasma. Binding of vWF to the Ib-IX-V complex regulates adhesion of platelets to the sub-endothelium at high shear flow, and initiates signal transduction leading to platelet activation. The GP Ib-IX-V complex also constitutes a binding site for α-thrombin, an interaction that facilitates thrombin-dependent platelet activation. Analysis of GP Ib-IX-V complex and vWF has identified discrete amino acid sequences that mediate their interaction. An anionic/sulfated tyrosine sequence of the GP Ib α-chain that is critical for binding of Ib-IX-V complex to both vWF and α-thrombin is analogous to sulfated anionic amino acid sequences mediating interactions of other adhesive proteins, including P-selectin binding to PSGL-1 and Factor VIII binding to vWF (Andrews et al. 1997).

27.2 P-Selectin (GMP-140, PADGEM, CD62): A Member of Selectin Adhesion Family

27.2.1 Platelets and Vascular Endothelium Express P-Selectin

Selectins are adhesion receptors that mediate leukocyte adhesion to platelets or endothelial cells through Ca^{2+}-dependent interactions with cell surface oligosaccharides.

The selectin family consists of three closely related cell-surface molecules with differential expression by leukocytes (L-selectin), platelets (P-selectin), and vascular endothelium (E- and P-selectin). The forms, functions and some clinical applications of L-selectins have been discussed in previous chapter (Chap. 26). P-selectin and E-selectin are adhesion receptors for monocytes and neutrophils and expressed by stimulated endothelial cells. P-selectin is stored in Weibel-Palade bodies, and rapidly delivered to plasma membrane upon exocytosis of these secretory granules. E-selectin is not stored, and its synthesis is induced by cytokines. The fate of the two proteins has been studied after their surface expression by following the intracellular routing of internalized antibodies to the selectins. After a brief surface exposure, internalized E-selectin is degraded in the lysosomes, whereas P-selectin returns to the storage granules from where it can be reused. E-selectin and P-selectin are two closely related vascular cell adhesion proteins. P-selectin is a 140-kDa glycoprotein expressed on endothelial cells and platelets and expressed on the surface of these cells in response to inflammatory stimuli. P-selectin mediates the tethering and rolling of leukocytes along the endothelium during early step of leukocyte extravasation. Each selectin has an amino-terminal C-type lectin domain that possesses carbohydrate binding site, which recognizes certain vascular mucin-type glycoproteins bearing the carbohydrate structure sialyl-Lewis^x (sLe^x or CD15s) (Neu5Acα2-3Galβ1-4(Fucα1-3) GlcNAc). In addition to sLex carbohydrate, P-selectin binds sulfated proteoglycan, 3-sulfated galactosyl ceramide (sulfatide), and heparin. The clinical prognosis and metastatic progression of many epithelial carcinomas has been correlated independently with production of tumor mucins and with enhanced expression of sialyl-Lewis^x.

P-selectin is also expressed on macrophages in the arterial wall after carotid denudation injury and spontaneous atherosclerosis in atherosclerosis-prone apoE-deficient (apoE$^{-/-}$) mice. Furthermore, P-selectin mRNA expression was readily detectable in macrophage-rich plaques of atherosclerotic innominate arteries and blood monocyte-derived macrophages from apoE$^{-/-}$ mice. Thus, macrophages in carotid injury-induced neointimal lesions and spontaneous atherosclerotic plaques of the innominate artery acquire the ability to express P-selectin, as does regenerating endothelium. These findings provide a potential new paradigm in macrophage-mediated vascular inflammation, atherosclerosis, and neointimal hyperplasia after arterial injury (Li et al. 2005) (Fig. 27.1).

27.2.2 Human Granule Membrane Protein 140 (GMP-140)/P-Selectin

A structurally and functionally related group of genes, lymph node homing receptor (LHR), granule membrane protein 140 (GMP-140), and endothelial leukocyte adhesion molecule 1 (ELAM-1) constitute a gene cluster on mouse and human chromosome 1. GMP-140 is an inducible granule membrane protein of the selectin family of cell surface receptors that mediate interactions of leukocytes with the blood vessel wall. After cellular activation, it is rapidly redistributed to the plasma membrane. The cDNA-derived primary structure of GMP-140 predicted a cysteine-rich protein with multiple domains, including a "lectin" region, an "EGF:" domain, nine tandem consensus repeats related to those in complement-binding proteins, a transmembrane domain, and a short cytoplasmic tail. Some cDNAs also predicted a soluble protein with a deleted transmembrane segment. The domain organization of GMP-140 is similar to that of ELAM-1, which binds neutrophils. This similarity suggests that GMP-140 belongs to a new family of inducible receptors with related structure and function on vascular cells. In situ hybridization mapped GMP-140 to human chromosome 1 bands 21–24 consistent with chromosomal localization of LHR. Gene linkage analysis in the mouse indicated that these genes and serum coagulation factor V (FV) all map to a region of distal mouse chromosome 1 that is syntenic with human chromosome 1, with no crossovers identified between these four genes in 428 meiotic events. Moreover, long range restriction site mapping demonstrated that these genes map to within 300 kb in both the human and mouse genomes. Studies suggest that LHR, ELAM-1, and GMP-140 comprise the selectins that arose by multiple gene duplication events before divergence of mouse and human. The location of these genes on mouse and human chromosome 1 is consistent with a close evolutionary relationship to the complement receptor-related genes, which also are positioned on the same chromosomes in both species and with which these genes share a region of sequence homology (Johnston et al. 1989; Watson et al. 1990).

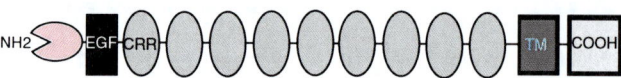

Fig. 27.1 Domain organization of selectin-P (CD62) also called granule membrane protein 140 (GMP-140). The extracellular portion of P-selectin contains an amino terminal domain homologous to C-type lectins and an adjacent epidermal growth factor (*EGF*)-like domain. The single EGF element that is found in P-selectin is followed by nine repetitive elements, each ~60 amino acids long, which resemble protein motives found in complement regulatory repeats (*CRR*) and a transmembrane (*TM*) sequence. A short cytoplasmic sequence is at the carboxyl terminus (*COOH*) of P-selectin. The number of amino acids present in the mature P-selectin as deduced from the cDNA sequences is 789

Two variant cDNAs for GMP-140 have been identified, one predicting a soluble form of the molecule lacking the transmembrane domain and the other predicting a molecule containing eight instead of nine consensus repeats. The human gene encoding GMP-140, which spans over 50 kbp, contains 17 exons. Almost all exons encode distinct structural domains, including the lectin-like domain, the EGF-like domain, each of the nine consensus repeats, and the transmembrane region. Each of the two deletions found in the variant cDNAs is precisely encoded by an exon, suggesting that these forms of GMP-140 are derived from alternative splicing of mRNA. Transcripts encoding the putative soluble form of GMP-140 can be amplified from both platelet and endothelial cell RNA by PCR (Johnston et al. 1990).

27.2.3 Murine P-Selectin

P-selectin contains a NH2-terminal carbohydrate-recognition domain, an epidermal growth factor motif, nine consensus repeats, a transmembrane domain, and a cytoplasmic tail. E-selectin (ELAM-1) and P-selectin (PADGEM) have both been described as human endothelial cell adhesion molecules for neutrophils and monocytes. The molecular cloning and sequencing of murine cDNA for P-selectin revealed the presence of lectin, EGF-like, transmembrane, and cytoplasmic domains, which are highly conserved between mouse and human, with an overall amino acid identity of 79% (Sanders et al. 1992; Weller et al. 1992). cDNA clones covering the full-length coding sequence of the homologous mouse proteins for both endothelial selectins were isolated by Weller et al. (1992). Rat P-selectin cDNA shows a significant nucleotide and amino-acid identity with human and mouse P-selectin. Similar to mouse P-selectin, the rat sequence lacks the equivalent of human complement regulatory protein-like repeat 2 (CR2). Seven potential N-linked glycosylation sites are conserved between the three species, suggesting that carbohydrate modification may play an important role in P-selectin function. P-selectin mRNA is undetectable in tissues of vehicle-treated animals.

Mouse P-selectin, like E-selectin, is transiently induced by TNF-α in several endothelioma cell lines derived from mouse tissues. The TNF induced, newly synthesized P-selectin protein is detected on the cell surface. Thus, P-selectin can be regulated on two different levels. The transport of stored P-selectin to the cell surface is controlled by regulated secretion of storage granules. Cell surface appearance of E an d P selectins on human umbilical vein endothelial cells (HUVEC) can be regulated by different mechanisms: E-selectin is transcriptionally induced, within hours, by TNF-α while P-selectin is transported from storage granules to the plasma membrane within minutes upon induction by various stimulating agents. After injection of endotoxin, the cellular response in vivo includes a rapid increase in the level of mRNA, presumably for new synthesis of P-selectin. After administration of endotoxin, the highest levels of mRNA expression were detected in liver, lung, kidney, thymus, spleen, and small intestine and heart (Sanders et al. 1992; Auchampach et al. 1994).

27.2.3.1 Rabbit P-Selectin

In rabbit platelets, P-selectin is an alpha-granule membrane protein that mediates leukocyte adhesion and thrombus propagation. The sequences of tryptic peptides of rabbit P-selectin showed an overall sequence identity of 74% with human P-selectin, and 69–77% identity with cow, dog, mouse, rat and sheep P-selectins. The apparent molecular mass of reduced rabbit P-selectin is 117 + 7 kDa which is approximately 8 kDa larger than the unreduced protein (109 + 5 kDa). Rabbit P-selectin appears smaller than human P-selectin, but is comparable to other species P-selectins, that have fewer 'complement regulatory protein' repeat domains. Rabbit P-selectin is nearly absent from the surface of platelets (290 + 30 molecules/cell). However, cellular activation with thrombin causes nearly a 30-fold increase in expression to 14,200 + 1,100 molecules/cell. P-selectin is expressed on the surface of rabbit platelets activated by other agonists like ADP, A23817 and epinephrine. P-selectin in rabbit platelets is similar in structure, cell localization and expression to human and other species P-selectins (Reed et al. 1998).

27.2.4 P-Selectin Promoter

27.2.4.1 Multiple Initiation Sites in 5′-Flanking Region of P-Selectin Gene

The 5′-flanking region of the human P-selectin gene shows multiple transcriptional initiation sites from -95 to -25 nucleotides relative to the start of protein-coding sequence. P-selectin gene is regulated by a combination of cis elements and their cognate transcription factors. Transfection experiments indicated that the sequence from -249 to -13 was sufficient to promote high level gene expression. Deletions to -197, -147, and -128 gradually reduced expression to basal levels, and further deletion to -100 abolished expression. The putative regulatory elements in short 5′-flanking sequence included a CACCC sequence, two inverted repeats similar to binding sites for the Ets and NF-kB/rel families, a GATA motif, and a sequence related to the GT-IIC element of the SV40 enhancer. The GATA element was functional, as it bound recombinant GATA-2 (Pan and McEver 1993).

27.2.4.2 Species-Specific Mechanisms for Transcriptional Regulation in Endothelial Cells

The 5′-flanking region of murine P-selectin gene was compared with the human gene. The murine and human genes shared conserved Stat-like, Hox, Ets, GATA, and GT-IIC elements. In the murine gene, a conserved GATA element bound to GATA-2 and functioned as a positive regulatory element, whereas a conserved Ets element bound to GA-binding protein and functioned as a negative regulatory element. However, the murine P-selectin gene had several features not found in the human gene. These included an insertion from -987 to -649 that contained tandem GATA and tandem AP1-like sequences, which resembled enhancers in β-globin locus control regions. Both tandem elements bound specifically to nuclear proteins. The murine gene lacked the unique κB site specific for p50 or p52 homodimers found in the human gene. Instead, it contained two tandem κB elements and a variant activating transcription factor/cAMP response element site, which closely resembled sites in the E-selectin gene that are required for TNF-α or LPS-inducible expression. Deletional analysis of the murine 5′-flanking region revealed several sequences that were required for either constitutive or inducible expression (Pan et al. 1998a).

27.2.4.3 kB Site in P-Selectin Promoter

Pan and McEver (1995) identified and characterized a kB site (-218GGGGGTGACCCC-207) in the promoter of the human P-selectin gene. The κB site is unique in that it binds constitutive nuclear protein complexes containing p50 or p52, but not inducible nuclear protein complexes containing p65. Furthermore, the element bound recombinant p50 or p52 homodimers, but not p65 homodimers. The p50 or p52 homodimers contacted the guanines at positions -218 to -214 on the coding strand and at -210 to -207 on the noncoding strand. Changes in the three central residues at -213 to -211 altered binding specificity for members of the NF-kB/Rel family. Data suggest that Bcl-3 differentially regulates the effects of p50 and p52 homodimers bound to kB site of the

P-selectin promoter. This site may be a prototype for kB elements in other genes that bind specifically to p50 and/or p52 homodimers.

kB Sites and an ATF/CREB for TNF-α- or LPS-Induced Expression of Murine P-Selectin Gene: The ATF-a$_o$ cDNA, a variant of the ATF/CREB transcription factor family contains a large in-frame deletion of 525 bp that removes the P/S/T-rich putative transactivation domain and binds E-selectin promoter, NF-ELAM1/δ A. The putative mRNAs for ATF-a$_o$ and ATF-a are present at varying ratios in different tissues. ATF-a$_o$ is an important member of the ATF family with a negative regulatory role in transactivation (Pescini et al. 1994). The TNF-α and LPS augment the expression of a reporter gene driven by murine, but not the human P-selectin promoter in transfected endothelial cells. The regions from -593 to -474 and from -229 to -13 in the murine P-selectin promoter are required for TNF-α or LPS to stimulate reporter gene expression. Within these regions, there are two tandem kB elements, a reverse-oriented kB site and a variant ATF/cAMP response element (ATF/CRE), that participate in TNF-α or LPS-induced expression. The tandem kB elements bound to NF-kB heterodimers and p65 homodimers, the reverse-oriented kB site bound to p65 homodimers, and the variant ATF/CRE bound to nuclear proteins that included ATF-2. Co-overexpression of p50 and p65 enhanced murine P-selectin promoter activity in a kB site-dependent manner. Results indicated that kB sites and the variant ATF/CRE are required for TNF-α and LPS to optimally induce the expression of murine P-selectin gene. The presence of these elements in the murine, but not the human, P-selectin gene may explain in part why TNF-α or LPS stimulates transcription of P-selectin in a species-specific manner. Thus, both species-specific and conserved mechanisms regulate transcription of the human and murine P-selectin genes (Pan et al. 1998b).

Interference of Sp1 and NF-kB: Gene activation by NF-kB/Rel transcription factors is modulated by synergistic or antagonistic interactions with other promoter-bound transcription factors. For example, Sp1 sites are often found in NF-kB-regulated genes, and Sp1 can activate certain promoters in synergism with NF-kB through nonoverlapping binding sites. It was found that Sp1 acts directly through a subset of NF-kB binding GC box sites. In contrast, NF-kB does not bind to a GC box Sp1 site. Sp1 can activate transcription through immunoglobulin k-chain enhancer or P-selectin promoter NF-kB sites. The p50 homodimers replace Sp1 from the P-selectin promoter by binding site competition and thereby either inhibit basal Sp1-driven expression or, in concert with Bcl-3, stimulate expression. The interaction of Sp1 with NF-kB sites thus provides a means to keep an elevated basal expression of NF-kB-dependent genes in the absence of activated nuclear NF-kB/Rel (Hirano et al. 1998).

27.2.4.4 Stat6 Activation Is Essential for IL-4 Induction of P-Selectin Transcription

Agonists of P-selectin expression fall into two categories: those that induce a very rapid, transient increase, lasting only hours, and those that induce prolonged upregulation lasting days. The latter group includes IL-4 that is a mediator of chronic P-selectin upregulation. The increase in P-selectin expression induced by IL-4 results from increased transcriptional activation of the P-selectin gene. Khew-Goodall et al. (1999) demonstrated the existence of two functional signal transducers and activator of transcription 6 (Stat6) binding sites on the P-selectin; binding of at least one site at activated Stat6 is essential for IL-4-induction in P-selectin transcription. Site 1 (nt -142) bound Stat6 with a higher affinity than did site 2 (nt -229). IL-4 also induced prolonged activation of Stat6, which was contingent on the continuous presence of IL-4. The sustained activation of Stat6 induced by IL-4 is likely to be a key factor leading to the prolonged activation of the P-selectin promoter, thereby resulting in prolonged P-selectin upregulation.

27.2.4.5 c-GMP Dependent Down-Regulation of P-Selectin Expression

The NO receptor, soluble guanylate cyclase (sGC), is expressed in endothelial cells. Fluorescence-activated cell sorting analysis in vitro and a specific P-selectin antibody in vivo revealed that selective down-regulation of P-selectin expression accounted the anti-adhesive effects of sGC activation. Thus, sGC plays a key anti-inflammatory role by inhibiting P-selectin expression and leukocyte recruitment (Ahluwalia et al. 2004).

27.2.5 Structure-Function Studies

27.2.5.1 Amino Acid Residue in E- and P-Selectin EGF Domains and Carbohydrate Specificity

Both E- and P-selectin have an EGF domain that is immediately adjacent to C-terminal lectin domain. The mutagenic substitution of single amino acid residues in either the P- or E-selectin EGF domain can dramatically alter selectin binding to sLe^x, heparin, or sulfatide. Substitution of E- and P-selectin EGF domain residue Ser128 with an arginine results in E- and P-selectin proteins that have lost the requirement for α1-3-linked fucose and are thus able to bind to sialyllactosamine (Graves et al. 1994). Additionally, conservative substitution of EGF domain residues 124 and 128 can alter E-selectin binding such that it is able to adhere to heparin or sulfatide and reduce P-selectin adherence to these ligands. The distance between the substituted EGF domain amino acid residues and the primary carbohydrate binding site within the lectin domain and their relative positioning as determined by crystal structure analysis of the E-selectin lectin and EGF domains suggest that there is little direct contact between the two domains. Selectin oligosaccharide binding may be modulated by both domains and that wild-type E- and P-selectin/sLe^x binding interactions may be significantly different from those previously hypothesized (Revelle et al. 1996a).

27.2.5.2 Monomeric Soluble- and Membrane P-Selectins

Ushiyama et al. (1993) expressed two soluble forms of P-selectin, one truncated after the ninth repeat (tPS) and the other lacking the transmembrane domain due to alternative RNA splicing (asPS). When visualized by electron microscopy, each was a monomeric rod-like structure with a globular domain at one end, whereas membrane P-selectin (mPS) from platelets formed rosettes with the globular domains facing outward. While tPS and asPS are asymmetric monomers, mPS is oligomeric. HL-60 cells adhered to immobilized tPS and asPS, although less efficiently than to mPS. 125I-Labeled tPS and asPS bound to approximately 25,000 sites/neutrophil and approximately 36,000 sites/HL-60 cell with an apparent K_D of 70 nM. Treatment of HL-60 cells with O-sialoglycoprotease eliminated the binding sites for asPS. Thus; (1) P-selectin is a rigid, asymmetric protein; (2) monomeric soluble P-selectin binds to high affinity ligands with sialylated O-linked oligosaccharides on leukocytes; and (3) oligomerization of mPS enhances its avidity for leukocytes (Ushiyama et al. 1993).

27.2.5.3 Cytoplasmic Domain of P-Selectin

The cytoplasmic domains of many membrane proteins have short sequences, usually including a tyrosine or a di-leucine, that function as sorting signals. The cytoplasmic domain of P-selectin was sufficient to confer rapid turnover on LDL-R. The 35-residue cytoplasmic domain of P-selectin contains signals for sorting into regulated secretory granules, for endocytosis, and for movement from endosomes to lysosomes. The domain has a membrane-distal sequence, YGVFTNAAF that resembles some tyrosine-based signals. Deletion of 10 amino acids from the cytoplasmic domain of P-selectin implicates this sequence as a necessary element of a lysosomal targeting signal. Sorting of P-selectin away from efficiently recycled proteins occurs in endosomes. This sorting event represents a constitutive equivalent of receptor down regulation, and may function to regulate the expression of P-selectin at the surface of activated endothelial cells (Green et al. 1994). However, mutations and deletions in the putative tyrosine-based motif in the cytoplasmic tail of human P-selectin did not clearly implicate these residues as critical components of a short internalization signal in transfected CHO cells, and it was not possible to identify a short internalization signal in the cytoplasmic tail of P-selectin. Residues throughout the cytoplasmic domain, and perhaps the transmembrane sequence to which the domain is attached, affect the efficiency of internalization (Setiadi et al. 1995).

Norcott et al. (1996) reported that P-selectin contains signals that target a chimera composed of horseradish peroxidase (HRP) and P-selectin, to the synaptic-like microves. Mutagenesis of the chimera followed by transient expression showed that at least two different sequences within the carboxy-terminal cytoplasmic tail of P-selectin are necessary, but that neither is sufficient for trafficking to the synaptic-like microvesicles (SLMV). One of these sequences is centred on the 10 amino acids of the membrane-proximal C1 exon that is also implicated in lysosomal targeting. The other sequence needed for trafficking to the SLMV includes the last four amino acids of the protein. Blagoveshchenskaya et al. (1998b) discovered a lysosomal targeting signal, KCPL, located within the C1 domain of the cytoplasmic tail. Alanine substitution of this tetrapeptide reduced lysosomal targeting to the level of a tailless HRP-P-selectin chimera, which was previously found to be deficient in both internalization and delivery to lysosomes. A proline residue within this lysosomal targeting signal makes a major contribution to the efficiency of lysosomal targeting. The sequence KCPL within the cytoplasmic tail of P-selectin is a structural element that mediates sorting from endosomes to lysosomes. A balance of positive and negative signals is required for proper lysosomal sorting of P-selectin. Within the sequence KCPL, Cys-766 plays a major role along with Pro-767, whereas Lys-765 and Leu-768 make no contribution to promoting lysosomal targeting. In addition, HRP-P-selectin chimeras were capable of acylation in vivo with ^{3H}palmitic acid at Cys-766, since no labeling of a chimera in which Cys-766 was replaced with Ala was detected (Blagoveshchenskaya et al. 1998a). The cytoplasmic domain is acylated at Cys766 through a thioester bond. Fatty acid acylation may regulate intracellular trafficking or other functions of P-selectin (Fujimoto et al. 1993).

P-selectin mutants expressed in AtT-20, a murine cell line with secretory granules containing the hormone corticotropin ('ACTH') revealed that wild-type P-selectin and mutants with alanine substitutions at 14 different positions in the cytoplasmic tail were concentrated in the tips of the cellular processes, which contained the majority of corticotropin granules. However, targeting to the cell tips was greatly decreased for Tyr^{777}—>Ala, Tyr^{777}—>Phe, Gly^{778}—>Ala, Phe^{780}—>Ala and Leu^{768}/Asn^{769}—>Ala/Ala mutants. Results indicated that Tyr^{777}, Gly^{778} and Phe^{780} form part of an atypical tyrosine-based motif, which also requires the presence of Leu768 and/or Asn769 to mediate sorting of P-selectin to secretory granules (Modderman et al. 1998).

Within the cytoplasmic domain of P-selectin, Tyr^{777} is needed for the appearance of P-selectin in immature and mature dense core granules (DCG), as well as for targeting to synaptic-like microvesicles (SLMV). The latter destination also requires additional sequences (Leu^{768} and $^{786}DPSP^{789}$) which are responsible for movement through endosomes en route to the SLMV. Leu^{768} also mediates transfer from early transferrin (Trn)-positive endosomes to the lysosomes; i.e., operates as a lysosomal targeting signal. Together, results are consistent with a model of SLMV biogenesis which involves an endosomal intermediate in neuroendocrine PC12 cells. In addition, the impairment of SLMV or DCG targeting results in a concomitant increase in lysosomal delivery, illustrating the entwined relationships between routes leading to regulated secretory organelles (RSO) and to lysosomes (Blagoveshchenskaya et al. 1999). The addition of the cytoplasmic domain of P-selectin to FIX modifies the cellular fate of the FIX molecule by directing the recombinant protein toward regulated-secretory granules without altering its coagulant activity (Plantier et al. 2003).

27.2.6 P-Selectin-Sialyl Lewis^x Binding Interactions

P-selectin adhesion to leukocytes is mediated by the amino-terminal lectin domain that binds sLe^x carbohydrate (Neu5Acα2-3Galβ1-4(Fucα1-3)GlcNAc). Using known binding interactions that occur between the rat MBP and its ligand (oligomannose) as a template (Weis et al. 1992), and substituting Ala-77 with lysine, it was observed that P-selectin-carbohydrate binding specificity changed from sLe^x to oligomannose. Results indicated that P-selectin binds sLe^x in a shallow cleft that is similar to the MBP saccharide-binding cleft. Additionally, Lys-113 has been implicated earlier in P-selectin binding to both sLe^x and 3-sulfated galactosylceramide (sulfatide). But Revelle et al. (1996b) demonstrated that Lys-113 is probably not involved in P-selectin binding to either sulfatide or sLe^x. Functionally, it appeared that P-selectin has retained a conserved carbohydrate and calcium coordination site that enables it to bind carbohydrate in a manner that is quite similar to that which has been determined for the rat mannose-binding protein (Revelle et al. 1996b). Several peptides corresponding to amino acid sequences within the lectin domains of selectins inhibit neutrophil (PMN) adhesion to P-selectin. One of the active regions, 109–118 aa contains residues, which are critical for E-selectin binding to sLe^x counter receptor. Peptide sequences which inhibited PMN binding to the fusion proteins were not necessarily those that inhibited fusion protein binding to sLe^x. In addition, various amino acid substitutions could be tolerated at the 111 and 113 positions without altering inhibitory activity. Modeling suggests that structural conformations of peptide analogues could explain the differences in biological activity of peptide analogues compared to mutants of the native protein (Tam et al. 1996).

Compared to weak binding of selectins to sLe^x-like glycans, selectins bind with high-affinity to specific glycoprotein counterreceptors, including PSGL-1. Somers et al. (2000) reported crystal structures of human P- and E-selectin constructs containing the lectin and EGF (LE) domains co-complexed with s-Le^x the crystal structure of P-selectin LE co-complexed with the N-terminal domain of human PSGL-1 modified by both tyrosine sulfation and s-Le^x. These structures reveal differences in how E- and P-selectin bind SLe^x and the molecular basis of the high-affinity interaction between P-selectin and PSGL-1.

27.2.7 Functions of P-Selectin

27.2.7.1 Mediation of Leukocyte and Platelet Rolling on the Vessel Wall

Selectins not only mediate leukocyte rolling, but also platelet rolling on the vessel wall. The functional significance of platelet rolling has not been established. The process could be important for hemostasis leading to firm platelet adhesion at sites of denuded endothelium and/or in inflammation. After activation, platelets may help in leukocyte recruitment as shown by studies of lymphocyte homing to peripheral lymph nodes. Surprisingly, work with the P-selectin mutant mice has also revealed an anti-inflammatory aspect of platelet P-selectin. P-selectin binding to leukocytes promoted the transcellular production of an anti-inflammatory mediator limiting the extent of acute glomeluronephritis. In addition, soluble P-selectin was shown to be shed from both activated platelets and endothelium and there are strong indications that it too could have an attenuating effect on inflammatory disease progression. The selectin-deficient mice demonstrated crucial role of P- and E-selectins in the homing of hematopoietic progenitor cells to the bone marrow. The selectin mutant mice have taught us a great deal about the role of selectins in normal physiology and in pathology. Further studies are needed to explore the regulation of shedding of the selectins and the function of soluble selectins in vivo. Exploring new territories of selectin-mediated interactions may provide a basis for developing new interventions and treatments for diseases in which the role of selectins has not yet been suspected (Hartwell and Wagner 1999).

27.2.7.2 Migration of T Lymphocyte Progenitors to Thymus

It has been shown that P-selectin is expressed by thymic endothelium and that lymphoid progenitors in bone marrow and thymus bind P-selectin. Parabiosis, competitive thymus reconstitution and short-term homing assays indicated that P-selectin and its ligand PSGL-1 are functionally important components of the thymic homing process. Accordingly, thymi of mice lacking PSGL-1 contained fewer early thymic progenitors and had increased empty niches for prothymocytes compared with wild-type mice. Furthermore, the number of resident thymic progenitors controling thymic expression of P-selectin suggested that regulation of P-selectin expression by a thymic 'niche occupancy sensor' may be used to direct progenitor access (Rossi et al. 2005).

27.2.7.3 Acute Emigration of Neutrophils

Whereas P-selectin is known to be involved in early stages of an inflammatory response, Johnson et al. (1995) indicated that it is additionally responsible for leukocyte rolling and macrophage recruitment in more prolonged tissue injury. P-Selectin also mediates rolling or slowing of neutrophils, while ICAM-1 contributes to the firm adhesion and emigration of neutrophils. Removing the function of either molecule partially prevents neutrophil emigration. P-selectin-deficient mice exhibit peripheral neutrophilia (Mayadas et al. 1993). This is not caused by changes in bone marrow precursors nor by a lack of neutrophil margination. Bullard et al. (1995) generated a line of mice with mutations in both of these molecules. While mice with either mutation alone show a 60–70% reduction in acute neutrophil emigration into the peritoneum during *S. pneumoniae*-induced peritonitis, double mutant mice show a complete loss of neutrophil emigration. In contrast, neutrophil emigration into the alveolar spaces during acute *S. pneumoniae*-induced pneumonia is normal in double mutant mice. These data demonstrate organ-specific differences, since emigration into the peritoneum requires both adhesion molecules while emigration into the lung requires neither. In the peritoneum, P-selectin-independent and ICAM-1-independent adhesive mechanisms permit reduced emigration when one of these molecules is deficient, but P-selectin-independent mechanisms cannot lead to ICAM-1-independent firm adhesion and emigration (Bullard et al. 1995).

27.2.7.4 P-Selectin Mediates Adhesion of Leukocytes, Platelets, and Cancer Cells

The P-selectin has long been known to support leukocyte rolling and emigration at sites of inflammation. Rolling under the hydrodynamic drag forces of blood flow is mediated by the interaction between selectins and their ligands across the leukocyte and endothelial cell surfaces. The high strength of binding combined with force-dependent rate constants and high molecular elasticity are tailored to support physiological leukocyte rolling (Fritz et al. 1998). Multiple-targeted deficiencies in selectins revealed a predominant role for P-selectin in leukocyte recruitment (Robinson et al. 1999). P-selectin is also a key molecule in hemostasis and thrombosis, mediating platelet rolling, generating procoagulant microparticles containing active tissue factor and enhancing fibrin deposition. Elevated levels of plasma P-selectin are indicative of thrombotic disorders and predictive of future cardiovascular events. Because the interaction between P-selectin and its receptor P-selectin glycoprotein ligand-1 (PSGL-1/CD162) represents an important mechanism by which P-selectin induces the formation of procoagulant microparticles and recruits the microparticles to thrombi, anti-thrombotic strategies are currently aimed at inhibiting this interaction. Recent developments also suggest that the procoagulant potential of P-selectin could be used to treat coagulation disorders such as hemophilia A (Cambien and Wagner 2004).

Stimulated endothelial cells and activated platelets express P-selectin, which interacts with PSGL-1 for leukocyte rolling on stimulated endothelial cells and heterotypic aggregation of activated platelets onto leukocytes. Crosslinking of PSGL-1 by P-selectin also primes leukocytes intracellularly for cytokine and chemoattractant-induced β2-integrin activation for firm adhesion of leukocytes. Furthermore, P-selectin mediates heterotypic aggregation of activated platelets to cancer cells and adhesion of cancer cells to stimulated endothelial cells. P-selectin blockade significantly inhibits inflammation and neointimal formation after arterial injury. In atherosclerosis, both platelet and endothelial cell P-selectins are important. Platelet P-selectin expression, but not endothelial P-selectin, plays a crucial role in the development of neointimal formation after arterial injury, and therapeutic strategies targeting leukocyte-platelet interactions could be effective in inhibiting restenosis (Wang et al. 2005). Chen and Geng (2006) provided a comprehensive summary of the functional roles and the biological importance of P-selectin-mediated cell adhesive interactions in the pathogeneses of inflammation, thrombosis, and the growth and metastasis of cancers (Chen and Geng 2006). Crystal structures of lectin and EGF–like domains of P-selectin show 'bent' and 'extended' conformations. An extended conformation would be 'favored' by forces exerted on a selectin bound at one end to a ligand and at the other end to a cell experiencing hydrodynamic drag force. The introduction of an N-glycosylation site to 'wedge open' the interface between the lectin and EGF-like domains of P-selectin increased the affinity of

P-selectin for its ligand PSGL-1, and thereby the strength of P-selectin-mediated rolling adhesion. Similarly, an asparagine-to-glycine substitution in the lectin-EGF-like domain interface of L-selectin enhanced rolling adhesion under shear flow. These results demonstrate that force, by 'favoring' an extended selectin conformation, can strengthen selectin-ligand bonds (Phan et al. 2006).

The selectins must resist applied forces to mediate leukocyte tethering and rolling along the endothelium and have 2 conformational states. Selectin–ligand bond dissociation increases only modestly with applied force, and exhibits catch bond behavior in a low-force regime where bond lifetimes counterintuitively increase with increasing force. Both allosteric and sliding–rebinding models have emerged to explain catch bonds. Waldron TT and Springer (2009) suggest a large residue into a cleft that opens within the lectin domain to stabilize the more extended, high-affinity selectin conformation. This mutation stabilizes the high-affinity state, but surprisingly makes rolling less stable. The position of the mutation in the lectin domain provides evidence for an allosteric pathway through the lectin domain, connecting changes at the lectin–EGF interface to the distal binding interface (Fig. 27.2).

Superoxide Anion Release by Monocytes and Neutrophils Through P-Selectin: The leukocytes are functionally modified by their adhesion to activated platelets. The levels of superoxide anion production were found to be markedly elevated when thrombin-activated platelets were used. The extent of this enhancement was much smaller when leukocytes were cultured with resting platelets than activated platelets. The membranes prepared from activated platelets also induced superoxide anion production, but the culture supernatant of activated platelets did not. This indicated that the adhesion of activated platelets to the leukocytes through P-selectin is a crucial step for the activation of leukocyte function, and support the idea that activated platelets are actively involved in inflammation processes (Nagata et al. 1993).

Two Sites on P-Selectin Are Involved in Adhesion of Monocytes to Thrombin-Activated EC: Peptides derived from both the lectin (residues 19–34 and 51–61) and EGF-like (residues 127–139) domains inhibit the adhesion of peripheral blood mononuclear cells (PBMC), elutriated monocytes and a monocytic cell line (U937) to thrombin-activated EC. All three peptides, when conjugated to BSA and coated on plastic plates, mediated U937 cell adhesion. This study shows that two sites on P-selectin, the lectin and EGF-like domains, are involved in the adhesion of monocytes to thrombin-activated EC (Murphy and McGregor 1994).

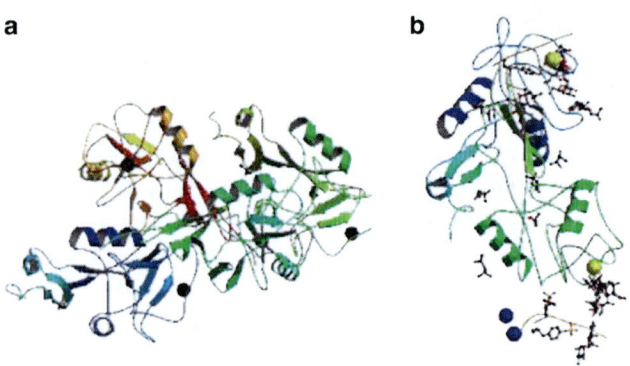

Fig. 27.2 (a) Crystal structure of human P-selectin lectin/EGF domains (PDB ID: 1G1Q); (b) crystal structure of P-selectin lectin/EGF domains complexed with PSGL-1 peptide (PDB ID:1G1S) (Somers et al. 2000)

27.2.7.5 P-Selectin Regulates MCP-1 and TNF-α Secretion

Adhesion of human monocytes to P-selectin, the most rapidly expressed endothelial tethering factor, increased the secretion of monocyte chemotactic protein-1 (MCP-1) and TNF-α by the leukocytes when they were stimulated with platelet-activating factor (PAF). Tethering by P-selectin specifically enhanced nuclear translocation of NF-kB required for expression of MCP-1, TNF-α, and other immediate-early genes. Results demonstrate that P-selectin, through its ligands on monocytes, may locally regulate cytokine secretion in inflamed tissues (Weyrich et al. 1995). P-selectin plays an important role in PAF-induced injury in mice, and the selectins and the integrin-ICAM-1 system work in concert to mediate the inflammatory response to PAF in vivo (Sun et al. 1997).

27.2.7.6 B Lymphocyte Binding to E- and P-Selectins

Activated but not resting B cells are able to interact with E and P selectins. This binding capacity of activated B cells paralleles the induction of different carbohydrate epitopes (Lewis^x, sialyl-Lewis^x, CD57 and CDw65) as well as other molecules bearing these or related epitopes in myeloid cells (L-selectin, αLβ2 and αXβ2 integrins, and CD35) involved in the interaction of different cell types with selectins. B cells infiltrating inflamed tissues like in Hashimoto's thyroiditis, also express these selectin-binding carbohydrates in parallel with the expression of E-selectin by surrounding follicular dendritic cells. Thus, in addition to the involvement of integrins, E- and P-selectins could play an important role in the interaction of B lymphocytes with the endothelium during B cell extravasation into lymphoid tissues and inflammatory foci as well as in their organization into lymphoid organs (Postigo et al. 1994).

27.2.7.7 Nonrolling Interaction Mediated by P-Selectin

The renal glomerulus is one of the few sites within the microvasculature in which leukocyte recruitment occurs in capillaries. Following infusion of anti-glomerular basement membrane (GBM) Ab, leukocytes became adherent in glomerular capillaries via a process of immediate arrest, without undergoing prior detectable rolling. However, despite the absence of rolling, this recruitment involved non-redundant roles for the P-selectin/P-selectin glycoprotein ligand-1 and β2 integrin/ICAM-1 pathways, suggesting that a novel form of the multistep leukocyte adhesion cascade occurs in these vessels. Perhaps, anti-GBM Ab-induced leukocyte adhesion in glomeruli occurs via a pathway that involvs a nonrolling interaction mediated by platelet-derived P-selectin (Kuligowski et al. 2006).

27.2.7.8 P-Selectin-Dependent, PSGL-1-Independent Rolling

Interestingly, antibodies and pharma-cological inhibitors (e.g., rPSGL-Ig) that target the N-terminus of PSGL-1 reduce but do not abolish P-selectin-dependent leukocyte rolling in vivo whereas PSGL-1-deficient mice have almost no P-selectin-dependent rolling. Therefore, Ridger et al. (2005) investigated mechanisms of P-selectin-dependent, PSGL-1-independent rolling. Data suggest that leukocytes can continue to roll in the absence of optimal P-selectin/PSGL-1 interaction using an alternative mechanism that involves P-selectin-, L-selectin-, and sialyl Le^x-bearing ligands.

27.3 P-Selectin Glycoprotein Ligand-1 (PSGL-1)

27.3.1 PSGL-1: The Major Ligand for P-Selectin

Based on site-directed mutagenesis, blocking monoclonal antibodies, and biochemical analyses, the extreme amino terminal extracellular domain of PSGL-1 is critical for interactions with selectins The current hypothesis is that for high affinity interactions with P-selectin, PSGL-1 must contain O-glycans with a core-2 branched motif containing the s-Lewis X antigen (NeuAc α2—>3Gal β1—>4[Fuc α1—>3]GlcNAc β1—>R). In addition, high affinity interactions require the co-expression of tyrosine sulfate on tyrosine residues near the critical O-glycan structure. This review addresses the biochemical evidence for this hypothesis and the evidence that PSGL-1 is an important in vivo ligand for cell adhesion (Cummings 1999; Moore 1998; Kum et al. 2009).

27.3.1.1 Human (PSGL-1, CD162)

Human P-selectin glycoprotein ligand-1 (PSGL-1, CD162) is expressed on the surface of myeloid cells and serves as the high affinity counter receptor for P-selectin. PSGL-1 is a disulfide-bonded, homodimeric mucin (approximately 250 kDa) on leukocytes that binds to P-selectin on platelets and endothelial cells during the initial steps in inflammation. PSGL-1 is a transmembrane glycoprotein, which is constitutively expressed on leukocytes, binds selectins. It mediates the initial tethering of leukocytes to activated platelets and endothelium. PSGL-1 is an essential adhesive molecule mediating the rolling of leukocytes on the endothelial cells and the recruitment of leukocytes to the inflamed tissue. In addition to its direct role in capture of leukocytes from the blood stream, PSGL-1 also functions as a signal-transducing receptor and initiates a series of intracellular signal events during the activation of leukocytes. PSGL-1 shares common features with platelet glycoprotein Ibα. A recently described polymorphism in this receptor that results in a variable number of tandem repeats (VNTR) sequence present either 16, 15 or 14 times (alleles A, B or C) could, similar to GP Ibα, be functionally relevant. PSGL-1 is a disulfide-bonded, homodimeric mucin (approximately 250 kDa) on leukocytes that binds to P-selectin on platelets and endothelial cells during the initial steps in inflammation. Genes involved in inflammatory processes are candidates for predisposition to prothrombotic syndromes. Moreover, adhesion to E-selectin inhibits the proliferation of human CD34$^+$ cells isolated either from blood, or steady-state bone marrow (Winkler et al. 2004). A dimeric sialoglycoprotein from myeloid cells with subunits of Mr of 120 kDa is selectively recognized by P-selectin. This P-selectin ligand carries α2-3-linked sialic acids and the sialyl-Lewis^x (sialyl Le^x) tetrasaccharide motif and contains < 1% of the total membrane-bound sialic acids and a very small fraction of the total sialyl Le^x on neutrophil membranes. The predominant form of sialic acid on the ligand is N-acetylneuraminic acid. The P-selectin ligand carries large numbers of closely spaced sialylated O-linked oligosaccharides. The 120-kDa ligand is the major determinant of P-selectin: myeloid cell interaction in vivo (Norgard et al. 1993). Platelet PSGL-1 expression is 25–100-fold lower than that of leukocytes. Presence of a functional PSGL-1 on platelets suggests one of the mechanism by which selectins and their ligands participate in inflammatory and/or hemostatic responses (Frenette et al. 2000).

The predicted amino acid sequence of a functional ligand for P-selectin from an HL-60 reveals a novel mucin-like transmembrane protein. Binding of transfected COS cells to P-selectin requires coexpression of both the protein ligand and a fucosyltransferase. This binding is calcium dependent and can be inhibited by mAb to P-selectin. Cotransfected COS cells express the ligand as a homodimer of 220 kDa. A soluble ligand construct, when coexpressed with fucosyltransferase in COS cells, also mediates P-selectin binding and is immunocrossreactive with the major HL-60 glycoprotein that specifically binds P-selectin (Sako et al. 1993). In vitro studies have suggested that PSGL-1 may also be a ligand for E- and L-selectins. However, P-selectin-PSGL-1 interaction alone is sufficient to mediate rolling in vivo and that E-selectin-PSGL-1 interaction supports slow rolling (Norman et al. 2000).

27.3.1.2 The Mouse Homolog of PSGL-1

The *mouse* homolog of PSGL-1, flanking the entire ORF of 397 amino acids is composed of a single exon. Mouse and human PSGL-1 show an overall similarity of 67% and an identity of 50% and contain a similar domain organization. However, there are 10 threonine/serine-rich decameric repeats in mouse PSGL-1 as compared to 15 threonine-rich repeats in human PSGL-1. The mouse PSGL-1 gene, *Selpl*, was mapped to a position on mouse chromosome 5 (Chr 5). Moderate expression of a PSGL-1 mRNA occurs in species in most tissues and high levels of expression in blood, bone marrow, brain, adipose tissue, spleen, and thymus. Whereas certain mouse myeloid cell lines including PU5-1.8, WEHI-3B, and 32 DC13 express high levels of PSGL-1 mRNA, only WEHI-3B and 32 DC13 bind to P-selectin. WEHI-3B cells bind significantly better to P-selectin than to E-selectin. Thus, mouse PSGL-1 has structural and functional homology to human PSGL-1 but is characterized by differences in the composition and number of the decameric repeats. PSGL-1 on mouse myeloid cells is critical for high-affinity binding to P-selectin but not E-selectin (Yang et al. 1996).

27.3.1.3 PSGL-1 in Equine

A high baseline level of P-selectin expression in circulating equine platelets suggests a primed state toward inflammation and thrombosis via P-selectin/PSGL-1 adhesion. Equine PSGL-1 (ePSGL-1) subunit is predicted to be 43 kDa and composed of 420 amino acids with a predicted 18-amino-acid signal sequence showing 78% homology to hPSGL-1. Though earlier work has shown that binding of P-selectin requires sulfation of at least one of three tyrosines and O-glycosylation of one threonine in the N-terminus of human PSGL-1, the corresponding domain in ePSGL-1, spanning residues 19–43, contains only one tyrosine in the vicinity of two threonines at positions 25 and 41. The ePSGL-1 contains 14 threonine/serine-rich decameric repeats as compared to hPSGL-1 which contains 14–16 threonine-rich decameric repeats. The transmembrane and cytoplasmic domains display 91% and 74% homology to corresponding human PSGL-1 domains, respectively. Overall, there is 71% homology between ORF of ePSGL-1 and hPSGL-1. The greatest homologies between species exist in the transmembrane domain and cytoplasmic tail while substantial differences exist in the extra-cellular domain (Xu et al. 2005).

27.3.1.4 Bovine PSGL-1

The ORF of bovine PSGL-1 (bPSGL-1) cDNA is 1,284 bp in length, predicting a protein of 427 amino acids including an 18-amino-acid signal peptide, an extra-cellular region with a mucin-like domain, and transmembrane and cytoplasmic domains. The amino acid sequence of bPSGL-1 demonstrated 52%, 49% and 40% overall homology to equine, human and mouse, respectively. A single extra-cellular cysteine, at the transmembrane and extracellular domain junction, suggests a disulfide-bonding pattern. Alignment of bovine with equine, human and mouse PSGL-1 demonstrates high conservation of transmembrane and cytoplasmic domains, but diversity of the extracellular domain, especially in the anionic NH_2-terminal of PSGL-1, the putative P-selectin binding domain. In the NH_2-terminal of bPSGL-1, there are three potential tyrosine sulfation sites and three potential threonine O-glycosylation sites, all of which are required for P-selectin binding in human PSGL-1. The bPSGL-1 shares only 57% homology in amino acid sequence with the corresponding epitope region in human PSGL-1 without cross-reactivity with bovine leukocytes (Xu et al. 2006).

27.3.1.5 Eosinophil Versus Neutrophil PSGL-1

Studies have indicated an important role for P-selectin in eosinophil adhesion. Eosinophils bound to twofold more blood vessels within the nasal polyp tissue than neutrophils. The eosinophil P-selectin ligand is a sialylated, homodimeric glycoprotein consistent with the known structure of PSGL-1. However, expression of PSGL-1 by eosinophils was greater than on neutrophils. The eosinophil ligand had a molecular mass of approximately 10 kDa greater than the neutrophil ligand, which was not due to differences in N-glycosylation. Eosinophils expressed the 15-decapeptide repeat form of PSGL-1 compared with neutrophils that have the 16-decapeptide repeat form. The increased binding of eosinophils, compared with neutrophils, to P-selectin in both an ex-vivo and in vitro assay suggests that P-selectin may have a role directing the specific migration of eosinophils in diseases such as asthma (Symon et al. 1996). While the roles of P- and E-selectin in neutrophil recruitment are well established, the mechanisms of L-selectin-mediated neutrophil recruitment remain elusive. One proposal is that tethering is mediated by L-selectin on flowing neutrophils interacting with PSGL-1 on adherent neutrophils. To clarify this point, Shigeta et al. (2008) examined the impact of L-selectin deficiency in PSGL-1-deficient mice. In L-selectin and PSGL-1 double-knockout mice, Shigeta et al. (2008) provided evidence for the existence of another L-selectin ligand distinct from PSGL-1 in inflammation and indicated that such a ligand is expressed on endothelial cells, that promotes neutrophil rolling in vivo.

27.3.2 Genomic Organization

27.3.2.1 The PSGL-1 Gene

The PSGL-1 gene from human placenta contains a single intron of approximately 9 kb in the 5′-untranslated region; the complete coding region resides in exon 2. The genomic clone differs from the cDNA clone isolated from HL-60 cells in that it encodes an extra copy of the decameric repeat located in the extracellular domain of PSGL-1. Further analysis indicated that the PSGL-1 genes of HL-60 and U-937 cells contain 15 repeats, whereas the PSGL-1 genes of polymorphonuclear leukocytes, monocytes, and several other cell lines contain 16 repeats. There was no functional difference among these two variants of PSGL-1. The organization of the PSGL-1 gene closely resembles those of CD43 and human platelet glycoprotein GPIb α, both of which have an intron in the 5′-noncoding region, a long second exon containing the complete coding region, and TATA-less promoters. The gene for human PSGL-1, designated SELPLG, was mapped to chromosome 12q24 (Veldman et al. 1995).

27.3.3 Specificity of PSGL-1 as Ligand for P-Selectin

P-, L, and E- Selectins Have Overlapping Leukocyte Ligand Specificities: Although, L-selectin binds multiple ligands expressed on endothelial cells and P-selectin is proposed to interact exclusively with PSGL-1 on leukocytes, one study showed that L-selectin binds leukocytes through PSGL-1, although at lower levels than P-selectin. L-selectin binding to PSGL-1 appeared to be specific since it was blocked by Abs to L-selectin or PSGL-1, and required appropriate glycosylation of PSGL-1. The lectin and EGF domains of L- and P-selectin contributed significantly to binding through similar, if not identical, regions of PSGL-1. The different chimeric selectins revealed that the lectin domain was the dominant determinant for ligand binding, while cooperative interactions between the lectin, EGF, and short consensus repeat domains of the selectins also modified ligand binding specificity. It seems that L-selectin binding to PSGL-1 expressed by leukocytes may mediate neutrophil rolling on stationary leukocytes bound to cytokine-induced endothelial cells, which is a L-selectin-dependent process (Tu et al. 1996). The proposed ligands for P- and E-selectin receptors contain the Le^x core and sialic acid. Since both E-selectin and P-selectin bind to sialylated Lex, Larsen et al. (1992) evaluated whether E-selectin and P-selectin recognize the same counter-receptor on leukocytes. Although sialic acid and Le^x are components of the P-selectin ligand and the E-selectin ligand, results indicated that the ligands are related, having overlapping specificities, but are structurally distinct. A protein component containing sialyl Le^x in proximity to sialyl-2,6 βGal structures on the P-selectin ligand may contribute to its specificity for P-selectin.

27.3.4 Structural Polymorphism in PSGL-1 and CAD Risk

27.3.4.1 PSGL-1 Is Highly Polymorphic and Shows Structural Polymorphism

P-selectin and P-selectin glycoprotein ligand constitutes a receptor/ligand complex that is likely to be involved in the development of atherosclerosis and its complications. PSGL-1 is highly polymorphic and contains a structural polymorphism that potentially indicates functional variation in the human population (Afshar-Kharghan et al. 2001; Tregouet et al. 2003). The functional relevance of a variable number of tandem repeats polymorphism affecting the PSGL-1 has been demonstrated. Platelet glycoprotein (GP) Ibα and leukocyte PSGL-1 are membrane mucins with a number of structural and functional similarities. Like GP Ibα, PSGL-1 is affected by a variable number of tandem repeat polymorphism in its mucin-like region. By PCR amplification of the genomic region encoding the PSGL-1 repeats, three allelic variants were identified in the human population. The three alleles-A, B, and C-from largest to smallest, contained 16, 15, and 14 decameric repeats, respectively, with the B variant lacking repeat 2 and the C variant retaining repeat 2 but lacking repeats 9 and 10. Allele frequencies were highest for the A variant and lowest for the C variant in the two populations studied (frequencies of 0.81, 0.17, and 0.02 in white persons and 0.65, 0.35, and 0.00 in Japanese). The coding and regulatory sequences of the PSGL-1 showed nine polymorphisms. The identified polymorphisms were genotyped in the AtheroGene study. Haplotype analysis revealed that two polymorphisms of PSGL-1, the M62I and the VNTR, independently influence plasma PSGL-1 levels. Conversely, haplotypes of PSGL-1 are not associated with CAD risk (Tregouet et al. 2003).

Hancer et al. (2005) determined the allele and genotype frequencies of polymorphisms GP Ibα Kozak and PSGL-1 in the Turkish population. Allele frequencies of T and C were calculated to be 0.873 and 0.127 for the GPIbα Kozak polymorphism and no significant difference was found between Turkish and French populations. In contrast, the difference between Turkish and Japanese population was highly significant. In the PSGL-1 group, allele frequencies of A, B and C were calculated as 0.818, 0.160, 0.022 respectively. Thus, for the PSGL-1, although the difference between Turkish and French populations was not significant, the difference between the Turkish and Japanese was extremely significant. A Turkish population database has been established for these gene polymorphisms (Hancer et al. 2005).

Short Alleles Protect Against Premature Myocardial Infarction: Neutrophils, carrying short alleles, exhibit a significantly lower capacity to bind activated platelets. These alleles consistently protect against transient ischemic attack, probably because of their lesser adhesive capacity. The role of this polymorphism in premature myocardial infarction was evaluated for the fact that genetic risk factors are more relevant in the development of disease in young patients. Hence, Roldan et al. (2004) genotyped 219 Caucasian patients who had suffered a premature myocardial infarction (MI) and 594 control subjects from Mediterranean area. The frequency of the short alleles (B and C) was significantly lower in patients than in controls. Multiple regression analysis revealed that B and C alleles had an independent protective effect on the development of premature MI.

In Cerebro-Vascular Disease, CHD, and Venous Thrombosis: The variable number of tandem repeat (VNTR) polymorphism in the PSGL-1 gene has been associated with ischemic cerebro-vascular disease (CVD) and with CHD. A recently described polymorphism in this receptor that results in VNTR sequence present 16, 15 or 14 times (alleles A, B or C) could, similar to GPIbα be functionally relevant. The allelic frequency of this polymorphism in 469 individuals from the south of Spain, was similar to that described in other Caucasian populations: 85% A, 14% B and 1% C alleles. Lozano et al. (2001) identified two new polymorphisms genetically linked to the C isoform, resulting in the Ser273Phe and Met274Val changes. Neutrophils carrying the shortest C allele and the amino acid variations in residues 273 and 274 exhibited a significantly lower capacity to bind activated platelets than A/B and A/A samples. The distribution of VNTR was analyzed in three case-control studies including 104 patients of CVD, 101 patients of CHD and 150 deep venous thrombosis (DVT) patients. Smaller (B and C) alleles seemed to be associated with a lower risk of developing CVD but not to be related to CHD or DVT (Lozano et al. 2001).

In another study involving low number of CHD patients (281) and 397 healthy blood donors, the prevalence of homozygous carriers of the PSGL-1 VNTR allele with 15 repeat units was significantly higher in the CHD patients (5.3% vs. 1.5%) than in controls, suggesting an effect of this marker in CHD. But genotyping, performed in a larger sample size including 2,578 CHD patients, 731 patients without CHD, and 1,084 healthy blood donors, failed to confirm the putative role of PSGL-1 VNTR polymorphism in CHD. Frequencies of the PSGL-1 VNTR 15 repeats for homozygous carriers were 2.2% in healthy blood donors, 2.3% in patients without CHD and 2.7%, in CHD cases, respectively. Based on these results, the PSGL-1 VNTR polymorphism is not a genetic risk factor for CHD (Bugert et al. 2003). Thus, polymorphisms of the PSGL-1 may influence the neutrophil-platelet binding, and may represent a risk factor for CVD

Multiple Sclerosis and PSGL-1: PSGL-1 represents a crucial step in the pathogenesis of multiple sclerosis (MS). Three hundred twenty-one MS patients and 342 controls were genotyped for the presence of a polymorphism in the PSGL-1 gene, consisting of a VNTR originating three alleles: A, B and C, in order to test whether they influence the susceptibility and the course of the disease. No significant differences among allelic frequencies of A, B and C alleles in MS as compared with controls were observed. However, the C allele of the VNTR polymorphism in PSGL-1 is likely to be associated with PP-MS. As this allele has been demonstrated to have a very low efficiency in mediating lymphocyte binding to brain endothelium during attacks, its high frequency in PP-MS could be related to the absence of exacerbations in such patients (Scalabrini et al. 2005).

27.3.5 Carbohydrate Structures on PSGL-1

27.3.5.1 Primitive Human Hematopoietic Progenitors Adhere to P-Selectin

P-selectin binds human hematopoietic progenitors (colony-forming unit-granulocyte-macrophage [CFU-GM] and burst-forming unit-erythroid [BFU-E]) as identified by their expression of CD34 antigen. In addition, P-selectin binds all precursors (pre-CFU) of committed myeloid progenitors. This suggested that PSGL-1 comprises a P-selectin ligand expressed by primitive hematopoietic cells, but did not preclude the existence of additional P-selectin ligands on these cells. It has been proposed that only covalently dimerized PSGL-1 can bind P-selectin. Epperson et al. (2000) examined whether covalent dimers of PSGL-1 are required for binding to P-selectin. Recombinant forms of PSGL-1, in which the single extracellular Cys (Cys-320) was replaced with either Ser (C320S-PSGL-1) or Ala (C320A-PSGL-1), suggested that Cys(320)-dependent dimerization of PSGL-1 is not required for binding to P-selectin and that a monomeric fragment of PSGL-1 is sufficient for P-selectin recognition (Epperson et al. 2000).

27.3.5.2 Structure Containing GalNAc-Lewis^x and Neu5Acα2-3Galβ1-3GalNAc Sequences on PSGL-1

The PSGL-1-P-selectin interaction is calcium dependent and requires presentation of sialyl-Le^x-type structures on the

O-linked glycans of PSGL-1. The selectins interact under normal and pathological situations with certain sialylated, fucosylated glycoconjugate ligands containing sialyl-Le^x (Neu5Acα2-3Galβ1-4(Fucα1-3)GlcN Ac). Much effort has gone into the synthesis of sialylated and sulfated Le^x analogs as competitive ligands for the selectins. Other studies have shown that sulfate esters can replace sialic acid in some selectin ligands (Yuen et al. 1992; Imai et al. 1993). Since the natural selectin ligands GlyCAM-1 and PSGL-1 carry sialyl-Le^x as part of a branched Core 2 O-linked structure, Koenig et al. (1997) synthesized Galβ1-4(Fucα1-3) GlcNAcβ1-6(SE-3Galβ1^{+++}-3)GalNAc1αOMe and found it to be a moderately superior ligand for L and P-selectin. Jain et al. (1998) synthesized Galβ1-4(Fucα1-3)GlcNAcβ1-6 (Neu5Acα2^{+++} 3Galβ1-3)-GalNAc α1-OB, which was found to be 2- to 3-fold better than sialyl Le^x for P and L selectin, respectively. The unusual structure GalNAcβ1-4(Fucα1-3)GlcNAcβ1-OMe (GalNAc-Le^x-O-methyl glycoside) also proved to be a better inhibitor of L- and P-selectin than sialyl-Le^x-OMe. In view of Core 2 branched structures, Jain et al. (1998) synthesized a molecule that is 5- to 6-fold better at inhibiting L- and P-selectin than sialyl-Le^x-OMe. By contrast to un-branched structures, substitution of a sulfate ester group for a sialic acid residue in such a molecule resulted in a considerable loss of inhibition ability. Thus, the combination of a sialic acid residue on the primary (β1-3) arm, and a modified Le^x unit on the branched (β1-6) arm on an O-linked Core 2 structure generated a monovalent synthetic oliogosaccharide inhibitor superior to Sialyl-Le^x for both L- and P-selectin (Jain et al. 1998). The density of α-Gal epitopes on PSGL-1 is dependent on the expression of O-linked glycans with core 2 structures and lactosamine extensions. Hence, the structural complexity of the terminal Gal-Gal expressing O-glycans with both neutral as well as sialic acid-containing structures is likely to contribute to the high adsorption efficacy (Liu et al. 2005). Moore et al. (1994) suggested that PSGL-1 from human neutrophils displays complex, sialylated, and fucosylated O-linked poly-N-acetyllactosamine that promote high affinity binding to P-selectin, but not to E-selectin.

27.3.5.3 Carbohydrate and Non-Carbohydrate Moieties of PSGL-1 in Binding P-Selectin

In order to define the modifications required for PSGL-1 to bind P- and E-selectin, CHO cells were transfected with cDNAs for PSGL-1 and specific glycosyltransferases. CHO cells synthesized only core 1 O-linked glycans (Galβ1-3GalNAc α1-Se r/Thr), which lacked core 2 O-linked glycans (Galβ1-3(Galβ1-4GlcNAcβ1-6)GalNAc α1 -Ser/Thr). PSGL-1 expressed on transfected CHO cells bound P- and E-selectin only when it was co-expressed with both C2GnT and a α1,3 fucosyltransferase (Fuc-TIII, Fuc-TIV, or Fuc-TVII). Results indicated that PSGL-1 requires core 2 O-linked glycans that are sialylated and fucosylated to bind P- and E-selectin. PSGL-1 also requires tyrosine sulfate to bind P-selectin but not E-selectin (Li et al. 1996). Sako et al. (1995) reported the identification of a non-carbohydrate component of the binding determinant that is critical for high affinity binding to P-selectin. Located within the first 19 amino acids, this anionic polypeptide segment contains at least one sulfated tyrosine residue. This sulfotyrosine-containing segment of PSGL-1, in conjunction with sLex presented on O-linked glycans, constitutes the high affinity P-selectin-binding site (Sako et al. 1995).

27.3.6 Role of PSGL-1

Selectin-dependent rolling is the earliest observable event in the recruitment of leukocytes to inflamed tissues. Several glycoproteins decorated with sialic acid, fucose, and/or sulfate, have been shown to bind the selectins. The best-characterized selectin ligand is PSGL-1 that supports P-selectin- dependent rolling in vitro and in vivo.

27.3.6.1 PSGL-1 Mediates Rolling of Neutrophils on P-Selectin

Neutrophils roll on P-selectin expressed by activated platelets or endothelial cells under the shear stresses in the microcirculation. PSGL-1 accounts for the high affinity binding sites for P-selectin on leukocytes, and it must interact with P-selectin for neutrophils to roll on P-selectin at physiological shear stresses (Moore et al. 1995). Activated T cells regulate inflammatory diseases in the intestinal tract. P-selectin and PSGL-1 are dominating molecules in supporting adhesive interactions of CD8 T cells in inflamed colonic venules and may be useful targets to protect against pathological inflammation in the large bowel (Asaduzzaman et al. 2009). Human PSGL-1 interacts with CCL27, a skin-associated chemokine that attracts skin-homing T lymphocytes. Hirata et al. (2004) suggest a role for PSGL-1 in regulating chemokine-mediated responses, in addition to its role as a selectin ligand. The aspartyl protease BACE1 cleaves the amyloid precursor protein and the sialyltransferase ST6Gal I and is important in the pathogenesis of Alzheimer's disease. It was shown that BACE1 acts on PSGL-1, which mediates leukocyte adhesion in inflammatory reactions and that PSGL-1 is an additional substrate for BACE1 (Lichtenthaler et al. 2003).

Memory T cells expressing Cutaneous lymphocyte-associated antigen (CLA) occur in humans and accumulate in normal and inflamed skin. These cells uniformly bind to E-selectin, yet only a subset binds to P-selectin. Findings indicate that unlike memory CLA$^+$ CD4$^+$ T cells, when activated these cells can broadly bind to P-selectin, suggesting a more diverse tissue trafficking capacity (Ni and Walcheck 2009).

27.3.6.2 Interaction of ERM Proteins with Syk Mediates Signaling by PSGL-1

In addition to its direct role in capture of leukocytes from the bloodstream, PSGL-1 also functions as a signal-transducing receptor and initiates a series of intracellular signal events during activation of leukocytes. Antibody engagement of PSGL-1 up-regulates the transcriptional activity of CSF-1 promotor and increases the endogenous expression of CSF-1 mRNA in Jurkat cell. The PSGL-1 associates with Syk. This association is mediated by the actin-linking proteins moesin and ezrin, which directly interact with Syk in an ITAM-dependent manner. PSGL-1 engagement induces tyrosine phosphorylation of Syk and SRE-dependent transcriptional activity. Study revealed a new role for ERMs (ezrin/radixin/moesin) as adaptor molecules in the interactions of adhesion receptors and intracellular tyrosine kinases and showed that PSGL-1 is a signaling molecule in leukocytes (Urzainqui et al. 2002). Studies suggest that signal transduced by PSGL-1 up-regulates the transcriptional activity of CSF-1, and non-receptor tyrosine kinase Syk participates in this pathway (Ba et al. 2005). The PSGL-1 peptide corresponding to the 18-residue juxtamembrane region from the mouse PSGL-1 cytoplasmic tail (residues 2–19 in Fig. 27.3) binds to the radixin FERM domain with a dissociation constant K_D of 201 nm.

Structural Basis of PSGL-1 Binding to ERM Proteins: Proteins involved in mediating cell migration and attachment have been identified using molecular genetic and biochemical approaches and many of these proteins contain highly conserved protein interaction domains. One such family of proteins is: which contains a FERM (Four.1 protein, ezrin, radixin, moesin) domain that functions as a protein docking surface with the cytosolic tail of transmembrane proteins such as CD44. On activation, ERM proteins mediate the redistribution of PSGL-1 on polarized cell surfaces to facilitate binding to target molecules. ERM proteins recognize a short binding motif, Motif-1, conserved in cytoplasmic tails of adhesion molecules, whereas PSGL-1 lacks Motif-1 residues important for binding to ERM proteins. The crystal structure of the complex between the radixin FERM domain and a PSGL-1 juxtamembrane peptide reveals that the peptide binds the groove of FERM subdomain C by forming a β-strand associated with strand β5C, followed by a loop flipped out towards the solvent (Takai et al. 2007) (Fig. 27.4).

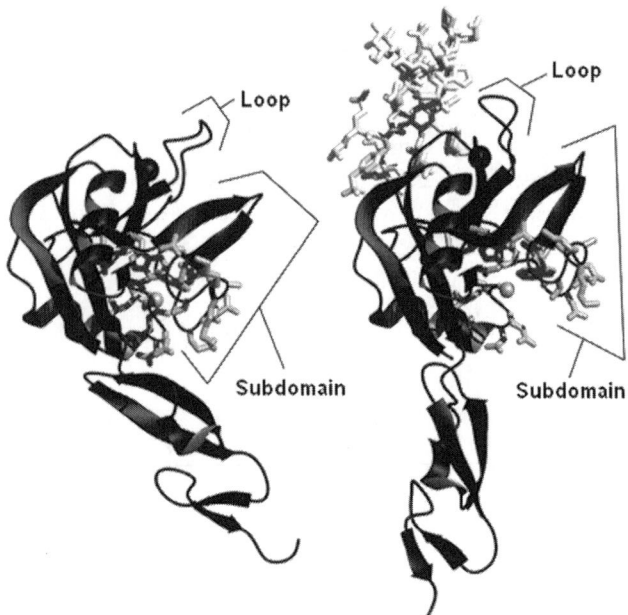

Fig. 27.3 Opening of a cleft in the lectin domain in the extended selectin conformation. (*Left*) P-selectin in the unliganded conformation (PDB ID code 1G1Q). (*Right*) P-selectin in the liganded, extended conformation (1G1S). In each structure, the Cβ atom of Ala-28 in the cleft is shown as a *gray sphere*, and other side chains that line the cleft are shown as *white sticks*. The divalent cation in the ligand binding site is shown as a larger *black sphere*. Molecules are shown as *ribbons* in identical orientations after superposition on the lectin domain (Adapted with permission from Waldron and Springer 2009 © National Academy of Sciences, USA)

27.4 Other Ligands of P-Selectin

CD24 as a Ligand for P-Selectin in Carcinoma Cells: P-selectin can also bind to breast and a small cell lung carcinoma cell line, which are negative for PSGL-1. But CD24, a mucin-type glycosyl phosphatidyl-inositol-linked cell surface molecule on human neutrophils, pre B lymphocytes, and many tumors can promote binding to P-selectin. Results establish a role of CD24 as a ligand for P-selectin on tumor cells. The CD24/P-selectin binding pathway could be important in the dissimination of tumor cells by facilitating the interaction with platelets or endothelial cells (Aigner et al. 1997).

CD44v as Ligands on Colon Carcinoma Cells: Variant isoforms of CD44 (CD44v) on LS174T colon carcinoma cells possess P-/L-/E-selectin binding activity, in contrast to the standard isoform of CD44 (CD44s) on hematopoietic-progenitor cells (HPCs), which is primarily

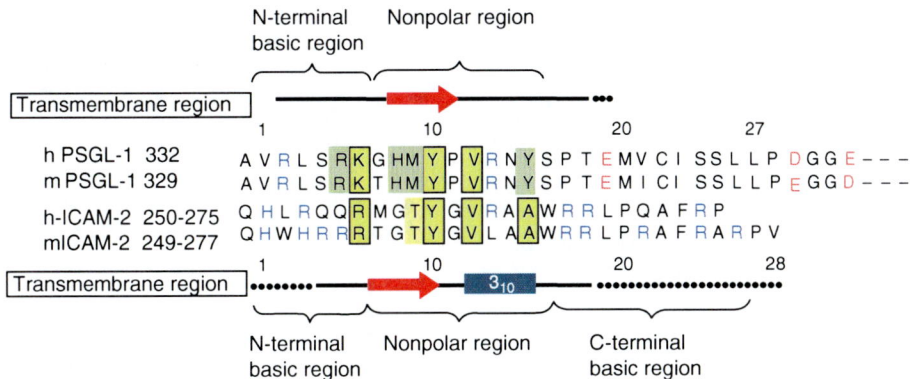

Fig. 27.4 Comparison of FERM-binding sequences in PSGL-1 and ICAM-2. The juxtamembrane region of the cytoplasmic tail of PSGL-1 is compared with that of ICAM-2, a member of the Ig superfamily of adhesion molecules that bind the FERM domain of ERM proteins. Basic and acidic residues are in *blue* and *red*, respectively. Compared with the ICAM-2 tail that displays three characteristic regions which include the N-terminal basic region, the middle nonpolar region and the C-terminal basic region, the PSGL-1 juxtamembrane region lacks the corresponding C-terminal basic region and instead contains acidic residues. PSGL-1-specific residues that are important in stabilizing conformation and binding are highlighted in *green*. The residue numbering for PSGL-1 follows that of ICAM-2. Part of the nonpolar region forms a short β-strand (residues 8–11; a *red arrow*) in FERM–PSGL-1 complex. Of the 18 PSGL-1 peptide residues, 17 (residues 2–18; indicated with *bold lines*) were defined in the current electron density map. The secondary structure of the mouse ICAM-2 cytoplasmic tail is shown at the *bottom* of the alignment. Conserved Motif-1 residues (RXXTYXVXXA) present in ICAM-2 are highlighted in *yellow* and the most important residues for binding to the FERM domain are *boxed*. PSGL-1 lacks Thr and Ala of Motif-1 (Reprinted with permission from Takai et al. 2007 © John Wiley and Sons)

an L-/E-selectin ligand. Moreover, the selectin-binding determinants on CD44v from LS174T cells are sialofucosylated structures displayed on O-linked glycans, akin to those on PSGL-1, but distinct from the HECA-452-reactive N-glycans on CD44s expressed on HPCs. Hanley et al. (2006) suggested that CD44v selectin ligands offer a unifying perspective on the apparent enhanced metastatic potential associated with tumor cell CD44v over-expression and the critical role of selectins in metastasis.

E-Selectin Promotes Growth Inhibition and Apoptosis of Hematopoietic Progenitor Cells Independent of PSGL-1: In human bone marrow, PSGL-1 is expressed by primitive bone marrow $CD34^+$ cells, mediates their adhesion to P-selectin, and inhibits their proliferation. Moreover, adhesion to E-selectin inhibits the proliferation of human $CD34^+$ cells isolated either from blood, or steady-state bone marrow. Furthermore, a subpopulation, which does not contain the most primitive hematopoietic progenitor cells, undergoes apoptosis following E-selectin-mediated adhesion. Studies suggest that an E-selectin ligand(s) other than PSGL-1 transduces growth inhibitory and pro-apoptotic signals and requires post-translational fucosylation to be functional (Winkler et al. 2004).

Lysosomal Membrane Glycoprotein-1 Can Present Ligands for E-Selectin: Lysosomal membrane glycoprotein (lamp)-1 and -2 are the most abundant glycoproteins in lysosomal membrane. A small amount of lamp-1 and lamp-2 molecules, however, can be present on the cell surface. The extent of adhesion to E-selectin and cell surface sialyl-Le^x determinants is proportional to the amount of cell surface lamp-1. Moreover, it was demonstrated that such adhesion can be inhibited by soluble lamp-1 generated from CHO cells expressing sialyl-Le^x structures. Results indicate that lamp-1 can efficiently present ligands for E-selectin and at the same time can be a useful reagent for inhibition of E-selectin (and possibly P-selectin)-mediated interaction (Sawada et al. 1993).

Sulfatides: Leukocytes may be important in the development of intimal hyperplasia. Sulfatides (3-sulfated galactosyl ceramides) are native ligands of L- and P-selectin. Findings suggest that neutrophil accumulation on the subendothelial matrix or adherence of platelets mediated by adhesive interactions between L- or P-selectin and sulfatides may contribute to the development of intimal hyperplasia. The neutrophil accumulation may be mediated by an increase in Mac-1 caused by the agonistic effects of sulfatides on the neutrophil membrane surface, or by an increase in L- and P-selectin ligands resulting from the binding of sulfatides onto the exposed subendothelial matrix (Shimazawa et al. 2005).

References

Afshar-Kharghan V, Diz-Kucukkaya R, Ludwig EH et al (2001) Human polymorphism of P-selectin glycoprotein ligand 1 attributable to variable numbers of tandem decameric repeats in the mucinlike region. Blood 97:3306–3307

Ahluwalia A, Foster P, Scotland RS et al (2004) Antiinflammatory activity of soluble guanylate cyclase: cGMP-dependent downregulation of P-selectin expression and leukocyte recruitment. Proc Natl Acad Sci USA 101:1386–1391

Aigner S, Sthoeger ZM, Fogel M et al (1997) CD24, a mucin-type glycoprotein, is a ligand for P-selectin on human tumor cells. Blood 89:3385–3395

Andrews RK, Lopez JA, Berndt MC (1997) Molecular mechanisms of platelet adhesion and activation. Int J Biochem Cell Biol 29:91–105

Asaduzzaman M, Mihaescu A, Wang Y et al (2009) P-selectin and P-selectin glycoprotein ligand 1 mediate rolling of activated CD8+ T cells in inflamed colonic venules. J Investig Med 57:765–768

Auchampach JA, Oliver MG, Anderson DC et al (1994) Cloning, sequence comparison and in vivo expression of the gene encoding rat P-selectin. Gene 145:251–255

Ba XQ, Chen CX, Xu T et al (2005) Engagement of PSGL-1 upregulates CSF-1 transcription via a mechanism that may involve Syk. Cell Immunol 237:1–6

Blagoveshchenskaya AD, Hewitt EW, Cutler DF (1998a) A balance of opposing signals within the cytoplasmic tail controls the lysosomal targeting of P-selectin. J Biol Chem 273:27896–27903

Blagoveshchenskaya AD, Hewitt EW, Cutler DF (1999) A complex web of signal-dependent trafficking underlies the triorganellar distribution of P-selectin in neuroendocrine PC12 cells. J Cell Biol 145:1419–1433

Blagoveshchenskaya AD, Norcott JP, Cutler DF (1998b) Lysosomal targeting of P-selectin is mediated by a novel sequence within its cytoplasmic tail. J Biol Chem 273:2729–2737

Bugert P, Hoffmann MM, Winkelmann BR et al (2003) The variable number of tandem repeat polymorphism in the P-selectin glycoprotein ligand-1 gene is not associated with coronary heart disease. J Mol Med 81:495–501

Bullard DC, Qin L, Lorenzo I et al (1995) P-selectin/ICAM-1 double mutant mice: acute emigration of neutrophils into the peritoneum is completely absent but is normal into pulmonary alveoli. J Clin Invest 95:1782–1788

Cambien B, Wagner DD (2004) A new role in hemostasis for the adhesion receptor P-selectin. Trends Mol Med 10:179–186

Chen M, Geng JG (2006) P-selectin mediates adhesion of leukocytes, platelets, and cancer cells in inflammation, thrombosis, and cancer growth and metastasis. Arch Immunol Ther Exp (Warsz) 54:75–84

Cummings RD (1999) Structure and function of the selectin ligand PSGL-1. Braz J Med Biol Res 32:519–528

Epperson TK, Patel KD, McEver RP, Cummings RD (2000) Noncovalent association of P-selectin glycoprotein ligand-1 and minimal determinants for binding to P-selectin. J Biol Chem 275:7839–7853

Frenette PS, Denis CV, Weiss L et al (2000) P-Selectin glycoprotein ligand 1 (PSGL-1) is expressed on platelets and can mediate platelet-endothelial interactions in vivo. J Exp Med 191:1413–1422

Fritz J, Katopodis AG, Kolbinger F, Anselmetti D (1998) Force-mediated kinetics of single P-selectin/ligand complexes observed by atomic force microscopy. Proc Natl Acad Sci USA 95:12283–12288

Fujimoto T, Stroud E, Whatley RE et al (1993) P-selectin is acylated with palmitic acid and stearic acid at cysteine 766 through a thioester linkage. J Biol Chem 268:11394–11400

Graves BJ, Crowther RL, Chandran C et al (1994) Insight into E-selectin/ligand interaction from the crystal structure and mutagenesis of the lec/EGF domains. Nature 367:532–538

Green SA, Setiadi H, McEver RP, Kelly RB (1994) The cytoplasmic domain of P-selectin contains a sorting determinant that mediates rapid degradation in lysosomes. J Cell Biol 124:435–448

Hancer VS, Diz-Kucukkaya R, Nalcaci M (2005) Turkish population data on the factor XIII Val34Leu, glycoprotein (GP)Ibα Kozak and P-selectin glycoprotein ligand 1 (PSGL-1) loci. Cell Biochem Funct 23:55–58

Hanley WD, Napier SL, Burdick MM et al (2006) Variant isoforms of CD44 are P- and L-selectin ligands on colon carcinoma cells. FASEB J 20:337–339

Hartwell DW, Wagner DD (1999) New discoveries with mice mutant in endothelial and platelet selectins. Thromb Haemost 82:850–857

Hirano F, Tanaka H, Hirano Y et al (1998) Functional interference of Sp1 and NF-kB through the same DNA binding site. Mol Cell Biol 18:1266–1274

Hirata T, Furukawa Y, Yang BG et al (2004) Human P-selectin glycoprotein ligand-1 (PSGL-1) interacts with the skin-associated chemokine CCL27 via sulfated tyrosines at the PSGL-1 amino terminus. J Biol Chem 279:51775–51782

Imai Y, Lasky LA, Rosen SD (1993) Sulphation requirement for GlyCAM-1, an endothelial ligand for L-selectin. Nature 361:555

Jain RK, Piskorz CF, Huang BG et al (1998) Inhibition of L- and P-selectin by a rationally synthesized novel core 2-like branched structure containing GalNAc-Lewisx and Neu5Ac α2-3Galβ1-3GalNAc sequences. Glycobiology 8:707–717

Johnson RC, Mayadas TN, Frenette PS et al (1995) Blood cell dynamics in P-selectin-deficient mice. Blood 86:1106–1114

Johnston GI, Bliss GA, Newman PJ, McEver RP (1990) Structure of the human gene encoding granule membrane protein-140, a member of the selectin family of adhesion receptors for leukocytes. J Biol Chem 265:21381–21385

Johnston GI, Cook RG, McEver RP (1989) Cloning of GMP-140, a granule membrane protein of platelets and endothelium: sequence similarity to proteins involved in cell adhesion and inflammation. Cell 24(56):1033–1044

Khew-Goodall Y, Wadham C, Stein BN et al (1999) Stat6 activation is essential for interleukin-4 induction of P-selectin transcription in human umbilical vein endothelial cells. Arterioscler Thromb Vasc Biol 19:1421–1429

Koenig A, Jain R, Vig R et al (1997) Evaluation of novel sialylated, sulfated and fucosylated oligosaccharides, including the major capping group of GlyCAM-1. Glycobiology 7:79–93

Kuligowski MP, Kitching AR, Hickey MJ (2006) Leukocyte recruitment to the inflamed glomerulus: a critical role for platelet-derived P-selectin in the absence of rolling. J Immunol 176:6991–6999

Kum WW, Lee S, Grassl GA et al (2009) Lack of functional P-selectin ligand exacerbates Salmonella serovar typhimurium infection. J Immunol 182:6550–6561

Larsen GR, Sako D, Ahern TJ et al (1992) P-selectin and E-selectin. Distinct but overlapping leukocyte ligand specificities. J Biol Chem 267:11104–11110

Li F, Wilkins PP, Crawley S et al (1996) Post-translational modifications of recombinant P-selectin glycoprotein ligand-1 required for binding to P- and E-selectin. J Biol Chem 271:3255–3264

Li G, Sanders JM, Phan ET, Ley K, Sarembock IJ (2005) Arterial macrophages and regenerating endothelial cells express P-selectin in atherosclerosis-prone apolipoprotein E-deficient mice. Am J Pathol 167:1511–1518

Lichtenthaler SF, Dominguez DI, Westmeyer GG et al (2003) The cell adhesion protein P-selectin glycoprotein ligand-1 is a substrate for the aspartyl protease BACE1. J Biol Chem 278:48713–48719

Liu J, Gustafsson A, Breimer ME et al (2005) Anti-pig antibody adsorption efficacy of α-Gal carrying recombinant P-selectin glycoprotein ligand-1/immunoglobulin chimeras increases with core2 β1,6-N-acetylglucosaminyltransferase expression. Glycobiology 15:571–583

Lozano ML, Gonzalez-Conejero R, Corral J et al (2001) Polymorphisms of P-selectin glycoprotein ligand-1 are associated with neutrophil-platelet adhesion and with ischaemic cerebrovascular disease. Br J Haematol 115:969–976

Mayadas TN, Johnson RC, Rayburn H et al (1993) Leukocyte rolling and extravasation are severely compromised in P selectin-deficient mice. Cell 74:541–554

Modderman PW, Beuling EA, Govers LA et al (1998) Determinants in the cytoplasmic domain of P-selectin required for sorting to secretory granules. Biochem J 336:153–161

Moore KL, Eaton SF, Lyons DE et al (1994) The P-selectin glycoprotein ligand from human neutrophils displays sialylated, fucosylated, O-linked poly-N-acetyllactosamine. J Biol Chem 269:23318–23327

Moore KL, Patel KD, Bruehl RE et al (1995) P-selectin glycoprotein ligand-1 mediates rolling of human neutrophils on P-selectin. J Cell Biol 128:661–671

Moore KL (1998) Structure and function of P-selectin glycoprotein ligand-1. Leuk Lymphoma 29:1–15

Murphy JF, McGregor JL (1994) Two sites on P-selectin (the lectin and epidermal growth factor-like domains) are involved in the adhesion of monocytes to thrombin-activated endothelial cells. Biochem J 303(Pt 2):619–624

Nagata K, Tsuji T, Todoroki N et al (1993) Activated platelets induce superoxide anion release by monocytes and neutrophils through P-selectin (CD62). J Immunol 151:3267–3273

Ni Z, Walcheck B (2009) Cutaneous lymphocyte-associated antigen (CLA) T cells up-regulate P-selectin ligand expression upon their activation. Clin Immunol 133:257

Norcott JP, Solari R, Cutler DF (1996) Targeting of P-selectin to two regulated secretory organelles in PC12 cells. J Cell Biol 134:1229–1240

Norgard KE, Moore KL, Diaz S et al (1993) Characterization of a specific ligand for P-selectin on myeloid cells. A minor glycoprotein with sialylated O-linked oligosaccharides. J Biol Chem 268:12764–12774

Norman KE, Katopodis AG, Thoma G et al (2000) P-selectin glycoprotein ligand-1 supports rolling on E- and P selectin in vivo. Blood 96:3585–3591

Pan J, McEver RP (1993) Characterization of the promoter for the human P-selectin gene. J Biol Chem 268:22600–22608

Pan J, McEver RP (1995) Regulation of the human P-selectin promoter by Bcl-3 and specific homodimeric members of the NF-kB/Rel family. J Biol Chem 270:23077–23083

Pan J, Xia L, McEver RP (1998a) Comparison of promoters for the murine and human P-selectin genes suggests species-specific and conserved mechanisms for transcriptional regulation in endothelial cells. J Biol Chem 273:10058–10067

Pan J, Xia L, Yao L, McEver RP (1998b) Tumor necrosis factor-α- or lipopolysaccharide-induced expression of the murine P-selectin gene in endothelial cells involves novel kB sites and a variant activating transcription factor/cAMP response element. J Biol Chem 273:10068–10077

Pescini R, Kaszubska W, Whelan J et al (1994) ATF-a0, a novel variant of the ATF/CREB transcription factor family, forms a dominant transcription inhibitor in ATF-a heterodimers. J Biol Chem 269:1159–1165

Phan UT, Waldron TT, Springer TA (2006) Remodeling of the lectin–EGF-like domain interface in P- and Lselectin increases adhesiveness and shear resistance under hydrodynamic force. Nat Immunol 7:883–889

Plantier JL, Enjolras N, Rodriguez MH et al (2003) The P-selectin cytoplasmic domain directs the cellular storage of a recombinant chimeric factor IX. J Thromb Haemost 1:292–299

Postigo AA, Marazuela M, Sanchez-Madrid F et al (1994) B lymphocyte binding to E- and P-selectins is mediated through the de novo expression of carbohydrates on in vitro and in vivo activated human B cells. J Clin Invest 94:1585–1596

Reed GL, Houng AK, Bianchi C (1998) Comparative biochemical and ultrastructural studies of P-selectin in rabbit platelets. Comp Biochem Physiol B Biochem Mol Biol 119:729–738

Revelle BM, Scott D, Beck PJ (1996a) Single amino acid residues in the E- and P-selectin epidermal growth factor domains can determine carbohydrate binding specificity. J Biol Chem 271:16160–16170

Revelle BM, Scott D, Kogan TP, Zheng J, Beck PJ (1996b) Structure-function analysis of P-selectin-sialyl LewisX binding interactions. Mutagenic alteration of ligand binding specificity. J Biol Chem 271:4289–4297

Ridger VC, Hellewell PG, Norman KE (2005) L- and P-selectins collaborate to support leukocyte rolling in vivo when high-affinity P-selectin-P-selectin glycoprotein ligand-1 interaction is inhibited. Am J Pathol 166:945–952

Robinson SD, Frenette PS, Raybur H et al (1999) Multiple, targeted deficiencies in selectins reveal a predominant role for P-selectin in leukocyte recruitment. Proc Natl Acad Sci USA 96:11452–11457

Roldan V, Gonzalez-Conejero R, Marin F et al (2004) Short alleles of P-selectin glycoprotein ligand-1 protect against premature myocardial infarction. Am Heart J 148:602–605

Rossi FM, Corbel SY, Merzaban JS et al (2005) Recruitment of adult thymic progenitors is regulated by P-selectin and its ligand PSGL-1. Nat Immunol 6:626–634

Sako D, Chang XJ, Barone KM et al (1993) Expression cloning of a functional glycoprotein ligand for P-selectin. Cell 75:1179–1186

Sako D, Comess KM, Barone KM et al (1995) A sulfated peptide segment at the amino terminus of PSGL-1 is critical for P-selectin binding. Cell 83:323–331

Sanders WE, Wilson RW, Ballantyne CM et al (1992) Molecular cloning and analysis of in vivo expression of murine P-selectin. Blood 80:795–800

Sawada R, Lowe JB, Fukuda M (1993) E-selectin-dependent adhesion efficiency of colonic carcinoma cells is increased by genetic manipulation of their cell surface lysosomal membrane glycoprotein-1 expression levels. J Biol Chem 268:12675–12681

Scalabrini D, Galimberti D, Fenoglio C et al (2005) P-selectin glycoprotein ligand-1 variable number of tandem repeats (VNTR) polymorphism in patients with multiple sclerosis. Neurosci Lett 388:149–152

Setiadi H, Disdier M, Green SA et al (1995) Residues throughout the cytoplasmic domain affect the internalization efficiency of P-selectin. J Biol Chem 270:26818–26826

Shigeta A, Matsumoto M, Tedder TF et al (2008) An L-selectin ligand distinct from P-selectin glycoprotein ligand-1 is expressed on endothelial cells and promotes neutrophil rolling in inflammation. Blood 112:4915–4923

Shimazawa M, Kondo K, Hara H et al (2005) Sulfatides, L- and P-selectin ligands, exacerbate the intimal hyperplasia occurring after endothelial injury. Eur J Pharmacol 520:118–126

Somers WS, Tang J, Shaw GD, Camphausen RT (2000) Insights into the molecular basis of leukocyte tethering and rolling revealed by structures of P- and E-selectin bound to SLe(X) and PSGL-1. Cell 103:467–479

Sun X, Rozenfeld RA, Qu X et al (1997) P-selectin-deficient mice are protected from PAF-induced shock, intestinal injury, and lethality. Am J Physiol 273(1 Pt 1):G56–G61

Symon FA, Lawrence MB, Williamson ML et al (1996) Functional and structural characterization of the eosinophil P-selectin ligand. J Immunol 157:1711–1719

Takai Y, Kitano K, Terawaki S, Maesaki R, Hakoshima T (2007) Structural basis of PSGL-1 binding to ERM proteins. Genes Cells 12:1329–1338

Tam SH, Nakada MT, Kruszynski M et al (1996) Structure-function studies on synthetic peptides derived from the 109-118 lectin domain of selectins. Biochem Biophys Res Commun 227:712–717

Tregouet DA, Barbaux S, Poirier O, AtheroGene Group et al (2003) SELPLG gene polymorphisms in relation to plasma SELPLG levels and coronary artery disease. Ann Hum Genet 67:504–511

Tu L, Chen A, Delahunty MD et al (1996) L-selectin binds to P-selectin glycoprotein ligand-1 on leukocytes: interactions between the lectin, epidermal growth factor, and consensus repeat domains of the selectins determine ligand binding specificity. J Immunol 157:3995–4004

Urzainqui A, Serrador JM, Viedma F et al (2002) ITAM-based interaction of ERM proteins with Syk mediates signaling by the leukocyte adhesion receptor PSGL-1. Immunity 17:401–412

Ushiyama S, Laue TM, Moore KL et al (1993) Structural and functional characterization of monomeric soluble P-selectin and comparison with membrane P-selectin. J Biol Chem 268:15229–15237

Veldman GM, Bean KM, Cumming DA et al (1995) Genomic organization and chromosomal localization of the gene encoding human P-selectin glycoprotein ligand. J Biol Chem 270:16470–16475

Waldron TT, Springer TA (2009) Transmission of allostery through the lectin domain in selectin-mediated cell adhesion. Proc Natl Acad Sci USA 106:85–90

Wang K, Zhou X, Zhou Z et al (2005) Platelet, not endothelial, P-selectin is required for neointimal formation after vascular injury. Arterioscler Thromb Vasc Biol 25:1584–1589

Watson ML, Kingsmore SF, Johnston GI et al (1990) Genomic organization of the selectin family of leukocyte adhesion molecules on human and mouse chromosome 1. J Exp Med 172:263–272

Weis WI, Drickamer K, Hendrickson WA (1992) Structure of a C-type mannose-binding protein complexed with an oligosaccharide. Nature 360:127–134

Weller A, Isenmann S, Vestweber D (1992) Cloning of the mouse endothelial selectins. Expression of both E- and P-selectin is inducible by tumor necrosis factor alpha. J Biol Chem 267:15176–15183

Weyrich AS, McIntyre TM, McEver RP et al (1995) Monocyte tethering by P-selectin regulates monocyte chemotactic protein-1 and tumor necrosis factor-alpha secretion. Signal integration and NF-kappa B translocation. J Clin Invest 95:2297–2303

Winkler IG, Snapp KR, Simmons PJ, Levesque JP (2004) Adhesion to E-selectin promotes growth inhibition and apoptosis of human and murine hematopoietic progenitor cells independent of PSGL-1. Blood 103:1685–1692

Xu J, Cai J, Barger BA, Peek S, Darien BJ (2006) Molecular cloning and characterization of bovine P-selectin glycoprotein ligand-1. Vet Immunol Immunopathol 110:155–161

Xu J, Lasry JB, Svaren J, Wagner B, Darien BJ (2005) Identification of equine P-selectin glycoprotein ligand-1 (CD162). Mamm Genome 16:66–71

Yang J, Galipeau J, Kozak CA et al (1996) Mouse P-selectin glycoprotein ligand-1: molecular cloning, chromosomal localization, and expression of a functional P-selectin receptor. Blood 87:4176–4186

Yuen C, Lawson AM, Chai W et al (1992) Novel sulfated ligands for the cell adhesion molecule E-selectin revealed by the neoglycolipid technology among O-linked oligosaccharides on an ovarian cystadenoma glycoprotein. Biochemistry 31:9126–9131

E-Selectin (CD62E) and Associated Adhesion Molecules

G.S. Gupta

28.1 E-Selectin (Endothelial Leukocyte Adhesion Molecule 1: ELAM-1)

28.1.1 Endothelial Cells Express E-Selectin (ELAM-1/CD62E)

E-selectin, also known as CD62 antigen-like family member E (CD62E), endothelial-leukocyte adhesion molecule 1 (ELAM-1), or leukocyte-endothelial cell adhesion molecule 2 (LECAM2), is a cell adhesion receptor expressed on endothelial cells activated by cytokines. Like other selectins, it plays an important part in inflammation. In humans, E-selectin is encoded by the *E-sel* gene. E-selectin or ELAM-1 is a member of the lectin/epidermal-growth-factor/complement-regulatory-protein-like cell-adhesion molecule family, which includes structurally related molecules referred to as selectins. Whereas P-selectin is stored in Weibel-Palade bodies, and it reaches the plasma membrane after exocytosis of these granules, E-selectin is not stored, and its synthesis is induced by cytokines. The fate of E-selectin has been studied after its surface expression following the intracellular routing of internalized antibodies to the selectin. After a brief surface exposure, internalized E-selectin is degraded in the lysosomes, whereas P-selectin returns to the storage granules from where it can be reused (Subramaniam et al. 1993). The presence of selectins on the cell surface is tightly regulated and abnormal appearance is associated with a number of inflammatory disease conditions. Multiple studies indicate that the selectins mediate neutrophil, monocyte, and lymphocyte rolling along the venular wall (Tedder et al. 1995).

28.1.1.1 E-Selectin Expression Is Not Confined to Endothelium

Proper differentiation and maturation of trophoblast contributes to the fetal-maternal vascular interface of the mature placenta and is required for all subsequent stages of embryogenesis. E-selectin is expressed in a unique pattern in secondary trophoblast giant cells, trophoblast lining the central artery, and a subpopulation of labyrinthine trophoblast all located at the fetal-maternal interface of the murine placenta. Placentae lacking E-selectin show increased trophoblast glycogen cells and fewer labyrinthine neutrophils compared with normal placentae, suggesting that recognition of E-selectin on trophoblast by counter-receptors on other cells contributes to placental development (Milstone et al. 2000). The E-selectin-specific monoclonal antibody H4/18 to keratinocytes in inflamed human oral mucosa, particularly gingival epithelium shows that E-selectin expression is not confined to endothelium (Pietrzak et al. 1996). The E-selectin and ICAM-1 are also localized to subfoveal choroidal neovascular membranes (CNVMs) in patients with age-related macular degeneration (AMD). Mesothelial cells can express a set of adhesion molecules overlapping with, but distinct from those expressed in vascular endothelium (ICAM-1, ICAM-2, VCAM-1, E-selectin), which are functionally relevant for interacting with mononuclear phagocytes (Jonjic et al. 1992).

28.2 E-Selectin Genomic DNA

28.2.1 Human E-Selectin

A full-length cDNA for E-selectin was isolated by transient expression in COS cells. The ELAM-1 clone express a surface structure recognized by E-selectin specific monoclonal antibodies and supported the adhesion of human neutrophils and the promyelocytic cell line HL-60. The primary sequence of E-selectin predicts an amino-terminal lectin-like domain, an EGF domain, and six tandem repetitive motifs (about 60 amino acids each) related to those found in complement regulatory proteins. A similar domain structure is also found in L-selectin homing receptor, and in P-selectin, a membrane glycoprotein of platelet and endothelial secretory granules (Bevilacqua et al. 1989) (Fig. 28.1).

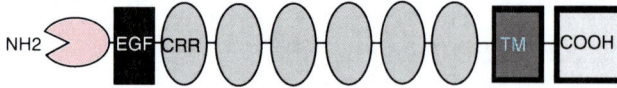

Fig. 28.1 Domain composition of human E-selectin. The extracellular portion of E-selectin contains an amino terminal domain homologous to C-type lectins and an adjacent epidermal growth factor (*EGF*)-like domain. The single EGF element, found in E-selectin is followed by six repetitive elements, each ~60 amino acids long, which resemble protein motives found in complement regulatory proteins (*CRR*), and a transmembrane (*TM*) sequence. A short cytoplasmic sequence is at the carboxyl terminus of E-selectin. The number of amino acids present in the mature E-selectin as deduced from the cDNA sequences is 589

The *E-sel* gene is present in a single copy in the human genome and contains 14 exons spanning about 13 kb of DNA. The positions of exon-intron boundaries correlate with the putative functional subdivisions of the protein. Introns are found at similar positions in all of the six complement regulatory repeats suggesting that these elements arose by internal gene duplication. A consensus TATAA element is located upstream of the transcriptional start site. The E-selectin promoter contains an inverted CCAAT box and consensus NF-kB- and AP-1-binding sites. The *E-sel* gene was assigned to the q12 greater than qter region of human chromosome. Two other members of the selectin gene family, the LAM-1/L-selectin and CD62P/P-selectin have been localized to the long arm of chromosome 1, as have the structurally related complement binding proteins, suggesting that these genes may share a common evolutionary history (Collins et al. 1995). Furthermore, the location of these genes on mouse and human chromosome 1 is consistent with a close evolutionary relationship to the complement receptor-related genes, which also are positioned on the same chromosomes in both species and with which these genes share a region of sequence homology (Watson et al. 1990).

28.2.2 E-Selectin Gene in Other Species

Murine E-selectin is encoded by a single-copy gene, spanning about 13 kb, which is structurally organized into 14 exons and 13 introns; very similar to its human counterpart. The exon/intron architecture exactly parallels the domain structure of the encoded protein. The nucleotide sequence of murine *E-sel* from heart tissue of an IL-1-treated mouse shows an overall similarity of 70% to human *E-sel* cDNA. Highest expression of murine *E-sel* gene occurs in heart and only low expression in lung tissue. Within the promoter, most of the identified regulatory elements are conserved. An exception is the NF-kB box sequence, which, in the murine *E-sel* promoter, does not correspond to the consensus NF-kB sequence (Becker-Andre et al. 1992). Transient expression in COS cells demonstrated that the lectin and EGF domains were sufficient to mediate the binding of mouse and human neutrophils as well as HL60 cells (Norton et al. 1993). The porcine gene comprises 12 exons and 11 introns. Two pseudoexons are contained within intron 4 and intron 6. These sequences are similar to the corresponding exons in the human E-selectin sequence; however, they are not present in the porcine E-selectin-encoding cDNA. Transcription starts at position −498 relative to the translation initiation site. The first ATG is located within exon 2. Translation stops in exon 11 leaving exon 12 untranslated in its entirety (Winkler et al. 1996).

28.3 E-Selectin Gene Regulation

28.3.1 Transcriptional Regulation of CAMs

The innate immune system plays an important role as a first defense against pathogens and involves the recognition of bacteria and viruses, and byproducts thereof, by Toll receptors on immunecompetent cells. Activated cells synthesize and secrete cytokines, which in turn activate systemic responses directed at clearing the pathogen. The "inflammatory triad" of IL-1, TNF-α and LPS are potent stimulators of the endothelial cell (EC) activation/adhesion molecules: ICAM-1, E-selectin (ELAM-1), and VCAM-1 (Bender et al. 1994; Fries et al. 1992). These inflammatory triad molecules, as well as the Toll receptors, initiate intracellular signaling cascades that activate nuclear factor κB (NF-κB), activator protein-1 (AP-1), cAMP responsive element binding proteins (CREBs), and various other transcription factors, which are essential for the regulation of numerous genes, many of which play important roles in immunological processes. The positive regulatory elements required for maximal levels of cytokine induction have been defined in the promoters of all three genes of cell adhesion molecules (CAMs). DNA binding studies reveal a requirement for NF-κB and other transcriptional activators. The organization of the cytokine-inducible element in the E-selectin promoter is remarkably similar to that of the virus-inducible promoter of the human IFN-β gene in that both promoters require NF-κB, ATF-2, and high mobility group protein I(Y) for induction. Based on this structural similarity, a model was proposed for the cytokine-induced E-selectin enhancer that is similar to the stereospecific complex proposed for the IFN-β gene promoter. In these models, multiple DNA bending proteins facilitate the assembly of higher order complexes of transcriptional activators that interact as a unit with the basal transcriptional machinery. The assembly of unique enhancer complexes from similar sets of transcriptional factors may provide the specificity required to regulate complex patterns of gene expression and correlate with the distinct patterns of expression of the leukocyte adhesion molecules. Not only IL-1 but also weakly oxidized LDL, glycated LDL, and hypoxia may be possible factors that cause the expression of ELAM-1 (Collins et al. 1995).

28.3.2 Induction of E-Selectin and Associated CAMs by TNF-α

28.3.2.1 Cell Surface Expression of CAMs Can be Regulated by Different Mechanisms

Expression of some of the endothelial-leukocyte adhesion molecules is dynamically regulated at sites of leukocyte recruitment. For example, endothelial expression of E-selectin and VCAM-1 is dramatically induced, and expression of ICAM-1 is substantially increased at sites of inflammation. These dynamic changes in surface proteins provide the endothelial cell with a mechanism of regulating cell-cell interactions. The three endothiehal cell-surface proteins have different patterns of expression, and are structurally and functionally distinct. Transcription of all three of these genes is substantially increased when endothelial cells are exposed to cytokines. Cell surface appearance of E-selectin and P-selectin on human umbilical vein endothelial (HUVE) cells can be regulated by different mechanisms. E-selectin is transcriptionally induced, within hours, by TNF-α while P-selectin is transported from storage granules to the plasma membrane within minutes upon induction by various stimulating agents (Weller et al. 1992).

PKC Pathway in Expression of CAMs: A protein kinase C agonist, 12-deoxyphorbol 13 phenylacetate 20-acetate (dPPA), selective for the β I isozyme, induces NF-kB-like binding activity and surface expression of E-selectin and VCAM-1 in human umbilical vein endothelial cells (HUVECs), similar to TNF-α. Induction of E-selectin and VCAM-1 expression by dPPA was completely inhibited by the PKC inhibitors staurosporine and Ro31-7549. The PKC inhibitors also reduced TNF-α-induced VCAM-1 expression. However, neither dPPA nor TNF translocated PKC from cytosolic to plasma/or nuclear membrane particulate fractions in HUVEC. These results indicated that activation of the βI PKC isozyme is sufficient for expression of E-selectin and VCAM-1, and suggested that PKC might mediate the effects of TNF-α and dPPA without requiring the translocation normally associated with activation of PKC (Deisher et al. 1993b). Inhibitors of topoisomerase II prevent cytokine-induced expression of VCAM-1, while augmenting the expression of ELAM-1 on human HUVECs. It appeared that topoisomerase II activity may differentially regulate the expression of adhesion molecules on HUVEC (Deisher et al. 1993a). The lower classic PKC activity on pretreatment with phorbol ester (phorbol 12-myristate 13-acetate (PMA)

markedly decreased IL-1β-induced E-selectin mRNA expression in HUVEC in the presence of fetal calf serum and bFGF, although the induction of ICAM-1 mRNA expression was only influenced a little by the PKC down-regulation. On the other hand, TNFα-induced gene expression of these adhesion molecules was unaffected by such PKC modulation. Promoter analysis of E-selectin indicated that the NF-ELAM1/ATF element is critical for the synergistic effect of the cotreatment with IL-1β and PMA (Tamaru and Narumi 1999).

Tyrosine Phosphorylation in Endothelial Adhesion Molecule Induction: Induction of endothelial adhesion molecules by TNFα can occur independently of PKC. Activation of a protein tyrosine kinase (PTK) has been implicated in the upregulation of VCAM-1 by IL-4 on endothelial cells. The PTK inhibitors herbimycin A or genistein suppress induction of endothelial VCAM-1 and E-selectin, as well as subsequent monocytic cell adhesion to endothelial cells stimulated by TNF. Inhibition studies indicated that specific tyrosine phosphorylation following PTK activation is involved in the mobilization of NF-kB, and VCAM-1 mRNA expression. This may have implications for pathophysiological conditions that involve the upregulation of these molecules (e.g. inflammation and atherosclerosis) (Weber 1996)

E-Selectin Expression by Double-Stranded RNA and TNF-α: Since the double stranded RNA (dsRNA)-activated kinase (PKR) mediates dsRNA induction of NF-kB, murine aortic endothelial (MuAE) cells from wild-type and PKR-null mice were investigated for the role of PKR in the induction of E-selectin expression by dsRNA (pIC) and TNF-α. Study indicated that PKR is required for full activation of E-selectin expression by pIC and TNF-α in primary MuAE cells identifying activating transcription factor 2 as a new target for PKR-dependent regulation and suggested a role for PKR in leukocyte adhesion (Bandyopadhyay et al. 2000).

TNF-α Activates Two Signaling Pathways, NF-kB and JNK/p38, Both Required for Expression of E-Selectin: Transcriptional regulation of the E-selectin promoter by TNFα requires multiple NF-kB binding sites and a cAMP-responsive element/activating transcription factor-like binding site designated positive domain II (PDII). Read et al. (1997) while studying the role of MAP kinases in induced expression of E-selectin, suggested that TNFα activates two signaling pathways, NF-kB and JNK/p38, which are both required for maximal expression of E-selectin (Xu et al. 1998). Moreover, transient overexpression of catalytically inactive JNK or truncated TRAF2 (TNF receptor-associated factor 2) partially inhibits endogenous E-selectin protein expression in human endothelial cells. Min and Pober (1997) suggest that TNF activates parallel TRAF-NF-kB and TRAF-RAC/CDC42-JNK-c-Jun/ATF2 pathways to initiate E-selectin transcription. Furthermore, endogenous inhibitors of the MAPK cascade, such as the dual-specificity phosphatases may be important for the postinduction repression of MAPK activity and E-selectin transcription in endothelial cells. These inhibitors may play an important role in limiting the inflammatory effects of TNF-α and IL-1β (Wadgaonkar et al. 2004).

IL-4 Suppression of TNFα-Stimulated E-Selectin Gene: IL-4, secreted by activated T-helper 2 lymphocytes, eosinophils, and mast cells, stimulates the expression of pyrogen-induced upregulation of ICAM-1, E-selectin and VCAM-1 in cultured HUVEC (Kapiotis et al. 1994). IL-4 stimulates a number of immune system genes via activation of STAT6. However, IL-4 can concomitantly suppress the expression of other immune-related gene products, including k light chain, FcγRI, IL-8, and E-selectin (Bennett et al. 1997).

Regulation by Cyclosporin A and Transcription Factor NFAT: Nuclear factor of activated T cells (NFAT) supports the activation of cytokine gene expression and mediates the immunoregulatory effects of cyclosporin A (CsA). Activated endothelial cells also express NFAT. CsA completely suppressed the induction of NFAT in endothelial cells and inhibited the activity of GM-CSF gene regulatory elements. CsA also suppressed E-selectin, but not VCAM-1 expression in endothelial cells, even though E-selectin promoter is activated by NF-kB rather than NFAT. Hence, induction of cell surface expression of E-selectin by TNF-α is reduced in presence of CsA, and this was reflected by a decrease in neutrophil adhesion. This suggests a mechanism by which CsA could function as an antiinflammatory agent (Cockerill et al. 1995).

28.3.2.2 TNF-α-Induced Cell Adhesion to Human Endothelial Cells Through TNF-Receptors

Mackay et al. (1993) reported that TNF-α triggers cell responses through two distinct membrane receptors. The HUVEC express both TNF receptor types, TNF-R55 and TNF-R75. But the TNF-α-induced cell adhesion to HUVEC is controlled almost exclusively by TNF-R55, but not through TNF-R75. This finding correlated with the exclusive activity of TNF-R55 in the TNF-α-dependent regulation of the expression of the ICAM-1, E-selectin, and VCAM-1. However, both TNF-R55 and TNF-R75 upregulate α2 integrin expression in HUVEC. The predominant role of TNF-R55 in TNF-α-induced adhesion in HUVEC may correlate with its specific control of NF-kB activation, since kB elements are known to be present in ICAM-1, E-selectin, and VCAM-1 gene regulatory sequences (Mackay et al. 1993).

Fish TNF-α Displays Different Sorts of Bioactivity to Their Mammalian Counterparts: In two complementary fish models, gilthead seabream and zebrafish, the proinflammatory effects of fish TNF-α are mediated through the activation of endothelial cells leading to the expression of E-selectin and different chemokines in endothelial cells, thus explaining the recruitment and activation of phagocytes observed in vivo in both species. Results indicate that fish TNF-α displays different sorts of bioactivity to their mammalian counterparts and point to the complexity of the evolution that has taken place in the regulation of innate immunity by cytokines (Roca et al. 2008).

28.3.3 IL-1-Mediated Expression of CAMs

Stimulation of E-selectin, VCAM-1, and ICAM-1 after treatment with IL-1β, TNF-α, and LPS, in a dose-dependent fashion was observed in cultured human intestinal microvascular endothelial cells (HIMEC). Each molecule displayed a time-related response comparable to those obtained with HUVEC (Haraldsen et al. 1996). Sphingomyelin hydrolysis to ceramide did not trigger, but rather enhanced cytokine-induced E-selectin expression. Sphingomyelin hydrolysis to ceramide did not mediate all the effects of IL-1β, although it might play important roles in IL-1β signal transduction in HUVECs (Masamune et al. 1996).

The infiltration of leukocytes into the CNS is associated with many pathologic conditions of the brain. The elevated levels of IL-1 in brain appear to accompany the pathogenesis. The evidences suggest that IL-1 can induce the expression of adhesion molecules for leukocytes on glial cells and a role for NF-kB in the induction process. Thus, the expression of adhesion molecules for leukocytes on glial cells in response to IL-1 may represent an important mechanism for retention of immune cells in the CNS that may be a prologue to inflammatory conditions in the brain (Moynagh et al. 1994; Couffinhal et al. 1994).

28.4 Factors in the Regulation of CAMs

28.4.1 NF-kB: A Dominant Regulator

A growing body of evidence demonstrates that nuclear transcription factor NF-kB acts as a dominant regulator of transcription of endothelial CAMs, E-selectin, VCAM-1 and

ICAM-1 genes. This molecular pathway may be involved in the pathogenesis of acute inflammatory diseases. Pharmacologic antagonism of transcription from these genes therefore represents a novel approach to the development of anti-inflammatory therapeutics and NF-kB may represent a prime target for therapeutic intervention in pathologic conditions associated with EC activation such as allo- and xenograft rejection, atherosclerosis, ischemic reperfusion injury and vasculitis.

NF-kB Is Required for TNF-α-Induced Expression of CAMs in Endothelial Cells: In response to inflammation stimuli, TNF-α induces expression of CAMs in ECs. Studies suggested that NF-kB and p38 MAP kinase (p38) signaling pathways play central roles in this process. However, subsequent analysis revealed that p38 activity is not essential for TNF-α-induced CAMs. Rajan et al. (2008) demonstrated that NF-kB, but not p38, is critical for TNF-α-induced CAM expression. In addition to TNF-α, Tie-1, an endothelial specific cell surface protein upregulates VCAM-1, E-selectin, and ICAM-1, partly through a p38-dependent mechanism (Chan et al. 2008). Similarly, LPS can activate p38 MAPK in equine ECs and that both neutrophil adhesion to LPS-activated EDVEC and prostacyclin (PGI$_2$) release are dependent upon p38 MAPK phosphorylation. Results reveal that inhibition of this kinase may reduce inflammatory events in the endotoxemic horse (Brooks et al. 2009).

NF-kB and IkBα: An Inducible Regulatory System in Endothelial Activation: Endothelial cells encode transcripts encoding the p50/p105 and p65 components of NF-kB and the rel-related proto-oncogene c-rel. These transcripts are transiently increased by TNF-α. Stimulation of endothelial cells with TNF-α resulted in nuclear accumulation of the p50 and p65 components of NF-kB and their binding to the E-selectin kB site. Endothelial cells also express I kB-α (MAD-3), an inhibitor of NF-kB activation. Level of I kB-α inhibitor falls rapidly after TNF-α stimulation. In parallel, p50 and p65 accumulate in the nucleus, where as RNA transcripts for IkB-α are dramatically upregulated. Studies suggest that NF-kB and IkB-α system may be an inducible regulatory mechanism in endothelial activation (Read et al. 1994, 1996). TNF causes persistent activation of NF-κB in human EC and that this may result from sustained reductions in I kB-β levels (Johnson et al. 1996).

Nuclear NF-kB DNA-binding activity was rapidly increased within lung and heart tissues of rats administered endotoxin, consistent with the translocation of NF-kB complexes from the cytoplasm to the nucleus. NF-kB activation preceded the transcriptional activation of E-selectin, VCAM-1, and ICAM-1 genes. These molecular events were temporally associated with the sequestration of leukocytes and the development of pulmonary inflammation. Report supported that NF-κB activation is the underlying molecular mechanism for constitutive expression of E-selectin, VCAM-1, and ICAM-1 on human B lymphocytes, plasma cells and in hepatic vascular lining cells though the hepatic parenchymal cells, despite NF-kB activation do not express E- selectin mRNA. This indicated that NF-kB activation alone is not sufficient for E-selectin gene transcription in vivo (Essani et al. 1996; Manning et al. 1995; Wang et al. 1995).

Platelet factor 4 (PF4), a platelet-specific chemokine, has been localized to atherosclerotic lesions, including macrophages and endothelium. The PF4, which is able to increase expression of E-selectin by endothelial cells, represents one of the potential mechanisms by which platelets may participate in atherosclerotic lesion progression (Yu et al. 2005). E-selectin and its RNA are up-regulated in HUVE cells exposed to PF4. It appeared that activation of NF-kB is critical for PF4-induced E-selectin expression. The LDL receptor-related protein as the cell surface receptor may mediate this effect (Yu et al. 2005). The POZ domain of FBI-1 (factor that binds to the inducer of short transcripts of HIV-1) interacts with the Rel homology domain of the p65 subunit of NF-kB in in vivo and in vitro protein-protein interactions. FBI-1 enhanced NF-kB-mediated transcription of E-selectin genes in HeLa cells upon stimulation and overcame gene repression (Lee et al. 2005).

Glucocorticoid-Mediated E-Selectin Repression: Induction of E-selectin expression by stimuli such as TNF-α or LPS is reduced markedly in the presence of dexamethasone, a potent anti-inflammatory agent. E-selectin promoter analysis revealed that induction by proinflammatory stimuli as well as repression by dexamethasone are mediated by the same promoter region containing three closely spaced binding sites for NF-kB and an element, NF-ELAM-1, constitutively occupied by ATF and c-Jun. NF-ELAM-1 contributes to maximal promoter activity, but does not confer glucocorticoid inhibition, as demonstrated by site-directed mutagenesis. In contrast, transcription directed by the E-selectin NF-kB elements is reduced strongly in the presence of dexamethasone, thus identifying NF-kB as the primary target for glucocorticoid-mediated E-selectin repression. The interference by glucocorticoids receptor (GR) with the transcriptional activation potential of DNA-bound NF-kB complexes might contribute to mechanisms underlying the anti-inflammatory effects of glucocorticoids (Brostjan et al. 1997; Ray et al. 1997).

28.4.2 Cyclic AMP Inhibits NF-kB-Mediated Transcription

Cytokines induce the expression of E-selectin, VCAM-1, and ICAM-1 on HUVECs. Expression of these surface proteins is differently affected by cAMP. Increased cAMP levels decrease E-selectin and VCAM-1 but increase ICAM-1 expression. The cAMP repression of E-selectin occurs at the transcription level. This effect is abolished by protein kinase A (PKA) inhibition, suggesting that repression is mediated by PKA-driven phosphorylation. A minimal E-selectin promoter sequence necessary to confer cytokine inducibility is also sufficient to mimic the cAMP effect in transfected HUVECs. There are two regions (NF-kB and NF-ELAM1) of the minimal promoter that bind transcription factors necessary for E-selectin induction. Increased cAMP did not alter the binding of the complexes formed on either the NF-kB or NF-ELAM1 site. In contrast, in IL-1-treated HUVECs transactivity due to an NF-kB site was reduced by elevated cAMP. Increased cAMP in HUVECs appears to induce a protein kinase activity that reduces the cytokine signal for E-selectin and VCAM-1 expression (Ghersa et al. 1994). Ollivier et al. (1996) examined the molecular mechanism by which agents that elevate intracellular cAMP inhibit the expression of TNFα, tissue factor, ELAM-1, and VCAM-1 genes. Both forskolin and dibutyryl cAMP, which elevate intracellular cAMP by independent mechanisms, inhibited TNFα and tissue factor expression at the level of transcription. Induction of NF-kB-dependent gene expression in transiently transfected human monocytic THP-1 cells and HUVEC was inhibited by elevated cAMP and by over-expression of the catalytic subunit of PKA. This indicated that activation of PKA reduced the induction of a distinct set of genes in monocytes and endothelial cells by inhibiting NF-kB-mediated transcription.

Cilostazol Inhibits Cytokine-Induced NF-kB Activation Via AMP-Activated Protein Kinase Activation: Cilostazol is a selective inhibitor of phosphodiesterase-3 that increases intracellular cAMP levels and activates PKA, thereby inhibiting platelet aggregation and inducing peripheral vasodilation. Cilostazol also inhibited TNFα-induced NF-kB activation and TNFα-induced I κB kinase activity. Furthermore, cilostazol attenuated the TNFα-induced gene expression of various pro-inflammatory and cell adhesion molecules, such as VCAM-1, E-selectin, ICAM-1, MCP-1, and PECAM-1 in HUVEC. Hattori et al. (2009) suggested that cilostazol might attenuate the cytokine-induced expression of adhesion molecule genes by inhibiting NF-κB following AMP-activated protein kinase.

28.4.3 Peroxisome Proliferator-Activated Receptors

Peroxisome proliferator-activated receptors (PPARs) are nuclear receptors which down-regulate inflammatory signaling pathways. Three peroxisome proliferator-activated receptors (PPARs) have been identified: PPARα, PPARβ/δ, and PPARγ, all of which have multiple biological effects, especially the inhibition of inflammation. The expressions of 3 PPAR isoforms and PPAR-responsive genes are markedly upregulated in spontaneously hypertensive rats (SHR) compared with those of Wistar-Kyoto rats (WKY). An enhanced inflammatory response in the organs of SHR might play a key role in pathogenesis of hypertension and secondary organ complications. Changes (increases) in PPARs expression may reflect a compensatory mechanism to the inflammatory status of hypertensive rats (Sun et al. 2008). PPAR-γ has an essential role in adipogenesis and glucose homeostasis. The PPAR-γ is expressed in vascular tissues including endothelial cells. Its activity is regulated by many pharmacological agonists. PPAR-γ might be involved in the control of inflammation and in modulating the expression of various cytokines. PPAR-γ activation suppresses the expression of vascular adhesion molecules in ECs and the ensuing leukocyte recruitment. Evidence shows that constitutive activation of PPAR-γ is sufficient to prevent ECs from converting into a pro-inflammatory phenotype (Wang et al. 2002).

The PPAR activators have been shown to inhibit the expression of E-selectin of HVECs in response to TNF-α. Troglitazone, pioglitazone, α-clofibrate, and 15-deoxy-δ 12,14-prostaglandin J2 inhibited the TNF-α-stimulated E-selectin gene transcription. The activators caused a significant induction of liver regenerating factor 1 (LRF1)/ATF3, which bound to NF-ELAM1 site and repressed the TNF-α-induced E-selectin gene expression. It appears that the effect of PPAR activators is mediated, in part, through the induction of LRF1/ATF3 (Nawa et al. 2000). In another study, pioglitazone inhibited the expression of VCAM-1 protein and mRNA on HUVEC after IL-1β stimulation, though it showed little effect on the expression of ICAM-1 and E-selectin. The study revealed that pioglitazone can influence monocyte-EC binding by inhibiting VCAM-1 expression on activated EC and neutrophil-EC binding by inhibiting upregulation of CD11b/CD18 on activated neutrophils. This might provide a molecular basis of antiinflammatory effect of PPAR activators (Nawa et al. 2000; Imamoto et al. 2004).

The role of PPAR-δ in endothelial activation remains poorly understood. In human umbilical vein endothelial cells (HUVECs), the synthetic PPAR-δ ligands GW0742 and GW501516 significantly inhibited TNF-α-induced expression of VCAM-1 and E-selectin, as well as ensuing endothelial-leukocyte adhesion. Activation of PPAR-δ upregulated the expression of antioxidant genes: SOD-1, catalase, and thioredoxin and decreased ROS production in ECs. This shows that ligand activation of PPAR-δ in ECs has a potent antiinflammatory effect, probably via a binary mechanism involving the induction of antioxidative genes and the release of nuclear corepressors (Fan et al. 2008).

PPAR-δ agonists may have a potential for treating inflammatory diseases such as atherosclerosis and diabetes. Alterations of PPAR functions can contribute to HIV-1-induced dysfunction of brain endothelial cells. Treatment with HIV-1 transactivator of transcription (Tat) protein decreased PPAR transactivation in brain endothelial cells. Tat-induced up-regulation of inflammatory mediators, such as IL-1β, TNF-α, CCL2, and E-selectin were markedly attenuated in hCMEC/D3 over-expressing PPARα or PPARγ. Data suggest that targeting PPAR signaling may provide a novel therapeutic approach to attenuate HIV-1-induced local inflammatory responses in brain endothelial cells (Huang et al. 2008).

28.4.4 Endogenous Factors Regulating CAM Genes

In order to understand the role of complement components as regulators of expression of endothelial adhesive molecules in response to immune complexes (ICs), Lozada et al. (1995) showed that ICs stimulate both the endothelial adhesiveness for leukocytes and expression of E-selectin, ICAM-1, and VCAM-1. ICs (BSA-anti-BSA) stimulated EC adhesiveness for added leukocytes in the presence of complement-sufficient human serum. Studies showed that ICs stimulate ECs to express adhesive proteins for leukocytes in presence of a heat-labile serum factor, C1q (Lozada et al. 1995).

Native LDL (n-LDL) increases HUVEC adherence of mononuclear cells. Such phenotypic changes suggest that n-LDL alters the usual expression of cell adhesion molecules to enhance the adhesive properties of the endothelium. n-LDL increased ICAM-1 protein expression corresponded with increased ICAM-1 mRNA levels. n-LDL also appeared to increase E-selectin and VCAM-1 message levels, but these changes were not statistically significant. n-LDL increase of ICAM-1 expression seems to enhance the adhesive properties of the endothelium. Such perturbations in HUVEC function likely represent a proinflammatory response to protracted n-LDL exposure and one of the early steps in atherogenesis (Smalley et al. 1996).

Adrenomedullin (AM) and corticotrophin (ACTH) are both vasoactive peptides produced by a variety of cell types, including ECs. Their role in inflammation and the immune response was studied by Hagi-Pavli et al. (2004). AM and ACTH induce cell surface expression of E-selectin, VCAM-1, and ICAM-1 on HUVEC. This effect appears to be mediated in part via elevation of cAMP, and that both peptides elevate cAMP. The effect of AM and ACTH is inhibited by the adenylyl cyclase inhibitor SQ-22528. This demonstrates a role for AM and ACTH in the regulation of the immune and inflammatory response through adhesion molecules (Hagi-Pavli et al. 2004).

Methylation May Play a Role in Blocking E-Selectin Expression in Non-Endothelial Cells: The E-selectin promoter in cultured endothelial cells is under-methylated in comparison with non-expressing HeLa cells. Plasmid constructs carrying a reporter driven by E-selectin promoter and methylated in vitro are not transcribed in either an in vitro transcription system or in transiently transfected cells. The NF-kB site in the promoter of E-selectin is the likely target for this methylation-mediated repression in a minimal promoter carrying only this and an associated element. Thus, methylation is likely to play a role in blocking E-selectin expression in non-endothelial cells (Smith et al. 1993).

28.4.5 Tat Protein Activates Human Endothelial Cells

Tat protein, an HIV gene product, functions as a transactivator for HIV replication and is known to be secreted extra-cellularly by infected cells. Tat stimulates endothelial cells. This is evidenced by the expression of E-selectin, critical for the initial binding of leukocytes to the blood vessel wall, and its effect on increased synthesis of IL-6, a cytokine known to enhance endothelial cell permeability. Furthermore, tat acts synergistically with low concentrations of TNF-α to enhance IL-6 secretion. Tat activates human CNS-derived endothelial cells (CNS-EC) by the increase in the expression of E-selectin, the synthesis of IL-6, and the secretion of plasminogen activator inhibitor-1. It was suggested that secreted tat protein may increase leukocyte binding, and alter the blood-brain barrier permeability to enhance dissemination of HIV-infected cells into the CNS (Hofman et al. 1994). Treatment of HUVEC with Tat induces the cell surface expression of ICAM-1, VCAM-1, and E-selectin-1 and the effect on expression of adhesion molecules was potentiated by TNF. Like TNF, Tat also enhanced the adhesion of human promyelomonocytic HL-60 cells to EC (Dhawan et al. 1997). Both Tat and TNF activated p65 translocation and binding to an oligonucleotide containing the E-selectin kB site 3 sequence. A super-repressor adenovirus (AdIkBαSR) that constitutively sequesters IkB in the cytoplasm as well as cycloheximide or actinomycin D inhibited Tat- or TNF-mediated kB translocation and E-selectin up-regulation (Cota-Gomez et al. 2002).

28.4.6 Reactive Oxygen Species (ROS)

ROS is critical signal in the activation of NF-kB and in E-selectin expression: Decreased NO activity, the formation of ROS and increased endothelial expression of the redox-sensitive VCAM-1 gene in the vessel wall are early and characteristic features of atherosclerosis. Through this mechanism, NO may function as an immunomodulator of the vessel wall and thus mediate inflammatory events involved in the pathogenesis of atherosclerosis (Khan et al. 1996). Reactive oxygen intermediates may act as second messengers in the activation of NF-kB.

Antioxidants and NO Inhibit Adhesion Molecule Expression: Using the antioxidant pyrrolidine dithiocarbamate (PDTC) for the study of the effect of NF-kB inhibition in TNF-α-induced EC activation in vitro, Ferran et al. (1995) showed that PDTC strongly reduces the TNF-α-mediated induction of E-selectin, VCAM-1, and ICAM-1. The N-acetyl-L-cysteine (NAC) is a thiol-containing antioxidant that inhibits agonist-induced monocytic cell adhesion to endothelial cells. Unlike PDTC, which specifically inhibits VCAM-1, NAC inhibits IL-1β -induced mRNA and cell surface expression of both E-selectin and VCAM-1. Although NAC reduced NF-kB activation in EC, the antioxidant had no appreciable effect when an oligomer corresponding to the consensus NF-kB binding site of the E-selectin gene was used. (Faruqi et al. 1997). Trans fatty acid intake could also affect biomarkers of inflammation and endothelial dysfunction including C-reactive protein (CRP), IL-6, sTNFR-2, E-selectin, sICAM-1 and sVCAM-1. Higher intake of trans fatty acids could adversely affect endothelial function, which might partially explain why the positive relation between trans fat and cardiovascular risk is greater than one would predict based solely on its adverse effects on lipids (Lopez-Garcia et al. 2005). Findings support that TNF-α induces NF-kB activation of the resultant E-selectin gene by a pathway that involves formation of ROS and that the E-selectin expression can be inhibited by the antioxidant action of NAC or PDTC. The results support the hypothesis that generation of ROS in endothelial cells induced by proinflammatory cytokines such as TNF-α is a critical signal mediating E-selectin expression (Rahman et al. 1998; Spiecker et al. 2000). The inhibitory effects of NO on adhesion molecule expression differ from that of antioxidants in terms of the mechanism by which NF-kB is inactivated (Spiecker et al. 2000). NO inhibited IL-1α-stimulated VCAM-1 expression in a concentration-dependent manner. NO also decreased the endothelial expression of E-selectin and to a lesser extent ICAM-1 and secretable cytokines (IL-6 and IL-8) (De Caterina et al. 1995).

28.4.7 Role of Hypoxia

Anoxia/reoxygenation (A/R) leads to a biphasic increase in neutrophil adhesion to HUVECs, with peak responses occurring at 30 min (phase 1) and 240 min (phase 2) after reoxygenation. It was indicated that A/R elicits a two-phase neutrophil-endothelial cell adhesion response that involves transcription-independent and transcription-dependent surface expression of different endothelial CAMs. Tissue injury that accompanies hypoxemia/reoxygenation shares features with the host response in inflammation, suggesting that cytokines, such as IL-1, may act as mediators in this setting. Human endothelial cells subjected to hypoxia elaborated IL-1 activity into conditioned media in a time-dependent manner. Production of IL-1 activity by hypoxic ECs was associated with an increase in the level of mRNA for IL-1 alpha, and was followed by induction of ELAM-1 and enhanced ICAM-1 during reoxygenation. This suggests that hypoxia is a stimulus, which induces EC synthesis and release of IL-1 alpha, resulting in an autocrine enhancement in the expression of adhesion molecules (Shreeniwas et al. 1992). Findings make the use of inducible enhancers a promising strategy for increasing tissue specific gene expression.

28.5 Binding Elements in E-Selectin Promoter

28.5.1 Transcription Factors Stimulating *E-sel* Gene

As seen, expression of E-selectin is cell and stimulus specific, as it is mainly expressed on endothelial cells in response to induction by IL-1 and TNF-α as well as LPS and phorbol myristate acetate. IL-1 and TNF-α initiate intracellular signaling cascades that activate NF-kB, activator protein-1 (AP-1), and cAMP responsive element binding proteins (CREBs), which are transcription factors essential for the regulation of numerous genes, many of which play important roles in immunological processes. In addition, E-selectin gene activity is transient. Expression is maximal 2–4 h following cytokine induction and returns to the basal level by 24 h. This tight regulation of gene activity is likely to require complex control mechanisms (Jensen and Whitehead 2003) .

Factors in addition to NF-kB appear to play a central role in determining the specific expression of ELEM-1 gene in response to cytokines. In support of this, Whelan et al. (1991) have shown that NF-kB alone, although essential, is not sufficient to mediate IL-1-induced activation of E-selectin gene. Hooft van Hujsduinen et al. (1992) identified two additional factors, which were referred to as NF-ELAM1 and NF-ELAM2 (binding at positions −153 to −144 and −104 to −100, respectively), that play critical roles in controlling cytokine-induced expression of the E-selectin gene. While neither of these elements alone is sufficient to confer enhancer activity on a heterologous promoter, NF-ELAM1 was shown to cooperate with NF-kB to augment cytokine-induced expression to levels

28.5 Binding Elements in E-Selectin Promoter

significantly above that observed with NF-kB alone. Hooft van Hujsduinen et al. (1992) demonstrated that NF-ELAM1 functionally cooperates with NF-kB in IL-1 induction of the E-selectin gene.

Multiple NF-kB Binding Sites in Human *E-sel* Gene: Promoter binding factors responsible for induced ELAM-1 gene expression include NF-kB, which binds at three sites within the E-selectin promoter; two of these NF-kB binding sites are partially overlapping, that have been reported to be required for full induction by cytokines. In band shift assays, Hooft van Huijsduijnen et al. (1993) detected two distinct NF-kB complexes in nuclear extracts from several cytokine-induced cells. A putative AP-1 site at position −499 to −493 within the promoter has been shown not to affect cytokine induction (Schindler and Baichwal 1994). The presence of a TNF-α-inducible element close to the transcriptional start site. 170 bp of upstream sequences was sufficient to confer TNF-α induction. Site-directed mutagenesis of this region revealed two regulatory elements (−129 to −110 and −99 to −80) that are essential for maximal promoter activity following cytokine treatment. Binding studies with crude nuclear extracts and recombinant proteins showed that the two elements correspond to three NF-kB binding sites (site 1, −126; site 2, 116; and site 3, −94). Thus induction of E-selectin gene requires the interaction of NF-kB proteins at multiple regulatory elements (Schindler and Baichwal 1994). Consequently, this promoter is often used to drive expression of luciferase in reporter assays and is frequently considered to be an "NF-kB-specific" promoter. Whitley et al. identified four positive regulatory domains (PDI to PDIV) in the E-selectin promoter that are required for maximal induction by TNF-α in endothelial cells. Two of these domains contain adjacent binding sites for NF-kB (PDIII and PDIV), where as the third corresponds to CRE/ATF site (PDII), and a fourth is a consensus NF-kB site (PDI) (Fig. 28.2). Mutations that decrease the binding of NF-kB to any one of the NF-kB binding sites in vitro abolish cytokine-induced E-selectin gene expression in vivo. A similar correlation was demonstrated between ATF binding to PDII and E-selectin gene expression (Whitley et al. 1994).

28.5.2 CRE/ATF Element or NF-ELAM1

Activating Transcription Factor (ATF) is a basic leucine zipper protein, whose DNA target sequence is widely distributed in cAMP response element (CRE). The CRE/ATF element of E-selectin promoter is necessary for full cytokine responsiveness. It differs from a consensus CRE by 1 nucleotide (G—>A conversion) and does not mediate

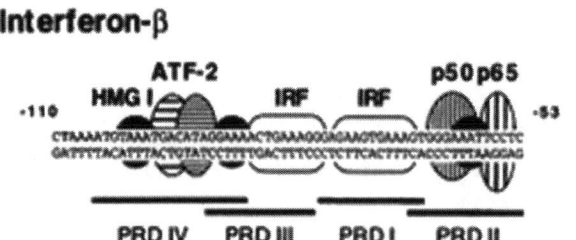

Fig. 28.2 Comparison of the human promoters for the IFN-β and E-selectin genes. The double-stranded DNA sequences of the IFN-β promoter from −110 to −53 and of the human E-selectin from −156 to −83, upstream of the transcriptional start site, are shown. Both regions contain four regulatory domains, designated PRDI through PRDIV, in the human IFN-P gene and PDI through PDIV in the E-selectin gene. Transcription factors that bind to each of the elements are shown for both genes. Those for the IFN-1 gene are as follows (reviewed in reference 19): NF-KB binds to PRDII, IFN regulatory factor 1 (*IRF*) binds to PRDI and PRDIII, and ATF-2 binds to PRDIV. Binding sites for HMG I(Y) are found in PRDII and PRDIV and are indicated. Transcription factors that bind to the E-selectin cytokine response region are as follows: NF-kB binds to PDI, PDIII, and PDIV, and ATF-a or ATF-2 homodimer or the corresponding c-Jun heterodimer binds to PDII. HMG I(Y) binds to a region within PDII as well as to the AT-rich regions in the kB sites contained in PDI, PDIII, and PDIV (Adapted with permission from Whitley et al. 1994 © American Society for Microbiology)

transcriptional activation in response to cAMP. The cAMP actually decreases E-selectin synthesis induced by TNF-α in aortic endothelial cells; the cAMP-mediated inhibition maps to the CRE/ATF element. Study suggests that a change in the composition of the proteins binding to the CRE/ATF promoter element contributes to the competing effects of TNF-α and cAMP on E-selectin gene expression (De Luca et al. 1994). The abnormalities in activating transcription factor-2 (ATF-2) mutant mice demonstrate its absolute requirement for skeletal and CNS development, and for maximal induction of select genes with CRE sites, such as E-selectin (Reimold et al. 1996).

The cAMP-independent members of the ATF family bind specifically to the NF-ELAM1 promoter element (Fig. 28.2). This sequence operates in a cAMP-independent manner to induce transcription and thus it was defined as a non-cAMP-responsive element (NCRE). It was found that ATFα is a component of the NF-ELAM1 complex and its over-expression activates the E-selectin promoter. Furthermore, the ability of over-expressed NF-kB to transactivate the E-selectin promoter in vivo was dependent on the NF-ELAM1 complex (Kaszubska et al. 1993). Hooft van Huijsduijnen et al. (1992) identified two proximal E-selectin promoter elements and their DNA-binding factors that are, in addition to NF-kB, essential for E-selectin transcription. This site, ACATCAT, is recognized by NF-ELAM-1. The site corresponds to NF-ELAM-1s preferential binding sequence (A/T)CA(G/T)CA(G/T). This element is identical to the T-cell δ A enhancer found in the T-cell receptor-α, -β, and CD3δ genes. Studies suggest that δ A/NF-ELAM-element can function as a modulator of NF-kB in endothelial cells both as well as a T-cell enhancer (Hooft van Huijsduijnen et al. 1992). It was shown that a novel protein-protein interaction occurs between the p50 and p65 subunits of NF-kB and certain ATFs. Furthermore, results demonstrate that NF-kB is dependent on NFELAM1 for strong induction of E-selectin expression. These studies demonstrate a novel mechanism operating to specifically control E-selectin gene transcription in response to cytokines.

28.5.3 HMG-I(Y) Mediates Binding of NF-kB Complex

The high-mobility-group protein I (Y) [HMG-I(Y)] also binds specifically to the E-selectin promoter and thereby enhances the binding of both ATF-2 and NF-kB to the E-selectin promoter in vitro. In addition to interaction of NF-kB and ATF with promotor elements, analysis of the E-selectin promoter revealed an additional region (−140 to −105 nt referred as −140/−105) which is essential in controlling promoter activation by cytokines. The HMG-I(Y) interacts specifically at two sites within this region. One of the HMG-I(Y)-binding sites overlaps a sequence element (−127/−118) diverging at only one position from the NF-kB consensus binding sequence. Lewis et al. (1994) searched whether −127/−118 element represents a second functional NF-kB-binding site within the E-selectin promoter. It was shown that p50, p65, and RelB are components of the complex interacting at this site. Mutations that interfere with HMG-I(Y) binding decrease the level of cytokine-induced E-selectin expression. Mutations at −127/−118 NF-kB site indicated that both NF-kB and HMG-I(Y) binding at this site are essential for IL-1 induction of the promoter. The binding affinity of the p50 subunit of NF-kB to both NF-kB sites within the E-selectin promoter was significantly enhanced by HMG-I(Y). Thus, HMG-I(Y) mediates binding of a distinct NF-kB complex at two sites within the E-selectin promoter (Fig. 28.2). Furthermore, a unique cooperativity between these NF-kB complexes is essential for induced E-selectin expression. These results suggest mechanisms by which NF-kB complexes are involved in specific gene activation (Lewis et al. 1994). Studies suggest that the organization of the TNF-α-inducible element of the E-selectin promoter is remarkably similar to that of the virus-inducible promoter of human β interferon gene in that both promoters require NF-kB, ATF-2, and HMG-I(Y). Thus, HMG-I(Y) functions as a key architectural component in the assembly of inducible transcription activation complexes on both promoters (Whitley et al. 1994). HMG-I(Y) binds to the A/T-rich core found at the centre of these binding sites. Distamycin, an antibiotic that also binds A/T-rich DNA, inhibits HMG-I(Y) DNA binding. The distamycin effect on transcription was mediated through one of the three NF-kB-HMG-I(Y) binding sites (NF-kBII) within the promoter. This suggests that the NF-kB-HMG-I(Y) complex interacting at the NF-kBII site plays a role not only in cytokine induction of E-selectin expression, but also in its down-modulation (Ghersa et al. 1997).

28.5.4 Phased-Bending of E-Selectin Promoter

As suggested, DNA elements required for IL-1 inducibility are located in the proximal promoter: an NF-ELAM site, two NF-kB sites (I and II), the NF-E-selectin 2 element and a TATA box. IL-1 induced promoter activity is exquisitely sensitive to the spatial arrangements of these elements. Phasing of the ATF and NF-kB II elements indicates that their relative helix orientation is more important than distance per se. This sensitivity is partly due to a requirement for correctly oriented, transcription factor-induced DNA-bending. (1) Band shift analyses with permuted ATF- and NF-kB elements showed that their associated factors all bend DNA. (2) One can functionally replace the NF-E-selectin/ATF element by a subset of a panel of DNA fragments that contain defined bends in various planes. It appears that the main role of the factors binding at the NF-E-selectin/ATF element is to alter the conformation of the E-selectin promoter, presumably looping distant enhancer elements into each other's proximity (Meacock et al. 1994).

28.6 E-Selectin Ligands

28.6.1 Carbohydrate Ligands (Lewis Antigens)

Various Lewis Antigens, Lewis A (Le^a), Lewis B (Le^b), and Sialyl Lewis A (sLe^a) have been studied in different biological contexts, for example in microbial adhesion and cancer. Adhesion molecules recognize carbohydrate moieties sLe^x, s-diLe^x, or sLe^a, though with different affinity. The potency of oligosaccharides sLe^x, sLe^a, HSO$_3$Le^x, and HSO$_3$Le^a, their conjugates with polyacrylamide (40 kDa), and other monomeric and polymeric selectin inhibitors was compared with that of the polysaccharide fucoidan. The monomeric sLe^x is more effective as a selectin blocker in vivo than in vitro, whereas the opposite is true for fucoidan and the bi-ligand neoglycoconjugate HSO$_3$LeA-PAA-sTyr (Ushakova et al. 2005). Melanoma cell lines show differential binding to different Lewis antigens. For example melanoma cell line NKI-4 binds to E-selectin, but not to P-selectin. Cell line NKI-4 does not express sLe^x, but was positive for sdiLe^x and sLe^a. In contrast, melanoma cell lines, MeWo and SK-MEL-28, expressing either sdiLe^x or sLe^a on the cell surface, bound neither E-selectin nor P-selectin (Kunzendorf et al. 1994). High metastatic colonic carcinoma cells express relatively more Lamp molecules and sLe^x structures on the cell surface than their corresponding low metastatic counterparts (Saitoh et al. 1992). High and low metastatic colonic carcinoma cells differ in their adhesion efficiency to E-selectin-expressing cells. The high metastatic cells, as compared to their low metastatic counterparts, bind more efficiently to activated human endothelial cells that express E-selectin. In addition, the high metastatic cells also adhere more efficiently to mouse endothelioma cells after activation with IL-1β. It was also shown that the adhesion can be inhibited by soluble Lamp-1 or soluble leukosialin that contain sLe^x termini (Sawada et al. 1994). Among O-glycans released from an ovarian cystadenoma glycoprotein using ethylamine, three variants of the sulfated Le$^{a/x}$ sequences were identified as ligands for E-selectin, one of which is based on the unusual backbone Gal-3/4GlcNAc-3Gal-3Gal (Chai et al. 1997).

28.6.1.1 Sialyl-Lewis^x

The sLe^x determinant (Neu5Ac α2 → 3Gal β1 → 4[Fuc α-1 → 3]GlcNAc) has been identified as a major ligand in the selectin-mediated adhesion of neutrophils and monocytes to activated endothelium or platelets. This carbohydrate epitope is formed by the sequential action of α3-sialyltransferase and α3-fucosyltransferase on N-acetyllactosamine (Gal β1 → 4 GlcNAc) disaccharide termini of glycoconjugates. In addition, the α3 fucosyltransferase of the cells can use sialylated acceptors. Characterization of the product obtained with a sialylated oligosaccharide indicated that the enzyme can catalyze the formation of the sLe^x structure. The enzyme studied appeared to be a myeloid-type and was involved in the synthesis of sLe^x in leukocytes provided that its expression is at a sufficiently high level (Easton et al. 1993). A single amino acid can determine the ligand specificity of E-selectin. Kogan et al. (1995) developed a model of E-selectin binding to the sLe^x tetrasaccharide, (Neu5Ac α2 → 3Gal β1 → 4(Fuc α 1 → 3)GlcNAc), using the E-selectin-bound solution conformation of sLe^x (Cooke et al.1994) and the crystallographic structures of E-selectin (Graves et al. 1994) (ligand unbound) and the MBP (Weis et al. 1992) (ligand bound). Analysis of this model indicated that the alteration of one E-selectin amino acid, alanine 77, to a lysine residue might shift binding specificity from sLe^x to mannose. The E-selectin mutant protein possessing this change displays preferential binding to mannose containing oligosaccharides and that further mutagenesis of this mannose-binding selectin confers galactose recognition in a predictable manner. These mutagenesis data support the presented model of the detailed interactions between E-selectin and the sLe^x oligosaccharide.

Evidences suggest for density-dependent binding of the membrane-associated E-selectin not only to 3′-sialyl-lacto-N-fucopentaose II (3′-S-LNFP-II) and 3′-sialyl-lacto-N-fucopentaose III (3′-S-LNFP-III), which express the sialyl Le^a and sLe^x antigens, respectively, but also to the nonsialylated analogue LNFP-II; there is a threshold density of E-selectin required for binding to these sialylated sequences, and binding to the nonsialylated analogue is a property only of cells with the highest density of E-selectin expression. The presence of fucose linked to subterminal rather than to an internal N-acetylglucosamine is a requirement for E-selectin binding. Moreover, the presence of sialic acid 3-linked to the terminal galactose of the LNFP-II or LNFP-III sequences substantially enhances E-selectin binding where as the presence of 6-linked sialic acid abolishes E-selectin binding (Larkin et al. 1992).

28.6.1.2 Sulfated Polysaccharide Ligands

A new class of oligosaccharide ligand—sulfate-containing—for the human E-selectin molecule from among oligosaccharides was identified on ovarian cystadenoma glycoprotein. Several components with strong E-selectin binding activity were revealed among acidic oligosaccharides. The smallest among these preparations showed equimolar mixture of the Le^a- and Le^x/SSEA-1-type fucotetrasaccharides sulfated at position 3 of outer galactose. The binding activity was substantially greater than those of lipid-linked Le^a and Le^x/SSEA-1 sequences and is at least equal to that of the 3′- sLe^x/SSEA-1 glycolipid analogue (Yuen et al. 1992). The sulfo derivative of sLe^x, GM 1998-016, which blocks the P- and E-selectins interaction with a ligand, showed a significant decrease in bacterial translocation, both local (MLN) and systemic, in association with the decrease in the neutrophil infiltration, the oxygen free radicals production and the cytokines. Thus, sulfo derivative of sLe^x shows a protective effect in experimental model of bacterial translocation, downregulating the inflammatory response and the leukocyte-endothelium interactions (Garcia-Criado et al. 2005).

Interactions with Fuco-Oligosaccharides: Sweeney et al. (2000) found that sulfated fucans, whether branched and linear, are capable of increasing mature white cells in the periphery and mobilizing stem/progenitor cells of all classes. Though the presence of sulfate groups was necessary, yet it was not sufficient. Significant mobilization of stem/progenitor cells and leukocytosis, elicited in selectin-deficient mice ($L^{-/-}$, $PE^{-/-}$ or $LPE^{-/-}$) similar to that of controls, suggested that the mode of action of sulfated fucans is not through blockade of known selectins. The influence of the location of fucose residue(s) was investigated using 14 structurally defined and variously fucosylated oligosaccharides. Results showed that the recognition motifs for E-selectin include 4-fucosyl-lacto (Le^a) and 3-fucosyl-neo-lacto (Le^x) sequences strictly at capping positions and not Le^x at an internal position as a part of VIM-2 antigen sequence. Additional fucose residues α1-2-linked to neighboring galactoses or α1-3-linked to inner N-acetyglucosamines or to reducing-terminal glucose residues of the tetrasaccharide backbone had little or no effect on the selectin binding (Martin et al. 2002).

Selectin/MBP Chimeras Bind Selectin Ligand sLe^x Through Ca^{2+} Dependent Subsite: Selectin/MBP chimeras created by transfer of key sequences from E-selectin into MBP bind the selectin ligand sLe^x through a Ca^{2+} dependent subsite, present in many C-type lectins, and an accessory site containing positively charged amino acid residues. These chimeras demonstrate selectin-like interaction with sLe^x and can be faithfully reproduced even though structural evidence indicates that the mechanisms of binding to E-selectin and the chimeras are different. Selectin-like binding to the nonfucosylated sulfatide and sulfoglucuronyl glycolipids can also be reproduced with selectin/MBP chimeras that contain the two subsites involved in sLe^x binding. Results indicate that binding of structurally distinct anionic glycans to C-type carbohydrate-recognition domains can be mediated by the Ca^2-dependent subsite in combination with a positively charged region that forms an ionic strength-sensitive subsite (Blanck et al. 1996; Bouyain et al. 2001).

E-Selectin Receptors on Human l Neutrophils: The human neutrophil receptor for E-selectin has not been established. Sialylated glycosphingolipids with 5N-acetyllactosamine (LacNAc, Galβ1-4GlcNAcβ1-3) repeats and 2–3 fucose residues are major functional E-selectin receptors on human neutrophils. E-selectin-expressing cells tethered and rolled on selected glycolipids, whereas P-selectin-expressing cells failed to interact. Results support that the glycosphingolipid NeuAcα2-3Galβ1-4GlcNAcβ1-3[Galβ1-4(Fucα1-3)GlcNAcβ1-3](2)[Galβ1-4GlcNAcβ1-3](2)Galβ1-4GlcβCer (and closely related structures) are functional E-selectin receptors on human neutrophils (Nimrichter et al. 2008; Gege et al. 2008).

Enzymes Involved in the Synthesis of Lewis Antigens: The enzymes involved in the synthesis of Lewis antigens have been identified. FucT-VI is an enzyme involved in the biosynthesis of sLe^x, in myeloid cells (Koszdin and Bowen 1992). Their biosynthesis is complex and involves β1 → 3-galactosyltransferases (β3Gal-Ts) and a combined action of α2- and/or α4-fucosyltransferases (Fuc-Ts). O-glycans with different core structures have been identified, and the ability of β3Gal-Ts and Fuc-Ts to use them as substrates has been tested (Holgersson and Lofling 2006). This knowledge enables us to engineer recombinant glycoproteins with glycan- and core chain-specific Lewis antigen substitution. Such tools may be important for investigations on the fine carbohydrate specificity of Le B-binding lectins, such as Helicobacter pylori adhesins and DC-SIGN, and may also prove useful as therapeutics (Holgersson and Lofling 2006).

28.6.1.3 Transcriptional Control of Expression of Carbohydrate Ligands

Cell adhesion mediated by selectins and their carbohydrate ligands is involved in the adhesion of cancer cells to endothelial cells during the course of hematogenous metastasis of cancer. Extravasation and tissue infiltration of malignant cells in patients with adult T-cell leukemia is mediated by the interaction of selectins and their carbohydrate ligand sialyl-Le^x, which is strongly and constitutively expressed on the leukemic cells. Constitutive expression of Lewis X in these cells is due to the transcriptional activation of Fuc-T VII, the rate-limiting enzyme in the sLe^x synthesis, induced by the Tax protein encoded by the human T-cell leukemia virus-1, the etiological virus for this leukemia. This transactivation is in clear contrast to the regulation of typical CRE-element found in various cellular genes in that it is independent of phosphorylation-dependent regulation. This must be the reason for the strong and constitutive expression of sialyl-Le^x, which exacerbates the tissue infiltration of leukemic cells. This is a good example corroborating the proposition that the abnormal expression of carbohydrate determinant at the surface of malignant cells is intimately associated with the genetic mechanism of malignant transformation of cells (Kannagi 2001).

28.6.2 E-Selectin Ligand-1

E-Selectin-Ligand-1 Is a Variant of FGF-R: A 150 kDa glycoprotein is the major ligand for E-selectin on plasma membranes of myeloid cells. Mouse c-DNA for this E-selectin ligand (ESL-1) predicted amino-acid sequence, which is 94% identical (over 1,078 amino acids) to the chicken cysteine-rich FGF-R, except for a unique 70-amino-acid aminoterminal domain of mature ESL-1. Fucosylation of ESL-1 is imperative for affinity isolation with E-selectin-IgG. ESL-1, with a structure essentially identical to that of a receptor functions as a cell adhesion ligand of E-selectin (Steegmaier et al. 1995). E-selectin binds to ESL-1 with a fast dissociation rate constant of 4.6 s^{-1} and a calculated association rate constant of 7.4×10^4 $m^{-1} s^{-1}$. A K_D of 62 µM resembles the affinity of L-selectin binding to glycosylation-dependent cell adhesion molecule-1. The affinity of the E-selectin-ESL-1 interaction did not change significantly from 5°C to 37°C, indicating that the enthalpic contribution to the binding is small at physiological temperatures, and that, in contrast to typical protein-carbohydrate interactions, binding is driven primarily by favorable entropic changes (Wild et al. 2001). MG160 is a conserved membrane sialoglycoprotein of the Golgi apparatus displaying over 90% amino acid sequence identities with two apparently unrelated molecules, namely, a chicken fibroblast growth factor receptor (CFR), and ESL-1 (Steegmaier et al. 1995). MG160 from rat brain membranes binds to b-FGF. The gene for MG160 has been assigned to human chromosome 16q22-23 (Mourelatos et al. 1996).

28.6.3 PSGL-1 and Relating Ligands

The selectin family of adhesion molecules mediates tethering and rolling of leukocytes to the vessel wall in the microcirculation. Selectins promote these interactions by binding to glycoconjugate ligands expressed on apposing cells. Selectin-mediated rolling is a prerequisite for firm adhesion and subsequent transendothelial migration of leukocytes into tissues. PSGL-1 interacts with selectins to support leukocyte rolling along vascular wall. PSGL-1 (CD162 P) is a dimeric mucin-like 120-kDa glycoprotein on leukocyte surfaces that binds to P-, L- and E- selectin and promotes cell adhesion in the inflammatory response (see Chap. 26 and 27). Present understanding on selectin functions reveals that selectins and their ligands have a complex role during inflammatory diseases. Among selectin ligands, mucin PSGL-1 has a well-documented role in organ targeting during inflammation in animal models. Although inhibition of selectins and their ligands in animal models of inflammatory diseases has proven the validity of this approach in vivo, only a limited number of anti-selectin drugs have been tested in humans. Results in clinical trials for asthma and psoriasis show that, although very challenging, the development of selectin antagonists holds concrete promise for the therapy of inflammatory diseases (Rossi and Constantin 2008).

PSGL-1 requires sialylated, fucosylated O-linked glycans and tyrosine sulfate to bind P-selectin. Less is known about the determinants that PSGL-1 requires to bind E-selectin. Tyrosine residues on PSGL-1 expressed in CHO cells were shown to be sulfated. Phenylalanine replacement of three tyrosines within a consensus sequence for tyrosine sulfation abolished binding to P-selectin but not to E-selectin. Results suggest that PSGL-1 requires core 2 O-linked glycans that are sialylated and fucosylated to bind P- and E-selectin. PSGL-1 also requires tyrosine sulfate to bind P-selectin but not E-selectin (Li et al. 1996). Studies using CHO-E and -P monolayers demonstrated that the first 19 amino acids of PSGL-1 are sufficient for attachment and rolling on both E- and P-selectin and that a sialyl Le^x-containing glycan at Threonine-16 was critical for this sequence of amino acids to mediate attachment to E- and P-selectin. However, a sulfated-anionic polypeptide segment within the amino terminus of PSGL-1 is necessary for PSGL-1-mediated attachment to P- but not to E-selectin. In addition, PSGL-1 had more than one binding site for E-selectin: one site located within the first 19 amino acids of PSGL-1 and one or more sites located between amino acids 19 through 148 (Goetz et al. 1997).

28.6.3.1 PSGL-1-Dependent and -Independent E-Selectin Rolling

PSGL-1 supports P-selectin-dependent rolling in vivo and in vitro. However, controversy exists regarding the importance of PSGL-1-dependent and -independent E-selectin rolling. Using antibodies against PSGL-1 and PSGL-1$^{-/-}$ mice, Zanardo et al. (2004) demonstrated abolition of P-selectin-dependent rolling but only partial inhibition of E-selectin-mediated rolling in the cremaster microcirculation following local administration of TNF-α. Data support an E-selectin ligand present on PSGL-1$^{-/-}$ neutrophils that is down-regulatable upon systemic but not local activation. In P- and E-selectin-dependent cutaneous contact hypersensitivity model, binding studies showed no E-selectin ligand down-regulation. In conclusion, it was suggested that E-selectin mediates PSGL-1-dependent and independent rolling and the latter can be down-regulated by systemic activation and can replace PSGL-1 to support the development of inflammation (Zanardo et al. 2004; Shigeta et al. 2008).

28.6.3.2 PSGL-1 Participates in E-Selectin-Mediated Progenitor Homing

The nature and exact function of selectin ligands involved in hematopoietic progenitor cell (HPC) homing to the bone marrow (BM) are not clear. Using murine progenitor homing assays, it was found that PSGL-1 plays a partial role in HPC homing to BM. Homing studies with PSGL-1-deficient HPCs pretreated with anti-α4 integrin antibody revealed that PSGL-1 contributes to approximately 60% of E-selectin ligand-mediated homing activity. Results thus underscore a major difference between mature myeloid cells and immature stem/progenitor cells in that E-selectin ligands cooperate with α4 integrin rather than P-selectin ligands (Katayama et al. 2003). PSGL-1, expressed by primitive human bone marrow CD34$^+$ cells, mediates their adhesion to P-selectin and inhibits their proliferation. Adhesion to E-selectin also inhibits the proliferation of human and mouse CD34$^+$ cells. Furthermore, a subpopulation, which does not contain the most primitive hematopoietic progenitor cells, undergoes apoptosis following E-selectin-mediated adhesion. This shows that PSGL-1 is not the ligand involved in E-selectin-mediated growth inhibition and apoptosis, but an E-selectin ligand(s) other than PSGL-1 transduces growth inhibitory and proapoptotic signals and suggests the requirement of posttranslational fucosylation (Winkler et al. 2004).

28.6.3.3 PSGL-1 Decameric Repeats Regulate Selectin-Dependent Rolling Under Flow Conditions

PSGL-1 regulates leukocyte rolling by binding P-selectin, and also by binding E- and L-selectins with lower affinity. L- and P-selectin bind to N-terminal tyrosine sulfate residues and to core-2 O-glycans attached to Thr-57, whereas tyrosine sulfation is not required for E-selectin binding. PSGL-1 extracellular domain contains decameric repeats, which extend L- and P-selectin binding sites far above the plasma membrane. It was suggested that decamers may play a role in regulating PSGL-1 interactions with selectins. Deletion of decamers abrogated sL-selectin binding and cell rolling on L-selectin, whereas their substitution partially reversed these diminutions. P-selectin-dependent interactions with PSGL-1 were less affected by decamer deletion. Analysis showed that decamers are required to stabilize L-selectin-dependent rolling. However, adhesion assays performed on recombinant decamers demonstrated that they directly bind to E-selectin and promote slow rolling. Therefore, the role of decamers is to extend PSGL-1N terminus far above the cell surface to support and stabilize leukocyte rolling on L- or P-selectin. In addition, they function as a cell adhesion receptor, which supports approximately 80% of E-selectin-dependent rolling (Tauxe et al. 2008). Tomita et al. (2009) investigated cutaneous wound healing in PSGL-1$^{-/-}$ mice in comparison with E-selectin$^{-/-}$, P-selectin$^{-/-}$, and P-selectin$^{-/-}$ mice treated with an anti-E-selectin antibody. PSGL-1 contributes to wound healing predominantly as a P-selectin ligand and partly as an E-selectin ligand by mediating infiltration of inflammatory cells.

E-selectin binding to PSGL-1 can activate β2 integrin lymphocyte function-associated antigen-1 by signaling through spleen tyrosine kinase (Syk). This signaling is independent of G αi-protein-coupled receptors, results in slow rolling, and promotes neutrophil recruitment to sites of inflammation. An ITAM-dependent pathway involving Src-family kinase Fgr and the ITAM-containing adaptor proteins DAP12 and FcRγ is involved in the initial signaling events downstream of PSGL-1 that are required to initiate neutrophil slow rolling (Zarbock et al. 2008).

28.6.3.4 Hematopoietic Cell E-/L-selectin Ligand (HCELL) (CD44) in Stem Cell Homing

Though a glycoform of PSGL-1 functions as the principal E-selectin ligand on human T lymphocytes, the E-selectin ligand(s) of human hematopoietic progenitor cells (HPCs) is not well defined. While the PSGL-1 expressed on human HPCs is an E-selectin ligand, HSCs express a novel glycoform of CD44 known as hematopoietic cell E-/L-selectin ligand (HCELL) that acts as ligand. Current understanding of the molecular basis of HSC homing describes the fundamental "roll" of HCELL in opening the avenues for efficient HSC trafficking to the bone marrow, the skin and other extramedullary sites (Sackstein 2004). The E-selectin ligand activity of CD44 is expressed on primitive CD34$^+$ human HPCs, but not on more mature hematopoietic cells. Under physiologic flow conditions, this molecule mediates E-selectin-dependent rolling interactions over a wider shear range than that of PSGL-1, and promotes human HPC rolling interactions on E-selectin expressed on human BM endothelial cells (Dimitroff et al. 2001). The LS174T colon carcinoma cell line expresses the CD44 glycoform known as hematopoietic cell E-/L-selectin ligand (HCELL), which functions as a high affinity E- and L-selectin ligand on these cells. Burdick et al. (2006) measured the binding of LS174T cells transduced with CD44 short interfering RNA (siRNA). It indicated that expression of HCELL confers robust and predominant tumor cell binding to E- and L-selectin, highlighting a central role for HCELL in promoting shear-resistant adhesive interactions essential for hematogenous cancer dissemination.

28.6.3.5 CD43 as a Ligand for E-Selectin on CLA$^+$ Human T Cells

Human memory T cells that infiltrate skin express the carbohydrate epitope cutaneous lymphocyte-associated antigen (CLA). Expression of the CLA epitope on T cells has been described on PSGL-1 and associated with the acquisition of both E-selectin and P-selectin ligand functions. The CD43, a sialomucin expressed constitutively on T cells, can also be decorated with the CLA epitope and serves as an E-selectin ligand. CLA expressed on CD43 is found exclusively on the 125 kDa glycoform bearing core-2-branched O-linked glycans. CLA$^+$ CD43 from human T cells supported tethering and rolling in shear flow via E-selectin but did not support binding of P-selectin. The identification and characterization of CD43 as a T-cell E-selectin ligand distinct from PSGL-1 expands the role of CD43 in the regulation of T-cell trafficking and provides new targets for the modulation of immune functions in skin (Fuhlbrigge et al. 2006). It has been reported that CD43 on activated T cells functions as an

E-selectin ligand and thereby mediates T cell migration to inflamed sites, in collaboration with PSGL-1, a major P- and E-selectin ligand. CD43 on neutrophils also functions as an E-selectin ligand. Observations suggest that CD43 generally serves as an antiadhesive molecule to attenuate neutrophil-endothelial interactions, but when E-selectin is expressed on endothelial cells, it also plays a proadhesive role as an E-selectin ligand (Matsumoto et al. 2008).

28.6.4 Endoglycan, a Ligand for Vascular Selectins

While several ligands have been characterized on human T cells, monocytes and neutrophils, there is limited information concerning ligands on B cells. Endoglycan (EG) together with CD34 and podocalyxin comprise the CD34 family of sialomucins. EG was previously implicated as an L-selectin ligand on endothelial cells. EG is present on human B cells, T cells and peripheral blood monocytes. Upon activation of B cells, EG increased with a concurrent decrease in PSGL-1. Expression of EG on T cells remained constant under the same conditions. Native EG from several sources (a B cell line, a monocyte line and human tonsils) was reactive with HECA-452, a mAb that recognizes sialyl Lewis X and related structures. Finally, an EG construct supported slow rolling of E- and P-selectin bearing cells in a sialic acid and fucose dependent manner, and the introduction of intact EG into a B cell line facilitated rolling interactions on a P-selectin substratum. These findings indicate that endoglycan can function as a ligand for vascular selectins (Kerr et al. 2008). Basigin (Bsg)/CD147 is a ligand for E-selectin that promotes renal inflammation in ischemia/reperfusion (I/R). Bsg is a physiologic ligand for E-selectin that plays a critical role in renal damage induced by I/R (Kato et al. 2009).

28.6.5 L-Selectin as E-Selectin Ligand

L-Selectin from Human, but Not from Mouse Neutrophils Binds E-Selectin: L-selectin has been suggested as a carbohydrate presenting ligand for E- and P-selectin. However, affinity experiments with an E-selectin-Ig failed to detect L-selectin in the isolated E-selectin ligands from mouse neutrophils. In contrast to mouse neutrophils, L-selectin from human neutrophils could be affinity-isolated as a major ligand from cell extracts using E-selectin-Ig as affinity probe. It was suggested that L-selectin on human neutrophils is a major glycoprotein ligand among very few glycoproteins that could be isolated by an E-selectin affinity matrix. The difference between human and mouse L-selectin suggests that E-selectin-binding carbohydrate moieties are attached to different protein scaffolds in different species (Zollner et al. 1997). L-selectin acquires E-selectin-binding activity following phorbol ester (PMA) treatment of the Jurkat T cell line and anti-CD3/IL-2-driven proliferation of human T lymphocytes in vitro. It appears that L-selectin on human T lymphoblasts is one of the several glycoproteins that interacts directly with E-selectin and contributes to rolling under flow (Jutila et al. 2002).

In addition to endogenous ligands for E-selectin, several proteins are found in cancer cell lines or solid tumors that act as ligands for E, L, and P selectins. E-selectin ligands present on cancer cells are: (1) Glycodelin A (GdA) is primarily produced in endometrial and decidual tissue and secreted to amniotic fluid Glycodelins is expressed in in ovarian cancer can act as an inhibitor of lymphocyte activation and/or adhesion (Jeschke et al. 2009); (2) The cysteine-rich fibroblast growth factor receptor (FGF-R) represents the main E-selectin ligand (ESL-1) on granulocytes. Hepatic stellate cells (HSC) are pericytes of liver sinusoidal endothelial cells, which are involved in the repair of liver tissue injury and angiogenesis of liver metastases. HSC express FGF-R together with FucT7 and exhibit a functional E-selectin binding activity on their cell surface (Antoine et al. 2009). (3) Although B-cell precursor acute lymphoblastic leukemia (BCP-ALL) cell lines do not express the ligand PSGL-1, a major proportion of carbohydrate selectin ligand was carried by another sialomucin, CD43, in NALL-1 cells. CD43 plays an important role in extravascular infiltration of NALL-1 cells and the degree of tissue engraftment of BCP-ALL cells may be controlled by manipulating CD43 expression (Nonomura et al. 2008). (4) Thomas et al. (2009a) identified podocalyxin-like protein (PCLP) as an alternative selectin ligand. PCLP on LS174T colon carcinoma cells possesses E-/L-, but not P-, selectin binding activity. PCLP functions as an alternative acceptor for selectin-binding glycans. The finding that PCLP is an E-/L-selectin ligand on carcinoma cells offers a unifying perspective on the apparent enhanced metastatic potential associated with tumor cell PCLP overexpression and the role of selectins in metastasis (Thomas et al. 2009b). (5) E-selectin has been shown to play a pivotal role in mediating cell-cell interactions between breast cancer cells and endothelial monolayers during tumor cell metastasis. The counterreceptor for E-selectin was found as CD44v4. However, CD44 variant (CD44v) isoforms was functional P-, but not E-/L- selectin ligands on colon carcinoma cells. Furthermore, a ~180-kDa sialofucosylated glycoprotein(s) mediated selectin binding in CD44-knockdown cells. This glycoprotein was identified as carcinoembryonic antigen (CEA). CEA serves as an auxiliary L-selectin ligand, which stabilizes L-selectin-dependent cell rolling against fluid shear (Thomas et al. 2009b). Zen et al. (2008) identified a ~170 kDa human

CD44 variant 4 (CD44v4) as E-selectin ligand. Purified CD44v4 has a high affinity for E-selectin via sLe^x moieties.

28.7 Structural Properties of E-Selectin

28.7.1 Soluble E-Selectin: An Asymmetric Monomer

The gene coding for a soluble form of human E-selectin (sE-selectin) has been expressed in CHO cells. The sE-selectin showed a broad band of M_r 75 kDa on nonreducing SDS-PAGE and eluted with M_r 310 kDa from size exclusion chromatography. Matrix-assisted laser-desorption MS gave a molecular weight of 80 kDa, while the minimum monomer molecular weight from the gene sequence should be 58.571 kDa, demonstrating that the monomeric molecule thus expressed had 27% carbohydrate. Equilibrium ultracentrifugation gave an average solution molecular weight of 81.6 kDa. Velocity ultracentrifugation gave a sedimentation coefficient of 4.3S and, from this, an apparent axial ratio of 10.5:1, assuming a prolate ellipsoid of revolution. An analysis of the NMR NOESY spectra of sE-selectin, sialyl-Lewis^x, and sE-selectin with sialyl-Lewis^x demonstrated that the recombinant protein binds sialyl-Lewis^x productively. Hence, in solution, sE-selectin is a functional elongated monomer (Hensley et al. 1994). E-selectin-expressed in COS cells bind the promyelocytic cell line HL-60 by a Ca^{2+} dependent mechanism. Although E-selectin is homologous to mammalian lectins, its interaction with HL-60 cells is not inhibited by simple carbohydrate structures. E-selectin-expressing COS cells also bind human neutrophils and the human colon carcinoma cell line HT-29, but not the B-cell line Ramos (Hession et al. 1990).

The three-dimensional structure of the ligand-binding region of human E-selectin reveals limited contact between the two domains and a coordination of Ca^{2+} not predicted from other C-type lectins. Structure/function analysis indicates a defined region and specific amino-acid side chains that might be involved in ligand binding. The features of E-selectin/ligand interaction have important implications for understanding the recruitment of leukocytes to sites of inflammation. The selectins bind weakly to sialyl Lewis-X (sLe^X)-like glycans, but with high-affinity to specific glycoprotein counterreceptors, including PSGL-1. The crystal structure of human E-selectin constructs containing the lectin and EGF (LE) domains co-complexed with sLe^X has been elucidated (Fig. 28.3) (Somers et al. 2000). Somers et al. (2000) also solved crystal structure of P-selectin-lectin and EGF domains co-complexed with the N-terminal

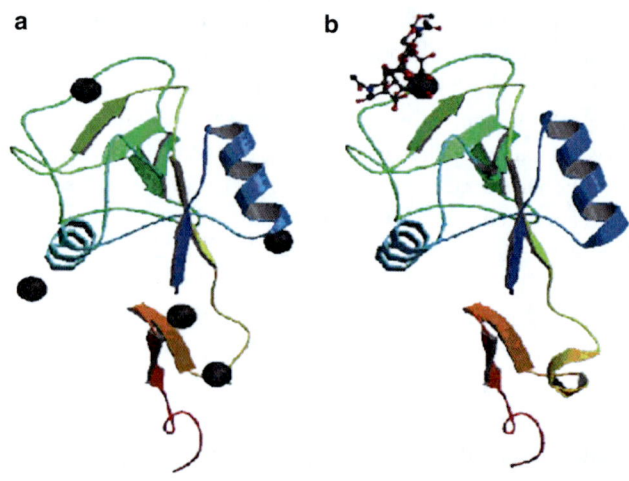

Fig. 28.3 (a) Crystal structure of E-selectin lectin-EGF domains (PDB ID: 1ESL) (Graves et al. 1994); (b) crystal structure of E-selectin lectin-EGF domains complexed with s-Le^x (PDB ID: 1G1T; Somers et al. 2000). Serum sE-selectin ranges from 0.13 to 2.8 ng/ml, suggesting that even in the absence of overt inflammatory processes E-selectin is being synthesized and released into the bloodstream. In addition, bacteremic patients with hypotension, but not those without, showed markedly elevated sE-selectin values. The blood-derived form of E-selectin is biologically active

domain of human PSGL-1 modified by both tyrosine sulfation and sLe^X. These structures reveal differences in binding of E- and P-selectin to sLe^X and the molecular basis of interaction between P-selectin and PSGL-1.

28.7.2 Complement Regulatory Domains

Studies have shown that different animal species express E-selectin mRNAs that encode different numbers of complement regulatory (CR) domains. The isolation of these two rat E-selectin cDNA fragments, which differ only for the presence of CR5, represents the first direct evidence for the existence of E-selectin CR-variant mRNAs in the same species. Moreover, the sequence of the CR5$^-$ cDNA is consistent with its origin from an mRNA splice variant of a CR5$^+$ mRNA. Billups et al. (1995) demonstrated the presence of the two predicted mRNA species in rat heart tissue and investigated their expression in response to LPS. Although both mRNA variants were greatly induced by LPS, the CR5$^-$ form was more abundant in both treated and control tissues. This difference in mRNA abundance indicates different levels of CR5 variant proteins that perform functionally distinct tasks in E-selectin dependent inflammatory processes (Billups et al. 1995).

Consensus Repeat Domains: To study the structural characteristics of E-selectin necessary for mediating cell adhesion, Li et al. (1994) examined the role of consensus repeat domains in E-selectin function. E-selectin containing all six consensus repeat domains (Lec-EGF-CR6) at its COOH terminus was the most potent in blocking neutrophil or HL-60 cell adhesion to either immobilized E-selectin or cytokine-stimulated HUVE cells. Therefore, although the lectin and EGF domains are necessary and sufficient for mediating cell adhesion, the additional six consensus repeat domains, present in native E-selectin, contribute to the enhanced binding of E-selectin to its ligand (Li et al. 1994).

Modulation by Metal Ions: Since E-selectin recognizes carbohydrate ligands in a Ca^{2+}-dependent manner, Anostario and Huang (1995) examined the E-selectin structure by limited proteolysis. Apo-Lec-EGF-CR6, a Ca^{2+}-free form of soluble E-selectin containing the entire extracellular domain, was sensitive to limited proteolysis by Glu-C endoproteinase. Amino-terminal sequencing analysis of the proteolytic fragments revealed that the major cleavage site is at Glu98, which is in the loop (residues 94–103) adjacent to the Ca^{2+} binding region of the lectin domain. Upon Ca^{2+} binding, Lec-EGF-CR6 was protected from proteolysis. This Ca^{2+}-dependent protection was augmented upon sialyl Lewis x (sLex) binding. This implied that Ca^{2+} binding to E-selectin induces a conformational change and perhaps facilitates ligand binding. Ba^{2+} was a potent antagonist in blocking Lec-EGF-CR6-mediated HL-60 cell adhesion. Sr^{2+} also bound to apo-Lec-EGF-CR6 tighter than Ca^{2+}. Thus, E-selectin function can be modulated by different metal ions (Anostario and Huang 1995).

28.7.3 Three-Dimensional Structure

The three-dimensional structure of the ligand-binding region of human E-selectin has been determined at 2.0 Å resolution. The structure reveals limited contact between the two domains and a coordination of Ca^{2+} not predicted from other C-type lectins. Structure/function analysis indicates a defined region and specific amino-acid side chains that may be involved in ligand binding. These features of the E-selectin/ligand interaction have important implications for understanding the recruitment of leukocytes to sites of inflammation (Graves et al. 1994). Surface presentation of adhesion receptors influences cell adhesion, although the mechanisms underlying these effects are not well understood. The orientation and length of an adhesion receptor influences its rate of encountering and binding a surface ligand but does not subsequently affect the stability of binding (Huang et al. 2004).

28.8 Functions of E-Selectin

28.8.1 Functions in Cell Trafficking

In antigen-challenged mice a significant increase in leukocyte rolling and adhesion, and a very dramatic increase in emigration have been observed for 24 h. Although rolling and adhesion was dramatically blunted in P-selectin- or P selectin/ICAM-1-deficient mice, emigrated cell number was similar to that observed in wild-type mice. Leukocyte rolling, adhesion and emigration were almost totally abrogated over in E/P-selectin-deficient mice, demonstrating that antigen-induced leukocyte recruitment can be entirely disrupted in the absence of both endothelial selectins (McCafferty et al. 2000). The absence of P-selectin, and not E-selectin, resulted in an altered adhesion environment with subsequent expansion of megakaryocyte progenitors and immature megakaryoblasts, enhanced secretion of TGF-β1, and apparent increased responsiveness to inflammatory cytokines (Banu et al. 2002; Khew-Goodall et al. 1993).

In vitro assays suggest that both E-selectin and chemokines can trigger arrest of rolling neutrophils, but E-selectin$^{-/-}$ mice have normal levels of adherent neutrophils in inflamed venules (Asaduzzaman et al. 2009). Smith et al. (2004) showed that E-selectin- and chemokine-mediated arrest mechanisms are overlapping and identified CXCR2 as an important neutrophil arrest chemokine in vivo.

28.8.1.1 Trafficking of Th1 Cells to Lung Depends on Selectins

Expression of CD62E and CD62P is induced in the lungs of mice primed and then challenged with intratracheal (i.t.) SRBC. In addition, endogenous lymphocytes accumulate in the lungs of E- and P-selectin-deficient ($E^{-/-}P^{-/-}$) mice after i.t. SRBC challenge. $E^{-/-}P^{-/-}$ mice showed a 85–95% decrease in CD8$^+$ T cells and B cells in the lungs at both early and late time points (Curtis et al. 2002). Both P- and E-selectin play an important role in Th1 lymphocyte migration to lung (Clark et al. 2004). Moreover, Yiming et al. (2005) suggested that interactions of circulating inflammatory cells with P-selectin critically determine proinflammatory endothelial activation during high tidal volume ventilation.

Changes in P-Selectin/ICAM-1 in Mutant Mice Are Organ-Specific: Mice homozygous for an E-selectin null mutation were viable and exhibited no obvious developmental alterations. Neutrophil emigration during an inflammatory response is mediated through interactions between adhesion molecules on endothelial cells and neutrophils.

P-Selectin mediates rolling or slowing of neutrophils, while ICAM-1 contributes to the firm adhesion and emigration of neutrophils. Removing the function of either molecule partially prevents neutrophil emigration. While mice with either mutation alone showed a 60–70% reduction in acute neutrophil emigration into the peritoneum during *Streptococcus pneumoniae*-induced peritonitis, double mutant mice showed a complete loss of neutrophil emigration. In contrast, neutrophil emigration into the alveolar spaces during acute *S. pneumoniae*-induced pneumonia was normal in double mutant mice. These results demonstrated organ-specific differences, since emigration into the peritoneum requires both adhesion molecules while emigration into the lung requires neither (Bullard et al. 1996).

28.8.1.2 Cooperation Between E-Selectin and CD18 Integrin

CD18-deficient mice (CD18$^{-/-}$ mice) suffer from a severe leukocyte recruitment defect in some organs, with no detectable defect in other models. Mice lacking E-selectin (CD62E$^{-/-}$ mice) have either no defect or a mild defect of neutrophil infiltration, depending on the model. The CD18$^{-/-}$CD62E$^{-/-}$, but not CD18$^{-/-}$CD62P$^{-/-}$, mice generated by cross-breeding failed to thrive, reaching a maximum body weight of 10–15 g. To explore the mechanisms underlying reduced viability, Forlow et al. (2000) suggested that the greatly reduced viability of CD18$^{-/-}$CD62E$^{-/-}$ mice appeared to result from an inability to mount an adequate inflammatory response. This study showed that cooperation between E-selectin and CD18 integrins is necessary for neutrophil recruitment and that alternative adhesion pathways cannot compensate for the loss of these molecules (Forlow et al. 2000).

28.8.1.3 Selectin-Independent Leukocyte Rolling and Adhesion

To study selectin-independent leukocyte recruitment and the role of ICAM-1, Forlow and Ley (2001) generated mice lacking all three selectins and ICAM-1 (E/P/L/I$^{-/-}$) by bone marrow transplantation. A striking similarity of leukocyte adhesion efficiency in E/P/L$^{-/-}$ and E/P/I$^{-/-}$ mice suggests a pathway in which leukocyte rolling through L-selectin required ICAM-1 for adhesion and recruitment. Comparison of data with mice lacking individual or other combinations of adhesion molecules revealed that elimination of more adhesion molecules further reduces leukocyte recruitment but the effect is less than additive.

28.8.2 E-Selectin in Neutrophil Activation

Neutrophil adherence to activated endothelium mediated by E-selectin is reversibly inhibited by hypothermia. The protective effect of hypothermia (e.g., in cardiopulmonary bypass) may, in part, be mediated by transiently inhibiting the expression of an endothelial cell activation phenotype (Haddix et al. 1996). In transgenic mice, the number of blood neutrophils was reduced, without any other obvious phenotype or tissue damage. These neutrophils, however, displayed two significant changes: first, an alteration in the levels of expression of two membrane receptors involved in neutrophil adhesion to endothelial cells, namely a marked increased in the Mac-1 antigen (CD11b/CD18) and a decrease in L-selectin; second, an increased oxidative activity when compared to blood neutrophils of non-transgenic mice. These observations indicated that the binding of E-selection with neutrophils bearing its ligands promotes neutrophil activation in vivo (Araki et al. 1996). Mac-1 (CD11b/CD18) is the major neutrophil glycoprotein decorated with sLe^x and ligation of these carbohydrate moieties significantly impairs neutrophil functions. Protein-binding assays indicate that sLe^x moieties on Mac-1 are critical for binding interaction of Mac-1 to E-selectin. Ligation of Mac-1 sLe^x by anti-sLe^x antibody induces a significant degranulation of neutrophil secondary granules at the absence of chemoattractant stimulation. This "dysregulated" degranulation induced by anti-sLe^x antibody strongly inhibits neutrophil transmigration in response to formyl-Met-Leu-Phe (fMLP). Thus, Mac-1 sLe^x moieties play a critical role in regulating β2 integrin functions during neutrophil transmigration and degranulation (Zen et al. 2007).

However, E-selectin-deficient mice displayed no significant change in the trafficking of neutrophils in several models of inflammation, although blocking both endothelial selectins by treatment of the E-selectin-deficient animals with an anti-murine P-selectin antibody significantly inhibited neutrophil emigration in two distinct models of inflammation. Labow et al. (1994) demonstrated that the majority of neutrophil migration in both models requires an endothelial selectin but that E-selectin and P-selectin are functionally redundant. Studies on the functional effects of mAbs against murine E-selectin on neutrophil recruitment in vivo, leukocyte rolling and circulating leukocyte concentrations in vivo, and adhesion of myeloid cells to E-selectin transfectants and recombinant E-selectin-IgG fusion protein in vitro indicated that E-selectin serves a function, other than rolling, that appears to be critically important for neutrophil recruitment to inflammatory sites in Balb/c mice (Ramos et al. 1997).

28.8.3 P- and E-Selectin in Differentiation of Hematopoietic Cells

The P- and E-selectins are critically important for adhesion and homing of hematopoietic progenitor cells (HPC) into the bone marrow. The most primitive HPC capable of long-term in vivo repopulation express PSGL-1, a receptor common to both P- and E-selectin. In addition, P-selectin delays the differentiation of HPC whereas E-selectin enhances their differentiation along the monocyte/granulocyte pathway, describing different roles for these selectins in the regulation of hematopoiesis. Thus, the two endothelial selectins, E-selectin and P-selectin, have very different effects on HPC. E-selectin accelerates the differentiation of maturing HPC towards granulocyte and monocyte lineages while maintaining the production of more immature CFU-S(12) in ex vivo liquid suspension culture. In marked contrast, P-selectin delays the differentiation of Lin$^-$ Sca-1$^+$ c-kit$^+$ cells, allowing enhanced ex vivo expansion of CFC and CFU-S(12) but not HSCs (Eto et al. 2005; Schweitzer et al. 1996).

Role in Loose Adhesion of Allogeneic Lymphocytes and in Structural and Functional Lung Alterations: The interaction between host lymphocytes and graft endothelial cells plays an important role in graft rejection. Isogeneic perfusion induced nonspecific endothelial cell activation, which was characterized by up-regulation of E-selectin, ICAM-1, and of TNF-α. It was revealed that E-selectin expression (1) is not a consequence of TNF-α triggering, (2) E-selectin expression up-regulates its own expression and expression of I-A, VCAM-1, TNF-α, and lymphotoxin-α mRNAs, and down-regulates expression of LFA-3 and ICAM-1 mRNAs. Study indicated that the E-selectin plays major role in the loose adhesion of allogeneic lymphocytes and in structural and functional lung alterations (Joucher et al. 2004).

Memory B Lymphocytes from Secondary Lymphoid Organs Interact with E-Selectin: A subset of human tonsillar B cells that interact with E-selectin but not with P-selectin have a phenotype of non-germinal center (CD10$^-$, CD38$^-$, CD44$^+$), memory (IgD$^-$) cells. Furthermore, FucT-VII is expressed selectively in CD44$^+$ E-selectin-adherent B lymphocytes. These results assigned resident memory B lymphocytes a adhesion function, the rolling on E-selectin that provides insights on the adhesion pathways involved in homing of memory B cells to tertiary sites (Montoya et al. 1999).

28.8.4 Transmembrane Signaling in Endothelial Cells

In addition to supporting rolling and stable arrest of leukocytes, there is increasing evidence that E-selectin functions in transmembrane signaling into endothelial cells during these adhesive interactions. Adhesion of HL-60 cells (which express ligands for E-selectin), or antibody-mediated cross-linking of E-selectin, results in the formation of a Ras/Raf-1/phospho-MEK macrocomplex, extracellular signal-regulated protein kinase (ERK1/2) activation, and c-fos up-regulation. These downstream events require an intact cytoplasmic domain of E-selectin. Tyrosine-603 plays an important role in mediating the association of E-selectin with SHP2, and the catalytic domain of SHP2, in turn, is critical for E-selectin-dependent ERK1/2 activation. Events suggest that cross-linking of E-selectin by engagement of ligands on adherent leukocytes can initiate a multifunctional signaling pathway in the activated endothelial cell at the sites of inflammation (Hu et al. 2001).

Src-Family Kinases Mediate an Outside-In Signal: E-selectin-ligand engagement differs between lymphocytes and PMN, and that these differences may be accentuated by the CR1 and CR2 domains in the E-selectin (Hammel et al. 2001). In cell suspensions subjected to high-shear rotatory motion, human PMN adhered to E-selectin-expressing CHO (CHO-E), and formed homotypic aggregates when challenged by E-selectin-IgG fusion protein, by a mechanism that involved β2 integrins. It appeared that Src-family kinases, and perhaps Pyk2, mediate a signal necessary for β2 integrin function in PMN tethered by E-selectin (Totani et al. 2006).

E-Selectin Is Unique in Clustering Sialylated Ligands and Transducing Signals: Two adhesive events critical to efficient recruitment of neutrophils at vascular sites of inflammation are up-regulation of endothelial selectins that bind sLe^x ligands and activation of β2-integrins that support neutrophil arrest by binding ICAM-1. Neutrophils rolling on E-selectin is sufficient for signaling cell arrest through β2-integrins binding of ICAM-1 in a process dependent upon ligation of L-selectin and PSGL-1. Spatial and temporal events showed that binding of E-selectin to sLe^x on L-selectin and PSGL-1 drives their co-localization into membrane caps at the trailing edge of neutrophils rolling on HUVECs and on an L-cell monolayer co-expressing E-selectin and ICAM-1. Inhibition of p38 and p42/44 mitogen-activated protein kinase blocked the cocapping of L-selectin and PSGL-1 and the subsequent clustering of high-affinity β2-integrins. Results suggest that E-selectin is unique among selectins in its capacity for clustering sialylated ligands and transducing signals leading to neutrophil arrest in shear flow (Green et al. 2004).

References

Anostario M Jr, Huang KS (1995) Modulation of E-selectin structure/function by metal ions. Studies on limited proteolysis and metal ion regeneration. J Biol Chem 270:8138–8144

Antoine M, Tag CG, Gressner AM et al (2009) Expression of E-selectin ligand-1 (CFR/ESL-1) on hepatic stellate cells: implications for leukocyte extravasation and liver metastasis. Oncol Rep 21:357–362

Araki M, Araki K, Miyazaki Y et al (1996) E-selectin binding promotes neutrophil activation in vivo in E-selectin transgenic mice. Biochem Biophys Res Commun 224:825–830

Asaduzzaman M, Mihaescu A, Wang Y et al (2009) P-selectin and P-selectin glycoprotein ligand 1 mediate rolling of activated CD8+ T cells in inflamed colonic venules. J Invest Med 57:765

Bandyopadhyay SK, de La Motte CA, Williams BR (2000) Induction of E-selectin expression by double-stranded RNA and TNF-α is attenuated in murine aortic endothelial cells derived from double-stranded RNA-activated kinase (PKR)-null mice. J Immunol 164:2077–2083

Banu N, Avraham S, Avraham HK (2002) P-selectin, and not E-selectin, negatively regulates murine megakaryocytopoiesis. J Immunol 169:4579–4585

Becker-Andre M, Hooft van Huijsduijnen R, Losberger C et al (1992) Murine endothelial leukocyte-adhesion molecule 1 is a close structural and functional homologue of the human protein. Eur J Biochem 206:401–411

Bender JR, Sadeghi MM, Watson C et al (1994) Heterogeneous activation thresholds to cytokines in genetically distinct endothelial cells: evidence for diverse transcriptional responses. Proc Natl Acad Sci USA 91:3994–3998

Bennett BL, Cruz R, Lacson RG et al (1997) Interleukin-4 suppression of tumor necrosis factor alpha-stimulated E-selectin gene transcription is mediated by STAT6 antagonism of NF-kappaB. J Biol Chem 272:10212–10219

Bevilacqua MP, Stengelin S, Gimbrone MA Jr, Seed B (1989) Endothelial leukocyte adhesion molecule 1: an inducible receptor for neutrophils related to complement regulatory proteins and lectins. Science 243(4895):1160–1165

Billups KL, Sherley JL, Palladino MA et al (1995) Evidence for E-selectin complement regulatory domain mRNA splice variants in the rat. J Lab Clin Med 126:580–587

Blanck O, Iobst ST, Gabel C, Drickamer K (1996) Introduction of selectin-like binding specificity into a homologous mannose-binding protein. J Biol Chem 271:7289–7292

Bouyain S, Rushton S, Drickamer K (2001) Minimal requirements for the binding of selectin ligands to a C-type carbohydrate-recognition domain. Glycobiology 11:989–996

Brooks AC, Menzies-Gow NJ, Wheeler-Jones C et al (2009) Endotoxin-induced activation of equine digital vein endothelial cells: role of p38 MAPK. Vet Immunol Immunopathol 129:174–180

Brostjan C, Anrather J, Csizmadia V et al (1997) Glucocorticoids inhibit E-selectin expression by targeting NF-kB and not ATF/c-Jun. J Immunol 158:3836–3844

Bullard DC, Kunkel EJ, Kubo H et al (1996) Infectious susceptibility and severe deficiency of leukocyte rolling and recruitment in E-selectin and P-selectin double mutant mice. J Exp Med 183:2329–2336

Burdick MM, Chu JT, Godar S, Sackstein R (2006) HCELL is the major E- and L-selectin ligand expressed on LS174T colon carcinoma cells. J Biol Chem 281:13899–13905

Chai W, Feizi T, Yuen CT et al (1997) Nonreductive release of O-linked oligosaccharides from mucin glycoproteins for structure/function assignments as neoglycolipids: application in the detection of novel ligands for E-selectin. Glycobiology 7:861–872

Chan B, Yuan HT, Ananth Karumanchi S et al (2008) Receptor tyrosine kinase Tie-1 overexpression in endothelial cells upregulates adhesion molecules. Biochem Biophys Res Commun 371:475–479

Clark JG, Mandac-Dy JB, Dixon AE et al (2004) Trafficking of Th1 cells to lung: a role for selectins and a P-selectin glycoprotein-1-independent ligand. Am J Respir Cell Mol Biol 30:220–227

Cockerill GW, Bert AG, Ryan GR et al (1995) Regulation of granulocyte-macrophage colony-stimulating factor and E-selectin expression in endothelial cells by cyclosporin A and the T-cell transcription factor NFAT. Blood 86:2689–2698

Collins T, Read MA, Neish AS (1995) Transcriptional regulation of endothelial cell adhesion molecules: NF-kB and cytokine-inducible enhancers. FASEB J 9:899–909

Cooke RM, Hale RS, Lister SG et al (1994) The conformation of the sialyl Lewis X ligand changes upon binding to E-selectin. Biochemistry 33:10591–10596

Cota-Gomez A, Flores NC, Cruz C et al (2002) The human immunodeficiency virus-1 Tat protein activates human umbilical vein endothelial cell E-selectin expression via an NF-kB-dependent mechanism. J Biol Chem 277:14390–14399

Couffinhal T, Duplaa C, Moreau C et al (1994) Regulation of vascular cell adhesion molecule-1 and intercellular adhesion molecule-1 in human vascular smooth muscle cells. Circ Res 74:225–234

Curtis JL, Sonstein J, Craig RA et al (2002) Subset-specific reductions in lung lymphocyte accumulation following intratracheal antigen challenge in endothelial selectin-deficient mice. J Immunol 169:2570–2579

De Caterina R, Libby P, Peng HB et al (1995) Nitric oxide decreases cytokine-induced endothelial activation. Nitric oxide selectively reduces endothelial expression of adhesion molecules and proinflammatory cytokines. J Clin Invest 96:60–68

De Luca LG, Johnson DR, Whitley MZ et al (1994) cAMP and tumor necrosis factor competitively regulate transcriptional activation through and nuclear factor binding to the cAMP-responsive element/activating transcription factor element of the endothelial leukocyte adhesion molecule-1 (E-selectin) promoter. J Biol Chem 269:19193–19196

Deisher TA, Kaushansky K, Harlan JM (1993a) Inhibitors of topoisomerase II prevent cytokine-induced expression of vascular cell adhesion molecule-1, while augmenting the expression of endothelial leukocyte adhesion molecule-1 on human umbilical vein endothelial cells. Cell Adhes Commun 1:133–142

Deisher TA, Sato TT, Pohlman TH et al (1993b) A protein kinase C agonist, selective for the β I isozyme, induces E-selectin and VCAM-1 expression on HUVEC but does not translocate PKC. Biochem Biophys Res Commun 193:1283–1290

Dhawan S, Puri RK, Kumar A et al (1997) Human immunodeficiency virus-1-tat protein induces the cell surface expression of endothelial leukocyte adhesion molecule-1, vascular cell adhesion molecule-1, and intercellular adhesion molecule-1 in human endothelial cells. Blood 90:1535–1544

Dimitroff CJ, Lee JY, Rafii S et al (2001) CD44 is a major E-selectin ligand on human hematopoietic progenitor cells. J Cell Biol 153:1277–1286

Easton EW, Schiphorst WE, van Drunen E et al (1993) Human myeloid alpha 3-fucosyltransferase is involved in the expression of the sialyl-Lewis(x) determinant, a ligand for E- and P-selectin. Blood 81:2978–2986

Essani NA, McGuire GM, Manning AM et al (1996) Endotoxin-induced activation of the nuclear transcription factor kB and expression of E-selectin messenger RNA in hepatocytes, Kupffer cells, and endothelial cells in vivo. J Immunol 156:2956–2963

Eto T, Winkler I, Purton LE et al (2005) Contrasting effects of P-selectin and E-selectin on the differentiation of murine hematopoietic progenitor cells. Exp Hematol 33:232–242

Fan Y, Wang Y, Tang Z et al (2008) Suppression of pro-inflammatory adhesion molecules by PPAR-δ in human vascular endothelial cells. Arterioscler Thromb Vasc Biol 28:315–321

Faruqi RM, Poptic EJ, Faruqi TR et al (1997) Distinct mechanisms for N-acetylcysteine inhibition of cytokine-induced E-selectin and VCAM-1 expression. Am J Physiol 273(2 Pt 2):H817–H826

Ferran C, Millan MT, Csizmadia V et al (1995) Inhibition of NF-kappa B by pyrrolidine dithiocarbamate blocks endothelial cell activation. Biochem Biophys Res Commun 214:212–223

Forlow SB, Ley K (2001) Selectin-independent leukocyte rolling and adhesion in mice deficient in E-, P-, and L-selectin and ICAM-1. Am J Physiol Heart Circ Physiol 280:H634–H641

Forlow SB, White EJ, Barlow SC et al (2000) Severe inflammatory defect and reduced viability in CD18 and E-selectin double-mutant mice. J Clin Invest 106:1457–1466

Fries JW, Williams AJ, Atkins RC et al (1993) Expression of VCAM-1 and E-selectin in an in vivo model of endothelial activation. Am J Pathol 143:725–737

Fuhlbrigge RC, King SL, Sackstein R et al (2006) CD43 is a ligand for E-selectin on CLA + human T cells. Blood 107:1421–1426

Garcia-Criado FJ, Lozano FS et al (2005) P- and E-selectin blockade can control bacterial translocation and modulate systemic inflammatory response. J Invest Surg 18:167–176

Gege C, Schumacher G, Rothe U et al (2008) Visualization of sialyl Lewis(X) glycosphingolipid microdomains in model membranes as selectin recognition motifs using a fluorescence label. Carbohydr Res 343:2361–2368

Ghersa P, Hooft van Huijsduijnen R, Whelan J et al (1994) Inhibition of E-selectin gene transcription through a cAMP-dependent protein kinase pathway. J Biol Chem 269:29129–29137

Ghersa P, Whelan J, Cambet Y et al (1997) Distamycin prolongs E-selectin expression by interacting with a specific NF-kB-HMG-I(Y) binding site in the promoter. Nucleic Acids Res 25:339–346

Goetz DJ, Greif DM, Ding H et al (1997) Isolated P-selectin glycoprotein ligand-1 dynamic adhesion to P- and E-selectin. J Cell Biol 137:509–519

Graves BJ, Crowther RL, Chandran C et al (1994) Insight into E-selectin/ligand interaction from the crystal structure and mutagenesis of the lec/EGF domains. Nature 367(6463):532–538

Green CE, Pearson DN, Camphausen RT et al (2004) Shear-dependent capping of L-selectin and P-selectin glycoprotein ligand 1 by E-selectin signals activation of high-avidity β2-integrin on neutrophils. J Immunol 172:7780–7790

Haddix TL, Pohlman TH, Noel RF et al (1996) Hypothermia inhibits human E-selectin transcription. J Surg Res 64:176–183

Hagi-Pavli E, Farthing PM, Kapas S (2004) Stimulation of adhesion molecule expression in human endothelial cells (HUVEC) by adrenomedullin and corticotrophin. Am J Physiol Cell Physiol 286:C239–C246

Hammel M, Weitz-Schmidt G, Krause A et al (2001) Species-specific and conserved epitopes on mouse and human E-selectin important for leukocyte adhesion. Exp Cell Res 269:266–274

Haraldsen G, Kvale D, Lien B et al (1996) Cytokine-regulated expression of E-selectin, intercellular adhesion molecule-1 (ICAM-1), and vascular cell adhesion molecule-1 (VCAM-1) in human microvascular endothelial cells. J Immunol 156:2558–2565

Hattori Y, Suzuki K, Tomizawa A et al (2009) Cilostazol inhibits cytokine-induced nuclear factor-kB activation via AMP-activated protein kinase activation in vascular endothelial cells. Cardiovasc Res 8:133–139

Hensley P, McDevitt PJ, Brooks I et al (1994) The soluble form of E-selectin is an asymmetric monomer. Expression, purification, and characterization of the recombinant protein. J Biol Chem 269:23949–23958

Hession C, Osborn L, Goff D et al (1990) Endothelial leukocyte adhesion molecule 1: direct expression cloning and functional interactions. Proc Natl Acad Sci USA 87:1673–1677

Hofman FM, Dohadwala MM, Wright AD et al (1994) Exogenous tat protein activates central nervous system-derived endothelial cells. J Neuroimmunol 54:19–28

Holgersson J, Lofling J (2006) Glycosyltransferases involved in type 1 chain and Lewis antigen biosynthesis exhibit glycan and core chain specificity. Glycobiology 16:584–593

Hooft van Huijsduijnen R, Pescini R, DeLamarter JF (1993) Two distinct NF-kB complexes differing in their larger subunit bind the E-selectin promoter kB element. Nucleic Acids Res 21:3711–3717

Hooft van Huijsduijnen R, Whelan J, Pescini R et al (1992) A T-cell enhancer cooperates with NF-kB to yield cytokine induction of E-selectin gene transcription in endothelial cells. J Biol Chem 267:22385–22391

Hu Y, Szente B, Kiely JM, Gimbrone MA Jr (2001) Molecular events in transmembrane signaling via E-selectin. SHP2 association, adaptor protein complex formation and ERK1/2 activation. J Biol Chem 276:48549–48553

Huang J, Chen J, Chesla SE et al (2004) Quantifying the effects of molecular orientation and length on two-dimensional receptor-ligand binding kinetics. J Biol Chem 279:44915–44923

Huang W, Rha GB, Han MJ et al (2008) PPARα and PPARγ effectively protect against HIV-induced inflammatory responses in brain endothelial cells. J Neurochem 107:497–509

Imamoto E, Yoshida N, Uchiyama K et al (2004) Inhibitory effect of pioglitazone on expression of adhesion molecules on neutrophils and endothelial cells. Biofactors 20:37–47

Jensen LE, Whitehead AS (2003) ELAM-1/E-selectin promoter contains an inducible AP-1/CREB site and is not NF-kappa B-specific. Biotechniques 35(54–6):58

Jeschke U, Mylonas I, Kunert-Keil C et al (2009) Immunohistochemistry, glycosylation and immunosuppression of glycodelin in human ovarian cancer. Histochem Cell Biol 131:283–295

Johnson DR, Douglas I, Jahnke A et al (1996) A sustained reduction in IkappaB-β may contribute to persistent NF-kappaB activation in human endothelial cells. J Biol Chem 271:16317–16322

Jonjic N, Peri G, Bernasconi S et al (1992) Expression of adhesion molecules and chemotactic cytokines in cultured human mesothelial cells. J Exp Med 176:1165–1174

Joucher F, Mazmanian GM, German-Fattal M (2004) E-selectin early overexpression induced by allogeneic activation in isolated mouse lung. Transplantation 78:1283–1289

Jutila MA, Kurk S, Jackiw L et al (2002) L-selectin serves as an E-selectin ligand on cultured human T lymphoblasts. J Immunol 169:1768–1773

Kannagi R (2001) Transcriptional regulation of expression of carbohydrate ligands for cell adhesion molecules in the selectin family. Adv Exp Med Biol 491:267–278

Kapiotis S, Quehenberger P, Sengoelge G et al (1994) Modulation of pyrogen-induced upregulation of endothelial cell adhesion molecules (CAMs) by interleukin-4: transcriptional mechanisms and CAM-shedding. Circ Shock 43:18–28

Kaszubska W, Hooft van Huijsduijnen R, Ghersa P et al (1993) Cyclic AMP-independent ATF family members interact with NF-kappa B and function in the activation of the E-selectin promoter in response to cytokines. Mol Cell Biol 13:7180–7190

Katayama Y, Hidalgo A, Furie BC et al (2003) PSGL-1 participates in E-selectin-mediated progenitor homing to bone marrow: evidence for cooperation between E-selectin ligands and alpha4 integrin. Blood 102:2060–2067

Kato N, Yuzawa Y, Kosugi T, Hobo A et al (2009) The E-selectin ligand basigin/CD147 is responsible for neutrophil recruitment in renal ischemia/reperfusion. J Am Soc Nephrol 20:1565–1576

Kerr SC, Fieger CB, Snapp KR, Rosen SD (2008) Endoglycan, a member of the CD34 family of sialomucins, is a ligand for the vascular selectins. J Immunol 181:1480–1490

Khan BV, Harrison DG, Olbrych MT et al (1996) Nitric oxide regulates vascular cell adhesion molecule 1 gene expression and redox-sensitive transcriptional events in human vascular endothelial cells. Proc Natl Acad Sci USA 93:9114–9119

Khew-Goodall Y, Gamble JR, Vadas MA (1993) Regulation of adhesion and adhesion molecules in endothelium by transforming growth factor-β. Curr Top Microbiol Immunol 184:187–199

Kogan TP, Revelle BM, Tapp S et al (1995) A single amino acid residue can determine the ligand specificity of E-selectin. J Biol Chem 270:14047–14055

Koszdin KL, Bowen BR (1992) The cloning and expression of a human alpha-1,3 fucosyltransferase capable of forming the E-selectin ligand. Biochem Biophys Res Commun 187:152–157

Kunzendorf U, Kruger-Krasagakes S, Notter M et al (1994) A sialyl-Le (x)-negative melanoma cell line binds to E-selectin but not to P-selectin. Cancer Res 54:1109–1112

Labow MA, Norton CR, Rumberger JM et al (1994) Characterization of E-selectin-deficient mice: demonstration of overlapping function of the endothelial selectins. Immunity 1:709–720

Larkin M, Ahern TJ, Stoll MS et al (1992) Spectrum of sialylated and nonsialylated fuco-oligosaccharides bound by the endothelial-leukocyte adhesion molecule E-selectin. Dependence of the carbohydrate binding activity on E-selectin density. J Biol Chem 267:13661–13668

Lee DK, Kang JE, Park HJ et al (2005) FBI-1 enhances transcription of the nuclear factor-kB (NF-kB)-responsive E-selectin gene by nuclear localization of the p65 subunit of NF-kB. J Biol Chem 280:27783–27791

Lewis H, Kaszubska W, DeLamarter JF, Whelan J (1994) Cooperativity between two NF-kB complexes, mediated by high-mobility-group protein I(Y), is essential for cytokine-induced expression of the E-selectin promoter. Mol Cell Biol 14:5701–5709

Li F, Wilkins PP, Crawley S et al (1996) Post-translational modifications of recombinant P-selectin glycoprotein ligand-1 required for binding to P- and E-selectin. J Biol Chem 271:3255–3264

Li SH, Burns DK, Rumberger JM et al (1994) Consensus repeat domains of E-selectin enhance ligand binding. J Biol Chem 269:4431–4437

Lopez-Garcia E, Schulze MB, Meigs JB et al (2005) Consumption of Trans Fatty Acids Is Related to Plasma Biomarkers of Inflammation and Endothelial Dysfunction. J Nutr 135:562–566

Lozada C, Levin RI, Huie M et al (1995) Identification of C1q as the heat-labile serum cofactor required for immune complexes to stimulate endothelial expression of the adhesion molecules E-selectin and intercellular and vascular cell adhesion molecules 1. Proc Natl Acad Sci USA 92:8378–8382

Mackay F, Loetscher H, Stueber D et al (1993) Tumor necrosis factor alpha (TNF-α)-induced cell adhesion to human endothelial cells is under dominant control of one TNF receptor type, TNF-R55. J Exp Med 177:1277–1286

Manning AM, Bell FP, Rosenbloom CL et al (1995) NF-kB is activated during acute inflammation in vivo in association with elevated endothelial cell adhesion molecule gene expression and leukocyte recruitment. J Inflamm 45:283–296

Martin MJ, Feizi T, Leteux C et al (2002) An investigation of the interactions of E-selectin with fuco-oligosaccharides of the blood group family. Glycobiology 12:829–835

Masamune A, Igarashi Y, Hakomori S (1996) Regulatory role of ceramide in interleukin (IL)-1 β-induced E-selectin expression in human umbilical vein endothelial cells. Ceramide enhances IL-1 β action, but is not sufficient for E-selectin expression. J Biol Chem 271:9368–9375

Matsumoto M, Shigeta A, Miyasaka M et al (2008) CD43 plays both antiadhesive and proadhesive roles in neutrophil rolling in a context-dependent manner. J Immunol 181:3628–3635

McCafferty DM, Kanwar S, Granger DN, Kubes P (2000) E/P-selectin-deficient mice: an optimal mutation for abrogating antigen but not tumor necrosis factor-alpha-induced immune responses. Eur J Immunol 30:2362–2371

Meacock S, Pescini-Gobert R, DeLamarter JF et al (1994) Transcription factor-induced, phased bending of the E-selectin promoter. J Biol Chem 269:31756–31762

Milstone DS, Redline RW, O'Donnell PE et al (2000) E-selectin expression and function in a unique placental trophoblast population at the fetal-maternal interface: regulation by a trophoblast-restricted transcriptional mechanism conserved between humans and mice. Dev Dyn 219:63–76

Min W, Pober JS (1997) TNF initiates E-selectin transcription in human endothelial cells through parallel TRAF-NF-kappa B and TRAF-RAC/CDC42-JNK-c-Jun/ATF2 pathways. J Immunol 159:3508–3518

Montoya MC, Holtmann K, Snapp KR et al (1999) Memory B lymphocytes from secondary lymphoid organs interact with E-selectin through a novel glycoprotein ligand. J Clin Invest 103:1317–1327

Mourelatos Z, Gonatas JO, Cinato E et al (1996) Cloning and sequence analysis of the human MG160, a fibroblast growth factor and E-selectin binding membrane sialoglycoprotein of the Golgi apparatus. DNA Cell Biol 15:1121–1128

Moynagh PN, Williams DC, O'Neill LA (1994) Activation of NF-kB and induction of vascular cell adhesion molecule-1 and intracellular adhesion molecule-1 expression in human glial cells by IL-1. Modulation by antioxidants. J Immunol 153:2681–2690

Nawa T, Nawa MT, Cai Y et al (2000) Repression of TNF-α-induced E-selectin expression by PPAR activators: involvement of transcriptional repressor LRF-1/ATF3. Biochem Biophys Res Commun 275:406–411

Nimrichter L, Burdick MM, Aoki K et al (2008) E-selectin receptors on human leukocytes. Blood 112:3744–3752

Nonomura C, Kikuchi J, Kiyokawa N et al (2008) CD43, but not P-selectin glycoprotein ligand-1, functions as an E-selectin counter-receptor in human pre-B-cell leukemia NALL-1. Cancer Res 68:790–799

Norton CR, Rumberger JM, Burns DK et al (1993) Characterization of murine E-selectin expression in vitro using novel anti-mouse E-selectin monoclonal antibodies. Biochem Biophys Res Commun 195:250–258

Ollivier V, Parry GC, Cobb RR et al (1996) Elevated cyclic AMP inhibits NF-kB-mediated transcription in human monocytic cells and endothelial cells. J Biol Chem 271:20828–20835

Pietrzak ER, Savage NW, Aldred MJ, Walsh LJ (1996) Expression of the E-selectin gene in human gingival epithelial tissue. J Oral Pathol Med 25:320–324

Rahman A, Kefer J, Bando M et al (1998) E-selectin expression in human endothelial cells by TNF-α-induced oxidant generation and NF-kB activation. Am J Physiol 275:L533–L544

Rajan S, Ye J, Bai S et al (2008) NF-kB, but not p38 MAP kinase, is required for TNF-α-induced expression of cell adhesion molecules in endothelial cells. J Cell Biochem 105:477–486

Ramos CL, Kunkel EJ, Lawrence MB et al (1997) Differential effect of E-selectin antibodies on neutrophil rolling and recruitment to inflammatory sites. Blood 89:3009–3018

Ray KP, Farrow S, Daly M et al (1997) Induction of the E-selectin promoter by interleukin 1 and tumor necrosis factor alpha, and inhibition by glucocorticoids. Biochem J 328:707–715

Read MA, Neish AS, Gerritsen ME, Collins T (1996) Postinduction transcriptional repression of E-selectin and vascular cell adhesion molecule-1. J Immunol 157:3472–3479

Read MA, Whitley MZ, Gupta S et al (1997) Tumor necrosis factor alpha-induced E-selectin expression is activated by the nuclear factor-kB and c-JUN N-terminal kinase/p38 mitogen-activated protein kinase pathways. J Biol Chem 272:2753–2761

Read MA, Whitley MZ, Williams AJ et al (1994) NF-k B and I kB α: an inducible regulatory system in endothelial activation. J Exp Med 179:503–512

Reimold AM, Grusby MJ, Kosaras B et al (1996) Chondrodysplasia and neurological abnormalities in ATF-2-deficient mice. Nature 379(6562):262–265

Roca FJ, Mulero I, López-Muñoz A et al (2008) Evolution of the inflammatory response in vertebrates: fish TNF-α is a powerful activator of endothelial cells but hardly activates phagocytes. J Immunol 181:5071–5081

Rossi B, Constantin G (2008) Anti-selectin therapy for the treatment of inflammatory diseases. Inflamm Allergy Drug Targets 7:85–93

Sackstein R (2004) The bone marrow is akin to skin: HCELL and the biology of hematopoietic stem cell homing. J Invest Dermatol 122:1061–1069

Saitoh O, Wang W-L, Lotan R, Fukuda M (1992) Differential glycosylation and cell surface expression of lysosomal membrane glycoproteins in sublines of a human colon cancer exhibiting distinct metastatic potentials. J Biol Chem 267:5700–5711

Sawada R, Tsuboi S, Fukuda M (1994) Differential E-selectin-dependent adhesion efficiency in sublines of a human colon cancer exhibiting distinct metastatic potentials. J Biol Chem 269:1425–1431

Schindler U, Baichwal VR (1994) Three NF-kB binding sites in the human E-selectin gene required for maximal tumor necrosis factor alpha-induced expression. Mol Cell Biol 14:5820–5831

Schweitzer KM, Drager AM, van der Valk P et al (1996) Constitutive expression of E-selectin and vascular cell adhesion molecule-1 on endothelial cells of hematopoietic tissues. Am J Pathol 148:165–175

Shigeta A, Matsumoto M, Tedder TF et al (2008) An L-selectin ligand distinct from P-selectin glycoprotein ligand-1 is expressed on endothelial cells and promotes neutrophil rolling in inflammation. Blood 112:4915–4923

Shreeniwas R, Koga S, Karakurum M et al (1992) Hypoxia-mediated induction of endothelial cell interleukin-1 alpha. An autocrine mechanism promoting expression of leukocyte adhesion molecules on the vessel surface. J Clin Invest 90:2333–2339

Smalley DM, Lin JH, Curtis ML et al (1996) Native LDL increases endothelial cell adhesiveness by inducing intercellular adhesion molecule-1. Arterioscler Thromb Vasc Biol 16:585–590

Smith GM, Whelan J, Pescini R et al (1993) DNA-methylation of the E-selectin promoter represses NF-kB transactivation. Biochem Biophys Res Commun 194:215–221

Smith ML, Olson TS, Ley K (2004) CXCR2- and E-selectin-induced neutrophil arrest during inflammation in vivo. J Exp Med 200:935–939

Somers WS, Tang J, Shaw GD, Camphausen RT (2000) Insights into the molecular basis of leukocyte tethering and rolling revealed by structures of P- and E-selectin bound to SLe(X) and PSGL-1. Cell 103:467–479

Spiecker M, Darius H, Liao JK (2000) A functional role of I κB-ε in endothelial cell activation. J Immunol 164:3316–3322

Steegmaier M, Levinovitz A, Isenmann S et al (1995) The E-selectin-ligand ESL-1 is a variant of a receptor for fibroblast growth factor. Nature 373(6515):615–620

Subramaniam M, Koedam JA, Wagner DD (1993) Divergent fates of P- and E-selectins after their expression on the plasma membrane. Mol Biol Cell 4:791–801

Sun L, Ke Y, Zhu CY et al (2008) Inflammatory reaction versus endogenous peroxisome proliferator-activated receptors expression, re-exploring secondary organ complications of spontaneously hypertensive rats. Chin Med J (Engl) 121:2305–2311

Sweeney EA, Priestley GV, Nakamoto B et al (2000) Mobilization of stem/progenitor cells by sulfated polysaccharides does not require selectin presence. Proc Natl Acad Sci USA 97:6544–6549

Edelstein LC, Pan A, Collins T et al (2005) T. Chromatin modification and the endothelial-specific activation of the E-selectin gene. J Biol Chem 280:11192–11202

Tamaru M, Narumi S (1999) E-selectin gene expression is induced synergistically with the coexistence of activated classic protein kinase C and signals elicited by interleukin-1β but not tumor necrosis factor-alpha. J Biol Chem 274:3753–3763

Tauxe C, Xie X, Joffraud M et al (2008) P-selectin glycoprotein ligand-1 decameric repeats regulate selectin-dependent rolling under flow conditions. J Biol Chem 283:28536–28545

Tedder TF, Steeber DA, Chen A et al (1995) The selectins: vascular adhesion molecules. FASEB J 9:866–873

Thomas SN, Schnaar RL, Konstantopoulos K (2009a) Podocalyxin-like protein is an E-/L-selectin ligand on colon carcinoma cells: comparative biochemical properties of selectin ligands in host and tumor cells. Am J Physiol Cell Physiol 296:C505–C513

Thomas SN, Zhu F, Zhang F et al (2009b) Different roles of galectin-9 isoforms in modulating E-selectin expression and adhesion function in LoVo colon carcinoma cells. Mol Biol Rep 36:823–830

Tomita H, Iwata Y, Ogawa F et al (2009) P-selectin glycoprotein ligand-1 contributes to wound healing predominantly as a p-selectin ligand and partly as an e-selectin ligand. J Invest Dermatol 129:2059–2067

Totani L, Piccoli A, Manarini S et al (2006) Src-family kinases mediate an outside-in signal necessary for β2 integrins to achieve full activation and sustain firm adhesion of polymorphonuclear leucocytes tethered on E-selectin. Biochem J 396:89–98

Ushakova NA, Preobrazhenskaya ME, Bird MI et al (2005) Monomeric and multimeric blockers of selectins: comparison of in vitro and in vivo activity. Biochemistry (Mosc) 70:432–439

Wadgaonkar R, Pierce JW, Somnay K et al (2004) Regulation of c-Jun N-terminal kinase and p38 kinase pathways in endothelial cells. Am J Respir Cell Mol Biol 31:423–431

Wang N, Verna L, Chen NG et al (2002) Constitutive activation of peroxisome proliferator-activated receptor-gamma suppresses pro-inflammatory adhesion molecules in human vascular endothelial cells. J Biol Chem 277:34176–34181

Wang X, Feuerstein GZ, Gu JL et al (1995) Interleukin-1 β induces expression of adhesion molecules in human vascular smooth muscle cells and enhances adhesion of leukocytes to smooth muscle cells. Atherosclerosis 115:89–98

Watson ML, Kingsmore SF, Johnston GI et al (1990) Genomic organization of the selectin family of leukocyte adhesion molecules on human and mouse chromosome 1. J Exp Med 172:263–272

Weber C (1996) Involvement of tyrosine phosphorylation in endothelial adhesion molecule induction. Immunol Res 15:30–37

Weis WI, Drickamer K, Hendrickson WA (1992) Structure of a C-type mannose-binding protein complexed with an oligosaccharide. Nature 360:127–134

Weller A, Isenmann S, Vestweber D (1992) Cloning of the mouse endothelial selectins. Expression of both E- and P-selectin is inducible by tumor necrosis factor alpha. J Biol Chem 267:15176–15183

Whelan J, Ghersa P, Hooft van Hujsduinen R et al (1991) An NF-kB like factor is essential but not sufficient for cytokine induction of endothelial leukocyte adhesion molecule (ELAM) gene transcription. Nucleic Acids Res 19:2645–2653

Whitley MZ, Thanos D, Read MA et al (1994) A striking similarity in the organization of the E-selectin and β interferon gene promoters. Mol Cell Biol 14:6464–6475

Wild MK, Huang MC, Schulze-Horsel U et al (2001) Affinity, kinetics, and thermodynamics of E-selectin binding to E-selectin ligand-1. J Biol Chem 276:31602–31612

Winkler H, Brostjan C, Csizmadia V et al (1996) The intron-exon structure of the porcine E-selectin-encoding gene. Gene 176:67–72

Winkler IG, Snapp KR, Simmons PJ et al (2004) Adhesion to E-selectin promotes growth inhibition and apoptosis of human and murine hematopoietic progenitor cells independent of PSGL-1. Blood 103:1685–1692

Xu XS, Vanderziel C, Bennett CF et al (1998) A role for c-Raf kinase and Ha-Ras in cytokine-mediated induction of cell adhesion molecules. J Biol Chem 273:33230–33238

Yiming MT, Parthasarathi K, Issekutz AC et al (2005) Sequence of endothelial signaling during lung expansion. Am J Respir Cell Mol Biol 33:549–554

Yu G, Rux AH, Ma P et al (2005) Endothelial expression of E-selectin is induced by the platelet-specific chemokine platelet factor 4 through LRP in an NF-kB-dependent manner. Blood 105: 3545–3551

Yuen CT, Lawson AM, Chai W et al (1992) Novel sulfated ligands for the cell adhesion molecule E-selectin revealed by the neoglycolipid technology among O-linked oligosaccharides on an ovarian cystadenoma glycoprotein. Biochemistry 31:9126–9131

Zanardo RC, Bonder CS, Hwang JM et al (2004) A down-regulatable E-selectin ligand is functionally important for PSGL-1-independent leukocyte-endothelial cell interactions. Blood 104: 3766–3773

Zarbock A, Abram CL, Hundt M et al (2008) PSGL-1 engagement by E-selectin signals through Src kinase FgR and ITAM adapters DAP12 and FcRγ to induce slow leukocyte rolling. J Exp Med 205: 2339–2347

Zen K, Cui LB, Zhang CY, Liu Y (2007) Critical role of mac-1 sialyl Lewis X moieties in regulating neutrophil degranulation and transmigration. J Mol Biol 374:54–63

Zen K, Liu DQ, Guo YL et al (2008) CD44v4 is a major E-selectin ligand that mediates breast cancer cell transendothelial migration. PLoS One 3:e1826

Zollner O, Lenter MC, Blanks JE et al (1997) L-selectin from human, but not from mouse neutrophils binds directly to E-selectin. J Cell Biol 136:707–716

Part X

C-Type Lectins: Lectin Receptors on NK Cells

KLRB Receptor Family and Human Early Activation Antigen (CD69)

Rajesh K. Gupta and G.S. Gupta

29.1 Lectin Receptors on NK Cell

29.1.1 NK Cell Receptors

Natural killer cells are important component of the innate immune system, providing protection against intracellular infection particularly viruses and also neoplasia through direct cytotoxic mechanisms and the secretion of cytokines. They mediate their effects through direct cytolysis, release of cytokines and regulation of subsequent adaptive immune responses. They are called 'natural' killers because, unlike cytotoxic T cells, they do not require a previous challenge and preactivation to become active. NK cells can be activated by a range of soluble factors, including type I interferons, IL-2, IL-12, IL-15 and IL-18, but also by direct cell to cell contact between NK cell receptors and target cell ligands. NK cells possess an elaborate array of receptors, which regulate NK cytotoxic and secretory functions upon interaction with target cell MHC class I proteins. Determination of structures of NK cell receptors and their ligand complexes has led to a fast growth in our understanding of the activation and ligand recognition by these receptors as well as their function in innate immunity. B and T cells significantly and differentially influence the homeostasis and the phenotype of NK cells. The function of NK cell is tightly regulated by a fine balance of inhibitory and activating signals that are delivered by a diverse array of cell surface receptors. A prerequisite for a NK cell attack is the presence on target cells of ligands for activating receptors and low level or absence of ligands for inhibitory receptors. It was believed that NK self-tolerance was achieved by expression on each NK cell of at least one self-MHC specific inhibitory receptor. However, this dogma has been challenged after identification of a NK cell population in normal mice that lack inhibitory receptors specific for self-MHC class I molecules (Kumar and McNerney 2005; Fernandez et al. 2005). Therefore, it was made clear that some additional surface receptors contribute to NK self-tolerance and to the modulation of NK cell responses. The characterization and the identification of their physiological ligands allow us a comprehensive understanding of NK cell function.

NK cell receptors (NKR) include: (1) non-HLA class I–specific receptors, such as CD56, CD57, or CD161 (Lee et al. 2003), (2) as well as receptors that recognize HLA class I molecules or their structural relatives. Later group comprises killer receptors belonging to the immunoglobulin family (KIR), such as KIR2DL1/S1 and KIR2DL2/3/S2 (Moretta and Moretta 2004) or immunoglobulin-like transcripts (ILTs), which are expressed mainly on B, T, and myeloid cells, although some members are also expressed on NK cells. The immunoglobulin-like transcripts (ILTs) include CD94/NKG2 heterodimers (also called KLRD/KLRC), rodent Ly49 (KLRA), NKG2D (KLRK), NKR-P1 (KLRB), and KLRG1 belonging to the C-type lectin receptor family, called killer cell lectin-like receptors (KLR approved symbol) (Borrego et al. 2001; Vetter et al. 2000). The superfamily of KLRs was originally discovered in NK cells in rats. Functionally, NK cell surface receptors are divided into two groups, the inhibitory and the activating receptors. Classically, inhibitory signals are mediated by receptors that recognize MHC class I molecules or their structural relatives such as MICA, ULBP, RAE-1, and H-60. The inhibitory form of NK receptors provides the protective immunity through recognizing class I MHC molecules with self-peptides on healthy host cells (Raulet et al. 2001; Yokoyama and Plougastel 2003).

Recent studies also suggest that MHC class I-independent inhibitory signals can also result in inhibition of cytotoxic cells and some NK cell specificities may be conveyed by orphan receptors expressed on NK cells (Lebbink and Meyaard 2007; Kumar and McNerney 2005). The recent identification of non-MHC class I ligands for inhibitory immune receptors, such as KLRG1, KLRB1 and LAIR-1, indicates that MHC class I-independent inhibitory immune receptors play crucial roles in inducing peripheral tolerance.

The presence of these receptors on many other immune cell types besides effector cells suggests that tight regulation of cell activation is necessary in all facets of the immune response in both normal and diseased tissue. Lebbink and Meyaard (2007) gave an overview of the known ligand-receptor pairs by grouping the ligands according to their properties and discussed implications of these interactions for the maintenance of immune balance and for the defense against tumors and pathogens (Lebbink and Meyaard 2007). Different mechanisms, through which NK cells can be activated, are shown in Fig. 29.1 (also see Fig. 31.1 in Chap. 31).

29.1.2 NKC Gene Locus

The human NK cell gene locus (NKC) is located on the short arm of chromosome 12 and contains a number of genes encoding C-type lectin receptors important for NK cell function. A part of human chromosome 12p12.3-p13.2 containing NKC has been sequenced. Genes among them are: CD94 and the five NKG2 genes. A detailed analysis shows that all six genes are found within a region of 100–200 kb proximal of the marker D12S77. The gene order established is D12S77—CD94—NKG2D—NKG2F—NKG2E—NKG2C—NKG2A. The NKG2 genes are of identical transcriptional orientation, whereas the CD94 gene is placed in opposite orientation. The tight genomic linkage of these genes and the identical orientation of the NKG2 genes suggest coordinate regulation of expression during differentiation of NK cells (Renedo et al. 2000; Sobanov et al. 1999). Renedo et al. (2000) localized 17 genes, 5 expressed sequence tags, and 49 STSs within this contig and established the order of the genes as tel-M6PR-MAFAL (HGMW-approved symbol KLRG1)-A2M-PZP-A2MP-NKRP1A (HGMW-approved symbol KLRB1)-CD69-AICL (HGMW-approved symbol CLECSF2)-KLRF1-OLR1-CD94 (HGMW-approved symbol KLRD1)-NKG2D (HGMW-approved symbol D12S2489E)-PGFL-NKG2F (HGMW-approved symbol KLRC4)-NKG2E (HGMW-approved symbol KLRC3)-NKG2A (HGMW-approved symbol KLRC1)-LY49L (HGMW-approved symbol KLRA1)-cen (Renedo et al. 2000).

In marmoset monkey, NKC is 1.5 times smaller than its human counterpart, but the genes are colinear and orthologous. One exception is the activating NKG2CE gene, which is probably an ancestral form of the NKG2C- and NKG2E-activating receptor genes of humans and great apes. Analyses of NKC genes in nine additional marmoset individuals revealed a moderate degree of polymorphism of the CD94, NKG2A, NKG2CE, and NKG2D genes. Furthermore, expression analyses identified several alternatively spliced transcripts, particularly of the CD94 gene. Several products of alternative splicing of NKC genes are highly conserved among primates (Averdam et al. 2007).

29.2 The Ever-Expanding Ly49 Receptor Gene Family

29.2.1 Activating and Inhibitory Receptors

The molecular mechanism that explains why NK cells do not kill indiscriminately has been elucidated. It is due to several specialized receptors that recognize MHC class I molecules expressed on normal cells. The lack of expression of one or more HLA class I alleles leads to NK-mediated target cell lysis. Different types of receptors specific for groups of HLA-C, HLA-B, and, later HLA-A alleles had been identified. While in most instances, they function as inhibitory receptors, an activatory form of the HLA-C-specific receptors has been identified in some donors. The mouse lectin-related Ly49 family and the human KIR family represent structurally distinct, yet functionally analogous, class I MHC receptors that are expressed on NK cells and some T cells. The functional similarity of these two families has been borne out by the demonstration of identical signal transduction pathways associated with each receptor family. The Ly49 family therefore provides a useful model system to study the role of this class of receptors in the regulation of the immune system. Reports relating to the Ly49 repertoire in several mouse strains have revealed an additional evolutionary parallel between KIR and Ly49 receptor families. There is now an appreciation of the variation in the number and type of Ly49s expressed in different mouse strains, similar to the previously demonstrated differences in the number of KIR genes found in humans. Different reviews have appeared describing properties of members of the Ly49 gene family, their MHC class I recognition and associated signal transduction pathways. The Ly49 receptor repertoire may be initially generated by a stochastic process that distributes receptors randomly to different cells and treats the two alleles of a given Ly49 gene independently. However, class I-MHC-dependent "education" processes shape the functional repertoire. The education processes silence potentially auto-aggressive NK cells, probably by ensuring that each NK cell expresses at least one self-specific Ly49 receptor. In addition, NK cell clones that express multiple self-specific Ly49 receptors are disfavored by the education processes, perhaps to confer greater discrimination on to individual NK cells (Raulet et al. 2001; Anderson et al. 2001).

29.2.1.1 Two Structurally Dissimilar NK Cell Receptor Families, Ly49 and KIR

In mice, C-type lectin receptors are represented by the Ly49 family of receptors, whereas in humans, NK cells express the distantly related CD94, which forms MHC class I-specific heterodimers with NKG2 family members. The Ly49 family of receptors is encoded by a highly polymorphic multigene

29.2 The Ever-Expanding Ly49 Receptor Gene Family

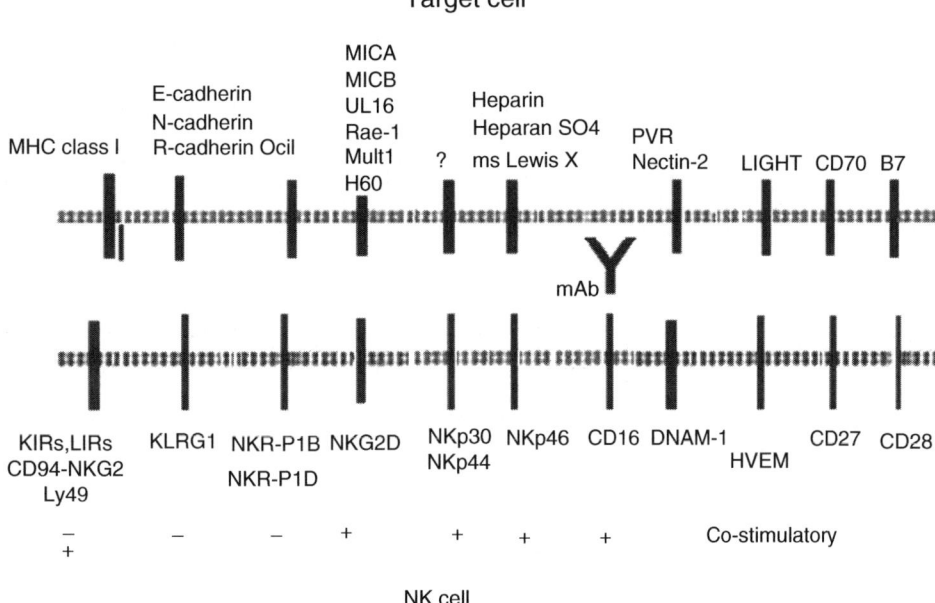

Fig. 29.1 The activating or the non-inhibitory NK receptors mediate the killing of tumor or virally infected cells through their specific ligand recognition. During target cell recognition, these cells receive both activating and inhibitory signals. Integration of opposing signals from two types of receptors determines the appropriate response: activation or inhibition of NK cells

family in mouse and is also present in multiple copies in rat. However, this gene exists as a single copy in primates and is mutated to non-function in humans. Humans also express MHC class I-specific p50/p58/p70 family of Ig-like receptors, and have been identified in mice. One Ly49-like gene, Ly49L, exists in humans but is incorrectly spliced and assumed to be nonfunctional. Mouse KIR-like genes have not been found, and the evidence suggests that the primate KIRs amplified after rodents and primates diverged. Thus, two structurally dissimilar families, Ly49 and KIR, had evolved to play similar roles in mouse and human NK cells respectively. It is not known, however, when the Ly49L gene became nonfunctional and if this event affected the functional evolution of the KIRs. The distribution of these gene families in different mammals has not been well studied, nor it is known if any species uses both types of receptors. However, the Ly49L gene shows evidence of conservation in other mammals and that the human gene likely became nonfunctional 6–10 million years ago. Furthermore, baboon lymphocytes express both full-length Ly49L transcripts and multiple KIR genes (Mager et al. 2001).

Interestingly, many activating and inhibitory receptors come in pairs, with a strong homology of the extracellular domains and partially overlapping ligand specificity but opposite signaling capacities. Examples for these paired receptors are KIRs, which comprise inhibitory and activating receptors, both recognizing the same major MHC class I ligand. Also in the family of KLRs can be found antithetic pairs that are specific for the same ligand but transmit opposing signals. For example, the NKG2A-CD94 heterodimer transmits an inhibitory signal and NKG2C-CD94 signals in an activating fashion while both receptors recognize the same ligand, HLA-E. The structures of activating and inhibitory NK cell surface receptors and their complexes with ligands and complexes with class I MHC homologs have been reviewed (Dimasi et al. 2004; Natarajan et al. 2002; Radaev and Sun 2003). Although CD94/NKG2 is expressed in both humans and rodents, KIRs are only expressed in humans and Ly49s only in rodents. Examples of other C-type lectin-like NK cell receptors that occur as individual genes are CD69, and activation-induced C-type lectin (AICL).

29.2.1.2 Ly49 Gene Expression in Different Inbred Mouse Strains

Mouse NK cells family of Ly49 contains at least 23 expressed members (A-w) and is further subdivided into activating and inhibitory subfamilies based on intracellular and transmembrane characteristics. The level of sequence identity between different members varies dramatically. However, comparison of the extracellular domain has revealed that several of the Ly49 molecules also form "pairs," where one member is activating and the other is inhibitory. Until recently, most Ly49 molecules described have come from the C57B1/6 strain of inbred mice. Using molecular cloning and immunochemical analysis it has been revealed that different mouse strains express novel Ly49 molecules. Comparison of the allelic forms of some Ly49 molecules has shown that the dividing line between different genes and different alleles is blurred (Deng and Mariuzza 2006; Deng et al. 2008).

29.2.1.3 Mouse Ly49 NK Receptors: Balancing Activation and Inhibition

In addition to lysis, a major consequence of triggering the murine activating NK receptor Ly49D is the expression of cytokines and chemokines. The activating Ly49D murine NK cell receptor can potently synergize during co-stimulation with IL-12 and IL-18 for selective production of IFN-γ. Activation both in vitro and in vivo and synergistic production of IFN-γ by Ly49D expressing NK cells results from cytokine stimulation combined with co-receptor ligation. Costimulation of the activating Ly49D murine NK cell receptor with IL-12 or IL-18 is capable of over-riding the inhibitory Ly49G2 receptor blockade for cytokine production both in vitro and in vivo. This synergy is mediated by and dependent upon Ly49D-expressing NK cells and results in significant systemic expression of IFN-γ. This would place NK cells and their activating Ly-49 receptors as important initiators of microbial, antiviral, and antitumor immunity and provide a mechanism for the release of activating Ly49 receptors from inhibitory receptor blockade (Ortaldo and Young 2003, 2005) (Fig. 29.2).

29.2.2 Crystal Analysis of CTLD of Ly49I and comparison with Ly29A, NKG2D and MBP-A

Dimasi and Biassoni (2005) reviewed the functions and X-ray crystal structure of the Ly49 NK cell receptors (Ly49A and Ly49I) (Fig. 29.3a, b), and the structural features of the Ly49/MHC class I interaction as revealed by the X-ray crystal structures of Ly49A/H-2D^d and the Ly49C/H-2K^b. The Ly49 monomer consists of two α-helices (α1 and α2), and two anti-parallel β-sheets formed by β-strands β0, β1, β5; β2, β2′, β3 and β4 (Fig. 29.3). To form the Ly49 homodimer, the monomers associate through strand β0, creating an extended anti-parallel β-sheet. In Ly49A, the C-terminal ends of the α2 helices pack against one another, creating a 'closed' dimer (Dimasi and Biassoni 2005). In Ly49C and Ly49I, however, the α2 helices are not juxtaposed in the interface, opening these dimers by ~20° compared with Ly49A. An important consequence of this variability in Ly49 dimerization geometry is to modulate the way these NK receptors bind MHC, as revealed by the Ly49A/H-2D^d and Ly49C/H-2K^b structures (Tormo et al. 1999; Dam et al. 2003, 2006). The complex structures provide a framework for understanding MHC-I recognition by NK receptors from both families and reveal striking

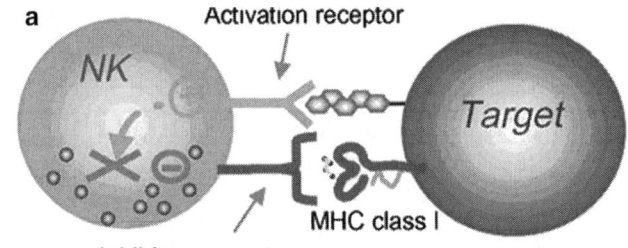

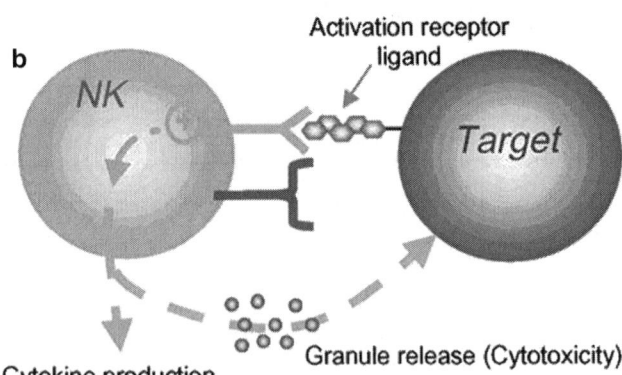

Fig. 29.2 NK cell activation is controlled by the integration of signals from activation and inhibitory receptors. (**a**) Inhibitory NK cell receptors recognize self MHC class I and restrain NK cell activation. (**b**) When unimpeded by the inhibitory receptors, binding of NK cell activation receptors to their ligands on target cells results in NK cell stimulation. In the absence or downregulation of self MHC class I on the target cells, these stimulatory signals are no longer suppressed, resulting in NK cell responses including cytokine production and granule release leading to cytotoxicity. Note that this model indicates that NK cells do not kill by default; that is, when MHC class I inhibition is absent, the NK cell must still be stimulated through activation receptors. Moreover, whether or not an individual NK cell is activated by a target is determined by this complex balance of receptors with opposing function and expression of the corresponding ligands. In general, however, inhibition dominates over activation. Finally, NK cells can be directly stimulated by cytokines such as interleukin-12 that trigger the production of other cytokines by NK cells (not shown). These direct cytokine-mediated responses are not affected by MHC class I expression (Reproduced from French and Yokoyama 2004)

differences in the nature of this recognition, despite the receptors' functional similarity (Sawicki et al. 2001). The crystal structure of inhibitory NK receptor Ly49I monomer, at 3.0 Å adopts a fold similar to that of other C-type lectin-like NK receptors, including Ly49A, NKG2D (Fig. 29.3) and CD69.

However, the Ly49I monomers associate in a manner distinct from that of these other NK receptors, forming a more open dimer. As a result, the putative MHC-binding surfaces of the Ly49I dimer are spatially more distant than the corresponding surfaces of Ly49A or NKG2D. These structural differences probably reflect the fundamentally different ways in which Ly49 and NKG2D receptors recognize their respective ligands. For example the single MICA binding site of NKG2D is formed by the precise

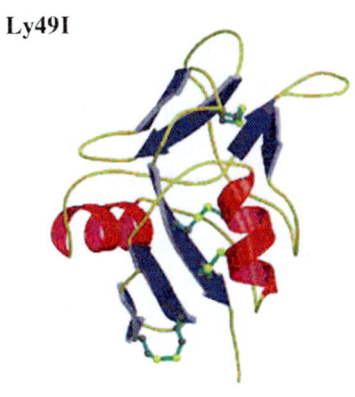

Fig. 29.3 Ribbon diagrams of C-type lectin-like domains of NK receptor Ly49A (PDB entry code 1QO3) (Tormo et al. 1999) and Ly49I (1JA3) (Dimasi et al. 2002). The secondary structural elements are colored as follows: β- strands *blue*, ℝ-helices *red*, and loop regions *gold*. The disulphide bonds are shown in green as ball-and-stick representation

juxtaposition of two monomers, each Ly49 monomer contains an independent binding site for MHC-I. Hence, the structural constraints on dimerization geometry may be relatively relaxed within the Ly49 family. Such variability may enable certain Ly49 receptors, like Ly49I, to bind MHC-I molecules bivalently, thereby stabilizing receptor-ligand interactions and enhancing signal transmission to the NK cell (Dimasi et al. 2002).

29.3 NK Cell Receptor Protein 1 (NKR-P1) or KLRB1

29.3.1 *Ly49* and *Nkrp1 (Klrb1)* Recognition Systems

Three classes of multigene family-encoded receptors enable NK cells to discriminate between polymorphic MHC class I molecules: Ly-49 homodimers, CD94/NKG2 heterodimers and the killer cell inhibitory receptors (KIR). Of these, CD94/NKG2 has been characterized in both rodents and humans (Chap. 31) and does not show lectin activity. In contrast, Ly-49 family members have hitherto been found only in rodents, and KIR molecules only in the human. Like *Ly49* (*Klra*), *Nkrp1(Klrb1)* locus encodes structurally and functionally related cell surface protein that regulates NK cell-mediated cytotoxicity and cytokine production. The Ly49 receptors are encoded in NKC that contains clusters of genes for other lectin-like receptors on NK cells and other hematopoietic cells. Though the NKR-P1 molecules were the first family of NK cell receptors identified, yet they remained enigmatic in their contribution to self-nonself discrimination until recently. The *Ly49* and *Nkrp1* genetically linked loci for receptor-ligand pairs suggest a genetic strategy to preserve this interaction and show several other contrasts with Ly49-MHC interactions. Despite their close relatedness and genetic linkage within NKC, these two multi-gene families have adopted dissimilar evolutionary strategies. While families, like *Ly49* are polygenic in rodents (Dissen et al. 1996), there is only a single family member (Ly49L) in humans (Westgaard et al. 1998). Ly49 genes are extremely polymorphic, with distinct gene numbers, remarkable allelic diversity, and varying MHC-I-ligand specificities and affinities among different murine haplotypes. In contrast, the *Nkrp1* genes have opted for overall conservation of genomic organization, sequences, and ligand specificities, with only limited allelic polymorphism (Carlyle et al. 2008).

29.3.2 NKRP1

The NKR-P1family of receptors has been of considerable interest due to its conservation in rats, mice, and humans (Lanier 2005; Brissette-Storkus et al. 1994). Rodents have several *Klrb1* genes encoding either activating or inhibitory NK receptors of C-type lectin superfamily, including Nkr-p1c, the NK1.1 Ag defining mouse NK cells (Plougastel et al. 2001a). In rats, the KLRs are encoded by the NK gene complex located on chromosome 4, and syntenic chromosomal regions have been identified in many other species. Several KLR gene families are clustered together in specific sub-regions of NKC, and there is considerable species-to-species variation in the number, expression patterns, and putative functions of the genes encoded by these different KLR families.

Initially thought as markers for NK cells, expression of NKR-P1 has now been documented in other cell types, including subsets of T cells as well as on granulocytes and spleen DCs. In mice, six NKR-P1 genes have been identified (Plougastel et al. 2001a), whereas in humans there appears to be only one gene, CD161 (NKR-P1A), expressed on most NK cells and subsets of thymocytes and fetal liver T cells

(Lanier 2005). Subsets of mature CD3$^+$ T cells, including CD4$^+$ and CD8$^+$ cells also express this receptor, particularly CD1d reactive NKT cells (Exley et al. 1998). This latter population is thought to be involved in the regulation of immune responses associated with a broad range of diseases, including autoimmunity, infectious diseases, and cancer (Godfrey et al. 2004). Rat NKR-P1A was shown to confer reactivity of an NK cell line to certain mouse tumor cell lines (Ryan et al. 1995), and some reports suggest a role for NKR-P1 receptors in the lysis of MHC allogeneic or semiallogeneic target cells (Dissen et al. 1996). Khalturin et al. (2003) identified a gene in the tunicate *Botryllus schlosseri* that encodes a lectin-like protein that is more similar in its extracellular domain to NKR-P1 than to other lectin-like genes (Mesci et al. 2006; Plougastel and Yokoyama 2006).

29.3.2.1 Rat NKR-P1

NKR-P1 is a 60-kDa homodimer expressed on rat NK cells. NKR-P1 may play a role in NK cell activation because antibody to NKR-P1 stimulates the release of granules from NK cells, and anti-NKR-P1 antibody causes redirected lysis by activated NK cells against targets that express FcR. NKR-P1 molecules in rodents may activate cytotoxicity by transducing biochemical signals. Structures of these molecules have not been elucidated. The murine NKR-P1 and Ly-49 molecules are encoded by members of polymorphic gene families that reside in the NK gene complex on the distal region of mouse chromosome 6. The rat NKR-P1 Ag shares several features with the mouse Ly-49 Ag, including selective cell surface expression on NK cells, homology to the C-type lectins, expression as a type II integral membrane protein, and disulfide-linked homodimeric structure. The mouse and rat NKR-P1-deduced polypeptide sequences are highly conserved, suggesting a similar tertiary structure. Although the deduced amino acid sequences of mNKR-P1 and Ly-49 reveal that these proteins are structurally similar, they are only 24% identical at the amino acid level and the cDNA sequences do not demonstrate significant nucleotide homology (Yokoyama and Plougastel 2003). NKR-P1 expresses at low levels on a small subset of rat T cells with an NKR-P1$_{dim}$/TCR-$\alpha\beta^+$ phenotype and on a small subset of cells with an NKR-P1$_{dim}$/TCR-$\alpha\beta^-$ phenotype ($\gamma\delta^+$ T cells?). The expression of functional NKR-P1 (i.e., ability to signal rADCC) correlates with and potentially contributes to MHC-unrestricted cytotoxicity (Brissette-Storkus et al. 1994; Chambers et al. 1992).

Peripheral blood neutrophils in normal animals express very low or undetectable levels of NKR-P1. Detectable level of NKR-P1 was induced as early as day 1 following small bowel transplantation in all allografted animals, whereas expression was only rarely detected in isografted animals (Webster et al. 1994). A major subset of non-alloreactive NK cells in PVG strain rats is generally low in Ly49 receptors, but expresses rat NKR-P1B(PVG) receptor (previously termed NKR-P1C). The NKR-P1B$^+$ NK subset is inhibited by a non-polymorphic target cell ligand, which is a C-type lectin-related molecule (Clr). Rat Clr molecules appear to be constitutively expressed by hematopoietic cells; expression in tumor cell lines is more variable (Kveberg et al. 2009).

Rat NK Activating and Inhibitory Receptors: While NKR-P1A is an activating receptor on rat NK cells (Giorda et al. 1992), the NKR-P1B receptor (Dissen et al. 1996), in contrast, has inhibitory structural features though it was not possible to prove NKR-P1B-specific inhibitory functions on primary cells since available anti-NKR-P1 mAb failed to distinguish NKR-P1A from NKR-P1B (Kveberg et al. 2006; Li et al. 2003). NKR-P1B functions as an inhibitory receptor because of the presence of an ITIM. Xu et al. (2005) showed that the frequency of CD161$^+$ (human NKR-P1A) NK cells in the epithelia of DA rats was greater than that of WKAH and F344 rats. DA rats have far stronger resistance in the colon to preneoplastic lesion than do other strains. Perhaps CD161$^+$ NK cells play an important role in immune-surveillance at the bottom of the crypt. These observations suggest some allelic variations among NKR-P1 molecules.

A Ly49-negative NK subset is known to selectively express an inhibitory rat NKR-P1 molecule termed NKR-P1C with unique functional characteristics. NKR-P1C$^+$ NK cells efficiently lyse certain tumor target cells, secrete cytokines upon stimulation. However, they specifically fail to kill MHC-mismatched lymphoblast target cells. The NKR-P1C$^+$ NK cell subset al.so appears earlier during development and shows a tissue distribution distinct from its complementary Ly49s3$^+$ subset, which expresses a wide range of Ly49 receptors (Kveberg et al. 2006). The NKR-P1C can act as an autonomous activation structure for NK cell cytotoxicity and cytokine secretion (Ryan et al. 1995). Thus, functionally distinct populations of rat NK cells possess different killer cell lectin-like receptor repertoires (Kveberg et al. 2006).

29.3.2.2 Mouse NKR-P1: Six Distinct Murine *Nkr-p1* Genes

An initial report described three distinct murine *Nkr-p1* genes: *Nkr-p1a*, *Nkr-p1b*, and *Nkr-p1c*. A fourth *Nkr-p1* gene, *Nkr-p1d*, was identified subsequently (Carlyle et al. 1999; Giorda et al. 1992). Later, genomic sequencing identified a fifth functional gene, *Nkr-p1f* (Plougastel et al. 2001a). Analysis of NK cells from different mouse strains revealed that the NK1.1 alloantigen epitope is shared by products of two distinct genes, *Nkr-p1b* and *Nkr-p1c* (Carlyle et al. 1999; Kung et al. 1999). This discovery led to the first demonstration of the inhibitory function of mouse NKR-P1B (Carlyle et al. 1999; Kung et al. 1999) and to the cDNA cloning of a related inhibitory receptor, the B6-derived NKR-P1D (Carlyle et al. 1999; Kung et al. 1999).

29.3 NK Cell Receptor Protein 1 (NKR-P1) or KLRB1

Since then, there have been a total of five NKR-P1 proteins identified to date: three stimulatory receptors, NKR-P1A, NKR-P1C, and NKR-P1F, which possess a charged transmembrane arginine (R) residue thought to be important for association with the FcRγ adaptor protein (Arase et al. 1997; Carlyle et al. 1999; Plougastel et al. 2001a; Ryan and Seaman 1997); and two inhibitory receptors, NKR-P1B and NKR-P1D (Carlyle et al. 2004; Iizuka et al. 2003; Plougastel et al. 2001b), which possess a consensus cytoplasmic ITIM (L/VxYxxL/I/V). The *Nkr-p1* gene cluster is located on NKC on chromosome 6 in mice, syntenic to a similar region on chromosome 4 in rats and chromosome 12 in humans (Yokoyama and Plougastel 2003; Ryan and Seaman 1997).

29.3.2.3 NK1.1 (NKR-P1C)

NK1.1 (mouse NKR-P1C) is a member of a family of NKR NKR-P1 family of disulfide-linked homodimeric type II transmembrane C-type lectin-like receptors initially defined in rats (Ryan and Seaman 1997). First described in 1977 (Glimcher et al. 1977), the original NK1.1 alloantigen has been widely used as a marker to identify NK cells and NKT cells in C57BL/6 (B6) mice. In mice, NK1.1 is expressed on almost all NK cells and on a subset of T cells, NK T cells. Almost all NK1.1$^+$ T cells recognize their TCR ligands in association with the Ag-presenting molecule CD1d, and the majority of NK1.1$^+$ T cells in the thymus, liver, and secondary lymphoid tissues express the invariant Vα14Jα18 TCR (Brigl and Brenner 2004; Godfrey et al. 2004; Kronenberg and Gapin 2002; Taniguchi et al. 2003). Subsets of NK1.1$^+$ T cells are CD4$^+$, CD8$^+$, or CD4$^-$CD8$^-$. In the thymus, blood, and secondary lymphoid tissues, NK1.1$^+$ T cells represent 1–2% of all T cells; in the liver and bone marrow, they represent 20–40% of all T cells (Hammond et al. 2001). Mouse NKR-P1C (NK1.1) is expressed on CD117$^+$ progenitor thymocytes capable of giving rise to cells of the T and NK lineages. NKR-P1C engagement with a mAb leads to IFN-γ production and the directed release of cytotoxic granules from NK cells. The NK2.1 antigen is likely to be anchored in the plasma membrane by a peptide moiety. In addition to be present on a splenic NK cell subset, the NK2.1 antigen was shown to be expressed by a small number of CD^4-CD^8-thymocytes and by a subset of CD^4-CD^8-IgG$^-$ lymph node cells. Finally, unlike NKR-P1, the rat homologue of the murine NK1.1 antigen, neither the NK2.1 nor the NK1.1 antigen is expressed by polymorphonuclear leukocytes.

In contrast to the NKR-P1C$^+$ T cells in mouse liver, the majority of NKR-P1A$^+$ T cells in the human liver are CD8$^+$. Almost all of the NKR-P1A$^+$ T cells in the human liver expressed CD69, suggesting that they were activated. Furthermore, the NKR-P1A$^+$ T cells in the human liver exhibited strong cytotoxicity against a variety of tumor cell lines including K562, Molt4 and some colonic adenocarcinoma cell lines (Ishihara et al. 1999).

Strain-Dependent NK1.1 Alloreactivity of Mouse NK Cells: Historically, NK cells from selected mouse strains have been phenotypically defined using NK-1 alloantigen-specific antisera (Glimcher et al. 1977) or the anti-NK1.1 mAb, PK136 (Koo and Peppard 1984). The NK1.1 alloantigen is now known to identify NK cells from CE, B6, NZB, C58, Ma/My, ST, SJL, FVB, and Swiss outbred mice, but not BALB/c, AKR, CBA, C3H, DBA, or 129 mice (Carlyle et al. 2006). All of *Nkr-p1* genes were reportedly identified in B6 mice, but it is now clear that *Nkr-p1b* is not expressed by B6 mice but is expressed in other strains (Iizuka et al. 2003). In B6 mice, the PK136 mAb binds only NKR-P1C, which can be considered the NK1.1 antigen (Glimcher et al. 1977), but the Ab also binds NKR-P1B in mouse strains that express it (Iizuka et al. 2003).

The underlying molecular basis for the lack of NK1.1 reactivity of NK cells from BALB/c and other mouse strains remains an enigma. The lack of NK1.1 reactivity could be due to deletion of *Nkrp1* genes, defective gene expression, or allelic polymorphism in BALB/c mice. Extreme variation in gene content between the BALB/c and B6 haplotypes has been observed previously for the related *Ly49* gene family (Anderson et al. 2005). This suggests that other NK gene complex regions, including the *Nkrp1-Ocil/Clr* region, also may be subject to rapid evolutionary divergence and/or polymorphism. Since cognate ligands for NKR-P1 have been identified (Carlyle et al. 2004; Iizuka et al. 2003; Kumar and McNerney 2005), a BALB/c defect in NKR-P1 expression could be functionally significant for NK cell function and innate immunity. Determination of the gene content of the BALB/c *Nkrp1-Ocil/Clr* region and the basis of the BALB/c defect in NK1.1 expression could have implications for the importance of the NKR-P1–Ocil/Clr system in self-nonself discrimination in mice and other species (Kumar and McNerney 2005).

To investigate the NK1.1$^-$ phenotype of BALB/c NK cells, Carlyle et al. (2006) studied NK1.1 epitope mapping and genomic analysis of the BALB/c *Nkrp1* region. Analysis reveals that, unlike the *Ly49* region, the *Nkrp1-Ocil/Clr* region displays limited genetic divergence between B6 and BALB/c mice. In fact, significant divergence is confined to the *Nkrp1b* and *Nkrp1c* genes. Strikingly, the B6 *Nkrp1d* gene appears to represent a divergent allele of the *Nkrp1b* gene in BALB/c mice and other strains. Allelic divergence of the *Nkrp1b/c* gene products and limited divergence of the BALB/c *Nkrp1-Ocil/Clr* region explain the strain-specific NK1.1 alloantigen reactivity of mouse NK cells (Carlyle et al. 2006).

Mouse NKR-P1B, a NK1.1 Antigen with Inhibitory Function: The mouse NK1.1 Ag originally defined as NKR-P1C (CD161) mediates NK cell activation. Another member of the mouse CD161 family, NKR-P1B, represents a novel NK1.1 Ag. In contrast to NKR-P1C, which functions as an activating receptor, NKR-P1B inhibits NK cell activation. Association of NKR-P1B with Src homology 2-containing protein tyrosine phosphatase-1 provides a molecular mechanism for this inhibition. The existence of these two NK1.1 Ags with opposite functions suggests a potential role for NKR-P1 molecules, such as those of the Ly-49 gene family, in regulating NK cell function (Carlyle et al. 1999).

29.4 Human NKR-P1A (CD161)

29.4.1 Cellular Localization

CD161 (human NKR-P1A) is a major phenotypic marker of NK and NKT cell types and is thought to be involved in the regulation of functions of these cells. As against six NKR-P1 genes in mice, there appears to be only one gene in humans. Human NKR-P1A is found on a subset of peripheral T cells, including $CD4^+$ and $CD8^+$ T cells, invariant NKT cells, and $\gamma\delta$-TCR^+ T cells, fetal liver T cells and on a subset of $CD3^+$ thymocytes. It is predominantly expressed on T cells with the memory phenotype ($CD45RO^+$) (Lanier 2005; Werwitzke et al. 2003). Comparison of the predicted amino acid of human NKR-P1A with rat and mouse NKR-P1 indicates 46% homology. NKR-P1A is located on human chromosome 12, the syntenic of mouse chromosome 6, where murine NKR-P1 genes are located.

CD161 is type II transmembrane C-type lectin-like receptor and expressed on the cell membrane as disulphide-linked homodimer. In contrast to rodents, no clear functional role has been ascribed to CD161 in humans. Engagement of CD161 on cultured NK cells can result in inhibition of cytotoxicity (Poggi et al. 1999). The functional consequences of treating human NK cells with anti-CD161 mAb are more complex, resulting in no effect, activation, or inhibition, depending on the NK cell population studied (Lanier 2005; Poggi et al. 1997a), although mAb to CD161 has also been reported to trigger proliferation of immature thymocytes (Poggi et al. 1999). These diverse responses elicited by anti-NKR-P1 mAb suggest that additional, functionally distinct, isoforms or alleles of NKR-P1 may exist in humans. Consistent with the latter observation, human NKR-P1A, though not involved in direct activation of NKT cells or their recognition of the CD1d molecule, is an important costimulatory molecule for this population of T cells (Exley et al. 1998). The molecular basis of such widely different functional activities of human CD161 is not clear. Although both human NK cells and a subset of human T cells express the CD56 marker, only a small minority of $CD161^+$ T cells also express CD56 (Loza et al. 2002). Gene encoding the NKR-P1 from cattle showed identity to human nucleotide sequences at 90% and 75%, respectively, and all structural residues of C-type lectin carbohydrate recognition domains were conserved. The identification of two of its members allows to hypothesise the existence of a bovine NK gene complex, prospectively located on chromosome 5 (Govaerts and Goddeeris 2001).

While CD161 is expressed on the minority of human blood T cells, $CD161^+$ T cells expressing either CD4 or CD8 represent the majority of T cells from the epithelial and lamina propria layers of the human duodenum and colon (Ishihara et al. 1999; O'Keeffe et al. 2004). The latter cells contain few, if any, $V\alpha24V\beta11$ invariant NK T cells and secrete IFN-γ and TNF-α without IL-4 after in vitro stimulation. An abundance of $CD161^+$ T cells was also found in the liver and intestinal epithelial cells of the jejunum; the majority of the latter cells were $CD161^+CD8^+$ T cells (Iiai et al. 2002). CD161 is also expressed on human monocytes and dendritic cells. Bone marrow-derived DC of rat uniformly expressed low levels of CD161 and expressed OX62 in a bimodal distribution (Brissette-Storkus et al. 2002). Engagement of CD161 on the cell surface using an appropriate mAb induced the production of IL-1β and IL-12 by monocytes and dendritic cells, respectively (Azzoni et al. 1998; Poggi et al. 1997b). CD161on $CD4^+$ T cells and $\gamma\delta$-TCR^+ T cells has been implicated in transendothelial migration (Poggi et al. 1999). Unexpectedly, the $CD8^+CD161^+$ cells contained an anergic $CD8\alpha^+CD8\beta^{low/-}CD161^{high}$ T cell subset that failed to proliferate, secrete cytokines, or mediate NK lytic activity (Takahashi et al. 2006).

29.4.2 Transcriptional Regulation

Although the 5'-flanking region of NKR-P1C from C57Bl/6 mouse is TATA-less, there is an initiator region and a downstream promoter element, which together form the principal minimal functional promoter. Analyses of the 10-kb 5'-flanking region revealed potential binding sites for regulatory factors, and a single hypersensitive site (HS) of about a 9-kb upstream of the transcriptional initiation site. This site, HS1, could act as a transcriptional enhancer element in a NK cell line. The minimal upstream cis-acting elements point to a complex regulatory mechanism involved in the lineage-specific control of NKR-P1C expression in NK lymphocytes (Ljutic et al. 2003). IL-12, in contrast to IL-2, strongly up-regulates the expression of NKRP1A on human NK cells. The NKRP1A in turn, can regulate the NK cell activation induced via different triggering pathways. This would imply that NKRP1A-mediated functions may be regulatd by the cytokine microenvironment that NK cells may encounter at inflammatory sites (Poggi et al. 1999).

The stochastic expression of individual members of NK cell receptor gene families on subsets of NK cells has attracted considerable interest in the transcriptional

29.4 Human NKR-P1A (CD161)

regulation of these genes. Each receptor gene can contain up to three separate promoters with distinct properties. The recent discovery that an upstream promoter can function as a probabilistic switch element in the Ly49 gene family has revealed a novel mechanism of variegated gene expression. An important question to be answered is whether or not the other NK cell receptor gene families contain probabilistic switches. The promoter elements currently identified in the Ly49, NKR-P1, CD94, NKG2A, and KIR gene families are described (Anderson 2006).

29.4.3 Ligands of CD161/ NKR-P1

29.4.3.1 Binding of Carbohydrates to NKR-P1

A diversity of high-affinity oligosaccharide ligands has been identified for NKR-P1 that contains an extracellular Ca^{2+}-dependent lectin domain. Dimerization of soluble recombinant rat NKR-P1 is predominantly dependent on the presence of an intact juxta-membrane stalk region and independent of N-glycosylation. Aminosugars have a good affinity for the NKR-P1A protein. NKR-P1 is a lectin with a preference order of GalNAc > GlcNAc >> Fuc >> Gal > Man. At neutral pH, Ca^{2+} is tightly associated with the protein. However, NKR-P1 can be decalcified at pH 10 with a total loss of carbohydrate binding. NKR-P1 differs from other calcium-dependent animal lectins in its pattern of monosaccharide recognition and in the tightness of Ca^{2+} binding. NK-resistant tumor cells are rendered susceptible by preincubation with liposomes expressing NKR-P1 ligands, suggesting that purging of tumor or virally infected cells in vivo may be a therapeutic possibility (Bezouska et al. 1994).

While N-acetylD-mannosamine was the best neutral monosaccharide ligand, its participation in the context of an extended oligosaccharide sequence was equally important. The IC_{50} value for the GalNAcβ1 —>ManNAc disaccharide was nearly 10^{-10} M with a further possible increase depending on the type of the glycosidic linkage and the aglycon nature (Krist et al. 2001). GlcNAc and chitooligomers were identified as strong activation ligands for NKR-P1 protein in vitro and in vivo. Their clustering brings about increase of their affinity to the NKR-P1 by 3–6 orders. In analogy with previous observations with GlcNAc clustered on protein or PAMAM backbones the synthetic chitooligomer clusters should provide considerably better ligands in the in vivo antitumor treatment (Semenuk et al. 2001). NKR-P1A also binds to the chitobiose core of uncompletely glycosylated N-linked glycans, and to linear chitooligomers (Bezouska et al. 1997).

Ligands for NKR-P1A include a fully sulphated disaccharide, sucrose octasulphate as observed by NMR spectroscopy and described for the screening of compound libraries for bioaffinities. These findings raise the possibility that NKR-P1A recognises sulphated natural ligands in common with certain other members of the C-type lectin family (Kogelberg et al. 2002). Uncharacterized receptors on human NK cells interact with ligands containing the terminal Galα(1,3)Gal xenoepitope. Among carbohydrate binding proteins from NK cells that bind αGal or N-acetyllactosamine (LacNAc), created by deletion of α1,3galactosyltransferase (GT) in animals is NKRPIA. Moreover, exposing LacNAc by removal of αGal resulted in an increase in binding. This may be relevant in later phases of xenotransplant rejection if $GT^{-/-}$ pigs, like $GT^{-/-}$ mice, display increased LacNAc expression (Christiansen et al. 2006). Structural studies suggested the preference of these receptors for either linear (NKR-P1) or branched (CD69) carbohydrate sequences (Pavlícek et al. 2004). Recombinant soluble form of rat NKR-P1 also recognized carbohydrate GalNac and GlcNac moieties. Ganglioside GM2 and heparin related-IS oligosaccharides representing the high affinity ligands for this receptor, increased the sensitivity of targets for killing by the rat effectors isolated from blood and spleen in vitro. Synthetic three mono- and bivalent LacdiNAc glycomimetics proved to be powerful ligands of NKR-P1 and CD69. A synthetic bivalent tethered di-LacdiNAc is the best currently known precipitation agent for both of these receptors and has promising potential for the development of immunoactive glycodrugs (Bojarová et al. 2009).

29.4.3.2 Ocil/Clr-b as Ligand in Rodents

While ligands for the stimulatory NKR-P1A/C receptors remain elusive, ligands for inhibitory NKR-P1B and NKR-P1D receptors (Carlyle et al. 2004) and activating NKR-P1F receptors are products of the *Ocil/Clr* family of genes (Carlyle et al. 2004; Iizuka et al. 2003; Zhou et al. 2001; Plougastel et al. 2001a), which are intermingled with the *Nkrp1* genes themselves in the NK gene complex. Ocil/Clr-b molecules have been shown to be the ligands for several members of the NKR-P1 family of receptors (Giorda et al. 1992; Ryan and Seaman 1997). The activating NKR-P1F receptor recognizes Clr-g (encoded by *Clec2i*) (Iizuka et al. 2003). The *Clr* genes are interspersed between the *Nkr-p1* genes. The cloning, expression, and function of Ocil/Clr-b as a ligand for the inhibitory NKR-P1B and NKR-P1D receptors have been reported. This suggests that specific receptor-ligand pairs are not inherited separately, but rather en bloc (Iizuka et al. 2003; Carlyle et al. 2004).

The NKR-P1 receptors have been shown to recognize other lectin-like molecules in contrast to other lectin-like NKRs, which recognize class I MHC molecules or relatives of class I MHC molecules. The inhibitory NKR-P1B/D orphan NK cell receptors functionally interact with Ocil/Clr-b ligand and can mediate missing self recognition of tumor cells by mouse NK cells. The findings of Clr ligands have created a renewed interest in the NKR-P1 family, because these ligands and their receptors represent a form of self-nonself discrimination that is independent of MHC class I, a principle that can now be extended to other NK cell receptors (Kumar and McNerney 2005). It is speculated that the NKR-P1 proteins and their Clr ligands may be descended from a common C-type lectin-like ancestral protein that engaged in homophilic interactions. The Ocil/Clr-b is expressed broadly on normal hematopoietic cells but is frequently down-regulated on tumor cells, indicating that NKR-P1B/D receptors play a role in "missing self-recognition" of Ocil/Clr-b. Clr-b is also named as osteoclast inhibitory lectin, because it was also identified as an osteoblast-derived glycoprotein, which inhibits in vitro osteoclastogenesis (Hu et al. 2004; Zhou et al. 2001).

29.4.3.3 LLT1 as Ligand in Humans

Although mice have multiple *Clr* family genes, only one ortholog, *CLEC2D* (also named lectin-like transcript-1 or LLT1), exists in humans. Like mouse Clr-b, human LLT1 blocks osteoclast differentiation (Hu et al. 2004; Zhou et al. 2001). The LLT1 (alternative name, CLEC2D) is a physiologic ligand for NKR-P1A in humans. The human LLT1 is similar to the mouse Clr molecules and is expressed on cells of lymphocytic origin (Boles et al. 1999; Carlyle et al. 2004; Aldemir et al. 2005; Rosen et al. 2005). Human LLT1 is expressed on NK, T, and B cells and localized to the NK gene complex within 100 kb of CD69. In addition to NK and T cells, LLT1 is expressed on TLR-activated plasmacytoid dendritic, TLR-activated monocyte-derived DCs, and on B cells stimulated through TLR9, surface Ig, or CD40. Interactions between NKR-P1A on NK cells and LLT1 on target cells inhibit NK cell-mediated cytotoxicity and cytokine production and can inhibit TNF-α production by TCR-activated NKR-P1A$^+$ CD8$^+$ T cells. In contrast, NKR-P1A failed to inhibit or augment the TCR-dependent activation of NKR-P1A-bearing CD4$^+$ T cells. Expression of LLT1 on activated DCs and B cells suggests that it might regulate the cross-talk between NK cells and APCs (Rosen et al. 2008).

The cDNA encodes a predicted protein of 191 amino acid residues with a transmembrane domain near the N-terminus and an extracellular domain of 132 amino acid residues with similarity to CRD of C-type lectins. The predicted protein of LLT1 shows 59% and 56% similarity to AICL and CD69, respectively. The predicted protein does not contain any intracellular ITIM motifs, suggesting that LLT1 may be involved in mediating activation signals. The human LLT1 shows 43–48% of homology to Clr at the amino acid level and seems to be primarily expressed by monocytes and B cells in peripheral blood (Mathew et al. 2004).

Engagement of CD161 on NK cells with LLT1 expressed on target cells inhibited NK cell-mediated cytotoxicity and IFN-γ secretion. Conversely, LLT1/CD161 interaction in the presence of a TCR signal enhanced IFN-γ production by T cells (Aldemir et al. 2005). The lytic activity of NK cells is negatively regulated via CD161 expression mediated by IL-12 (Azzoni et al. 1998). However, LLT1 is induced rapidly in PMA-stimulated PBMCs (Eichler et al. 2001) and in IL-2-activated NK cells or T cells (Boles et al. 1999). Cross-linking of LLT1 with an Ab induced production of IFN-γ by NK cells (Mathew et al. 2004). LLT1-containing liposomes bind to NKR-P1A$^+$ cells, and binding is inhibited by anti-NKR-P1A mAb. Moreover, LLT1 on target cells can inhibit NK cytotoxicity via interactions with NKR-P1A (Rosen et al. 2005). This interaction inhibits NK cell-mediated cytotoxicity and IFN-γ production while enhancing CD3-triggered IFN-γ production by T cells. It appears that LLT1, expressed by gliomas, contributes to tumor-associated immunosuppression by affecting the lytic activity of NK cells (Roth et al. 2006, 2007).

Oosteoclast Inhibitory Lectin: Osteoclast inhibitory lectin (Ocil) is an inhibitor of osteoclast formation and shows promise as an antiresorptive protein. Murine, rat, human Ocils (mOcil, rOcil, and hOcil respectively) are type II membrane C-type lectins expressed by osteoblasts and other extraskeletal tissues, with the extracellular domain of each, are able to inhibit in vitro osteoclast formation. The hOcil is highly conserved with mOcil in its C-lectin domain, genomic structure, and activity to inhibit osteoclastogenesis (Hu et al. 2004). The hOcil gene predicts a 191 amino acid membrane protein, with the 112 amino acid C-type lectin region in the extracellular domain, having 53% identity with rOcil and mOcil. The hOcil gene is 25 kb in length, comprising five exons, and is a member of a superfamily of NK cell receptors encoded by the NK gene complex located on chromosome 12. Human Ocil mRNA expression is upregulated by IL-1α and prostaglandin E2 in human osteogenic sarcoma MG63 cells, but not by dexamethasone or 1,25 dihydroxyvitamin D3. In addition, Ocil can also inhibit bone resorption by mature, giant-cell tumor-derived osteoclasts (Hu et al. 2004). Ocil is notably localized in bone, skin, and other connective tissues, binds a range of physiologically important glycosaminoglycans, and this property may modulate Ocil actions upon other cells (Gange et al. 2004).

29.4.4 Signaling Pathways

The mechanisms of signaling pathways of CD161 are poorly understood. Different rodent NKR-P1 molecules use different signal transduction pathways to achieve their different functions. The cytoplasmic tail of rodent NKR-P1s has been reported to associate with *Src* family kinases such as p56^{lck}

and various heterotrimeric G proteins (Ljutic et al. 2005), whereas the activating NKR-P1C molecule associates with the γ-chain of the FcR (Arase et al. 1997) and the inhibitory isoform NKR-P1B (expressed by NK cells of SJL/J mice) recruits Src homology protein 1 on cross-linking (Carlyle et al. 1999; Kung et al. 1999).

Target cells expressing class I MHC bound to NKR-P1-expressing NK cells can generate a great variety of intracellular signals that are largely resistant to PTX treatment (Maghazachi et al. 1996). The functional homodimeric form of NKR-P1 selectively binds and activates G_z, G_s, $G_{q/11}$, and G_{i3}, as revealed in [^{35}S]GTP γS binding (Ho and Wong 2001) and to $G\alpha_o$ and $G\alpha_z$ which are also activated in the process of NK cell lysis of allogeneic and tumor target cells (Maghazachi et al. 1996). The activation of NKR-P1 induces IP_3 production, Ca^{2+} flux, interferon-γ secretion, degranulation and cytotoxicity of NK cells (Ho and Wong 2001).

Humans: The cytoplasmic tail of human CD161 (NKR-P1A) does not contain the CxCP/S/T lck-binding motif found in CD4, CD8, and rodent NKR-P1 (Exley et al. 1998; Ljutic et al. 2005). Although it has been reported that human NKR-P1A can be found in association with *Src* family kinases, including p56^{lck} (Cerný et al. 1997), others report a failure to reproduce these observations (Exley et al. 1998). To identify molecules that can interact with the cytoplasmic tail of human CD161 (NKR-P1A), Pozo et al. identified acid sphingomyelinase as a novel intracellular signaling pathway linked to CD161. The mAb-mediated cross-linking of CD161, in both transfectants and primary human NK cells, triggers the activation of acid, but not neutral sphingomyelinase. The sphingomyelinases represent the catabolic pathway for *N*-acyl-sphingosine (ceramide) generation, an emerging second messenger with key roles in the induction of apoptosis, proliferation, and differentiation and define a novel signal transduction pathway for the CD161 in NK and NKT cell biology (Pozo et al. 2006).

Rodents: Cross-linking of the stimulatory rat NKR-P1A or mouse NKR-P1C molecules stimulates phosphatidylinositol turnover and Ca^{2+} flux (Ryan and Seaman 1997), as well as NK cell-mediated cytotoxicity and cytokine production (Arase et al. 1997). Sequence analysis revealed that all murine NKR-P1 proteins possess the Cys-X-Cys-Pro (CxCP) motif also found in the cytoplasmic domains of CD4 and CD8 that mediates association with the Src-related nonreceptor protein tyrosine kinase, p56^{lck}. Campbell and Giorda (1997) demonstrated a physical association between the rat NKR-P1A cytoplasmic CxCP motif and p56^{lck}; however, no functional requirements for this association were shown. A feature of the inhibitory NKR-P1 receptors is the presence of an ITIM in the cytoplasmic domains of mouse NKR-P1B/D. Studies have shown that, like other ITIM-bearing receptors expressed by NK cells, mouse NKR-P1B binds Src homology 2 (SH2)-containing protein tyrosine phosphatase-1 (SHP-1) in a phosphorylation-dependent manner, suggesting a molecular mechanism for the inhibition of NK cell cytotoxicity through NKR-P1B (Carlyle et al. 1999; Kung et al. 1999). The requirement for SHP-1 recruitment to the cytoplasmic ITIM in mediating NKR-P1 inhibition was substantiated by Ljutic et al. (2005). During NKR-P1 signaling in mice, both NKR-P1B and NKR-P1C functionally associate with the tyrosine kinase, p56(lck) p56^{lck}. Mutation at putative Lck-recruitment CxCP motif abolished signal transduction through NKR-P1B/C. Lck appears to be involved in the initiation of NKR-P1 signaling, and SHP-1 in effector function of NKR-P1 receptor (Ljutic et al. 2005).

Ljutic et al. (2005) proposed a model for the initiation and effector signaling through the stimulatory and inhibitory mouse NKR-P1 receptors (Fig. 29.4). Model suggests that the initial phosphorylation of ITIM tyrosine in NKR-P1B is mediated by Lck, providing a docking site for the SH2 domain of the SHP-1 phosphatase, leading in turn to dephosphorylation and inhibition of proximal kinases. In contrast, association of Lck with the stimulatory NKR-P1C receptor leads, upon cross-linking, to transphosphorylation of the cytoplasmic ITAM tyrosines in the FcRγ adaptor protein, in turn leading to recruitment of Syk kinase and activation of downstream second messengers. Thus, findings support a stepwise model for the signaling requirements of the stimulatory and inhibitory NKR-P1 receptors (Fig. 29.4). The identification of cognate ligands for the NKR-P1 receptors (Carlyle et al. 2004; Iizuka et al. 2003) leads to new insight into the physiology of NKR-P1-mediated recognition of target cells by this functionally dichotomous receptor family (Ljutic et al. 2005).

29.4.5 Functions of NKR-P1

The biological function of human CD161 is still insufficiently understood; probably it is involved in regulation of the cytotoxic functions of the cells and in regulation of cytokine production. NKR-P1 molecules are involved in natural killing of certain tumor targets. The loss during differentiation of NKR-P1 and CD2, which are involved in target adhesion and triggering of NK cells, is consistent with the poor cytolytic capacity reported for these cells (Head et al. 1994). The NK1.1 (NKR-P1C) molecule is the most specific serological marker on murine NK cells in C57BL/6 mice. Studies of NKR-P1 have indicated that anti-NKR-P1 mAb induced NK cells to kill otherwise insensitive targets, NK cell phosphoinositol turnover and Ca^{2+} flux. Results demonstrate that immobilized anti-NK1.1 triggers only a subpopulation of NK cells

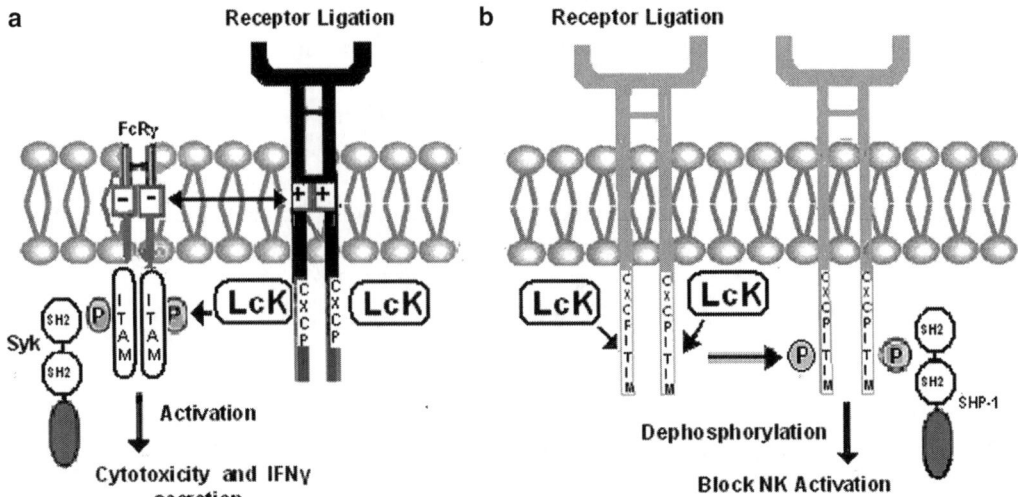

Fig. 29.4 Signaling mechanisms of stimulatory and inhibitory NKR-P1 receptors. (a) Upon receptor cross-linking, Lck associated with cytoplasmic tail of stimulatory NKR-P1C phosphorylates ITAM tyrosines in the FcRγ adaptor protein, in turn leading to recruitment of Syk and activation of downstream second messengers. (b) Phosphorylation of the cytoplasmic ITIM tyrosine in NKR-P1B is mediated by Lck, providing a docking site for the SH2 domain of SHP-1, leading to dephosphorylation and inhibition of proximal kinases

(Yokoyama and Plougastel 2003) and that NK cell responses can be greatly diminished after in vivo treatment with these mAbs (Levik et al. 2001).

The sequencing of the chicken MHC led to the identification of two ORFs, designated B-NK and B-lec that were predicted to encode C-type lectin domains. The genes for B-NK and B-lec are located next to each other in opposite orientations in the chicken MHC. The B-NK is an inhibitory receptor expressed on NK cells, whereas B-lec is an activation-induced receptor with a broader expression pattern. The B-NK and B-lec genes share greatest homology with CTLD receptors encoded by human NK gene complex, in particular NKR-P1 and lectin-like transcript 1 (LLT1), respectively. Like human NKR-P1, B-NK has a functional inhibitory signaling motif in the cytoplasmic tail and is expressed in NK cells. In contrast, B-lec contains an endocytosis motif in the cytoplasmic tail, and like LLT1, is an early activation Ag. B-NK and B-lec are potential candidate genes for the MHC-mediated resistance to MDV. Both glycosylated type II membrane proteins form disulphide-linked homodimers (Rogers and Kaufman 2008; Rogers et al. 2005).

29.5 NKR-P1A in Clinical Disorders

29.5.1 Autoimmune Reactions

Experimental Autoimmune Neuritis: NK cells are implicated in T cell-mediated autoimmune diseases. Experimental autoimmune neuritis (EAN) is a $CD4^+$ T cell-mediated animal model of the Guillain-Barré syndrome in human. Lymphoid cells co-cultured with IL-12 and IL-18 cytokines transferred aggressive clinical and histological EAN into all recipients. NKR-P1$^+$ cells (including NK and NKT cells) played an immunosuppressive function in passive transfer EAN and depletion of NKR-P1$^+$ cells by anti-NKR-P1 Ab and complement induced a more serious form of EAN. Nevertheless, lymphoid cells co-cultured with both IL-12 and IL-18 induced high levels of IFNγ and promoted Th1 differentiation partially through NKR-P1$^+$ cells and to some extent, NKR-P1$^+$ cell depletion inhibited the autoreactivity of lymphoid cells treated with IL-12 and IL-18 (Sun et al. 2007; Yu et al. 2002).

Autoimmune Uveoretinitis: Kitaichi et al. (2002) treated experimental autoimmune uveoretinitis (EAU)-susceptible mice with anti-CD161c antibodies to deplete natural NK cells. Results indicated that the severity of EAU is augmented by NK1.1$^+$ NK cells.

Rheumatic Diseases: The number of both $CD4^-CD8^-$ and $CD4^+$ NKT cells are selectively decreased in the peripheral blood of patients with rheumatic diseases. In addition, both the frequency and the absolute number of $CD161^+CD8^+$ T cells were decreased in the peripheral blood of patients suffering from SLE, MCTD, SSc and PM/DM. Thus there is also an abnormality of NKT cells in the $CD8^+$ population (Mitsuo et al. 2006).

Graves' Disease: To clarify changes in the intra-thyroidal NKT cell subset in patients with Graves' disease (GD),

Watanabe et al. (2008) examined intrathyroidal and peripheral lymphocytes patients and peripheral lymphocytes in healthy volunteers. The proportion of CD161$^+$ T cell receptor Vα24$^+$ Vβ11$^+$ cells was lower in the thyroid of patients than in the peripheral blood of the same patients and in the peripheral blood of healthy subjects. These results indicated that the proportion of intra-thyroidal NKT cells is decreased in patients with GD and that this decrease may contribute to incomplete regulation of autoreactive T cells in GD (Watanabe et al. 2008).

Type 1 Diabetes Mellitus: Invariant NKT cells are considered to be important in some autoimmune diseases including Type 1 diabetes mellitus (T1DM). The published reports are contradictory in regard to the role of iNKT cells in T1DM. Kis et al. (2007) studied iNKT cell frequency and the function of different iNKT cell subgroups in T1DM and suggested that the decrease in the CD4$^+$ population among the iNKT cells and their Th1 shift indicates dysfunction of these potentially important regulatory cells in T1DM.

Asthma: In humans, T cells expressing NKR-P1A constitute around 20% of the circulating CD3$^+$ cells and are potentially immunoregulatory in several diseases. González-Hernández et al. (2007) suggested that during an asthma attack, IFN-γ produced by CD161$^+$ T cells could help to reestablish the Th1/Th2 equilibrium and can be related to asthama.

Atopic Dermatitis: Atopic dermatitis is a chronic inflammatory skin disease associated with cutaneous hyperreactivity to environmental triggers and is often the first step in the atopic march that results in asthma and allergic rhinitis. Helper T cells and their cytokines, in addition to IgE and eosinophils, play a major role in the pathogenesis of atopic dermatitis. NKT cells may play a role in atopic dermatitis status. The reduction of Vα24$^+$CD161$^+$ NKT cells subtypes may be involved in the immunopathogenesis of atopic dermatitis (Ilhan et al. 2007).

29.5.2 Other Diseases

Evidence suggests that NK and NKT cells contribute to inflammation and mortality during septic shock caused by cecal ligation and puncture (CLP). Further studies indicated that NK but not CD1-restricted NKT cells contribute to acute CLP-induced inflammation. NK cells appear to mediate their proinflammatory functions during septic shock, in part, by migration into the peritoneal cavity and amplification of the proinflammatory activities of specific myeloid cell populations. This study provides a new insight into the mechanisms used by NK cells to facilitate acute inflammation during septic shock (Etogo et al. 2008). Selective reduction in the population of colonic mucosal NKR$^+$ T cells may contribute to the development of intestinal inflammation in ulcerative colitis (Shimamoto et al. 2007).

Hepatitis C virus (HCV) causes chronic infection accompanied by a high risk of liver failure. A subset of CD161$^+$CD56$^{+/-}$ NKT cells can recognize glycolipids presented by CD1d and positively or negatively regulate inflammatory responses, implicated in several models of hepatitis. CD1d is expressed at very low levels in the healthy liver, but there is a large fraction of CD161$^+$CD56$^+$ NKT cells. There are high levels of non-classical proinflammatory hepatic CD1d-reactive T cells in HCV infection. Durante-Mangoni et al. (2004) confirmed large numbers of hepatic CD161$^+$ T cells, lower levels of CD56$^+$ T cells, and small numbers of iNKT cells in HCV infection. However, hepatic CD1d-reactivity was not restricted to any of these populations (Durante-Mangoni et al. 2004). CD161 was significantly expressed on HCV-specific cells but not on CD8$^+$ T cells specific for human immunodeficiency virus (3.3%), cytomegalovirus (3.4%), or influenza (3.4%). Northfield et al. (2008) proposed that expression of CD161 indicates a unique pattern of T cell differentiation that might help elucidate the mechanisms of HCV immunity and pathogenesis. Functional capacities and coexpression patterns of lectin-like receptors on lymphocytes are differentially affected in HIV patients depending on the state of therapy or the cell type (NK or T cells), respectively (Jacobs et al. 2004). Reports identify a novel population of human T cells that could contribute to destructive as well as protective immune responses in the liver. It is likely that CD1d-reactive T cells may have distinct roles in different tissues (Exley et al. 2002).

Cytomegaloviruses are known to encode several gene products that function to subvert MHC-dependent immune recognition. Voigt et al. (2007) characterized a rat cytomegalovirus (RCMV) C-type lectin-like (RCTL) gene product with homology to the Clr ligands for the NKR-P1 receptors. RCMV infection rapidly extinguished host Clr-b expression, thereby sensitizing infected cells to killing by NK cells. However, the RCTL protein functioned as a decoy ligand to protect infected cells from NK killing via direct interaction with the NKR-P1B inhibitory receptor. Findings indicate a strategy adopted by cytomegaloviruses to evade MHC-independent self-nonself discrimination. The existence of lectin-like genes in several poxviruses suggests that this may be a common theme for viral evasion of innate immunity (Voigt et al. 2007). The Vδ2$^+$ TCR$\gamma\delta^+$ T lymphocyte subset, expressing the NKRP1a, is expanded in patients with relapsing-remitting multiple sclerosis and uses this molecule to migrate through endothelium. Poggi et al. (2002) suggested that subsets of $\gamma\delta$T lymphocytes may migrate to the site of lesion in multiple sclerosis using two different signaling pathways to extravasate. The mechanism involved in the spontaneous acceptance of liver allografts in some rat strain combinations remains unclear. Immunoregulatory NKR-P1TCRαβT (NKT) cells primarily produce IL-4 and IFN-γ, and enhance the polarization of immune responses to Th2 and Th1, respectively. The role of graft-derived NKT cells in inducing the spontaneous acceptance of rat orthotopic liver transplantation (OLTx) has suggested that graft-derived NKT cells might be responsible for spontaneous acceptance in the rat OLTx (Kiyomoto et al. 2005).

29.5.3 NKR-P1A Receptor (CD161) in Cancer Cells

NK cells were originally defined by their ability to spontaneously eliminate rare cells lacking expression of MHC-I self molecules, which is commonly referred to as "missing self" recognition. The molecular basis for missing self recognition emerges from the expression of MHC-I-specific inhibitory receptors on the NK cell surface that tolerize NK cells toward normal MHC-I-expressing cells. By lacking inhibitory receptor ligands, tumor cells or virus-infected cells that have down-modulated surface MHC-I expression become susceptible to attack by NK cells. Killer cell Ig-like receptors (KIR; CD158) and Ig-like transcripts (ILTs), CD94/NKG2 heterodimers, NKG2D, NKR-P1, and KLRG1 belonging to the C-type lectin receptor family, called killer cell lectin-like receptors (KLR approved symbol) (Borrego et al. 2001; Vetter et al. 2000). constitute a family of MHC-I binding receptors that play major roles in regulating the activation thresholds of NK cells and some T cells in humans. Although the function of NKR on T cells infiltrating tumors and their potential effect on antitumor immunity has been investigated, little is known about T cells expressing CD161 in cancer patients. CD4$^+$CD161$^+$CD56$^-$ cells represent a distinct memory T-cell population significantly increased in cancer patients. Depending on the type of signals provided by the tumor microenvironment, CD4$^+$CD161$^+$ cells may regulate the immune response (Iliopoulou et al. 2006).

Cutaneous malignant melanoma is one of the most immunogenic tumors. CD8 T-lymphocytes and NK cells are believed to be important effector cells involved in eliciting a protection against melanoma (Lozupone et al. 2004). The early stages of cutaneous primary melanoma are evidenced by changes in the number of CD8$^+$DR$^+$ or CD8$^+$CD161$^+$ T-cells with respect to healthy individuals. In addition, the number of CD56 NK cells was also increased at early disease stages. Campillo et al. (2006) provided evidence of immune activation in early stages of cutaneous melanoma, together with an increase of cells expressing CD158a in patients bearing the corresponding HLA-C ligand. Low expression of CD161 and activating NKG2D receptors is associated with a significant impairment in NK cell activity in metastatic melanoma (MM) patients (Konjević et al. 2007).

Glioblastoma is the most malignant intrinsic brain tumor. The activating NK and costimulatory T-cell ligands that are expressed by glioma cells are overridden by several inhibitory signals, including the immunosuppressive molecule TGF-ß (Friese et al. 2004). Glioma cells showing high surface expression of classic MHC class I molecules and a significant expression of the nonclassic MHC molecules HLA-E and HLA-G immune escape phenotype by interacting with different NK cell receptors (Lanier 2005). A tolerance factor is another molecule that may help immune escape of gliomas by inhibition of NK and T cells. Glioma cells express LLT1 mRNA and protein in vitro and in vivo, whereas expression levels in normal brain are low. It appears that LLT1, expressed by gliomas, contributes to tumor-associated immunosuppression by affecting the lytic activity of NK cells (Roth et al. 2006, 2007). LLT1 expression in human gliomas increases with the WHO grade of malignancy. TGF-β up-regulates the expression of LLT1 in glioma cells. Small iRNA-mediated down-regulation of LLT1 in LNT-229 and LN-428 cells promotes their lysis by NK cells. Thus, LLT1 acts as a mediator of immune escape and contributes to the immunosuppressive properties of glioma cells (Roth et al. 2007). Co-culture of NK cells and DCs results in their reciprocal co-activation, and an enhancement of lysis of tumor target cells (Yang et al. 2006).

In rat 9L gliosarcoma, a substantial number of cells express NKR-P1, a marker expressed only on rat lymphocytes capable of non-MHC-restricted cytotoxicity. Previous investigations have determined the existence of three populations of NKR-P1$^+$ lymphocytes in normal rats, including NKR-P1$_{bright}$/T-cell receptor (TCR)$^-$/CD3-/CD5$^-$ (~5–15%), NKR-P1$_{dim}$/TCRβ$^+$/CD3$^+$/CD5$^+$ (1–5%), and NKR-P1$_{dim}$/TCR γδ/CD3$^+$/CD5$^+$ (~0.5–2%). It was suggested that there is selective localization of cells capable of mediating antitumor responses in 9L, but that tumor-associated factors may down-regulate their function and expression of NKR-P1 (Chambers et al. 1996).

Because the activity of cytotoxic cells is suppressed in most cancer patients, it was suggested that the *Klrb1* expression might be suppressed in cancerous cells. The transcription of the KLRB1 was suppressed in tumor tissues in 68% patients with nonsmall-lung-cancer and 57% patients with esophageal squamous-cell carcinoma. This parameter can be a marker of lung and esophageal cancers (Pleshkan et al. 2007).

29.6 Human Early Activation Antigen (CD69)

CD69 belongs to a subfamily of CTLDs, which are referred to as NK-cell domains (NKDs). Its gene maps in the NK gene complex, close to other genes coding for NK receptors.

29.6 Human Early Activation Antigen (CD69)

CD69 is a dimeric glycoprotein of 33 and 27 kDa. The CD69 has a structural homology with other type II lectin cell surface receptors, such as T cell antigen Ly49, the low avidity IgE receptor (CD23), and the hepatic asialoglycoprotein receptors. The CD69, one of the first described members of the NKC family of receptors (López-Cabrera et al. 1993; Ziegler et al. 1994), is present at the cell surface as a disulfide linked homodimer, with subunits of 28 and 32 kDa resulting from the differential glycosylation at a single extracellular N-linked glycosylation site. Contrary to other NKC gene products, whose expression is restricted to NK cells, CD69 has been found on the surface of most hematopoietic lineages. It is one of the earliest markers induced upon activation in T and B lymphocytes, NK cells, macrophages, neutrophils, and eosinophils. In addition, it is constitutively expressed on monocytes, platelets, Langerhans cells, and a small percentage of resident lymphocytes in thymus and secondary lymphoid tissues (Sánchez-Mateos et al. 1989). The activation of T lymphocytes induces the expression of CD69. It functions as a signal transmitting receptor in lymphocytes, NK cells, and platelets. Although CD69 is absent from peripheral blood resting lymphocytes, it is expressed by in vivo activated lymphocytes infiltrating sites of chronic inflammation in several pathologies, as well as by lymphocytes after in vitro activation with different stimuli. The TNF-α gene expression and protein secretion could be induced in peripheral blood T cells through CD69 molecule. CD69-deficient mice revealed its modulatory role on B cell development and antibody synthesis (Lauzurica et al. 2000).

29.6.1 Organization of CD69 Gene

The CD69 gene mapped on chromosome 12 p13-p12. The cDNA coding for CD69 exhibited a single open reading frame of 597 bp coding and predicted a 199-amino acid protein of type II membrane topology, with extracellular (COOH-terminal), transmembrane, and intracellular domains. The CD69 clone hybridized to a 1.7-kb mRNA species, which was rapidly induced and degraded after lymphocyte stimulation, consistent with the presence of rapid degradation signals at the 3′ untranslated region. The CD69 protein shares functional characteristics with other members of this superfamily, which act as transmembrane signaling receptors in early phases of cellular activation (López-Cabrera et al. 1993)

The genomic structure of the human gene encoding CD69 sequence is divided into five exons separated by four introns. The first two exons corresponded to separate functional domains: cytoplasmic tail and the transmembrane region. The final three exons encoded the CRD. The major transcription initiation site has been located 30 nt downstream of a consensus TATA box (Santis et al. 1994). The conserved intron position between the exons encoding the CRD indicated that this protein is closely related to other type-II C-type lectins, such as the asialoglycoprotein receptor, the CD23, and NKR-P1 and Ly49. In contrast to the broad NKR-P1 and Ly-49 gene families, CD69 is a single-copy gene. The mouse CD69 gene has phorbol ester-inducible promoter element within the first 700 bp upstream of the start of transcription. Chromosomal mapping placed the mouse CD69 gene on the long arm of chromosome 6 near the NK gene complex that contains the related NKR-P1 and Ly-49 gene families. The human CD69 gene mapped to chromosome 12p13 near the related NKG2 gene cluster and in a region associated with rearrangements in approximately 10% of cases of childhood acute lymphocytic leukemia (Ziegler et al. 1994). Recombinant CD69 protein exists as a disulfide-linked homodimer on the cell surface and crystallizes as a symmetrical dimer, similar to those formed by the related NK cell receptors Ly49A and CD94.

The 5′-flanking region of the promoter of CD69 gene contains a potential TATA element 30 bp upstream of the major transcription initiation site and several putative binding sequences for inducible transcription factors (NF-kB, Egr-1, AP-1), which might mediate the inducible expression of this gene. The proximal promoter region spans positions −78 to +16 containing the cis-acting sequences necessary for basal and phorbol 12-myristate 13-acetate-inducible transcription of CD69 gene. Removal of the upstream sequences located between positions −78 and −38 resulted in decreased promoter strength and abolished the response to phorbol 12-myristate 13-acetate. The TNF-α is capable of inducing the surface expression of CD69 molecule as well as the promoter activity of fusion plasmids that contain 5′-flanking sequences of *CD69* gene, suggesting that TNF may regulate in vivo the expression of CD69 (López-Cabrera et al. 1995).

29.6.2 Src-Dependent Syk Activation of CD69-Mediated Signaling

CD69 engagement leads to the rapid and selective activation of the tyrosine kinase Syk, but not of the closely related member of the same family, ZAP70, in IL-2-activated human NK cells. The requirement for Src family kinases in the CD69-triggered activation of Syk suggests a role for Lck in this event. It was demonstrated that Syk and Src family tyrosine kinases control the CD69-triggered tyrosine phosphorylation and activation of phospholipase Cγ2 and the Rho family-specific exchange factor Vav1 and are responsible for CD69-triggered cytotoxicity of activated NK cells. Thus CD69 receptor functionally couples to the activation of Src family tyrosine kinases, which, by inducing Syk activation, initiate downstream signaling pathways, induce rise in intracellular Ca^{2+}. Synthesis of different cytokines and/or proliferation regulates CD69-triggered functions on human NK cells (Pisegna et al. 2002; Llera et al. 2001). The receptor-proximal signaling pathways activated by CD69 cross-linking are involved in CD69-mediated cytotoxic activity. CD69 engagement leads to the activation of ERK enzymes belonging to the MAPK family, and that this event is required for CD69-mediated cell degranulation. The co-engagement of CD94/NKG2-A inhibitory receptor effectively suppressed both CD69-triggered cell degranulation in RBL transfectants, through the inhibition of ERK activation, and CD69-induced cytotoxicity in human NK cells. Thus, CD69-initiated signaling pathways and functional activity are negatively regulated by CD94/NKG2-A inhibitory complex (Llera et al. 2001). The wide distribution of CD69, along with its activating signal-transducing properties, suggest an important role of CD69 in the physiology of leukocyte activation (Llera et al. 2001)

29.6.3 Crystal Analysis of CD69

The structure of extracellular portion of NK cell receptors are divergent from true C-type lectins and are referred to as NK-cell domains (NKDs). CD69 NKD adopts the canonical CTLD fold but lacks the features involved in Ca^{2+} and carbohydrate binding by C-type lectins. The NKD of human CD69 at 2.27 A resolution reveals conservation of the C-type lectin-like fold, including preservation of the two α-helical regions found in Ly49A and MBP. Using comparative computer modeling, a 3-D model of the extracellular domain of CD69 based on the crystal structure of the MBL was generated. The sequence of CD69 appears to be highly compatible with the C-type lectin fold. Compared with MBL and selectins, CD69 displays significant deletions in loop regions. The conserved calcium binding sites found in the C-type lectin family are not conserved in CD69; only one of the nine residues coordinated to Ca^{2+} in MBP is conserved in CD69 and no bound Ca^{2+} is evident in the crystal structure. In this respect, CD69 departs from some of the conserved motifs seen in crystal structures of MBL and selectins. The CD69 model shows cavity-shaped hydrophobic regions surrounded by charged residues. One of these cavities is proximal to a potential low affinity calcium binding site and may be implicated in specific interactions with ligands. Surprisingly, electron density suggestive of a puckered six-membered ring was discovered at a site structurally analogous to the ligand-binding sites of MBP and Ly49A. This sugar-like density may represent, or mimic, part of the natural ligand recognized by CD69 (Bajorath and Aruffo 1994; Natarajan et al. 2000). CD69 NKD dimerizes noncovalently, both in solution and in crystalline state. The dimer interface consists of a hydrophobic, loosely packed core, surrounded by polar interactions, including an interdomain β-sheet. The intersubunit core shows certain structural plasticity that may facilitate conformational rearrangements for binding to ligands. The surface equivalent to the binding site of other members of the CTLD superfamily reveals a hydrophobic patch surrounded by conserved charged residues that probably constitutes the CD69 ligand-binding site (Llera et al. 2001).

References

Aldemir H, Prod'homme V, Dumaurier MJ et al (2005) Cutting edge: lectin-like transcript 1 is a ligand for the CD161 receptor. J Immunol 175:7791–7795

Anderson SK, Dewar K, Goulet ML et al (2005) Complete elucidation of a minimal class I MHC natural killer cell receptor haplotype. Genes Immun 6:481–492

Anderson SK, Ortaldo JR, McVicar DW (2001) The ever-expanding Ly49 gene family: repertoire and signaling. Immunol Rev 181:79–89

Anderson SK (2006) Transcriptional regulation of NK cell receptors. Curr Top Microbiol Immunol 298:59–75

Arase N, Arase H, Park S et al (1997) Association with FcRγ is essential for activation signal through NKR-P1 (CD161) in natural killer (NK) cells and NK. 1.1^+ T cells. J Exp Med 186: 1957–1963

Averdam A, Kuhl H, Sontag M et al (2007) Genomics and diversity of the common marmoset monkey NK complex. J Immunol 178:7151–7161

Azzoni L, Zatsepina O, Abebe B et al (1998) Differential transcriptional regulation of CD161 and a novel gene, 197/15a, by IL-2, IL-15, and IL-12 in NK and T cells. J Immunol 161:3493–3500

Bajorath J, Aruffo A (1994) Molecular model of the extracellular lectin-like domain in CD69. J Biol Chem 269:32457–32463

Bezouska K, Sklenár J, Dvoráková J et al (1997) NKR-P1A protein, an activating receptor of rat natural killer cells, binds to the chitobiose core of uncompletely glycosylated N-linked glycans, and to linear chitooligomers. Biochem Biophys Res Commun 238:149–153

Bezouska K, Yuen CT, O'Brien J et al (1994) Oligosaccharide ligands for NKR-P1 protein activate NK cells and cytotoxicity. Nature 372 (6502):150–157

Bojarová P, Krenek K, Wetjen K et al (2009) Synthesis of LacdiNAc-terminated glycoconjugates by mutant galactosyltransferase-a way to new glycodrugs and materials. Glycobiology 19:509–519

Boles KS, Barten R, Kumaresan PR et al (1999) Cloning of a new lectin-like receptor expressed on human NK cells. Immunogenetics 50:1–7

Borrego F, Kabat J, Kim D-K et al (2001) Structure and function of major histocompatibility complex (MHC) class I specific receptors expressed on human natural killer (NK) cells. Mol Immunol 38:637–660

Brigl M, Brenner MB (2004) CD1: antigen presentation and T cell function. Annu Rev Immunol 22:817–890

Brissette-Storkus C, Kaufman CL, Pasewicz L et al (1994) Characterization and function of the NKR-P1dim/T cell receptor-α β+ subset of rat T cells. J Immunol 152:388–396

Brissette-Storkus CS, Kettel JC, Whitham TF et al (2002) Flt-3 ligand (FL) drives differentiation of rat bone marrow-derived dendritic cells expressing OX62 and/or CD161 (NKR-P1). J Leukoc Biol 71:941–949

Campbell KS, Giorda R (1997) The cytoplamic domain of rat NKR-P1 receptor interacts with the N-terminal domain of p56^{lck} via cysteine residues. Eur J Immunol 27:72–77

Campillo JA, Martínez-Escribano JA, Muro M et al (2006) HLA class I and class II frequencies in patients with cutaneous malignat melanoma from southeastern Spain: the role of HLA-C in disease prognosis. Immunogenetics 57:926–933

Carlyle J, Mesci A, Ljutic B et al (2006) Molecular and genetic basis for strain-dependent NK1.1 alloreactivity of mouse NK cells. J Immunol 176:7511–7524

Carlyle J, Martin A, Mehra A et al (1999) Mouse NKR-P1B, a novel NK1.1 antigen with inhibitory function. J Immunol 162:5917–5923

Carlyle JR, Jamieson AM, Gasser S et al (2004) Missing self-recognition of Ocil/Clr-b by inhibitory NKR-P1 natural killer cell receptors. Proc Natl Acad Sci USA 101:3527–3532

Carlyle JR, Mesci A, Fine JH (2008) Evolution of the Ly49 and Nkrp1 recognition systems. Semin Immunol 20:321–330

Cerný J, Fiserová A, Horváth O et al (1997) Association of human NK cell surface receptors NKR-P1 and CD94 with Src-family protein kinases. Immunogenetics 46:231–236

Chambers WH, Bozik ME, Brissette-Storkus SC et al (1996) NKR-P1$^+$ cells localize selectively in Rat 9 L gliosarcomas but have reduced cytolytic function. Cancer Res 56:3516–3525

Chambers WH, Brumfield AM, Hanley-Yanez K et al (1992) Functional heterogeneity between NKR-P1bright/Lycopersicon esculentum lectin (L.E.)bright and NKR-P1bright/L.E.dim subpopulations of rat natural killer cells. J Immunol 148:3658–3665

Christiansen D, Mouhtouris E, Milland J et al (2006) Recognition of a carbohydrate xenoepitope by human NKRP1A (CD161). Xenotransplantation 13:440–446

Dam J, Baber J, Grishaev A et al (2006) Variable dimerization of the Ly49A natural killer cell receptor results in differential engagement of its MHC class I ligand. J Mol Biol 362:102–113

Dam J, Guan R, Natarajan K et al (2003) Variable MHC class I engagement by Ly49 natural killer cell receptors demonstrated by the crystal structure of Ly49C bound to H-2K^b. Nat Immunol 4:1213–1222

Deng L, Cho S, Malchiodi EL et al (2008) Molecular architecture of the major histocompatibility complex class I-binding site of Ly49 natural killer cell receptors. J Biol Chem 283:16840–16849

Deng L, Mariuzza RA (2006) Structural basis for recognition of MHC and MHC-like ligands by natural killer cell receptors. Semin Immunol 18:159–166

Dimasi N, Biassoni R (2005) Structural and functional aspects of the Ly49 natural killer cell receptors. Immunol Cell Biol 83:1–8

Dimasi N, Moretta L, Biassoni R (2004) Structure of the Ly49 family of natural killer (NK) cell receptors and their interaction with MHC class I molecules. Immunol Res 30:95–104

Dimasi N, Sawicki MW, Reineck LA et al (2002) Crystal structure of the Ly49I natural killer cell receptor reveals variability in dimerization mode within the Ly49 family. J Mol Biol 320:573–585

Dissen E, Ryan JC, Seaman WE, Fossum S (1996) An autosomal dominant locus, Nka, mapping to the Ly-49 region of a rat natural killer (NK) gene complex, controls NK cell lysis of allogeneic lymphocytes. J Exp Med 183:2197–2207

Durante-Mangoni E, Wang R, Shaulov A et al (2004) Hepatic CD1d expression in hepatitis C virus infection and recognition by resident proinflammatory CD1d-reactive T cells. J Immunol 173:2159–2166

Eichler W, Ruschpler P, Wobus M, Drossler K (2001) Differentially induced expression of C-type lectins in activated lymphocytes. J Cell Biochem 81:201–208

Etogo AO, Nunez J, Lin CY et al (2008) NK but not CD1-restricted NKT cells facilitate systemic inflammation during polymicrobial intra-abdominal sepsis. J Immunol 180:6334–6345

Exley M, Porcelli S, Furman M et al (1998) CD161 (NKR-P1A) costimulation of CD1d-dependent activation of human T cells expressing invariant Vα 24 Jα Q T cell receptor αchains. J Exp Med 188:867–876

Exley MA, He Q, Cheng O et al (2002) Cutting edge: compartmentalization of Th1-like noninvariant CD1d-reactive T cells in hepatitis C virus-infected liver. J Immunol 168:1519–1523

Fernandez NC, Treiner E, Vance RE et al (2005) A subset of natural killer cells achieves self-tolerance without expressing inhibitory receptors specific for self-MHC molecules. Blood 105:4416–4423

French AR, Yokoyama WM (2004) Natural killer cells and autoimmunity. Arthritis Res Ther 6:8–14

Friese MA, Wischhusen J, Wick W et al (2004) RNA interference targeting transforming growth factor-ß enhances NKG2D-mediated antiglioma immune response, inhibits glioma cell migration and invasiveness, and abrogates tumorigenicity in vivo. Cancer Res 64:7596–7603

Gange CT, Quinn JM, Zhou H et al (2004) Characterization of sugar binding by osteoclast inhibitory lectin. J Biol Chem 279:29043–29049

Giorda R, Weisberg EP, Ip TK, Trucco M (1992) Genomic structure and strain-specific expression of the natural killer cell receptor NKR-P1. J Immunol 149:1957–1963

Glimcher L, Shen FW, Cantor H (1977) Identification of a cell-surface antigen selectively expressed on the natural killer cell. J Exp Med 145:1–9

Godfrey DI, MacDonald HR, Kronenberg M et al (2004) NKT cells: what's in a name? Nat Rev Immunol 4:231–237

Godfrey DI, Kronenberg M (2004) Going both ways: immune regulation via CD1d-dependent NKT cells. J Clin Invest 114:1379–1388

González-Hernández Y, Pedraza-Sánchez S, Blandón-Vijil V et al (2007) Peripheral blood CD161$^+$ T cells from asthmatic patients are activated during asthma attack and predominantly produce IFN-$^\gamma$. Scand J Immunol 65:368–375

Govaerts MM, Goddeeris BM (2001) Homologues of natural killer cell receptors NKG2-D and NKR-P1 expressed in cattle. Vet Immunol Immunopathol 80:339–344

Hammond KJ, Pellicci DG, Poulton LD et al (2001) CD1d-restricted NKT cells: an interstrain comparison. J Immunol 167:1164–1173

Head JR, Kresge CK, Young JD, Hiserodt JC (1994) NKR-P1$^+$ cells in the rat uterus: granulated metrial gland cells are of the natural killer cell lineage. Biol Reprod 51:509–523

Ho MK, Wong YH (2001) G_z signaling: emerging divergence from G_i signaling. Oncogene 20:1615–1625

Hu YS, Zhou H, Myers D et al (2004) Isolation of a human homolog of osteoclast inhibitory lectin that inhibits the formation and function of osteoclasts. J Bone Miner Res 19:89–99

Iiai T, Watanabe H, Suda T et al (2002) CD161$^+$ T (NT) cells exist predominantly in human intestinal epithelium as well as in liver. Clin Exp Immunol 129:92–98

Iizuka K, Naidenko OV, Plougastel BFM et al (2003) Genetically linked C-type lectin-related ligands for the NKRP1 family of natural killer cell receptors. Nat Immunol 4:801–807

Ilhan F, Kandi B, Akbulut H et al (2007) Atopic dermatitis and Vα24$^+$ natural killer T cells. Skinmed 6:218–220

Iliopoulou EG, Karamouzis MV, Missitzis I et al (2006) Increased frequency of CD4$^+$ cells expressing CD161 in cancer patients. Clin Cancer Res 12:6901–6909

Ishihara S, Nieda M, Kitayama J et al (1999) CD8$^+$NKR-P1A$^+$ T cells preferentially accumulate in human liver. Eur J Immunol 29:2406–2413

Jacobs R, Weber K, Wendt K et al (2004) Altered coexpression of lectin-like receptors CD94 and CD161 on NK and T cells in HIV patients. J Clin Immunol 24:281–286

Khalturin K, Becker M, Rinkevich B, Bosch TC (2003) Urochordates and the origin of natural killer cells: identification of a CD94/NKR-P1-related receptor in blood cells of Botryllus. Proc Natl Acad Sci USA 100:622–627

Kis J, Engelmann P, Farkas K et al (2007) Reduced $CD4^+$ subset and Th1 bias of the human iNKT cells in Type 1 diabetes mellitus. J Leukoc Biol 81:654–662

Kitaichi N, Kotake S, Morohashi T et al (2002) Diminution of experimental autoimmune uveoretinitis (EAU) in mice depleted of NK cells. J Leukoc Biol 72:1117–1121

Kiyomoto T, Ito T, Uchikoshi F et al (2005) The potent role of graft-derived $NKR-P1^+TCR\alpha\beta^+$ T (NKT) cells in the spontaneous acceptance of rat liver allografts. Transplantation 80: 1749–1755

Kogelberg H, Frenkiel TA, Birdsall B et al (2002) Binding of sucrose octasulphate to the C-type lectin-like domain of the recombinant natural killer cell receptor NKR-P1A observed by NMR spectroscopy. Chembiochem 3:1072–1077

Konjević G, Mirjačić Martinović K, Vuletić A et al (2007) Low expression of CD161 and NKG2D activating NK receptor is associated with impaired NK cell cytotoxicity in metastatic melanoma patients. Clin Exp Metastasis 24:1–11

Koo GC, Peppard JR (1984) Establishment of monoclonal anti-NK-1.1 antibody. Hybridoma 3:301–303

Krist P, Herkommerová-Rajnochová E, Rauvolfová J et al (2001) Toward an optimal oligosaccharide ligand for rat natural killer cell activation receptor NKR-P1. Biochem Biophys Res Commun 287:11–20

Kronenberg M, Gapin L (2002) The unconventional life style of NKT cells. Nat Rev Immunol 2:557–568

Kumar V, McNerney ME (2005) A new self: MHC-class-I-independent natural-killer-cell self-tolerance. Nat Rev Immunol 5:363–374

Kung SKP, Su R-C, Shannon J, Miller RG (1999) The NKR-P1B gene product is an inhibitory receptor on SJL/J NK cells. J Immunol 162:5876–5887

Kveberg L, Bäck CJ, Dai KZ et al (2006) The novel inhibitory NKR-P1C receptor and Ly49s3 identify two complementary, functionally distinct NK cell subsets in rats. J Immunol 176:4133–4140

Kveberg L, Dai KZ, Westgaard IH et al (2009) Two major groups of rat NKR-P1 receptors can be distinguished based on chromosomal localization, phylogenetic analysis and Clr ligand binding. Eur J Immunol 39:541–551

Lanier LL (2005) NK cell recognition. Annu Rev Immunol 23:225–274

Lauzurica P, Sancho D, Torres M et al (2000) Phenotypic and functional characteristics of hematopoietic cell lineages in CD69-deficient mice. Blood 95:2312–2320

Lebbink RJ, Meyaard L (2007) Non-MHC ligands for inhibitory immune receptors: novel insights and implications for immune regulation. Mol Immunol 44:2153–2164

Lee SH, Zafer A, de Repentigny Y et al (2003) Transgenic expression of the activating natural killer receptor Ly49H confers resistance to cytomegalovirus in genetically susceptible mice. J Exp Med 197:515–526

Levik G, Vaage JT, Rolstad B, Naper C (2001) The effect of in vivo depletion of $NKR-P1^+$ or $CD8^+$ lymphocytes on the acute rejection of allogeneic lymphocytes (ALC) in the rat. Scand J Immunol 54:341–347

Li J, Rabinovich BA, Hurren R et al (2003) Expression cloning and function of the rat NK activating and inhibitory receptors NKR-P1A and -P1B. Int Immunol 15:411–416

Ljutic B, Carlyle JR, Filipp D et al (2005) Functional requirements for signaling through the stimulatory and inhibitory mouse NKR-P1 (CD161) NK cell receptors. J Immunol 174:4789–4796

Ljutic B, Carlyle JR, Zúñiga-Pflücker JC (2003) Identification of upstream cis-acting regulatory elements controlling lineage-specific expression of the mouse NK cell activation receptor, NKR-P1C. J Biol Chem 278:31909–31919

Llera AS, Viedma F, Sánchez-Madrid F, Tormo J (2001) Crystal structure of the C-type lectin-like domain from the human hematopoietic cell receptor CD69. J Biol Chem 276:7312–7319

López-Cabrera M, Muñoz E, Blázquez MV et al (1995) Transcriptional regulation of the gene encoding the human C-type lectin leukocyte receptor AIM/CD69 and functional characterization of its tumor necrosis factor-alpha-responsive elements. J Biol Chem 270:21545–21551

Lopez-Cabrera M, Santis AG, Fernandez-Ruiz E et al (1993) Molecular cloning, expression, and chromosomal localization of the human earliest lymphocyte activation antigen AIM/CD69, a new member of the C-type animal lectin superfamily of signal-transmitting receptors. J Exp Med 178:537–547

Loza MJ, Metelitsa LS, Perussia B (2002) NKT and T cells: coordinate regulation of NK-like phenotype and cytokine production. Eur J Immunol 32:3453–3462

Lozupone F, Pende D, Burgio VL et al (2004) Effect of human natural killer and $\gamma\delta$T-cells on the growth of human autologous melanoma xenografts in SCID mice. Cancer Res 64:378–385

Mager DL, McQueen KL, Wee V, Freeman JD (2001) Evolution of natural killer cell receptors: coexistence of functional Ly49 and KIR genes in baboons. Curr Biol 11:626–630

Maghazachi AA, Al-Aoukaty A, Naper C et al (1996) Preferential involvement of Go and Gz proteins in mediating rat natural killer cell lysis of allogeneic and tumor target cells. J Immunol 157:5308–5314

Mathew PA, Chuang SS, Vaidya SV et al (2004) The LLT1 receptor induces IFN-γ production by human natural killer cells. Mol Immunol 40:1157–1163

Mesci A, Ljutic B, Makrigiannis AP, Carlyle JR (2006) NKR-P1 biology: from prototype to missing self. Immunol Res 35:13–26

Mitsuo A, Morimoto S, Nakiri Y et al (2006) Decreased $CD161^+CD8^+$ T cells in the peripheral blood of patients suffering from rheumatic diseases. Rheumatology (Oxford) 45:1477–1484

Moretta L, Moretta A (2004) Killer immunoglobulin-like receptors. Curr Opin Immunol 16:626–633

Natarajan K, Dimasi N, Wang J et al (2002) Structure and function of natural killer cell receptors: multiple molecular solutions to self, nonself discrimination. Annu Rev Immunol 20:853–885

Natarajan K, Sawicki MW, Margulies DH, Mariuzza RA (2000) Crystal structure of human CD69: a C-type lectin-like activation marker of hematopoietic cells. Biochemistry 39:14779–14786

Northfield JW, Kasprowicz V, Lucas M et al (2008) CD161 expression on hepatitis C virus-specific $CD8^+$ T cells suggests a distinct pathway of T cell differentiation. Hepatology 47:396–406

O'Keeffe J, Doherty DG, Kenna T et al (2004) Diverse populations of T cells with NK cell receptors accumulate in the human intestine in health and in colorectal cancer. Eur J Immunol 34:2110–2119

Ortaldo JR, Young HA (2003) Expression of IFN-γ upon triggering of activating Ly49D NK receptors in vitro and in vivo: costimulation with IL-12 or IL-18 overrides inhibitory receptors. J Immunol 170:1763–1769

Ortaldo JR, Young HA (2005) Mouse Ly49 NK receptors: balancing activation and inhibition. Mol Immunol 42:445–450

Pavlícek J, Kavan D, Pompach P et al (2004) Lymphocyte activation receptors: new structural paradigms in group V of C-type animal lectins. Biochem Soc Trans 32:1124–1126

Pisegna S, Zingoni A, Pirozzi G et al (2002) Src-dependent Syk activation controls CD69-mediated signaling and function on human NK cells. J Immunol 169:68–74

Pleshkan VV, Zinov'eva MV, Vinogradova TV, Sverdlov ED (2007) Transcription of the KLRB1 gene is suppressed in human cancer tissues. Mol Gen Mikrobiol Virusol 4:3–7, Article in Russian

Plougastel B, Dubbelde C, Yokoyama WM (2001a) Cloning of Clr, a new family of lectin-like genes localized between mouse Nkrp1a and Cd69. Immunogenetics 53:209–214

Plougastel B, Matsumoto K, Dubbelde C, Yokoyama WM (2001b) Analysis of a 1-Mb BAC contig overlapping the mouse Nkrp1 cluster of genes: cloning of three new Nkrp1 members, Nkrp1d, Nkrp1e, and Nkrp1f. Immunogenetics 53:592–598

Plougastel BF, Yokoyama WM (2006) Extending missing-self? Functional interactions between lectin-like NKrp1 receptors on NK cells with lectin-like ligands. Curr Top Microbiol Immunol 298:77–89

Poggi A, Zocchi MR, Costa P et al (1999) IL-12-mediated NKRP1A up-regulation and consequent enhancement of endothelial transmigration of V$\delta 2^+$ TCR $\gamma\delta^+$ T lymphocytes from healthy donors and multiple sclerosis patients. J Immunol 162:4349–4354

Poggi A, Costa P, Zocchi MR, Moretta L (1997a) NKRP1A molecule is involved in transendothelial migration of $CD4^+$ human T lymphocytes. Immunol Lett 57:121–123

Poggi A, Rubartelli A, Moretta L, Zocchi MR (1997b) Expression and function of NKRP1A molecule on human monocytes and dendritic cells. Eur J Immunol 27:2965–2970

Poggi A, Zocchi MR, Carosio R et al (2002) Transendothelial migratory pathways of V $\delta 1^+$TCR $\gamma\delta^+$ and V $\delta 2^+$TCR $\gamma\delta^+$ T lymphocytes from healthy donors and multiple sclerosis patients: involvement of phosphatidylinositol 3 kinase and calcium calmodulin-dependent kinase II. J Immunol 168:6071–6077

Pozo D, Valés-Gómez M, Mavaddat N et al (2006) CD161 (Human NKR-P1A) signaling in NK cells involves the activation of acid sphingomyelinase. J Immunol 176:2397–2406

Radaev S, Sun PD (2003) Structure and function of natural killer cell surface receptors. Annu Rev Biophys Biomol Struct 32:93–114

Raulet DH, Vance RE, McMahon CW (2001) Regulation of the natural killer cell receptor repertoire. Annu Rev Immunol 19:291–330

Renedo M, Arce I, Montgomery K et al (2000) A sequence-ready physical map of the region containing the human natural killer gene complex on chromosome 12p12.3-p13.2. Genomics 65:129–136

Rogers SL, Göbel TW, Viertlboeck BC et al (2005) Characterization of the chicken C-type lectin-like receptors B-NK and B-lec suggests that the NK complex and the MHC share a common ancestral region. J Immunol 174:3475–3483

Rogers SL, Kaufman J (2008) High allelic polymorphism, moderate sequence diversity and diversifying selection for B-NK but not B-lec, the pair of lectin-like receptor genes in the chicken MHC. Immunogenetics 60:461–475

Rosen DB, Bettadapura J, Alsharifi M et al (2005) Cutting edge: lectin-like transcript-1 is a ligand for the inhibitory human NKR-P1A receptor. J Immunol 175:7796–7799

Rosen DB, Cao W, Avery DT et al (2008) Functional consequences of interactions between human NKR-P1A and its ligand LLT1 expressed on activated dendritic cells and B cells. J Immunol 180:6508–6519

Roth P, Aulwurm S, Gekel I et al (2006) Regeneration and tolerance factor: a novel mediator of glioblastoma-associated immunosuppression. Cancer Res 66:3852–3858

Roth P, Mittelbronn M, Wick W et al (2007) Malignant glioma cells counteract antitumor immune responses through expression of lectin-like transcript-1. Cancer Res 67:3540–3544

Ryan JC, Niemi EC, Nakamura MC et al (1995) NKR-P1A is a target-specific receptor that activates natural killer cell cytotoxicity. J Exp Med 18:1911–1915

Ryan JC, Seaman WE (1997) Divergent functions of lectin-like receptors on NK cells. Immunol Rev 155:79–89

Sánchez-Mateos P, Cebrián M, Acevedo A et al (1989) Expression of a gp33/27,000 MW activation inducer molecule (AIM) on human lymphoid tissues. Induction of cell proliferation on thymocytes and B lymphocytes by anti-AIM antibodies. Immunology 68:72–79

Santis AG, López-Cabrera M, Hamann J et al (1994) Structure of the gene coding for the human early lymphocyte activation antigen CD69: a C-type lectin receptor evolutionarily related with the gene families of natural killer cell-specific receptors. Eur J Immunol 24:1692–1697

Sawicki MW, Dimasi N, Natarajan K et al (2001) Structural basis of MHC class I recognition by natural killer cell receptors. Immunol Rev 181:52–65

Semenuk T, Krist P, PavlÃcek J, Bezouska K, Kuzma M, Novák P, Kren V (2001) Synthesis of chitooligomer-based glycoconjugates and their binding to the rat natural killer cell activation receptor. NKR-P1. Glycoconj J 18:817–826

Shimamoto M, Ueno Y, Tanaka S et al (2007) Selective decrease in colonic $CD56^+$ T and $CD161^+$ T cells in the inflamed mucosa of patients with ulcerative colitis. World J Gastroenterol 13:5995–6002

Sobanov Y, Glienke J, Brostjan C et al (1999) Linkage of the NKG2 and CD94 receptor genes to D12S77 in the human natural killer gene complex. Immunogenetics 49:99–105

Sun B, Li HL, Wang JH, Wang GY et al (2007) Passive transfer of experimental autoimmune neuritis by IL-12 and IL-18 synergistically potentiated lymphoid cells is regulated by NKR-P1$^+$ cells. Scand J Immunol 65:412–420

Takahashi T, Dejbakhsh-Jones S, Strober S (2006) Expression of CD161 (NKR-P1A) defines subsets of human CD4 and CD8 T cells with different functional activities. J Immunol 176:211–216

Taniguchi M, Harada M, Kojo S et al (2003) The regulatory role of Vα14 NKT cells in innate and acquired immune response. Annu Rev Immunol 21:483–513

Tormo J, Natarajan K, Margulies DH, Mariuzza RA (1999) Crystal structure of a lectin-like natural killer cell receptor bound to its MHC class I ligand. Nature 402:623–631

Vetter CS, Straten PT, Terheyden P et al (2000) Expression of CD94/NKG2 subtypes on tumor-infiltrating lymphocytes in primary and metastatic melanoma. J Invest Dermatol 114:941–947

Voigt S, Mesci A, Ettinger J et al (2007) Cytomegalovirus evasion of innate immunity by subversion of the NKR-P1B:Clr-b missing-self axis. Immunity 26:617–627

Watanabe M, Nakamura Y, Matsuzuka F et al (2008) Decrease of intrathyroidal $CD161^+V\alpha 24^+V\beta 11^+$ NKT cells in Graves' disease. Endocr J 55:199–203

Webster GA, Bowles MJ, Karim MS et al (1994) Activation antigen expression on peripheral blood neutrophils following rat small bowel transplantation. NKR-P1 is a novel antigen preferentially expressed during allograft rejection. Transplantation 58:707–712

Werwitzke S, Tiede A, Drescher BE et al (2003) CD8 β/CD28 expression defines functionally distinct populations of peripheral blood T lymphocytes. Clin Exp Immunol 133:334–343

Westgaard IH, Berg SF, Ørstavik S et al (1998) Identification of a human member of the Ly-49 multigene family. Eur J Immunol 28:1839–1846

Xu H, Imanishi S, Yamada K et al (2005) Strain and age-related changes in the localization of intestinal $CD161^+$ natural killer cells and $CD8^+$ intraepithelial lymphocytes along the longitudinal crypt axis in inbred rats. Biosci Biotechnol Biochem 69:567–574

Yang T, Flint MS, Webb KM, Chambers WH (2006) CD161B:ClrB interactions mediate activation of enhanced lysis of tumor target cells following NK cell:DC co-culture. Immunol Res 36:43–50

Yokoyama WM, Plougastel BFM (2003) Immune functions encoded by the natural killer gene complex. Nat Rev Immunol 3:304–316

Yu S, Zhu Y, Chen Z et al (2002) Initiation and development of experimental autoimmune neuritis in Lewis rats is independent of the cytotoxic capacity of NKR-P1A$^+$ cells. J Neurosci Res 67:823–828

Zhou H, Kartsogiannis V, Hu YS et al (2001) A novel osteoblast-derived C-type lectin that inhibits osteoclast formation. J Biol Chem 276:14916–14923

Ziegler SF, Levin SD, Johnson L et al (1994) The mouse CD69 gene. Structure, expression, and mapping to the NK gene complex. J Immunol 152:1228–1236

NKG2 Subfamily C (KLRC)

Rajesh K. Gupta and G.S. Gupta

30.1 NKG2 Subfamily C (KLRC)

30.1.1 NKG2 Gene Family and Structural Organization

NKG2 receptors are type II C-type, lectin-like, integral membrane glycoproteins, which are expressed on the cell surface as heterodimers with CD94, which is an invariant type II C-type, lectin-like polypeptide. CD94 lacks a cytoplasmic tail and therefore, cannot transduce signals. It is however essential for the expression of NKG2 receptors. Four distinct genes, A/B, C, E/H, and F, encode the NKG2 receptors. Of these receptors, CD94/NKG2A is an inhibitory one, as it contains a long cytoplasmic tail with two ITIMs. Others have short cytoplasmic tails, and each associates noncovalently with a homodimer of DAP-12, as in the case of activating KIRs. The NKG2 family of genes (HGMW-approved symbol KLRC) contains at least six members (NKG2-A, -B, -C, -E, -F and -H) which are localized to human chromosome 12p12.3-p13.2, in the same region where CD69 genes have been mapped. In addition, the human CD94 and NKR-P1A genes map to the short arm of chromosome 12. The physical distance spanned by NK gene complex (NKC) in humans ranges between 0.7 and 2.4 megabases (Renedo et al. 1997). The NKG2 and CD94 genes are localized in a small region (< 350 kb) and mapped in the following order: (NKG2-C/NKG2-A)/NKG2-E/NKG2-F/NKG2-D/CD94. Sequence analysis of 62 kb spanning the NKG2-A, -E, -F, and -D loci allowed the identification of two LINE elements that could have been involved in the duplication of the NKG2 genes. Presence of one MIR and one L1ME2 element at homologous positions in the NKG2-A and NKG2-F genes is consistent with the existence of rodent NKG2 gene(s). The 5′-ends of the NKG2-A transcripts were mapped into two separate regions showing the existence of two separate transcriptional control regions upstream of the NKG2-A locus and defining putative promoter elements for these genes (Plougastel and Trowsdale 1998). Restriction mapping and sequencing revealed the NKG2-C, -D, -E, and -F genes to be closely linked to one another, and of the same transcriptional orientation. The NKG2-C, -E, and -F genes, despite being highly similar, are variable at their 3′ ends. It was found that NKG2-C consists of six exons, whereas NKG2-E has seven, and the splice acceptor site for the seventh exon occurs in an Alu repeat. NKG2-F consists of only four exons and part of exon IV is in some cases spliced to the 5′ end of the NKG2-D transcript. NKG2-D has only a low similarity to the other NKG2 genes Glienke et al. (1998). The murine NKG2-D-like sequence also maps to the murine NK complex near CD94 and Ly49 family members.

30.1.1.1 Association of NKG2 Family Proteins with CD94

NKG2 family proteins have been shown to be covalently associated with CD94. Structural heterogeneity in NKG2 gene family and the formation of heterodimers with CD94 provides the creation of a diverse class of NK cell repertoire (Houchins et al. 1997; Lazetic et al. 1996). The human CD94 glycoprotein forms disulfide-bonded heterodimers with the NKG2A/B, NKG2C, and NKG2E glycoproteins. Human NKG2-F gene is localized 25 kb from NKG2-A, and related to NKG2-D cDNA. Despite the similarities with other NKG2 genes, NKG2-F encodes a putative protein which does not contain any lectin domain. However, a conserved 24-amino acid sequence, present in all members of NKG2 family, suggests that NKG2-F is also able to form heterodimers with CD94 (Plougastel and Trowsdale 1997). NK cells inhibited upon HLA recognition express the CD94/p43 dimer, whose specificity for HLA molecules partially overlaps the Ig-SF receptor system (López-Botet et al. 1997). In addition to NK cells, CD94-NKG2-A/B heterodimers are also expressed by TCR-γ/δ cells, and a subset of TCR-α/β cells. The human NKG2A-CD94 (KLRC1) and NKG2C-CD94 (KLRC2) heterodimers recognize the nonclassic MHC class I molecule, HLA-E (Qa-1 in mice) as ligand, which primarily displays peptides derived from the signal peptides of classic MHC class I molecules. NKG2D (KLRK1) molecules are also expressed by T cells, mediating co-stimulatory functions dependent on the availability of the adaptor protein DAP10 (Fig. 30.1).

30.1.2 Human NKG2-A, -B, and -C

The NKG2 cDNA clone is expressed in all NK cells. The original isolate, from a CD3- NK cell clone, was found to cross-hybridize with a family of transcripts that fell into four distinct groups designated NKG2-A, -B, -C, and -D. All four human transcripts encode type II membrane proteins of 215–233 amino acids. NKG2-A and -B peptides appear to be alternative splicing products of a single gene *Nkg2a*, which encodes two isoforms, NKG2A and NKG2B, with the latter lacking the stem region. NKG2-C is highly homologous with group A, having 94% homology in the external (C-terminal) domain and 56% homology throughout the internal and transmembrane regions. Cell surface expression of NKG2A is dependent on the association with CD94 since glycosylation patterns characteristic of mature proteins are found only in NKG2A that is associated with CD94. The induction of an inhibitory signal is consistent with the presence of two ITIM (V/LXYXXL) on the cytoplasmic domain of NKG2A. Similar motifs are found on Ly49 and KIR receptors, which also transmit negative signals to NK cells (Brooks et al. 1997). Brostjan et al. (2002) found differential expression of inhibitory (NKG2-A/B) versus triggering (NKG2-C and potentially -E, -F, -H) NK receptor chains. The generation of the splice variants NKG2-E and -H seemed to occur at a constant ratio. NKG2-D is distantly but significantly related (21% amino acid homology) to the first three groups. Large panel of NK clones indicated that NKG2-A$^-$ P25$^+$ NK clones express the NKG2-C transcript (Cantoni et al. 1998).

30.1.2.1 Promoter Sequences
Brostjan et al. (2000) established app. 3 kb of upstream promoter sequences of the human NKG2-C, -E and -F genes and compared with available NKG2-A sequences. Extended regions of homology contain numerous putative transcription factor binding sites conserved in the NKG2 genes. However, variation in Alu insertion among family members has led to promoter structures unique to the respective family members, which could contribute to differences in transcriptional initiation as well as gene-specific regulation.

NKG2C gene is deleted in Japanese population. The location of the break-point was determined to be 1.5–1.8 kb telomeric from the 3′ end of NKG2A. The frequency of NKG2C deletion haplotype was 30.2% in Japanese and 30.0% in Dutch populations. The frequency of homozygous deletion was 4.1% in Japanese and 3.8% in Dutch. *Nkg2c* deletion in Japanese and Dutch suggests that NKG2C is not essential for survival and reproduction, and is not associated with rheumatic diseases (Miyashita et al. 2004). Each of the NKG2 molecules is paired with CD94 for expression and recognizes HLA-E and Qa-1b as a ligand.

30.1.2.2 NKG2 Receptors on Fetal NK Cells and T Lymphocytes
Using mAb against NKG2A/C/E, Van Beneden et al. precipitated the NKG2A as a 38-kDa protein that decreased drastically from 2 week after birth. Phenotypic analysis showed that ~ 90% of fetal NK cells and ~ 50% of adult NK cells express high levels of CD94/NKG2. The remaining 50% of adult NK cells expressed low surface levels of CD94/NKG2. Expression of CD94/NKG2 was not restricted to NK cells, but was also observed on NK T and memory T cells. Functional analysis showed that CD94/NKG2$^+$ fetal NK cells could discriminate between MHC class I-positive and MHC class I-negative tumor cells. The expression levels of CD94/NKG2 were similar in wild-type compared with β2m$^{-/-}$ mice (Stevenaert et al. 2003). CD94/NKG2 receptors are also expressed on a subpopulation of peripheral CD8 memory TCR αβ lymphocytes and on mature TCR Vγ3$^+$ cells in the fetal thymus. Skin-located Vγ3 T cells, the progeny of fetal thymic Vγ3 cells, also expressed CD94/NKG2 and Ly49E but not the other members of the Ly49 family. This suggested that Vγ3 T cells expressing CD94/NKG2 receptors are mature and display a memory phenotype, and that CD94/NKG2 functions as an inhibitory receptor on these T lymphocytes (Van Beneden et al. 2002).

30.1.3 Murine NKG2A, -B, -C

The genes coding for *NKG2A,-B, -C* are clustered on the distal mouse chromosome 6 and on the rat chromosome 4 in a region designated the NK gene complex. The deduced amino acid sequence of mouse NKG2-A contains only one consensus cytoplasmic ITIM. NKG2-A from B6 and BALB/c mice differ by six amino acid residues in the extracellular domain. Murine NKG2B, like its human conterpart, appears to be a splice variant of NKG2-A, and lacks a large portion of the stalk region. Murine NKG2-C lacks an ITIM in its cytoplasmic domain, a feature shared by human and rat NKG2-C. However, unlike the human counterpart, the transmembrane domain of mouse NKG2-C does not contain a charged amino acid residue. Mouse NKG2-A mRNA was detected in IL-2-activated NK cells and spleen cells but not in other tissues (Lohwasser et al. 1999). The identification of a rat CD94 orthologue implied that NK cell receptors equivalent to NKG2/CD94 also exist in rat. The rat NKG2A

30.1 NKG2 Subfamily C (KLRC)

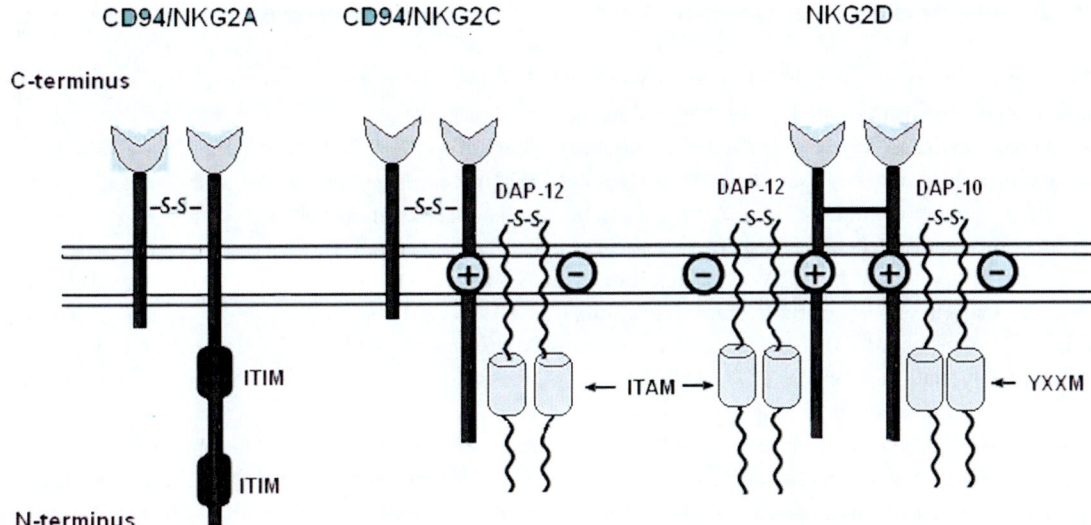

Fig. 30.1 How inhibitory and activating C-lectin receptors operate in NK cells. The C-type lectin receptors are disulfide-linked heterodimers of CD94 and NKG2 family members, either the inhibitory NKG2A or the activating NKG2C, and recognize the nonclassical MHC class I molecule HLA-E. Similar to KIRs with long cytoplasmic tails, ITIM-containing NKG2A signals through SHP-1/2 that mediate inhibitory signals. Likewise, NKG2C has a positively charged transmembrane domain that interacts with DAP-12 and transduces activating signals through Syk family members. NKG2D exists as a homodimer on the surface of NK cells. The dimmer is formed as an extension of an anti parallel β-sheet across the interface with a twofold crystallographic axis perpendicular to the extended sheet (Chap. 31) (Call et al. 2010). The NKG2D dimmer interface is primarily composed of a central hydrophobic core encased by a hydrogen bond and a salt bridge network. In contrast to NKG2C, NKG2D is only distantly related to the other NKG2 family members, does not associate with CD94, and binds to the MHC-like ligands MICA, MICB, and ULBP family. Through its positively charged transmembrane domain NKG2D associates with the adaptor molecule DAP-10 that contains an YXNK motif to bind PI-3 kinase (PI3K) and sends activating signals through this alternative pathway (see Chap. 31 for details). As the PI3K cascade is not inhibited by SHP-1/2, NKG2D may be able to mediate a dominant activation

expression is strain dependent, with high expression in DA and low in PVG NK cells, correlating with the expression of rat CD94. Presence of Ly-49 genes and the existence of rat NKG2 genes in addition to a CD94 orthologue suggest that NK cells utilize different C-type lectin receptors for MHC class I molecules in parallel.

30.1.4 NKG2 Receptors in Monkey

Labonte and Letvin (2004) demonstrated a significant variability in NKG2 mRNA expression in peripheral blood mononuclear cells (PBMCs) from rhesus monkeys. In the absence of DAP12, rhesus monkey NKG2A is preferentially expressed at the cell surface with CD94 due to a single amino acid difference in the transmembrane of NKG2A and NKG2C. In the presence of DAP12, the ability of NKG2C to compete for cell surface CD94 heterodimerization is enhanced and approaches that of NKG2A (LaBonte et al. 2004). Both *M. fascicularis* and *M. mulatta* NK cells express NKp80, NKG2D, and NKG2C molecules, which display a high degree of sequence homology with their human counterpart and reduced surface expression of selected NK cell-triggering receptors associated with a decreased NK cell function in some animals (Biassoni et al. 2005). The full-length cDNAs for CD94, NKG2A, and NKG2CE, in three unrelated squirrel monkeys showed three alternatively spliced forms of CD94 in which part of intron 4 was included in the mature transcript, suggesting evolutionary pressure for changes in the corresponding loop 3 region of the lectin domain (Andersen et al. 2004). Squirrel monkey NKG2A contains a three-nucleotide indel (insertions or deletions) that results in an additional amino acid in the predicted NKG2A protein compared to NKG2A in other species. Transmembrane-deleted forms of CD94 and NKG2CE were also expressed in the squirrel monkey (LaBonte et al. 2007). The NKC of marmoset is 1.5 times smaller than its human counterpart, but the genes are colinear and orthologous. Analyses of NKC genes in additional marmoset individuals revealed a moderate degree of polymorphism of the CD94, NKG2A, NKG2CE, and NKG2D genes. Expression analyses identified several alternatively spliced transcripts, particularly of the CD94 gene (LaBonte et al. 2007).

30.1.5 NKG2 Receptors in Other Species

In contrast to human, the cattle have multiple distinct NKG2A genes, some of which show minor allelic variation. All of the sequences designated NKG2A have two ITIM in the cytoplasmic domain and one putative gene has, in addition, a charged residue in the transmembrane domain. NKG2C appears to be essentially monomorphic in cattle. All of the NKG2A sequences are similar apart from NKG2A-01, which, in contrast, shares the majority of its carbohydrate recognition domain with NKG2-C. Most of the genes appear to generate multiple alternatively spliced forms. The CD94/NKG2A heterodimers in cattle, in contrast to other species, bind several ligands. Because NKG2C is not polymorphic, this raises questions as to the combined functional capacity of the CD94/NKG2 gene families in cattle (Birch and Ellis 2007).

The NKC is not known in chicken. Instead, NK receptor genes were found in the MHC region. In chicken, two C-type lectin-like receptor genes were identified in a region on chromosome 1 that is syntenic to mammalian NKC region. Based on 3D structure and sequence homology, one receptor is the orthologue of mammalian CD69, and the other is highly homologous to CD94 and NKG2. Like CD94/NKG2 gene found in teleostean fishes, chicken CD94/NKG2 has the features of both human CD94 and NKG2A. The arrangement of several other genes that are located outside the mammalian NKC is conserved among chicken, human, and mouse. The chicken NK C-type lectin-like receptors in the NKC syntenic region indicate that this chromosomal region existed before the divergence between mammals and apes (Chiang et al. 2007). A cDNA clone derived from the bony fish *Paralabidochromis chilotes* encodes a protein related to the CD94/NKG2 subfamily of the NK cell receptors. The gene encoding this receptor in a related species, Oreochromis niloticus, has a similar structure to the human CD94/NKG2 genes and is a member of a multigene cluster that resembles the mammalian NK cell gene complex (Sato et al. 2003).

The 1,176 bp cDNA of a C-type lectin gene, from a homozygous rainbow trout contains a 714 bp ORF predicting a 238-amino-acid (27 kDa) protein. The predicted sequence contains a 48 aa cytoplasmic domain, a 20 aa transmembrane domain (TM), a 46 aa stalk region and a 124 aa CRD. Sequence alignment and phylogenetic analysis of the CRD indicated that the protein had similarity with human dendritic cell immunoreceptor (DCIR), gp120 binding C-type lectin (gp120BCL) and mammalian hepatic lectins. The N-terminus (aa 4-183) has similarity with NKG2, important in human NK cell function (Zhang et al. 2000). This indicated that the CD94/NKG2 subfamily of NK cell receptors must have arisen before the divergence of fish and tetrapods and may have retained its function for >400 million years

30.1.6 Inhibitory and Activatory Signals

The induction of an inhibitory signal is consistent with the presence of two ITIM (V/LXYXXL) on the cytoplasmic domain of NKG2A. Similar motifs are found on Ly49 and KIR receptors, which also transmit negative signals to NK cells (Brooks et al. 1997).

Association of DAP12 with Activating CD94/NKG2C NK Cell Receptors: While inhibitory NK cell receptors for MHC class I express ITIM that recruit intracellular tyrosine phosphatases and prevent NK cell effector function, the activating NK cell receptors lack intrinsic sequences required for cellular stimulation. In contrast, the NKG2C and -E/H associate with DAP12 via a positively charged residue in their transmembrane domains and function as activation receptors. Therefore, the NKG2A/-2B contain an ITIM sequences in its cytoplasmic domain, which may be responsible for the inhibitory function of these receptors, whereas NKG2C and -E/H, lack ITIMs and associate with DAP12 via a positively charged residue in their transmembrane domains function as activation receptors and potentially transmit positive signals. Efficient expression of CD94/NKG2C on the cell surface requires the presence of DAP12. The charged residues in the transmembrane domains of DAP12 and NKG2C are necessary for this interaction (Lanier et al. 1998) (Fig. 30.1).

30.2 CD94 (KLRD1)

30.2.1 Human CD94 in Multiple Transcripts

CD94 is a receptor for human leukocyte antigen class I molecules. The CD94 is a type II membrane protein encoded by a unique gene of the C-type lectin superfamily. The primary sequences of NKG2 molecules A/B, C, and E share 27–32% identity with CD94 within the C-type lectin domain with minimal insertions or deletions, suggesting a strong structural similarity between CD94 and the NKG2 molecules. However, homodimeric CD94 has been found on the surface of certain transfected cell lines in which the expression of NKG2 is absent. Cell surface expression of NKG2A is dependent on the association with CD94 as glycosylation patterns are found only in NKG2A that is associated with CD94. Three new alternative transcripts of the *cd94* gene were identified by Lieto et al. (2006) in addition to the originally described canonical CD94Full. One of the transcripts, termed CD94-T4, lacks the portion that encodes the stem region. CD94-T4 associates with both NKG2A and NKG2B, but preferentially associates with the latter. This is probably due to the absence of a stem region in both CD94-T4 and NKG2B.

CD94 Binding with NKG2: Functionally, CD94 exists in a heterodimeric disulfide-linked complex with NKG2A/B, -C, or -E (Lazetic et al. 1996; Brooks et al. 1997; Cantoni et al. 1998) and has been shown to bind to a 26-kD NKG2A or a 43-kD protein on the surface of NK cells. The cloned CD94 molecule also covalently assembles with at least two different glycoproteins to form functional receptors. The human CD94 glycoprotein forms disulfide-bonded heterodimers with the NKG2A/B, NKG2C, and NKG2E glycoproteins. Structural heterogeneity in the NKG2 gene family and the formation of heterodimers with CD94 provides the creation of a diverse class of NK cell repertoire (Lazetic et al. 1996). Homology of CD94 with the NK cell-associated NKR-P1 and NKG2 C-type lectin genes is limited to the structural motifs conserved in the carbohydrate recognition domain. An unexpected feature of CD94 is the essential absence of a cytoplasmic domain, implying that association with other receptors may be necessary for the function of this molecule (LaBonte et al. 2001; Boyington et al. 2000).

CD94 Promoter Genes: CD94 gene expression is regulated by distal and proximal promoters that transcribe unique initial exons specific to each promoter. This results in two species of transcripts; the CD94 mRNA and a CD94C mRNA. All NK cells and CD94$^+$, CD8$^+$αβ T cells transcribe CD94 mRNA. Stimulation of NK and CD8$^+$αβ T cells with IL-2 or IL-15 induced the transcription of CD94C mRNA. The distal and proximal promoters both contain elements with IFN-γ-activated and Ets binding sites, known as GAS/EBS. Additionally, an unknown element, termed site A, was identified in the proximal promoter. EMSA analyses showed that constitutive factors could bind to oligonucleotide probes containing each element. After treatment of primary NK cells with IL-2 or IL-15, separate inducible complexes could be detected with oligonucleotide probes containing either the proximal or distal GAS/EBS elements. These elements are highly conserved between mice and humans, which suggests that both species regulate CD94 gene expression via mechanisms that predate their evolutionary divergence (Lieto et al. 2003).

30.2.2 Mouse CD94

Mouse CD94 is 54% identical and 66% similar to human CD94, and is a member of the C-type lectin superfamily. Mouse CD94 is expressed efficiently on the cell surface of cells transiently transfected with the corresponding cDNA, but surface CD94 was unable to mediate detectable binding to MHC class I-expressing Con-A blasts. Notably, mouse CD94, like human CD94, has a very short cytoplasmic tail, suggesting the existence of partner chains that may play a role in ligand binding and signaling. Like its human counterpart, the mouse CD94 protein associates with different NKG2 isoforms and recognizes the atypical MHC class I molecule Qa-1b.

Like many other C-type lectins expressed by NK cells, mouse *Cd94* maps to the NK complex on distal chromosome 6, synteneic to human CD94. The mouse *Cd94* is highly expressed specifically by mouse NK cells, raising the possibility that mice, like humans, express multiple families of MHC class I-specific receptors on their NK cells. The murine NKG2D-like sequence also maps to the murine NK complex near *Cd94* and Ly49 family members. The genomic organization of the mouse *Cd94* gene contains six exons separated by five introns. Exons I and II encode the 5′ untranslated region (UTR) and the transmembrane domain. Exon III encodes the stalk region and exons IV-VI encode the carbohydrate recognition domain (CRD). The *Cd94* promoter region and putative regulatory DNA elements have been identified (Lohwasser et al. 2000).

The murine *Cd94* gene has two promoters. Lymphoid cell types use these two promoters differentially. The promoter usage seen in adult cells is established during fetal development. The differential promoter usage by NK cells appears to be susceptible to perturbation, as both the murine NK cell line LNK, as well as cultured C57BL/6 NK cells showed altered promoter usage relative to fresh NK cells. Since, the promoter activity observed in transfection assays did not correlate with expression of the endogenous *Cd94* gene, it suggested the involvement of chromatin structure/methylation in transcriptional regulation (Wilhelm et al. 2003). The identification of two of NK cell receptors with residues of C-type lectin CRDs indicated the existence of a bovine NK gene complex, prospectively located on chromosome 5 (Govaerts and Goddeeris 2001). Bovine CD94, named KLRJ1, is most similar to Ly-49 (KLRA) and mapped to the bovine NK gene complex on chromosome 5. It suggests that KIR multigene families with divergent signaling motifs exist also in bovine species and that the primate and bovine KIR multigene families have evolved independently (Storset et al. 2003).

30.3 Cellular Sources of NKG2/CD94

NKG2 family proteins have been shown to be covalently associated with CD94. Structural heterogeneity in NKG2 gene family and the formation of heterodimers with CD94 provides the creation of a diverse class of NK cell repertoire (Houchins et al. 1997; Lazetic et al. 1996). The NKG2 cDNA clone is expressed on all NK cells. The proportion of CD94$^+$NKG2C$^-$ (NKG2A$^+$) NK cells and the level of expression of NKG2D, NKp30 and NKp46 decreases with age (Sundström et al. 2007). NK cells derived from human umbilical cord blood (CB) are the major effector cells involved in graft-versus-host disease (GVHD) and graft-versus-leukemia (GVL). The expression of NKG2A/CD94 was significantly higher on CB NK cells. The high expression of NKG2A/CD94 and low expression of granzyme B may be related with the reduced activity of CB NK cells (Tanaka et al. 2007; Wang et al. 2007). CD94-NKG2-A/B heterodimers are also expressed by TCR-γ/δ cells, and a subset of TCR-α/β cells. NKG2A, commonly expressed on cytotoxic T cells, has been found on activated T helper (Th) cells. In identifying markers differentiating between Th1 and Th2 lymphocytes, Freishtat et al. (2005) identified co-induction of NKG2A and CD56 on activation of Th2 cells.

As in NK cells, CD94 is expressed on Th1 cells together with members of the NKG2 family of molecules, including NKG2A, C, and E. Meyers et al. (2002) proposed that CD94/NKG2 heterodimers may costimulate effector functions of differentiated Th1 cells. Tanaka et al. (2004) found increased expression of CD94/NKG2A on $CD3^+/CD8^+$ T cells from G-CSF mobilized peripheral blood mononuclear cells (G-PBMC) after stimulation with immobilized anti-CD3 mAb with or without cytokines. A large majority of multiple myeloma (MM) $CD8^+$ cells do not express a functional CD94 receptor. It seemed that their ability to 'fine-tune' an appropriate immune response against tumor cells is impaired (Besostri et al. 2000). Activating and inhibitory CD94/NKG2 receptors regulate CTL responses by altering TCR signaling, thus modifying antigen activation thresholds set during thymic selection. Reports suggest that TCR antigenic specificity dictates NKG2A commitment, which critically regulates subsequent activation of CTL (Jabri et al. 2002). NK cells are the most abundant lymphocyte population at the maternal-fetal interface. They are considered to be important during placentation by controlling trophoblast invasion.

Bellon et al. (1999) studied the expression of CD94 heterodimers in different αβ or γ/δ T cell clones. While most of the $CD94^+NKG2A^-$ T cells have a low to intermediate expression of CD94, the cross-linking of the CD94/NKG2 heterodimer in one of these CD8αβ $CD94^+NKG2A^-$ T cell clones (K14B06) resulted in to induce the up-regulation of CD25 and the secretion of IFN-γ and to trigger the redirected cytotoxicity in a TCR-independent manner. The CD94 heterodimer showed a 39-kDa band with a similar activating heterodimer found on other NK cells and led to identification of NKG2H in K14B06 T cells. The NKG2H is an alternative spliced form of *NKG2E* gene and displayed a charged residue in the transmembrane portion and a cytoplasmic tail, which lacks ITIM. The expression of NKG2H forms part of the activating CD94/Kp39 heterodimer present on K14B06 cells.

Melanocyte differentiation antigen MART-1 specific T cells express CD94/NKG2 receptors. Detailed analysis revealed the exclusive presence of inhibitory NKG2-A/B receptors in the vitiligo-like leukoderma, whereas both the inhibitory receptors and the activating NKG2-C/E isoforms were present within the tumor (Pedersen et al. 2002).

Expression of CD94/NKG2 Receptors by Viral- and Bacterial-Specific CD8 T Cells Subsets of CD8 T cells express receptors that are critical in regulating the activity of NK cells. NKG2A/CD94 receptor expression is up-regulated by antiviral $CD8^+$ T cells during acute polyoma infection; this is responsible for down-regulating their antigen-specific cytotoxicity during both viral clearance and virus-induced oncogenesis. Miller et al. (2003) suggested that CD94/NKG2 expression is not correlated with inhibition of T cell function. This was shown following acute infection, with lymphocytic choriomeningitis virus (LCMV), of C57BL/6 and BALB/cJ mice, which expressed Ag-specific CD94/NKG2 $CD8^+$ T cells, and hence CD94/NKG2 expression is not necessarily correlated with inhibition of T cell function (c/r Miller et al. 2003). Wojtasiak et al. (2004) established that while Ag-stimulated gB-specific CD8 T cells primarily express inhibitory isoforms of CD94/NKG2 receptors, these cells remained capable of producing IFNγ upon peptide stimulation.

30.4 The Crystal Analysis of CD94

30.4.1 An Intriguing Model for CD94/NKG2 Heterodimer

Functionally, CD94 exists in a heterodimeric disulfide-linked complex with NKG2A/B, -C, or -E. A twofold related crystallographic dimer was observed in CD94 crystals in which two monomers hydrogen bond through their respective first β strands, creating an extended six-stranded antiparallel β sheet. The interface of this elongated dimer is relatively flat and contains a central hydrophobic region (residues Val66, Tyr68, Ile75, Phe107, and Met108) (Fig. 30.2a), surrounded by hydrophilic residues. A significant part of the interface observed in this dimer involves loop 3, where the helix-to-loop transformation has occurred in CD94, suggesting that this loop conformation may be partly stabilized through hydrophobic interactions across the dimer interface (Boyington et al. 1999). The extracellular ligand-binding domain of human CD94, corresponding to residues 34–179, was expressed in a bacterial expression system and refolded in vitro. The crystal structure of this fragment at 2.6 Å resolution showed that final model comprises residues Cys59 to Ile179. The molecule, with overall dimensions of approximately $42 \times 37 \times 33$ Å, consists of a three stranded antiparallel β sheet (strands 1,

30.4 The Crystal Analysis of CD94

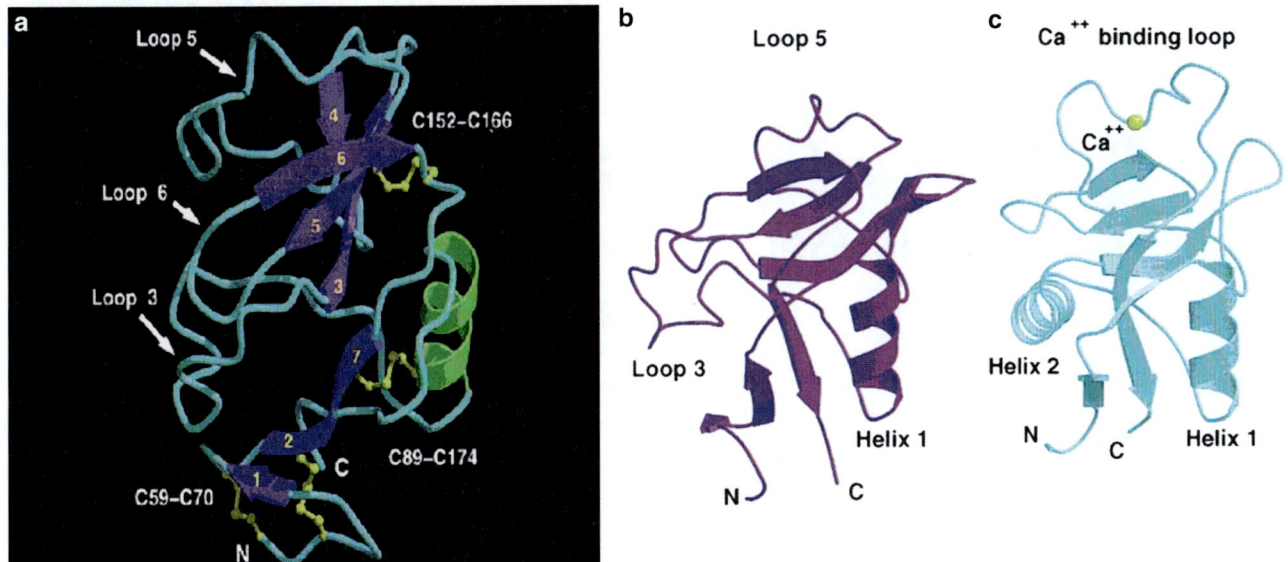

Fig. 30.2 (**a**). Ribbon diagram of the structure of CD94. β strands are shown in *purple* and are numbered according to their order in the sequence; the α helix is shown in *green*. The disulfide bonds are represented by *yellow ball-and-stick models* and labeled. Secondary structure assignments are as follows: β strands 1–7 corresponding to residues 66–68, 71–75, 115–122, 127–129, 151–155, 161–165, and 171–176; α helix 1 corresponding to residues 82–91. Comparison of Human CD94 (**b**) and Rat MBP-A (**c**) Superimposed and separated ribbon diagrams of C-type lectin domains from human CD94 and MBP-A (Reprinted with permission from Boyington et al. 1999 © Elsevier)

2, and 7), a four stranded antiparallel β sheet (strands 4, 3, 5, and 6) and an α helix after strand 2 (Fig. 30.2). There are four intrachain disulfide bonds in CD94, three of which are the characteristic invariant disulfides (Cys61–Cys72, Cys89–Cys174, and Cys152–Cys166) found in long-form C-type lectins. The fourth disulfide, Cys59–Cys70, which forms a looped structure with N-terminal β strands, is unique to the structure of CD94. The only extracellular cysteine not involved in intrachain disulfide pairing, Cys58, is expected to pair with the equivalent cysteine of NKG2 (e.g., Cys116 in NKG2A) to form the interchain disulfide in the CD94/NKG2 heterodimer (Boyington et al. 1999).

Unlike the canonical C-type lectin fold, CD94 lacks one of two major α helices present in other C-type lectin structures known till this date. Specifically, the region from residues 102 to 112 (within loop 3), corresponding to the second helix of the consensus C-type lectin fold, adopts a loop conformation in CD94. The putative Ca^{2+}-dependent carbohydrate-binding loop (loop 5, residues 142–152) is 2–5 residues shorter in CD94 compared to other C-type lectins of known structure and displays a markedly different conformation as well as a different sequence. Apart from the missing helix and the conformation of the Ca^{2+}-binding loop, the rest of the CD94 structure is quite similar to the classical C-type lectin fold even though the average sequence identity is only 20% between CD94 and C-type lectins of known structure.

The primary sequences of NKG2-A/-B, -C, and -E suggest a strong structural similarity between CD94 and the NKG2 molecules. The dimerization of CD94 brings the two N-terminal α carbons to within 7.4 Å of each other, which is consistent with having a disulfide bond between the two chains of the receptor. Modeling of CD94 homodimer structure reveals that of the five hydrophobic core residues at the dimer interface of CD94, three are completely conserved in the NKG2 sequences, and two are replaced with other nonpolar residues. This distinctly hydrophobic patch forms the largest contiguous nonpolar surface on both the CD94 monomer and the model of NKG2. While the CD94 homodimer has no interchain salt bridges, there are two potential regions of charge complementarity across the modeled CD94/NKG2 interface: one between Asp106 of CD94 and Lys135 of NKG2A and the other between Arg69 of CD94 and Glu122 of NKG2A. Lys135 and Glu122 are each conserved throughout NKG2A/B, -C, and -E sequences, but are replaced by Ser and Lys, respectively, in CD94, creating an unfavorable Arg69–Lys64 interaction across the CD94/CD94 interface. This helps to explain the favorable interaction between CD94 and NKG2 compared to CD94 homodimerization (Boyington et al. 1999).

The areas of greatest sequence divergence between human CD94 and NKG2 occur outside the dimer interface in the C-terminal half of the molecule, forming a contiguous surface at one end of CD94 (Fig. 30.3). These variable regions border each other to form a flat, uninterrupted surface across one face of the molecule opposite from the N- and C-termini. Based on this structure and sequence alignments, it was postulated that the putative ligand-binding site resides within this variable region including

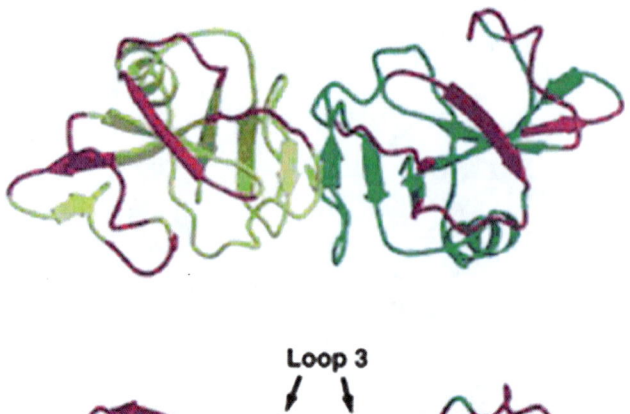
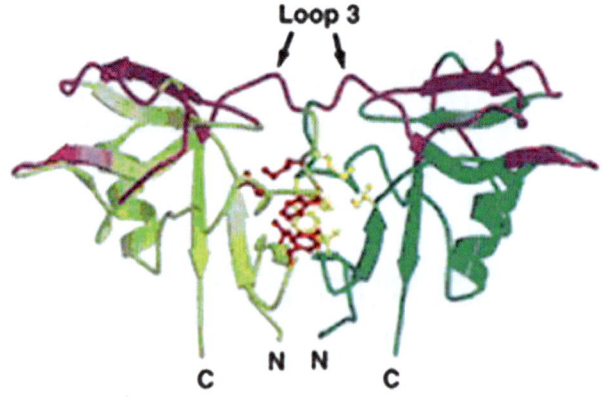

Fig. 30.3 Ribbon models showing two views of the CD94 dimer: each monomer is colored with a different *green shade*. Regions that have low sequence identity with the NKG2 sequences are colored *purple*. The *top view* is rotated 90° from the *bottom view* along the *horizontal axis*. In the *bottom view*, residues in the hydrophobic core of the dimer interface (V66, Y68, I75, F107, and M108) are represented by *ball-and-stick models*. These residues are colored *red* in one monomer and *yellow* in the other (Reprinted with permission from Boyington et al. 1999 © Elsevier)

residues 110–115, 120–124, 137–142, 144–150, 159–165, and 167–171 of human CD94 and the equivalent residues of human NKG2. The variable region of CD94 has an overall net negative charge, with seven acidic residues and only three basic residues. In contrast, the corresponding regions of the NKG2A sequence is considerably more positively charged than CD94, with six basic residues and four acidic residues (Boyington et al. 1999).

30.5 CD94/NKG2 Complex

30.5.1 CD94 and NKG2-A Form a Complex for NK Cells

The CD94/NKG2 family of receptors play an important role in regulating responses against infected and tumorigenic cells. While soluble NKG2A, -B and -C lectin domains interact with CD94 lectin domains to form complexes, NKG2D and human NKR-P1 lectin domains do not (Ding et al. 1999). The NKG2-B protein, which is an alternatively spliced product of the NKG2-A gene, can also assemble with CD94. Both NKG2-A and NKG2-B proteins contain ITIM. This provides the molecular basis of the inhibitory function mediated by the CD94/NKG2-A receptor complexes. Soluble NKG2C, -D and CD94 lectin domains bind solubilized purified HLA class I antigens independently, whereas NKG2A and -B require association of CD94 lectin domains for binding. Through differential CD94/NKG2 gene expression, human NK cells generate diverse repertoires, each cell having an inhibitory receptor for autologous HLA class I. The lytic capacity of a NK cell is regulated, in part, by the balance in cell surface expression between inhibitory CD94/NKG2A and activating CD94/NKG2C heterodimers (Lazetic et al. 1996; Borrego et al. 2006; Carretero et al. 1997).

The heterodimeric CD94/NKG2A receptor, expressed by mouse NK cells, transduces inhibitory signals upon ligation of its ligand, Qa-1b. The CD94/NKG2C and CD94/NKG2E also bind to Qa-1b. Within their extracellular CRDs, NKG2C and NKG2E share extensive homology with NKG2A (93–95% amino acid similarity). CD94/NKG2 molecules are the only Qa-1b binding receptors on NK cells. However, NKG2C/E receptors differ from NKG2A in their cytoplasmic domains (only 33% similarity) and contain features that suggest that CD94/NKG2C and CD94/NKG2E may be activating receptors (Vance et al. 1998, 1999).

30.6 Acquisition of NK Cell Receptors

Coordinated Acquisition of Inhibitory and Activating Receptors: The stages of human NK cell differentiation are not well established. Culturing CD34$^+$ progenitors, Grzywacz et al. (2006) identified 2 non-overlapping subsets of differentiating CD56$^+$ cells based on CD117 and CD94 [CD117^{high}CD94$^-$ and CD117$^{low/-}$CD94$^+$ cells]. Both populations expressed CD161 and NKp44, but differed with respect to NKp30, NKp46, NKG2A, NKG2C, NKG2D, CD8, CD16, and KIR. These two subsets represent distinct stages of NK cell differentiation, since purified CD117^{high} CD94$^-$ cells gave rise to CD117$^{low/-}$CD94$^+$ cells. The identified stages of NK-cell differentiation suggested an evidence for coordinated acquisition of HLA-specific inhibitory receptors (i.e., CD94/NKG2A) and function in developing human NK cells. However, CD94/NKG2 receptor acquisition by NK cells does not require lymphotoxin-β receptor expression (Stevenaert et al. 2005).

Hierarchy of NK Cell Response Is Determined by Class and Quantity of Inhibitory Receptors for Self-HLA-B and HLA-C Ligands: Yu et al. (2007) evaluated the resting NK repertoire analyzing the responsiveness of NK subgroups expressing discrete combinations of non-KIR and KIR class I-specific receptors. CD94:NKG2A and

ILT2-expressing cells have a modest response to class I-negative target cells, but NK cells expressing inhibitory KIRs to self-MHC class I ligands, both HLA-B and HLA-C ligands, achieve significantly higher effector capacity. These findings defined how inhibitory receptor and autologous HLA interactions impact single-cell function and demonstrated how hierarchy of NK cell response is determined by class and quantity of inhibitory receptors for self-HLA-B and HLA-C ligands. These findings have important implications for the resting NK response to viral pathogens and malignancy and for donor selection in allogeneic hemopoietic cell transplantation (Yu et al. 2007; Salmon-Divon et al. 2005).

30.7 Regulation of CD94/NKG2

30.7.1 Transcriptional Regulation of NK Cell Receptors

The stochastic expression of individual members of NK cell receptor gene families on subsets of NK cells has attracted considerable interest in the transcriptional regulation of these genes. The discovery that an upstream promoter can function as a probabilistic switch element in the Ly49 gene family has revealed a mechanism of variegated gene expression. The other NK cell receptor gene families contain probabilistic switches. The promoter elements identified in the Ly49, NKR-P1, CD94, NKG2A, and KIR gene families have been reviewed. In the human population, there is a wide variation in the NK cell repertoire of KIRs and CD94:NKG2A expression. Variation is principally due to KIR gene variation and polymorphism, with a smaller effect due to MHC class I.

The human NKG2A gene of 3.9-kb genomic fragment contains a 1.65-kb region upstream of the exon 1, as well as exon 1 (untranslated), intron 1 and exon 2. A region immediately upstream from the most upstream transcriptional initiation site led to increased transcriptional activity. Within a DNase I hypersensitivity site to this region is 80-bp segment that shows two GATA (a zinc-finger transcription factor) binding sites. Mutation of GATA binding site II (−2,302 bp) but not GATA binding site I (−2,332 bp) led to decreased transcriptional activity. GATA-3 specifically binds to the NKG2A promoter in situ in NKL and primary NK cells, but not in Jurkat T cells. Thus GATA-3 is an important transcription factor for regulating NKG2A gene expression (Marusina et al. 2005; Muzzioli et al. (2007). The cAMP induced the upregulation of NKG2A at the mRNA level in human CD8$^+$ T lymphocytes. The PGE2/cAMP/PKA type I pathway appears to be involved in the expression of CD94/NKG2A on human CD8$^+$ T lymphocytes (Zeddou et al. 2005).

30.7.2 Regulation by Cytokines

Members of NKG2 family, NKG2A, -C, -D, and -E are sequentially expressed on CD4$^+$ cells. This expression was tightly regulated by cytokines, among which TGFβ1 and IL-10 are the main factors that positively contribute to the expression of CD94 and NKG2A (Romero et al. (2001). Ortega et al. (2004) found a constitutive expression of NKG2E in CD94-depleted resting human peripheral CD4$^+$ cells, whereas inductions of NKG2A (day 15) and NKG2C (day 20) after CD3-mediated stimulation required chronic cell activation and occurred after expression of CD94 (day 5). The mRNA induction of NKG2-A and NKG2-C genes was influenced by the presence of cytokines (IL-10 and TGF-β). This indicates that there are strict gene regulatory mechanisms for CD94 and NKG2 gene expression on CD4$^+$ cells, which are different from the mechanisms governing the expression of these same genes in CD8$^+$ cells (Gunturi et al. 2005). IL-12 modulates the expression of the CD94/NKG2-A inhibitory receptor by CD8$^+$ T cells in culture. Expression of the CD94/NKG2-A was also induced by IL-12 during T cell Ag stimulation. This implies the role of IL-12 in the modulation of immune responses through NKR induction (Derre et al. 2002; Gays et al. 2005).

Regulation by IL-15: The IL-15 provides an appropriate stimulus to the expression of CD94/NKG2A, but not of other class I-specific NK receptors in the process of maturation of NK cells from thymocytes (Mingari et al. 1997). Among different cytokines, only IL-2 or IL-15 induced cell proliferation when used alone. IL-15Rα-deficient (IL-15Rα$^{-/-}$) mice lack NK cells. However, when bone marrow (BM) progenitors from IL-15Rα$^{-/-}$ mice were cultured with IL-7, stem cell factor and flt3 ligand, followed by IL-15, they were able to differentiate into functional NK cells, indicating that IL-15Rα is not critical for NK cell development (Kawamura et al. 2003).

Autocrine IL-15 Mediates Intestinal Epithelial Cell Death: Intestinal intraepithelial lymphocytes (IELs), which reside between the basolateral faces of intestinal epithelial cells (IECs), provide a first-line defense against pathogens. IL-15 mediates the reciprocal interaction between IELs and IECs, an important interaction for the regulation of appropriate mucosal immunohomeostasis. These intraepithelial NK cells expressed Ly-49 antigens, NKG2 receptors, and perforin. It suggests the possibility that the apoptosis of IECs could be regulated by self-produced IL-15 through the activation of intraepithelial NK cells. The IELs, T-cell receptor αβCD8⁺ T cells located between epithelial cells, were thought to contribute to Fas ligand (FasL)-mediated epithelial cell death in coeliac disease, a condition characterized by excess IL-15.

IELs express CD94 in celiac disease (CD), which is characterized by IL-15 cytokine response of IELs to gluten-containing diet with concomitant increase in expression of IFN-β and down-regulatory IL-10 without increase of TNF-α or TGF-β1. CD is characterized by the presence of gliadin-specific CD4⁺ T cells in the lamina propria and by the intraepithelial T-cell infiltration of unknown mechanism. In normal intestine, different proportions of IELs, which were mainly T cells, express NK receptors including CD94/NKG2. During active CD, the frequency of CD94⁺ IELs, which were mostly αβT cells, conspicuously increased, while the expression of other NK markers was not modified. It appeared that the gut epithelium favors the development of T cells that express NK receptors. The specific and selective increase of IELs expressing CD94 may be related to T-cell receptor activation and/or IL-15 secretion (Jabri et al. 2000). In three major IEL-subsets γδIELs, CD4⁺αβIELs and CD8⁺αβELs, as well as CD94⁺CD8⁺αβIELs, all of which selectively expanded in active CD, CD8⁺αβ IELs showed a significant increase in expression of both IFN-γ and IL-10. Production of IL-10 may be a common feature of IELs producing pro-inflammatory cytokines, thereby attempting to limit inflammation in an autocrine fashion (Ebert 2005; Forsberg et al. 2007; Kinoshita et al. 2002).

IL-21 Induces the Functional Maturation of Murine NK Cells: IL-21 stimulates mouse NK cell effector functions in vitro. The IL-21 is produced by CD4⁺ T and NKT cells and mediates potent effects on a variety of immune cells including NK, -T, and -B cells. It achieves its stimulatory effect by inducing the development of mature NK cells into a large granular lymphocyte phenotype with heightened effector function. IL-21 treatment results in increased cell size and granularity and a corresponding decrease in cell viability and proliferative potential. These cells up-regulate the expression of the inhibitory CD94-NKG2A receptor complex and the activation markers CD154 and lectin-like-receptor G1. Surprisingly, IL-21 treatment also results in down-regulation of the pan-NK marker, NK1.1. These developmental changes suggested that IL-21 functions to induce the terminal differentiation of mouse NK cells, resulting in heightened NK cell-mediated cytotoxicity and immune surveillance (Brady et al. 2004).

IL-21 increases the proliferation of NKT cells in combination with IL-2 or IL-15, and particularly with the CD1d-restricted glycosphingolipid Ag α-galactosylceramide. Similar to its effects on NK cells, IL-21 enhances NKT cell granular morphology, granzyme B expression, and some inhibitory NK receptors, including Ly49C/I and CD94. This study suggests that NKT cells are potentially a major source of IL-21, and that IL-21 may be an important factor in NKT cell-mediated immune regulation (Coquet et al. 2007).

IFN-γ-Mediated Negative Feedback Regulation of NKT-Cell Function by CD94/NKG2: Activation of invariant NK T (iNKT) cells with CD1d-restricted T-cell receptor (TCR) ligands is a powerful means to modulate various immune responses. The CD94/NKG2A inhibitory receptor plays a critical role in down-regulating iNKT-cell responses. The iNKT-cell response is of limited duration and iNKT cells appear refractory to secondary stimulation. Both TCR and NK-cell receptors expressed by iNKT cells were rapidly down-modulated by priming with α-galactosylceramide (α-GalCer) or its analog OCH. TCR and CD28 were re-expressed more rapidly than the receptors CD94/NKG2A and Ly49, temporally rendering the primed iNKT cells hyperreactive to ligand restimulation. Blockade of the CD94/NKG2-Qa-1b interaction markedly augmented recall and primary responses of iNKT cells. This shows the critical role for NK-cell receptors in controlling iNKT-cell responses and provides a novel strategy to augment the therapeutic effect of iNKT cells by priming with OCH or blocking of the CD94/NKG2A inhibitory pathway in clinical applications (Kaiser et al. 2005; Ota et al. 2005).

30.8 Functions of CD94/NKG2

NK cell receptors create a delicate balance of activating and inhibitory signals. These receptors are present on different immune cells but play a major role for the innate immune system. For example, the NKG2A-CD94 heterodimer transmits an inhibitory signal and NKG2C-CD94 signals in an activating fashion while both receptors recognize the same ligand, HLA-E. CD94-NKG2-A/B heterodimers are expressed by NK cells, TCR-γ/δ cells, and a subset of TCR-α/β cells. Internalization of CD94/NKG2A is independent of ligand cross-linking or the presence of functional ITIM (Borrego et al. 2002). Thus, the mechanisms that control cell surface homeostasis of CD94/NKG2A are independent of functional signaling. Masilamani et al. (2008) indicated that CD94/NKG2A utilizes a novel endocytic mechanism coupled with an abbreviated trafficking pattern, perhaps to insure surface expression.

30.8.1 CD94/NKG2 in Innate and Adaptive Immunity

CD94/NKG2 receptor varies in function as an inhibitor or activator depending on which isoform of NKG2 is expressed. The lytic capacity of a NK cell is regulated, in part, by the balance in cell surface expression between inhibitory CD94/NKG2A and activating CD94/NKG2C heterodimers. The ligand for CD94/NKG2 is HLA-E in human and its homolog, Qa1 in mouse, which are both nonclassical class I molecules that bind leader peptides from other class I molecules. Although <5% of CD8 T cells express the receptor in a naive mouse, its expression is upregulated upon specific recognition of antigen. Similar to NK cells, most CD8 T cells that express high levels of CD94 co-express NKG2A, the inhibitory isoform. The engagement of this receptor can lead to a blocking of cytotoxicity. However, these receptors have also been implicated in the cell survival of both NK and CD8 T cells. It was found that the extent of apoptosis in $CD8^+$ T and NK cells was inversely related to the expression of CD94, with lower levels of apoptosis seen in $CD94^{high}$ cells after 1–3 days of culture. The expression of CD94/NKG2 is correlated with a lower level of apoptosis and may play an important role in the maintenance of $CD8^+$ T and NK cells. Thus, CD94/NKG2 receptors may regulate effector functions and cell survival of NK cells and CD8 T cells, thereby playing a crucial role in the innate and adaptive immune response to a pathogen (Gunturi et al. 2004, 2003; Lohwasser et al. 2001). However, NK cells whose inhibitory receptors lack any apparent self-ligand can also be found in healthy individuals. On examining these NK cells, Grau et al. (2004) detected NK cells whose sole inhibitory receptors were CD94/NKG2-A and that had no affinity for autologous HLA-C molecules. Findings demonstrated the presence of potentially autoreactive NK cells in otherwise healthy individuals.

CD94 Participates in Qa-1-Mediated Self Recognition by NK Cells: In comparison to human CD94/NKG2 heterodimer, mouse Ly-49, CD94/NKG2 homologues and CD94 have been observed on all NK and NK T cells as well as small fractions of T cells in all mouse strains tested. Two distinct populations of CD94 were identified among NK and NK T cells, $CD94^{bright}$ and $CD94^{dim}$ cells, independent of Ly-49 expression. Importantly, $CD94^{bright}$ but not $CD94^{dim}$ cells were found to be functional in the Qa-1/Qdm-mediated inhibition. Toyama-Sorimachi et al. (2001) suggested that mouse CD94 participates in the protection of self cells from NK cytotoxicity through the Qa-1 recognition, independent of inhibitory receptors for classical MHC class I such as Ly-49.

Qa-1b in Association with CD94/NKG2A on CD8 T Cells Regulates Cytotoxic T Cells: The CD94/NKG2A recognizes the non-classical MHC class I molecule Qa-1b and inhibits NK cytotoxicity. Qa-1b presents a peptide derived from the leader sequence of classical MHC class I molecules. During action of CD94/NKG2A in T cell-mediated cytotoxicity, tetrameric Qa-1b binds to almost all $CD8^+$, but not $CD4^+$ T cells. Most murine $CD8^+$ T cells constitutively express CD94 and NKG2A transcripts. Co-expression of Qa-1b and D(k) on target cells significantly inhibited cytotoxicity of D(k)-specific cytotoxic T lymphocytes generated by MLR, indicating that Qa-1b on antigen-presenting cells interacts with CD94/NKG2A on CD8 T cells and regulates classical MHC class I-restricted cytotoxic T cells (Lohwasser et al. 2001). The Qa-1-NKG2A interaction protected activated CD4+ T cells from lysis by a subset of $NKG2A^+$ NK cells and was essential for T cell expansion and development of immunologic memory. Findings suggest a new clinical strategy for elimination of antigen-activated T cells in the context of autoimmune disease and transplantation (Lu et al. 2007).

30.8.2 Modulation of Anti-Viral and Anti-Tumoral Responses of γ/δ T Cells

Viral, bacterial, protozoal, and cancer-associated Ags elicit strong responses in human γ/δ T lymphocytes. Vγ9Vδ2 T cells stimulated with nonpeptidic mycobacterial antigens produce IFN-γ and TNF-α. Signaling through the CD94/NKG2 receptor interferes with the synthesis of these cytokines. The CD94/HLA class I interaction is also involved in the cytotoxic activity of Vγ9Vδ2 T cells. The Vγ9Vδ2 T cell regulation through the CD94 receptor may be important for the potentially dual function in innate immunity, i.e., (1) NK-like and (2) TCR ligand-induced cytolytic activities (Poccia et al. 1997).

Different Mechanisms of CD94/NKG2-A Expression on γ/δ T Lymphocytes: Most adult peripheral blood γ/δ T cells express Vγ9Vδ2-encoded TCR that recognize a restricted set of nonpeptidic phosphorylated compounds, referred to as phosphoantigens. They also express MHC class I-specific inhibitory receptors, in particular CD94/NKG2-A heterodimers, which participate in the fine tuning of their TCR-mediated activation threshold. Most mature Vγ9Vδ2 T cells express surface CD94 receptors, unlike cord blood or thymus-derived Vγ9Vδ2 clones, thus suggesting a role for the microenvironment in inhibitory receptors expression. Most $CD94^-$ Vγ9Vδ2 peripheral blood lymphocytes (PBL), ex vivo, express an intracellular pool of CD94/

NKG2-A receptors that is translocated to the cell surface upon activation by phosphoantigens or IL-2. In sharp contrast, intracellular CD94/NKG2-A complexes are undetectable in CD94⁻ thymus or PBL-derived mature Vδ2 T cell clones, and no surface induction is observed following phosphoantigen activation of T cell clones. These results provide the existence of distinct mechanisms controlling in vivo and in vitro induction of inhibitory receptors on Vγ9Vδ2 T cells (Boullier et al. 1998).

CD94/NKG2 Complex or KIRs Exert Inhibitory Effect on HLA-I-Mediated NK Cell Apoptosis: The $CD8^{dull}$, $CD8^{intermediate}$, and $CD8^{bright}$ NK cell clones can be identified. Triggering of CD8 with its natural ligand(s), such as soluble HLA class I (sHLA-I) leads to NK cell apoptosis. The magnitude of apoptosis directly correlated with the level of CD8 expression. The sHLA-I-induced apoptosis depends on the interaction with CD8, as it was inhibited by masking this molecule with specific mAbs. However, inhibitory receptor, such as CD94/NKG2 complex or KIRs exerted an inhibitory effect on sHLA-I-mediated apoptosis and secretion of FasL. Thus the interaction between sHLA-I and CD8 evoking an apoptotic signal is down-regulated by inhibitory receptor superfamily that functions as survival receptors in NK cells (Spaggiari et al. 2002).

30.9 NK Cells in Female Reproductive Tract and Pregnancy

NK cells are present in the various female reproductive tract (FRT) tissues; their regulation is largely dependent upon the FRT tissue where they reside. NK cells in the Fallopian tube, endometrium, cervix, and ectocervix expressed CD9 while blood NK cells did not. Unique subsets of NK cells are found in specific locations of the FRT. The NK cells in the lower reproductive tract did not express CD94, but they did express CD16. In contrast, NK cells in the upper FRT express high amounts of CD94 and CD69, but few NK cells expressed CD16. All FRT NK cells were able to produce IFN-γ upon stimulation with cytokines. These unique characteristics of the tissues may account of specific localization of different NK cell subsets (Mselle et al. 2007).

30.9.1 Decidual NK Cell Receptors

NK cells are the most abundant lymphocyte population at the maternal-fetal interface. They are considered to be important during placentation by controlling trophoblast invasion. In early human pregnancy, uterine decidual NK cells (dNK) are abundant (50–90% of lymphocytes) and considered as cytokine producers but poorly cytotoxic despite their cytolytic granule content, suggesting a negative control of this latter effector function. There is no clear evidence that dNK cells kill trophoblast cells. Instead they are able to secrete cytokines which are likely to be beneficial for the placental development, maternal uterine spiral arteries remodeling, and the antiviral immune response (Rabot et al. 2005). To investigate the basis of a negative control of effector function, El Costa et al. (2008) examined the relative contribution to the cytotoxic function of different activating receptors expressed by dNK. In fresh dNK, mAb-specific engagement of NKp46- and to a lesser extent NKG2C-, but not NKp30-activating receptors induced intracellular calcium mobilization, perforin polarization, granule exocytosis and efficient target cell lysis. It was found that in dNK, mAb-specific engagement of NKp30, but not NKp46, triggered the production of IFN-γ, TNF-α, MIP-1α MIP-1β, and GM-CSF pro-inflammatory molecules.

The balance of inhibitory and activating NK receptors on maternal dNK cells, most of which are $CD56^{bright}$, is thought to be crucial for the proper growth of trophoblasts in placenta and maintenance of pregnancy. The CD94/NKG2, a receptor for HLA-E, is expressed on trophoblasts. Women with alloimmune abortions have a limited inhibiting KIR repertoire and such miscarriages may occur because trophoblastic HLA class I molecules are recognized by dNK cells lacking the appropriate inhibitory KIRs (Varla-Leftherioti et al. 2003). dNK cells express normal surface levels of certain activating receptors, including NKp46, NKG2D, and 2B4, as well as KIRs and CD94/NKG2A inhibitory receptor. In addition, they are characterized by high levels of cytoplasmic granules despite their $CD56^{bright}$ CD16⁻ surface phenotype. Moreover, in dNK cells, activating NK receptors display normal triggering capability whereas 2B4 functions as an inhibitory receptor. This might suggest that dNK cells, although potentially capable of killing, are inhibited in their function when interacting with cells expressing CD48 (Eidukaite et al. 2004; Vacca et al. 2006).

Kusumi et al. (2006) compared the expression patterns of NK receptor, CD94/NKG2A, and CD94/NKG2C, on dNK cells in early stage of normal pregnancy with those on peripheral NK cells, most of which are $CD56^{dim}$. The rate of NKG2A-positive cells was significantly higher for decidual $CD56^{bright}$ NK cells than for peripheral $CD56^{dim}$ NK cells, but the rates of NKG2C-positive cells were comparable between the two cell types. Interestingly, peripheral $CD56^{dim}$ NK cells reciprocally expressed inhibitory NKG2A and activating NKG2C, but decidual $CD56^{bright}$ NK cells that expressed activating NKG2C simultaneously expressed inhibitory NKG2A. The co-expression of inhibitory and activating NKG2 receptors may fine-tune the immunoregulatory functions of the dNK cells to control the trophoblast invasion in constructing placenta (Kusumi et al. 2006).

Pre-eclamptic women had higher number of CD56$^+$ and CD94$^+$ cells in the decidua, indicating an altered receptor expression of dNK cells. The villous trophoblasts from women suffering from pre-eclampsia had significantly less IL-12 in placentae and significantly elevated IL-12 and IL-15 levels in serum. The altered receptor expression of dNK cells together with diminished placental IL-12 expression could implicate an altered NK cell-regulation in pre-eclampsia (Bachmayer et al. 2006).

Potentially cytotoxic Vδ$^+$ T lymphocytes recognize HLA-E on the trophoblast via their CD94/NKG2A receptors. While the percentage of viable Vδ2$^+$ T cells was higher, the percentage of Vδ1$^+$ T cells was lower in women at risk of premature pregnancy termination than in healthy pregnant women. Nonetheless, the percentage of NKG2A$^+$ Vδ2$^+$ T cells was significantly lower in pregnant women at risk of premature pregnancy termination. This indicates the involvement of γδ T lymphocytes in the pathogenesis of premature pregnancy termination (Szereday et al. 2003).

30.10 Signal Transduction by CD94/NKG2

30.10.1 Engagement of CD94/NKG2-A by HLA-E and Recruitment of Phosphatases

Many receptors share common signaling motifs to transmit their signal into the cell. Activating receptors usually signal via the immunoreceptor tyrosine-based activation motif (ITAM) containing tyrosine residues that can be phosphorylated upon receptor engagement and can recruit Syk family kinases. Inhibitory receptors often possess an Immunoreceptor Tyrosine-based Inhibition Motif (ITIM), which can recruit phosphatases upon phosphorylation. These phosphatases, like SHP-1 or SHP-2, are able to de-phosphorylate and therefore inhibit intracellular factors that otherwise would promote cellular activation. NKG2C and NKG2D, the C-type lectin-like receptors, play important roles for the activation of NK cells. These receptors have no intracellular signaling domain but instead pair with the adaptor molecules DAP10 and/or DAP12 to mediate an activating signal. The src-family tyrosine kinases include SH2-domain-containing protein tyrosine phosphatase 1 (SHP-1), SHP-2, and SH2-domain-containing inositol polyphosphate 5′ phosphatase (SHIP1). SHP-1, in particular, has been demonstrated to associate with phosphorylated ITIMs and to mediate inhibition of NK cell cytotoxicity (Fig. 30.1). A selective engagement of the CD94/NKG2A receptor with a specific mAb was sufficient to induce tyrosine phosphorylation of NKG2A subunit and SHP-1 recruitment. Furthermore, mAb cross-linking of the CD94/NKG2A receptor, segregated from other NK-associated molecules by transfection into a leukemia cell line (RBL-2H3), promoted tyrosine phosphorylation of NKG2A and co-precipitation of SHP-1, together with an inhibition of secretory events triggered via FcεRI (Carretero et al. 1998).

Specific engagement of the receptor complex expressed on the surface of an NK clone induced the phosphorylation of MAPK. It was demonstrated that the MAPK pathway participates in the CD94-dependent TNF-α production and cytotoxicity. Cross-linking of the receptor induced calcium mobilization, serotonin release and phosphorylation of MAPK (Carretero et al. 2000). Palmieri et al. (1999) indicated that CD94/NKG2-A inhibits the CD16-triggered activation of two signaling pathways involved in the cytotoxic activity of NK cells. They thus provide molecular evidence to explain the inhibitory function of CD94/NKG2-A receptor on NK effector functions.

NK Cell inhibitory receptors can interfere with tyrosine phosphorylation of 2B4 (CD244), which is an NK cell activation receptor that can provide a co-stimulatory signal to other activation receptors. Cross-linking of 2B4 on NK cells and ligation of 2B4 in the context of an NK cell-target cell interaction leads to 2B4 tyrosine phosphorylation, target cell lysis, and IFN-γ release. Coligation of 2B4 with CD94/NKG2 completely blocks NK cell activation. The rapid tyrosine phosphorylation of 2B4 after contact of NK cells with sensitive target cells is abolished when CD94/NKG2 are engaged by their cognate MHC class I ligand on resistant target cells. These results demonstrate that NK inhibitory receptors can interfere with a step as proximal as phosphorylation of an activation receptor (Watzl et al. 2000; Zingoni et al. 2000.

30.11 Ligands for CD94/NKG2

30.11.1 HLA-E as Ligand for CD94/NKG2A

The protein HLA-E is a non-classical MHC molecule of limited sequence variability. Human HLA-E, transcribed in most tissues, is the primary ligand for CD94/NKG2A-inhibitory receptors. The apparent CD94-mediated specific recognition of different HLA class Ia allotypes, transfected into the HLA-defective cell line 730.221, depends on their selective ability to concomitantly stabilize the surface expression of endogenous HLA-E molecules, which confer protection against CD94/NKG2A$^+$ effector cells. Further studies on CD94/NKG2$^+$ NK cell-mediated recognition of .221 cells transfected with different HLA class I allotypes confirmed that the inhibitory interaction was mediated by CD94/NKG2A recognizing the surface HLA-E molecule, because only antibodies directed against either HLA-E, CD94, or CD94/NKG2A specifically restored lysis.

Consistent with the prediction that the ligand for CD94/NKG2A is expressed ubiquitously, the HLA-E distribution indicated that it is detectable on the surface of a wide variety of cell types (Lee et al. 1998; Brooks et al. 1999).

The inhibitory CD94/NKG2-A receptor has a higher binding affinity for HLA-E than the activating CD94/NKG2-C receptor and, that the recognition of HLA-E by both CD94/NKG2-A and CD94/NKG2-C is peptide dependent. There appeared to be a strong, direct correlation between the binding affinity of the peptide-HLA-E complexes for the CD94/NKG2 receptors and the triggering of a response by the NK cell (Braud et al. 1998; Vales-Gomez et al. 1999; Wada et al. 2004).

HLA-E-Bound Peptides Influence Recognition by CD94/NKG2 Receptors: Reports support the notion that the primary structure of HLA-E-bound peptides influences CD94/NKG2-mediated recognition, beyond their ability to stabilize surface HLA-E. Further, CD94/NKG2A$^+$ NK clones appeared more sensitive to the interaction with most HLA-E-peptide complexes than did effector cells expressing the activating CD94/NKG2C receptor. However, a significant exception to this pattern was HLA-E loaded with the HLA-G-derived nonamer. This complex triggered cytotoxicity very efficiently over a wide range of peptide concentrations, suggesting that the HLA-E/G-nonamer complex interacts with the CD94/NKG2 triggering receptor with a significantly higher affinity. Study raises the possibility that CD94/NKG2-mediated recognition of HLA-E expressed on extravillous cytotrophoblasts plays an important role in maternal-fetal cellular interactions (Llano et al. 1998). HLA-E interacts with CD94/NKG2A receptors on NK cells and this inhibits NK cell lysis of the cell displaying HLA-E (LaBonte et al. 2001; Boyington et al. 2000).

The uterine mucosa in early pregnancy (decidua) is infiltrated by large numbers of NK cells, which are closely associated with placental trophoblast cells. Trophoblast cells express HLA-E on their cell surface in addition to the expression of HLA-G and HLA-C. The vast majority of decidual NK cells bind to HLA-E tetrameric complexes and this binding is inhibited by mAb to CD94. This shows that recognition of fetal HLA-E by decidual NK cells may play a key role in regulation of placentation. The functional consequences of decidual NK cell interaction between CD94/NKG2 and HLA-E is the inhibition of cytotoxicity by decidual NK cells. It was suggested that HLA-E interaction with CD94/NKG2 receptors may regulate other functions besides cytolysis during implantation (King et al. 2000).

A Peptide from Leader Sequences of HSP 60 Binds Also to HLA-E: HLA-E also presents a peptide derived from the leader sequence of human heat shock protein 60 (HSP60). This peptide gains access to HLA-E intracellularly, resulting in up-regulated HLA-E/HSP60 signal peptide cell-surface levels on stressed cells. Notably, HLA-E molecules in complex with the HSP60 signal peptide are no longer recognized by CD94/NKG2A inhibitory receptors. Thus, during cellular stress an increased proportion of HLA-E molecules may bind the nonprotective HSP60 signal peptide, leading to a reduced capacity to inhibit a major NK cell population. Such stress induced peptide interference would gradually uncouple CD94/NKG2A inhibitory recognition and provide a mechanism for NK cells to detect stressed cells in a peptide-dependent manner (Crew et al. 2005; Michaelsson et al. 2002; Wooden et al. 2005).

30.11.2 CD94/NKG2-A Recognises HLA-G1

There is no evidence that p58 and p70 KIRs may interact with HLA-G1. By contrast, NK recognition of cells expressing HLA-G1 involves at least two non-overlapping receptor-ligand systems: (1) the direct engagement of the ILT2 (LIR1) receptor by HLA-G1; and (2) the interaction of CD94/NKG2A and CD94/NKG2C receptors with the nonclassical class I molecule HLA-E, co-expressed on the surface upon binding to a nonamer (VMAPRTLFL) from the HLA-G leader sequence (López-Botet et al. 1999 ; Navarro et al. 1999). Perez-Villar et al. (1997) supported the idea that the CD94/NKG2 receptor is involved in the recognition of cells expressing HLA-G1.

CD94/NKG2 Is Predominant Receptor in Recognition of HLA-G at Maternal-Fetal Interface: NK cells isolated from adult peripheral blood kill the HLA-A-, HLA-B-, and HLA-C-deficient B lymphoblastoid cell line 730.221, but many are unable to kill 730.221 cells transfected with HLA-G, a molecule expressed preferentially on fetal cytotrophoblasts. The nonclassical human HLA-G is selectively expressed on fetal trophoblast tissue at the maternal-fetal interface in pregnancy. It seems that HLA-G may inhibit maternal NK cells through interaction with particular NK cell receptors. Recognition of HLA-G by NK cells was prevented in the presence of anti-CD94 mAb, implicating CD94/NKG2 as the predominant inhibitory NK cell receptor for HLA-G used by dNK cells (Soderstrom et al. 1997). Allan et al. (1999) suggested that the primary role of HLA-G may be the modulation of myelomonocytic cell behavior in pregnancy.

HLA-E and HLA-G Expression on Porcine Endothelial Cells Inhibit Xenoreactive Human NK Cells: Studies showed that peptides derived from the leader sequence of HLA-G binds and up-regulates the surface expression of HLA-E molecules, which was considered to consequently provide negative signals to human NK cells. HLA-G protects porcine cells from human NK cells through a CD94/NKG2-independent pathway. These results demonstrated that both HLA-E and HLA-G could directly inhibit human NK cells in the absence of other endogenous HLA class I molecules (Sasaki et al. 1999). These results have implications in preventing xenograft rejection mediated by human NK cells.

30.11.3 Non-Classical MHC-I Molecule Qa-1b as Ligand

Vance et al. (1998) provided evidence that mouse NK cell CD94/NKG2A heterodimer recognizes Qa-1b. The NK recognition of Qa-1b results in the inhibition of target cell lysis. Inhibition appears to depend on the presence of Qdm (Qa-1 determinant modifier), a Qa-1b-binding peptide derived from the signal sequences of some classical class I molecules. Qa-1 predominantly assembles with a single Qdm. The Qa-1/Qdm complex is the primary ligand for CD94/NKG2A inhibitory receptors expressed on a major fraction of NK cells. Cells become susceptible to killing by NK cells under conditions where surface expression of the Qa-1/Qdm inhibitory ligand is reduced (Borrego et al. 1998).

Qdm (Qa-1 Determinant Modifier): Qdm peptide, derived from the signal/leader sequence of many MHC class Ia protein molecules, has the sequence (AMAPRTLLL). This peptide binds with- and accounts for almost all of the peptides associated with this molecule. The Qa-1b was found to bind related ligands representing peptides derived from the leaders of class I molecules from several mammalian species, indicating a conservation of this "Qdm-like" epitope throughout mammalian evolution (Kurepa et al. 1998). Human HLA-E, the homologue of Qa-1b, binds similar peptides derived from human class Ia molecules and interacts with CD94/NKG2 receptors on NK cells. All of the peptides, which bind HLA-E bound readily to Qa-1b. However, the exact circumstances under which Qa1 protects cells from NK lysis and, in particular the role of Qdm, were addressed by Gays et al. (2001). Results obtained with a series of substituted Qdm peptides suggest that residues at positions 3, 4, 5, and 8 of the Qdm sequence, AMAPRTLLL, are important for recognition of Qa1-Qdm complexes by inhibitory CD94/NKG2 receptors (Jensen et al. 2004). The CD94/NKG2A receptor, expressed by mouse NK cells, transduces inhibitory signals upon recognition of its ligand, Qa-1b. The CD94/NKG2C and CD94/NKG2E also bind to Qa-1b. Within their extracellular CRDs, NKG2C and NKG2E share extensive homology with NKG2A (93–95% amino acid similarity) and may bind Qa-1b.

The Qa-1b/Qdm Tetramer Binds to CD94/NKG2 Complex Expressed on CD94^{high} Murine NK Cells: Although very few CD8$^+$ T cells from naive mice express CD94/NKG2 receptors, approximately 50% of CD8$^+$ T cells taken from mice undergoing a secondary response against *Listeria monocytogenes* (LM) are CD94^{high} and bind the Qa-1b/Qdm tetramer. Also, CD94^{int} NK cells do not bind the tetramer, where as CD94^{int} CD8$^+$ T cells do, and this binding is dependent on the CD8 co-receptor (Gunturi et al. 2003).

A Peptide from HSP60 Binds to Qa-1 in Absence of Qdm: In the light of stress induced peptide interference by HSP-60 peptide to uncouple CD94/NKG2A inhibitory recognition and to provide a mechanism for NK cells to detect stressed cells in a peptide-dependent manner (Michaelsson et al. 2002), Davies et al. (2003) reported the isolation and sequencing of a HSP-60-derived peptide (GMKFDRGYI) from Qa-1. This peptide is the dominant peptide bound to Qa-1 in the absence of Qdm. A Qa-1-restricted CTL clone recognizes this HSP60 peptide, further verifying that it binds to Qa-1 and a peptide from the homologous *Salmonella typhimurium* protein GroEL (GMQFDRGYL). These observations have implications for how Qa-1 can influence NK cell and T cell effector function via the TCR and CD94/NKG2 family members, and how this effect can change under conditions that cause the peptides bound to Qa-1 to change (Davies et al. 2003).

30.12 Structure Analysis of CD94-NKG2 Complex

30.12.1 Crystal Analysis of NKG2A/CD94: HLA-E Complex

HLA-E binds peptides derived from the leader sequences of other HLA class I molecules. NK cell recognition of HLA-E molecules, via the CD94-NKG2 NK family, represents a central innate mechanism for monitoring MHC expression levels within a cell. Kaiser et al. (2008) determined the crystal structure of the NKG2A/CD94/HLA-E complex at 4.4-Å resolution, revealing two critical aspects of this interaction. First, the C-terminal region of the peptide, which displays the most variability among class I leader sequences, interacts entirely with CD94, the invariant component of these receptors. Second, residues 167–170 of NKG2A/C account for the approximately sixfold-higher affinity of the inhibitory NKG2A/CD94 receptor compared to its activating NKG2C/CD94 counterpart. These residues do not contact HLA-E or peptide directly but instead form part of the heterodimer interface with CD94. An evolutionary analysis across primates reveals that whereas CD94 is evolving under purifying selection, both NKG2A and NKG2C are evolving under positive selection. It seemed that the evolution of the NKG2x/CD94 family of receptors has been shaped both by the need to bind the invariant HLA-E ligand and the need to avoid subversion by pathogen-derived decoys (Kaiser et al. 2008).

Changes in Peptide Conformation Affect Recognition of HLA-E by CD94-NKG2: The leader sequence-derived peptides bound to HLA-E exhibit very limited polymorphism, yet subtle differences affect the recognition of HLA-E by the CD94-NKG2 receptors. In order to prove this point, the structure of HLA-E was determined in complex with two leader peptides, namely, HLA-Cw*07 (VMAPRALLL), which is poorly recognised by CD94-NKG2 receptors, and HLA-G*01 (VMAPRTLFL), a high-affinity ligand of CD94-NKG2 receptors. A comparison of these structures revealed that allotypic variations in the bound leader sequences do not result in conformational changes in HLA-E heavy chain, although subtle changes in the conformation of the peptide within the binding groove of HLA-E were evident. Accordingly, results indicate that the CD94-NKG2 receptors interact with HLA-E in a manner that maximises the ability of the receptors to discriminate between subtle changes in both the sequence and conformation of peptides bound to HLA-E (Hoare et al. 2008).

30.12.2 CD94-NKG2A Binding to HLA-E

Petrie et al. (2008) described the crystal structure of CD94-NKG2A in complex with HLA-E bound to a peptide derived from the leader sequence of HLA-G. The CD94 subunit dominated the interaction with HLA-E, whereas the NKG2A subunit was more peripheral to the interface. Moreover, the invariant CD94 subunit dominated the peptide-mediated contacts, albeit with poor surface and chemical complementarity. There were few conformational changes in either CD94-NKG2A or HLA-E upon ligation, and such a "lock and key" interaction is typical of many receptor-ligand interactions. The structure provided insight into how this interaction can be modulated by subtle changes in the peptide ligand or by the pairing of CD94 with other members of NKG2 family (Petrie et al. 2008) (Figs. 30.4 and 30.5).

Results showed that the CD94-NKG2A docked toward the C-terminal end of the HLA-E antigen-binding cleft, binding at an angle of ~70° (Fig. 30.4a, b) in a manner that permitted CD94-NKG2A to sit across both the α1 and α2 helices of HLA-E (Fig. 30.4). A comparison of the CD94-NKG2A–HLA-E^{VMAPRTLFL} complex with the nonligated CD94-NKG2A and HLA-E^{VMAPRTLFL} structures revealed no significant conformational change in either HLA-E or CD94-NKG2A upon complex formation. The one disordered loop of the nonligated CD94-NKG2A heterodimer (residues 199–204 in NKG2A) became ordered in the complex, although this observation was attributable to crystal-packing effects, as this loop did not contact HLA-E^{VMAPRTLFL}. Only one residue, Gln 112 of CD94, was re-orientated upon ligation to maximize the complementarity at the interface. Accordingly, this "lock and key" engagement between HLA-E^{VMAPRTLFL} and CD94-NKG2A exemplified the "innate characteristic" of this interaction.

Analysis of the electrostatic surfaces of HLA-E and CD94-NKG2A highlighted a role for charge complementarity at the CD94-NKG2A–HLA-E interface. Namely, a basic region on the α1 helix of HLA-E interacted with an acidic region on CD94 and, conversely, an acidic region on the HLA-E α2 helix docked with a distinct patch of basic charge on NKG2A. The CD94 footprint on HLA-E was broad, with residues within loops 2, 3, and 5, and β strands 6 and 7 from CD94 interacting with a region spanning residues 65–89 from α1 helix of HLA-E (Petrie et al. 2008). When compared with that of CD94, the footprint of NKG2A on HLA-E was markedly smaller and more focused with loop 3 and β strands 2, 5, and 6 interacting with residues 151–162 of the α2 helix of HLA-E. Nevertheless, analogous to the CD94–HLA-E interactions, the NKG2A–HLA-E contacts were dominated by polar residues. Accordingly, the large and predominantly polar network of interactions between CD94-NKG2A and HLA-E underscored the specificity of this interaction and highlighted the dominant role of the CD94 subunit with respect to the NKG2A subunit (Petrie et al. 2008).

The location of the CD94-NKG2A footprint on HLA-E^{VMAPRTLFL} was analogous to the positioning of the footprint of human NKG2D on its MHC-like ligands, MICA and, to a lesser extent, ULBP3 (Fig. 30.5). Nevertheless, there were notable differences between these footprints, which are attributable to the narrower cleft between the α helices of MICA and ULBP3 compared with HLA-E, and the structural differences between CD94-NKG2A and NKG2D. CD94-NKG2A possessed a flat interacting surface with HLA-E, whereas that of NKG2D was more "saddle-like," which enabled it to clamp around both α helices of the MHC-like ligands (Fig. 30.5) (Petrie et al. 2008). Accordingly, the innate NK receptors and the αβTCR focus on a similar region of the HLA-E or MHC-like ligand, regardless of whether the ligand presents peptide or not. However, the characteristics of the footprint deviate between these receptors, thereby providing a basis for the NKG2D promiscuity versus the CD94-NKG2A specificity.

30.12 Structure Analysis of CD94-NKG2 Complex

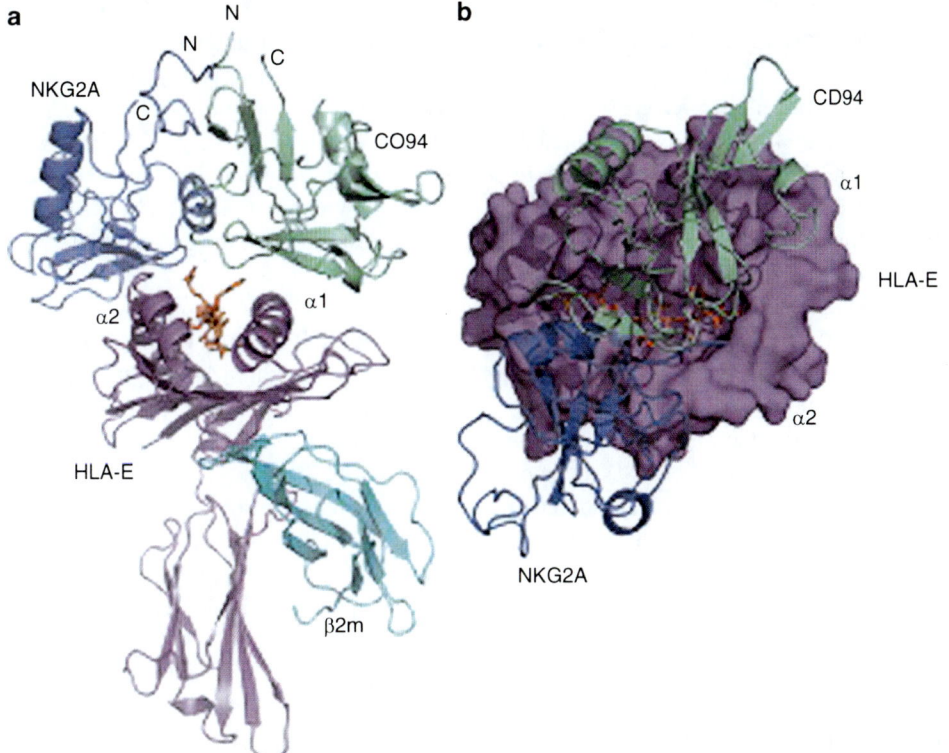

Fig. 30.4 The CD94-NKG2A – HLA-E VMAPRTLFL complex: NKG2A and CD94 are represented as *blue* and *pale green* ribbon structures, respectively. The heavy chain of HLA-E and β 2 m are shown as *violet* and *cyan* ribbons, respectively, with the VMAPRTLFL peptide in *orange sticks*. (**a**) Side view of CD94-NKG2A docking onto HLA-E VMAPRTLFL. (**b**) *Top view* of CD94-NKG2A docking onto the surface of HLA-E VMAPRTLFL (Adapted with permission from Petrie et al. 2008 © Rockefeller University Press)

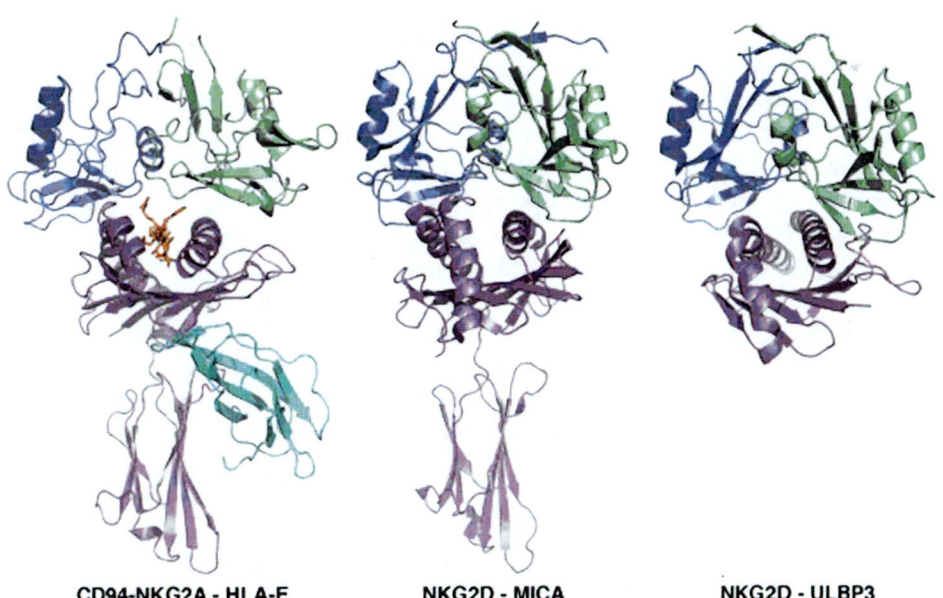

Fig. 30.5 Comparison of polar versus nonpolar interactions by CD94-NKG2A and NKG2D homodimer: Nonpolar interactions are represented in *purple*, and polar interactions are in *blue*. In situations where a residue makes both nonpolar interactions as well either a salt bridge or hydrogen bond, it was represented as making a polar interaction. Ribbon representation of CD94-NKG2A in complex with HLA-E VMAPRTLFL and NKG2D homodimers binding to MICA and ULBP3. The subunit of the NKG2D homodimer equivalent to CD94 is represented in *pale green*, and the subunit equivalent to NKG2A is in *blue*. The MHC-like molecules, MICA and ULBP3, not associated with β2m (represented in *cyan*), are represented in *violet* (Adapted with permission from Petrie et al. 2008 © The Rockefeller University Press)

30.13 Inhibitory Receptors in Viral Infection

Virus Associated Innate Immunity in Liver: The expression of inhibitory receptor CD94/NKG2A is up-regulated on NK cells in patients with chronic hepatitis C. HLA-E, a ligand for NKG2A, was expressed in all human hepatoma cell lines as well as in nontransformed hepatocytes, but not in K562 cells, a classic NK-sensitive target. Aberrant expression of CD94/NKG2A should have negative impact on innate resistance and subsequent adaptive immunity toward HCV-infected or transformed cells in chronic hepatitis C (Takehara and Hayashi 2005).

Kanto (2008) sought to clarify the role of innate immune system in the pathogenesis of HCV infection. DCs are known to sense virus infection via toll-like receptors (TLR) or retinoic acid inducible gene-I (RIG-I), resulting in the secretion of type-I IFN and inflammatory cytokines. Blood DCs consist of two subsets; myeloid DC (MDC) and plasmacytoid DC (PDC). In MDC from HCV-infected patients, the levels of TLR/RIG-I-mediated IFN-β or TNF-α induction are lower than those in uninfected donors suggesting that their signal transduction in MDC is impaired. In response to IFN-α, DC are able to express MHC class-I related chain A/B (MICA/B) and activate NK cells following ligation of NKG2D.

Interestingly, DC from HCV-infected patients are unresponsive to exogenous IFN-α to enhance MICA/B expression and fail to activate NK cells. Alternatively, NK cells from HCV-infected patients down-regulate DC functions in the presence of HLA-E-expressing hepatocytes by secreting IL-10 and TGF-β1. Such functional alteration of NK cells in HCV infection is ascribed to the enhanced expression of inhibitory receptor NKG2A/CD94 compared to the healthy counterparts. Invariant NKT cells activated by CD1d-positive DC secrete both Th1 and Th2 cytokines, serving as immune regulators. The frequency of NKT cells in chronic HCV infection does not differ from those in healthy donors. Activated NKT cells produce higher levels of IL-13 but comparable levels of IFN-γ with those from healthy subjects, showing that NKT cells are biased to Th2-type in chronic HCV infection. It is evident that the cross-talks among DC, NK cells and NKT cells are critical in shaping subsequent adaptive immune response against HCV (Kanto and Hayashi 2007; Kanto 2008).

Qa-1b-Dependent Modulation of DC and NK Cell Cross-Talk In Vivo: DC trigger activation and IFN-γ release by NK cells in lymphoid tissues, a process important for the polarization of Th1 responses. Colmenero et al. (2007) showed that the interaction between Qa-1b expressed on DC and its CD94/NKG2A receptor on NK cells is important in the regulation of DC-induced NK cell IFN-γ synthesis. NK cells from CD94/NKG2A-deficient mice displaying higher IFN-γ production upon DC stimulation along with other experiments demonstrated that Qa-1b is critically involved in regulating IFN-γ synthesis by NK cells in vivo through its interaction with CD94/NKG2A inhibitory receptors. This receptor-ligand interaction may be essential to prevent unabated cytokine production by NK cells during an inflammatory response (Colmenero et al. 2007).

CD94/NKG2 Receptors During HIV-1 Infection: NK HIV has evolved several strategies to evade recognition by the host immune system including down-regulation of MHC class I molecules. However, reduced expression of MHC class I molecules may stimulate NK cell lysis in cells of haematopoietic lineage. Dysfunction of cytotoxic activity of T and NK lymphocytes is main feature in patients with AIDS. The augmented expression of CD94 observed in HIV-infected individuals could be related to high levels of IL-10 in HIV-1-infected individuals (Galiani et al. 1999). The functional capacities and coexpression patterns of NK receptors (CD94 and CD161) are differentially affected depending on the state of therapy [HAART or HAART naïve] or the cell type (NK or T cells), respectively (Jacobs et al. 2004). However, Mela et al. (2005) suggested that changes in the NK cell repertoire during HIV-1 infection were not the result of HIV-1 viraemia alone but resembled those associated with concomitant infections (Mela et al. 2005).

Increased NK Cell Activity in Viremic HIV-1 Infection: Abnormal upregulation of CD94/NKG2A inhibitory NKR on CTLs could be responsible for a failure of immunosurveillance in cancer or HIV infection. Despite reduced NK cell numbers in subjects with ongoing viral replication, NK cells were significantly more active in secreting both IFN-γ and TNF-α than NK cells from aviremic subjects or HIV-1-negative controls and expressed significantly higher levels of CD107a, while the numbers of $CD3^-/CD56^+/CD94^+$ and $CD3^-/CD56^+/CD161^+$ NK cells were reduced. Therefore, viremic HIV-1 infection is associated with a reduction in NK cell numbers and a perturbation of NK cell subsets, but increased overall NK cell activity (Alter et al. 2004). However, Zeddou et al. (2007) showed that the chronic stimulation with HIV antigens in viraemic patients leads to a decreased rather than increased CD94/NKG2A expression on $CD8^+$ T lymphocytes and NK cells. Ballan et al. (2007) observed an increased frequency of NK cells expressing inhibitory KIRs in infected children. Moreover, increased expression of KIR2DL3, NKG2C, and NKp46 on NK cells correlated with decreased $CD4^+$ T-lymphocyte percentage, an indicator of disease severity in HIV-1-infected children (Ballan et al. 2007; Goodier et al. 2007)

NKG2C is a Triggering Receptor in Vδ1 T Cell-Mediated Cytotoxicity Against HIV-Infected CD4 T Cells: γδT cells share with NK cells many cell-surface proteins, including the NKG2 receptor. A subset of γδT cells that express the variable Vδ1 region plays a critical role in immune regulation, tumor surveillance and viral infection. Dramatic expansion of Vδ1 T cells has been observed in HIV infection. Results raise the possibility that induction of NKG2C expression on Vδ1 T cells plays a key role in the destruction of HIV-infected CD4 T cells during HIV disease (Fausther-Bovendo et al. 2008). Nattermann et al. (2005) demonstrated that HIV-mediated up-regulation of HLA-E is an additional evasion strategy targeting the antiviral activities of NK cells, which may help the virus in establishing chronic infection (Nattermann et al. 2005; Martini et al. 2005).

SIV-Infected Rhesus Monkeys
The NK cells from simian immunodeficiency virus (SIV)-infected rhesus monkeys are significantly impaired in their ability to secrete IFN-γ, TNF-α, and IL-2, while NK cell function in SIV-infected long-term non-progressor monkeys is similar to that of normal monkeys. The activating molecules NKG2C and NKG2C2 are significantly down-regulated in peripheral blood mononuclear cells of SIV-infected rhesus monkeys, suggesting that the dysregulation of these molecules may contribute to the abnormal NK cell function observed in the setting of infection (LaBonte et al. 2006).

Regulation of Cytolytic Activity of HSV-Specific Memory CD8$^+$ T Cells: Evidences suggest that sensory neurons regulate the effector functions and phenotype of CD8$^+$ T cells during active immunosurveillance of herpes simplex virus (HSV) latency. After ocular infection, HSV-specific CD8$^+$ T cells migrate to and retained in the ophthalmic branch of the trigeminal ganglia (TG). Virus-specific CD8$^+$ T cells maintain an activation phenotype and secrete IFN-γ in the latent TG. The activated virus-specific memory CD8$^+$ T cells, although potentially cytolytic, also express the CD94-NK cell receptor subfamily G2a inhibitory molecule and are unable to exert cytotoxicity when engaged by Qa-1b expressing targets. Suvas et al. (2006) indicated that the Qa-1b and CD94-NKG2a interaction regulate cytolytic activity of HSV-specific memory CD8$^+$ T cells in the latently infected TG and serves to protect irreplaceable neurons from destruction by the immune system.

CD94/NKG2A Expression During Polyoma Virus Infection: Memory CD8$^+$ T cells form critical component of durable immunity because of their capacity to rapidly proliferate and exert effector activity upon Ag rechallenge. In mice infected with polyoma virus, the majority of persistence-phase antiviral CD8$^+$ T cells expressed the inhibitory NK cell receptor CD94/NKG2A. The CD94/NKG2A expression was associated with Ag-specific recall of polyoma virus-specific CD8$^+$T cells. Polyoma virus-specific CD8$^+$ T cells that expressed CD94/NKG2A were found to preferentially proliferate; this proliferation was dependent on cognate Ag both in vitro and in vivo. In addition, CD94/NKG2A$^+$ polyoma-specific CD8$^+$ T cells had an enhanced capacity to produce IL-2 upon ex vivo Ag stimulation compared with CD94/NKG2A$^-$ polyoma-specific CD8$^+$ cells (Byers et al. 2006).

NK Cell Receptors in Response to HCMV/MCMV Infection: Human cytomegalovirus (HCMV) infection is a pattern of complexity shown by host-pathogen interactions. To avoid recognition by CTLs, HCMV inhibits the expression of HLA class I molecules. As a consequence, engagement of inhibitory NKRs specific for HLA class I molecules is impaired, and infected cells become vulnerable to an NK cell response driven by activating receptors. In addition to the well-defined role of the NKG2D, the involvement of other triggering receptors (i.e., activating KIR, CD94/NKG2C, NKp46, NKp44, and NKp30) in response to HCMV is under investigation. (Gumá et al. 2006a).

During HCMV infection, the CD94/NKG2C$^+$ cells express lower levels of NKp30 and NKp46 and higher proportions of KIR$^+$ and CD85j$^+$ cells than CD94/NKG2A$^+$ cells. The CD8$^+$ T cell compartment of human CMV-seropositive individuals characteristically contains a high proportion of cells that express NKRs which may contribute to the surveillance of virus-infected cells. Like other NKR, CD94/NKG2C is predominantly expressed by a CD8$^+$ T cell subset, though TCRγδ$^+$ NKG2C$^+$ and rare CD4$^+$ NKG2C$^+$ cells are also detected in some individuals. Studies support the notion that CD94/NKG2C may constitute an alternative T cell activation pathway capable of driving the expansion and triggering the effector functions of a CTL subset in response to HCMV (Guma et al. 2004, 2005).

The increase of CD94/NKG2C$^+$ NK cells in healthy individuals infected with human CMV suggested that HCMV infection may shape the NK cell receptor repertoire (Gumá et al. 2006c). The CD94/NKG2C$^+$ cells outnumbered the CD94/NKG2A$^+$ subset. Results support that the interaction of CD94/NKG2C with HCMV-infected fibroblasts, concomitant to the inhibition of HLA class I expression, promotes an outgrowth of CD94/NKG2C$^+$ NK cells (Gumá et al. 2006b). In addition to NKG2C, to escape from NK cell-mediated surveillance, HCMV interferes with the expression of NKG2D ligands in infected cells. Saez-Borderias et al. (2006) suggested that NKG2D functions as a prototypic costimulatory receptor in a subset of HCMV-specific CD4$^+$ T lymphocytes and thus might have a role in the response against infected HLA class II$^+$ cells displaying NKG2D ligands. In the acute phase of infection, a strong induction of CD94 on CD3$^+$ T cells was observed with surface expression of activating CD94^{dim} NKG2C dimers appearing before inhibitory CD94^{bright} NKG2A ones (van Stijn et al. 2008).

NK Cell Recognition of MCMV-Infected Cells: Studies on mouse model have shown that NK cells deploy multiple mechanisms to deal with mouse cytomegalovirus (MCMV) infection, which involve receptors of the C-lectin type superfamily. While reviewing these attack-counterattack measures, Vidal and Lanier (2006) pointed to the central role of NK cells in host resistance to infection. Following acute infection, MCMV replicates persistently in the salivary glands, despite the vigorous response of activated CD8 T cells that infiltrate this gland. Virus-specific CD8 T lymphocytes from this organ were found to express the CD94/NKG2A receptor that confers an inhibitory response to CTLs. In response to MCMV infection, expression of the CD94/NKG2A ligand, Qa-1b, increased dramatically in the sub-mandibular gland (SMG) prior to up-regulation of H-2D^d. However, there was no net negative impact on virus-specific T-cell function. Thus, the expression of inhibitory CD94/NKG2A receptors does not account for the failure of MCMV-specific CTLs to clear the SMG of infection (Cavanaugh et al. 2007).

Epstein-Barr Virus Differs with HCMV: Following activation of Epstein-Barr virus (EBV)-infected B cells from latent to productive (lytic) infection, there is a concomitant reduction in the level of MHC class I molecules and an impaired antigen-presenting function that may facilitate evasion from EBV-specific CD8$^+$ cytotoxic T cells. In some other herpesviruses studied, such as HCMV, evasion of virus-specific CD8$^+$ effector responses via down-regulation of surface MHC class I molecules is supplemented with specific mechanisms for evading NK cells. Pappworth et al. (2007) reported that EBV differs from HCMV in matters such as: resistance to lysis by NK cell lines, down-regulation of HLA-A, -B, and -C molecules that bind to the KIR family of inhibitory receptors and also the down-regulation of HLA-E molecules that bind CD94/NKG2A inhibitory receptors. Conversely the ligands for NK cell-activating receptors NKG2D and DNAM-1, respectively, were elevated. These results highlight a fundamental difference between EBV and HCMV with regards to evasion of innate immunity.

Induction of CD94 and NKG2 in CD4$^+$ T Cells by Influenza A Virus (IAV) Infection: Graham et al. (2007) investigated CD94 and NKG2 gene expression in memory CD4 T-cell clones established from the spleens of C57BL/10 (H-2^b) and BALB/c (H-2^d) mice infected with IAV. The in-vivo-generated CD4 Th1 cells, but not Th2 cells, expressed full-length CD94 and NKG2A following activation with viral peptide. A truncated isoform of NKG2A was detectable in a Th2 cell clone. NKG2D, but not NKG2C or E, was also differentially expressed in Th1 cells. Graham et al. (2007) suggested that CD94 and NKG2A may exist in multiple isoforms and help to distinguish helper T-cell subsets.

30.14 Pathophysiological Role of CD94/NKG2 Complex

30.14.1 Polymorphism in NKG2 Genes

CD94 is highly conserved, while the NKG2 genes exhibit some polymorphism. For all the genes, alternative mRNA splicing variants were frequent among the human clones. Similar alternative splicing occurs in human and chimpanzee to produce the CD94B variant from the CD94 gene and the NKG2B variant from the NKG2A gene. Whereas single chimpanzee orthologs for CD94, NKG2A, NKG2E, and NKG2F were identified, two chimpanzee paralogs of the human NKG2C gene were defined. The chimpanzee Pt-NKG2CI gene encodes a protein similar to human NKG2C, whereas in the chimpanzee Pt-NKG2CII gene the translation frame changes near the beginning of the carbohydrate recognition domain, causing premature termination. Genomic DNA from 80 individuals representing six primate species was typed for the presence of CD94 and NKG2. Each species gave distinctive typing patterns, with NKG2A and CD94 being most conserved. Seven different NK complex genotypes within the panel of 48 common chimpanzees were due to differences in Pt-NKG2C and Pt-NKG2D genes (Shum et al. 2002).

30.14.2 Phenotypes Associated with Leukemia

The polymorphic nature of NK cell receptor genes generates diverse repertoires in the human population, and displays specificity in the innate immune response. This was substantiated by Verheyden et al. (2004) who suggested that an important percentage of leukemic patients express a KIR phenotype in favor of escape from NK cell immunity. Studies have revealed the existence of a distinct type of NK cell leukaemia of the juvenile type, which shows with hypersensitivity to mosquito bites (HMB) as a clinical manifestation and is infected with clonal Epstein-Barr virus (EBV). This disorder is thus called HMB-EBV-NK disease and has been reported mostly from Japan. More than 98% of NK cells from the patients with HMB-EBV-NK disease exhibited CD94 at a higher level than did normal NK cells, whereas p70 or NKAT2, belonging to Ig-like receptor, was not expressed in those NK cells. Leukaemic NK cells transcribed mRNA for CD94-associated molecule NKG2C at an abnormally high level, and upon stimulation with IL-2 and/or IL-12 they expressed NKG2A as well. The expression of these receptors provides insights into phenotypic markers for the diagnosis of this type of NK cell leukaemia (Seo et al. 2000).

B-Cell Chronic Lymphocytic Leukaemia: Reduced reactivity of cytotoxic T cells (CTL) towards tumor cells can result into tumor progression and loss of tumor control. In B-cell chronic lymphocytic leukaemia (B-CLL), the reactivity of tumor-reactive CTL seems to relate inversely to disease stage. Analysis of $CD8^+$ T from patients with advanced disease (Binet stage C) had a significantly greater percentage of CTL expressing CD158b, CD158e, and CD94 than patients with non-progressive disease (Binet stage A) and healthy controls. The increased expression of KIR and CD94 on CTL in advanced stage B-CLL may significantly contribute to the impaired anti-tumor immune response in these patients (Junevik et al. 2007).

Heat Shock Protein 70: A Link Between NK Cells and Tissue Response: Stress-inducible heat shock protein 70-kDa (Hsp70) provides a molecular link between inflammatory responses and tissue repair. In addition to molecular chaperoning, Hsp70 exerts modulatory effects on endothelial cells and leukocytes involved in inflammatory networks. The 14 amino acid sequence (aa-450-463) TKDNNLLGRFELSG (TKD) of Hsp70 has been identified as a tumor-selective recognition structure for NK cells. The receptor CD94 participates in the interaction of NK cells with Hsp70-protein and Hsp70-peptide TKD. Hsp70 reactivity could be associated with elevated cell surface densities of CD94 after TKD stimulation independent of MHC class I molecules (Gross et al. 2003). Incubation of peripheral blood lymphocytes with TKD plus low-dose IL-2 enhances the cytolytic activity of NK cells against Hsp70 membrane-positive tumors, in vitro and in vivo. From clinical Phase I trial, reinfusion of Hsp70-activated autologous NK cells is safe and warrant additional studies in patients with lower tumor burden (Krause et al. 2004). Further studies of Gross et al. (2008) revealed that Hsp70 peptide initiates NK cell killing of leukemic blasts after stem cell transplantation and that Hsp70 is the target structure for TKD-activated NK cells. Studies indicated that Hsp70 (as an activatory molecule) and HLA-E (as an inhibitory ligand) expression influence the susceptibility of leukemic cells to the cytolytic activities of cytokine/TKD-activated NK cells (Stangl et al. 2008).

30.14.3 CD94/NKG2A on NK Cells in T Cell Lymphomas

The NK and cytotoxic T-cell lymphomas and noncytotoxic nodal T-cell lymphoma controls are stained for variety of NKR molecules including CD94, NKG2A, and p70. The NK-cell lymphomas expressed at least the CD94/NKG2A complex. Detection of CD94/NKG2A may be a useful tool to the diagnosis of NK-cell lymphomas and to delineate the subgroup of cytotoxic T-cell lymphomas (Haedicke et al. 2000; Kamarashev et al. 2001). However, a restricted KIR repertoire without monoclonal T-cell receptor rearrangement (mTCR-R) supports a NK lineage in nasal-type extranodal NK/T-cell lymphoma (NTENL) but does not correlate with clinical findings. Developing NK cells express first CD94, then NKG2A, NKG2E, and finally NKG2C. This sequence suggests an immature $CD94^-$ and a mature $CD94^+$ subtype of NTENL (Lin et al. 2003).

30.14.4 CD94/NKG2 Subtypes on Lymphocytes in Melanoma Lesions

Vetter et al. (2000) characterized the expression of CD94/NKG2 on tumor-infiltrating lymphocytes in melanoma lesions. 5–10% of the tumor-infiltrating lymphocytes, both in primary and metastatic melanoma lesions, expressed CD94. More than 95% of these $CD94^+$ cells coexpressed CD8 and the percentage of CD94 expression within the $CD8^+$ cell population ranged from 5% to 20% with a higher expression in metastatic lesions. RT-PCR revealed the presence of NKG2-C/E in all primary and metastatic lesions. In contrast, the inhibitory NKG2-A/B was only present in 50% of primary tumors, whereas 80% of tumor-infiltrating lymphocytes in metastatic lesions expressed these transcripts. In healthy humans, the number of inhibitory NK cell receptors is higher than that of activating receptors, but the opposite was true for tumor-infiltrating lymphocytes in melanoma (Casado et al. 2005; Naumova et al. 2005).

Becker et al. (2000) demonstrated that the expression of CD94/NKG2 on T cells depends on the state of differentiation during the immune response to solid tumors. To this end Becker et al. identified clonally expanded T cells which were present both in the sentinel lymph node of primary melanoma, as well as in the tumor itself. Within the early stages of T cell activation, i.e. priming in the lymph node, T cells did not express CD94/NKG2 whereas the same T cell clones expressed high levels of CD94/NKG2 having reached the effector state at the tumor site. It is likely that NK cell receptors are involved in peripheral regulatory mechanisms avoiding overwhelming immune responses and immunopathology, particularly in situations of long-lasting immune activation. However, DBA/2J mice are naturally CD94-deficient and do not express cell-surface CD94/NKG2A receptors, even on neonatal NK cells. Thus, self-tolerance of neonatal NK cells cannot be attributed to CD94/NKG2A expression. The results of Vance et al. (2002) lead to the reconsideration of current models of NK cell development and self-tolerance.

30.14.5 Cancers of Female Reproductive Tract

Very often, up-regulation of inhibitory NKRs has been linked to the modulation of virus- and/or tumor-specific immune responses in animal models. However, in human cervical cancer, the percentage expression of Ig-like NKR^+CD8^+ T lymphocytes was similar in gated $CD8^+$-autologous tumor-infiltrating lymphocytes (TILs) and peripheral blood mononuclear cells (Sheu et al. 2005). On the contrary, cervical cancer-infiltrating $CD8^+$ T lymphocytes expressed up-regulated CD94/NKG2A compared with peripheral blood $CD8^+$ T cells and/or normal cervix-infiltrating $CD8^+$ T lymphocytes, where as CD94 and NKG2A mainly expressed on $CD56^-$ $CD161^-$ $CD8^+$ TILs within the cancer milieu. Cytotoxicity experiments showed that up-regulated expressions of CD94/NKG2A restrain $CD8^+$ T lymphocyte cytotoxicity. This study indicated that cervical cancer cells could promote the expression of inhibitory NKRs via IL-15- and possibly TGF-β-mediated mechanism and abrogate the antitumor cytotoxicity of TILs. $CD8^+$ T cells express higher ratios of CD94 and NKG2A in TILs than in peripheral blood mononuclear cells (PBMCs) in human endometrial carcinoma (EC). Increased expression of CD94/NKG2A restricted to tumor-infiltrating $CD8^+$ T cell subsets may shape the cytotoxic responses, which indicate a possible role of tumor escape from host immunity in human EC (Chang et al. 2005).

30.14.6 Disorders of Immune System

CD94 in Pathogenenesis of Behcet's Disease: Behçet's disease (BD) is a multisystemic disorder with a possible underlying pathology of immune-mediated vasculitis. An imbalance in cytotoxic activity and cytokine production and genetic susceptibility associated with HLA-B*51 and B*2702 has been implicated in its pathogenesis. The BD patients show increased $CD3^+$ T cells, with no change in NK cells. Increased expression of CD94 in BD was observed on $CD16^+CD56^+$ and on $CD3^+$ and $CD3^+CD56^+$ T cells. Considering the defined regulatory mechanisms of NK cells through HLA class I binding receptors, the interactions of NK and T cells through the NK receptors may be important in the pathogenesis of BD. Although the effects of SNPs remain unclear, results indicate that the SNPs of CD94/NKG2A and its haplotypes, as well as its ligand HLA-E, are associated with BD immune systems (Seo et al. 2007).

Behcet's Uveitis: Behçet's uveitis, characterized by chronic recurrent uveitis and obliterating retinal vasculitis, frequently causes bilateral blindness. Intraocular infiltration of $TCR\alpha\beta^+$ $CD8^{bright}$ $CD56^+$ cells is a distinct feature in Behçet's uveitis. Interestingly, $CD45RA^{dim}$ $CD45RO^-$ phenotypes were expanded, and CD94 expression was markedly up-regulated in contrast to the down-regulation of NKG2D. Furthermore, these subsets were polarized to produce IFN-γ and contained high amounts of preformed intracellular perforin while exclusively expressing surface FasL upon PI stimulation. Hence, $CD8^{bright}$ $CD56^+$ T cells in Behçet's uveitis are characterized by cytotoxic effector phenotypes with functional NK receptors and function as strong cytotoxic effectors through both FasL-dependent and perforin-dependent pathways. The combined low dose cyclosporine and prednisone treatments in active Behçet uveitis may down-regulate the NK-like effector functions of $CD8^{bright}$ $CD56^+$ T cells (Ahn et al. 2005).

$CD8^+$ Regulatory T Cells in Multiple Sclerosis: To investigate $CD8^+$ regulatory T cell influence on multiple sclerosis development, peripheral blood and cerebrospinal fluid (CSF) $CD8^+$ T cell clones (TCCs) recognizing MBP(83–102) and MOG(63–87)-specific $CD4^+$ T cells were analysed. During exacerbations, CD94/NKG2A expression was significantly higher in $CD8^+$ TCCs, limiting their cytotoxic activity. Moreover, IL-15 and IFN-γ significantly increased CD94 and NKG2A expression. Evidence suggests that CD94/NKG2A receptors play an important role in regulating T cell activity during the course of MS (Correale and Villa 2008).

Negative Regulation of CD8+ TNF-α by CD94/NKG2A: Lung injury due to influenza pneumonia is mediated after Ag recognition by $CD8^+$ T cell in the distal airways and alveoli. TNF-α produced by Ag-specific $CD8^+$ T cells appears primarily responsible for this immunopathology. Zhou et al. (2008) suggested negative regulation of $CD8^+$ TNF-α by CD94/NKG2A when engaged with its receptor, Qa-1b. TNF-α production by $CD8^+$ T cells was enhanced by NKG2A blockade in vitro, and mice deficient in Qa-1b manifested greater pulmonary pathology upon $CD8^+$ T cell-mediated clearance in influenza pneumonia. It appears that CD94/NKG2A transduces a biologically important signal in vivo to activated $CD8^+$ T cells that limits immunopathology in severe influenza infection.

Susceptibility for Rheumatoid Arthritis: To examine a possible association between variations in CD94 and NKG2 genes and genetic susceptibility to rheumatoid arthritis (RA) and systemic lupus erythematosus (SLE), Hikami et al. (2003) carried out a systematic polymorphism screening of NKG2A, NKG2C and CD94 genes on a population basis. Although human NKG2-A, -C and CD94 are generally conserved with respect to amino acid sequences, NKG2A is polymorphic in the noncoding region, and that the number of genes encoded in the human NKC is variable among individuals, as shown for leukocyte receptor complex (LRC), HLA and Fcγ receptor (FCGR) regions (Hikami et al. 2003). Park et al. (2008) revealed that the major NKG2A c.338-90*A/*A, NKG2C102*Ser/*Ser, and NKG2D72*Ala/*Ala genotypes in RA were significantly associated compared with controls. The minor NKG2A c.338-90*G/*G, NKG2C102*Phe/*Phe, and NKG2D72*Thr/*Thr genotypes showed a risk of RA compared with controls.

Synovial NK cells were similar to the well-characterized CD56^{bright} peripheral blood (PB) NK-cell subset present in healthy individuals. However, compared to peripheral blood subset the synovial NK cells expressed a higher degree of activation receptors including CD69 and NKp44, the latter being up-regulated also on CD56(bright) NK cells in the peripheral blood of patients. Activated synovial NK cells produced IFN-γ and TNF, which were further up-regulated by antibody masking CD94/NKG2A, and down-regulated by target cells expressing HLA-E in complex with peptides known to engage CD94/NKG2A. It suggested that synovial NK cells have an activated phenotype and that CD94/NKG2A is the key regulator of synovial NK-cell cytokine synthesis. Human eosinophils express several inhibitory receptors IRp60, LIR3/ILT5, FcγRIIB, and p75/AIRM but not LIR1/ILT2, p58.1, p58.2, p70, or NKG2A/CD94 (Munitz et al. 2006; de Matos et al. 2007).

CD94/NKG2A$^+$ T Cells in Patients with Psoriasis: Psoriasis is a common inflammatory cutaneous disorder characterized by activated T-cell infiltration. T lymphocytes bearing NKRs have been suggested to play an important role in the pathogenesis of psoriasis. Liao et al. (2006) demonstrated increased proportions of particular subsets of inhibitory CD158b$^+$ and/or CD94/NKG2A$^+$ T cells in patients with psoriasis. The elevation of these inhibitory NKR-expressing T cells was correlated with disease severity, which may signify the unregulated cytokine production in the pathogenesis of psoriasis.

CD94/NKG2A in Sarcoidosis Patients: The majority of the lung NK cell subpopulation expressed CD56(bright). In contrast, there was a predominant CD56(dim) subset in the blood of both patients and healthy controls. Most lung NK cells expressed HLA-E-specific inhibitory receptor (i.e. CD94/NKG2A), but only a few lung NK cells expressed KIRs specific for HLA-A, -B or -C molecules. After in vitro stimulation, both lung NK and CD56$^+$ T-cells produced considerable amounts of IFN-γ and TNF-α. Thus, patients with pulmonary sarcoidosisproduce a distinct phenotype of NK cells with the capacity to produce cytokines (Katchar et al. 2005).

Frequency of CD94/NKG2A in Peritoneal NK Cells in Patients with Endometriosis: In women with endometriosis, the percentage of CD94/NKG2A-positive peritoneal NK cells was significantly higher than in the control group. It was associated with high levels mRNA transcripts of HLA-E, the ligand of CD94/NKG2A. The increased expression of CD94/NKG2A in peritoneal NK cells may mediate the resistance of endometriotic tissue to NK cell-mediated lysis leading to the progression of the disease (Galandrini et al. 2008).

Lymphoproliferative Disease of Granular Lymphocytes: Mitsui et al. (2004) analysed the chemokine receptors and NKRs in addition to NK-cell markers in patients with lymphoproliferative disease of granular lymphocytes (LDGL). There were no marked differences in the expression patterns of chemokine receptors between NK- and T-LDGL patients. Although restricted NKR subsets were expressed on both NK- and T-large granular lymphocytes (LGLs), CD94 was the most widely expressed marker.

CD94 and KIRs After Transplantation: The effect of NK cell alloreactivity on the outcome of haploidentical hematopoietic stem cell transplantation (HSCT), with or without in vitro T cell depletion, remains controversial. KIRs recognize HLA-C and -B epitopes on target cells, thereby regulating NK cell activity. The occurrence of a GVHD or the receipt of high doses of T cells in the allograft altered KIR reconstitution. The high levels of CD94 expression in donors or in recipients by day 60 might be a good predictor for poor prognosis (Zhao et al. 2007). While hepatic CD56$^+$ T cells are not expanded in malignancy, downregulation of KIR and CD94 expression may be a mechanism by which the hepatic immune system can be activated to facilitate tumor rejection (Norris et al. 2003).

NK cells mediate bone marrow allograft rejection. It was shown that murine NK cells recognize allogeneic target cells through Ly-49s and CD94/NKG2 heterodimers. NKG2I is the activating receptor mediating recognition and rejection of allogeneic target cells (Koike et al. 2004). NKG2I was composed of 226 amino acids, showed ~40% homology to the murine NKG2D and CD94 The expression of NKG2I was largely confined to NK and NKT cells. The cross-linking of NKG2I enhanced interleukin 2– and interleukin 12–dependent interferon- production (Koike et al. 2004).

References

Ahn JK, Chung H, Lee DS et al (2005) CD8brightCD56+ T cells are cytotoxic effectors in patients with active Behcet's uveitis. J Immunol 175:6133–6142

Allan DS, Colonna M, Lanier LL et al (1999) Tetrameric complexes of human histocompatibility leukocyte antigen (HLA)-G bind to peripheral blood myelomonocytic cells. J Exp Med 189:1149–1156

Alter G, Malenfant JM, Delabre RM et al (2004) Increased natural killer cell activity in viremic HIV-1 infection. J Immunol 173:5305–5311

Andersen H, Rossio JL, Coalter V et al (2004) Characterization of rhesus macaque natural killer activity against a rhesus-derived target cell line at the single-cell level. Cell Immunol 231:85–95

Bachmayer N, Rafik Hamad R, Liszka L et al (2006) Aberrant uterine natural killer (NK)-cell expression and altered placental and serum levels of the NK-cell promoting cytokine interleukin-12 in pre-eclampsia. Am J Reprod Immunol 56:292–301

Ballan WM, Vu BA, Long BR et al (2007) Natural killer cells in perinatally HIV-1-infected children exhibit less degranulation compared to HIV-1-exposed uninfected children and their expression of KIR2DL3, NKG2C, and NKp46 correlates with disease severity. J Immunol 179:3362–3370

Becker JC, Vetter CS, Schrama D et al (2000) Differential expression of CD28 and CD94/NKG2 on T cells with identical TCR β variable regions in primary melanoma and sentinel lymph node. Eur J Immunol 30:3699–3706

Bellon T, Heredia AB, Llano M et al (1999) Triggering of effector functions on a $CD8^+$ T cell clone upon the aggregation of an activatory CD94/kp39 heterodimer. J Immunol 162:3996–4002

Besostri B, Beggiato E, Bianchi A et al (2000) Increased expression of non-functional killer inhibitory receptor CD94 in CD8+ cells of myeloma patients. Br J Haematol 109:46–53

Birch J, Ellis SA (2007) Complexity in the cattle CD94/NKG2 gene families. Immunogenetics 59:273–280

Biassoni R, Fogli M, Cantoni C et al (2005) Molecular and functional characterization of NKG2D, NKp80, and NKG2C triggering NK cell receptors in rhesus and cynomolgus macaques: monitoring of NK cell function during simian HIV infection. J Immunol. 174:5695–705

Borrego F, Kabat J, Sanni TB, Coligan JE (2002) NK cell CD94/NKG2A inhibitory receptors are internalized and recycle independently of inhibitory signaling processes. J Immunol 169:6102–6111

Borrego F, Masilamani M, Marusina AI et al (2006) The CD94/NKG2 family of receptors: from molecules and cells to clinical relevance. Immunol Res 35:263–278

Borrego F, Ulbrecht M, Weiss EH et al (1998) Recognition of human histocompatibility leukocyte antigen (HLA)-E complexed with HLA class I signal sequence-derived peptides by CD94/NKG2 confers protection from natural killer cell-mediated lysis. J Exp Med 187:813–818

Boullier S, Poquet Y, Halary F et al (1998) Phosphoantigen activation induces surface translocation of intracellular CD94/NKG2A class I receptor on CD94- peripheral Vγ9Vδ2 T cells but not on CD94- thymic or mature γδ T cell clones. Eur J Immunol 28:3399–3410

Boyington JC, Raiz AN, Brooks AG, Patamawenu A, Sun PD (2000) Reconstitution of bacterial expressed human CD94: the importance of the stem region for dimer formation. Protein Expr Purif 18:235–241

Boyington JC, Riaz AN, Patamawenu A et al (1999) Structure of CD94 reveals a novel C-type lectin fold: implications for the NK cell-associated CD94/NKG2 receptors. Immunity 10:75–82

Brady J, Hayakawa Y, Smyth MJ, Nutt SL (2004) IL-21 induces the functional maturation of murine NK cells. J Immunol 172:2048–2058

Braud VM, Allan DS, O'Callaghan CA et al (1998) HLA-E binds to natural killer cell receptors CD94/NKG2A, B and C. Nature 391(6669):795–799

Brooks AG, Borrego F, Posch PE et al (1999) Specific recognition of HLA-E, but not classical, HLA class I molecules by soluble CD94/NKG2A and NK cells. J Immunol 162:305–313

Brooks AG, Posch PE, Scorzelli CJ et al (1997) NKG2A complexed with CD94 defines a novel inhibitory natural killer cell receptor. J Exp Med 185:795–800

Brostjan C, Bellon T, Sobanov Y et al (2002) Differential expression of inhibitory and activating CD94/NKG2 receptors on NK cell clones. J Immunol Methods 264:109–119

Brostjan C, Sobanov Y, Glienke J et al (2000) The NKG2 natural killer cell receptor family: comparative analysis of promoter sequences. Genes Immun 1:504–508

Byers AM, Andrews NP, Lukacher AE (2006) CD94/NKG2A expression is associated with proliferative potential of CD8 T cells during persistent polyoma virus infection. J Immunol 176:6121–6129

Call ME, Wucherpfennig K, Chou JJ (2010) The structural basis for intramembrane assembly of an activating immunoreceptor complex. Nature Immunol 11:1023–1029

Cantoni C, Biassoni R, Pende D et al (1998) The activating form of CD94 receptor complex: CD94 covalently associates with the Kp39 protein that represents the product of the NKG2-C gene. Eur J Immunol 28:327–338

Carretero M, Cantoni C, Bellon T et al (1997) The CD94 and NKG2-A C-type lectins covalently assemble to form a natural killer cell inhibitory receptor for HLA class I molecules. Eur J Immunol 27:563–567

Carretero M, Llano M, Navarro F et al (2000) Mitogen-activated protein kinase activity is involved in effector functions triggered by the CD94/NKG2-C NK receptor specific for HLA-E. Eur J Immunol 30:2842–2848

Carretero M, Palmieri G, Llano M et al (1998) Specific engagement of the CD94/NKG2-A killer inhibitory receptor by the HLA-E class Ib molecule induces SHP-1 phosphatase recruitment to tyrosine-phosphorylated NKG2-A: evidence for receptor function in heterologous transfectants. Eur J Immunol 28:1280–1291

Casado JG, Soto R, DelaRosa O et al (2005) CD8 T cells expressing NK associated receptors are increased in melanoma patients and display an effector phenotype. Cancer Immunol Immunother 54:1162–1171

Cavanaugh VJ, Raulet DH, Campbell AE (2007) Upregulation of CD94/NKG2A receptors and Qa-1b ligand during murine cytomegalovirus infection of salivary glands. J Gen Virol 88:1440–1445

Chang WC, Huang SC, Torng PL et al (2005) Expression of inhibitory natural killer receptors on tumor-infiltrating CD8+ T lymphocyte lineage in human endometrial carcinoma. Int J Gynecol Cancer 15:1073–1080

Chiang HI, Zhou H, Raudsepp T et al (2007) Chicken CD69 and CD94/NKG2-like genes in a chromosomal region syntenic to mammalian natural killer gene complex. Immunogenetics 59:603–611

Colmenero P, Zhang AL, Qian T et al (2007) Qa-1(b)-dependent modulation of dendritic cell and NK cell cross-talk in vivo. J Immunol 179:4608–4615

Coquet JM, Kyparissoudis K, Pellicci DG et al (2007) IL-21 is produced by NKT cells and modulates NKT cell activation and cytokine production. J Immunol 178:2827–2834

Correale J, Villa A (2008) Isolation and characterization of CD8+ regulatory T cells in multiple sclerosis. J Neuroimmunol 195:121–134

Crew MD, Cannon MJ, Phanavanh B et al (2005) An HLA-E single chain trimer inhibits human NK cell reactivity towards porcine cells. Mol Immunol 42:1205–1214

Davies A, Kalb S, Liang B et al (2003) A peptide from heat shock protein 60 is the dominant peptide bound to Qa-1 in the absence of the MHC class Ia leader sequence peptide Qdm. J Immunol 170:5027–5033

de Matos CT, Berg L, Michaëlsson J et al (2007) Activating and inhibitory receptors on synovial fluid natural killer cells of arthritis patients: role of CD94/NKG2A in control of cytokine secretion. Immunology 122:291–301

Derre L, Corvaisier M, Pandolfino MC et al (2002) Expression of CD94/NKG2-A on human T lymphocytes is induced by IL-12: implications for adoptive immunotherapy. J Immunol 168:4864–4870

Ding Y, Sumitran S, Holgersson J (1999) Direct binding of purified HLA class I antigens by soluble NKG2/CD94 C-type lectins from natural killer cells. Scand J Immunol 49:459–465

Dissen E, Berg SF, Westgaard IH, Fossum S (1997) Molecular characterization of a gene in the rat homologous to human CD94. Eur J Immunol 27:2080–2086

Ebert EC (2005) IL-15 converts human intestinal intraepithelial lymphocytes to CD94 producers of IFN-gamma and IL-10, the latter promoting Fas ligand-mediated cytotoxicity. Immunology 115:118–126

Eidukaite A, Siaurys A, Tamosiunas V (2004) Differential expression of KIR/NKAT2 and CD94 molecules on decidual and peripheral blood CD56bright and CD56dim natural killer cell subsets. Fertil Steril 81:863–868

El Costa H, Casemayou A, Aguerre-Girr M et al (2008) Critical and differential roles of NKp46- and NKp30-activating receptors expressed by uterine NK cells in early pregnancy. J Immunol 181:3009–3017

Fausther-Bovendo H, Wauquier N, Cherfils-Vicini J et al (2008) NKG2C is a major triggering receptor involved in the Vδ1 T cell-mediated cytotoxicity against HIV-infected CD4 T cells. AIDS 22:217–226

Forsberg G, Hernell O, Hammarström S, Hammarström ML (2007) Concomitant increase of IL-10 and pro-inflammatory cytokines in intraepithelial lymphocyte subsets in celiac disease. Int Immunol 19:993–1001

Freishtat RJ, Mitchell LW, Ghimbovschi SD et al (2005) NKG2A and CD56 are coexpressed on activated TH2 but not TH1 lymphocytes. Hum Immunol 66:1223–1234

Galandrini R, Porpora MG, Stoppacciaro A et al (2008) Increased frequency of human leukocyte antigen-E inhibitory receptor CD94/NKG2A-expressing peritoneal natural killer cells in patients with endometriosis. Fertil Steril 89(5 Suppl):1490–1496

Galiani MD, Aguado E, Tarazona R et al (1999) Expression of killer inhibitory receptors on cytotoxic cells from HIV-1-infected individuals. Clin Exp Immunol 115:472–476

Gays F, Fraser KP, Toomey JA et al (2001) Functional analysis of the molecular factors controlling Qa1-mediated protection of target cells from NK lysis. J Immunol 166:1601–1610

Gays F, Martin K, Kenefeck R, Aust JG, Brooks CG (2005) Multiple cytokines regulate the NK gene complex-encoded receptor repertoire of mature NK cells and T cells. J Immunol 175:2938–2947

Glienke J, Sobanov Y, Brostjan C et al (1998) The genomic organization of NKG2C, E, F, and D receptor genes in the human natural killer gene complex. Immunogenetics 48:163–173

Goodier MR, Mela CM, Steel A et al (2007) NKG2C$^+$ NK cells are enriched in AIDS patients with advanced-stage Kaposi's sarcoma. J Virol 81:430–433

Govaerts MM, Goddeeris BM (2001) Homologues of natural killer cell receptors NKG2-D and NKR-P1 expressed in cattle. Vet Immunol Immunopathol 80:339–344

Graham CM, Christensen JR, Thomas DB (2007) Differential induction of CD94 and NKG2 in CD4 helper T cells. A consequence of influenza virus infection and interferon-γ? Immunology 121:238–247

Grau R, Lang KS, Wernet D et al (2004) Cytotoxic activity of natural killer cells lacking killer-inhibitory receptors for self-HLA class I molecules against autologous hematopoietic stem cells in healthy individuals. Exp Mol Pathol 76:90–98

Gross C, Holler E, Stangl S, Dickinson A et al (2008) An Hsp70 peptide initiates NK cell killing of leukemic blasts after stem cell transplantation. Leuk Res 32:527–534

Gross C, Schmidt-Wolf IG, Nagaraj S et al (2003) Heat shock protein 70-reactivity is associated with increased cell surface density of CD94/CD56 on primary natural killer cells. Cell Stress Chaperones. Winter 8:348–360

Grzywacz B, Kataria N, Sikora M et al (2006) Coordinated acquisition of inhibitory and activating receptors and functional properties by developing human natural killer cells. Blood 108:3824–3833

Gumá M, Angulo A, López-Botet M (2006a) NK cell receptors involved in the response to human cytomegalovirus infection. Curr Top Microbiol Immunol 298:207–223

Guma M, Angulo A, Vilches C et al (2004) Imprint of human cytomegalovirus infection on the NK cell receptor repertoire. Blood 104:3664–3671

Gumá M, Budt M, Sáez A et al (2006b) Expansion of CD94/NKG2C$^+$ NK cells in response to human cytomegalovirus-infected fibroblasts. Blood 107:3624–3631

Gumá M, Busch LK, Salazar-Fontana LI et al (2005) The CD94/NKG2C killer lectin-like receptor constitutes an alternative activation pathway for a subset of CD8+ T cells. Eur J Immunol 35:2071–2080

Gumá M, Cabrera C, Erkizia I et al (2006c) Human cytomegalovirus infection is associated with increased proportions of NK cells that express the CD94/NKG2C receptor in aviremic HIV-1-positive patients. J Infect Dis 194:38–41

Gunturi A, Berg RE, Crossley E et al (2005) The role of TCR stimulation and TGF-β in controlling the expression of CD94/NKG2A receptors on CD8 T cells. Eur J Immunol 35:766–775

Gunturi A, Berg RE, Forman J (2003) Preferential survival of CD8 T and NK cells expressing high levels of CD94. J Immunol 170:1737–1745

Gunturi A, Berg RE, Forman J (2004) The role of CD94/NKG2 in innate and adaptive immunity. Immunol Res 30:29–34

Haedicke W, Ho FC, Chott A, Moretta L, Rudiger T, Ott G, Muller-Hermelink HK (2000) Expression of CD94/NKG2A and killer immunoglobulin-like receptors in NK cells and a subset of extranodal cytotoxic T-cell lymphomas. Blood 95:3628–3630

Hikami K, Tsuchiya N, Yabe T, Tokunaga K (2003) Variations of human killer cell lectin-like receptors: common occurrence of NKG2-C deletion in the general population. Genes Immun 4:160–167

Hoare HL, Sullivan LC, Clements CS et al (2008) Subtle changes in peptide conformation profoundly affect recognition of the non-classical MHC class I molecule HLA-E by the CD94-NKG2 natural killer cell receptors. J Mol Biol 377:1297–1303

Houchins JP, Lanier LL, Niemi EC et al (1997) Natural killer cell cytolytic activity is inhibited by NKG2-A and activated by NKG2-C. J Immunol 158:3603–3609

Jabri B, de Serre NP, Cellier C et al (2000) Selective expansion of intraepithelial lymphocytes expressing the HLA-E-specific natural killer receptor CD94 in celiac disease. Gastroenterology 118:867–879

Jabri B, Selby JM, Negulescu H et al (2002) TCR specificity dictates CD94/NKG2A expression by human CTL. Immunity 17:487–499

Jacobs R, Weber K, Wendt K et al (2004) Altered coexpression of lectin-like receptors CD94 and CD161 on NK and T cells in HIV patients. J Clin Immunol 24:281–286

Jensen PE, Sullivan BA, Reed-Loisel LM et al (2004) Qa-1, a nonclassical class I histocompatibility molecule with roles in innate and adaptive immunity. Immunol Res 29:81–92

Junevik K, Werlenius O, Hasselblom S et al (2007) The expression of NK cell inhibitory receptors on cytotoxic T cells in B-cell chronic lymphocytic leukaemia (B-CLL). Ann Hematol 86:89–94

Kaiser BK, Barahmand-Pour F, Paulsene W et al (2005) IFN-γ-mediated negative feedback regulation of NKT-cell function by CD94/NKG2. Blood 106:184–192

Kaiser BK, Pizarro JC, Kerns J, Strong RK (2008) Structural basis for NKG2A/CD94 recognition of HLA-E. Proc Natl Acad Sci USA 105:6696–6701

Kamarashev J, Burg G, Mingari MC et al (2001) Differential expression of cytotoxic molecules and killer cell inhibitory receptors in CD8+ and CD56+ cutaneous lymphomas. Am J Pathol 158:1593–1598

Kanto T, Hayashi N (2007) Innate immunity in hepatitis C virus infection: interplay among dendritic cells, natural killer cells and natural killer T cells. Hepatol Res 37(Suppl 3):S319–S326

Kanto T (2008) Virus associated innate immunity in liver. Front Biosci 13:6183–6192

Katchar K, Söderström K, Wahlstrom J et al (2005) Characterisation of natural killer cells and CD56+ T-cells in sarcoidosis patients. Eur Respir J 26:77–85

Kawamura T, Koka R, Ma A, Kumar V (2003) Differential roles for IL-15R α-chain in NK cell development and Ly-49 induction. J Immunol 171:5085–5090

King A, Allan DS, Bowen M et al (2000) HLA-E is expressed on trophoblast and interacts with CD94/NKG2 receptors on decidual NK cells. Eur J Immunol 30:1623–1631

Kinoshita N, Hiroi T, Ohta N et al (2002) Autocrine IL-15 mediates intestinal epithelial cell death via the activation of neighboring intraepithelial NK cells. J Immunol 169:6187–6192

Koike J, Wakao H, Ishizuka Y et al (2004) Bone marrow allograft rejection mediated by a novel murine NK receptor, NKG2I. J Exp Med 199:137–144

Krause SW, Gastpar R, Andreesen R et al (2004) Treatment of colon and lung cancer patients with ex vivo heat shock protein 70-peptide-activated, autologous natural killer cells: a clinical phase i trial. Clin Cancer Res 10:3699–3707

Kurepa Z, Hasemann CA, Forman J (1998) Qa-1b binds conserved class I leader peptides derived from several mammalian species. J Exp Med 188:973–978

Kusumi M, Yamashita T, Fujii T et al (2006) Expression patterns of lectin-like natural killer receptors, inhibitory CD94/NKG2A, and activating CD94/NKG2C on decidual CD56bright natural killer cells differ from those on peripheral CD56dim natural killer cells. J Reprod Immunol 70:33–42

LaBonte ML, Hershberger KL, Korber B et al (2001) The KIR and CD94/NKG2 families of molecules in the rhesus monkey. Immunol Rev 183:25–40

LaBonte ML, Choi EI, Letvin NL (2004) Molecular determinants regulating the pairing of NKG2 molecules with CD94 for cell surface heterodimer expression. J Immunol 172:6902–6912

Labonte ML, Letvin NL (2004) Variable NKG2 expression in the peripheral blood lymphocytes of rhesus monkeys. Clin Exp Immunol 138:205–212

LaBonte ML, McKay PF, Letvin NL (2006) Evidence of NK cell dysfunction in SIV-infected rhesus monkeys: impairment of cytokine secretion and NKG2C/C2 expression. Eur J Immunol 36:2424–2433

LaBonte ML, Russo J, Freitas S, Keighley D (2007) Variation in the ligand binding domains of the CD94/NKG2 family of receptors in the squirrel monkey. Immunogenetics 59:799–811

Lanier LL, Corliss B, Wu J, Phillips JH (1998) Association of DAP12 with activating CD94/NKG2C NK cell receptors. Immunity 8:693–701

Lazetic S, Chang C, Houchins JP et al (1996) Human natural killer cell receptors involved in MHC class I recognition are disulfide-linked heterodimers of CD94 and NKG2 subunits. J Immunol 157:4741–4745

Lee N, Llano M, Carretero M et al (1998) HLA-E is a major ligand for the natural killer inhibitory receptor CD94/NKG2A. Proc Natl Acad Sci USA 95:5199–5204

Liao YH, Jee SH, Sheu BC et al (2006) Increased expression of the natural killer cell inhibitory receptor CD94/NKG2A and CD158b on circulating and lesional T cells in patients with chronic plaque psoriasis. Br J Dermatol 155:318–324

Lieto LD, Borrego F, You CH, Coligan JE (2003) Human CD94 gene expression: dual promoters differing in responsiveness to IL-2 or IL-15. J Immunol 171:5277–5286

Lieto LD, Maasho K, West D et al (2006) The human CD94 gene encodes multiple, expressible transcripts including a new partner of NKG2A/B. Genes Immun 7:36–43

Lin CW, Chen YH, Chuang YC et al (2003) CD94 transcripts imply a better prognosis in nasal-type extranodal NK/T-cell lymphoma. Blood 102:2623–2631

Llano M, Lee N, Navarro F et al (1998) HLA-E-bound peptides influence recognition by inhibitory and triggering CD94/NKG2 receptors: preferential response to an HLA-G-derived nonamer. Eur J Immunol 28:2854–2863

Lohwasser S, Hande P, Mager DL, Takei F (1999) Cloning of murine NKG2A, B and C: second family of C-type lectin receptors on murine NK cells. Eur J Immunol 29:755–761

Lohwasser S, Kubota A, Salcedo M et al (2001) The non-classical MHC class I molecule Qa-1^b inhibits classical MHC class I-restricted cytotoxicity of cytotoxic T lymphocytes. Int Immunol 13:321–327

Lohwasser S, Wilhelm B, Mager DL, Takei F (2000) The genomic organization of the mouse CD94 C-type lectin gene. Eur J Immunogenet 27:149–151

López-Botet M, Carretero M, Pérez-Villar J, Bellón T, Llano M, Navarro F (1997a) The CD94/NKG2 C-type lectin receptor complex: involvement in NK cell-mediated recognition of HLA class I molecules. Immunol Res 16:175–185

López-Botet M, Navarro F, Llano M (1999) How do NK cells sense the expression of HLA-G class Ib molecules? Semin Cancer Biol 9:19–26

López-Botet M, Pérez-Villar JJ, Carretero M et al (1997b) Structure and function of the CD94 C-type lectin receptor complex involved in recognition of HLA class I molecules. Immunol Rev 155:165–174

Lu L, Ikizawa K, Hu D, Werneck MB, Wucherpfennig KW, Cantor H (2007) Regulation of activated CD4+ T cells by NK cells via the Qa-1-NKG2A inhibitory pathway. Immunity 26:593–604

Martini F, Agrati C, D'Offizi G, Poccia F (2005) HLA-E up-regulation induced by HIV infection may directly contribute to CD94-mediated impairment of NK cells. Int J Immunopathol Pharmacol 18:269–276

Marusina AI, Kim DK, Lieto LD et al (2005) GATA-3 is an important transcription factor for regulating human NKG2A gene expression. J Immunol 174:2152–2159

Masilamani M, Narayanan S, Prieto M et al (2008) Uncommon endocytic and trafficking pathway of the natural killer cell CD94/NKG2A inhibitory receptor. Traffic 9:1019–1034

Mela CM, Burton CT, Imami N et al (2005) Switch from inhibitory to activating NKG2 receptor expression in HIV-1 infection: lack of reversion with highly active antiretroviral therapy. AIDS 19:1761–1769

Meyers JH, Ryu A, Monney L et al (2002) CD94/NKG2 is expressed on Th1 but not Th2 cells and costimulates Th1 effector functions. J Immunol 169:5382–5386

Michaelsson J, Teixeira de Matos C et al (2002) A signal peptide derived from hsp60 binds HLA-E and interferes with CD94/NKG2A recognition. J Exp Med 196:1403–1414

Miller JD, Weber DA, Ibegbu C et al (2003) Analysis of HLA-E peptide-binding specificity and contact residues in bound peptide required for recognition by CD94/NKG2. J Immunol 171:1369–1375

Mingari MC, Vitale C, Cantoni C et al (1997) Interleukin-15-induced maturation of human natural killer cells from early thymic precursors: selective expression of CD94/NKG2-A as the only HLA class I-specific inhibitory receptor. Eur J Immunol 27:1374–1380

Mitsui T, Maekawa I, Yamane A et al (2004) Characteristic expansion of CD45RA CD27 CD28 CCR7 lymphocytes with stable natural killer (NK) receptor expression in NK- and T-cell type lymphoproliferative disease of granular lymphocytes. Br J Haematol 126:55–62

Miyashita R, Tsuchiya N, Hikami K et al (2004) Molecular genetic analyses of human NKG2C (KLRC2) gene deletion. Int Immunol 16:163–168

Mselle TF, Meadows SK, Eriksson M et al (2007) Unique characteristics of NK cells throughout the human female reproductive tract. Clin Immunol 124:69–76

Munitz A, Bachelet I, Eliashar R et al (2006) The inhibitory receptor IRp60 (CD300a) suppresses the effects of IL-5, GM-CSF, and eotaxin on human peripheral blood eosinophils. Blood 107:1996–2003

Muzzioli M, Stecconi R, Donnini A, Re F, Provinciali M (2007) Zinc improves the development of human CD34+ cell progenitors towards natural killer cells and induces the expression of GATA-3 transcription factor. Int J Biochem Cell Biol 39:955–965

Nattermann J, Nischalke HD, Hofmeister V et al (2005) HIV-1 infection leads to increased HLA-E expression resulting in impaired function of natural killer cells. Antivir Ther 10:95–107

Naumova E, Mihaylova A, Stoitchkov K et al (2005) Genetic polymorphism of NK receptors and their ligands in melanoma patients: prevalence of inhibitory over activating signals. Cancer Immunol Immunother 54:172–178

Navarro F, Llano M, Bellon T et al (1999) The ILT2(LIR1) and CD94/NKG2A NK cell receptors respectively recognize HLA-G1 and HLA-E molecules co-expressed on target cells. Eur J Immunol 29:277–283

Norris S, Doherty DG, Curry M et al (2003) Selective reduction of natural killer cells and T cells expressing inhibitory receptors for MHC class I in the livers of patients with hepatic malignancy. Cancer Immunol Immunother 52:53–58

Ortega C, Romero P, Palma A et al (2004) Role for NKG2-A and NKG2-C surface receptors in chronic CD4+ T-cell responses. Immunol Cell Biol 82:587–595

Ota T, Takeda K, Akiba H et al (2005) IFN-γ-mediated negative feedback regulation of NKT-cell function by CD94/NKG2. Blood 106:184–192

Palmieri G, Tullio V, Zingoni A et al (1999) CD94/NKG2-A inhibitory complex blocks CD16-triggered Syk and extracellular regulated kinase activation, leading to cytotoxic function of human NK cells. J Immunol 162:7181–7188

Pappworth IY, Wang EC, Rowe M (2007) The switch from latent to productive infection in epstein-barr virus-infected B cells is associated with sensitization to NK cell killing. J Virol 81:474–482

Park KS, Park JH, Song YW (2008) Inhibitory NKG2A and activating NKG2D and NKG2C natural killer cell receptor genes: susceptibility for rheumatoid arthritis. Tissue Antigens 72:342

Pedersen LO, Vetter CS, Mingari MC et al (2002) Differential expression of inhibitory or activating CD94/NKG2 subtypes on MART-1-reactive T cells in vitiligo versus melanoma: a case report. J Invest Dermatol 118:595–599

Perez-Villar JJ, Melero I, Navarro F et al (1997) The CD94/NKG2-A inhibitory receptor complex is involved in natural killer cell-mediated recognition of cells expressing HLA-G1. J Immunol 158:5736–5743

Petrie EJ, Clements CS, Lin J et al (2008) CD94-NKG2A recognition of human leukocyte antigen (HLA)-E bound to an HLA class I leader sequence. J Exp Med 205:725–735

Plougastel B, Trowsdale J (1997) Cloning of NKG2-F, A new member of the NKG2 family of human natural killer cell receptor genes. Eur J Immunol 27:2835–2839

Plougastel B, Trowsdale J (1998) Sequence analysis of a 62-kb region overlapping the human KLRC cluster of genes. Genomics 49:193–199

Poccia F, Cipriani B, Vendetti S et al (1997) CD94/NKG2 inhibitory receptor complex modulates both anti-viral and anti-tumoral responses of polyclonal phosphoantigen-reactive Vγ 9Vδ 2T lymphocytes. J Immunol 159:6009–6017

Rabot M, Tabiasco J, Polgar B et al (2005) HLA class I/NK cell receptor interaction in early human decidua basalis: possible functional consequences. Chem Immunol Allergy 89:72–83

Renedo M, Arce I, Rodríguez A et al (1997) The human natural killer gene complex is located on chromosome 12p12-p13. Immunogenetics 46:307–311

Romero P, Ortega C, Palma A et al (2001) Expression of CD94 and NKG2 molecules on human CD4$^+$ T cells in response to CD3-mediated stimulation. J Leukoc Biol 70:219–224

Saez-Borderias A, Guma M, Angulo A et al (2006) Expression and function of NKG2D in CD4+ T cells specific for human cytomegalovirus. Eur J Immunol 36:3198–3206

Salmon-Divon M, Höglund P, Johansson MH et al (2005) Computational modeling of human natural killer cell develop- ment suggests a selection process regulating coexpression of KIR with CD94/NKG2A. Mol Immunol 42:397–403

Sasaki H, Xu XC, Mohanakumar T (1999) HLA-E and HLA-G expression on porcine endothelial cells inhibit xenoreactive human NK cells through CD94/NKG2-dependent and -independent pathways. J Immunol 163:6301–6305

Sato A, Mayer WE, Overath P, Klein J (2003) Genes encoding putative natural killer cell C-type lectin receptors in teleostean fishes. Proc Natl Acad Sci USA 100:7779–7784

Seo J, Park JS, Nam JH et al (2007) Association of CD94/NKG2A, CD94/NKG2C, and its ligand HLA-E polymorphisms with Behcet's disease. Tissue Antigens 70:307–313

Seo N, Tokura Y, Ishihara S, Takeoka Y et al (2000) Disordered expression of inhibitory receptors on the NK1-type natural killer (NK) leukaemic cells from patients with hypersensitivity to mosquito bites. Clin Exp Immunol 120:413–419

Sheu BC, Chiou SH, Lin HH et al (2005) Up-regulation of inhibitory natural killer receptors CD94/NKG2A with suppressed intracellular perforin expression of tumor-infiltrating CD8+ T lymphocytes in human cervical carcinoma. Cancer Res 65:2921–2929

Shum BP, Flodin LR, Muir DG et al (2002) Conservation and variation in human and common chimpanzee CD94 and NKG2 genes. J Immunol 168:240–252

Soderstrom K, Corliss B, Lanier LL, Phillips JH (1997) CD94/NKG2 is the predominant inhibitory receptor involved in recognition of HLA-G by decidual and peripheral blood NK cells. J Immunol 159:1072–1075

Spaggiari GM, Contini P, Carosio R et al (2002) Soluble HLA class I molecules induce natural killer cell apoptosis through the engagement of CD8: evidence for a negative regulation exerted by members of the inhibitory receptor superfamily. Blood 99:1706–1714

Stangl S, Gross C, Pockley AG et al (2008) Influence of Hsp70 and HLA-E on the killing of leukemic blasts by cytokine/Hsp70 peptide-activated human natural killer (NK) cells. Cell Stress Chaperon 13:221–230

Stevenaert F, Van Beneden K, De Colvenaer V et al (2005) Ly49 and CD94/NKG2 receptor acquisition by NK cells does not require lymphotoxin-β receptor expression. Blood 106:956–962

Stevenaert F, Van Beneden K, De Creus A et al (2003) Ly49E expression points toward overlapping, but distinct, natural killer (NK) cell differentiation kinetics and potential of fetal versus adult lymphoid progenitors. J Leukoc Biol 73:731–738

Storset AK, Slettedal IO, Williams JL et al (2003) Natural killer cell receptors in cattle: a bovine killer cell immunoglobulin-like receptor multigene family contains members with divergent signaling motifs. Eur J Immunol 33:980–990

Sundström Y, Nilsson C, Lilja G et al (2007) The expression of human natural killer cell receptors in early life. Scand J Immunol 66:335–344

Suvas S, Azkur AK, Rouse BT (2006) Qa-1b and CD94-NKG2a interaction regulate cytolytic activity of herpes simplex virus-specific memory CD8+ T cells in the latently infected trigeminal ganglia. J Immunol 176:1703–1711

Szereday L, Barakonyi A, Miko E et al (2003) γδT-cell subsets, NKG2A expression and apoptosis of Vδ2+ T cells in pregnant women with or without risk of premature pregnancy termination. Am J Reprod Immunol 50:490–496

Takehara T, Hayashi N (2005) Natural killer cells in hepatitis C virus infection: from innate immunity to adaptive immunity. Clin Gastroenterol Hepatol 3(10 Suppl 2):S78–S81

Tanaka J, Sugita J, Kato N et al (2007) Expansion of natural killer cell receptor (CD94/NKG2A)-expressing cytolytic CD8 T cells and CD4 + CD25+ regulatory T cells from the same cord blood unit. Exp Hematol 35:1562–1566

Tanaka J, Toubai T, Miura Y et al (2004) Differential expression of natural killer cell receptors (CD94/NKG2A) on T cells by the stimulation of G-CSF-mobilized peripheral blood mononuclear cells with anti-CD3 monoclonal antibody and cytokines: a study in stem cell donors. Transplant Proc 36:2511–2512

Toyama-Sorimachi N, Taguchi Y, Yagita H et al (2001) Mouse CD94 participates in Qa-1-mediated self recognition by NK cells and delivers inhibitory signals independent of Ly-49. J Immunol 166:3771–3779

Vacca P, Pietra G, Falco M et al (2006) Analysis of natural killer cells isolated from human decidua: evidence that 2B4 (CD244) functions as an inhibitory receptor and blocks NK-cell function. Blood 108:4078–4085

Vales-Gomez M, Reyburn HT, Erskine RA et al (1999) Kinetics and peptide dependency of the binding of the inhibitory NK receptor CD94/NKG2-A and the activating receptor CD94/NKG2-C to HLA-E. EMBO J 18:4250–4260

Van Beneden K, De Creus A, Stevenaert F et al (2002) Expression of inhibitory receptors Ly49E and CD94/NKG2 on fetal thymic and adult epidermal TCR Vγ3 lymphocytes. J Immunol 168:3295–3302

van Stijn A, Rowshani AT, Yong SL et al (2008) Human cytomegalovirus infection induces a rapid and sustained change in the expression of NK cell receptors on CD8+ T cells. J Immunol 180:4550–4560

Vance RE, Jamieson AM, Cado D, Raulet DH (2002) Implications of CD94 deficiency and monoallelic NKG2A expression for natural killer cell development and repertoire formation. Proc Natl Acad Sci USA 99:868–873

Vance RE, Jamieson AM, Raulet DH (1999) Recognition of the class Ib molecule Qa-1(b) by putative activating receptors CD94/NKG2C and CD94/NKG2E on mouse natural killer cells. J Exp Med 190:1801–1812

Vance RE, Kraft JR, Altman JD et al (1998) Mouse CD94/NKG2A is a natural killer cell receptor for the nonclassical major histocompatibility complex (MHC) class I molecule Qa-1b. J Exp Med 188:1841–1848

Varla-Leftherioti M, Spyropoulou-Vlachou M, Papadimitropoulos M, Lepage V, Balafoutas C, Stavropoulos-Giokas C et al (2003) Natural killer (NK) cell receptors' repertoire in couples with recurrent spontaneous abortions. Am J Reprod Immunol 49:183–191

Verheyden S, Bernier M, Demanet C (2004) Identification of natural killer cell receptor phenotypes associated with leukemia. Leukemia 18:2002–2007

Vetter CS, Straten PT, Terheyden P et al (2000) Expression of CD94/NKG2 subtypes on tumor-infiltrating lymphocytes in primary and metastatic melanoma. J Invest Dermatol 114:941–947

Vidal SM, Lanier LL (2006) NK cell recognition of mouse cytomegalovirus-infected cells. Curr Top Microbiol Immunol 298:183–206

Wada H, Matsumoto N, Maenaka K et al (2004) The inhibitory NK cell receptor CD94/NKG2A and the activating receptor CD94/NKG2C bind the top of HLA-E through mostly shared but partly distinct sets of HLA-E residues. Eur J Immunol 34:81–90

Wang Y, Xu H, Zheng X et al (2007) High expression of NKG2A/CD94 and low expression of granzyme B are associated with reduced cord blood NK cell activity. Cell Mol Immunol 4:377–382

Watzl C, Stebbins CC, Long EO (2000) NK cell inhibitory receptors prevent tyrosine phosphorylation of the activation receptor 2B4 (CD244). J Immunol 165:3545–3548

Wilhelm BT, Landry JR, Takei F, Mager DL (2003) Transcriptional control of murine CD94 gene: differential usage of dual promoters by lymphoid cell types. J Immunol 171:4219–4226

Wojtasiak M, Jones CM, Sullivan LC et al (2004) Persistent expression of CD94/NKG2 receptors by virus-specific CD8 T cells is initiated by TCR-mediated signals. Int Immunol 16:1333–1341

Wooden SL, Kalb SR, Cotter RJ et al (2005) HLA-E binds a peptide derived from the ATP-binding cassette transporter multidrug resistance-associated protein 7 and inhibits NK cell-mediated lysis. J Immunol 175:1383–1387

Yu J, Heller G, Chewning J et al (2007) Hierarchy of the human natural killer cell response is determined by class and quantity of inhibitory receptors for self-HLA-B and HLA-C ligands. J Immunol 179:5977–5989

Zeddou M, Greimers R, de Valensart N et al (2005) Prostaglandin E2 induces the expression of functional inhibitory CD94/NKG2A receptors in human CD8+ T lymphocytes by a cAMP-dependent protein kinase A type I pathway. Biochem Pharmacol 70:714–724

Zeddou M, Rahmouni S, Vandamme A et al (2007) Downregulation of CD94/NKG2A inhibitory receptors on CD8$^+$ T cells in HIV infection is more pronounced in subjects with detected viral load than in their aviraemic counterparts. Retrovirology 4:72

Zhang H, Robison B, Thorgaard GH, Ristow SS (2000) Cloning, mapping and genomic organization of a fish C-type lectin gene from homozygous clones of rainbow trout (Oncorhynchus mykiss). Biochim Biophys Acta 1494:14–22

Zhao XY, Huang XJ, Liu KY et al (2007) Reconstitution of natural killer cell receptor repertoires after unmanipulated HLA-mismatched/haploidentical blood and marrow transplantation: analyses of CD94:NKG2A and killer immunoglobulin-like receptor expression and their associations with clinical outcome. Biol Blood Marrow Transplant 13:734–744

Zhou J, Matsuoka M, Cantor H et al (2008) Cutting edge: engagement of NKG2A on CD8+ effector T cells limits immunopathology in influenza pneumonia. J Immunol 180:25–29

Zingoni A, Palmieri G, Morrone S et al (2000) CD69-triggered ERK activation and functions are negatively regulated by CD94/NKG2-A inhibitory receptor. Eur J Immunol 30:644–651

NKG2D Activating Receptor

Rajesh K. Gupta and G.S. Gupta

31.1 NKG2D Activating Receptor (CD314, Synonyms KLRK1)

Information on receptor ligand systems used by NK cells to specifically detect transformed cells has been accumulating rapidly. Killer cell lectin-like receptor subfamily K, member 1, also known as KLRK1, is the product of human gene. The KLRK1 has been designated as CD314 and contains a C-type lectin-like domain (CTLD). KLRK1 is also known as: KLR; NKG2D; NKG2-D; FLJ17759; FLJ75772; D12S2489E. Human NKG2D was originally identified in 1991 as an orphan receptor on NK cells (Houchins et al. 1991). Although genetically mapping near the C-type lectin receptors CD94 and NKG2A-E, the NKG2D activating NK cell receptor has little sequence homology with these receptors and is expressed as a homodimer that signals through DAP10 rather than CD94 (Chap. 30). NKG2D binds to two distinct families of ligands, the MHC class I chain-related peptides (MICA and MICB) and the UL-16 binding proteins (ULBP). These ligands are upregulated in cells that have undergone neoplastic transformation, and NK cytotoxicity on tumor cells correlates with tumor expression of MICA and ULBP. The NKG2D differs from other members of the NKG2 family in significant ways. They do not form heterodimers with CD94 on the cell surface. Instead, they are expressed as homodimers, and each homodimer associates noncovalently with a homodimer of the adaptor protein DAP-10. The cytoplasmic tail of DAP-10 carries a YxxM motif, which can recruit the regulatory subunit p85 of phosphatidylinositol-3 kinase and Grb2 (see also Chap. 30).

The NKG2D (KLRK1) is also widely expressed on T cells and other immune system cells, providing stimulatory or co-stimulatory signals. NKG2D drives target cell killing following engagement of diverse, conditionally expressed MHC class I-like protein ligands whose expression can signal cellular distress due to infection or transformation. The NKG2D ligand-binding site recognition is highly degenerate that demonstrates its ability to simultaneously accommodate multiple non-conservative allelic or isoform substitutions in the ligands. The NKG2D degeneracy is achieved using distinct interaction mechanisms at each rigid interface: recognition degeneracy by "rigid adaptation." While forming similar complexes with their ligand (HLA-E), other NKG2x NKR family members do not require such recognition degeneracy. However, NK cells are known to efficiently kill target cells that do not express HLA class I molecules, thus implying the existence of triggering receptors for non-HLA ligands (Zhang et al. 2005).

31.2 Characteristics of NKG2D

31.2.1 The Protein

NKG2D is encoded by the *KLRK1* (killer cell lectin-like receptor subfamily member 1) gene. NKG2D is a type II homodimeric transmembrane protein with an extracellular C-type (i.e. Ca^{2+}-binding) lectin-like domain. The NKG2D gene exists within the "NK complex" on human chromosome 12 and mouse chromosome 6 (Ho et al. 1998). The physical location of the *Cd94* (centromeric) and *Nkg2d* (telomeric) genes were mapped between *Cd69* and the *Ly49* cluster in the NK complex. Genes within this complex encode structurally similar type II lectin-like receptors, and many of these genes are expressed primarily in NK cells. In mice, alternative DNA splicing generates two isoforms of NKG2D that differ in the length of their cytoplasmic domains (NKG2D-S (short) and NKG2D-L (long)). Their ability to induce cellular activation is mediated via association with two membrane-bound, signaling adaptor molecules, DAP10 and DAP12. Two NKG2D isoforms, NKG2D-S and NKG2D-L, are known to associate differentially with DAP10 and DAP12 adaptor proteins. The mouse (Ho et al. 1998), rat (Berg et al. 1998), and porcine (Yim et al. 2001) homologues of NKG2D have been identified. Interspecies amino acid sequence identities range from 52% to 78% for the entire protein (mouse and rat are the most closely related sequences) and from 72% to 90% within the lectin domain. Function of NKG2D was first described in 1999 by two separate groups investigating MICA/MICB ligands (Bauer et al. 1999) or signal transduction through the DAP10 adapter protein (Wu et al. 1999). Ho et al. (1998) cloned C57BL/6-derived cDNAs homologs to human NKG2-D and CD94. Among normal tissues, murine NKG2-D transcripts are highly expressed only in activated NK cells, including both Ly49A$^+$ and Ly49A$^-$ subpopulations. Additionally, mNKG2-D is expressed in murine NK cell clones KY-1 and KY-2. NKG2D is present on the surface of natural killer (NK) cells, some NKT cells, CD8$^+$ cytotoxic T cells, subsets of γδ T cells, and under certain conditions CD4$^+$ T cells. Present in both humans and mice, NKG2D binds to a surprisingly diverse family of ligands that are distant relatives of MHC-class-I molecules. There is increasing evidence that ligand expression can result in both immune activation (tumor clearance, viral immunity, autoimmunity, and transplantation) and immune silencing (tumor evasion).

Tolerance of NK cells toward normal cells is mediated through their expression of inhibitory receptors that detect the normal expression of self in the form of MHC-I molecules on target cells. These MHC-I-binding inhibitory receptors recruit tyrosine phosphatases, which are believed to counteract activating receptor-stimulated tyrosine kinases. The perpetual balance between signals derived from inhibitory and activating receptors controls NK cell responsiveness and provides an interesting paradigm of signaling cross talk. MacFarlane and Campbell (2006) and Champsaur and Lanier (2010) reviewed the knowledge of the intracellular mechanisms by which cell surface receptors influence biological responses by NK cells with a special emphasis on the dynamic signaling events at the NK immune synapse and the unique signaling characteristics of specific receptors, such as NKG2D.

31.2.2 Orthologues to Human NKG2D

The CLEC12B is a lectin-like NK cell receptor on myeloid cells. The extracellular domain of CLEC12B shows a considerable homology to activating NKG2D, but unlike NKG2D, CLEC12B contains an ITIM in its intracellular domain. Despite the homology, CLEC12B does not appear to bind NKG2D ligands and therefore does not represent the inhibitory counterpart of NKG2D. However, CLEC12B has the ability to counteract NKG2D-mediated signaling, and this function is dependent on the ITIM and the recruitment of the phosphatases SHP-1 and SHP-2. This receptor seems to play a role in myeloid cell function (Hoffmann et al. 2007).

The rat orthologue to human NKG2D (hNKG2D) represents a separate, single gene NKLLR family, named NKR-P2. Rat NKR-P2 is expressed in NK cells and strongly expressed in resting CD81 and CD41 T cells. Sequence analysis showed that the two molecules form a distinct NKLLR family, not related to NKG2A/B, -C or -E than to other NKLLR families. *Nkrp2* is a single-copy gene containing seven introns, mapping to the rat NK gene complex. Rat NKR-P2 differs from the human orthologue in that its cytoplasmic tail contains 13 additional amino acids, encoded by a separate exon (Berg et al. 1998). NKR-P2 is a potent activation receptor on rat DCs (Srivastava et al. 2008)

31.3 NKG2D Ligands

31.3.1 The Diversity of NKG2D Ligands

The NKG2D drives target cell killing following engagement of diverse, conditionally expressed MHC class I-like protein ligands whose expression can signal cellular distress due to infection or transformation. NKG2D recognizes some self ligands such as MICA/B in human and RAE-1 in mice. The interaction of NKG2D with its ligands plays an important role in immunosurveillance of tumors and infectious pathogens, but dysregulation of this system may lead to autoimmunity. Several human NKG2D ligands (NKG2DLs), UL16-binding proteins (ULBP) 1, 2, 3 and 4, retinoic acid early transcript 1G (RAET1G) and MHC class I-related chains A and B have been reported. However, the list of ligands for NKG2D, often up-regulated during cellular distress, is growing. Moreover, NKG2D ligands have an intrinsic ability to regulate tissue-associated immune compartments. Certain cytokines and specifically the IFNs regulate expression of specific NKG2D ligands on murine tumors. Why there is such diversity of NKG2D ligands is not known but one hypothesis is that they are differentially expressed in different tissues in response to different stresses. The expression of diverse NKG2D-binding molecules in different tissues and different properties is consistent with multiple modes of infection- or stress-induced activation.

Rodent homologs of MICA and MICB have not been identified, but other molecules with weak homology to MHC class I, notably RAE-I and H-60, serve as ligands for mouse NKG2D. Like MICA and MICB, RAE-1 and H-60 are frequently expressed on tumor but not normal cells, suggesting that NKG2D functions as a receptor for tumor surveillance by NK cells. In addition, HCMV glycoprotein UL16 binding proteins (ULBP) and mouse UL16-binding protein–like transcript 1 (MULT1) have been identified as ligand for human and mouse NKG2D respectively. UL16 is believed to act as a decoy receptor for NKG2D ligands, facilitating viral evasion of the immune system. Although the homodimeric NKG2D ligands are distant structural homologs of MHC class I, including MICA, MICB, ULBP3, and RAE-1β, unlike true MHC class I molecules, NKG2D ligands neither bind peptides (or other small molecules) nor β$_2$m, and ULBP3 and RAE-1β even lack the heavy chain α3 domain, existing on the cell surface as glycophosphatidylinositol-linked α1/α2 platform domains. In humans, MICA and MICB are minimally expressed in normal tissues, but are upregulated in stressed cells and epithelial tumors. The interaction of mouse NKG2D with RAE-1, H-60 and MULT1 is disrupted by the HCMV-encoded MHC class I homologs m152, m155, m157 and m144. By preventing NK cell activation, this disruption promotes viral survival (Cerwenka 2009; Deng and Mariuzza 2006; Eagle et al. 2006 Eagle and Trowsdale 2007; Mans et al. 2007).

31.3.2 MHC Class I Chain Related (MIC) Proteins

31.3.2.1 A Second Lineage of Mammalian MHC Class I Genes

MHC class I genes typically encode polymorphic peptide-binding chains which are ubiquitously expressed and mediate the recognition of intracellular antigens by cytotoxic T cells. They constitute diverse gene families in different species and include the numerous so-called non-classical genes in the mouse H-2 complex, of which some have been adapted to variously modified functions. Bahram et al. (1994, 2005) identified a distinct family of five related sequences in human MHC which are distantly homologous to class I chains. These MIC genes (*M*HC class *I c*hain-related genes) evolved in parallel with the human class I genes and with those of most if not all mammalian orders. The MIC-A gene is located near HLA-B and is by far the most divergent mammalian MHC class I gene known. It is further distinguished by its unusual exon-intron organization and preferential expression in fibroblasts and epithelial cells. However, the presence of diagnostic residues in the MIC-A amino acid sequence translated from cDNA suggests that the putative MIC-A chain folds similarly to typical class I chains and may have the capacity to bind peptide or other short ligands. These results define a second lineage of evolutionarily conserved MHC class I genes. Major structural deviations of MIC-A from HLA-A2 (Bjorkman et al. 1987) include a deletion of amino acid positions 45–49 and an insertion of 6 amino acids at position 147. This implies that MIC-A and possibly other members in this family have been selected for specialized functions that are either ancient or derived from those of typical MHC class I genes, in analogy to some of the nonclassical mouse H-2 genes (Bahram et al. 1994).

The MIC-A and MIC-B are inducible by stress on cell surface and recognized by immunocytes bearing the receptor NKG2D, including intestinal epithelial Vδ1 γδ T cells, which may play a role in immunological reaction in intestinal mucosa. NKG2D ligands are expressed by infected and transformed cells. They transmit danger signals to NKG2D-expressing immune cells, leading to lysis of NKG2D ligand-expressing cells. Expression of MIC-A and MIC-B has been proposed to play an important role in the immunosurveillance of tumors Groh et al. (1998). Proteolytic shedding of NKG2D ligands from cancer cells therefore constitutes an immune escape mechanism impairing anti-tumor reactivity by NKG2D-bearing cytotoxic lymphocytes. Serum levels of sMICA have been shown to be of diagnostic significance in malignant diseases of various origins (Sutherland et al. 2006). NKG2D homodimers form stable complexes with monomeric MICA in solution and also with cell surface MICB, which has structural and functional properties similar to those of MICA. Comparison of allelic variants of MICA revealed large differences in NKG2D binding that were associated with a single amino acid substitution at position 129 in α2 domain. Varying affinities of MICA alleles for NKG2D may affect thresholds of NK-cell triggering and T-cell modulation (Steinle et al. 2001). MICA has no function in antigen presentation.

31.3.3 Retinoic Acid Early (RAE) Transcripts

31.3.3.1 A Cluster of MHC Class I Related Genes on Human Chromosome 6

In addition to human MIC-A and MIC-B, human ligands also include retinoic acid early transcript-1 proteins (RAET-1, originally called unique long 16) [UL-16-binding proteins (ULBP)]. Radosavljevic et al. (2002) identified a cluster of ten human MHC class I related genes, among which 6 encode potentially functional glycoproteins. The 180-kb cluster containing these genes is not located within the MHC on 6p44.3, but near the tip of its long arm at q24.2-q31.3, close to the human equivalent of the mouse H2-linked t-complex, a sub-chromosomal region syntenic to a segment of mouse chromosome 10. Mouse chromosome 10 harbors the orthologous MHC class I related transcript loci, *Raet1a-d*. The human identified loci were called *RAET1E-N*. Human RAET1 products are all devoid of membrane-proximal Ig-like α3 domain. Most, but not all RAET1 products, are predicted to remain membrane-anchored via GPI-linkage and are known to display an atypical pattern of polymorphism. *RAET1* transcripts are absent from hematopoietic tissues, but largely expressed in tumors.

Mouse NKG2D ligands include the family of retinoic acid inducible genes-1 (RAE-1α), the minor histocompatibility antigen H60, two H60 variants (H60b and H60c), and mouse UL16-binding protein–like transcript 1 (MULT1) (Diefenbach et al. 2003; Carayannopoulos et al. 2002). Ligands for mouse NKG2D were identified by their capacity to bind dimeric NKG2D-Fc fusion proteins or tetrameric NKG2D-streptavidin complexes (Cerwenka et al. 2000; Diefenbach et al. 2000). Members of RAET1 family (RAE-1α) are MHC class I-related genes located within a 180-kb cluster on chromosome 6q24.2-q31.3. RAET1 proteins contain MHC class I-like α-1 and α-2 domains. RAET1E and RAET1G differ from the other RAET1 proteins (e.g., RAET1I, or ULBP1) in that they have type I membrane-spanning sequences at their C termini rather than GPI anchor sequences (Radosavljevic et al. 2002).

The mouse NKG2D ligands are expressed at appreciable levels on thymocytes but at low levels or not at all on most normal tissues (Cerwenka et al. 2000; Diefenbach et al. 2000). Expression is up-regulated, however, on many tumor cells. RAE-1 mRNA was detected throughout early embryos and in the brain/head region of day 10–14 embryos, but transcripts were not observed in day 18 embryos (Cerwenka et al. 2000; Nomura et al. 1996). Very low levels of *Rae-1* transcripts were detected in adult spleen and liver, and transcripts were absent from adult brain and kidney. *Rae-1* transcripts were present in several transformed cell lines. Expression of NKG2D ligands RAE-1, MULT1, and H60 within the CNS following JHM strain of mouse hepatitis virus (JHMV) infection demonstrated a functional role for NKG2D in host defense during acute viral encephalitis by selectively enhancing CTL activity by infiltrating virus-specific CD8[+]T cells (Walsh et al. 2008).

The cDNAs encoding NKG2D ligands, RAE-1ß and RAE-1 (Cerwenka et al. 2000; Diefenbach et al. 2000) had been described earlier as products of retinoic acid early inducible (Rae-1) transcripts (Zou et al. 1996; Nomura et al. 1996). This family is known to contain at least four distinct loci encoding polypeptides that share 92–95% amino acid identity. All RAE-1 family members were shown to bind NKG2D (Zou et al. 1996; Savithri and Khar 2006). Savithri and Khar (2006) described the cloning of a ligand for NKR-P2, the rat ortholog of human and mouse NKG2D, termed as rat RAE-1-like transcript (RRLT). The RRLT ligand is homologous to mouse RAE-1 family of proteins, but differs from them in being transmembrane anchored. Strid et al. (2008) reported that epidermis-specific upregulation of RAE-1 induces changes in the organization of tissue-resident Vγ5Vδ1 TCRγδ[+] intraepithelial T cells and Langerhans cells, followed by epithelial infiltration by unconventional αβT cells. Whereas local Vγ5Vδ1[+] T cells attempted to limit carcinogenesis, Langerhans cells unexpectedly promoted it.

31.3.3.2 The UL16-Binding Proteins

The UL16-binding proteins (ULBPs, also termed as retinoic acid early transcripts, encoded by RAET1 genes) are a family of human, MHC class I-related, cell surface proteins. ULBP1 was identified based on its ability to bind to human CMV (HCMV) glycoprotein, UL16. ULBP2 and ULBP3 were subsequently discovered as expressed sequence tags with high homology to ULBP1. Unlike traditional MHC class I proteins, ULBPs are GPI-linked, lack an α$_3$ domain, and do not associate with β$_2$-microglobulin (Cosman et al. 2001). ULPB transcripts were found in various tissues in healthy individuals, including heart, lung, liver, and testis (Cosman et al. 2001). UL16 also binds to a member of another family of nonclassical MHC class I proteins, MIC-B (Groh et al. 1996). Presently, the human RAET1/ULBP gene family comprises of ten members (RAET1E to N) with six loci encoding for potentially functional proteins. These are ULBP1 or RAET1I, ULBP2 or RAET1H, ULBP3 or RAET1N, and RAET1L, which are glycosylinositol phospholipid (GPI)-linked glycoproteins and ULBP4 or RAET1E and ULBP5 or RAET1G, which are transmembrane glycoproteins. The RAET1 products contain α1 and α2 domains but lack α3 domain and do not associate with β2-microglobulin. RAET1/ULBPs have tissue-specific expressions, and some of them are also polymorphic. The MIC-B and the closely related MIC-A protein share some similar properties with ULBPs (Sutherland et al. 2002).

ULBPs are important activators of NK cells. Soluble recombinant forms of ULBPs bind to human NK cells and stimulate NK cytotoxicity against tumor targets (Kubin et al. 2001). Soluble ULBPs also induce production of the cytokines IFN-γ, GM-CSF, TNF-α, and TNF-β, and the chemokines macrophage-inflammatory protein (MIP)-1α, MIP-1β, and I-309. In all cases, costimulation with IL-12 has a superadditive effect on the production of these factors (Cosman et al. 2001; Kubin et al. 2001).

UL16-binding proteins are frequently expressed by malignant transformed cells, a variety of leukemias, carcinomas, melanomas, and tumor cell lines and mediate cytotoxicity of NKG2D-positive NK cells, CD8[+] αβ T cells and γδ T cells in vitro and in vivo to tumor cells. ULBPs 1, 2, 3 and 4 are functional ligands of the NKG2D/DAP10 receptor complex on human NK cells. ULBP1, −2, −3, −4 and RAET1L are linked to membrane through GPI. Two more members of the RAET1/ULBP gene cluster, RAET1E and RAET1G have been characterized by Bacon et al. (2004). The encoded proteins RAET1E and RAET1G were similar to the ULBP in their class I-like α1 and α2 domains, but differed in one aspect: instead of being GPI-anchored, their sequences were like type 1 membrane-spanning molecule and contain transmembrane and cytoplasmic domains. Both proteins were capable of being expressed at the cell surface. Both proteins bound the activating receptor NKG2D, and RAET1G bound the human CMV protein UL16. Tissue expression of ULBP4 differs from other members of the family, in that it is expressed predominantly in the skin (Chalupny et al. 2003). ULBP-1 can be detected on mature DCs both in situ in the T cell areas of lymph nodes as well as in vitro after artificial maturation (Schrama et al. 2006). Human gastric cancer cell lines, which expressed ULBP are susceptible to NK cells in a NKG2D-dependent manner. However, cancer cells which had no ULBP on their surface were resistant to NK cells. ULBP 1, 2, and 3 being GPI-anchored proteins are sensitive to phosphatidylinositol-specific phospholipase C (PI-PLC). Down-regulation of NKG2D by soluble ULBP provides a potential mechanism by which gastric cancer cells escape NKG2D-mediated attack by the immune cells (Song et al. 2006). The ULBP4 binds to TCRγ9/δ2 and induces cytotoxicity to tumor cells through both TCRγδ and NKG2D (Kong et al. 2009).

In the NKG2D/ULBP3 complex, the structure of ULBP3 resembles α1 and α2 domains of classical MHC molecules without a bound peptide. The lack of α3 and β2m domains is compensated by replacing two hydrophobic patches at the underside of the class I MHC-like β sheet floor with a group of hydrophilic and charged residues in ULBP3. NKG2D binds diagonally across the ULBP3 α helices, creating a complementary interface, an asymmetrical subunit orientation, and local conformational adjustments in the receptor. The interface is stabilized primarily by hydrogen bonds and hydrophobic interactions. Unlike the KIR receptors that recognize a conserved HLA region by a lock-and-key mechanism, NKG2D recognizes diverse ligands by an induced-fit mechanism (Radaev et al. 2001).

Monocyte-derived DCs (mDCs) and NK cells are reciprocally activated via cytokines and cell-cell contact. ULBPs have been detected on human mDCs. While ULBP1 was upregulated on mDCs in response to LPS or infection with human respiratory syncytial virus (RSV), the expression of ULBP2 was induced by LPS and poly I:C, indicating that the TLR-containing adapter molecule-1, IFN inducing pathway is associated with ULBP2 (Ebihara et al. 2007).

Proteolytic cleavage of ULBP2 produces truncated and soluble forms that may counteract NKG2D-mediated tumor immune surveillance. It is suggested that RAET1E can produce a soluble, 35-kDa protein (termed as RAET1E2) lacking the transmembrane region by selective splicing in tumor cells. The expressions of both RAET1E2 transcripts and protein can be found in different tumor cells and tissues. Incubation of NK-92 cells with RAET1E2 protein decreases the surface expression of NKG2D and in marked reduction in cytotoxiccity to tumor cells. These results suggest for an immune escape mechanism of tumors via alternative splicing of ULBP RNA to generate RAET1E2 that may impair NKG2D-mediated NK cell cytotoxicity to tumors (Cao et al. 2007).

31.3.4 Role of NKG2D Ligands

As NKG2D plays an important role in the immunosurveillance of tumors, studies suggest that release of MIC-A from cancer cells constitutes an immune escape mechanism that systemically impairs antitumor immunity. The interaction of human NK cells with MIC-A-positive human cancer cells in an in vivo setting showed that MIC-A overexpression can function as NK cell-mediated immunotherapy in experimental lung cancer. Moreover, low-dose application of the proteasome inhibitor bortezomib enhances expression of human NKG2D ligands MIC-A and MIC-B on hepatocellular carcinoma cells (Armeanu et al. 2008; Busche et al. 2006).

An understanding of effects of hyperthermia on NK cell cytotoxicity is important for maximizing the clinical benefits in cancer therapy. At temperatures above 40°C, (normally achieved during fever or exercise), both enhancing and inhibitory effects on cytotoxic activity of NK cells against tumor cells have been reported. Dayanc et al. (2008) have shown that thermal stress (using a temperature of 39.5°C) enhances human NK cell cytotoxicity against tumor target cells. This effect requires function of the NKG2D receptor of NK cells, and is maximal when both NK and tumor cell targets are heated. Heat sensitive cellular targets affected by hyperthermia on tumor cells include HSPs, MIC-A and MHC Class I. The expressions of NKG2D ligands were induced by both heat shock and ionizing radiation in various cell lines with peaks at 2 h and 9 h after treatment, respectively, although inducibility of each NKG2D ligand was dependent on cell lines (Kim et al. 2006). The available studies indicate a strong potential for heat-induced enhancement of NK cell activity in mediating, at least in part, the improved clinical responses seen when hyperthermia is used in combination with other therapies (Dayanc et al. 2008).

31.3.5 Regulation of Ligands

The three main mechanisms by which NKG2D ligand transcription can be induced are: DNA damage, TLR stimulation, and cytokine exposure. The DNA damage response pathway is involved in maintaining the integrity of the genome. The PI3K-related protein kinases ATM (ataxia telangiectasia, mutated) and ATR (ATM and Rad3 related) sense DNA lesions, specifically double-strand breaks and stalled DNA replication, respectively. This sensing results in cell-cycle arrest and DNA repair, or cell apoptosis if the DNA damage is too extensive to be repaired. This pathway has been shown to be constitutively active in human cancer cells (Bartkova et al. 2005; Gorgoulis et al. 2005; Stephan and David 2006). The DNA damage response, previously known to arrest the cell cycle and enhance DNA repair functions, or to trigger apoptosis, may also participate in alerting the immune system to the presence of potentially dangerous cells (Gasser et al. 2005). Gasser et al. provided evidence that this pathway actively regulates NKG2D ligand transcription (Gasser et al. 2005). Both mouse and human cells upregulated NKG2D ligands following treatment with DNA-damaging agents. This effect was dependent on ATR function, as inhibitors of ATR and ATM kinases prevented ligand upregulation in a dose-dependent fashion. These findings provide a link between the constitutive activity of the DNA damage response in tumors (Bartkova et al. 2005; Gorgoulis et al. 2005) and the frequent upregulation of NKG2D ligands by these transformed cells. The exact mechanism linking the ATR/ATM-dependent recognition of DNA damage and the transcription of NKG2D ligands remain elusive (Champsaur and Lanier 2010). Toll-like receptor (TLR) signaling and Cytokines can also influence NKG2D ligand expression. In particular, interferons have pleiotropic effects on NKG2D ligand expression. Transforming growth factor (TGF-β) also decreases the transcription of MICA, ULBP2, and ULPB4 on human malignant gliomas. Therefore, cytokines and interferons can differentially affect NKG2D ligand expression in different cell types and environments. Other stimuli have also been reported to induce NKG2D ligand transcription.

Various mechanisms are responsible for the post-transcriptional regulation of NKG2D ligands. A group of endogenous cellular microRNAs (miRNAs) that bound to the 3′-UTR (untranslated region) of MICA and MICB, repressed their translation. Nice et al. showed that MULT1 protein undergoes ubiquitination dependent on the lysines in its cytoplasmic tail, resulting in its rapid degradation. Thomas et al. have recently described the capacity of the KSHV (Kaposi's sarcoma-associated herpesvirus)-encoded E3 ubiquitin ligase K5 to down-regulate cell surface expression of MICA and MICB. In this case, ubiquitination resulted in the redistribution of MICA to the plasma membrane, rather than its targeting to degradation as observed with MULT1 (Champsaur and Lanier 2010).

31.4 Crystal Structure of NKG2D

The basic folding pattern of murine NKG2D is similar to other C-type lectin domains such as CD94, Ly49A, rat MBP-A, and CD69 despite relatively low sequence homology (Boyington et al. 1999; Natarajan et al. 2000; Tormo et al. 1999; Weis et al. 1991). However, the precise mode of dimeric assembly varies among these natural killer receptors, as well as their surface topography and electrostatic properties.

31.4.1 Structures of NKG2D-Ligand Complexes

The NKG2D structure provides the first structural insights into the role and ligand specificity of this stimulatory receptor in the innate and adaptive immune system (Wolan et al. 2001) (Fig. 31.1). Crystal structures have been determined for mouse and human NKG2Ds in complex with MICA (Li et al. 2001). The structure of MICA - NKG2D complex reveals an NKG2D homodimer bound to a MICA monomer in an interaction that is analogous to that seen in αß-T cell receptor-MHC-I protein complexes. Similar surfaces on each NKG2D monomer interact with different surfaces on either the α1 or α2 domains of MICA. The binding interactions are large in area and highly complementary. The central section of the α2-domain helix, disordered in the structure of MICA alone, is ordered in the complex and forms part of the NKG2D interface. The extensive flexibility of the interdomain linker of MICA is shown by its altered conformation when crystallized alone or in complex with NKG2D (Li et al. 2001). Although there is little obvious sequence similarity between the human and mouse NKG2D ligands, there appears to be considerable structural similarity as evidenced by the ability of mouse NKG2D to react with human ligands, MICB, ULBP1, and ULBP2 (Kubin et al. 2001).

In NKG2D/ULBP3 complex, the structure of ULBP3 resembles the α1 and α2 domains of classical MHC molecules without a bound peptide. The lack of α3 and β2m domains is compensated by replacing two hydrophobic patches at the underside of the class I MHC-like β-sheet floor with a group of hydrophilic and charged residues in ULBP3. NKG2D binds diagonally across the ULBP3 α-helices, creating a complementary interface, an asymmetrical subunit orientation, and local conformational adjustments in the receptor. The interface is stabilized primarily by hydrogen bonds and hydrophobic interactions. Unlike the KIR receptors that recognize a conserved HLA region by a lock-and-key mechanism, NKG2D recognizes diverse ligands by an induced-fit mechanism (Radaev et al. 2001).

Rodent RAE-1 proteins (α, β, γ and δ) are distant MHC class I homologs, comprising isolated α1α2 platform domains. The crystal structure of RAE-1β was distorted from other MHC homologs and displayed noncanonical disulfide bonds. The loss of any remnant of a peptide binding groove was facilitated by the close approach of the groove-defining helices through a hydrophobic, leucine-rich interface. The RAE-1β-murine NKG2D complex structure resembled the human NKG2D-MICA receptor-ligand complex and further demonstrated the promiscuity of the NKG2D ligand binding site (Li et al. 2002).

The interaction of NKG2D receptor with several MHC I-like ligands has been analyzed thermodynamically. NKG2D ligand-binding site recognition is highly degenerate, as demonstrated by NKG2D's ability to simultaneously accommodate multiple non-conservative allelic or isoform substitutions in the ligands. In TCRs, "induced-fit" recognition explains cross-reactivity, but analyses of multiple NKG2D-ligand pairs showed that rather than classical "induced-fit" binding, NKG2D degeneracy is achieved using distinct interaction mechanisms at each rigid interface: recognition

31.4 Crystal Structure of NKG2D

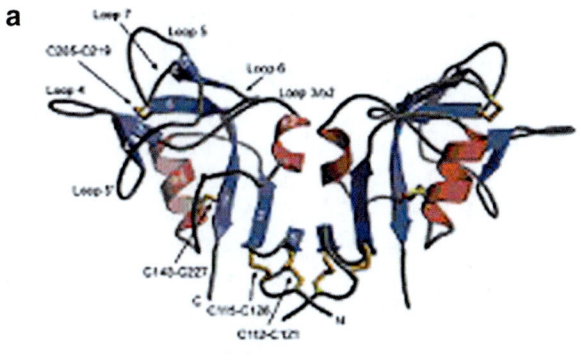

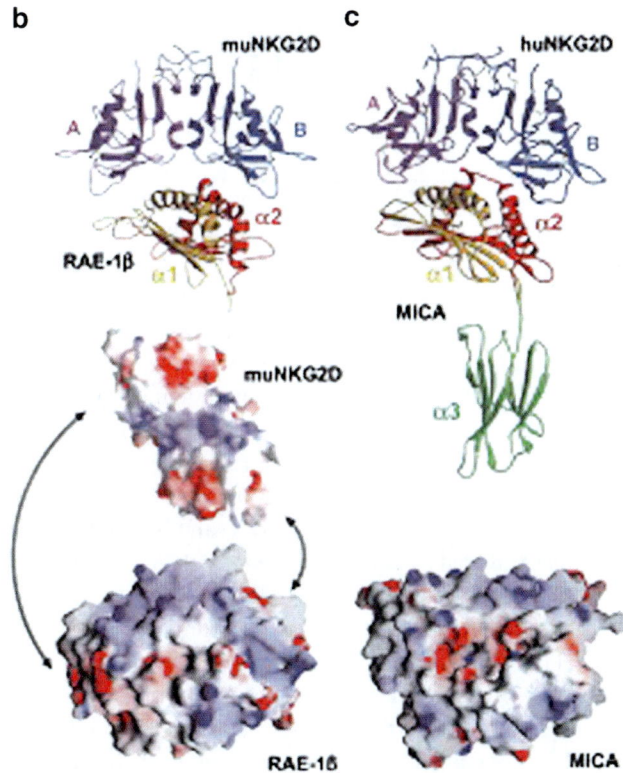

Fig. 31.1 (a) **Anatomy of killer cell lectin-like receptor subfamily K, member 1 (KLRK1) or NKG2D.** Structure of murine NKG2D homodimer shows side view showing the NH_2 terminus at the *bottom* of the structure that would connect to the membrane at the cell surface by a stalk comprised of 16 residues. The dimer is formed primarily of an extension of the first β-sheet (β1) across the interface. The NKG2D structure has a C-type lectin fold like the one reported in CD94 (Boyington et al. 1999), Ly49A (Tormo et al. 1999) and rat MBP-A (Weis et al. 1991). However, loop 5 in NKG2D is not a Ca^{2+}-dependent carbohydrate-binding loop like rat MBP-A. Only CD94 does not contain an α2-helix in loop 3 (Figure not shown) (Adapted with permission from Boyington et al. 1999 © Elsevier). (b) and (c) **Structures of murine NKD NK cell receptor-ligand complexes:** ribbon representations (*top*) and GRASP molecular surfaces (*bottom*) are shown for the structures of (a) the muNKG2D–RAE-1β, and (b) huNKG2D–MICA. Ribbons of the ligands are colored by domain: α1, *yellow*; α2, *red*; α3 (when present), *green*; and β₂-m (when present), *cyan*; ribbons of the receptors are colored by chain: *blue* or *purple*. Molecular surfaces of the platform domains are oriented such that the view is looking down onto the NKG2D binding surface of RAE-1 and MICA. In (a), the molecular surface of muNKG2D was included in an orientation looking down onto the RAE-1 binding surface, as if the receptor had been peeled away from the complex. Molecular surfaces are colored by electrostatic potential, with positively charged areas in blue and negatively charged areas in red. The RAE-1β-murine NKG2D complex structure resembles the human NKG2D-MICA receptor-ligand complex and demonstrates the promiscuity of the NKG2D ligand binding site (Adapted with permission from Li et al. 2002 © Elsevier)

Table 31.1 Activating homodimeric receptors and co-receptors with two basic TM residues

	Receptor	Signaling polypeptide	Ligand	Protein family
Human	NKG2D	DAP10	MIC-A/MIC-B; ULBPs(RAET1)	C-type lectin
	NKG2F (NKp80)	DAP12	AICL	C-type lectin
	NKp46	ξ/γ	Heparin, lewis-x	Ig-SF
	NKp30	ξ/γ	HLA-B-associated transcript 3;B7h6	Ig-SF
	NKp44	DAP12	Heparan sulfate epitope(s)	Ig-SF
Mouse	NKG2D	DAP10	MIC-A/MIC-B; ULBPs (RAET1)	C-type lectin
	Ly49D	DAP12	H2-D	C-type lectin
	Ly49H	DAP12	m157	C-type lectin
	NKRP1C	ξ/γ	unknown	C-type lectin
	NKRP1F	?	Clr-g	C-type lectin

TM transmembrane residues

degeneracy by "rigid adaptation." While likely forming similar complexes with their ligand (HLA-E), other NKG2x NKR family members do not require such recognition degeneracy (Strong and McFarland 2004).

31.5 DAP10/12 Adapter Proteins

31.5.1 DAP10

A number of receptors that are primarily expressed by human and/or murine NK cells represent homodimers with a basic residue in each transmembrane (TM) domain and may thus form a hexameric structure (Table 31.1). The human NKp46 receptor can assemble with Fcγ or ξ-ξ signaling dimers (Pessino et al. 1998), and experiments demonstrated that it forms a homodimer (Garrity et al. 2005). The human NKp80 and NKp40 and mouse Ly49H and Ly49D receptors assemble with DAP12 (Smith et al. 1998), whereas the mouse NKR-P1C receptor assembles with Fcγ (Lanier 2005). The stoichiometry described for NKG2D–DAP10 complex may be relevant for several receptors from the C-type lectin and immunoglobulin families.

In the recognition of self from non-self, activating and inhibitory receptors signal immune cells via adapter molecules, which determine the outcome of immune response. NKG2D lacks signal elements in its cytoplasmic domain. However, DAP10, the transmembrane adapter protein expressed broadly in hematopoietic cells, associates with NKG2D activating receptor forming a multisubunit complex, which recognizes self-proteins upregulated during tumorigenesis, infection, and autoimmune response. NKG2D can deliver stimulatory signals only in association with DAP10 or DAP12. Human NKG2D transduces activation signals exclusively via an associated DAP10 adaptor containing a YxNM motif, whereas murine NKG2D can signal through either DAP10 or the DAP12 adaptor, which contains an ITAM sequence. DAP10 signaling is thought to be mediated, at least in part, by PI3K and is independent of Syk/Zap-70 kinases. NKG2D couples to the non-ITAM-containing DAP10 and initiates at least two signaling branches that are both required for cytotoxicity. The NKG2D-mediated response can be modulated by factors such as cytokine milieu and possibly the particular ligand expressed. This nontraditional NKG2D-DAP10 initiated signal pathway enables lymphocytes to differentially respond to ligands based on cellular environment (Upshaw and Leibson 2006).

31.5.2 KARAP (DAP12 or TYROBP)

The signaling adaptor protein KARAP/DAP12/TYROBP (killer cell activating receptor-associated protein/DNAX activating protein of 12 kDa/tyrosine kinase binding protein) belongs to the family of transmembrane polypeptides bearing an intracytoplasmic immunoreceptor tyrosine-based activation motif (ITAM). This adaptor, initially characterized in NK cells, is associated with multiple cell-surface activating receptors expressed in both lymphoid and myeloid lineages. The main features of KARAP/DAP12 describe its involvement in a broad array of biological functions. KARAP/DAP12 is a wiring component for NK cell anti-viral function (e.g. mouse cytomegalovirus via its association with mouse Ly49H) and NK cell anti-tumoral function (e.g. via its association with mouse NKG2D or human NKp44). KARAP/DAP12 is also involved in inflammatory reactions via its coupling to myeloid receptors, such as triggering receptors expressed by myeloid cells (TREM) displayed by neutrophils, monocytes/macrophages and dendritic cells. Moreover, bone remodeling and brain function are also dependent upon the integrity of KARAP/DAP12 signals, as shown by the analysis of KARAP/DAP12-deficient mice and KARAP/DAP12-deficient Nasu-Hakola patients (Tomasello and Vivier 2005).

31.5.3 Characterization of DAP10 and DAP12

Human, mouse, pig, and the bovine DAP10 cDNA clones codes for a phosphatidyl-inositol-3 (PI-3) kinase-binding site (YxxM) in its cytoplasmic region. Similar to human, mouse, pig, the bovine DAP12 demonstrates one ITAM in its cytoplasmic domain (Fig. 31.2). The short mouse NKG2D-S and the bovine NKG2D could associate with either DAP10 or DAP12 adaptor protein for its cell surface expression. The advances in our understanding of the structural properties and signaling pathways of the inhibitory

31.5 DAP10/12 Adapter Proteins

and activating NK cell receptors, with a particular focus on the ITAM-dependent activating receptors, the NKG2D-DAP10 receptor complexes system have been reviewed (Lanier 2008, 2009).

NKG2D is required for cytolysis of tumor cells and both activated and expanded $CD8^+$ T cells and NK cells use DAP10. In addition, direct killing was partially dependent on the DAP12 signaling pathway. Silencing human NKG2D, DAP10, and DAP12 reduces cytotoxicity of activated $CD8^+$ T cells and NK cells in human effector cell function (Karimi et al. 2005), although NKG2D function was normal in patients lacking functional DAP12, indicating that DAP10 is sufficient for human NKG2D signal transduction (Rosen et al. 2004).

31.5.3.1 NKG2D Receptor Assembles in the Membrane with Two Signaling Dimers into a Hexameric Structure

The human receptor NKG2D assembles with the DAP10 signaling dimer and it was thought that one NKG2D homodimer pairs with a single DAP10 dimer by formation of two salt bridges between charged transmembrane (TM) residues. However, stoichiometry analysis demonstrated that one NKG2D homodimer assembles with four DAP10 chains giving hexameric structure (Fig. 31.2). Selective mutation of one of the basic TM residues of NKG2D resulted in loss of two DAP10 chains, indicating that each TM arginine serves as an interaction site for a DAP10 dimer. Assembly of the hexameric structure was cooperative because this mutation also significantly reduced NKG2D dimerization. A monomeric NKG2D TM peptide was sufficient for assembly with a DAP10 dimer, indicating that the interaction between these proteins occurs in the membrane environment. Formation of a three-helix interface among the TM domains involved ionizable residues from all three chains: the TM arginine of NKG2D and both TM aspartic acids of the DAP10 dimer. The organization of the TM domains thus shows similarities to the T cell antigen receptor-CD3 complex, in particular to the six-chain assembly intermediate between T cell antigen receptor and the CD3δε and CD3γε dimers. Binding of a single ligand can thus result in phosphorylation of four DAP10 chains, which may be relevant for the sensitivity of NKG2D receptor signaling, in particular in situations of low ligand density (Garrity et al. 2005) (Fig. 31.2). DAP10 signaling is involved in adjusting the activation threshold and generation of NKT cells and T_{reg} to avoid autoreactivity, but also modulates antitumor mechanisms (Hyka-Nouspikel and Phillips 2006).

31.5.3.2 Association of Adapter Proteins with NKG2D Splice Variants

It has been reported that the long form of murine NKG2D (NKG2D-L) associates exclusively with DAP10, whereas the short form of NKG2D (NKG2D-S) can pair with either DAP10 or DAP12. The short isoform was almost undetectable in naive NK cells. However, using two distinct cell types, it was observed that like the short isoform, the long variant of NKG2D also associates not only with DAP10 but also with DAP12. Cross-linking either isoform of NKG2D induces a calcium flux when associated exclusively with DAP10 or DAP12 (Rabinovich et al. 2006). The differential adapter pairing has functional consequences, as the different adapters trigger distinct signaling cascades. The cytoplasmic YINM motif of DAP10 recruits the p85 subunit of phosphoinositide kinase-3 (PI3K) and growth factor receptor-bound protein 2 (Grb2). DAP12 carries an immunoreceptor tyrosine-based activation motif (ITAM) whose phosphorylation leads to the recruitment of the zeta-chain associated protein kinase 70 (Zap70) and spleen tyrosine kinase (Syk). Thus, NKG2D engagement can result in both the PI3K and Grb2 and Syk and Zap70 signaling cascades (Fig. 31.2). It appears that signaling via mouse NKG2D isoforms is more complex than originally presented (Champsaur and Lanier 2010).

31.5.4 NKG2D Receptor Complex and Signaling

Each NKG2D homodimer associates with two homodimers of DAP10, forming a hexameric structure. Whether a single NKG2D homodimer can pair with both DAP10 and DAP12 homodimers has not yet been determined, but one could imagine this scenario to be beneficial to induce both signaling cascades upon triggering a single receptor. Crystal structures of both mouse and human NKG2D receptors in the soluble form and bound to ligands suggest that NKG2D binds to its ligands through "rigid adaptation" recognition, allowing binding to a wide variety of ligands (Champsaur and Lanier 2010).

The NK cell signaling pathway illustrates the signals produced by the activating cell surface receptors that initiate PTK (protein tyrosine kinase)-dependent pathways through their noncovalent association with transmembrane signaling adaptors that harbor ITAMs. Vivier et al. (2004) described the mechanism by which these positive pathways are antagonized by intracytoplasmic PTPs (protein tyrosine phosphatases) that are activated upon engagement of cell surface receptors with intracytoplasmic ITIMs. The tyrosine phosphorylation status of several signaling components that are substrates for both PTKs and PTPs is thus key to the propagation of the NK cell effector pathways. Additional cell surface receptors that are not directly coupled to ITAMs also participate in NK cell activation, such as NKG2D (which is noncovalently associated with the DAP10 transmembrane signaling adaptor), adhesion molecules, and cytokine receptors (Vivier et al. 2004).

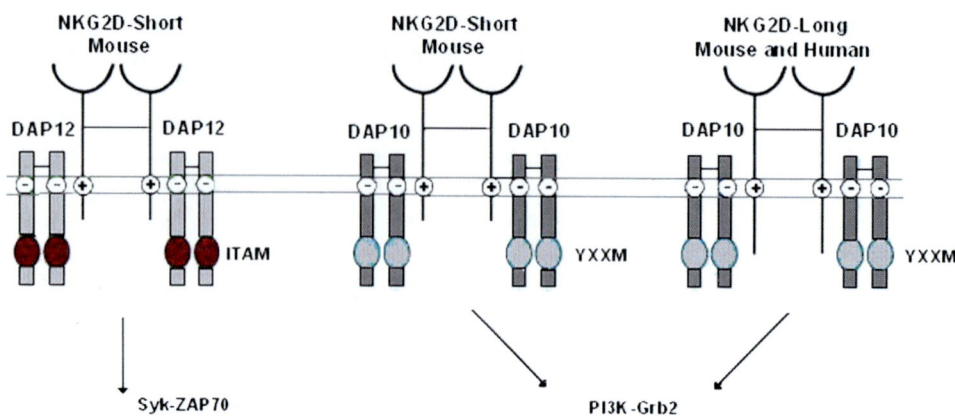

Fig. 31.2 Schematic representation of the NKG2D receptor complexes and ITAM-mediated signaling in NK cells: Hexameric NKG2D–DAP10 or NKG2D–DAP12 receptor complexes: Murine NKG2D exists as a homodimer on the surface of NK cells (Fig. 31.1a). The side view shows that the dimmer is formed as an extension of an anti parallel β-sheet across the interface with a two-fold crystallographic axis perpendicular to the extended sheet. The NKG2D dimmer interface is primarily composed of a central hydrophobic core encased by a hydrogen bond and a salt bridge network (Call et al. 2010). Each TM domain of human NKG2D assembles with one DAP10 dimer. This interaction involves both TM aspartic acids of DAP10 and the TM arginine (Garrity et al. 2005). NKG2D signals via association with adapter molecules through charged residues in the transmembrane domain. Mouse NKG2D associates with both DAP10 and DAP12 signaling molecules, whereas human NKG2D associates with DAP10 only. The pairing of NKG2D with DAP12 results in the phosphorylation of immunoreceptor tyrosine-based activation motif (ITAM) and triggering of the Syk and/or Zap70 cascade. Association with DAP10 leads to tyrosine phosphorylation on the YINM motif and triggering of the PI3K and Grb2 signaling cascade. ITAM-bearing signaling subunits are phosphorylated, probably by Src family kinases, after receptor engagement. Syk and/or ZAP-70 (both of which are expressed by human and mouse NK cells) are recruited to the phosphorylated ITAMs, initiating a cascade of downstream signaling as depicted. The signaling pathways depicted are hypothetical and were deduced by synthesizing results from many studies investigating ITAM-coupled receptor signaling in human and mouse NK cells (Reproduced with permission from Champsaur and Lanier 2010 © John Wiley and Sons)

Findings suggest how NKG2D-DAP10 mediates cytotoxicity and provides a framework for evaluating activation by other receptor complexes that lack ITIM (Upshaw et al. 2006).

Isoforms of PLC-γ in NK Cell Cytotoxicity and Innate Immunity: The two isoforms of PLC-γ couple immune recognition receptors to important calcium- and PKC-dependent cellular functions. It has been assumed that PLC-γ 1 and PLC-γ2 have redundant functions and that the receptors can use whichever PLC-γ isoform is preferentially expressed in a cell of a given hemopoietic lineage. The ITAM-containing immune recognition receptors can use either PLC- γ1 or PLC-γ2, whereas the NK cell-activating receptor NKG2D preferentially couples to PLC-γ2. Experimental models evaluating signals from either endogenous receptors (FcR verus NKG2D-DAP10) or ectopically expressed chimeric receptors (with ITAM-containing cytoplasmic tails vs DAP10-containing cytoplasmic tails) showed that PLC-γ1 and PLC-γ 2 both regulate the functions of ITAM-containing receptors, whereas only PLC-γ2 regulates the function of DAP10-coupled receptors. These data suggest that specific immune recognition receptors can differentially couple to the two isoforms of PLC-γ (Upshaw et al. 2005). Cytotoxicity was completely abrogated in PLC-γ2-deficient cells, regardless of whether targets expressed NKG2D ligands or others. It was suggested that exocytosis of cytotoxic granules, but not cellular polarization toward targets, depends on intracellular Ca^{2+} rise during NK cell cytotoxicity. In vivo, PLC-γ2 regulates selective facets of innate immunity because it is essential for NK cell responses to malignant and virally infected cells but not to bacterial infections (Caraux et al. 2006). Hence PLCγ1 and PLCγ2 play non-redundant and obligatory roles in NK cell ontogeny and in its effector functions (Regunathan et al. 2006).

Phosphatidylinositol 3-kinase Activation in the Formation of NKG2D Immunological Synapse: In human cells, NKG2D signaling is mediated through the associated DAP10 adapter. However, Giurisato et al. (2007) showed that engagement of NKG2D by itself is sufficient to stimulate the formation of the NK immunological synapse (NKIS), with recruitment of NKG2D to the center synapse. Mutagenesis studies of DAP10 revealed that the phosphatidylinositol 3-kinase binding site, but not the Grb2 binding site, was required and sufficient for recruitment of DAP10 to the NKIS. Surprisingly, in the absence of the Grb2 binding site, Grb2 was still recruited to the NKIS. Since the recruitment of Grb2 was dependent on phosphatidylinositol-trisphosphate (PIP3), the possibility that recruitment to the NKIS is mediated by a pleckstrin homology (PH) domain-containing binding partner for Grb2 was explored. It was found that the PH domain of SOS1, but not that of

Vav1, was able to be recruited by PIP3. These results demonstrate how multiple mechanisms can be used to recruit the same signaling proteins to the plasma membrane (Giurisato et al. 2007).

Many NK Cell Receptors Activate ERK2 and JNK1 to Trigger Microtubule Organizing Center: The NK activating receptors on NK cell signal the polarization of the microtubule organizing center (MTOC) together with cytolytic granules to the synapse with target cells. After ligation of any one of these receptors, Src family kinases initiate activation of two signal pathways, the phosphoinositide-3 kinase —>ERK2 and the phospholipase Cγ —>JNK1 pathways. Both are required for polarization of the MTOC and cytolytic granules, a prerequisite for killing the targets. Crosslinking of CD28, NKG2D, NKp30, NKp46, NKG2C/CD94, or 2B4 leads to the phosphorylation of both ERK2 and JNK1, although they use different proximal signaling modules. Thus, many, if not all, activating receptors stimulate these two distal pathways, independent of the proximal signaling module used. By contrast, CD2, DNAM-1, and β1-integrin crosslinking do not activate either pathway; they may be costimulatory molecules or have another function in the synapse (Chen et al. 2007).

In search of signaling pathway involved after NKG2D ligation, Li et al. either incubated NKG2D-bearing human NKL tumor cell line with K562 target cells or cross-linked with NKG2D mAb induced strong activation of the MAP kinases. Selective inhibition of JNK MAP kinase greatly reduced NKG2D-mediated cytotoxicity toward target cells and blocked the movement of the MTOC, granzyme B, and paxillin (a scaffold protein) to the immune synapse. NKG2D-induced activation of JNK kinase was also blocked by inhibitors of Src protein tyrosine kinases and phospholipase PLCγ, upstream of JNK. A second MAP kinase pathway through ERK was also shown to be required for NK cell cytotoxicity. Thus, activation of two MAP kinase pathways is required for cytotoxic granule and MTOC polarization and for cytotoxicity of human NK cells when NKG2D is ligated (Li et al. 2008).

Regulation of NK Cell-Mediated Cytotoxicity by the Tyrosine Kinase Itk: In activated human NK cells, the tyrosine kinase IL-2-inducible T cell kinase (Itk) differentially regulates distinct NK-activating receptors. Itk plays a complex role in regulating the functions initiated by distinct NK cell-activating receptors. Enhanced expression of Itk leads to increases in calcium mobilization, granule release, and cytotoxicity upon stimulation of the ITAM-containing FcR, suggesting that Itk positively regulates FcR-initiated cytotoxicity (Khurana et al. 2007). Gross et al. (2008) identified a molecular basis for the segregation of NK cell receptor-induced signals for cytokine release and target cell killing and suggested the roles for CARD-protein/Bcl10/Malt1 complexes in ITAM receptor signaling in innate and adaptive immune cells (Gross et al. 2008).

31.6 Functions of NKG2D

31.6.1 Engagement of NKG2D on γδ T Cells and Cytolytic Activity

Engagement of NKG2D on NK cells and γδ T cells can trigger cytolytic activity. The number of γδ T cells has been found to increase in some tumor tissues. This was demonstrated in transfectants that express the MIC proteins (Bauer et al. 1999; Cosman et al. 2001), the ULBPs (Cosman et al. 2001), the Rae-1 proteins (Cerwenka et al. 2000; Diefenbach et al. 2000) and H-60 (Cerwenka et al. 2000; Diefenbach et al. 2000). The parent cell lines were resistant to NK lysis while the transfectants were readily susceptible to NK lytic activity. The results also indicated that the NKG2D activating signal can overcome the inhibitory signals resulting from MHC class I recognition (Bauer et al. 1999; Cosman et al. 2001). Populations of γδ T cells that occur in tumors and in the intestinal epithelium also killed target cells, which expressed MIC proteins (Girardi et al. 2001). Most tumor cells express comparable levels of MICA and MICB as well as ULBP (Groh et al. 1999). Most epithelial tumors were susceptible to allogeneic γδT-cell lysis and in the case of an established ovarian carcinoma to autologous γδT cell killing. The NKG2D pathway appears to be involved in the lysis of different melanomas, pancreatic adenocarcinomas, squamous cell carcinomas of the head and neck, and lung carcinoma (Wrobel et al. 2007). The CD56$^+$ γδ T cells are potent anti-tumor effectors capable of killing squamous cell carcinoma and may play an important therapeutic role in patients with head and neck cancer and other malignancies (Alexander et al. 2008). Despite the increase in γδ T cells in some cancers, the levels of NKG2D receptors, responsible for the cytolytic effect of γδ T cells, were lower in cancer patients than in healthy adults (Bilgi et al. 2008) (see Section 31.7.2 on Immune Evasion Mechanisms). Based on the surface expression of CD56 and CD16 or inhibitory and activatory receptors, NK cells from cord blood (CB) could be divided into four subsets. Interestingly, CB NK cells, similar to the naïve T cells, express CD45RA but not CD45RO molecules. Moreover, CD27, a memory T cell marker, highly expressed on CD56^{hi}CD16$^-$ NK cells. The killing-associated molecules, NKG2A, NKG2D, CD95 and the intracellular granzyme B and perforin are heterogeneously expressed among the 4 subsets. Addition of IL-12 into cultures resulted in the induction of IFN-γ expression by

CD56^{hi} CD16$^-$ and CD56^{lo} CD16$^-$ subsets and the enhancement of NK cytolytic activity (Fan et al. 2008).

Activation of Vγ 9Vδ 2T Cells by NKG2D: Human γδT cells expressing a Vγ 9Vδ2 T-cell receptor (TCR) kill various tumor cells including autologous tumors. In addition to TCR-dependent recognition, activation of NKG2D-positive γδT cells by tumor cell-expressed NKG2D ligands can also trigger cytotoxic effector function. Human Vγ9 Vδ2 T cells recognize phosphoantigens, certain tumor cells, and cells treated with aminobisphosphonates. Evidences suggest that Vγ9 Vδ2T cells may also be directly activated by NKG2D. Culture of PBMC with immobilized NKG2D-mAb or NKG2D ligand MICA induces the up-regulation of CD69 and CD25 in NK and Vγ9 Vδ2 but not in CD8 T cells. Furthermore, NKG2D triggers the production of TNF-α but not of IFNγ, as well as the release of cytolytic granules by Vγ9 Vδ2 T cells. Due to remarkable similarities in NKG2D function in NK and Vγ9 Vδ2 T cells, it suggests a new perspective for Vγ9 Vδ2 T cell-based immunotherapy based on NKG2D ligand-expressing tumors (Rincon-Orozco et al. 2005). Trichet et al. (2006) reported that activating NKRs have less affinity for their MHC ligands than homologous inhibitory NKRs. Despite this, activating NKRs recognizing MHC class I molecules play an important role in the increased killing by Vγ9Vδ2 T cells of tumor cells with down-regulated MHC class I molecule expression, and suggest that these T cells will best lyse tumor cells combining MHC class I molecule expression down-regulation with up-regulated NKG2D ligand expression (Trichet et al. 2006).

31.6.2 NKG2D: A Co-stimulatory Receptor for Naive CD8$^+$ T Cells

CD8$^+$ T cells require a signal through a co-stimulatory receptor in addition to TCR to become activated. The role of CD28 in co-stimulating T cell activation is well established. The CD28 has a cytoplasmic YXXM motif that activates PI3-kinase via the DAP10 adapter protein (Wu et al. 1999) suggesting a similar co-stimulatory capacity of NKG2D. It is likely that NKG2D acts as a co-stimulatory molecule only under restricted conditions or requires additional cofactors. Whereas NKG2D can only function as a co-stimulatory receptor on CD8$^+$ αβ T cells under the domination of αβ TCR in spite of deficiency of co-stimulatory molecule CD28, NKG2D can directly activate NK cells even in the presence of inhibitory signals from MHC-I and corresponding receptor complexes. Results indicate that NKG2D can regulate the priming of human naive CD8$^+$ T cells, which may provide an alternative mechanism for potentiating and channeling the immune response (Maasho et al. 2005). In mice, however, two alternative splicing products NKG2D polypeptides associate differentially with the DAP10 and DAP12 signaling subunits and that differential expression of these isoforms and of signaling proteins determines whether NKG2D only functions as a co-stimulatory receptor in the adaptive immune system (CD8$^+$ T cells) or as both a primary recognition unit and a co-stimulatory receptor in the innate immune system (NK cells and macrophages) (Cao and He 2004; Diefenbach et al. 2002; Ogasawara and Lanier 2005). The 4-1BB co-stimulatory molecule expressed by resting CD8$^+$ T cells is an essential regulator of NKG2D in CD8$^+$ T cells. Kim et al. (2008) proposed that 4-1BB plays a critical role in protecting NKG2D from TGF-β1-mediated down-modulation.

Experiments with blocking antibodies revealed that NKG2D engagement contributed to cytolytic activity only during later stages of the infection when MHC class I expression was down-regulated and the MIC proteins were up-regulated. Lytic activity absolutely required the T cell receptor, and NKG2D provided a co-stimulatory signal. Same T cell clones secreted increased levels of cytokines in response to their target antigens when simultaneous NKG2D engagement occurred. Co-stimulation was also observed in mouse macrophages (Diefenbach et al. 2000; Groh et al. 2001). Transfectants expressing H-60 or Rae-1ß were used to stimulate activated macrophages. TNF-α transcription and nitric oxide production were enhanced in response to the signal, but only in the presence of LPS, which also provided an activating signal to macrophages (Ehrlich et al. 2005). NKG2D functions as a prototypic costimulatory receptor in a subset of human cytomegalovirus (HCMV)-specific CD4$^+$ T lymphocytes and thus may have a role in the response against infected HLA class II$^+$ cells displaying NKG2D ligands (Saez-Borderias et al. 2006).

31.6.3 NKG2D in Cytokine Production

The activating signal of NKG2D can also elicit cytokine production. NK cell production of MIP-1ß, TNF-α and IFN-γ was enhanced upon treatment with soluble ULBP (Kubin et al. 2001). A much more striking effect of ULBP was observed when IL-12 was included in the culture. Similarly, HCMV-specific γδ T cells produced IFN-γ, TNF-α, IL-2, and IL-4 when exposed to target antigen presented by C1R cells. Cytokine secretion was strongly enhanced if the C1R cells were transfected with MICA (Groh et al. 2001).

It was shown that dendritic epidermal T cells (DETCs), also known to express NKG2D, constitutively express NKG2D-S, NKG2D-L, DAP10, and DAP12 transcripts as well as cell surface NKG2D protein. Blocking of NKG2D inhibited DETC-mediated cytotoxicity against target cells that do not express T cell receptor ligands. Cross-linking of NKG2D on DETCs induced IFNγ production. Thus, DETCs constitutively express NKG2D that acts as a primary activating receptor, and plays important role in cutaneous immune surveillance (Nitahara et al. 2006). Chan et al. (2006) described a third DC lineage, termed interferon-producing killer DCs (IKDCs), distinct from conventional DCs and plasmacytoid DCs. They produce substantial amounts of type I IFN and IL-12 or IFNγ, depending on activation stimuli. Upon stimulation with CpG oligodeoxynucleotides, ligands for Toll-like receptor (TLR)-9, IKDCs kill typical NK target cells using NK-activating receptors. Their cytolytic capacity subsequently diminishes, associated with the loss of NKG2D receptor and its adaptors, DAP10 and DAP12. By virtue of their capacity to kill target cells, followed by antigen presentation, IKDCs provide a link between innate and adaptive immunity (Chan et al. 2006).

31.6.4 Heterogeneity of NK Cells in Umbilical Cord Blood

NK cells from umbilical cord blood (CB) play an important role in allogeneic stem cell transplantation and defending infections of newborn. The proportion of NK cells is high in CB, but the IFN-γ production low compared to later part in life. In contrast, the proportion of T cells is low in CB, an observation that indicates a deviation of the regulatory function of NK cells in CB compared to later in life. Additionally, the level of expression of NKG2D, NKp30 and NKp46 decreased with age. These age related changes in NK cell populations defined by the expression of activating and inhibitory receptors may be the result of pathogen exposure and/or a continuation of the maturation process that begins in the bone marrow (Sundström et al. 2007). Human NK cells from cord blood are the key effector cells involved in graft-versus-host disease (GVHD) and graft-versus-leukemia (GVL). The activity of CB NK cells was shown to be lower than that of adult peripheral blood (PB) NK cells. Though, the expression of activating NK receptors, CD16, NKG2D and NKp46 did not show significant difference between CB and PB NK cells, the expression of inhibitory receptor NKG2A/CD94 was significantly higher on CB NK cells. The high expression of NKG2A/CD94 and low expression of granzyme B may be related with the reduced activity of CB NK cells (Wang et al. 2007).

31.7 Cytotoxic Effector Function and Tumor Immune Surveillance

31.7.1 Anti-Tumor Activity

Recent advances clearly implicate that NKG2D recognition plays an important role in tumor immune surveillance and that NKG2D primarily acts to trigger perforin-mediated apoptosis. NKG2D-mediated tumor rejection was demonstrated in a murine system using the tumor-forming RMA cell line (Cerwenka et al. 2001). RMA transfectants expressing Rae-1 proteins were rejected. In vivo depletion experiments demonstrated that tumor rejection was mediated by NK cells, and studies in CD1 deficient mice demonstrated that CD1-restricted NKT cells were not involved in the rejection. A second report using similar methods confirmed that ectopic expression of the NKG2D ligands, Rae-1ß or H60, in several tumor cell lines resulted in potent rejection of the tumor cells by syngeneic mice (Diefenbach et al. 2001). Cell depletion experiments demonstrate that both NK cells and $CD8^+$ T cells are involved in this rejection. These authors also report a potent immune memory in contrast to observation of Cerwenka et al. (2001). NKG2D ligands might be useful in the design of tumor vaccines. Though the ligands for the NKG2D stimulatory receptor are frequently upregulated on tumors, rendering them sensitive to NK cells killing, Guerra et al. (2008) provided the evidence in NKG2D-deficient mice that NKG2D was not necessary for NK cell development but was critical for immunosurveillance of epithelial and lymphoid malignancies in transgenic models. These findings provided important genetic evidence for surveillance of primary tumors by an NK cell receptor.

The importance of the NKG2D pathway was further illustrated in mice deficient for either IFNγ or TNF-related apoptosis-inducing ligand, whereas mice depleted of NK cells, T cells, or deficient for perforin did not display any detectable NKG2D phenotype. Furthermore, IL-12 therapy preventing MCA-induced sarcoma formation was also largely dependent on the NKG2D pathway. The data begin to place the NKG2D pathway into the context of other recognition-effector systems used by NK cells (Hayakawa and Smyth 2006a, b). Ewing sarcoma (EWS) cells are potentially susceptible to NK cell cytotoxicity due to the expression of activating NK cell receptor ligands. The use of cytokine-activated NK cells rather than resting NK cells in immunotherapy may be instrumental to optimize NK cell reactivity to EWS (Verhoeven et al. 2008; von Strandmann et al. 2006).

Tumor-Derived Exosomes Down-Modulate NKG2D Expression: Soluble NKG2D ligands and growth factors, such as TGFβ1 emanating from tumors, are mechanisms for down-regulating NKG2D expression. Cancers thereby impair the capacity of lymphocytes to recognize and destroy them (Groh et al. 2002). Exosomes derived from cancer cells express ligands for NKG2D and express TGF β1. Exosomes produced by various cancer cell lines, or by mesothelioma patients triggered down-regulation of surface NKG2D expression by NK cells and CD8+ T cells. Other markers remained unchanged, indicating the selectivity and non-activatory nature of the response. Exosomal NKG2D ligands were partially responsible for this effect. Exosomally expressed TGF β1 was the principal mechanism. NKG2D is a likely physiological target for exosome-mediated immune evasion in cancer (Clayton et al. 2008).

NKR-P2, the rat orthologue of human NKG2D, is a potent activation receptor on rat DCs. A potential agonistic anti-NKR-P2 mAb (1A6) mimics the NKR-P2 ligand and induces activation and maturation of DCs. Interaction of DCs with 1A6 enhances nitric oxide-mediated apoptosis in tumor cells. Cross-linking of NKR-P2 with 1A6 up-regulates MHC II and decreases endocytic activity of DC, thus suggesting a pivotal role in adaptive immune responses. Blocking of 1A6-mediated activation and maturation with inhibitors of PI3K, p38K and ERK1/2K suggests the involvement of MAP kinase in signal transduction. 1A6 cross-linking activates NF-kB, which acts as key executioner of DC activation (Srivastava et al. 2007).

NK Cell-Mediated Cytotoxicity in Human Lung Adenocarcinoma: Although adenocarcinoma (ADC) cells express heterogeneous levels of NKG2DLs, they are often resistant to NK cell-mediated killing. Resistance of ADC-Coco to allogeneic polyclonal NK cells and autologous NK cell clones are correlated with shedding of NKG2DLs resulting from a matrix metalloproteinase (MMP) production. Treatment of ADC-Coco cells with a MMP inhibitor (MMPI) combined with IL-15 stimulation of autologous NK cell clones lead to a potentiation of NK cell-mediated cytotoxicity. This lysis is mainly NKG2D mediated, since it is abrogated by anti-NKG2D-neutralizing mAb. These results suggest that MMPIs, in combination with IL-15, may be useful for overcoming tumor cell escape from the innate immune response (Le Maux Chansac et al. 2008).

31.7.2 Immune Evasion Mechanisms

Acute Myeloid Leukemia (AML): Little is known regarding the regulation of NKp46 and NKG2D expression in AML. Mononuclear and polymorphonuclear phagocytes down-regulate the cell surface density of NKp46 and NKG2D on NK cells with CD56^{dim} phenotype in vitro by a mechanism that is dependent on the availability of phagocyte-derived reactive oxygen species (ROS). Histamine maintained NKp46 and NKG2D expression despite the presence of inhibitory phagocytes by targeting an H2 receptor on phagocytes. By contrast, NKp46 and NKG2D expression by the CD56^{bright} subset of NK cells was resistant to inhibition by phagocytes. These findings are suggestive of a mechanism of relevance to the regulation of NKp46/NKG2D receptor expression (Romero et al. 2006).

Malignant blood cells in AML display low levels of ligands for NKG2D and can thus evade NK immunosurveillance. Rohner et al. (2007) attempted to up-regulate NKG2D-L ULBP using anti-neoplastic compounds with myeloid differentiation potential. Combinations of 5-aza-2′-deoxycytidine, trichostatin A, vitamin D3, bryostatin-1, and all-trans-retinoic acid, used together with myeloid growth factors and IFNγ, increased cell surface ULBP expression in the AML cell line HL60 and in primary AML blasts. Up-regulation of ULBP ligands was associated with induction of myelomonocytic differentiation of AML cells. Higher ULBP expression increased NKG2D-dependent sensitivity of HL60 cells to NK-mediated killing (Rohner et al. 2007).

Myelodysplastic Syndromes: (MDS) are characterized by ineffective hematopoiesis with potential for progression to acute myeloid leukemia (AML). Findings suggest that impairment of NK cytolytic function derives in part from reduced activating NK receptors such as NKG2D in association with disease progression. Evasion of NK immunosurveillance may have importance for MDS disease progression (Epling-Burnette et al. 2007).

Human Melanomas: Several human melanomas (cell lines and freshly isolated metastases) do not express MICA on the cell surface but have intracellular deposits of this NKG2DL. Fuertes et al. (2008) identified a strategy developed by melanoma cells to evade NK cell-mediated immune surveillance based on the intracellular sequestration of immature forms of MICA in the endoplasmic reticulum. The low expression of CD161 and activating NKG2D receptors, without increased expression of KIR receptors CD158a and CD158b, as well as a decrease in the cytotoxic, CD16^{bright} NK cell subset, is associated with a significant impairment in NK cell activity in metastatic melanoma patients. Furthermore, the predictive pretherapy finding that IL-2, IFN, IFN and RA, unlike RA alone, can enhance NK cell activity of metastatic melanoma patients against FemX melanoma tumor cell line can be of help in the design and development of therapeutic regimens (Konjević et al. 2007).

Risk of Colorectal Cancer with the High NK Cell Activity NKG2D Genotype: Cancer patients with advanced disease display signs of immune suppression, which constitute a major obstacle for effective immunotherapy. Both T cells and NK cells are affected by a multitude of mechanisms of which the generation of reactive oxygen species is of major importance. Two weeks of high-dose treatment with the anti-oxidant vitamin E may enhance NK cell function in cancer patients by protecting from oxidative stress. Vitamin E treatment was associated with a minor, but consistent, induction of NKG2D expression in patients with colorectal cancer (Hanson et al. 2007; Melum et al. 2008).

It is reported that there are two haplotypes of NKG2D, HNK1 (high NK activity) and LNK1 (low NK activity). The HNK1 is reported to reduce the overall cancer risk. To elucidate its impact on colorectal cancer (CRC), Furue et al. (2008) found a reduced risk of CRC with the NKG2D HNK1. It was suggested that the HNK1 genotype, associated with high NK cell activity, might be an independent protective factor for CRC among the Japanese population.

Differing Phenotypes Between Intraepithelial and Stromal Lymphocytesin Tongue Cancer: The significance of tumor-infiltrating lymphocytes (TIL) has attracted much attention in relation to the prognosis of patients. The tumor nest-infiltrating (intraepithelial) $CD8^+$ T cells frequently expressed PD-1, an inhibitory receptor, in sharp contrast to those in the stroma or in the lichen planus. Conversely, the intraepithelial $CD8^+$ T cells only infrequently expressed NKG2D, an activating receptor, in contrast to those in the stroma or in the lichen planus. Furthermore, the intraepithelial NK cells expressed NKG2A, an inhibitory receptor, more frequently than those in the stroma or the lichen planus. Collectively, the intraepithelial $CD8^+$ T cells and NK cells are phenotypically inactivated, whereas stromal counterparts are phenotypically just as active as those in the lichen planus. These results suggest the first-step occurrence of an immune evasion mechanism in the tumor nest of oral squamous cell carcinoma (Katou et al. 2007). The expression of NKG2D on $CD8^+$ T lymphocytes from esophageal cancer patients was significantly lower than in those of normal controls. The decreased NKG2D expression on $CD8^+$ T cells was correlated with disease severity (Osaki et al. 2009).

Cholangiocarcinoma in Primary Sclerosing Cholangitis: Melum et al. (2008) investigated the influence of genetic variations in the NKG2D-MICA receptor-ligand pair on the risk of cholangiocarcinoma (CCA) in patients with primary sclerosing cholangitis (PSC). Seven SNPs covering the NKG2D gene were genotyped in Scandinavian PSC patients and controls. This suggested that genetic variants of the NKG2D receptor are associated with development of CCA in PSC patients and the interaction between NKG2D and MICA is involved in protection against CCA in PSC. Patients who are homozygous for the nonrisk alleles are unlikely to develop CCA; this finding could be helpful in identifying PSC patients with a low CCA risk.

Gastric Cancer: Some studies suggest that the NKG2D expression on $CD8^+$ T cells is down-regulated and this reduction may be involved in immune evasion in cancer patients. In gastric cancer patients, NKG2D expression on circulating $CD8^+$ T cells was down-regulated and significantly correlated with IFN-γ production in gastric cancer patients. There was no difference in soluble MICA between gastric cancer patients and normal controls. Decreased NKG2D expression may be one of the key factors responsible for immune evasion by tumors in gastric cancer (Osaki et al. 2007).

MIF in the Immune Escape of Ovarian Cancer: The proinflammatory cytokine macrophage migration inhibitory factor (MIF) stimulates tumor cell proliferation, migration, and metastasis; promotes tumor angiogenesis; suppresses p53-mediated apoptosis; and inhibits antitumor immunity by unknown mechanisms. Functionally, MIF may contribute to the immune escape of ovarian carcinoma by transcriptionally down-regulating NKG2D in vitro and in vivo which impairs NK cell cytotoxicity toward tumor cells. Perhaps inhibitors of MIF can be used for the treatment of MIF-secreting cancers (Krockenberger et al. 2008).

31.8 NKG2D in Immune Protection and Inflammatory Disorders

31.8.1 NKG2D Response to HCMV

Human cytomegalovirus (HCMV) infection is a paradigm of the complexity of host-pathogen interactions. To avoid recognition by cytotoxic T lymphocytes (CTL) HCMV inhibits the expression of HLA-I molecules. As a consequence, engagement of KIR, CD94/NKG2A, and CD85j (ILT2 or LIR-1) NKR specific for HLA-I molecules is impaired, and infected cells become vulnerable to an NK cell response driven by activating receptors. In addition to the well-defined role of the NKG2D, the involvement of other triggering receptors (i.e., activating KIR, CD94/NKG2C, NKp46, NKp44, and NKp30) in the response to HCMV is being explored (Gumá et al. 2006). To escape from NK cell-mediated surveillance, HCMV interferes with the expression of NKG2D ligands in infected cells. In addition, the virus may keep NK inhibitory receptors engaged preserving HLA-I molecules with a limited role in antigen presentation (i.e., HLA-E) or, alternatively, displaying class I surrogates. Despite considerable progress in the field, a number of issues regarding the involvement of NKR in the innate immune response to HCMV remain uncertain.

31.8.2 HTLV-1-Associated Myelopathy

Human T cell lymphotropic virus type I (HTLV-1)-associated myelopathy/tropical spastic paraparesis (HAM/TSP) is a chronic inflammatory disease of the CNS that resembles multiple sclerosis. Disease progression involves production of IL-15 and its receptor through transactivation. Substantial proportions of HAM/TSP patient CD4 T and CD8 cells are positive for NKG2D and both subsets express MIC. Engagement of MIC by NKG2D promotes spontaneous HAM/TSP T cell proliferation and, apparently, CTL activities against HTLV-1-infected T cells. Perhaps it is the viral strategy that exploits immune stimulatory mechanisms to maintain a balance between promotion and limitation of infected host T cell expansions (Azimi et al. 2006).

NKG2D in Clearance of Picornavirus from Infected Murine Brain: The role of NKG2D-mediated augmentation in the clearance of viral infections from CNS was explored using Theiler's murine encephalomyelitis virus model. Studies suggest that NKG2D-positive $CD8^+$ cytotoxic T cells enter the brain, and that NKG2D ligands are expressed in the brain during acute infection. The interruption of NKG2D ligand recognition via treatment with a blocking antibody attenuates the efficacy of viral clearance from the CNS (Deb and Howe 2008).

31.8.3 Protection Against Bacteria

The NKG2D-activating receptor interacts with ligands expressed on the surface of cells stressed by pathogenic and nonpathogenic stimuli. Lysis of bronchial epithelial cells is not MHC I restricted but depends on NKG2D signaling. It appears that respiratory epithelium has antigen presenting function and directly alloactivates cytotoxic $CD8^+$ T cells, which show nonclassical effector function (Kraetzel et al. 2008). In response to acute pulmonary *Pseudomonas aeruginosa* infection, the expression of mouse NKG2D ligands (Rae1) increased in airway epithelial cells and alveolar macrophages in vivo and also increased the cell surface expression of human ULBP2 on airway epithelial cells in vitro. NKG2D receptor blockade inhibited the pulmonary clearance of *P. aeruginosa* and also resulted in decreased production of Th1 cytokines and nitric oxide in the lungs of *P. aeruginosa*-infected mice. These results suggest the importance of NKG2D-mediated immune activation in the clearance of acute bacterial infection and that epithelial cell-lymphocyte interactions mediate pulmonary cytokine production and bacterial clearance (Borchers et al. 2006).

Expression of the activating receptors NKp30, NKp46, and NKG2D is enhanced on NK cells after exposure to *M. tuberculosis*-infected monocytes, whereas expression of DNAX accessory molecule-1 and 2B4 was not. Anti-NKG2D and anti-NKp46 inhibited NK cell lysis of M. tuberculosis-infected monocytes, but Abs to NKp30, DNAX accessory molecule-1, and 2B4 had no effect. Following infection of monocytes, the expression of the ULBP1, but not of ULBP2, ULBP3, or MICA/B was upregulated. Therefore, NKp46 and NKG2D are the principal receptors involved in lysis of *M. tuberculosis*-infected mononuclear phagocytes, and that ULBP1 on infected cells is the major ligand for NKG2D. It appears that TLR2 contributes to up-regulation of ULBP1 expression (Vankayalapati et al. 2005).

31.8.4 Autoimmune Disorders

Studies have shown that NK-DC interaction plays an important role in the induction of immune response against tumors and certain viruses. The effect of this interaction is bidirectional. Coculture with NK cells causes several fold increase in IL-12 production by *Toxoplasma gondii* lysate Ag-pulsed DC. In vitro blockade of NKG2D neutralizes the NK cell-induced up-regulation of DC response. Results emphasize the critical role played by NKG2D in the NK-DC interaction (Guan et al. 2007). A role for NKG2D has been indicated in several autoimmune diseases in humans and in animal models of type 1 diabetes (T1DM) and multiple sclerosis, and treatment with monoclonal antibodies to NKG2D attenuated disease severity in these models.

NKG2D in Progression of Autoimmune Diabetes: RAE-1 is present in prediabetic pancreas islets of NOD mice and the autoreactive $CD8^+$ T cells infiltrating the pancreas express NKG2D. Treatment with a nondepleting anti-NKG2D monoclonal antibody (mAb) during the prediabetic stage completely prevented disease by impairing the expansion and function of autoreactive $CD8^+$ T cells. These results suggest that NKG2D is associated with disease progression and suggest a new therapeutic target for autoimmune type I diabetes (Ogasawara et al. 2004).

Rheumatoid Arthritis: Stimulation of T cell autoreactivity by anomalous expression of NKG2D and its MIC ligands in rheumatoid arthritis has been suggested by Groh et al. (2003). With the reported expansion of peripheral $CD4^+CD28^-NKG2D^+$ T cells in rheumatoid arthritis (RA), this interaction is important in triggering autoimmunity. But in studies of Schrambach et al. (2007), despite occasional and indiscriminate expansion of incriminated T cell subpopulation, no correlation could be observed between the $CD4^+CD28^-NKG2D^+$ T cells and auto-immunity.

Moreover, in situ, the presence of NKG2D matched that of CD8$^+$, but not that of CD4$^+$ T cells (Schrambach et al. 2007).

NKG2D Does Not Differ in Systemic Lupus Erythematosus: Schepis et al. (2009) studied NK cells in SLE, a B-cell-driven systemic autoimmune disease and found that CD56^{bright} NK cells increase in SLE patients, regardless of disease activity, in association with some increased expression of NKp46/CD335 on NK cells. Other receptors on lymphocytes or NKRs, including CD94, NKG2C/CD159c, NKG2D/CD314, NKp30/CD337, NKp44/CD336, CD69 did not differ between patient groups.

Sjögren's Syndrome: Izumi et al. (2006) found that NK cell number, NK cell killing activity, and NKG2D were significantly decreased, and the expression of NKp46, as well as the percentage of apoptotic NK cells, were significantly increased in primary Sjögren's syndrome (SS) patients compared with healthy controls. Data suggested that reduced NK cell numbers, probably a result of apoptotic death, might contribute to impaired NK cell activity in patients with primary SS.

NKG2D and Its Ligands in Allograft Transplant Rejection: The early "danger" signals associated with transplantation lead to rapid up-regulation of NKG2D ligands. A second wave of NKG2D ligand up-regulation is mediated by the adaptive immune response to allografts. Treatment with an Ab to NKG2D was highly effective in preventing CD28-independent rejection of cardiac allografts. Notably, NKG2D blockade did not deplete CD8$^+$ T cells or NK1.1$^+$ cells nor affect their migration to the allografts. Results suggest a functional role of NKG2D and its ligands in the rejection of solid organ transplants (Kim et al. 2007; Zhang and Stastny 2006).

Role of CD45 for Cytokine and Chemokine Production from Killing in Primary NK Cells: Engagement of receptors on the surface of NK cells initiates a biochemical cascade ultimately triggering cytokine production and cytotoxicity. Huntington et al. (2005) investigated the role of cell surface phosphatase CD45 in NK cell development and intracellular signaling from activating receptors. Stimulation via MHC class I-binding receptor, Ly49D on CD45$^{-/-}$ primary NK cells resulted in the activation of phosphoinositide-3-kinase and normal cytotoxicity but failed to elicit a range of cytokines and chemokines. This blockage was associated with impaired phosphorylation of Syk, Vav1, JNK, and p38, and mimics results obtained using inhibitors of the src-family kinases (SFK). These data, supported by analogous findings after CD16 and NKG2D stimulation of CD45$^{-/-}$ primary NK cells, place CD45 upstream of SFK in NK cells after stimulation via immunoreceptor tyrosine-based activation motif-containing receptors. Thus CD45 is a pivotal enzyme in eliciting a subset of NK cell responses.

Interaction Between Human NK Cells and Stromal Cells: It appears that HLA-I molecules do not protect bone marrow stromal cells (BMSC) from NK cell-mediated injury. But NK cells, activated upon binding with BMSC, may regulate BMSC survival (Poggi et al. 2005). While the NK receptors NKp30 and NKp46 are responsible for the delivery of lethal hit to APC, the NKG2D-activating receptor, the MICA, and the ULBP are involved in stromal cell killing. These events are dependent on the activation of phosphoinositol 3-kinase and consequent release of perforins and granzymes. Altogether, studies support the notion that NK cells can recognize self-cells possibly affecting both APC function and interaction between lymphocytes and microenvironment leading to autoreactivity (Poggi et al. 2007). Activated T cells become susceptible to autologous NK lysis via NKG2D/NKG2DLs interaction and granule exocytosis, suggesting that NK lysis of T lymphocytes via NKG2D may be an additional mechanism to limit T-cell responses (Cerboni et al. 2007; Coudert and Held 2006; Poggi et al. 2005).

31.9 Decidual/Placental NK Cell Receptors in Pregnancy

Inhibitory and Activating Receptors on Decidual/Placenta NK Cells: The interaction of decidual KIRs with on HLA-C trophoblast cells appears to block NK cytotoxicity against trophoblast cells. NK cell receptors protect the trophoblasts and the outcome of pregnancy depends on women's NKRs. Peripheral blood NK cells express CD56 surface antigen in low (CD56^{dim}) or high (CD56^{bright}) density. In contrast to CD56^{bright} NK cells, CD56^{dim} cells express KIR such as CD158a, CD158b, and NKB1. Evidences indicate that CD56^{bright} cells are specialized NK cells that regulate immunological response mechanisms rather by cytokine supply than by their cytotoxic potential. The poor cytolytic capacity of CD56^{bright} NK cells can be explained by weak ability in forming conjugates with target cells and low contents of perforin and granzyme A in their granules (Jacobs et al. 2001). The balance of inhibitory and activating NKRs on maternal decidual NK cells, most of which are CD56^{bright}, is thought to be crucial for the proper growth of trophoblasts in placenta.

In Recurrent Spontaneous Abortions (RSA): Women with alloimmune abortions have a limited inhibiting KIR repertoire and such miscarriages could occur because trophoblastic HLA class I molecules are recognized by decidual NK cells lacking the appropriate inhibitory KIRs (Varla-Leftherioti et al. 2003). Ntrivalas et al. (2005) suggested that CD158a and CD158b inhibitory receptor expression by $CD56^{dim}/CD16^+$ and $CD56^{bright}/CD16^-$ NK cells were significantly decreased, and CD161-activating receptor expression by $CD56^+/CD3^+$ NKT cells was significantly increased in women with implantation failures when compared with normal controls, suggesting an imbalance between inhibitory and activating receptor expression on NK cells of women with implantation failures. The expression of CD94/NKG2, the receptor for HLA-E, on trophoblasts, motivated Kusumi et al. (2006) to investigate the expression patterns of inhibitory receptor CD94/NKG2A, and activating receptor, CD94/NKG2C, on decidual NK cells in early stage of normal pregnancy and to compare them with those on peripheral NK cells, most of which are $CD56^{dim}$. The rate of NKG2A-positive cells was significantly higher for decidual $CD56^{bright}$ NK cells than for peripheral $CD56^{dim}$ NK cells, but the rates of NKG2C-positive cells were comparable between the two cell types. Interestingly, peripheral $CD56^{dim}$ NK cells reciprocally expressed inhibitory NKG2A and activating NKG2C, but decidual $CD56^{bright}$ NK cells that expressed activating NKG2C simultaneously expressed inhibitory NKG2A. The co-expression of inhibitory and activating NKG2 receptors may fine-tune the immunoregulatory functions of the decidual NK cells to control the trophoblast invasion in constructing placenta (Kusumi et al. 2006).

Down-Regulation of NKG2D on PBMC by MICA and MICB: A novel mechanism for fetal evasion of maternal immune attack, based on the down-regulation of the activating NKR NKG2D on peripheral blood mononuclear cells (PBMC) by soluble MHC class I chain-related proteins A and B has been proposed. A similar immune escape pathway has been described in tumors. The MIC mRNA is constitutively expressed by human placenta and restricted to the syncytiotrophoblast on the apical and basal cell membrane and in cytoplasmic vacuoles as MIC-loaded microvesicles/exosomes. Simultaneously, the cell surface NKG2D expression on maternal PBMC was down-regulated compared with nonpregnant controls. The soluble MIC molecules in pregnancy serum were shown to interact with NKG2D and down-regulate the receptor on PBMC from healthy donors, with the consequent inhibition of the NKG2D-dependent cytotoxic response. These results suggested a new physiological mechanism of silencing the maternal immune system that promotes fetal allograft immune escape and supports the view of the placenta as an immunoregulatory organ (Mincheva-Nilsson et al. 2006).

Decidual NK Cells Are Not Cytotoxic: In early pregnancy, invading fetal trophoblasts encounter abundant maternal decidual (d) dNK cells. dNK express perforin, granzymes A and B and the activating receptors NKp30, NKp44, NKp46, NKG2D, and 2B4 as well as LFA-1. Even though they express essential molecules required for lysis, fresh dNK displayed very low lytic activity on classical MHC I negative targets K562 and 721.221. dNK formed conjugates and activating immune synapses with 721.221 and K562 cells in which CD2, LFA-1 and actin were polarized toward the contact site. However, in contrast to peripheral NK cells, they failed to polarize their microtubule organizing centers and perforin-containing granules to the synapse, accounting for their lack of cytotoxicity (Kopcow et al. 2005).

31.10 Regulation of NKG2D Functions

31.10.1 NKG2D Induction by Chronic Exposure to NKG2D Ligand

It seems that signaling capacity of NKG2D receptor complexed with ligands is altered. The prolonged encounter with tumor cell-bound, but not soluble, ligand can completely uncouple the receptor from mobilization of $(Ca^{2+})_i$ and the exertion of cell-mediated cytolysis. However, cytolytic effector function is intact since NKG2D ligand-exposed NK cells can be activated via Ly49D receptor. While NKG2D-dependent cytotoxicity is impaired, prolonged ligand exposure results in constitutive IFNγ production, suggesting sustained signaling. The functional changes are associated with a reduced presence of the relevant signal transducing adaptors DAP-10 and KARAP/DAP-12 (Coudert et al. 2005). Coudert et al. (2008) found that sustained NKG2D engagement induces cross-tolerization of several unrelated NK cell activation receptors. It was observed that receptors that activate NK cells via the DAP12/KARAP and DAP10 signaling adaptors cross-tolerize preferentially NK cell activation pathways that function independent of DAP10/12, such as antibody-dependent CMC and missing-self recognition. These results identify NK cell activation receptors that can tolerize mature NK cells (Coudert et al. 2008).

Inhibitory Role of Ly49 Receptors on NKG2D: The inhibitory role of Ly49 receptors on NKG2D-mediated activation has only started emerging. Evidence suggests that NKG2D-mediated cytotoxicity and cytokine production result from the fine balance between activating and inhibitory receptors, thereby defining the NK cell-mediated immune responses (Regunathan et al. 2005). It is hypothesised that NKG2D-mediated activation is a function of 'altering the balance' in the signaling strength between the activating NKG2D and inhibiting Ly49 receptors. Balance in the signaling strength depends on the expression levels of activating ligands on the target cells. Qualitative and quantitative variations of MHC class I molecules expressed on the target cells also plays a major role in determining this 'altered-balance'. Consequently, the nature of Ly49 receptors expressed on specific NK subsets determines the level of NKG2D-mediated NK cell activation. These observations provide a firm basis of 'altered-balance' in NK signaling and describe an active interplay between inhibitory Ly49 and activating NKG2D receptors (Malarkannan 2006; Regunathan et al. 2005)

31.10.2 Effects of Cytokines and Other Factors

Elevated TGF-β1 Down-Regulates NKG2D Surface Expression: Lee et al. (2004) showed that TGF-β1 secreted by tumors is responsible for down-regulating of NKG2D and poor NK lytic activity. The elevated plasma level of TGF-β1 in human lung and colorectal cancer patients compared with in normal controls was inversely related with surface expression of NKG2D on NK cells in these patients. In a murine model of head and neck squamous cell carcinoma (SCC VII/SF), SCC VII/SF tumors expressed high levels of TGF-β1, which were down-modulated by vaccination with rvv-IL-2 (Dasgupta et al. 2005). TGF-β is known to play a major role for the reduced expression of NKG2D in cancer patients. The protected expression of NKG2D by IL-2/IL-18 provides insight into the mechanism of NKG2D regulation (Song et al. 2006). An inverse correlation does exist between NK cell activation and $CD4^+CD25^+$ regulatory T (T_{reg}) cells expansion in tumor-bearing patients. Results indicated that T_{reg} cells expressed membrane-bound TGF-β, which directly inhibited NK cell effector functions and down-regulated NKG2D receptors on the NK cell surface. Thus T_{reg} cells generated during tumor growth can blunt the NK cell functions of the innate immune system (Ghiringhelli et al. 2005).

Interleukin-15 stimulates NKG2D: IL-15 is the major physiologic growth factor responsible for NK cell differentiation, survival and cytolytic activity of mature NK cells. The IL-15 treatment increases NKG2D transcripts and surface expression in NK cells. The up-regulatory effect of IL-15 on NK cytolysis is partly dependent on the interaction of NKG2D and MICA (Zhang et al. 2008b). IL-15 activates peripheral blood mononuclear cells to lyse *Cryptosporidium*-infected epithelial cells. Flow cytometry revealed that IL-15 increased expression of NKG2D on NK cells. Thus, IL-15 has an important role in activating an NK cell-mediated pathway that leads to the elimination of intracellular protozoans from the intestines (Dann et al. 2005). Results also indicate that the pro-inflammatory cytokines (IL-2, IL-15 and IL-21) can modify the peripheral repertoire of NK cells (de Rham et al. 2007).

Interleukin-12, 18 and 21: IL-12 improves cytotoxicity of NK cells and expression of NKG2D. IL-12, produced primarily by APCs, plays an important role in the interaction between the innate and adaptive arms of immunity acting upon T and NK cells to generate CTLs. Ortaldo et al. (2006) demonstrated that cross-linking of NKG2D and NK1.1 results in a synergistic NK IFN- IFNγ response when combined with IL-12 or IL-18. Multiple convergent signals maximize the innate immune response by triggering complementary biochemical signaling pathways (Ortaldo et al. 2006; Zhang et al. 2008a).

IL-21 mediates its biological effects via the IL-21R in conjunction with the common receptor γ-chain that is also shared by members of the IL-2 family. Human primary NK and $CD8^+$ T cells with IL-21 in combination with IL-2- results in significant reduction of surface expression of NKG2D, compared with that in cells treated with IL-2 alone. This was attributed to a dramatic reduction in DAP10 promoter activity. In contrast to NKG2D expression, IL-21 was able to induce the expression of the NK activation receptors NKp30 and 2B4 as well as the co-stimulatory receptor CD28 on $CD8^+$ T cells (Burgess et al. 2006). IL-21 has a synergistic effect by increasing the numbers of NK cells on a large scale. IL-2 and IL-15 may induce the expression of KIRs, the NKG2D and NKp31. Findings indicated that the pro-inflammatory cytokines IL-21 and others (IL-2 and IL-15) can modify the peripheral repertoire of NK cells (de Rham et al. 2007). IL-21 can promote the anti-tumor responses of the innate and adaptive immune system. Mice treated with IL-21 reject tumor cells more efficiently, and a higher percentage of mice remain tumor-free compared with untreated controls. In certain tumor models, IL-21-enhanced tumor rejection is NKG2D dependent. IL-21 therapy may work optimally against tumors that can elicit a NKG2D-mediated immune response (Takaki et al. 2005).

Other Factors: NKG2D, expressed on γδ T cells participates in demyelination since blockade of NKG2D interactions in vivo resulted in 60% reduction in demyelination (Dandekar et al. 2005). All-trans retinoic acid (ATRA), a metabolite of vitamin A plays an important role in regulating immune responses. ATRA suppressed cytotoxic activities of NK92 cells without affecting their proliferation. It suppressed NF-kB activity and prevented IkBα degradation in a dose-dependent way, inhibited IFN-γ production and gene expression of granzyme B and NKp46 (Li et al. 2007). Activation of NK cells is triggered by multiple receptors. SLP-76 is required for CD16- and NKG2D-mediated NK cell cytotoxicity, while MIST negatively regulates these responses in an SLP-76-dependent manner (Hidano et al. 2008). Tryptophan (Trp) catabolism mediated by indoleamine 2,3-dioxygenase (IDO) plays a central role in the regulation of T-cell-mediated immune responses and NK-cell functions. The effect of l-kynurenine appears to be restricted to NKp46 and NKG2D, while it does not affect other surface receptors such as NKp30 or CD16 (Della Chiesa et al. 2006).

31.11 Role in Immunotherapy

Requirement of DNAM-1, NKG2D, and NKp46 in NK Cell-Mediated Killing of Myeloma Cells: Ovarian carcinoma cells display ubiquitous expression of DNAX accessory molecule-1 (DNAM-1) ligand PVR and sparse/heterogeneous expression of MICA/MICB and ULBP1, ULBP2, and ULBP3. In line with the NK receptor ligand expression profiles, antibody-mediated blockade of activating receptor pathways revealed a dominant role for DNAM-1 and a complementary contribution of NKG2D signaling in tumor cell recognition. Thus, resting NK cells are capable of directly recognizing ovarian carcinoma as a potential target for adoptive NK cell-based immunotherapy (Carlsten et al. 2007). DNAM-1 is also important in the NK cell-mediated killing of myeloma cells expressing the cognate ligands (El-Sherbiny et al. 2007). Heat shock proteins (HSPs) are known activate both adaptive and innate immune responses. Vaccination with autologous tumor-derived HSP96 of colorectal cancer patients, radically resected for liver metastases, induced a significant boost of NK activity detected as cytokine secretion and cytotoxicity in the presence of NK-sensitive targets. Increased NK activity was associated with an enhanced expression of NKG2D and/or NKp46 receptors (Pilla et al. 2005).

Immunotherapy with Chimeric NKG2D Receptors: Adoptive transfer of tumor-reactive T cells is a promising antitumor therapy for many cancers. Adoptive transfer of selected populations of alloreactive HLA class I-mismatched NK cells in combination with pharmacologic induction of NKG2D-ligands merits clinical evaluation as a new approach to immunotherapy of human acute myeloid leukemia (AML) (Diermayr et al. 2008). Barber et al. designed a chimeric receptor linking NKG2D to the CD3ζ chain of the T-cell receptor to target ovarian tumor cells. More than 80% of primary human ovarian cancer samples express ligands for NKG2D on the cell surface. Engagement of chimeric NKG2D receptors (chNKG2D) with ligands for NKG2D leads to T-cell secretion of proinflammatory cytokines and tumor cytotoxicity. ChNKG2D-expressing T cells lysed ovarian cancer cell lines. T cells from ovarian cancer patients that express chNKG2D secrete proinflammatory cytokines when cultured with autologous tumor cells. chNKG2D T cells can be used therapeutically in a murine model of ovarian cancer and offer potential application as immunotherapy for ovarian cancer (Barber et al. 2007). Most myeloma tumor cells express NKG2D ligands and respond to human chNKG2D T cells. The chNKG2D T cells can be generated with serum-free media, and maintained effector functions on cryopreservation (Barber et al. 2008a). The chNKG2D T-cell-mediated therapeutic effects were mediated by both cytokine-dependent and cytotoxic mechanisms in vivo (Zhang et al. 2007). Specifically, chNKG2D T cell expression of perforin, GM-CSF, and IFN-γ were essential for complete antitumor efficacy (Barber et al. 2008b).

Regulatory T Cells Suppress NK Cell-Mediated Immunotherapy: $CD4^+CD25^+$ regulatory T cells (Treg) that suppress T cell-mediated immune responses may also regulate other arms of an effective immune response. In particular, T_{reg} directly inhibit NKG2D-mediated NK cell cytotoxicity in vitro and in vivo, effectively suppressing NK cell-mediated tumor rejection. In vitro, T_{reg} were shown to inhibit NKG2D-mediated cytolysis largely by a TGF-β-dependent mechanism and independently of IL-10. Adoptively transferred T_{reg} suppressed NK cell antimetastatic function in RAG-1-deficient mice. Depletion of T_{reg} before NK cell activation via NKG2D and the activating IL-12 cytokine dramatically enhanced NK cell-mediated suppression of tumor growth and metastases. The study shows at least one mechanism by which T_{reg} can suppress NK cell antitumor activity and requirement of subsequent NK cell activation to promote strong innate anti-tumor immunity (Smyth et al. 2006; Roy et al. 2008).

References

Alexander AAZ, Maniar A, Cummings J-S et al (2008) Isopentenyl pyrophosphate activated $CD56^+$ γδ T lymphocytes display potent anti-tumor activity towards human squamous cell carcinoma. Clin Cancer Res 14:4232–4240

Armeanu S, Krusch M, Baltz KM et al (2008) Direct and natural killer cell-mediated antitumor effects of low-dose bortezomib in hepatocellular carcinoma. Clin Cancer Res 14:3520–3528

Azimi N, Jacobson S, Tanaka Y et al (2006) Immunostimulation by induced expression of NKG2D and its MIC ligands in HTLV-1-associated neurologic disease. Immunogenetics 58:252–258

Bacon L, Eagle RA, Meyer M et al (2004) Two human ULBP/RAET1 molecules with transmembrane regions are ligands for NKG2D. J Immunol 173:1078–1084

Bahram S, Bresnahan M, Geraghty DE, Spies T (1994) A second lineage of mammalian major histocompatibility complex class I genes. Proc Natl Acad Sci USA 91:6259–6263

Bahram S, Inoko H, Shiina T, Radosavljevic M (2005) MIC and other NKG2D ligands: from none to too many. Curr Opin Immunol 17:505–509

Barber A, Zhang T, DeMars LR et al (2007) Chimeric NKG2D receptor-bearing T cells as immunotherapy for ovarian cancer. Cancer Res 67:5003–5008

Barber A, Zhang T, Megli CJ et al (2008a) Chimeric NKG2D receptor-expressing T cells as an immunotherapy for multiple myeloma. Exp Hematol 36:1318–1328

Barber A, Zhang T, Sentman CL (2008b) Immunotherapy with chimeric NKG2D receptors leads to long-term tumor-free survival and development of host antitumor immunity in murine ovarian cancer. J Immunol 180:72–78

Barber A, Rynda A, Sentman CL (2009) Chimeric NKG2D expressing T cells eliminate immunosuppression and activate immunity within the ovarian tumor microenvironment. J Immunol 183:6939–6947

Bartkova J et al (2005) DNA damage response as a candidate anticancer barrier in early human tumorigenesis. Nature 434:864–870

Bauer S, Groh V, Wu J et al (1999) Activation of NK cells and T cells by NKG2D, a receptor for stress-inducible MICA. Science 285:727–729

Berg SF, Dissen E, Westgaard IH, Fossum S (1998) Molecular characterization of rat NKR-P2, a lectin-like receptor expressed by NK cells and resting T cells. Int Immunol 10:379–385

Bilgi O, Karagoz B, Turken O et al (2008) Peripheral blood γδT cells in advanced-stage cancer patients. Adv Ther 25:218–224

Bjorkman PJ, Saper MA, Samraoui B et al (1987) Structure of the human class I histocompatibility antigen, HLA-A2. Nature 329(6139):506–512

Borchers MT, Harris NL, Wesselkamper SC et al (2006) The NKG2D-activating receptor mediates pulmonary clearance of *Pseudomonas aeruginosa*. Infect Immun 74:2578–2586

Boyington JC, Riaz AN, Patamawenu A, Coligan JE, Brooks AG, Sun PD et al (1999) Structure of CD94 reveals a novel C-type lectin fold: implications for the NK cellassociated CD94/NKG2 receptors. Immunity 10:75–82

Burgess SJ, Marusina AI, Pathmanathan I et al (2006) IL-21 down-regulates NKG2D/DAP10 expression on human NK and CD8+ T cells. J Immunol 176:1490–1497

Busche A, Goldmann T, Naumann U, Steinle A, Brandau S (2006) Natural killer cell-mediated rejection of experimental human lung cancer by genetic overexpression of major histocompatibility complex class I chain-related gene A. Hum Gene Ther 171:35–46

Cao W, He W (2004) UL16 binding proteins. Immunobiology 209:283–290

Cao W, Xi X, Hao Z et al (2007) RAET1E2, a soluble isoform of the UL16-binding protein RAET1E produced by tumor cells, inhibits NKG2D-mediated NK cytotoxicity. J Biol Chem 282:18922–18928

Caraux A, Kim N, Bell SE et al (2006) Phospholipase C-γ2 is essential for NK cell cytotoxicity and innate immunity to malignant and virally infected cells. Blood 107:994–1002

Carayannopoulos LN, Naidenko OV, Fremont DV, Yokoyama WM (2002) Cutting Edge: murine UL16-binding protein-like transcript 1: a newly described transcript encoding a high-affinity ligand for murine NKG2D. J Immunol 169:4079–4083

Carlsten M, Björkström NK, Norell H et al (2007) DNAX accessory molecule-1 mediated recognition of freshly isolated ovarian carcinoma by resting natural killer cells. Cancer Res 67:1317–1325

Cerboni C, Zingoni A, Cippitelli M et al (2007) Antigen-activated human T lymphocytes express cell-surface NKG2D ligands via an ATM/ATR-dependent mechanism and become susceptible to autologous NK- cell lysis. Blood 110:606–615

Cerwenka A (2009) New twist on the regulation of NKG2D ligand expression. J Exp Med 206:265–268

Cerwenka A, Bakker AB, McClanahan T (2000) Retinoic acid early inducible genes define a ligand family for the activating NKG2D receptor in mice. Immunity 12:721–727

Cerwenka A, Baron JL, Lewis L, Lanier LL (2001) Ectopic expression of retinoic acid early inducible-1 gene (RAE-1) permits natural killer cell-mediated rejection of a MHC class I-bearing tumor *in vivo*. Proc Natl Acad Sci USA 98:11521–11526

Chalupny NJ, Sutherland CL, Lawrence WA et al (2003) ULBP4 is a novel ligand for human NKG2D. Biochem Biophys Res Commun 305:129–135

Champsaur M, Lanier LL (2010) Effect of NKG2D ligand expression on host immune responses. Immunol Rev 235:267–285

Chan CW, Crafton E, Fan HN et al (2006) Interferon-producing killer dendritic cells provide a link between innate and adaptive immunity. Nat Med 12:207–213

Chen X, Trivedi PP, Ge B et al (2007) Many NK cell receptors activate ERK2 and JNK1 to trigger microtubule organizing center and granule polarization and cytotoxicity. Proc Natl Acad Sci USA 104:6329–6334

Clayton A, Mitchell JP, Court J et al (2008) Human tumor-derived exosomes down-modulate NKG2D expression. J Immunol 180:7249–7258

Cosman D, Mullberg J, Sutherland CL et al (2001) ULBPs, novel MHC class I-related molecules, bind to CMV glycoprotein UL16 and stimulate NK cytotoxicity through the NKG2D receptor. Immunity 14:123–133

Coudert JD, Held W (2006) The role of the NKG2D receptor for tumor immunity. Semin Cancer Biol 16:333–343

Coudert JD, Scarpellino L, Gros F et al (2008) Sustained NKG2D engagement induces cross tolerance of multiple distinct NK cell activation pathways. Blood 11:3571–3578

Coudert JD, Zimmer J, Tomasello E et al (2005) Altered NKG2D function in NK cells induced by chronic exposure to NKG2D ligand-expressing tumor cells. Blood 106:1711–1717

Dandekar AA, O'Malley K, Perlman S (2005) Important roles for γ interferon and NKG2D in γδ T-cell-induced demyelination in T-cell receptor β-deficient mice infected with a coronavirus. J Virol 79:9388–9396

Dann SM, Wang HC, Gambarin KJ et al (2005) Interleukin-15 activates human natural killer cells to clear the intestinal protozoan cryptosporidium. J Infect Dis 192:1294–1302

Dasgupta S, Bhattacharya-Chatterjee M et al (2005) Inhibition of NK cell activity through TGF-β1 by down-regulation of NKG2D in a murine model of head and neck cancer. J Immunol 175:5541–5550

Dayanc BE, Beachy SH, Ostberg JR, Repasky EA (2008) Dissecting the role of hyperthermia in natural killer cell mediated anti-tumor responses. Int J Hyperthermia 24:41–56

de Rham C, Ferrari-Lacraz S, Jendly S et al (2007) The proinflammatory cytokines IL-2, IL-15 and IL-21 modulate the repertoire of mature human natural killer cell receptors. Arthritis Res Ther 9:R125

Deb C, Howe CL (2008) NKG2D contributes to efficient clearance of picornavirus from the acutely infected murine brain. J Neurovirol 14:261–266

Della Chiesa M, Carlomagno S, Frumento G et al (2006) The tryptophan catabolite L-kynurenine inhibits the surface expression of NKp46- and NKG2D-activating receptors and regulates NK-cell function. Blood 108:4118–4125

Deng L, Mariuzza RA (2006) Structural basis for recognition of MHC and MHC-like ligands by natural killer cell receptors. Semin Immunol 18:159–166

Diefenbach A, Hsia JK, Hsiung MY, Raulet DH (2003) A novel ligand for the NKG2D receptor activates NK cells and macrophages and induces tumor immunity. Eur J Immunol 33:381–391

Diefenbach A, Jamieson AM, Liu SD, Shastri N, Raulet DH (2000) Ligands for the murine NKG2D receptor: expression by tumor cells and activation of NK cells and macrophages. Nat Immunol 1:119–126

Diefenbach A, Jensen ER, Jamieson AM et al (2001) Rae1 and H60 ligands of the NKG2D receptor stimulate tumour immunity. Nature 413:165–171

Diefenbach A, Tomasello E, Lucas M et al (2002) Selective associations with signaling proteins determine stimulatory versus costimulatory activity of NKG2D. Nat Immunol 3:1142–1149

Diermayr S, Himmelreich H, Durovic B et al (2008) NKG2D ligand expression in AML increases in response to HDAC inhibitor valproic acid and contributes to allorecognition by NK-cell lines with single KIR-HLA class I specificities. Blood 111:1428–1436

Eagle RA, Traherne JA, Ashiru O, Wills MR, Trowsdale J (2006) Regulation of NKG2D ligand gene expression. Hum Immunol 67:159–169

Eagle RA, Trowsdale J (2007) Promiscuity and the single receptor: NKG2D. Nat Rev Immunol 7:737–744

Ebihara T, Masuda H, Akazawa T et al (2007) Induction of NKG2D ligands on human dendritic cells by TLR ligand stimulation and RNA virus infection. Int Immunol 19:1145–1155

Ehrlich LI, Ogasawara K, Hamerman JA et al (2005) Engagement of NKG2D by cognate ligand or antibody alone is insufficient to mediate costimulation of human and mouse CD8+ T cells. J Immunol 174:1922–1931

El-Sherbiny YM, Meade JL, Holmes TD et al (2007) The requirement for DNAM-1, NKG2D, and NKp46 in the natural killer cell-mediated killing of myeloma cells. Cancer Res 67:8444–8449

Epling-Burnette PK, Bai F, Painter JS et al (2007) Reduced natural killer (NK) function associated with high-risk myelodysplastic syndrome (MDS) and reduced expression of activating NK receptors. Blood 109:4816–4824

Fan YY, Yang BY, Wu CY (2008) Phenotypic and functional heterogeneity of natural killer cells from umbilical cord blood mononuclear cells. Immunol Invest 37:79–96

Fuertes MB, Girart MV, Molinero LL, Domaica CI, Rossi LE, Barrio MM, Mordoh J, Rabinovich GA, Zwirner NW (2008) Intracellular retention of the NKG2D ligand MHC class I chain-related gene A in human melanomas confers immune privilege and prevents NK cell-mediated cytotoxicity. J Immunol 180:4606–4614

Furue H, Matsuo K, Kumimoto H et al (2008) Decreased risk of colorectal cancer with the high natural killer cell activity NKG2D genotype in Japanese. Carcinogenesis 29:316–322

Garrity D, Call ME, Feng J, Wucherpfennig KW (2005) The activating NKG2D receptor assembles in the membrane with two signaling dimers into a hexameric structure. Proc Natl Acad Sci USA 102:7641–7646

Gasser S, Orsulic S, Brown EJ, Raulet DH (2005) The DNA damage pathway regulates innate immune system ligands of the NKG2D receptor. Nature 436(7054):1186–1190

Ghiringhelli F, Ménard C, Terme M et al (2005) $CD4^+CD25^+$ regulatory T cells inhibit natural killer cell functions in a transforming growth factor-β-dependent manner. J Exp Med 202:1075–1085

Girardi M, Oppenheim DE, Steele CR et al (2001) Regulation of cutaneous malignancy by γδ T cells. Science 294(5542):605–609

Giurisato E, Cella M, Takai T (2007) Phosphatidylinositol 3-kinase activation is required to form the NKG2D immunological synapse. Mol Cell Biol 27:8583–8599

Gorgoulis VG et al (2005) Activation of the DNA damage checkpoint and genomic instability in human precancerous lesions. Nature 434:907–913

Groh V, Bahram S, Bauer S et al (1996) Cell stress-regulated human major histocompatibility complex class I gene expressed in gastrointestinal epithelium. Proc Natl Acad Sci USA 93:12445–12450

Groh V, Brühl A, El-Gabalawy H et al (2003) Stimulation of T cell autoreactivity by anomalous expression of NKG2D and its MIC ligands in rheumatoid arthritis. Proc Natl Acad Sci USA 100:9452–9457

Groh V, Rhinehart R, Randolph-Habecker J et al (2001) Co-stimulation of CD8 αβT cells by NKG2D via engagement by MIC induced on virus-infected cells. Nat Immunol 2:255–260

Groh V, Rhinehart R, Secrist H et al (1999) Broad tumor-associated expression and recognition by tumor-derived γδ T cells of MICA and MICB. Proc Natl Acad Sci USA 96:6879–6884

Groh V, Steinle A, Bauer S, Spies T (1998) Recognition of stress-induced MHC molecules by intestinal epithelial γδ T cells. Science 279:1737–1740

Groh V, Wu J, Yee C, Spies T (2002) Tumour-derived soluble MIC ligands impair expression of NKG2D and T-cell activation. Nature 419:734–738

Gross O, Grupp C, Steinberg C et al (2008) Multiple ITAM-coupled NK cell receptors engage the Bcl10/Malt1 complex via Carma1 for NF-κB and MAPK activation to selectively control cytokine production. Blood 112(6):2421–2428

Guan H, Moretto M, Bzik DJ, Gigley J, Khan IA (2007) NK cells enhance dendritic cell response against parasite antigens via NKG2D pathway. J Immunol 179:590–596

Guerra N, Tan YX, Joncker NT et al (2008) NKG2D-deficient mice are defective in tumor surveillance in models of spontaneous malignancy. Immunity 28:571–580

Gumá M, Angulo A, López-Botet M (2006) NK cell receptors involved in the response to human cytomegalovirus infection. Curr Top Microbiol Immunol 298:207–222

Hanson MG, Ozenci V, Carlsten MC et al (2007) A short-term dietary supplementation with high doses of vitamin E increases NK cell cytolytic activity in advanced colorectal cancer patients. Cancer Immunol Immunother 56:973–984

Hayakawa Y, Smyth MJ (2006a) Innate immune recognition and suppression of tumors. Adv Cancer Res 95:293–322

Hayakawa Y, Smyth MJ (2006b) NKG2D and cytotoxic effector function in tumor immune surveillance. Semin Immunol 18:176–185

Hidano S, Sasanuma H, Ohshima K et al (2008) Distinct regulatory functions of SLP-76 and MIST in NK cell cytotoxicity and IFN-γ production. Int Immunol 20:345–352

Ho EL, Heusel JW, Brown MG et al (1998) Murine Nkg2d and Cd94 are clustered within the natural killer complex and are expressed independently in natural killer cells. Proc Natl Acad Sci USA 95:6320–6325

Hoffmann SC, Schellack C, Textor S et al (2007) Identification of CLEC12B, an inhibitory receptor on myeloid cells. J Biol Chem 282:22370–22375

Houchins JP, Yabe T, McSherry C, Bach FH (1991) DNA sequence analysis of NKG2, a family of related cDNA clones encoding type II integral membrane proteins on human natural killer cells. J Exp Med 173:1017–1022

Huntington ND, Xu Y, Nutt SL, Tarlinton DM (2005) A requirement for CD45 distinguishes Ly49D-mediated cytokine and chemokine production from killing in primary natural killer cells. J Exp Med 201:1421–1433

Hyka-Nouspikel N, Phillips JH (2006) Physiological roles of murine DAP10 adapter protein in tumor immunity and autoimmunity. Immunol Rev 214:106–117

Izumi Y, Ida H, Huang M et al (2006) Characterization of peripheral natural killer cells in primary Sjögren's syndrome: impaired NK cell activity and low NK cell number. J Lab Clin Med 147:242–249

Jacobs R, Hintzen G, Kemper A et al (2001) $CD56^{bright}$ cells differ in their KIR repertoire and cytotoxic features from $CD56^{dim}$ NK cells. Eur J Immunol 31:3121–3127

Karimi M, Cao TM, Baker JA et al (2005) Silencing human NKG2D, DAP10, and DAP12 reduces cytotoxicity of activated CD8+ T cells and NK cells. J Immunol 175:7819–7828

Katou F, Ohtani H, Watanabe Y et al (2007) Differing phenotypes between intraepithelial and stromal lymphocytes in early-stage tongue cancer. Cancer Res 67:11195–11201

Khurana D, Arneson LN, Schoon RA et al (2007) Differential regulation of human NK cell-mediated cytotoxicity by the tyrosine kinase Itk. J Immunol 178:3575–3582

Kim J, Chang CK, Hayden T et al (2007) The activating immunoreceptor NKG2D and its ligands are involved in allograft transplant rejection. J Immunol 179:6416–6422

Kim JY, Son YO, Park SW et al (2006) Increase of NKG2D ligands and sensitivity to NK cell-mediated cytotoxicity of tumor cells by heat shock and ionizing radiation. Exp Mol Med 38:474–484

Kim YJ, Han MK, Broxmeyer HE (2008) 4-1BB regulates NKG2D costimulation in human cord blood CD8+ T cells. Blood 111:1378–1386

Kong Y, Cao W, Xi, X et al (2009) The NKG2D ligand ULBP4 binds to TCRγ9/δ2 and induces cytotoxicity to tumor cells through both TCRγδ and NKG2D. Blood 114:310–317

Konjević G, Jović V, Vuletić A et al (2007) CD69 on CD56+ NK cells and response to chemoimmunotherapy in metastatic melanoma. Eur J Clin Invest 37:887–896

Kopcow HD, Allan DS, Chen X et al (2005) Human decidual NK cells form immature activating synapses and are not cytotoxic. Proc Natl Acad Sci USA 102:15563–15568

Kraetzel K, Stoelcker B, Eissner G et al (2008) NKG2D-dependent effector function of bronchial epithelium activated alloreactive T cells. Eur Respir J 32:563–570

Krockenberger M, Dombrowski Y, Weidler C et al (2008) Macrophage migration inhibitory factor contributes to the immune escape of ovarian cancer by down-regulating NKG2D. J Immunol 180:7338–7348

Kubin M, Cassiano L, Chalupny J et al (2001) ULBP1, 2, 3: novel MHC class I-related molecules that bind to human cytomegalovirus glycoprotein UL16, activate NK cells. Eur J Immunol 31:1428–1437

Kusumi M, Yamashita T, Fujii T et al (2006) Expression patterns of lectin-like natural killer receptors, inhibitory CD94/NKG2A, and activating CD94/NKG2C on decidual CD56^{bright} natural killer cells differ from those on peripheral CD56^{dim} natural killer cells. J Reprod Immunol 70:33–42

Lanier LL (2009) DAP10- and DAP12-associated receptors in innate immunity. Immunol Rev 227:150–160

Lanier LL (2005) NK cell recognition. Annu Rev Immunol 23:225–274

Lanier LL (2008) Up on the tightrope: natural killer cell activation and inhibition. Nat Immunol 9:495–502

Le Maux Chansac B, Missé D et al (2008) Potentiation of NK cell-mediated cytotoxicity in human lung adenocarcinoma: role of NKG2D-dependent pathway. Int Immunol 20:801–810

Lee JC, Lee KM, Kim DW, Heo DS (2004) Elevated TGF-β1 secretion and down-modulation of NKG2D underlies impaired NK cytotoxicity in cancer patients. J Immunol 172:7335–7340

Levin SD, Brandt CS, Yi EC et al (2009) Identification of a cell surface ligand for NKp30. J Immunol 182:134.20

Li A, He M, Wang H et al (2007) All-trans retinoic acid negatively regulates cytotoxic activities of nature killer cell line 92. Biochem Biophys Res Commun 352:42–47

Li C, Ge B, Nicotra M, Stern JN et al (2008) JNK MAP kinase activation is required for MTOC and granule polarization in NKG2D-mediated NK cell cytotoxicity. Proc Natl Acad Sci USA 105:3017–3022

Li P, McDermott G, Strong RK (2002) Crystal structures of RAE-1β and its complex with the activating immunoreceptor NKG2D. Immunity 16:77–86

Li P, Morris DL, Willcox BE et al (2001) Complex structure of the activating immunoreceptor NKG2D and its MHC class I-like ligand MICA. Nat Immunol 2:443–451

Maasho K, Opoku-Anane J, Marusina AI et al (2005) NKG2D is a costimulatory receptor for human naive CD8$^+$ T cells. J Immunol 174:4480–4484

MacFarlane AW 4th, Campbell KS (2006) Signal transduction in natural killer cells. Curr Top Microbiol Immunol 298:23–57

Malarkannan S (2006) The balancing act: inhibitory Ly49 regulate NKG2D-mediated NK cell functions. Semin Immunol 18:186–192

Mans J, Natarajan K, Balbo A et al (2007) Cellular expression and crystal structure of the murine cytomegalovirus major histocompatibility complex class I-like glycoprotein, m153. J Biol Chem 282:35247–35258

Call ME, Wucherpfennig K, Chou JJ (2010) The structural basis for intramembrane assembly of an activating immunoreceptor complex. Nat Immunol 11:1023–1029

Melum E, Buch S, Schafmayer C et al (2008) Investigation of cholangiocarcinoma associated NKG2D polymorphisms in colorectal carcinoma. Int J Cancer 123:241–242

Mincheva-Nilsson L, Nagaeva O et al (2006) Placenta-derived soluble MHC class I chain-related molecules down-regulate NKG2D receptor on peripheral blood mononuclear cells during human pregnancy: a possible novel immune escape mechanism for fetal survival. J Immunol 176:3585–3592

Natarajan K, Sawicki MW, Margulies DH, Mariuzza RA (2000) Crystal structure of human CD69: a C-type lectin-like activation marker of hematopoietic cells. Biochemistry 39:14779–14786

Nitahara A, Shimura H, Ito A et al (2006) NKG2D ligation without T cell receptor engagement triggers both cytotoxicity and cytokine production in dendritic epidermal T cells. J Invest Dermatol 126:1052–1058

Nomura M, Zou Z, Joh T et al (1996) Genomic structures and characterization of Rae1 family members encoding GPI-anchored cell surface proteins and expressed predominantly in embryonic mouse brain. J Biochem 120:987–995

Ntrivalas EI, Bowser CR, Kwak-Kim J, Beaman KD, Gilman-Sachs A (2005) Expression of killer immunoglobulin-like receptors on peripheral blood NK cell subsets of women with recurrent spontaneous abortions or implantation failures. Am J Reprod Immunol 53:215–221

Ogasawara K, Hamerman JA, Ehrlich LR et al (2004) NKG2D blockade prevents autoimmune diabetes in NOD mice. Immunity 20:757–767, Comment in: Immunity. 2004; 21: 303–4

Ogasawara K, Lanier LL (2005) NKG2D in NK and T cell-mediated immunity. J Clin Immunol 25:534–540

Ortaldo JR, Winkler-Pickett R, Wigginton J et al (2006) Regulation of ITAM-positive receptors: role of IL-12 and IL-18. Blood 107:1468–1475

Osaki T, Saito H, Fukumoto Y et al (2009) Inverse correlation between NKG2D expression on CD8+ T cells and the frequency of CD4 + CD25+ regulatory T cells in patients with esophageal cancer. Dis Esophagus 22:49–54

Osaki T, Saito H, Tetal Y (2007) Decreased NKG2D expression on CD8+ T cell is involved in immune evasion in patient with gastric cancer. Clin Cancer Res 13:382–387

Pessino A, Sivori S, Bottino C et al (1998) Molecular cloning of NKp46: a novel member of the immunoglobulin superfamily involved in triggering of natural cytotoxicity. J Exp Med 188:953–960

Pilla L, Squarcina P, Coppa J et al (2005) Natural killer and NK-Like T-cell activation in colorectal carcinoma patients treated with autologous tumor-derived heat shock protein 96. Cancer Res 65:3942–3949

Poggi A, Prevosto C et al (2007) NKG2D and natural cytotoxicity receptors are involved in natural killer cell interaction with self-antigen presenting cells and stromal cells. Ann N Y Acad Sci 1109:47–57

Poggi A, Prevosto C, Massaro AM et al (2005) Interaction between human NK cells and bone marrow stromal cells induces NK cell triggering: role of NKp30 and NKG2D receptors. J Immunol 175:6352–6360

Rabinovich B, Li J, Wolfson M et al (2006) NKG2D splice variants: a reexamination of adaptor molecule associations. Immunogenetics 58:81–88

Radaev S, Rostro B, Brooks AG, Colonna M, Sun PD (2001) Conformational plasticity revealed by the cocrystal structure of

NKG2D and its class I MHC-like ligand ULBP3. Immunity 15:1039–1049

Radosavljevic M, Cuillerier B, Wilson MJ et al (2002) A cluster of ten novel MHC class I related genes on human chromosome 6q24.2-q25.3. Genomics 79:114–123

Regunathan J, Chen Y, Kutlesa S et al (2006) Differential and nonredundant roles of phospholipase Cγ2 and phospholipase Cγ1 in the terminal maturation of NK cells. J Immunol 177:5365–5376

Regunathan J, Chen Y, Wang D et al (2005) NKG2D receptor-mediated NK cell function is regulated by inhibitory Ly49 receptors. Blood 105:233–240

Rincon-Orozco B, Kunzmann V, Wrobel P et al (2005) Activation of Vγ 9Vδ 2 T cells by NKG2D. J Immunol 175:2144–2151

Rohner A, Langenkamp U, Siegler U et al (2007) Differentiation - promoting drugs up-regulate NKG2D ligand expression and enhance the susceptibility of acute myeloid leukemia cells to natural killer cell-mediated lysis. Leuk Res 3:1393–1402

Romero AI, Thorén FB, Brune M, Hellstrand K (2006) NKp46 and NKG2D receptor expression in NK cells with CD56dim and CD56bright phenotype: regulation by histamine and reactive oxygen species. Br J Haematol 132:91–98

Rosen DB, Araki M, Hamerman JA et al (2004) A structural basis for the association of DAP12 with mouse, but not human, NKG2D. J Immunol 173:2470–2478

Roy S, Barnes PF, Garg A, Wu S, Cosman D, Vankayalapati R (2008) NK cells lyse T regulatory cells that expand in response to an intracellular pathogen. J Immunol 180:1729–1736

Saez-Borderias A, Guma M, Angulo A et al (2006) Expression and function of NKG2D in CD4+ T cells specific for human cytomegalovirus. Eur J Immunol 36:3198–3206

Savithri B, Khar A (2006) A transmembrane-anchored rat RAE-1-like transcript as a ligand for NKR-P2, the rat ortholog of human and mouse NKG2D. Eur J Immunol 36:107–117

Schepis D, Gunnarsson I, Eloranta ML et al (2009) Increased proportion of CD56^{bright} natural killer cells in active and inactive systemic lupus erythematosus. Immunology 126:140–146

Schrama D, Terheyden P, Otto K et al (2006) Expression of the NKG2D ligand UL16 binding protein-1 (ULBP-1) on dendritic cells. Eur J Immunol 36:65–72

Schrambach S, Ardizzone M, Leymarie V et al (2007) In vivo expression pattern of MICA and MICB and its relevance to auto-immunity and cancer. PLoS One 2:e518

Smith KM, Wu J, Bakker AB et al (1998) Ly-49D and Ly-49H associate with mouse DAP12 and form activating receptors. Immunol 161:7–10

Smyth MJ, Teng MW, Swann J et al (2006) CD4+CD25; T regulatory cells suppress NK cell mediated immunotherapy of cancer. J Immunol 176:1582–1587

Song H, Hur DY, Kim KE et al (2006) IL-2/IL-18 prevent the down-modulation of NKG2D by TGF-β in NK cells via the c-Jun N-terminal kinase (JNK) pathway. Cell Immunol 242:39–45

Srivastava RM, Varalakshmi C, Khar A (2008) The ischemia-responsive protein 94 (Irp94) activates dendritic cells through NK cell receptor protein-2/NK group 2 member D (NKR-P2/NKG2D) leading to their maturation. J Immunol 180:1117–1130

Srivastava RM, Varalakshmi Ch, Khar A (2007) Cross-linking a mAb to NKR-P2/NKG2D on dendritic cells induces their activation and maturation leading to enhanced anti-tumor immune response. Int Immunol 19:591–607

Steinle A, Li P, Morris DL et al (2001) Interactions of human NKG2D with its ligands MIC-A, MIC-B, and homologs of the mouse RAE-1 protein family. Immunogenetics 53:279–287

Stephan G, David HR (2006) Activation and self-tolerance of natural killer cells. Immunol Rev 214:130–142

Strid J, Roberts SJ, Filler RB et al (2008) Acute upregulation of an NKG2D ligand promotes rapid reorganization of a local immune compartment with pleiotropic effects on carcinogenesis. Nat Immunol 9:146–154

Strong RK, McFarland BJ (2004) NKG2D and related immunoreceptors. Adv Protein Chem 68:281–312

Sundström Y, Nilsson C, Lilja G et al (2007) The expression of human natural killer cell receptors in early life. Scand J Immunol 66:335–344

Sutherland CL, Chalupny NJ, Schooley K et al (2002) UL16-binding proteins, novel MHC class I-related proteins, bind to NKG2D and activate multiple signaling pathways in primary NK cells. J Immunol 168:671–679

Sutherland CL, Rabinovich B, Chalupny NJ, Brawand P, Miller R, Cosman D (2006) ULBPs, human ligands of the NKG2D receptor, stimulate tumor immunity with enhancement by IL-15. Blood 108:1313–1319

Takaki R, Hayakawa Y, Nelson A, Sivakumar PV, Hughes S, Smyth MJ, Lanier LL (2005) IL-21 enhances tumor rejection through a NKG2D-dependent mechanism. J Immunol 175:2167–2173

Tomasello E, Vivier E (2005) KARAP/DAP12/TYROBP: three names and a multiplicity of biological functions. Eur J Immunol 35:1670–1677

Tormo J, Natarajan K, Margulies DH et al (1999) Crystal structure of a lectin-like natural killer cell receptor bound to its MHC class I ligand. Nature 402:623–631

Trichet V, Benezech C, Dousset C, Gesnel MC, Bonneville M, Breathnach R (2006) Complex interplay of activating and inhibitory signals received by Vγ9Vδ2 T cells revealed by target cell β2-microglobulin knockdown. J Immunol 177:6129–6136

Upshaw JL, Arneson LN, Schoon RA et al (2006) NKG2D-mediated signaling requires a DAP10-bound Grb2-Vav1 intermediate and phosphatidylinositol-3-kinase in human natural killer cells. Nat Immunol 7:524–532

Upshaw JL, Leibson PJ (2006) NKG2D-mediated activation of cytotoxic lymphocytes: unique signaling pathways and distinct functional outcomes. Semin Immunol 18:167–175

Upshaw JL, Schoon RA, Dick CJ et al (2005) The isoforms of phospholipase C-γ are differentially used by distinct human NK activating receptors. J Immunol 175:213–218

Vankayalapati R, Garg A, Porgador A et al (2005) Role of NK cell-activating receptors and their ligands in the lysis of mononuclear phagocytes infected with an intracellular bacterium. J Immunol 175:4611–4617

Varla-Leftherioti M, Spyropoulou-Vlachou M, Niokou D et al (2003) Natural killer (NK) cell receptors' repertoire in couples with recurrent spontaneous abortions. Am J Reprod Immunol 49:183–191

Verhoeven DH, de Hooge AS, Mooiman EC et al (2008) NK cells recognize and lyse Ewing sarcoma cells through NKG2D and DNAM-1 receptor dependent pathways. Mol Immunol 45(15):3917–3925

Vivier E, Nunès JA, Vély F (2004) Natural killer cell signaling pathways. Science 306(5701):1517–1519

von Strandmann EP, Hansen HP, Reiners KS et al (2006) A novel bispecific protein (ULBP2-BB4) targeting the NKG2D receptor on natural killer (NK) cells and CD138 activates NK cells and has potent antitumor activity against human multiple myeloma in vitro and in vivo. Blood 107:1955–1962

Walsh KB, Lodoen MB, Edwards RA, Lanier LL, Lane TE (2008) Evidence for differential roles for NKG2D receptor signaling in innate host defense against coronavirus-induced neurological and liver disease. J Virol 82:3021–3030

Wang Y, Xu H, Zheng X et al (2007) High expression of NKG2A/CD94 and low expression of granzyme B are associated with reduced cord blood NK cell activity. Cell Mol Immunol 4:377–382

Weis WI, Kahn R, Fourme R, Drickamer K, Hendrickson WA (1991) Structure of the calcium-dependent lectin domain from a rat mannose-binding protein determined by MAD phasing. Science 254:1608–1615

Wolan DW, Teyton L, Rudolph MG et al (2001) Crystal structure of the murine NK cell-activating receptor NKG2D at 1.95 A. Nat Immunol 2:248–254

Wrobel P, Shojaei H, Schittek B et al (2007) Lysis of a broad range of epithelial tumour cells by human gamma delta T cells: involvement of NKG2D ligands and T-cell receptor- versus NKG2D-dependent recognition. Scand J Immunol 66:320–328

Wu J, Song Y, Bakker AB, Bauer S et al (1999) An activating immunoreceptor complex formed by NKG2D and DAP10. Science 285:730–732

Yim D, Jie HB, Sotiriadis J et al (2001) Molecular cloning and characterization of pig immunoreceptor DAP10 and NKG2D. Immunogenetics 53:243–249

Zhang C, Zhang J, Niu J et al (2008a) Interleukin-12 improves cytotoxicity of natural killer cells via upregulated expression of NKG2D. Hum Immunol 69:490–500

Zhang C, Zhang J, Niu J, Zhang J, Tian Z (2008b) Interleukin-15 improves cytotoxicity of natural killer cells via up-regulating NKG2D and cytotoxic effector molecule expression as well as STAT1 and ERK1/2 phosphorylation. Cytokine 42:128–136

Zhang C, Zhang J, Wei H, Tian Z (2005) Imbalance of NKG2D and its inhibitory counterparts: how does tumor escape from innate immunity? Int Immunopharmacol 5:1099–1111

Zhang T, Barber A, Sentman CL (2007) Chimeric NKG2D modified T cells inhibit systemic T-cell lymphoma growth in a manner involving multiple cytokines and cytotoxic pathways. Cancer Res 67:11029–11036

Zhang Y, Stastny P (2006) MIC-A antigens stimulate T cell proliferation and cell-mediated cytotoxicity. Hum Immunol 67: 215–222

Zou Z, Nomura M, Takihara Y, Yasunaga T et al (1996) Isolation and characterization of retinoic acid-inducible cDNA. J Biochem 119:319–328

KLRC4, KLRG1, and Natural Cytotoxicity Receptors

Rajesh K. Gupta and G.S. Gupta

In earlier Chapters we discussed mouse and human NK cell lectin receptors which can deliver inhibitory or activatory NK cell cytotoxicity in presence or absence of MHC-I molecules. Among non-lectin inhibitory receptors, killer immunoglobulin (Ig)-like receptors (KIRs) recognize different allelic groups of HLA-A, -B or -C molecules. Immunoglobulin-like transcript 2 (ILT2 or LIR1 or CD85j) receptors are more 'promiscuous', as they recognize a large number of HLA class I alleles, while CRD containing CD94-NKG2A recognizes HLA-E, an HLA class I molecule with a limited polymorphism. It is well known that various HLA class I alleles provide signal sequence peptides that bind HLA-E and enable it to be expressed on the cell surface. Importantly, each type of KIR is expressed only by a subset of NK cells (Braud et al. 1998). A common characteristic of the various HLA class I-specific inhibitory receptors is the presence, in their cytoplasmic tail, of ITIM that enable them to recruit and activate SHP-1 and SHP-2-phosphatases (Moretta et al. 2001; Lanier 1998). In turn, these phosphatases switch off the activating signaling cascade initiated by the various activating receptors (Moretta and Moretta 2004). In this chapter we continue discussing cytotoxic receptors which do not depend on HLA antigens for their action and do not contain ITIM motif in their cytoplasmic tail as well as activating NK cell receptors which are of recent origin but do not contain lectin-like domain till date.

32.1 Killer Cell Lectin-Like Receptor F-1 (NKG2F/KLRC4/*CLEC5C*/NKp80)

32.1.1 NKp80 or Killer Cell Lectin-Like Receptor Subfamily F-Member 1 (KLRF1)

Killer cell lectin-like receptor subfamily F-member 1 (KLRF1), also known as KLRC4 or NKp80, is a human gene which is localized 25 kb from NKG2-A (Roda-Navarro et al. 2000) and related to NKG2-D cDNA on the high-resolution physical map of chromosome 12p. It encodes a putative type II transmembrane glycoprotein. The 3′ end of NKp80/KLRC4/KLRF1 transcript includes the first non-coding exon found at the 5′ end of the adjacent D12S2489E gene transcript (Renedo et al. 2000). The genomic structure of the NKp80/KLRF1 gene and the existence of one spliced variant have been described. It is expressed at the mRNA level in peripheral blood leukocytes, activated NK cells, monocytes and NK and myeloid cell lines. NKp80, a C-type lectin-like receptor, stimulates the cytotoxicity of NK cells and releases cytokine. Roda-Navarro et al. described the existence of one alternative spliced form, lacking the stalk region of the extracellular domain and two novel KLRF1 alternative spliced variants coding for truncated proteins lacking the C-type lectin-like domain (Plougastel and Trowsdale 1998a, b; Roda-Navarro et al. 2001). A conserved 24-amino acid sequence, present in all members of the NKG2 family, suggests that NKG2-F is also able to form heterodimers with CD94 (Plougastel and Trowsdale 1998a, b).

NKp80 is a type II transmembrane protein of 231 amino acids identical to the putative protein encoded by a cDNA termed KLRF1 (Vitale et al. 2001) (Fig. 32.1). NKp80 is protein with a C-type lectin domain in its extracellular region. NKp80 is expressed at the cell surface as a dimer of ~80 kDa on virtually all NK cells and by a minor subset of T cells characterized by the CD56 surface antigen and shown to function as a co-receptor rather than as true receptor with natural cytotoxicity receptors (NCR) to induce activation of NK cell-mediated cytotoxicity. NKp80 surface expression was also detected in all CD3⁻ and in 6/10 CD3⁺ large granular lymphocyte expansions derived from patients with lympho-proliferative disease of granular lymphocytes. mAb-mediated cross-linking of NKp80 resulted in induction of cytolytic activity and Ca^{2+} mobilization. The NKp80 recognizes a ligand on normal T cells that may be down-regulated during tumor transformation. Indeed, it induces natural cytotoxicity only when co-engaged with a triggering receptor (Vitale et al. 2001). NK cells in monkeys were generally identified by negative selection of peripheral

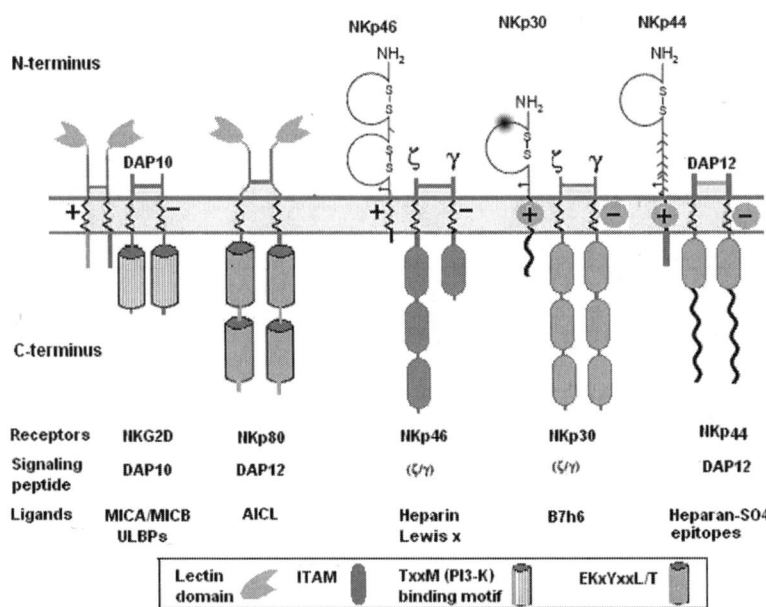

Fig. 32.1 Activating NK cell receptors and co-receptors and their cellular ligands. The Figure illustrates the molecular structure of the NK receptors NKG2D, NKp80, NKp46, NKp30, NKp44. Their interaction with signaling polypeptides or with relevant cytoplasmic molecules is also shown

blood mononuclear cells (PBMCs) for the absence of T-cell, B-cell, and monocyte markers.

32.1.1.1 Triggering Receptors in Sub-Human Primate NK Cells

Mavilio et al. (2005) indicated that mAb-mediated ligation of NKp80 induced NK cell cytotoxicity, while NKG2A inhibited the lysis of target cells by NK cells from macaques, as well as from humans. This phenotypic and functional characterization of NKG2A and NKp80 in rhesus and pig-tailed macaque NK cells is useful in the analysis of their innate immune system. Both M. fascicularis and M. mulatta NK cells express NKp80, which displayed a high degree of sequence homology with their human counterpart. Analysis of NK cells in simian HIV-infected M. fascicularis revealed reduced surface expression of selected NK cell-triggering receptors associated with a decreased NK cell function only in some animals (Biassoni et al. 2005). Functional characterization for NKp46, NKp30, NKp80, and NKG2D on Pan troglodytes NK cells revealed that, in this AIDS-resistant species, relevant differences to human NK cells involve NKp80 and particularly NKp30, which is primarily involved in NK-DC interactions (Rutjens et al. 2007).

32.1.2 NKp80/KLRF1 Associates with DAP12

The presence of two ITIM within the cytoplasmic tail of NKp80/KLRF1 suggests an inhibitory role in NK cell and monocyte activity. NKG2F/KLRF1 is an orphan gene within the NKG2 family whose translated product would contain both a positively charged residue in its transmembrane region, an intracellular ITIM-like sequence (YxxL) and an extracellular domain (62 residues) that is truncated relative to other NKG2 molecules. Expression of NKp80/KLRF1 appears to be confined to intracellular compartments probably due to its inability to associate with CD94. It can, however, associate with DAP12 thereby providing activation signaling potential. Since it was not possible to demonstrate phosphorylation of Tyr residue in the ITIM-like motif, it was suggested that it is a mock ITIM. NKG2F, a receptor component with an unidentified partner(s), can regulate cell activation through competition for DAP12 with other receptors, such as NKG2C and -E/H, or it could simply be a vestigial gene product (Kim et al. 2004). Using the NK cell line NK92MI, Dennehy et al. (2011) demonstrated that NKp80, but not NKp80 mutated at tyrosine 7 (NKp80/Y7F), is tyrosine phosphorylated. Accordingly, NKp80/Y7F, but not NKp80/Y30F or NKp80/Y37F, failed to induce cytotoxicity. Altogether, report suggests that NKp80 uses an atypical hemi-ITAM and Syk kinase to trigger cellular cytotoxicity (Fig. 32.1).

32.2 Activation-Induced C-Type Lectin (CLECSF2): Ligand for NKp80

Thomas et al. (2008) demonstrated that the ligand for NKp80 is the myeloid-specific CTLR activation-induced C-type lectin (AICL), which is encoded in the natural killer gene complex (NKC) adjacent to NKp80. Hamann et al. (1997) cloned a cDNA of a molecule, designated activation-induced C-type lectin (AICL), whose gene maps to the human NKC proximal to the CD69 gene. AICL is a myeloid-specific activating orphan receptor that is upregulated by Toll-like receptor stimulation. AICL is a 149-amino acid polypeptide with a short cytoplasmic tail of seven amino acids and a C-type lectin domain separated from the transmembrane region by only nine amino acids. The AICL shows highest sequence similarity to the domains of CD69 and the 17.5 chicken-like lectin. NKp80 binds to the genetically linked receptor AICL, which, like NKp80, is absent from rodents. AICL-NKp80 interactions promote NK cell-mediated cytolysis of malignant myeloid cells. In addition, during crosstalk between NK cells and monocytes, NKp80 stimulated the release of proinflammatory cytokines from both cell types. Thus, NKp80-AICL interactions may contribute to the initiation and maintenance of immune responses at sites of inflammation (Welte et al. 2006). The Kaposi's sarcoma-associated herpesvirus (KSHV) immune evasion gene, K5, reduces cell surface expression of AICL. Down-regulation of AICL requires the ubiquitin E3 ligase activity of K5 to target substrate cytoplasmic tail lysine residues. The down-regulation of ligands for both the NKG2D and NKp80 activation pathways provides KSHV with a powerful mechanism for evasion of NK cell antiviral functions (Thomas et al. 2008).

The AICL is also expressed on a minor fraction of human $CD8^+$ T cells that exhibit a high responsiveness and an effector memory phenotype. The $NKp80^+$ T-cell subset is characterized by the co-expression of other NK receptors and NKp80 ligation augmentes CD3-stimulated degranulation and IFNγ secretion by effector memory T cells. Hence, NKp80 may enable effector memory CD8 T cells to interact functionally with cells of myeloid origin at sites of inflammation (Kuttruff et al. 2009).

32.3 KLRG1 (Rat MAFA/CLEC15A or Mouse 2F1-Ag)

32.3.1 Mast Cell Function-Associated Antigen (MAFA)

The mAb G63 identified a membrane component of mast cells, called as mast cell function-associated antigen (MAFA). Mast cell antigen is a glycoprotein with an M_r of 28–40 kDa, and is present on the surface of rat mucosal and serosal mast cells. Its density on cells of the mucosal mast cell line RBL-2H3 is $1–2 \times 10^4$ copies per cell. Cross-linking of this protein by mAb G63 results in the inhibition of Fc ε RI-mediated secretion of RBL-2H3 cells and inhibits biochemical processes initiated by FcεRI aggregation, such as the hydrolysis of phosphatidylinositides, the influx of Ca^{2+} ions, and the synthesis and release of inflammatory mediators and in redistribution of the membrane component recognized by G63 (Ortega et al. 1991). Based on MAFA density the rat mucosal mast cell line, RBL-2H3, can be classified into two subpopulations: cells carrying either high or low density of a MAFA (Cohen-Dayag et al. 1992). The MAFA is expressed at a ratio of approximately 1:30 with respect to the Type I FcεRI on rat mast cell line. Despite this stoichiometry, clustering MAFA by its specific mAb G63 substantially inhibits secretion of both granular and de novo synthesized mediators induced upon FcεRI aggregation. Since the FcεRIs apparently signal from within raft micro-environments, Barisas et al. (2007) investigated co-localization of MAFA within these membrane compartments containing aggregated FcεRI and suggested for constitutive interactions between FcεRI and MAFA.

32.3.2 MAFA Is a Lectin

The recombinant MAFA (rMAFA) was expressed as a monomeric and disulfide-linked homodimeric glycoprotein in the membrane of insect cells. The rMAFA binds specifically the terminal mannose residues in a Ca^{2+}-dependent manner. Results support the notion that the extracellular domain of the MAFA is able to bind ligands, which may be modulatory for the mast cell response (Binsack and Pecht 1997). The MAFA interferes with the coupling cascade of the type 1 FcεR upstream to PLCγ1 activation by protein-tyrosine kinases. The MAFA cDNA shows that MAFA contains a single ORF, encoding a 188-amino acid-long type II integral membrane protein. The 114C-terminal amino acids display sequence homology with the CRD of calcium-dependent animal lectins. The cytoplasmic tail of MAFA contains a YXXL (YSTL) motif, which is conserved among related C-type lectins and is an ITIM. Changes in the MAFA tyrosyl- and seryl-phosphorylation levels are observed in response to binding of mAb G63, antigenic stimulation, and a combination of both treatments (Guthmann et al. 1995b). The high conservation of cysteinyl residues suggests an important role for intrachain disulfide bonds in attaining its structure and biological activity. The MAFA clustering by G63 inhibits the de novo synthesis and secretion of IL-6 induced by the FcεRI stimulus. Experiments suggested that MAFA might have a role in cell adhesion in addition to its immunomodulatory capacity (Guthmann et al. 1995a).

32.3.3 Interactions of MAFA with FcεR

The nature of MAFA-FcεRI interactions giving rise to this inhibition remains unclear. By time-resolved phosphorescence anisotropy, Song et al. (2002) studied the rotational behavior of both MAFA and FcεRI as ligated by various reagents involved in FcεRI -induced degranulation and MAFA-mediated inhibition there of. Clustering the FcεRI-IgE complex by antigen or by anti-IgE increases the phosphorescence anisotropy of G63 Fab and slows its rotational relaxation. Studies indicated that unperturbed MAFA associates with clustered FcεRI (Song et al. 2002).

Clustering of MAFA Induces Tyrosyl Phosphorylation of the FcεRI-β Subunit: Clustering of MAFA on the surface of rat mucosal mast cells RBL-2H3 leads to suppression of the secretory response induced by FcεRI. The MAFA is located in the vicinity of the FcεRI on resting cells, and clustering of Fcε RI leads to no significant change in the proximity of the two molecular species. Reports suggest that the secretory response inhibition by MAFA interferes with the signal transduction cascade initiated via the FcεRI (Jürgens et al. 1996). The inhibition was found to take place upstream to the production of inositol phosphates and the transient increase in free cytosolic Ca^{2+} ion concentration. Hence it probably interferes with the cascade at the level of the protein tyrosyl kinases (PTK) activity (Schweitzer-Stenner et al. 1999).

Studies indicate that MAFA's inhibitory action involves at least two different enzymes: Following the tyrosyl-phosphorylation of MAFA ITIM by the PTK Lyn, two phosphatases SHIP and SHP2 are recruited at the plasma membrane where they propagate the inhibitory signals. The inhibition has been shown to take place upstream to the production step of inositol phosphates in the FcεRI coupling cascade. Syk tyrosine phosphorylation and activation, as well as LAT (linker for activation of T cells) tyrosine phosphorylation, both induced by FcεRI clustering, were found to be reduced upon MAFA clustering. In contrast, the activity of the SH2-containing protein tyrosine phosphatase (SHP-2) increased. Reports suggest that one possible mechanism by which MAFA affects the FcεRI stimulation cascade is suppression of Syk activity, i.e. MAFA clustering leads SHP-2 to act on Syk, thereby reducing its tyrosine phosphorylation and its activity (Abramson et al. 2002; Xu et al. 2001). MAFA has also a capacity of modulating the cell cycle of RBL-2H3 line and cell proliferation rate by inhibiting RasGTP formation in the Ras signaling pathway (Abramson et al. 2002; Licht et al. 2006).

Inhibition of FcεRI-Induced de Novo Synthesis of Mediators by MAFA: Aggregation of the type 1 Fc-ε-RI on mast cells initiates a network of biochemical processes culminating in secretion of both granule-stored and de novo-synthesized inflammatory mediators. A strict control of this response is obviously a necessity. MAFA selectively regulates the FcεRI stimulus–response coupling network and the subsequent de novo production and secretion of inflammatory mediators. Specifically, MAFA suppresses the PLC-γ2-$[Ca^{2+}]i$, Raf-1-Erk1/2, and PKC-p38 coupling pathways, while the Fyn-Gab2-mediated activation of PKB and Jnk is essentially unaffected. Hence, the activities of several transcription/nuclear factors for inflammatory mediators (NF-kB, NFAT) are markedly reduced, while those of others (Jun, Fos, Fra, p90rsk) are unaltered. This results in a selective inhibition of gene transcription of cytokines including IL-1beta, IL-4, IL-8, and IL-10, while that of TNF-alpha, MCP-1, IL-3, IL-5, or IL-13 remains unaffected. These results illustrate the capacity of an ITIF-containing receptor to cause tight and specific control of the production and secretion of inflammatory mediators by mast cells (Abramson et al. 2006).

32.3.4 MAFA Gene

MAFA is encoded by a single-copy gene that spans 13 kb in the rat genome and is composed of five exons. Three separate exons encode the CRD of the MAFA, defining its close homology to the genes of CD23, CD69, CD72, NKR-P1, and Ly49. Analysis of the 5′ flanking region of the gene reveals that a cell type-specific promoter is located within the first 664 bp upstream of the transcription origin. The promoter lacks any obvious TATA box and drives gene transcription originating from multiple start sites. Examination for possible polymorphism of the MAFA transcripts revealed two transcripts, generated by alternative splicing. The transcription of the MAFA gene was detected in normal rat lungs, where both transmembranal and soluble MAFA appear to be expressed. Lung immunohistochemical analysis further suggests that MAFA expression is restricted to mast cells.

32.4 KLRG1 (OR 2F1-AG): A Mouse Homologue of MAFA

32.4.1 KLRG1: A Mouse Homologue of MAFA

The KLRG1or formally known as 2F1-AG or MAFA is expressed by NK cells and memory T cells in man and mice. A mAb (2F1), specific for mouse NK cells, recognized the mouse homolog of MAFA that is expressed on rat mast cells. The 2F1 antigen (2F1-Ag) and rat MAFA are highly conserved and contain a cytoplasmic ITIM used in inhibitory signaling. The human homologue is closely related to the rodent MAFA/2F1-Ag proteins. Like rat MAFA, 2F1-Ag is probably encoded by a single gene, which exhibits relatively little polymorphism. While rat MAFA is considered a mast cell antigen, it was not possible to detect cell surface expression of 2F1-Ag by mouse mast cell lines, bone marrow-derived mast cells, or peritoneal mast cells. However, mouse bone marrow-derived mast cells were devoid of 2F1-Ag mRNA. Instead, 40% of mouse NK cells express 2F1-Ag (Hanke et al. 1998).

The mouse MAFA (mMAFA) gene expression is strongly induced in effector CD8$^+$ T cells and lymphokine-activated NK cells but not in effector CD4$^+$ T cells and in mouse mast cells. Moreover, mMAFA gene expression was only found in effector CD8$^+$ T cells that had been primed in vivo with live virus because in vitro activated CD8$^+$ T cells did not express mMAFA. Sequence comparison revealed a high degree of conservation (89% similarity) between rat MAFA and mMAFA (Blaser et al. 1998). Using a mAb (2F1), Corral et al. (2000) characterized a mouse NK cell KLRG. KLRG1 is expressed on 30–60% of murine NK cells, and a small fraction of T cells, and is composed of a homodimer of glycosylated 30–38-kDa subunits. Cell surface expression of KLRG1 by NK cells is down-regulated in mice deficient for expression of class I molecules, in contrast to the Ly49 lectin-like NK receptors, which are up-regulated in class I-deficient mice. KLRG1 does not bind class I molecules in a cell-cell adhesion assay. Transgenic expression of KLRG1 was unaffected by class I deficiency, indicating that class I molecules do not affect the KLRG1 protein directly, and suggesting that regulation is at the level of expression of the endogenous KLRG1 gene. Evidence showed that class I molecules regulate KLRG1 via interactions with inhibitory Ly49 molecules and SHP-1 signaling (Corral et al. 2000). The KLRG1 is an inhibitory C-type lectin expressed on NK cells and activated CD8 T cells. The mouse gene spans about 13 kb and consists of five exons. Short interspersed repeats of the B1 and B2 family, a LINE-1-like element, and a (CTT)170 triplet repeat were found in intron sequences. In contrast to human KLRG1 and to the murine KLR family members, mouse KLRG1 locates outside the NK complex on Chromosome 6 between the genes encoding CD9 and CD4 (Voehringer et al. 2001).

32.4.1.1 KLRG1 on Effector and Memory CD8 T Cells

KLRG1 was present on 1–3% of adult splenic CD8 cells that expressed CD8αβ heterodimers as well as a polyclonal TCR Vβ repertoire indicative of conventional CD8 cells and CD8$^+$ effector/memory cells that can secrete cytokines but have poor proliferative capacity. Spontaneous IFN-γ production by approximately 20% of KLRG1$^+$ CD8 cells identified them as pro-inflammatory effector cells. In contrast to NK cells, Ly49 and KLRG1 expression on CD8 cells was found to be mutually exclusive. KLRG1 triggering interferes with TCR αβ -mediated Ca^{2+} mobilization and cytotoxicity, raising the possibility that KLRG1 functionally participates in down-regulation of CD8 T cell responses (Beyersdorf et al. 2001). Furthermore, CD8$^+$ cells expressing CD57, a marker of replicative senescence, also expressed KLRG1 and a population of CD57$^-$KLRG1$^+$ cells has also been identified. The combination of KLRG1 and CD57 expression might thus aid in refining functional characterization of CD8$^+$ T cell subsets (Ibegbu et al. 2005). KLRG1 expression is dramatically induced on CD8$^+$ T cells during both a viral and a parasitic infection. The engagement of KLRG1 on a transfected NK cell line inhibits both cytokine production and NK cell-mediated cytotoxicity (Robbins et al. 2002, 2003).

The acquisition of the NK cell inhibitory markers NK1.1 and KLRG on T cells exposed to high numbers of DCs suggests a role for these molecules in the protection of antigen-responsive T cells from exhaustion by overstimulation. Using a mAb for human KLRG1, Voehringer et al. (2002) identified human T cells that are capable of secreting cytokines but fail to proliferate after stimulation. Furthermore, the lack of proliferative capacity of CD8 T cells correlates better with KLRG1 expression than with absence of the CD28 marker.

32.4.1.2 Characterization of Mouse CD4 T Cell Subsets Defined by Expression of KLRG1

In normal mice, while a polyclonal TCR repertoire suggests thymic origin of KLRG1$^+$ CD4$^+$ cells, KLRG1 expression was found to be restricted to peripheral CD4$^+$ T cells. Based on phenotypic analyses, a minority of KLRG1$^+$ CD4$^+$ cells are effector/memory cells with a proliferative history. The majority of KLRG1$^+$ CD4$^+$ cells are, however, bonafide Treg cells that depend on IL-2 and/or CD28 and express both FoxP3 and high levels of intracellular CD152. KLRG1 expression represents a distinctive subset of senescent effector/memory and potent regulatory CD4$^+$ T cells (Beyersdorf et al. 2007).

CCR7$^+$ central memory (T$_{CM}$) CD4$^+$ T cells play a central role in long-term immunological memory. Reports indicate that a proportion of CD4$^+$ T$_{CM}$ is able to produce effector cytokines. Stubbe et al. (2008) characterized cytokine-producing human CD4$^+$ T$_{CM}$ specific for cleared protein and persistent viral Ag. The type of Ag stimulation is the major determinant of CD4$^+$ T$_{CM}$ differentiation. CMV-specific T$_{CM}$ were significantly more differentiated than protein Ag-specific T$_{CM}$ and included higher proportions of IFN-γ-producing cells. The expression of KLRG1 by protein Ag- and CMV-specific T$_{CM}$ was associated with increased production of effector cytokines. KLRG1$^+$ T$_{CM}$ expressing high levels of CD127 suggests that they can survive long term under the influence of IL-7 (Stubbe et al. 2008). Studies with hepatitis C virus (HCV)-specific CD8$^+$ T also demonstrated central memory (CCR7$^+$) and early-differentiated phenotypes of HCV-specific CD8$^+$ T cells (Bengsch et al. 2007).

32.4.1.3 KLRG1 in Cord Blood

Umbilical cord blood T cells in humans represent a homogenous pool of naive T cells. However, analysis of KLRG1 expression in cord blood revealed an unexpected heterogeneity of human T cells in newborns. A substantial subsets of

CD4 (30%) and CD8 (20%) αβ T cells in cord blood expressed KLRG1. In contrast to T cells in adult, KLRG1⁺ T cells in cord blood exhibited predominantly a naive CCR7⁺CD45RA⁺ and CD11a^{low} phenotype. After birth, KLRG1 expression in T cells from peripheral blood decreased rapidly to reappear in effector/memory T cells in adults. KLRG1⁺ T cells in cord blood expressed a diverse T cell receptor β repertoire and the cells proliferated normally, in contrast to KLRG1⁺ T cells from adults (Marcolino et al. 2004). In naive C57BL/6 mice KLRG1 is expressed on a subset of CD44^{high}CD62L^{low} T cells. KLRG1 expression can be detected on a small number of Vα14i NK T cells but not on CD8αα⁺ intraepithelial T cells that are either TCRγδ⁺ or TCRαβ⁺ (Eberl et al. 2005).

32.4.1.4 ER Stress Regulator XBP-1 Contributes to Effector CD8+ T Cell Differentiation During Acute Infection

The transcription factor X-box-binding protein-1 (XBP-1) plays an essential role in activating the unfolded protein response in the ER. The IRE-1/XBP-1 pathway is activated in effector CD8⁺ T cells during acute infection. An XBP-1 splicing reporter was enriched in terminal effector cells expressing high levels of KLRG1. Over-expression of the spliced form of XBP-1 in CD8⁺ T cells enhanced KLRG1 expression during infection, whereas XBP-1$^{-/-}$ CD8⁺ T cells or cells expressing a dominant-negative form of XBP-1 showed a decreased proportion of KLRG1^{high} effector cells. These results suggest that, in the response to pathogen, activation of ER stress sensors and XBP-1 splicing contribute to the differentiation of end-stage effector CD8⁺ T cells (Kamimura and Bevan 2008).

32.4.1.5 Viral and Bacterial Infections Induce Expression of Multiple NK Cell Receptors in Responding CD8⁺ T Cells

Viral infections are often accompanied by extensive proliferation of reactive CD8 T cells. NK cell receptors are also expressed by certain memory phenotype CD8⁺ T cells, and in some cases are up-regulated in T cells responding to viral infection. The majority of pathogen-specific CD8⁺ T cells initiated expression of the inhibitory CD94/NKG2A heterodimer, the KLRG1 receptor, and a murine NK cell marker (10D7); very few Ag-specific T cells expressed Ly49 family members. The up-regulation of these receptors was independent of IL-15 and persisted long after clearance of the pathogen. Thus, CD94/NKG2A expression is a common consequence of CD8⁺ T cell activation (McMahon et al. 2002).

After a defined number of divisions, normal somatic cells enter a nonreplicative stage termed senescence. The KLRG1 is a unique marker for replicative senescence of murine CD8 T cells. KLRG1 expression is induced in a substantial portion (30–60%) of CD8 T cells in C57BL/6 mice infected with lymphocytic choriomeningitis virus (LCMV), vesicular stomatitis virus, or vaccinia virus. Similarly, KLRG1 was found on a large fraction of LCMV gp33 peptide-specific TCR-transgenic (tg) effector and memory cells activated in vivo using an adoptive transfer model. Thus study demonstrates that senescent CD8 T cells are induced in abundant numbers during viral infections in vivo (Voehringer et al. 2001).

The frequency of CD8⁺ T cells expressing KLRG1, unable to undergo further clonal expansion, was markedly elevated in CD8⁺ T cells from old donors. Moreover, the elevated frequency of CMV-specific CD8⁺ T cells in the elderly was due to an accumulation of cells bearing this dominant negative receptor. The fraction of CMV-specific T cells able to secrete IFN-γ after specific antigenic stimulation was significantly lower in the elderly than in the young, although the total number of functional cells was comparable. Therefore, the majority of the clonally expanded virus-specific CD8⁺ cells in the elderly was dysfunctional. Thus, T cell responses are altered in the aged by an accumulation of replicatively senescent dysfunctional T cells carrying receptors for persistent herpes viruses (Ouyang et al. 2003).

The repetitive and persistent antigen stimulation leads to an increase in KLRG1 expression of virus-specific CD8⁺ T cells in mice and that virus-specific CD8⁺ T cells are mostly KLRG1⁺ in chronic human viral infections (HIV, cytomegalovirus, and Epstein-Barr virus) while CD8⁺ T cells targeting resolved viral antigens (influenza virus) typically display high CD127 and low KLRG1 expressions. Thus, by using KLRG1 as a T-cell marker, results suggest that the differentiation status and function of virus-specific CD8⁺ T cells are directly influenced by persistent antigen stimulation (Thimme et al. 2005).

As a result of clonal expansion T lymphocytes, some T lymphocytes acquire a senescent phenotype, fail to replicate in response to further antigenic stimulation, and express KLRG1 and/or CD57. The T lymphocytes with a senescent phenotype are mobilized and subsequently removed from the bloodstream in response to acute high-intensity exercise (Simpson et al. 2007). There is a greater proportion of senescent CD3⁺/CD8⁺ T-lymphocytes in the blood of older adults compared to young at rest and immediately after exhaustive exercise, indicating that the greater frequency of KLRG1⁺/CD8+ T-lymphocytes in older humans is ubiquitous and not localised to the peripheral blood (Simpson et al. 2008).

32.4.2 NK Cell Maturation and Homeostasis Is Linked to KLRG1 Up-Regulation

KLRG1 expression is acquired during periods of NK cell division such as development and homeostatic proliferation. KLRG1⁺ NK cells are mature in phenotype, and these cells have a slower in vivo turnover rate, reduced proliferative response to IL-15, and poorer homeostatic expansion potential compared with mature NK cells lacking KLRG1. Results indicate that NK cells acquire KLRG1 on their surface during development, and this expression correlates with functional distinctions from other peripheral NK cells in vivo (Huntington et al. 2007). While both IL-7 and IL-15 induced proliferation of KLRG1^{low} CD8$^+$ T cells, KLRG1^{high} cells exhibited an extraordinarily high level of resistance to cytokine-driven proliferation in vivo despite their dramatic accumulation upon IL-15 administration. Thus, IL-15 and IL-2 greatly improve the survival of KLRG1^{high} CD8$^+$ T cells, which are usually destined to perish during contraction, in the absence of proliferation (Rubinstein et al. 2008).

32.4.3 Cadherins as Ligands of KLRG1

Cadherins have been identified as ligands for mouse and human KLRG1, which binds three of the classical cadherins (E-, N-, and R-). Cadherins are ubiquitously expressed in vertebrates and mediate cell–cell adhesion by homotypic or heterotypic interactions. By expression cloning using the mouse KLRG1 tetramer as a probe, human E-cadherin was found to be a xenogeneic ligand with a syngeneic interaction between mouse KLRG1 and mouse E-cadherin. KLRG1 also binds N- and R-cadherins. E-cadherin binding of KLRG1 prevents the lysis of E-cadherin–expressing targets by KLRG1$^+$ NK cells (Ito et al. 2006; Tessmer et al. 2007). E-cadherin is expressed by normal epithelial cells, Langerhans cells, and keratinocytes and is usually down-regulated on metastatic cancer cells. KLRG1 ligation by E-cadherin in healthy tissue may thus exert an inhibitory effect on primed T cells (Gründemann et al. 2006). E-cadherin function is often inactivated during development of human carcinomas and splice-site mutations resulting in in-frame loss of exon 8 or 9 occur frequently in diffuse type gastric carcinomas. KLRG1 ligation by E-, N-, or R-cadherins may regulate the cytotoxicity of killer cells to prevent damage to tissues expressing the cadherins (Ito et al. 2006; Schwartzkopff et al. 2007; Tessmer et al. 2007).

Li et al. (2009) determined the structure of KLRG1 in complex with E-cadherin. KLRG1 mediates missing self recognition by binding to a highly conserved site on classical cadherins, enabling it to monitor expression of several cadherins (E-, N- and R-) on target cells (Fig. 32.2). This site overlaps the site responsible for cell–cell adhesion, but is distinct from the integrin $\alpha_E\beta_7$ binding site. Li et al. (2009) proposed that E-cadherin might co-engage KLRG1 and $\alpha_E\beta_7$, and that KLRG1 overcomes its exceptionally weak affinity for cadherins through multipoint attachment to target cells, resulting in inhibitory signaling.

32.5 MAFA-Like Receptor in Human NK Cell (MAFA-L)

32.5.1 Characteristics and Biological Properties

Butcher et al. (1998) identified a receptor gene in NK cell gene complex on human chromosome 12p12-13 encoding a putative type II transmembrane glycoprotein. The product was 54% identical to the rat MAFA, which inhibits mast cell activation by IgE. The human MAFA-like receptor (MAFA-L) and the rat MAFA protein are expressed by basophils and both have an ITIM in the cytoplasmic tail, consistent with an inhibitory role in basophil activation. Like other genes in the NK cell gene complex, MAFA-L is also expressed by NK cells as well as the monocyte-like cell-line U937. Expression in NK cells is restricted to peripheral blood NK cells, where as decidual NK cells did not express MAFA-L. While MAFA-L and rat MAFA might have a similar role in basophils, the expression of MAFA-L in other cell types implies additional functions for this molecule.

Human MAFA cDNA product is similar to the rat MAFA having an intracellular domain containing a putative ITIM and an extracellular C type lectin-like domain. However, in contrast to rat MAFA, the amino acid sequence suggests the presence of two additional extracellular N-linked glycosylation sites. In addition, alternative mRNA transcripts are observed that differ substantially from those found in the rat (Lamers et al. 1998).

32.6 Killer Cell Lectin Like Receptor Subfamily E, Member 1 (KLRE1)

A novel NK receptor, killer cell lectin like receptor subfamily E, member 1 (KLRE1) (also known as NKG2I), belonging to KLRs family has been characterized (Koike et al. 2004; Westgaard et al. 2003; Wilhelm and Mager 2003). A series of experiments using anti–KLRE-1 mAbs indicate that this receptor plays a role in the cytotoxicity mediated by NK cells in vitro (Koike et al. 2004; Westgaard et al. 2003). KLRE1 receptor is a type II transmembrane protein with a C-terminal lectin-like domain and belongs to the NKG2 family. Rat *Klre1* gene was mapped to the NK gene complex. KLRE1 was shown to be expressed by NK cells and a subpopulation of CD3$^+$ cells, with pronounced inter-strain variation. KLRE1 can be expressed on the NK cell surface as a disulphide-linked dimer. The predicted proteins do not contain ITIMs or a positively charged amino acid in the transmembrane domain. Despite absence of a signaling motif, KLRE1 has nevertheless been shown to regulate NK cell-mediated cytotoxicity. Thus, KLRE1 may form a functional heterodimer with ITIM-

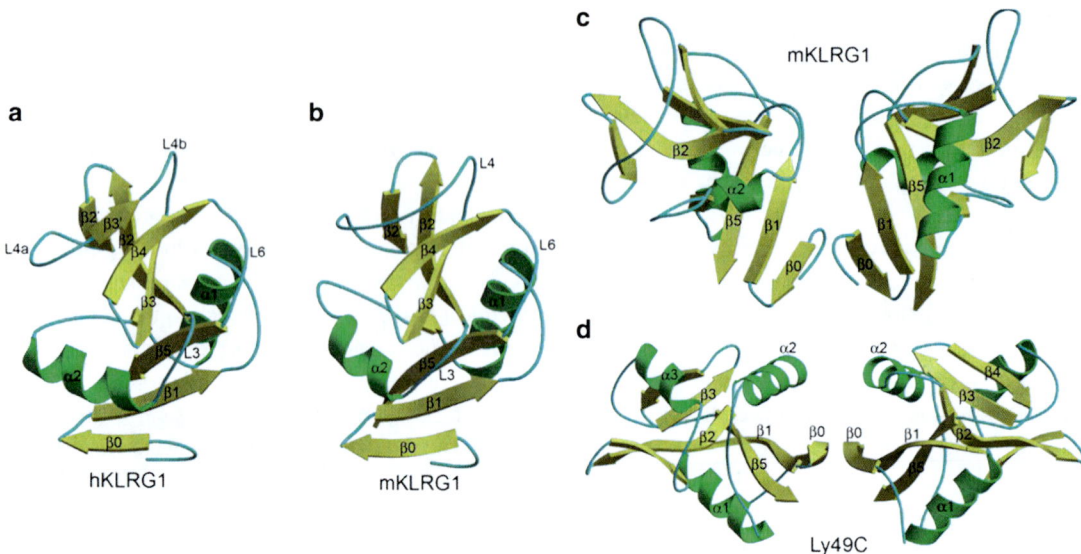

Fig. 32.2 Structure of KLRG1 (**a**) Ribbon diagram of human KLRG1, as observed in the hKLRG1–hEC1 complex. Secondary structure elements are labeled. α-helices are colored in *green*, β-stands in *yellow*, and loops in *cyan*. (**b**) Structure of mouse KLRG1 in unbound form. (**c**) Mouse KLRG1 homodimer, as observed in the mKLRG1–hEC1 complex. This dimer was not observed in structures of mKLRG1 alone or in the hKLRG1–hEC1 complex. (**d**) Structure of the Ly49C homodimer (PDB ID: 3C8J). Secondary structure elements are colored as in (**a**) (Adapted by permission from Li et al. 2009 © Elsevier)

bearing partner that recruits SHP-1 to generate an inhibitory receptor complex (Westgaard et al. 2003). Murine NK receptor, NKG2I, which was composed of 226 amino acids, showed ~40% homology to the murine NKG2D and CD94 in the C-type lectin domain. The expression of NKG2I was largely confined to NK and NKT cells, but was not seen in T cells. Furthermore, anti-NKG2I mAb inhibited NK cell–mediated cytotoxicity, whereas cross-linking of NKG2I enhanced IL-2 and IL-12–dependent IFN-γ production. NKG2I is an activating receptor mediating recognition and rejection of allogeneic target cells (Koike et al. 2004).

KLRE/I1 and KLRE/I2: A Pair of Heterodimeric Receptors: Saether et al. (2008) demonstrated that KLRE1 forms functional heterodimers with either KLRI1 or KLRI2. Cotransfection with KLRE1 was necessary for surface expression of the NKR chains KLRI1 and KLRI2 in 293T cells and can be co-precipitated with KLRI1 or KLRI2 from transfected NK cell lines. Unlike other KLRs, KLRE1/KLRI1 and KLRE1/KLRI2 heterodimers predominantly migrate as single chains in SDS-PAGE, indicating noncovalent association. KLRI1 showed activity of SH-2 domain-containing phosphatase 1. Ab to KLRE1 induced inhibition in KLRI1-transfected cells but increased cytotoxicity in KLRI2 transfectants, demonstrating that KLRE/I1 is a functional inhibitory heterodimer in NK cells, whereas KLRE/I2 is an activating heterodimeric receptor (Saether et al. 2008).

KLRE-1 in BM Allograft Rejection: KLRE-1-deficient mice were born at an expected frequency and showed no aberrant phenotype on growth and lymphoid development. However, KLRE-1-deficient cells showed a severely compromised allogeneic cytotoxic activity compared with the wild-type cells. Allogeneic bone marrow (BM) transfer culminated in colony formation in the spleen of KLRE-1-deficient mice, whereas no colony formation was observed in wild-type recipient mice. Therefore, KLRE-1 is a receptor mediating recognition and rejection of allogeneic target cells in the host immune system (Shimizu et al. 2004). These results indicate that NKG2I on NK cells recognizes putative ligands present on allogeneic BM cells and induces signals leading to the rejection of allografts. It should be mentioned that administration of anti-NK1.1, anti–asialo GM1, or anti–Ly-49D mAb also abrogated the rejection of allogeneic BM grafts, but these effects were primarily due to the depletion of NK cells expressing these molecules

32.7 KLRL1 from Human and Mouse DCs

KLRL1 belongs to the KLR family and is a novel inhibitory NK cell receptor. The KLRL1 from human and mouse DCs is a type II transmembrane protein with an ITIM and a C-type lectin-like domain. The KLRL1 gene is located in the central region of the NK gene complex in both humans and mice, on human chromosome 12p13 and mouse chromosome 6F3, adjacent to the other KLR genes. KLRL1 is preferentially expressed in lymphoid tissues and immune cells, including NK cells, T cells, DCs, and monocytes or macrophages. KLRL1 forms a heterodimer with an as yet unidentified partner. Human and mouse KLRL1 both contain a putative ITIMs, and associates with the tyrosine phosphatases SHP-1 and SHP-2 (Han et al. 2004).

32.8 Natural Cytotoxicity Receptors

The function of NK cells are finely regulated by a series of inhibitory or activating receptors. The inhibitory receptors, specific for MHC class I molecules, allow NK cells to discriminate between normal cells and cells that have lost the expression of MHC class I (e.g., tumor cells). The general molecular strategies that allow NK cells to spare normal cells and kill tumor or virally infected cells have been clarified. The major receptors responsible for NK cell triggering are NKp46, NKp30, NKp44 and NKG2D. The NK-mediated lysis of tumor cells involves several such receptors, while killing of DCs involves only NKp30. The target-cell ligands recognized by some receptors have been identified, but those to which major receptors bind are not yet known. Nevertheless, functional data suggest that they are primarily expressed on cells upon activation, proliferation or tumor transformation. Thus, the ability of NK cells to lyse target cells requires both the lack of surface MHC class I molecules and the expression of appropriate ligands that trigger NK receptors.

32.8.1 NK Cell Triggering: The Activating Receptors

Provided that turning NK cells 'off' represents the major fail-safe device to prevent the NK-mediated attack of normal HLA class I$^+$ autologous cells, an 'on' signal must be generated upon interaction of NK cells with potential target cells. This signal is extinguished whenever appropriate interactions occur between inhibitory receptors and MHC class I molecules. On the other hand, the 'on' signal can be readily detected when NK cells interact with target cells that lack MHC class I molecules. The receptors responsible for NK cell activation in the process of natural cytotoxicity were identified recently and characterized. Collectively termed natural cytotoxicity receptors (NCRs), NKp46 (Sivori et al. 1997; Pessino et al. 1998), NKp44 (Vitale et al. 1998; Cantoni et al. 1999) and NKp30 (Pende et al. 1999) possess limited homology with known human molecules and no homology to each other (Moretta et al. 2000, 2001) (Fig. 32.1). As their expression is restricted to NK cells, they represent the most accurate surface markers for human NK cell identification. NCRs play a major role in the NK-mediated killing of most tumor cell lines (Moretta et al. 2001). Moreover, their surface density on NK cells correlates with the magnitude of the cytolytic activity against NK-susceptible target cells (Sivori et al. 2000). The ligands of NCRs are still not defined. However, as revealed by cytolytic assays, they are expressed by cells belonging to different histotypes (Moretta et al. 2001; Sivori et al. 2000; Costello et al. 2002; Pende et al. 2002). While NKp46 and NKp30 enable a precise identification of all NK cells, regardless of whether these cells are resting or activated (which is not true for other widely used NK cell markers including CD56 and CD16), NKp44 is selectively expressed by activated NK cells (Vitale et al. 1998; Cantoni et al. 1999; Moretta et al. 2001). This might explain, at least in part, the higher levels of cytolytic activity of activated NK cells cultured in IL-2.

NKG2D, a surface receptor of the NKG2 family, also plays a role in NK-mediated cytolysis. NKG2D is not restricted to NK cells, but is also expressed by cytolytic T lymphocytes. NKG2D is specific for the stress-inducible MICA and MICB or ULBP proteins. These ligands are expressed predominantly, but not exclusively, by cells of epithelial origin (Chap. 31).

32.8.2 Activating Receptors and Their Ligands

Natural cytotoxicity receptors (NCRs) (Moretta et al. 1999) are involved in NK cell triggering upon recognition of non-HLA ligands. These receptors appear to complement each other in the induction of target cell lysis by NK cells. Other triggering surface molecules expressed by NK cells (but shared by other leukocyte types) appear to function primarily as co-receptors (Fig. 32.1). That is, their ability to signal depends on the simultaneous co-engagement of one or another triggering receptor (Moretta et al. 2001). They may function to amplify signaling by true receptors. Other molecules that function as a triggering co-receptor in NK cells were found to be 2B4, NTB-A, the Poliovirus receptor (PVR, CD155) and Nectin-2 (CD112); the last two are members of the nectin family.

32.8.2.1 NKp46, NKp44, and NKp30

Sivori et al. (1997) identified a surface molecule (p46), which induces strong NK cell triggering and, unlike other NK cell antigens, is expressed exclusively by NK cells. NKp46 is expressed by all resting or activated NK cells. Unlike other NK cell antigens, the expression of p46 was strictly confined to NK cells. The p46-mediated induction of Ca^{2+} increases or triggering of cytolytic activity was downregulated by the simultaneous engagement of inhibitory receptors including p58, p70, and CD94/NKG2A. Both the unique cellular distribution and functional capability of p46 molecules suggest a possible role in the mechanisms of non-MHC-restricted cytolysis mediated by human NK cells.

Vitale et al. (1998) identified a 44-kD surface molecule (NKp44) that is absent in freshly isolated peripheral blood lymphocytes but is progressively expressed by all NK cells in vitro after culture in IL-2. Different from other markers of cell activation, NKp44 is absent in activated T lymphocytes or T cell clones. NKp44 was not detected in any of the other cell lineages analyzed. The mAb-mediated cross-linking of NKp44 in cloned NK cells resulted in strong activation of target cell lysis in a redirected killing assay. NKp44 functions as a triggering receptor selectively expressed by activated NK cells and that together with p46, may be involved in the process of non-MHC-restricted lysis. Both p46 and NKp44 are coupled to the intracytoplasmic transduction machinery via the association with CD3 or KARAP/DAP12, respectively; and these associated molecules are tyrosine phosphorylated upon NK cell stimulation (Fig. 32.1).

Pende et al. (1999) identified NKp30, a 30-kD triggering receptor selectively expressed by all resting and activated human NK cells. Although mAb-mediated cross-linking of NKp30 induces strong NK cell activation, mAb-mediated masking inhibits the NK cytotoxicity against normal or tumor target cells. NKp30 cooperates with NKp46 and/or NKp44 in the induction of NK-mediated cytotoxicity against the majority of target cells, whereas it represents the major triggering receptor in the killing of certain tumors. This receptor is associated with CD3 ζ chains that become tyrosine phosphorylated upon sodium pervanadate treatment of NK cells. Molecular cloning of NKp30 cDNA revealed a member of the immunoglobulin superfamily, characterized by a single V-type domain and a charged residue in the transmembrane portion. Moreover, NKp30 is encoded by the 1C7 gene. Although the identification of different NCRs constitutes a major step forward in our understanding of the NK cell physiology, both the nature and the distribution of the NCR ligands on target cells remain to be determined. The occupational exposure to mixture of toxics is one of the important factors in the diminution of the NK cell receptor expression (De Celis et al. 2008). Following exposure of women to toxic compounds, NKp30 and NKG2D receptor expression was diminished.

Similar to NKp46, NKp30 is selectively expressed by all NK cells, both freshly isolated and cultured in IL-2, thus representing an optimal marker for NK cell identification. Although both belong to the Ig-SF, NKp30 does not display any substantial homology with previously identified NK receptors. In many respects, NKp30 appeared similar to NKp42. Indeed, their parallel expression on all NK cells (including the rare CD16⁻ cells), the existence for both of a high or low density pattern of surface expression, together with their similar functional characteristics, led to the thought that the surface molecule recognized by the new mAbs could be identical or strictly related to NKp42. However, NKp30 and NKp46 displayed different molecular masses and, functionally, appeared to play a complementary role in the induction of natural cytotoxicity. Moreover, molecular cloning revealed that NKp30 is a protein with very limited homology with NKp46, as the two molecules display only 13% identity and 15% similarity, and are encoded by genes located on different chromosomes.

Both NKp46 and NKp44 receptors are involved in recognition and lysis of a variety of tumor targets, but neither displays significant identity. NKp46 and NKp44 were shown to cooperate in the process of tumor cell lysis by human NK cells. However, studies indicated the existence of additional receptor(s) cooperating with NKp46 and NKp44. Indeed, NKp30 represents a receptor that may cooperate with NKp46 and NKp44 in the induction of cytotoxicity against a variety of target cells. Perhaps, NKp30 represents the major receptor in inducing NK-mediated killing of certain tumor target cells, the lysis of which is largely NKp46/NKp44 independent. Molecular cloning revealed that NKp30 is the product of 1C7, a gene mapped on human chromosome 6 in the HLA class III region. Persistent high risk Human papillomavirus (HPV) infection can lead to cervical cancer. NK cells play a crucial role against tumors and virus-infected cells through a fine balance between activating and inhibitory receptors. Expression of triggering receptors NKp30, NKp44, NKp46 and NKG2D on NK cells correlates with cytolytic activity against tumor cells and their down-regulation represents an evasion mechanism associated to low NK cell activity, HPV-16 infection and cervical cancer progression (Garcia-Iglesias et al. 2009).

mAbs to NCR block to differing extents the NK-mediated lysis of various tumors. Moreover, lysis of certain tumors can be virtually abrogated by the simultaneous masking of the three NCRs. There is a coordinated surface expression of the three NCRs, their surface density varying in different individuals and also in the NK cells isolated from a given individual. A direct correlation exists between the surface density of NCR and the ability of NK cells to kill various tumors. NKp46 is the only NCR involved in human NK-mediated killing of murine target cells. Accordingly, a homologue of NKp46 has been detected in mouse. Molecular cloning of NCR revealed novel members of the Ig superfamily displaying a low degree of similarity to each other and to known human molecules. NCRs are coupled to different signal transducing adaptor proteins, including CD3 ζ, Fc εRI γ, and KARAP/DAP12. Besides different NCRs, several other surface molecules that can mediate NK cell triggering have been identified in humans and rodents. These include CD2, CD69, CD28, 2B4, and NKR-P1 (Pende et al. 1999).

32.8.2.2 Ligands of Activating Receptors

Although many core questions regarding NK cell receptors have been answered in recent years, there are still relevant issues that remain to be tackled in NK cell physiology. One major problem is with regard to the identification of the cellular ligands for the major triggering receptors (NKp46, NKp30 and NKp44) as well as for other receptors or co-receptors (Fig. 32.1). Available information is compatible with the concept that, similar to MICA/B, PVR or Nectin-2, they may also be represented by molecules primarily expressed by cells that have been 'stressed' by cytokine activation, proliferation, high temperature, viral infection or tumor transformation. NKp46 ligands are expressed in reticular dermis, and their expression is enhanced in the active proliferation zone (dermoepidermal junction) of nevi and melanomas. The physiological role of NKp46 ligands in the progression of malignancy is unknown (Cagnano et al. 2008). Myeloid DCs express ligands for NKp46 and NKp30 natural cytotoxicity receptors, which are partially reduced on HCMV infection; yet. The studies stress the importance of the dynamics of viral immune evasion mechanisms (Magri et al. 2011). NKp46 binds to heparin and heparan sulfate (HS); and also to multimeric sialyl Lewis X expressing transferrin secreted by human hepatoma HepG2 cells (HepTF). NKp46 binds to sulfate- and 2,3-NeuAc-containing glycans mainly via ionic interactions. The HS GAG are not ligands for NKp30 (Ito et al. 2011).

References

Abramson J, Xu R, Pecht I (2002) An unusual inhibitory receptor–the mast cell function-associated antigen (MAFA). Mol Immunol 38:1307–1313

Abramson J, Licht A, Pecht I (2006) Selective inhibition of the FcεRI-induced de novo synthesis of mediators by an inhibitory receptor. EMBO J 25:323–334

Barisas BG, Smith SM, Liu J et al (2007) Compartmentalization of the Type I Fc ε receptor and MAFA on mast cell membranes. Biophys Chem 126:209–217

Bengsch B, Spangenberg HC, Kersting N et al (2007) Analysis of CD127 and KLRG1 expression on hepatitis C virus-specific CD8+ T cells reveals the existence of different memory T-cell subsets in the peripheral blood and liver. J Virol 81:945–953

Beyersdorf NB, Ding X, Karp K, Hanke T (2001) Expression of inhibitory "killer cell lectin-like receptor G1" identifies unique subpopulations of effector and memory CD8 T cells. Eur J Immunol 31:3443–3452

Beyersdorf N, Ding X, Tietze JK et al (2007) Characterization of mouse CD4 T cell subsets defined by expression of KLRG1. Eur J Immunol 37:3445–3454

Biassoni R, Cantoni C, Pende D et al (2001) Human natural killer cell receptors and co-receptors. Immunol Rev 181:203–214

Biassoni R, Fogli M, Cantoni C et al (2005) Molecular and functional characterization of NKG2D, NKp80, and NKG2C triggering NK cell receptors in rhesus and cynomolgus macaques: monitoring of NK cell function during simian HIV infection. J Immunol 174:5695–5705

Binsack R, Pecht I (1997) The mast cell function-associated antigen exhibits saccharide binding capacity. Eur J Immunol 27:2557–2561

Blaser C, Kaufmann M, Pircher H (1998) Virus-activated CD8 T cells and lymphokine-activated NK cells express the mast cell function-associated antigen, an inhibitory C-type lectin. J Immunol 161:6451–6454

Bocek P Jr, Guthmann MD, Pecht I (1997) Analysis of the genes encoding the mast cell function-associated antigen and its alternatively spliced transcripts. J Immunol 158:3235–3243

Braud VM, Allan DS, O'Callaghan CA et al (1998) HLA-E binds to natural killer cell receptors CD94/NKG2A, B and C. Nature 391:795–799

Butcher S, Arney KL, Cook GP (1998) MAFA-L, an ITIM-containing receptor encoded by the human NK cell gene complex and expressed by basophils and NK cells. Eur J Immunol 28:3755–3762

Cagnano E, Hershkovitz O, Zilka A et al (2008) Expression of ligands to NKp46 in benign and malignant melanocytes. J Invest Dermatol 128:972–979

Cantoni C, Bottino C, Vitale M et al (1999) NKp44, a triggering receptor involved in tumor cell lysis by activated human natural killer cells, is a novel member of the immunoglobulin superfamily. J Exp Med 189:787–796

Cohen-Dayag A, Schneider H, Pecht I (1992) Variants of the mucosal mast cell line (RBL-2H3) deficient in a functional membrane glycoprotein. Immunobiology 185:124–149

Corral L, Hanke T, Vance RE et al (2000) NK cell expression of the killer cell lectin-like receptor G1 (KLRG1), the mouse homolog of MAFA, is modulated by MHC class I molecules. Eur J Immunol 30:920–930

Costello RT, Sivori S, Marcenaro E et al (2002) Defective expression and function of natural killer cell-triggering receptors in patients with acute myeloid leukemia. Blood 99:3661–3667

De Celis R, Feria-Velasco A, Bravo-Cuellar A et al (2008) Expression of NK cells activation receptors after occupational exposure to toxics: a preliminary study. Immunol Lett 118:125–131

Dennehy KM, Klimosch SN, Steinle A (2011) NKp80 uses an atypical hemi-ITAM to trigger NK cytotoxicity. J Immunol 186:657–661

Eberl M, Engel R, Aberle S et al (2005) Human Vγ9/Vδ2 effector memory T cells express the killer cell lectin-like receptor G1 (KLRG1). J Leukoc Biol 77:67–70

Garcia-Iglesias T, Del Toro-Arreola A et al (2009) Low NKp30, NKp46 and NKG2D expression and reduced cytotoxic activity on NK cells in cervical cancer and precursor lesions. BMC Cancer 9:186

Gründemann C, Bauer M, Schweier O et al (2006) Cutting edge: identification of E-cadherin as a ligand for the murine killer cell lectin-like receptor G1. J Immunol 176:1311–1315

Guthmann MD, Tal M, Pecht I (1995a) A new member of the C-type lectin family is a modulator of the mast cell secretory response. Int Arch Allergy Immunol 107:82–86

Guthmann MD, Tal M, Pecht I (1995b) A secretion inhibitory signal transduction molecule on mast cells is another C-type lectin. Proc Natl Acad Sci USA 92:9397–9401

Hamann J, Montgomery KT, Lau S et al (1997) AICL: a new activation-induced antigen encoded by the human NK gene complex. Immunogenetics 45:295–300

Han Y, Zhang M, Li N et al (2004) KLRL1, a novel killer cell lectinlike receptor, inhibits natural killer cell cytotoxicity. Blood 104:2858–2866, KLR

Hanke T, Corral L, Vance RE, Raulet DH (1998) 2F1 antigen, the mouse homolog of the rat "mast cell function-associated antigen", is a lectin-like type II transmembrane receptor expressed by natural killer cells. Eur J Immunol 28:4409–4417

Huntington ND, Tabarias H, Fairfax K et al (2007) NK cell maturation and peripheral homeostasis is associated with KLRG1 up-regulation. J Immunol 178:4764–4770

Ibegbu CC, Xu Y-X, Harris W et al (2005) Expression of killer cell lectin-like receptor G1 on antigen-specific human CD8$^+$ T lymphocytes during active, latent, and resolved infection and its relation with CD57. J Immunol 174:6088–6094

Ito M, Maruyama T, Saito N et al (2006) Killer cell lectin-like receptor G1 binds three members of the classical cadherin family to inhibit NK cell cytotoxicity. J Expt Med 203:289–295

Ito K, Higai K, Sakurai M et al (2011) Binding of natural cytotoxicity receptor NKp46 to sulfate- and α2,3-NeuAc-containing glycans and its mutagenesis. Biochem Biophys Res Commun 406:377–382

Jürgens L, Arndt-Jovin D, Pecht I, Jovin TM (1996) Proximity relationships between the type I receptor for Fc ε (FcεRI) and the mast cell function-associated antigen (MAFA) studied by donor photobleaching fluorescence resonance energy transfer microscopy. Eur J Immunol 26:84–91

Kamimura D, Bevan MJ (2008) Endoplasmic reticulum stress regulator XBP-1 contributes to effector CD8+ T cell differentiation during acute infection. J Immunol 181:5433–5441

Kim DK, Kabat J, Borrego F et al (2004) Human NKG2F is expressed and can associate with DAP12. Mol Immunol 41:53–62

Koike J, Wakao H, Ishizuka Y et al (2004) Bone marrow allograft rejection mediated by a novel murine NK receptor, NKG2I. J Exp Med 199:137–144

Kuttruff S, Koch S, Kelp A et al (2009) NKp80 defines and stimulates a reactive subset of CD8 T cells. Blood 113:358–369

Lamers MBAC, Lamont AG, Williams DH (1998) Human MAFA has alternatively spliced variants. Biochim Biophys Acta 1399:209–212. doi:dx.doi.org

Lanier LL (1998) NK receptors. Annu Rev Immunol 16:359–393

Li Y, Hofmann M, Wang Q et al (2009) Structure of natural killer cell receptor KLRG1 bound to E-cadherin reveals basis for MHC-independent missing self recognition. Immunity 31:35–46

Licht A, Pecht I, Schweitzer-Stenner R (2005) Regulation of mast cells' secretory response by co-clustering the Type 1 Fcε receptor with the mast cell function-associated antigen. Eur J Immunol 35:1621–1633

Licht A, Abramson J, Pecht I (2006) Co-clustering activating and inhibitory receptors: impact at varying expression levels of the latter. Immunol Lett 104:166–170

Magri G, Muntasell A, Romo N et al (2011) NKp46 and DNAM-1 NK-cell receptors drive the response to human cytomegalovirus-infected myeloid dendritic cells overcoming viral immune evasion strategies. Blood 117:848–856

Marcolino I, Przybylski GK, Koschella M et al (2004) Frequent expression of the natural killer cell receptor KLRG1 in human cord blood T cells: correlation with replicative history. Eur J Immunol 34:2672–2680

Mavilio D, Benjamin J, Kim D et al (2005) Identification of NKG2A and NKp80 as specific natural killer cell markers in rhesus and pigtailed monkeys. Blood 106:1718–1725

McMahon CW, Zajac AJ, Jamieson AM et al (2002) Viral and bacterial infections induce expression of multiple NK cell receptors in responding CD8+ T cells. J Immunol 169:1444–1452

Moretta A (2002) Natural killer cells and dendritic cells: rendezvous in abused tissues. Nat Rev Immunol 2:957–965

Moretta A, Bottino C, Millo R, Biassoni R (1999) HLA-specific and non-HLA-specific human NK receptors. Curr Top Microbiol Immunol 244:69–84

Moretta L, Moretta A (2004) Unravelling natural killer cell function: triggering and inhibitory human NK receptors. EMBO J 23:255–259

Moretta A, Biassoni R, Bottino C et al (2000) Natural cytotoxicity receptors that trigger human NK-mediated cytolysis. Immunol Today 21:228–234

Moretta A, Bottino C, Vitale M et al (2001) Activating receptors and co-receptors involved in human natural killer cell-mediated cytolysis. Annu Rev Immunol 19:197–223

Ortega E, Schneider H, Pecht I (1991) Possible interactions between the Fc ε receptor and a novel mast cell function-associated antigen. Int Immunol 3:333–342

Ouyang Q, Wagner WM, Voehringer D et al (2003) Age-associated accumulation of CMV-specific CD8+ T cells expressing the inhibitory killer cell lectin-like receptor G1 (KLRG1). Exp Gerontol 38:911–920

Pende D, Parolini S, Pessino A et al (1999) Identification and molecular characterization of NKp30, a novel triggering receptor involved in natural cytotoxicity mediated by human natural killer cells. J Exp Med 190:1505–1516

Pende D, Rivera P, Marcenaro S et al (2002) Major histocompatibility complex class I-related chain A and UL16-binding protein expression on tumor cell lines of different histotypes: analysis of tumor susceptibility to NKG2D-dependent natural killer cell citotoxicity. Cancer Res 62:6178–6186

Pessino A, Sivori S, Bottino C et al (1998) Molecular cloning of NKp46: a novel member of the immunoglobulin superfamily involved in triggering of natural cytotoxicity. J Exp Med 188:953–960

Plougastel B, Trowsdale J (1998a) Cloning of NKG2-F, A new member of the NKG2 family of human natural killer cell receptor genes. Eur J Immunol 27:2835–2839

Plougastel B, Trowsdale J (1998b) Sequence analysis of a 62-kb region overlapping the human KLRC cluster of genes. Genomics 49:193–199

Renedo M, Arce I, Montgomery K et al (2000) A sequence-ready physical map of the region containing the human natural killer gene complex on chromosome 12p12.3-p13.2. Genomics 65:129–136

Robbins SH, Nguyen KB, Takahashi N et al (2002) Cutting edge: Inhibitory functions of the killer cell lectin-like receptor G1 molecule during the activation of mouse NK cells. J Immunol 168:2585–2589

Robbins SH, Terrizzi SC, Sydora BC et al (2003) Differential regulation of killer cell lectin-like receptor G1 expression on T cells. J Immunol 170:5876–5885

Roda-Navarro P, Arce I, Renedo M et al (2000) Human KLRF1, a novel member of the killer cell lectin-like receptor gene family: molecular characterization, genomic structure, physical mapping to the NK gene complex and expression analysis. Eur J Immunol 30:568–576

Roda-Navarro P, Hernanz-Falcón P, Arce I et al (2001) Molecular characterization of two novel alternative spliced variants of the KLRF1 gene and subcellular distribution of KLRF1 isoforms. Biochim Biophys Acta 1520:141–146

Rubinstein MP, Lind NA, Purton JF et al (2008) IL-7 and IL-15 differentially regulate CD8+ T cell subsets during contraction of the immune response. Blood 112:3704–3712

Rutjens E, Mazza S, Biassoni R et al (2007) Differential NKp30 inducibility in chimpanzee NK cells and conserved NK cell phenotype and function in long-term HIV-1-infected animals. J Immunol 178:1702–1712

Saether PC, Westgaard IH, Hoelsbrekken SE et al (2008) KLRE/I1 and KLRE/I2: a novel pair of heterodimeric receptors that inversely regulate NK cell cytotoxicity. J Immunol 181:3177–3182

Schwartzkopff S, Gründemann C, Schweier O et al (2007) Tumor-associated E-cadherin mutations affect binding to the killer cell lectin-like receptor G1 in humans. J Immunol 179:1022–1029

Schweitzer-Stenner R, Engelke M, Licht A, Pecht I (1999) Mast cell stimulation by co-clustering the type I Fc ε-receptors with mast cell function-associated antigens. Immunol Lett 68:71–78

Shimizu E, Koike J, Wakao H et al (2004) Role of a NK receptor, KLRE-1, in bone marrow allograft rejection: analysis with KLRE-1-deficient mice. Blood 104:781–783

Simpson RJ, Florida-James GD, Cosgrove C et al (2007) High-intensity exercise elicits the mobilization of senescent T lymphocytes into the peripheral blood compartment in human subjects. J Appl Physiol 103:396–401

Simpson RJ, Cosgrove C, Ingram LA et al (2008) Senescent T-lymphocytes are mobilised into the peripheral blood compartment in young and older humans after exhaustive exercise. Brain Behav Immun 22:544–551

Sivori S, Vitale M, Morelli L et al (1997) p46, a novel natural killer cell-specific. J Exp Med 186:1129–1136

Sivori S, Parolini S, Falco M et al (2000) 2B4 functions as a co-receptor in human natural killer cell activation. Eur J Immunol 30:787–793

Song J, Hagen GM, Roess DA et al (2002) The mast cell function-associated antigen and its interactions with the type I Fcε receptor. Biochemistry 41:881–889

Stubbe M, Vanderheyde N, Pircher H et al (2008) Characterization of a subset of antigen-specific human central memory CD4+ T lymphocytes producing effector cytokines. Eur J Immunol 38:273–282

Tessmer MS, Fugere C, Stevenaert F et al (2007) KLRG1 binds cadherins and preferentially associates with SHIP-1. Int Immunol 19:391–400

Thimme R, Appay V, Koschella M et al (2005) Increased expression of the NK cell receptor KLRG1 by virus-specific CD8 T cells during persistent antigen stimulation. J Virol 79:12112–12116

Thomas M, Boname JM, Field S et al (2008) Down-regulation of NKG2D and NKp80 ligands by Kaposi's sarcoma-associated herpesvirus K5 protects against NK cell cytotoxicity. Proc Natl Acad Sci USA 105:1656–1661

Vitale M, Bottino C, Sivori S et al (1998) NKp44, a novel triggering surface molecule specifically expressed by activated natural killer cells is involved in non-MHC restricted tumor cell lysis. J Exp Med 187:2065–2072

Vitale M, Falco M, Castriconi R et al (2001) Identification of NKp80, a novel triggering molecule expressed by human NK cells. Eur J Immunol 31:233–242

Voehringer D, Blaser C, Brawand P et al (2001) Viral infections induce abundant numbers of senescent CD8 T cells. J Immunol 167:4838–4843

Voehringer D, Koschella M, Pircher H (2002) Lack of proliferative capacity of human effector and memory T cells expressing killer cell lectin like receptor G1 (KLRG1). Blood 100:3698–3702

Warren HS, Jones AL, Freeman C, Bettadapura J, Parish CR (2005) Evidence that the cellular ligand for the human NK cell activation receptor NKp30 is not a heparan sulfate glycosaminoglycan. J Immunol 175:207–212

Welte S, Kuttruff S, Waldhauer I, Steinle A (2006) Mutual activation of natural killer cells and monocytes mediated by NKp80-AICL interaction. Nat Immunol 7:1334–1342

Westgaard IH, Dissen E, Torgersen KM et al (2003) The lectin-like receptor KLRE1 inhibits natural killer cell cytotoxicity. J Expt Med 197:1551–1561

Wilhelm BT, Mager DL (2003) Identification of a new murine lectin-like gene in close proximity to CD94. Immunogenetics 55:53–56

Xu R, Abramson J, Fridkin M, Pecht I (2001) SH2 domain-containing inositol polyphosphate 5′-phosphatase is the main mediator of the inhibitory action of the mast cell function-associated antigen. J Immunol 167:6394–6402

Part XI
C-Type Lectins: Endocytic Receptors

Asialoglycoprotein Receptor and the Macrophage Galactose-Type Lectin

Anita Gupta

33.1 Asialoglycoprotein Receptor: The First Animal Lectin Discovered

Mammalian liver expresses endocytic cell surface receptors that specifically bind natural or synthetic molecules containing terminal galactosyl or N-acetylgalactosaminyl sugars. One of these hepatocyte receptors is the asialoglycoprotein receptor (ASGP-R), which mediates the endocytosis and subsequent lysosomal degradation of these glyco-molecules. The ASGP-R was the first mammalian lectin discovered by Morell et al. (1968) and is a member of the C-type lectin family, which require Ca^{2+} for ligand binding and to contain disulfide bridges in the CRD (Zelensky and Gready 2005). ASGP-R is an integral membrane protein, predominately expressed on the sinusoidal surface of mammalian hepatocytes. Morphometric analysis estimated that approximately 90% of the receptor molecules were at the sinusoidal face, approximately 10% at the lateral face, and approximately 1% at the bile canalicular face. Nonhepatic cells such as endothelial and Kupffer cells had no receptor specific for asialoglycoproteins (Matsuura et al. 1982). At cellular level, the basic molecular function of the receptor is to mediate the uptake and ultimate degradation of galactosyl/N-acetylgalactosaminyl-containing molecules (ligands). At the organism level, however, the physiological function is uncertain. Weigel (1994) proposed physiological role of ASGP-R in regulating the dynamic flux of galactosyl/N-acetylgalactosaminyl glycoconjugates in mammals. The ASGP-R is one of the best characterized systems for receptor-mediated endocytosis via the clathrin-coated pit pathway (Breitfeld et al. 1985; Weigel and Yik 2002; Stockert 1995).

33.2 Rat Asialoglycoprotein Receptor

33.2.1 Characteristics

Hepatic plasma membrane receptors mediate the specific binding and uptake of partially deglycosylated glycoproteins. These receptors, referred to as hepatic lectins, have been isolated from several mammalian liver, peritoneal macrophages (Ashwell and Harford 1982) and avian species (Baenziger and Maynard 1980). While the mammalian (rabbit, rat, and human) receptors recognize galactose residues exposed upon removal of terminal sialic acid, the avian (chicken) receptor recognizes terminal N-acetylglucosamine exposed upon further removal of the galactose. In spite of this difference in binding specificity, the mammalian and avian receptors share a number of properties, such as a strict Ca^{2+} requirement for ligand binding and similar pH profiles for binding and release of ligand (Brown et al. 1983). The complete amino acid sequence of the chicken hepatic lectin (Drickamer and Mamon 1982) suggests a unique organization for this membrane protein with its NH2 terminus in the cytoplasm and COOH terminus outside the cell.

Rat ASGP-R contains three subunits, designated rat hepatic lectins (RHL) 1, 2, and 3 where as human ASGP-R contains two subunits, HHL1 and HHL2. Peptide mapping experiments suggest that the three polypeptides of rat RHL may be structurally related to each other (Schwartz et al. 1981; Warren and Doyle 1981). The predominant polypeptide RHL-1 has an M_r of 41.5 kDa, while two less abundant species appeared to be of higher molecular weight (RHL-2 of 49.0 kDa and RHL-3 of 54 kDa); the RHL-1 represents 70–80% of the total mass of the receptor. Each of the RHL-1 and RHL-2/3 polypeptides is associated into homo-oligomers but are physically unlinked to each other. The complete sequence of the major RHL species (RHL-1) has been established (Drickamer et al. 1984; Schwartz 1984). The complete sequence of RHL-1 is 283 residues long, although 20% of the protein as isolated is missing the first two residues at NH2 terminus. The overall arrangement of the polypeptide, similar to chicken ASGP-R, consists of an NH2-terminal stretch of hydrophilic amino acids, a segment of about 30 uncharged residues, and a COOH-terminal portion which contains three oligosaccharide attachment sites. The COOH terminus of the rat and chicken receptors showed 28% identity. Evidence shows that there must be at least two distinct genes for RHL receptor polypeptides. ASGP-Rs are covalently modified by fatty acylation (Zeng et al. 1995; Zeng and Weigel 1996). The single Cys residue in the cytoplasmic domain of each RHL or HHL subunit is fatty acylated. Deacylation of ASGP-Rs with hydroxylamine results in the spontaneous formation of dimers through reversible disulfide bonds, indicating that deacylation concomitantly generates free thiol groups. Cys^{57} within the transmembrane domain of HHL1 is not normally palmitoylated. In conclusion, Cys^{35} in RHL1, Cys^{54} in RHL2 and RHL3, and Cys^{36} in HHL1 are fatty acylated. Cys^{57} in HHL1 and probably Cys^{56} in RHL1 are not palmitoylated.

The second subunit, RHL-2/3, consists of two species that are differentially glycosylated forms of a second, homologous polypeptide. The structure of RHL-2/3 polypeptide reveals that this protein is homologous to RHL-1 throughout its length but contains one major insertion of 18 amino acids near its NH2 terminus. RHL-1 and RHL-2/3 bind carbohydrate ligands through COOH-terminal, Ca^{2+}-dependent CRDs. Results indicate that RHL-1 and RHL-2/3 polypeptides are self-associated into two distinct molecules, each of which has galactose-binding activity (Halberg et al. 1987).

Structural analysis of wild-type and mutant forms of homologous C-type CRD from serum MBP suggests that carbohydrate ligands bind to the CRDs of RHL-1 at a conserved Ca^{2+} designated site two. The primary interaction with sugar is through hydroxyl groups that form coordination bonds with the Ca^{2+} and hydrogen bonds with protein side chains that also ligate the Ca^{2+} (Weis et al. 1992). The molecular basis for galactose binding to C-type CRDs has been studied in a mutant CRD from serum MBP which has been engineered to bind galactose. Three single amino acid replacements ($E^{185}Q$, $N^{187}D$, and $H^{189}W$) and insertion of a glycine-rich loop result in a modified CRD designated QPDWG, which has affinity and selectivity similar to RHL-1 (Iobst and Drickamer 1994). The crystal structures of this modified CRD and of a further mutant that binds N-acetylgalactosamine with high affinity have been solved (Kolatkar and Weis 1996; Kolatkar et al. 1998). These structures provide a basis for modeling the CRD of RHL-1. Murine ASGP-R cDNA exhibits homology with rat and human receptors. The membrane-bound M (mouse) HL polypeptide does not contain a cleavable N-terminal signal sequence and is probably anchored to the membrane via an internal insertion sequence (Sanford and Doyle 1990).

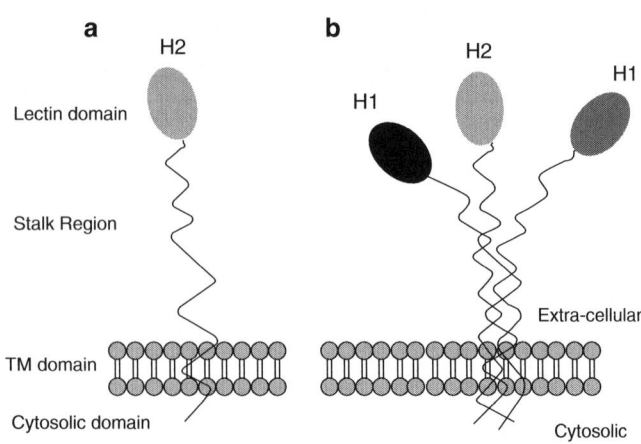

Fig. 33.1 Schematic presentation of asialoglycoprotein receptor (ASGP-R). (**a**) Each subunit, H1 and H2, consists of four domains; a N-terminal cytoplasmic domain, a single-pass transmembrane domain, an extracellular stalk segment and a C-type lectin carbohydrate recognition domain. (**b**) A hetero-oligomeric complex of two H1 and one H2 subunits has been proposed as the minimum size of an active ASGP-R

33.3 Human Asialoglycoprotein Receptor

33.3.1 Structural Characteristics

The human liver ASGP-R is a multichain heterooligomer composed of two homologous subunits, the major H1 and the minor H2, sharing 58% sequence identity. The ASGP-R from human liver and from the human hepatoma cell line HepG2 migrates as a single species of 45 kDa. The receptor encoding a cDNA clones has been isolated in H1 and H2 forms, with a protein sequence homology of 58%. Each subunit consists of four domains, a N-terminal cytoplasmic domain, a single-pass transmembrane domain, an extracellular stalk segment and a CRD, the latter being responsible for the ligand binding (Fig. 33.1a) (Weigel and Yik 2002; Chiacchia and Drickamer 1984; Shia and Lodish 1989). The H1 is a single-spanning membrane protein with amino terminus facing the cytoplasm and the carboxy terminus exposed on the exoplasmic side of plasma membrane. The transmembrane segment, residues 38–65, functions as an internal signal directing protein synthesis to the endoplasmic reticulum and initiating membrane insertion (Spiess 1987; Spiess and Handschin 1987). A notable difference between H1 and H2 is an 18-amino acid insert present in the cytoplasmic domain of only H2. Comparison of sequences of rat RHL-1 and RHL-2 indicates that H1 is more homologous to R1 than to H2, and H2 is more similar to R2 than to H1 (Weigel and Yik 2002; Spiess and Lodish 1985).

Furthermore, while H1 only occurs as one protein isoform, H2 transcripts exist in multiple variants both in HepG2 cells and in the normal human liver. Some of these appear to be the result of alternative splicing events. Three splice variants of H2, designated H2a, H2b and H2c, have been isolated from human liver and HepG2 cells. However, H2a does not occur in native ASGP-R complexes. In contrast, both H2b and H2c associate with H1 in functional ASGP-Rs, but never together in the same receptor complex (Yik et al. 2002).

Though, H1 and H2 are highly homologous, the major H1 subunit is stably expressed by itself, whereas in the absence of H1 most of the H2 subunits are degraded in ER. Since the internal pool of H2 was also able to traffick to the cell surface, it was suggested that H2 recycles between the surface and intracellular compartments, similar to the constitutive recycling of hetero-oligomeric ASGP-R complexes. However, unlike H1, which can bind the ligand asialo-orosomucoid (ASOR), H2 failed to bind or endocytose ASOR. It seems that the H2 subunit of the human ASGP-R contains functional, although weak, signal(s) for endocytosis and recycling and has the ability to oligomerize. H2 homo-oligomers, however, do not create binding sites for desialylated glycoproteins, such as ASOR, that contain tri- and tetra-antennary N-linked oligosaccharides (Saxena et al. 2002).

The X-ray crystal structure of the carbohydrate recognition domain of the major subunit H1 was studied at 2.3 Å resolution (Meier et al. 2000). While the overall fold of this and other known C-type lectin structures are well conserved, the positions of the bound calcium ions are not, indicating that the fold is stabilised by alternative mechanisms in different branches of the C-type lectin family. It was the first CRD structure where three calcium ions form an intergral part of the structure. In addition, the structure provided direct confirmation for the conversion of the ligand-binding site of the mannose-binding protein to an asialoglycoprotein receptor-like specificity suggested by Drickamer and colleagues. In agreement with the prediction that the coiled-coil domain of the ASGP-R is separated from the CRD and its N-terminal disulphide bridge by several residues, these residues are indeed not alpha-helical, while in tetranectin they form an alpha-helical coiled-coil (Drickamer 1988; Meier et al. 2000).

33.3.1.1 Position of Cysteine Is Critical for Palmitoylation of H1

The two subunits, H1 and H2 ASGP-R are palmitoylated at the cytoplasmic Cys residues near their transmembrane domains (TMD). The cytoplasmic Cys^{36} in H1 is located at a position that is five amino acids from the transmembrane junction. Neither the native amino acid sequence surrounding Cys^{36} nor the majority of the cytoplasmic domain sequence is critical for palmitoylation. Palmitoylation was also not dependent on the native TMD of H1. In contrast, the attachment of palmitate was abolished if the Cys residue was transposed to a position that was 30 amino acids away from the transmembrane border. Thus, the spacing of a Cys residue relative to the TMD in the primary protein sequence of H1 is the major determinant for successful palmitoylation (Yik and Weigel 2002).

33.3.1.2 Macrophage ASGP-Binding Protein

The primary structure of macrophage asialoglycoprotein-binding protein, M-ASGP-BP consists of 306 amino acid residues with a molecular mass of 34,242 Da. The sequence was highly homologous with that of the rat liver asialoglycoprotein receptor (RHL), particularly that of RHL-1 (the major form of RHL), throughout its whole length, and especially so in its putative membrane-spanning region and CRD. There were two N-glycosylation sites in M-ASGP-BP, the location of which were identical to those in RHL-1 (Ii et al. 1990). However, M-ASGP-BP was characteristic in having a shorter cytoplasmic tail, and an inserted segment of 24 amino acids containing an Arg-Gly-Asp sequence between the membrane-spanning region and carbohydrate recognition domain (Ii et al. 1990).

Ligand specificity and binding affinity depend on the arrangement of the subunits in the complex. Presence of both subunits is a prerequisite for the ASGP-R to reach full functionality. Although the subunit composition of the native receptor remains unspecified (Weigel and Yik 2002), a stoichiometry of 2:1 of H1 and H2 has been proposed as the minimum size of an active ASGP-R complex, exhibiting high affinity ligand binding (Fig. 33.1b) (Hardy et al. 1985; Schwartz 1984). Results suggest that the stalk segments of the receptor subunits oligomerize to constitute an α-helical coiled coil stalk on top of which the carbohydrate binding domains are exposed for ligand binding. This subunit ratio was partially supported by another study in which the functional receptor was suggested to be a 2:2 heterotetramer. Extracellular stalk segments of H1 and H2, responsible for the subunit oligomerization, were expressed and shown to associate in homo- and herooligomers in vitro (Bider et al. 1996). A study of e binding of three different ^{125}I-labeled, galactose-terminated ligands to the hepatic galactose/N-acetylgalactosamine-specific lectin revealed that the different ligands manifest different physical parameters of binding. For example, there were many more total binding sites for di-tris-lac on the surface of rabbit hepatocytes than there were for asialoorosomucoid, although the dissociation constants were similar for these ligands (Hardy et al. 1985).

33.4 Ligand Binding Properties of ASGP-R

The ASGP-Rs specifically recognize galactose- or N-acetylgalactosamine-terminated oligosaccharides, present on desialylated glycoproteins. The ASGP-R recognizes with high affinity (K_D in nM range) tri- and tetra-antennary N-linked sugar side chains with terminal galactose residues. Glycoproteins with such glycosylation patterns are rapidly endocytosed by the ASGP-R via clathrin-coated pits and vesicles. The specificity and affinity of ligand binding of ASGP-R depends on the number and spatial arrangement of several galactose-binding sites within the receptor complex. A single Gal residue exhibits only modest binding affinity with a K_D in the range of 1 mM. Additional sugars, from a mono- to a triantennary structure, result in a significant increase in affinity with K_D ~10 nM for the native receptor. Hence, high affinity binding is achieved through multiple interactions between the CRDs in a receptor complex and multiple sugar residues (Westerlind et al. 2004). Furthermore, the sugar spacing has been shown to be important for optimal receptor recognition. A spacer length of >20 Å, separating the sugar moieties from the branching point of the ligand, results in high affinity binding to the ASGP-R of Gal containing ligand constructs (Westerlind et al. 2004; Rensen et al. 2001). In addition, the ASGP-R clearly favors binding of GalNAc over Gal, showing an approximate 50-fold higher affinity for the former (Rensen et al. 2001; Baenziger and Maynard 1980). Studies indicate that both subunits are required for high-affinity ligand binding, i.e. for the simultaneous interaction with three galactose residues within an N-linked glycan. However, asialoorosomucoid (ASOR) and asialofetuin (ASF) bind to transfected COS-7 cells expressing subunit H1 in absence of the second subunit H2. It was found that, at a sufficiently high density of H1 on the cell surface, high-affinity binding of ASOR and ASF is the result of two or more glycans interacting with H1 oligomers with low affinity in a bivalent manner. ASOR displays a K_D of <10 nM for the native receptor. It has also been observed to bind H1, overexpressed in transfected COS-7 cells, in the absence of H2 with a K_D of <40 nM (Bider et al. 1996). In order to develop the non-viral Bioplex vector system for targeted delivery of genes to hepatocytes, Westerlind et al. (2004) evaluated the structure-function relationship for a number of synthetic ligands designed for specific interaction with the hepatic lectin ASGP-R. Uptake efficiency increased with number of displayed GalNAc units per ligand, in a receptor dependent manner. Thus, a derivative displaying six GalNAc units showed the highest uptake efficacy both in terms of number of internalizing cells and increased amount of material taken up by each cell. However, this higher efficiency was shown to be due not so much to higher number of sugar units, but to higher accessibility of the sugar units for interaction with the receptor (longer spacer) (Westerlind et al. 2004).

Analogous binding specificity can be engineered into the homologous rat MBP-A by changing three amino acids and inserting a glycine-rich loop (Iobst and Drickamer 1994). Crystal structures of this mutant complexed with β-methyl galactoside and N-acetylgalactosamine (GalNAc) revealed that as with wild-type MBPs, the 3- and 4-OH groups of the sugar directly coordinate Ca^{2+} and form hydrogen bonds with amino acids that also serve as Ca^{2+} ligands. The different stereochemistry of the 3- and 4-OH groups in mannose and galactose, combined with a fixed Ca^{2+} coordination geometry, leads to different pyranose ring locations in the two cases. The glycine-rich loop provides selectivity against mannose by holding a critical tryptophan in a position optimal for packing with the apolar face of galactose but incompatible with mannose binding. The 2-acetamido substituent of GalNAc is in the vicinity of amino acid positions identified by site-directed mutagenesis (Iobst and Drickamer 1996) as being important for the formation of a GalNAc-selective binding site (Kolatkar and Weis 1996). ASGP-R was also postulated to account for the low density lipoprotein (LDL) receptor-independent clearance of lipoproteins including chylomicron remnants (Ishibashi et al. 1996). Later, immunoglobulin A (Rifai et al. 2000) and fibronectin (Rotundo et al. 1998) emerged as likely candidates of natural ligands for ASGP-R. The clearance of apoptotic cells or a subpopulation of lymphocytes in the liver has been also attributed to ASGP-R (Stockert 1995).

33.4.1 Interaction with Viruses

It is particularly noteworthy that ASGP-R has also been proposed to be utilized as entry sites into hepatocytes by several hepatotropic viruses including hepatitis B virus (Treichel et al. 1994), Marburg virus (Becker et al. 1995), and hepatitis A virus (Dotzauer et al. 2000). The liver is one of the main target organs of Marburg virus (MBG), a filovirus causing severe haemorrhagic fever with a high fatality rate in humans and non-human primates. MBG grown in certain cells does not contain neuraminic acid, but has terminal galactose on its surface glycoprotein. The lack of neuraminic acid together with the marked hepatotropism of the MBG infection led to suggest that the ASGP-R of hepatocytes may serve as a receptor for MBG in the liver. MBG lacking sialic acid specifically binds to the ASGP-R. This observation may at least in part explain why the liver is a central target for MBG and may thus contribute to an understanding of the pathogenesis of this infection (Becker et al. 1995). Structural proteins of hepatitis C virus (HCV) genotype 1a bind to cultured human hepatic cells. This binding was decreased by calcium depletion and was partially prevented by ligands of ASGP-R, thyroglobulin, asialothyroglobulin, and antibody against a peptide in CRD of ASGP-R but not preimmune antibody (Saunier et al. 2003).

33.5 The Crystal Structure of H1-CRD

The structure of ASGP-R H1-CRD was solved in 2000 by X-ray crystallography (Meier et al. 2000). It is a globular protein, containing two α-helices and eight β-strands. The β-strands are arranged in a bent plane and form the core of the protein, while the α-helices are positioned on either side of the plane. Three calcium ions can be seen in the structure, forming an integral part of the protein as they pin together several loops. The two calcium ions at site 1 and 2 are seen in close proximity, both coordinated by Glu^{252}. Site 2 is of particular interest as it is essential for sugar binding and is present in all C-type lectins. Calcium binding site 3 is found close to the N- and C-terminus of the protein (Fig. 33.2).

C-type CRDs can be classified into short-form CRDs, with two disulfide bridges, or long-form CRDs, containing a conserved extension with an additional disulfide bond at the N-terminus. H1-CRD falls into the latter category, containing seven cysteines out of which six are engaged in three disulfide bridges. One bridge constitutes part of the sugar binding site, seen in the crystal structure between Cys^{254} and Cys^{268}. A second bond is formed between Cys^{181} and Cys^{276}, bringing the C- and N-terminus close together and contributes to the tertiary structure of the subunit. Finally, a third typical long-form CRD disulfide bridge is found at the N-terminus between Cys^{153} and Cys^{164}. The N-terminal residues 147–152, including the seventh odd cysteine (Cys^{152}), and the C-terminal residues 281–290 could not be positioned into the electron density and hence, cannot be seen in the crystal structure (Meier et al. 2000; Johansson 2007).

33.5.1 The Sugar Binding Site of H1-CRD

The specificity and affinity of ligand binding of ASGP-R depends on the number and spatial arrangement of several galactose-binding sites within the receptor complex. Studies indicate that both subunits are required for high-affinity ligand binding, i.e. for the simultaneous interaction with three galactose residues within an N-linked glycan. However, asialoorosomucoid (ASOR) and asialofetuin (ASF) bind to transfected COS-7 cells expressing subunit H1 in the absence of the second subunit H2. It was found that, at a sufficiently high density of H1 on the cell surface, high-affinity binding of ASOR and ASF is the result of two or more glycans interacting with H1 oligomers with low affinity in a bivalent manner. The region around the sugar binding site in H1-CRD is formed by one continuous stretch of the polypeptide chain from Arg^{236} to Cys^{268} (Johansson 2007).

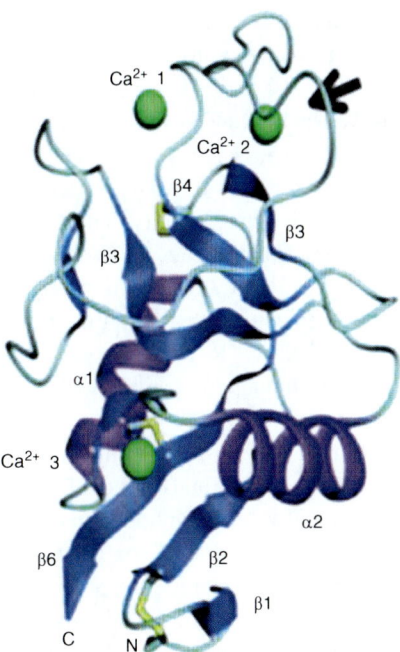

Fig. 33.2 Ribbon diagram of the carbohydrate recognition domain of the H1 subunit. The three calcium ions can be seen in *green*. The second calcium binding site, which is part of the sugar binding site, is denoted by a *black arrow*. The disulfide bridges are marked in *yellow* (PDB ID: 1DV8) (Adapted with permission from Meier et al. 2000 © Elsevier)

Mutagenesis studies have been carried out with closely related protein RHL-1, aiming to deduce the residues giving rise to it's ligand binding specificity. RHL-1 has been used as a reference in attempts to mimic its Gal- and GalNAc-binding properties in other lectins, e.g. the MBP (Iobst and Drickamer 1994) and the macrophage galactose receptor (MGR) (Iobst and Drickamer 1996). Such studies form the basis for explanation of H1-CRDs sugar preference as well as its distinction between high- and low-affinity ligands. Introduction of two point mutations in MBP, $E^{185}Q$ and $N^{187}D$, corresponding to Q^{239} and D^{241} in RHL-1, was sufficient to achieve galactose binding in the protein, which is normally binding mannose (Iobst and Drickamer 1994). However, as the Gal affinity of the MBP mutant, designated QPD, was rather low it was concluded that other residues in H1-CRD also must contribute to the ligand binding. An additional mutation, $H^{189}W$, showed to increase the affinity significantly, making the galactose binding of mutant QPDW comparable to that of RHL-1. Selectivity for galactose over mannose was achieved by incorporating a glycine-rich stretch following the QPDW sequence, referred to as the QPDWG mutant. Stepwise mutations within the glycine rich loop, corresponding to residues Y^{244}, G^{245} and H^{246} in RHL-1, showed that they all contribute to the selectivity for Gal (Iobst and Drickamer 1994). However, the influence of the amino acids on the selectivity could equally be attributed to stabilization and support of the protein structure as to an actual effect on the ligand binding.

RHL-1, as well as H1-CRD, shows preferential binding to GalNAc over Gal. The selectivity for GalNAc has been probed by mutations studies of MGR, which binds Gal and GalNAc with roughly equal affinity. The CRDs of RHL-1 and MGR are highly homologues (77%) and the difference in selectivity is likely to stem from divergences in the sequences. It was concluded that substitution of four amino acids in MGR, $V^{230}N$, $A^{258}R$, $K^{260}G$ and $S^{281}T$, is sufficient to induce GalNAc binding comparable with that of RHL-1. N^{230} increased the selectivity 20-fold, while the contributions by R^{258} and G^{260} were less prominent. The latter two residues only showed marginal or no effect at all when substituted individually, but a twofold increase when present together. It is possible that glycine contributes to the affinity by positioning arginine, which has the potential for forming hydrogen bonds with the ligand. T^{281} significantly increased the selectivity for GalNAc when inserted into MGR simultaneously with R^{258} and G^{260}, but is more likely to play an indirect role for ligand binding as it is predicted to be positioned at a considerable distance (10 Å). The residue appearing most important for achieving GalNAc binding is a histidine (His, H), found in both RHL-1 and MGR at position 256 and 278 respectively. Initial studies exchanging H^{278} in MGR for an alanine (Ala, A) resulted in an almost complete loss of GalNAc selectivity, but without any apparent effect on the galactose binding. The importance of the histidine was further investigated by repeating the substitution in RHL-1, causing a 25-fold loss in affinity for GalNAc, but without affecting that for Gal (Iobst and Drickamer 1996). Corresponding residue in MBP, T^{202}, was also substituted by histidine and resulted in a 14-fold increase in the relative affinity for GalNAc of the protein (referred to as the QPDWGH mutant) (Kolatkar et al. 1998).

33.5.2 Sugar Binding to H1-CRD

A model for sugar binding to H1-CRD has been proposed based on the mutagenesis experiments with MBP (Kolatkar and Weis 1996; Kolatkar et al. 1998). Crystallographic data of the QPDWG mutant could show that Gal and GalNAc bind directly to the Ca^{2+} in the second calcium binding site. The 3-OH and 4-OH groups of the sugar replaces two water molecules, normally coordinated by the calcium. In addition, the same OH groups also forms hydrogen bonds with amino acid chains that are Ca^{2+} site 2 ligands (Q^{239}, D^{241}, E^{252}, N^{264} in H1-CRD). Further interactions between the ligand and the protein are formed by stacking of the apolar face of the sugar against the side-chain of a tryptophan (W^{243} in H1-CRD) (Kolatkar and Weis 1996). The GalNAc-specific MBP mutant QPDWGH was also analyzed by crystallography.

Regions of the CRD of the receptor believed to be important in preferential binding to N acetylgalactosamine were inserted into the homologous CRD of a MBP mutant that was previously altered to bind galactose. Introduction of a single histidine residue corresponding to residue 256 of the hepatic ASGP-R was found to cause a 14-fold increase in the relative affinity for N-acetylgalactosamine compared with galactose. The relative ability of various acyl derivatives of galactosamine to compete for binding to this modified CRD suggest that it is a good model for the natural N-acetylgalactosamine binding site of the ASGP-R. Crystallographic analysis of this mutant CRD in complex with N-acetylgalactosamine reveals a direct interaction between the inserted histidine residue and the methyl group of the N-acetyl substituent of the sugar. The structure of this mutant reveals that the beta-branched valine side chain interacts directly with the histidine side chain, resulting in an altered imidazole ring orientation (Kolatkar et al. 1998).

33.5.2.1 Histidine 256: Responsible for pH Dependent Ligand Binding

Efficient release of ligands from CRD of the hepatic ASGP-R at endosomal pH requires a small set of conserved amino acids that includes a critical histidine residue. Mutagenesis studies of RHL-1 have shown that histidine 256 plays an important role in pH dependent ligand binding exhibited by the subunit (Wragg and Drickamer 1999; Feinberg et al. 2000). When these residues are incorporated at corresponding positions in an homologous galactose-binding derivative of serum MBP, the pH dependence of ligand binding becomes more like that of the receptor. The modified CRD displays 40-fold preferential binding to N-acetylgalactosamine compared with galactose, making it a good functional mimic of ASGP-R. In the crystal structure of the modified CRD bound to N-acetylgalactosamine, the His^{202} contacts the 2-acetamido methyl group and also participates in a network of interactions involving Asp^{212}, Arg^{216}, and Tyr^{218} that positions a water molecule in a hydrogen bond with the sugar amide group. These interactions appear to produce the preference for N-acetylgalactosamine over galactose and are also likely to influence the pK_a of His^{202}. Protonation of His^{202} would disrupt its interaction with an asparagine that serves as a ligand for Ca^{2+} and sugar. The structure of the modified CRD without sugar displays several different conformations that may represent structures of intermediates in the release of Ca^{2+} and sugar ligands caused by protonation of His^{202} (Feinberg et al. 2000).

In another work (Johansson 2007), His^{256} of H1-CRD was substituted by glutamate, the corresponding residue in H2-CRD. Binding of GalNAc to mutant $H^{256}E$ and WT H1-CRD was investigated at different pH ranging from 7.4 to 5. Results showed that binding of the GalNAc-polymer decreased at pH 6 or lower. $Hist^{256}$ clearly renders H1-CRD more sensitive to low pH compared to a glutamate in the same position. His^{256} has been proposed to interact with Asn^{264}, which in turn is involved in both calcium- and ligand binding. The histidine stabilizes Asn^{264} by a hydrogen bond formed between the imidazole and the amide. Protonation of imidazole, caused by a drop in pH during endocytosis, will disrupt the hydrogen bond. As a consequence, the asparagine is destabilized and both ligand and calcium binding is disturbed.

33.6 Physiological Functions

A number of diverse physiological roles have been proposed for ASGP-R over the years. Among them, hepatic clearance of the desialylated and senescent serum proteins was most originally proposed. Later physiological and pathophysiological functions to this lectin include: the removal of apoptotic cells, clearance of lipoproteins, and the sites of entry for hepatotropic viruses. The primary function of the ASGP-R has been considered to be the removal and degradation of desialylated glycoproteins from the circulation. Normally, many oligosaccharide chains on glycoproteins carry terminal sialic acid residues. Upon removal of the sialic acid, caused by the action of neuraminidases, penultimate galactose residues are exposed and recognized by ASGP-R (Tozawa et al. 2001). The assembly of two homologous subunits, H1 and H2, is required to form functional, high affinity receptors on the cell surface. However, the importance of the individual subunits for receptor transport to the cell surface is controversial. To explore the significance of the minor H2 subunit for receptor expression and function in vivo, homozygous H2-deficient mice (MHL-$2^{-/-}$) are superficially normal. However, H1 expression in the liver is greatly reduced, indicating that H2 may H1 is strictly required for the stable expression of H2. Although these mice are completely unable to clear asialoorosomucoid, a high affinity ligand for asialoglycoprotein receptor, they do not accumulate desialylated glycoproteins or lipoproteins in their circulation (Ishibashi et al. 1994; Tozawa et al. 2001). Results suggest that ASGP-R is not ultimately responsible for the clearance of plasma glycoproteins, but is likely to possess other functions.

ASGP-R has been proposed to be involved in the metabolism of plasma lipoproteins and cellular fibronectin (Rotundo et al. 1998). However, studies with ASGP-R knock-out mice could not confirm this result, as the plasma levels of fibronectin and lipoprotein appeared unaffected in the absence of the receptor (Tozawa et al. 2001). ASGP-R has also been implicated in the clearance of apoptotic cells by the liver. Studies showed that the uptake of apoptotic bodies was blocked in the presence of an ASGP-R specific antibody or following the addition of receptor specific ligands such as ASF or Gal$N$Ac (Dini et al. 1992). Immunoglobulin A has also been proposed as a ligand for the receptor (Rifai et al. 2000).

Finally, ASGP-R is thought to act as an entry point for a few specified pathogenic viruses to the hepatocytes. Experimental data indicate that the Marburg virus (Becker et al. 1995), the hepatitis B virus (Owada et al. 2006; Treichel et al. 1994) and the hepatitis C virus (Saunier et al. 2003) are capable of binding to the receptor, followed by infection of the host cell. Results on the infectivity of HBV in vitro showed that ASGP-R may be a specific HBV receptor once viral particles are desialylated (Owada et al. 2006).

Geuze et al. (1984) compared the endocytotic pathways of the receptors for ASGP-R, mannose-6-phosphate ligands (M6PR), and polymeric IgA (IgA-R). All three were found within the Golgi complex, along the entire plasma membrane, in coated pits and vesicles, and within a compartment of uncoupling of receptors and ligand (CURL). The receptors occurred randomly at the cell surface, in coated pits and vesicles. Within CURL tubules ASGP-R and M6PR were co-localized, but IgA-R and ASGP-R displayed microheterogeneity. Thus, in addition to its role in uncoupling and sorting recycling receptor from ligand, CURL serves as a compartment to segregate recycling receptor (e.g. ASGP-R) from receptor involved in transcytosis (e.g. IgA-R).

33.6.1 Impact of ASGP-R Deficiency on the Development of Liver Injury

Function of ASGP receptor is impaired in pathological states such as liver disease among alcoholics. The knockout ASGP-receptor$^{-/-}$ mice, after various toxic challenges (alcohol, anti-Fas, CCl$_4$ and LPS/galactosamine), consistently showed more liver injury than the wild-type animals. This suggested that ASGP receptor functions as a protective agent. Thus, receptor-mediated endocytosis may be a novel mechanism that may be involved in the induction of toxin-induced liver injury. However, it is likely that impaired clearance of apoptotic bodies, perturbations in extracellular matrix deposition, oxidative stress, and cytokine dysregulation may play roles in the progression of disease (Lee et al. 2009). Overall, results suggest a possible link between hepatic receptors and liver injury. In particular, adequate function and content of the ASGP receptor may provide protection against various toxin-mediated liver diseases.

33.7 ASGP-R: A Marker for Autoimmune Hepatitis and Liver Damage

33.7.1 Autoimmune Hepatitis

Antibodies to ASGP-R have prognostic value. Circulating anti-ASGP-R autoantibodies in autoimmune hepatitis is independent of geographic or ethnic criteria. Results in different ethnic groups suggest highest frequency (76%) of anti-human ASGP-R in autoimmune hepatitis patients (11/24 U.S.; 21/25 European; 28/30 Japanese), particularly in those with active disease before treatment (53/62, 85%), and decreased in titer with response to immunosuppressive therapy.

The research has been focused on the hepatocytes by examining endocytosis using the ASGP-R pathway as a model and identified multiple ethanol-induced impairments in receptor function. The altered uptake of apoptotic cells via ASGP-R may result in the release of proinflammatory mediators, the introduction of autoimmune responses, and inflammatory injury to the tissue. It was found that uptake of apoptotic bodies is impaired in hepatocytes isolated from ethanol-fed animals compared to controls, and that this impairment is linked to altered ASGP-R function. There is an attempt to examine a link between ethanol-impaired ASGP-R function, apoptotic body accumulation, and inflammation in the liver (Casey et al. 2008; McVicker et al. 2002). Hemolysis in patients with advanced alcoholic liver disease is a common clinical problem and indicates an unfavorable prognosis. All patients with liver disease have a soluble variant of the human s-ASGP-R in their serum, as well as high titers of autoantibodies against this receptor. Examination of patients with alcoholic liver disease reveal a high incidence for s-ASGP-R (36%) and anti-ASGP-R (27%) in patients with alcoholic liver cirrhosis compared to patients with cirrhosis due to viral hepatitis. This suggests a non-immunological mechanism for hemolysis in patients with alcoholic liver disease, mediated through agglutination by a soluble variant of the human ASGP-R and mechanical shear stress (Hilgard et al. 2004).

33.7.2 Hepatocellular Carcinoma

ASGP-Rs expression is decreased in various chronic liver diseases and hepatoma. Although the active shedding of the receptors clearly occurs in vitro, the significance of these shed receptors in blood will require further studies. The expression of ASGP-R on human hepatocellular carcinoma (HCC) cells might be exploited to reduce the extrahepatic toxicity of DNA synthesis inhibitors by their conjugation with galactosyl-terminating peptides. The results clearly demonstrated that DNA synthesizing cancer cells expressed the ASGP-R on their surface. The presence of ASGP-R on cell plasma membrane in the majority of differentiated HCCs and its maintenance on proliferating cells encourages studies in order to restrict the action of the inhibitors of DNA synthesis of HCC cells by their conjugation with galactosyl-terminating carriers internalized through this receptor (Trerè et al. 1999). Both acute and chronic phenobarbital administration decreased the number of ASGP-Rs per cell. Partial hepatectomy had a similar effect on the number of receptors per cell (Evarts et al. 1985).

33.7.3 Extra-Hepatic ASGP-R

In humans, ASGP-R is predominantly expressed by hepatocytes, at an estimated density of 100,000–500,000 binding sites per cell (Wall and Hubbard 1981; Matsuura et al. 1982) but does occur extrahepatically in thyroid, in small and large intestines, and in the testis. There is evidence of expression of the receptor in the T-cell line Jurkat. Moreover, Tera-1 cells derived from human testis gave rise to a weak signal, indicating presence of ASGP-R (Park et al. 1998). In the kidney, there has been evidence both for and against its existence in mesangial cells. Studies support ASGP-R expression in cells originating from human bone intestine and kidney (Park et al. 2006; Seow et al. 2002). Primary renal proximal tubular epithelial cells have a functional ASGP-R, consisting of the H1 and H2 subunit which are capable of specific ligand binding and uptake (Seow et al. 2002).

33.8 ASGP-R: A Model Protein for Endocytosis

Transport of macromolecules into the cell by receptor-mediated endocytosis follows a complex series of intracellular transfers, passing through distinct environments. The ASGP-R constitutively enters cells via coated pits and delivers ligand to these intracellular compartments. In addition to being a model of receptor-mediated endocytosis, the presence of the receptor on hepatocytes provides a membrane-bound active site for cell-to-cell interactions, has made possible the selective targeting of chemotherapeutic agents and foreign genes, and has also been implicated as a site mediating hepatitis B virus uptake. Regulated expression of receptor subunits and their intracellular trafficking during biosynthesis and endocytosis has provided insights into the relationship of receptor structure to its overall function. As a marker of hepatocellular differentiation, its study has uncovered a unique response to intracellular guanosine $3'$, $5'$-cyclic monophosphate and translational regulation of the receptor.

Receptor-mediated endocytosis serves as a mechanism by which cells can internalize macromolecules like peptides and proteins. The ASGP-R has been the focus of several studies aiming to understand endocytosis via the clathrin-coated pit pathway. Following binding of ligand to the ASGP-R at the cell surface, the receptor-ligand complex is endocytosed via clathrin-coated pits and directed to endosomes, where the complex dissociates. The carbohydrate ligand is targeted for degradation in lysosomes while the receptor recycles to the cell surface with a vacant binding site (Spiess 1990). Upon ligand binding, the ASGP-Rs cluster into clathrin-coated domains of the membrane, which in turn invaginates. Clathrin-coated pits are then formed and subsequently turned into coated vesicles, which are internalized by the cell. With few exceptions, receptor-mediated endocytosis of specific ligands is mediated through clustering of receptor-ligand complexes in coated pits on the cell surface, followed by internalization of the complex into endocytic vesicles. During this process, ligand-receptor dissociation occurs, most probably in a low pH prelysosomal compartment. In most cases the ligand is ultimately directed to the lysosomes, wherein it is degraded, while the receptor recycles to the cell surface. In a human hepatoma cell line, Ciechanover et al. (1983) gained some insight into the complex mechanisms which govern receptor recycling as well as ligand sorting and targeting and explained why transferrin is exocytosed intact from the cells, while asialoglycoproteins are degraded in lysosomes. The receptor appears to be a transmembrane protein and is localized both to the cell surface as well as to several membranous intracellular compartments (Schwartz 1984). The ASGP-Rs meet a less grim fate as they are recycled and transported back to the plasma membrane (Geffen and Spiess 1992; Spiess 1990).

Recycling of the ASGP-R is a continuous process and occurs several hundred times during the life span of an individual receptor, the time of which is estimated to approximately 30 h (Schwartz 1984). Internalization of the ASGP-R takes place independently of ligand binding, but was shown to increase 2-fold upon binding of ASOR. Speculations attribute this enhanced internalization rate to a possible conformational change induced by ligand binding (Bider and Spiess 1998). Furthermore, a tyrosine residue in H1 has been shown to be of critical importance for efficient endocytosis of the ASGP-R. The corresponding residue in H2, a phenylalanine, does not appear to contribute to the internalization of hetero-oligomeric receptor complexes (Fuhrer et al. 1991, 1994).

Endosomal pH is an important determinant of recycling of the asialoglycoprotein receptor as well as other endocytic receptors (Mellman et al. 1986). The dissociation process that occurs in endosomes can be mimicked in vitro by ligand release at pH 5.4. In the case of the chicken hepatic lectin, an homologous endocytic receptor, pH modulates structural transitions between several distinct states of the CRD (Loeb and Drickamer 1988). At endosomal pH, the structural change causes an approximately 10-fold reduction in affinity for Ca^{2+} with concomitant loss of ligand binding activity. Site-directed mutagenesis of the CRD to residues His^{256}, Asp^{266}, and Arg^{270} singly and in combination indicate that these residues reduce the affinity of the CRD for Ca^{2+}, so that ligands are released at physiological Ca^{2+} concentrations. The proximity of these three residues to the ligand-binding site at Ca^{2+} site 2 of the domain suggests that they form a pH-sensitive switch for Ca^{2+} and ligand binding. Introduction of histidine and aspartic acid residues into the mannose-binding protein CRD at positions equivalent to His^{256} and Asp^{266} raises the pH for half-maximal binding of ligand to 6.1. The results, as well as sequence comparisons with other C-type CRDs, confirm the importance of these residues in conferring appropriate pH dependence in this family of domains (Wragg and Drickamer 1999).

Glycans as Endocytosis Signals Animal cells internalize specific extracellular macromolecules (ligands) by using specialized cell surface receptors that operate through a complex and highly regulated process known as receptor-mediated endocytosis, which involves the binding, internalization, and transfer of ligands through a series of distinct intracellular compartments. For the uptake of a variety of carbohydrate-containing macromolecules, such as glycoproteins, animal cells use specialized membrane-bound lectins as endocytic receptors that recognize different sugar residues or carbohydrate structures present on various ligands. Studies of how the asialoglycoprotein receptor functions have led to the discovery of two functionally distinct, parallel pathways of clathrin-mediated endocytosis (called the State 1 and State 2 pathways), which may also be utilized by all the other endocytic recycling receptor systems. Weigel and Yik (2002) discussed the characteristics and physiological importance of ASGP-R as an example of how lectins can function as endocytic receptors.

33.9 ASGP-R for Targeting Hepatocytes

The ASGP-R on mammalian hepatocytes provides a unique means for the development of liver-specific carriers, such as liposomes, recombinant lipoproteins, and polymers for drug or gene delivery to the liver, especially to hepatocytes (Wu et al. 1998). The abundant receptors on the cells specifically recognize ligands with terminal galactose or N-acetylgalactosamine residues, and endocytose the ligands for an intracellular degradation process. The use of its natural ligand, i.e. asialofetuin, or synthetic ligands with galactosylated or lactosylated residues, such as galactosylated cholesterol, glycolipids, or galactosylated polymers has achieved significant targeting efficacy to the liver. There are several examples of successful targeted therapy for acute liver injury with asialofetuin-labeled and vitamin E-associated liposomes or with a caspase inhibitor loaded in sugar-carrying polymer particles, as well as for the delivery of an antiviral agent, 9-(2-phosphonylmethoxyethyl) adenine. Liposome-mediated gene delivery to the liver is more difficult than to other organs, such as to lungs. Galactosylated polymers are promising for gene delivery, but require further studies to verify their potential applications (Wu et al. 2002) (see Chap. 46).

Labeling conventional liposomes with asialofetuin was seen to result in a significant increase of liver uptake, compared to unlabeled liposomes, after intravenous injection into mice. The enhanced uptake was most likely mediated by the ASGP-R (Wu et al. 1998). Another study used glycolipids containing a cluster galactoside moiety for targeting to ASGP-R. The liver uptake of the glycolipid-liposomes was estimated to exceed 80% compared to less than 10% for conventional liposomes after injection. DNA-galactosylated cationic liposome complexes show higher DNA uptake and gene expression in the liver parenchymal cells in vitro than DNA complexes with bare cationic liposomes. In the in vitro gene transfer experiment, galactosylated liposome complexes are more efficient than DNA-galactosylated poly(amino acids) complexes but they have some difficulties in their biodistribution control. On the other hand, introduction of mannose residues to carriers resulted in specific delivery of genes to non-parenchymal liver cells. These results suggest advantages of these glycosylated carriers in cell-specific targeted delivery of genes (Hashida et al. 2001).

In order to reduce the extrahepatic side-effects of antiviral nucleoside analogues in the treatment of chronic viral hepatitis, these drugs were conjugated with galactosyl-terminating macromolecules. The conjugates selectively enter hepatocytes after interaction of the carrier galactose residues with the ASGP-R present in large amounts and high affinity only on these cells. The validity of this chemotherapeutic strategy has been endorsed by a clinical study (Fiume et al. 1997). Receptors like ASGP-R provide unique opportunity to target liver parenchymal cells. The results obtained so far reveal tremendous promise and offer enormous options to develop novel DNA based pharmaceuticals for liver disorders in near future. The ^{99m}Tc-labeled asialoglycoprotein analog, TcGSA (galactosyl-human serum albumin-diethylenetriamine-pentaacetic acid) has been applied to human hepatic receptor imaging. This method is unique and provides information that is totally independent of the ICG test or Child-Turcotte Score (Kokudo et al. 2003; Pathak et al. 2008).

33.10 Macrophage Galactose-Type Lectin (MGL) (CD301)

33.10.1 Human MGL (CD 301)

Serial analysis of gene expression in monocyte-derived DCs, monocytes, and macrophages revealed that 7 of the 19 C-type lectin mRNAs are present in immature DCs. Two of these, the macrophage mannose receptor (MMR) and the macrophage lectin specific for galactose/N-acetylgalactosamine (MGL), were found only in immature DCs (Steinman et al. 2003). Subcloning and sequencing the amplified mRNA, revealed nucleotide sequences encoding seven different human MGL (hMGL) subtypes, which were apparently derived from alternatively spliced mRNA. In addition, the hMGL gene locus on human chromosome 17p13 contains one gene. A single nucleotide polymorphism was identified at a position in exon 3 that corresponds to the cytoplasmic region proximal to the transmembrane domain. Of all the splicing variants, the hMGL variant 6 C was expressed at the highest levels on immature DCs from all donors tested.

Macrophage galactose-type lectins are a family of type II C-type lectins expressed in connective tissue macrophages and in bone marrow-derived DCs (Mizuochi et al. 1997; Denda-Nagai et al. 2002). The human macrophage C-type lectin was homologous to galactose- and N- acetylgalactosamine-specific C-type macrophage lectins of rodents. In the putative CRD, deduced amino acid sequence revealed 60 and 63% homology to galactose- and N- acetylgalactosamine-specific C-type macrophage lectins of mice and rats, respectively. The *Mgl* gene consists of ten exons and has been mapped to mouse chromosome 11 (Tsuiji et al. 1999). The *MGL* gene has been mapped to human chromosome 17p13.2 and contains one gene. The chromosome localization of MGL is distinct from that of many C-type lectins, MMR, DEC-205, DCIR, langerin, DC-SIGN, and Dectin-1, claimed to be specific for DCs. A single nucleotide polymorphism was identified at a position in exon 3 that corresponds to the cytoplasmic region proximal to the transmembrane domain. Of all the splicing variants, the hMGL variant 6 C was expressed at the highest levels on immature DCs. It was proposed that hMGL is a marker of immature DCs and that it functions as an endocytic receptor for glycosylated antigens (Higashi et al. 2002a). MGL is also called DC-asialoglycoprotein receptor (DC-ASGP-R) or human macrophage lectin (HML) (Higashi et al. 2002b; Valladeau et al. 2001; Suzuki et al. 1996). The liver-specific ASGP-R is closest homolog of MGL.

33.10.2 Murine MGL1/MGL2 (CD301)

Murine bone marrow–derived immature DCs also bind and internalize α-*N* acetylgalactosaminides conjugated to soluble polyacrylamide (α-GalNAc polymers), whereas mature DCs and bone marrow cells did not. It was suggested that mMGL is transiently expressed on bone marrow–derived DCs during their development and maturation and seemed to be involved in the uptake of glycosylated antigens for presentation (Denda-Nagai et al. 2002).

There are two MGL genes in mice: *Mgl1*, and *Mgl2* (Tsuiji et al. 2002; Onami et al. 2002; Dupasquier et al. 2006). The *Mgl* family is known to have two homologous genes in mice, *Mgl1* and *Mgl2*, and these two lectins have distinct carbohydrate recognition specificities, although their distinct roles have not yet been defined (Tsuiji et al. 2002; Oo-Puthinan et al. 2008). MGL1 and/or MGL2 are mainly expressed on macrophages and immature DCs, and that these cells were observed mainly in the connective tissue of various organs, especially in skin, large intestines, and lymph nodes (Mizuochi et al. 1997). MGL'CD301 recognizes terminal galactose and (GalNAc) residues as monosaccharides in a calcium-dependent manner (Imai and Irimura 1994; Sato et al. 1992; Yamamoto et al. 1994; Denda-Nagai et al. 2010). These lectins were found to be involved in the uptake of mucin-like GalNAc-conjugated polymers by murine bone marrow-derived and human monocyte-derived DCs (Higashi et al. 2002b; Denda-Nagai et al. 2002), which was thought to be an important process of antigen processing. *Mgl1*-deficient mice did not show obvious defects in lymphoid and erythroid homeostasis (Onami et al. 2002). In an in vivo study with mouse embryos, MGL1 was shown to function as an endocytic receptor for X-irradiation-induced apoptotic cells, whereas *Mgl1*-deficient mice showed retarded clearance of apoptotic cells in neural tubes (Yuita et al. 2005). It is also suggested that MGL1 regulates trafficking of MGL1-expressing cells from skin to lymph nodes (Chun et al. 2000a, b; Kumamoto et al. 2004). Antigen-induced inflammatory tissue formation in skin was abrogated in *Mgl1*-deficient mice (Sato et al. 2005a), suggesting that MGL1 functioned under inflammatory conditions. MGL2 expresses in the skin and the cutaneous LNs. It was highly restricted to DDCs in the skin and the LNs. Evidence indicates that hapten-incorporated DDCs are sufficient to induce CHS in vivo. The availability of MGL2 as a marker for DDCs suggested the contribution of MGL2$^+$ DDCs for initiating contact hypersensitivity (Kumamoto et al. 2009).

The sequence of mMGL2 is highly homologous to the mMGL, which should now be called mMGL1. The ORF of mMGL2 contains a sequence corresponding to a type II transmembrane protein with 332 amino acids having a single extracellular C-type lectin domain. The 3′-untranslated region included long terminal repeats of mouse early transposon. The *Mgl2* gene spans 7,136 base pairs and consists of ten exons, similar to the genomic organization of mMGL1. The mMGL2 mRNA was also detected in mMGL1-positive cells. The soluble recombinant proteins of mMGL2 exhibited carbohydrate specificity for α- and β-GalNAc-conjugated soluble polyacrylamides, whereas mMGL1 preferentially bound Lewis X-conjugated soluble polyacrylamides (Tsuiji et al. 2002).

The MGL, expressed by DC and macrophages, mediates binding to glycoproteins and lipids that contain terminal GalNAc moieties. MGL represents an exclusive marker for myeloid-type APC. Dexamethasone increased MGL expression on DC in a time- and dose-dependent manner. In contrast, DC generated in the presence of IL-10 did not display enhanced MGL levels. Furthermore, dexamethasone and IL-10 also differentially regulated expression of other C-type lectins, such as DC-SIGN and mannose receptor. Results indicate that depending on the local microenvironment, DC can adopt different C-type lectin profiles, which could have major influences on cell-cell interactions, antigen uptake and presentation (van Vliet et al. 2006b).

33.10.3 Ligands of MGL

Ligands of Human MGL: Van Die et al. (2003) identified an exclusive specificity for terminal α- and β-linked GalNAc residues that naturally occur as parts of glycoproteins or glycosphingolipids. Specific glycan structures containing terminal GalNAc moieties, expressed by the human helminth parasite *Schistosoma mansoni* as well as tumor antigens and a subset of gangliosides, were identified as ligands for MGL. The dendritic cell-specific DC-SIGN is a receptor for Schistosoma 8 MGL recognition of terminal GalNAc residues (Van Die et al. 2003).

Exclusive specificity of human MGL for rare terminal GalNAc structures was revealed on the tumor-associated mucin MUC1 and CD45 on effector T cells. Tumor glycoproteins, such as carcinoembryonic antigen and MUC-1/ MUC1-Tn are known to interact with MGL on APCs. In vivo studies in mice demonstrated the potency of targeting antigens to C-type lectins on antigen-presenting cells for anti-tumor vaccination strategies (Aarnoudse et al. 2006). Glycosylation changes during malignant transformation create tumor-specific carbohydrate structures that interact with C-type lectins on DCs. The detection of MGL positive cells in situ at the tumor site together with the modified glycosylation status of MUC1 to target MGL on DC suggests that MGL positive antigen presenting cells may play a role in tumor progression (Saeland et al. 2007). Tumor-associated Tn-MUC1 glycoform is internalized through the MGL and delivered to the HLA class I and II compartments in dendritic cells (Napoletano et al. 2007). Soluble model antigens are efficiently internalized by MGL and subsequently presented to responder CD4+ T cells. The tyrosine-5 residue in the YENF motif, present in the MGL cytoplasmic domain, was essential for the MGL-mediated endocytosis in CHO cells. MGL contributes to the antigen processing and presentation capacities of DC and may provide a suitable target for the initiation of anti-tumor immune responses (van Vliet et al. 2007). Findings implicate MGL in the homeostatic control of adaptive immunity. van Vliet et al. (2008b) discussed the functional similarities and differences between MGL orthologs and compared MGL to its closest homolog, the liver-specific ASGP-R (van Vliet et al. 2008). Evidence indicates that both N-linked glycoproteins and distinct lipooligosaccharide glycoforms of *C. jejuni* are ligands for the human MGL and that the *C. jejuni* N-glycosylation machinery can be exploited to target recombinant bacteria to MGL-expressing eukaryotic cells (van Sorge et al. 2009).

Studies indicate the involvement of a C-type lectin, the MMR and the macrophage surface molecule MGL that binds influenza virus, both of which are known to be endocytic. Binding of influenza virus to MMR and MGL occurred independently of sialic acid through Ca^{2+}-dependent recognition of viral glycans by the CRDs of the two lectins. Thus, lectin-mediated interactions of influenza virus with the MMR or the MGL are required for the endocytic uptake of the virus into macrophages, and these lectins can thus be considered secondary or coreceptors with sialic acid for infection of this cell type (Upham et al. 2010).

Ligands of Murine MGLs: Despite the high similarity between the primary sequences of MGL1 and MGL2, they have different carbohydrate specificities, respectively, for Lewis X and a/b-GalNAc structures (Tsuiji et al. 2002). Earlier studies on COS-1 transfectants of MGL suggested a specificity for galactose and *N*-acetylgalactosamine as monosaccharides (Suzuki et al. 1996). In contrast, recombinant MGL displayed restricted binding to GalNAc (Iida et al. 1999). MGL1 shows high affinity for the Lewis-X trisaccharide among 111 oligosaccharides tested, whereas MGL2 preferentially bound globoside Gb4. Molecular modeling illustrated potential direct molecular interactions of Leu^{61}, Arg^{89}, and His^{109} in MGL2 CRD with GalNAc (Oo-Puthinan et al. 2008). NMR analyses of the MGL1-Lewis-X complex presented a Lewis-X binding mode on MGL1 where the galactose moiety is bound to the primary sugar binding site, including Asp^{94}, Trp^{96}, and Asp^{118}, and the fucose moiety interacts with the secondary sugar binding site, including Ala^{89} and Thr^{111}. Ala^{89} and Thr^{111} in MGL1 are replaced with arginine and serine in MGL2, respectively. The hydrophobic environment formed by a small side chain of Ala^{89} and a methyl group of Thr^{111} is a requisite for the accommodation of the fucose moiety of the Lewis-X trisaccharide within the sugar binding site of MGL1 (Sakakura et al. 2008; Oo-Puthinan et al. 2008). Using a glycan array, murine MGL1 was highly specific for Lewis-X and Lewis-A structures, whereas mMGL2, more similar to the human MGL, recognized GalNAc and galactose, including the O-linked Tn-antigen, TF-antigen and core 2. Strikingly, MGL2 interacted strongly to adenocarcinoma cells, suggesting a potential role in tumor immunity (Singh et al. 2009).

Specific glycan structures containing terminal GalNAc moieties, expressed by the human helminth parasite *Schistosoma mansoni* as well as tumor antigens and a subset of gangliosides, were identified as ligands for MGL. Results indicated an endogenous function for DC-expressed MGL in the clearance and tolerance to self-gangliosides, and in the pattern recognition of tumor antigens and foreign glycoproteins derived from helminth parasites (van Vliet et al. 2005).

33.10.4 Functions of MGL

The MGL on immature DCs is involved in mediating down-regulation of effector T cell function and T cell death by interaction with CD45 avoiding potentially harmful T cell activation (van Vliet et al. 2006a). Several lines of evidence suggest that MGL may also be involved in the trafficking of APC that express this α/β-GalNAc-specific lectin (van Vliet et al. 2005; Dupasquier et al. 2006; Chun et al. 2000a, b). MGL1/2-positive cells represent a distinct sub-population of macrophages, having unique functions in the generation and maintenance of granulation tissue induced by antigenic stimuli (Sato et al. 2005a). It is highly likely that MGL1-positive cells are not involved in tissue remodeling when inflammation is driven by nonspecific stimuli (Sato et al. 2005b). Mice DCs have two MGL genes, *Mgl1* and *Mgl2*. A report demonstrates the involvement of GalNAc residues in antigen uptake and presentation by DCs that lead to $CD4^+$ T cell activation.

Pregnant mice with $Mgl1^{+/-}$ genotype were mated with $Mgl1^{+/-}$ or $Mgl1^{-/-}$ genotype males, and the embryos were used to assess a hypothesis that this molecule plays an important role in the clearance of apoptotic cells. After X-ray irradiation at 1 Gy of developing embryos at 10.5 days post coitus (d.p.c.), the number of $Mgl1^{-/-}$ pups was significantly reduced as compared with $Mgl1^{+/+}$ pups. Results strongly suggest that MGL1 is involved in the clearance of apoptotic cells (Yuita et al. 2005).

MGL induction in Activated Macrophages by Parasitic Infections: The expression of the two members of the mouse mMGL1 and mMGL2 is induced in diverse populations of alternatively activated macrophages (aaMF), including peritoneal macrophages elicited during infection with the protozoan *Trypanosoma brucei brucei* or the Helminth *Taenia crassiceps* and alveolar macrophages elicited in a mouse model of allergic asthma. In addition, interleukin-4 (IL-4) and IL-13 up-regulate mMGL1 and mMGL2 expression in vitro, and that in vivo, induction of mMGL1 and mMGL2 is dependent on IL-4 receptor signaling. Moreover, expression of MGL on human monocytes is also up-regulated by IL-4 (Raes et al. 2005).

Initiation of Contact Hypersensitivity In Vivo: The MGL, expressed in immature DCs, mediates binding to glycoproteins carrying GalNAc moieties. MGL ligands are present on the sinusoidal and lymphatic endothelium of lymph node and thymus, respectively. MGL binding strongly correlated with the expression of the preferred MGL ligand, α-GalNAc-containing glycan structures. MGL$^+$ cells were localized in close proximity of the endothelial structures that express the MGL ligand. Strikingly, instead of inducing migration, MGL mediated retention of human immature DCs, as blockade of MGL interactions enhanced DC trafficking and migration. Thus, MGL$^+$ DCs are hampered in their migratory responses and only upon maturation, when MGL expression is abolished; these DCs will be released from their MGL-mediated restraints (van Vliet et al. 2008a).

Anti-Inflammatory Role in Murine Experimental Colitis: Inflammatory bowel disease is caused by abnormal inflammatory and immune responses to harmless substances, such as commensal bacteria, in the large bowel. MGL1 expressed on intestinal lamina propria macrophages functions through its inter-action with commensal bacteria by magnifying the IL-10 production by these cells. Results in $Mgl1^{-/-}$ mice and their wild-type littermates strongly suggest that MGL1/CD301a plays a protective role against colitis by effectively inducing IL-10 production by colonic lamina propria macrophages in response to invading commensal bacteria (Saba et al. 2009). Mgl1 is not required for the trafficking of type 2 Adipose tissue macrophages to adipose tissue. But MGL1 is a novel regulator of inflammatory monocyte trafficking to adipose tissue in response to diet-induced obesity (Westcott et al. 2009).

References

Aarnoudse CA, Garcia Vallejo JJ et al (2006) Recognition of tumor glycans by antigen-presenting cells. Curr Opin Immunol 18:105–111

Ashwell G, Harford J (1982) Carbohydrate-specific receptors of the liver. Annu Rev Biochem 51:531–554

Baenziger JU, Maynard Y (1980) Human hepatic lectin. Physiochemical properties and specificity. J Biol Chem 255:4607–4613

Becker S, Spiess M, Klenk H-D (1995) The asialoglycoprotein receptor is a potential liver-specific receptor for Marburg virus. J Gen Virol 76:393–399

Bider MD, Spiess M (1998) Ligand-induced endocytosis of the asialoglycoprotein receptor: evidence for heterogeneity in subunit oligomerization. FEBS Lett 434:37–41

Bider MD, Wahlberg JM, Kammerer RA et al (1996) The oligomerization domain of the asialoglycoprotein receptor preferentially forms 2:2 heterotetramers in vitro. J Biol Chem 271:31996–32001

Breitfeld PP, Simmons CF Jr, Strous GJ et al (1985) Cell biology of the asialoglycoprotein receptor system: a model of receptor-mediated endocytosis. Int Rev Cytol 97:47–95

Brown MS, Anderson RG, Goldstein JL (1983) Recycling receptors: the round-trip itinerary of migrant membrane proteins. Cell 32:663–667

Casey CA, Lee SM, Aziz-Seible R, McVicker BL (2008) Impaired receptor-mediated endocytosis: its role in alcohol-induced apoptosis. J Gastroenterol Hepatol 23(Suppl 1):S46–S49

Chiacchia KB, Drickamer K (1984) Direct evidence for the transmembrane orientation of the hepatic glycoprotein receptors. J Biol Chem 259:15440–15446

Chun KH, Imai Y, Higashi N, Irimur T (2000a) Migration of dermal cells expressing a macrophage C-type lectin during the sensitization phase of delayed-type hypersensitivity. J Leukocyte Biol 68:471–478

Chun KH, Imai Y, Higashi N, Irimura T (2000b) Involvement of cytokines in the skin-to-lymph node trafficking of cells of the monocyte-macrophage lineage expressing a C-type lectin. Int Immunol 12:1695–1703

Ciechanover A, Schwartz AL, Lodish HF (1983) Sorting and recycling of cell surface receptors and endocytosed ligands: the asialoglycoprotein and transferrin receptors. J Cell Biochem 23:107–130

Denda-Nagai K, Kubota N, Tsuiji M et al (2002) Macrophage C-type lectin on bone marrow–derived immature dendritic cells is involved in the internalization of glycosylated antigens. Glycobiology 7:443–450

Denda-Nagai K, Aida S, Saba K et al (2010) Distribution and function of macrophage galactose-type C-type lectin 2 (MGL2/CD301b): efficient uptake and presentation of glycosylated antigens by dendritic cells. J Biol Chem 285:19193–19204

Dini L, Autuori F, Lentini A et al (1992) The clearance of apoptotic cells in the liver is mediated by the asialoglycoprotein receptor. FEBS Lett 296:174–178

Dotzauer A, Gebhardt U, Bieback K et al (2000) Hepatitis A virus-specific immunoglobulin A mediates infection of hepatocytes with hepatitis A virus via the asialoglycoprotein receptor. J Virol 74:10950–10957

Drickamer K (1988) Two distinct classes of carbohydrate-recognition domains in animal lectins. J Biol Chem 263:9557–9560

Drickamer K, Mamon JF (1982) Phosphorylation of a membrane receptor for glycoproteins. Possible transmembrane orientation of the chicken hepatic lectin. J Biol Chem 257:15156–15161

Drickamer K, Mamon JF, Binns G, Leung JO (1984) Primary structure of the rat liver asialoglycoprotein receptor. Structural evidence for multiple polypeptide species. J Biol Chem 259:770–778

Dupasquier M, Stoitzner P, Wan H et al (2006) The dermal microenvironment induces the expression of the alternative activation marker CD301/mMGL in mononuclear phagocytes, independent of IL-4/IL-13 signaling. J Leukoc Biol 80:838–849

Evarts RP, Marsden ER, Thorgeirsson SS (1985) Modulation of asialoglycoprotein receptor levels in rat liver by phenobarbital treatment. Carcinogenesis 6:1767–1773

Feinberg H, Torgersen D, Drickamer K et al (2000) Mechanism of pH-dependent N-acetylgalactosamine binding by a functional mimic of the hepatocyte asialoglycoprotein receptor. J Biol Chem 275:35176–35184

Fiume L, Di Stefano G, Busi C et al (1997) Liver targeting of antiviral nucleoside analogues through the asialoglycoprotein receptor. J Viral Hepat 4:363–370

Fuhrer C, Geffen I, Spiess M (1991) Endocytosis of the ASGP receptor H1 is reduced by mutation of tyrosine-5 but still occurs via coated pits. J Cell Biol 114:423–431

Fuhrer C, Geffen I, Huggel K, Spiess M (1994) The two subunits of the asialoglycoprotein receptor contain different sorting information. J Biol Chem 269:3277–3282

Geffen I, Spiess M (1992) Asialoglycoprotein receptor. Int Rev Cytol 137B:181–219

Geuze HJ, Slot JW, Strous GJ et al (1984) Intracellular receptor sorting during endocytosis: comparative immunoelectron microscopy of multiple receptors in rat liver. Cell 37:195–204

Halberg DF, Wager RE, Farrell DC et al (1987) Major and minor forms of the rat liver asialoglycoprotein receptor are independent galactose-binding proteins. Primary structure and glycosylation heterogeneity of minor receptor forms. J Biol Chem 262:9828–9838

Hardy MR, Townsend RR, Parkhurst SM, Lee YC (1985) Different modes of ligand binding to the hepatic galactose/N-acetylgalactosamine lectin on the surface of rabbit hepatocytes. Biochemistry 24:22–28

Hashida M, Nishikawa M, Yamashita F et al (2001) Cell-specific delivery of genes with glycosylated carriers. Adv Drug Deliv Rev 52:187–196

Higashi N, Fujioka K, Denda-Nagai K et al (2002a) The macrophage C-type lectin specific for galactose/N-acetylgalactosamine is an endocytic receptor expressed on monocyte-derived immature dendritic cells. J Biol Chem 277:20686–20693

Higashi N, Morikawa A, Fujioka K et al (2002b) Human macrophage lectin specific for galactose/N-acetylgalactosamine is a marker for cells at an intermediate stage in their differentiation from monocytes into macrophages. Int Immunol 14:545–554

Hilgard P, Schreiter T, Stockert RJ et al (2004) Asialoglycoprotein receptor facilitates hemolysis in patients with alcoholic liver cirrhosis. Hepatology 39:1398–1407

Ii M, Kurata H, Itoh N et al (1990) Molecular cloning and sequence analysis of cDNA encoding the macrophage lectin specific for galactose and N-acetylgalactosamine. J Biol Chem 265:11295–11298

Iida S, Yamamoto K, Irimura T (1999) Interaction of human macrophage C-type lectin with O-linked N-acetylgalactosamine residues on mucin glycopeptides. J Biol Chem 274:10697

Imai Y, Irimura T (1994) Quantitative measurement of carbohydrate binding activity of mouse macrophage lectin. J Immunol Methods 171:23–31

Iobst ST, Drickamer K (1994) Binding of sugar ligands to Ca(2+)-dependent animal lectins. II. Generation of high-affinity galactose binding by site-directed mutagenesis. J Biol Chem 269:15512–15519

Iobst ST, Drickamer K (1996) Selective sugar binding to the carbohydrate recognition domains of the rat hepatic and macrophage asialoglycoprotein receptors. J Biol Chem 271:6686–6693

Ishibashi S, Hammer RE, Herz J (1994) Asialoglycoprotein receptor deficiency in mice lacking the minor receptor subunit. J Biol Chem 269:27803–27806

Ishibashi S, Perrey S, Chen Z et al (1996) Role of the low density lipoprotein (LDL) receptor pathway in the metabolism of chylomicron remnants. A quantitative study in knockout mice lacking the LDL receptor, apolipoprotein E, or both. J Biol Chem 271:22422–22427

Johansson AK (2007) Linking structure and function of the asialoglycoprotein receptor H1-CRD using site-directed mutagenesis and isotope labeling, Inauguraldissertation, zur Erlangung der Würde eines Doktors der Philosophie, vorgelegt der Philosophisch-Naturwissenschaftlichen. Fakultät der Universität, Basel

Kokudo N, Vera DR, Makuuchi M (2003) Clinical application of TcGSA. Nucl Med Biol 30:845–849

Kolatkar AR, Weis WI (1996) Structural basis of galactose recognition by C-type animal lectins. J Biol Chem 271:6679–6685

Kolatkar AR, Leung AK, Isecke R et al (1998) Mechanism of N-acetylgalactosamine binding to a C-type animal lectin carbohydrate-recognition domain. J Biol Chem 273:19502–19508

Kumamoto Y, Higashi N, Denda-Nagai K et al (2004) Identification of sialoadhesin as a dominant lymph node counter-receptor for mouse macrophage galactose-type C-type lectin 1. J Biol Chem 279:49274–49280

Kumamoto Y, Denda-Nagai K, Aida S et al (2009) MGL2$^+$ dermal dendritic cells are sufficient to initiate contact hypersensitivity in vivo. PLoS One 4:e5619

Lee SM, Casey CA, McVicker BL (2009) Impact of asialoglycoprotein receptor deficiency on the development of liver injury. World J Gastroenterol 15:1194–1200

Loeb JA, Drickamer K (1988) Conformational changes in the chicken receptor for endocytosis of glycoproteins. Modulation of ligand-binding activity by Ca2+ and pH. J Biol Chem 263:9752–9760

Matsuura S, Nakada H, Sawamura T, Tashiro Y (1982) Distribution of an asialoglycoprotein receptor on rat hepatocyte cell surface. J Cell Biol 95:864–875

McVicker BL, Tuma DJ, Kubik JA et al (2002) The effect of ethanol on asialoglycoprotein receptor-mediated phagocytosis of apoptotic cells by rat hepatocytes. Hepatology 36:1478–1487

Meier M, Bider MD, Malashkevich VN et al (2000) Crystal structure of the carbohydrate recognition domain of the H1 subunit of the asialoglycoprotein receptor. J Mol Biol 300:857–865

Mellman I, Fuchs R, Helenius A (1986) Acidification of the endocytic and exocytic pathways. Annu Rev Biochem 55:663–700

Mizuochi S, Akimoto Y, Imai Y et al (1997) Unique tissue distribution of a mouse macrophage C-type lectin. Glycobiology 7:137–146

Morell AG, Irvine RA, Sternlieb I, Scheinberg IH, Ashwell G (1968) Physical and chemical studies on ceruloplasmin. V. Metabolic studies on sialic acid-free ceruloplasmin in vivo. J Biol Chem 243:155–159

Napoletano C, Rughetti A, Agervig Tarp MP et al (2007) Tumor-associated Tn-MUC1 glycoform is internalized through the macrophage galactose-type C-type lectin and delivered to the HLA class I and II compartments in dendritic cells. Cancer Res 67:8358–8367

Onami TM, Lin MY, Page DM et al (2002) Generation of mice deficient for macrophage galactose- and N-acetylgalactosamine-specific lectin: limited role in lymphoid and erythroid homeostasis and evidence for multiple lectins. Mol Cell Biol 22:5173–5181

Oo-Puthinan S, Maenuma K, Sakakura M et al (2008) The amino acids involved in the distinct carbohydrate specificities between macrophage galactose-type C-type lectins 1 and 2 (CD301a and b) of mice. Biochim Biophys Acta 1780:89–100

Owada T, Matsubayashi K, Sakata H et al (2006) Interaction between desialylated hepatitis B virus and asialoglycoprotein receptor on hepatocytes may be indispensable for viral binding and entry. J Viral Hepat 13:11–18

Park JH, Cho EW, Shin SY et al (1998) Detection of the asialoglycoprotein receptor on cell lines of extrahepatic origin. Biochem Biophys Res Commun 244:304–311

Park JH, Kim KL, Cho EW (2006) Detection of surface asialoglycoprotein receptor expression in hepatic and extra-hepatic cells using a novel monoclonal antibody. Biotechnol Lett 28:1061–1069

Pathak A, Vyas SP, Gupta KC (2008) Nano-vectors for efficient liver specific gene transfer. Int J Nanomedicine 3:31–49

Raes G, Brys L, Dahal BK et al (2005) Macrophage galactose-type C-type lectins as novel markers for alternatively activated macrophages elicited by parasitic infections and allergic airway inflammation. J Leukoc Biol 77:321–327

Rensen PC, Sliedregt LA, Ferns M et al (2001) Determination of the upper size limit for uptake and processing of ligands by the asialoglycoprotein receptor on hepatocytes in vitro and in vivo. J Biol Chem 276:37577–37584

Rifai A, Fadden K, Morrison SL et al (2000) The N-glycans determine the differential blood clearance and hepatic uptake of human immunoglobulin (Ig)A1 and IgA2 isotypes. J Exp Med 191:2171–2182

Rotundo RF, Rebres RA, Mckeown-Longo PJ et al (1998) Circulating cellular fibronectin may be a natural ligand for the hepatic asialoglycoprotein receptor: possible pathway for fibronectin deposition and turnover in the rat liver. Hepatology 28:475–485

Saba K, Denda-Nagai K, Irimura T (2009) A C-Type Lectin MGL1/CD301a plays an anti-inflammatory role in murine experimental colitis. Am J Pathol 174:144–152

Saeland E, van Vliet SJ, Bäckström M et al (2007) The C-type lectin MGL expressed by dendritic cells detects glycan changes on MUC1 in colon carcinoma. Cancer Immunol Immunother 56:1225–1236

Sakakura M, Oo-Puthinan S, Moriyama C et al (2008) Carbohydrate binding mechanism of the macrophage galactose-type C-type lectin 1revealed by saturation transfer experiments. J Biol Chem 283:33665–33673

Sanford JP, Doyle D (1990) Mouse asialoglycoprotein receptor cDNA sequence: conservation of receptor genes during mammalian evolution. Biochim Biophys Acta 1087:259–261

Sato M, Kawakami K, Osawa T, Toyoshima S (1992) Molecular cloning and expression of cDNA encoding a galactose/N-acetylgalactosamine-specific lectin on mouse tumoricidal macrophages. J Biochem (Tokyo) 111:331–336

Sato K, Imai Y, Higashi N et al (2005a) Lack of antigen-specific tissue remodeling in mice deficient in the macrophage galactose-type calcium-type lectin 1/CD301a. Blood 106:207–215

Sato K, Imai Y, Higashi N et al (2005b) Redistributions of macrophages expressing the macrophage galactose-type C-type lectin (MGL) during antigen-induced chronic granulation tissue formation. Int Immunol 17:559–568

Saunier B, Triyatni M, Ulianich L et al (2003) Role of the asialoglycoprotein receptor in binding and entry of hepatitis C virus structural proteins in cultured human hepatocytes. J Virol 77:546–559

Saxena A, Yik JH, Weigel PH (2002) H2, the minor subunit of the human asialoglycoprotein receptor, trafficks intracellularly and forms homo-oligomers, but does not bind asialo-orosomucoid. J Biol Chem 277:35297–35304

Schwartz AL (1984) The hepatic asialoglycoprotein receptor. CRC Crit Rev Biochem 16:207–233

Schwartz AL, Marshak-Rothstein A, Rup D, Lodish HF (1981) Identification and quantification of the rat hepatocyte asialoglycoprotein receptor. Proc Natl Acad Sci USA 78:3348–3352

Seow YY, Tan MG, Woo KT (2002) Expression of a functional asialoglycoprotein receptor in human renal proximal tubular epithelial cells. Nephron 91:431–438

Shia MA, Lodish HF (1989) The two subunits of the human asialoglycoprotein receptor have different fates when expressed alone in fibroblasts. Proc Natl Acad Sci USA 86:1158–1162

Singh SK, Streng-Ouwehand I, Litjens M et al (2009) Characterization of murine MGL1 and MGL2 C-type lectins: distinct glycan specificities and tumor binding properties. Mol Immunol 46:1240–1249

Spiess M (1990) The asialoglycoprotein receptor: a model for endocytic transport receptors. Biochemistry 29:10009–10018

Spiess M, Handschin C (1987) Deletion analysis of the internal signal-anchor domain of the human asialoglycoprotein receptor H1. EMBO J 6:2683–2691

Spiess M, Lodish HF (1985) Sequence of a second human asialoglycoprotein receptor: conservation of two receptor genes during evolution. Proc Natl Acad Sci USA 82:6465–6469

Steinman RM (2003) The control of immunity and tolerance by dendritic cell. Pathol Biol 51:59–60

Steinman R, Hawiger MD, Nussenzweig MC (2003) Tolerogenic dendritic cells. Annu Rev Immunol 21:685–711

Stockert RJ (1995) The asialoglycoprotein receptor: relationships between structure, function, and expression. Physiol Rev 75:591–609

Suzuki N, Yamamoto K, Toyoshima S et al (1996) Molecular cloning and expression of cDNA encoding human macrophage C- type lectin. Its unique carbohydrate binding specificity for Tn antigen. J Immunol 156:128–135

Tozawa R, Ishibashi S, Osuga J et al (2001) Asialoglycoprotein receptor deficiency in mice lacking the major receptor subunit. Its obligate requirement for the stable expression of oligomeric receptor. J Biol Chem 276:12624–12628

Treichel U, McFarlane BM, Seki T et al (1994) Demographics of anti-asialoglycoprotein receptor autoantibodies in autoimmune hepatitis. Gastroenterology 107:799–804

Trerè D, Fiume L, De Giorgi LB et al (1999) The asialoglycoprotein receptor in human hepatocellular carcinomas: its expression on proliferating cells. Br J Cancer 81:404–408

Tsuiji M, Fujimori M, Seldin MF et al (1999) Genomic structure and chromosomal location of the mouse macrophage C-type lectin gene. Immunogenetics 50:67–70

Tsuiji M, Fujimori M, Ohashi Y et al (2002) Molecular cloning and characterization of a novel mouse macrophage C-type lectin, mMGL2, which has a distinct carbohydrate specificity from mMGL1. J Biol Chem 277:28892–28901

Upham JP, Pickett D, Irimura T, Anders EM, Reading PC (2010) Macrophage receptors for influenza A virus: role of the macrophage galactose-type lectin and mannose receptor in viral entry. J Virol 84:3730–3737

Valladeau J, Duvert-Frances V, Pin JJ et al (2001) Immature human dendritic cells express asialoglycoprotein receptor isoforms for efficient receptor-mediated endocytosis. J Immunol 167:5767–5774

Van Die I, van Vliet SJ, Nyame AK et al (2003) The dendritic cell-specific C-type lectin DC-SIGN is a receptor for Schistosoma mansoni egg antigens and recognizes the glycan antigen Lewis x. Glycobiology 13:471

van Sorge NM, Bleumink NM, van Vliet SJ et al (2009) N-glycosylated proteins and distinct lipooligosaccharide glycoforms of Campylobacter jejuni target the human C-type lectin receptor MGL. Cell Microbiol 11:1768–1781

van Vliet SJ, van Liempt E, Saeland E et al (2005) Carbohydrate profiling reveals a distinctive role for the C-type lectin MGL in the recognition of helminth parasites and tumor antigens by dendritic cells. Int Immunol 17:661–669

van Vliet SJ, Gringhuis SI, Geijtenbeek TB, van Kooyk Y (2006a) Regulation of effector T cells by antigen-presenting cells via interaction of the C-type lectin MGL with CD45. Nat Immunol 7:1200–1208

van Vliet SJ, van Liempt E, Geijtenbeek TB, van Kooyk Y (2006b) Differential regulation of C-type lectin expression on tolerogenic dendritic cell subsets. Immunobiology 211:577–585

van Vliet SJ, Aarnoudse CA, Broks-van den Berg VC et al (2007) MGL-mediated internalization and antigen presentation by dendritic cells: a role for tyrosine-5. Eur J Immunol 37:2075–2081

van Vliet SJ, Paessens LC, Broks-van den Berg VC et al (2008a) The C-type lectin macrophage galactose-type lectin impedes migration of immature APCs. J Immunol 18:3148–3155

van Vliet SJ, Saeland E, van Kooyk Y (2008b) Sweet preferences of MGL: carbohydrate specificity and function. Trends Immunol 29:83–90

Wall DA, Hubbard AL (1981) Galactose-specific recognition system of mammalian liver: receptor distribution on the hepatocyte cell surface. J Cell Biol 90:687–696

Warren R, Doyle D (1981) Turnover of the surface proteins and the receptor for serum asialoglycoproteins in primary cultures of rat hepatocytes. J Biol Chem 256:1346–1355

Weigel PH (1994) Galactosyl and N-acetylgalactosaminyl homeostasis: a function for mammalian asialoglycoprotein receptors. Bioessays 16:519–524

Weigel PH, Yik JH (2002) Glycans as endocytosis signals: the cases of the asialoglycoprotein and hyaluronan/chondroitin sulfate receptors. Biochim Biophys Acta 1572:341–363

Weis WI, Drickamer K, Hendrickson WA (1992) Structure of a C-type mannose-binding protein complexed with an oligosaccharide. Nature 360(6400):127–134

Westcott DJ, Delproposto JB, Geletka LM et al (2009) MGL1 promotes adipose tissue inflammation and insulin resistance by regulating 7/4hi monocytes in obesity. J Exp Med 206:3143–3156

Westerlind U, Westman J, Törnquist E et al (2004) Ligands of the asialoglycoprotein receptor for targeted gene delivery, part 1: Synthesis of and binding studies with biotinylated cluster glycosides containing N- acetylgalactosamine. Glycoconj J 21: 227–241

Wragg S, Drickamer K (1999) Identification of amino acid residues that determine pH dependence of ligand binding to the asialoglycoprotein receptor during endocytosis. J Biol Chem 274: 35400–35406

Wu J, Liu P, Zhu JL, Maddukuri S, Zern MA (1998) Increased liver uptake of liposomes and improved targeting efficacy by labeling with asialofetuin in rodents. Hepatology 27:772–778

Wu J, Nantz MH, Zern MA (2002) Targeting hepatocytes for drug and gene delivery: emerging novel approaches and applications. Front Biosci 7:d717–d725

Yamamoto K, Ishida C, Shinohara Y et al (1994) Interaction of immobilized recombinant mouse C-type macrophage lectin with glycopeptides and oligosaccharides. Biochemistry 33:8159–8166

Yik JH, Weigel PH (2002) The position of cysteine relative to the transmembrane domain is critical for palmitoylation of H1, the major subunit of the human asialoglycoprotein receptor. J Biol Chem 277:47305–47312

Yik JHN, Saxena A, Weigel PH (2002) The minor subunit splice variants, H2b and H2c, of the human asialoglycoprotein receptor are present with the major subunit H1 in different hetero-oligomeric receptor complexes. J Biol Chem 277:23076–23083

Yuita H, Tsuiji M, Tajika Y et al (2005) Retardation of removal of radiation-induced apoptotic cells in developing neural tubes in macrophage galactose-type C-type lectin-1-deficient mouse embryos. Glycobiology 15:1368–1375

Zelensky AN, Gready JE (2005) The C-type lectin-like domain superfamily. FEBS J 272:6179–6217

Zeng FY, Weigel PH (1996) Fatty acylation of the rat and human asialoglycoprotein receptors. A conserved cytoplasmic cysteine residue is acylated in all receptor subunits. J Biol Chem 271:32454–32460

Zeng FY, Kaphalia BS, Ansari GAS, Weigel PH (1995) Fatty acylation of the rat asialoglycoprotein receptor. The three subunits from active receptors contain covalently bound palmitate and stearate. J Biol Chem 270:21382–21387

Dectin-1 Receptor Family

Rajesh K. Gupta and G.S. Gupta

34.1 Natural Killer Gene Complex (NKC)

Natural killer (NK) cell receptors belong to two unrelated, but functionally analogous gene families: the immunoglobulin superfamily, situated in the leukocyte receptor complex (LRC) and the C-type lectin receptors (CLRs) superfamily, located in the natural killer gene complex (NKC). Wong et al. (2009) described the largest NK receptor gene expansion seen to date and identified 213 putative C-type lectin NK receptor homologs in the genome of the platypus. Many have arisen as the result of a lineage-specific expansion. Orthologs of OLR1, CD69, KLRE, CLEC12B, and CLEC16p genes were also identified. The NKC is split into at least two regions of the genome: 34 genes map to chromosome 7, two map to a small autosome, and the remainder are unanchored in the current genome assembly. No NK receptor genes from the LRC were identified. The massive C-type lectin expansion and lack of Ig-domain-containing NK receptors represents the most extreme polarization of NK receptors found to date. This new data from platypus was utilized to trace the possible evolutionary history of the NK receptor clusters.

The myeloid cluster within NK gene complex comprises several CLRs genes of diverse and highly important functions in the immune system such as LOX-1 and DECTIN-1. The type II transmembrane CLRs are best known for their involvement in the detection of virally infected or transformed cells, through the recognition of endogenous (or self) proteinacious ligands. However, certain CLR families within the NKC, particularly those expressed by myeloid cells, recognize structurally diverse ligands and perform a variety of other immune and homoeostatic functions. One such family is the 'Dectin-1 cluster' of CLRs, which includes MICL (CLEC12A), CLEC-2 (CLEC1B), CLEC12B, CLEC9A, CLEC-1(CLEC-1A), in addition to Dectin-1 (CLEC7A), and LOX-1 (Fig. 34.1). Current understanding of these CLRs, high-lighting their ligands, functions and new insights into the underlying mechanisms of immunity and homeostasis has been reviewed by Huysamen and Brown (2009) and given in Table 34.1. The arrangement of genes within the primate cluster differs from the order and orientation of the corresponding genes in the rodent complex which can be explained by evolutionary duplication and inversion events. Analysis of individual genes revealed a high sequence conservation supporting the prime importance of the encoded proteins (Sattler et al. 2010; Sobanov et al. 2001).

34.2 β-Glucan Receptor (Dectin 1) (CLEC7A or CLECSF12)

β-1—>3-D-Glucans are biological response modifiers with potent effects on the immune system. A number of receptors are thought to play a role in mediating these responses, including murine Dectin-1, which was identified as a β-glucan receptor in humans. The receptor possessed a single C-type lectin-like CRD connected to the transmembrane region by a stalk and a cytoplasmic tail possessing an immunoreceptor tyrosine-based activation motif (ITAM). Dectin-1 is widely expressed in mouse tissues and acts as a pattern recognition receptor, recognizing a variety of carbohydrates containing β-1—>3- and/or β-1—>6-glucan linkages and intact *Saccharomyces cerevisiae* and *Candida albicans* (Brown and Gordon 2001). Dectin-2 and its isoforms, together with dectin-1, represent a unique subfamily of DC-associated C-type lectins (Ariizumi et al. 2000a).

34.2.1 Characterization of β-Glucan Receptor (Dectin 1)

Dectin-1 gene encodes a type II membrane-integrated polypeptide of 244 amino acids containing a single carbohydrate recognition domain motif at the COOH-terminal end. This molecule was expressed abundantly at both mRNA and protein levels by the XS52 DC line, but not by non-DC lines (Taylor et al. 2002). Dectin-1 mRNA was detected predominantly in spleen and thymus and in skin-resident DC, i.e. Langerhans cells. Dectin-1 was identified a 43-kDa glycoprotein in membrane fractions of XS52 DC line and from dectin-1 cDNA-transfected COS-1 cells. In vitro results suggest that dectin-1 on DC may bind to a ligand(s) on T cells, thereby delivering T cell co-stimulatory signals (Ariizumi et al. 2000b). Mouse Dectin 1 has a CRD consisting of six cysteine residues, which are highly conserved in C-type lectins. The human homologue of Dectin 1 is structurally and functionally similar to the mouse receptor. The human β-glucan receptor is a type II transmembrane receptor with a single extracellular CRD and an ITAM in its cytoplasmic tail.

Dectin-1 is regulated by different immune stimuli; GM-CSF and IL-4 up-regulate surface expression, whereas IL-10 and LPS down-regulate the expression (Willment et al. 2003). Murine dectin-1 has been suggested to have an alternative splice form that lacks the stalk region (Yokota et al. 2001). In humans, eight isoforms of the homologue of dectin-1 have been described (Willment et al. 2003). The most common isoforms, BGR-A and BGR-B (β-glucan receptor/dectin-1), represent full-length and stalkless isoforms, respectively, and both mediate the recognition of yeast particles in a β-glucan-dependent manner. The human homologue is expressed by cells similar to those in the mouse as well as by peripheral B cells and eosinophils (Willment et al. 2005). Human dectin-1 undergoes cell-specific isoform expression during monocyte maturation. Monocytes express both BGR-A and BGR-B, but during maturation to Mφ, the expression levels of BGR-A decrease with time. Immature DCs express high levels of both isoforms; however, DCs lose the expression of both isoforms when matured with LPS (Willment et al. 2005). RT-PCR analysis revealed that mice have at least two splice forms of dectin-1, generated by differential usage of exon 3, encoding the full-length dectin-1A and a stalkless MΦ dectin-1B. MΦ from BALB/c mice and genetically related mice expressed both isoforms in similar amounts, whereas MΦ from C57BL/6 and related mice mainly expressed the smaller isoform. Evidence suggests that dectin-1 isoforms are functionally distinct and indicate that differential isoform usage may represent a mechanism of regulating cellular responses to β-glucans (Heinsbroek et al. 2006). The Dectin-1 gene was

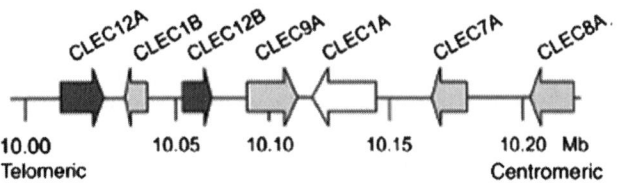

Fig. 34.1 Genomic organization of 'Dectin-1 cluster' in NKC on human chromosome 12. Activation receptors are shown in *light grey*, inhibitory receptors in *dark grey*, and those whose function is unclear are shown in *white* (Adapted with permission from Huysamen and Brown 2009 © John Wiley and Sons)

Table 34.1 Selected ligands and expression profiles of the Dectin-1 cluster of C-type lectin receptors (Adapted and modified from Huysamen and Brown 2009 © John Wiley and Sons)

Official name	Alternative name	Cell expression	Exogenous ligands	Endogenous ligands	Functions
CLEC7A	Dectin 1	Myeloid cells, B cells, Mast cells, T cell subsets, eosinophils	β-Glucon, mycobacterial ligands	T cells	Activation, apoptoic cells
CLEC12A	MICL, DCAL2, CLL1, KLRL1	Myeloid cells	?	Yes, identity unknown	Inhibition
CLEC1B	CLEC2	Platelets, myeloid cells, B cells, CD8$^+$ T cells, NK cells in BM[a]	Rhodocytin, HIV	Podoplanin RACK1[b]	Activation, modulation of platelet activity
CLEC12B	Macrophage Antigen H	Macrophages	?	?	Inhibition
CLEC9A	DNGR1	BDCA3$^+$ DC, monocyte subsets, B-cells	?	?	Activation
CLEC1A	CLEC-1	DC	?	?	?
CLEC8A	Lox-1	Endothelium, smooth muscle, platelets, fibroblasts, macrophgages	Gram positive and Gram negative bacteria	ox-LDL, modified lipoproteins aged/apoptotic cells, AGE-products, HSP70	Activation

'?' refers to unknown
BM bone marrow, *RACK1* the receptor for activated C-kinase-1, *AGE-products* advanced glycation end products

localized in NKC on human Chromosome 12p12.3-p13.2, between OLR1 and CD94 (position 21.8 cM on genetic map). The Dectin-1 gene is highly expressed at the mRNA level in DCs and is not further up-regulated during the maturation of DCs (Hermanz-Falcón et al. 2001). The amino acid sequence Try^{221}-Ile^{222}-His^{223} seemed critical for formation of a β-glucan binding site in CRD of dectin 1 (Adachi et al. 2004).

34.2.2 Crystal Structure of Dectin-1

Brown et al. (2007) reported dectin-1 crystal structures, including a short soaked natural β-glucan, trapped in the crystal lattice. In vitro characterization of dectin-1 in presence of its natural ligand indicates higher-order complex between dectin-1 and β-glucans. These combined structural and biophysical results considerably extend the current knowledge of dectin-1 structure and function, and suggest potential mechanisms of defense against fungal pathogens. Amino acid analysis indicates that dectin-1 is a 28-kDa type II membrane protein. An extracellular C-type lectin-like domain (CTLD) is connected by a stalk to a transmembrane region, followed by a cytoplasmic tail containing an ITAM-like motif (Ariizumi et al. 2000b). First recognized as a calcium-dependent carbohydrate-binding domain, the CTLD fold is also seen in non-calcium dependent protein recognition interactions. Few of the residues required for calcium coordination in classical CTLDs are conserved in dectin-1. In classical CTLDs, the long loop region (LLR) contains residues responsible for calcium binding, but these residues are absent in dectin-1 and no metal ions stabilize the LLR. As expected from sequence analysis, the dectin-1 fold is similar to that of other long-form CTLDs (Brown et al. 2007) (Fig. 34.2), comprising two antiparallel β-sheets and two α-helices.

The N and C termini are close together with domain integrity maintained by three disulphide bridges—Cys^{119}-Cys^{130}, Cys^{147}-Cys^{240}, and Cys^{219}-Cys^{232}—stabilizing the LLR. For their CTLDs alone, the residue identity between murine dectin-1 and its human homolog β-glucon receptor is 59.5%, implying that the structure of murine dectin-1 provides a reasonably good model for the structure of the human protein. Among several hydrophobic side chains exposed to solvent, Trp^{221} and His^{223} deserve attention and are in agreement with results of Adachi et al. (2004), who suggested that side groups of these amino acids participate in ligand binding and are conserved in all Dectin-1 homologues. Xray analysis of Dectin-1 structure revealed a shallow surface groove running between the Trp^{221} and His^{223} side chains (Fig. 34.2).

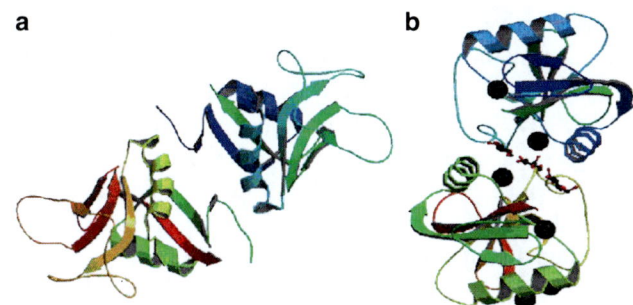

Fig. 34.2 Crystal structure of murine Dectin-1 (known as β–glucon receptor in humans). Two dectin-1 monomers (**a**) form a dimer into which a short β-glucan binds (**b**). (**a**) A cartoon diagram of the dectin-1 dimer (PDB: 2BPD), with each monomer from N terminus to C terminus. (**b**) Dectin-1 in complex with β-glucan (PDB: 2CL8). Disulphide linkages, the metal ion, and the bound β-glucan are shown

34.2.3 Interactions of Dectin-1 with Natural or Synthetic Glucans

Dectin-1 (CLEC7A) specifically recognizes β-D-(1—>3)-glucan, a polysaccharide and a component of fungal cell wall. As a result, dectin-1 is involved in recognition of fungi such as *C. albicans* and *Aspergillus fumigatus*. Due to the presence of (1—>3)-β-D-glucans on fungal cell walls, dectin 1 is considered important for recognizing fungal invasion. Although Dectin-1 is highly specific for β-D-(1—>3)-glucans, it does not recognize all glucans equally and interacts with β-D-(1—>3)-glucan over a very wide range of binding affinities (2.6 mM-2.2 pM). Among different polysachharides, Dectin-1 binding was detected exclusively to (1—>3)-linked glucose oligomers, the minimum length required for detectable binding being a 10- or 11-mer (Palma et al. 2006; Ujita et al. 2009). Dectin-1 can differentiate between glucan ligands based on structural determinants and can interact with both natural product and synthetic glucan ligands (Adams et al. 2008). Dectin-1 lacks residues involved in calcium ligation that mediates carbohydrate-binding by classical C-type lectins; nevertheless, it binds zymosan, a particulate β-glucan-rich extract of *Saccharomyces cerevisiae*, and binding is inhibited by polysaccharides rich in (1—>3)-β or both (1—>3)-β/(1—>6)-linked glucose (Palma et al. 2006). Dectin-1 is responsible for β-glucan-dependent, nonopsonic recognition of zymosan by primary macrophages and is an important target for examining the immunomodulatory properties of β-glucans for therapeutic drug design (Brown et al. 2002).

34.2.4 Regulation of Dectin-1

In mice, the expression of Dectin-1 can be influenced by various cytokines, steroids and microbial stimuli.

Interleukin-4 and IL-13, for example, which are associated with the alternative activation of macrophages, markedly increase the expression of Dectin-1 at the cell surface, whereas LPS and dexamethasone repress Dectin-1 expression (Willment et al. 2003). However, in *Pneumocystis*-infected mice, the mRNA levels of Dectin-1 gene decreased in AMΦs. This was associated with the decreased expression of Dectin-1 on the surface of these cells and reduced expression of mRNA of transcription factor PU.1 in AMΦs from *Pneumocystis*-infected mice. Down-regulation of PU.1 during *Pneumocystis* pneumonia appears to decrease the expression of Dectin-1 in AMΦs (Zhang et al. 2010). Leukocyte Dectin-1 levels are modulated in response to infections of fungal and nonfungal origin (Ozment-Skelton et al. 2009). Glucan phosphate (GP) resulted in a significant reduction in peripheral leukocyte membrane-associated Dectin-1 positivity. The systemic administration of GP has a specific and prolonged effect on loss of leukocyte membrane Dectin-1 positivity (Ozment-Skelton et al. 2006).

34.2.5 Signaling Pathways by Dectin-1

34.2.5.1 Raf-1 and CARD9 Dependent Pathways

Dectin-1-induced-signaling leads to the production of cytokines and non-opsonic phagocytosis of yeast by murine macrophages (LeibundGut-Landmann et al. 2007; Herre et al. 2004). Studies indicate that the alveolar macrophage inflammatory response, specifically the production of TNF-α, IL-1α, IL-1β, IL-6, CXCL2, CCL3-CSF, and GM-CSF, to live *A. fumigatus* is dependent on recognition via Dectin-1. The Dectin-1 is involved in calcium-independent recognition of β-(1—>3)-glucans exposed on particles such as zymosan, or many fungal species, including *Saccharomyces, Pneumocystis, Aspergillus* and *Candida* (Brown 2006; Brown et al. 2003; Heinsbroek et al. 2008). Signaling pathway activation of Dectin-1 depends on its ITAM the phosphorylation of which by Src kinase leads to the recruitment of spleen tyrosine kinase Syk in macrophages. Studies suggest that the increase in Dectin-1 expression by IL-4 involved the STAT signaling pathway (Murray 2007). Moreover, genetic models of macrophage specific peroxisome proliferator-activated receptor γ (PPARγ) or STAT-6 knockout mice showed that IL-4/IL-13/STAT-6/PPARγ axis is required for the maturation of alternatively activated macrophages (Odegaard et al. 2007; Ricote et al. 1998; Galès et al. 2010). In vitro and in vivo Dectin-1 is essential both to trigger the phagocytosis of non-opsonized *C. albicans* and the respiratory burst after yeast challenge and to control fungal gastrointestinal infection (Galès et al. 2010); the ManR alone is not sufficient to trigger antifungal functions during macrophage alternative activation.

Interestingly, Dectin-1 and ManR were increased by IL-13 through the activation of the nuclear receptor PPARγ, suggesting that PPARγ could be a therapeutic target to eliminate fungal infection.

The intracellular signaling of Dectin-1 has been demonstrated to be mediated mainly by Raf-1 and Syk-adaptor molecule *C*aspase *A*ctivating *R*ecruitment *D*omain 9 (CARD9) dependent pathways to induce production of pro-inflammatory cytokines and reactive oxygen species (Underhill et al. 2005; Gringhuis et al. 2009; Rogers et al. 2005; Gross et al. 2006; Hara et al. 2007). Upon activation, dectin-1 recruits spleen Syk which in turn activates NF-κB, requiring the CARD9, a key adaptor for non-TLR signal transduction (Gross et al. 2006) and also has a critical function in NOD2-mediated activation of kinases p38 and Jnk, required for the production of pro-inflammatory cytokines in innate immune responses to intracellular pathogens (Hsu et al. 2007). LeibundGut-Landmann et al. (2007) showed that Dectin-1-Syk-CARD9 signaling induces DC maturation and secretion of pro-inflammatory cytokines like IL-6, TNF-α, IL-17 and IL-23 (Fig. 34.3). Several studies suggest that dectin-1 converges with Toll Like Receptor (TLR) signaling (Ferwerda et al. 2008; Dennehy et al. 2008) for the induction of cytokine responses and is able to promote Th17 and cytotoxic T-cell responses through activation of DCs (Gerosa et al. 2008; Leibundgut-Landmann et al. 2007; Gow et al. 2007). There is also evidence of Syk-dependent, but CARD9-independent, pathways, such as those leading to the induction of ERK, a MAP kinase regulating the Dectin-1-mediated production of cytokines, particularly IL-10 and IL-2 (Slack et al. 2007; Dillon et al. 2006).

Dectin-1 can also induce intracellular signaling through Syk-independent pathways. Phagocytosis in macrophages, for example, does not require Syk, although this response still involves the ITAM-like motif of the receptor (Brown 2006). These pathways are still largely uncharacterised, but Dectin-1 was found to induce a Syk-independent pathway involving the serine–threonine kinase Raf-1 (Gringhuis et al. 2009). This pathway was shown to integrate with the Syk pathway, at the level of NF-κB, and to be involved in controlling Dectin-1 mediated cytokine production. Thus, dectin-1 activates two independent signaling pathways, one through Syk and one through Raf-1, to induce immune responses.

Dectin-1 signaling can also directly modulate gene expression via activation of NFAT. Dectin-1-triggered NFAT activation plays a role in the induction of early growth response transcription factors, and cyclooxygenase-2. Furthermore, NFAT activation regulates IL-2, IL-10 and IL-12 p70 production by zymosan-stimulated DCs. This study establishes NFAT activation in myeloid cells as a novel mechanism of regulation of innate antimicrobial

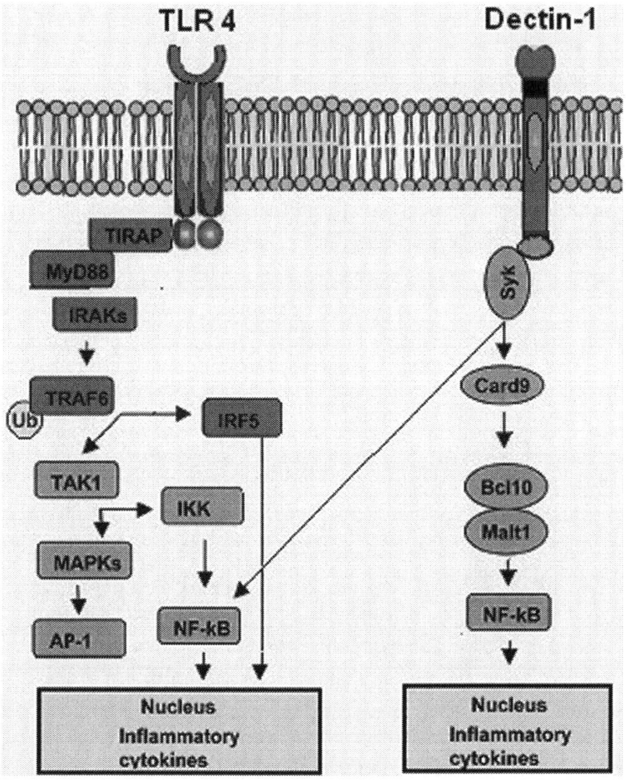

34.2 β-Glucan Receptor (Dectin 1) (CLEC7A or CLECSF12)

Fig. 34.3 Signaling pathway downstream of TLR4 and Dectin-1. Antigen presenting cells (APCs), including monocytes, macrophages and DCs, engage pathogens (fungi) and activate host responses via several PRRs including the Toll-like receptors (*TLR-4*) and Dectin-1. Dectin-1 is alternatively spliced into two functional isoforms, which differ by the presence or absence of a stalk region (shown in *dark black*). Dectin-1 recognises linear or branched 1—>3-linked β-glucan, which triggers intracellular signaling through at least two pathways, involving Syk kinase (shown in Figure) and Raf-1(not shown in Figure). Upon zymosan recognition, Dectin-1 can recruit Syk, and subsequently, Card9 relays the signal to the Bcl10-Malt1 complex to activate NF-κB, thus leading to the expression of inflammatory cytokines (IL-6, IL-10, IL-12, IL-17, IL-23, TNF). TLRs, on the contrary, which recognise various mannosylated and other fungal cell wall structures, signal through the MyD88-Mal mediated NF-κB pathway and induce the production of both pro and anti-inflammatory cytokines, including TNF, IL-10, IL-12 and TGFβ. Costimulation of both receptors can amplify the production of cytokines, including TNF, IL-23, IL-10 and IL-6 while downregulating the production IL-12, influencing the resultant generation of adaptive immunity (Lee and Kim 2007; Reid et al. 2009)

34.2.5.2 PLCγ2 Is Critical for Dectin-1-Mediated Ca²⁺ Flux and Cytokine Production in DCs

Stimulation of DCs with zymosan triggers an intracellular Ca^{2+} flux that can be attenuated by a blocking anti-Dectin-1 antibody or by pre-treatment of cells with PLCγ-inhibitor U73122. This suggests that Dectin-1 could elicit Ca^{2+} signaling through PLCγ2 in Dectin-1 signal transduction pathway (Xu et al. 2009b). Lipid rafts are plasma membrane microdomains that are enriched in cholesterol, glycosphingolipids, and glycosylphosphatidylinositol-anchored proteins and play an important role in the signaling of ITAM-bearing lymphocyte antigen receptors. It was demonstrated that Dectin-1 translocates to lipid rafts upon stimulation of DCs with β-glucan. In addition, two key signaling molecules, Syk and PLCγ2 are also recruited to lipid rafts upon the activation of Dectin-1, suggesting that lipid raft microdomains facilitate Dectin-1 signaling. Xu et al. (2009a) indicated that Dectin-1 and perhaps also other CTLDs are recruited to lipid rafts upon activation and that the integrity of lipid rafts is important for the signaling and cellular functions initiated by this class of innate receptors.

34.2.5.3 Activation and Regulation of Phospholipase A2

Dectin-1 also mediates respiratory burst (Gantner et al. 2003) and its involvement has been suggested in the activation and regulation of phospholipase A2 (PLA2) and cyclooxygenase-2 (COX-2) (Suram et al. 2006). Secretory PLA_2 ($sPLA_2$) translocates from Golgi and recycling endosomes of mouse peritoneal macrophages to newly formed phagosomes and regulates the phagocytosis of zymosan, suggesting a role in innate immunity. The $sPLA_2$ regulates phagocytosis and contributes to the innate immune response against *C. albicans* through a mechanism that is likely dependent on phagolysosome fusion (Balestrieri et al. 2009). Co-stimulation of dectin-1 and DC-SIGN triggers the arachidonic acid cascade in human monocyte-derived DCs through both opsonic and nonopsonic receptors. The FcγR route depends on ITAM/Syk/cytosolic phospholipase A_2 axis, whereas the response to zymosan involves the interaction with dectin-1 and DC-SIGN (Valera et al. 2008).

34.2.6 Functions of Dectin-1

34.2.6.1 Recognition of Pathogens and Antifungal Defense

Dectin-1 has been shown to be involved in AMΦ recognition, nonopsonic phagocytosis, and killing of *Pneumocystis* organisms both in vitro and in vivo. Dectin-1 is a major β-glucan receptor on the surface of macrophages, DCs, neutrophils and it is also expressed on certain lymphocytes

response (Goodridge et al. 2007). DCs stimulated through Dectin-1 can generate efficient Th, CTL and B cell responses and can therefore be used as effective mucosal and systemic adjuvants in humans (Agrawal et al. 2010). Engagement of Dectin-1 with β-glucan on the surface of murine primary microglia results also in an increase in tyrosine phosphorylation of Syk, a feature of Dectin-1 signaling pathway. Dectin-1 pathway may play an important role in antifungal immunity in the CNS (Shah et al. 2008).

(Taylor et al. 2002), subpopulations of MΦs in splenic red and white pulp, Kupffer cells, and MΦs and DCs in the lamina propria of gut villi (Reid et al. 2004). This is consistent with its role in pathogen surveillance. Tissue localization thus revealed potential roles of Dectin-1 in leukocyte interactions during innate immune responses and T cell development. Dectin-1 functionally interacts with leukocyte-specific tetraspanin CD37. Dectin-1 and CD37 colocalize on the surface of human APCs. Tetraspanin CD37 is important for dectin-1 stabilization in APC membranes and controls dectin-1-mediated IL-6 production (Meyer-Wentrup et al. 2007).

RAW 264.7 macrophages overexpressing Dectin-1 bind *Pneumocystis* organisms (Steele et al. 2003) and increases macrophage-dependent killing of *Pneumocystis* (Gross et al. 2006; Rapaka et al. 2007). Dectin-1-deficient mice have compromised clearance of *Pneumocystis* (Saijo et al. 2007) and *Candida albicans* (Taylor et al. 2007), as well as attenuated macrophage inflammatory responses to these organisms. Dectin-1 is centrally required for the generation of alveolar macrophage proinflammatory responses to *A. fumigatus* and provides in vivo evidence for the role of dectin-1 in fungal innate defense (Leal et al. 2010; Steele et al. 2005). Naive mice lacking Dectin-1 (Dectin-1$^{-/-}$) are more sensitive to intratracheal challenge with *A. fumigatus* than control mice, exhibiting >80% mortality within 5 days, ultimately attributed to a compromise in respiratory mechanics (Werner et al. 2009). A fusion protein consisting of Dectin-1 extracellular domain linked to the Fc portion of murine IgG1 augmented alveolar macrophage killing of *A. fumigatus* and shifted mortality associated with IPA via attenuation of *A. fumigatus* growth in the lung (Mattila et al. 2008). A role for Dectin-1 in promoting M. tuberculosis-induced IL-12p40 production by DC has been suggested by Rothfuchs et al. (2007) in which the receptor augments bacterial-host cell interaction and enhances the subsequent cytokine response through an unknown mechanism involving Syk signaling. Non-typeable *Haemophilus influenzae* (NTHi) can induce an innate inflammatory response in eosinophils that is mainly mediated via β-glucan receptors (Ahren et al. 2003; Weck et al. 2008). Dectin-1 is a potential targeting molecule for immunization and has implications for the specialization of DC subpopulations (Carter et al. 2006).

34.2.6.2 Collaborative Responses Mediated by Dectin-1 and TLR2

During fungal infection, a variety of receptors initiates immune responses, including TLR and the Dectin-1. TLR recognition of fungal ligands and subsequent signaling through myeloid differentiation factor 88 (MyD88) pathway (Fig. 34.3) was thought to be the most important interactions required for the control of fungal infection. However, recent studies have highlighted the role of Dectin-1 in induction of cytokine responses and the respiratory burst. Mitogen-activated protein kinase (MAPK) activation and TNF-α production in macrophage infected with *Mycobacterium avium* or *M smegmatis* is dependent on MyD88 and TLR2 but not TLR4, ManR, or CR3. Interestingly, the TLR2-mediated production of TNF-α by macrophages infected with *M smegmatis* requires dectin-1. A similar requirement for Dectin-1 in TNF-α production was observed for macrophages infected with *M bovis* Bacillus Calmette-Guerin (BCG), *M. phlei*, *M avium 2151*-rough, and *M tuberculosis H37Ra*. Studies established a significant role for Dectin-1, in cooperation with TLR2, to activate a macrophage's proinflammatory response to a mycobacterial infection (Yadav and Schorey 2006)

Saijo et al. (2007) argue that Dectin-1 plays a minor role in control of *Pneumocystis carinii* by direct killing and that TLR-mediated cytokine production controls *P. carinii* and *C. albicans*. By contrast, Taylor et al. (2007) argue that Dectin-1-mediated cytokine and chemokine production, leading to efficient recruitment of inflammatory cells, is required for control of fungal infection. Dennehy and Brown (2007) argued that collaborative responses induced during infection may partially explain these apparently contradictory results. It appears that Dectin-1 can mediate their own signaling, as well as synergize with TLR to initiate specific responses to infectious agents (Dennehy et al. 2009).

The DCs activated via Dectin can convert T_{reg} to IL17 producing cells (Osorio et al. 2008). Furthermore, they also prime cytotoxic T-lymphocyte (CTL) and mount potent CTL responses (LeibundGut-Landmann et al. 2008). Dectin-1 also induces antibody production in rodents (Kumar et al. 2009). Thus, DCs stimulated through Dectin-1 can generate efficient Th, CTL and B cell responses and can therefore be used as effective mucosal and systemic adjuvants in humans (Agrawal et al. 2010).

34.2.7 Genetic Polymorphism in Relation to Pathology

CARD9 is a susceptibility locus for inflammatory bowel disease (IBD) (Zhernakova et al. 2008). Dectin-1 polymorphism c.714T>G on chromosome 12p13 has been described with a transition from a tyrosine to an early stop codon on amino acid position 238 (p.Y238X) (Ferwerda et al. 2009). The functional consequence of this polymorphism is a complete loss of function, and immune cells expressing this truncated protein produce significantly less cytokines, including TNF-α, IL-1β and IL-17, upon in vitro stimulation with β-glucan or *C. albicans* (Plantinga et al. 2009). Th17 responses are considered to be involved in the pathogenesis of auto-immune diseases. Interestingly, both NOD2 [Nucleotide-binding oligomerization domain-containing protein 2 (also known as caspase recruitment domain-containing protein 15 (CARD15)] and Dectin-1 are shown to be capable of inducing Th17 responses after activation (LeibundGut-Landmann et al. 2007; van Beelen et al. 2007). In this respect, the Dectin-1 c.714T>G polymorphism could influence the Th17 response towards fungi such as *C. albicans* in gastrointestinal tract. de Vries et al. (2009) demonstrated that Dectin-1 expression is elevated on macrophages, neutrophils, and other immune cells involved in the inflammatory reaction in IBD. The Dectin-1 c.714T>G polymorphism however, is not a major susceptibility factor for developing IBD.

It has been well established that fungal particles, either intact yeast or fungal cell wall components that can be recognized by dectin-1, such as zymosan, can act as adjuvants in several experimental models of RA (Leibundgut-Landmann et al.. 2007; Frasnelli et al. 2005; Hida et al. 2007). In addition, Yoshitomi et al. (2005) revealed that β-glucan induced autoimmune arthritis in genetically susceptible SKG mice could be prevented by blocking the dectin-1 receptor. These studies imply that dectin-1 plays a pivotal role in the innate immune system and is able to modulate adaptive immune responses, of which, especially Th17 responses are implicated in immunopathology. Furthermore, Dectin-1 is involved in the induction of arthritis in mouse models through induction of intracellular signaling on recognition of fungal components. As a consequence, dectin-1 mediated inflammatory responses could contribute to the aetiology or disease severity of RA. An early stop codon polymorphism Y238X (c.714T>G, rs16910526) in Dectin-1 (Veerdonk et al. 2009) has resulted in a complete loss of function of the protein. Cytokine production capacity of peripheral blood mononuclear cells (PBMCs) from individuals homozygous for the Dectin-1 Y238X polymorphism on β-glucan or *C. albicans* exposure are impaired, including TNF-α, interleukin (IL-)1β, IL-6, and IL-17 responses. In the same assays, individuals heterozygous for Dectin-1 Y238X polymorphism exhibited intermediate cytokine responses compared with wild-type individuals (Ferwerda et al. 2009). Considering both the involvement of Dectin-1 in pro-inflammatory responses and the significant consequences of the Y238X polymorphism for Dectin-1 function, Plantinga et al. (2010) suggest that Dectin-1 Y238X polymorphism does not play a role in the pathogenesis of RA. Although expression of dectin-1 was high in synovial tissue of RA patients, and reduced cytokine production was observed in macrophages of individuals bearing Dectin-1 Y238X polymorphism, loss of one functional allele of Dectin-1 was not associated with either susceptibility to or severity of RA (Plantinga et al. 2010).

34.3 The C-Type Lectin-Like Protein-1 (CLEC-1) or CLEC-1A

34.3.1 *CLEC-1* Gene

The region telomeric of CD94 contains, in addition to the LOX-1 and DECTIN-1, genes: the CLEC-1 and CLEC-2 genes within about 100 kb. Sequence similarities and chromosomal arrangement suggest that these genes form a separate subfamily of lectin-like genes within the NK gene complex. Human CLEC-1 displays a single CRD and a cytoplasmic tyrosine-based motif. It is homologous to the NK cell receptors NKG2s and CD94 and also to LOX-1and preferentially transcribes in DCs (Colonna et al. 2000; Sobanov et al. 2001). CLEC-1 is over-expressed in a model of rat allograft tolerance. CLEC-1 is expressed by myeloid cells and specifically by endothelial cells in tolerated allografts. CLEC-1 expression can be induced in endothelial cells by alloantigen-specific regulatory $CD4^+CD25^+T$ cells. Expression of CLEC-1 is down-regulated by inflammatory stimuli but increased by the immunoregulators IL-10 or TGFβ. Interestingly, inhibition of CLEC-1 expression in rat DCs increases the subsequent differentiation of allogeneic Th17 T cells and decreases the regulatory $Foxp3^+$ T cell pool in vitro. In chronically rejected allograft, the decreased expression of CLEC-1 is associated with a higher production of IL-17. It is suggested that CLEC-1, expressed by myeloid cells and endothelial cells, is enhanced by regulatory mediators and moderates Th17 differentiation (Thebault et al. 2009).

34.4 CLEC-18 or the C-Type Lectin-Like Protein-2 (CLEC-2) or CLEC1B

34.4.1 Characterization

CLEC-1 and CLEC-18 (CLEC-2) possess a single CRD and a cytoplasmic tyrosine-based motif. Both are homologous to the NK cell receptors NKG2s and CD94 and also to LOX-1. CLEC-2 is expressed on the surface of transfected cells as a protein of approximately 33 kDa. The CLEC-2 is expressed on platelets and signaling through CLEC-2 is sufficient to mediate platelet aggregation (Suzuki-Inoue et al. 2006). The gene encoding CLEC-2 is located in the human NK complex on chromosome 12, along with the C-type lectin-like receptors NKG2D, LOX-1, and Dectin-1 (Sobanov et al. 2001). CLEC-2 is a type II transmembrane receptor, and its transcripts have been identified in immune cells of myeloid origin, including monocytes, DCs, and granulocytes, and in liver (Colonna et al. 2000). Human CLEC-2 has been reported to facilitate the capture of HIV-1. Xie et al.

(2008) identified two novel splicing variants of mCLEC-2 derived from omission of exon 2 and 2/4, respectively. These two variants had different expression profiles and subcellular localization from full-length mCLEC-2. Moreover, the full-length mCLEC-2 could be cleaved probably by proteases sensitive to aprotinin and PMSF into a soluble form that partially existed as a disulfide-linked homodimer. The interacting partner of the cytoplasmic region of CLEC-2 is RACK1, the receptor for activated C-kinase 1. Moreover, over-expression of RACK1 decreased the stability of CLEC-2 through promoting its ubiquitin-proteasome degradation, suggesting that RACK1 as a novel modulator of CLEC-2 expression (Ruan et al. 2009).

34.4.2 Ligands for CLEC-2

CLEC-2 has been shown to be a receptor on the surface of platelets for the snake venom toxin rhodocytin, which is produced by the Malayan pit viper *Calloselasma rhodostoma* (Shin and Morita 1998; Suzuki-Inoue et al. 2006) and the endogenous sialoglycoprotein podoplanin and functions as stimulator in platelet activation.

34.4.2.1 Rhodocytin/Aggretin

Aggretin, also known as rhodocytin, is a C-type lectin from *Calloselasma rhodostoma* snake venom. It is a potent activator of platelets, resulting in a collagen-like response by binding and clustering platelet receptor CLEC-2 (Suzuki-Inoue et al. 2006). The rhodocytin has been reported to bind to integrin $\alpha 2\beta 1$ and glycoprotein (GP) Ibα on platelets, but it is also able to induce activation independent of two receptors and of GPVI. CLEC-2 in platelets confers signaling responses to rhodocytin when expressed in a cell line. CLEC-2 has a single tyrosine residue in a YXXL motif in its cytosolic tail, which undergoes tyrosine phosphorylation upon platelet activation by rhodocytin or an antibody to CLEC-2. The pathway in platelets included activation of CLEC-2 by rhodocytin, binding of Syk, initiation of downstream tyrosine phosphorylation, and the activation of PLCγ2. CLEC-2 is the first C-type lectin receptor on platelets which signals through this novel pathway (Suzuki-Inoue et al. 2006). Rhodocytin may induce cytokine TNF-α/IL-6 release after interaction with CLEC-2 and the subsequent MAPK and NF-kB activation in monocytes/macrophages (Chang et al. 2010).

The Rhodocytin/Aggretin at 1.7 A is an unique tetrameric quaternary structure. The two $\alpha\beta$ heterodimers are arranged through twofold rotational symmetry, resulting in an antiparallel side-by-side arrangement. The Rhodocytin thus presents two ligand binding sites on one surface and can therefore cluster ligands in a manner reminiscent of convulxin and flavocetin. Molecular interaction of the rhodocytin with CLEC-2 was studied using molecular modeling after docking the aggretin $\alpha\beta$ structure with the CLEC-2 N-terminal domain (CLEC-2N). This model positions the CLEC-2N structure face down in the "saddle"-shaped binding site which lies between the rhodocytin α and β lectin-like domains. A twofold rotation of this complex to generate the rhodocytin tetramer reveals dimer contacts for CLEC-2N which bring the N- and C-termini into the proximity of each other, and a series of contacts involving two interlocking β-strands close to N-terminus. A comparison with homologous lectin-like domains from the immunoreceptor family reveals a similar but not identical dimerization mode, suggesting this structure may represent the clustered form of CLEC-2 capable of signaling across the platelet membrane (Hooley et al. 2008). Rhodocytin displays a concave binding surface, which is highly complementary to the experimentally determined binding interface on CLEC-2. Using computational dynamic methods, surface electrostatic charge and hydrophobicity analyses, and protein-protein docking predictions, it was proposed that the $(\alpha\beta)_2$ rhodocytin tetramer induces clustering of CLEC-2 receptors on the platelet surface, which will trigger major signaling events resulting in platelet activation and aggregation (Watson et al. 2008).

34.4.2.2 Podoplanin as Ligand

Podoplanin (aggrus), a transmembrane mucin-like sialoglycoprotein, is involved in tumor cell-induced platelet aggregation, tumor metastasis, and lymphatic vessel formation. Podoplanin is an endogenous ligand for CLEC-2 and facilitates tumor metastasis by inducing platelet aggregation. Podoplanin induces platelet aggregation with a long lag phase, which is dependent upon Src and phospholipase Cγ2 activation. However, it does not bind to glycoprotein VI. This mode of platelet activation was reminiscent of the snake rhodocytin, the receptor of CLEC-2 (Suzuki-Inoue et al. 2006). Results suggest that CLEC-2 is a physiological target protein of podoplanin and imply that it is involved in podoplanin-induced platelet aggregation, tumor metastasis, and other cellular responses related to podoplanin (Kato et al. 2008; Suzuki-Inoue et al. 2007). In addition, indirect evidence indicates for an endogenous ligand for CLEC-2 in renal cells expressing HIV-1. Podoplanin is expressed on renal cells (podocytes) and acts as a ligand for CLEC-2 on renal cells (Christou et al. 2008). Incorporation of podoplanin into HIV released from HEK-293T cells is required for efficient binding to the attachment factor CLEC-2 (Chaipan et al. 2010). Mice, deficient in podoplanin, which is also expressed on the surface of lymphatic endothelial cells, show abnormal patterns of lymphatic vessel formation.

34.4 CLEC-18 or the C-Type Lectin-Like Protein-2 (CLEC-2) or CLEC1B

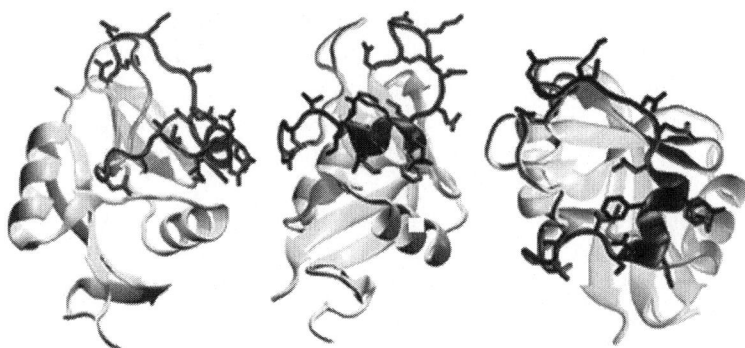

Fig. 34.4 **Extracellular domain of the platelet-activating receptor CLEC-2: the semi-helical long loop region.** Semi-helical long loop region dominates the upper surface of CLEC-2. The loop region is shown in *black* with the side chains illustrated. With respect to the *left panel*, the central panel is rotated 270° anticlockwise about the *y*-axis, and the *right panel* 90° anticlockwise about the *x*-axis. *B*, CLEC-2 is the only structurally characterized C-type lectin-like molecule with a formal helix in the long loop region (Adapted with permission from Watson et al. 2007 © American Society for Biochemistry and Molecular Biology)

Laminin VWF bridges exposed collagen, at damaged vessels, to GPIb. Subsequently, GPVI binds to collagen, leading to integrin α2β1 activation. Platelets also adhere to laminin, another major ECM component, through integrin α6β1, and are activated through GPVI. Laminin also interacts with VWF, leading to platelet adhesion via GPIb under sheer stress (Ozaki et al. 2009).

34.4.3 Crystal Structure of CLEC-2

The crystals of CLEC-2 belongs to the orthorhombic space group $P2_12_12_1$, with unit-cell parameters a = 35.407, b = 55.143, c = 56.078 A. The presence of one molecule per asymmetric unit is consistent with a crystal volume per unit weight (V_M) of 1.82 $A^3 Da^{-1}$ and a solvent content of 32.6% (Watson and O'Callaghan 2005). Watson et al. (2007) solved the crystal structure of the extracellular domain of CLEC-2 and identified the key structural features involved in ligand binding. A semi-helical loop region and flanking residues (Fig. 34.4) dominate the surface that is available for ligand binding. The precise distribution of hydrophobic and electrostatic features in this loop will determine the nature of any endogenous ligand with which it can interact. Major ligand-induced conformational change in CLEC-2 is unlikely as its overall fold is compact and robust. However, ligand binding could induce a tilt of 3–10 helical portion of long loop region in CLEC-2. Mutational analysis and surface plasmon resonance binding studies support these observations (Watson et al. 2007).

34.4.4 Functions of CLEC-2

34.4.4.1 Platelet Activation

CLEC-2 has been described to play crucial roles in thrombosis/hemostasis, tumor metastasis, and lymphangiogenesis. Platelets play an essential role in wound healing by forming thrombi that plug holes in the walls of damaged blood vessels. To achieve this, platelets express a diverse array of cell surface receptors and signaling proteins that induce rapid platelet activation. Platelet glycoprotein receptors that signal via an ITAM or an ITAM-like domain, namely collagen receptor complex GPVI-FcR γ-chain and CLEC-2, respectively, support constitutive (i.e. agonist-independent) signaling. The CLEC-2 mediates powerful platelet activation through Src and Syk kinases, but regulates Syk through a novel dimerization mechanism via a single YXXL motif known as a hemITAM. Inhibition of constitutive signaling through Src and Syk tyrosine kinases by G6b-B may help to prevent unwanted platelet activation (Mori et al. 2008; O'Callaghan et al. 2009). CLEC-2 is a receptor for podoplanin, which is expressed at high levels in several tissues (Section 34.4.2.2) but is absent from vascular endothelial cells and platelets. Platelet activation at sites of vascular injury is critical for primary hemostasis, but can also trigger arterial thrombosis in vascular disease. The ability of rhodocytin or CLEC-2-specific antibodies to trigger platelet aggregation in the absence of other stimuli indicates the potency with which CLEC-2 can modulate platelet activity (Suzuki-Inoue et al. 2006). Therefore, CLEC-2 is a potential therapeutic target in thrombotic cardiovascular disease. In addition, a molecular understanding of the effects of rhodocytin could lead to more effective therapy of snake envenomation (Chaipan et al. 2006). Phenotypes in CLEC-2-deficient mice are lethal at the embryonic/neonatal stages associated with disorganized and blood-filled lymphatic vessels and severe edema. Transplantation of fetal liver cells from Clec-2$^{-/-}$ demonstrated that CLEC-2 is involved in thrombus stabilization in vitro and in vivo, possibly through homophilic interactions without apparent increase in bleeding tendency. These results reveal an essential function of CLEC-2 in hemostasis and thrombosis (May et al. 2009;

Kerrigan et al. 2009; Watson et al. 2009). CLEC-2 could be an ideal novel target protein for an anti-platelet drug, which inhibits pathological thrombus formation but not physiological hemostasis (Suzuki-Inoue et al. 2010; Spalton et al. 2009).

34.4.4.2 Lymphatic Vascular Development Through CLEC-2-SLP-76 Signaling

Although platelets appear by embryonic day 10.5 in the developing mouse, an embryonic role for these cells has not been identified. The SYK-SLP-76 signaling pathway is required in blood cells to regulate embryonic blood-lymphatic vascular separation, but the cell type and molecular mechanism underlying this regulatory pathway are not known. Bertozzi et al. (2010) demonstrated that platelets regulate lymphatic vascular development by directly interacting with lymphatic endothelial cells through CLEC-2 receptors. Podoplanin, expressed on surface of lymphatic endothelial cells, is required in nonhematopoietic cells for blood-lymphatic separation. These studies identify a nonhemostatic pathway in which platelet CLEC-2 receptors bind lymphatic endothelial podoplanin and activate SLP-76 signaling to regulate embryonic vascular development.

CLEC-2, originally thought to be restricted to platelets, is also expressed by peripheral blood neutrophils, and only weakly by bone marrow or elicited inflammatory neutrophils. On circulating neutrophils, CLEC-2 can mediate phagocytosis of Ab-coated beads and the production of proinflammatory cytokines, including TNF-α, in response to CLEC-2 ligand, rhodocytin. Like dectin-1, CLEC-2 can recruit the signaling kinase Syk in myeloid cells, however, stimulation of this pathway does not induce the respiratory burst. Thus, CLEC-2 also functions as an activation receptor on neutrophils (Kerrigan et al. 2009).

34.4.5 CLEC-2 Signaling

The CLEC-2 and Dectin-1 have been shown to signal through a Syk-dependent pathway, despite the presence of only a single YXXL in their cytosolic tails. Stimulation of CLEC-2 in platelets and in two mutant cell lines is dependent on YXXL motif and on proteins that participate in signaling by ITAM receptors, including Src, Syk, and Tec family kinases, and on phospholipase Cγ. Results demonstrate that CLEC-2 signals through a single YXXL motif that requires the tandem SH2 domains of Syk but is only partially dependent on the SLP-76/BLNK family of adapters (Fuller et al. 2007). Ligand binding by CLEC-2 promotes phosphorylation of a tyrosine in the cytoplasmic domain YXXL motif of CLEC-2 by Src kinases and further downstream signaling events trigger platelet activation and aggregation. The snake venom protein rhodocytin and the endogenous protein podoplanin have been identified as ligands. The structures of CLEC-2 and rhodocytin suggest that ligand binding could cluster CLEC-2 molecules at the platelet surface and initiate signaling.

The signaling pathway used by CLEC-2 shares many similarities with that used by receptors that have 1 or more copies of an ITAM, defined by the sequence YXX(L/I)X(6–12)YXX(L/I), in their cytosolic tails or associated receptor chains. Evidence suggests that Syk activation by CLEC-2 is mediated by the cross-linking through the tandem SH2 domains with a stoichiometry of 2:1. Cross-linking and electron microscopy demonstrated that CLEC-2 is present as a dimer in resting platelets and converted into larger complexes on activation. This is a unique mode of activation of Syk by a single YXXL-containing receptor (Hughes et al. 2010a, b). Pollitt et al. (2010) demonstrated that CLEC-2 translocates to lipid rafts upon ligand engagement and that translocation is essential for hemITAM phosphorylation and signal initiation. HemITAM phosphorylation, but not translocation, is also critically dependent on actin polymerization, Rac1 activation, and release of ADP and thromboxane A_2 (TxA_2). This reveals a unique series of proximal events in CLEC-2 phosphorylation involving actin polymerization, secondary mediators, and Rac activation (Pollitt et al. 2010).

34.5 CLEC9A (DNGR-1)

CLEC9A is a group V C-type lectin-like receptor located in the "Dectin-1 cluster" of related receptors, which are encoded within the NK-gene complex. Expression of human CLEC9A is highly restricted in peripheral blood, being detected only on BDCA3$^+$ dendritic cells and on a small subset of CD14$^+$CD16$^-$ monocytes. CLEC9A is expressed at the cell surface as a glycosylated dimer and can mediate endocytosis, but not phagocytosis. CLEC9A possesses a cytoplasmic ITAM that can recruit Syk kinase. This receptor can induce proinflammatory cytokine production and can function as an activating receptor (Huysamen et al. 2008). Mouse *Clec9A* encodes a type II membrane protein with a single extracellular C-type lectin domain. Surface staining of mAbs against mouse and humans revealed that *Clec9A* was selective for mouse DCs and was restricted to the CD8$^+$ conventional DC and plasmacytoid DC subtypes. A subset of human blood DCs also expressed CLEC9A (Caminschi et al. 2008).

Sancho et al. (2008) characterized a C-type lectin of the NK cell receptor group that was named DC, NK lectin group receptor-1 (DNGR-1). DNGR-1 is expressed in mice at high levels by CD8$^+$ DCs and at low levels by plasmacytoid DCs but not by other hematopoietic cells. Human DNGR-1 was also restricted in expression to a small subset of blood DCs that bear similarities to mouse CD8α$^+$ DCs. The selective expression pattern and observed endocytic activity of DNGR-1 suggested that it could be used for antigen targeting to DCs. DNGR-1 is a highly specific marker of mouse and human DC subsets that can be exploited for CTL cross-priming and tumor therapy.

In the mouse, the CD8α$^+$ subset of dendritic cells phagocytose dead cell remnants and cross-primes CD8$^+$ T cells against cell-associated antigens. Sancho et al. (2009) showed that CD8α$^+$ DCs use CLEC9A (DNGR-1), to recognize a preformed signal that is exposed on necrotic cells. Loss or blockade of CLEC9A does not impair the uptake of necrotic cell material by CD8$^+$ DCs, but specifically reduces cross-presentation of dead-cell-associated antigens in vitro and decreases the immunogenicity of necrotic cells in vivo. The function of CLEC9A requires a key tyrosine residue in its intracellular tail that allows the recruitment and activation of the tyrosine kinase SYK, which is also essential for cross-presentation of dead-cell-associated antigens. Thus, CLEC9A functions as a SYK-coupled C-type lectin receptor to mediate sensing of necrosis by the principal DC subset involved in regulating cross-priming to cell-associated antigens. In addition, DNGR-1 is a target for selective in vivo delivery of antigens to DC and the induction of CD8$^+$ T-cell and Ab responses (Joffre et al. 2010; Poulin et al. 2010).

34.6 Myeloid Inhibitory C-Type Lectin-Like Receptor (MICL/CLEC12A)

Marshall et al. (2004) reported the characterization of a human myeloid inhibitory C-type lectin-like receptor (MICL) which is primarily restricted to myeloid cells, including granulocytes, monocytes, macrophages, and dendritic cells. MICL contains a single C-type lectin-like domain and a cytoplasmic ITIM, and is variably spliced and highly N-glycosylated. It preferentially associates with the signaling phosphatases SHP-1 and SHP-2, but not with SHIP (Huysamen and Brown 2009). Chimeric construct combining MICL and the β-glucan receptor showed that MICL can inhibit cellular activation through its cytoplasmic ITIM. MICL is a negative regulator of granulocyte and monocyte function (Marshall et al. 2004). Although MICL was highly N-glycosylated in primary cells, the level of glycosylation varied between cell types. MICL surface expression was down-regulated during inflammatory/activation conditions in vitro, as well as during an in vivo model of acute inflammation. This suggests that human MICL may be involved in the control of myeloid cell activation during inflammation (Marshall et al. 2006). The murine homolog of MICL (mMICL) is structurally and functionally similar to the human orthologue (hMICL), although there are some notable differences. mMICL is expressed as a dimer and is not heavily glycosylated; however, like hMICL, the receptor can recruit inhibitory phosphatases upon activation, and is down-regulated on leukocytes following stimulation with selected TLR agonists. Like the human receptor, mMICL is predominantly expressed by myeloid cells. However, mMICL is also expressed by B cells and CD8$^+$ T cells in peripheral blood, and NK cells in the bone marrow. The mMICL recognises an endogenous ligand in a variety of murine tissues, suggesting that the receptor plays a role in homeostasis (Pyz et al. 2008).

34.7 CLEC12B or Macrophage Antigen H

Hoffmann et al. (2007) described a new member of CTLD family, CLEC12B. The extracellular domain of CLEC12B shows considerable homology to the activating natural killer cell receptor NKG2D, but unlike NKG2D, CLEC12B contains an ITIM (VxYxxL) within its cytoplasmic tail in its intracellular domain. Human CLEC12B is widely expressed at low levels in various human tissues, except the brain, and that the receptor is alternatively spliced, generating at least two isoforms, one of which lacks part of CRD and predicted to be nonfunctional (Hoffmann et al. 2007). Despite the homology of CLEC12B to NKG2D, CLEC12B does not appear to bind NKG2D ligands and therefore does not represent the inhibitory counterpart of NKG2D. However, CLEC12B has the ability to counteract NKG2D-mediated signaling, and that this function is dependent on the ITIM and the recruitment of the phosphatases SHP-1 and SHP-2. This receptor plays a role in myeloid cell function.

34.8 CLECSF7 (CD303)

C-type lectin domain superfamily member 7 was previously called CLECSF11 (C-type lectin domain superfamily member 11). The new designation is CLEC4C (C-type lectin domain family 4 member C). It is also named as blood dendritic cell antigen 2 (BDCA-2) or dendritic lectin, CD303 and contains one C-type lectin domain. It functions in antigen-capturing and targets ligand into antigen processing and peptide-loading compartments for presentation to T-cells. It may mediate potent inhibition of induction of IFN- α/β. Its expression in plasmacytoid dendritic cells may act as a signaling receptor that activates protein-tyrosine kinases and mobilizes intracellular Ca^{+2}. CLECSF7 does not seem to bind mannose (Fernandes et al. 2000). *HECL* (HGMW-approved symbol *CLECSF7*) gene maps close to the NK gene complex on human chromosome 12p47. Sequence analysis revealed a complete ORF of

549 bp comprising several putative glycosylation and phosphorylation sites as well as a C-terminal C-type CRD. *HECL* exhibits a significant degree of divergence from the NKCs that comprise the NK gene complex. The NKC receptors all belong to group V of the C-type lectin superfamily, where as HECL, is most closely related to the sole group II C-type lectins reported to map near this region of the genome, the murine *Nkcl* and *Mpcl* genes. Like Nkcl, HECL is expressed in a variety of hematopoietic cell types and has a complete Ca^{2+}-binding site 2. Despite the presence of critical amino acids for sugar binding in Ca^{2+} binding site 2, HECL does not seem to bind carbohydrate. Moreover, HECL is the first non-receptor-like C-type lectin to map near the NK gene complex (Fernandes et al. 2000).

34.9 Lectin-Like Oxidized LDL Receptor (LOX-1) (CLEC8A)

Endothelial activation, accumulation of oxidatively modified low density lipoprotein (Ox-LDL) and intense inflammation characterize atherosclerotic plaque in acute myocardial ischemia. Endothelial dysfunction, or activation, elicited by Ox-LDL and its lipid constituents has been shown to play a key role in the pathogenesis of atherosclerosis. Ox-LDL induces expression of lectin-like receptor (LOX-1) on endothelial cells and leads to the expression of matrix metalloproteinases (MMPs), which destabilize atherosclerotic plaque. LOX-1 and scavenger receptor for phosphatidylserine and oxidized lipoprotein (SR-PSOX) are type II and I membrane glycoproteins, respectively, both of which can act as cell-surface endocytosis receptors for Ox-LDL. In vivo, endothelial cells that cover early atherosclerotic lesions, and intimal macrophages and activated vascular smooth muscle cells (VSMC) in advanced atherosclerotic plaques dominantly express LOX-1 (Adachi and Tsujimoto 2006). In association with Ox-LDL, LOX-1 is a marker of atherosclerosis that induces vascular endothelial cell activation and dysfunction, resulting in pro-inflammatory responses, pro-oxidative conditions and apoptosis. In addition to binding Ox-LDL, it acts as a receptor for HSP70 protein involved in antigen cross-presentation to naive T-cells in DCs, thereby participating in cell-mediated antigen cross-presentation. Soluble LOX-1 concentration in human blood also has been shown to be elevated in coronary heart diseases especially in acute coronary syndrome (Kume and Kita 2004). In addition to endothelial cells, LOX-1 is expressed in macrophages and smooth muscle cells accumulated in the intima of advanced atherosclerotic plaques in vivo. LOX-1 is critical in foam cell formation of macrophages (Mφ) and smooth muscle cells (SMC). Inhibition of LOX-1 expression reduces foam cell formation and might influence lipid core formation in atherosclerotic lesions. In contrast to LOX-1 expressed by a variety of cell types, SR-PSOX expression appeared relatively confined to macrophages in atherogenesis.

34.9.1 General Features of LOX-1

LOX-1, the lectin-like oxidized LDL receptor is the receptor for Ox-LDL and is abundantly expressed in endothelial cells; it might mediate some of the actions of Ox-LDL in the endothelium. LOX-1 is initially synthesized as a 40-kDa precursor protein with N-linked high mannose-type carbohydrate, which is further glycosylated and processed into a mature form of 52-kD. The mature form of LOX-1 is a lectin-like receptor and is a homodimer linked through disulfide bond. It may form a hexamer composed of three homodimers. LOX-1 belongs to the same family as NKR and functionally undergoes dimerization. The LOX-1 is a type II membrane protein, and acts as a cell-surface endocytosis receptor for atherogenic Ox-LDL, though secretory form of LOX-1 also exists. Its mRNA was shown to be expressed in human atheromatous lesions. LOX-1 can support binding, internalization and proteolytic degradation of Ox-LDL, but not of significant amounts of acetylated LDL, which is a well-known high-affinity ligand for class A scavenger receptors and scavenger receptor expressed by endothelial cells (Sawamura et al. 1997). Cell-surface LOX-1 can be cleaved through some protease activities that are associated with the plasma membrane, and released into the culture media. The N-terminal amino-acid sequencing identified two cleavage sites (Arg86-Ser87 and Lys89-Ser90), both of which are located in the membrane proximal extracellular domain of LOX-1. LOX-1 is often co-localized with apoptotic cells and implicated in vascular inflammation and atherosclerotic plaque initiation, progression, and destabilization (Kume and Kita 2001).

Mouse LOX-1 is composed of 363 amino acids and has a C-type lectin domain type II membrane protein structure. Mouse LOX-1 has triple repeats of the sequence in extracellular "Neck domain," which is unlike human and bovine LOX-1. The LOX-1 binds Ox-LDL with two affinity constants in the presence of serum. The binding component with higher affinity showed the lowest value of K_D among the known receptors for Ox-LDL (Hoshikawa et al. 1998). Measurement of soluble LOX-1 in vivo may provide a novel diagnostic tool for the evaluation and prediction of atherosclerosis and vascular disease (Kume and Kita 2001). Soluble forms of LOX-1 are present in conditioned media of cultured bovine aortic endothelial cells (BAECs) and CHO-K1 cells transfected with LOX-1 cDNA. In TNF-α-activated BAECs, cell-surface expression of LOX-1 precedes soluble LOX-1 production. The soluble LOX-1 in cell-conditioned media is derived from LOX-1 expressed on the cell surface (Murase et al. 2000).

LOX-1 expression is not constitutive, but can be induced by proinflammatory stimuli, such as TNF-α, TGF-β and bacterial endotoxin, and fluid shear stress. It is up-regulated in atherosclerotic lesions, by Ox-LDL, reactive oxygen species, suggesting that it may participate in amplification of Ox-LDL-induced vascular dysfunction. This receptor is upregulated by angiotensin II, endothelin, cytokines and all participants in atherosclerosis and is up-regulated in the arteries of hypertensive, dyslipidemic, and diabetic animals. Defects in LOX-1 gene (HGMW-approved symbol *OLR1*) may be a cause of susceptibility to myocardial infarction. Defects in *OLR1* may also be associated with susceptibility to Alzheimer disease (AD). Upregulation of LOX-1 has been identified in atherosclerotic arteries of several animal species and humans, not only on the endothelial lining, but also in choroidal neovasculature of the atherosclerotic plaque (Mehta et al. 2006; Mukai et al. 2004; Vohra et al. 2006).

The extracellular domains of LOX-1 are post-translationally modified by N-linked glycosylation. LOX-1 is synthesized as a 40-kDa precursor protein with N-linked high mannose carbohydrate chains (pre-LOX-1), which is subsequently further glycosylated and processed into the 48-kDa mature form. Both TNF-α-activated bovine aortic endothelial cells and CHO-K1 cells stably expressing bovine LOX-1 (BLOX-1-CHO) exclusively produced a 32-kDa deglycosylated form of LOX-, follwing treatment with an N-glycosylation inhibitor, tunicamycin. The deglycosylated form of LOX-1 is not efficiently transported to the cell surface, but is retained in the ER or Golgi apparatus in TNF-α-activated bovine aortic endothelial cells, but not in BLOX-1-CHO cells. It shows that N-linked glycosylation plays key roles in the cell-surface expression and ligand binding of LOX-1 (Kataoka et al. 2000).

34.9.2 LOX-1 Gene in Human

The natural killer (NK) gene complex is a genomic region containing lectin-type receptor genes located on human chromosome 12 that contains several families of lectin-like genes including the CD94 and NKG2 NK receptor genes (Fig. 34.1). Bull et al. (2000) established a contig of PAC and BAC clones comprising about 1 Mb of the centromeric part of the NK gene complex. This region extending from LOX-1 gene and found within 100 kb telomeric of STS marker D12S77, contains CD94 and NKG2 NK receptor genes and reaches beyond D12S852 on the proximal side.

The human LOX-1 has the highest homology with C-type lectin receptors expressed on NK cells. The human LOX-1 gene is a single-copy gene and assigned to p12.3-p13.2 region of chromosome 12. In contrast, the cellular expression pattern of LOX-1 is different from that of NK cell receptors. A 1,753-bp fragment of 5′ flanking region of LOX-1 gene had a functional promoter activity and contains binding sites for several transcription factors, including STAT family and NF-IL6. The 5′-regulatory region contained several potential cis-regulatory elements, such as GATA-2 binding element, c-ets-1 binding element, 12-O-tetradecanoylphorbol 13-acetate-responsive element and shear-stress-responsive elements, which may mediate the endothelium-specific and inducible expression of LOX-1. The major transcription-initiation site was found to be located 29 nt downstream of the TATA box and 61 nt upstream from the translation-initiation codon. The minor initiation site was found to be 5 bp downstream from the major site. Most of the promoter activity of the LOX-1 gene was ascribed to the region (−150 to −90) containing the GC and CAAT boxes. The coding sequence was divided into six exons by five introns. The first three exons corresponded to the different functional domains of the protein (cytoplasmic, transmembrane and neck domains), and the residual three exons encoded the carbohydrate-recognition domain similar to other C-type lectin genes. Since the locus for a familial hypertension has been mapped to the overlapping region, LOX-1 might be the gene responsible for the hypertension. The expression of LOX-1 was upregulated by several cytokines (Aoyama et al. 1999; Yamanaka et al. (1998)).

34.9.2.1 Rat LOX-1 Gene

The genomic organization of rat LOX-1 shows several consensus sequences in 59-flanking region. Rat LOX-1 cDNA clone encodes a single-transmembrane protein with its N terminus in the cytoplasm (Nagase et al. 1997). The extracellular region consists of a spacer, 46-amino acid triple repeats, and C-type lectin-like domains. The rat LOX-1 gene, encoded by a single copy gene spanning over 19 kb, consists of eight exons (Fig. 34.5). The promoter region contained putative TATA and CAAT boxes and multiple cis-elements such as NF-kB, AP-1 and AP-2 sites, and a shear stress response element. Exon boundaries correlated well with the functional domain boundaries of the receptor protein (Nagase et al. 1998b). The rat endothelial receptor LOX-1 cDNA encodes a protein of 364 amino acids that showed approximately 60% similarity to its bovine and human counterparts. The protein consisted of intracellular N-terminal, transmembrane and extracellular lectin-like domains. Rat LOX-1 is unique in having three repeats of a 46-amino-acid motif between the transmembrane and lectin-like regions. Two isoforms of mRNA were found to be generated by alternative use of two polyadenylation signals in a tissue-specific manner. The 3′-untranslated region contained multiple A + U-rich elements for rapid degradation of mRNA. Northern-blot analysis revealed that LOX-1 mRNA was expressed predominantly in the lung. Nagase et al. (1998b) reported an unexpected blood-pressure-

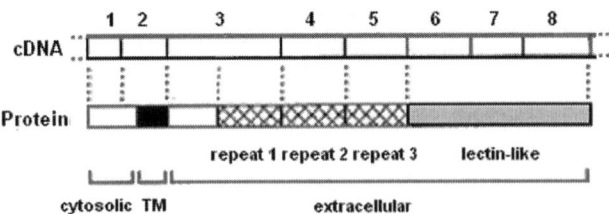

Fig. 34.5 Genomic organization of rat LOX-1. Schematic representation of the cDNA and protein. Exons in the cDNA and functional domains in the protein are indicated by *boxes*. The *solid box* represents the transmembrane (*TM*) domain. *Hatched boxes* represent the 46-amino acid repeat unit (*repeats 1–3*). The *dotted box* represents the C-type lectin-like domain (Adapted from Nagase et al. 1998; © American Society for Biochemistry and Molecular Biology)

associated regulation of LOX-1 expression, a new relation between lipid metabolism and blood-pressure control. The expression of LOX-1 was dramatically up-regulated in the aorta in hypertensive SHR-SP/Izm rats compared with very low levels in control WKY/Izm rats, suggesting a potential role for LOX-1 in the pathogenesis of hypertension as well as atherosclerosis.

34.9.3 Ligands for LOX-1

As a scavenger receptor, LOX-1 is capable of binding to a variety of structurally unrelated ligands. Ox-LDLs appear to play key roles in atherosclerotic progression and plaque rupture (Li et al. 2002b). LOX-1 is a major receptor for Ox-LDL in activated vascular smooth muscle cells (VSMC) as in endothelial cells. Human recombinant LOX-1 protein showed that the protein had specific Ox-LDL-binding activity (Huang et al. 2005). LOX-1 binds multiple classes of ligands that are implicated in the pathogenesis of atherosclerosis. Besides Ox-LDL, LOX-1 can recognize apoptotic/aged cells, activated platelets, and bacteria, implying versatile physiological functions (Chen et al. 2002; Shimaoka et al. 2001b). LOX-1 from bovine aortic endothelial cells (BAE) is the major binding protein for Ox-LDL on the surface of BAE and is expressed in atheromatous intima of human carotid artery as well as in intima of normal bovine aorta. In addition, human LOX-1 recognizes a key cellular phospholipid, PS (phosphatidylserine), in a Ca^{2+}-dependent manner, both in vitro and in cultured cells. A recombinant, folded and glycosylated LOX-1 molecule binds PS, but not other phospholipids. LOX-1 recognition of PS was maximal in presence of mM Ca^{2+} levels (Murphy et al. 2006). C-reactive protein (CRP), a risk factor for cardiovascular events and an amplifier of vascular inflammation through promoting endothelial dysfunction, can act as a novel ligand for LOX-1. This interaction could be disrupted with known LOX-1 ligands, such as Ox-LDL and carrageenan. In vivo assays revealed that in LOX-1 the basic spine arginine residues are important for binding, which is lost upon mutation of Trp^{150} with alanine. Simulations of the wild-type LOX-1 and of the Trp^{150}Ala mutant C-type lectin-like domains revealed that the mutation does not alter the dimer stability, but a different dynamical behaviour differentiates the two proteins. The symmetrical motion of monomers is completely damped by the structural rearrangement caused by the Trp^{150}Ala mutation. An improper dynamical coupling of the monomers and different fluctuations of the basic spine residues are observed, with a consequent altered binding affinity (Falconi et al. 2007). Mutagenesis studies demonstrated that the arginine residues forming the basic spine structure on LOX-1 ligand-binding interface were dispensable for CRP binding, suggesting a novel ligand-binding mechanism for LOX-1, distinct from that used for Ox-LDL binding. Study suggests that CRP- LOX-1 interaction may mediate CRP-induced endothelial dysfunction (Shih et al. 2009). Advanced glycation end product (AGE) is recognized by LOX-1. Cellular binding experiments revealed that AGE-bovine serum albumin (AGE-BSA) specifically binds CHO cells over-expressing bovine LOX-1 (bLOX-1). Cultured bovine aortic endothelial cells also showed specific binding for AGE-BSA (Jono et al. 2002). Interaction of AGE with AGE receptors induces several cellular phenomena potentially relating to diabetic complications. Study indicates that CD36-mediated interaction of AGE-modified proteins with adipocytes might play a pathological role in obesity or insulin-resistance (Kuniyasu et al. 2003).

Adhesion of bacteria to vascular endothelial cells as well as mucosal cells and epithelial cells appears to be one of the initial steps in the process of bacterial infection, including infective endocarditis. LOX-1 with C-type lectin-like structure, can support adhesion of bacteria and work as a cell surface receptor for Gram-positive and Gram-negative bacteria, such as *S. aureus* and *E. coli*, however, other unknown molecules may also be involved in the adhesion of *E. coli* to cultured bovine aortic endothelial cells, which is enhanced by poly(I) (Shimaoka et al. 2001a). Development of antagonists for LOX-1 might be a good therapeutic approach to vascular diseases (Chen et al. 2002)

34.9.4 Structural Analysis

34.9.4.1 Carbohydrate Recognition Domain

LOX-1 from human aortic ECs has a sequence identical to that from human lung. Human LOX-1 can recognize modified LDL, apoptotic cells and bacteria. The CRD is the ligand-binding domain of human LOX-1. LOX-1s carrying a mutation on each of six Cys in CRD resulted in a variety of N-glycosylation and failed to be transported to cell surface. This gave an evidence for the involvement of all six Cys in the intrachain disulfide bonds required for proper folding, processing and transport of LOX-1. The C-terminal sequence

(KANLRAQ) was also essential for protein folding and transport, while the final residues (LRAQ) were involved in maintaining receptor function. In addition, positively charged (R^{208}, R^{209}, H^{226}, R^{229} and R^{231}) and uncharged hydrophilic (Q^{193}, S^{198}, S^{199} and N^{210}) residues are involved in ligand binding, suggesting that ligand recognition of LOX-1 is not merely dependent on the interaction of positively charged residues with negatively charged ligands (Shi et al. 2001). Truncation of the lectin domain of LOX-1 abrogated oxLDL-binding activity (Chen et al. 2001). Deletion of the utmost C-terminal ten amino acid residues (261–270) was enough to disrupt the Ox-LDL-binding activity. Substitutions of Lys-262 and/or Lys-263 with Ala additively attenuated the activity. Serial-deletion analysis showed that residues up to 265 are required for the expression of minimal binding, although deletion of C-terminal three residues (268–270) still retained full binding activity. These results demonstrate distinct role of lectin domain as functional domain recognizing LOX-1 ligand.

To understand the interaction between hLOX-1 and its ligand Ox-LDL, Xie et al. (2003) reconstituted the functional C-type lectin-like domain (CTLD) of LOX-1 from inactive aggregates in *E. coli*. The CD spectra of the domain suggested that the domain has α-helical structure where as the blue shift of Trp residues indicated refolding of the domain. Like wild-type hLOX-1, the refolded CTLD was able to bind modified LDL. Thus, even though CTLD contains six Cys residues that form disulfide bonds, it recovered its specific binding ability on refolding. This suggests that the correct disulfide bonds in CTLD were formed by the artificial chaperone technique. Although the domain lacked N-glycosylation, it showed high affinity for its ligand in surface plasmon resonance experiments. Thus, unglycosylated CTLD is sufficient for binding modified LDL (Xie et al. 2003).

Though, the lectin-like domain of LOX-1 is essential for ligand binding, the neck domain is not. In particular, the large loop between the third and fourth cysteine of lectin-like domain plays a critical role for Ox-LDL binding as well as C-terminal end residues. Alanine-directed mutagenesis of the basic amino acid residues around this region revealed that all the basic residues are involved in Ox-LDL binding. Therefore, electrostatic interaction between basic residues in lectin-like domain of LOX-1 and negatively charged Ox-LDL is critical for the binding activity of LOX-1 (Chen et al. 2001). Chen et al. (2005) suggested that Ox-LDL/LOX-1 plays a role in signaling pathway in rat cardiac fibroblasts, which naturally express low levels of LOX-1. It was suggested that in cardiac fibroblasts, Ox-LDL binds to LOX-1 and activates p38 MAPK, followed by the expression of ICAM-1, VCAM-1, and MMP-1. Thus, fibroblasts transform into an endothelial phenotype on transfection with CMV-LOX-1wild-type (WT) and subsequent exposure to Ox-LDL.

Role of Cytoplasmic Sequences in Cell-Surface Sorting of LOX-1: The ability of LOX-1 binding to Ox-LDL can be distinguished from other NK receptors. Domain swapping of the lectin-like domain between LOX-1 and the NK cell receptors CD94, NKG2D, and LY-49A demonstrated the crucial role of this domain for recognition of Ox-LDL by LOX-1, but not for the correct cell-surface sorting of LOX-1. Using N-terminal deletions Chen and Sawamura (2005) found that the correct cell-surface localization is dependent on a positively charged motif present in the cytosolic juxtamembrane region of LOX-1. Furthermore, the extracellular localization of LOX-1 C-terminus is disrupted when cytoplasmic basic amino acids, Lys-22, Lys-23 and Lys-25 were mutated to Glu. These results indicated that the N-terminal cytoplasmic domain of LOX-1 determines the correct expression of lectin domain on the cell-surface.

Dominant-Negative LOX-1 Blocks Homodimerization of Wild-Type LOX-1-Induced Cell Proliferation: LOX-1 is thought to function as a monomer. Yet, Xie et al. (2004) suggest that human LOX-1 (hLOX-1) forms constitutive homo-interactions in vivo. Site-directed mutagenesis studies indicated that Cys^{140} has a key role in the formation of disulfide-linked hLOX-1 dimers. Eliminating this intermolecular disulfide bond markedly impairs the recognition of *E. coli* by hLOX-1. These dimers can act as a "structural unit" to form noncovalently associated oligomers. These results provided evidence for the existence of hLOX-1 dimers/oligomers (Xie et al. 2004).

Although Lys^{262} and Lys^{263} in C-terminus of bovine bLOX-1 play important roles in the uptake of Ox-LDL, mutation of these residues did not suggest as potential source of the dominant-negative property of bLOX-1. It is possible that dominant-negative human hLOX-1 forms a heterodimer with LOX-1-wild-type (WT) and blocks LOX-1-WT-induced cell signaling. Homodimerization of hLOX-1-WT was localized in cell membrane, and Ox-LDL activated ERK-1/2 without the translocation of hLOX-1-WT. Tanigawa et al. (2006a) gave an evidence that blocking cell-proliferative pathways at the receptor level could be useful for impairing LOX-1-induced cell proliferation.

34.9.4.2 Crystal Structure

The X-ray crystallography of 136–270 subunit revealed the presence of disulfide bonds. The 1.4 Å crystal structure of the extracellular C-type lectin-like domain of human Lox-1 reveals a heart-shaped homodimer with a ridge of six basic amino acids extending diagonally across the apolar top of Lox-1, a central hydrophobic tunnel that extends through the entire molecule, and an electrostatically neutral patch of 12 charged residues that resides next to the tunnel at each opening. Based on the arrangement of critical binding residues on the LOX-1 structure, a binding mode for the ligand binding. Single amino acid substitution in the dimer interface caused a severe reduction in LOX-1 binding activity, suggesting that the correct dimer arrangement is crucial for binding to Ox-LDL. Based on the LDL model structure, possible binding modes of LOX-1 to Ox-LDL are proposed (Ohki et al. 2005).

The crystal of the monomeric ligand-binding domain of LOX-1 belongs to space group $P2_12_12_1$, with unit-cell parameters a = 56.79, b = 67.57, c = 79.02 Å. The crystal of the dimeric form belongs to space group C2, with unit-cell parameters a = 70.86, b = 49.56, c = 76.73 Å, beta = 98.59°. Data for the dimeric form of the LOX-1 ligand-binding domain have been collected to 2.4 Å. For the monomeric form of the ligand-binding domain, native, heavy-atom derivative and SeMet-derivative crystals have been obtained; their diffraction data have been measured to 3.0, 2.4 and 1.8 Å resolution, respectively (Ishigaki et al. 2005).

The NECK domain of LOX-1 displays sequence similarity to the coiled-coil region of myosin, having been suggested it adopts a rod-like structure. The structural analysis of human LOX-1 reveals a unique structural feature of LOX-1 NECK. Despite significant sequence similarity with the myosin coiled-coil, LOX-1 NECK does not form a uniform rod-like structure. Although not random, one-third of the N-terminal NECK is less structured than the remainder of the protein and is highly sensitive to cleavage by a variety of proteases. The coiled-coil structure is localized at the C-terminal part of the NECK, but is in dynamic equilibrium among multiple conformational states on a mus-ms time scale. This chimeric structural property of the NECK region may enable clustered LOX-1 on the cell surface to recognize Ox-LDL (Ishigaki et al. 2007).

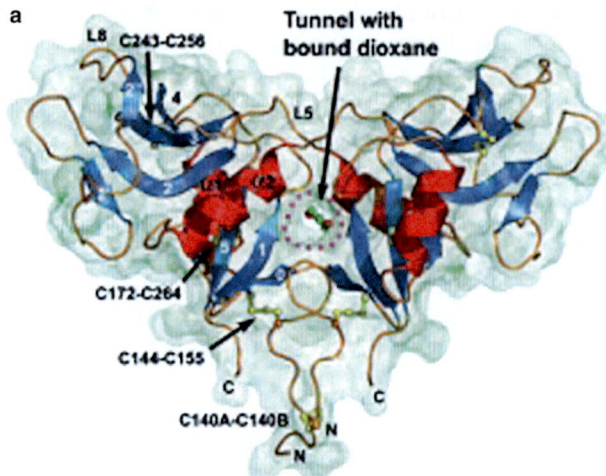

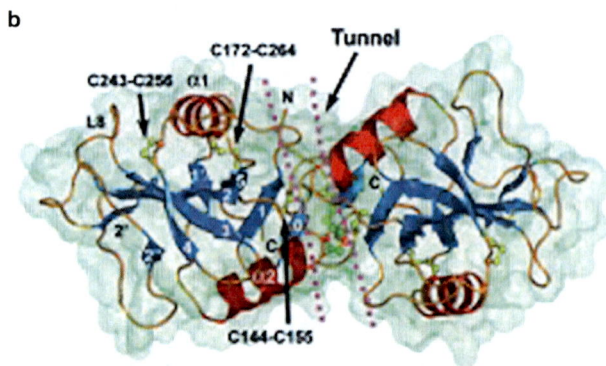

Fig. 34.6 Ribbon structure and transparent surface of human Lox-1 CTLD homodimer. *Side view* (**a**) and *top view* (**b**) (opposite the N and C termini) of Lox-1 from the high resolution monoclinic crystal. The central tunnel containing bound dioxane is outlined by a *dotted line*. α-Helices are in *red*; β-strands are *blue*. Secondary structures are *numbered* as proposed for CTLDs by Zelensky and Gready (2003). Loops are *numbered sequentially*. A semitransparent molecular surface representation is shown in *light green*. The disulfide bonds and dioxane are represented by *yellow* and *green ball-and-stick models*, respectively (Adapted with permission from Park et al. 2005a © American Society for Biochemistry and Molecular Biology)

recognition of modified LDL and other Lox-1 ligands has been proposed (Fig. 34.6) (Park et al. 2005a). The central hydrophobic tunnel that extends through the entire LOX-1 molecule is a key functional domain of the receptor and is critical for the recognition of modified LDL (Francone et al. 2009).

Ohki et al. (2005) determined the crystal structure of the ligand binding domain of 143–271 subunit (and mutagenesis of Trp[150]; Arg[208]; Arg[209]; His[226]; Arg[229]; Arg[231] and Arg[248]) of LOX-1 at 2.4 Å and revealed a short stalk region connecting the domain to the membrane-spanning region, as a homodimer linked by an interchain disulfide bond. In vivo assays with LOX-1 mutants revealed that the "basic spine," consisting of linearly aligned arginine residues spanning over the dimer surface, is responsible for

34.9.5 Functions of LOX-1

34.9.5.1 In Support of Cell Adhesion

LOX-1 can support adhesion of mononuclear leukocytes. Under a static condition, CHO-K1 cells stably expressing LOX-1 showed more prominent adhesion of human peripheral blood mononuclear leukocytes and human monocytic cell line than untransfected CHO-K1 cells. Mononuclear leukocytes also adhered to plastic plates precoated with recombinant soluble LOX-1 extracellular domain (Hayashida et al. 2002). Kakutani et al. (2000) showed the possibility that LOX-1 is involved in the platelet-endothelium interaction and hence directly in endothelial dysfunction. More importantly, the binding of platelets to LOX-1 enhanced the release of endothelin-1 from endothelial cells, supporting the induction of endothelial dysfunction, which would, in turn, promote the atherogenic process. LOX-1 may initiate and promote atherosclerosis, binding not only Ox-LDL but also platelets. LOX-1 can support cell adhesion to fibronectin (FN) in a divalent cation-independent fashion. CHO-K1 cells expressing bovine LOX-1, but not untransfected CHO-K1 cells, can adhere to FN-coated plates, but not to collagen-coated plates, in presence of EDTA (Shimaoka et al. 2001a).

LOX-1 on Endothelial Cells Binds with Bacterial Ligands: Adhesion of bacteria to vascular endothelial cells as well as mucosal cells and epithelial cells appears to be one of the initial steps in the process of bacterial infection, including infective endocarditis. LOX-1 with C-type lectin-like structure, can support adhesion of bacteria and works as a cell surface receptor for Gram-positive and Gram-negative bacteria, such as *S. aureus* and *E. coli*, in a mechanism similar to that of class A scavenger receptors; however, other unknown molecules may also be involved in the adhesion of *E. coli* to cultured bovine aortic endothelial cells (BAEC), which is enhanced by poly-I (Palaniyar et al. 2002; Shimaoka et al. 2001b). Thus, LOX-1 is an adhesion molecule involved in leukocyte recruitment and represents an attractive target for modulation of endotoxin-induced inflammation (Honjo et al. 2003).

34.9.5.2 Role in Apoptosis

LOX-1 mediates Ox-LDL-induced apoptosis of endothelial cells, monocyte adhesion to endothelium, smooth muscle cells (SMCs), and phagocytosis of aged cells. LOX-1 activation by Ox-LDL causes endothelial changes that are characterized by activation of NF-kB through an increased ROSs, subsequent induction of adhesion molecules, and endothelial apoptosis. Ox-LDL also induced apoptosis of HCAECs and LOX-1 mediates in this function. Findings suggest that Ox-LDL through its receptor LOX-1 decreases the expression of anti-apoptotic proteins Bcl-2 and c-IAP-1. This is followed by activation of apoptotic signaling pathway, involving release of cytochrome c and Smac and activation of caspase-9 and then caspase-3 (Chen et al. 2004). Duerrschmidt et al. (2006) reported the presence of LOX-1 in granulosa cells from patients under in vitro fertilization therapy after Ox-LDL induced expression of LOX-1 mRNA. The Ox-LDL treatment caused autophagy form of programmed cell death in association with reorganization of the actin cytoskeleton. It suggested that follicular atresia is not under the exclusive control of apoptosis. The LOX-1-dependent autophagy represents an alternate form of programmed cell death. Obese women with high blood levels of Ox-LDL may display an increased rate of autophagic granulosa cell death.

34.9.5.3 In Organ Transplantation

The pathogenesis of posttransplant CAD, which is a major form of chronic rejection after cardiac transplantation, is not fully understood. Suga et al. (2004) investigated the expression of LOX-1 mRNA in murine allografted hearts that develop diffuse coronary obstruction. Results indicate that alloimmune responses induce up-regulation of LOX-1 mRNA in transplanted hearts and hence increased LOX-1 may be involved in the progression of obstructive vascular changes (Suga et al. 2004). In human atherosclerotic lesions, the expression of LOX-1 is significantly increased in atherosclerotic grafted vein and carotid artery specimens compared with that in normal arteries. LOX-1 expression was colocalized with apoptotic cells. The apoptotic cells were present mostly in the rupture-prone regions of the atherosclerotic plaque. These observations indicated that LOX-1 is extensively expressed in the proliferated intima of grafted veins and in advanced atherosclerotic carotid arteries (Li et al. 2002a, b; Jin et al. 2004).

Older-age renal allografts are associated with inferior survival, though the mechanisms are not clear. Since, ROS participate in aging and in chronic vascular disease, Brasen et al. (2005) reported that induction of LOX-1-related oxidation pathways and increased susceptibility to oxidative stress plays significant role in promoting vascular injury in old transplants independent of age of recipient (Brasen et al. 2005).

34.9.6 LOX-1 and Pathophysiology

LOX-1 in Atherogenesis: The expression of LOX-1 receptor activates a variety of intracellular processes that leads to expression of adhesion molecules and endothelial activation. The experimental evidence suggested that LOX-1 may contribute to the development of vascular injury. For example, LOX-1 is up-regulated in aorta from hypertensive, diabetic, and hyperlipidemic animal models. Also, LOX-1 over-expression is observed in atherosclerotic regions and damaged kidneys. In vivo, endothelial cells covering early atherosclerotic lesions and macrophages and smooth muscle cells accumulated in the intima of advanced atherosclerotic plaques express LOX-1. LOX-1 is cleaved at membrane proximal extracellular domain by some protease activities and released from the cell surface. Measurement of soluble LOX-1 in vivo may provide novel diagnostic strategy for the evaluation and prediction of atherosclerosis and vascular diseases (Kume and Kita 2001).

LOX-1 in Hypertensive Rats: LOX-1 is expressed in hypertensive state. LOX-1 expression was low in the aorta of salt-resistant (DR: 0.3% salt) and of Dahl salt-sensitive (DS: 8% salt) rats on a control diet, whereas it was elevated in salt-loaded DS rats. Results indicated that LOX-1 expression in aorta and vein was upregulated in hypertensive rats, which may be involved in impaired endothelium-dependent vasodilation in these rats (Nagase et al. 1997). In kidneys and glomeruli of hypertensive DS rats, LOX-1 gene expression was markedly elevated compared with those of normotensive DR and DS rats. Prolonged salt loading further increased the renal LOX-1 expression in DS rats. Nagase et al. (2000) suggested a possible pathogenetic role for renal LOX-1 in the progression to hypertensive glomerulosclerosis.

LOX-1 and monocyte chemoattractant protein-1 (MCP-1) are involved in the initiation and progression of atherosclerosis. In order to examine differential role of LOX-1 and MCP-1 on the severity of early stage of atherosclerosis, Hamakawa et al. (2004) investigated atherosclerotic changes by exposure to hypertension and hyperlipidemia in common carotid arteries (CCAs) of stroke-prone spontaneously hypertensive rat (SHR-SP). Three groups were examined as: control [Wistar Kyoto rat (WKY) group], hypertension (SHR-SP group) and hypertension + hyperlipidemia [SHR-SP + high fat and cholesterol (HFC) group]. LOX-1 and MCP-1 expressions were coordinately up-regulated at mRNA and protein levels in an early stage of sclerosis depending on the severity of atherosclerotic stress. Thus, activations of LOX-1 and MCP-1 were collectively involved in the early stage of atherosclerosis (Hamakawa et al. 2004).

LOX-1 is critical in foam cell formation of macrophages (Mφ) and SMC. Inhibition of LOX-1 expression reduces foam cell formation and might influence lipid core formation in atherosclerotic lesions. Balloon-injury model of rabbit aorta revealed that LOX-1 mRNA is expressed 2 days after injury, and remained elevated until 24 weeks after injury. However, LOX-1 was not detected in media of non-injured aorta but expressed in both medial and neointimal SMC weeks after injury. Thus, LOX-1 mediates Ox-LDL-induced SMC proliferation and plays a role in neointimal formation after vascular injury (Eto et al. 2006) and confirmed in a rat model of balloon injury (Hinagata et al. 2006).

Platelet-Endothelium Interaction Mediated by LOX-1 Reduces NO in ECs: The role of LOX-1 in the thrombotic system was addressed by Cominacini et al. (2003b) following platelet interaction with LOX-1. The incubation of Bovine aortic endothelial cells (BAECs) and CHO-K1 cells stably expressing bovine LOX-1 (BLOX-1-CHO) cells with human platelets induced a sharp and dose-dependent increase in intracellular concentration of ROS and O_2^-. The increase in intracellular concentration of O_2^- was followed by a dose-dependent reduction in basal and bradykinin-induced intracellular NO concentration. The increase in O_2^- and the reduction of NO were inhibited by the presence of vitamin C and anti-LOX-1 mAb. The results show that one of the pathophysiologic consequences of platelet binding to LOX-1 may be the inactivation of NO through an increased cellular production of O_2^-.

Remnant lipoprotein particles (RLPs), products of lipolytic degradation of triglyceride-rich lipoprotein derived from VLDL, exert atherogenesis. RLPs stimulated superoxide formation and induction of cytokines in HUVECs via activation of LOX-1, consequently leading to reduction in cell viability with DNA fragmentation. The cilostazol exerts a cell-protective effect by suppressing these variables. Upon incubation of HUVECs with RLPs, adherent monocytes increased along with increased cell surface expression of adhesion molecules and increased expression of LOX-1 receptor protein. Cilostazol repressed these variables and hence it was suggested that cilostazol suppresses RLP-stimulated increased monocyte adhesion to HUVECs by suppression of LOX-1 receptor-coupled NF-kB-dependent nuclear transcription (Park et al. 2005b; Shin et al. 2004).

LOX-1 Contributes to Vein Graft Atherosclerosis: Ge et al. (2004) designed a study to examine the expression of LOX-1 in vein grafts atherosclerosis and the modulating effect of losartan (an AT-1 receptor antagonist) on it in male rabbits. After surgery, rabbits were fed with high cholesterol diet (HC), high cholesterol diet plus losartan (LHC) or regular chow (control) for 12 weeks. Study revealed that LOX-1 was expressed in endothelium and neointima of autologous vein grafts of rabbits. Increased LOX-1 expression was associated with vein grafts atherosclerosis development. Downregulation of LOX-1 by losartan might contribute to its attenuating effect on vein grafts atherosclerosis.

34.9.7 Macrophage Differentiation to Foam Cells

Foam cell formation from macrophages with subsequent fatty streak formation plays a key role in early atherogenesis. LOX-1 may play an important role in Ox-LDL uptake by macrophages and subsequent foam cell formation in this cell type (Yoshida et al. 1998). This role of LOX-1 was substantiated by studies on monocyte adhesion to human coronary artery endothelial cells (HCAECs) and endothelial injury in response to Ox-LDL in presence of antisense oligodeoxynucleotides to 5′-coding sequence of human LOX-1 gene. Studies indicated that LOX-1 is a key factor in Ox-LDL-mediated monocyte adhesion to HCAECs (Li and Mehta 2000). Depression of scavenger receptor function in monocytes by TGF-β1 at low concentrations reduced foam cell formation (Draude and Lorenz 2000). The expression of many genes is likely modulated during macrophage transformation into a foam cell. Functional consequences of modulation of three groups of genes: Scavenger Receptors including LOX-1, the PPAR family of nuclear receptors, and a number of genes involved in eicosanoid biosynthesis have been described (Shashkin et al. 2005). Scavenger receptors appear to play a key role in uptake of Ox-LDL, while minimally-modified LDL appears to interact with CD14/TLR4. The regulation of scavenger receptors is, in part, mediated by the PPAR family of nuclear receptors (Shashkin et al. 2005).

References

Adachi H, Tsujimoto M (2006) Endothelial scavenger receptors. Prog Lipid Res 45:379–404

Adachi Y, Ishii T, Ikeda Y et al (2004) Characterization of beta-glucan recognition site on C-type lectin, dectin 1. Infect Immun 72:4159–4171

Adams EL, Rice PJ, Graves B et al (2008) Differential high-affinity interaction of dectin-1 with natural or synthetic glucans is dependent upon primary structure and is influenced by polymer chain length and side-chain branching. J Pharmacol Exp Ther 325:115–123

Agrawal S, Gupta S, Agrawal A (2010) Human dendritic cells activated via dectin-1 are efficient at priming Th17, cytotoxic CD8 T and B cell responses. PLoS One 5:e13418

Ahren IL, Eriksson E, Egesten A, Riesbeck K (2003) Nontypeable Haemophilus influenzae activates human eosinophils through β-glucan receptors. Am J Respir Cell Mol Biol 29:598–605

Aoyama T, Sawamura T, Furutani Y et al (1999) Structure and chromosomal assignment of the human lectin-like oxidized low-density-lipoprotein receptor-1 (LOX-1) gene. Biochem J 339:177–184

Ariizumi K, Shen G-L, Shikano S et al (2000a) Cloning of a second dendritic cell-associated c-type lectin (Dectin-2) and its alternatively spliced isoforms. J Biol Chem 275:11957–11963

Ariizumi K, Shen G-L, Shikano S et al (2000b) Identification of a novel, dendritic cell-associated molecule, dectin-1, by subtractive cDNA cloning. J Biol Chem 275:20157–20167

Balestrieri B, Maekawa A, Xing W et al (2009) Group V secretory phospholipase A2 modulates phagosome maturation and regulates the innate immune response against Candida albicans. J Immunol 182:4891–4898

Bertozzi CC, Schmaier AA, Mericko P et al (2010) Platelets regulate lymphatic vascular development through CLEC-2-SLP-76 signaling. Blood 116:661–670

Brasen JH, Nieminen-Kelha M, Markmann D et al (2005) Lectin-like oxidized low-density lipoprotein (LDL) receptor (LOX-1)-mediated pathway and vascular oxidative injury in older-age rat renal transplants. Kidney Int 67:1583–1594

Brown GD (2006) Dectin-1: a signaling non-TLR pattern-recognition receptor. Nat Rev Immunol 6:33–43

Brown GD, Gordon S (2001) Immune recognition: a new receptor for β-glucans. Nature 413:36–37

Brown GD, Taylor PR, Reid DM et al (2002) Dectin-1 is a major β-glucan receptor on macrophages. J Exp Med 196:407–412

Brown GD, Herre J, Williams DL et al (2003) Dectin-1 mediates the biological effects of β-glucans. J Exp Med 197:1119–1124

Brown J, O'Callaghan CA, Marshall AS et al (2007) Structure of the fungal β-glucan-binding immune receptor dectin-1: implications for function. Protein Sci 16:1042–1052

Bull C, Sobanov Y, Rohrdanz B et al (2000) The centromeric part of the human NK gene complex: linkage of LOX-1 and LY49L with the CD94/NKG2 region. Genes Immun 1:280–287

Caminschi I, Proietto AI et al (2008) The dendritic cell subtype-restricted C-type lectin Clec9A is a target for vaccine enhancement. Blood 112:3264–3273

Carter RW, Thompson C, Reid DM et al (2006) Preferential induction of CD4+ T cell responses through in vivo targeting ofantigen to dendritic cell-associated C-type lectin-1. J Immunol 177:2276–2284

Chaipan C, Soilleux EJ, Simpson P et al (2006) DC-SIGN and CLEC-2 mediate human immunodeficiency virus type 1 capture by platelets. J Virol 80:8951–8960

Chaipan C, Steffen I, Tsegaye TS et al (2010) Incorporation of podoplanin into HIV released from HEK-293T cells, but not PBMC, is required for efficient binding to the attachment factor CLEC-2. Retrovirology 7:47

Chang CH, Chung CH, Hsu CC, Huang TY, Huang TF (2010) A novel mechanism of cytokine release in phagocytes induced by aggretin, a snake venom C-type lectin protein, through CLEC-2 ligation. J Thromb Haemost 8:2563

Chen M, Sawamura T (2005) Essential role of cytoplasmic sequences for cell-surface sorting of the lectin-like oxidized LDL receptor-1 (LOX-1). J Mol Cell Cardiol 39:553–561

Chen M, Nagase M, Fujita T et al (2001) Diabetes enhances lectin-like oxidized LDL receptor-1 (LOX-1) expression in the vascular endothelium: possible role of LOX-1 ligand and AGE. Biochem Biophys Res Commun 287:962–968

Chen M, Masaki T, Sawamura T (2002) LOX-1, the receptor for oxidized low-density lipoprotein identified from endothelial cells: implications in endothelial dysfunction and atherosclerosis. Pharmacol Ther 95:89–100

Chen J, Mehta JL, Haider N et al (2004) Role of caspases in Ox-LDL-induced apoptotic cascade in human coronary artery endothelial cells. Circ Res 94:370–376

Chen K, Chen J, Liu Y et al (2005) Adhesion molecule expression in fibroblasts: alteration in fibroblast biology after transfection with LOX-1 plasmids. Hypertension 46:622–627

Christou CM, Pearce AC, Watson AA et al (2008) Renal cells activate the platelet receptor CLEC-2 through podoplanin. Biochem J 411:133–140

Colonna M, Samaridis J, Angman L (2000) Molecular characterization of two novel C-type lectin-like receptors, one of which is selectively expressed in human dendritic cells. Eur J Immunol 30:697–704

Cominacini L, Pasini AF, Garbin U et al (2003) The platelet-endothelium interaction mediated by lectin-like oxidized low-density lipoprotein receptor-1 reduces the intracellular concentration of nitric oxide in endothelial cells. J Am Coll Cardiol 41:499–507

de Vries HS, Plantinga TS, van Krieken JH et al (2009) Genetic association analysis of the functional c.714T>G polymorphism and mucosal expression of dectin-1 in inflammatory bowel disease. PLoS One 4:e7818

Dennehy KM, Brown GD (2007) The role of the β-glucan receptor dectin-1 in control of fungal infection. J Leukoc Biol 82:253–258

Dennehy KM, Ferwerda G, Faro-Trindade I et al (2008) Syk kinase is required for collaborative cytokine production induced through dectin-1 and toll-like receptors. Eur J Immunol 38:500–506

Dennehy KM, Willment JA, Williams DL, Brown GD (2009) Reciprocal regulation of IL-23 and IL-12 following co-activation of dectin-1 and TLR signaling pathways. Eur J Immunol 39:1379–1386

Dillon S, Agrawal S, Banerjee K, van Dyke T et al (2006) Yeast zymosan, a stimulus for TLR2 and dectin-1, induces regulatory antigenpresenting cells and immunological tolerance. J Clin Invest 116:916–928

Draude G, Lorenz RL (2000) TGF-β1 downregulates CD36 and scavenger receptor A but upregulates LOX-1 in human macrophages. Am J Physiol Heart Circ Physiol 278:H1042–H1048

Duerrschmidt N, Zabirnyk O, Nowicki M et al (2006) Lectin-like oxidized low-density lipoprotein receptor-1-mediated autophagy in human granulosa cells as an alternative of programmed cell death. Endocrinology 147:3851–3860

Eto H, Miyata M, Kume N et al (2006) Expression of lectin-like oxidized LDL receptor-1 in smooth muscle cells after vascular injury. Biochem Biophys Res Commun 341:591–598

Falconi M, Biocca S, Novelli G, Desideri A (2007) Molecular dynamics simulation of human LOX-1 provides an explanation for the lack of OxLDL binding to the Trp150Ala mutant. BMC Struct Biol 7:73

Fernandes MJ, Iscove NN, Gingras G, Calabretta B (2000) Identification and characterization of the gene for a novel C-type lectin (CLECSF7) that maps near the natural killer gene complex on human chromosome. Genomics 69:263–270

Ferwerda G, Meyer-Wentrup F et al (2008) Dectin-1 synergizes with TLR2 and TLR4 for cytokine production in human primary monocytes and macrophages. Cell Microbiol 10:2058–2066

Ferwerda B, Ferwerda G, Plantinga TS et al (2009) Human dectin-1 deficiency and mucocutaneous fungal infections. N Engl J Med 361:1760–1767

Francone OL, Tu M, Royer LJ et al (2009) The hydrophobic tunnel present in LOX-1 is essential for oxidized LDL recognition and binding. J Lipid Res 50:546–555

Frasnelli ME, Tarussio D, Chobaz-Peclat V et al (2005) TLR2 modulates inflammation in zymosan-induced arthritis in mice. Arthritis Res Ther 7:R370–R379

Fuller GL, Williams JA, Tomlinson MG et al (2007) The C-type lectin receptors CLEC-2 and dectin-1, but not DC-SIGN, signal via a novel YXXL-dependent signaling cascade. J Biol Chem 282:12397–12409

Galès A, Conduché A, Bernad J et al (2010) PPARγ controls dectin-1 expression required for host antifungal defense against *Candida albicans*. PLoS Pathog 6:e1000714

Gantner BN, Simmons RM, Canavera SJ et al (2003) Collaborative induction of inflammatory responses by dectin-1 and Toll-like receptor 2. J Exp Med 197:1107–1117

Ge J, Huang D, Liang C et al (2004) Upregulation of lectinlike oxidized low-density lipoprotein receptor-1 expression contributes to the vein graft atherosclerosis: modulation by losartan. Atherosclerosis 177:263–268

Gerosa F, Baldani-Guerra B, Lyakh LA et al (2008) Differential regulation of interleukin 12 and interleukin 23 production in human dendritic cells. J Exp Med 205:1447–1461

Goodridge HS, Simmons RM, Underhill DM (2007) Dectin-1 stimulation by *Candida albicans* yeast or zymosan triggers NFAT activation in macrophages and dendritic cells. J Immunol 178:3107–3115

Gow NA, Netea MG, Munro CA et al (2007) Immune recognition of *Candida albicans* β-glucan by dectin-1. J Infect Dis 196:1565–1571

Gringhuis SI, den Dunnen J, Litjens M et al (2009) Dectin-1 directs T helper cell differentiation by controlling noncanonical NF-kappaB activation through Raf-1 and Syk. Nat Immunol 10:203–213

Gross O, Gewies A, Finger K et al (2006) Card9 controls a non-TLR signaling pathway for innate anti-fungal immunity. Nature 442:651–656

Hamakawa Y, Omori N, Ouchida M et al (2004) Severity dependent up-regulations of LOX-1 and MCP-1 in early sclerotic changes of common carotid arteries in spontaneously hypertensive rats. Neurol Res 26:767–773

Hara H, Ishihara C, Takeuchi A et al (2007) The adaptor protein CARD9 is essential for the activation of myeloid cells through ITAM-associated and Toll-like receptors. Nat Immunol 8:619–629

Hayashida K, Kume N, Minami M, Kita T (2002) Lectin-like oxidized LDL receptor-1 (LOX-1) supports adhesion of mononuclear leukocytes and a monocyte-like cell line THP-1 cells under static and flow conditions. FEBS Lett 511:133–138

Heinsbroek SE, Taylor PR, Rosas M et al (2006) Expression of functionally different dectin-1 isoforms by murine macrophages. J Immunol 176:5513–5518

Heinsbroek SE, Taylor PR, Martinez FO et al (2008) Stage-specific sampling by pattern recognition receptors during *Candida albicans* phagocytosis. PLoS Pathog 4:e1000218

Hermanz-Falcón P, Arce I, Roda-Navarro P et al (2001) Cloning of human DECTIN-1, a novel C-type lectin-like receptor gene expressed on dendritic cells. Immunogenetics 53:288–295

Herre J, Marshall AS, Caron E et al (2004) Dectin-1 uses novel mechanisms for yeast phagocytosis in macrophages. Blood 104:4038–4045

Hida S, Miura NN, Adachi Y, Ohno N (2007) Cell wall β-glucan derived from *Candida albicans* acts as a trigger for autoimmune arthritis in SKG mice. Biol Pharm Bull 30:1589–1592

Hinagata J, Kakutani M, Fujii T et al (2006) Oxidized LDL receptor LOX-1 is involved in neointimal hyperplasia after balloon arterial injury in a rat model. Cardiovasc Res 69:263–271

Hoffmann SC, Schellack C, Textor S et al (2007) Identification of CLEC12B, an inhibitory receptor on myeloid cells. J Biol Chem 282(31):22370–22375

Honjo M, Nakamura K, Yamashiro K et al (2003) Lectin-like oxidized LDL receptor-1 is a cell-adhesion molecule involved in endotoxin-induced inflammation. Proc Natl Acad Sci USA 100:1274–1279

Hooley E, Papagrigoriou E, Navdaev A et al (2008) The crystal structure of the platelet activator aggretin reveals a novel (αβ)2 dimeric structure. Biochemistry 47:7831–7837

Hoshikawa H, Sawamura T, Kakutani M et al (1998) High affinity binding of oxidized LDL to mouse lectin-like oxidized LDL receptor (LOX-1). Biochem Biophys Res Commun 245:841–846

Hsu YM, Zhang Y, You Y et al (2007) The adaptor protein CARD9 is required for innate immune responses to intracellular pathogens. Nat Immunol 8:198–205

Huang Z, Zhang T, Yang J et al (2005) Cloning and expression of human lectin-like oxidized low density lipoprotein receptor-1 in Pichia pastoris. Biotechnol Lett 27:49–52

Hughes CE, Navarro-Nunez L, Finney BA et al (2010a) CLEC-2 is not required for platelet aggregation at arteriolar shear. J Thromb Haemost 8:2328–2332

Hughes CE, Pollitt AY, Mori J et al (2010b) CLEC-2 activates Syk through dimerization. Blood 115:2947–2955

Huysamen C, Brown GD (2009) The fungal pattern recognition receptor, dectin-1, and the associated cluster of C-type lectin-like receptors. FEMS Microbiol Lett 290:121–128

Huysamen C, Willment JA, Dennehy KM, Brown GD (2008) CLEC9A is a novel activation C-type lectin-like receptor expressed on BDCA3+ dendritic cells and a subset of monocytes. J Biol Chem 283:16693–16701

Ishigaki T, Ohki I, Oyama T et al (2005) Purification, crystallization and preliminary X-ray analysis of the ligand-binding domain of human lectin-like oxidized low-density lipoprotein receptor 1 (LOX-1). Acta Crystallograph Sect F Struct Biol Cryst Commun 61:524–527

Ishigaki T, Ohki I, Utsunomiya-Tate N, Tate SI (2007) Chimeric structural stabilities in the coiled-coil structure of the NECK domain in human lectin-like oxidized low-density lipoprotein receptor 1 (LOX-1). J Biochem 141:855–866

Jin S, Mathis AS, Rosenblatt J et al (2004) Insights into cyclosporine A-induced atherosclerotic risk in transplant recipients: macrophage scavenger receptor regulation. Transplantation 77:497–504

Joffre OP, Sancho D, Zelenay S et al (2010) Efficient and versatile manipulation of the peripheral CD4+ T-cell compartment by antigen targeting to DNGR-1/CLEC9A. Eur J Immunol 40:1255–1265

Jono T, Miyazaki A, Nagai R et al (2002) Lectin-like oxidized low density lipoprotein receptor-1 (LOX-1) serves as an endothelial receptor for advanced glycation end products (AGE). FEBS Lett 511:170–174

Kakutani M, Masaki T, Sawamura T (2000) A platelet-endothelium interaction mediated by lectin-like oxidized low-density lipoprotein receptor-1. Proc Natl Acad Sci USA 97:360–364

Kataoka H, Kume N, Miyamoto S et al (2000) Biosynthesis and post-translational processing of lectin-like oxidized low density lipoprotein receptor-1 (LOX-1). N-linked glycosylation affects cell-surface expression and ligand binding. J Biol Chem 275:6573–6579

Kato Y, Kaneko MK, Kunita A et al (2008) Molecular analysis of the pathophysiological binding of the platelet aggregation-inducing factor podoplanin to the C-type lectin-like receptor CLEC-2. Cancer Sci 99:54–61

Kerrigan AM, Dennehy KM et al (2009) CLEC-2 is a phagocytic activation receptor expressed on murine peripheral blood neutrophils. J Immunol 182:4150–4157

Kumar H, Kumagai Y, Tsuchida T et al (2009) Involvement of the NLRP3 inflammasome in innate and humoral adaptive immune responses to fungal β-glucan. J Immunol 183:8061–8067

Kume N, Kita T (2001) Roles of lectin-like oxidized LDL receptor-1 and its soluble forms in atherogenesis. Curr Opin Lipidol 12:419–423

Kume N, Kita T (2004) Apoptosis of vascular cells by oxidized LDL: involvement of caspases and LOX-1 and its implication in atherosclerotic plaque rupture. Circ Res 94:370–376

Kuniyasu A, Ohgami N, Hayashi S et al (2003) CD36-mediated endocytic uptake of advanced glycation end products (AGE) in mouse 3T3-L1 and human subcutaneous adipocytes. FEBS Lett 537:85–90

Leal SM Jr, Cowden S, Hsia YC et al (2010) Distinct roles for dectin-1 and TLR4 in the pathogenesis of Aspergillus fumigatus keratitis. PLoS Pathog 6:e1000976

Lee MS, Kim Y-J (2007) Pattern-recognition receptor signaling initiated from extracellular, membrane, and cytoplasmic space. Mol Cells 23:1–10

Leibundgut-Landmann S, Gross O, Robinson MJ et al (2007) Syk- and CARD9-dependent coupling of innate immunity to the induction of T helper cells that produce interleukin 17. Nat Immunol 8:630–638

Leibundgut-Landmann S, Osorio F, Brown GD et al (2008) Stimulation of dendritic cells via the dectin-1/Syk pathway allows priming of cytotoxic T-cell responses. Blood 112:4971–4980

Li D, Mehta JL (2000) Antisense to LOX-1 inhibits oxidized LDL-mediated upregulation of monocyte chemoattractant protein-1 and monocyte adhesion to human coronary artery endothelial cells. Circulation 101:2889–2895

Li D, Chen H, Romeo F, Sawamura T, Saldeen T, Mehta JL (2002a) Statins modulate oxidized low-density lipoprotein-mediated adhesion molecule expression in human coronary artery endothelial cells: role of LOX-1. J Pharmacol Exp Ther 302:601–605

Li DY, Chen HJ, Staples ED et al (2002b) Oxidized low-density lipoprotein receptor LOX-1 and apoptosis in human atherosclerotic lesions. J Cardiovasc Pharmacol Ther 7:147–153

Marshall AS, Willment JA, Lin HH et al (2004) Identification and characterization of a novel human myeloid inhibitory C-type lectin-like receptor (MICL) that is predominantly expressed on granulocytes and monocytes. J Biol Chem 279:14792–14802

Marshall AS, Willment JA, Pyz E et al (2006) Human MICL (CLEC12A) is differentially glycosylated and is down-regulated following cellular activation. Eur J Immunol 36:2159–2169

Mattila PE, Metz AE, Rapaka RR et al (2008) Dectin-1 Fc targeting of Aspergillus fumigatus β-glucans augments innate defense against invasive pulmonary aspergillosis. Antimicrob Agents Chemother 52:1171–1172

May F, Hagedorn I, Pleines I et al (2009) CLEC-2 is an essential platelet-activating receptor in hemostasis and thrombosis. Blood 114:3464–3472

Mehta JL, Chen J, Hermonat PL et al (2006) Lectin-like, oxidized low-density lipoprotein receptor-1 (LOX-1): a critical player in the development of atherosclerosis and related disorders. Cardiovasc Res 69:36–45

Meyer-Wentrup F, Figdor CG, Ansems M et al (2007) Dectin-1 interaction with tetraspanin CD37 inhibits IL-6 production. J Immunol 178:154–162

Mori J, Pearce AC, Spalton JC et al (2008) G6b-B inhibits constitutive and agonist-induced signaling by glycoprotein VI and CLEC-2. J Biol Chem 283:35419–35427

Mukai E, Kume N, Hayashida K et al (2004) Heparin-binding EGF-like growth factor induces expression of lectin-like oxidized LDL receptor-1 in vascular smooth muscle cells. Atherosclerosis 176:289–296

Murase T, Kume N, Kataoka H et al (2000) Identification of soluble forms of lectin-like oxidized LDL receptor-1. Arterioscler Thromb Vasc Biol 20:715–720

Murphy JE, Tacon D et al (2006) LOX-1 scavenger receptor mediates calcium-dependent recognition of phosphatidylserine and apoptotic cells. Biochem J 393:107–115

Murray PJ (2007) The JAK-STAT signaling pathway: input and output integration. J Immunol 178:2623–2629

Nagase M, Abe J, Takahashi K et al (1998a) Genomic organization and regulation of expression of the lectin-like oxidized low-density lipoprotein receptor (LOX-1) gene. J Biol Chem 273:33702–33707

Nagase M, Hirose S, Sawamura T, Masaki T, Fujita T (1997) Enhanced expression of endothelial oxidized low-density lipoprotein receptor (LOX-1) in hypertensive rats. Biochem Biophys Res Commun 237:496–498

Nagase M, Hirose S, Fujita T (1998b) Unique repetitive sequence and unexpected regulation of expression of rat endothelial receptor for oxidized low-density lipoprotein (LOX-1). Biochem J 330:1417–1422

Nagase M, Kaname S, Nagase T et al (2000) Expression of LOX-1, an oxidized low-density lipoprotein receptor, in experimental hypertensive glomerulosclerosis. J Am Soc Nephrol 11:1826–1836

O'Callaghan CA (2009) Thrombomodulation via CLEC-2 targeting. Curr Opin Pharmacol 9:90–95

Odegaard JI, Ricardo-Gonzalez RR et al (2007) Macrophage-specific PPARγ controls alternative activation and improves insulin resistance. Nature 447:1116–1120

Ohki I, Ishigaki T, Oyama T et al (2005) Crystal structure of human lectin-like, oxidized low-density lipoprotein receptor 1 ligand binding domain and its ligand recognition mode to OxLDL. Structure 13:905–917

Osorio F, LeibundGut-Landmann S et al (2008) DC activated via dectin-1 convert Treg into IL-17 producers. Eur J Immunol 38:3274–3281

Ozaki Y, Suzuki-Inoue K, Inoue O (2009) Novel interactions in platelet biology: CLEC-2/podoplanin and laminin/GPVI. J Thromb Haemost 7(Suppl 1):191–194

Ozment-Skelton TR, Goldman MP, Gordon S et al (2006) Prolonged reduction of leukocyte membrane-associated dectin-1 levels following β-glucan administration. J Pharmacol Exp Ther 318:540–546

Ozment-Skelton TR, deFluiter EA, Ha T et al (2009) Leukocyte dectin-1 expression is differentially regulated in fungal versus polymicrobial sepsis. Crit Care Med 37:1038–1045

Palaniyar N, Nadesalingam J, Reid KB (2002) Pulmonary innate immune proteins and receptors that interact with gram-positive bacterial ligands. Immunobiology 205:575–594

Palma AS, Feizi T, Zhang Y et al (2006) Ligands for the β-glucan receptor, dectin-1, assigned using "designer" microarrays of oligosaccharide probes (neoglycolipids) generated from glucan polysaccharides. J Biol Chem 281:5771–5779

Park H, Adsit FG, Boyington JC (2005a) The 1.4 angstrom crystal structure of the human oxidized low density lipoprotein receptor lox-1. J Biol Chem 280:13593–13599

Park SY, Lee JH, Kim YK et al (2005b) Cilostazol prevents remnant lipoprotein particle-induced monocyte adhesion to endothelial cells by suppression of adhesion molecules and monocyte chemoattractant protein-1 expression via lectin-like receptor for oxidized low-density lipoprotein receptor activation. J Pharmacol Exp Ther 312:1241–1248

Plantinga TS, van der Velden WJ et al (2009) Early stop polymorphism in human dectin-1 is associated with increased candida colonization in hematopoietic stem cell transplant recipients. Clin Infect Dis 49:724–732

Plantinga TS, Fransen J, Takahashi N et al (2010) Functional consequences of DECTIN-1 early stop codon polymorphism Y238X in rheumatoid arthritis. Arthritis Res Ther 12:R26

Pollitt AY, Grygielska B, Leblond B et al (2010) Phosphorylation of CLEC-2 is dependent on lipid rafts, actin polymerization, secondary mediators, and Rac. Blood 115:2938–2946

Poulin LF, Salio M, Griessinger E et al (2010) Characterization of human DNGR-1+ BDCA3+ leukocytes as putative equivalents of mouse CD8α+dendritic cells. J Exp Med 207:1261–1271

Pyz E, Huysamen C, Marshall AS et al (2008) Characterisation of murine MICL (CLEC12A) and evidence for an endogenous ligand. Eur J Immunol 38:1157–1163

Rapaka RR, Goetzman ES, Zheng M et al (2007) Enhanced defense against *Pneumocystis carinii* mediated by a novel dectin-1 receptor Fc fusion protein. J Immunol 178:3702–3712

Reid DM, Montoya M, Taylor PR et al (2004) Expression of the β-glucan receptor, dectin-1, on murine leukocytes in situ correlates with its function in pathogen recognition and reveals potential roles in leukocyte interactions. J Leukoc Biol 76:86–94

Reid DM, Gow NAR, Brown GD (2009) Pattern recognition: recent insights from Dectin-1. Curr Opin Immunol 21:30–37

Ricote M, Li AC, Willson TM et al (1998) The peroxisome proliferator-activated receptor-γ is a negative regulator of macrophage activation. Nature 391:79–82

Rogers NC, Slack EC, Edwards AD et al (2005) Syk-dependent cytokine induction by Dectin-1 reveals a novel pattern recognition pathway for C type lectins. Immunity 22:507–517

Rothfuchs AG, Bafica A, Feng CG et al (2007) Dectin-1 interaction with Mycobacterium tuberculosis leads to enhanced IL-12p40 production by splenic dendritic cells. J Immunol 179:3463–3471

Ruan Y, Guo L et al (2009) RACK1 associates with CLEC-2 and promotes its ubiquitin-proteasome degradation. Biochem Biophys Res Commun 390:217–222

Saijo S, Fujikado N, Furuta T et al (2007) Dectin-1 is required for host defense against *Pneumocystis carinii* but not against *Candida albicans*. Nat Immunol 8:39–46

Sancho D, Mourão-Sá D, Joffre OP et al (2008) Tumor therapy in mice via antigen targeting to a novel, DC-restricted C-type lectin. J Clin Invest 118:2098–2110

Sancho D, Joffre OP, Keller AM et al (2009) Identification of a dendritic cell receptor that couples sensing of necrosis to immunity. Nature 458:899–903

Sattler S, Ghadially H, Reiche D et al (2010) Evolutionary development and expression pattern of the myeloid lectin-like receptor gene family encoded within the NK gene complex. Scand J Immunol 72:309–318

Sawamura T, Kume N, Aoyama T et al (1997) An endothelial receptor for oxidized low-density lipoprotein. Nature 386(6620):73–77

Shah VB, Huang Y, Keshwara R et al (2008) B-glucan activates microglia without inducing cytokine production in Dectin-1-dependent manner. J Immunol 180:2777–2785

Shashkin P, Dragulev B, Ley K (2005) Macrophage differentiation to foam cells. Curr Pharm Des 11:3061–3072

Shi X, Niimi S, Ohtani T, Machida S (2001) Characterization of residues and sequences of the carbohydrate recognition domain required for cell surface localization and ligand binding of human lectin-like oxidized LDL receptor. J Cell Sci 114(Pt 7):1273–1282

Shih HH, Zhang S, Cao W et al (2009) CRP is a novel ligand for the oxidized LDL receptor LOX-1. Am J Physiol Heart Circ Physiol 296:H1643–H1650

Shimaoka T, Kume N, Minami M et al (2001a) Lectin-like oxidized low density lipoprotein receptor-1 (LOX-1) supports cell adhesion to fibronectin. FEBS Lett 5042:65–68

Shimaoka T, Kume N, Minami M et al (2001b) LOX-1 supports adhesion of gram-positive and gram-negative bacteria. J Immunol 166:5108–5114

Shin Y, Morita T (1998) Rhodocytin, a functional novel platelet agonist belonging to the heterodimeric C-type lectin family, induces platelet aggregation independently of glycoprotein Ib. Biochem Biophys Res Commun 245:741–745

Shin HK, Kim YK, Kim KY et al (2004) Remnant lipoprotein particles induce apoptosis in endothelial cells by NAD(P)H oxidase-mediated production of superoxide and cytokines via lectin-like oxidized low-density lipoprotein receptor-1 activation: prevention by cilostazol. Circulation 109:1022–1028

Slack EC, Robinson MJ, Hernanz-Falcon P et al (2007) Syk-dependent ERK activation regulates IL-2 and IL-10 production by DC stimulated with zymosan. Eur J Immunol 37:1600–1612

Sobanov Y, Bernreiter A, Derdak S et al (2001) A novel cluster of lectin-like receptor genes expressed in monocytic, dendritic and endothelial cells maps close to the NK receptor genes in the human NK gene complex. Eur J Immunol 31:3493–3503

Spalton JC, Mori J, Pollitt AY et al (2009) The novel Syk inhibitor R406 reveals mechanistic differences in the initiation of GPVI and CLEC-2 signaling in platelets. J Thromb Haemost 7:1192–1199

Steele C, Marrero L, Swain S et al (2003) Alveolar macrophage-mediated killing of *Pneumocystis carinii f. sp. Muris* involves molecular recognition by the Dectin-1 β-glucan receptor. J Exp Med 198:1677–1688

Steele C, Rapaka RR, Metz A et al (2005) The β-glucan receptor dectin-1 recognizes specific morphologies of *Aspergillus fumigatus*. PLoS Pathog 1:e42

Suga M, Sawamura T, Nakatani T et al (2004) Expression of lectin-like oxidized low-density lipoprotein receptor-1 in allografted hearts. Transplant Proc 36:2440–2442

Suram S, Brown GD, Ghosh M et al (2006) Regulation of cytosolic phospholipase A2 activation and cyclooxygenase 2 expression in macrophages by the β-glucan receptor. J Biol Chem 281:5506–5514

Suzuki-Inoue K, Fuller GL, Garcĩa A, Tybulewicz VL, Ozaki Y, Watson SP et al (2006) A novel Syk-dependent mechanism of platelet activation by the C-type lectin receptor CLEC-2. Blood 107:542–549

Suzuki-Inoue K, Kato Y, Inoue O et al (2007) Involvement of the snake toxin receptor CLEC-2, in podoplanin-mediated platelet activation, by cancer cells. J Biol Chem 282:25993–26001

Suzuki-Inoue K, Inoue O, Ding G et al (2010) Essential in vivo roles of the C-type lectin receptor CLEC-2: embryonic/neonatal lethality of CLEC-2-deficient mice by blood/lymphatic misconnections and impaired thrombus formation of CLEC-2-deficient platelets. J Biol Chem 285:24494–24507

Tanigawa H, Miura S, Matsuo Y et al (2006) Dominant-negative Lox-1 blocks homodimerization of wild-type Lox-1-induced cell proliferation through extracellular signal regulated kinase 1/2 activation. Hypertension 48:294–300

Taylor PR, Brown GD, Reid DM et al (2002) The β-glucan receptor, dectin-1, is predominantly expressed on the surface of cells of the monocyte/macrophage and neutrophil lineages. J Immunol 169:3876–3882

Taylor PR, Tsoni SV, Willment JA et al (2007) Dectin-1 is required for β-glucan recognition and control of fungal infection. Nat Immunol 8:31–38

Thebault P, Lhermite N, Tilly G et al (2009) The C-type lectin-like receptor CLEC-1, expressed by myeloid cells and endothelial cells, is up-regulated by immunoregulatory mediators and moderates T cell activation. J Immunol 183:3099–3108

Ujita M, Nagayama H, Kanie S et al (2009) Carbohydrate binding specificity of recombinant human macrophage β-glucan receptor dectin-1. Biosci Biotechnol Biochem 73:237–240

Underhill DM, Rossnagle E, Lowell CA et al (2005) Dectin-1 activates Syk tyrosine kinase in a dynamic subset of macrophages for reactive oxygen production. Blood 106:2543–2550

Valera I, Fernández N, Trinidad AG et al (2008) Costimulation of dectin-1 and DC-SIGN triggers the arachidonic acid cascade in human monocyte-derived dendritic cells. J Immunol 180:5727–5736

van Beelen AJ, Zelinkova Z, Taanman-Kueter EW et al (2007) Stimulation of the intracellular bacterial sensor NOD2 programs dendritic cells to promote interleukin-17 production in human memory T cells. Immunity 27:660–669

van de Veerdonk FL, Marijnissen RJ et al (2009) The macrophage mannose receptor induces IL-17 in response to *Candida albicans*. Cell Host Microbe 5:329–340

Vohra RS, Murphy JE, Walker JH et al (2006) Atherosclerosis and the lectin-like OXidized low-density lipoprotein scavenger receptor. Trends Cardiovasc Med 16:60–64

Watson AA, O'Callaghan CA (2005) Crystallization and X-ray diffraction analysis of human CLEC-2. Acta Crystallogr Sect F Struct Biol Cryst Commun 61(Pt 12):1094–1096

Watson AA, Brown J, Harlos K et al (2007) The crystal structure and mutational binding analysis of the extracellular domain of the platelet-activating receptor CLEC-2. J Biol Chem 282:3165–3172

Watson AA, Eble JA, O'Callaghan CA (2008) Crystal structure of rhodocytin, a ligand for the platelet-activating receptor CLEC-2. Protein Sci 17:1611–1616

Watson AA, Christou CM, James JR et al (2009) The platelet receptor CLEC-2 is active as a dimer. Biochemistry 48:10988–10996

Weck MM, Appel S, Werth D et al (2008) hDectin-1 is involved in uptake and cross present-ation of cellular antigens. Blood 111:4264–4272

Werner JL, Metz AE, Horn D et al (2009) Requisite role for the dectin-1 β-glucan receptor in pulmonary defense against *Aspergillus fumigatus*. J Immunol 182:4938–4946

Willment JA, Lin HH, Reid DM et al (2003) Dectin-1 expression and function are enhanced on alternatively activated and GM-CSF-treated macrophages and are negatively regulated by IL-10, dexamethasone, and lipopolysaccharide. J Immunol 171: 4569–4573

Willment JA, Marshall AS, Reid DM et al (2005) The human β-glucan receptor is widely expressed and functionally equivalent to murine dectin-1 on primary cells. Eur J Immunol 35:1539–1547

Wong ES, Sanderson CE, Deakin JE et al (2009) Identification of natural killer cell receptor clusters in the platypus genome reveals an expansion of C-type lectin genes. Immunogenetics 61:565–579

Xie Q, Matsunaga S, Shi X et al (2003) Refolding and characterization of the functional ligand-binding domain of human lectin-like oxidized LDL receptor. Protein Expr Purif 32:68–74

Xie Q, Matsunaga S, Niimi S et al (2004) Human lectin-like oxidized low-density lipoprotein receptor-1 functions as a dimer in living cells. DNA Cell Biol 23:111–117

Xie J, Wu T, Guo L et al (2008) Molecular characterization of two novel isoforms and a soluble form of mouse CLEC-2. Biochem Biophys Res Commun 371:180–184

Xu S, Huo J, Gunawan M, Su IH, Lam KP (2009a) Activated dectin-1 localizes to lipid raft microdomains for signaling and activation of phagocytosis and cytokine production in dendritic cells. J Biol Chem 284:22005–22011

Xu S, Huo J, Lee KG et al (2009b) Phospholipase Cγ2 is critical for Dectin-1-mediated Ca2+ flux and cytokine production in dendritic cells. J Biol Chem 284:7038–7046

Yadav M, Schorey JS (2006) The β-glucan receptor dectin-1 functions together with TLR2 to mediate macrophage activation by mycobacteria. Blood 108:3168–3175

Yamanaka S, Zhang XY, Miura K et al (1998) The human gene encoding the lectin-type oxidized LDL receptor (OLR1) is a novel member of the natural killer gene complex with a unique expression profile. Genomics 54:191–199

Yokota KA, Takashima A, Bergstresser PR et al (2001) Identification of a human homologue of the dendritic cell-associated C-type lectin-1, dectin-1. Gene 272:51–60

Yoshida H, Kondratenko N, Green S et al (1998) Identification of the lectin-like receptor for oxidized low-density lipoprotein in human macrophages and its potential role as a scavenger receptor. Biochem J 334:9–13

Yoshitomi H, Sakaguchi N, Kobayashi K et al (2005) A role for fungal {β}-glucans and their receptor dectin-1 in the induction of autoimmune arthritis in genetically susceptible mice. J Exp Med 201:949–960

Zelensky AN, Gready JE (2003) Comparative analysis of structural properties of the C-type-lectin-like domain (CTLD). Proteins 52:466–477

Zhang C, Wang SH, Liao CP et al (2010) Downregulation of PU.1 leads to decreased expression of dectin-1 in alveolar macrophages during *Pneumocystis pneumonia*. Infect Immun 78:1058–1065

Zhernakova A, Festen EM, Franke L et al (2008) Genetic analysis of innate immunity in Crohn's disease and ulcerative colitis identifies two susceptibility loci harboring CARD9 and IL18RAP. Am J Hum Genet 82:1202–1210

35 Dendritic Cell Lectin Receptors (Dectin-2 Receptors Family)

Rajesh K. Gupta and G.S. Gupta

35.1 Dendritic Cells

Dendritic cells (DCs) are special subsets of antigen presenting cells characterized by their potent capacity to activate immunologically naive T cells to induce initial immune responses. DCs are far more potent than macrophages and B cells in their capacity to activate immunologically naive T cells. Infact, DCs are responsible for initiating T cell-mediated immune responses to a variety of antigens. Members of the DC family are distributed to virtually all the organs (except the brain), where they serve as tissue resident APCs, playing critical roles in presenting environmental, microbial, and tumor-associated antigens to the immune system. DCs operate at the interface of innate and acquired immunity by recognizing pathogens and presenting pathogen-derived peptides to T lymphocytes. As a component of the innate immune system, DCs organize and transfer information from the outside world to the cells of the adaptive immune system. DCs can induce such contrasting states as active immune responsiveness or immunological tolerance. Recent years have brought a wealth of information regarding DC biology and pathophysiology that shows the complexity of this cell system. Presentation of an antigen by immature (non-activated) DC leads to tolerance, whereas mature, antigen-loaded DCs are geared towards the launching of antigen-specific immunity.

DCs Form a Major Resident Leukocyte Population in Human Skin Two main types of DCs are found in noninflamed skin: epidermal Langerhans cells (LCs) and dermal DCs. LCs express Langerin/CD207 that localizes to and forms Birbeck granules, as well as the CD1a class I–like molecule that presents glycolipids. Plasmacytoid dendritic cells (PDCs), which are also known as plasmacytoid T cells, plasmacytoid monocytes, natural IFN-α/ß–producing cells (natural IPCs), and type 2 predendritic cells (pDC2), constitute a subset of immature DCs, which is capable of differentiating in vitro into mature DCs with typical dendritic cell morphology and potent T cell stimulatory function when exposed to IL-3 alone, IL-3 and CD40L, viruses, and bacterial DNA containing unmethylated CpG motifs (CpG-DNA) (Dzionek et al. 2001; Ebner et al. 2004). Dzionek et al. (2000) identified two novel markers of PDCs, blood dendritic cell antigen-2 (BDCA-2) and BDCA-4, which enable direct identification of PDCs in human blood. BDCA-2 is presumably involved in ligand internalization, processing and presentation, as well as in inhibition of IFN-α/ß synthesis in PDCs.

Langerhans Cells Langerhans cells (LCs) are a unique population of dendritic cells found in the epidermis, where they capture and process antigens and subsequently migrate to the draining lymph nodes to activate naive T cells. About 70% of all dermal cells represent CD45$^+$ leukocytes. The vast majority of these cells (60% of total) express the mononuclear phagocyte markers murine macrophage galactose/N-acetylgalactosamine-specific C-type lectin (mMGL), F4/80 and CD11b. Studies suggest that mononuclear phagocyte populations form the majority of dermal cells underestimated earlier (Guironnet et al. 2002; Dupasquier et al. 2004). Evidences suggest that fresh human LCs are the only cells in the epidermis to express a fucose-mannose receptor on their surface (Condaminet et al. 1998). Freshly isolated LCs in epidermal cell suspensions phagocytose the yeast cell wall derivative zymosan, intact *S. cerevisiae*, *Corynebacterium parvum* and *S. aureus*, as well as 0.5–3.5 μ latex microspheres. Zymosan uptake by LCs is mediated by a mannose/β-glucan receptor(s) that is differentially expressed in mice and that is down-regulated during maturation of LCs in culture. In epidermis, they can be identified by expression of E-cadherin and cytoplasmic Birbeck granules (BGs) as their hallmark; the presence of Birbeck granules is the unique feature of LCs. BGs are disks of two limiting membranes, separated by leaflets with periodic "zipperlike" striations. Birbeck granules are unusual rod-shaped structures specific to epidermal LCs, whose origin and function remain undetermined. The normal dermis contains typical immunostimulatory myeloid DCs identified by CD11c and BDCA-1, as well as an additional population of poorly stimulatory macrophages marked by CD163 and FXIIIA (Zaba et al. 2007). Using Langerin as marker of LCs revealed the heterogeneity in the phenotype of gingival LC population (Seguier et al. 2003).

35.2 Dendritic Cell-Associated Lectins

35.2.1 Type-I and Type-II Surface Lectins on DC

The identification of DC-associated lectins is of particular interest, because one of the characteristic features of DC is the expression of many C-type lectins. DCs have a number of receptors for adsorptive uptake of antigens. Some are shared with other cells, such as Fcγ receptors, DEC-205, a type I membrane-integrated glycoprotein and the macrophage mannose receptor (MMR). Other receptors are more DC restricted, e.g., Langerin/CD207, DC-SIGN/CD209, asialoglycoprotein receptor or hepatic lectins (HL), and dendritic cell lectin (DLEC; also referred to as BDCA-2) (Abner et al. 2004; Dzionek et al. 2001) and C-type lectin receptor 1 (CLEC-1). Of the type II surface lectins, DCs express CD23, CD69, DCIR, dectin-1, and dectin-2 α, β, and γ isoforms) that form the list of DC-associated C-type lectin family. It was proposed that dectin-2 and its isoforms, together with dectin-1, represent a unique subfamily of DC-associated C-type lectins (Graham and Brown 2009). The cytoplasmic domains of the C-type lectins are diverse and contain several conserved motifs that are important for antigen uptake: a tyrosine-containing coated-pit intracellular targeting motif, a triad of acidic amino acids and a dileucine motif. Some type II C-type lectins contain potential signaling motifs (ITIM, ITAM, and proline-rich regions) (Table 35.1) (Fig. 35.1). Immature human dendritic cells express asialoglycoprotein receptor isoforms for efficient receptor-mediated endocytosis (Valladeau et al. 2001).

35.3 Macrophage Mannose Receptor (CD206) on DC

35.3.1 Expression and Characteristics

Humans express two types of mannose receptors, each encoded by its own gene: Macrophage mannose receptor 1 (CD206, MMR) and Macrophage mannose receptor 2 (C-type mannose receptor 2, Urokinase-type plasminogen activator receptor-associated protein, CD280). The macrophage mannose receptor (CD206) is a 175-kDa transmembrane glycoprotein that appears to be expressed on the surface of terminally differentiated macrophages and endothelial cell subsets whose natural ligands include both self glycoproteins and microbial glycans. The expression of the mannose receptor CD206 is regarded a differentiation hallmark of immature DCs, whereas monocytes, mature DCs, and epidermal LCs do not express CD206. In immature cultured DC, MMR mediates high efficiency uptake of glycosylated antigens. The expression of MMR by MHC class II positive APC in non-lymphoid organs of the mouse is also described. The MMR positive APC have been identified in peripheral organs: skin, liver, cardiac and skeletal muscle and tongue. The mannose receptor positive cells in salivary gland, thyroid and pancreas co-express MHC class II and the myeloid markers macrosialin and sialoadhesin, but not the DC markers CD11c or DEC-205 (Linehan 2005). Being an R-type lectin, structural details of mannose receptors have been discussed in Chap 14 and Chap 15 and hence readers are advised to consult these Chapters for more details.

The mannose receptor family consists of four members [the mannose receptor, the M-type phospholipase A_2 receptor, DEC-205 and Endo180 (or urokinase plasminogen activator receptor-associated protein)] all of which share a common extracellular arrangement of an amino-terminal cysteine-rich domain related to R-type domain, followed by a fibronectin type II domain and 8–10 C-type lectin-like CRD domains within a single polypeptide. In addition, all have a short cytoplasmic domain, which mediates their constitutive recycling between the plasma membrane and the endosomal apparatus, suggesting that these receptors function to internalize ligands for intracellular delivery. However, despite the common presence of multiple lectin-like domains, these four endocytic receptors have divergent ligand binding activities, and it is clear that the majority of these domains do not bind sugars. All of the MR family members except DEC-205 recycle back to the cell surface from early endosomes, but DEC-205 recycles from late endosomes. However, each receptor has evolved to have distinct functions and distributions.

Understanding the molecular basis of cell surface ligand recognition and endosomal release by the MR requires information about how individual domains interact with sugars as well as the structural arrangement of the multiple domains. The NH2-terminal cysteine-rich domain and the fibronectin type II repeat are not necessary for endocytosis of mannose-terminated glycoproteins. The CRDs 1–3 have at most very weak affinity for carbohydrate, where as, of the eight C-type CRDs, CRDs 4–8 are required for binding and endocytosis of mannose/GlcNAc/fucose-terminated ligands, but only CRD-4 has demonstrable sugar binding activity in isolation. CRD 4 shows the highest affinity binding and has multi-specificity for a variety of monosaccharides. As the main mannose-recognition domain of MR (CRD4) is the central ligand binding domain of the receptor, analysis of this domain suggests ways in which multiple CRDs in whole

Table 35.1 Characteristics of C-type lectins produced by dendritic cells (DCs) and Langerhans cells (LCs) (Adapted and modified from Figdor et al. 2002; Feinberg et al. 2011; Graham and Brown 2009; Shrimpton et al. 2009; Tateno et al. 2010)

C-type lectin	Type	Amino acids (human/mouse)	Chromosome localization (human)	Types of cells produce	Ligands	Functions
MMR (CD206)	I	1,456	10p13	DCs, LCs, Mo, MΦs, DMECs	Mannose, fucose, sLe^x	Antigen uptake
DEC-205 (CD205)	I	1,722	2q24	DCs, LCs, Act DCs, Thymic ECs	?	Antigen uptake
Dectin 1	II	247	12p13	DCs, LCs	β-Glucan	T cell interaction
Dectin 2	II	209	2p13	DCs, LCs	High mannose structures; ligands on CD4$^+$CD25$^+$ T cells ?	Antigen uptake
Langerin (CD207)	II	328	2p13	LCs	Galactose-6- SO4 oligosaccharides; oligomannose; Bd Gr-B antigen; Glcβ1-3Glcβ13Glc: A fr of β-glucan	BGs formation
DC-SIGN (CD209)	II	404	19p13	DCs	Mannan, ICAM2, ICAM3, HIV-1 (gp120), simion virus	T cell interaction, antigen uptake, migration, HIV-1 pathology
BDCA-2	II	?	12	pDCs	?	Antigen uptake
DCIR (LLIR)	II	237	12p13	DCs, Mo, MΦ, B, PMN	HIV-1	Autoimmunity
DLEC	?	231	12p13			
CLEC-1	II	280/229	12	DCs	?	?
CLEC9A	II	241/238	12	DCs in mouse and pDCs in humans	Necrotic cells	Activation
DCAR		?/209	?	DCs in mouse	?	Activation

The CLEC-1 and CLEC9A have been discussed with dectin-1 family in Chap. 34
? indicates unknown data
acDCs activated DCs, *pDCs* plasmacytoid DCs, *B* B cells, *Bd* blood, *BG* Birbeck granules, *DMECs* dermal microvascular endothelial cells, *Mo* monocytes, *MΦ* macrophage, *PMN* polymorphic nuclear cells, *sLeX* sialyl Lewis X, *Thymic ECs* thymic endothelial cells

receptor might interact with each other (Feinberg et al. 2000; Mullin et al. 1997). However, CRD 4 alone cannot account for the binding of the receptor to glycoproteins. At least 3 CRDs (4, 5, and 7) are required for high affinity binding and endocytosis of multivalent glycoconjugates. In this respect, the MR is like other carbohydrate-binding proteins, in which several CRDs, each with weak affinity for single sugars, are clustered to achieve high affinity binding to oligosaccharides (Taylor et al. 1992). The overall structure of CRD-4 (Fig. 15.1a) is similar to other C-type CRDs, containing two α helices and two small antiparallel β sheets. MBP (Fig. 15.1b). The core region of the CRD-4 domain, consisting of β strands 1–5 and the two α helices, superimposes on the equivalent residues of the rat MBP-A CRD. The principal difference resides in the position of helix α2, which is the most variable element of secondary structure among the C-type lectin-like folds.

The ectodomain of mannose receptor recognizes the patterns of sugars that adorn a wide array of bacteria, parasites, yeast, fungi, and mannosylated ligands. The capability to take up mannosylated protein antigens is important for the biologic function of DCs, as many glycoproteins derived from bacteria and fungi are mannosylated. The mannose receptor is also expressed by immature cultured DCs, where it mediates high efficiency uptake of glycosylated antigens, though its role in antigen handling in vivo is not clear. Endocytosis of mannose receptor -antigen complexes takes place via small coated vesicles, while non-mannosylated antigens were mainly present in larger vesicles. Shortly after internalization the mannose receptor and its ligand appear in the larger vesicles. Within 10 min, the mannosylated and non-mannosylated antigens co-localize with typical markers for MHC class II-enriched compartments and lysosomes (Tan et al. 1997). Peripheral blood DCs produce IFN-α in response to challenge by many enveloped viruses including herpes simplex virus (HSV) and HIV. The mannose receptor is an important receptor for the nonspecific recognition of enveloped viruses by DCs and stimulates the production of IFN-α by these viruses (Milone and Fitzgerald-Bocarsly 1998). The inflammatory dendritic-epidermal cells expressing MMR/CD206 in situ use CD206

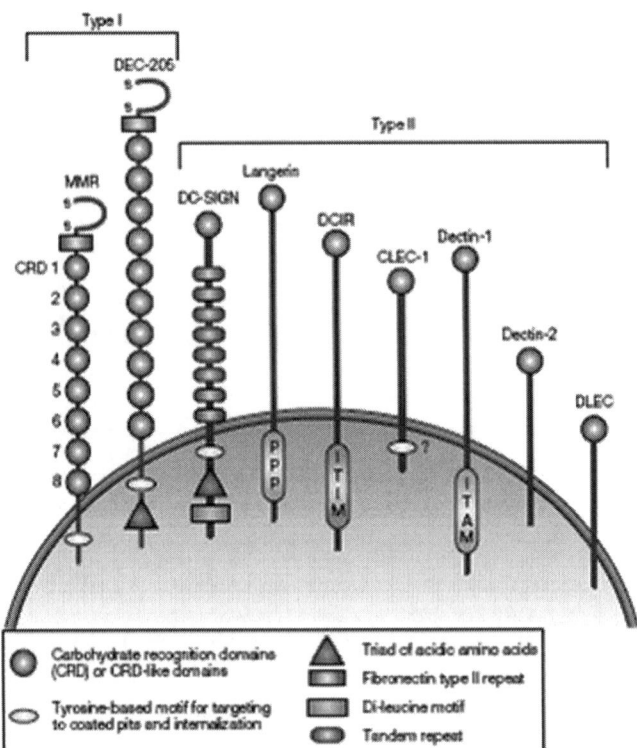

Fig. 35.1 Two types of C-type lectins or lectin-like molecules are produced by dendritic cells and Langerhans cells. Type I C-type lectins (MMR and DEC-205) contain an amino-terminal cysteine-rich repeat (S–S), a fibronectin type II repeat (FN) and 8–10 carbohydrate recognition domains (CRDs), which bind ligand in a Ca^{2+}-dependent manner. MMR binds ligand through CRD4 and CRD5. Type II C-type lectins contain only one CRD at their carboxy-terminal extracellular domain. The cytoplasmic domains of the C-type lectins are diverse and contain several conserved motifs that are important for antigen uptake: a tyrosine-containing coated-pit intracellular targeting motif, a triad of acidic amino acids and a dileucine motif. Other type II Ctype lectins contain other potential signaling motifs (ITIM, ITAM, proline-rich regions (P)). CLEC-1, C-type lectin receptor 1; DCIR, dendritic cell immunoreceptor; DC-SIGN, dendritic-cell specific ICAM-3 grabbing non-integrin; DLEC, dendritic cell lectin; ITAM, immunoreceptor tyrosine-based activation motif; ITIM, immunoreceptor tyrosine-based inhibitory motif; MMR, macrophage Mannose receptor (Reprinted by permission from Macmillan Publishers Ltd: Nature Rev Immunol., Figdor et al. © 2002)

for receptor-mediated endocytosis (Wollenberg et al. 2002). The MMR and DEC-205 belonging to R type lectins, which are characterized by the presence of trefoil group. Hence these lectins have been discussed along with R type lectins (Chap. 14 and Chap. 15).

35.3.2 Functions of Mannose Receptor in DC

Endocytosis Through Mannose Receptors: One of the major functions proposed for mannose receptor found on DCs as well as on macrophages and hepatic endothelial cells is in enhancing the uptake and processing of glycoprotein antigens for presentation by MHC class II molecules. Targeting recycling endocytic receptors with specific Abs provides a means for introducing a variety of tumor-associated Ags into human DCs, culminating in their efficient presentation to T cells. The specific targeting of soluble exogenous tumor Ag to the DC MR (CD206) directly contributes to the generation of multiple HLA-restricted Ag-specific T cell responses (Ramakrishna et al. 2004).

Cell surface-bound receptors are suitable sites for gene delivery into cells by receptor-mediated endocytosis. Presence of mannose receptor on DCs can be exploited for targeted gene transfer by employing mannosylated conjugates (Bonifaz et al. 2002). It was demonstrated that DCs transfected with ManPEI/DNA complexes containing adenovirus particles are effective in activating T cells of T cell receptor transgenic mice in an antigen-specific fashion (Diebold et al. 1999).

Langerin Mediates Efficient Antigen Presentation: The LCs show poor endocytic capacity and do not exploit MR-mediated endocytosis pathways. This may serve to avoid hyper-responsiveness to harmless protein antigens that are likely to be frequently encountered in the skin due to skin damage (Mommaas et al. 1999). This seems to be related to low level of expression of the MR (CD206) by epidermal LCs. However, some epidermal DCs may express CD206 under inflammatory skin conditions such as atopic dermatitis and psoriasis. The inflammatory dendritic epidermal cells expressing CD206 in situ can be used for receptor-mediated endocytosis (Wollenberg et al. 2002; Idoyaga et al. 2008).

Neonatal LCs preferentially utilize a wortmannin-sensitive, fluid-phase pathway, rather than receptor-mediated endocytosis, to internalize antigen (Bellette et al. 2003).

The mannose receptor mediates induction of IFN-α in peripheral blood DCs. Suppressed stimulation of IFN-α synthesis by herpes simplex virus (HSV) by monosaccharides as well as the yeast polysaccharide mannan, supported a role for lectin(s) in the IFN-α stimulation. Blood DC and IFN-α-producing cells responding to HSV stimulation also express mannose receptor. The mannose receptor is an important receptor for the nonspecific recognition of enveloped viruses by DCs and the subsequent stimulation of IFN-α production by these viruses (Milone et al. 1998).

35.4 DEC-205

DEC-205 is an endocytic receptor with 10 membrane-external, contiguous C-type lectin domains. DEC-205 is expressed at high levels on DCs in the T cell areas of lymphoid organs (Witmer-Pack et al. 1995). αDEC-205 antibodies selectively target these DCs in mice (Hawiger et al. 2001). Small amounts of injected antigen, targeted to DCs by the DEC-205 adsorptive pathway, are able to induce solid peripheral CD8$^+$ T cell tolerance. The antigen-presentation function of DCs is associated with the high-level expression of DEC-205, which mediates the efficient processing and presentation of antigens in vivo (Hawiger et al. 2001; Jiang et al. 1995; Mahnke et al. 2000; Maruyam et al. 2002). The DEC-205 is rapidly taken up by means of coated pits and vesicles, and is delivered to a multi-vesicular endosomal compartment that resembles the MHC class II-containing vesicles implicated in antigen presentation (refer Chap. 15). Results of Shrimpton et al. (2009) demonstrate that two areas of the CD205 molecule, within C-type lectin-like domains (CTLDs) 3 + 4 and 9 + 10, recognise ligands expressed during apoptosis and necrosis of multiple cell types. Thus, CD205 acts as a recognition receptor for dying cells, potentially providing an important pathway for the uptake of self-antigen in intrathymic and peripheral tolerance.

35.5 Langerin: A C-Type Lectin on Langerhans Cells

35.5.1 Human Langerin (CD207)

Use of mAbs restricted to human DCs has enabled the identification of Langerin (CD207), a Ca^{2+}-dependent type II the mannose-type lectin receptor. Langerin reactivity was strongly stimulated by LCs differentiation factor TGF-β and down-regulated by CD40 ligation. The monoclonal antibody, DCGM4, selectively stained Langerhans Cells; hence the antigen was termed Langerin. Langerin neither co-localized with MHC class II rich compartments nor with lysosomal LAMP-1 markers. The DCGM4 was rapidly internalized at 37°C. The Langerin was found as a 40-kDa protein with a pI of 5.2–5.5 (Valladeau et al. 1999) and only expressed by LCs. Langerin is responsible for Birbeck granule (BG) formation by membrane superimposition and zippering. Langerin is a mannose specific lectin expressed by LCs in epidermis and epithelia. Remarkably, transfection of Langerin cDNA into fibroblasts created a compact network of membrane structures with typical features of BG. Langerin is thus a potent inducer of membrane superimposition and zippering leading to BG formation. The induction of BG is a consequence of the antigen-capture function of Langerin, allowing routing into these organelles and providing access to a non-classical antigen-processing pathway (Valladeau et al. 2000, 2003).

The gene encoding Langerin is localized on chromosome 2p13, a region that does not contain any known genetic complex. The human Langerin gene spans over 5.6 kbp and is organized in six exons. Cloning of a cDNA for langerin shows that langerin gene has an open reading frame 987 bp predicting a polypeptide of 328 amino acids with a molecular weight of 37.5 kDa. The characteristics of different types of lectins and organization of various domains are schematically shown in Fig 35.1. Langerin is a type II transmembrane cell surface receptor with an intracytoplasmic portion of 43 amino acids with proline rich motif (WPREPPP) and an extracellular lectin domain. The extracellular domain of langerin consists of a neck region containing a series of heptad repeats and a C-terminal C-type carbohydrate-recognition domain (CRD). The CRD of Langerin contains a EPN (Glu-Pro-Asn) motif characteristic of lectins with mannose specificity. As in other lectins having a single CRD, Langerin features a particular domain located between the CRD and the transmembrane domain that is rich in leucine zipper sequences. The extracellular region of Langerin exists as a stable trimer held together by a coiled-coil of α-helices formed by the neck region, a structure found in mannose binding proteins. Finally Langerin has two potential sites of N-glycosylation at positions 87–89 and 180-18 amino acids. Trimer formation is essential for binding to oligosaccharide ligands because, as is typical for C-type CRDs, the CRD of langerin has only low affinity for monosaccharides. Oligomerization of C-type lectins is also important for determining selectivity for particular oligosaccharide structures.

35.5.1.1 The CRD of Human Langerin

Langerin CRD crystals at 1.5 Å resolution belonged to the tetragonal space group P4$_2$, with unit-cell parameters a = b = 79.55, c = 90.14 Å (Thépaut et al. 2008). The CRD of human Langerin, expressed in the periplasm of *E. coli*, was crystallized and analysed for Xray crystallography for apo-Langerin and for the complexes with mannose and maltose, respectively. The Langerin CRD (dubbed LangA) fold resembles that of DC-SIGN (Chap. 36). However, especially in the long loop region (LLR), which is responsible for carbohydrate-binding, two additional secondary structure elements were present: a 3_{10} helix and a small β-sheet arising from the extended β-strand 2, which enters into a hairpin and a new strand β2'. Unexpectedly, the crystal structures in the presence of maltose and mannose revealed two sugar-binding sites. One was calcium-dependent and structurally conserved in the C-type lectin family whereas the second one represented a calcium-independent type. Based on these data, a model for the binding of mannan was proposed and the differences in binding behavior between Langerin and DC-SIGN with respect to the Lewis X carbohydrate antigen and its derivatives were explained (Chatwell et al. 2008).

Crystal structures of the CRD of human langerin in complex with mannose or maltose show that it binds monosaccharides by ligation to a bound Ca^{2+} at a site that is conserved in all C-type CRDs (Chatwell et al. 2008). Interestingly, the co-crystals also show the presence of a second sugar-binding site that has not been seen in many C-type lectins. Both monosaccharide residues of maltose, or the monosaccharide mannose, are bound in this second site largely via polar interactions with backbone residues in a cleft formed between two of the large loop regions in the top half of the domain (Chatwell et al. 2008). This cleft is wider in langerin than in other C-type CRDs, most likely due to the absence of auxiliary Ca^{2+} sites present in many other CRDs, including those of mannose-binding protein and DC-SIGN (Feinberg et al. 2001). Like the full-length protein, truncated langerin exists as a stable trimer in solution. Glycan array screening with the trimeric fragment shows that high mannose oligosaccharides are the best ligands for langerin. Structural analysis of the trimeric fragment of langerin confirms that the neck region forms a coiled-coil of α-helices. Multiple interactions between the neck region and the CRDs make the trimer a rigid unit with the three CRDs in fixed positions and the primary sugar-binding sites separated by a distance of 42 Å. The fixed orientation of the sugar-binding sites in the trimer is likely to place constraints on the ligands that can be bound by langerin (Feinberg et al. 2010).

35.5.1.2 Polymorphisms in Human Langerin

Analysis of human genome has identified three SNPs that result in amino acid changes in the CRD of langerin. Expression of full-length versions of the four common langerin haplotypes in fibroblasts revealed that all of these forms can mediate endocytosis of neoglycoprotein ligands. However, fragments from the extracellular domain showed that two of the amino acid changes reduced the affinity of the CRD for mannose and decreased the stability of the extracellular domain. In addition, analysis of sugar binding by langerin containing the rare W264R mutation, previously identified in an individual lacking Birbeck granules, showed that this mutation could abolish sugar binding activity. It suggests that certain langerin haplotypes may differ in their binding to pathogens and thus might be associated with susceptibility to infection (Ward et al. 2006).

35.5.2 Mouse Langerin: Homology to Human Langerin

Mouse Langerin (m-Langerin) displays 65% and 74% homologies in total amino acid and lectin domains with those of human (h)-Langerin. The cognate mouse and rat genes were assigned to chromosome 6D1-D2 and chromosome 4q33 distal-q34.1 proximal respectively, syntenic to the h-Langerin gene on chromosome 2p35. m-Langerin transcripts were as expected detected in MHC class II$^+$, but not MHC class II$^-$, cells from epidermis. However, m-Langerin transcripts were also expressed in spleen, lymph nodes (LN), thymus, liver, lung and even heart, but not gut-associated lymphoid tissues. In single-cell lymphoid suspensions, m-Langerin transcripts were mainly detected in the CD11c$^+$ DC, especially the CD11b^{low}/CD8^{high} fraction of spleen and LN. Unexpectedly, significant amounts of m-Langerin transcripts were detected in skin and LN of TGF-β1-deficient mice, although in much lower amounts than littermate controls. It indicates that Langerin expression is regulated at several levels: by TGF-β1, DC subsets, DC maturation and the tissue environment (Takahara et al. 2002). The organization of human and mouse Langerin genes are similar, consisting of six exons, three of which encode the CRD. The m-Langerin, detected as a 48-kDa species, is abundant in epidermal LC in situ and is down-regulated upon culture (Valladeau et al. 2002).

Riedl et al (2004) isolated m-Langerin cDNA from murine fetal skin-derived DCs (FSDDC). In vitro-generated FSDDC and epidermal LC expressed both full-length and δE3Langerin mRNA, but tissue expression was not restricted to skin. Mouse Langerin protein isoforms were readily detected in fibroblasts transfected with cDNAs encoding epitope-tagged Langerin and δE3Langerin. Full-length m-Langerin bound mannan, whereas δE3Langerin and soluble bacterial recombinant Langerin protein lacking the neck domain did not. Fibroblasts transfected with m-Langerin cDNA contained typical BG and cored tubules, whereas δE3Langerin cDNA did not induce BG or cored tubule formation in transfected fibroblasts. Developmentally regulated expression of Langerin isoforms provides a mechanism by which Langerin involvement in antigen uptake and processing could be regulated (Riedl et al. 2004).

35.5.3 Ligands of Langarin

Mannoside-binding capacity was detected in normal epithelial cells. The Langerin CRD shows specificity for mannose, GlcNAc, and fucose, but only the trimeric extracellular domain fragment binds to glycoprotein ligands. Langerin extracellular domain binds mammalian high mannose oligosaccharides, as well mannose-containing structures on yeast invertase but does not bind complex glycan structures. Full-length Langerin, stably expressed in rat fibroblast transfectants, mediates efficient uptake and degradation of a mannosylated neoglycoprotein ligand. The pH-dependent ligand release appears to involve interactions between the CRDs or between the CRDs and the neck region in the trimer (Stambach and Taylor 2003). A role for langerin in processing of glycoprotein antigens has been proposed. Langerin binds HIV through its CRD and plays a protective role against its propagation by the internalization of virions in Birbeck granules. Access to the CRD of Langerin appeared to be impaired in proliferatively active environments (malignancies, hair follicles), indicating presence of an endogenous ligand with high affinity to saturate the C-type lectin under these conditions (Plzák et al. 2002).

Using glycoconjugate microarray and other analyses, Langerin showed outstanding affinity to galactose-6-sulfated oligosaccharides, including keratan sulfate, while it preserved binding activity to mannose, as a common feature of the C-type lectins with an EPN motif. Mutagenesis study showed that Lys^{299} and Lys^{313} form extended binding sites for sulfated glycans. Consistent with the former observation, the sulfated Langerin ligands were found to be expressed in brain and spleen, where the transcript of keratan sulfate 6-O-sulfotransferase is expressed. Langerin also recognized pathogenic fungi, such as *Candida* and *Malassezia*, expressing heavily mannosylated glycans. Observations provide strong evidence that Langerin mediates diverse functions on Langerhans cells through dual recognition of sulfated as well as mannosylated glycans by its uniquely evolved C-type CRD (Tateno et al. 2010). For endogenous ligands, fibroblast membrane fractions seemed to contain 140 and 240-kDa proteins, which bind Langerin. Though mass spectrometry suggested types I and III procollagen and fibronectin as candidate ligands, results indicated that Langerin selectively interacts with at least one ligand in extracellular matrix (type I procollagen) and may have an unanticipated role in cell-matrix interactions to modulate LC development (Tada et al. 2006). Crystal structures of the carbohydrate-recognition domain from human langerin bound to a series of oligomannose compounds, the blood group B antigen, and a fragment of β-glucan reveal binding to mannose, fucose, and glucose residues by Ca^{2+} coordination of vicinal hydroxyl groups with similar stereochemistry. Likewise, a β-glucan fragment, Glcβ1-3Glcβ1-3Glc, binds to langerin through the interaction of a single glucose residue with the Ca^{2+} site. The fucose moiety of the blood group B trisaccharide Galα1-3 (Fucα1-2)Gal also binds to the Ca^{2+} site (Feinberg et al. 2011). Langerin binds to an unusually diverse number of endogenous and pathogenic cell surface carbohydrates, including mannose-containing O-specific polysaccharides derived from bacterial lipopolysaccharides.

35.5.4 Functions of Langerin

From absence of classical macrophage mannose receptor (MMR) from LCs, it could be thought that Langerin might be involved in the endocytosis of microorganisms through mannosylated components. Experiments showed that Langerin is implicated in recognition and internalization of *Myobacteria, HIV, Leishmania* or *Candida* components (Valladeau et al. 2003).

Birbeck Granule (BG) Formation: Birbeck granules are unusual rod-shaped structures specific to epidermal LCs, whose origin and function remain undetermined (Fig. 35.2). The presence of Birbeck granules is the unique feature of LCs. The BGs are disks of two limiting membranes, separated by leaflets with periodic "zipperlike" striations. The identification of Langerin/CD207 has allowed researchers to decipher the mechanism of BG formation and approach an understanding of their function. Remarkably, transfection of Langerin cDNA into fibroblasts creates a dense network of membrane structures with features typical of BGs. Furthermore, mutated and deleted forms of Langerin have been engineered to map the functional domains essential for BG formation. Langerin is a potent LC-specific regulator of membrane superimposition and zippering, representing a key molecule to trace LCs and to probe BG function (McDermott et al. 2002; Valladeau et al. 2003).

In the steady state, Langerin is predominantly found in the endosomal recycling compartment and in BGs. Langerin internalizes by classical receptor-mediated endocytosis and the first BGs accessible to endocytosed Langerin are those connected to recycling endosomes in the pericentriolar area, where Langerin accumulates. Drug-induced inhibition of endocytosis results in the appearance of abundant open-ended Birbeck granule-like structures appended to the plasma membrane, whereas inhibition of recycling induces BGs to merge with a tubular endosomal network. In mature Langerhans cells, Langerin traffic is abolished and the loss of internal Langerin is associated with a concomitant depletion of BGs. An exchange of Langerin occurs between early endosomal compartments and the plasma membrane, with dynamic retention in the endosomal recycling compartment. Mc Dermott et al. (2002) suggested that BGs are not endocytotic structures, rather they

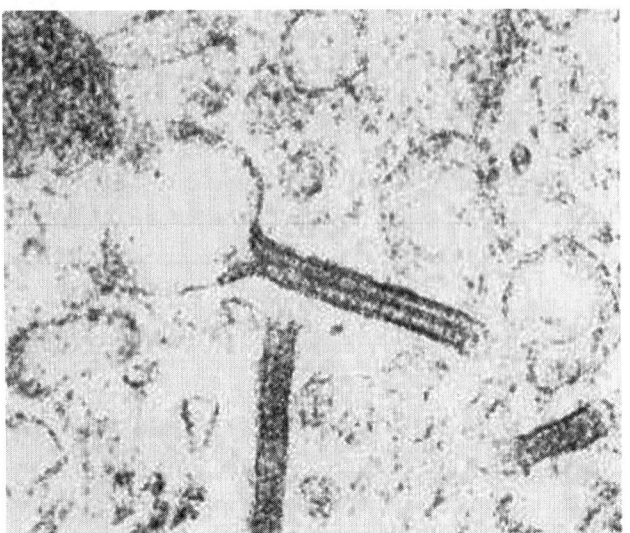

Fig. 35.2 Birbeck granules (BGs) are "tennis-racket" or rod shaped cytoplasmic organelles with a central linear density and a striated appearance. Birbeck granules were discovered by Michael Stanley Clive Birbeck (1925–2005), a British scientist and electron microscopist. Figure shows Birbeck granules, EM—Tennis racket shaped structures found in Langerhan's cells in histiocytosis X (Langerhan's cell histiocytosis) which includes eosinophilic granuloma, Letterer-Siwe disease and Hand-Schuller-Christian disease. The origin and function of BGs remain under question, but one theory is that they migrate to the periphery of the Langerhans cells and release its contents into the extracellular matrix. Another theory is that the Birbeck granule functions in receptor-mediated endocytosis, similar to clathrin-coated pits

are subdomains of the endosomal recycling compartment that form where Langerin accumulates. ADP-ribosylation factor proteins are implicated in Langerin trafficking and the exchange between BGs and other endosomal membranes (Mc Dermott et al. 2002).

Rab11A in Biogenesis of BG by Regulating Langerin Recycling and Stability: The extent to which Rab GTPases, Rab-interacting proteins, and cargo molecules cooperate in the dynamic organization of membrane architecture remains to be clarified. Langerin, accumulating in the Rab11-positive compartments of LCs, induces the formation of BGs, which are membrane subdomains of the endosomal recycling network. Results show that Rab11A and Langerin are required for BG biogenesis, and they illustrate the role played by a Rab GTPase in the formation of a specialized subcompartment within the endocytic-recycling system (Uzan-Gafsou et al. 2007).

Contact Hypersensitivity and LCs: The general role of LCs in skin immune responses is not clear because distinct models of LC depletion resulted in opposite conclusions about their role in contact hypersensitivity (CHS) responses. For example, LCs not only dispensable for CHS, but they regulate the response as well (Kaplan et al. 2005). While comparing various models, Bursch et al. (2007) suggested that dermal LCs could mediate CHS and provided an explanation for previous differences observed in the two-model systems (Bursch et al. 2007). Bennett and Clausen (2007) reviewed the impact of CD11c-DTR and Langerin-DTR mice on DC immunobiology, and highlighted the problems while interpreting data from these models. In another study Wang et al. (2008) found that acute depletion of mouse LC reduced CHS, but the timing of toxin administration was critical: toxin administration 3 days before priming did not impair CHS, whereas toxin administration 1 day before priming did. Moreover, LC elimination reduced the T cell response to epicutaneous immunization with OVA Ag. However, this reduction was only observed when OVA was applied on the flank skin, and not on the ear. Additionally, peptide immunization was not blocked by depletion, regardless of the site. Finally it was shown that conditions which eliminate epidermal LC but spare other Langerin$^+$ DC do not impair the epicutaneous immunization response to OVA. Overall, these results reconciled with previous conflicting data, and suggested that Langerin$^+$ cells do promote T cell responses to skin Ags, but only under defined conditions (Wang et al. 2008).

Langerin: A Natural Barrier to HIV-1 Transmission: The protective function of LCs is dependent on the function of Langerin and is thought to mediate HIV-1 transmission sexually. In the genital tissues, two different DC subsets are present: the LCs and the DC-SIGN$^+$-DCs. Although DC-SIGN$^+$-DCs mediate HIV-1 transmission, studies demonstrate that LCs prevent HIV-1 transmission by clearing invading HIV-1 particles. This protective function of LCs is dependent on the function of Langerin. Thus Langerin is a natural barrier to HIV-1 infection, and strategies to combat infection must enhance, preserve or, at the very least, not interfere with Langerin expression and function (de Witte et al. 2007, 2008).

35.6 DC-SIGN and DC-SIGNR on DCs

Additional C-type lectins with specificity for mannose, which have been characterized, are DC-SIGN (dendritic cell-specific ICAM-3 grabbing non-integrin where ICAM is intercellular adhesion molecule) and DC-SIGNR (DC-SIGN-related; L-SIGN, Liver-SIGN) (Geijtenbeek et al. 2004). DC-SIGN or CD209 is a mannose-specific C-type lectin expressed by DCs and plays an important role in the activation of T-lymphocytes. Evidences suggest that DC-SIGN can function both as an adhesion receptor and as a phagocytic pathogen-recognition receptor, similar to the Toll-like receptors. Although major differences in the cytoplasmic domains of these receptors might predict their function, findings show that differences in glycosylation of ligands can dramatically alter C-type lectin-like receptor usage (Cambi and Figdor 2003).

DC-SIGN and DC-SIGNR may function in DC migration and DC/T-cell synapse formation through interaction with ICAM-2 and ICAM-3, respectively, as well as playing roles in pattern recognition by binding carbohydrates on the surface of viruses and other microbes in a manner similar to other C-type lectin receptors (Figdor et al. 2002). The DC-SIGN can also bind HIV gp120 in a calcium-dependent manner. DC-SIGN may deliver bound HIV to permissive cell types, mediating infection with high efficiency. Determination of the crystal structure of DC-SIGN/DC-SIGNR has revealed high-affinity binding to an internal feature of high-mannose oligosaccharides (Feinberg et al. 2001). This is in contrast to MBL and the MR, which have preferred specificity for terminal mannose residues. This, taken with evidence that DC-SIGN forms tetramers at the cell surface (Mitchell et al. 2001), provides a mechanism for lectins to exhibit different specificity, whilst all being loosely classified as 'mannose specific'. Recent experiments with model oligosaccharide ligands demonstrated distinct ligand preferences for both MR and DC-SIGN (Frison et al. 2003). The macrophage-expressed murine homolog of DC-SIGNR (SIGNR1) appears to function analogously to its human counterpart (Geijtenbeek et al. 2002; Kang et al. 2003) but has distinct activities compared to other murine DC-SIGN homologs.

35.7 Dectin-2 Cluster in Natural Killer Gene Complex (NKC)

35.7.1 Natural Killer Gene Complex

A number of genes encoding C-type lectin molecules have been mapped to the natural killer gene complex (NKC) at the distal region of mouse chromosome 6 and to a syntenic region on human chromosome 12p12-p13. In addition to those receptors which regulate NK cell function, related structures expressed on other cells types have also been localized to this chromosomal region. Although many receptors of the NKC are expressed primarily by NK and T cells, a growing number have been found to be expressed on myeloid cells (Pyz et al. 2006). In contrast to the NK and T cell specific receptors which function mostly in detection of tumorous or virally infected cells, largely by means of MHC class I recognition, the myeloid expressed receptors seem to have a more diverse repertoire of ligands and cellular functions, including pathogen recognition and maintenance of homeostasis (Yokoyama and Plougastel 2003). The Dectin-1 cluster of receptors is one such example, which includes Dectin-1, LOX-1, C-type lectin-like receptor-1 (CLEC-1), CLEC-2, CLEC12B, CLEC9A and myeloid inhibitory C-type lectin-like receptor (MICL) and form part of the Group V C-type lectin-like receptors (Huysamen and Brown 2009) and discussed in Chap. 34. Dectin-1 specifically recognizes fungal (1,3)-linked β-glucans, while CLEC9A recognizes a ligand in necrotic cells (Brown and Gordon 2001; Brown 2006). Another cluster of receptors of interest is the Dectin-2 family of C-type lectins. These receptors are clustered in the telomeric region of the NKC, in close proximity to the Dectin-1 family (Figs. 35.3 and 35.4) and also appear to have diverse functions in both immunity and homeostasis (Graham and Brown 2009). Dectin-1 is selectively expressed in DCs and to a lower extent in monocytes and macrophages. mRNA forms with and without a stalk exon are observed. This family of lectin-like genes encodes receptors with important immune and/or scavenger functions in monocytic, dendritic and endothelial cells (Sobanov et al. 2001).

The Dectin-2 cluster comprises Dectin-2, DCIR, DCAR, BDCA-2, MCL, Mincle and Clecsf8, which are members of the Group II C-type lectin family. These type II receptors are expressed on myeloid and non-myeloid cells and contain a single extracellular CRD and have diverse functions in both immunity and homeostasis. DCIR is the only member of the family which contains a cytoplasmic signaling motif (ITIM) and has been shown to act as an inhibitory receptor, while BDCA-2, Dectin-2, DCAR and Mincle all associate with FcRγ chain to induce cellular activation, including phagocytosis and cytokine production. Dectin-2 and Mincle have been shown to act as pattern recognition receptors for fungi, while DCIR acts as an attachment factor for HIV. In addition to pathogen recognition, DCIR has been shown to be pivotal in preventing autoimmune disease by controlling dendritic cell proliferation, whereas Mincle recognizes a nuclear protein released by necrotic cells (Graham and Brown 2009).

The amino acid sequences comprising the single C-type lectin domains of MCL, Mincle, DCIR and Dectin-2 are closely related to each other. These molecules show overall similarity to groups of animal C-type lectins, which demonstrate type II transmembrane topology. Sequence analysis suggests that MCL, Mincle, DCIR and Dectin-2 represent a subset of group II-related C-type lectins which may participate in analogous recognition events on macrophages and DCs. The genomic organization of the MCL gene and the sequence of the promoter region, with putative regulatory elements, have been described (Balch et al. 2002). Flornes et al (2004) reported the cDNA cloning and positional arrangement of C-type lectin superfamily (CLSF) receptor genes, which represent rat orthologues to human Mincle and

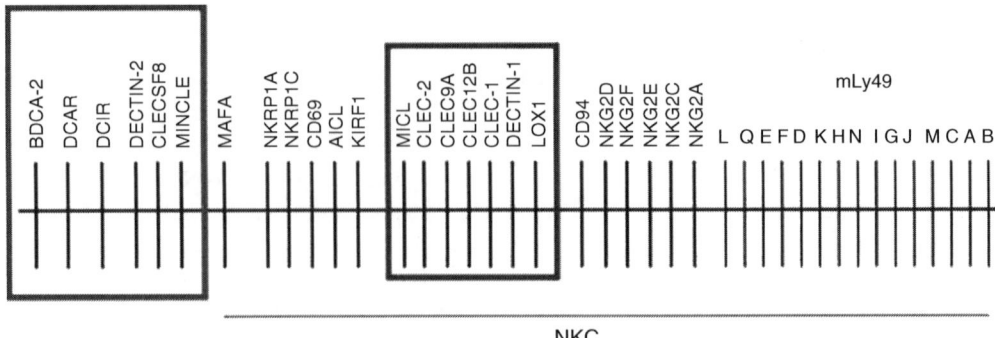

Fig. 35.3 The dectin-2 family genes form a cluster in the telomeric region of the NKC. The dectin-2 gene family includes BDCA-2, DCAR, DCIR, dectin-2, Clecsf8 and mincle, and form a cluster (*red square*) in the telomeric region of the NKC, close to the Dectin-1 cluster (*blue square*), on mouse chromosome 6 and human chromosome 12 (Adapted with permission from Graham and Brown 2009 © Elsevier)

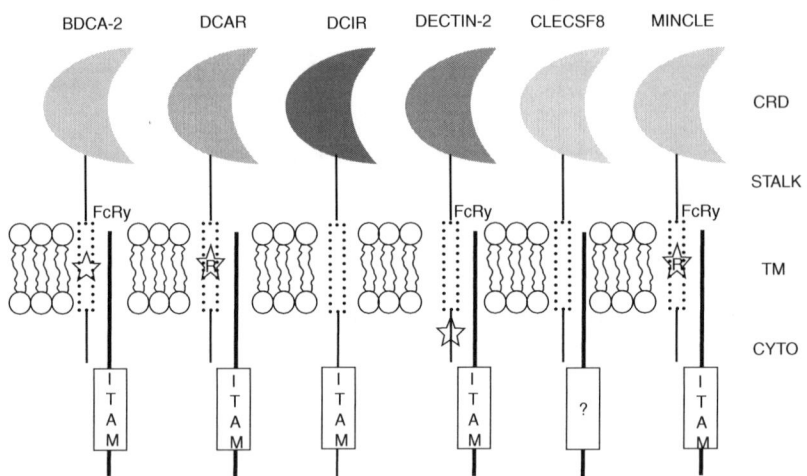

Fig. 35.4 Cartoon representation of the dectin-2 family of C-type lectins. Members of the dectin-2 family of C-type lectins are type II proteins with a single C-terminal extracellular carbohydrate recognition domain (CRD), a stalk region, a transmembrane region (TM), and a cytoplasmic domain (cyto). DCIR contains an immunoreceptor tyrosine-based signalling motif (ITIM) in its cytoplasmic domain, while BDCA-2, DCAR, dectin-2 and mincle associate with FcRγ chain which contains an ITAM. It is not yet known whether Clecsf8 associates with an adaptor molecule (?). The *star* represents the region responsible for association with the adaptor. R, arginine (Reprinted with permission from Graham and Brown 2009 © Elsevier)

DCIR and to mouse MCL and Dectin-2, as well as four other receptors DCIR2, DCIR3, DCIR4 and DCAR1, not reported in other species. Flornes et al (2004) also reported the cDNA cloning of human Dectin-2 and MCL, and of the mouse orthologues to the rat receptors. Similar to KLR some of these receptors exhibit structural features, which suggest that they regulate leukocyte reactivity; e.g., human DCIR and rodent DCIR1 and DCIR2 carry an ITIM, predicting inhibitory function. Conversely, Mincle has a positively charged amino acid in the transmembrane region, which suggests activating function in all three species. Sequence comparisons show that the receptors form a discrete family, more closely related to group II CLSF receptors than to the group V KLR. Most of the genes are preferentially expressed by professional APCs (DCs, macrophages and B cells) and neutrophils. In all three species, the genes map together, forming an evolutionary conserved gene complex, which is called as the antigen presenting lectin-like receptor complex (APLEC) (Flornes et al. 2004). Arce et al. (2001) identified a protein designated DLEC (dendritic cell lectin), which is a type II membrane glycoprotein of 213 amino acids and belongs to the human C-type lectin family. The cytoplasmic tail of DLEC lacks consensus signaling motifs and its extracellular region shows a single CRD, closest in homology to DCIR CRD. The DLEC gene has been localized linked to DCIR on the telomeric region of the NK gene complex.

DLEC mRNA is preferentially expressed in monocyte-derived DCs.

35.7.2 Antigen Presenting Lectin-Like Receptor Complex (APLEC)

In an experimental autoimmune rat model, a quantitative trait locus (QTL) conferring susceptibility to experimentally induced arthritis, a region on distal part of Chromosome 4 (Ribbhammar et al. 2003) was predicted to contain genes that encode proteins related to the human and mouse C-type lectin superfamily (CLSF) receptors: dendritic cell immunoreceptor (DCIR) (Bates et al. 1999), dendritic cell immunoactivating receptor (DCAR) (Kanazawa et al. 2003), dendritic cell lectin (DLEC; also referred to as BDCA-2) (Arce et al. 2001; Dzionek et al. 2001), dendritic cell associated C-type lectin 2 (Dectin-2) (Ariizumi et al. 2000; Fernandes et al. 1999), macrophage C-type lectin (MCL) (Balch et al. 1998) and macrophage inducible C-type lectin (Mincle) (Matsumoto et al. 1999). An important subgroup of CLSF receptors belongs to the opposing regulatory leukocyte receptors that come in two functional forms: activating and inhibitory, mediating their effects via protein tyrosine kinases and phosphatases, respectively. As a rule, the inhibitory variants have cytoplasmic tails containing immunoreceptor tyrosinebased inhibitory motifs (ITIMs) (Ravetch and Lanier 2000). The activating receptors lack such motifs, but contain instead a positively charged residue in the transmembrane (TM) region, mediating non-covalent association with chains containing immunoreceptor tyrosine- based activating motifs (ITAMs) (Lanier 2001). In the human and the mouse, regulatory roles for DCIR are suggested by the presence of an ITIM in the cytoplasmic tail, and for Mincle by a positively charged amino acid in the TM region. Functional studies of DCIR showed ITIM-dependent inhibition of B-cell receptor-mediated Ca^{2+} mobilization and protein tyrosine phosphorylation (Kanazawa et al. 2002), whereas antibody crosslinking of DLEC/BDCA-2 (Dzionek et al. 2001) and DCAR (Kanazawa et al. 2003) induced Ca^{2+} mobilization and protein tyrosine phosphorylation in dendritic cells. DCIR, DCAR and DLEC were reported to be preferentially expressed by professional antigen-presenting cells (APC), in particular by dendritic cells (Bates et al. 1999; Dzionek et al. 2001). Their suggested role as quantitative regulators of APC reactivity makes these receptors interesting candidates for being QTL associated with autoimmune diseases. Future positional cloning of QTL associated genes relies on exact definition of gene content and arrangement. Flornes et al. (2004) described cDNA cloning of the corresponding receptors in rat and for related receptors in rat, mouse and human. Flornes et al. (2004) described the cDNA cloning and positional arrangement of these receptor genes, which represent rat orthologues to human Mincle and DCIR and to mouse MCL and Dectin-2, as well as four novel receptors DCIR2, DCIR3, DCIR4 and DCAR1, not reported in other species. These genes are expressed mainly by professional APC (DCs, macrophages and B cells) and neutrophils and. map together, close to, but separate from NKC (Yokoyama and Plougastel 2003), with positions and orientations largely conserved between the human, the mouse and the rat.

35.7.2.1 Impact of APLEC on Arthritis and Autoimmunity

Congenic DA rats were resistant to oil-induced arthritis (OIA) when they carried PVG alleles for APLEC, which encodes immunoregulatory C-type lectin-like receptors. Five corresponding human APLEC genes were identified and targeted. The SNP rs1133104 in *Dcir* and a haplotype including that marker and 4 other SNPs in *Dcir* and its vicinity gave an indication of allelic association with susceptibility to RA in patients who were negative for antibodies to cyclic citrullinated peptide. Results supported that rat APLEC is associated with susceptibility to polyarthritis, and human APLEC and *Dcir* may be associated with susceptibility to anti-CCP-negative RA (Lorentzen et al. 2007). The APLEC locus is a major locus regulating the severity of experimentally induced arthritis in rats (Rintisch et al. 2010). Rat APLEC regulates autoimmunity and multiple phenotypes in several types of arthritis. However, delineating the genetic impact may require stratification for sex or mode of arthritis induction. This pathogenetic complexity should be considered when evaluating APLEC in inflammatory and autoimmune diseases, including RA (Guo et al. 2008a; Bäckdahl et al. 2009). In vivo and in vitro phenotypes in DA rats and APLEC-congenic rats suggested that both strains had a notably dichotomous expression of genes, with general down-regulation of all four *Dcir* genes and upregulation of *Mincle* and *Mcl*. It is suggested that human APLEC genes may similarly regulate infectious diseases, differential delayed type hypersensitivity (DTH) and status of general macrophage activation (Guo et al. 2009).

35.8 Dectin-2 (CLECF4N)

35.8.1 Characteristics

Dendritic cell-associated lectin-2 (Dectin-2) was discovered in an experimental murine model of acute myeloid leukemia as a receptor over-expressed in the spleen (Fernandes et al. 1999). It is the best characterized member of this receptor family and was identified as an over-expressed transcript in a myeloid leukaemia mouse model and in macrophages, neutrophils and pleuripotent myeloid precursors (Fernandes et al. 1999). Tissue distribution of Dectin-2 is tightly restricted to myeloid cells of the macrophage and DC lineages. Dectin-2 surface expression on these cells was also found to be low but exhibited enhanced expression on inflammatory monocytes during the acute phase of an inflammatory reaction (Taylor et al. 2005). Dectin-2 is a C-type lectin-like receptor, which is associated to DC and involved in the initiation and maintenance of UV-induced tolerance. In naïve mice Dectin-2 is predominantly expressed by a wide variety of tissue macrophages and has a novel distribution pattern compared with other myeloid markers. Its expression is limited to DCs and notably absent from brain microglia and choroid plexus or meningeal macrophages. On peripheral blood monocytes, Dectin-2 expression is very low on the surface but is transiently and markedly up-regulated on induction of inflammation in vivo using a variety of stimuli. This change in Dectin-2 expression occurs on 'inflammatory' monocytes after arrival at the inflammatory lesion as demonstrated by adoptive cell-transfer studies, and is independent of whether the macrophages elicited by the stimuli ultimately expressed Dectin-2. Observations suggest Dectin-2 expression to be characteristic of monocyte activation/maturation at an inflammatory lesion and provide a new interpretation of Dectin-2 function in vivo (Taylor et al. 2005). Dectin-2 is encoded in the NK cell complex of C-type lectin-like genes and shares features with other classical C-type lectins in that it has conserved motifs for the recognition of mannose in a Ca^{2+}-dependent manner (Fernandes et al. 1999). Despite this, reports of a lectin activity of Dectin-2 have been contradictory (Fernandes et al. 1999; Ariizumi et al. 2000).

Clone 1B12 gene is expressed in a DC-specific manner as a type II membrane-integrated polypeptide of 209 amino acids containing a single CRD motif in the COOH terminus. The expression pattern of dectin-2 was almost indistinguishable from that of dectin-1; that is, both were expressed abundantly at mRNA and protein levels by the XS52 DC line, but not by non-DC lines, and both were detected in spleen and thymus, as well as in skin resident DC (i.e. Langerhans cells). One of the striking findings was the identification of two truncated isoforms of dectin-2, i.e. the β isoform with 34 aa deletions in the neck domain and the γ isoform with 41 aa deletions within the CRD domain. Genomic analyses indicated that a full-length dectin-2 (α isoform) is encoded by 6 exons, whereas truncated isoforms (β and γ) are produced by alternative splicing (Ariizumi et al. 2000a). RT-PCR and immunoblotting of multiple bands of dectin-2 transcripts and proteins confirmed molecular heterogeneity.

The human Dectin-2 (hDectin-2) (Kanazawa et al. 2004a; Gavino et al. 2005) transcripts have been localized in lung, spleen, lymph node, leukocytes, bone marrow and tonsils, but unlike mouse, Dectin-2 was not expressed in the human thymus. In peripheral blood cells, hDectin-2- transcripts were preferentially expressed in plasmacytoid, rather than myeloid DCs and constitutively expressed in $CD14^+$ monocytes and B cells. The hDectin-2 could be induced in $CD4^+$ T cells upon activation with Con A (Kanazawa et al. 2004a; Gavino et al. 2005). Similar to mouse Dectin-2, hDectin-2 appears to be upregulated in inflammatory settings, as gene expression in $CD14^+$ monocytes could be upregulated by treatment with GM-CSF, TGF-β1 and TNF-α and downregulated with the addition of IL-4 and IL-10 (Gavino et al. 2005). The hDectin-2 was also expressed on Langerhans cells.

A truncated isoform of hDectin-2 has been identified by Gavino et al. (2005). Truncated isoform of hDectin-2 lacks part of the intracellular domain and most of the transmembrane domain of the receptor. The lack of transmembrane region has been proposed to encode a secreted protein which may act as an antagonist to full-length Dectin-2 (Gavino et al. 2005). Human Dectin-1 isoform E also lacks a transmembrane region. This isoform is retained intracellularly where it interacts with Ran-binding protein, which is presumed to act as a scaffold protein to coordinate signals from cell surface receptors with intracellular signaling pathways (Xie et al. 2006). The Dectin-2 promoter was also defined as a Langerhans cell specific regulatory element and, while numerous splice forms of Dectin-2 mRNA were shown to be highly expressed in a Langerhans cell-like skin-derived cell line compared to other cell lines, these transcript were also found to be abundant in spleen and thymus (Ariizumi et al. 2000a; Bonkobara et al. 2001).

35.8.2 Ligands of Dectin-2

35.8.2.1 Dectin-2 Shows Specificity for High Mannose and Fucose

The CRD of Dectin-2 functions as a C-type lectin with specificity for glycoconjugates bearing high-mannose sugars, which was predicted to be due to the presence of an EPN motif in the CRD (Fernandes et al. 1999). The CRD of Dectin-2 exhibited cation-dependent mannose/fucose-like lectin activity, with an IC_{50} for mannose of approximately 20 mM compared to an IC_{50} of 1.5 mM for the MMR when assayed by similar methodology. The extracellular domain of Dectin-2 exhibited binding to live Candida albicans and the Saccharomyces-derived particle zymosan. Binding of C. albicans hyphae by RAW cells transduced with Dectin-2 resulted in tyrosine phosphorylation of intracellular proteins (Sato et al. 2006). Though, both mannose receptor and SIGNR1 were able to bind bacterial capsular polysaccharides derived from *Streptococcus pneumoniae*, the Dectin-2 CRD exhibited only weak interactions to these capsular polysaccharides, indicative of different structural or affinity requirements for binding, when compared with the other two lectins. Glycan array analysis of the carbohydrate recognition by Dectin-2 indicated specific recognition of high-mannose structures ($Man_9GlcNAc_2$). The differences in the specificity of these three mannose-specific lectins indicated that mannose recognition is mediated by distinct receptors, with unique specificity, and are expressed by discrete subpopulations of cells. This highlights the complex nature of carbohydrate recognition by immune cells (McGreal et al. 2006; Sato et al. 2006).

Use of a soluble form of the CRD of Dectin-2 as a probe, revealed that the receptor could recognize zymosan and numerous pathogens including *S. cerevisiae*, *C. albicans*, *M. tuberculosis*, *Microsporum audounii*, *Trichophyton rubrum Paracoccoides brasiliensis*, *Histoplasma capsulatum* and capsule-deficient *Cryptococcus neoformans*. Although the binding to these pathogens differed greatly, binding could be inhibited by chelation of Ca^{2+} or in presence of mannose (McGreal et al. 2006; Sato et al. 2006). A glycan microarray showed that the receptor had specificity for high-mannose structures (McGreal et al. 2006). Dectin-2 has been shown to play a role in response to allergens. Dectin-2 on bone-marrow-derived DCs (BMDCs) was able to bind to extracts from house dust mite (*Dermatophagoides farinae* and *Dermatophagoides pteronyssinus*) and *Aspergillus fumigatus* in a mannose-dependent manner. Stimulation of mast cells co-expressing Dectin-2 and FcRγ chain with these extracts resulted in production of proinflammatory lipid mediators which are not produced by untransfected cells (Barrett et al. 2009).

Experiments with a soluble recombinant form of Dectin-2 have suggested the presence of a ligand on $CD4^+CD25^+$ T cells, and in vivo administration of this protein in mice impaired the development and maintenance of ultraviolet (UV)-induced tolerance (Aragane et al. 2003). It has been speculated that prevention of the interaction between endogenous Dectin-2 and T cells causes this defect, although the nature of this ligand was yet uncharacterized. UV-B irradiation was shown to increase Dectin-2 expression in Langerhans cells of the skin at both mRNA and protein levels (Bonkobara et al. 2005). It is possible that Dectin-2 recognises an endogenous ligand that is not a carbohydrate, perhaps via an alternative binding site to that which recognises fungi, as has been reported for other C-type lectins, such as Dectin-1 (Willment et al. 2001). However, while the expression of Dectin-2 was upregulated upon UV-B radiation in mice, it was downregulated in human cells. The discrepency may be due to use of $CD14^+$ monocytes as a surrogate model for epidermal Langerhans cells in the human experiment (Gavino et al. 2005).

35.8.2.2 Dectin-1 and -2, Share Several Important Features

DC-associated, dectin-2, shared several important features with dectin-1. First, both dectin-2 and dectin-1 exhibited a common domain structure, consisting of a relatively short cytoplasmic domain, a transmembrane domain, an extracellular neck domain, and a single CRD in the COOH termini. Second, dectin-1 mRNA and dectin-2 mRNA expression profiles were indistinguishable, i.e. both were expressed: (a) at relatively high levels in the XS52 DC line but not in other tested non-DC lines, (b) most abundantly in spleen and thymus, and (c) constitutively in the Ia^+ epidermal cell population (i.e. Langerhans cells). Third, expression of dectin-2 and dectin-1 proteins also occurred exclusively in the XS52 DC line among the tested cell lines. Despite these similarities, the degree of sequence homology between dectin-2 and dectin-1 was relatively low (19.6% identity in the overall sequence and 24.8% within the CRD motif). Thus, dectin-2 and dectin-1 represent two structurally independent, DC-associated C-type lectins.

35.8.2.3 Dectin-1 and Dectin-2 Receptors in Control of Fungal Infection

Both, dectin-1 and dectin-2 are type II- transmembrane proteins with extracellular domains containing a carbohydrate recognition domain highly conserved among C-type lectins (Ariizumi et al. 2000a, b). Dectin-1 is expressed widely by APC (Ariizumi et al. 2000a) and is a PRR for β-glucan in yeasts (Brown et al. 2003). Dectin-2 is constitutively expressed at very high levels by mature DC and can be induced on macrophages after activation (Ariizumi et al. 2000b; Taylor et al. 2005). Dectin-2 is a PRR for fungi that employ Fc receptor (FcR) chain signaling to induce internalization, activate NF-κB, and up-regulate production of TNFα and IL-1ra (Sato et al. 2006).

During fungal infection, a variety of receptors, including TLR and Dectin-1 initiate immune responses. TLR recognition of fungal ligands and subsequent signaling through the MyD88 pathway were thought to be the most important interactions required for the control of fungal infection. However, Dectin-1-deficient mice, address the role of Dectin-1 in control of fungal infection. Saijo et al. (2007) argue that Dectin-1 plays a minor role in control of *Pneumocystis carinii* by direct killing and that TLR-mediated cytokine production controls *P. carinii* and *Candida albicans*. By contrast, Taylor et al. (2005) argue that Dectin-1-mediated cytokine and chemokine production, leading to efficient recruitment of inflammatory cells, is required for control of fungal infection.

Binding assays using soluble dectin-2 receptors showed the extracellular domain to bind preferentially to hyphal (rather than yeast/conidial) components of *Candida albicans*, *Microsporum audouinii*, and *Trichophyton rubrum*. Selective binding for hyphae was also observed using RAW macrophages expressing dectin-2, the ligation of which by hyphae or cross-linking with dectin-2-specific antibody led to protein tyrosine phosphorylation. Because dectin-2 lacks an intracellular signaling motif, Sato et al. (2006) searched for a signal adaptor that permits it to transduce intracellular signals. First, it was found that the Fc receptor (FcR) chain can bind to dectin-2. Second, ligation of dectin-2 on RAW cells induced tyrosine phosphorylation of FcR, activation of NF-κB, internalization of a surrogate ligand, and up-regulated secretion of tumor necrosis factor α and interleukin-1 receptor antagonist. Finally, these dectin-2-induced events were blocked by PP2, an inhibitor of Src kinases that are mediators for FcR chain-dependent signaling. It is suggested that dectin-2 is a PRR for fungi that employs signaling through FcR to induce innate immune responses (Sato et al. 2006). Dennehy and Brown (2007) argue that collaborative responses induced during infection may partially explain some apparently contradictory results. It seems that Dectin-1 is the first of many pattern recognition receptors that can mediate their own signaling, as well as synergize with TLR to initiate specific responses to infectious agents (Dennehy and Brown 2007).

35.9 The DC Immunoreceptor and DC-Immunoactivating Receptor

The dendritic cell (DC) immunoreceptors (DCIR) and DC-immunoactivating receptors (DCAR) represent a subfamily of cell surface C-type lectin receptors, whose multifunctional capacities range from classical Ag uptake and immunoregulatory mechanisms to the involvement in DC ontogeny.

35.9.1 Dendritic Cell Immunoreceptor (DCIR) (CLECsF6): Characterization

DCIR was identified by screening a nucleotide database for molecules homologous to the Group II C-type lectin hepatic asialoglycoprotein receptors, which also contain a single CRD at the C-terminal end (Bates et al. 1999). The DCIR (official gene symbol *Clec4a2*, called *Dcir*) is a type II C-type lectin expressed mainly in DCs that has a CRD in its extracellular region of the protein. DCIR mRNA was found to be highly expressed in human peripheral blood leukocytes, and at lower levels in lymph node, spleen, bone marrow and thymus, while mouse DCIR was found to be expressed at highest levels in spleen and lymph node. The term "DC immunoreceptors" is applied to a distinct set of signaling pattern-recognition receptors (PRR) described by Kanazawa et al. (2004b) and Kanazawa (2007), who reviewed their signaling mechanisms, carbohydrate recognition, and other features that contribute to the function of DC to control immunity.

DCIR is a type II membrane glycoprotein of 237 aa with a single CRD, closest in homology to those of the macrophage lectin and hepatic asialoglycoprotein receptors. The intracellular domain of DCIR contains a consensus ITIM. The gene encoding human DCIR was localized to chromosome 12p13, in a region close to the NK gene complex. The DCIR mRNA is predominantly transcribed in hematopoietic tissues. A closer look at protein expression on cells, revealed that DCIR was expressed on APCs such as $CD14^+$ monocytes, $CD19^+$ B cells, macrophages, neutrophils as well as myeloid and plasmacytoid DCs (pDCs), but not on $CD3^+$ T cells nor on $CD56^+$ or $CD16^+$NK cells (Bates et al. 1999; Kanazawa et al. 2002; Richard et al. 2002). DCIR acts as an inhibitory receptor via an ITIM in its cytoplasmic tail that transduces negative signals into cells. In contrast, DCAR is a molecule that forms a putative pair with DCIR. While both molecules share the highly homologous extracellular lectin domain, DCAR lacks the ITIM in its short cytoplasmic tail and acts as an activating receptor through association with the Fc receptor γ chain. Dectin-2 and BDCA-2 are highly related to DCAR by similarities of their amino acid sequence, molecular structure and chromosomal localization.

The DCIR was strongly expressed by DC derived from blood monocytes cultured with GM-CSF and IL-4. DCIR was mostly expressed by monocyte-related rather than Langerhans cell related DCs obtained from $CD34^+$ progenitor cells. The DCIR expression was down-regulated by signals inducing DC maturation such as CD40 ligand, LPS, or TNF-α. Thus, DCIR is differentially expressed on DC depending on their origin and stage of maturation/activation. DCIR represents a novel surface molecule expressed by Ag presenting cells, and of potential importance in regulation of DC function (Bates et al. 1999). Expression of mouse DCIR mRNA was observed specifically in spleen and lymph node, slightly increased with DC maturation during in vitro culture of bone marrow cells, and was not detected in cultured NK cells. Surface expression of mouse DCIR protein is observed in splenic APCs including B cells, monocytes/macrophages, and DCs, but not in T cells. The dectin-2 α isoform showed the highest degree of homology (33.5% identity in the overall sequence and 44.8% within the CRD motif) to murine DCIR. Dectin-2 differs from DCIR in ITIM motif found in the intracellular domain of DICR. ITIM was absent from short intracellular domain (14 aa) of dectin-2. Interestingly, dectin-2 also lacked an ITIM, which was identified in the intracellular domain of dectin-1. Thus, a simplified scenario would be that dectin-1 and DCIR deliver counteracting signals into DC, whereas dectin-2 has no apparent signaling potential. Unfortunately, information with respect to the natural ligands that are recognized by CD69, DCIR, dectin-1, and/or dectin-2 is not available.

Duck DCIR encodes an inhibitory receptor that features an ITIM in the cytoplasmic domain. While DCAR1 is a pseudogene, DCAR2 encodes an activating receptor with a positively charged residue in transmembrane region. Study suggests the presence of full-length and alternatively spliced forms of both DCIR and DCAR2. Duck DCIR and DCAR transcripts are preferentially expressed in immune and mucosal tissues including spleen, bursa of Fabricius, intestine and lung. Targeting these receptors on DCs holds promise for enhancing immune responses relevant for hepatitis B and vaccination against avian influenza (Guo et al. 2008b). Recombinant duck β-defensin, expressed in HEK293T cells significantly down-regulated DCIR mRNA, without changing the expression of TLR-7, DCAR, CD44, CD58 and cytokines (Soman et al. 2009).

DCIR as Inhibitory Receptor with an ITIM Motif: After coligation with a chimeric Fcγ receptor IIB containing the cytoplasmic portion of mouse DCIR, Kanazawa et al. (2002) detected two distinct inhibitory effects of cytoplasmic DCIR (-) inhibition of B-cell-receptor-mediated Ca^{2+} mobilization and protein tyrosine phosphorylation (-); both these effects required the tyrosine residue inside the ITIM. This report suggested that mouse DCIR expressed on APCs can exert two distinct inhibitory signals depending on its ITIM tyrosine residue (Kanazawa et al. 2002). This inhibition was completely abolished when the tyrosine of the DCIR ITIM motif was mutated to a phenylalanine (Kanazawa et al. 2002).

DCIR also called CLECSF6, is dominantly present in neutrophil membranes. The expression of this protein was down-regulated in neutrophils treated with GM-CSF, TNF-α, LPS, and IL-1α, where as anti-inflammatory stimuli, including IL-4, IL-10 and IL-13, did not affect expression, that suggests that DCIR may be down-regulated during inflammation (Richard et al. 2002). Interestingly, GM-CSF, IL-3, IL-4 and IL-13 stimulation of neutrophils resulted in accumulation of a short form of DCIR mRNA, which encodes a putative non-functional protein which has been proposed to act as an antagonist to the full-length receptor (Richard et al. 2002, 2003).

During neutrophil signaling, the peptide bearing the ITIM in its phosphorylated form associates with both SHP-1 and SHP-2. Phosphorylated SHP-1 binds the ITIM whereas phosphorylated SHP-2 does not. In addition, GM-CSF reduces the binding of SHP-2 to the ITIM of CLECSF6 while enhancing the phosphorylation level of SHP-2. GM-CSF is known to recruit SHP-2 to its receptor. Results suggest that the phosphorylation of SHP-2 by GM-CSF promotes the binding of SHP-2 to the GM-CSF receptor to the disadvantage of CLECSF6. Therefore, upon a treatment with GM-CSF, SHP-2 could move from a CLECSF6 associated signalosome with a repressor function to a GM-CSF receptor associated signalosome with an activator function. This work supports the hypothesis that CLECSF6 is involved in the control of inflammation in neutrophils (Richard et al. 2003, 2006).

DCIR Inhibits TLR8-Mediated Cytokine Production: Targeting DCIR on PDCs not only results in efficient antigen presentation but also affects TLR9-induced IFN-α production. It suggests that targeting of DCIR can modulate human PDC (Meyer-Wentrup et al. 2008). The endogenous DCIR is internalized efficiently into human moDC after triggering with DCIR-specific mAb. Collectively, DCIR acts as an APC receptor that is endocytosed efficiently in a clathrin-dependent manner and negatively affects TLR8-mediated cytokine production (Meyer-Wentrup et al. 2009). Cross-priming $CD8^+$ T cells by targeting antigens to human DCs through DCIR allows activation of specific $CD8^+$ T-cell immunity (Klechevsky et al. 2010).

DCIR As an Attachment Factor for HIV-1: DCIR also plays a role in the capture and transmission of HIV-1 by DCs (Lambert et al. 2008). Although DC-SIGN was responsible for the trans-infection function of the virus, subsequent studies demonstrated that trans-infection of $CD4^+$ T cells with HIV-1 can also occur through DC-SIGN-independent mechanisms as well (Lambert et al. 2008). It was proposed that DCIR interaction with HIV could allow the virus to gain access to nondegradative endosomal organelles and lead to fusion of viral and endosomal membranes, allowing productive infection of the cells (Lambert et al. 2008). Thus, DCIR acts as a ligand for HIV-1 and is involved in events leading to productive virus infection. In this process, the neck domain of DCIR is important for DCIR-mediated effect on virus binding and infection. Therefore, DCIR plays a role in HIV-1 pathogenesis by supporting the productive infection of DCs and promoting virus propagation (Lambert et al. 2008).

DCIR Deficiency and Autoimmune Disorders: In addition to acting as a PRR for HIV, DCIR has been shown to play a role in controlling autoimmune disease. Aged DCIR deficient mice were found to spontaneously develop joint abnormalities, had elevated levels of autoantibodies and showed higher levels of $CD11c^+$ DCs and a proportional expansion of T cell populations in their lymph nodes (Fujikado et al. 2008). The DCIR deficiency was related to development of autoimmune disorders and DCIR gene polymorphisms were associated with rheumatoid arthritis (RA). When studying DCIR expression in patients with rheumatoid arthritis, the receptor was found to be abundantly expressed in synovial fluid but was not found in healthy controls (Eklöw et al. 2008). In these samples, the DCIR was expressed on numerous cell types and surprisingly also on $CD56^+$ NK cells and $CD4^+$ and $CD8^+$ T cells. $DCIR^+$ T cells in the synovial fluid were activated, as well as much more abundant, than those found in peripheral blood (Eklöw et al. 2008). *Dcir* expression was also high in joints of two mouse RA models. These observations suggested that DCIR has an essential role in maintaining homeostasis of the immune system by controlling DC expansion and the development of autoimmune disease. Ronninger et al. (2008) analyzed the mRNA expression from the four known transcripts of DCIR in IFN-γ-treated human leukocytes together with fine mapping across the locus. RA patients and healthy controls were genotyped for several single nucleotide polymorphisms (SNPs) in *Dcir* and flanking regions. Results revealed that IFN-γ significantly downregulates the average expression of transcripts DCIR_v1, DCIR_v2, DCIR_v3 and DCIR_v4. The expression of *Dcir* showed significant association with variations in the gene. Cells with the RA-associated allele rs2024301 exhibited a significant increase in the expression of DCIR_v4. A fifth isoform, lacking exons 2, 3 and 4, illustrated that common genetic variations might influence *Dcir* mRNA expression. It also showed that the expression is regulated by the inflammatory mediator IFN-γ, affecting all four transcripts and that this was independent of genotype (Ronninger et al. 2008).

Since the *Dcir* may have an immune regulatory role, and since autoimmune-related genes are mapped to the DCIR locus in humans, Fujikado et al. (2008) found that aged $Dcir^{-/-}$ mice spontaneously develop sialadenitis and enthesitis associated with elevated serum autoantibodies. $Dcir^{-/-}$ mice showed a markedly exacerbated response to collagen-induced arthritis. The DC population was expanded excessively in aged and type II collagen-immunized $Dcir^{-/-}$ mice. Upon treatment with GMC-SF, $Dcir^{-/-}$ mouse-derived bone marrow cells (BMCs) differentiated into DCs more efficiently than did wild-type BMCs, owing to enhanced signal transducer and activator of transcription-5 phosphorylation. These observations indicated that Dcir is a negative regulator of DC expansion and has a crucial role in maintaining the homeostasis of the immune system.

35.9.2 Dendritic Cell Immunoactivating Receptor (DCAR)

In contrast to DCIR that induces negative signals through an ITIM in its cytoplasmic tail, Kanazawa et al. (2003) identified a C-type lectin receptor, dendritic cell immunoactivating receptor (DCAR), whose extracellular lectin domain is highly homologous to that of DCIR. The DCAR is expressed in tissues similar to DCIR, but its short cytoplasmic portion lacks signaling motifs like ITIM. However, presence of a positively charged arginine residue in the transmembrane region of the DCAR may explain its association with FcRγ chain and its stable expression on the cell surface. In A20 cells co-transfected with the FcRγ chain, cross-linking a chimeric receptor consisting of the extracellular region of FcγRIIB demonstrated that signaling from DCAR takes place via ITAM motif of the adaptor. The FcRγ chain was also required for surface expression of DCAR, as it enhanced receptor expression in transduced in 293T cells. Two isoforms of DCAR have been identified, one of which lacks the stalk region, but the ligands and biological functions of these isoforms still remain undefined. Thus, DCAR introduces activating signals into APCs through its physical and functional association with ITAM-bearing γ chain and provides an example of signaling via a DC-expressed C-type lectin receptor (Kanazawa et al. 2003). The gene of the DCAR has been identified next to the DCIR gene, and this acts as a putative activating pair of DCIR through association with an ITAM-bearing FcRγ chain. Kaden et al. (2009) functionally characterized mouse DCAR1 (mDCAR1) whose expression was strongly tissue dependent. The mDCAR1 expression was restricted to the $CD8^+$ DC subset in spleen and thymus and on subpopulations of $CD11b^+$ myeloid cells in bone marrow and spleen, and not detectable on both cell types in lymph nodes and peripheral blood. The Ag delivered via mDCAR1 was internalized, trafficked to early and late endosomes/lysosomes and, as a consequence, induced cellular and humoral responses in vivo even in absence of CD40 stimulation. Results indicated that mDCAR1 is a functional receptor on cells of the immune system and provides further insights into the regulation of immune responses by CLRs.

35.10 Macrophage-Inducible C-Type Lectin (Mincle)

Matsumoto et al. (1999) reported a macrophage-inducible C-type lectin (Mincle), as a downstream target of NF-IL6 in macrophages. NF-IL6 belongs to the CCAAT/enhancer binding protein (C/EBP) of transcription factors and plays a crucial role in activated macrophages. Mincle exhibits the highest homology to the members of group II C-type lectins. Mincle mRNA expression was strongly induced in response to several inflammatory stimuli, such as LPS, TNF-α, IL-6, and IFN-γ in wild-type macrophages. In contrast, NF-IL6-deficient macrophages displayed a much lower level of

Mincle mRNA induction following treatment with these inflammatory reagents. The mouse Mincle proximal promoter region contains an indispensable NF-IL6 binding element, demonstrating that Mincle is a direct target of NF-IL6. The Mincle gene locus was mapped at 0.6 centiMorgans proximal to CD4 on mouse chromosome 6. Mincle (also called as Clec4e and Clecsf9) is an FcRγ-associated activating receptor that senses damaged cells and involved in sensing necrosis (Brown 2008). Mincle selectively associates with the Fc receptor common γ-chain and activated macrophages to produce inflammatory cytokines and chemokines. Mincle-expressing cells are activated in the presence of dead cells. SAP130, a component of small nuclear ribonucloprotein, acts as a Mincle ligand that is released from dead cells. Thus, Mincle is a receptor that senses nonhomeostatic cell death and thereby induces the production of inflammatory cytokines to drive the infiltration of neutrophils into damaged tissue (Yamasaki et al. 2008).

35.10.1 Recognition of Pathogens

Yeast *Candida albicans* is a causative agent in mycoses of the skin, oral cavity, and gastrointestinal tract (Flores-Langarica et al. 2005). Identification of receptors, and their respective ligands, that are engaged by immune cells when in contact with *C. albicans* is crucial for understanding inflammatory responses leading to invasive candidiasis. Mincle, expressed predominantly on macrophages, was shown to play a role in macrophage responses to *Candida albicans*. The carbohydrate-recognition domain of human and mouse Mincle demonstrated the recognition of whole *C. albicans* yeast cells. After exposure to the yeast in vitro, Mincle localized to the phagocytic cup. However, it was not essential for phagocytosis. In absence of Mincle, production of TNF-α by macrophages was reduced, both in vivo and in vitro. In addition, mice lacking Mincle showed a significantly increased susceptibility to systemic candidiasis. Thus, Mincle plays a novel and non-redundant role in the induction of inflammatory signaling in response to *C. albicans* infection (Bugarcic et al. 2008; Wells et al. 2008). Mincle specifically recognizes *Malassezia* species among 50 different fungal species tested. *Malassezia* is a pathogenic fungus that causes skin diseases, such as tinea versicolor and atopic dermatitis, and fatal sepsis. Analyses of glycoconjugate revealed that Mincle selectively binds to α-mannose but not mannan. *Malassezia* activated macrophages to produce inflammatory cytokines/chemokines. *Malassezia*-induced cytokine/chemokine production by macrophages from Mincle$^{-/-}$ mice was significantly impaired. Results indicate that Mincle is the specific receptor for *Malassezia* species that plays a crucial role in immune responses to this fungus (Yamasaki et al. 2009).

35.10.2 Recognition of Mycobacterial Glycolipid, Trehalose Dimycolate

Trehalose-6,6′-dimycolate (TDM; also called cord factor) is a mycobacterial cell wall glycolipid that is the most studied immunostimulatory component of *M. tuberculosis*. However, its host receptor has not been clearly identified. Mincle is an essential receptor for TDM, which acts as a mincle ligand. TDM activated macrophages produce inflammatory cytokines and nitric oxide, which are completely suppressed in Mincle-deficient macrophages. The TDM and its synthetic analog trehalose-6,6-dibehenate (TDB) are potent adjuvants for Th1/Th17 vaccination that activate Syk-Card9 signaling in APCs. Further studies established that Mincle is a key receptor for the mycobacterial cord factor and controls the Th1/Th17 adjuvanticity of TDM and TDB (Geijtenbeek et al. 2003; Ishikawa et al. 2009; Matsunaga and Moody 2009; Schoenen et al. 2010).

35.11 Blood Dendritic Cell Antigen-2 (BDCA-2) (CLEC-4C)

35.11.1 BDCA-2: A Plasmacytoid DCs (PDCs)-Specific Lectin

Dzionek et al. (2000) identified three human blood DC Ags: BDCA-2, BDCA-3, and BDCA-4 which were expressed by PDCs. BDCA-2 and BDCA-4 were expressed on CD11c$^-$ CD123^{bright} PDCs, whereas BDCA-3 was expressed on small population of CD11c$^+$ CD123$^-$ DCs. All three Ags were not detectable on a third blood DC population, which corresponded to CD1c$^+$ CD11c^{bright} CD123^{dim}. Expression of all three Ags dramatically changes once blood DCs undergo in vitro maturation. For example, BDCA-2 was completely down-regulated on plasmacytoid CD11c$^-$ CD123^{bright} DCs where as expression of BDCA-3 and BDCA-4 was up-regulated on sub-types of DCs on which they were localized. BDCA-2 was rapidly internalized at 37°C after mAb labeling. The three Ags serve as markers for respective subpopulations of blood DCs in fresh blood (Dzionek et al. 2000). Thus, BDCA-2 is a marker for human PDC.

35.11.2 Characterization

Molecular cloning of BDCA-2 (CD303) revealed that BDCA-2 is a type II C-type lectin, which shows 50.7% sequence identity at the amino acid level to its putative murine ortholog, the murine DC-associated C-type lectin 2. Anti-BDCA-2 mAbs are rapidly internalized and efficiently presented to T cells, indicating that BDCA-2 could play a role in ligand internalization and presentation. Furthermore, ligation of BDCA-2 potently suppresses induction of interferon α/β production in plasmacytoid DCs, presumably by a mechanism dependent on protein-tyrosine phosphorylation by protein-tyrosine kinases (Dzionek et al. 2001). The BDCA-2 receptor is involved in the down-regulation of virus triggered interferon-α/β production in PDC (Dzionek et al. 2001).

BDCA-2 transcripts were weakly detected in tonsils, bone marrow, pancreas, lymph nodes, peripheral blood leukocytes, testis and ovary, (Dzionek et al. 2001; Fernandes et al. 2000). While testes from patients with testicular cancer, neoplastic and normal epithelium were negative for BDCA-2, CD123$^+$ PDCs were associated with lymphoid aggregates in tumors. It appears that expression of BDCA-2 in tissues is restricted to PDCs (Dzionek et al. 2001). The murine homolog(s) of BDCA-2 are not yet known, whereas, at least five truncated BDCA-2 mRNA species have been detected in humans (Dzionek et al. 2001; Fernandes et al. 2000). BDCA-2 contains an EPN motif in the CRD but ligand(s) for this receptor have not been identified. BDCA-2 from cell lysates was shown not to recognize mannose as carbohydrate (Fernandes et al. 2000; Arce et al. 2001). However, transfection studies on 293T and Jurkat cells and freshly isolated PDCs revealed that BDCA-2 couples with FcRγ chain. This was substantiated in Jurkat cells co-transfected with BDCA-2 and FcRγ chain, resulting into stimulation of BDCA-2 induced intracellular protein phosphorylation and Ca^{2+} influx which was not possible when BDCA-2 was expressed alone (Dzionek et al. 2001; Cao et al. 2007).

35.11.3 BDCA-2 Signals in PDC via a BCR-Like Signalosome

Human BDCA2 protein lacks an identifiable signaling motif. Signaling through BDCA-2 is dependent on the ITAM motif in FcRγ chain and downstream pathways involve Syk, SH-2 domain-containing leukocyte protein of 65 kDa (Slp65), Vav1, phospholipase C-γ (PLCγ2) and Erk1/2. It forms a complex with the transmembrane adapter Fc εRI γ. The signaling machinery in human PDCs is similar to that which operates downstream of BCR, which is distinct from the system involved in TCR signaling. BDCA2 crosslinking resulted in the activation of the BCR-like cascade, which potently suppressed the ability of PDCs to produce type I interferon and other cytokines in response to Toll-like receptor ligands (Cao et al. 2007; Rock et al. 2007).

By associating with Fc εRI γ, BDCA2 activates a BCR-like signaling pathway to regulate the immune functions of PDCs. BDCA-2 signaling induces tyrosine phosphorylation and Src kinase dependent Ca^{2+} influx. Cross-linking BDCA2 results in the inhibition of IFN-I production in stimulated PDC. PDCs express a signalosome similar to the BCR signalosome, consisting of Lyn, Syk, Btk, Slp65 (Blnk) and PLCγ2. BDCA2 associates with the signaling adapter FcR γ-chain. Triggering BDCA2 leads to tyrosine phosphorylation of Syk, Slp65, PLCγ2 and cytoskeletal proteins. Analogous to BCR signaling, BDCA2 signaling is likely linked with its internalization by clathrin-mediated endocytosis. The inhibition of IFN-I production by stimulated PDC is at least partially regulated at the transcriptional level. These results support a possible therapeutic value of an anti- BDCA2 mAb strategy, since the production of IFN-I by PDCs is considered to be a major pathophysiological factor in SLE patients (Röck et al. 2007).

Engagement of BDCA-2 Blocks TRAIL-Mediated Cytotoxic Activity of PDC: PDCs express Toll-like receptor (TLR) 9, which mediates recognition of microbial DNA during infection or self-DNA in autoimmune diseases. BDCA-2-induced signaling in PDCs inhibits up-regulation of CD86 and CD40 molecules in CpG-activated PDCs, but not in CD40L-activated PDCs. Furthermore, triggering of BDCA-2 diminished the ability of CpG- and CD40L-stimulated PDCs to process and present antigen to antigen-specific autologous memory T cells. Jähn et al. (2010) suggest that BDCA-2 represents an attractive target for clinical immunotherapy of IFN-I dependent autoimmune diseases influencing both, IFN-I production and antigen-specific T-cell stimulation by PDCs.

TLR7 and TLR9 ligands can induce the secretion of biologically active TNF-related apoptosis-inducing ligand (TRAIL) by PDCs. Accordingly PDC supernatant is endowed with TRAIL-mediated cytotoxic activity when tested on a TRAIL-sensitive Jurkat cell line. Importantly, both TRAIL secretion and cytotoxic activity of PDC supernatants are completely abolished by BDCA2 ligation. These results document a negative regulatory pathway of PDC cytotoxic activity that may be relevant in pathological situations such as tumors and autoimmune diseases (Riboldi et al. 2009).

35.11.4 Functions of BDCA-2

Regulation of IFN-α Production: PDCc are the natural type I IFN-producing cells that produce large amounts of IFN-α in response to viral stimulation. Cross-linking BDCA-2, BDCA-4, and CD4 on PDC regulates IFN-α production at the level of IRF-7, while the decrease in IFN-α production after CD123 cross-linking is due to stimulation of the IL-3R and induction of PDC maturation (Fanning et al. 2006).

Pregnancy: Dendritic cells are involved in the immune regulation during physiological pregnancy. $CD1c^+$ and $BDCA-2^+$ cells can influence the Th2 phenomenon which is observed during physiological pregnancy. It seems possible that lower $BDCA-2^+$ cells percentage and higher $CD1c^+$: $BDCA-2^+$ ratio can be associated with increased Th1-type immunity in patients with pre-eclampsia (Darmochwal-Kolarz et al. 2003). DCs within the decidua have been implicated in pregnancy maintenance. Ban et al. (2008) identified three DC subsets in normal human first-trimester decidua: $BDCA-1^+$ $CD19^-$ $CD14^-$ myeloid DC type 1 (MDC1), $BDCA-3^+$ $CD14^-$ myeloid DC type 2 (MDC2) and $BDCA-2^+$ $CD123^+$ plasmacytoid DC (PDC).

$BDCA-2^+$ in SLE: Type 1 IFN is thought to be implicated in the autoimmune process of SLE. Plasmacytoid DC, which are natural IFN-α producing cells, play a pivotal epipathogenic role in SLE. The phenotypic characteristics of peripheral blood DC in SLE patients show a reduced number of both $BDCA-2^+$ plasmacytoid DC and $CD11c^+$ myeloid DC. These alternations of the DC subset may drive the autoimmune response in SLE (Blomberg et al. 2003; Migita et al. 2005). BDCA-2-expressing pDCs are termed natural IFN α-producing cells, IFN α production could be inhibited by anti-BDCA-2/4 mAb (Blomberg et al. 2003; Wu et al. 2008).

Hematopoietic Malignancies: $CD4^+CD56^+$ hematodermic neoplasms are rare, aggressive hematopoietic malignancies usually presenting with cutaneous masses followed by a leukemic phase. Accumulating evidence suggests that these neoplasms represent malignant counterparts to the plasmacytoid DCs. BDCA-2, expressed predominantly on plasmacytoid DCs in $CD7^+$ subset of hematodermic neoplasms, and similar to non-neoplastic plasmacytoid dendritic cells, indicates a relatively more mature differentiation state. Clinical follow-up data confirm the aggressiveness of these tumors and suggests that BDCA-2 immunoreactivity may herald a significant reduction in survival (Jaye et al. 2006).

35.12 CLECSF8

Murine Clecsf8 was first identified through a differential display PCR screen of numerous cell lines for macrophage-specific genes (Balch et al. 1998). The molecule, named mouse macrophage C-type lectin, is a 219-amino acid, type II transmembrane protein with a single extracellular C-type lectin domain and expressed in cell lines and normal mouse tissues in a macrophage-restricted manner. The expression of this receptor was upregulated on these cells by IL-6, IL-10, TNF-α or IFN-γ, but was downregulated with LPS. The cDNA and genomic sequences of mouse macrophage C-type lectin indicate that it is related to the Group II animal C-type lectins. The mcl gene locus has been mapped between the genes for the interleukin-17 receptor and CD4 on mouse chromosome 6, the same chromosome as the mouse NK cell gene complex (Arce et al. 2004).

Arce et al. (2004) characterized the human orthologue of the mouse Mcl/Clecsf8. Human CLECSF8 codes for a type II membrane glycoprotein of 215 amino acids that belongs to human C-type lectin family. The cytoplasmic tail of CLECSF8 lacks consensus signaling motifs and its extracellular region shows a single CRD. The CLECSF8 gene has been localized on the telomeric region of the NK gene complex on chromosome 12p13 close to MINCLE. CLECSF8 mRNA shows a monocyte/macrophage expression pattern. Biochemical analysis of CLECSF8 on transiently transfected cells showed a glycoprotein of 30 kDa. Cross-linking of the receptor leads to a rapid internalization suggesting that CLECSF8 constitutes an endocytic receptor. Preliminary studies suggest that Clecsf8 could be upregulated in proinflammatory settings, as has been described for Mincle and Dectin-2 (Arce et al. 2004).

35.13 Macrophage Galactose/N-acetylgalactosamine Lectin (MGL)

The hMGL is a marker of immature DCs and it functions as an endocytic receptor for glycosylated antigens (Higashi et al. 2002). Van Die et al. (2003) identified an exclusive specificity for terminal α- and β-linked GalNAc residues that naturally occur as parts of glycoproteins or glycosphingolipids. Specific glycan structures containing terminal GalNAc moieties, expressed by helminth parasite *Schistosoma mansoni* as well as tumor antigens and a subset of gangliosides, were found as ligands for MGL. Studies indicate an endogenous function for DC-expressed MGL in the clearance and tolerance to self-gangliosides, and in the pattern recognition of tumor antigens and foreign glycoproteins derived from helminth parasites. The dendritic cell-specific C-type lectin DC-SIGN is a receptor for Schistosoma 8 MGL recognition of terminal GalNAc residues (Van Die et al. 2003). The MGL has been discussed in greater details in Chap. 33

References

Aragane Y, Maeda A, Schwarz A et al (2003) Involvement of dectin-2 in ultraviolet radiation-induced tolerance. J Immunol 171(3):801–807

Arce I, Martínez-Muñoz L, Roda-Navarro P et al (2004) The human C-type lectin CLECSF8 is a novel monocyte/macrophage endocytic receptor. Eur J Immunol 34:210–220

Arce I, Roda-Navarro P, Montoya MC et al (2001) Molecular and genomic characterization of human DLEC, a novel member of the C-type lectin receptor gene family preferentially expressed on monocyte-derived dendritic cells. Eur J Immunol 31:2733–2740

Ariizumi K, Shen G-L, Shikano S et al (2000a) Cloning of a second dendritic cell-associated c-type lectin (dectin-2) and its alternatively spliced isoforms. J Biol Chem 275:11957–11963

Ariizumi K, Shen G-L, Shikano S et al (2000b) Identification of a novel, dendritic cell-associated molecule, dectin-1, by subtractive cDNA cloning. J Biol Chem 275:20157–20167

Bäckdahl L, Guo JP, Jagodic M et al (2009) Definition of arthritis candidate risk genes by combining rat linkage-mapping results with human case-control association data. Ann Rheum Dis 68:1925–1932

Balch SG, Greaves DR, Gordon S, McKnight AJ (2002) Organization of the mouse macrophage C-type lectin (Mcl) gene and identification of a subgroup of related lectin molecules. Eur J Immunogenet 29:61–64

Balch SG, McKnight AJ, Seldin MF, Gordon S (1998) Cloning of a novel C-type lectin expressed by murine macrophages. J Biol Chem 273:18656–18664

Ban YL, Kong BH, Qu X, Yang QF (2008) Ma YY BDCA-1$^+$, BDCA-2 + and BDCA-3$^+$ dendritic cells in early human pregnancy decidua. Clin Exp Immunol 151:399–406

Barrett NA, Maekawa A, Rahman OM et al (2009) Dectin-2 recognition of house dust mite triggers cysteinyl leukotriene generation by dendritic cells. J Immunol 182:1119–1128

Bates EE, Fournier N, Garcia E et al (1999) APCs express DCIR, a novel C-type lectin surface receptor containing an immunoreceptor tyrosine-based inhibitory motif. J Immunol 163:1973–1983

Bellette BM, Woods GM, Wozniak T et al (2003) DEC-205lo Langerinlo neonatal Langerhans' cells preferentially utilize a wortmannin-sensitive, fluid-phase pathway to internalize exogenous antigen. Immunology 110:466–473

Bennett CL, Clausen BE (2007) DC ablation in mice: promises, pitfalls, and challenges. Trends Immunol 28:525–531

Blomberg S, Eloranta ML, Magnusson M et al (2003) Expression of the markers BDCA-2 and BDCA-4 and production of interferon-α by plasmacytoid dendritic cells in systemic lupus erythematosus. Arthritis Rheum 48:2524–2532, Comment in: Arthritis Rheum. 2003; 48: 2396–401

Bonifaz L, Bonnyay D, Mahnke K et al (2002) Efficient targeting of protein antigen to the dendritic cell receptor DEC-205 in the steady state leads to antigen presentation on major histocompatibility complex class I products and peripheral CD8+ T cell tolerance. J Exp Med 196:1627–1638

Bonkobara M, Yagihara H, Yudate T et al (2005) Ultraviolet-B radiation upregulates expression of dectin-2 on epidermal Langerhans cells by activating the gene promoter. Photochem Photobiol 81:944–948

Bonkobara M, Zukas PK et al (2001) Epidermal Langerhans cell-targeted gene expression by a dectin-2 promoter. J Immunol 167:6893–6900

Brown GD (2008) Sensing necrosis with Mincle. Nat Immunol 9:1099–1100

Brown GD, Gordon S (2001) Immune recognition. A new receptor for beta-glucans. Nature 413:36–37

Brown GD, Herre J, Williams DL et al (2003) Dectin-1 mediates the biological effects of β-glucans. J Exp Med 197:1119–1124

Brown GD (2006) Dectin-1: a signaling non-TLR pattern-recognition receptor. Nat Rev Immunol 6:33–43

Bugarcic A, Hitchens K, Beckhouse AG, Blanchard H et al (2008) Human and mouse macrophage-inducible C-type lectin (Mincle) bind *Candida albicans*. Glycobiology 18:679–685

Bursch LS, Wang L, Igyarto B et al (2007) Identification of a novel population of Langerin + dendritic cells. J Exp Med 204:3147–3156

Cambi A, Figdor CG (2003) Dual function of C-type lectin-like receptors in the immune system. Curr Opin Cell Biol 15:539–546

Cao W, Zhang L, Rosen DB et al (2007) BDCA2/Fc epsilon RI gamma complex signals through a novel BCR-like pathway in human plasmacytoid dendritic cells. PLoS Biol 5:e248

Chatwell L, Holla A, Kaufer BB, Skerra A (2008) The carbohydrate recognition domain of Langerin reveals high structural similarity with the one of DC-SIGN but an additional, calcium-independent sugar-binding site. Mol Immunol 45:1981–1994

Condaminet B, Peguet-Navarro J, Stahl PD et al (1998) Human epidermal Langerhans cells express the mannose-fucose binding receptor. Eur J Immunol 28:3541–3551

de Witte L, Nabatov A, Geijtenbeek TB (2008) Distinct roles for DC-SIGN + -dendritic cells and Langerhans cells in HIV-1 transmission. Trends Mol Med 14:12–19

de Witte L, Nabatov A, Pion M et al (2007) Langerin is a natural barrier to HIV-1 transmission by Langerhans cells. Nat Med 13:367–371

Dennehy KM, Brown GD (2007) The role of the β-glucan receptor dectin-1 in control of fungal infection. J Leukoc Biol 82:253–258

Diebold SS, Kursa M, Wagner E et al (1999) Mannose polyethylenimine conjugates for targeted DNA delivery into dendritic cells. J Biol Chem 274:19087–19094

Dupasquier M, Stoitzner P, van Oudenaren A et al (2004) Macrophages and dendritic cells constitute a major subpopulation of cells in the mouse dermis. J Invest Dermatol 123:876–879

Dzionek A, Fuchs A, Schmidt P et al (2000) BDCA-2, BDCA-3, and BDCA-4: three markers for distinct subsets of dendritic cells in human peripheral blood. J Immunol 165:6037–6046

Dzionek A, Sohma Y, Nagafune J et al (2001) BDCA-2, a novel plasmacytoid dendritic cell-specific type II C-type lectin, mediates antigen capture and is a potent inhibitor of interferon alpha/beta induction. J Exp Med 194:1823–1834

Ebner S, Ehammer Z, Holzmann S et al (2004) Expression of C-type lectin receptors by subsets of dendritic cells in human skin. Int Immunol 16:877–887

Eklöw C, Makrygiannakis D, Bäckdahl L et al (2008) Cellular distribution of the C-type II lectin DCIR and its expression in the rheumatic joint- identification of a subpopulation of DCIR + T cells. Ann Rheum Dis 67:1742–1749

Fanning SL, George TC, Feng D et al (2006) Receptor cross-linking on human plasmacytoid dendritic cells leads to the regulation of IFN-α production. J Immunol 177:5829–5839

Feinberg H, Park-Snyder S, Kolatkar AR et al (2000) Structure of a C-type carbohydrate recognition domain from the macrophage mannose receptor. J Biol Chem 275:21539–48

Feinberg H, Mitchell DA, Drickamer K, Weis WI (2001) Structural basis for selective recognition of oligosaccharides by DC-SIGN and DC-SIGNR. Science 294(5549):2163–2166

Feinberg H, Powlesland AS, Taylor ME, Weis WI (2010) Trimeric structure of Langerin. J Biol Chem 285:13285–13293

Feinberg H, Taylor ME, Razi N et al (2011) Structural basis for Langerin recognition of diverse pathogen and mammalian glycans through a single binding site. J Mol Biol 405:1027–1039

Fernandes MJ, Finnegan AA, Siracusa LD et al (1999) Characterization of a novel receptor that maps near the natural killer gene complex: demonstration of carbohydrate binding and expression in hematopoietic cells. Cancer Res 59:2709–2717

Fernandes MJ, Iscove NN, Gingras G et al (2000) Identification and characterization of the gene for a novel C-type lectin (CLECSF7) that maps near the natural killer gene complex on human chromosome 12. Genomics 69:263–270

Figdor CG, van Kooyk Y, Adema GJ (2002) C-type lectin receptors on dendritic cells and Langerhans cells. Nat Rev Immunol 2:77–84

Flores-Langarica A, Meza-Perez S, Calderon-Amador J et al (2005) Network of dendritic cells within the muscular layer of the mouse intestine. Proc Natl Acad Sci USA 102:19039–19044

Flornes LM, Bryceson YT, Spurkland A et al (2004) Identification of lectin-like receptors expressed by antigen presenting cells and neutrophils and their mapping to a novel gene complex. Immunogenetics 56:506–517

Frison N, Taylor ME, Soilleux E et al (2003) Oligolysine-based oligosaccharide clusters: selective recognition and endocytosis by the mannose receptor and dendritic cell-specific intercellular adhesion molecule 3 (ICAM-3)-grabbing nonintegrin. J Biol Chem 278:23922–23929

Fujikado N, Saijo S, Yonezawa T et al (2008) Dcir deficiency causes development of autoimmune diseases in mice due to excess expansion of dendritic cells. Nat Med 14:176–180

Geijtenbeek TB, Groot PC, Nolte MA et al (2002) Marginal zone macrophages express a murine homologue of DC-SIGN that captures blood-borne antigens in vivo. Blood 100:2908–2916

Gavino AC, Chung JS, Sato K et al (2005) Identification and expression profiling of a human C-type lectin, structurally homologous to mouse dectin-2. Exp Dermatol 14:281–288

Geijtenbeek TB, van Vliet SJ, Engering A et al (2004) Self- and nonself-recognition by C-type lectins on dendritic cells. Annu Rev Immunol 22:33–54

Geijtenbeek TB, van Vliet SJ, Koppel EA et al (2003) *Mycoba- cteria* target DC-SIGN to suppress dendritic cell function. J Exp Med 197:7–17

Graham LM, Brown GD (2009) The dectin-2 family of C-type lectins in immunity and homeostasis. Cytokine 48:148–155

Guironnet G, Dezutter-Dambuyant C, Vincent C et al (2002) Antagonistic effects of IL-4 and TGF-beta1 on Langerhans cell-related antigen expression by human monocytes. J Leukoc Biol 71:845–853

Guo JP, Bäckdahl L, Marta M et al (2008a) Profound and paradoxical impact on arthritis and autoimmunity of the rat antigen-presenting lectin-like receptor complex. Arthritis Rheum 58:1343–1353

Guo JP, Verdrengh M, Tarkowski A et al (2009) The rat antigen-presenting lectin-like receptor complex influences innate immunity and development of infectious diseases. Genes Immun 10:227–236

Guo X, Branton WG, Moon DA et al (2008b) Dendritic cell inhibitory and activating immunoreceptors (DCIR and DCAR) in duck: genomic organization and expression. Mol Immunol 45:3942–3946

Hawiger D, Inaba K, Dorsett Y et al (2001) Dendritic cells induce peripheral T cell unresponsiveness under steady state conditions in vivo. J Exp Med 194:769–780

Higashi N, Fujioka K, Denda-Nagai K et al (2002) The macrophage C-type lectin specific for galactose/N-acetylgalactosamine is an endocytic receptor expressed on monocyte-derived immature dendritic cells. J Biol Chem 277:20686–20693

Huysamen C, Brown GD (2009) The fungal pattern recognition receptor, dectin-1, and the associated cluster of C-type lectin-like receptors. FEMS Microbiol Lett 290:121–128

Idoyaga J, Cheong C, Suda K et al (2008) Cutting edge: Langerin/CD207 receptor on dendritic cells mediates efficient antigen presentation on MHC I and II products in vivo. J Immunol 180:3647–3650

Ishikawa E, Ishikawa T, Morita YS et al (2009) Direct recognition of the mycobacterial glycolipid, trehalose dimycolate, by C-type lectin Mincle. J Exp Med 206:2879–2888

Jähn PS, Zänker KS, Schmitz J, Dzionek A (2010) BDCA-2 signaling inhibits TLR-9-agonist-induced plasmacytoid dendritic cell activation and antigen presentation. Cell Immunol 265:15–22

Janssens S, Beyaert R (2003) Role of toll-like receptors in pathogen recognition. Clin Microbiol Rev 16:637–646

Jaye DL, Geigerman CM, Herling M et al (2006) Expression of the plasmacytoid dendritic cell marker BDCA-2 supports a spectrum of maturation among CD4+ CD56+ hematodermic neoplasms. Mod Pathol 19:1555–1562

Jiang W, Swiggard WJ, Heufler C et al (1995) The receptor DEC-205 expressed by dendritic cells and thymic epithelial cells is involved in antigen processing. Nature 375:151–155

Kaden SA, Kurig S, Vasters K et al (2009) Enhanced dendritic cell-induced immune responses mediated by the novel C-type lectin receptor mDCAR1. J Immunol 183:5069–5078

Kanazawa N, Okazaki T, Nishimura H et al (2002) DCIR acts as an inhibitory receptor depending on its immunoreceptor tyrosine-based inhibitory motif. J Invest Dermatol 118:261–266

Kanazawa N, Tashiro K, Inaba K et al (2004a) Molecular cloning of human dectin-2. J Invest Dermatol 122:1522–1524

Kanazawa N, Tashiro K, Inaba K, Miyachi Y (2003) Dendritic cell immunoactivating receptor, a novel C-type lectin immunoreceptor, acts as an activating receptor through association with Fc receptor γ chain. J Biol Chem 278:32645–32652

Kanazawa N, Tashiro K, Miyachi Y (2004b) Signaling and immune regulatory role of the dendritic cell immunoreceptor (DCIR) family lectins: DCIR, DCAR, dectin-2 and BDCA-2. Immunobiology 209:179–190

Kanazawa N (2007) Dendritic cell immunoreceptors: C-type lectin receptors for pattern-recognition and signaling on antigen-presenting cells. J Dermatol Sci 45:77–86

Kang YS, Yamazaki S, Iyoda T et al (2003) SIGN-R1, a novel C-type lectin expressed by marginal zone macrophages in spleen, mediates uptake of the polysaccharide dextran. Int Immunol 15:177–186

Kaplan DH, Jenison MC, Saeland S et al (2005) Epidermal Langerhans cell-deficient mice develop enhanced contact hypersensitivity. Immunity 23:611–620

Kato M, Neil TK et al (1998) cDNA cloning of human DEC-205, a putative antigen-uptake receptor on dendritic cells. Immunogenetics 47:442–450

Klechevsky E, Flamar AL et al (2010) Cross-priming CD8+ T cells by targeting antigens to human dendritic cells through DCIR. Blood 116:1685–1697

Lambert AA, Gilbert C, Richard M et al (2008) The C-type lectin surface receptor DCIR acts as a new attachment factor for HIV-1 in dendritic cells and contributes to trans- and cis-infection pathways. Blood 112:1299–1307

Lanier LL (2001) On guard—activating NK cell receptors. Nat Immunol 2:23–27

Linehan SA (2005) The mannose receptor is expressed by subsets of APC in non-lymphoid organs. BMC Immunol 6:4

Lorentzen JC, Flornes L, Eklöw C et al (2007) Association of arthritis with a gene complex encoding C-type lectin-like receptors. Arthritis Rheum 56:2620–2632

Mahnke K, Guo M, Lee S et al (2000) The dendritic cell receptor for endocytosis, DEC-205, can recycle and enhance antigen presentation via major histocompatibility complex class II-positive lysosomal compartments. J Cell Biol 151:673–684

Maruyama K, Akiyama Y, Cheng J et al (2002) Hamster DEC-205, its primary structure, tissue and cellular distribution. Cancer Lett 181:223–232

Matsumoto M, Tanaka T, Kaisho T et al (1999) A novel LPS-inducible C-type lectin is a transcriptional target of NF-IL6 in macrophages. J Immunol 163:5039–5048

Matsunaga I, Moody DB (2009) Mincle is a long sought receptor for mycobacterial cord factor. J Exp Med 206:2865–2868

McDermott R, Ziylan U, Spehner D et al (2002) Birbeck granules are subdomains of endosomal recycling compartment in human epidermal Langerhans cells, which form where Langerin accumulates. Mol Biol Cell 13:317–335

McGreal EP, Rosas M, Brown GD et al (2006) The carbohydrate-recognition domain of dectin-2 is a C-type lectin with specificity for high mannose. Glycobiology 16:422–430

Meyer-Wentrup F, Benitez-Ribas D, Tacken PJ et al (2008) Targeting DCIR on human plasmacytoid dendritic cells results in antigen presentation and inhibits IFN-alpha production. Blood 111:4245–4253

Meyer-Wentrup F, Cambi A, Joosten B et al (2009) DCIR is endocytosed into human dendritic cells and inhibits TLR8-mediated cytokine production. J Leukoc Biol 85:518–525

Migita K, Miyashita T, Maeda Y et al (2005) Reduced blood BDCA-2$^+$ (lymphoid) and CD11c$^+$ (myeloid) dendritic cells in systemic lupus erythematosus. Clin Exp Immunol 142:84–91

Mitchell DA, Fadden AJ, Drickamer K (2001) A novel mechanism of carbohydrate recognition by the C-type lectins DC-SIGN and DC-SIGNR. Subunit organization and binding to multivalent ligands. J Biol Chem 276:28939–28945

Mommaas AM, Mulder AA, Jordens R et al (1999) Human epidermal Langerhans cells lack functional mannose receptors and a fully developed endosomal/lysosomal compartment for loading of HLA class II molecules. Eur J Immunol 29:571–580

Mullin NP, Hitchen PG, Taylor ME (1997) Mechanism of Ca^{2+} and monosaccharide binding to a C-type carbohydrate-recognition domain of the macrophage mannose receptor. J Biol Chem 272:5668–81

Plzák J, Holíková Z, Dvořánková B et al (2002) Analysis of binding of mannosides in relation to Langerin (CD207) in Langerhans cells of normal and transformed epithelia. Histochem J 34:247–253

Pyz E, Marshall AS, Gordon S, Brown GD (2006) C-type lectin-like receptors on myeloid cells. Ann Med 38:242–251

Ramakrishna V, Treml JF, Vitale L et al (2004) Mannose receptor targeting of tumor antigen pmel17 to human dendritic cells directs anti-melanoma T cell responses via multiple HLA molecules. J Immunol 172:2845–2852

Ravetch JV, Lanier LL (2000) Immune inhibitory receptors. Science 290:84–89

Ribbhammar U, Flornes L, Backdahl L et al (2003) High resolution mapping of an arthritis susceptibility locus on rat chromosome 4, and characterization of regulated phenotypes. Hum Mol Genet 12:2087–2096

Riboldi E, Daniele R, Cassatella MA et al (2009) Engagement of BDCA-2 blocks TRAIL-mediated cytotoxic activity of plasmacytoid dendritic cells. Immunobiology 214:868–876

Richard M, Thibault N, Veilleux P et al (2006) Granulocyte macrophage-colony stimulating factor reduces the affinity of SHP-2 for the ITIM of CLECSF6 in neutrophils: a new mechanism of action for SHP-2. Mol Immunol 43:1716–1721

Richard M, Thibault N, Veilleux P et al (2003) The ITIM-bearing CLECSF6 (DCIR) is down-modulated in neutrophils by neutrophil activating agents. Biochem Biophys Res Commun 310:767–773

Richard M, Veilleux P, Rouleau M et al (2002) The expression pattern of the ITIM-bearing lectin CLECSF6 in neutrophils suggests a key role in the control of inflammation. J Leukoc Biol 71:871–880

Riedl E, Tada Y, Udey MC (2004) Identification and characterization of an alternatively spliced isoform of mouse Langerin/CD207. J Invest Dermatol 123:78–86

Rintisch C, Kelkka T, Norin U et al (2010) Finemapping of the arthritis QTL Pia7 reveals co-localization with Oia2 and the APLEC locus. Genes Immun 11:239–245

Röck J, Schneider E, Grün JR et al (2007) CD303 (BDCA-2) signals in plasmacytoid dendritic cells via a BCR-like signalosome involving Syk, Slp65 and PLCgamma2. Eur J Immunol 37:3564–3575

Ronninger M, Eklöw C, Lorentzen JC et al (2008) Differential expression of transcripts for the autoimmunity-related human dendritic cell immunoreceptor. Genes Immun 9:412–418

Rossi G, Heveker N, Thiele B et al (1992) Development of a Langerhans cell phenotype from peripheral blood monocytes. Immunol Lett 31:189–197

Saijo S, Fujikado N, Furuta T et al (2007) Dectin-1 is required for host defense against *Pneumocystis carinii* but not against *Candida albicans*. Nat Immunol 8:39–46

Sato K, X-li Y, Yudate T et al (2006) Dectin-2 is a pattern recognition receptor for fungi that couples with the fc receptor γchain to induce innate immune responses. J Biol Chem 281:38854–38866

Schoenen H, Bodendorfer B, Hitchens K et al (2010) Cutting edge: mincle is essential for recognition and adjuvanticity of the mycobacterial cord factor and its synthetic analog trehalosedibehenate. J Immunol 184:2756–2760

Seguier S, Bodineau A, Godeau G et al (2003) Langerin + versus CD1a + Langerhans cells in human gingival tissue: a comparative and quantitative immunohistochemical study. Arch Oral Biol 48:255–262

Shrimpton RE, Butler M, Morel A-S, Eren E, Hue SS, Ritter MA et al (2009) CD205 (DEC-205): a recognition receptor for apoptotic and necrotic self. Mol Immunol 46:1229–1239

Sobanov Y, Bernreiter A, Derdak S et al (2001) A novel cluster of lectin-like receptor genes expressed in monocytic, dendritic and endothelial cells maps close to the NK receptor genes in the human NK gene complex. Eur J Immunol 31:3493–3503

Soman SS, Nair S, Issac A et al (2009) Immunomodulation by duck defensin, Apl_AvBD2: in vitro dendritic cell immunoreceptor (DCIR) mRNA suppression, and B- and T-lymphocyte chemotaxis. Mol Immunol 46:3070–3075

Stambach NS, Taylor ME (2003) Characterization of carbohydrate recognition by Langerin, a C-type lectin of Langerhans cells. Glycobiology 13:401–410

Tada Y, Riedl E, Lowenthal MS et al (2006) Identification and characterization of endogenous Langerin ligands in murine extracellular matrix. J Invest Dermatol 126:1549–1558

Takahara K, Omatsu Y, Yashima Y et al (2002) Identification and expression of mouse Langerin (CD207) in dendritic cells. Int Immunol 14:433–444, 1

Tan MC, Mommaas AM, Drijfhout JW et al (1997) Mannose receptor-mediated uptake of antigens strongly enhances HLA class II-restricted antigen presentation by cultured dendritic cells. Eur J Immunol 27:2426–2435

Tateno H, Ohnishi K, Yabe R et al (2010) Dual specificity of Langerin to sulfated and mannosylated glycans via a single c-type carbohydrate recognition domain. J Biol Chem 285:6390–6400

Taylor ME, Bezouska K, Drickamer K (1992) Contribution to ligand binding by multiple carbohydrate-recognition domains in the macrophage mannose receptor. J Biol Chem 267:1719–26

Taylor PR, Reid DM et al (2005) Dectin-2 is predominantly myeloid restricted and exhibits unique activation-dependent expression on maturing inflammatory monocytes elicited in vivo. Eur J Immunol 35:2163–2174

Thépaut M, Vivès C, Pompidor G et al (2008) Overproduction, purification and preliminary crystallographic analysis of the carbohydrate-recognition domain of human Langerin. Acta Crystallogr Sect F Struct Biol Cryst Commun 64:115–118

Uzan-Gafsou S, Bausinger H, Proamer F et al (2007) Rab11A controls the biogenesis of Birbeck granules by regulating Langerin recycling and stability. Mol Biol Cell 18:3169–3179

Valladeau J, Clair-Moninot V, Dezutter-Dambuyant C et al (2002) Identification of mouse Langerin/CD207 in Langerhans cells and some dendritic cells of lymphoid tissues. J Immunol 168:782–792

Valladeau J, Dezutter-Dambuyant C, Saeland S (2003) Langerin/CD207 sheds light on formation of birbeck granules and their possible function in Langerhans cells. Immunol Res 28:93–107

Valladeau J, Duvert-Frances V, Pin JJ et al (1999) The monoclonal antibody DCGM4 recognizes Langerin, a protein specific of

Langerhans cells, and is rapidly internalized from the cell surface. Eur J Immunol 29:2695–2704

Valladeau J, Duvert-Frances V, Pin J-J, Kleijmeer MJ, Ait-Yahia S, Ravel O, Vincent C, Vega F, Helms A Jr et al (2001) Immature human dendritic cells express asialoglycoprotein receptor isoforms for efficient receptor-mediated endocytosis. J Immunol 167:5767–5774

Valladeau J, Ravel O, Dezutter-Dambuyant C et al (2000) Langerin, a novel C-type lectin specific to Langerhans cells, is an endocytic receptor that induces the formation of Birbeck granules. Immunity 12:71–81

Van Die I, van Vliet SJ, Nyame AK et al (2003) The dendritic cell-specific C-type lectin DC-SIGN is a receptor for Schistosoma 8 MGL recognition of terminal GalNAc residues (Van Die et al. 2003) mansoni egg antigens and recognizes the glycan antigen Lewis x. Glycobiology 13:471

Wang L, Bursch LS, Kissenpfennig A et al (2008) Langerin expressing cells promote skin immune responses under defined conditions. J Immunol 180:4722–4727

Ward EM, Stambach NS, Drickamer K, Taylor ME (2006) Polymorphisms in human Langerin affect stability and sugar binding activity. J Biol Chem 281:15450–15456

Wells CA, Salvage-Jones JA, Li X et al (2008) The macrophage-inducible C-type lectin, mincle, is an essential component of the innate immune response to *Candida albicans*. J Immunol 180:7404–7413

Witmer-Pack MD, Swiggard WJ, Mirza A et al (1995) Tissue distribution of the DEC-205 protein that is detected by the monoclonal antibody NLDC-145. II. Expression in situ in lymphoid and nonlymphoid tissues. Cell Immunol 163:157–162

Willment JA, Gordon S, Brown GD (2001) Characterization of the human beta -glucan receptor and its alternatively spliced isoforms. J Biol Chem 276:43818–43823

Wollenberg A, Mommaas M, Oppel T et al (2002) Expression and function of the mannose receptor CD206 on epidermal dendritic cells in inflammatory skin diseases. J Invest Dermatol 118:327–334

Wu P, Wu J, Liu S et al (2008) TLR9/TLR7-triggered downregulation of BDCA2 expression on human plasmacytoid dendritic cells from healthy individuals and lupus patients. Clin Immunol 129:40–48

Xie J, Sun M, Guo L, Liu W et al (2006) Human Dectin-1 isoform E is a cytoplasmic protein and interacts with RanBPM. Biochem Biophys Res Commun 347:1067–1073

Yamasaki S, Ishikawa E, Sakuma M et al (2008) Mincle is an ITAM-coupled activating receptor that senses damaged cells. Nat Immunol 9:1179–1188

Yamasaki S, Matsumoto M, Takeuchi O et al (2009) C-type lectin mincle is an activating receptor for pathogenic fungus, *Malassezia*. Proc Natl Acad Sci USA 106:1897–1902

Yokoyama WM, Plougastel BF (2003) Immune functions encoded by the natural killer gene complex. Nat Rev Immunol 3: 304–316

Zaba LC, Fuentes-Duculan J, Steinman RM et al (2007) Normal human dermis contains distinct populations of CD11c + BDCA-1+ dendritic cells and CD163 + FXIIIA + macrophages. J Clin Invest 117:2517–2525

DC-SIGN Family of Receptors

Rajesh K. Gupta and G.S. Gupta

36.1 DC-SIGN (CD209) Family of Receptors

In the immune system, C-type lectins and CTLDs have been shown to act both as adhesion and as pathogen recognition receptors. The **D**endritic **c**ell-specific ICAM-3 **g**rabbing **n**on-integrin (DC-SIGN) and its homologs in human and mouse represent an important C-type lectin family. DC-SIGN contains a lectin domain that recognizes in a Ca^{2+}-dependent manner carbohydrates such as mannose-containing structures present on glycoproteins such as ICAM-2 and ICAM-3. DC-SIGN is a prototype C-type lectin organized in microdomains, which have their role as pathogen recognition receptors in sensing microbes. Although the integrin LFA-1 is a counter-receptor for both ICAM-2 and ICAM-3 on DC, DC-SIGN is the high affinity adhesion receptor for ICAM-2/-3. While cell–cell contact is a primary function of selectins, collectins are specialized in recognition of pathogens. Interestingly, DC-SIGN is a cell adhesion receptor as well as a pathogen recognition receptor. As adhesion receptor, DC-SIGN mediates the contact between dendritic cells (DCs) and T lymphocytes, by binding to ICAM-3, and mediates rolling of DCs on endothelium, by interacting with ICAM-2. As pathogen receptor, DC-SIGN recognizes a variety of microorganisms, including viruses, bacteria, fungi and several parasites (Cambi et al. 2005). The natural ligands of DC-SIGN consist of mannose oligosaccharides or fucose-containing Lewis-type determinants. In this chapter, we shall focus on the structure and functions of DC-SIGN and related CTLDs in the recognition of pathogens, the molecular and structural determinants that regulate the interaction with pathogen-associated molecular patterns. The heterogeneity of carbohydrate residues exposed on cellular proteins and pathogens regulates specific binding of DC-expressed C-type lectins that contribute to the diversity of immune responses created by DCs (van Kooyk et al. 2003a; Cambi et al. 2005).

The DC-SIGN, originally described in 1992 as a C-type lectin able to bind the HIV surface protein, gp120 (Curtis et al. 1992), is important for efficient infection with HIV (Geijtenbeek et al. 2000c). Recent advances on a broad perspective concerning DC-SIGN structure, signaling and immune function have appeared in excellent reviews (den Dunnen et al. 2009; Gringhuis and Geijtenbeek 2010; Svajger et al. 2010). Interaction of DC-SIGN with the viral envelope glycoproteins may evoke cellular signal transduction implicated in viral pathogenesis.

36.1.1 CD209 Family Genes in Sub-Human Primates

Two *CD209* family genes identified in humans, *CD209* (*DC-SIGN*) and *CD209L* (*DC-SIGNR/L-SIGN/LSECtin*), encode C-type lectins that serve as adhesion receptors for ICAM-2 and ICAM-3 and participate in the transmission of HIV and SIV respectively to target cells in vitro. The *CD209* gene family that encodes C-type lectins in primates includes CD209 (DC-SIGN), CD209L (L-SIGN) and CD209L2. The *CD209* gene family in sub-human primates showed evolutionary alterations that occurred in this family across primate species. All of the primate species, specifically, Old World monkeys (OWM) and apes, have orthologues of human *CD209*. In contrast, *CD209L* is missing in OWM but present in apes. A third family member, that has been named *CD209L2*, was cloned from rhesus monkey cDNA and subsequently identified in OWM and apes but not in humans. Rhesus *CD209L2* mRNA was prominently expressed in the liver and axillary lymph nodes. Despite a high level of sequence similarity to both human and rhesus CD209, rhesus CD209L2 was substantially less effective at binding ICAM-3 and poorly transmitted HIV type 1 and SIV to target cells relative to CD209. The Toll-like receptor (TLR) gene family shares with CD209 genes a common

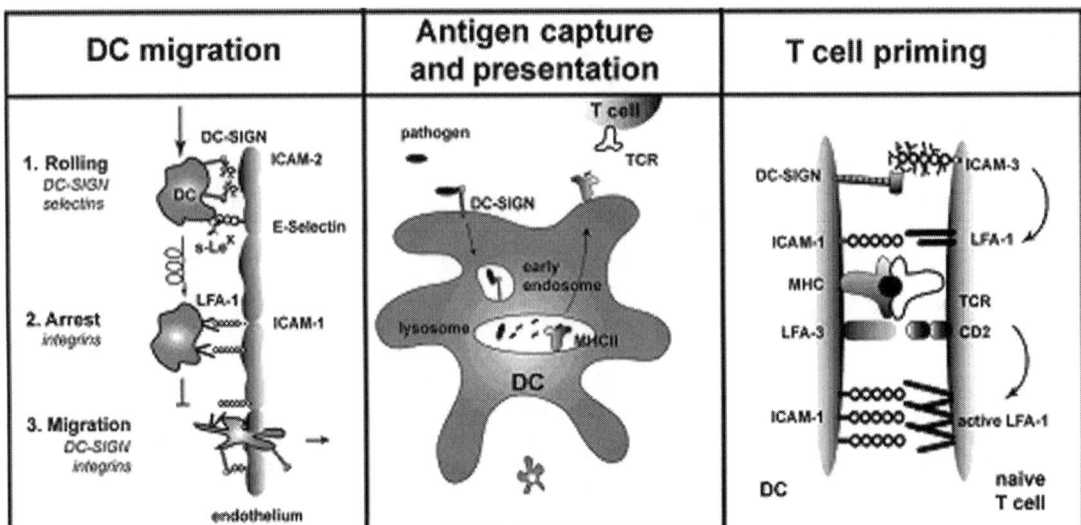

Fig. 36.1 DC-SIGN controls many functions of DC to elicit immune responses. The egress of precursor DC from blood into tissues is mediated partly by DC-SIGN. DC-SIGN facilitates rolling and transendothelial migration of DC-SIGN_ precursor DC, whereas arrest is mediated by integrin-mediated interactions. DC-SIGN also functions as an antigen receptor. DC-SIGN internalizes rapidly upon binding soluble ligand and is targeted to late endosomes/lysosomes, where antigens are processed and presented by MHC class II molecules. Moreover, initial DC-T-cell clustering, necessary for an efficent immune response, is mediated by transient interactions between DC-SIGN and ICAM-3. This interaction facilitates the formation of low-avidity LFA-1/ICAM-1 interaction and scanning of the antigen-MHC repertoire. It is becoming clear that other C-type lectins also participate in these processes. Selectins and the MR may regulate DC migration. Many other C-type lectins on DC, such as MR, DEC205, DC-ASPGR, BCDA-2, and dectin-1, are antigen receptors that recognize various distinct carbohydrate-containing antigens. It has been postulated that dectin-1 regulates T-cell priming, however its interaction with T cells is not carbohydrate-dependent (Reprinted with permission from Geijtenbeek et al. 2002a © Journal of Leukocyte Biology)

profile of evolutionary constraint (Bashirov et al. 2003b; Ortiz et al. 2008).

36.2 DC-SIGN (CD209): An Adhesion Molecule on Dendritic Cells

DC-SIGN (dendritic cell-specific ICAM-grabbing non-integrin, where ICAM is intercellular adhesion molecule) or CD209 is a type II C-type lectin expressed by DCs. DC-SIGN is specifically expressed on DCs and has been identified on monocyte-derived DCs in vitro and on DC subsets of skin, mucosal tissues, tonsils, lymph nodes, and spleen in vivo (Geijtenbeek et al. 2000a, c). DC-SIGN on DC binds the intercellular adhesion molecule (ICAM)-3 (CD50) with very high affinity. Although ICAM-3, a member of the immunoglobulin (Ig) superfamily, is known to be a ligand for β2 integrins lymphocyte function-associated antigen-1 (LFA-1; αLβ2) and $α_Dβ_2$ (Bleijs et al. 2001; Geijtenbeek et al. 2002a; van Kooyk and Geijtenbeek 2002) (Fig. 36.1), these receptors did not contribute to the binding activity of ICAM-3 by DC. Many C-type lectins have been identified on DCs: (1) Type I multi-CRD lectins are represented by the mannose receptor (Sallusto et al. 1995) and DEC-205 (Kato et al. 2000); and (2) type II single-CRD lectins by DC-SIGN, dectin-1, dectin-2, Langerin, BCDA-2, DCIR, DLEC, CLEC-1, and DC-ASGPR (Geijtenbeek et al. 2000a) (Chap. 35). Within the CRD, the highly conserved EPN or QPD sequences are essential in recognizing mannose- and galactose-containing structures, respectively.

DC-SIGN contains a short, cytoplasmic N-terminal domain with several intracellular sorting motifs, an extracellular stalk of seven complete and one partial tandem repeat, and a terminal lectin or CRD (Geijtenbeek et al. 2000a; Curtis et al. 1992). The DC-SIGN has three conserved cytoplasmic tail motifs: the tyrosine (Y)-based, dileucine (LL), and triacidic cluster (EEE), which are believed to regulate ligand binding, uptake, and trafficking. The full-length porcine DC-SIGN cDNA encodes a type II transmembrane protein of 240 amino acids. Phylogenetic analysis revealed that porcine DC-SIGN, together with bovine, canis and equine DC-SIGN, is more closely related to mouse SIGNR7 and SIGNR8 than to human DC-SIGN. Porcine DC-SIGN has the same gene structure as bovine, canis DC-SIGN and mouse SIGNR8 with eight exons. Porcine DC-SIGN mRNA expression was detected in spleen, thymus, lymph node, lung, bone marrow and muscles. Porcine DC-SIGN protein was found to express on the surface of monocyte-derived macrophages and dendritic cells, alveolar macrophages, lymph node sinusoidal macrophage-like, dendritic-like and endothelial cells but not of monocytes, peripheral blood lymphocytes or lymph node lymphocytes (Huang et al. 2009). Bovine ortholog of human DC-SIGN, within the bovine genome, exists as a single copy with a sequence similar to that of SIGNR7 (Yamakawa et al. 2008).

36.3 Ligands of DC-SIGN

36.3.1 Carbohydrates as Ligands of DC-SIGN

DC-SIGN and its close relative DC-SIGNR recognize various oligosaccharide ligands found on human tissues as well as on pathogens including viruses, bacteria, and parasites through the receptor lectin domain-mediated carbohydrate recognition. The DC-SIGN and DC-SIGNR bind to high-mannose carbohydrates on a variety of viruses. Studies have shown that these receptors bind the outer trimannose branch Manα1-3(Manα1-6)Manα present in high mannose structures. Although the trimannoside binds to DC-SIGN or DC-SIGNR more strongly than mannose, additional affinity enhancements are observed in presence of one or more Manα1-2Manα moieties on nonreducing termini of oligomannose structures. The molecular basis of this enhancement was investigated in crytstals of DC-SIGN bound to a synthetic six-mannose fragment of a high mannose N-linked oligosaccharide, Manα1-2Manα1-3 (Manα1-2Manα1-6)Manα1-6Man and to the disaccharide Manα1-2Man. The structures revealed mixtures of two binding modes in each case. Each mode features typical C-type lectin binding at main Ca^{2+}-binding site by one mannose residue. In addition, other sugar residues form contacts unique to each binding mode. Thus the affinity enhancement of DC-SIGN toward oligosaccharides decorated with Manα1-2Manα structure is due to multiple binding modes at the primary Ca^{2+} site, which provides both additional contacts and a statistical (entropic) enhancement of binding (Feinberg et al. 2007).

36.3.1.1 Lewis Antigen as Ligand

In addition to high-mannose moieties, DC-SIGN recognizes nonsialylated Lewis^X (Le^x) and Le^Y glycans and binds to Le^X-expressing pathogens such as *Schistosoma mansoni* and *Helicobacter pylori* (Appelmelk et al. 2003; van Die et al. 2003). Mouse homolog of human DC-SIGN has similar carbohydrate specificity for high mannose-containing ligands present on both cellular and pathogen ligands and called mSIGNR1or SIGNR1. However, mSIGNR1 interacts not only with Le$^{x/y}$ and Le$^{a/b}$ antigens similar to DC-SIGN, but also with sialylated Le^x, a ligand for selectins. The differential recognition of Lewis antigens suggests differences between mSIGNR1 and DC-SIGN in the recognition of cellular ligands and pathogens that express Lewis epitopes (Koppel et al. 2005a). Using the known 3D structure of the Lewis-x trisaccharide, Timpano et al. (2008) identified some monovalent α-fucosylamides that bind to DC-SIGN. α-fucosylamides work as functional mimics of chemically and enzymatically unstable α-fucosides and describes interesting candidates for the preparation of multivalent systems able to block DC-SIGN with high affinity and with potential biomedical applications.

The binding partner of DC-SIGN on endothelial cells is the glycan epitope Le^Y, expressed on ICAM-2. The interaction between DC-SIGN on DCs and ICAM-2 on endothelial cells is strictly glycan-specific. ICAM-2 expressed on CHO cells only served as a ligand for DC-SIGN when properly glycosylated, underscoring its function as a scaffolding protein (García-Vallejo et al. 2008). Oligosaccharide ligands expressed on SW1116, a typical human colorectal carcinoma are recognized by DC-SIGN, and has similar carbohydrate-recognition specificities as MBP. These tumor-specific oligosaccharide ligands comprise clusters of tandem repeats of Le^a/Le^b glycans on carcinoembryonic Ag (CEA) and CEA-related cell adhesion molecule 1 (CEACAM1). DC-SIGN ligands containing Le^a/Le^b glycans are also highly expressed on primary cancer colon epithelia but not on normal colon epithelia (Nonaka et al. 2008). Fucosylated glycans similar to pathogens are also found in a variety of allergens, but their functional significance remains unclear. Results suggest that allergens are able to interact with DC-SIGN and induce TNF-α expression in monocyte-derived DCs (MDDCs) via, in part, Raf-1 signaling pathways (Hsu et al. 2010).

Structural characterization of glycolipids, in combination with solid phase and cellular binding studies revealed that DC-SIGN binds to carbohydrate moieties of both glycosphingolipid species with Galβ1-4(Fucα1-3)GlcNAc (Le^x) and Fucα1-3Galβ1-4(Fucα1-3)GlcNAc (pseudo-Le^y) determinants. These data indicate that surveying DCs in the skin may encounter schistosome-derived glycolipids immediately after infection. Crystal structure of the CRD of DC-SIGN bound to Le^x provided insight into the ability of DC-SIGN to bind fucosylated ligands. The observed binding of schistosome-specific pseudo-Le^y to DC-SIGN is not directly compatible with the model described (Meyer et al. 2005).

36.4 Structure of DC-SIGN

36.4.1 Neck-Domains

Human DC-SIGN is a type II membrane protein which contains 404 amino acids and is of 44 kDa in molecular weight. DC-SIGN consists of extracellular domain, transmembrane region and cytoplasmic region (Fig. 36.2). The extracellular portion of each receptor contains a membrane-distal CRD and forms tetramers stabilized by an extended neck region consisting of 23 amino acid repeats. Cross-linking analysis of full-length receptors expressed in

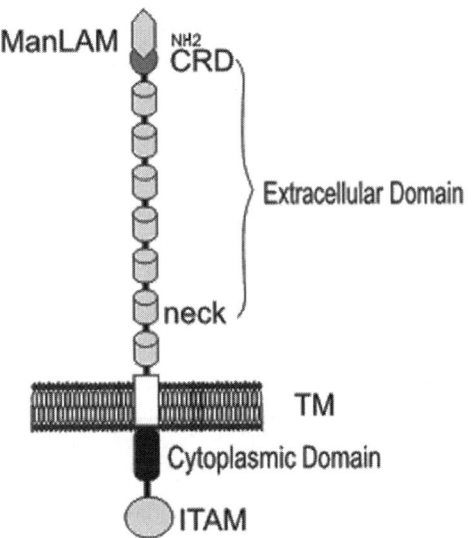

Fig. 36.2 Structure of DC-SIGN. Cytoplasmic domain, transmembrane region (TM) and extracellular domain are the three parts of DC-SIGN. The extracellular domain contains carbohydrate recognition domain (CRD) and neck domain. Cytoplasmic domain contains LL (di-leucine), EEE (tri-acidic clusters) and other internalization motifs and is connected to an incomplete ITAM. CRD recognizes certain carbohydrate-contained antigens like ManLAM and Lewis^X by four amino acids (Glu^{347}, Asn^{349}, Glu^{354} and Asn^{365}) and one Ca^{2+}-binding site in it (Zhou et al. 2006)

fibroblasts confirmed the tetrameric state of the intact receptors. Alternative splicing and genomic polymorphism generate DC-SIGN mRNA variants, which were detected at the sites of pathogen entrance and transmission. Naturally occurring DC-SIGN neck variants differ in multimerization competence in the cell membrane, exhibit altered sugar binding ability, and retain pathogen-interacting capacity, implying that pathogen-induced cluster formation predominates over the basal multimerization capability. Reports highlight the central role of the neck domain in the pH-sensitive control of oligomerization state, in the extended conformation of the protein, and in CRD organization and presentation (Guo et al. 2006; Tabarani et al. 2009). Analysis of DC-SIGN neck polymorphisms indicated that the number of allelic variants is higher than previously thought and that the multimerization of the prototypic molecule is modulated in presence of allelic variants with a different neck structure. Serrano-Gómez et al. (2008) demonstrated that the presence of allelic variants or a high level of expression of neck domain splicing isoforms might influence the presence and stability of DC-SIGN multimers on the cell surface, thus providing a molecular explanation for the correlation between DC-SIGN polymorphisms and altered susceptibility to HIV-1 and other pathogens. Neck domains of DC-SIGN and DC-SIGNR are shown to form tetramers in the absence of the CRDs. Analysis indicates that interactions between the neck domains account full stability of the tetrameric extracellular portions of the receptors. The neck domains are ~40% α-helical based on circular dichroism analysis. However, in contrast to other glycan-binding receptors in which fully helical neck regions are intimately associated with C-terminal C-type CRDs, the neck domains in DC-SIGN and DC-SIGNR act as autonomous tetramerization domains and the neck domains and CRDs are organized independently. Neck domains from polymorphic forms of DC-SIGNR that lack some of the repeat sequences show modestly reduced stability, but differences near the C-terminal end of the neck domains lead to significantly enhanced stability of DC-SIGNR tetramers compared to DC-SIGN (Yu et al. 2009). The length of the neck region shows variable levels of polymorphism, and can critically influence the pathogen binding properties of these two receptors. In Colored South African population of 711 individuals, including 351 tuberculosis patients and 360 healthy controls, Barreiro et al. (2007) revealed that none of the DC-SIGN and L-SIGN neck-region variants or genotypes seems to influence the individual susceptibility to develop tuberculosis.

Surface force measurements between apposed lipid bilayers displaying the extracellular domain of DC-SIGN and a neoglycolipid bearing an oligosaccharide ligand provide evidence that the receptor is in an extended conformation and that glycan docking is associated with a conformational change that repositions the carbohydrate-recognition domains during ligand binding. The results further show that the lateral mobility of membrane-bound ligands enhances the engagement of multiple carbohydrate-recognition domains in the receptor oligomer with appropriately spaced ligands. These studies highlight differences between pathogen targeting by DC-SIGN and receptors in which binding sites at fixed spacing bind to simple molecular patterns (Menon et al. 2009).

36.4.2 Crystal Structure of DC-SIGN (CD209) and DC-SIGNR (CD299)

To understand the tetramer-based ligand binding avidity, the crystal structure of DC-SIGNR was determined with its last repeat region. Compared to the carbohydrate-bound CRD structure, the structure revealed conformational changes in the calcium and carbohydrate coordination loops of CRD, an additional disulfide bond between the N and the C termini of the CRD, and a helical conformation for the last repeat. On the basis of the current crystal structure and other published structures with sequence homology to the repeat domain, Snyder et al. (2005) generated a tetramer model for DC-SIGN/DC-SIGNR using homology modeling and proposed a ligand-recognition index to identify potential receptor ligands.

The CRD of DC-SIGN is a globular structure consisting of 2 α-helices, 12 β-strands, and 3 disulphide bridges. A loop protrudes from the protein surface and forms part of two Ca^{2+}- binding sites. One of such sites is essential for the conformation of CRD, and the other is essential for direct coordination of the carbohydrate structures. Four amino acids (Glu^{347}, Asn^{349}, Glu^{354} and Asn^{365}) interact with Ca^{2+} at this site and dictate the recognition of specific carbohydrate structures. CRD can recognize certain carbohydrate containing antigens like ManLAM and Le^x. Crystal structures of CRDs of DC-SIGN and of DC-SIGNR bound to oligosaccharide revealed that these receptors selectively recognize endogenous high-mannose oligosaccharides and represent a new avenue for developing HIV prophylactics (Feinberg et al. 2001). Hydrodynamic studies on truncated receptors demonstrated that the portion of the neck of each protein adjacent to the CRD was sufficient to mediate the formation of dimers, whereas regions near the N terminus were needed to stabilize the tetramers. Some of the intervening repeats are missing from polymorphic forms of DC-SIGNR. Two different crystal forms of truncated DC-SIGNR comprising two neck repeats and the CRD revealed that the CRDs are flexibly linked to the neck, which contains α-helical segments interspersed with non-helical regions. Differential scanning calorimetry measurements indicated that the neck and CRDs were independently folded domains (Feinberg et al. 2005).

The neck domain contains 7 or 8 complete tandem repeats of 23 amino acids each and 1 incomplete repetitive sequence. It is required for oligomerization, which regulates carbohydrate specificity. Transmembrane region is essential in localization of DC-SIGN on cell surface. The cytoplasmic region contains internalization motifs, such as di-leucine (LL) motif, tri-acidic (EEE) clusters and an incomplete immunoreceptor tyrosine (Y) based activation motif (ITAM) which are believed to regulate ligand binding, uptake, and trafficking. The LL motif participates in antigen internalization and EEE clusters participate in signal transduction (Guo et al. 2004; Feinberg et al. 2001; van Kooyk and Geijtenbeek 2003). DC-SIGNR is homologous to DC-SIGN and is also denoted DC-related protein (L-SIGN), which has a similar structure to DC-SIGN. The genes encoding DC-SIGNR are similar to those of DC-SIGN (Pohlmann et al. 2001). Based on crystal structures and hydrodynamic data, models for the full extracellular domains of the receptors have been generated. The observed flexibility of the CRDs in the tetramer, combined with previous data on the specificity of these receptors, suggests an important role for oligomerization in the recognition of endogenous glycans, in particular those present on the surfaces of enveloped viruses recognized by these proteins (Feinberg et al. 2005). Azad et al. (2008) mutated each of the three conserved cytoplasmic tail motifs of DC-SIGN by alanine substitution and tested their roles in phagocytosis and receptor-mediated endocytosis of highly mannosylated ligands, *M. tuberculosis* ManLAM and HIV-1 surface gp120, respectively, in transfected human myeloid K-562 cells. Azad et al. (2008) indicated a dual role for EEE motif as a sorting signal in the secretory pathway and a lysosomal targeting signal in the endocytic pathway. The DC-SIGN and L-SIGN have been shown to interact with a vast range of infectious agents, including *M. tuberculosis*.

36.5 DC-SIGN versus DC-SIGN-RELATED RECEPTOR [DC-SIGNR or L-SIGN (CD 209)/LSEctin] (Refer Section 36.9)

36.5.1 DC-SIGN Similarities with DC-SIGNR/L-SIGN/LSEctin

DC-SIGNR is homologous to DC-SIGN and is also denoted as DC- related protein (L-SIGN), which has a similar structure to DC-SIGN. The genes encoding DC-SIGNR are similar to those of DC-SIGN (Pohlmann et al. 2001). The mRNA of L-SIGN shows about 90% similarity with DC-SIGN, which has a similar binding specificity to L-SIGN. L-SIGN like DC-SIGN binds to the cellular ligands i.e. ICAM-2 and ICAM-3 (Bashirova et al. 2001). The sequences of CRDs of DC-SIGN and DC-SIGNR show greatest identity to human asialoglycoprotein receptors (41% and 34% at amino acid level, respectively) and rat CD23 (both 33% at amino acid level). Consistent with previous reports (Curtis et al. 1992; Geijtenbeek et al. 2000a, b), DC-SIGN shows features of a mannose binding lectin, as opposed to the features of a protein-binding NK cell lectin (Weis et al. 1998). DC-SIGNR shows 77% identity to DC-SIGN at the amino acid level and also possesses all the residues shown to be required for the binding of mannose (Weis et al. 1998). The closely linked gene, DC-SIGNR, shows 73% identity to DC-SIGN at RNA level and a similar genomic organization.

The DC-SIGN and DC-SIGNR/L-SIGN (or CLEC4M) directly recognize a wide range of micro-organisms of major impact on public health. Both genes have long been considered to share similar overall structure and ligand-binding characteristics. Both DC-SIGN and DC-SIGNR efficiently bind HIV-1 surface glycoproteins of viruses and other viral as well as nonviral pathogens by interacting with high mannose oligosaccharides and assist either cis or trans infection. DC-SIGNR/L-SIGN (Lozach et al. 2004) is specifically expressed by liver sinusoidal endothelial cells (LSEC), a liver-resident APC, by endothelial cells in lymph nodes (Bashirova et al. 2001; Pohlmann et al. 2001) and by placenta (Soilleux 2003b). Similar *trans* activity for other viruses has been reported, and the receptors can also directly mediate infection of cells in *cis* (Soilleux 2003; Alvarez et al. 2002; Gardner et al. 2003). There is also evidence that these receptors interact with bacterial pathogens and with parasites (Appelmelk et al. 2003). The L-SIGN/DC-SIGNR functions as a HIV-1 trans-receptor similar to DC-SIGN (Bashirova et al. 2001). Moreover, L-SIGN interacts with other pathogens such as *Ebola virus* (Alvarez et al. 2002), to the envelope glycoproteins from *HIV-1, Hepatitis C virus* and *cytomegalovirus* (Alvarez et al. 2002; Bashirova et al. 2001; Bovin et al. 2003; Gardner et al. 2003; Halary et al. 2002), similar to DC-SIGN. L-SIGN is a liver-specific capture receptor for hepatitis C virus (Bovin et al. 2003).

Both DC-SIGN and DC-SIGNR bind ligands bearing mannose and related sugars through CRDs. The CRDs of DC-SIGN and DC-SIGNR bind $Man_9GlcNAc_2$ oligosaccharide 130- and 17-fold more tightly than mannose (Mitchell et al. 2001). Both DC-SIGN and DC-SIGNR possess a neck region, made up of multiple repeats, which supports the ligand-binding domain. Cross-linking analysis of full-length receptors expressed in fibroblasts confirms the tetrameric state of the intact receptors. The extra-cellular domain of DC-SIGN and DC-SIGNR, each comprises seven 23-residue tandem repeats, encoded by a single exon to form a coiled coil neck region. There is very high sequence identity between the repeat units, within each protein, and between DC-SIGN and DC-SIGNR. By analogy to other lectin receptors, such as the asialoglycoprotein receptors and CD23 (Bates et al. 1999; Beavil et al. 1995), it was suggested that this domain could mediate oligomerization, forming an α-helical coiled coil.

A subset of B cells in the blood and tonsils of normal donors also express DC-SIGN, which increased after stimulation in vitro with IL-4 and CD40 ligand, with enhanced expression of activation and co-stimulatory molecules CD23, CD58, CD80, and CD86, and CD22. The activated B cells captured and internalized X4 and R5 tropic strains of HIV-1, and mediated trans- infection of T cells. DC-SIGN serves as a portal on B cells for HIV-1 infection of T cells in trans. Transmission of HIV-1 from B cells to T cells through DC-SIGN pathway could be important in the pathogenesis of HIV-1 infection (Gupta and Rinaldo 2006).

36.5.2 Domain Organization of DC-SIGN and DC-SIGNR

DC-SIGN and DC-SIGNR consist of an N-terminal cytoplasmic domain, a repeat region consisting of seven 23-amino-acid tandem repeats, and a C-terminal C-type CRD that binds mannose-enriched carbohydrate modifications of host and pathogen proteins. They bind with highest affinity to larger glycans that contain 8 or 9 mannose residues (Mitchell et al. 2001; Guo et al. 2004). In addition, DC-SIGN, but not DC-SIGNR, binds to fucose-containing glycans, such as those present on the surfaces of nematode parasites (Appelmelk et al. 2003; Guo et al. 2004). The sugar-binding activity of each protein is conferred by a Ca^{2+}-dependent CRD that is located at the C terminus of the receptor polypeptide. This domain is separated from the membrane anchor by a neck region consisting of multiple 23-amino acid repeats. The neck forms an extended structure that associates to create a tetramer at the cell surface (Feinberg et al. 2001) (Fig. 36.3). Signals in the N-terminal cytoplasmic domain of DC-SIGN direct internalization of the receptor, which can thus mediate endocytosis and degradation of glycoproteins (Guo et al. 2004). DC-SIGNR lacks such signals and seems not to be a recycling receptor. Although these lectins on primary sinusoidal cells support HCV E2 binding, they are unable to support HCV entry. Lai et al. (2006) provided evidence for binding of circulating HCV with DC-SIGN and DC-SIGNR on sinusoidal endothelium within the liver allowing subsequent transfer of the virus to underlying hepatocytes in a manner analogous to DC-SIGN presentation of HIV on DCs (Lai et al. 2006).

36.5.3 Differences Between DC-SIGN and DC-SIGNR/L-SIGN

DC-SIGN and L-SIGN/DC-SIGNR genes have long been considered to share similar overall structure and ligand-binding characteristics. However, biochemical and structural studies show that they have distinct ligand-binding properties and different physiological functions. Of importance in both these genes is the presence of an extra-cellular domain consisting of an extended neck region encoded by tandem repeats that support the CRD, which plays a crucial role in influencing the pathogen-binding properties of these receptors. The notable difference between these two genes is in the extra-cellular domain. Whilst the tandem-neck-repeat region remains relatively constant in size in DC-SIGN, there is considerable polymorphism in L-SIGN. Homo-oligomerization of the neck region of L-SIGN has been shown to be important for high-affinity ligand binding, and heterozygous expression of the polymorphic variants of L-SIGN in which neck lengths differ could thus affect ligand-binding affinity. Despite DC-SIGN and DC-SIGNR bind HIV and enhance infection, comparison of these receptors reveals that they have very different physiological functions.

Screening an extensive glycan array demonstrated that DC-SIGN and DC-SIGNR have distinct ligand-binding properties. Structural and mutagenesis studies explain how both receptors bind high-mannose oligosaccharides on enveloped viruses and why only DC-SIGN binds blood group antigens, including those present on microorganisms. DC-SIGN mediates endocytosis, trafficking as a recycling receptor and releasing ligand at endosomal pH, whereas

36.5 DC-SIGN versus DC-SIGN-RELATED RECEPTOR [DC-SIGNR or L-SIGN (CD 209)/LSEctin]

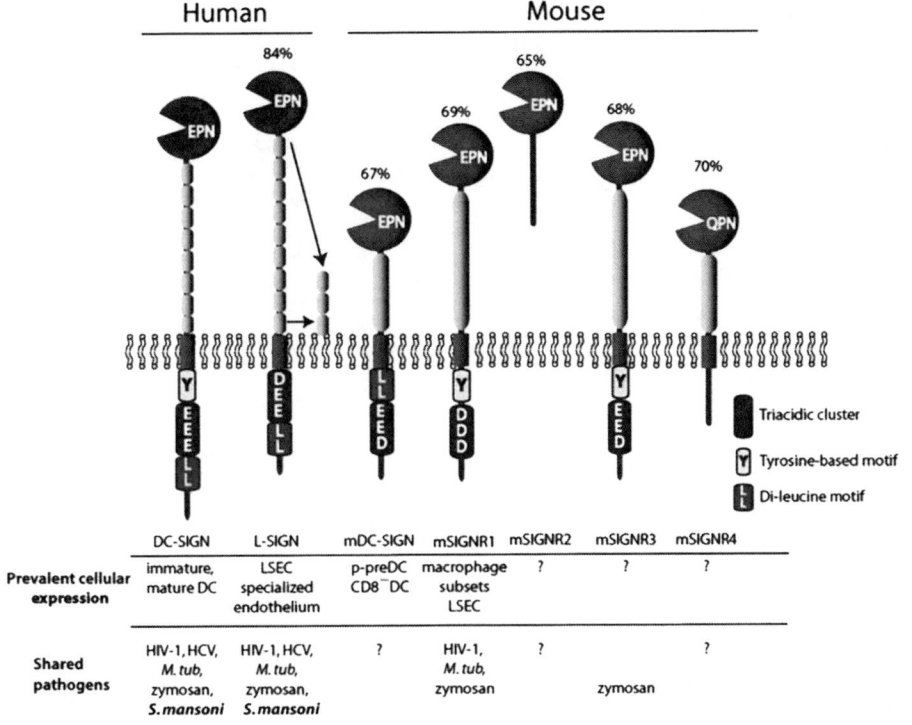

Fig. 36.3 Schematic representation of the structure, expression and binding specificities of DC-SIGN and its human and murine homologs. All SIGN homologs are transmembrane receptors, except for mSIGNR2, which is a soluble receptor. The percentage amino-acid-homology of CRDs of different homologs compared to DC-SIGN is depicted above the CRDs. Within the CRDs the highly conserved EPN sequence is essential for recognizing mannose-containing structures. All SIGN homologs contain the EPN motif, except for mSIGNR4, which has a QPN motif, indicating that mSIGNR4 may have another ligand specificity compared to the other SIGN molecules. In contrast to DC-SIGN that contains seven complete and one incomplete repeats, the number of repeats of L-SIGN is variable and varies between three and nine which is indicated with arrows. Within the cytoplasmic tail several internalization motifs are found. The di-leucin (LL) motif is thought to be important for internalization of DC-SIGN. The tyrosine-based motif and the tri-acidic cluster are also involved in internalization. However, the internalization capacities of the homologs of DC-SIGN have not been extensively explored. Based on its expression pattern, mSIGNR1 seems more similar to the homolog of L-SIGN rather than DC-SIGN whereas the expression of mDC-SIGN seems more homologous to DC-SIGN, although they are expressed by different DC subtypes. Strikingly, DC-SIGN, L-SIGN and mSIGNR1 share the ability to bind a number of pathogens (Reprinted with permission from Koppel et al. 2005 © John Wiley and Sons)

DC-SIGNR does not release ligand at low pH or mediate endocytosis. Thus, whereas DC-SIGN has dual ligand-binding properties and functions both in adhesion and in endocytosis of pathogens, DC-SIGNR binds a restricted set of ligands and has only the properties of an adhesion receptor (Guo et al. 2004). Functional studies on the effect of tandem-neck-repeat region on pathogen-binding, as well as genetic association studies for various infectious diseases and among different populations, have been reported. Worldwide demographic data of the tandem-neck-repeat region showing distinct differences in the neck-region allele and genotype distribution among different ethnic groups have been presented. These findings support the neck region as an excellent candidate acting as a functional target for selective pressures exerted by pathogens (Khoo et al. 2008). Chung et al (2010) identified Trp-258 in the DC-SIGN CRD to be essential for HIV-1 transmission. Although introduction of a K270W mutation at the same position in L-SIGN was insufficient for HIV-1 binding, an L-SIGN mutant molecule with K270W and a C-terminal DC-SIGN CRD subdomain transmitted HIV-1 (Chung et al. 2010).

36.5.4 Recognition of Oligosaccharides by DC-SIGN and DC-SIGNR

Both DC-SIGN and DC-SIGNR bind mannose bearing ligands and related sugars through CRDs. The CRDs of DC-SIGN and DC-SIGNR bind $Man_9GlcNAc_2$ oligosaccharide 130- and 17-fold more tightly than mannose. Results indicate that CRDs contain extended or secondary oligosaccharide binding sites that accommodate mammalian-type glycan structures. When the CRDs are clustered in the tetrameric extracellular domain, their arrangement provides a means of amplifying specificity for multiple glycans on host molecules targeted by DC-SIGN and DC-SIGNR. Binding to clustered oligosaccharides may also explain the interaction of these receptors with the gp120 envelope protein of HIV-1, which contributes to virus infection (Mitchell et al. 2001). Crystal structures of carbohydrate-recognition domains of DC-SIGN and of DC-SIGNR bound to oligosaccharide, in combination with binding studies, revealed that these receptors selectively recognize endogenous high-mannose oligosaccharides and may represent a new avenue for developing HIV prophylactics (Feinberg et al. 2001) (Fig. 36.4).

Similar to DC-SIGN, DC-SIGNR/L-SIGN can recognize high-mannose type N-glycans and the fucosylated glycan epitopes Lewis[A] (Le[a], Galß1-3(Fuc1-4)GlcNAc-), Lewis [B] (Le[b], Fuc1-2Galß1-3(Fuc1-4)GlcNAc-) and Lewis Y (Le[Y], Fuc1-2Galß1-4(Fuc1-3)GlcNAc-) (Geijtenbeek et al. 2003; Guo et al. 2004; van Liempt et al. 2004). L-SIGN, however, does not bind to the Le[X] epitope, which is one of the major ligands of DC-SIGN, although the formation of crystals between L-SIGN and Le[X] indicates that a weak interaction is possible (Guo et al. 2004). The inability of L-SIGN to bind to Le[X] epitopes is mainly due to the presence of a single amino acid in the CRD of L-SIGN, Ser[363] that prevents interaction with the Fuc(1-3)GlcNAc unit in Le[X], but supports binding of the Fuc1-4GlcNAc moiety present in Le[A] and Le[B] antigens. The equivalent amino acid residue Val[351] in DC-SIGN creates a hydrophobic pocket that strongly interacts with the Fuc(1-3/4)GlcNAc moiety of Le[X], other Lewis antigens, and probably LDN-F (Guo et al. 2004; van Liempt et al. 2004; van Liempt et al. 2006). The interaction of L-SIGN with *S. mansoni* egg glycoproteins and its location on liver endothelial cells suggest that L-SIGN may function in the recognition of glycan antigens of eggs that are trapped in the liver, thus contributing to glycan-specific immune responses and/or the immunopathology of schistosomiasis.

36.5.5 Extended Neck Regions of DC-SIGN and DC-SIGNR

Two different crystal forms of truncated DC-SIGNR comprising two neck repeats and the CRD reveal that the CRDs are flexibly linked to the neck, which contains α-helical segments interspersed with non-helical regions. Differential scanning calorimetry measurements indicated that the neck and CRDs are independently folded domains. Based on the crystal structures and hydrodynamic data, models for the full extracellular domains of the receptors have been generated. The observed flexibility of the CRDs in the tetramer, combined with reported data on the specificity of these receptors, suggests an important role for oligomerization in the recognition of endogenous glycans, in particular those present on the surfaces of enveloped viruses recognized by these proteins (Feinberg et al. 2005, 2009). To understand the tetramer-based ligand binding avidity, Snyder et al. (2005)

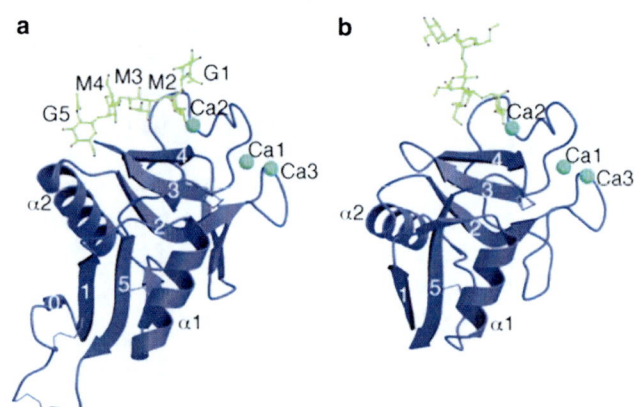

Fig. 36.4 Structure of CRD of DC-SIGN bound to GlcNAc$_2$Man$_3$. (**a**) Ribbon diagram of the DC-SIGN CRD (*blue*), with the bound oligosaccharide shown in a *ball-and-stick* representation (*yellow-green*, bonds and carbon atoms; *red*, oxygen; *blue*, nitrogen). Oligosaccharide residues are shown with the single letter code G for GlcNAc and M for mannose. Large cyan spheres are three Ca^{2+} ions. Disulfide bonds are shown in *pink*. The DC-SIGNR complex is very similar, except that the fourth disulfide connecting the NH$_2$ and COOH termini is not visible in either copy. (**b**) Rat serum mannose-binding protein bound to a high-mannose oligosaccharide. The color scheme is same as in (**a**) (Adapted from Feinberg et al. 2001 © American Association for the Advancement of Science)

determined the crystal structure of DC-SIGNR with its last repeat region and showed that compared to the carbohydrate-bound CRD structure, there are conformational differences in the calcium and carbohydrate coordination loops of CRD, an additional disulfide bond between the N and the C termini of the CRD, and a helical conformation for the last repeat. Snyder et al. (2005) generated a tetramer model for DC-SIGN/R using homology modeling and proposed a ligand-recognition index to identify potential receptor ligands. Polymorphisms associated with the length of the extracellular neck region of DC-SIGNR have been linked to differences in susceptibility to infection by enveloped viruses. The heterotetramers provide a molecular basis for interpreting the way polymorphisms affect interactions with viruses (Guo et al. 2006).

CRDs in DC-SIGN and DC-SIGNR are projected from the membrane surface by extended neck domains containing multiple repeats of a largely conserved 23-amino-acid sequence motif. The repeats are largely α-helical. Based on the structure and arrangement of the repeats in the crystal, the neck region can be described as a series of four-helix bundles connected by short, non-helical linkers. Combining the structure of the isolated neck domain with a overlapping structure of the distal end of the neck region with the CRDs attached provides a model of the almost-complete extracellular portion of the receptor. The organization of the neck suggests how CRDs may be disposed differently in DC-SIGN compared with DC-SIGNR and in variant forms of DC-SIGNR assembled from polypeptides with different numbers of repeats in the neck domain (Feinberg et al, 2009).

CRD of Langerin with Structural Similarity with DC-SIGN: Though DCs are thought to mediate HIV-1 transmission, yet it is becoming evident that different DC subsets at the sites of infection have distinct roles. In the genital tissues, two different DC subsets are present: the LCs and the DC-SIGN$^+$-DCs. Although DC-SIGN$^+$-DCs mediate HIV-1 transmission, recent data demonstrate that LCs prevent HIV-1 transmission by clearing invading HIV-1 particles. However, this protective function of LCs is dependent on the function of Langerin: blocking Langerin function by high virus concentrations enables HIV-1 transmission by LCs. A better understanding of the mechanism of these processes is crucial to understand and develop strategies to combat transmission (de Witte et al. 2008b). The CRD of human Langerin was examined by X-ray analyses for apo-Langerin and for complexes with mannose and maltose. The fold of the Langerin CRD resembles that of DC-SIGN. However, especially in the long loop region (LLR), which is responsible for carbohydrate-binding, two additional secondary structure elements are present: a 3(10) helix and a small β-sheet arising from the extended β-strand 2, which enters into a hairpin and a new strand β2'. However, the crystal structures in presence of maltose and mannose revealed two sugar-binding sites. One is calcium-dependent and structurally conserved in the C-type lectin family whereas the second one represents a calcium-independent type. Based on these data, the differences in binding behavior between Langerin and DC-SIGN with respect to the Lewis X carbohydrate antigen and its derivatives has been explained (Chatwell et al. 2008).

36.5.6 Signaling by DC-SIGN through Raf-1

Adaptive immune responses by DCs are critically controlled by Toll-like receptor (TLR) function. Little is known about modulation of TLR-specific signaling by other pathogen receptors. DC-SIGN has gained an exponential increase in attention because of its involvement in multiple aspects of immune function. Besides being an adhesion molecule, particularly in binding ICAM-2 and ICAM-3, it is also crucial in recognizing several endogenous and exogenous antigens. Additionally, the intracellular domain of DC-SIGN includes molecular motifs, which enable the activation of signal transduction pathways involving serine and threonine kinase Raf-1 and subsequent modulation of DC-maturation status, through direct modification of nuclear factor Nf-kB in DCs. DC-SIGN modulates TLR signaling at the level of the transcription factor NF-kB.

The DC-SIGN has emerged as a key player in the induction of immune responses against numerous pathogens by modulating TLR-induced activation. Upon DC-SIGN engagement by mannose- or fucose-containing oligosaccharides, the latter leads to a tailored Toll-like receptor signaling, resulting in an altered DC-cytokine profile and skewing of Th1/Th2 responses. Gringhuis et al. (2007) demonstrated that pathogens trigger DC-SIGN on human DCs to activate Raf-1, which subsequently leads to acetylation of the NF-kB subunit p65, but only after TLR-induced activation of NF-kB. Acetylation of p65 both prolonged and increased IL10 transcription to enhance anti-inflammatory cytokine responses. Different pathogens such as *M. tuberculosis, M. leprae, Candida albicans, measles virus,* and *HIV-1* interacted with DC-SIGN to activate the Raf-1-acetylation-dependent signaling pathway to modulate signaling by different TLRs. Thus, this pathway is involved in regulation of adaptive immunity by DCs to bacterial, fungal, and viral pathogens (Gringhuis et al. 2007). In addition, other DC-SIGN-ligands induce different signaling pathways downstream of Raf-1, indicating that DC-SIGN-signaling is tailored to the pathogen.

36.6 Functions of DC-SIGN

36.6.1 DC-SIGN Supports Immune Response

Immature DCs are recruited from blood into tissues to patrol for foreign antigens. Cells expressing DC-SIGN stable transfectants were able to mediate phagocytosis of *E. coli*. Ca^{2+} binding sites in the CRD of DC-SIGN were involved in phagocytosis of bacteria as well as multimerization of DC-SIGN, and neck region played a role in efficiency of binding to microbes as well as multimerization of the protein (Iyori et al. 2008; Valera et al. 2008). While analyzing early stages of DC-SIGN-mediated endocytosis, Cambi et al. (2009) demonstrated that both membrane cholesterol and dynamin are required and that DC-SIGN-mediated internalization occurs via clathrin-coated pits. Electron microscopy studies confirmed the involvement of DC-SIGN in clathrin-dependent HIV-1 internalization by DCs. Recent studies showed that some functions of decidual dendritic cells appear to be essential for pregnancy. In humans, decidual dendritic cells are identifiable by their expression of DC-SIGN. In normal decidua, DC-SIGN$^+$ cells expressed antigens associated with immature myeloid dendritic cells. In samples from spontaneous abortions, the decidual DC-SIGN$^+$ cells at a significantly lower proportion compared to normal pregnancies seem to play a role in pathological pregnancy outcomes (Tirado-González et al. 2010).

The dendritic cell-specific DC-SIGN internalizes antigen for presentation to T cells. After antigen uptake and processing, DCs mature and migrate to the secondary lymphoid organs where they initiate immune responses. As an adhesion molecule, DC-SIGN is able to mediate rolling and adhesion over endothelial cells under shear flow. The normal functions of DC-SIGN and DC-SIGNR include binding to ICAM-2 and ICAM-3. Binding of DC-SIGN to ICAM-2 on endothelial cells facilitates chemokine-induced DC extravasation; binding to ICAM-3 on T lymphocytes provides the initial step for establishing cell-mediated immunity (Liu and Zhu 2005). DCs could activate arrest T cells in the lymph node, but the mechanism is poorly understood. Studies showed that binding of DC-SIGN to both carcinoembryonic antigen (CEA)-related cell adhesion molecule 1 (CEACAM1) and Mac1 was required to establish cellular interaction between DCs and neutrophils, and such interaction promoted T cell proliferation and transformation to Th1 cells (van Gisbergen et al. 2005a, c). Subpopulations of human macrophages express DC-SIGN, to which Le^x-carrying CEACAMs may modulate the immune response in normal tissues such as the human placenta or in malignant tumors, for example in colorectal, pancreatic or lung carcinomas (Samsen et al. 2010). Reports suggest that DCs participate in the contact between itself and resting T cells, and also in T cell activation, and such effect is related to its cytoplasmic ITAM signal transduction (Engering et al. 2002). Martinez et al. discovered that DC-SIGN could promote CD3-activated T cells to produce IL-2 and receive a strong TCR signal, thus strengthening TCR-APC interaction and enhancing immune response (Martinez et al. 2005). Inhibiting DC-SIGN on DCs could reduce T cell proliferation and inhibit co-stimulator CD11c, CD83, CD80 and CD86 expression. Such effects are achieved by NF-κB signaling pathway (Zhou et al. 2006a). Recent evidences suggest that there is a cross-talk between DC-SIGN together with its CLRs and TLRs, and such crosstalk could lead to immune activation or T cells depression (Geijtenbeek et al. 2004; Gantner et al. 2003; Zhou et al. 2006b).

36.6.2 DC-SIGN Recognizes Pathogens

Viral and Bacterial Antigens DC-SIGN plays an important role in recognizing and capturing pathogens, DC migration and initiation of T cell responses. The role of DC-SIGN as a broad pathogen receptor has been well established (Geijtenbeek et al. 2000a; Alvarez et al. 2002; Colmenares et al. 2002; Cambi et al. 2003; Geijtenbeek et al. 2003; Lozach et al. 2003). In addition, DC-SIGN functions as a cell adhesion receptor mediating the interaction between DCs and resting T cells by binding to ICAM-3, and the transendothelial migration of DCs by binding to ICAM-2 (Geijtenbeek et al. 2000b). DC-SIGN has been identified as a receptor for HIV-1, HCV, Ebola virus, CMV, dengue virus, and the SARS coronavirus. Evidences suggest that DC-SIGN can function both as an adhesion receptor and as a phagocytic pathogen-recognition receptor, similar to the Toll-like receptors. Although major differences in the cytoplasmic domains of these receptors might predict their function, findings show that differences in glycosylation of ligands can dramatically alter C-type lectin-like receptor usage (Cambi and Figdor 2003; Engering et al. 2002). As a pathogen receptor, DC-SIGN displays affinity for high mannose moieties and functions as an internalization receptor for *HIV-1, hepatitis C virus, Mycobacterium tuberculosis, Helicobacter pylori,* and *Schistosoma manson, Leishmania,* and *Candida albicans* (Appelmelk et al. 2003; Bashirova et al. 2003a; Feinberg et al. 2001; Jack et al. 2001; Colmenares et al. 2004; Geijtenbeek et al. 2000a; Lozach et al. 2003; van Kooyk and Geijtenbeek 2003) and other pathogens that express mannose-containing carbohydrates (Alvarez et al. 2002; Halary et al. 2002; Tassaneetrithep et al. 2003; Zhou et al. 2008). The DC-SIGN molecule is used by HIV to attach to DCs in the genitourinary tract and rectum. Geitjenbeek et al. (2000a, c) suggest that DCs then carry HIV particles to lymph nodes, where the infection of T lymphocytes via receptors such as CD4 and CCR5 may occur. The virus may remain bound to DC-SIGN for protracted periods. DC-SIGN may deliver bound HIV to permissive cell types, mediating infection with high efficiency.

In vitro, DC-SIGN specifically interacts with *S. pneumoniae* serotype 3 and 14 in contrast to other serotypes such as 19F. While DC-SIGN interacts with *S. pneumoniae* serotype 14 through a ligand expressed by the capsular polysaccharide, the binding to *S. pneumoniae* serotype 3 appeared to depend on an as yet unidentified ligand (Koppel et al. 2005b). Leptospirosis is a global zoonotic disease, caused by pathogenic Leptospira species including *Leptospira interrogans* that causes public health and livestock problems. *L. interrogans* binds DC-SIGN and induces DCs maturation and cytokine production, which should provide new insights into cellular immune processes during leptospirosis (Gaudart et al. 2008).

Schistosoma Mansoni Antigens: Schistosomiasis is a human parasitic disease caused by helminths of the genus *Schistosoma. Schistosoma mansoni* synthesizes a multitude of carbohydrate complexes, which include both parasite-specific glycan antigens, as well as glycan antigens that are shared with the host. One example for a host-like glycan is the Le^X epitope Galß1-4(Fuc1-3)GlcNAc, which is expressed in all schistosomal life stages, but also on human leukocytes. Glycan antigens expressed by schistosomes induce strong humoral and cellular immune responses in their host (Meyer et al. 2007). The recognition of carbohydrates is mediated by DC-SIGN, and DC-SIGNR bind to glycans of *S. mansoni* soluble egg antigens (SEA) (van Die et al. 2003; van Liempt et al. 2004; Meyer et al. 2005; Saunders et al. 2009; van Die and Cummings 2006, 2010). L-SIGN binds both SEA and egg glycosphingolipids, and can mediate internalization of SEA by L-SIGN expressing cells. L-SIGN predominantly interacts with oligomannosidic *N*-glycans of SEA. L-SIGN binds to a glycosphingolipid fraction containing fucosylated species with compositions of $Hex_1HexNAc_{5-7}dHex_{3-6}Cer$. Results indicate that L-SIGN recognizes both oligomannosidic *N*-glycans and multiply fucosylated carbohydrate motifs within *Schistosoma* egg antigens, which demonstrate that L-SIGN has a broad but specific glycan recognition profile (Meyer et al. 2007).

DC-SIGN Captures Also Fungi: The DC-SIGN is located in the submucosa of tissues, where they mediate HIV-1 entry. Interestingly, the pathogen *C. albicans*, the major cause of hospital-acquired fungal infections, penetrates at similar submucosal sites. The DC-SIGN is able to bind *C. albicans* both in DC-SIGN-transfected cell lines and in human monocyte-derived DC. Moreover, in immature DC, DC-SIGN was able to internalize *C. albicans* in specific DC-SIGN-enriched vesicles, distinct from those containing the ManR. These results demonstrated that DC-SIGN is an exquisite pathogen-uptake receptor that captures not only viruses but also fungi (Cambi et al. 2003). However, DC-SIGN regulation in monocyte-derived macrophages does not singly predict the transmission potential of this cell type (Bashirova et al. 2003; Chehimi et al. 2003).

Microbial Uptake Capacities of DC-SIGN, SIGNR1, SIGNR3 and Langerin: Using transfected non-macrophage cell lines, Takahara et al. (2004) compared the polysaccharide and microbial uptake capacities of three lectins—DC-SIGN, SIGNR1 and SIGNR3—to another homolog mLangerin. Each molecule shares a potential mannose-recognition EPN-motif in its carbohydrate recognition domain. Using an anti-Tag antibody, it was found that each molecule could be internalized, although the rates differed. However, mDC-SIGN was unable to take up FITC-dextran, FITC-ovalbumin, zymosan or heat-killed *C. albicans*. The other three lectins showed distinct carbohydrate recognition properties. Furthermore, only SIGNR1 was efficient in mediating the capture by transfected cells of Gram-negative bacteria, such as *E. coli and S. typhimurium*, while none of the lectins tested were competent to capture Gram-positive bacteria, *S. aureus*. Therefore, these homologous C-type lectins have distinct recognition patters for microbes despite similarities in the carbohydrate recognition domains (Takahara et al. 2004).

36.6.3 DC-SIGN as Receptor for Viruses

DC-SIGN and L-SIGN can function as attachment receptors for *Sindbis (SB) virus*, an arbovirus of the Alphavirus genus. DC-SIGN is a universal pathogen receptor and can be used by *hepatitis C virus* (HCV) and other viral pathogens including *Ebola* virus, *cytomegalovirus* (CMV), *and Dengue virus* to facilitate infection by a mechanism that is distinct from that of HIV-1, leading to inhibition of the immunostimulatory function of DC and, hence, promotion of pathogen survival. Reports show that DC-SIGN not only plays a role in entry into DC, but HCV E2 interaction with DC-SIGN might also be detrimental to the interaction of DC with T cells during antigen presentation (Zhou et al. 2008). The surface membrane glycoprotein of *Borna disease virus* (BDV) is a polypeptide of 57 kDa and N-glycosylated to a precursor glycoprotein (GP) of about 94 kDa. Analysis showed that the precursor GP contains only mannose-rich N-glycans (Kiermayer et al. 2002) and has the potential to bind with DC-SIGN, L-SIGN in addition to other mannose spectic lectins discussed in different Chapters.

Reduced expression of DC-SIGN in spleen specifically characterizes pathogenic forms of simian immunodeficiency virus (SIV) infection, correlates with disease progression, and may contribute to SIV pathogenesis (Yearley et al. 2008).

N-glycan Status Modifies Virus Interaction: The virus-producing cell type is an important factor in dictating both N-glycan status and virus interaction with DC-SIGN/DC-SIGNR (Lin et al. 2003). In contrast, viruses bearing Ebola (Zaire strain) and Marburg (Musoke strain) envelope glycoproteins bind at significantly higher levels to immobilized MBL compared with virus particles pseudo typed with vesicular stomatitis virus glycoprotein or with no virus glycoprotein. Importantly, the tetrameric complexes, in contrast to DC-SIGN monomers, bind with high affinity to high mannose glycoproteins such as mannan or HIV gp120 suggesting that such an assembly is required for high affinity binding of glycoproteins to DC-SIGN (Bernhard et al. 2004).

36.6.4 HIV-1 gp120 and Other Viral Envelope Glycoproteins

DC-SIGN has been described as an attachment molecule for human HIV-1 with the potential to mediate its transmission. About half of the carbohydrates on gp120 are terminally mannosylated, a pattern common to many pathogens. The DC-SIGN binds to HIV and SIV gp120 and mediates the binding and transfer of HIV from monocyte-derived dendritic cells (MDDCs) to permissive

T cells (Geijtenbeek et al. 2002c). However, DC-SIGN binding to HIV gp120 may also be carbohydrate independent. Hong et al. (2002) formally demonstrated that gp120 binding to DC-SIGN and MDDCs is largely if not wholly carbohydrate dependent.

DC-SIGN was mainly expressed in tubular epithelial cells and DC-SIGN$^+$ DCs were primarily distributed in renal tubulo-interstitial areas during the early stage of nephritis. In a rat model of chronic renal interstitial fibrosis, there was a significant correlation of DC-SIGN expression with DC-SIGN$^+$ DC distribution and the degree of tubulo-interstitial lesion. DC-SIGN plays an important role in DC-mediated renal tubular interstitial lesions induced by immuno-inflammatory responses (Zhou et al. 2009). HIV-1 infection of renal cells has been proposed to play a role in HIV-1-associated nephropathy. The HIV-1 internalization was DC-SIGN receptor mediated. It appeared that HIV-1 routing occurred through nonacid vesicular compartments and the clathrin-coated vesicles and caveosomes may not be contributing to HIV-1-associated membrane traffic (Mikulak et al. 2010). However, Hatsukari et al. (2007) showed no expression of DC-SIGN, or mannose receptors in tubular cells and suggested that DEC-205 (Chap.15 and Chap. 35) acts as an HIV-1 receptor that mediates internalization of the virus into renal tubular cells (HK2), from which the virus can be rescued and disseminated by encountering immune cells.

Herpes Simplex Virus and Human Herpes Virus 8: Of the two Herpes simplex virus (HSV) subtypes described, HSV-1 causes mainly oral-facial lesions, whilst HSV-2 is associated with genital herpes. HSV-1 and -2 both interact with DC-SIGN. Analyses demonstrated that DC-SIGN interacts with the HSV glycoproteins gB and gC. In another setting, human herpesvirus 8 (HHV-8) is the etiological agent of Kaposi's sarcoma, primary effusion lymphoma, and some forms of multicentric Castleman's disease. DC-SIGN is an entry receptor for HHV-8 on DC and macrophages. The infection of B cells with HHV-8 resulted in increased expression of DC-SIGN and a decrease in the expression of CD20 and MHC-I. It was indicated that the expression of DC-SIGN is essential for productive HHV-8 infection of and replication in B cells (de Jong et al. 2008; Rappocciolo et al. 2006, 2008).

West Nile Virus: West Nile virus (WNV), a mosquito-borne flavivirus, has recently emerged in North America. The elderly are particularly susceptible to severe neurological disease and death from infection with this virus. DC-SIGN enhances infection of cells by West Nile virus (WNV) glycosylated strains, which may at least in part explain the higher pathogenicity of glycosylated L1 strains versus most non-glycosylated L2 strains (Martina et al. 2008). Kong et al. (2008) found that the binding of the glycosylated WNV envelope protein to DC-SIGN leads to a reduction in the expression of TLR3 in macrophages from young donors via signal transducer and activator of transcription 1 (STAT1)-mediated pathway. This signaling is impaired in the elderly, and the elevated levels of TLR3 result in an elevation of cytokine levels. This alteration of the innate immune response with aging may contribute to the permeability of blood–brain barrier and a possible mechanism for the increased severity of WNV infection in aged individuals.

Corona Virus Infection and SARS: The MBL deficiency is a susceptibility factor for acquisition of severe acute respiratory syndrome (SARS) corona virus (CoV) (Ip et al. 2005). On similar pattern, entry of the serotype II feline corona virus strains feline infectious peritonitis virus (FIPV) and DF2 into nonpermissive mouse 3T3 cells could be rescued by the expression of human DC-SIGN and the infection of a permissive feline cell line (Crandall-Reese feline kidney) was markedly enhanced by the overexpression of DC-SIGN. Treatment with mannan considerably reduced infection of feline monocyte-derived cells expressing DC-SIGN, indicating a role for FIPV infection in vivo (Regan and Whittaker 2008).

Dengue Virus Envelope Glycoprotein and DC-SIGN: Dengue virus (DV) primarily targets immature DCs after a bite by an infected mosquito vector. Navarro-Sanchez et al. (2003) showed that DC-SIGN is a binding receptor for DV that recognizes N-glycosylation sites on the viral E-glycoprotein and allows viral replication. Mosquito-cell-derived DVs may have differential infectivity for DC-SIGN-expressing cells. There is evidence that infection of immature myeloid DCs plays a crucial role in dengue pathogenesis and that the interaction of the viral envelope E glycoprotein with DC-SIGN is a key element for their productive infection (Kwan et al. 2008).

Measles Virus Targets DC-SIGN: DCs are involved in the pathogenesis of measles virus (MV) infection by inducing immune suppression and possibly spreading the virus from the respiratory tract to lymphatic tissues. The DC-SIGN is the receptor for laboratory-adapted and wild-type MV strains. The ligands for DC-SIGN are both MV glycoproteins F and H. DC-SIGN was found important for the infection of immature DCs with MV, since both attachment and infection of immature DCs with MV were blocked in the presence of DC-SIGN inhibitors. Moreover, MV might not only target DC-SIGN to infect DCs but may also use DC-SIGN for viral transmission and immune suppression (de Witte et al. 2006). Thus, DCs play a prominent role during the initiation, dissemination, and clearance of MV infection (de Witte et al. 2008a).

36.7 Subversion and Immune Escape Activities of DC-SIGN

36.7.1 The entry and dissemination of viruses can be mediated by DC-SIGN

Dendritic cells are likely the first cells to encounter invading pathogens. The entry and dissemination of viruses in several families can be mediated by DC-SIGN. Terminal mannoses at positions 2 or 3 in the trisaccharides are the most important moiety and present the strongest contact with the binding site of DC-SIGN (Reina et al. 2008). Although, the mechanism of DC-SIGN and HIV-1 interaction remains unclear, Smith et al. (2007) identified a cellular protein that binds specifically to the cytoplasmic region of DC-SIGN and directs internalized virus to the proteasome degradation. This cellular protein, leukocyte-specific protein 1 (LSP1) is an F-actin binding protein involved in leukocyte motility and found on the cytoplasmic surface of the plasma membrane. LSP1 interacted specifically with DC-SIGN and other C-type lectins, but not the inactive mutant DC-SIGNΔ35, which lacks a cytoplasmic domain and shows altered virus transport in DCs. Thus, LSP1 protein facilitates virus transport into the proteasome after its interaction with DC-SIGN through its interaction with cytoskeletal proteins. Thus, it has been proposed that attachment of HIV to DC-SIGN enables the virus to hijack cellular transport processes to ensure its transmission to adjacent T cells.

Interestingly, not all interactions between DC-SIGN receptors and pathogenic ligands have beneficial results. During the interaction between body and pathogens or tumors, the latter could escape immune surveillance and survive. Such mechanism is related to suppressions of DCs by DC-SIGN. Thus, though DCs are vital in the defense against pathogens, it is becoming clear that some pathogens subvert DC functions to escape immune surveillance. It appears that some pathogens have evolved immunoevasive or immunosuppressive activities through receptors such as DC-SIGN. For example, HIV-1 targets the DC-SIGN to hijack DCs for viral dissemination. Binding to DC-SIGN protects HIV-1 from antigen processing and facilitates its transport to lymphoid tissues, where DC-SIGN promotes HIV-1 infection of T cells. Studies have also demonstrated that different ligand binding and/or sensing receptors collaborate for full and effective immune responses (McGreal et al. 2005; van Kooyk et al. 2003; Wang et al. 2004). It was also demonstrated that DC-SIGN efficiently reduces the amount of gp120 present on the cell plasma membrane, and completely strips off gp120 from the virions produced by the host cells, suggesting that blockage of HIV budding is due to internalization of gp120 by DC-SIGN (Wang and Pang 2008; Solomon Tsegaye and Pohlmann 2010).

Perhaps, binding of DC-SIGN to gp120 may facilitate or stabilize these transitions. Further studies demonstrated that HIV-1 would lose its activity if it is kept in vitro for 24 h, but DC-SIGN-bound HIV-1 could be kept within DCs for more than 4 days, allowing incoming virus to persist for 25 days before infecting target (Zhou et al. 2004). The mechanism of DC-SIGN prolonging viral infectivity is still poorly understood (Lozach et al. 2005). Thus, HIV sequestration by and stimulation of DC-SIGN helps HIV evade immune responses and spread to cells (Hodges et al. 2007; Marzi et al. 2007).

Many Pathogens Target DC-SIGN to Escape Host Immunity: For Hepatitis C Virus, both DC-SIGN and L-SIGN are known to bind envelope glycoproteins E1 and E2. Soluble DC-SIGN and L-SIGN specifically bound HCV virus-like particles. It is also speculated that HBV exploits mannose trimming as a way to escape recognition by DC-SIGN and thereby subvert a possible immune activation response (Op den Brouw et al. 2008). Reduced expression of DC-SIGN in spleen specifically characterizes pathogenic forms of SIV infection, correlates with disease progression, and may contribute to SIV pathogenesis (Yearley et al. 2008). DC-SIGN acts as a capture or attachment molecule for avian H5N1 virus, and mediates infections in cis and in trans (Wang et al. 2008b). Mittal et al. (2009) demonstrated that *Enterobacter sakazakii* (ES) targets DC-specific DC-SIGN to survive in myeloid DCs for which outer membrane protein A expression in ES is critical, although it is not required for uptake. ES interaction with DC-SIGN seems to subvert the host immune responses by disarming MAPK pathway in DCs (Mittal et al. 2009). *Yersinia pestis* is the etiologic agent of bubonic and pneumonic plagues. It is speculated that *Y. pestis* hijacks DCs and alveolar macrophages, in order to be delivered to lymph nodes. DC-SIGN is a receptor for *Y. pestis* that promotes phagocytosis by DCs in vitro (Zhang et al. 2008b). Accumulating evidence supports that certain pathogens target DC-SIGN to escape host immunity. Unlike certain other host pathogen interactions, activation of DCs by *B. pseudomallei* is not dependent on DC-SIGN. Evidence also indicates that the LPS mutant that binds DC-SIGN has a suppressive effect on DC cytokine production (Charoensap et al. 2008).

36.7.2 DC-SIGN and Escape of Tumors

Recognition of Tumor Glycans: Dendritic cells play an important role in the induction of antitumor immune responses. Glycosylation changes during malignant transformation create tumor-specific carbohydrate structures that interact with C-type lectins on DCs. Studies suggest that tumor glycoproteins, such as carcinoembryonic antigen (CEA) and MUC-1, indeed interact with DC-SIGN and macrophage galactose-type lectin on APCs. DC-SIGN has been detected on immature DCs that were associated with melanoma and myxofibrosarcoma (Soilleux et al. 2003; Vermi et al. 2003). The consequences for anti-cancer immunity or tolerance induction can be extrapolated from the function of C-type lectins in pathogen recognition and antigen presentation. In addition, in vivo studies in mice demonstrated the potency of targeting antigens to C-type lectins on APCs for anti-tumor vaccination strategies (Aarnoudse et al. 2006).

DC-SIGN and Escape of Tumors: DC-SIGN has been related to immune escape of tumors (van Gisbergen et al. 2005b; Gijzen et al. 2008). Immature DCs are located intratumorally within colorectal cancer and intimately interact with tumor cells, whereas mature DCs are present peripheral to the tumor. The majority of colorectal cancers over-express CEA and malignant transformation changes the glycosylation of CEA on colon epithelial cells, resulting in higher levels of Le^X and de novo expression of Le^Y on tumor-associated CEA. Since DC-SIGN has high affinity for nonsialylated Lewis antigens, it is possible that DC-SIGN is involved in recognition of colorectal cancer cells by DCs. It was shown that immature DCs within colorectal cancer express DC-SIGN, which mediates these interactions through binding of Le^X/Le^Y carbohydrates on CEA of colorectal cancer cells. In contrast, DC-SIGN does not bind CEA expressed on normal colon epithelium due to low levels of Lewis antigens. This indicates that DCs may recognize colorectal cancer cells through binding of DC-SIGN to tumor-specific glycosylation on CEA. Similar to pathogens that target DC-SIGN to escape immunosurveillance, tumor cells may interact with DC-SIGN to suppress DC functions (van Gisbergen et al. 2005b). At the same time, tumor cells can suppress DC maturation by DC-SIGN and escape immune surveillance. Bogoevska et al. (2006) showed that CEA-related cell adhesion molecule 1 (CEACAM1) selectively attaches and specifically interacts with DC-SIGN, and participates in cancer development. DC-SIGN is involved in the interaction of DCs with colorectal tumor SW1116 cells through the recognition of aberrantly glycosylated forms of Le^a/Le^b glycans on CEA and CEACAM. DC-SIGN ligands containing Le^a/Le^b glycans are also highly expressed on primary cancer colon epithelia but not on normal colon epithelia, and DC-SIGN is suggested to be involved in the association between DCs and colorectal cancer cells in situ by DC-SIGN recognizing these cancer-related Le glycan ligands. Observations imply that colorectal carcinomas affecting DC function and differentiation through interactions between DC-SIGN and colorectal tumor-associated Le glycans may induce generalized failure of a host to mount an effective antitumor response (Nonaka et al. 2008). In acute lymphoblastic leukemia (ALL), aberrant glycosylation of blast cells can alter their interaction with DC-SIGN and L-SIGN, thereby affecting their immunological elimination. High binding of B-ALL peripheral blood cells to DC-SIGN and L-SIGN correlates with poor prognosis. Apparently, when B-ALL cells enter the blood circulation and are able to interact with DC-SIGN and L-SIGN the immune response is shifted toward tolerance (Gijzen et al. 2008).

The role of DCs in progression of primary cutaneous T-cell lymphoma (CTCL) is not established. Schlapbach et al. (2010) found a significant infiltration of CTCL lesions by immature DC-SIGN$^+$ DCs with close contact to tumor cells. Matured and activated DCs were only rarely detected in lesions of CTCL. The preponderance of immature DC-DIGN$^+$ DCs in contact with regulatory T cells in lesions of CTCL points to an important role of this subset in the host's immune reaction to the malignant T cells. Since these immature DCs are known to induce immunotolerance, they might play a role in the mediation of immune escape of the proliferating clone (Schlapbach et al. 2010). In contrast, studies showed that there are CEA specific T cells in colorectal cancer patients which have anti-tumor effects (Nagorsen et al. 2000). Such results suggest that DCs could recognize and bind to colorectal tumor cells by DC-SIGN and participate in anti-tumor immune response.

36.7.3 *Mycobacterial* Carbohydrates as Ligands of DC-SIGN, L-SIGN and SIGNR1

Mycobacteria, including *M. tuberculosis* (Mtb), are surrounded by a loosely attached capsule that is mainly composed of proteins and polysaccharides. Although the chemical composition of the capsule is relatively well studied, its biological function is only poorly understood. *M. tuberculosis*, the causative agent of tuberculosis (TB), is recognized by pattern recognition receptors on macrophages and DCs, thereby triggering phagocytosis, antigen presentation to T cells and cytokine secretion. Mtb spreads through aerosol carrying them deep into the lungs, where they are internalized by phagocytic cells, such as neutrophils (PMNs), DCs, and macrophages. The DC- Mtb manipulates cells of the innate immune system to provide the bacteria with a sustainable intracellular niche (Schaefer et al. 2008). Mannosylated moieties of the mycobacterial cell wall, such as mannose-capped lipoarabinomannan (ManLAM) or higher-order phosphatidylinositol-mannosides (PIMs) of Mtb, were shown to bind to DC-SIGN on immature DCs and macrophage subpopulations. This interaction reportedly impaired dendritic cell maturation, modulated cytokine secretion by phagocytes and dendritic cells and was postulated to cause suppression of protective immunity to TB. However, experimental Mtb infections in mice transgenic for human DC-SIGN revealed that, instead of favoring immune evasion of mycobacteria, DC-SIGN may promote host protection by limiting tissue pathology. Furthermore, infection studies with mycobacterial strains genetically engineered to lack ManLAM or PIMs demonstrated that manLAM/PIM-DC-SIGN interaction was not critical for cytokine secretion in vitro and protective immunity in vivo (Ehlers 2010).

Reports suggest that *M. tuberculosis* targets DC-SIGN to inhibit the immuno-stimulatory function of DC through the interaction of the mycobacterial ManLAM to DC-SIGN, which prevents DC maturation and induces the formation of immuno-suppressive cytokine IL-10 that helps in the survival and persistence of *M. tuberculosis* (Fig. 36.2). The pathogen-derived carbohydrate structure on ManLAM that is recognized by DC-SIGN has been identified. The synthetic mannose-cap oligosaccharides manara, (Man)2-ara and (Man)3-ara specifically bound by DC-SIGN. The human and murine DC-SIGN homolog L-SIGN and SIGNR1, respectively, also interact with mycobacteria through ManLAM. Both homologs have the highest affinity for the (Man)3-ara structure, similar to DC-SIGN. The identification of SIGNR1 as a receptor for ManLAM enabled in vivo studies to investigate the role of DC-SIGN in *M. tuberculosis* pathogenesis (Geijtenbeek et al. 2003; Tailleux et al. 2003; Koppel et al. 2004). In addition to manLAM, Mtb α-glucan is another ligand for DC-SIGN. The recognition of α-glucans by DC-SIGN is a general feature and the interaction is mediated by internal glucosyl residues. As for manLAM, an abundant mycobacterial cell wall-associated glycolipid, binding of α-glucan to DC-SIGN stimulated the production of immunosuppressive IL-10 by LPS-activated monocyte-derived DCs. This IL-10 induction was DC-SIGN-dependent and also required acetylation of NF-kB (Geurtsen et al. 2009).

The mannose cap of LAM is a crucial factor in mycobacterial virulence. Appelmelk et al. (2008) evaluated the biological properties of capless mutants of *M. marinum* and *M. bovis* BCG, made by inactivating homologs of Rv1635c and showed that its gene product is an undecaprenyl phospho-mannose-dependent mannosyltransferase. Compared with parent strain, capless *M. marinum* induced slightly less uptake by and slightly more phagolysosome fusion in infected macrophages but this did not lead to decreased survival of the bacteria in vitro, nor in vivo in zebra fish. Appelmelk et al. (2008) contradicted the current paradigm and demonstrated that mannose-capped LAM does not dominate the M.-host interaction. Although the mannose caps of the mycobacterial surface (ManLAM) are essential for the binding to DC-SIGN, genetic removal of these caps did not diminish the interaction of whole mycobacteria with DC-SIGN and DCs. Like ManLAM, Hexamannosylated PIM (6), which contains terminal α(1—>2)-linked mannosyl residues identical to the mannose cap on ManLAM showed high affinity and represents a bonafide DC-SIGN ligand but

that other, as-yet-unknown, ligands dominate in the interaction between *mycobacteria* and DCs (Driessen et al. 2009) (Fig. 36.1, 36.2).

36.7.4 Decreased Pathology of Human DC-SIGN Transgenic Mice During Mycobacterial Infection

Although, the *M. tuberculosis*, the causative agent of pulmonary tuberculosis, interacts with DC-SIGN to evade the immune system, transgenic mice after high dose aerosol infection with the strain *Mtb*-H37Rv, showed massive accumulation of DC-SIGN+ cells in infected lungs, reduced tissue damage and prolonged survival. Based on these results, it was proposed that instead of favoring the immune evasion of mycobacteria, human DC-SIGN may have evolved as a pathogen receptor promoting protection by limiting tuberculosis-induced pathology (Schaefer et al. 2008). Hedlund et al. (2010) demonstrated that DCs can distinguish between normal and infected apoptotic PMNs via cellular crosstalk, where the DCs can sense the presence of danger on the Mtb-infected PMNs and modulate their response accordingly. Balboa et al. (2010) showed that early interaction of Î-irradiated *M. tuberculosis* with Mo subverts DC differentiation in vitro and suggested that *M. tuberculosis* escapes from acquired immune response in tuberculosis may be caused by an altered differentiation into DC leading to a poor Mtb-specific T-cell response. Since, *M. tuberculosis* interacts with DC-SIGN to evade the immune system, the dominant Mtb-derived ligands for DC-SIGN are presently unknown, and a major role of DC-SIGN in the immune response to Mtb infection may lie in its capacity to maintain a balanced inflammatory state during chronic TB (Ehlers 2010).

36.7.5 Genomic Polymorphism of DC-SIGN (CD209) and Consequences

36.7.5.1 CD209 Genetic Polymorphism and HIV Infectivity

DC-SIGN and DC-SIGNR have been thought to play an important role in establishing HIV infection by enhancing trans-infection of CD4+ T cells in the regional lymph nodes. The variation of DC-SIGNR genotypes affects the efficacy of trans-infection by affecting the amounts of the protein expressed on cell surface and augmenting the infection. A potential association of DC-SIGN and DC-SIGNR neck domain repeat polymorphism and risk of HIV-1 infection is currently under debate. Rathore et al. (2008a) showed that polymorphism in DC-SIGN neck repeats region was rare and not associated with HIV-1 susceptibility among North Indians. But sequencing analysis of DC-SIGN gene confirmed four novel genetic variants in intronic region flanking exon 4 coding region. A total of 13 genotypes were found in DC-SIGNR neck repeat region polymorphism. Among all the genotypes, only 5/5 homozygous showed significant reduced risk of HIV-1 infection in HIV-1-exposed seronegative individuals. A unique genotype 8/5 heterozygous was also found in HIV-1 seropositive individual, which is not reported elsewhere (Rathore et al. 2008a, b).

The association of polymorphism of homolog of DC-SIGNR gene with susceptibility to virus infection suggests that the tandem-repeat polymorphisms of the DC-SIGNR gene in the Chinese Han population exhibit unique genetic characteristics not recognized earlier in the Caucasian population. Genotype 9/5 seems to be a risk factor for HIV-1 infection in the Chinese population (Wang et al. 2008a). To understand the role of DC-SIGN neck-region length variation in HIV-1 transmission, Zhang et al. (2008a) studied 530 HIV-1-positive and 341 HIV-1-negative individuals in China. The carrier frequency of a DC-SIGN allele with <5 repeat units in the neck-region was 0.9% in HIV-1-positive and 3.8% in HIV-1-negative individuals. This observation suggests that DC-SIGN variation plays a role in HIV-1 transmission. These naturally occurring DC-SIGN neck-region variants were significantly more frequent in the Chinese population than in the US population and in a worldwide population. Several transcripts of DC-SIGN have been identified, some of which code for putative soluble proteins. However, little is known about the regulation and the functional properties of such putative sDC-SIGN variants. Based on the analysis of the cytokine/chemokine content of sDC-SIGN culture supernatants, results confirmed that sDC-SIGN, like membrane DC-SIGN counterpart, may play a pivotal role in CMV-mediated pathogenesis (Plazolles et al. 2011). Variations in genes encoding virus recognition and reactivation and patients following allogeneic stem-cell transplantation suggested that two SNP (rs735240, G > A; rs2287886, C > T) in the promoter region of DC-SIGN are significantly associated with an increased risk of development of hCMV reactivation and disease. These genetic markers influence the expression levels of DC-SIGN on immature DCs, as well as infection efficiency of immature DCs by hCMV and might help to predict the individual risk of hCMV reactivation and the disease (Mezger et al. 2008).

Variations in the number of repeats in the neck region of DC-SIGN and DC-SIGNR possibly influence host susceptibility to HIV-1 infection. Chaudhary et al. (2008a) examined the SNP of DC-SIGN and DC SIGNR in healthy HIV seronegative individuals, high risk STD patients seronegative for HIV, and HIV-1 seropositive patients from northern India. DC-SIGN polymorphism was rare and genotype 7/7 was predominant in all groups studied. DC-SIGNR was highly polymorphic and 11 genotypes were observed among the different study groups. The precise role of the polymorphic variants of DC-SIGNR needs to be elucidated in the population (Chaudhary et al. 2008a, b). GG genotype of SDF-1α 3′UTR polymorphism may be associated with susceptibility to PTB in HIV-1 infected patients in south India. Genotype frequencies of DC-SIGN polymorphisms did not differ significantly between HIV patients with or without TB. A better understanding of genetic factors that are associated with TB could help target preventive strategies to those HIV patients likely to develop tuberculosis (Alagarasu et al. 2009). Whether variants in the DC-SIGN encoding CD209 gene are associated with susceptibility to or protection against HIV-1 infection or development of TB among HIV-1 infected south Indian patients, study suggests that -336G/G genotype while associated with protection against HIV-1 infection the same genotype is also associated with susceptibility to HIV-TB among south Indians (Selvaraj et al. 2009). Olesen et al. (2007) support the report that vitamin D receptor (VDR) gene SNPs modulate the risk for TB in West Africans and suggest that variation within DC-SIGN and PTX3 also affect the disease outcome.

Reports suggest that CD209 promoter SNP-336A/G exerts an effect on CD209 expression and is associated with human susceptibility to dengue, HIV-1 and tuberculosis in humans. The CD209 -336G variant allele is also associated with significant protection against tuberculosis in individuals from sub-Saharan Africa and, cases with -336GG were significantly less prone to develop tuberculosis-induced lung cavitation. Previous in vitro work demonstrated that the promoter variant -336G allele causes down-regulation of CD209 mRNA expression. This report suggests that decreased levels of the DC-SIGN receptor may be protective against both clinical tuberculosis in general and cavitory tuberculosis disease in particular. This is consistent with evidence that *Mycobacteria* can utilize DC-SIGN binding to suppress the protective pro-inflammatory immune response (Vannberg et al. 2008).

Wichukchinda et al. (2007) genotyped two SNPs in DC-SIGN promoter (-139A/G and 336A/G), a repeat number of 69 bp in Exon 4 of DC-SIGN and DC-SIGNR, and one SNP in Exon 5 of DC-SIGNR and showed that the proportion of individuals possessing a heterozygous 7/5 and 9/5 repeat and A allele at rs2277998 of DC-SIGNR in HIV-seronegative individuals of HIV-seropositive spouses was significantly higher than HIV-seropositive individuals. These associations were observed only in females but not in males. The proportion of individuals possessing the 5A haplotype in HIV-seronegative females was significantly higher than HIV-seropositive females. These associations suggested that DC-SIGNR might affect susceptibility to HIV infection by a mechanism that is different in females and males (Wichukchinda et al. 2007; Zhu et al. 2010).

36.7.5.2 Human T-Cell Lymphotropic Virus Type 1 (HTLV) Infection

DC-SIGN plays a critical role in HTLV-1 binding, transmission, and infection, thereby providing an attractive target for the development of antiretroviral therapeutics and microbicides (Jain et al. 2009). Kashima et al. (2009) evaluated four polymorphisms located in the DC-SIGN gene promoter region (positions -336, -332 -201 and -139) in DNA samples from Brazilian ethnic groups (Caucasians, Afro-Brazilian, Asians and Amerindians) to establish the population distribution of these SNPs and correlated DC-SIGN polymorphisms and infection in samples from human T-cell lymphotropic virus type 1 (HTLV-1)-infected individuals. The -336A and -139A SNPs were quite common in Asians and that the -201T allele was not observed in Caucasians, Asians or Amerindians. No significant differences were observed between individuals with HTLV-1 disease and asymptomatic patients. However, the -336A variant was more frequent in HTLV-1-infected patients. In addition, the -139A allele was found to be associated with protection against HTLV-1 infection when the HTLV-1-infected patients as a whole were compared with the healthy-control group. Kashima et al. (2009) suggested that the -139A allele might be associated with HTLV-1 infection, although no significant association was observed among asymptomatic and HAM/TSP patients. Koizumi et al. (2007) showed that RANTES -28G was associated with delayed AIDS progression, while DC-SIGN -139C was associated with accelerated AIDS progression in HIV-1-infected Japanese hemophiliacs. While analyzing DC-SIGN and DC-SIGNR polymorphisms in Caucasian Canadian and indigenous African populations, Boily-Larouche et al. (2007) found several novel nucleotide variants within regulatory 5'- and 3'-untranslated regions of the genes that could affect their transcription and translation. Study demonstrated that Africans show greater genetic diversity at these two closely-related immune loci than observed in other major population groups.

36.7.5.3 CD209 Genetic Polymorphism and Tuberculosis Disease

DC-SIGN is a receptor capable of binding and internalizing *M. tuberculosis*. The CD209 promoter single SNP-336A/G exerts an effect on CD209 expression and is associated with human susceptibility to dengue, HIV-1 and tuberculosis in humans. In vitro studies confirmed that this SNP modulates gene promoter activity. An association study was performed in Tunisian patients comprising tuberculosis and healthy controls. Sequencing of the DC-SIGN promoter region detected four polymorphisms (-939, -871, -601, and -336), but no differences in their allelic distribution were observed between the two groups. In addition, the analysis of length variation in the DC-SIGN neck region indicated extremely low levels of polymorphisms and, again, no differences between patients and controls. Results suggested neither promoter variants nor length variation in the neck region of DC-SIGN is associated with susceptibility to tuberculosis in Tunisian patients (Ben-Ali et al. 2007).

Among Caucasians patients suffering from pulmonary tuberculosis, DC-SIGN revealed no significant differences in loci -336A/G and -871A/G with controls. Analysis of MIRU-VNTR patterns identified 50 unique profiles, among which there were genotypes of the families Beijing, T. LAM, Haarlem, "Ural" (Haarlem 4) and X. Among 90 MIRU-VNTR genotypes, 42 profiles belonged to the Beijing family. Moreover, the minimum spanning tree (MST) test revealed a number of Beijing-like strains. The genotypes of the subjects affected with Beijing and Beijing-like strains and those affected with the strains of other families (non-Beijing) were compared. A significance reduction was found in the incidence of the -336G genotype among the subjects affected with Beijing strains versus those infected with non-Beijing strains at a frequency of 0.09 and 0.24, respectively (Ogarkov et al. 2007). CD209 facilitates severe acute respiratory syndrome (SARS)-coronavirus spike protein-bearing pseudotype driven infection of permissive cells in vitro. Genetic association analysis of SNP with clinico-pathologic outcomes in 824 serologic confirmed that the -336AG/GG genotype SARS patients were associated with lower LDH levels compared with the -336AA patients. High LDH levels are known to be an independent predictor for poor clinical outcome, probably related to tissue destruction from immune hyperactivity. Hence, SARS patients with the CD209 -336 AA genotype carry a 60% chance of having a poorer prognosis. This association is in keeping with the role of CD209 in modulating immune response to viral infection. The relevance of these findings for other infectious diseases and inflammatory conditions would be worth investigating (Chan et al. 2010).

36.8 SIGNR1 (CD209b): The Murine Homologues of DC-SIGN

36.8.1 Characterization

The mouse (m) DC-SIGN family consists of several homologous type II transmembrane proteins located in close proximity on chromosome 8 and having a single carboxyl terminal carbohydrate recognition domain. Initial screening of mouse cDNA libraries led to the identification of multiple mouse homologs of DC-SIGN and DC-SIGNR, designated DC-SIGN and SIGNR1 through SIGNR4 by Park et al. (2001). More murine homologs of human SIGNs have been identified, and the biochemical and cell biological properties of all murine SIGNs have been compared (Fig. 36.3). The SIGNR1 is a C-type lectin domain of murine homolog of DC-SIGN and functions in vivo as a pathogen recognition receptor on macrophages that captures blood-born antigens, which are rapidly internalized and targeted to lysosomes for processing (Geijtenbeek et al. 2002b). In addition to five SIGNR proteins, a pseudogene, encoding a hypothetical SIGNR6, and a further two expressed proteins, SIGNR7 and SIGNR8, have been identified. Screening of a glycan array demonstrated that only mouse SIGNR3 shares with human DC-SIGN the ability to bind both high mannose and fucose-terminated glycans in this format and to mediate endocytosis. SIGNR3 is a differentiation marker for myeloid mononuclear cells and that some DCs, especially in sLNs, are possibly replenished by Ly6C(high) monocytes (Nagaoka et al. 2010). The mouse homologs of DC-SIGN have a diverse set of ligand-binding and intracellular trafficking properties, some of which are distinct from the properties of any of the human receptors (Powlesland et al. 2006).

SIGN Related 1 (SIGNR1) or CD209b is expressed at high levels on macrophages in lymphoid tissues, especially within the marginal zone of the spleen. SIGNR1 can bind and mediate the uptake of various microbial polysaccharides, including dextrans, lipopolysaccharides and pneumococcal capsular polysaccharides. SIGNR1 mediates the clearance of encapsulated pneumococcus, complement fixation via binding C1q independent of antibody and innate resistance to pneumococcal infection. Recently, SIGNR1 has also been demonstrated to bind sialylated antibody and mediate its activity to suppress autoimmunity (Silva-Martin et al. 2009). The CRD of SIGNR1 has been cloned and overexpressed in a soluble secretory form in CHO cells. The single crystal of CRD protein SIGNR1 belonged to the monoclinic space group C2 with unit-cell parameters $a = 146.72, b = 92.77, c = 77.06$ A, $\beta = 121.66°$, allowed the collection of a full X-ray data set to a maximum resolution of 1.87 A (Silva-Martin et al. 2009).

36.8.2 Functions

SIGNR1 receptor, involved in the uptake of capsular polysaccharides (caps-PS) by APCs, is necessary for the antibody response to pneumococcal caps-PS and phosphorylcholine (PC). Moens et al. (2007) found that SIGNR1 is not involved in the IgM antibody production to PC and caps-PS serotype 3 or 14 and the IgG immune response to PC and caps-PS serotype 14. There is no direct relation between capture and uptake of caps-PS serotype 14 by SIGNR1 and the initiation of the anti-caps-PS antibody production in mice. Resident peritoneal macrophages (PEMs) express SIGNR1 on the cell surface as a major mannose receptor. These cells also ingest oligomannose-coated liposomes (OMLs) in an oligomannose-dependent manner following intraperitoneal administration. SIGNR1 on macrophages acts as a receptor for recognition of OMLs under physiological conditions (Takagi et al. 2009). Sialylated Fc from IgG require SIGNR1 which preferentially binds to 2,6-sialylated Fc compared with similarly sialylated, biantennary glycoproteins, suggesting that a specific binding site is created by sialylation of IgG Fc. A human DC-SIGN displays a binding specificity similar to SIGNR1 but differs in its cellular distribution. These studies thus identify an antibody receptor specific for sialylated Fc, and present the initial step that is triggered by intravenous Ig to suppress inflammation (Anthony et al. 2008).

SIGNR1 also interacts with *M. tuberculosis* similar to DC-SIGN. Peritoneal macrophages from SIGNR1 deficient (KO) mice produce less IL-10 upon stimulation with ManLAM than those from wild-type mice, suggesting that the interaction of ManLAM with SIGNR1 can result in immuno-suppression similar to its human homolog. Studies suggest that although SIGNR1 has a similar binding specificity as DC-SIGN, its role is limited during murine M. tuberculosis infection (Wieland et al. 2007). Resistance to *M. tuberculosis* was impaired only in SIGNR3-deficient animals. SIGNR3 was expressed in lung phagocytes during infection, and interacted with *M. tuberculosis* bacilli and mycobacterial surface glycoconjugates to induce secretion of critical host defense inflammatory cytokines, including TNF. SIGNR3 signaling was dependent on an intracellular tyrosine-based motif and the tyrosine kinase Syk. Thus, the mouse DC-SIGN homolog SIGNR3 makes a unique contribution to protection of the host against a pulmonary bacterial pathogen (Tanne et al. 2009). The rat CD209b mediates the uptake of dextran or CPS14 within the rat splenic marginal zone, similar to SIGNR1. On microglia, rat CD209b also mediates the uptake of CPS14 of *S. pneumoniae*. Findings suggest that both rat CD209b and SIGNR1 on microglia mediate the SIGNR1 complement activation pathway against *S. pneumoniae*, and thereby plays an important role in the pathogenesis of pneumococcal meningitis (Park et al. 2009).

Marginal zone macrophages in the murine spleen play an important role in the capture of blood-borne pathogens and are viewed as an essential component of host defense against the development of pneumococcal sepsis. However, reports described the loss of marginal zone macrophages associated with the splenomegaly that follows a variety of viral and protozoal infections; this finding raises the question of whether these infected mice would become more susceptible to secondary pneumococcal infection. Contrary to expectations, Kirby et al. (2009) demonstrated that the normal requirement for SIGNR1[+] marginal zone macrophages to protect against a primary pneumococcal infection can be readily compensated for by activated red pulp macrophages under conditions of splenomegaly. SIGNR1 is crucial for the capture of *S.*

pneumoniae from blood. SIGNR1 is able to interact in vitro with the juxtaposing marginal zone B cell population, which is responsible for the production of early IgM response against the *S. pneumoniae*-epitope phosphorylcholine. Strikingly, SIGNR1-deficient mice display a reduction in the marginal zone B cell population. In addition, ex vivo B cell stimulation assays demonstrate a decrease in phosphorylcholine specificity in the splenic B cell population derived from SIGNR1-deficient mice, whereas the total IgM response was unaffected. Therefore, interaction of SIGNR1 expressed by marginal zone macrophages with marginal zone B cells is essential to early IgM responses against *S. pneumoniae* (Koppel et al. 2008; Saunders et al. 2009).

Inflammatory Bowel Disease: In context of the etiology of inflammatory bowel disease, there is a less defined function for C-type lectins. The CD209 gene is located in a region linked to inflammatory bowel disease (IBD). Though the CD209 functional polymorphism (rs4804803) has been associated to other inflammatory conditions, it does not seem to be influencing Crohn's disease susceptibility. However, it could be involved in the etiology or pathology of Ulcerative Colitis in HLA-DR3-positive individuals (Núñez et al. 2007). Saunders et al. (2010) demonstrated that mice deficient in SIGNR1 have reduced susceptibility to experimental colitis, with a reduction in the disease severity, colon damage, and levels of the proinflammatory cytokines IL-1β, TNF-α, and IL-6. SIGNR1$^{-/-}$ peritoneal macrophages, but not bone marrow-derived macrophages, had a specific defect in IL-1β and IL-18 production, but not other cytokines, in response to TLR4 ligand LPS. SIGNR1 was associated in the regulation of inflammation in a model of experimental colitis and is a critical innate factor in response to LPS.

36.9 Liver and Lymph Node Sinusoidal Endothelial Cell C-Type Lectin (LSECtin) (or CLEC4G or L-SIGN or CD209L)

36.9.1 Characterization and Localization

The liver is an organ with paradoxic immunologic properties and is known for its tolerant microenvironment, which holds important implications for hepatic diseases. Liver and lymph node sinusoidal endothelial cell C-Type lectin (LSECtin) (or CLEC4G or L-SIGN or CD209L) displays 77% amino acid identity with DC-SIGN, and is expressed on endothelial cells in lymph node sinuses, capillary endothelial cells in the placenta and on liver sinusoidal cells (LSECs) (Soilleux et al. 2000; Bashirova et al. 2001; Pohlmann et al. 2001; Engering et al. 2004). The LSECtin (CD209L), gene encodes a protein of 293 amino acids and maps to chromosome 19p13.3 adjacent to C-type lectin genes, CD23, DC-SIGN, and DC-SIGNR. The four genes form a tight cluster in an insert size of 105 kb and have analogous genomic structures. The LSECtin is a type II integral membrane protein of approximately 40 kDa in size with a single C-type lectin-like domain at the C-terminus, close in homology to DC-SIGNR, DC-SIGN, and CD23. LSECtin mRNA was expressed in liver and lymph node among 15 human tissues tested, intriguingly neither expressed on hematopoietic cell lines nor on monocyte-derived DCs. Colmenares et al. (2007) detected LSECtin expression in human peripheral blood and thymic dendritic cells. LSECtin is also detected in monocyte-derived macrophages and dendritic cells at RNA and protein level (Liu et al. 2004). LSECtin could also be detected in the MUTZ-3DC cell line at mRNA and protein level. Human liver revealed its presence in Kupffer cells coexpressing the myeloid marker CD68 (Domínguez-Soto et al. 2009). In vitro, IL-4 induces the expression of 3 LSECtin alternatively spliced isoforms, including a potentially soluble form (Δ2 isoform) and a shorter version of the prototypic molecule (Δ3/4 isoform).

Full-length porcine (p) CLEC4G (L-SIGN) cDNA encodes a type II transmembrane protein of 290 amino acids. The pCLEC4G gene has same gene structure as human and the predicted bovine, canis, mouse and rat CLEC4G genes with nine exons. The pCLEC4G mRNA expresses in liver, lymph node and spleen tissues. A series of sequential intermediate products of pCLEC4G pre-mRNA were also identified during splicing from pig liver. The chromosomal regions syntenic to the human cluster of genes CD23/CLEC4G/DC-SIGN/L-SIGN have been compared in mammalian species including primates, domesticated animal, rodents and opossum. The L-SIGN homologs do not exist in non-primates mammals (Huang and Peng 2009).

In the liver LSECs functions as liver-resident antigen presenting cells (Knolle and Gerken 2000) and is important in tolerance induction (Knolle and Limmer 2001). LSECs may mediate the clearance of antigens from the circulation in same manner as DCs (Bashirova et al. 2001; Karrar et al. 2007). DC-SIGN and L-SIGN (LSECtin) share a di-leucine motif and a cluster of three acidic amino acids in their cytoplasmic tails, which are known to be essential for antigen uptake (Bashirova et al. 2001; Engering et al. 2002). Recent studies with Ebola virus, Severe Acute Respiratory Syndrome (SARS) virus or antibodies against L-SIGN, clearly demonstrated that L-SIGN indeed is able to internalize antigens (Klimstra et al. 2003; Liu et al. 2004; Jeffers et al. 2004; Ludwig et al. 2004; Dakappagari et al. 2006). DC-SIGN transient expression in HEK293T is a useful model for investigating p38 MAPK pathway triggered by hepatitis C virus glycoprotein E2, which may provide information for understanding cellular receptors-mediated signaling events and the viral pathogenesis (Chen et al. 2010).

36.9.2 Ligands of LSECtin

LSECtin binds to mannose, GlcNAc, and fucose in a Ca^{2+}-dependent manner but not to galactose (Liu et al. 2004). The DC-SIGN and DC-SIGNR (DC-SIGN/R) bind to high-mannose carbohydrates on a variety of viruses. In contrast, the related lectin LSECtin does not recognize mannose-rich glycans and interacts with a more restricted spectrum of viruses. LSECtin and DC-SIGNR, which are co-expressed by liver, lymph node and bone marrow sinusoidal endothelial cells, bind to soluble Ebola virus glycoprotein (EBOV-GP) with comparable affinities. Similarly, LSECtin, DC-SIGN and Langerin readily bound to soluble HIV-1 GP. However, only DC-SIGN captured HIV-1 particles, indicating that binding to soluble GP is not necessarily predictive of binding to virion-associated GP. Results reveal important differences between pathogen capture by DC-SIGN/R and LSECtin and hint towards different biological functions of these lectins (Gramberg et al. 2008). To compare the sugar and pathogen binding properties of LSECtin with those of related but more extensively characterized receptors, such as DC-SIGN, a soluble fragment of LSECtin consisting of the C-terminal CRD was expressed in bacteria and used to probe a glycan array and to characterize binding to oligosaccharide and glycoprotein ligands. LSECtin binds with high selectivity to glycoproteins terminating in GlcNAcβ1-2Man. Glycan analysis of the surface glycoprotein of Ebola virus reveals the presence of such truncated glycans, explaining the ability of LSECtin to facilitate infection by Ebola virus (Powlesland et al. 2008). A systematic study of DC-SIGN, DC-SIGNR and LSECtin suggested that 'agalactosylated N-glycans' are candidate ligands common to these lectins (Dominguez-Soto et al. 2010).

Polymorphisms of *CLEC4M* have been associated with predisposition for infection by severe acute respiratory syndrome coronavirus (SARS-CoV). LSECtin not only acts as an attachment factor for pathogens, but also recognizes "endogenous" activated T cells. The CD44 on Jurkat T cells is a candidate ligand of LSECtin. Moreover, LSECtin selectively bound CD44s, CD44v4 and CD44v8-10 by screening a series of typical CD44 isoforms. The interaction between CD44 and LSECtin is dependent on protein-glycan recognition. Findings indicate that CD44 is the first endogenous ligand of LSECtin, and that LSECtin is a ligand of CD44 (Tang et al. 2010).

36.9.3 Functions

LSECtin functions as a pathogen receptor, because its expression confers Ebola virus-binding capacity to leukemic cells. Sugar-binding studies indicate that LSECtin specifically recognizes N-acetyl-glucosamine, whereas no LSECtin binding to Mannan- or N-acetyl-galactosamine-containing matrices are observed. Antibody or ligand-mediated engagement triggers a rapid internalization of LSECtin, which is dependent on tyrosine and diglutamic-containing motifs within the cytoplasmic tail. Therefore, LSECtin is a pathogen-associated molecular pattern receptor in human myeloid cells. In addition, LSECtin participates in antigen uptake and internalization, and might be a suitable target in vaccination strategies (Colmenares et al. 2007). In liver, LSECtin specifically recognized activated T cells and negatively regulated their immune responses. In mice with T-cell-mediated acute liver injury, the lack of LSECtin accelerated the disease owing to an increased T-cell immune response, whereas the exogenous administration of recombinant LSECtin protein or plasmid ameliorated the disease via down-regulation of T-cell immunity. Results reveal that LSECtin is a novel regulator of T cells and expose a crucial mechanism for hepatic T-cell immune suppression, perhaps opening up a new approach for treatment of inflammatory diseases in the liver (Tang et al. 2009).

The L-SIGN (or LSECtin) is also expressed in human lung in type II alveolar cells and endothelial cells, both potential targets for SARS-CoV. Since, several other enveloped viruses including *Ebola* and *Sindbis* use CD209L as a portal of entry, and HIV and *hepatitis C virus* can bind to L-SIGN on cell membranes but do not use it to mediate virus entry, it appears that the large S glycoprotein of SARS-CoV may use L-SIGN, in addition to ACE2 in infection and pathogenesis (Jeffers et al. 2004). Capture of *Hepatitis C virus* (HCV) by L-SIGN results in *trans*-infection of hepatoma cells. L-SIGN polymorphism could influence the establishment and progression of HCV infection (Falkowska et al. 2006). There is no significant correlation between the genetic polymorphism of DC-SIGN's exon 4 and HCV infection susceptibility. 9/5 genotype distribution frequency of DC-SIGNR's exon 4 in patients with hepatitis C is significantly higher and may be associated with HCV infection susceptibility (Wang et al. 2007). The L-SIGN binds mycobacterial ManLAM but not AraLAM, suggesting that L-SIGN may bind *M. tuberculosis*. Binding assays suggest that L-SIGN interacts strongly with the (Man)2-ara and (Man)3-ara, but not with the man-ara, similar to DC-SIGN. It indicates that L-SIGN may be involved in the pathogenesis of *M. tuberculosis* infection and that the L-SIGN captures the infection through ManLAM and rapidly internalizes it to lysosomes. This shows that L-SIGN may be involved in the clearance of mycobacteria since L-SIGN is expressed on those sites in lymph nodes and liver which are ideally suited for antigen capture and clearance. However, mycobacteria may target L-SIGN to invade those tissues. More research is necessary to investigate the specific role of L-SIGN in these infections.

36.9.4 Role in Pathology

LSECtin enhances infection driven by filovirus glycoproteins (GP) and the S protein of SARS coronavirus, but does not interact with HIV-1 and hepatitis C virus envelope proteins. Ligand binding to LSECtin was inhibited by EGTA but not by mannan, suggesting that LSECtin unlike DC-SIGN/R does not recognize high-mannose glycans on viral glycoproteins. LSECtin is N-linked glycosylated and glycosylation is required for cell surface expression. In nut-shell, LSECtin is an attachment factor that in conjunction with DC-SIGNR might concentrate viral pathogens in liver and lymph nodes (Gramberg et al. 2005). Li et al. (2008) genotyped 23 tagSNPs in 181 SARS patients and reported no significant association with disease predisposition. Genetic variations in this cluster also did not predict disease prognosis. However, Li et al. (2008) detected a population stratification of the VNTR alleles in a sample of 1145 Han Chinese collected from different parts of China. Li et al. (2008) indicated that the genetic predisposition allele was not found in this lectin gene cluster and population stratification might have caused the previous positive association.

References

Aarnoudse CA, Garcia Vallejo JJ et al (2006) Recognition of tumor glycans by antigen-presenting cells. Curr Opin Immunol 18:105–111

Alagarasu K, Selvaraj P, Swaminathan S et al (2009) CCR2, MCP-1, SDF-1a & DC-SIGN gene polymorphisms in HIV-1 infected patients with & without tuberculosis. Indian J Med Res 130:444–450

Alvarez CP, Lasala F, Carrillo J et al (2002) C-type lectins DC-SIGN and L-SIGN mediate cellular entry by Ebola virus in cis and in trans. J Virol 76:6841–6844

Anthony RM, Wermeling F et al (2008) Identification of a receptor required for the anti-inflammatory activity of IVIG. Proc Natl Acad Sci USA 105:19571–19578

Appelmelk BJ, Van Die I et al (2003) Cutting edge: carbohydrate profiling identifies new pathogens that interact with dendritic cell-specific ICAM-3- grabbing nonintegrin on dendritic cells. J Immunol 170:1635–1639

Appelmelk BJ, den Dunnen J, Driessen NN et al (2008) The mannose cap of mycobacterial lipoarabinomannan does not dominate the M.-host interaction. Cell Microbiol 10:930–944

Azad AK, Torrelles JB, Schlesinger LS (2008) Mutation in the DC-SIGN cytoplasmic triacidic cluster motif markedly attenuates receptor activity for phagocytosis and endocytosis of mannose-containing ligands by human myeloid cells. J Leukoc Biol 84:1594–1603

Balboa L, Romero MM, Yokobori N et al (2010) M. tuberculosis impairs dendritic cell response by altering CD1b, DC-SIGN and MR profile. Immunol Cell Biol 88:716–726

Barreiro LB, Neyrolles O, Babb CL et al (2007) Length variation of DC-SIGN and L-SIGN neck-region has no impact on tuberculosis susceptibility. Hum Immunol 68:106–112

Bashirova AA, Geijtenbeek TBH, van Duijnhoven GCF et al (2001) A dendritic cell-specific intercellular adhesion molecule 3-grabbing nonintegrin (DC-SIGN)-related protein is highly expressed on human liver sinusoidal endothelial cells and promotes HIV-1 infection. J Exp Med 193:671–678

Bashirova AA, Geijtenbeek TB, van Duijnhoven GC et al (2003a) The C-type lectin DC-SIGN (CD209) is an antigen-uptake receptor for Candida albicans on dendritic cells. Eur J Immunol 33:532–538

Bashirova AA, Wu L, Cheng J et al (2003b) Novel member of the CD209 (DC-SIGN) gene family in primates. J Virol 77:217–227

Bates EE, Fournier N, Garcia E et al (1999) APCs express DCIR, a novel C-type lectin surface receptor containing an immunoreceptor tyrosine-based inhibitory motif. J Immunol 163:1973–1983

Beavil RL, Graber P, Aubonney N et al (1995) CD23/Fc epsilon RII and its soluble fragments can form oligomers on the cell surface and in solution. Immunology 84:202–206

Ben-Ali M, Barreiro LB, Chabbou A et al (2007) Promoter and neck region length variation of DC-SIGN is not associated with susceptibility to tuberculosis in Tunisian patients. Hum Immunol 68:908–912

Bernhard OK, Lai J, Wilkinson J, Sheil MM, Cunningham AL (2004) Proteomic analysis of DC-SIGN on dendritic cells detects tetramers required for ligand binding but no association with CD4. J Biol Chem 279:51828–51835

Bleijs DA, Geijtenbeek TBH, Figdor CG et al (2001) DC-SIGN and LFA-1: a battle for ligand. Trends Immunol 22:457–463

Bogoevska V, Horst A, Klampe B et al (2006) CEACAM1, An adhesion molecule of human granulocytes, is fucosylated by fucosyltransferase IX and interacts with DC-SIGN of dendritic cells via Lewis x residues. Glycobiology 16:197–209

Boily-Larouche G, Zijenah LS, Mbizvo M et al (2007) DC-SIGN and DC-SIGNR genetic diversity among different ethnic populations: potential implications for pathogen recognition and disease susceptibility. Hum Immunol 68:523–530

Bovin NV, Korchagina EY et al (2003) L-SIGN (CD 209L) is a liver-specific capture receptor for hepatitis C virus. Proc Natl Acad Sci USA 100:4498–4503

Cambi A, Figdor CG (2003) Dual function of C-type lectin-like receptors in the immune system. Curr Opin Cell Biol 15:539–546

Cambi A, Gijzen K, de Vries JM et al (2003) The C-type lectin DC-SIGN (CD209) is an antigen-uptake receptor for Candida albicans on dendritic cells. Eur J Immunol 33:532–538

Cambi A, Koopman M, Figdor CG (2005) How C-type lectins detect pathogens. Cell Microbiol 7:481–488

Cambi A, Beeren I, Joosten B, Figdor CG et al (2009) The C-type lectin DC-SIGN internalizes soluble antigens and HIV-1 virions via a clathrin-dependent mechanism. Eur J Immunol 39:1923–1928

Chan KY, Xu MS, Ching JC et al (2010) CD209 (DC-SIGN) -336A > G promoter polymorphism and severe acute respiratory syndrome in Hong Kong Chinese. Hum Immunol 71:702–707

Charoensap J, Engering A, Utaisincharoen P et al (2008) Activation of human monocyte-derived dendritic cells by Burkholderia pseudomallei does not require binding to the C-type lectin DC-SIGN. Trans R Soc Trop Med Hyg 102(Suppl 1):S76–S81

Chatwell L, Holla A, Kaufer BB, Skerra A (2008) The carbohydrate recognition domain of langerin reveals high structural similarity with the one of DC-SIGN but an additional, calcium-independent sugar-binding site. Mol Immunol 45:1981–1994

Chaudhary O, Bhasin R, Luthra K (2008a) DC-SIGN and DC-SIGNR polymorphic variants in Northern Asian Indians. Int J Immunogenet 35:475–479

Chaudhary O, Rajsekar K, Ahmed I et al (2008b) Polymorphic variants in DC-SIGN, DC-SIGNR and SDF-1 in high risk seronegative and HIV-1 patients in Northern Asian Indians. J Clin Virol 43:196–201

Chehimi J, Luo Q, Azzoni L et al (2003) HIV-1 transmission and cytokine-induced expression of DC-SIGN in human monocyte-derived macrophages. J Leukoc Biol 74:757–763

Chen QL, Zhu SY, Bian ZQ et al (2010) Activation of p38 MAPK pathway by hepatitis C virus E2 in cells transiently expressing DC-SIGN. Cell Biochem Biophys 56:49–58

Chung NP, Breun SK, Bashirova A et al (2010) HIV-1 transmission by dendritic cell-specific ICAM-3-grabbing nonintegrin (DC-SIGN) is regulated by determinants in the carbohydrate recognition domain that are absent in liver/lymph node-SIGN (L-SIGN). J Biol Chem 285:2100–2112

Colmenares M, Puig-Kroger A, Pello OM et al (2002) Dendritic cell (DC)-specific intercellular adhesion molecule 3 (ICAM-3)-grabbing nonintegrin (DC-SIGN, CD209), a C-type surface lectin in human DCs, is a receptor for *Leishmania* amastigotes. J Biol Chem 277:36766–36769

Colmenares M, Corbi AL, Turco SJ, Rivas L (2004) The dendritic cell receptor DC-SIGN discriminates among species and life cycle forms of *Leishmania*. J Immunol 172:1186–1190

Colmenares M, Naranjo-Gomez M, Borras FE et al (2007) The DC-SIGN-related lectin LSECtin mediates antigen capture and pathogen binding by human myeloid cells. Blood 109: 5337–5345

Curtis BM, Scharnowske S, Watson AJ (1992) Sequence and expression of a membrane-associated C-type lectin that exhibits CD4-independent binding of human immunodeficiency virus envelope glycoprotein gp120. Proc Natl Acad Sci USA 89: 8356–8360

Dakappagari N, Maruyama T, Renshaw M et al (2006) Internalizing antibodies to the C-type lectins, L-SIGN and DC-SIGN, inhibit viral glycoprotein binding and deliver antigen to human dendritic cells for the induction of T cell responses. J Immunol 176:426–440

de Jong MA, de Witte L, Bolmstedt A et al (2008) Dendritic cells mediate herpes simplex virus infection and transmission through the C-type lectin DC-SIGN. J Gen Virol 89:2398–2409

de Witte L, Abt M, Schneider-Schaulies S et al (2006) Measles virus targets DC-SIGN to enhance dendritic cell infection. J Virol 80:3477–3486

de Witte L, de Vries RD, van der Vlist M et al (2008a) DC-SIGN and CD150 have distinct roles in transmission of measles virus from dendritic cells to T-lymphocytes. PLoS Pathog 4:e1000049

de Witte L, Nabatov A, Geijtenbeek TB (2008b) Distinct roles for DC-SIGN$^+$-dendritic cells and langerhans cells in HIV-1 transmission. Trends Mol Med 14:12–19

den Dunnen J, Gringhuis SI, Geijtenbeek TB (2009) Innate signaling by the C-type lectin DC-SIGN dictates immune responses. Cancer Immunol Immunother 58:1149–1157

Domínguez-Soto A, Aragoneses-Fenoll L, Gómez-Aguado F et al (2009) The pathogen receptor liver and lymph node sinusoidal endotelial cell C-type lectin is expressed in human Kupffer cells and regulated by PU.1. Hepatology 49:287–296

Dominguez-Soto A, Aragoneses-Fenoll L, Martin-Gayo E et al (2010) Frontal affinity chromatography analysis of constructs of DC-SIGN, DC-SIGNR and LSECtin extend evidence for affinity to agalactosylated N-glycans. FEBS J 277:4010–4026

Driessen NN, Ummels R, Maaskant JJ et al (2009) Role of phosphatidylinositol mannosides in the interaction between mycobacteria and DC-SIGN. Infect Immun 77:4538–4547

Ehlers S (2010) DC-SIGN and mannosylated surface structures of *M. tuberculosis*: a deceptive liaison. Eur J Cell Biol 89:95–101

Engering A, Geijtenbeek TB, van Vliet SJ et al (2002) The dendritic cell-specific adhesion receptor DC-SIGN internalizes antigen for presentation to T cells. J Immunol 168:2118–2126

Engering A, van Vliet SJ, Hebeda K et al (2004) Dynamic populations of dendritic cell-specific ICAM-3 grabbing nonintegrin-positive immature dendritic cells and liver/lymph node-specific ICAM-3 grabbing nonintegrin-positive endothelial cells in the outer zones of the paracortex of human lymph nodes. Am J Pathol 164:1587–1595

Falkowska E, Durso RJ, Gardner JP et al (2006) L-SIGN (CD209L) isoforms differently mediate *trans*-infection of hepatoma cells by hepatitis C virus pseudoparticles. J Gen Virol 87:2571–2576

Feinberg H, Mitchell DA, Drickamer K et al (2001) Structural basis for selective recognition of oligosaccharides by DC-SIGN and DC-SIGNR. Science 294:2163–2166

Feinberg H, Guo Y, Mitchell DA, Drickamer K, Weis WI (2005) Extended neck regions stabilize tetramers of the receptors DC-SIGN and DC-SIGNR. J Biol Chem 280:1327–1335

Feinberg H, Castelli R, Drickamer K et al (2007a) Multiple modes of binding enhance the affinity of DC-SIGN for high mannose N-linked glycans found on viral glycoproteins. J Biol Chem 282:4202–4209

Feinberg H, Taylor ME, Weis WI (2007b) Scavenger receptor C-type lectin binds to the leukocyte cell surface glycan Lewis(x) by a novel mechanism. J Biol Chem 282:17250–17258

Feinberg H, Tso CK, Taylor ME et al (2009) Segmented helical structure of the neck region of the glycan-binding receptor DC-SIGNR. J Mol Biol 394:613–620

Gantner BN, Simmons RM, Canavera SJ et al (2003) Collaborative induction of inflammatory responses by dectin-1 and toll-like receptor 2. J Exp Med 197:1107–1117

García-Vallejo JJ, van Liempt E, da Costa MP et al (2008) DC-SIGN mediates adhesion and rolling of dendritic cells on primary human umbilical vein endothelial cells through LewisY antigen expressed on ICAM-2. Mol Immunol 45:2359–2369

Gardner JP, Durso RJ et al (2003) L-SIGN (CD 209L) is a liver-specific capture receptor for hepatitis C virus. Proc Natl Acad Sci USA 100:4498–4503

Gaudart N, Ekpo P, Pattanapanyasat K et al (2008) Leptospira interrogans is recognized through DC-SIGN and induces maturation and cytokine production by human dendritic cells. FEMS Immunol Med Microbiol 53:359–367

Geijtenbeek TB, Krooshoop DJ, Bleijs DA et al (2000a) DC-SIGN-ICAM-2 interaction mediates dendritic cell trafficking. Nat Immunol 1:353–357

Geijtenbeek TBH, Kwon DS et al (2000b) DC-SIGN, a dendritic cellspecific HIV-1-binding protein that enhances trans-infection of T cells. Cell 100:587–597

Geijtenbeek TBH, Torensma R, van Vliet SJ et al (2000c) Identification of DC-SIGN, a novel dendritic cell-specific ICAM-3 receptor that supports primary immune responses. Cell 100:575–585

Geijtenbeek TBH, Engering A et al (2002a) DC-SIGN, a C-type lectin on dendritic cells that unveils many aspects of dendritic cell biology. J Leukoc Biol 71:921–931

Geijtenbeek TB, Groot PC, Nolte MA et al (2002b) Marginal zone macrophages express a murine homolog of DCSIGN that captures blood-borne antigens in vivo. Blood 100:2908–2916

Geijtenbeek TB, van Duijnhoven GC, van Vliet SJ et al (2002c) Identification of different binding sites in the dendritic cell-specific receptor DC-SIGN for intercellular adhesion molecule 3 and HIV-1. J Biol Chem 277:11314–11320

Geijtenbeek TB, van Vliet SJ, Koppel EA et al (2003) Mycobacteria target DC-SIGN to suppress dendritic cell function. J Exp Med 197:7–17

Geijtenbeek TB, van Vliet SJ, Emgering A et al (2004) Self- and nonself-recognition by C-type lectins on dendritic cells. Annu Rev Immunol 22:33–54

Geurtsen J, Chedammi S, Mesters J et al (2009) Identification of mycobacterial α-glucan as a novel ligand for DC-SIGN: involvement of mycobacterial capsular polysaccharides in host immune modulation. J Immunol 183:5221–5231

Gijzen K, Raymakers RA et al (2008) Interaction of acute lymphopblastic leukemia cells with C-type lectins DC-SIGN and L-SIGN. Exp Hematol 36:860–870

Gramberg T, Hofmann H, Möller P et al (2005) LSECtin interacts with filovirus glycoproteins and the spike protein of SARS coronavirus. Virology 340:224–236

Gramberg T, Soilleux E, Fisch T et al (2008) Interactions of LSECtin and DC-SIGN/DC-SIGNR with viral ligands: differential pH dependence, internalization and virion binding. Virology 373:189–201

Gringhuis SI, Geijtenbeek TB (2010) Carbohydrate signaling by C-type lectin DC-SIGN affects NF-kB activity. Methods Enzymol 480:151–164

Gringhuis SI, den Dunnen J, Litjens M et al (2007) C-type lectin DC-SIGN modulates toll-like receptor signaling via Raf-1 kinase-dependent acetylation of transcription factor NF-kB. Immunity 26:605–616

Guo Y, Feinberg H, Conroy E et al (2004) Structural basis for distinct ligand-binding and targeting properties of the receptors DC-SIGN and DC-SIGNR. Nat Struct Mol Biol 11:591–598

Guo Y, Atkinson CE, Taylor ME, Drickamer K (2006) All but the shortest polymorphic forms of the viral receptor dc-signr assemble into stable homo- and heterotetramers. J Biol Chem 281:16794–16798

Gupta P, Rinaldo CR (2006) DC-SIGN on B lymphocytes is required for transmission of HIV-1 to T lymphocytes. PLoS Pathog 2:e70

Halary F, Amara A, Lortat-Jacob H et al (2002) Human cytomegalovirus binding to DC-SIGN is required for dendritic cell infection and target cell trans-infection. Immunity 17:653–664

Hatsukari I, Singh P, Hitosugi N, Messmer D et al (2007) DEC-205-mediated internalization of HIV-1 results in the establishment of silent infection in renal tubular cells. J Am Soc Nephrol 18:780–787

Hedlund S, Persson A et al (2010) Dendritic cell activation by sensing M. tuberculosis-induced apoptotic neutrophils via DC-SIGN. Hum Immunol 71:535–540

Hodges A, Sharrocks K, Edelmann M et al (2007) Activation of the lectin DC-SIGN induces an immature dendritic cell phenotype triggering Rho-GTPase activity required for HIV-1 replication. Nat Immunol 8:569–577

Hong PW, Flummerfelt KB, de Parseval A et al (2002) Human immunodeficiency virus envelope (gp120) binding to DC-SIGN and primary dendritic cells is carbohydrate dependent but does not involve 2G12 or cyanovirin binding sites: implications for structural analyses of gp120-DC-SIGN binding. J Virol 76:12855–12865

Hsu SC, Chen CH, Tsai SH et al (2010) Functional interaction of common allergens and a C-type lectin receptor, dendritic cell-specific ICAM3-grabbing non-integrin (DC-SIGN), on human dendritic cells. J Biol Chem 285:7903–7910

Huang YW, Meng XJ (2009) Identification of a porcine DC-SIGN-related C-type lectin, porcine CLEC4G (LSECtin), and its order of intron removal during splicing: comparative genomic analyses of the cluster of genes CD23/CLEC4G/DC-SIGN among mammalian species. Dev Comp Immunol 33:747–760

Huang YW, Dryman BA, Li W, Porcine MXJ (2009) DC-SIGN, molecular cloning, gene structure, tissue distribution and binding characteristics. Dev Comp Immunol 33:464–480

Iyori M, Ohtani M, Hasebe A et al (2008) A role of the Ca2+ binding site of DC-SIGN in the phagocytosis of E. coli. Biochem Biophys Res Commun 377:367–372

Ip WK, Chan KH, Law HK et al (2005) Mannose-binding lectin in severe acute respiratory syndrome coronavirus infection. J Infect Dis 191:1697–704

Jack DL, Klein NJ et al (2001) Mannose-binding lectin: targeting the microbial world for complement attack and opsonophagocytosis. Immunol Rev 180:86–99

Jain P, Manuel SL, Khan ZK et al (2009) DC-SIGN mediates cell-free infection and transmission of human T-cell lymphotropic virus type 1 by dendritic cells. J Virol 83:10908–10921

Jeffers SA, Tusell SM et al (2004) CD209L (L-SIGN) is a receptor for severe acute respiratory syndrome coronavirus. Proc Natl Acad Sci USA 101:15748–15753

Kashima S, Rodrigues ES, Azevedo R et al (2009) DC-SIGN (CD209) gene promoter polymorphisms in a Brazilian population and their association with human T-cell lymphotropic virus type 1 infection. J Gen Virol 90:927–934

Karrar A, Broome U, Uzunel M et al (2007) Human liver sinusoidal endothelial cells induce apoptosis in activated T cells: a role in tolerance induction. Gut 56:243–252

Kato M, Neil TK et al (2000) Expression of multilectin receptors and comparative FITCdextran uptake by human dendritic cells. Int Immunol 12:1511–1519

Khoo US, Chan KY, Chan VS, Lin CL (2008) DC-SIGN and L-SIGN: the SIGNs for infection. J Mol Med 86:861–874

Kiermayer S, Kraus I, Richt JA (2002) Identification of the amino terminal subunit of the glycoprotein of Borna disease virus. FEBS Lett 531:255–258

Kirby AC, Beattie L, Maroof A (2009) SIGNR1-negative red pulp macrophages protect against acute streptococcal sepsis after Leishmania donovani-induced loss of marginal zone macrophages. Am J Pathol 175:1107–1115, SIGNR1

Klimstra WB, Nangle EM, Smith MS et al (2003) DC-SIGN and L-SIGN can act as attachment receptors for alphaviruses and distinguish between mosquito cell- and mammalian cell-derived viruses. J Virol 77:12022–12032

Knolle PA, Gerken G (2000) Local control of the immune response in the liver. Immunol Rev 174:21–34

Knolle PA, Limmer A (2001) Neighborhood politics: the immunoregulatory function of organ-resident liver endothelial cells. Trends Immunol 22:432–437

Koizumi Y, Kageyama S, Fujiyama Y et al (2007) RANTES -28G delays and DC-SIGN - 139C enhances AIDS progression in HIV type 1-infected Japanese hemophiliacs. AIDS Res Hum Retroviruses 23:713–719

Kong KF, Delroux K, Wang X et al (2008) Dysregulation of TLR3 impairs the innate immune response to West Nile virus in the elderly. J Virol 82:7613–7623

Koppel EA, Ludwiga IS, Hernandeza MS et al (2004) Identification of the mycobacterial carbohydrate structure that binds the C-type lectins DC-SIGN, L-SIGN and SIGNR1. Immunobiology 209:117–127

Koppel EA, Ludwig IS, Appelmelk BJ, van Kooyk Y, Geijtenbeek TB (2005a) Carbohydrate specificities of the murine DC-SIGN homolog mSIGNR1. Immunobiology 210:195–201

Koppel EA, Saeland E, de Cooker DJM et al (2005b) DC-SIGN specifically recognizes Streptococcus pneumoniae serotypes 3 and 16. Immunobiology 210:203–210

Koppel EA, van Gisbergen KPJM, Geijtenbeek TBH, van Kooyk Y (2005c) Distinct functions of DC-SIGN and its homologs L-SIGN (DC-SIGNR) and mSIGNR1 in pathogen recognition and immune regulation. Cell Microbiol 7:157–165

Koppel EA, Litjens M, van den Berg VC et al (2008) Interaction of SIGNR1 expressed by marginal zone macrophages with marginal zone B cells is essential to early IgM responses against Streptococcus pneumoniae. Mol Immunol 45:2881–2887

Kwan WH, Navarro-Sanchez E, Dumortier H et al (2008) Dermal-type macrophages expressing CD209/DC-SIGN show inherent resistance to dengue virus growth. PLoS Negl Trop Dis 2:e311

Lai WK, Sun PJ, Zhang J et al (2006) Expression of DC-SIGN and DC-SIGNR on human sinusoidal endothelium A. Role for capturing hepatitis C virus particles. Am J Pathol 169:200–208

Li H, Tang NL, Chan PK et al (2008) Polymorphisms in the C type lectin genes cluster in chromosome 19 and predisposition to severe acute respiratory syndrome coronavirus (SARS-CoV) infection. J Med Genet 45:752–758

Lin G, Simmons G, Pohlmann S et al (2003) Differential N-linked glycosylation of human immunodeficiency virus and Ebola virus envelope glycoproteins modulates interactions with DC-SIGN and DC-SIGNR. J Virol 77:1337–1346

Liu H, Zhu T (2005) Determination of DC-SIGN and DC-SIGNR repeat region variations. Methods Mol Biol 304:471–481

Liu W, Tang L, Zhang G et al (2004) Characterization of a novel C-type lectin-like gene, LSECtin: demonstration of carbohydrate binding and expression in sinusoidal endothelial cells of liver and lymph node. J Biol Chem 279:18748–18758

Lozach PY, Lortat-Jacob H, de Lacroix de Lavalette A et al (2003) DC-SIGN and L-SIGN are high affinity binding receptors for hepatitis C virus glycoprotein E2. J Biol Chem 278:20358–20366

Lozach PY, Amara A, Bartosch B et al (2004) C-type lectins L-SIGN and DC-SIGN capture and transmit infectious hepatitis C virus pseudotype particles. J Biol Chem 279:32035–32045

Lozach PY, Burleigh L, Staropoli I et al (2005) Dendritic cell-specific intercellular adhesion molecule 3-grabbing non-integrin (DCSIGN)-mediated enhancement of dengue virus infection is independent of DC-SIGN internalization signals. J Biol Chem 280:23698–23708

Ludwig IS, Lekkerkerker AN, Depla E et al (2004) Hepatitis C virus targets DC-SIGN and L-SIGN to escape lysosomal degradation. J Virol 78:8322–8332

Martina BE, Koraka P, van den Doel P et al (2008) DC-SIGN enhances infection of cells with glycosylated West Nile virus in vitro and virus replication in human dendritic cells induces production of IFN-α and TNF-α. Virus Res 135:64–71

Martinez O, Brackenridge S, M-A El-Idrissi, Prabhakar BS (2005) DC-SIGN, but not sDC-SIGN, can modulate IL-2 production from PMA- and anti-CD3-stimulated primary human CD4 T cells. Int Immunol 17:769–778

Marzi A, Mitchell DA et al (2007) Modulation of HIV and SIV neutralization sensitivity by DC-SIGN and mannose-binding lectin. Virology 368:322–330

McGreal EP, Miller JL, Gordon S (2005) Ligand recognition by antigen-presenting cell C-type lectin receptors. Curr Opin Immunol 17:18–24

Menon S, Rosenberg K, Graham SA et al (2009) Binding-site geometry and flexibility in DC-SIGN demonstrated with surface force measurements. Proc Natl Acad Sci USA 106:11524–11529

Meyer S, van Liempt E, Imberty A et al (2005) DC-SIGN mediates binding of dendritic cells to authentic pseudo-LewisY glycolipids of Schistosoma mansoni cercariae, the first parasite-specific ligand of DC-SIGN. J Biol Chem 280:7349–7359

Meyer S, Tefsen B, Imberty A et al (2007) The C-type lectin L-SIGN differentially recognizes glycan antigens on egg glycosphingolipids and soluble egg glycoproteins from Schistosoma mansoni. Glycobiology 17:1104–1119

Mezger M, Steffens M, Semmler C et al (2008) Investigation of promoter variations in dendritic cell-specific ICAM3-grabbing nonintegrin (DC-SIGN) (CD209) and their relevance for human cytomegalovirus reactivation and disease after allogeneic stem-cell transplantation. Clin Microbiol Infect 14:228–234

Mikulak J, Teichberg S, Arora S et al (2010) DC-specific ICAM-3-grabbing nonintegrin mediates internalization of HIV-1 into human podocytes. Am J Physiol Renal Physiol 299:F664–F673

Mitchell DA, Fadden AJ, Drickamer K (2001) A novel mechanism of carbohydrate recognition by the C-type lectins DC-SIGN and DC-SIGNR. Subunit organization and binding to multivalent ligands. J Biol Chem 276:28939–28945

Mittal R, Bulgheresi S, Emami C, Prasadarao NV (2009) Enterobacter sakazakii targets DC-SIGN to induce immunosuppressive responses in dendritic cells by modulating MAPKs. J Immunol 183:6588–6599

Moens L, Jeurissen A, Wuyts G et al (2007) Specific intracellular adhesion molecule-grabbing nonintegrin R1 is not involved in the murine antibody response to pneumococcal polysaccharides. Infect Immun 75:5748–5752

Nagaoka K, Takahara K et al (2010) Expression of C-type lectin, SIGNR3, on subsets of dendritic cells, macrophages, and monocytes. J Leukoc Biol 88(5):913–924

Nagorsen D, Keiholz U, Rivoltini L et al (2000) Natural T-cell response against MHC class I epitopes of epithelial cell adhesion molecule, her-2/neu, and carcinoembryonic antigen in patients with colorectal cancer. Cancer Res 60:4850–4854

Navarro-Sanchez E, Altmeyer R, Amara A et al (2003) Dendritic-cell-specific ICAM3-grabbing non-integrin is essential for the productive infection of human dendritic cells by mosquito-cell-derived dengue viruses. EMBO Rep 4:723–728

Nonaka M, Ma BY, Mura R et al (2008) Glycosylation-dependent interactions of C-type lectin DC-SIGN with colorectal tumor-associated Lewis glycans impair the function and differentiation of monocyte-derived dendritic cells. J Immunol 180:3347–3356

Núñez C, Oliver J, Mendoza JL et al (2007) CD209 in inflammatory bowel disease: a case-control study in the Spanish population. BMC Med Genet 8:75

Ogarkov OB, Medvedeva TV, Nekipelov OM et al (2007) Study of DC-SIGN gene polymorphism in patients infected with Mycobacterium tuberculosis strains of different genotypes in the Irkutsk region. Probl Tuberk Bolezn Legk 11:37–42 [Article in Russian]

Olesen R, Wejse C, Velez DR et al (2007) DC-SIGN (CD209), pentraxin 3 and vitamin D receptor gene variants associate with pulmonary tuberculosis risk in West Africans. Genes Immun 8:456–467

Op den Brouw ML, de Jong MA, Ludwig IS et al (2008) Branched oligosaccharide structures on HBV prevent interaction with both DC-SIGN and L-SIGN. J Viral Hepat 15:675–683

Ortiz M, Kaessmann H, Zhang K, Bashirova A et al (2008) The evolutionary history of the CD209 (DC-SIGN) family in humans and non-human primates. Genes Immun 9:483–492

Park CG, Takahara K, Umemoto E et al (2001) Five mouse homologs of the human dendritic cell C-type lectin DC-SIGN. Int Immunol 13:1283–1290

Park JY, Choi HJ, Prabagar MG et al (2009) The C-type lectin CD209b is expressed on microglia and it mediates the uptake of capsular polysaccharides of Streptococcus pneumoniae. Neurosci Lett 450:246–51

Plazolles N, Humbert JM, Vachot L et al (2011) Pivotal Advance: the promotion of soluble DC-SIGN release by inflammatory signals and its enhancement of cytomegalovirus-mediated cis-infection of myeloid dendritic cells. J Leukoc Biol 83(3):329–342

Pohlmann S, Soilleux EJ, Baribaud F et al (2001) DC-SIGNR, a DC-SIGN homolog expressed in endothelial cells, binds to human and simian immunodeficiency viruses and activates infection in trans. Proc Natl Acad Sci USA 98:2670–2675

Powlesland AS, Ward EM et al (2006) Widely divergent biochemical properties of the complete set of mouse DC-SIGN-related proteins. J Biol Chem 281:20440–20449

Powlesland AS, Fisch T, Taylor ME et al (2008) A novel mechanism for LSECtin binding to Ebola virus surface glycoprotein through truncated glycans. J Biol Chem 283:593–602

Rappocciolo G, Jenkins FJ, Hensler HR et al (2006) DC-SIGN is a receptor for human herpesvirus 8 on dendritic cells and macrophages. J Immunol 176:1741–1749

Rappocciolo G, Hensler HR, Jais M et al (2008) Human herpesvirus 8 infects and replicates in primary cultures of activated B lymphocytes through DC-SIGN. J Virol 82:4793–4806

Rathore A, Chatterjee A, Sivarama P et al (2008a) Role of homozygous DC-SIGNR 5/5 tandem repeat polymorphism in HIV-1 exposed seronegative North Indian individuals. J Clin Immunol 28:50–57

Rathore A, Chatterjee A, Sood V et al (2008b) Risk for HIV-1 infection is not associated with repeat-region polymorphism in the DC-SIGN neck domain and novel genetic DC-SIGN variants among North Indians. Clin Chim Acta 391:1–5

Regan AD, Whittaker GR (2008) Utilization of DC-SIGN for entry of feline coronaviruses into host cells. J Virol 82:11992–11996

Reina JJ, Díaz I, Nieto PM et al (2008) Docking, synthesis, and NMR studies of mannosyl trisaccharide ligands for DC-SIGN lectin. Org Biomol Chem 6:2743–2754

Sallusto F, Cella M, Danieli C et al (1995) Dendritic cells use macropinocytosis and the mannose receptor to concentrate macromolecules in the major histocompatibility complex class II compartment: downregulation by cytokines and bacterial products. J Exp Med 182:389–400

Samsen A, Bogoevska V, Klampe B et al (2010) DC-SIGN and SRCL bind glycans of carcinoembryonic antigen (CEA) and CEA-related cell adhesion molecule 1 (CEACAM1): recombinant human glycan-binding receptors as analytical tools. Eur J Cell Biol 89:87–94

Saunders SP, Walsh CM, Barlow JL et al (2009) The C-type lectin signr1 binds Schistosoma mansoni antigens in vitro, but SIGNR1-deficient mice have normal responses during schistosome infection. Infect Immun 77:399–404

Saunders SP, Barlow JL et al (2010) C-type lectin SIGNR1 has a role in experimental colitis and responsiveness to lipopolysaccharide. J Immunol 184:2627–2637

Schaefer M, Reiling N, Fessler C et al (2008) Decreased pathology and prolonged survival of human DC-SIGN transgenic mice during mycobacterial infection. J Immunol 180:6836–6845

Schlapbach C, Ochsenbein A, Kaelin U et al (2010) High numbers of DC-SIGN + dendritic cells in lesional skin of cutaneous T-cell lymphoma. J Am Acad Dermatol 62:995–1004

Selvaraj P, Alagarasu K, Swaminathan S et al (2009) CD209 gene polymorphisms in South Indian HIV and HIV-TB patients. Infect Genet Evol 9:256–262

Serrano-Gómez D, Sierra-Filardi E, Martínez-Nuñez RT et al (2008) Structural requirements for multimerization of the pathogen receptor dendritic cell-specific ICAM3-grabbing non-integrin (CD209) on the cell surface. J Biol Chem 283:3889–3903

Silva-Martin N, Schauer JD et al (2009) Crystallization and preliminary X-ray diffraction studies of the carbohydrate-recognition domain of SIGNR1, a receptor for microbial polysaccharides and sialylated antibody on splenic marginal zone macrophages. Acta Crystallogr Sect F Struct Biol Cryst Commun 65:1264–1266

Smith AL, Ganesh L, Leung K et al (2007) Leukocyte-specific protein 1 interacts with DC-SIGN and mediates transport of HIV to the proteasome in dendritic cells. J Exp Med 204:421–430

Snyder GA, Colonna M, Sun PD (2005) The structure of DC-SIGNR with a portion of its repeat domain lends insights to modeling of the receptor tetramer. J Mol Biol 347:979–989

Soilleux EJ (2003) DC-SIGN (dendritic cell-specific ICAM-grabbing non-integrin) and DC-SIGN-related(DC-SIGNR): friend or foe? Clin Sci (Lond) 104:437–446

Soilleux EJ, Barten R, Trowsdale J (2000) Cutting edge: DC-SIGN; a related gene, DC-SIGNR; and CD23 form a cluster on 19p13. J Immunol 165:2937–2942

Soilleux EJ, Rous B, Love K et al (2003) Myxofibrosarcomas contain large numbers of infiltrating immature dendritic cells. Am J Clin Pathol 119:540–545

Solomon Tsegaye T, Pohlmann S (2010) The multiple facets of HIV attachment to dendritic cell lectins. Cell Microbiol 12:1553–1561

Svajger U, Anderluh M, Jeras M, Obermajer N (2010) C-type lectin DC-SIGN: an adhesion, signaling and antigen-uptake molecule that guides dendritic cells in immunity. Cell Signal 22:1397–1405

Tabarani G, Thépaut M, Stroebel D et al (2009) DC-SIGN neck domain is a pH-sensor controlling oligomerization: SAXS and hydrodynamic studies of extracellular domain. J Biol Chem 284:21229–21240

Tailleux L, Schwartz O, Herrmann JL et al (2003) DC-SIGN is the major Mycobacterium tuberculosis receptor on human dendritic cells. J Exp Med 197:121–148

Takagi H, Numazaki M, Kajiwara T et al (2009) Cooperation of specific ICAM-3 grabbing nonintegrin-related 1 (SIGNR1) and complement receptor type 3 (CR3) in the uptake of oligomannose-coated liposomes by macrophages. Glycobiology 19:258–266

Takahara K, Yashima Y et al (2004) Functional comparison of the mouse DC-SIGN, SIGNR1, SIGNR3 and langerin, C-type lectins. Int Immunol 16:819–829

Tang L, Yang J, Liu W et al (2009) Liver sinusoidal endothelial cell lectin, LSECtin, negatively regulates hepatic T-cell immune response. Gastroenterology 137:1498–1508, e1-5

Tang L, Yang J, Tang X et al (2010) The DC-SIGN family member LSECtin is a novel ligand of CD44 on activated T cells. Eur J Immunol 40:1185–1191

Tanne A, Ma B, Boudou F et al (2009) A murine DC-SIGN homolog contributes to early host defense against M. tuberculosis. J Exp Med 206:2205–2220

Tassaneetrithep B, Burgess TH, Granelli-Piperno A et al (2003) DC-SIGN (CD209) mediates dengue virus infection of human dendritic cells. J Exp Med 197:823–829

Timpano G, Tabarani G, Anderluh M et al (2008) Synthesis of novel DC-SIGN ligands with an α-fucosylamide anchor. Chembiochem 9:1921–1930

Tirado-González I, Muñoz-Fernández R, Blanco O et al (2010) Reduced proportion of decidual DC-SIGN + cells in human spontaneous abortion. Placenta 31:1019–1022

Valera I, Fernández N, Trinidad AG et al (2008) Costimulation of dectin-1 and DC-SIGN triggers the arachidonic acid cascade in human monocyte-derived dendritic cells. J Immunol 180:5727–5736

van Die I, Cummings RD (2006) Glycans modulate immune responses in helminth infections and allergy. Chem Immunol Allergy 90:91–112

van Die I, Cummings RD (2010) Glycan gimmickry by parasitic helminths: a strategy for modulating the host immune response? Glycobiology 20:2–12

van Die I, van Vliet SJ, Nyame AK et al (2003) The dendritic cell-specific C-type lectin DC-SIGN is a receptor for Schistosoma mansoni egg antigens and recognizes the glycan antigen Lewis x. Glycobiology 13:471–478

van Gisbergen KP, Ludwig IS, Geijtenbeek TB et al (2005a) Interactions of DC-SIGN with Mac-1 and CEACAM1 regulate contact between dendritic cells and neutrophils. FEBS Lett 579:6159–6168

van Gisbergen KPJM, Aarnoudse CA, Meijer GA et al (2005b) Dendritic cells recognize tumor-specific glycosylation of carcinoembryonic antigen on colorectal cancer cells through dendritic cell–specific intercellular adhesion molecule-3–grabbing nonintegrin. Cancer Res 65:5935–5944

van Gisbergen KPJM, Sanchez-Hernandez M, Geijtenbeek TBH, van Kooyk Y (2005c) Neutrophils mediate immune modulation of dendritic cells through glycosylation- dependent interactions between Mac-1 and DC-SIGN. J Exp Med 201:1281–1292

van Kooyk Y, Geijtenbeek TB (2002) A novel adhesion pathway that regulates dendritic cell trafficking and T cell interactions. Immunol Rev 186:47–56

van Kooyk Y, Geijtenbeek TBH (2003) DC-SIGN: escape mechanism for pathogens. Nat Rev Immunol 3:697–709

van Kooyk Y, Appelmelk B, Geijtenbeek TB (2003) A fatal attraction: *Mycobacterium tuberculosis* and HIV-1 target DC-SIGN to escape immune surveillance. Trends Mol Med 9:153–159

van Liempt E, Imberty A, Bank CM et al (2004) Molecular basis of the differences in binding properties of the highly related C-type lectins DC-SIGN and L-SIGN to Lewis X trisaccharide and *Schistosoma mansoni* egg antigens. J Biol Chem 279:33161–33167

van Liempt E, Bank CM, Mehta P et al (2006) Specificity of DC-SIGN for mannose- and fucose-containing glycans. FEBS Lett 580:6123–6131

Vannberg FO, Chapman SJ, Khor CC et al (2008) CD209 genetic polymorphism and tuberculosis disease. PLoS One 3:e1388

Vermi W, Bonecchi R, Facchetti F et al (2003) Recruitment of immature plasmacytoid dendritic cells (plasmacytoid monocytes) and myeloid dendritic cells in primary cutaneous melanomas. J Pathol 200:255–268

Wang Q, Pang S (2008) An intercellular adhesion molecule-3 (ICAM-3) -grabbing nonintegrin (DC-SIGN) efficiently blocks HIV viral budding. FASEB J 22:1055–1064

Wang QC, Feng ZH, Nie QH, Zhou YX (2004) DC-SIGN: binding receptors for hepatitis C virus. Chin Med J (Engl) 117:1395–1400

Wang M, Wang H, Jiang XL et al (2007) Genetic polymorphism of dendritic cell-specific ICAM-3 grabbing nonintegrin and DC-SIGNR's exon 4 in Chinese hepatitis C patients. Zhonghua Gan Zang Bing Za Zh 15:889–892 [Article in Chinese]

Wang H, Wang C, Feng T et al (2008a) DC-SIGNR polymorphisms and its association with HIV-1 infection. Zhonghua Yi Xue Yi Chuan Xue Za Zhi 25:542–545

Wang SF, Huang JC, Lee YM et al (2008b) DC-SIGN mediates avian H5N1 influenza virus infection in cis and in trans. Biochem Biophys Res Commun 373:561–566

Weis WI, Taylor ME, Drickamer K (1998) The C-type lectin superfamily in the immune system. Immunol Rev 163:19–34

Wichukchinda N, Kitamura Y et al (2007) The polymorphisms in DC-SIGNR affect susceptibility to HIV type 1 infection. AIDS Res Hum Retroviruses 23:686–692

Wieland CW, Koppel EA, den Dunnen J et al (2007) Mice lacking SIGNR1 have stronger T helper 1 responses to *M. tuberculosis*. Microbes Infect 9:134–141

Yamakawa Y, Pennelegion C, Willcocks S et al (2008) Identification and functional characterization of a bovine orthologue to DC-SIGN. J Leukoc Biol 83:1396–1403

Yearley JH, Kanagy S, Anderson DC et al (2008) Tissue-specific reduction in DC-SIGN expression correlates with progression of pathogenic simian immunodeficiency virus infection. Dev Comp Immunol 32:1510–1521

Yu QD, Oldring AP, Powlesland AS et al (2009) Autonomous tetramerization domains in the glycan-binding receptors DC-SIGN and DC-SIGNR. J Mol Biol 387:1075–1080

Zhang J, Zhang X, Fu J et al (2008a) Protective role of DC-SIGN (CD209) neck-region alleles with <5 repeat units in HIV-1 transmission. J Infect Dis 198:68–71

Zhang P, Skurnik M, Zhang SS et al (2008b) Human dendritic cell-specific intercellular adhesion molecule-grabbing nonintegrin (CD209) is a receptor for *Yersinia pestis* that promotes phagocytosis by dendritic cells. Infect Immun 76:2070–2079

Zhou T, Sun GZ, Li X et al (2004) Role of P-selectin and dendritic cells in human renal tubulointerstitial lesions. Shanghai Di 2 Yi Ke Da Xue Xue Bao 24:501–504

Zhou T, Chen Y, Hao L, Zhang Y (2006a) DC-SIGN and immunoregulation. Cell Mol Immunol 3:279–283

Zhou T, Sun GZ, Zhang YM et al (2006b) The inhibitory effects of anti-P-selectin lectin EGF monoclonal antibody on maturation and function of human dendritic cells. Xi Bao Sheng Wu Xue Za Zhi 28:201–205

Zhou T, Zhang Y, Sun G et al (2008) PsL-EGFmAb inhibits the stimulatory functions of human dendritic cells via DC-SIGN. Front Biosci 13:7269–7276

Zhou T, Li X, Zou J et al (2009) Effects of DC-SIGN expression on renal tubulointerstitial fibrosis in nephritis. Front Biosci 14:3814–3824

Zhu D, Kawana-Tachikawa A et al (2010) Influence of polymorphism in dendritic cell-specific intercellular adhesion molecule-3-grabbing nonintegrin-related (DC-SIGNR) gene on HIV-1 trans-infection. Biochem Biophys Res Commun 393:598–602

Part XII

C-Type Lectins: Proteoglycans

Lectican Protein Family

G.S. Gupta

37.1 Proteoglycans

Proteoglycans (PGs) are a diverse group of glycoproteins, which are formed through covalent and noncovalent aggregation of proteins and glycosaminoglycans (GAGs) in the extracellular matrix. GAGs are linear polysaccharides, whose building blocks (disaccharides) consist of an amino sugar (either GlcNAc or GalNAc) and an uronic acid (GlcA and IdoA). Virtually all mammalian cells produce proteoglycans and either secrete them into the ECM, insert them into the plasma membrane, or store them in secretory granules. The matrix proteoglycans include small interstitial proteoglycans (decorin, biglycan, fibromodulin), a proteoglycan form of type IX collagen, and one or more members of the aggrecan family of proteoglycans (aggrecan, brevican, neurocan, or versican). Some of these proteoglycans contain only one GAG chain (e.g., decorin), whereas others have more than 100 chains (e.g., aggrecan). The matrix proteoglycans typically contain the GAGs known as chondroitin sulfate (CS) or dermatan sulfate (DS). Exceptions to this generalization exist, since the HS proteoglycans perlecan and agrin are major species found in basement membranes. The size and shape of PGs vary widely that suggests numerous functions for them. The protein components of PGs are synthesized by ribosomes and translocated into the lumen of the rough endoplasmic reticulum. Glycosylation of the proteoglycan occurs in the Golgi apparatus in multiple enzymatic steps. The point of attachment is a Serine (Ser) residue to which the glycosaminoglycan is joined through a tetrasaccharide bridge (For example: chondroitin sulfate-GlcA-Gal-Gal-Xyl-PROTEIN). The Ser residue is generally in the sequence -Ser-Gly-X-Gly- (where X can be any amino acid residue), although not every protein with this sequence has an attached glycosaminoglycan. Thus, a PG contains a core protein and one to many glycosaminoglycans or carbohydrate chains. Along with glycoproteins and glycolipids proteoglycans on the noncytosolic side of the cell membrane form the carbohydrate layer. This layer on top of the bilipid layer serves to protect the cell from both chemical and mechanical damage. These components absorb water, giving the cell a slimy surface. Another main function of PGs is to regulate the movement of molecules through the matrix. Proteoglycans also contribute to the activity and stability of proteins within the matrix.

Proteoglycans being the major component of the animal extracellular matrix, are called the "filler" substances between cells in an organism. Here they form large complexes, both to other proteoglycans, to hyaluronan and to fibrous matrix proteins (such as collagen). They are also involved in binding cations (such as sodium, potassium and calcium) and water, and also regulating the movement of molecules through the matrix. Evidence also shows that they can affect the activity and stability of proteins and signalling molecules within the matrix. Individual functions of proteoglycans can be attributed to either the protein core or the attached GAG chain and serve as lubricants.

37.1.1 Nomenclature of PGs

Nomenclature for PGs is based on the source of glycosylation or the functions of PG. Many PGs can be classified on the basis of the size of core protein. Size is also used for classification of PGs (Table 37.1). Large PGs are considered aggrecans, which are a major part of cartilage (Fig. 37.1). Certain members are considered members of the "small leucine-rich proteoglycan family". These include decorin, biglycan, fibromodulin and lumican. Small interstitial PGs have a high degree of homology in their protein core sequence. Each has between 10 and 12 highly conserved leucine rich tandem repeats which make up the central portion of the core protein. The fibromodulin and lumican are keratan sulfate substituted where as decorin and biglycan are both chondroitin sulfate substituted.

Decorins are small extracellular PGs usually within 90–140 kDa in size. Decorins have a core protein of ~42 kDa in size and one GAG chain consisting of either CS or DS. They were named so because they appeared decorating collagen fibers under EM. Biglycans are

Table 37.1 Classification of proteoglycans on the basis of their localization and type of core protein

Localization	GAG-chain	M_r of core protein (kDa)	Principal members
ECM	hyaluronic acid, chondroitin sulfate, keratan sulfate	225–250	Aggrecan (the major proteoglycan in cartilage), versican (present in many adult tissues including blood vessels and skin)
Collagen-associated	chondroitin sulfate, keratan sulfate, dermatan sulfate	40	decorin, biglycan fibromodulin[c]
Basement membrane	hyaluronic acid	120	perlecan
Cell-surface	hyaluronic acid, chondroitin sulfate	33[a]; 60[b]; 92[c]	syndecans[a]; glypican[b]; betaglycan[c], CD44E, cerebroglycan
Intracellular granules	heparin, chondroitin sulfate	17–19	serglycin

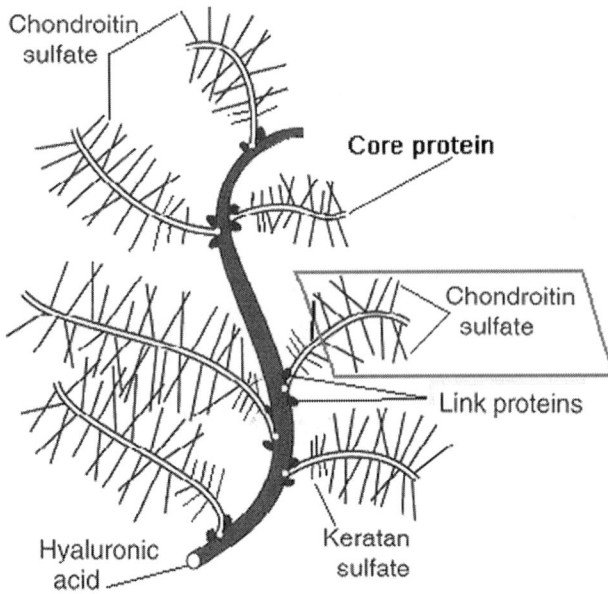

Fig. 37.1 Proteoglycan Complex. In the cartilage matrix, individual proteoglycans (in *box*) are linked to a nonsulfated GAG, hyaluronic acid, to form a giant complex with a molecular mass of about 3,000,000. The *box* indicates one of the proteoglycans of the type shown in figure

also small extracellular PGs that are similar to decorins. They are about 150–240 kDa in size and are generally found on the surface of cells, bones, and cartilage. Among small PGs, fibromodulin is the most abundant. Fibromodulin is about 60 kDa in size with a core protein of ~42 kDa. Fibromodulin is very similiar to decorins and biglycans and contains 4 keratan sulfate (KS) chains.

Large proteoglycans are versicans, neurocans, brevicans, and collagens. Chondroitin sulfate proteoglycan or versican is a sulfated glycosaminoglycan attributed for possessing good hydration potential. This property allows for interaction with surrounding cells in cell adhesion, proliferation, and migration. Another type of chondroitin sulfate proteoglycan called neurocan is located in the extracellular matrix within the brain. Proteoglycans can be a main contributor for all types of amyloidosis. High density proteoglycans are also linked to joint autoimmune responses and arthritis in mice. Joint problems such as these may improve with glucosamine and chondroitin added to the diet (Baeurle et al. 2009). Heparan sulfate proteoglycans (HSPGs) interact with many proteins and protein receptors that influence growth factors and structural proteins. Mutations in genes relating to HSPGs are associated with increased cancer risk. HSPGs are involved in several human diseases and affect the cartilage and musculoskeletal systems like hereditary multiple exostoses (HME).

37.1.2 Glycosaminoglycans

Glycosaminoglycans (GAGs) or mucopolysaccharides are long unbranched polysaccharides consisting of a repeating disaccharide unit as a building block. The disaccharides consist of an amino sugar (either GlcNAc or GalNAc) and an uronic acid (GlcA and IdoA) (Figs. 37.2 and 37.3). Protein cores made in the rough endoplasmic reticulum are posttranslationally modified by glycosyltransferases in the Golgi apparatus, where GAG disaccharides are added to protein cores to yield proteoglycans; the exception is hyaluronan, which is uniquely synthesized without a protein core and is "spun out" by enzymes at cell surfaces directly into the extracellular space. This family of carbohydrates is essential or important for life. GAGs form an important component of connective tissues and may be covalently linked to a protein to form proteoglycans. Water sticks to GAGs; this is where the resistance to pressure comes from. The density of sugar molecules and the net negative charge attract salts. Water does not compress, unlike gas. Some examples of glycosaminoglycan uses in nature include heparin as an anticoagulant, hyaluronate as a component in the synovial fluid lubricant in body joints, and chondroitins which can be found in connective tissues, cartilage and tendons. Classification and examples of GAGs are given in Table 37.1.

37.1 Proteoglycans

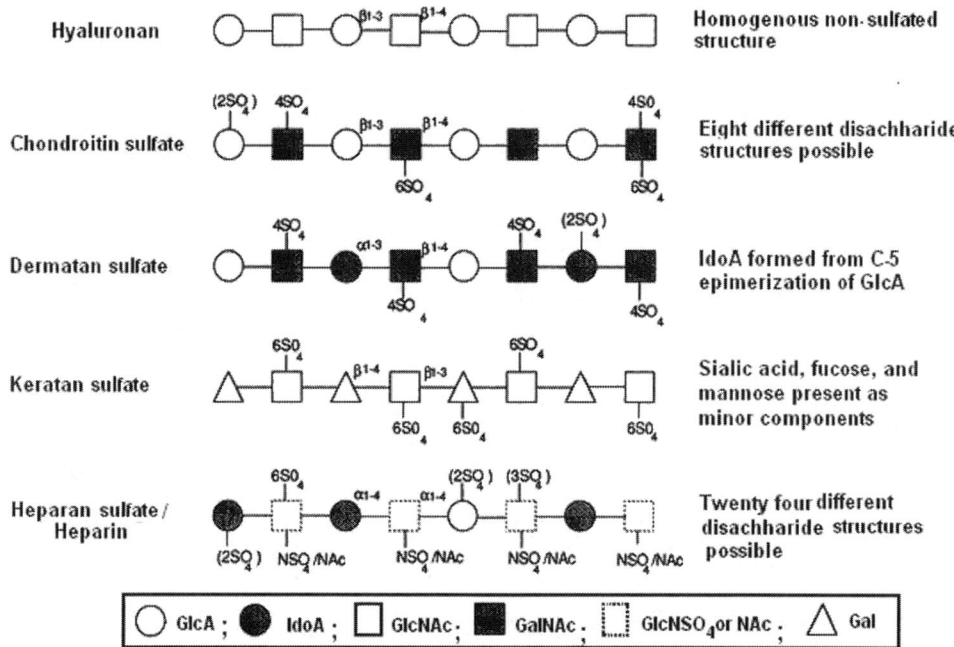

Fig. 37.2 Repeating disaccharide structures of glycosaminoglycans. Glycosaminoglycans, consisting repeating disaccharide units, are composed of an N-acetylated or N-sulfated hexosamine and either a uronic acid (glucuronic acid or iduronic acid) or galactose. Hyaluronan is synthesized without a covalent link to any protein. Hyaluronan lacks sulfate groups, but the rest of the glycosaminoglycans contain sulfates at various positions. Dermatan sulfate is formed from chondroitin sulfate by intracellular eprimenzation of glucuronate (GlcA) to iduronate (IdoA) and distinguished from chondroitn sulfate by the presence of iduronic acid. 2-sulfated hexuronate residues are more common in dermatan sulfate than in chondroitin-sulfate. Keratan sulfates lack uronic acids and instead consist of sulfated galactose and N-acetylglucosamine residues. Heparin is a more extensively epimerized and sulfated form of heparan sulfate. Heparan sulfate frequently contains some chain segments with little or no epimerization or sulfation. Heparin is synthesized only on a mast cell granule proteoglycan, whereas heparan sulfate is found on many cell surfaces and on some matrix proteoglycans. Abbreviations: GlcNAc, N-acetyl glucosamine; GalNAc, N-acetyl galactosamine; GlcA, glucuronic acid; Xyl, xylose; Gal, galactose; (Figure adapted with permission from Hardingham and Fosang 1992 © Federation of American Societies for Experimental Biology)

37.1.3 Chondroitin

Chondroitin is a chondrin derivative without "sulfate" and has been used to describe a fraction with no sulfation. Types of chondroitin include: (1) Chondroitin sulfate and (2) Dermatan sulfate. Chondroitin's functions depend largely on the properties of overall proteoglycan of which it is a part. These functions can be broadly divided into structural and regulatory roles. However, this division is not absolute, and some proteoglycans have both structural and regulatory roles (see versican). Chondroitin is an ingredient found commonly in dietary supplements and used as an alternative medicine to treat osteoarthritis and also approved and regulated as a symptomatic slow-acting drug for this disease. It is commonly sold together with glucosamine.

Chondroitin Sulfate: Chondroitin sulfate is a major component of extracellular matrix, and is important in maintaining the structural integrity of the tissue. Chondroitin sulfate is a sulfated GAG composed of a chain of alternating sugars (N-acetylgalactosamine and glucuronic acid). It is usually found attached to proteins as part of a proteoglycan. A chondroitin chain can have over 100 individual sugars, each of which can be sulfated in variable positions and quantities. Chondroitin sulfate is an important structural component of cartilage and provides much of its resistance to compression. Members of the glycosaminoglycan family vary in the type of hexosamine, hexose or hexuronic acid unit they contain (e.g. glucuronic acid, iduronic acid, galactose, galactosamine, glucosamine). They also vary in the geometry of the glycosidic linkage (Figs. 37.2 and 37.3).

Chondroitin sulfate is important in maintaining the structural integrity of the tissue. This function is typical of the large aggregating proteoglycans: aggrecan, versican, brevican, and neurocan, collectively termed the lecticans. As part of aggrecan, chondroitin sulfate is a major component of cartilage. The tightly packed and highly charged sulfate groups of chondroitin sulfate generate electrostatic repulsion that provides much of the resistance of cartilage to compression. Loss of chondroitin sulfate from the cartilage is a major cause of osteoarthritis. Chondroitin sulfate readily interacts with proteins in the extracellular matrix due to its negative charges.

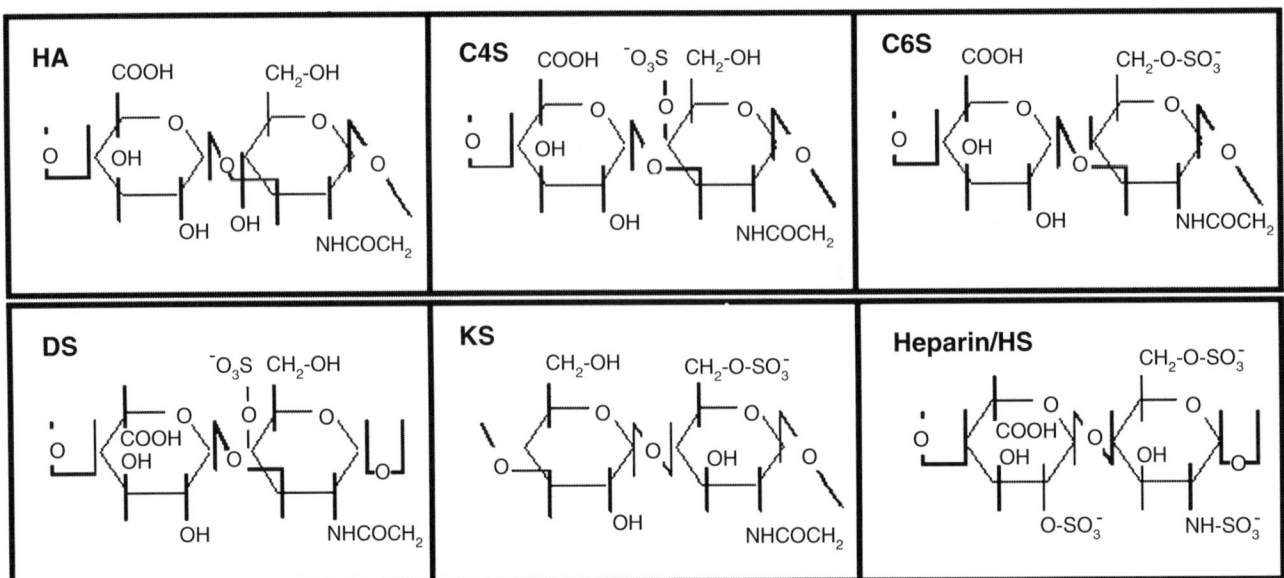

Fig. 37.3 Glycosidic bonds in disachharides present in GAGs. Heparin/HS and hyaluronic acid (HA) are glycosaminoglycans. Heparin has highest negative charge density among any known biological molecule and the only GAG that is exclusively non-sulfated; HS: Highly similar in structure to heparin, however heparan sulfates disaccharide units are organised into distinct sulfated and non-sulfated domains. Since heparin/HS structures are highly heterogeneous, IdoA (2-OSO_3)-GlcNSO_3(6-OSO_3), the most abundant disaccharide unit of HS, is shown here; Chondroitin-4-sulfate (C4S) and chondroitin-6-sulfate (C6S) are most prevalent GAG; C6S and dermatan sulfate (DS) are galactosaminoglycans. DS can be distinguished from chondroitin sulfate by the presence of iduronic acid, although some hexuronic acid monosaccharides may be glucuronic acid; Keratan sulfate (KS) is a sulfated polylactosamine

These interactions are important for regulating a diverse array of cellular activities. The lecticans are a major part of the brain extracellular matix, where the chondroitin sugar chains function to stabilize normal brain synapses as part of perineuronal nets. The levels of chondroitin sulfate proteoglycans are vastly increased after injury to the CNS where they act to prevent regeneration of damaged nerve endings. Although these functions are not as well characterized as those of heparan sulfate, new roles continue to be discovered for the chondroitin sulfate proteoglycans. The dosage of oral chondroitin used in human clinical trials is 800–1,200 mg per day. Most chondroitin appears to be made from extracts of cartilaginous cow and pig tissues (cow trachea and pig ear and nose), but other sources such as shark, fish, and bird cartilage are also used. CSPG is secreted by human B cell lines that closely resembles in its structure the serum-derived C1q inhibitor (C1qI). Strong binding of B cell CSPG to C1q, its inhibition of C1q activity, and its structural similarities to a human serum C1qI indicate that B cells produce a soluble CSPG, which may act as C1qI under physiologic conditions (Kirschfink et al. 1997).

The benefit of chondroitin sulfate in patients with osteoarthritis is likely the result of a number of effects including its anti-inflammatory activity, the stimulation of the synthesis of proteoglycans and hyaluronic acid, and the decrease in catabolic activity of chondrocytes inhibiting the synthesis of proteolytic enzymes, nitric oxide, and other substances that contribute to damage cartilage matrix and cause of death of articular chondrocytes (Monfort et al. 2008). The rationale behind the use of chondroitin sulfate is based on the belief that osteoarthritis is associated with a local deficiency in some natural substances, including chondroitin sulfate. In several prospective controlled trials, chondroitin sulfate had been thought to decrease pain, improve functional disability, reduce NSAID or acetaminophen consumption, and provide good tolerability with an additional carry-over effect. Bruyere and Reginster (2007) suggest that glucosamine and chondroitin sulfate act as valuable symptomatic therapies for osteoarthritis disease with some potential structure-modifying effects.

Structure of Chondroitin Sulfate: Chondroitin sulfate chains are unbranched polysaccharides of variable length containing two alternating monosaccharides: D-glucuronic acid (GlcA) and N-acetyl-D-galactosamine (GalNAc). Some GlcA residues are epimerized into L-iduronic acid (IdoA); the resulting disaccharide is then referred to as dermatan sulfate (Fig. 37.3). Chondroitin sulfate chains are linked to hydroxyl groups on serine residues of certain proteins. Glycosylated serines are often followed by a glycine and have neighboring acidic residues, but this motif does not always predict glycosylation. Attachment of GAG chain begins with four monosaccharides in a fixed pattern: Xyl – Gal – Gal – GlcA. Each sugar is attached by a specific enzyme, allowing for multiple levels of control over GAG synthesis. Xylose begins to be attached to proteins in endoplasmic reticulum, while rest of the sugars is attached in Golgi apparatus. Each monosaccharide may be left unsulfated, sulfated once, or sulfated twice. In the most common scenario, the hydroxyls of the four and six positions of the N-acetyl-galactosamine are sulfated, with some chains having the two position of glucuronic acid.

Dermatan Sulfate: Dermatan sulfate is a glycosaminoglycan (formerly called a mucopolysaccharide) found mostly in skin, but also in blood vessels, heart valves, tendons, and lungs. Chondroitin sulfate B is an old name for dermatan sulfate, and is no longer classified as a form of chondroitin sulfate. Dermatan sulfate may have roles in coagulation, cardiovascular disease, carcinogenesis, infection, wound repair, and fibrosis. Dermatan sulfate accumulates abnormally in several of the mucopolysaccharidosis disorders. An excess of dermatan sulfate in the Mitral Valve is characteristic of myxomatous degeneration of the leaflets leading to redundancy of valve tissue and ultimately, Mitral Valve Prolapse (into the Left atrium) and insufficiency. This chronic prolapse occurs mainly in women over the age of 60, and can predispose the patient to mitral annular calcification. Mitral valve insufficiency can lead to eccentric (volume dependent or dilated) hypertrophy and eventually Left Heart failure if untreated (Table 37.1).

37.1.4 Heparan Sulfate Proteoglycans (HSPG)

PGs can be classified on the basis of their localization, GAG-chain composition, and on the type of the core protein. As shown in Table 37.1, HSPGs are localized mainly in the basement membrane and on the cell surface (Figs. 37.2 and 37.3). The final structure of heparin/HS depends upon the incompleteness of the reactions that occur during the biosynthetic process. The modification process is more complete in heparin where the final disaccharide IdoA (2-OSO_3)-GlcNSO_3(6-OSO_3) represents up to 70% of the chain, leading to a heavily O-sulfated polysaccharide with a high IdoA/GlcA ratio.

Typical concentrations of HSPGs on the cell surface are in the range of 10^5–10^6 molecules/cell. HSPGs can link to plasma membrane through a hydrophobic transmembrane domain of their core protein or through a glycosylphosphatidylinositol (GPI) anchor covalently bound to core protein (transmembrane HSPGs). Also, HSPGs can interact with the cell by non-covalent linkage to different cell-surface macromolecules (peripheral membrane HSPGs). Transmembrane HSPGs are glypican, cerebroglycan, betaglycan, CD44, and the members of the syndecan family: syndecan 1, fibroglycan (syndecan 2), N-syndecan (syndecan 3) and ryudocan (syndecan 4). It is important to know that cell-associated HSPGs can be internalized via endocytosis and metabolized in the lysosomal compartment. In some cell types oligosaccharides originated during intracellular degradation appear to be delivered specifically to the nucleus (Gallagher 2001).

Glypican and cerebroglycan are typical GPI-anchored HSPGs. Syndecans and betaglycan are typical transmembrane HSPGs characterized by a core protein composed of an extracellular domain, a single membrane-spanning domain and a short cytoplasmic domain. In the extracellular domain are present the consensus sequences for glycosylation and a conserved putative proteolytic cleavage site. The cytoplasmic domain of syndecans can interact with the cytoskeleton and contains four conserved tyrosine residues, one of them being a substrate for enzymatic phosphorylation. Perlecan is a typical peripheral membrane HSPG that interacts with the cell surface through its core protein. The cell-adhesion motif Arg-Gly-Asp within the core protein of perlecan binds integrins β_1 or β_3 present on endothelial cell surface. However, HSPGs may associate to the cell surface and/or ECM also through their GAG-chain. In contrast, the modifications that occur during the biosynthesis of HS are less extensive, leading to HS molecules characterized by lower IdoA content and a lower overall degree of O-sulfation and resulting in high heterogeneity of distribution of the sulfate groups along the chain. Eventually, disaccharides containing GlcNAc or GlcNSO$_3$ may form clusters ranging from 2 to 20 adjacent GlcNAc-containing disaccharides and from 2 to 10 adjacent GlcNSO$_3$-containing disaccharides. However, about 20–30% of the chain contains alternate GlcNAc- and GlcNSO$_3$-disaccharides units. HSPGs are necessary for structural and functional integrity of endothelium.

37.2 Proteoglycans in Tissues

Skin Proteoglycans: Expression of versican is observed in various adult tissues such as blood vessels, skin, and developing heart. Smooth muscle cells of blood vessels, epithelial cells of skin, and the cells of central and peripheral nervous system are a few examples of cell types that express versican physiologically. Dramatic changes occur in skin as a function of age, including changes in morphology, physiology, and mechanical properties. The major proteoglycans detected in extracts of human skin are dermaton sulfate (DS) proteoglycons, decorin and versican. Studies indicate that human fetal skin is structurally different from adult skin in terms of both the distribution and the composition of the large, aggregating chondroitin sulfate proteoglycan versican (Sorrell et al. 1999). Versican is a significant component of the interstitial extracellular matrix of skin development. An apparent codistribution of versican with the various fiber forms of the elastic network of the dermis suggested an association of versican with microfibrils. Both dermal fibroblasts and keratinocytes express versican in culture during active cell proliferation. General function of versican is in cell proliferation processes that may not solely be confined to the skin (Zimmermann et al. 1994; Carrino et al. 2003). Human skin fibroblasts, in addition to versican, express a second large chondroitin sulfate/dermatan sulfate proteoglycan, which is produced in cultures of fetal fibroblasts. Proteoglycans have been described in adult human hair follicles. Versican may play an essential role both in mesenchymal condensation and in hair induction (Kishimoto et al. 1999). Glycosaminoglycan (GAG) and PG expression along the human anagen hair follicle has been characterized by Malgouries et al. (2008).

Brain: Many CSPGs have been shown to influence CNS axon growth in vitro and in vivo. These interactions can be mediated through the core protein or through CS GAG side chains. Degrading CS GAG side chains using chondroitinase ABC enhances dopaminergic nigrostriatal axon regeneration in vivo and interfering with complete CSPGs limit axon growth in vivo. Neurocan, versican, aggrecan, and brevican CSPGs may be anchored within extracellular matrix through hyaluronan glycosaminoglycan. Partial degradation of hyaluronan and chondroitin sulfate and depletion of hyaluronan-binding CSPGs enhances local sprouting of cut CNS axons, but long-distance regeneration fails in regions containing residual hyaluronan-binding CSPGs. Hyaluronan, chondroitin sulfate and hyaluronan-binding CSPGs may, therefore, contribute toward the failure of spontaneous axon regeneration in the injured adult mammalian brain and spinal cord (Moon et al. 2003) (see Chap. 38).

Human Eye: Apart from significant amounts of collagen, hyaluronan and sialylated glycoproteins, the human vitreous gel also contains low amounts of versican-like proteoglycan with a molecular mass of 380 kDa. Its core protein is substituted by CS side chains (37 kDa), in which 6-sulfated disaccharides predominated. Versican, which is able to bind lectins via its C-terminal region, may bridge or interconnect various constituents of the extracellular matrix via its terminal domains in order to stabilize large supramolecular complexes at the vitreous, contributing towards the integrity and specific properties of the tissue (Theocharis et al. 2002).

Glomerular Mesangial Cells Synthesize Versican: Mesangial cells derived from human adult glomeruli synthesize a number of proteoglycans including a large chondroitin sulfate proteoglycan (CSPG), two DS proteoglycans (biglycan and decorin) and two HS proteoglycans. Experiments with ^{3}H-labelled mesangial-cell proteoglycans showed that only the large CSPG, with core protein molecular masses of 400 kDa and 500 kDa, interacted with HA. Thus human mesangial large CSPG is a member of the versican family of proteoglycans. The interaction of CSPG and HA within the glomerulus may be important in glomerular cell migration and proliferation (Thomas et al. 1994). Both basement membrane and interstitial PGs are secreted by Madin-Darby canine kidney (MDCK) cells. HSPGs expressed by MDCK cells are perlecan, agrin, and collagen XVIII. Various CSPG core proteins are made by MDCK cells and have been identified as biglycan, bamacan, and versican/PG-M. These PGs are also associated with mammalian kidney tubules in vivo (Erickson and Couchman 2001).

Human Follicular Fluid and Umbilical Cord Vein: Two proteoglycans differing in size and composition are present in human follicular fluid. The larger one of high density had a molecular mass of 3.0×10^6 Da, and was substituted with 15–20 CS chains (Mr 60–65 kDa). EM of CS proteoglycan revealed a versican-like structure, with one globular domain at each end of a long extended segment substituted with CS side chains. The smaller proteoglycan had a molecular mass of 1.1×10^6 Da that showed a globular-protein core structure. The protein core was found to be heterogeneous, with bands occurring at 215, 330 and 400-kDa after enzymic degradation of the glycosaminoglycan chains (Eriksen et al. 1999). PGs from human umbilical cord arteries (UCAs) are especially enriched in CS/DS PGs. The predominant PG fraction included small PGs with molecular mass of 160–200 kDa and 90–150 kDa, i.e. typical for biglycan and decorin, respectively (Gogiel et al. 2007). Wharton's jelly contains mainly small CS/DS proteoglycans, with decorin strongly predominating over biglycan (Gogiel et al. 2003).

Tendons/Ligaments: The major proteoglycans of tendons are decorin and versican. Other species that were detected were biglycan and the large proteoglycans versican and aggrecan. Majority of the large proteoglycans present in the matrix of tendon are degraded and did not contain the G1 globular domain (Samiric et al. 2004a, b). Sulfated glycosaminoglycan content increases in pathologic tendons compared to normal (Tom et al. 2009). Bovine collateral ligament synthesizes a large proteoglycan which contained only chondroitin sulfate chains with a M_r 32 kDa and core proteins with a range of molecular masses above 200 kDa. Findings indicate that the large CS proteoglycan synthesized by bovine collateral ligament may be a versican-like proteoglycan which exhibited the potential to form like protein-stabilized complexes. Approximately 90% of the total proteoglycans in fresh ligament was decorin. Other species that were detected were biglycan and the large proteoglycans versican and aggrecan (Ilic et al. 2005).

Human Articular Cartilage, Intervertebral Disc, and Bone: Splicing variation of versican and size heterogeneity of versican core protein has been observed in human articular cartilage and intervertebral disc. All articular cartilage extracts from the fetus to the mature adult contained multiple core protein sizes of greater than 200 kDa. The adult cartilage extracts tended to have an increased proportion of the smaller sized core proteins and osteoarthritic cartilage possessed similar core protein sizes to the normal adult. The increased presence of versican in the disc relative to articular cartilage may suggest a more pronounced functional role for this PG, particularly in the nucleus pulposus (Sztrolovics et al. 2002). Large hyaluronate-binding proteoglycans were characrerized in developing mandible of fetal rats at embryonic day 15 (E15) to E18. The large proteoglycan having smaller molecular weight is preferentially localized to bone nodules and may correlate with bone matrix mineralization (Lee et al. 1998). The Versican has been found in osteoarthritic cartilage. Though the control cartilage showed no staining, but in osteoarthritic cartilage there was strong staining of the cytoplasm of chondrocytes with abnormal morphology (Nishida et al. 1994). Cleavage of aggrecan by aggrecanase in articular cartilage characterizes cartilage degeneration in rheumatoid arthritis (Milz et al. 2002).

Tooth: Biglycan, decorin, versican, and link protein mRNAs are expressed in human dental pulp cells. Assuming expression of link protein and versican in vivo, the larger proteoglycans in the dental pulp are capable of forming large proteoglycan aggregates (Yamada et al. 1997; Abiko et al. 2001). Cementum is believed to play a regulatory role in periodontal regeneration through a variety of macromolecules including PGs present in its ECM. Immunoreactivity to versican, decorin, biglycan and lumican was evident at the borders and lumina of a proportion of lacunae and canaliculi surrounding cementocytes in cellular cementum. Versican, decorin, biglycan and lumican are components of the ECM of cellular, but not of acellular cementum. The distribution of PG epitopes around a proportion of cementocytes suggests the existence of different cementocyte subpopulations, or a differential response of these cells to yet undefined stimuli (Ababneh et al. 1999).

37.3 Hyaluronan: Proteoglycan Binding Link Proteins

Hyaluronan (HA) is a ubiquitous component of extracellular matrices, and in several systems it plays a central role in regulating cellular proliferation and differentiation. In the mammalian CNS, HA is present throughout development and into adulthood. The HA plays an important role in tissue reorganization in response to injury. Cell or tissue-specific functions of HA are likely to be mediated by cell or tissue-specific HA-binding proteins. Link proteins are glycoproteins in cartilage that are involved in the stabilization of aggregates of proteoglycans and hyaluronic acid. Link protein (LP), an extracellular matrix protein in cartilage, stabilizes aggregates of aggrecan and hyaluronan, giving cartilage its tensile strength and elasticity. Targeted mutations in mice in the gene encoding LP (Crtl1) showed defects in cartilage development and delayed bone formation with short limbs and craniofacial anomalies. Thue LP is important for the formation of proteoglycan aggregates and normal organization of hypertrophic chondrocytes (Watanabe et al. 1999) (Fig. 37.1).

37.4 Lecticans (Hyalectans)

The term *hyalectans* (or *lecticans*) defines a family of large hyaluronan-binding proteoglycans whose members include versican, aggrecan, neurocan, and brevican. Lecticans are a family of chondroitin sulfate proteoglycans, encompassing aggrecan (abundant in cartilage), brevican and neurocan (nervous system proteoglycans) and versican (also known as chondroitin sulfate proteoglycan core protein 2 or chondroitin sulfate proteoglycan 2). These proteoglycans are characterized by the presence of a hyaluronan-binding domain and a C-type lectin domain in their core proteins. Through these domains, lecticans interact with carbohydrate and protein ligands in the extracellular matrix and act as linkers of these extracellular matrix molecules. Molecular cloning has allowed identification of genes encoding core proteins of various proteoglycans, leading to a better understanding of the diversity of proteoglycan structure and function, as well as the evolution of a classification of proteoglycans. The postulated functions of proteoglycans in basements include: a structural role in maintaining tissue histoarchitecture, or aid in selective filtration processes; sequestration of growth factors; and regulation of cellular differentiation. Furthermore, expression of PGs has been found to vary in several disease states (Yamaguchi 2000). Lecticans contain N-terminal G1 domains and C-terminal G3 domains. Only aggrecan contains the G2 domain. The G1 domain consists of an Ig-like loop and two link modules, whereas the G2 domain consists only of two link modules. The G3 domain consists of one or two EGF repeats, a C-type lectin domain and CRP-like domain (Fig. 37.4). Lecticans share a number of structural features. The amino-terminal domains mediate the interaction with hyaluronic acid (HA) as well as with the link protein, a relatively small polypeptide involved in the aggregation of proteoglycans. The link protein consists basically of a HA binding domain. BEHAB, a protein deduced from its cDNA sequence, has been suggested to constitute a brain-specific link protein. Members of the aggrecan/versican family are soluble chondroitin sulfate proteoglycans whereas most of the membrane-spanning or glycosylphosphatidylinositol (GPI)-

Cbfa1 (Takeda et al. 2001). Presently, the best candidate for a general chondrogenic transcription factor, Sox9, is implicated in the control of several cartilage-specific genes (Bi et al. 2001). Aggrecan has been shown to interact with versican (Matsumoto et al. 2003). There is association of aggrecan and tenascin in tissues. The expression of aggrecan is mainly restricted to cartilages while tenascin mRNA is present at variable levels in most of the tissues. In the newborn mouse skeleton tenascin and aggrecan mRNAs were expressed essentially in a mutually exclusive manner, tenascin transcripts being present in osteoblasts, periosteal and perichondrial cells, and in cells at articular surfaces. None of these cells expressed the cartilage specific collagen or aggrecan genes. Patterns of gene expression depend on the location of chondrocytes in different cartilages (Glumoff et al. 1994).

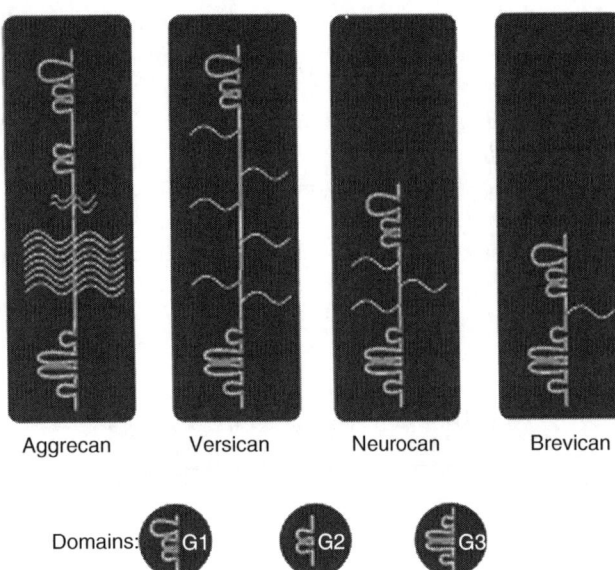

Fig. 37.4 Domain structures of lecticans. Lecticans contain N-terminal G1 domains and C-terminal G3 domains. Only aggrecan contains the G2 domain. The G1 domain consists of an Ig-like loop and two link modules, whereas the G2 domain consists only of two link modules. The G3 domain consists of one or two EGF repeats, a C-type lectin domain and CRP-like domain. All lecticans contain chondroitin sulfate chains (*yellow*) in the central domain. Aggrecan also contains keratan sulfate chains (*pink*) in the N-terminal part of the central domain (Reprinted with permission from Yamaguchi 2000 © Birkhauser Verlag). The number of chondroitin sulfate side chains present in brain aggrecan is not known but appears to be lower than the 100 GAG chains attached to cartilage aggrecan. Because the versicans, neurocan, and brevican form an almost perfect array of functionally related but differently sized molecules, it seems conceivable that the spatiotemporally regulated expression of these hyalectans (lecticans) fine tunes the inhibitory capacity of specialized hyaluronan-rich matrices in the developing and mature CNS (Bandtlow and Zimmermann 2000)

anchored proteoglycans carry heparan sulfate or keratan sulfate side chains.

37.5 Cartilage Proteoglycan: Aggrecan

37.5.1 Skeletogenesis

Skeletogenesis proceeds through a complex process involving patterning signals, cell differentiation, and growth. Most of the skeletal elements including those of limbs, as well as many parts of the craniofacial and axial skeleton, develop through the specialized process of endochondral ossification, in which a cartilaginous rudiment is first formed, then converted to bone. The growth and differentiation of cartilage into bone is controlled in part by interactions between the Indian hedgehog and parathyroid hormone-related peptide signaling pathways in the growth plate, and the induction of the bone- and cartilage-specific transcription factor

37.5.2 Human Aggrecan

37.5.2.1 Domain Organization of Core Protein

The aggregating group of CSPGs is characterized by N-terminal and C-terminal globular (or selectin-like) domains, known as G1, and G3 domains, respectively (Fig. 37.4). The proteoglycans bind hyaluronan through their N-terminal G1 domains, and other ECM proteins through C-type lectin repeat in their C-terminal G3 domains. Members of CSPG family exhibit structural similarity: a G1 domain at the N-terminus and a G3 domain at the C-terminus, with a central sequence for modification by CS chains. However, a unique feature of aggrecan is the insertion of three additional domains, an inter-globular domain (IGD), a G2 domain and a keratan sulfate (KS) domain (sequence modified by KS chains), between the G1 domain and the CS domain (sequence modified by CS chains). The G1 region at the amino terminus of the core protein can be further sub-divided into three functional domains, termed A, B1 and B2, with the B-type domains being responsible for the interaction with HA (Watanabe et al. 1997). The G2 region also possesses two B-type domains, but does not appear to interact with HA (Fosang and Hardingham, 1989) (Fig. 37.4). The domain structure of core proteins of aggrecan/versican family members is reflected at the genomic level. The genes for rat and mouse aggrecan as well as human versican have a very similar exon/intron organization that resembles the arrangement of functional domains of the corresponding proteins (Schwartz et al. 1999). The human form of aggrecan is 2,316 amino acids long and can be expressed in multiple isoforms due to alternative splicing (Doege et al. 1994). Along with Type-II collagen, aggrecan forms a major structural component of cartilage, particularly articular cartilage. Aggrecan is detected as an early event in

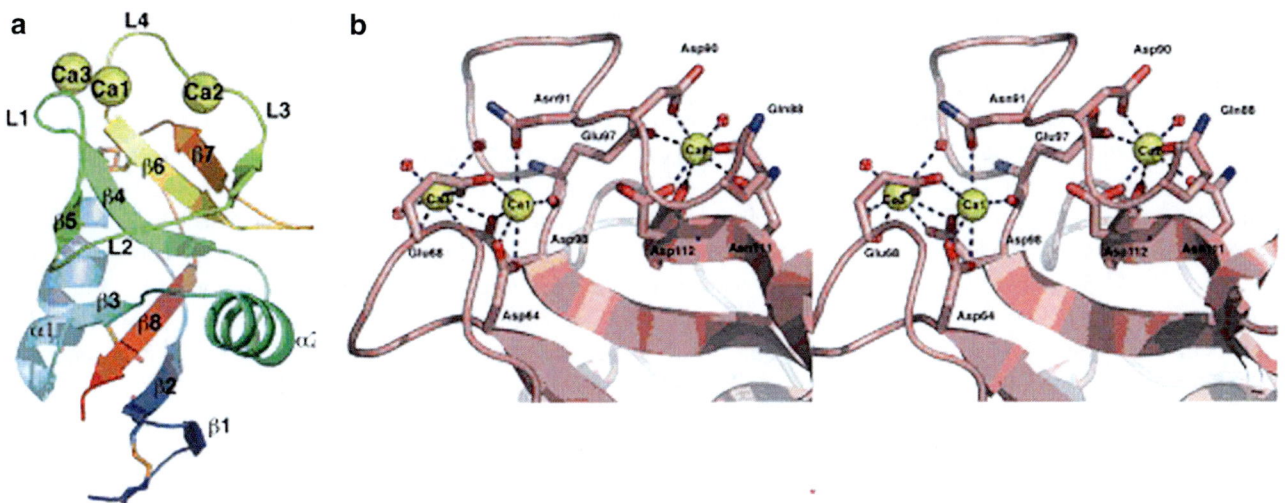

Fig. 37.5 The Aggrecan CLD fold and comparison to other C-type lectin domains (CLD) and Ca^{2+} binding. (a) The aggrecan CLD: The elements are colored from *blue* at the N terminus to *red* at the C terminus. The disulfide bridges are highlighted, and the three Ca^{2+} ions are shown as *yellow* spheres. (b) A stereoview of Ca^{2+} coordination in the aggrecan CLD. Water molecules are drawn as solid crosses (Reprinted with permission from Lundell et al. 2004 © Elsevier)

chondrocyte differentiation (Sandell 1994). Together with the CLDs of the other lecticans brevican, neurocan, and versican, the aggrecan CLD forms group I of the C-type lectin family. The overall structure is highly canonical for this family (Fig. 37.5). Aggrecan contains a long-form CLD (126 residues) with three disulfide bridges, like, e.g., tetranectin and asialoglycoprotein receptor H1 subunit (ASPGR). A search revealed strong similarity to a variety of other Ca^{2+} binding and non-Ca^{2+} binding CTLDs, including lithostatine, mannose binding protein MBP, and E-selectin (Lundell et al. 2004).

37.5.2.2 The Human Aggrecan Gene

The aggrecan gene has been characterized in several species including rat, human, mouse, and chicken (Schwartz et al. 1999). The basal promoter has been defined and shown to be active in various cell types. The human aggrecan gene, which resides at chromosome 15q26, consists of 19 exons (Doege et al. 2002; Valhmu et al. 1995), with each exon encoding a distinct structural domain of the core protein. Exons 3–6 encode the G1 region and exons 8–10 encode the G2 region. The CS-rich domain and much of the KS-rich domain are encoded by large exon 12. The G3 region is encoded by exons 13–18. It can give rise to different aggrecan transcripts due to alternative splicing (Fülöp et al. 1996), though it was not clear whether this is of any functional consequence. The promoter activity has been shown to be influenced by first exon sequences (Valhmu et al. 1998a), biomechanical stimuli (Valhmu et al. 1998b), sox9 (Sekiya et al. 2000), and a region that inhibits expression in chondrocytes (Pirok et al. 2001).

Glycosaminoglycan (GAG) Chain Attachment Region: The KS-attachment domain residing adjacent to G2 region is composed largely of repeat motifs whose number varies between species (Barry et al. 1994; Funderburgh 2000). The neighbouring CS-attachment domain is divided into two subdomains – the CS1 and CS2 domains. The CS1 domain lies adjacent to the KS-rich domain and is also composed largely of repeat motifs whose number varies between species. In addition, the human CS1 domain exhibits size polymorphism between individuals due to a variable number of 19 amino acid repeats (Doege et al. 1997). This results in the aggrecan molecules of different individuals being able to bear different numbers of CS chains. Irrespective of the number of CS chains present, their structure varies throughout life due to changes in length and sulphation pattern (Roughley and White 1989), though the functional consequence of this change is not clear. The GAG-attachment region also possesses sites for the attachment of O-linked oligosaccharides (Nilsson et al. 1982), which with age may become substituted with KS (Santer et al. 1982). KS may also be present within the G1 region, the IGD and the G2 region, attached to either O-linked or N-linked oligosaccharides (Barry et al. 1995). The KS chains also show age-related changes in structure (Brown et al. 1998).

The GAG-attachment region provides the high anionic charge density needed for unique osmotic properties of aggrecan. Normal cartilage function depends on a high aggrecan content, high GAG substitution and large aggregate size. Loss of cartilage integrity in arthritis is associated with impaired aggrecan function due either to proteolytic cleavage of the aggrecan core protein, which decreases aggrecan charge, or to cleavage of the HA, which decreases aggregate size. It has also been suggested that aggrecan charge and hence function could be affected by size polymorphism within the CS1 domain, as those individuals with the shortest core protein length would possess aggrecan with the lowest CS substitution. Such individuals might be at risk for premature cartilage degeneration. CS2 domain processing by aggrecanases would result in aggrecan fragments enriched in the CS1 domain and therefore enhance any influence that CS1 domain polymorphism may have on aggrecan function. While size polymorphism in the aggrecan CS1 domain has been associated with both articular cartilage and intervertebral disc degeneration (Horton et al. 1998; Kawaguchi et al. 1999), the reason for the linkage is not clear as it is not always the shorter CS1 domains that have been associated with disease. It is possible that the presence of one short aggrecan allele may be of little functional consequence, and only individuals with two short alleles would be at risk. Such individuals represent less than 1% of the population and have been of little focus.

The G1 and G3 domains have been implicated in product secretion, but G2, although structurally similar to the tandem repeats of G1, performs an unknown function. Each aggrecan domain has been cloned and expressed in various combinations in COS-7 cells (Kiani et al. 2001). Yang et al. (2000) indicated that the presence of G1 domain was sufficient to inhibit product secretion, while the G3 domain enhanced this process. The inhibition of secretion by G1 was mediated by its two tandem repeats, while G3's promotion of glycosaminoglycan chain attachment was apparently dependent on G3's complement-binding protein (CBP)-like motif. The modulatory effects of these two molecular domains contribute to versican's biological activities. Thus, G3 domain enhances product secretion, alone or in combination with the KS or CS domain, and promotes GAG chain attachment (Kiani et al. 2001; Yang et al. 2000; Zheng et al. 1998) (Fig. 37.6).

Effect of Age on Content of C-Terminal Region of Aggrecan: The C-terminal region of aggrecan contains G3 domain which is removed from aggrecan in mature cartilage, by proteolytic cleavage. The content of the C-terminal region decreases with age relative to G1 domain content. The content of this region of the molecule shows a fall of 92% from newborn to 65 years of age. Hence, the content of C-terminal region gives a measure of abundance of newly synthesized aggrecan. The loss of the C-terminal region is not direct part of the process of aggrecan turnover, but it is a slow independent matrix process that occurs more extensively with aging as turnover rates become slower. Young cartilage with the fastest turnover contains least molecules lacking the C-terminal region, whereas in old tissue with slow turnover few molecules retain this region. An increase in the cleavage of this region with age may also contribute to this change (Dudhia et al. 1996; Trowbridge and Gallo 2002).

37.5.3 Rat Aggrecan

Overlapping cDNA clones for coding sequence of rat cartilage proteoglycan core protein from rat chondrosarcoma hybridize to two sizes of RNA transcripts of 8.2 and 8.9 kb pairs, which contain large 3′-untranslated sequences. The entire contiguous cDNA is 6.55 kb pairs in size, and codes for a 2124-residue protein, including a 19-residue signal peptide. The sequence forms a series of eight structural domains including two globules, ($Mr = 37,000$ and 22,000) at the NH2 terminus of the molecule, one a complete and one a partial copy of the cartilage link protein. The deduced sequence is a 1,104-residue protein, containing 117 Ser-Gly sequences, possible CS attachment sites. These are arranged in three domains of 428, 503, and 173 amino acids. The first domain contains 11 complete or partial repeats of a 40-residue unit, and the second domain is composed of six copies of a 100-residue repeating sequence. The first pattern is the more highly conserved, and may have given rise to the second. The carboxyl-terminal domain is a third globule which has homology with animal lectins (Doege et al. 1987). Two sets of cDNAs encoding isoforms of aggrecan/versican core proteins expressed in rat brain have been characterized. One isoform constitutes most likely the rat homolog of bovine soluble brevican, whereas another group of isoforms represents the first example of GPI-anchored proteins of the aggrecan/versican family. GPI-anchored brevican isoforms are up-regulated late during postnatal development (see Chap. 38).

Nucleotide sequencing of 2 kb of coding sequences for the human protein in comparison with same region in rat and chick indicated that domain 8, the lectin-like domain, is highly conserved among species. In contrast, domain 7 is poorly conserved among species. Some of the cDNA clones also contained an additional structural domain between domains 7 and 8 which was not described in the rat or chick sequences. The additional domain of 38 amino acids was highly homologous to EGF-like sequences seen in other proteins. Because some cDNA clones contained codons for the EGF-like domain and some did not, the results suggested that the EGF-like domain underwent alternative RNA splicing (Baldwin et al. 1989).

37.5.3.1 Rat Aggrecan Gene
Aggrecan DNA sequencing shows 18 exons, most of which encode structural or functional modules; exceptions are domains G1-B and G2-B, which are split into two exons and the G3 lectin domain, which is encoded by three exons. There is one expressed EGF-like exon and in addition a non-

37.5 Cartilage Proteoglycan: Aggrecan

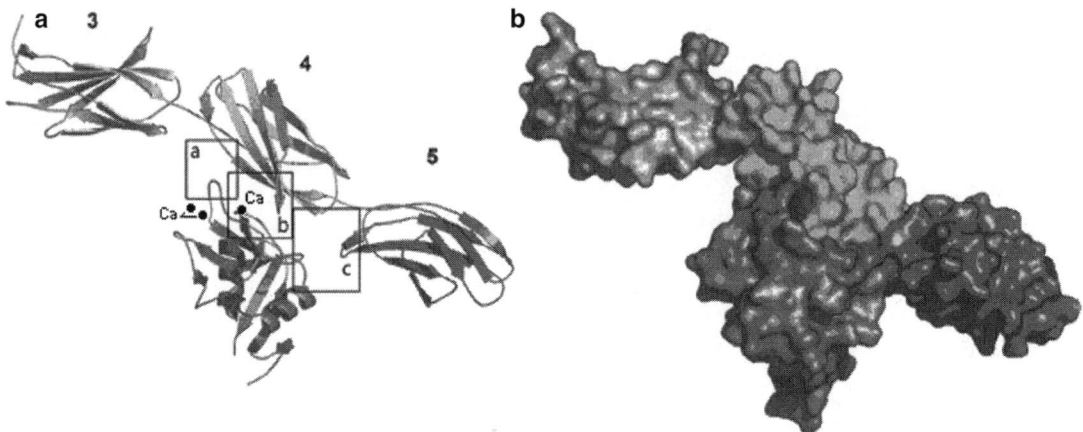

Fig. 37.6 (a) Overall structure of the complex between the aggrecan CTLD and FnIII repeats 3–5 of Tenascin-R, TN3–5. A consistent coloring scheme is used throughout for the different domains. Surfaces, cartoon elements, and carbon atoms from domain 3 are colored beige, those in domain 4 are colored cyan, and those in domain 5 are purple. The three Ca^{2+} ions are shown as yellow spheres. The boxes marked *a–c* denote the interaction areas. (b) Surface representation colored by domain. The aggrecan CLD amino acid residues mediating interaction with tenascin-R form a surface corresponding to the sulfopeptide binding surface of P-selectin in complex with PSGL-1 (Somers et al. 2000) (Adapted with permission from Lundell et al. 2004 © Elsevier)

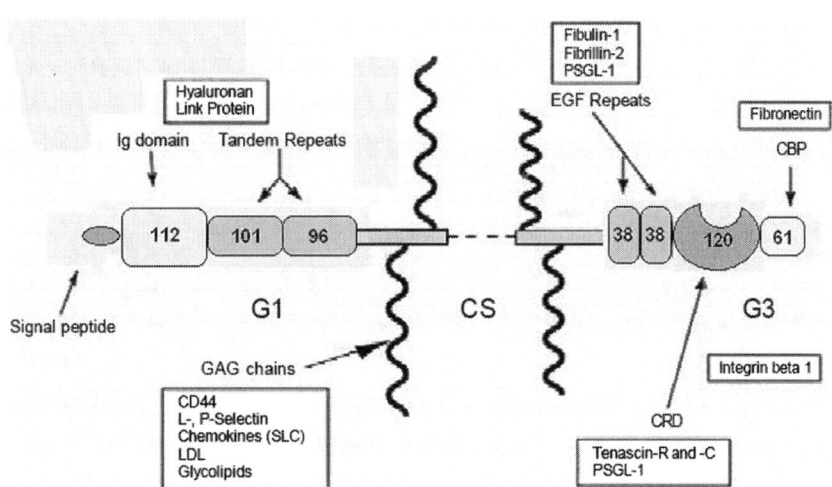

Fig. 37.7 The interaction of versican with various molecules. The locations of versican motifs that interact with other molecules are shown in the figure. Structure shows presence of G1 domain comprising of an Ig domain and two tandom repeats; a proteoglycan with Chondroitin sulfate (CS) region; and G3 domain comprising of an EGF repeats, a carbohydrate recognition domain (CRD) and a complementary binding protein (CBP) region. Molecules interactiong with various domains are boxed (Wu et al. Cell Res 2005)

expressed "pseudo-exon" encoding a heavily mutated EGF-like domain. A 30-kb intron separates exons 1 and 2. Exon 1 encodes 381 bp of 5′-untranslated sequence. There is a minor promoter which initiates transcription of an additional 68 bp 5′ of the major promoter start site. DNA sequence is reported for a 529-bp fragment encompassing exon 1, including 120 bp of 5′-flanking DNA comprising the promoter. This promoter is lacking the TATAA or CCAAT elements but has several putative binding sites for transcription factors. A 922-bp DNA fragment with 640-bp 5′-flanking DNA and 282-bp exon 1 sequence showed higher promoter activity in transfected chondrocytes, is completely inactive in the reverse orientation, and is strongly enhanceable in the forward direction by the SV40 enhancer (Doege et al. 1994).

DNA comprising aggrecan gene (83 kb including the 30-kb first intron) was surveyed for active elements in rat. A 4.7-kb DNA fragment (P3) with cell-specific enhancer activity was discovered ~12 kb upstream of the transcription start site; this active DNA fragment strongly stimulates aggrecan promoter expression in chondrocytes and weakly suppressing transcription in fibroblasts. Most of this activity has been localized to P3-7, a 2.3-kb internal fragment of P3. Another enhancer element (A23), which is not tissue-specific, was discovered about 70 kb downstream of the transcription start site. Several lines of transgenic mice revealed that neither a short (900 bp) nor a long (3.7 kb) promoter alone showed detectable expression in 14.5-day embryos, whereas placing the P3 tissue-specific enhancer together with P0 gave strong expression restricted to embryonic cartilage of transgenic mice. The A23 downstream enhancer in conjunction with P0 did not confer expression. This gene confers authentic tissue-specific regulation of aggrecan in vitro or in vivo (Doege et al. 2002).

37.5.4 Mouse Aggrecan

Overlapping cDNA clones encoding the entire core protein of aggrecan localized the aggrecan gene in mouse chromosome 7. Walcz et al. (1994) determined 7386 bp of the cDNA sequence, including 132 and 854 nt of 5′ and 3′ untranslated regions, respectively. The core protein precursor is 2,132 amino acids long (M_r 222,008), including a 19-residue secretory signal peptide. The overall amino acid sequence of the mouse aggrecan shows 91.6% identity to rat and 72.5% to human aggrecan. Comparison of the amino acid sequences of various domains and subdomain structures of mouse aggrecan to known sequences of other species and related proteins (versican, neurocan, link protein, and lymphocyte homing receptor CD44) revealed high levels of identity of the G1, G2, and G3 globular domains and relatively less conserved structures in the interglobular and glycosaminoglycan-attachment regions (Walcz et al. 1994).

The mouse aggrecan gene spans at least 61 kb and contains 18 exons. Exon 1 encodes 5′-untranslated sequence and exon 2 contains a translation start codon, methionine. The coding sequence is 6545 bp for a 2,132-amino-acid protein with $M_r = 259,131$ including an 18-amino-acid signal peptide. The chondroitin sulfate domain consisting of 1,161 amino acids is encoded by a single exon of 3.6 kb. Although link protein has similar structural domains and subdomains, the sequence identity and the organization of exons encoding the subdomains B and B′ of G1 and G2 domains revealed a strong similarity of mouse aggrecan to both human versican and rat neurocan. Primer extension analysis identified four transcription start sites which are close together. The promoter sequence showed high G/C content and contained several consensus binding motifs for transcription factors including Sp-1 and the glucocorticoid receptor. There are stretches of sequences similar to the promoter region of both the type-II collagen and link protein genes. These sequences may be important for cartilage gene expression (Watanabe et al. 1995, 2006).

37.5.5 Chick Aggrecan

Pirok et al. (1997) cloned and sequenced the 1.8-kb genomic 5′ flanking sequence of the chick aggrecan gene. Sequence analysis revealed potential Sp1, AP2, and NF-I related sites, as well as several putative transcription factor binding sites, including the cartilage-associated silencers CIIS1 and CIIS2. A number of these transcription factor binding motifs are embedded in a sequence flanked by prominent inverted repeats. Although lacking a classic TATA box, there are two instances in the 1.8-kb genomic fragment of TATA-like TCTAA sequences, as have been defined previously in other promoter regions. Primer extension and S1 protection analyses revealed three major transcription start sites, also located between the inverted repeats. Expression of aggrecan is both cell-specific and developmentally regulated. Seven functionally defined cis elements in aggrecan promoter region are known to repress aggrecan gene expression. A functional repressor cis element, (T/C)TCCCCT(A/C)RRC occurs at multiple locations within the chick aggrecan regulatory region (Pirok et al. 2005).

37.6 Versican or Chondroitin Sulfate Proteoglycan Core Protein 2

Versican, also known as chondroitin sulfate proteoglycan core protein 2 (or chondroitin sulfate proteoglycan 2 (CSPG2), or PG-M), is a large extracellular matrix proteoglycan that is present in a variety of human tissues. It is encoded by the VCAN gene. It is a large chondroitin sulfate proteoglycan with M_r of more than 1,000-kDa. Versican is encoded by a single gene, which is located on chromosome 5q12-14 in the human genome. Versican occurs in four isoforms: V0, V1, V2, V3. The central domain of versican V0 contains both the GAG-α and GAG-β domains. V1 isoforms has the GAG-β domain, V2 has the GAG-α domain, and V3 is void of any GAG attachment domains. The GAGs, being composed of repeating disaccharide units, contribute to the negative charge and many other properties of proteoglycans. These proteoglycans share a homologous globular N-terminal, C-terminal, and glycosaminoglycan (GAG) binding regions. The N-terminal (G1) globular domain consists of Ig-like loop and two link modules, and has Hyaluronan binding properties. The C-terminal (G3) globular domain consists of one or two EGF repeats, a C-type lectin domain and complement regulatory protein (CRP)-like domain. The C-terminal domain binds a variety of ligands in ECM which contribute significantly to the functions of lecticans (Fig. 37.1).

37.6.1 Expression

Expression of versican is observed in various adult tissues such as blood vessels, skin, and developing heart. Smooth muscle cells of blood vessels, epithelial cells of skin, and the cells of central and peripheral nervous system are a few examples of cell types that express versican physiologically. In adult CNS, versican is found in perineuronal nets, where it may stabilize synaptic connections. Also, there is differential temporal and spatial expression of versican by multiple cell types and in different developmental and pathological time frames. To fully appreciate the functional roles of versican as it relates to changing patterns of expression in development and disease, an in depth knowledge of versican's biosynthetic processing is necessary. Rahmani et al. (2006) evaluated the current status of our knowledge regarding the transcriptional control of versican gene regulation with a focuss on the signal transduction pathways, promoter regions, cis-acting elements, and trans-factors that have been characterized.

cDNA Clones: The cDNA clones of versican expressed by human fibroblasts code for 2,389 amino acid long core protein and the 20-residue signal peptide. The sequence predicts a potential hyaluronic acid-binding domain in the amino-terminal portion. This domain contains sequences virtually identical to partial peptide sequences from a glial hyaluronate-binding protein. Putative glycosaminoglycan attachment sites are located in the middle of the protein. The carboxy-terminal portion includes two (EGF) epidermal growth factor-like repeats, a lectin-like sequence and a complement regulatory protein-like domain. Amino- and carboxy-terminal portions of the fibroblast core protein are closely related to the core protein of a large chondroitin sulfate proteoglycan of chondrosarcoma cells. However, the glycosaminoglycan attachment regions in the middle of the core proteins are different and only the fibroblast core protein contains EGF-like repeats. Based on the similarities of its domains with various binding elements of other proteins, it was suggested that the large fibroblast proteoglycan, herein be referred to as versican (Zimmermann and Ruoslahti 1989).

Alternative Splice Variants of Human Versican: Alternative splicing of versican generates at least four isoforms named V0, V1, V2, and V3. An alternatively spliced glycosaminoglycan attachment domain (GAG-α) of human versican was cloned from cDNA libraries. Inserted carboxyl-terminal of the hyaluronan-binding region, this domain adds another 987 amino acids to the original versican (V1) core protein giving rise to the large V0 isoform with 3,396 amino acids and 17–23 putative glycosaminoglycan attachment sites. The GAG-α domain is encoded by exon 7 of the human versican gene. Sequence comparisons revealed a slight similarity to the alternative splice domain of PG-M, further supporting the notion that PG-M is the chicken homologue of versican. Both V0 and V1 isoforms were detected in cerebral cortex, aorta, intervertebral disc, liver, myometrium, and prostate, whereas keratinocytes exclusively expressed versican V1. In brain tissue, a short versican variant (V2) including only the GAG-α domain was identified (Dours-Zimmermann and Zimmermann 1994).

Multiple Forms of Mouse Versican: Versicon from chick limb buds, named as PG-M, was shown to be expressed in the prechondrogenic condensation area of the developing chick limb buds. PG-M shows the presence of a hyaluronic acid binding domain at the amino-terminal side and two EGF-like domains, a lectin-like domain, and a complement regulatory protein-like domain at the carboxyl-terminal side. These domains show high homology to corresponding domains of a human fibroblast versican. The chondroitin sulfate attachment domain of PG-M core protein is about 100 kDa larger than that of versican core protein. The finding of alternatively spliced forms of PG-M core protein suggests that versican might be one of the multiple forms of PG-M (Shinomura et al. 1993). Ito et al. (1995) sequenced cDNA clones that encode the core protein of PG-M-like proteoglycan produced by cultured mouse aortic endothelial cells. A homology search of cDNA revealed that the core protein is a mouse equivalent of chick PG-M-V1, one of the alternatively spliced forms of the PG-M core protein, which may correspond to human versican. Northern blot analysis revealed three mRNA species of 10, 9, and 8 kb in size. The analysis of PG-M mRNA species in embryonic limb buds and adult brain indicated the presence of other mRNA species with different sizes; the one with the largest size (12 kb) was found in embryonic limb buds, and the ones with smaller sizes of 7.5 and 6.5 kb were present in adult brain. Sequencing of cDNA for smaller forms from adult brain showed that they were different from PG-M-V1 in encoding the second chondroitin sulfate attachment domain (CSα) alone. The mRNA of 12 kb in size corresponded to a transcript without the alternative splicing (V0), PG-M-V0. It is likely that multiforms of the PG-M core protein may be generated by alternative usage of either or both of the two different CS attachment domains (α and β) and that molecular forms of PG-M may vary from tissue to tissue by such an alternative splicing. The ectopic expression of versican V1 isoform induced mesenchymal-epithelial transition (MET) in NIH3T3 fibroblasts, and inhibition of endogenous versican expression abolished the MET in metanephric mesenchyme. Reports indicate the involvement of versican in MET: expression of versican is sufficient to induce MET in NIH3T3 fibroblasts and reduction of versican expression decreased MET in metanephric mesenchyme (Sheng et al. 2006).

Versican Gene: The human VCAN gene is divided into 15 exons over 90–100 kb (Naso et al. 1994). The entire primary structures of versican have been generated from human, murine, bovine and chick cDNA clones; the chick form was originally named PG-M (Shinomura et al. 1993). The exon organization corresponds to the protein subdomains encoded by homologous proteins, with a remarkable conservation of exon size and intron phase. An additional exon is present just proximal to the glycosaminoglycan-binding region that is identical to a splice variant of versican (Dours-Zimmermann and Zimmermann 1994). The versican promoter harbors a typical TATA box located approximately 16 bp upstream of the transcription start site and binding sites for a number of transcription factors involved in regulated gene expression. Stepwise 5′ deletions identified a strong enhancer element between −209 and −445 bp and a strong negative element between −445 and −632 bp (Naso et al. 1994).

37.6.2 Structure

Versican is a large (1–2 × 10^6 Da) chondroitin-sulfate proteoglycan that can form large aggregates by means of interaction with hyaluronan and also binds to a series of other extracellular matrix proteins, chemokines and cell-surface molecules. A recombinant construct corresponding to the C-type lectin domain of versican demonstrates a calcium-dependent self-association of this region. This binding reaction could contribute to the ability of versican to organize formation of the proteoglycan extracellular matrix by inducing binding of individual versican molecules or by modulating binding reactions to other matrix components (Ney et al. 2006) (Fig. 37.4). The large aggregating chondroitin sulfate proteoglycans are characterized by N-terminal and C-terminal globular (or selectin-like) domains known as the G1 and G3 domains, respectively. The modulatory effects of G1 and G3 molecular domains may contribute to versican's biological activities (Yang et al. 2000). Versican, like other members of its family, has unique N- and C-terminal globular regions, each with multiple motifs. A large glycosaminoglycan-binding region lies between them. These structures have been discussed in the context of ECM proteoglycans. All isoforms of versican have homologous N-terminal (HA binding) and C-terminal (lectin-like) domains. The central domain of versican V0 contains both the GAG-α and GAG-β domains. V1 isoforms has the GAG-β domain, V2 has the GAG-α domain, and V3 is void of any GAG attachment domains, and only consists of the N-terminal and C-terminal globular domains.

N-Terminus: The N-terminal (G1) globular domain consists of Ig-like loop and two link modules, and has hyaluronan (HA) binding properties. The N-terminal of versican has an important role in maintaining the integrity of the ECM by interacting with hyaluronan. Its interactions with link protein have also been observed.

C-Terminus: The C-terminal (G3) globular domain consists of one or two EGF repeats, a C-type lectin domain and complement regulatory protein (CRP)-like domain. The C-terminal domain binds a variety of ligands in ECM which contribute significantly to the functions of lecticans. One important family of ligands is the tenascin family. For example, The C-lectin domain of versican interacts with Tenascin R through its fibronectin type III (FnIII) repeat 3–5 domain in a calcium dependent manner, in vivo. Different tenascin domains interact with a wide range of cellular receptors, including integrins, cell adhesion molecules and members of the syndecan and glypican proteoglycan families (Fig. 37.3B). Versican's C-terminal domain interacts with Fibulin-2, a protein whose expression is associated with that of versican in the developing heart. The EGF domain of the C-terminal of Versican also binds the EGF-receptor molecule, in vivo.

GAG Binding Region: The central domain of Versican is decorated with glycosaminoglycans. The structural and functional diversity of Versican is increased by variations in GAG sulfation patterns and the type of GAG chains bound to the core protein. The central domain of versican V0 contains both the GAG-α and GAG-β domains. V1 isoforms has the GAG-β domain, V2 has the GAG-α domain, and V3 is void of any GAG attachment domains. The GAGs, being composed of repeating disaccharide units, contribute to the negative charge and many other properties of proteoglycans.

37.6.3 Interactions of Versican

The diverse binding partners afforded to versican by virtue of its modular design include ECM components, such as hyaluronan, type I collagen, tenascin-R, fibulin-1, and -2, fibrillin-1, fibronectin, P- and L-selectins, and chemokines.

Versican also binds to the cell surface proteins CD44, integrin β1, epidermal growth factor receptor, and PSGL-1. These multiple interactors play important roles in cell behaviour. Roles of versican in modulating such processes have been discussed.. Fibulin-1 is strongly expressed in tissues where versican is expressed, and also expressed in developing cartilage and bone. The interactions with fibulin-1 are Ca^{2+}-dependent with K_D values in the low nM range. The binding site for aggrecan and versican lectin domains was mapped to the EGF-like repeats in domain II of fibulin-1. Solid phase assays with known ligands demonstrated that fibulin-2 and tenascin-R bind the same site on the proteoglycan lectin domains (Wu et al. 2005) (Fig. 37.7).

The interactions of the CS/DS chains of versican with L- and P-selectin and chemokines are sulfation-dependent but the interaction with CD44 is sulfation-independent. The binding of versican to L- and P-selectin was inhibited by CS and HS. On the other hand, the binding to CD44 was inhibited by hyaluronic acid, chondroitin (CH), CS A, CS B, CS C, CS D, and CS E but not by HS or keratan sulfate. A cross-blocking study indicated that L- and P-selectin recognize close or overlapping sites on versican, whereas CD44 recognizes separate sites. Structural analysis showed that versican is modified with at least CS B and CS C (Kawashima et al. 2000). Studies suggest that oversulfated CS/DS chains containing GlcAβ1/IdoAα1-3GalNAc(4,6-O-disulfate) are recognized by L- and P-selectin and chemokines, and imply that these chains are important in selectin- and/or chemokine-mediated cellular responses (Kawashima et al. 2002).

37.6.4 Functions

Versican is a multifunctional molecule with roles in cell adhesion, matrix assembly, cell migration proliferation and angiogenesis, and hence plays a central role in tissue morphogenesis and maintenance. Versican is involved in development, guiding embryonic cell migration important in the formation of the heart and outlining the path for neural crest cell migration. Versican is a key factor in inflammation through interactions with adhesion molecules on the surfaces of inflammatory leukocytes and interactions with chemokines that are involved in recruiting inflammatory cells. Versican is often considered an anti-adhesive molecule. Considering the large size (>1,000 kDa) and hydration capability of versican, it is possible that the interaction of integrins with their cell surface receptors is sterically hindered. In addition, because of negatively charged sulfates or carboxyl groups, chondroitin sulfate chains are attracted to various positively charged molecules such as certain growth factors, cytokines, and chemokines. This interaction in the ECM or on the cell surface is important in the formation of immobilized gradients of these factors, their protection from proteolytic cleavage, and their presentation to specific cell-surface receptors. The binding of versican with leukocyte adhesion molecules L-selectin, P-selectin, and CD44 is also mediated by the interaction of CS chains of versican with the carbohydrate-binding domain of these molecules. Both CD44 and L-selectin have been implicated in leukocyte trafficking. The ability of versican to bind a large panel of chemokines and the biological consequences of such binding has also been examined. Versican can bind specific chemokines through its CS chains and this interaction down-regulates the chemokines function. In light of results that V1 and V2 isoforms of versican have opposite effects on cell proliferation, glycosaminoglycan domain GAG-β has been implicated in versican-enhanced cell proliferation and versican-induced reduction of cell apoptosis.

Versican Guides Migratory Neural Crest Cells: Selective expression of versican V0 and V1 in barrier tissues impede the migration of neural crest cells during embryonic trunk development (Landolt et al. 1995). Pure preparations of either a mixture of versican V0/V1 or V1 alone strongly inhibit the migration of multipotent Sox10/p75NTR-positive early neural crest stem cells. This inhibition is largely core glycoprotein-dependent. Findings support the notion that versican variants V0 and V1 act, possibly in concert with other inhibitory molecules such as aggrecan and ephrins, in directing the migratory streams of neural crest cells to their appropriate target tissues (Dutt et al. 2006).

The versican accumulates in lesions of atherosclerosis and restenosis. The unique structural features make this molecule to bind growth factors, enzymes, lipoproteins, and a variety of other ECM components to influence fundamental events involved in vascular diseases. Versican is upregulated after vascular injury and is a prominent component in stented and nonstented restenotic lesions. The synthesis of versican is highly regulated by specific growth factors and cytokines and the principal source of versican is the smooth muscle cell. Versican interacts with hyaluronan to create expanded viscoelastic pericellular matrices that are required for arterial smooth muscle cell (ASMC) proliferation and migration. Versican is also prominent in advanced lesions of atherosclerosis, at the borders of lipid-filled necrotic cores as well as at the plaque-thrombus interface, suggesting roles in lipid accumulation, inflammation, and thrombosis. Versican influences the assembly of ECM and controls elastic fiber fibrillogenesis, which is of fundamental importance in ECM remodeling during vascular disease. Collectively, studies highlight the importance of this specific ECM component in atherosclerosis and restenosis (Wight and Merrilees 2004).

37.7 Pathologies Associated with PGS

37.7.1 Proteoglycans Facilitate Lipid Accumulation in Arterial Wall

The accumulation of extracellular matrix components such as proteoglycans is a hallmark of an atherosclerotic lesion. Proteoglycans are considered to facilitate lipid accumulation in the arterial wall, as part of the injury and repair process in atherogenesis. Srinivasan et al. (1995) determined (1) characteristics of arterial tissue CS monomers of versican type that vary in binding affinity to low-density lipoproteins (LDL), and (2) the ability of these variants to modulate LDL metabolism by macrophages. A large CS devoid of DS was identified and purified from bovine aorta intima-media under dissociative conditions. Variations in LDL-binding affinity of CS could modulate the lipid accumulation in atherogenesis. LDL enhances mesangial cell (MC) matrix deposition by modulating the production of PG and hyaluronan (HA) in the culture medium and to a lesser extent in the cell layer. The GAG chains did not show difference in either their size or charge. The synthesis of both CS and HS was enhanced as was the production of versican, perlecan and to a lesser extent decorin. An increase in HA synthesis was also demonstrated following LDL stimulation (Chana et al. 2000).

Heparan sulfate in ECM of artery wall possesses anti-atherogenic properties and interferes in lipoprotein retention, by suppression of inflammation, and by inhibition of smooth muscle cell (SMC) growth. The amount of HS in atherosclerotic lesions from humans and animals is known to be reduced. The atherosclerotic lesions retrieved from patients undergoing surgery for symptomatic carotid stenosis showed a selective reduction in perlecan gene expression, whereas, expression of other HS proteoglycans in the artery wall, agrin and collagen XVIII, the large CSPG and the versican remained unchanged. This suggests a reduction of perlecan mRNA-expression and protein deposition in human atherosclerosis, which in part explains the low levels of HS in this disease (Tran et al. 2007).

Fatty Acids Modulate the Composition of ECM in Cultured Human Arterial SMCs: In diabetes-associated angiopathies and atherosclerosis, there are alterations in the intima of small and large arteries. High concentrations of nonesterified fatty acids (NEFAs) might alter the basement membrane composition of endothelial cells. In arteries, SMCs are the major producers of proteoglycans and glycoproteins in the intima, which is the site of lipoprotein deposition and modification during atherogenesis. Exposure of human arterial SMCs (ASMCs) to albumin-bound linoleic acid lowered their rate of proliferation and altered cell morphology. ASMCs expressed 2–10 times more mRNA for the core proteins of versican, decorin, and syndecan 4. Studies suggest that some of the NEFA effects are mediated by PPAR-γ. These effects of NEFAs, in vivo, could contribute to changes of the matrix of arterial intima associated with micro- and macroangiopathies (Olsson et al. 1999).

Human hASMC matrix PGs contribute to the retention of apoB lipoproteins in the intima, a possible key step in atherogenesis. The response of ASMC to NEFA could induce ECM alterations favoring apoB lipoprotein deposition and atherogenesis (Rodríguez-Lee et al. 2006). Extracellular matrix changes occur in heart valve pathologies. Decreased expression of the hyaluronan receptor for endocytosis in myxomatous mittral valves suggested that hyaluronan metabolism could be altered in myxomatous mitral valve disease (Gupta et al. 2009).

Perlecan in Atherosclerosis-Prone and Atherosclerosis-Resistant Human Arteries: A large HSPG, perlecan, dramatically increases in the advanced lesion, and vascular SMC are the cell type responsible for the accumulation. In cultured human coronary SMCs indicate that thrombin increases the cell layer-associated HSPG with a core protein size of approximately 400 kDa without any change in the length of the glycosaminoglycan chains when the cell density is high. The HSPG was identified as perlecan. In addition, thrombin elevated the steady-state level of perlecan mRNA but not that of versican, decorin, and syndecan-1 mRNAs, although biglycan mRNA was moderately elevated. It is suggested that thrombin induces the perlecan core protein synthesis without influencing the formation of the heparan sulfate chains in human coronary SMCs at a high cell density. The regulation of proteoglycan synthesis by thrombin may be involved in the accumulation of perlecan in advanced lesions of atherosclerosis (Yamamoto et al. 2005).

A proteomics-based approach indicated that intimal proteoglycan composition in preatherosclerotic lesions was more complex than previously appreciated with up to eight distinct core proteins present, including the large versican and aggrecan, the basement membrane proteoglycan perlecan, the class I small leucine-rich proteoglycans biglycan and decorin, and the class II small leucine-rich proteoglycans lumican, fibromodulin, and prolargin/PRELP (proline arginine-rich end leucine-rich repeat protein). Although most of these proteoglycans seem to be present in similar amounts at the two locations, there was a selective enhanced deposition of lumican in the intima of the atherosclerosis-prone internal carotid artery compared with the intima of the atherosclerosis-resistant internal thoracic artery. The enhanced deposition of lumican in the intima of an atherosclerosis prone artery has important implications for the pathogenesis of atherosclerosis (Talusan et al. 2005).

Molecular Structure and Organization of Versicon in Human Aneurysmal Abdominal Aortas: Differences have been noted in matrix metalloproteinase expression patterns in cultures of medial smooth muscle cells from tissue affected by abdominal aortic aneurysm (AAA) or atherosclerotic occlusive disease and from normal arterial tissue. Versican, perlecan, and biglycan levels were significantly elevated in AAA smooth muscle cell cultures. Two populations of smooth muscle cell versican were identified. Because heparan sulfate proteoglycans can bind growth factors, their elevated synthesis by AAA smooth muscle cells in combination with an increased expression of matrix metalloproteinases may at least partly explain the differential proliferative capacity of the AAA smooth muscle cells and may govern the pattern of abnormal cellular proliferation and matrix protein synthesis observed in the pathogenesis of vascular disease (Melrose et al. 1998).

Versican in aortic wall participates in various biological functions of the tissue. In human aorta and aneurysmal aortic tissue versican is exclusively substituted with chondroitin sulfate chains, in contrast to other human tissues where both chondroitin and dermatan sulfate chains are attached onto versican core proteins. Except for the significant decrease in the concentration of versican in the aneurysmal tissue, this PG undergoes specific alterations in the aneurysmal tissue. The molecular size of versican isolated from diseased tissue is decreased with a simultaneous increase in the ratio of glycosaminoglycan to protein in this tissue. Although the size of chondroitin sulfate chains was identical in both versican preparations, a significant increase in the percentage of 6-sulfated disaccharides was observed in chondroitin sulfate chains of versican in aneurysmal aortas, which is accompanied by decrease in 4-sulfated and non-sulfated units (Theocharis et al. 2003a). Mature human aorta contains a 70-kDa versican fragment, which appears to represent G1 domain of versican V1 and has been generated in vivo by proteolytic cleavage at the Glu(441)-Ala(442) bond, within the sequence DPEAAE (441)-A(442)RRGQ (Sandy et al. 2001). Increased versican expression is often observed in tumor growth in tissues such as breast, brain, ovary, gastrointestinal tract, prostate, and melanoma, Sarcoma, and mesothelioma. In addition, versican contributes to the development of a number of pathologic processes including atherosclerotic vascular diseases, tendon remodeling, hair follicle cycling, CNS injury, and neurite outgrowth.

37.8 Diseases of Aggrecan Insufficiency

37.8.1 Aggrecanase-Mediated Cartilage Degradation

The linker domain between the N-terminal globular domains, called the interglobular domain, is highly sensitive to proteolysis. Such degradation has been associated with the development of arthritis. Proteases capable of degrading aggrecans are called aggrecanases, and they are members of the ADAM (A Disintegrin And Metalloprotease) protein family. Aggrecanases and matrix metalloproteinases (MMPs) are associated with aggrecan proteolysis (East et al. 2007). The aggrecanases are of particular interest because of their selectivity for aggrecan. Five aggrecanase cleavage sites have been described in aggrecan (Tortorella et al. 2000), with one residing in the IGD domain and four in the CS2 domain. Increasing evidence is accumulating for the importance of the aggrecanases ADAMTS-4 and ADAMTS-5 in cartilage degradation in arthritis. A number of laboratories have provided insight into the regulation of the expression and activity of these proteins and the molecular basis of their role in aggrecan catabolism. Recombinant ADAMTS-4 and ADAMTS-5 cleave aggrecan at five distinct sites along the core protein and aggrecan fragments generated by cleavage at all of these sites have been identified in cartilage explants undergoing matrix degradation. This proteolytic activity of the aggrecanases can be modulated by several means, including altered expression, activation by proteolytic cleavage at a furin-sensitive site, binding to the aggrecan substrate through the C-terminal thrombospondin motif, activation through post-translational processing of a portion of the C-terminus and inhibition of activity by the endogenous inhibitor TIMP-3. ADAMTS-4 and ADAMTS-5 activity is detected in joint capsule and synovium in addition to cartilage, and may be upregulated in arthritic synovium at either the message level or through post-translational processing. Additional substrates include the CSPG brevican and versican. Advances are occurring in the development of selective aggrecanase inhibitors designed to serve as therapeutics for the treatment of arthritis (Arner 2002)

Site of Catabolism of Aggrecan: Characterization of aggrecan core protein peptides present in the matrix of adult human articular cartilage showed that at least 11 aggrecan core proteins were present with M_r between 300,000 and 43,000. All these core proteins were found to have the same N-terminal sequences as that observed in human aggrecan. Articular cartilage in explant culture in medium containing 10^{-6} M retinoic acid showed a 3.5-fold increase in the loss of aggrecan into the medium compared to tissue maintained in absence of retinoic acid. Analysis showed the presence of N-terminal sequence ARGS-, which starts at residue 393 of the human aggrecan core protein, located within the interglobular region between the G1 and G2 domains and is the site of aggrecan catabolism by the putative protease aggrecanase. The G1 regions may accumulate in the cartilage matrix for many years (Ilic et al. 1995; Maroudas et al. 1998). The structure of aggrecan fragments in human synovial fluid suggests the involvement of a proteinase in osteoarthritis which cleaves the Glu^{373}-Ala^{374} bond of the interglobular domain (Sandy et al. 1992; Fosang et al. 1995).

Induction of Arthritis in BALB/c Mice by Cartilage Link Protein: Both type II collagen and the proteoglycan aggrecan are capable of inducing an erosive inflammatory polyarthritis in mice. The link protein (LP) from bovine cartilage can produce a persistent, erosive, inflammatory polyarthritis when injected repeatedly intraperitoneally into BALB/c mice. Another cartilage protein, LP, like type II collagen and the proteoglycan aggrecan, is capable of inducing an erosive inflammatory arthritis in mice and the immunity to LP involves recognition of both T- and B-cell epitopes. This immunity may be of importance in the pathogenesis of inflammatory joint diseases, such as juvenile rheumatoid arthritis, in which cellular immunity to LP has been demonstrated (Zhang et al. 1998).

37.8.2 A Mutation in Aggrecan Gene Causes Spondyloepiphyseal Dysplasia

Spondyloepiphyseal dysplasia (SED) encompasses a heterogeneous group of disorders characterized by shortening of the trunk and limbs. Mutations in the aggrecan gene leading to chondrodyplasias have been described in the human, mouse and chicken. The autosomal dominant SED type Kimberley (SEDK) is associated with premature degenerative arthropathy and has been mapped in a multigenerational family to a novel locus on 15q26.1. This locus contains the gene AGC1, which encodes aggrecan, the core protein of proteoglycan of cartilage. In the human, A single-base-pair insertion, within the variable repeat region of exon 12 affected individuals from the family with SEDK, that introduces a frameshift of 212 amino acids, including 22 cysteine residues, followed by a premature stop codon and results in a form of spondyloepiphyseal dysplasia (Gleghorn et al. 2005). Roughley et al. (2006) suggested the involvement of aggrecan polymorphism in degeneration of human intervertebral disc and articular cartilage.

In the mouse, a 7 bp deletion in exon 5 results in a premature termination codon arising in exon 6 (Watanabe et al. 1994, 1998). In the chicken, a premature stop codon is present within exon 10 encoding the CS-attachment region (Li et al. 1993), resulting in decreased message accumulation and under-production of a truncated aggrecan. It is likely that the absence of a G3 region impairs secretion of the mutant aggrecan molecules. In the human, chondrodystrophic phenotypes have also been associated with the under-sulphation of aggrecan due to gene defects in a sulfate transporter (Superti-Furga et al. 1996). These disorders illustrate the importance of aggrecan content and charge in embryonic cartilage development and growth.

37.9 Expression of Proteoglycans in Carcinogenesis

Glycosaminoglycans (GAGs) in PG forms or as free GAGs are implicated in the growth and progression of malignant tumors. Human gastric carcinoma (HGC) contained about twofold increased amounts of GAGs in comparison to normal gastric mucosa (HNG). Specifically, increase in chondroitin sulfate (CS) and hyaluronan (HA) contents was associated with HGC. In HGC, the amounts of versican and decorin were significantly increased. Analysis of Δ-disaccharide of versican and decorin from HGC showed an increase of 6-sulfated Δ-disaccharides (Δ di-6 S) and non-sulfated Δ-disaccharides (Δ di-0 S) with a parallel decrease of 4-sulfated Delta-disaccharides (Δdi-4 S) as compared to HNG. In addition, the accumulation of core proteins of versican and decorin in HGC was also associated with many post-translational modifications, referring to the number, size, degree and patterns of sulphation and epimerization of CS/DS chains (Theocharis et al. 2003b). Prostate tumor cells induce host stromal cells to secrete increased versican levels via a paracrine mechanism mediated by transforming growth factor beta1 (Sakko et al. 2001).

Although, versican expression is not an independent prognostic factor in pharyngeal squamous cell carcinoma (Pukkila et al. 2004), Skandalis et al. (2006) provided direct evidence for a significant and stage-related accumulation of versican and decorin in the tumor-associated stroma of laryngeal squamous cell carcinoma (LSCC) in comparison to normal larynx. The accumulated versican and decorin were markedly modified on both protein core and glycosaminoglycan (GAG) levels. Versican was found to undergo stage-related structural modifications. The modified chemical structure of both PGs could be associated with the degree of aggressiveness of laryngeal squamous cell carcinomas (Skandalis et al. 2006).

37.10 Regulation of Proteoglycans

Dex stimulates cell-associated and soluble chondroitin 4-sulfate proteoglycans. During differentiation into adipocytes, there was a 1.68-fold increase in the ^{35}S in medium of 3 T3-L1 preadipocytes, whereas cell-associated proteoglycan showed no increase. Analyses indicated that all ^{35}S label was recovered as two major species of chondroitin 4-sulfate proteoglycans (CSPG-I and CSPG-II) and 7% as heparan sulfate proteoglycan. Cell differentiation was associated with a specific increase in CSPG-I (Calvo et al. 1991). Platelet-derived growth factor (PDGF) and TGF-β1 increase ^{35}Sulfate incorporation into PG by monkey arterial smooth muscle cells but have opposite effects on cell proliferation. Although both of these growth factors increase the net synthesis of a large versican like CSPG, they differ in their effects on the structure of the glycosaminoglycan chains (Schönherr et al. 1991). TGF-β1 markedly enhanced the expression of biglycan and versican mRNAs and the enhancement of biglycan expression was coordinate with elevated type I procollagen gene expression in the same cultures. In contrast, the expression of decorin mRNA was markedly inhibited by TGF-β1. Results indicate differential regulation of PG gene expression in fibroblasts by TGF-β1 (Kähäri et al. 1991; Barry et al. 2001). Changes have been observed in synthesis of proteoglycans by vascular SMCs in response to controlled mechanical strains. Following strain mRNA and protein for versican, biglycan, and perlecan increased. Thus, mechanical deformation increases specific vascular SMC proteoglycan synthesis and aggregation, indicating a highly coordinated extracellular matrix response to biomechanical stimulation (Lee et al. 2001).

37.11 CD44: A Major Hyaluronan Receptor

37.11.1 CD44: A Hyaluronan Receptor

The CD44 is a transmembrane surface glycoprotein widely distributed in both epithelial and nonepithelial normal tissues (Ponta et al. 2003). This protein participates in a wide variety of cellular functions including lymphocyte activation, recirculation and homing, hematopoiesis, and tumor metastasis. CD44 functions as a major hyaluronan receptor, which mediates cell adhesion and migration in a variety of physiological processes (Sillanpää et al. 2003). It can also interact with other ligands, such as osteopontin, collagens, and matrix metalloproteinases (MMPs). A specialized sialofucosylated glycoform of CD44 called HCELL is found natively on human hematopoietic stem cells, and is a highly potent E-selectin and L-selectin ligand. CD44 is encoded by a single gene located on human chromosome 11. However, alternative splicing of mRNA produces several larger CD44 isoforms in addition to the standard isoform, CD44s. The extracellular region of CD44 gives rise to multiple CD44 isoforms. Alternative splicing is the basis for the structural and functional diversity of this protein, and may be related to tumor metastasis All isoforms contain an amino-terminal domain, which is homologous to cartilage link proteins. The cartilage link protein-like domain of CD44 is important for hyaluronan binding. Splice variants of CD44 on colon cancer cells display the HCELL glycoform, which mediates binding to vascular E-selectin under hemodynamic flow conditions, a critical step in colon cancer metastasis.

CD44 is a lectin cell adhesion molecule that is also expressed in keratinocytes. Amino sugars such as NAG may competitively bind to CD44, modulating keratinocyte cellular adhesion. These amino sugars may modulate keratinocyte cellular adhesion and differentiation, leading to the normalization of stratum corneum exfoliation (Mammone et al. 2009). CD44, along with CD25, is used to track early T cell development in the thymus. Its expression is an indicative marker for effector-memory T-cells. It is tracked with CFSE chemical tagging.

37.11.2 Hyaluronan Binding Sites in CD44

CD44 is a receptor for hyaluronic acid and can also interact with other ligands, such as osteopontin, collagens, and matrix metalloproteinases (MMPs). CD44 function is controlled by its posttranslational modifications. CD44 binds hyaluronan ligand via a lectin-like fold termed the Link module, but only after appropriate functional activation. Among twenty-four point mutants, eight residues were identified as critical for binding or to support the interaction. These residues form a coherent surface, the location of which approximately corresponds to the carbohydrate binding sites in MBP and E-selectin (CD62E) (Bajorath et al. 1998). The interaction with hyaluronan is dominated by shape and hydrogen-bonding complementarity. Banerji et al. (2007) identified two conformational forms of the murine Cd44 that differ in orientation of a crucial hyaluronan-binding residue (Arg^{45}, equivalent to Arg^{41} in human CD44). NMR studies indicated that the conformational transition can be induced by hyaluronan binding, the possible mechanisms for regulation of Cd44.

Although full activity of CD44 requires binding to ERM (ezrin/radixin/moesin) proteins, the CD44 cytoplasmic region, consisting of 72 amino acid residues, lacks the Motif-1 consensus sequence for ERM binding found in ICAM-2 and other adhesion molecules of the Ig superfamily. CD44 exists as an extended monomeric form of the cytoplasmic peptide in solution. The crystal structure of the radixin FERM domain complexed with a CD44 cytoplasmic peptide reveals that the KKKLVIN sequence of the peptide forms a β strand followed by a short loop structure that binds subdomain C of the FERM domain. Like Motif-1 binding, the CD44 β strand binds the shallow groove between strand β5C and helix α1C and augments the β sheet β5C- β7C from subdomain C. Two hydrophobic CD44 residues, Leu and Ile, are docked into a hydrophobic pocket with the formation of hydrogen bonds between Asn of the CD44 short loop and loop β4C- β5C from subdomain C. This binding mode resembles that of neutral endopeptidase 24.11 (NEP) rather than ICAM-2. These results reveal a characteristic versatility of peptide recognition by the FERM domains from ERM proteins, a possible mechanism by which the CD44 tail is released from the cytoskeleton for nuclear translocation by regulated intra-membrane proteolysis, and provide a structural basis for Smad1 interactions with activated CD44 bound to ERM protein (Mori et al. 2008).

37.11.3 CD44 in Cancer

CD44 is involved in cell proliferation, cell differentiation, cell migration, angiogenesis, and as growth factors to the corresponding receptors, and docking of proteases at the cell membrane, as well as in signaling for cell survival. All these biological properties are essential for physiological activities of normal cells, but they are also associated with the pathologic activities of cancer cells. CD44 is remarkable for its ability to generate alternatively spliced forms, many of which differ in their activities. This remarkable flexibility has led to speculation that CD44, via its changing nature, plays a role in some of the methods that tumor cells use to progress successfully through growth and metastasis. Experiments in animals have shown that targeting of CD44 by antibodies, antisense, and CD44-soluble proteins markedly reduces the malignant activities of various neoplasms, stressing the therapeutic potential of anti-CD44 agents (Naor et al. 2002). Variations in CD44 are reported as cell surface markers for some breast and prostate cancer stem cells (Li et al. 2007) and has been seen as an indicator of increased survival time in epithelial ovarian cancer patients (Martin et al. 2004; Sillanpää et al. 2003; Yasuda et al. 2003). The role of CD44 in prostate cancer development and progression is controversial with studies showing both tumor-promoting and tumor-inhibiting effects. However, Patrawala et al. (2006) suggest that the CD44$^+$ prostate cancer cell population is enriched in tumorigenic and metastatic progenitor cells. CD44$^+$ cells displayed clustered growth within colorectal cancer. Knockdown of CD44, strongly prevented clonal formation and inhibited tumorigenicity in xenograft model. Results indicated that CD44 is a robust marker and is of functional importance for colorectal CSC for cancer initiation (Du et al. 2008).

References

Ababneh KT, Hall RC, Embery G (1999) The proteoglycans of human cementum: immunohistochemical localization in healthy, periodontally involved and ageing teeth. J Periodontal Res 34:87–96

Abiko Y, Nishimura M, Rahemtulla F et al (2001) Immunohistochemical localization of large chondroitin sulphate proteoglycan in porcine gingival epithelia. Eur J Morphol 39:99–104

Arner EC (2002) Aggrecanase-mediated cartilage degradation. Curr Opin Pharmacol 2:322–329

Baeurle SA, Kiselev MG, Makarova ES et al (2009) Effect of the counterion behavior on the frictional-compressive properties of chondroitin sulfate solutions. Polymer 50:1805–1813

Bajorath J, Greenfield B, Munro SB et al (1998) Identification of CD44 residues important for hyaluronan binding and delineation of the binding site. J Biol Chem 273:338–343

Baldwin CT, Reginato AM, Prockop DJ (1989) A new epidermal growth factor-like domain in the human core protein for the large cartilage-specific proteoglycan. Evidence for alternative splicing of the domain. J Biol Chem 264:15747–15750

Bandtlow CE, Zimmermann DR (2000) Proteoglycans in the developing brain: new conceptual insights for old proteins. Physiol Rev 80:1267–1290

Banerji S, Wright AJ, Noble M, Mahoney DJ, Campbell ID, Day AJ, Jackson DG (2007) Structures of the Cd44-hyaluronan complex provide insight into a fundamental carbohydrate-protein interaction. Nat Struct Mol Biol 14:234–239

Barry FP, Neame PJ, Sasse J, Pearson D (1994) Length variation in the keratan sulfate domain of mammalian aggrecan. Matrix 14:323–328

Barry FP, Rosenberg LC, Gaw JU, Koob TJ, Neame PJ (1995) N- and O-linked keratin sulfate on the hyaluronan binding region of aggrecan from mature and immature bovine cartilage. J Biol Chem 270:20516–20524

Barry F, Boynton RE, Liu B, Murphy JM (2001) Chondrogenic differentiation of mesenchymal stem cells from bone marrow: differentiation-dependent gene expression of matrix components. Exp Cell Res 268:189–200

Bi W, Huang W, Whitworth DJ et al (2001) Haploinsufficiency of Sox9 results in defective cartilage primordia and premature skeletal mineralization. Proc Natl Acad Sci USA 98:6698–6703

Brown GM, Huckerby TN, Bayliss MT, Nieduszynski IA (1998) Human aggrecan keratan sulfate undergoes structural changes during adolescent development. J Biol Chem 273:26408–26414

Bruyere O, Reginster JY (2007) Glucosamine and chondroitin sulfate as therapeutic agents for knee and hip osteoarthritis. Drugs Aging 24:573–580

Calvo JC, Rodbard D, Katki A et al (1991) Differentiation of 3 T3-L1 preadipocytes with 3-isobutyl-1-methylxanthine and dexamethasone stimulates cell-associated and soluble chondroitin 4-sulfate proteoglycans. J Biol Chem 266:11237–11244

Carrino DA, Onnerfjord P, Sandy JD et al (2003) Age-related changes in the proteoglycans of human skin. Specific cleavage of decorin to yield a major catabolic fragment in adult skin. J Biol Chem 278:17566–17572

Chana RS, Wheeler DC, Thomas GJ, Williams JD, Davies M (2000) Low-density lipoprotein stimulates mesangial cell proteoglycan and hyaluronan synthesis. Nephrol Dial Transplant 15:167–172

Doege KJ, Coulter SN, Meek LM, Maslen K, Wood JG (1997) A human-specific polymorphism in the coding region of the aggrecan gene – variable number of tandem repeats produce a range of core protein sizes in the general population. J Biol Chem 272:13974–13979

Doege K, Sasaki M, Horigan E et al (1987) Complete primary structure of the rat cartilage proteoglycan core protein deduced from cDNA clones. J Biol Chem 262:17757–17767

Doege KJ, Garrison K, Coulter SN, Yamada Y (1994) The structure of the rat aggrecan gene and preliminary characterization of its promoter. J Biol Chem 269:29232–29240

Doege K, Hall LB, McKinnon W et al (2002) A remote upstream element regulates tissue-specific expression of the rat aggrecan gene. J Biol Chem 277:13989–13997

Dours-Zimmermann MT, Zimmermann DR (1994) A novel glycosaminoglycan attachment domain identified in two alternative splice variants of human versican. J Biol Chem 269:32992–32998

Du L, Wang H, He L et al (2008) CD44 Is of functional importance for colorectal cancer stem cells. Clin Cancer Res 14:6751–6760

Dudhia J, Davidson CM, Wells TM et al (1996) Age-related changes in the content of the C-terminal region of aggrecan in human articular cartilage. Biochem J 313:933–940

Dutt S, Kléber M, Matasci M et al (2006) Versican V0 and V1 guide migratory neural crest cells. J Biol Chem 281:12123–12131

East CJ, Stanton H, Golub SB et al (2007) ADAMTS-5 deficiency does not block aggrecanolysis at preferred cleavage sites in the chondroitin sulfate-rich region of aggrecan. J Biol Chem 282:8632–8640

Erickson AC, Couchman JR (2001) Basement membrane and interstitial proteoglycans produced by MDCK cells correspond to those expressed in the kidney cortex. Matrix Biol 19:769–778

Eriksen GV, Carlstedt I, Mörgelin M et al (1999) Isolation and characterization of proteoglycans from human follicular fluid. Biochem J 340:613–620

Fosang AJ, Hardingham TE (1989) Isolation of the N-terminal globular protein domains from cartilage proteoglycans. Identification of G2 domain and its lack of interaction with hyaluronate and link protein. Biochem J 261:801–809

Fosang AJ, Last K, Neame PJ et al (1995) Neutrophil collagenase (MMP-8) cleaves at the aggrecanase site E373-A374 in the interglobular domain of cartilage aggrecan. Biochem J 304:347–351

Fülöp C, Cs-Szabó G, Glant TT (1996) Species-specific alternative splicing of the epidermal growth factor-like domain 1 of cartilage aggrecan. Biochem J 319:935–940

Funderburgh JL (2000) Keratan sulfate: structure, biosynthesis, and function. Glycobiology 10:951–958

Gallagher JT (2001) Heparan sulfate: growth control with a restricted sequence menu. J Clin Invest 108:357–361

Gleghorn L, Ramesar R, Beighton P, Wallis G (2005) A mutation in the variable repeat region of the aggrecan gene (AGC1) causes a form of spondyloepiphyseal dysplasia associated with severe, premature osteoarthritis. Am J Hum Genet 77:484–490

Glumoff V, Savontaus M, Vehanen J, Vuorio E (1994) Analysis of aggrecan and tenascin gene expression in mouse skeletal tissues by northern and in situ hybridization using species specific cDNA probes. Biochim Biophys Acta 1219:613–622

Gogiel T, Bańkowski E, Jaworski S (2003) Proteoglycans of Wharton's jelly. Int J Biochem Cell Biol 35:1461–1469

Gogiel T, Galewska Z, Romanowicz L et al (2007) Pre-eclampsia-associated alterations in decorin, biglycan and versican of the umbilical cord vein wall. Eur J Obstet Gynecol Reprod Biol 134:51–56

Gupta V, Barzilla JE, Mendez JS et al (2009) Abundance and location of proteoglycans and hyaluronan within normal and myxomatous mitral valves. Cardiovasc Pathol 18:191–197

Hardingham TE, Fosang AJ (1992) Proteoglycans: many forms and many functions. FASEB J 6:861–870

Horton WE, Lethbridge-Çejku M, Hochberg MC, Balakir R, Precht P, Plato CC, Tobin JD, Meek L, Doege K (1998) An association between an aggrecan polymorphic allele and bilateral hand osteoarthritis in elderly white men: data from the Baltimore Longitudinal Study of Aging (BLSA). Osteoarthritis Cartilage 6:245–251

Ilic MZ, Mok MT, Williamson OD et al (1995) Catabolism of aggrecan by explant cultures of human articular cartilage in the presence of retinoic acid. Arch Biochem Biophys 322:22–30

Ilic MZ, Carter P, Tyndall A et al (2005) Proteoglycans and catabolic products of proteoglycans present in ligament. Biochem J 385:381–388

Ito K, Shinomura T, Zako M et al (1995) Multiple forms of mouse PG-M, a large chondroitin sulfate proteoglycan generated by alternative splicing. J Biol Chem 270:958–965

Jomphe C, Gabriac M, Hale TM et al (2008) Chondroitin sulfate inhibits the nuclear translocation of nuclear factor-kB in Interleukin-1β-stimulated chondrocytes. Basic Clin Pharmacol Toxicol 102 (1):59–65

Kähäri VM, Larjava H, Uitto J (1991) Differential regulation of extracellular matrix proteoglycan (PG) gene expression. Transforming growth factor-beta 1 up-regulates biglycan (PGI), and versican (large fibroblast PG) but down-regulates decorin (PGII) mRNA levels in human fibroblasts in culture. J Biol Chem 266:10608–10615

Kawaguchi Y, Osada R, Kanamori M, Ishihara H, Ohmori K, Matsui H, Kimura T (1999) Association between an aggrecan gene polymorphism and lumbar disc degeneration. Spine 24:2456–2460

Kawashima H, Hirose M, Hirose J et al (2000) Binding of a large chondroitin sulfate/dermatan sulfate proteoglycan, versican, to L-selectin, P-selectin, and CD44. J Biol Chem 275:35448–35456

Kawashima H, Atarashi K, Hirose M et al (2002) Oversulfated chondroitin/dermatan sulfates containing GlcAbeta1/IdoAalpha1-3GalNAc(4,6-O-disulfate) interact with L- and P-selectin and chemokines. J Biol Chem 277:12921–12930

Kiani C, Lee V, Cao L et al (2001) Roles of aggrecan domains in biosynthesis, modification by glycosaminoglycans and product secretion. Biochem J 354:199–207

Kirschfink M, Blase L, Engelmann S et al (1997) Secreted chondroitin sulfate proteoglycan of human B cell lines binds to the complement protein C1q and inhibits complex formation of C1. J Immunol 158:1324–1331

Kishimoto J, Ehama R, Wu L et al (1999) Selective activation of the versican promoter by epithelial- mesenchymal interactions during hair follicle development. Proc Natl Acad Sci USA 96:7336–7341

Landolt RM, Vaughan L, Winterhalter KH et al (1995) Versican is selectively expressed in embryonic tissues that act as barriers to neural crest cell migration and axon outgrowth. Development 212:2303–2312

Lee I, Ono Y, Lee A et al (1998) Immunocytochemical localization and biochemical characterization of large proteoglycans in developing rat bone. J Oral Sci 40:77–87

Lee RT, Yamamoto C, Feng Y et al (2001) Mechanical strain induces specific changes in the synthesis and organization of proteoglycans by vascular smooth muscle cells. J Biol Chem 276:13847–13851

Li F, Tiede B, Joan Massagué J, Yibin Kang Y (2007) Beyond tumorigenesis: cancer stem cells in metastasis. Cell Res 17:3–14

Li H, Schwartz NB, Vertel BM (1993) cDNA cloning of chick cartilage chondroitin sulfate (aggrecan) core protein and identification of a stop codon in the aggrecan gene associated with the chondrodystrophy, nanomelia. J Biol Chem 268:23504–23511

Lundell A, Olin AI, Mörgelin M et al (2004) Structural basis for interactions between tenascins and lectican C-type lectin domains: evidence for a crosslinking role for tenascins. Structure 12:1495–1506

Malgouries S, Thibaut S, Bernard BA (2008) Proteoglycan expression patterns in human hair follicle. Br J Dermatol 158:234–242

Mammone T, Gan D, Fthenakis C, Marenus K (2009) The effect of N-acetyl-glucosamine on stratum corneum desquamation and water content in human skin. J Soc Cosmetic Chemists 60:423–428

Maroudas A, Bayliss MT, Uchitel-Kaushansky N, Schneiderman R, Gilav E (1998) Aggrecan turnover in human articular cartilage: use of aspartic acid racemization as a marker of molecular age. Arch Biochem Biophys 350:61–71

Martin TA, Harrison G, Mansel RE, Jiang WG (2004) The role of the CD44/ezrin complex in cancer metastasis. Crit Rev Oncol Hematol 46:165–186

Matsumoto K, Shionyu M, Go M et al (2003) Distinct interaction of versican/PG-M with hyaluronan and link protein. J Biol Chem 278:41205–41212

Melrose J, Whitelock J, Xu Q, Ghosh P (1998) Pathogenesis of abdominal aortic aneurysms: possible role of differential production of proteoglycans by smooth muscle cells. J Vasc Surg 28:676–686

Milz S, Regner F, Putz R, Benjamin M (2002) Expression of a wide range of extracellular matrix molecules in the tendon and trochlea of the human superior oblique muscle. Invest Ophthalmol Vis Sci 43:1330–1334

Monfort J, Pelletier J-P, Garcia-Giralt N et al (2008) Biochemical basis of the effect of chondroitin sulfate on osteoarthritis articular tissues. Ann Rheum Dis 67(6):735–740

Moon LD, Asher RA, Fawcett JW (2003) Limited growth of severed CNS axons after treatment of adult rat brain with hyaluronidase. J Neurosci Res 71:23–37

Mori T, Kitano K, Terawaki SI et al (2008) Structural basis for CD44 recognition by ERM proteins. J Biol Chem 283:29602–29612

Naor D, Nedvetzki S, Golan I et al (2002) CD44 in cancer. Crit Rev Clin Lab Sci 39:527–579

Naso MF, Zimmermann DR, Iozzo RV (1994) Characterization of the complete genomic structure of the human versican gene and functional analysis of its promoter. J Biol Chem 269:32999–33008

Ney A, Booms P, Epple G et al (2006) Calcium-dependent self-association of the C-type lectin domain of versican. Int J Biochem Cell Biol 38:23–29

Nilsson B, De Luca S, Lohmander S, Hascall VC (1982) Structures of N-linked and O-linked oligosaccharides on proteoglycan monomer isolated from the Swarm rat chondrosarcoma. J Biol Chem 257:10920–10927

Nishida Y, Shinomura T, Iwata H et al (1994) Abnormal occurrence of a large chondroitin sulfate proteoglycan, PG-M/versican in osteoarthritic cartilage. Osteoarthritis Cartilage 2:43–49

Olsson U, Bondjers G, Camejo G (1999) Fatty acids modulate the composition of extracellular matrix in cultured human arterial smooth muscle cells by altering the expression of genes for proteoglycan core proteins. Diabetes 48:616–622

Patrawala L, Calhoun T, Schneider-Broussard R et al (2006) Highly purified CD44+ prostate cancer cells from xenograft human tumors are enriched in tumorigenic and metastatic progenitor cells. Oncogene 25:1696–1708

Pirok EW III, Li H, Mensch JR Jr, Henry J, Schwartz NB (1997) Structural and functional analysis of the chick chondroitin sulfate proteoglycan (aggrecan) promoter and enhancer region. J Biol Chem 272:11566–11574

Pirok EWIII, Henry J, Schwartz NB (2001) Cis elements that control the expression of chick aggrecan. J Biol Chem 276:16894–16903

Pirok PEWIII, Domowicz MS, Henry J et al (2005) APBP-1, a DNA/RNA-binding protein, interacts with the chick aggrecan regulatory region. J Biol Chem 280:35606–35616

Ponta H, Sherman L, Herrlich PA (2003) CD44: from adhesion molecules to signalling regulators. Nat Rev Mol Cell Biol 4:33–45

Pukkila MJ, Kosunen AS, Virtaniemi JA et al (2004) Versican expression in pharyngeal squamous cell carcinoma: an immunohistochemical study. J Clin Pathol 57:735–739

Rahmani M, Wong BW, Ang L et al (2006) Versican: signaling to transcriptional control pathways. Can J Physiol Pharmacol 84:77–92

Rodríguez-Lee M, Ostergren-Lundén G, Wallin B et al (2006) Fatty acids cause alterations of human arterial smooth muscle cell proteoglycans that increase the affinity for low-density lipoprotein. Arterioscler Thromb Vasc Biol 26:130–135

Roughley PJ, White RJ (1989) Dermatan sulphate proteoglycans of human articular cartilage. The properties of dermatan sulphate proteoglycans I and II. Biochem J 262:823–827

Roughley P, Martens D, Rantakokko J et al (2006) The involvement of aggrecan polymorphism in degeneration of human intervertebral disc and articular cartilage. Eur cells Mater 11:1–7

Sakko AJ, Ricciardelli C, Mayne K et al (2001) Versican accumulation in human prostatic fibroblast cultures is enhanced by prostate cancer cell-derived transforming growth factor beta1. Cancer Res 61:926–930

Samiric T, Ilic MZ, Handley CJ (2004a) Characterisation of proteoglycans and their catabolic products in tendon and explant cultures of tendon. Matrix Biol 23:127–140

Samiric T, Ilic MZ, Handley CJ (2004b) Large aggregating and small leucine-rich proteoglycans are degraded by different pathways and at different rates in tendon. Eur J Biochem 271:3612–3620

Sandell L (2004) In situ expression of collagen and proteoglycan genes in notochord and during skeletal development and growth. J Microsc Res Tech 28:470–482

Sandy JD, Flannery CR, Neame PJ, Lohmander LS (1992) The structure of aggrecan fragments in human synovial fluid. Evidence for the involvement in osteoarthritis of a novel proteinase which cleaves the Glu 373-Ala 374 bond of the interglobular domain. J Clin Invest 89:1512–1516

Sandy JD, Westling J, Kenagy RD et al (2001) Versican V1 proteolysis in human aorta in vivo occurs at the Glu441-Ala442 bond, a site that is cleaved by recombinant ADAMTS-1 and ADAMTS-4. J Biol Chem 276:13372–13378

Santer V, White RJ, Roughley PJ (1982) O-linked oligosaccharides of human articular cartilage proteo-glycans. Biochim Biophys Acta 716:277–282

Schönherr E, Järveläinen HT, Sandell LJ, Wight TN (1991) Effects of platelet-derived growth factor and transforming growth factor-beta 1 on the synthesis of a large versican-like chondroitin sulfate proteoglycan by arterial smooth muscle cells. J Biol Chem 266:17640–17647

Schwartz NB, Pirok EW 3rd, Mensch JR Jr et al (1999) Domain organization, genomic structure, evolution, and regulation of expression of the aggrecan gene family. Prog Nucleic Acid Res Mol Biol 62:177–225

Sekiya I, Tsuji K, Koopman P et al (2000) SOX9 Enhances aggrecan gene promoter/enhancer activity and is up-regulated by retinoic acid in a cartilage-derived cell line, TC6. J Biol Chem 275:10738–10744

Sheng W, Wang G, La Pierre DP et al (2006) Versican mediates mesenchymal-epithelial transition. Mol Biol Cell 17:2009–2020

Shinomura T, Nishida Y, Ito K et al (1993) CDNA cloning of PG-M, a large chondroitin sulfate proteoglycan expressed during chondrogenesis in chick limb buds. Alternative spliced multiforms of PG-M and their relationships to versican. J Biol Chem 268:14461–14469

Sillanpää S, Anttila MA, Voutilainen K et al (2003) CD44 Expression indicates favorable prognosis in epithelial ovarian cancer. Clin Cancer Res 9:5318–5324

Skandalis SS, Theocharis AD, Papageorgakopoulou N et al (2006) The increased accumulation of structurally modified versican and decorin is related with the progression of laryngeal cancer. Biochimie 88:1135–1143

Somers WS, Tang J, Shaw GD, Camphausen RT (2000) Insights into the molecular basis of leukocyte tethering and rolling revealed by structures of P- and E-selectin bound to SLe(X) and PSGL-1. Cell 103:467–479

Sorrell JM, Carrino DA, Baber MA, Caplan AI (1999) Versican in human fetal skin development. Anat Embryol (Berl) 199:45–56

Srinivasan SR, Xu JH, Vijayagopal P et al (1995) Low-density lipoprotein binding affinity of arterial chondroitin sulfate proteoglycan variants modulates cholesteryl ester accumulation in macrophages. Biochim Biophys Acta 1272:61–67

Superti-Furga A, Rossi A, Steinmann B, Gitzelmann R (1996) A chondrodysplasia family produced by mutations in the diastrophic dysplasia sulfate transporter gene: genotype/phenotype correlations. Am J Med Genet 63:144–147

Sztrolovics R, Grover J, Cs-Szabo G et al (2002) The characterization of versican and its message in human articular cartilage and intervertebral disc. J Orthop Res 20:257–266

Takeda S, Bonnamy JP, Owen MJ et al (2001) Continuous expression of Cbfa1 in nonhypertrophic chondrocytes uncovers its ability to induce hypertrophic chondrocyte differentiation and partially rescues Cbfa1-deficient mice. Genes Dev 15:467–481

Talusan P, Bedri S, Yang S et al (2005) Analysis of intimal proteoglycans in atherosclerosis-prone and atherosclerosis-resistant human arteries by mass spectrometry. Mol Cell Proteomics 4:1350–1357

Theocharis AD, Papageorgakopoulou N, Feretis E et al (2002) Occurrence and structural characterization of versican-like proteoglycan in human vitreous. Biochimie 84:1237–1243

Theocharis AD, Tsolakis I, Hjerpe A et al (2003a) Versican undergoes specific alterations in the fine molecular structure and organization in human aneurysmal abdominal aortas. Biomed Chromatogr 17:411–416

Theocharis AD, Vynios DH, Papageorgakopoulou N et al (2003b) Altered content composition and structure of glycosaminoglycans and proteoglycans in gastric carcinoma. Int J Biochem Cell Biol 35:376–390

Thomas GJ, Bayliss MT, Harper K et al (1994) Glomerular mesangial cells in vitro synthesize an aggregating proteoglycan immunologically related to versican. Biochem J 302:49–56

Tom S, Parkinson J, Ilic MZ et al (2009) Changes in the composition of the extracellular matrix in patellar tendinopathy. Matrix Biol 28:230–236

Tortorella MD, Pratta M, Liu RQ, Austin J, Ross OH, Abbaszade I, Burn T, Arner E (2000) Sites of aggrecan cleavage by recombinant human aggrecanase – 1 (ADAMTS-4). J Biol Chem 275:18566–18573

Tran PK, Agardh HE, Tran-Lundmark K, Hedin U et al (2007) Reduced perlecan expression and accumulation in human carotid atherosclerotic lesions. Atherosclerosis 190:264–270

Trowbridge JM, Gallo RL (2002) Dermatan sulfate: new functions from an old glycosaminoglycan. Glycobiology 12:117R–125R

Valhmu WB, Palmer GD, Rivers PA, Ebara S, Cheng JF, Fischer S, Ratcliffe A (1995) Structure of the human aggrecan gene: exon-intron organization and association with the protein domains. Biochem J 309:535–542

Valhmu WB, Palmer GD, Dobson J et al (1998a) Regulatory activities of the 5′- and 3′-untranslated regions and promoter of the human aggrecan gene. J Biol Chem 273:6196–6202

Valhmu WB, Stazzone EJ, Bachrach NM et al (1998b) Load-controlled compression of articular cartilage induces a transient stimulation of aggrecan gene expression. Arch Biochem Biophys 353:29–36

Walcz E, Deák F, Erhardt P et al (1994) Complete coding sequence, deduced primary structure, chromosomal localization, and structural analysis of murine aggrecan. Genomics 22:364–371

Watanabe H, Kimata K, Line S, Strong D et al (1994) Mouse cartilage matrix deficiency (cmd) caused by a 7 bp deletion in the aggrecan gene. Nat Genet 7:154–157

Watanabe H, Gao L, Sugiyama S, Doege K (1995) Mouse aggrecan, a large cartilage proteoglycan: protein sequence, gene structure and promoter sequence. Biochem J 308:433–440

Watanabe H, Cheung SC, Itano N, Kimata K, Yamada Y (1997) Identification of hyaluronan-binding domains of aggrecan. J Biol Chem 272:28057–28065

Watanabe H, Yamada Y, Kimata K (1998) Roles of aggrecan, a large chondroitin sulfate proteoglycan, in cartilage structure and function. J Biochem 124:687–693

Watanabe H, Yamada Y (1999) Mice lacking link protein develop dwarfism and craniofacial abnormalities. Nat Genet 21:225–229

Watanabe H, Watanabe H, Kimata K (2006) The roles of proteoglycans for cartilage. Clin Calcium 16:1029–1033

Wight TN, Merrilees MJ (2004) Proteoglycans in atherosclerosis and restenosis: key roles for versican. Circ Res 94:1158–1167

Wu YJ, La Pierre DP, Wu J et al (2005) The interaction of versican with its binding partners. Cell Res 15:483–494

Yamada K, Yamada T, Sasaki T et al (1997) Light and electron microscopical immunohistochemical localization of large proteoglycans in human tooth germs at the bell stage. Histochem J 29:167–175

Yamaguchi Y (2000) Lecticans: organizers of the brain extracellular matrix. Cell Mol Life Sci 57:276–289

Yamamoto C, Wakata T, Fujiwara Y, Kaji T (2005) Induction of synthesis of a large heparan sulfate proteoglycan, perlecan, by thrombin in cultured human coronary smooth muscle cells. Biochim Biophys Acta 1722:92–102

Yang BL, Cao L, Kiani C et al (2000) Tandem repeats are involved in G1 domain inhibition of versican expression and secretion and the G3 domain enhances glycosaminoglycan modification and product secretion via the complement-binding protein-like motif. J Biol Chem 275:21255–21261

Yasuda M, Nakano K, Yasumoto K, Tanaka Y (2003) CD44: functional relevance to inflammation and malignancy. Histol Histopathol 17:945–950

Zhang Y, Guerassimov A, Leroux JY et al (1998) Induction of arthritis in BALB/c mice by cartilage link protein: involvement of distinct regions recognized by T and B lymphocytes. Am J Pathol 153:1283–1291

Zheng J, Luo W, Tanzer ML (1998) Aggrecan synthesis and secretion – a paradigm for molecular and cellular coordination of multiglobular protein folding and intracellular trafficking. J Biol Chem 273:12999–13006

Zimmermann DR, Ruoslahti E (1989) Multiple domains of the large fibroblast proteoglycan, versican. EMBO J 8:2975–2981

Zimmermann DR, Dours-Zimmermann MT, Schubert M et al (1994) Versican is expressed in the proliferating zone in the epidermis and in association with the elastic network of the dermis. J Cell Biol 124:817–825

Proteoglycans of the Central Nervous System

G.S. Gupta

38.1 Proteoglycans in Central Nervous System

The extracellular matrix (ECM) is a complex molecular framework that provides physical support to cells and tissues, while also providing signals for cell growth, migration, differentiation and survival. Proteoglycans, as part of the extracellular or cell-surface milieu of most tissues and organ systems, play important roles in morphogenesis by modulating cell-matrix or cell-cell interactions, cell adhesiveness, or by binding and presenting growth and differentiation factors. The basic concept, that specialized extracellular matrices rich in hyaluronan (HA), chondroitin sulfate proteoglycans (CS-PG: aggrecan, versican, neurocan, brevican, phosphacan), link proteins and tenascins (TN-R and TN-C) can regulate cellular migration and axonal growth and thus, actively participate in the development and maturation of nervous system, has gained rapidly expanding experimental support (Zimmermann and Dours-Zimmermann 2008). The distribution of ECM molecules displays area-specific differences along the dorso-ventral axis, delimiting functionally and developmentally distinct areas. In gray matter, laminae I and II lack perineuronal nets (PNN) of extracellular matrix and contain low levels of chondroitin sulfate glycosaminoglycans (CS-GAGs), brevican, and tenascin-R, possibly favoring the maintenance of local neuroplastic properties. Conversely, CS-GAGs, brevican, and phosphacan were abundant, with numerous thick PNNs, in laminae III-VIII and X. Motor neurons (lamina IX), surrounded by PNNs, contained various amounts of CS-GAGs (Vitellaro-Zuccarello et al. 2007).

38.1.1 Large Proteoglycans in Brain

Proteoglycans (PGs) of aggrecan family include extracellular PGs: aggrecan, versican (and avian homologue PG-M), neurocan, brevacan and cell surface specific receptor CD44. They are modular PGs containing combinations of structural motifs such as EGF-like domains, lectin like domains, complement regulatory like domains, immunoglobulin folds, and proteoglycan tandem repeats that are found in other proteins (Fig. 38.1). The protein-hyaluronate aggregates in brain ECM contain versican. The bovine versican splice variant, versican V2 is together with brevican, a major component of the mature brain ECM. Versicans V0 and V1 are only present in relatively small amounts (Schmalfeldt et al. 1998). Several other proteins are related to this family of molecules including link proteins, BEHAV and TSG-6 and contain some of the same highly conserved motifs but lack GAG side chains. Aggrecan and CD44 have been characterized and discussed in general in Chap. 37. In this chapter brain specific PGs, related to aggrecan have been discussed. In adult brain, lecticans are thought to interact with hyaluronan and tenascin-R to form a ternary complex. It is proposed that hyaluronan-lectican-tenascin-R complex constitutes the core assembly of adult brain ECM, which is found mainly in pericellular spaces of neurons as 'perineuronal nets'. Brevican is a member of the aggrecan/versican family of proteoglycans, containing a hyaluronic acid-binding domain in its N-terminus and a lectin-like domain in its C-terminus.

PGs in Developing Brain: A diverse set of proteoglycans (phosphocan, neurocan, versican, aggrecan, and neuronglia antigen 2 (NG2) proteoglycan) is expressed in developing and adult brain. In developing rat brain from embryonic day 14 (E14), the concentration of aggrecan increased steadily up to 5 months of age, when it reached a level that was 18-fold higher than at E14, where as spliced versican isoforms were present at a relatively low level during the late embryonic and early postnatal period, decreased by approximately 50% between 1 and 2 weeks postnatal, and then increased steadily to a maximum at 100 days (Milev et al. 1998a). In contrast, versican isoforms containing β domain doubled between E14 and birth, after which

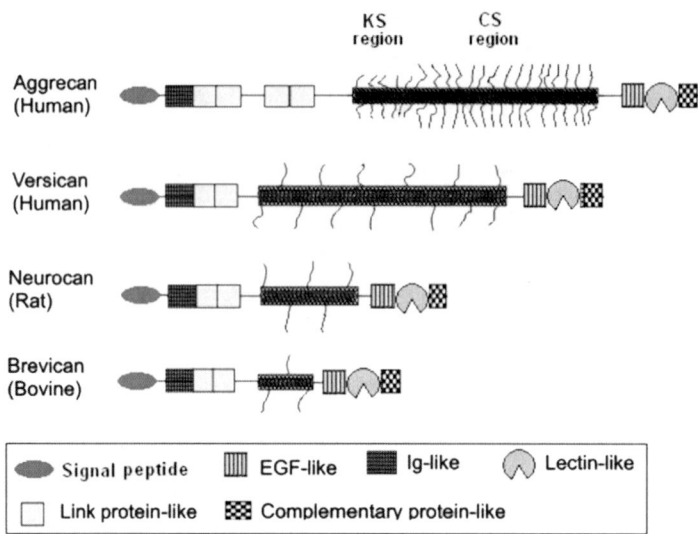

Fig. 38.1 The modular structures of the proteoglycans aggrecan, versican 2, neurocan, and brevican. Aggrecan shows both keratin sulfate (KS) and chondritin sulfate (CS) regions where as versican, neurocan and brevican have CS. Boxes represent putative functional domains shared by all the members of the aggrecan/versican family (Yamada et al. 1994; Fosang and Hardingham 1996)

versican isoforms decreased to reach a low "mature" level that remained unchanged between 2 and 8 months. As in aggrecan, brevican was detected in traces in embryonic brain and increased steadily after birth to reach an adult level that was ~14-fold higher than that in neonatal brain.

Astrocytes as primary site for synthesis of PGs: The CS-PG is primarily synthesized by astrocytes and is believed to influence astroglial motility during development and under certain pathological conditions. The CS-PGs expressed by reactive astrocytes may contribute to the axon growth-inhibitory environment of the injured central nervous system (CNS). In the glial scar, after cortical injury, neurocan and phosphacan can be localized to reactive astrocytes 30 days after CNS injury, whereas brevican and versican are not expressed in the chronic glial scar. Because these CS-PGs are capable of inhibiting neurite outgrowth in vitro, reports suggest that phosphacan and neurocan in areas of reactive gliosis may contribute to axonal regenerative failure after CNS injury. Reports also indicate that brevican is upregulated in areas of brain damage as well as in areas denervated by a lesion and involved in the synaptic reorganization of denervated brain areas (McKeon et al. 1999; Thon et al. 2000). Oligodendrocyte (OL) and their progenitors secrete brevican prior to and during active developmental myelination whereas NG2 is a transmembrane CS-PG produced by early OL progenitors (Table 38.1).

Perineuronal Nets (PNN): Perineuronal nets (PNN) are specialized ECM structures enwrapping CNS neurons, which are important regulators for neuronal and synaptic functions. The decrease in plasticity in CNS during postnatal development is accompanied by appearance of PNNs around the cell body and dendrites of many classes of neuron. These structures are composed of ECM molecules, such as CS-PGs, HA, tenascin-R, and link proteins. In rat deep cerebellar nuclei, only large excitatory neurons were surrounded by nets, which contained the CS-PGs aggrecan, neurocan, brevican, versican, and phosphacan, along with tenascin-R and HA. In the cerebellar cortex, Golgi neurons possessed PNNs and also synthesized HA synthases (HASs), cartilage link protein, and BRAL2 mRNAs. Carulli et al. (2007) proposed that HASs, which can retain HA on the cell surface, may act as a link between PNNs and neurons. Thus, HAS and link proteins might be key molecules for PNN formation and stability. During postnatal development the expression of link protein and aggrecan mRNA is upregulated at the time of PNN formation, and these molecules may therefore trigger their formation (Galtrey et al. 2008).

38.1.2 Ligands of Lecticans

Interaction Between Tenascin-R and Phosphacan and Neurocan: CS PGs in perineuronal nets include neurocan and phosphacan. Morphological alterations were noticed in Wisteria floribunda labelled nets around cortical interneurones both in tenascin-R knockout and tenascin-R/parvalbumin double knockout mice. This alteration reflects the loss of phosphacan and neurocan from cortical nets in mice deficient in tenascin-R. No effect on the membrane related cytoskeleton, as revealed by ankyrin(R), was observed in any of the mice. These results on mice lacking tenascin-R substantiate in vitro interactions between tenascin-R and phosphacan and neurocan (Haunsø et al. 2000) (Table 38.2).

Table 38.1 Structure and expression of lecticans (Hyalectans) in central nervous system (CNS)[a]

Name	Type	Core protein size (kD) calculated/SDS-PAGE	GAG type number		Cell origin in CNS	CNS specificity	Sources out side CNS
Brevican	ECM-PG	97	145	CS 0–5	Glial cell/neurons	yes	
GPI-Brevican	GPI-PG	64	140/125	CS 0–5	Glial cells		
Neurocans	ECM-PG	133	245	CS 3	Neurons/Astrocytes	yes	
Versican V0	ECM-PG	370	~550	CS 17–23	Astrocytes	No	Blood vessels; mesenchymes, etc.
Versican V1	ECM-PG	262	500	CS 12–15	Astrocytes	No	Blood vessels; mesenchymes, etc.
Versican V2	ECM-PG	180	400	CS 5–8	Oligodendrocytes	Yes	
Aggrecan	ECM-PG	234	370	CS ?	Neurons	No	Cartilage, notochord

[a]Bbandtlow and Zzimmermann (2000)

Table 38.2 Ligands of proteoglycans in central nervous system[a]

Name	Growth factors	ECM molecules	Cell surface molecules
Brevican		Hyaluronan; tenacin-R	Sulfatides (glycolipids)
Neurocan	FGF2	Hyaluronan; tenacin-C; Tenacin-R	N-CAM; Ng-CAM/L1 Tag-1/axonin, sulfatides
Versicans		Hyaluronan; tenacin-R; Fibronectin	Sulfatides
Aggrecan		Hyaluronan	Sulfatides

FGF fibroblast growth factor, *NCAM* neural cell adhesion molecule, *L1* cell adhesion molecule highly enriched on axon, *Ng-CAM* neuron-glia cell adhesion molecule (a chick homolog of L1)

[a]Bbandtlow and Zzimmermann (2000)

Sulfoglucuronylglycolipids: Two classes of sulfated glycolipids: sulfatides and HNK-1-reactive sulfoglucuronylglycolipids (SGGLs) act as cell surface receptors for brevican. The lectin domain of brevican binds sulfatides and SGGLs in a calcium-dependent manner as expected of a CTLD. The interaction between the lectin domains of lecticans and sulfated glycolipids suggests that lecticans in ECMs serve as substrate for adhesion and migration of cells expressing these glycolipids in vivo (Miura et al. 1999). The Ig chimera of the brevican lectin domain binds to the surface of SGGL-expressing rat hippocampal neurons. The substrate of the brevican chimera promotes adhesion and neurite outgrowth of hippocampal neurons. The full-length brevican also promotes neuronal cell adhesion and neurite outgrowth. Miura et al. (2001) suggest that the interaction between the lectin domain of brevican and cell surface SGGLs acts as a cell recognition system that promotes neuronal adhesion and neurite outgrowth.

Tenascin-C and Tenascin-R as Ligands of Brevican: Tenascin-C (TN-C) is a ligand for brain lecticans and binding site maps on the tenascin molecule to fibronectin type III repeats, which corresponds to the proteoglycan lectin-binding site on tenascin-R (TN-R). In the G3 domain, the C-type lectin is flanked by EGF repeats and a complement regulatory protein-like motif. In aggrecan, these are subject to alternative splicing. The mRNA for this splice variant was shown to be expressed in human chondrocytes. The alternative splicing in the aggrecan G3 domain may be a mechanism for modulating interactions and ECM assembly (Day et al. 2004).

The TN-R and the proteoglycans of the lectican family show an overlapping distribution in developing brain. In TN-R-deficient animals, the perineuronal nets tend to show a granular component within their lattice-like structure at early stages of development and the staining intensity for brevican was reduced in perineuronal nets, extremely low for hyaluronan and neurocan, and virtually no for phosphacan. It was indicated that the lack of TN-R initially and continuously disturbs the molecular scaffolding of extracellular matrix components in perineuronal nets (Brückner et al. 2000). Differential patterns of co-expression of proteoglycans, hyaluronan and TN-R in the individual regions and laminae of the hippocampal formation and the inhomogeneous composition of these components suggest that the extracellular matrix is specifically adapted to the functional domains of intrahippocampal connections and afferent fibre systems (Brückner et al. 2003). Zacharias and Rauch (2006) suggest that a complex network of protein-protein interactions within the brain extracellular matrix, as shown for TN-R and lecticans, is important for the fine-regulation of developmental processes such as microprocess formation along the neurite and neurite outgrowth.

Though the C-type lectin domain of brevican also binds TN-R, this interaction is mediated by a protein-protein interaction through the fibronectin type III domains 3–5 of TN-R, independent of any carbohydrate or sulfated amino acid. Interestingly, the lectin domains of versican and other lecticans also bind the same domain of TN-R by protein-protein interactions. In earlier study, the C-type lectin domain of versican was shown to bind TN-R and the interaction was mediated by a carbohydrate-protein interaction. Surface plasmon resonance analysis revealed that brevican lectin has at least a tenfold higher affinity than the other lectican lectins. Reports suggest that C-type lectin domain can interact with fibronectin type III domains through protein-protein interactions, and that brevican is a physiological TN-R ligand in the adult brain (Aspberg et al. 1997; Hagihara et al. 1999; Lundell et al. 2004).

38.2 Neurocan

Neurocan is a multidomain hyaluronan-binding CS-PG with a 136-kDa core protein that is synthesized by neurons and binds to hyaluronic acid, whereas the astroglial proteoglycan phosphacan is an mRNA splice variant representing the entire extracellular portion of a receptor-type protein tyrosine phosphatase. The concentration of neurocan in brain increases during late embryonic development followed by a steep decline during early postnatal period together with hyaluronan. Neurocan also undergoes extensive proteolytic processing during the course of brain development (Miller et al. 1995). Mouse brain neurocan, but not brevican, is retained on a heparin affinity matrix (Feng et al. 2000). Due to its inhibition of neuronal adhesion and outgrowth in vitro and its expression pattern in vivo it was suggested to play an important role in axon guidance and neurite growth. Neurocan of brain is developmentally regulated with respect to its concentration, carbohydrate composition, sulfation, and localization. Although the possible neuroprotective involvement of CS-PG remains to be investigated, Mizuno et al. (2008) suggested that both the reactive astrocytes and the differential accumulation of CS-PGs may create a nonpermissive microenvironment for neural regeneration in neurodegenerative diseases such as amyotrophic lateral sclerosis (ALS).

38.2.1 Cellular Sites of Synthesis

Neurocan mRNA was evident in neurons, including cerebellar granule cells and Purkinje cells, and in neurons of the hippocampal formation and cerebellar nuclei. The distribution of neurocan message is more wide spread, extending to the cortex, hippocampal formation, caudate putamen, and basal telencephalic neuroepithelium. Neurocan mRNA is present in both the ependymal and mantle layers of the spinal cord but not in the roof plate. The presence of neurocan mRNA in areas where the proteoglycan is not expressed suggests that the short open reading frame in the 5′-leader of neurocan may function as a cis-acting regulatory signal for the modulation of neurocan expression in the developing CNS (Engel et al. 1996). Ihara's epileptic rat (IER) is an animal model of temporal lobe epilepsy with mycrodysgenesis that exhibits abnormal migration of hippocampal neurons and recurrent spontaneous seizures. It was suggested that the insufficient expression of neurocan may affect the development of neuronal organization in the hippocampus, and that the remodeling of ECM in the dentate gyrus may contribute to the mossy fiber sprouting into the inner molecular layer (Kurazono et al. 2001).

38.2.2 Characterization

The composite sequence of overlapping cDNA clones is 5.2 kb long, including 1.3 kb of 3′-untranslated sequence and 76 bp of 5′-untranslated sequence. An ORF of 1257 amino acids encodes a protein with a molecular mass of 136 kDa containing ten peptide sequences present in the adult and/or early postnatal brain proteoglycans. The deduced amino acid sequence revealed a 22-amino acid signal peptide followed by an immunoglobulin domain, tandem repeats characteristic of the hyaluronic acid-binding region of aggregating proteoglycans, and an RGDS sequence. The C-terminal portion (amino acids 951–1215) has approximately 60% identity to regions in the C termini of the fibroblast and cartilage proteoglycans, versican and aggrecan, including two EGF-like domains, a lectin-like domain, and a complement regulatory protein-like sequence. The proteoglycan contains six potential N-glycosylation sites and 25 potential threonine O-glycosylation sites. In the adult form of the proteoglycan (which represents the C-terminal half of neurocan) a single 32-kDa chondroitin 4-sulfate chain is linked at serine-944, whereas three additional potential chondroitin sulfate attachment sites are present in the larger proteoglycan species. The 1D1 (one of the CS-PG) proteoglycan of adult brain, containing a 68-kDa core protein, is generated by a developmentally regulated in vivo proteolytic processing of 136-kDa species which is predominant in early postnatal brain (Rauch et al. 1992).

Neurocan gene of 25 kb contains the coding sequence for the mRNA on 15 exons. The exon-intron structure reflected the structural organization of neurocan, which is a multidomain protein belonging to proteoglycan family. All introns between conserved modular domains are phase I introns. Primer extension experiments indicate a transcriptional start point 28 bases downstream of a consensus TATA sequence. Analysis of 1 kb of 5′ flanking sequence revealed in addition to AP1, AP2, and SP1 consensus binding sites multiple E-box elements and a glucocorticoid responsive element. Single-strand conformation polymorphism mapped neurocan to chromosome 8 between the microsatellite markers D8Mit29 and D8Mit78. The multidomain structure and the preferential expression of neurocan in the brain suggest its involvement in neurological disorders (Rauch et al. 1995).

The coding sequence of human neurocan mRNA known as CS-PG3 was described by Prange et al. (1998). Sequence homology searches indicated close homology to the mouse and rat neurocan. Neurocan mRNA is a brain-specific transcript of approx. 7.5 kb. A longer cDNA clone, GT-5 was mapped to the physical map of chromosome 19. Full coding sequence of the mRNA indicates a 3963 bp ORF corresponding to a 1321 amino acid protein, similar to protein in mouse and rat. The amino acid sequence of human neurocan shows 63% identity with both the mouse and rat sequences. The genomic structure of the gene spans approx. 41 kb, and is transcribed in the telomere to centromere orientation.

Two Proteoglycan Fragments of Neurocan: In adult rat brain, neurocan is completely cleaved into some proteoglycan fragments including the C-terminal half known as neurocan-C and an N-terminal fragment with a 130 kDa core glycoprotein (neurocan-130). Results suggested that not only the C-terminal half (CSPG-150) but also the N-terminal half (CSPG-130) of CSPG-220 existed in a CSPG form in rat brain. The contents of CSPG-130 and CSPG-150 in the rat brain reached maximum levels around the time of birth. Both CSPG-130 and 150 were observed when CSPG-220 was hardly detectable in extracts from the adult rat brain (Matsui et al. 1994). The two neurocan-derived CS-PGs are distributed in adult rat cerebrum. Neurocan-130 exhibits pericellular localization around a subset of neurons in addition to diffuse distribution in the neuropil where as, neurocan-C was distributed only diffusely in the neuropil. The neurocan-130 was mainly localized in the cytoplasm of glial cell processes, the so-called glial perineuronal net, encompassing the cell bodies of certain neurons. The presence of neurocan-130 in a limited number of glial cells may reflect some functional heterogeneity of the glia (Matsui et al. 1998). Under electron microscopy neurocan shows two globular domains interconnected by an extended flexible filament of 60–90 nm. Electron microscopic shapes indicated a mucin-like character for the central neurocan region (Retzler et al. 1996b). Neurocan has been shown to bind to the neural cell adhesion molecule N-CAM and to inhibit its homophilic interaction (Retzler et al. 1996a).

The full-length neurocan is detected in juvenile brains but not in adult brains. However, full-length neurocan transiently appeared in adult rat hippocampus when it was lesioned by kainate-induced seizures. The full-length neurocan was detectable also in the adult brain when it was exposed to mechanical incision or epileptic stimulation. The full-length neurocan transiently appeared in the peri-ischemic region of transient middle cerebral artery occlusion (tMCAO) in adult rat. The induction of neurocan expression by reactive astrocytes suggested that juvenile-type neurocan plays some roles in brain repair (Matsui et al. 2002; Deguchi et al. 2005).

Neurocan-Like and 6B4 Proteoglycan: The neurocan-like and 6B4 proteoglycan-like reactivities in rat embryo were investigated from E10.5–15.5. It is likely that neurocan serves as a barrier molecule to regulate the direction of axonal growth from the dorsal root ganglia. By contrast, results suggested possible functions of 6B4 proteoglycan in rat embryo (Katoh-Semba et al. 1998). Within first week postnatally, neurocan expression is strongly downregulated. But it is re-expressed in areas of axonal growth (sprouting) after brain injury as found in the denervated fascia dentata of the rat after entorhinal cortex lesion. In the denervated zone, neurocan-positive astrocytes were confirmed by electron microscopy. In contrast to the situation during development, astrocytes, but not neurons, express neurocan and enrich the ECM with this molecule. Similar to the situation during development it is suggested that neurocan acts to maintain the boundaries of the denervated fascia dentata after entorhinal cortex lesion (Haas et al. 1999).

38.2.3 Ligand Interactions

Neurocan can bind to various structural extracellular matrix components, such as hyaluronan, heparin, tenascin-C and tenascin-R, and the growth and mobility factors FGF-2, HB-GAM, and amphoterin. Neurocan can also interact with several cell surface molecules, such as N-CAM, L1/Ng-CAM, TAG-1/axonin-1, and an N-cadherin-binding N-acetyl-galactosamine-phosphoryl-transferase. In vitro studies have shown that neurocan is able to modulate the cell-binding and neurite outgrowth promoting activites of these molecules. Binding of phosphacan and neurocan to intact tenascin-C, and of phosphacan to the fibrinogen globe, is significantly increased in presence of calcium. These interactions seem to be mediated by the proteoglycan core proteins rather than through the glycosaminoglycan chains (Milev et al. 1997). A divalent cation-dependent interaction between the COOH-terminal domain of neurocan and those fibronectin type III repeats is substantially involved in the binding of neurocan to tenascin-C (Rauch et al. 1997). Neurocan like phosphacan is also high affinity ligand of amphoterin and HB-GAM (K_D = 0.3–8 nM), two heparin-binding proteins that are developmentally regulated in brain and functionally involved in neurite outgrowth (Milev et al. 1998a). Analysis of the molecular structures and substructures involved in homophilic and heterophilic interactions of these molecules and complementary loss-of-function mutations might shed some light on the roles played by neurocan and interacting molecules in the fine tuning of the nervous system (Rauch et al. 2001; Rauch 2004). Treatment with phosphacan and neurocan on poly-L-lysine (PL) significantly impaired both neuronal attachment and neurite extension, and both phosphacan and neurocan are repulsive substrata for adhesion and neurite regeneration of adult rat dorsal root ganglion (DRG) neurons in vitro (Sango et al. 2003). Li et al. (2005) examined the formation of thalamocortical pathway in the cerebral neocortex of normal and reeler mutant mice. Results support the hypothesis that a heterophilic molecular interaction between L1 and neurocan is involved in determining the thalamocortical pathway within the neocortical anlage.

38.3 Brevican

Brevican has the smallest core protein among aggrecan/versican/neurocan family and is one of the most abundant CS-PG in the adult brain. Expression of brevican is highly specific in the brain and increases as the brain develops. Brevican may play a role in maintaining the extracellular environment of mature brain as a major constituent of the adult brain ECM. Because of smallest core protein, this novel proteoglycan has been named "brevican" (from the Latin word *brevis* meaning "short") (Fig. 38.1).

38.3.1 Characterization

Brevican carries chondroitin sulfate chains and, like other members of the family, contains a hyaluronic acid-binding domain in its N-terminal region, an EGF-like repeat, a lectin-like and a complement regulatory protein-like domain in its C-terminal region (see Fig. 37.4). In contrast, the central region of brevican is much shorter than that of aggrecan, versican, or neurocan, and shows little homology with these proteoglycans. Brevican core protein exists as a 145 kDa full-length form and a 80 kDa N terminally truncated form. A significant amount of brevican devoid of any glycosaminoglycan chains is present in the brain, indicating that brevican is a "part-time" proteoglycan (Yamada et al. 1994).

Two sets of overlapping Brevican cDNA clones from rat brain have been characterized that differ in their 3′-terminal regions. Each set was found to hybridize with two brain-specific transcripts of 3.3 and 3.6 kb. The 3.6-kb transcript encodes a polypeptide that exhibits 82% sequence identity with bovine brevican and is thought to be the rat ortholog of brevican. The polypeptide deduced from ORF of 3.3-kb transcript is truncated just carboxyl-terminal of the central domain of brevican and instead contains a putative glypiation signal. Both soluble and membrane-bound brevican isoforms exist. Isoforms of brevican are indeed glycosylphosphatidylinositol-anchored to the plasma membrane. Brevican is widely distributed in the brain and is localized extracellularly. During postnatal development, amounts of both soluble and phosphatidylinositol-specific phospholipase C-sensitive isoforms increase, suggesting a role for brevican in the terminally differentiating and the adult nervous system (Seidenbecher et al. 1995).

38.3.2 Murine Brevican Gene

The mouse genomic sequence of 13,700 nt revealed that the murine gene has a size of approximately 13 kb and contains the sequence of mRNA for the secreted brevican isoform on 14 exons. The exon-intron structure reflected the structural organization of the multidomain protein brevican. No consensus TATA sequence was found upstream of the first exon, and RNase protection experiments revealed multiple transcriptional start sites for the brevican gene. The first part of the sequence of intron 8 corresponded to an alternative brevican cDNA, coding for a GPI-linked isoform. Single strand conformation polymorphism analysis mapped the brevican gene (*Bcan*) to chromosome 3 between the microsatellite markers D3Mit22 and D3Mit11 (Rauch et al. 1997).

Site of Expression: In rat cerebellar cortex, brevican occurs predominantly in the protoplasmic islet in the internal granular layer after the third postnatal week. Brevican is localized in close association with the surface of astrocytes that form neuroglial sheaths of cerebellar glomeruli where incoming mossy fibers interact with dendrites and axons from resident neurons. Studies indicate that brevican is synthesized by astrocytes and retained on their surface by an interaction involving its core protein. Purified brevican inhibits neurite outgrowth from cerebellar granule neurons in vitro, an activity that requires chondroitin sulfate chains. The brevican present on the surface of neuroglial sheaths may be controlling the infiltration of axons and dendrites into maturing glomeruli (Yamada et al. 1997).

Ogawa et al. (2001) demonstrated that brevican is expressed by both oligodendrocytes and white matter astrocytes in the fimbria, but the expression of brevican in these two glial cell types is differently regulated during development. At P14, brevican immunoreactivity was observed throughout the fimbria, with particularly strong immunoreactivity in the developing interfascicular glial rows. In contrast, the expression in astrocytes started around P21 as oligodendrocytes began to down-regulate the expression. In the adult fimbria, brevican expression was restricted to astrocytes. The secreted brevican transcript was detectable in the pituitary of both male and female adult rats. In posterior lobe of pituitary, pituicytes were heavily labelled. In anterior and intermediate lobes of pituitary, signals for brevican transcripts were observed in cells of various sizes (Dong et al. 2004).

Brevican is expressed by neuronal and glial cells, and as a component of the perineuronal nets it decorates the surface of large neuronal somata and primary dendrites. One brevican isoform harbors a glycosylphosphatidylinositol anchor attachment site and is indeed glypiated in stably transfected HEK293 cells as well as in oligodendrocyte precursor Oli-neu cells. The major isoform is secreted into the extracellular space, although a significant amount appears to be tightly attached to the cell membrane. Brevican is most prominent in the microsomal, light membrane and synaptosomal fractions of rat brain membrane preparations. The association with the particulate fraction is in part sensitive to chondroitinase ABC and phosphatidylinositol-specific phospholipase C treatment. Furthermore, brevican staining on the surface of hippocampal neurons in culture is diminished after hyaluronidase or chondroitinase ABC treatment. Results provide a mechanism by which perineuronal nets are anchored on neuronal surfaces (Seidenbecher et al. 2002). Brevican, TN-R and phosphacan are present at the nodes of Ranvier on myelinated axons with a particularly large diameter in CNS of rat brain.

Transcripts for Secreted and GPI-anchored Brevican Are Differentially Distributed in Rat Brain: In contrast to the other family members, brevican occurs both as soluble isoforms secreted into the extracellular space and membrane-bound isoforms which are anchored to the cell surface via a glycosylphosphatidylinositol (GPI) moiety. Expression of both variants, which are encoded by two differentially processed transcripts from the same gene, is confined to the nervous system. Whereas the 3.6-kb transcript encoding secreted brevican displays a widespread distribution in grey matter structures, including cerebellar and cerebral cortex, hippocampus and thalamic nuclei with silver grains accumulating over neuronal cell bodies, the smaller transcript (3.3 kb) encoding GPI-anchored isoforms appears to be largely confined to white matter tracts and diffusely distributed glial cells. During ontogenetic development, both brevican transcripts are generally up-regulated. However, the expression of glypiated brevican is delayed by about 1 week, compared with the expression of the secreted isoform (Seidenbecher et al. 1998).

38.4 Hyaluronan: Proteoglycan Binding Link Protein

Hyaluronan (HA) is a ubiquitous component of extracellular matrices, and in several systems it plays a central role in regulating cellular proliferation and differentiation. Cell or tissue-specific functions of HA are likely to be mediated by cell or tissue-specific HA-binding proteins (Meyer-Puttlitz et al. 1995). The HA plays an important role in tissue reorganization in response to injury. In the mammalian CNS, HA is present throughout development and into adulthood. Link proteins are glycoproteins in cartilage that are involved in the stabilization of aggregates of proteoglycans and hyaluronic acid Link protein (LP), an extracellular matrix protein in cartilage, stabilizes aggregates of aggrecan and hyaluronan, giving cartilage its tensile strength and elasticity. targeted mutations in mice in the gene encoding LP (Crtl1) showed defects in cartilage development and delayed bone formation with short limbs and craniofacial anomalies. Thus LP is important for the formation of proteoglycan aggregates and normal organization of hypertrophic chondrocytes (Watanabe et al. 1999) (Fig. 37.1).

38.4.1 Cartilage link protein 1 and Brain Link Proteins

Spicer et al. (2003) described a vertebrate hyaluronan and proteoglycan binding link protein gene family (HAPLN), consisting of four members including cartilage link protein CRTL1. Human and mouse link proteins share 81–96% amino acid sequence identity. Two of the four link protein genes (HAPLN2 and HAPLN4) are restricted to the brain/CNS, while one HAPLN3 is widely expressed. Genomic structures revealed that all four HAPLN genes are similar in exon-intron organization and also similar in genomic organization to the 5′ exons for the CS-PG core protein genes. The HAPLN genes expressed by most tissues suggest the fundamental importance of the hyaluronan-dependent extracellular matrix to tissue architecture and function in vertebrates.

The CRTL1 stabilizes aggregates of aggrecan and hyaluronan in cartilage (Czipri et al. 2003). The aggrecan forms link protein-stabilized complexes with HA, via its N-terminal G1-domain that provides cartilage with its load bearing properties. Similar aggregates, in which other CS-PGs (i.e. versican, brevican, and neurocan) substitute for aggrecan, may contribute to the structural integrity of many other tissues including skin and brain (Seyfried et al. 2005). A human brain link protein-1 (BRAL1) is predominantly expressed in brain (Hirakawa et al. 2000). The predicted ORF of human *Bral1* encodes a polypeptide of 340 amino acids containing three protein modules: Ig-like fold and proteoglycan tandem repeat 1 and 2 domains, with an estimated mass of 38 kDa. Brain link protein-1 functions to stabilize the binding between hyaluronan and brevican. The deduced amino acid sequence of human BRAL1 exhibited 45% identity with human CRTL1. The immunoreactivity of BRAL1 was predominantly observed in myelinated fiber tracts in the adult brain and could be detected at P20 in the white matter of the developing cerebellum. Furthermore, BRAL1 colocalized with the versican V2 isoform at the nodes of Ranvier. BRAL1 may play a pivotal role in the formation of the hyaluronan-associated matrix in the CNS that facilitates neuronal conduction by forming an ion diffusion barrier at the nodes (Oohashi et al. 2002).

The *Bral2* gene is predominantly expressed in brain. The *Bral2* mRNA expression is first detected at P20 and continued through adulthood. In situ hybridization revealed that BRAL2 is synthesized by these neurons themselves, especially by the GABAergic neurons in the cerebellar cortex. The colocalization and synergic importance of BRAL2 and brevican in the perineuronal nets is indicated by the analysis using wild-type and brevican-deficient mouse brain, which suggests that BRAL2 is involved in the formation of ECM contributing to perineuronal nets and facilitate the understanding of a functional role of these ECMs (Bekku et al. 2003).

38.4.2 Brain Enriched Hyaluronan Binding (BEHAB)/Brevican

While the functions of HA are mediated by HA-binding proteins, Hockfield S and co-workers reported the characterization of a cDNA with a high degree of sequence homology to members of the proteoglycan tandem repeat (PTR) family of HA-binding proteins. Unlike other HA-binding proteins, the expression of this cDNA is restricted to the CNS. Hockfield S and co-workers proposed the name BEHAB, Brain Enriched Hyaluronan Binding protein, for this gene. The expression of BEHAB mRNA is developmentally regulated. In the embryo, BEHAB is expressed at highest levels in mitotically active cells. The sequence of BEHAB suggests that the encoded protein is functionally important. The size and sequence of BEHAB are consistent with the possibility that it could serve a function like link protein, stabilizing interactions between HA and brain proteoglycans. In rat and cat brain, BEHAB is expressed at very high levels in ventricular zones throughout the neuraxis. Expression is first detected at embryonic day 15 (E15) in the spinal cord, and is detected at progressively more rostral levels at later ages (Jaworski et al. 1996). There exist several BEHAB/brevican isoforms, each of which may mediate different functions. BEHAB/brevican has been cloned from bovine, mouse, rat and human. Two isoforms have been reported: a full-length isoform that is secreted into ECM and a shorter isoform with a sequence that predicts a glycophosphatidylinositol (GPI) anchor. The BEHAB/brevican gene maps to human chromosome 1q31 (Gary et al. 2000).

Glial tumors, gliomas, are the most common primary intracranial tumors. Their distinct ability to invade the normal surrounding tissue makes them difficult to control and nearly impossible to completely remove surgically. The ECM can modulate, in part, the permissiveness of a tissue to cell movement. One ECM molecule that shows dramatic upregulation in gliomas is BEHAB/brevican. The BEHAB/brevican gene is consistently expressed by human glioma and is not expressed by tumors of nonglial origin (Gary and Hockfield 2000; Jaworski et al. 1996; Zhang et al. 1998). BEHAB/brevican expression is also upregulated during periods of increased glial cell motility in development and following brain injury. Experimental evidence suggests that in glioma, in addition to upregulation of BEHAB/brevican, proteolytic processing of the full-length protein also may contribute to invasion. Reports suggest that up-regulation and proteolytic cleavage of BEHAB/brevican increases significantly the aggressiveness of glial tumors. It will be important to determine the therapeutic potential of inhibiting BEHAB/brevican cleavage in gliomas (Nutt et al. 2001).

38.5 Proteolytic Cleavage of Brevican

BEHAB/brevican can be cleaved into an N-terminal fragment that contains a hyaluronan-binding domain (HABD) and a C-terminal fragment (Yamada et al. 1995). The BEHAB/brevican protein is cleaved in invasive human and rodent gliomas. Matthews et al. (2000) suggested a function for ADAMTS family members in BEHAB/brevican cleavage and glioma and indicated that inhibition of ADAMTS in glioma might provide a novel therapeutic strategy (Viapiano et al. 2008). ADAMTS (a disintegrin and metalloproteinase with thrombo- spondin motifs) proteases are complex secreted enzymes containing a prometalloprotease domain of the reprolysin type attached to an ancillary domain with a highly conserved structure that includes at least one thrombospondin type 1 repeat (Lemons et al. 2001). ADAMTS proteases are synthesized as zymogens, with constitutive proprotein convertase removal of the propeptide occurring prior to secretion. Their enzymatic specificity is heavily influenced by their ancillary domain, which plays a critical role in directing these enzymes to their substrates, the cell surface and the extracellular matrix. Known functions of ADAMTS proteases include (1) processing of procollagens N-proteinase; (2) cleavage of the matrix proteoglycans aggrecan, versican and brevican; (3) inhibition of angiogenesis; and (4) blood coagulation homoeostasis as the von Willebrand factor cleaving protease. ADAMTS-1, -4 and -5 are present in CNS, which are able to cleave the aggregating CS-PG, aggrecan, phosphacan, neurocan and brevican. ADAMTS-4 mRNA and protein expression was decreased. Changes in ADAMTS expression during the course of CNS inflammation may contribute to ECM degradation and disease progression (Cross et al. 2006). ADAMTS4 that selectively cleaves lecticans was detected in cultures of neurons, astrocytes and microglia (Hamel et al. 2005). The ADAMTS-induced cleavage of brevican in the dentate outer molecular layer is closely associated with diminished synaptic density, and may, therefore, contribute to synaptic loss and/or reorganization in this region (Mayer et al. 2005; Porter et al. 2005).

In human glioblastomas ADAMTS4 and ADAMTS5 (aggrecanases 1 and 2) are expressed in considerable amounts. ADAMTS5 expression is confined to proliferating glioblastoma cells of surgical tumor sections and with lower intensity to astroglial cells in normal brain sections, as opposed to brevican. Selective cleavage of the Glu^{395}-Ser^{396} bond of brevican in adult brain tissues is thought to be important for glioma cell invasion. ADAMTS-4 has such an activity (Nakada et al. 2005). In vitro, glioblastoma-derived ADAMTS5 degrades recombinant human brevican to several smaller fragments. Results show that ADAMTS4 and 5 are upregulated on proliferating glioblastoma cells and these proteases may contribute to their invasive potential (Held-Feindt et al. 2006). ADAMTS-5 is capable of degrading brevican and is overexpressed in glioblastoma cells, and that ADAMTS-5 may play a role in glioma cell invasion through the cleavage of brevican (Nakada et al. 2005; Porter et al. 2005).

38.6 Proteoglycans in Sensory Organs

Various proteoglycans are expressed in ocular tissues. The TGF-β2 regulates the expression of proteoglycans in aqueous humor. The expression of CS in stroma was upregulated at early postnatal stages and reduced during development in rat eyes. Decorin is expressed in trabecular meshwork tissues. In retinal tissues, neurocan and phosphacan were expressed mainly in nerve fiber-rich layers during rat postnatal stages. Soluble extracellular proteoglycans in corneal and trabecular meshwork tissues contribute to the stromal transparency in the corneal tissues and the resistance of the aqueous humor outflow in trabecular meshwork tissues (Inatani and Tanihara 2002).

CS and HS proteoglycans are the major constituents expressed in and secreted by retinal tissue. Soluble HS proteoglycans are found in extracellular matrices of the basement membrane, whereas HS proteoglycans with their membrane-binding domain are localized primarily in the neurites of retinal neuronal cells, indicating their role as receptors for cytokines. The expression in the nerve fiber-rich layers of several CS proteoglycans, such as neurocan and phosphacan, is restricted in the nervous tissues, and is upregulated as retinal development proceeds, then decreases after maturation of the retina. In vitro data suggest that these proteoglycans regulate axon guidance and synapse formation during the development of nervous tissue. In contrast, in adult vertebrate retina, the IPM is a rich source of CS proteoglycans (Inatani and Tanihara 2002). Co-localization of the molecule with TN-R in the retina and optic nerve suggested a functional relationship between TN-R and phosphacan-related CS proteoglycan in vivo (Inatani et al. 2001; Xiao et al. 1997).

Aggrecan and versican were first seen at E16 in the optic nerve and retina of rat, whereas neurocan was not detected in the embryonic eye. At postnatal day P0, β-versican is largely confined to the inner plexiform layer whereas α-versican is also apparent in the neuroblastic layer. Both aggrecan and, much more weakly, neurocan immunoreactivity is present throughout the neonatal retina. Aggrecan and α-versican are also present throughout the optic nerve and disk, whereas β-versican and neurocan are confined to the laminar beams of the optic nerve (Popp et al. 2004). In mouse retinofugal pathway, phosphacan but not neurocan is likely the major carrier of the CS glycosaminoglycans that play crucial functions in axon divergence and age-related axon ordering in the mouse optic pathway. These two proteoglycans regulate axon growth and patterning not only through the sulfated sugars but also by interactions of the protein parts with guidance molecules on the optic axons (Leung et al. 2004).

Accumulation of neurocan in association with the retinal vasculature does not correlate with photoreceptor cell loss, because it was not observed in the rhodopsin mutant rats. During the earliest stages of the disease, accumulation of debris in the subretinal space in RCS rats with dystrophic retinas may be sufficient per se to initiate a cascade of metabolic changes that result in accumulation of neurocan. With time, the neurocan accumulated perivascularly may, by interaction with other matrix molecules, modulate at least some of the vascular alterations observed in rat animal model (Zhang et al. 2003). In transient retinal ischemia the intensity of the 220-kDa band as well as the 150-kDa band increased markedly after reperfusion. The neurocan expression by Müller cells suggests that this proteoglycan plays a role in the damage and repair processes in diseased retina (Inatani et al. 2000).

Neurocan was conspicuous in the medial and lateral superior olivary nuclei and much less stained in the cochlear nucleus and posterior colliculus. Different locations of CS-PG, including neurocan, may be associated with focal sites composed of neuronal surface, terminal boutons and ECM in the lower auditory tract of the adult dog (Atoji et al. 1997).

Neurocan was first detected in primary olfactory neurons at E11.5. Neurocan was expressed by primary olfactory axons as they extended toward the rostral pole of the telencephalon as well as by their arbors in glomeruli after they contacted the olfactory bulb. Being a strong promoter for neurite outgrowth neurocan supports the growth of primary olfactory axons through ECM. Phosphacan, unlike neurocan, was present within the mesenchyme surrounding the E11.5 and E12.5 nasal cavity. This expression decreased at E13.5, concomitant with a transient appearance of phosphacan in nerve fascicles. Within the embryonic olfactory bulb, phosphacan was localised to the external and internal plexiform layers. The spatiotemporal expression patterns of neurocan and phosphacan indicate that these CS-PGs have diverse in situ roles, which are dependent on context-specific interactions with extracellular and cell adhesion molecules within the developing olfactory nerve pathway (Clarris et al. 2000). Glycosaminoglycan chains, such as CS and KS, and core proteins of the CS proteoglycan, neurocan and phosphacan, were barely detected in the migratory pathway from the olfactory placode. N-syndecan mRNA was localized in virtually all of migrating neurons as well as in cells of the olfactory epithelium and the vomeronasal organ. It is likely that a heterophilic interaction between NCAM and N-syndecan participates in the neuronal migration from the rat olfactory placode (Toba et al. 2002).

38.7 Functions of Proteoglycans in CNS

38.7.1 Functions of Chondroitin Sulphate Proteoglycans

In the adult, CS-PGs are generally thought to be inhibitory for tissue plasticity and to have a barrier function during regeneration (Bradbury et al. 2002). However, in the developing CNS, CS-PGs show a functional dualism: depending on the neuronal cell type and the way they are presented to neurons, both growth inhibitory and growth promoting effects have been described (Emerling and Lander 1996; Akita et al. 2004; Wu et al. 2004). The inhibitory properties of CS-PGs have mostly been attributed to their glycosaminoglycan (GAG) chains (Asher et al. 2002; Sivasankaran et al. 2004), however, the core protein may also act as an inhibitor (Monnier et al. 2003; Ughrin et al. 2003). In regions containing low levels of adhesion molecules, various CS-PGs may act as barriers to cell migration and axonal growth. In regions containing high levels of adhesion proteins, brain CS-PGs may still act to maintain certain boundaries while allowing selective axonal extension to proceed. There are numerous regions of overlap in the expression patterns of CS-PGs and adhesion molecules in vivo, and the relative levels of these molecules as well as the organization of the ECM may be important factors that regulate the rate of axonal growth locally. Differential expression of CS-PGs may be important for modulating cell adhesion as well as axonal growth and guidance during neural development, and continued expression may prevent these processes in normal mature nervous system as well as following brain injury (Margolis et al. 1996). The swift assembly and remodeling of ECMs have been associated with axonal guidance functions in the periphery and with the structural stabilization of myelinated fiber tracts and synaptic contacts in the maturating CNS. Particular interest has been focused on the putative role of CS-PG in suppressing CNS regeneration after lesions. The axon growth inhibitory properties of several of these CS-PGs in vitro and the partial recovery of structural plasticity in lesioned animals treated with chondroitin sulfate degrading enzymes in vivo have significantly contributed to the increased awareness of this long time neglected structure.

Little is known about the contribution of CS-PGs to regeneration failure in vivo at dorsal root entry zone (DREZ), a CNS region that blocks regeneration of sensory fibers following dorsal root injury without glial scar formation. It is suggested that the proteoglycans (Brevican, Neurocan, Versican V1, and Versican V2) are abundant in the DREZ at the time regenerating sensory fibers reach the PNS/CNS border and may therefore participate in growth-inhibition in this region (Beggah et al. 2005). Quaglia et al. (2008) demonstrated that (1) neurocan and/or brevican contribute to the non-permissive environment of the DREZ several weeks after lesion and that (2) delayed stimulation of the growth program of sensory neurons can facilitate regeneration across DREZ provided its growth-inhibitory properties are attenuated. Post-injury enhancement of the intrinsic growth capacity of sensory neurons combined with removal of inhibitory CS-PG may therefore help to restore sensory function and thus attenuate the chronic pain resulting from human brachial plexus injury.

In areas of denervation the role of CS-PGs is less clear, since they are enriched in zones of sprouting, i.e. zones of axonal growth. However, results reveal a temporal pattern in CS-PG mRNA expression in the denervated fascia dentata. This suggests specific biological functions for CS-PGs during the denervation-induced reorganization process: whereas the early increase in CS-PGs in the denervated zone could influence the pattern of sprouting, the late increase of aggrecan mRNA suggests a different role during the late phase of reorganization (Schäfer et al. 2008).

Functions of Sciatic Nerve-Derived Versican- and Other CS-PGs: Human sciatic nerves contained two chondroitin sulphate proteoglycans of M_r of 130 and 900 kDa. Following chondroitinase ABC treatment of 130 kDa proteoglycan, a core protein of ~45 kDa was shown to react with polyclonal antibodies against the chondroitin-dermatan sulphate proteoglycan decorin from human fibroblasts. Chondroitinase ABC treatment of 900 kDa proteoglycan yielded a core protein with a M_r of ~400 kDa. Observations suggest two proteoglycans from human sciatic nerve are immunochemically related to chondroitin sulphate proteoglycans versican and decorin that may contribute to successful regeneration in the peripheral nervous system of mammals. The versican- and decorin-like molecules are involved in cell-extracellular matrix interactions and inhibit adhesion of several cell lines, neonatal dorsal root ganglion neurons and Schwann cells.

Observations suggest that binding of the two proteoglycans to fibronectin is involved in the modulation of adhesion of cells to fibronectin (Braunewell et al. 1995).

38.7.2 Functions of Brevican and Neurocan in CNS

Perineuronal nets (PNN) are specialized ECM structures enwrapping CNS neurons, which are important regulators for neuronal and synaptic functions. Brevican is an integral component of PNN, which appear in hippocampal primary cultures. Brevican is primarily synthesized by co-cultured glial fibrillary acidic protein (GFAP-)-positive astrocytes and co-assembles with its interaction partners in PNN-like structures on neuronal somata and neurites. Both excitatory and inhibitory synapses are embedded into PNN. Furthermore, axon initial segments are strongly covered by a dense brevican coat. Altogether, mature primary cultures can form PNN (John et al. 2006). Digestions with chondroitinase ABC and hyaluronidase indicated that aggrecan, versican, neurocan, brevican, and phosphacan are retained in PNNs through binding to HA. A comparison of the brain and spinal cord ECM with respect to CS-PGs indicated that the PNNs in both parts of the CNS have the same composition (Deepa et al. 2006).

Mice lacking brevican gene are viable and fertile and have a normal life span. Brain anatomy was normal, although alterations in the expression of neurocan were detected. The PNN formed but appeared to be less prominent in mutant than in wild-type mice. Brevican-deficient mice showed significant deficits in the maintenance of hippocampal long-term potentiation (LTP). Detailed behavioral analysis revealed no significant deficits in learning and memory. Results indicated that brevican is not crucial for brain development but has restricted structural and functional roles (Brakebusch et al. 2002). A brevican deficiency resulted in a reorganization of the nodal matrices, which was characterized by the shift of tenascin-R (TN-R), and concomitantly phosphacan, from an axonal diameter-dependent association with nodes to an axonal diameter independent association. Bekku et al. (2009) revealed that brevican plays a crucial role in determining the specialization of the hyaluronan-binding nodal matrix assemblies in large diameter nodes. Brevican associates with astrocytes ensheathing cerebellar glomeruli and inhibits neurite outgrowth from granule neurons.

At molecular level, brevican promotes EGFR activation, increases the expression of cell adhesion molecules, and promotes the secretion of fibronectin and accumulation of fibronectin microfibrils on the cell surface. Moreover, the N-terminal cleavage product of brevican, but not the full-length protein, associates with fibronectin in cultured cells and in surgical samples of glioma. Results provide evidence of the cellular and molecular mechanisms that may underlie the motility-promoting role of brevican in primary brain tumors. In addition, these results underscore the important functional implications of brevican processing in glioma progression (Hu et al. 2008).

Neurocan-deficient mice are viable and fertile and have no obvious deficits in reproduction and general performance. Brain anatomy, morphology, and ultrastructure are similar to those of wild-type mice. Perineuronal nets surrounding neurons appear largely normal. Mild deficits in synaptic plasticity may exist, as maintenance of late-phase hippocampal long-term potentiation is reduced. Results indicate that neurocan has either a redundant or a more subtle function in the development of the brain (Zhou et al. 2001; Okamoto et al. 2001). The distal transected cords of infant rats are more permissive for axon extension than those of adults. Neurocan mRNA was up-regulated in the distal cord of adult rats shortly after transection, followed by a longer wide distribution of neurocan activity in both neurons and astrocytes. By contrast, upregulation of neurocan mRNA was not seen in infant rats, although transient expression of neurocan immunoreactivity was seen in neurons. Combined with the different regenerative capacity of infant and adult rats, it indicates that neurocan inhibits spinal cord regeneration (Qi et al. 2003).

Extracellular Matrix Alterations: During hippocampal development, neurocan is expressed throughout the alveus, neuropil layers, and parts of the dentate gyrus from E16 to P2. The CS-PG phosphacan is expressed primarily in the neuropil layers at postnatal stages. After E18, intense labeling of neurocan was observed in regions of the alveus surrounding L1-expressing axon fascicles. In vitro, axons from brain regions that project through the alveus during development would not grow across CS-PG substrata, in a concentration-dependent manner. In addition, hippocampal axons from dissociated neuron cultures only traveled across CS-PG substrata as fasciculated axon bundles. These results implicate CS-PG in the regulation of axon trajectory and fasciculation during hippocampal axon tract formation (Wilson and Snow 2000).

Watanabe et al. (1995) revealed that neurocan was distributed throughout the cerebral cortex of rat during early postnatal development but was absent from the centres of cortical barrels at the time of entry and arborization of thalamocortical axons. At this developmental stage, expression of neurocan mRNA was shown by in situ hybridization to be down-regulated in the barrel centres. It appears that neuronal stimuli through early thalamocortical fibres from the sensory periphery cause reduced expression of neurocan mRNA in neurocan-producing cells in the presumptive barrel centres.

Further, mice lacking neurocan, brevican, tenascin-R, and tenascin-C brain extracellular matrix molecules were found to be viable and fertile. However, the brains of 1-month-old quadruple KO mice revealed increased levels of fibulin-1 and fibulin-2 with an unusual parenchymal deposition of these molecules. The quadruple KO mice also displayed obvious changes in the pattern of deposition of hyaluronan (Rauch et al. 2005).

Regulation of Cadherin and Integrin Function by Neurocan: Li et al. (2000) showed that the interaction of neurocan with its GalNAcPTase receptor coordinately inhibits both N-cadherin- and β1-integrin-mediated adhesion and neurite outgrowth. The inhibitory activity is localized to an NH_2-terminal fragment of neurocan containing an Ig loop and an HA-binding domain. The coordinate inhibition of cadherin and integrin function on interaction of neurocan with its receptor may prevent cell and neurite migration across boundaries.

The L1/Ng-CAM adhesion molecule of the Ig superfamily is implicated in neural processes including neuronal cell migration, axon outgrowth, learning, and memory formation. It has been suggested that neurocan can block homophilic binding. The sequences involved in neurocan binding are localized on the surface of the first Ig domain and largely overlap with the G-F-C β-strands proposed to interact with the fourth Ig domain during homophilic binding. It was found that the C-terminal portion of neurocan is sufficient to mediate binding to the first Ig domain of L1, and that the sushi domain cooperates with a glycosaminoglycan side chain in forming the binding site for L1 (Oleszewski et al. 2000).

38.7.3 Neurocan in Embryonic Chick Brain

Neurocan has been found in the embryonic avian heart and vasculature. In stage 11 quail embryos, neurocan was prominently expressed in the myocardium, dorsal mesocardium, heart-forming fields, splanchnic mesoderm, and vicinity of the extraembryonic vaculature, and at lower levels in the endocardium. Results suggest that neurocan may function as a barrier that regulates vascular patterning during development (Mishima and Hoffman 2003). Zanin et al. (1999) revealed the presence of aggrecan and versican in stages 12–21 chicken embryo hearts in distinctive spatial and temporal patterns. Versican is found in the myocardium and the myocardial basement membrane. In contrast, aggrecan is specifically colocalized with several groups of migrating cells including endocardial cushion tissue cells, epicardial cells, a mesenchymal cell population in the outflow tract that may be of neural crest origin, and a mesenchymal cell population in the inflow tract. The observations indicated that versican and aggrecan are expressed in unique patterns and that they play different roles in development.

In the chick embryo, aggrecan has a regionally specific and developmentally regulated expression profile during brain development. Aggrecan expression is first detected in chick brain on E7, increases from E7 to E13, declines markedly after E16, and is not evident in hatchling brains. The time course and pattern of aggrecan expression observed in ventricular zone cells suggested that it might play a role in gliogenesis. In aggrecan-deficient model (nanomelic chicks) expression and levels of neurocan and brevican are not affected, indicating a non-redundant role for these members of the aggrecan gene family. Results suggest that aggrecan functions in specification of a sub-set of glia precursors that might give rise to astrocytes in vivo (Schwartz and Domowicz 2004).

38.8 Chondroitin Sulfate Proteoglycans in CNS Injury Response

38.8.1 CS-PGs in CNS in Response to Injury

Injury to the CNS leads to permanent loss of function due to the inability of severed nerve fibers to regenerate back to their targets. Lack of CNS repair has been attributed to several causes, including to extrinsic growth-inhibitory molecules associated with myelin, and to ECM CS-PGs produced by activated glial cells post-injury (Filbin; 2003; Silver and Miller 2004; Zurn and Bandtlow 2006). Treatment with antibodies against the myelin inhibitor Nogo (Schnell and Schwab 1990), with a Nogo-66 receptor antagonist peptide (GrandPre et al. 2002), or with enzymes degrading CS-PGs (Bradbury et al. 2002; Barritt et al. 2006), has provided evidence for the growth-inhibitory function of these molecules. However, the limited ability to regenerate following injury is also due to an intrinsic low capacity of adult axons to grow. Interestingly, the growth capacity of central branches of sensory neurons located in the spinal cord can be increased if their peripheral branches are previously cut (conditioning lesion), i.e. if their axonal growth program is activated before spinal cord lesion (Richardson and Issa 1984; Neumann and Woolf 1999). Yet the timing of this peripheral lesion is critical since peripheral nerve priming delayed by 2 weeks post-lesion fails to promote growth of central sensory axons (Neumann and Woolf 1999). Only the combination of two priming lesions, one at the time of spinal cord injury, and one after 1 week, has been shown to facilitate growth in the same model (Neumann et al. 2005).

Experimental data now indicate that the expression of a number of different CS-PGs is increased following CNS injury. After spinal cord injury, neurocan, brevican, and versican increased within days in injured spinal cord parenchyma surrounding the lesion site and peaked at 2 weeks. Neurocan and versican were persistently elevated for 4 weeks postinjury, and brevican expression persisted for at least 2 months. On the other hand, phosphacan labeling decreased in the same region immediately following injury but later recovered and then peaked after 2 months. Thus, the production of several CS-PG family members is differentially affected by spinal cord injury, overall establishing a CS-PG-rich matrix that persists for up to 2 months following injury. Optimization of strategies to reduce CS-PG expression to enhance regeneration may need to target several different family members over an extended period following injury (Jones et al. 2003a; b). Upregulation of hyalectans (lecticans) neurocan, versican and brevican, plus NG2 and phosphacan following injury have been shown to exhibit inhibitory effects on neurite outgrowth in vitro. It is likely therefore that the increased expression of these molecules contributes to the non-permissive nature of the glial scar. It is important to remember also that not only does the glial scar contain many different inhibitory molecules, but that these are the products of a number of different cells, including not just astrocytes, but also oligodendrocyte progenitor and meningeal cells. It is arguable that the latter two cell types make a greater contribution than astrocytes to the inhibitory environment of the injured CNS. Evaluation of total soluble CS-PGs 2 weeks after dorsal column lesion in the rat expressed after spinal cord injury (SC-I) demonstrated that NG2 is highly upregulated and is a major CS-PG species. NG2 expression is upregulated within 24 h of injury, peaks at 1 week, and remains elevated for at least an additional 7 weeks (Jones et al. 2002). Attempts have been made to alter the CS-PG component of the glial scar in the hope that this will facilitate improved axonal regeneration. Studies have reported an improved regenerative response following treatment of the injured CNS with chondroitinase ABC. The CS-PGs representing a significant source of inhibition within the injured CNS, studies indicate that successful CNS regeneration may be brought about by interventions which target these molecules and/or the cells which produce them(Morgenstern et al. 2002; Chan et al. 2007).

Axonal Regeneration Through Regions of CS-PG Deposition After Spinal Cord Injury: Jones et al. (2003b). examined growth responses of several classes of axons to this inhibitory environment in the presence of a cellular fibroblast bridge in a spinal cord lesion site and after a growth factor stimulus at the lesion site. Analysis showed dense labeling of NG2, brevican, neurocan, versican, and phosphacan at the host-lesion interface after SC-I. NG2 expression also increased after sciatic nerve injury, wherein axons successfully regenerate. Cellular sources of NG2 in SC-I and peripheral nerve lesion sites included Schwann cells and endothelial cells. Notably, these cellular lesion sites produced the cell adhesion molecules L1 and laminin. Thus, axons grow along substrates coexpressing both inhibitory and permissive molecules, suggesting that regeneration is successful when local permissive signals balance and exceed inhibitory signals (Jones et al. 2003b; Tang et al. 2003). Versican V2 protein levels, however, displayed an opposite trend, dropping below unlesioned spinal cord values at all time points studied (Tang et al. 2003).

Haddock et al. (2007) examined changes in brevican and phosphacan, following transient middle cerebral artery occlusion, a model of stroke in rat and their spatial relationship with ADAMTS-4. The co-localization of ADAMTS or its activity indicated a functional role for this matrix-protease pair in degeneration/regeneration processes that occur in stroke.

Effect of Hypoxic-ischemic Brain Injury in Perinatal Rats: CS-PG are involved in the pathologic process of hypoxia-ischemia (H-I) in the neonatal brain. Brevican is secreted by OLs and their progenitors prior to and during active myelination whereas NG2 is produced by early OL progenitors. In neonatal hippocampus of rat after H-I of unilateral carotid artery ligation and exposure to hypoxia, a cavitary infarct involving the ipsilateral parietal and temporal regions of cerebral cortex, hippocampus, and striatum of most rat pups was evident after H-I. The abundance of brevican was reduced in ipsilateral hippocampus after H-I and the total G1 proteolytic fragment of brevican was lower in the ipsilateral hippocampus and the level of a protease-generated brevican fragment was significantly diminished in OL-rich hippocampal fimbria. Hippocampal NG2 levels were also lower after H-I, but were not different from contralateral side at 14 days. The early events in the process could be involved in apoptotic cell death and/or tissue injury (Aya-ay et al. 2005). In contrast to injured adult CNS, the amount of neurocan was reduced after hypoxia in the neonatal hypoxic-ischemic cerebral hemisphere. The amounts of phosphacan and neuroglycan C were also reduced significantly 24 h after hypoxia at the right injured cortex compared to those at the left cortex. But phosphacan was conversely intensified both at 24 h and 8 days after hypoxia at the infarcted area. Hypoxic-ischemic insult may unmask phosphacan epitopes at the injured sites, resulting in intensified immunostaining. Because intensified immunostaining for neurocan and neuroglycan C was not observed, unmasking seems to be specific to phosphacan among these three CS-PG (Matsui et al. 2005).

Versican and brevican are expressed with distinct pathology in neonatal hypoxic-ischemic injury. Unique expression profiles for lecticans after neonatal H-I suggested deposition of brevican at elevated rates in response to progressive gray matter injury, whereas diminished versican expression may be associated with deep cerebral white matter injury (Leonardo et al. 2008).

38.8.2 Glial Scar and CNS Repair

Injury to CNS results in the formation of the glial scar, a primarily astrocytic structure that represents an obstacle to regrowing axons. CS-PG are greatly upregulated in the glial scar. The glial scar, a primarily astrocytic structure bordering the infarct tissue inhibits axonal regeneration after stroke. In this environment, axon regeneration fails, and remyelination may also be unsuccessful. The glial reaction to injury recruits microglia, oligodendrocyte precursors, meningeal cells, astrocytes and stem cells. Damaged CNS also contains oligodendrocytes and myelin debris. Most of these cell types produce molecules that have been shown to be inhibitory to axon regeneration. Oligodendrocytes produce NI250, myelin-associated glycoprotein (MAG), and tenascin-R, oligodendrocyte precursors produce NG2 DSD-1/phosphacan and versican, astrocytes produce tenascin, brevican, and neurocan, and can be stimulated to produce NG2, meningeal cells produce NG2 and other proteoglycans, and activated microglia produce free radicals, nitric oxide, and arachidonic acid derivatives. Most of these molecules participate in rendering the damaged CNS inhibitory for axon regeneration. Demyelinated plaques in multiple sclerosis consist mostly of scar-type astrocytes and naked axons. The extent to which the astrocytosis is responsible for blocking remyelination is not established, but astrocytes inhibit the migration of both oligodendrocyte precursors and Schwann cells which must restrict their access to demyelinated axon (Fawcett and Asher 1999; Buss et al. 2009).

Various reports indicate that Neurocan is up-regulated in the scar region after stroke. Neurocan, expressed in the CNS, exerts a repulsive effect on growing cerebellar axons. Immunocytochemistry revealed neurocan to be deposited on the substrate around and under astrocytes but not on the cells. Astrocytes therefore lack the means to retain neurocan at the cell surface. Findings raise the possibility that neurocan interferes with axonal regeneration after CNS injury. In the spinal cord under various types of injury, reactive gliosis emerges in the lesion accompanied by CS-PG up-regulation. Several types of CS-PG core proteins and their side chains have been shown to inhibit axonal regeneration in vitro and in vivo. The bone morphogenetic proteins (BMPs) stimulate glial scar formation in demyelinating lesions of adult spinal cord. BMP4 and BMP7 increase rapidly at the site of demyelination, and astrocytes surrounding the lesion increase expression of phosphorylated Smad1/5/8. Thus, local increase in BMPs at the site of a demyelinating lesion causes upregulation of gliosis, glial scar formation, and heightened expression of CS-PGs such as neurocan and aggrecan that may inhibit remyelination (Fuller et al. 2007). Bone marrow stromal cells (BMSCs) reduce the thickness of glial scar wall and facilitate axonal remodeling in the ischemic boundary zone. Analysis showed that reactive astrocytes were the primary source of neurocan, and BMSC-treated animals had significantly lower neurocan and higher growth associated protein 43 expression in the penumbral region compared with control rats. Neurocan gene expression was significantly down-regulated in rats receiving BMSC transplantation. Shen et al. (2008) suggest that BMSCs promote axonal regeneration by reducing neurocan expression in peri-infarct astrocytes.

38.9 Other Proteoglycans in CNS

In addition to lecticans, phosphacan, NG2, and testicans as proteoglycans are expressed in CNS (Schnepp et al. 2005). The NG2 chondroitin sulfate proteoglycan, a structurally unique, integral membrane proteoglycan, is found on the surfaces of several different types of immature cells. NG2 is associated with multipotential glial precursor cells (O2A progenitor cells), chondroblasts of the developing cartilage, brain capillary endothelial cells, aortic smooth muscle cells, skeletal myoblasts and human melanoma cells (Levine and Nishiyama 1996). Neuroglycan C (NGC), a brain-specific transmembrane proteoglycan, is thought to bear not only CS but also N- and O-linked oligosaccharides on its core protein. The structure of carbohydrate moiety of NGC is developmentally regulated, and differs from those of neurocan and phosphacan. The developmentally-regulated structural change of carbohydrates on NGC may be partly implicated in the modulation of neuronal cell recognition during brain development (Shuo et al. 2004). Brain-specific CS-PGs, including neurocan, phosphacan/receptor-type protein-tyrosine phosphatase beta, and neuroglycan C, have been detected in the CNS. These CS-PGs are involved in the proliferation of neural stem cells as a group of cell microenvironmental factors (Ida et al. 2006). Xiao et al. (1997) isolated a CS-PG from mouse brain with a fragment of ECM tenascin-R that comprises the amino-terminal cysteine-rich stretch and the 4.5 EGF-like repeats. The 173-kDa core protein of phosphacan is synthesized by glia and represents an extracellular variant of the receptor-type protein tyrosine phosphatase RPTP ζ/β. Keratan sulfate-containing glycoforms of phosphacan (designated phosphacan-KS) are also present in brain. In early postnatal rat cerebellum, neurocan, phosphacan, and phosphacan-KS show the overlapping localization with tenascin, an extracellular matrix protein that modulates cell adhesion and migration. Phosphacan and neurocan are involved in the modulation of cell adhesion and neurite outgrowth during neural development and regeneration (Alliel et al. 2004; Meyer-Puttlitz et al. 1996).

38.10 Regulation of Proteoglycans in CNS

38.10.1 Growth Factors and Cytokine Regulate CS-PGs by Astrocytes

After injury to adult CNS, numerous cytokines and growth factors are released that contribute to reactive gliosis and ECM production. The presence of TGF-β1 and EGF greatly increase the production of several CS-PGs by astrocytes. Treatment of astrocytes with other EGF-receptor (ErbB1) ligands produced increases in CS-PG production similar to those observed with EGF. Treatment of astrocytes, however, with heregulin, which signals through other members of the EGF-receptor family (ErbB2, ErbB3, ErbB4), did not induce CS-PG upregulation. The specificity of activation through the ErbB1 receptor indicated the presence of multiple core proteins containing 4-sulfated or 6-sulfated chondroitin. Further analyses showed that treatment of astrocytes with EGF increased phosphacan expression, whereas treatment with TGF-β1 increased neurocan expression. These results elucidate some of the injury-induced growth factors that regulate the expression of CS-PGs which could be targeted in modulate CS-PG production after injury to the CNS (Smith and Strunz 2005).

ECM alterations in CNS of multiple sclerosis (MS) patients result from blood–brain barrier breakdown, release and activation of proteases, and synthesis of ECM components. In active MS plaque edges, CNS lecticans (versican, aggrecan, and neurocan) and dermatan sulfate PG were increased in association with astrocytosis; in active plaque centers they were decreased in the ECM and accumulated in foamy macrophages, suggesting that these ECM PGs are injured and phagocytosed along with myelin. Results indicate that ECM PG alterations are specific, temporally dynamic, and widespread in MS patients and may play critical roles in lesion pathogenesis and CNS dysfunction (Sobel and Ahmed 2001).

Glutamate activation of excitatory amino acid receptors induces the synthesis and release of PGs with neurite-promoting activity from hippocampal neurones. Both cerebroglycan (CBG), a glycosylphatidylinositol-anchored heparan sulphate PG, and neurocan are expressed in hippocampal neurones. Exposure of hippocampal neurones to 100 μM glutamate resulted in an increase in CBG mRNA levels and an increase in axonal and dendritic length. The increase in CBG mRNA levels following glutamate exposure was mediated via both N-methyl-D-aspartate and metabotropic receptor activation (Wang and Dow 1997).

38.10.2 Decorin Suppresses PGs Expression

Decorin transcripts express in brain on postnatal day 3 followed by a slow decline to the lower adult level. Postnatal decorin expression was observed in the grey matter of neocortex, hippocampus and thalamus, in myelinated fibre tracts and in several mesenchymal tissues (blood vessels, pia mater and the choroid plexus). In the neocortex, decorin is expressed in a specific laminar pattern with intense staining of the cortical plate and its derivatives, which differs remarkably from the distributions observed for other proteoglycans (Miller et al. 1995). Decorin seems to serve yet unknown functions in the developing rat brain parenchyma in addition to its well-established role as a constituent of the mesenchymal ECM (Kappler et al. 1998).

Though the inhibitory CS-PGs and myelin-associated molecules are major impediments to axon regeneration within CNS, decorin infusion can suppress the levels of multiple inhibitory CS-PGs and promote axon growth across spinal cord injuries. Decorin treatment of acute spinal cord injury and cultured adult spinal cord microglia can increase plasminogen/plasmin synthesis and induced 10- and 17-fold increases in plasminogen and plasmin protein levels, respectively, within sites of injury. In addition to potentially degrading multiple axon growth inhibitory components of the glial scar, plasmin is known to play major roles in activating neurotrophins and promoting CNS plasticity (Davies et al. 2004, 2006). The formation of misaligned scar tissue by a variety of cell types expressing multiple axon growth inhibitory proteoglycans presents a physical and molecular barrier to axon regeneration after adult spinal cord injuries. Decorin reduces astrogliosis and basal lamina formation in acute cerebral cortex stab injuries. Decorin pretreatment of meningial fibroblasts in vitro resulted in increase in neurite outgrowth from co-cultured adult sensory neurons and suppression of NG2 immunoreactivity. In addition to suppressing inhibitory scar formation, decorin can directly boost the ability of neurons to extend axons within CS-PG or myelin rich environments. The ability of decorin to promote axon growth across acute spinal cord injuries via a coordinated suppression of inflammation, CS-PG expression and astroglial scar formation make decorin treatment a promising component of future spinal cord regeneration strategies (Minor et al. 2008).

References

Akita K, Toda M, Hosoki Y et al (2004) Heparan sulphate proteoglycans interact with neurocan and promote neurite outgrowth from cerebellar granule cells. Biochem J 383:129–138

Alliel PM, Perin J-P, Jollès P, Bonnet FJ (2004) Testican, a multidomain testicular proteoglycan resembling modulators of cell social behaviour. Eur J Biochem 214:347–350

Asher RA, Morgenstern DA, Shearer MC et al (2002) Versican is upregulated in CNS injury and is a product of oligodendrocyte lineage cells. J Neurosci 22:2225–2236

Aspberg A, Miura R, Bourdoulous S et al (1997) The C-type lectin domains of lecticans, a family of aggregating chondroitin sulfate proteoglycans, bind tenascin-R by protein-protein interactions independent of carbohydrate moiety. Proc Natl Acad Sci USA 94:10116–10121

Atoji Y, Yamamoto Y, Suzuki Y et al (1997) Immunohistochemical localization of neurocan in the lower auditory nuclei of the dog. Hear Res 110:200–208

Aya-ay J, Mayer J, Eakin AK et al (2005) The effect of hypoxic-ischemic brain injury in perinatal rats on the abundance and proteolysis of brevican and NG2. Exp Neurol 193:149–162

Barritt AW, Davies M, Marchand F et al (2006) Chondroitinase ABC promotes sprouting of intact and injured spinal systems after spinal cord injury. J Neurosci 26:10856–10867

Bbandtlow CE, Zzimmermann DR (2000) Proteoglycans in the developing brain: new conceptual insights for old proteins. Physiol Rev 80:1267–1290

Beggah AT, Dours-Zimmermann MT, Barras FM et al (2005) Lesion-induced differential expression and cell association of neurocan, brevican, versican V1 and V2 in the mouse dorsal root entry zone. Neuroscience 133:749–762

Bekku Y, Su WD, Hirakawa S et al (2003) Molecular cloning of Bral2, a novel brain-specific link protein, and immunohistochemical colocalization with brevican in perineuronal nets. Mol Cell Neurosci 24:148–159

Bekku Y, Rauch U, Ninomiya Y, Oohashi T (2009) Brevican distinctively assembles extracellular components at the large diameter nodes of Ranvier in the CNS. J Neurochem 108:1266–1276

Bradbury EJ, Moon LD, Popat RJ et al (2002) Chondroitinase ABC promotes functional recovery after spinal cord injury. Nature 416:636–640

Brakebusch C, Seidenbecher CI, Asztely F et al (2002) Brevican-deficient mice display impaired hippocampal CA1 long-term potentiation but show no obvious deficits in learning and memory. Mol Cell Biol 22:7417–7427

Braunewell KH, Pesheva P, McCarthy JB et al (1995) Functional involvement of sciatic nerve-derived versican- and decorin-like molecules and other chondroitin sulphate proteoglycans in ECM-mediated cell adhesion and neurite outgrowth. Eur J Neurosci 7:805–814

Brückner G, Grosche J, Schmidt S et al (2000) Postnatal development of perineuronal nets in wild-type mice and in a mutant deficient in tenascin-R. J Comp Neurol 428:616–629

Brückner G, Grosche J, Hartlage-Rübsamen M et al (2003) Region and lamina-specific distribution of extracellular matrix proteoglycans, hyaluronan and tenascin-R in the mouse hippocampal formation. J Chem Neuroanat 26:37–50

Buss A, Pech K, Kakulas BA et al (2009) NG2 and phosphacan are present in the astroglial scar after human traumatic spinal cord injury. BMC Neurol 9:32

Carulli D, Rhodes KE, Fawcett JW (2007) Upregulation of aggrecan, link protein 1, and hyaluronan synthases during formation of perineuronal nets in the rat cerebellum. J Comp Neurol 501:83–94

Chan CC, Wong AK, Liu J et al (2007) ROCK inhibition with Y27632 activates astrocytes and increases their expression of neurite growth-inhibitory chondroitin sulfate proteoglycans. Glia 55:369–384

Clarris HJ, Rauch U, Key B (2000) Dynamic spatiotemporal expression patterns of neurocan and phosphacan indicate diverse roles in the developing and adult mouse olfactory system. J Comp Neurol 423:99–111

Cross AK, Haddock G, Surr J et al (2006) Differential expression of ADAMTS-1, -4, -5 and TIMP-3 in rat spinal cord at different stages of acute experimental autoimmune encephalomyelitis. J Autoimmun 26:16–23

Czipri M, Otto JM, Cs-Szabó G et al (2003) Genetic rescue of chondrodysplasia and the perinatal lethal effect of cartilage link protein deficiency. J Biol Chem 278:39214–39223

Davies JE, Tang X, Denning JW et al (2004) Decorin suppresses neurocan, brevican, phosphacan and NG2 expression and promotes axon growth across adult rat spinal cord injuries. Eur J Neurosci 19:1226–1242

Davies JE, Tang X, Bournat JC, Davies SJ (2006) Decorin promotes plasminogen/plasmin expression within acute spinal cord injuries and by adult microglia in vitro. J Neurotrauma 23:397–408

Day JM, Olin AI, Murdoch AD et al (2004) Alternative splicing in the aggrecan G3 domain influences binding interactions with tenascin-C and other extracellular matrix proteins. J Biol Chem 279:12511–12518

Deepa SS, Carulli D, Galtrey C et al (2006) Composition of perineuronal net extracellular matrix in rat brain: a different disaccharide composition for the net-associated proteoglycans. J Biol Chem 281:17789–17800

Deguchi K, Takaishi M, Hayashi T et al (2005) Expression of neurocan after transient middle cerebral artery occlusion in adult rat brain. Brain Res 1037:194–199

Dong Y, Han X, Xue Y et al (2004) Secreted brevican mRNA is expressed in the adult rat pituitary. Biochem Biophys Res Commun 314:745–748

Emerling DE, Lander AD (1996) Inhibitors and promoters of thalamic neuron adhesion and outgrowth in embryonic neocortex: functional association with chondroitin sulfate. Neuron 17:1089–1100

Engel M, Maurel P, Margolis RU, Margolis RK (1996) Chondroitin sulfate proteoglycans in the developing central nervous system. I. cellular sites of synthesis of neurocan and phosphacan. J Comp Neurol 366:34–43

Fawcett JW, Asher RA (1999) The glial scar and central nervous system repair. Brain Res Bull 49:377–391

Feng K, Arnold-Ammer I, Rauch U (2000) Neurocan is a heparin binding proteoglycan. Biochem Biophys Res Commun 272:449–455

Filbin MT (2003) Myelin-associated inhibitors of axonal regeneration in the adult mammalian CNS. Nat Rev Neurosci 4:703–713

Fosang AJ, Hardingham TE (1996) In: Comper WD (ed) Extracellular matrix, vol 1. Harwood Academic Publishers, Amsterdam

Fuller ML, DeChant AK, Rothstein B et al (2007) Bone morphogenetic proteins promote gliosis in demyelinating spinal cord lesions. Ann Neurol 62:288–300

Galtrey CM, Kwok JC, Carulli D et al (2008) Distribution and synthesis of extracellular matrix proteoglycans, hyaluronan, link proteins and tenascin-R in the rat spinal cord. Eur J Neurosci 27:1373–1390

Gary SC, Hockfield S (2000) BEHAB/brevican: an extracellular matrix component associated with invasive glioma. Clin Neurosurg 47:72–82

Gary SC, Zerillo CA, Chiang VL et al (2000) cDNA cloning, chromosomal localization, and expression analysis of human BEHAB/brevican, a brain specific proteoglycan regulated during cortical development and in glioma. Gene 256:139–147

GrandPre T, Li S, Strittmatter SM (2002) Nogo-66 receptor antagonist peptide promotes axonal regeneration. Nature 417:547–551

Haas CA, Rauch U, Thon N et al (1999) Entorhinal cortex lesion in adult rats induces the expression of the neuronal chondroitin sulfate proteoglycan neurocan in reactive astrocytes. J Neurosci 19:9953–9963

Haddock G, Cross AK, Allan S et al (2007) Brevican and phosphacan expression and localization following transient middle cerebral artery occlusion in the rat. Biochem Soc Trans 35:692–694

Hagihara K, Miura R, Kosaki R et al (1999) Immunohistochemical evidence for the brevican-tenascin-R interaction: colocalization in perineuronal nets suggests a physiological role for the interaction in the adult rat brain. J Comp Neurol 410:256–264

Hamel MG, Mayer J, Gottschall PE (2005) Altered production and proteolytic processing of brevican by transforming growth factor beta in cultured astrocytes. J Neurochem 93:1533–1541

Haunsø A, Ibrahim M, Bartsch U et al (2000) Morphology of perineuronal nets in tenascin-R and parvalbumin single and double knockout mice. Brain Res 864:142–145

Held-Feindt J, Paredes EB, Blömer U et al (2006) Matrix-degrading proteases ADAMTS4 and ADAMTS5 (disintegrins and metalloproteinases with thrombospondin motifs 4 and 5) are expressed in human glioblastomas. Int J Cancer 118:55–61

Hirakawa S, Oohashi T, Su WD et al (2000) The brain link protein-1 (BRAL1): cDNA cloning, genomic structure, and characterization as a novel link protein expressed in adult brain. Biochem Biophys Res Commun 276:982–989

Hu B, Kong LL, Matthews RT, Viapiano MS (2008) The proteoglycan brevican binds to fibronectin after proteolytic cleavage and promotes glioma cell motility. J Biol Chem 283:24848–24859

Ida M, Shuo T, Hirano K et al (2006) Identification and functions of chondroitin sulfate in the milieu of neural stem cells. J Biol Chem 281:5982–5991

Inatani M, Tanihara H (2002) Proteoglycans in retina. Prog Retin Eye Res 21:429–447

Inatani M, Tanihara H, Oohira A et al (2000) Upregulated expression of neurocan, a nervous tissue specific proteoglycan, in transient retinal ischemia. Invest Ophthalmol Vis Sci 41:2748–2754

Inatani M, Honjo M, Otori Y et al (2001) Inhibitory effects of neurocan and phosphacan on neurite outgrowth from retinal ganglion cells in culture. Invest Ophthalmol Vis Sci 42:1930–1938

Jaworski DM, Kelly GM, Piepmeier JM et al (1996) BEHAB (brain enriched hyaluronan binding) is expressed in surgical samples of glioma and in intracranial grafts of invasive glioma cell lines. Cancer Res 56:2293–2298

John N, Krügel H, Frischknecht R et al (2006) Brevican-containing perineuronal nets of extracellular matrix in dissociated hippocampal primary cultures. Mol Cell Neurosci 31:774–784

Jones LL, Yamaguchi Y, Stallcup WB, Tuszynski MH (2002) NG2 is a major chondroitin sulfate proteoglycan produced after spinal cord injury and is expressed by macrophages and oligodendrocyte progenitors. J Neurosci 22:2792–2803

Jones LL, Margolis RU, Tuszynski MH (2003a) The chondroitin sulfate proteoglycans neurocan, brevican, phosphacan, and versican are differentially regulated following spinal cord injury. Exp Neurol 182:399–411

Jones LL, Sajed D, Tuszynski MH (2003b) Axonal regeneration through regions of chondroitin sulfate proteoglycan deposition after spinal cord injury: a balance of permissiveness and inhibition. J Neurosci 23:9276–9288

Kappler J, Stichel CC, Gleichmann M et al (1998) Developmental regulation of decorin expression in postnatal rat brain. Brain Res 793:328–332

Katoh-Semba R, Matsuda M, Watanabe E et al (1998) Two types of brain chondroitin sulfate proteoglycan: their distribution and possible functions in the rat embryo. Neurosci Res 31:273–282

Kurazono S, Okamoto M, Sakiyama J et al (2001) Expression of brain specific chondroitin sulfate proteoglycans, neurocan and phosphacan, in the developing and adult hippocampus of Ihara's epileptic rats. Brain Res 898:36–48

Lemons ML, Sandy JD, Anderson DK, Howland DR (2001) Intact aggrecan and fragments generated by both aggrecanse and metalloproteinase-like activities are present in the developing and adult rat spinal cord and their relative abundance is altered by injury. J Neurosci 21:4772–4781

Leonardo CC, Eakin AK, Ajmo JM, Gottschall PE (2008) Versican and brevican are expressed with distinct pathology in neonatal hypoxic-ischemic injury. J Neurosci Res 86:1106–1114

Leung KM, Margolis RU, Chan SO (2004) Expression of phosphacan and neurocan during early development of mouse retinofugal pathway. Brain Res Dev Brain Res 152:1–10

Levine JM, Nishiyama A (1996) The NG2 chondroitin sulfate proteoglycan: a multifunctional proteoglycan associated with immature cells. Perspect Dev Neurobiol 3:245–259

Li H, Leung TC, Hoffman S, Balsamo J, Lilien J (2000) Coordinate regulation of cadherin and integrin function by the chondroitin sulfate proteoglycan neurocan. J Cell Biol 149:1275–1288

Li HP, Oohira A, Ogawa M et al (2005) Aberrant trajectory of thalamocortical axons associated with abnormal localization of neurocan immunoreactivity in the cerebral neocortex of reeler mutant mice. Eur J Neurosci 22:2689–2696

Lundell A, Olin AI, Mörgelin M, al-Karadaghi S, Aspberg A, Logan DT (2004) Structural basis for interactions between tenascins and lectican C-type lectin domains: evidence for a crosslinking role for tenascins. Structure 12:1495–1506

Margolis RK, Rauch U, Maurel P, Margolis RU (1996) Neurocan and phosphacan: two major nervous tissue-specific chondroitin sulfate proteoglycans. Perspect Dev Neurobiol 3:273–290

Matsui F, Watanabe E, Oohira A (1994) Immunological identification of two proteoglycan fragments derived from neurocan, a brain-specific chondroitin sulfate proteoglycan. Neurochem Int 25:425–431

Matsui F, Nishizuka M, Yasuda Y et al (1998) Occurrence of a N-terminal proteolytic fragment of neurocan, not a C-terminal half, in a perineuronal net in the adult rat cerebrum. Brain Res 790:45–51

Matsui F, Kawashima S, Shuo T et al (2002) Transient expression of juvenile-type neurocan by reactive astrocytes in adult rat brains injured by kainate-induced seizures as well as surgical incision. Neuroscience 112:773–781

Matsui F, Kakizawa H, Nishizuka M et al (2005) Changes in the amounts of chondroitin sulfate proteoglycans in rat brain after neonatal hypoxia-ischemia. J Neurosci Res 81:837–845

Matthews RT, Gary SC, Zerillo C et al (2000) Brain-enriched hyaluronan binding (BEHAB)/brevican cleavage in a glioma cell line is mediated by a disintegrin and metalloproteinase with thrombospondin motifs (ADAMTS) family member. J Biol Chem 275:22695–22703

Mayer J, Hamel MG, Gottschall PE (2005) Evidence for proteolytic cleavage of brevican by the ADAMTSs in the dentate gyrus after excitotoxic lesion of the mouse entorhinal cortex. BMC Neurosci 6:52

McKeon RJ, Jurynec MJ, Buck CR (1999) The chondroitin sulfate proteoglycans neurocan and phosphacan are expressed by reactive astrocytes in the chronic CNS glial scar. J Neurosci 19:10778–10788

Meyer-Puttlitz B, Milev P, Junker E et al (1995) Chondroitin sulfate and chondroitin/keratan sulfate proteoglycans of nervous tissue: developmental changes of neurocan and phosphacan. J Neurochem 65:2327–2337

Meyer-Puttlitz B, Junker E, Margolis RU, Margolis RK (1996) Chondroitin sulfate proteoglycans in the developing central nervous system. II. Immunocytochemical localization of neurocan and phosphacan. J Comp Neurol 366:44–54

Milev P, Maurel P, Häring M, Margolis RK, Margolis RU (1996) TAG-1/axonin-1 is a high-affinity ligand of neurocan, phosphacan/protein-tyrosine phosphatase- ζ/β, and N-CAM. J Biol Chem 271:15716–15723

Milev P, Fischer D, Häring M et al (1997) The fibrinogen-like globe of tenascin-C mediates its interactions with neurocan and phosphacan/protein-tyrosine phosphatase- ζ/β. J Biol Chem 272:15501–15509

Milev P, Chiba A, Häring M et al (1998a) High affinity binding and overlapping localization of neurocan and phosphacan/protein-tyrosine phosphatase- ζ/β with tenascin-R, amphoterin, and the heparin-binding growth-associated molecule. J Biol Chem 273:6998–7005

Milev P, Maurel P, Chiba A et al (1998b) Differential regulation of expression of hyaluronan-binding proteoglycans in developing brain: aggrecan, versican, neurocan, and brevican. Biochem Biophys Res Commun 247:207–212

Miller B, Sheppard AM, Bicknese AR, Pearlman AL (1995) Chondroitin sulfate proteoglycans in the developing cerebral cortex: the distribution of neurocan distinguishes forming afferent and efferent axonal pathways. J Comp Neurol 355:615–628

Minor K, Tang X, Kahrilas G et al (2008) Decorin promotes robust axon growth on inhibitory CS-PGs and myelin via a direct effect on neurons. Neurobiol Dis 32:88–95

Mishima N, Hoffman S (2003) Neurocan in the embryonic avian heart and vasculature. Anat Rec A Discov Mol Cell Evol Biol 272:556–562

Miura R, Aspberg A, Ethell IM et al (1999) The proteoglycan lectin domain binds sulfated cell surface glycolipids and promotes cell adhesion. J Biol Chem 274:11431–11438

Miura R, Ethell IM, Yamaguchi Y (2001) Carbohydrate-protein interactions between HNK-1-reactive sulfoglucuronyl glycolipids and the proteoglycan lectin domain mediate neuronal cell adhesion and neurite outgrowth. J Neurochem 76:413–432

Mizuno H, Warita H, Aoki M, Itoyama Y (2008) Accumulation of chondroitin sulfate proteoglycans in the microenvironment of spinal motor neurons in amyotrophic lateral sclerosis transgenic rats. J Neurosci Res 86:2512–2523

Monnier PP, Sierra A, Schwab JM et al (2003) The Rho/ROCK pathway mediates neurite growth-inhibitory activity associated with the chondroitin sulfate proteoglycans of the CNS glial scar. Mol Cell Neurosci 22:319–332

Morgenstern DA, Asher RA, Fawcett JW (2002) Chondroitin sulphate proteoglycans in the CNS injury response. Prog Brain Res 137:313–332

Nakada M, Miyamori H, Kita D et al (2005) Human glioblastomas overexpress ADAMTS-5 that degrades brevican. Acta Neuropathol 110:239–246

Neumann H, Woolf CJ (1999) Regeneration of dorsal column fibers into and beyond the lesion site following adult spinal cord injury. Neuron 23:83–91

Neumann S, Skinner K, Basbaum AI (2005) Sustaining intrinsic growth capacity of adult neurons promotes spinal cord regeneration. Proc Natl Acad Sci USA 102:16848–16852

Nutt CL, Zerillo CA, Kelly GM, Hockfield S (2001) Brain enriched hyaluronan binding (BEHAB)/brevican increases aggressiveness of CNS-1 gliomas in Lewis rats. Cancer Res 61:7056–7059

Ogawa T, Hagihara K, Suzuki M et al (2001) Brevican in the developing hippocampal fimbria: differential expression in myelinating oligodendrocytes and adult astrocytes suggests a dual role for brevican in central nervous system fiber tract development. J Comp Neurol 432:285–295

Okamoto M, Sakiyama J, Kurazono S et al (2001) Developmentally regulated expression of brain-specific chondroitin sulfate proteoglycans, neurocan and phosphacan, in the postnatal rat hippocampus. Cell Tissue Res 306:217–229

Oleszewski M, Gutwein P, von der Lieth W et al (2000) Characterization of the L1-neurocan-binding site. Implications for L1-L1 homophilic binding. J Biol Chem 275:34478–34485

Oohashi T, Hirakawa S, Bekku Y et al (2002) Bral1, a brain-specific link protein, colocalizing with the versican V2 isoform at the nodes of Ranvier in developing and adult mouse central nervous systems. Mol Cell Neurosci 19:43–57

Popp S, Maurel P, Andersen JS, Margolis RU (2004) Developmental changes of aggrecan, versican and neurocan in the retina and optic nerve. Exp Eye Res 79:351–356

Porter S, Clark IM, Kevorkian L, Edwards DR (2005) The ADAMTS metalloproteinases. Biochem J 386:15–27

Prange CK, Pennacchio LA, Lieuallen K et al (1998) Characterization of the human neurocan gene, CS-PG3. Gene 221:199–205

Qi ML, Wakabayashi Y, Enomoto M, Shinomiya K (2003) Changes in neurocan expression in the distal spinal cord stump following complete cord transection: a comparison between infant and adult rats. Neurosci Res 45:181–188

Quaglia X, Beggah AT, Seidenbecher C, Zurn AD (2008) Delayed priming promotes CNS regeneration post-rhizotomy in neurocan and brevican-deficient mice. Brain 131:240–249

Rauch U (2004) Extracellular matrix components associated with remodeling processes in brain. Cell Mol Life Sci 61:2031–2045

Rauch U, Karthikeyan L, Maurel P et al (1992) Cloning and primary structure of neurocan, a developmentally regulated, aggregating chondroitin sulfate proteoglycan of brain. J Biol Chem 267:19536–19547

Rauch U, Grimpe B, Kulbe G et al (1995) Structure and chromosomal localization of the mouse neurocan gene. Genomics 28:405–410

Rauch U, Clement A, Retzler C, Fröhlich L et al (1997a) Mapping of a defined neurocan binding site to distinct domains of tenascin-C. J Biol Chem 272:26905–26912

Rauch U, Meyer H, Brakebusch C et al (1997b) Sequence and chromosomal localization of the mouse brevican gene. Genomics 44:15–21

Rauch U, Feng K, Zhou XH (2001) Neurocan: a brain chondroitin sulfate proteoglycan. Cell Mol Life Sci 58:1842–1856

Rauch U, Zhou XH, Roos G (2005) Extracellular matrix alterations in brains lacking four of its components. Biochem Biophys Res Commun 328:608–617

Retzler C, Göhring W, Rauch U (1996b) Analysis of neurocan structures interacting with the neural cell adhesion molecule N-CAM. J Biol Chem 271:27304–27310

Retzler C, Wiedemann H, Kulbe G, Rauch U (1996a) Structural and electron microscopic analysis of neurocan and recombinant neurocan fragments. J Biol Chem 271:17107–17113

Richardson PM, Issa VM (1984) Peripheral injury enhances central regeneration of primary sensory neurones. Nature 309:791–793

Sango K, Oohira A, Ajiki K et al (2003) Phosphacan and neurocan are repulsive substrata for adhesion and neurite extension of adult rat dorsal root ganglion neurons in vitro. Exp Neurol 182:1–11

Schäfer R, Dehn D, Burbach GJ, Deller T (2008) Differential regulation of chondroitin sulfate proteoglycan mRNAs in the denervated rat fascia dentata after unilateral entorhinal cortex lesion. Neurosci Lett 439:61–65

Schmalfeldt M, Dours-Zimmermann MT, Winterhalter KH, Zimmermann DR (1998) Versican V2 is a major extracellular matrix component of the mature bovine brain. J Biol Chem 273:15758–15764

Schnell L, Schwab ME (1990) Axonal regeneration in the rat spinal cord produced by an antibody against myelin-associated neurite growth inhibitors. Nature 343:269–272

Schnepp A, Lindgren PK, Hülsmann H et al (2005) Mouse Testican-2: Eexpression, glycosylation, and effects on neurite outgrowth. J Biol Chem 280(12):11274–11280

Schwartz NB, Domowicz M (2004) Proteoglycans in brain development. Glycoconj J 21:329–341

Seidenbecher CI, Richter K, Rauch U et al (1995) Brevican, a chondroitin sulfate proteoglycan of rat brain, occurs as secreted and cell surface glycosylphosphatidylinositol-anchored isoforms. J Biol Chem 270:27206–27212

Seidenbecher CI, Gundelfinger ED, Böckers TM et al (1998) Transcripts for secreted and GPI-anchored brevican are differentially distributed in rat brain. Eur J Neurosci 10:1621–1632

Seidenbecher CI, Smalla KH, Fischer N et al (2002) Brevican isoforms associate with neural membranes. J Neurochem 83:738–746

Seyfried NT, McVey GF, Almond A et al (2005) Expression and purification of functionally active hyaluronan-binding domains from human cartilage link protein, aggrecan and versican: formation of ternary complexes with defined hyaluronan oligosaccharides. J Biol Chem 280:5435–5448

Shen LH, Li Y, Gao Q et al (2008) Down-regulation of neurocan expression in reactive astrocytes promotes axonal regeneration and facilitates the neurorestorative effects of bone marrow stromal cells in the ischemic rat brain. Glia 56:1747–1754

Shuo T, Aono S, Matsui F et al (2004) Developmental changes in the biochemical and immunological characters of the carbohydrate moiety of neuroglycan C, a brain-specific chondroitin sulfate proteoglycan. Glycoconj J 20:267–278

Silver J, Miller JH (2004) Regeneration beyond the glial scar. Nat Rev Neurosci 5:146–156

Sivasankaran R, Pei J, Wang KC et al (2004) PKC mediates inhibitory effects of myelin and chondroitin sulfate proteoglycans on axonal regeneration. Nat Neurosci 7:261–268

Smith GM, Strunz C (2005) Growth factor and cytokine regulation of chondroitin sulfate proteoglycans by astrocytes. Glia 52:209–218

Sobel RA, Ahmed AS (2001) White matter extracellular matrix chondroitin sulfate/dermatan sulfate proteoglycans in multiple sclerosis. J Neuropathol Exp Neurol 60:1198–1207

Spicer AP, Joo A, Bowling RA Jr (2003) A hyaluronan binding link protein gene family whose members are physically linked adjacent to chondroitin sulfate proteoglycan core protein genes: the missing links. J Biol Chem 278:21083–21091

Tang X, Davies JE, Davies SJ (2003) Changes in distribution, cell associations, and protein expression levels of NG2, neurocan, phosphacan, brevican, versican V2, and tenascin-C during acute to chronic maturation of spinal cord scar tissue. J Neurosci Res 71:427–444

Thon N, Haas CA, Rauch U et al (2000) The chondroitin sulphate proteoglycan brevican is upregulated by astrocytes after entorhinal cortex lesions in adult rats. Eur J Neurosci 12:2547–2558

Toba Y, Horie M, Sango K et al (2002) Expression and immunohistochemical localization of heparan sulphate proteoglycan N-syndecan in the migratory pathway from the rat olfactory placode. Eur J Neurosci 15:1461–1473

Ughrin YM, Chen ZJ, Levine JM (2003) Multiple regions of the NG2 proteoglycan inhibit neurite growth and induce growth cone collapse. J Neurosci 23:175–186

Viapiano MS, Hockfield S, Matthews RT (2008) BEHAB/brevican requires ADAMTS-mediated proteolytic cleavage to promote glioma invasion. J Neurooncol 88:261–272

Vitellaro-Zuccarello L, Bosisio P, Mazzetti S et al (2007) Differential expression of several molecules of the extracellular matrix in functionally and developmentally distinct regions of rat spinal cord. Cell Tissue Res 327:433–447

Wang W, Dow KE (1997) Differential regulation of neuronal proteoglycans by activation of excitatory amino acid receptors. Neuroreport 8:659–663

Watanabe E, Aono S, Matsui F et al (1995) Distribution of a brain-specific proteoglycan, neurocan, and the corresponding mRNA during the formation of barrels in the rat somatosensory cortex. Eur J Neurosci 7:547–554

Wilson MT, Snow DM (2000) Chondroitin sulfate proteoglycan expression pattern in hippocampal development: potential regulation of axon tract formation. J Comp Neurol 424:532–546

Wu Y, Sheng W, Chen L et al (2004) Versican V1 isoform induces neuronal differentiation and promotes neurite outgrowth. Mol Biol Cell 15:2093–2104

Xiao ZC, Bartsch U, Margolis RK et al (1997) Isolation of a tenascin-R binding protein from mouse brain membranes. A phosphacan-related chondroitin sulfate proteoglycan. J Biol Chem 272:32092–32101

Yamada H, Watanabe K, Shimonaka M, Yamaguchi Y (1994) Molecular cloning of brevican, a novel brain proteoglycan of the aggrecan/versican family. J Biol Chem 269:10119–10126

Yamada H, Watanabe K, Shimonaka M et al (1995) cDNA cloning and the identification of an aggrecanase-like cleavage site in rat brevican. Biochem Biophys Res Commun 216:957–963

Yamada H, Fredette B, Shitara K et al (1997) The brain chondroitin sulfate proteoglycan brevican associates with astrocytes ensheathing cerebellar glomeruli and inhibits neurite outgrowth from granule neurons. J Neurosci 17:7784–7795

Zacharias U, Rauch U (2006) Competition and cooperation between tenascin-R, lecticans and contactin 1 regulate neurite growth and morphology. J Cell Sci 119:3456–3466

Zanin MK, Bundy J, Ernst H et al (1999) Distinct spatial and temporal distributions of aggrecan and versican in the embryonic chick heart. Anat Rec 256:366–380

Zhang H, Kelly G, Zerillo C et al (1998) Expression of a cleaved brain-specific extracellular matrix protein mediates glioma cell invasion In vivo. J Neurosci 18:2370–2376

Zhang Y, Rauch U, Perez MT (2003) Accumulation of neurocan, a brain chondroitin sulfate proteoglycan, in association with the retinal vasculature in RCS rats. Invest Ophthalmol Vis Sci 44:1252–1261

Zhou XH, Brakebusch C, Matthies H et al (2001) Neurocan is dispensable for brain development. Mol Cell Biol 21:5970–5978

Zimmermann DR, Dours-Zimmermann MT (2008) Extracellular matrix of the central nervous system: from neglect to challenge. Histochem Cell Biol 130:635–653

Zurn AD, Bandtlow CE (2006) Regeneration failure in the CNS: cellular and molecular mechanisms. Adv Exp Med Biol 557:54–76

Part XIII

C-Type Lectins: Emerging Groups of C-Type Lectins

Regenerating (Reg) Gene Family

G.S. Gupta

39.1 The Regenerating Gene Family

Regenerating (Reg) gene is a multigene super-family grouped into four subclasses, types I, II, III, and IV based on the primary structures of the encoded proteins and represents the products of small secretory proteins, which can function as acute phase reactants, lectins, antiapoptotic factors or growth factors for pancreatic β-cells, neural cells and epithelial cells in the digestive system. Among C-type lectin gene superfamily, the Reg-related genes are classified in group VII and encode small secreted proteins with a single carbohydrate recognition domain linked to a short N-terminal peptide (Drickamer 1999; Zhang et al. 2003a). The genes coding these proteins are tandemly clustered on chromosome 2p12 in humans and may have arisen from the same ancestral gene by gene duplication. The expression of this group of proteins is controlled by complex mechanisms, some members being constitutively expressed in certain tissues while, in others, they require activation by several factors. These members have several apparently unrelated biological effects, depending on the member studied and the target cell. insoluble fibrils at physiological pH (Iovanna and Dagorn 2005).

39.2 Regenerating (*Reg*) Gene Products

Reg and Reg-related genes constitute a multigene family, the Reg family. Identified members of the human *Reg* gene family include Reg Iα, Reg Iβ, Reg IIIα (HIP/PAP), Reg IIIβ and Reg IV (Chakraborty et al. 1995; Dieckgraefe et al. 2002; Hartupee et al. 2001; Lasserre et al. 1994; Laurine et al. 2005). A glycoprotein expressed in exocrine pancreas (where it has been called PSP/LIT) and endocrine pancreas (where it has been called the Reg protein) is encoded by *Reg* gene which maps to 2p12. The *Reg* gene characterized in the exocrine pancreas has been found to be expressed in regenerating islets of 90% depancreatized rats and not in normal islets. In humans, it was identified only in the exocrine pancreas. The Reg gene is known to be involved in the growth of not only pancreatic β-cells, but also epithelial cells of the GI tract and carcinoma of its lineage. The expression of the REG gene is closely related to the infiltrating property of gastric carcinoma, and may be a prognostic indicator of differentiated adenocarcinoma of the stomach (Rosty et al. 2002; Yonemura et al. 2003).

Reg gene, first isolated from a rat regenerating islet cDNA library, (Terazono et al. 1988) and Reg-related genes constitute a multigene family (Reg Family), which consists of four subtypes (type I, II, III, and IV) based on the primary structures of the encoded proteins (Watanabe et al. 1994) that ameliorates the diabetes of 90% depancreatized rats and non-obese diabetic mice. The Reg protein family consists solely of a M_r 16-kDa CTLD and N-terminal secretion signal. Despite having a canonical C-type lectin fold, Reg proteins lack conserved sequences that support Ca^{2+}-dependent carbohydrate binding in other C-type lectins (Drickamer 1999). The human REG family genes have a common gene structure with 6 exons and 5 introns, and encode homologous 158–175-aa secretory proteins. The human REG family genes on chromosome 2, except for REG IV on chromosome 1, were mapped to a contiguous 140 kb region of the human chromosome 2p12. The gene order from centromere to telomere was 5′ HIP/PAP-III′-5′ RS 3′-3′ REG Iα 5′-5′ REG Iβ 3′-3′ REG-III 5′. The human REG gene family is constituted from an ancestor gene by gene duplication and forms a gene cluster on the region (Nata et al. 2004). Okamoto grouped the members of the family, Reg and Reg-related genes from human, rat and mouse, into three subclasses, types I, II, and III (Okamoto 1999). Stephanova et al. determined that the three rat PAP genes and the related Reg gene (REGL, regenerating islet-derived-like/pancreatic stone protein-like/pancreatic thread protein-like) were all located at 4q33–q34 (Stephanova et al. 1996). The mouse Reg family genes were mapped to a contiguous 75 kb region in chromosome 6, including Reg I, Reg II, Reg IIIα, Reg IIIβ, Reg IIIγ, and Reg IIIδ (Abe et al. 2000). Reg IIIδ was expressed predominantly in exocrine pancreas, whereas both Reg I and Reg II

Table 39.1 Reg proteins, length of amino acids and chromosome localization in different species (Zhang et al. 2003a)

Super family memeber	Species	Length of amino acids	Chromosome localization	Reference(s)[a]
Reg I	Mouse Reg I	165	6	Abe et al. 2000; Unno et al. 1993
			12	
	Rat Reg	165	4q33-q34	Stephanova et al. 1996
	Human Reg I and II	166	2p12	Miyashita et al. 1995; Perfetti et al. 1994
	Human PSP/PTP	166	2p12	Miyashita et al. 1995; Dusetti et al. 1994a
Reg II	Mouse Reg II	173	6	Abe et al. 2000; Unno et al. 1993
			3	
Reg III	Rat PAP-I, PAP-II, Peptide-23	175	4q33-q34	Stephanova et al. 1996
	Mouse Reg-III-α; Reg IIIβ	175	6	Abe et al. 2000; Narushima et al. 1997
	Mouse Reg-III-γ	174	6	Abe et al. 2000; Narushima et al. 1997
	Human HIP	175	2p12	Miyashita et al. 1995
	Bovine PTP	175	11	NCBI[b]
Reg IV	Human Reg IV	158	1q12-q21	Hartupee et al. 2001
	Mouse Reg IV	157	3	NCBI[c]

[a]References represent chromosomal localization
[b]Accession NP_991356
[c]Accession NP_080604

were expressed in hyperplastic islets and Reg IIIα, Reg IIIβ and Reg IIIγ were expressed strongly in the intestinal tract and weakly in pancreas (Table 39.1).

Although Reg IV was not found in the same chromosome as other members of human Reg gene and Reg-related gene (Table 39.1), it shared some common features with other members such as: sequence homology, tissue expression profiles, and exon-intron junction genomic organization (Hartupee et al. 2001). Several Reg III proteins are highly expressed in the pancreas and small intestine, including mouse Reg-IIIγ and human HIP/PAP (hepatointestinal pancreatic/pancreatitis associated protein) (Cash et al. 2006a; Christa et al. 1996) and in the pancreas and GI tract in pathological conditions (Iovanna et al. 1991; Okamoto 1999; Abe et al. 2000). Pancreatitis-associated protein (PAP)-I and -II are members of Reg-III subclass and encoded by gene PAP-I and PAP-II, respectively. PAP-I is also known as peptide 23 in the rat, and regenerating (Reg) islet-derived Reg-IIIβ in mouse, and PAP-II known as Reg-III in the rat and Reg-III-α in the mouse (Narushima et al. 1997). Thus, four types of Reg gene family have been identified. Among these, a total of 17 members have been discovered in mammals across human, pig, mouse, bovine and rat species. Table 39.1 lists some important members of Reg family (Schiesser et al. 2001; Zenilman et al. 1996). Hartupee et al. also suggested that a mouse homologue of Reg IV likely existed (Hartupee et al. 2001) and substantiated (Bishnupuri et al. 2010). While rat Reg cDNA had a single open reading frame that encoded a 165-amino acid protein with a 21-amino acid signal peptide, the human REG cDNA encoded a 166-amino acid protein with a 22-amino acid signal peptide and showed 68% homology to that of rat Reg protein. While some members of the family (Reg 1 and islet neogenesis-associated protein, i.e. INGAP) have been implicated in β-cell replication and/or neogenesis, the roles of the other members have yet to be characterized (Planas et al. 2006).

39.3 Reg 1

39.3.1 Tissue Expression

Reports suggest that the expression of Reg mRNAs and of the corresponding protein(s) was restricted to exocrine tissue irrespective of age, sex, and presence of insulitis and/or diabetes. Moreover, Reg remains localized in acinar cells in the two opposite situations of (a) cyclophosphamide-treated males in a prediabetic stage presenting a high level of both insulin and Reg mRNAs, and (b) the overtly diabetic females with no insulin but a high level of Reg mRNA (Sanchez et al. 2000). The Reg gene was more significantly increased in female mice than in male mice, and in both cases, the expression was not influenced by age. Nondiabetic female mice had a significantly higher expression of the gene than diabetic female mice. Overexpression of the Reg gene was found in male mice treated with cyclophosphamide, an agent known to be a potent inducer of diabetes in male non-obese diabetic (NOD) mice (Baeza et al. 1997; Baeza et al. 2001). Bimmler et al. (2004) supported the concept that PSP/Reg and PAP are coordinately regulated secretory stress proteins (SSP). The cystic fibrosis (CF) mouse pancreas had constitutively elevated expression of the Reg/PAP cell stress genes, which are suggested to be involved in protection or recovery from pancreatic injury. The severity of caerulein-induced pancreatitis was not ameliorated in the CF mouse even though the Reg/PAP stress genes were already highly upregulated. While Reg/PAP may be protective they may also have a negative effect during pancreatitis due to their anti-apoptotic activity, which has been shown to increase the severity of pancreatitis (Norkina et al. 2006).

The two nonallelic Reg genes and the two insulin genes are expressed differentially during early embryogenesis. The differential expression of Reg-I and -II suggests that they may be induced by different and independent stimuli and have distinct functions (Perfetti et al. 1996). Correlation between Reg and insulin gene expression does not exist in the fetal pancreas during the developmental period but, on the contrary, such a correlation was present in the adult pancreas (Moriscot et al. 1996). Scattered distribution of Reg protein is observed in pancreatic islet cells of streptozotocin (STZ)-treated rats. The increased Reg gene expression in neonatal STZ-treated rat pancreas is a useful model for studying the relationship between Type II (Non-Insulin Dependent) Diabetes (NIDDM) and β cell Regeneration or Reg gene protein.

39.3.2 Gastric Mucosal Cells

Reg mRNA and its product are distributed in the basal part of the oxyntic mucosa and expressed mainly in enterochromaffin-like (ECL) cells. Levels of both Reg mRNA and its product are markedly increased during the healing process of water immersion-induced gastric lesions (Asahara et al. 1996). Although Reg protein is reported to have a trophic effect on gastric epithelial cells, its involvement in human gastric diseases was studied by Fukui et al. (2003). Both gastrin and gastric mucosal inflammation enhance Reg gene expression in the fundic mucosa in rats. The Reg gene is associated with hypergastrinemia and fundic mucosal inflammation and may be involved in *H. pylori*-induced gastritis. The expression of Reg I is controlled through separate promoter elements by gastrin and *Helicobacter* (Bernard-Perrone et al. 1999; Steele et al. 2007). A link between Reg protein and *H. pylori* infection may help explain the molecular mechanisms underlying *H. pylori*-associated diseases, including gastric cancer (Yoshino et al. 2005). During the healing course of gastric erosion, Reg expression is highly increased in ECL cells surrounding the ulcer crater, suggesting its role as a regulator of gastric mucosal regeneration. During healing, the gene expression of several proinflammatory cytokines and Reg was markedly augmented. Among the proinflammatory cytokines, CINC-2 β is the only cytokine in which augmented expression preceded the increase of Reg gene expression. CINC-2 β, expressed in damaged gastric mucosa, stimulates the production of Reg protein in ECL cells via CXCR-2 and may be involved in the accelerated healing of injured gastric mucosa (Kazumori et al. 2000; Fukuhara et al. 2010). REG expression has been observed in various tumors including gastric carcinoma (Dhar et al. 2004; Miyaoka et al. 2004). Gastrin regulates Reg mRNA abundance in human corpus. Mutations of Reg that prevent secretion are associated with ECL cell carcinoids, suggesting a function as an autocrine or paracrine tumor suppressor (Higham et al. 1999).

39.3.3 Ectopic Expression

Ectopic expression of the *Reg* gene occurs in some human colonic and rectal tumors, suggesting that enhanced Reg expression may be related to the proliferative state of tumor cells. Regenerating protein may act not only as a regulator of gastric epithelial cell proliferation but also as a modifier of other multiple physiologic functions (Kazumori et al. 2002). Reg-1 protein is expressed in human hearts obtained from autopsied patients who died of myocardial infarction. In view of emerging evidence of Reg for tissue regeneration in a variety of tissues/organs, it is proposed that the damaged heart may be a target for Reg action and that Reg may protect against acute heart stress (Kiji et al. 2005). While hepatocyte and cholangiocyte proliferation was suppressed, hepatic stem cells and/or hepatic progenitor cells were activated. Reg I was significantly upregulated in liver of the 2-AAF/PH rat model, accompanied by the formation of bile ductules during liver regeneration. The presence of Reg I in normal testis is weak. But its strong expression in the testis cancer suggests its potential role in normal and neoplastic germ cell proliferation (Mauro et al. 2008). Reg I may act through IL-6 to exert effects on squamous esophageal cancer cell biology. Transfecting TE-5 and TE-9 cells with Reg Iα and -Iβ led to significantly increased expression of IL-6 mRNA and protein, but with little or no effect on expression of IL-2, IL-4, IL-5, IL-10, IL-12, IL-13, IL-17A, IFN-γ, TNF-α, GC- or TGF-β1. The elevated IL-6 expression in REG Iα transfectants was silenced by siRNA-mediated knockdown (Usami et al. 2010).

Ectopic expression of Reg also was observed in ductal cell carcinomas. In ductal cell carcinomas, expression of Reg immunoreactivity was considered as one of phenotypic heterogeneity, as seen in AAT, lysozyme, and CMG immunoreactivity (Kimura et al. 1992). The diminution in pancreatic β-cell mass caused by subcutaneous implantation of an insulinoma is associated with reduced Reg gene expression and that the increase in β-cell replication after resection of the tumor is preceded by return of Reg gene expression toward normal. *Reg I* expression in untreated endoscopic biopsy specimens may provide a basis for new treatments of locally advanced thoracic squamous cell esophageal cancers (Motoyama et al. 2006). *Reg I*-knockout mice reveal the role of Reg-1 in the regulation of cell growth that is required in generation and maintenance of the villous structure of small intestine (Ose et al. 2007). Reg I appears to enhance the chemo- and radiosensitivity of squamous esophageal cancer cells, which suggests that it may be a useful target for improved and more individualized treatments for patients with esophageal squamous cell carcinoma (Hayashi et al. 2008).

39.3.4 Reg Protein

Unno et al. (1993) isolated two mouse distinct cDNAs and genes, one of which was a mouse homologue to rat and human Reg gene, the other a novel type of Reg gene. They were designated Reg I and Reg II, respectively. The two proteins encoded by these genes share 76% amino acid sequence identity with each other. Both genes span about 3 kbp, and the genomic organization of six exons and five introns is conserved between them. Chromosomal mapping studies indicated that the Reg I gene is localized on mouse chromosome 12, whereas the Reg II gene is localized on chromosome 3. Both Reg I and Reg II mRNAs are detected in the normal pancreas and hyperplastic islets of aurothioglucose-treated mice, but not in the normal islets. It is remarkable that in the gallbladder Reg I is expressed, but Reg II is not (Unno et al. 1993).

Moriizumi et al. (1994) isolated a novel human gene and cDNA encoding a member of the Reg I proteins, Reg Iβ. The gene encodes a 166-amino acid protein which has 22 amino acid substitutions in comparison with previously isolated human Reg protein, Reg Iα. While Reg Iβ was expressed only in pancreas, RegI α was expressed in kidney and stomach as well as in pancreas. The human REG gene has a high degree of similarity to the rat Reg gene. A REG-related sequence (REGL) is also located in 2p12 and expressed in the pancreas and physically mapped within a 100-kb genomic region (Bartoli et al. 1995; Perfetti et al. 1994). Reg may be involved in the expansion of β-cell mass during regeneration as well as in the maintenance of normal β-cell function. Sequence comparisons of Reg and Reg 1 suggested similar exon-intron organisation. The proteins encoded by Reg and Reg 1 comprise 166 amino acids and differ by 22 amino acids only (Bartoli et al. 1995). The 5′-Regulatory region −304/−237 of rat *Reg I* gene encoding a growth stimulating factor for pancreatic β-cells contained positive cis-acting elements. Gel shift assays showed the formation of a specific complex with the −256/−237 oligonucleotide (Miyashita et al. 1995).

39.3.5 *REGIA (Reg Iα)* Gene in Human Pancreas

[Synonym names for Reg1A protein are: pancreatic stone protein; pancreatic thread protein; Islet of Langerhans regenerating protein; regenerating protein I α; Islet cells regeneration factor; ICRF; regenerating islet-derived 1 α].

In human, four REG family genes, i.e., REG 1α, Reg Iβ, REG-related sequence (RS) and HIP/PAP, have been isolated and characterized but only Reg Iα protein has been isolated from human pancreatic secretion. A discoordinate expression of the two *REG* genes was found with a higher level of Reg Iα mRNA in fetus and a higher level of Reg Iβ in adult. In addition, while Reg Iα mRNA level was correlated with the expression of genes encoding exocrine proteins in adults, Reg Iβ mRNA level presented no correlation with any ductular, endocrine, or exocrine gene expression. In human pancreatic cell lines only Reg Iβ gene and protein expressed. These results suggest that two Reg genes and proteins play different roles in human pancreas (Sanchez et al. 2001). Reg Iα is a growth factor known to affect pancreatic islet β cells. Since tropical calcific pancreatitis (TCP) is known to have a variable genetic basis, reports do not suggest the interaction between mutations in the susceptibility genes: serine protease inhibitor Kazal 1 (*SPINK1*) and *CTSB* (a gene for cathepsin B) with REG 1α polymorphisms (Mahurkar et al. 2007).

Function of REGIA (Reg Iα) in Pancreas: Acinar cells can transdifferentiate into other pancreatic-derived cells. In these cells, Reg I overexpression is linked to acinar cell differentiation, whereas inhibition of Reg I leads to β cell and possibly ductal phenotype. Human and rat Reg I proteins, and recombinant protein are mitogenic to primary cultures of β- and ductal cells. Reg I expression in acinar cells is important in maintaining pancreatic cell lineage, and when decreased, cells can dedifferentiate and move toward becoming other pancreatic cells (Sanchez et al. 2004, 2009). Pancreatic-derived cells exposed to Reg I grow by activation of signal transduction pathways involving the mitogen-activated protein kinase phosphatases and cyclins, with concomitant induction of mitogen-activated protein kinase phosphatase (MKP-1). However, high intracellular levels of Reg I lead to decreased growth, likely via a binding to and inactivation of MKP-1. Inhibition of cell growth, and possible induction of apoptosis, may lead to differentiation of these cells to other cell types (Mueller et al. 2008; Zenilman et al. 1998; Levine et al. 2000). Induction of Reg I and its receptor may be important for recovery from acute pancreatitis (Bluth et al. 2006)

The administration of Reg protein can be used as a therapeutic approach for diabetes mellitus (Watanabe et al. 1994), although the serum Reg protein level is not a marker for progression of type I diabetes (Christofilis et al. 1999). Infact, Pdx1-Cre-mediated pancreas inactivation of IGF-I gene [in pancreatic-specific IGF-I gene-deficient (PID) mice] results in increased β-cell mass and significant protection against both type 1 and type 2 diabetes, where multiple Reg family genes (*Reg-2, -3α,* and *-3β,*) were significantly upregulated in pancreas. Interestingly, Reg family genes were also activated after streptozotocin-induced β-cell damage and diabetes (wild-type T1D mice) when islet cells were undergoing regeneration and Reg proteins increased in exocrine as well as endocrine pancreas, suggesting their potential role in β-cell neogenesis in PID or T1D mice. Reg proteins (Reg 1 and islet neogenesis-associated protein) were also shown to promote islet cell replication and neogenesis (Lu et al. 2006). The administration of poly (ADP-ribose) synthetase/polymerase (PARP) inhibitors such as nicotinamide to 90% depancreatized rats induces islet regeneration. Reg protein induces β-cell replication via the Reg receptor and ameliorates experimental diabetes. Studies suggest that poly(ADP-ribose) polymerase binds Reg promoter and regulates the transcription by autopoly (ADP-ribosyl)ation (Akiyama et al. 2001). Depletion of Reg I, associated with the pathogenesis of impaired glucose tolerance of pancreatitis-associated diabetes, Bluth et al. (2008) support that replacement therapy could be useful in such patients.

REG Iα gene expression and its promoter activity were enhanced by IFN-γ and IL-6. Reg Iα protein promoted cell growth and cell resistance to H_2O_2-induced apoptosis in AGS cells. REG Iα gene is inducible by cytokine stimulation and its gene product may function as a mitogenic and/or an antiapoptotic factor in the development of early gastric cancer (Sekikawa et al. 2005). The IL-6-responsive element was located within the sequence from −142 to −134 of the REG Iα promoter region. Reg Iα protein mediated the anti-apoptotic effects of STAT3 signaling in gastric cancer cells by enhancing Akt activation, Bad phosphorylation and Bcl-xL expression. The expression of Reg Iα protein was significantly correlated with that of p-STAT3 in gastric cancer tissues. It appeared that Reg Iα protein plays a pivotal role in anti-apoptosis in gastric tumorigenesis under STAT3 activation (Sekikawa et al. 2008). The enhanced expression of IL-22 in infiltrating inflammatory cells and concommitant enhancement of Reg Iα in the inflamed epithelium indicated the role of IL-22/REG Iα axis in ulcerative colitis (UC). The IL-22-responsive element was located between −142 and −134 in the REG Iα promoter region. REG Iα protein may have a pathophysiological role as a biological mediator for immune cell-derived IL-22 in the UC mucosa (Sekikawa et al. 2010).

Role in Pathology: PAP and Reg Iα are up-regulated during the pancreas regeneration. The PAP expression represents a subset of low-grade, low-stage human hepatocellular carcinomas (HCC) with frequent β-catenin mutation and hence more favorable prognosis, whereas further genetic or epigenetic alterations, such as p53 mutation and Reg 1α expression, lead to more advanced HCCs (Yuan et al. 2005). Three genes encoding related proteins, PAP, Reg Iα and Reg Iβ, are over-expressed in cancer (Rechreche et al. 1999). Human Reg I protein is ectopically expressed in colorectal mucosa at the transition zone of colorectal cancer, and occasionally within the tumor itself. Although ectopic Reg I expression in colorectal epithelia is not a marker for the presence of carcinoma, it may be a sensitive marker for mucosa at risk for development of neoplasia (Zenilman et al. 1997). High levels of REG 1α expression within tumors are an independent predictor of poor prognosis in patients with breast cancer (Sasaki et al. 2008). Reg 1A is a molecular marker of prognostic value and is associated with peritoneal carcinomatosis in colorectal cancer (Astrosini et al. 2008). High levels of REG 1A expression by tumor cells are an independent predictor of a poor prognosis in patients with NSCLC (Minamiya et al. 2008). Stage Ta/T1 urothelial carcinoma of the bladder (Ta/T1 BC) has a marked tendency to reoccur. Expression of REG 1A is an independent predictor of recurrence in Ta/T1 BC (Geng et al. 2009). Reg Iα protein may play a role in the development of gastric cancers (Fukui et al. 2004).

Reg I-expressing cells are present in the bile ductules and increased during regeneration. Reg I is significantly upregulated in the liver of the 2-acetylaminofluorene-PH rat model, accompanied by the formation of bile ductules during liver regeneration (Wang et al. 2009). Reg Iα protein, which was rarely expressed in ductal epithelial cells of normal minor salivary gland (MSG), was overexpressed in patients with Sjögren's syndrome (SS). Reg Iα protein may play a role in the regeneration of ductal epithelial cells in the MSGs of patients with SS (Kimura et al. 2009). REG expresses in Barrett's esophagus. Expression of Reg Iα was more frequently observed in patients who showed squamous re-epithelialization of Barrett's esophagus at biopsy sites (Chinuki et al. 2008). Yamauchi et al. (2009) demonstrated that PPARγ-agonist thiazolidinediones (TZDs) inhibited cell proliferation and Reg Iα protein/mRNA expression in gastrointestinal cancer through a PPARγ-dependent pathway. TZDs may, therefore, be a candidate for novel anti-cancer drugs for patients with gastrointestinal cancer expressing both REG Iα and PPARγ.

39.3.6 REGIB (Reg Iβ) Gene in Human Pancreas

Human REG Iβ gene encodes a protein secreted by the exocrine pancreas that is highly similar to the Reg1A (Reg 1α) protein. Human Reg I β, also known as secretory pancreatic stone protein 2 and lithostathine 1 β, is a type I subclass member of Reg family. The Reg 1β protein is associated with islet cell regeneration and diabetogenesis, and may be involved in pancreatic lithogenesis. Reg family members REG1A, REGL, PAP and this gene are tandemly clustered on chromosome 2p12 and may have arisen from the same ancestral gene by gene duplication.

Pancreatic juice in vertebrates contains a group of 16-kDa proteins which is composed of the protein species such as pancreatic stone protein (PSP). PSP is a 16-kDa acidic protein with an isoelectric point in the range of pH 5.5–6. A truncated form of this protein was originally isolated from calcium carbonate stones surgically removed from the main pancreatic duct of humans with chronic pancreatitis. PSP is a secretory protein that is related to C-type lectins.

39.4 Pancreatic Stone Protein/Lithostathine (PSP/LIT)

[Alternative names: Lithostathine-1-β; Regenerating islet-derived protein 1-β/Regenerating protein I β (REG-1β); Pancreatic stone protein 2 (PSP-2)]

39.4.1 Characterization

Reg protein was first found in pancreatic stones. It was named Pancreatic Stone Protein (PSP) and later renamed lithostathine (LIT), as it was assumed to prevent stone formation. The protein was found in regenerating endocrine pancreas and was named Reg (for Regenerating) protein. The functional human Reg gene is a single copy gene, spans approximately 3.0 kbp, and is composed of six exons and five introns. TATA box and CCAAT box-like sequences are located at 27 and 100 bp upstream from the transcriptional initiation site. The human Reg mRNA was detected predominantly in the pancreas, and at lower levels in the gastric mucosa and kidney. Furthermore, the Reg gene was found to be expressed ectopically in colon and rectal tumors. Immunoblot analysis demonstrated several molecular forms (15–18 kDa) of the Reg protein in the pancreas. The 166-amino acid sequence encoded by the human Reg gene contains the 144-amino acid sequence of pancreatic stone protein determined by De Caro et al. (1989) and the partially determined 45-amino acid sequence of pancreatic thread protein (Gross et al. 1985), indicating that the Reg protein, pancreatic stone protein, and pancreatic thread protein are simply different names for a single protein deriving from the Reg gene (Watanabe et al. 1990). The 144 amino acid protein is O-glycosylated on Thr-5. The glycan chain is variable in length and in charge. The PSP/LIT 3-D structural organization is of the C-lectin type, even though CRD of PSP'LIT is unlikely to have any functional calcium-binding site. However, the PSP/LIT binds Ca^{2+} with 1:1 stoichiometry (Lee et al. 2003). EPR studies, using divalent vanadyl (VO^{2+}) ion as a paramagnetic substitute for Ca^{2+} also showed that VO^{2+} binds to PSP/LIT with a metal:protein binding stoichiometry of 1:1 and that VO^{2+} competes with Ca^{2+} in binding to PSP/LIT. Mutations of a cluster of acidic residues on the molecular surface resulted in almost complete loss (95–100%) of binding of Ca^{2+} and VO^{2+}, showing that these residues are critical for calcium binding by PSP/LIT (Lee et al. 2003). The Arg^{11}-Ile^{12} bond is readily cleaved by trypsin and the resulting C-terminal polypeptide precipitates at physiological pH and tends to form fibrils.

The rat PSP/LIT gene was characterized over 2.7 kbp of gene sequence and 2.43 kbp of 5′-flanking sequence. The lithosathine sequence spanned over six exons. The promoter region contained the TATAAA and CCAAT consensus sequences 30 and 107 bp upstream of the cap site, respectively. Furthermore, a tract of (TG)22 repeat, with potential Z-DNA conformation, was found at position-1081 (Dusetti et al. 1993). The rat cDNA encoding a protein homologous to Reg encodes a protein, designated Reg-2 and shows 60%, 78% and 61% similarities with the reported amino acid sequences of the rat, bovine and human proteins, respectively (Kamimura et al. 1992). Using in situ hybridization, the three rat PAP, and the related REG genes mappped in the same chromosomes region, namely 4q33—>q39. This rat chromosome region is thus homologous to the human 2p12 region, which also contains the PAP gene, the REG1A gene, and a REG-related gene (REGL) suggesting the tandem organization of Reg/lithostathin genes (Gharib et al. 1993; Stephanova et al. 1996).

Reg-I (PSP) and Reg-III (PAP) are induced after the onset of acute pancreatitis, and both have been proposed as potential markers of pancreatitis. After induction of pancreatitis, serum levels of Reg I and III protein differ significantly. PSP/Reg is up-regulated in blood after trauma and its level is related to the severity of inflammation. The serum PSP/Reg-protein concentration may reflect pancreatic damage, especially in acute pancreatitis, and may be a sensitive a marker for such damage as elastase-1, although false positivity was apparent inrenal failure and in some patients with hepatic

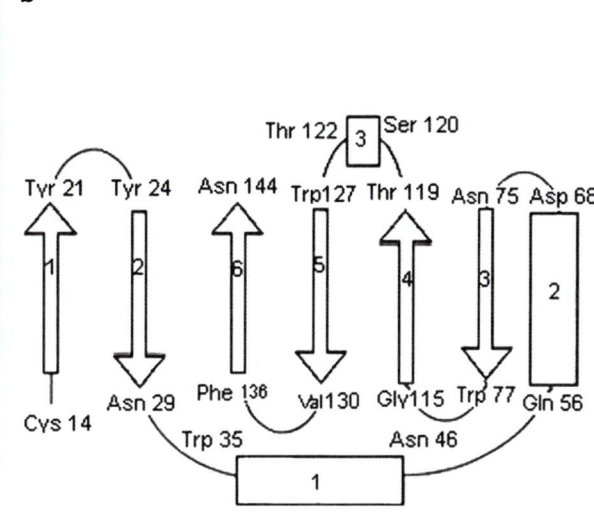

Fig. 39.1 Structure of hPSP/hLIT. The protein folds into one globular domain that consists of three α helices (*Boxed*) and six β-strands, the latter forming two antiparallel β-sheets (**a**). The sequential connectivity follows the scheme β1-β2-α1-α2-β3-β4-α3-β5-β6 (**b**). (**b**) shows schematic diagram of hLIT polypeptide topology. The β-strands are depicted as *arrows*, with the *arrowheads* indicating the direction of the chain. α-Helices are shown as *rectangles* (Reprinted by permission from Macmillan Publishers Ltd: EMBO J, Bertrand et al. © 1996)

dysfunction or digestive system malignancies (Satomura et al. 1995). Furthermore, PSP/Reg binds to and activates neutrophils. Therefore, PSP/Reg and Reg-III/PAP are acute-phase proteins that could serve as a marker for disease severity and posttraumatic complications (Keel et al. 2009; Zenilman et al. 2000). PSP/LIT/Reg mRNA was strongly induced by serum from rats with acute pancreatitis (SAP) in AR-42J cells. Treatment with interleukins (IL) IL-1 or IL-6 or dexamethasone alone was ineffective. Combination of IL-6 with dexamethasone resulted in strong induction of the PSP/LIT/Reg gene, but the further addition of IL-1 to the mixture reduced induction. Treatment with tumor TNFα or IFNγ induced PSP/LIT/Reg mRNA expression. Findings suggest that expression of the PSP/LIT/Reg mRNA during acute pancreatitis could be mediated by specific combinations of cytokines and/or glucocorticoids.

39.4.2 3D-Structure

Crystals of human PSP/LIT grown in PEG 4000 as the precipitating agent, belong to the hexagonal space group $P6_1$ (or its enantiomorph $P6_5$) and diffract to 1.55 A resolution. There is one molecule in the asymmetric unit and the crystals have 39% solvent (Pignol et al. 1995).

The C-terminal part of hPSP/hLIT is homologous with the animal C-type lectins showing the highest percentage of homology with the rat cartilage proteoglycan and the human IgE receptor (Petersen 1988). Amino acid similarities are distributed over regions corresponding to the consensus CRD and two of the three disulfide bridges in hPSP/hLIT are conserved in the C-type lectin sequence motif (Drickamer 1999). Studies on the three-dimensional structure of hPSP/hLIT confirmed that hPSP/hLIT belongs to the family of C-type lectins. Overall protein structure of hPSP/hLIT has the shape of a heart with dimensions 45 × 30 × 25 A (Fig. 39.1a) (Bertrand et al. 1996). The protein folds into one globular domain that consists of three α helices and six 5-strands, the latter forming two antiparallel β-sheets. The sequential connectivity follows the scheme β1-β2-α1-α2-β3-β4-α3-β5-β6 (Fig. 39.1b). The well-defined N-terminal residue of the structure participates in an intramolecular disulfide bond (Cys14-Cys25) forming a loop. This loop contains a short β-strand (β1; residues 19–21) which forms an anti-parallel β1-sheet with strands β2 and β6. The two major helices α1 and α2, of respectively three and four helical turns, are oriented perpendicular to one another framing the three β-strands (β1, β2 and β6) of the lower half of the molecule (Fig. 39.1a). The Cys42- Cys140 disulfide bridge connects α1 to α6. The upper half of the molecule includes the other three β-strands (β3, β4 and β5). Two strands of this second antiparallel β-sheet, β4 and β5, are separated by a single helical turn (α3) and are connected at one extremity by the Cys 115-Cys 132 disulfide bridge. In addition to the secondary structure elements, the structure also contains several extended loops and stretches of non-regular secondary structure which represent >60% of the molecule.

39.4.3 Secretory Forms of PSP

Secretory forms of pancreatic stone protein (sPSP, Mr 17,500-22,000) have been purified from human pancreatic juice. Secretory forms of PSP are inhibitors of $CaCO_3$ crystal growth. Proteins homologous to human sPSP are present in other mammalian species and may act as stabilizers of Ca^{2+}-supersaturated pancreatic juice (Bernard et al. 1991). In mature pancreas, expression of Reg/sPSP gene occurs primarily in acinar cells. The gene product, which encodes a secretory protein inhibiting CaCO3 crystal growth in juice, is unlikely to play a specific role in islet regeneration (Rouquier et al. 1991). A cDNA encoding pre-sPSP revealed that it comprised all but the 5' end of sPSP mRNA, which was obtained by sequencing the first exon of the sPSP gene. The complete mRNA sequence is 775 nt long, including 5'- and 3'-noncoding regions of 80 and 197 nucleotides, respectively, attached to a poly(A) tail of approximately 125 nt. It encodes a preprotein of 166 amino acids, including a prepeptide of 22 amino acids. PSP-S gene expression is specifically reduced in chronic calcifying pancreatitis (CCP) patients (Giorgi et al. 1989). Ovocleidin 17, a major protein of the chicken eggshell calcified layer shows a C-type lectin domain with a sequence similariy of 30% to that of PSP and lectins and anticoagulant proteins from snake venom (Mann and Siedler 1999).

Serum PSP showed an isoelectric point (pH 9) similar to that reported for the pancreatic thread protein. Schmiegel et al. (1990) did not support the etiopathogenic role postulated for pancreatic stone protein in chronic pancreatitis and chronic calcifying pancreatitis by other investigators. The PSP in human calculi derives from human pancreatic juice (PSP S2-5) through the tryptic cleavage of the Arg-^{11}-Ile-^{12} bond (De Caro et al. 1989; Rouimi et al. 1987). EM studies revealed that the stone protein was markedly present in the zymogen granules and condensing vacuoles of the normal pancreatic acinar cells, the label was found in the acinar and ductal lumen. In chronic pancreatitis, the localization of PSP/LIT, when it occurred, was extremely weak in the acinar cells. No PSP/LIT was specifically characterized in hepatocytes, gastric mucosa, and enterocytes. However, a weak but specific reaction was found in the secretory granules of Paneth cells (Lechene de la Porte et al. 1986).

De Reggi et al. (1995) isolated the major oligosaccharide chains of human pancreatic PSP/LIT and determined their sequences by means of NMR. There were 11 different glycoforms and seven of them were sequenced. They all were from the same site of glycosylation (Thr5) and displayed the same core 2 structure: GlcNAc(β 1–6)[Gal(β 1–3)]GalNAc α-. They ranged in size from 4 to 9 sugar residues. Elongation was found to proceed from a common tetrasaccharidic core: Gal(β 1–4)GlcNAc(β 1–6)[Gal(β 1–3)]GalNAc-ol through N-acetyllactosamine units. The non-reducing ends of some oligosaccharides carry the antigenic determinant H, with presence of external Fuc linked only in (α 1–2) to Gal. All the glycans, except one, carry a sialic acid in (α 2–3) linkage to Gal, with one disialylated form which displays a supplementary (α 2–6) linkage. These findings are consistent with the polymorphism of the protein, either in its native form or after enzymic processing.

39.4.4 Functions of PSP/LIT

39.4.4.1 As Inhibitor of Crystal Growth

Calcium carbonate crystals are observed in the juice of patients suffering with chronic calcifying pancreatitis (CCP), a lithogenic disorder characterized by the presence of stones, which obstruct pancreatic ducts and remain scarce or absent in juice from healthy individuals (De Caro et al. 1988). Several lines of evidence indicate that the inhibition of calcium carbonate crystal nucleation and growth in human pancreatic juice is exerted by human lithostathine (hPSP/hLIT); patients suffering from CCP have an abnormally low concentration of hPSP/hLIT (Bernard et al. 1995). The hPSP/hLIT binds to the surface of calcium carbonate crystals, and provides inhibitory effects (Bernard et al. 1992). Inhibitory effects of hPSP/hLIT can be seen with the entire protein, which shows an activity 100-fold greater (Geider et al. 1996; Gerbaud et al. 2000) than that of 11-residue N-terminal peptide obtained after cleavage with trypsin (Bernard et al. 1992). Consequently, it is likely that both the N-terminal peptide and the C-terminal domain bind to calcium carbonate crystals; a conclusion reinforced by the fact that C-terminal proteolytic fragment (residues 12–144) is a major component of pancreatic stones (De Caro et al. 1988).

As pancreatic juice is supersaturated with respect to calcium carbonate, it was hypothesized that PSP/LIT stabilizes pancreatic juice and calcite precipitation is prevented by PSP/LIT; the decreased PSP levels could be a key factor in the growth of calcium carbonate crystals and stone development during the course of chronic calcifying pancreatitis (CCP) (Multigner et al. 1985). The secretary PSP (sPSP) could play that role, since an activity inhibiting the nucleation and growth in vitro of CaCO3 crystals was found in pancreatic juice, associated with these proteins. Moreover, sPSP concentration was significantly lower in the pancreatic juice of patients with CCP than in control patients. While aiming at discovering how peptides inhibit calcium salt crystal growth, Gerbaud et al. (2000) showed that the peptide backbone governed the binding more than did the lateral chains. The ability of peptides to inhibit crystal growth is essentially based on backbone flexibility. Proteins homologous to sPSP were also found in the dog, rat, swine, monkey and ox. They constitute a new family of pancreatic secretory proteins, whose biological role would be to maintain pancreatic juice in a stable state towards CaCO3 (De Caro et al. 1988).

The inhibition property of crystal nucleation and growth property of hPSP/hLIT is shared by antifreeze proteins (AFPs) in fish. The AFPs are synthesized and secreted in the liver of marine fishes to provide protection from freezing environments and are classified into three groups (Davies and Hew 1990); type I (helical peptides, M_r 3–5 K), type II (cysteine-rich, M_r 14 K), and type III (predominantly β-sheet, M_r 6–7 K). The type II AFPs are homologous to hPSP/hLIT, showing sequence similarities distributed over both the N- and C-terminal domains (Ewart et al. 1992; Ng and Hew 1992). PSP/LIT is cleaved by trace amounts of trypsin, resulting in a C-terminal polypeptide and an N-terminal undecapeptide. The N-terminal undecapeptide has been identified as the active site of PSP/LIT regarding crystal inhibition. Thus, it has been assumed to inhibit calcium carbonate precipitation and therefore to prevent stone formation in the pancreatic ducts. This function is, however, debatable (Bimmler et al. 1997) since most of these studies have been carried out in vitro. De Reggi (1998) showed conclusively that PSP/LIT does not inhibit calcium carbonate nucleation and crystal growth. Based on the findings it seemed unlikely that PSP/LIT is a physiologically relevant calcite crystal inhibitor (De Reggi et al. 1998).

39.4.4.2 Role in Regeneration of Islets

For several years it was believed that PSP served as an inhibitor of calcium carbonate precipitation in pancreatic juice, and it was proposed that its name should be changed to "lithostathine" (LIT) (Sarles et al. 1990). However, it was later shown that PSP has no more crystal inhibitory activity than several of the pancreatic digestive enzymes (Bimmler et al. 1997; De Reggi et al. 1998). Studies demonstrated that the expression of PSP/LIT protein is increased during regeneration of islets after nicotinamide treatment and partial pancreatectomy (Terazono et al. 1990; Unno et al. 1993). These observations led to suggest that PSP/LIT may be a protein involved in regeneration (Watanabe et al. 1994) and may act as a growth mediator stimulating the proliferation of β-cells. Tissue culture studies implied a mitogenic activity of PSP/LIT on the growth of various cell types (Zenilman et al. 1996; Fukui et al. 1998), and application of PSP/LIT was observed to partially ameliorate diabetes in NOD mice (Gross et al. 1998; Kobayashi et al. 2000). PSP/LIT is a bifunctional protein which might be involved in the control of the bacterial ecosystem in the intestine. PSP/LIT is not expected to present sugar- or calcium-binding properties (Patard et al. 1996).

39.4.4.3 PSP/LIT in Alzheimer's Disease

PSP/LIT is a found precipitated in the form of fibrils in Alzheimer's disease (AD). Recombinant PSP/LIT is essentially monomeric at acidic pH while it aggregates at physiological pH. Electron microscopic studies of aggregates showed an apparently unorganized structure of numerous monomers which tend to precipitate forming regular unbranched fibrils. Aggregated forms seemed to occur prior to the apparition of fibrils. In addition, these fibrils resulted from a proteolytic mechanism due to a specific cleavage of the Arg^{11}-Ile^{12} peptide bond. It is deduced that the NH_2-terminal undecapeptide of PSP/LIT normally impedes fiber formation but not aggregation. A theoretical model explaining the formation of amyloid plaques in neurodegenerative diseases or stones in lithiasis starting from PSP/LIT has been described (Cerini et al. 1999). PSP/LIT and PAP were significantly increased in the cerebrospinal fluid of patients with AD at the very early stages of AD, and their level remained elevated during the course of the AD. Because PSP/LIT undergoes an autolytic cleavage leading to its precipitation and the formation of fibrils, it is likely that it may be involved in amyloidosis and tangles by allowing heterogeneous precipitation of other proteins (Duplan et al. 2001).

Symptoms of pancreaticobiliary maljunction/choledochal cysts are caused by the obstruction of bile and pancreatic ducts due to protein plugs compacted in the common channel. PSP is reported to be a key protein to form protein plugs in chronic pancreatitis (Chen et al. 2005; Ochiai et al. 2004). PSP expression is not increased because of a protein-deficient diet. After a temporary increase in PSP levels due to a carbohydrate-deficient high-protein diet, there were no signs of a diet-dependent regulation of this protein, which may differ from that of other pancreatic secretory proteins. These findings contradict earlier reports that had drawn conclusions based solely on mRNA levels (Bimmler et al. 1999). Both chronic ethanol consumption and dietary protein deficiency increase the capacity of the pancreatic acinar cell to synthesize PSP/LIT (Apte et al. 1996). Alternatively, activated trypsin cleaves soluble PSP and creates insoluble PSP.

39.5 Pancreatic Thread Protein

39.5.1 Pancreatic Thread Proteins (PTPs)

PTPs are acinar cell products and members of the regenerating gene (Reg) family. PTP/Reg protein was mitogenic to both β-cell and ductal cell lines but not to mature, nondividing islets. This supports the hypothesis that PTP/Reg protein is mediator of β-cell growth and may be involved in modulating the duct-to-islet axis (Zenilman et al. 1996). PTP forms double helical threads in the neutral pH range after purification, undergoing freely reversible, pH-dependent globule-fibril transformation. PTP forms filamentous bundles reminiscent of the paired helical filaments of Alzheimer's disease in vitro. A major 0.9-kb as well as several minor transcripts of PTP have been identified in human pancreas. An excess of serum PTP may filter to CSF through blood brain barrier (Blennow et al. 1995). There were significantly higher levels of PTP mRNA in brains with AD compared with aged controls, with increased amounts of 1.2-, 0.6-, and 0.4-kb transcripts (de la Monte et al. 1990; Ozturk et al. 1989). The experimentally derived secondary structure of human PTP (HPTP) consists of a significant proportion of β-sheets and β-turns and lesser amounts of α-helical structures. The β-sheet component presumably plays an important role in the pH-dependent globule-fibril transformation of HPTP leading to antiparallel β-sheet structure in the aggregated state. The secondary structure of HPTP and its globule-fibril transformation lend credence to the belief that AD may be viewed as a conformational disease (Renugopalakrishnan et al. 1999).

39.5.2 Neuronal Thread Proteins

Neuronal thread proteins (NTP) are phosphoproteins which are expressed during neuritic sprouting. The 15–18-kDa NTP cluster is associated with development and neuronal differentiation, whereas the 21-kDa and 39–42-kDa species are overexpressed in AD, correlating with neurodegenerative sprouting and synaptic disconnection. Both human and experimental (rat) focal cerebral infarcts revealed up-regulation of NTP gene expression in perifocal neurons (de la Monte et al. 1996a). In AD, high levels of NTP immunoreactivity were detected in neuronal perikarya, neuropil fibers, and white matter fibers (axons). NTP accumulates in cortical neurons and colocalizes with phospho-tau-immunoreactive cytoskeletal lesions that correlate with dementia. NTPs overexpression in relation to paired helical filament-associated neurodegenerative lesions in AD has been indicated (de la Monte et al. 1997). In view of the rapid phosphorylation, the accumulation of NTP in AD cortical neuronal perikarya suggests a problem related to post-translational processing and transport of NTP molecules in AD neurodegeneration (de la Monte et al. 1996b). Primary human primitive neuroectodermal tumor (PNETs), malignant astrocytomas, and several human PNET and glioblastoma cell lines also express thread protein immunoreactivity. However, in addition to the 21-kDa species, there are ~8, 17 and ~14-kDa thread protein-immunoreactive molecules expressed in both PNET and glioblastoma cell lines. Glycosylated residues were not detected in either the PNET- or glioblastoma-derived thread proteins. Studies suggest that there are several distinct neuronal and glial derived thread proteins expressed in the central nervous system and that their levels of expression may be modulated with cell growth (Xu et al. 1993).

Insulin, insulin-like growth factor, type 1 (IGF-1), and nerve growth factor modulate NTP gene expression during neuronal differentiation in PNET cell lines. Insulin effected neuronal differentiation and modulation of NTP gene expression in PNET cells utilizes a signal transduction cascade that requires tyrosyl phosphorylation of IRS-1 (Xu et al. 1995b). Ethanol may inhibit NTP expression associated with CNS neuronal differentiation by uncoupling the IRS-1-mediated insulin signal transduction pathway (Xu et al. 1995a). Studies indicate a functional role for NTP in relation to the turnover or processing of neuronal cytoskeletal proteins, attributes that may be modulated by insulin/IGF-1-mediated signaling (de la Monte et al. 2003). Further studies suggest that reduced survival in neurons that over-express AD7c-NTP may be mediated by impaired insulin/IGF-1 (de la Monte and Wands 2004). Brains with AD, AD + Parkinson's disease (PD), and AD + Down's syndrome (DN) contain significantly higher densities of NTP immunoreactive neurons and more frequent immunostaining of neuropil and white matter fibers compared with PD dementia (PDD) and aged controls which had few or no AD lesions.

39.5.3 Pancreatic Proteins Form Fibrillar Structures upon Tryptic Activation

Graf et al. (2001) studied the structural/functional consequences of trypsin activation on 16-kDa proteins with respect to the kinetics of conversion from soluble to insoluble protein forms and the kinetics of assembly of protein subunits into polymerized thread structures. Trypsin activation of recombinant stress proteins or counterparts (PSP/Reg, PAP-I and PAP-III) resulted in conversion of 16-kDa soluble proteins into 14-kDa soluble isoforms [called pancreatic thread protein and pancreatitis-associated thread protein (PATP), respectively] that rapidly polymerize into sedimenting structures. Activated thread proteins show long lived resistance to a wide spectrum of proteases contained in pancreatic juice, including serine proteases and metalloproteinases. In contrast, PAP-II, following activation with trypsin or pancreatic juice, did not form insoluble structures and was rapidly digested by pancreatic proteases. SEM and TEM demonstrated that activated thread proteins polymerized into highly organized fibrillar structures with helical configurations. Through bundling, branching, and extension processes, these fibrillar structures formed dense matrices that span large topological surfaces (Fig. 39.2) (Graf et al. 2001). Such studies suggested that PSP/Reg and PAP-I and III isoforms consist of a family of highly regulated soluble secretory stress proteins, which, upon trypsin activation, convert into a family of insoluble helical thread proteins. Dense extracellular matrices, composed of helical thread proteins organized into higher ordered matrix structures, may serve physiological functions within luminal compartments in the exocrine pancreas.

39.5.3.1 PSP/LIT in Alzheimer's Disease

The term 'cerebral proteopathies' has been proposed to designate all brain diseases the hallmarks of which are the misfolding and subsequent aggregation of proteins (Walker and LeVineH 2000). The most characteristic feature of many neuron-degenerative diseases is the formation of fibrillar

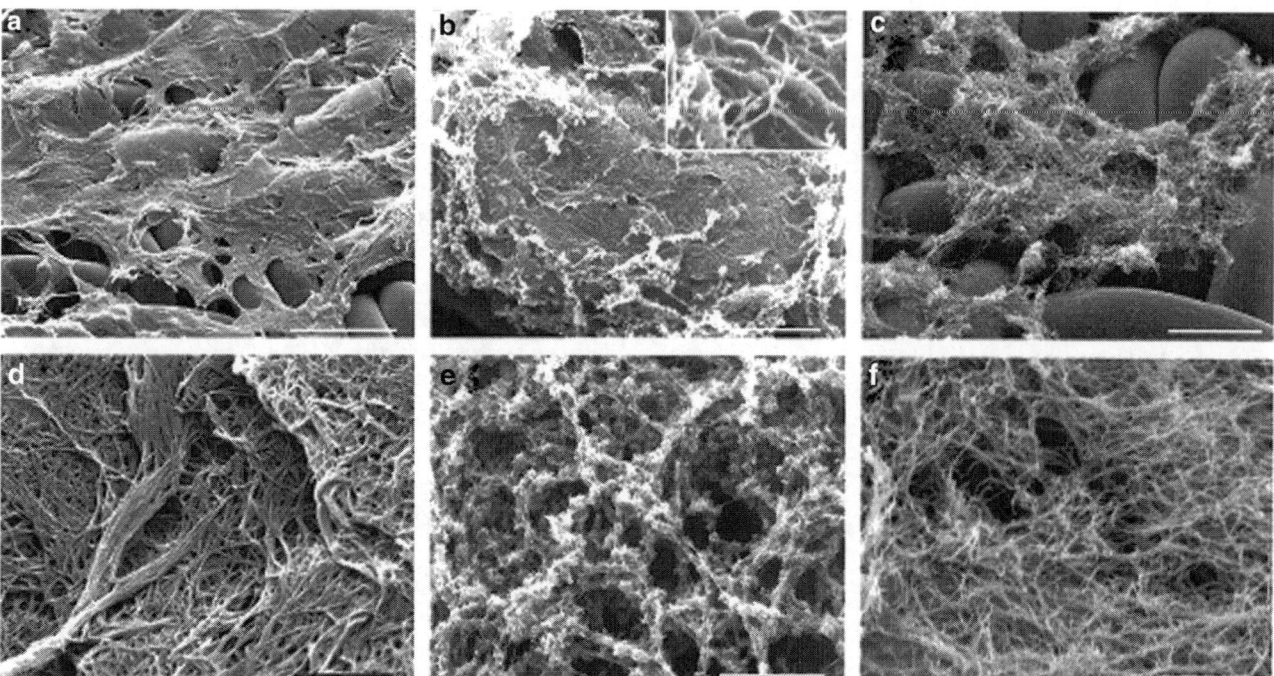

Fig. 39.2 Trypsin activation of purified recombinant secretory stress proteins generates a matrix of highly organized fibrils. Scanning electron microscopy of PTP and PATP generated in vitro. The micrographs demonstrate fibrous networks following activation of PAP-I, PAP-III and PSP/reg with trypsin. (**a**) and (**d**) show the matrix obtained with PATP I (**a**, bar 70 mm; **d**, 3 mm). (**b**) and (**e**) show the matrix obtained with PATP III (**b**, 20 mm; **e**, 7 mm). (**c**) and (**f**) show the matrix obtained with PTP (**c**, 40 mm; **f**, 4 mm). Inset in (**b**) gives a higher magnification micrograph showing individual fibrils that emerge from the dense matrix. (**a**) and (**c**), part of the plastic mesh of the pouch can be seen (Reprinted with permission from Graf et al. 2001© American Society for Biochemistry and Molecular Biology)

aggregates rather than the type of protein involved. Several neuron-degenerative diseases, such as AD are characterized by fibrillar deposits of proteins or peptides in brain. Many of these deposits are characterized by properties similar to those of starch and hence were called amyloid structures. Among proteins that form clinical fibril deposits, PSP/LIT readily polymerizes into fibrils after self-proteolysis of its N-terminal undecapeptide (Cerini et al. 1999). Autocatalytic cleavage leads PSP/LIT to the formation of quadruple-helical fibrils (QHF-litho) that are present in AD, and overexpressed during the very early stages of the disease before clinical signs appear (Duplan et al. 2001). PSP/LIT does not appear to change its native-like, globular structure during fibril formation. Therefore, it was thought that PSP/LIT constitutes an important protein in the deposition of polypeptides in vivo in relation to neurodegenerative diseases

Grégoire et al. (2001) studied 3D structure of hPSP/hLIT protofibrils using atomic force microscopy. These aggregates consisted of a network of protofibrils, each of which had a twisted appearance. Electron microscopy analysis showed that this twisted protofibril has a quadruple helical structure. Three-dimensional X-ray structural data and the results of biochemical experiments demonstrated that when forming a protofibril, hPSP/hLIT was first assembled via lateral hydrophobic interactions into a tetramer. Each tetramer then linked up with another tetramer as the result of longitudinal electrostatic interactions. All these results were used to build a structural model for the PSP/LIT protofibril called the quadruple-helical filament (QHF-litho) (Laurine et al. 2003). Ho et al. (2006) investigated the structural properties of recombinant hPAP lacking N-terminal propeptide, both in the soluble form and as fibrillar aggregates. Based on the solution structure analysis and the TANGO prediction, peptide hPAP84–116 was synthesized and found to form fibrils. The mechanism of fibril formation by hPAP has been compared with that of hPSP/hLIT. Over-all, studies support the idea that PSP/LIT may play a role in the ethiology of AD and hPSP/hLIT strongly resembles the prion protein in its dramatic proteolysis and amyloid proteins in its ability to form fibrils.

39.6 REG-II/REG-2/PAP-I

The 16 kDa Regenerating islet derived-2 (REG-2) has also been referred to as Islet of Langerhans regenerating protein 2, Lithostathine-2, Pancreatic stone protein 2, or Pancreatic thread protein 2. The rat cDNA for this gene has been cloned by Kamimura et al. (1992). Reg-2 is the murine homologue of the human secreted HIP/PAP C-type lectin. Reg-2, related to the protein encoded by the REG gene, is identical with PAP-1 (pancreatitis-associated protein-1) and a β-cell-derived autoantigen in non-obese diabetic mice. The autoimmune response against this protein may convert a regenerative into an islet-destructive process accelerating development of type 1 diabetes (Gurr et al. 2007; Liu et al. 2008). Reg-2 is a likely mouse exocrine pancreas cytoprotective protein whose expression is regulated by keratin filament organization and phosphorylation (Zhong et al. 2007).

Motor neurons are the only adult mammalian neurons of the CNS to regenerate after injury. This ability is dependent on the environment of the peripheral nerve and an intrinsic capacity of motor neurons for regrowth. The 16 kDa Reg-2, that is expressed solely in regenerating and developing rat motor and sensory neurons, is a potent Schwann cell mitogen in vitro. In vivo, Reg-2 protein is transported along regrowing axons and inhibition of Reg-2 signaling significantly retards the regeneration of Reg-2-containing axons. During development, Reg-2 production by motor and sensory neurons is regulated by contact with peripheral targets. Strong candidates for peripheral factors regulating Reg-2 production are cytokines of leukemia inhibitory factor (LIF)/CNTF family, because Reg-2 is not expressed in developing motor or sensory neurons of mice carrying a targeted disruption of LIF receptor gene (Livesey et al. 1997). Purified Reg-2 can itself act as an autocrine/paracrine neurotrophic factor for a subpopulation of motoneurons, by stimulating a survival pathway involving phosphatidylinositol-3-kinase, Akt kinase and NF-kB and therefore Reg-2 expression is a necessary step in the CNTF survival pathway in late embryonic spinal cord (Nishimune et al. 2000; Fang et al. 2010).

Reg-2 is normally not expressed by dorsal root ganglion (DRG) cells but Reg-2 is rapidly upregulated in DRG cells after sciatic nerve transection and after 24 h recovery is expressed almost exclusively in small-diameter neurons (Averill et al. 2002). Subsequently, expression shifts from small to large neurons (Nishimune et al. 2000). Reg-2 is also upregulated by DRG neurons in inflammation with a very unusual expression pattern. In a rat model of monoarthritis, Reg-2 immunoreactivity was detected in DRG neurons at 1 day. In addition, the presence of Reg-2 in central axon terminals implicates Reg-2 as a possible modulator of second order dorsal horn cells (Averill et al. 2008).

Analysis on gene-expression made in cardiomyocytes during myocarditis, revealed that the Reg-2/PAP-I mRNA level is most markedly increased in cardiomyocytes rather than in noncardiomyocytes. Reg-2/PAP-I mRNA was approximately 2000-fold greater in cardiomyocytes under active myocarditis than in normal. Moreover, Reg-2/PAP-I protein and other Reg/PAP family gene expressions were remarkably increased in EAM hearts. In addition, IL-6 expression was significantly related to Reg-2/PAP-I. Therefore, the Reg/PAP family, which was found to dramatically increase is suspected to play an important role in myocarditis (Watanabe et al. 2008).

To assess the role of Reg-2, Lieu et al. (2006) used Reg-$2^{-/-}$ mice in a model of fulminant hepatitis induced by Fas and in the post-hepatectomy Regeneration and demonstrated that Reg-2 deficiency enhanced liver sensitivity to Fas-induced oxidative stress and delayed liver regeneration with persistent TNFα/IL6/STAT3 signaling. In contrast, overexpression of human HIP/PAP promoted liver resistance to Fas and accelerated liver regeneration with early activation/deactivation of STAT3. Reg-2/HIP/PAP-I is therefore a critical mitogenic and antiapoptotic factor for the liver (Lieu et al. 2006).

Reg-2 expression was found to be upregulated in pancreatic islets both during diabetes development and as a result of adjuvant treatment in diabetic NOD mice and in C57BL/6 mice made diabetic by streptozotocin treatment. The upregulation of Reg-2 by adjuvant treatment was independent of signaling through MyD88 and IL-6. Upregulation of Reg-2 was also observed in the pancreas of diabetic mice undergoing β cell regenerative therapy with exendin-4 or with islet neogenesis-associated protein. Adjuvant immunotherapy regulates T1D in diabetic mice and induces Reg-2-mediated regeneration of β cells (Huszarik et al. 2010). In contrast to pancreatic islet neogenesis-associated protein [INGAP, Reg3δ] promote the growth or regeneration of the endocrine islet cells, Reg-2 overexpression did not protect acinar cells against caerulein-induced acute pancreatitis, indicating clear subtype specificities of the Reg family of proteins (Li et al. 2010). Pancreas-specific ablation of IGF-I in mice induced an over-expression of Reg-2 and Reg3β in the pancreas and protected them from streptozotocin (Stz)-induced β-cell damage. This indicated that Reg-2 protects insulin-producing cells against Stz-induced apoptosis by interfering with its cytotoxic signaling upstream of the intrinsic proapoptotic events by preventing its ability to inactivate JNK (Liu et al. 2010).

39.7 Reg-III

39.7.1 Murine *Reg-IIIα*, *Reg-IIIβ*, *Reg-III γ*, and *Reg-IIIδ* Genes

Human type III REG gene is divided into six exons spanning about 3 kb, and encodes a 175 aa protein with 85% homology with HIP/PAP. REG-III was expressed predominantly in pancreas and testis, but not in small intestine, whereas HIP/PAP was expressed strongly in pancreas and small intestine.

IL-6 responsive elements existed in the 5′-upstream region of human REG-III gene indicating that human REG-III gene might be induced during acute pancreatitis (Nata et al. 2004). The murine cDNAs type III Reg, Reg-IIIα, Reg-IIIβ and Reg-IIIγ encode 175-, 175- and 174- aa proteins, respectively, with 60–70% homology. All three genes are composed of six exons and five introns spanning approx. 3 kb, and exhibit distinctive structural features unique for members of the Reg gene family. All mouse Reg genes, Reg-IIIα, Reg-IIIβ, Reg-IIIγ, Reg-I and Reg-II, are assigned to the adjacent site of chromosome 6C. The Reg family genes were mapped to a contiguous 75 kb region of mouse genome according to the following order: 5′-Reg-IIIβ-Reg-IIIα-Reg II-Reg I-Reg-IIIγ-3′. Reg-IIIα, Reg-IIIβ and Reg-IIIγ were expressed weakly in pancreas, strongly in intestinal tract, but not in hyperplastic islets, whereas both Reg-I and Reg-II were expressed in hyperplastic islets (Narushima et al. 1997). REG3α is normally expressed in pancreatic acinar and endocrine cells. Transfection assays suggest that REG3α stimulates β-cell replication, by activating Akt kinase and increasing the levels of cyclin D1/CDK4 (Cui et al. 2009).

Abe et al. (2000) sequenced the 6.8 kb interspace fragment between Reg-IIIβ and Reg-IIIα and encountered a novel type III Reg gene, Reg-IIIδ. Reg-IIIδ gene is divided into six exons spanning about 3 kb, and encodes a 175 amino acid protein with 40–52% identity with the other five mouse Reg proteins. Reg-IIIδ was expressed predominantly in exocrine pancreas, but not in normal islets, hyperplastic islets, intestine or colon, whereas both Reg I and Reg II were expressed in hyperplastic islets and Reg-IIIα, Reg-IIIβ and Reg-IIIγ expressed strongly in the intestinal tract. In mouse, type I, type II and type III Reg genes (i.e. *Reg-I, Reg-II*, and *Reg-III* gene) have also been isolated.

The rat PAP-III gene spans over 2.5 kb sequences along with a 1.7 kb of 5′-flanking sequence. The PAP-III coding sequence spanned over six exons. There were striking similarities between PAP-III and PAP-I and II genes, in genomic organization as well as in promoter sequences. Moreover, the rat PAP-III gene was mapped to chromosome 4 which coincides with that of PAP-I and II genes. The three genes are derived from same ancestral gene by duplication. Expression of the PAP-III gene was compared with that of PAPs I and II (Dusetti et al. 1995). Rat cDNA and a regeneration-promoting gene encodes a 174-amino-acid (aa) Reg-III protein with a 25-aa signal peptide. Reg-III was expressed in regenerating pancreatic islets, but not in normal islets (Suzuki et al. 1994). Analysis of the open reading frame of a cDNA indicated that the deduced protein from the mRNA was a polypeptide of 174 amino acids, unexpectedly similar to that of rat PAP-II/Reg-III (Honda et al. 2002). However, the length of the identified mRNA (1,467 bp) was longer than that of rat PAP-II mRNA (885 bp), because the elongated mRNA was generated through a different polyadenylation site in the same gene. The elongated mRNA after acute pancreatitis was strongly induced in the restricted early phase, in comparison with the original mRNA. It was suggested that the elongated mRNA affects the function of PAP-II/Reg-III protein because the elongated mRNA with long three; untranslated regions is known to be involved in the translation efficiency. The identified mRNA may play an important role in the progression of pancreatitis (Honda et al. 2002). Proteomic approach identified that rat Reg-III could be functionally associated with Reg I (Shin et al. 2005).

39.7.2 Human Reg-III or HIP/PAP

Reg-III was strongly induced in gut epithelial cells following bacterial reconstitution, as well as in the colitis initiated by DSS. The mRNA expression of HIP/PAP was enhanced in colonic epithelial cells of patients with inflammatory bowel disease (IBD). Reg-III mRNA expression was localized in the epithelial cells including goblet cells and columnar cells in mice; on the other hand, HIP/PAP-expressing cells were correlated with Paneth cell metaplasia in human colon. Epithelial expression of Reg-III or HIP/PAP was induced under mucosal inflammation initiated by exposure to commensal bacteria or DSS as well as inflamed IBD colon (Dieckgraefe et al. 2002). *Salmonella* infection increased ileal mucosal PAP/Reg-III protein levels in enterocytes located at the crypt-villus junction. Increased colonization and translocation of *Salmonella* was associated with higher ileal mucosal PAP/Reg-III levels and secretion of this protein in feces. PAP/Reg-III protein is increased in enterocytes of the ileal mucosa during *Salmonella* infection and is associated with infection severity. Fecal PAP/Reg-III might be used as a new and non-invasive infection marker (van Ampting et al. 2009).

REG-IIIα and REG-1α are activated in many human hepatocellular carcinomas (HCC), REG-IIIα is down-regulated in most primary human gastric cancer cells. In gastric cancer REG-IIIα might be useful in the diagnosis of cancer (Choi et al. 2007). Two genes involved in HCC belong to REG family. They encode the regenerating islet-derived 3 α (REG-IIIα/HIP/PAP/REG-III) and 1α (REG-1α) proteins, both involved in liver and pancreatic regeneration and proliferation. REG-IIIα is a target of β-catenin signaling in Huh7 hepatoma cells. The upregulation of REG-IIIα and REG-Iα expression is significantly correlated to the β-catenin status in HCC and hepatoblastomas. The Wnt/β-catenin signaling pathway is activated in HCC. Evidence shows that both genes are downstream targets of the Wnt pathway during liver tumorigenesis (Cavard et al. 2006). Results predict that the REGα and REGγ inserts play virtually no role in oligomerization or in proteasome activation. By contrast, removal of REGβ insert reduces binding of this subunit and REGα/REGβ oligomers to proteasomes. However, findings showed that REG inserts are not required for binding and activating the proteasome (Zhang et al. 1998).

Reg-III proteins bind their bacterial targets via interactions with cell wall peptidoglycan but lack the canonical sequences that support CRD in other C-type lectins. HIP/PAP recognizes the peptidoglycan carbohydrate backbone in a calcium-independent manner via a conserved "EPN" motif that is critical for bacterial killing. While EPN sequences govern calcium-dependent carbohydrate recognition in other C-type lectins, the unusual location and calcium-independent functionality of the HIP/PAP EPN motif suggest that this sequence is a versatile functional module that can support both calcium-dependent and calcium-independent carbohydrate binding. Further, HIP/PAP binding affinity for carbohydrate ligands depends on carbohydrate chain length, supporting a binding model in which HIP/PAP molecules "bind and jump" along the extended polysaccharide chains of peptidoglycan, reducing dissociation rates and increasing binding affinity. The dynamic recognition of highly clustered carbohydrate epitopes in native peptidoglycan is an essential mechanism governing high-affinity interactions between HIP/PAP and the bacterial cell wall.

39.8 Rat PAPs

39.8.1 Three Forms of PAP in Rat

Rat PAP-I, PAP-II, and PAP-III are members of a multigenic family of proteins expressed in several tissues. Rat PAP appears in pancreatic juice after induction of prancreatic inflammation. In acute pancreatitis in rats, PAP was first observed 6 h after induction of pancreatitis with taurocholate or cerulein, reached maximal levels in zymogen granules and in pancreatic tissue during the acute phase (48 h), and disappeared during recovery (day 5). It was never detected in spleen, liver, kidney, heart, or lung. The PAP-I is therefore an acute-phase protein that differs from other proteins of that family because of its exocrine nature. PAP-I is expressed in the pancreas in relation to the severity of cerulein-induced pancreatitis (Keim et al. 1991; Keim et al. 1994). PAP-I was shown to be antiapoptotic, mitogenic, and anti-inflammatory and can promote cell adhesion to the extracellular matrix. PSP/LIT/RegIα can be mitogenic. Because polymerization might regulate activity, the ability of rat PAP-I was examined to interact with itself (homodimerization), PAP-II, PAP-III, and PSP/LIT/RegIα (heterodimerization). PAP-I interacted significantly with all members of the PAP protein family, homodimerization showing the strongest interaction as judged by the β-galactosidase test (Bodeker et al. 1999). A change in gene expression of PAP reflects acute inflammatory changes in the pancreas most sensitively (Funakoshi et al. 1995).

In vitro, dexamethasone and IL-6 induced a marked transcription of PAP-I, II and III genes in AR42J cells. In vivo, pancreas mRNA levels of PAP-I, II or III increased by 2.6-fold, 1.9-fold, and 1.3-fold respectively after dexamethasone treatment. Histopathologic evaluation revealed less inflammation and necrosis in pancreata obtained from dexamethasone treated animals. Thus, dexamethasone significantly decreases the severity of pancreatitis. The protective mechanism of dexamethasone may be via upregulating PAP gene expression during injury (Kandil et al. 2006).

The PAP-I and III, but not PAP-II mRNAs are constitutively expressed in the small intestine of rats. Between day 20 of gestation and day 21 of age, PAP mRNAs could barely be detected. Their concentrations increased dramatically from day 21 to day 45 of age and remained constant thereafter. Rats adapted to a diet with low carbohydrate content showed a significant decrease in PAP mRNA concentrations. Gene expression of PAP-I and III mRNAs is regulated in a coordinate manner in the rat small intestine during development and on nutritional and hormonal manipulations (Sansonetti et al. 1995).

Rat PAP mRNA is barely detectable in normal pancreas and overexpressed during acute pancreatitis (Iovanna et al. 1991). Rat PAP mRNA was constitutively expressed in duodenum, jejunum, and ileum, at similar levels as in pancreas during the acute phase of pancreatitis. A weak expression was also detected in several other tissues. Serum from rats with acute pancreatitis (SAP) induced the expression of PAP mRNA in AR-42J cells. in a dose-dependent manner. It suggests that SAP contains factors responsible for the PAP mRNA expression and that the cis-acting elements are localized within the 1.2 kbp upstream region of the transcription initiation site (Dusetti et al. 1994b). Free radicals are involved in the pathogenesis of acute pancreatitis, during which PAP-I is overexpressed. PAP-I expression could be induced by oxidative stress and could affect apoptosis. It appears that during oxidative stress, PAP-I might be part of a mechanism of pancreatic cell protection against apoptosis (Ortiz et al. 1998).

The mRNA level of a PAP-like protein was found to be elevated in the ileal Peyer's patch of lambs during the early phase of scrapie infection. The ovine PAP-like protein cDNA encodes a putative 178 amino acid protein with a signal peptide and a C-lectin binding domain. REG/PAP protein's deduced amino acid sequences were conserved. The overall amino acid identity between the ovine PAP-like protein and bovine, human and rat REG/PAP proteins varied from 23% to 85%. The expression of the ovine PAP-like protein mRNA was restricted to the ileal and jejunal Peyer's patches. The data provided will offer the possibility to search for a link between this PAP-like protein and early events in the development of scrapie (Skretting et al. 2006).

Although induction of PAP family genes has been reported in peripheral nerve injury models, the expression of PAP in CNS after traumatic brain injury (TBI) induced by weight drop was examined by Ampo et al. (2009). There was a significant upregulation of PAP-I and PAP-III mRNA in the injured cortex beginning at 1 day after TBI. PAP-I and PAP-III staining was localized in a subpopulation of neurons in the peri-injured region. Expression of both PAP-I and PAP-III mRNA was observed following a transient increase in inflammatory cytokines, including TNF-α, IL-6, and IL-1β mRNA. It appears that expression of PAP family members in response to traumatic and inflammatory stimuli are not restricted to the pancreas, intestine, and peripheral nervous system, but are likely a more general cellular response, including the CNS in rat (Ampo et al. 2009). Studies suggest that the Reg family of proteins is protective in acute pancreatitis (Viterbo et al. 2009).

39.8.2 PAP-I/Reg-2 Protein (or HIP, p23)

The rat PAP gene was characterized over 3.2 kb of gene sequence and 1.2 kb of 5′-flanking sequence. Several potential regulatory elements were identified in the promoter region, including a pancreas-specific consensus sequence. The PAP coding sequence spanned over six exons. The first three exons encoded the 5′-untranslated region of the mRNA, the signal peptide, and 39 amino acids of the NH2-terminal end of the mature protein, respectively. The other three exons encoded a domain of the protein with significant homology to the CRD of animal lectins (Dusetti et al. 1993). Analysis of the rat PAP-I promoter indicated that the region between nt −180 and −81 possessed silencer activity in cells that did not express PAP-I. Transient transfection assays revealed that the sequence with silencer activity was located within the rep27 region (position −180/−153). Suppressor activity was observed when rep27 was inserted upstream from the core PAP-I promoter, in both orientations. Results suggest that the rep27 cis-acting element contributes to the tissue specific expression of PAP-I gene (Ortiz et al. 1997). The promoter of the PAP-I gene represents a potential candidate to drive expression of therapeutic molecules to the diseased pancreas. Studies show that (1) a recombinant adenovirus containing a fragment of the PAP-I promoter allows specific targeting of a reporter gene to the mouse pancreas and (2) expression of the reporter gene in pancreas is induced during acute pancreatitis (Dusetti et al. 1997).

A correlation between PAP-I mRNA levels and glutathione levels seems to exist in the mouse pancreas (Fu et al. 1996). In experimental pancreatitis a correlation was found between the severity of pancreatitis and the amount of PAP in pancreatic homogenates (Keim and Löffler 1986). Consumption of alcohol for short term does not alter serum PAP values during 56 h after drinking. However, longterm drinking induced at least a 10-fold increase in serum PAP. Nordback et al. (1995) support the suggestion that heavy long term drinking often induces subclinical pancreatic damage, but not clinical pancreatitis (Nordback et al. 1995).

Oxidative stress has an important role in the pathogenesis of pancreatitis. PAP-1 is a protein secreted upon induction of acute pancreatitis. As a result, oxidative stress induced PAP-1 mRNA expression in AR42J cells in a time-dependent manner. Cell viability decreased with the concentration of glucose oxides delivered to the cells that had received glucose. Oxidative stress-induced PAP-1 expression was augmented in cells transfected with PAP-1S cDNA. PAP-1 induction by oxidative stress decreased in the cells transfected with PAP-1 AS cDNA. PAP-1 may be a defensive gene for oxidative stress-induced cell death of pancreatic acinar cells (Lim et al. 2009).

Induction of PAP-I gene was also described in liver during hepatocarcinogenesis. It was noted that expression of PAP-I in hepatocarcinoma (HCC) occurred through inactivation of its silencer element and was not concomitant in all malignant cells. On that basis, PAP-I was assayed in serum from patients with chronic hepatitis, liver cirrhosis or hepatocarcinoma. PAP-I levels were normal in chronic active or persistent hepatitis, significantly higher in cirrhosis and strongly elevated in hepatocarcinoma. Because those clinical entities often develop in that sequence, serum PAP-I appeared as a potential marker of hepatocarcinoma development (Dusetti et al. 1996). However PAP assay can only be recommended in cases of justified suspicion of HCC with negative α-fetoprotein (Montalto et al. 1998). PAP-I was not found in normal melanocytes, melanoma tumors, and melanoma cell lines, even after stress induction. Exogenous PAP-I can modify the adhesion and motility of normal and transformed melanocytes, suggesting a potential interaction with melanoma invasivity (Valery et al. 2001).

39.8.3 PAP-II/PAP2

Frigerio et al. (1993b) cloned two overlapping cDNAs encoding a protein structurally related to PAP. The second PAP, which was called PAP-II, was of same size as the original PAP (PAP-I) and showed 74.3% amino acid homology. PAP-II mRNA concentration increased within 6 h following induction of pancreatitis, reached maximal levels

(>200 times control values) at 24–48 h, and decreased thereafter, similar to PAP-I. However, PAP-II mRNA could not be detected in the intestinal tract or in other tissues. PAP-II genomic DNA fragment was characterized over 2.7 kb of gene sequence and 1.9 kb of 5′ flanking sequence. The PAP-II coding sequence spanned six exons separated by five introns. Several potential Regulatory elements were identified in the promoter region, including two glucocorticoid-response elements and one IL-6-response element. Grønborg et al. (2004) reported a protein that is 85% identical to HIP/PAP and was designated as PAP-II. Two transcripts are generated from PAP-II gene in rat pancreas by alternative splicing in the 5′ untranslated region as shown by the existence of two forms of PAP-II mRNA with identical coding sequence but a different 5′-untranslated region. We demonstrate that this is the result of a differential splicing. (Vasseur et al. 1995).

NR8383 macrophages which were cultured in the presence of PAP2 aggregated and exhibited increased expression of IL-1, IL-6, TNF-α, and IL-10. Chemical inhibition of the NFkB pathway abolished cytokine production and PAP facilitated nuclear translocation of NF-kB and phosphorylation of IkBα inhibitory protein suggesting that PAP2 signaling involves this pathway. Similar findings were observed with primary macrophages derived from lung, peritoneum, and blood but not spleen. Furthermore, PAP2 activity was inhibited by the presence of serum, inhibition which was overcome with increased PAP2. These results demonstrate a new function for PAP2: it stimulates macrophage activity and likely modulates the inflammatory environment of pancreatitis (Viterbo et al. 2008b).

In rats, each of the three PAP-I isoforms has independent immunologic function on macrophages. PAP2 up-regulates inflammatory cytokines in macrophages in a dose-dependent manner and acts through NF-kB mechanisms. Truncation of the first 25 residues on the N terminus of PAP2 did not affect protein activity whereas truncation of the last 30 residues of the C terminus of PAP2 completely inactivated the function of PAP2. Additionally, reduction of three disulfide bonds proved to be important for the activity of this protein. Further investigation revealed two invariant disulfide bonds were important for activity of PAP2 while the disulfide bond that is observed in long-form C-type lectin proteins was not essential for activity. Further, preincubation with select rPAP2 mutant proteins affect translocation of this transcription factor into the nucleus (Viterbo et al. 2008a).

39.8.4 PAP-III

A third member of the rat PAP gene family was described by Frigerio et al. (1993a). The encoded protein, designated PAP-III, shows 66% and 63% identity with the rat PAP-I and II, respectively. The PAP-III gene is constitutively expressed in the small intestine and in the pancreas with acute pancreatitis, but not in the healthy pancreas. In vitro experiments revealed that PAP-III possessed a strong macrophage chemoattractant activity that was comparable with that of monocyte chemoattractant protein-1. PAP-III is involved in peripheral nerve regeneration and provides new insights into Schwann cell-macrophage interactions and therapeutic interventions (Namikawa et al. 2006). The ovulatory process in mammals involves gross physiological events in the ovary that cause transient deterioration of the ovarian connective tissue and rupture of the apical walls of mature follicles. A study revealed that the ovulatory events included induction of mRNA for PAP-III. In situ hybridization indicated that PAP-III mRNA expression was limited mainly to the hilar Region of the ovarian stroma, with most of the signal emanating from endothelial cells that lined the inner walls of blood vessels, and from small secondary follicles. Ovarian transcription of PAP-III mRNA was moderately dependent on ovarian progesterone synthesis. In conclusion, the present evidence of an increase in PAP-III gene expression in gonadotropin-stimulated ovaries provides further evidence that the ovulatory process is comparable to an inflammatory reaction (Yoshioka et al. 2002).

PAP-III is expressed with increased frequency in the bladder urothelium in a rat cystitis model and associated with bladder inflammation and implicates PAP-I in the abnormal sensation in cystitis (Takahara et al. 2008). Reg-III kills Gram-positive bacteria and plays a vital role in antimicrobial protection of the mammalian gut. Reg-III proteins bind their bacterial targets via interactions with cell wall peptidoglycan but lack the canonical sequences that support calcium-dependent carbohydrate binding in other C-type lectins.

Reg-IIIβ is found in many areas of the body and shown to play an important role in both the development and regeneration of subsets of motor neurons. The Reg-IIIβ expressed by motor neurons is both an obligatory intermediate in the downstream signaling of the leukemia inhibitory factor/ciliary neurotrophic factor (CNTF) family of cytokines, maintaining the integrity of motor neurons during development, as well as a powerful influence on Schwann cell growth during regeneration of the peripheral nerve. In mice, Reg-IIIβ positive motor neurons are concentrated in cranial motor nuclei that are involved in the patterning of swallowing and suckling. Suckling was impaired in Reg-IIIβ KO mice and correlated this with a significant delay in myelination of the hypoglossal nerve. In summary, Reg-IIIβ has an important role in the developmental fine-tuning of neonatal motor behaviors mediating the response to peripherally derived cytokines and growth factors and regulating the myelination of motor axons (Tebar et al. 2008).

39.9 Functions of PAP

39.9.1 PAP: A Multifunctional Protein

The fact that PAP is secreted by pancreatic acinar cells into the pancreatic juice initially suggested a role for this protein in pancreatic juice homeostasis. Pancreatic juice is supersaturated in $CaCO_3$ and, in the absence of physiological inhibitors this salt will precipitate in crystal formation. In vitro, the rat HIP/PAP has been reported to prevent cell death in neuronal primary cultures (Nishimune et al. 2000) and pancreatic cells (Malka et al. 2000), promote the growth of epithelial intestinal cells (Moucadel et al. 2001), and stimulate DNA synthesis in Schwann cells (Livesey et al. 1997). The hamster homologue to HIP/PAP, called INGAP peptide (section 39.12), has been shown to enhance nerve outgrowth from explanted dorsal root ganglia (Tam et al. 2004). In vivo, HIP/PAP can function as an acute phase reactant in human pancreatitis (Orelle et al. 1992). Reports also suggest that HIP/PAP could regulate both viability and proliferation in hepatocytes. Human HIP/PAP expressed in hepatocytes exhibits mitogenic and an antiapoptotic properties in primary cultures and liver regeneration after partial hepatectomy is stimulated in HIP/PAP transgenic mice (Simon et al. 2003). HIP/PAP accelerates liver regeneration and protects against acetaminophen injury in mice (Lieu et al. 2005). Despite the obvious interest of these observations, it is difficult to link these effects with the enormous amount of PAP released by pancreas during acute pancreatitis or cellular stress.

39.9.2 PAP in Bacterial Aggregation

Evidence from a large number of studies seemed to indicate that PAP had a variety of activities, which included anti-inflammatory, anti-apoptotic, proliferative, and antibacterial effects (Closa et al. 2007; De Reggi and Gharib 2001; Graf et al. 2006; Iovanna and Dagorn 2005). With respect to the antimicrobial activity, PAP and PSP were shown to induce bacterial aggregation without inhibiting growth, and trypsin appeared to stimulate the effect of PSP (Iovanna et al. 1991, 1993). Cash et al. (2006a) demonstrated that PAP can bind to the peptidoglycan layer of Gram positive bacteria and exert a direct bactericidal effect. PAP also bound to chitin and the mannose polymer mannan, indicating that PAP recognizes carbohydrate patterns (Cash et al. 2006a, b). Cash et al. (2006a) proposed that PAP-I is part of the innate immune system and its intestinal expression plays an important role in the maintenance of the intestinal bacterial flora (Cash et al. 2006a). By analogy, in the pancreas PAP-I is probably upregulated during acute panceatitis to prevent bacterial infection.

PAP and its paralog PSP induce bacterial aggregation. It is proposed that insoluble PAP might be the biologically active form. PAP has been shown to bind to the peptidoglycan of Gram positive bacteria and to exert a direct bactericidal effect. Medveczky et al. (2009) showed that N-terminal cleavage of PAP by trypsin at the Arg^{37}-Ile^{38} peptide bond or by elastase at the Ser^{35}-Ala^{36} peptide bond is a prerequisite for binding to the peptidoglycan of the Gram positive bacterium *Bacillus subtilis*. Trypsin-mediated processing of PAP resulted in the formation of the characteristic insoluble PAP species, whereas elastase-processed PAP remained soluble. N-terminally processed PAP induced rapid aggregation of *Bacillus subtilis* without significant bacterial killing. Thus, N-terminal processing is necessary for the peptidoglycan binding and bacterial aggregating activity of PAP and that trypsin-processed and elastase-processed forms are functionally equivalent. The observations also extend the complement of proteases capable of PAP processing, which now includes trypsins, pancreatic elastases and bacterial zinc metalloproteases of the thermolysin type.

It has been demonstrated that Reg-IIIγ and HIP/PAP are directly bactericidal for Gram-positive bacteria at μM concentrations (Mukherjee et al. 2009), revealing a unique biological function for mammalian lectins. The bactericidal activities of Reg-IIIγ and HIP/PAP depend on binding to cell wall peptidoglycan, a polymer of alternating N-acetylglucosamine (GlcNAc) and N-acetylmuramic acid (MurNAc) cross-linked by short peptides. This finding identified Reg-III lectins as a distinct class of peptidoglycan binding proteins. However, given the lack of canonical carbohydrate binding motifs in Reg-III proteins, the molecular basis for lectin-mediated peptidoglycan recognition remains unknown. Furthermore, Reg-III lectins are secreted into the intestinal lumen where there are abundant soluble peptidoglycan fragments derived from the resident microbiota through shedding or enzymatic degradation of the bacterial cell wall. Thus, it remains unclear how Reg-III lectins selectively bind to bacterial surfaces without competitive inhibition by soluble peptidoglycan fragments in the luminal environment.

39.9.3 PAP-I: An Anti-inflammatory Cytokine

PAP-I has been implicated in the endogenous regulation of inflammation. In addition to its role in acute pancreatitis, PAP-I has also been associated with inflammatory diseases, such as Crohn's disease. During acute pancreatitis, PAP-I may contribute to stress response to control bacterial proliferation (Iovanna et al. 1991). In pancreatic AR4-2J cells, PAP-I is one of the effectors for the TNF-α-induced apoptosis inhibition (Malka et al. 2000). Although PAP-II is first isolated as a pancreatic secretory protein that contributes to pancreatic regeneration, it is up-regulated during the acute phase of pancreatitis and likely modulates the inflammatory environment of pancreatitis (Iovanna et al. 1991; Cavard et al. 2006; Choi et al. 2007; Dusetti et al. 1994; Lasserre et al. 1992; Viterbo et al. 2008). PAP-II is found to inhibit TNF-α-mediated inflammatory responses (Vasseur et al. 2004; Gironella et al. 2005). In the exocrine pancreas, PAP-I is associated with pancreatic acinar cell and protects cells from oxidative stress and TNF-α-induced pancreatic stress (Malka et al. 2000; Ortiz et al. 1998). Moreover, expression of PAP-II is also increased in gut epithelial cells in human inflammatory bowel disease (Ogawa et al. 2003).

PAP2 mediates the expression of inflammatory cytokines in macrophages through the NF-kB pathway (Viterbo et al. 2008). It was shown that both antisense gene knockdown and Ab neutralization of PAP2 in rats with experimental acute pancreatitis caused a significant increase in disease severity (Zhang et al. 2004). These findings corroborate other studies which showed protective roles served by PAP proteins during tissue injury (Vasseur et al. 2004; Kandil et al. 2006; Gironella et al. 2005). Therefore, PAP proteins are key regulators of inflammation and their absence causes a dysregulated inflammatory process. Truncation of the first 25 residues on the N terminus of PAP2 did not affect protein activity whereas truncation of the last 30 residues of the C terminus of PAP2 completely inactivated the function of PAP2. In addition, reduction of three disulfide bonds proved to be important for the activity of this protein (Viterbo et al. 2008).

Several reports suggest that Reg proteins could also be functional in the nervous system and expressed in the brain during development and Alzheimer's disease (Watanabe et al. 1994; de la Monte et al. 1990) (section 39.5.3.1). PAP-I may contribute to the signaling pathway of ciliary neurotrophic factor and is involved in the regeneration and survival of motor neurons (Livesey et al. 1997; Nishimune et al. 2000). The expression of PAP-I has been found to be induced in urinary tract afferent neurons following cyclophosphamide-induced cystitis (Takahara et al. 2008), suggesting its potential role in the abnormal sensation in cystitis. Moreover, the PAP-I expression in isolectin B4 (IB4)-positive small DRG neurons is up-regulated and followed by a dynamic shift from small to large DRG neurons after peripheral nerve injury (Averill et al. 2002). Increased expression of PAP-I also occurs in neurons following traumatic brain injury (Ampo et al. 2009). These data indicate that PAP-I expression in response to injury and inflammation could be a general response in the pancreas, intestine, and both peripheral and central nervous systems. However, it remains unclear whether PAP-II is involved in the response of primary sensory neurons to the stimulations of peripheral inflammation and nerve injury. A recent study shows that the expression of PAP-II is strongly induced in DRG neurons following peripheral tissue inflammation and nerve injury, suggesting an involvement of PAP-II in the signal processing of the spinal sensory pathways in chronic pain states. PAP-II may play potential roles in the modulation of spinal sensory pathways in pathological pain states (He et al. 2010).

While PAP has been shown to be anti-bacterial and anti-apoptotic in vitro, its definitive biological function in vivo is not clear. In vivo evidence indicate that PAP mediates significant protection against pancreatic injury. PAP may exert its protective function by suppressing local pancreatic as well as systemic inflammation during acute pancreatitis (Heller et al. 1999; Zhang et al. 2004). During pancreatitis, PAP released by the pancreas could mediate lung inflammation through induction of hepatic TNFα expression and subsequent increase in circulating TNFα (Folch-Puy et al. 2003). Since, PAP-I and IL-10 responses share several features, Folch-Puy et al. (2006) assessed their expression and involvement of JAK/STAT and NF-kB signaling pathways in the suppression of inflammation mediated by PAP using pancreatic acinar cell line (AR42J). PAP-I inhibits the inflammatory response by blocking NF-kB activation through a STAT3-dependent mechanism. Important functional similarities to the anti-inflammatory cytokine IL-10 suggested that PAP-I could play a role similar to that of IL-10 in epithelial cells (Folch-Puy et al. 2006).

PAP-I is able to activate the expression of the anti-inflammatory factor: suppressor of cytokine signaling (SOCS)3 through the JAK/STAT3-dependent pathway. The JAK/STAT3/SOCS3 pathway seems to be a common point between PAP and several cytokines. Therefore, it is reasonable to propose that PAP-I is an anti-inflammatory cytokine (Closa et al. 2007). A model of caerulein-induced pancreatitis was used to compare the outcome of pancreatitis in PAP/HIP$^{-/-}$ and wild-type mice. PAP can be strongly induced by IL-6 and IL-10 and IL-10-related cytokines through a STAT3-mediated pathway. The expression of PAP itself appears to be induced in pancreatic acinar cells by the presence of PAP in the medium. This is also related to the activation of STAT3 pathway since at least two functional STAT-responsible elements have been reported in the promoter of the PAP gene (Dusetti et al. 1995b). This self-induction suggests the existence of a positive feedback mechanism in pancreatic acinar cells via a PAP receptor and a cross-talk with other cytokines (Closa et al. 2007).

Arginine-induced pancreatic acinar cell injury has been reported in vivo. Arginine inhibits the proliferation of pancreatic aciner AR4-2J cells. The anti-proliferation by arginine was due to an increase in apoptosis. Studies suggest that arginine induces apoptosis and PAP gene expression in pancreatic acinar cells and that PAP might inhibit the induction of apoptosis (Motoo et al. 2000). The anti-apoptotic and anti-inflammatory functions described in vitro for PAP/HIP have physiological relevance in the pancreas in vivo during caerulein-induced pancreatitis (Gironella et al. 2007). Assuming a protective role of PAP, Li et al. (2009) showed that PAP had no significant effect on proliferation and migration on human pancreatic stellate cell (PSCs). Cell-associated fibrillar collagen types I and III and fibronectin increased after addition of PAP to PSCs. PAP diminished the expression of MMP-1 and −2 and TIMP-1 and −2 and their concentrations in PSC supernatants. Studies offer new insights into the biological functions of PAP, which may play an important role in wound healing response and cell-matrix interactions (Li et al. 2009). The RII α regulatory subunit of cAMP-dependent protein kinase (PKA) is a binding partner of HIP/PAP. The HIP/PAP co-immunoprecipitates with RIIα in HIP/PAP expressing cells. Increase in basal PKA activity in HIP/PAP expressing cells suggests that HIP/PAP may alter PKA signaling (Demaugre et al. 2004).

Ferrés-Masó et al. (2009) described the intracellular pathways triggered by PAP-I in a pancreatic acinar cells and showed that PAP-I increased the transactivation activity of *PAP-I* and the binding on its promoter of the nuclear factors C/EBPβ, P-CREB, P-ELK1, EGR1, STAT3, and ETS2, which are downstream targets of MAPK signaling. p44/42, p38, and JNK MAPKs activity increased after PAP-I treatment. In addition, pharmacological inhibition of these kinases markedly inhibited the induction of PAP-I mRNA. These results indicated that the mechanism of PAP-I action involves the activation of the MAPK superfamily (Ferrés-Masó et al. 2009).

39.10 Panceatitis Associated Protein(PAP)/Hepatocarcinoma-Intestine Pancreas (HIP)

39.10.1 HIP Similarity to PAP

Lasserre et al. (1994) identified a gene (named HIP) the expression of which is markedly increased in 25% of human primary liver cancers. HIP mRNA expression is tissue specific since it is restricted to pancreas and small intestine. Pleiotropic biological activities have been ascribed to this protein, but little is known about the function of HIP/PAP in the liver. HIP protein consists in a signal peptide linked to a CRD, typical of C-type lectins without other binding domains. The analysis of HIP/PAP gene indicates that the HIP/PAP CRD is encoded by four exons, a pattern shared with all members of this group of proteins. This common intron-exon organization indicates an ancient divergence of the free CRD-lectin group from other groups of C-type lectins. Lasserre et al. (1994) provided evidence for the localization of HIP/PAP on chromosome 2, suggesting previous duplication of HIP/PAP and the related Reg Iα and Reg Iβ genes from the same ancestral gene. The sequence of the 5′ upstream region of the HIP gene shows several potential regulatory elements which might account for the enhanced expression of the gene during pancreatic inflammation and liver carcinogenesis (Lasserre et al. 1994). In normal liver, the protein is undetectable in normal mature hepatocytes and found only in some ductular cells, representing potential hepatic progenitor cells. Itoh et al. (1995) especially focussed on the 5′-flanking region, which spans about 3 kb and is composed of six exon. Exon 1 encodes the 5′-noncoding sequence and exon 2 consists of three miniexons, 2a, 2b and 2c; the common exon 2c encodes the sequence including the start codon. Analysis revealed the presence of at least three different types 5′-ends of human PAP/HIP transcripts which were derived from alternative use of 5′-exons. Although all three types of transcripts were expressed in both normal small intestine and pancreas, their gene expression was increased ectopically in gastric cancer, hepatocellular cancer and pancreatic acinar cell carcinoma. Furthermore, significant differences among the transcript types were detected between normal and tumor tissues, and especially between gastric and hepatocellular cancers, suggesting that PAP/HIP expression may vary with differences in 5′-alternative splicing.

Simon et al. (2003) proposed that HIP/PAP acts as a hepatic cytokine that combines mitogenic and anti-apoptotic functions through the PKA signaling pathway and consequently acts as a growth factor in vivo to enhance liver regeneration. The HIP/PAP (=human Reg-2) encoding gene is activated in primary liver cancers. The involvement of HIP/PAP has been suggested in the pathophysiology of interstitial cystitis (IC) patients, because, the urinary HIP/PAP levels were significantly higher and the HIP/PAP expression in the bladder urothelium was more frequently apparent in IC patients (Makino et al. 2010). The rHIP regulatory sequence is a potent liver tumor-specific promoter for the transfer of therapeutic genes, and AdrHIP-NIS-mediated ^{131}I therapy is a valuable option for the treatment of multinodular HCC (Herve et al. 2008).

To test whether HIP/PAP-I is a target of islet-directed autoimmunity, Gurr et al. (2002) measured splenic T-cell responses against HIP/PAP in NOD mice. Differential cloning of Reg from islets of a type 1 diabetic patient and the response of Reg to the cytokine IL-6 suggests that HIP/PAP becomes overexpressed in human diabetic islets because of the local inflammatory response and. HIP/PAP acts as a T-cell autoantigen in NOD mice. Therefore, autoimmunity to HIP/PAP might create a vicious cycle, accelerating the immune process leading to diabetes (Gurr et al. 2002; Shervani et al. 2004).

Laurine et al. (2005) focused on two proteins, REG Iα and newly identified PAP-IB, belonging to more closely related FI and FII families, respectively. REG Iα and PAP-IB share 50% sequence identity. PAP-IB was expressed almost only in pancreas, unlike REG Iα, whose expression was ubiquitous. In addition, the two proteins displayed distinctive surface charge distribution, which may lead to different ligands binding. In spite of their common fold that should result in closely related functions, REG Iα and PAP-IB are a good example of duplication and divergence, probably with the acquisition of new functions, thus participating in the evolution of the protein repertoire.

39.10.2 Characterization of HIP/PAP

Pancreatitis-associated protein hepatocarcinoma-intestine-pancreas (PAP/HIP) is 16-kDa basic protein with an isoelectric point in the range of pH 6.5–7.6. Although most species contain a single PAP form, rat contains three isoforms, PAP-I, PAP-II, and PAP-III, transcribed from three separate genes (Iovanna et al. 1991; Frigerio et al. 1993a, b). PAP levels increase in pancreatic juice during experimental and clinical pancreatitis (Keim et al. 1992). Although showing an acute phase response under conditions of pancreatic disease, the function of PAP is not clear. Several studies have demonstrated that PSP and PAP both act as acute phase reactants in pancreatic juice under a variety of conditions including acute pancreatitis (Keim et al. 1994), post-weaning period (Bimmler et al. 1999). Trypsin cleavage of PSP and PAP has resulted in the appearance of precipitated proteins believed to represent insoluble thread structures in humans (Gross et al. 1985a) and cows (Gross et al. 1985b). It is difficult to understand how precipitation properties could serve a useful function in pancreatic physiology.

The PAP coding sequence spans over six exons and the putative CRD is encoded by exons IV, V, and VI. This gene organization suggests that PAP belongs to C-type lectins, which have evolved from the same carbohydrate-recognition domain ancestral precursor through a different process. It is interesting to note that PAP is the smallest protein reported among the C-type lectins. In fact, it comprises a single CRD linked to a signal peptide whereas other C-type lectins contain the sugar-binding consensus combined with a variety of other protein domains which confer specific functions of lectins. In contrast, PAP does not have additional functional domains. One PAP gene (HIP/PAP) was described by Keim et al. as a 16 kDa secretory protein in rat pancreatic juice which appeared upon induction of pancreatic inflammation and represented up to 5% of total protein (Keim et al. 1991). The human ortholog was isolated in 1992 from pancreatic juice of diabetic patients who underwent combined kidney and pancreas transplantation (Keim et al. 1992). In these patients the donor pancreas developed acute pancreatitis and levels of PAP-I in the juice reached up to 7.5% of the total secretory protein. Because PAP was hardly detectable in the normal pancreatic juice, but became upregulated in inflammation, it was considered a unique acute phase protein, which was targeted to the exocrine secretions. Subsequent cDNA cloning revealed that PAP exhibited homology to the carbohydrate binding region of Ca^{2+}-dependent (C-type) lectins and thus PAP belongs to a larger secretory protein family which contain only a single C-type lectin domain (Iovanna et al. 1991; Orelle et al. 1992). Members of this protein family, which includes pancreatic stone protein (PSP), have been described in excellent reviews (Closa et al. 2007; De Reggi and Gharib 2001; Graf et al. 2006; Iovanna and Dagorn 2005). PAP expression was also detected in the intestines and other extrapancreatic tissues, but the physiological function of PAP in the pancreas or elsewhere has remained contentious.

Graf et al. (2001) demonstrated that trypsin cleaved all three isoforms of PAP at N-terminus after Arg^{37} (Arg^{11} in the mature PAP protein), which in turn forms insoluble fibrils and the processed PAP became resistant to proteases and formed insoluble, fibrillar structures (Graf et al. 2001; Schiesser et al. 2001). Specific PAP-I isoforms have been shown to be normal constituents of intestinal paneth cells (McKie et al. 1996) and, more importantly, serve to maintain gut microbial integrity. PAP proteins also serve important roles within the nervous system, being involved in motor neuron regeneration (Namikawa et al. 2005). Coupling its diffuse expression pattern with the interspecies conservation of this group of proteins supports important functional roles for this network of proteins. In addition, HIP/PAP-I is a promising candidate for the prevention and treatment of liver failure (Christa et al. 2000; Lieu et al. 2005).

The secretory human PAP was structurally related to rat PAP and had the same size as rat PAP. It showed 71% amino acid identity, the six half-cystines being in identical positions. Domains of the proteins showing homologies with calcium-dependent lectins were also conserved. In addition, expression in pancreas of the genes encoding the human protein and rat PAP showed similar characteristics (Orelle et al. 1992). The human PAP synthesis increases during inflammation and its use as biological marker of acute pancreatitis has been suggested. The human PAP gene spans 2748 bp and contains six exons interrupted by five introns. The gene has a typical promoter containing the sequences TATAAA and CCAAT 28 and 52 bp upstream of the cap site, respectively. The human PAP gene was mapped to chromosome 2p12. This localization coincides with that of the Reg/PSP/LIT gene, which encodes a pancreatic secretory protein structurally related to PAP (Dusetti et al. 1994a).

39.10.3 PAP Ligands (PAP Interactions)

HIP/PAP protein has a very high affinity for D-lactose, but insignificantly to α-D-glucopyroanose, α-L-fucose, α-D-galactopyranose, or N-acetyl-β-D-glucosamine (Christa et al. 1994). It promotes rat hepatocyte adhesion in primary culture, and interacts with the proteins of the extracellular matrix (Christa et al. 1996). NMR studies showed that HIP/PAP recognizes the peptidoglycan carbohydrate backbone in a calcium-independent manner via a conserved "EPN" motif that is critical for bacterial killing. While EPN sequences govern calcium-dependent carbohydrate recognition in other C-type lectins (Drickamer 1999), the unusual location and calcium-independent functionality of the HIP/PAP EPN motif suggest that this sequence is a versatile functional module that can support both calcium-dependent and calcium-independent carbohydrate binding. Further, it was shown that HIP/PAP binding affinity for carbohydrate ligands depends on carbohydrate chain length, supporting a binding model in which HIP/PAP molecules "bind and jump" along the extended polysaccharide chains of peptidoglycan, reducing dissociation rates and increasing binding affinity (Fig. 39.3). Lehotzky et al. (2010) proposed that dynamic recognition of highly clustered carbohydrate epitopes in native peptidoglycan is an essential mechanism governing high-affinity interactions between HIP/PAP and the bacterial cell wall. HIP/PAP protein binds laminin and acts as an adhesion molecule for hepatocytes (Christa et al. 1999).

39.10.4 Crystal Structue of human PAP (hPAP)

Crystals of hPAP belong to the orthorhombic space group P2(1)2(1)2(1), with unit-cell parameters a = 30.73, b = 49.35, c = 92.15 A and one molecule in the asymmetric unit (Abergel et al. 1999). hPAP folds into an α + β structure and consists of two α-helices and eight β-strands cross-linked by three disulfide bridges packed into a heart-like shape, similar to that of the C-type lectin domain (Fig. 39.4). However, hPAP was found to have a unique distribution of charged residues on the surface. Nine of the eleven acidic residues (Asp^{37}, Asp^{39}, Glu^{58}, Asp^{82}, Glu^{88}, Glu^{92}, Glu^{95}, Asp^{100}, and Glu^{149}) were found to be clustered on one side, whereas 12 of the 13 basic residues (Arg^{13}, Lys^{16}, Lys^{19}, Lys^{33}, Lys^{44}, Arg^{45}, Lys^{67}, Arg^{109}, Arg^{125}, Arg^{131}, Lys^{133}, and Arg^{140}) were located on the other side. This surface charge distribution results in a highly polarized protein molecule, suggesting strong electrostatic interactions between molecules. In addition, several hydrophobic residues, such as residues Leu^{28}–Pro^{32} and Val^{101}–Trp^{107}, were found to be highly exposed and to constitute a continuous hydrophobic surface. The hydrophobic loop Val^{101}–Trp^{107} is located between the negatively charged and positively charged surfaces and is especially interesting, because the TANGO program predicts that this region in hPAP shows high propensity for β aggregation. Because hPAP forms fibrils with a native-like structure, it is possible that some conformational rearrangements in the loop regions that do not perturb the native secondary structure may drive the assembly of protein into fibril (Ho et al. 2006).

39.10.5 Expression of PAP

HIP/PAP is rapidly overexpressed during the acute phase of pancreatitis (Iovanna et al. 1994). Motor neurons are the only adult mammalian neurons of the CNS that regenerate after injury. The gene corresponding to human HIP/PAP gene in rat, called Reg-2 or rat PAP-I, has been found to be expressed in regenerating motor neurons, and its product exhibits mitogen-like activity in vitro on Schwann cells (Livesey et al. 1997). In the exocrine pancreas, HIP/PAP expression pattern is consistent with that of an acute phase reactant; however, this does not account for its expression in the islets of Langerhans of endocrine pancreas. HIP/PAP was expressed in normal subjects in the intestine (Paneth and neuroendocrine cells), and the pancreas (acinar pancreatic cells and islets of Langerhans), whereas in the liver, its expression was triggered in the event of primary liver cancer (Christa et al. 1996). Lasserre et al. (1992) discovered HIP/PAP by differential screening of a human hepatocarcinoma, and showed its frequent overexpression in liver tumors and regenerative hepatic/ductular diseases (Christa et al. 1999). In human, HIP/PAP is expressed in HCC and increased in serum, suggesting its potential importance in human liver carcinogenesis (Dusetti et al. 1996). During early stages of chemical hepatocarcinogenesis in rat (Petersen et al. 1998), oval cell proliferation was observed during hepatocarcinogenesis in woodchuck virus carriers and in human livers from patients with different pathological conditions (Christa et al. 1999). HIP/PAP mRNA is abundantly expressed in tumoral but not in nontumoral/normal livers. HIP/PAP gene expression is increased in 25% of HCC. HIP/PAP expression is not restricted to hepatocellular carcinoma, but also detected in cholangiocarcinoma cells as well as in reactive non-malignant bile ductules. Altogether, reports suggest that HIP/PAP protein may be implicated in hepatocytic and cholangiolar differentiation and proliferation (Christa et al. 1999).

PAP-I Is Upregulated During Acute and Chronic Pancreatitis: Early after its discovery it became clear that acinar cells of the pancreas are the main source of hPAP in pathological situations. In experimental model of acute pancreatitis in rat, Morisset et al. (1997) found the induction of PAP and its localization in zymogen granules. Bodeker and co-

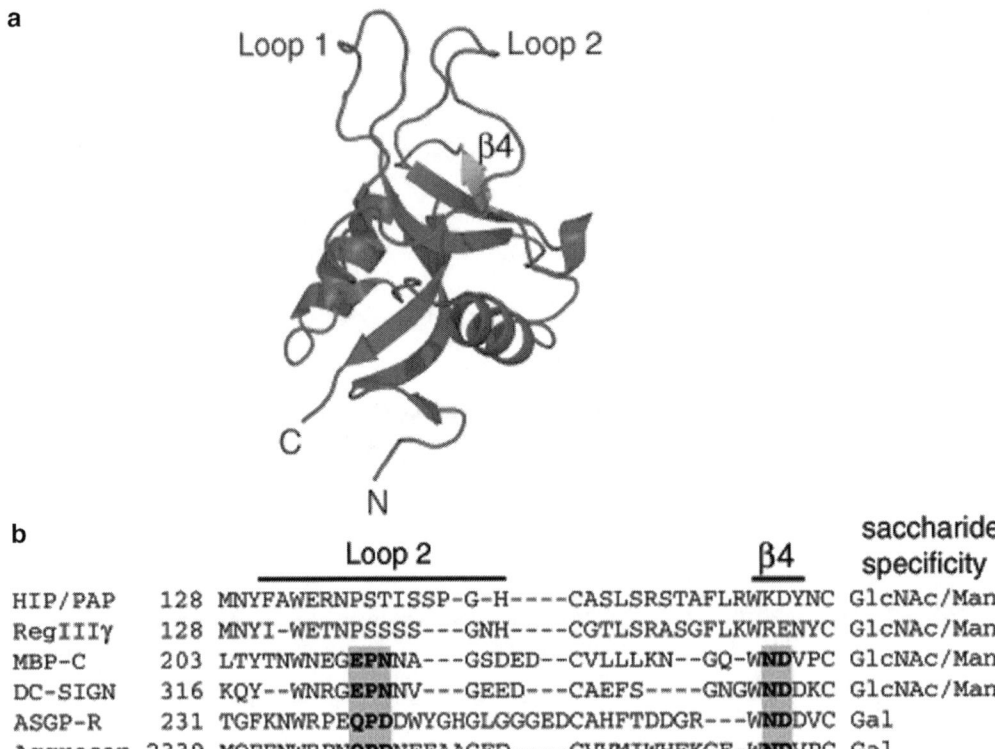

Fig. 39.3 Absence of canonical C-type lectin-carbohydrate binding motifs in Reg-III and HIP/PAP lectins: (**a**) NMR structure of HIP/PAP (PDB code 2GO0) (Ho et al. 2006: Fig. 39.4), with *Loops 1* and *2* of the *long loop* region and the β4 strand marked. (**b**) Alignment of the *long loop* region of Reg-III family members with other C-type lectin family members. MBP-C and DC-SIGN harbor *Loop 2* EPN motifs that govern sugar ligand binding and confer selectivity for mannose (Man) and GlcNAc (Drickamer 1992). Asialoglycoprotein receptor (ASGP-R) and aggrecan contain *Loop 2* QPD motifs that confer selectivity for Gal and GalNAc. Despite their selective binding to GlcNAc and Man polysaccharides RegIII lectins lack the *Loop 2* EPN motif (Cash et al. 2006a) (Adapted with permission from Lehotzky et al. 2010 © National Academy of Sciences)

workers (1998) described the pattern of PAP up-regulation in exocrine pancreas during the progression of the disease. The acute-phase response of pancreas seems to be a powerful emergency defense mechanism against further pancreatic aggression, as shown by the improved survival of the animals (Fiedler et al. 1998). PAP, induced in acute pancreatitis, was decreased by antisense-mediated gene knockdown of PAP gene expression and worsened pancreatitis (Lin et al. 2008).

PAP-I is not only expressed during acute pancreatitis but also in pancreatic adenocarcinoma, gastric carcinoma, hepatocellular carcinoma, and colorectal carcinoma. Expression in carcinoma might be another characteristic of PAP. Elevation of PAP in patients with pancreatic cancer is not merely explained by concomitant pancreatitis, but seems to be due to increased PAP production by the cancer cells and is also correlated to tumor load as expressed by the UICC stages (Cerwenka et al. 2001). PAP was overexpressed in 79% of pancreatic ductal adenocarcinoma, 19% of chronic pancreatitis, and 29% of mucinous cystadenoma. PAP was found in malignant ductular structures in pancreatic carcinomas as well as in benign proliferating ductules and acinar cells in chronic pancreatitis. It was not expressed in normal pancreas. The incidence of PAP overexpression was significantly higher in pancreatic cancer than in the other pancreatic diseases. PAP overexpression was significantly correlated with nodal involvement, distant metastasis, and short survival in pancreatic cancer. The overexpression of PAP in human pancreatic ductal adenocarcinoma indicates tumor aggressiveness (Xie et al. 2003). Functional differences between arteries and veins are based upon differences in gene expression. The most prominent difference was PAP-I, expressed 64-fold higher in vena cava versus aorta (Szasz et al. 2009).

TNF-α contributes to the development of acute pancreatitis. Because TNF-α is involved in the control of apoptosis and antiapoptotic mechanisms are mediated by NF-kB and MAP kinases, PAP-I is one of the effectors of apoptosis inhibition (Malka et al. 2000). The endotoxemia, which is caused by a bacterial infection, can exacerbate acute pancreatitis, whereas PAP has the ability to induce bacterial aggregation and supposed to protect the tissue from infection during inflammation. Studies suggest that PAP-I mRNA might be modulated by endotoxemia, independent of

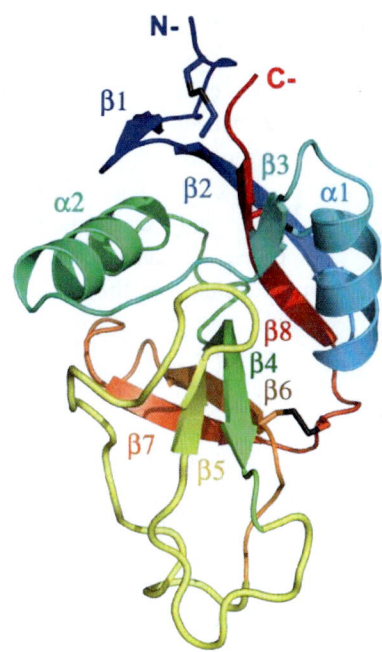

Fig. 39.4 NMR solution structure of hPAP at pH 4.0. Ribbon representation shows native hPAP. The tertiary fold of hPAP resembles that of the C-type lectin domain, which consists of two α-helices and eight β-strands cross-linked by three disulfide bridges (colored in *black*). The N and C termini are indicated. The region from Thr^{84} to Ser^{116} are highlighted in *yellow* (Adapted with permission from Ho et al. 2006 © The American Society for Biochemistry and Molecular Biology)

cerulein-pancreatitis. There were no strong correlations between PAP-I mRNA expression and the severity of pancreatitis (Wang et al. 2001).

PAP in Human Intestinal Tract: Members of the Reg family play regulatory roles in various endocrine cell populations and that their expression in endocrine cells is lineage-specific Analysis shows the presence of a transcript in the jejunum that has the same electrophoretic mobility as the pancreatic mRNA. No signal was detected in colon, however. In addition, PAP revealed the presence of a protein of 16,000 Da (as in pancreatic juice) in the homogenate of jejunum, but not of the colon. Positive immunoreactivity was observed on Paneth cells and in some goblet cells located in jejunum at the bottom of the crypts (Masciotra et al. 1995). In both adult and fetal normal tissues, HIP/PAP expression was detected only in endocrine cells of the small intestine, ascending colon, and pancreas (Hervieu et al. 2006).

Expression of PAP-I is altered in patients with celiac disease, where active phase of the disease is accompanied by an increased serum concentration of PAP (Carroccio et al. 1997). Increased serum level of PAP was diagnosed at ileal location in active CD with a sensitivity of 60%, a specificity of 94%, a positive predictive value of 84% and a negative predictive value of 81%. Elevated serum PAP is significantly associated with disease activity and ileal location (Desjeux et al. 2002). Clinical data suggest that pancreatitis could be an extraintestinal manifestation of inflammatory bowel disease. However, no experimental data support such a clinical relationship. Increased PAP mRNA has been reported in active inflammatory bowel disease (IBD). PAP-I is synthesised by Paneth cells and is overexpressed in colonic tissue of active IBD. PAP inhibits NF-kB activation and downregulates cytokine production and adhesion molecule expression in inflamed tissue. It may represent an anti-inflammatory mechanism and new therapeutic strategy in IBD (Gironella et al. 2005). PAP overexpression in pancreas demonstrates that inflammatory stress early occurs in the mouse pancreas during the course of TNBS-induced colitis. The concomitant pancreatic overexpression of IL-1B supported a pancreatic inflammatory mechanism mediated by cytokines (Barthet et al. 2003). Morphological pancreatic damage is also induced by the inhalation of cigarette smoke, which is likely to be mediated by alterations of acinar cell function (Wittel et al. 2006). PAP-I is also known as a marker for cystic fibrosis. PAP-I is increased in most neonates with cystic fibrosis and could be used for cystic fibrosis screening. Its combination with immunoreactive trypsinogen looks promising (Sarles et al. 1999). There was no evidence for polymorphism of the PAP gene in patients with hereditary or alcoholic pancreatitis. But the expression of the PAP in both groups of patients is related to the degree of cellular damage of the pancreas (Keim et al. 1999).

39.10.6 Similarity of PAP to Peptide 23 from Pituitary

Peptide 23 is a protein secreted by rat pituitary cells in primary culture. Although the secretion of this protein is stimulated by GH-releasing hormone and inhibited by somatostatin, the N-terminal amino acid sequence of peptide 23 shows no homology to rat GH. The cDNA of peptide 23 contained 777 nt and encoded a 175-amino acid protein with a 26-amino acid putative signal peptide with a calculated mol wt of mature protein (16,613 Da). Northern blot analysis revealed a major mRNA species of about 0.9 kb and a minor species of about 1.7 kb in cultured rat anterior pituitary cells. In rats, peptide 23 was most abundant in pancreas and GI tract. Sequence search revealed complete sequence identity between peptide 23 cDNA and PAP cDNA, 73% homology with HHC cDNA, and 55% homology with rat and human Reg cDNAs, expressed in regenerating pancreatic islets. Cloning of peptide-23 c-DNA revealed that it is identical to PAP. A dramatic increase in the expression of both genes was seen at the time of weaning in the third week postpartum. Peptide-23/PAP mRNA is most abundant in the ileum, whereas PSP/Reg is maximally expressed in the pancreas and duodenum. GRF modulates peptide-23/PAP expression in GI tract in a similar fashion to that previously reported for pituitary cells in primary culture (Chakraborty et al. 1995c; Katsumata et al. 1995). Later studies demonstrated that peptide 23/PAP, previously thought to be of pituitary origin, is widely expressed in GI tract and that it is rapidly removed from the circulation by the kidney and by tissues which express peptide 23/PAP. Peptide 23/PAP is also expressed in rat uterus where estrogen may act as a physiological regulator of peptide 23 (Chakraborty et al. 1995a). Human fetal and adult tissues express at least one of the different transcripts of the PAP/Reg family, suggesting that the regulation of these homologous genes is coordinately controlled (Bartoli et al. 1998). The protein, present in very high concentration in pancreatic secretion, has been detected in brain lesions in AD and Down syndrome and in regenerating rat pancreatic islets (Cai et al. 1990).

39.10.7 Serum PAP: An Indicator of Pancreatic Function

Measuring serum PAP-I in acute pancreatitis has proved valuable in monitoring the course of the disease and the recovery of the patient. However, at admission PAP did not distinguish severe from mild acute pancreatitis better than C reactive protein; the measurement of PAP did not give appreciable diagnostic advantage in the early phase of acute pancreatitis (Kemppainen et al. 1996; Pezzilli et al. 1997). The serum PAP levels were significantly elevated in the patients with gastric, colorectal, biliary tract, hepatocellular, or pancreatic cancers compared with the healthy subjects. The increase of serum PAP levels in patients with gastrointestinal cancers reflects an ectopic expression of PAP-I. In cancer cells, the increased serum levels of PAP are correlated with the disease severity (Motoo et al. 1998, 1999). The PAP mRNA expression and serum PAP levels are closely related to neoplastic proliferative activity in patients with colorectal carcinoma (Cao et al. 2009). Serum PAP levels strongly correlate with creatinine clearance measurements. In patients with a pancreas-kidney transplantation, PAP may prove a useful biological and histological marker of pancreatic graft rejection (van der Pijl et al. 1997). Human PAP could be detected in pancreatic juice from patients with pancreatic diseases. Determination of PAP in pancreatic juice might be helpful for early detection of pancreatic injury (Motoo et al. 2001).

39.11 Reg IV (RELP)

Hartupee et al. (2001) isolated two cDNAs which encode a novel member of this multigene family. Based on primary sequence homology, tissue expression profiles, and shared exon-intron junction genomic organization, gene was assigned to the regenerating gene family as *Reg IV*. *Reg IV* has a highly restricted tissue expression pattern, with prominent expression in mucosal cells of GI tract. Namikawa et al. (2005) identified and sequenced a type-IV Reg gene in rats. Reg family members are mediators among injured neurons and glial cells that may play pivotal roles during nerve regeneration. Unlike other C-type lectins, human Reg IV binds to polysaccharides, mannan, and heparin in the absence of calcium. To elucidate the structural basis for carbohydrate recognition by NMR, mutant protein with Pro91 replaced by Ser (hReg IV-P91S) showed that the structural property and carbohydrate binding ability of hRegIV-P91S are almost identical with those of wild-type protein. The solution structure of hRegIV-P91S adopts a typical fold of C-type lectin. Based on NMR studies, two calcium-independent mannan-binding sites were proposed. One site is similar to the calcium-independent sugar-binding site on human Reg-III and Langerin. Interestingly, the other site is adjacent to the conserved calcium-dependent site 2 at position typical of C-type lectins (Ho et al. 2010)

39.11.1 Tissue Expression

Reg IV gene expresses in human colon, small intestine, stomach and pancreas, in rat spleen and colon and constitutively expressed in neuroendocrine cells of the intestinal mucosa. A potential role for Reg IV protein has been indicated in spleen and colon cell growth (StarceviĂć-Klasan et al. 2008). Heiskala et al. (2006) reported a robust de novo expression of Reg IV (RELP) in the neoplastic goblet cells of appendiceal mucinous cystadenomas and in the epithelial implants of pseudomyxoma peritonei (PMP). Reg IV serves as a marker for appendiceal mucinous cystadenomas and PMP, and may contribute to the pathogenesis of these disorders (Heiskala et al. 2006). Reg IV mRNA expression is significantly up-regulated by mucosal injury from active Crohn's disease or ulcerative colitis (UC). Reg IV mRNA was strongly expressed in inflamed epithelium and in dysplasias and cancerous lesions in UC tissues. Reg IV gene induction promoted cell growth and conferred resistance to H_2O_2-induced apoptosis in human colon cancer (DLD1) cells. Therefore, Reg IV gene is inducible by growth factors and may function as a growth promoting and/or an antiapoptotic factor in the pathophysiology of UC (Nanakin et al. 2007).

Intestinal neuroendocrine neoplasms showed co-expression of Reg IV and Hath1. Lung small-cell carcinoma and gastric mucocellular carcinoma expressed only Reg IV. Pancreatic islet-derived tumors, pheochromocytomas, and paragangliomas expressed only Hath1. The dissimilar expression patterns suggest that the proteins belong to different signaling pathways and are activated at different stages of neuroendocrine differentiation. Local Reg IV expression may be influenced by the growth factors bFGF and HGF and/or their receptors CD138 and c-met, which were found to co-localize with Reg IV in intestinal neuroendocrine tumors (Heiskala et al. 2010). Reg IV might accelerate cell growth and disease progression of adenoid cystic carcinomas (ACCs) in salivary glands s (Sasahira et al. 2008). Reg IV expression is associated with intestinal and neuroendocrine differentiation in gastric adenocarcinoma (Oue et al. 2005). Bishnupuri et al. (2006a) examined Reg gene expression and associated changes in anti-apoptotic genes in an animal model of GI tumorigenesis and showed that dys-regulation of Reg genes occurs early in tumorigenesis. Furthermore, increased expression of Reg genes, specifically Reg IV contribute to adenoma formation and lead to increased resistance to apoptotic cell death in CRC (Bishnupuri et al. 2006a). Serum olfactomedin 4 (GW112, hGC-1) in combination with Reg IV is a highly sensitive biomarker for gastric cancer patients (Yoshida and Yasui 2009).

Levels of Reg IV protein in peritoneal lavage fluids increased in Reg IV-transfectants inoculated mice, but decreased in Reg IV-knockdown cell inoculated mice. It appears that Reg IV might accelerate peritoneal metastasis in gastric cancer and that Reg IV in lavage might be a good marker for peritoneal metastasis (Kuniyasu et al. 2009). The serum Reg IV concentration in presurgical gastric cancer patients is significantly elevated even at stage I and represents a novel biomarker for gastric cancer (Mitani et al. 2007). Reg IV is involved in gallbladder carcinoma carcinogenesis through intestinal metaplasia and is associated with relatively favorable prognosis in patients after surgery. The serum level of Reg IV may be of use or indicative of neoplasia (Tamura et al. 2009).

39.11.1.1 Colorectal Cancer (CRC)

Regenerating IV gene is overexpressed in HT-29 drug-resistant cells and more strongly expressed in 71% of colorectal tumors (in particular in mucinous carcinomas) than in normal colon tissues. Reg IV overexpression may be an early event in CRC carcinogenesis. A high level of RegIV and RegIV CRD expression was demonstrated in RegIV and RegIV CRD-transfected cells. RegIV enhances LoVo cell migration and invasion, and its CRD domain is critical for these effects (Guo et al. 2010). The comparison of Reg IV expression with that of other REG genes, regenerating Iα (Reg Iα), Reg Iβ and PAP, highlights its predominant expression in colorectal tumors (Violette et al. 2003). The over-expression of Reg IV in CRCs was significantly lower and inversely correlated with poor differentiation and venous invasion. Serum levels of Reg IV may predict CRC recurrence in the liver (Li et al. 2010; Oue et al. 2007). The expression of Reg IV in CRCs was positively linked to MUC2 and phosphorylation of the EGFR on Tyr^{1068}, suggesting that Reg IV may be a useful marker for intestinal type mucinous carcinoma and a good candidate as a molecular therapeutic target for CRCs. Reg IV is a potent activator of the EGF receptor/Akt/AP-1 signaling pathway in CRC (Bishnupuri et al. 2006b). Detection of Reg IV overexpression may be useful in the early diagnosis of carcinomatous transformation of adenoma (Zhang et al. 2003b). Reg IV is an important modulator of gastrointestinal cell susceptibility to irradiation; hence, it is a potential target for adjunctive treatments for human CRC and other gastrointestinal malignancies (Bishnupuri et al. 2010).

To clarify the role of Reg IV in gastric carcinogenesis and subsequent progression, Zheng et al. (2010) indicated that Reg IV expression experienced up-regulation in gastric intestinal metaplasia and adenoma and then down-regulation with malignant transformation of gastric epithelial cells. It was suggested that Reg IV expression is a good biomarker for gastric precancerous lesions and is especially related to the histogenic pathway of signet ring cell carcinoma (Zheng et al. 2010).

39.11.1.2 Marker for Hormone Refractory Metastatic Prostate Cancer

Gene, Reg IV is differentially expressed in the LAPC-9 hormone refractory xenograft. Consistent with its up-regulation in a hormone refractory xenograft, it is expressed in several prostate tumors after neoadjuvant hormone ablation therapy. As predicted by its sequence homology, it is secreted from transiently transfected cells. It is also expressed strongly in a majority of hormone refractory metastases. In comparison, it is not expressed by any normal prostate specimens and only at low levels in approximately 40% of primary tumors. Reports support Reg IV is a candidate marker for hormone refractory metastatic prostate cancer (Gu et al. 2005; Ohara et al. 2008). Serum Reg IV represents a novel biomarker for prostate cancer (Hayashi et al. 2009). Reg IV participates in 5-fluorouracil (5-FU) resistance and peritoneal metastasis, and its expression was associated with an intestinal phenotype of gastric cancer and with endocrine differentiation (Yamagishi et al. 2009; Yasui et al. 2009). Signet-ring cell carcinoma (SRCC) is a unique subtype of adenocarcinoma that is characterized by abundant intracellular mucin accumulation. Reg IV staining and claudin-18 staining can aid in diagnosis of gastrointestinal SRCC (Oue et al. 2008).

39.12 Islet Neogenesis-Associated Protein (INGAP) in Hamster

39.12.1 Characterization

Islet neogenesis-associated protein (INGAP) is expressed in islets and in exocrine pancreatic cells of normal hamsters. INGAP is a 175-amino-acid pancreatic acinar protein that stimulates pancreatic duct cell proliferation in vitro and islet neogenesis in vivo. The INGAP gene is a novel pancreatic gene expressed during islet neogenesis whose protein product is a constituent of Ilotropin and is capable of initiating duct cell proliferation, a prerequisite for islet neogenesis. Induction of islet neogenesis by cellophane wrapping (CW) reverses streptozotocin-induced (STZ) diabetes. Administration of Ilotropin, a protein extract isolated from CW pancreata, causes recapitulation of normal islet ontogeny and reverses STZ diabetes. Although INGAP shows homology to the PAP and Reg/PSP families of genes, the increased expression of INGAP in CW is unlikely to be a result of acute pancreatitis (Rafaeloff et al. 1997). Sasahara et al. (2000) reported a INGAP-related protein (INGAPrP). While INGAP expressed in cellophane-wrapped pancreas and not in normal pancreas, the INGAPrP was abundantly expressed in normal pancreas. INGAP could be a potentially useful tool to treat conditions in which there is a decrease in β-cell mass (Del Zotto et al. 2000). A 6 kb of hamster genomic INGAP has identified a 3 kb 5-prime region with core promoter elements that is rich in transcription factor binding sites and six exons for the coding region. Analysis of promoter activity reveals stimulus-responsive DNA elements which have been identified though deletion analysis. Comparison of transcription factor binding sites in INGAP to the related gene Reg-IIIδ exposes potential sites for differential gene regulation (Taylor-Fishwick et al. 2003).

The ubiquitous localization of INGAP suggests its possible role in the physiological process of islet growth and its protective effect upon streptozotocin-induced diabetes (Flores et al. 2003). It participates in the regulation of islet neogenesis, and PDX-1/INGAP-positive cells represent a new stem cell subpopulation at an early stage of development, highly activateable in neogenesis (Flores et al. 2003; Gagliardino et al. 2003). Findings implicate PDX-1 in a possible feedback loop to block unbridled islet expansion (Taylor-Fishwick et al. 2006). INGAP expression is probably restricted to pancreas cells exerting its effect in a paracrine fashion. INGAP would be released and circulate bound to a serum protein from where it is bound and inactivated by the liver. Tissue binding could also explain INGAP's immunocytochemical presence in small intestine, where it could affect epithelial cell turnover (Borelli et al. 2007). INGAP stimulates cells in the pancreatic duct epithelium of healthy dogs (putative islet progenitor cells) to develop along a neuroendocrine pathway and form new islets in response to INGAP peptide. The INGAP might be an effective therapy for diabetes (Pittenger et al. 2007). INGAP is localized to the pancreatic endocrine cells in mouse. INGAP- and insulin-immunoreactive cells are mutually exclusive, with INGAP-immunoreactive cells being preserved after streptozotocin-mediated destruction of β-cells.

Findings reveal that INGAP and/or related group 3 Reg proteins have a conserved expression in the pancreatic islet (Taylor-Fishwick et al. 2008).

39.12.2 Properties of Pentadecapeptide from INGAP

A pentadecapeptide having the 104–118 amino acid sequence of INGAP (INGAP-PP) affects insulin secretion and transcript profile expression in cultured normal pancreatic neonatal rat islets. Islets cultured with INGAP-PP released significantly more insulin in response to glucose than those cultured without the peptide. INGAP-PP enhances specifically the secretion of insulin and the transcription of several islet genes, many of them directly or indirectly involved in the control of islet metabolism, β-cell mass and islet neogenesis. Studies strongly indicate an important role of INGAP-PP, and possibly of INGAP, in the regulation of islet function and development (Barbosa et al. 2006; Barbosa et al. 2008). Whilst the peptides with scrambled sequences showed no definite prevalent structure in solution, INGAP-PP maintained a notably stable tertiary fold, namely, a conformer with a central β-sheet and closed C-terminal. Such structure resembles the one corresponding to the amino acid sequence of human PAP-1, which presents 85% sequence homology with INGAP. These results explain why the two scrambled sequences tested showed no biological activity, while INGAP-PP significantly increases β-cells function and mass both in vitro and in vivo conditions. These data can help to obtain more potent INGAP-PP analogs, suitable for the prevention and treatment of diabetes (McCarthy et al. 2009)

Administration of the peptide fragment of INGAP (INGAP peptide) has been demonstrated to reverse chemically induced diabetes as well as improve glycemic control and survival in an animal model of type 1 diabetes. Cultured human pancreatic tissue has also been shown to be responsive to INGAP peptide, producing islet-like structures with function, architecture and gene expression matching that of freshly isolated islets. Studies in normoglycemic animals show evidence of islet neogenesis. These clinical studies suggest an effect of INGAP peptide to improve insulin production in type 1diabetes and glycemic control in type 2 diabetes (Lipsett et al. 2007). The INGAP peptide acts as a mitogen in the peripheral nervous system (PNS), and enhances neurite outgrowth from DRGs in vitro. The neuritogenic action of INGAP peptide correlates with an increase in ^{3}H-thymidine incorporation and mitochondrial activity and promotes Schwann cell proliferation in the DRG which releases trophic factors that promote neurite outgrowth (Tam et al. 2002).

References

Abe M, Nata K, Akiyama T et al (2000) Identification of a novel Reg family gene, Reg-IIIdelta, and mapping of all three types of Reg family gene in a 75 kilobase mouse genomic region. Gene 246:111–122

Abergel C, Chenivesse S, Stinnakre MG et al (1999) Crystallization and preliminary crystallographic study of HIP/PAP, a human C-lectin overexpressed in primary liver cancers. Acta Crystallogr D Biol Crystallogr 55:1487–1489

Akiyama T, Takasawa S, Nata K et al (2001) Activation of Reg gene, a gene for insulin-producing β -cell regeneration: poly(ADP-ribose) polymerase binds Reg promoter and regulates the transcription by autopoly(ADP-ribosyl)ation. Proc Natl Acad Sci USA 98:48–53

Ampo K, Suzuki A, Konishi H et al (2009) Induction of pancreatitis-associated protein (PAP) family members in neurons after traumatic brain injury. J Neurotrauma 26:1683–1693

Apte MV, Norton ID et al (1996) Both ethanol and protein deficiency increase messenger RNA levels for pancreatic lithostathine. Life Sci 58:485–492

Asahara M, Mushiake S, Shimada S et al (1996) Reg gene expression is increased in rat gastric enterochromaffin-like cells following water immersion stress. Gastroenterology 111:45–55

Astrosini C, Roeefzaad C, Dai YY et al (2008) REG 1A expression is a prognostic marker in colorectal cancer and associated with peritoneal carcinomatosis. Int J Cancer 123:409–413

Averill S, Davis DR et al (2002) Dynamic pattern of Reg-2 expression in rat sensory neurons after peripheral nerve injury. J Neurosci 22:7493–7501

Averill S, Inglis JJ et al (2008) Reg-2 expression in dorsal root ganglion neurons after adjuvant-induced monoarthritis. Neuroscience 155:1227–1236

Baeza N, Sanchez D, Vialettes B et al (1997) Specific Reg II gene overexpression in the non-obese diabetic mouse pancreas during active diabetogenesis. FEBS Lett 416:364–368

Baeza N, Sanchez D, Christa L et al (2001) Pancreatitis-associated protein (HIP/PAP) gene expression is upregulated in NOD mice pancreas and localized in exocrine tissue during diabetes. Digestion 64:233–9

Barbosa H, Bordin S, Stoppiglia L et al (2006) Islet neogenesis associated protein (INGAP) modulates gene expression in cultured neonatal rat islets. Regul Pept 136:78–84

Barbosa HC, Bordin S, Anhê G et al (2008) Islet neogenesis-associated protein signaling in neonatal pancreatic rat islets: involvement of the cholinergic pathway. J Endocrinol 199:299–306

Barthet M, Dubucquoy L, Garcia S et al (2003) Pancreatic changes in TNBS-induced colitis in mice. Gastroenterol Clin Biol 27:895–900

Bartoli C, Dagorn JC, Fontes M et al (1995) A limited genomic region contains the human REG and REG-related genes. Eur J Hum Genet 3:344–350

Bartoli C, Baeza N, Figarella C et al (1998) Expression of peptide-23/pancreatitis-associated protein and Reg genes in human pituitary and adenomas: comparison with other fetal and adult human tissues. J Clin Endocrinol Metab 83:4041–4046

Bernard JP, Adrich Z, Montalto G et al (1991) Immunoreactive forms of pancreatic stone protein in six mammalian species. Pancreas 6:162–167

Bernard JP, Adrich Z, Montalto G et al (1992) Inhibition of nucleation and crystal growth of calcium carbonate by human lithostathine. Gastroenterology 103:1277–1284

Bernard JP, Barthet M, Gharib B et al (1995) Quantification of human lithostathine by high performance liquid chromatography. Gut 36:630–636

Bernard-Perrone FR, Renaud WP, Guy-Crotte OM et al (1999) Expression of REG protein during cell growth and differentiation of two human colon carcinoma cell lines. J Histochem Cytochem 47:863–870

Bertrand JA, Pignol D, Bernard JP et al (1996) Crystal structure of human lithostathine, the pancreatic inhibitor of stone formation. EMBO J 15:2678–2684

Bimmler D, Graf R, Scheele GA et al (1997) Pancreatic stone protein (lithostathine), a physiologically relevant pancreatic calcium carbonate crystal inhibitor? J Biol Chem 272:3073–3082

Bimmler D, Angst E et al (1999) Regulation of PSP/Reg in rat pancreas: immediate and steady-state adaptation to different diets. Pancreas 19:255–267

Bimmler D, Schiesser M, Perren A et al (2004) Coordinate regulation of PSP/Reg and PAP-Isoforms as a family of secretory stress proteins in an animal model of chronic pancreatitis. J Surg Res 118:122–135

Bishnupuri KS, Luo Q, Korzenik JR et al (2006a) Dysregulation of Reg gene expression occurs early in gastrointestinal tumorigenesis and regulates anti-apoptotic genes. Cancer Biol Ther 5:1714–1720

Bishnupuri KS, Luo Q, Murmu N et al (2006b) Reg IV activates the epidermal growth factor receptor/Akt/AP-1 signaling pathway in colon adenocarcinomas. Gastroenterology 130:137–149

Bishnupuri KS, Luo Q, Sainathan SK et al (2010) Reg IV regulates normal intestinal and colorectal cancer cell susceptibility to radiation-induced apoptosis. Gastroenterology 138:616–626

Blennow K, Wallin A, Chong JK (1995) Cerebrospinal fluid 'neuronal thread protein' comes from serum by passage over the blood–brain barrier. Neurodegeneration 4:187–193

Bluth MH, Patel SA, Dieckgraefe BK et al (2006) Pancreatic regenerating protein (Reg I) and Reg I receptor mRNA are upregulated in rat pancreas after induction of acute pancreatitis. World J Gastroenterol 12:4511–4516

Bluth M, Mueller CM, Pierre J et al (2008) Pancreatic regenerating protein I in chronic pancreatitis and aging: implications for new therapeutic approaches to diabetes. Pancreas 37:386–395

Bodeker H, Fiedler F et al (1998) Pancreatitis-associated protein is upregulated in mouse pancreas during acute pancreatitis. Digestion 59:186–191

Bodeker H, Keim V, Fiedler F et al (1999) PAP-I interacts with itself, PAP-II, PAP-III, and lithostathine/RegIα. Mol Cell Biol Res Commun 2:150–154

Borelli MI, Del Zotto H, Flores LE et al (2007) Transcription, expression and tissue binding in vivo of INGAP and INGAP-related peptide in normal hamsters. Regul Pept 140:192–197

Cai L, Harris WR, Marshak DR et al (1990) Structural analysis of bovine pancreatic thread protein. J Protein Chem 9:623–632

Cao G, Ma J, Zhang Y et al (2009) Pancreatitis-associated protein is related closely to neoplastic proliferative activity in patients with colorectal carcinoma. Anat Rec (Hoboken) 292:249–253

Carroccio A, Iovanna JL, Iacono G et al (1997) Pancreatitis-associated protein in patients with celiac disease: serum levels and immunocytochemical localization in small intestine. Digestion 58:98–103

Cash HL, Whitham CV, Behrendt CL et al (2006a) Symbiotic bacteria direct expression of an intestinal bactericidal lectin. Science 313:1126–1130

Cash HL, Whitham CV, Hooper LV (2006b) Refolding, purification, and characterization of human and murine Reg-III proteins expressed in *Escherichia coli*. Protein Expr Purif 48:151–159

Cavard C, Terris B, Grimber G et al (2006) Overexpression of regenerating islet-derived 1 α and 3 α genes in human primary liver tumors with β-catenin mutations. Oncogene 25:599–608

Cerini C, Peyrot V, Garnier C et al (1999) Biophysical characterization of lithostathine. Evidences for a polymeric structure at physiological pH and a proteolysis mechanism leading to the formation of fibrils. J Biol Chem 274:22266–22274

Cerwenka H, Aigner R, Bacher H et al (2001) Pancreatitis-associated protein (PAP) in patients with pancreatic cancer. Anticancer Res 21:1471–1474

Chakraborty C, Katsumata N, Myal Y et al (1995a) Age-related changes in peptide-23/pancreatitis-associated protein and pancreatic stone protein/Reg gene expression in the rat and regulation by growth hormone-releasing hormone. Endocrinology 136: 1843–1849

Chakraborty C, Sharma S, Katsumata N et al (1995b) Plasma clearance, tissue uptake and expression of pituitary peptide 23/pancreatitis-associated protein in the rat. J Endocrinol 145:461–469

Chakraborty C, Vrontakis M, Molnar P et al (1995c) Expression of pituitary peptide 23 in the rat uterus: regulation by estradiol. Mol Cell Endocrinol 108:149–154

Chen CY, Lin XZ, Wu HC et al (2005) The value of biliary amylase and hepatocarcinoma-intestine-pancreas/pancreatitis-associated protein I (HIP/PAP-I) in diagnosing biliary malignancies. Clin Biochem 38:520–525

Chinuki D, Amano Y, Ishihara S et al (2008) REG Iα protein expression in Barrett's esophagus. J Gastroenterol Hepatol 23:296–302

Choi B, Suh Y, Kim WH et al (2007) Downregulation of regenerating islet-derived 3 α (REG3ς) in primary human gastric adenocarcinomas. Exp Mol Med 39:796–804

Christa L, Felin M, Morali O et al (1994) The human HIP gene, overexpressed in primary liver cancer encodes for a C-type carbohydrate binding protein with lactose binding activity. FEBS Lett 337:114–118

Christa L, Carnot F, Simon MT et al (1996) HIP/PAP-is an adhesive protein expressed in hepatocarcinoma, normal Paneth, and pancreatic cells. Am J Physiol 271:G993–G1002

Christa L, Simon M-T, Brezault-Bonnet C et al (1999) Hepatocarcinoma-intestine-pancreas/pancreatic associated protein (HIP/PAP) is expressed and secreted by proliferating ductules as well as by hepatocarcinoma and cholangiocarcinoma cells. Am J Pathol 155:1525–1533

Christa L, Pauloin A et al (2000) High expression of the human hepatocarcinoma-intestine-pancreas/pancreatic-associated protein (HIP/PAP) gene in the mammary gland of lactating transgenic mice. Secretion into the milk and purification of the HIP/PAP lectin. Eur J Biochem 267:1665–1671

Christofilis MA, Carrere J, Atlan-Gepner C et al (1999) Serum Reg protein level is not related to the β cell destruction/regeneration process during early phases of diabetogenesis in type I diabetes. Eur J Endocrinol 141:368–373

Closa D, Motoo Y, Iovanna JL (2007) Pancreatitis-associated protein: from a lectin to an anti-inflammatory cytokine. World J Gastroenterol 13:170–174

Cui W, De Jesus K, Zhao H et al (2009) Overexpression of REG3α increases cell growth and the levels of cyclin D1 and CDK4 in insulinoma cells. Growth Factors 27:195–202

Davies PL, Hew CL (1990) Biochemistry of fish antifreeze proteins. FASEB J 4:2460–2468

De Caro A, Multigner L, Dagorn JC, Sarles H (1988) The human pancreatic stone protein. Biochimie 70:1209–1214

De Caro AM, Adrich Z, Fournet B et al (1989) N-terminal sequence extension in the glycosylated forms of human pancreatic stone protein. The 5-oxoproline N-terminal chain is O-glycosylated on the 5th amino acid residue. Biochim Biophys Acta 994:281–284

de la Monte SM, Wands JR (2004) Alzheimer-associated neuronal thread protein mediated cell death is linked to impaired insulin signaling. J Alzheimers Dis 6:231–242

de la Monte SM, Ozturk M, Wands JR (1990) Enhanced expression of an exocrine pancreatic protein in Alzheimer's disease and the developing human brain. J Clin Invest 86:1004–1013

de La Monte SM, Carlson RI, Brown NV, Wands JR (1996a) Profiles of neuronal thread protein expression in Alzheimer's disease. J Neuropathol Exp Neurol 55:1038–1050

de la Monte SM, Xu YY, Wands JR (1996b) Modulation of neuronal thread protein expression with neuritic sprouting: relevance to Alzheimer's disease. J Neurol Sci 138:26–35

de la Monte SM, Garner W, Wands JR (1997) Neuronal thread protein gene modulation with cerebral infarction. J Cereb Blood Flow Metab 17:623–635

de la Monte SM, Chen GJ, Rivera E, Wands JR (2003) Neuronal thread protein regulation and interaction with microtubule-associated proteins in SH-Sy5y neuronal cells. Cell Mol Life Sci 60:2679–2691

De Reggi M, Gharib B (2001) Protein-X, pancreatic stone-, pancreatic thread-, reg-protein, P19, lithostathine, and now what? Characterization, structural analysis and putative function(s) of the major non-enzymatic protein of pancreatic secretions. Curr Protein Pept Sci 2:19–42

De Reggi M, Capon C, Gharib B et al (1995) The glycan moiety of human pancreatic lithostathine. Structure characterization and possible pathophysiological implications. Eur J Biochem 230:503–510

De Reggi M, Gharib B, Patard L, Stoven V (1998) Lithostathine, the presumed pancreatic stone inhibitor, does not interact specifically with calcium carbonate crystals. J Biol Chem 273:4967–4971

Del Zotto H, Massa L, Rafaeloff R et al (2000) Possible relationship between changes in islet neogenesis and islet neogenesis-associated protein-positive cell mass induced by sucrose administration to normal hamsters. J Endocrinol 165:725–733

Demaugre F, Philippe Y, Sar S et al (2004) HIP/PAP, a C-type lectin overexpressed in hepatocellular carcinoma, binds the RII α regulatory subunit of cAMP-dependent protein kinase and alters the cAMP-dependent protein kinase signaling. Eur J Biochem 271: 3812–3820

Desjeux A, Barthet M, Barthellemy S et al (2002) Serum measurements of pancreatitis associated protein in active Crohn's disease with ileal location. Gastroenterol Clin Biol 26:23–28

Dhar DK, Udagawa J, Ishihara S et al (2004) Expression of regenerating gene I in gastric adenocarcinomas: correlation with tumor differentiation status and patient survival. Cancer 100:1130–1136

Dieckgraefe BK, Crimmins DL, Landt V et al (2002) Expression of the regenerating gene family in inflammatory bowel disease mucosa: Reg Iα upregulation, processing, and antiapoptotic activity. J Investig Med 50:421–434

Drickamer K (1992) Engineering galactose-binding activity into a C-type mannose binding protein. Nature 360:183–186

Drickamer K (1999) C-type lectin-like domains. Curr Opin Struct Biol 9:585–590

Duplan L, Michel B, Boucraut J et al (2001) Lithostathine and pancreatitis-associated protein are involved in the very early stages of Alzheimer's disease. Neurobiol Aging 22:79–88

Dusetti NJ, Frigerio JM, Dagorn JC et al (1993a) Rapid PCR cloning and sequence determination of the rat lithostathine gene. Biochim Biophys Acta 1174:99–102

Dusetti NJ, Frigerio JM, Keim V et al (1993b) Structural organization of the gene encoding the rat pancreatitis-associated protein. Analysis of its evolutionary history reveals an ancient divergence from the other carbohydrate-recognition domain-containing genes. J Biol Chem 268:14470–14475

Dusetti NJ, Frigerio JM, Fox MF et al (1994a) Molecular cloning, genomic organization, and chromosomal localization of the human pancreatitis-associated protein (PAP) gene. Genomics 19:108–114

Dusetti NJ, Mallo G, Dagorn JC, Iovanna JL (1994b) Serum from rats with acute pancreatitis induces expression of the PAP mRNA in the pancreatic acinar cell line AR-42J. Biochem Biophys Res Commun 204:238–243

Dusetti NJ, Frigerio JM, Szpirer C et al (1995a) Cloning, expression and chromosomal localization of the rat pancreatitis-associated protein III gene. Biochem J 307:9–16

Dusetti NJ, Ortiz EM, Mallo GV et al (1995b) Pancreatitis-associated protein I (PAP I), an acute phase protein induced by cytokines. Identification of two functional interleukin-6 response elements in the rat PAP I promoter region. J Biol Chem 270:22417–22421

Dusetti NJ, Mallo GV, Ortiz EM et al (1996a) Induction of lithostathine/Reg mRNA expression by serum from rats with acute pancreatitis and cytokines in pancreatic acinar AR-42 J cells. Arch Biochem Biophys 330:129–132

Dusetti NJ, Montalto G et al (1996b) Mechanism of PAP-I gene induction during hepatocarcinogenesis: clinical implications. Br J Cancer 74:1767–1775

Dusetti NJ, Vasseur S, Ortiz EM et al (1997) The pancreatitis-associated protein I promoter allows targeting to the pancreas of a foreign gene, whose expression is up-regulated during pancreatic inflammation. J Biol Chem 272:5800–5804

Ewart KV, Rubinsky B, Fletcher GL (1992) Structural and functional similarity between fish antifreeze proteins and calciumdependant lectins. Biochem Biophys Res Comm 185:335–340

Fang M, Huang JY, Ling SC et al (2010) Effects of Reg-2 on survival of spinal cord neurons in vitro. Anat Rec (Hoboken) 293:464–476

Ferrés-Masó M, Sacilotto N, López-Rodas G et al (2009) PAP-I signaling involves MAPK signal transduction. Cell Mol Life Sci 66:2195–2204

Fiedler F, Croissant N, Rehbein C et al (1998) Acute-phase response of the rat pancreas protects against further aggression with severe necrotizing pancreatitis. Crit Care Med 26:887–894

Flores LE, GarcÃa ME, Borelli MI et al (2003) Expression of islet neogenesis-associated protein in islets of normal hamsters. J Endocrinol 177:243–248

Folch-Puy E, GarcÃa-Movtero A et al (2003) The pancreatitis-associated protein induces lung inflammation in the rat through activation of TNFα expression in hepatocytes. J Pathol 199:398–408

Folch-Puy E, Granell S, Dagorn JC et al (2006) Pancreatitis-associated protein I suppresses NF-kappa B activation through a JAK/STAT-mediated mechanism in epithelial cells. J Immunol 176:3774–3779

Frigerio JM, Dusetti NJ, Garrido P et al (1993a) The pancreatitis associated protein III (PAP-III), a new member of the PAP gene family. Biochim Biophys Acta 1216:329–331

Frigerio JM, Dusetti NJ, Keim V et al (1993b) Identification of a second rat pancreatitis-associated protein. Messenger RNA cloning, gene structure, and expression during acute pancreatitis. Biochemistry 32:9236–9241

Fu K, Sarras MP Jr, De Lisle RC, Andrews GK (1996) Regulation of mouse pancreatitis-associated protein-I gene expression during caerulein-induced acute pancreatitis. Digestion 57:333–340

Fukuhara H, Kadowaki Y, Ose T et al (2010) In vivo evidence for the role of RegI in gastric regeneration: transgenic overexpression of RegI accelerates the healing of experimental gastric ulcers. Lab Invest 90:556–565

Fukui H, Kinoshita Y, Maekawa T et al (1998) Regenerating gene protein may mediate gastric mucosal proliferation induced by hypergastrinemia in rats. Gastroenterology 115:1483–1493

Fukui H, Franceschi F, Penland RL et al (2003) Effects of *Helicobacter pylori* infection on the link between regenerating gene expression and serum gastrin levels in Mongolian gerbils. Lab Invest 83:1777–1786

Fukui H, Fujii S, Takeda J et al (2004) Expression of Reg Iα protein in human gastric cancers. Digestion 69:177–184

Funakoshi A, Miyasaka K, Jimi A et al (1995) Changes in gene expression of pancreatitis-associated protein and pancreatic secretory trypsin inhibitors in experimental pancreatitis produced by pancreatic duct occlusion in rats: comparison with gene expression of cholecystokinin and secretin. Pancreas 11:147–153

Gagliardino JJ, Del Zotto H, Massa L et al (2003) Pancreatic duodenal homeobox-1 and islet neogenesis-associated protein: a possible combined marker of activateable pancreatic cell precursors. J Endocrinol 177:249–259

Geider S, Baronnet A, Cerini C et al (1996) Pancreatic lithostathine as a calcite habit modifier. J Biol Chem 271:26302–26306

Geng J, Fan J, Wang P et al (2009) REG 1A predicts recurrence in stage Ta/T1 bladder cancer. Eur J Surg Oncol 35:852–857

Gerbaud V, Pignol D, Loret E et al (2000) Mechanism of calcite crystal growth inhibition by the N-terminal undecapeptide of lithostathine. J Biol Chem 275:1057–1064

Gharib B, Fox MF, Bartoli C et al (1993) Human regeneration protein/lithostathine genes map to chromosome 2p12. Ann Hum Genet 57:9–16

Giorgi D, Bernard JP, Rouquier S et al (1989) Secretory pancreatic stone protein messenger RNA. Nucleotide sequence and expression in chronic calcifying pancreatitis. J Clin Invest 84:100–106

Gironella M, Iovanna JL et al (2005) Anti-inflammatory effects of pancreatitis associated protein in inflammatory bowel disease. Gut 54:1244–1253

Gironella M, Folch-Puy E, LeGoffic A et al (2007) Experimental acute pancreatitis in PAP/HIP knock-out mice. Gut 56:1091–1097

Graf R, Schiesser M, Scheele GA et al (2001) A family of 16-kDa pancreatic secretory stress proteins form highly organized fibrillar structures upon tryptic activation. J Biol Chem 276:21028–21038

Graf R, Schiesser M, Reding T et al (2006) Exocrine meets endocrine: pancreatic stone protein and regenerating protein–two sides of the same coin. J Surg Res 133:113–120

Grégoire C, Marco S, Thimonier J et al (2001) Three-dimensional structure of the lithostathine protofibril, a protein involved in Alzheimer's disease. EMBO J 20:3313–3321

Grønborg M, Bunkenborg J, Kristiansen TZ et al (2004) Comprehensive proteomic analysis of human pancreatic juice. J Proteome Res 3:1042–1055

Gross J, Brauer AW, Bringhurst RF et al (1985a) An unusual bovine pancreatic protein exhibiting pH-dependent globule-fibril transformation and unique amino acid sequence. Proc Natl Acad Sci USA 82:5627–5631

Gross J, Carlson RI, Brauer AW et al (1985b) Isolation, characterization, and distribution of an unusual pancreatic human secretory protein. J Clin Invest 76:2115–2126

Gross DJ, Weiss L, Reibstein I et al (1998) Amelioration of diabetes in nonobese diabetic mice with advanced disease by linomide-induced immunoregulation combined with Reg protein treatment. Endocrinology 139:2369–2374

Gu Z, Rubin MA, Yang Y et al (2005) Reg IV: a promising marker of hormone refractory metastatic prostate cancer. Clin Cancer Res 1:2237–2243

Guo Y, Xu J, Li N, Gao F, Huang P (2010) RegIV potentiates colorectal carcinoma cell migration and invasion via its CRD domain. Cancer Genet Cytogenet 199:38–44

Gurr W, Yavari R, Wen L et al (2002) A Reg family protein is overexpressed in islets from a patient with new-onset type 1 diabetes and acts as T-cell autoantigen in NOD mice. Diabetes 51:339–346

Gurr W, Shaw M, Li Y, Sherwin R (2007) RegII is a β-cell protein and autoantigen in diabetes of NOD mice. Diabetes 56:34–40

Hartupee JC, Zhang H, Bonaldo MF et al (2001) Isolation and characterization of a cDNA encoding a novel member of the human regenerating protein family: Reg IV. Biochim Biophys Acta 1518:287–293

Hayashi K, Motoyama S, Koyota S et al (2008a) REG I enhances chemo- and radiosensitivity in squamous cell esophageal cancer cells. Cancer Sci 99:2491–2495

Hayashi K, Motoyama S, Sugiyama T et al (2008b) REG Iα is a reliable marker of chemoradiosensitivity in squamous cell esophageal cancer patients. Ann Surg Oncol 15:1224–1231

Hayashi T, Matsubara A, Ohara S et al (2009) Immunohistochemical analysis of Reg IV in urogenital organs: frequent expression of Reg IV in prostate cancer and potential utility as serum tumor marker. Oncol Rep 21:95–100

He SQ, Yao JR, Zhang FX et al (2010) Inflammation and nerve injury induce expression of pancreatitis-associated protein-II in primary sensory neurons. Mol Pain 6:23

Heiskala K, Giles-Komar J, Heiskala M, Andersson LC (2006) High expression of RELP (Reg IV) in neoplastic goblet cells of appendiceal mucinous cystadenoma and pseudomyxoma peritonei. Virchows Arch 448:295–300

Heiskala K, Arola J, Heiskala M et al (2010) Expression of Reg IV and Hath1 in neuroendocrine neoplasms. Histol Histopathol 25:63–72

Heller A, Fiedler F, Schmeck J et al (1999) Pancreatitis-associated protein protects the lung from leukocyte-induced injury. Anesthesiology 91:1408–1414

Herve J, Sa Cunha A, Liu B et al (2008) Internal radiotherapy of liver cancer with rat hepatocarcinoma-intestine-pancreas gene as a liver tumor-specific promoter. Hum Gene Ther 19:915–926

Hervieu V, Christa L, Gouysse G et al (2006) HIP/PAP, a member of the Reg family, is expressed in glucagon-producing enteropancreatic endocrine cells and tumors. Hum Pathol 37:1066–1075

Higham AD, Bishop LA, Dimaline R et al (1999) Mutations of RegIα are associated with enterochromaffin-like cell tumor development in patients with hypergastrinemia. Gastroenterology 116:1310–1318

Ho MR, Lou YC, Lin WC et al (2006) Human pancreatitis-associated protein forms fibrillar aggregates with a native-like conformation. J Biol Chem 281:33566–33576

Ho MR, Lou YC, Wei SY et al (2010) Human RegIV protein adopts a typical C-Type lectin fold but binds mannan with two calcium-independent site\s. J Mol Biol 402:682–695

Honda H, Nakamura H, Otsuki M (2002) The elongated PAP-II/Reg-III mRNA is upregulated in rat pancreas during acute experimental pancreatitis. Pancreas 25:192–197

Huszarik K, Wright B, Keller C et al (2010) Adjuvant immunotherapy increases beta cell regenerative factor Reg-2 in the pancreas of diabetic mice. J Immunol 185:5120–5129

Iovanna JL, Dagorn JC (2005) The multifunctional family of secreted proteins containing a C-type lectin-like domain linked to a short N-terminal peptide. Biochim Biophys Acta 1723:8–18

Iovanna J, Orelle B, Keim V et al (1991) Messenger RNA sequence and expression of rat pancreatitis-associated protein, a lectin-related protein overexpressed during acute experimental pancreatitis. J Biol Chem 266:24664–24669

Iovanna JL, Keim V, Bosshard A et al (1993) PAP, a pancreatic secretory protein induced during acute pancreatitis, is expressed in rat intestine. Am J Physiol 265:G611–G618

Iovanna JL, Keim V, Nordback I et al (1994) Serum levels of pancreatitis-associated protein as indicators of the course of acute pancreatitis. Gastroenterology 106:728–734

Itoh T, Sawabu N, Motoo Y et al (1995) The human pancreatitis-associated protein (PAP)-encoding gene generates multiple transcripts through alternative use of 5′ exons. Gene 155:283–287

Kamimura T, West C, Beutler E (1992) Sequence of a cDNA clone encoding a rat Reg-2 protein. Gene 118:299–300

Kandil E, Lin YY, Bluth MH et al (2006) Dexamethasone mediates protection against acute pancreatitis via upregulation of pancreatitis-associated proteins. World J Gastroenterol 12:6806–6811

Katsumata N, Chakraborty C, Myal Y et al (1995) Molecular cloning and expression of peptide 23, a growth hormone-releasing hormone-inducible pituitary protein. Endocrinology 136:1332–1339

Kazumori H, Ishihara S, Hoshino E et al (2000) Neutrophil chemoattractant 2 β regulates expression of the Reg gene in injured gastric mucosa in rats. Gastroenterology 119:1610–1622

Kazumori H, Ishihara S, Fukuda R, Kinoshita Y (2002) Localization of Reg receptor in rat fundic mucosa. J Lab Clin Med 139:101–108

Keel M, Härter L, Reding T et al (2009) Pancreatic stone protein is highly increased during posttraumatic sepsis and activates neutrophil granulocytes. Crit Care Med 37:1642–1648

Keim V, Löffler HG (1986) Pancreatitis-associated protein in bile acid-induced pancreatitis of the rat. Clin Physiol Biochem 4:136–142

Keim V, Iovanna JL, Rohr G et al (1991) Characterization of a rat pancreatic secretory protein associated with pancreatitis. Gastroenterology 100:775–782

Keim V, Iovanna JL, Orelle B et al (1992) A novel exocrine protein associated with pancreas transplantation in humans. Gastroenterology 103:248–254

Keim V, Iovanna JL, Dagorn JC (1994a) The acute phase reaction of the exocrine pancreas. Gene expression and synthesis of pancreatitis-associated proteins. Digestion 55:65–72

Keim V, Willemer S, Iovanna JL et al (1994b) Rat pancreatitis-associated protein is expressed in relation to severity of experimental pancreatitis. Pancreas 9:606–612

Keim V, Hoffmeister A, Teich N et al (1999) The pancreatitis-associated protein in hereditary and chronic alcoholic pancreatitis. Pancreas 19:248–254

Kemppainen E, Sand J, Puolakkainen P et al (1996) Pancreatitis associated protein as an early marker of acute pancreatitis. Gut 39:675–678

Kiji T, Dohi Y, Takasawa S et al (2005) Activation of regenerating gene Reg in rat and human hearts in response to acute stress. Am J Physiol Heart Circ Physiol 289:H277–H284

Kimura N, Yonekura H, Okamoto H, Nagura H (1992) Expression of human regenerating gene mRNA and its product in normal and neoplastic human pancreas. Cancer 70:1857–1863

Kimura T, Fukui H, Sekikawa A et al (2009) Involvement of REG Iα protein in the regeneration of ductal epithelial cells in the minor salivary glands of patients with Sjögren's syndrome. Clin Exp Immunol 155:16–20

Kobayashi S, Akiyama T, Nata K et al (2000) Identification of a receptor for Reg (regenerating gene) protein, a pancreatic β-cell regeneration factor. J Biol Chem 275:10723–10726

Kuniyasu H, Oue N, Sasahira T et al (2009) Reg IV enhances peritoneal metastasis in gastric carcinomas. Cell Prolif 42:110–121

Lasserre C, Christa L, Simon MT et al (1992) A novel gene (HIP) activated in human primary liver cancer. Cancer Res 52:5089–5095

Lasserre C, Simon MT, Ishikawa H et al (1994) Structural organization and chromosomal localization of a human gene (HIP/PAP) encoding a C-type lectin overexpressed in primary liver cancer. Eur J Biochem 224:29–38

Laurine E, Grégoire C, Fändrich M et al (2003) Lithostathine quadruple-helical filaments form proteinase K-resistant deposits in Creutzfeldt-Jakob disease. J Biol Chem 278:51770–51778

Laurine E, Manival X, Montgelard C et al (2005) PAP-IB, a new member of the Reg gene family: cloning, expression, structural properties, and evolution by gene duplication. Biochim Biophys Acta 1727:177–187

Lechene de la Porte P, de Caro A, Lafont H, Sarles H (1986) Immunocytochemical localization of pancreatic stone protein in the human digestive tract. Pancreas 1:301–308

Lee BI, Mustafi D, Cho W, Nakagawa Y (2003) Characterization of calcium binding properties of lithostathine. J Biol Inorg Chem 8:341–347

Lehotzky RE, Partch CL, Mukherjee S et al (2010) Molecular basis for peptidoglycan recognition by a bactericidal lectin. Proc Natl Acad Sci USA 107:7722–7727

Levine JL, Patel KJ, Zheng Q et al (2000) A recombinant rat regenerating protein is mitogenic to pancreatic derived cells. J Surg Res 89:60–65

Li L, Bachem MG, Zhou S et al (2009) Pancreatitis-associated protein inhibits human pancreatic stellate cell MMP-1 and −2, TIMP-1 and −2 secretion and RECK expression. Pancreatology 9:99–110

Li B, Wang X, Liu JL (2010a) Pancreatic acinar-specific overexpression of Reg-2 gene offered no protection against either experimental diabetes or pancreatitis in mice. Am J Physiol Gastrointest Liver Physiol 299:G413–G421

Li XH, Zheng Y, Zheng HC et al (2010b) REG IV overexpression in an early stage of colorectal carcinogenesis: an immunohistochemical study. Histol Histopathol 25:473–484

Lieu HT, Batteux F, Simon MT et al (2005) HIP/PAP accelerates liver regeneration and protects against acetaminophen injury in mice. Hepatology 42:618–626

Lieu HT, Simon MT, Nguyen-Khoa T et al (2006) Reg-2 inactivation increases sensitivity to Fas hepatotoxicity and delays liver regeneration post-hepatectomy in mice. Hepatology 44:1452–1464

Lim JW, Song JY, Seo JY et al (2009) Role of pancreatitis-associated protein 1 on oxidative stress-induced cell death of pancreatic acinar cells. Ann N Y Acad Sci 1171:545–548

Lin YY, Viterbo D, Mueller CM et al (2008) Small-interference RNA gene knockdown of pancreatitis-associated proteins in rat acute pancreatitis. Pancreas 36:402–410

Lipsett M, Hanley S, Castellarin M, Austin E, Suarez-Pinzon WL, Rabinovitch A, Rosenberg L (2007) The role of islet neogenesis-associated protein (INGAP) in islet neogenesis. Cell Biochem Biophys 48:127–137

Liu JL, Cui W, Li B, Lu Y (2008) Possible roles of Reg family proteins in pancreatic islet cell growth. Endocr Metab Immune Disord Drug Targets 8:1–10

Liu L, Liu JL, Srikant CB (2010) Reg-2 protects mouse insulinoma cells from streptozotocin-induced mitochondrial disruption and apoptosis. Growth Factors 28:370–378

Livesey FJ, O'Brien JA, Li M et al (1997) A Schwann cell mitogen accompanying regeneration of motor neurons. Nature 390:614–618

Lu Y, Ponton A, Okamoto H et al (2006) Activation of the Reg family genes by pancreatic-specific IGF-I gene deficiency and after streptozotocin-induced diabetes in mouse pancreas. Am J Physiol Endocrinol Metab 291:E50–E58

Mahurkar S, Bhaskar S, Reddy DN et al (2007) Comprehensive screening for Reg 1α gene rules out association with tropical calcific pancreatitis. World J Gastroenterol 13:5938–5943

Makino T, Kawashima H, Konishi H et al (2010) Elevated urinary levels and urothelial expression of hepatocarcinoma-intestine-pancreas/pancreatitis-associated protein in patients with interstitial cystitis. Urology 75:933–937

Malka D, Vasseur S, Bodeker H et al (2000) Tumor necrosis factor α triggers antiapoptotic mechanisms in rat pancreatic cells through pancreatitis-associated protein I activation. Gastroenterology 119:816–828

Mann K, Siedler F (1999) The amino acid sequence of ovocleidin 17, a major protein of the avian eggshell calcified layer. Biochem Mol Biol Int 47:997–1007

Masciotra L, Lechâne de la Porte P et al (1995) Immunocytochemical localization of pancreatitis-associated protein in human small intestine. Dig Dis Sci 40:519–524

Mauro V, Carette D, Chevallier D et al (2008) Reg I protein in healthy and seminoma human testis. Histol Histopathol 23:1195–1203

McCarthy AN, Mogilner IG, Grigera JR et al (2009) Islet neogenesis associated protein (INGAP): structural and dynamical properties of its active pentadecapeptide. J Mol Graph Model 27:701–705

McKie AT, Simpson RJ, Ghosh S, Peters TJ, Farzaneh F (1996) Regulation of pancreatitis-associated protein (HIP/PAP) mRNA levels in mouse pancreas and small intestine. Clin Sci 91:213–218

Medveczky P, Szmola R, Sahin-Tóth M (2009) Proteolytic activation of human pancreatitis associated protein is required for peptidoglycan binding and bacterial aggregation. Biochem J 420:335–343

Minamiya Y, Kawai H, Saito H et al (2008) REG 1A expression is an independent factor predictive of poor prognosis in patients with non-small cell lung cancer. Lung Cancer 60:98–104

Mitani Y, Oue N, Matsumura S et al (2007) Reg IV is a serum biomarker for gastric cancer patients and predicts response to 5-fluorouracil-based chemotherapy. Oncogene 26:4383–4393

Miyaoka Y, Kadowaki Y, Ishihara S et al (2004) Transgenic overexpression of Reg protein caused gastric cell proliferation and differentiation along parietal cell and chief cell lineages. Oncogene 23:3572–3579

Miyashita H, Nakagawara K, Mori M et al (1995) Human REG family genes are tandemly ordered in a 95-kilobase region of chromosome 2p12. FEBS Lett 377:429–433

Montalto G, Iovanna JL, Soresi M et al (1998) Clinical evaluation of pancreatitis-associated protein as a serum marker of hepatocellular carcinoma: comparison with α-fetoprotein. Oncology 55:421–425

Moriizumi S, Watanabe T, Unno M et al (1994) Isolation, structural determination and expression of a novel Reg gene, human RegI β. Biochim Biophys Acta 1217:199–202

Moriscot C, Renaud W, Bouvier R et al (1996) Absence of correlation between Reg and insulin gene expression in pancreas during fetal development. Pediatr Res 39:349–353

Morisset J, Iovanna J, Grondin G (1997) Localization of rat pancreatitis-associated protein during bile salt-induced pancreatitis. Gastroenterology 112:543–550

Motoo Y, Itoh T, Su SB et al (1998) Expression of pancreatitis-associated protein (PAP) mRNA in gastrointestinal cancers. Int J Pancreatol 23:11–16

Motoo Y, Satomura Y, Mouri I et al (1999) Serum levels of pancreatitis-associated protein in digestive diseases with special reference to gastrointestinal cancers. Dig Dis Sci 44:1142–1147

Motoo Y, Taga K, Su SB et al (2000) Arginine induces apoptosis and gene expression of pancreatitis-associated protein (PAP) in rat pancreatic acinar AR4-2J cells. Pancreas 20:61–66

Motoo Y, Watanabe H, Yamaguchi Y et al (2001) Pancreatitis-associated protein levels in pancreatic juice from patients with pancreatic diseases. Pancreatology 1:43–47

Motoyama S, Sugiyama T, Ueno Y et al (2006) REG I expression predicts long-term survival among locally advanced thoracic squamous cell esophageal cancer patients treated with neoadjuvant chemoradiotherapy followed by esophagectomy. Ann Surg Oncol 13:1724–1731

Moucadel V, Soubeyran P, Vasseur S et al (2001) Cdx1 promotes cellular growth of epithelial intestinal cells through induction of the secretory protein PAP-I. Eur J Cell Biol 80:156–163

Mueller CM, Zhang H, Zenilman ME (2008) Pancreatic Reg I binds MKP-1 and regulates cyclin D in pancreatic-derived cells. J Surg Res 150:137–143

Mukherjee S, Partch CL, Lehotzky RE et al (2009) Regulation of C-type lectin antimicrobial activity by a flexible N-terminal prosegment. J Biol Chem 284:4881–4888

Multigner L, Sarles H, Lombardo D, De Caro A (1985) Pancreatic stone protein. II. Implication in stone formation during the course of chronic calcifying pancreatitis. Gastroenterology 89:387–391

Namikawa K, Fukushima M, Murakami K et al (2005) Expression of Reg/PAP family members during motor nerve regeneration in rat. Biochem Biophys Res Commun 332:126–134

Namikawa K, Okamoto T, Suzuki A et al (2006) Pancreatitis-associated protein-III is a novel macrophage chemoattractant implicated in nerve regeneration. J Neurosci 26:7460–7467

Nanakin A, Fukui H, Fujii S et al (2007) Expression of the REG IV gene in ulcerative colitis. Lab Invest 87:304–314

Narushima Y, Unno M, Nakagawara K et al (1997) Structure, chromosomal localization and expression of mouse genes encoding type III Reg, Reg-III α, Reg-III β, Reg-III γ. Gene 185:159–168

Nata K, Liu Y, Xu L et al (2004) Molecular cloning, expression and chromosomal localization of a novel human REG family gene, REG-III. Gene 340:161–170

Ng NFL, Hew CL (1992) Structure of an antifreeze polypeptide from the sea raven. J Biol Chem 267:16069–16075

Nishimune H, Vasseur S, Wiese S et al (2000) Reg-2 is a motoneuron neurotrophic factor and a signaling intermediate in the CNTF survival pathway. Nat Cell Biol 2:906–914

Nordback I, Jaakkola M, Iovanna JL, Dagorn JC (1995) Increased serum pancreatitis associated protein (PAP) concentration after longterm alcohol consumption: further evidence for regular subclinical pancreatic damage after heavy drinking? Gut 36:117–120

Norkina O, Graf R, Appenzeller P, De Lisle RC (2006) Caerulein-induced acute pancreatitis in mice that constitutively overexpress Reg/PAP genes. BMC Gastroenterol 6:16

Ochiai K, Kaneko K, Kitagawa M et al (2004) Activated pancreatic enzyme and pancreatic stone protein (PSP/Reg) in bile of patients with pancreaticobiliary maljunction/choledochal cysts. Dig Dis 49:1953–1956

Ogawa H, Fukushima K, Naito H et al (2003) Increased expression of HIP/PAP and regenerating gene III in human inflammatory bowel disease and a murine bacterial reconstitution model. Inflamm Bowel Dis 9:162–170

Ohara S, Oue N, Matsubara A et al (2008) Reg IV is an independent prognostic factor for relapse in patients with clinically localized prostate cancer. Cancer Sci 99:1570–1577

Okamoto H (1999) The Reg gene family and Reg proteins: with special attention to the regeneration of pancreatic β-cells. J Hepatobiliary Pancreat Surg 6:254–262

Orelle B, Keim V, Masciotra L, Dagorn JC, Iovanna JL (1992) Human pancreatitis-associated protein. Messenger RNA cloning and expression in pancreatic diseases. J Clin Invest 90:2284–2291

Ortiz EM, Dusetti NJ, Dagorn JC, Iovanna JL (1997) Characterization of a silencer regulatory element in the rat PAP-I gene which confers tissue-specific expression and is promoter-dependent. Arch Biochem Biophys 340:111–116

Ortiz EM, Dusetti NJ, Vasseur S et al (1998) The pancreatitis-associated protein is induced by free radicals in AR4-2J cells and confers cell resistance to apoptosis. Gastroenterology 114:808–816

Ose T, Kadowaki Y, Fukuhara H et al (2007) Reg I-knockout mice reveal its role in regulation of cell growth that is required in generation and maintenance of the villous structure of small intestine. Oncogene 26:349–359

Oue N, Mitani Y, Aung PP et al (2005) Expression and localization of Reg IV in human neoplastic and non-neoplastic tissues: Reg IV expression is associated with intestinal and neuroendocrine differentiation in gastric adenocarcinoma. J Pathol 207:185–198

Oue N, Kuniyasu H, Noguchi T et al (2007) Serum concentration of Reg IV in patients with colorectal cancer: overexpression and high serum levels of Reg IV are associated with liver metastasis. Oncology 72:371–380

Oue N, Sentani K, Noguchi T et al (2008) Immunohistochemical staining of Reg IV and claudin-18 is useful in the diagnosis of gastrointestinal signet ring cell carcinoma. Am J Surg Pathol 32:1182–1189

Ozturk M, de la Monte SM, Gross J, Wands JR (1989) Elevated levels of an exocrine pancreatic secretory protein in Alzheimer disease brain. Proc Natl Acad Sci USA 86:419–423

Patard L, Stoven V, Gharib B et al (1996) What function for human lithostathine?: structural investigations by three-dimensional structure modeling and high-resolution NMR spectroscopy. Protein Eng 9:949–957

Perfetti R, Hawkins AL et al (1994) Assignment of the human pancreatic regenerating (REG) gene to chromosome 2p12. Genomics 20:305–307

Perfetti R, Raygada M et al (1996) Regenerating (Reg) and insulin genes are expressed in prepancreatic mouse embryos. J Mol Endocrinol 17:79–88

Petersen TE (1988) The amino-terminal domain of thrombomodulin and pancreatic stone protein are homologous with lectins. FEBS Lett 231:51–53

Petersen BE, Zajac VF, Michalopoulos GK (1998) Hepatic oval cell activation in response to injury following chemically induced periportal or pericentral damage in rats. Hepatology 27:1030–1038

Pezzilli R, Billi P, Migliori M, Gullo L (1997) Clinical value of pancreatitis-associated protein in acute pancreatitis. Am J Gastroenterol 92:1887–1890

Pignol D, Bertrand JA, Bernard JP et al (1995) Crystallization and preliminary crystallographic study of human lithostathine. Proteins 23:604–606

Pittenger GL, Taylor-Fishwick DA et al (2007) Intramuscular injection of islet neogenesis-associated protein peptide stimulates pancreatic islet neogenesis in healthy dogs. Pancreas 34:103–111

Planas R, Alba A, Carrillo J et al (2006) Reg (regenerating) gene overexpression in islets from non-obese diabetic mice with accelerated diabetes: role of IFNβ. Diabetologia 49:2379–2387

Rafaeloff R, Pittenger GL, Barlow SW et al (1997) Cloning and sequencing of the pancreatic islet neogenesis associated protein (INGAP) gene and its expression in islet neogenesis in hamsters. J Clin Invest 99:2100–2109

Rechreche H, Montalto G, Mallo GV et al (1999) PAP, Reg Iα and Reg Iβ mRNAs are concomitantly up-regulated during human colorectal carcinogenesis. Int J Cancer 81:688–694

Renugopalakrishnan V, Dobbs JC, Collette TW et al (1999) Human pancreatic thread protein, an exocrine thread protein with possible implications to Alzheimer's disease: secondary structure in solution at acid pH. Biochem Biophys Res Commun 258:653–656

Rosty C, Christa L, Kuzdzal S et al (2002) Identification of hepatocarcinoma-intestine pancreas/pancreatitis-associated protein I as a biomarker for pancreatic ductal adenocarcinoma by protein biochip technology. Cancer Res 62:1868–1875

Rouimi P, Bonicel J, Rovery M, De Caro A (1987) Cleavage of the Arg-Ile bond in the native polypeptide chain of human pancreatic stone protein. FEBS Lett 216:195–199

Rouquier S, Verdier JM, Iovanna J et al (1991) Rat pancreatic stone protein messenger RNA. Abundant expression in mature exocrine cells, regulation by food content, and sequence identity with the endocrine Reg transcript. J Biol Chem 266:786–791

Sanchez D, Baeza N, Blouin R et al (2000) Over-expression of the Reg gene in non-obese diabetic mouse pancreas during active diabetogenesis is restricted to exocrine tissue. J Histochem Cytochem 48:1401–1410

Sanchez D, Figarella C, Marchand-Pinatel S et al (2001) Preferential expression of Reg Iβ gene in human adult pancreas. Biochem Biophys Res Commun 284:729–737

Sanchez D, Gmyr V, Kerr-Conte J et al (2004) Implication of Reg I in human pancreatic duct-like cells in vivo in the pathological pancreas and in vitro during exocrine dedifferentiation. Pancreas 29:14–21

Sanchez D, Mueller CM, Zenilman ME (2009) Pancreatic regenerating gene I and acinar cell differentiation: influence on cellular lineage. Pancreas 38:572–577

Sansonetti A, Romeo H, Berthezene P et al (1995) Developmental, nutritional, and hormonal regulation of the pancreatitis-associated protein I and III gene expression in the rat small intestine. Scand J Gastroenterol 30:664–669

Sarles H, Dagorn JC, Giorgi D, Bernard JP (1990) Renaming pancreatic stone protein as 'lithostathine'. Gastroenterology 99:900–901

Sarles J, Barthellemy S, Férec C et al (1999) Blood concentrations of pancreatitis associated protein in neonates: relevance to neonatal screening for cystic fibrosis. Arch Dis Child Fetal Neonatal Ed 80: F118–F122

Sasahara K, Yamaoka T, Moritani M et al (2000) Molecular cloning and tissue-specific expression of a new member of the regenerating protein family, islet neogenesis-associated protein-related protein. Biochim Biophys Acta 1500:142–146

Sasahira T, Oue N, Kirita T et al (2008) Reg IV expression is associated with cell growth and prognosis of adenoid cystic carcinoma in the salivary gland. Histopathology 53:667–675

Sasaki Y, Minamiya Y, Takahashi N et al (2008) REG 1A expression is an independent factor predictive of poor prognosis in patients with breast cancer. Ann Surg Oncol 15:3244–3251

Satomura Y, Sawabu N, Mouri I et al (1995) Measurement of serum PSP/Reg-protein concentration in various diseases with a newly developed enzyme-linked immunosorbent assay. J Gastroenterol 30:643–650

Schiesser M, Bimmler D, Frick TW, Graf R (2001) Conformational changes of pancreatitis-associated protein (PAP) activated by trypsin lead to insoluble protein aggregates. Pancreas 22:186–192

Schmiegel W, Burchert M, Kalthoff H et al (1990) Immunochemical characterization and quantitative distribution of pancreatic stone protein in sera and pancreatic secretions in pancreatic disorders. Gastroenterology 99:1421–1430

Sekikawa A, Fukui H, Fujii S et al (2005) REG Iα protein may function as a trophic and/or anti-apoptotic factor in the development of gastric cancer. Gastroenterology 128:642–653

Sekikawa A, Fukui H, Fujii S et al (2008) REG Iα protein mediates an anti-apoptotic effect of STAT3 signaling in gastric cancer cells. Carcinogenesis 29:76–83

Sekikawa A, Fukui H, Suzuki K et al (2010) Involvement of the IL-22/REG Iα axis in ulcerative colitis. Lab Invest 90:496–505

Shervani NJ, Takasawa S, Uchigata Y et al (2004) Autoantibodies to REG, a β-cell regeneration factor, in diabetic patients. Eur J Clin Invest 34:752–758

Shin JS, Lee JJ, Lee EJ et al (2005) Proteome analysis of rat pancreas induced by pancreatectomy. Biochim Biophys Acta 1749:23–32

Simon MT, Pauloin A, Normand G et al (2003) HIP/PAP stimulates liver regeneration after partial hepatectomy and combines mitogenic and anti-apoptotic functions through the PKA signaling pathway. FASEB J 17:1441–1450

Skretting G, Austbø L, Olsaker I, Espenes A (2006) Cloning and expression analysis of an ovine PAP-like protein cDNA, a gene differentially expressed in scrapie. Gene 376:116–122

Starcević-Klasan G, Azman J, Picard A et al (2008) Reg IV protein is expressed in normal rat tissues. Coll Antropol 32(2):89–93

Steele IA, Dimaline R, Pritchard DM et al (2007) Helicobacter and gastrin stimulate Reg 1 expression in gastric epithelial cells through distinct promoter elements. Am J Physiol Gastrointest Liver Physiol 293:G347–G354

Stephanova E, Tissir F, Dusetti N et al (1996) The rat genes encoding the pancreatitis-associated proteins I, II and III (PAP-I, Pap2, Pap3), and the lithostathin/pancreatic stone protein/regeneration protein (Reg) colocalize at 4q33– > q34. Cytogenet Cell Genet 72:83–85

Suzuki Y, Yonekura H, Watanabe T et al (1994) Structure and expression of a novel rat Reg-III gene. Gene 144:315–316

Szasz T, Eddy S, Paulauskis J et al (2009) Differential expression of pancreatitis-associated protein and thrombospondins in arterial versus venous tissues. J Vasc Res 46:551–560

Takahara Y, Suzuki A, Maeda M et al (2008) Expression of pancreatitis associated proteins in urothelium and urinary afferent neurons following cyclophosphamide induced cystitis. J Urol 179:1603–1609

Tam J, Rosenberg L, Maysinger D (2002) Islet-neogenesis-associated protein enhances neurite outgrowth from DRG neurons. Biochem Biophys Res Commun 291:649–654

Tam J, Rosenberg L, Maysinger D (2004) INGAP peptide improves nerve function and enhances regeneration in streptozotocin-induced diabetic C57BL/6 mice. FASEB J 18:1767–1769

Tamura H, Ohtsuka M, Washiro M et al (2009) Reg IV expression and clinicopathologic features of gallbladder carcinoma. Hum Pathol 40:1686–1692

Taylor-Fishwick DA, Rittman S, Kendall H, et al (2003) Cloning genomic INGAP: a Reg-related family member with distinct transcriptional regulation sites. Biochim Biophys Acta 1638:83–89

Taylor-Fishwick DA, Shi W, Pittenger GL et al (2006) PDX-1 can repress stimulus-induced activation of the INGAP promoter. J Endocrinol 188:611–621

Taylor-Fishwick DA, Bowman A, Korngiebel-Rosique M et al (2008) Pancreatic islet immunoreactivity to the Reg protein INGAP. J Histochem Cytochem 56:183–191

Tebar LA, Géranton SM, Parsons-Perez C et al (2008) Deletion of the mouse Reg-IIIβ (Reg-2) gene disrupts ciliary neurotrophic factor signaling and delays myelination of mouse cranial motor neurons. Natl Acad Sci USA 105:11400–11405

Terazono K, Yamamoto H, Takasawa S et al (1988) A novel gene activated in regenerating islets. J Biol Chem 263:2111–2114

Terazono K, Uchiyama Y, Ide M et al (1990) Expression of reg protein in rat regenerating islets and its co-localization with insulin in the β cell secretory granules. Diabetologia 33:250–252

Unno M, Yonekura H, Nakagawara K et al (1993) Structure, chromosomal localization, and expression of mouse Reg genes, Reg I and Reg II. A novel type of Reg gene, Reg II, exists in the mouse genome. J Biol Chem 268:15974–15982

Usami S, Motoyama S, Koyota S et al (2010) Regenerating gene I regulates interleukin-6 production in squamous esophageal cancer cells. Biochem Biophys Res Commun 392:4–8

Valery C, Vasseur S, Sabatier F et al (2001) Pancreatitis associated protein I (PAP-I) alters adhesion and motility of human melanocytes and melanoma cells. J Invest Dermatol 116:426–433

van Ampting MT, Rodenburg W, Vink C et al (2009) Ileal mucosal and fecal pancreatitis associated protein levels reflect severity of Salmonella infection in rats. Dig Dis Sci 54:2588–2597

van der Pijl JW, Boonstra JG et al (1997) Pancreatitis-associated protein: a putative marker for pancreas graft rejection. Transplantation 63:995–1003

Vasseur S, Frigerio JM, Dusetti NJ et al (1995) Two transcripts are generated from the pancreatitis associated protein II gene by alternative splicing in the 5′ untranslated region. Biochim Biophys Acta 1261:272–274

Vasseur S, Folch-Puy E, Hlouschek V, Iovanna JL et al (2004) p8 improves pancreatic response to acute pancreatitis by enhancing the expression of the anti-inflammatory protein pancreatitis-associated protein I. J Bio Chem 279:7199–7207

Violette S, Festor E, Pandrea-Vasile I et al (2003) Reg IV, a new member of the Regenerating gene family, is overexpressed in colorectal carcinomas. Int J Cancer 103:185–193

Viterbo D, Bluth MH, Lin YY et al (2008a) Pancreatitis-associated protein 2 modulates inflammatory responses in macrophages. J Immunol 181:1948–1958

Viterbo D, Bluth MH, Mueller CM et al (2008b) Mutational characterization of pancreatitis-associated protein 2 domains involved in mediating cytokine secretion in macrophages and the NF-kB pathway. J Immunol 181:1959–1968

Viterbo D, Callender GE, DiMaio T et al (2009) Administration of anti-Reg I and anti-PAP-II antibodies worsens pancreatitis. JOP 10:15–23

Walker LC, LeVineH III (2000) The cerebral proteopathies. Neurobiol Aging 21:559–561

Wang X, Wang B, Wu J (2001) Pancreatitis-associated protein-I mRNA expression in mouse pancreas is upregulated by lipopolysaccharide independent of cerulein-pancreatitis. J Gastroenterol Hepatol 16:79–86

Wang J, Koyota S, Zhou X et al (2009) Expression and localization of regenerating gene I in a rat liver regeneration model. Biochem Biophys Res Commun 380:472–477

Watanabe T, Yonekura H, Terazono K et al (1990) Complete nucleotide sequence of human Reg gene and its expression in normal and tumoral tissues. The Reg protein, pancreatic stone protein, and pancreatic thread protein are one and the same product of the gene. J Biol Chem 265:7432–7439

Watanabe T, Yonemura Y, Yonekura H et al (1994) Pancreatic β-cell replication and amelioration of surgical diabetes by Reg protein. Proc Natl Acad Sci USA 91:3589–3592

Watanabe R, Hanawa H, Yoshida T et al (2008) Gene expression profiles of cardiomyocytes in rat autoimmune myocarditis by DNA microarray and increase of regenerating gene family. Transl Res 152:119–127

Wittel UA, Pandey KK, Andrianifahanana M et al (2006) Chronic pancreatic inflammation induced by environmental tobacco smoke inhalation in rats. Am J Gastroenterol 101:148–159

Xie MJ, Motoo Y, Iovanna JL et al (2003) Overexpression of pancreatitis-associated protein (PAP) in human pancreatic ductal adenocarcinoma. Dig Dis Sci 48:459–464

Xu YY, Wands JR, de la Monte SM (1993) Characterization of thread proteins expressed in neuroectodermal tumors. Cancer Res 53:3823–3829

Xu YY, Bhavani K, Wands JR et al (1995a) Ethanol inhibits insulin receptor substrate-1 tyrosine phosphorylation and insulin-stimulated neuronal thread protein gene expression. Biochem J 310:125–132

Xu YY, Bhavani K, Wands JR et al (1995b) Insulin-induced differentiation and modulation of neuronal thread protein expression in primitive neuroectodermal tumor cells is linked to phosphorylation of insulin receptor substrate-1. J Mol Neurosci 6:91–108

Yamagishi H, Fukui H, Sekikawa A et al (2009) Expression profile of REG family proteins REG Iα and REG IV in advanced gastric cancer: comparison with mucin phenotype and prognostic markers. Mod Pathol 22:906–913

Yamauchi A, Takahashi I, Takasawa S et al (2009) Thiazolidinediones inhibit REG Iα gene transcription in gastrointestinal cancer cells. Biochem Biophys Res Commun 379:743–748

Yasui W, Oue N, Sentani K et al (2009) Transcriptome dissection of gastric cancer: identification of novel diagnostic and therapeutic targets from pathology specimens. Pathol Int 59:121–136

Yonemura Y, Sakurai S et al (2003) REG gene expression is associated with the infiltrating growth of gastric carcinoma. Cancer 98:1394–1400

Yoshida K, Yasui W (2009) Serum olfactomedin 4 (GW112, hGC-1) in combination with Reg IV is a highly sensitive biomarker for gastric cancer patients. Int J Cancer 125:2383–2392

Yoshino N, Ishihara S, Rumi MA et al (2005) Interleukin-8 regulates expression of Reg protein in *Helicobacter pylori*-infected gastric mucosa. Am J Gastroenterol 100:2157–2166

Yoshioka S, Fujii S, Richards JS, Espey LL (2002) Gonadotropin-induced expression of pancreatitis-associated protein-III mRNA in the rat ovary at the time of ovulation. J Endocrinol 174:485–492

Yuan RH, Jeng YM, Chen HL et al (2005) Opposite roles of human pancreatitis-associated protein and REG 1A expression in hepatocellular carcinoma: association of pancreatitis-associated protein expression with low-stage hepatocellular carcinoma, β-catenin mutation, and favorable prognosis. Clin Cancer Res 11:2568–2575

Zenilman ME, Magnuson TH et al (1996) Pancreatic thread protein is mitogenic to pancreatic-derived cells in culture. Gastroenterology 110:1208–1214

Zenilman ME, Kim S, Levine BA et al (1997) Ectopic expression of Reg protein: a marker of colorectal mucosa at risk for neoplasia. J Gastrointest Surg 1:194–201

Zenilman ME, Chen J, Magnuson TH (1998) Effect of Reg protein on rat pancreatic ductal cells. Pancreas 17:256–261

Zenilman ME, Tuchman D, Zheng Q et al (2000) Comparison of Reg I and Reg-III levels during acute pancreatitis in the rat. Ann Surg 232:646–652

Zhang Z, Realini C, Clawson A et al (1998) Proteasome activation by REG molecules lacking homolog-specific inserts. J Biol Chem 273:9501–9509

Zhang Y, Lai M, Lv B et al (2003a) Overexpression of Reg IV in colorectal adenoma. Cancer Lett 200:69–76

Zhang YW, Ding LS, Lai MD (2003b) Reg gene family and human diseases. World J Gastroenterol 9:2635–2641

Zhang H, Kandil E, Lin YY et al (2004) Targeted inhibition of gene expression of pancreatitis-associated proteins exacerbates the severity of acute pancreatitis in rats. Scand J Gastroenterol 39:870–881

Zheng HC, Xu XY, Yu M et al (2010) The role of Reg IV gene and its encoding product in gastric carcinogenesis. Hum Pathol 41:59–69

Zhong B, Pavel Strnad P, Diana M, Toivola DM et al (2007) Reg-II is an exocrine pancreas injury-response product that is up-regulated by keratin absence or mutation. Mol Biol Cell 18:4969–4978

Emerging Groups of C-Type Lectins

G.S. Gupta

40.1 Layilin Group of C-Type Lectins (Group VIII of CTLDS)

40.1.1 Layilin: A Hyaluronan Receptor

Layilin, a ~55-kDa lectin, is a widely expressed integral membrane hyaluronan receptor, originally identified as a binding partner of talin located in membrane ruffles. It is recruited to membrane ruffles in cells induced to migrate in in vitro wounding experiments and in peripheral ruffles in spreading cells. Layilin is a transmembrane C-type lectin (Fig. 40.1) and binds specifically to hyaluronan (HA) but not to other tested glycosaminoglycans and belongs to group VIII of CTLDs. The other member of this group is chondrolectin. Layilin's ability to bind hyaluronan reveals an interesting parallel between layilin and CD44, because both can bind to cytoskeleton-membrane linker proteins through their cytoplasmic domains and to hyaluronan through their extracellular domains. This parallelism suggests a role for layilin in cell adhesion and motility (Banerji et al. 1999; Bono et al. 2001). There is no sequence homology with known hyaluronan receptors, including CD44, RHAMM (receptor for HA-mediated motility), or LYVE-1 (lymphatic vessel endothelial HA receptor-1). Thus, by binding to HA, layilin may facilitate cell migration by mediating early interactions between spreading cells and the extracellular matrix (ECM). Since, hyaluronan is involved in invasion of a variety of tumor cells and layilin was detected in the human lung cell line A549, layilin might play crucial roles in lymphatic metastasis of lung carcinoma. Suppression of layilin expression by iRNA significantly increased survival of tumor-bearing mice. The suppression of layilin expression might be a promising strategy for treatment of human lung carcinoma (Chen et al. 2008).

40.1.2 Interactions of Layilin

The actin cytoskeleton plays a significant role in change of cell shape and motility. Several adaptor proteins, including talin, maintain the cytoskeleton-membrane linkage by binding to integral membrane proteins and to the cytoskeleton. The talin binds to integrins, vinculin, actin and layilin. A ten-amino acid motif in layilin cytoplasmic domain is sufficient for talin binding. Layilin colocalizes with talin in ruffles and binds to talin's ~50-kDa head domain (amino acids 280–435). This region overlaps a binding site for focal adhesion kinase (Borowsky and Hynes 1998). The other known binding partners of layilin include other members of the talin/band 4.1/ERM (ezrin, radixin, and moesin) family of cytoskeletal-membrane linker molecules. The neurofibromatosis 2 tumor suppressor protein, merlin, is commonly mutated in human benign brain tumors. Merlin and radixin interact with the carboxy-terminal domain of layilin. This suggests that layilin may mediate signals from extracellular matrix to the cell cytoskeleton via interaction with different intracellular binding partners and thereby be involved in the modulation of cortical structures in the cell (Bono et al. 2005; Scoles 2008). The talin F3 domain binds to short sequences in the layilin cytoplasmic domain, integrins, and PIPK1γ (Wegener et al. 2008).

40.1.3 Functions of Layilin

The colocalization of layilin with talin and the binding interaction between them make layilin as a good candidate for a membrane-binding site for talin in ruffles. The colocalization of both epitope-tagged and endogenous layilin with talin in actin-rich membrane ruffles and significant homology to proteins with C-type lectin activity of layilin leads to suggest two general models for layilin

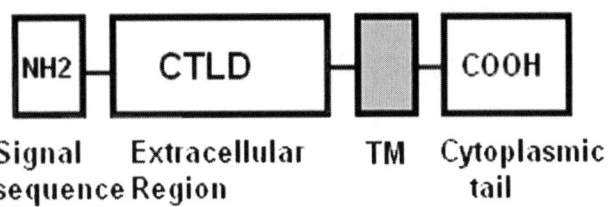

Fig. 40.1 The protein domains of layilin. Sig (NH$_2$-terminal signal sequence), Lectin (layilin's extracellular part which is homologous to C-type lectins) and TM (transmembrane) domain

function. In the first model, layilin acts in cell migration by anchoring to the membrane talin, which in turn binds F-actin. This chain of interactions may transmit force from actin to the membrane, resulting in its characteristic deformation into ruffles. In this scenario, layilin may serve simply as a talin docking site, or it could be an early-acting adhesion molecule that binds extracellular matrix, nucleating the formation of focal contact precursors at the cell periphery. In this case, layilin would be functioning analogously to selectins, which mediate transient adhesion between rolling leukocytes and the endothelium followed by tight integrin-mediated adhesion. However, whereas selectins mediate cell adhesion in specialized cells of immune system, it was hypothesized that layilin performs fundamental cell adhesion tasks common to most cells because layilin and talin are present in many different tissues. Layilin would thus form transient adhesion sites between ruffles and extracellular matrix that are refined into focal contacts after integrin recruitment. This could occur in both spreading and migrating cells. Talin, which can bind both layilin and integrins, may provide continuity between the two types of cell–matrix linkages as transient layilin-containing structures mature into integrin-containing focal adhesions. Integrin extracellular matrix receptors are known to be signaling as well as adhesion receptors, and if layilin encounters matrix early in the process of cell adhesion it may also signal. The three internally repeated motifs in the layilin cytoplasmic domain are potential binding sites for cytoskeletal or signaling molecules in addition to talin. Conservation of these motifs in hamster and human layilins suggests that they are of importance, and two of these repeats form a binding-site for talin (Borowsky and Hynes 1998). In the second model, layilin may function as an endo- or phagocytic receptor, similar to some other C-type lectins such as the macrophage mannose receptor. This model is suggested by the observation that talin is present in phagocytic cups in macrophages and at the sites of uptake of some bacterial pathogens (Young et al. 1992). In addition, the layilin cytoplasmic domain contains two YXXΦ motifs (YNVI, YDNM). These motifs are similar to sequences that mediate clathrin-dependent endocytosis of other membrane proteins by binding to the μ$_2$ chain of the AP-2 adaptor complex (Ohno et al. 1995; Marks et al. 1997; Borowsky and Hynes 1998).

40.1.4 Chondrolectin (CHODL)

The *Chondrolectin (Chodl)*, a novel human gene encoding chondrolectin is localized at chromosome 21q21 and consists of six exons and five introns. The ORF of *Chodl* encodes a type I transmembrane protein containing a single CRD of C-type lectins in its extracellular portion. CHODL expresses in testis, prostate, and spleen and the expression of CHODL is mainly limited to vascular muscle of testis, smooth muscle of prostate stroma, heart muscle, skeletal muscle, crypts of small intestine, and red pulp of spleen. CHODL is an N-glycosylated protein with a molecular weight of ~36 kDa and shows a predominantly perinuclear localization in transiently transfected COS1 cells. Although the predicted CHODL protein shares a significant homology (45% overall and 60% within the CRD) with layilin, a hyaluronan receptor, there was no specific interaction between CHODL and hyaluronan using cetylpyridinium chloride precipitation (Weng et al. 2002).

The mouse orthologue of human chondrolectin gene, *Chodl* is located at chromosome 16C3 and consists of six exons and five introns. The putative full-length mouse cDNA of *Chodl* consists of 2393 bp, with an ORF of 839 bp, 243 bp of 5′ untranslated region and 1310 bp of 3′ untranslated region. The predicted CHODL protein containing one CRD of C-type lectin in its extracellular portion shares a similarity (45%) with human layilin. In adult mice, CHODL is preferentially expressed in skeletal muscle, testis, brain, and lung. During gestation, E 7–15 its expression is up-regulated (Weng et al. 2003).

40.2 Tetranectin Group of Lectins (Group IX of CTLDs)

The tetranectin group of CTLD-containing secreted proteins has three members in both human and mouse: tetranectin (TN), cartilage-derived C-type lectin, and stem cell growth factor (SCGF). Each polypeptide has an N-terminal tail rich in charged residues, a central neck region, and a C-terminal CTLD. The neck region mediates homotrimerization through the formation of a coiled-coil of α-helices. It is not known if carbohydrate recognition plays a role in the function of TN group of proteins. The CTLDs in these proteins resemble CTLDs of the galactose-binding subtype, but attempts to demonstrate sugar-binding activity have not been successful.

40.2.1 Tetranectin

By stimulating plasminogen activation, Tetranectin (TN) may regulate proteolysis of extracellular matrix components in development, tissue remodelling and cancer, and influence the proteolytic activation of proteases and growth factors. Tetranectin is a member of the C-type lectin family and can be detected in serum and the extracellular matrix. It is composed of three identical, non-covalently linked 20 kDa subunits, and binds to the fourth kringle domain of plasminogen, which can stimulate plasminogen activation in vitro (Høgdall 1998). Each of the mature TN monomer of 181 amino acids consists of three functional domains encoded by separate exons (Fuhlendorff et al. 1987; Wewer and Albrechtsen 1992). A short lysine-rich region at the N terminus binds to heparin (Lorentsen et al. 2000). This domain is followed by an α-helical domain responsible for multimerization by forming a triple coiled-coil α-helix (Nielsen et al. 1997). The final 132 amino acids comprise a C-type lectin-like domain, homologous to the CRD of calcium-dependent animal lectins (Drickamer 1999) and contain binding sites for calcium and plasminogen (Graversen et al. 1998). The three intrachain disulfide bonds connect Cys residues 50–60, 77–176, and 152–168. The TN is homologous (17–24% identical positions) with those parts of the asialoglycoprotein receptor family that are considered to be extracellular. The TN has no structures corresponding to those parts of the receptors considered to be intracellular and membrane anchoring. The sequence of TN is also homologous (22–23% identical positions) with the C-terminal globular domain of the core protein of the cartilage proteoglycan (Fuhlendorff et al. 1987). In addition, O-linked glycosylation of the TN molecule at threonine 4 has been described (Fuhlendorff et al. 1987; Jaquinod et al. 1999). Tetranectin has a distinct binding site in its N-terminus that mediates binding to complex sulfated carbohydrates (e.g., heparin) (Lorentsen et al. 2000). The N-terminus of TN could, therefore, mediate binding to ECM components. The crystal structure of TN is similar to that of mannose binding protein, a member of the collectin group of the C-type lectin superfamily, except that the presence of an extra disulfide bridge in TN may restrict the flexibility of the C-type lectin-like domain (Nielsen et al. 1997). The CTLD of TN undergoes structural rearrangements upon calcium binding and in its calcium-free form binds to proteins with kringle domains, such as plasminogen and to isolated kringle-4 (Clemmensen et al. 1986), apparently to its lysine-binding site. Recombinant tetranectin- as well as natural tetranectin from human plasma was shown to be a homo-trimer in solution as are other lectins in collectin family of C-type lectins.

The gene for human TN is about 12 kbp and contains two intervening sequences. The gene encodes a protein of 202 amino acid residues, with a signal peptide of 21 amino acid residues, followed by the TN sequence of 181 amino acid residues ($M_r = 20,169$ Da) (Wewer and Albrechtsen 1992). The 3′ noncoding region contained a single polyadenylation signal and a 26-residue poly A tail. This protein is produced locally by cells of the stromal compartment of tumors and is deposited into the extracellular matrix. Since TN binds to plasminogen, it could function as an anchor and/or reservoir for plasminogen and similar substances that regulate tumor invasion and metastasis as well as tumor angiogenesis (Wewer and Albrechtsen 1992).

By screening a human chromosome 3 somatic cell hybrid mapping panel, Durkin et al. (1997) localized the human TN gene to 3p22–p21.3, which is distinct from the loci of two human connective tissue disorders that map to the short arm of chromosome 3, MFS2 and LRS. The human TN gene consists of three exons (Durkin et al. 1997). Each separate exon encodes specific functional domains. The short exon 1 encodes the heparin binding site of TN. The second exon encodes an α helix protein, which governs the trimerisation of TN monomers by assembling them into a triple helical coiled structural element. The final exon encodes a C-type lectin-like domain, homologous to the carbohydrate recognition domains of calcium dependent animal lectins (Nielsen et al. 2000). Biochemical evidence shows that an N-terminal domain encoded within exons 1 and 2 of the tetranectin gene is necessary and sufficient to govern subunit trimerization (Holtet et al. 1997). The human, mouse, and chicken TN cDNA sequences and gene structures are very similar (Berglund and Petersen 1992; Ibaraki et al. 1995; Sørensen et al. 1995, 1997; Wewer and Albrechtsen 1992; Xu et al. 2001). Neame et al. (1992) isolated a protein from reef shark (Carcharhinus springeri) cartilage that bears a striking resemblance to TN monomer originally described by Clemmensen et al. (1986). The amino acid sequence had 166 amino acids and a calculated molecular weight of 18,430. The shark protein was 45% identical to human TN, indicating that it was in the family of mammalian C-type lectins and an analog of human TN (Neame et al. 1992).

Murine Tetranectin: The mouse TN 992-bp cDNA with an open reading frame of 606 bp is identical in length to the human TN cDNA encoding a protein of 202 aa including a signal peptide of 21 aa. An overall identity of 79–87% in amino acids between human and murine TN revealed a high evolutionary conservation of the protein. The highest expression of mouse TN was found in lung and skeletal muscle. The sequence analysis revealed a difference in both sequence and size of the non-coding regions between mouse and human cDNAs. The cloned murine TN was mapped to region F1-F3 on mouse chromosome 9. Mapping of the transcription start point (tsp) suggests that murine TN has more than one of these. In addition, no consensus TATA-box was found upteam for the putative tsp(s) in the 5′-flanking region of the gene, indicating that the murine TN promoter belongs to the TATA-less class of genes (Ibaraki et al. 1995; Sørensen et al. 1997).

40.2.2 Cell and Tissue Distribution

Tetranectin shows a wide tissue distribution. Predominantly, it was found in secretory cells of endocrine tissue like pituitary, thyroid, parathyroid glands, and the liver, pancreas, and adrenal medulla. Endocrine cells with a known protein or glycoprotein hormonal production such as chromophils (pituitary), follicular and parafollicular cells (thyroid), chief cells (parathyroid), hepatocytes (liver), islet cells (pancreas) and ganglion cells of the adrenal medulla displayed a convincing, positive staining reaction for TN, which varied from cell to cell within the different tissues. The liver showed a distinct and universal reaction within almost all hepatocytes, thus raising suspicion of producing the bulk of TN to the blood. Suspicion indicates that this protein might have a dual function, serving both as a regulator in the secretion of certain hormones and as a participant in the regulation of the limited proteolysis, which is considered important for the activation of prohormones (Christensen et al. 1987; Jensen et al. 1987). TN was found in endothelial and epithelial tissues, particularly in cells with a high turn-over or storage function such as gastric parietal and zymogenic cells, absorptive surface epithelium of the small intestine, ducts of exocrine glands and pseudostratified respiratory epithelium. Also mesenchymal cells produced a TN positive staining reaction, which was most conspicuous in mast cells, but also present in some lymphocytes, plasma cells, macrophages, granulocytes, striated and smooth muscle cells and fibroblasts (Christensen et al. 1989). Hermann et al. (2005) described the in vivo and in vitro localisation of TN in human and murine islet cells. The amount and localization of TN is influenced by different culture conditions. Tetranectin may play an important role in the survival of islets in the liver after islet transplantation.

A distinct accumulation of TN was observed in the surrounding extracellular matrix of various carcinomas where it colocalized with plasminogen (De Vries et al. 1996). Plasma levels of TN are approximately 100 nM in healthly adults (Jensen and Clemmensen 1988). Tetranectin is also found in a mobilizable set of granules in neutrophils (Borregaard et al. 1990), in monocytes (Nielsen et al. 1993), fibroblasts (Clemmensen et al. 1991), platelets (Christensen 1992), and in various tissue locations like cartilage and the extracellular matrix of developing or regenerating muscle (Wewer et al. 1994; Clemmensen et al. 1991; Wewer et al. 1998). Furthermore, TN is a crucial component of the extracellular matrix during muscle cell development and regeneration. During muscle cell development and regeneration, TN expression marks active myogenesis in vivo and in vitro (Wewer et al. 1998a). The decrease in concentration of TN in serum after myocardial infarction (Kamper et al. 1998a) suggests consumption of circulating soluble TN during myocardial repair. A role in skeletal formation during development was evidenced since targeted deletion of the protein results in spinal deformity (Iba et al. 2001a). Iba et al. (2009) also supported that TN might play a role in the wound healing process.

Crab-eating macaque serum showed the strongest reaction for TN, followed by horse and cat. Serum from cow, goat, pig, mouse and chicken reacted weakly, while dog, trout, and the amphibian and the reptile species did not react (Thougaard et al. 2001). TN has been demonstrated in normal brain and CSF. In amniotic fluid a significant, positive correlation exists between TN concentration and gestational age. Fetal TN has been correlated to fetal maturation (Høgdall et al. 1998). Tetranectin is enriched in the cartilage of the shark. During osteogenesis, in the newborn mouse, strong TN immunoreactivity was found in the newly formed woven bone around the cartilage anlage in the future bone marrow and along the periosteum forming the cortex. TN may play an important direct and/or indirect role during osteogenesis and my have a potential role in mineralization in vivo and in vitro (Wewer et al. 1994).

40.2.3 Interactions of Tetranectin

Interactions with Complex Sulfated Polysaccharides
Tetranectin interacts with complex sulfated polysaccharides (Clemmensen 1989; Lorentsen et al. 2000), in a lysine sensitive way, specifically with apolipoprotein A (Kluft et al. 1989a), and in a calcium dependent fashion with fibrin (Kluft et al. 1989b). The distinct binding properties of TN to plasminogen and its expression in normal tissue development point to an important physiological function of TN in tissue formation and repair. Fourth kringle domain of plasminogen can stimulate plasminogen activation in vitro. In the search for new ligands for the plasminogen kringle-4 binding-protein TN, it has been found that TN specifically bound to the plasminogen-like hepatocyte growth factor and tissue-type plasminogen activator. The dissociation constants of these complexes were found to be within the same order of magnitude as the one for the plasminogen-tetranectin complex. TN did not interact with the kindred proteins: macrophage-stimulating protein, urokinase-type plasminogen activator and prothrombin (Westergaard et al. 2003).

Kluft et al. (1989a) demonstrated reversible binding of apolipoprotein-A to kringle-4-binding with an apparent K_D of 0.013 µM/l, where as Lys- and Glu-plasminogen showed an apparent K_D of 0.5 µM/l. It suggested that plasminogen and lipoprotein-A show functional analogy in their binding to TN, but TN primarily targets at lipoprotein-A (Kluft et al. 1989a). TN binds to fibrin and the amount of TN bound to fibrin varies with TN in plasma. Normally, it constitutes a constant percentage of 13–17% of plasma TN. The TN, which is reduced in serum after blood clotting can, upon coagulation, be released from platelets and become partially bound to fibrin (Kluft et al. 1989b). TN may interact with angiostatin, which is formed in cancer tissues by proteolytic degradation of plasminogen. The predominant form of angiostatin produced in cancer tissues is AST^{K1-4} (O'Reilly et al. 1994; Gately et al. 1997; Richardson et al. 2002), which inhibits cancer progression and metastasis by inhibiting cancer-related angiogenesis. Mogues et al. (2004) demonstrated that TN binds to the kringle 1–4 form of angiostatin (AST^{K1-4}) and reduces its ability to bind to ECM of endothelial cells or to inhibit endothelial cell growth.

Plasminogen and Heparin-Binding Sites in Tetranectin: TN, related to corresponding regions of the mannose-binding proteins, is known specifically to bind the plasminogen kringle-4 protein domain, an interaction sensitive to lysine. Surface plasmon resonance and isothermal calorimetry binding analyses using single-residue and deletion mutant TN derivatives showed that the kringle-4 binding site resides in the CRD and includes residues of the putative carbohydrate binding site. Furthermore, the interaction is sensitive to calcium in addition to lysine (Graversen et al. 1998). The heparin-binding site in TN resides not in the CRD but within the N-terminal region, comprising the 16 amino acid residues encoded by exon 1. In particular, the lysine residues in the decapeptide segment KPKKIVNAKK (TN residues 6–15) at the N terminus are shown to be of primary importance in heparin binding (Lorentsen et al. 2000).

Tetranectin-Binding Site on Plasminogen Kringle-4: It was reported that all amino acid residues of plasminogen kringle-4 were involved in binding to TN. Notably, one amino acid residue of plasminogen kringle-4, Arg^{32}, not involved in binding trans-aminomethyl-cyclohexanoic acid (t-AMCHA), is critical for binding TN. Moreover Asp^{57} and As^{55} of plasminogen kringle-4, which both were found to interact with low molecular weight ligand with an almost identical geometry in the crystal of the complex, are not of equal functional importance in t-AMCHA binding. Mutating Asp^{57} to an Asn totally eliminates binding, whereas the Asp^{55} to Asn, like the Arg^{71} to Gln mutation, was found only to decrease the affinity (Graversen et al. 2000a).

A site, involving Lys^{148}, Glu^{150}, and Asp^{165} in TN is known to mediate calcium-sensitive binding to plasminogen kringle-4. Substitution of Thr^{149} in TN with a tyrosine residue considerably increases the affinity for plasminogen kringle-4, and, in addition, confers affinity for plasminogen kringle 2. This new interaction is stronger than the binding of wild-type TN to plasminogen kringle-4. The insight into molecular determinants is important for binding selectivity and affinity of C-type lectin kringle interactions (Graversen et al. 2000b).

40.2.4 Crystal Structure of Tetranectin

Tetranectin is a homotrimeric protein (TN3) which contains a C-type lectin-like domain. The TN3 domain can bind calcium. In absence of calcium, the TN3 domain binds a number of kringle-type protein ligands. The structure of calcium free-form of TN3 (apoTN3) has been studied by NMR. Compared to the structure of the calcium-bound form of TN3 (holoTN3), the core region of secondary structural elements is conserved, while large displacements occur in the loops involved in calcium or K4 binding. A conserved proline, present in *cis* conformation in holoTN3, is in apoTN3 predominantly in the *trans* conformation. Backbone dynamics indicated that, in apoTN3 especially, two of the three calcium-binding loops and two of the three K4-binding residues exhibit increased flexibility, whereas no such flexibility was observed in holoTN3. In apoTN3, the residues critical for K4 binding, are spanned over a large conformational space. Study indicated that the K4-ligand-binding site in apoTN3 is not preformed (Nielbo et al. 2004).

The recombinant human TN and the CRD of TN3 have been crystallized. TN3 crystallizes in the tetragonal space group $P4_22_12$ with cell dimensions a = b = 64.0, c = 75.7 Å and with one molecule per asymmetric unit. The crystals of TN are rhombohedral, space group R3, with the hexagonal axes a = b = 89.1, c = 75.8 Å, and diffract to at least 2.5 Å. The asymmetric unit contains one monomer of TN. The rhombohedral space group indicates that trimers of TN are formed in accordance with the observation of trimerization in solution (Kastrup et al. 1997). The two C-terminal domains, TN23 (residues 17–181) of human recombinant TN have been crystallized in two different space groups. Crystals belonging to the monoclinic space group C2 showed unit-cell parameters: a = 160.4, b = 44.7, c = 107.5 Å, β = 127.6°. TN23 crystallizes in a rhombohedral space group, with unit-cell parameters a = b = c = 107.4 Å, α = β = γ = 78.3°. A full data set to 4.5 Å, collected from the monoclinic crystals, shows that TN23 crystallizes as a trimer, with one trimer in the asymmetric unit (Nielsen et al. 2000).

Fig. 40.2 (**a** and **b**): The overall structure of the TN trimer viewed (**a**) along and (**b**) perpendicular to the threefold axis (Reprinted by permission from Macmillan Publishers Ltd: FEBS Letters, Nielsen et al. © 1997)

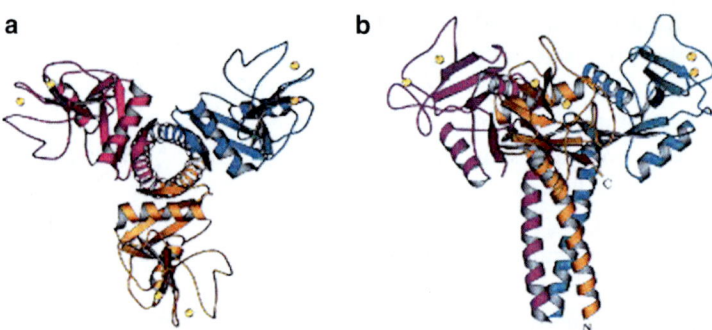

Human tetranectin is a homotrimer forming a triple α-helical coiled coil. Each monomer consists of a CRD connected to a long α-helix. TN has been classified in a distinct group of the C-type lectin superfamily but has structural similarity to the proteins in the group of collectins. TN has three intramolecular disulfide bridges. Two of these are conserved in the C-type lectin superfamily, whereas the third is present only in long-form CRDs. In TN, the third disulfide bridge tethers the CRD to the long helix in the coiled-coil. The residues 26–52 of TN form a long α-helix (E2), which is connected to the CRD. The CRD of TN has characteristic topology similar to other proteins of the C-type lectin superfamily. However, structural differences occur at insertions and deletions in TN as compared to human lithostathine, human MBP, and human E-selectin (Nielsen et al. 1997). Two calcium ions have been located at same positions as calcium-binding sites 1 and 2 in structures of MBPs. In TN, the ligands are Asp^{116}, Glu^{120}, Gly^{147}, Glu^{150}, and Asn^{151} at calcium site 1 and Gln^{143}, Asp^{145}, Glu^{150}, and Asp^{165} at calcium site 2. TN, as well as lithostathine, contains a long-form CRD with three disulphide bridges in contrast to MBPs and E-selectin, which have short-form CRDs with two disulfide bridges. In TN, the disulfide bridge Cys^{50}-Cys^{60} tethers the E2 helix to a short L-strand of the CRD, whereas this disulfide bridge is part of a loop in lithostathine.

The long E2 helix provides trimerization of TN by the formation of a triple α-helical coiled coil (Fig. 40.2a, b). Only few contacts were observed among the CRDs of the trimer, in agreement with the observation that the isolated CRD of TN is a monomer in solution. The trimeric structure of TN (TN3) resembles those of rat MBP and human MBP, in which the neck region forms a long α-helix. Each monomer is built of two distinct regions, one region consisting of six β-strands and two α-helices, and the other region is composed of four loops harboring two calcium ions. The calcium ion at site 1 forms an eightfold coordinated complex and has Asp^{116}, Glu^{120}, Gly^{147}, Glu^{150}, Asn^{151}, and one water molecule as ligands. The calcium ion at site 2, which is believed to be involved in recognition and binding of oligosaccharides, is sevenfold coordinated with ligands Gln^{143}, Asp^{145}, Glu^{150}, Asp^{165}, and two water molecules. One sulfate ion was located at the surface of TN, forming contacts to Glu^{120}, Lys^{148}, Asn^{106} of a symmetry-related molecule, and to an ethanol molecule (Kastrup et al. 1998). The orientations of the CRDs are determined by the coiled-coil structure, by intramolecular interactions between the helix and its CRD, and also by intermolecular interactions between each of the helices of the coiled-coil and a neighboring CRD. The trimerization of TN as well as the fixation of the CRDs relative to the helices in the coiled-coil indicates a demand for high specificity in the recognition and binding of ligands (Nielsen et al. 1997).

40.2.5 Functions of Tetranectin

Tissue Remodeling

The function(s) of TN in postnatal life have not been elucidated although there is evidence for roles in tissue remodeling, coagulation, and cancer. Tetranectin may affect coagulation and angiogenesis through interaction with fibrin and angiostatin (a plasminogen fragment), respectively. During development, TN stimulates muscle cell differentiation and fibre formation, and promotes bone mineralization. Tetranectin-deficient mice develop a spinal deformity. TN was originally isolated as a plasminogen-binding protein that can enhance plasminogen activation in presence of tissue plasminogen activator (Clemmensen et al. 1986). Tetranectin binds to plasminogen through a calcium-sensitive interaction of its C-terminal domain with kringle-4 domain of plasminogen (Graversen et al. 1998; Graversen et al. 2000b). Because of the distinct binding properties of TN and its dynamic expression in development and disease, one may expect that TN plays a role in tissue remodeling. The precise function of TN in these processes, however, is not known, and no human genetic disorders have yet been associated with mutations in the single-copy TN gene (Durkin et al. 1997).

Osteoblastic Activity

TN is also expressed during bone development, and transfection studies suggest that TN can induce osteogenesis (Wewer et al. 1994). Studies indicated that the expression of TN in osteoblastic cells is regulated by Dex and TGF-β 1 and that the TN expression is tightly linked to the process of mineralization. Expression of TN is completely inhibited by treatment with retinoic acid, irrespective of the stage of cell differentiation. Retinoic acid downregulates the TN expression in human osteoblastic cells independent of the stage of cell differentiation, and is correlated with inhibition of mineralization (Iba et al. 2001a). Targeted loss of TN expression interferes with proper postnatal development of the vertebral bodies and results in a mild spine deformity. The TN-deficient mice exhibit kyphosis, a type of spinal deformity characterized by an increased curvature of the thoracic spine. The spines of these mice revealed an apparently asymmetric development of the growth plate and of the intervertebral disks of the vertebrae. TN-null mice had a normal peak bone mass density and were not more susceptible to ovariectomy-induced osteoporosis than were their littermates (Iba et al. 2001b).

40.2.6 Association of TN with Diseases

The concentration of TN in human serum decreases in pathological conditions such as cancer (Høgdall 1998), myocardial infarction (Kamper et al. 1998a), and rheumatoid arthritis (RA) (Jensen and Clemmensen 1988; Christensen 1992; Kamper et al. 1998b). TN decreases with the increase of RA activity from group to group. Study points to the usefulness of TN assessment as a specific fibrinolytic marker in the evaluation of disease activity in patients with RA. Reports suggest the implication of TN in the impaired regulation of fibrinolysis associated with the inflammatory process (Kamper et al. 1998b). Acute coronary syndrome (ACS) is triggered by the occlusion of a coronary artery usually due to the thrombosis caused by an atherosclerotic plaque (Bugge et al. 1995). Some of the proteins identified have been associated with ACS whereas others (such as ç-1-B-glycoprotein, Hakata antigen, Tetranectin, Tropomyosin-4) constitute novel proteins that are altered in this pathology (Darde et al. 2010). Altered abundance of TN among few proteins suggests that it may be potential biomarker for alcohol abuse (Lai et al. 2009).

In Prognosis of Cancer
Tetranectin is significantly reduced in patients with various malignancies and may be related to the pathogenesis of cancer spread and metastasis (Jensen and Clemmensen 1988). TN has been detected in tumors of the breast, colon, stomach, and ovary, whereas no TN immunoreactivity was seen in the corresponding normal epithelium of the breast, colon, or ovary (Christensen and Clemmensen 1991; Arvanitis et al. 2002; Obrist et al. 2004). Notably, TN was colocalised together with plasminogen at the invasion front of cutaneous melanomas (De Vries et al. 1996). In case of ovarian cancer, decreased plasma levels of TN were a stronger predictor of adverse prognosis than cancer stage (Høgdall 1998). Furthermore, TN is present in the stroma of various cancers (e.g., breast, ovary, colon), whereas it is not present in normal tissue from which the cancers arose (Wewer and Albrechtsen 1992; Christensen and Clemmensen 1991; Tuxen et al. 1995). Positive staining for TN in cancer stroma has also been strongly correlated with cancer progression (Høgdall 1998). While several investigators have reported that plasma and serum TN concentrations in patients with malignant tumors correlate with disease stage and survival (Blaakaer et al. 1995; Høgdall 1998; Jensen and Clemmensen 1988; Høgdall 1998; Obrist et al. 2004), others found that TN expression in tumor tissue correlated with tumor histological grading (Høgdall 1998). A significantly shortened survival was found for patients with low serum TN values compared to patients with serum TN levels above one of the cutoff levels. Thus, serum TN determination may be valuable in the selection of patients with relapse of ovarian cancer for new treatment strategies in future studies (Deng et al. 2000; Høgdall et al. 2002b). While TN was a strong prognostic variable in patients with advanced ovarian cancer, CA125 was a strong prognostic factor in patients with a localised ovarian cancer and of no prognostic value in patients with advanced cancer (Høgdall et al. 2002a). Serum TN seems to be useful prognostic factor in metastatic breast cancer (Høgdall 1998; Begum et al. 2009; 2010). A slight, but significant reduction in serum TN was found in pelvic inflammatory disease patients. The finding is important in the assessment of TN used as a potential screening marker for ovarian cancer, or as a diagnostic tool for pelvic tumors (Høgdall 1998). Low serum levels of TN are associated with high serum levels of CASA or YKL-40 with increased risk of second-line chemoresistance in patients with ovarian cancer (Gronlund et al. 2006).

In patients with multiple myeloma, either untreated or previously treated, serum levels were found to be significantly reduced (Nielsen et al. 1990). The TN, in association with the circulating intercellular adhesive molecule-1 and IL-10, may be involved in the metastatic cascade of B-chronic lymphocytic leukemia (B-CLL). The findings and good performance characteristics of TN and cICAM-1 in B-CLL suggest the potential usefulness of these adhesive molecules as prognostic markers in B-CLL (Kamper et al. 1999). Colonic neoplasia is associated with a decrease in the tissue distribution of TN, without an obvious change in the tissue level, and a low plasma TN level (Verspaget et al. 1994). Høgdall et al. (2002a) confirmed that TN is a strong prognostic factor in patients with colorectal cancer and may be valuable as a prognostic variable in future studies evaluating new treatment strategies for colorectal cancer. TN is expressed in a subgroup of bladder cancer patients with a higher risk of recurrence who may take benefit from a closer follow-up (Brunner et al. 2007).

Neurological Diseases
TN may play a role in neurological diseases and may serve as a diagnostic aid in multiple sclerosis (MS). TN appears in CSF of MS patients (Stoevring et al. 2005; 2006; Hammack et al. 2004). Wang et al. (2010a) suggest that CSF-TN and serum-TN are potential biomarkers in epilepsy and drug-refractory epilepsy and would be useful for diagnosis. The application of biomarkers may potentially improve the efficiency of the diagnosis for Parkinson's disease (PD), since no reliable biomarker has been identified to date. Preliminary results of Wang et al. (2010b) suggest that TN and apoA-I may serve as potential biomarkers for PD, though further validation is needed.

40.2.7 Cartilage-Derived C-Type Lectin (CLECSF1)

Cartilage-derived C-type lectin is expressed in cartilage, where it may have a role in organizing the extracellular matrix. Cartilage is a tissue that is primarily extracellular matrix, the bulk of which consists of proteoglycan aggregates constrained within a collagen framework. Candidate components that organize the extracellular assembly of the matrix consist of collagens, proteoglycans and multimeric glycoproteins. Neame et al. (1999) described the human gene structure of a potential organizing factor, a cartilage-derived member of the C-type lectin superfamily (CLECSF1; C-type lectin superfamily) related to tetranectin. The CLECSF1 is restricted to cartilage and the gene is located on chromosome 16q23. The 10.9 kb of sequence upstream of the first exon was characterized. Similarly to human TN, there are three exons. The residues that are conserved between CLECSF1 and TN suggest that the cartilage-derived protein forms a trimeric structure similar to that of TN, with three N-terminal α-helical domains aggregating through hydrophobic faces. The globular, C-terminal domain that has been shown to bind carbohydrate in some members of the family and plasminogen in TN, is likely to have a similar overall structure to that of TN.

40.2.8 Stem Cell Growth Factor (SCGF): Tetranectin Homologous Protein

The SCGF is expressed in skeletal and connective tissues, where it stimulates proliferation and differentiation of haematopoietic precursor cells. The cDNA encoding stem cell growth factor (SCGF; 245 aa), a human growth factor for primitive hematopoietic progenitor cells, has been reported by Hiraoka et al. (1997). This protein consists of 323, 328 and 328 aa in the human, murine and rat forms, the latter two of which share 85.1% and 83.3% aa identity, and 90.4% and 90.4% aa similarity to the human protein, respectively. Because the newly identified human clone encodes the protein longer by 78 aa than that identified earlier, the longer clone was termed as hSCGF-α and the shorter one as hSCGF-β. SCGF is a new member of the C-type lectin superfamily, and shows the greatest homology to TN among the members of the family (27.2–33.7% aa identity and 46.0–40.6% aa similarity). SCGF transcripts are detected in spleen, thymus, appendix, bone marrow and fetal liver. The SCGF gene is located on chromosome 19 at position q13.3 for human form, and on chromosome 7 at position B3-B5 for murine form, which are close to flk-2/flt3 ligand and interleukin-11 genes of both human and murine species (Mio et al. 1998).

40.3 Attractin Group of CTLDs (Group XI)

The autosomal recessive gene of attractin has received great attention within last few decades. The attractin group of CTLD-containing type 1 transmembrane proteins has two widely-expressed members in both human and mouse: attractin (ATRN) and attractin-like protein (ATRNL1).

40.3.1 Secreted and Membrane Attractins

Attractin, initially identified as a soluble human plasma protein with dipeptidyl peptidase IV (DPPIV) activity that is expressed and released by activated T lymphocytes, has also been identified as the product of the murine mahogany gene with connections to control of pigmentation and energy metabolism. The mahogany gene product, however, is a transmembrane protein, raising the possibility of a human membrane attractin in addition to the secreted form. The genomic structure of human attractin reveals that soluble attractin arises from transcription of 25 sequential exons on human chromosome 20p13, where the 3′ terminal exon contains sequence from a long interspersed nuclear element-1 (LINE-1) retrotransposon element that includes a stop codon and a polyadenylation signal. The mRNA isoform for membrane attractin splices over the LINE-1 exon and includes five exons encoding transmembrane and cytoplasmic domains with organization and coding potential almost identical to that of the mouse gene. The relative abundance of soluble and transmembrane isoforms is differentially regulated in lymphoid tissues. Since activation of peripheral blood leukocytes with phytohemagglutinin induces strong expression of cell surface attractin followed by release of soluble attractin, results suggest that a genomic event unique to mammals, LINE-1 insertion, has provided an evolutionary mechanism for regulating cell interactions during an inflammatory reaction (Tang and Duke-Cohan 2000).

40.3.2 Attractin has Dipeptidyl Peptidase IV Activity?

Dipeptidyl peptidase IV (DPPIV) is a serine type protease with an important modulatory activity on a number of chemokines, neuropeptides and peptide hormones. It is also known as CD26 or adenosine deaminase (ADA; EC 3.5.4.4) binding protein. DPPIV has been demonstrated on the plasma membranes of T cells and activated natural killer or B cells as well as on a number of endothelial and differentiated epithelial cells. A soluble form of CD26/DPPIV has been described in serum. Attractin, the human orthologue of mouse mahogany protein, was postulated to be responsible for the majority of DPPIV-like activity in serum. 95% of the serum dipeptidyl peptidase activity is associated with a protein with ADA-binding properties. The natural DPPIV serum enzyme was confirmed as CD26 (Durinx et al. 2000). However, the attractin acts as a receptor or adhesion protein rather than a protease (Friedrich et al. 2007). Human peripheral blood monocytes express a DPPIV-like enzyme activity, which could be inhibited completely by the synthetic DPPIV inhibitor. Drugs directed to DPPIV-like enzyme activity can affect monocyte function via attractin inhibition (Wrenger et al. 2006). The expression of DPPV and attractin in circulating blood monocytes of human subjects are influenced by metabolic abnormalities with obesity being an important factor (Laudes et al. 2010). In the CNS, attractin has been detected in neurons and in glioma cell lines at different degree of transformation. In human U373 and U87 glioma cells, membrane-bound attractin displays hydrolytic activity between 5% and 25% of total cellular DPPIV-like enzyme activity, respectively. Attractin presence in glioma, but not in normal glial cells, together with its differential enzymatic activity, suggests its role in growth properties of tumors of glial cell origin (MalÃk et al. 2001).

Attractin is a normal human serum glycoprotein of 175 kDa that was found to be rapidly expressed on activated T cells and released extracellularly after 48–72 h. There are two mRNA species with hematopoietic tissue-specific expression that code for a 134-kDa protein with a putative serine protease catalytic serine. Both have large extracellular regions consisting of a membrane-distal epidermal growth factor (EGF) domain followed by a CUB domain; two EGF domains; five Kelch domains; a cysteine-rich/plexin domain; a CTLD; a second cysteine-rich/plexin domain; and two laminin EGF domains. Except for the latter two domains, the overall structure shares high homology with the *Caenorhabditis elegans* F33C8.1 protein, suggesting that attractin has evolved new domains and functions in parallel with the development of cell-mediated immunity (Duke-Cohan et al. 1998, 2000). Attractin is found in both membrane-bound and secreted forms as a result of alternative splicing.

40.3.3 Attractin-Like Protein

The gene dosage effect of melanocortin-4 receptor (MC4R) on obesity suggests that regulation of MC4-R expression is critically important to the central control of energy homoeostasis. In order to identify putative MC4-R regulatory proteins, Haqq et al. (2003) identified a positive clone that shares 63% amino acid identity with the C-terminal part of the mouse attractin gene product, a single-transmembrane-domain protein characterized as being required for agouti signaling through MC4R. A direct interaction between this (ATRNL1) and the C-terminus of the mouse MC4-R was confirmed and the regions in this interaction involved residues 303–313 in MC4-R and residues 1280–1317 in ATRNL1. ATRNL1 is highly expressed in brain, but also in heart, lung, kidney and liver. Furthermore, co-localization analyses in mice showed co-expression of ATRNL1 in cells expressing MC4-R in a number of regions known to be important in the regulation of energy homoeostasis by melanocortins, such as the paraventricular nucleus of hypothalamus and the dorsal motor nucleus of the vagus (Haqq et al. 2003).

Attractin and Attractin-like 1 are highly similar type I transmembrane proteins. *Atrn* null mutant mice have a pleiotropic phenotype including dark fur, juvenile-onset spongiform neurodegeneration, hypomyelination, tremor, and reduced body weight and adiposity, implicating ATRN in numerous biological processes. Bioinformatic analysis indicated that Atrn and Atrnl1 arose from a common ancestral gene early in vertebrate evolution. Characterization of Atrnl1 loss- and gain function mutant mice suggested that *Atrnl1* mutant mice were grossly normal with no alterations of pigmentation, CNS pathology or body weight. *Atrn* null mutant mice carrying a β-actin promoter-driven *Atrnl1* transgene had normal, agouti-banded hairs and significantly delayed onset of spongiform neurodegeneration, indicating that overexpression of ATRNL1 compensates for loss of ATRN. Thus, two genes are redundant from the perspective of gain-of-function but not loss-of-function mutations (Walker et al. 2007). The cytoplasmic region of ATRNL1 interacts with the cytoplasmic tail of the MC4R, a G-protein-coupled receptor in the brain which regulates appetite and metabolism, but the function of ATRNL1 is not understood. The CTLDs in attractin and ATRNL1 contain some of the residues which in other CTLDs are involved in binding mannose-type sugars. It is not known if carbohydrate recognition by the CTLDs is involved in the functions of these proteins.

A heterozygous deletion of the *Attractin-like* (*ATRNL1*) gene in a patient presented with a distinctive phenotype comprising dysmorphic facial appearance, ventricular septal defect, toe syndactyly, radioulnar synostosis, postnatal growth retardation, cognitive impairment with autistic features, and ataxia. The phenotype of mice with homozygous *Atrn* mutations overlaps considerably with the features observed in this patient. It was postulated that patient's phenotype is caused by the deletion of ATRNL1 (Stark et al. 2010).

40.3.4 Functions of Attractin

Multiple Functions: Attractin is known to play multiple roles in regulating physiological processes that are involved in monocyte-T cell interaction, agouti-related hair pigmentation, and control of energy homeostasis. Attractin affects the balance between agonist and antagonist at receptors on melanocytes, modifies behaviour and basal metabolic rate, intervenes in the development of CNS and its functions. Attractin modulates an interaction between activated T cells and macrophages. The loss of murine membrane attractin due to mahogany mutation results with severe repercussions upon skin pigmentation and control of energy metabolism. In each of these latter instances, there is a strong likelihood that attractin is moderating the interaction of cytokines with their respective receptors. Attractin is performing a similar function in the immune system through capture and proteolytic modification of the N-terminals of several cytokines and chemokines. This regulatory activity allows cells to interact and form immunoregulatory clusters and subsequently aids in downregulating chemokine/cytokine activity once a response has been initiated. These two properties are likely to be affected by the balance of membrane-expressed to soluble attractin (Duke-Cohan et al. 2000). The attractin protein is involved in the suppression of diet-induced obesity. Its expression in brain has a significant relationship with obesity and indicates the therapeutic potential of attractin in the treatment of obesity. Furthermore, the murine attractin locus is located in a region harboring several QTL for body weight and fatness.

Membrane-Bound Attractin Is a Co-Receptor for Agouti and Regulates Pigmentation in Skin: Melanocortin-1 receptor (MC1R) and its ligands, α-melanocyte stimulating hormone (αMSH) and agouti signaling protein (ASIP), regulate switching between eumelanin and pheomelanin synthesis in melanocytes. ASIP-MC1R signaling includes a cAMP-independent pathway through attractin and mahogunin, while the known cAMP-dependent component requires neither attractin nor mahogunin. Attractin may be a component of a pathway for regulated protein turnover that also involves mahogunin, a widely-expressed E3 ubiquitin ligase found at particularly high levels in brain. Membrane-bound attractin is a co-receptor for Agouti, the paracrine signaling molecule in the skin which regulates hair pigmentation by antagonizing the MC1R (Hida et al. 2009; Jackson 1999). Genetic, biochemical and pharmacological studies in humans and rodents have established that signaling through the G-protein-coupled melanocortin-4 receptor (MC4R) by pro-opiomelano- cortin (POMC)-derived ligands plays a critical role in the central suppression of appetite. As a consequence, malfunction of this signaling system leads to the development of obesity. It has been shown that melanocortin signaling can be modulated by attractin, apparently acting as a co-receptor for the inhibitory ligand agouti. Haqq et al. (2003) demonstrated that the cytosolic tail of an attractin-like protein binds directly and specifically to the C-terminal region of MC4R, raising the possibility that proteins of the attractin family influence melanocortin receptor function through multiple mechanisms (Carlson and Moore 1999; Yeo and Siddle 2003).

Agouti is expressed ubiquitously, such as lethal yellow, have pleiotropic effects that include a yellow coat, obesity, increased linear growth, and immune defects. The *mahogany* mutation suppresses the effects of lethal yellow on pigmentation and body weight. Results of genetic studies place mahogany downstream of transcription of *Agouti* but upstream of melanocortin receptors. Positional cloning of *mahogany* gene *Mgca* predicted a protein of 1,428 aa with a single-transmembrane-domain that is expressed in many tissues, including pigment cells and the hypothalamus. The extracellular domain of the Mgca protein is the orthologue of human attractin, a molecule produced by activated T cells that has been implicated in immune-cell interactions (Gunn et al. 1999).

Mahogunin Ring Finger-1 and Attractin Act Through Common Pathway: Oxidative stress, ubiquitination defects and mitochondrial dysfunction are common factors associated with neurodegeneration. Mice lacking mahogunin ring finger-1 (MGRN1) or attractin (ATRN) develop age-dependent spongiform neurodegeneration through an unknown mechanism. It has been suggested that they act through a common pathway. As MGRN1 is an E3 ubiquitin ligase, it was suggested that many mitochondrial proteins were reduced in *Mgrn1* mutants. Mitochondrial dysfunction was obvious many months before onset of vacuolation, implicating this as a causative factor. Compatible with the hypothesis that ATRN and MGRN1 act in the same pathway, mitochondrial dysfunction and increased oxidative stress were also observed in the brains of Atrn mutants. Study of *Mgrn1* and *Atrn* mutant mice can provide insight into molecular mechanism common to many neurodegenerative disorders (Sun et al. 2007; Kadowaki et al. 2007; Walker and Gunn 2010).

40.3 Attractin Group of CTLDs (Group XI)

Cerebral Spongiform Changes: A mutation characterized by mahogany coat color, sprawling gait, tremors, and severe vacuolization of cerebrum, brainstem, granular layer of cerebellum and spinal cord was discovered in a stock of Mus castaneus mice. Histopathological analysis of brains from Mice homozygous for 2 known mahogany attractin (*Atrnmg*) mutants showed that they also have severe spongiform changes. This surprising finding raises questions about the mechanism by which mahogany controls appetite and metabolic rate, as reported (Bronson et al. 2001). A null mutation for *mahoganoid* causes a similar age-dependent neuropathology that includes many features of prion diseases but without accumulation of protease-resistant prion protein. The gene mutated in *mahoganoid* encodes a RING-containing protein with E3 ubiquitin ligase activity in vitro. Similarities in phenotype, expression, and genetic interactions suggest that *mahoganoid* and *Atrn* genes are part of a conserved pathway for regulated protein turnover whose function is essential for neuronal viability (He et al. 2003b).

Secreted Attractin Disrupts Neurite Formation in Differentiating Cortical Neural Cells In Vitro: Mutations at *Atrn* locus affects the secreted form mRNA, which is down-regulated in discrete regions of human brain while membrane attractin mRNA is well represented, resulting in the apparent absence of secreted attractin protein in CSF. In vitro, transcription of secreted form of attractin mRNA is strongly down-regulated upon differentiation of a human cortical neuron-derived cell line (HCN-1A) to a mature neuron phenotype in response to nerve growth factor. Tang and Duke-Cohan (2002) proposed that inappropriate expression of secreted attractin in the CNS blocks membrane attractin function and that its presence, either by leakage from the periphery, aberrant transcription, or release from inflammatory foci may affect neuron extracellular interactions leading to neurodegeneration in the human.

Attractin/Mahogany Protein in Rodent CNS: Both the secreted and membrane-bound forms of ATRN may be involved in the development and maintenance of the CNS. ATRN was intensely expressed in most neurons and dendrites of large neurons, such as cortical pyramidal neurons and cerebellar Purkinje neurons in CNS of rat and mice. Intense ATRN expression was also observed in the neuropil of gray matter in many regions of the CNS and indicates that ATRN is more widely expressed throughout the CNS. Expression of ATRN by various cell types suggests that ATRN serves multiple functions in the CNS (Nakadate et al. 2008). The zitter (zi/zi) rat, a loss-of-function mutant of the glycosylated transmembrane ATRN exhibits widespread age-dependent spongiform degeneration, hypomyelination, and abnormal metabolism of reactive oxygen species (ROS) in the brain. The onset of the impairment of oligodendrocyte differentiation occurs in a non-cell autonomous manner in zi/zi rats (Nakadate et al. 2008; Sakakibara et al. 2008; Ueda et al. 2008).

The Mahogany Is a Receptor Involved in Suppression of Obesity: The pathogenesis of obesity is multifactorial and involves the interaction of genetic and environmental factors. A number of important signaling molecules play important roles in obesity. One family of these molecules is the melanocortin system, which consists of several components: (1) melanocortin peptides; (2) the five seven-transmembrane G-protein coupled melanocortin receptors (MCRs); (3) the endogenous MCR antagonists, agouti and agouti-related protein; (4) the endogenous melanocortin mediators, mahogany, and syndecan. This system plays a key role in the central nervous system control of feeding behaviour and energy expenditure (Carlson and Moore 1999; Yang and Harmon 2003).

Expression of *mahogany* gene is broad; in situ hybridization analysis emphasizes the importance of its expression in the ventromedial hypothalamic nucleus, a region that is intimately involved in the regulation of body weight and feeding. Genetic studies indicate that the *mahogany* locus does not suppress the obese phenotype of the MC4-R null allele or those of the monogenic obese models (Lep^{db}, tub and Cpe^{fat}). However, mahogany can suppress diet-induced obesity, the mechanism of which is likely to have implications for therapeutic intervention in common human obesity. The amino-acid sequence of the mahogany protein suggests that it is a large, single-transmembrane-domain receptor-like molecule, with a short cytoplasmic tail containing a site that is conserved between *Caenorhabditis elegans* and mammals. It was suggested that mahogany can act as a signaling receptor (Nagle et al. 2000).

Attractin in Testis: The ATRN is expressed in Leydig cells, primitive spermatogonia, primary spermato- cytes, spermatids, Sertoli cells and peritubular myoid cells. The attractin protein was mainly located on the cell membrane and cytoplasm, and its mRNA distributed in the nucleus and cytoplasm. The rat testis has the ability of synthesizing the ATRN protein throughout sexual development. The loss of ATRN in mice showed no significantly different pathological changes from the control mice in 3-month-old ATRN (*mg-3J*) mice. But age-related *Atrn* gene progressively lost its function and caused testis vacuolization and impaired sperm function, which may be responsible for the impairment of male reproductive ability (Li et al. 2009). Human sperm-associated antigen 11 (SPAG11) iso-proteins, closely related to β-defensins in structure and function, are predominantly expressed in male reproductive tract, where their best-known roles are in innate host defense and reproduction. Human SPAG11B isoform D (SPAG11B/D) interacts with ATRN. SPAG11B/D and the ATRN interacting proteins are expressed in the proximal epididymis, and function in immunity and fertility control. Radhakrishnan et al. (2009) showed that SPAG11B/D is both a substrate and a potent inhibitor of TPSAB1 activity. Llike SPAG11B/D, ATRN is associated with spermatozoa.

40.3.5 Genetic and Phenotypic Studies of *Mahogany/Attractin* Gene

Pigment-Type Switching: The mouse *mahogany* mutation affects melanocortin signaling pathways that regulate energy homeostasis and hair color. The gene mutated in *mahogany* mice encodes attractin that is broadly expressed and conserved among multicellular animals. The mouse mutations *mahogany (mg)* and *mahoganoid (md)* are negative modifiers of the *Agouti* coat color gene, which encodes a paracrine signaling molecule that induces a switch in melanin synthesis from eumelanin to pheomelanin. Results suggest that *md* and *mg* interfere directly with Agouti signaling, possibly at the level of protein production or receptor regulation (Miller et al. 1997). The mouse *Attractin (Atrn)* (formerly *mahogany*) gene has been proposed as a downstream mediator of Agouti signaling because yellow hair color and obesity in lethal yellow (A^y) mice are suppressed by the *mahogany* ($Atrn^{mg}$) mutation. Atrn mRNA was found widely distributed throughout the CNS, with high levels in regions of the olfactory system, some limbic structures, regions of the brainstem, cerebellum and spinal cord. In the hypothalamus, ATRN mRNA was found in specific nuclei including the suprachiasmatic nucleus, the supraoptic nucleus, the medial preoptic nucleus, the paraventricular hypothalamic nucleus, the ventromedial hypothalamic nucleus, and the arcuate nucleus. Results suggest a broad spectrum of physiological functions for *Atrn* gene product (Lu et al. 1999).

Mutations of the mouse *Attractin* gene were originally recognized because they suppress Agouti pigment type switching. Agouti protein, a paracrine signaling molecule normally limited to skin, is ectopically expressed in lethal yellow (A^y) mice, and causes obesity by mimicking agouti-related protein (Agrp), found primarily in the hypothalamus. Mouse attractin is a widely expressed protein whose loss of function in *mahogany* ($Atrn^{mg-3J}/Atrn^{mg-3J}$) mutant mice blocks the pleiotropic effects of A^y. Results showed that attractin is a low-affinity receptor for agouti protein, but not Agrp, in vitro and in vivo. Histopathologic abnormalities in $Atrn^{mg-3J}/Atrn^{mg-3J}$ mice and cross-species genomic comparisons indicate that Atrn has multiple functions distinct from both a physiologic and an evolutionary perspective (He et al. 2001).

Characterization of two additional *Atrn* alleles, $Atrn^{mg}$ and $Atrn^{mg-L}$ by Gunn et al. (2001), who examined in parallel the phenotypes of homozygous and compound heterozygous animals showed that the three alleles have similar effects on pigmentation and neurodegeneration, with a relative severity of $Atrn^{mg-3J} > Atrn^{mg} > Atrn^{mg-L}$, which also corresponds to the effects of three alleles on levels of normal Atrn mRNA. Animals homozygous for $Atrn^{mg-3J}$ or $Atrn^{mg}$, but not $Atrn^{mg-L}$, showed reduced body weight, reduced adiposity, and increased locomotor activity, all in the presence of normal food intake. These results confirm that the mechanism responsible for the neuropathological alteration is a loss—rather than gain—of function and indicate that abnormal body weight in *Atrn* mutant mice is caused by a central process leading to increased energy expenditure, and that the pigmentation is more sensitive to levels of *Atrn* mRNA than the nonpigmentary phenotypes (Gunn et al. 2001).

Pleiotropic phenotype in *dark-like (dal)* mutant mouse includes dark dorsal hairs and reproductive degeneration. Their pigmentation phenotype is similar to *Atrn* mutants, which also develop vacuoles throughout brain. Testicular degenerated *dal* mutant males showed reduced serum testosterone and developed vacuoles in their testes. Genetic crosses placed *dal* upstream of the *melanocortin 1 receptor (Mc1r)* and downstream of *agouti*, although *dal* suppressed the effect of agouti on pigmentation but not body weight. $Atrn^{mg-3J}$ and dal showed additive effects on pigmentation, testicular vacuolation, and spongiform neurodegeneration, but transgenic over-expression of *Attractin-like-1 (Atrnl1)*, which compensates for loss of ATRN, did not rescue *dal* mutant phenotypes. Results suggest that *dal* and *Atrn* function in the same pathway and that identification of *dal* gene will provide insight into molecular mechanisms of vacuolation in multiple cell types (Cota et al. 2008).

The Mouse Coat Color Mutant Mahoganoid (md): *Mahoganoid* is a mouse coat-color mutation whose pigmentary phenotype and genetic interactions resemble those of *Atrn*. The mouse coat color mutant *mahoganoid (md)* darkens coat color and decreases the obesity of A^y mice that ectopically over-express agouti-signaling protein. The phenotypic effects of *md* are similar to those of coat color mutant mahogany ($Atrn^{mg}$). The mahoganoid encodes a 494-amino acid protein containing a C3HC4 RING (really interesting new gene) domain that may function as an E3 ubiquitin ligase. The mutations in mahoganoid allelic series (md, md^{2J}, md^{5J}) are all due to large retroviral insertions. In md and md^{2J}, the result is minimal expression of the normal size transcripts in all tissues examined. Unlike $Atrn^{mg}/Atrn^{mg}$ animals, there was no evidence of neurological deficit or neuropathology in *md/md* mice. Body weight and body mass index (a surrogate for adiposity) measurements of B6.C3H-*md*-A *md/+* and *md/md* animals on 9% and 45% kcal fat diets indicate that *mahoganoid* does not suppress body weight in B6.C3H animals in a gene dose-dependent fashion. *Mahoganoid* effects on energy homeostasis are, therefore, most evident in the circumstances of epistasis to hypothalamic over-expression of ASP in A^y and possible other obesity-causing mutations (Phan et al. 2002).

Genes Controlling the Synthesis of Eumelanin and Pheomelanin: Mutations that affect the balance between the synthesis of eumelanin and pheomelanin provide a powerful set of tools with which to understand general aspects of cell signaling. It has been demonstrated that pheomelanin synthesis is triggered by the ability of Agouti protein to inhibit signaling through the *Melanocortin 1 receptor (Mc1r)*. The pigmentary effects of Agouti are suppressed by previously existing coat-color mutations *mg*, *md*, and *Umbrous (U)*. Double mutant studies, with animals deficient for *Mc1r* or those which carry A^y, indicated that *mg* and *md* are genetically upstream of the Mc1r, and can suppress the effects of A^y on both pigmentation and body weight. Positional cloning has identified the gene mutated in mahogany as a single transmembrane-spanning protein whose ectodomain is orthologous to human ATRN (Barsh et al. 2000). The coat color mutations *mg* and *md* prevent hair follicle melanocytes from responding to Agouti protein. The gene mutated in *mg*, encodes ATRN that functions as an accessory receptor for Agouti protein. The gene mutated in *mahoganoid*, which is also known as *Mahogunin (Mgrn1)*, encodes an E3 ubiquitin ligase. Like Attractin, Mahogunin is conserved in invertebrate genomes, and its absence causes a pleiotropic phenotype that includes spongiform neurodegeneration (He et al. 2003a).

Control of Metabolic Rate: Mouse attractin is likely to have additional roles outside melanocortin signaling (Gunn and Barsh 2000). The *mahogany (mg) or Attractin (Atrn))* locus was originally identified as a recessive suppressor of *agouti*, a locus encoding a skin peptide that modifies coat color by antagonizing the MSH-R or MC1-R. Certain dominant alleles of *agouti* cause an obesity syndrome when ectopic expression of the peptide aberrantly antagonizes the MC4-R, a related melanocyte-stimulating hormone receptor expressed in hypothalamic circuitry and involved in the regulation of feeding behavior and metabolism. Reports indicate that *mg*, when homozygous, blocks not only the ability of *agouti* to induce a yellow coat color when expressed in the skin of the lethal yellow mouse (A^Y), but also the obesity resulting from ectopic expression of *agouti* in the brain. Detailed analysis of *mg/mg* A^Y/a animals, demonstrates that *mg/mg* blocks the obesity, hyperinsulinemia, and increased linear growth induced by ectopic expression of the agouti peptide. Remarkably, however, *mg/mg* did not reduce hyperphagia in the A^Y/a mouse. Furthermore, *mg/mg* induced hyperphagia and an increase in basal metabolic rate in the C57BL/6J mouse in the absence of A^Y. Consequently, although mahogany is broadly required for agouti peptide action, it also appears to be involved in the control of metabolic rate and feeding behavior independent of its suppression of *agouti* (Dinulescu et al. 1998).

Hypomyelination in Rat Zitter: In absence of attractin there is a decline in plasma membrane glycolipid-enriched rafts. The structural integrity of lipid rafts depends upon cholesterol and sphingomyelin, and can be identified by partitioning of ganglioside GM1. Despite a significant fall in cellular cholesterol with maturity, and a lesser fall in both membrane and total cellular GM1, these parameters lag behind raft loss, and are normal when hypomyelination/neurodegeneration has already begun thus supporting consequence rather than cause. These findings can be recapitulated in Atrn-deficient cell lines propagated in vitro (Azouz et al. 2007) Histopathological analysis of brains from Mice homozygous for 2 known mahogany attractin (*Atrnmg*) mutants showed that they also have severe spongiform changes. This surprising finding raises questions about the mechanism by which mahogany controls appetite and metabolic rate, as recently reported (Bronson et al. 2001).

The rat zitter (zi) mutation induces hypomyelination and vacuolation in CNS, which results in early-onset tremor and progressive flaccid paresis. Kuramoto et al. (2001) found a marked decrease in ATRN mRNA in the brain of the zi/zi rat and identified zi as an 8-bp deletion at a splice donor site of *Atrn*. Rat *Atrn* gene encoded two isoforms, a secreted and a membrane form, as a result of alternative splicing. The zi mutation at the *Atrn* locus darkened coat color when introduced into agouti rats, as also described in *mahogany (mg)* mice, carrying the homozygous mutation at *Atrn* locus. Transgenic rescue rats showed that the membrane-type ATRN complemented both neurological alteration and abnormal pigmentation in zi/zi rats, but that the secreted-type ATRN complemented neither mutant phenotype. Furthermore, *mg* mice exhibited hypomyelination and vacuolation in the CNS associated with body tremor (Kuramoto et al. 2001). Ultrastructurally these vacuoles mainly consisted of splitting of myelin lamella both in the periaxonal and intermyelinic spaces. Linkage analysis using intercross progeny between the myelin vacuolation (mv) rat, named after the pathologic characteristics, and normal control rat strains showed that the mv phenotypes were cosegregated with polymorphic markers adjacent to the *Atrn* (Attractin, formerly zi [zitter]) locus on rat chromosome 3. Discovery of the rat null mutation *Atrn(mv)*, different from *Atrn(zi)*, provides a new animal model for studying the functions of the attractin protein (Kuwamura et al. 2002). Studies suggest that lack of *Atrn* gene expression induced neurodegeneration by a decrease in active ERK through an intracellular signaling via oxidative stress (Muto and Sato 2003). These studies indicate that membrane ATRN has a critical role in normal myelination in CNS and provide insights into the physiology of myelination as well as the etiology of myelin diseases. Izawa et al. (2008) indicated that the attractin defect results in oligodendrocyte dysfunction, and is associated with astrogliosis and microglial activation in mv rats. Attractin may be directly involved in the function of oligodendro-cytes in CNS myelination. Results indicate that myelinogenesis but not oligodendrogenesis is severely altered both in the white and gray matter of mv rats (Izawa et al. 2010).

The Hamster Black Tremor (bt) Mutation: The hamster black tremor (bt) mutation induces a black coat color and a defective myelination in CNS that manifests as a tremor. On the other hand, loss-of-function mutations of the *Atrn* gene, such as *Atrnmg*, *Atrnmg-L*, and *Atrnmg-3J* in mice, and *Atrnzi* in rats, induce both darkening of coat color and hypomyelination and vacuolation in CNS. The close resemblance of mutant phenotypes led to postulate that the bt/bt hamster also might harbor a mutation in *Atrn*. While the human and rat *Atrn* genes encode both membrane- and secreted-type proteins, the hamster *Atrn* gene encoded only membrane-type protein with 1,427 amino acids, as in the case of the mouse. Hamster ATRN had 93.6%, 96.8%, and 96.8% identities with human, rat, and mouse membrane-type Attractin. In the brain of the bt/bt hamster, aberrant transcripts with more than three size species were observed, and the most predominant transcript encoded the truncated Attractin without transmembrane domain. The hamster bt mutation was the approximately 10-kb retrotransposon-like insertion into the *Atrn* gene, which resulted in aberrant transcripts (Kuramoto et al. 2002).

40.3.6 Therapeutic Applications of ATRN/ *Mahogany* Gene Products

The C-terminal peptide product encoded by *mahogany* gene crosses the blood–brain barrier (BBB) by a transport mechanism that is saturable. The ability of this system indicates the therapeutic potential of mahogany (1377–1428) in the treatment of obesity (Kastin and Akerstrom 2000). Avy/agouti (A^{vy}) mice have late onset obesity related to over-expression of agouti signaling protein (ASP) in the hypothalamus. As mahogany modulates the actions of ASP, Pan and Kastin (2007) tested the transport of mahogany peptide across BBB. The brain uptake of mahogany peptide was significantly higher in young A^{vy} mice, and it preceded the surge of fat mass quantified by NMR. The results suggest a role of accelerated BBB transport in the epigenetics of A^{vy} mice.

The proliferative responses of T lymphocytes of a subset of patients with CVID are abnormally low. This may be due to abnormalities in extracellular interactions or signaling defects downstream from membrane-associated receptors. Attractin is a rapidly expressed T cell activation antigen involved in forming an association between T cells and monocytes. Due to the likely role of attractin in cell guidance and amplification of immune response, results indicate that the lack of up-regulation of the molecule in patients with CVID may in turn affect any further step of productive immune response. This finding may imply a potential therapeutic role for this novel molecule (Pozzi et al. 2001). Attractin is a reliable secreted marker for high-grade gliomas. Additionally, it may be an important mediator of tumor invasiveness, and thus, a potential target in future therapies (Khwaja et al. 2006). ATRN may play a protective role against environmental toxins that implies a potential therapeutic effect of ATRN for neurodegenerative diseases (Paz et al. 2007).

40.4 Eosinophil Major Basic Protein 1 (EMBP1) (Group XII of CTLD)

40.4.1 Characterization of EMBP

Eosinophils are implicated in the combat of infections caused by helminth parasites and viruses and are associated with tissue damage in a variety of diseases (Gleich 2000). A distinctive group of proteins form the eosinophil granule. Among these, the eosinophil major basic protein 1 (EMBP1) is the most abundant eosinophil granule protein that forms the crystalline core of the granule and belongs to Group XII in CTLD classificatiion. Proteoglycan 2, bone marrow (natural killer cell activator, eosinophil granule major basic protein), also known as PRG2, is the protein which in humans is encoded by the *PRG2* gene. EMBP is a 14 kDa (117 amino acid residues) intensely basic protein whose complement of basic residues consists of arginines, giving rise to a calculated pI of 11.4 and a net charge of 15.0 at pH 7 (Barker et al. 1988; McGrogan et al. 1988). However, pro-EMBP is almost neutral with a calculated pI of 6.2, because of the presence of a large number of acidic residues in the 90 residue long pro-portion (pI 3.9). It has been suggested that the pro-portion of EMBP serves predominantly to protect cells from the extreme basicity of EMBP during transport from the Golgi to the eosinophil granule (Barker et al. 1991).

EMBP is a monomer under physiological conditions that readily polymerizes in solution forming insoluble aggregates because of the presence of five free thiol groups (in addition to the four cysteines involved in disulfide bond formation) in the protein (Oxvig et al. 1994a). It has been shown to be non-glycosylated, contrasting with the pro-protein, which is heavily glycosylated with N-glycans, O-glycans, and glycosaminoglycans, raising the molecular weight to between 30 and 50 kDa (Oxvig et al. 1994b). During eosinophilopoiesis, EMBP is initially transcribed as a promolecule containing a markedly acidic pro-piece. The promolecule can be identified in eosinophil granules, and current evidence suggests that it is processed here into EMBP, which then condenses to form the eosinophil granule core (Popken-Harris et al. 1998). A EMBP homologue (EMBPH) similar to EMBP has also been identified in human eosinophil granule lysates (Plager et al. 1999).

40.4.2 Functions of EMBP

EMBP is expressed in eosinophils, basophils, and placental X cells. In eosinophils, the protective acidic region is removed and the CTLD covalently polymerizes to form insoluble aggregates that make up the crystalline core of eosinophil granules. EMBP is a cytotoxin deployed against helminth parasites, and possibly also against bacteria, fungi and cancerous or infected cells. EMBP triggers release of histamine from mast cells and basophils, and activates neutrophils and alveolar macrophages. Release of EMBP and its deposition onto tissues is linked to hypersensitivity reactions and causes tissue dysfunction in inflammatory and allergic conditions, including asthma. Towards the end of pregnancy, pro-EMBP circulates in a complex with pregnancy-associated plasma protein A, angiotensin and C3dg. EMBPH is expressed in eosinophils. It is less basic and less potent than EMBP, but has similar biological activities. The CTLDs in these proteins do not contain the sequence motifs associated with sugar-binding activity. However, EMBP binds heparin through a novel calcium-independent carbohydrate-recognition mechanism.

EMBP is a potent helminthotoxin and is cytotoxic toward bacteria and mammalian cells in vitro (Swaminathan et al. (2001) and may have important roles in allergic and inflammatory reactions. The EMBP causes release of histamine from mast cells and basophils, activates neutrophils and alveolar macrophages, and is directly implicated in epithelial cell damage, exfoliation and bronchospasm in asthma (Swaminathan et al. 2001). The exact mechanism of EMBP action is not known, although it has been speculated that the highly basic nature of the protein contributes to its toxicity.

The concentration of EMBP is elevated in biological fluids (e.g. sputa and bronchoalveolar lavage fluids) from patients with asthma and other eosinophil-associated diseases. Deposition of EMBP onto damaged tissues in various human diseases is strongly associated with dysfunction of those tissues, and instillation of EMBP into the lung of monkeys causes a tenfold increase in bronchial hyperreactivity, as well as marked bronchospasm (Gundel et al. 1991). EMBP appears to damage cells by disrupting the bilayer lipid membrane (Gleich et al. 1993) or altering the activity of enzymes within tissues (Hastie et al. 1987). Evidence also implicates pro-EMBP as an inhibitor of human pregnancy-associated plasma protein A (Overgaard et al. 2000).

40.4.3 Crystal Structure of EMBP

EMBP does not exhibit high sequence similarity to any known proteins, but has weak sequence identity (23–28%) with C-type lectin domains of mannose-binding protein, human lithostathine (LIT), and the low affinity IgE receptor

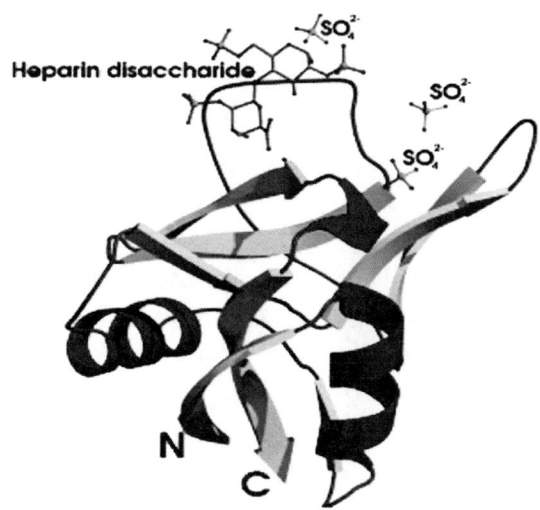

Fig. 40.3 Crystal structure of EMBP in complex with heparin disaccharide (PDB: 2BRS). The sugars recognized by EMBP are likely to be proteoglycans such as heparin, leading to new interpretations for EMBP function (Printed with permission from Swaminathan et al. 2005 © American Chemical Society)

FceRII (Patthy 1989). Swaminathan et al. (2001) reported the crystal structure of EMBP at 1.8 Å resolution. This protein is structurally related to members of the CTLD family, but is significantly different from known CTLDs, lacking the potential to bind carbohydrates in a calcium dependent manner. EMBP are likely to be proteoglycans such as heparin, leading to new interpretations for EMBP function (Swaminathan et al. 2005). Evidence of binding of EMBP to heparin and heparin disaccharide has been presented. The crystal structure of EMBP in complex with a heparin disaccharide and in the absence of Ca^{2+} is shown in Fig. 40.3, a report of any C-type lectin with this sugar. Structural analysis shows that the potential carbohydrate ligands recognized by EMBP are likely to be different from those in other CTLDs.

40.4.4 DEC-205-Associated C-Type Lectin (DCL)-1

DEC-205-associated C-type lectin (DCL)-1 was also formerly classified as a member of group XII. However, DCL-1 is a type 1 transmembrane protein with an extracellular CTLD and neck region and may be classified as belonging to a new group of CTLD-containing proteins. DCL-1 was discovered in cells associated with the development of Hodgkin's lymphoma. Intergenic splicing of DCL-1 and DEC-205 (a CTLD-containing type 1 transmembrane protein of the mannose receptor family (Chaps 15 and 35) results in production of a fusion protein containing both DCL-1 and DEC-205 extracellular domains, plus the transmembrane and cytoplasmic regions of DCL-1. Non-intergenically spliced DCL-1 transcripts are present in myeloid and B lymphocyte cell lines.

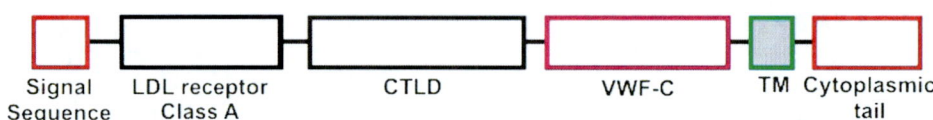

Fig. 40.4 Domaim organization of Integral membrane protein (IDD/DGCR2) (Group XIII of CTLD)

40.5 Integral Membrane Protein, Deleted in Digeorge Syndrome (IDD) (Group XIII of CTLDs)

Integral membrane protein, deleted in DiGeorge syndrome (IDD) is a unique CTLD-containing type 1 transmembrane protein expressed in a range of cell types and notably in developing nervous system. The extracellular region consists of a CTLD sandwiched between a membrane-distal low density lipoprotein (LDL)-receptor class A domain and a membrane-proximal von Willebrand Factor C domain (Fig. 40.4). IDD also has a sizeable intracellular region.

40.5.1 DiGeorge Syndrome Critical Region 2 (DGCR2)

IDD may function as an adhesion receptor in cell-cell or cell-matrix interactions during cell differentiation and migration, particularly in the nervous system, or may be involved in signaling. The IDD gene lies within the DiGeorge syndrome critical region – the minimal region of chromosome 22. Deletions of the 22q11.2 have been associated with a wide range of developmental defects (notably DiGeorge syndrome, velocardiofacial syndrome, conotruncal anomaly face syndrome and isolated conotruncal cardiac defects) classified under the acronym CATCH 22. The DGCR2 gene encodes a novel putative adhesion receptor protein, which could play a role in neural crest cells migration, a process which has been proposed to be altered in DiGeorge syndrome (Taylor et al. 1997). The CTLD in IDD is not known to have sugar-binding activity. Alternative splicing of gene results into multiple transcript variants. The DGCR2 gene is conserved in chimpanzee, dog, cow, mouse, rat, chicken, and zebrafish.

SEZ-12 is one of the seizure-related cDNAs which was isolated from primary cultured neurons from mouse cerebral cortex with or without pentylene tetrazol (PTZ). SEZ-12 expression is transiently down-regulated in mouse brain by injection of PTZ. The deduced amino acid sequence of SEZ-12 revealed that it encodes membrane-bound C-type lectin and has a significant homology to that of human cDNA of DGCR2 and IDD, associated with the DiGeorge syndrome. The message was expressed ubiquitously in various organs with low-abundance. The cloned transmembrane protein was probably involved in cell-cell interaction.

Findings suggest that transmembrane signaling in neuronal cells may have an important role in PTZ-induced seizure (Kajiwara et al. 1996). Several lines of evidence have established the presence of an association between a 3-Mb deletion in chromosome 22q11 and schizophrenia with a reduced expression of DGCR2 (Iida et al. 2001; Shifman et al. 2006). However, Georgi et al. (2009) did not support a significant role of DGCR2 in the aetiology of schizophrenia in the German population.

References

Layilin group of C-type Lectins (Group VIII of CTLDs)

Banerji S, Ni J, Wang SX et al (1999) LYVE-1, a new homologue of the CD44 glycoprotein, is a lymph-specific receptor for hyaluronan. J Cell Biol 144:789–801

Bono P, Cordero E, Johnson K et al (2005) Layilin, a cell surface hyaluronan receptor, interacts with merlin and radixin. Exp Cell Res 308:177–187

Bono P, Rubin K, Higgins JM, Hynes RO (2001) Layilin, a novel integral membrane protein, is a hyaluronan receptor. Mol Biol Cell 12:891–900

Borowsky ML, Hynes RO (1998) Layilin, a novel talin-binding transmembrane protein homologous with C-type lectins, is localized in membrane ruffles. J Cell Biol 143:429–442

Chen Z, Zhuo W, Wang Y, Ao X, An J (2008) Down-regulation of layilin, a novel hyaluronan receptor, via RNA interference, inhibits invasion and lymphatic metastasis of human lung A549 cells. Biotechnol Appl Biochem 50:89–96

Marks MS, Ohno H, Kirchhausen T, Bonifacino JS (1997) Protein sorting by tyrosine-based signals: adapting to the Ys and wherefores. Trends Cell Biol 7:124–128

Ohno H, Stewart J, Fournier M-C et al (1995) Interaction of tyrosine-based sorting signals with clathrin-associated proteins. Science 269:1872–1875

Scoles DR (2008) The merlin interacting proteins reveal multiple targets for NF2 therapy. Biochim Biophys Acta 1785:32–54

Wegener KL, Basran J, Bagshaw CR et al (2008) Structural basis for the interaction between the cytoplasmic domain of the hyaluronate receptor layilin and the talin F3 subdomain. J Mol Biol 382:112–126

Weng L, Hübner R, Claessens A, Smits P et al (2003) Isolation and characterization of chondrolectin (Chodl), a novel C-type lectin predominantly expressed in muscle cells. Gene 308:21–29

Weng L, Smits P, Wauters J et al (2002) Molecular cloning and characterization of human chondrolectin, a novel type I transmembrane protein homologous to C-type lectins. Genomics 80:62–70

Young VB, Falkow S, Schoolnik GK (1992) The invasin protein of *Yersinia enterocolitica*: internalization of invasin-bearing bacteria by eukaryotic cells is associated with reorganization of the cytoskeleton. J Cell Biol 116:197–207

Tetranectin Group of Lectins (Group IX of CTLDs)

Arvanitis DL, Kamper EF, Kopeikina L et al (2002) Tetranectin expression in gastric adenocarcinomas. Histol Histopathol 17:471–475

Begum FD, Høgdall E, Christensen IJ et al (2010) Serum tetranectin is a significant prognostic marker in ovarian cancer patients. Acta Obstet Gynecol Scand 89:190–198

Begum FD, Høgdall E, Kjaer SK et al (2009) Preoperative serum tetranectin, CA125 and menopausal status used as single markers in screening and in a risk assessment index (RAI) in discriminating between benign and malignant ovarian tumors. Gynecol Oncol 113:221–227

Berglund L, Petersen TE (1992) The gene structure of tetranectin, a plasminogen binding protein. FEBS Lett 309:15–19

Blaakaer J, Høgdall CK, Micic S et al (1995) Ovarian carcinoma serum markers and ovarian steroid activity—is there a link in ovarian cancer? A correlation of inhibin, tetranectin and CA-125 to ovarian activity and the gonadotropin levels. Eur J Obstet Gynecol Reprod Biol 59:53–56

Borregaard N, Christensen L, Bejerrum OW et al (1990) Identification of a highly mobilizable subset of human neutrophil intracellular vesicles that contains tetranectin and latent alkaline phosphatase. J Clin Invest 85:408–416

Brunner A, Ensinger C, Christiansen M et al (2007) Expression and prognostic significance of Tetranectin in invasive and non-invasive bladder cancer. Virchows Arch 450:659–664

Bugge TH, Flick MJ, Daugherty CC, Degen JL (1995) Plasminogen deficiency causes severe thrombosis but is compatible with development and reproduction. Genes Dev 9:794–807

Christensen L, Clemmensen I (1989) Tetranectin immunoreactivity in normal human tissues. An immunohistochemical study of exocrine epithelia and mesenchyme. Histochemistry 92:29–35

Christensen L, Clemmensen I (1991) Differences in tetranectin immunoreactivity between benign and malignant breast tissue. Histochemistry 95:427–433

Christensen L, Johansen N, Jensen BA et al (1987) Immunohistochemical localization of a novel, human plasma protein, tetranectin, in human endocrine tissues. Histochemistry 87:195–199

Christensen L (1992) The distribution of fibronectin, laminin and tetranectin in human breast cancer with special attention to the extracellular matrix. APMIS Suppl 26:1–39

Clemmensen I, Lund LR, Christensen L et al (1991) A tetranectin-related protein is produced and deposited in extracellular matrix by human embryonal fibroblasts. Eur J Biochem 195:735–741

Clemmensen I, Petersen LC, Kluft C (1986) Purification and characterization of a novel, oligomeric, plasminogen kringle-4 binding protein from human plasma: tetranectin. Eur J Biochem 156:327–333

Clemmensen I (1989) Interaction of tetranectin with sulphated polysaccharides and trypan blue. Scand J Clin Lab Invest 49:719–725

Darde ÌVM, de la Cuesta F, Dones FG et al (2010) Analysis of the plasma proteome associated with acute coronary ayndrome: does a permanent protein signature exist in the plasma of ACS patients? J Proteome Res 9(9):4420–4432

De Vries TJ, De Wit PE et al (1996) Tetranectin and plasmin/plasminogen are similarly distributed at the invasive front of cutaneous melanoma lesions. J Pathol 179:260–265

Deng X, Høgdall EV, Høgdall CK et al (2000) The prognostic value of pretherapeutic tetranectin and CA-125 in patients with relapse of ovarian cancer. Gynecol Oncol 79:416–419

Drickamer K (1999) C-type lectin-like domains. Curr Opin Struct Biol 9:585–590

Durkin ME, Naylor SL et al (1997) Assignment of the gene for human tetranectin (TNA) to chromosome 3p22 → p21.3 by somatic cell hybrid mapping. Cytogenet Cell Genet 76:39–40

Fuhlendorff J, Clemmensen I, Magnusson S (1987) Primary structure of tetranectin, a plasminogen kringle-4 binding plasma protein: homology with asialoglycoprotein receptors and cartilage proteoglycan core protein. Biochemistry 26:6757–6764

Gately S, Twardowski P, Stack MS et al (1997) The mechanism of cancer-mediated conversion of plasminogen to the angiogenesis inhibitor angiostatin. Proc Natl Acad Sci USA 94:10868–10872

Graversen JH, Lorentsen RH, Jacobsen C et al (1998) The plasminogen binding site of the C-type lectin tetranectin is located in the carbohydrate recognition domain, and binding is sensitive to both calcium and lysine. J Biol Chem 273:29241–29246

Graversen JH, Jacobsen C, Sigurskjold BW et al (2000a) Mutational analysis of affinity and selectivity of kringle-tetranectin interaction. Grafting novel kringle affinity ontp the trtranectin lectin scaffold. J Biol Chem 275:37390–37396

Graversen JH, Sigurskjold BW, ThøgersenHC EM (2000b) Tetranectin-binding site on plasminogen kringle-4 involves the lysine-binding pocket and at least one additional amino acid residue. Biochemistry 39:7414–7419

Gronlund B, Høgdall EV, Christensen IJ et al (2006) Pre-treatment prediction of chemoresistance in second-line chemotherapy of ovarian carcinoma: value of serological tumor marker determination (tetranectin, YKL-40, CASA, CA 125). Int J Biol Markers 2:141–148

Gunn TM, Barsh GS (2000) Mahogany/attractin: en route from phenotype to function. Trends Cardiovasc Med 10(2):76–81

Hammack BN, Fung KY, Hunsucker SW et al (2004) Proteomic analysis of multiple sclerosis cerebrospinal fluid. Mult Scler 10:245–260

Hermann M, Pirkebner D, Draxl A et al (2005) In the search of potential human islet stem cells: is tetranectin showing us the way? Transplant Proc 37:1322–1325

Hiraoka A, Sugimura A, Seki T et al (1997) Cloning, expression, and characterization of a cDNA encoding a novel human growth factor for primitive hematopoietic progenitor cells. Proc Natl Acad Sci USA 94:7577–7582

Høgdall CK, Christensen IJ, Stephens RW et al (2002a) Serum tetranectin is an independent prognostic marker in colorectal cancer and weakly correlated with plasma suPAR, plasma PAI-1 and serum CEA. APMIS 110:630–638

Høgdall CK, Norgaard-Pedersen B, Mogensen O (2002b) The prognostic value of pre-operative serum tetranectin, CA-125 and a combined index in women with primary ovarian cancer. Anticancer Res 22:1765–1768

Høgdall CK (1998) Human tetranectin: methodological and clinical studies. APMIS Suppl 86:1–31

Holtet TL, Graversen JH, Clemmensen I, Thøgersen HC, Etzerodt M (1997) Tetranectin, a trimeric plasminogen-binding C-type lectin. Protein Sci 6:1511–1515

Iba K, Chiba H, Yamashita T, Ishii S, Sawada N (2001a) Phase-independent inhibition by retinoic acid of mineralization correlated with loss of tetranectin expression in a human osteoblastic cell line. Cell Struct Funct 26:227–233

Iba K, Durkin ME, Johnsen L et al (2001b) Mice with a targeted deletion of the tetranectin gene exhibit a spinal deformity. Mol Cell Biol 21:7817–7825

Iba K, Hatakeyama N, Kojima T et al (2009) Impaired cutaneous wound healing in mice lacking tetranectin. Wound Repair Regen 17:108–112

Ibaraki K, Kozak CA, Wewer UM et al (1995) Mouse tetranectin: cDNA sequence, tissue-specific expression, and chromosomal mapping. Mamm Genome 6:693–696

Jaquinod M, Holtet TL, Etzerodt M et al (1999) Mass spectrometric characterisation of post-translational modification and genetic variation in human tetranectin. J Biol Chem 380:1307–1314

Jensen BA, Clemmensen I (1988) Plasma tetranectin is reduced in cancer and related to metastasia. Cancer 62:869–872

Jensen BA, McNair P, Hyldstrup L, Clemmensen I (1987) Plasma tetranectin in healthy male and female individuals, measured by enzyme-linked immunosorbent assay. J Lab Clin Med 110:612–617

Kamper EF, Kopeikina L, Mantas A et al (1998a) Tetranectin levels in patients with acute myocardial infarction and their alterations during thrombolytic treatment. Ann Clin Biochem 35:400–407

Kamper EF, Kopeikina LT, Trontzas P et al (1998b) Comparative study of tetranectin levels in serum and synovial fluid of patients with rheumatoid arthritis, seronegative spondylarthritis and osteoarthritis. Clin Rheumatol 17:318–324

Kamper EF, Papaphilis AD, Angelopoulou MK et al (1999) Serum levels of tetranectin, intercellular adhesion molecule-1 and interleukin-10 in B-chronic lymphocytic leukemia. Clin Biochem 32:639–645

Kastrup JS, Nielsen BB, Rasmussen H et al (1998) Structure of the C-type lectin carbohydrate recognition domain of human tetranectin. Acta Crystallogr D Biol Crystallogr 54(Pt 5):757–766

Kastrup JS, Rasmussen H, Nielsen BB et al (1997) Human plasminogen binding protein tetranectin: crystallization and preliminary X-ray analysis of the C-type lectin CRD and the full-length protein. Acta Crystallogr D Biol Crystallogr 53(Pt 1):108–111

Kluft C, Jie AF, Los P et al (1989a) Functional analogy between lipoprotein(a) and plasminogen in the binding to the kringle-4 binding protein, tetranectin. Biochem Biophys Res Commun 161:427–433

Kluft C, Los P, Clemmensen I et al (1989b) Quantitation of plasma levels of tetranectin—effects of oral contraceptives, pregnancy, treatment with L-asparaginase and liver cirrhosis. Thromb Haemost 62:792–796

Lai X, Liangpunsakul S, Crabb DW, Ringham HN, Witzmann FA (2009) A proteomic workflow for discovery of serum carrier protein-bound biomarker candidates of alcohol abuse using LC-MS/MS. Electrophoresis 30:2207–2214

Lorentsen RH, Graversen JH, Caterer NR, Thøgersen HC, Etzerodt M (2000) The heparin-binding site in tetranectin is located in the N-terminal region and binding does not involve the carbohydrate recognition domain. Biochem J 347:83–87

Mio H, Kagami N, Yokokawa S et al (1998) Isolation and characterization of a cDNA for human mouse, and rat full-length stem cell growth factor, a new member of C-type lectin superfamily. Biochem Biophys Res Commun 249:124–130

Mogues T, Etzerodt M, Hall C et al (2004) Tetranectin binds to the kringle 1–4 form of angiostatin and modifies its functional activity. J Biomed Biotechnol 2004:73–78

Neame PJ, Tapp H, Grimm DR (1999) The cartilage-derived, C-type lectin (CLECSF1): structure of the gene and chromosomal location. Biochim Biophys Acta 1446:193–202

Neame PJ, Young CN, Treep JT (1992) Primary structure of a protein isolated from reef shark (Carcharhinus springeri) cartilage that is similar to the mammalian C-type lectin homolog, tetranectin. Protein Sci 1:161–168

Nielbo S, Thomsen JK, Graversen JH et al (2004) Structure of the plasminogen kringle-4 binding calcium-free form of the C-type lectin-like domain of tetranectin. Biochemistry 43:8636–8643

Nielsen BB, Kastrup JS, Rasmussen H et al (2000) Crystallization and molecular-replacement solution of a truncated form of human recombinant tetranectin. Acta Crystallogr D Biol Crystallogr 56:637–639

Nielsen BB, Kastrup JS, Rasmussen H et al (1997) Crystal structure of tetranectin, a trimeric plasminogen-binding protein with an α-helical coiled coil. FEBS Lett 412:388–396

Nielsen H, Clemmensen I, Kharazmi A (1993) Tetranectin: a novel secretory protein from human monocytes. Scand J Immunol 37:39–42

Nielsen H, Clemmensen I, Nielsen HJ, Drivsholm A (1990) Decreased tetranectin in multiple myeloma. Am J Hematol 33:142–144

Obrist P, Spizzo G et al (2004) Aberrant tetranectin expression in human breast carcinomas as a predictor of survival. J Clin Pathol 57:417–421

O'Reilly MS, Holmgren L, Shing Y et al (1994) Angiostatin: a novel angiogenesis inhibitor that mediates the suppression of metastases by a Lewis lung carcinoma. Cell 79:315–328

Richardson M, Gunawan J, Hatton MW et al (2002) Malignant ascites fluid (MAF), including ovarian-cancer-associated MAF, contains angiostatin and other factor(s) which inhibit angiogenesis. Gynecol Oncol 86:279–287

Sørensen CB, Berglund L, Petersen TE (1997) Cloning of the murine tetranectin gene and 5′-flanking region. Gene 201:199–202

Sørensen CB, Berglund L, Petersen TE (1995) Cloning of a cDNA encoding murine tetranectin. Gene 152:243–245

Stoevring B, Jaliashvili I et al (2006) Tetranectin in cerebrospinal fluid of patients with multiple sclerosis. Scand J Clin Lab Invest 66:577–583

Stoevring B, Jaliashvili I, Thougaard AV et al (2005) Tetranectin in cerebrospinal fluid: biochemical characterisation and evidence of intrathecal synthesis or selective uptake into CSF. Clin Chim Acta 359:65–71

Thougaard AV, Jaliashvili I, Christiansen M (2001) Tetranectin-like protein in vertebrate serum: a comparative immunochemical analysis. Comp Biochem Physiol B Biochem Mol Biol 128:625–634

Tuxen MK, Sölétormos G, Dombernowsky P (1995) Tumor markers in the management of patients with ovarian cancer. Cancer Treat Rev 21:215–245

Verspaget HW, Clemmensen I, Ganesh S et al (1994) Tetranectin expression in human colonic neoplasia. Histopathology 25:463–467

Wang E-S, Sun Y, Guo J-G et al (2010a) Tetranectin and apolipoprotein A-I in cerebrospinal fluid as potential biomarkers for Parkinson's disease. Acta Neurol Scand 122:350–359

Wang L, Pan Y et al (2010b) Tetranectin is a potential biomarker in cerebrospinal fluid and serum of patients with epilepsy. Clin Chim Acta 411:581–583

Westergaard UB, Andersen MH, Heegaard CW, Fedosov SN, Petersen TE (2003) Tetranectin binds hepatocyte growth factor and tissue-type plasminogen activator. Eur J Biochem 270:1850–1854

Wewer UM, Albrechtsen R (1992) Tetranectin, a plasminogen kringle-4-binding protein. Cloning and gene expression pattern in human colon cancer. Lab Investig 67:253–262

Wewer UM, Iba K, Durkin ME et al (1998) Tetranectin is a novel marker for myogenesis during embryonic development, muscle regeneration, and muscle cell differentiation in vitro. Dev Biol 200:247–259

Wewer UM, Ibaraki K, Schjørring P et al (1994) A potential role for tetranectin in mineralization during osteogenesis. J Cell Biol 127:1767–1775

Xu X, Gilpin B, Iba K et al (2001) Tetranectin in slow intra- and extrafusal chicken muscle fibers. J Muscle Res Cell Motil 22:121–132

Attractin Group of CTLDs (Group XI)

Azouz A, Gunn TM, Duke-Cohan JS (2007) Juvenile-onset loss of lipid-raft domains in attractin-deficient mice. Exp Cell Res 313:761–771

Barsh G, Gunn T, He L et al (2000) Biochemical and genetic studies of pigment-type switching. Pigment Cell Res 13(Suppl 8):48–53

Bronson RT, Donahue LR, Samples R et al (2001) Mice with mutations in the mahogany gene Atrn have cerebral spongiform changes. J Neuropathol Exp Neurol 60:724–730

Carlson GA, Moore KJ (1999) The mahogany protein is a receptor involved in suppression of obesity. Nature 398(6723):148–152

Cota CD, Liu RR, Sumberac TM et al (2008) Genetic and phenotypic studies of the dark-like mutant mouse. Genesis 46:562–573

Dinulescu DM, Fan W, Boston BA et al (1998) Mahogany (mg) stimulates feeding and increases basal metabolic rate independent of its suppression of agouti. Proc Natl Acad Sci USA 95:12707–12712

Duke-Cohan JS, Gu J, McLaughlin DF et al (1998) Attractin (DPPT-L), a member of the CUB family of cell adhesion and guidance proteins, is secreted by activated human T lymphocytes and modulates immune cell interactions. Proc Natl Acad Sci USA 95:11336–11341

Duke-Cohan JS, Tang W, Schlossman SF (2000) Attractin: a cub-family protease involved in T cell-monocyte/macrophage interactions. Adv Exp Med Biol 477:173–185

Durinx C, Lambeir AM, Bosmans E et al (2000) Molecular characterization of dipeptidyl peptidase activity in serum: soluble CD26/dipeptidyl peptidase IV is responsible for the release of X-Pro dipeptides. Eur J Biochem 267:5608–5613

Friedrich D, Hoffmann T, Bär J et al (2007) Does human attractin have DP4 activity? J Biol Chem 388:155–162

Graphodatskaya D, Perelman P, Serdukova N et al (2003) Assignment of the bovine attractin (ATRN) gene to chromosome 13q21q22 by in situ hybridization. Cytogenet Genome Res 103:204K

Gunn TM, Barsh GS (2000) Mahogany/attractin: en route from phenotype to function. Trends Cardiovasc Med 10:76–81

Gunn TM, Inui T, Kitada K et al (2001) Molecular and phenotypic analysis of attractin mutant mice. Genetics 158:1683–1695

Gunn TM, Miller KA, He L et al (1999) The mouse mahogany locus encodes a transmembrane form of human attractin. Nature 398(6723):152–156

Haqq AM, René P, Kishi T et al (2003) Characterization of a novel binding partner of the melanocortin-4 receptor: attractin-like protein. Biochem J 376:595–605

He L, Eldridge AG, Jackson PK et al (2003a) Accessory proteins for melanocortin signaling: attractin and mahogunin. Ann N Y Acad Sci 994:288–298

He L, Gunn TM, Bouley DM et al (2001) A biochemical function for attractin in agouti-induced pigmentation and obesity. Nat Genet 27:40–47

He L, Lu XY, Jolly AF et al (2003b) Spongiform degeneration in mahoganoid mutant mice. Science 299(5607):710–712

Hida T, Wakamatsu K, Sviderskaya EV et al (2009) Agouti protein, mahogunin, and attractin in pheomelanogenesis and melanoblast-like alteration of melanocytes: a cAMP-independent pathway. Pigment Cell Melanoma Res 22:623–634

Izawa T, Takenaka S, Ihara H et al (2008) Cellular responses in the spinal cord during development of hypomyelination in the mv rat. Brain Res 1195:120–129

Izawa T, Yamate J et al (2010) Abnormal myelinogenesis both in the white and gray matter of the attractin-deficient mv rat. Brain Res 1312:145–155

Jackson IJ (1999) The mahogany mouse mutation: further links between pigmentation, obesity and the immune system. Trends Genet 15:429–431

Kadowaki T, Nakadate K, Sakakibara S et al (2007) Expression of Iba1 protein in microglial cells of zitter mutant rat. Neurosci Lett 411:26–31

Kastin AJ, Akerstrom V (2000) Mahogany (1377–1428) enters brain by a saturable transport system. J Pharmacol Exp Ther 294:633–636

Khwaja FW, Duke-Cohan JS, Brat DJ, Van Meir EG (2006) Attractin is elevated in the cerebrospinal fluid of patients with malignant astrocytoma and mediates glioma cell migration. Clin Cancer Res 12:6331–6336

Kuramoto T, Kitada K, Inui T et al (2001) Attractin/mahogany/zitter plays a critical role in myelination of the central nervous system. Proc Natl Acad Sci USA 98:559–564

Kuramoto T, Nomoto T, Fujiwara A et al (2002) Insertional mutation of the attractin gene in the black tremor hamster. Mamm Genome 13:36–40

Kuwamura M, Maeda M, Kuramoto T et al (2002) The myelin vacuolation (mv) rat with a null mutation in the attractin gene. Lab Invest 82:1279–1286

Laudes M, Oberhauser F, Schulte DM et al (2010) Dipeptidyl-peptidase 4 and attractin expression is increased in circulating blood monocytes of obese human subjects. Exp Clin Endocrinol Diabetes 118:473–477

Li J, Wang S, Huang S et al (2009) Attractin gene deficiency contributes to testis vacuolization and sperm dysfunction in male mice. J Huazhong Univ Sci Technolog Med Sci 29:750–754

Lu X, Gunn TM, Shieh K et al (1999) Distribution of mahogany/attractin mRNA in the rat central nervous system. FEBS Lett 462:101–107

Malãk R, Mares V, Kleibl Z et al (2001) Expression of attractin and its differential enzyme activity in glioma cells. Biochem Biophys Res Commun 284:289–294

Miller KA, Gunn TM, Carrasquillo MM et al (1997) Genetic studies of the mouse mutations mahogany and mahoganoid. Genetics 146:1407–1415

Muto Y, Sato K (2003) Pivotal role of attractin in cell survival under oxidative stress in the zitter rat brain with genetic spongiform encephalopathy. Brain Res Mol Brain Res 111:111–122

Nagle DL, McGrail SH, Vitale J et al (2000) Mahogany/attractin: en route from phenotype to function. Trends Cardiovasc Med 10:76–81

Nakadate K, Sakakibara S, Ueda S (2008) Attractin/mahogany protein expression in the rodent central nervous system. J Comp Neurol 508:94–111

Pan W, Kastin AJ (2007) Mahogany, blood–brain barrier, and fat mass surge in A^{vy} mice. Int J Obes (Lond) 31:1030–1032

Paz J, Yao H, Lim HS et al (2007) The neuroprotective role of attractin in neurodegeneration. Neurobiol Aging 28:1446–1456

Phan LK, Lin F et al (2002) The mouse mahoganoid coat color mutation disrupts a novel C3HC4 RING domain protein. J Clin Invest 110:1449–1459

Pozzi N, Gaetaniello L, Martire B, De Mattia D et al (2001) Defective surface expression of attractin on T cells in patients with common variable immunodeficiency (CVID). Clin Exp Immunol 123:99–104

Radhakrishnan Y, Hamil KG, Tan JA et al (2009) Novel partners of SPAG11B isoform D in the human male reproductive tract. Biol Reprod 81:647–656

Sakakibara S, Nakadate K, Ookawara S, Ueda S (2008) Non-cell autonomous impairment of oligodendrocyte differentiation precedes CNS degeneration in the Zitter rat: implications of macrophage/microglial activation in the pathogenesis. BMC Neurosci 9:35

Stark Z, Bruno DL, Mountford H et al (2010) De novo 325 kb microdeletion in chromosome band 10q25.3 including ATRNL1 in a boy with cognitive impairment, autism and dysmorphic features. Eur J Med Genet 53:337–339

Sun K, Johnson BS, Gunn TM (2007) Mitochondrial dysfunction precedes neurodegeneration in mahogunin (Mgrn1) mutant mice. Neurobiol Aging 28:1840–1852

Tang W, Duke-Cohan JS (2002) Human secreted attractin disrupts neurite formation in differentiating cortical neural cells in vitro. J Neuropathol Exp Neurol 61:767–777

Tang W, Gunn TM, McLaughlin DF et al (2000) Secreted and membrane attractin result from alternative splicing of the human ATRN gene. Proc Natl Acad Sci USA 97:6025–6030

Ueda S, Sakakibara S, Kadowaki T, Naitoh T, Hirata K, Yoshimoto K (2008) Chronic treatment with melatonin attenuates serotonergic degeneration in the striatum and olfactory tubercle of zitter mutant rats. Neurosci Lett 448:212–216

Walker WP, Aradhya S, Hu CL et al (2007) Genetic analysis of attractin homologs. Genesis 45:744–756

Walker WP, Gunn TM (2010) Shades of meaning: the pigment-type switching system as a tool for discovery. Pigment Cell Melanoma Res 23:485–495

Wrenger S, Faust J, Friedrich D, Hoffmann T et al (2006) Attractin, a dipeptidyl peptidase IV/CD26-like enzyme, is expressed on human peripheral blood monocytes and potentially influences monocyte function. J Leukoc Biol 80:621–629

Yang YK, Harmon CM (2003) Recent developments in our understanding of melanocortin system in the regulation of food intake. Obes Rev 4:239–248

Yeo GS, Siddle K (2003) Attractin' more attention - new pieces in the obesity puzzle? Biochem J 376:e7–e8

Eosinophil Major Basic Protein 1 (EMBP1) (Group XII of CTLDs)

Barker RL, Gleich GJ, Pease LR (1988) Acidic precursor revealed in human eosinophil granule major basic protein cDNA. J Exp Med 168:1493–1498

Barker RL, Loegering DA, Arakawa KC et al (1990) Cloning and sequence analysis of the human gene encoding eosinophil major basic protein. Gene 86:285–289

Barker RL, Gundel RH, Gleich GJ et al (1991) Acidic polyamino acids inhibit human eosinophil granule major basic protein toxicity. Evidence of a functional role for ProMBP. J Clin Invest 88:798–805

Gleich GJ (2000) Mechanisms of eosinophil-associated inflammation. J Allergy Clin Immunol 105:651–663

Gleich GJ, Adolphson CR, Leiferman KM (1993) The biology of the eosinophilic leukocyte. Annu Rev Med 44:85–101

Gundel RH, Letts LG, Gleich GJ (1991) Human eosinophil major basic protein induces airway constriction and airway hyperresponsiveness in primates. J Clin Invest 87:1470–1473

Hastie AT, Loegering DA, Gleich GJ, Kueppers F (1987) The effect of purified human eosinophil major basic protein on mammalian ciliary activity. Am Rev Respir Dis 135:848–853

McGrogan M, Simonsen C, Scott R et al (1988) Isolation of a complementary DNA clone encoding a precursor to human eosinophil major basic protein. J Exp Med 168:2295–2308

Overgaard MT, Haaning J, Boldt HB et al (2000) Expression of recombinant human pregnancy-associated plasma protein-A and identification of the proform of eosinophil major basic protein as its physiological inhibitor. J Biol Chem 275:31128–31133

Oxvig C, Gleich GJ, Sottrup-Jensen L (1994a) Localization of disulfide bridges and free sulfhydryl groups in human eosinophil granule major basic protein. FEBS Lett 341:213–217

Oxvig C, Haaning J, Højrup P, Sottrup-Jensen L (1994b) Location and nature of carbohydrate groups in proform of human major basic protein isolated from pregnancy serum. Biochem Mol Biol Int 33:329–336

Patthy L (1989) Homology of cytotoxic protein of eosinophilic leukocytes with IgE receptor Fc epsilon RII: implications for its structure and function. Mol Immunol 26:1151–1154

Plager DA, Loegering DA, Weiler DA et al (1999) A novel and highly divergent homolog of human eosinophil granule major basic protein. J Biol Chem 274:14464–14473

Popken-Harris P, Checkel J, Loegering D et al (1998) Regulation and processing of a precursor form of eosinophil granule major basic protein (ProMBP) in differentiating eosinophils. Blood 92:623–631

Swaminathan GJ, Arthur J, Weaver AJ et al (2001) Crystal structure of the eosinophil major basic protein at 1.8 Å: an atypical lectin with a paradigm shift in specificity. J Biol Chem 276:26197–26203

Swaminathan GJ, Myszka DG, Katsamba PS et al (2005) Major eosinophil-granule basic protein, a C-type lectin, binds heparin. Biochemistry 44:14152–14158

Integral Membrane Protein, Deleted in Digeorge Syndrome (IDD) (Group XIII of CTLDs)

Georgi A, Schumacher J, Leon CA et al (2009) No association between genetic variants at the DGCR2 gene and schizophrenia in a German sample. Psychiatr Genet 19:104

Iida A, Ohnishi Y, Ozaki K, Ariji Y et al (2001) High-density single-nucleotide polymorphism (SNP) map in the 96-kb region containing the entire human DiGeorge syndrome critical region 2 (DGCR2) gene at 22q11.2. J Hum Genet 46:604–608

Kajiwara K, Nagasawa H, Shimizu-Nishikawa K et al (1996) Cloning of SEZ-12 encoding seizure-related and membrane-bound adhesion protein. Biochem Biophys Res Commun 222:144–148

Shifman S, Levit A, Chen ML et al (2006) A complete genetic association scan of the 22q11 deletion region and functional evidence reveal an association between DGCR2 and schizophrenia. Hum Genet 120:160–170

Taylor C, Wadey R, O'Donnell H et al (1997) Cloning and mapping of murine Dgcr2 and its homology to the Sez-12 seizure-related protein. Mamm Genome 8:371–375

Family of CD93 and Recently Discovered Groups of CTLDs

G.S. Gupta

41.1 Family of CD93

CD93 is a ~120 kDa O-sialoglycoprotein that within the hematopoietic system is selectively expressed on cells of the myeloid lineage. CD93 in humans is encoded by the *CD93* gene (Nepomuceno et al. 1997; Webster et al. 2000). CD93 is a C-type lectin transmembrane receptor which plays a role not only in cell–cell adhesion processes but also in host defence. CD93 belongs to the Group XIV C-Type lectin family, a group containing two other members, endosialin (CD248) and thrombomodulin (TM, CD141), a well characterized anticoagulant. All of them contain a C-type lectin domain, a series of epidermal growth factor like domains (EGF), a highly glycosylated mucin-like domain, a unique transmembrane domain and a short cytoplasmic tail. Due to their strong homology and their close proximity on chromosome 20, CD93 has been suggested to have arisen from thrombomodulin gene through a duplication event. Sequence analysis revealed that CD93 is identical to a protein on human phagocytes termed C1q receptor (C1qRp). The mouse homologue of human C1qRp is known as AA4. The AA4 and TM are derived from a common ancestor since both genes are co-localized to the same region of the chromosome 2 and also because they share similar domain composition and organization.

Endosialin (tumor endothelial marker 1) is expressed preferentially by tumor endothelial cells but not by normal endothelium. Its protein domain architecture is homologous to that of CD93 and TM (CD141), suggesting a similar function in mediating cell-cell interactions. Endosialin colocalization with thrombomodulin suggests that these proteins may have complementary functions in tumor progression (Brady et al. 2004). The complement C1q receptor (C1qRp) and TM are involved in cell-cell interactions and innate immune host defense. However, not much is known on the lectin 14A, a member of CD93 family of C-type lectins.

41.2 CD93 or C1q Receptor (C1qRp): A Receptor for Complement C1q

41.2.1 CD93 Is Identical to C1qRp

The first component of complement, C1, is a multi-molecular complex comprising C1q and the Ca^{2+}-dependent tetramer $C1r_2$-$C1s_2$. The C1q is an essential component of innate complement (C) system and is crucial to ward off infection and clear toxic cell debris (e.g. amyloid fibrils, apoptotic cells) in natural way. Another important function of C1q is its ability to bind to a wide range of cell types resulting in the induction of cell-specific biological responses. These cells include polymorphonuclear leukocytes, monocytes, lymphocytes, dendritic cells, endothelial cells and platelets. Interaction of C1q with endothelial cells and platelets, for example, leads to cellular activation followed by release of biological mediators and/or expression of adhesion molecules, all of which contribute, directly or indirectly to the inflammatory process. These specific responses are mediated by the interaction of C1q with C1q binding proteins or receptors on the cell surfaces. Several candidate C1q receptors [C1q receptor for phagocytosis enhancement (C1qRp/cC1qR/CD93, a 120 kDa, O-sialoglycoprotein), complement receptor (CR) 1, calreticulin (CRT: Chap. 2), and binding protein for globular head of C1q (gC1qR/p33, a 33 kDa homotrimeric protein)] have been studied and reviewed for their structure-function relationships (McGreal and Gasque 2002). Cell-surface molecule, C1qRp, has emerged as a defence collagen receptor for C1q, as well as MBL and surfactant protein A (Chaps. 23–25). C1qRp (known as AA4 antigen in rodents) is the antigen recognized by a pro-adhesive monoclonal antibody (mNI-11) and antibodies against CD93. Although the specific role of each of these receptors in a given C1q-mediated cellular response is yet to be known, all of them may participate in the inflammatory processes associated with vascular or atherosclerotic lesions, autoimmune diseases, or infections. Study of structure of cC1qR/CD93 and gC1qR/p33, both of which have been characterized on the basis of their ability to bind C1q, and their role in infection and inflammation is the subject of recent and future investigations (Ghebrehiwet and Peerschke 2004). Within the hematopoietic system CD93 is selectively expressed on cells of the myeloid lineage. CD93 is identical C1qRp, which was shown to mediate enhancement of phagocytosis in monocytes and suggested to be a receptor of C1q and two other structurally related molecules. Cells expressing CD93 have enhanced capacity to bind C1q. Furthermore, immature DCs express CD93/C1qRp, and mature DCs, known to have reduced capacity for antigen uptake and to have lost the ability to phagocytose, show weak-to-negative CD93/C1qRp expression (Steinberger et al. 2002). Narayanan et al. (1997) investigated the distribution of receptors for C1q-collagen domain (cC1qR) and globular domain (gC1qR) in adult human lung fibroblasts. The cC1qR was expressed predominantly by a population of human lung fibroblasts, while the 38 kD gC1qR was produced by all cells. Therefore, two lung fibroblast subtypes may be distinguished based on production of cC1qR and another protein which binds to C1q-globular domain.

41.2.2 Characterization of CD93/C1qR$_P$

The CD93/C1qR$_P$ is a type I membrane protein of 631 amino acid containing a region homologous to C-type lectin carbohydrate recognition domain. The C1qR$_P$ is a 100 kDa M$_r$ protein (126,000 M$_r$ under reducing conditions). Monoclonal antibodies that immunoprecipitate 126 kDa M$_r$ polypeptide inhibit the enhancement of phagocytosis triggered not only by C1q but also by MBL and pulmonary SP-A providing evidence that this polypeptide is a functional receptor that mediates the enhancement of phagocytosis. Specifically, the *Cd93* gene is located at 84 cM on murine chromosome 2 (Kim et al. 2000) and encodes a type I O-glycosylated transmembrane protein whose domain structure includes an amino-terminal C-type lectin domain, a tandem array of five EGF-like repeats, a single hydrophobic transmembrane region, and a short cytoplasmic domain that contains a PDZ binding domain and a moesin interaction site (Bohlson et al. 2005; Kim et al. 2000; Zhang et al. 2005). This domain structure bears a unique resemblance to the selectin family of adhesion molecules (Dean et al. 2001; Kim et al. 2000; Rosen 2004). Both post-translational glycosylation and the nature of the amino acid sequence of the protein explain to the difference between its predicted molecular weight and its migration on SDS-PAGE. The mature C1qRp contains a relatively high degree of O-linked glycoslyation, and engaged in cell surface ligation of SP-A, as well as C1q and MBL, and enhances phagocytosis. The O-glycosylation is important in the stable cell surface expression of C1qR$_P$/CD93 (Park and Tenner 2003). While the cDNA for C1qRp encodes a 631 amino acid membrane protein, CHO cells transfected with the cDNA of the C1qR$_P$ coding region express a surface glycoprotein with the identical 126,000 M$_r$ as the native C1qR$_P$ (Nepomuceno et al. 1999). CD93, similar to C1qR$_P$, is expressed by endothelial cells, cells of myeloid lineage, platelets, and early hematopoietic stem cells and is an important lineage-specific marker of early B-cell developmental stages (Cancro 2004; Danet et al. 2002; Fonseca et al. 2001; Nepomuceno and Tenner 1998). Normally, CD93 is expressed at high levels on pro-, pre- and immature bone marrow (BM) B-cell progenitors as well as transitional (TR) B cells in the periphery (Cancro 2004). Additionally, CD93 is subject to metalloprotease-mediated ecto-domain cleavage or shedding, which is characteristic of several inflammatory mediators and adhesion molecules including TNF-α, TGF-α, TGF-β, EGF, CD44, and L-selectin (Bohlson et al. 2005a). In phagocytic cells, it has been characterized as contributing to the enhancement of FcR- and CR1-induced phagocytosis triggered by innate immune system defense collagens such as C1q and MBL (Park and Tenner 2003). There seems a novel mechanism of regulated expression of CD93 that may have implications in stem cell development and inflammation. CD93 is susceptible to ectodomain shedding. Bohlson et al. (2005a) identified multiple stimuli that trigger shedding, and identified both a soluble form of CD93 in human plasma and intracellular domain containing cleavage products within cells that may contribute to the physiologic role of CD93 (Bohlson et al. 2005a). The C1qR from U937 cells and human tonsil lymphocytes interacts with C1q, MBP, conglutinin and SP-A (Malhotra et al. 1990). The interaction of SP-A with U937 cells triggers the expression of an intracellular pool of C1qR (Malhotra et al. 1992).

41.2.3 Two Types of Cell Surface C1q-Binding Proteins (C1qR)

Synonyms of C1qR are: C1q/MBL/SPA receptor, C1qR(p), C1qr1, C1qrp, C1qRp, cell surface antigen AA4 and CD93. Selective expression of C1qR$_P$ in specific cell types supports the hypothesis that there is more than one C1q receptor mediating the diverse responses triggered by C1q (Nepomuceno and Tenner 1998; Guan et al. 1994; Tenner 1998). A 60-kDa calreticulin-homologue which binds to the collagen-like "stalk" of C1q (called cC1qR) and a 33-kDa protein with affinity for the globular "heads" of the molecule (called gC1qR), have been described. The two molecules are also secreted by Raji cells and peripheral blood lymphocytes and can be isolated in soluble form from serum-free culture supernatant. The two purified soluble proteins had immunochemical and physical characteristics similar to their membrane counterparts in the sense that both bound to intact C1q and to their respective C1q ligands, cC1q and gC1q. In addition, N-terminal amino acid sequence analyses of the soluble cC1qR and gC1qR were found to be identical to the reported sequences of the respective membrane-isolated proteins. Ligand blot analyses using biotinylated membrane or soluble cC1qR and gC1qR showed that both bind to the denatured and nondenatured A-chain and moderately to the C-chain of C1q. Moreover, like their membrane counterparts, the soluble proteins were found to inhibit serum C1q hemolytic activity. Although cC1qR was released when both peripheral blood lymphocytes and Raji cells were incubated in phosphate-buffered saline for 1 h under tissue culture conditions, gC1qR was releasable only from Raji cells, suggesting that perhaps activation or transformation leading to immortalization is required for gC1qR release (Peterson et al. 1997).

gC1q-binding Proteins (gC1qR): Differences from cC1qR

Human neutrophils have multiple C1q-binding proteins. Ligand-binding studies with the globular domain of C1q revealed two gC1q-binding proteins (gC1qR): a 33 kDa protein (pI 4.5) mainly in the neutrophil plasma membrane and an 80–90 kDa protein (pI 4.1–4.2) located mainly in the granules. Direct binding studies showed that C1q bound to this higher molecular weight protein under physiological conditions. In contrast, anti-cC1qR antibody, which recognizes a protein binding to collagenous tails of C1q, detected only a 68,000 M_r protein in the plasma membrane. Both the 33,000 and 68,000 M_r receptors appear early on the surface of differentiating HL-60 cells. Phorbol myristate acetate treatment of neutrophils down-regulated both the receptors from cell surface, and significant amounts of soluble gC1qR were in cell media supernatants, suggesting receptor shedding or secretion. gC1qR, unlike cC1qR, did not bind to other C1q-like ligands, namely mannose binding protein, surfactant protein-A, surfactant protein-D, or conglutinin under normal ionic conditions, suggesting a greater specificity for C1q than the "collectin" type receptor (cC1qR). Rather, gC1qR only bound purified C1q, and the binding was enhanced under low ionic conditions and in the absence of calcium. C1qR/CD93 is also proteolytically cleaved from the surface of activated human monocytes and neutrophils in response to inflammatory signals in vitro and a soluble form of CD93 (sCD93) exists in human plasma. Inflammation triggers release of sCD93 in vivo and identifies the inflammatory macrophage as a source of sCD93 (Greenlee et al. 2008, 2009).

Ghebrehiwet et al. (1997) showed that biotinylated cC1qR binds to recombinant as well as native gC1qR and the binding sites for cC1qR are located within N-terminal residues 76 through 93 of the mature form of gC1qR and within residues 204 through 218. The evidence suggests that cC1qR is able to form a complex with gC1qR and may associate with gC1qR on the cell surface (Ghebrehiwet et al. 1997).

C1q receptor (C1qR/collectin receptor/cC1qR) with an almost complete amino acid sequence identity with calreticulin (CRT) is located on the surface of many cell types. C1qR also interacts with the collectins SP-A, MBL, CL43 and conglutinin via a cluster of charged residues on the collagen tails of the ligands. The C1q and collection binding site on C1qR/CRT is localised to the residues 160–283 (Stuart et al. 1997).

Collectins and C1q Enhance Mononuclear Phagocytosis Through a Common Receptor: Enhanced monocyte FcR- and CR1-mediated phagocytosis by MBP, C1q, and SP-A suggests acommon structural feature of the collagen-like domains that may provide a basis for this immunologically important function. MBP also enhances FcR-mediated phagocytosis by both monocytes and macrophages, and stimulated CR1-mediated phagocytosis in human culture-derived macrophages and in phorbol ester-activated monocytes. The mAb that recognizes a 126 kDa cell surface protein and inhibits C1q-enhanced phagocytosis, also inhibited the MBP-mediated enhancement of phagocytosis. This common feature among defense collagen receptors suggests that the enhancement of phagocytosis by MBP and C1q share at least one critical functional component, the C1qR_P (Nepomuceno et al. 1997; Tenner et al. 1995). mAbs that recognize a cell surface C1qR_P inhibit C1q-, MBL- and SP-A-mediated enhancement of phagocytosis. Similar inhibition of enhancement of phagocytosis by these mAbs suggests that C1qR_P is a common component of a receptor for these macromolecules. The C1qR_P is cross-linked directly by monoclonal anti-C1qR_P or engaged as a result of cell surface ligation of SP-A, C1q and MBL resulting in enhanced phagocytosis (Nepomuceno et al. 1999).

41.2.4 Tissue Expression and Regulation of CD93/C1qRp

CD93/C1qR_P was originally identified in mice as an early B cell marker through the use of AA4.1 monoclonal antibody. Then this molecule was shown to be expressed on an early population of hematopoietic stem cells, which give rise to the entire spectrum of mature cells in the blood. Now CD93 is known to be expressed by a wide variety of cells such as platelets, monocytes, microglia and endothelial cells. In the immune system CD93 is also expressed on neutrophils, activated macrophages, B cell precursors until the T2 stage in the spleen, a subset of DCs and of natural killer cells.

Though C1qRp is identical to CD93 protein, a study failed to demonstrate a direct interaction between CD93 and C1q under physiological conditions. Radioiodinated C1qR and CRT bind to C1q with identical characteristics (McGreal et al. 2002).

Studies show that human monocyte-derived DCs express gC1qR and cC1qR; their expression on the cell surface is maturation dependent, and immature DCs secrete C1q (Vegh et al. 2003). Immature DCs express C1qR, gC1qR and cC1qR/CR and, accordingly, display a vigorous migratory response to soluble C1q with maximal cell movement. In contrast, mature DCs neither express C1qR nor do move to a gradient of soluble C1q. C1q functions as a chemotactic factor for immature DC, and migration is mediated through ligation of both gC1qR and cC1qR/CR (Vegh et al. 2006).

The TGF-β and TNF-α upregulate the mRNA levels of cC1qR and collagen but do not affect gC1qR mRNA levels significantly. It was also indicated that different subsets of human lung fibroblasts respond differently to inflammatory mediators (Lurton et al. 1999). Protein kinases regulate the expression of CD93 on the cell surface. The CD93 expression on human monocyte-like cell line (U937), a human NK-like cell line (KHYG-1), and a human umbilical vein endothelial cell line (HUV-EC-C) involves the PKC isoenzymes (Ikewaki et al. 2006). PKC activator phorbol myristate acetate (PMA) effectively up-regulated CD93 expression on cultured cell lines and its regulation was mainly controlled by a PKCδ-isoenzyme (Ikewaki et al. 2007).

41.2.5 Functions

41.2.5.1 Adhesion, Migration and Phagocytosis

The CD93 was initially thought to be a receptor for C1q, but now is thought to instead be involved in intercellular adhesion and in the clearance of apoptotic cells. The intracellular cytoplasmic tail of this protein contains two highly conserved domains which may be involved in CD93 function. Indeed, the highly charged juxtamembrane (JX) domain has been found to interact with moesin, a protein known to play a role in linking transmembrane proteins to the cytoskeleton and in the remodelling of the cytoskeleton (Fig. 41.1). The interaction of moesin with CD93 cytoplasmic domain is modulated by binding of other intracellular molecules to the C11 region and implies that a PIP_2 signaling pathway is involved in CD93 function (Norsworthy et al. 2004; Zhang and Bohlson 2005). This process appears crucial for adhesion, migration and phagocytosis, three functions in which CD93 may be involved. CD93 is required for efficient engulfment of apoptotic cells via an unknown mechanism. Neonatal rat microglia are known to express C1qRp. Interaction of these cells with substrate-bound C1q was shown to enhance both FcR-and CR1-mediated phagocytosis. Results suggest that C1q in areas of active degeneration may promote the phagocytic capacity of microglia via interaction with microglial C1qRp (Webster et al. 2000). Reports also indicate that besides such-well-known complement regulatory molecules as CD55 (DAF), CD46 (MCP), CD35 (CR1) and CD59(HRF), C1qR too is able to regulate complement activity (van den Berg et al. 1995).

41.2.5.2 CD93 in Maintenance of Plasma Cells in Bone Marrow

In the context of late B cell differentiation, CD93 has been shown to be important for the maintenance of high antibody titres after immunization and in the survival of long-lived plasma cells in the bone marrow. Indeed, CD93 deficient mice failed to maintain high antibody level upon immunization and present a lower amount of antigen specific plasma cells in the bone marrow. CD93, expressed during early B-cell development, appeared to be restimulated during plasma-cell differentiation. High CD93/CD138 expression was restricted to antibody-secreting cells both in T-dependent and T-independent responses as naive, memory, and germinal-center B cells remained CD93-negative. CD93 was expressed on (pre)plasmablasts/plasma cells, including long-lived plasma cells. Strikingly, while humoral immune responses initially proceeded normally, CD93-deficient mice were unable to maintain antibody secretion and bone-marrow plasma-cell numbers, demonstrating that CD93 is important for the maintenance of plasma cells in bone marrow niches (Chevrier et al. 2009).

41.2.5.3 Role in Fertilization

Complement components might play a role in fertilization. C1q has the ability to promote sperm agglutination in a capacitation-dependent manner as well as an effect on sperm-oolemma binding and fusion. The gC1qR is present on the surface of capacitated sperm. Confocal and immunofluorescence microscopy revealed an increase in receptor expression over the rostral portion of the sperm head after capacitation. It appears that gC1qR may play a role in human fertilization (Grace et al. 2002).

41.2.6 CD93 in Pathogenesis of SLE

The association between C1q and autoimmune diseases such as rheumatoid arthritis and systemic lupus erythematosus (SLE) is well established. The SLE is characterized by the presence of multiple autoantibodies and high levels of circulating immune complexes. In a cross-sectional study it was found that higher titres of antibodies against cC1qR/CRT are present in sera of SLE patients compared with normal donors. Deficiency in C1q is considered to be a

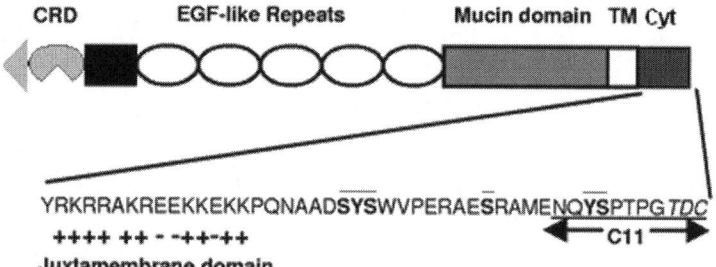

Fig. 41.1 Schematic representation of C1QRp/CD93. CD93 has been shown to influence the phagocytic activity in response to defense collagens such as C1q. CD93 has a 47 amino acid cytoplasmic tail and modulates phagocytosis. The cytoplasmic tail contains a highly charged juxtamembrane (Jx) region, and charge (+ or −) is indicated below the single-letter amino acid residue. Potential sites of serine or tyrosine phosphorylation are marked by horizontal line. Cytoplasmic (*underlined*) 11 amino acids region has been implicated in signal transduction; this region contains a class I PDZ-binding domain (*italics*). Abbreviations: CRD, Carbohydrate recognition domain; EGF, epidermal growth factor; TM, transmembrane domain; Cyt, cytosolic region (Bohlson et al. 2005b)

strong susceptibility factor and is corroborated by the fact that ~92% of the known cases of hereditary deficiency in C1q develop rheumatic disease. C1q-deficient mice have shown the presence of glomerulonephritis with immune deposits and a large number of apoptotic bodies in the diseased glomeruli suggesting a defect in the clearance of apoptotic cell by macrophages and DCs. Although these data are consistent with the hypothesis that C1q deficiency may induce a generalized failure to clear immune complexes and apoptotic cells, this concept alone cannot wholly explain why individuals with C1q deficiency are prone to develop SLE. Therefore, C1q in conjunction with other surface molecules must play a more fundamental role in immunoregulation, especially those processes that regulate T cell function and tolerance. In support of this hypothesis is the finding that C1q causes inhibition of mitogen-induced T cell-proliferative response by interaction with C1q receptors. Furthermore, macrophages and possibly DCs not only synthesize but also display C1q as a type II cell surface molecule, especially at sites of inflammation. Although it is not yet known what role the surface-expressed C1q plays, it is tempting to assume that it plays a role in the priming of naive T cells by DCs (Ghebrehiwet and Peerschke 2004; van den Berg et al. 1998). However, in a contradiction, C1qRp expression and regulation was not altered on peripheral blood monocytes of SLE patients. Possible relations with disease activity and medication need further investigations (Moosig et al. 2006)

41.3 Murine Homologue of Human C1qRp (AA4)

A murine genomic clone is 73% identical in sequence with the coding region for human C1qR$_P$ cDNA. Chromosomal localization of the human and murine genes demonstrated that these genes are syntenic. Murine cell lines of diverse myeloid origins are shown to respond to interaction of C1q with the enhancement of phagocytosis similar to that seen in human peripheral blood monocytes. mC1qR$_P$ is expressed in murine myeloid cell lines, but not in a mouse epithelial cell line, similar to the cell type expression of the human gene product (Kim et al. 2000). A 110–130-kDa membrane glycoprotein, expressed by rat NK cells, is a structural orthologue of the phagocytosis-stimulating receptor for complement factor C1q and mannose-binding lectin on human macrophages, C1qRp (receptor for C1q, regulating phagocytosis). Rat C1qRp is a monomeric type I integral membrane protein consisting of 643 amino acids with an N-terminal lectin-like domain, five epidermal growth factor-like domains, a transmembrane domain and a 45-residue cytoplasmic domain. It is encoded by a single gene on rat chromosome 3q41–q42 and is 67% and 87.5% identical at the amino acid level to human and mouse C1qRp, respectively. Rat C1qRp is expressed by resting and by activated NK cells, on subpopulations of NKR-P1$^+$ T cells (NK/T cells), dendritic cells, macrophages, neutrophils and granulocytes, but not by B cells or NKR-P1$^-$ T cells. Expression of this innate immune receptor is therefore not restricted to hematopoietic cells of the myeloid lineage, but is also expressed on subsets of cells of lymphoid origin (Løvik et al. 2000) The C1qRp, present on macrophages and neutrophils, is presumed to stimulate phagocytosis in these cells. In rat NK cells, an intracellular Ca^{2+}-response is induced upon stimulation of C1qRp with mAbs, LOV3 and LOV8. The response induced by the antibodies was owing to the Ca^{2+} mobilization from intracellular calcium stores (Løvik et al. 2001). Murine mast cells express specific receptors for C1q. Both cC1qR and gC1qR are present on these cells. C1qRs may play a significant role in mast cell function and regulation by providing an important signal through which mast cells can be recruited to inflammatory sites of increased C1q concentration (Ghebrehiwet et al. 1995).

The major site of AA4 expression in all tissue tested is on endothelial cells and that despite the apparent involvement of AA4 in the phagocytic response, it is not expressed by tissue macrophages. Although AA4 has been described on all hematopoietic progenitors, only circulating immature B cells, monocytes and NK cells but not T cells and neutrophils expressed AA4 (Dean et al. 2001). The CD93-mRNA is highly induced after transient focal cerebral ischemia. CD93 protein is upregulated in endothelial cells, but also in selected macrophages and microglia. Harhausen et al. (2010) speculated that CD93-neuroprotection is mediated via suppression of the neuroinflammatory response through downregulation of CCL21 (Harhausen et al. 2010).

The murine fetal stem cell marker AA4 is a known homologue of the human phagocytic C1q receptor involved in host defense. The rat AA4 displays similar domain composition and organization to thrombomodulin. The rat AA4 was encoded by a single transcript of 7 kb expressed constitutively in all tissues and was expressed predominantly by pneumocytes and vascular endothelial cells. AA4 was identified as a glycosylated protein of 100 kDa expressed by endothelial cells > platelets > NK cells and monocytes (ED1+ cells). The staining was associated to the cell surface and intracytoplasmic vesicles. Interestingly, alveolar epithelial cells (lung) but not other epithelia (e.g. uterus) were strongly stained for AA4. Although AA4 has been described on all hematopoietic progenitors, only circulating immature B cells, monocytes and NK cells but not T cells and neutrophils expressed AA4. These results support the idea that AA4/C1qRp is involved in some cell-cell interactions (Dean et al. 2000, 2001).

The hematopoietic system of mice is established during the early to midgestational stage of development. Although AA4.1$^+$ and AA4.1$^-$ cells had equivalent potency to generate myeloid cell lineages, the lymphoid potential in ES-cell-derived cells was largely restricted to the cells expressing AA4.1. The same cell type was present abundantly in the early yolk sac and in fewer numbers (~5% of that in the yolk sac) in the caudal half of the developing embryos. These results suggest that AA4.1 is a cell surface marker that can identify the earliest lymphohematopoietic progenitors in mouse development (Yamane et al. 2009).

Zekavat et al. (2010) identified a point mutation in the NOD *Cd93* gene, which maps to the *Idd13* locus, a region encoding a high degree of penetrance for diabetes susceptibility in non-obese diabetic (NOD) mice (Kim et al. 2000; Serreze et al. 1998). In addition, they also identified this *Cd93* polymorphism in NZB/W F1 mice, to which the lupus susceptibility loci, *Wbw1* and *Nkt2*, are tightly linked (Rahman et al. 2002; Esteban et al. 2003). This point mutation is associated with aberrant expression of CD93 on NOD and NZB/W F1 B cells in the pro-/pre-, immature, and TR subsets as compared with non-autoimmune B6 mice. The *Idd13* and *Wbw1* and *Nkt2* loci are known to play a role in the regulation of the invariant natural killer T (iNKT) cell compartment (Chen et al. 2007; Esteban et al. 2003; Jordan et al. 2004; Rahman et al. 2002).

This allele carries a coding polymorphism in the first EGF-like domain of CD93, which results in an amino acid substitution from Asn→His at position 264. This polymorphism does not appear to influence protein translation or ecto-domain cleavage, as CD93 is detectable in bone-marrow-derived macrophage and B-cell precursor lysates and in soluble form in the serum. The NOD CD93 isoform causes a phenotypic aberrancy in the early B-cell developmental stages (i.e., pro-, pre-, immature, and transitional), likely related to a conformational variation. Consistent with genetic linkage, B6 CD93$^{-/-}$ and B6.NOD^{Idd13} mice were found susceptible to a profound CD4$^+$ NKT cell deficient state. Data suggest that *Cd93* may be an autoimmune susceptibility gene residing within the *Idd13* locus, which plays a role in regulating absolute numbers of CD4$^+$ NKT cells (Zekavat et al. 2010).

41.4 The gC1qR (p33, p32, C1qBP, TAP)

41.4.1 A Multi-Multifunctional Protein - Binding with C1q

The gC1qR (also named as gC1qBP) is a 33 kDa single chain, highly acidic glycoprotein that binds to the globular 'heads' of C1q. Comparison of the cDNA-derived amino acid sequences of gC1qBP reveals that either of the rodents, rat and mouse, sequences is 89.9% identical to human sequence. Recombinant rat gC1qBP binds avidly to human C1q. gC1qBP mRNA is abundantly expressed in every rat and mouse tissue analysed. Rat mesangial cells synthesise gC1qBP, but do not express gC1qBP on the cell surface. In rat serum, gC1qBP is present at low levels (Lynch et al. 1997).

Tha gC1qR was first isolated from membrane preparation of Raji cells and now appears to be ubiquitously distributed. Although, gC1qR was originally identified as a protein which binds to the globular "heads" of C1q, evidence suggests that the molecule is in fact a multi-ligand binding, multifunctional protein with affinity for diverse ligands which at best are functionally related. These molecules include: thrombin, vitronectin, and high molecular weight kininogen. The gC1qR molecule, which is identical to the transcription factors SF2 and the Tat-associated protein, or TAP, is the product of a single gene localized on chromosome 17p13.3 in human, and chromosome 11 in mouse, and is encoded by an approximately 1.5–1.6 kb mRNA. The full length cDNA encodes a primary translation protein of 282 residues and the 'mature' or membrane form of the protein isolated from Raji cells corresponds to residues 74–282 and is presumed to be generated by a site-specific cleavage and removal of the highly basic, 73-residues long, N-terminal segment during post-translational processing. The translated amino acid sequence does not predict for the presence of a conventional sequence motif compatible with a

41.4 The gC1qR (p33, p32, C1qBP, TAP)

transmembrane segment and does not have a consensus site for a GPI anchor. However, there is strong evidence which indicates that gC1qR is expressed both inside the cell and on the membrane. In addition, the membrane expression of gC1qR can be upregulated with inflammatory cytokines such as INF-gamma, TNF-α, or LPS (Ghebrehiwet and Peerschke 1998).

Although expressed on the surface of cells, an intriguing feature of the membrane-associated form of gC1qR is that its translated amino acid sequence does not predict the presence of either a sequence motif compatible with a transmembrane segment or a consensus site for a glycosylphosphatidylinositol anchor. Moreover, the N-terminal sequence of the pre-pro-protein gC1qR contains a motif that targets the molecule to the mitochondria and as such was deemed unlikely to be expressed on the surface. However, several lines of experimental evidence clearly show that gC1qR is present in all compartments of the cell, including the extracellular cell surface. First, surface labeling of B lymphocytes with the membrane-impermeable reagent sulfosuccinimidyl 6-(biotinamido) hexanoate shows specific biotin incorporation into the surface-expressed but not the intracellular form of gC1qR. Second, FACS and confocal laser scanning microscopic analyses demonstrated specific staining of Raji cells surface. Third, endothelial gC1qR, which is associated with the urokinase plasminogen activator receptor, and cytokeratin 1 bind 125I-high molecular weight kininogen in a specific manner. Fourth, native gC1qR purified from Raji cell membranes but not intracellular gC1qR is glycosylated, as evidenced by a positive periodic acid Schiff stain as well as sensitivity to digestion with endoglycosidase H and F. Finally, cross-linking experiments using C1q as a ligand indicate that both cC1qR and gC1qR are co immunoprecipitated with anti-C1q. Taken together, the evidence accumulated to date supports the concept that in addition to its intracellular localization, gC1qR is expressed on the cell surface and can serve as a binding site for plasma and microbial proteins, but also challenges the existing paradigm that mitochondrial proteins never leave their designated compartment. It is therefore proposed that gC1qR belongs to a growing list of a class of proteins initially targeted to the mitochondria but then exported to different compartments of the cell through specific mechanisms which have yet to be identified. The designation 'multifunctional and multicompartmental cellular proteins' is proposed for this class of proteins (Ghebrehiwet et al. 2001).

Platelets are involved in the development of many types of vascular lesions. In addition to their role in primary hemostasis, they participate in inflammatory processes that may contribute to the development of thrombosis, atherosclerosis and vasculitis. C1q has been shown to modulate platelet interactions with collagen and immune complexes, and has been identified at the sites of vascular injury and inflammation, as well as in atherosclerotic lesions. Platelets express a variety of C1q binding sites, including gC1qR/p33 (gC1qR) (Peerschke and Ghebrehiwet 2001).

The gC1qR (or P32) protein is a binding protein for nuclear pre-mRNA splicing factor SF2/ASF and numerous other nuclear and cell surface proteins, yet is targeted to the mitochondrial matrix compartment where these proteins are not present (Soltys et al. 2000). The endothelial protein p33/gC1qR is thought to mediate the assembly of components of the kinin-forming and complement-activating pathways on the surface of cardiovascular cells. Fractionation studies demonstrated that the vesicular but not the membrane fraction of EA.hy926 cells is rich in p33. It seems that externalization of p33 must precede its complex formation with target proteins on the endothelial cell surface (Dedio and Muller-Esterl 1996).

The binding of the ligand C1q by recombinant gC1qR was indistinguishable from binding shown by gC1qR isolated from Raji cells. gC1qR as a novel vitronectin-binding protein that may participate in the clearance of vitronectin-containing complexes or opsonized particles or cooperate with vitronectin in the inhibition of complement-mediated cytolysis (Ghebrehiwet et al. 1994; Lim et al. 1996).

41.4.2 Human gC1qR/p32 (C1qBP) Gene

The 7.8-kbp human gC1qR/p32 (C1qBP) gene was cloned and found to consist of 6 exons and 5 introns. Analysis of a 1.3-kb DNA fragment at the 5′-flanking region of this gene revealed the presence of multiple TATA, CCAAT, and Sp1 binding sites. Subsequent 5′ and 3′ deletion of this fragment confined promoter elements to within 400 bp upstream of the translational start site. Because the removal of the 8-bp consensus TATATATA at -399 to -406 and CCAAT at -410 to -414 did not significantly affect the transcription efficiency of the promoter, GC-rich sequences between this TATA box and the translation start site may be very important for the promoter activity of the C1qBP gene. One of seven GC-rich sequences in this region binds specifically to PANC-1 nuclear extracts and the transcription factor Sp1. Primer extension analysis mapped three major transcription start regions. The farthest transcription start site is 49 bp upstream of the ATG translation initiation codon and is in close proximity of the specific SP1 binding site (Tye et al. 2001).

41.4.3 Ligand Interactions

41.4.3.1 A Receptor for C1q

Although originally isolated as a receptor for C1q by virtue of its specificity for the globular heads of that molecule, a large body of evidence has now been accumulated which shows that in addition to C1q, gC1qR can serve as a receptor for diverse ligands including proteins of the intrinsic coagulation/bradykinin forming cascade, as well as antigens of cellular, bacterial, and viral origin. Furthermore, since gC1qR has been shown to regulate the functions of protein kinase C (PKC), it is postulated that gC1qR-induced signaling cascade may involve activation of PKC. Collective reports suggest that gC1qR plays an important role in blood coagulation, inflammation, and infection. It is still unclear as to how the molecule is anchored on the membrane since its sequence is devoid of a classical transmembrane domain or a glycosylphosphatidylinositol (GPI) anchor. Furthermore, while recombinant gC1qR can bind to cell surfaces suggesting that it may bind directly to the phospholipid bilayer, experiments showed that, at least in vitro, gC1qR does not bind to unilamellar vesicle preparations of either phosphatidylcholine (PC) or phosphatidylserine:phosphatidylcholine. The three-dimensional structure of gC1qR.

41.4.3.2 Interaction of High Molecular Weight Kininogen

Kininogens, the precursor proteins of the vasoactive kinins, bind specifically, reversibly, and saturably to platelets, neutrophils, and endothelial cells. Two domains of the kininogens expose major cell binding sites: domain D3 that is shared by H- and L-kininogen and domain D5H that is exclusively present in H-kininogen. The kininogen cell binding sites were mapped to 27 residues of D3 ("LDC27") and 20 residues of D5H ("HKH20'''", respectively) (Herwald et al. 1995; Hasan et al. 1995). An endothelial binding protein of 33 kDa similar to gC1qR has been indicated to mediates the assembly of critical components of the kinin-generating pathway on the surface of endothelial cells, thereby linking the early events of kinin formation and complement activation (Herwald et al. 1996).

To further probe cellular trafficking routes of gC1qR, Dedio et al. (1999) over-expressed human gC1qR in a mammalian cell and monitored cell surface exposure of recombinant gC1qR by virtue of its capacity to bind labeled H-kininogen. Transient transfection of COS1 cells with the full-length cDNA of human gC1qR resulted in a high level of recombinant protein that matched the pool of endogenous gC1qR present in these cells. Overexpression of gC1qR did not significantly increase the number of H-kininogen binding sites exposed by the transfected cells thus denying the possibility that alternative routing of gC1qR to the surface of COS1 cells occurs at significant levels. Hence gC1qR has the capacity to tightly bind H-kininogen, but because gC1qR is routed to mitochondria it cannot fulfill the postulated functions as a cell docking site for kininogens and complement factors (Dedio et al. 1999).

The gC1qR has been shown to bind a number of plasma proteins involved in the coagulation and kinin systems. Biotinylated gC1qR was found to bind to fibrinogen in a manner which was specific and inhibited by excess soluble fibrinogen or polyclonal antibodies directed against either gC1qR or fibrinogen. Observations may suggest a potential role for gC1qR in modulating fibrin formation particularly at local sites of immune injury or inflammation (Lu et al. 1999).

High molecular weight kininogen (HK) and factor XII bind to human umbilical vein endothelial cells (HUVEC) in a zinc-dependent and saturable manner indicating that HUVEC express specific binding site(s) for those proteins. The binding of HK and factor XII to HUVECs occurs via a 33-kDa cell surface glycoprotein that appears to be identical to gC1qR but with a site on gC1qR distinct from that which binds C1q (Joseph et al. 1996; 2001). However, unlike C1q, whose interaction with gC1qR does not require divalent ions, the binding of HK to gC1qR is absolutely dependent on the presence of zinc. Zinc can induce the exposure of hydrophobic sites in the C-terminal domain of gC1qR involved in binding to HK/FXII (Kumar et al. 2002).

Cell surface proteins reported to participate in the binding and activation of the plasma kinin-forming cascade includes gC1qR, cytokeratin 1 and u-PAR. Each of these proteins binds high molecular weight kininogen (HK) as well as Factor XII. The studies suggest that formation of HK (and Factor XII) binding sites along endothelial cell membranes consists of bimolecular complexes of gC1qR-cytokeratin 1 and u-PAR-cytokeratin 1, with gC1qR binding being favored (Joseph et al. 2004).

Human p32 (also known as SF2-associated p32, p32/TAP, and gC1qR) localizes predominantly in the mitochondrial matrix. It is thought to be involved in mitochondrial oxidative phosphorylation and in nucleus-mitochondrion interactions. The three-dimensional structure of gC1qR identified unique structural features that may serve not only to anchor the protein but also to explain its affinity for such a diversity of plasma as well as microbial and viral ligands (Ghebrehiwet et al. 2002; Jiang et al. 1999). The crystal structure reveals that p32 adopts a novel fold with seven consecutive antiparallel β-strands flanked by one N-terminal and two C-terminal α-helices. Three monomers form a doughnut-shaped quaternary structure with an unusually asymmetric charge distribution on the surface. The implications of the structure on previously proposed functions of p32 are discussed and new specific functional properties are suggested (Jiang et al. 1999).

41.4.3.3 Adrenergic Receptor

gC1qR was found to bind with the carboxyl-terminal cytoplasmic domain of the α1B-adrenergic receptor (amino acids 344–516) in a yeast two-hybrid screen of a cDNA library prepared from the rat liver. The gC1qR interaction with an arginine-rich motif in the C-tail of hamster α1B-adrenergic receptors seems to control their expression and subcellular localization. Studies suggest that gC1qR interacts specifically with α1B- and α1D-, but not α1A-adrenergic, and this interaction depends on the presence of an intact C-tail. This suggests a role for gC1qR in regulating the cellular localization and expression of adrenergic receptors (Pupo and Minneman 2003; Xu et al. 1999).

41.4.3.4 Interaction with Bacteria and Viruses

On the surface of activated platelets, gC1qR has been shown to serve as a binding site for *Staphylococcus* aureus and this binding is mediated by protein A. Since the binding of *S. aureus* to platelets is postulated to play a major role in the pathogenesis of endocarditis, gC1qR may provide a suitable surface for the initial adhesion of the bacterium. Recent data also demonstrate that the exosporium of Bacillus cereus, a member of a genus of aerobic, Gram-positive, spore-forming rod-like bacilli, which includes the deadly Bacillus anthracis, contains a binding site for gC1qR. Therefore, by virtue of its ability to recognize plasma proteins such as C1q and HK, as well as bacterial and viral antigens, cell-surface gC1qR not only is able to generate proinflammatory byproducts from the complement and kinin/kallikrein systems, but also can be an efficient vehicle and platform for a plethora of pathogenic microorganisms (Peerschke and Ghebrehiwet 2007).

InlB is a *Listeria monocytogenes* protein that promotes entry of the bacterium into mammalian cells by stimulating tyrosine phosphorylation of the adaptor proteins Gab1, Cbl and Shc, and activation of phosphatidyl-inositol (PI) 3-kinase. However, Braun et al. (2000) demonstrated a direct interaction between InlB and the gC1qR. The platelet gC1qR is a cellular binding site for *Staphylococcal* protein A which suggests an additional mechanism for bacterial cell adhesion to sites of vascular injury and thrombosis (Nguyen et al. 2000).

Circulating, nonenveloped Hepatitis C virus (HCV) core protein interacts with immunocytes through gC1qR. The HCV core protein is the protein expressed during the early phase of HCV infection and suppresses host immune responses, including anti-viral cytotoxic T-lymphocyte responses in a murine model. Like C1q, HCV core protein can inhibit T-cell proliferative responses in vitro. Biochemical analysis of the interaction between core and gC1qR indicates that HCV core binds the region spanning amino acids 188–259, a site distinct from the binding region of C1q (Kittlesen et al. 2000). Binding of HCV core to gC1qR on T cells leads to impaired Lck/Akt activation and T-cell function (Yao et al. 2004). Up-regulation of gC1qR expression is a distinctive feature of MC, and dysregulated shedding of C1qR molecules contributes to vascular cryoglobulin-induced damage via the classic complement-mediated pathway (Sansonno et al. 2009). The gC1qR is the main cellular partner of the hepatitis B virus P22 which was colocalized with the endogenous gC1qR in both the cytoplasm and the nucleus but never in the mitochondria (Lainé et al. 2003). The gC1qR is an effective inhibitor of HIV-1 infection, which prevents viral entry by blocking the interaction between CD4 and gp120. Since gC1qR is a human protein, it is most probably not antigenic in humans. It would seem logical, therefore, to consider gC1qR or its fragments involved in the CD4 binding as potential therapeutic agents (Szabo et al. 2011).

41.4.4 Role in Pathology

The localization of gC1qR and its ligands C1q and HK in atherosclerotic lesions, and the ability of gC1qR to modulate complement, kinin, and coagulation cascades, suggests that gC1qR may play an important role in promoting inflammation and thrombosis in atherosclerotic lesions (Peerschke et al. 2004).

Sequencing of the gene encoding tumor-associated gC1qR did not reveal any consistent tumor-specific mutations. However, histochemical staining demonstrated marked differential expression of gC1qR in thyroid, colon, pancreatic, gastric, esophageal and lung adenocarcinomas compared to their nonmalignant histologic counterparts. In contrast, differential expression was not seen in endometrial, renal and prostate carcinomas. Despite high expression in breast carcinoma, gC1qR was also expressed in nonmalignant breast tissue. Although the precise relation of gC1qR to carcinogenesis remains unclear, the finding of tumor overexpression and the known multivalent binding of gC1qR to a variety of circulating plasma proteins as well as its involvement in cell-to-cell interactions suggest that gC1qR may have a role in tumor metastases (Rubinstein et al. 2004).

Membrane type 1 matrixmetallo-proteinase (MT1-MMP), a key enzyme in cell locomotion, is known to be primarily recruited to the leading edge of migrating cells. MT1-MMP via its cytoplasmic tail directly associates with a compartment-specific regulator gC1qR. Although a direct functional link between these two proteins remains uncertain, observations suggest that the transient associations of gC1qR with the cytoplasmic tail of MT1-MMP are likely to be involved in the mechanisms regulating presentation of the protease at the tumor cell surface (Rozanov et al. 2002).

41.5 Thrombomodulin

The thrombomodulin group of CTLD-containing cell surface proteins has three members in both human and mouse: thrombomodulin, CD93, endosialin, and C-type lectin 14A. The extracellular region of each of these type 1 transmembrane proteins consists of a CTLD projected from the cell surface by a Sushi domain, a number of epidermal growth factor (EGF) domains, and a region rich in proline, serine and threonine that carries multiple O-linked glycans. The short cytoplasmic tail often contains tyrosine-based motifs.

41.5.1 Localization

Thrombomodulin (TM) (CD141) or BDCA-3 is an integral membrane glycoprotein, initially described on an endothelial cell surface. Most of the vascular TM is contained in capillaries which comprise greater than 99% of endothelial surface area. Immunohistochemical analysis of human tissues demonstrated TM on endothelia of blood vessels and lymphatics except for in the CNS (Maruyama et al. 1985; Ishii et al. 1986; DeBault et al. 1986). The syncytiotrophoblast of the placenta contains TM, while hepatic sinusoids and the postcapillary venules of lymph nodes are devoid of it. The TM has been identified on the squamous epithelium of the epidermis (Yonezawa et al. 1987) and in human platelets where their number per endothelial cell is low (Suzuki et al. 1988). TM has been found in a variety of cultured cells, including NIH 3T3 cells, A549 small cell lung cancer cells and CHO cells. Some endothelial neoplasms express TM. Choriocarcinoma but not choriocarcinoma cell lines have been found to express TM (Dittman and Majerus 1990). The antigen described as BDCA-3 has turned out to be identical to TM (Dzionek et al. 2000, 2002). Thus it was revealed that this molecule also occurs on a very rare (0.02%) subset of human dendritic cells called MDC2. Its function on these cells is unknown at present, but apparently, TM has at least one other ligand apart from thrombin, because anticoagulation is a common place function, in contrast to the rarity of MDC2 cells. In humans, TM is encoded by the THBD gene (Wen et al. 1987).

41.5.2 Characteristics

The human TM is a 75 kDa protein when analyzed on SDS-PAGE. After reduction of disulfide bonds, the Mr was 105 kDa, indicating secondary structure involving multiple cystine bridges. This secondary structure renders TM stable under extremes of pH and under denaturing conditions. The cDNAs for human (Wen et al. 1987; Suzuki et al. 1987) and murine TM (Dittman and Majerus 1989) have been sequenced. The structure resembles the low density lipoprotein (LDL) receptor (Goldstein et al. 1985) (Fig. 41.2). The initiating methionine at amino terminal is followed by an 18 amino acid hydrophobic leader sequence. The mature protein starts with sequences distantly homologous to CTLDs such as the asialoglycoprotein receptor (residues 19–179) (Patthy 1988). The large area of identity between the mouse and human cDNAs in the lectin domain implies that the conserved sequence has some unknown function. Human TM consists of ten structural elements: an N-terminal domain homologous to the family of C-type lectins (residues 1–226), six tandemly repeated epidermal growth factor (EGF)-like1 domains (residues 227–462), a Ser/Thr-rich region (residues 463–497), a transmembrane domain (residues 498–521), and a short cytoplasmic tail (residues 522–557) (Jackman et al. 1986; Suzuki et al. 1987; Wen et al. 1987) (Fig. 41.2). The short cytoplasmic tail contains several potential sites for phosphorylation, and phosphorylation in cells has been associated with increased endocytosis and degradation of TM. Thrombomodulin is glycosylated on asparagine residues (Stearns et al. 1989).

The TM gene (THBD) maps to chromosome 20p11.2, contains a single exon and no introns, and spans 4 kb (Ireland et al. 1996; Wen et al. 1987). Both the human (Jackman et al. 1987; Shira et al. 1988) and mouse genes are intronless, for which no special function is known. The TM transcribes a single mRNA of approximately 3,700 bases, consistent with the size of cDNAs for human (3,693 bases) and mouse (3,658 bases) TM. In humans the 5′ untranslated region is 150 bases, the coding region is 1,725 bases encoding 575 amino acids, and the 3′ untranslated region is 1,779 bases. The 3′ untranslated region is highly conserved between species and contains the sequence TTATTTAT, which has been associated with short mRNA half-lives (Shaw and Kamen 1986). Several potential polyadenylation signals, AATAAA are also present in the cDNAs of three species.

41.5.3 Regulation of TM Activity

TM is subject to regulation by internalization of the cell-surface molecule, with loss of protein C activating activity. Thrombin binding to TM can induce internalization of thrombin-TM complexes, with transport to lysosomes, release and degradation of thrombin, and subsequent return of TM to the cell surface (Dittman and Majerus 1990). Cytokines also affect the expression of TM. Tumor necrosis factor (TNF) rapidly alters the anticoagulant properties of the endothelial cell surface by inducing expression of the procoagulant, tissue factor, and reducing TM activity. The reduction in activity is associated with endocytosis and lysosomal degradation of TM. TNF also inhibits transcription of TM mRNA; the subsequent decrease in TM synthesis may also contribute to the decline in TM activity. Other cytokines, interleukin-1 (Naworth et al. 1986) and endotoxin (Moore et al. 1987) also cause decreased TM activity on endothelial cell surfaces. In fact, thrombin increases TM transcription rates, mRNA levels, and total TM protein in cultured mouse hemangioma cells (Dittman et al. 1989). Infusion of TNF into humans increases the production of fibrin, indicating activation of coagulation. Protein C activation was also increased in contrast to what might have been predicted by the in vitro effect of TNF on TM activity and protein C activation. This observation is consistent with additional regulation of TM in vivo by components of the hemostatic system, perhaps by thrombin as discussed above (Dittman and Majerus 1990).

41.5.4 Structure-Function Relations - Binding to Thrombin

Structural requirements for thrombin binding to TM and cofactor activity have been studied by mutagenesis of recombinant human TM. Deletion of the fourth EGF-like domain abolished cofactor activity but did not affect thrombin binding. Deletion of either the fifth or the sixth EGF-like domain markedly reduced both thrombin binding affinity and cofactor activity. Studies suggest that the anticoagulant function of TM is mediated by EGF-like domains 4, 5, and 6 (EGF456) (Zushi et al. 1989). While TM binds to EGF-like domains 5 and 6 (Kurosawa et al. 1988), EGF-like domain 4 is not required for thrombin binding but is essential for accelerating protein C activation (Tsiang et al. 1992; Parkinson et al. 1992). Membrane association is not necessary for cofactor activity (Kurosawa et al. 1988; Suzuki et al. 1989), and the smallest fully active soluble TM fragment consists of EGF456 (Zushi et al. 1989; Hayashi et al. 1990). The EGF456 contains the region required for thrombin binding and protein C activation. Neither mouse nor bovine TM contains the serine-glycine-X-glycine signal sequence required for heparin-like glycosoaminoglycan attachment 58; human TM has an unfavored site for such modification.

Thrombin binding sequences were also localized by assaying the ability of synthetic peptides derived from TM. The two most active peptides corresponded to (a) the entire third loop of the fifth EGF-like domain (Kp = 85 ± 6 μM) and (b) parts of the second and third loops of the sixth EGF-like domain (Kp = 117 ± 9 μM). Studies suggest that thrombin interacts with two discrete elements in TM. Deletion of the Ser/Thr-rich domain dramatically decreased both thrombin binding affinity and cofactor activity and also prevented the formation of a high molecular weight TM species containing chondroitin sulfate. Substitutions of this domain with polypeptide segments of decreasing length and devoid of glycosylation sites progressively decreased both cofactor activity and thrombin binding affinity. This correlation suggests that increased proximity of the membrane surface to the thrombin binding site may hinder efficient thrombin binding and the subsequent activation of protein C. Membrane-bound TM therefore requires the Ser/Thr-rich domain as an important spacer, in addition to EGF-like domains 4–6 (EGF456), for efficient protein C activation. The EGF456 appears to have anticoagulant properties which make it more suitable anticoagulant in extracorporeal circulation in monkeys (Suzuki et al. 1998).

41.5.4.1 The Disulfide Bonding Pattern

The disulfide bonding pattern of the fourth and fifth EGF-like domains within the smallest active fragment of TM has been determined. The disulfide bonding pattern of the fourth EGF-like domain was (1–3, 2–4, 5–6). Surprisingly, the disulfide bonding pattern of the fifth domain was (1–2, 3–4, 5–6), which is different from EGF or any other EGF-like domain analyzed so far. The observation that not all EGF-like domains have an EGF-like disulfide bonding pattern reveals an additional element of diversity in the structure of EGF-like domains (White et al. 1996).

41.5.4.2 Critical Amino Acids in TM

Studies suggest that the region 333–350 in EGF3-4 is critical for protein C activation by the thrombin-TM complex and the region 447–462 in EGF6 is critical for thrombin binding (Parkinson et al. 1992). The 80–90% of TM activity is lost by oxidation of Met^{388}, located within the short interdomain loop between EGF-like domains 4 and 5. The interdomain loop is critical in the biological anticoagulant properties of TM (Clarke et al. 1993). Site-directed mutagenesis suggested that amino acid Asp^{349} of EGF456, in a recombinant protein is essential for retaining full protein C-activating cofactor activity. Thus, Asp^{349} in the fourth EGF-like structure of TM plays a role in its Ca^{2+}-mediated binding to protein C (Zushi et al. 1991).

Although, the fourth EGF-like domain is unnecessary for high affinity thrombin binding, yet it is required for cofactor activity. Two regions of the fourth EGF-like domain were identified as essential for cofactor activity: (1) the sequence consisting of amino acids Glu^{357}, Tyr^{358}, and Gln^{359} shared by the overlapping first and second disulfide loops, and (2) the amino-terminal region of the third disulfide loop containing amino acids Glu^{374}, Gly^{375}, and Phe^{376} (Lentz et al. 1993). In another study, mutants between Cys^{333} and Cys^{462} of TM were expressed in E. coli. In EGF4, which is essential for protein C activation by the thrombin-TM complex, critical residues were: Asp^{349}, Glu^{357}, Tyr^{358}, and Phe^{376}. In EGF5-EGF6, critical residues within a proposed

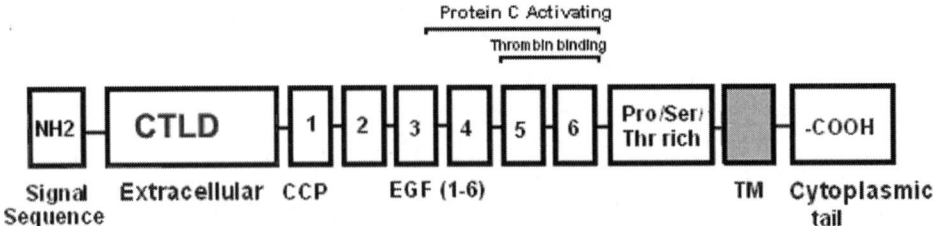

Fig. 41.2 Domain organization of Thrombomodulin: TM is composed of five structural domains. Extending from a short cytoplasmic tail and transmembrane domain is a serine/threonine–rich region to which a chondroitin sulfate moiety that optimizes anticoagulant function is attached. Next is a domain that consists of six epidermal growth factor (EGF)-like repeats, four of which are responsible for the protein's anticoagulant and antifibrinolytic functions. The NH$_2$-terminal domain has two modules. The first, adjacent to the EGF-like domain, is a ~70–amino acid residue hydrophobic region. The second, which is ~155–amino acid residues long, has homology to C-type lectins (Conway et al. 2002)

acidic thrombin-binding region were: Glu^{408}, Tyr^{413}, Ile^{414}, Leu^{415}, Asp^{416}, Asp^{417}, Asp^{423}, Ile^{424}, Asp^{425}, and Glu^{426}. A potential Ca^{2+}-binding site, which is comprised of residues Asp^{423}, Asp^{425}, Glu^{426}, Asn^{439}, Leu^{440}, and Phe^{444}, overlaps the thrombin-binding region. Asp^{461}, in the C-loop of EGF6 shown to be critical for thrombin and residues Asp^{398}, Asp^{400}, Asn^{402}, and Asn^{429} in EGF5 were critical for binding. Thus, 22 residues are critical in the region comprising EGF4-6, which is essential for thrombin binding and protein C activation by the thrombin-TM complex (Nagashima et al. 1993). These results suggest that amino acids critical for TM cofactor activity are located near the junction between the two subdomains of the fourth EGF-like domain.

41.5.4.3 Methionine Oxidation Is Deleterious for TM

Although TM is a large protein, only EGF4,5-like domains are required for anticoagulant function. These two domains must work together, and the linker between the two domains contains a single methionine residue, Met^{388}. Oxidation of Met^{388} is deleterious for TM activity. Structural studies, both X-ray and NMR, of wild type and variants at position 388 showed that Met^{388} provides a key linkage between the two domains. Oxidation of the methionine has consequences for the structure of the fifth domain, which binds to thrombin. The functional consequences of oxidation of Met^{388} include decreased anticoagulant activity. Oxidative stress from several causes is reflected in lower serum levels of activated protein C and a higher thrombotic tendency, and this is thought to be linked to the oxidation of Met^{388} in TM. Thus, TM structure and function are altered in a subtle but functionally critical way upon oxidation of Met^{388} (Wood et al. 2005).

41.5.4.4 Role of Exosites 1 and 2 in Thrombin Reaction

Thrombin has three main binding sites: the catalytic or active site, where its primary plasma inhibitor, antithrombin, binds and inhibits its coagulation activity; and two anion-binding sites, exosite 1 and exosite 2. Exosite 1 binds to fibrinogen, TM, hirudin, and the amino terminus of heparin cofactor II (HCII) (Becker et al. 1999). Exosite 2 mediates binding of thrombin to heparin and to platelet surface glycoprotein Ib-α (GPIbα). Ligand binding to either exosite 1 or exosite 2 may also influence reactivity of the active site of thrombin (Fredenburgh et al. 1997). The cofactors heparin, vitronectin (VN), and TM modulate the reactivity of α-thrombin with plasminogen activator inhibitor (PAI-1). While heparin and VN accelerate the reaction by two orders of magnitude, TM protects α-thrombin from rapid inactivation by PAI-1 in presence of VN. Kinetic studies suggest that (1) heparin binds to exosite 2 of α-thrombin to accelerate the reaction by a template mechanism, (2) VN accelerates PAI-1 inactivation of α-thrombin by lowering the K_D for initial complex formation by an unknown mechanism that does not require binding to either exosite 1 or exosite 2 of α-thrombin, (3) α-thrombin may have a binding site for PAI-1 within or near exosite 1, and (4) TM occupancy of exosite 1 partially accounts for the protection of thrombin from rapid inactivation by PAI-1 in presence of vitronectin (Rezaie 1999).

Exosite 1 of thrombin consists of a cluster of basic residues (Arg^{35}, Lys^{36}, Arg^{67}, Lys^{70}, Arg^{73}, Arg^{75}, and Arg^{77} in chymotrypsinogen numbering) that play key roles in the function of thrombin. Structure suggests that the side chain of Arg^{35} projects toward the active site pocket of thrombin, but all other residues are poised to interact with TM. While studying the role of these residues in TM-mediated protein C activation by thrombin, results suggest that Arg^{35} is responsible for the Ca^{2+} dependence of protein C activation by the thrombin–TM complex and that a function for TM in the activation complex is the allosteric alleviation of the inhibitory interaction of Arg^{35} with the substrate. The results suggest that TM modulates the conformation of Arg^{35}, and a cofactor function of TM is to alleviate the inhibitory interaction of Arg^{35} with protein C in the activation complex (Rezaie and Yang 2003).

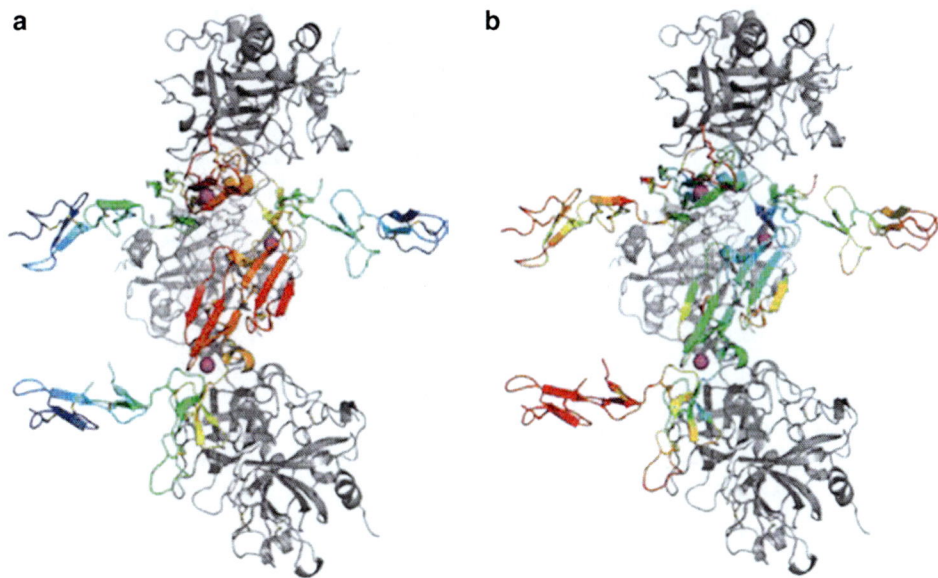

Fig. 41.3 Crystal structure of Na-free thrombin in complex with thrombomodulin: Structure of the thrombin-TM domains 4, 5, 6 (TM456) complex. (a) Ribbon diagram of the three complexes in the asymmetric unit, with thrombin coloured *grey* and TM456 coloured from N-to-C terminus from *blue* (EGF4) to *red* (EGF6), with *magenta* balls indicating Ca2 + ions. (b) The same image as in (a), but coloured according to B-factor to illustrate the inherent flexibility of the 4th EGF domains (Adapted with permission from Adams et al. 2009 © John Wiley and Sons)

Structural and mutagenesis data indicated that the interaction of basic residues of the heparin-binding exosite of protein C with the acidic residues of EGF4 is partially responsible for the efficient activation of the substrate by the thrombin-EGF456 complex. Similar to protein C, protein C inhibitor has a basic exosite (H-helix) that constitutes the heparin-binding site of the serpin. It was suggested that TM enhances the reactivity of protein C inhibitor with thrombin by providing both a binding site for the serpin and a conformational modulation of the extended binding pocket of thrombin (Yang et al. 2003).

41.5.5 The Crystal Structure

The crystal structure of the complex between human thrombin and the minimal cofactor fragment of TM, EGF456 revealed the features of TM-thrombin interaction and corroborated with earlier biochemical information (Fuentes-Prior et al. 2000) (Fig. 41.3). A reasonably small contact interface of ~900 Å is seen between exosite I of thrombin and EGF domains 5 and 6 of TM. Although the surfaces at the interface displayed charge complementarity, which may aid in steering the two fragments together (Myles et al. 2001) only a single salt bridge exists between lysine 110 of thrombin and aspartate^{461} of TM. The TM-thrombin complex crystal structure revealed no obvious conformational changes within the active site of thrombin, but the presence of a bound active-site inhibitor clouds this issue. However, allostery may play a role in switching the substrate preference of thrombin, as several biochemical studies suggest a tightening of the active site in response to TM binding (Koeppe et al. 2005). The cofactor effect of TM may also be partially attributable to an improved accessibility of the activation peptide of TM-bound protein C (Vindigni et al. 1997; Yang et al. 2006; Adams and Huntington 2006). The structure of EGF-like domain 4 has been determined by NMR spectroscopy. These structures are small steps toward an understanding of how TM regulates thrombin (Sadler 1997). The interaction of thrombin with a 28-residue peptide, hTM422-449, corresponding to the N-terminal subdomain of the sixth EGF-like repeat of human TM plus the junction between the fifth and the sixth EGF-like domains, characterized by NMR spectroscopy, is conformationally flexible in absence of thrombin. Upon addition of thrombin, differential resonance perturbations and transferred nuclear Overhauser effects (NOEs) are observed for TM peptide, suggesting specific and rapidly reversible binding and structuring of hTM422-449 in complex with thrombin (Tolkatchev et al. 2000).

Molecular modeling of the protein C activation based on the crystal structure of thrombin in complex with the EGF456 predicts that the binding of EGF56 to exosite 1 of thrombin positions EGF4 so that a negatively charged region on this domain juxtaposes a positively charged region of protein C. It has been hypothesized that electrostatic interactions between these oppositely charged residues of EGF4 and protein C facilitate a proper docking of the substrate into the catalytic pocket of thrombin. To test this hypothesis, Yang and Rezaie

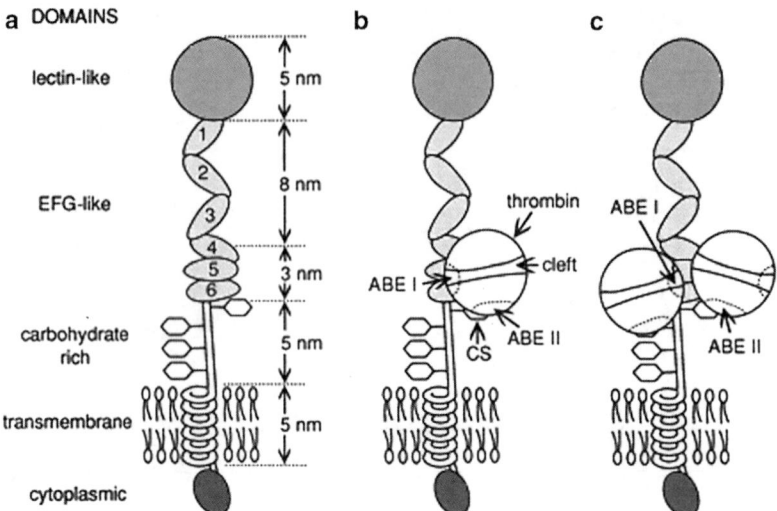

Fig. 41.4 Schematic diagram of Thrombomodulin and TM bound with thrombin. (**a**) schematic diagram of TM based on the electron microscope images of soluble TM. Domains are identified on the *left*, while their sizes are given on the *right*. EGF modules are numbered and carbohydrate moieties are represented as hexagons. The membrane is represented as a lipid bilayer, and the transmembrane region is shown as a typical single helix. (**b**) TM with one thrombin molecule bound. The locations of the catalytic cleft, anion-binding exosite I (ABE I), and anion-binding exosite II (ABE II) of thrombin are shown. Anion-binding exosite I is shown interacting with EGF-like modules 5 and 6 of TM, and anion-binding exosite II is shown interacting with the chondroitin sulfate (CS) residue on TM. (**c**) TM with two thrombin molecules bound, one via interactions of its anion-binding exosite I with EGF modules 5 and 6 and the other via interactions of its anion-binding exosite II with the chondroitin sulfate (Adapted with permission from Weisel et al. 1996 © The American Society for Biochemistry and Molecular Biology)

(2003) constructed several mutants of EGF456 and protein C in which charges of the putative interacting residues on both EGF4 (Asp/Glu) and protein C (Lys/Arg) were reversed. Results of TM-dependent protein C activation studies by such a compensatory mutagenesis approach support the molecular model that EGF4 interacts with the basic exosite of protein C (Yang and Rezaie 2003).

41.5.5.1 TM-Thrombin Complex Under EM

Weisel et al. (1996) determined the shape of SolulinTM, a soluble recombinant form of human TM missing the transmembrane and cytoplasmic domains, by electron microscopy. Solulin appears to be an elongated molecule about 20 nm long that has a large nodule at one end and a smaller nodule near the other end from which extends a thin strand. About half of the molecules form bipolar dimers apparently via interactions between these thin strands. Electron microscopy of complexes formed between Solulin and human α-thrombin revealed that a single thrombin molecule appears to bind to the smaller nodule of Solulin, suggesting that this region contains the EGF41. Epidermal growth factor-like domains 1–4 comprise the connector between the small and large nodule, which is the lectin-like domain; the thin strand at the other end of the molecule is the carbohydrate-rich region. With chondroitin sulfate-containing soluble TM produced from either human melanoma cells or CHO cells, a higher percentage of molecules bound thrombin and, in some cases, two thrombin molecules were attached to one soluble TM in approximately the same region (Weisel et al. 1996) (Fig. 41.4).

41.5.6 Functions

TM is considered a protein involved in coagulation, cancer and embryogenesis. The altered coagulation process leads to a precocious placental failure, but some studies suggest that TM has an anti-inflammatory ability through both protein C dependent and independent pathways.

41.5.6.1 TM Initiates Anticoagulant Process

Thrombomodulin is an endocytic receptor expressed primarily on endothelial cells. It exerts an anticoagulant effect through the sequestration of thrombin and activation of protein C, which degrades coagulation factors. Thrombin sequestration inhibits activation of the G-protein-coupled protease-activated receptor (PAR)-1. TM improves the catalytic efficiency of thrombin toward protein C in the presence of Ca^{2+} by ~3 orders of magnitude by improving both the Km and the kcat of the protein C activation reaction. TM forms a complex with thrombin which is responsible of converting protein C to activated protein C (Maruyama et al. 1985; Rezaie and Yang 2003; Wouwer et al. 2004; Dahlbäck and Villoutreix 2005). The TM-thrombin complex balances the substrate specificity of thrombin, erasing its procoagulant properties, and improving the conversion of protein C into its activated form (Esmon 2000; Fuentes-Prior et al. 2000). Protein C acts on a variety of targets where as the TM-protein C system is essential for maintaining the placenta during pregnancy. TM influences inflammation and tumor progression and TM polymorphisms have been linked to thrombotic conditions such as heart attack. The activity of TM initiates an essential physiological anticoagulant process (Hofsteenge et al. 1986) which by binding to thrombin inhibits the procoagulant functions of thrombin and acts as a cofactor for thrombin catalyzed activation of protein C. This raises the speed of protein C activation 1,000-fold. Activated protein C degrades clotting factors Va and VIIIa and thereby inhibits blood clotting. TM-bound thrombin has no procoagulant effect. When bound to TM, thrombin also activates the carboxypeptidase B-like enzyme thrombin activatable fibrinolysis inhibitor (TAFI) (Bajzar et al. 1996). The binding site for thrombin has been localized by alanine-scanning mutagenesis and peptide-binding studies to the final 2 EGF domains, EGF56 (Jakubowski and Owen 1989; Kurosawa et al. 1988; Nagashima et al. 1993; Adams and Huntington 2006).

Primarily, TM modifies the substrate specificity of thrombin, apparently by an allosteric mechanism. The thrombin-TM-complex also inhibits fibrinolysis by cleaving thrombin-activatable fibrinolysis inhibitor (TAFI) into its active form, initiating an essential anticoagulant pathway. The cofactor function of membrane-associated TM requires EGF456, as well as a Ser/Thr-rich spacer between EGF-like domain 6 and the transmembrane domain. The Ser/Thr-rich domain is variably modified with a chondroitin sulfate chain that influences the affinity of thrombin binding and the calcium ion dependence of cofactor function. Deletion of the N-terminal lectin-like domain and EGF-like domains 1 and 2 had no effect on TAFI or protein C activation, whereas deletions including EGF-like domain 3 selectively abolished TM cofactor activity for TAFI activation. Thus, the anticoagulant and antifibrinolytic cofactor activities of TM have distinct structural requirements: protein C binding to the thrombin-TM complex requires EGF-like domain 4, whereas TAFI binding requires EGF-like domain 3 (Kokame et al. 1998) (Fig. 41.5).

Deletion and point mutants of soluble TM were used to compare and contrast elements of primary structure required for the activation of TAFI and protein C. The smallest mutant capable of efficiently promoting TAFI activation contained residues including the c-loop of EGF3 through EGF6. This mutant is 13 residues longer than the smallest mutant that functioned well with protein C; the latter consisted of residues from the interdomain loop connecting EGF3 and EGF4 through EGF6. Wang et al. (2000) suggest that structural elements in the thrombin-binding domain are needed for the activation of both protein C and TAFI, but more of the primary structure is needed for TAFI activation. In addition, some residues are needed for one of the reactions but not the other (Wang et al. 2000).

Single-chain urokinase-type plasminogen activator (scu-PA) can be cleaved by thrombin into a virtually inactive form called thrombin-cleaved two-chain urokinase-type plasminogen activator (tcu-PA/T), a process accelerated by TM. TM-EGF56 bound to thrombin but did not accelerate the activation of protein C. In contrast, the inactivation of scu-PA by thrombin was accelerated to the same extent as that induced by TM-LEO and TM-Ei4-6. This study demonstrated that, in addition to the chondroitin sulfate moiety, only EGF-like domains 5 and 6 are essential for the acceleration of the inactivation of scu-PA by thrombin. This differs from the domains that are critical for activation of protein C (EGF-like domains i4–6) and thrombin activatable fibrinolysis inhibitor (EGF-like domains 3–6) (Schenk-Braat et al. 2001).

Molecular Basis of TM Activation of Slow Thrombin
Thrombin is critical enzyme for producing and stabilizing a clot, but when complexed with TM it is converted to a powerful anticoagulant. Thrombin function is also controlled by Na^+. Its apparent affinity suggests that half of the thrombin generated is in a Na^+-free "slow" state and half is in a Na^+-coordinated "fast" state. While slow thrombin is a poor procoagulant enzyme, when complexed to TM it is an effective anticoagulant. A structure of thrombin complexed with EGF456 of TM in absence of Na^+ and other cofactors or inhibitors suggests that TM binds thrombin (Adams et al. 2009). As a result the thrombin component resembles structures of fast form. The Na^+ binding loop in thrombin in thrombin-TM complex is in a conformation identical to the Na^+-bound form, with conserved water molecules compensating for the missing ion. It was shown that activation of slow thrombin by TM principally involves the opening of the primary specificity pocket. Therefore, TM binding alters the conformation of thrombin in a similar manner as Na^+ coordination, resulting in an ordering of the Na^+ binding loop and an opening of the adjacent S1 pocket (Adams et al. 2009) (Fig. 41.5).

41.5.6.2 Regulator of Inflammation
Although, the TM is best characterized as a natural anticoagulant, reports have illuminated the importance of TM in the regulation of inflammation. The antigen described as BDCA-3 is identical to TM (Chap. 35) (Dzionek et al. 2000, 2002). This molecule also occurs on a very rare (0.02%) subset of human dendritic cells called MDC2. Its function on these cells is unknown at present, but apparently, TM has at least one other ligand apart from thrombin, because anticoaglulation is a common place function,

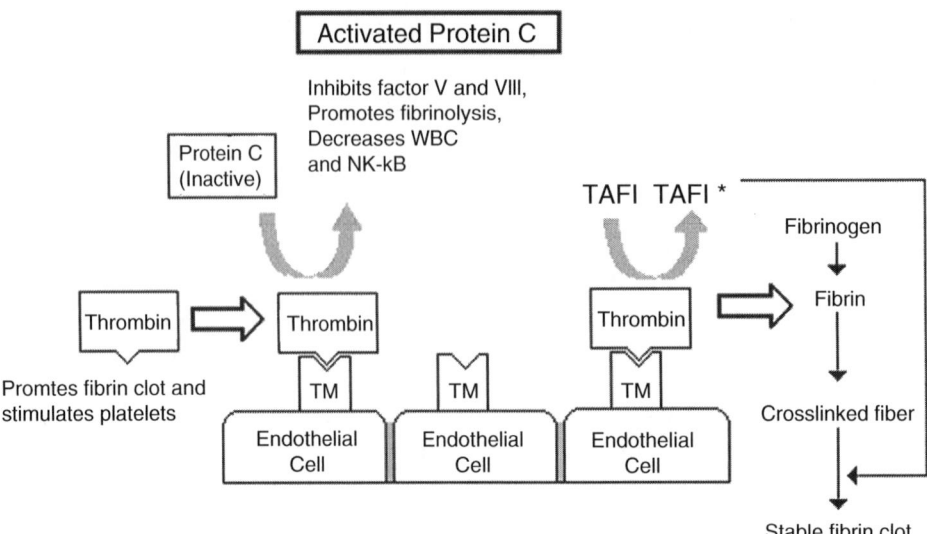

Fig. 41.5 Functions of membrane bound thrombomodulin (TM): (1) TM exerts an anticoagulant effect through the sequestration of thrombin and activation of protein C, which degrades coagulation factors. (2) TM also accelerates the proteolytic activation of a plasma procarboxypeptidase referred to as thrombin-activable fibrinolysis inhibitor (TAFI*). Activated TAFI* can be generated from the binary thrombin-TAFI complex (T-TAFI) or the ternary thrombin-thrombomodulin-TAFI complex (T-TM-TAFI) (Nesheim 2003). The anticoagulant and antifibrinolytic cofactor activities of TM have distinct structural requirements: protein C binding to the thrombin-TM complex requires EGF-like domain 4, whereas TAFI binding also requires EGF-like domain 3 (Kokame et al. 1998)

in contrast to the rarity of MDC2 cells. Thrombomodulin expression on DCs may be involved in the pathogenesis of atopy and asthma (Yerkovich et al. 2009). Thrombomodulin inhibits allergic bronchial asthma by inducing tolerogenic DCs. Treatment of bronchial asthma with TM improved lung function and reduced IgE and cells in alveolar lavage fluid by inducing tolerogenic dendritic dells. These were characterized by high expression of surface CD141/TM$^+$ and low expression of maturation markers and possessed reduced T-cell co-stimulatory activity. The TM effect was independent of its role in coagulation. Rather, it was mediated via the TM lectin domain directly interacting with DCs. Results showed that TM is a modulator of DC immunostimulatory properties and a novel candidate drug for the prevention of bronchial asthma in atopic patients (Takagi et al. 2011; Yerkovich et al. 2009). CD141$^+$ DC subset is an important functionally distinct human DC subtype with characteristics similar to those of the mouse CD8α^+ DC subset. CD141$^+$ DCs play a role in the induction of cytotoxic T lymphocyte and may be the most relevant targets for vaccination against cancers, viruses, and other pathogens (Jongbloed et al. 2010). Thrombomodulin was present in all keratinizing squamous carcinomas and the great majority (87%) of nonkeratinizing tumors in a membrane-staining pattern. It was moderately common in small cell (27%) and large cell carcinomas (25%) but relatively rare in adenocarcinomas (13%) (Miettinen and Sarlomo-Rikala 2003).

41.5.6.3 Regulator of Placental Functions

Thrombomodulin is expressed mainly on the endothelial surface of blood vessels and in the placental syncytiotrophoblast (Maruyama et al. 1985; Dittman and Majerus 1990). The functions of TM are exerted in two distinct tissues: in non-endothelial extra-embryonic tissues, required for proper function of the early placenta, and in embryonic blood vessel endothelium whose absence causes lethal consumptive coagulopathy (Bicheng et al. 2005). Disruption of the mouse gene encoding the TM leads to embryonic lethality caused by an unknown defect in the placenta. Isermann et al. (2003) showed that the abortion of TM-deficient embryos is caused by tissue factor–initiated activation of the blood coagulation cascade at the feto-maternal interface. These findings show a new function for the TM–protein C system in controlling the growth and survival of trophoblast cells in the placenta. This function is essential for the maintenance of pregnancy (Isermann et al. 2003).

41.5.7 Abnormalities Associated with Thrombomodulin Deficiency

41.5.7.1 Polymorphism and the Risk of Heart Disease

The genes encoding proteins in the TM-protein C pathway are promising candidate genes for stroke susceptibility because of their importance in thrombosis regulation and inflammatory response. Thrombosis is a dynamic balance between factors that promote clot formation, antithrombotic mechanisms, and fibrinolysis. Central to this balance is the TM-protein C antithrombotic mechanism. TM forms a 1:1 complex with thrombin on the vascular endothelium, thereby inhibiting the procoagulant actions of thrombin and converting protein C to activated protein C. Activated protein C promotes fibrinolysis, inhibits thrombosis by inactivating clotting factors Va and VIIIa, and reduces inflammation by decreasing white blood cell and NF-kB activation (Esmon 1987, 1995, 2001; Barnes and Karin 1997). These relationships are demonstrated in Fig. 41.5. Because of the central role that the TM-protein C pathway plays in thrombosis regulation and inflammatory response, the genes encoding these pathway proteins are promising candidate genes regarding stroke susceptibility.

Eosinophil (EO) specific major basic protein accumulates on endocardial surfaces in the course of hypereosinophilic heart disease and promotes thrombosis by binding to the anionic TM and impairing its anticoagulant activities. Major basic protein potently inhibits the capacity of cell surface TM to generate the natural anticoagulant activated protein C. Thus, EO cationic proteins potently inhibit anticoagulant activities of the glycosylated form of TM, thereby suggesting a potential mechanism for thromboembolism in hypereosinophilic heart disease (Slungaard et al. 1993).

The TM gene (*THBD*) maps to chromosome 20p11.2, and spans 4 kb (Ireland et al. 1996). A single nucleotide polymorphism (C→T) at position + 1418 (C1418T) encodes for an amino acid change from alanine to valine at protein position 455 (Ala^{455}Val) (van der Velden et al. 1991). The location of this amino acid variation corresponds to the sixth EGF region of the TM protein (Fig. 41.2). This location has been shown to be responsible for the high-affinity binding of thrombin and for the suspension of thrombin at a specific position above the endothelial surface in relation to other cofactors, thereby producing optimal protein C activation by thrombin (Esmon 1995; Sadler 1997).

Studies have shown that the *THBD* Ala^{455}Val polymorphism is associated with ischemic heart disease (Norlund et al. 1997; Wu et al. 2001). Based on results of Kittner et al. (1998), Cole et al. (2004) determined the association between the *THBD* Ala^{455}Val polymorphism and the occurrence of ischemic stroke in young women. Schulz et al. (2004) indicated that among women aged 15–44 years, the AA genotype is more prevalent among blacks than whites and is associated with increased risk of early-onset ischemic stroke. Removing strokes potentially related to cardioembolic phenomena increased this association. Whether this association is due to population stratification, needs to be confirmed.

41.5.7.2 Spontaneous Recurrent Abortion

Reduced expression of TM in placental tissue of women were associated with spontaneous recurrent abortion and the loss of TM caused early post-implantation embryonic lethality (Isermann et al. 2001; Kaare et al. 2007; Stortoni et al. 2010). This suggested that TM expression in non-endothelial placental cells is required for a normal function of the early placenta. Further studies are needed to elucidate that the TM expression in non-endothelial placental cells is required for a normal function of the early placenta.

41.6 Endosialin (Tumor Endothelial Marker-1 or CD248)

41.6.1 A Marker of Tumor Endothelium

Endosialin (Tumor Endothelial Marker-1 or CD248) was originally described as a marker of tumor endothelium, and later as a marker of fibroblasts (Rettig et al. 1992; St Croix et al. 2000). Endosialin is a 165-kDa cell surface glycoprotein expressed by tumor blood vessel endothelium in a broad range of human cancers but not detected in blood vessels or other cell types in many normal tissues. A full-length endosialin cDNA with an open reading frame of 2274 bp encodes a type I membrane protein of 757 amino acids with a predicted molecular mass of 80.9 kDa. The sequence matches with an expressed sequence tag of unknown function in public data bases, named Tumor Endothelial Marker-1(TEM-1). Endosialin is classified as a C-type lectin-like protein, composed of a signal leader peptide, five globular extracellular domains (including a C-type lectin domain, one domain with similarity to the Sushi/ccp/scr pattern, and three EGF repeats), followed by a mucin-like region, a transmembrane segment, and a short cytoplasmic tail (Fig. 41.6). The endosialin core protein carries abundantly sialylated, O-linked oligosaccharides and is sensitive to O-sialoglycoprotein endopeptidase, placing it in the group of sialomucin-like molecules. The N-terminal 360 amino acids of endosialin show homology to thrombomodulin, a receptor involved in regulation of blood coagulation (Section 41.5), and to complement receptor C1qRp (Section 41.4) (Christian et al. 2001; Teicher 2007). Endosialin is cell surface receptor of unknown function, with a distinctive pattern of endothelial expression in newly formed blood vessels in human cancers. It is not known if carbohydrate recognition is involved in the function of proteins in the endosialin group. The CTLDs in these proteins do not contain the sequence motifs connected with sugar-binding activity.

The single copy endosialin gene was mapped to chromosome 19. Endosialin gene is intronless and encodes a 92-kDa protein that has 77.5% overall homology to the human protein. This gene is ubiquitously expressed in normal human and mouse somatic tissues and during development.

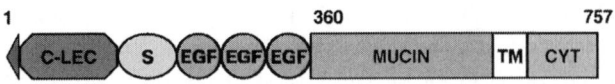

Fig. 41.6 Domain architecture of endosialin. The diagram shows the arrangement of the N-terminal signal leader peptide (*triangle*), the C-type lectin domain, the domain similar to a Sushi/SCR/CCP domain (*S*), three EGF-like repeats (*EGF*), and the transmembrane region (*TM*), flanked by two segments with low complexity regions, namely a sialomucin-like sequence (*MUCIN*) and a short, putative cytoplasmic tail (*CYT*) (Adapted from Christian et al. 2001© American Society for Biochemistry and Molecular Biology)

The murine orthologue of endosialin has been identified, opening up the analysis of developmental regulation in the embryo and in aberrant tissue remodeling. Its expression at the mRNA level is density-dependent and up-regulated in serum-starved cells. In vitro, its expression is limited to cells of embryonic, endothelial, and preadipocyte origin, suggesting that the wide distribution of its expression in vivo is due to the presence of vascular endothelial cells in all the tissues. The ubiquitous expression in vivo is in contrast to previously reported expression limited to corpus luteum and highly angiogenic tissues such as tumors and wound tissue (Opavsky et al. 2001). Further analyses confirmed selective *TEM-1/endosialin* expression in tumor endothelium, pericytes and a subset of fibroblast-like cells of tumor stroma in breast carcinoma, anaplastic astrocytoma and GBM (Simonavicius et al. 2008; Davies et al. 2004; Brady et al. 2004; MacFadyen et al. 2005; Carson-Walter et al. 2001; Christian et al. 2008).

Findings on mouse endosialin suggest that: (1) endosialin protein (also termed as TEM-1) is restricted to vascular endothelium and fibroblast-like cells in developing organs, and largely disappears during adulthood, (2) endothelial expression varies markedly between organs regarding spatial and temporal patterns, (3), circumscribed mesenchymal expression in fibroblast-like cells was evident throughout development, most pronounced adjacent to certain budding epithelia, as in the lung and kidney glomeruli, but unrelated to the endothelial expression. The endosialin protein persists in the stromal fibroblasts of the adult uterus (Rupp et al. 2006). Brain pericytes in culture had higher levels of endosialin/TEM 1 than TEMs-2, -3, -4, -5, -7, and -8. Endosialin was present in the cytoplasmic body and in the elongated extensions essential to pericyte function. It seems that endosialin is strongly expressed by pericytes during periods of active angiogenesis during embryonic and tumor development. With the appropriate agent, targeting endosialin may interfere with blood vessel growth during tumor development (Bagley et al. 2008a).

41.6.1.1 Upregulation of Endosialin Gene by Hypoxia

Angiogenesis is well known to be principally driven by hypoxia in growing solid tumors as well as in developing embryos (Liao and Johnson 2007). The endosialin gene transcription is induced by hypoxia predominantly through a mechanism involving hypoxia-inducible factor-2 (HIF-2) cooperating with the Ets-1 transcription factor. It was shown that HIF-2 activates the endosialin promoter both directly, through binding to a hypoxia-response element adjacent to an Ets-binding site in the distal part of the upstream regulatory region, and indirectly, through Ets-1 and its two cognate elements in the proximal promoter. Results also suggested that the SP1 transcription factor mediates responsiveness of endosialin promoter to high cell density (Ohradanova et al. 2008).

41.6.2 Interactions with Ligands

It is not clear if carbohydrate recognition is involved in the function of proteins in the endosialin group. The CTLDs in these proteins do not contain the sequence motifs connected with sugar-binding activity. Recent evidence suggests that endosialin may interact with extracellular matrix components including fibronectin (FN) and collagen types I and IV as well as Mac-2 BP/90 K in promoting vascular migration and invasion (Tomkowicz et al. 2007; Becker et al. 2008). More importantly, cells expressing endosialin exhibit enhanced adhesion to FN as well as enhanced migration through matrigel and increased adhesive properties (Tomkowicz et al. 2007). Subsequently, endosialin/collagen IV co-localization was confirmed by confocal microscopy. Other studies have demonstrated the importance of endosialin in regulating proliferation tube formation on matrigel (Bagley et al. 2008b). However, the exact mechanism(s) as to how endosialin exerts its pleiotropic cellular effects has not been addressed. Becker et al. (2008) provided evidence that PDGF receptor and the MAP kinase ERK-1/2 are involved with the endosialin-regulated signaling pathways that control proliferation of pericytes. Mac-2 BP/90 K has been identified as a specific interacting partner of endosialin. C-terminal fragment of Mac-2 BP/90 K, a binding partner for galectin-3, and collagens were responsible for endosialin binding. Intriguingly, the expression patterns of Mac-2 BP/90 K and endosialin were mutually exclusive in all human tissues. These studies demonstrate a novel repulsive interaction between endosialin on stromal fibroblasts and Mac-2 BP/90 K on tumor cells (Becker et al. 2008). Tomkowicz et al. (2010) showed that normal pericytes expressing high levels of endosialin were able to proliferate, respond to PDGF-BB stimulation by phosphorylating both the PDGF receptor and the MAP-kinase ERK-1/2, and induce the expression of immediate early transcription factor *c-Fos*. However, when endosialin expression was knocked-down, PDGF-BB-induced proliferation, ERK-1/2-phosphorylation, and *c-Fos* expression were significantly impaired. These results provide evidence for a endosialin-dependent signal pathway that controls proliferation of human pericytes and suggest targeting this pathway for future strategies aimed at mitigating tumor angiogenesis (Tomkowicz et al. 2010).

41.6.3 Functions

Abundant endosialin expression was found in the vasculature and fibroblast-like cells of the developing mouse and human embryos (Rupp et al. 2006; MacFadyen et al. 2007; Virgintino et al. 2007). Nanda et al. (2006) have shown that endosialin is dispensable for normal development and subcutaneous tumor growth, but can modulate invasiveness and metastatic progression in an orthotopic Xenograft model of colorectal cancer. All these data support the functional involvement of endosialin in angiogenesis, a complex process of vascular branching and sprouting that plays a key role in tumor expansion and progression, and therefore represents an opportunity for therapeutic intervention against cancer. MacFadyen et al. demonstrated that endosialin was expressed by fibroblasts and pericytes associated with tumor vessels, but not on HUVECs or endothelial cells of normal or diseased tissue (MacFadyen et al. 2005, 2007). Moreover, endosialin expression was upregulated on tumor-derived pericytes and confirmed by Virgintino et al. (2007).

The functional significance of endosialin expression on pericytes and stromal cells, its upregulation during cancer, and its role in angiogenesis has largely remained unknown. The knockout (KO) mouse model showed the absence of endosialin expression, reduced growth, invasion, and metastasis of human tumor Xenografts. In addition, lack of endosialin led to an increase in small immature blood vessels and decreased numbers of medium and large tumor vessels. This abnormal angiogenic response could be responsible for the reduced tumor growth and invasion observed in endosialin KO mice, suggesting a role for endosialin in controlling the interaction among tumor cells, endothelia, and stromal matrix (Nanda et al. 2006). Data indicated that the stroma can control tumor aggressiveness and that this control varies with anatomic site. Endosialin is required for complete popliteal LN (pLN) expansion but not for coordination of B and T cell compartmentalisation or antibody production following 4-hydroxy-3-nitrophenyl)acetyl chicken γ-globulin immunisation. In vitro, Endosialin expression in human MG63 stromal cells and mouse embryonic fibroblasts leads to a pro-proliferative and pro-migratory phenotype. This correlates with a proliferating $CD248^+$ population observed in vivo during pLN expansion (Lax et al. 2010).

41.6.3.1 Endosialin in Cancer

Several studies confirmed that endosialin is upregulated in blood vessels and activated stromal fibroblasts in human colorectal, brain and breast tumors as well as in mouse Xenograft models (Carson-Walter et al. 2001; Brady et al. 2004; Davies et al. 2004; Madden et al. 2004; Dolznig et al. 2005; MacFadyen et al. 2005, 2007; Huber et al. 2006; Rupp et al. 2006). Endosialin expression is restricted to capillaries and shows a heterogeneous pattern typical for molecules involved in vascular reorganization. Human colon and ovarian Xenografts showed that endosialin expression was largely confined to NG2-expressing perivascular cells and not CD31-positive endothelial cells (Bagley et al. 2008b).

Endosialin is a stromal marker that is differentially expressed on fibroblasts and pericytes in the thymus, lymph node and spleen. Expression is high during LT development but largely disappears in the adult. Endosialin is re-expressed in a Salmonella-induced model of splenic enlargement; peak expression corresponding to the peak of splenic enlargement. These results suggest that endosialin expression helps define a subset of LT stromal cells which play a role in remodelling during tissue development, infection and repair (Dolznig et al. 2005; Lax et al. 2007; MacFadyen et al. 2005; Teicher 2007). Cancer-associated fibroblasts (CAF) have been implicated in promoting tumor development and have been associated with mesenchymal stem cells (MSC). MSC can form networks in a tube formation assay that is inhibited by an anti-endosialin antibody. MSC are a potential target for anticancer therapeutic intervention and endosialin expression offers a tool for the identification of MSC. Endosialin expression by both CAF and MSC further implies the potential contribution of MSC to tumor stroma via differentiation into tumor stromal fibroblasts (Bagley et al. 2009; Demoulin 2010).

In an effort to better characterize the desmoplastic response to human skin tumors, Huber et al. (2006) evaluated the expression pattern of three stromal cell markers, fibroblast-activation protein (FAP), endoglyx-1, and endosialin, in a series of melanocytic and epithelial skin tumors. FAP-positive fibroblasts were detected in all tumor tissues tested, including cases of melanocytic nevi, melanoma metastases, basal cell carcinomas, and squamous cell carcinomas. Endoglyx-1 expression was confined to normal and tumor blood vessel endothelium including 'hot spots' of neoangiogenesis within the cutaneous melanoma metastases. Endosialin was selectively induced in subsets of small- and medium-sized tumor blood vessels in melanoma metastases and squamous cell carcinomas. These data describe novel aspects of targets for novel therapeutic strategies aimed at the tumor stroma (Huber et al. 2006). Christian et al. (2008) claim that endosialin is expressed by tumor-associated myofibroblasts and mural cells and not by endothelial cells. Endosialin expression is barely detectable in normal human tissues with moderate expression only detectable in the stroma of the colon and the prostate. Corresponding cellular experiments confirmed endosialin expression by mesenchymal cells and indicated that it may in fact be a marker of mesenchymal stem cells. Collectively, experiments validate endosialin as a marker of tumor-associated myofibroblasts and tumor vessel-associated mural cells (Christian et al. 2008).

Brain Tumors

Human brain tumors express endosialin in a heterogeneous manner. Gliomas are the most frequent primary tumors of the central nervous system in adults. The most prevalent and aggressive subclass of these is glioblastoma multiforme, which is characterized by massive neovascularization. The largest proportion of endosialin-expressing tumors was found in highly invasive glioblastoma multiforme, anaplastic astrocytomas, and metastatic carcinomas. Endosialin is not expressed in normal human adult brain but is strongly upregulated in the angiogenic vasculature of all high-grade glioma specimens. Endosialin was localized to the endothelium of small and large CD31 stained vessels and also expressed by Thy-1-positive fibroblast-like cells close to the meninges and α-smooth muscle actin-positive cells in some vessels. Endosialin colocalized with thrombomodulin, suggesting these proteins may have complementary functions in tumor progression (Brady et al. 2004). However, Simonavicius et al. (2008) demonstrated unambiguously that endosialin is not expressed by the glioma endothelial cells but on closely associated perivascular cells. Carson-Walter et al. (2009) characterized the expression pattern of endosialin in astrocytic and metastatic brain tumors and investigated its role as a therapeutic target in human endothelial cells and mouse xenograft models. Eendosialin was induced in the vasculature of high-grade brain tumors where its expression was inversely correlated with patient age. Although lack of endosialin did not suppress growth of intracranial GBM xenografts, it did increase tumor vascularity (Carson-Walter et al. 2009). The recognition of endosialin as an antigen that is selectively over-expressed in human tumor tissues offers new strategies for treating cancer. Not only do the tumor vasculature and stromal compartments upregulate endosialin but, the malignant cells of sarcomas also strongly express endosialin. Thus, endosialin holds potential value both as a therapeutic target and as a biomarker for certain human cancers (Bagley 2009; Rouleau et al. 2008; Simonavicius et al. 2008). The function of C-type lectin 14A is not known.

41.7 Bimlec (DEC 205 Associated C-type lectin-1 or DCL-1) or (CD302) (Group XV of CTLD)

BIMLEC is synonymous with C-type lectin domain family 13 member A (CLEC13A), KIAA0022 and DEC-205-associated C-type lectin-1 (DCL-1). BIMLEC is a Type I transmembrane protein with neck region and CTLD in extracellular region. This new group of C-type lectins was predicted by Zelensky and Gready (2004) in Fugu whole genome and supported by a database cDNA sequences. Group was named Bimlec, linked to DEC-205. Function of Bimlec in humans is unknown, but it is expressed as a fusion protein with DEC-205 in Hodgkin's lymphoma cells (Kato et al. 2003). The cDNA encoding DCL-1 has been identified in Reed-Sternberg cells associated with the development of classic Hodgkin's lymphoma as part of fusion protein between DEC-205 (CD205) and a C-type lectin DCL-1 (Kato et al. 2003). Although the 7.5-kb DEC-205 and 4.2-kb DCL-1 mRNA were expressed independently in myeloid and B lymphoid cell lines, the DEC-205/DCL-1 fusion mRNA (9.5 kb) predominated in the HRS cell lines. The DEC-205 and DCL-1 genes comprising 35 and 6 exons, respectively, are juxtaposed on chromosome band 2q24 and separated by only 5.4 kb. The fusion mRNA predominates in these cells, but both proteins are expressed independently also in myeloid cells and B-lymphoid cell lines. DCL-1 protein is highly conserved among the human, mouse, and rat orthologs. DCL-1 expression in leukocytes is restricted to monocytes, macrophages, granulocytes, and dendritic cells, although DCL-1 mRNA is present in many tissues. These results imply an unusual transcriptional control mechanism in HRS cells, which cotranscribe an mRNA containing DEC-205 and DCL-1 prior to generating the intergenically spliced mRNA to produce a DEC-205/DCL-1 fusion protein. Kato et al. (2007) have suggested that DCL-1 is an unconventional lectin receptor that plays roles in endocytosis and phagocytosis but also in cell adhesion and migration. In nomenclature of CD antigens this protein has been given the designation CD302.

41.8 Proteins Containing SCP, EGF, EGF and CTLD (SEEC) (Group XVI of CTLD)

Souble protein containing SCP, EGF, EGF and CTLD (SEEC) domains is a group of CTLD which has been predicted during Fugu whole-genome analysis and numbered as group XVI of CTLD. Group XVI of CTLD is well conserved between human and fish and supported by available cDNAs. The sperm-coating glycoprotein (SCP) domain, which is present in organisms from yeast and plants to mammals, but whose function is unknown (Szyperski et al. 1998), is rarely observed in combination with other domains in proteins. This SCP/CTLD combination is observed in one known protein – Nowa from hydra (Engel et al. 2002). CTLD has potential Ca-carbohydrate-binding motif (QPD) characteristic of galactose specificity

41.8.1 Nematocyst Outer Wall Antigen (NOWA)

The nematocyst is a unique extrusive organelle involved in the defense and capture of prey in cnidarians. The nematocyst outer wall antigen (NOWA) was identified in nematocysts, explosive organelles of *Hydra*, jellyfish, corals and other *Cnidaria*. Minicollagens along with glycoprotein NOWA form the major components of nematocyst capsule outer wall, which resists osmotic pressure of 15 MPa. NOWA spontaneously assembles to globular macromolecular particles that are sensitive to reduction as the native wall structure. NOWA, the major component of the outer wall, is formed very early in morphogenesis. NOWA is a 90-kDa glycoprotein which comprises a C-terminal eightfold repeat of the minicollagen cysteine-rich domain, suggesting a possible disulfide-dependent hetero-assembly of minicollagens and NOWA protein (Ozbek et al. 2004). NOWA has a modular structure with an N-terminal sperm coating glycoprotein domain, a central C-type lectin-like domain, and an eightfold repeated cysteine-rich domain at the C-terminus. Interestingly, the cysteine-rich domains are homologous to the cysteine-rich domains of minicollagens. The cysteines of these minicollagen cysteine-rich domains undergo an isomerization process from intra- to intermolecular disulfide bonds, which mediates the crosslinking of minicollagens to networks in the inner wall of the capsule. The minicollagen cysteine-rich domains present in both proteins provide a potential link between NOWA in the outer wall and minicollagens in the inner wall. Engel et al. (2002) proposed a model that integrates the role of microtubule cytoskeleton and the interaction of NOWA and minicollagen in forming the nematocyst wall. Data suggest a continuous suprastructure of the nematocyst wall, assembled from wall proteins that share a common oligomerization motif (Meier et al. 2007).

41.8.2 The Cysteine-Rich Secretory Proteins (CRISP) Super-Family

The cysteine-rich secretory proteins (CRISP), antigen 5, and pathogenesis-related 1 proteins (CAP) superfamily members are found in a remarkable range of organisms spanning each of the animal kingdoms. Within humans and mice, there are 31 and 33 individual family members, respectively, and although many are poorly characterized, a majority of them shows a notable expression bias to the reproductive tract (Gupta 2005) and immune tissues or are deregulated in cancers. CAP superfamily proteins are most often secreted and have an extracellular endocrine or paracrine function and are involved in processes including the regulation of extracellular matrix and branching morphogenesis, potentially as either proteases or protease inhibitors; in ion channel regulation in fertility; as tumor suppressor or prooncogenic genes in tissues including prostate; and in cell-cell adhesion during fertilization. Several reviews describing mammalian CAP superfamily gene expression profiles, phylogenetic relationships, protein structural properties, and biological functions draw into focus their potential role in health and disease. The nine subfamilies of the mammalian CAP superfamily include: the human glioma pathogenesis-related 1 (GLIPR1), Golgi associated pathogenesis related-1 (GAPR1) proteins, peptidase inhibitor 15 (PI15), peptidase inhibitor 16 (PI16), cysteine-rich secretory proteins (CRISPs), CRISP LCCL domain containing 1 (CRISPLD1), CRISP LCCL domain containing 2 (CRISPLD2), mannose receptor like and the R3H domain containing like proteins. The overall protein structural conservation within the CAP super-family results in fundamentally similar functions for the CAP domain in all members, yet the diversity outside this core region dramatically alters target specificity and, therefore, the biological consequences (Gibbs et al. 2008).

41.8.2.1 Human Glioma Pathogenesis-Related Protein

Specifically, the human glioma pathogenesis-related protein (GliPR) is highly expressed in the brain tumor glioblastoma multiforme and exhibits 35% amino acid sequence identity with the tomato pathogenesis-related (PR) protein P14a, which has an important role for the plant defense system. GliPR is homologous to group 1 plant pathogenesis-related proteins (PR-1) that are implicated in plant defense responses to viral, bacterial, and fungal infection (Klessig et al. 2000; van Loon et al. 2006). Since GliPR shows structural similarities with its homologous plant PR-1 proteins, mammalian testis proteins (TPX1) and the insect venom Ag-5 protein, which are secretory proteins (van Loon et al. 2006; Foster and Gerton 1996), it has been suspected that GliPR is also secreted. Comparison of the GliPR model with the P14a structure resulted into identification of a common partially solvent-exposed spatial cluster of four amino acid residues, His-69, Glu-88, Glu-110, and His-127 in the GliPR numeration. This cluster is conserved in all known plant PR proteins of class 1, indicating a common putative active site for GliPR and PR-1 proteins and thus a functional link between the human immune system and a plant defense system (Szyperski et al. 1998).

There isTGliPR is related to testes-specific, vespid and pathogenesis protein 1 (RTVP-1) (Rich et al. 1996). Increased expression of GliPR was associated with myelomonocytic differentiation in macrophages (Gingras and Margolin 2000). Whereas GliPR has been reported to act as a tumor suppressor gene inducing apoptosis in prostate cancer (Naruishi et al. 2006; Ren et al. 2004), it appears to be an oncogene in glioblastomas (Rosenzweig et al. 2006) and

Wilms tumors (Chilukamarri et al. 2007). RTVP-1 protein is reported to contain a N-terminal signal peptide sequence and a transmembrane domain (Szyperski et al. 1998). Furthermore, homology studies revealed a putative active enzymatic center in GliPR (Szyperski et al. 1998). GliPR's homology with plant PR-1 proteins raises the question whether GliPR has an evolutionarily conserved role in innate immune response and human host defense of viral infection including HIV-1. There is an early up-regulation of GliPR expression (fivefold) in CEM T cells infected with HIV-1 (Scheuring et al. 1998). Alternatively or additionally, HIV-1 may induce and exploit GliPR for viral replication. The up-regulation of GliPR by HIV-1 and the early significant inhibition of HIV-1 replication mediated by knockdown of GliPR reveal GliPR as an important HIV-1 dependency factor (HDF), which may be exploited for HIV-1 inhibition (Capalbo et al. 2010).

Data from a number of systems suggests that sequences within the C-terminal CAP domain of CAP proteins have the ability to promote cell-cell adhesion. Cloned mouse *Glipr1l1* has a testis-enriched expression profile where GliPR1L1 is post-translationally modified by N-linked glycosylation during spermatogenesis and ultimately becomes localized to the connecting piece of elongated spermatids and sperm. After sperm capacitation, however, GliPR1L1 is also localized to the anterior regions of the sperm head. Zona pellucida binding assays indicate that GliPR1L1 has a role in the binding of sperm to the zona pellucida surrounding the oocyte. Studies suggest that, along with other members of the CAP superfamily and several other proteins, GliPR1L1 is involved in the binding of sperm to the oocyte complex. Results strengthen the role of CAP domain-containing proteins in cellular adhesion and propose a mechanism whereby CAP proteins show overlapping functional significance during fertilization (Gibbs et al. 2010).

41.8.2.2 C9orf19

A human transcript, C9orf19, mapped to the genomic region involved in hereditary inclusion body myopathy (IBM2) at chromosome 9p12-p13, has been characterized. Genomic characterization of the C9orf19 gene identified five exons extending over 27.2 kb of genomic DNA, located 12 kb centromeric to the tumor suppressor RECK gene. C9orf19 mRNA is expressed in a wide range of adult tissues as a single transcript, most abundantly in lung and peripheral blood leukocytes. The predicted protein contains SCP-like extracellular protein signature classified to IPR001283, a family of evolutionary related proteins with extracellular domains, which includes the human GliPR, the human testis specific glycoprotein (TPX-1), and several other extracellular proteins from rodents (SCP), insects venom allergens (Ag5, Ag3), plants pathogenesis proteins (PR-1) and yeast hypothetical proteins (Eisenberg et al. 2002).

41.8.2.3 A 28 kDa Glycoprotein from Human Neutrophils

A 28 kDa glycoprotein has been purified from exocytosed material from human neutrophils. The deduced 245 amino acid sequence of the 2124 bp full-length cDNA showed high degree of similarity to the deduced sequences of human gene TPX-1 and of sperm-coating glycoprotein from rat and mouse. Subcellular fractionation of human neutrophils indicated that the protein is localized in specific granules, hence named SGP28 (specific granule protein of 28 kDa) (Kjeldsen et al. 1996).

41.9 Calx-*b* and CTLD Containing Protein (CBCP) (Group XVII of CTLD)

41.9.1 CBCP/Frem1/QBRICK

CBCP/Frem1/QBRICK large proteoglycan (~2,100 residues) contains a set of chondroitin sulphate proteoglycan (CSPG) repeats (homologous to the NG2 ectodomain) (Nishiyama et al. 1991), a calcium-binding Calx-β domain and a CTLD. The emerging family of extracellular matrix proteins characterized by 12 consecutive CSPG repeats and the presence of Calx-β motif(s) includes Fras1, QBRICK/Frem1, Frem2 and Frem3. Frem1 belongs to a family of structurally related extracellular matrix proteins of which Fras1 is the founding member. Fras1/Frem proteins have been shown to be strictly co-localized in the sublamina densa of mouse epithelial basement membranes during development. Frem3 was present in a broad range of epithelial basement membranes where Fras1, Frem1 and Frem2 were missing. Fras1 and Frem2 were colocalized with Frem3 in the basement membrane of certain skin parts, underlying the thin-layer, of rapidly proliferating keratinocytes, whereas Frem1 was detected only in the basement membrane of the tail (Pavlakis et al. 2008).

41.9.1.1 Fras1 and Frem1 in Sublamina Densa of Epithelial Basement Membranes

Fras1 co-localizes with the markers of epithelial basement membranes and is ultrastructurally detected underneath the lamina densa of embryonic mouse epithelia. The loss of Fras1 mainly affects the cohesiveness of the embryonic skin basement membrane with its underlying mesenchyme. Fras1 could serve as a direct link between the sublamina densa and mesenchyme. The localization of Fras1 is consistent with previous results indicating that Fras1 exerts its function below the lamina densa and that Fras1 displays the same localization pattern in all epithelial basement membranes (Dalezios et al. 2007). It was shown that basement membrane levels of collagen VII rise at late embryonic life, concomitant with descending Fras1/Frem immunolabeling (Chiotaki et al. 2007).

Mutations in Fras1 and Frem1 have been identified in mouse models for Fraser syndrome, which display a strikingly similar embryonic skin blistering phenotype due to impaired dermal-epidermal adhesion. Frem1 originates from both epithelial and mesenchymal cells, in contrast to Fras1 that is exclusively derived from epithelia. However, both proteins are localized in an absolutely overlapping fashion in diverse epithelial basement membranes. At the ultrastructural level, Frem1 exhibits a clustered arrangement in the sublamina densa coinciding with fibrillar structures reminiscent of anchoring fibrils. Furthermore, in addition to its extracellular deposition, around E16, Frem1 displays an intracellular distribution in distinct epidermal cell types such as the periderm layer and basal keratinocytes. Since periderm cells are known to participate in temporary epithelial fusions like embryonic eyelid closure, defective function of Frem1 in these cells could provide a molecular explanation for the "eyes open at birth" phenotype, a feature unique for Frem1 deficient mouse mutants. Finally, Frem1 localization in the basement membrane is lost but not in periderm cells in the skin of Fras1$^{-/-}$ embryos. Reports indicate that besides a cooperative function with Fras1 in embryonic basement membranes, Frem1 can also act independently in processes related to epidermal differentiation (Petrou et al. 2007a).

FREM1 Mutations Cause Bifid Nose, Renal Agenesis, and Anorectal Malformations Syndrome

An autosomal-recessive syndrome of bifid nose and anorectal and renal anomalies (BNAR) was reported in a consanguineous Egyptian sibship. Alazami et al. (2009) identified a shared region of homozygosity on chromosome 9p22.2-p23 that revealed homozygous frameshift and missense mutations in FREM1, which encodes an extracellular matrix component of basement membranes. The phenotypic variability reported for different Frem1 mouse mutants suggests that the apparently distinct phenotype of BNAR in humans may represent an unrecognized variant of Fraser syndrome (Alazami et al. 2009).

41.9.1.2 Association of Fras1/Frem Family with Fraser Syndrome

Fras1 and the structurally related proteins are involved in the structural adhesion of the skin epithelium to its underlying mesenchyme. Deficiency in the individual murine Fras1/Frem genes gives rise to the bleb phenotype, which is equivalent to the human hereditary disorder Fraser syndrome, characterized by cryptophthalmos (hidden eyes), embryonic skin blistering, renal agenesis, and syndactyly. Recent studies revealed a functional cooperation between the Fras1/Frem gene products, in which Fras1, Frem1 and Frem2 are simultaneously stabilized at the lower most region of the basement membrane by forming a macromolecular ternary complex. Loss of any of these proteins results in the collapse of the protein assembly, thus providing a molecular explanation for the highly similar phenotypic defects displayed by the respective mutant mice. Petrou et al. (2008) reviewed the current knowledge regarding the structure, function, and interplay between the proteins of the Fras1/Frem family and proposed a possible scenario for the evolution of the corresponding genes.

QBRICK (Frem 1) in Hair Morphogenesis

During mouse hair morphogenesis, gene predominantly expressed at the tip of developing hair follicles encodes a protein characterized by the presence of 12 tandem repeats of ~120 amino acids and a novel N-terminal domain containing an Arg-Gly-Asp cell-adhesive motif. The protein encoded by this gene, named QBRICK, was localized at the basement membrane zone of embryonic epidermis and hair follicles, in which it was more enriched at the tip rather than the stalk region. QBRICK was active in mediating cell-substratum adhesion through integrins containing α_v or α_8 chain, but not integrin $\alpha_5\beta_1$. QBRICK colocalized with α_v-containing integrins in the interfollicular region, but with the α_8-containing integrin at the tip region of developing hair follicles. These results indicate that QBRICK is an adhesive ligand of basement membrane distinctively recognized by cells in the embryonic skin and hair follicles through different types of integrins directed to the Arg-Gly-Asp motif (Kiyozumi et al. 2005).

41.9.1.3 Genetics of Frem1 in Fraser Syndrome and Blebs Mouse Mutants

Fraser syndrome is a recessive multi-system disorder characterized by embryonic epidermal blistering, cryptophthalmos, syndactyly, renal defects and a range of other developmental abnormalities. The family of four mapped mouse blebs mutants has been proposed as models of this disorder, given their striking phenotypic overlaps. These loci have been cloned, uncovering a family of three large extracellular matrix proteins and an intracellular adapter protein which are required for normal epidermal adhesion early in development. These proteins have been shown to play a crucial role in the development and homeostasis of the kidney. The cloning and characterization of these genes and the consequences of their loss have been explored (Smyth et al. 2004). Smyth et al. (2004) reported the mutations in Frem1 in both the classic head blebs mutant and in an N-ethyl-N-nitrosourea-induced allele and showed that inactivation of the gene results in the formation of in utero epidermal blisters beneath the lamina densa of the basement membrane and also in renal agenesis. Frem1 is expressed widely in the developing embryo in regions of epithelial/mesenchymal interaction and epidermal remodeling. Furthermore, Frem1 appears to act as a dermal mediator of basement membrane adhesion, apparently independently of the other known "blebs" proteins Fras1 and GRIP1. Since the collagen VI and Fras1 deposition in the basement membrane was normal, it indicated that the protein plays an independent role in epidermal differentiation and is required for epidermal adhesion during embryonic development (Smyth et al. 2004).

Jadeja et al. (2005) mapped myelencephalic blebs to Frem2, a gene related to Fras1 and Frem1, and showed that a Frem2 gene-trap mutation was allelic to myelencephalic blebs. Expression of Frem2 in adult kidneys correlated with cyst formation in myelencephalic blebs homozygotes, indicating that the gene is required for maintaining the differentiated state of renal epithelia. Two individuals with Fraser syndrome were homozygous with respect to the same missense mutation of FREM2, confirming genetic heterogeneity. This is the only missense mutation reported in any blebbing mutant or individual with Fraser syndrome, suggesting that calcium binding in the CALXβ-cadherin motif is important for normal functioning of FREM2.

In Frem2 mutant mice, not only Frem2 but Fras1 and QBRICK/Frem1 were depleted from the basement membrane zone. This coordinated reduction in basement membrane deposition was also observed in another Fraser syndrome model mouse, in which GRIP1, a Fras1- and Frem2-interacting adaptor protein, is primarily affected. Targeted disruption of Qbrick/Frem1 also resulted in diminished expression of Fras1 and Frem2 at the epidermal basement membrane, confirming the reciprocal stabilization of QBRICK/Frem1, Fras1, and Frem2 in this location. These proteins formed a ternary complex, raising the possibility that their reciprocal stabilization at the basement membrane is due to complex formation. The coordinated assembly of three Fraser syndrome-associated proteins at the basement membrane appears to be instrumental in epidermal-dermal interactions during morphogenetic processes (Kiyozumi et al. 2006, 2007).

41.9.2 Frem3

The Fraser syndrome protein Fras1 and the structurally related proteins Frem1, Frem2 and Frem3 comprise a novel family of extracellular matrix proteins implicated in the structural adhesion of the embryonic epidermis to the underlying mesenchyme. Fras1, Frem1 and Frem2 have been shown to be simultaneously and interdependently stabilized in the basement membrane by forming a ternary complex located underneath the lamina densa. Although Fras1, QBRICK/Frem1 and Frem2 have been shown to localize at the basement membrane through reciprocal stabilization, Frem3 localized at the basement membrane with tissue distribution patterns clearly distinct from those of other 12 CSPG repeats-containing proteins (12-CSPG proteins). In adult mice, Frem3 was present at the basement membrane underlying ductal cells of the salivary gland, retinal ganglion cells, basal cells of epidermis and hair follicles, where other 12-CSPG proteins were barely expressed. Frem3 is distinct from other 12-CSPG proteins in its tissue distribution and competence to assemble into the basement membrane (Kiyozumi et al. 2007). In absence of Fras1 the basement membrane localization of Frem3 remains unaffected in contrast to Frem1 and Frem2 which are completely abolished from the basement membrane (Petrou et al. 2007b).

41.9.3 Zebrafish Orthologues of FRAS1, FREM1, or FREM2

Using forward genetics, Carney et al. (2010) identified the genes mutated in two classes of zebrafish fin mutants. The mutants of these genes display characteristic blistering underneath the basement membrane of the fin epidermis. Three of them are due to mutations in zebrafish orthologues of FRAS1, FREM1, or FREM2, large basement membrane protein encoding genes that are mutated in mouse bleb mutants and in human patients suffering from Fraser Syndrome. In addition to mutations in Hemicentin1 (Hmcn1) Carney et al. (2010) identified the extracellular matrix protein Fibrillin2 as an indispensable interaction partner of Hmcn1 (Carney et al. 2010).

References

CD93/gC1qR/Throbomodulin/Endosialin

Adams TE, Huntington JA (2006) Thrombin-cofactor interactions: structural insights into regulatory mechanisms. Arterioscler Thromb Vasc Biol 26:1738–1745

Adams TE, Li W, Huntington JA (2009) Molecular basis of thrombomodulin activation of slow thrombin. J Thromb Haemost 7:1688–1695

Bagley RG (2009) Endosialin: from vascular target to biomarker for human sarcomas. Biomark Med 3:589–604

Bagley RG, Honma N, Weber W, Boutin P, Rouleau C, Shankara S, Kataoka S, Ishida I, Roberts BL, Teicher BA (2008a) Endosialin/TEM 1/CD248 is a pericyte marker of embryonic and tumor neovascularization. Microvasc Res 76:180–188

Bagley RG, Rouleau C, St Martin T et al (2008b) Human endothelial precursor cells express tumor endothelial marker 1/endosialin/CD248. Mol Cancer Ther 8:2536–2546

Bagley RG, Weber W, Rouleau C et al (2009) Human mesenchymal stem cells from bone marrow express tumor endothelial and stromal markers. Int J Oncol 34:619–627

Bajzar L, Morser J, Nesheim M (1996) TAFI, or plasma procarboxypeptidase B, couples the coagulation and fibrinolytic cascades through the thrombin- thrombomodulin complex. J Biol Chem 271:16603–16608

References

Barnes PJ, Karin M (1997) Nuclear factor-kappa b: a pivotal transcription factor in chronic inflammatory disease. N Engl J Med 336:1066–1071

Becker DL, Fredenburgh JC, Stafford AR, Weitz JI (1999) Exosites 1 and 2 are essential for protection of fibrin-bound thrombin from heparin-catalyzed inhibition by antithrombin and heparin cofactor II. J Biol Chem 274:6226–6233

Becker R, Lenter MC, Vollkommer T et al (2008) Tumor stroma marker endosialin (TEM-1) is a binding partner of metastasis-related protein Mac-2 BP/90 K. FASEB J 22:3059–3067

Bicheng N, Hui Y, Shaoyu Y et al (2005) C-reactive protein decreases expression of thrombomodulin and endothelial protein C receptor in human endothelial cells. Surgery 138:212–222

Bohlson SS, Silva R, Fonseca MI, Tenner AJ (2005a) CD93 is rapidly shed from the surface of human myeloid cells and the soluble form is detected in human plasma. J Immunol 175:1239–1247

Bohlson SS, Zhang M, Ortiz CE et al (2005b) CD93 interacts with the PDZ domain-containing adaptor protein GIPC: implications in the modulation of phagocytosis. J Leukoc Biol 77:80–89

Brady J, Neal J, Sadakar N, Gasque P (2004) Human endosialin (tumor endothelial marker 1) is abundantly expressed in highly malignant and invasive brain tumors. J Neuropathol Exp Neurol 63: 1274–1283

Braun L, Ghebrehiwet B, Cossart P (2000) gC1qR/p32, a C1q-binding protein, is a receptor for the InlB invasion protein of Listeria monocytogenes. EMBO J 19:1458–1466

Cancro MP (2004) Peripheral B-cell maturation: the intersection of selection and homeostasis. Immunol Rev 197:89–101

Carson-Walter EB, Watkins DN et al (2001) Cell surface tumor endothelial markers are conserved in mice and humans. Cancer Res 61:6649–6655

Carson-Walter EB, Winans BN et al (2009) Characterization of TEM-1/endosialin in human and murine brain tumors. BMC Cancer 9:417

Chen YG, Driver JP, Silveira PA, Serreze DV (2007) Subcongenic analysis of genetic basis for impaired development of invariant NKT cells in NOD mice. Immunogenetics 59:705–712

Chevrier S, Genton C, Kallies A et al (2009) CD93 is required for maintenance of antibody secretion and persistence of plasma cells in the bone marrow niche. Proc Natl Acad Sci USA 106:3895–3900

Christian S, Ahorn H, Koehler A et al (2001) Molecular cloning and characterization of endosialin, a C-type lectin-like cell surface receptor of tumor endothelium. J Biol Chem 276:7408–7414

Christian S, Winkler R, Helfrich I et al (2008) Endosialin (TEM-1) is a marker of tumor-associated myofibroblasts and tumor vessel-associated mural cells. Am J Pathol 172:486–494

Clarke JH, Light DR, Blasko E et al (1993) The short loop between epidermal growth factor-like domains 4 and 5 is critical for human thrombomodulin function. J Biol Chem 268:6309–6315

Conway EM, Van de Wouwer M, Pollefeyt S et al (2002) The lectin-like domain of thrombomodulin confers protection from neutrophil-mediated tissue damage by suppressing adhesion molecule expression via nuclear factor kappaB and mitogen-activated protein kinase pathways. J Exp Med 196:565–577

Cole JW, Roberts SC, Gallagher M et al (2004) Thrombomodulin Ala455Val Polymorphism and the risk of cerebral infarction in a biracial population: the Stroke Prevention in Young women study. BMC Neurol 4:21

Dahlbäck B, Villoutreix BO (2005) The anticoagulant protein C pathway. FEBS 579:3310–3316

Danet GH, Luongo JL, Butler G et al (2002) C1qRp defines a new human stem cell population with hematopoietic and hepatic potential. Proc Natl Acad Sci USA 99:10441–10445

Davies G, Cunnick GH, Mansel RE et al (2004) Levels of expression of endothelial markers specific to tumor-associated endothelial cells and their correlation with prognosis in patients with breast cancer. Clin Exp Metastasis 21:31–37

Dean YD, McGreal EP, Akatsu H et al (2000) Molecular and cellular properties of the rat AA4 antigen, a C-type lectin-like receptor with structural homology to thrombomodulin. J Biol Chem 275:34382–34392

Dean YD, McGreal EP, Gasque P (2001) Endothelial cells, megakaryoblasts, platelets and alveolar epithelial cells express abundant levels of the mouse AA4 antigen, a C-type lectin-like receptor involved in homing activities and innate immune host defense. Eur J Immunol 31:1370–1381

DeBault LE, Esmon NL, Olson JR, Esmon CT (1986) Distribution of the TM antigen in the rabbit vasculature. Lab Invest 54:172–178

Dedio J, Müller-Esterl W (1996) Kininogen binding protein p33/gC1qR is localized in the vesicular fraction of endothelial cells. FEBS Lett 399:255–258

Dedio J, Renné T, Weisser M et al (1999) Subcellular targeting of multiligand-binding protein gC1qR. Immunopharmacology 45:1–5

Demoulin JB (2010) No PDGF receptor signal in pericytes without endosialin? Cancer Biol Ther 9:916–918

Dittman WA, Kumada T, Majerus PW (1989) Transcription of thrombomodulin mRNA in mouse hemagioma cells is enhanced by cycloheximide and thrombin. Proc Natl Acad Sci USA 86:7179–7182

Dittman WA, Majerus PW (1990) Structure and function of thrombomodulin: a natural anticoagulant. Blood 75:329–336

Dittman WA, Majerus PW (1989) Structure of a cDNA for mouse thrombomodulin and comparison of the predicted mouse and human amino acid sequences. Nucleic Acids Res 172:302

Dolznig H, Schweifer N, Puri C et al (2005) Characterization of cancer stroma markers: in silico analysis of an mRNA expression database for fibroblast activation protein and endosialin. Cancer Immun 5:10

Dzionek A, Fuchs A, Schmidt P et al (2000) BDCA-2, BDCA-3, and BDCA-4: three markers for distinct subsets of dendritic cells in human peripheral blood. J Immunol 165:6037–6046

Dzionek A, Inagaki Y, Okawa K et al (2002) Plasmacytoid dendritic cells: from specific surface markers to specific cellular functions. Hum Immunol 63:1133–1148

Eggleton P, Ghebrehiwet B, Sastry KN et al (1995) Identification of a gC1q-binding protein (gC1qR) on the surface of human neutrophils. Subcellular localization and binding properties in comparison with the cC1qR. J Clin Invest 95:1569–1578

Esmon CT (2000) Regulation of blood coagulation. Biochim Biophys Acta 1477:349–360

Esmon CT (2001) Role of coagulation inhibitors in inflammation. Thromb Haemost 86:51–56

Esmon CT (1987) The regulation of natural anticoagulant pathways. Science 235(4794):1348–1352

Esmon CT (1995) Thrombomodulin as a model of molecular mechanism that modulates protease specificity and function at the vessel surface. FASEB J 9:946–955

Esteban LM, Tsoutsman T, Jordan MA et al (2003) Genetic control of NKT cell numbers maps to major diabetes and lupus loci. J Immunol 171:2873–2878

Fonseca MI, Carpenter PM, Park M et al (2001) C1qR(P), a myeloid cell receptor in blood, is predominantly expressed on endothelial cells in human tissue. J Leukoc Biol 70:793–800

Fredenburgh JC, Stafford AR, Weitz JI (1997) Evidence for allosteric linkage between exosites 1 and 2 of thrombin. J Biol Chem 272:25493–25499

Fuentes-Prior P, Iwanaga Y et al (2000) Structural basis for the anticoagulant activity of the thrombin thrombomodulin complex. Nature 404:518–525

Ghebrehiwet B, Jesty J, Peerschke EI (2002) gC1qR/p33: structure-function predictions from the crystal structure. Immunobiology 205:421–432

Ghebrehiwet B, Kew RR, Gruber B et al (1995) Murine mast cells express two types of C1q receptors that are involved in the induction of chemotaxis and chemokinesis. J Immunol 155:2614–2619

Ghebrehiwet B, Lim BL, Kumar R et al (2001) gC1qR/p33, a member of a new class of multifunctional and multicompartmental cellular proteins, is involved in inflammation and infection. Immunol Rev 180:65–77

Ghebrehiwet B, Lim B-L, Peerschke EIB et al (1994) Isolation, cDNA cloning, and overexpression of a 33-kD cell surface glycoprotein that binds to the globular "heads" of C1q. J Exp Med 179:1809–1821

Ghebrehiwet B, Lu PD, Zhang W et al (1997) Evidence that the two C1q binding membrane proteins, gC1qR and cC1qR, associate to form a complex. J Immunol 159:1429–1436

Ghebrehiwet B, Peerschke EI (2004a) cC1qR (calreticulin) and gC1qR/p33: ubiquitously expressed multi-ligand binding cellular proteins involved in inflammation and infection. Mol Immunol 41:173–183

Ghebrehiwet B, Peerschke EI (2004b) Role of C1q and C1q receptors in the pathogenesis of systemic lupus erythematosus. Curr Dir Autoimmun 7:87–97

Ghebrehiwet B, Peerschke EI (1998) Structure and function of gC1qR: a multiligand binding cellular protein. Immunobiology 199:225–238

Grace KS, Bronson RA, Ghebrehiwet B (2002) Surface expression of complement receptor gC1qR/p33 is increased on the plasma membrane of human spermatozoa after capacitation. Biol Reprod 66:823–829

Greenlee MC, Sullivan SA, Bohlson SS (2008) CD93 and related family members: their role in innate immunity. Curr Drug Targets 9:130–138

Greenlee MC, Sullivan SA, Bohlson SS (2009) Detection and characterization of soluble CD93 released during inflammation. Inflamm Res 58:909–919

Goldstein JL, Brown MS, Anderson RG, Russell DW, Schneider WJ (1985) Receptor-mediated endocytosis: concepts emerging from the LDL receptor system. Annu Rev Cell Biol 1:1–39

Guan E, Robinson SL, Goodman EB, Tenner AJ (1994) Cell-surface protein identified on phagocytic cells modulates the C1q-mediated enhancement of phagocytosis. J Immunol 152:4005–4016

Harhausen D, Prinz V et al (2010) CD93/AA4.1: a novel regulator of inflammation in murine focal cerebral ischemia. J Immunol 184:6407–6417

Hasan AA, Cines DB, Herwald H, Schmaier AH, Müller-Esterl W (1995) Mapping the cell binding site on high molecular weight kininogen domain 5. J Biol Chem 270:19256–19261

Hayashi T, Zushi M, Yamamoto S et al (1990) Further localization of binding sites for thrombin and protein C in human thrombomodulin. J Biol Chem 265:20156–20159

Herwald H, Dedio J, Kellner R et al (1996) Isolation and characterization of the kininogen-binding protein p33 from endothelial cells. Identity with the gC1q receptor. J Biol Chem 271:13040–13047

Herwald H, Hasan AA, Godovac-Zimmermann J et al (1995) Identification of an endothelial cell binding site on kininogen domain D3. J Biol Chem 270:14634–14642

Hofsteenge J, Taguchi H, Stone SR (1986) Effect of thrombomodulin on the kinetics of the interaction of thrombin with substrates and inhibitors. Biochem J 237:243–251

Huber MA, Kraut N et al (2006) Expression of stromal cell markers in distinct compartments of human skin cancers. J Cutan Pathol 33:145–155

Ikewaki N, Kulski JK, Inoko H (2006) Regulation of CD93 cell surface expression by protein kinase C isoenzymes. Microbiol Immunol 50:93–103

Ikewaki N, Tamauchi H, Inoko H (2007) Decrease in CD93 (C1qRp) expression in a human monocyte-like cell line (U937) treated with various apoptosis-inducing chemical substances. Microbiol Immunol 51:1189–1200

Ireland H, Kyriakoulis K, Kunz G, Lane DA (1996) Directed search for thrombomodulin gene mutations. Haemostasis 26(Suppl 4):227–232

Isermann B, Hendrickson SB, Hutley K et al (2001) Tissue restricted expression of TM in the placenta rescues thrombomodulin deficient mice from early lethality and reveals a secondary developmental block. Development 28:827–838

Isermann B, Sood R, Pawlinski R et al (2003) The thrombomodulin–protein C system is essential for the maintenance of pregnancy. Nat Med 9:331–337

Ishii H, Salem HH, Bell CE et al (1986) Thrombomodulin, an endothelial anticoagulant protein, is absent from the human brain. Blood 67:362–365

Jackman RW, Beeler DL, Fritze L et al (1987) Human thrombomodulin gene is intron depleted: nucleic acid sequences of the cDNA and gene predict protein structure and suggest sites of regulatory control. Proc Natl Acad Sci USA 84:6425–6429

Jackman RW, Beeler DL, van de Water L et al (1986) Characterization of a thrombomodulin cDNA reveals structural similarity to the low density lipoprotein receptor. Proc Natl Acad Sci USA 83:8834–8838

Jakubowski HV, Owen WG (1989) Macromolecular specificity determinants on thrombin for fibrinogen and thrombomodulin. J Biol Chem 264:11117–11121

Jiang J, Zhang Y, Krainer AR, Xu RM (1999) Crystal structure of human p32, a doughnut-shaped acidic mitochondrial matrix protein. Proc Natl Acad Sci USA 96:3572–3577

Jordan MA, Fletcher J, Baxter AG (2004) Genetic control of NKT cell numbers. Immunol Cell Biol 82:276–284

Jongbloed SL, Kassianos AJ, McDonald KJ, Clark GJ, Ju X, Angel CE, Chen CJ, Dunbar PR, Wadley RB, Jeet V, Vulink AJ, Hart DN, Radford KJ (2010) Human CD141[+] (BDCA-3)[+] dendritic cells (DCs) represent a unique myeloid DC subset that cross-presents necrotic cell antigens. J Exp Med 207:1247–1260

Joseph K, Ghebrehiwet B, Peerschke EI et al (1996) Identification of the zinc-dependent endothelial cell binding protein for high molecular weight kininogen and factor XII: identity with the receptor that binds to the globular "heads" of C1q (gC1qR). Proc Natl Acad Sci USA 93:8552–8557

Joseph K, Shibayama Y, Ghebrehiwet B, Kaplan AP (2001) Factor XII-dependent contact activation on endothelial cells and binding proteins gC1qR and cytokeratin 1. Thromb Haemost 85:119–124

Joseph K, Tholanikunnel BG, Ghebrehiwet B, Kaplan AP (2004) Interaction of high molecular weight kininogen binding proteins on endothelial cells. Thromb Haemost 91:61–70

Kaare M, Ulander VM, Painter JN et al (2007) Variations in the thrombomodulin and endothelial protein C receptor genes in couples with recurrent miscarriage. Hum Reprod 22:864–868

Kim TS, Park M, Nepomuceno RR et al (2000) Characterization of the murine homolog of C1qR(P): identical cellular expression pattern, chromosomal location and functional activity of the human and murine C1qR(P). Mol Immunol 37:377–389

Kittlesen DJ, Chianese-Bullock KA, Yao ZQ et al (2000) Interaction between complement receptor gC1qR and hepatitis C virus core protein inhibits T-lymphocyte proliferation. J Clin Invest 106:1239–1249

Kittner SJ, Stern BJ, Wozniak M et al (1998) Cerebral infarction in young adults: the Baltimore-Washington cooperative young stroke study. Neurology 50:890–894

Koeppe JR, Seitova A et al (2005) Thrombomodulin tightens the thrombin active site loops to promote protein C activation. Biochemistry 44:14784–14791

Kokame K, Zheng X, Evan Sadler J (1998) Activation of thrombin-activatable fibrinolysis inhibitor requires epidermal growth factor-like domain 3 of thrombomodulin and is inhibited competitively by protein C. J Biol Chem 273:12135–12139

Kumar R, Peerschke EI, Ghebrehiwet B (2002) Zinc induces exposure of hydrophobic sites in the C-terminal domain of gC1qR/p33. Mol Immunol 39:69–75

Kurosawa S, Stearns DJ, Jackson KW, Esmon CT (1988) A 10-kDa cyanogen bromide fragment from the epidermal growth factor homology domain of rabbit thrombomodulin contains the primary thrombin binding site. J Biol Chem 263:5993–5996

Løvik G, Larsen Sand K, Iversen JG et al (2001) C1qRp elicits a Ca++ response in rat NK cells but does not influence NK-mediated cytotoxicity. Scand J Immunol 53:410–415

Løvik G, Vaage JT, Dissen E et al (2000) Characterization and molecular cloning of rat C1qRp, a receptor on NK cells. Eur J Immunol 30:3355–3362

Lainé S, Thouard A, Derancourt J et al (2003) In vitro and in vivo interactions between the hepatitis B virus protein P22 and the cellular protein gC1qR. J Virol 77:12875–12880

Lax S, Hardie DL, Wilson A et al (2010) The pericyte and stromal cell marker CD248 (endosialin) is required for efficient lymph node expansion. Eur J Immunol 40:1884–1889

Lax S, Hou TZ, Jenkinson E et al (2007) CD248/Endosialin is dynamically expressed on a subset of stromal cells during lymphoid tissue development, splenic remodeling and repair. FEBS Lett 581:3550–3556

Lentz SR, Chen Y, Sadler JE (1993) Sequences required for thrombomodulin cofactor activity within the fourth epidermal growth factor-like domain of human thrombomodulin. J Biol Chem 268:15312–15317

Liao D, Johnson RS (2007) Hypoxia: a key regulator of angiogenesis in cancer. Cancer Metastasis Rev 26:281–290

Lim BL, Reid KB, Ghebrehiwet B et al (1996) The binding protein for globular heads of complement C1q, gC1qR. Functional expression and characterization as a novel vitronectin binding factor. J Biol Chem 271:26739–26744

Lu PD, Galanakis DK, Ghebrehiwet B et al (1999) The receptor for the globular "heads" of C1q, gC1qR, binds to fibrinogen/fibrin and impairs its polymerization. Clin Immunol 90:360–367

Lurton J, Soto H, Narayanan AS, Raghu G (1999) Regulation of human lung fibroblast C1qReceptors by transforming growth factor-β and tumor necrosis factor-α. Exp Lung Res 25:151–164

Lynch NJ, Reid KB, van den Berg RH et al (1997) Characterisation of the rat and mouse homologs of gC1qBP, a 33 kDa glycoprotein that binds to the globular 'heads' of C1q. FEBS Lett 418:111–114

MacFadyen J, Savage K, Wienke D, Isacke CM (2007) Endosialin is expressed on stromal fibroblasts and CNS pericytes in mouse embryos and is downregulated during development. Gene Expr Patterns 7:363–369

MacFadyen JR, Haworth O, Roberston D et al (2005) Endosialin (TEM-1, CD248) is a marker of stromal fibroblasts and is not selectively expressed on tumor endothelium. FEBS Lett 579:2569–2575

Madden SL, Cook BP, Nacht M et al (2004) Vascular gene expression in nonneoplastic and malignant brain. Am J Pathol 165:601–608

Malhotra R, Haurum J, Thiel S, Sim RB (1992) Interaction of C1q receptor with lung surfactant protein A. Eur J Immunol 22:1437–1445

Malhotra R, Thiel S, Reid KB, Sim RB (1990) Human leukocyte C1q receptor binds other soluble proteins with collagen domains. J Exp Med 172:955–959

Maruyama I, Elliott Bell C, Majerus PW (1985) Thrombomodulin is found on endothelium of arteries, veins, capillaries, and lymphatics, and on syncytiotrophoblast of human placenta. J Cell Biol 101:363–371

McGreal E, Gasque P (2002) Structure-function studies of the receptors for complement C1q. Biochem Soc Trans 30:1010–1014

McGreal EP, Ikewaki N, Akatsu H et al (2002) Human C1qRp is identical with CD93 and the mNI-11 antigen but does not bind C1q. J Immunol 168:5222–5232

Miettinen M, Sarlomo-Rikala M (2003) Expression of calretinin, thrombomodulin, keratin 5, and mesothelin in lung carcinomas of different types: an immunohistochemical analysis of 596 tumors in comparison with epithelioid mesotheliomas of the pleura. Am J Surg Pathol 27:150–158

Moore KL, Andreoli SP, Esmon NL et al (1987) Endotoxin enhances tissue factor and suppresses thrombomodulin expression of human vascular endothelium in vitro. J Clin Invest 79:124–130

Moosig F, Fähndrich E, Knorr-Spahr A et al (2006) C1qRP (CD93) expression on peripheral blood monocytes in patients with systemic lupus erythematosus. Rheumatol Int 26:1109–1112

Myles T, Le Bonniec BF, Betz A, Stone SR (2001) Electrostatic steering and ionic tethering in the formation of thrombin-hirudin complexes: the role of the thrombin anion-binding exosite-I. Biochemistry 40:4972–4979

Nagashima M, Lundh E, Leonard JC et al (1993) Alanine-scanning mutagenesis of the epidermal growth factor-like domains of human thrombomodulin identifies critical residues for its cofactor activity. J Biol Chem 268:2888–2892

Nanda A, Karim B, Peng Z et al (2006) Tumor endothelial marker 1 (TEM-1) functions in the growth and progression of abdominal tumors. Proc Natl Acad Sci USA 103:3351–3356

Narayanan AS, Lurton J, Raghu G (1997) Distribution of receptors of collagen and globular domains of C1q in human lung fibroblasts. Am J Respir Cell Mol Biol 17:84–90

Naworth PP, Handley DA, Esmon CT, Stern DM (1986) Interleukin1 induces endothelial cell procoagulant while suppressing cellsurface anticoagulant activity. Proc Natl Acad Sci USA 83:3460–3464

Nepomuceno RR, Henschen-Edman AH, Burgess WH et al (1997) cDNA cloning and primary structure analysis of C1qR(P), the human C1q/MBL/SPA receptor that mediates enhanced phagocytosis in vitro. Immunity 6:119–129

Nepomuceno RR, Ruiz S, Park M, Tenner AJ (1999) C1qRP is a heavily O-glycosylated cell surface protein involved in the regulation of phagocytic activity. J Immunol 162:3583–3589

Nepomuceno RR, Tenner AJ (1998) C1qRP, the C1q receptor that enhances phagocytosis, is detected specifically in human cells of myeloid lineage, endothelial cells, and platelets. J Immunol 160:1929–1935

Nesheim M (2003) Thrombin and fibrinolysis. Chest 124:S33–S95

Nguyen T, Ghebrehiwet B, Peerschke EI (2000) Staphylococcus aureus protein A recognizes platelet gC1qR/p33: a novel mechanism for staphylococcal interactions with platelets. Infect Immun 68:2061–2068

Norlund L, Holm J, Zoller B, Ohlin AK (1997) A common thrombomodulin amino acid dimorphism is associated with myocardial infarction. Thromb Haemost 77:248–251

Norsworthy PJ, Fossati-Jimack L, Cortes-Hernandez J et al (2004) Murine CD93 (C1qRp) contributes to the removal of apoptotic cells in vivo but is not required for C1q-mediated enhancement of phagocytosis. J Immunol 172:3406–3414

Ohradanova A, Gradin K, Barathova M et al (2008) Hypoxia upregulates expression of human endosialin gene via hypoxia-inducible factor 2. Br J Cancer 99:1348–1356

Opavsky R, Haviernik P, Jurkovicova D et al (2001) Molecular characterization of the mouse TEM-1/endosialin gene regulated by cell density in vitro and expressed in normal tissues in vivo. J Biol Chem 276:38795–38807

Park M, Tenner AJ (2003) Cell surface expression of C1qRP/CD93 is stabilized by O-glycosylation. J Cell Physiol 196:512–522

Parkinson JF, Nagashima M, Kuhn I et al (1992) Structure-function studies of the epidermal growth factor domains of human thrombomodulin. Biochem Biophys Res Commun 185: 567–576

Patthy L (1988) Detecting distant homologies of mosaic proteins: analysis of the sequences of thrombomodulin, thrombospondin, complement components C9, C8 α, and C8 β, vitronectin and plasma cell membrane glycoprotein PC-1. J Mol Biol 202: 689–696

Peerschke EI, Ghebrehiwet B (2001) Human blood platelet gC1qR/p33. Immunol Rev 180:56–64

Peerschke EI, Ghebrehiwet B (2007) The contribution of gC1qR/p33 in infection and inflammation. Immunobiology 212:333–342

Peerschke EI, Minta JO, Zhou SZ et al (2004) Expression of gC1qR/p33 and its major ligands in human atherosclerotic lesions. Mol Immunol 41:759–766

Peterson KL, Zhang W, Lu PD et al (1997) The C1q-binding cell membrane proteins cC1qR and gC1qR are released from activated cells: subcellular distribution and immunochemical characterization. Clin Immunol Immunopathol 84:17–26

Pupo AS, Minneman KP (2003) Specific interactions between gC1qR and α1-adrenoceptor subtypes. J Recept Signal Transduct Res 23:185–195

Rahman ZS, Tin SK, Buenaventura PN et al (2002) A novel susceptibility locus on chromosome 2 in the (New Zealand Black x New Zealand White)F1 hybrid mouse model of systemic lupus erythematosus. J Immunol 168:3042–3049

Rettig WJ, Garin-Chesa P, Healey JH et al (1992) Identification of endosialin, a cell surface glycoprotein of vascular endothelial cells in human cancer. Proc Natl Acad Sci USA 89:10832–10836

Rezaie AR, Yang L (2003) Thrombomodulin allosterically modulates the activity of the anticoagulant thrombin. Proc Natl Acad Sci 100:12051–12056

Rezaie AR (1999) Role of exosites 1 and 2 in thrombin reaction with plasminogen activator inhibitor-1 in the absence and presence of cofactors. Biochemistry 38:14592–14599

Rosen SD (2004) Ligands for L-selectin: homing, inflammation, and beyond. Annu Rev Immunol 22:129–156

Rouleau C, Curiel M, Weber W et al (2008) Endosialin protein expression and therapeutic target potential in human solid tumors: sarcoma versus carcinoma. Clin Cancer Res 14:7223–7236

Rozanov DV, Ghebrehiwet B, Ratnikov B et al (2002) The cytoplasmic tail peptide sequence of membrane type-1 matrix metalloproteinase (MT1-MMP) directly binds to gC1qR, a compartment-specific chaperone-like regulatory protein. FEBS Lett 527:51–57

Rubinstein DB, Stortchevoi A, Boosalis M et al (2004) Receptor for the globular heads of C1q (gC1qR, p33, hyaluronan-binding protein) is preferentially expressed by adenocarcinoma cells. Int J Cancer 110:741–750

Rupp C, Dolznig H, Puri C et al (2006) Mouse endosialin, a C-type lectin-like cell surface receptor: expression during embryonic development and induction in experimental cancer neoangiogenesis. Cancer Immun 6:10

Sadler JE (1997) TM structure and function. Thromb Haemost 78:392–395

Sansonno D, Tucci FA, Ghebrehiwet B et al (2009) Role of the receptor for the globular domain of C1q protein in the pathogenesis of hepatitis C virus-related cryoglobulin vascular damage. J Immunol 183:6013–6020

Schenk-Braat EA, Morser J, Rijken DC (2001) Identification of the epidermal growth factor-like domains of thrombomodulin essential for the acceleration of thrombin-mediated inactivation of single-chain urokinase-type plasminogen activator. Eur J Biochem 268:5562–5569

Schulz UGR, Flossmann E, Rothwell PM (2004) Heritability of ischemic stroke in relation to age, vascular risk factors, and subtypes of incident stroke in population-based studies. Stroke 35: 819–824

Serreze DV, Bridgett M, Chapman HD et al (1998) Subcongenic analysis of the Idd13 locus in NOD/Lt mice: evidence for several susceptibility genes including a possible diabetogenic role for β 2-microglobulin. J Immunol 160:1472–1478

Shaw G, Kamen R (1986) A conserved AU sequence from the 3′untranslated region of GM-CSF mRNA mediates selective mRNA degradation. Cell 46:659–667

Shira T, Shiojiri S, Ito H et al (1988) Gene structure of human thrombomodulin, a cofactor for thrombin catalyzed activation of protein C. J Biochem 103:281–285

Simonavicius N, Robertson D et al (2008) Endosialin (CD248) is a marker of tumor-associated pericytes in high-grade glioma. Mod Pathol 21:308–315

Slungaard A, Vercellotti GM, Tran T et al (1993) Eosinophil cationic granule proteins impair thrombomodulin function. A potential mechanism for in hypereosinophilic heart disease. J Clin Invest 91:1721–1730

Soltys BJ, Kang D, Gupta RS (2000) Localization of P32 protein (gC1qR) in mitochondria and at specific extramitochondrial locations in normal tissues. Histochem Cell Biol 114:245–255

St Croix B, Rago C, Velculescu V et al (2000) Genes expressed in human tumor endothelium. Science 289:1197–1202

Stearns DJ, Kurosawa S, Esmon CT (1989) Micro thrombomodulin. Residues 3 10-486 from the epidermal growth factor precursor homology domain of TM will accelerate protein C activation. J Biol Chem 264:3352–3356

Steinberger P, Szekeres A, Wille S et al (2002) Identification of human CD93 as the phagocytic C1q receptor (C1qRp) by expression cloning. J Leukoc Biol 71:133–140

Stortoni P, Cecati M, Giannubilo SR et al (2010) Placental thrombomodulin expression in recurrent miscarriage. Reprod Biol Endocrinol 8:1

Stuart GR, Lynch NJ et al (1997) The C1q and collectin binding site within C1q receptor (cell surface calreticulin). Immunopharmacology 38:73–80

Stuart GR, Lynch NJ, Lu J et al (1996) Localisation of the C1q binding site within C1q receptor/calreticulin. FEBS Lett 397:245–249

Suzuki K, Hayashi T, Nishioka J et al (1989) A domain composed of epidermal growth factor-like structures of human thrombomodulin is essential for thrombin binding and for protein C activation. J Biol Chem 264:4872–4876

Suzuki K, Kusumoto H, Deyashiki Y et al (1987) Structure and expression of human thrombomodulin, a thrombin receptor on endothelium acting as a cofactor for protein C activation. EMBO J 6:1891–1897

Suzuki K, Nichioka J, Hayaoni T, Kosaka Y (1988) Functionally active thrombomodulin is present in human platelets. J Biochem 104:628

Suzuki M, Mohri M, Yamamoto S (1998) In vitro anticoagulant properties of a minimum functional fragment of human thrombomodulin and in vivo demonstration of its benefit as an anticoagulant in extracorporeal circulation using a monkey model. Thromb Haemost 79:417–422

Szabo J, Cervenak L, Toth FD et al (2011) Soluble gC1qR/p33, a cell protein that binds to the globular "heads" of C1q, effectively inhibits the growth of HIV-1 strains in cell cultures. Clin Immunol 99:222–231

Takagi T, Taguchi O, Toda M, Ruiz DB et al (2011) Inhibition of allergic bronchial asthma by thrombomodulin is mediated by dendritic cells. Am J Respir Crit Care Med 183:31–42

Teicher BA (2007) Newer vascular targets: endosialin (review). Int J Oncol 30:305–312

Tenner AJ, Robinson SL, Ezekowitz RA (1995) Mannose binding protein (MBP) enhances mononuclear phagocyte function via a receptor that contains the 126,000 Mr component of the C1q receptor. Immunity 3:485–493

Tenner AJ (1998) C1q receptors: regulating specific functions of phagocytic cells. Immunobiology 199:250–264

Tolkatchev D, Ng A, Zhu B, Ni F (2000) Identification of a thrombin-binding region in the sixth epidermal growth factor-like repeat of human thrombomodulin. Biochemistry 39:10365–10372

Tomkowicz B, Rybinski K, Foley B et al (2007) Interaction of endosialin/TEM-1 with extracellular matrix proteins mediates cell adhesion and migration. Proc Natl Acad Sci USA 104:17965–17970

Tomkowicz B, Rybinski K, Sebeck D et al (2010) Endosialin/TEM-1/CD248 regulates pericyte proliferation through PDGF receptor signaling. Cancer Biol Ther 9:11

Tsiang M, Lentz SR, Sadler JE (1992) Functional domains of membrane-bound human thrombomodulin. EGF-like domains four to six and the serine/threonine-rich domain are required for cofactor activity. J Biol Chem 267:6164–6170

Tye AJ, Ghebrehiwet B, Guo N et al (2001) The human gC1qR/p32 gene, C1qBP. Genomic organization and promoter analysis. J Biol Chem 276:17069–17075

van den Berg RH, Faber-Krol M, van Es LA, Daha MR (1995) Regulation of the function of the first component of complement by human C1q receptor. Eur J Immunol 25:2206–2210

van den Berg RH, Siegert CE, Faber-Krol MC et al (1998) Anti-C1q receptor/calreticulin autoantibodies in patients with systemic lupus erythematosus (SLE). Clin Exp Immunol 111:359–364

van der Velden PA, Krommenhoek-Van Es T, Allaart CF et al (1991) A frequent thrombomodulin amino acid dimorphism is not associated with thrombophilia. Thromb Haemost 65:511–513

Vegh Z, Goyarts EC, Rozengarten K et al (2003) Maturation-dependent expression of C1q binding proteins on the cell surface of human monocyte-derived dendritic cells. Int Immunopharmacol 3:39–51

Vegh Z, Kew RR, Gruber BL, Ghebrehiwet B (2006) Chemotaxis of human monocyte-derived dendritic cells to complement component C1q is mediated by the receptors gC1qR and cC1qR. Mol Immunol 43:1402–1407

Vindigni A, White CE, Komives EA, Di Cera E (1997) Energetics of thrombin- thrombomodulin interaction. Biochemistry 36: 6674–6681

Virgintino D, Girolamo F, Errede M et al (2007) An intimate interplay between precocious, migrating pericytes and endothelial cells governs human fetal brain angiogenesis. Angiogenesis 10:35–45

Wang W, Nagashima M, Schneider M et al (2000) Elements of the primary structure of thrombomodulin required for efficient thrombin-activable fibrinolysis inhibitor activation. J Biol Chem 275:22942–22947

Webster SD, Park M, Fonseca MI, Tenner AJ (2000) Structural and functional evidence for microglial expression of C1qR(P), the C1q receptor that enhances phagocytosis. J Leukoc Biol 67:109–116

Weisel JW, Nagaswami C, Young TA, Light DR (1996) The shape of thrombomodulin and interactions with thrombin as determined by electron microscopy. J Biol Chem 271:31485–31490

Wen D, Dittman WA, Ye RD et al (1987) Human thrombomodulin: complete cDNA sequence and chromosome localization of the gene. Biochemistry 26:4350–4357

White CE, Hunter MJ, Meininger DP et al (1996) The fifth epidermal growth factor-like domain of thrombomodulin does not have an epidermal growth factor-like disulfide bonding pattern. Proc Natl Acad Sci USA 93:10177–10182

Wood MJ, Helena Prieto J, Komives EA (2005) Structural and functional consequences of methionine oxidation in thrombomodulin. Biochim Biophys Acta 1703:141–147

Wouwer M, Collen D, Conway EM (2004) TM protein C-EPCR system: integrated to regulate coagulation and inflammation. Arterioscler Thromb Vasc Biol 24:1374–1383

Wu KK, Aleksic N, Ahn C et al (2001) Atherosclerosis risk in communities study (ARIC) investigators. Thrombomodulin Ala455Val polymorphism and risk of coronary heart disease. Circulation 103:1386–1389

Xu Z, Hirasawa A, Shinoura H et al (1999) Interaction of the $\alpha(1B)$-adrenergic receptor with gC1qR, a multifunctional protein. J Biol Chem 274:21149–21154

Yamane T, Hosen N, Yamazaki H et al (2009) Expression of AA4.1 marks lymphohematopoietic progenitors in early mouse development. Proc Natl Acad Sci USA 106:8953–8958

Yang L, Manithody C, Rezaie AR (2006) Activation of protein C by the thrombin-thrombomodulin complex: cooperative roles of Arg-35 of thrombin and Arg-67 of protein C. Proc Natl Acad Sci USA 103:879–884

Yang L, Manithody C, Walston TD et al (2003) Thrombomodulin enhances the reactivity of thrombin with protein C inhibitor by providing both a binding site for the serpin and allosterically modulating the activity of thrombin. J Biol Chem 278:37465–37470

Yang L, Rezaie AR (2003) The fourth epidermal growth factor-like domain of thrombomodulin interacts with the basic exosite of protein C. J Biol Chem 278:10484–10490

Yao ZQ, Eisen-Vandervelde A, Waggoner SN et al (2004) Direct binding of hepatitis C virus core to gC1qR on CD4+ and CD8+ T cells leads to impaired activation of Lck and Akt. J Virol 78:6409–6419

Yonezawa S, Maruyama I, Sakae K et al (1987) Thrombomodulin as a marker for vascular tumors: comparative study with factor VI11 and UZex europaes I lectin. Am J Clin Pathol 88:405–411

Zekavat G, Mozaffari R, Arias VJ et al (2010) A novel CD93 polymorphism in non-obese diabetic (NOD) and NZB/W F1 mice is linked to a CD4+ iNKT cell deficient state. Immunogenetics 62:397–407

Zhang M, Bohlson SS, Dy M, Tenner AJ (2005) Modulated interaction of the ERM protein, moesin, with CD93. Immunology 115:63–73

Zushi M, Gomi K, Honda G et al (1991) Aspartic acid 349 in the fourth epidermal growth factor-like structure of human thrombomodulin plays a role in its Ca^{2+}-mediated binding to protein C. J Biol Chem 266:19886–19889

Zushi M, Gomi K, Yamamoto S et al (1989) The last three consecutive epidermal growth factor-like structures of human thrombomodulin comprise the minimum functional domain for protein C-activating cofactor activity and anticoagulant activity. J Biol Chem 264:10351–10353

CTLDs of Group XV–XVII

Alazami AM, Shaheen R, Alzahrani F et al (2009) FREM1 mutations cause bifid nose, renal agenesis, and anorectal malformations syndrome. Am J Hum Genet 85:414–418

Capalbo G, Müller-Kuller T, Dietrich U et al (2010) Inhibition of HIV-1 replication by small interfering RNAs directed against glioma pathogenesis related protein (GliPR) expression. Retrovirology 7:26

Carney TJ, Feitosa NM, Sonntag C et al (2010) Genetic analysis of fin development in zebrafish identifies furin and hemicentin1 as potential novel fraser syndrome disease genes. PLoS Genet 6:e1000907

Chilukamarri L, Hancock AL, Malik S et al (2007) Hypomethylation and aberrant expression of the glioma pathogenesis-related 1 gene in Wilms tumors. Neoplasia 9:970–978

Chiotaki R, Petrou P, Giakoumaki E et al (2007) Spatiotemporal distribution of Fras1/Frem proteins during mouse embryonic development. Gene Expr Patterns 7:381–388

Dalezios Y, Papasozomenos B, Petrou P, Chalepakis G (2007) Ultrastructural localization of Fras1 in the sublamina densa of embryonic epithelial basement membranes. Arch Dermatol Res 299:337–343

Eisenberg I, Barash M, Kahan T et al (2002) Cloning and characterization of a human novel gene C9orf19 encoding a conserved putative protein with an SCP-like extracellular protein domain. Gene 293:141–148

Engel U, Ozbek S, Streitwolf-Engel R et al (2002) Nowa, a novel protein with minicollagen Cys-rich domains, is involved in nematocyst formation in Hydra. J Cell Sci 115:3923–3934

Foster JA, Gerton GL (1996) Autoantigen 1 of the guinea pig sperm acrosome is the homologue of mouse Tpx-1 and human TPX1 and is a member of the cysteine-rich secretory protein (CRISP) family. Mol Reprod Dev 44:221–229

Gibbs GM, Lo JC, Nixon B et al (2010) Glioma pathogenesis-related 1-like 1 is testis enriched, dynamically modified, and redistributed during male germ cell maturation and has a potential role in sperm-oocyte binding. Endocrinology 151:2331–2342

Gibbs GM, Roelants K, O'Bryan MK (2008) The CAP superfamily: cysteine-rich secretory proteins, antigen 5, and pathogenesis-related 1 proteins–roles in reproduction, cancer, and immune defense. Endocr Rev 29:865–897

Gingras MC, Margolin JF (2000) Differential expression of multiple unexpected genes during U937 cell and macrophage differentiation detected by suppressive subtractive hybridization. Exp Hematol 28:65–76

Gupta GS (2005) Sperm maturation in epididymis. In: Gupta GS (ed) Proteomics of spermatogenesis. Springer, New York, pp 811–837

Jadeja S, Smyth I, Pitera JE et al (2005) Identification of a new gene mutated in Fraser syndrome and mouse myelencephalic blebs. Nat Genet 37:520–525

Kato M, Khan S, d'Aniello E et al (2007) The novel endocytic and phagocytic C-Type lectin receptor DCL-1/CD302 on macrophages is colocalized with F-actin, suggesting a role in cell adhesion and migration. J Immunol 179:6052–6063

Kato M, Khan S, Gonzalez N et al (2003) Hodgkin's lymphoma cell lines express a fusion protein encoded by intergenically spliced mRNA for the multilectin receptor DEC-205 (CD205) and a novel C-type lectin receptor DCL-1. J Biol Chem 278: 34035–34041

Kiyozumi D, Osada A, Sugimoto N et al (2005) Identification of a novel cell-adhesive protein spatiotemporally expressed in the basement membrane of mouse developing hair follicle. Exp Cell Res 306:9–23

Kiyozumi D, Sugimoto N, Nakano I, Sekiguchi K (2007) Frem3, a member of the 12 CSPG repeats-containing extracellular matrix protein family, is a basement membrane protein with tissue distribution patterns distinct from those of Fras1, Frem2, and QBRICK/Frem1. Matrix Biol 26:456–462

Kiyozumi D, Sugimoto N, Sekiguchi K (2006) Breakdown of the reciprocal stabilization of QBRICK/Frem1, Fras1, and Frem2 at the basement membrane provokes Fraser syndrome-like defects. Proc Natl Acad Sci USA 103:11981–11986

Kjeldsen L, Cowland JB, Johnsen AH et al (1996) SGP28, a novel matrix glycoprotein in specific granules of human neutrophils with similarity to a human testis-specific gene product and a rodent sperm-coating glycoprotein. FEBS Lett 380:246–250

Klessig DF, Durner J, Noad R et al (2000) Nitric oxide and salicylic acid signaling in plant defense. Proc Natl Acad Sci USA 97:8849–8855

Meier S, Jensen PR, Adamczyk P et al (2007) Sequence-structure and structure-function analysis in cysteine-rich domains forming the ultrastable nematocyst wall. J Mol Biol 368:718–728

Naruishi K, Timme TL, Kusaka N et al (2006) Adenoviral vector-mediated RTVP-1 gene-modified tumor cell-based vaccine suppresses the development of experimental prostate cancer. Cancer Gene Ther 13:658–663

Nishiyama A, Dahlin KJ, Prince JT et al (1991) The primary structure of NG2, a novel membrane-spanning proteoglycan. J Cell Biol 114:359–371

Ozbek S, Pokidysheva E, Schwager M et al (2004) The glycoprotein NOWA and minicollagens are part of a disulfidelinked polymer that forms the cnidarian nematocyst wall. J Biol Chem 279:52016–52023

Pavlakis E, Makrygiannis AK, Chiotaki R et al (2008) Differential localization profile of Fras1/Frem proteins in epithelial basement membranes of newborn and adult mice. Histochem Cell Biol 130:785–793

Petrou P, Chiotaki R, Dalezios Y, Chalepakis G (2007a) Overlapping and divergent localization of Frem1 and Fras1 and its functional implications during mouse embryonic development. Exp Cell Res 313:910–920

Petrou P, Makrygiannis AK, Chalepakis G (2008) The Fras1/Frem family of extracellular matrix proteins: structure, function, and association with Fraser syndrome and the mouse bleb phenotype. Connect Tissue Res 49:277–282

Petrou P, Pavlakis E, Dalezios Y, Chalepakis G (2007b) Basement membrane localization of Frem3 is independent of the Fras1/Frem1/Frem2 protein complex within the sublamina densa. Matrix Biol 26:652–658

Ren C, Li L, Yang G, Timme TL et al (2004) RTVP-1, a tumor suppressor inactivated by methylation in prostate cancer. Cancer Res 64:969–976

Rich T, Chen P, Furman F, Huynh N, Israel MA (1996) RTVP-1, a novel human gene with sequence similarity to genes of diverse species, is expressed in tumor cell lines of glial but not neuronal origin. Gene 180:125–130

Rosenzweig T, Ziv-Av A, Xiang C et al (2006) Related to testes-specific, vespid, and pathogenesis protein-1 (RTVP-1) is overexpressed in gliomas and regulates the growth, survival, and invasion of glioma cells. Cancer Res 66:4139–4148

Scheuring UJ, Corbeil J, Mosier DE, Theofilopoulos AN (1998) Early modification of host cell gene expression induced by HIV-1. AIDS 12:563–570

Smyth I, Du X, Taylor MS et al (2004) The extracellular matrix gene Frem1 is essential for the normal adhesion of the embryonic epidermis. Proc Natl Acad Sci USA 101:13560–13565

Szyperski T, Fernández C et al (1998) Structure comparison of human glioma pathogenesis-related protein GliPR and the plant pathogenesis-related protein P14a indicates a functional link between the human immune system and a plant defense system. Proc Natl Acad Sci USA 95:2262–2266

van Loon LC, Rep M, Pieterse CM (2006) Significance of inducible defense-related proteins in infected plants. Annu Rev Phytopathol 44:135–162

Yerkovich ST, Roponen M, Smith ME et al (2009) Allergen-enhanced thrombomodulin (blood dendritic cell antigen 3, CD141) expression on dendritic cells is associated with a TH2-skewed immune response. J Allergy Clin Immunol 123:209–216.e4

Zelensky AN, Gready JE (2004) C-type lectin-like domains in Fugu rubripes. BMC Genomics 5:51

Part XIV

Clinical Significance of Animal Lectins

MBL Deficiency as Risk of Infection and Autoimmunity

Anita Gupta

42.1 Pathogen Recognition

In pathogen recognition by C-type lectins, several levels of complexity can be distinguished; these might modulate the immune response in different ways. Firstly, the pathogen-associated molecular pattern repertoire expressed at the microbial surface determines the interactions with specific receptors (Fig. 42.1). Secondly, each immune cell type possesses a specific set of pathogen-recognition receptors. Thirdly, changes in the cell-surface distribution of C-type lectins regulate carbohydrate binding by modulating receptor affinity for different ligands. Crosstalk between these receptors results in a network of multimolecular complexes, adding a further level of complexity in pathogen recognition (Cambi and Figdor 2005; Jack et al. 2001; Thiel et al. 2006) (see Chap. 23). MBL deficiency is genetically determined and predisposes to recurrent infections and chronic inflammatory diseases. MBL deficiency has been implicated in susceptibility and course of viral, bacterial, fungal, and protozoan infection. More than 10% of the general population may, depending on definition, be classified as MBL deficient, underlining the redundancy of the immune system. MBL-disease association studies have been a fruitful area of research, which implicates a role for MBL in infective, inflammatory and autoimmune disease processes. MBL deficiency predisposes both to infection by extra-cellular pathogens and to autoimmune disease.

42.1.1 MBL Characteristics

Mannose-binding lectin is a C-type serum lectin and is primarily produced by the liver (Bouwman et al. 2005) in response to infection, and is part of many other factors termed "http://en.wikipedia.org/wiki/Acute_phase_protein" \o "Acute phase protein" acute phase proteins. MBL is made up of 96-kDa structural units, which in turn are composed of three identical 32-kDa primary subunits. The subunits consist of an N-terminal cross-linking region, a collagenlike domain, and a C-terminal carbohydrate-recognition domain (CRD) (Chap. 23; Turner and Hamvas 2000). Circulating MBL is composed of higher-order oligomeric structures, which include dimers, trimers, tetramers, pentamers, and hexamers of the structural homotrimeric unit. The oligomeric configuration of the structural units allows the MBL molecule to have multiple CRDs, facilitating multivalent ligand binding. Each CRD of MBL is structurally identical and is able to bind a range of oligosaccharides including N-acetylglucosamine D-mannose, Nacetylmannosamine, and L-fucose (Turner 1996). Although the various sugars are bound with different affinities, the cluster-like array of multiple binding sites allows activation of complement to be most effective. MBL is considered to play a major role in innate defense against pathogens, involving recognition of arrays of MBLbinding carbohydrates on microbial surfaces. However, recent studies have shown that MBL is also involved in the recognition of self-targets, such as apoptotic and necrotic cells (Nauta et al. 2003). In plasma, MBL is associated with MBL-associated serine proteases (MASPs). Currently, three MASPs have been identified, MASP-1, MASP-2, and MASP-3 (Chap. 23).

42.1.2 Pathogen Recognition and Role in Innate Immunity

MBL belongs to the class of collectins in the C-type lectin superfamily, whose function appears to be pattern recognition in the first line of defense in the pre-immune host. MBL recognizes carbohydrate patterns, found on the surface of a large number of pathogenic micro-organisms, including bacteria, viruses, protozoa and fungi. Mannose-binding lectin binds to the repeating sugar arrays on surfaces of pathogens through multiple lectin domains and, following binding, is able to activate the complement system via an associated serum protease, MASP-2. Importantly, MBL activates the complement system through a distinctive third pathway, independent of

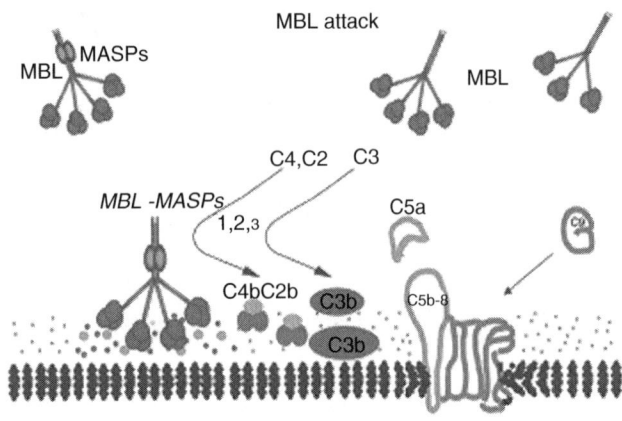

Fig. 42.1 *MBL-mediated complement attack.* Mannose-binding lectin (MBL) complexed with the MASPs binds to sugar arrays on a microorganism and mediates a complement attack through MASP2. MASPs denote MBL-associated serine proteases (Reprinted by permission from Macmillan Publishers Ltd: Genes Immun. Garred et al. © 2006). Mannan-binding lectin binds to patterns of carbohydrate groups in the correct spatial orientation. MBL is composed of two to six clusters of carbohydrate-binding lectin domains that interact with each other via a collagen-like domain (Chap. 23). Within each cluster are three separate binding sites that have a fixed orientation relative to each other; all three sites can therefore only bind when their ligands – mannose and fucose residues in bacterial cell-wall polysaccharides – have the appropriate spacing

Table 42.1 Some clinically relevant microorganisms recognized by MBL

Bacteria	Viruses
Staphylococcus aureus	HIV-1 and HIV-2
Staphylococcus pneumoniae	Herpes simplex-2
Staphylococcus pyrogenes	Influenza A
Enterococcus spp.	Hepatitis B virus
Listeria monocytogenes	Hepatitis C virus
Haemophilus influenzae	**Fungi**
Neisseria meningitidis	*Aspergillus fumigatus*
Neisseria gonorrhoeae	*Candida albicans*
Escherichia coli	*Cryptococcus neoformans*
Klebsiella spp.	*Saccharomyces cerevisiae*
Pseudomonas aeroginosae	**Protozoa**
Salmonella montevideo	*Plasmodium falciparum*
Salmonella typhimurium	*Cryptosporidium parvum*
H. pylori	*Trypanosoma*
Chlamydia trachomatis	
Chlamydia pneumonia	
Proprionibacterium acnes	
Mycobacterium avium	
Mycobacterium tuberculosis	
Mycobacterium leprae	
Leishmania chagasi	

both antibody and the C1 complex (Table 42.1). The MBL binds to neutral carbohydrates on microbial surfaces and recognises carbohydrates such as mannose, glucose, L-fucose, N-acetyl-mannosamine (ManNAc), and N-acetyl-glucosamine (GlcNAc). Oligomerisation of MBL enables high avidity binding to repetitive carbohydrate ligands, such as those present on microbial surfaces, including *E. coli, Klebsiella aerogenes, Neisseria meningitides, Staphylococcus aureus, S. pneumonia, A. fumigatus* and *C. albicans* (c/r Kerrigan and Brown 2009). However, there is also a great variation in the binding of MBL to various organisms; *Candida albicans*, β-haemolytic group A *Streptococci* and *Staphylococcus aureus* bind with high affinity, while *Clostridium* sp. *Pseudomonas aeruginosa, Staphylococcus epidermidis*, b-haemolytic streptococci and *Streptococcus pneumoniae* exhibit low or no binding (Santos et al. 2001). It is also observed that some organisms (e.g. *Klebsiella* sp. and *Escherichia coli*) show a variable pattern of binding. Later, it was shown that the absence of sialic acid from the lipo-oligosaccharide (LOS) of *Neisseria meningitidis* serogroup B, serogroup C and *Neisseria gonorrhoeae* permits MBL binding and presence of sialic acid on LOS results in poor or no MBL binding. In a similar study on *Salmonella* sp, it was found that MBL binds to rough chemotype but exhibits low or no binding with smooth chemotype (Ambrosio and De Messias-Reason 2005). In order to activate the complement system after MBL binds to its target (for example, mannose on the surface of a bacterium), the MASP protein functions to cleave the blood protein C4 into C4a and C4b. The C4b fragments can then bind to the surface of the bacterium, and initiate the formation of a C3 convertase. The subsequent complement cascade, catalyzed by C3 convertase, results in creating a membrane attack complex, which causes the lysis of the pathogen to which MBL is bound (Worthley et al. 2005, 2006) (Fig. 42.1). Being an important component of innate immunity, acting as an ante-antibody and/or as a disease modifier, MBL is thought to influence disorders as diverse as meningococcal disease, rheumatoid arthritis, cystic fibrosis and recurrent miscarriage. Vulvovaginal candidiasis is a yeast infection of vulva and vagina; millions of women suffer from vulvovaginal candidiasis 5 worldwide. Women bearing MBL variant allele are at a higher risk for vulvovaginal candidiasis syndrome 343. The cervicovaginal lavage (CVL) MBL levels and gene mutation frequency were both higher in women suffering from vulvovaginal candidiasis than in controls (Liu et al. 2006b). On the other hand, MBL levels were low (0.30 ng/mL) in women with recurrent vulvovaginal candidiasis and were associated with a high gene mutation frequency compared to controls (1.28 ng/mL) (Ip and Lau 2004). Parenteral administration of MBL increased resistance of mice to hematogenously disseminated candidiasis. Thus, MBL plays an important role in innate resistance to candidiasis, suggesting a protective role of lectin in female genital tract infection (Pellis et al. 2005).

The autologous function for MBL, perhaps, is to perform a regulatory role within the immune system. The MBL interacts with human peripheral blood cells such as B lymphocytes and natural killer cells. The MBL is capable of binding to differently glycosylated ligands on several autologous cell types via its carbohydrate-recognition domain. It was speculated that this could have functional significance at extravascular sites, but perhaps only in individuals possessing MBL genotypes conferring MBL sufficiency (Downing et al. 2003, 2005).

Nevertheless, MBL genotyping of various populations has led to the suggestion that there may be some biological advantage associated with absence of the protein. In addition, the protein also modulates disease severity, at least in part through a complex, dose-dependent influence on cytokine production. Moreover, there appears to be a genetic balance in which individuals generally benefit from high levels of the protein. These findings suggest that the concept of MBL as a protein involved solely in first line defense is an over-simplification and the protein should rather be viewed as having a range of activities including disease modulation (Turner and Hamvas 2000; Dommett et al. 2006; Worthley et al. 2006). The mechanisms and signaling pathways involved in such processes remain to be elucidated. Though the deficiency of MBL is associated with increased susceptibility to infections, Roos et al. (2004) indicated that antibody-mediated classical pathway activation can compensate for impaired target opsonization via MBL pathway in MBL-deficient individuals. Lack of MBL may be most relevant in the context of a co-existing secondary immune deficiency. Replacement therapy appears promising. The development of a recombinant product should permit the extension of MBL therapy to randomized clinical trials of sufficient size to provide clear evidence about the physiological significance of this intriguing glycoprotein. The MBL has attracted great interest as a potential candidate for passive immunotherapy to prevent infection (Gupta et al. 2008).

42.2 MBL Deficiency as Risk of Infection and Autoimmunity

42.2.1 MBL Deficiency and Genotyping

The single point mutations in exon 1 of human *MBL-2* gene appear to impair the generation of functional oligomers leading to the secondary structural abnormalities of the collagenous triple helix and a failure to form biologically functional higher order oligomers. Such deficiencies of the functional protein are common in certain populations, e.g. in sub-Saharan Africa, but virtually absent in others, e.g. indigenous Australians. There is an increased incidence of infections in individuals with such mutations and an association with the autoimmune disorders such as SLE and rheumatoid arthritis. Thus, the MBL is a potential candidate for passive immunotherapy to prevent infection.

42.2.1.1 Polymorphism in MBL Gene Is Associated in Exon 1 at Codon 52, 54, and 57

The concentration of MBL in plasma is determined genetically, primarily by the genetic polymorphism of the first exon of the structural gene and promoter region. The gene encoding MBL, *MBL2* (*MBL1* is a pseudogene), is located on long arm chromosome 10q11.2–q21 and contains four exons. A number of single nucleotide polymorphisms (SNPs) have been characterized in the gene. Exon 1 harbours three missense SNPs, giving rise to amino acid exchanges in the first part of the collagenous region. Two of these SNPs: $Gly^{54}Asp$, named 'B' and $Gly^{57}Glu$, named 'C' exchange glycine with an acetic amino acid. The third ($Arg^{52}Cys$, 'D') introduces a cysteine in the collagen region (the residue numbers includes the leader sequence of 20 residues) (Fig. 42.2). The wild type is denoted 'A'. Heterozygous individuals for D, B, and C mutations have a substantial decrease in MBL serum concentration, whereas MBL is undetectable in the serum of homozygous individuals. Three structural mutations within exon 1 at codons 52, 54, and 57 have been invariably referred to as the D, B and C variants. In addition to these three variant structural alleles, the promoter region also shows a number of SNPs as well, some of which influences the expression of MBL. The three SNPs in the coding region of the MBL2 gene those are associated with abnormal polymerization of the MBL molecule, decreased serum concentrations of MBL and strongly impaired function. These MBL SNPs are associated with increased susceptibility to infections, especially in immune-compromised persons, as well as with accelerated progression of chronic diseases. Normal serum levels of MBL range from 800 to 1,000 ng/mL in healthy Caucasians, however, wide variations can occur due to point mutations in codons 52, 54 and 57 of exon 1 and in the promotor region of the MBL gene (Turner 2003). Separate point mutations in the collagenous domain of human MBL are associated with immune-deficiency, caused by reduced complement activation by the variant MBLs as well as by lower serum MBL concentrations.

MBL deficiency with B mutation is associated with 26% of Caucasian populations (Steffensen et al. 2000). In a cohort of 236 Australian blood donors, 5 MBL promoter and coding SNPs were genotyped. Significant associations were found between both coding and promoter polymorphisms and MBL antigenic and functional levels (Minchinton et al. 2002). Point mutations in exon 1 at codons 52, 54 and 57

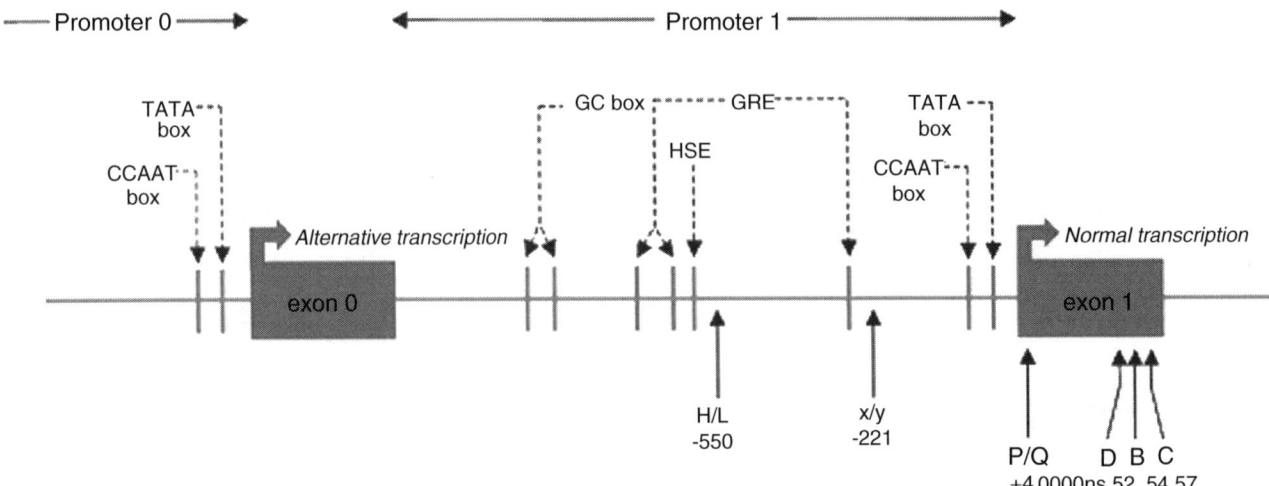

Fig. 42.2 *MBL2 polymorphisms.* Two promoters, promoter 1 and promoter 0 regulate the transcription of the human MBL2 gene. Similar to promoter 1 promoter 0 also includes a TATA box for transcription initiation, and transcription factor binding DNA sequences. Six DNA polymorphisms in the MBL2 gene are known to be associated with variation in quantity and/or function of MBL in serum. Three variants affect the expression of the MBL2 gene. They are localized in the promoter 1 (position −550, H/L variant and −221, X/Y variant) and in the 5′-untranslated region (position þ4, P/Q variant) of the MBL2 gene. Three base substitutions in exon 1 in codons 52 (D), 54 (B) and 57 (C) result in amino-acid changes (arginine to cysteine, glycine to aspartic acid and glycine to glutamic acid, respectively) and decreased level and function of MBL. The normal allele is named A proteases (Reprinted by permission from Macmillan Publishers Ltd: Genes Immun. Garred et al. © 2006)

and a promoter polymorphism at −221 bp of MBL gene were associated with increased susceptibility to various infectious diseases (Steffensen et al. 2003; Roos et al. 2006). The codon 54 mutation was frequent in both a British Caucasian and a Hong Kong Chinese population. The replacement of glycine-54 with an aspartic acid residue disrupts the fifth Gly-Xaa-Yaa repeat in the collagen-like domain of each 32 kDa MBP peptide chain and prevents the formation of the normal triple helix (Sumiya et al. 1991). Super et al. (1992) suggested that this genotype occurs in 5% of the population and encodes a functional protein. Super et al. (1992) also suggested that the Gly^{54}Asp allele does not account for a deficiency state, but instead suggested that MBP may have two predominant allelic forms that have overlapping function and differ only in their ability to activate the classical pathway of complement.

Of 123 healthy Danish individuals investigated, 93 were homozygous (743.6%) for GGC, 28 heterozygous (22.8%), and 2 homozygous for GAC (1.6%). The gene frequency of the GAC allele was found to be 0.13. DNA sequencing of the cloned exon 1 from one GAC homozygous individual revealed no other substitution. The median MBP concentration in the group containing the GAC allele was 43.4 times lower than in the GGC homozygous group. However, the range of MBP in plasma was wide and overlapping between the groups. MBP protein was detected in both the GAC homozygotes. This study suggested that the GAC allele is able to produce a functional MBP protein which may be detected in serum at low concentrations (Garred et al. 1992).

The point mutation (GGA to GAA) involving codon 57 of exon 1 has been reported in Gambians from West Africa. In the Gambians the codon 57 mutation was remarkably common whereas the codon 54 mutation was very rare. In contrast, the codon 54 mutation was frequent in both the British Caucasian and the Hong Kong Chinese population. It was predicted that both homozygous and heterozygous individuals would have profoundly reduced serum levels of the protein and this was confirmed by immunoassay complement activation through lectin pathway. These two mutations have arisen independently because of the divergence of African and non-African populations.

Codon 52 Polymorphism Increases Risk of Premature Delivery: MBL2 gene polymorphisms are associated with an increased risk of neonatal infections. A relation between the maternal *MBL2* genotype and the risk of premature delivery has been indicated. Bodamer et al. (2006) suggested that the frequency of the codon 52 polymorphism was higher in the pre-term group compared to the term group (10.8% vs. 4.9%), while the frequency of codon 54 polymorphism was equal in both groups (11.3% vs. 11.8%). Data suggest that the fetal MBL2 genotype might be an additional genetic factor contributing to the risk of premature delivery.

SNPs at Codon 54 in MBL Are Associated with Increased Prevalence of Respiratory Infection: The SNPs of the innate immunity receptors CD14, MBL, and Toll-like receptor-2 with clinical phenotype in critically ill patients with systemic inflammatory response syndrome are associated with increased prevalence of positive bacterial cultures and sepsis but not with altered prevalence of septic shock or decreased 28-day survival. Furthermore, CD14 SNPs were associated with Gram-negative bacteria and Toll-like receptor-2 with Gram-positive bacteria, whereas MBL was not associated with a particular organism class. The prevalence of the codon 54 mutation of the MBL gene in patients having repeated respiratory infections as well as the prevalence of the MBL mutant genotype among patients with diffuse panbronchiolitis was further supported (Gomi et al. 2004).

MBL Gene Polymorphisms in Gestational Diabetes Mellitus: Insulin resistance is a feature of gestational diabetes mellitus (GDM). A genetic predisposition to a pro-inflammatory state could favor the appearance of GDM during pregnancy. An association has been found between G54D and GDM. GDM patients carrying the G54D mutation require insulin therapy more frequently and have heavier infants than GDM women homozygous for the wild-type allele. An inverse correlation in GDM patients between neonatal weight and plasma MBL levels has been reported. Thus, pregnant women bearing the G54D MBL allele have a greater risk for developing GDM and having heavier infants (Megia et al. 2004).

MBL Genotypes in Acute Lymphoblastic Leukemia: Epidemiological studies show that acute lymphoblastic leukemia (ALL) can be induced by interactions between the immune system and early childhood infections. Since, certain types of childhood acute lymphoblastic leukemia develop as a multiple step process involving both pre- and postnatal genetic events, the MBL may play a critical role in the immune response in early childhood before specific immune protection develops. Schmiegelow et al. (2002) indicated that low-level MBL genotypes is associated with an increased risk of childhood ALL, particularly with early age at onset. Childhood ALL may often be initiated in utero. The prenatal origin of childhood leukemia was ascertained in children with B-precursor acute lymphoblastic leukemia carrying the chromosomal translocation t(12;21), the most common subtype of all childhood ALL. Study provided evidence that the development of t(12;21) B-precursor ALL may be initiated in utero. However, age at leukemia may be inversely correlated with the burden of cells with leukemia clonal markers, i.e. leukemia predisposed cells at birth (Hjalgrim et al. 2002).

MBL Genotypes in Viral Hepatitis: The prevalence of mutations in MBL gene was assessed in patients of hepatitis B. A mutation in codon 52 of the MBP gene was present in two (11%) of 19 Caucasian patients with acute hepatitis B and nine (27%) of 33 Caucasian patients with chronic hepatitis B, compared with four (4%) of 98 Caucasian controls.

By contrast the prevalence of the mutation was similar in Asian patients with chronic hepatitis B and in Asian controls (one [5%] of 20 vs. two [2%] of 117). Mutations in codon 54 and codon 57 were found in similar proportions of patients and controls. These findings showed in Caucasian, but not Asian, patients an association of the codon 52 mutation of the MBL gene with persistent hepatitis B virus (HBV) infection. They suggest an important role for this gene, or a gene in linkage disequilibrium with MBL, in determining outcome after HBV infection in adult but not neonatal life (Thomas et al. 1996).

Bellamy et al. (1998) investigated the association between variant MBP alleles and malaria, tuberculosis, and HBV in adults and children in Gambia. Of the 2,041 Gambians screened for MBP mutations, 944 (46%) were homozygous for the wild-type allele, 922 (45%) were carriers of a single variant allele, and 175 (8.6%) possessed 2 mutant alleles. The most common mutation in Africans – the codon 57 variant allele – was weakly associated with resistance to tuberculosis in both patients and controls. Although MBP deficiency may predispose to recurrent infections, this study failed to provide evidence that such a deficiency is a major risk factor for infectious diseases (Bellamy et al. 1998).

The hepatitis C virus (HCV) envelope glycoprotein E2 binds the DC-SIGN and the related liver endothelial cell lectin through high-mannose N-glycans. Several high-mannose N-glycans in a structurally defined cluster on E2 bind to several subunits of the oligomeric lectin CRD and represent a strategy by which HCV targets to and concentrates in the liver and infects dendritic cells (Lozach et al. 2003). To determine the relevance of MBL polymorphism to hepatitis C virus infection, Matsushita et al. (1998) determined the MBL genotypes in 159 hepatitis C virus-infected chronic hepatitis patients and 218 healthy controls in Japan by looking at 4 polymorphic loci: 2 (H/L and X/Y) within the promoter region and 2 (P/Q and A/B) within exon-1 of the *MBL* gene. It indicated that the MBL-related innate immune system plays an important role in elimination of hepatitis C virus during interferon therapy (Matsushita et al. 1998). Matsushita et al. (2001) demonstrated that the MBL genotype can be significantly associated with increased risk for to primary biliary cirrhosis (PBC), and that increased production of MBL plays a critical role in immunopathogenesis.

42.2.1.2 Loss of Carbohydrate Binding and MASP-2 Auto-Activation in Mutated MBL

The mutations Gly25→Asp and Gly^{28}→Glu disrupt the disulfide-bonding arrangement of the protein and cause at least a fivefold increase in the half-time of secretion of MBP compared with wild-type rat serum MBP. A similar phenotype, including a threefold increase in the half-time of secretion, disruption of the disulfide bonding arrangement, and inefficient complement fixation, was observed when nearby glucosylgalactosyl hydroxylysine residues at positions 27 and 30 were replaced with arginine residues. The results suggest that defective secretion resulting from structural changes in the collagen-like domain is likely to be a contributory factor for MBP immunodeficiency (Heise et al. 2000).

To investigate the molecular defects associated with heterozygosity, rat serum MBP polypeptides (MBP-A: 56% identical in sequence to human MBP) and rat MBP polypeptides containing mutations associated with human immunodeficiency were co-expressed in mammalian expression system. The resulting proteins are secreted almost exclusively as heterooligomers that were defective in activating the complement cascade. Functional defects were caused by structural changes to the N-terminal collagenous and cysteine-rich domains of MBP, disrupting interactions with associated serine proteases. These mutations demonstrated how a SNP gives rise to the molecular defects that lead to the disease phenotype in heterozygous individuals (Wallis 2002). Wallis et al. (2005) analyzed the molecular and functional defects associated with two variant proteins of lectin pathway. Mutations Gly^{25}→Asp and Gly^{28}→Glu created comparable structural changes in rat MBL but the G28E variant activated complement >10-fold less efficiently than the G25D variant, which in turn had approximately sevenfold lower activity than wild-type MBL. Analysis of mutant MBL-MASP-2 complexes formed from recombinant proteins showed that reduced complement activation by both mutant MBLs was due to failure of activation of MASP-2 efficiently on binding to a mannan-coated surface. Disruption of MBL-MASP-2 interactions as well as to changes in oligomeric structure and reduced binding to carbohydrate ligands compared with wild-type MBL probably account for the intermediate phenotype of the G25D variant. However, carbohydrate binding and -MASP-2 activation are ostensibly completely decoupled in complexes assembled from the G28E mutant, such that the rate of MASP-2 activation is no greater than the basal rate of zymogen MASP-2 autoactivation. Analogous molecular defects in human MBL probably combine to create the mutant phenotypes of immunodeficient individuals (Wallis et al. 2005).

Since it is difficult to evaluate MBL in patients blood on the only basis of protein contents, or in combination with MBL genotyping due to possible association of altered oligomeric state of MBL, Dumestre-Perard et al. (2002) purified MBL from human plasma and showed the presence of MBL in two different oligomeric forms. Data on the specific activity of these forms showed that the higher oligomeric forms of MBL had the ability to induce C4 cleavage more efficiently than the corresponding lower oligomers (Dumestre-Perard et al. 2002).

How to Define Abnormal MBL Pathway and Disease Associations: Although, the association between MBL deficiency and risk of infection with other common diseases and death during years of follow-up has established the role of MBL in innate immune system, yet, in a large ethnically homogeneous Caucasian population, there was no evidence for significant differences in infectious disease or mortality in MBL-deficient individuals versus controls and suggested that MBL deficiency is not a major risk factor for morbidity or death in the adult Caucasian population (Dahl et al. 2004; Eisen and Minchinton 2003; Kilpatrick 2002; Summerfield et al. 1997). While addressing possible correlation between MBL levels and clinical conditions an issue is how to define MBL deficiency. The physiologically relevant MBL level leading to clinical manifestations is likely to differ in different diseases. In the examples given below, a number of different levels have been used as cut-off values defining MBL deficiency. Judged from clinical trials it appears that at least 200 ng MBL/mL plasma is needed for reconstituting in vitro functional activity (C4b deposition) after MBL infusion in MBL deficient individuals (Valdimarsson et al. 2004). On the other hand, in leukaemic patients a cut-off level of 500 ng/mL or more has been suggested (Peterslund et al. 2001; Neth et al. 2001), and in the cases of obstetric problems even lower levels (100 ng/mL) have been indicated.

42.2.2 MBL and Viral Infections

42.2.2.1 MBL and HIV Interaction

A broad range of proteins binds high-mannose carbohydrates found on the surface of the envelope protein gp120 of the HIV and thus interfere with the viral life cycle, providing a new method of controlling HIV infection. While glycosylation of HIV gp120/gp41 provides a formidable barrier for development of strong antibody responses to the virus, it also provides a potential site of attack by the innate immune system through MBL/MBP. The MBL that binds to HIV depends on the high-mannose glycans present on gp120 while host cell glycans incorporated into virions do not contribute substantially to this interaction. The MBL has been shown to interact with all tested HIV strains. However, drugs that alter processing of carbohydrates enhance neutralization of HIV primary isolates by MBL. Complement activation on gp120 and opsonization of HIV due to MBL binding have also been observed but these immune mechanisms have not been studied in detail. MBL has also been shown to block the interaction between HIV and DC-SIGN. Clinical studies show that levels of MBL, an acute-phase protein, increase during HIV disease. Because of apparently universal reactivity with HIV strains, MBL clearly represents an important mechanism for recognition of HIV by the immune system. However, further studies are needed to define the in vivo contribution of MBL to clearance and destruction of HIV (Botos and Wlodawer 2005; Ji et al. 2005). MBL that binds to high mannose glycans on HIV-1 (gp120), has been shown to neutralize the cell line-adapted strain HIV(IIIB). But HIV primary isolates (PI) are generally more resistant to neutralization by antibodies. Considering that PI are produced in primary cells that could alter the number of high mannose glycans on HIV relative to cell lines, Ying et al. (2004) showed that both PI and cell line-adapted HIV, despite binding of MBL, are relatively resistant to neutralization by levels of MBL normally present in serum. However, binding and opsonization of HIV by MBL may alter virus trafficking and viral-antigen presentation during HIV infection (Ying et al. 2004).

In search of the effect of MBL-2 polymorphisms on susceptibility and progression of HIV-1 infection in children, Dzwonek et al. (2006) observed MBL deficiency more frequently in patients with severe disease. The study suggested that MBL-2 variants may be less frequent in children classified as long-term non-progressors (LTNPs) and hence MBL analysis can be useful in identifying children with slow disease progression and, consequently, may not require immediate antiretroviral treatment (Dzwonek et al. 2006). Production of HIV in the presence of the mannosidase I inhibitor deoxymannojirimycin (dMM) significantly enhanced binding of HIV to MBL and increased MBL neutralization of an M-tropic HIV primary isolate. In contrast, HIV cultured in presence of alpha-glucosidase I and II inhibitors, castanospermine and deoxynojirimycin showed only slight effect on virus binding and neutralization by MBL. The study suggested that specific alterations of the N-linked carbohydrates on HIV gp120/gp41 can enhance MBL-mediated neutralization of virus by strengthening the interaction of HIV-1 with MBL (Hart et al. 2003).

42.2.2.2 Susceptibility to RSV, CoV and HTLV

Respiratory syncytial virus (RSV) is the most important microbiological cause of lower respiratory tract infection (LRTI) in infants. MBL deficiency is the most common immunodeficiency on the African Continent. MBL deficiency has an impact on the hospitalization for LRTI caused by RSV in infants from Soweto, South Africa. But in contrary to expectations, results suggested no association between low levels of MBL or carriage of variant alleles and LRTI caused by RSV (Kristensen et al. 2004).

Corona Virus Infection and Acute Respiratory Syndrome: Little is known about the innate immune response to severe acute respiratory syndrome (SARS) corona virus (CoV) infection. The MBL plays an important role in SARS-CoV infection. The distribution of MBL gene polymorphisms was significantly different between patients with SARS and normal subjects, with a higher frequency of haplotypes associated with low or deficient serum levels of MBL in patients with SARS. There was, however, no association between MBL genotypes, which are associated with low or deficient serum levels of MBL, and mortality related to SARS. MBL could bind SARS-CoV in calcium-dependent and mannan-inhibitable fashion in vitro, suggesting that binding is through the carbohydrate recognition domains of MBL. Furthermore, deposition of complement C4 on SARS-CoV was enhanced by MBL. Inhibition of the infectivity of SARS-CoV by MBL in fetal rhesus kidney cells suggested that MBL contributes to the first-line host defense against SARS-CoV and that MBL deficiency is a susceptibility factor for acquisition of SARS (Ip et al. 2005).

Susceptibility to T-cell Lymphotropic Virus: Pontes et al. (2005) investigated the association between MBL gene polymorphism and the susceptibility to human T-cell lymphotropic virus (HTLV) infection in 83 HTLV-infected asymptomatic subjects. Detection of MBL*A, MBL*B, and MBL*C was performed by amplifying a fragment of 349 bp (exon 1) and restriction fragment length polymorphism analysis with BanI and MboII endonucleases. A strong association has been demonstrated between MBL polymorphism and HTLV infection. Presence of genotype BB may be associated with the susceptibility to HTLV. Though further studies, with a larger number of individuals, are necessary, MBL polymorphism could have a possible impact on diseases associated with HTLV infection (Pontes et al. 2005).

42.3 Autoimmune and Inflammatory Diseases

Studies demonstrating the binding of MBL to the endothelium and causing excessive complement activation and subsequent tissue damage are known (Jordan et al. 2003), where as MBL deficiency may be advantageous in some circumstances since MBL may lead to an increased cytokine secretion by macrophages (Takahashi et al. 2002). There seems to be a delicate balance as to when MBL levels may be involved in harmful or in beneficial inflammation such as in the cardiovascular and other systems.

42.3.1 MBL Gene in Rheumatoid Arthritis

The etiology of autoimmune diseases is largely unknown. Studies on associations between MBL deficiency and rheumatoid arthritis (RA) have been discussed in reviews (Graudal 2004; Barton et al. 2004). Results depend on ethnic groups, type of patients and the symptoms studied. Low levels of serum MBL are associated with a higher erythrocyte sedimentation rate (ESR), joint swelling score, limitation of joint motion score, and annual increase in radiographic destruction score. Despite this, indications are that low MBL levels may be linked with symptoms indicating a poor prognosis as well as an earlier debut. Whether variant alleles of the MBL gene causing low serum concentrations of MBL are associated with increased susceptibility to RA and erosive outcome in an inception cohort of patients with early polyarthritis was studied by Jacobsen et al. (2001). Jacobsen et al. (2001) suggested that MBL variant alleles appear to be weak susceptibility markers for RA, and patients with early polyarthritis and homozygous for MBL structural variant alleles have a higher risk of developing early erosive RA. These findings, together with the positive association between MBL variant alleles and the increased serum levels of IgM RF and CRP, point at the MBL gene as a relevant locus in the pathophysiology of RA. Graudal (2004) indicated that MBL-deficient patients have a relative risk of a severe radiographic event of 3.1 compared with the MBL competent group. The relative risk (RR) of early IL-1α auto-antibodies positive patients developing serious radiographic joint destruction was significantly lower than for IL-1α auto-antibodies-negative patients. Perhaps MBL and IL-1α auto-antibodies are predictors of prognosis of RA and play important roles in the pathogenesis of RA (Graudal 2004; Lee et al. 2005). Low levels of the protein have been related to a poor prognosis in rheumatoid arthritis perhaps due to the modulatory action that MBL exerts on the secretion of tumor necrosis factor α, a central molecule in the pathogenesis of rheumatoid arthritis (Graudal et al. 2000). Low MBL levels have also been associated with adult dermatomyositis and are probably related to a reduced clearance of apoptotic keratinocytes (Werth et al. 2002). Genotypes related to a lower production of MBL have also been linked to the development of systemic lupus erythematosus (Villarreal et al. 2001; Davies et al. 1995).

42.3.2 Systemic Lupus Erythematosus

Genetic factors play a major role in the development of SLE. More than 5% of cases are familial and the concordance rate between identical twins is 40%. Genetic studies in mice suggest a complex mechanism of transmission involving interactions among several susceptibility genes and, probably, protective genes. Genetic studies in humans have identified nearly 50 chromosomal areas possibly involved in lupus transmission. Significant linkage has been found for at least six regions, two on chromosome 1, one near the HLA region on chromosome 6, and three on chromosomes 2, 4, and 16, respectively. The genetic polymorphism of cytokines and, perhaps, of the T-cell receptor (TCR) may contribute to deregulate lymphocyte activity. The polymorphism of the Fc receptors of immunoglobulins may affect immune complex clearance, thereby promoting tissue damage. Further genetic studies are needed to enrich the fund of knowledge on lupus and to identify new targets for treatment (Perdriger et al. 2003).

From the several investigations on MBL and SLE the consensus is emerging that low levels of MBL predisposes to development of the disease. However, the connection is not like for C1q where SLE develops in almost all of the rare cases of deficiency. Rather, it seems that MBL deficiency aggravates the disease or the development (Ohlenschlaeger et al. 2004; Takahashi et al. 2005). In SLE patients, MBL deficiency increase the risk for respiratory tract infections (Garred et al. 2001; Takahashi et al. 2005) as well as the risk of developing arterial thromboses (Ohlenschlaeger et al. 2004).

As with infectious disease, there is some evidence that the risk of pathology increases if there is another co-existing immune defect. For example, in a cohort of Spanish patients, the odds ratio for developing SLE was 2.4 for individuals with MBL deficiency, but this increased to 3.2 when there was also a co-existing partial C4 deficiency (Davies et al. 1997). Studies in patients with SLE have reported that MBL deficiency also influences their risk of developing certain complications, which include arterial thromboses (Ohlenschlaeger et al. 2004) and respiratory tract infections (Garred et al. 2001; Takahashi et al. 2005).

42.3.2.1 MBL Polymorphisms in SLE

Whether dysfunctional or deficient MBP variants are found with increased frequency in black patients with SLE was determined (Sullivan et al. 1996). Two structural polymorphisms of MBP, associated with low serum levels of MBP, were found with significantly increased frequency in the SLE patient population. In contrast, a promoter haplotype associated with particularly high serum levels of MBP was negatively associated with SLE. Thus, it seemed that deficiencies of MBP predispose individuals to SLE (Sullivan et al. 1996). The distribution of promoter variants of the MBL gene and correlations between the promoter variants and serum MBL concentrations in Chinese patients with SLE were investigated. Significant differences in the distribution of the two pairs of promoter polymorphisms, H/L and Y/X, between SLE patients and controls, were observed. Analysis of the correlation between promoter haplotypes and serum MBL levels revealed HY as the highest-producing, LY as the intermediate-producing, and LX as the lowest-producing haplotypes. The LX haplotype was present at a frequency of 0.259 in SLE patients and 0.154 in controls and was significantly associated with SLE. The low-producing promoter polymorphism of the MBL gene is associated with SLE, and a low serum MBL level is a risk factor for SLE (Ip et al. 1998). Whether occurrence, characteristics, and progression of SLE are associated with polymorphism of the MBL gene and with serum MBL concentration, Takahashi et al. (2005) reported that MBL gene polymorphism influences susceptibility to SLE, but has no direct effect on disease characteristics. Serum MBL levels fluctuate during the course of SLE in individual patients. MBL genotyping may be useful in assessing the risk of infection during treatment of SLE (Takahashi et al. 2005). MBL and FcγRII (CD32) polymorphisms have both been implicated as candidate susceptibility genes in SLE. These patients carried MBL codon 54 mutant allele more frequently than controls and the haplotype HY W52 W54 W57 was significantly lower in cases compared with controls. The MBL gene codon 54 mutant allele appears to be a risk factor for SLE, whilst haplotypes encoding for high levels of MBL are protective against the disease. However, differences between controls and patients were not significant when FcγRIIa polymorphisms were considered (Villarreal et al. 2001).

42.3.2.2 Lectin Pathway in Murine Lupus Nephritis

In SLE, hypocomplementaemia and complement deposition have been described both in man and in experimental models. In mice, MBL is expressed in two forms, MBL-A and MBL-C. In young and old MRL-lpr and control MRL+/+, the declining levels of MBL-A and MBL-C showed a high degree of correlation. In aged MRL-lpr mice in which autoimmunity is most pronounced, high auto-Ab titers and strong deposition of glomerular immune complexes were associated with deposition of C1q, C3, MBL-A and MBL-C. Thus, in addition to the classical pathway and the alternative pathway, the lectin pathway of complement activation is also involved in murine lupus nephritis (Trouw et al. 2005). SLE patients were associated with a reduced functional activity of the MBL pathway of complement, in relation to expression of MBL variant alleles, increased levels of autoantibodies against cardiolipin and C1q, but not against MBL. The enhanced production of autoantibodies may be related to disturbed clearance of apoptotic material due to impaired MBL function (Seelen et al. 2005). Cohorts of MRL-lpr mice, which are known to develop age-dependent SLE-like disease showed that at 2 months of age all mice already had elevated levels of anti-C1q autoantibodies, and elution of kidneys confirmed the presence of these antibodies in renal immune deposits in MRL-lpr mice and not in control MRL+/+ mice. Thus, anti-C1q antibodies are already present in serum and immune deposits of the kidney early in life and therefore can play a role in nephritis during experimental SLE-like disease in mice (Trouw et al. 2004).

Effect of (S)-Armepavine on Autoimmune Disease of MRL/MpJ-lpr/lpr Mice: (S)-Armepavine (C19H23O3N; MW313) from *Nelumbo nucifera* suppresses T cells proliferation. (S)-armepavine prevents lymphadenopathy and elongated life span of MRL/MpJ-lpr/lpr mice, which have disease features similar to human SLE. The action seemed to be mediated by inhibition of splenocytes proliferation, suppression of IL-2, IL-4, IL-10, and IFN-γ gene expressions, reduction of glomerular hypercellularity and immune complexes deposition, and decrease of urinary protein and anti-double stranded DNA autoantibody production. It has been suggested that (S)-armepavine is an immunomodulator for the management of autoimmune diseases like SLE (Liu et al. 2006a).

42.3.2.3 Autoantibodies to C1q and MBL in SLE

In SLE patients, there is an association between the occurrence of autoantibodies to C1q and MBL and renal involvement. The presence of autoantibodies to MBL, analogous to autoantibodies to C1q in patients with SLE, contributes to the disease development. Anti-MBL autoantibodies were of the IgG isotype and the binding site of IgG anti-MBL was located in the F(ab')2 portion. Anti-MBL are present in sera from SLE patients and influence the functional activity of MBL (Seelen et al. 2003). More SLE patients have IgG anti-MBL antibodies than normal controls. However, in SLE, these antibodies are neither sensitive nor specific for this condition. They occur more frequently in (proliferative) lupus nephritis, particularly during active disease. Furthermore, levels of anti-C1q rise, in many cases, prior to a relapse of lupus nephritis, suggesting a pathogenic role for the autoantibodies in immune complex-mediated renal disease. In addition, anti-C1q may interfere with the clearance of apoptotic cells, so influencing induction and expression of autoimmunity (Kallenberg 2008; Mok et al. 2004).

Cardiovascular Disease and SLE: Cardiovascular disease is an important complication in patients with SLE. Variant alleles of the MBL gene are associated with SLE as well as with severe atherosclerosis. Ohlenschlaeger et al. (2004) determined whether *MBL* variant alleles were associated with an increased risk of arterial thrombosis among Danish patients with SLE and suggested that among patients with SLE, homozygosity for *MBL* variant alleles is associated with an increased risk of arterial thrombosis. The risk of venous thrombosis is not increased, indicating that MBL has a specific role in providing protection against arterial thrombosis (Ohlenschlaeger et al. 2004).

42.3.3 Systemic Inflammatory Response Syndrome/Sepsis

A systemic inflammatory response syndrome (SIRS) is well known in patients after major surgery (Bone 1992). Clinical studies of critically ill patients requiring intensive care management have shown that individuals who are MBL deficient are more likely to develop the SIRS and progress to septic shock and death (Garred et al. 2003b; Fidler et al. 2004), findings which may well relate to the proinflammatory cytokine response. In some cases, SIRS occurs in response to infection and "sepsis" is then used to describe the symptoms. More severely a septic shock may develop with multi-organ dysfunctions (MOD). The MBL levels and genotypes were investigated in 272 adults (197 with sepsis) prospectively admitted to the ICU. No difference was seen between genotype frequencies in patients with SIRS as compared to healthy controls. But the frequency of MBL variant genotypes was significantly higher in patients with sepsis compared with the patients without sepsis, and the risk ratios for the development of "severe sepsis" and "septic shock" ranged from 1.3 to 3.2 times higher in patients with A/O or O/O versus A/A genotype. MBL levels were inversely related to the severity of sepsis (Garred et al. 2003a, b). The SNP of MBL with clinical phenotype in critically ill Caucasians with SIRS are associated with increased prevalence of positive bacterial cultures at admission to the ICU. Patients in low MBL haplotype group did not have significantly increased rates of sepsis or septic shock at admission to the ICU. Survival at day 28 did not differ significantly between the low MBL haplotype and high MBL haplotype groups. Furthermore, MBL was not associated with a particular organism class. The prevalence of the codon 54 mutation of MBL gene in patients having repeated respiratory infections as well as the prevalence of MBL mutant genotype among patients with diffuse panbronchiolitis was further supported (Gomi et al. 2004).

In contrast, the prevalence of the MBL mutant genotype among patients with nontuberculous mycobacteria or Aspergillus chronic infection was not different from that in control subjects. Thus, SNPs in innate immunity receptors may alter recognition and clearance of bacteria without changing outcomes of critically ill adults with systemic inflammatory response syndrome (Sutherland et al. 2005). Polymorphisms of both codon 54 allele and promoter variants of the mannose MBL gene in patients with primary Sjogren's syndrome (SS) was suggested as one of the genetic factors that determines susceptibility to SS (Wang et al. 2001).

Fidler et al. (2004) analyzed the MBL levels and genotypes levels in 100 critically ill children admitted to ICU. A sevenfold greater risk of developing SIRS within 48 h of admission (60% of the patients) was observed for those carrying MBL variant alleles than those with wild type alleles (A/O + O/O vs. A/A). A significant relation was also found between severity of the systemic response to infection and the presence of an MBL mutation. If the severity of illness among the patients admitted with infections was divided into localized infection, sepsis, and septic shock, the median MBL levels were inversely related to severity, and the children with MBL levels below 1,000 ng/mL had a greater chance of developing SIRS. A study of the frequency of sepsis in very low birth weight infants did not reveal statistical significance in clinical data between infants with and without specific mutations in a number of genes, including MBL genotypes (Ahrens et al. 2004). Following surgery of 156 patients undergoing major elective gastrointestinal surgery for malignant disease, Siassi et al. (2003) reported that patients who developed sepsis or SIRS showed significantly lower mean post-operative MBL levels. In colorectal cancer patients, postoperative infection is associated with poor prognosis. Ytting et al. (2005) have reported significantly increased frequency of pneumonia after primary operation in colorectal carcinoma patients who were having low MBL levels. MBL deficiency appears to play an important role in susceptibility of critical ill patients to the development and progression of sepsis and septic shock, and confers a substantially increased risk of fatal outcome. There is clearly a need for improvement in defining which patient groups and which clinical data are relevant to examine.

42.3.4 MBL and Inflammatory Bowel Diseases

Inflammatory bowel disease (IBD) is a pathological spectrum encompassing ulcerative colitis (UC), Crohn's disease (CD), and indeterminate colitis. The resultant IBD phenotype is the consequence of multiple interactions between environmental factors, particularly enteric flora, and the host response to this environment, determined by immunogenetic, epithelial, and other non-immune genetic factors. MBL, as an important component of innate immunity, has engendered considerable research interest. In an early study of 340 unrelated patients with IBD genotyped for *MBL2* exon 1 coding mutations, the frequency of deficient alleles was significantly lower in patients with UC than either the control group, or those with CD (Rector et al. 2001). The study by Rector et al. (2001) suggests that MBL deficiency could be protective against UC; alternatively, it could be interpreted that MBL deficiency, in individuals otherwise predisposed to IBD, may skew the phenotype away from the UC spectrum of disease towards CD. This concept is supported by another study that genotyped *MBL2* in patients with CD, UC, or healthy controls (Seibold et al. 2004). The study also assessed anti-*Saccharomyces cerevisiae* antibody (ASCA) and MBL levels within the same subsets of patients, albeit slightly different numbers within each group. CD patients with MBL deficiency were significantly more likely to be positive for ASCA and for their lymphocytes to proliferate in response to mannan. Thus, it appears that MBL deficiency could impair normal processing of mannan-expressing microbial antigens, such as those found on the cell surface of many common microorganisms. The accumulated antigens could then stimulate the immune system, and contribute to the production of ASCA and possibly the pathogenesis of Crohn's disease. Thus, MBL deficiency might act primarily to influence IBD-specific phenotype in these patients. It should be noted, however, that a follow-up study, testing a larger cohort of CD patients (n = 241), failed to confirm the significant association between variant MBL genotypes and ASCA positivity (Joossens et al. 2006). Seibold et al. (2004) suggested that the frequency of homozygous and compound heterozygous for variant exon 1 alleles differed significantly between patients suffering from CD or UC and the healthy controls.

42.3.4.1 Celiac Disease and MBL

Celiac disease is a multi-factorial/auto-immune disorder caused in genetically susceptible patients, by the ingestion of dietary gluten. Though very little is known about the genetic factors, but there is a strong association of two HLA haplotypes (DQ2 or α1*05, β1*02 and DQ8 or α1*0301, β1*0302) with the disease. Boniotto et al. (2002) indicated an association between celiac disease and the presence of variant MBL alleles. Boniotto et al. (2005) and Iltanen et al. (2003) reported the frequency of homozygosity for variant MBL alleles to be higher in the patients. Among 149 Italian celiac patients, 116 showed the presence of DQ2 or DQ8. MBL2 allele and genotype frequencies varied significantly between celiac patients and healthy humans. It is likely that in those rare cases of celiac disease that are negative for DQ2 and DQ8, the non-HLA susceptibility genotypes would exert a greater effect. The study also analyzed apoptosis within small intestinal biopsy specimens, and showed that MBL tended to aggregate to areas of apoptosis within the epithelium. MBL has been implicated in the normal clearance of apoptotic bodies (Nauta et al. 2003; Ogden et al. 2001). The authors postulated that the association between MBL and celiac disease, and indeed other autoimmune conditions, could relate to impaired apoptosis, whereby MBL deficiency impairs the normal removal and clearance of apoptotic cells that may subsequently reveal previously hidden self-antigen, causing loss of self-tolerance, and spreading of autoimmunity (Boniotto et al. 2005). The association between variant MBL2 alleles and coeliac disease has also been confirmed within the Finnish population (Iltanen et al. 2003) (Worthley et al. 2006). The low MBL genotypes were strongly associated with more celiac disease symptoms as well with increased frequency of secondary autoimmune diseases. By immunohistochemistry MBL was found to be present, together with apoptotic cells, in the basal lamina under the intestinal epithelium, where they had previously found mRNA for MBL. Boniotto et al. suggested that impaired removal of apoptotic cells due to MBL deficiency might predispose to the development of autoimmune symptoms. Mice lacking MBL have been shown to be less efficient in removal of apoptotic cells (Stuart et al. 2005). In vitro studies have also implicated MBL in removal of apoptotic cells (Ogden et al. 2001; Nauta et al. 2004). Alternative explanation could be that increased susceptibility to intestinal infections and diarrhea, associated with low MBL, may change the intestinal epithelia thus allowing for abnormal stimulation of anti-gliadin immune responses and triggering of the cascade leading to celiac disease. A role for MBL like in celiac disease could be easily applied to other autoimmune disorders.

42.3.4.2 MBL and Gastrointestinal Infection

Despite the well-established role of MBL in innate immunity, there have been relatively few studies describing the clinical effect of MBL deficiency in enteric infections. One notable exception is the association between MBL deficiency and risk of *Cryptosporidium parvum* enteritis. This study indicated that patients with biallelic coding mutations (O/O) had a significantly greater chance of cryptosporidiosis compared to those who were either wild-type or heterozygous for *MBL2* mutation (Kelly et al. 2000). The association between MBL deficiency and cryptosporidiosis was confirmed in young Haitian children by Kirkpatrick et al. (2006). However, the two combined studies present compelling evidence for the role of MBL in the host defense against *Cryptosporidium spp.* infection. However, MBL deficiency was not associated with an increased risk of *Escherichia coli* 0157: H7 colitis nor the complication of HUS (Proulx et al. 2003). *H. pylori* is one of the most common human bacterial infections, affecting approximately 50% of humans. Several immunogenetic polymorphisms are associated with clinical outcomes in *H. pylori* infection, as well as with the risk of infection itself. *H pylori* activates MBL in vitro (c/r Worthley et al. 2006). Studies demonstrated that *H. pylori*-related chronic gastritis causes an increase in gastric mucosal MBL expression, but no association was found between *MBL2* genotype and risk of chronic gastritis (Bak-Romaniszyn et al. 2006; Worthley et al. 2007).

MBL has been implicated in mediating gastrointestinal ischemia/reperfusion injury in mice (Hart et al. 2005). But MBL-null mice (deficient in the murine genes encoding MBL) developed only minor gut injury after induced ischemia/reperfusion insult compared to the wild-type mice. On the contrary, MBL has been implicated as a mediator of ischemia/reperfusion injury in both the myocardium (Walsh et al. 2005) and the kidney (Møller-Kristensen et al. 2005).

42.3.5 Rheumatic Heart Disease

Whereas MBL deficiency has been associated with rheumatic disorders, high MBL levels associated to disease has been reported by Hansen et al. (2003). Rheumatic fever (RF) is the most common cause of acquired valvular disease in children and young adults. The pathogenic mechanisms responsible for the development of RF/RHD are associated to an abnormal host immune response (both at humoral and cellular level) to cross-reactive streptococcal antigens. The significantly elevated circulating MBL levels in patients with RHD together with the greater prevalence of MBL deficiency in controls suggest that MBL may cause undesirable complement activation contributing to the pathogenesis of RHD (Schafranski et al. 2004). Probably, MBL deficiency may represent an advantage against the development of rheumatic mitral stenosis and that increased MBL levels may be related to the development of RHD. Under normal conditions, MBL does not bind to the organism's own tissues, but in situations of cellular hypoxia, glycosylation of cell surfaces may occur, leading to the deposition of MBL followed by complement activation. The significantly elevated levels of MBL observed in chronic RHD suggest that MBL may represent a pathogenic factor in the complex physiopathology of the disease, whereas MBL deficient individuals might be less susceptible to develop chronic RHD. Studies demonstrating the binding of MBL to the endothelium and causing excessive complement activation and subsequent tissue damage are known (Jordan et al. 2003), where as MBL deficiency may be advantageous in some circumstances since MBL may lead to an increased cytokine secretion by macrophages (Takahashi et al. 2002). Under some other conditions, MBL is associated to disease severity in both infectious and autoimmune disease (Garred et al. 1997, 1999). Although MBL is an acute-phase protein produced by the liver, its level only shows a moderate increase and determined genetically in inflammatory diseases. The elevated MBL levels in patients with chronic RHD might corroborate the chronic inflammatory activity present in these individuals and contribute to valve injury through complement activation. In addition, MBL may act as an immunomodulatory molecule, inducing a higher secretion of cytokines by macrophages (Turner and Hamvas 2000).

42.3.6 MBL in Cardio-Vascular Complications

There seems to be a delicate balance as to when MBL levels may be involved in harmful or in beneficial inflammation in the cardiovascular system. For example, Kawasaki disease is a systemic vasculitis in childhood possibly caused by infections and in the developed world Kawasaki disease is the most common cause of acquired heart disease in children (Royle et al. 2005). MBL as an initiator of inflammation, Biezeveld et al. (2003) while studying the frequency of MBL genotypes with Kawasaki disease in Dutch patients found a higher frequency of MBL mutations as compared to the genotypes in controls. Children younger than 1 year were at higher risk of development of CAD. Kawasaki disease occurs more frequent in oriental children (10 times more frequent than in Caucasians) (Royle et al. 2005). In Chinese, Cheung et al. (2004) did not see a difference between the MBL genotypes of patients and controls. The recent observation of an association between a human coronavirus and Kawasaki disease (Esper et al. 2005) fits with the many indications of MBL having antiviral activity. Plaque material may be removed from inside of the carotid artery (e.g., by endarterectomy) to avoid cerebral attack. Rugonfalvi-Kiss et al. (2005) indicated that female patients with genotypes associated with lower MBL levels had a slower rate to early restenosis, suggesting that a high level of MBL may be part of the pathophysiology of this condition.

In a study of 76 patients with severe atherosclerosis, Madsen et al. (1998) found that there were more patients with myocardial infarcts among Norwegians with low MBL allotypes than in controls. Saevarsdottir et al. (2005) found in a cohort study in Iceland that the risk of developing myocardial infarction was higher in MBL deficient individuals. The relationship between markers of innate immunity and clinical outcomes in patients with heart failure (HF) after acute myocardial infarction (AMI) suggested that atherogenesis and heart failure are associated with the altered control of inflammation by innate immune defenses, which include TLRs and MBL. Circulating levels of MBL and sTLR2 may reflect different aspects of the innate immune response and the involvement of innate immunity responses in the pathogenesis of post-MI heart failure (Ueland et al. 2006).

Lectin Pathway in Myocardial Ischemia-Reperfusion: MBL plays a dual role in modifying inflammatory responses to sterile and infectious injury. Although complement is widely accepted as participating in the pathophysiology of ischemia-reperfusion injury, the specific role of the lectin pathway has been addressed by Jordan et al. (2001). Since, blockade of the lectin pathway with inhibitory mAbs protects the heart from ischemia-reperfusion by reducing neutrophil infiltration and attenuating pro-inflammatory gene expression, it appears that the lectin complement pathway is activated after myocardial ischemia-reperfusion and leads to tissue injury.

Mice devoid of MBL-dependent lectin pathway activation but fully active for alternative and classical complement pathways, are protected from cardiac reperfusion injury with resultant preservation of cardiac function. Significantly, mice that lack a major component of the classical complement pathway initiation complex (C1q) but have an intact MBL complement pathway are not protected from injury. Thus, the MBL-dependent pathway of complement activation is the key regulator of myocardial reperfusion ischemic injury (Walsh et al. 2005).

Diabetic Patients at High Risk for Diabetic-Nephropathy: Whether lectin pathway of complement activation plays a role in the pathogenesis of human glomerulonephritis, is not well known. It has been proposed that MBL may bind to altered self components, which may possibly be found in diabetic patients, and MBL could thus be a potential pathogenic factor for diabetic cardiovascular complications, e.g., nephropathy (Hansen et al. 2005; Saevarsdottir et al. 2005). Normo-albuminuric type 1 diabetics have been found to have higher MBL levels than non-diabetic controls, with a stepwise increase in circulating MBL levels with increasing levels of urinary albumin excretion (Hansen et al. 2003). In another study, a significantly larger proportion of patients with diabetic nephropathy presented a MBL genotype associated with higher MBL level, when compared to the group with MBL genotypes associated with low MBL levels (Hansen et al. 2004). The elevated serum MBL levels in type 1 diabetic patients with diabetic nephropathy were confirmed by Saraheimo et al. (2005). In a long follow up study on 270 type 1 diabetic patients, it was found that for patients with type 1 diabetes and MBL-levels below the median of 1.6 µg/mL the risk of developing micro or macro-albuminuria was 26%, while the patients with MBL-levels above the median had a risk of 41% of developing micro or macro-albuminuria (Hovind et al. 2005). These studies suggested that a high MBL geno- and phenotype is associated with an increased risk of developing diabetic kidney disease and that assessing MBL status may prove beneficial in identifying patients at risk for micro- and macro-vascular complications (c/r Thiel et al. 2006).

MBL has been detected in the glomeruli of patients with lupus nephropathy, membranous nephropathy, membranoproliferative glomerulonephritis type I and anti-GBM nephritis. It was proposed that MBL binds to agalactosyl oligosaccharides of IgG that terminate in N-acetylglucosamine (Lhotta et al. 1999). Elevated serum MBL levels of were implicated in the pathogenesis of renal manifestations of Henoch-Schönlein purpura (Endo et al. 2000), in IgA nephropathy (Endo et al. 1998), in other forms of human glomerulonephritis (Lhotta et al. 1999), and in vascular complications of diabetes mellitus type 1 (Hansen et al. 2003).

42.3.7 Other Inflammatory Disorders

42.3.7.1 Sarcoidosis and MBL Variants

Sarcoidosis is a chronic granulomatous disease of unknown aetiology. The causative agent may be an infectious microorganism. MBL variants predisposed to sarcoidosis by increasing their susceptibility to micro-organisms among sarcoidosis patients showed that the frequencies of variants were similar regardless of severity of disease outcome. MBL gene variants did not indicate to influence susceptibility to sarcoidosis, age of disease onset, or severity of disease. The average patient ages at time of diagnosis were similar for all MBL genotypes (Foley et al. 2000). However, a 57-year-old woman patient with pulmonary sarcoidosis suggested interstitial nephritis without proteinuria and hematuria, whereas a renal biopsy showed granulomatous interstitial nephritis and mild mesangial proliferative glomerulonephritis. From this case of renal sarcoidosis, it was hypothesized that *P. acnes* might be involved in pathogenesis of granulomatous interstitial nephritis and that it plays a role in glomerular complement activation via the lectin pathway (Hagiwara et al. 2005).

42.3.7.2 Behcet's Disease

Behcet's disease (BD) is a multisystemic, recurrent inflammatory disease caused by the combination of genetic and environmental factors. The haplotypes of the MBL2 gene can influence therapeutic response in BD, thus affecting the clinical symptoms in BD patients. The promoter region, MBL2-550*C/*C (L/L) homozygote was found to have a lower frequency in BD patients than that in controls. No difference was observed in the allele frequencies of G-221 C (Y/X), C + 4 T (P/Q) or Gly54Asp (A/B) of the MBL2 gene in BD patients and in controls. The HYPA haplotype contributed to BD occurrence, whereas the LYPA haplotype was negatively associated with BD. BD patients with several symptoms and with an earlier disease-onset age had a higher HYPA haplotype frequency. BD patients showing poor response (S) to therapy had a higher HYPA frequency than those showing good response (M). It appeared that possessing HYPA increases the risk of BD and that the MBL2 HYPA haplotype plays a role in MBL levels and increases the susceptibility to BD (Park et al. 2005).

42.3.7.3 Cystic Fibrosis (CF) Lung Disease

The MBL deficiency has been associated with poor outcome in cystic fibrosis (CF) lung disease. A mutation in MASP-2 and higher serum levels of MBL than healthy controls belonging to the MBL pathway in serum have been described. Thus, MBL pathway function is affected both by MBL and by MASP-2 genotypes (Carlsson et al. 2005). Patients undergoing abdominal aortic aneurysm (AAA) repair are exposed to an ischaemia-reperfusion injury (IRI), which is in part mediated by complement activation. During IRI, the patients undergoing AAA repair experience a mean decrease in plasma MBL level of 41% representing significant lectin pathway activation. This indicated that consumption of MBL occurs during AAA repair, which suggested an important role for lectin pathway in IRI. Hence, specific transient inhibition of lectin pathway activity can be of significant therapeutic value in patients undergoing open surgical AAA repair (Norwood et al. 2006).

In Guillain-Barre syndrome (GBS), complement activation plays a crucial role in the induction and extent of the post-infectious immune-mediated peripheral nerve damage. The MBL2 genotype, serum MBL level, and MBL complex activity are associated with the development and severity of GBS. The MBL2 B allele was associated with functional deficiency and relatively mild weakness. Studies support the hypothesis that complement activation mediated by MBL contributes to the extent of nerve damage in GBS, which is codetermined by the MBL2 haplotype (Geleijns et al. 2006).

42.3.7.4 Experimental Polymicrobial Peritonitis

Peritonitis is the most common and major complication in the treatment modality of peritoneal dialysis (PD) for uraemic patients. The contribution of the different complement activation pathways was studied in the host defense against experimental polymicrobial peritonitis induced by cecal ligation and puncture by using mice deficient in either C1q or factors B and C2. The C1q-deficient mice lack the classical complement activation pathway. Mice with a deficiency of both factors B and C2 lack complement activation via the classical, the alternative, and the lectin pathways and exhibited the maximum mortality of 92%, indicating a significant contribution of the lectin and alternative pathways of complement activation to survival (Windbichler et al. 2004). While examining the role of serum MBL concentration and point mutations in MBL gene in PD-related peritonitis, Lam et al. (2005) found that both homozygous and heterozygous patients had profoundly reduced serum level of MBL. Thus, dialysis patients having lower MBL levels may increase the susceptibility of infection.

42.4 Significance of MBL Gene in Transplantation

MBL Replacement Therapy Following Stem Cell Transplantation: Life-threatening complications such as graft versus host disease and infection remain major barriers to the success of allogeneic hemopoietic stem cell transplantation (SCT). Among various factors, MBL deficiency is a risk factor for infection in other situations where immunity is compromised. MBL2 coding mutations were associated with an increased risk of major infection following transplantation. MBL2 promoter variants were also associated with major infection. The high-producing haplotype HYA was associated with a markedly reduced risk of infection. Donor MBL2 coding mutations and recipient HYA haplotype were independently associated with infection in multivariate analysis. Thus, these results suggest that MBL2 genotype influences the risk of infection following allogeneic SCT and that both donor and recipient MBL2 genotype are important (Mullighan et al. 2002).

A retrospective study examining associations between polymorphisms in the gene encoding MBL, MBL2 and risk of major infection post-SCT was conducted in 96 related myeloablative transplants. The study showed that "low-producing" MBL2 coding alleles, when present in the donor, were significantly associated with increased risk of major infection in the recipient following neutrophil count recovery. Furthermore, a "high-producing" MBL2 haplotype, HYA, when present in the recipient, was protective against infection. Since MBL is under development as a therapeutic agent, findings suggest that administration of MBL may reduce the risk of infection post-transplant. Further work is required to confirm these results. These results indicate a report of a genetic determinant of risk of infection post-SCT, and highlight the importance of non-HLA genetic factors in determining the risk of transplant complications (Mullighan and Bardy 2004).

Genetic MBL variants are frequent and associated with low MBL serum levels. Higher MBL levels may be associated with more complement-mediated damage resulting in inferior graft survival. Berger et al. (2005) showed no significant difference in incidence of delayed graft function in recipients with a low MBL level (< or = 400 ng/mL) compared to those with a higher MBL level (>400 ng/mL). If these data can be confirmed, pre-transplant MBL levels may provide additional information for risk stratification prior to kidney transplantation (Berger et al. 2005).

Complement Activation Is Harmful for the Allograft Endothelium: In heart transplant recipients, Fiane et al. (2005) recorded transplant-associated coronary artery disease and observed an association with MBL deficiency. They also recorded that acute rejection of the transplant was seen in 6 out of 6 with MBL deficiency as compared to 15 out of 32 with higher MBL levels. Assuming that MBL may interact with the transplanted tissue and initiate complement activation, this study added to the list of studies, which suggested that complement activation is harmful for the endothelium in general, and possibly for the allograft endothelium in particular.

MBL Pathway and SPKT Graft Survival: Simultaneous pancreas-kidney transplantation (SPKT) is the treatment of choice for patients with type 1 diabetes and renal failure. However, this procedure is characterized by a high rate of postoperative infections, acute rejection episodes, and cardiovascular mortality. The lectin pathway of complement activation contributes to cardiovascular disease in diabetes and may play an important role in inflammatory damage after organ transplantation. MBL serum levels and MBL genotypes in patients who received an SPKT from 1990 through 2000 and related graft survivals revealed that survival was significantly better in recipients with MBL gene polymorphisms associated with low MBL levels. Thus,

MBL is a potential risk factor for graft and patient survival in SPKT. It is hypothesized that MBL contributes to the pathogenesis of inflammation-induced vascular damage both in the transplanted organs and in the recipient's native blood vessels (Berger et al. 2007). To address further the role of MBL deficiency, Verschuren et al. (2008) showed that high levels of serum MBL are associated with protection against urinary tract infections and, more specifically, against urosepsis after SPKT. These results indicate an important role for the lectin pathway of complement activation in antimicrobial defense in these transplant recipients (Verschuren et al. 2008).

42.5 MBL in Tumorigenesis

42.5.1 Polymorphisms in the Promoter

In paediatric oncology patients with febrile neutropenia, MBL levels are correlated to clinical and laboratory parameters. Structural exon-1 MBL2 mutations and the LX promoter polymorphism were related to deficient MBL levels. The capacity to increase MBL concentrations during febrile neutropenia was associated with MBL2 genotype. Infectious parameters did not differ between MBL-deficient and MBL-sufficient neutropenic children. However, most patients (61%) were severely neutropenic, compromising the opsonophagocytic effector function of MBL. MBL substitution might still be beneficial in patients with phagocytic activity (Frakking et al. 2003).

Five polymorphisms in the promoter and first exon of the MBL2 gene alter the expression and function of MBL in humans and are associated with inflammation-related disease susceptibility. These five polymorphisms create six well-characterized haplotypes that result in lower (i.e., LYB, LYC, HYD, and LXA) or higher (i.e., HYA and LYA) serum MBL concentrations. A statistically significant association was found between the X allele of the promoter Y/X polymorphism (which results in a lower serum MBL concentration) and improved lung cancer survival among white patients but not among African American patients. The functional Y/X polymorphism of the innate-immunity gene MBL2 and MBL2 haplotypes and diplotypes appear to be associated with lung cancer survival among white patients (Pine et al. 2007). A significantly higher incidence of MBL deficiency/insufficiency-associated genotypes was found among patients with malignant disease. Findings reflecting anti-tumorigenic activity of MBL protein suggest potential therapeutic application. However, it cannot be excluded that mbl-2 mutant alleles may be in linkage disequilibrium with an unidentified tumor susceptibility gene(s) (Swierzko et al. 2007).

The genetic variations in key genes of MBL2 could influence the risk for breast cancer. A preliminary analysis of SNPs in MBL2 in breast cancer [166 African-American (AA) patients vs. 180 controls and 127 Caucasian (CAU) patients vs. 137 controls] presents evidence that common genetic variants in the 3′-UTR of MBL2 might influence the risk for breast cancer in AA women, probably in interaction with the 5′ secretor haplotypes that are associated with high concentrations of MBL (Bernig et al. 2007).

42.5.2 MBL Binding with Tumor Cells

Changes in cell surface structures during oncogenic transformation appear to promote binding of MBL to cancer cells (Hakomori 2001) where the protein can mediate cytotoxic effects including MBL-dependent cell mediated cytotoxicity (Ma et al. 1999; Nakagawa et al. 2003). Experimental studies suggest that MBL (both wild-type and the mutant B allele) may possess anti-colorectal cancer tumor activity (Ma et al. 1999). The MBP/MBL binds specifically to oligosaccharides expressed on the surface of human colorectal carcinoma cells, SW1116. MBL binding occurs in colon adenocarcinoma cell lines (Colo205, Colo201 and DLD-1), but not in any of the leukemic cell lines. The binding of MBL to these cell lines was sugar-specific and calcium-dependent. The degree of MBL binding was correlated with the expression of Lewis A and Lewis B antigens on these cell lines (Muto et al. 1999). Intra-tumoral administration of the recombinant vaccinia virus carrying a MBL2 gene significantly reduced tumor size as compared to controls, along with prolonged survival of mice (Ma et al. 1999). However, these results were not reflected in clinical trials. In fact, patients with colorectal cancer have increased activation of the lectin-complement pathway and increased levels of serum MBL (Ytting et al. 2004). In patients undergoing surgery for colorectal cancer, however, low pre-operative levels of serum MBL have been linked to an increased risk of developing post-colectomy pneumonia (Ytting et al. 2005). Further studies may clarify the role of the lectin-complement pathway in colorectal cancer.

42.6 Complications Associated with Chemotherapy

42.6.1 Neutropenia

Secondary immunodeficiencies due to disease or treatment have provided interesting patient populations within which to study the role of MBL. One such group comprises those receiving chemotherapy for malignancy. In these patients, chemotherapy induces neutropenia and increased risk of infection. Studies have thus attempted to analyze the correlation between MBL deficiency and infections in such patients. It is difficult to compare these studies since these studies include patients with a variety of underlying malignancies and variety of chemotherapy, different combinations of antimicrobial agents and various other factors (Klein and Kilpatrick 2004). However, there is clear evidence of the importance of MBL for protection in leukemia patients. MBL genotypes and MBL levels were correlated to the causes, frequency and duration of febrile neutropenic periods in children receiving chemotherapy (Neth et al. 2001). The majority of children were patients of acute lymphoblastic leukemia (ALL). Children with variant MBL alleles exhibited twice as many days of febrile neutropenia as children with wild type genotypes. Analysis by MBL quantification supported this as children with less than 1 μg MBL/mL had a higher number of days with febrile neutropenia. Peterslund et al. (2001) described infections defined as bacteremia, pneumonia or both in hematological malignancies of 54 adults treated with chemotherapy. All patients with the infections, except one, showed MBL levels below 0.5 μg/mL. Vekemans et al. (2005) conducted a prospective observational study focusing on assessment of MBL as a risk factor for infection during chemotherapy induced neutropenia in adult hematological cancer patients. They included 255 patients and determined MBL levels as well as MBL genotypes. A higher rate of severe infection was seen in MBL deficient patients. The impact was further increased when acute leukaemic patients were excluded. In a contrasting study, Bergmann et al. (2003) followed 80 adults undergoing therapy for acute myeloid leukaemia (AML), which involves intense highly myelosuppressive treatment. They found no effect of MBL deficiency on frequency, severity or duration of fever and suggested that the severe immunosuppression induced by the combination of the myeloid cancer and chemotherapy may obscure the normal effector functions of MBL, though, Kilpatrick et al. (2003) failed to see anything but a modest effect of MBL levels below 100 ng/mL in a retrospective study on 128 patients, most of whom were prepared for bone marrow transfer and more than half presented with AML. Results cast doubt on the potential value of MBL replacement therapy in this clinical context (Kilpatrick et al. 2003). A growing body of evidence indicates that genetic factors are involved in an increased risk of infection. MBL gene polymorphisms that cause low levels of MBL are associated with the occurrence of major infections in patients, mainly bearing hematological malignancies, after high-dose chemotherapy (HDT) rescued by autologous peripheral blood stem cell transplantation (auto-PBSCT). A retrospective examination of 113 patients treated with HDT and auto-PBSCT revealed that the low-producing genotypes, B/B and B/LXA, were associated with major bacterial infection. A nation-wide study, conducted to assess the allele frequency of the MBL coding mutation in a total of 2,623 healthy individuals in Japan, revealed the frequency of allele B as 0.2, almost the same in seven different areas of Japan. This common occurrence suggested that MBL deficiency may play an important role in the clinical settings of immune-suppression (Horiuchi et al. 2005). Studies by Aittoniemi et al. (1999) in patients with chronic lymphocytic leukemia did not observe any effect of MBL on infections. A possible association between MBL genotypes and severe infections in patients with multiple myeloma receiving moderate strength induction chemotherapy has been studied. From the MBP genotypes, identified in bone marrow biopsies, the study concluded that during induction chemotherapy in patients with multiple myeloma, a general protective effect of wild-type MBL2 against chemotherapy-related infections was not apparent. However, indications were there of a reduced occurrence of septicaemia in patients with wild-type compared with variant MBL2. Further studies in larger cohorts of patients are relevant (Molle et al. 2006). Thus, further studies are required to describe the patients particularly at risk when being MBL deficient.

42.6.2 Animal Studies

The MBL knock-out mice have made possible experimental investigations of the effect of MBL deficiency. The mouse has two genes encoding different MBL molecules (MBL-A and -C) compared to one in humans. Both MBLs in mice are able to bind to carbohydrate surfaces and activate the complement system. A slight difference in carbohydrate specificity has been reported for the two mouse MBLs. Mice with only MBL-A knocked-out were first produced, but only mice with both MBL-A and -C knocked out (MBL DKO) are suitable as animal model of human MBL deficiency. In 2004, Shi et al. demonstrated that MBL DKO sepsis model mice were highly susceptible to intravenous inoculation via tail vein with *Staphylococcus aureus*, all dying within 48 h, compared with 55% survival of MBL wild-type mice. Infusion of recombinant MBL reversed the phenotype. No difference was seen when the bacteria were injected intra-peritoneally. However, if the mice were treated with cyclophosphamide, simulating chemotherapy-induced neutropenia, before the intra peritoneal infection, the MBL DKO had more abscesses than the wild type. The MBL DKO mice were also more susceptible to challenge with herpes simplex virus type 2 (Gadjeva et al. 2004). In line with the suggested involvement of MBL in autoimmune diseases the MBL DKO mice were examined for autoimmune symptoms when 18-month-old (Stuart et al. 2005). No such signs were observed. On the other hand, the ability to clear apoptotic cells was less efficient in the MBL knock-outs. It has been hypothesized that while MBL does not bind significantly to healthy tissue, changes due to abnormal conditions might reveal MBL ligands. Indeed, MBL is expressed by some tumor cell lines, and gene therapy with an MBL-vaccinia construct was found protective in nude mice transplanted with a human colorectal cancer cell line (Ma et al. 1999). In vitro studies have indicated binding of MBL to cells exposed to hypoxia-reoxygenation (simulating ischemia/reperfusion) and subsequently it was shown that infusion of a blocking anti-MBL antibody would protect against myocardial destruction following ischemia/reperfusion in a rat model (Jordan et al. 2001). Using MBL DKO mice Møller-Kristensen et al. (2005) found, in a model of kidney ischemia reperfusion (I/R) injury, that the MBL DKO were partially protected as evidenced by a better kidney function in these mice after ischemia/reperfusion. Increased deposition of the complement factor C3 was seen in wild type mice, and binding of MBL to sections of kidney could be inhibited with mannose. In agreement with this, deVries et al. (2004) found MBL-A and -C deposited in the kidneys after ischemia/reperfusion in MBL wild type mice. The recombinant vaccinia virus carrying human MBP gene possesses a potent growth-inhibiting activity against human colorectal carcinoma cells transplanted in KSN nude mice. The treatment resulted into a prolonged life span of tumor-bearing mice. Local production of MBP had a cytotoxic activity, which was proposed as MBP dependent cell-mediated cytotoxicity (MDCC). This study offers a model for the development of an effective and specific host defense factor for cancer gene therapy (Ma et al. 1999; Thiel et al. 2006).

42.7 MBL: A Reconstitution Therapy

Since genetically determined MBL deficiency is very common and can be associated with increased susceptibility to a variety of infections, the potential benefits of MBL reconstitution therapy need to be evaluated. In a phase I trial on 20 MBL-deficient healthy adult volunteers receiving a total of 18 mg of MBL in three 6 mg doses given (i.v.), once a week for a period of 3 weeks did not show adverse clinical changes or any sign of infusion-associated complement activation. Study suggested that infusion of purified MBL as prepared by Statens Serum Institut (SSI) is safe. However, adults have to be given at least 6 mg twice or thrice weekly for maintaining protective MBL levels assumed to be about 1,000 ng/mL (Valdimarsson et al. 2004). Considerations of MBL genotyping and association with infection opens the possibility of producing clinical grade recombinant MBL that resulted to establishing a company having this aim (Jensenius et al. 2003). The treatment of chronic disorders may possibly also be considered on the longer term. The invention led to use of at least MBL oligomer comprising at least one MBL subunit, for the manufacture of a medicament for prophylaxis and/or treatment of infection (Thiel and Jensenius 2007).

References

Ahrens P, Kattner E, Kohler B et al (2004) Mutations of genes involved in the innate immune system as predictors of sepsis in very low birth weight infants. Pediatr Res 55:652–656

Aittoniemi J, Miettinen A, Laine S et al (1999) Opsonising immunoglobulins and mannanbinding lectin in chronic lymphocytic leukemia. Leuk. Lymphoma 34:381–385

Ambrosio AR, De Messias-Reason IJ (2005) *Leishmania* (Viannia) *braziliensis*: interaction of mannose-binding lectin with surface glycoconjugates and complement activation. An antibodyindependent defence mechanism. Parasite Immunol 2:333–340

Bak-Romaniszyn L, Cedzyński M, Szemraj J et al (2006) Mannan-binding lectin in children with chronic gastritis. Scand J Immunol 63:131–5

Barton A, Platt H, Salway F et al (2004) Polymorphisms in the mannose binding lectin (MBL) gene are not associated with radiographic erosions in rheumatoid or inflammatory polyarthritis. J Rheumatol 31:442–447

Bellamy R, Ruwende C, McAdam KP et al (1998) Mannose binding protein deficiency is not associated with malaria, hepatitis B carriage nor tuberculosis in Africans. Q J Med 91:13–18

Berger SP, Roos A, Mallat MJ et al (2005) Association between mannose-binding lectin levels and graft survival in kidney transplantation. Am J Transplant 5:1361–1366

Berger SP, Roos A, Mallat MJ et al (2007) Low pretransplantation mannose-binding lectin levels predict superior patient and graft survival after simultaneous pancreas-kidney transplantation. J Am Soc Nephrol 18:2416–2422

Bergmann OJ, Christiansen M, Laursen I et al (2003) Low levels of mannose-binding lectin do not affect occurrence of severe infections or duration of fever in acute myeloid leukaemia during remission induction therapy. Eur J Haematol 70:91–97

Bernig T, Boersma BJ, Howe TM et al (2007) The mannose-binding lectin (MBL2) haplotype and breast cancer: an association study in African-American and Caucasian women. Carcinogenesis 28:828–836

Biezeveld MH, Kuipers IM, Geissler J et al (2003) Association of mannose-binding lectin genotype with cardiovascular abnormalities in Kawasaki disease. Lancet 361:1268–1270

Bodamer OA, Mitterer G, Maurer W et al (2006) Evidence for an association between mannose-binding lectin 2 (MBL2) gene polymorphisms and pre-term birth. Genet Med 8:518–524

Bone RC, (1992) Toward an epidemiology and natural history of SIRS (systemic inflammatory response syndrome). JAMA 268: 3452–3455

Boniotto M, Braida L, Spano A et al (2002) Variant mannose-binding lectin alleles are associated with celiac disease. Immunogenetics 54:596–598

Boniotto M, Braida L, Baldas V et al (2005) Evidence of a correlation between mannose binding lectin and celiac disease: a model for other autoimmune diseases. J Mol Med 83:308–316

Botos I, Wlodawer A (2005) Proteins that bind high-mannose sugars of the HIV envelope. Prog Biophys Mol Biol 88:233–282

Bouwman LH, Roos A, Terpstra OT, de Knijff P et al (2005) Mannose binding lectin gene polymorphisms confer a major risk for severe infections after liver transplantation. Gastroenterology 129:408

Cambi A, Figdor CG (2005) Levels of complexity in pathogen recognition by C-type lectins. Curr Opin Immunol 17:345–351

Carlsson M, Sjoholm AG, Eriksson L et al (2005) Deficiency of the mannan-binding lectin pathway of complement and poor outcome in cystic fibrosis: bacterial colonization may be decisive for a relationship. Clin Exp Immunol 139:306–313

Cheung YF, Ho MH, Ip WK et al (2004) Modulating effects of mannose binding lectin genotype on arterial stiffness in children after Kawasaki disease. Pediatr Res 56:591–596

Dahl M, Tybjaerg-Hansen A, Schnohr P et al (2004) A population-based study of morbidity and mortality in mannose-binding lectin deficiency. J Exp Med 199:1391–1399

Davies EJ, Snowden N, Hillarby MC et al (1995) Mannose-binding protein gene polymorphism in systemic lupus erythematosus. Arthritis Rheum 38:110–114

Davies EJ, Teh LS, Ordi-Ros J et al (1997) A dysfunctional allele of the mannose binding protein gene associates with systemic lupus erythematosus in a Spanish population. J Rheumatol 24:485–488

deVries B, Walter SJ, Peutz-Kootstra CJ et al (2004) The mannose-binding lectin-pathway is involved in complement activation in the course of renal ischemia-reperfusion injury. Am J Pathol 165:1677–1688

Dommett RM, Klein N, Turner MW (2006) Mannose-binding lectin in innate immunity: past, present and future. Tissue Antigens 68:193–209

Downing I, Koch C, Kilpatrick DC (2003) Immature dendritic cells possess a sugar-sensitive receptor for human mannan-binding lectin. Immunology 109:360–364

Downing I, MacDonald SL, Turner ML et al (2005) Detection of an autologous ligand for mannan-binding lectin on human B lymphocytes. Scand J Immunol 62:507–514

Dumestre-Perard C, Ponard D, Arlaud GJ et al (2002) Evaluation and clinical interest of mannan binding lectin function in human plasma. Mol Immunol 39(7–8):465–473

Dzwonek A, Novelli V, Bajaj-Elliott M et al (2006) Mannose-binding lectin in susceptibility and progression of HIV-1 infection in children. Antivir Ther 11:499–506

Eisen DP, Minchinton RM (2003) Impact of mannose-binding lectin on susceptibility to infectious diseases. Clin Infect Dis 37:1496–1506

Endo M, Ohi H, Ohsawa I et al (1998) Glomerular deposition of mannose-binding lectin (MBL) indicates a novel mechanism of complement activation in IgA nephropathy. Nephrol Dial Transplant 13:1984–1990

Endo M, Ohi H, Ohsawa I et al (2000) Complement activation through the lectin pathway in patients with Henoch-Schönlein purpura nephritis. Am J Kidney Dis 35:401–719

Esper F, Shapiro ED, Weibel C et al (2005) Association between a novel human coronavirus and Kawasaki disease. J Infect Dis 191:499–502

Fiane AE, Ueland T, Simonsen S et al (2005) Low mannose-binding lectin and increased complement activation correlate to allograft vasculopathy, ischaemia, and rejection after human heart transplantation. Eur Heart J 26(16):1660–1665

Fidler KJ, Wilson P, Davies JC et al (2004) Increased incidence and severity of the systemic inflammatory response syndrome in patients deficient in mannose-binding lectin. Int Care Med 30:1438–1446

Foley PJ, Mullighan CG, McGrath DS et al (2000) Mannose-binding lectin promoter and structural gene variants in sarcoidosis. 30:549–552

Frakking F, van de Wetering M et al (2003) The role of mannose-binding lectin (MBL) in paediatric oncology patients with febrile neutropenia. Eur J Cancer 42:909–916

Gadjeva M, Paludan SR, Thiel S et al (2004) Mannan-binding lectin modulates the response to HSV-2 infection. Clin Exp Immunol 138:304–311

Garred P, Thiel S, Madsen HO et al (1992) Gene frequency and partial protein characterization of an allelic variant of mannan binding protein associated with low serum concentrations. Clin Exp Immunol 90:517–521

Garred P, Madsen HO, Baslev U et al (1997) Susceptibility to HIV infection and progression of AIDS in relation to variant alleles of mannose binding lectin. Lancet 349:236–240

Garred P, Madsen HO, Halberg P et al (1999) Mannose-binding lectin polymorphism and susceptibility to infection in systemic lupus erythematosus. Arthritis Rheum 42:2145–2152

Garred P, Voss A, Madsen HO, Junker P (2001) Association of mannose-binding lectin gene variation with disease severity and infections in a population-based cohort of systemic lupus erythematosus patients. Genes Immun 2:442–450

Garred P, Larsen F, Madsen HO, Koch C (2003a) Mannose-binding lectin deficiency-revisited. Mol Immunol 40:73–84

Garred P, Strom JJ, Quist L et al (2003b) Association of mannose-binding lectin polymorphisms with sepsis and fatal outcome, in patients with systemic inflammatory response syndrome. J Infect Dis 188:1394–1403

Garred P, Larsen F, Seyfarth J, Fujita R, Madsen HO (2006) Mannose-binding lectin and its genetic variants. Genes Immun 7:85–94

Geleijns K, Roos A, Houwing-Duistermaat JJ et al (2006) Mannose-binding lectin contributes to the severity of Guillain-Barre syndrome. J Immunol 177:4211–4217

Gomi K, Tokue Y, Kobayashi T et al (2004) Mannose-binding lectin gene polymorphism is a modulating factor in repeated respiratory infections. Chest 126:95–99

Graudal N (2004) The natural history and prognosis of rheumatoid arthritis: association of radiographic outcome with process variables, joint motion and immune proteins. Scand J Rheumatol Suppl 118:1–38

Graudal NA, Madsen HO, Tarp U et al (2000) The association of variant mannose-binding lectin genotypes with radiographic outcome in rheumatoid arthritis. Arthritis Rheum 43:515–521

Gupta K, Gupta RK, Hajela K (2008) Disease associations of mannose-binding lectin & potential of replacement therapy. Indian J Med Res 127:431–440

Hagiwara S, Ohi H, Eishi Y et al (2005) A case of renal sarcoidosis with complement activation via the lectin pathway. Am J Kidney Dis 45:580–587

Hakomori S (2001) Tumor-associated carbohydrate antigens defining tumor malignancy: basis for development of anti-cancer vaccines. Adv Exp Med Biol 491:369–402

Hansen TK (2005) Mannose-binding lectin (MBL) and vascular complications in diabetis. Horm Metab Res 37(suppl 1):1–4

Hansen TK, Thiel S, Knudsen ST et al (2003) Elevated levels of mannan-binding lectin in patients with type 1 diabetes. J Clin Endocrinol Metab 88:4857–4861

Hansen TK, Tarnow L, Thiel S, Parving HH, Flyvbjerg A et al (2004) Association between mannose-binding lectin and vascular complications in type 1 diabetes. Diabetes 53:1570–1576

Hart ML, Saifuddin M, Spear GT (2003) Glycosylation inhibitors and neuraminidase enhance human immunodeficiency virus type 1 binding and neutralization by mannose-binding lectin. J Gen Virol 84(Pt 2):353–360

Hart ML, Ceonzo KA, Shaffer LA et al (2005) Gastrointestinal ischemia-reperfusion injury is lectin complement pathway dependent without involving C1q. J Immunol 174:6373–6380

Heise CT, Nicholls JR, Leamy CE, Wallis R (2000) Impaired secretion of rat mannose-binding protein resulting from mutations in the collagen-like domain. J Immunol 165:1403–1409

Hjalgrim LL, Madsen HO, Melbye M et al (2002) Presence of clone-specific markers at birth in children with acute lymphoblastic leukaemia. Br J Cancer 87:994–999

Horiuchi T, Gondo H, Miyagawa H et al (2005) Association of MBL gene polymorphisms with major bacterial infection in patients treated with high-dose chemotherapy and autologous PBSCT. Genes Immun 6:162–166

Hovind P, Hansen TK, Tarnow L et al (2005) Mannose-binding lectin as a predictor of microalbuminuria in type 1 diabetes: an inception cohort study. Diabetes 54:1523–1527

Iltanen S, Maki M, Collin P et al (2003) The association between mannan-binding lectin gene alleles and celiac disease. Am J Gastroenterol 98:2808–2809

Ip WK, Lau YL (2004) Role of mannose-binding lectin in the innate defense against *Candida albicans*: enhancement of complement activation, but lack of opsonic function, in phagocytosis by human dendritic cells. J Infect Dis 190:632–640

Ip WK, Chan SY, Lau CS, Lau YL (1998) Association of systemic lupus erythematosus with promoter polymorphisms of the mannose-binding lectin gene. Arthritis Rheum 41:1663–1668

Ip WK, Chan KH, Law HK et al (2005) Mannose-binding lectin in severe acute respiratory syndrome coronavirus infection. J Infect Dis 191:1697–1704

Jack DL, Klein NJ, Turner MW (2001) Mannose-binding lectin: targeting the microbial world for complement attack and opsonophagocytosis. Immunol Rev 180:86–99

Jacobsen S, Madsen HO, Klarlund M, TIRA Group et al (2001) The influence of mannose binding lectin polymorphisms on disease outcome in early polyarthritis. J Rheumatol 28:935–942

Jensenius JC, Jensen PH, McGuire K et al (2003) Recombinant mannan-binding lectin (MBL) for therapy. Biochem Soc Trans 31:763–767

Ji X, Gewurz H, Spear GT (2005) Mannose binding lectin (MBL) and HIV. Mol Immunol 42:145–152

Joossens S, Pierik M, Rector A et al (2006) Mannan binding lectin (MBL) gene polymorphisms are not associated with anti-*Saccharomyces cerevisiae* (ASCA) in patients with Crohn's disease. Gut 55:746

Jordan JE, Montalto MC, Stahl GL (2003) Inhibition of mannose-binding lectin reduces postischemic myocardial reperfusion injury. Circulation 104:1413–1418

Kallenberg CG (2008) Anti-C1q autoantibodies. Autoimmun Rev 7:612–616

Kelly P, Jack DL, Naeem A et al (2000) Mannose-binding lectin is a component of innate mucosal defense against *Cryptosporidium parvum* in AIDS. Gastroenterology 119:1236–1242

Kerrigan AM, Brown GD (2009) C-type lectins and phagocytosis. Immunobiology 214:562–576

Kilpatrick DC (2002) Mannan-binding lectin: clinical significance and applications. Biochim Biophys Acta 1572:401–413

Kilpatrick DC, McLintock LA, Allan EK et al (2003) No strong relationship between mannan binding lectin or plasma ficolins and chemotherapy-related infections. Clin Exp Immunol 134:279–284

Kirkpatrick BD, Huston CD, Wagner D et al (2006) Serum mannose-binding lectin deficiency is associated with cryptosporidiosis in young Haitian children. Clin Infect Dis 43:289–294

Klein NJ, Kilpatrick DC (2004) Is there a role for mannan/mannose-binding lectin (MBL) in defence against infection following chemotherapy for cancer? Clin Exp Immunol 138:202–204

Kristensen IA, Thiel S, Steffensen R et al (2004) Mannan-binding lectin and RSV lower respiratory tract infection leading to hospitalization in children: a case-control study from Soweto, South Africa. Scand J Immunol 60:184–188

Lam MF, Leung JC, Tang CC et al (2005) Mannose binding lectin level and polymorphism in patients on long-term peritoneal dialysis. Nephrol Dial Transplant 20:2489–2496

Lee YH, Witte T, Momot T et al (2005) The mannose-binding lectin gene polymorphisms and systemic lupus erythematosus: two case-control studies and a meta-analysis. Arthritis Rheum 52:3966–3974

Lhotta K, Wurzner R, Konig P (1999) Glomerular deposition of mannose-binding lectin in human glomerulonephritis. Nephrol Dial Transplant 14:881–886

Liu CP, Tsai WJ, Shen CC et al (2006a) Inhibition of (S)-armepavine from *Nelumbo nucifera* on autoimmune disease of MRL/MpJ-lpr/lpr mice. Eur J Pharmacol 531:270–279

Liu F, Liao Q, Liu Z (2006b) Mannose-binding lectin and Vulvovaginal candidiasis. Int J Gynaecol Obstet 92:43–47

Lozach PY, Lortat-Jacob H, de Lacroix de Lavalette A et al (2003) DC-SIGN and L-SIGN are high affinity binding receptors for hepatitis C virus glycoprotein E2. J Biol Chem 278:20358–66

Ma Y, Uemura K, Oka S, Kozutsumi Y et al (1999) Antitumor activity of mannan-binding protein in vivo as revealed by a virus expression system: mannan-binding protein dependent cell-mediated cytotoxicity. Proc Natl Acad Sci USA 96:371–376

Madsen HO, Videm V, Svejgaard A et al (1998) Association of mannose-binding-lectin deficiency with severe atherosclerosis. Lancet 352:959–960

Matsushita M, Hijikata M, Matsushita M et al (1998) Association of mannose-binding lectin gene haplotype LXPA and LYPB with interferon-resistant hepatitis C virus infection in Japanese patients. J Hepatol 29:695–700

Matsushita M, Miyakawa H, Tanaka A et al (2001) Single nucleotide polymorphisms of the mannose-binding lectin are associated with susceptibility to primary biliary cirrhosis. J Autoimmun 17:251–257

Megia A, Gallart L, Fernandez-Real JM et al (2004) Mannose-binding lectin gene polymorphisms are associated with gestational diabetes mellitus. J Clin Endocrinol Metab 89:5081–5087

Minchinton RM, Dean MM, Clark TR et al (2002) Analysis of the relationship between mannose-binding lectin (MBL) genotype, MBL levels and function in an Australian blood donor population. Scand J Immunol 56:630–641

Mok MY, Jack DL, Lau CS et al (2004) Antibodies to mannose binding lectin in patients with systemic lupus erythematosus. Lupus 13:522–528

Molle I, Steffensen R, Thiel S et al (2006) Chemotherapy-related infections in patients with multiple myeloma: associations with mannan-binding lectin genotypes. Eur J Haematol 77:19–26

Møller-Kristensen M, Wang W, Ruseva M et al (2005) Mannan-binding lectin recognizes structures on ischemic reperfused mouse kidneys and is implicated in tissue injury. Scand J Immunol 61:426–434

Mullighan CG, Bardy PG (2004) Mannose-binding lectin and infection following allogeneic hemopoietic stem cell transplantation. Leuk Lymphoma 45:247–256

Mullighan CG, Heatley S, Doherty K et al (2002) Mannose-binding lectin gene polymorphisms are associated with major infection following allogeneic hemopoietic stem cell transplantation. Blood 99:3524–3529

Muto S, Sakuma K, Taniguchi A et al (1999) Human mannose-binding lectin preferentially binds to human colon adenocarcinoma cell lines expressing high amount of Lewis A and Lewis B antigens. Biol Pharm Bull 22:347–352

Nakagawa T, Kawasaki N, Ma Y et al (2003) Antitumor activity of mannan-binding protein. Methods Enzymol 363:26–33

Nauta AJ, Raaschou-Jensen N, Roos A et al (2003) Mannose-binding lectin engagement with late apoptotic and necrotic cells. Eur J Immunol 33:2853–2863

Nauta AJ, Castellano G, Xu W et al (2004) Opsonization with C1q and mannose-binding lectin targets apoptotic cells to dendritic cells. J Immunol 173:3044–3050

Neth O, Hann I, Turner MW, Klein NJ (2001) Deficiency of mannose-binding lectin and burden of infection in children with malignancy: a prospective study. Lancet 358:614–618

Norwood MG, Sayers RD, Roscher S et al (2006) Consumption of mannan-binding lectin during abdominal aortic aneurysm repair. Eur J Vasc Endovasc Surg 31:239–243

Ogden CA, deCathelineau A, Hoffmann PR et al (2001) C1q and mannose binding lectin engagement of cell surface calreticulin and CD91 initiates macropinocytosis and uptake of apoptotic cells. J Exp Med 194:781–796

Ohlenschlaeger T, Garred P, Madsen HO, Jacobsen S (2004) Mannose-binding lectin variant alleles and the risk of arterial thrombosis in systemic lupus erythematosus. N Engl J Med 351:260–267

Park KS, Min K, Nam JH et al (2005) Association of HYPA haplotype in the mannose-binding lectin gene-2 with Behcet's disease. Tissue Antigens 65:260–266

Pellis V, De Seta F, Crovella S et al (2005) Mannose binding lectin and C3 act as recognition molecules for infectious agents in the vagina. Clin Exp Immunol 139:120–126

Perdriger A, Werner-Leyval S, Rollot-Elamrani K (2003) The genetic basis for systemic lupus erythematosus. Joint Bone Spine 70:103–108

Peterslund NA, Koch C, Jensenius JC, Thiel S (2001) Association between deficiency of mannose-binding lectin and severe infections after chemotherapy. Lancet 358:637–638

Pine SR, Mechanic LE, Ambs S et al (2007) Lung cancer survival and functional polymorphisms in MBL2, an innate-immunity gene. J Natl Cancer Inst 99:1401–1409

Pontes GS, Tamegao-Lopes B, Machado LF et al (2005) Characterization of mannose-binding lectin gene polymorphism among human T-cell lymphotropic virus 1 and 2-infected asymptomatic subjects. Hum Immunol 66:892–896

Proulx F, Wagner E, Toledano B et al (2003) Mannan-binding lectin in children with Escherichia coli O157:H7 haemmorrhagic colitis and haemolytic uraemic syndrome. Clin Exp Immunol 133:360–363

Rector A, Lemey P, Laffut W et al (2001) Mannan-binding lectin (MBL) gene polymorphisms in ulcerative colitis and Crohn's disease. Genes Immun 2:323–328

Roos A, Garred P, Wildenberg ME et al (2004) Antibody-mediated activation of the classical pathway of complement may compensate for mannose-binding lectin deficiency. Eur J Immunol 34:2589–2598

Roos A, Dieltjes P, Vossen RH et al (2006) Detection of three single nucleotide polymorphisms in the gene encoding mannose-binding lectin in a single pyrosequencing reaction. J Immunol Methods 309:108–114

Royle J, Burgner D, Curtis N (2005) The diagnosis and management of Kawasaki disease. J Paediatr Child Health 41:87–93

Rugonfalvi-Kiss S, Dosa E, Madsen HO et al (2005) High rate of early restenosis after carotid eversion endarterectomy in homozygous carriers of the normal mannose-binding lectin genotype. Stroke 36:944–948

Saevarsdottir S, Oskarsson OO, Aspelund T et al (2005) Mannan binding lectin as an adjunct to risk assessment for myocardial infarction in individuals with enhanced risk. J Exp Med 201:117–126

Santos IK, Costa CH, Krieger H et al (2001) Mannan-binding lectin enhances susceptibility to visceral leishmaniasis. Infect Immun 69:5212–5216

Saraheimo M, Forsblom C, Hansen TK et al (2005) On behalf of the Finn Diane Study Group, Increased levels of mannan-binding lectin in type 1 diabetic patients with incipient and overt nephropathy. Diabetologia 48:198–202

Schafranski MD, Stier A, Nisihara R et al (2004) Significantly increased levels of mannose-binding lectin (MBL) in rheumatic heart disease: a beneficial role for MBL deficiency. Clin Exp Immunol 138:521–526

Schmiegelow K, Garred P, Lausen B et al (2002) Increased frequency of mannose-binding lectin insufficiency among children with acute lymphoblastic leukemia. Blood 100:3757–3760

Seelen MA, Trouw LA, van der Hoorn JW et al (2003) Autoantibodies against mannose-binding lectin in systemic lupus erythematosus. Clin Exp Immunol 134:335–343

Seelen MA, van der Bijl EA, Trouw LA et al (2005) A role for mannose-binding lectin dysfunction in generation of autoantibodies in systemic lupus erythematosus. Rheumatology (Oxford) 44:111–119

Seibold F, Konrad A, Flogerzi B et al (2004) Genetic variants of the mannan-binding lectin are associated with immune reactivity to mannans in Crohn's disease. Gastroenterology 127:1076–1084

Shi L, Takahashi K, Dundee J et al (2004) Mannose-binding lectin-deficient mice are susceptible to infection with Staphylococcus aureus. J Exp Med 199:1379–1390

Siassi M, Hohenberger W, Riese J (2003) Mannan-binding lectin (MBL) serum levels and post-operative infections. Biochem Soc Trans 31:774–776

Steffensen R, Thiel S, Varming K et al (2000) Detection of gene mutations and promoter polymorphisms in the mannan-binding lectin (MBL) gene by polymerase chain reaction with sequence specific primers. J Immunol Methods 241:33–42

Steffensen R, Hoffmann K, Varming K (2003) Rapid genotyping of MBL2 gene mutations using real-time PCR with fluorescent hybridisation probes. J Immunol Methods 278:191–199

Stuart LM, Takahashi K, Shi L et al (2005) Mannose-binding lectin-deficient mice display defective apoptotic cell clearance but no autoimmune phenotype. J Immunol 174:3220–3226

Sullivan KE, Wooten C, Goldman D, Petri M (1996) Mannose-binding protein genetic polymorphisms in black patients with systemic lupus erythematosus. Arthritis Rheum 39:2046–2051

Sumiya M, Super M, Tabona P et al (1991) Molecular basis of opsonic defect in immunodeficient children. Lancet 337(8757):1569–1570

Summerfield JA, Sumiya M, Levin M, Turner MW (1997) Association of mutations in mannose binding protein gene with childhood infection in consecutive hospital series. Br Med J 314:1229–1231

Super M, Gillies SD, Foley S et al (1992) Distinct and overlapping functions of allelic forms of human mannose binding protein. Nat Genet 2:50–56

Sutherland AM, Walley KR, Russell JA (2005) Polymorphisms in CD14, mannose-binding lectin, and Toll-like receptor-2 are associated with increased prevalence of infection in critically ill adults. Crit Care Med 33:638–644, Comment in: Crit Care Med. 2005; 33: 695-6

Swierzko AS, Florczak K, Cedzyński M et al (2007) Mannan-binding lectin (MBL) in women with tumors of the reproductive system. Cancer Immunol Immunother 56:959–971

Takahashi K, Gordon J, Liu H et al (2002) Lack of mannose-binding lectin-A enhances survival in a mouse model of acute septic peritonitis. Microbes Infect 4:773–784

Takahashi R, Tsutsumi A, Ohtani K et al (2005) Association of mannose binding lectin (MBL) gene polymorphism and serum MBL concentration with characteristics and progression of systemic lupus erythematosus. Ann Rheum Dis 64:311–314

Thiel S, Jensenius C (2007) Indications of mannan-binding lectin (MBL) in the treatment of immunocompromised individuals. US Patent Issued on 10 Apr 2007

Thiel S, Frederiksen PD, Jensenius JC (2006) Clinical manifestations of mannan-binding lectin deficiency. Mol Immunol 43:86–96

Thomas HC, Foster GR, Sumiya M et al (1996) Mutation of gene of mannose-binding protein associated with chronic hepatitis B viral infection. Lancet 348(9039):1417–1419

Trouw LA, Seelen MA, Visseren R et al (2004) Anti-C1q autoantibodies in murine lupus nephritis. Clin Exp Immunol 135:41–48

Trouw LA, Seelen MA, Duijs JM, Wagner S, Loos M, Bajema IM, van Kooten C, Roos A, Daha MR (2005) Activation of the lectin pathway in murine lupus nephritis. Mol Immunol 42:731–740

Turner MW (2003) The role of mannose-binding lectin in health and disease. Mol Immunol 40:423–429

Turner MW (1996) Mannose-binding lectin: the pluripotent molecule of the innate immune system. Immunol Today 17:532

References

Turner MW, Hamvas RM (2000) Mannose-binding lectin: structure, function, genetics and disease associations. Rev Immunogenet 2:305–322

Ueland T, Espevik T, Kjekshus J et al (2006) Mannose binding lectin and soluble Toll-like receptor 2 in heart failure following acute myocardial infarction. J Card Fail 12:659–663

Valdimarsson H, Vikingsdottir T, Bang P et al (2004) Human plasma-derived mannose-binding lectin: a phase I safety and pharmacokinetic study. Scand J Immunol 59:97–102

Vekemans M, Georgala A, Heymans C et al (2005) Influence of mannan binding lectin serum levels on the risk of infection during chemotherapy-induced neutropenia in adult haematological cancer patients. Clin Microbiol Infect 11(suppl 2):20

Verschuren JJ, Roos A, Schaapherder AF et al (2008) Infectious complications after simultaneous pancreas-kidney transplantation: a role for the lectin pathway of complement activation. Transplantation 85:75–80

Villarreal J, Crosdale D, Ollier W et al (2001) Mannose binding lectin and FcγRIIa (CD32) polymorphism in Spanish systemic lupus erythematosus patients. Rheumatology 40:1009–1012

Wallis R (2002) Dominant effects of mutations in the collagenous domain of mannose-binding protein. J Immunol 168:4553–4558

Wallis R, Lynch NJ, Roscher S et al (2005) Decoupling of carbohydrate binding and MASP-2 autoactivation in variant mannose-binding lectins associated with immunodeficiency. J Immunol 175:6846–6851

Walsh MC, Bourcier T, Takahashi K et al (2005) Mannose-binding lectin is a regulator of inflammation that accompanies myocardial ischemia and reperfusion injury. J Immunol 175:541–546

Wang ZY, Morinobu A, Kanagawa S et al (2001) Polymorphisms of the mannose binding lectin gene in patients with Sjogren's syndrome. Ann Rheum Dis 60:483–486

Werth VP, Berlin JA, Callen JP et al (2002) Mannose binding lectin (MBL) polymorphisms associated with low MBL production in patients with dermatomyositis. J Invest Dermatol 119:1394–1399

Windbichler M, Echtenacher B, Hehlgans T et al (2004) Involvement of the lectin pathway of complement activation in antimicrobial immune defense during experimental septic peritonitis. Infect Immun 72:5247–5252

Worthley DL, Bardy PG, Mullighan CG (2005) Mannose-binding lectin: biology and clinical implications. Intern Med J 35:548–556

Worthley DL, Bardy PG, Gordon DL et al (2006) Mannose-binding lectin and maladies of the bowel and liver. World J Gastroenterol 12:6420–6428

Worthley DL, Mullighan CG, Dean MM et al (2007) Mannose-binding lectin deficiency does not increase the prevalence of *Helicobacter pylori* seropositivity. Eur J Gastroenterol Hepatol 19:147–152

Ying H, Ji X, Hart ML, Gupta K et al (2004) Interaction of mannose-binding lectin with HIV type 1 is sufficient for virus opsonization but not neutralization. AIDS Res Hum Retroviruses 20:327–336

Ytting H, Jensenius JC, Christensen IJ, Thiel S, Nielsen HJ (2004) Increased activity of the mannan-binding lectin complement activation pathway in patients with colorectal cancer. Scand J Gastroenterol 39:674–679

Ytting H, Christensen IJ, Jensenius JC et al (2005) Preoperative mannan-binding lectin pathway and prognosis in colorectal cancer. Cancer Immunol Immunother 54:265–272

43 Pulmonary Collectins in Diagnosis and Prevention of Lung Diseases

Anita Gupta

43.1 Pulmonary Surfactant

Pulmonary surfactant is a complex mixture of lipids and proteins, and is synthesized and secreted by alveolar type II epithelial cells and bronchiolar Clara cells. It acts to keep alveoli from collapsing during the expiratory phase of the respiratory cycle. After its secretion, lung surfactant forms a lattice structure on the alveolar surface, known as tubular myelin. Surfactant proteins (SP)-A, B, C and D make up to 10% of the total surfactant. SP-B and SPC are relatively small hydrophobic proteins, and are involved in the reduction of surface-tension at the air-liquid interface. SP-A and SP-D, on the other hand, are large oligomeric, hydrophilic proteins that belong to the collagenous Ca^{2+}-dependent C-type lectin family (known as "Collectins"), and play an important role in host defense and in the recycling and transport of lung surfactant (Awasthi 2010) (Fig. 43.1). In particular, there is increasing evidence that surfactant-associated proteins A and -D (SP-A and SP-D, respectively) contribute to the host defense against inhaled microorganisms (see Chaps. 24 and 25). Based on their ability to recognize pathogens and to regulate the host defense, SP-A and SP-D have been recently categorized as "Secretory Pathogen Recognition Receptors". While SP-A and SP-D were first identified in the lung, the expression of these proteins has also been observed at other mucosal surfaces, such as lacrimal glands, gastrointestinal mucosa, genitourinary epithelium and periodontal surfaces. SP-A is the most prominent among four proteins in the pulmonary surfactant-system. The expression of both SP-A and SP-D is complexly regulated on the transcriptional and the chromosomal level. SP-A is a major player in the pulmonary cytokine-network and has been described to act in the pulmonary host defense. This chapter gives an overview on the understanding of role of SP-A and SP-D in for human pulmonary disorders and points out the importance for pathology-orientated research to further elucidate the role of these molecules in adult lung diseases. As an outlook, it will become an issue of pulmonary pathology which might provide promising perspectives for applications in research, diagnosis and therapy (Awasthi 2010).

43.2 SP-A and SP-D in Interstitial Lung Disease

SP-A and SP-D appear in the circulation in specific lung diseases. Interstitial lung disease (ILD), also known as diffuse parenchymal lung disease (DPLD), refers to a group of lung diseases affecting the interstitium of lung: alveolar epithelium, pulmonary capillary endothelium, basement membrane, perivascular and perilymphatic tissues. The term ILD is used to distinguish these diseases from obstructive airways diseases. Most types of ILD involve fibrosis, but this is not essential; indeed fibrosis is often a later feature. The phrase "pulmonary fibrosis" is no longer considered a synonym, but the term is still used to denote ILD involving fibrosis. The term is commonly combined with idiopathic in "idiopathic pulmonary fibrosis", denoting fibrotic ILD that cannot be ascribed to a distinct primary cause.

43.2.1 Pneumonitis

Chronic hypersensitivity pneumonitis (HP) eventually ensues to extensive lung fibrosis when exposure to causative antigen continues. Klebs von den Lungen (KL)-6, a mucin-like glycoprotein and SP-D are elevated in most cases. Correct diagnosis in the early stage is crucial, since chronic summer-type HP can result in a fatal outcome after continuous exposure to the causative antigen (Inase et al. 2007). In pulmonary tissues of collagen vascular disease-associated interstitial pneumonia (CVD-IP) and hypersensitivity pneumonitis (HP), SP-D can be a marker for maturity of regenerating epithelial cells. SP-A along with KL-6 is detected in intimate relationship to the stage of regeneration

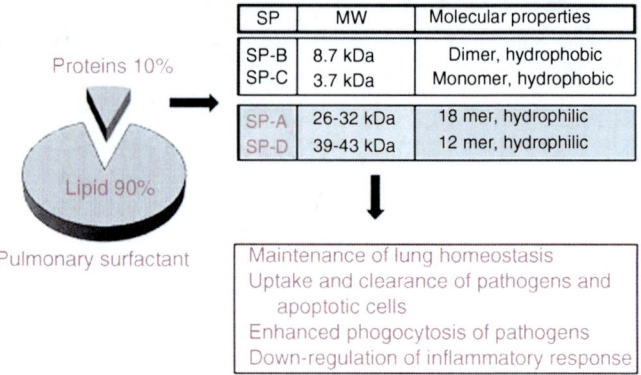

Fig. 43.1 Presence of surfactant proteins (SP) in lung surfactant, their properties and major functions of SP-A and SP-D

of alveolar epithelial cells and expressed before SP-D (Ohtsuki et al. 2007). Radiation pneumonitis (RP) is most common complication of radiotherapy for thoracic tumors. Both SP-A and SP-D concentrations in sera from patients with RP were significantly higher than those from patients without RP. Serum SP-A and SP-D may be of diagnostic value for detection of RP, even when radiographic change is faint (Takahashi et al. 2001). Despite the rise of SP-D and KL-6 in serum in adult patients with various types of interstitial pneumonia (IP) and collagen diseases with interstitial pneumonia, KL-6 may be superior in sensitivity of IP, whereas SP-D may be more specific for IP than KL-6. Early decrease of SP-D contrasts with the transient increase of KL-6 levels after prednisolone pulse therapy (Arai et al. 2001). High serum KL-6 value is an indicator of ILD of Wilson-Mikity syndrome and better than SP-D and LDH levels (Takami et al. 2003). Thus serum SP-A and SP-D monitoring along with KL-6 is useful indicator for estimating RP (Matsuno et al. 2006).

43.2.2 Interstitial Pneumonia (IP)

SP-A and SP-D in BAL as Indicator of Pneumonia in Children

SP-A and SP-D in serum significantly increase in patients with pulmonary alveolar proteinosis (PAP), idiopathic pulmonary fibrosis (IPF) and interstitial pneumonia with collagen vascular diseases (IPCD) (Kuroki et al. 1998; Takahashi et al. 2006b). The concentrations of SP-A and SP-D in BAL fluids from patients with IPF and IPCD are rather lower than those in healthy controls; and the SP-A/phospholipid ratio may be a useful marker of survival prediction. SP-D-deficient patients have more frequently pneumonias and their long-term outcome is worse than those with detectable SP-D. Among children with recurrent bronchitis and SP-D detectable in bronchoalveolar lavage (BAL), patients with allergic asthma had threefold levels of SP-D compared with controls. In contrast, SP-D deficiency due to consumption or failure to up-regulate SP-D may be linked to pulmonary morbidity in children (Griese et al. 2008).

SP-A Levels can Differentiate Usual Interstitial Pneumonia with Non-Specific Interstitial Pneumonia (NSIP)

There is a need to use serum markers for differentiating usual interstitial pneumonia (UIP) from other ILD. Serum levels of SP-A and SP-D in patients with UIP and nonspecific interstitial pneumonia (NSIP) are significantly higher than in healthy volunteers. In particular, serum SP-A levels in patients with UIP are significantly higher than in patients with NSIP, where as SP-D in BAL fluid in UIP patients were significantly lower than in patients with NSIP. Thus, serum SP-A level seems useful marker to differentiate UIP from NSIP (Ishii et al. 2003).

Abnormal tracheal aspirate surfactant phospholipids and SP-A are noted in children with bacterial pneumonia, viral pneumonitis, and ARDS, but not in children on cardiopulmonary bypass (Baughman et al. 1993; LeVine et al. 1996). SP-A in pneumonia group is significantly reduced and the reduction was better indicator in the Gm^+-pneumonia group than in Gm^--pneumonia group patients (Baughman et al. 1993). Fulminant early-onset neonatal pneumonia is associated with ascending intrauterine infection (IUI) and alveolar M showed significantly less nitric oxide synthase 2 (NOS2) isoform than in the controls. In the airway samples, the infants with fulminant pneumonia after birth had low intracellular NOS2 and significantly low IL-1β and SP-A than noninfected IUI infants (Aikio et al. 2000).

Foster et al. (2002) suggested that signaling of EGF axis and differential regulation of SPs persist during postnatal lung development, and SP-A and SP-D may modulate postpneumonectomy (PNX) lung growth in dogs. SP-D in patients, hospitalized for community-acquired pneumonia of suspected bacterial origin, indicates significant changes during pulmonary infection (Daimon et al. 2005; Leth-Larsen et al. 2003). The SA-A and SP-D in sera are useful for identification of the clinical condition of horses with bacterial pneumonia (Hobo et al. 2007).

43.2.3 ILD Due to Inhaled Substances

Cigarette smoke may alter component and function of pulmonary surfactant. Alterations in serum levels of SP-A may reflect smoking habits since serum SP-A was higher in active smokers than in nonsmokers (Nomori et al. 1998). However, SP-A is not a sensitive discriminating factor to separate smokers from nonsmokers. The contents of SP-A and SP-D in BAL fluids were significantly decreased in smokers compared to those in nonsmokers, although there was no significant difference of total phospholipid content between two groups (Honda et al. 1996). SP-A may decrease due to the cumulative effects of long-term smoking and development of emphysema, while SP-D decreases due to long-term smoking (Betsuyaku et al. 2004; Shijubo et al. 1998). Emphysema can be induced in mice by chronic cigarette smoke exposure with increase of SP-D in emphysema lungs. While accumulation of foamy alveolar macrophages may play a key role in the development of smoking-induced emphysema, increased SP-D may play a protective role in the development of smoking-induced emphysema, in part by preventing alveolar cell death (Hirama et al. 2007).

Although effects of maternal smoking on fetal growth and viability are overwhelmingly negative, there is a paradoxical enhancement of lung maturation as evidenced, in part, by a lower incidence of RDS in infants of smoking mothers. Epidemiologic and experimental evidence further support the view that a tobacco smoke constituent, possibly nicotine, affects the development of the lung in utero. The murine embryonic lungs explanted at 11 days gestation showed a 32% increase in branching after 4 days in culture in presence of 1 μM nicotine and 7–15-fold increases in mRNAs encoding SP-A and SP-C after 11 days. The nicotine-induced stimulation of surfactant gene expression could, in part, account for the effect of maternal smoking on the incidence of RDS (Wuenschell et al. 1998).

Intratracheal administration of crystalline silica to rats elicits a marked increase in alveolar accumulation of surfactant lipids and SP-A. The extracellular accumulation of SP-D is markedly increased in silica-induced lipoproteinosis, and that SP-D is associated with amorphous components identified by electron microscopy. SP-D may be useful biomarkers for early diagnosis and serum SP-D concentration may associate with the pathogenesis of silicosis (Barbaro et al. 2002; Wang et al. 2007b). Alcohol consumption at high levels during pregnancy is associated with immuno-modulation and premature birth. Chronic maternal ethanol consumption during the third trimester of pregnancy alters SP-A gene expression in fetal lung. These alterations may underlie increased susceptibility of preterm infants, exposed to ethanol in utero, to RSV and other microbial agents (Lazic et al. 2007). The exposure to moderate and high occupational levels of Diesel exhaust (DE) causes an increase in lung injury and inflammation, and a decrease in host defense molecules, which could result in increased severity of infectious and allergic lung disease. Several inflammatory and immune cytokines are upregulated at various time points and concentrations, in contrast to SP-A and SP-D which were significantly decreased at protein level. (Gowdy et al. 2008).

43.2.4 Idiopathic Pulmonary Fibrosis

Idiopathic pulmonary fibrosis (IPF) is a progressive disease of lung characterized by an inflammatory infiltrate, alveolar type II cell hypertrophy and hyperplasia, and ultimate parenchymal scarring. The phospholipid composition of the surface-active material recovered by BAL is abnormal in this disease. The content of SP-A in lavage was reduced, even when normalized for the total amount of surface-active material (SP-A/total phospholipids (PL)) recovered. The reduction in SP-A was not specific to IPF but also occurred in other interstitial lung diseases. Despite this, SP-A/PL in BAL is a biochemical marker that predicts survival in patients with IPF (McCormack et al. 1995; Phelps et al. 2004).

The serum SP-A and SP-D levels are significantly elevated in patients with IPF and systemic sclerosis compared to sarcoidosis, beryllium disease and normal controls, and correlated with radiographic abnormalities in patients with IPF. Dohmoto et al. (2000) hypothesized that regenerated or premature bronchoepithelial cells may circulate in the blood in patients with IPF. RT-PCR for cytokeratin 19 (CK19) and pulmonary SP-A in peripheral blood in patients with IPF and pulmonary fibrosis (PF) associated with collagen vascular disorders suggests that there were some circulating bronchoepithelial cells expressing mRNA for SP-A in peripheral blood of patients associated with collagen vascular disorders. Thus, both serum SP-A and SP-D levels are highly predictive of survival in patients with IPF (Greene et al. 2002; Takahashi et al. 2006b) and the measurement of SP-D in sera can provide an easily identifiable and useful clinical marker for the diagnosis of IPF, IPCD, and PAP, and can predict the disease activity of IPF and IPCD and the disease severity of PAP (Honda et al. 1995). However, KL-6 is the best serum marker for ILD (Ohnishi et al. 2002). Serum KL-6 and SP-D were also prognostic markers in acute exacerbation of IPF after treatment with Sivelestat (Endo et al. 2006; Nakamura et al. 2007). High levels of SP-D in BAL fluids are associated in patients with PAP, but not with IPF and IPCD.

Selman et al. (2003) examined associations between IPF and genetic polymorphic variants of SP-A1, SP-A2, SP-B, SP-C, and SP-D. One SP-A1 (6A^4) allele and SNPs that characterize the 6A^4 allele and one SP-B (B1580_C) were found with higher in nonsmoker and smoker IPF subgroups, respectively, compared with healthy controls. To explore whether a tryptophan (in 6A^4) or an arginine (in other SP-A1 alleles and in all SP-A2 alleles) at amino acid 219 alters protein behavior, two truncated proteins that varied only at amino acid 219 were oxidized by exposure to ozone. Differences in the absorption spectra (310–350 nm) between the two truncated rSP-A proteins, before and after protein oxidation, suggested allele-specific aggregation attributable to amino acid 2143. The SP-B SNP B1580_C, to be a risk factor for IPF smokers, was also shown to be a risk factor for other pulmonary diseases. The SP-C and SP-D SNPs and SP-B-linked microsatellite markers did not associate with IPF. These findings indicated that surfactant protein variants may serve as markers to identify subgroups of patients at risk. The observed alleles of SP-A and SP-D in association with various diseases are summarized in Table 43.1. Different alleles of these genes seem to predispose the individuals to various diseases. A logical explanation seems to be that different SNPs lead to different alterations in function or expression. However, common SNPs predispose Caucasians to RDS and Mexicans to TB. Similarly, common SNPs predispose the Indian population to ABPA and TB. Furthermore, Met11 SP-D allele is predisposing Mexicans to TB and Finns to RSV infection. It is also interesting to note that some of the alleles of SP-A interact with other alleles of SP-A and SP-B and thus increase the susceptibility of subjects to a disease (Kishore et al. 2005).

43.2.5 Cystic Fibrosis

Cystic fibrosis (CF) is an inherited disorder of CFTR gene, a chloride ion channel. The lack of this channel causes reduced water content of secretions. This affects the mucus secreted as part of the lung's defence and creates sticky, viscous mucus. In patients with CF, neutrophils are recruited in excess to the airways yet pathogens are not cleared and the patients suffer from chronic infections. In CF, the disease-causing gene has been clearly identified as the CF transmembrane conductance regulator gene, but genetic variants of the MBP and SP-A have been associated with disease severity in CF.

Table 43.1 Broad range of pathogens interacting with surfactant protein (SP)-A and SP-D

Type	Name of pathogen	Surfactant protein	Reference(s)
Bacteria	*E. coli*	SP-D	Kuan et al. (1992)
	Salmonela minnesota	SP-D	Kuan et al. (1992)
	H. pylori	SP-D	Appelmelk et al. (2005)
	Klebsiella pneumoniae	SP-D	Keisari et al. (2001), Ofek et al. (2001)
	Mycoplasma pneumoniae and *Histoplasma capsulatum*	SP-A and SP-D	Ernst (1998), Chiba et al. (2002), Gaynor et al. (1995)
	Haemophilus influenzae	Minimal effects of SP-D	Tino and Wright (1996), Restrepo et al. (1999)
	Pseudomonas aeruginosa Stenotrophomonas maltophilia	SP-D, not SP-A	Malloy et al. (2005), Giannoni et al. (2006), Bufler et al. (2004)
	Mycobacterium tuberculosis	Virulent and attenuated M. tuberculosis strains bind best	Ferguson et al. (1999, 2002), Hall-Stoodley et al. (2006)
	Mycobacterium avium	SP-A and SP-D	Kudo et al. (2004)
	Group B streptococcus' (*Streptococcus agalactiae*) and *S. pneumoniae*	SP-A and SP-D	Jounblat et al. (2004), Kuronuma et al. (2004)
	B. bronchiseptica (LPS); Ruminant bronchopneumonia	SP-D	Schaeffer et al. (2004), Grubor et al. (2004)
	Alloiococcus otitidis	SP-A	Konishi et al. (2006)
Yeast and fungi	*Cryptococcus neoformans*	SP-A and SP-D	Schelenz et al. (1995), Walenkamp et al. (1999), van de Wetering et al. (2004)
	Aspergillus fumigatus	SP-A and SP-D	Allen et al. (1999), Madan et al. (1997a, b)
	Coccidioides posadasii	SP-A and SP-D	Awasthi et al. (2004), Awasthi (2010)
	Candida albicans	SP-D	Van Rozendaal et al. (2000)
	Pneumocystis carinii	SP-D	O'Riordan et al. (1995), Vuk-Pavlovic et al. (2001), Atochina et al. (2004a), Yong et al. (2003)
Viruses	Influenza A virus	SP-A and SP-D	Malhotra et al. (1994), Hartshorn et al. (1994, 1997), Levine et al. (2001), Tecle et al. (2007b)
	HIV	SP-D binds HIV – gp120	Meschi et al. (2005)
	Respiratory syncytial virus	SP-A and SP-D	Ghildyal et al. (1999), Hickling et al. (1999), Griese (2002)

Allele associations and allele interaction of surfactant protein genes in relation to RDS have been discussed (Floros and Fan 2001). Studies have shown a deficiency of SP-A in airway fluids from patients with CF and other inflammatory pulmonary conditions. Findings suggest that the neutrophil serine proteases cathepsin G and/or elastase and/or proteinase-3 may contribute to degradation of SP-A and SP-D, thereby diminishing innate pulmonary antimicrobial defence (Rubio et al. 2004; von Bredow et al. 2001, 2003).

The dramatic decrease of SP-A and SP-D in the presence of normal surfactant phospholipid may be a mechanism underlying the relative ineffectiveness of cellular inflammatory response in killing invading bacteria in lungs of patients with CF. In bronchoalveolar lavage fluids (BALFs), although SP-A levels tend to decline in CF patients compared with non-CF, and the decline was only significant in presence of bacterial infection. Among CF patients, SP-A concentrations in BALF were inversely related to inflammation and age (Hull et al. 1997; Noah et al. 2003). Reports suggest that decreasing protease activity and increasing collectin activity may be beneficial in early CF (Alexis et al. 2006; Baker et al. 1999).

However, both, SP-D and TNF-α, are significantly increased in CF patients compared with patients of allergic fungal rhinosinusitis (AFRS), suggesting activation of both innate immunity and Th1-mediated inflammation and potential correlation between SPs and downstream adaptive immune responses (Skinner et al. 2007). Rat SP-D is highly resistant to degradation by a wide range of proteolytic enzymes. Patients with CF and chronic rhinosinusitis (CRS) with nasal polyposis demonstrated elevated SP-A1, -A2, and -D. While in patients with AFS, SP-A1, SP-A2, and SP-D, were not significantly different, these proteins are up-regulated in various forms of CRS, particularly in CF-CRS (Woodworth et al. 2007).

43.2.6 Familial Interstitial Lung Disease

Amin et al. (2001) studied the development of chronic lung injury in a familial form of ILD. An 11-year-old girl, her sister, and their mother who were diagnosed with chronic ILD were negative for SP-C and decreased levels of SP-A and SP-B in BALF. Lung biopsy from both children demonstrated a marked decrease of pro-SP-C in the alveolar epithelial cells but strong staining for pro-SP-B, SP-B, SP-A, and SP-D. The apparent absence of SP-C and a decrease in the levels of SP-A and SP-B were related to familial ILD. Several linkage and association studies have been done using SPs genes as markers to locate pulmonary disease susceptibility genes, but few have studied markers systematically in different ethnic groups.

43.3 Connective Tissue Disorders

43.3.1 Systemic Sclerosis

Significant progress is being made in terms of understanding the pathogenesis and various options for therapy of systemic sclerosis patients whose disease course is complicated by ILD. The significance of serum SP-A, SP-D and KL-6 for diagnosis and treatment of ILD in connective tissue disorders has been evaluated by different workers. Serum KL-6 and SP-D levels are more specific and useful markers for diagnosis and evaluation of ILD compared with serum LDH in connective tissue disorders (Ogawa et al. 2003; Suematsu et al. 2003). Characteristics or disease activity of early ILD has been evaluated in subjects. In abnormal group, curvilinear subpleural lines or thickened interlobular and intralobular lines were observed more frequently in lower lung fields and SP-A and SP-D were higher in true abnormalities group than in control group. True parenchymal abnormalities in posterior subpleural aspect of lung may indicate early ILD activity (Al-Salmi et al. 2005; Kashiwabara 2006). Since higher levels of SP-A and SP-D are associated with more severe lung function impairment at presentation, and better recovery over time, Janssen et al. (2005) suggested that SP-A, SP-D and KL-6 are especial markers of disease activity. Nevertheless, serum pulmonary and activation-regulated chemokine (PARC) levels may be more useful marker for active PF in systemic sclerosis (SSc) (Kodera et al. 2005) since elevated PARC values correlated more sensitively reflecting the PF activity than serum KL-6 or SP-D levels.

In lung fibrosis in patients with SSc and inflammatory myopathies, KL-6, von Willebrandt factor (vWF), soluble E-selectin (sES), SP-D are good surrogate factors of PF but cannot replace conventional diagnostic procedures. However, these markers are suitable for the assessment of progression and severity of PF in systemic autoimmune disorders once the diagnosis is established (Kumánovics et al. 2008). Takahashi et al. (2006b) indicated that elevated levels of serum SP-A and SP-D reflect the presence of ILD and the combination of SP-D and X-ray contributes to reduce the risk of clinicians overlooking ILD complicated by SSc (Highland and Silver 2005; Yanaba et al. 2004).

Maeda et al. (2001) compared serum SP-D in collagen diseases such as systemic scleroderma (SSd), scleroderma spectrum disorders (SSD), systemic lupus erythematodes (SLE), Sjogren syndrome (Sjs), dermatomyositis (DM), rheumatoid arthritis (RA), and dermatitis (DE) as a control. Patients with SSc possess higher levels of SP-D than those with other collagen diseases and dermatitis, which may correspond to severity of pulmonary fibrosis (Maeda et al. 2001). The basic and clinical studies of SSc patients with ILD are yielding promising data that may be translated in to more effective diagnostic and therapeutic strategies Although the SP-D level in sera of patients with polymyositis/dermatomyositis (PM/DM) is significantly elevated, the serum SP-D in patients with ILD was still higher than those without ILD, suggesting that serum SP-D level is a useful marker for ILD in patients with PM/DM (Ihn et al. 2002). However, there is a need to investigate whether another connective tissue disease has developed when laboratory findings cannot be explained by usual clinical course of an existing connective tissue disease (Ishiguro et al. 2007).

43.3.2 Sarcoidosis

Sarcoidosis also called sarcoid, Besnier-Boeck disease or Besnier-Boeck-Schaumann disease, is a disease in which abnormal collections of chronic inflammatory cells form as nodules in multiple organs. KL-6, SP-A and SP-D levels in BALF were increased in pulmonary sarcoidosis. Since these markers are specifically derived from epithelial cells, it is considered that KL-6 and SP-D levels are reflecting damage or release of these markers from epithelial cells due to the inflammatory response. Among serum Clara cell 16 (CC16), KL-6, and SP-D as markers of ILD, and their ability to reflect pulmonary disease severity and prognosis in sarcoidosis, KL-6 is the best marker in differentiating patients from healthy controls (Günther et al. 1999; Hamm et al. 1994; Janssen et al. 2003; Kunitake et al. 2001). The median amounts of SP-A in BAL fluid in control subjects was 2.82 mg/L (range, 0.92–5.17 mg/L). In comparison to control, SP-A in patients with asthma had a lower value of SP-A, which remained unchanged in patients with pulmonary sarcoidosis (van de Graaf et al. 1992). In contrast, SP-A levels in BAL fluids from patients with sarcoidosis were markedly higher than in control subjects and it was comparable with patients of hypersensitivity pneumonitis (HP). In both conditions, SP-A$^+$ alveolar macrophages were increased (Günther et al. 1999; Hamm et al. 1994).

The serum levels of SP-A in patients with IPF (205 ± 23 ng/mL) and PAP (285 ± 23 ng/mL) were significantly higher than those in healthy controls (45 ± 3 ng/mL). In patients of sarcoidosis, pneumonia, and tuberculosis SP-A values were 52 ± 27 ng/mL, 65 ± 11 ng/mL, and 49 ± 23 ng/mL, respectively. The SP-A appears to circulate in the bloodstream as a complex with Ig in IPF and in PAP (Kuroki et al. 1993).

43.4 Pulmonary Alveolar Proteinosis

A diffuse lung process of unknown etiology is characterized by the presence of alveolar spaces filled with amorphous eosinophilic (but sometimes basophilic) PAS-positive material of predominantly phospholipid nature in alveolar lumina. It is generally regarded as type of response to alveolar injury and results from accumulation of surfactant apoprotein through either: increased secretion by granular pneumocytes, or abnormal uptake and handling by alveolar macrophages. The prominent increase of SP-A and SP-D in BAL fluids and sputum is diagnostic for pulmonary alveolar proteinosis (PAP) (Kuroki et al. 1998; Brasch and Müller 2004; Takahashi et al. 2006a). There are reports about polymorphisms and mutations on the surfactant protein genes, especially SP-B that may be associated with congenital alveolar proteinosis.

43.4.1 Idiopathic Pulmonary Alveolar Proteinosis

SP-A in BALF of PAPs patients is significantly increased in comparison to normal volunteers and hence can be used as a diagnostic tool in the clinical laboratory (Brasch et al. 2004; Honda et al. 1996). PAP is a rare lung disorder and can be caused by inactivation of either granulocyte-macrophage colony-stimulating factor (GM-CSF) or GM receptor common β-chain (βc) genes in mice [GM$^{-/-}$, βc$^{-/-}$], demonstrating a critical role of GM-CSF signaling in surfactant homeostasis. Studies demonstrate abnormal accumulation of SP-A and SP-D in air spaces of patients with PAP (Crouch et al. 1993) and the precursors of SP-B, SP-B and SP-C. Although lung histology in βc$^{-/-}$ and GM$^{-/-}$ mice was indistinguishable, distinct differences were observed in surfactant phospholipid and surfactant protein concentrations in lungs of βc$^{-/-}$ and GM$^{-/-}$ mice. The defect in clearance was significantly more severe in GM$^{-/-}$ than in βc$^{-/-}$ mice. GM-CSF concentrations, increased in BALF but not in serum of βc$^{-/-}$ mice, were consistent with a pulmonary response to the lack of GM-CSF signaling. The observed differences in surfactant metabolism suggest the presence of alternative clearance mechanisms regulating surfactant homeostasis in mice and may provide a molecular basis for the range in severity of PAP symptoms (Reed et al. 2000). In a young patient with idiopathic PAP, the enhanced serum anti-GM-CSF antibody level demonstrated a striking difference in the distribution of SP-A and SP-D in intraalveolar substance with idiopathic PAP (Ohtsuki et al. 2008; Kobayashi et al. 2008b).

Evidence suggests that not only an impairment of surfactant clearance by alveolar macrophages, but also an abnormal secretion of transport vesicles containing precursors of SP-B (but not SP-C) and an insufficient palmitoylation of SP-C, which may lead to the formation of di- and oligomeric SP-C forms, play a role in the pathogenesis of pulmonary alveolar proteinosis.

43.4.2 Structural Changes in SPs in PAP

The primary structures of human pulmonary SPs isolated from lung lavage of patients with alveolar proteinosis demonstrate significant differences from lung surfactant proteins isolated from lungs of healthy individuals. In contrast to SP-A from normal lungs, PAP-SP-A was shown to contain large amounts of non-reducable cross-linked β chains, where as proteinosis SP-B showed a significantly increased molecular weight by approx. 500 Da for the unreduced protein dimer. In contrast, SP-C from proteinosis patients was modified by (1) partial or even complete removal of palmitate residues and (2) additional N-terminal proteolytic degradation (Voss et al. 1992).

Pathophysiological structural modifications in SP-A seemed to occur in the alveolar space, and may lead to a reduced surfactant function (Voss et al. 1992). Multimerized form of SP-A oligomer (alveolar proteinosis protein-I, APP-I) has been detected besides the normal-sized octadecamer (APP-II) in SP-As isolated from PAP patients. Analysis of APP revealed that it was composed of two proteins. The M_r of APP-I and APP-II were 1.65 MDa and 0.93 MDa, respectively. APP-I and APP-II showed almost identical amino acid compositions. Electron microscopy revealed that APP-II was a hexameric particle, presumably consisting mainly of octadecamers whose diameter was approximately 30 nm. In contrast, APP-I was made of multimerized larger aggregates whose diameter appeared to be about 70–90 nm. Both APP-I and APP-II retained the abilities to bind DPPC. Reconstitution experiments with porcine SP-B and phospholipids revealed that multilamellated membranes in structures formed from APP-I consisted of several layers of doubled unit membranes. APP-I failed to form tubular myelin structures. In contrast, APP-II formed well-formed lattice structures seen in tubular myelin The multimerized form of human SP-A oligomer exhibits the reduced capacity to regulate phospholipid secretion from type II cells, and lower affinity to bind to type II cells. It is to be reminded that the integrity of a flower-bouquet-like octadecameric structure of SP-A oligomer is important for the expression of full activity of this protein, indicating the importance of the oligomeric structure of mammalian lectins with collagenous domains. Thus there exists an abnormal multimerized form of SP-A oligomer in the alveoli of patients with PAP that exhibits abnormal function on phospholipid membrane organization (Hattori et al. 1996a, b).

In alveolar proteinosis, cholesterol/disaturated phospholipid ratios (CHOL/DSP) are invariably elevated, whereas the SP-A/DSP and SP-B/DSP ratios are generally elevated. Because the SP-B/SP-A ratio was normal in all cases, it was suggested that structural changes to the proteins occurred secondarily and that caution must be used in comparing functional data derived using SP-A obtained from patients with PAP (Doyle et al. 1998). The major part of SP-A from a proteinosis patient consisted of SP-A2 gene product while SP-A1 gene product was present in only a small amount. The disulfide bridges in the carbohydrate recognition domain were identified to be in the 1–4, 2–3 pattern common for collectins. Interchain disulfide bridges were discovered between two Cys-48 residues and cysteine residues in the N-terminal region. However, the exact disulfide bridge connections within the bouquet-like ultrastructure could not be established (Berg et al. 2000).

43.5 Respiratory-Distress Syndrome and Acute Lung Injury

43.5.1 ARDS and Acute Lung Injury

Acute respiratory distress syndrome (ARDS), also known as respiratory distress syndrome (RDS) or adult respiratory distress syndrome (in contrast with IRDS) is a serious reaction to various forms of injuries to lung. ARDS is caused by a variety of direct and indirect issues. It is characterized by inflammation of lung parenchyma leading to impaired gas exchange with concomitant systemic release of inflammatory mediators causing inflammation, hypoxemia and frequently resulting in multiple organ failure. A less severe form is called acute lung injury (ALI). Clinical and biochemical evidences suggest that the etiology of RDS is multifactorial with a significant genetic component. There are reports about polymorphisms and mutations on the surfactant protein genes, especially surfactant proteins-B that may be associated with RDS, ARDS, and congenital alveolar proteinosis. The measurement of SP-A and SP-D in amniotic fluids and tracheal aspirates reflects lung maturity and the production level of the lung surfactant in infants with RDS. The SP-A concentrations in BAL fluids are significantly reduced in patients with ARDS and also in patients at risk to develop ARDS (Kuroki et al. 1998; Takahashi et al. 2006a). Patients with low concentrations of SP-A and SP-B in the BAL are at risk for ARDS before onset of clinically defined lung injury, though the SP-D concentrations remain in normal range. Thus, SP abnormalities occur before and after the onset of ARDS, and the responses of SP-A, SP-B, and SP-D differ in important ways. However, plasma SP-D is a valuable biomarker in ALI/ARDS and SP-A increases during the early phase of ARDS, including some molecular alteration followed by decrease during the late phase (Endo et al. 2002; Kuroki et al. 1998; Takahashi et al. 2006b; Zhu et al. 2001).

Elevated level of SP-A has also been reported in the sera of patients with acute cardiogenic pulmonary edema (APE) and in patients with ARDS relative to healthy subjects and ventilated patients with no cardio-respiratory disease. Serum SP-A was inversely related to blood oxygenation and to static respiratory system compliance both at the time of patient's entry into the study and during the course of admission. Since SP-B is synthesized as a precursor smaller than alveolar SP-A, Doyle et al. (1995, 1997) suggested that immunoreactive SP-B that enters more readily than SP-A, is cleared acutely, and provides a better indicator of lung trauma (Shimura et al. 1996).

Prematurely born infants can develop the neonatal RDS because of a deficiency of pulmonary surfactant. At autopsy RDS lungs lacked tubular myelin and had decreased immunoreactivity for antisera to SP-A, an important component of tubular myelin. Therefore, a role for SP-A in the conversion of lamellar bodies to tubular myelin and in the pathogenesis of RDS was proposed. It was postulated that if SP-A is indeed necessary for the conversion of lamellar bodies to tubular myelin, in RDS either there is a deficiency of adequate amounts of functional SP-A, or some other important component of surfactant is missing (deMello et al. 1993). Mechanical ventilation is the main modality of treatment of ARDS. On mechanical ventilation, there is a progressive increase in SP-A levels in patients with ARDS, and may be one of the contributors for recovery in ARDS. A significant increase within the first 4 days was found in those infants who survived, whereas no such change was found in those infants who died (Balamugesh et al. 2003; Stevens et al. 1992). Intratracheal aerosolization of LPS in rats produces typical features of human ARDS. The SP-D binds inhaled LPS-endotoxin in vivo, which may help to protect the lung from endotoxin-induced disease (van Rozendaal et al. 1999). The SP-D was reduced in lung of young rats following ALI at early stage and early administration of Dex could reverse the SP-D content (Shu et al. 2007). SP-A in sera of cord blood from infants born at gestational ages <32 weeks with RDS was 15.1 ng/mL compared to without RDS (5.8 ng/mL) and significantly related to the non-RDS outcome (Cho et al. 2000). Shimoya et al. (2000) suggested that IL-6 elevation in fetuses with chorioamnionitis promotes fetal lung maturation by inducing SP-A synthesis, thereby decreasing the incidence of RDS in the preterm neonates.

Acute Lung Injury (ALI): Plasma SP-A, but not SP-D, was higher in patients with fewer days of unassisted ventilation and in patients with an absence of intact alveolar fluid clearance. In contrast, pulmonary edema fluid SP-D, but not SP-A, was lower in patients with worse oxygenation. Reduced pulmonary edema fluid SP-D and elevated plasma SP-A concentrations at the onset of ALI may be associated with more severe disease and worse clinical outcome and may serve as valuable biochemical markers of prognosis (Cheng et al. 2003). The BALF proteome analysis showed the presence of several isoforms of SP-A, in which an N-non-glycosylierte form and several proline hydroxylations were identified (Bai et al. 2007). In the plasma and edema fluid, protein profile of ALI patients showed multiple qualitative changes. Nearly all ALI patients also had protein spots that indicated truncation or other posttranslational modifications (Bowler et al. 2004).

43.5.2 Bronchopulmonary Dysplasia (BPD)

The pathophysiology of bronchopulmonary dysplasia (BPD) as an inflammatory disorder, secondary to neonatal RDS represents a major complication of prematurity. Maximum SP-A and anti-SP-A antibodies (SAS) immune complex values between 2 and 4 weeks after birth correlate with subsequent development of BPD independently and may be useful in analyzing the course and outcome of neonatal RDS, in particular the likelihood of subsequent development of BPD (Strayer et al. 1995). Weber et al. (2000) investigated an association of polymorphisms of SP-A1 and SP-A2 encoding genes and the risk of BPD in Caucasian preterm infants below 32 weeks of gestation matched for immaturity and year of birth. BPD was defined as oxygen dependency or need for mechanical ventilation at day 243. A significantly increased frequency of SP-A1 polymorphism $6A^6$ in infants was associated with BPD compared with controls. In addition to established risk factors for BPD, $6A^6$ polymorphism for SP-A1 gene is an independent co-factor.

BPD_28D (O2 dependency at 28 days of life) and BPD_36W (O2 dependency at 36 week post-menstrual age) are diseases of prematurely born infants exposed to mechanical ventilation and/or oxygen supplementation. Genetic variants of SP-A, B, C, and D and SP-B-linked microsatellite markers are risk factors in BPD. Significant associations were observed for alleles of SP-B and SP-B-linked microsatellite markers, and haplotypes of SP-A, SP-D, and SP-B. Unlike SP-A, SP-D does not contribute to lowering surface tension. SP-D-deficient mice have no respiratory abnormalities at birth, but it causes development of emphysema and predisposition to specific infections. No human infant or child with respiratory distress and mutation in the SP-D gene has been identified (Yurdakök 2004). Studies in larger sample size are warranted to confirm these observations and delineate genetic background of BPD subgroups (Pavlovic et al. 2006).

43.5.2.1 SP-A Deficiency in Primate Model of BPD with Infection

In a baboon model of hyperoxia-induced BPD and superimposed infection, animals constituting a group- pro re nata (PRN) were delivered by hysterotomy at 140 days gestational age and ventilated on clinically appropriate oxygen for a 16-day experimental period and served as controls. Immunostaining with SP-A, SP-B, and SP-C antibodies showed variable staining patterns. The study demonstrated that a deficiency of SP-A mRNA expression persists in chronic lung injury and variable protein staining patterns are manifested depending upon the underlying pathology (Coalson et al. 1995; King et al. 1995).

Awasthi et al. (1999) measured SP-A and SP-D levels and their mRNAs in three groups of animals: (1) nonventilated premature baboon fetuses; (2) neonatal baboons delivered prematurely at 140 d gestation age (ga) and ventilated with PRN O_2; (3) animals of same age ventilated with 100% O_2 to induce chronic lung injury. In chronic lung injury, SP-A is significantly reduced in alveolar space. SP-D concentration in lavage was nearly equal to that in normal adults, but the total collectin pool in lavage was still significantly reduced. Because these collectins may bind and opsonize bacteria and viruses, decrements in their amounts may present additional risk to those premature infants who require prolonged periods of ventilatory support (Awasthi et al. 1999). Reduced SP-D expression in BAL fluid was associated with progression of bronchial dysplasia in heavy smokers. SP-D levels in BAL fluid may serve a potential biomarker to identify smokers who are at risk of early lung cancer (Sin et al. 2008b). Cheng et al. (2003) proved the hypothesis that reduced pulmonary edema fluid SP-D and elevated plasma SP-A concentrations at onset of ALI may be associated with more severe disease and worse clinical outcome and may serve as valuable biochemical markers of prognosis (Cheng et al. 2003).

43.6 Chronic Obstructive Pulmonary Disease (COPD)

43.6.1 COPD as a Group of Diseases

Obstructive lung disease is a category of respiratory disease characterized by airway obstruction. Chronic obstructive pulmonary disease (COPD), also known as chronic obstructive airways disease (COAD) or chronic airflow limitation (CAL) is a group of illnesses characterised by airflow limitation that is not fully reversible. The flow of air into and out of the lungs is impaired. The COPD is characterized by chronic inflammation. It is most likely the result of complex interactions of environmental and genetic factors. Term COPD includes the conditions of emphysema and chronic bronchitis although most patients with COPD have characteristics of both conditions to varying degrees. Asthma being a reversible obstruction of airways is often considered separately, but many COPD patients also have some degree of reversibility in their airways. The most common cause of COPD is cigarette smoking. COPD may also be caused by breathing in other particles and gases.

Diagnosis of COPD is established through spirometry and chest X-ray although other pulmonary function tests can be helpful. Emphysema can only be seen on CT scan. COPD is generally irreversible although lung function can partially recover if the patient stops smoking. α1-antitrypsin deficiency is a rare genetic condition that results in COPD (particularly emphysema) due to lack of antitrypsin protein which protects fragile alveolar walls from protease enzymes released by inflammatory processes.

The prevalence of COPD is age-dependent, suggesting an intimate relationship between the pathogenesis of COPD and aging. Genetic polymorphism in SP-A is associated with the development of COPD in Chinese Hans. The genotypes of patients with COPD and healthy smoking subjects as controls for SP-A gene showed that in COPD group, the frequencies of +186 locus genotypes AA, AG and GG were 86.4%, 12.5% and 1.1%i respectively; compared to 66.7%, 27.6% and 5.7% in control group. The frequencies of polymorphic genotypes at +655 locus and +667 loci showed no significant difference between the COPD group and control group (Xie et al. 2005).

43.6.1.1 Serum SP-A in COPD and Its Relation to Smoking

SP-A occurs physiologically in small amounts in blood. Tobacco smoke induces increased alveolo-capillary leakage of SPs into blood and its level in blood may help in the assessment of lung injury caused by smoke. SP-A is occasionally elevated in non-ILD pulmonary patients. Serum SP-A increased in current smokers than in never- or ex-smokers and in COPD and pulmonary thromboembolism than in other diseases. Serum SP-D and KL-6 were unaffected by smoking. Therefore, different baseline levels of serum SP-A need to be established for smokers and non-smokers. Serum SP-A may be a useful marker for predicting COPD in the preclinical stage (Behera et al. 2005; Kobayashi et al. 2008a). Different alleles of SP-A and SP-D associated with various diseases have been summarized by Kishore et al. (2005) and given in Table 43.2. Analysis between COPD and smokers revealed several COPD susceptibility alleles (AA62_A, B1580_C, D2S388_5), based on an odds ratio (OR > 2.5). Results indicate that surfactant protein alleles may be useful in COPD by either predicting the disease in a subgroup and/or by identifying disease subgroups that may be used for therapeutic intervention (Guo et al. 2001).

Proteome research revealed increased levels of SP-A in COPD but not in normal or fibrotic lung. Furthermore, elevated SP-A protein levels were detected from the induced sputum supernatants of COPD patients. The levels of other surfactant proteins (SP-B, SP-C, SP-D) were not altered. It is suggested that SP-A is linked to the pathogenesis of COPD and can be considered as a potential COPD biomarker (Ohlmeier et al. 2008). Toxic metals and transition elements are detectable in exhaled breath condensate (EBC) of studied subjects (Mutti et al. 2006).

43.6.1.2 SP-D Is an Ideal Biomarker in COPD
In COPD, SP-D is an ideal biomarker that is produced mostly in lungs and can be measured in the peripheral circulation. It changes with the clinical status of the patient and has inherent functional attributes that suggest a possible causal role in pathogenesis of disease (Sin et al. 2008b, c).

In a multivariable linear regression model, COPD was independently associated with lower SP-D levels. Given the importance of this molecule in lung, low levels may play a role in the pathogenesis and/or progression of COPD (Sims et al. 2008). Inhaled corticosteroids alone or in combination exhibited partial systemic anti-inflammatory effects, reducing significantly only SP-D serum levels. ICS in conjunction with long-acting β2-adrenergic agonist significantly reduced serum SP-D levels. These drugs reduce lung-specific but not generalized biomarkers of systemic inflammation in COPD. Hydrofluoroalkane-beclomethasone dipropionate (HFA-BDP) controls eosinophilic inflammation, including in distal airways, more effectively than fluticasone propionate (FP) Diskus (Ohbayashi and Adachi 2008; Sin et al. 2008a).

43.6.2 Emphysema

Emphysema is a chronic pulmonary disease marked by an abnormal increase in size of air spaces. Pulmonary emphysema, a major component of COPD, is pathologically characterized by destructive alterations in pulmonary architectures as a result of persistent inflammation. Emphysema may be a dynamic disease process in which alveolar wall cell death and proliferation are repeated. The decrease of surfactant protein secreted by the alveolar type II cell is one of the important causes of limiting air of pulmonary emphysema and the changes of SP-A may be related to emphysematous changes in the lung. Cigarette smoke and LPS alter lung SP-A gene activity and protein homeostasis (Hu et al. 2008). Mice deficient in SP-D$^{-/-}$ develop progressive emphysema with age. SP-D gene-targeted mice develop severe pulmonary lipidosis, and foamy macrophage infiltrations. By lowering surface tension at the air-water interface in the surfactant deficient premature lung, exogenous surfactant replacement therapy for neonatal RDS has been highly successful in decreasing mortality after preterm birth. It has emerged that SP-A and SP-D have additional roles in host defence distinct from the surface tension lowering effects of surfactant. Recombinant forms of SP-D could be useful therapeutically in attenuating inflammatory processes in neonatal chronic lung disease, cystic fibrosis, and emphysema (Clark and Reid 2003).

43.6.3 Allergic Disorders

43.6.3.1 Allergic Inflammation in Asthma
Asthma is an obstructive lung disease where the bronchial tubes (airways) are extra sensitive (hyperresponsive). The airways become inflamed and produce excess mucus and muscles around the airways tighten making the airways narrower. Asthma is usually triggered by breathing in things present in air such as dust or pollen that produces an allergic reaction. It may be triggered by other things such as an upper respiratory tract infection, cold air, exercise or smoke. Asthma is diagnosed by the characteristic pattern of symptoms. A peak flow meter can record variations in the severity of asthma over time. Spirometry can provide an assessment of the severity, reversibility, and variability of airflow limitation, and help confirm the diagnosis of asthma. Significant changes occur in levels of SP-A and SP-D during the asthmatic response in animal models as well as in asthmatic patients. The impact of the SP-A and SP-D on asthmatic allergic inflammation and vice versa has been reviewed (Hohlfeld et al. 2002). Serum SP-D concentrations are affected in allergic patients and correlate with changes in allergic airway inflammation. Serum SP-D levels may give additional information, beside bronchial hyperresponsiveness (BHR) and sputum eosinophils, about the degree of bronchial inflammation in allergic patients (Koopmans et al. 2004).

Immunoregulatory Roles of SP-A and SP-D
Studies on allergen-sensitized murine models and asthmatic patients show that SP-A and SP-D can: specifically bind to aero-allergens; inhibit mast cell degranulation and histamine release; and modulate the activation of alveolar macrophages and DCs during the acute hypersensitive phase of allergic response (Erpenbeck et al. 2005; Wang et al. 1998). They also can alleviate chronic allergic inflammation by inhibiting T-lymphocyte proliferation as well as increasing phagocytosis of DNA fragments and clearance of apoptotic cell debris. Furthermore, it has emerged, from the studies on SP-D-deficient mice, that, when these mice are challenged with allergen, they develop increased eosinophil infiltration, and abnormal activation of lymphocytes, leading to the production of Th2 cytokines. Intranasal administration of SP-D significantly attenuated the asthmatic-like symptoms seen in allergen-sensitized wild-type, and SP-D-deficient, mice. These findings provide a new insight of role that surfactant proteins play in handling environmental stimuli and in their immunoregulation of airway inflammatory disease (Wang and Reid 2007).

Both SP-A and SP-D can inhibit histamine release in the early phase of allergen provocation and suppress lymphocyte proliferation in the late phase of bronchial inflammation, the two essential steps in the development of asthmatic symptoms (Wang et al. 1998). Studies suggest that the increased levels of SP-A and D may play a protective role in an allergic inflammation in the pathogenesis of bronchial asthma. Structural remodelling of airways in asthma that follows inflammation may be affected by SP-D-mediated effects on immune response. SP-D accumulation is increased in this model of allergen-induced eosinophilia, both in upper and lower airways (Cheng et al. 2000; Kasper et al. 2002). SP-D gene-deficient mice ($Sftpd^{-/-}$) have an impaired systemic Th-2 response at baseline and reduced inflammation and airway responses after allergen exposure. Translational studies revealed that a polymorphism in $SFTPD$ gene was associated with lower atopy and possibly asthma susceptibility. Thus, SP-D-dependent innate immunity influences atopy and asthma (Brandt et al. 2008). Dex significantly down-regulates SP-D in allergic airways and lavage fluid. In addition, Dex promoted airway expression of vitamin D-binding protein, heptoglobin and α1-antitrypsin (Zhao et al. 2007).

Serum SP-D is increased in acute and chronic inflammation in mice. Profiles of SP-A and SP-D in acute and chronic inflammation indicated that serum SP-D can serve as a biomarker of lung inflammation in both acute and chronic lung injury in mice (Fujita et al. 2005). Because of their capability to directly inhibit T-cell activation and T-cell-dependent allergic inflammatory events, SP-A and SP-D may be significant contributors to the local control of Th-2 type inflammation in the airways. SP-D is able to reduce the immediate allergen-induced mediator release and the early bronchial obstruction in addition to its effects on airway inflammation and bronchial hyperresponsiveness in an *A. fumigatus* mouse asthma model. Thus, SP-D not only reduces allergen-induced eosinophilic inflammation and airway hyper-responsiveness but also provides protection against early airway obstruction by inhibition of early mediator release (Erpenbeck et al. 2006; Takeda et al. 2003). However, mice sensitized and challenged with either *A. fumigatus* or OVA increased SP-D levels in their lung. Allergen exposure induced elevation in SP-D protein levels in an IL-4/IL-13-dependent manner, which in turn, prevents further activation of sensitized T cells. This negative feedback regulatory circuit could be essential in protecting the airways from inflammatory damage after allergen inhalation (Haczku et al. 2006). Haczku (2006) support the hypothesis that SP-A and SP-D have a role in regulation of allergic airway sensitization.

Murine Model of Asthma

Dust mite allergens can directly activate alveolar macrophages (AΦs), induce inflammatory cytokines, and enhance T-helper type 2 cytokine production. The SP-D is able to bind mite allergens and alleviates allergen-induced airway inflammation and may be an important modulator of allergen-induced pulmonary inflammation (Liu et al. 2005a). There is marked reduction in SP-A and SP-D levels in the BALF of dust mite (*Dermatophagoides pteronyssinus*, Der p)-sensitized BALB/c mice after allergen challenge. Both SP-A and SP-D were able to suppress Der p-stimulated intrapulmonary lymphocyte proliferation of naïve mice with saline or allergen challenge, or of Der p-sensitized mice with saline challenge. On the contrary, this suppressive effect was mild on lymphocytes from sensitized mice after allergen challenge. These results indicated the involvement of pulmonary surfactant proteins in the allergic bronchial inflammation of sensitized mice (Wang et al. 1996, 2001). Both SP-A and SP-D down-regulate the eosinophilic inflammation in murine asthma models and shift the cytokine profile towards a T helper cell type 1 response. In addition, they are effective at alleviating bronchial hyperresponsiveness. There is evidence of activation of innate immune system in asthma which results in the production of pro-inflammatory cytokines and may contribute to the pathogenesis of neutrophilic asthma (Simpson et al. 2007).

43.6.3.2 Chronic Sialadenitis and Chronic Rhinosinusitis

SP-A and mRNA and protein were detected in glands of patients with chronic sialadenitis. The expression in salivary glands of patients with chronic sialadenitis was significantly higher than from healthy salivary glands. SP-A immunoreactivity, localized in the epithelial cells and submucosal glands of paranasal sinus mucosa in normal and chronic sinusitis patients, was enhanced in chronic rhinosinusitis mucosa as compared with normal paranasal sinus mucosa (Lee et al. 2004, 2006). SP-A expression in human nasal tissue was correlated with symptoms suggestive of allergic rhinitis. (Wootten et al. 2006).

43.6.4 Interactions of SP-A and SP-D with Pathogens and Infectious Diseases

Microbial targets for SP-D include both Gram-positive and Gram-negative respiratory pathogens, influenza, and respiratory syncytial viruses, *Cryptococcus neoformans*, *Pneumocystis carinii*, and *Aspergillus fumigatus*. Both monocytes/macrophages and neutrophils express surface receptors that can interact with SP-D. The interactions between SP-D and microorganisms and in many instances immune cells promote both microbial aggregation and enhanced phagocytosis. SP-D has been shown to bind to a variety of bacteria, including rough strains of *Salmonella Minnesota* and *E. coli* as well as *Klebsiella pneumoniae* and *Pseudomonas aeruginosa* (Lim et al. 1994). SP-D also stimulates the phagocytosis of *Pseudomonas aeruginosa* (Restrepo et al. 1999). The interaction of SP-D with bacteria often results in CRD-dependent bacterial aggregation or agglutination. Unlike SP-A (van Iwaarden et al. 1994), SP-D does not bind to lipid A. It interacts with *E. coli* through the core polysaccharides and/or the O-specific antigens. The core region of the LPS of other gram-negative bacteria is broadly recognized by SP-D as well (Kuan et al. 1992). SP-D can be used as a biomarker for chronic periodontitis. As no significant associations of SFTPD gene polymorphisms could be detected, other mechanisms influencing SP-D serum/plasma expression might exist (Glas et al. 2008).

SP-D has been shown to bind to the influenza A virus, resulting in aggregation of the target (Hartshorn et al. 1996a). The binding and inhibition of hemagglutination was inhibited by chelation of calcium and by carbohydrates, suggesting that the interaction of SP-D with the virus was mediated via the CRD. SP-D also enhances the neutrophil uptake of the virus in a calcium-dependent manner (Hartshorn et al. 1997). Further enhanced antiviral and opsonic activity for influenza A virus was obtained by making a human MBP and SP-D chimera (White et al. 2000) (Table 43.1). The degree of multimerization of SP-D also appears to be important for its interactions with viruses (Brown-Augsburger et al. 1996; Hartshorn et al. 1996b). SP-D induces massive aggregation of influenza A virus particles (Hartshorn et al. 1996a). This massive agglutination of organisms could contribute to lung host defence by promoting airway mucociliary clearance, but it could also promote internalization by phagocytic cells. Recombinant SP-D inhibited RSV infectivity both in vitro and in vivo (Hickling et al. 1999; Le Vine et al. 2004), and reduced SP-D protein levels have been detected in RSV infection (Kerr and Paton 1999). A direct interaction between the yeast *Candida albicans* and SP-D confirms the importance of SP-D in innate immunity (van Rozendaal et al. 2000).

43.6.4.1 Distinct Effects of SP-A or -D Deficiency During Bacterial Infection

Surfactant proteins A and D expressed in respiratory tract bind bacterial, fungal and viral pathogens, enhancing their opsonization and killing by phagocytic cells. Clearance of bacterial pathogens including group *B streptococci*, *Haemophilus influenza*, *Pseudomonas aeruginosa* and viral pathogens, *respiratory syncytial virus, adenovirus and influenza A virus*, was deficient in SP-A$^{-/-}$ mice (Table 43.1). Mice lacking SP-A (SP-A$^{-/-}$) or SP-D (SP-D$^{-/-}$) and wild-type mice, infected with group *B streptococcus* or *Haemophilus influenzae*, are associated with increased inflammation and inflammatory cell recruitment in lung after infection. Although, decreased killing of group B *streptococcus* and *H. influenzae* was observed only in SP-A$^{-/-}$ mice but not in SP-D$^{-/-}$ mice, bacterial uptake by alveolar macrophages was reduced in both SP-A- and SP-D-deficient mice. Isolated alveolar macrophages from SP-A$^{-/-}$ mice generated significantly less, whereas those from SP-D$^{-/-}$ mice generated significantly greater superoxide and H_2O_2 compared with wild-type alveolar macrophages.

In SP-D$^{-/-}$ mice, bacterial killing was associated with increased lung inflammation and increased oxidant production. Where as, bacterial killing was decreased and associated with increased lung inflammation and decreased oxidant production in SP-A$^{-/-}$, macrophage phagocytosis was decreased in both SP-A and SP-D deficient mice. SP-A deficiency was associated with enhanced inflammation and synthesis of pro-inflammatory cytokines. SP-D$^{-/-}$ mice cleared these bacteria as efficiently as wild-type mice; however, clearance of viral pathogens was deficient in SP-D$^{-/-}$ mice and associated with increased inflammation. Study suggests that SP-A and SP-D play distinct roles during bacterial infection of lung (LeVine et al. 2000, 2001).

Alloiococcus otitidis has been found to be associated with otitis media with effusion. SP-A and MBL interact with *A. otitidis* in Ca^{2+}-dependent manner. Results demonstrate that *A. otitidis* is a ligand for SP-A and TLR2, and that the collectins enhance the phagocytosis of *A. otitidis* by macrophages, suggesting important roles of collectins and TLR2 in the innate immunity of the middle ear against *A. otitidis* infection (Konishi et al. 2006). Meningococcal disease occurs after colonization of nasopharynx with *Neisseria meningitidis*. Variation in genes of surfactant proteins affects the expression and function of SPs. Gene polymorphism resulting in substitution of glutamine with lysine at residue 223 in the CRD of SP-A2 increases susceptibility to meningococcal disease, as well as the risk of death (Jack et al. 2006). In contrast to defensive function, SP-D in BALF binds β-glucan onB. *Dermatitidis and*, blocks BAM access to β-glucan, thereby inhibiting TNF-α production. Thus, whereas BALF constituents commonly mediate antimicrobial activity, *B. dermatitidis* may utilize BALF constituents, such as SP-D, to blunt the host defensive reaction; this effect could reduce inflammation and tissue destruction but could also promote disease (Lekkala et al. 2006)

43.7 Pulmonary Tuberculosis

43.7.1 Enhanced Phagocytosis of *M. tuberculosis* by SP-A

During initial infection with *M. tuberculosis*, bacteria that reach the distal airspaces of lung are phagocytosed by AMΦs in presence of pulmonary surfactant. Studies indicated a direct interaction between SP-A and macrophage in mediating enhanced adherence of *M. tuberculosis* (Gaynor et al. 1995). Since, SP-A binds mannose, it was hypothesized that SP-A attaches to *M. tuberculosis* and serves as a ligand between *M. tuberculosis* and AΦs. Stokes et al. (1998) demonstrated that explanted alveolar AΦs do not efficiently bind *M. tuberculosis* in a serum-free system, although a small subpopulation of these AΦs could bind mycobacteria. In contrast, almost 100% of peritoneal AΦs bind mycobacteria under similar conditions. Evidence suggests that opsonic binding of *M. tuberculosis* by differentiated alveolar Ms is mediated by complement and CR3, and that the poor binding by resident alveolar AΦs is due to their poor expression of CR3. Thus, attachment of *M. tuberculosis* to AΦs is an essential early event in primary pulmonary tuberculosis and SP-A helps in early capture and phagocytosis of *M. tuberculosis* by AΦs. Ferguson et al. (2002) provided evidence for specific binding of SP-D to *M. tuberculosis* and indicated that SP-D and SP-A serve different roles in the innate host response to this pathogen in lung.

43.7.1.1 Lipomannan and ManLAM are Major Mycobacterial Lipoglycans as Potential Ligands

The SP-A binds to *M. bovis* Bacillus Calmette-Guerin (BCG), the vaccinating strain of pathogenic mycobacteria, and also to a lesser extent to *M. smegmatis*, which indicates that SP-A does not discriminate virulent from nonpathogenic strains. Lipomannan and mannosylated lipoarabinomannan (ManLAM) are two major mycobacterial cell-wall lipoglycans, which act as potential ligands for binding of SP-A. Both the terminal mannose residues and the fatty acids are critical for binding. It appears that recognition of carbohydrate epitopes on lipoglycans by SP-A is dependent on the presence of fatty acids (Sidobre et al. 2000, 2002).

Rivière et al, (2004) claim that the hydrophobic aglycon part of ManLAM is associated to a supra-molecular organization of these complex molecules. Furthermore, the deacylated ManLAMs or the lipid-free mannosylated arabinomannans, which do not exhibit characteristic ManLAM activities, do not display this supra-molecularorganization. These observations suggest that the ManLAMs immunomodulatory activities might be associated to their particular organization. The critical micellar concentration of ManLAMs obviously supports the notion that this supra-molecularorganization may be responsible for the specific biological activities of these complex molecules (Rivière et al. 2004).

As indicated, the molecular recognition of ManLAM terminal mannose units by CRDs of SP-A depends on the presence of lipid moiety of ManLAMs associated to a characteristic supra-molecular organization of ManLAM complex. On the other hand, the deacylated ManLAM or the lipid-free mannosylated arabinomannans, which do not exhibit characteristic ManLAM activities, do not display this supra-molecular organization. Therefore the ManLAM immunomodulatory activities might be associated to their particular organization. The critical micellar concentration of ManLAM supports the notion that this supra-molecular organization is responsible for specific biological activities of these complex molecules.

Apa Glycoprotein on *M. tuberculosis*: A Potential Adhesion to SP-A: Although lipoglycan ManLAM is considered as the major C-type lectin target on mycobacterial surface, Ragas et al. (2007) identified Apa (alanine- and proline-rich antigenic) glycoprotein as new potential target for SP-A, which binds to purified Apa. Apa is associated to the cell wall for a long time to aid in the attachment of SP-A. Because, Apa seems to be restricted to the *M. tuberculosis* complex strains, it was proposed that it may account for selective recognition of complex strains by SP-A containing homologous functional domains.

SP-A Enhances *M. avium* Ingestion by Macrophages: Tuberculosis leads to immune activation and increased HIV-1 replication in lung. SP-A promotes attachment of *M. tuberculosis* to AΦs during infection with HIV. SP-A levels and attachment of *M. tuberculosis* to AΦs inversely correlate with peripheral blood CD4 lymphocyte counts (Downing et al. 1995). *M. avium* complex (MAC) is a significant cause of opportunistic infection in patients with AIDS. Once in lung, MAC can interact with SP-A. Work on pulmonary pathogens including *M. bovis* BCG suggests that SP-A participates in promoting efficient clearance of these organisms by AMs. Lopez et al. (2003) reported that SP-A can bind to and enhance the uptake of MAC by AΦs, similar to BCG and *M. tuberculosis*. However, unlike BCG and other pulmonary pathogens that are cleared in presence of SP-A via a NO-dependent pathway, macrophage-mediated clearance of MAC is not enhanced by SP-A.

Suppression of Reactive Nitrogen Intermediates by SP-A in AMs in Response to *M. tuberculosis*: Reactive nitrogen intermediates (RNIs) play a significant role in the killing of *mycobacteria*. RNI levels generated by AΦs were significantly increased when IFNγ-primed AΦs were incubated with *M. tuberculosis*. However, the RNI levels were significantly suppressed in presence of SP-A. Furthermore, incubation of deglycosylated SP-A with *M. tuberculosis* failed to suppress RNI by AΦs, suggesting that the oligosaccharide of SP-A, which binds to *M. tuberculosis*, is necessary for this effect. Pasula et al. (1999) showed that SP-A-mediated binding of *M. tuberculosis* to AΦs and decreased RNI levels may be one mechanism by which *M. tuberculosis* diminishes the cytotoxic response of activated AΦs.

43.7.2 SP-A Modulates Inflammatory Response in AΦs During Tuberculosis

There is a severe reduction in SP-A levels in BAL during tuberculosis only in the radiographically involved lung segments, and the levels returned to normal after 1 month of treatment. The SP-A levels were inversely correlated with the percentage of neutrophils in BAL fluid, suggesting that low SP-A levels were associated with increased inflammation in the lung. SP-A has pleiotropic effects even at low concentrations found in tuberculosis patients. This protein augments inflammation in presence of infection and inhibits inflammation in uninfected macrophages, protecting uninvolved lung segments from the deleterious effects of inflammation (Gold et al. 2004).

SP-A modulates phenotypic and functional properties of cells of adaptive immune response such as DCs and lymphocytes. Bone marrow-derived DCs generated in presence of SP-A fail to increase LPS-induced up-regulation of MHC class II and CD86 co-stimulatory molecule on DCs surface and behaves like "tolerogenic DCs". SP-A may also induce tolerance by suppressing the proliferation of activated T lymphocytes (Hussain 2004). SP-A suppresses lymphocyte proliferation and IL-2 secretion, in part, by binding to its receptor, SP-R210. However, the mechanisms underlying this effect are not well understood. The effects of antibodies against the SP-A-binding (neck) domain (α-SP-R210n) or nonbinding C-terminal domain (α-SP-R210ct) of SP-R210 on human peripheral blood T cell immune responses against *M. tuberculosis* support the hypothesis that SP-A, via SP-R210, suppresses cell-mediated immunity against *M. tuberculosis* via a mechanism that up-regulates secretion of IL-10 and TGF-β1 (Samten et al. 2008). Role of SP-A and SP-D in linking innate and adaptive immunity to regulate host defense has been suggested by Wright (2005). Although both SP-A and SP-D can bind to T cells and directly inhibit proliferation, SP-A can also indirectly inhibit T-cell proliferation via suppression of dendritic cell (DC) maturation. SP-D has been shown to enhance antigen uptake and presentation. Taken together, these in vitro results suggest that the combined role of SP-A and SP-D is to modulate the immunologic environment of the lung so as to protect the host, yet thwart an overzealous inflammatory response that could potentially damage the lung and impair gas exchange (Wright 2005)

43.7.3 Marker Alleles in *M. tuberculosis*

Regression analyses of tuberculosis and tuberculin-skin test positive groups, on the basis of odds ratios, revealed tuberculosis susceptibility (DA11_C and GATA_3) and protective (AAGG_2) marker alleles. Similarly, between tuberculosis patients and general population control subjects, susceptibility $1A^3$, $6A^4$, and B1013_A and protective AAGG_1, and AAGG_7 marker alleles were observed. Moreover, interactions were seen between alleles $6A^2$ and $1A^3$ and between $1A^3$ and B1013_A. Studies indicate a possible involvement of SP alleles in tuberculosis pathogenesis (Floros et al. 2000). Malik et al. (2006) investigated polymorphisms in the *SFTPA1* and *SFTPA2* genes for association with tuberculosis in 181 Ethiopian families comprising 226 tuberculosis cases. Four polymorphisms, SFTPA1 307A, SFTPA1 776T, SFTPA2 355C, and SFTPA2 751C, were associated with tuberculosis. Additional subgroup analysis in male, female and more severely affected patients provided evidence for SFTPA1/2-covariate interaction. Among five intragenic haplotypes identified in SFTPA1 gene and nine identified in SFTPA2 gene, $1A^3$ was most significantly associated with tuberculosis susceptibility (Table 43.2).

SNPs in Collagen Region of SP-A2 as a Contributing Factor: Relation exists between polymorphisms in the collagen regions of SP-A2 genes and pulmonary tuberculosis. Seven SNPs (4 exonic and 3 intronic) were identified in collagen regions of SP-A1 and SP-A2 genes in Indian population. Two intronic polymorphisms, SP-A1C1416T and SP-A2C1382G showed significant association with pulmonary tuberculosis. A redundant SNPA1660G of SP-A2 gene showed significant association with pulmonary tuberculosis. This polymorphism, when existing along with a nonredundant polymorphism, SP-A2G1649C (Ala91Pro) resulted in a stronger association with pulmonary tuberculosis. The SNPs in collagen region of SP-A2 may be one of the contributing factors to the genetic predisposition to pulmonary tuberculosis (Madan et al. 2002).

43.7.4 Interaction of SP-D with *M. tuberculosis*

Since many mycobacteria are facultative intracellular pathogens, their ability to cause disease involves entry, survival and replication within host cells. Although much progress has been made in our understanding of entry by mycobacteria, we anticipate that clarification of role of entry in pathogenesis will require further application of newly developed molecular tools to dissect each of the proposed mechanisms.

SP-D is known to bind *M. tuberculosis*. Binding of SP-D to *M. tuberculosis* is calcium dependent, and carbohydrate inhibitable. The binding of SP-D to Erdman lipoarabinomannan is mediated by terminal mannosyl oligosaccharides of this lipoglycan. Incubation of *M. tuberculosis* with subagglutinating concentrations of SP-D leads to reduced adherence of bacteria to macrophages, whereas incubation of

Table 43.2 SP-A and SP-D alleles associated with various diseases (Kishore et al. 2005).

Polymorphism		Disease association, population, type of study
Gene	Allele	
SP-A2	1A0	Susceptibility, RDS, Caucasian
SP-A1	6A2	Susceptibility, RDS, Caucasian
SP-A1	6A3	Protection, RDS, Caucasian
SP-A2	1A2	Protection, RDS, Caucasian
SP-A1	6A3	Protection, RDS, Negroids
SP-A1, SP-A2	6A2-1A0	Susceptibility, RDS, Caucasian, family
SP-A1	6A2	Protection, RDS, Caucasian, twins
SP-A1, SP-A2	6A2-1A0	Protection, RDS, Caucasian, twins
SP-A1	6A6	Susceptibility, BPD, Caucasian
SP-A	AA62_A	Susceptibility, COPD, Mexican
SP-D	D2S388_5	Susceptibility, COPD, Mexican
SP-A1	6A1	Susceptibility, IPF, Mexican
SP-D	DA11_C	Susceptibility, TB, Mexican
SP-A2	1A3	Susceptibility, TB, Mexican
SP-A1	6A4	Susceptibility, TB, Mexican
SP-A1, SP-A2	6A2-1A0	Susceptibility, TB, Mexican
SP-A1	C1416T	Susceptibility, TB, Indian
SP-A2	C1382G	Susceptibility, TB, Indian
SP-A2	A1660G-G1649C	Susceptibility, TB, Indian
SP-D	G459A	Susceptibility, TB, Indian
SP-D	T3130G	Susceptibility, TB, Indian
SP-D	Met 11	Susceptibility, RSV, Finnish
SP-A2	A1660G-G1649C	Susceptibility, ABPA, Indian

RDS respiratory distress syndrome, *BPD* bronchopulmonary dysplasia, *COPD* chronic obstructive pulmonary disease, *IPF* idiopathic pulmonary fibrosis, *TB* tuberculosis, *RSV* respiratory syncy–tial virus, *ABPA* allergic bronchopulmonary aspergillosis (Adapted with permission from Kishore et al. 2005 © Springer)

bacteria with SP-A leads to significantly increased adherence to monocyte-derived macrophages. Ferguson et al. (2002) provided evidence for specific binding of SP-D to *M. tuberculosis* and indicated that SP-D and SP-A serve different roles in the innate host response to this pathogen in lung. Further studies provide direct evidence that inhibition of phagocytosis of *M. tuberculosis* affected by SP-D occurs independently of aggregation process. SP-D limits the intracellular growth of bacilli in macrophages by increasing phagosome-lysosome fusion but not by generating a respiratory burst (Ferguson et al. 2006). Results also provide evidence that SP-A and SP-D enhance mannose receptor-mediated phagocytosis of *M. avium* by macrophages (Kudo et al. 2004). Virulent and attenuated *M. tuberculosis* strains bind best to immobilized SP-A (Hall-Stoodley et al. 2006). *Mycobacterium avium* has developed numerous mechanisms for entering mononuclear phagocytes. The SP-A, and SP-D, exhibit a concentration-dependent binding to *M. avium*. Studies provide evidence that SP-A and SP-D enhance mannose receptor-mediated phagocytosis of *M. avium* by macrophages (Kudo et al. 2004).

43.7.5 Association of SPs with Diabetes

Insulin decreased SP-A gene transcription in human lung epithelial cells (Miakotina et al. 2002). Alveolar type II cells and nonciliated bronchiolar epithelial (Clara) cells in lungs of rats with diabetes have decreased SP-A, but increased mRNA. This is on account of differential expression in the level of SP-A, SP-B, and SP-C mRNAs in both alveolar and bronchiolar epithelial cells from diabetic lungs in comparison to control lungs (Sugahara et al. 1994). Nonetheless, Fernández-Real et al. (2008) reported circulating SP-A significantly higher among patients with glucose intolerance and type 2 diabetes than in subjects with normal glucose tolerance, even after adjustment for BMI, age, and smoking status. In amniotic fluid from diabetic women, SP-A levels were significantly less than in nondiabetic pregnancies. Hypertension did not modify SP-A in diabetic women. Although Snyder et al. (1988) suggested that the concentration of amniotic fluid SP-A is decreased in diabetic pregnancies, McMahan et al. (1987) concluded that in well controlled diabetic pregnancies fetal lung maturation is not adversely affected. SP-A and SP-B were significantly elevated in amniotic fluid from black mothers and in amniotic fluid from mothers who smoked during pregnancy (Pryhuber et al. 1991).

43.8 Expression of SPs in Lung Cancer

43.8.1 Non-Small-Cell Lung Carcinoma (NSCLC)

Molecular mechanisms underlying carcinogenesis of non-small cell lung cancer(NSCLC) may provide gene targets in critical pathways valuable for improving the efficacy of therapy and survival of patients with NSCLC (Chong et al. 2006). SP-A is described for a portion of NSCLC facilitating a diagnostic marker for these carcinomas (Goldmann et al. 2009). Studies in human lung carcinoma reported positive staining of tumor cells for SP-A, especially in peripheral airway cell carcinoma, which include bronchioloalveolar carcinoma and in some reports also papillary subtypes. The SP-A gene is expressed at higher levels in hyperplastic cells; the expression occurs predominantly, but not exclusively, in adenocarcinomas (Broers et al. 1992; Linnoila et al. 1992). The determination of SP-A in malignant effusions may help in distinguishing primary lung adenocarcinoma from adenocarcinomas of miscellaneous origin. Analysis of SP-A gene transcript in pleural effusion is useful for diagnosis of primary lung adenocarcinoma (Saitoh et al. 1997; Shijubo et al. 1992). Gene expression of SP-A and SP-C was restricted to metastatic pulmonary adenocarcinomas (Betz et al. 1995). Camilo et al. (2006) suggested that all adenocarcinomas were negative for p63 where as 4 (26.6%) of 15 were positive for SP-A.

Uzaslan et al. (2005) studied 169 primary adenocarcinomas of lung (109 acinar, 32 solid with mucin, 24 papillary and 4 mucinous) for SP-A expression. Twenty-five percent of acinar, 38% of papillary and 3% of solid adenocarcinoma with mucin showed a positive intracytoplasmic SP-A reaction of the tumor cells. Results support the theory that SP-A-producing cells may generate not only bronchioloalveolar and papillary carcinoma, but also other subtypes of lung adenocarcinoma (Stoffers et al. 2004; Uzaslan et al. 2005). Tsutsumida et al. (2007) advocate that high MUC1 expression on the surface is an important characteristic of a micropapillary pattern, where as reduced surfactant apoprotein A expression in the micropapillary pattern may be an excellent indicator for poor prognosis in small-size lung adenocarcinoma.

43.8.1.1 Genetic Factors as Lung Cancer Risk

Deletions of the SP-A gene are specific genomic aberrations in bronchial epithelial cells adjacent to and within NSCLC, and are associated with tumor progression and a history of smoking. SP-A deletions might be a useful biomarker to identify poor prognoses in patients with NSCLC who might therefore benefit from adjuvant treatment (Jiang et al. 2005). Seifart et al. (2005) genotyped for SP-A1, -A2, -B, and -D marker alleles in lung cancer subgroups, which included 99 patients with small cell lung carcinoma (SCLC), or non-SCLC (NSCLC, n = 68) consisting of squamous cell carcinoma (SCC), and adenocarcinoma (AC); controls and healthy individuals (population control). Seifart et al. (2005) found (a) no significant marker associations with SCLC, (b) rare SP-A2 ($1A^9$) and SP-A1 ($6A^{11}$) alleles associate with NSCLC risk when compared with population control, (c) the same alleles ($1A9, 6A^{11}$) associate with risk for AC when compared with population ($6A^{11}$) or clinical control ($1A^9$), and (d) the SP-A1-$6A^4$ allele (found in ~10% of the population) associates with SCC, when compared with control. A correlation between SP-A variants and lung cancer susceptibility appears to exist, indicating that SP-A alleles may be useful markers of lung cancer risk.

The SP mRNAs with SP-A, B, and C were coexpressed in 10/12 (83%) of adenomas and 4/5 (80%) of carcinomas in both solid and tubulopapillary areas. SP-D mRNA signals were not noted in normal or neoplastic lung. ISH for SP A, B, or C mRNA was a helpful aid in the diagnosis of proliferative lesions of the murine lung (Pilling et al. 1999). In ovine pulmonary adenocarcinoma, caused by jaagsiekte sheep retrovirus, SP-A and C were expressed in 70% and 80% of tumor cells, respectively, whereas Clara cell 10-kDa protein was expressed in 17% of tumor cells (Platt et al. 2002).

43.8.1.2 TTF-1 and SP-A in Differential Diagnosis

Results suggest that TTF-1 can play an important role for the maintenance and/or differentiation process in bronchiolar and alveolar cells (Nakamura et al. 2002). TTF-1 is frequently expressed in human lung cancer, especially in adenocarcinoma and small cell lung cancer, and TTF-1 expression is closely related to the expression of surfactant protein. Zamecnik and Kodet (2002) described positive results for TTF-1 and SP-A in 75% and 46% of pulmonary adenocarcinomas and in 50% and 25% of pulmonary non-neuroendocrine large cell carcinomas (LCCs), respectively. Small cell lung carcinomas were TTF-1 positive in 89% of cases and completely negative for SP-A. Squamous cell carcinomas and carcinoid tumors were negative for both proteins. The frequency of TTF-1 expression in the nucleus was very low in human lung cancer cell lines; however, their cytoplasmic positivities should be further investigated (Fujita et al. 2003. Rossi et al. (2003) (1) support the metaplastic histogenetic theory for pulmonary carcinomas group of tumors; (2) show that cytokeratin 7 and TTF-1, but not SP-A, are useful immunohistochemical markers in this setting, and (3) suggest that this group of tumors has a worse prognosis than conventional NSCL carcinoma at surgically curable stages I, justifying their segregation as an independent histologic type. Lu et al. (2006) suggested that nuclear inclusions positive for SP-A antibody staining in adenocarcinomas of lung were derived from accumulated content in the perinuclear cistern resembling pseudoinclusion processes and composed of proteins antigenically cross-reactive with SP-A. Because of its diagnostic utility TTF-1 should be added to a panel of antibodies used for assessing tumors of unknown origin. The combination of anti-TTF-1 with anti-SP-A does not increase the diagnostic usefulness of TTF-1 alone (Lu et al. 2006). Suzuki et al. (2005) and Ueno et al. (2003) reported that Napsin is better marker than SP-A for diagnosis of lung adenocarcinoma. Napsin A is an aspartic proteinase expressed in lung and kidney. Napsin A is expressed in type II pneumocytes and in adenocarcinomas of lung.

43.8.1.3 SPs as a Tool for Diagnosis of Lung Tumors

Most bronchioloalveolar carcinomas of lung react positively for SP-A. Positive SP-A staining of large cell carcinoma of the lung could indicate that at least part of these tumors have the same cellular origin or differentiation as bronchioloalveolar carcinoma. Twenty of 63 (32%) tumors stained positive for SP-A. This may imply that about one third of large cell carcinomas of lung have a similar cellular origin or differentiation as bronchioloalveolar carcinoma (Uzaslan et al. 2006).

SP-A, a marker for lung adenocarcinomas, can be used to differentiate lung adenocarcinomas from other types and metastatic cancers of other origins (Kuroki et al. 1998; Takahashi et al. 2006b). RT-PCR and primers specific for SP-A, SP-B, SP-C and SP-D genes were used to detect nodal metastases and occult tumor spread of pulmonary adenocarcinomas. A combination of SP-A and SP-D may help to establish a differential prognosis in patients with gefitinib-induced ILD (Kitajima et al. 2006). ILD is a serious adverse event in lung cancer patients treated with gefitinib, an epidermal growth factor receptor tyrosine kinase inhibitor (EGFR-TKI). Pretreatment with gefitinib exacerbated LPS-induced lung EGFR-TKI by reducing SP-A expression in lung. EGFR-TKI may reduce SP-A expression in lungs of lung cancer patients and thus patients treated with EGFR tyrosine kinase inhibitor may be susceptible to pathogens (Inoue et al. 2008).

SP mRNAs were present in all lung tumors, with SPs A, B, and C being co-expressed in 83% of adenomas and 80% of carcinomas in both solid and tubulopapillary areas. No signals for SP D mRNA were noted in normal or neoplastic lung. Additionally, no staining for any SP transcript was observed in the hepatocellular carcinoma metastases. In situ hybridization for SP A, B, or C mRNA was helpful in diagnosis of proliferative lesions of the murine lung, enabling differentiation from hepatocellular metastases (Pilling et al. 1999; Qi et al. 2002).

43.9 Other Inflammatory Disorders

43.9.1 Airway Inflammation in Children with Tracheostomy

The long-term tracheostomy in infants and children may perpetuate chronic airway inflammation and airway remodeling due to easier access to the lungs for microorganisms. The SP-A and SP-D may directly interact with invading microorganisms and also modulate the activity of local immune cells. Children with tracheostomy had an increased total number of cells, increased neutrophils, and more frequently bacteria, but no viruses were recovered. SP-D concentration was reduced to half, though SP-A, SP-B, and SP-C were not different from controls. SP-D was inversely correlated to neutrophils, and high numbers of bacteria were associated with lower SP-D concentrations. It was suggested that bacteria and low SP-D support neutrophilic inflammation in the lower respiratory tract of nonsymptomatic with children with tracheostomy (Griese et al. 2004). BAL fluids from patients carrying a chronic tracheostoma agglutinated *P. aeruginosa*, which was completely inhibited by maltose. The agglutination of *P. aeruginosa* by BAL fluid was related in part to the concentration of SP-D. Additional factors, such as the multimeric organization of SP-D, are likely to contribute to the agglutination of microorganisms by BAL or other body fluids (Griese and Starosta 2005).

Pulmonary Alveolar Microlithiasis (PAM): Pulmonary alveolar microlithiasis (PAM) is an uncommon chronic disease characterized by calcifications within the alveoli and a paucity of symptoms in contrast to image findings. PAM occurs in the absence of any known disorder of calcium metabolism. Takahashi et al. (2006a) reported two cases of PAM, with markedly elevated sera concentrations of SP-A and SP-D, which showed a tendency to increase as the disease progressed. Therefore, SP-A and SP-D may function as serum markers to monitor the disease activity and progression of PAM.

Gastroesophageal Reflux Disease: Children with gastroesophageal reflux often suffer from chronic, severe lung damage and recurrent infections. The mechanisms may involve reflux induced lung injury with alterations of the SP-A and SP-D, which bind specifically to various microbes and increase their elimination by granular leukocytes and macrophages. In children with gastroesophageal reflux disease (GERD), the macromolecular organization of SP-A and SP-D were significantly reduced. The more active SP-A and especially those of SP-D were diminished, whereas the smaller sized forms of SP-D were markedly increased. Reduced amounts of SP-A and SP-D and an altered structural organization of the surfactant proteins may contribute to pathogenesis of chronic lung disease commonly observed in these children (Griese et al. 2002).

43.9.2 Surfactant Proteins in Non-ILD Pulmonary Conditions

Infants with increased pulmonary blood flow secondary to congenital heart disease suffer from tachypnea, dyspnea, and recurrent pulmonary infections. In congenital heart disease with pulmonary hypertension secondary to increased pulmonary blood flow, there is a decrease in SP-A gene expression as well as a decrease in SP-A and SP-B protein contents (Gutierrez et al. 2001). In an experiment involving 4-week-old lambs with pulmonary hypertension secondary to increased pulmonary blood flow following an in utero placement of an aortopulmonary vascular graft, Lee et al. (2004) found a decrease in SP-A gene expression as well as a decrease in SP-A and SP-B protein contents. But in a lamb model of congenital heart disease with pulmonary hypertension and increased pulmonary blood flow, the effect of the shunt on SP gene expression and protein content was not apparent within first week of life (Lee et al. 2004). No significant association between the common genetic variants of SP-A and SP-D and victims of sudden infant death syndrome (SIDS) was disclosed by Stray-Pedersen et al. (2009). However, low SP-A protein expression may possibly be determined by the 6A2/1A0 SP-A haplotype, which should be a subject for further investigation.

The SP-A level decreases significantly in acute pulmonary embolism, which may play an important role in hypoxemia in pulmonary embolism (Xie et al. 2005). Although an immunohistochemical investigation of pulmonary SP-A suggested a characteristic increase in fatal asphyxiation, no particular change was observed in the total amount of SP-A mRNA. The analysis of the SP-A1/A2 ratio may assist interpretation of the molecular alterations of SP-A related to acute asphyxial death (Ishida et al. 2002).

In hyperpnea there is a significant increase in lamellar bodies (LB) SP-A, lysozyme, and phospholipid (PL) but no change in the protein-to-prolonged hyperpnea ratios. It was suggested that (1) surfactant-associated lysozyme is secreted with LB, (2) the majority of SP-A is linked to lipid secretion but not necessarily with LB, and (3) the majority of SP-B secretion is independent of PL secretion. (4) Hyperpnea did not alter the mRNA expression of SP-A, SP-B, SP-C, or lysozyme in alveolar type II cells, but expression of SP-A and SP-B mRNA was significantly increased in lung tissue (Yogalingam et al. 1996).

43.10 DNA Polymorphisms in SPs and Pulmonary Diseases

Though the genes underlying susceptibility to RDS are insufficiently known, genes coding for SP-A and B have been assigned as the most likely genes in the etiology of RDS. Acute-RDS (ARDS) develops in association with many serious medical disorders. Mortality is at least 40%, and there is no specific therapy. The deficiency in SP-A level has been implicated in the pathophysiology of ARDS. Associations between single nucleotide polymorphisms (SNPs) of human gene coding *SFTPA1*, *SFTPA2*, and *SFTPD* and infectious pulmonary diseases have been established by several groups.

43.10.1 Association Between SP-A Gene Polymorphisms and RDS

Evidences suggest that the etiology of RDS is multifactorial with a significant genetic component. There are reports about polymorphisms and mutations on the surfactant protein genes, especially surfactant proteins-B, that may be associated with RDS, ARDS, and congenital alveolar proteinosis. The human SP-A gene locus includes two functional genes, *SFTPA1* and *SFTPA2* which are expressed independently, and a pseudo gene. SP-A polymorphisms play a role in respiratory distress syndrome, allergic bronchopulmonary aspergillosis and idiopathic pulmonary fibrosis. The levels of SP-A are decreased in lungs of patients with CF, RDS and chronic lung diseases (Heinrich et al. 2006).

Both low levels of SP-A and SP-A alleles have been associated with RDS. Floros et al. characterized four allelic variants of SP-A1 gene (6A, $6A^2$, $6A^3$, and $6A^4$) and five allelic variants of the SP-A2 gene (1A, $1A^0$, $1A^1$, $1A^2$, and $1A^3$) and hypothesized that specific SP-A alleles/genotypes are associated with increased risk of RDS. Because race, gestational age (**GA**), and sex are risk factors for RDS, Kala et al. (1998) studied the distribution and frequencies of SP-A alleles/genotypes while adjusting for these factors as confounders or effect modifiers in control and RDS populations with GAs ranging from 24 week to term. Although the odds ratios of several alleles and genotypes were in opposite directions for black and white subjects, the homogeneity of odds ratio reached statistical significance only in case of $6A^3/6A^3$. Although differences were observed in subgroups with different GAs of RDS white population, definitive conclusions could not be made regarding the effect of modification by GA or as a function of sex. Study suggested that (1) the genetic analyses of RDS and SP-A locus should be performed separately for black and white populations and (2) SP-A alleles/genotypes and SP-B variant may contribute to the etiology of RDS and/or may serve as markers for disease subgroups. In a genetically homogeneous Finnish population, Rämet et al. (2000) showed that certain SP-A1 alleles ($6A^2$ and $6A^3$) and an SP-A1/SP-A2 haplotype ($6A^2/1A^0$) were associated with RDS. The $6A^2$ allele was over-represented and the $6A^3$ allele was under-represented in infants with RDS. According to results, diseases associated with premature birth did not explain the association between the odds of a particular homozygous SP-A1 genotype ($6A^2/6A^2$ and $6A^3/6A^3$) and RDS. In the population evaluated, SP-B intron 4 variant frequencies were low and had no association with RDS. Thus, SP-A gene locus is an important determinant for predisposition to RDS in premature infants.

Floros et al. (2001b), in family-based linkage studies to discern linkage of SP-A to RDS, showed a link between SP-A and RDS; certain SP-A alleles/haplotypes are susceptibil ($1A^0$, $6A^2$, $1A^0/6A^2$) or protective ($1A^5$, $6A^4$, $1A^5/6A^4$) for RDS. Some differences between blacks and whites with regard to SP-A alleles may exist. In a 107 father-mother-offspring trios, divided into two sets according to proband's phenotype, Haataja et al. (2001) evaluated familial segregation of candidate gene polymorphisms by the transmission disequilibrium test. A set of 76 trios were analyzed for transmission disequilibrium from parents to affected offspring. Another set of

31 trios were studied for allele transmission from parents to hypernormal offspring born very prematurely before GA of 32 weeks. SP-A1-A2 haplotype $6A^2$-$1A^0$ showed significant excess transmission to affected infants and SP-A1 allele $6A^2$ decreased transmission to the hypernormals. Study provides a support for a role of SP-A alleles as genetic predisposers to RDS in premature infants.

43.10.2 SP-A and SP-B as Interactive Genetic Determinants of Neonatal RDS

Haataja et al. (2000) investigated if SP-B gene or interaction between SP-A and SP-B genes has a role in genetic susceptibility to RDS. Of the two SP-B polymorphisms genotyped, the Ile131Thr variation, a putative N-terminal N:-linked glycosylation site of proSP-B and length variation of intron 4 have been suggested to associate with RDS. Neither of the two SP-B polymorphisms associated directly with RDS or with prematurity. Instead, results showed that known association between SP-A alleles and RDS was dependent on the SP-B Ile131Thr genotype. Hence, the SP-B Ile131Thr polymorphism is a determinant for certain SP-A alleles as factors causing genetic susceptibility to RDS ($6A^2$, $1A^0$) or protection against it ($6A^3$, $1A^2$).

Floros et al. (2001a) studied genotypes for SP-B intron 4 size variants and for four SNPs [-18 (A/C), 1013 (A/C), 1580 (C/T), 9306 (A/G)] in SP-B in black and white subjects. Based on odds ratio: (1) the SP-B intron 4 deletion variant in white subjects is more of an RDS risk factor for males and for subjects of 28 weeks $<$ gestational age (GA) $<$33 weeks; (2) the SP-B intron 4 insertion variant in black subjects is more of an RDS risk factor in females; (3) in white subjects, SP-A1 ($6A^2/6A^2$) or SP-A2 ($1A^0/1A^0$ or $1A^0/*$) genotypes in subjects of certain GA and with a specific SP-B genotype (9306 (A/G) or deletion/*) are associated with an enhanced risk for RDS; (4) in black subjects, SP-A1 ($6A^3/6A^3$ or $6A^3/*$) genotypes in subjects of 31 weeks $<$ or $=$ GA $<$ or $=$ 35 weeks and with the SP-B (1580 (T/T)) genotype are associated with a reduced risk for RDS. The SP-B polymorphisms are important determinants for RDS. These may identify differences between black and white subjects, as well as, between males and females regarding the risk for RDS. Moreover, SP-A susceptibility or protective alleles, in specific SP-B background, are associated with an increased or reduced risk for RDS.

43.10.3 RDS in Premature Infants

DNA samples from 441 premature singleton infants and 480 twin or multiple infants were genotyped for SP-A1, SP-A2, and SP-B exon 4 polymorphisms and intron 4 size variants in a homogeneous white population. Distribution of SP-A and SP-B gene variants between RDS and no-RDS infants were determined alone and in combination. The SP-A1 allele 6A2 and homozygous genotype 6A2/6A2 are over-represented in RDS of singletons when SP-B exon 4 genotype was Thr/Thr, and under represented in RDS of multiples when the SP-B genotype was Ile/Thr. The SP-A 6A2 allele in SP-B Thr131 background predisposed the smallest singleton infants to RDS, whereas near-term multiples were protected from RDS. There was a continuous association between fetal mass and risk of RDS, defined by SP-A and SP-B variants. Labeled lung explants with the Thr/Thr genotype showed proSP-B amino-terminal glycosylation, which was absent in Ile/Ile samples. Hence, Genetic and environmental variation may influence intracellular processing of surfactant complex and the susceptibility to RDS (Marttila et al. 2003b). However, the association between SP-A polymorphisms and RDS may not be applicable to entire population of premature infants. In twins, the association between SP-A polymorphism and RDS is different from that seen in premature singleton infants. The factor associated with SP-A genotype-specific susceptibility to RDS appears to be related to the size of uterus and length of gestation at birth (Marttila et al. 2003a). Zhai et al. (2008) reported that the frequency of SP-A1 allele $6A^2$ and $6A^3$ expression of SP-A in Chinese premature infants was low in neonatal RDS. In contrast, the frequency of SP-A2 allele $1A^0$ and $1A^1$ was high in normal Chinese premature infants. It supports that SP-A1 allele $6A^2$ may be a susceptible gene for RDS.

43.10.4 Gene Polymorphism in Patients of High-altitude Pulmonary Edema

A pathogenetic cofactor for development of high-altitude pulmonary edema (HAPE) is an increase in capillary permeability, which could occur as a result of an inflammatory reaction and/or free-radical-mediated injury to lung. Pulmonary SP-A has potent antioxidant properties and protects unsaturated phospholipids and growing cells from oxidative injury (Swenson et al. 2002). In view of protective role of SP-A against oxidative damage, Saxena et al. (2005) examined the association of constitutional susceptibility to HAPE with polymorphisms in SP-A1 and SP-A2. Allele frequencies of three loci in SP-A1 and one in SP-A2 were significantly different between low-altitude native (LAN) HAPE patients and LAN control subjects. Heterozygous individuals, with respect to SP-A1 C1101T and SP-A2 A3265C, showed less severity in oxidative damage in comparison with homozygous subjects (SP-A1 T1101 and SP-A2 C3265). The polymorphisms in SP-A1 might be one of the genetic factors contributing to susceptibility to HAPE (Saxena et al. 2005).

43.10.5 SNPs in Pulmonary Diseases

Four validated SNPs were genotyped with sequence-specific probes (TaqMan 7000) in 284 newborn infants below 32 weeks of GA. The finding of an association of a variant of the *Sftpd* gene, that has previously been shown to be associated with increased SP-D serum levels in adult patients with RDS in preterm infants, may provide a basis for the initial risk assessment of RDS and modification of surfactant treatment strategies. A role for SP-D in neonatal pulmonary adaptation has to be postulated. Genotyping for three SNP altering amino acids in the mature protein in codon 11 (Met^{11}Thr), 160 (Ala^{160}Thr), and 270 (Ser^{270}Thr) of the SP-D gene was performed and related to the SP-D levels in serum. Individuals with Thr/Thr-11-encoding genotype had significantly lower SP-D serum levels than individuals with Met/Met (11) genotype. Polymorphic variation in the N-terminal domain of the SP-D molecule influences oligomerization, function, and the concentration of the molecule in serum (Hilgendorff et al. 2009; Leth-Larsen et al. 2005; Sorensen et al. 2007).

Studies on twins indicated very strong genetic dependence for serum levels of SP-D. Sequencing of 5′ untranslated region (5′UTR), the coding region and the 3′ region of *Sftpd* gene of 32 randomly selected blood donors indicated one single *Sftpd* haplotype (allele frequency 13.53%) that showed a negative association with serum SP-D levels. The discovery of a frequent negative variant of *Sftpd* gene provides a basis for genetic analysis of function of SP-D in resistance against pulmonary infections and inflammatory disorders in humans (Heidinger et al. 2005). The presence of SP-D in non-pulmonary tissues, such as gastrointestinal tract and genital organs, suggest additional functions located to other mucosal surfaces. Sorensen et al. (2007) summarized studies on genetic polymorphisms, structural variants, and serum levels of human SP-A and SP-D and their associations with human pulmonary disease.

Polymorphisms of genes are transmitted together in haplotypes, which can be used in study of development of complex diseases such as RDS. Genetic haplotypes of these SP genes are associated with the development of RDS. Studies identify protective haplotypes against RDS and support findings related to SP genetic differences in children who develop RDS. An allele association study of 19 polymorphisms in SP-A1, SP-A2, SP-B, and SP-D genes in ARDS was carried out. Analysis revealed differences in frequency of alleles for some of the microsatellite markers flanking SP-B, and for one polymorphism (C/T) at nucleotide 1580 [C/T (1580)], within codon 131 (Thr^{131}Ile) of the SP-B gene. The latter determines the presence or absence of a potential N-linked glycosylation site. Based on the odds ratio, the C allele may be viewed as a susceptibility factor for ARDS. These data suggest that SP-B or a linked gene contributes to susceptibility to ARDS (Lin et al. 2000; Thomas et al. 2007).

Amino Acid Variants in SP-D Are Not Associated with Bronchial Asthma: As SP-D binds and neutralizes common allergens like house dust mites it is especially important in allergic asthma. Levels of SP-D are elevated in serum and alveolar lavage of asthmatic patients. Three common amino acid variants have been identified in SP-D and association of first variant has been described to severe infection with respiratory syncytial virus. The three polymorphisms leading to amino acid exchanges (Met^{11}Thr, Ala^{160}Thr, and Ser^{270}Thr) were typed in 322 asthmatic children and none of these polymorphisms was associated with bronchial asthma. Haplotype analyses revealed four major haplotypes all of which were evenly distributed between the populations. Functional amino acid variants in SP-D do not seem to play a major role in the genetic pre-disposition to bronchial asthma in children (Krueger et al. 2006).

Following allergen exposure in vivo, SP-D$^{-/-}$ mice expressed higher bronchoalveolar lavage (BAL) eosinophils and IL-13 and lower FN-γ expression at early time points compared with wild mice. IL-10 expression was increased at early time points in SP-D$^{-/-}$ compared with wild mice. SP-D may be critical for the modulation of early stages of allergic inflammation in vivo (Schaub et al. 2004).

Pettigrew et al. (2006, 2007) evaluated gene polymorphisms in loci encoding SP-A and risk of otitis media during first year of life among a cohort of infants at risk for developing asthma in white infants. Polymorphisms at codons 19, 62, and 133 in SP-A1, and 223 in SP-A2 were associated with race/ethnicity. In regression models incorporating estimates of uncertainty in haplotype assignment, the 6A^4/1A^5 haplotype was protective for otitis media among white infants. On similar line, analyses suggested that polymorphisms within *SFTPA* loci may be associated with wheeze and persistent cough in white infants at risk for asthma. These associations require replication and exploration in other ethnic/racial groups.

43.10.6 Allergic Bronchopulmonary Aspergillosis and Chronic Cavitary Pulmonary Aspergillosis (CCPA)

Individuals with any structural or functional defects in SP-A and SP-D due to genetic variations might be susceptible to aspergillosis. Single nucleotide polymorphism in genes of collagen region of SP-A1 and SP-A2 has been associated with allergic bronchopulmonary aspergillosis (ABPA) and its clinical markers. SP-A2 G1649C and SP-A2 A1660G, polymorphisms in the collagen region of SP-A2, might be one of the contributing factors to genetic predisposition and severity of clinical markers of ABPA. SNPs in SP-A2 and MBL genes showed significant associations with patients of ABPA in an Indian population. Patients carrying either one or both of GCT and AGG alleles of SP-A2 and patients with A allele at position 1011 of MBL had markedly higher eosinophilia, total IgE antibodies and lower FEV1. Therapeutic administration of SP-D and MBL proteins in a murine model of pulmonary invasive aspergillosis rescued mice from death. In mice mimicking human ABPA, SP-A and SP-D suppressed IgE levels, eosinophilia, pulmonary cellular infiltration and cause a marked shift from a pathogenic Th2 to a protective Th1 cytokine profile. Thus, collectins play an important role in Aspergillus mediated allergies and infections (Madan et al. 2005; Saxena et al. 2003).

Patients with CCPA or ABPA of Caucasian origin were screened for SNPs in collagen region of SP-A1 and SP-A2 and MBL. The T allele at T1492C and G allele at G1649C of SP-A2 were observed at slightly higher frequencies in ABPA patients (86% and 93%) than in controls (63% and 83%), and the C alleles at position 1492 and 1649 were found in higher frequencies in CCPA patients (33% and 25%) than in ABPA patients (14% and 7%). However, the CC genotype at position 1649 of SP-A2 was significantly associated with CCPA. Similarly, ABPA patients showed a higher frequency of TT genotype (71%) at 1492 of SP-A2 than controls (43%) and CCPA patients (41%). In case of MBL, the T allele and CT genotype at position 868 (codon 52) were significantly associated with CCPA, but not with ABPA. Further analysis of genotype combinations at position 1649 of SP-A2 and at 868 of MBL between patient groups showed that both CC/CC and CC/CT SP-A2/MBL were found only in CCPA patients, while GG/CT SP-A2/MBL was significantly higher in CCPA patients in comparison to ABPA patients. SNPs in SP-A1 did not differ between patients and controls. Distinct alleles, genotypes and genotype combinations of SP-A2 and MBL may contribute to differential susceptibility of the host to CCPA or ABPA (Vaid et al. 2007).

Allergic Airway Inflammation: The SP-A has potent immunomodulatory activities. SP-A protein levels in the BAL fluid showed a rapid, transient decline that reached the lowest values (25% of controls) 12 h after intranasal Af provocation of sensitized mice. It was speculated that a transient lack of SP-A following allergen exposure of airways may significantly contribute to the development of a T-cell dependent allergic immune response (Scanlon et al. 2005). After acute ovalbumin-induced allergic airway inflammation (1) alveolar epithelial type II cells (AEII) but not Clara cells show a significantly higher expression of SP-A and SP-D in rats leading also to higher amounts of both SPs in BALF and (2) macrophages gather predominantly SP-A (Schmiedl et al. 2008).

43.10.7 Autoreactivity Against SP-A and Rheumatoid Arthritis

Circulating SP-D is decreased in early rheumatoid arthritis and SP-A and SP-D levels in synovial fluid from patients correlated with rheumatoid factor, CRP, IgA, IgM, and IgG, and total lipid content. SP-A and SP-D seem to participate in initiation of immune system and joint inflammation within the joint (Kankavi 2006) and may be an additional RA disease modifier like MBL. The Met^{11}Thr polymorphism in the N-terminal part of SP-D is important determinant in serum SP-D. But this polymorphism is also essential to the function and assembly into oligomers. SP-D levels did not correlate with traditional disease activity measures. The Thr^{11}/Thr^{11} genotype and the Thr^{11} allele tended to be more frequent in RA patients. Therefore, the low serum level of SP-D and the lack of correlation with traditional disease activity measures indicate that SP-D reflects a distinctive aspect in the RA pathogenesis (Hoegh et al. 2008; Miyata et al. 2002). Trinder et al. (2000) were able to show autoreactivity to SP-A, as expressed by IgG and IgM autoantibodies, and present in synovial fluid (SF) from patients with RA. There was no cross-reactivity between autoantibodies reactive with type II collagen (CII) and those reactive with SP-A or C1q; However, autoantibodies reacted with polymeric (dimers and larger) SP-A, but not with monomeric SP-A subunits, indicating that a degree of quaternary structure is required for antibody binding.

43.11 Inhibition of SP-A Function by Oxidation Intermediates of Nitrite

43.11.1 Protein Oxidation by Chronic Pulmonary Diseases

The oxidation of proteins may play an important role in the pathogenesis of chronic inflammatory lung diseases, and may contribute to lung damage. Higher levels of protein oxidation than in healthy controls were observed in patients with interstitial lung disease, gastro-esophageal reflux disease, and PAP. The proteins most sensitive to oxidation were serum albumin, SP-A, and α1-antitrypsin. Abundance of reactive oxygen species produced during neutrophilic inflammation may be a deleterious factor that leads to pulmonary damage in these patients (Starosta and Griese 2006). Primary chain and quaternary structure of SP-D in BALFs showed significant changes under oxidative conditions in vitro and in vivo and functional capacity to agglutinate bacteria was impaired by oxidation. Free radicals generated in lungs resulting in oxidation of SP-D may impair host defense and may contribute to the suppurative lung diseases like cystic fibrosis (Starosta and Griese 2006).

43.11.2 Oxidation Intermediates of Nitrite

Nitration of protein tyrosine residues by peroxynitrite ($ONOO^-$) has been implicated in a variety of inflammatory diseases such as ARDS. A mixture of hypochlorous acid (HOCl) and nitrite (NO_2^-) induces nitration, oxidation, and chlorination of tyrosine residues in human SP-A, and inhibits SP-A's ability to aggregate lipids and bind mannose. Nitration and oxidation of SP-A was not altered by the presence of lipids, suggesting that proteins are preferred targets in lipid-rich mixtures such as pulmonary surfactant. Moreover, both horseradish peroxidase and myeloperoxidase (MPO) can utilize NO_2^- and H_2O_2 as substrates to catalyze tyrosine nitration in SP-A, and inhibit its lipid aggregation function. SP-A nitration and oxidation by MPO is markedly enhanced in presence of Cl^- and the lipid aggregation function of SP-A is completely abolished. Studies suggest that MPO released by activated neutrophils during inflammation utilizes physiological or pathological levels of NO_2^- to nitrate proteins, and may provide an additional mechanism in addition to ONOO – formation, for tissue injury in ARDS and other inflammatory diseases associated with upregulated NO* and oxidant production. The oxidant-mediated tissue injury is likely to be important in the pathogenesis of ARDS/ALI (Davis et al. 2002; Lang et al. 2002; Narasaraju et al. 2003).

In vitro and in vivo data suggest that NO alters surfactant protein gene expression. The role of NO in ALI remains controversial. Although inhaled NO increases oxygenation in clinical trials, inhibiting NOS can be protective. However, inhalation of NO may not be indicated in sepsis because of excessive NO production. Aikio et al. (2003) indicated that inhaled NO is effective in a select group of small premature infants and that the responsiveness to NO is associated with low NOS2 enzyme. Very low birth-weight infants (birthweight <1,500 g), infants with progressive respiratory failure and infection at birth have deficient pulmonary NOS2 and cytokine response. After surfactant therapy, these infants responded strikingly to inhaled NO. An acute pulmonary inflammatory response may contribute to respiratory adaptation in early-onset pneumonia. In intact lambs inhaled NO increased SP-A and SP-B mRNA and protein content with no change in DNA content. The mechanisms and physiological effects of these findings warrant further investigation (Stuart et al. 2003; Hu et al. 2007). Exposure of rats to NO_2 showed impairment of SP-A and a higher alveolar pool size after in vivo exposure. The NO_2-induced alterations of SP-A may contribute to the pulmonary toxicity of this oxidant (Müller et al. 1992). NO production from NOS2 expressed in lung parenchymal cells in a murine model of ARDS correlates with abnormal surfactant function and reduced SP-B expression. $NOS2^{-/-}$ null mice exhibit significantly less physiologic lung dysfunction and loss of SP-B expression. Study indicated that the expression of NOS2 in lung epithelial cells is critical for the development of lung injury and mediates surfactant dysfunction independent of NOS2 inflammatory cell expression and cytokine production (Baron et al. 2004).

43.11.3 BPD Treatment with Inhaled NO

Inhaled NO is used to treat a number of disease processes. BPD is characterized by arrested alveolar and vascular development of immature lung. The increased expression of SP-A mRNA under hyperoxia can be attributed, at least in part, to an induction of mRNA and protein expression in bronchial Clara cells. The expanded role of Clara cells in the defence against hyperoxic injury suggests that they support alveolar type 2 cell function and may play an important role in the supply of surfactant proteins to the lower airways (ter Horst et al. 2006). The inhaled nitric oxide treatment of premature infants at risk for bronchopulmonary dysplasia does not adversely affect endogenous surfactant function or composition and may improve surfactant function transiently (Ballard et al. 2007). Chorioamnionitis is a risk factor for the development of bronchopulmonary dysplasia. Endotoxin-induced oxidative stress to the fetus in the uniquely hypoxic intrauterine environment has been reported. SP-A and B mRNAs were highest at Day 2, suggesting that oxidative stress did not contribute to the lung maturation response. A modest lung oxidative stress in chorioamnionitis could contribute to bronchopulmonary dysplasia (Cheah et al. 2008).

43.12 Congenital Diaphragmatic Hernia

Pulmonary hypoplasia is one of the main causes for high mortality rate in patients with congenital diaphragmatic hernia (CDH). The expression of SP-A in hypoplastic CDH lung is reduced, and its concentration is decreased in amniotic fluid of pregnancies complicated by CDH. In animal models, surfactant deficiency contributes to the pathophysiology of the disease. In humans surfactant disaturated phosphatidylcholine (DSPC) synthesis and SP-A were significantly lower in infants with CDH than in control subjects (Cogo et al. 2002).

SP-A is altered in developing lungs from rat fetuses with CDH induced by maternal ingestion of Nitrofen on Day 9 of gestation. There is decreased expression of SP-A in rat fetuses with CDH secondary to Nitrofen exposure (Mysore et al. 1998). In rat CDH model, induced in pregnant rats following administration of nitrofen, SP-A, SP-B, and SP-D mRNA expression in CDH lung were significantly decreased compared to controls at birth and 6 h after ventilation. The inability of O2 to increase SP mRNA expression in hypoplastic CDH lung suggests that the hypoplastic lung is not responsive to increased oxygenation for synthesis of SP (Shima et al. 2000). Though, SP's deficiency appears to be a common feature among various CDH models, TTF-1 expression was not altered in surgical model in contrast to nitrofen model, indicating different molecular mechanisms in two models (Benachi et al. 2002).

43.13 Protective Effects of SP-A and SP-D on Transplants

Surfactant treatment has been shown to improve lung transplant function, but the effect is variable. Erasmus et al. (2002) indicated that SP-A enrichment of surfactant improves the efficacy of surfactant in lung transplantation. After instillation of SP-A-enriched surfactant, P_{O2} values were reached to control values, whereas after SP-A-deficient surfactant treatment, the P_{O2} values did not improve (Erasmus et al. 2002). The impairment of surfactant adsorption from transplanted lungs may be correlated with decreased levels of SP-A, and increased levels of serum acute-phase protein C-reactive protein (CRP). The elevated levels of CRP in BAL can be a very sensitive marker of lung injury (Casals et al. 1998).

Bronchiolitis obliterans syndrome (BOS) affects long-term survival of lung transplant recipients (LTRs). Among 11 differentially expressed proteins in BALF, peroxiredoxin 2 (Prdx2) exclusively expressed in BOS; and SP-A expressed consistently less in BOS patients than in stable LTRs. The reduction of SP-A in BALF was detectable early after lung transplant, preceding BOS onset in four of five patients and indicated that SP-A levels in BALF could predict LTR patients who are at higher risk of BOS development (Meloni et al. 2007) BOS and IPS cause high mortality and impaired survival after allogeneic hematopoietic stem-cell transplantation (allo-HSCT). The pretransplant serum SP-D levels but not SP-A, KL-6 in BOS/IPS patients were lower than those in non-BOS/IPS patients. However, the patients with lower pretransplant serum SP-D level had a trend toward frequent development of BOS/IPS. Constitutive serum SP-D level before allo-HSCT may be a useful, noninvasive predictor for the development of BOS/IPS (Nakane et al. 2008).

Keratinocyte growth factor (KGF) given before bone marrow transplantation (BMT) can prevent allogeneic T cell-dependent lung inflammation, but the antiinflammatory effects of KGF were impaired in mice injected with both T cells and conditioning regimen of cyclophosphamide. Yang et al. (2000, 2002) demonstrated that addition of cyclophosphamide interferes with the ability of KGF to enhance SP-A production. The systemic pre-BMT injection of KGF in recipients of allogeneic T cells up-regulates SP-A, which may contribute to the early antiinflammatory effects of KGF. Exogenous and basal endogenous SP-A can suppress donor T-cell-dependent inflammation that occurs during the generation of idiopathic pneumonia syndrome after BMT. Wild-type and SP-A-deficient mice, given allogeneic donor bone marrow plus inflammation-inducing spleen T cells, showed that basal endogenous SP-A, and enhanced alveolar SP-A level modulate donor T-cell-dependent immune responses and prolong survival after allogeneic BMT.

43.14 Therapeutic Effects of SP-A, SP-D and Their Chimeras

43.14.1 SP-A Effects on Inflammation of Mite-sensitized Mice

SP-A and SP-D interact with a wide range of inhaled allergens, competing for their binding to cell-sequestered IgE resulting in inhibition of mast cell degranulation. SP-D interacts with glycoprotein allergens of house dust mite (*Dermatophagoides pteronyssinus*, Derp) via its CRDs and thus inhibits specific IgE, isolated from mite-sensitive asthmatic patients, from binding these allergens, and blocking subsequent histamine release from sensitized basophils. Exogenous administration of SP-A and SP-D diminishes allergic hypersensitivity in vivo. A fragment of recombinant human SP-D (rfh SP-D) has a therapeutic effect on allergen-induced bronchial inflammation through its inhibitory effect on NO and TNF-α production by AΦs, and thus preventing the development of Th-2 type cytokine response (Liu et al. 2005b; Singh et al. 2003). The rfhSP-D that is effective in diminishing allergic hypersensitivity in mouse models of dust mite allergy was more susceptible to degradation than the native full-length protein. The degradation and consequent inactivation of SP-A and SP-D may be a mechanism to account for the potent allergenicity of these common dust mite allergens (Deb et al. 2007). Evidence suggests for an antiinflammatory role for SP-D in response to noninfectious, subacute lung injury via modulation of oxidative-nitrative stress (Casey et al. 2005).

43.14.2 SP-D Increases Apoptosis in Eosinophils of Asthmatics

The effect of exogenous rfhSP-D on protection of adult mouse lung from LPS-induced and lipoteichoic acid (LTA)-induced injury was assessed in *Sftpd*$^{+/+}$ and *Sftpd*$^{-/-}$

mice. Intratracheal rhSP-D inhibited inflammation induced by intratracheal LPS and LTA instillation in lung. The antiinflammatory effects of rhSP-D were enhanced by addition of pulmonary surfactant, providing a potential therapy for the treatment of lung inflammation (Ikegami et al. 2007). In view of therapeutic effects of exogenous SP-D or rfhSP-D (composed of eight Gly-X-Y collagen repeat sequences, homotrimeric neck and lectin domains) in murine models of lung allergy and hypereosinophilic SP-D gene-deficient mice, Mahajan et al. (2008) suggested that rfhSP-D mediated preferential increase of apoptosis of primed eosinophils while not affecting the normal eosinophils. The increased phagocytosis of apoptotic eosinophils may be important mechanisms of rfhSP-D and SP-D-mediated resolution of allergic eosinophilic inflammation in vivo.

43.14.3 Targeting of Pathogens to Neutrophils Via Chimeric SP-D/Anti-CD89 Protein

Intratracheal rfhSP-D prevents shock caused by endotoxin released from the lung during ventilation in the premature newborn (Ikegami et al. 2006). In lambs, preterm infants experience enhanced susceptibility and severity to respiratory syncytial virus (RSV) infection. This was observed when SP-A, -D and TLR4 mRNA expression increased from late gestation to term birth, where as in preterm lungs, studies showed reduced SP-A, -D, and TLR4 expression and enhanced RSV susceptibility (Meyerholz et al. 2006).

A chimeric protein, consisting of a recombinant fragment of human SP-D coupled to a Fab' fragment directed against human Fc α receptor (CD89) (chimeric rfSP-D/anti-Fc), effectively targets pathogens recognized by SP-D to human neutrophils. A recombinant trimeric fragment of SP-D (rfSP-D), consisting of CRD and neck domain of human SP-D, cross-linked to the Fab' of an Ab directed against the human Fc α RI (CD89) (chimeric rfSP-D/anti-CD89 protein) enhanced uptake of *E. coli*, *C. albicans*, and *influenza A virus* by human neutrophils (Tacken et al. 2004). Both chimeric rfSP-D/anti-Fc receptor proteins increased internalization of *E. coli* by human promonocytic cell line U937, but only after induction of monocytic differentiation. Both CD64 and CD89 on U937 cells proved suitable for targeting by rfSP-D/anti-Fc receptor proteins (Tacken and Batenburg 2006). Collectin-based chimeric proteins may thus offer promise for therapy of infectious disease.

43.14.4 Anti-IAV and Opsonic Activity of Multimerized Chimeras of rSP-D

A recombinant human SP-D, consisting of a short collagen region (two repeats of Gly-Xaa-Yaa amino acid sequences), the neck domain and CRD can form a trimeric structure owing to neck domain and exhibits sugar-binding activity and specificity similar to those of native human SP-D. Though the truncated SP-D could bind to IAV, like native SP-D, but the truncated human SP-D was less effective in agglutinating bacteria than the native structure and failed to inhibit haemagglutination by IAV (Eda et al. 1997). On the other hand, chimeric collectin containing N-terminus and collagen domain of human SP-D and CRD of MBL showed greater anti-IAV activity than similarly multimerized preparations of SP-D or incompletely oligomerized preparations of the chimera. Highly multimerized preparations of chimera also caused greater increases in uptake of IAV by neutrophils. These studies may be useful for development of collectins as therapeutic agents against IAV infection (Hartshorn et al. 2000b; White et al. 2000).

Bovine serum conglutinin has greater ability to inhibit IAV infectivity than other collectins. Altering the carbohydrate binding properties of SP-D [e.g., by replacing its CRD with that of either MBL or conglutinin] can increase its activity against IAV. Hence, recombinant conglutinin and a chimeric protein containing NH_2 terminus and collagen domain of rat SP-D (rSP-D) fused to neck region and CRD of conglutinin (termed SP-D/Cong(neck + CRD)) have markedly greater ability to inhibit infectivity of IAV than wild-type recombinant rSP-D, confirming that potent IAV-neutralizing activity of conglutinin resides in its neck region and CRD. Furthermore, SP-D/Cong(neck + CRD) also caused substantially greater enhancement of neutrophil binding and H_2O_2 responses to IAV than r-conglutinin or rSP-D. Hence, chimeric SP-D/Cong(neck + CRD) protein showed favorable antiviral and opsonic properties of conglutinin and SP-D (Hartshorn et al. 2000a). Thus, the SP-D N-terminal and/or collagen domains contribute to the enhanced bacterial binding and aggregating activities of SP-D. Although replacement of neck recognition domains and CRDs of SP-D with those of MBL and conglutinin confer increased viral binding activity, it does not favorably affect bacterial binding activity, suggesting that requirements for optimal collectin binding to influenza virus and bacteria differ (Hartshorn et al. 2007).

Chimera of Trimeric Neck + CRDs of Human SP-D: The recombinant trimeric neck + CRDs of human SP-D (NCRD) retains binding activity for some ligands and mediates some functional activities. In comparison to strong neutralizing activity of lung SP-D for IAVs in vitro and in vivo, the NCRD derived from SP-D has weak viral-binding ability and lacks neutralizing activity. Using a panel of mAbs against NCRD, Tecle et al. (2008) showed that antiviral activities of SP-D can be reproduced without the N-terminal and collagen domains and that cross-linking of NCRDs is essential for antiviral activity of SP-D with respect to IAV (Tecle et al. 2008).

Incubation of native SP-D or NCRDs with peroxynitrite results into nitration and nondisulfide cross-linking. Modifications could be blocked by peroxynitrite scavengers or pH inactivation of peroxynitrite. Abnormal cross-linking leads to defective aggregation. Thus, modification of SP-D by reactive oxygen-nitrogen species could contribute to alterations in the structure and function of SP-D at sites of inflammation in vivo (Matalon et al. 2009). In contrast, a trimeric neck and CRD construct of bovine serum collectin CL-46 induces aggregation of IAV and potently increases IAV uptake by neutrophils. CL-46-NCRD showed calcium-dependent and sugar-sensitive binding to both neutrophils and IAV. Results indicate that collectins can act as opsonins for IAV even in the absence of the collagen domain or higher order multimerization. This may involve increased affinity of individual CRDs for glycoconjugates displayed on host cells or the viral envelope (Hartshorn et al. 2010).

Insertion of Arg-Ala-Lys in NCRD Increases Inhibitory Activity: Arg-Ala-Lys (RAK) (immediately N-terminal to the first motif) in CL-43 contributes to differences in saccharide selectivity and host defense function. Insertion of CL-43 RAK sequence or a control Ala-Ala-Ala sequence (AAA) into corresponding position in NCRD increased the efficiency of binding to mannan and changed the inhibitory potencies of competing saccharides to more closely resemble those of CL-43. In addition, RAK resembled CL-43 in its greater capacity to inhibit infectivity of IAV and to increase uptake of IAV by neutrophils (Crouch et al. 2005).

43.15 Lessons from SP-A and SP-D Deficient Mice

SP-D deficient (SP-D$^{-/-}$) mice exhibit an increase in the number and size of airway macrophages, peribronchiolar inflammation, increases in metalloproteinase activity, and development of emphysema. Mice deficient in SP-D$^{-/-}$ develop progressive emphysema with age, associated with loss of parenchymal tissue, subpleural fibrosis, and accumulation of abnormal elastin fibers. The changes in lung structure in SP-D$^{-/-}$ mice are reflected in the mechanical properties of both airway and lung parenchyma measured in vivo (Yoshida and Whitsett 2006).

Gene-targeted mice deficient in SP-D develop abnormalities in surfactant homeostasis, hyperplasia of alveolar epithelial type II cells, and emphysema-like pathology. Alveolar and tissue phosphatidylcholine pool sizes are markedly increased in SP-D$^{-/-}$ mice. The pulmonary lipidosis in SP-D$^{-/-}$ mice was not associated with accumulation of SP-B or C, or their mRNAs, distinguishing the disorder from alveolar proteinosis syndromes. Surfactant protein A mRNA was reduced and, SP-A protein appeared to be reduced in SP-D$^{-/-}$ compared with wild type mice. Targeting of mouse SP-D gene caused accumulation of surfactant lipid and altered phospholipid structures, demonstrating a unsuspected role for SP-D in surfactant lipid homeostasis in vivo (Botas et al. 1998; Korfhagen et al. 1998; Ikegami et al. 2005). HDL cholesterol was significantly elevated in SP-D$^{-/-}$ mice while treatment of SP-D$^{-/-}$ mice with rhSP-D resulted in decreases of HDL-cholesterol as well as total cholesterol, and LDL cholesterol along with reduced plasma TNF-α in SP-D$^{-/-}$ mice. It shows that SP-D regulates atherogenesis in mouse model (Sorensen et al. 2006). SP-D plays a critical role in the suppression of alveolar macrophage activation, which may contribute to the pathogenesis of chronic inflammation and emphysema (Wert et al. 2000). Oxidant production and reactive oxygen species were increased in lungs of SP-D$^{-/-}$ mice, in turn activate NF-kB and MMP expression. SP-D plays an unexpected inhibitory role in the regulation of NF-kB in AΦs (Yoshida et al. 2001).

Studies indicate that GM-CSF-dependent macrophage activity is not necessary for emphysema development in SP-D-deficient mice, but that type II cell metabolism and proliferation are, either directly or indirectly, regulated by GM-CSF in this model (Hawgood et al. 2001; Ochs et al. 2004). SP-D and GM-CSF play distinct roles in the regulation of surfactant homeostasis and lung structure (Ikegami et al. 2001).

SP-A and SP-D Double Deficient Mice SP-A and SP-D proteins have overlapping as well as distinct functions. Mice singly deficient in SP-A and SP-D have distinct phenotypes and produce altered inflammatory responses to microbial challenges. Adult mice deficient in both SP-A and SP-D (A$^-$D$^-$) show fewer and larger alveoli, an increase in the number and size of type II cells, as well as more numerous and larger alveolar macrophages. Chronic deficiency of SP-A and SP-D in mice leads to parenchymal remodeling, type II cell hyperplasia and hypertrophy, and disturbed intracellular surfactant metabolism (Jung et al. 2005) In double deficient mice, there is a progressive increase in bronchoalveolar lavage phospholipid, protein, and macrophage content through 24 week of age. The macrophages from doubly deficient mice express high levels of the MMP-12 and develop intense but patchy lung inflammation. Qualitative changes resemble the lung pathology seen in SP-D-deficient mice (Hawgood et al. 2002).

Treatment of SP-D deficient mice with a truncated recombinant fragment of human SP-D (rfhSP-D) decreased lipidosis and alveolar macrophage accumulation as well as production of proinflammatory chemokines. The rfhSP-D treatment reduced the structural abnormalities in parenchymal architecture and type II cells characteristic of SP-D deficiency and reduced degree of emphysema and a corrected type II cell hyperplasia and hypertrophy. This suggests that rfhSP-D might become a therapeutic option in diseases that are characterized by decreased SP-D levels in the lung (Knudsen et al. 2007; Zhang et al. 2002).

Treatment with a recombinant fragment of human SP-D consisting of a short collagen-like stalk (but not the entire collagen-like domain of native SP-D), neck, and CRD inhibited development of emphysema-like pathology in SP-D deficient mice. On the other hand, the entire collagen-like domain was necessary for preventing SP-D knockout mice from pulmonary emphysema development. The fragment of SP-D lacking the short collagen-like stalk failed to correct pulmonary emphysematous alterations demonstrating the importance of the short collagen-like stalk for the biological activity of the recombinant fragment of human SP-D (Knudsen et al. 2009; Breij and Batenburg 2008).

NO Production and S-Nitrosylation of SP-D Controls Inflammatory Function SP-D$^{-/-}$ mice exhibit an increase in the number and size of airway macrophages, peribronchiolar inflammation, increases in metalloproteinase activity, and development of emphysema. SP-A inhibited production of NO and inducible nitric oxide synthase (iNOS) in rat AΦ stimulated with smooth LPS. In contrast, SP-A enhanced production of NO and iNOS in cells stimulated with IFN-γ or IFN-γ plus LPS. SP-A contributes to the lung inflammatory response by exerting differential effects on the responses of immune cells, depending on their state and mechanism of activation (Stamme et al. 2000). NO is involved in a variety of signaling processes, and because altered NO metabolism has been observed in inflammation, it is predicted that alterations in its metabolism would underlie the proinflammatory state observed in SP-D deficiency (Atochina et al. 2004a, c). Treatment with the iNOS inhibitor 1,400 W can inhibit inflammatory phenotype and can attenuate inflammatory processes within SP-D deficiency. Mice treated with 1,400 W reduced total lung NO synthase activity (Atochina-Vasserman et al. 2007). Guo et al. (2008) suggest that NO controls the dichotomous nature of SP-D and that posttranslational modification by S-nitrosylation causes quaternary structural alterations in SP-D causing it to switch its inflammatory signaling role. This represents new insight into both the regulation of protein function by S-nitrosylation and NO's role in innate immunity (Guo et al. 2008). Thus, inflammation that occurs in SP-D deficiency is due to an increase in NO production and a shift in the chemistry and targets of NO from a disruption of NO-mediated signaling within the innate immune system. However, purified preparations of SPs often contain endotoxin and the functions of SP-A and SP-D are affected by endotoxin. Therefore, the monitoring of SP preparations for endotoxin contamination is important (Wright et al. 1999).

References

Aikio O, Vuopala K, Pokela ML, Hallman M (2000) Diminished inducible nitric oxide synthase expression in fulminant early-onset neonatal pneumonia. Pediatrics 105:1013–1019

Aikio O, Saarela T, Pokela ML et al (2003) Nitric oxide treatment and acute pulmonary inflammatory response in very premature infants with intractable respiratory failure shortly after birth. Acta Paediatr 92:65–69

Alexis NE, Muhlebach MS, Peden DB, Noah TL (2006) Attenuation of host defense function of lung phagocytes in young cystic fibrosis patients. J Cyst Fibros 5:17–25

Allen MJ, Harbeck R, Smith B et al (1999) Binding of rat and human surfactant proteins A and D to *Aspergillus fumigatus* conidia. Infect Immun 67:4563–4569

Al-Salmi QA, Walter JN, Colasurdo GN et al (2005) Serum KL-6 and surfactant proteins A and D in pediatric interstitial lung disease. Chest 127:403–407

Amin RS, Wert SE, Baughman RP et al (2001) Surfactant protein deficiency in familial interstitial lung disease. J Pediatr 139:85–92

Appelmelk BJ, Eggleton P, Reid KB et al (2005) Variations in *Helicobacter pylori* lipopolysaccharide to evade the innate immune component surfactant protein D. Infect Immun 73:7677–7686

Arai Y, Obinata K, Sato Y, Hisata K et al (2001) Clinical significance of the serum surfactant protein D and KL-6 levels in patients with measles complicated by interstitial pneumonia. Eur J Pediatr 160:425–429

Atochina EN, Beck JM, Preston AM et al (2004a) Enhanced lung injury and delayed clearance of *Pneumocystis carinii* in surfactant protein A-deficient mice: attenuation of cytokine responses and reactive oxygen-nitrogen species. Infect Immun 72:6002–6011

Atochina EN, Beers MF, Hawgood S et al (2004b) Surfactant protein-D, a mediator of innate lung immunity, alters the products of nitric oxide metabolism. Am J Respir Cell Mol Biol 30:271–279

Atochina EN, Gow AJ, Beck JM et al (2004c) Delayed clearance of *Pneumocystis carinii* infection, increased inflammation, and altered nitric oxide metabolism in lungs of surfactant protein-D knockout mice. J Infect Dis 189:1528–1539

Atochina-Vasserman EN, Beers MF et al (2007) Selective inhibition of inducible NO synthase activity in vivo reverses inflammatory abnormalities in surfactant protein D-deficient mice. J Immunol 179:8090–8097

Awasthi S (2010) Surfactant protein (SP)-A and SP-D as antimicrobial and immunotherapeutic agents. Recent Pat Antiinfect Drug Discov 5:115–123

Awasthi S, Coalson JJ, Crouch E, Yang F, King RJ (1999) Surfactant proteins A and D in premature baboons with chronic lung injury (Bronchopulmonary dysplasia). Evidence for an inhibition of secretion. Am J Respir Crit Care Med 160:942–949

Awasthi S, Magee DM, Coalson JJ (2004) *Coccidioides posadasii* infection alters the expression of pulmonary surfactant proteins (SP)-A and SP-D. Respir Res 5:28

Bai Y, Galetskiy D, Damoc E et al (2007) Lung alveolar proteomics of bronchoalveolar lavage from a pulmonary alveolar proteinosis patient using high-resolution FTICR mass spectrometry. Anal Bioanal Chem 389:1075–1085

Baker CS, Evans TW, Randle BJ et al (1999) Damage to surfactant-specific protein in acute respiratory distress syndrome. Lancet 353 (9160):1232–1237

Balamugesh T, Kaur S, Majumdar S, Behera D (2003) Surfactant protein-A levels in patients with acute respiratory distress syndrome. Indian J Med Res 117:129–133

Ballard PL, Merrill JD, Truog WE et al (2007) Surfactant function and composition in premature infants treated with inhaled nitric oxide. Pediatrics 120:346–353

Barbaro M, Cutroneo G, Costa C, Sciorio S, et al (2002) Early events of experimental exposure to amorphous and crystalline silica in the rat: time course of surfactant protein D. Ital J Anat Embryol 107:243–256

Baron RM, Carvajal IM, Fredenburgh LE et al (2004) Nitric oxide synthase-2 down-regulates surfactant protein-B expression and enhances endotoxin-induced lung injury in mice. FASEB J 18:1276–1279

Baughman RP, Sternberg RI, Hull W et al (1993) Decreased surfactant protein A in patients with bacterial pneumonia. Am Rev Respir Dis 147:653–657

Behera D, Balamugesh T, Venkateswarlu D et al (2005) Serum surfactant protein-A levels in chronic bronchitis and its relation to smoking. Indian J Chest Dis Allied Sci 47:13–17

Benachi A, Chailley-Heu B, Barlier-Mur AM et al (2002) Expression of surfactant proteins and thyroid transcription factor 1 in an ovine model of congenital diaphragmatic hernia. J Pediatr Surg 37:1393–1399

Berg T, Leth-Larsen R, Holmskov U et al (2000) Structural characterisation of human proteinosis surfactant protein A. Biochim Biophys Acta 1543:159–173

Betsuyaku T, Kuroki Y, Nagai K, Nasuhara Y, Nishimura M (2004) Effects of ageing and smoking on SP-A and SP-D levels in bronchoalveolar lavage fluid. Eur Respir J 24:964–70

Betz C, Papadopoulos T, Buchwald J et al (1995) Surfactant protein gene expression in metastatic and micrometastatic pulmonary adenocarcinomas and other non-small cell lung carcinomas: detection by reverse transcriptase-polymerase chain reaction. Cancer Res 55:4283–4286

Botas C, Poulain F, Akiyama J et al (1998) Altered surfactant homeostasis and alveolar type II cell morphology in mice lacking surfactant protein D. Proc Natl Acad Sci USA 95:11869–11874

Bowler RP, Duda B, Chan ED et al (2004) Proteomic analysis of pulmonary edema fluid and plasma in patients with acute lung injury. Am J Physiol Lung Cell Mol Physiol 286:L1095–L1104

Brandt EB, Mingler MK, Stevenson MD et al (2008) Surfactant protein D alters allergic lung responses in mice and human subjects. J Allergy Clin Immunol 121:1140–1147, e2

Brasch F, Müller KM (2004) Classification of pulmonary alveolar proteinosis in newborns, infants, and children. Pathologe 25: 299–309

Brasch F, Birzele J, Ochs M et al (2004) Surfactant proteins in pulmonary alveolar proteinosis in adults. Eur Respir J 24:426–435

Breij EC, Batenburg JJ (2008) Surfactant protein D/anti-Fc receptor bifunctional proteins as a tool to enhance host defence. Expert Opin Biol Ther 8:409–419

Broers JL, Jensen SM, Travis WD et al (1992) Expression of surfactant associated protein-A and Clara cell 10 kilodalton mRNA in neoplastic and non-neoplastic human lung tissue as detected by in situ hybridization. Lab Invest 66:337–346

Brown-Augsburger P, Hartshorn K, Chang D et al (1996) Site-directed mutagenesis of Cys-15 and Cys-20 of pulmonary surfactant protein D. Expression of a trimeric protein with altered anti-viral properties. J Biol Chem 271:13724–13730

Bufler P, Schikor D, Schmidt B, Griese M (2004) Cytokine stimulation by *Pseudomonas aeruginosa* – strain variation and modulation by pulmonary surfactant. Exp Lung Res 30:163–179

Camilo R, Capelozzi VL, Siqueira SA et al (2006) Expression of p63, keratin 5/6, keratin 7, and surfactant-A in non-small cell lung carcinomas. Hum Pathol 37:542–546

Casals C, Varela A, Ruano ML et al (1998) Increase of C-reactive protein and decrease of surfactant protein A in surfactant after lung transplantation. Am J Respir Crit Care Med 157:43–49

Casey J, Kaplan J, Atochina-Vasserman EN, Gow AJ (2005) Alveolar Surfactant Protein D Content Modulates Bleomycin-induced Lung Injury. Am J Respir Crit Care Med 172: 869–877

Cheah FC, Jobe AH, Moss TJ et al (2008) Oxidative stress in fetal lambs exposed to intra-amniotic endotoxin in a chorioamnionitis model. Pediatr Res 63:274–279

Cheng G, Ueda T, Numao T et al (2000) Increased levels of surfactant protein A and D in bronchoalveolar lavage fluids in patients with bronchial asthma. Eur Respir J 16:831–835

Cheng IW, Ware LB, Greene KE et al (2003) Prognostic value of surfactant proteins A and D in patients with acute lung injury. Crit Care Med 31:20–27

Chiba H, Pattanajitvilai S, Evans AJ et al (2002) Human surfactant protein D (SP-D) binds *Mycoplasma pneumoniae* by high affinity interactions with lipids. J Biol Chem 277:20379–20385

Cho K, Matsuda T, Okajima S et al (2000) Prediction of respiratory distress syndrome by the level of pulmonary surfactant protein A in cord blood sera. Biol Neonate 77:83–87

Chong IW, Chang MY, Chang HC et al (2006) Great potential of a panel of multiple hMTH1, SPD, ITGA11 and COL11A1 markers for diagnosis of patients with non-small cell lung cancer. Oncol Rep 16:981–988

Clark H, Reid K (2003) The potential of recombinant surfactant protein D therapy to reduce inflammation in neonatal chronic lung disease, cystic fibrosis, and emphysema. Arch Dis Child 88:981–984

Coalson JJ, King RJ, Yang F et al (1995) SP-A deficiency in primate model of bronchopulmonary dysplasia with infection. In situ mRNA and immunostains. Am J Respir Crit Care Med 151:854–866

Cogo PE, Zimmermann LJ, Rosso F et al (2002) Surfactant synthesis and kinetics in infants with congenital diaphragmatic hernia. Am J Respir Crit Care Med 166:154–159

Crouch E, Persson A, Chang D (1993) Accumulation of surfactant protein D in human pulmonary alveolar proteinosis. Am J Pathol 142:241–248

Crouch E, Tu Y, Briner D et al (2005) Ligand specificity of human surfactant protein D: expression of a mutant trimeric collectin that shows enhanced interactions with influenza A virus. J Biol Chem 280:17046–17056

Daimon T, Tajima S, Oshikawa K et al (2005) KL-6 and surfactant proteins A and D in serum and bronchoalveolar lavage fluid in patients with acute eosinophilic pneumonia. Intern Med 44:811–817

Davis IC, Zhu S, Sampson JB et al (2002) Inhibition of human surfactant protein A function by oxidation intermediates of nitrite. Free Radic Biol Med 33:1703–1713

Deb R, Shakib F, Reid K, Clark H (2007) Major house dust mite allergens Dermatophagoides pteronyssinus 1 and Dermatophagoides farinae 1 degrade and inactivate lung surfactant proteins A and D. J Biol Chem 282:36808–19

deMello DE, Heyman S, Phelps DS, Floros J (1993) Immunogold localization of SP-A in lungs of infants dying from respiratory distress syndrome. Am J Pathol 142:1631–1640

Dohmoto K, Hojo S, Fujita J et al (2000) Circulating bronchoepithelial cells expressing mRNA for surfactant protein A in patients with pulmonary fibrosis. Respir Med 94:475–481

Downing JF, Pasula R, Wright JR et al (1995) Surfactant protein A promotes attachment of *Mycobacterium tuberculosis* to alveolar macrophages during infection with human immunodeficiency virus. Proc Natl Acad Sci USA 92:4848–4852

Doyle IR, Nicholas TE, Bersten AD (1995) Serum surfactant protein-A levels in patients with acute cardiogenic pulmonary edema and adult respiratory distress syndrome. Am J Respir Crit Care Med 152:307–317

Doyle IR, Bersten AD, Nicholas TE (1997) Surfactant proteins-A and -B are elevated in plasma of patients with acute respiratory failure. Am J Respir Crit Care Med 156:1217–1229

Doyle IR, Davidson KG, Barr HA et al (1998) Quantity and structure of surfactant proteins vary among patients with alveolar proteinosis. Am J Respir Crit Care Med 157:658–664

Eda S, Suzuki Y, Kawai T et al (1997) Structure of a truncated human surfactant protein D is less effective in agglutinating bacteria than the native structure and fails to inhibit haemagglutination by influenza A virus. Biochem J 323:393–399

Endo S, Sato N, Nakae H, Yamada Y et al (2002) Surfactant protein A and D (SP-A, AP-D) levels in patients with septic ARDS. Res Commun Mol Pathol Pharmacol 111:245–251

Endo S, Sato N, Yaegashi Y et al (2006) Sivelestat sodium hydrate improves septic acute lung injury by reducing alveolar dysfunction. Res Commun Mol Pathol Pharmacol 119:53–65

Erasmus ME, Hofstede GJ, Petersen AH et al (2002) SP-A-enriched surfactant for treatment of rat lung transplants with SP-A deficiency after storage and reperfusion. Transplantation 73:348–352

Ernst JD (1998) Macrophage receptors for *Mycobacterium tuberculosis*. Infect Immun 66:1277–1281

Erpenbeck VJ, Malherbe DC, Sommer S et al (2005) Surfactant protein D increases phagocytosis and aggregation of pollen-allergen starch granules. Am J Physiol Lung Cell Mol Physiol 288:L692–L698

Erpenbeck VJ, Ziegert M, Cavalet-Blanco D et al (2006) Surfactant protein D inhibits early airway response in *Aspergillus fumigatus*-sensitized mice. Clin Exp Allergy 36:930–940

Ferguson JS, Voelker DR, McCormack FX, Schlesinger LS (1999) Surfactant protein D binds to *Mycobacterium tuberculosis* bacilli and lipoarabinomannan via carbohydrate-lectin interactions resulting in reduced phagocytosis of the bacteria by macrophages. J Immunol 163:312–321

Ferguson JS, Voelker DR, Ufnar JA et al (2002) Surfactant protein D inhibition of human macrophage uptake of *Mycobacterium tuberculosis* is independent of bacterial agglutination. J Immunol 168:1309–1314

Ferguson JS, Martin JL, Azad AK et al (2006) Surfactant protein D increases fusion of *Mycobacterium tuberculosis*-containing phagosomes with lysosomes in human macrophages. Infect Immun 74:7005–7009

Fernández-Real JM, Chico B, Shiratori M et al (2008) Circulating surfactant protein A (SP-A), a marker of lung injury, is associated with insulin resistance. Diabetes Care 31:958–963

Floros J, Fan R (2001) Surfactant protein A and B genetic variants and respiratory distress syndrome: allele interactions. Biol Neonate 80 (Suppl 1):22–25

Floros J, Lin HM, García A et al (2000) Surfactant protein genetic marker alleles identify a subgroup of tuberculosis in a Mexican population. J Infect Dis 182:1473–1479

Floros J, Fan R, Diangelo S et al (2001a) Surfactant protein (SP) B associations and interactions with SP-A in white and black subjects with respiratory distress syndrome. Pediatr Int 43:567–576

Floros J, Fan R, Matthews A et al (2001b) Family-based transmission disequilibrium test (TDT) and case–control association studies reveal surfactant protein A (SP-A) susceptibility alleles for respiratory distress syndrome (RDS) and possible race differences. Clin Genet 60:178–187

Foster DJ, Yan X, Bellotto DJ et al (2002) Expression of epidermal growth factor and surfactant proteins during postnatal and compensatory lung growth. Am J Physiol Lung Cell Mol Physiol 283:L981–L990

Fujita J, Ohtsuki Y, Bandoh S et al (2003) Expression of thyroid transcription factor-1 in 16 human lung cancer cell lines. Lung Cancer 39:31–36

Fujita M, Shannon JM, Ouchi H et al (2005) Serum surfactant protein D is increased in acute and chronic inflammation in mice. Cytokine 31:25–33

Gaynor CD, McCormack FX, Voelker DR et al (1995) Pulmonary surfactant protein A mediates enhanced phagocytosis of *Mycobacterium tuberculosis* by a direct interaction with human macrophages. J Immunol 155:5343–5351

Ghildyal R, Hartley C, Varrasso A et al (1999) Surfactant protein A binds to the fusion glycoprotein of respiratory syncytial virus and neutralizes virion infectivity. J Infect Dis 180:2009–2013

Giannoni E, Sawa T, Allen L et al (2006) Surfactant proteins A and D enhance pulmonary clearance of *Pseudomonas aeruginosa*. Am J Respir Cell Mol Biol 34:704–710

Glas J, Beynon V, Bachstein B et al (2008) Increased plasma concentration of surfactant protein D in chronic periodontitis independent of SFTPD genotype: potential role as a biomarker. Tissue Antigens 72:21–28

Gold JA, Hoshino Y, Tanaka N et al (2004) Surfactant protein A modulates the inflammatory response in macrophages during tuberculosis. Infect Immun 72:645–650

Goldmann T, Kähler D, Schultz H et al (2009) On the significance of surfactant protein-A within the human lungs. Diagn Pathol 4:9

Gowdy K, Krantz QT, Daniels M, et al (2008) Modulation of pulmonary inflammatory responses and antimicrobial defenses in mice exposed to diesel exhaust. Toxicol Appl Pharmacol 229:310–9

Greene KE, King TE Jr, Kuroki Y et al (2002) Serum surfactant proteins-A and -D as biomarkers in idiopathic pulmonary fibrosis. Eur Respir J 19:439–446

Griese M (2002) Respiratory syncytial virus and pulmonary surfactant. Viral Immunol 15:357–363

Griese M, Starosta V (2005) Agglutination of *Pseudomonas aeruginosa* by surfactant protein D. Pediatr Pulmonol 40:378–384

Griese M, Maderlechner N, Ahrens P, Kitz R (2002) Surfactant proteins A and D in children with pulmonary disease due to gastroesophageal reflux. Am J Respir Crit Care Med 165:1546–1550

Griese M, Felber J, Reiter K, Strong P et al (2004) Airway inflammation in children with tracheostomy. Pediatr Pulmonol 37:356–361

Griese M, Steinecker M, Schumacher S et al (2008) Children with absent surfactant protein D in bronchoalveolar lavage have more frequently pneumonia. Pediatr Allergy Immunol 19:639–647

Grubor B, Gallup JM, Meyerholz DK et al (2004) Enhanced surfactant protein and defensin mRNA levels and reduced viral replication during parainfluenza virus type 3 pneumonia in neonatal lambs. Clin Diagn Lab Immunol 11:599–607

Günther A, Schmidt R, Nix F et al (1999) Surfactant abnormalities in idiopathic pulmonary fibrosis, hypersensitivity pneumonitis and sarcoidosis. Eur Respir J 14:565–573

Guo X, Lin HM, Lin Z et al (2001) Surfactant protein gene A, B, and D marker alleles in chronic obstructive pulmonary disease of a Mexican population. Eur Respir J 18:482–490

Guo CJ, Atochina-Vasserman EN, Abramova E et al (2008) S-nitrosylation of surfactant protein-D controls inflammatory function. PLoS Biol 6:e266

Gutierrez JA, Parry AJ, McMullan DM et al (2001) Decreased surfactant proteins in lambs with pulmonary hypertension secondary to increased blood flow. Am J Physiol Lung Cell Mol Physiol 281:L1264–L1270

Haataja R, Rämet M, Marttila R et al (2000) Surfactant proteins A and B as interactive genetic determinants of neonatal respiratory distress syndrome. Hum Mol Genet 9:2751–2760

Haataja R, Marttila R, Uimari P et al (2001) Respiratory distress syndrome: evaluation of genetic susceptibility and protection by transmission disequilibrium test. Hum Genet 109:351–355

Haczku A (2006) Role and regulation of lung collectins in allergic airway sensitization. Pharmacol Ther 110:14–34

Haczku A, Cao Y, Vass G et al (2006) IL-4 and IL-13 form a negative feedback circuit with surfactant protein-D in the allergic airway response. J Immunol 176:3557–3565

Hall-Stoodley L, Watts G, Crowther JE et al (2006) *Mycobacterium tuberculosis* binding to human surfactant proteins A and D, fibronectin, and small airway epithelial cells under shear conditions. Infect Immun 74:3587–3596

Hamm H, Lührs J, Guzman Y, Rotaeche J et al (1994) Elevated surfactant protein A in bronchoalveolar lavage fluids from sarcoidosis and hypersensitivity pneumonitis patients. Chest 106:1766–1770

Hartshorn KL, Crouch EC, White MR et al (1994) Evidence for a protective role of pulmonary surfactant protein D (SP-D) against influenza A viruses. J Clin Invest 94:311–319

Hartshorn K, Chang D, Rust K et al (1996a) Interactions of recombinant human pulmonary surfactant protein D and SP-D multimers with influenza A. Am J Physiol 271:L753–L762

Hartshorn KL, Reid KB, White MR et al (1996b) Neutrophil deactivation by influenza A viruses: mechanisms of protection after viral opsonization with collectins and hemagglutination-inhibiting antibodies. Blood 87:3450–3461

Hartshorn KL, White MR, Shepherd V et al (1997) Mechanisms of anti-influenza activity of surfactant proteins A and D: comparison with serum collectins. Am J Physiol 273:L1156–L1166

Hartshorn KL, Sastry KN, Chang D et al (2000a) Enhanced anti-influenza activity of a surfactant protein D and serum conglutinin fusion protein. Am J Physiol Lung Cell Mol Physiol 278:L90–L98

Hartshorn KL, White MR, Voelker DR et al (2000b) Mechanism of binding of surfactant protein D to influenza A viruses: importance of binding to haemagglutinin to antiviral activity. Biochem J 351:449–458

Hartshorn KL, White MR, Tecle T et al (2007) Reduced influenza viral neutralizing activity of natural human trimers of surfactant protein D. Respir Res 8:9

Hartshorn KL, White MR, Tecle T et al (2010) Viral aggregating and opsonizing activity in collectin trimers. Am J Physiol Lung Cell Mol Physiol 298:L79–L88

Hattori A, Kuroki Y, Katoh T et al (1996a) Surfactant protein A accumulating in the alveoli of patients with pulmonary alveolar proteinosis: oligomeric structure and interaction with lipids. Am J Respir Cell Mol Biol 14:608–619

Hattori A, Kuroki Y, Sohma H et al (1996b) Human surfactant protein A with two distinct oligomeric structures which exhibit different capacities to interact with alveolar type II cells. Biochem J 317:939–944

Hawgood S, Akiyama J, Brown C et al (2001) GM-CSF mediates alveolar macrophage proliferation and type II cell hypertrophy in SP-D gene-targeted mice. Am J Physiol Lung Cell Mol Physiol 280:L1148–L1156

Hawgood S, Ochs M, Jung A et al (2002) Sequential targeted deficiency of SP-A and -D leads to progressive alveolar lipoproteinosis and emphysema. Am J Physiol Lung Cell Mol Physiol 283:L1002–L1010

Heidinger K, König IR, Bohnert A et al (2005) Polymorphisms in the human surfactant protein-D (SFTPD) gene: strong evidence that serum levels of surfactant protein-D (SP-D) are genetically influenced. Immunogenetics 57:1–7

Heinrich S, Hartl D, Griese M (2006) Surfactant protein A – from genes to human lung diseases. Curr Med Chem 13:3239–3252

Hickling TP, Bright H, Wing K et al (1999) A recombinant trimeric surfactant protein D carbohydrate recognition domain inhibits respiratory syncytial virus infection in vitro and in vivo. Eur J Immunol 29:3478–3484

Highland KB, Silver RM (2005) New developments in scleroderma interstitial lung disease. Curr Opin Rheumatol 17:737–745

Hilgendorff A, Heidinger K, Bohnert A et al (2009) Association of polymorphisms in the human surfactant protein-D (SFTPD) gene and postnatal pulmonary adaptation in the preterm infant. Acta Paediatr 98:112–117

Hirama N, Shibata Y, Otake K, et al (2007) Increased surfactant protein-D and foamy macrophages in smoking-induced mouse emphysema. Respirology 12:191–201

Hobo S, Niwa H, Anzai T (2007) Evaluation of serum amyloid A and surfactant protein D in sera for identification of the clinical condition of horses with bacterial pneumonia. J Vet Med Sci 69:827–830

Hoegh SV, Lindegaard HM, Sorensen GL et al (2008) Circulating surfactant protein D is decreased in early rheumatoid arthritis: a 1-year prospective study. Scand J Immunol 67:71–76

Hohlfeld JM, Erpenbeck VJ, Krug N (2002) Surfactant proteins SP-A and SP-D as modulators of the allergic inflammation in asthma. Pathobiology 70:287–292

Honda Y, Kuroki Y, Matsuura E et al (1995) Pulmonary surfactant protein D in sera and bronchoalveolar lavage fluids. Am J Respir Crit Care Med 152:1860–1866

Honda Y, Takahashi H, Kuroki Y et al (1996) Decreased contents of surfactant proteins A and D in BAL fluids of healthy smokers. Chest 109:1006–1009

Hu X, Guo C, Sun B (2007) Inhaled nitric oxide attenuates hyperoxic and inflammatory injury without alteration of phosphatidylcholine synthesis in rat lungs. Pulm Pharmacol Ther 20:75–84

Hu QJ, Xiong SD, Zhang HL et al (2008) Altered surfactant protein A gene expression and protein homeostasis in rats with emphysematous changes. Chin Med J (Engl) 121:1177–1182

Hull J, South M, Phelan P, Grimwood K (1997) Surfactant composition in infants and young children with cystic fibrosis. Am J Respir Crit Care Med 156:161–165

Hussain S (2004) Role of surfactant protein a in the innate host defense and autoimmunity. Autoimmunity 37:125–130

Ihn H, Asano Y, Kubo M et al (2002) Clinical significance of serum surfactant protein D (SP-D) in patients with polymyositis/dermatomyositis: correlation with interstitial lung disease. Rheumatology (Oxford) 41:1268–1272

Ikegami M, Hull WM, Yoshida M et al (2001) SP-D and GM-CSF regulate surfactant homeostasis via distinct mechanisms. Am J Physiol Lung Cell Mol Physiol 281:L697–L703

Ikegami M, Na CL, Korfhagen TR, Whitsett JA (2005) Surfactant protein D influences surfactant ultrastructure and uptake by alveolar type II cells. Am J Physiol Lung Cell Mol Physiol 288:L552–L561

Ikegami M, Carter K, Bishop K et al (2006) Intratracheal recombinant surfactant protein d prevents endotoxin shock in the newborn preterm lamb. Am J Respir Crit Care Med 173:1342–1347

Ikegami M, Scoville EA, Grant S et al (2007) Surfactant protein-D and surfactant inhibit endotoxin-induced pulmonary inflammation. Chest 132:1447–1454

Inase N, Ohtani Y, Usui Y et al (2007) Chronic summer-type hypersensitivity pneumonitis: clinical similarities to idiopathic pulmonary fibrosis. Sarcoidosis Vasc Diffuse Lung Dis 24:141–147

Inoue A, Xin H, Suzuki T et al (2008) Suppression of surfactant protein A by an epidermal growth factor receptor tyrosine kinase inhibitor exacerbates lung inflammation. Cancer Sci 99:1679–1684

Ishida K, Zhu BL, Maeda H (2002) A quantitative RT-PCR assay of surfactant-associated protein A1 and A2 mRNA transcripts as a diagnostic tool for acute asphyxial death. Leg Med (Tokyo) 4:7–12

Ishiguro T, Yasui M, Takato H et al (2007) Progression of interstitial lung disease upon overlapping of systemic sclerosis with polymyositis. Intern Med 46:1237–1241

Ishii H, Mukae H, Kadota J et al (2003) High serum concentrations of surfactant protein A in usual interstitial pneumonia compared with non-specific interstitial pneumonia. Thorax 58:52–57

Jack DL, Cole J, Naylor SC et al (2006) Genetic polymorphism of the binding domain of surfactant protein-A2 increases susceptibility to meningococcal disease. Clin Infect Dis 43:1426–1433

Janssen R, Sato H, Grutters JC et al (2003) Study of Clara cell 16, KL-6, and surfactant protein-D in serum as disease markers in pulmonary sarcoidosis. Chest 124:2119–2125

Janssen R, Grutters JC, Sato H et al (2005) Analysis of KL-6 and SP-D as disease markers in bird fancier's lung. Sarcoidosis Vasc Diffuse Lung Dis 22:51–57

Jiang F, Caraway NP, Nebiyou Bekele B et al (2005) Surfactant protein A gene deletion and prognostics for patients with stage I non-small cell lung cancer. Clin Cancer Res 11:5417–5424

Jounblat R, Kadioglu A, Iannelli F et al (2004) Binding and agglutination of Streptococcus pneumoniae by human surfactant protein D (SP-D) vary between strains, but SP-D fails to enhance killing by neutrophils. Infect Immun 72:709–716

Jung A, Allen L, Nyengaard JR et al (2005) Design-based stereological analysis of the lung parenchymal architecture and alveolar type II cells in surfactant protein A and D double deficient mice. Anat Rec A Discov Mol Cell Evol Biol 286:885–890

Kala P, Ten Have T, Nielsen H et al (1998) Association of pulmonary surfactant protein A (SP-A) gene and respiratory distress syndrome: interaction with SP-B. Pediatr Res 43:169–177

Kankavi O (2006) Increased expression of surfactant protein A and D in rheumatoid arthritic synovial fluid (RASF). Croat Med J 47:155–161

Kashiwabara K (2006) Characteristics and disease activity of early interstitial lung disease in subjects with true parenchymal abnormalities in the posterior subpleural aspect of the lung. Chest 129:402–406

Kasper M, Sims G, Koslowski R et al (2002) Increased surfactant protein D in rat airway goblet and Clara cells during ovalbumin-induced allergic airway inflammation. Clin Exp Allergy 32:1251–1258

Kerr MH, Paton JY (1999) Surfactant protein levels in severe respiratory syncytial virus infection. Am J Respir Crit Care Med 159:1115–1118

Keisari Y, Wang H, Mesika A et al (2001) Surfactant protein D-coated Klebsiella pneumoniae stimulates cytokine production in mononuclear phagocytes. J Leukoc Biol 70:135–141

King RJ, Coalson JJ, deLemos RA et al (1995) Surfactant protein-A deficiency in a primate model of bronchopulmonary dysplasia. Am J Respir Crit Care Med 151:1989–1997

Kishore U, Bernal AL, Kamran MF et al (2005) Surfactant proteins SP-A and SP-D in human health and disease. Arch Immunol Ther Exp 53:399–417

Kitajima H, Takahashi H, Harada K et al (2006) Gefitinib-induced interstitial lung disease showing improvement after cessation: disassociation of serum markers. Respirology 11:217–220

Knudsen L, Ochs M, Mackay R et al (2007) Truncated recombinant human SP-D attenuates emphysema and type II cell changes in SP-D deficient mice. Respir Res 8:70

Knudsen L, Wucherpfennig K, MackaY R-M et al (2009) A recombinant fragment of human surfactant protein D lacking the short collagen-like stalk fails to correct morphological alterations in lungs of SP-D deficient mice. Anat Rec 292:183–189

Kobayashi H, Kanoh S, Motoyoshi K (2008a) Serum surfactant protein-A, but not surfactant protein-D or KL-6, can predict preclinical lung damage induced by smoking. Biomarkers 13:385–392

Kobayashi M, Takeuchi T, Ohtsuki Y (2008b) Differences in the immunolocalization of surfactant protein (SP)-A, SP-D, and KL-6 in pulmonary alveolar proteinosis. Pathol Int 58:203–207

Kodera M, Hasegawa M, Komura K et al (2005) Serum pulmonary and activation-regulated chemokine/CCL18 levels in patients with systemic sclerosis: a sensitive indicator of active pulmonary fibrosis. Arthritis Rheum 52:2889–2896

Konishi M, Nishitani C, Mitsuzawa H et al (2006) Alloiococcus otitidis is a ligand for collectins and Toll-like receptor 2, and its phagocytosis is enhanced by collectins. Eur J Immunol 36:1527–1536

Koopmans JG, van der Zee JS, Krop EJ et al (2004) Serum surfactant protein D is elevated in allergic patients. Clin Exp Allergy 34:1827–1833

Korfhagen TR, Sheftelyevich V, Burhans MS et al (1998) Surfactant protein-D regulates surfactant phospholipid homeostasis in vivo. J Biol Chem 273:28438–28443

Krueger M, Puthothu B, Gropp E et al (2006) Amino acid variants in surfactant protein D are not associated with bronchial asthma. Pediatr Allergy Immunol 17:77–81

Kuan SF, Rust K, Crouch E (1992) Interactions of surfactant protein D with bacterial lipopolysaccharides. Surfactant protein D is an Escherichia coli- binding protein in bronchoalveolar lavage. J Clin Invest 90:97–106

Kudo K, Sano H, Takahashi H et al (2004) Pulmonary collectins enhance phagocytosis of Mycobacterium avium through increased activity of mannose receptor. J Immunol 172:7592–7602

Kumánovics G, Minier T, Radics J et al (2008) Comprehensive investigation of novel serum markers of pulmonary fibrosis associated with systemic sclerosis and dermato/polymyositis. Clin Exp Rheumatol 26:414–420

Kunitake R, Kuwano K, Yoshida K et al (2001) KL-6, surfactant protein A and D in bronchoalveolar lavage fluid from patients with pulmonary sarcoidosis. Respiration 68:488–495

Kuroki Y, Tsutahara S, Shijubo N et al (1993) Elevated levels of lung surfactant protein A in sera from patients with idiopathic pulmonary fibrosis and pulmonary alveolar proteinosis. Am Rev Respir Dis 147:723–729

Kuroki Y, Takahashi H, Chiba H, Akino T (1998) Surfactant proteins A and D: disease markers. Biochim Biophys Acta 1408:334–345

Kuronuma K, Sano H, Kato K et al (2004) Pulmonary surfactant protein A augments the phagocytosis of Streptococcus pneumoniae by alveolar macrophages through a casein kinase 2-dependent increase of cell surface localization of scavenger receptor A. J Biol Chem 279:21421–21430

Lang JD, McArdle PJ, O'Reilly PJ et al (2002) Oxidant-antioxidant balance in acute lung injury. Chest 122(6 Suppl):314S–320S

Lazic T, Wyatt TA, Matic M, Meyerholz DK et al (2007) Maternal alcohol ingestion reduces surfactant protein A expression by preterm fetal lung epithelia. Alcohol 41:347–55

Lee JW, Ovadia B, Azakie A et al (2004) Increased pulmonary blood flow does not alter surfactant protein gene expression in lambs within the first week of life. Am J Physiol Lung Cell Mol Physiol 286:L1237–L1243

Lee HM, Kang HJ, Woo JS et al (2006) Upregulation of surfactant protein A in chronic rhinosinusitis. Laryngoscope 116:328–330

Lekkala M, LeVine AM, Linke MJ et al (2006) Effect of lung surfactant collectins on bronchoalveolar macrophage interaction with Blastomyces dermatitidis: inhibition of tumor necrosis factor alpha production by surfactant protein D. Infect Immun 74:4549–4556

Leth-Larsen R, Nordenbaek C, Tornoe I et al (2003) Surfactant protein D (SP-D) serum levels in patients with community-acquired pneumonia small star, filled. Clin Immunol 108:29–37

Leth-Larsen R, Garred P, Jensenius H et al (2005) A common polymorphism in the SFTPD gene influences assembly, function, and concentration of surfactant protein D. J Immunol 174:1532–1538

LeVine AM, Lotze A, Stanley S et al (1996) Surfactant content in children with inflammatory lung disease. Crit Care Med 246:1062–1067

LeVine AM, Whitsett JA, Gwozdz JA et al (2000) Distinct effects of surfactant protein A or D deficiency during bacterial infection on the lung. J Immunol 165:3934–3940

LeVine AM, Whitsett JA, Hartshorn KL et al (2001) Surfactant protein D enhances clearance of influenza A virus from the lung in vivo. J Immunol 167:5868–5873

LeVine AM, Elliott J, Whitsett JA et al (2004) Surfactant protein-D enhances phagocytosis and pulmonary clearance of respiratory syncytial virus. Am J Respir Cell Mol Biol 31:193–199

Lim BL, Wang JY, Holmskov U et al (1994) Expression of the carbohydrate recognition domain of lung surfactant protein D and demonstration of its binding to lipopolysaccharides of gram-negative bacteria. Biochem Biophys Res Commun 202:1674–1680

Lin Z, Pearson C, Chinchilli V et al (2000) Polymorphisms of human SP-A, SP-B, and SP-D genes: association of SP-B Thr131Ile with ARDS. Clin Genet 58:181–191

Linnoila RI, Mulshine JL, Steinberg SM, Gazdar AF (1992) Expression of surfactant-associated protein in non-small-cell lung cancer: a discriminant between biologic subsets. J Natl Cancer Inst Monogr 13:61–66

Liu CF, Chen YL, Chang WT et al (2005a) Mite allergen induces nitric oxide production in alveolar macrophage cell lines via CD14/toll-like receptor 4, and is inhibited by surfactant protein D. Clin Exp Allergy 35:1615–1624

Liu CF, Chen YL, Shieh CC et al (2005) Therapeutic effect of surfactant protein D in allergic inflammation of mite-sensitized mice. Clin Exp Allergy 35:515–521

Lopez JP, Clark E, Shepherd VL (2003) Surfactant protein A enhances *Mycobacterium avium* ingestion but not killing by rat macrophages. J Leukoc Biol 74:523–530

Lu SH, Ohtsuki Y, Nonami Y et al (2006) Ultrastructural study of nuclear inclusions immunohistochemically positive for surfactant protein A in pulmonary adenocarcinoma with special reference to their morphogenesis. Med Mol Morphol 39:214–220

Madan T, Eggleton P, Kishore U et al (1997a) Binding of pulmonary surfactant proteins A and D to *Aspergillus fumigatus* conidia enhances phagocytosis and killing by human neutrophils and alveolar macrophages. Infect Immun 65:3171–3179

Madan T, Kishore U, Shah A et al (1997b) Lung surfactant proteins A and D can inhibit specific IgE binding to the allergens of *Aspergillus fumigatus* and block allergen-induced histamine release from human basophils. Clin Exp Immunol 110:241–249

Madan T, Saxena S, Murthy KJ et al (2002) Association of polymorphisms in the collagen region of human SP-A1 and SP-A2 genes with pulmonary tuberculosis in Indian population. Clin Chem Lab Med 40:1002–1009

Madan T, Kaur S, Saxena S et al (2005) Role of collectins in innate immunity against aspergillosis. Med Mycol 43(Suppl 1):S155–S163

Maeda M, Ichiki Y, Aoyama Y et al (2001) Surfactant protein D (SP-D) and systemic scleroderma (SSc). J Dermatol 28:467–474

Mahajan L, Madan T, Kamal N et al (2008) Recombinant surfactant protein-D selectively increases apoptosis in eosinophils of allergic asthmatics and enhances uptake of apoptotic eosinophils by macrophages. Int Immunol 20:993–1007

Malhotra R, Haurum JS, Thiel S, Sim RB (1994) Binding of human collectins (SP-A and MBP) to influenza virus. Biochem J 304:455–461

Malik S, Greenwood CM, Eguale T et al (2006) Variants of the SFTPA1 and SFTPA2 genes and susceptibility to tuberculosis in Ethiopia. Hum Genet 118:752–759

Malloy JL, Veldhuizen RA, Thibodeaux BA et al (2005) *Pseudomonas aeruginosa* protease IV degrades surfactant proteins and inhibits surfactant host defense and biophysical functions. Am J Physiol Lung Cell Mol Physiol 288:L409–L418

Marttila R, Haataja R, Guttentag S et al (2003a) Surfactant protein A and B genetic variants in respiratory distress syndrome in singletons and twins. Am J Respir Crit Care Med 168:1216–1222

Marttila R, Haataja R, Rämet M et al (2003b) Surfactant protein A gene locus and respiratory distress syndrome in Finnish premature twin pairs. Ann Med 35:344–352

Matalon S, Shrestha K, Kirk M et al (2009) Modification of surfactant protein D by reactive oxygen-nitrogen intermediates is accompanied by loss of aggregating activity, in vitro and in vivo. FASEB J 23(5):1415–1430

Matsuno Y, Satoh H, Ishikawa H et al (2006) Simultaneous measurements of KL-6 and SP-D in patients undergoing thoracic radiotherapy. Med Oncol 23:75–82

McCormack FX, King TE Jr, Bucher BL et al (1995) Surfactant protein A predicts survival in idiopathic pulmonary fibrosis. Am J Respir Crit Care Med 152:751–759

McMahan MJ, Mimouni F, Miodovnik M et al (1987) Surfactant associated protein (SAP-35) in amniotic fluid from diabetic and nondiabetic pregnancies. Obstet Gynecol 70:94–8

Meloni F, Salvini R, Bardoni AM et al (2007) Bronchoalveolar lavage fluid proteome in bronchiolitis obliterans syndrome: possible role for surfactant protein A in disease onset. J Heart Lung Transplant 26:1135–1143

Meschi J, Crouch EC, Skolnik P et al (2005) Surfactant protein D binds to human immunodeficiency virus (HIV) envelope protein gp120 and inhibits HIV replication. J Gen Virol 86:3097–3107

Meyerholz DK, Kawashima K, Gallup JM, Grubor B, Ackermann MR (2006) Expression of select immune genes (surfactant proteins A and D, sheep beta defensin 1, and Toll-like receptor 4) by respiratory epithelia is developmentally regulated in the preterm neonatal lamb. Dev Comp Immunol 30:1060–1069

Miakotina OL, Goss KL, Snyder JM (2002) Insulin utilizes the PI 3-kinase pathway to inhibit SP-A gene expression in lung epithelial cells. Respir Res 3:27

Miyata M, Sakuma F, Fukaya E et al (2002) Detection and monitoring of methotrexate-associated lung injury using serum markers KL-6 and SP-D in rheumatoid arthritis. Intern Med 41:467–473

Müller B, Barth P, von Wichert P (1992) Structural and functional impairment of surfactant protein A after exposure to nitrogen dioxide in rats. Am J Physiol 263(2 Pt 1):L177–84

Mutti A, Corradi M, Goldoni M et al (2006) Exhaled metallic elements and serum pneumoproteins in asymptomatic smokers and patients with COPD or asthma. Chest 129:1288–1297

Mysore MR, Margraf LR, Jaramillo MA et al (1998) Surfactant protein A is decreased in a rat model of congenital diaphragmatic hernia. Am J Respir Crit Care Med 157:654–657

Nakamura N, Miyagi E, Murata S et al (2002) Expression of thyroid transcription factor-1 in normal and neoplastic lung tissues. Mod Pathol 15:1058–67

Nakamura M, Ogura T, Miyazawa N et al (2007) Outcome of patients with acute exacerbation of idiopathic interstitial fibrosis (IPF) treated with sivelestat and the prognostic value of serum KL-6 and surfactant protein D. Nihon Kokyuki Gakkai Zasshi 45:455–459

Nakane T, Nakamae H, Kamoi H et al (2008) Prognostic value of serum surfactant protein D level prior to transplant for the development of bronchiolitis obliterans syndrome and idiopathic pneumonia syndrome following allogeneic hematopoietic stem cell transplantation. Bone Marrow Transplant 42:43–49

Narasaraju TA, Jin N, Narendranath CR et al (2003) Protein nitration in rat lungs during hyperoxia exposure: a possible role of myeloperoxidase. Am J Physiol Lung Cell Mol Physiol 285:L1037–L1045

Noah TL, Murphy PC, Alink JJ et al (2003) Bronchoalveolar lavage fluid surfactant protein-A and surfactant protein-D are inversely related to inflammation in early cystic fibrosis. Am J Respir Crit Care Med 168:685–691

Nomori H, Horio H, Fuyuno G et al (1998) Serum surfactant protein A levels in healthy individuals are increased in smokers. Lung 176:355–361

Ochs M, Knudsen L, Allen L, Stumbaugh A, Levitt S, Nyengaard JR, Hawgood S (2004) GM-CSF mediates alveolar epithelial type II cell changes, but not emphysema-like pathology, in SP-D-deficient mice. Am J Physiol Lung Cell Mol Physiol 287:L1333–L1341

Ofek I, Mesika A, Kalina M et al (2001) Surfactant protein D enhances phagocytosis and killing of unencapsulated phase variants of Klebsiella pneumoniae. Infect Immun 69:24–33

Ogawa N, Shimoyama K, Kawabata H et al (2003) Clinical significance of serum KL-6 and SP-D for the diagnosis and treatment of interstitial lung disease in patients with diffuse connective tissue disorders. Ryumachi 43:19–28

Ohbayashi H, Adachi M (2008) Hydrofluoroalkane-beclomethasone dipropionate effectively improves airway eosinophilic inflammation including the distal airways of patients with mild to moderate persistent asthma as compared with fluticasone propionate in a randomized open double-cross study. Allergol Int 57:231–239

Ohlmeier S, Vuolanto M, Toljamo T et al (2008) Proteomics of human lung tissue identifies surfactant protein A as a marker of chronic obstructive pulmonary disease. J Proteome Res 7:5125–5132

Ohnishi H, Yokoyama A, Kondo K et al (2002) Comparative study of KL-6, surfactant protein-A, surfactant protein-D, and monocyte chemoattractant protein-1 as serum markers for interstitial lung diseases. Am J Respir Crit Care Med 165:378–381

Ohtsuki Y, Nakanishi N, Fujita J et al (2007) Immunohistochemical distribution of SP-D, compared with that of SP-A and KL-6, in interstitial pneumonias. Med Mol Morphol 40:163–167

Ohtsuki Y, Kobayashi M, Yoshida S et al (2008) Immunohistochemical localisation of surfactant proteins A and D, and KL-6 in pulmonary alveolar proteinosis. Pathology 40:536–539

O'Riordan DM, Standing JE, Kwon KY et al (1995) Surfactant protein D interacts with *Pneumocystis carinii* and mediates organism adherence to alveolar macrophages. J Clin Invest 95:2699–2710

Pasula R, Wright JR, Kachel DL et al (1999) Surfactant protein A suppresses reactive nitrogen intermediates by alveolar macrophages in response to *Mycobacterium tuberculosis*. J Clin Invest 103:483–490

Pavlovic J, Papagaroufalis C, Xanthou M et al (2006) Genetic variants of surfactant proteins A, B, C, and D in bronchopulmonary dysplasia. Dis Markers 22:277–291

Pettigrew MM, Gent JF, Zhu Y et al (2006) Association of surfactant protein A polymorphisms with otitis media in infants at risk for asthma. BMC Med Genet 7:69

Pettigrew MM, Gent JF, Zhu Y et al (2007) Respiratory symptoms among infants at risk for asthma: association with surfactant protein A haplotypes. BMC Med Genet 8:15

Phelps DS, Umstead TM, Mejia M et al (2004) Increased surfactant protein-A levels in patients with newly diagnosed idiopathic pulmonary fibrosis. Chest 125:617–625

Pilling AM, Mifsud NA, Jones SA et al (1999) Expression of surfactant protein mRNA in normal and neoplastic lung of B6C3F1 mice as demonstrated by in situ hybridization. Vet Pathol 36:57–63

Platt JA, Kraipowich N, Villafane F et al (2002) Alveolar type II cells expressing jaagsiekte sheep retrovirus capsid protein and surfactant proteins are the predominant neoplastic cell type in ovine pulmonary adenocarcinoma. Vet Pathol 39:341–352

Pryhuber GS, Hull WM, Fink I et al (1991) Ontogeny of surfactant proteins A and B in human amniotic fluid as indices of fetal lung maturity. Pediatr Res 30:597–605

Qi ZL, Xiao L, Gao YT et al (2002) Expression and clinical significance of surfactant protein D mRNA in peripheral blood of lung cancer patients. Ai Zheng 21:772–775, Abstract

Ragas A, Roussel L, Puzo G et al (2007) The *Mycobacterium tuberculosis* cell-surface glycoprotein Apa as a potential adhesin to colonize target cells via the innate immune system pulmonary C-type lectin surfactant protein A. J Biol Chem 282:5133–5142

Rämet M, Haataja R, Marttila R et al (2000) Association between the surfactant protein A (SP-A) gene locus and respiratory-distress syndrome in the Finnish population. Am J Hum Genet 66:1569–1579

Reed JA, Ikegami M, Robb L et al (2000) Distinct changes in pulmonary surfactant homeostasis in common β-chain- and GM-CSF-deficient mice. Am J Physiol Lung Cell Mol Physiol 278:L1164–L1171

Restrepo CI, Dong Q, Savov J et al (1999) Surfactant protein D stimulates phagocytosis of *Pseudomonas aeruginosa* by alveolar macrophages. Am J Respir Cell Mol Biol 21:576–585

Rivière M, Moisand A, Lopez A, Puzo G (2004) Highly ordered supramolecular organization of the mycobacterial lipoarabinomannans in solution. Evidence of a relationship between supra-molecular organization and biological activity. J Mol Biol 344:907–919

Rossi G, Cavazza A, Sturm N et al (2003) Pulmonary carcinomas with pleomorphic, sarcomatoid, or sarcomatous elements: a clinicopathologic and immunohistochemical study of 75 cases. Am J Surg Pathol 27:311–324

Rubio F, Cooley J, Accurso FJ et al (2004) Linkage of neutrophil serine proteases and decreased surfactant protein-A (SP-A) levels in inflammatory lung disease. Thorax 59:318–323

Saitoh H, Shimura S, Fushimi T et al (1997) Detection of surfactant protein-A gene transcript in the cells from pleural effusion for the diagnosis of lung adenocarcinoma. Am J Med 103:400–404

Samten B, Townsend JC, Sever-Chroneos Z et al (2008) An antibody against the surfactant protein A (SP-A)-binding domain of the SP-A receptor inhibits T cell-mediated immune responses to *Mycobacterium tuberculosis*. J Leukoc Biol 84:115–123

Saxena S, Madan T, Shah A et al (2003) Association of polymorphisms in the collagen region of SP-A2 with increased levels of total IgE antibodies and eosinophilia in patients with allergic bronchopulmonary aspergillosis. J Allergy Clin Immunol 111:1001–1007

Saxena S, Kumar R, Madan T et al (2005) Association of polymorphisms in pulmonary surfactant protein A1 and A2 genes with high-altitude pulmonary edema. Chest 128:1611–1619

Scanlon ST, Milovanova T, Kierstein S et al (2005) Surfactant protein-A inhibits *Aspergillus fumigatus*-induced allergic T-cell responses. Respir Res 6:97

Schaeffer LM, McCormack FX, Wu H, Weiss AA (2004) Interactions of pulmonary collectins with *Bordetella bronchiseptica* and *Bordetella pertussis* lipopolysaccharide elucidate the structural basis of their antimicrobial activities. Infect Immun 72:7124–7130

Schaub B, Westlake RM, He H et al (2004) Surfactant protein D deficiency influences allergic immune responses. Clin Exp Allergy 34:1819–1826

Schelenz S, Malhotra R, Sim RB et al (1995) Binding of host collectins to the pathogenic yeast Cryptococcus neoformans: human surfactant protein D acts as an agglutinin for acapsular yeast cells. Infect Immun 63:3360–3366

Schmiedl A, Lührmann A, Pabst R, Koslowski R (2008) Increased surfactant protein A and D expression in acute ovalbumin-induced allergic airway inflammation in Brown Norway rats. Int Arch Allergy Immunol 148:118–126

Seifart C, Lin HM, Seifart U et al (2005) Rare SP-A alleles and the SP-A1-6A[4] allele associate with risk for lung carcinoma. Clin Genet 68:128–136

Selman M, Lin HM, Montaño M et al (2003) Surfactant protein A and B genetic variants predispose to idiopathic pulmonary fibrosis. Hum Genet 113:542–550

Shijubo N, Tsutahara S, Hirasawa M, Takahashi H, Honda Y, Suzuki A, Kuroki Y, Akino T (1992) Pulmonary surfactant protein A in pleural effusions. Cancer 69:2905–2909

Shijubo N, Honda Y, Itoh Y et al (1998) BAL surfactant protein A and Clara cell 10-kDa protein levels in healthy subjects. Lung 176:257–265

Shima H, Guarino N, Puri P (2000) Effect of hyperoxia on surfactant protein gene expression in hypoplastic lung in nitrofen-induced diaphragmatic hernia in rats. Pediatr Surg Int 16:473–477

Shimoya K, Taniguchi T, Matsuzaki N et al (2000) Chorioamnionitis decreased incidence of respiratory distress syndrome by elevating fetal interleukin-6 serum concentration. Hum Reprod 15: 2234–2240

Shimura S, Masuda T, Takishima T et al (1996) Surfactant protein-A concentration in airway secretions for the detection of pulmonary oedema. Eur Respir J 9:2525–2530

Shu LH, Xue XD, Shu LH et al (2007) Effect of dexamethasone on the content of pulmonary surfactant protein D in young rats with acute lung injury induced by lipopolysaccharide. Zhongguo Dang Dai Er Ke Za Zhi 9:155–158

Sidobre S, Nigou J, Puzo G, Rivière M (2000) Lipoglycans are putative ligands for the human pulmonary surfactant protein A attachment to mycobacteria. Critical role of the lipids for lectin-carbohydrate recognition. J Biol Chem 275:2415–2422

Sidobre S, Puzo G, Rivière M (2002) Lipid-restricted recognition of mycobacterial lipoglycans by human pulmonary surfactant protein A: a surface-plasmon-resonance study. Biochem J 365:89–97

Simpson JL, Grissell TV, Douwes J et al (2007) Innate immune activation in neutrophilic asthma and bronchiectasis. Thorax 62:211–219

Sims MW, Tal-Singer RM, Kierstein S et al (2008) Chronic obstructive pulmonary disease and inhaled steroids alter surfactant protein D (SP-D) levels: a cross-sectional study. Respir Res 9:13

Sin DD, Man SF, Marciniuk DD et al (2008a) The effects of fluticasone with or without salmeterol on systemic biomarkers of inflammation in chronic obstructive pulmonary disease. Am J Respir Crit Care Med 177:1207–1214

Sin DD, Man SF, McWilliams A, Lam S (2008b) Surfactant protein D and bronchial dysplasia in smokers at high risk of lung cancer. Chest 134:582–588

Sin DD, Pahlavan PS, Man SF (2008c) Surfactant protein D: a lung specific biomarker in COPD? Ther Adv Respir Dis 2:65–74

Singh M, Madan T, Waters P et al (2003) Protective effects of a recombinant fragment of human surfactant protein D in a murine model of pulmonary hypersensitivity induced by dust mite allergens. Immunol Lett 86:299–307

Skinner ML, Schlosser RJ, Lathers D et al (2007) Innate and adaptive mediators in cystic fibrosis and allergic fungal rhinosinusitis. Am J Rhinol 21:538–541

Snyder JM, Kwun JE, O'Brien JA et al (1988) The concentration of the 35-kDa surfactant apoprotein in amniotic fluid from normal and diabetic pregnancies. Pediatr Res 24:728–734

Sorensen GL, Madsen J, Kejling K et al (2006) Surfactant protein D is proatherogenic in mice. Am J Physiol Heart Circ Physiol 290: H2286–H2294

Sorensen GL, Husby S, Holmskov U (2007) Surfactant protein A and surfactant protein D variation in pulmonary disease. Immunobiology 212:381–416

Stamme C, Walsh E, Wright JR (2000) Surfactant protein A differentially regulates IFN-γ- and LPS-induced nitrite production by rat alveolar macrophages. Am J Respir Cell Mol Biol 23:772–779

Starosta V, Griese M (2006) Oxidative damage to surfactant protein D in pulmonary diseases. Free Radic Res 40:419–425

Stevens PA, Schadow B, Bartholain S et al (1992) Surfactant protein A in the course of respiratory distress syndrome. Eur J Pediatr 151:596–600

Stoffers M, Goldmann T, Branscheid D et al (2004) Transcriptional activity of surfactant-apoproteins A1 and A2 in non small cell lung carcinomas and tumor-free lung tissues. Pneumologie 58:395–399 [Article in German]

Stokes RW, Thorson LM, Speert DP (1998) Nonopsonic and opsonic association of *Mycobacterium tuberculosis* with resident alveolar macrophages is inefficient. J Immunol 160:5514–5521

Strayer DS, Merritt TA, Hallman M (1995) Levels of SP-A-anti-SP-A immune complexes in neonatal respiratory distress syndrome correlate with subsequent development of bronchopulmonary dysplasia. Acta Paediatr 84:128–131

Stray-Pedersen A, Vege A, Opdal SH et al (2009) Surfactant protein A and D gene polymorphisms and protein expression in victims of sudden infant death. Acta Paediatr 98:62–69

Stuart RB, Ovadia B, Suzara VV et al (2003) Inhaled nitric oxide increases surfactant protein gene expression in the intact lamb. Am J Physiol Lung Cell Mol Physiol 285:L628–L633

Suematsu E, Miyamura T, Shimada H et al (2003) Assessment of serum markers KL-6 and SP-D for interstitial pneumonia associated with connective tissue diseases. Ryumachi 43:11–18

Sugahara K, Iyama K, Sano K et al (1994) Differential expressions of surfactant protein SP-A, SP-B, and SP-C mRNAs in rats with streptozotocin-induced diabetes demonstrated by in situ hybridization. Am J Respir Cell Mol Biol 11:397–404

Suzuki A, Shijubo N, Yamada G et al (2005) Napsin A is useful to distinguish primary lung adenocarcinoma from adenocarcinomas of other organs. Pathol Res Pract 20:579–586

Swenson ER, Maggiorini M, Mongovin S et al (2002) Pathogenesis of high-altitude pulmonary edema: inflammation is not an etiologic factor. JAMA 287:2228–2235

Tacken PJ, Hartshorn KL, White MR et al (2004) Effective targeting of pathogens to neutrophils via chimeric surfactant protein D/anti-CD89 protein. J Immunol 172:4934–4940

Tacken PJ, Batenburg JJ (2006) Monocyte CD64 or CD89 targeting by surfactant protein D/anti-Fc receptor mediates bacterial uptake. Immunology. 117:494–501

Takahashi H, Fujishima T, Koba H et al (2000) Serum surfactant proteins A and D as prognostic factors in idiopathic pulmonary fibrosis and their relationship to disease extent. Am J Respir Crit Care Med 162:1109–1114

Takahashi H, Imai Y, Fujishima T et al (2001) Diagnostic significance of surfactant proteins A and D in sera from patients with radiation pneumonitis. Eur Respir J 17:481–487

Takahashi H, Chiba H, Shiratori M et al (2006a) Elevated serum surfactant protein A and D in pulmonary alveolar microlithiasis. Respirology 11:330–333

Takahashi H, Shiratori M, Kanai A et al (2006b) Monitoring markers of disease activity for interstitial lung diseases with serum surfactant proteins A and D. Respirology 11:S51–S54

Takami T, Kumada A, Takei Y et al (2003) A case of Wilson-Mikity syndrome with high serum KL-6 levels. J Perinatol 23:56–58

Takeda K, Miyahara N, Rha YH, Gelfand EW et al (2003) Surfactant protein D regulates airway function and allergic inflammation through modulation of macrophage function. Am J Respir Crit Care Med 168:783–789

Tecle T, White MR, Gantz D et al (2007) Human neutrophil defensins increase neutrophil uptake of influenza A virus and bacteria and modify virus-induced respiratory burst responses. J Immunol 178:8046–8052

Tecle T, White MR, Sorensen G et al (2008) Critical role for cross-linking of trimeric lectin domains of surfactant protein D in antiviral activity against influenza A virus. Biochem J 412:323–329 ter Horst SA, Fijlstra M, Sengupta S et al (2006) Spatial and temporal expression of surfactant proteins in hyperoxia-induced neonatal rat lung injury. BMC Pulm Med 6:9

Thomas NJ, Fan R, Diangelo S et al (2007) Haplotypes of the surfactant protein genes A and D as susceptibility factors for the development of respiratory distress syndrome. Acta Paediatr 96:985–989

Trinder PK, Hickling TP, Sim RB et al (2000) Humoral autoreactivity directed against surfactant protein-A (SP-A) in rheumatoid arthritis synovial fluids. Clin Exp Immunol 120:183–187

Tsutsumida H, Nomoto M, Goto M et al (2007) A micropapillary pattern is predictive of a poor prognosis in lung adenocarcinoma, and reduced surfactant apoprotein A expression in the micropapillary pattern is an excellent indicator of a poor prognosis. Mod Pathol 20:638–647

Ueno T, Linder S, Elmberger G (2003) Aspartic proteinase napsin is a useful marker for diagnosis of primary lung adenocarcinoma. Br J Cancer 88:1229–1233

Uzaslan E, Stuempel T, Ebsen M et al (2005) Surfactant protein A detection in primary pulmonary adenocarcinoma without bronchioloalveolar pattern. Respiration 72:249–253

Uzaslan E, Ebsen M, Stuempel T et al (2006) Surfactant protein A detection in large cell carcinoma of the lung. Appl Immunohistochem Mol Morphol 14:88–90

Vaid M, Kaur S, Sambatakou H et al (2007) Distinct alleles of mannose-binding lectin (MBL) and surfactant proteins A (SP-A) in patients with chronic cavitary pulmonary aspergillosis and allergic bronchopulmonary aspergillosis. Clin Chem Lab Med 45:183–186

van de Graaf EA, Jansen HM, Lutter R et al (1992) Surfactant protein A in bronchoalveolar lavage fluid. J Lab Clin Med 120:252–263

van de Wetering JK, Coenjaerts FE et al (2004) Aggregation of Cryptococcus neoformans by surfactant protein D is inhibited by its capsular component glucuronoxylomannan. Infect Immun 72:145–153

van Iwaarden JF, Pikaar JC, Storm J et al (1994) Binding of surfactant protein A to the lipid A moiety of bacterial lipopolysaccharides. Biochem J 303 (Pt 2):407–411

van Rozendaal BA, van de Lest CH, van Eijk M et al (1999) Aerosolized endotoxin is immediately bound by pulmonary surfactant protein D in vivo. Biochim Biophys Acta 1454:261–269

van Rozendaal BA, van Spriel AB, van De Winkel JG, Haagsman HP (2000) Role of pulmonary surfactant protein D in innate defense against Candida albicans. J Infect Dis 182:917–922

von Bredow C, Birrer P, Griese M (2001) Surfactant protein A and other bronchoalveolar lavage fluid proteins are altered in cystic fibrosis. Eur Respir J 17:716–722

von Bredow C, Wiesener A, Griese M (2003) Proteolysis of surfactant protein D by cystic fibrosis relevant proteases. Lung 181:79–88

Voss T, Schäfer KP, Nielsen PF et al (1992) Primary structure differences of human surfactant-associated proteins isolated from normal and proteinosis lung. Biochim Biophys Acta 1138:261–267

Vuk-Pavlovic Z, Standing JE, Crouch EC et al (2001) Carbohydrate recognition domain of surfactant protein D mediates interactions with Pneumocystis carinii glycoprotein A. Am J Respir Cell Mol Biol 24:475–484

Walenkamp AM, Verheul AF, Scharringa J, Hoepelman IM (1999) Pulmonary surfactant protein A binds to Cryptococcus neoformans without promoting phagocytosis. Eur J Clin Invest 29:83–92

Wang JY, Reid KB (2007) The immunoregulatory roles of lung surfactant collectins SP-A, and SP-D, in allergen-induced airway inflammation. Immunobiology 212:417–425

Wang JY, Kishore U, Lim BL et al (1996) Interaction of human lung surfactant proteins A and D with mite (Dermatophagoides pteronyssinus) allergens. Clin Exp Immunol 106:367–373

Wang JY, Shieh CC, You PF et al (1998) Inhibitory effect of pulmonary surfactant proteins A and D on allergen-induced lymphocyte proliferation and histamine release in children with asthma. Am J Respir Crit Care Med 158:510–518

Wang JY, Shieh CC, Yu CK, Lei HY (2001) Allergen-induced bronchial inflammation is associated with decreased levels of surfactant proteins A and D in a murine model of asthma. Clin Exp Allergy 31:652–662

Wang SX, Liu P, Wei MT et al (2007) Roles of serum clara cell protein 16 and surfactant protein-D in the early diagnosis and progression of silicosis. J Occup Environ Med 49:834–839

Weber B, Borkhardt A, Stoll-Becker S et al (2000) Polymorphisms of surfactant protein A genes and the risk of bronchopulmonary dysplasia in preterm infants. Turk J Pediatr 42:181–185

Wert SE, Yoshida M, LeVine AM et al (2000) Increased metalloproteinase activity, oxidant production, and emphysema in surfactant protein D gene-inactivated mice. Proc Natl Acad Sci USA 97:5972–5977

White MR, Crouch E, Chang D et al (2000) Enhanced antiviral and opsonic activity of a human mannose-binding lectin and surfactant protein D chimera. J Immunol 165:2108–2115

Woodworth BA, Wood R, Baatz JE et al (2007) Sinonasal surfactant protein A1, A2, and D gene expression in cystic fibrosis: a preliminary report. Otolaryngol Head Neck Surg 137:34–38

Wootten CT, Labadie RF, Chen A, Lane KF (2006) Differential expression of surfactant protein A in the nasal mucosa of patients with allergy symptoms. Arch Otolaryngol Head Neck Surg 132:1001–1007

Wright JR (2005) Immunoregulatory functions of surfactant proteins. Nat Rev Immunol 5:58–69

Wright JR, Zlogar DF, Taylor JC et al (1999) Effects of endotoxin on surfactant protein A and D stimulation of NO production by alveolar macrophages. Am J Physiol 276:L650–L658

Wuenschell CW, Zhao J, Tefft JD et al (1998) Nicotine stimulates branching and expression of SP-A and SP-C mRNAs in embryonic mouse lung culture. Am J Physiol 274:L165–L170

Xie JG, Xu YJ, Zhang ZX et al (2005) Surfactant protein A gene polymorphisms in chronic obstructive pulmonary disease. Zhonghua Yi Xue Yi Chuan Xue Za Zhi 22:91–93, Article in Chinese

Yanaba K, Hasegawa M, Takehara K, Sato S (2004) Comparative study of serum surfactant protein-D and KL-6 concentrations in patients with systemic sclerosis as markers for monitoring the activity of pulmonary fibrosis. J Rheumatol 31:1112–1120

Yang S, Panoskaltsis-Mortari A, Ingbar DH et al (2000) Cyclophosphamide prevents systemic keratinocyte growth factor-induced up-regulation of surfactant protein A after allogeneic transplant in mice. Am J Respir Crit Care Med 162:1884–1890

Yang S, Milla C, Panoskaltsis-Mortari A et al (2002) Surfactant protein A decreases lung injury and mortality after murine marrow transplantation. Am J Respir Cell Mol Biol 27:297–305

Yogalingam G, Doyle IR, Power JH (1996) Expression and distribution of surfactant proteins and lysozyme after prolonged hyperpnea. Am J Physiol 270:L320–L330

Yong SJ, Vuk-Pavlovic Z, Standing JE et al (2003) Surfactant protein D-mediated aggregation of Pneumocystis carinii impairs phagocytosis by alveolar macrophages. Infect Immun 71:1662–1671

Yoshida M, Whitsett JA (2006) Alveolar macrophages and emphysema in surfactant protein-D-deficient mice. Respirology 11(Suppl):S37–S40

Yoshida M, Korfhagen TR, Whitsett JA (2001) Surfactant protein D regulates NF-kB and matrix metalloproteinase production in alveolar macrophages via oxidant-sensitive pathways. J Immunol 166:7514–7519

Yurdakök M (2004) Inherited disorders of neonatal lung diseases. Turk J Pediatr 46:105–114

Zamecnik J, Kodet R (2002) Value of thyroid transcription factor-1 and surfactant apoprotein A in the differential diagnosis of pulmonary carcinomas: a study of 109 cases. Virchows Arch 440: 353–61

Zhai L, Wu HM, Wei KL et al (2008) Genetic polymorphism of surfactant protein A in neonatal respiratory distress syndrome. Zhongguo Dang Dai Er Ke Za Zhi 10:295–298, Article in Chinese

Zhang L, Ikegami M, Dey CR et al (2002) Reversibility of pulmonary abnormalities by conditional replacement of surfactant protein D (SP-D) in vivo. J Biol Chem 277:38709–38713

Zhang F, Pao W, Umphress SM et al (2003) Serum levels of surfactant protein D are increased in mice with lung tumors. Cancer Res 63:5889–5894

Zhao J, Yeong LH, Wong WS (2007) Dexamethasone alters bronchoalveolar lavage fluid proteome in a mouse asthma model. Int Arch Allergy Immunol 142:219–229

Zhu BL, Ishida K, Quan L et al (2001) Immunohistochemistry of pulmonary surfactant-associated protein A in acute respiratory distress syndrome. Leg Med (Tokyo) 3:134–140

Selectins and Associated Adhesion Proteins in Inflammatory disorders

G.S. Gupta

44.1 Inflammation

Inflammation is defined as the normal response of living tissue to injury or infection. It is important to emphasize two components of this definition. First, that inflammation is a normal response and, as such, is expected to occur when tissue is damaged. Infact, if injured tissue does not exhibit signs of inflammation this would be considered abnormal and wounds and infections would never heal without inflammation. Secondly, inflammation occurs in living tissue, hence there is need for an adequate blood supply to the tissues in order to exhibit an inflammatory response. The inflammatory response may be triggered by mechanical injury, chemical toxins, and invasion by microorganisms, and hypersensitivity reactions. Three major events occur during the inflammatory response: the blood supply to the affected area is increased substantially, capillary permeability is increased, and leucocytes migrate from the capillary vessels into the surrounding interstitial spaces to the site of inflammation or injury. The inflammatory response represents a complex biological and biochemical process involving cells of the immune system and a plethora of biological mediators. Cell-to-cell communication molecules such as cytokines play an extremely important role in mediating the process of inflammation. Inflammation and platelet activation are critical phenomena in the setting of acute coronary syndromes. An extensive exposition of this complex phenomenon is beyond the scope of this article (Rankin 2004).

Inflammation can be classified as either acute or chronic. Acute inflammation is the initial response of the body to harmful stimuli and is achieved by the increased movement of plasma and leukocytes (especially granulocytes) from the blood into the injured tissues. A cascade of biochemical events propagates and matures the inflammatory response, involving the local vascular system, the immune system, and various cells within the injured tissue. Prolonged inflammation, known as chronic inflammation, leads to a progressive shift in the type of cells present at the site of inflammation and is characterized by simultaneous destruction and healing of the tissue from the inflammatory process. However, chronic inflammation can also lead to a host of diseases, such as hay fever, atherosclerosis, rheumatoid arthritis, and even cancer (e.g., gallbladder carcinoma). It is for that reason that inflammation is normally closely regulated by the body. Acute and chronic inflammation differ in matter of causative agent, major cells involved, primary mediators, onset, duration and final outcomes. Generally speaking, acute inflammation is mediated by granulocytes, while chronic inflammation is mediated by mononuclear cells such as monocytes and lymphocytes.

44.2 Cell Adhesion Molecules

Cell adhesion molecules are glycoproteins expressed on the cell surface and play an important role in inflammatory as well as neoplastic diseases. There are four main groups: the integrin family, the immunoglobulin superfamily, selectins, and cadherins. The integrin family has eight subfamilies, designated as β1 through β8. The immunoglobulin superfamily includes leukocyte function antigen-2 (LFA-2 or CD2), leukocyte function antigen-3 (LFA-3 or CD58), intercellular adhesion molecules (ICAMs), vascular adhesion molecule-1 (VCAM-1), platelet-endothelial cell adhesion molecule-1 (PE-CAM-1), and mucosal addressin cell adhesion molecule-1 (MAdCAM-1). The selectin family includes L-selectin (CD62L), P-selectin (CD62P), and E-selectin (CD62E). Cadherins are major cell-cell adhesion molecules and include epithelial (E), placental (P), and neural (N) subclasses. The binding sites (ligands/receptors) are different for each of these cell adhesion molecules (e.g., ICAM binds to CD11/CD18; VCAM-1 binds to VLA-4). The specific cell adhesion molecules and their ligands that may be involved in pathologic conditions and potential therapeutic strategies by modulating the expression of these molecules have been discussed (Elangbam et al. 1997). Most adhesion molecules play fairly broad roles in the generation of immune responses. The three selectins act in concert with other cell adhesion molecules e.g., intracellular adhesion molecule (ICAM-1), vascular cell adhesion molecule-1 (VCAM-1), and leukocyte integrins to effect adhesive interactions of leukocytes, platelets, and endothelial cells. The structure and functions of selectins, which belong to C-type lectins family, have been reviewed in Chaps. 26, 27, and 28.

44.2.1 Selectins

The selectin family of lectins consists of three closely related cell-surface molecules with differential expression by leukocytes (L-selectin), platelets (P-selectin), and vascular endothelium (E- and P-selectin). Structural identity of a selectins resides in its unique domain composition (Chap. 26). E-, P-, and L-selectin are >60 % identical in their NH2 terminus of 120 amino acids, which represent the lectin domain (Chaps. 26, 27, and 28). The ligands (counter structures) of selectins are sialylated and fucosylated carbohydrate molecules which, in most cases, decorate mucin-like glycoprotein membrane receptors. Their common structure consists of an N-terminal Ca^{2+}-dependent lectin-type domain, an epidermal growth factor (EGF)-like domain, multiple short consensus repeat (SCR) domains similar to those found in complement regulatory proteins, a transmembrane region, and a short cytoplasmic C-terminal domain. Together this arrangement results in an elongated structure which projects from the cell surface, ideal for initiating interactions with circulating leucocytes. The lectin domain forms the main ligand binding site, interacting with a carbohydrate determinant typified by fucosylated, sialylated, and usually sulphated glycans such as sialyl Lewis X (s-Le^X). The EGF domain may also play a role in ligand recognition. The short consensus repeat (SCR) domains (two for L-selectin, six for E-selectin, and nine for P-selectin) probably act as spacer elements, ensuring optimum positioning of the lectin and EGF domains for ligand interaction. The EGF repeats have comparable sequence similarity. Each complement regulatory-like module is 60 amino acids in length and contains six cysteinyl residues capable of disulfide bond formation. This feature distinguishes the selectin modules from those found in complement binding proteins, such as complement receptors 1 and 2, which contain four cysteines (Chap. 26).

The selectins cell-surface receptors play a key role in the initial adhesive interaction between leukocytes and endothelial cells at sites of inflammation. Selectins (P, E and L) and their ligands (mainly P-selectin ligand) are involved in the rolling and tethering of leukocytes on the vascular wall. Activation of endothelial cells (EC) with different stimuli induces the expression of E- and P-selectins, and other adhesion molecules (ICAM-1, VCAM-1), involved in their interaction with circulating cells. Lymphocytes home to peripheral lymph nodes (PLNs) via high endothelial venules (HEVs) in the subcortex and incrementally larger collecting venules in the medulla. HEVs express ligands for L-selectin, which mediates lymphocyte rolling (Horstman et al. 2004). For structure and functions of selectins, the readers are advised to consult Chaps 26–28. In this chapter we will emphasize mainly on the role of selectins in inflammatory disorders including cancer.

44.3 Atherothrombosis

Atherothrombosis, defined as atherosclerotic plaque disruption with superimposed thrombosis, is the leading cause of mortality in the Western world. Atherosclerosis is a diffuse process that starts early in childhood and progresses asymptomatically through adult life. Later in life, it is clinically manifested as coronary artery disease (CAD), stroke, transient ischaemic attack (TIA), and peripheral arterial disease. From the clinical point of view, we should envision this disease as a single pathologic entity that affects different vascular territories. A suggestive analogy is that TIA and intermittent claudication are the unstable angina of the brain and lower limbs, respectively; and stroke and gangrene are the myocardial infarction. Circulating platelets display reversible interactions with atherosclerotic lesions. Atherosclerotic arterial disease is associated with an increased share of platelets unable to express P-selectin and an increased fraction of platelets that microaggregate in citrate anticoagulant. These platelet alterations are not completely explained by either focal arterial injury or abnormal rheology associated with arterial stenosis but appear to be an effect of the atherosclerotic process (McBane et al. 2004). The pathogenesis of arterial thrombotic disease involves multiple genetic and environmental factors related to atherosclerosis and thrombosis.

44.3.1 Venous Thrombosis

Venous Thrombosis is a world wide health problem in the general population. Injury to the endothelium leads to dysfunction. The causes of injury include lipids, immune complexes, microorganisms, smoking, hypertension, aging, diabetes mellitus and trauma. The selectins are thought to be largely responsible for the initial attachment and rolling of leukocytes on stimulated vascular endothelium. Platelet activation is an important process in the pathogenesis of atherothrombosis. Platelet adhesion, activation, and aggregation at the sites of vascular endothelial disruption caused by atherosclerosis are key events in arterial thrombus formation. Platelet tethering and adhesion to the arterial wall, particularly under high shear forces, are achieved through multiple high-affinity interactions between platelet membrane receptors (integrins) and ligands within the exposed subendothelium, most notably collagen and von Willebrand factor (vWF). Platelet adhesion to collagen occurs both indirectly, via binding of the platelet glycoprotein (GP) Ib-V-IX receptor to circulating vWF, which binds to exposed collagen, and directly, via interaction with platelet receptors GP VI and GP Ia/IIb. Platelet activation, initiated by exposed collagen and locally generated soluble platelet agonists (primarily thrombin, ADP, and thromboxane A2), provides the stimulus for the release of platelet-derived growth factors, adhesion molecules and coagulation factors, activation of adjacent platelets, and conformational changes in the platelet $\alpha(IIb)\beta3$ integrin (GP IIb/IIIa receptor). Platelet aggregation, mediated primarily by interaction between the activated platelet GP IIb/IIIa receptor and its ligands, fibrinogen and vWF, results in the formation of a platelet-rich thrombus (Steinhubl and Moliterno 2005).

44.3.2 Arterial Thrombosis

P-selectin expression in platelets is elevated in disorders associated with arterial thrombosis such as coronary artery disease, acute myocardial infarction, stroke, and peripheral artery disease. During thrombosis, P-selectin is expressed on the surface of activated endothelial cells and platelets. P-selectin mediates rolling of platelets and leukocytes on activated endothelial cells as well as interactions of platelets with leukocytes. Platelet P-selectin interacts with PSGL-1 on leukocytes to form platelet-leukocyte aggregates. Furthermore, this interaction of P-selectin with PSGL-1 induces the upregulation of tissue factor, several cytokines in leukocytes and the production of procoagulant microparticles, thereby contributing to a prothrombotic state. P-selectin is also involved in platelet-platelet interactions, i. e. platelet aggregation which is a major factor in arterial thrombosis. P-selectin interacts with platelet sulfatides, thereby stabilizing initial platelet aggregates formed by GPIIb/IIIa-fibrinogen bridges. Inhibtion of the P-selectin-sulfatide interaction leads to a reversal of platelet aggregation. Thus, P-selectin plays a significant role in platelet aggregation and platelet- leukocyte interactions, both important mechanisms in the development of arterial thrombosis. Following activation, P-selectin is rapidly translocated to the cell surface (Merten and Thiagarajan 2004; Wang et al. 2005).

44.3.3 Thrombogenesis in Atrial Fibrillation

Platelet activation occurs in peripheral blood of patients with rheumatic mitral stenosis (MS). The plasma levels of soluble P-selectin are elevated in permanent atrial fibrillation (AF) patients; the plasma levels of soluble P-selectin in the left atrium do not significantly differ from those in the right atrium, femoral vein, or femoral artery. The venous plasma levels of sP-selectin in patients with moderate-to-severe MS are significantly higher than those in healthy volunteers or patients with lone AF. In addition, in patients with MS, there was no difference in the plasma levels of sP-selectin between the left and right atrial blood and between peripheral and atrial blood. Moreover, there was no change in sP-selectin levels as a result of percutaneous transluminal mitral valvuloplasty (PTMV) (Chen et al. 2004). Lip et al. (2005) studied the relations of plasma vWf (an index of endothelial damage and dysfunction) and sP-selectin levels in relation to the presence and onset of clinical congestive heart failure (CHF) and degree of left ventricular dysfunction in patients taking part in SPAF (stroke prevention in AF). While plasma vWf was higher among patients with AF and CHF, plasma P-selectin concentrations were not affected by presence, onset, or severity of heart failure.

44.3.4 Atherosclerosis

Atherosclerosis is a complex chronic inflammatory disease of the arterial wall. Though the inflammatory nature of atherosclerosis has been established, the initial events that trigger this response in the arterial intima remain obscure. Studies reveal a significant rate of genomic alterations in human atheromas. The accumulation of genomic rearrangements in vascular endothelium and smooth muscle cells are important for disease development. It is well accepted that the induction of EC adhesion molecules is a critical component in acute inflammatory responses as well as allogeneic interactions in vascularized allografts and, possibly, atherogenesis. Inflammation and genetics are both prominent mechanisms in the pathogenesis of atherosclerosis and arterial thrombosis. Accordingly, population studies have explored the association of ischaemic heart disease with gene polymorphisms of the inflammatory molecules: tumor necrosis factors (TNF) α and β, transforming growth factors (TGF) $\beta1$ and 2, P and E selectins, and platelet endothelial cell adhesion molecule (PECAM) 1. The partly conflicting data provide some evidence that alterations in the genetics of the inflammatory system may modify the risk of ischaemic heart disease.

44.3.4.1 Formation of Reactive Oxygen Species (ROS) as an Initial Event

In recent years, reactive oxygen species (ROS) are considered as initial event in causing atherosclerosis. ROS are a family of molecules including molecular oxygen and its derivatives produced in all aerobic cells. Excessive production of ROS, outstripping endogenous antioxidant defense mechanisms, has been implicated in processes in which they oxidize biological macromolecules, such as DNA, protein, carbohydrates, and lipids. Many ROS possess unpaired electrons and thus are free radicals. These include molecules such as superoxide anion (O_2^-), hydroxyl radical (HO•), nitric oxide (NO•), and lipid radicals. Other reactive oxygen species, such as hydrogen peroxide ($H2O2$), peroxynitrite ($ONOO^-$), and hypochlorous acid (HOCl), are not free radicals per se but have oxidizing effects that contribute to oxidant stress. The cellular production of one ROS may lead to the production of several others via radical chain reactions. For example, reactions between radicals and polyunsaturated fatty acids within cell membrane may result in a fatty acid peroxyl radical (R-COO•) that can attack adjacent fatty acid side chains and initiate production of other lipid radicals. Lipid radicals produced in this chain reaction accumulate in the cell membrane and may have a myriad of effects on cellular function, including leakage of the plasmolemma and dysfunction of membrane-bound receptors. Of note, end products of lipid peroxidation, including unsaturated aldehydes and other metabolites, have cytotoxic and mutagenic properties. A decline in NO bioavailability may be caused by decreased expression of the endothelial cell NO synthase (eNOS), a lack of substrate or cofactors for eNOS (Fig. 44.1 and 44.2).

In mammalian cells, potential enzymatic sources of ROS include the mitochondrial respiration, arachidonic acid pathway enzymes lipoxygenase and cyclooxygenase, cytochrome p450s, xanthine oxidase, NADH/NADPH oxidases, NO synthase, peroxidases, and other hemoproteins. Although many of these sources could potentially produce ROS that inactivate NO•, 3 sources have been studied extensively in cardiovascular system. These include xanthine oxidase, NADH/NADPH oxidase, and NO synthase (Cai and Harrison 2000; Hamilton et al. 2004; Vijya Lakshmi et al. 2009).

44.3.4.2 CAMs as Predicators of Atherosclerosis

During initial step in atherosclerosis, there is rapid targeting of monocytes to the sites of inflammation and endothelial injury; the adhesion of leukocytes to activated endothelial cells is mediated by ICAM-1. The induction of EC adhesion molecules is a critical component in acute inflammatory responses as well as allogeneic interactions in vascularized allografts and, possibly, atherogenesis. The "inflammatory triad" of IL-1, TNF, and LPS are potent stimulators of the EC activation and adhesion molecules E-selectin or ELAM-1 (or also known as CD62E), ICAM-1 and VCAM-1. PECAM-1 plays also a key role in the transendothelial migration of circulating leukocytes (diapedesis) during vascular inflammation. ICAM-1 and VCAM-1 are inflammatory predicators of adverse prognosis in patients with acute coronary syndromes (ACS) (Postadzhiyan et al. 2008) (Fig. 44.2).

Levels of P-selectin are increased in the blood of patients with familial hypercholesterolemia (FH) in spite of long-term intensive extracorporeal LDL-elimination, documenting the activity of atherosclerosis. Low levels of P-selectin and MCP-1 after hypolidemic procedure can be used as a marker showing the effectivity of the extracorporeal LDL-cholesterol elimination (Blaha et al. 2004). In an extended study, the levels of expression of tissue factor, ICAM-1, P- and E-selectin, and PAI-1 were found low, whereas those of endothelial protein C receptor and VCAM-1 were high (Merlini et al. 2004).

44.3.4.3 Gene Polymorphisms in E-Selectin

Polymorphisms in the E-selectin gene are associated with accelerated atherosclerosis in young (age <40 years) patients, further suggesting a role of inflammation in atherosclerosis. A further change in endothelial physiology is an increase in the surface expression of E-selectin, which regulate adhesive interactions between certain blood cells and endothelium. Intravascular fibrinolysis induced by tissue-type plasminogen activator or urokinase may contribute to the initiation of atherosclerosis by inducing P-selectin and platelet activating factor as well as to plaque rupture, either directly or indirectly, by activating metalloproteinases. As E-selectin is only expressed on activated endothelium, it provides an opportunity to study pathophysiological aspects of this cell in cardiovascular and other disease. However, sE-selectin can be found in the plasma, which has potential role in the pathogenesis of cardiovascular disease as raised levels have been found in hypertension, diabetes and hyperlipidemia, although its association in established atherosclerosis disease and its value as a prognostic factor is more controversial (Holvoet and Collen 1997).

Polymorphisms for three genes, P-selectin, L-selectin, and E-selectin (genes *P-sel, L-sel,* and *E-sel,* respectively) showed that the selectin cluster is linked to markers at chromosome 1q23 (Vora et al. 1994). Significant genomic alterations were found on 1q22-q25 in *Sel-L* gene. The message indicated somatic DNA rearrangements, on loci associated to leukocyte adhesion, vascular smooth muscle cells growth, differentiation and migration, to atherosclerosis development as an inflammatory condition (Arvanitis et al. 2005). Wenzel et al. (1999) and Yoshida et al. (2003) described an adenine to cytosine (A/C) substitution for cDNA position 561 resulting in an amino acid exchange from serine to arginine at position 128 (S/R or Ser^{128}Arg) was detected in the epidermal growth factor (EGF) domain. A higher mutation frequency was observed in patients aged 50 years or less with proven severe atherosclerosis as well as in patients aged 40 years or less. If Ser^{128}Arg substitution

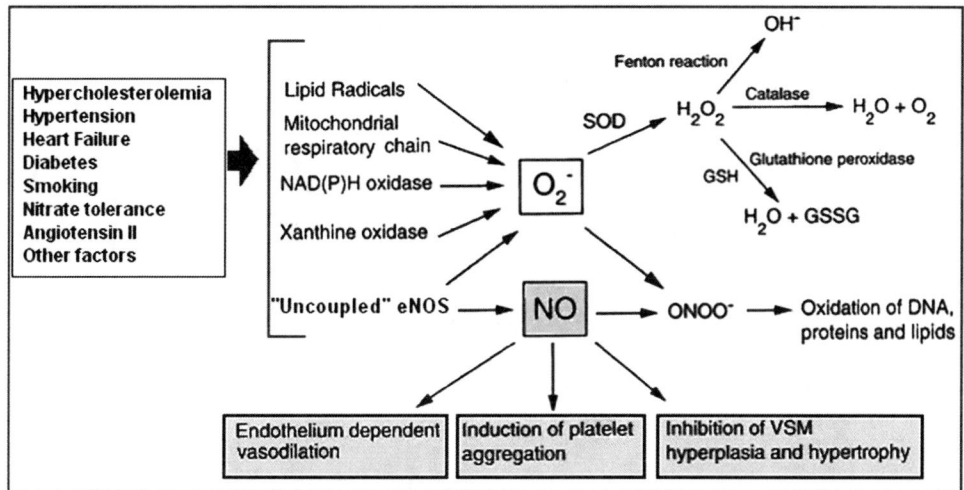

Fig. 44.1 *Mechanisms for oxidant stress-induced endothelial dysfunction in cardiovascular diseases* (Hamilton et al. 2004). Excessive production of ROS has been implicated in processes in which they oxidize biological macromolecules, such as DNA, protein, carbohydrates, and lipids. Many ROS possess unpaired electrons and thus are free radicals. These include molecules such as superoxide anion (O_2^-), hydroxyl racial (HO·), nitric oxide (NO·), and lipid radicals. The cellular production of one ROS may lead to the production of several others via radical chain reactions. A decline in NO bioavailability may be caused by decreased expression of the endothelial cell NO synthase (eNOS), a lack of substrate or cofactors required for eNOS action. Low-density lipoprotein (LDL) is oxidized to oxidized form of LDL (ox-LDL) and initiates the atherosclerotic process in the vessel wall (see Fig. 44.2). Abbreviations: O2−, superoxide; NO, nitric oxide; ONOO−, peroxynitrite; H2O2, hydrogen peroxide; OH−, hydroxyl radical; SOD, superoxide dismutase; GSH, reduced glutathione; GSSG, oxidised glutathione; VSM, vascular smooth muscle

had an effect on the adhesion of blood cells to the endothelium, the polymorphism could be of interest with respect to association studies in a number of pathological conditions, such as cardiovascular diseases. The Ser^{128}Arg polymorphism is associated with a higher risk for early severe atherosclerosis. Yoshida et al. (2003) suggested that the E-selectin Ser^{128}Arg polymorphism could functionally alter leukocyte-endothelial interactions as well as biochemical and biological consequences, which may account for the pathogenesis of myocardial infarction (Li et al. 2005).

Leu^{554}Phe E-selectin mutations in Hypertension and CAD

Wenzel et al. (1996, 1999) detected 17 mutations, five of which resulted in an amino acid substitution. In E-selectin, exchange at Ser^{128}Arg in EGF domain and Leu^{554}Phe in membrane domain, and a DNA mutation from guanine to thymine (position 98) presented different allele frequencies in young patients with severe atherosclerosis, compared with an unselected population. The bi-allelic A/C polymorphism in the E-selectin gene may be implicated in the clinical expression of erythema nodosum (EN) secondary to sarcoidosis (Amoli et al. 2004). However, the E-selectin polymorphism may be associated with severity of atherosclerotic disease, but it is unknown if it is actually a risk factor for atherosclerosis (Ghilardi et al. 2004). A strong relationship was confirmed between 561A > C and 98G > T polymorphisms of E-selectin gene and susceptibility to CAD by Zak et al. (2008). A body mass index (BMI)-specific effect of Leu^{554}Phe polymorphism of E-selectin gene on blood pressure has been reported by Marteau et al. (2004) who strengthened the view that E-selectin is implicated in hypertension (Marteau et al. 2004). Serum levels of E- and P-selectin in patients with essential hypertension (EH) are significantly higher than in controls, where as differences in serum levels of soluble L-selectin, VCAM-1, or ICAM-1 between the patients with EH and the controls were not different (Sanada et al. 2005).

44.3.4.4 Genomic Arrangement of P-Selectin Gene

P-Selectin Thr^{715}Pro (A/C) Polymorphism

Genetic analyses of P-selectin in the progression of atherosclerosis have provided conflicting results regarding the role of variation within the P-selectin gene and risk for heart disease. Miller et al. (2004b) suggested that the Thr^{715}Pro C allele was rare in blacks (0.8 %) and intermediate in South Asians (3.0 %) compared to whites (11.2 %). sP-selectin levels were significantly lower in the individuals with the AC or CC compared to the AA genotype in both whites and South Asians. Thus, in whites and South Asians the C allele of the Thr^{715}Pro P-selectin polymorphism is associated with

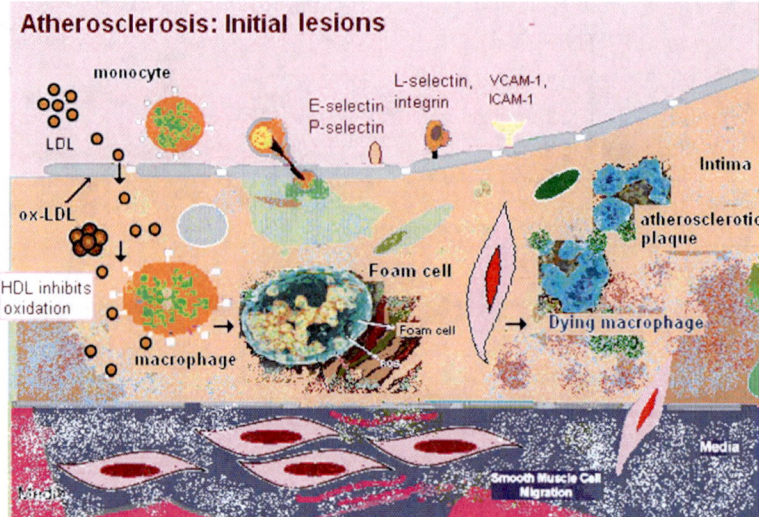

Fig. 44.2 Oxidation of LDL initiates the atherosclerotic process in the vessel wall by acting as a potent stimulus for the induction of inflammatory gene products in vascular endothelial cells. By activating the NF-kB transcription factor, oxidized LDL (ox-LDL) stimulates increased expression of cellular adhesion molecules. There are several different types of adhesion molecules with specific functions in the endothelial leukocyte interaction: The selectins tether and trap monocytes and other leukocytes. Importantly, VCAMs and ICAMs mediate firm attachment of these leukocytes to the endothelial layer. Ox-LDL also augments expression of monocyte chemoattractant protein 1 (MCP-1) and macrophage colony stimulating factor (M-CSF). MCP-1 mediates the attraction of monocytes and leukocytes and their diapedesis through the endothelium into the intima. M-CSF plays an important role in the transformation of monocytes to macrophage foam cells. Macrophages express scavenger receptors, which internalize oxLDL in their transformation into foam cells. Migration of smooth muscle cells (SMCs) from the intima into the media is another early event initiating a sequence that leads to formation of a fibrous atheroma

lower sP-selectin levels (Miller et al. 2004b). The P-selectin Thr^{715}Pro polymorphism is not associated with incident CHD or ischemic stroke in either whites or African-Americans (Volcik et al. 2006).

Leu^{125}Val polymorphism of PECAM-1 gene and elevated soluble PECAM-1 were related to severe coronary artery stenosis in CAD patients.

44.3.4.5 Predictive Value of sE-Selectin in CAD

In CAD, inflammatory biomarkers have been extensively investigated; more evidence exists for C-reactive protein (CRP; Chap. 8). Fatty acid (FA) composition in serum has been associated with CRP and E-selectin but not with other inflammatory markers (Petersson et al. 2009). Studies suggest that, besides CRP, other inflammatory biomarkers such as cytokines, s-CD40 ligand, serum amyloid A (SAA), selectins (E-selectin, P-selectin), ICAM-1, VCAM-1, and several others may have a potential role for the prediction of risk for developing CAD and may correlate with severity of CAD (Eikemo et al. 2004; Fang et al. 2004; Potapov et al. 2005; Zakynthinos and Pappa 2009). The combination of natriuretic peptide (BNP) and E-selectin offers increased predictive value. Plasma levels of adhesive molecules are correlated in patients with stable IHD. Cigarette smoke condensate (CSC)-induced surface expression of ICAM-1, E-selectin, and VCAM-1 in HUVEC. Fang et al. (2005) reported a significant decrease in C allele frequency of PECAM-1 gene and showed that

44.3.5 Myocardial Infarction

sP-Selectin is Associated with Myocardial Damage: Platelets are known to be activated during myocardial infarction (MI). Though, the levels of sP-selectin, sE-selectin and sPECAM-1 did not differ significantly in the pathogenesis of atherosclerosis, sP-selectin was substantially increased in patients with acute myocardial infarction (AMI). Yip et al. (2006) tested the hypothesis that platelet activity shown by CD62P is enhanced and predictive of both the extent of myocardial damage and 30-day clinical outcome in patients with ST-se AMI undergoing primary coronary stenting. Xu et al. (2006) suggested that activated-platelets play an important role in the process of myocardial ischemia-reperfusion injury, and platelet-derived P-selectin is a critical mediator. P-selectin expression, along with CD40 ligand and tissue factor is significantly increased in infarcted rabbits with respect to controls. Clopidogrel administration reduced P-selectin expression and CD40 ligand (Molero et al. 2005).

Hyperhomocysteinemia and Selecins in MI: Hyperhomocysteinemia is regarded as an independent risk factor for vascular diseases, and homocysteine is supposed to contribute to oxidative stress and endothelial damage. Hyperhomocysteinemia is significantly associated with MI in comparison with controls with an odd ratio of 6.26 (Khare et al. 2005). Folic acid corrected and reduced hyperhomocysteinemia in a large majority of the cases. Although the levels of sP-selectin, sE-selectin and sPECAM-1 decrease after folic acid therapy, it was only sE-selectin which was significantly reduced. Apart from their lipid-lowering capacity, statins also exert anti-inflammatory and antioxidant effects.

DNA Polymorphism in MI: Some polymorphisms may increase the risk of MI within specific ethnic groups or in certain populations. P-selectin expression is increased in atherosclerotic plaques, and high plasma levels of this molecule have been observed in patients with unstable angina. DNA polymorphisms in P-selectin gene may be a possible candidate for MI. The P-selectin gene is situated on chromosome 1q21-q24, spans >50 kb and contains 17 exons. Four polymorphisms (Ser^{290}Asn, Asn^{562}Asp, Leu^{599}Val and Thr^{715}Pro) predicted a change in the amino acid sequence of P-selectin. In patients with MI from four regions of France and Northern Ireland (the ECTIM study) the P-selectin polymorphisms provided a heterozygosity of 91 %. The polymorphisms were tightly associated with one another and displayed patterns of linkage disequilibrium suggesting the existence of highly conserved ancestral haplotypes. Study illustrates the complexity of the relationship between gene variability and disease and the necessity to explore in detail the polymorphisms of candidate genes (Herrmann et al. 1998; Tregouet et al. 2002).

The E-selectin gene Arg128, 98 T, and Phe554 alleles and PECAM1 Leu1^{25}Val and Ser^{563}Asn polymorphisms may increase the risk of atherosclerosis, but not necessarily the risk of MI. This association seems to be more pronounced in younger patients and may be especially important in patients with a low risk for developing atherosclerosis. Reports indicated that screening for CD14-260 C/T genotypes is unlikely to be a useful tool for risk assessment and it remains unclear whether CD14 polymorphisms significantly increase the risk of MI. The A^{252}G polymorphism of lymphotoxin-α (LTA) gene, a member of the TNF family, is strongly related with the onset of AMI (Auer et al. 2003). Quantitative real-time RT-PCR confirmed that LTA increased the expressions of E-Selectin and VCAM1 both in HUVEC and HCAEC, suggesting the roles of LTA in the development of atherosclerosis. Aminian et al. (2007) determined the possible role of Gly^{241}Arg and Lys^{469}Glu polymorphisms in development of CAD and acute or chronic MI. Although the frequency of Gly-Arg and Arg-Arg genotypes were higher in the control group compared to the CHD patients, no strong corelation was found between Gly^{241}Arg and Lys^{469}Glu polymorphisms and occurrence of CHD and MI in population from Iran.

44.3.6 Atherosclerotic Ischemic Stroke

In ischemic event in patients with atherosclerotic ischemic stroke, though the platelet aggregability was decreased after day 3 compared to that at day 1 of stroke onset, platelet CD63 and P-selectin/CD62P expression remained high even 90 days after the events. This suggested that platelet hyperactivation in atherosclerotic ischemic stroke might be sustained for a considerable period (Cha et al. 2004; Nadar et al. 2004b; Yip et al. 2006). Blood levels of ICAM-1 and CD62P expression in different typing of patients with ischemic stroke are different. Evidences suggest that MPS (Meridian-phlegm stagnancy) group of patients is the key pathogenic factor of ischemic stroke.

Mucosal tolerance to E-selectin after booster tolerization can relieve cerebral ischemia-reperfusion injury and induce ischemia tolerance in rats. The mechanisms may involve decreased frequencies of CD8$^+$ T cells, heightened mRNA expression of IL-10 and lowered mRNA expression of E-selectin in the ischemic hemisphere (Yun et al. 2008). Selakovic et al. (2009) defined changes of soluble CAMS in cerebrospinal fluid and plasma in the patients with the acute brain infarction, in which significant increase in the level of soluble adhesion molecules occurs within the first seven days. Studies show that hypoxia/reoxygenation stimulates ICAM-1 and apoptosis (Antonova et al. 2009).

Cerebral arteriovenous malformations (AVMs) showed significant upregulation of E-selectin, VCAM-1 and ICAM-1 (Storer et al. 2008; Chan and Sukhatme 2008; Tuttolomondo et al. 2009). Li et al. (2008) showed that ICAM-1 Lys^{469}Glu polymorphism was involved in the causation of ischemic stroke, especially in female but not in male (Rodrigues et al. 2008). Two allelic variants were related to ischemic stroke. Multivariable regression analysis after adjustment for vascular risk factors demonstrated that alleles Arg of Ser^{128}Arg and Phe of Leu^{554}Phe polymorphisms are independent risk factors for ischemic stroke. The combination of two minor alleles of E-selectin genes appeared to be the strongest susceptibility factor for ischemic stroke (Haidari et al. 2009). Sarecka-Hujar et al. (2010) could not confirm the relationship between the 98 G > T polymorphism of the E-selectin gene and childhood ischemic stroke. The G allele of the E- selectin 98 G > T polymorphism was more frequently transmitted to the children after stroke compared to the T allele. There is a need for further studies in these areas.

44.3.7 Hypertension

The association between blood pressure and different adhesion molecules appeared to be present in women younger than 50 years, who were likely to be pre-menopausal (Miller et al. 2004a). Serum levels of E- and P-selectin in patients with essential hypertension (EH) are significantly higher than in the controls (Sanada et al. 2005). After adjustment for age, only sE-selectin concentrations were significantly associated with blood pressure. Higher levels of plasma sP-selectin were confirmed in hypertensive patients alone with VEGF (Nadar et al. 2004a). It is stated that decrease in blood pressure may reduce the rate of progression of atherosclerosis by affecting the expression of E- and P-selectin in the endothelium, the platelets, or both.

44.3.8 Reperfusion Injury

In vitro studies indicate that complement activation regulates the expression of P-selectin on endothelial cells. This suggests that in disorders such as ischemia/reperfusion injury, in which both complement and P-selectin have been shown to play a role, complement activation is a primary event and the effects of P-selectin are secondary. In mouse kidney model of I/R injury, results indicated that complement and P-selectin-mediated pathways of renal reperfusion injury are mutually independent (Farrar et al. 2004). Induction of circulating polymorphonuclear neutrophils (PMNs) might contribute to the superior outcome following stenting and early intervention compared to conventional balloon angioplasty (PTCA). A substantial increase in sE-selectin levels early after PTCA and stent implantation may predict development of restenosis (Heider et al. 2006; Kilickap et al. 2004). After reperfusion of myocardial vessels, P-selectin expressed on majority of vessels (77 %) though the expression decreased during subsequent remaining duration of reperfusion (Chukwuemeka et al. 2005). In rats, the mRNA expression for several genes was associated with inflammation after transient middle cerebral artery occlusion (MCAO). Gene expression increased in the injured hemisphere for IL-1β, IL-6 and ICAM-1. TNF-α mRNA was upregulated in the injured versus uninjured hemisphere, while E-selectin mRNA showed a significant increase from 6 to 24 h after MCAO (Berti et al. 2002). Both P-selectin and LFA-1 may be important targets to control pathologic inflammation in I/R-induced tissue injury in the colon (Riaz et al. 2002). The study in intestinal ischemia and reperfusion injury (IR/I) using murine models demonstrated the importance of P-selectin in warm and cold IR/I. The blockade of P-selectin using rPSGL1-lg or the absence of P-selectin KO mice confers a survival advantage and reduction in tissue injury. The mechanism appears to be independent of neutrophil infiltration (Carmody et al. 2004). Enterocyte apoptosis is increased following intestinal I/R injury. Hyperoxia following intestinal I/R in rat increased E-selectin expression in the jejunum and ileum and a concomitant increase in neutrophil recruitment in the ileum, accompanied by increased cell apoptosis (Braun et al. 2004; Sukhotnik et al. 2008). Germ cell-specific apoptosis that occurs after I/R of murine testis is dependent on neutrophil recruitment to the testis and is dependent on E-selectin. Blockage of E-selectin may be a strategy to treat postischaemic testis (Celebi and Paul 2008).

44.4 CAMS in Allergic Inflammation

Allergic inflammation is characterized by recruitment of specific leukocyte subpopulations from blood into tissue and requires a series of cell adhesion-molecule-mediated interactions between postcapillary vascular endothelium and the leukocyte cell surface. Three major groups are involved: selectins, integrins, and the immunoglobulin gene superfamily. P- and E-selectin mediate initial leukocyte adhesion, whereas beta 2-integrin/ICAM-1 and VLA-4/VCAM-1 pathways mediate leukocyte arrest and transendothelial migration. Because VLA-4 expression is restricted to eosinophils and lymphocytes, VCAM-1 has been implicated in selective eosinophil recruitment characterizing allergic inflammation. However, additional factors such as profile of cytokine release are likely to operate since tissue eosinophilia has been observed in the absence of VCAM-1 expression (Smith et al. 1993a).

44.4.1 Dermal Disorders

E-selectin is highly expressed on vascular endothelium in atopic dermatitis and psoriasis, and in patients with measles. The cutaneous lymphocyte-associated antigen (CLA), which is expressed on peripheral skin-homing helper memory T cells in healthy persons, is at least partly the sialyl 6-sulfo Le^x determinant (Ohmori et al. 2006) and a ligand for selectins. The differential polyadenylation of E-selectin transcripts may provide the molecular basis for the observed chronic expression of E-selectin in human dermal disorders. In atopic dermatitis, patients express ICAM-1 and ICAM-3, E-selectin and L-selectin (60 %) in the dermis, without expression of E- and L-selectins in the epidermis. A high expression of adhesion molecules in the skin lesions of atopic dermatitis patients may play an important role in the pathogenesis of atopic dermatitis (Lugovic et al. 2006). The blood markers for atopic dermatitis, including soluble forms of E-selectin, VCAM-1 and ICAM-1 were reduced after treatment with cetiridine (Izu and Tokura 2005). The extracts from dust mites, *Dermatophagoides farinae*, *D. pteronyssinus* and *Euroglyphus maynei* with and without endotoxin (LPS) stimulated endothelial cells to express ICAM-1, VCAM-1, and E-selectin and to secrete IL-6, IL-8, MCP-1, and GM-CSF. Serum levels of sE-selectin are higher in children with measles than in children with atopic dermatitis, atopic asthma and healthy controls. But it was not correlated with measles. There was no correlation between sE-selectin and TNF-α level (Park et al. 2008). Pollinosis from Parietaria judaica is one of the main causes of allergy in the Mediterranean area. The treatment of endothelial cells with pollen extract causes an increase of E-selectin and VCAM-1 protein levels as well as an increase of IL-8 production. The stimulation of cell adhesion molecules was paralleled by an increase of adhesion of polymorphonuclear cells (PMNs) to HMVEC-L monolayer (Taverna et al. 2008).

44.4.2 Rhinitis and Nasal Polyposis

Allergic rhinitis is an inflammatory disease of the nasal mucosa, caused by an IgE-mediated reaction after exposure to the allergen. Persistent inflammation is induced by the presence of an inflammatory cell infiltrate, together with ICAM-1 expression in the epithelial cells of the mucosa exposed to the allergen to which they are sensitized, in the absence of clinical symptoms (Montoro et al. 2007). Nasal polyposis is a chronic non-infectious inflammatory disease of the nasal and paranasal cavity mucosa. Eosinophil migration from blood stream to nasal polyps involves different molecules such as ICAM-1, VCAM-1, and L-, P- and E-selectins. Patients with nasal polyposis exhibit a higher expression of VCAM-1, E-selectin, and L-selectin compared to healthy controls (Corsi et al. 2008). *Staphylococcal enterotoxin A* (SEA) and staphylococcal enterotoxin B (SEB) infection increased ICAM-1 expression and cytokine secretion (Wang et al. 2007).

44.4.3 Lung Injury

Excessive leukocyte accumulation is involved in the pathogenesis of the sepsis-induced acute lung injury. Studies suggest that P-selectin has a substantial role in the pathogenesis of the lung injury induced by LPS (Ohnishi et al. 1999). In bleomycin-induced fibrosis in mice, the L-selectin and/or ICAM-1 deficiency inhibited skin and lung fibrosis with decreased Th2 and Th17 cytokines and increased Th1 cytokines. In contrast, P-selectin deficiency, E-selectin deficiency with or without P-selectin blockade, or PSGL-1 deficiency augmented the fibrosis in parallel with increased Th2 and Th17 cytokines and decreased Th1 cytokines. Yoshizaki et al. (2010) suggest that L-selectin and ICAM-1 regulate Th2 and Th17 cell accumulation in skin and the lung, leading to the development of fibrosis, and that P-selectin, E-selectin, and PSGL-1 regulate Th1 cell infiltration, resulting in the inhibition of fibrosis (Yoshizaki et al. 2010). Adult respiratory distress syndrome (ARDS) appears to develop as the acute lung injury in the course of many severe diseases, as the result of damage of alveolar-capillary barrier. Clinical observations suggest that analysis of E-, P-selectin and ICAM-1 concentrations in the serum of patients with ARDS may be helpful in monitoring the course and treatment of the disease (Skiba-Choińska and Rogowski 1996). In the pathogenesis of paracoccidioidomycosis, Gonzalez et al. (2005) suggest that during early stages, up-regulation of ICAM-1, VCAM-1, CD18 and Mac-1 expression may participate in the inflammatory process.

44.4.4 Bronchial Asthma and Human Rhinovirus

The house dust mite (HDM) is the common indoor allergen associated with bronchial asthma. ICAM-1, VCAM-1, and E-selectin are newly synthesized prior to spontaneous asthma attacks, and their expression may play a key role in eosinophil infiltration into the airway (Ohkawara et al. 1995). Crude extract of *D. farinae* induces ICAM-1 expression in EoL-1 cells through signaling pathways involving both NF-kB and JNK (Kwon et al. 2007). Kirchberger et al. (2006) demonstrated that signaling via ICAM-1 induces adhesiveness of mononuclear phagocytes, which critically involves PECAM-1 and is mediated via LFA-1/ICAM-3. The most common acute infection in humans, Human Rhinovirus (HRV) is a leading cause of exacerbations of asthma and chronic obstruction pulmonary disease. ICAM-1 is a critical target-docking molecule on epithelial cells for 90 % HRV serotypes. ICAM-1 regulates not only viral entry and replication but also signaling pathways that lead to inflammatory mediator production (Lau et al. 2008; Lee et al. 2008). The sICAM-1 but not sE-selectin from patients with asthma is significantly higher than healthy controls. Although serum levels of sICAM-1 are higher in asthmatics, it may be necessary to establish individual baseline values for serial estimation to evaluate their clinical relevance (Bijanzadeh et al. 2009). The serum levels of sICAM-1 were significantly higher in obese nonasthmatic and obese asthmatic children versus control and lean asthmatic children (Huang et al. 2008). P-selectin is an important controller of the inflammation by mediating selective eosinophil cell influx to the lung. It can be used as a sensitive marker in mild asthma (Sjosward et al. 2004).

44.5 Autoimmune Diseases

Adhesion molecule expression and interactions are involved in initiation and propagation of autoimmune diseases including rheumatoid arthritis (RA), systemic lupus erythematosus, Sjögren's syndrome, autoimmune thyroid disease, multiple sclerosis, systemic sclerosis (SSc) and diabetes mellitus. Increased adhesion molecule expression and avidity changes occurring with cellular activation are the principal methods regulating leukocyte adhesion. Although differences between specific autoimmune diseases exist, key interactions facilitating the development of autoimmune inflammation appear to include L-selectin/P-selectin/E-selectin, LFA-1/ICAM-1, very late antigen-4 (VLA-4)/VCAM-1, and α4B7/MadCAM or VCAM-1 adhesion. A vast array of adhesive interactions occurs between immunocompetent cells, endothelium, extracellular matrix, and target tissues during the evolution of an autoimmune disease. Dermatitis herpetiformis (DH) and bullous pemphigoid (BP), the autoimmune diseases, are characterized by destruction of the basement membrane zone (BMZ) and anchoring fibres by autoantibodies and infiltration. Skin biopsies from patients with DH, with BP, and from healthy subjects showed the expression of E and L selectins mainly in the skin leukocytes in all samples where as β1, β3 integrins was detected mainly in basal keratinocytes. Integrins and selectins seem to play an important role in the destruction of BMZ in DH and BP (Erkiert-Polguj et al. 2009). P-selectin levels were significantly higher than normal in RA and SSc, but not in SLE. In contrast, mean L-selectin levels were significantly higher than normal in SLE, but not in RA or SSc. Where as soluble IL-2 receptors in patients with active RA, SSc and SLE were almost double the normal level, showing a strong positive correlation only between L-selectin and sIL-2R, and only in patients with SLE. These findings indicated a distinct pattern of immune cell activation in chronic diseases that share an over-activation of T-lymphocytes (Sfikakis and Mavrikakis 1999).

44.5.1 Endothelial Dysfunction in Diabetes (Type 1 Diabetes)

Adhesion molecules have been implicated in the development and progression of cardiovascular disease, particularly in people with diabetes. Diabetes mellitus type 1 (Type 1 diabetes or T1DM, also called insulin-dependent diabetes mellitus—IDDM, or, formerly, juvenile diabetes) is a form of diabetes mellitus that results from autoimmune destruction of insulin-producing β cells of the pancreas. The chronic hyperglycemic state in T1DM patients produces an aggression to vascular endothelium leading to a premature development of atherosclerosis. In both boys and girls, sE-selectin is an early marker of endothelial dysfunction and a probable risk marker of atherosclerosis in children with T1DM (Carrizo et al. 2008). The levels of C-reactive protein, E-selectin, and cytokines in association with severity index were significantly increased in T1DM and type 1 diabetic patients with microvascular complications (T1DM-MV patients) compared with control subjects (Devaraj et al. 2007). Nerve microvasculitis and ischemic injury appear to be the primary and important pathogenic alterations in lumbosacral radiculoplexus neuropathy (LRPN) of patients with diabetes mellitus (DLRPN) and without diabetes mellitus (LRPN). The up-regulation of inflammatory mediators target different cells at different disease stages and that these mediators may be sequentially involved in an immune-mediated inflammatory process that is shared by both DLRPN and LRPN (Kawamura et al. 2008).

Adhesion molecules are upregulated in endothelial cells of the placental bed in pregnancies complicated by T1DM in association with increased adherence of peripheral blood monocytes. The increase in monocyte adhesion to decidual endothelial cells from diabetic pregnancies was associated with increased endothelial cell expression of ICAM-1, but not VCAM-1. ICAM-1 expression in normal decidual endothelial cells was stimulated by pro-atherogenic and pro-inflammatory stimuli (Xie et al. 2008; Telejko et al. 2009).

Type 2 Diabetes: In contrast to T1DM, type 2 diabetes mellitus (T2DM) results from insulin resistance, a condition in which cells fail to use insulin properly, sometimes combined with an absolute insulin deficiency (Formerly referred to as non-insulin-dependent diabetes mellitus, NIDDM for short). Endothelial dysfunction in type 2 diabetic patients is associated with inflammation, increased levels of circulating soluble adhesion molecules (VCAM-1 and E-selectin), and inducing production of ROS, and urinary albumin excretion (Potenza et al. 2009). Diabetic patients have increased susceptibility to infection, which may be related to impaired inflammatory response observed in experimental models of diabetes, and restored by insulin treatment (Riad et al. 2008; West et al. 2008).

Serum Levels of CAMs in Diabetic Patients: Abnormal levels some of serum ICAM-1, VICAM-1, E-selectin, P-selectin, L-selectin have been detected in T2DM. High-fat load and glucose alone produce an increase of nitrotyrosine, ICAM-1, VCAM-1, and E-selectin plasma levels in normal and diabetic subjects. A decrease in neutrophil surface CD62L expression and significantly higher concentrations of sICAM-1, sVCAM-1, sE-selectin, vWF, hsCRP, IL-6 and fibrinogen in patients with diabetic microangiopathy in comparison with diabetic group without microangiopathic complications and healthy controls suggested that: (1) diabetic microangiopathy is accompanied by increase in CD11b expression and decrease in CD62L (L-selectin) expression on peripheral blood neutrophils; (2) neutrophil activation and intensified adhesion; (3) the development of diabetic microangiopathy is accompanied by an increase in soluble adhesion molecules and inflammatory markers concentrations in the blood (Lim et al. 2004; Mastej and Adamiec; 2008). Levels of E-selectin positively correlated with high triglyceride levels in type 2 diabetic subjects with silent ischemia (Adamikova et al. 2008; Okapcova and Gabor 2004; Rubio-Guerra et al. 2008). Though, baseline plasma levels of vascular markers (hsCRP, sICAM-1, sVCAM-1, E-selectin and P-selectin) were significantly elevated, they did not improve after aerobic exercise. The sE-selectin and vWF are elevated in chronic heart failure patients with DM but not in those without DM. High sE-selectin levels may be associated with ischaemic events in patients with DM (Kistorp et al. 2008).

Diabetic Nephropathy (DN) and Diabetic Heart: Diabetic nephropathy (DN) is the leading cause of end stage renal disease (ESRD) (Malatino et al. 2007). Although the pathogenesis of DN is multifactorial, local inflammatory stress may result from both the metabolic and hemodynamic derangements observed in DN. The current evidence supporting the role of inflammation in the early phases of clinical and experimental DN has been reviewed (Fornoni et al. 2008). Inflammatory markers such as IL-18 and TNF-α are increased in the serum of patients with diabetes and DN. This occurs at an early stage of disease, and correlates with the degree of albuminuria. The pharmacologic interventions for DN by angiotensin converting enzyme inhibitors, angiotensin receptor blockers and aldosterone antagonists may have anti-inflammatory effects, which are independent of their hemodynamic effect.

Diabetic Retinopathy: The association between soluble adhesion molecules levels and retinopathy in type 2 diabetic patients has been clarified. sE-selectin levels are elevated in diabetic patients compared to control subjects, with no significant difference in sICAM-1 and sVCAM-1 levels. The progression of retinopathy was not associated with an increase in soluble adhesion molecules. However, Nowak et al. (2008) observed that serum levels of sICAM-1 and sELAM-1 were significantly elevated and the concentration sVCAM-1 was elevated but not significantly in diabetic patients. Increase in sICAM-1 and sVCAM-1 levels, as well as their correlation with high vitreous IL-6 and TNF-α concentrations in patients with diabetic retinopathy seems to confirm the inflammatory-immune nature of this process. Significantly increased TNF-α concentration in the vitreous body was related to the rise of VCAM-1 (Adamiec-Mroczek and Oficjalska-Młyńczak 2008; Leal et al. 2008; Khalfaoui et al. 2008). Intravitreal injection of corticosteroid has been used to treat diabetic macular edema.

44.5.2 Rheumatic Diseases

Plasma levels of vWF and sP-selectin (but not sE-selectin) are significantly higher among Rheumatoid disease (RD) patients compared to controls. Levels of vWF progressively rise with increasing cardiovascular risk (Bhatia et al. 2009). Serum levels of ICAM-1, ICAM-3, VCAM-1, L-selectin, and E-selectin have been detemined in children with a variety of pediatric rheumatic diseases. A trend toward higher levels of sE-selectin was found in vasculitis vs other diagnoses. The sICAM-1 was higher in patients with active vs inactive disease across all diagnoses. Report suggests that (1) elevated E-selectin levels in vasculitis likely reflect the high degree of endothelial activation and possibly overt vascular damage in those conditions. (2) The correlation of sL-selectin with C4 in SLE may indicate that downregulation of shedding of cell surface L-selectin is involved in continued adherence of leukocytes to endothelium, possibly causing further damage and immune complex deposition in this condition. (3) The trend toward inverse correlation between sE-selectin and vWF:Ag in diabetes mellitus is interesting. (4) Levels of sICAM- I may be a useful marker of active vs quiescent disease in general in the pediatric rheumatic diseases, although lack of correlation with disease activity indices indicates that it is too insensitive to smaller differences in disease activity to be recommended for routine clinical use (Bloom et al. 2002).

44.5.3 Rheumatoid Arthritis

Considerable evidence indicates that patients with Rheumatoid arthritis (RA) are at greater risk of developing atherosclerosis and cardiovascular disease. Atherosclerotic cardiovascular mortality is increased in RA patients. The markers proposed for assessing RA activity include rheumatoid factor, anti-citrullinated protein/peptide antibodies, IgM anti-IgG advanced glycation end products, markers of bone/cartilage metabolism, mannose-binding lectin, E-selectin, IL-6, and leptin. Various studies have investigated the correlation between some of these markers and other variables that might indicate disease activity, e.g., inflammatory activity tests and disease activity scores. However, there is as yet insufficient evidence that any of these markers, in isolation or in combination, are useful in the assessment of RA activity. Many numerous endothelial cells become positive for E-selectin and E-selectin mRNA in RA synovial membranes and the E-selectin expression appeared to correlate with inflammatory activity. P-selectin deficiency in mice resulted in accelerated onset of joint inflammation in the murine collagen-immunized arthritis model. Mice deficient either in E-selectin or in E- selectin and P-selectin (E/P-selectin mutant) also exhibit accelerated development of arthritis compared with wild type mice in CIA model. The strong vascular expression of E-selectin indicates an activation of endothelial cells in the recruitment of cells associated with the chronic inflammation of RA (Foster et al. 2009; da Mota et al. 2009). E-Selectin and ICAM-1 are upregulated on the synovial endothelium, while VCAM-1 plays an important role in synovial lining layer cells and within the synovial stroma. The expression of CAMs may be blocked by mAbs and modified by nonsteroidal anti-inflammatory drugs and disease-modifying antirheumatic drugs. (Cobankara et al. 2004). Serum soluble adhesion molecules concentrations are down-regulated following anti-TNF-α antibody therapy combined with methotrexate (MTX) (Klimiuk et al. 2004, 2007; Levälampi et al. 2007; Bosello et al. 2008).

In comparison with osteoarthritis (OA), patients with early RA are characterized by high serum concentrations of sICAM-1, sVCAM-1, and sE-selectin (Yildirim et al. 2005), while LDL-cholesterol was decreased in all RA patients (Pemberton et al. 2009). P-selectin deficiency in mice results in accelerated onset of joint inflammation in the murine collagen-immunized arthritis model. Mice deficient either in E-selectin or in E-selectin and P-selectin (E/P-selectin mutant) also exhibit accelerated development of arthritis compared with wild type mice, suggesting that these adhesion molecules perform overlapping functions in regulating joint disease. Ruth et al. (2005) suggested that E-selectin and P-selectin expression can significantly influence cytokine and chemokine production in joint tissue, and that these adhesion molecules play important regulatory roles in the development of RA in E/P-selectin mutant mice (Singh et al. 2008).

In RA patients, P-selectin expression, PMC and sCD40L levels were increased when compared with controls. The increase in markers of active platelets, P-selectin and sCD40L, and platelet-monocyte levels might be associated with the increased cardiovascular mortality in RA. Psoriatic arthritis (PsA) is associated with the development of endothelial dysfunction and increased atherosclerotic complications. Endothelial activation might have a role in the pathogenesis of both psoriasis and PsA. Among parameters of platelet activation, only PMC might play a role in the pathogenesis of PsA (Pamuk et al. 2008, 2009). Though, sE-selectin correlated with severity of joint disease, further follow-up studies should evaluate if sE-selectin is useful as prognosis marker for progression of articular damage (Corona-Sanchez et al. 2009).

44.5.4 Other Autoimmune Disorders

Systemic Lupus Erythematosus (SLE): Elevated serum concentrations of ET-1, s-thrombomodulin (TM), and sE-selectin reflect persisting endothelial cell activation in SLE, and point to an important role of ET-1 in the pathogenesis of internal organ involvement (Kuryliszyn-Moskal et al. 2008). The s-E-selectin, TM and s-VCAM-1 are significantly elevated in lupus nephritis (LN) with renal vascular lesions (VLS) than in LN without VLS. A positive correlation was found between TM and serum creatinine in patients with vascular lesions. Therefore, serum TM and s-VCAM-1 can be biomarkers of VLS in LN patients (Yao et al. 2008, Rho et al. 2008).

Autoimmune Thyroiditis: Autoimmune thyroiditis is multifactorial in etiology with genetic and environmental factors contributions. Patients with untreated Graves' disease (GD) show high serum level of sE-selectin, which correlated with the activity of the disease. The expression of ICAM-1 and VCAM-1 was increased in EC from patients from Graves' disease (GD) and Hashimoto's thyroiditis (HT). Results suggest that both the LFA-1/ICAM-1, ICAM-3 and VLA-4/VCAM-1 pathways could play a relevant role in autoimmune thyroid disorders (Marazuela et al. 1994). In patients with GD, the 721 G-A polymorphism was associated with an earlier age of GD onset (before age 40) and that the 1405A-G polymorphism could predispose to Graves ophthalmopathy. It was concluded that G241R and K469E amino acid substitutions in the ICAM1 molecule could influence the intensity/duration of the autoimmunity process and the infiltration of orbital tissues (Kretowski et al. 2003). Chen et al. (2008) suggested that common *SELE* variants may be associated with susceptibility to GD in Chinese population, though the limitation of sample size and multiple test problems exists (Chen et al. 2008).

Sjogren's Syndrome: CAMs are involved in the lymphoid cell infiltration of the salivary and lacrimal glands in Sjogren's syndrome (SS) patients. Biopsies from SS patients showed a marked expression of VCAM-1 and ICAM-1 in the venules surrounded by infiltrated $CD4^+$ $CD45RO^+$ T cells. E-selectin was expressed on vascular endothelium with weak intensity (Saito et al. 1993). Pisella et al. (2000) reported that a significant increase of HLA-DR and ICAM1 expression by epithelial cells was consistently found in patients with keratoconjunctivitis sicca (Sjogren syndrome). These markers were well correlated with each other and correlated inversely with tear break-up time and tear production. Cytokine-mediated up-regulation of VCAM-1 and ICAM-1 that facilitates the recruitment of VLA-4 and LFA-1 expressing T cells might contribute to lymphoid cell infiltration in the salivary and lacrimal glands in SS

44.6 CAMs in System Related Disorders

44.6.1 Gastric Diseases

CAMs mediate the extravasation of leukocytes and their accumulation in inflamed intestinal mucosa. Eosinophilic inflammation is a common feature of numerous eosinophil-associated gastrointestinal (EGID) diseases. Increased intestinal expression of E-selectin has been associated with multiple organ failure and an adverse outcome. VCAM-1 is not altered in in mucosa of patients with inflammatory bowel disease (IBD) regardless of the activity of the inflammatory process. In contrast, E-selectin was not detected in normal colonic mucosa or in colonic mucosa of patients with IBD. However, high levels of E-selectin were consistently found on endothelial surfaces in association with active inflammation in affected areas of colonic mucosa in patients with either ulcerative colitis or Crohn's colitis. In addition, E-selectin appeared to be present within neutrophils which had migrated into crypt abscesses in affected mucosa. Thus E-selectin may play an important role in facilitating leukocyte migration into sites of active IBD involvement (Koizumi et al. 1992)

ICAM-1 was expressed to a greater degree in ulcerative colitis (UC) specimens. Serum ICAM-1 levels in UC patients showed lower levels than those in the control group and were found to vary according to degree of clinical severity (Ogawa et al. 2008). Characterization of integrin expression on colonic eosinophils revealed that colonic CC chemokine receptor 3^+ eosinophils express ICAM-1 counter-receptor integrins αL, αM, and β2. It appears that β2-integrin/ICAM-1-dependent pathways are integral to eosinophil recruitment in colon during GI inflammation associated with colonic injury (Forbes et al. 2006).

McCafferty et al. (1999) examined the role of P-selectin in intestinal inflammation in P-selectin deficient mice alone or in combination with either ICAM-1 or E-selectin and suggested that anti-adhesion therapy might play only a limited, beneficial role and often a detrimental role in intestinal inflammation. The sE-selectin levels of Crohn's disease patients with active disease are higher than those with remission of the disease. L-selectin does not change in patients with active disease compared to those with remission. Thus, determination of sE-selectin in children with Crohn's disease is of significance in estimation of inflammation activity (Adamska et al. 2007).

Khazen et al. (2009) investigated mutations in CAM genes in Tunisian patients, implicated in determining susceptibility to ulcerative colitis (UC) and Crohn's disease (CD). A significant increase in allele frequencies of 206 L of L-selectin and the associated genotype F/L was observed in patients with UC and CD compared with controls; the L206 allele and F/L206 genotype frequencies were significantly increased in UC patients with left-sided type; whereas, the F/L206 genotype was significant in CD patients with ileocolonic location. No significant differences in allele or genotype frequencies were observed for ICAM-1 K469E, E-selectin, and PECAM-1 polymorphisms between UC patients, CD patients, and controls. Khazen et al. (2009) suggest an association of inflammatory bowel disease with allele L206 of L-selectin gene, whereas genotype L/F was associated with a subgroup of UC (left-sided type) and CD patients with more extensive location of disease and stricturing behavior.

However, Vischer et al. (2008) did not reveal any difference in mRNA and protein expression levels for any construct or a major impact of missense variants on ICAM-1 biological function. Pulse-chase experiments showed that two variants, K469E and arg478 to trp (R478W), had a prolonged half-life compared with wildtype ICAM1, whereas two other variants, G241R and pro352 to leu (P352L), had a decreased half-life, implying differences in protein degradation.

Celiac Disease: Celiac disease is a chronic intestinal inflammatory disease that develops in genetically susceptible individuals after gluten ingestion. The *ICAM-1* gene, located in the Celiac disease linkage region 19p13, encodes ICAM-1 involved in inflammatory processes. Increased levels of ICAM-1 were observed in intestinal biopsies and in sera of Celiac disease patients. In addition, an association between the ICAM1 polymorphism G241R and Celiac disease patients has been described in a French population.though, in spanish population results discard the importance of ICAM1 G241R in celiac disease (Dema et al. 2008).

Behçet Syndrome: Behçet's disease/syndrome (BD/BS) is a multisystemic inflammatory disorder of which oral aphthous ulceration is a major feature. CD3 and γδ T-cell expression and other adhesion molecules including VCAM-1 and ICAM-1 were upregulated, whereas CD40 showed little change in BD. The changes in cell-cell and cell-extracellular matrix interactions may affect cell homeostasis and participate in the formation of oral ulcers in BD (Kose et al. 2008). However, Demirkesen et al. (2008) found no significant differences between the BS and control groups in regard to E-selectin, P-selectin, VCAM-1, PNCAM-1 except for ICAM-1.

Systemic Sclerosis or Systemic Scleroderma: Systemic sclerosis or systemic scleroderma is a systemic autoimmune disease or systemic connective tissue disease that is a subtype of scleroderma. Severe fibrosis and increased expression of profibrotic cytokines are important hallmarks in the gastric wall of patients with systemic sclerosis (SSc; scleroderma). The $CD4^+/CD8^+$ T cell ratio is significantly increased in SSc specimens. T cells strongly express the activation markers

VLA-4, LFA-1, and ICAM-1. Endothelial cells showed corresponding surface activation with strong expression of VCAM-1 and ICAM-1. These results provide the evidence that endothelial/lymphocyte activation leading to prominent CD4+ T cell infiltration may play a key pathogenetic role within the gastric wall of patients with SSc (Manetti et al. 2008). In patients with SSc with and without pulmonary arterial hypertension (PAH), serum sICAM-1, sVCAM-1, sP-selectin and sPECAM-1 levels were higher than in healthy donors (HD) at baseline and fell to normal values after 12 months of bosentan therapy. Endothelial activation occurs in SSc, and that changes in the T cell/endothelium interplay take place in SSc-associated PAH. Bosentan seems to be able to hamper these changes and restore T cell functions in these patients (Iannone et al. 2008).

44.6.2 Liver Diseases

Soluble adhesion molecules play a significant role in hepatitis. Biliary atresia (BA) is a congenital or acquired liver disease and one of the principle forms of chronic rejection of a transplanted liver allograft. In the congenital form, the common bile duct between the liver and the small intestine is blocked or absent. The acquired type most often occurs in the setting of autoimmune disease, and is one of the principle forms of chronic rejection of a transplanted liver allograft. The serum sE-selectin of BA patients was higher than that of controls. Subgroup analysis showed that there was an increase in sE-selectin levels of BA patients with jaundice compared to those without jaundice. Also, sE-selectin was positively correlated with serum alanine transferase (ALT), a marker for liver injury, but not with serum gamma glutamyl transpeptidase (GGT) (Vejchapipat et al. 2008).

Cholangitis without a modifier—from Greek *chol-*, bile + *ang-*, vessel + *itis-*, inflammation) is an infection of the bile duct (cholangitis). In secondary cholangitis, ICAM-1 expression is increased along with de novo VCAM-1 and E-selectin appearance on the endothelium of microvessels in chronic exacerbated cholangitis (Gulubova et al. 2008).

Primary biliary cirrhosis (PBC) is an autoimmune disease of the liver marked by the slow progressive destruction of the small bile ducts (bile canaliculi) within the liver. Patients with PBC, primary sclerosing cholangitis and chronic active hepatitis (autoimmune) show significant increase in sICAM-1 compared with normal healthy subjects. Significant elevation in sICAM-1 is also detected in patients with inactive alcoholic cirrhosis, suggesting that impaired liver may, in part, account for the increased serum level in patients with autoimmune liver disease. In contrast, sE-selectin did not differ significantly from healthy controls. Although, peripheral blood mononuclear cells (PBMC) may be a source of sICAM-1, Thomson et al. (1994) suggested that PBMC may not be a significant source of sICAM-1 in this disease. The differential expression of CAMs in the liver is consistent with the suggestion of selectins involvement in neutrophil rolling in the vasculature and ICAM-1 in transendothelial migration and adherence to parenchymal cells (Essani et al. 1995).

Wu et al. (2009) investigated the relationships between the polymorphisms of E-selectin gene and plasma sE-selectin levels in relation to disease progression in a hepatitis B virus (HBV)-infected Chinese Han population. The frequency of C allele (AC or CC) of the $A^{561}C$ polymorphism was significantly increased in patients with liver cirrhosis (LC) compared to normal population. There was no difference in allele distribution of the $G^{98}T$ polymorphism. The $A^{561}C$ polymorphism of E-selectin gene may be associated with disease progression in patients with chronic HBV infection and control the expression of plasma soluble levels, while the $G^{98}T$ polymorphism may be related to fibrotic severity in Chinese population (Wu et al. 2009).

Mice with targeted deletion of the P-selectin gene developed unpolarized type 1/type 2 cytokine responses and severely aggravated liver pathology following infection pathogen *Schistosoma mansoni*. Liver fibrosis increased 6 fold, despite simultaneous induction of IFN-γ and increase in inflammation in absence of P-selectin. This suggested a critical role of P-selectin in the progression of chronic liver disease caused by schistosome parasites (Wynn et al. 2004).

44.6.3 Neuro/Muscular Disorders

Axonal degeneration was confirmed as the major pathological feature of critical illness polyneuropathy (CIP). Expression of E-selectin was significantly increased in endothelium of epineurial and endoneurial vessels, suggesting endothelial cell activation (Fenzi et al. 2003). Increasing evidence indicates that inflammatory responses are implicated in the pathogenesis of cerebral vasospasm after aneurismal subarachnoid hemorrhage (SAH). Murine SAH model provided the evidence of effective prevention of SAH-induced vasospasm by a mAb implied the possible role of E selectin in the pathogenesis of vasospasm after SAH (Lin et al. 2005).

Neuroinflammation is present in the substantia nigra (SN) of patients of Parkinson disease (PD). A large number of ICAM-1-positive reactive astrocytes have been observed in the SN of patients with neuropathologically confirmed PD, including three of familial origin. The ICAM-1-positive reactive astrocytes were mainly concentrated around residual neurons in areas of heavy neuronal loss and extracellular melanin accumulation (Miklossy et al. 2006). The sVCAM-1 plasma levels were higher in late onset Alzheimer's disease (LOAD) and vascular dementia (VD) compared with controls. Among patients (LOAD, VD, and not-dementia

(CDND), sE-selectin levels were higher in individuals with most severe cerebrovascular disease on CT scan. Increased sVCAM-1 plasma levels in LOAD and VD suggest the existence of endothelial dysfunction in both types of dementia. Results support the possible role of E-selectin in the pathogenesis of cerebrovascular disease (Zuliani et al. 2008).

44.6.4 Acute Pancreatitis

Upregulation of ICAM-1, LFA-1, Mac-1 and subsequent leukocyte infiltration appears to be significant events of pancreatic and pulmonary injuries in Acute pancreatitis (AP) (Sun et al. 2006). Proinflammatory cytokines and oxidative stress seem to be involved in the development of local and particularly systemic complications in AP patients. Acute pancreatitis patients show VCAM-1 and P-selectin concentrations significantly lower and L-selectin concentrations significantly higher than the healthy subjects. Only E-selectin was significantly higher in severe than in mild disease (Pezzilli et al. 2008). Kleinhans et al. (2009) showed that the endothelial cell expression of PECAM-1, VCAM, E-selectin, and P-selectin was upregulated in severe porcine pancreatitis. In acute pancreatitis, plasma levels of sE-selectin and soluble thrombomodulin (sTM) serve as endothelial markers; the former is an endothelial activation marker, while the latter is an endothelial injury marker (Chooklin 2009; Ida et al. 2009).

44.6.5 Renal Failure

In patients affected by microscopic polyangiitis (MPA) and associated with myeloperoxidase (MPO)-anti-neutrophil cytoplasmic antibodies (ANCA), higher sICAM-1 and sE-selectin levels during active phase and their slower decline during the treatment period, could be a prognostic risk factor for chronic renal failure development (Di Lorenzo et al. 2004; Musial et al. 2005). An increased level of sE-selectin in patients susceptible to restenosis supports a role for white blood cell/endothelial interaction in restenosis after angioplasty (Sainani and Maru 2005). The impairment of vascular endothelial function was obvious in uremic patients with maintaining hemodialysis (MHD). The changes of ICAM-1 and E-selectin could be accepted as biochemical criterions of vascular endothelial injury (Li et al. 2005). Diuresis, serum creatinine, urea, and enzyme elimination are pathological among patients with acute renal failure (ARF). Higher elimination rates of sICAM-1 and higher values of sE-selectin compared to patients without ARF indicated additional parameters for early signs of kidney damage (Dehne et al. 2008). Both circulating and urinary TNF-α levels are increased in inflammatory chronic renal diseases. TNF-α appeared to play a crucial role in the immunopathogenesis of nephritis by the induction of chemokine, ICAM-1 and VCAM-1 expression via the activation of the intracellular MAPK signaling pathway, which may contribute to macrophage and lymphocyte infiltration (Ho et al. 2008; Li et al. 2008).

SNPs in Selectin genes and IgA Nephropathy: Although intensive efforts have been made to elucidate the genetic basis of Ig A nephropathy (IgAN), genetic factors associated with the pathogenesis of this disease are not well understood. A case–control study, based on linkage disequilibrium among SNPs in selectin gene cluster on chromosome 1q24-25 revealed two SNPs in the E-selectin gene (*SELE8* and *SELE13*) and six SNPs in the L-selectin gene (*SELL1*, *SELL4*, *SELL5*, *SELL6*, *SELL10*, and *SELL11*), that were significantly associated with IgAN in Japanese patients. *SELE8* and *SELL10* caused amino acid substitutions from His to Tyr and from Pro to Ser for His-to-Tyr substitutions; and *SELL1* could affect promoter activity of the L-selectin gene. The TGT haplotype at these three loci was associated significantly with IgAN. These SNPs in selectin genes may be useful for screening populations susceptible to the IgAN phenotype. (Takei et al. 2002)

Transplant Rejection: Soluble adhesion molecules are not valuable markers for stable kidney graft (STx) rejection reaction. However, patients with chronic renal failure showed increased levels of adhesion molecules, which could reflect an impaired elimination (Alcalde et al. 1995). The expression levels of ICAM-1 and VCAM-1 show positive correlation with the severity of graft rejection and can provide evidence for early diagnosis and prevention of CR. Chronic allograft failure (CAF) is the major cause for late graft loss in renal transplantation. ICAM-1 polymorphisms may represent a predetermined genetic risk factor for CAF. This was substantiated by the polymorphism in exon 4 at the Mac-1 binding site and in exon 6 at fifth Ig-like domain (McLaren et al. 1999). Khazen et al. (2007) found no evidence for an association of any polymorphism with acute rejection in E- and L-selectin. During kidney reperfusion, E-selectin, ICAM-1, and VCAM-1 concentrations correlated positively with hypoxanthine concentrations during reperfusion, whereas concentrations of ICAM-1 correlated negatively with xanthine concentrations, indicating metabolic changes in renal tissue (Domanski et al. 2009).

44.6.6 Other Inflammatory Disorders

Serum CAM levels have been analyzed in many organ diseases, including diseases of nervous system, endocrine disorders and others. Immune dysfunction has been proposed as a mechanism for pathophysiology of autistic-spectrum disorders. Levels of sP-selectin and sL-selectin were significantly lower in patients than in controls. Furthermore, sP-selectin levels were negatively correlated with impaired social development during early childhood (Iwata et al. 2008). In multiple sclerosis and in its animal model experimental autoimmune encephalomyelitis (EAE), inflammatory cells migrate across the endothelial blood–brain barrier (BBB) and gain access to the CNS. The role of E- and P-selectin in this process has been controversial. Döring et al. (2007) suggest that absence of E- and P-selectin did neither influence the activation of myelin-specific T cells nor the composition of the cellular infiltrates in the CNS during EAE. Thus, E- and P-selectin are not required for leukocyte recruitment across BBB and the development of EAE in C57BL/6 and in SJL mice (Döring et al. 2007). No significant differences in allelic or genotypic frequency in all the SNPs (rs6133, rs4987310 and rs5368 substitutions) tested were found in the Italian population (Fenoglio et al. 2009).

The pathophysiology of cluster headache (CH) is supposed to involve the lower posterior part of the hypothalamus, the trigeminal nerve, autonomic nerves and vessels in the orbital/retro-orbital region. Remahl et al. (2008) compared serum levels of sICAM-1, sVCAM-1 and sE-selectin in patients with episodic CH and in patients with biopsy-positive giant cell arteritis (GCA), a vasculitic disorder of large and medium-sized arteries. Within the CH group, sICAM-1, sVCAM-1 and sE-selectin showed an increasing trend in remission compared with active cluster headache period, but sE-selectin only was significant. Remahl et al. (2008) suggest that cluster headache is not a vasculitic disorder of medium-sized arteries, but CH patients may have an immune response that reacts differently from that of healthy volunteers.

Adhesion molecules have a role in many vasculitic disorders. Compared to controls, Takayasu's arteritis (TA) patients had elevated levels of sE-selectin, sVCAM-1, and sICAM-1. Compared to controls, patients with inactive TA also had elevated levels of sE-selectin, sVCAM-1, and sICAM-1. There was no difference between active TA and controls. The sE-selectin had a trend towards increased levels in inactive versus active TA, but there was no difference in sVCAM-1 and sICAM-1 levels between the groups. Patients with inactive TA had elevated levels of sE-selectin, sVCAM-1, and sICAM-1 that might indicate persistent vasculopathy in clinically inactive disease (Tripathy et al. 2008).

44.6.7 Inflammation in Hereditary Diseases

Serum levels of sVCAM-1, sICAM-1, sTM, P-selectin, E-selectin and CRP levels as inflammation markers are increased in patients of β-thalassemia intermedia and not influenced by treatment (Kanavaki et al. 2009). Pseudoxanthoma elasticum (PXE) is a hereditary disorder predominantly affecting the skin, retina and vascular system. P-selectin concentrations were increased in male and female PXE patients and levels correlated with the ABCC6 gene status of the patients. Patients harboring two mutant ABCC6 alleles had 1.5-fold increased P-selectin concentrations in comparison to patients with at least one wild-type allele. E- and L-selectin levels were within normal range and the allelic frequencies did not differ between from controls. Elevated P-selectin levels in PXE patients are potentially due to oxidative stress and elevated protease activity in PXE (Götting et al. 2008).

Fabry disease, an X-linked systemic vasculopathy, is caused by a deficiency of α-galactosidase A resulting in globotriaosylceramide (Gb_3) storage in cells. Accumulation of Gb_3 in the vascular endothelium of Fabry disease is associated with increased production of reactive oxygen species (ROS) and increased expression of CAMs. Increased Gb_3 induces expression of ICAM-1, VCAM-1, and E-selectin. Reduction of endogenous Gb_3 by treatment of the cells with an inhibitor of glycosphingolipid synthase or α-galactosidase A led to decreased expression of adhesion molecules. This study indicates that excess intracellular Gb_3 induces oxidative stress and up-regulates the expression of CAMs in vascular endothelial cells (Shen et al. 2008).

44.7 Role of CAMs in Cancer

Recent reports have expanded the concept that inflammation is a critical component of tumor progression. Many cancers arise from sites of infection, chronic irritation and inflammation. It is now becoming clear that the tumor microenvironment, which is largely orchestrated by inflammatory cells, is an indispensable participant in the neoplastic process, fostering proliferation, survival and migration. In addition, tumor cells have co-opted some of the signaling molecules of the innate immune system, such as selectins, chemokines and their receptors for invasion, migration and metastasis. These insights are fostering new anti-inflammatory therapeutic approaches to cancer development

44.7.1 Selectin Ligands in Cancer cells

Sialosyl Lewis[a] in Adhesion of Colon and other Cancers: The complexity of the tumor microenvironment has been revealed in the past decade. The CAMs in the process of inflammation are responsible for recruiting leukocytes onto the vascular endothelium before extravasation to the injured tissues. Some circulating cancer cells have been shown to extravasate to a secondary site using a process similar to inflammatory cells. The most studied ligands for CAMs expressed on cancer cells, s-Lewis[a] and s-Lewis[x] antigens, are shown to be involved in adhesion to endothelial cells by binding to E-selectin. This process, shared by inflammatory cells and cancer cells, may partially explain the link between inflammation and tumorigenesis. The adhesion of colon cancer cells to E-selectin can be directly affected by changes in the expression level of sialosyl Le[a] antigen. The specific lack of expression of sialosyl Le[a] carbohydrate structure on the surface of colon cancer cells completely abolished their adhesion to E-selectin. It is proposed that glycoproteins as well as gangliosides carrying sialosyl Le[a] structures, when properly exposed and present in high density on surface of cancer cells, can effectively support the adhesion of cancer cells to E-selectin (Klopocki et al. 1998; Kobayashi et al. 2007). In addition to endogenous ligands for L-, P-, and E-selectins (Chap. 26, 27, and 28), several proteins are found in cancer cell lines or solid tumors that act as ligands for E, L, and P selectins. Selectin ligands present in cancers are:

(1) Glycodelin A (GdA) is primarily produced in endometrial and decidual tissue and secreted to amniotic fluid. GdA is expressed in ovarian cancer where it can act as an inhibitor of lymphocyte activation and/or adhesion (Jeschke et al. 2009); (2) The cysteine-rich fibroblast growth factor receptor (FGF-R) represents the main E-selectin ligand (ESL-1) on granulocytes. Hepatic stellate cells (HSC) are pericytes of liver sinusoidal endothelial cells, which are involved in the repair of liver tissue injury and angiogenesis of liver metastases. HSC express FGF-R together with FucT7 and exhibit a functional E-selectin binding activity on their cell surface (Antoine et al. 2009). (3) Although B-cell precursor acute lymphoblastic leukemia (BCP-ALL) cell lines do not express the ligand PSGL-1, a major proportion of carbohydrate selectin ligand was carried by another sialomucin, CD43, in NALL-1 cells. CD43 plays an important role in extravascular infiltration of NALL-1 cells and the degree of tissue engraftment of BCP-ALL cells may be controlled by manipulating CD43 expression (Nonomura et al. 2008). (4) Thomas et al. (2009a) identified podocalyxin-like protein (PCLP) as an alternative selectin ligand. PCLP on LS174T colon carcinoma cells possesses E-/L-, but not P-, selectin binding activity. PCLP functions as an alternative acceptor for selectin-binding glycans. The finding that PCLP is an E-/L-selectin ligand on carcinoma cells offers a unifying perspective on the apparent enhanced metastatic potential associated with tumor cell PCLP overexpression and the role of selectins in metastasis (Thomas et al. 2009b). (5) E-selectin has been shown to play a pivotal role in mediating cell-cell interactions between breast cancer cells and endothelial monolayers during tumor cell metastasis. The counterreceptor for E-selectin was found as CD44v4. However, CD44 variant (CD44v) isoforms was functional P-, but not E-/L- selectin ligands on colon carcinoma cells. Furthermore, a ~180-kDa sialofucosylated glycoprotein(s) mediated selectin binding in CD44-knockdown cells. This glycoprotein was identified as carcinoembryonic antigen (CEA). CEA serves as an auxiliary L-selectin ligand, which stabilizes L-selectin-dependent cell rolling against fluid shear (Thomas et al. 2009b). Zen et al. (2008) identified a ~170 kDa human CD44 variant 4 (CD44v4) as E-selectin ligand, which has a high affinity for E-selectin via sLe[x] moieties.

44.7.2 E-Selectin-Induced Angiogenesis

Angiogenesis plays an important role in a variety of pathophysiologic processes, including tumor growth and rheumatoid arthritis. Studies on capillary morphogenesis and angiogenesis in vitro have suggested a role for E-selectin in the process of differentiation into tube-like structures. Soluble E-selectin is a potent mediator of human dermal microvascular endothelial cell (HMVEC) chemotaxis, which is predominantly mediated through the Src and the phosphatidylinositiol 3-kinase (PI3K) pathways (Kumar et al. 2003). Gastrin-17 (G17) has marked proangiogenic effects in vivo on experimental gliomas and in vitro on HUVECs and transiently decreased the expression of E-selectin, but not P-selectin, whereas IL-8 increased the expression of E-selectin. Specific antisense oligonucleotides against E- and P-selectin decreased HUVEC tubulogenesis processes in vitro. This showed that gastrin has marked proangiogenic effects in vivo on experimental gliomas and in vitro on HUVECs. This effect depends in part on the level of E-selectin activation, but not on IL-8 expression/release by HUVECs (Lefranc et al. 2004).

44.7.3 E-Selectin in Cancer Cells

Adhesion molecules are thought to have a role in the host defense against carcinogenesis. Significantly increased P-selectin, s-VCAM-I and s-ICAM-I levels were observed in patients with bladder cancer, and s-VCAM-I levels correlated with tumor stage (Coskun et al. 2006). Selectins mediate attachment of leukocytes to activated endothelium as well as the adhesion reaction of tumor cells during malignancy (Borsig 2007). In a breast tumor xenograft model, the effect of combined TNF-α and IFN-γ therapy involved the selective destruction of the tumor vasculature and death of tumor cells. Concomitant with these changes RT-PCR analysis revealed the increase of stromal mRNA levels for a series of stromal cytokines, cytokine receptors including TNF-α, sICAM-1, VCAM-1, P-selectin, which could be implicated in the observed events (de Kossodo et al. 1995).

Squamous Cell Carcinomas: In order to evaluate the risk of postoperative haematogenic recurrence of esophageal squamous cell carcinoma (SCC) patients, Shimada et al. (2003) examined the preoperative serum levels of sE-selectin and pathological status of the patients. The patients with a high serum soluble E-selectin concomitant with expression of s-Lewis antigens had a significant risk of postoperative haematogenic recurrence. SCCs of sun-induced skin cancers are particularly numerous in patients on T cell immunosuppression. Blood vessels in SCCs did not express E-selectin, and tumors contained few cutaneous lymphocyte antigen (CLA$^+$) T cells, the cell type thought to provide cutaneous immunosur-veillance. Clark et al. (2008) found that SCCs evade the immune response at least in part by down-regulating E-selectin and recruiting T$_{reg}$ cells.

Cutaneous T-Cell Lymphoma (CTCL): The CTCL is characterized by accumulation of malignant CD4$^+$ T cells in the skin. In malignant T cells from Sezary syndrome (SS), a leukemic variant of CTCL, in dermal microvessels in mouse skin, Hoeller et al. (2009) found that SS cells rolled along dermal venules in a P-selectin- and E-selectin-dependent manner at ratios similar to CD4$^+$ memory T cells from normal donors. Chemokine CCL17/TARC was sufficient to induce the arrest of SS cells in the microvasculature. Together, experiments suggested molecular adhesion cascade operant in SS cell homing to the skin in vivo. Patients with CTCL showed increased levels of sICAM-1 and sICAM-3 when compared with healthy individuals and patients with inflammatory dermatosis. The sE-selectin and sVCAM-1 levels were not affected (López-Lerma and Estrach 2009).

Hodgkin's Disease: Increased sICAM-1 and sE-selectin have been observed in Hodgkin's Disease/lymphoma (HD/HL) patients at diagnosis and sVCAM-1 at diagnosis correlated with both sICAM-1 and sE-selectin levels. Chemotherapy resulted in a significant decrease of sICAM-1 and sE-selectin (Syrigos et al. 2004). Serum sICAM-1 level increases at advanced stages of untreated multiple myeloma (MM) patients, but did not differ significantly from controls.

A positive correlation of IL-6 appeared with sICAM-1 and sE-selectin (Uchihara et al. 2006). Epstein-Barr virus (EBV)-positive NK/T cells showed affinity to vascular components. EBV-positive NK lymphoma cells express ICAM-1 and VCAM-1 at much higher levels than those in EBV-negative T cell lines. Furthermore, NK lymphoma cell lines exhibited increased adhesion to cultured endothelial cells stimulated with TNF-α or IL-1β. The up-regulated expression of VCAM-1 on cytokine-stimulated endothelial cells can be important to initiate the vascular lesions (Kanno et al. 2008).

Non-small Cell Lung Cancer: Serum levels of ICAM-1 increased in advanced stage non-small cell lung cancer (NSCLC) patients, whereas sE-selectin levels were not significantly different from healthy controls Reports suggest that higher serum ICAM-1 can be useful for diagnosis while E-selectin levels have prognostic significance and could be a potential prognostic factor in NSCLC patients (Dowlati et al. 2008; Guney et al. 2008). The Cyfra 21–1 and sE-selectin showed good performance in detecting lung cancer from normal groups. However, Cyfra 21–1 was superior to sE-selectin in discriminating lung cancer from benign lung diseases (Swellam et al. 2008).

44.7.3.1 Thyroid Cancer

Maspin, a serine protease inhibitor belonging to serpin family, is known as a tumor-suppressor protein and also exhibits an inhibitor effect on angiogenesis. Positive correlations were found for maspin positivity and lymph node metastases; E-selectin positivity and lymph node metastases, and P-selectin positivity and lymph node metastases and lymphovascular invasion. Correlations do exist between maspin, E- and P-selectin expressions with each other and with tumor stage. Inactive cytoplasmic maspin cannot act as a tumor suppressor. Expression of E- and P-selectins in tumor cells facilitates the occurrence of metastases, lymphovascular invasion, and perithyroidal soft tissue invasion. Further studies are needed to reveal detailed interactions between maspin, E-selectin, and P-selectin expression (Bal et al. 2008).

Primary Hyperparathyroidism: Patients with primary hyperparathyroidism (PHPT) have impaired vasodilation. Based on small number of patients, a study suggested that classic cardiovascular risk factors seem to be the main determinants for the high plasma levels of sE-selectin and vWF in PHPT. Together with unaltered thrombomodulin and sE-selectin levels, the vWF decrease in plasma after parathyroidectomy reflects a specific mechanism of its endothelial calcium- and/or PTH-stimulated secretion in some PHPT patients without risk factors (Fallo et al. 2006).

Colorectal Cancer: Plasma level of sP-selectin, sE-selectin and ICAM-1 were significantly higher in colorectal cancer (CRC) patients. The highest levels of sE-selectin and ICAM-1 were observed in patients with liver metastasis. There was no correlation between sP-selectin and sE-selectin, but a significant correlation was seen between sE-selectin and ICAM-1 in all patients. Plasma concentration of E-selectin and ICAM-1 may indicate tumor progression and liver metastasis (Dymicka-Piekarska and Kemona 2009; Sato et al. 2010).

The interaction between rE-selectin and CRC cells alters the gene expression profile of cancer cells. A DNA microarry analysis indicated that E-selectin-mediated alterations were significantly more pronounced in metastatic CRC variants SW620 and KM12SM than in the corresponding non-metastatic local SW480 and KM12C variants. The number of genes altered by E-selectin in metastatic variants was 10-fold higher than the number of genes altered in the corresponding local variants. Analysis indicated that E-selectin down regulated (at least by 1.6-folds) the expression of seven genes in a similar fashion, in both metastatic cells. Confocal microscopy indicated that E-selectin down-regulated the cellular expression of HMGB1 protein and enhanced the release of HMGB1 into the culture medium. The released HMGB1 in turn, activated endothelial cells to express E-selectin (Aychek et al. 2008).

The entrapment of malignant cells within the hepatic sinusoids and their interactions with resident non-parenchymal cells are considered very important for the whole metastatic sequence. In the sinusoids, cell connection and signaling is mediated by multiple cell adhesion molecules, such as the selectins. The three members of the selectin family, E-, P- and L-selectin, in conjunction with sialylated Lewis ligands and CD44 variants, regulate colorectal cell communication and adhesion with platelets, leucocytes, sinusoidal endothelial cells and stellate cells. Therefore, trials have already commenced aiming to exploit selectins and their ligands in the treatment of benign and malignant diseases. Multiple pharmacological agents have been developed that are being tested for potential therapeutic applications (Schnaar et al. 2008; Paschos et al. 2010; Zigler et al. 2010).

44.7.4 Metastatic Spreading

The degree of selectin ligand expression by cancer cells is well correlated with metastasis and poor prognosis for cancer patients. Initial adhesion events of cancer cells facilitated by selectins result in activation of integrins, release of chemokines and are possibly associated with the formation of permissive metastatic microenvironment. While E-selectin is one of the initiating adhesion events during metastasis, it is becoming apparent that P-selectin and L-selectin-mediated interactions significantly contribute to this process as well (Gout et al. 2008; Läubli and Borsig 2010).

E-Selectin in Progression of Metastasis of Breast Cancer: Extravasation of cancer cells is a pivotal step in the formation of hematogenous metastasis. Extravasation is initiated by the loose adhesion of cancer cells to endothelial cells via an interaction between endothelial selectins and selectin ligands expressed by the tumor cells. Metastatic spreading is a dreadful complication of neoplastic diseases that is responsible for most deaths due to cancer. It consists in the formation of secondary neoplasms from cancer cells that have detached from the primary site. Leukocytes and tumor cells use E-selectin binding ligands to attach to activated endothelial cells expressing E-selectin during inflammation or metastasis. The formation of these secondary sites is not random and several clinical observations indicate that the metastatic colonization exhibits organ selectivity. This organ tropism relies mostly on the complementary adhesive interactions between cancer cells and their microenvironment. E-selectin and sLewis antigens might play important role in breast tumor, lymph node and liver metastasis. High levels of sE-selectin have been reported in melanoma and some epithelial tumors, especially in colorectal carcinoma. But sE-selectin may not be used as a predictive marker of metastasis in colorectal carcinoma, though high levels of sE-selectin may support diagnosis of liver metastasis (Uner et al. 2004; Eichbaum et al. 2004). It appeared that serum levels of sE-selectin are associated with the clinical course of liver metastases from breast cancer. Eichbaum et al. (2004) observed a possible trend for certain unfavorable prognostic parameters (e.g., young women, low-graded tumors, human epidermal growth factor receptor 2 over-expression) that could be related to higher serum levels of sE-selectin.

Role of E-Selectin in Diapedesis of Cancer Cells: Diapedesis is a vital part of tumor metastasis, whereby tumor cells attach to and cross the endothelium to enter the circulation. E-selectin was found to regulate initial attachment and rolling of colon cancer cells and also the subsequent diapedesis through the endothelium. Evidence indicates that E-selectin-dependent paracellular extravasation is independent of ICAM and VCAM and that it requires the activation of extracellular signal-regulated kinase (ERK) mitogen-activated protein kinase downstream of E-selectin. Studies establish the role of E-selectin in diapedesis of circulating cancer cells (Tremblay et al. 2008; Woodward 2008).

Polymorphisms within E-selectin gene, especially the $S^{128}R$ polymorphism, may increase the risk of metastases by facilitating adhesion of tumor cells to endothelium. Blood DNA from patients treated for stage II or III colorectal cancer (CRC) and from healthy controls was assessed for three polymorphisms within E-selectin gene ($S^{128}R$, $G^{98}T$ and $L^{554}F$) and one within the P-selectin gene ($V^{640}L$). The $S^{128}R$ polymorphism was detected in 22.3 % patients and was correlated with $G^{98}T$ polymorphism. In multivariate analysis, the $S^{128}R$ polymorphism was associated with shorter event-free survival (EFS) and overall survival (OS) in whole population, in patients with stage II CRC, and in patients with stage III CRC. $L^{554}F$ and $V^{640}L$ polymorphisms had no prognostic value. The $S^{128}R$ polymorphism is a constitutional factor associated with a higher risk of relapse and death in patients treated for CRC and its detection may permit better selection of patients suitable for adjuvant therapy, especially among those with stage II disease (Hebbar et al. 2009).

P-selectin Deficiency Attenuates Tumor Growth and Metastasis: Metastasis is thought to involve the formation of tumor-platelet-leukocyte emboli and their interactions with the endothelium of distant organs. A link between these observations shows that P-selectin, which normally binds leukocyte ligands, can promote tumor growth and facilitate the metastatic seeding of a mucin-producing carcinoma. P-selectin-deficient (P-sele$^{-/-}$) mice showed three potential pathophysiological mechanisms: (1) intravenously injected tumor cells home to the lungs of P-sele$^{-/-}$ mice at a lower rate; (2) P-sele$^{-/-}$ mouse platelets fail to adhere to tumor cell-surface mucins; and (3) tumor cells lodged in lung vasculature after intravenous injection often are decorated with platelet clumps, and these are markedly diminished in P-selectin$^{-/-}$ animals (Kim et al. 1998). However, the surgical procedure did not totally eliminate the factors responsible for platelet activation and did not normalize platelet activation (Dymicka-Piekarska et al. 2005; Hanley et al. 2006).

Role of Sialyl-Lewis Antigens: During inflammation, E- and P-selectins appear on activated endothelial cells to interact with leukocytes through sialyl-Lewis X (sLe^x) and sialyl-Lewis A (sLe^a). These selectins can also interact with tumor cells in a sialyl-Lewis-dependent manner and hence, they are thought to play a key role in metastasis. Diverting the biosynthesis of sialyl-Lewis antigens toward nonadhesive structures is an attractive gene therapy for preventing the hematogenous metastatic spread of cancers. The transduced α1,2-fucosyltransferase-1 (FUT1) efficiently fucosylated the P-selectin ligand PSGL-1 without altering P-selectin binding (Mathieu et al. 2004).

The metastasis of cancer cells and leukocyte extravasation into inflamed tissues share common features. Carbohydrate antigen sLe^a (CA19-9) is the most frequently applied serum tumor marker for diagnosis of cancers in the digestive organs. The normal counterpart of the determinant, namely disialyl-Le^a is predominantly expressed in non-malignant epithelial cells of the digestive organs. The disialyl-Le^a determinant carries one extra sialic residue attached through a $2 \rightarrow 6$ linkage to GlcNAc moiety compared to cancer-associated sLe^a, which carries only one $2 \rightarrow 3$ linked sialic acid residue (monosialyl Lewis A) (Fig. 44.3). Disialyl-Le^a in normal epithelial cells serves as a ligand for immunosuppressive receptors such as sialic acid binding Ig-like lectins (siglec-7 and -9) expressed on resident monocytes/macrophages and maintains immunological homeostasis of mucosal membranes in digestive organs. Epigenetic silencing of a gene for a $2 \rightarrow 6$ sialyl-transferase in the early stages of carcinogenesis results in impairment of $2 \rightarrow 6$ sialylation, leading to incomplete synthesis and accumulation of sLe^a, which lacks the $2 \rightarrow 6$ linked sialic acid residue, in cancer cells. Simultaneous determination of serum levels of sLe^a and disialyl-Le^a, and calculation of the sLe^a/disialyl-Le^a ratio provides information useful for excluding a false-positive serum diagnosis. During cancer progression in locally advanced cancers, tumor hypoxia induces transcription of several glyco genes involved in sLe^a synthesis. Expression of the determinant, consequently, is further accelerated in more malignant hypoxia-resistant cancer cell clones, which become predominant clones in advanced stage cancers and frequently develop hematogenous metastasis. sLe^a, as well as its positional isomer sLe^a, serves as a ligand for vascular E-selectin and facilitates hematogenous metastasis through mediating adhesion of circulating cancer cells to vascular endothelium. Patients having both strong sLe^a expression on cancer cells and enhanced E-selectin expression on vascular beds are at a greater risk of developing distant hematogenous metastasis (Kannagi 2007). In a human-mouse model, the selectin ligand s-Le^a is involved in in vivo extravasation of colorectal carcinoma (CRC) cells. Highly metastatic CRC cells expressing high levels of s-Le^a extravasate more efficiently than non-metastatic CRC cells expressing low levels of s-Le^a. Down-regulating the expression of s-Le^a in CRC cells by genetic manipulations, significantly reduced CRC extravasation. The arrest and adhesion of CRC cells, and possibly of other types of cancer cells as well, to endothelium depend on the expression of the selectin ligand sLe^a by the tumor cells (Ben-David et al. 2008). 3'-Sulfo-Le^a is known to be the potent ligand of E-selectin which is important in cell adhesion and migration. The serum 3'-sulfo-Le^a can provide important information in patients with primary gastric cancer, which might be useful as a predictive marker especially for the detection of tumor metastasis (Zheng et al. 2009).

Specialized carbohydrates modified with sLe^x antigens on leukocyte membranes are ligands for selectin adhesion molecules on activated vascular endothelial cells at

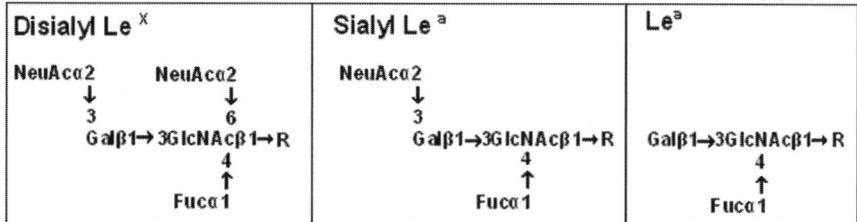

Fig. 44.3 Structures of three carbohydrate determinants, disialyl Lewis x (Le^x), sialyl Lewis a and Lewis a (Le^a). In panel, note that the only difference between the three determinants is the linkage of sialic acid residues. The α(2 → 6) sialic acid residue in disialyl Le^a is synthesized by a sialyltransferase ST6GalNAcVI, which shows a significant decrease in its mRNA level upon malignant transformation

inflammatory sites. The sLe^x expression of invasive micropapillary carcinoma was higher than that of invasive ductal carcinoma, which was also associated with lymph node metastasis. E-selectin combined with sLe^x might play an important role in lymph node metastasis in invasive micropapillary carcinoma. The expression pattern of sLe^x in invasive micropapillary carcinoma suggested that the reversal of cell polarity of invasive micropapillary carcinoma might be as an important factor for the morphogenesis and possibly the pathogenesis, especially their higher rates of lymph node metastasis (Wei et al. 2010). The activity of core 2 β1,6 N-acetylglucosaminyltransferase (C2GnT1) in leukocytes greatly increases their ability to bind to endothelial selectins. C2GnT1 is essential for the synthesis of core 2-branched O-linked carbohydrates terminated with sLe^x (C2-O-sLe^x). E-selectin and its ligand-sLe^x are closely correlated with the metastasis of hepatocellular carcinoma. C2-O-sLe^x is a potentially useful early predictor of metastasis (Zhang et al. 2002). The expression profiles of C2-O-sLe^x in the malignant progression and metastasis of colorectal adenocarcinomas is upregulated in colorectal adenocarcinomas and metastatic liver tumors (St Hill et al. 2009).

44.7.5 Survival Benefits of Heparin

Endothelial P-Selectin as a Target of Heparin Action: Metastasis can be effectively inhibited by the anticoagulant heparin in different tumor models. At the cellular level, many of the antimetastatic effects of heparin in vivo are due to its action on P-selectin-mediated binding. Ludwig et al. (2007) addressed the potential contribution of endothelial P-selectin expression to adhesive events between the microvasculature and melanoma cells in vivo. Heparin not only inhibits P-selectin-mediated melanoma cell rolling but also attenuates melanoma metastasis formation in vivo, supporting the concept that endothelial P-selectin expression may represent an additional target of heparin in experimental melanoma lung metastasis (Ludwig et al. 2007). The low molecular weight heparin (LMWH) significantly improved colonic inflammation in rats with trinitrobenzene sulphonic acid (TNBS) induced colitis. The effect is possibly related to inhibition of proinflammatory cytokine IL-8, but not involved platelet surface P-selectin expression (Xia et al. 2004). The survival benefits in patients with cancer treated with LMWH may result from a LMWH-mediated effect on the immune system or on the cross-talk between platelets and tumor cells. However, survival observed with LMWH in patients with cancer apparently cannot be explained by a LMWH effect on these circulating markers (Di Nisio et al. 2005). Nonetheless, in vivo antimetastatic effects of heparins reflect their action on P-selectin-mediated binding. Therefore, these commonly used anticoagulants widely differ in their potential to interfere with P-selectin mediated cell binding. Importantly, the superior inhibitory capacity on P-selectin function of unfractionated heparin and LMWH nadroparin as opposed to LMWH enoxaparin and synthetic heparin pentasaccharide fondaparinux strongly correlated to the inhibitory potency of each in inhibiting experimental lung metastasis in vivo. Hence, P-selectin inhibition constitutes a valuable feature to identify anticoagulants that are suitable for anticancer therapy (Ludwig et al. 2006). Stevenson et al. (2005) studied metastasis inhibition by clinically relevant levels of various heparins and investigated the structural basis for selectin inhibition differences. Five clinically approved heparins were evaluated for inhibition of P-selectin and L-selectin binding to carcinoma cells and showed differing abilities to inhibit selectins, likely explained by size distribution. It should be possible to size fractionate heparins and inhibit selectins at concentrations that do not have a large effect on coagulation. Gao et al. (2005) prepared periodate-oxidized, borohydride-reduced heparin (RO-heparin) and tested its anticoagulant and anti-inflammatory activities. Compared with heparin, RO-heparin had greatly reduced anticoagulant activity. Intravenous administration of this compound led to reduction in the peritoneal infiltration of neutrophils in a mouse acute inflammation model. In vitro studies showed that the effect of RO-heparin on inflammatory responses was mainly due to inhibiting the interaction of P-selectin with its ligands. These results indicate that RO-heparin may be a safer treatment for inflammation than heparin, especially when selectin is targeted.

To clarify the mechanism of heparin antimetastatic activity, several biological effects are being investigated. Cancer progression and metastasis are associated with enhanced expression of heparanase, which is inhibited efficiently by heparin. Heparin is also a potent inhibitor of selectin-mediated interactions. P- and L-selectin were shown to contribute to the early stages of metastasis, which is associated with platelet-tumor cell thrombi formation. Low anticoagulant heparin preparations still inhibited metastasis efficiently indicating that anticoagulation is not a necessary component for heparin attenuation of metastasis. Modified heparins characterized for heparanase inhibitory activity are also potential inhibitors of selectins. Selectin inhibition is a clear component of heparin inhibition of metastasis. The contribution of selectin or heparanase inhibition by heparin can provide evidence about its antimetastatic activity (Borsig 2007). One of the mechanisms by which heparin inhibits metastasis is by blocking the P-selectin-based interaction of platelets with tumor cell. The sulfate groups at C6/N and especially C6, but not C2 and C3, of heparin play a critical role in P-selectin recognition and that 2-O,3-O-desulfated heparin can block P-selectin-mediated A375 human melanoma cell adhesion. Thus chemical modification of heparin, especially 2-O,3-O-desulfation, may result in a therapeutic agent that is anti-metastatic because it blocks unwanted P-selectin-dependent adhesion but that lacks dose-limiting anticoagulant effects (Wei et al. 2005).

Heparin-Induced Thrombocytopenia: The pathophysiology of heparin-induced thrombocytopenia (HIT) is a complex process which involves platelets, vascular endothelium, and leukocytes. The activation products from these sites also contribute to the activation of coagulation and to the fibrinolytic deficit. Many of the markers of hemostatic activation processes have been found to be at increased levels during acute phases of the HIT syndromes. Since the pathophysiology of HIT involves the activation of platelets, endothelium, and leukocytes, it is expected that activation products related to these hemostatic systems, including soluble selectins, will also be increased in circulating blood. These alterations may provide an index of the pathophysiologic process. Fareed et al. (1999) reviewed on the circulating levels of P-, E-, and L-selectins in HIT patients and their modulation after therapeutic intervention. With the availability of recombinant hirudin, it is now possible to provide alternate anticoagulants to HIT patients. However, Fareed et al. (1999) suggest that the immunoactivation of platelets and other cells may require additional adjunct therapeutic approaches.

44.8 Adhesion Proteins in Transplantation

Activated protein C (APC) is the major physiological anticoagulant with concomitant anti-inflammatory properties. Turunen et al. (2005) suggest that APC has an anti-inflammatory role in I/R injury in clinical renal transplantation (Turunen et al. 2005). Bimosiamose prolongs survival of kidney allografts. Binding of the P-, L-, and E-selectins to sLe X retards circulating leukocytes, thereby facilitating their attachment to the blood vessels of allografts. Selectin inhibitor bimosiamose (BIMO) inhibits the rejection process of kidney allografts in a rat model in association with reduced intragraft expression of P-selectin glycoprotein ligand-1, CX(3)CL1, CCL19, CCL20, and CCL2. Thus, BIMO blocks allograft rejection by reduction of intragraft expression of cytokines and chemokines (Langer et al. 2004).

Brain death (BD), a significant antigen-independent process, the donor-related injury up-regulates variety of inflammatory mediators in peripheral organs. One of the immediate responses is the expression of selectins by endothelial cells of the transplanted tissues, which in turn trigger a cascade of nonspecific events that may enhance host alloresponses. Using a rat model in which donor BD accentuates subsequent renal allograft injury, Gasser et al. (2005) tested the effects of therapy with rPSGL-Ig alone, or in combination with sirolimus (SRL) and cyclosporin A. It was found that in contrast to the effects of standard doses of SRL or cyclosporine, rPSGL-Ig decreased inflammation in the early posttransplant period such that lower doses of maintenance immunosuppression were sufficient to maintain long-term graft function.

Intestinal transplantation (ITx) is severely limited by ischemia-reperfusion (I/R) injury. T lymphocyte is an important regulatory cell in this inflammatory process (Farmer et al. 2005a). rPSGL-Ig treatment leads to marked improvement in the outcome. The mechanism of action seems to involve the blockade of neutrophil and lymphocyte infiltration that leads to a decreased inflammatory response possibly driven by Th2 cytokines (Farmer et al. 2005b).

It was suggested that liver transplantation and liver resection, together with portal clamping time, might be a potential stimulus for platelet activation. Becker et al. (2004) indicated that neither liver transplantation nor liver resection influences GPIIb/IIIa and P-selectin expression on circulating platelets (Becker et al. 2004).

44.9 Inflammation During Infection

44.9.1 Microbial Pathogens

Endothelial activation contributes significantly to the systemic inflammatory response to bacteraemia. Release of soluble endothelial markers into the circulation has been demonstrated together with elevated plasma levels of CAMs and has been reported in bacteraemic patients. It has been proposed that the infection of endothelial cells with *Staphylococcus aureus, Streptococcus sanguis,* or *Staphylococcus epidermidis* induces surface expression of ICAM-1 and VCAM-1 and monocyte adhesion. In general, leukocyte/endothelial cell interactions such as capture, rolling, and firm adhesion should be viewed as a series of overlapping synergistic interactions among adhesion molecules resulting in an adhesion cascade. These cascades thereby direct leukocyte migration, which is essential for the generation of effective inflammatory responses and the development of rapid immune responses (Golias et al. 2007). *Helicobacter pylori* is a common bacterial pathogen that infects world's population up to 50 %. Carbohydrate components on *H. pylori* (sequences related to Le^x or Le^a antigens) are responsible for the persistent inflammation through interactions with leukocyte-endothelial adhesion molecules of the host. *H. pylori* isolates from patients with chronic gastritis, duodenal ulcer and gastric cancer interact with E- and L-selectins (Galustian et al. 2003). Expression of E-selectin was specifically upregulated in *H. pylori*-induced gastritis but not in gastritis induced by acetylsalicylic acid or pouchitis. The upregulated E-selectin expression was localized to the gastric mucosa rather than being a systemic response to the infection (Svensson et al. 2009).

Although mice with mutations in individual selectins showed no spontaneous disease and had a mild or negligible deficiencies of inflammatory responses, Bullard et al. (1996), in contrast, found that mice with null mutations in both endothelial selectins (P and E) develop a phenotype of leukocyte adhesion deficiency characterized by mucocutaneous infections in response to intraperitoneal *S. pneumoniae* peritonitis. These mice provide strong evidence for the functional importance of selectins in vivo (Bullard et al. 1996). Anthrax lethal toxin (LT), a key virulence factor of *Bacillus anthracis*, enhanced VCAM-1 expression on primary human endothelial cells suggesting a causative link between dysregulated adhesion molecule expression and the poor immune response and vasculitis associated with anthrax. Results suggest that LT can differentially modulate NF-kB target genes and highlight the importance of VCAM-1 enhancement (Warfel and D'Agnillo 2008). Vascular endothelium stimulation in vitro that lead to the upregulation of CAMs is known for the pathogenic spirochaetes, including rLIC10365 of *Leptospira interrogans*. The recombinant proteins of *L. interrogans* in *E. coli* as a host were capable to promote the upregulation of ICAM-1 and E-selectin on monolayers of HUVECS. In addition, pathogenic and non-pathogenic Leptospira are both capable to stimulate endothelium E-selectin and ICAM-1, but the pathogenic *L. interrogans serovar Copenhageni* strain promoted a higher activation than the non-pathogenic *L. biflexa serovar Patoc* (Atzingen et al. 2009; Gómez et al. 2008). *Chlamydia pneumoniae* has been associated with cardiovascular disease and atherosclerosis. To determine the ability of *C. pneumoniae* to elicit inflammation, Högdahl et al. (2008) infected human coronary artery endothelial cells (HCAEC) with *C. pneumoniae*. Secretion of IL-8, MCP-1, and ICAM-1 was significantly increased after *C. pneumoniae* infection of HCAEC in comparison with uninfected controls, where as release of E-selectin or MMP-1did not change. This suggested that *C. pneumoniae* initiates and propagates vascular inflammation in ways that contribute to coronary artery disease (Högdahl et al. 2008).

CAMs in Gingival Crevicular Fluid: The sICAM-1, sVCAM-1, and sE-Selectin are present in gingival crevicular fluid (GCF) and changes in their levels may be a sensitive indicator to differentiate healthy sites from those with periodontitis (Hannigan et al. 2004; Tamai et al. 2007). *Porphyromonas gingivalis* is a Gram-negative bacterium that is an important etiologic agent of human adult periodontitis. *E. coli* LPS and isoforms of *P. gingivalis* LPS were potent in stimulating the expression of inflammatory markers, with *E. coli* LPS being more potent (Liu et al. 2008). DNA samples from blood of periodontitis patients genotyped for E-selectin Ser^{128}Arg and L-selectin Phe^{206}Leu revealed a significant difference in the Ser^{128}Arg polymorphism of E-selectin, but not in L-selectin, between periodontal patients and controls; the 128Arg allele was present more frequently in patients. Houshmand et al. (2009) suggested that Ser^{128}Arg polymorphism of E-selectin might contribute to the susceptibility of Iranian individuals to periodontitis.

CAMs in Subjects with HIV Disease: Swingler et al. (2003) suggested that while both soluble CD23 and ICAM1 promote resting cell HIV1 infection, productive infection of cycling cells requires soluble ICAM1. Swingler et al. (2003) noted that these results may explain in part the existence of a resting T-cell reservoir infected with HIV-1. Subjects with HIV disease have multiple risk factors for cardiovascular disease, including elevated levels of ICAM-1 and VCAM-1. Many of the variables associated with

ICAM-1 and VCAM-1 levels can be related to their impact on inflammation (Melendez et al. 2008). The LFA-1, ICAM-1, and ICAM-3 are enriched at virological synapse (VS). The cognate adhesion molecule interactions at VS are important for HIV-1 spread between T cells (Jolly et al. 2007).

44.9.2 Yeasts and Fungi

Zuccarello et al. (2002) described a distinct form of familial chronic mucocutaneous candidiasis characterized by early-onset infections by different species of Candida, restricted to the nails of the hands and feet and associated with low serum concentration of ICAM-1. Phan and Filler (2009) measured the effects of *C. albicans* on the endothelial cell production of E-selectin and TNF-α in vitro. During invasive pulmonary aspergillosis, *A. fumigatus* hyphae invade the abluminal endothelial cell surface, whereas they invade the luminal endothelial cell surface during haematogenous dissemination. Infection with hyphae stimulates endothelial cells to synthesize E-selectin, VCAM-1, IL-8, and TNF-α in vitro. In neutropenic mice infected with wild-type *A. fumigatus*, increased pulmonary expression of E-selectin and TNF-α occurred only when neutropenia had resolved. In nonneutropenic mice immunosuppressed with corticosteroids, *A. fumigatus* stimulated earlier pulmonary expression of E-selectin and VCAM-1, while expression of ICAM-1 and TNF-α was suppressed. In both mouse models, expression of E-selectin was associated with high pulmonary fungal burden, angioinvasion, and neutrophil adherence to endothelial cells (Chiang et al. 2008; Kamai et al. 2009).

44.9.3 Parasites and Amoeba

44.9.3.1 Falciparum Malaria

Significant differences are observed between falciparum malaria patients and the healthy people in term of levels of both sE-selectin and thrombomodulin (TM). The levels of both sE-selectin and TM correlated positively with temperature, levels of IFN-γ and levels of TNF-α; and negatively with hemoglobin levels. Trends of positive correlations were observed between sP-selectin or vWF and temperature (Matondo et al. 2008). Evidence from autopsy and in vitro binding studies suggests that adhesion of erythrocytes infected with *Plasmodium falciparum* to the human host ICAM-1 receptor is important in the pathogenesis of severe malaria. Fernandez-Reyes et al. (1997) identified a mutation (K29M) in the ICAM1 gene, which they designated 'ICAM1 Kilifi,' that was associated with susceptibility to cerebral malaria with relative risks of 2.23 and 1.39 for homozygotes and heterozygotes, respectively. The available epidemiological, population genetic and functional evidence link ICAM-1(Kilifi) to severe malaria susceptibility (Fry et al. 2008; Cojean et al. 2008).

Increased serum concentrations of soluble sICAM-1, CD54 and of soluble E-, but not soluble P- and L-selectins were detected in Malagasy patients living in hyperendemic focus of Schistosoma mansoni. Serum levels of ICAM-1 were significantly correlated with the disease severity (Esterre et al. 1998). Studies in several models of inflammation have underscored the importance of P- and E-selectins in the migration of T cells to inflamed tissues. $CD4^+$ T cells recruited to the cutaneous compartment during infection with Leishmania major express P- and E-selectin ligands. Results suggest that by blocking P- and E-selectins, the immune pathology associated with cutaneous leishmaniasis might be ameliorated without compromising immunity to infection (Zaph and Scott 2003). Invasive amebiasis offers a new model that poses an inadequate immune response leading to a continuous and prolonged activation of endothelial cells (ECs) by amebas, amebic molecules and cytokines, leading to necrosis. Hyperactivated endothelial cells continuously express ICAM-1 and E-selectin, pro-coagulant molecules (tissue factor, vWF, and the plasminogen activator inhibitor), resulting in ever greater inflammation and thrombosis (Campos-Rodríguez et al. 2009)

44.9.3.2 Sepsis

Sepsis is a multifactorial, and often fatal, disorder typically characterized by widespread inflammation and immune activation with resultant endothelial activation. Though bacterial sepsis is most common, sepsis occurs with fungal, parasitic and mycobacterial organisms. During bacterial sepsis in vivo, in wild-type mice and mice with E-, P-, or E-/P-selectin deficiencies, a phenotypic abnormality in E-selectin-deficient mice suggested that E- and P-selectin are important in the host defense against *S. pneumoniae* infection (Munoz et al. 1997). P-selectin is an important mediator of eosinophil recruitment to the cornea from limbal vessels to the corneal stroma, suggesting that P-selectin interactions may be potential targets for immunotherapy in eosinophil-mediated ocular inflammation (Kaifi et al. 2000).

Staphylococcus aureus is one of the most significant pathogens in human sepsis and endocarditis. Peptidoglycan induced surface expression of EC inflammation markers ICAM-1 and VCAM-1, which supported the adhesion of monocytes to these ECs (Mattsson et al. 2008). Teoh et al. (2008) assigned adiponectin as a modulator of survival and endothelial inflammation in experimental sepsis and a potential mechanistic link between adiposity and increased sepsis. Newborn infants with clinical diagnosis of sepsis demonstrated significantly higher plasma sE-selectin levels in infected infants. Infants with gram-negative sepsis had higher sE-selectin levels than did those with gram-positive sepsis. C-reactive protein was the best test for diagnosis of neonatal sepsis (Zaki and el-Sayed 2009).

Hofer et al. (2008) compared two different models of sepsis LPS-induced endotoxemia and cecal ligation perforation (CLP) bacteremia in rats with respect to changes in endothelial expression of CAMS as a marker for capillary breakdown of the blood brain barrier. Increased ICAM-1 expression might be an early factor involved in these pathogenic events. Although the role of PECAM-1 could not be determined, it was possible to show its expression on cerebral endothelium in all groups (Hofer et al. 2008). In mouse models of sepsis, Shapiro et al. (2009) demonstrated increased circulating levels of sE-selectin, sICAM-1, sVCAM-1 and sP-selectin at 24 h, while CLP was associated with increased levels of sE-selectin alone. In real-time PCR, mRNA levels for P-selectin, ICAM-1 and PAI-1 were increased in skin from endotoxemic mice. In CLP, mRNA levels for P-selectin, ICAM-1, E-selectin and PAI-1 were elevated, while VCAM-1 expression was reduced in skin. Most, but not all of these changes correlated with alterations in immunohistochemical staining (Shapiro et al. 2009).

44.10 Action of Drugs and Physical Factors on CAMS

The field of selectin inhibition has matured significantly in recent years in the ability to inhibit selectin/ligand interactions with drug-like molecules and to demonstrate disease modification in human trials. A comprehensive review of new developments in the field of selectin inhibition through discussion of patents/patent applications from 2003 to August 2009 has been reported by Bedard and Kaila (2010)

44.10.1 Inhibitors of Gene Transcription

Treatment of human endothelial cells with cytokines such as IL-1, TNF-α or IFN-γ induces the expression of specific leukocyte adhesion molecules on the endothelial cell surface. Interfering with either leukocyte adhesion or upregulation of adhesion protein is an important therapeutic target as evidenced by the potent anti-inflammatory actions of neutralizing antibodies to these ligands in various animal models and in patients. The induction of E-selectin, VCAM-1, and ICAM-1 genes requires the transcription factor NF-κB. Pharmaceutical agents, which prevent the induced expression of one or more of cell adhesion molecules on endothelium, might be expected to provide a novel mechanism to attenuate the inflammatory responses associated with chronic inflammatory diseases. E-selectin expression is induced on the endothelial cell surface of vessels in response to inflammatory stimuli but is absent in the normal vessels. Thus, E-selectin is an attractive molecular target, and high affinity ligands for E-selectin could be powerful tools for the delivery of therapeutics and/or imaging agents to inflamed vessels. Zimmerman and Blanco (2008) reviewed the structure and regulation of LFA-1 and different classes of inhibitors that interfere LFA-1/ICAM-1 interactions. Alicaforsen (ISIS 2302), an antisense to ICAM-1, designed to inhibit ICAM-1 expression did not reveal significant effect in Crohn's disease. However, topical enemas for ulcerative colitis demonstrated some effect in secondary outcomes, and initial studies in pouchitis are promising (Philpott and Miner 2008). ICAM-1 antibody (UV3) was highly effective at slowing the growth of tumors and/or prolonging survival in SCID mice xenografted with human multiple myeloma, lymphoma, melanoma and other cell lines (Brooks et al. 2008). A structurally diverse collection of small molecule inhibitors has been characterized and developed either to bind the IDAS site of α_L I-domain or to the MIDAS of the β_2 I-like domain.

44.10.2 Anti-NF-kB Reagents

CAMs play important roles in a critical step of tumor metastasis and arrest of tumor cells onto the venous or capillary bed of the target organ. In this process, IL-1β induces nuclear translocation of NF-kB in HUVE cells, followed by induction of cell surface expression of E-selectin, ICAM-1, and VAM-1, and subsequent adhesion of those cancer cells expressing sialyl Le^X antigen, which is a ligand to E-selectin. The adhesion of tumor cells to IL-1β-treated HUVE cells can be inhibited by anti-NF-kB reagents such as N-acetyl L-cysteine, aspirin, or pentoxifylline. These observations indicate the involvement of NF-kB in cancer metastasis and the feasibility of using anti-NF-kB reagents in preventing metastasis (Tozawa et al. 1995). Incubation of HUVEC with N,N,N-trimethylsphingosine (TMS) resulted in a dose-dependent inhibition of IL-1β-induced E-selectin expression. Sphingosine or N,N-dimethylsphingosine had no effects on the expression. This inhibitory effect of TMS on IL-1β-dependent endothelial cell activation may partly explain the known anti-inflammatory or anti-metastatic effect of TMS in vivo (Masamune et al. 1995). Cimetidine inhibits the expression of E-selectin on vascular endothelial cells in gastric- and colorectal cancer patients, treated for chemotherapy (Kawase et al. 2005). Since the expression of E-selectin and Mac-1 is regulated either directly or indirectly by NF-kB, studies provide in vivo evidence that tepoxalin is a potent inhibitor of NF-kB mediated events in animal models and this novel molecular mechanism clearly defines it as a new class of anti-inflammatory compounds. E-selectin transcription requires binding of transcription factors, NF-kB, ATF-2, and HMG-I(Y). HUVE cells treated with TNF-α showed E-selectin surface expression, which peaked at 4 h and then declined. However, ATF-2 binding was unchanged after stimulation with TNF-α. The termination of E-selectin expression is controlled at the level of transcription, with loss of protein-DNA interactions at only one of three NF-kB-binding sites in the E-selectin promoter (Boyle et al. 1999).

E-selectin is synthesized following X-ray exposure to doses as low as 0.5 Gy. X-ray-induced expression of E-selectin and ICAM-1 has been proposed to contribute to radiation injury in normal tissues. E-selectin expression does not require cytokine synthesis, but involves NF-kB activation (Hallahan et al. 1995). NFκB inhibition using NFκB inhibitors abrogates X-ray induced inflammatory mediators (Hallahan et al. 1998). Andrographolide, the principal component of medicinal plant Andrographis paniculata, has been shown to inhibit NF-kB activity and may attenuate allergic asthma via inhibition of the NF-kB signaling pathway. Andrographolide attenuated OVA-induced lung tissue eosinophilia and airway mucus production, mRNA expression of E-selectin, and inducible NOS in lung tissues. Findings implicate a potential therapeutic value of andrographolide in the treatment of asthma (Bao et al. 2009; Jiang et al. 2007).

Effects of TGF-β and IFN-γ on E-Selectin Expression: Transforming growth factor (TGF-β) has been shown to decrease the adhesiveness of endothelial cells for neutrophils, lymphocytes, and tumor cells. TGF-β inhibits the basal E-selectin expression and TNF-stimulated expression. While TGF-β had no effect on the expression of VCAM-1 and ICAM-1, the effect was additive with IL-4 in inhibiting the expression of E-selectin. Thus, perivascular TGF-β appears to act as an inhibitor of inflammatory responses involving neutrophils and a subset of lymphocytes (Gamble et al. 1993). IFN-γ down-regulates the induction by a viral mimetic, polyinosinic-polycytidylic acid [poly-(I:C)], of E-selectin. The inhibitory effect of IFN-γ on poly(I:C)-induced E-selectin was specific for dsRNA. Results indicated the role for IFN-γ in the regulation of E-selectin gene expression in response to dsRNA by a transcriptional mechanism independent of NF-kB, as well as by a minor decrease in message stability (Faruqi and DiCorleto 1997).

Retinoic Acid Inhibits the Expression of VCAM-1 but not E-Selectin: Several genes are regulated by tocopherols which can be categorized, based on their function. Genes that are related to inflammation, cell adhesion and platelet aggregation include E-selectin, ICAM-1, and others (Azzi et al. 2004). Retinoic acid and synthetic derivatives are known to exert anti-inflammatory effects in cutaneous diseases. Pretreatment with all-trans-retinoic acid (t-RA) specifically prevented TNFα-induced VCAM-1 expression, but not ICAM-1 and E-selectin induction (Gille et al. 1997). The TNFα-mediated activation of the human VCAM-1 promoter was also inhibited after t-RA treatment, while the ICAM-1 promoter activation was unaffected, indicating that the selective inhibition of CAM expression is regulated in part at the level of gene transcription. Furthermore, the transcriptional inhibition by t-RA appears to be mediated by its effects upon the activation of NF-kB-dependent complex formation. The specific inhibition of cytokine-mediated VCAM-1 gene expression in vitro provides a potential basis by which retinoids exert their biological effects at sites of inflammation in vivo (Gille et al. 1997). Radiation-induced expression of E-selectin was also blocked by t-RA, whereas 9-cis retinoic acid was ineffective. Application of statins and t-RA might have clinical impact in protecting against E-selectin-promoted metastasis, which might arise as an unwanted side effect from radiation treatment (Holler et al. 2009; Nubel et al. 2004).

Methylation of E-Selectin Promoter Gene Represses NF-kB Transactivation: The E-selectin promoter in cultured endothelial cells is under-methylated in comparison with non-expressing HeLa cells. Thus, methylation is likely to play a role in blocking E-selectin expression in non-endothelial cells (Smith et al. 1993). In intestine, MUC2 is the main mucin carrying s-Le^x, which interacts with E-selectin. This interaction may contribute to the extravasation of tumor cells and thus to the metastases. In several colorectal carcinoma cell lines the methylation of the 5'-flanking region of MUC2 correlated with the suppression of the MUC2 gene. The increase in MUC2 expression after the inhibition of the methylation with 5-aza-2' deoxycytidine strongly supports the notion that the suppression of MUC2 gene is related to the methylation of the promoter (Riede et al. 1998).

44.10.3 Strategies to Combat Atherogenesis and Venous Thrombosis

The advances in the development of adhesion molecule blocking agents, as well as an insight into the potential of these molecules in cardiovascular therapy have been reviewed from time to time (Lutters et al. 2004). Prophylactic dosing of a recombinant P-selectin ligand decreases venous thrombosis in a dose-dependent fashion in both feline and nonhuman primate animal models. Additionally, treatment of 2-day iliac thrombi with a recombinant protein, P-selectin inhibitor, significantly improves vein reopening in nonhuman primates (Register 2009). It is interesting to note that P-selectin inhibition decreases thrombosis without adverse anticoagulation. Myers et al. (2005) evaluated an orally bioavailable inhibitor of P-selectin (PSI-697), which decreased thrombosis. Since, P-selectin is expressed on the surface of activated endothelial cells and platelets during thrombosis, targeting the plasminogen activator (PA) to P-selectin would enhance local thrombolysis and reduce bleeding risk. A urokinase (uPA)/anti-P-selectin antibody (HuSZ51) fusion protein is known to increase fibrinolysis in a hamster pulmonary embolism (Dong et al. 2004).

Aspirin reduces risks of myocardial infarction, stroke and cardiovascular death (Serebruany et al. 2004). The impact of cyclooxygenase (COX)-2 antagonist treatment on acute coronary risk is controversial. Prolonged COX-2 inhibition attenuates CRP and IL-6, does not modify P-selectin and MMP-9, and has no deleterious effect on endothelial function in stable patients with a history of recurrent acute coronary events and raised C-reactive protein (CRP) (Bogaty et al. 2004). Statins used in the control of hypercholesterolemia exert a protective effect on the endothelium reflected by a reduced level of circulating adhesion molecules. Statins exert a beneficial effects on endothelial function and atherosclerotic plaque, modulating oxidative stress and inflammation, with subsequent, well documented, primary and secondary prevention of CAD. Following statin treatment, sP-selectin, and ICAM-1 and highly sensitive CRP decreased compared to baseline levels. Other proteins (sVCAM-1, sE-selectin and platelet ECAM-1) did not show significant changes. In contrast to CRP, the reduction of sP-selectin concentrations correlated directly with the lowering of total cholesterol and inversely with the progression of CAD (Marschang et al. 2006)

44.10.4 Anti-inflammatory Drugs

While diclofenac is capable of inhibiting the expression of E-selectin, ICAM-1 and VCAM-1, the SJC13 is selective in inhibiting the expression of E-selectin and VCAM-1, but not ICAM-1 in endothelial cells. Nonsteroidal anti-inflammatory agents, such as sodium salicylate and aspirin, inhibit NF-κB-dependent gene activation. Salicylate blocked the TNF-α-induced increase in mRNA levels of adhesion molecules and gave a dose-dependent inhibition of TNF-α-induced surface expression of VCAM-1 and ICAM-1 with higher doses required to inhibit E-selectin expression. Ibuprofen appeared a potent inhibitor of IL-1α and TNF-α-induced surface expression of VCAM-1 and a less potent inhibitor of ICAM-1. Indomethacin, a nonsalicylate cyclooxygenase inhibitor, had no effect on surface expression of adhesion molecules, suggesting that the effects were not due to inhibition of cyclooxygenase (Pierce et al. 1996). Methimazole, used in treating autoimmune diseases, may also diminish pathological inflammation by suppressing E-selectin expression. The phenyl methimazole can also reduce cytokine-induced E-selectin expression and consequent leukocyte adhesion. Compound 10, which dramatically inhibits TNF-α-induced VCAM-1 mRNA and protein expression in human aortic endothelial cells, has a modest inhibitory effect on TNF-α induced E-selectin expression and has no effect on ICAM-1 expression (Dagia et al. 2004).

A thieno(2,3-d)pyrimidine, A-155918 inhibits the TNFα-induced expression of E-selectin, ICAM-1, or VCAM-1 on HEVCs (Stewart et al. 2001). Co-treatment of human endothelial cells with certain hydroxyflavones and flavanols blocks cytokine-induced ICAM-1, VCAM-1, and E-selectin expression on human endothelial cells. One of the potent flavones, apigenin, exhibited a dose- and time-dependent, reversible effect on adhesion protein expression as well as inhibiting adhesion protein upregulation at the transcriptional level (Gerritsen et al. 1996). Enalapril and losartan but not placebo induced a small but stable decrease of cardiovascular ICAM-1 and VCAM-1, while E-selectin and leukocyte expression of ICAM-1 remained unchanged. The lowering of plasma adhesion molecules may indicate an antiatherogenic effect of angiotensin II blockade in hypercholesterolemia (Graninger et al. 2004).

Carbohydrates, Synthetic Oligopeptides and Steroids

Targeting interaction of selectins and appropriate carbohydrate ligand is a promising approach to treat chronic inflammation. β-1,3-glucan sulfate (PS3) has inhibitory activity toward L and P-selectins under static conditions (Alban et al. 2009). Access to synthetic carbohydrates is an urgent need for the development of carbohydrate-based drugs, vaccines, adjuvants as well as novel drug delivery systems. Besides traditional synthesis in solution, synthetic carbohydrates have been generated by chemoenzymatic methods as well as automated solid-phase synthesis. Synthetic oligosaccharides have proven to be useful for identifying ligands of carbohydrate-binding proteins such as C-type lectins and siglecs using glycan arrays. Furthermore, glyconanoparticles and glycodendrimers have been used for specific targeting of lectins of the immune system such as selectins, DC-SIGN, and CD22 (Lepenies et al. 2010).

Compounds that target both heparanase and selectins offer a promising approach for cancer therapy. Borsig et al.

(2011) reported semisynthetic sulfated tri mannose C-C-linked dimers (STMCs) which are endowed with heparanase and selectin inhibitory activity. STMC hexasaccharide is an effective inhibitor of P-selectin in vivo. P-selectin-specific STMC attenuated metastasis in animal models, indicating that inhibition of tumor cell interaction with the vascular endothelium is critical for cancer dissemination. The small size, the stability of the C-C bond, and the chemically defined structure of STMCs make them superior to heparin derivatives and signify STMCs as valuable candidates for further evaluation.

Steroids down-regulate the expression of CAMs in endothelial cells stimulated by LPS in vitro. Low-dose hydrocortisone is a new treatment of patients with septic shock, a state that is characterized by an endothelial injury. Treatment with glucocorticoids differently affected the pattern of evolution of sCAMs, with sE-selectin being decreased and sICAM-1 being increased. Expression of sP-selectin and sVCAM-1 was not affected (Leone et al. 2004). Methotrexate (MTX) markedly reduces the expression of vascular E-selectin. A positive correlation between disease severity and the frequency of cutaneous lymphocyte-associated antigen (CLA)-positive T cells in the blood of untreated patients with psoriasis has been observed. It is suggested that MTX decreases the expression of CLA and E-selectin and that this may be a major mechanism for the therapeutic effect of MTX on psoriatic skin lesions (Sigmundsdottir et al. 2004).

References

Adamiec-Mroczek J, Oficjalska-Młyńczak J (2008) Assessment of selected adhesion molecule and proinflammatory cytokine levels in the vitreous body of patients with type 2 diabetes—role of the inflammatory-immune process in the pathogenesis of proliferative diabetic retinopathy. Graefes Arch Clin Exp Ophthalmol 246:1665–1670

Adamikova A, Kojecky V, Rybka J, Svacina S (2008) Levels of adhesion molecules bear a relationship to triglyceride levels in type 2 diabetic subjects with proven silent ischemia. Int Angiol 27:307–312

Adamska I, Czerwionka-Szaflarska M, Kulwas A et al (2007) Value of E-selectin and L-selectin determination in children and youth with inflammatory bowel disease. Med Wieku Rozwoj 11:413–418

Alban S, Ludwig RJ, Bendas G et al (2009) PS3, a semisynthetic β-1,3-glucan sulfate, diminishes contact hypersensitivity responses through inhibition of L- and P-selectin functions. J Invest Dermatol 129:1192–1202

Alcalde G, Merino J, Sanz S et al (1995) Circulating adhesion molecules during kidney allograft rejection. Transplantation 59:1695–1699

Aminian B, Abdi Ardekani AR, Arandi N (2007) ICAM-1 polymorphisms (G241R, K469E), in coronary artery disease and myocardial infarction. Iran J Immunol 4:227–235

Amoli MM, Llorca J, Gomez-Gigirey A et al (2004) E-selectin polymorphism in erythema nodosum secondary to sarcoidosis. Clin Exp Rheumatol 22:230–232

Antoine M, Tag CG, Gressner AM et al (2009) Expression of E-selectin ligand-1 (CFR/ESL-1) on hepatic stellate cells: implications for leukocyte extravasation and liver metastasis. Oncol Rep 21:357–362

Antonova OA, Loktionova SA, Romanov YA et al (2009) Activation and damage of endothelial cells upon hypoxia/reoxygenation. Effect of extracellular pH. Biochemistry (Mosc) 74:605–612

Arvanitis DA, Flouris GA, Spandidos DA (2005) Genomic rearrangements on VCAM1, SELE, APEG1and AIF1 loci in atherosclerosis. J Cell Mol Med 9:153–159

Atzingen MV, Gómez RM, Schattner M et al (2009) Lp95, a novel leptospiral protein that binds extracellular matrix components and activates E-selectin on endothelial cells. J Infect 59:264–276

Auer J, Weber T, Berent R et al (2003) Genetic polymorphisms in cytokine and adhesion molecule genes in coronary artery disease. Am J Pharmacogenomics 3:317–331

Aychek T, Miller K, Sagi-Assif O et al (2008) E-selectin regulates gene expression in metastatic colorectal carcinoma cells and enhances HMGB1 release. Int J Cancer 123:1741–1750

Azzi A, Gysin R, Kempna P et al (2004) Regulation of gene expression by α-tocopherol. J Biol Chem 385:585–591

Bal N, Kocer NE, Ertorer ME et al (2008) E-selectin, and P-selectin expressions in papillary thyroid carcinomas and their correlation with prognostic parameters. Pathol Res Pract 204:743–750

Bao Z, Guan S, Cheng C et al (2009) A novel antiinflammatory role for andrographolide in asthma via inhibition of the nuclear factor-kB pathway. Am J Respir Crit Care Med 179:657–665

Becker T, Juttner B, Elsner HA et al (2004) Platelet P-selectin and GPIIb/IIIa expression after liver transplantation and resection. Transpl Int 17:442–448

Bedard PW, Kaila N (2010) Selectin inhibitors: a patent review. Expert Opin Ther Pat 20:781–793

Ben-David T, Sagi-Assif O, Meshel T et al (2008) The involvement of the sLe-a selectin ligand in the extravasation of human colorectal carcinoma cells. Immunol Lett 116:218–224

Berti R, Williams AJ, Moffett JR et al (2002) Quantitative real-time RT-PCR analysis of inflammatory gene expression associated with ischemia-reperfusion brain injury. J Cereb Blood Flow Metab 22:1068–1079

Bhatia GS, Sosin MD, Patel JV et al (2009) Plasma indices of endothelial and platelet activation in Rheumatoid Disease: relationship to cardiovascular co-morbidity. Int J Cardiol 134:97–103

Bijanzadeh M, Ramachandra NB, Mahesh PA et al (2009) Soluble intercellular adhesion molecule-1 and E-selectin in patients with asthma exacerbation. Lung 187:315–320

Blaha M, Krejsek J, Blaha V et al (2004) Selectins and monocyte chemotactic peptide as the markers of atherosclerosis activity. Physiol Res 53:273–278

Bloom BJ, Miller LC, Blier PR (2002) Soluble adhesion molecules in pediatric rheumatic diseases. J Rheumatol 29:832–836

Bogaty P, Brophy JM, Noel M et al (2004) Impact of prolonged cyclooxygenase-2 inhibition on inflammatory markers and endothelial function in patients with ischemic heart disease and raised C-reactive protein: a randomized placebo-controlled study. Circulation 110:934–939

Borsig L (2007) Antimetastatic activities of modified heparins: selectin inhibition by heparin attenuates metastasis. Semin Thromb Hemost 33:540–546

Borsig L, Vlodavsky I, Ishai-Michaeli R et al (2011) Sulfated hexasaccharides attenuate metastasis by inhibition of P-selectin and heparanase. Neoplasia 13:445–452

Bosello S, Santoliquido A, Zoli A et al (2008) TNF-α blockade induces a reversible but transient effect on endothelial dysfunction in patients with long-standing severe rheumatoid arthritis. Clin Rheumatol 27:833–839

Boyle EM Jr, Sato TT, Noel RF Jr et al (1999) Transcriptional arrest of the human E-selectin gene. J Surg Res 82:194–200

Braun F, Hosseini M, Wieland E et al (2004) Expression of E-selectin and its transcripts during intestinal ischemia-reperfusion injury in pigs. Transplant Proc 36:265–266

Brooks KJ, Coleman EJ, Vitetta ES (2008) The antitumor activity of an anti-CD54 antibody in SCID mice xenografted with human breast, prostate, non-small cell lung, and pancreatic tumor cell lines. Int J Cancer 123:2438–2445

Bullard DC, Kunkel EJ, Kubo H, Hicks MJ et al (1996) Infectious susceptibility and severe deficiency of leukocyte rolling and recruitment in E-selectin and P-selectin double mutant mice. J Exp Med 183:2329–2336

Cai H, Harrison DG (2000) Endothelial dysfunction in cardiovascular diseases: the role of oxidant stress. Circ Res 87:840–844

Campos-Rodríguez R, Jarillo-Luna RA, Larsen BA et al (2009) Invasive amebiasis: A microcirculatory disorder? Med Hypotheses 73:687–697

Carmody IC, Meng L, Shen XD et al (2004) P-selectin knockout mice have improved outcomes with both warm ischemia and small bowel transplantation. Transplant Proc 36:263–264

Carrizo Tdel R, Prado MM, Velarde MS et al (2008) Soluble E- selectin in children and adolescents with type 1 diabetes. Medicina (B Aires) 68:193–197

Celebi M, Paul AG (2008) Blocking E-selectin inhibits ischaemia-reperfusion-induced neutrophil recruitment to the murine testis. Andrologia 40:235–239

Cha JK, Jo WS, Shin HC et al (2004) Increased platelet CD63 and P-selectin expression persist in atherosclerotic ischemic stroke. Platelets 15:3–7

Chan B, Sukhatme VP (2009) Receptor tyrosine kinase EphA2 mediates thrombin-induced upregulation of ICAM-1 in endothelial cells in vitro. Thromb Res 123:745–752

Chen MC, Chang HW, Juang SS et al (2004) Increased plasma levels of soluble P-selectin in rheumatic mitral stenosis. Chest 126:54–58

Chen H, Cui B, Wang S et al (2008) The common variants of E-selectin gene in Graves' disease. Genes Immun 9:182–186

Chiang LY, Sheppard DC, Gravelat FN et al (2008) Aspergillus fumigatus stimulates leukocyte adhesion molecules and cytokine production by endothelial cells in vitro and during invasive pulmonary disease. Infect Immun 76:3429–3438

Chooklin S (2009) Pathogenic aspects of pulmonary complications in acute pancreatitis patients. Hepatobiliary Pancreat Dis Int 8:186–192

Chu JW, Abbasi F, Lamendola C et al (2005) Effect of rosiglitazone treatment on circulating vascular and inflammatory markers in insulin-resistant subjects. Diab Vasc Dis Res 2:37–41

Chukwuemeka AO, Brown KA, Venn GE et al (2005) Changes in P-selectin expression on cardiac microvessels in blood-perfused rat hearts subjected to ischemia-reperfusion. Ann Thorac Surg 79:204–211

Clark RA, Huang SJ, Murphy GF et al (2008) Human squamous cell carcinomas evade the immune response by down-regulation of vascular E-selectin and recruitment of regulatory T cells. J Exp Med 205:2221–2234

Cobankara V, Ozatli D, Kiraz S et al (2004) Successful treatment of rheumatoid arthritis is associated with a reduction in serum sE-selectin and thrombomodulin level. Clin Rheumatol 23:430–434

Cojean S, Jafari-Guemouri S, Le Bras J, Durand R (2008) Cytoadherence characteristics to endothelial receptors ICAM-1 and CD36 of Plasmodium falciparum populations from severe and uncomplicated malaria cases. Parasite 15:163–169

Collins RG, Velji R, Guevara NV et al (2000) P-Selectin or intercellular adhesion molecule (ICAM)-1 deficiency substantially protects against atherosclerosis in apolipoprotein E-deficient mice. J Exp Med 191:189–194

Corona-Sanchez EG, Gonzalez-Lopez L, Muñoz-Valle JF et al (2009) Circulating E-selectin and tumor necrosis factor-alpha in extraarticular involvement and joint disease activity in rheumatoid arthritis. Rheumatol Int 29:281–286

Corsi MM, Pagani D, Dogliotti G, Perona F, Sambataro G, Pignataro L (2008) Protein biochip array of adhesion molecule expression in peripheral blood of patients with nasal polyposis. Int J Biol Markers 23:115–120

Coskun U, Sancak B, Sen I et al (2006) Serum P-selectin, soluble vascular cell adhesion molecule-I (s-VCAM-I) and soluble intercellular adhesion molecule-I (s-ICAM-I) levels in bladder carcinoma patients with different stages. Int Immunopharmacol 6:672–677

da Mota LM, dos Santos Neto LL, de Carvalho JF (2009) Autoantibodies and other serological markers in rheumatoid arthritis: predictors of disease activity? Clin Rheumatol 28:1127–1134

Dagia NM, Harii N, Meli AE et al (2004) Phenyl methimazole inhibits TNF-α-induced VCAM-1 expression in an IFN regulatory factor-1-dependent manner and reduces monocytic cell adhesion to endothelial cells. J Immunol 173:2041–2049

de Kossodo S, Moore R, Gschmeissner S et al (1995) Changes in endogenous cytokines, adhesion molecules and platelets during cytokine-induced tumor necrosis. Br J Cancer 72:1165–1172

Dehne MG, Sablotzki A, Mühling J et al (2008) Evaluation of sE-Selectin and sICAM-1 as parameters for renal function. Ren Fail 30:675–684

Dema B, Martínez A, Polanco I et al (2008) ICAM1 R241 is not associated with celiac disease in the Spanish population. Hum Immunol 69:675–678

Demirkesen C, Tüzüner N, Senocak M et al (2008) Comparative study of adhesion molecule expression in nodular lesions of Behçet syndrome and other forms of panniculitis. Am J Clin Pathol 130:28–33

Devaraj S, Cheung AT, Jialal I et al (2007) Evidence of increased inflammation and microcirculatory abnormalities in patients with type 1 diabetes and their role in microvascular complications. Diabetes 56:2790–2796

Di Lorenzo G, Pacor ML, Mansueto P et al (2004) Circulating levels of soluble adhesion molecules in patients with ANCA-associated vasculitis. J Nephrol 17:800–807

Di Nisio M, Niers TM, Reitsma PH, Buller HR (2005) Plasma cytokine and P-selectin levels in advanced malignancy: prognostic value and impact of low-molecular weight heparin administration. Cancer 104:2275–2281

Domanski L, Pawlik A, Safranow K et al (2009) Circulating adhesion molecules and purine nucleotides during kidney allograft reperfusion. Transplant Proc 41:40–43

Dong N, Da Cunha V, Citkowicz A et al (2004) P-selectin-targeting of the fibrin selective thrombolytic Desmodus rotundus salivary plasminogen activator alpha1. Thromb Haemost 92:956–965

Döring A, Wild M, Vestweber D et al (2007) E- and P-selectin are not required for the development of experimental autoimmune encephalomyelitis in C57BL/6 and SJL mice. J Immunol 179:8470–8479

Dowlati A, Gray R, Sandler AB et al (2008) Cell adhesion molecules, vascular endothelial growth factor, and basic fibroblast growth factor in patients with non-small cell lung cancer treated with chemotherapy with or without bevacizumab–an Eastern Cooperative Oncology Group Study. Clin Cancer Res 14:1407–1412

Dymicka-Piekarska V, Kemona H (2009) Does colorectal cancer clinical advancement affect adhesion molecules (sP-selectin, sE-selectin and ICAM-1) concentration? Thromb Res 124:80–83

Dymicka-Piekarska V, Butkiewicz A, Matowicka-Karna J et al (2005) Soluble P-selectin concentration in patients with colorectal cancer. Neoplasma 52:297–301

Eichbaum MH, de Rossi TM, Kaul S, Bastert G (2004) Serum levels of soluble E-selectin are associated with the clinical course of metastatic disease in patients with liver metastases from breast cancer. Oncol Res 14:603–610

Eikemo H, Sellevold OF, Videm V (2004) Markers for endothelial activation during open heart surgery. Ann Thorac Surg 77:214–219

Elangbam CS, Qualls CW Jr, Dahlgren RR (1997) Cell adhesion molecules–update. Vet Pathol 34:61–73

Erkiert-Polguj A, Pawliczak R, Sysa-Jedrzejowska A (2009) Expression of selected adhesion molecules in dermatitis herpetiformis and bullous pemphigoid. Pol J Pathol 60:26–34

Essani NA, McGuire GM, Manning AM, Jaeschke H (1995) Differential induction of mRNA for ICAM-1 and selectins in hepatocytes, Kupffer cells and endothelial cells during endotoxemia. Biochem Biophys Res Commun 211:74–82

Esterre P, Raobelison A, Ramarokoto CE et al (1998) Serum concentrations of sICAM-1, sE-, sP- and sL-selectins in patients with Schistosoma mansoni infection and association with disease severity. Parasite Immunol 20:369–376

Fallo F, Cella G, Casonato A et al (2006) Biochemical markers of endothelial activation in primary hyperparathyroidism. Horm Metab Res 38:125–129

Fang L, Wei H, Mak KH et al (2004) Markers of low-grade inflammation and soluble cell adhesion molecules in Chinese patients with coronary artery disease. Can J Cardiol 20:1433–1438

Fang L, Wei H, Chowdhury SH et al (2005) Association of Leu125Val polymorphism of platelet endothelial cell adhesion molecule-1 (PECAM-1) gene & soluble level of PECAM-1 with coronary artery disease in Asian Indians. Indian J Med Res 121:92–99

Fareed J, Walenga JM, Hoppensteadt DA et al (1999) Selectins in the HIT syndrome: pathophysiologic role and therapeutic modulation. Semin Thromb Hemost 25(Suppl 1):37–42

Farmer DG, Anselmo D, Da Shen X et al (2005a) Disruption of P-selectin signaling modulates cell trafficking and results in improved outcomes after mouse warm intestinal ischemia and reperfusion injury. Transplantation 80:828–835

Farmer DG, Shen XD, Amersi F et al (2005b) CD62 blockade with P-Selectin glycoprotein ligand-immunoglobulin fusion protein reduces ischemia-reperfusion injury after rat intestinal transplantation. Transplantation 79:44–51

Farrar CA, Wang Y, Sacks SH, Zhou W (2004) Independent pathways of P-selectin and complement-mediated renal ischemia/reperfusion injury. Am J Pathol 164:133–141

Faruqi TR, DiCorleto PE (1997) IFN-γ inhibits double-stranded RNA-induced E-selectin expression in human endothelial cells. J Immunol 159:3989–3994

Fenoglio C, Scalabrini D, Piccio L et al (2009) Candidate gene analysis of selectin cluster in patients with multiple sclerosis. J Neurol 256:832–833

Fenzi F, Latronico N, Refatti N, Rizzuto N (2003) Enhanced expression of E-selectin on the vascular endothelium of peripheral nerve in critically ill patients with neuromuscular disorders. Acta Neuropathol (Berl) 106:75–82

Fernandez-Reyes D, Craig AG, Kyes SA et al (1997) A high frequency African coding polymorphism in the N-terminal domain of ICAM-1 predisposing to cerebral malaria in Kenya. Hum Molec Genet 6:1357–1360

Forbes E, Hulett M, Ahrens R et al (2006) ICAM-1-dependent pathways regulate colonic eosinophilic inflammation. J Leukoc Biol 80:330–341

Fornoni A, Ijaz A, Tejada T, Lenz O (2008) Role of inflammation in diabetic nephropathy. Curr Diabetes Rev 4:10–17

Foster W, Shantsila E, Carruthers D, Lip GY, Blann AD (2009) Circulating endothelial cells and rheumatoid arthritis: relationship with plasma markers of endothelial damage/dysfunction. Rheumatology (Oxford) 48:285–288

Fry AE, Auburn S, Diakite M et al (2008) Variation in the ICAM1 gene is not associated with severe malaria phenotypes. Genes Immun 9:462–469

Galustian C, Elviss N, Chart H et al (2003) Interactions of the gastrotropic bacterium Helicobacter pylori with the leukocyte-endothelium adhesion molecules, the selectins–A preliminary report. FEMS Immunol Med Microbiol 36:127–134

Gamble JR, Khew-Goodall Y, Vadas MA (1993) Transforming growth factor-beta inhibits E-selectin expression on human endothelial cells. J Immunol 150:4494–4503

Gao Y, Li N, Fei R et al (2005) P-Selectin-mediated acute inflammation can be blocked by chemically modified heparin, RO-heparin. Mol Cells 19:350–355

Gasser M, Waaga-Gasser AM, Grimm MW et al (2005) Selectin blockade plus therapy with low-dose sirolimus and cyclosporin A prevent brain death-induced renal allograft dysfunction. Am J Transplant 5:662–670

Gerritsen ME, Shen CP, Atkinson WJ et al (1996) Microvascular endothelial cells from E-selectin-deficient mice form tubes in vitro. Lab Invest 75:175–184

Ghilardi G, Biondi ML, Turri O et al (2004) Ser128Arg gene polymorphism for E-selectin and severity of atherosclerotic arterial disease. J Cardiovasc Surg (Torino) 45:143–147

Gille J, Paxton LL, Lawley TJ et al (1997) Retinoic acid inhibits the regulated expression of vascular cell adhesion molecule-1 by cultured dermal microvascular endothelial cells. J Clin Invest 99:492–500

Golias C, Tsoutsi E, Matziridis A et al (2007) Leukocyte and endothelial cell adhesion molecules in inflammation focusing on inflammatory heart disease. In Vivo 21:757–769

Gómez RM, Vieira ML, Schattner M et al (2008) Putative outer membrane proteins of Leptospira interrogans stimulate human umbilical vein endothelial cells (HUVECS) and express during infection. Microb Pathog 45:315–322

Gonzalez A, Lenzi HL, Motta EM et al (2005) Expression of adhesion molecules in lungs of mice infected with Paracoccidioides brasiliensis conidia. Microbes Infect 7:666–673

Götting C, Adam A, Szliska C, Kleesiek K (2008) Circulating P-,L- and E-selectins in pseudoxanthoma elasticum patients. Clin Biochem 41:368–374

Gout S, Tremblay PL, Huot J (2008) Selectins and selectin ligands in extravasation of cancer cells and organ selectivity of metastasis. Clin Exp Metastasis 25:335–344

Graninger M, Reiter R, Drucker C et al (2004) Angiotensin receptor blockade decreases markers of vascular inflammation. J Cardiovasc Pharmacol 44:335–339

Gulubova M, Vlaykova T, Manolova I et al (2008) Implication of adhesion molecules in inflammation of the common bile duct in patients with secondary cholangitis due to biliary obstruction. Hepatogastroenterology 55:836–841

Guney N, Soydinc HO, Derin D et al (2008) Serum levels of intercellular adhesion molecule ICAM-1 and E-selectin in advanced stage non-small cell lung cancer. Med Oncol 25:194–200

Haidari M, Hajilooi M, Rafiei AR et al (2009) E-selectin genetic variation as a susceptibility factor for ischemic stroke. Cerebrovasc Dis 28:26–32

Hallahan D, Clark ET, Kuchibhotla J et al (1995) E-selectin gene induction by ionizing radiation is independent of cytokine induction. Biochem Biophys Res Commun 217:784–795

Hallahan DE, Virudachalam S, Kuchibhotla J (1998) Nuclear factor kappaB dominant negative genetic constructs inhibit X-ray induction of cell adhesion molecules in the vascular endothelium. Cancer Res 58:5484–5488

Hamilton CA, Miller WH, Al-Benna S, Brosnan MJ et al (2004) Strategies to reduce oxidative stress in cardiovascular disease. Clin Sci 106:219–234

Hanley WD, Napier SL, Burdick MM et al (2006) Variant isoforms of CD44 are P- and L-selectin ligands on colon carcinoma cells. FASEB J 20:337–339

Hannigan E, O'Connell DP, Hannigan A, Buckley LA (2004) Soluble cell adhesion molecules in gingival crevicular fluid in periodontal health and disease. J Periodontol 75:546–550

Hebbar M, Adenis A, Révillion F et al (2009) E-selectin gene S128R polymorphism is associated with poor prognosis in patients with stage II or III colorectal cancer. Eur J Cancer 45:1871–1876

Heider P, Wildgruber MG, Weiss W et al (2006) Role of adhesion molecules in the induction of restenosis after angioplasty in the lower limb. J Vasc Surg 43:969–977

Herrmann SM, Ricard S, Nicaud V et al (1998) The P-selectin gene is highly polymorphic: reduced frequency of the Pro715 allele carriers in patients with myocardial infarction. Hum Mol Genet 7:1277–1284

Ho AW, Wong CK, Lam CW (2008) Tumor necrosis factor-alpha up-regulates the expression of CCL2 and adhesion molecules of human proximal tubular epithelial cells through MAPK signaling pathways. Immunobiology 213:533–544

Hoeller C, Richardson SK, Ng LG et al (2009) In vivo imaging of cutaneous T-cell lymphoma migration to the skin. Cancer Res 69:2704–2708

Hofer S, Bopp C, Hoerner C et al (2008) Injury of the blood brain barrier and up-regulation of ICAM-1 in polymicrobial sepsis. J Surg Res 146:276–281

Högdahl M, Söderlund G, Kihlström E (2008) Expression of chemokines and adhesion molecules in human coronary artery endothelial cells infected with Chlamydia (Chlamydophila) pneumoniae. APMIS 116:1082–1088

Holler V, Buard V, Gaugler MH et al (2009) Pravastatin limits radiation-induced vascular dysfunction in the skin. J Invest Dermatol 129:1280–1291

Holvoet P, Collen D (1997) Thrombosis and atherosclerosis. Curr Opin Lipidol 8:320–328

Horstman LL, Jy W, Jimenez JJ, Ahn YS (2004) Endothelial microparticles as markers of endothelial dysfunction. Front Biosci 9:1118–1135

Houshmand B, Rafiei A, Hajilooi M et al (2009) E-selectin and L-selectin polymorphisms in patients with periodontitis. J Periodontal Res 44:88–93

Huang F, del-Río-Navarro BE, Monge JJ et al (2008) Endothelial activation and systemic inflammation in obese asthmatic children. Allergy Asthma Proc 29:453–460

Iannone F, Riccardi MT, Guiducci S et al (2008) Bosentan regulates the expression of adhesion molecules on circulating T cells and serum soluble adhesion molecules in systemic sclerosis-associated pulmonary arterial hypertension. Ann Rheum Dis 67:1121–1126

Ida S, Fujimura Y, Hirota M, Imamura Y et al (2009) Significance of endothelial molecular markers in the evaluation of the severity of acute pancreatitis. Surg Today 39:314–319

Iwata Y, Tsuchiya KJ, Mikawa S et al (2008) Serum levels of P-selectin in men with high-functioning autism. Br J Psychiatry 193:338–339

Izu K, Tokura Y (2005) The various effects of four H1-antagonists on serum substance P levels in patients with atopic dermatitis. J Dermatol 32:776–781

Jeschke U, Mylonas I, Kunert-Keil C et al (2009) Immunohistochemistry, glycosylation and immunosuppression of glycodelin in human ovarian cancer. Histochem Cell Biol 131:283–295

Jiang CG, Li JB, Liu FR et al (2007) Andrographolide inhibits the adhesion of gastric cancer cells to endothelial cells by blocking E-selectin expression. Anticancer Res 27:2439–2447

Jolly C, Mitar I, Sattentau QJ (2007) Adhesion molecule interactions facilitate human immunodeficiency virus type 1-induced virological synapse formation between T cells. J Virol 81:13916–13921

Kaifi JT, Hall LR, Diaz C et al (2000) Impaired eosinophil recruitment to the cornea in P-selectin-deficient mice in Onchocerca volvulus keratitis (River blindness). Invest Ophthalmol Vis Sci 41:3856–3861

Kamai Y, Lossinsky AS, Liu H et al (2009) Polarized response of endothelial cells to invasion by Aspergillus fumigatus. Cell Microbiol 11:170–182

Kanavaki I, Makrythanasis P, Lazaropoulou C et al (2009) Soluble endothelial adhesion molecules and inflammation markers in patients with β-thalassemia intermedia. Blood Cells Mol Dis 43:230–234

Kannagi R (2007) Carbohydrate antigen sialyl Lewis A–its pathophysiological significance and induction mechanism in cancer progression. Chang Gung Med J 30:189–209

Kanno H, Watabe D, Shimizu N, Sawai T (2008) Adhesion of Epstein-Barr virus-positive natural killer cell lines to cultured endothelial cells stimulated with inflammatory cytokines. Clin Exp Immunol 151:519–527

Kawamura N, Dyck PJ, Schmeichel AM et al (2008) Inflammatory mediators in diabetic and non-diabetic lumbosacral radiculoplexus neuropathy. Acta Neuropathol 115:231–239

Kawase J, Ozawa S, Kobayashi K et al (2009) Increase in E-selectin expression in umbilical vein endothelial cells by anticancer drugs and inhibition by cimetidine. Oncol Rep 22(6):1293–1297

Khalfaoui T, Lizard G, Ouertani-Meddeb A (2008) Adhesion molecules (ICAM-1 and VCAM-1) and diabetic retinopathy in type 2 diabetes. J Mol Histol 39:243–249

Khare A, Shetty S, Ghosh K et al (2005) Evaluation of markers of endothelial damage in cases of young myocardial infarction. Atherosclerosis 180:375–380

Khazen D, Jendoubi-Ayed S, Gorgi Y et al (2007) Adhesion molecule polymorphisms in acute renal allograft rejection. Transplant Proc 39:2563–2564

Khazen D, Jendoubi-Ayed S, Aleya WB et al (2009) Polymorphism in ICAM-1, PECAM-1, E-selectin, and L-selectin genes in Tunisian patients with inflammatory bowel disease. Eur J Gastroenterol Hepatol 21:167–175

Kilickap M, Tutar E, Aydintug O et al (2004) Increase in soluble E-selectin level after PTCA and stent implantation: a potential marker of restenosis. Int J Cardiol 93:13–18

Kim YJ, Borsig L, Varki NM, Varki A (1998) P-selectin deficiency attenuates tumor growth and metastasis. Proc Natl Acad Sci USA 95:9325–9330

Kirchberger S, Vetr H, Majdic O et al (2006) Engagement of ICAM-1 by major group rhinoviruses activates the LFA-1/ICAM-3 cell adhesion pathway in mononuclear phagocytes. Immunobiology 211:537–547

Kistorp C, Chong AY, Gustafsson F et al (2008) Biomarkers of endothelial dysfunction are elevated and related to prognosis in chronic heart failure patients with diabetes but not in those without diabetes. Eur J Heart Fail 10:380–387

Kleinhans H, Kaifi JT, Mann O, Reinknecht F et al (2009) The role of vascular adhesion molecules PECAM-1 (CD 31), VCAM-1 (CD 106), E-selectin (CD62E) and P-selectin (CD62P) in severe porcine pancreatitis. Histol Histopathol 24:551–557

Klimiuk PA, Sierakowski S, Domyslawska I et al (2004) Reduction of soluble adhesion molecules (sICAM-1, sVCAM-1, and sE-selectin) and vascular endothelial growth factor levels in serum of rheumatoid arthritis patients following multiple intravenous infusions of infliximab. Arch Immunol Ther Exp (Warsz) 52:36–42

Klimiuk PA, Fiedorczyk M, Sierakowski S, Chwiecko J (2007) Soluble cell adhesion molecules (sICAM-1, sVCAM-1, and sE-selectin) in patients with early rheumatoid arthritis. Scand J Rheumatol 36:345–350

Klopocki AG, Laskowska A, Antoniewicz-Papis J et al (1998) Role of sialosyl Lewis(a) in adhesion of colon cancer cells–the antisense RNA approach. Eur J Biochem 253:309–318

Kobayashi H, Boelte KC, Lin PC (2007) Endothelial cell adhesion molecules and cancer progression. Curr Med Chem 14:377–386

Koizumi M, King N, Lobb R et al (1992) Expression of vascular adhesion molecules in inflammatory bowel disease. Gastroenterology 103:840–847

Kose O, Stewart J, Waseem A et al (2008) Expression of cytokeratins, adhesion and activation molecules in oral ulcers of Behçet's disease. Clin Exp Dermatol 33:62–69

Kretowski A, Wawrusiewicz N, Mironczuk K et al (2003) Intercellular adhesion molecule 1 gene polymorphisms in Graves' disease. J Clin Endocr Metab 88:4945–4949

Kumar P, Amin MA, Harlow LA et al (2003) Src and phosphatidylinositol 3-kinase mediate soluble E-selectin-induced angiogenesis. Blood 101:3960–3968

Kuryliszyn-Moskal A, Klimiuk PA, Ciolkiewicz M, Sierakowski S (2008) Clinical significance of selected endothelial activation markers in patients with systemic lupus erythematosus. J Rheumatol 35:1307–1313

Kwon BC, Sohn MH, Kim KW et al (2007) House dust mite induces expression of intercellular adhesion molecule-1 in EoL-1 human eosinophilic leukemic cells. J Korean Med Sci 22:815–819

Langer R, Wang M, Stepkowski SM et al (2004) Selectin inhibitor bimosiamose prolongs survival of kidney allografts by reduction in intragraft production of cytokines and chemokines. J Am Soc Nephrol 15:2893–2901

Lau C, Wang X, Song L et al (2008) Syk associates with clathrin and mediates phosphatidylinositol 3-kinase activation during human rhinovirus internalization. J Immunol 180:870–880

Läubli H, Borsig L (2010) Selectins promote tumor metastasis. Semin Cancer Biol 20:169–177

Leal EC, Aveleira CA, Castilho AF et al (2008) Müller cells do not influence leukocyte adhesion to retinal endothelial cells. Ocul Immunol Inflamm 16:173–179

Lee HM, Kim HJ, Won KJ et al (2008) Contribution of soluble intercellular adhesion molecule-1 to the migration of vascular smooth muscle cells. Eur J Pharmacol 579:260–268

Lefranc F, Mijatovic T, Mathieu V et al (2004) Characterization of gastrin-induced proangiogenic effects in vivo in orthotopic U373 experimental human glioblastomas and in vitro in human umbilical vein endothelial cells. Clin Cancer Res 10:8250–8265

Leone M, Boutiere-Albanese B, Valette S et al (2004) Cell adhesion molecules as a marker reflecting the reduction of endothelial activation induced by glucocorticoids. Shock 21:311–314

Lepenies B, Yin J, Seeberger PH (2010) Applications of synthetic carbohydrates to chemical biology. Curr Opin Chem Biol 14:404–411

Levälampi T, Honkanen V, Lahdenne P et al (2007) Effects of infliximab on cytokines, myeloperoxidase, and soluble adhesion molecules in patients with juvenile idiopathic arthritis. Scand J Rheumatol 36:189–193

Li Y, Wei YS, Wang M et al (2005) Association between the Ser128Arg variant of the E-selectin and risk of coronary artery disease in the central China. Int J Cardiol 103:33–36

Li XX, Liu JP, Cheng JQ et al (2008) Intercellular adhesion molecule-1 gene K469E polymorphism and ischemic stroke: a case–control study in a Chinese population. Mol Biol Rep 36:1565–1571

Lim HS, Blann AD, Lip GY (2004) Soluble CD40 ligand, soluble P-selectin, interleukin-6, and tissue factor in diabetes mellitus: relationships to cardiovascular disease and risk factor intervention. Circulation 109:2524–2528

Lin CL, Dumont AS, Calisaneller T et al (2005) Monoclonal antibody against E selectin attenuates subarachnoid hemorrhage-induced cerebral vasospasm. Surg Neurol 64:201–205

Lip GY, Pearce LA, Chin BS et al (2005) Effects of congestive heart failure on plasma von Willebrand factor and soluble P-selectin concentrations in patients with non-valvar atrial fibrillation. Heart 91:759–763

Liu R, Desta T, Raptis M et al (2008) P. gingivalis and E. coli lipopolysaccharides exhibit different systemic but similar local induction of inflammatory markers. J Periodontol 79:1241–1247

López-Lerma I, Estrach MT (2009) A distinct profile of serum levels of soluble intercellular adhesion molecule-1 and intercellular adhesion molecule-3 in mycosis fungoides and Sézary syndrome. J Am Acad Dermatol 61:263–270

Ludwig RJ, Alban S, Bistrian R, Boehncke WH et al (2006) The ability of different forms of heparins to suppress P-selectin function in vitro correlates to their inhibitory capacity on bloodborne metastasis in vivo. Thromb Haemost 95:535–540

Ludwig RJ, Schön MP, Boehncke WH (2007) P-selectin: a common therapeutic target for cardiovascular disorders, inflammation and tumour metastasis. Expert Opin Ther Targets 11:1103–1117

Lugovic L, Cupic H, Lipozencic J et al (2006) The role of adhesion molecules in atopic dermatitis. Acta Dermatovenerol Croat 14:2–7

Lutters BC, Leeuwenburgh MA, Appeldoorn CC et al (2004) Blocking endothelial adhesion molecules: a potential therapeutic strategy to combat atherogenesis. Curr Opin Lipidol 15:545–552

Malatino LS, Stancanelli B, Cataliotti A et al (2007) Circulating E-selectin as a risk marker in patients with end-stage renal disease. J Intern Med 262:479–487

Manetti M, Neumann E, Müller A, Schmeiser T et al (2008) Endothelial/lymphocyte activation leads to prominent CD4+ T cell infiltration in the gastric mucosa of patients with systemic sclerosis. Arthritis Rheum 58:2866–2873

Marazuela M, Postigo AA, Acevedo A et al (1994) Adhesion molecules from the LFA-1/ICAM-1,3 and VLA-4/VCAM-1 pathways on T lymphocytes and vascular endothelium in Graves' and Hashimoto's thyroid glands. Eur J Immunol 24:2483–2490

Marschang P, Friedrich GJ, Ditlbacher H et al (2006) Reduction of soluble P-selectin by statins is inversely correlated with the progression of coronary artery disease. Int J Cardiol 106:183–190

Marteau JB, Sass C, Pfister M et al (2004) The Leu554Phe polymorphism in the E-selectin gene is associated with blood pressure in overweight people. J Hypertens 22:305–311

Masamune A, Hakomori S, Igarashi Y (1995) N,N,N-trimethylsphingosine inhibits interleukin-1 beta-induced NF-kB activation and consequent E-selectin expression in human umbilical vein endothelial cells. FEBS Lett 367:205–209

Mastej K, Adamiec R (2008) Neutrophil surface expression of CD11b and CD62L in diabetic microangiopathy. Acta Diabetol 45:183–190

Mathieu S, Prorok M, Benoliel AM et al (2004) Transgene expression of α(1,2)-fucosyltransferase-I (FUT1) in tumor cells selectively inhibits sialyl-Lewis x expression and binding to E-selectin without affecting synthesis of sialyl-Lewis a or binding to P-selectin. Am J Pathol 164:371–383

Matondo Maya DW, Mewono L, Nkoma AM et al (2008) Markers of vascular endothelial cell damage and P. falciparum malaria: association between levels of both sE-selectin and thrombomodulin, and cytokines, hemoglobin and clinical presentation. Eur Cytokine Netw 19:123–130

Mattsson E, Heying R, van de Gevel JS et al (2008) Staphylococcal peptidoglycan initiates an inflammatory response and procoagulant activity in human vascular endothelial cells: a comparison with highly purified lipoteichoic acid and TSST-1. FEMS Immunol Med Microbiol 52:110–117

McBane RD 2nd, Karnicki K, Miller RS, Owen WG (2004) The impact of peripheral arterial disease on circulating platelets. Thromb Res 113:137–145

McCafferty DM, Smith CW, Granger DN, Kubes P (1999) Intestinal inflammation in adhesion molecule-deficient mice: an assessment of P-selectin alone and in combination with ICAM-1 or E-selectin. J Leukoc Biol 66:67–74

McLaren AJ, Marshall SE, Haldar NA et al (1999) Adhesion molecule polymorphisms in chronic renal allograft failure. Kidney Int 55:1977–1982

Melendez MM, McNurlan MA, Mynarcik DC et al (2008) Endothelial adhesion molecules are associated with inflammation in subjects with HIV disease. Clin Infect Dis 46:775–780

Mel-S Z, el-Sayed H (2009) Evaluation of microbiologic and hematologic parameters and E-selectin as early predictors for outcome of neonatal sepsis. Arch Pathol Lab Med 133:1291–1296

Merlini PA, Rossi ML, Faioni EM et al (2004) Expression of endothelial protein C receptor and thrombomodulin in human coronary atherosclerotic plaques. Ital Heart J 5:42–47

Merten M, Thiagarajan P (2004) P-selectin in arterial thrombosis. Z Kardiol 93:855–863

Miklossy J, Doudet DD, Schwab C et al (2006) Role of ICAM-1 in persisting inflammation in Parkinson disease and MPTP monkeys. Exp Neurol 197:275–283

Miller MA, Kerry SM, Cook DG et al (2004a) Cellular adhesion molecules and blood pressure: interaction with sex in a multi-ethnic population. J Hypertens 22:705–711

Miller MA, Kerry SM, Dong Y et al (2004b) Association between the Thr715Pro P-selectin gene polymorphism and soluble P-selectin levels in a multiethnic population in South London. Thromb Haemost 92:1060–1065

Molero L, Lopez-Farre A, Mateos-Caceres PJ et al (2005) Effect of clopidogrel on the expression of inflammatory markers in rabbit ischemic coronary artery. Br J Pharmacol 146:419–427

Montoro J, Sastre J, Jáuregui I et al (2007) Allergic rhinitis: continuous or on demand antihistamine therapy? J Investig Allergol Clin Immunol 17(Suppl 2):21–27

Munoz FM, Hawkins EP, Bullard DC et al (1997) Host defense against systemic infection with Streptococcus pneumoniae is impaired in E-, P-, and E-/P-selectin-deficient mice. J Clin Invest 100:2099–2106

Musial K, Zwolinska D, Polak-Jonkisz D et al (2005) Serum VCAM-1, ICAM-1, and L-selectin levels in children and young adults with chronic renal failure. Pediatr Nephrol 20:52–55

Myers DD Jr, Rectenwald JE, Bedard PW et al (2005) Decreased venous thrombosis with an oral inhibitor of P selectin. J Vasc Surg 42:329–336

Nadar SK, Blann AD, Lip GY (2004a) Plasma and platelet-derived vascular endothelial growth factor and angiopoietin-1 in hypertension: effects of antihypertensive therapy. J Intern Med 256:331–337

Nadar SK, Lip GY, Blann AD (2004b) Platelet morphology, soluble P selectin and platelet P-selectin in acute ischaemic stroke. The West Birmingham Stroke Project. Thromb Haemost 92:1342–1348

Nonomura C, Kikuchi J, Kiyokawa N et al (2008) CD43, but not P-selectin glycoprotein ligand-1, functions as an E-selectin counter-receptor in human pre-B-cell leukemia NALL-1. Cancer Res 68:790–799

Nowak M, Wielkoszyński T, Marek B et al (2008) Blood serum levels of vascular cell adhesion molecule (sVCAM-1), intercellular adhesion molecule (sICAM-1) and endothelial leucocyte adhesion molecule-1 (ELAM-1) in diabetic retinopathy. Clin Exp Med 8:159–164

Nubel T, Dippold W, Kaina B, Fritz G (2004) Ionizing radiation-induced E-selectin gene expression and tumor cell adhesion is inhibited by lovastatin and all-trans retinoic acid. Carcinogenesis 25:1335–1344

Ogawa N, Saito N, Kameoka S et al (2008) Clinical significance of intercellular adhesion molecule-1 in ulcerative colitis. Int Surg 93:37–44

Ohkawara Y, Yamauchi K, Maruyama N et al (1995) In situ expression of the cell adhesion molecules in bronchial tissues from asthmatics with air flow limitation: in vivo evidence of VCAM-1/VLA-4 interaction in selective eosinophil infiltration. Am J Respir Cell Mol Biol 12:4–12

Ohmori K, Fukui F, Kiso M et al (2006) Identification of cutaneous lymphocyte-associated antigen as sialyl 6-sulfo Lewis X, a selectin ligand expressed on a subset of skin-homing helper memory T cells. Blood 107:3197–3204

Ohnishi M, Imanishi N, Tojo SJ (1999) Protective effect of anti-P-selectin monoclonal antibody in lipopolysaccharide-induced lung hemorrhage. Inflammation 23:461–469

Okapcova J, Gabor D (2004) The levels of soluble adhesion molecules in diabetic and nondiabetic patients with combined hyperlipoproteinemia and the effect of ciprofibrate therapy. Angiology 55:629–639

Pamuk GE, Vural O, Turgut B et al (2008) Increased platelet activation markers in rheumatoid arthritis: are they related with subclinical atherosclerosis? Platelets 19:146–154

Pamuk GE, Nurı Pamuk O, Orum H et al (2009) Elevated platelet-monocyte complexes in patients with psoriatic arthritis. Platelets 14:1–5

Park EY, Shim JY, Kim DS et al (2008) Elevated serum soluble E-selectin levels in Korean children with measles. Pediatr Int 50:519–522

Paschos KA, Canovas D, Bird NC (2010) The engagement of selectins and their ligands in colorectal cancer liver metastases. J Cell Mol Med 14:165–174

Pemberton PW, Ahmad Y, Bodill H et al (2009) Biomarkers of oxidant stress, insulin sensitivity and endothelial activation in rheumatoid arthritis: a cross-sectional study of their association with accelerated atherosclerosis. BMC Res Notes 2:83

Petersson H, Lind L, Hulthe J et al (2009) Relationships between serum fatty acid composition and multiple markers of inflammation and endothelial function in an elderly population. Atherosclerosis 203(1):298–303

Pezzilli R, Corsi MM, Barassi A et al (2008) Serum adhesion molecules in acute pancreatitis: time course and early assessment of disease severity. Pancreas 37:36–41

Phan QT, Filler SG (2009) Endothelial cell stimulation by Candida albicans. Methods Mol Biol 470:313–326

Philpott JR, Miner PB Jr (2008) Antisense inhibition of ICAM-1 expression as therapy provides insight into basic inflammatory pathways through early experiences in IBD. Expert Opin Biol Ther 8:1627–1632

Pierce JW, Read MA, Ding H et al (1996) Salicylates inhibit I κB-α phosphorylation, endothelial-leukocyte adhesion molecule expression, and neutrophil transmigration. J Immunol 156:3961–3969

Pisella P-J, Brignole F, Debbasch C et al (2000) Flow cytometric analysis of conjunctival epithelium in ocular rosacea and keratoconjunctivitis sicca. Ophthalmology 107:1841–1849

Postadzhiyan AS, Tzontcheva AV, Kehayov I, Finkov B (2008) Circulating soluble adhesion molecules ICAM-1 and VCAM-1 and their association with clinical outcome, troponin T and C-reactive protein in patients with acute coronary syndromes. Clin Biochem 41:126–133

Potapov EV, Hennig F, Wagner FD et al (2005) Natriuretic peptides and E-selectin as predictors of acute deterioration in patients with inotrope-dependent heart failure. Eur J Cardiothorac Surg 27:899–905

Potenza MA, Gagliardi S, Nacci C et al (2009) Endothelial dysfunction in diabetes: from mechanisms to therapeutic targets. Curr Med Chem 16:94–112

Rankin JA (2004) Biological mediators of acute inflammation. AACN Clin Issues 15:3–17

Register TC (2009) Primate models in women's health: inflammation and atherogenesis in female cynomolgus macaques (Macaca fascicularis). Am J Primatol 71:766–775

Remahl AI, Bratt J, Möllby H et al (2008) Comparison of soluble ICAM-1, VCAM-1 and E-selectin levels in patients with episodic cluster headache and giant cell arteritis. Cephalalgia 28:157–163

Rho YH, Chung CP, Oeser A et al (2008) Novel cardiovascular risk factors in premature coronary atherosclerosis associated with systemic lupus erythematosus. J Rheumatol 35:1789–1794

Riad A, Westermann D, Linthout SV et al (2008) Enhancement of endothelial nitric oxide synthase production reverses vascular dysfunction and inflammation in the hindlimbs of a rat model of diabetes. Diabetologia 51:2325–2332

Riaz AA, Wan MX, Schaefer T et al (2002) Fundamental and distinct roles of P-selectin and LFA-1 in ischemia/reperfusion-induced leukocyte-endothelium interactions in the mouse colon. Ann Surg 236:777–784

Riede E, Gratchev A, Foss HD et al (1998) Increased methylation of promotor region suppresses expression of MUC2 gene in colon carcinoma cells. Langenbecks Arch Chir Suppl Kongressbd 115 (Suppl I):299–302

Rodrigues SF, de Oliveira MA, dos Santos RA et al (2008) Hydralazine reduces leukocyte migration through different mechanisms in spontaneously hypertensive and normotensive rats. Eur J Pharmacol 589:206–214

Rubio-Guerra AF, Vargas-Robles H, Vargas-Ayala G et al (2008) The effect of trandolapril and its fixed-dose combination with verapamil on circulating adhesion molecules levels in hypertensive patients with type 2 diabetes. Clin Exp Hypertens 30:682–688

Ruth JH, Amin MA, Woods JM et al (2005) Accelerated development of arthritis in mice lacking endothelial selectins. Arthritis Res Ther 7:R959–R970

Sainani GS, Maru VG (2005) The endothelial leukocyte adhesion molecule. Role in coronary artery disease. Acta Cardiol 60:501–507

Saito I, Terauchi K, Shimuta M et al (1993) Expression of cell adhesion molecules in the salivary and lacrimal glands of Sjogren's syndrome. J Clin Lab Anal 7:180–187

Sanada H, Midorikawa S, Yatabe J et al (2005) Elevation of serum soluble E- and P-selectin in patients with hypertension is reversed by benidipine, a long-acting calcium channel blocker. Hypertens Res 28:871–878

Sarecka-Hujar B, Zak I, Emich-Widera E et al (2010) Association analysis of the E-selectin 98 G > T polymorphism and the risk of childhood ischemic stroke. Cell Biochem Funct 28:591–596

Sato H, Usuda N, Kuroda M et al (2010) CA19-9 and E-selectin as markers of hematogenous metastases and as predictors of prognosis in colorectal cancer. Jpn J Clin Oncol 40:1073–1080

Schnaar RL, Alves CS, Konstantopoulos K (2008) Carcinoembryonic antigen and CD44 variant isoforms cooperate to mediate colon carcinoma cell adhesion to E- and L-selectin in shear flow. J Biol Chem 283:15647–15655

Selakovic V, Raicevic R, Radenovic L (2009) Temporal patterns of soluble adhesion molecules in cerebrospinal fluid and plasma in patients with the acute brain infraction. Dis Markers 26:65–74

Serebruany VL, Malinin AI, Oshrine BR et al (2004) Lack of uniform platelet activation in patients after ischemic stroke and choice of antiplatelet therapy. Thromb Res 113:197–204

Sfikakis PP, Mavrikakis M (1999) Adhesion and lymphocyte costimulatory molecules in systemic rheumatic diseases. Clin Rheumatol 18:317–327

Shapiro NI, Yano K, Sorasaki M et al (2009) Skin biopsies demonstrate site-specific endothelial activation in mouse models of sepsis. J Vasc Res 46:495–502

Shen JS, Meng XL, Moore DF et al (2008) Globotriaosylceramide induces oxidative stress and up-regulates cell adhesion molecule expression in Fabry disease endothelial cells. Mol Genet Metab 95:163–168

Shimada Y, Maeda M, Watanabe G, Imamura M (2003) High serum soluble E-selectin levels are associated with postoperative haematogenic recurrence in esophageal squamous cell carcinoma patients. Oncol Rep 10:991–995

Sigmundsdottir H, Johnston A, Gudjonsson JE et al (2004) Methotrexate markedly reduces the expression of vascular E-selectin, cutaneous lymphocyte-associated antigen and the numbers of mononuclear leucocytes in psoriatic skin. Exp Dermatol 13:426–434

Singh K, Colmegna I, He X et al (2008) Synoviocyte stimulation by the LFA-1-intercellular adhesion molecule-2-Ezrin-Akt pathway in rheumatoid arthritis. J Immunol 180:1971–1978

Sjöswärd KN, Uppugunduri S, Schmekel B (2004) Decreased serum levels of P-selectin and eosinophil cationic protein in patients with mild asthma after inhaled salbutamol. Respiration 71:241–245

Skiba-Choińska I, Rogowski F (1996) Adhesion molecules and their role in pathogenesis of ARDS. Przegl Lek 53:627–630

Smith CH, Barker JN, Lee TH (1993a) Adhesion molecules in allergic inflammation. Am Rev Respir Dis 148:S75–S78

Smith GM, Whelan J, Pescini R et al (1993b) DNA-methylation of the E-selectin promoter represses NF-kB transactivation. Biochem Biophys Res Commun 194:215–221

St Hill CA, Farooqui M, Mitcheltree G et al (2009) The high affinity selectin glycan ligand C2-O-sLex and mRNA transcripts of the core 2 β-1,6-N acetylglucosaminyl-transferase (C2GnT1) gene are highly expressed in human colorectal adenocarcinomas. BMC Cancer 9:79

Steinhubl SR, Moliterno DJ (2005) The role of the platelet in the pathogenesis of atherothrombosis. Am J Cardiovasc Drugs 5:399–408

Stevenson JL, Choi SH, Varki A (2005) Differential metastasis inhibition by clinically relevant levels of heparins–correlation with selectin inhibition, not antithrombotic activity. Clin Cancer Res 11:7003–7011

Stewart AO, Bhatia PA, McCarty CM et al (2001) Discovery of inhibitors of cell adhesion molecule expression in human endothelial cells. 1. Selective inhibition of ICAM-1 and E-selectin expression. J Med Chem 44:988–1002

Storer KP, Tu J, Karunanayaka A et al (2008) Inflammatory molecule expression in cerebral arteriovenous malformations. J Clin Neurosci 15:179–184

Sukhotnik I, Coran AG, Greenblatt R et al (2008) Effect of 100 % oxygen on E-selectin expression, recruitment of neutrophils and enterocyte apoptosis following intestinal ischemia-reperfusion in a rat. Pediatr Surg Int 24:29–35

Sun W, Watanabe Y, Wang ZQ (2006) Expression and significance of ICAM-1 and its counter receptors LFA-1 and Mac-1 in experimental acute pancreatitis of rats. World J Gastroenterol 12:5005–5009

Svensson H, Hansson M, Kilhamn J et al (2009) Selective upregulation of endothelial E-selectin in response to Helicobacter pylori-induced gastritis. Infect Immun 77:3109–3116

Swellam M, Ragab HM, Abdalla NA, El-Asmar AB (2008) Soluble cytokeratin-19 and E-selectin biomarkers: their relevance for lung cancer detection when tested independently or in combinations. Cancer Biomark 4:43–54

Swingler S, Brichacek B, Jacque J-M et al (2003) HIV-1 Nef intersects the macrophage CD40L signaling pathway to promote resting-cell infection. Nature 424:213–219

Syrigos KN, Salgami E, Karayiannakis AJ et al (2004) Prognostic significance of soluble adhesion molecules in Hodgkin's disease. Anticancer Res 24:1243–1247

Takei T, Iida A, Nitta K et al (2002) Association between single-nucleotide polymorphisms in selectin genes and immunoglobulin A nephropathy. Am J Hum Genet 70:781–786

Tamai R, Asai Y, Kawabata A et al (2007) Possible requirement of intercellular adhesion molecule-1 for invasion of gingival epithelial cells by Treponema medium. Can J Microbiol 53:1232–1238

Taverna S, Flugy A, Colomba P et al (2008) Effects of Parietaria judaica pollen extract on human microvascular endothelial cells. Biochem Biophys Res Commun 372:644–649

Telejko B, Zonenberg A, Kuzmicki M et al (2009) Circulating asymmetric dimethylarginine, endothelin-1 and cell adhesion molecules in women with gestational diabetes. Acta Diabetol 46:303–308

Teoh H, Quan A, Bang KW et al (2008) Adiponectin deficiency promotes endothelial activation and profoundly exacerbates sepsis-related mortality. Am J Physiol Endocrinol Metab 295:E658–E664

Thomas SN, Schnaar RL, Konstantopoulos K (2009a) Podocalyxin-like protein is an E-/L-selectin ligand on colon carcinoma cells: comparative biochemical properties of selectin ligands in host and tumor cells. Am J Physiol Cell Physiol 296:C505–C513

Thomas SN, Zhu F, Zhang F et al (2009b) Different roles of galectin-9 isoforms in modulating E-selectin expression and adhesion function in LoVo colon carcinoma cells. Mol Biol Rep 36:823–830

Thomson AW, Satoh S, Nussler AK et al (1994) Circulating intercellular adhesion molecule-1 (ICAM-1) in autoimmune liver disease and evidence for the production of ICAM-1 by cytokine-stimulated human hepatocytes. Clin Exp Immunol 95:83–90

Tozawa K, Sakurada S, Kohri K, Okamoto T (1995) Effects of antinuclear factor kB reagents in blocking adhesion of human cancer cells to vascular endothelial cells. Cancer Res 55:4162–4167

Tregouet DA, Barbaux S, Escolano S et al (2002) Specific haplotypes of the P-selectin gene are associated with myocardial infarction. Hum Mol Genet 11:2015–2023

Tremblay PL, Huot J, Auger FA (2008) Mechanisms by which E-selectin regulates diapedesis of colon cancer cells under flow conditions. Cancer Res 68:5167–5176

Tripathy NK, Chandran V, Garg NK et al (2008) Soluble endothelial cell adhesion molecules and their relationship to disease activity in Takayasu's arteritis. J Rheumatol 35:1842–1845

Turunen AJ, Fernandez JA, Lindgren L et al (2005) Activated protein C reduces graft neutrophil activation in clinical renal transplantation. Am J Transplant 5:2204–2212

Tuttolomondo A, Pinto A, Corrao S et al (2009) Immuno-inflammatory and thrombotic/fibrinolytic variables associated with acute ischemic stroke diagnosis. Atherosclerosis 203:503–508

Uchihara JN, Matsuda T, Okudaira T et al (2006) Transactivation of the ICAM-1 gene by CD30 in Hodgkin's lymphoma. Int J Cancer 118:1098–1107

Uner A, Akcali Z, Unsal D (2004) Serum levels of soluble E-selectin in colorectal cancer. Neoplasma 51:269–274

Vejchapipat P, Sookpotarom P, Theamboonlers A et al (2008) Elevated serum soluble E-selectin is associated with poor outcome and correlated with serum ALT in biliary atresia. Eur J Pediatr Surg 18:254–257

Vijya Lakshmi SV, Padmaja G, Kuppusamy P, Kutala VK (2009) Oxidative stress in cardiovascular disease. Ind J Biochem Biophys 46:421–440

Vischer P, Telgmann R, Schmitz B et al (2008) Molecular investigation of the functional relevance of missense variants of ICAM-1. Pharmacogenet Genomics 18:1017–1019

Volcik KA, Ballantyne CM, Coresh J et al (2006) P-selectin Thr715Pro polymorphism predicts P-selectin levels but not risk of incident coronary heart disease or ischemic stroke in a cohort of 14595 participants: the Atherosclerosis Risk in Communities Study. Atherosclerosis 186:74–79

Vora DK, Rosenbloom CL, Beaudet AL, Cottingham RW (1994) Polymorphisms and linkage analysis for ICAM-1 and the selectin gene cluster. Genomics 21:473–477

Wang K, Zhou X, Zhou Z et al (2005) Platelet, not endothelial, P-selectin is required for neointimal formation after vascular injury. Arterioscler Thromb Vasc Biol 25:1584–1589

Wang JH, Kwon HJ, Lee BJ, Jang YJ (2007) Staphylococcal enterotoxins A and B enhance rhinovirus replication in A549 cells. Am J Rhinol 21:670–674

Warfel JM, D'Agnillo F (2008) Anthrax lethal toxin enhances TNF-induced endothelial VCAM-1 expression via an IFN regulatory factor-1-dependent mechanism. J Immunol 180:7516–7524

Wei M, Gao Y, Tian M et al (2005) Selectively desulfated heparin inhibits P-selectin-mediated adhesion of human melanoma cells. Cancer Lett 229:123–126

Wei J, Cui L, Liu F et al (2010) E-selectin and Sialyl Lewis X expression is associated with lymph node metastasis of invasive micropapillary carcinoma of the breast. Int J Surg Pathol 18:193–200

Wenzel K, Ernst M, Rohde K et al (1996) DNA polymorphisms in adhesion molecule genes–a new risk factor for early atherosclerosis. Hum Genet 97:15–20

Wenzel K, Stahn R, Speer A et al (1999) Functional characterization of atherosclerosis-associated Ser128Arg and Leu554Phe E-selectin mutations. J Biol Chem 380:661–667

West MB, Ramana KV, Kaiserova K et al (2008) L-Arginine prevents metabolic effects of high glucose in diabetic mice. FEBS Lett 582:2609–2614

Woodward J (2008) Crossing the endothelium: E-selectin regulates tumor cell migration under flow conditions. Cell Adh Migr 2:151–152

Wu S, Zhou X, Yang H et al (2009) Polymorphisms and plasma soluble levels of E-selectin in patients with chronic hepatitis B virus infection. Clin Chem Lab Med 47:159–164

Wynn TA, Hesse M, Sandler NG et al (2004) P-selectin suppresses hepatic inflammation and fibrosis in mice by regulating interferon-γ and the IL-13 decoy receptor. Hepatology 39:676–687

Xia B, Han H, Zhang KJ et al (2004) Effects of low molecular weight heparin on platelet surface P-selectin expression and serum interleukin-8 production in rats with trinitrobenzene sulphonic acid-induced colitis. World J Gastroenterol 10:729–732

Xie L, Galettis A, Morris J et al (2008) Intercellular adhesion molecule-1 (ICAM-1) expression is necessary for monocyte adhesion to the placental bed endothelium and is increased in type 1 diabetic human pregnancy. Diabetes Metab Res Rev 24:294–300

Xu Y, Huo Y, Toufektsian MC et al (2006) Activated platelets contribute importantly to myocardial reperfusion injury. Am J Physiol Heart Circ Physiol 290:H692–H699

Yao GH, Liu ZH, Zhang X et al (2008) Circulating thrombomodulin and vascular cell adhesion molecule-1 and renal vascular lesion in patients with lupus nephritis. Lupus 17:720–726

Yildirim K, Senel K, Karatay S et al (2005) Serum E-selectin and erythrocyte membrane Na + K + ATPase levels in patients with rheumatoid arthritis. Cell Biochem Funct 23:285–289

Yip HK, Chang LT, Sun CK et al (2006) Platelet activity is a biomarker of cardiac necrosis and predictive of untoward clinical outcomes in patients with acute myocardial infarction undergoing primary coronary stenting. Circ J 70:31–36

Yoshida M, Takano Y, Sasaoka T et al (2003) E-selectin polymorphism associated with myocardial infarction causes enhanced leukocyte-endothelial Interactions under flow conditions. Arterioscler Thromb Vasc Biol 23:783–788

Yoshizaki A, Yanaba K, Iwata Y et al (2010) Cell adhesion molecules regulate fibrotic process via Th1/Th2/Th17 cell balance in a bleomycin-induced scleroderma model. J Immunol 185:2502–2515

Yun W, Qing-Cheng L, Lei Y, Jia-Yin M (2008) Mucosal tolerance to E-selectin provides protection against cerebral ischemia-reperfusion injury in rats. J Neuroimmunol 205:73–79

Zak I, Sarecka B, Krauze J (2008) Synergistic effects between 561A > C and 98 G > T polymorphisms of E-selectin gene and hypercholesterolemia in determining the susceptibility to coronary artery disease. Heart Vessels 23:257–263

Zakynthinos E, Pappa N (2009) Inflammatory biomarkers in coronary artery disease. J Cardiol 53:317–333

Zaph C, Scott P (2003) Th1 cell-mediated resistance to cutaneous infection with Leishmania major is independent of P- and E-selectins. J Immunol 171:4726–4732

Zen K, Liu DQ, Guo YL et al (2008) CD44v4 is a major E-selectin ligand that mediates breast cancer cell transendothelial migration. PLoS One 3:e1826

Zhang BH, Chen H, Yao XP et al (2002) E-selectin and its ligand-sLeX in the metastasis of hepatocellular carcinoma. Hepatobiliary Pancreat Dis Int 1:80–82

Zheng J, Bao WQ, Sheng WQ et al (2009) Serum 3'-sulfo-Le^a indication of gastric cancer metastasis. Clin Chim Acta 405:119–126

Zigler M, Dobroff AS, Bar-Eli M (2010) Cell adhesion: implication in tumor progression. Minerva Med 101:149–162

Zimmerman T, Blanco FJ (2008) Inhibitors targeting the LFA-1/ICAM-1 cell-adhesion interaction: design and mechanism of action. Curr Pharm Des 14:2128–2139

Zuccarello D, Salpietro DC, Gangemi S et al (2002) Familial chronic nail candidiasis with ICAM-1 deficiency: a new form of chronic mucocutaneous candidiasis. J Med Genet 39:671–675

Zuliani G, Cavalieri M, Galvani M et al (2008) Markers of endothelial dysfunction in older subjects with late onset Alzheimer's disease or vascular dementia. J Neurol Sci 272:164–170

Polycystins and Autosomal Polycystic Kidney Disease

G.S. Gupta

45.1 Polycystic Kidney Disease Genes

Autosomal dominant polycystic kidney disease (ADPKD) is one of the most common monogenic disorders, and globally is the third most common cause of end-stage kidney disease. Although cystic renal disease is the major cause of morbidity, the occurrence of nonrenal cysts, most notably in the liver (occasionally resulting in clinically significant polycystic liver disease) and the increased prevalence of other abnormalities including intracranial aneurysms, indicate that ADPKD is a systemic disorder. Approximately 85% of ADPKD cases are attributable to mutations in polycystic kidney disease (PKD) gene 1 (*PKD1*) on chromosome 16, while mutations in *PKD2* gene on chromosome 4 account for almost all of the remaining cases. These two diseases are phenotypically almost identical, differing only by the higher age of diagnosis with *PKD2*, and its slower progression to end-stage renal disease (Rapoport 2007). The product of *PKD1*, polycystin-1 (PC-1), is a very large protein (4,303 amino acids), and is a membrane glycoprotein widely expressed in epithelial cells. It is also expressed in tight junctions, adherens junctions, desmosomes, apical junctions and primary cilia. Polycystin-2 (PC-2), the product of *PKD2*, is a smaller protein (968 amino acids) mainly present in the endoplasmic reticulum, but also in the cell plasma membrane. PC-1 and -2 are joined via a domain in the carboxy-tail of PC-1, and appear to act in concert (Ong and Harris 2005). PC-2 acts as a Ca^{2+} channel. It appears that PKD1 and PKD2 proteins associate physically in vivo and may be partners of a common signaling cascade involved in tubular morphogenesis (Qian et al. 1997). Qian et al. (1997) defined naturally occurring pathogenic mutations of *PKD1* and *PKD2* that disrupt their associations. Portions of the cellular populations of PC-1 and PC-2 localize to the primary cilium. The ADPKD is the founding member of the "ciliopathies," a recently defined class of genetic disorders that result from mutations in genes encoding cilia-associated proteins.

The human *PKD1* gene encodes an ~14-kb transcript, but full characterization was complicated, because most of the gene lies in a genomic region that is duplicated elsewhere on chromosome 16; the duplicate area encodes three genes with substantial homology to *PKD*. The *PKD1* has been identified in the chromosome region 16p13.3. Other *PKD1*-like loci on chromosome 16 are approximately 97% identical to *PKD1*. The 14.5 kb *PKD1* transcript encodes a 4303/4 amino acid protein with a calculated mass of ~460 kDa with a novel domain architecture. The PKD1 gene covers ~52 kb of genomic DNA and is divided into 46 exons. The amino-terminal half of the protein consists of a mosaic of domains, including leucine-rich repeats flanked by characteristic cysteine-rich structures, LDL-A and C-type lectin domains, and 16 units of a novel 80–85 amino acid domain. The presence of these domains suggests that the PKD1 protein is involved in adhesive protein-protein and protein-carbohydrate interactions in the extracellular compartment. The C-terminal third of the protein has multiple hydrophobic regions, and modeling of this region suggests the presence of many transmembrane domains and a cytoplasmic C terminus. The ADPKD phenotype suggests that polycystin may play a role in cell-matrix communication, which is important for normal basement membrane production and for controlling cellular differentiation (Harris et al. 1995; International Polycystic Kidney Disease Consortium 1995). Mutations in PKD1 gene are the most common cause of ADPKD.

Mouse gene transcript homologous to human PC-1 predicted 79% protein identity to human PC-1 and showed the presence of most of the domains identified in the human sequence. The mouse homolog is transcribed from a unique gene and there are no transcribed, closely related copies as observed in human *PKD1*. In mouse at the junction of exons 12 and 13, several different splicing variants lead to a predicted protein that would be secreted. These forms are predominantly found in newborn brain, while in kidney the transcript homologous to human RNA predominates (Löhning et al 1997). The murine homolog of human *PKD2* gene, *Pkd2* is localized on mouse Chromosome 5 proximal to anchor marker D5Mit175, spans at least 35 kb of the mouse genome, and consists of 15 exons. Its translation product consists of 966 amino acids, and the peptide shows a 95% homology to human PC-2. Functional domains are particularly well conserved in the mouse homolog. The expression of mouse PC-2 in the developing embryo at day 12.5 post conception is localized in mesenchymally derived structures. In adult mouse, the protein is mostly expressed in kidney, which suggests its functional relevance for this organ (Pennekamp et al. 1998).

45.1.1 Regulatory Elements in Promoter Regions

The *PKD1* and *PKD2* genes are developmentally regulated and their aberrant expression leads to cystogenesis. The 5′-flanking regions of the murine and canine *PKD1* genes have been characterized and compared with sequences from human and Fugu rubripes orthologues as well as the *PKD2* promoters from mouse and human (Lantinga-van Leeuwen et al. 2005). Sequences revealed a variety of conserved putative binding sites for transcription factors and no TATA-box element. Nine elements were conserved in the mammalian *PKD1* promoters: AP2, E2F, E-Box, EGRF, ETS, MINI, MZF1, SP1, and ZBP-89; six of these elements were also found in the mammalian *PKD2* promoters. Deletion studies with the mouse *PKD1* promoter defined a functional promoter region for *PKD1* and implied that E2F, EGRF, Ets, MZF1, Sp1, and ZBP-89 are potential key regulators of *PKD1* and *PKD2* gene products in mammals. The proximal *PKD1* promoter region is a potential target of Ets family transcription factors, which regulate the polycystic kidney disease-1 promoter (Puri et al. 2006).

45.2 Polycystins: The Products of PKD Genes

45.2.1 Polycystins

Polycystins are transmembrane proteins that form a distinct subgroup of the transient receptor potential (TRP) superfamily of channels (Montell 2005; Clapham 2003). The polycystin family is divided structurally and functionally into two subfamilies, the polycystin 1 (PKD1)-like or TRPP1 and polycystin 2 (PKD2)-like or TRPP2 proteins, both of which have a modest degree of sequence similarity between subfamilies (Delmas 2004; Igarashi and Somlo 2002; Nauli and Zhou 2004). In humans, the PKD2-like subgroup contains three homologous proteins—PKD2, PKD2L1 and PKD2L2—which are now referred to as TRPP2, TRPP3 and TRPP5. The PKD1-like subgroup contains five homologous proteins—all with an 11-transmembrane topology and by virtue of their structure are not considered as members of the TRP superfamily. Although channel activity has not been demonstrated for PKD1, it has 11 transmembrane segments of which the last 6 are similar to both PKD2 and other voltage-gated calcium channels (Hughes et al. 1995; Mochizuki et al. 1996). PKD2 assembles with PKD1 to form a functional Ca^{2+}-permeable cation channel complex in the plasma membrane (Delmas et al. 2004; Hanaoka et al. 2000) and the primary cilium (Nauli et al. 2003). In addition to currents observed at the plasma membrane in conjunction with PKD1, PKD2-mediated channel activity has also been shown in other sub-cellular membranes where it acts as an intracellular Ca^{2+} release channel (Koulen et al. 2002). Interactions between PKD1 and PKD2 are driven by their cytoplasmic carboxyl-terminal regions (Hanaoka et al. 2000; Newby et al. 2002). PKD1–PKD2 interactions are critical to kidney architecture and function because complex-disrupting mutations in either partner lead to development of autosomal dominant polycystic kidney disease (ADPKD), which affects 1 in 1,000 individuals (Wilson 2004; Igarashi and Somlo 2002; Molland et al. 2010).

45.2.2 Polycystin-1 (TRPP1) with a C-type Lectin Domain

Polycystin-1 (PC-1) is a large transmembrane protein with an estimated molecular weight of ~500 kDa having 4303 amino acids (Hughes et al. 1995; Burn et al. 1995). The large extracellular N-terminal region contains several specific motifs including leucine-rich repeats (LRRs), C-type lectin domain, LDL-A region, multiple Ig-like domains (or PKD domains), REJ domain and GPS domain. It has 11 transmembrane domains, with a PLAT domain located in the first cytoplasmic loop and a small cytoplasmic tail with a G-protein-binding motif and coiled-coil region (Figs. 45.1a and 45.2). The 16 Ig-like domains are segmented such that the first Ig-like domain is localized between the LRRs and the C-type lectin domain, while the remaining 15 Ig-like domains are clustered together between LDL-A and REJ domains. This Ig-like domain cluster forms strong homophilic interactions that are important for cell-cell adhesion (Bycroft et al. 1999; Ibraghimov-Beskrovnaya et al. 2000; Streets et al. 2003). Polycystin is likely a multifunctional protein with important roles in cell-cell/matrix adhesion and ciliary functions (Hildebrandt and Otto 2005; Wilson 2004). Polycystin-1 undergoes partial cleavage at the GPS domain such that N-terminal and C-terminal polypeptides remain non-covalently linked (Qian et al.

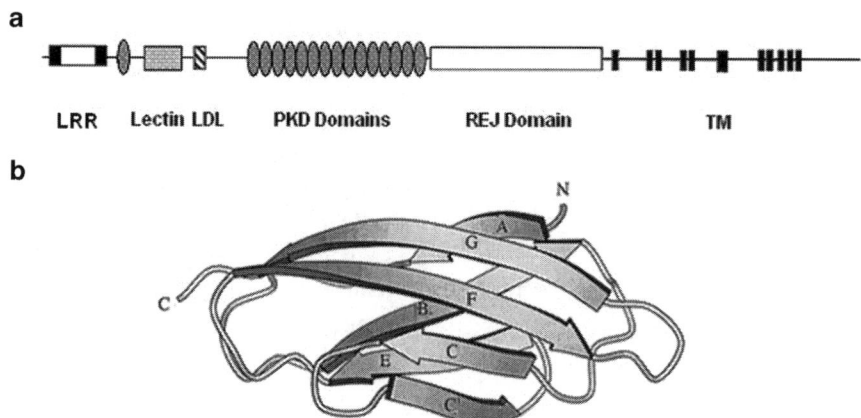

Fig. 45.1 (**a**) A schematic representation of the domains in polycystin-1. *LRR* leucine-rich repeat domain with flanking cysteine-rich regions, *Lectin* C-type lectin domain, *LDL* LDL-A module, *PKD* PKD (Ig-like) domains, *REJ domain* similar to sea urchin REJ protein, *TM* transmembrane domains. (**b**) PKD (Ig-like)-1 showing the elements of secondary structure. The β-strands are labeled (Reprinted with permission from Macmillan Publishers Ltd: EMBO J, Bycroft et al. ©1999)

2002; Wei et al. 2007). It is subsequently cleaved at the second site, which releases its C-terminal tail. The cytoplasmic tail of PC-1 enters the nucleus and regulates cell signaling events. This signaling function of PC-1 is regulated by PC-2 and may possibly be initiated by mechanical stimuli (Chauvet et al. 2004).

Analysis of the C-terminal cytosolic domain of human and mouse PC-1 has identified a number of conserved protein motifs, including a 20-amino-acid heterotrimeric G-protein activation sequence, suggesting that PC-1 may function as a heterotrimeric G-protein coupled receptor (Parnell et al. 1998). The *Xenopus* homologue of human *PKD1* (*xPKD1*) gene predicts the sequence of the putative protein, homologous to human PKD1 with a high level of expression in the kidney. A similar analysis in developing embryos and in an in vitro nephrogenic system suggests that *xPKD1* is associated with development of the amphibian pronephros (Burtey et al. 2005). The C-terminal intracellular segments of PC-1 have been extensively studied, mainly with respect to their putative involvement in cell signaling. Mutations in PC-1 result in ADPKD which is characterised by perturbation of transport resulting in fluid accumulation, cell proliferation and modification of the extracellular matrix (Weston et al. 2003). Cloned and expressed *PKD1* C-type lectin domain demonstrated that PC-1 may be involved in protein–carbohydrate interactions in vivo, and that Ca^{2+} is required for this interaction (Weston et al. 2001). Mutations in *PKD1* are implicated in human autosomal dominant polycystic disease.

Pletnev et al. (2007) made a model on a 3D structure and derived function of individual domains of PC-1. A three dimensional model of CTLD of PC-1 (sequence region 405–534) complexed with galactose and a Ca^{2+} was developed. The model suggests a αβ structural organization, which is composed of eight β strands and three α helices, and includes three disulfide bridges. It is consistent with the observed Ca^{2+} dependence of sugar binding to CLD and identifies the amino acid side chains (E^{499}, H^{501}, E^{506}, N^{518}, T^{519} and D^{520}) that are likely to bind the ligand (Pletnev et al. 2007). A coiled-coil domain within the C terminus of polycystin has been described to bind specifically to the C terminus of PKD2.

45.2.2.1 PKD (or Ig-like) Domains

The extracellular portion of PC-1 is modular in nature, and has a β sandwich-Ig-like protein module called PKD domain and comprises approximately 40% of the structure (Hughes et al. 1995). PC-1 consists of 16 copies of the PKD domain. Although the β-sandwich fold is common to a number of cell-surface modules, the PKD domain represents a distinct protein family. The tenth PKD domain of human and Fugu polycystin-1 showed extensive conservation of surface residues suggesting that this region could be a ligand-binding site. This structure will allow the likely effects of missense mutations in a large part of PKD1 gene to be determined (Bycroft et al. 1999) (Fig. 45.1b).The mechanical properties of PC-1 PKD domains (Forman et al. 2005; Qian et al. 2005) were first investigated for their importance in PKD. Atomic force microscopy experiments, applying force at the N and C termini of PKD domains, showed that PKD domains resist unfolding under significant force, a requirement for their function as mechanosensors (Forman et al. 2005; Qian et al. 2005). Hence, PC-1 was proposed to act as a mechanosensor, transducing fluid flow detected by the cilia of kidney epithelial cells into changes in intracellular calcium levels (Nauli et al. 2003). On the basis of predicted domain structure PC-1 seems to be involved in protein-protein and protein-carbohydrate interactions.

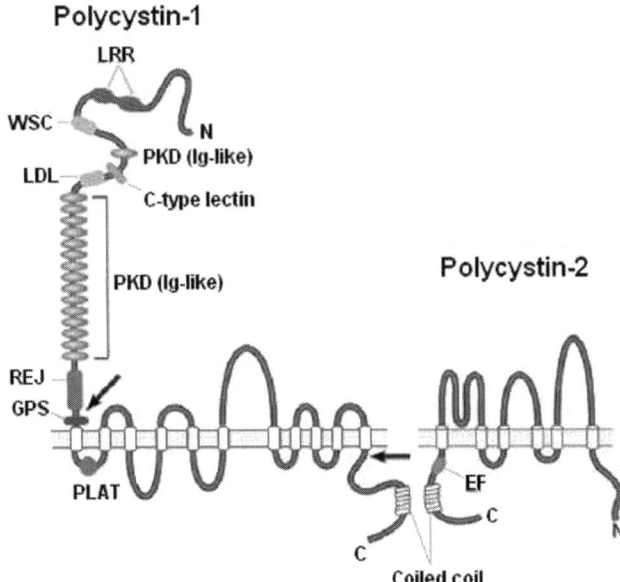

Fig. 45.2 The structure of polycystin-1 and polycystin-2 (*LRR* leucine-rich repeats, *WSC* cell wall integrity and stress response component 1, *PKD* (Ig-like), Ig-like domains, *LDL* low density lipoprotein domain, *REJ* receptor for egg jelly, *GPS* proteolytic G protein-coupled receptor proteolytic site, *PLAT* lipoxygenase domain, *EF* EF hand (helix-loop-helix) domain). Polycystin-1 undergoes cleavage at site shown by the *arrow* (Adapted with permission from Ibraghimov-Beskrovnaya and Bukanov 2008 © Springer)

45.2.2.2 Amino Acid Polymorphism

Many disease-causing mutations have been characterized in *PKD1* gene, most of them resulting in premature protein termination. A few intragenic polymorphisms have been described that are also useful for linkage studies. A new diallelic polymorphism is described for amino acid residue 4058, Ala/Val4058, with allelic frequencies of 0.88 and 0.12, respectively, and a heterozygosity of 0.23, in the Greek and Greek-Cypriot populations. Interestingly, this polymorphism and Ala4091-A/G, which were described in Caucasians, were not detected in DNA from 44 Japanese. This is particularly important when allelic frequencies in a particular population are used for linkage analysis of families of different ethnic origin. Also, observation of the two polymorphisms together as haplotypes suggests that the Ala/Val4058 polymorphism occurred more recently than the establishment of the Ala4091-A/G polymorphism, and specifically on the G allele (Constantinides et al. 1997).

45.2.3 Polycystin-2 (TRPP2)

Polycystin-2 (or TRPP2) is a 968-amino acid containing protein of ~110 kDa with six transmembrane domains and cytoplasmic N- and C-terminal domains (Fig. 45.2) (Mochizuki et al. 1996). Polycystin-2 is thought to be a new member of the transient receptor potential (TRP) family of ion channels. It was shown to be a cation channel with some selectivity for Ca^{2+} (Gonzalez-Perrett et al. 2001) and functions in multiple subcellular locations including plasma membrane (Hanaoka et al. 2000; Babich et al. 2004), endoplasmic reticulum (Koulen et al. 2002) and the primary cilia (Nauli et al. 2003). Several domains present in N and C termini of PC-2 are responsible for PC-2's protein–protein interactions and Ca^{2+} sensitivity. At least two domains, one in each cytoplasmic tail, contribute to PC-2 oligomerization. Immediately distal to PC-2's last transmembrane domain is a functionally complex region of the C terminus that includes coiled-coil, EF-hand (helix-loop-helix), and ER retention domains. A calcium-binding EF hand domain begins upstream of and extends into the PC-1-interacting coiled-coil region (Mochizuki et al. 1996; Qian et al. 1997; Celić et al. 2008). The helix-loop-helix structure of the EF-hand binds Ca^{2+}, permitting the protein to sense or to buffer changes in Ca^{2+} (Gifford et al. 2007). The PC-2 EF-hand has a single Ca^{2+}-binding site with micromolar affinity (Celić et al. 2008). Slightly overlapping with both the coiled-coil and the EF-hand is the sequence that is required for maintaining PC-2's ER and Golgi localization (Cai et al. 1999). A naturally occurring truncation mutation that removes this C-terminal domain, and thus presumably abrogates all of its interactions and regulatory potential, is sufficient to cause ADPKD (Mochizuki et al. 1996).

Polycystin-2 (or TRPP2) has been implicated in various biological functions including cell proliferation, sperm fertilization, mating behaviour, mechanosensation and asymmetric gene expression. Polycystin-1 and -2 can function together as a complex as well as independently in a variety of subcellular compartments. Direct interaction between the cytoplasmic tails of the polycystins has been shown using yeast two-hybrid assay (Qian et al. 1997; Tsiokas et al. 1997). Being a calcium-activated channel PC-2 releases calcium from intracellular stores in response to local increases in calcium concentrations. The calcium-conducting pore of PC-2 is likely formed by the loop between the fifth and sixth transmembrane domains, with some involvement of the third transmembrane domain (Koulen et al. 2002). A missense mutation that perturbs this putative conducting pore (D511V) is causative of ADPKD (Koulen et al. 2002). PC2 also indirectly regulates cytoplasmic calcium levels through interactions with two major intracellular Ca^{2+} channels: the ryanodine receptor and the inositol 1,4,5-trisphosphate receptor (IP3R). The ryanodine receptor mediates calcium-induced calcium release, and PC-2 inhibits its function by binding the channel in its open state and decreasing its conductance (Anyatonwu et al. 2007). PC-2 also modifies IP3-induced Ca^{2+} flux through direct binding between the PC-2C terminus and the IP3R (Li et al. 2005b, 2009).

45.2.3.1 Polycystin-L

Polycystin-L (PC-L), the third member of the polycystin family of proteins, functions as a Ca^{2+}-modulated nonselective cation channel when expressed in Xenopus oocytes. PC-L is predominantly expressed in adult mouse tissues and has a more restricted pattern of expression than either PC-1 or -2. In kidney, PC-L expression was first detected at E16, and levels increased into adulthood. Localization of PC-L was predominantly found in the apical region of the principal cells of inner medullary collecting ducts. PC-L was also found in discrete cell types of the retina, testis, liver, pancreas, heart, and spleen, but not in the lung. The expression pattern of PC-L suggests that it is unlikely to be a candidate gene for ADPKD (Basora et al. 2002). Results raise the possibility that PKD2L1 represents the third genetic locus that is responsible for PKD. However, its CTLD status remains to be investigated.

45.2.4 Interactions of Polycystins

PC-1 and PC-2 interact through their C-terminal cytoplasmic tails. This interaction results in an up-regulation of PC-1 but not PC-2. Furthermore, the cytoplasmic tail of PC-2 but not PC-1 formed homodimers through a coiled-coil domain distinct from the region required for interaction with PC-1. These interactions indicate that PC-1 and PC-2 may function through a common signaling pathway that is necessary for normal tubulogenesis and that PC-1 may require the presence of PC-2 for stable expression (Tsiokas et al. 1997). The intermediate filament (IF) protein vimentin is a strong PC-1-interacting partner. Cytokeratins K8 and K18 and desmin were also found to interact with PC-1. These interactions were mediated by coiled-coil motifs in PC-1 and IF proteins. Polycystin-1 may utilize this association for structural, storage, or signaling functions (Xu et al. 2001). Fibrocystin, a ~450 kDa protein, may form a complex with PC-2 to regulate calcium responses in kidney epithelia, but its exact role in normal and cystic epithelia is unknown (Wang et al. 2007).

A potential target for PC-1 signal transduction is the β catenin signal transduction pathway. This cytoplasmic protein is a key player in the regulation of cell polarity, proliferation, and morphogenesis, processes that are all affected in ADPKD. It was found that polycystin-1 interacts with the E-cadherin–catenin complex containing β catenin and suggested that polycystin-1 modulates signaling by β catenin. Several lines of evidence implicate β catenin in the pathogenesis of polycystic kidney disease. It appeared that aberrant β catenin signaling is a common feature of polycystic kidney diseases. Cytoplasmic β catenin interacts with members of the LEF/TCF family of transcription factors (van Adelsberg 2000).

45.2.5 Tissue and Sub-Cellular Distribution of Polycystins

Both proteins localize to primary cilia of renal epithelial cells, where they are implicated in mechanosensitive transduction signals (Yoder et al. 2002; Nauli et al. 2003). Polycystin 2/TRPP2 has been also documented at the plasma membrane (PM) and the ER—but it is still disputed whether it functions as an intracellular or a plasmalemmal channel. The finding that TRPP2 is retained in the ER of most cell systems has supported the view that TRPP2 might function as a reticular Ca^{2+}-release channel (Koulen et al. 2002). Conversely, TRPP2 has been shown to reside and act at the PM, notably in Madine–Darby canine kidney (MDCK) cells derived from cortical collecting ducts (Luo et al. 2003; Scheffers et al. 2002). These apparently incongruent views might be reconciled by the demonstration that subcellular transport and localization of TRPP2 are controlled by many interactions with adaptor proteins and enzymes (Köttgen et al. 2005; Streets et al. 2006; Geng et al. 2006). Such varied transport behaviour provides a mechanism for the dynamic regulation of TRPP2 channel density at the ER, PM and cilial localizations, and for different subcellular TRPP2 signaling functions. Polycystin is localized in membranes of renal and endothelial cells. When cultured cells made cell-cell contact, polycystin was localized to the lateral membranes of cells in contact (Ibraghimov-Beskrovnaya et al. 1997). Subcellular localization studies found PKD-1 to be a component of various cell junctional complexes and to be associated with the cytoskeleton (Xu et al. 2001). Scheffers et al. (2002) suggest that polycystin-2 can move freely in certain regions of the membrane where it probably functions as a channel, activated by, or in complex with, polycystin-1. PC-1 and PC-2 are functionally expressed in B-lymphoblastoid cells (LCLs), easily obtainable from ADPKD patients, to study PKD gene expression and function (Aguiari et al. 2004).

In fetal kidney, polycystin was localized to the plasma membranes of ureteric bud and comma and S-shaped bodies.

However, in more mature tubules in fetal kidney, in adult kidney, and in polycystic kidney, the majority of polycystin staining was intracellular. The temporal and spatial regulation of polycystin during renal development indicates that polycystin may play a role in nephrogenesis (van Adelsberg et al. 1997; van Adelsberg 1999a). Thus, the knowledge of the compartment-specific regulations of TRPP2 is of crucial importance for the understanding of its roles in health and disease (Giamarchi et al. 2006). The pattern of polycystin expression changed with gestational age in kidney. The widespread distribution of polycystin is consistent with the systemic nature of ADPKD and the role of epithelial cells in the disease (Weston et al. 2001).

45.3 Functions of Polycystin-1 and Polycystin-2

45.3.1 Polycystin-1 and Polycystin-2 Function Together

Polycystin-1 and -2 are thought to function together as part of a multiprotein membrane-spanning complex involved in cell-cell or cell-matrix interactions. Polycystin-1 and -2 interact to produce new calcium-permeable non-selective cation currents. Neither polycystin-1 nor -2 alone is capable of producing currents. Moreover, disease-associated mutant forms of either polycystin protein that are incapable of heterodimerization do not result in new channel activity. Thus, polycystin-1 and -2 co-assemble at the plasma membrane to produce a new channel and to regulate renal tubular morphology and function (Hanaoka et al. 2000). Polycystin-1 and -2 co-assembly can initiate signal transduction, leading to the activation of a number of downstream effectors, including heterotrimeric G-proteins, protein kinase C, mitogen-activated protein kinases, β-catenin, and the AP-1 transcription factor. In addition, PC-2 may function in mediating calcium flux.

The overlapping expression and localization patterns of polycystin-1 and -2 support their role as a complex in regulating multiple processes in tubular epithelia (Ong 2000). Both proteins are found in basolateral membranes and the primary cilium, where they may act together to regulate cellular adhesion and Ca^{2+} signaling. On the other hand, PC-2 is mainly expressed in ER, where it functions as a Ca^{2+} release channel (Koulen et al. 2002). In addition, PC-1 is highly expressed during development, with significant down-regulation of its expression in adult tissues. In contrast, expression of PC-2 seems to persist into adult life (Ong 2000). Experimental evidence from several groups has established an important role for polycystins in epithelial cell morphogenesis, including differentiation and maturation in vivo (Kim et al. 2000; Lu et al. 1997). Studies using MDCK cells demonstrated that expression of polycystin-1 at cell-cell junctions at controlled levels is critical for proper tubular differentiation (Bukanov et al. 2002). It has been shown that PC-1 is directly involved in intercellular adhesion via formation of strong homophilic interactions of its PKD (Ig-like) domains (Ibraghimov-Beskrovnaya et al. 2000). A direct role for Ig-like domains in cell-cell adhesion was demonstrated by specific perturbation of intercellular adhesion using antibodies against Ig-like domains in cell cultures (Ibraghimov-Beskrovnaya et al. 2000; Streets et al. 2003). Polycystin-1 was localized to the cell-cell adhesion complexes with adherens junctions and desmosomal junctions in epithelial cells of different origin (Bukanov et al. 2002; Huan and van Adelsberg 1999; Scheffers et al. 2000). Because alterations in PC-1-mediated adhesion may cause the abnormal epithelial cell phenotype observed in ADPKD cells, including dedifferentiation and loss of epithelial polarity, several studies examined cell-cell adhesion junctions in primary cells derived from ADPKD kidneys (Streets et al. 2003; Roitbak et al. 2004; Russo et al. 2005). Abnormal adherens and desmosomal junctions are found in ADPKD: intracellular junctions are devoid of desmosomal cadherins and associated proteins, which were sequestered to the cytoplasmic pools, and adherens junctions appeared disrupted, accompanied by a great reduction of E-cadherin expression and partial compensatory expression of N-cadherin (Roitbak et al. 2004) Streets et al. (2003) demonstrated that one of the primary functions of polycystin-1 is to mediate cell-cell adhesion in renal epithelial cells, probably via homophilic or heterophilic interactions of the PKD domains. Polycystin-1 localizes with desmosomal junctions (DJs) and adherens junctions (AJs). AJs and DJs are disrupted in ADPKD cells, while tight junctions (TJ) remain intact. In normal epithelial cells, PC-1 is found in a complex with talin (TAL), paxillin (PAX), vinculin (VINC), focal adhesion kinase (FAK), c-src (SRC), p130-cas (CAS), nephrocystin (NPH1) and tensin (TEN). In ADPKD cells, expression of FAK is lost from the focal adhesion complex (Ibraghimov-Beskrovnaya and Bukanov 2008; Chapin and Caplan 2010).

45.3.2 Cell-Cell and Cell-Matrix Adhesion

The *PKD* genes are required for normal fetal development. Consistent with its manifestations, PC-1 is widely expressed in both epithelial and nonepithelial tissues during embryological development. Mice with targeted mutations of either the *PKD1* or the *PKD2* genes die during embryogenesis. The observation, that loss of polycystin-1 or -2 function causes death during embryogenesis, suggests that *PKD1* and *PKD2* might be part of a morphoregulatory pathway (van Adelsberg 1999b). The cystein-flanked leucine-rich repeats (LRR) of PC-1 act as mediators of the PC-1 interaction with the ECM. The observed suppression effect of the LRR on cell proliferation suggests a functional role of the LRR-mediated PC-1 involvement in cell-matrix and cell-cell interactions. These interactions may result in the enhanced cell proliferation that is a characteristic feature of ADPKD (Malhas et al. 2002; Chapin and Caplan 2010).

45.3.3 Role in Ciliary Signaling

The study of the polycystins has revealed some entirely novel insights into fundamental cell biology but these have not yet been satisfactorily integrated into a verified pathogenetic pathway for the development of ADPKD (Sutters 2006). The majority of epithelial cells along the nephron, except intercalated cells, possess a primary cilium, an organelle projecting from the cell's apical surface into the luminal space. Recent studies indicate that renal cilia have a sensory function. Many of the molecular players, which should help solve the mystery of how the renal cilium senses fluid flow, have been studied. Several proteins implicated in the pathogenesis of polycystic kidney disease localize to cilia. The role(s) of the polycystin signaling complex in mediating mechanosensory function by the primary cilium of renal epithelium as well as of the embryonic node have been reviewed (Nauli and Zhou 2004; Chapin and Caplan 2010). It is likely that the central pathogenetic pathway for cystogenesis stems from de-differentiation of tubular epithelial cells. Available evidence indicates that loss of polycystin activity leads to subtle derangements of cell calcium regulation through several possible pathways. Abnormal cell calcium homeostasis might then lead to altered differentiation in affected cells. The precise mechanism by which PC-1 functions, however, remains unclear. Polycystin-1 undergoes a proteolytic cleavage that releases its C-terminal tail (CTT), which enters the nucleus and initiates signaling processes. The cleavage occurs in vivo in association with alterations in mechanical stimuli. Polycystin-2 modulates the signaling properties of the PC-1 CTT.

Results show also a novel pathway by which PC-1 transmits messages directly to the nucleus (Chauvet et al. 2004). Low et al. (2006) showed that PC-1 tail interacts with the transcription factor STAT6 and the coactivator P100, and it stimulates STAT6-dependent gene expression. Under normal conditions, STAT6 localizes to primary cilia of renal epithelial cells. Cessation of apical fluid flow results in nuclear translocation of STAT6. Cyst-lining cells in ADPKD exhibit elevated levels of nuclear STAT6, P100, and the PC-1 tail. Exogenous expression of human PC-1 tail results in renal cyst formation in zebrafish embryos. This is a novel mechanism of cilia function in transduction of a mechanical signal to changes of gene expression involving PC-1 and shows that this pathway is inappropriately activated in ADPKD.

Primary Cilia of inv/inv Mouse Renal Epithelial Cells Sense Physiological Fluid Flow: Anomalous structure of primary cilia and/or impairment of increases in intracellular Ca^{2+} in response to fluid flow are thought to result in renal cyst formation in conditional kif3a knockout, Tg737 and pkd1/pkd2 mutant mice. The mutant inv/inv mouse develops multiple renal cysts like kif3a, Tg737 and pkd1/pkd2 mutants. Inv proteins have been shown to be localized in the renal primary cilia (Shiba et al. 2005).

Polycystins have been suggested to form mechanosensory transduction complexes involved in a variety of biological functions including sperm fertilization, mating behavior, and asymmetric gene expression in different species. Furthermore, their dysfunction is the cause of cyst formation in human kidney disease. The extracellular region of polycystin-1, which has a number of putative binding domains, may act as a mechanosensor. New evidence shows that a mechanosensitive signal, cilia bending, activates the PC-1-PC-2 channel complex. When working properly, this functional complex elicits a transient Ca^{2+} influx, which is coupled to the release of Ca^{2+} from intracellular stores (Cantiello 2003). Delmas (2004) focused on the pros and cons of their candidacy as mechanically gated channels and on recent findings that have significantly advanced our physiological insight. Nauli et al. (2006) proposed that calcium response to fluid-flow shear stress can be used as a read out of polycystin function and that loss of mechanosensation in the renal tubular epithelia is a feature of PKD cysts. Report supports a two-hit hypothesis as a mechanism of cystogenesis.

A large proportion of the extracellular region of polycystin-1 consists of β-sandwich PKD domains in tandem array. Using atomic force microscopy, it was shown that these domains, despite having a low thermodynamic stability, exhibit a remarkable mechanical strength, similar to that of Ig domains in the giant muscle protein titin. The simulations suggest that the basis for this mechanical stability is the formation of aforce-stabilised intermediate. Results suggest that these domains will remain folded under external force supporting the hypothesis that polycystin-1 could act as a mechanosensor, detecting changes in fluid flow in the kidney tubule (Forman et al. 2005). Both polycystins are localized to motile oviduct cilia and this localization is greatly increased upon ovulatory gonadotropic stimulation. It was suggested that PC-1 and -2 play an important role in granulosa cell differentiation and in development and maturation of ovarian follicles. In the oviduct both TRPV4 and polycystins could be important in relaying physiochemical changes in the oviduct upon ovulation (Nauli et al. 2003; Teilmann et al. 2005).

45.3.4 Cilia and Cell Cycle

There is an intimate link between cilia and the cell cycle. The basal bodies/centrosomes of the cilia act as organizers of the mitotic spindle poles during cell division, directly connecting ciliogenesis with cell cycle regulation, and cilia are resorbed when cells enter the cell cycle. Because a number of cystoproteins causing PKD in humans and animals are expressed, at least partially, in cilia or the basal body of the cilia, polycystins and other cystoproteins may play an important role in connecting the mechanosensory function of the cilia to the centrosome and thus influence cell cycle control. Disruption of cystoproteins associated with cilia or basal bodies could, therefore, lead to dysregulation of the cell cycle and proliferation, resulting in cystic disease. Several lines of evidence support this hypothesis. Li et al. (2005b) showed that PC-2 regulates the cell cycle through direct interaction with Id2, a member of the helix-loop-helix (HLH) protein family that is known to regulate cell proliferation and differentiation. Id2 expression suppresses the induction of a cyclin-dependent kinase inhibitor, p21, by either PC-1 or PC-2. It was proposed that Id2 has a crucial role in cell-cycle regulation that is mediated by PC-1 and PC-2 (Chapin and Caplan 2010).

45.3.5 Polycystins and Sperm Physiology

Polycystins also play a significant role in sperm development and function. Drosophila PC-2 is associated with the head and tail of mature sperm. Targeted disruption of the PKD2 homologue results in nearly complete male sterility without disrupting spermatogenesis. Mutant sperm are motile but are unable to reach the female storage organs (seminal receptacles and spermathecae). The sea urchin PC-1-equivalent suPC-2 colocalizes with the PC-1 homolog REJ3 to plasma membrane over the acrosomal vesicle (Galindo et al. 2004; Kierszenbaum 2004). Like other PC-2 family members, suPC-2 is a six-pass transmembrane protein containing C-terminal cytoplasmic EF hand and coiled-coil domains. The location of suPC-2 suggests that it may function as a cation channel mediating the sperm acrosome reaction. The low cation selectivity of PC-2 channels would explain data indicating that Na^+ and Ca^{2+} may enter sea urchin sperm through same channel during the acrosome reaction (Neill et al. 2004). This localization also suggests that the suPC-2-REJ3 complex may function as a cation channel mediating acrosome reaction when sperm contact the jelly layer surrounding the egg at fertilization. Future studies leading to the identification of specific ligands for polycystins, including the signaling pathways, might define the puzzling relationship between renal tubular morphogenesis and sperm development and function.

45.4 Autosomal Dominant Polycystic Kidney Disease

45.4.1 Mutations in *PKD1* and *PKD2* and Association of Polycystic Kidney Disease

Autosomal dominant polycystic kidney disease (ADPKD) is a common inherited nephropathy affecting over 1:1000 of worldwide population. It is a systemic condition with frequent hepatic and cardiovascular manifestations in addition to the progressive development of fluid-filled renal cysts that eventually result in loss of renal function in the majority of affected individuals. Mutations in either PC-1 or PC-2 account for the majority of autosomal dominant PKD. The cysts that grow in the kidney are the result of any number of mutations within the genes *PKD1* and *PKD2* that code for PC-1 and PC-2, respectively. Over 90 mutations have been discovered in *PKD1* alone, which lead to the disruption of proper function. These mutations have been found to be missense, nonsense, and frame-shift (Eo et al. 2002). The diagnosis of ADPKD is typically made using renal imaging despite the identification of mutations in *PKD1* and *PKD2* that account for virtually all cases. Most *PKD* gene mutations are loss of function and a 'two-hit' mechanism has been demonstrated underlying focal cyst formation. The protein products of the *PKD* genes, the polycystins, form a calcium-permeable ion channel complex that regulates the cell cycle and the function of the renal primary cilium. Abnormal cilial function is now thought to be the primary defect in several types of PKD including autosomal recessive polycystic kidney disease and represents a novel and exciting mechanism underlying a range of human diseases (Boucher and Sandford 2004; Chapin and Caplan 2010).

Russo et al. (2005) indicated that polycystin-1 is involved in cell proliferation and morphogenesis. Mutated PC-1 affects intercellular adhesion as shown by desmosomal junctions in primary cells derived from ADPKD cysts. While, primary epithelial cells from normal kidney showed co-localization of PC-1 and desmosomal proteins at cell-cell contacts, a striking difference was seen in ADPKD cells, where PC-1 and desmosomal proteins were lost from the intercellular junction membrane, despite unchanged protein expression levels. Results demonstrated that, in the absence of functional PC-1, desmosomal junctions cannot be properly assembled and remain sequestered in cytoplasmic compartments. Thus, PC-1 is crucial for formation of intercellular contacts. The abnormal expression of PC-1 causes disregulation of cellular adhesion complexes leading to increased proliferation, loss of polarity and, ultimately cystogenesis.

45.4.2 Proliferation and Branching Morphogenesis in Kidney Epithelial Cells

Grimm et al. (2006) showed that the basal and EGF-stimulated rate of cell proliferation is higher in cells that do not express PC-2 versus those that do, indicating that PC-2 acts as a negative regulator of cell growth. In addition, cells not expressing PC-2 exhibited significantly more branching morphogenesis and multicellular tubule formation under basal and Hepatocyte GF-stimulated conditions than their PC-2-expressing counterparts, suggesting that PC-2 may also play an important role in the regulation of tubulogenesis. Cells expressing a channel mutant of PC-2 proliferated faster than those expressing the wild-type protein, but exhibited blunted tubule formation. Thus, the channel activity of PC-2 may be an important component of its regulatory machinery. Finally, PC-2 regulation of cell proliferation appears to be dependent on its ability to prevent phosphorylated extracellular-related kinase from entering the nucleus. These results indicate that PC-2 is necessary for the proper growth and differentiation of kidney epithelial cells (Grimm et al. 2006). Polycystin-2 is thought to function with PC-1, as part of a multiprotein complex involved in transducing Ca^{2+}-dependent information. Although its function as a Ca^{2+}-permeable cation channel is well established, its precise role in the plasma membrane, the endoplasmic reticulum and the cilium is controversial. Studies suggest that PC-2 (TRPP2) function is highly dependent on the subcellular compartment of expression, and is regulated by many interactions with adaptor proteins (Giamarchi et al. 2006).

Although the PKD genes were identified a decade ago, the pathway(s) leading from mutation to disease remain the subject of intense investigation. The pathogenesis of cyst formation is thought to involve increased cell proliferation, fluid accumulation, and basement membrane remodeling. It appears that cAMP metabolism is a central component of cyst formation, stimulating apical chloride secretion and driving the accumulation of cyst fluid. Evidence has shown that ADPKD cells also have an altered responsiveness to cAMP. In contrast to normal kidney cells whose cell proliferation is inhibited by cAMP, ADPKD cells are stimulated to proliferate. Thus, it is likely that an alteration in polycystin function transforms the normal cellular phenotype to one that responds to elevated cAMP by an increased rate of cell proliferation and that the enlarging cyst expands by an increased rate of cAMP-driven fluid secretion. Cyclic AMP and growth factors, including EGF, have complementary effects to accelerate the enlargement of ADPKD cysts, and thereby to contribute to the progression of the disease. This knowledge should facilitate the discovery of inhibitors of signal transduction cascades that can be used in the treatment of ADPKD (Calvet and Grantham 2001). Li et al. (2005) suggested that PC-2 and IP3R functionally interact and modulate intracellular Ca^{2+} signaling. Therefore, mutations in either PC-1 or PC-2 could result in the misregulation of intracellular Ca^{2+} signaling, which in turn could contribute to the pathology of ADPKD (Li et al. 2005).

Involvement of G-proteins was demonstrated by Delmas et al. (2004) who found that full-length PC-1 functions as a constitutive activator of $G_{i/o}$-type but not G_q-type G-proteins and modulates the activity of Ca^{2+} and K^+ channels via release of $G_{\beta\gamma}$ subunits. PC-1 lacking N-terminal 1,811 residues replicated the effects of full-length PC-1. Evidence indicates that full-length PC-1 acts as an untraditional G-protein-coupled receptor, activity of which is physically regulated by PC-2. Thus, it suggests that mutations in PC-1 or PC-2 that distort the polycystin complex would initiate abnormal G-protein signaling in ADPKD (Delmas et al. 2002). Regulation of intracellular Ca^{2+} mobilization has been associated with the functions of PC-1 and PC-2. PC-1 can activate the calcineurin/NFAT (nuclear factor of activated T-cells) signaling pathway (Puri et al. 2004). Puri et al. (2006) suggested a model in which PC-1 signaling leads to a sustained elevation of intracellular Ca^{2+} mediated by PC-1 activation of $G_{\alpha q}$ followed by PLC activation, release of Ca^{2+} from intracellular stores, and activation of store-operated Ca^{2+} entry, thus activating calcineurin and NFAT.

References

Adelsberg JV (2000) Polycystin-1 interacts with E-cadherin and the catenins—clues to the pathogenesis of cyst formation in ADPKD? Nephrol Dial Transplant 15:1–2

Aguiari G, Banzi M, Gessi S et al (2004) Deficiency of polycystin-2 reduces Ca^{2+} channel activity and cell proliferation in ADPKD lymphoblastoid cells. FASEB J 18:884–886

Anyatonwu GI, Estrada M, Tian X et al (2007) Regulation of ryanodine receptor-dependent calcium signaling by polycystin-2. Proc Natl Acad Sci USA 104:6454–6459

Babich V, Zeng WZ et al (2004) The N-terminal extracellular domain is required for polycystin-1-dependent channel activity. J Biol Chem 279:25582–25589

Basora N, Nomura H, Berger UV et al (2002) Tissue and cellular localization of a novel polycystic kidney disease-like gene product, polycystin-L. J Am Soc Nephrol 13:293–301

Boucher C, Sandford R (2004) Autosomal dominant polycystic kidney disease (ADPKD, MIM 173900, PKD1 and PKD2 genes, protein products known as polycystin-1 and polycystin-2). Eur J Hum Genet 12:347–354

Bukanov NO, Husson H, Dackowski WR et al (2002) Functional polycystin-1 expression is developmentally regulated during epithelial morphogenesis in vitro: downregulation and loss of membrane localization during cystogenesis. Hum Mol Genet 11:923–936

Burn TC, Connors TD, Dackowski WR et al (1995) Analysis of the genomic sequence for the autosomal dominant polycystic kidney disease (PKD1) gene predicts the presence of a leucine-rich repeat. The American PKD1 Consortium (APKD1 Consortium). Hum Mol Genet 4:575–582

Burtey S, Leclerc C, Nabais E et al (2005) Cloning and expression of the amphibian homologue of the human PKD1 gene. Gene 357:29–36

Bycroft M, Bateman A et al (1999) The structure of a PKD domain from polycystin-1: implications for polycystic kidney disease. EMBO J 18:297–305

Cai Y, Maeda Y, Cedzich A et al (1999) Identification and characterization of polycystin-2, the PKD2 gene product. J Biol Chem 274:28557–28565

Calvet JP, Grantham JJ (2001) The genetics and physiology of polycystic kidney disease. Semin Nephrol 21:107–123

Cantiello HF (2003) A tale of two tails: ciliary mechanotransduction in ADPKD. Trends Mol Med 9:234–236

Celić A, Petri ET, Demeler B et al (2008) Domain mapping of the polycystin-2 C-terminal tail using de novo molecular modeling and biophysical analysis. J Biol Chem 283:28305–28312

Chapin HC, Caplan MJ (2010) The cell biology of polycystic kidney disease. J Cell Biol 191:701–710

Chauvet V, Tian X et al (2004) Mechanical stimuli induce cleavage and nuclear translocation of the polycystin-1 C terminus. J Clin Invest 114:1433–1443

Clapham DE (2003) TRP channels as cellular sensors. Nature 426:517–524

Constantinides R, Xenophontos S, Neophytou P et al (1997) New amino acid polymorphism, Ala/Val4058, in exon 45 of the polycystic kidney disease 1 gene: evolution of alleles. Hum Genet 99:644–647

Delmas P (2004) Polycystins: from mechanosensation to gene regulation. Cell 118:145–148

Delmas P, Padilla F, Osorio N, Coste B (2004) Polycystins, calcium signaling, and human diseases. Biochem Biophys Res Commun 322:1374–1383

Delmas P, Nomura H, Li X et al (2002) Constitutive activation of G-proteins by polycystin-1 is antagonized by polycystin-2. J Biol Chem 277:11276–11283

Eo HS, Lee JG, Ahn C et al (2002) Three novel mutations of the PKD1 gene in Korean patients with autosomal dominant polycystic kidney disease. Clin Genet 62:169–174

Forman JR, Qamar S, Paci E et al (2005) The remarkable mechanical strength of polycystin-1 supports a direct role in mechanotransduction. J Mol Biol 349:861–871

Galindo BE, Moy GW, Vacquier VD (2004) A third sea urchin sperm receptor for egg jelly module protein, suREJ2, concentrates in the plasma membrane over the sperm mitochondrion. Dev Growth Differ 46:53–60

Geng L, Okuhara D et al (2006) Polycystin-2 traffics to cilia independently of polycystin-1 by using an N-terminal RVxP motif. J Cell Sci 119:1383–1395

Giamarchi A, Padilla F, Coste B et al (2006) The versatile nature of the calcium-permeable cation channel TRPP2. EMBO Rep 7:787–793

Gifford JL, Walsh MP, Vogel HJ (2007) Structures and metal-ion-binding properties of the Ca^{2+}–binding helix-loop-helix EF-hand motifs. Biochem J 405:199–221

Gonzalez-Perrett S, Kim K, Ibarra C et al (2001) Polycystin-2, the protein mutated in autosomal dominant polycystic kidney disease (ADPKD), is a Ca^{2+}-permeable nonselective cation channel. Proc Natl Acad Sci USA 98:1182–1187

Grimm DH, Karihaloo A, Cai Y et al (2006) Polycystin-2 regulates proliferation and branching morphogenesis in kidney epithelial cells. J Biol Chem 281:137–144

Hanaoka K, Qian F, Boletta A et al (2000) Co-assembly of polycystin-1 and -2 produces unique cation-permeable currents. Nature 408(6815):990–994

Harris PC, Ward CJ, Peral B, Hughes J (1995) Polycystic kidney disease. 1: identification and analysis of the primary defect. J Am Soc Nephrol 6:1125–1133

Hildebrandt F, Otto E (2005) Cilia and centrosomes: a unifying pathogenic concept for cystic kidney disease? Nat Rev Genet 6:928–940

Huan Y, van Adelsberg J (1999) Polycystin-1, the PKD1 gene product, is in a complex containing E-cadherin and the catenins. J Clin Invest 104:1459–1468

Hughes J, Ward CJ, Peral B et al (1995) The polycystic kidney disease 1 (PKD1) gene encodes a novel protein with multiple cell recognition domains. Nat Genet 10:151–160

Ibraghimov-Beskrovnaya O, Bukanov N (2008) Polycystic kidney diseases: from molecular discoveries to targeted therapeutic strategies. Cell Mol Life Sci 65:605–619

Ibraghimov-Beskrovnaya O, Dackowski WR, Foggensteiner L et al (1997) Polycystin: in vitro synthesis, in vivo tissue expression, and subcellular localization identifies a large membrane-associated protein. Proc Natl Acad Sci USA 94:6397–6402

Ibraghimov-Beskrovnaya O, Bukanov NO, Donohue LC et al (2000) Strong homophilic interactions of the Ig-like domains of polycystin-1, the protein product of an autosomal dominant polycystic kidney disease gene, PKD1. Hum Mol Genet 9:1641–1649

Igarashi P, Somlo S (2002) Genetics and pathogenesis of polycystic kidney disease. J Am Soc Nephrol 13:2384–2398

International Polycystic Kidney Disease Consortium (1995) Polycystic kidney disease: the complete structure of the PKD1 gene and its protein. Cell 81:289–298

Kierszenbaum AL (2004) Polycystins: what polycystic kidney disease tells us about sperm. Mol Reprod Dev 67:385–388

Kim K, Drummond I et al (2000) Polycystin 1 is required for the structural integrity of blood vessels. Proc Natl Acad Sci USA 97:1731–1736

Köttgen M, Benzing T, Simmen T et al (2005) Trafficking of TRPP2 by PACS proteins represents a novel mechanism of ion channel regulation. EMBO J 24:705–716

Koulen P, Cai Y, Geng L et al (2002) Polycystin-2 is an intracellular calcium release channel. Nat Cell Biol 4:191–197

Lantinga-van Leeuwen IS, Leonhard WN, Dauwerse H et al (2005) Common regulatory elements in the polycystic kidney disease 1 and 2 promoter regions. Eur J Hum Genet 13:649–659

Li X, Luo Y et al (2005a) Polycystin-1 and polycystin-2 regulate the cell cycle through the helix-loop-helix inhibitor Id2. Nat Cell Biol 7:1202–1212

Li Y, Wright JM, Qian F et al (2005b) Polycystin 2 interacts with type I inositol 1,4,5-trisphosphate receptor to modulate intracellular Ca^{2+} signaling. J Biol Chem 280:41298–41306

Li Y, Santoso NG, Yu S et al (2009) Polycystin-1 interacts with inositol 1,4,5-trisphosphate receptor to modulate intracellular Ca^{2+} signaling with implications for polycystic kidney disease. J Biol Chem 284:36431–36441

Löhning C, Nowicka U, Frischauf AM (1997) The mouse homolog of PKD1: sequence analysis and alternative splicing. Mamm Genome 8:307–311

Low SH, Vasanth S, Larson CH et al (2006) Polycystin-1, STAT6, and P100 function in a pathway that transduces ciliary mechanosensation and is activated in polycystic kidney disease. Dev Cell 10:57–69

Lu W, Peissel B et al (1997) Perinatal lethality with kidney and pancreas defects in mice with a targeted PKD1 mutation. Nat Genet 17:179–181

Luo Y, Vassilev PM, Li X et al (2003) Native polycystin 2 functions as a plasma membrane Ca^{2+}-permeable cation channel in renal epithelia. Mol Cell Biol 23:2600–2607

Malhas AN, Abuknesha RA, Price RG (2002) Interaction of the leucine-rich repeats of polycystin-1 with extracellular matrix proteins: possible role in cell proliferation. J Am Soc Nephrol 13:19–26

Mochizuki T, Wu G et al (1996) PKD2, a gene for polycystic kidney disease that encodes an integral membrane protein. Science 272:1339–1342

Molland KL, Narayanan A, Burgner JW et al (2010) Identification of the structural motif responsible for trimeric assembly of the carboxyl-terminal regulatory domains of polycystin channels PKD2L1 and PKD2. Biochem J 429:171–183

Montell C (2005) The TRP superfamily of cation channels. Sci STKE 2005(272):re3

Nauli SM, Zhou J (2004) Polycystins and mechanosensation in renal and nodal cilia. Bioessays 26:844–856

Nauli SM, Alenghat FJ, Luo Y et al (2003) Polycystins 1 and 2 mediate mechanosensation in the primary cilium of kidney cells. Nat Genet 33:129–137

Nauli SM, Rossetti S, Kolb RJ et al (2006) Loss of polycystin-1 in human cyst-lining epithelia leads to ciliary dysfunction. J Am Soc Nephrol 17:1015–1025

Neill AT, Moy GW, Vacquier VD (2004) Polycystin-2 associates with the polycystin-1 homolog, suREJ3, and localizes to the acrosomal region of sea urchin spermatozoa. Mol Reprod Dev 67:472–477

Newby LJ, Streets AJ, Zhao Y et al (2002) Identification, characterization, and localization of a novel kidney polycystin-1-polycystin-2 comple44. J Biol Chem 277:20763–20773

Ong AC (2000) Polycystin expression in the kidney and other tissues: complexity, consensus and controversy. Exp Nephrol 8:208–214

Ong ACM, Harris PC (2005) Molecular pathogenesis of ADPKD: the polycystin complex gets complex. Kidney Int 67:1234–1247

Parnell SC, Magenheimer BS, Maser RL et al (1998) The polycystic kidney disease-1 protein, polycystin-1, binds and activates heterotrimeric G-proteins in vitro. Biochem Biophys Res Commun 251:625–631

Pennekamp P, Bogdanova N, Wilda M et al (1998) Characterization of the murine polycystic kidney disease (Pkd2) gene. Mamm Genome 9:749–752

Pletnev V, Huether R, Habegger L et al (2007) Rational proteomics of PKD1. I. Modeling the three dimensional structure and ligand specificity of the C-lectin binding domain of polycystin-1. J Mol Model 13:891–896

Puri S, Magenheimer BS, Maser RL et al (2004) Polycystin-1 activates the calcineurin/NFAT (nuclear factor of activated T-cells) signaling pathway. J Biol Chem 279:55455–55464

Puri S, Rodova M, Islam MR et al (2006) Ets factors regulate the polycystic kidney disease-1 promoter. Biochem Biophys Res Commun 342:1005–1013

Qian F, Germino FJ, Cai Y et al (1997) PKD1 interacts with PKD2 through a probable coiled-coil domain. Nat Genet 16:179–183

Qian F, Boletta A, Bhunia AK et al (2002) Cleavage of polycystin-1 requires the receptor for egg jelly domain and is disrupted by human autosomal-dominant polycystic kidney disease 1-associated mutations. Proc Natl Acad Sci USA 99:16981–16986

Qian F, Wei W, Germino G, Oberhauser A (2005) The nanomechanics of polycystin-1 extracellular region. J Biol Chem 280:40723–40730

Rapoport J (2007) Autosomal dominant polycystic kidney disease: pathophysiology and treatment. Q J Med 100:1–9

Roitbak T, Ward CJ, Harris PC et al (2004) A polycystin-1 multiprotein complex is disrupted in polycystic kidney disease cells. Mol Biol Cell 15:1334–1346

Russo RJ, Husson H, Joly D et al (2005) Impaired formation of desmosomal junctions in ADPKD epithelia. Histochem Cell Biol 124:487–497

Scheffers MS, van der Bent P, Prins F et al (2000) Polycystin-1, the product of the polycystic kidney disease 1 gene, co-localizes with desmosomes in MDCK cells. Hum Mol Genet 9:2743–2744

Scheffers MS, Le H, van der Bent P et al (2002) Distinct subcellular expression of endogenous polycystin-2 in the plasma membrane and Golgi apparatus of MDCK cells. Hum Mol Genet 11:59–67

Shiba D, Takamatsu T, Yokoyama T (2005) Primary cilia of inv/inv mouse renal epithelial cells sense physiological fluid flow: bending of primary cilia and Ca^{2+} influ44. Cell Struct Funct 30:93–100

Streets AJ, Newby LJ, O'Hare MJ et al (2003) Functional analysis of PKD1 transgenic lines reveals a direct role for polycystin-1 in mediating cell-cell adhesion. J Am Soc Nephrol 14:1804–1815

Streets AJ, Moon DJ, Kane ME et al (2006) Identification of an N-terminal glycogen synthase kinase 3 phosphorylation site which regulates the functional localization of polycystin-2 in vivo and in vitro. Hum Mol Genet 15:1465–1473

Sutters M (2006) The pathogenesis of autosomal dominant polycystic kidney disease. Nephron Exp Nephrol 103:e149–e155

Teilmann SC, Byskov AG, Pedersen PA et al (2005) Localization of transient receptor potential ion channels in primary and motile cilia of the female murine reproductive organs. Mol Reprod Dev 71:444–452

Tsiokas L, Kim E et al (1997) Homo- and heterodimeric interactions between the gene products of PKD1 and PKD2. Proc Natl Acad Sci USA 94:6965–6970

van Adelsberg J (1999a) Peptides from the PKD repeats of polycystin, the PKD1 gene product, modulate pattern formation in the developing kidney. Dev Genet 24:299–308

van Adelsberg JS (1999b) The role of the polycystins in kidney development. Pediatr Nephrol 13:454–459

van Adelsberg J, Chamberlain S, D'Agati V (1997) Polycystin expression is temporally and spatially regulated during renal development. Am J Physiol 272:F602–F609

Wang S, Zhang J, Nauli SM et al (2007) Fibrocystin/polyductin, found in the same protein complex with polycystin-2, regulates calcium responses in kidney epithelia. Mol Cell Biol 27:3241–3252

Wei W, Hackmann K, Xu H et al (2007) Characterization of cis-autoproteolysis of polycystin-1, the product of human polycystic kidney disease 1 gene. J Biol Chem 282:21729–21737

Weston BS, Bagnéris C, Price RG, Stirling JL (2001) The polycystin-1C-type lectin domain binds carbohydrate in a calcium-dependent manner, and interacts with extracellular matrix proteins in vitro. Biochim Biophys Acta 1536:161–176

Weston BS, Malhas AN, Price RG (2003) Structure-function relationships of the extracellular domain of the autosomal dominant polycystic kidney disease-associated protein, polycystin-1. FEBS Lett 538:8–13

Wilson PD (2004) Polycystic kidney disease. N Engl J Med 350:151–164

Xu GM, Sikaneta T, Sullivan BM et al (2001) Polycystin-1 interacts with intermediate filaments. J Biol Chem 276:46544–46552

Yoder BK, Hou X, Guay-Woodford LM (2002) The polycystic kidney disease proteins, polycystin-1, polycystin-2, polaris, and cystin, are co-localized in renal cilia. J Am Soc Nephrol 13:2508–2516

Endogenous Lectins as Drug Targets

Rajesh K. Gupta and Anita Gupta

46.1 Targeting of Mannose-6-Phosphate Receptors and Applications in Human Diseases

46.1.1 Lysosomal Storage Diseases

The lysosome is an intracytoplasmic acidic vacuole containing more than 60 hydrolytic enzymes for digestion of macromolecules, such as nucleic acids, proteins, lipids and complex carbohydrates. Expression of lysosomal enzyme activities is regulated by various intracellular environmental factors. Mutation of a gene coding for a lysosomal enzyme results in a specific genetic disease, often involving the central nervous system in children. Three groups of functional proteins are known at present for regulation of the expressed enzyme activity in lysosomes. Targeting of a newly synthesized protein is achieved by the mannose 6-phosphate receptor (M6PR) system (Chaps. 3–5), which was revealed in the course of I-cell disease research. Many lysosomal enzymes are excessively secreted in the extracellular compartment in the absence of this regulatory system in patients with this disease (Kornfeld and Sly 2001). Lysosomal enzymes are also components of cell type-specific compartments referred to as lysosome-related organelles which include melanosomes, lytic granules, MHC class II compartments, platelet-dense granules, and synaptic-like microvesicles. Lysosomal storage diseases (LSDs) are inherited metabolic disorders caused by deficient activity of a single lysosomal enzyme or other defects resulting in deficient catabolism of large substrates in lysosomes. There are more than 40 forms of inherited LSDs known to occur in humans, with an aggregate incidence estimated at 1 in 7,000 live births. Clinical signs result from the inability of lysosomes to degrade large substrates; because most lysosomal enzymes are ubiquitously expressed, a deficiency in a single enzyme can affect multiple organ systems. Thus LSDs are associated with high morbidity and mortality and represent a significant burden on patients and society. Because lysosomal enzymes are trafficked by M6PR mechanism, normal enzyme provided to deficient cells can be localized to the lysosome to reduce and prevent storage. However, many LSDs remain untreatable, and gene therapy holds the promise for effective therapy. Other therapies for some LSDs do exist, or are under evaluation, including heterologous bone marrow or cord blood transplantation, enzyme replacement therapy (ERT), and substrate reduction therapy, but these treatments are associated with significant concerns, including high morbidity and mortality, limited positive outcomes, incomplete response to therapy the cost (Haskins 2009), life-long therapy, and the cost (Haskins 2009). Moreover, proteins, vesicles, nano-particles or other polymers could be functionalized with mannose 6-phosphate or its analogues in order to be used as carriers of bioactive molecules for the treatments of different diseases (Gary-Bobo et al. 2007).

Transport of lysosomal enzymes is mediated by two M6PRs: a cation dependent (CD-MPR) and a cation independent receptor (CI-MPR) (see Chaps. 3–5 for structures and functions). In hepatocytes of MPR-deficient neonatal mice lysosomal storage occurs when both MPRs are lacking, whereas deficiency of CI-MPR only has no effect on the ultrastructure of the lysosomal system (Schellens et al. 2003). Some structural features have been shown to be crucial for the binding of M6P to CI-MPR. The hydroxyl group at 2-position of pyranose ring must be axial to allow a strong binding to the CI-MPR. The analogues synthesized to target CI-MPR must be isosteric to M6P to efficiently bind to the receptor. Moreover, a single negative charge is sufficient to allow the binding to the receptor while the phosphorus atom is not necessary to ensure recognition. However, the best ligands for CI-MPR contain two negative charges like in the malonate or phosphonate isosteric analogues of M6P.

46.1.2 Enzyme Replacement Therapy (ERT)

Enzyme replacement therapy (ERT) represents a major breakthrough in the treatment of LSDs, initially used for the treatment of Gaucher disease and now available for several other LSDs. For example ERT with recombinant human acid α-glucosidase (GAA) (rhGAA) is presently the only approach for the treatment of Pompe disease patients. To date, several studies have pointed to the role of a variety of structural and biochemical responses triggered by intracellular storage, which are considered to be responsible for the pathogenetic manifestation of LSDs. Abnormal intracellular trafficking of lipids and proteins may affect the function of membrane-bound proteins, such as receptors, and ligands. The CI-MPR is of particular interest as it is a key player in the internalization of exogenous enzymes, and for the possible consequences of deranged CI-MPR function on ERT efficacy. At the steady state, a fraction (~10%) of CI-MPR is located on cell surface, where it mediates the uptake of lysosomal enzymes. CI-MPR being an integral membrane glycoprotein follows a complex and finely regulated itinerary from the trans-Golgi network (TGN), where it binds newly synthesized lysosomal hydrolases, travels through the early endosomes towards the late endosomal compartments, and recycles back to the TGN. CI-MPR at the cell surface mediates the uptake of lysosomal enzymes, internalize in the early endocytic compartment and then recycles again to the plasma membrane through the endocytic recycling compartment (ERC). Some of these advances in LSDs therapies have been enumerated in sections to follow. However many of these studies are at experimental stages.

Pompe Disease (Glycogen Storage Disease Type II)
Pompe disease also called Glycogen storage disease type II (or acid maltase deficiency) is an autosomal recessive metabolic disorder which damages muscle and nerve cells throughout the body. It is caused by an accumulation of glycogen in the lysosome due to deficiency of the lysosomal acid α-glucosidase (GAA). The GAA enzyme is responsible for breaking down glycogen in the lysosome. In individuals afflicted with Pompe Disease, the GAA enzyme is either absent or present in a very low quantity and glycogen builds in the lysosome, resulting in symptoms associated with this disorder. The glycogen storage disease with a defect in lysosomal metabolism is the first glycogen storage disease to be identified, in 1932. The build-up of glycogen causes progressive muscle weakness (myopathy) throughout the body and affects various body tissues, particularly in the heart, skeletal muscles, liver and nervous system. There are four primary forms of Pompe Disease: The classic infantile form, the non-classic infantile, the juvenile form, and the adult form. The classic infantile form is the most severe. Pompe Disease is estimated to occur in about one in 40,000 births worldwide.

One of the major therapeutic applications of M6P derivatives could be the enzyme replacement therapy (ERT) for lysosomal diseases. To address the consequences of abnormalities of cellular morphology and function on CI-MPR subcellular localization, fibroblasts from Pompe disease patients with different genotypes and phenotypes have been studied. In these cells, which showed abnormalities of cellular morphology, CI-MPR is mislocalized and its availability at the plasma membrane is reduced. These abnormalities in CI-MPR distribution result in a less efficient uptake of rhGAA by Pompe disease fibroblasts. CI-MPR-mediated endocytosis of rhGAA is an important pathway by which the enzyme is delivered to the affected lysosomes of Pompe muscle cells. Hence, the generation of rhGAA containing high affinity ligands for the CI-MPR represents a strategy by which the potency of rhGAA and the clinical efficacy of enzyme replacement therapy for Pompe disease may be improved. To enhance the delivery of rhGAA to the affected muscles in Pompe disease, the carbohydrate moieties on the enzyme were remodelled to exhibit a high affinity ligand for CI-MPR. This was achieved by chemically conjugating on to rhGAA, a synthetic oligosaccharide ligand bearing M6P residues in the optimal configuration for binding the receptor (Zhu et al. 2004, 2005). This approach allowed to decrease the effective dose of enzyme and to treat either less accessible tissues, such as skeletal muscles, or tissues with a relatively low abundance of CI-MPR (Gary-Bobo et al. 2007).

The ERT for Pompe disease was recently approved in Europe, the U.S., Canada, and Japan using a recombinant human GAA (Myozyme, alglucosidase alfa) produced in CHO cells (CHO-GAA). The ERT with rhGAA is at present the only approved treatment for Pompe disease, in addition to supportive and physical therapies. However, ERT shows limited efficacy in some patients and does not completely correct the disease phenotype. Recently, an improved knowledge of Pompe disease pathophysiology has provided clues to explain the limitations of ERT. A mechanical effect of lysosomal inclusions on muscle contractility has been proposed as a key factor of disease resulting in a severe loss of contractility. In addition, it has been shown that secondary abnormalities of housekeeping cellular functions, such as autophagy, have an important role in the pathogenesis of cell damage in Pompe disease. Abnormalities of intracellular trafficking of vesicles and membrane-bound proteins, such as the CI-MPR, may be deleterious for the efficacy of ERT. Other approaches, also in a pre-clinical stage, include substrate reduction and gene therapy (Parenti and Andria 2011; McVie-Wylie et al. 2008).

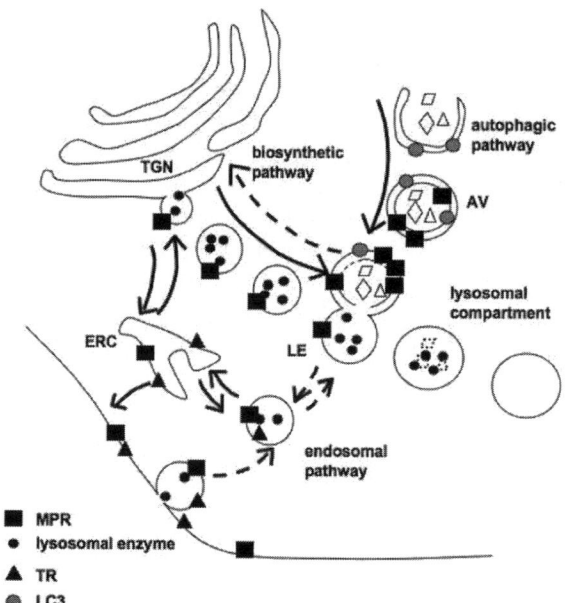

Fig. 46.1 Schematic representation of CI-MPR trafficking in the cells. CI-MPR follows different routes, including the biosynthetic pathway from the trans-Golgi network (*TGN*), where it binds and delivers newly synthesized lysosomal hydrolases to the late endosomal compartment. A fraction (~ 10%) of CI-MPR is located on the cell surface, where it mediates the uptake of lysosomal enzymes, is internalized in the endocytic pathway and recycles through the endocytic recycling compartment (*ERC*), again to the plasma membrane. Increased co-localization of CI-MPR with microtubule-associated protein 1 light chain 3 (*LC3*) in Pompe disease fibroblasts indicates that the CI-MPR in these cells is diverted from its normal routes and is sequestered in autophagosomes and autophagolysosomes. Since transferrin and transferring receptor trafficking is not affected in Pompe disease fibroblasts, it is likely that the disruption of CI-MPR trafficking occurs at the late endosomal compartment and in the retrograde route to the TGN (Cardone et al. 2008)

To investigate the mechanisms underlying the variable response to ERT, Cardone et al. (2008) studied cell morphology of Pompe disease fibroblasts, the distribution and trafficking of the CI-MPR that mediates rhGAA uptake, and rhGAA uptake itself. Immunofluorescence analysis showed abnormal intracellular distribution of CI-MPR in Pompe disease fibroblasts, increased co-localization with microtubule-associated protein 1 light chain 3 (LC3) and reduced availability of the receptor at the plasma membrane. The recycling of CI-MPR from the plasma membrane to the trans-Golgi network was also impaired. All these abnormalities were more prominent in severe and intermediate Pompe disease fibroblasts, correlating with disease severity. In severe and intermediate Pompe disease cells rhGAA uptake and processing were less efficient and correction of GAA activity was reduced. Results indicate a role for disrupted CI-MPR trafficking in the variable response to ERT in Pompe disease with implications for ERT efficacy and optimization of treatment protocols (Fig. 46.1).

Although ERT with acid α-glucosidase has become available for Pompe disease, the response of skeletal muscle, as opposed to the heart, has been attenuated. The poor response of skeletal muscle was attributed to the low abundance of the CI-MPR in skeletal muscle compared to heart. In CI-MPR-KO/GAA-KO (double KO) mice, the role of CI-MPR was emphasized by the lack of efficacy of ERT as demonstrated by markedly reduced biochemical correction of GAA deficiency and of glycogen accumulations in double KO mice, in comparison with the administration of the same doses in GAA-KO mice. Clenbuterol, a selective β_2-agonist, enhanced the CI-MPR expression in skeletal tissue and also increased efficacy from GAA therapy, thereby confirming the key role of CI-MPR with regard to ERT in Pompe disease. Biochemical correction improved in both muscle and non-muscle tissues, indicating that therapy could be similarly enhanced in other lysosomal storage disorders. In summary, enhanced CI-MPR expression might improve the efficacy of ERT in Pompe disease through enhancing receptor-mediated uptake of GAA (Koeberl et al. 2011).

Aspartylglycosaminuria

Aspartylglucosaminuria also called aspartylglycosaminuria (AGU), a severe lysosomal storage disease, is caused by the deficiency of the lysosomal glycosylasparaginase (GA), and accumulation of aspartylglucosamine (GlcNAc-Asn) in tissues. Aspartylglucosaminuria is caused by mutations in the gene encoding for a soluble aspartylglucosaminidase (AGA). Kyttälä et al. (1998) indicated that endocytotic capability of cultured telencephalic neurons for recombinant AGA was mediated by M6PRs. Human leukocyte glycosylasparaginase can correct the metabolic defect in Epstein-Barr virus (EBV)-transformed AGU lymphocytes rapidly and effectively by M6PR-mediated endocytosis or by contact-mediated cell-to-cell transfer from normal EBV-transformed lymphocytes, and that 2–7% of normal activity is sufficient to correct the GlcNAc-Asn metabolism in the cells. The combined evidence indicates that cell-to-cell transfer of GA plays a main role in enzyme replacement therapy of AGU by normal lymphocytes (Dunder and Mononen 2001).

Niemann-Pick Disease (NPD) Type-A, B and C

Acid sphingomyelinase (ASM), a member of the saposin-like protein family, is a lysosomal hydrolase that converts sphingomyelin to ceramide. Deficiency of ASM causes a variant form of Niemann-Pick disease. Progressive accumulation of lipid-laden macrophages is the hallmark of the ASM-deficient forms of Niemann-Pick disease (i.e. Types A and B NPD). Studies using receptor-specific ligands to inhibit enzyme uptake revealed that in normal cells rhASM was taken up by a combination of mannose receptor (MR) and M6PR, respectively. Whereas in the ASM deficient (ASM-KO) cells the M6PR had a minimal role in rhASM uptake. Expression of M6PR mRNA was normal in the ASM-KO cells, although it was hypothesized that lipid accumulation in ASM-KO macrophages led to abnormalities in M6PR trafficking and/or degradation, resulting in reduced enzyme uptake. Consistent with this hypothesis, it was also found that, when rhASM was modified to expose terminal mannose residues and target mannose receptors, the uptake of this modified enzyme by ASM-KO cells was tenfold greater when compared with the "complex" type rhASM. These findings have important implications for NPD enzyme replacement therapy, particularly in the lung (Dhami and Schuchman 2004). Further studies suggest that ASM uses in part the M6P-R and Sortilin, a type I transmembrane glycoprotein that belongs to a family of receptor proteins, involved in lysosomal targeting (Ni and Morales 2006).

Niemann-Pick disease type C (NPC) is an autosomal recessive disorder that leads to massive accumulation of cholesterol and glycosphingolipids in late endosomes and lysosomes. NPC, caused by mutations in the *NPC1* gene or the *NPC2* gene, is characterized by the accumulation of unesterified cholesterol and other lipids in lysosomal compartment. NPC2 is a small lysosomal protein that is targeted to this compartment via a M6P-inhibitable pathway. Either of M6PRs alone is sufficient to transport NPC2 to the endo/lysosomal compartment, although M6PR-300 seems to be more efficient than M6PR-47. In the absence of both MPRs, NPC2 is secreted into the culture medium, and only a small amount of intracellular NPC2 can be detected, mainly in the ER. This leads to massive accumulation of unesterified cholesterol in the endo/lysosomal compartment of the MPR46/300-deficient fibroblasts, a phenotype similar to that of the NPC patient fibroblasts. The lysosomal targeting of NPC2 is strictly dependent on M6PRs in fibroblasts. In another concept, Rab9 is likely sequestered in an inactive form on Niemann-Pick type C membranes, as cation-dependent M6PRs were missorted to the lysosome for degradation, a process that was reversed by over-expression of Rab9. It seems that cholesterol contributes directly to the sequestration of Rab9 on Niemann-Pick type C cell membranes, which in turn, disrupts M6PR trafficking (Naureckiene et al. 2000; Ganley and Pfeffer 2006).

Mucolipidosis Type II and III
Mucolipidosis (ML) II and III are rare autosomal recessive inherited diseases characterized by deficiency of multiple lysosomal enzymes resulting in a generalized storage of macromolecules in lysosomes of cells of mesenchymal origin. The M6P lysosomal targeting signal on acid hydrolases is synthesized by the sequential action of uridine 5′-diphosphate-N-acetylglucosamine: lysosomal enzyme N-acetylglucosamine-1-phosphotransferase (GlcNAc-1-phosphotransferase) and GlcNAc-1-phosphodiester α-N-acetylglucosaminidase ("uncovering enzyme" or UCE). Mutations in the two genes that encode GlcNAc-1-phosphotransferase give rise to lysosomal storage diseases (ML type II and III). In ML II and ML III fibroblasts, most of the newly synthesized lysosomal enzymes are secreted into the medium instead of being targeted correctly to lysosomes. Studies indicate that phospholipases, like most other lysosomal enzymes in these diseases, are secreted into the blood instead of being targeted specifically to lysosomes. The M6PR pathway is needed for proper delivery of lysosomal phospholipases to lysosomes (Jansen et al. 1999). Boonen et al. (2009) demonstrated that UCE accounts for all the uncovering activity in the Golgi. In the absence of UCE, the weak binding of the acid hydrolases to the CI-MPR allows sufficient sorting to lysosomes to prevent the tissue abnormalities seen with GlcNAc-1-phosphotranferase deficiency (Boonen et al. 2009).

46.1.3 M6PR-Mediated Transport Across Blood-Brain Barrier

Enzyme replacement therapy has been used successfully in many lysosomal storage diseases. However, correction of brain storage has been limited by the inability of infused enzyme to cross the blood-brain barrier (BBB). Delivering therapeutic levels of lysosomal enzymes across the BBB is pivotal issue in treating CNS diseases, including the mucopolysaccharidoses. The newborn mouse is an exception because recombinant enzyme is delivered to neonatal brain after M6PR-mediated transcytosis. Access to this route is limited for 2 weeks of age, after which this transporter is lost with maturation.

Glial and Neuronal Cells Express CI-MPR
Enzyme replacement therapy for lysosomal storage disorders depends on efficient uptake of recombinant enzyme into the tissues of patients. This uptake is mediated by oligosaccharide receptors including the CI-MPR and the mannose receptor (MR). Studies on the uptake of recombinant α-(L)-iduronidase into glial and neuronal cells, produced by retrovirally transduced NIH3T3 fibroblasts indicate that: (1) neuronal and glial cells take up α-(L)-iduronidase released into the medium by retrovirally transduced fibroblasts expressing high levels of α-(L)-iduronidase; (2) both glial and neuronal cells express the CI-MPR responsible for lysosomal enzyme uptake; and (3) uptake of the lysosomal enzyme could be blocked by excess free M6P, but not glucose-6-phosphate. Thus, various brain cells take up α-(L)-iduronidase, possibly through a CI-MPR mediated pathway, and this uptake is higher in actively dividing or immature brain cells (Stewart et al. 1997).

GM2 Gangliosidosis

Intralysosomal stability of β-galactosidase is regulated by a multifunctional protein that interacts with two lysosomal enzymes, β-galactosidase and sialidase, and also exerts catalytic activities as carboxypeptidase, esterase and deamidase under various pH conditions. It is encoded by a gene on chromosome 20, and its mutation results in a neurodegenerative disease in children and adults (galactosialidosis). For digestion of lipid substrates, lysosomal enzymes need specific activator proteins as natural detergents for molecular interaction with these nonpolar compounds. Two different groups of proteins have been revealed. A protein encoded by a gene on chromosome 5 interacts with ganglioside GM2 and its asialo derivative, for their catalytic hydrolysis by β-hexosaminidase A. Another protein encoded by a gene on chromosome 10 is expressed as a precursor (prosaposin) which is then processed to four small proteins (saposins) with heterogeneous functions. They are essential for hydrolysis of sphingolipid substrates, and genetic deficiency of each protein results in various lipid storage diseases.

Sandhoff disease is an autosomal recessive lysosomal storage disease caused by a defect of the β-subunit gene (*HEXB*) associated with simultaneous deficiencies of β-hexosaminidase A (HexA; αβ) and B (HexB; ββ), and excessive accumulation of GM2 ganglioside (GM2) and oligosaccharides with N-acetylglucosamine (GlcNAc) residues at their non-reducing termini. From the neonatal brains of Sandhoff disease model mice (SD mice) produced by disruption of the murine Hex β-subunit gene allele (Hexb$^{-/-}$) recombinant Hex isozyme sub-units were found to be incorporated into the SD microglia via cell surface CI-MPR and mannose receptor to degrade the intracellularly accumulated GM2 and GlcNAc-oligosaccharides (Tsuji et al. 2005).

Tay-Sachs disease is a severe neurodegenerative disorder due to mutations in the *HEXA* gene coding for the α-chain of the αβ heterodimeric lysosomal enzyme β-hexosaminidase A (HexA). Guidotti et al. (1998) corrected HexA-deficient cells by *HEXA* gene transfer. Murine HexA-deficient fibroblasts derived from HexA-/- mice were transduced with the G.HEXA vector. Transduced cells overexpressed the α-chain, resulting in the synthesis of interspecific HexA (human α-chain/murine β-chain) and in a total correction of HexA deficiency. The α-chain was secreted in the culture medium and taken up by HexA-deficient cells via M6PR binding, allowing for the restoration of intracellular HexA activity in non-transduced cells.

To develop a novel enzyme replacement therapy for Tay-Sachs disease (TSD) and Sandhoff disease (SD), which are caused by deficiency of Hex A, Matsuoka et al. (2011) designed a genetically engineered *HEXB* encoding the chimeric human β-subunit containing partial amino acid sequence of the α-subunit and succeeded in producing the modified HexB by a CHO cell line stably expressing the chimeric HexB, which can degrade artificial anionic substrates and GM2 ganglioside in vitro, and also retain the wild-type (WT) HexB-like thermostability in presence of plasma. The modified HexB was efficiently incorporated via CI-MPR into fibroblasts derived from Tay-Sachs patients, and reduced the GM2 ganglioside accumulated in the cultured cells. The intracerebroventricular enzyme replacement therapy involving the modified HexB should be more effective for Tay-Sachs and Sandhoff than that utilizing the HexA, especially as a low-antigenic enzyme replacement therapy for Tay-Sachs patients who have endogenous WT HexB.

Mucopolysaccharidosis (MPS)

Recently, several studies showed that multiple infusions of high doses of enzyme partially cleared storage in adult brain. For example, the neonate uses the M6PR to transport phosphorylated β-glucuronidase (P- GUS) across the BBB. These results raised the question if correction of brain storage by repeated high doses of enzyme depends on M6P-mediated uptake or whether enzyme gains access to brain storage by another route when brain capillaries are exposed to prolonged, high levels of circulating enzyme (Grubb et al. 2008). Pharmacological manipulation with epinephrine restores functional transport of P-GUS across the adult BBB. The effect of epinephrine on the transport of P-GUS was ligand specific (Urayama et al. 2007).

The rat amniotic epithelial cells (AEC) over-express and secrete human β-glucuronidase (GUS) following transduction with an adenoviral vector encoding human GUS. The AEC were used as donor cells for cell-mediated gene therapy of CNS lesions in mice with mucopolysaccharidosis type VII (MPSVII), a lysosomal storage disorder caused by an inherited deficiency of GUS activity. After confirmation that the secreted GUS was taken up mainly via M6PRs in primary cultured neurons, the AECs were transplanted into the brains of adult MPSVII mice. Results suggest that intracerebral transplantation of genetically engineered AEC has therapeutic potential for the treatment of CNS lesions in lysosomal storage disorders (Kosuga et al. 2001).

The availability of both MR$^{+/+}$ and MR$^{-/-}$ mice led to study the effects of eliminating the MR on MR- and MPR-mediated plasma clearance and tissue distribution of infused phosphorylated (P) and nonphosphorylated (NP) forms of human β-glucuronidase. In MR$^{+/+}$ MPS VII mice, the MR clearance system predominated at doses up to 6.4 mg/kg P-GUS. Genetically eliminating the MR slowed plasma clearance of both P- and NP-GUS and enhanced the effectiveness of P-GUS in clearing storage in kidney, bone, and retina. Saturating the MR clearance system by high doses of enzyme also improved targeting to MPR-containing tissues such as muscle, kidney, heart, and hepatocytes. Although ablating the MR clearance system genetically is not practical clinically, blocking the MR-mediated clearance system with high doses of enzyme is feasible. This approach delivers a larger fraction of enzyme to MPR-expressing tissues, thus enhancing the effectiveness of MPR-targeted ERT (Sly et al. 2006).

Mucopolysaccharidosis type VI (MPS VI) is an autosomal recessive disease caused by the deficiency of N-acetylgalactosamine 4-sulfatase (4S) leading to the lysosomal accumulation and urinary excretion of dermatan sulfate. MPS VI has also been described in the Siamese cat. Endocytosis of recombinant feline 4S (rf4S) by cultured feline MPS VI myoblasts was predominantly mediated by a M6PR and resulted in the correction of dermatan sulfate storage. The mutation causing feline MPS VI was identified as a base substitution at codon 476, altering a leucine codon to a proline (L476P) (Yogalingam et al. 1996). Differences in glycosylation of lysosomal enzymes can be an important factor in altering enzyme uptake by different cell types. In alternative approaches, carbohydrate modification variants may be useful for altering the distribution of exogenous enzyme in vivo (Fuller et al. 1998). Reports also suggest that muscle-mediated gene replacement therapy may be a viable method for achieving circulating levels of recombinant f4S (rf4S) in the MPS VI cat (Yogalingam et al. 1997).

Mucopolysaccharidosis Type IIIA (MPS IIIA) is a lysosomal storage disorder caused by a deficiency in the lysosomal enzyme sulfamidase, which is required for the degradation of heparan sulfate. The disease is characterized by neurological dysfunction but relatively mild somatic manifestations. In a naturally occurring mouse model to MPS IIIA, recombinant murine sulfamidase was able to correct the storage phenotype of MPS IIIA fibroblasts after endocytosis via M6PR (Bielicki et al. 1998; Gliddon et al. 2004; Urayama et al. 2008).

Mucopolysaccharidosis type IIIB (MPS-IIIB, Sanfilippo type B Syndrome) is a heterosomal, recessive lysosomal storage disorder resulting from a deficiency of α-N-acetylglucosaminidase (NAGLU). The use of secreted NAGLU in future enzyme and gene replacement therapy protocols will be with limited success due to its small degree of mannose-6-phosphorylation (Weber et al. 2001). Mucopolysaccharidosis type IIID or Sanfilippo D syndrome, a lysosomal storage disorder, is caused by the deficiency of N-acetylglucosamine-6-sulphatase (Glc6S). In addition to human patients, a Nubian goat with this disorder has been described. The r-caprine Glc6S was endocytosed by fibroblasts from patients with mucopolysaccharidosis type IIID via the M6PR-mediated pathway resulting in correction of the storage phenotype of these cells (Litjens et al. 1997). Mucopolysaccharidosis IVA (MPS IVA; Morquio A syndrome) is a lysosomal storage disorder caused by deficiency of N-acetylgalactosamine-6-sulfatase (GALNS), an enzyme that degrades keratan sulfate. The recombinant phosphorylated enzyme was dose-dependently taken up by M6PR thereby restoring enzyme activity in MPS IVA fibroblasts. Penetration of the therapeutic enzyme throughout poorly vascularized, but clinically relevant tissues, as well as macrophages and hepatocytes in wild-type mouse, supports development of rhGALNS as enzyme replacement therapy for MPS IVA (Dvorak-Ewell et al. 2010).

Palmitoyl-Protein Thioesterase (PPT) is a lysosomal long-chain fatty acyl hydrolase that removes fatty acyl groups from modified cysteine residues in proteins. Mutations in palmitoyl-protein thioesterase were found to cause the neurodegenerative disorder infantile neuronal ceroid lipofuscinosis, a disease characterized by accumulation of amorphous granular deposits in cortical neurons, leading to blindness, seizures, and brain death by the age of 3. The accumulation in cultured cells is reversed by the addition of recombinant palmitoyl-protein thioesterase that is competent for lysosomal uptake through M6PR (Lu et al. 1996). PPT expressed in COS-1 cells is recognized by M6PR and is routed to lysosome, but a substantial fraction of PPT is secreted. The PPT has a role outside the lysosomes in the brain and may be associated with synaptic functioning (Lehtovirta et al. 2001).

MPR as Target in MPS VII in Mice: Although ERT is an established method for treating lysosomal storage diseases, an alternative strategy to rectify lysosomal storage diseases depends on the interaction of a fragment of IGF2, with the IGF2 binding site on the IGF2/CI-MPR as tested in a murine mucopolysaccharidosis type VII (MPS VII) model. A chimeric protein containing a portion of mature human IGF2 fused to the C terminus of human β-glucuronidase was taken up by MPS VII fibroblasts in a M6P-independent manner. The tagged enzyme was delivered effectively to clinically significant tissues in MPS VII mice and effective in reversing the storage pathology. The peptide-based, glycosylation-independent lysosomal targeting system may enhance enzyme-replacement therapy for certain human lysosomal storage diseases (LeBowitz et al. 2004).

Lysosomal storage diseases may be treated by the transplantation of cells that secrete the enzyme which is deficient in patients. Secretion of lysosomal enzymes can be enhanced by reducing the MPR involved in the lysosomal sorting of newly synthesized lysosomal enzymes. Hammerhead ribozymes targeting the mRNA of murine CI-MPR cleave

RNA fragments efficiently and reduce the levels of murine MPR mRNA in transient transfection experiments. Thus, the reduction in MPR is sufficient to increase a lysosomal enzyme secretion (Yaghootfam and Gieselmann 2003).

46.1.4 Other Approaches using CI-MPR as Target

CI-MPR Binds and Internalizes Leukemia Inhibitory Factor: Leukemia inhibitory factor (LIF) is a multifunctional, highly glycosylated soluble protein belonging to the interleukin-6 (IL-6) subfamily of helical cytokines. LIF exerts an important role in neuronal, platelet and bone formation. Blanchard and co-workers have reported that glycosylated LIF and the macrophage-colony-stimulating factor were able to bind to CI-MPR in a M6P-sensitive manner (Blanchard et al. 1999). The M6P-containing cytokine LIF is rapidly internalized and degraded by cells expressing CI-MPR (Blanchard et al. 1999). Thus, CI-MPR is a candidate natural molecule, like other inhibitors of the IGF2 pathway, for the development of novel therapeutic strategies in treatment of haematopoietic disorders in which IL-6-type cytokines play a role, particularly multiple myeloma. CI-MPR can also bind and internalize renin and pro-renin (Admiraal et al. 1999). After internalization, pro-renin is swiftly activated by proteolytic cleavage (Saris et al. 2002). CI-MPR in myocytes may participate in the cardiac (pro)renin uptake by binding to the fraction of human (pro) renin, that is characterized by the presence of the M6P recognition marker (Saris et al. 2002).

Lysosomal Targeting for Cancer Therapy: CI-MPR has been reported to be a potential tumor suppressor in 70% of hepatocarcinomas and 15–30% of breast cancers (Hankins et al. 1996; Chappell et al. 1997; Oates et al. 1998). Moreover, over-expression of CI-MPR induced regression of tumors in mice and growth inhibition in cancer cells. These effects are probably due to the different functions of this receptor that indirectly controls cell growth: (1) internalization and degradation of various growth-promoting factors, such as IGF2 (Kiess et al. 1988), glycosylated LIF, and other M6P-containing cytokines, such as the macrophage colony-stimulating factor (Blanchard et al. 1999); (2) binding and uptake of granzyme B, an essential factor for T cell-mediated apoptosis (Motyka et al. 2000), and (3) regulation of secreted lysosomal enzymes that are responsible for extracellular matrix degradation and tumor dissemination. Attempts to target the acidic extracellular compartment of solid tumors have been performed by loading doxorubicin into pH-sensitive poly/PEG/folate micelles (Lee et al. 2005) or liposomes (Storm et al. 1987).

Inhibitors of PI3-Kinase Activity in Late Endocytic Pathway: Addition of wortmannin to normal rat kidney cells causes redistribution of the lysosomal type I integral membrane proteins Igp110 and Igp120 to a swollen vacuolar compartment. Wortmannin does not show gross morphological effect on TGN or lysosomes, or any effect on the delivery to the TGN of endocytosed antibodies against the type I membrane protein TGN38. The observed effects of wortmannin were due to inhibition of membrane traffic between CI-MPR-positive late endosomes and the TGN and to inhibition of membrane traffic between a novel Igp120-positive, CI-MPR-negative late endosomal compartment and lysosomes. The effects of wortmannin suggest a function for a PI3-kinase(s) in regulating membrane traffic in the late endocytic pathway (Reaves et al. 1996). 3-methyladenine (3-MA), a well-known inhibitor of autophagic sequestration, can also inhibit PI3-kinase activity, which is required for many processes in endosomal membrane trafficking. The treatment with 3-MA results in a specific redistribution of CI-MPR from the TGN to early/recycling endosomal compartments containing internalized transferrin. However, in contrast to wortmannin and LY294002, 3-MA did not cause the enlargement of late endosomal/lysosomal compartments. Hence, the effect of 3-MA is restricted to the retrieval of CI-MPR from early/recycling endosomes (Hirosako et al. 2004).

46.2 Cell Targeting Based on Mannan-Lectin Interactions

Carbohydrates exhibit properties of potential interest when developing drug-delivery mechanisms, such as specificity in their interaction with their receptors and the varied nature of potentially targetable receptors available. Macromolecular glycoconjugates, in particular, have shown some promise. Synthetic glycopolymers and glycoproteins have been used as carriers of covalently conjugated drugs, bearing carbohydrate ligands that provide delivery specificity. However, these systems commonly rely on endogenous mechanisms, such as lysosomal degradation, for release of the active drug, and so the unwanted release of the drug at sites other than the desired site of action is possible. Glycotargeting is being followed through two approaches: (1) relying on use of oligosaccharide moiety or (2) using the lectin as a component of the drug delivery system (Gao et al. 2007; Gupta et al. 2009; Minko 2004; Smart 2004). In the first approach, oligosaccharides or neoglycoconjugates form part of the drug delivery system, whereas, in the second approach using lectins towards glycotargeting, the principle is reversed of the first approach.

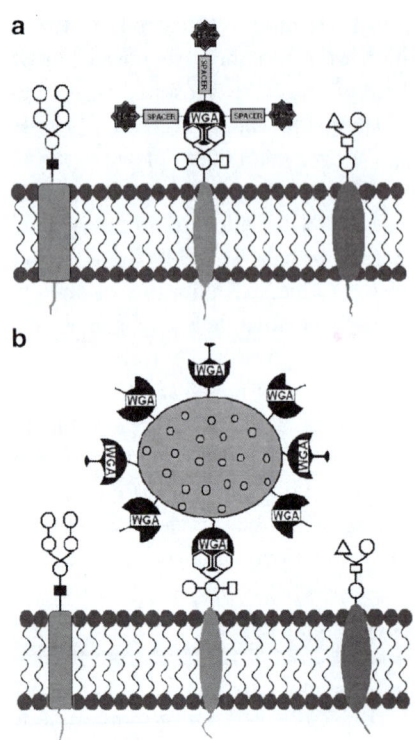

Fig. 46.2 Lectin-grafted formulations. (a) Lectin-grafted prodrug, (b) Lectin-grafted carrier system. ○ Galactose, ◆ N-acetylgalactosamine, □ N-acetylglucosamine, ▬ Spacer, △ Fructose, ✦ Drug, ○ Mannose

The delivery of therapeutic agents to, or via, the oral cavity is limited by the efficient removal mechanisms that exist in this area. The bio-recognition between lectinised drug delivery systems and glycosylated structures in the intestine can be exploited for improved peroral therapy. Lectins as proteins bind to specific sugar residues, and can interact with the glycoconjugates present on cell surfaces or salivary mucins. Endogenous lectins could also be used as points of attachment for carbohydrate-containing delivery systems. As lectins are multifunctional molecules, the possibility of using them is considered as a potential innovation for targeted and prolonged therapy within the oral cavity but considerations such as toxicity needs to be addressed before their routine use becomes a reality (Gao et al. 2007; Gupta et al. 2009; Minko 2004; Smart 2004).

The lectin-carbohydrate interaction can be made use of by the development of nanoparticles containing carbohydrate moieties that are directed to certain lectins (direct lectin targeting) as well as incorporating lectins into nanoparticles that are directed to cell surface carbohydrates (reverse lectin targeting) (Gabor et al. 2004) (Fig. 46.2). Treatment of glioblastoma multiforme (GBM), a primary malignant tumor of the brain, is one of the most challenging problems as no currently available treatment is curative.

Surgery remains the basic treatment in which the bulk of the tumor is removed and the peripheral infiltrating part is the target of supplementary treatments. Many of the technologies based on nanotechnology can be applied in the improvement of drug delivery to GBM (Jain 2007).

46.2.1 Receptor-Mediated Uptake of Mannan-Coated Particles (Direct Targeting)

Cell surface-bound receptors represent suitable entry sites for delivery of macromolecules or supramolecular structures into cells by receptor-mediated endocytosis. Carrying the carbohydrate-tag, the drug delivery system can be recognized by the cells/tissues and internalised by endogenous lectins at the cell surface. Mammalian mannose/fucose/galactose specific cell surface receptors are expressed on macrophages and other antigen presenting cells such as dendritic cells (DCs) in skin and M-cells in intestine. Macrophages and DCs play an important role in host immune functions such as antigen presentation. Attempts have been made to modulate the function of macrophages for the treatment of genetic metabolic diseases. The mannose receptor (ManR) on macrophages mediates the internalization of a wide range of molecules or microorganisms in a pattern recognition manner (see Chaps. 15 and 35 for structure and functions of ManR). Therefore, it represents an attractive entry for specific drug, gene, or antigen delivery to macrophages and antigen-presenting DCs. Particles coated with carbohydrate ligands offer potential future in the site directed delivery of macromolecules and therapeutic drugs. Based on this principle delivery systems containing asialofetuin, galactose, mannose, or N-acetyl-galactosamine were developed and tested for endocytois by macrophages, DCs, and liver cells (Chiu et al. 1994). The use of carbohydrate-modified HPMA or liposomes gave improved results (Andre et al. 2000; Dasi et al. 2001). Liver or the colon, macrophages and mouse brain have been shown to be targeted by mannosylated liposomes (Man-liposome) (Umezawa and Eto 1988).

46.2.2 Polymeric Glyco-Conjugates as Carriers

Mannosylated Conjugates as Cell-Specific Carrier: Hepatic uptake characteristics of mannosylated bovine serum albumin (Man-BSA) were assessed as a liver-specific carrier system (Ogawara et al. 1999). The ^{111}In-Man18-BSA accumulated in the liver up to 70%; the endothelial cells and Kupffer cells contributed major fraction. These results provide useful information in designing drug targeting systems to the liver nonparenchymal cells via mannose receptors (Ogawara et al. 1999).

Mannosylpolyethylenimine (ManPEI) Conjugates: Presence of mannose receptors on DCs can be exploited for targeted gene transfer by employing mannosylpolyethylenimine (ManPEI) conjugates. Several ManPEI conjugates have been used for formation of ManPEI/DNA transfection complexes. Results demonstrated that DCs transfected with ManPEI/DNA complexes containing adenovirus particles are effective in activating T cells of T cell receptor transgenic mice in an antigen-specific fashion (Diebold et al. 1999).

Poly-(L-Lysine Citramide Imide): Commercially available quinic and shikimic acids appear as stable mannose bioisosteres, which should prove valuable tools for specific cell delivery (Grandjean et al. 2001). With the aim of promoting the targeting of macrophage mannose receptors and the internalization of the norfloxacin antibiotic, which is active against some intracellular bacteria, was coupled to a polymeric carrier, namely poly-(L-lysine citramide imide). This carrier, derived from two metabolites, citric acid and L-lysine, is known to be biocompatible and slowly degradable under slight acidic conditions. Prodrug macromolecules compete effectively with glucose oxidase and thus should be able to bring the drug up to the mannosyl receptor-bearing membranes of macrophages infected by intracellular bacteria (Gac et al. 2000).

Poly[N-p-Vinylbenzyl-O-β-Mannopyranosyl-(1-4)-D-Gluconamide (PV-Man): PV-Man is a polystyrene derivative that contains mannose moieties and interacts with ManR-carrying cell line. The PV-Man strongly binds to macrophage cells, probably due to a specific interaction mediated by ManRs on the cell membrane. Using a PV-Man glycopolymer, receptor-mediated gene transfer via ManR is another method for targeted gene delivery into macrophages (Park et al. 2005). Polymeric nanospheres (NS) with a polystyrene core and a glucosyloxyethyl methacrylate (GEMA) oligomer corona nanosphere proved to be a useful material for studying sugar-biomolecule recognition and offered a potential for using a multi-lectin nanoparticle array in glycoprotein mapping (Fromell et al. 2005; Serizawa et al. 2001).

Cyclodextrin Conjugates: A structure based saccharide-directed molecular delivery system through biological receptors was studied by Benito et al. (2004). The dendritic β-cyclodextrin (βCD) derivatives bearing multivalent mannosyl ligands were assessed for their binding efficiency towards the ConA and mammalian mannose/fucose specific cell surface receptor from macrophages. This new type of βCD-dendrimer construct showed high drug solubilization capability. A subtle change in the structure of the conjugate may have important consequences on receptor affinity (Benito et al. 2004).

McNicholas et al. (2007) synthesized amphiphilic β-cyclodextrins bearing hexylthio, dodecylthio, and hexadecylthio chains at the 6-positions and glycosylthiocarbamoyl-oligo(ethylene glycol) units at the 2-positions. The glycosyl residues (α-D-mannosyl and β-L-fucosyl) were intended for cell-targeting. These amphiphilic glycosylated cyclodextrins form vesicles in water. Hexylthio assemblies exhibited selective binding to Lens culinaris lectin. A bioeliminable amphiphilic poly(ethylene oxide)-b-poly(ε-caprolactone) (PEO-b-PCL) diblock copolymer end-capped by a mannose residue, synthesized by sequential controlled polymerization of ethylene oxide and epsilon-caprolactone, followed by the coupling of a reactive mannose derivative to the PEO chain end showed that these colloidal systems have great potential for drug targeting and vaccine delivery systems (Rieger et al. 2007).

Delivery of Antisense Oligonucleotides: Antisense oligonucleotides (ONs) are useful for selective inhibition of gene expression. However, their effective use is limited by inefficient cellular uptake and lack of cellular targeting. A drug targeting system which utilizes ManR-mediated endocytosis to enhance cellular uptake of ONs in alveolar macrophages (AMs), employs a molecular complex consisting of partially substituted mannosylated poly(L-lysine) (MPL), linked to ON. Upon recognition by the macrophage ManRs, the MPL was internalized by the receptor-mediated pathway, co-transporting the ON. The AMs treated with the MPL:ON complex exhibited a significant increase in ON uptake over free ON-treated controls. The ON uptake was shown to require the recognition of the mannose moiety since unmodified polylysine was much less effective in promoting ON uptake. Following cellular internalization, the ON largely accumulated in endocytic vesicles (Liang et al. 1996).

46.2.3 Mannosylated Liposomes in Gene Delivery

Mannosylated Liposomes: Introduction of ligands for cell-surface receptors into liposomes has been continuing to improve transfection efficiency: as far as macrophages are concerned (Simões et al. 1999). In earlier studies, liposomes were coated with macromolecular ligands such as transferrin, immunoglobulins and asialoglycoproteins. Direct respiratory delivery via inhalation of mannose modified liposomal carriers to alveolar macrophages is of great interest. The success of targeting systems to alveolar macrophages depends on internalization into these cells for pharmacological intervention. Mannose grafted liposome intercalated Benzyl derivative of an antibiotic MT81 (Bz2MT81) eliminated intracellular amastigotes of *Leishmania donovani* within splenic macrophages more efficiently than the liposome intercalated Bz2MT81 or free Bz2MT81. Both Man-liposomes and Bz2MT81 appeared to be non-toxic to the host peritoneal macrophages. Liver and kidney function tests (SGPT, alkaline phosphatase, creatinine and urea in blood plasma) showed that the toxicity of Bz2MT81 was reduced up to normal level when mannose grafted liposomal Bz2MT81 were administered (Mitra et al. 2005).

Using cell receptors on the surface of mononuclear phagocyte cells, which are important hosts for HIV, stavudine-loaded Man-liposomal formulations have been developed for targeting HIV-infected cells. Using Con A as a model system for in-vitro ligand-binding capacity, Man-liposomes showed potential applications for the site-specific and ligand-directed delivery systems with better pharmacological activity (Garg et al. 2006). Perhaps, clustering of mannose residues on liposomal surfaces is important in determining the binding affinity of Man-liposomes with MBP and related to the mannose density of Man-liposomes (Terada et al. 2006; Wijagkanalan et al. 2008).

Mannosylated Cationic Liposomes: Several strategies have been developed to transfer genes directly into macrophages. In recent years, complexes of polylysine linked to ligands such as mannose (Ferkol et al. 1996; Erbacher et al. 1996) with DNA have been reported to enhance gene expression in macrophages. But, the transfection efficiency of many of these vectors is handicapped due to endosomal or lysosomal degradation. A promising nonviral gene delivery system developed involves cationic liposomes. Various kinds of cationic lipids have been synthesized and shown to be able to deliver genes into cells both in vitro and in vivo. DC-Chol liposomes have been used in gene therapy applications in clinical settings (Gao and Huang 1991; Nabel et al. 1993). A galactosylated cholesterol derivative in combination with dioleoylphosphatidylethanolamine (DOPE) efficiently transferred a plasmid DNA into human hepatoma cells (HepG2) via an asialoglycoprotein receptor-mediated mechanism (Kawakami et al. 1998). However, cationic liposomes do not exhibit any cell specificity in vivo. Kawakami et al. (2004) developed a low-molecular weight lipidic ligand, a mannosylated cholesterol derivative, cholesten-5-yloxy-N-(4-(1-imino-2-β-D-thiomannosylethyl)amino)butyl)-formamide (Man-C4-Chol), for gene delivery to hepatocytes and compared with other types of liposomes prepared with various molar ratios of Man-C4-Chol and particle size of about 200 nm for transfection assays in hepatocytes and mouse peritoneal macrophages (Kawakami et al. 2000).

The gene expression with Man-C4-Chol/DOPE (6:4) liposome/DNA complexes in the liver was observed preferentially in the non-parenchymal cells and was significantly reduced by predosing with Man-BSA. The gene expression in the liver was greater following intraportal injection. These results suggested that plasmid DNA complexed with Man-liposomes exhibits high transfection activity due to recognition by mannose receptors both in vitro and in vivo. Intravenous injection of DNAcationic liposome complexes resulted in gene expression in many tissues including the heart, lung, liver, kidney and spleen (Mahato et al. 1995) through participation of mannose receptors in liver Kuppfer and or endothelial cells. The splenic macrophages may also be targeted by this approach.

As far as the design of carriers for active targeting using receptor-mediated endocytosis was concerned, the density and stereochemistry of the ligand seemed to be important. It was demonstrated that galactosylated protein is recognized by liver cells in a manner directly related to the estimated surface density of the galactose residues. Similar strategy applies to Man-liposomes. The chemical structure and physicochemical characteristics of Man-C4-Chol seemed to satisfy the conditions for transfection in macrophages by offering a cationic charge and being recognized by the mannose structure on the liposomal surface.

Mannosylated-Emulsions: Carbohydrate grafted emulsions are one of the most promising cell-specific targeting systems for lipophilic drugs. Man-emulsions composed of soybean oil, EggPC and Man-C4-Chol with a ratio of 70:25:5 were significantly delivered to liver non-parenchymal cells (NPC) via mannose receptor-mediated mechanism after intravenous administration in mice and supported the design of pDNA/ligands-grafted cationic liposome complexes for cell-specific gene delivery (Kawakami et al. 2004). The in vitro study showed increased internalization of Man-5.0- and Man-7.5-emulsions and significant inhibition of uptake in the presence of mannan. The enhanced uptake of Man-emulsions was related to the increasing of Man-C4-Chol content. This suggested that the mannose density of Man-emulsions plays an important role in both cellular recognition and internalization via a mannose receptor-mediated mechanism (Yeeprae et al. 2006).

Wijagkanalan et al. (2008) demonstrated the efficient targeting to alveolar macrophages by the intratracheally administered Man-liposomes via ManR-mediated endocytosis in rats. The study involved Man-liposomes, with various ratio of mannosylated cholesterol derivatives, cholesten-5-yloxy-N-(4-((1-imino-2-D-thiomannosylethyl)-amino)alkyl)-formamide (Man-C4-Chol) and suggested that in vitro uptake of Man-liposomes occurs in a concentration-dependent manner. Through intratracheal route of administration of Man-7.5 and Man-5.0-liposomes, internalization was enhanced and selective to alveolar macrophages.

46.2.4 DC-Targeted Vaccines

A workshop on "Dendritic Cells: Biology and Therapeutic Applications," brought together basic and clinical research scientists to discuss the mechanisms underlying the control of immune responses and tolerance by DCs, as well as research in cancer immunotherapy based on DC vaccination (Ardavin et al. 2004).

Mannose Receptor Targeting Vaccine
DCs have a number of receptors for adsorptive uptake of antigens. Some are shared with other cells, such as Fcγ receptors, DEC-205, a type I membrane-integrated glycoprotein and the macrophage mannose receptor (MMR) (Chap. 35). Other receptors are more DC restricted, e.g., Langerin/CD207 (Chap. 35), DC-SIGN/CD209 (Chap. 36), asialoglycoprotein receptor or hepatic lectins (HL) (Chap. 33), and dendritic cell lectin (DLEC; also referred to as BDCA-2). Targeting antigens to endocytic receptors on professional APCs represents an attractive strategy to enhance the efficacy of vaccines. Such APC-targeted vaccines have the ability to guide exogenous protein antigens into vesicles that efficiently process the antigen for MHC class I and class II presentation. Efficient targeting not only requires high specificity for the receptor that is abundantly expressed on the surface of APCs, but also the ability to be rapidly internalised and loaded into compartments that contain elements of the antigen-processing machinery. The ManR and related C-type lectin receptors are especially designed to sample antigens, much like pattern recognition receptors, to integrate the innate with adaptive immune responses. A variety of approaches involving delivery of antigens to the ManR have demonstrated effective induction of potent cellular and humoral immune responses. ManR-targeted vaccines are likely to be most efficacious in vivo when combined with agents that elicit complementary activation signals. A better understanding of the mechanism associated with the induction of immune responses as a result of targeting antigens to the ManR, will be important in exploiting ManR-targeted vaccines not only for mounting immune defenses against cancer and infectious disease, but also for specific induction of tolerance in the treatment of autoimmune disease (Keler et al. 2004).

DEC-205 Mediated Cancer Immunotherapy
Anti α-DEC-205 antibodies target to the DEC-205 receptor that mediates antigen presentation to T cells by DC. DEC-205 is a mannose specific receptor, present on DC. One of the major functions of DEC-205 is to internalize the antigens and present to naïve T lymphocytes for development of T cell dependent immunity. To assess the potential of antigen targeting to DC, Bonifaz et al. (2004) incorporated ovalbumin into a mAb to the DEC-205 receptor, which is expressed on these cells in lymphoid tissues. A single low dose of antibody-conjugated ovalbumin initiated immunity from the naive $CD4^+$ and $CD8^+$ T cell repertoire. Unexpectedly, the αDEC-205 antigen conjugates targeted to DCs for long periods and ovalbumin peptide was presented on MHC class I. This was associated with stronger $CD8^+$ T cell-mediated immunity relative to other forms of antigen delivery, even when the latter was given at a 1,000 times higher doses. In parallel, the mice showed enhanced resistance to an established rapidly growing tumor and to viral infection at a mucosal site. By antibody-mediated antigen targeting via the DEC-205 receptor increased the efficiency of vaccination for T cell immunity, including systemic and mucosal resistance in disease models (Bonifaz et al. 2004; Mahnke et al. 2005). In preclinical studies, Badiee et al. (2007) prepared anti-human DEC-205 immunoliposomes (anti-hDEC-205 iLPSM) and compared their uptake by monocyte-derived DC and blood DC (BDC) with conventional liposomes (cLPSM). Confocal microscopy confirmed that the anti-hDEC-205 iLPSM were phagocytosed by DC and available for antigen processing. Thus, DEC-205 is one of effective targets for delivering liposomes to human DCs (Badiee et al. 2007).

Since, anti α-DEC-205 antibodies target to the DEC-205 receptor that mediates antigen presentation to T cells by DCs, these properties were exploited for immunization strategies by conjugating the melanoma antigen tyrosinase-related protein (TRP)-2 to α-DEC-205 antibodies and immunization of mice with these conjugates together with DC-activating oligonucleotides (CpG). Upon grafting of the melanoma cell line B16, α-DEC-TRP immunized mice were protected against tumor growth or showed substantially slow growth of implanted B16 cells into tumor bearing hosts. Approximately 70% of the animals were cured from existing tumors by treatment with α-DEC conjugates carrying two different melanoma antigens (TRP-2 and gp100). Thus, targeting of DCs in situ by antibody-antigen conjugates may be a novel approach to induce long-lasting antitumor immunity.

Mannan-Coated Nanoparticles in Vaccination: Genetic immunization using "naked" plasmid DNA (pDNA) has been used to elicit humoral and cellular immune responses. In search of a cell-targeted delivery system, cationic nanoparticles coated with pDNA for genetic immunization have been explored. Plasmid DNA-coated nanoparticles, especially with both an endosomolytic lipid and DC-targeting ligand, resulted in 16-fold increase in IgG titer and threefold release in Th1-type cytokine over "naked" pDNA, indicating that the engineered pDNA-coated nanoparticles could enhance in vitro cell transfection and enhanced in vivo immune responses (Cui and Mumper 2002a). Furthermore, pDNA-coated nanoparticles, especially the mannan-coated pDNA-nanoparticles with DOPE, resulted in significant enhancement in both antigen-specific IgG titers and splenocyte proliferation over 'naked' pDNA alone (Cui and Mumper 2002b). The freeze-dried nanoparticles (prior to pDNA coating) showed potential application for cell-specific targeting of macrophages. Moreover, incubation of nanoparticles with ManR positive mouse macrophage cell line (J774E) showed that the uptake of mannan-coated nanoparticles by the cells was 50% higher than that of uncoated nanoparticles (Cui et al. 2003).

DC-SIGN-Mediated Targeting
Targeting DC surface proteins to deliver liposomes carrying antigens has demonstrated potential for eliciting antigen-specific immune responses. Myeloid dendritic cells (MyDCs) can serve as one of the major reservoirs for HIV-1. Using monocyte-derived MyDCs, Gieseler et al. (2004) presented the evidence for liposomal compound delivery to these cells by specifically addressing. DC-SIGN (CD209), a MyDC-associated C-type lectin implicated in the transmission of HIV-1 to T helper cells. The DC-SIGN was demonstrated as a superior target as compared with other MyDC markers (CD1a, CD4, CD45R0, and CD83). This study implied that liposomal targeting to DC-SIGN (CD209) and related C-type lectins may afford therapeutic intracellular drug delivery to MyDCs and other reservoir and nonreservoir cells susceptible to infection with HIV-1.

Since DCs are a central element in the development of antigen-specific immune responses, current DC-based vaccines are based on ex vivo-generated autologous DCs loaded with antigen prior to re-administration into patients. A more direct and less laborious strategy is to target antigens to DCs in vivo via specific surface receptors. Therefore, a humanized antibody, hD1V1G2/G4 (hD1), directed against the DC-SIGN was explored for its capacity to serve as a target receptor for vaccination purpose. hD1 was cross-linked to a model antigen, keyhole limpet hemocyanin (KLH). The chimeric antibody-protein complex (hD1-KLH) bound specifically to DC-SIGN and was rapidly internalized and translocated to the lysosomal compartment. Antibody-mediated targeting of antigen to DCs via DC-SIGN effectively induces antigen-specific naive as well as recall T-cell responses. This identifies DC-SIGN as a promising target molecule for DC-based vaccination strategies (Gupta et al. 2009; Tacken et al. 2005). Optimal HIV vaccines should elicit $CD8^+$ T cells specific for HIV proteins presented on MHC class I products, because these T cells contribute to host resistance to viruses. Based on humans, studies with highly polymorphic MHC products reveal that DCs and DEC-205 can cross-present several different peptides from a single protein. Because of the consistency in eliciting $CD8^+$ T cell responses these data support the testing of αDEC-205 fusion mAb as a protein-based vaccine (Bozzacco et al. 2007).

The efficacy of adenoviral (Ad) vectors can be enhanced through alterations in vector tropism such that DC-targeted transduction is achieved. The efficiency of DC transduction by Ad vectors retargeted to DC-specific DC-SIGN was studied and compared to that of Ad vectors retargeted through CD40 (Korokhov et al. 2005). A comparable and significant enhancement of gene transfer to monocyte derived DCs (MDDCs) was accomplished by means of an Ad vector harboring the Fc-binding domain of *S. aureus* protein A in combination with antibodies to DC-SIGN or to CD40 or with fused complexes of human Ig-Fc with their natural ligands, i.e., ICAM-3 or CD40L, respectively. Results demonstrated the usefulness of DC-SIGN as a DC-restricted targeting motif for Ad-mediated vaccination strategies.

Targeted Oral Vaccine Delivery to M Cells: A strategy for mucosal vaccination and drug delivery: In the intestine, the delivery of antigens across the epithelial barrier to the underlying lymphoid tissue is accomplished by M cells, a specialized epithelial cell type that occurs only in the lymphoid follicle-associated epithelium. Selective and efficient transport of antigen by M cells is considered an essential requirement for effective mucosal vaccines. Therefore, particulate antigen formulations are currently being developed to take advantage of the capacity of M cells to endocytose particles. Delivery may be achieved using synthetic particulate delivery vehicles including poly(DL-lactide-co-glycolide) microparticles and liposomes. M cell interaction of these delivery vehicles is highly variable, and is determined by the physical properties of both particles and M cells. Delivery may be enhanced by coating with reagents including appropriate lectins, microbial adhesins and immunoglobulins which selectively bind to M cell surfaces (Brayden 2001; Clark et al. 2001). In an alternative approach, antigens are coupled to or encapsulated in particulate synthetic carriers. To enhance binding and uptake of such nonviable vectors, ligands are being attached which direct the vaccine particle to receptors on the M cell surface. While binding and uptake of M cell-targeted latex particles and stable liposomes by mouse M cells has been shown using the mouse M cell-specific lectin, Ulex europaeus 1 (UEA-1), a direct relationship between M cell particle uptake and immune outcome was reported by Brayden (2001).

Since various lectins and lectin containing pathogens bind specifically to oligosaccharides including mannose on intestinal cells, exploiting this specificity, lectins have been used as a ligand for targeted oral vaccine delivery to M cells (antigen-presenting cells) in follicle-associated epithelium. The antigen-sampling M cells offer a portal for absorption of colloidal and particulate delivery vehicles, including bacteria, viruses and inert microparticles. Consideration is also given to lectin-mediated targeting in non-intestinal sites and to the potential application of other bioadhesins to enhance M cell transport (Jepson et al. 2004). While ManR is found on lymphatic endothelial cells of small intestine, the intestinal serosa revealed a regular, dense, planar network of cells with prominent dendritic morphology within the external muscular layer and with increasing frequency along the length of the intestine (Flores-Langarica et al. 2005). The serosal-disposed layers show a significant fraction of DCs that express DEC-205, Langerin, CD14 and various other molecules. In vivo, these DCs responded to two microbial stimuli, systemic LPS and oral live bacteria, by up-regulating DEC-205, and Langerin within 12 h. This network of DCs, representing a unrecognized APC system in the intestine, needs to be explored for drug targeting and mucosal vaccination through ManR, DEC-205, and langerin present on intestinal cells (Flores-Langarica et al. 2005).

46.3 Asialoglycoprotein Receptor (ASGP-R) for Targeted Drug Delivery

46.3.1 Targeting Hepatocytes

Asialoglycoprotein receptor (ASGP-R) is predominantly expressed on the sinusoidal surface of mammalian hepatocytes and is responsible for the clearance of glycoproteins with desialylated galactose or acetylgalactosamine residues from the circulation by receptor-mediated endocytosis. The ASGP-R provides a unique means for the development of liver-specific carriers, such as liposomes, recombinant lipoproteins, and polymers for drug or gene delivery to the liver, especially to hepatocytes (Cawley et al. 1981; Wu et al. 1998) (Chap. 33). Another study used glycolipids containing a cluster galactoside moiety for targeting to ASGP-R. The liver uptake of the glycolipid-liposomes exceeds 80% compared to less than 10% for conventional liposomes after injection. The abundant receptors on the cells specifically recognize ligands with terminal galactose or N-acetylgalactosamine residues, and endocytose the ligands for an intracellular degradation process. The use of its natural ligand, i.e. asialofetuin, or synthetic ligands with galactosylated or lactosylated residues, such as galactosylated cholesterol, glycolipids, or galactosylated polymers has achieved significant targeting efficacy to the liver. There are several examples of successful targeted therapy for acute liver injury with asialofetuin-labeled and vitamin E-associated liposomes or with a caspase inhibitor loaded in sugar-carrying polymer particles, as well as for the delivery of an antiviral agent, 9-(2-phosphonylmethoxyethyl)adenine. Liposome-mediated gene delivery to the liver is still in its infancy due to difficulties in solving general issues, such as stability of liposome-DNA complexes in circulation, and lysosomal or endosomal degradation of plasmid DNA. Although, galactosylated polymers are promising for gene delivery, but require further studies to verify their potential applications (Wu et al. 2002).

Cationic liposomes and polymers have been accepted as effective non-viral vectors for gene delivery with low immunogenicity unlike viral vectors. Galactosylated liposomes and poly(amino acids) are selectively taken up by the ASGP-R-positive liver parenchymal cells in vitro and in vivo after intravenous injection. DNA-galactosylated cationic liposome complexes show higher DNA uptake and gene expression in the liver parenchymal cells in vitro than DNA complexes with bare cationic liposomes. In the in vitro gene transfer experiment, galactosylated liposome complexes are more efficient than DNA-galactosylated poly(amino acids) complexes but they have some difficulties in their biodistribution control. On the other hand, introduction of mannose residues to carriers resulted in specific delivery of genes to non-parenchymal liver cells (Hashida et al. 2001). Drugs conjugated with galactosyl-terminating macromolecules selectively enter hepatocytes after interaction of the carrier galactose residues with the ASGP-R present in large amounts and high affinity on hepatocytes. Within hepatocytes the conjugates are delivered to lysosomes where enzymes split the bond between the carrier and the drug, allowing the latter to become concentrated in the liver. The validity of this chemotherapeutic strategy has been endorsed by a clinical study (Fiume et al. 1997).

The results obtained reveal tremendous promise and offer enormous options to develop novel DNA based pharmaceuticals for liver disorders in near future. The ^{99m}Tc-labeled asialoglycoprotein analog, TcGSA (galactosyl-human serum albumin-diethylenetriamine - pentaacetic acid) has been applied to human hepatic receptor imaging. This method is unique and provides information that is totally independent of the ICG test or Child-Turcotte Score (Kokudo et al. 2003; Pathak et al. 2008).

The Specificity of ASGP-R for D-Galactose: The specificity of the receptor for D-galactose or D-mannose is accomplished by specific hydrogen bonding of the 3 and 4-hydroxyl groups with carboxylate and amide side-chains. Therefore, mutation of the amino acid sequence in the CRD results in a conversion of its specificity (Feinberg et al. 2000; Meier et al. 2000; Wu et al. 2002). The crystal structure provides a direct confirmation for the conversion of the ligand-binding site of mannose-binding protein to an ASGP-R-like specificity (Meier et al. 2000). A number of functional mimics for the CRDs of these lectins have been developed by modification of the domain amino acid residues. The modified CRD displayed 40-fold preferential binding to N- acetylgalactosamine compared with galactose, making it a good functional mimic for ASGP-R (Feinberg et al. 2000). Mannose-labeling shifted the ratio to more non-parenchymal cell incorporation (the majority to Kupffer cells) (Feinberg et al. 2000). Therefore, alternative approaches are needed to target liposomes to hepatocytes via ASGP-R (Wu et al. 2002; Yamazaki et al. 2000).

46.4 Siglecs as Targets for Immunotherapy

46.4.1 Anti-CD33-Antibody-Based Therapy of Human Leukemia

Targeting CD33 (Siglec-3) or CD45 is currently exploited for immunotherapy of acute myeloid leukemia (AML). In normal myelopoiesis, expression of CD33 is restricted to advanced stages of differentiation, whereas primitive stem cells do not express CD33. Leukaemic stem cells in patients with CD33$^+$ AML express CD33 (Hauswirth et al. 2007). Antibody-targeted chemotherapy is a therapeutic strategy in cancer therapy that involves a monoclonal antibody specific for a tumor-associated antigen, covalently linked via a suitable linker to a potent cytotoxic agent. The restricted expression of several siglecs (Chaps. 16 and 17) to one or a few cell types makes them attractive targets for cell-directed therapies. The anti-CD33 (Siglec-3) antibody gemtuzumab (Mylotarg) is approved for treatment of acute myeloid leukemia, and antibodies targeting CD22 (Siglec-2) are currently in clinical trials for treatment of B cell non-Hodgkins lymphomas and autoimmune diseases. Since siglecs are endocytic receptors, they are very well suited for a 'Trojan horse' strategy, whereby therapeutic agents conjugated to an antibody, or multimeric glycan ligand, bind to the siglec and are efficiently carried into the cell. Although the rapid internalization of unmodified siglec antibodies reduces their utility for induction of antibody-dependent cellular cytotoxicity or complement-mediated cytotoxicity, antibody binding of Siglec-8, Siglec-9 and CD22 (Siglec-2) has been demonstrated to induce apoptosis of eosinophils, neutrophils and depletion of B cells, respectively. The properties of siglecs that make them attractive for cell-targeted therapies have been reviewed in Chaps. 16 and 17.

Gemtuzumab Ozogamicin (GO): Anti-CD33 antibodies have been used alone-and more effectively, attached to chemotherapy agents or radioisotopes-to treat those with AML (Nemecek and Matthews 2002). Antibody-targeted chemotherapy with gemtuzumab ozogamicin (GO) (CMA-676, a CD33-targeted immunoconjugate of N-acetyl-γ-calicheamicin dimethyl hydrazide [CalichDMH], a potent DNA-binding cytotoxic antitumor antibiotic) is a clinically validated therapeutic option for patients with AML. Calicheamicin is a cytotoxic natural product isolated from Micromonospora echinospora that is at least 1,000-fold more potent than conventional cytotoxic chemotherapeutics. Calicheamicin binds DNA and causes double-strand DNA breaks, leading to cell death. Gemtuzumab ozogamicin is the first clinically validated cytotoxic immunoconjugate in which a humanised anti-CD33 antibody is linked to a derivative of calicheamicin. A similar conjugate, inotuzumab ozogamicin, is being evaluated in Phase I clinical trials in patients with non-Hodgkin's lymphoma. GO is part of clinical practice for AML, but is frequently associated with severe side effects. A number of tumor-targeted immunoconjugates of calicheamicin are being explored preclinically at present for their therapeutic applications (Damle 2004).

Immunoconjugates of calicheamicin targeted against tumor-associated antigens exhibit tumor-specific cytotoxic effects and cause regression of established human tumor xenografts in nude mice. CD33-specific binding triggers internalization of GO and subsequent hydrolytic release of calicheamicin. The histone deacetylase inhibitor valproic acid potently augments gemtuzumab ozogamicin-induced apoptosis in AML cells. The synergistic proapoptotic activity of cotreatment of AML cells with VPA and GO indicates the potential value of this strategy for AML (Ten Cate et al. 2007). Simultaneous targeting of CD45 could improve GO cytotoxicity against AML cell lines and primary AML cells. Further study of this antibody combination for clinical use in AML is warranted (Walter et al. 2008).

Lintuzumab: Lintuzumab (HuM195) is an unconjugated humanized murine mAb directed against cell surface myelomonocytic CD33. The efficacy of lintuzumab in combination with induction chemotherapy has been compared with chemotherapy alone in adults with first relapsed or primary refractory AML. The percent CR plus CRp with MEC plus lintuzumab was 36% vs. 28% in patients treated with MEC alone. The overall median survival was 156 days and was not different in the two arms of the study. The addition of lintuzumab to salvage induction chemotherapy was safe, but did not result in significant improvement in response rate or survival in patients with refractory/relapsed AML (Feldman et al. 2005).

46.4.2 CD22 Antibodies as Carrier of Drugs

The CD22 antigen is a viable target for therapeutic intervention for B-cell lymphomas. Several therapeutic anti-CD22 antibodies and an anti-CD22-based immunotoxin (HA22) are currently under investigation in clinical research. Coupling of anti-CD22 reagents with a nano-drug delivery vehicle is projected to significantly improve treatment efficacies. A mutant of the targeting segment of HA22 (a CD22 scFv), mut-HA22 conjugated to the surface of sonicated liposomes to generate immunoliposomes (mut-HA22-liposomes) may serve as promising carriers for targeted drug delivery to treat patients suffering from B-cell lymphoma (Loomis et al. 2010).

46.4.3 Immunogenic Peptides

Identification of immunogenic peptides for the generation of cytotoxic T lymphocytes (CTLs) may lead to the development of novel cellular therapies to treat disease relapse in AML patients. Bae et al. (2004) identified immunogeneic HLA-A2.1-specific CD33(65–73) peptide (AIISGDSPV) that was capable of inducing CTLs targeted to AML cells. The CD33-CTLs displayed HLA-A2.1-restricted cytotoxicity against both mononuclear cells from AML patients and the AML cell line. The peptide- was specific to CD33-CTLs that secreted IFN-γ in response to CD33(65–73) peptide stimulation. Alteration of native CD33(65–73) peptide at first amino acid residue from alanine (A) to tyrosine (Y) (YIISGDSPV) enhanced the HLA-A2.1 affinity/stability of the peptide and induced CTLs with increased cytotoxicity against AML cells. These results demonstrate the potential application of immunogenic HLA-A2.1-specific CD33 peptides in developing a cellular immunotherapy for treatment of AML patients (Bae et al. 2004). Bakker et al. (2004) selected a single chain Fv fragment that broadly reacted with AML samples and with myeloid cell lineages within peripheral blood. Expression cloning identified an antigen recognized as C-type lectin-like molecule-1 (CLL-1), and a transmembrane glycoprotein. CLL-1 showed variable expression in $CD34^+$ cells in chronic myelogenous leukemia and myelodysplastic syndrome but was absent in 12 of 13 cases of acute lymphoblastic leukemia. The AML reactivity combined with restricted expression on normal cells suggests CLL-1 as a novel potential target for AML treatment.

46.4.4 Blocking of CD33 Responses by SOCS3

CD33-related Siglecs 5–11 are inhibitory receptors that contain a membrane proximal ITIM, which can recruit SHP-1/2. The suppressor of cytokine signaling (SOCS) proteins, particularly SOCS1, are essential for regulating the inflammatory process. Gene-targeting studies revealed that SOCS play a nonredundant role in limiting the inflammatory response. SOCS expression is induced by cytokines, infective pathogen-associated molecular patterns, and other stimuli. They regulate cytokine signal transduction via a negative feedback loop. They are characterized by a phosphotyrosine binding SH2 domain and the SOCS box motif (Elliott and Johnston 2004; Krebs and Hilton 2001). SOCS3 binds phosphorylated ITIM of Siglec 7 and targets it for proteasomal-mediated degradation, suggesting that Siglec 7 is a SOCS target. SOCS3 can interact with a number of phosphorylated receptors and appears to potently inhibit JAKs in the presence of these receptors (Fujimoto and Naka 2003). Following ligation, the ECS E3 ligase is recruited by SOCS3 to target Siglec 7 for proteasomal degradation, and SOCS3 expression is decreased concomitantly. In addition, SOCS3 expression blocks Siglec 7-mediated inhibition of cytokine-induced proliferation. This may be a mechanism by which the inflammatory response is potentiated during infection (Orr et al. 2007a, b).

References

Admiraal PJ, van Kesteren CA, Danser AH et al (1999) Uptake and proteolytic activation of prorenin by cultured human endothelial cells. J Hypertens 17:621–629

Andre S, Frisch B, Kaltner H, Desouza DL et al (2000) Lectin-mediated drug targeting: selection of valency, sugar type (Gal/Lac), and spacer length for cluster glycosides as parameters to distinguish ligand binding to C-type asialoglycoprotein receptors and galectins. Pharm Res 17:985–990

Ardavin C, Amigorena S, Reise Sousa C (2004) Dendritic cells: immunobiology and cancer immunotherapy. Immunity 20:17–23

Badiee A, Davies N, McDonald K et al (2007) Enhanced delivery of immunoliposomes to human dendritic cells by targeting the multilectin receptor DEC-205. Vaccine 25:4757–4766

Bae J, Martinson JA, Klingemann HG (2004) Identification of novel CD33 antigen-specific peptides for the generation of cytotoxic T lymphocytes against acute myeloid leukemia. Cell Immunol 227:38–50

Bakker ABH, van den Oudenrijn S, Bakker AQ (2004) C-Type Lectin-Like Molecule-1: A Novel Myeloid Cell Surface Marker Associated with Acute Myeloid Leukemia. Cancer Res 64:8443–8450

Benito JM, Gomez-Garcia M, Ortiz Mellet C et al (2004) Optimizing saccharide-directed molecular delivery to biological receptors: design, synthesis, and biological evaluation of glycodendrimer-cyclodextrin conjugates. J Am Chem Soc 126:10355–10363

Bielicki J, Hopwood JJ, Melville EL, Anson DS (1998) Recombinant human sulphamidase: expression, amplification, purification and characterization. Biochem J 329:145–150

Blanchard F, Duplomb L, Raher S et al (1999) Mannose 6-Phosphate/Insulin-like growth factor ii receptor mediates internalization and degradation of leukemia inhibitory factor but not signal transduction. J Biol Chem 274:24685–24693

Bonifaz LC, Bonnyay DP, Charalambous A et al (2004) In vivo targeting of antigens to maturing dendritic cells via the DEC-205 receptor improves T cell vaccination. J Exp Med 199:815–824

Boonen M, Vogel P, Platt KA et al (2009) Mice lacking mannose 6-phosphate uncovering enzyme activity have a milder phenotype than mice deficient for N-acetylglucosamine-1-phospho transferase activity. Mol Biol Cell 20:4381–4389

Bozzacco L, Trumpfheller C, Siegal FP et al (2007) DEC-205 receptor on dendritic cells mediates presentation of HIV gag protein to CD8+ T cells in a spectrum of human MHC I haplotypes. Proc Natl Acad Sci USA 104:1289–1294

Brayden DJ (2001) Oral vaccination in man using antigens in particles: current status. Eur J Pharm Sci 14:183–189

Cardone M, Porto C, Tarallo A et al (2008) Abnormal mannose-6-phosphate receptor trafficking impairs recombinant α-glucosidase uptake in Pompe disease fibroblasts. Pathogenetics 1:6

Cawley DB, Simpson DL, Herschman HR (1981) Asialoglycoprotein receptor mediates the toxic effects of an asialofetuin-diphtheria toxin fragment A conjugate on cultured rat hepatocytes. Proc Natl Acad Sci USA 78:3383–3387

Chappell SA, Walsh T, Walker RA et al (1997) Loss of heterozygosity at the mannose 6-phosphate insulin-like growth factor 2 receptor gene correlates with poor differentiation in early breast carcinomas. Br J Cancer 76:1558–1561

Chiu M, Tamura T, Wadhwa MS, Rice KG (1994) In vivo targeting function of N-linked oligosaccharides with terminating galactose and N-acetyl-galactosamine residues. J Biol Chem 269:16195–16202

Clark MA, Jepson MA, Hirst BH (2001) Exploiting M cells for drug and vaccine delivery. Adv Drug Deliv Rev 50:81–106

Cui Z, Hsu CH, Mumper RJ (2003) Physical characterization and macrophage cell uptake of mannan-coated nanoparticles. Drug Dev Ind Pharm 29:689–700

Cui Z, Mumper RJ (2002a) Topical immunization using nanoengineered genetic vaccines. J Control Release 81:173–184

Cui Z, Mumper RJ (2002b) Genetic immunization using nanoparticles engineered from microemulsion precursors. Pharm Res 19:939–946

Damle NK (2004) Tumor-targeted chemotherapy with immunoconjugates of calicheamicin. Expert Opin Biol Ther 4:1445–1452

Dasi F, Benet M, Crespo J, Alino SF (2001) Asialofetuin liposome-mediated human alpha1-antitrypsin gene transfer in vivo results in stationary long-term gene expression. J Mol Med 79:205–212

Dhami R, Schuchman EH (2004) Mannose 6-phosphate receptor-mediated uptake is defective in acid sphingomyelinase-deficient macrophages: implications for Niemann-Pick disease enzyme replacement therapy. J Biol Chem 279:1526–1532

Diebold SS, Kursa M, Wagner E et al (1999) Mannose polyethylenimine conjugates for targeted DNA delivery into dendritic cells. J Biol Chem 274:19087–19094

Dunder U, Mononen I (2001) Human leukocyte glycosylasparaginase: cell-to-cell transfer and properties in correction of aspartylglycosaminuria. FEBS Lett 499:77–81

Dvorak-Ewell M, Wendt D, Hague C et al (2010) Enzyme replacement in a human model of mucopolysaccharidosis IVA in vitro and its biodistribution in the cartilage of wild type mice. PLoS One 5:e12194

Elliott J, Johnston JA (2004) SOCS: role in inflammation, allergy and homeostasis. Trends Immunol 25:434–444

Erbacher P, Bousser MT, Raimond J et al (1996) Gene transfer by DNA/ glycosylated polylysine complexes into human blood monocyte-derived macrophages. Hum Gene Ther 7:721–729

Feinberg H, Torgersen D, Drickaer K, Weis WI (2000) Mechanism of pH-dependent N-acetylgalactosamine binding by a functional mimic of the hepatocyte asialoglycoprotein receptor. J Biol Chem 275:35176–35184

Feldman EJ, Brandwein J, Stone R et al (2005) Phase III randomized multicenter study of a humanized anti-CD33 monoclonal antibody, lintuzumab, in combination with chemotherapy, versus chemotherapy alone in patients with refractory or first-relapsed acute myeloid leukemia. J Clin Oncol 23:4110–4116

Ferkol T, Perales JC, Mularo F, Hanson R (1996) Receptor-mediated gene transfer into macrophages. Proc Natl Acad Sci USA 93:101–105

Fiume L, Di Stefano G, Busi C et al (1997) Liver targeting of antiviral nucleoside analogues through the asialoglycoprotein receptor. J Viral Hepat 4:363–370

Flores-Langarica A, Meza-Perez S, Calderon-Amador J et al (2005) Network of dendritic cells within the muscular layer of the mouse intestine. Proc Natl Acad Sci USA 102:19039–19044

Fromell K, Andersson M, Elihn K, Caldwell KD (2005) Nanoparticle decorated surfaces with potential use in glycosylation analysis. Colloids Surf B Biointerfaces 46:84–91

Fujimoto M, Naka T (2003) Regulation of cytokine signaling by SOCS family molecules. Trends Immunol 24:659–666

Fuller M, Hopwood JJ, Anson DS (1998) Receptor mediated binding of two glycosylation forms of N-acetylgalactosamine-4-sulphatase. Biochim Biophys Acta 1406:283–290

Gabor F, Bogner E, Weissenboeck A, Wirth M (2004) The lectin-cell interaction and its implications to intestinal lectin-mediated drug delivery. Adv Drug Deliver Rev 56:459–480

Gac S, Coudane J, Boustta M et al (2000) Synthesis, characterisation and in vivo behaviour of a norfloxacin-poly(L-lysine citramide imide) conjugate bearing mannosyl residues. J Drug Target 7:393–406

Ganley IG, Pfeffer SR (2006) Cholesterol accumulation sequesters Rab9 and disrupts late endosome function in NPC1-deficient cells. J Biol Chem 28:17890–17899

Gao X, Chen J, Tao W et al (2007) UEA I-bearing nanoparticles for brain delivery following intranasal administration. Int J Pharm 340:207–215

Gao X, Huang L (1991) A novel cationic liposome reagent for efficient transfection of mammalian cells. Biochem Biophys Res Commun 179:280–285

Garg M, Asthana A, Agashe HB et al (2006) Stavudine-loaded mannosylated liposomes: in-vitro anti-HIV-I activity, tissue distribution and pharmacokinetics. J Pharm Pharmacol 58:605–616

Gary-Bobo M, Nirdé P, Jeanjean A et al (2007) Mannose 6-phosphate receptor targeting and its applications in human diseases. Curr Med Chem 14:2945–2953

Gieseler RK, Marquitan G, Hahn MJ et al (2004) DC-SIGN-specific liposomal targeting and selective intracellular compound delivery to human myeloid dendritic cells: implications for HIV disease. Scand J Immunol 59:415–424

Gliddon BL, Yogalingam G, Hopwood JJ (2004) Purification and characterization of recombinant murine sulfamidase. Mol Genet Metab 83:239–245

Grandjean C, Angyalosi G, Loing E et al (2001) Novel hyperbranched glycomimetics recognized by the human mannose receptor: quinic or shikimic acid derivatives as mannose bioisosteres. Chembiochem 2:747–757

Grubb JH, Vogler C, Levy B et al (2008) Chemically modified β-glucuronidase crosses blood-brain barrier and clears neuronal storage in murine mucopolysaccharidosis VII. Proc Natl Acad Sci USA 105:2616–2621

Guidotti J, Akli S, Castelnau-Ptakhine L et al (1998) Retrovirus-mediated enzymatic correction of Tay-Sachs defect in transduced and non-transduced cells. Hum Mol Genet 7:831–838

Gupta A, Gupta RK, Gupta GS (2009) Targeting cells for drug and gene delivery: emerging applications of mannans and mannan binding lectins. J Sci Ind Res 68:465–483

Hankins GR, De Souza AT, Bentley RC et al (1996) M6P/IGF2 receptor: a candidate breast tumor suppressor gene. Oncogene 12:2003–2009

Hashida M, Nishikawa M, Yamashita F, Takakura Y (2001) Cell-specific delivery of genes with glycosylated carriers. Adv Drug Deliv Rev 52:187–196

Haskins M (2009) Gene therapy for lysosomal storage diseases (LSDs) in large animal models. ILAR J 50:112–121

Hauswirth AW, Florian S, Printz D, Sotlar K, Krauth MT, Fritsch G, Schernthaner GH, Wacheck V, Selzer E, Sperr WR, Valent P (2007) Expression of the target receptor CD33 in CD34/CD38/CD123 AML stem cells. Eur J Clin Invest 37:73–82

Hirosako K, Imasato H, Hirota Y et al (2004) 3-Methyladenine specifically inhibits retrograde transport of cation-independent mannose 6-phosphate/insulin-like growth factor II receptor from the early endosome to the TGN. Biochem Biophys Res Commun 316:845–852

Jain KK (2007) Use of nanoparticles for drug delivery in glioblastoma multiforme Expert Rev Neurother 7:363–72

Jansen SM, Groener JE, Poorthuis BJ (1999) Lysosomal phospholipase activity is decreased in mucolipidosis II and III fibroblasts. Biochim Biophys Acta 1436:363–369

Jepson MA, Clark MA, Hirst BH (2004) M cell targeting by lectins: a strategy for mucosal vaccination and drug delivery. Adv Drug Deliv Rev 56:511–525

Kawakami S, Hattori Y, Lu Y et al (2004) Effect of cationic charge on receptor-mediated transfection using mannosylated cationic liposome/plasmid DNA complexes following the intravenous administration in mice. Pharmazie 59:405–408

Kawakami S, Sato A, Nishikawa M et al (2000) Mannose receptor-mediated gene transfer into macrophages using novel mannosylated cationic liposomes. Gene Ther 7:292–299

Kawakami S, Yamashita F, Nishikawa M et al (1998) Asialoglycoprotein receptor-mediated gene transfer using novel galactosylated cationic liposomes. Biochem Biophys Res Commun 252:78–83

Keler T, Ramakrishna V, Fanger MW (2004) Mannose receptor-targeted vaccines. Expert Opin Biol Ther 4:1953–1962

Kiess W, Blickenstaff GD, Sklar MM et al (1988) Biochemical evidence that the type II insulin-like growth factor receptor is identical to the cation-independent mannose 6-phosphate receptor. J Biol Chem 263:9339–9344

Koeberl DD, Luo X, Sun B et al (2011) Enhanced efficacy of enzyme replacement therapy in Pompe disease through mannose-6-phosphate receptor expression in skeletal muscle. Mol Genet Metab 103:107–112

Kokudo N, Vera DR, Makuuchi M (2003) Clinical application of TcGSA. Nucl Med Biol 30:845–849

Kornfeld S, Sly WS (2001) I cell disease and pseudo-Hurler polydystrophy: disorders of lysosomal enzyme phosphorylation and localization. In: Scriver CR, Beaudet AL, Sly WS, Valle D (eds) Metabolic and molecular bases of inherited diseases. McGraw-Hill, New York, pp 3469–3482

Korokhov N, de Gruijl TD, Aldrich WA et al (2005) High efficiency transduction of dendritic cells by adenoviral vectors targeted to DC-SIGN. Cancer Biol Ther 4:289–294

Kosuga M, Sasaki K, Tanabe A, Li XK et al (2001) Engraftment of genetically engineered amniotic epithelial cells corrects lysosomal storage in multiple areas of the brain in mucopolysaccharidosis type VII mice. Mol Ther 3:139–148

Krebs DL, Hilton DJ (2001) SOCS proteins: negative regulators of cytokine signaling. Stem Cells 19:378–387

Kyttälä A, Heinonen O, Peltonen L, Jalanko A (1998) Expression and endocytosis of lysosomal aspartylglucosaminidase in mouse primary neurons. J Neurosci 18:7750–7756

LeBowitz JH, Grubb JH, Maga JA et al (2004) Glycosylation-independent targeting enhances enzyme delivery to lysosomes and decreases storage in mucopolysaccharidosis type VII mice. Proc Natl Acad Sci USA 10:3083–3088

Lee ES, Na K, Bae YH (2005) Doxorubicin loaded pH-sensitive polymeric micelles for reversal of resistant MCF-7 tumor. J Control Release 103:405–418

Lehtovirta M, Kyttälä A, Eskelinen EL et al (2001) Palmitoyl protein thioesterase (PPT) localizes into synaptosomes and synaptic vesicles in neurons: implications for infantile neuronal ceroid lipofuscinosis (INCL). Hum Mol Genet 10:69–75

Liang WW, Shi X, Deshpande D et al (1996) Oligonucleotide targeting to alveolar macrophages by mannose receptor-mediated endocytosis. Biochim Biophys Acta 1279:227–234

Litjens T, Bielicki J, Anson DS et al (1997) Expression, purification and characterization of recombinant caprine N-acetylglucosamine-6-sulphatase. Biochem J 327:89–94

Loomis K, Smith B, Feng Y et al (2010) Specific targeting to B cells by lipid-based nanoparticles conjugated with a novel CD22-ScFv. Exp Mol Pathol 88:238–249

Lu JY, Verkruyse LA, Hofmann SL (1996) Lipid thioesters derived from acylated proteins accumulate in infantile neuronal ceroid lipofuscinosis: correction of the defect in lymphoblasts by recombinant palmitoyl-protein thioesterase. Proc Natl Acad Sci USA 93:10046–10050

Mahato RI, Kawabata K, Takakura Y, Hashida M (1995) In vivo disposition characteristics of plasmid DNA complexed with cationic liposomes. J Drug Target 3:149–157

Mahnke K, Qian Y, Fondel S et al (2005) Targeting of antigens to activated dendritic cells in vivo cures metastatic melanoma in mice. Cancer Res 65:7007–7012

Matsuoka K, Tamura T, Tsuji D (2011) Therapeutic potential of intracerebroventricular replacement of modified human β-hexosaminidase B for GM2 gangliosidosis. Mol Ther 19:1017

McNicholas S, Rencurosi A, Lay L et al (2007) Amphiphilic N-glycosyl-thiocarbamoyl cyclodextrins: synthesis, self-assembly, and fluorimetry of recognition by Lens culinaris lectin. Biomacromolecules 8:1851–1857

McVie-Wylie AJ, Lee KL et al (2008) Biochemical and pharmacological characterization of different recombinant acid α-glucosidase preparations evaluated for the treatment of Pompe disease. Mol Genet Metab 94:448–455

Meier M, Bider MD, Malashkevich VN et al (2000) Crystal structure of the carbohydrate recognition domain of the H1 subunit of the asialoglycoprotein receptor. J Mol Biol 300:857–865

Minko T (2004) Drug targeting to the colon with lectins and neoglycoconjugates. Adv Drug Deliv Rev 56:491–509

Mitra M, Mandal AK, Chatterjee TK, Das N (2005) Targeting of mannosylated liposome incorporated benzyl derivative of Penicillium nigricans derived compound MT81 to reticuloendothelial systems for the treatment of visceral leishmaniasis. J Drug Target 13:285–293

Motyka B, Korbutt G, Pinkoski MJ et al (2000) Mannose 6-phosphate/insulin-like growth factor II receptor is a death receptor for granzyme B during cytotoxic T cell-induced apoptosis. Cell 103:491–500

Nabel GJ, Nabel EG, Yang ZY et al (1993) Direct gene transfer with DNA–liposome complexes in melanoma: expression, biologic activity, and lack of toxicity in humans. Proc Natl Acad Sci USA 90:11307–11311

Naureckiene S, Sleat DE, Lackland H et al (2000) Identification of HE1 as the second gene of Niemann-Pick C disease. Science 290(5500): 2298–2301

Nemecek ER, Matthews DC (2002) Antibody-based therapy of human leukemia. Curr Opin Hematol 9:316–321

Ni X, Morales CR (2006) The lysosomal trafficking of acid sphingomyelinase is mediated by sortilin and mannose 6-phosphate receptor. Traffic 7:889–902

Oates AJ, Schumaker LM, Jenkins SB et al (1998) The mannose 6-phosphate/insulin-like growth factor 2 receptor (M6P/IGF2R), a putative breast tumor suppressor gene. Breast Cancer Res Treat 47:269–281

Ogawara K, Hasegawa S, Nishikawa M et al (1999) Pharmacokinetic evaluation of mannosylated bovine serum albumin as a liver cell-specific carrier: quantitative comparison with other hepatotropic ligands. J Drug Target 6:349–360

Orr SJ, Morgan NM, Buick RJ et al (2007a) SOCS3 targets Siglec 7 for proteasomal degradation and blocks Siglec 7-mediated responses. J Biol Chem 282:3418–3422

Orr SJ, Morgan NM, Elliott J et al (2007b) CD33 responses are blocked by SOCS3 through accelerated proteasomal-mediated turnover. Blood 109:1061–1068

Parenti G, Andria G (2011) Pompe disease: from new views on pathophysiology to innovative therapeutic strategies. Curr Pharm Biotechnol 12:902

Park KH, Sung WJ, Kim S et al (2005) Specific interaction of mannosylated glycopolymers with macrophage cells mediated by mannose receptor. J Biosci Bioeng 99:285–289

Pathak A, Vyas SP, Gupta KC (2008) Nano-vectors for efficient liver specific gene transfer. Int J Nanomedicine 3:31–49

Reaves BJ, Bright NA, Mullock BM, Luzio JP (1996) The effect of wortmannin on the localisation of lysosomal type I integral membrane glycoproteins suggests a role for phosphoinositide 3-kinase activity in regulating membrane traffic late in the endocytic pathway. J Cell Sci 109:749–762

Rieger J, Stoffelbach F, Cui D et al (2007) Mannosylated poly(ethylene oxide)-b-poly(epsilon-caprolactone) diblock copolymers: synthesis, characterization, and interaction with a bacterial lectin. Biomacromolecules 8:2717–2725

Saris JJ, van den Eijnden MM, Lamers JM, Saxena PR, Schalekamp MA, Danser AH (2002) Prorenin-induced myocyte proliferation: no role for intracellular angiotensin II. Hypertension 39:573–577

Schellens JP, Saftig P, von Figura K, Everts V (2003) Deficiency of mannose 6-phosphate receptors and lysosomal storage: a morphometric analysis of hepatocytes of neonatal mice. Cell Biol Int 27:897–902

Serizawa T, Yasunaga S, Akashi M (2001) Synthesis and lectin recognition of polystyrene core-glycopolymer corona nanospheres. Biomacromolecules 2(Summer):469–475

Simões S, Slepushkin V, Pretzer E et al (1999) Transfection of human macrophages by lipoplexes via the combined use of transferrin and pH-sensitive peptides. J Leukocyte Biol 65:270–279

Sly WS, Vogler C, Grubb JH et al (2006) Enzyme therapy in mannose receptor-null mucopolysaccharidosis VII mice defines roles for the mannose 6-phosphate and mannose receptors. Proc Natl Acad Sci USA 103:15172–15177

Smart JD (2004) Lectin-mediated drug delivery in the oral cavity. Adv Drug Deliv Rev 56:481–489

Stewart K, Brown OA, Morelli AE et al (1997) Uptake of α-(L)-iduronidase produced by retrovirally transduced fibroblasts into neuronal and glial cells in vitro. Gene Ther 4:63–75

Storm G, Roerdink FH, Steerenberg PA et al (1987) Influence of lipid composition on the antitumor activity exerted by doxorubicin-containing liposomes in a rat solid tumor model. Cancer Res 47:3366–3372

Tacken PJ, de Vries IJ, Gijzen K et al (2005) Effective induction of naive and recall T-cell responses by targeting antigen to human dendritic cells via a humanized anti-DC-SIGN antibody. Blood 106:1278–1285

Ten Cate B, Samplonius DF, Bijma T et al (2007) The histone deacetylase inhibitor valproic acid potently augments gemtuzumab ozogamicin-induced apoptosis in acute myeloid leukemic cells. Leukemia 21:248–252

Terada T, Nishikawa M, Yamashita F, Hashida M (2006) Influence of cholesterol composition on the association of serum mannan-binding proteins with mannosylated liposomes. Biol Pharm Bull 29:613–618

Tsuji D, Kuroki A, Ishibashi Y et al (2005) Metabolic correction in microglia derived from Sandhoff disease model mice. J Neurochem 94:1631–1638

Umezawa F, Eto Y (1988) Liposome targeting to mouse brain: mannose as recognition marker. Biochem Biophys Res Commun 153:1038–1044

Urayama A, Grubb JH, Banks WA, Sly WS (2007) Epinephrine enhances lysosomal enzyme delivery across the blood brain barrier by up-regulation of the mannose 6-phosphate receptor. Proc Natl Acad Sci USA 104:12873–12878

Urayama A, Grubb JH, Sly WS, Banks WA (2008) Mannose 6-phosphate receptor-mediated transport of sulfamidase across the blood-brain barrier in the newborn mouse. Mol Ther 16:1261–1266

Walter RB, Boyle KM, Appelbaum FR et al (2008) Simultaneously targeting CD45 significantly increases cytotoxicity of the anti-CD33 immunoconjugate, gemtuzumab ozogamicin, against acute myeloid leukemia (AML) cells and improves survival of mice bearing human AML xenografts. Blood 111:4813–4816

Weber B, Hopwood JJ, Yogalingam G (2001) Expression and characterization of human recombinant and α-N-acetylglucosaminidase. Protein Expr Purif 21:251–259

Wijagkanalan W, Kawakami S, Takenaga M et al (2008) Efficient targeting to alveolar macrophages by intratracheal administration of mannosylated liposomes in rats. J Control Release 125:121–130

Wu J, Liu P, Zhu JL, Maddukuri S, Zern MA (1998) Increased liver uptake of liposomes and improved targeting efficacy by labeling with asialofetuin in rodents. Hepatology 27:772–778

Wu J, Nantz MH, Zern MA (2002) Targeting hepatocytes for drug and gene delivery: emerging novel approaches and applications. Front Biosci 7:717–725

Yaghootfam A, Gieselmann V (2003) Specific hammerhead ribozymes reduce synthesis of cation-independent mannose 6-phosphate receptor mRNA and protein. Gene Ther 10:1567–1574

Yamazaki N, Kojima S, Bovin NV et al (2000) Endogenous lectins as targets for drug delivery. Adv Drug Deliver Rev 43:225–244

Yeeprae W, Kawakami S, Yamashita F (2006) Effect of mannose density on mannose receptor-mediated cellular uptake of mannosylated O/W emulsions by macrophages. J Control Release 114:193–201

Yogalingam G, Bielicki J, Hopwood JJ, Anson DS (1997) Feline mucopolysaccharidosis type VI: correction of glycosaminoglycan storage in myoblasts by retrovirus-mediated transfer of the feline N-acetylgalactosa mine 4-sulfatase gene. DNA Cell Biol 16:1189–1194

References

Yogalingam G, Litjens T, Bielicki J et al (1996) Feline mucopolysaccharidosis type VI. Characterization of recombinant N-acetylgalactosamine 4-sulfatase and identification of a mutation causing the disease. J Biol Chem 271:27259–27265

Zhu Y, Li X, Kyazike J et al (2004) Conjugation of mannose 6-phosphate-containing oligosaccharides to acid α-glucosidase improves the clearance of glycogen in pompe mice. J Biol Chem 279:50336–50341

Zhu Y, Li X, McVie-Wylie A et al (2005) Carbohydrate-remodelled acid α-glucosidase with higher affinity for the cation-independent mannose 6-phosphate receptor demonstrates improved delivery to muscles of Pompe mice. Biochem J 389:619–628

About the Author

Dr. Gupta is a former Professor and Chairman of the Department of Biophysics, Panjab University, Chandigarh, India. His primary areas of research are Molecular and Cell Biology, Enzymology and Protein Chemistry, and Radiation Biology. As a visiting researcher, he has worked at many institutions, including Northwestern University in Evanston, IL, the Center of Immunopathology and Experimental Immunology, INSERM, Paris and the Center of Cytogenetics and Immunogenetics, INSERM, Villejuif (France). He has been honored by Indian agencies as Emeritus Scientist by CSIR, Emeritus Professor by University Grants Commission (UGC), and Emeritus Medical Scientist by Indian Council of Medical Research (ICMR). Professor Gupta has made significant research contribution in scientific research and published 155 original research articles and reviews in Books and international journals of repute. He is the recipient of several awards and international fellowships including a WHO fellowship, the INSERM French Government fellowship, and the fellowship under Indo-French Exchange programme. Indian Council of Medical Research has honored him by conferring Swaran Kanta Dingley Oration Award of 1993 for his research contribution and extending knowledge in male reproduction. Dr. Gupta is the author of Proteomics of Spermatogenesis published by Springer (New York, USA) in 2005 and has contributed 34 reviews on the subject, as a single author. Associated with national and international scientific societies, he has chaired scientific sessions and delivered invited lectures at national and International conferences.

Professor G.S. Gupta holds master's degrees in Physical Chemistry and Biochemistry from Lucknow University, Lucknow, and a Ph.D. in Biophysics from Panjab University, Chandigarh (India).

Index

A

AA4, 901
AAA-ATPase Cdc48p/p97, 118
AAA F-type CRD, 440, 442
AA4.1$^+$ and AA4.1 cells, 906
AA4/C1qRp, 906
AAL: overall fold and organization, 447
    crystal structure, 447
    ribbon diagram of monomer A of AAL, 447
α-Amylase, 157
α-Amyloid deposits, 64
α4β1 (VLA-4), α5β1 (VLA-5), and α4β7 integrins, 234
Abdominal aortic aneurysm (AAA) or atherosclerotic occlusive
    disease, 817, 945
Abnormal G-protein signaling, 1035
Abnormal MBL pathway and disease associations, 938
ABPA in an Indian population, 975
Absence of C-type lectin-carbohydrate binding motifs in Reg-III, 868
Acidic-cluster-dileucine (ACLL), 95–97
Acidic mammalian chitinase (AMCase), 421, 429
Acidiphilium cryptum JF-5, 446
Acid maltase deficiency, 1040
Acid mammalian chitinase activity, 429
Acid sphingomyelinase (ASM), 225, 1041
Acinar cells, 848
ACTH adenomas, 301
Actin, 456, 881
Actinomycin D, 599
Action of drugs and physical factors on CAMS, 1015
Activated human neutrophils, 225
Activated MASP-2, 413
Activated protein C (APC), 916, 1012
Activating adapter protein DAP12, 383
Activating and inhibitory receptors, 620
Activating CD94/NKG2C, 642
Activating enzyme (E1), 124
Activating or inhibitory signalling, 362
Activating or the non-inhibitory NK receptors, 621
Activation-induced C-type lectin (AICL), 621, 694
Activation-induced C-type lectin (CLECSF2), 694
Activation of endothelial cells (EC), 992
Activation of E-selectin, VCAM-1, and ICAM-1 genes, 597
    cyclic AMP inhibits NF-kB-mediated transcription, 597
    ICAM-1 genes, 597
    I kBα (MAD-3), 597
    NF-kB activation, 597
    NF-kB: a dominant regulator, 596, 598
    nuclear NF-kB DNA-binding activity, 597
    platelet factor 4 (PF4), 597
    p38 MAPK in equine ECs, 597
Activation of NFAT, 224
Activation of NF-kB and kinases, 277

Activation of PI3K, 516
Activation of SPAR, 516
Activation of Src family tyrosine kinases, 633
Activation rafts, 366
Activator inhibitor-1, and ICAM-1, 167
Activator of the EGF receptor/Akt/AP-1 signaling pathway, 871
Activatory and inhibitory signals, 383
Active site of thrombin, 912
Active smokers, 957
Acute cardiogenic pulmonary edema (APE), 961
Acute coronary syndrome (ACS), 887
Acute familial mediterranean fever (FMF), 568
Acute heart stress, 849
Acute lung injury (ALI), 162, 961, 999
Acute myeloid leukemia (AML), 381, 399, 400, 680, 686, 1052
Acute myocardial infarction (AMI), 944, 996
Acute pancreatitis (SAP), 851, 853, 860, 861, 863, 867, 868, 1005
Acute-phase proteins, 164, 166
Acute phase reactant, 863
Acute-phase response (APR), 341
Acute respiratory distress syndrome (ARDS), 961, 972
Adamantinomatous, 251
ADAMTS, 837
ADAMTS-4, 817, 833, 837
ADAMTS5, 817, 832, 833
ADAMTS-1,-4 and-5v, 832
Adapter protein-1 (AP-1), 64
Adaptive immune responses, 543
Adaptor protein complexes, 61
Adaptor protein DAP-10, 667
Adaptors DAP-10 and KARAP/DAP-12, 667, 684
Adders, 71
AD + Down's syndrome (DN), 856
Addressin MAdCAM-1, 562
Adenocarcinoma, 847, 970
Adenocarcinoma (ADC) cells, 680
Adenoid cystic carcinoma (ACC), 298, 871
Adenomas, 871, 970
Adenosine deaminase, 889
Adenoviral (Ad) vectors, 1050
Adenovirus, 13
Adenylyl cyclase, 260
Adenylyl cyclase inhibitor SQ-22528, 599
Adherence lectin, 8
Adherens junctions (AJs), 1027, 1032
Adhesion, migration and phagocytosis, 904
Adhesion of endothelial cells, 428
Adhesion proteins in transplantation, 1012
Adipocyte differentiation, 226
A disintegrin and metalloprotease (ADAM), 817
AD neurodegeneration, 856
Adoptive transfer, 686

AD + Parkinson's disease (PD), 856
ADP-ribosylating toxin, 323
ADP-ribosylation, 322
ADP-ribosylation factor ARF-1, 72, 756
ADP ribosylation factor (Arf)-interacting (GGA), 95, 96
ADP-ribosyltransferase, 322–324
Adrenergic receptor, 909
Adrenomedullin (AM), 599
Adult respiratory distress syndrome (ARDS), 956, 961, 999
Advanced glycation end product (AGE), 269, 279, 726, 738
Advanced lipoxidation endproducts, 280
Affinity for GlcNAc, GalNAc and sialic acid, 411
A. fumigates, 182, 730, 934, 1014
A. fumigatus, 489
A. fumigatus mouse asthma model, 965
Age-related macular degeneration (AMD), 593
AGE-R3/Gal-3 and receptors for AGE (RAGE), 279
Agglutination, 539
Aggrecan, 476, 477, 802, 807, 808, 810, 812, 825, 827, 833
Aggrecanases 1 and 2, 833
Aggrecanases and matrix metalloproteinases (MMPs), 817
Aggrecan catabolism, 817
Aggrecan CLD, 809, 811
Aggrecan gene, 810
Aggrecan insufficiency, 817
Aggrecans, 801
Aggregation of phospholipid vesicles, 515
Aggresomes, 36
Agouti, 891, 893
Agouti coat color gene, 892
Agouti protein, 893
Agouti-related protein, 891
A/G polymorphism, 430
Ah, Bh and H determinants, 441
α-helical coiled-coil, 474, 778
AICL 628, 758
Air-liquid interface, 955
Air/liquid interface of the lung, 539
Airway inflammation, 971
Akt kinase, 858
Akt/protein kinase B, 43
Akt-signaling, 432
Alanine-and proline-rich antigenic (Apa) glycoprotein, 967
Alcohol consumption, 957
Algal lectins, 4
ALG-2 interacting protein-1 (AIP-1), 268
ALG-2 linked protein X (Alix), 268
α2-3-linked sialic acids, 570
Alix/AIP-1, 268
Alleles of SP-A and SP-D, 963
Allelic variants of SP-A1 gene, 972
Allelic variants of the SP-A2 gene, 972
Allergic airway inflammation, 421, 429, 975
Allergic asthma, 956
Allergic bronchial asthma, 429, 916
Allergic bronchopulmonary aspergillosis (ABPA), 542, 974
Allergic disorders, 964
Allergic encephalitis, inflammatory alveolitis, nephrotoxic, 167
Allergic fungal rhinosinusitis (AFRS), 959
Allergic inflammation in asthma, 964
Allergic rhinitis, 965, 999
Alligator hepatic lectin (AHL), 495
Allograft rejection, 700
Allosteric mechanism, 915
All-trans retinoic acid (ATRA), 685
α3β1, 197
α4β7, 1000
α-coiled neck regions, 486
Alternate polyadenylation signals, 65
Alternative DNA splicing, 667
Alternative pathway, 415
Alternative splicing, 245, 251, 254, 889
5'-Alternative splicing, 865
Alveolar human Mφ (AMφ), 517
Alveolar macrophage chemotaxis, 515
Alveolar macrophages, 539
Alveolar proteinosis protein-I (APP-I), 961
Alveolar type II cells, 539, 955
Alzheimer's disease (AD), 15, 64, 101, 174, 430, 544, 587, 737, 856, 864, 1004
AMCase, 422, 429
Amino-terminal Ig-like V-set domain, 362
A missense mutation, 1030
    (D511V) is causative of ADPKD, 1030
AMφ populations, 516
Amniotic fluid/the maturing fetal lung, 519
AMP-activated protein kinase activation, 598
Amphiphilic β-cyclodextrins, 1047
Amphisomes, 63
Amphoterin, 8, 829
Amyloidosis, 175
Amyloid P component, 175
Amyloid plaques, 171
Amyotrophic lateral sclerosis (ALS), 230, 828.
α-N-Acetylglucosaminidase (NAGLU), 1044
Anaphase-promoting complex or cyclosome (APC/C), 126
Anaplasma, 561
Anaplastic astrocytoma, 918
Anaplastic carcinomas, 292
Anchored proteoglycans, 808
Androgen, 45
Aneurismal subarachnoid hemorrhage (SAH), 1004
Angiogenesis, 43, 279, 428, 918
Angiogenic vasculature, 920
Angiostatin, 456, 885
    AST^{K1-4}, 885
Angiotensin, 895
Angiotensin II, 47
Anguilla Anguilla Agglutinin (AAA), 439, 440
Animal experiments, 359
Animal galectins, 191
Animal studies, 948
    MBL DKO, 948
    MBL knock-out mice, 948
Annexin 1, 274
Annexin 5, 466
Annexin A2 (p36), 459
    three distinct functional regions, 459
Annexin A4, 462
    lung, intestine, stomach, trachea, and kidney, 462
    in secretory epithelia, 462
Annexin A4 (or endonexin, protein I), 461
Annexin A4 (p33/41), 461
Annexin A6 (Annexin VI), 467
    annexin A6 by electron microscopy, 467
    crystal structure of annexin A6, 467
    functions, 467
    KFERQ-like sequences, 467
    a receptor for CS chains, 467
Annexin A4a, 462
Annexin A5/Annexin V, 464
Annexin A5 assay, 467

Annexin A5-mediated pathogenic mechanisms, 466
    antiphospholipid syndrome, 466
    prevention of atherothrombosis in SLE, 466
    SLE-associated cardiovascular disease (CVD), 466
Annexin A5 regulates IFN-γ signaling, 465
Annexin A4 trimerize, 463
Annexin core, 465
    structure for annexin A5, 465
Annexin family proteins and lectin activity, 458
Annexin interacting proteins, 456
Annexin IV 5, 09
Annexin IV, V, VI, 6
Annexins, 6
Annexins A1, A2, A4, A5, and A11, 456
Annexins A2, A4, A5, and A6, 459
    GAG binding properties, 459
    recognition elements for GAGs, 459
Annexins A1, A2, A4 and A11 in primary tumors, 463
Annexins A7, A8 and A10, 456
Annexins bind, 455
    phosphatidylethanolamine, 455
    phosphatidylinositol, 455
    to phosphatidylserine, 455
Annexins (Lipocortins): classification and nomenclature, 456
Annexins in plants, slime molds, metazoans, insects, birds, 457
Annexins in tissues, 457
Annexin 2 tetramer (A2t), 460, 461
    binds Ca^{2+}, 460
    GAGs (including heparin and heparan sulphate), 460
    heterotetramer, 460
    phospholipid and membranes, 460
Anorectal malformations syndrome, 923
Anoxia/reoxygenation, 600
Anti-adhesion therapy, 3
Antiallergic effect, 258
Antiapoptotic, 864
Antiapoptotic and/or pro-apoptotic factor, 272
Antiapoptotic factor, 858
Anti-bacterial, 864
Antibody-independent pathway of complement system, 493
Anti-CD33 (Siglec-3) antibody, 400
Anti-CD33-antibody-based therapy of human leukemia, 1052
Anti-CD33 (Siglec-3) antibody gemtuzumab (Mylotarg), 1052
Anticoagulant, 915
Anticoagulant proteins from snake venom, 854
Antifibrinolytic cofactor activities of TM, 915
Antifreeze glycoproteins, 474
Antifreeze polypeptide (AFP), 478
Antifreeze proteins (AFPs), 478, 854
Anti-galectin compounds as anti-cancer drugs, 305
Antigen presenting cells, 1046
Antigen presenting lectin-like receptor, 759
Antigen presenting lectin-like receptor complex (APLEC), 758, 759
Anti-inflammatory, 167, 168, 250
Anti-inflammatory drugs, 1017
Anti-inflammatory protein, 543
Anti-inflammatory role, 432
Anti-inflammatory role of SP-A, 516
Anti-metastatic potential in breast cancer, 259
Antimicrobial activity, 272
Anti-NF-kB reagents, 1015
Antinuclear autoimmunity, 174
Antioxidant enzymatic scavenger systems, 543
Antiphospholipid syndrome, 461
Anti-Ro/SSA, 48

Anti-*Saccharomyces cerevisiae* antibody (ASCA), 942
Antisense oligonucleotides, 1047
Antithy-1 glomerulonephritis, 342
Anti-tumor activity, 679
Antitumor effects of L-selectin, 565
Anti-tumor mechanisms by L-selectin, 565
Anti-viral and anti-tumoral responses of γ/δ T cells, 649
ANXA1, 456
ANXA2, 456
ANXA3, 456
ANXA4, 456
ANXA5, 456
ANXA6, 456
ANXA7, 456
ANXA8, 456
ANXA9, 456
ANXA10, 456
ANXA11, 456
ANXA13, 456
$Anx^{4-/-}$ mouse model, 462
AP-1, 72, 73, 90, 94, 224, 1032
AP-2, 72, 90, 94
AP-3, 73, 90, 94
AP-4, 73, 90, 94
AP-3 adaptor complex, 73
AP1, AP2, and SP1 consensus binding sites, 828
Apical-and basolateral routes, 71
Apical junctions, 1027
APLEC on arthritis and autoimmunity, 759
APLT, 43
ApoE-deficient mouse model of atherosclerosis, 280
$ApoE^{-/-}$ mice, 576
Apo form 1 of galectin-9 NCRD, 256
Apolipoprotein-A, 885
Apolipoprotein E, 164
Apoptosis, 35, 65, 93, 223, 225, 415, 544, 998
Apoptosis in eosinophils of asthmatics, 977
Apoptosis inhibition, 864
Apoptotic cells, 182, 183, 257, 726
Apoptotic neutrophil clearance, 274
Apoptotic signal, 215
Aporiina, 322–324
Appendiceal mucinous cystadenomas, 870
Appendix, 568
Appiadina, 322–324
APP-I and APP-II, 961
Arabidopsis, 133
Arabidopsis, Drosophila, Caenorhabditis, Danio, Xenopus, 207
Arabidopsis thaliana, 40, 127
Arabinose, ribose, and deoxyribose, 533
Arachidonic acid, 461
ARDS or respiratory distress syndrome (RDS), 535, 536
ARF-binding proteins (GGAs), 90
ARF-family GTPase Sar1, 60
Arf-GTP, 96
Arginine, 401
Arginine critical for sialic acid recognition, 383
Armenian hamster *(Cricetulus migratorius)*, 176, 177
Aromatic nucleophilic substitution reaction, 306
Arrangement of disulfide bonds in CTLDs, 478
Arterial stenosis, 992
Arterial thrombosis, 941, 993
Arthritis, 428, 429
Arthritis in Balb/c mice, 818
α-series gangliosides, 398
ASGP-R, 715, 716

ASGP-R: a marker for autoimmune hepatitis, 715
ASGP-R for targeting hepatocytes, 717
ASGP-R-like specificity, 1052
ASGR group, 479
Asialofetuin (ASF), 198, 712, 713
Asialoglycoprotein, 480
Asialoglycoprotein receptor (ASGPR), 9, 11, 476, 477, 479, 490, 633, 709, 717, 750, 1051
    macrophage galactose-type lectin, 709
Asialoglycoprotein receptor or hepatic lectins (HL), 1049
Asialo-GM2, 511
Asialomucin, 275
Asialoorosomucoid (ASOR), 712, 713
ASM-KO cells, 1042
Aspartylglucosamine (GlcNAc-Asn), 1041
Aspartylglycosaminuria (AGU), 1041
Aspergillus, 728
Aspergillus fumigatus, 527, 532, 934
Aspergillus oryzae, 114
Aspirin, 1017
Assay for apoptosis, 466
Association of Fras1/Frem family with Fraser Syndrome, 923
Association of SPs with diabetes, 969
Association of tetranectin with diseases, 887
    B-chronic lymphocytic leukemia (B-CLL), 887
    multiple myeloma, 887
    pelvic inflammatory disease, 887
    in prognosis of cancer, 887
    in tumors of the breast, colon, stomach, and ovary, 887
    various malignancies, 887
Asthma, 258, 272, 421, 428, 916, 964
Asthma pathogenesis, 429
Astrocytes, 826, 830, 838
Astrocytes and microglia, 335
Astroglial scar formation, 839
Asymmetric unit of apo form1, 255
Ataxia telangiectasia, mutated (ATM), 671
Atherogenesis, 169
Atherosclerosis, 169, 171, 176, 181, 364, 430, 944, 991–993, 1001, 1016
Atherosclerosis and restenosis, 815
Atherosclerosis-prone and atherosclerosis-resistant human arteries, 816
Atherosclerosis-prone apoE-deficient mice, 575
Atherosclerotic grafted vein and carotid artery, 741
Atherosclerotic ischemic stroke, 997
Atherosclerotic occlusive disease, 817
Atherosclerotic plaques, 429, 430
Atherothrombosis, 992, 997
ATL cells, 258
Atomic force microscopy (AFM), 507
Atopic dermatitis, 998
Atopic dermatitis and psoriasis, 998
ATPBD3 (ATB3) genes, 401
ATR (ATM and Rad3 related), 671
ATR/ATM-dependent recognition, 672
Atrial fibrillation (AF), 993
Atrn alleles, 892
Atrn^{mg}, 892
Atrn^{mg-3J} and dal, 892
Atrn^{mg-L}, 892
Attachment factor for HIV-1, 763
Attractin (Atrn) (formerly mahogany), 478–480, 892
Attractin, formerly zi [zitter] locus on rat chromosome 3, 893
Attractin group of CTLDs, 888
Attractin-like protein (ATRNL1), 888, 889

Autoantibodies to, 941
    MBL in SLE, 941
Autocrine cytokine, 517
Autocrine regulator of mast cell, 258
Autoimmune and inflammatory diseases, 682, 763, 939
Autoimmune diseases, 199, 400, 415, 428, 566, 731, 1000
Autoimmune hepatitis, 715
Autoimmune neuroinflammation, 234
Autoimmune reactions, 630
Autoimmune thyroiditis, 1002
Autoimmune uveoretinitis, 630
Autoimmunity, 192
Autologous peripheral blood stem cell transplantation (auto-PBSCT), 948
Autophagosomes, 63
Autopoly (ADP-ribosyl)ation, 851
Autosomal dominant polycystic kidney disease (ADPKD), 1027, 1034
Autosomal dominant polycystic liver disease (PCLD), 36
Autosomal recessive gene of attractin, 888
Autosomal recessive lysosomal storage disease, 1043
Autosomal recessive metabolic disorder, 1040
Autosomal X-linked genes, 82
αVβ3 receptor (CD51/CD61), 280
Avy/agouti (Avy) mice, 894
Axonal regeneration, 231, 371, 372
Axonal regeneration-promoting factor, 228
5-Azacytidine, 217, 218

B

2B4, 701, 702
Bacillus anthracis, 1013
Bacillus cereus, 909
Bacillus sphaericus, 321
Bacterial and parasitic pathogens, 272, 282
Bacterial pneumonia, 956
Bacterial zinc metalloproteases, 863
Bactericidal for gram-positive bacteria, 863
Balancing activation and inhibition, 622
BALB/c *Nkrp1-Ocil/Clr* region, 625
BALF of dust mite, 965
BALF proteome analysis, 962
BALF YKL-40 levels in IPF, 430
Balloon angioplasty, 998
Band shift analyses, 602
BanLec, 21
Bark lectins from the elderberry plant, *Sambucus sieboldiana* lectin (SSA), 314
    Sambucus nigra agglutinin, 314
β-Barrel structure, 89, 136, 440
(β/α)$_8$ Barrel topology, 424
Barrett's esophagus, 851
Basal telencephalic neuroepithelium, 828
Basigin (Bsg)/CD147, 607
Basolateral endocytic pathway, 91
Batten disease, 101
B cell activation, 363
B-cell chronic lymphocytic leukaemia (B-CLL), 659
B cell non-Hodgkins lymphomas, 367, 400
Bcl-2, 268
BDCA-1, 750
BDCA-3, 765, 910, 915
BDCA-4, 749, 765
BDCA-2: A plasmacytoid DCs (PDCs)-specific lectin, 14, 750, 765
BDCA-3$^+$ CD14-myeloid DC type 2 (MDC2), 767
BDCA-2$^+$ CD123$^+$ plasmacytoid DC (PDC), 767

BDCA-2+ cells, 767
BDCA3+DC, 726
BDCA-2+ in SLE, 767
BDCA-2 signals, 766
B. dermatitidis, 966
BDNF in astrocytes, 230
BEHAB/brevican, 832
BEHAV, 825
Behcet's disease (BD), 660, 945
Behçet syndrome, 1003
Benzodiazepine, 217
Beryllium disease, 957
β-cell-derived autoantigen in non-obese diabetic mice, 858
β cell regeneration or Reg gene protein, 849
β-sandwich structure, 149, 255
BeWo cells, 225
bFGF and HGF and/or their receptors CD138 and c-met, 871
BGR-A and BGR-B (β-glucan receptor/dectin-1), 726
b-haemolytic streptococci, 934
Biantennary analogs, 270
bi-CRD *Branchiostoma belcheri tsingtauense* galectin (BbtGal)-L, 208
bi-CRD galectins, 192
Bifid nose, 923
Biglycan, 533, 801, 802
Biliary atresia (BA), 1004
Bilobal, barbell-like structure, 314
BIMLEC, 478–480, 920
Bimlec (DEC 205) associated C-type lectin-1, 920
Bimosiamose (BIMO), 1012
Binary tandem domain F-Lectin from striped base, 442
Binding affinity of CI-MPR, 71
Binding characteristics of siglecs, 353
Binding domain for sPLA2, 340
Binding of bivalent oligosaccharides to Gal-3, 270
Binding of carbohydrates to NKR-P1, 627
Binding of MAG, SMP, and sialoadhesin, 369
Binding of SP-A to glycolipids, 511
Binding site for *Staphylococcus aureus*, 909
Binding site of Siglec-7 (Arg124), Siglec-9 (Arg120), 389
Binding sites (ligands/receptors), 991
Binding to CD45, 363
Binding to glycosphingolipids, 248
Binding to thrombin, 911
Binding with glycoprotein-340, 532
Binding with nucleic acids, 533
Bindin in invertebrate sperm, 445
β-inhibitor of the influenza virus, 492
Biogenesis of BG, 756
Biomarkers for asthma, 429
BiP, 36, 46, 156
Birbeck granule (BG), 751, 755, 756
Blebs mouse mutants, 923
Blood-brain barrier (BBB), 894, 1042
Blood-clotting factors, 40
Blood coagulation factors V (FV) and VIII (FVIII), 148, 155, 158, 324
Blood dendritic cell antigen-2 (BDCA-2) (CD303), 749, 757, 758, 762, 765, 766
Blood dendritic cell antigen-2 (BDCA-2)(CLEC-4C), 765
Blood group antigens H and Le^a, 270, 441
Blood group-O-sulfated Le^x, 335
B lymphocyte binding to E-and P-selectins, 582
B lymphocyte-induced maturation protein-1 (Blimp-1), 217
β2m, 655
β2m$^{-/-}$, 640
β2-microglobulin (β2-m), 42, 51
BMP4, 444

Body mass index, 544
Bone-marrow-derived DCs (BMDCs), 761
Bone marrow stromal cells (BMSCs), 838
Bone marrow (BM) transfer, 700
Bone marrow transplantation (BMT), 977
Bone morphogenetic proteins (BMPs), 838
Borna disease virus (BDV), 783
BOS/IPS, 977
Botrocetin, 478
Bouquet, 7, 416, 480
Bouquet arrangement, 416, 491
Bouquet-like, 7, 416, 483–487, 491, 502, 503, 961
Bovine CD94, named KLRJ1, 643
Bovine conglutinin, 481, 483
Bovine intelectins (bITLN1 and bITLN2), 450
Bovine LOX-1 (βLOX-1), 738
Bovine L-selectin, 556
Bovine NK gene complex, 643
Bovine serum conglutinin, 978
Bovine SP-A, 506
Bovine SP-D, 531
BPD treatment with inhaled NO, 976
Bradycardia, 9
Brain death (BD), 1012
Brain derived neurotrophic factor (BDNF), 230
Brain during development, 64
Brain enriched hyaluronan binding (BEHAB), 832
Brain link protein-1 (BRAL1), 831
Brain tumor, 20
BRAL2, 832
Bral2 gene, 32
BRAL2 mRNA, 26
Branched carbohydrate, 39
BRCA1-associated protein (BRAP), 201
Breast cancer, 94
Breast carcinomas, 294, 429
Brevican, 802, 806–808, 810, 825–827, 830, 834, 835, 837, 838
Brevican-deficient mice, 835
Brevican isoforms, 32
Broad range of pathogens interacting with surfactant protein (SP)-A and SP-D, 958
Bronchial asthma, 916, 999
Bronchial dysplasia in heavy smokers, 963
Bronchiolar Clara cells, 955
Bronchiolitis obliterans syndrome (BOS), 977
Bronchoalveolar lavage (BAL), 956, 959
Bronchoalveolar lavage fluid (BALF), 426
Bronchopulmonary dysplasia (BPD), 962, 976
Bronchus-associated lymphoid tissue (BALT), 543
Brucellosis, 564
Bryostatin, 566
β-sheet, 475
β-sheet-rich barrels, 248
(S1–S6/S6) β-sheets, 215
β-TrCP, 135
β-trefoil fold, 327
β-trefoil structure, 314, 321, 331
Budesonide, 202
Bullous pemphigoid, 1000
Bunch of flowers, 530
Butyrate, 219

C
CAAT box, 214, 737
CAAT promoter element, 172

Cabbage butterfly *Pieris rapae*, 321
Ca^{2+}-binding sites in CLRD/CTLD, 463, 476
$CaCO_3$ crystal, 853
Cadherin and integrin function, 836
Cadherins, 352
Cadherins as ligands of KLRG1, 699
Cadherins subclasses, 991
    epithelial (E), 991
    neural (N), 991
    placental (P), 991
Caenorhabditis elegans, 148, 213, 445, 477, 891
Calcineurin, 1035
Calcium-dependent neutrophil uptake of bacteria, 540
Calcium-responsive heat stable protein-28 (CRHSP-28), 456
Calicheamicin, 400, 1052
Calmegin, 30, 39, 43
Calnexin (Cnx), 6, 7, 29, 43, 147
Calnexin-t, 30
Calreticulin (Crt), 6, 7, 29, 36, 37, 40, 109, 147, 413, 901
Calreticulin and cardiac pathology, 48
Calreticulin and cell adhesiveness, 43
Calreticulin and steroid-sensitive gene, 45
Calreticulin/calnexin cycle, 29
Calreticulin in signal transduction, 45
Calsequestrin, 46
Calx-b, 922
Calx-b domain, 479, 922
cAMP metabolism, 1035
cAMP responsive element binding proteins (CREBs), 594
Campylobacter jejuni, 365, 399
CAMs as predicators of atherosclerosis, 994
CAMs in allergic inflammation, 998
CAMs in cancer, 1006
CAMs in diabetic patients, 1000
CAMs in gingival crevicular fluid, 1013
Canavalia ensifomist, 5
Cancers of female reproductive tract, 660
Candida albicans, 33, 272, 283, 360, 489, 725, 728, 782, 783, 934
Candida albicans glycans, 277
Canine surfactant SP-A, 507
CANLUC, 43
CA125 ovarian cancer antigen, 222
CAP domain of CAP proteins, 922
Capsular polysaccharide (CPS), 399, 415
CAP superfamily includes
    CRISP LCCL domain containing 1 (CRISPLD1), 921
    CRISP LCCL domain containing 2 (CRISPLD2), 921
    cysteine-rich secretory proteins (CRISPs), 921
    glioma pathogenesis-related protein (GliPR), 921
    Golgi associated pathogenesis related-1 (GAPR1) proteins, 921
    human glioma pathogenesis-related 1 (GLIPR1), 921
    mannose receptor like, 921
    peptidase inhibitor 15 (PI15), 921
    peptidase inhibitor 16 (PI16), 921
    R3H domain containing like proteins, 921
    tomato pathogenesis-related (PR) protein P14a, 921
carbohydrate binding module, 32, 47, 314, 321, 325, 331, 346
Carbohydrate-bound CRD structure, 780
Carbohydrate-independent binding, 271
Carbohydrate recognition domain (CRD), 3, 8, 149, 340, 474, 477, 483, 487, 489, 604, 737, 753, 755, 776, 866, 933
Carbohydrate recognition domain of LOX, 738
Carcinoembryonic antigen (CEA), 607, 780, 785, 1007
Carcinoembryonic antigen (CEA)-related cell adhesion molecule 1 (CEACAM1), 782
Carcinoid tumors, 251

Carcinomas, 970
CARD9, 17, 730
Cardiac (pro)renin uptake, 1045
CARD9-independent, pathways, 728
Cardiomyocytes, 47
Cardiopulmonary bypass, 956
Cardiovascular diseases, 181, 200, 429, 430, 1001
Cardiovascular risk, 163
CARD-protein/Bcl10/Malt1 complexes, 677
Cargo protein proteins of cytosolc sorting (Pacs), 33
Cargo receptor, 32, 57
Cargo receptor complex, 154
Cartilage-derived C-type lectin (CLECSF1), 888
Cartilage link protein CRTL1, 826, 831
Cartilage link proteins, 477
Cartilage proteoglycan, 808
Cartilaginous regions of the embryo, 338
Casein kinase II (CK-II), 73
Caspase activating recruitment domain 9 (CARD9), 17, 728, 730
Caspase independent neutrophil death, 392
Caspase recruitment domain-containing protein 15 (CARD15) and Dectin-1, 731
Catalytic domain, 421
Catalytic domains of triosephosphate isomerase (TIM barrel) fold, 423
Catalytic unit (Gal/GalNAc-T motif), 318
β-Catenin, 44, 47, 278, 851, 859, 1031, 1032
Catfish IntL1, 450
Catfish IntL2, 450
Cathepsins, 296
Cation channel, 1034
    mediating the sperm acrosome reaction, 1034
Cation-dependent mannose 6-phosphate receptor, 65
Cation dependent M6P receptor (CD-MPR), 61, 62, 1040
Cationic liposomes, 1048
Cation independent M6P receptor (CI-MPR), 61, 62, 81
Cation independent receptor (CI-MPR), 1040
Caudate putamen, 828
Caveolin 1, 456
CAZy database (carbohydrate-active enzymes database), 314
C3b, 415
C5b6, 174
C4b-binding protein (C4bp), 167, 174
CBCP, 478, 480, 922
CBCP/Frem1/QBRICK, 478, 480, 922
CBCP/Frem1/QBRICK large proteoglycan, 922
CBM14, 321
CBM13 family, 314
CBP70, 268
CCAAT, 907
CCAAT box-like, 852
CCAAT elements, 811
CCAAT/enhancer binding protein-α (C/EBPα), 220
CCAAT/enhancer-binding proteins (C/EBP), 529, 764
C1 complex, 934
C3 convertase C4b2a, 415
C3 convertase C3bBb, 415, 934
CCP and an EGF domain, 445
CCP (Sushi) domain, 479, 480
cC1qR/CD93, 902, 903
cC1qR/Crt, 415, 496
$CCR7^+$ $CD45RA^+$ and CD11a low phenotype, 698
$CCR7^+$ central memory (TCM) $CD4^+$ T cells, 697
CCV, 94
$CD3^+$, 660
CD7, 199, 223, 224
CD8, 646

Index

CD14, 1051
CD16, 621, 646
CD18, 999
CD19, 219
CD22 (Siglec-2), 15, 352, 354, 361, 366, 778, 1052
CD23, 219, 633, 750, 778, 791
CD25, 544
CD28, 677, 702
CD33 (Siglec-3), 15, 17, 353, 354, 362, 381–383, 386, 399, 1052
CD34, 562
CD35 (CR1), 904
CD40, 628
CD43, 199, 223, 224, 300, 362, 397, 607, 1007
CD44, 477, 567, 792, 805, 815, 819, 820, 825, 881
CD44 (HCELL), 819
CD45, 199, 219, 223, 224, 275, 336, 363, 489, 683, 1052
CD46 (MCP), 904
CD55 (DAF), 904
CD56, 619
CD56(dim), 661
CD58, 778
CD59(HRF), 904
CD66, 268
CD69, 476, 544, 621, 625, 628, 632, 702, 750, 758
CD80, 257, 778
CD83, 257, 778
CD86, 766, 778, 782
CD91, 544
CD93, 901
CD94, 476, 620, 639, 641, 758
CD94 (KLRD1), 642
CD98, 268
CD161, 626
CD161$^+$ (human NKR-P1A), 624
CD205, 343
CD209 (DCSIGN), 773
CD248, 917
CD303, 735
CD158a, 684
CD44: a hyaluronan receptor, 819
CD56$^+$ and CD94$^+$ cells, 651
CD43 and CD45 co-cluster on MDDCs, 234
CD33 (Siglec 3) and CD33-related siglecs, 17
CD4$^+$ and CD8$^+$ T cells, 761, 764
CD94 and NKG2 in CD4$^+$ T cells, 658
CD45 and sIgM, 362
CD22 antibodies as carrier of drugs, 1053
CD43 as a ligand for E-selectin, 606
CD22 as target for therapy, 367
CD43 as T cell counter-receptor for Sn, 356–357
CD158b, 684
CD44 binding sites in, 819
CD94^{bright} and CD94^{dim} cells, 649
CD56^{bright} CD16$^-$ surface phenotype, 650
CD8^{bright} CD56$^+$ T cells, 660
CD8 burst, 224
CD1c$^+$, 767
CD11c, 750
Cdc20, 126
Cdc2/cyclin b1, 45
CD11/CD18, 991
CD16$^+$ CD56$^+$, 660
CD27 CD28, 621
CD40, CD54, CD80, CD83, CD86, and HLA-DR, 257
CD3$^+$/CD4$^+$ cell function, 543
CD18$^{-/-}$CD62E$^{-/-}$ mice, 610
CD4$^+$ CD28$^-$NKG2D$^+$ T cells, 682
CD4$^+$ CD25$^+$ regulatory T (Treg) cells, 234, 558, 567, 685, 686
CD3$^+$ CD56$^+$ T cells, 660
CD4$^+$ CD8$^+$ thymocytes, 226
Cdc4, Grr1, and Met30, 135
CD18-deficient mice (CD18$^{-/-}$ mice), 610
CD34-deficient mice, 563
CD93-deficient mice, 904
CD56^{dim}/CD16$^+$ and CD56^{bright}/CD16$^-$ NK cells, 684
CD94 dimer, 646
CD94^{dim} NKG2C dimers, 658
CD56^{dim} or CD56^{bright} cells, 683
CD2, DNAM-1, 677, 702
CD26/DPPIV, 889
CD1d-reactive, 631
CD1d-restricted T-cell receptor (TCR), 648
CD44E, 802
CD22 expression level on B cells, 361
CD209 family genes in sub-human primates, 773
C3dg, 895
CD69 gene, 633
Cd93 gene, 902
CD209 genetic polymorphism; HIV infectivity, 787
CD44 glycoform, 566
    hematopoietic cell E-/L-selectin ligand (HCELL), 566
CD117^{high} CD94$^-$cells, 646
CD44 in cancer, 820
CD71 in cell death, 275, 276
CD44 in prostate cancer, 820
CD85j$^+$ cells, 657
CD57–KLRG1$^+$ cells mouse CD4 T cell subsets, 697
CD40L, 749
CD209L (DC-SIGNR/L-SIGN/LSECtin), 773
CD62L-deficient mice, 568
CD62L expression, 567
CD40 ligand, 762, 996
CD117^{low} CD94$^+$cells, 646
CD62L$^+$ sub-population of regulatory T, 558
CD22, MAG, and CD33, 355
CD56 marker, 626
CD22$^{-/-}$ mice, 365
CD-MPR, 30, 61, 62, 65
cDNA clones of versican, 813
CD45 negatively regulates galectin-1, 222
CD56$^+$ NK cells, 764
CD94/NKG2, 619, 621, 623, 640, 646
CD94/NKG2A, 639, 648, 650, 652, 657, 658, 661
CD94-NKG2-A/B, 639
CD94-NKG2-A/B heterodimers, 648
CD94/NKG2A-deficient mice, 656
CD94-NKG2A–HLA-E^{VMAPRTLFL} complex, 654
CD94-NKG2A in complex with HLA-E^{VMAPRTLFL}, 654
CD94/NKG2C, 646, 650, 652, 657
CD94, NKG2C/CD159c, 683
CD4$^+$ NKG2C$^+$ cells, 657
CD94/NKG2E, 646
CD94/NKG2 heterodimers, 619, 632, 644
CD94/NKG2-independent pathway, 653
CD94/NKG2 in innate and adaptive immunity, 649
CD94/NKG2 receptors during HIV-1 infection, 656
CD94/NKG2 subtypes on lymphocytes, 659
CD93 or C1q RECEPTOR (C1qRp), 901
CD3 or KARAP/DAP12, 199, 702
Cd93 polymorphism, 906
CD62P/P-selectin, 594
CD94 promoter genes, 643

CD8+ regulatory T cells, 660
CD33-related siglecs 5-11, 1053
CD45RO, 364, 488, 1002
CD33rSiglec binding, 351
CD33-rSiglec gene clusters, 394
CD33rSiglecs clusters, 401
CD44s, and CD44v8-10, 792
CD44 short interfering RNA (siRNA), 606
CD22/Siglec-2, 353
CD14 SNPs, 937
CD spectra of A2t, 460
CD94-T4, 642
CD4 T cells, 402
CD14/TLR4, 743
CD8+ TNF-α, 660
CD45 tyrosine phosphatase, 222
CD44v4, 607, 792, 1007
CD44 variant (CD44v), 1007
CEACAM-18 (CE18), 401
CEA-related cell adhesion molecule 1(CEACAM1), 775, 786
C/EBPb, P-CREB, P-ELK1, EGR1, STAT3, and C/EBPα, C/EBPβ, or C/EBPδ, 529
C/EBP proteins, 164
Cecal ligation and puncture (CLP), 631
C. elegans, 135, 317
CEL-I, 11, 478
Celiac disease (CD), 648
Celiac disease and MBL, 943
Cel-II, 8
Cell-bound siglecs-1,-5,-7, and-9, 399
Cell-cell and cell-matrix adhesion, 1032
Cell cycle arrest, 275
Cell distribution and regulation, 422
Cell homeostasis by galectins, 193
Cell-mediated cytotoxicity against HIV-Infected CD4 T cells
Cell migration, 196, 226
Cell motility, 226
Cell surface expression of CAMs, 595
Cell-surface glycoprotein organization, 196
Cell surface receptors for SP-A, 508
Cell-surface sorting of LOX-1, 739
Cell trafficking, 609
Cellular expression of sialoadhesin, 355
Cellular modulator of immune recognition (MIR), 456
Cellular sources of NKG2/CD94, 643
Central and effector memory T cells, 568
Centromere, 829
Centrosome, 1034
Ceramide, 225
Cerebral arteriovenous malformations (AVMs), 997
Cerebral proteopathies, 856
Cerebroglycan (CBG), 802, 805, 839
Cerebrospinal fluid, 997
Cerebro-vascular disease, CHD, and venous thrombosis, 586
Cervicovaginal lavage (CVL), 934
Cestode *Echinococcus granulosus* (Eg-ppGalNAc-T1), 319
CFTR gene, a chloride ion channel, 958
C. gigas, 445
Chagas disease, 360
Characteristics of annexins, 455
Characteristics of MAG, 368
Characteristics of NKG2D, 667
Characterization of CD93/C1qRP, 902
Characterization of HIP/PAP, 866
Characterization of Sialoadhesin/Siglec-1, 355
Charcot-Leyden crystal (CLC), 203, 205

Charged transmembrane arginine (R), 625
Chemical modification of SP-A, 504
Chemoattractant (FMLP), 568
Chemoattractant activity, 258
Chemoattraction of monocytes/macrophages, 272
Chemokines, 814
Chemotactic function, 513
Chemotaxis, 196, 226, 272, 542
Chemotaxis of AMφ and neutrophils, 515
Chemotaxis of eosinophils, 257
Chemotaxis of phagocytes, 539
Chemotherapy, 948
Chick aggrecan, 812
Chicken cysteine-rich FGF-R, 605
Chicken galectin-14, 206
Chicken galectin-16, 206
Chicken receptor, 83
CHI3L1, 6, 422
    expressed by: synovial cells, chondrocytes and smooth muscle cells, 422
    hepatic stellate cells, 422
    mammalian chitinaselike proteins (CLPs)
        AMCase, oviduct-specific glycoprotein, human cartilage, 422
        chitotriosidase, YKL-39, Ym1, 422
        glycoprotein 39 (HC-gp39) and stabilin-1-interacting (SI)-CLP, 422
    neutrophils, macrophages, fibroblast-like synovial cells, 422
CHI3L4, 6
Chi-lectin (chitinase 3-like-1), 250
Chi-lectins, 6, 8, 421
Chi-lectins as markers of pathogenesis, 428
CHI3L1 or YKL-40, 430
CHI3L1 or YKL-40 as biomarker of solid tumors
    breast cancer, 431
    colorectal cancer, 431
    endometrial cancer, 431
    extracellular, 431
    glioblastoma, 431
    glioma, 431
    hepatocellular carcinoma, 431
    Hodgkin's lymphoma, 431
    kidney tumor, 431
    lung cancer (small cell carcinoma), 431
    malignant melanoma, 431
    metastatic prostate cancer, 431
    multiple myeloma, 431
    myxoidchondrosarcoma, 431
    myxoid liposarcoma, 431
    oligodendroglioma, 431
    ovarian tumor, 431
    ovary, 431
    papillary thyroid carcinoma, 431
    primary prostate cancer, 431
    squamous cell carcinoma of the head and neck, 431
Chimeric galectin, 265
Chimeric NKG2D receptors, 686
Chimeric SP-D/anti-CD89 protein, 978
Chimpanzee brain microglia, 393
Chimpanzee Siglec-L1, 401
CHIT1, CHI3L1, and CHIA, 429
Chitin, 422
Chitin and chito oligosaccharides, 423
Chitinase insertion domain (CID), 425
Chitinase-like lectin (Chilectin), 8, 421–423
Chitinase 3-like-1 protein (CHI3L1) in clinical practice, 432
Chitinases, 421, 422

Chitlac, 236
Chitosan, 236
Chitotriosidase, 421, 422, 425, 426, 429
Chitotriosidase enzyme activity, 429
Chlamydial pathogens, 538
Chlamydia trachomatis, 934
ChNKG2D-expressing T cells, 686
Cholangiocarcinoma (CCA), 681
Cholangiocarcinoma cells, 867
Cholangitis, 1004
Cholesteatoma, 280
Cholesten-5-yloxy-N-(4-(1-imino-2-β-D-thiomannosylethyl)amino) butyl)-formamide (Man-C4-Chol), 1048
Cholesterol 3-sulfate, 249
Cholin-ERGIC neurons, 63
Chondrocyte mineralization, 465
Chondrocytes, 278, 338, 421
Chondroitin, 456, 803
Chondroitinase ABC, 428, 831, 834, 837
Chondroitin sulfate (CS), 455, 801–804
Chondroitin-4-sulfate (C4S), 335, 804
Chondroitin-6-sulfate (C6S), 804
Chondroitin sulfate B, 232, 805
Chondroitin sulfate glycosaminoglycans (CS-GAGs), 825
Chondroitin sulfate proteoglycan 2 (CSPG2) or PG-M or versican, 812, 813
Chondroitin sulfate proteoglycans (CS-PG), 825, 922
Chondroitin4-sulfate proteoglycans (CSPG-I and CSPG-II), 819
Chondrolectin (CHODL), 480, 882
Chondrolectin group, 479
CHOP, 36
Chorioamnionitis, 976
Choroidal neovascular membranes (CNVMs), 593
Chromatin, 167, 173, 174
Chromobindin, 4
Chromosomal mapping, 82
Chromosomal translocation t(12;21), 937
Chromosome 1, 847
Chromosome 1 (1q21-q23), 164, 174
Chromosome 2, 901
Chromosome 3, 830, 850, 883
Chromosome 4, 625
Chromosome 5, 643, 1043
Chromosome 6, 625
Chromosome 7, 725, 888
Chromosome 8, 828
Chromosome 10, 1043
Chromosome 12, 440, 625
Chromosome 14, 530
Chromosome 16, 1027
Chromosome 19, 361, 888, 917
Chromosome 19 (19q12–q13.2), 369
Chromosome 22, 896
Chromosome 25, 528
Chromosome 10 at 10q11.2-q21, 489
Chromosome 6C, 859
Chromosome 16C3, 882
Chromosome 6D1-D2, 754
Chromosome 11 in mouse, 906
Chromosome 2p13, 753
Chromosome 2p35, 754
Chromosome 12p13, 633, 730
Chromosome 20p11.2, 910, 917
Chromosome 17p13.3 in human, 906
Chromosome 2p12 in humans, 847
Chromosome 9p12-p13, 922

Chromosome 12 p13-p12, 633
Chromosome 1q23, 994
Chromosome 1q23-25, 554
Chromosome 1q24-25, 1005
Chromosome 1q24–25, 564
Chromosome 6q26, 70
Chromosome 16q23, 888
Chromosome 19q13.3, 392
Chromosome 19q13.4, 384, 393
Chromosome 19q13.33-41, 390
Chromosome 19q13.41-43, 385
Chromosome 21q21, 882
Chromosome 22q11, 896
Chromosome 22q12, 214
Chromosome 4q33 distal-q34.1, 754
Chromosome 1q21-q24, 997
Chromosome 6q24.2-q31.3, 670
Chromosome 10q11.2-q21, 935
Chromosomes 1, 2, 4, 5, 8, 9, 10 and 15, 457
Chronic allergic inflammation, 964
Chronic allograft failure, 1005
Chronic calcifying pancreatitis (CCP), 854
Chronic cavitary pulmonary aspergillosis (CCPA), 974
Chronic cholitis, 432
Chronic cigarette smoke exposure, 957
Chronic heart failure (CHF), 430
Chronic hepatitis B, 937
Chronic hypersensitivity pneumonitis, 955
Chronic infection, 192
Chronic maternal ethanol consumption, 957
Chronic obstructive pulmonary disease (COPD), 284, 430, 963
    Chronic obstructive airway disease (COAD) or chronic airflow limitation (CAL), 963
    also known as chronic obstructive, 963
Chronic pancreatitis, 854, 855, 867, 868
Chronic rhinosinusitis (CRS), 334, 542, 959, 965
Chronic sialadenitis, 965
Chrp, 268
Cigarette smoke, 957
Cilia and cell cycle, 1034
Ciliary neurotrophic factor (CNTF), 862
Ciliary roles, 278
Ciliary signaling, 1033
Ciliopathies, 1027
Cilostazol, 598, 742
CI-MPR, 81–101, 110, 111, 1040
CI-MPR binds and internalizes LIF, 1045
CI-MPR from opossum liver, 84
CI-MPR in bone cells, 64
CI-MPR in thyrocytes, 64
CI-MPRKO/GAA-KO (double KO) mice, 1041
CI-MPR trafficking, 97, 1041
CINC-2 b, 849
Ciona intestinalis galectin, 192
Circulating L-selectin, 558
Circulating MBL, 934
    higher-order oligomeric structures, 934
Cirrhosis, 861
Cis-acting element, 48
Cis-Golgi network (CGN), 57
Cis-medial-and trans-Golgi, 59
"cis" or "trans," 331
Cisternae, 59
Cisternal/tubular ER
Cis-trans isomerization, 413
C. jejuni LOS, 388, 389

CL-43, 527, 979
CL-46, 527
Clara cells, 539
Clara cell specific protein (CCSP), 529
Classical complement pathway, 170, 173
Classic Hodgkin's lymphoma, 920
Classification of CLRD/CTLD-containing proteins, 477
Classification of C-type lectins, 480
Classification of lectins, 9
Classification of proteoglycans, 802
Clathrin, 73
Clathrin adaptor proteins, 90
Clathrin and dynamin independent, 396
Clathrin-coated pits, 88, 96, 756
Clathrin-coated pits and vesicles, 712
Clathrin coated vesicles (CCV), 59, 60, 90
Clathrin-dependent endocytosis, 882
Clathrin/dynamin-dependent endocytosis, 92
Clathrin-mediated endocytosis, 717
Clearance of alveolar DPPC, 515
Clearance of pathogens, 512
Cleavage of brevican, 832
CLEC-1, 14
CLEC-2, 14, 726, 758
CLEC12, 14
CLEC-18 (CLEC-2), 726, 731
CLEC9A, 14, 758
CLEC9A (DNGR-1), 726, 734
CLEC1A CLEC-1 DC ?, 726
CLEC8A Lox-1, 726
CLEC12A MICL, 726
CLEC7A or CLECSF12, 726
CLEC12B, 668, 758
CLEC12B or macrophage antigen H, 726, 735
CLEC2D (also named lectin-like transcript-1 or LLT1), 628
CLEC-2-deficient mice, 733
CLEC4G (L-SIGN), 791
CLEC-1 gene, 731
CLEC-18 or C-type lectin-like protein-2 (CLEC-2) or CLEC1B, 731
CLEC16p genes, 725
CLECSF7 (CD303), 735
CLECSF8, 14, 757, 758, 767
CLEC-2 signaling, 734
Clenbuterol, 1041
Clinical pancreatitis, 861
Clinical syndrome of preeclampsia, 415
CLL1, 726
CL-46-NCRD, 979
α-Clofibrate, 598
Clone 1B12 gene, 760
Cloning of pierisin-1, 323
Clopidogrel, 996
Closed-bouquet, 507
Clostridium sp. *Pseudomonas aeruginosa*, 934
Clr-b, 628
CLR/CLEC, 474
CLRD fold, 9
Cluster headache, 1006
Cluster of genes on chromosome 19q13.3-4, 392
Cluster of MHC class I related genes, 669
CMC-544, 367
CMP-Neu5Ac hydroxylase, 351
CMV (HCMV) glycoprotein, UL16, ULBP2 and ULBP3, 670
C. neoformans, 336
CNS lecticans, 839

CNX, 109
Coactivator P100, 1033
Coagulation and kinin systems, 908
Coagulation factor G, 331
Coagulation factors, 154
Coagulation factor V and factor VIII, 154, 155
Coagulation factor V/VIII domain in, 446
Coagulation factor XIII, 413
Coagulation process, 462
Coatomer protein complex I (COPI), 59, 60
Coatomer protein complex II (COPII), 59, 60
Cochlear nucleus, 833
Co-clusters CD43/CD45, 234
Cofactor for thrombin, 915
Colitis, 200
Collagen, 335, 344
Collagen clearance cell matrix adhesion, 335
Collagen deposition, 280
Collagen diseases with interstitial pneumonia, 956
Collagen domain (CD), 489, 502
Collagen domain of SP-D, 540
Collagen-induced arthritis (CIA), 281
Collagen-like domain, 483, 490, 903
Collagen-like domain of human SP-A, 503
Collagen like region, 483, 490
Collagen-like triple helical domain, 496
Collagenous domain of C1q (cC1qR), 413
Collagenous region, 485
Collagenous triple helix, 484, 503
Collagenous type of lectins, 480, 483
Collagen receptor complex GPVI-FcR γ-chain, 733
Collagen triple helices, 423, 485, 496
Collagen type 2, 344, 423, 456
Collagen type 1, 2 and 3, 423
Collagen type II binding protein (anchorin CII), 466
Collagen type IV, 344
Collagen types I and IV, 918
Collagen V, 344
Collagen vascular disease-associated interstitial pneumonia (CVD-IP), 955
Collagen vascular disorders, 957
Collectin-based chimeric proteins, 978
Collectin from kidney (CL-K1), 480, 483
Collectin gene locus, 530
Collectin liver 1 (CL-L1), 480, 483
Collectin of 43 kDa (CL-43), 480, 483, 527, 533
Collectin of 46 kDa (CL-46), 480, 483
Collectin placenta 1 (CL-P1), 480, 483
Collectins, 5, 7, 9, 411, 477, 479, 480, 483, 527, 934, 955
Collectins as therapeutic agents, 978
Collectins bind DNA, 510
Collectins family, 473
Colon and other cancer, 1007
Colon carcinogenesis, 226, 251
Colonic lamina propria fibroblasts (CLPFs), 282
Colon neoplastic lesions, 299
Colorectal cancer (CRC), 304, 681, 871, 1009
Colorectal cancer (CRC) cell lines, 220
Colorectal carcinoma, 868, 870
Colorectal CSC, 820
Colorectal tumor cells, 786
Combination of markers, 293
Comparison of AAA with MBL and UEA-1, 441
Competitive inhibitors, 342
Complement-activating component (CRaRF), 494
Complement activation, 411, 946

Complement C1q, 483
Complement factor H 8 complement factor H as a ligand for
    L-selectin, 562
Complement receptor (CR) 1, 901
Complement receptor three (CR3), 14
Complement regulatory protein (H protein), 351
Complement regulatory protein (CRP)-like domain, 812
Complement regulatory protein-like repeat 2 (CR2), 576
Complement regulatory proteins (CRP domain), 556
Complements, 302
Complement system, 493
Complexes with two sialylated carbohydrates, 385
Complex N-glycans, 270
Complex sulphated carbohydrates (e.g., heparin), 883
Concanavalin A, 4, 6, 7, 165
Conceptus, 207
Congenic DA rats, 759
Congenital diaphragmatic hernia (CDH), 545, 976
Congenital disorders of glycosylation (CDG), 36
Conglutinin, 7, 480, 483, 533
Conglutinin gene, 531
    Bos taurus chromosome, 28, 531
Connective tissue disorders, 959
Conserved arginine residue, 383
Contact allergy, 200
Contact hypersensitivity and LCs, 756
Coordinated acquisition of inhibitory and activating
    receptors, 646
COPD as a group of diseases, 963
COPII, 58, 150
COPI vesicles, 58
Core domain and long loop region, 474
Core-fucosylated biantennary, 397
Core protein, 802, 808
C9orf19, 922
Corneal or epidermal wounds, 202
Coronary artery disease (CAD), 992
Coronary artery smooth muscle (CASM) cells, 283
Coronary artery stenosis, 996
Corona virus (CoV), 784
Corona virus infection and SARS, 784
Cortical barrels, 835
Cortical granule lectin (xCGL), 284
Corticotrophin (ACTH), 599
COTRAN 198
Cotreatment of AML cells with VPA and GO, 400
C. pneumoniae, 1013
C1q, 411, 486, 903
C1q and the collectins, 413
C1q-deficient mice, 905
C1q inhibitor (C1qI), 804
C1q-mediated complement activation, 183
C1qR(P), 491
C1q receptor (C1qRp), 901, 905
C1q receptor for phagocytosis enhancement, 901
C1qR for phagocytosis (C1qRp) (or CD93), 508
C1qRO2, 508
C1qRp/cC1qR/CD93; *O*-sialoglycoprotein, 901
    synonyms of C1qR are: C1q/MBL/SPA receptor C1qRp, C1qr1,
        C1qrp, C1qRp, AA4 and CD93, 902
    tissue expression and regulation of CD93/C1qRp, 903
C1qRP/CD93, 902, 905
CR3, 730
Crab-eating macaque serum, 884
Craniopharyngiomas, 251
CRD-4 monomer in MR, 332

CRD of a rat mannose-binding protein, 491
CRD of human Langerin, 753
    in complex with mannose or maltose, 754
CRDs 4–8, 332
CRDs in DC-SIGN and DC-SIGNR, 780
C-reactive protein (CRP), 163–165, 409, 415, 516, 738, 870, 996, 1000
C3 receptors, 413
CRM1 (exportin), 47
Crohan's disease (CD), 282, 428, 870, 1003
Cross-liking of HIV-1 infection, 235
Cross-linking of GM1 ganglioside by Gal-1, 235
Cross-linking properties of galectins, 215
CRP in BAL, 977
Cruciform, 8, 481, 483
C. rynchops, 411
Cryptococcal mannoproteins (MP), 336
Cryptococcus neoformans, 934
Cryptophthalmos, 923
Cryptosporidium parvum, 934
Cryptosporidium parvum enteritis, 943
Crystal analysis of, 461, 622
    annexin A2 as a heparin-binding protein, 461
    CD69, 622
    CD94, 644
    CTLD of Ly49I, 622
    Ly29A, NKG2D and MBP-A, 622
    NKG2A/CD94: HLA-E complex, 653
    sugar-annexin 2 complex, 461
    surface plasmon resonance (SPR), 461
Crystal forms of truncated DCSIGNR, 777
Crystallographic analysis of Ig-V like domain of p75/AIRM1, 388
Crystallographic and *in silico* analysis of sialoadhesin, 357
Crystallographic studies, 532, 535
    trimeric human SP-D neck + CRD domains, 532, 535
Crystals of human PSP/LIT, 853
Crystal structue of human PAP (hPAP), 867
Crystal structure, 173, 447, 491, 883, 1051
Crystal structure for domains of human MAdCAM-1, 564
Crystal structure of CEL-III from *Cucumaria echinata*, 320
Crystal structure of CLEC-2, 733
Crystal structure of CRD1of mouse Gal-4, 246
Crystal structure of DC-SIGN (CD209) and DC-SIGNR (CD299),
    776, 777
Crystal structure of EMBP, 895
    in complex with a heparin disaccharide, 895
Crystal structure of galectin-9, 255
Crystal structure of human annexin A5, 466
Crystal structure of human Siglec-5, 385
Crystal structure of mosquitocidal toxin, 322
Crystal structure of MS-FBP32, 442
Crystal structure of murine Dectin-1, 727
Crystal structure of murine ppGalNAc-T-T1, 318, 320
Crystal structure of NKG2D, 672
Crystal structure of ricin, 315
Crystal structure of siglec-7, 388
Crystal structure of tetranectin, 885
Crystal structure of the extracellular C-type lectin-like domain of
    human Lox-1, 739
Crystal structure of TL5A, 416, 417
Crystal structure of trimeric fragment of hSP-D, 535
Crystal structure of YKL-40, 423
Crystal structure of Ym1, 427
Crystal structures of *B. arenarum* Gal-1, 216
Crystal structures of CRD of M-ficolin, 412
Crystal structures of : EGF-like domains of P-selectin, 581, 582
    bent and extended conformations, 581

Crystal structures of human asialoglycoprotein receptor, 711, 712
    histidine[256], 714
        structure of H1-CRD, 713, 714
        sugar binding site of H1-CRD, 713, 714
Crystal structures of maltose-bound rfhSP-D, 536
Crystal structures of sialoadhesin in complex with sialyllactose, 465
Crystal structures of the CRD of human Langerin, 754, 755
C-set domains, 393
C2-set Ig domains, 383
CSPG, 479, 804, 808, 826, 828, 834, 836, 838, 839
C-terminal GRIP domains, 95
C-terminal region of aggrecan, 810
CTLD (canonical), 479
CTLD/CLRD, 8, 474, 477
CTLDs of MR mediate binding to, 335
    fucose, 335
    mannose, 335
    N-acetylglucosamine, 335
CTLD structure, 476
CTSB, 850
C-type lectin (mMGL), 749
C-type lectin 14A, 910, 920
C-type lectin domain (CLRD), 315, 473, 475, 485
C-type lectin domain family 13 member A (CLEC13A), 920
C-type lectin domains (CLD), 809
C-type lectin fold, 475
C-type lectin-like domains (CTLDs), 3, 8, 315, 473, 727, 753
C-type lectin-like (RCTL) gene, 431
C-type lectin-like molecule-1 (CLL-1), 1053
C-type lectin-like protein-1(CLEC-1) or CLEC-1A, 731
C-type lectin receptor 1 (CLEC-1), 750, 752, 758
C-type lectin-related molecule (Clr), 624
C-type lectins (CLEC), 6
C-type lectins: endocytic receptors, 717, 750, 752
C-type lectins family, 483
C-type lectin superfamily (CLSF) receptors, 757, 759
CUB domain, 479
Cubilin, 91
Cucumaria echinata, 478
Cullin1/Cdc53, 125
Cumulus matrix, 183
Cumulus oophorus, 183
Cushing's syndrome and bone cancer
Cutaneous lymphocyte-associated antigen (CLA), 606, 998
Cutaneous malignant melanoma, 632
Cutaneous T-cell lymphoma (CTCL), 786, 1008
CVID, 894
Cyanobacteria, 4, 1
Cyanovirin-N, 21
Cyclin-dependent kinase inhibitor p21, 253
Cyclin domains, 134
Cyclin D1 promoter activity, 275
Cyclins, 850
Cyclins Cln1p and Cln2p, 127
Cyclodextrin conjugates, 1047
Cycloheximide, 599
Cyclooxygenase (COX)-2 antagonist, 729, 1017
Cyclophosphamide, 217, 219
Cyclosporin A, 1012
Cyclosporin A and transcription factor NFAT, 596
CysR, 344
Cystatin C, 229
Cysteine-rich (CR) domain, 315, 335
Cysteine-rich fibroblast growth factor receptor (FGF-R), 607, 1007
Cysteine rich region (CRR), 489
Cysteine-rich R-type domain, 344

Cysteine-rich secretory proteins (CRISP) antigen 5, 921
Cystein-flanked leucine-rich repeats (LRR)
Cysteinyl isoforms, 504
Cyst formation, 1034
Cystic fibrosis (CF), 35, 124, 849, 869, 934, 958, 964
    lung disease, 945
Cystic renal disease, 1027
Cystogenesis, 1035
Cytokeratin 1, 907
Cytokeratin 19 (CK19), 957
Cytokeratins CBP70, 268
Cytokeratins K8, 1031
Cytokine and chemokine production, 683
Cytokines, 839, 996, 1000
Cytokines and other factors, 685
Cytokines IL-10, 496
Cytolytic activity of HSV-specific memory CD8+ T cells, 657
Cytomegalovirus (RCMV), 631, 778
Cytoplasmic domain of siglec-7, 387
Cytoplasmic ITIM (L/VxYxxL/I/V), 625
Cytosolic ITIMs, 402
Cytotoxic and apoptotic activity in Pierisin, 322
Cytotoxicity of butterflies extracts from, 323
    extracts from: *Appias lyncida, Leptosia nina, Anthocharis scolymus, Eurema hecabe, Catopsilia Pomona, Catopsilia Scylla, Colias erate*, 323
    family Pieridae, Papilionidae, Nymphalidae, Lycaenidae, *Celastrina argiolus, Lycaena phlaeas*, 323
Cytotoxic T cells, 649

D
DAMP, 396
DAP10, 640, 651, 674
DAP-12, 362, 396, 641, 642, 651, 674, 694
DAP10 and DAP12, 667, 674, 675
DAP10-coupled receptors, 676
DA rats, 624
Datura stramonium agglutinin (DSA), 445
DCAL2, 726
DCAR, 14
DC-immunoactivating receptor (DCAR), 758, 762, 764
DC immunoreceptor (DCIR), 750
Dcir[-/-], 764
DCIR (CLECSF6), 14, 750, 758, 762, 763
DCIR and to mouse MCL and Dectin-2, 758
DCIR2, DCIR3, DCIR4 and DCAR1, 758
DCIR deficiency, 763
DCIR, dendritic cell immunoreceptor, 752
DCIR+ T cells, 764
DCL-1 or (CD302), 920
DC maturation, 513
DC-Mtb, 786
DC receptors, 480
DCs form a major resident leukocyte population in human skin, 749
DC-SIGN, 13, 16, 17, 22
DC-SIGN (CD209), 750, 754, 756, 774, 785, 787, 789, 1049
DC-SIGN and DC-SIGNR, 476, 756, 757
DC-SIGN and escape of tumors, 785
DC-SIGN as receptor for viruses, 783
DC-SIGN+ DCs, 781, 784
DC-SIGN-mediated targeting, 1050
DC-SIGN polymorphisms, 788
DC-SIGN recognizes pathogens, 782
DC-SIGN related, 756
DC-SIGNR/L-SIGN (or CLEC4M), 13, 777

DC-SIGN, SIGNR1, SIGNR3 and Langerin, 783
DC-SIGN supports immune response, 781
DC-SIGN vs. DC-SIGN-related receptor, 777
DDR1, 325, 327
DDR2, 327
DDR1b, 325
Deacylated ManLAMs, 967
DEC-205 (CD205), 331, 342, 750, 753, 774, 1049
DEC-205-associated C-type lectin (DCL)-1, 895, 920
Decidua, 651
Decidua: BDCA-1$^+$ CD19$^-$CD14$^-$myeloid DC type 1(MDC1), 767
Decidual dendritic cells, 781
Decidual NK cell receptors, 650
Decidual/placental NK cell receptors in pregnancy, 683
DEC-205 mediated cancer immunotherapy, 1049
DEC-205/MR6-gp200, 315
Decorin, 341, 533, 801, 802, 839
Decorin-like molecules, 834
Dec oy receptor for NKG2D ligands, 668
Dectin-1 (CLEC7A), 14, 725–727, 729, 750, 757, 758, 761, 762, 774
Dectin-2, 14, 774
Dectin-2 α, β and γ isoforms, 750
Dectin-1 and TLR2, 730
Dectin-1 cluster, 14, 734
Dectin-2 cluster in natural killer gene complex (NKC), 757
Dectin-1 c.714TG polymorphism, 731
Dectin-2, DCIR, DCAR, 757
Dectin-1-deficient mice, 730
Dectin-1early stop codon polymorphism Y238X (c.714TG, rs16910526, 731
Dectin-1 in leukocyte interactions, 730
Dectin-1-mediated Ca^{2+} flux, 729
Dectin-1 polymorphism c.714TG, 730
Dectin-1 receptor family 34 Dectin-1 Y238X polymorphism, 731
    CLEC-1, CLEC-2, CLEC12B, CLEC9A and myeloid inhibitory C-type lectin-like receptor (MICL), 757
    Dectin-1, LOX-1, C-type lectin-like receptor-1, 757
Dectin-1-Syk-CARD9, 728
Defense collagens, 413, 486
Degradation, 362
Degranulation of human eosinophils, 542
Degranulation of porcine neutrophils, 233
Dementia, 1005
Demethylating stimulus, 219
Demyelinating brain lesions, 372
Denatured and aggregated forms of CRP (neo-CRP or modified CRP), 170
Dendritic cell-associated lectin-2 (Dectin-2) (CLECF4N), 258, 759
Dendritic cell-associated lectins, 750
Dendritic cell immunoreceptor (DCIR), 642, 757
Dendritic cell lectin (DLEC), 752
Dendritic cell lectin (DLEC; BDCA-2), 1049
Dendritic cells (DCs), 332, 749, 1046
Dendritic cell-specific ICAM-grabbing nonintegrin (DC-SIGN), 774
Dendritic cell-specific ICAM-3 grabbing non-integrin (DC-SIGN), 473
Dendritic cell-specific transmembrane protein (DC-STAMP), 117
Dengue virus, 182
Dengue virus envelope glycoprotein, 784
Densa of epithelial basement membranes, 922
15-Deoxy-δ12,14-prostaglandin, 598
Dermal dendritic cells, 342
Dermal disorders, 998
Dermal eosinophilia, 258
Dermatan sulfate (formerly called a mucopolysaccharide), 805, 810, 1044
Dermatitis (DE), 959

Dermatitis herpetiformis (DH), 1000
Dermatomyositis (DM), 959
Dermatophagoides farina, 999
Dermatophagoides pteronyssinus (Der p)-sensitized BALB/c mice, 965
DES-induced renal tumors, 301
Desmosomal cadherins, 1032
Desmosomal junctions (DJs), 1032
Desmosomes, 1027
Detergent solubilized MAG or L-MAG, 368
Deuterostomes, 433
Developmental expression, 530
Dexamethasone, 283, 819, 853, 860, 965
Dexamethasone (Dex), 426
Dexamethasone-treated, 334
δFosB, 230
d-galactopyranose configuration, 252
DGCR2, 479, 480
Diabetes, 169, 272, 430
Diabetes mellitus type 1 (type 1diabetes or T1DM), 1000
Diabetic heart, 1001
Diabetic kidney disease, 945
Diabetic-nephropathy, 945
Diabetic NOD mice, 858
Diabetic retinopathy, 1001
Diabetogenesis, 281
Diacylglycerol production, 219
Diallelic polymorphism, 1030
Diapedesis, 1009
Dicentrarchus labrax, 442
Dictyocaulus filaria, 450
Dictyostelium discoideum, 324–326, 445
Diester linkage (PDE), 67
Diethylstilbestrol (DES), 47
Differences between DC-SIGN and DC-SIGNR/L-SIGN, 778
Differentiation of hematopoietic lineage, 226
Differentiation of myogenic lineage, 226
Diffuse large B-cell lymphoma (DLBCL), 294
Diffuse panbronchiolitis, 937
Diffuse parenchymal lung disease (DPLD), 955
DiGeorge syndrome critical region gene 2 (DGCR2), 478, 896
1,2-Dihexadecyl-sn-glycero-3-phosphocholine (DPPC-ether), 515
Di-leucine (LL) motif, 71–74, 777
Di-Mannoside clusters, 339
Dimer formation, 255
Dimeric CG-16, 228
Dimeric form of Gal-1, 225
Dimeric protein, 232
Dimmers, 191
Di-N-acetylchitobiase, 422
Dioleylphosphatidylglycerol (DOPG), 515
Dipalmitoylphosphatidylcholine (DPPC), 511, 515, 517
Dipalmitoylphosphatidylcholine-egg phosphatidyl-choline, 507
Dipeptidyl peptidase IV (DPPIV) activity, 888, 889
Diphtheria toxin, 323
Direct targeting, 1046
Disaturated phosphatidylcholine (DSPC), 976
Discoidin domain (ds) (or f5/8 type C domain, or C2-like domain), 324
Discoidin domain and carbohydrate-binding module, 324, 325
Discoidin domain receptor proteins, 324
Discoidin domain receptors (DDR1 and DDR2), 326
Discoidin domains, 324, 326
    barrel-like structure, 324
    3-strand antiparallel β-sheet, 324
    5-strand antiparallel β-sheet, 324
Discoidin I-like domain (DLD, or D), 326

Discoidins, 6
Discoidins from *Dictyostelium discoideum* (DD), 325
Discoidins I and II, 325
Disease-causing mutations, 1030
    in PKD1 gene, 1030
Disorders of glycosylation (CDG), 114
Disorders of immune system, 660
Dissemination of T-lymphoma cells, 566
Distant metastasis, 868
Distearoylphosphatidylcholine (DSPC), 511
Distribution of polycystins, 1031
Disulfide bonding pattern, 911
Disulfide bonds, 480
Disulfide bonds in lectins and secondary structure, 478
Diubiquitin (UBD), 274
Divergence among galectins, 193
Diversity of NKG2D ligands, 668
DNAM-1, 621, 677, 686
DNA (cytosine-5-) methyltransferase 1 (DNMT1), 456
DNA polymorphism in ERGIC-53/LMAN1 genes, 155
DNA polymorphism in MI, 997
DNA polymorphisms in SPs, 972
DNAX accessory molecule-1 and 2B4, 682
DNAX activation protein (DAP)12 and DAP10, 373
DOCK180, 99
Dock and lock (DNL) method, 367
Dodecamers, 530
Dolichos biflorus, 21
Domain architecture of endosialin, 918
Domain architecture of vertebrate CTLD, 479
Domain composition of human E-selectin, 594
Domain of human Gal-4, 247
Domain organization of DC-SIGN and DC-SIGNR, 778
Domain organization of selectin-P, 576
Domain organization of thrombomodulin, 912
Domains in polycystin-1, 1029
$(\beta/\alpha)_8$ Domain structure, 423
Domain structures of lecticans, 808
Domain structures of M-ficolin, 412
Domain-swapped dimer structure, 333
Donor compound-7 (NOC-7) and NOC-8, 247
Dorin M, 416
Dorsal rhizotomy, 231
Dorsal root entry zone (DREZ), 834
Dorsal root ganglia, 229
Dorsal root ganglion (DRG), 830, 858
Dorsal root ganglion (DRG) neurons, 372
Double-stranded RNA (dsRNA), 254
Doublet p33/41 protein, 463
Doxorubicin, 283, 1045
DPPC, 506, 510, 512
DPPC liposomes, 511
1D1 (one of the CS-PG) proteoglycan, 598
D. pteronyssinus, 999
DRG neurons, 864
DrGRIFIN, 216
Drosophila lysosomes, 119
Drosophila melanogaster, 432, 477
Drosophila PC-2, 1034
Drug ezetimibe (Zetia), 170
DS47, 422
3-D structure of SP-A trimer, 305
d-talopyranose configuration, 252
Duck DCIR, 763
Dukes' C and D tumors, 299
Dust mite allergens, 965

E

E. aerogenes, 540
Early embryo development, 427
Early endosomal recycling exit, 98
Early endosomes (EE), 64, 95, 147
Early-onset ischemic stroke, 917
Early secretoty pathway (ESP), 57, 150
Early vertebrate, 433
Earth worm (EW)29 lectin, 327
Ebola infections, 233
Ebola virus, 778, 791, 792
Ebola virus glycoprotein (EBOV-GP), 792
E-box elements, 828
E-cadherin, 621
Ecalectin (galectin-9), 254, 258
ECM tenascin-R, 838
Ectocytosis, 267
Ectopic expression of *Reg* gene, 849
EDEM 1, 23, 137
EDEM 2, 138
EDEM 3, 138
EDEM 3', 32
EDEM variants, 110
Eel, 3
Eel aggutinins, 6
Effect of hyperoxia, 220, 545
Effects of exogenous SP-D or rfhSP-D, 978
Effects of O3, 545
Effects of SP-A and SP-D on transplants, 977
Effects of SP-A or-D deficiency, 966
Effects of TGF-β and IFN-γ on E-selectin expression, 1016
Efficiency of the diagnosis for Parkinson's disease, 888
Efficiency of vaccination, 1049
EGF domain, 479, 480
EGF-receptor family: ErbB1, ErbB2, ErbB3, ErbB4, 839
EGF receptors (EGFRs), 97
EGF repeats, 554
Eglectin (XL35), 6, 450, 451
ELAM-1/CD62E, 593
Elastase and/or proteinase-3, 959
Electron microscope images of soluble TM, 914
Electron microscopy, 357
Electrophoretic mobility shift assay (EMSA), 196, 224
Electrostatic potential, 485
Elegans and Xenopus, 148
Elements conserved in PKD1 promoters: AP2, E2F,E-Box, EGRF, ETS, MINI, MZF1, SP1, and ZBP-89, 1028
Elevated triglycerides, 281
EMBP, 476, 479
EMBP homologue (EMBPH), 894
Embryogenesis, 6, 65, 227
Embryonal carcinoma, 217
Embryonic epidermal blistering, 923
Emphysema, 957, 964
Emphysema-like pathology, 980
Emphysematous alveolar destruction, 430
Emp46/47p, 157
Enalapril and losartan, 1017
Endo180, 331, 335
Endo180 (or urokinase plasminogen activator receptor-associated protein), 315, 750
Endo-β-N-acetylglucosaminidase F, 65
Endo-β-N-acetylglucosaminidase H, 65
Endo-β-1,4-xylanase 10A (Xyn10A), 321
Endocrine pancreas, 847
Endocytic capacity, 233

Endocytic machinery, 88
Endocytic mechanisms of CD22, 366
Endocytic properties, 342
Endocytic receptor CD91, 396, 413
Endocytic receptor on immature dendritic cells, 343
Endocytic receptors, 9, 91, 316, 477
Endocytic recycling compartment (ERC), 1041
Endocytosis, 65, 66, 353, 362, 392, 396, 399
Endocytosis motif, 73
Endocytosis of dipalmitoylphosphatidylcholine (DPPC), 516
Endocytosis of glycoproteins, 340
Endocytosis targeting, 86
Endogenous factors regulating CAM genes, 599
Endogenous lectins, 1046
Endoglycan (EG), 607
Endometrial epithelial marker, 257
Endometriosis, 661
Endosialin (CD248), 901, 910
Endosialin (tumor endothelial marker-1), 901, 917
Endosialin gene, 917
Endosialin in astrocytic and metastatic brain tumors, 920
Endosialin in cancer, 919
Endosialin interactions with ligands, 918
Endosomal carrier vesicles, 147
Endosomal exit site, 73, 94
Endosomal-lysosomal (EL), 81
Endosomal pH, 717
Endostatin fold, 475
Endothelial cell adhesion molecule (ELAM), 555
Endothelial cell motility, 279
Endothelial heparan sulfate, 560
Endothelial lectin, 6
Endothelial-leukocyte adhesion, 595
Endothelial-leukocyte adhesion molecule 1(ELAM-1) or leukocyte-endothelial cell adhesion molecule 2 (LECAM2), 576, 593
Endothelial nitric oxide synthase 1, 67
Endothelial sLe^a and sLe^x during cardiac transplant rejection, 560
Endothelial surface of blood vessels, 916
Endotoxemia, 868
Endotoxic shock, 167
ENDO 180 (CD280)/uPARAP, 343
End stage renal disease (ESRD), 1001
Energy metabolism, 888
Engagement of CD94/NKG2-A by HLA-E, 651
Enhanced phagocytosis of *M. tuberculosis* by SP-A, 966
Ensheathing cells, 247
Enterochromaffin-like (ECL) cells, 849
Enterococcus spp., 934
Enterocyte like HT-29 cells, 250
Enterocytes, 246
Enthalpy (□H), 21
Entropy (□S), 21
Envelope glycoprotein E2, 937
Environmental influences on SP-D genes, 545
Enzymatic properties of pierisin-1, 324
Enzymatic sources of ROS, 994
    cyclooxygenase, 994
    cytochrome p450s, 994
    hemoproteins, 994
    lipoxygenase, 994
    NADH/NADPH oxidases, 994
    NO synthase, 994
    peroxidases, 994
    xanthine oxidase, 994
Enzyme replacement therapy (ERT), 1040
Eosinophil, 203, 894

Eosinophil associated gastrointestinal (EGID) diseases, 1003
Eosinophil chemoattractant (ECA), 206, 252, 258
Eosinophil granule major basic protein 2 (MBP2), 11, 479
Eosinophilia, 965
Eosinophilic mucus (EMCRS), 542
Eosinophil major basic protein (EMBP), 479, 480, 894
Eosinophil protein, 192
Eosinophils IL-8, 514
E-, P-, and L-selectin, 992
Ependymomas, 15
E/P/I$^{-/-}$, 610
Epidermal eosinophil, 258
Epidermal growth factor (EGF)-like domain, 553
Epidermal Langerhans cells, 342
Epithelial growth factor receptor, 456
Epithelial implants of pseudomyxoma peritonei (PMP), 870
Epithelial mucin MUC-1, 397
Epithelial tumor cell lines (MCF7 and BeWo), 223
Epithelium and connective tissue, 202
E/P/L/I$^{-/-}$, 610
Epratuzumab (e-mab), 367
Eps15 homology domain-containing protein-1 (EHD1), 99
Epstein-Barr virus (EBV), 259
Epstein-Barr virus differs with HCMV, 658
Epstein-Barr virus (EBV)-infected B cells, 658
ERAD pathway, 133
ERAD substrates, 115
ER α-mannosidase, 138
ER α1,2-mannosidase I (ERManI), 137
ER-associated degradation (ERAD), 51, 109, 123, 124, 128, 133
ER-associated degradation-enhancing α-mannosidase-like proteins, 135
ER degradation enhancing α-mannosidase-like protein (EDEM), 123, 135, 136
ER exit motif (KRFY), 158
ER-exit sites (ERES), 60
ERGIC-53, 6, 12, 30, 32, 57, 59
ERGIC-53 like (ERGL), 145, 147
ER glucosidase II β-subunit, 110, 111
ER-Golgi intermediates (ERGIC), 57, 59
ERK-1/2, 509
ERK2, 677
ERK and phosphatidylinositol 3-kinase signaling cascades, 252
ERK1/2, p38, and JNK, 277
ER mannosidase I (ER Man I), 147
ERp57, 29, 31, 33, 39, 40, 46, 51, 57, 72
ER retention signal (HDEL), 113
ER-retrieval signal, 151
Erythema nodosum (EN), 995
Escherichia coli, 5, 489, 540, 934
E-selectin, 593, 594, 1000, 1005
E-selectin$^{-/-}$, 606, 609
    consensus repeat domains, 608
    consensus repeat domains (Lec-EGF-CR6), 609
    crystal structure of E-selectin lectin-EGF, 608
    E-selectin: an asymmetric monomer, 608
    ligand-binding region of human E-selectin, 609
    modulation by metal Ions, 609
    sE-selectin complement regulatory domains, 608
    three-dimensional structure, 609
E-selectin (CD62E), 28
E-selectin also binds 562PSGL-1, 562
E-selectin and CD18 Integrin, 610
E-selectin fragment, 475
E-selectin gene regulation, 594
    activator protein-1 (AP-1), 594

E-selectin gene regulation (cont.)
  cAMP responsive element binding proteins, 594
  by double-stranded RNA, 595
  induction of E-selectin and associated CAMs by TNF-α, 595
  inflammatory triad of IL-1, TNF-α and LPS, 594
  nuclear factor kB (NF-kB), 594
  by phorbol ester (phorbol 12-myristate 13-acetate (PMA), 595
  transcriptional regulation of CAMs, 594
E-selectin genomic DNA, 593
  4 inverted CCAAT box, 594
  LAM-1/L-selectin, 594
  murine E-selectin, 594
  NF-ELAM1/ATF element, 595
  NF-kB 594AP-1-binding sites, 594
  porcine gene, 594
  TATAA, 59
E-selectin in cancer cells, 1007
E-selectin in diapedesis of cancer cells, 1009
E-selectin-induced angiogenesis, 1007
E-selectin ligand-1, 604
E-selectin ligands, 602
  carbohydrate ligands (Lewis antigens), 602
  4-fucosyl-lacto (Le^a), 604
  3-fucosyl-neo-lacto (Le^x), 604
  Lewis A (Le^a), Lewis B (Le^b), and sialyl Lewis A (sLe^a), sialyl-Lewis^x (sLe^x) s-diLe^x, 602, 603, 608
  neoglycoconjugate HSO3LeAPAA-sTyr, 603
  3'-sialyllacto-N-fucopentaose II (3'-S-LNFP-II), 603
  3'-sialyllacto-N-fucopentaose III (3'-S-LNFP-III), 603
  sulfated polysaccharide ligands, 603
  transcriptional control of expression of carbohydrate ligands, 604
E-selectin-mediated progenitor homing, 605
E-selectin or ELAM-1 (CD62E), 994
  binding elements in E-selectin promoter, 600
  E-selectin promoter phased-bending of E-selectin promoter, 602
  transcription factors stimulating *E-sel* gene, 600
    activating transcription factor-2 (ATF-2), 601
    activator protein-1 (AP-1), 600
    CRE/ATF element or NF-ELAM1, 601, 604
    CREBs, 600
    NF-ELAM1 and NF-kB, 600
    NF-k-binding sites in human *E-sel*, 601
    NF-k-specific promoter, 601
E-selectin promotes growth inhibition and apoptosis of hematopoietic progenitor cells, 589
E-selectin receptors on human l neutrophils, 604
ESL-1, 605, 1007
Esophageal squamous cell carcinoma (ESCC), 298, 850, 1008
Estrogen, 47, 463
Estrogen receptor α (ERα), 47
ET-1, 1002
Ethanol in utero, 957
Etiology of RDS, 972
ETS2, 865
E3 Ub ligase, 362
Eumelanin, 893
Euroglyphus maynei, 999
Ever-expanding Ly49 receptor gene family, 620
Evolution of galectins, 207
Evolution of mammalian chitinases (-like) of GH18 family, 432
Evolution of Siglec family, 401
Evolution of the C-type lectin-like domain, 477
EW29, 315, 327
Exocrine pancreas, 847
Exocytosis, 462
Exosite 1 and exosite 2, 912

Exosites 1 and 2 in thrombin reaction, 912
Experimental abdominal aortic aneurysm (AAA), 568
Experimental autoimmune encephalomyelitis (EAE), 235, 1006
Experimental autoimmune neuritis, 630
Experimental autoimmune uveoretinitis (EAU), 359
Experimental polymicrobial peritonitis, 946
Export signal, 151
Expression of CD62L, 566
Expression of E-selectin and associated CAMs by IL-4, 596
Expression of mahogany gene, 891
Expression of PAP, 867
Expression of SP-A, -B, and -C in lung explants byIL-1 and glucocorticoid, 539
Extended loops, 333
Extended neck regions of DC-SIGN and DC-SIGNR, 780
Extracellular matrix (ECM), 801, 825, 835
Extracellular signal-regulated protein kinase (ERK1/2), 611
Extracytoplasmic region, 89
Extra-hepatic ASGP-R, 716
Extra-hepatic sources of CRP, 171
Extrasynaptic and perisynaptic regions of the muscle, 232
Extravillous trophoblast (EVT), 236
Ezrin, radixin, and moesin (ERM), 819, 881

F

FA5/8, 326
Fabry disease, 1006
Factors V and-VIII, 325, 326
Factor VIII binding to vWF, 575
Factor XII, 908
2F1-AG *M. fascicularis* and *M. Mulatta* NK cells, 701, 702
Falciparum malaria, 1014
Familial hypercholesterolemia (FH), 994
Familial interstitial lung disease, 959
Families 18 and 19 of glycosyl hydrolases, 421
Family 18 glycosyl, 421, 423
Family of CD93, 901
Family 18 of glycosyl hydrolases, 421
F4/80 and CD11b, 749
Farnesyl cystein carboxymethylester, 227
Fas (CD95)/caspase-8-mediated cell death, 224
Fasciculation, 247
Fatal asphyxiation, 972
FBG1, 132
FBG3, 129, 130
FBG4 and FBG5, 129, 130
FBG domains, 410, 411
Fbl (or FBXL), 126
F-box lectins, 6, 8, 30
F box motif, 126
F-box protein, 125
F-box protein 2 (FBXO2), 132
  D-mannose, 440
  FBP affinity, 440
  for Fuc-BSA, 440
  galactosyl-BSA, 440
  L-fucose, 440
  N-acetylglucosaminyl-BSA, 440
FBP32, 442
FBP from European seabass, 442
Fbs, 32
Fbs1, 6, 123, 125, 129
Fbs2, 129, 130
Fbs1, Fbs2, Fbg3, 6
Fbs1 neural F-box protein (NFB42), 132

Fbw (or FBXW), 126
Fbx2, 132
Fbx17, 129
Fbx (or FBXO), 126
FcεRIs, 695
FcγR, 175
Fcγ receptors, 750, 1049
FcγRI, 168, 174
FcγRI-dependent phagocytosis, 233
FcγRII, 168
FcγRIII, 174
FcγR route, 729
FCN1, FCN2 and FCN3 genes, 410
FcRγ adaptor protein, 625
FcRγ chain, 14
FD1 (M-ficolin FBG domain), 412
Features of inhibitory receptors, 352
Fecal PAP/Reg-III, 859
Female fertility, 183
Female protein (FP), 163, 176, 177
FERM domains, 820
Ferric nitrilotriacetate (Fe-NTA), 35
Fertilization, 6, 13, 427
Fetal development, 1032
Fetal NK cells, 640
Fetuses with chorioamnionitis, 962
FGF-2, 829
FGF-R E-selectin ligand, 607
Fibrillar aggregates, 856
Fibrillar collagen types I and III, 865
Fibrillar structures upon tryptic activation, 856
Fibrillin-1, 814
Fibrin, 916
Fibrinogen, 916
Fibrinogen β/γ (homology) (FBG) domain, 409
Fibrinogen C domain containing 1 (FIBCD1), 413
Fibrinogen domain, 14, 451
Fibrinogen-like domain (FBG), 409
Fibrinogen-like domain (FD1), 409
Fibrinogen-like (FBG)-domains, 409
Fibrinogen-related domain (FReD), 413
Fibrinogen type lectins, 14, 409
Fibrinogen-type lectins, 6, 14
Fibrinolysis, 917
Fibrinolytic marker, 887
Fibroblastoid cells, 15
Fibroblasts and chondrocytes, 344
Fibroblasts MPRs, 63
Fibrocystin, 1031
Fibromodulin, 801, 802
Fibronectin (FN), 167, 222, 253, 479, 814, 827, 835, 865, 918
Fibronectin fibrillogenesis, 303
Fibronectin type II domain, 315, 316, 332, 340, 480
Fibronectin type II repeat, 334
Fibrous atheroma, 996
Fibulin-1, 814, 836
Fibulin-2, 814, 836
Ficolin-1 (M-ficolin), 410
Ficolin-2 (L-ficolin), 410
Ficolin-3 (H-ficolin or Hakata antigen), 410
Ficolins (FCNs), 6, 8, 14, 409, 486
Ficolins vs. collectins, 409
Fine-needle aspiration biopsy (FNAB), 292
Fish and amphibians, 442
Fish Siglec-4, 373
Fivefold symmetric β-propeller protein, 416

Five-stranded (F1–F5) and six-stranded (S1–S6a/S6b), 215
Five stranded (F1–F5) β-sheets, 255, 256
Flower-bouquet-like octadecameric structure, 483
5-Fluorouracil (5-FU) resistance, 871
FnIII repeats 3–5 of tenascin-R, TN3–5, 811
Foam cells, 742
Focal PTC-like immunophenotypic changes in HT, 294
Folic acid, 997
Follicular adenomas, 292, 293
Follicular carcinomas, 292, 293
Formation of folds and vesicles by DPPC monolayers, 506
Forms of the CI-MPR, 87
Formyl peptide receptor, 456
Forssman pentasaccharide, 255
Four-domain tandem F-type lectins, 439
Fragments of neurocan, 829
Fras1, 922, 924
Fraser syndrome, 923
Fraser syndrome-associated proteins, 924
Frem1, 480, 922, 924
Frem2, 924
Frem3, 922, 924
Frem2, a gene related to *Fras1* and *Frem1*, 924
Frog oocyte cortical granule lectins, 450, 451
Frog pentraxin, 440.
F-type domain, 442, 446
    coagulation factor V/VIII domain in, 446
    Saccharophagus degradans 2–40, 446
    from *Solibacter usitatus* Ellin6076, 446
F-type domains in copies of, 443
    1 (*X. laevis* X-epilectin), 443
    2 (*X. tropicalis* IIFBPL), 443
    3 (*X. tropicalis* III-FBPL), 443
    4 (*X. laevis* II-FBPL), 443
    5 F-type domains, 443
F-type lectin (DlFBL), 442
F-type lectin fold, 439
F-type lectin from the Japanese horseshoe crab *Tachypleus tridentatus*, 444
F-type lectin-like proteins in, 439
    invertebrates (e.g. the horseshoe crab *Tachypleus tridentatus*), 439
    vertebrates (e.g. the Japanese eel *Anguilla japonica*), 439
F-type lectins (Fuco-lectins), 6, 8, 20, 439
    in bacteria, 446
    from *Drosophila melanogaster*, 445
    in fish and amphibians, 441
    in fungi, 447
    in invertebrates, 444
    in mammalian vertebrates, 439
    in plants, 445
    in sea urchin, 445
Fucoidan, a sulfated fucopolysaccharide, 460
Fuco-lectin from *Aleuria aurantia* (AAL), 6, 447
Fucose-binding proteins (FBP), 440
    in boar spermatozoa, 440
    in mammamlian spermatozoa, 440
    oviductal sperm in cattle (Bos Taurus), 440
Fucose binding proteins/lectins in fertilization, 440
Fucose binding sites, 440
α1-3(4) Fucosylation, 570
Fucosyltransferase-7, 565
α1-3 Fucosyltransferase, 9, 17, 560
α3-Fucosyltransferase, 603
Fugu and zebrafish, 373
Functional overlap/divergence among galectins, 193
Functional polymorphism (rs4804803), 791

Function of REGIA (Reg Iα) in Pancreas, 850
Functions of annexin A2, 461
Functions of annexins, 458
Functions of attractin, 890
    multiple functions, 890
Functions of BDCA-2, 766
    IFN-α production, 766
Functions of brevican and neurocan, 835
Functions of CD22, 365
Functions of CD94/NKG2, 648
Functions of CHI3L1 (YKL-40), 428
    anti-catabolic effect, 428
    cancer cell proliferation, 428
    fibrosis, 428
    growth factor for fibroblasts, 428
    inflammation, 428
    regulation of cell-matrix interactions, 428
    solid carcinomas, 428
    stimulates migration, 428
    tissue remodelling, 428
Functions of chondroitin sulphate proteoglycans, 834
Functions of CLEC-2, 733
Functions of collectins, 486
Functions of DC-SIGN, 781
Functions of DEC-205, 343
Functions of Dectin-1, 729
Functions of EMBP, 895
Functions of Langerin, 755
Functions of layilin, 881
Functions of LOX-1, 740
    adhesion of mononuclear leukocytes, 740
    binds with bacterial ligands, 741
    cell adhesion, 740
    platelet endothelium interaction, 740
    promote atherosclerosis, 740
    release of endothelin-1, 740
    role in apoptosis, 741
Functions of MAG, 370
Functions of mannose receptor, 336
    adaptive immunity, 336
    antigen processing, 336
    cell migration, 336
    intracellular signalling, 336
    pattern recognition receptors, 336
    phagocytosis, 336
    presentation, 336
    role in immunity, 336
Functions of MBL, 492
Functions of M-type sPLA2 receptor, 342
Functions of NKR-P1, 628
Functions of PAP, 863
Functions of P-selectin, 580
    emigration of neutrophils, 581
    leukocyte and platelet rolling, 580
    migration of T lymphocyte progenitors, 581
    peripheral neutrophilia, 581
    P-selectin-deficient mice, 581
    P-selectin mediates adhesion of leukocytes, platelets, and cancer cells, 581
Functions of PSP/LIT, 854
Functions of sialoadhesin, 358
Functions of Siglec-7, 389
Functions of tetranectin, 886
Functions of TM: coagulation, embryogenesis, anticoagulant process as endocytic receptor, protease-activated receptor (PAR)-1, protein C activation, 914
Furin, 64

G
GAA-KO mice, 1041
GABAergic neurons, 832
G-actin helicase, 456
γ-adaptin ear domain, 95
GAG binding region, 814
Galactose, 192, 198
Galactose-specific C-type lectins, 479
Galactose-specific proteins, 198
β-Galactosidase, 1043
β-Galactoside, 232, 2221
Galactosides and lactosides as inhibitors of Gal-1 and-3, 305
Galactosylated liposomes, 1051
Galactosylceramide, 510
Galactosylsphingosine, 511
Galactosyltransferases, 19
Gal-3/AGE-receptor 3, 280
Gal-3 : a molecular switch from N-Ras to K-Ras, 305
Gal-3 and CD44v6, 293
Gal-3 and HBME-1 markers, 293
Gal-3 and obesity, 281
Gal-1 and the immune system, 232
Gal-3 as anticancer drug, 283
Gal-3 as a pattern recognition receptor, 283
Gal-3 as marker of cancer, 291
    an adjuvant marker for follicular carcinoma, 292
    anaplastic astrocytomas (grade 3), 295
    anti-apoptotic activity, 291
    anti-apoptotic effects, 274
    astrocytic tumors, 295
    benign tumors, 295
    B-small lymphocytic lymphoma (B-SLL), 297
    Burkitt lymphoma (BL), 296
    diffuse large B-cell lymphoma (DLBCL), 204, 206
    distinction between follicular carcinoma and adenoma, 291
    follicular lymphoma (FL), 296
    galectins and glioma cell migration, 296
    Gal-3 ligands in meningiomas, 296
    glioblastomas (grade 4 astrocytomas), 295
    higher in papillary carcinomas with metastases, 291
    Hodgkin's lymphomas, 297
    human gliomas, 295
    immune-cell neoplasms, 296
    lower in lymph node metastases of medullary thyroid carcinoma, 291
    MALT lymphoma, 297
    marginal zone lymphoma (MZL), 297
    marker for thyroid cancer 291
    multiple myeloma (MM), 296
    oncocytic cell adenomas (OCAs), 293
    oncocytic cell carcinomas (OCCs), 293
    oncocytic cell tumors (OCTs), 293
    primary effusion lymphoma (PEL), 296
    pro-apoptotic, 274
    RNA processing and cell cycle, 291
    tumors of CNS, 294, 295
Gal-1 binds to laminin cellular fibronectin thrombospondin plasma fibronectin vitronectin osteopontin, 222
Galectin-4, 191
Galectin-7, 202
Galectin-8, 251, 253
Galectin-9, 251, 255, 257
Galectin-10, 203
Galectin-11, 192
Galectin-13 (placental protein-13), 192, 205
Galectin 14, 206
Galectin-15, 206

Galectin-9 as a ligand for TIM-3, 256
Galectin-4 as lipid raft stabilizer, 249
Galectin-9, a tandem-repeat type galectin, 254
Galectin-2-binding protein, 201
Galectin-1 gene, 213
Galectin-glycan lattices, 196
Galectin-1-induced apoptosis, 224
Galectin-4 induced apoptosis, 250
Galectin-1 in fetomaternal tolerance, 235
Galectin-related inter-fiber protein (GRIFIN), 192, 204, 205
    DrGRIFIN, 205
    grifin/galectin-11, 204
    GRIFIN in zebrafish (Danio rerio)(DrGRIFIN), 205
    guinea pig lens GRIFIN, 205
    lens crystallin protein GRIFIN, 205
Galectin-related protein (GRP), 216
Galectins, 6, 251
    hinge region (short linker peptide sequence), 245
    prototype galectins (galectin-1,-2,-5,-7,-10,-11,-13,-14, and -15), 245
    single chimera-type galectin (galectin-3), 245
    tandem-repeat type galectins (galectin-4,-6,-8,-9, and -12), 245
Galectins-3 (formerly known as CBP35, Mac2, L-29, L-34, IgEBP, and LBP), 265
Galectins have multiple functions, 192, 193
    angiogenesis, 197
    cell-extracellular matrix interactions, 192, 193
    cell-to-cell interactions, 192, 193
    intracellular signaling, 192, 193
    posttranscriptional splicing, 192, 193
    receptor crosslinking or lattice formation, 192, 193
    regulate endothelial cell motility, 197
    regulate inflammation, 192
    role in chemotaxis, 196
    transcription processes, 192
Galectins, integrins and cell migration, 297
Galectins 3-4, intelectin at brush border of small intestine, 249
Galectin-3 structure, 265
Gal-4 from rat intestine, 247
Gal 1 from toad (*Bufo arenarum Hensel*) ovary, 216
Gal-3 in breast epithelial-endothelial interactions, 294
Gal-3 in CNS, 279
Gal-3 in distinguishing gliomas, 295
Gal-1-induced apoptosis, 232
Gal-3 inhibits anticancer drug-induced apoptosis, 300
Gal-3 inhibits apoptosis in bladder carcinoma cells, 301
Gal-1 in IMP1 deficient mice, 216
Gal-1 in its oxidized form, 231
Gal-3 in melanomas, 297
Gal-3 in metastasis, 302
Gallbladder carcinoma, 991
Gal-1 receptors: CD45, CD7, CD43, CD2, CD3, CD4, CD107, CEA, 222
Gal-3 regulates JNK gene in mast cells, 277
Gal-3 thyrotest, 292
Gal-3 up-regulates MUC2 transcription, 299
Gamma glutamyl transpeptidase, 1004
Gangliocytomas, 15
Ganglioside GM1, 893
Gangliosides, GD1α and GT1β, 370
Gangliosides GT1b, GQ1b, GD3, GM2, GM3 and GD1a, 398
Gangliotriaosylceramide, 510
GANL, 443
Gastric adenocarcinoma, 871
Gastric cancer, 301, 304, 681, 849, 851, 871
Gastric carcinoma, 847, 868

Gastric diseases, 1003
Gastric intestinal metaplasia, 871
Gastric mucosa, 199
Gastric mucosal cells, 849
Gastrin, 849
Gastrin-17 (G17), 1007
Gastroesophageal reflux disease, 971
Gastrointestinal mucosa, 955
Gastrointestinal SRCC, 871
GATA-2, 737
GATA (a zinc-finger transcription factor), 647
GBM, 918
GBS, 399
gC1q-binding proteins (gC1qR), 903
gC1qR, 902, 908
gC1qR (gC1qBP), 906
gC1qR (p33, p32, C1qBP, TAP), 902, 906, 909
g.-247C/T polymorphism, 429
GD3 ganglioside, 398
GD3, GQ1b and GT1b, 398
γδT cells, 657
Gefitinib, 971
Gemin-2, 268
Gemin-3, 268
Gemin4, 223, 268
Gemtuzumab, 400
Gemtuzumab (Mylotarg), 367
Gemtuzumab ozogamicin (GO), 400, 1052
Gene encoding human annexin A5, 464
Gene knock-out mice models, 527
Gene order from centromere to telomere: of regenerating genes, 847
Gene polymorphism in L-selectin, 555, 564
Gene polymorphisms in E-selectin, 994
Gene polymorphisms of
    P and selectins, 993
    platelet endothelial cell adhesion molecule (PECAM) 1, 993
    transforming growth factors (TGF) β1 and 2, 993
    tumor necrosis factors (TNF) α and β, 993
Generalized functions of collectins, 486
Gene structure of MBP, 489
Gene therapy, 1040, 1051
Genetic and phenotypic studies of *mahogany/attractin* gene, 892
Genetic factors, 170
Genetic factors as lung cancer risk, 970
Genetic polymorphic variants of, 958
    SNPs, 958
    SP-A1, SP-A2, SP-B,SP-C, and SP-D, 958
Genetic polymorphism, 789
Genetic polymorphism in SP-A: associated with development of COPD, 963
Genetics of collagenous lectins, 486
Genetics of Frem1 in Fraser syndrome, 923
Gene variations in autoimmune diseases, 361
Genistein, 275
Genitourinary epithelium, 955
Genomic arrangement of P-selectin gene, 995
Genomic organization of human SP-D, 528
Genomic polymorphism of DC-SIGN (CD209), 787
Germ cells, 65
Gestation, 227
Gestational diabetes mellitus, 937
GGA1, 2, 3, 95
G1, G2, and G3 globular domains, 808, 812, 818
GH47 proteins, 135
Giant cell arteritis (GCA), 1006
Giant mucin-like glycoprotein, 223

GI tumorigenesis, 871
Glc1Man9GlcNAc2 (G1M9), 112
GlcNAc-1-phosphotransferase, 1042
Glial and neuronal cells express CI-MPR, 1042
Glial fibrillary acidic protein (GFAP), 835
Glial scar and CNS repair, 826, 838
Glial tumors, 832
Glioblastoma cells, 43, 833, 856
Glioblastoma multiforme (GBM), 1046
Glioblastomas, 15
Gliomas, 832
GliPR, 922
Glipr1l1, 922
Globotetraosylceramide (Gb4), 322
Globotriaosylceramide (Gb3), 322
Globular conformation of the mannose receptor, 344
Globular head of C1q (gC1qR/p33), 901
Glomerulonephritis, 171, 174, 195
1, 3-β-D-D-Glucan, 505
Glucanase-like domain, 315
Glucan phosphate (GP), 728
β-Glucan receptor (Dectin 1), 725
Glucocorticoid receptor, 46
Glucocorticoid responsive element, 828, 862
Glucocorticoids (GCs), 45, 47, 490
β-Glucon, 726
α-D-Glucopyranoside, 5
Glucose-regulated protein 78 (Grp78), 37
Glucose transporters, 47
α-Glucosidase (GAA) (rhGAA), 1040
Glucosidase I (GI), 32, 109
Glucosidase II (GII), 32, 109, 147
Glucosylceramide, 511
Glucuronic acid, 803, 804
Glutamate, 229
Glutamate activation, 839
GlyCAM-1, 559, 561, 570
β Glycan, 802, 805
Glycan-dependent leukocyte adhesion, 559
Glycan inhibitors of MAG, 372
Glycans as endocytosis signals, 717
Glycans interacting AAA, 441
Glycan specificity of MAG, 369
Glycan structures, 220
Glycated hemoglobin, 281
Glycocalyx, 167
Glycoconjugate binding specificities of siglecs, 397
Glycodelin A (GdA), 607, 1007
Glycoform of CD34, 562
Glycogen synthase kinase 3, 178, 278
Glycolipids, 3, 631
Glyco-peptidolipids (GPLs), 336
Glycophosphatidylinositol (GPI) anchor, 832
Glycoprotein 2, 456
Glycoprotein-340, 508, 509
Glycoprotein D 93 glycoprotein ligands, 199
Glycoprotein (GP) Ibα, 732
Glycoprotein 90 K/MAC-2BP, 222
Glycoprotein transferase (UGGT), 32
Glycosaminoglycan chain attachment region, 809
Glycosaminoglycan network, 477
Glycosaminoglycans (GAGs), 6, 59, 222, 455, 801–803
Glycosidases, 19
Glycoside hydrolase 18 (GH18), 421
Glycoside hydrolase family 18 proteins, 421
Glycosidic activity, 425

Glycosidic bond, 804
Glycosphingolipid/cholesterol-enriched membrane microdomains, 249
Glycosphingolipid fraction, 783
Glycosylasparaginase (GA), 1041
Glycosylphosphatidylinositol (GPI), 805, 831
Glycosylphosphatidylinositol anchor, 830, 907
Glycosyltransferase 1 (GT1) motif, 318
Glycosyltransferases, 19
Glypiated brevican, 831
Glypiation signal, 830
Glypican, 341, 802, 805
GM-CSF, 726, 763
GM-CSF + IL-4, 335
GM1 ganglioside, 222, 223, 235
GM2 ganglioside (GM2), 1043
GM2 gangliosidosis, 1043
GM2/GD2 synthase, 370
Gm$^+$-pneumonia group, 956
GM receptor common β-chain (βc) genes in mice [GM-/-, βc-/-], 960
Goblet cells, 246, 869
Golgi α-mannosidase I (GolgiManI), 135
Golgi-associated retrograde protein (GARP) complex, 95
Golgi complex, 59
Golgi glycosyltransferases, 147
Golgi-localising, gamma-adaptin (GGA), 61
Golgi-localising, gamma-adaptin ear domain homology, ARF-binding proteins, 61, 74
Golgi neurons, 826
Golgi stacks, 57
Gonadotropin-stimulated ovaries, 862
GOX0967, 446
GOX0982, 446
GP-2, 462
GP39 (HC-gp39/YKL-40/CHI3L1), 421
GPI-anchored, 810
GPIb, 733
GPIb-IX-V complex, 575
GPIIb/IIIa, 1012
GPIIb/IIIa-fibrinogen bridges, 993
G-protein-coupled mechanism, 81
G-protein-coupled melanocortin-4 receptor (MC4R), 890
GPVI, 733
GQ1b, 398
Graft-*vs.*-host disease (GVHD), 679
Graft-*vs.*-leukemia (GVL), 679
Gram-negative bacteria, 416
Gram-positive and gram-negative respiratory pathogens, 965
Gram-positive and-negative bacteria in humans, 410
Gram-positive bacteria, 862
Granular gland cells, 248
Granule major basic protein, 894
Granule membrane protein 140 (GMP-140), 576
Granulocyte-macrophage colony-stimulating factor (GM-CSF), 960
Granzyme B, 62, 87, 100
Graves' disease (GD), 630, 1002
Grb2, 667, 676
Great apes (chimpanzees, bonobos, gorillas and orangutans), 402
Griffithsin, 21
GRIFIN (galectin-related inter fiber protein), 204
GRIFIN homologue in zebrafish, 216
GRIP1, 924
Group B Streptococcal capsular sialic acids, 399
Group B *Streptococci*, 365, 399
Group V C-type lectin-like receptors, 757
Growth and mobility factors, 829

Growth factor (EGF), 62, 97, 839
Growth factors βFGF, 871
Growth hormone (GH), 491
Growth stimulating effect, 428
Grp94, 37
γ-secretase complex, 32
GTPase activating proteins, 96
Guillain-Barre syndrome (GBS), 946
Guinea pig CRP, 172
Guinea pig pulmonary SP-A, 507
Guinea pig SAP, 172

H
H60, 621, 668, 669
HA-binding proteins, 831
Haemophilus influenza, 168, 934
Haemophilus influenzae (NTHi), 730
Hairpin, 40
Hamster FP, 177
Hamster homologue to HIP/PAP, 863
HAM/TSP patients, 789
HAPLN2, 831
HAPLN3, 831
HAPLN4, 831
Haplotypes of NKG2D, HNK1 (high NK activity) and LNK1 (low NK activity), 681
Hashimoto's thyroiditis (HT), 292, 294, 1002
HA synthases (HASs), 826
HB-GAM, 829
hCD33, mCD33, 395
HC gp-39, 425, 426
HCMV glycoprotein, 668
hDectin-2, 760
H2-deficient mice (MHL-2$^{-/-}$), 715
Head and neck carcinoma (HNSCC), 298
Head and neck domains, 540
Heart attack mechanisms, 164
Heart disease, stroke, hypertension, 169
Heart failure (HF), 944
Heart transplant recipients, 946
Heat shock, 153, 671
Heat shock protein 70 (Hsp70), 659
HECL (HGMW-approved symbol CLECSF7), 735
Hedgehoglike assembly, 322
Helical filaments of Alzheimer's disease, 855
Helicobacter pylori, 5, 282, 283, 561, 604, 782, 1013
Helicobacter SP-D-deficient (SP-D$^{-/-}$) mouse model, 538
Helminth parasites, 894
Hematopoietic cell E-/L-selectin ligand (HCELL) (CD44), 606
Hematopoietic progenitor cells (HPC), 610
Hematopoietic stem cell precursor (HSPC159), 216
Hemicentin1 (Hmcn1), 924
Hemitripterus americanus, 478
Hemocytes and hemolymph plasma, 416
Hemodialysis, 1005
Hemodynamic effect, 1001
Heparanase, 89, 91, 94
Heparanase, a β-D-endoglucuronidase, 94
Heparan sulfate, 423, 455, 621, 804, 808, 816, 828, 829
Heparan sulfate proteoglycans (HSPG), 802, 805, 839
Heparin, 455, 575, 621, 802, 804, 912
Heparin-binding site in TN, 885
Heparin-induced thrombocytopenia, 1012
Hepatic asialoglycoprotein receptor, 473
Hepatic endothelial cells, 332

Hepatic lectins, 642
Hepatic stellate cells (HSC), 282, 1007
Hepatitis, 861, 1004
Hepatitis and liver damage, 715
Hepatitis B, 763
Hepatitis B virus, 934
Hepatitis B virus (HBV)-infected, 1004
Hepatitis C virus (HCV), 631, 778, 782, 792, 909, 934, 937
Hepatitis C virus (HCV) and other viral pathogens, 783
Hepatoblastomas, 859
Hepatocarcinoma (HCC), 851, 859, 861, 867, 868
Hepatocarcinoma-intestine pancreas (HIP), 865
Hepatocellular carcinoma, 301, 716
Hepatocystin (β-subunit of glucosidase II), 114
Hereditary inclusion body myopathy (IBM2), 922
Hereditary multiple exostoses (HME), 802
Herpes simplex-2, 934
Herpes simplex virus (HSV), 62, 751, 753, 784
Herpes simplex virus 1 (HSV-1) UL9 protein, 132
Heterophilic protein-protein recognition, 352
Heterotetrameric adaptor protein 60, 3
Heterotrimeric G-proteins, 1032
Hexameric NKG2D–DAP10 receptor complex, 676
Hexameric NKG2D-DAP12 receptor complex, 676
β-Hexosaminidase A (HexA), 1043
β-Hexosaminidase A (HexA; αβ) and B (HexB; ββ), 1043
H-ficolin (Ficolin-3), 409–413
HGMW-approved symbol, 620
HHL1, 11
HIF-1α or C/EBPα, 220
High-altitude pulmonary edema (HAPE), 973
High endothelial venules (HEVs), 554, 992
High-mannose moieties, 775
High mannose oligosaccharides, 133
High-mobility group box-1 protein, 35
High molecular weight Kininogen, 908
High-resolution crystal structures, 135
HIPK2-and p53-induced apoptosis, 276
HIP/PAP, 850, 860
Hippocampal neurones, 839
Hippocampal NG2, 837
HIP similarity to PAP, 865
Histiocytosis X (Langerhan's cell histiocytosis), 756
Histones, and small nuclear ribonucleoproteins, 173
Histoplasma capsulatum, 545
hITLN-1, 449
HIV, 84, 785, 7751
HIV-1, 360, 599, 731, 934
HIV-2, 934
HIV and hepatitis C virus, 778, 792
HIV envelope glycoprotein, gp160, 49
HIV gene, 599
HIV-1 gp120, 360
HIV-1 gp120 and other viral envelope glycoproteins, 783
HIV-1, HCV, Ebola virus, CMV, dengue virus, SARS coronavirus, 782
HIV infection, 492
HIV-1 infection, 788
HIV-1 transmission, 756
HKH20, 908
HL-1, 6, 449
HL-2, 449, 450
HLA and Fcγ receptor (FCGR)
HLA-C, HLA-B, HLA-A alleles, 620
    lead to NK-mediated target cell lysis, 620
HLA class I heavy chain (HC), 51

HLA class I molecules, 42
HLA-DR3-positive individuals, 791
HLA-E, 640, 648, 651–654, 693
HLA-E (Qa-1), 639
HLA-E as ligand, 651
HLA-E/G-nonamer complex, 652
HLA-G1, 652
HL-60 cells and human placenta, 214
HMGB1 protein, 35, 1009
HMG I(Y), 601
HMVEC chemotaxis, 1007
Hodgkin's disease, 1008
Holotoxins, 313
Homodimeric Gal-2, 265
Homodimeric human galectin-1, 215, 265
Homologous to mannosidase I (HTM1), 138
Homology to thrombomodulin, 917
Homotrimeric structure, 414
Horseshoe crab, 3
    host defense, 416
    tachylectins, 415
Horseshoe crab *Limulus polyphemus*, 176
Host defense against *P. aeruginosa*, 541
House dust mite (HDM), 999
H. pylori, 934
    infection, 849
    related chronic gastritis, 943
H. pylori-associated diseases, 849
H. pylori isolates, 561
    interact with L-, E-and P-selectins, 561
H-Ras-Gal-1 interactions, 227
H-Ras-guanosine triphosphate (H-Ras-GTP), 227
Hrd3p, 118
Hrd1p/Hrd3p ubiquitin-ligase complex, 138
Hrd1p ligase complex, 118
Hrs and ESCRT proteins, 99
hSiglec-8, 398
hSiglec-3 and hSiglec-9, 402
hSiglec-10 ortholog, 394
hSiglecs-5,-7, and-9, 399
HSP70, 726, 736
Hsp47 procollagen chaperon, 216
HSV 1, 93
HTLV-1-associated myelopathy, 682
Htm1p, 32
Human activated regulatory CD4$^+$ CD25^{hi} T cells (Treg), 274
Human aggrecan, 808
Human aggrecan gene, 809
Human ANXA1 and ANXA2, 457
Human articular cartilage, 425
Human ASGP-R, 709
    H2 transcripts: H1 and H2, 711
    two subunits, HHL1 and HHL2, 709
Human autosomal recessive polycystic kidney disease, 115, 278
Human β-glucuronidase (GUS), 1043
Human breast carcinoma, 294
Human cartilage glycoprotein-39 (HUMGP39), 424
Human cartilage 39-kDa glycoprotein (or YKL-39)/(CHI3L2), 425
Human CD33 (Siglec-3), 381
Human CD94, 645
Human CD161 (NKR-P1A), 628
Human CD33 (Siglec-3): a myeloidspecific inhibitotry receptor, 381
Human CD62L, 558
Human CD44 link domain, 474
Human CD-MPR, 65
Human cervical cancer, 660

Human chi-lectin YKL-39 (CHI3L2), 421
Human cholesteatomas, 218
Human chromosome 3, 883
Human chromosome 6, 669
Human chromosome 10, 486, 502
Human chromosome 19, 368, 393 , 401
Human chromosome 12 a, 667
Human chromosome 1 bands 21–24, 576
Human chromosome 1p13, 426
Human chromosome 2p12, 847, 866
Human chromosome 12 p13, 620
Human chromosome 12p47, 735
Human chromosome 20p13, 355
Human chromosome 12p12.3-p13.2, 620, 727
Human chromosome 12p12-p13, 757
Human chromosome 1q31, 426 ., 832
Human chromosome 16q22-23, 605
Human chromosome 19q13.1–3, 355
Human chromosome 22q12, 214
Human chromosome 1q31–q32, 422
Human CLECSF8, 767
Human colon and ovarian Xenografts, 919
Human colonic and rectal tumors, 849
Human colorectal carcinoma, 493
Human coronary artery SMCs (HCASMCs), 543
Human cytomegalovirus (HCMV), 657, 681
Human Dectin-2 (hDectin-2), 760
Human diploid fibroblasts (HDF), 35
Human early activation antigen (CD69), 619, 632
Human endometrial carcinoma (EC), 660
Human E-selectin, 886
Human eye: versican, 806
Human fertilization, 904
Human fibrinogen γ fragment, 413
Human ficolins (L-, M and H-ficolins), 409, 414
Human galectin-4 by NMR, 247
Human gastric cancer, 323
Human gastric diseases, 849
Human gC1qR/p32 (C1qBP) gene, 907
Human gene TPX-1, 922
Human glioblastomas, 833
Human glioma cells, 225
Human hepatic carcinomas, 217
Human hepatocellular carcinoma, 251
Human hepatointestinal pancreatic/pancreatitis associated protein (HIP/PAP), 848
Human herpes virus 8, 784
Human HL-1/intelectin and HL-2, 451
Human ITLN-1 (hITLN), 449
Human keratinocytes, 336
Human Langerin (CD207), 753
Human lithostathine, 886
Human locus 14q21-22, 266
Human Lox-1 CTLD homodimer, 740
Human lung adenocarcinoma, 680
Human lung carcinoma, 881
Human lymphocyte proliferation, 513
Human macrophage mannose receptor (MMR) CD206, 332
    clathrin-dependent endocytosis, 332
    CTLDs 4 and 5, 332
    C-type lectin-like domains (CTLD), 332
    cysteine rich (CR) domain, 332
    fibronectin-type II (FNII) domain, 332
    macrophage MR (MMR), 332
    mannose recognition, 332
    phagocytosis of nonopsinized, 332

Human MAdCAM-1 gene, 563
Human mannan-binding protein, 487
Human MBP, 886
Human MBP1, MBP2, 480
Human melanomas, 680
Human MOLT-4, 223
Human muscular dystrophy, 226
Human nasal polyps, 202
Human neuronal pentraxin 2 (NPTX2), 178
Human NK cell gene locus (NKC), 620
Human NK complex on chromosome 12, 731
Human NKG2-A,-B, and-C, 640
Human NKR-P1A (CD161), 626
    cellular localization, 626
      human chromosome 12, 626
      mouse chromosome 6, 626
Human pancreatic juice (PSP S2-5), 854
Human parturition, 519
Human ppGalNAc-T isoform, GalNAc-T3, 317
Human PTX3 gene, 180
Human pulmonary SP-D, 528
Human Reg gene family, 847, 859
Human Reg protein, 850
Human rhinovirus (HRV), 999
Human siglec-3, 394
Human siglec-5, 383, 391
Human siglec-7, 395
Human siglec-9, 395
Human siglec-10 (mSiglec-G), 396
Human siglec-14, 383
Human siglec-like molecule (Siglec-L1), 355
Human SP-A: domain structure, 502
Human SPAG11B isoform D (SPAG11B/D), 891
Human sperm-associated antigen 11 (SPAG11), 891
Human T-cell leukemia virus type I (HTLV-I), 258
Human T cell lymphotropic virus type 1 (HTLV-1), 682, 789
Human testis specific glycoprotein (TPX-1), 922
Human T leukemia cells, 217
Human T lymphoblasts, 607
Human TN, 883
Human type II transmembrane, 509
Human U373 and U87 glioma cells, 889
Human ULBP2, 682
Human umbilical vein endothelial (HUVE) cells, 595
Human visceral fat, 449
Humoral immune responses L-Sel$^{-/-}$ mice, 557
Huntingtin-interacting protein-related (HIP1R), 61
Hurthle cell carcinomas, 292
HUVEC, 599, 908
H60 variants (H60b and H60c), 669
HVEM, 621
Hyalectans (lecticans), 9, 807, 825, 827, 838
Hyaluronan (hyaluronic acid), 423, 477, 802–804, 807, 825–827, 829, 837
Hyaluronan and proteoglycan binding link protein gene family (HAPLN), 831
Hyaluronan, type I collagen, 814
Hyaluronic acid (HA), 423, 477, 802, 804
Hyaluronic acid link protein (LP), 831
Hydrated polyhydroxylated glycan, 20
Hydrocortisone, 283, 1018
Hydrodynamic studies, 777
Hydrofluoroalkane-beclomethasone dipropionate (HFABDP), 964
Hydrolases, 61
Hydrolytic enzymes, 1040
Hydrophilic proteins, 955

Hydrophobic proteins, 955
Hydroxy fatty acids, 511
Hydroxyflavones and flavanols, 1017
Hypercholesterolemia, 281
Hypergastrinemia, 849
Hyperhomocysteinemia, 997
Hyperinsulinemia, 893
Hyperoxia-induced injury, 545
Hyperplastic nodules, 293
Hyperpnea, 972
Hypersensitivity pneumonitis (HP), 955
Hypertension, 998
Hypertensive glomerulosclerosis, 742
Hypertensive SHR-SP/Izm rats, 738
Hyperthermia, 671
Hypocretin neurons, 178
Hypomyelination, 889, 891
Hypomyelination in rat zitter, 893
Hypoxia-inducible factor 1 (HIF-1), 117, 126
Hypoxia-like tissue injury, 372
Hypoxia-like white matter damage in brain, 371
Hypoxic-ischemic brain injury in perinatal rats, 837
Hypoxic neuronal injury, 178

I

IBD, 731
(GP) Ib-IX-V complex, 575
ICAM-1$^{-/-}$, 567
ICAM-1 (CD54), 170, 554, 564, 567, 594, 611, 774, 994, 996, 998–1000, 1002, 1005
ICAM-1(Kilifi), 1014
ICAM-2, 593, 594
ICAM-3, 774, 998, 999, 1002
ICAM-2 and ICAM-3, 773
ICAM-1 genes, 597
ICAM-3 genes (p13.2–p13.3), 563
ICAM1 G241R in celiac disease, 1003
ICAM-1 mRNA, 599
ICAMs, 996
ICRF; regenerating islet-derived 1α, 850
Idd13 locus, 906
Ideal biomarker in COPD, 964
Idiopathic pulmonary alveolar proteinosis, 960
Idiopathic pulmonary fibrosis (IPF), 430, 955, 957
α-(L)-Iduronidase, 1042
Iduronate (IdoA), 803
IFN, 362, 680
IFN-γ, 281, 335, 358, 542, 648, 764, 767, 849
IFN-γ-and LPS-induced, 514
IgA, 494
IgA nephropathy, 415, 494
IgE, 268
IgE receptor (CD23), 633
Igf2 and Igf2r genes, 82
IGF-I (IGF1), 81
IGF-II (IGF2), 81
IGF-II/CIMPR, 30
IGF2/MPR, 81
IgG, IgM, IgE and secretory IgA, 543
Ig-like domain, 479
Ig-like transcripts (ILTs), 632
Ig receptors (FcγR), 167, 268
IgSF, 352
Ihara's epileptic rat (IER), 828
IL-1, 97

IL-2, 513, 543, 643, 680, 685, 728
IL-3, 749, 763
IL-4, 335, 338, 426, 542, 726, 760, 763
IL-5, 542
IL-6, 97, 496, 728, 764, 767, 791, 849, 852, 853, 858, 860, 862
IL-8, 170, 334, 515, 543
IL-10, 234, 281, 392, 517, 648, 686, 726, 728, 760, 763, 767, 781, 849, 862, 864
IL-11, 360
IL-12, 651, 728
IL-13, 542, 720, 728, 763
IL-15, 643, 651, 685, 699
IL-17, 281, 728
IL-18, 97
IL-21, 648, 685
IL-23, 728
IL-27, 234
IL-1α, 496, 763
IL-1β, 496, 791
ILD due to inhaled substances, 957
Ileal Peyer's patch of lambs, 860
IL-2, IL-4, IL-5, IL-10, IL-12, IL-13, IL-17A, 849
IL-4/IL-13/STAT-6/PPARγ axis, 728
IL-1 receptor, 496
IL-22/REG Iα axis, 851
IL-6-response element, 851, 862
IL-1, TNF, and LPS, 994
Imaginal disc cells, 428
Imaginal disc growth factors, 422
Immature thymocytes, 224
Immune complexes (IC), 567
Immune escape activities of DC-SIGN, 785
Immune evasion mechanisms, 680
Immune functions of SP-A and SP-D, 512
Immunity in liver, 656
Immunodeficiency virus, 631
Immunogenic peptides, 1053
Immunoglobulin binding protein (BiP), 156
Immunoglobulin-like transcript 2 (ILT2 or LIR1 orCD85j), 693
Immunoglobulin-like transcripts (ILTs), 619
Immunoglobulin superfamily (IgSF), 352, 991
    intercellular adhesion molecules (ICAMs), 991
    leukocyte function antigen-2 (LFA-2 or CD2), 991
    leukocyte function antigen-3 (LFA-3 or CD58), 778, 991
    mucosal addressin cell adhesion molecule-1 (MAdCAM-1), 991
    platelet-endothelial cell adhesion molecule-1(PE-CAM-1), 991
    vascular adhesion molecule-1(VCAM-1), 991
Immunological synapse, 676
Immunoreceptor tyrosine-based activation motif (ITAM), 167, 651, 674
Immunoreceptor tyrosine based inhibitory motif (ITIM), 167, 353, 381, 651
Immunoreceptor tyrosine-based switch-like motif (ITSM), 384
Immunoregulatory roles of galectins, 250
Immunoregulatory roles of SP-A and SP-D, 964
Impaired primary T cell responses in L-Sel$^{-/-}$, 557
Imprinting of mouse Igf2/Mpr or Igf2r gene, 82
Inactivating clotting factors Va and VIIIa, 917
Indoleamine 2,3-dioxygenase (IDO), 685
Indomethacin, 1017
Inducible transcription factors (NF-kB, Egr-1, AP-1), 633
Induction of p27, 227
Inflammation in hereditary diseases, 1006
Inflammatory bowel disease (IBD), 200, 250, 277, 428, 568, 730, 791, 864, 869, 1003
Inflammatory conditions, 428, 1013

Inflammatory cytokines, 729
Inflammatory diseases, 364
Inflammatory triad, 994
Influenza A, 934
Influenza A virus (IAV) infection, 13, 175, 658
Influenza infection, 631, 660
Infrared spectroscopy, 172
INGAP gene, 872
INGAP peptide, 863, 872
INGAP-related protein, 872
Inherited demyelinating diseases, 364
Inherited neuropathies, 360
Inhibitor of crystal growth, 854
Inhibitor of p38 signaling, 257
Inhibitors of gene transcription, 1015
Inhibitors of PI3-kinase activity, 1045
Inhibitors of regeneration of myelin, 372
Inhibitory and activatory signals, 642
Inhibitory immune receptors 619KLRG1, KLRB1 and LAIR-1, 619
Inhibitory NK cell receptors recognize self MHC class I, 622
Inhibitory potency of monosaccharides, 339
Inhibitory receptors in viral infection, 656
Inhibitory receptors, NKR-P1B and NKR-P1D, 625
Innate and adaptive immunity, 233
Innate immunity, 182, 539
Inositol phosphatase (SHIP), 384
Inositol phosphate production, 219
Inotuzumab ozogamicin, 367, 1052
Insects venom allergens (Ag5, Ag3), 922
Insect venom Ag-5 protein, 921
Insulin, 81
    receptors, 456
Insulin-dependent diabetes mellitus (IDDM), 1000
Insulinlike growth factor-I (IGF-I), 428, 491
Insulin-like growth factor II mRNA-binding protein 1, 216
Insulin-like growth factor 2 receptor (IGF2R), 81
Insulin-like growth factors, 81
Insulin-like growth factor, type 1 (IGF-1), 856
Integral membrane protein, deleted in DiGeorge syndrome (IDD), 896
Integrin A4, 456
Integrin B5, 456
Integrin β1, epidermal growth factor receptor, and PSGL-1, 815
Integrin LFA-1, 773
Integrins, 196, 197, 222, 253, 268, 352, 881
    trafficking of, 197
Integrins αL, αM, and β2, 1003
Intelectin (ITLN), 249, 448
Intelectin-1 (endothelial lectin HL-1 or Xenopus oocyte lectin), 448
    human intelectin (hITLN-1), 448
    intelectin-1 (ITLN-1), 448
    lactoferrin (LF) receptor (R) (LFR), 448
    omentin of adipose tissue Xenopus oocyte lectin XL-35, 448
Intelectin-2, 448
Intelectin-2 (endothelial lectin HL-2), 449
Intelectin-3, 449
Intelectins, 6, 8, 448
Intelectins in fish, 450
Interaction between human NK cells and stromal cells, 683
Interaction of A2t with heparin, 460
Interaction of calreticulin with C1q, 44
Interaction of SP-D with *M. tuberculosis*, 968
Interactions between galectins and integrins
Interactions of Annexin A5 with, 464
    collagen, 464
    type II and X collagen, 464
Interactions of Dectin-1, 727

Interactions of Endo180, 344
Interactions of layilin, 881
Interactions of MAFA with FcεR, 695
Interactions of polycystins, 1031
Interactions of siglec-7, 388
Interactions of SP-A, 509
Interactions of SP-A and SP-D, 965
   Aspergillus fumigates, 965
   Cryptococcus neoforman, 965
   E. coli, 965
   influenza, and respiratory syncytial viruses, 965
   Klebsiella pneumoniae, 965
   with pathogens and infectious diseases, 965
   Pneumocystis carinii, 965
   Pseudomonas aeruginosa, 965
   Salmonella Minnesota, 965
Interactions of SP-D, 531
   interactions with carbohydrates, 531
Interactions of tetranectin, 884
   with complex sulfated polysaccharides, 884
Interaction with bacteria and viruses, 909
Interaction with LPS, 532
Interaction with phospholipid liposomes, 511
Intercellular adhesion molecule (ICAM), 774
Interface of innate and adaptive immunity, 272
Interleukin, 8
Interleukin 2, 661
Interleukin-4 (IL-4), 720, 728
Interleukin-6 (IL-6), 1045
Interleukin 12, 661, 685
Interleukin (IL)-1, 852, 853, 862, 911
Interleukin-18 and 21, 685
Interleukin-1β, 6, 8
Interleukin-17 receptor, 767
Interleukin-15 stimulates NKG2D, 685
Intermediate compartment (IC), 58
Intermediate filament (IF) protein vimentin, 1031
Internalization of the IGF2R, 92
International Polycystic Kidney Disease Consortium, 1027
Interspersed nuclear element-1 (LINE-1), 888
Interstitial lung disease (ILD), 955
Interstitial pneumonia (IP), 956
Interstitium of lung, 955
Intestinal inflammation, 250
Intestinal ischemia, 998
Intestinal wound-healing, 250
Intestine lipid rafts, 246
Intracellular adhesion molecule (ICAM-1), 992
Intracellular C-terminal domain, 340
Intracellular MBP (I-MBP), 488
Intracerebroventricular enzyme, 1043
Intravascular fibrinolysis, 994
Intravitreal macrophage activation, 231
Invariant NK T (iNKT) cells, 648
Invertebrate C-type lectin CEL-I, 476
Ionizing radiation, 671
Ionomycin, 566
IPF and systemic sclerosis, 957
Ischaemia-reperfusion injury (IRI), 945
Ischaemic attack (TIA), 992
Ischemic heart disease, 917, 993
Islet neogenesis-associated protein (INGAP), 848, 871
Islet of Langerhans regenerating protein, 850
Isoforms of NKG2D, 667
   NKG2D-L (long), 667
   NKG2D-S (short), 667

Isoforms of PLC-γ, 676
Isolectin A from *Lotus tetragonolobus* (LTL-A), 446
Isothermal calorimetry, 885
ITAM, 675, 726, 750, 752, 777, 782
ITAM-bearing adaptor protein, DAP-12, 383
ITAM-dependent pathway, 606
ITAM motif, 384
ITIM, 383, 384, 386, 393, 644, 693, 750, 752, 757, 758, 762, 763
ITIM (V/LXYXXL), 640, 642
ITIM (VxYxxL), 735
ITIM and ITIM-like motifs, 366
ITIM-bearing inhibitory immunoreceptors, 400
ITIM-like motifs, 385, 389
ITIM of MIS, 396
I-type lectins, 6, 351, 352
I-type lectins MAG/Siglec-4, 353

J
JAK2, 272
JAK/STAT3/SOCS3 pathway, 864
Japanese eel *(Anguilla japonica)* fucolectins, 440
Japanese horseshoe crab, 416
Japanese horseshoe crab *T. tridentatus*, 417
Jejunal Peyer's patches, 861
Jelly-roll fold, 6, 163, 214, 272
Jellyroll topology, 165, 197
JNK1, 677
JNK, p38, ERK, and NF-kB, p65, 277
Jun and Fos family proteins, 229
JUPITER trial, 169
Jurkat T cells, 223, 224
Juvenile diabetes, 1000

K
K18 anddesmin, 1031
Kaposi's sarcoma (KS), 300, 565
KARAP (DAP12 or TYROBP), 674
KARAP/DAP12-deficient mice, 674
KARAP/DAP12-deficient Nasu-Hakola, 674
Karpas 299 T-lymphoma cells, 297
Kawasaki disease, 944
Kazal 1 (SPINK1), 850
28 kDa glycoprotein from human neutrophils, 922
KDEL sequence, 37, 58
Keratan sulfate, 802, 804
Keratan sulfate side chains, 808
Keratinocytes, 191, 203, 218, 235, 282
Keyhole limpet hemocyanin (KLH), 1050
KGF, 977
KIAA0022, 920
Kif3a knockout, Tg737 and pkd1/pkd2 mutant mice, 1033
Kifunensin and 1-deoxymannojirimycin, 138
Killer cell Ig-like receptors (KIR; CD158), 619, 632
Killer cell lectin like receptor subfamily E, member 1 (KLRE1), 699
Killer cell lectin-like receptor subfamily K, member 1 (KLRK1), 673
Killer immunoglobulin (Ig)-like receptors (KIRs), 693
Killer receptors, 619, 621
   immunoglobulin family (KIR), 619
   KIR2DL1/S1 and KIR2DL2/3/S2, 619
Kininogen, 906, 908
KIR, 646
KIR and Ly49 receptor families., 620
KIR2DL3, 656
KIRF1, 758

KIR gene, 647
KL-6, 90, 955, 957, 959
Klebisella aerogenes, 934
Klebsiella pneumonia, 337, 540
Klebsiella pneumoniae (KpOmpA), 181
Klebsiella spp. *Saccharomyces cerevisiae*, 934
Klebs von den Lungen (KL)-6, 955
KLR, 758
KLRB receptor family, 619
KLRE/I1 and KLRE/I2, 700
KLR; FLJ17759; FLJ75772; D12S2489E, 667
KLRG1, 521, 632, 697, 699
KLRG1 (or 2F1-AG), 696
    mouse homologue of MAFA, 696
KLRG1 (rat MAFA/CLEC15A or mouse 2F1-Ag), 695
KLRG1 in cord blood, 697
KLRK1 (killer cell lectin-like receptor subfamily member 1) gene, 667
KLRL1, 726
KLRL1 from human and mouse DCs, 700
90 K/Mac-2 binding protein (M2BP), 303
Knock-out (KO) mice, 359, 527, 542, 1041
KpOmpA, 182
Kremen1 and 2 (Krm1/2), 119
Kringle-4, 883
Kringle-4-binding, 885
Kringle domain, 883, 884
Kringle-4 domain of plasminogen, 886
Kupffer cells, 63, 332
KXFF(K/R)R synthetic peptide, 43

L

Lacrimal glands, 955
Lactodifucotetraose (LDFT), 442
Lactoferrin receptor, 249
αβ-Lactosamine glycolipid, 222
Lactose (Lac), 269
Lactose binding lectin in retina, 248
Lactosylceramide, 510
L-α-dipalmitoylphosphatidylethanolamine (PE), 455
LAM-1 and L-selectin are homologous, 554
Lamellar bodies (LB), 501, 972
Laminae III-VIII and X, 825
Lamina IX, 825
Lamina propria T lymphocytes, 200
Laminin, 167, 171, 222, 275, 302
Laminin α2 chain, 180
LAMP-1, 235, 268
LAMP-2, 268
LAMP-4, 73
L- and H-ficolins, 413, 414
L and P domains, 459
Langerhans cells, 670, 726
Langerhans cells (LCs) dermal DCs, 749, 760
Langerin: a natural barrier, 756
Langerin, BCDA-2, DCIR, DLEC, CLEC-1, 774
Langerin/CD207, 749, 750, 753, 755, 756, 774, 1049
L-29, another S-Lac lectin, 265
Large needle aspiration biopsy (LNAB), 292
Large proteoglycans, 802, 825
Laryngeal squamous cell carcinoma (LSCC), 818
Late endocytic pathway, 1045
Late endosomal compartment, 1041
Late endosomes (LE), 64, 147
Late endosomes/lysosomes, 343

Latent membrane protein 1 (LMP1), 259
Latent TGFβ precursor, 62
Layilin, 480, 881
Layilin: a hyaluronan receptor, 881
Layilin group of C-type lectins, 881
L1 cell adhesion molecule, 827
LDC27, 908
LDH, 956
LDL-A and C-type lectin domains, 1027
LDL class A domain, 479
Leader sequences of HSP, 60, 652
Le^a/Le^b glycans, 775
Lectican genes, 477
Lectican protein family, 37
Lecticans, 6, 477, 479
Lecticans (hyalectans), 477, 480, 807, 825, 827, 838
Lectin domain of GalNAc-T1, 318
Lectin from gill of bighead carp (*Aristichthys nobilis*), 443
Lectin-grafted carrier system, 1046
Lectin-grafted formulations, 1046
Lectin-grafted prodrug, 1046
Lectin-like oxidized ldl receptor (LOX-1) (CLEC8A), 736
Lectin-like transcript 1 (LLT1), 630
Lectin pathway in murine lupus nephritis, 941
Lectin pathway in myocardial ischemia-reperfusion, 944
Lectin pathway of complement activation, 413, 416
Lectin receptors on NK cells, 619
Lectin-related molecule (Clr), 624
Leech intramolecular transsialidase, 146
Legume lectins, 197
Leishmania, 336, 782
Leishmania chagasi, 934
Leishmania donovani, 336
Leishmania major, 175, 272, 282
Leishmaniasis, 255
Lens crystalline protein, 216
Lepidopteran, 433
Leptin-binding ability of siglec-6, 387
Leptin-binding protein of IgSF, 386
Leptospirosis, 782
Leucine-rich repeats (LRRs), 100, 126, 341
Leucine zipper protein, 601
Leucine zippers, 134
Leucocyte traffic, 257
Leukemia inhibitory factor (LIF)/CNTF family, 858, 862
Leukocyte activation, 556
Leukocyte adhesion molecule-1 (LAM-1), 555
Leukocyte-endothelial cell adhesion molecule 1, 554
    LECAM-1, 554
    Leu-8, 554
    leukocyte adhesion molecule-1 (LAM-1), 554
    L-selectin/CD62L, 554
    MEL-14, 554
    PLN homing receptor, 554
    TQ1, 554
Leukocyte extravasation, 575
Leukocyte integrins, 992
Leukocyte receptor complex (LRC), 725
Leukocyte tethering and rolling, 556
Leukodystrophic disorders, 360
Leukosialin (CD43), 567
Lewis antigen as ligand, 775
Lewis clusters target DC-SIGN, 339
Le^x, 777
LFA-1, 774, 999, 1002, 1004, 1005
L-ficolin, 409–413, 415

L-ficolin (synonymous with ficolin-2 or Ficolin/P35), 410
L-ficolin/MASP complex, 415
L-fucose-binding lectin from Nile tilapia
    (*Oreochromis niloticus L.*), 442
LG/LNS domains, 180
Ligand binding, 477
Ligand binding properties of ASGP-R, 711
Ligand-binding site of ficolins, 413
Ligand clustering, 358
Ligand for siglec-F, 396
Ligand interactions, 908
Ligand on colon carcinoma cells, 566
Ligand Qa-1b, 646
Ligands for CD94/NKG2, 651
Ligands for CLEC-2, 732
Ligands for cysteine-rich (CR) sialoadhesin, 338
Ligands for LOX-1, 738
Ligands for MBL, 488
Ligands for NKR-P1A, 627
Ligands for sialoadhesin, 356
Ligands for siglec-8, 390
Ligands of CD22, 362
Ligands of CD161/ NKR-P1, 627
Ligands of DC-SIGN, 775
Ligands of dectin-2, 760
Ligands of ficolins, 411
Ligands of langerin, 755
Ligands of lecticans, 826
Ligands of LSECtin, 792
Ligands of MAG, 368
Ligands of YKL-40, 423
LIGHT CD70, 621
Limax flavus, 8
Limulus horseshoe crab coagulation factor G, 315
Limulus polyphemus, 176
Lineage (primarily in echinoderms and vertebrates), 351
Linear (NKR-P1) or branched (CD69), 627
Linear tetrasaccharides, 248
LINE-1 insertion, 889
Link domains, 480
Link or protein tandem repeat (PTR) domains, 475
Link peptide of variable length, 245
Link protein (LP), 807
Link protein-encoding genes, 477
Link proteins, 477, 825, 826
L. interrogans, 1013
Lintuzumab (HuM195), 400, 1052
Lipid bilayers, 466, 507
Lipid-free mannosylated arabinomannans, 967
Lipid raft domains, 200
Lipid rafts, 249, 250
Lipid storage diseases, 1043
Lipomannan, 336
Lipomannan and ManLAM, 967
    mycobacterial lipoglycans as ligands, 967
Lipopolysaccharide (LPS), 388, 493, 510
Liposomes, 1046
Lipoxin A4 (Annexin A4), 463
LIRs, 621
Listeria, 182
Listeria monocytogenes, 934
Listeria monocytogenes protein, 909
Lithogenic disorder, 854
Lithostathine (hPSP/hLIT), 852, 854, 855
Liver and lymph node sinusoidal endothelial cell C-type lectin
    (LSECtin), 791

Liver cirrhosis, 428, 1004
Liver diseases, 1004
Liver metastasis, 1009
Liver-SIGN (L-SIGN), 756
Liver sinusoidal endothelial cells (LSEC), 777
Lizard *Podarcis hispanica*, 207
LLT1 as ligand, 628
LLT1/CD161 interaction, 628
LLT1 mRNA, 632
L-MAG, 369
L. major poly-β-galactosyl epitopes, 255
LMAN-2, 156
LMAN1/ERGIC53, 156
L1ME2 element, 639
LNCaP cells, 47
L1/Ng-CAM adhesion, 829, 836
Longest linker peptide, 252
Longevity of human neutrophils, 273
Long loop region (LLR), 754, 781
Long pentraxins, 163, 178
Loss of polarity, 1035
Lotus tetragonolobus agglutinin (LTA1), 446
Lovastatin (Mevacor), simvastatin (Zocor), rosuvastatin (Crestor),
    LoVo cell migration and invasion, 871
Low density lipoprotein (LDL), 712, 816
Low density lipoprotein (LDL) receptor, 910
Low molecular weight heparin (LMWH), 1011
LOX-1, 14, 725, 736, 737, 758
LOX-1 and pathophysiology, 741
LOX from bovine aortic endothelial cells (BAE), 738
LOX-1 gene (symbol OLR1), 737
LOX-1 gene in human, 737
LOX-1 in atherogenesis, 741
LOX-1 in hypertensive rats, 741
LPS, 277, 358, 416, 726, 762–764
LRF/ATF3, 598
LSECtin (or CLEC4G or L-SIGN or CD209L), 791, 793
L-selectin$^{-/-}$, 567
L-selectin (CD62L), 555, 567, 994
L-selectin as E-selectin ligand, 607
L-selectin: carbohydrate interactions, 559
L-selectin counter-receptors, 560
L-selectin deficiency, 567
    reduces immediate-type hypersensitivity, 567
L-selectin facilitate metastasis, 565
L-selectin for early neutrophil extravasation, 567
L-selectin/ICAM-1$^{-/-}$, 567
L-selectin in autoimmune diabetes, 566
L-selectin in mouse CNS during EAE, 567
L-selectin in pathological states, 564
L-selectin interactions with glycosaminoglycans, 559
L-selectin ligands, 560
    glycolipids, and proteoglycans, 560
    heparan sulfates, 560
    sulfated Lewis-type carbohydrates, 560
L-selectin (CD62L) ligands, 555
L-selectin-mediated neutrophil trafficking, 560
L-selectin/P-selectin/E-selectin, LFA-1, 1000
L-selectin shedding from antigen-activated T cells, 558
L-selectin towards sulphated oligosaccharides, 559
L-sel$^{-/-}$ mice, 556, 557
L-SIGN, 777, 793
LSTb oligosaccharide, 398
LT2, a smooth type strain of *S. typhimurium*, 410
LT stromal cells, 919
L-type lectin fold 145, 163

L-type lectins, 6, 7
Lumbosacral radiculoplexus neuropathy (LRPN) with diabetes mellitus (DLRPN) and without diabetes mellitus (LRPN), 1000
Lumican, 216, 801
Lung injury, 999
Lung (type II) pneumocytes, 501
Lung small-cell carcinoma, 870
Lung SP-D, 527
Lung squamous cell carcinomas, 253
Lung surfactant protein A (SP-A), 473
Lung surfactant protein D (SP-D), 473, 475, 476
Lupus nephritis, 941
Lutropin, 332, 335
Ly-49, 621, 633, 648
Ly-49 (KLRA), 643
LY294002, 1045
Ly49A, 474, 476
Ly49 and Nkrp1 (Klrb1) recognition systems, 623
Ly49 gene in different inbred mouse strains, 621
Ly49I, 476
Ly49L, 621
Ly49-like gene, 621
Lymphatic vascular development, 734
Lymph node carcinoma of prostate (LNCaP), 102
Lymph node homing receptor (LHR), 576
Lymphocyte homing and leukocyte rolling and migration, 556
Lymphocyte IgE receptor, 473
Lymphocytic choriomeningitis virus (LCMV), 568, 644
Lymphoproliferative disease of granular lymphocytes (LDGL), 661
Lymphotoxin-α, 200, 201
Ly49 (Klra), Nkrp1(Klrb1) locus, 623
Ly49 receptors on NKG2D, 684
Lysobisphosphatidic acid, 93
Lysosomal acid α-glucosidase (GAA), 1040
Lysosomal enzyme N-acetylglucosamine-1-phosphotransferase (GlcNAc-1-phosphotransferase) and GlcNAc-1-phosphodiester α-N acetylglucosaminidase ("uncovering enzyme" or UCE), 1042
Lysosomal enzymes in osteoclasts, 64
Lysosomal integral membrane protein (LIMP), 71
Lysosomal LAMP-1 markers, 753
Lysosomal storage diseases (LSDs), 1040
Lysosome associated membrane protein (LAMP), 71, 73, 94, 302, 753
Lysosome biogenesis, 90
Lysosome maturation, 63
Lysosome-related organelles, 1040
Lysosomes, 61, 62, 64, 1040
Lytic granules, 1040
LYVE-1 (lymphatic vessel endothelial HA receptor-1), 477, 881

M
3-MA, 1045
Mac-1, 999, 1005
Mac-1 antigen (CD11b/CD18), 610
Mac-2 BP/90, 918
Macrophage ASGP-binding protein, 711
Macrophage binding protein, 303
Macrophage chitotriosidase, 426
Macrophage colony stimulating factor (M-CSF), 996
Macrophage galactose-type lectin (MGL) (CD301), 718
    DC-asialoglycoprotein receptor (DC-ASGP-R) or human macrophage lectin (HML), 718
    functions of MGL, 720
    human chromosome 17p13.2, 718
    human MGL (CD 301), 718
    lectin specific for galactose/N-acetylgalactosamine, 718
    ligands of human MGL, 719
    macrophage mannose receptor (MMR), 718
    murine MGL1/MGL2 (CD301), 718
Macrophage-inducible C-type lectin (Mincle), 757, 758, 764, 765
Macrophage mannose receptor (MMR), 6, 479, 750, 752, 755, 1049
    internalization of Myobacteria, HIV, Leishmania, Candida components, 755
Macrophage mannose receptor (MMR) family, 473
Macrophage migration inhibitory factor (MIF), 218
Macrophage YM-1, 426
Macrosialin, 334
MadCAM, 1000
MAdCAM-1 gene, 563
Madine-Darby canine kidney (MDCK) cells, 156, 1031
MAFA, 758
MAFA gene, 696
MAFA is a lectin, 695
MAFA ITIM, 696
MAFA-L, 699
MAFA-like receptor, 699
MAG, 361
MAG (myelin-associated glycoprotein) (Siglec-4a), 352, 354
MAG and SMP, 398
MAG in demyelinating disorders, 371
MAG inhibits neurite outgrowth, 370
MAG is a bifunctional molecule, 370
MAG isoforms, 368
MAG knockout mice, 370
MAG recognizes only NeuAc α2→ 3Gal β1→ 3GalNAc, 354
Mahoganoid, 891, 892
Mahoganoid allelic series (md, md2J, md5J), 892
Mahogany (mg), 892
Mahogany is a receptor, 891
Mahogany (mg) mice, 893
Mahogunin, 890
Mahogunin ring finger-1 (MGRN1) or attractin, 890
Maintenance of plasma cells in bone marrow, 904
Major forms of MBL in human, 488
    intracellular MBP (I-MBP), 488
    serum MBP (S-MBP), 488
Major transcript of annexin A4, 462
Malayan pit viper *Calloselasma rhodostoma*, 732
Male germ cell specific homologue, 30
Malignant astrocytomas, 856
Maltose-binding protein, 485
Mammalian chitinases, 421
Mammalian testis proteins (TPX1), 921
Mammary gland protein (MGP-40), 424
Man1B1, Man1A1, Man1A2 and Man1C1, 135
Man-C4-Chol, 1048
ManLAM, 777
ManLAM or PIMs, 777, 787
ManLAM/PIM-DC-SIGN interaction, 787
Man-liposomes, 1048
Mannan-binding lectin (MBL), 480, 483, 487
Mannan-binding lectin pathway, 492
Mannan-coated nanoparticles in vaccination, 1049
Mannan-lectin interactions, 1045
Mannose-binding lectin-associated serine proteases (MASPs), 413
Mannose binding protein (MBP), 487
Mannosecapped lipoarabinomannan (ManLAM), 336, 786, 787
Mannose 6-phosphate (M6P), 61
Mannose 6-phosphate/insulin-like growth factor II receptor, 61

Mannose-6-phosphate receptor homology (MRH or M6PRH) domains, 111
Mannose-6-phosphate receptors and applications in human diseases, 1040
Mannose receptor (MR), 344
Mannose receptor CD 206 (MR/ManR), 14, 21, 315, 331, 332, 335, 1046
Mannose receptor family, 21, 315, 332
Mannose receptor-targeted drugs and vaccines, 339
Mannose receptor targeting vaccine, 1049
α-D-Mannopyranoside, 5
α-Mannosidase, 65, 135
α1,2-Mannosidase (ERManI), 135
Mannosidase(s), 109
Mannosidase I and II, 147
Mannosidase I inhibitor deoxymannojirimycin (dMM), 939
Mannosidase like proteins, 32
Mannosylated cationic liposomes, 1048
Mannosylated cholesterol derivative, 1048
Mannosylated conjugates, 1046
Mannosylated-emulsions, 1048
Mannosylated lipoarabinomannan, 967
Mannosylated liposomes (Man-liposome), 1046
Mannosylpolyethylenimine (ManPEI) conjugates, 1047
ManR, 730
ManR and TLRs expression by Mφ s, 517
MAp19 (19-kDa MBL-associated protein), 413
MAPK and NF-kB activation, 732
MAP kinase, 680
MAP kinase ERK-1/2, 918
MAPK p38 and ERK1/2, 257
MAPK signaling, 865
Marker alleles in *M. tuberculosis*, 968
Marker for hormone refractory metastatic prostate cancer, 871
Marker of ER-cargo exit site, 152
Marker of myeloid progenitor cells, 381
Marker of tumor endothelium, 917
Markers of ILD, 960
MASP-1, 14, 413, 494
MASP-2 (MBL-associated serine protease), 14, 415, 491, 494
MASP-3, 14, 413
MASP-2 auto-activation in mutated MBL, 937
MASP-binding sites in MBP, 491
MASP-2 efficiently, 938
Maspin, 1008
MASPs, 415, 492
Mast cell function-associated antigen (MAFA), 695
Maternal decidual (d) dNK cells, 684
Matrix metalloprotease-13 (MMP-13), 344
Matrix metalloproteinase (MMP), 49, 680, 736
Matrix of highly organized fibrils, 857
Mature myeloid cells, 381
M. avium, 517
M. avium complex (MAC), 967
M. avium 2151-rough, 730
MBL, 409, 486, 945, 9003
    nephropathy, 945
    pathogenic factor for diabetic cardiovascular complications, 945
MBL2, 495, 935
MBL and FcγRII (CD32) polymorphisms, 941
MBL and gastrointestinal infection, 943
MBL and HIV interaction, 938
MBL and inflammatory bowel diseases, 942
    Crohn's disease (CD), 942
    indeterminate colitis, 942
    ulcerative colitis (UC), 942

MBL and L-ficolin, 415
MBL and viral infections, 938
MBL: a reconstitution therapy, 949
MBL-associated protein (MAp19), 14
MBL-associated serine protease (MASP-2), 494
MBL-associated serine proteases (MASPs), 14, 410, 492–494, 934
    MASP-1, MASP-2, and MASP-3, 934
MBL2 B allele, 946
MBL binding with tumor cells, 947
MBL characteristics, 934
MBL deficiency, 945
    cystic fibrosis (CF) lung disease, 945
MBL deficiency and genotyping, 935
    to autoimmune disease, 934
    human MBL-2 gene, 935
    to infection by extra-cellular pathogens, 934
    MBL deficiency predisposes, 934
    polymorphism in MBL gene in Exon 1 at Codon 52, 54, and 57, 935
MBL-deficient individuals, 935
MBL gene in rheumatoid arthritis, 939
MBL gene in transplantation, 946
MBL genotypes, 490
MBL genotypes in acute lymphoblastic leukemia, 937
MBL genotypes in viral hepatitis, 937
MBL in cardio-vascular complications, 944
MBL in tumorigenesis, 947
    in lower (i.e., LYB, LYC, HYD, and LXA), 947
    or higher (i.e., HYA and LYA) serum MBL, 947
    polymorphisms in the promoter, 947
MBL knock-outs, 949
MBL levels, 490
MBL/MASP complex, 415
MBL-mediated complement attack, 934
MBL2 polymorphisms, 936
    codon 52 polymorphism, 936
    prevalence of respiratory infection, 936
    risk of premature delivery, 936
    SNPs at codon 54 in MBL, 936
    two promoters: promoter 1 and promoter 0, 936
MBL replacement therapy, 946
    following stem cell transplantation, 946
MBL-vaccinia, 949
M. bovis Bacillus Calmette-Guerin (BCG), 967
MBP, 480
MBP-A, 476
MBP alleles and malaria, tuberculosis, and HBV, 93
MBP dependent cell-mediated cytotoxicity (MDCC), 493, 949
MBP polypeptides, 938
M-cells, 1046, 1050
MCFD2 genes, 155
MCL or murine Clecsf8, 767
MCMV-infected cells, 658
MCR antagonists, 891
mCRP isoform, 171
MD-2 (lymphocyte antigen 96), 509, 532
mDCAR1, 764
MDDCs, 234
Measles, 998
Measles virus targets DC-SIGN, 784
MECA-79, 562, 563, 569
Mechanical ventilation, 962
Mechanosensory function of the cilia, 1034
Medial and lateral superior olivary nuclei, 833
Mediation of the migration and angiogenesis, 86
Megalin, 91
MEL-14, 556

Melanocortin 1 receptor (Mc1r), 892
Melanocortin-4 receptor (MC4R), 889
Melanocytes, 861
Melanoma tumors, 861
Melanosomes, 1040
Membrane bound thrombomodulin, 916
Membrane lattice structures, 507
Membrane type 1 matrixmetallo-proteinase (MT1-MMP), 909
Memory B lymphocytes, 611
Meningiomas, 15
Meningitis, 415
Meningococcal disease, 934
Meningotheliomatous meningiomas, 15
Menstrual cycle, 227
Meprin β, 116
Merlin and radixin, 881
Mesenchymal cells, 226
Metabolic acidosis, 278
Metabolic syndrome, 169
Metalloproteinases (MMPs), 296, 819
Metastases from breast cancer, 1009
Metastasis, 1009
Metastasis in lymph nodes, 565
Metastatic cancer, 430
Metastatic melanoma (MM), 631, 632
Metastatic prostate cancer, 871
Metastatic spreading, 1009
Methimazole, 1017
Methionine oxidation, 912
Methotrexate (MTX), 1002
M-ficolin (Ficolin-1), 409–412
M-ficolin FD1, 412
MG160, 605
MGP40 and HumGP39, 424
MG2 saccharides function, 561
MHC class I, 619
    assembly pathway, 51
    MICA, ULBP, RAE-1, and H-60, 619
    proteins, 51, 382, 619, 621
    structural relatives, 619
MHC class I chain related (MIC) proteins, 669
MHC class II compartments, 1040
MHC class I related chain A/B (MICA/B), 656
MHC-like ligands MICA, MICB, and ULBP, 641
MICA, 621, 668, 669, 672, 677, 682, 684
MICA and MICB or ULBP proteins, 701
MICA/B, PVR or Nectin-2, 703
MICA-NKG2D complex, 672
MIC-A-positive human cancer cells, 671
MICB, 621, 668, 669, 677, 682, 684
Mice deficient in E-selectin or in E-selectin and P-selectin
    (E/P-selectin mutant), 1002
Mice deficient in P-or L-selectins, 560
MICL/CLEC12A, 14, 735, 758
Microbial carbohydrates, 486
Microbial pathogens, 175, 1013
Microbial uptake capacities, 783
Microbiota, 863
Microcystis viridis, 4, 21
Microfibrilassociated protein 4 (MFAP4), 509
Microglia, 279, 356, 359
Microparticles and liposomes, 1050
MicroRNA (ribonucleoprotein (RNP)), 223
MicroRNAs (miRNAs), 672
Microsatellite markers, 830, 958, 962
Microsatellite markers D3Mit22 and D3Mit11, 830

Microsatellite unstable (MSI-H) tumors, 156
Microscopic polyangiitis (MPA), 1005
Microtubule organizing center (MTOC), 677
Microvillar "superrafts," 449
MIF in the immune escape of ovarian cancer, 681
Migratory neural crest cells, 815
Milk fat globule protein, 324
Mincle, 14
Mincle (also called as Clec4e and Clecsf9), 765
Mincle$^{-/-}$ mice, 765
Minor salivary gland (MSG), 851
MIR, 639
miRNP particle, 223
Mitogen-activated protein kinase (MAPK), 428, 651, 730, 791, 1032
Mitogen-activated protein kinase phosphatase (MKP-1), 850
m-Langerin, 754
MMBP, 258
MMP-2, 49
MMP-9, 49, 544
MMP-12, 544
Mmp9$^{-/-}$Spd$^{-/-}$mice, 517
Mmp12$^{-/-}$Spd$^{-/-}$mice, 517
Mobility on laminin, 222
Model invertebrates, 477
Model of I/R injury, 998
Model protein for endocytosis, 716
Modified CRP, 170
Modified lipoproteins, 726
Modular arrangement of SpGH98, 446
Modular structures of the proteoglycans structural motifs such as
    EGF-like domains, lectin like domains, complement regulatory
    like domains, immunoglobulin folds, and proteoglycan tandem
    repeats, 825
Modulation of adaptive immune responses by SPs, 513
Modulation of associated serine proteases, 494
Modulators of inflammation, 542
Moesin, 904
Molecular basis of TM activation, 915
Molecular phylogenetic analyses, 433
Molecular structure of annexin A5, 465
Monoclonal antibody, DCGM4, 753
Monocyte chemoattractant protein 1 (MCP-1), 170, 742, 862, 996
Monocyte derived dendritic cells (MDDCs), 233, 385, 1050
Monocytogens, 182
Monomeric and multimeric blockers of selectins, 569
Monomeric CRP (mCRP), 170
Monomeric GGA adapters, 61, 228
Monomeric structure of human MBP, 487
Monosaccharide binding at CRD-4 in MR, 334
Monosaccharide ligands, 441
Morone saxatilis, 442
Morphine treatment, 274
Morphogenesis in kidney epithelial cells, 1035
Mosquitocidal toxin (MTX), 321
Mosquitocidal toxin from *Bacillus sphaericus* SSII-1, 324
Motor neuron protein (SMN) complex, 223
Motor neurons (lamina IX), 825
Mouse breast regression protein 39 (BRP-39; CHI3l1), 427
Mouse CD94, 643
Mouse CD-MPR, 66
Mouse CD33/siglec-3, 394
Mouse chromosome 1, 426
Mouse chromosome 3, 426
Mouse chromosome 5, 1028
Mouse chromosome 6, 667, 847
Mouse chromosome 7, 355, 368, 394, 401

Mouse chromosome 10, 669
Mouse chromosome 12, 850
Mouse coat color mutant *mahoganoid (md)*, 892
Mouse GlyCAM-1, 561
Mouse hepatitis virus (JHMV), 670
Mouse intelectin, 451
Mouse Langerin, 754
    mouse and rat genes, 754
    mouse chromosome, 6, 757, 765, 767
Mouse lymph node homing receptor core protein (mLHR), 555
Mouse Ly49 NK receptors, 622
Mouse MAdCAM-1 gene on chromosome, 10, 563
Mouse mannose receptor, 337
Mouse Mcl/Clecsf8, 767
Mouse myoblastic cells, 217
Mouse NKG2D ligands, 670
Mouse NKR-P1, 624
Mouse NKR-P1B, 625
    NK1.1 antigen with inhibitory function, 625
Mouse NKR-P1B binds Src homology 2 (SH2), 628
Mouse NKR-P1C stimulates phosphatidylinositol, 628
Mouse orthologue of human Chodl, 882
Mouse Reg-IIIg, 848
Mouse retinoic acid inducible genes-1 (RAE-1a), 669
Mouse siglec (mSiglec-F), 394
Mouse siglec E (mSiglec-E), 395
Mouse siglec-F: convergent paralogs, 396
Mouse siglecs, 394
Mouse siglecs (mSiglecs) orthologous to hSiglec-1, 354
Mouse SP-D, 530
Mouse SP-D gene (Sftpd) genomic organization, 530
Mouse UL16-binding protein-like transcript 1 (MULT1), 662, 668, 669
MP20, 268
M6P-glycoforms, 91
M6P glycoforms of α-mannosidase B1, cathepsin D, 91
M6P-independent pathways, 92
MPR46, 66
MPR300, 66
M6PR-H, 111, 112
MPR homology (MRH) domains, 62
M6PR-mediated transport, 1042
MPRs in CNS, 63
MPRs in liver, 63
MPS (Meridian-phlegm stagnancy), 997
MPS IVA 1044
$MR^{-/-}$, 1043
MR as endocytic receptor, 337
MR endogenous ligands, 335
    lutropin, 335
    myeloperoxidase, 335
    thyroglobulin, 335
MRL1, 120
Mrl1, LERP, GlcNAc-1-phosphotransferase g-subunit, 110
MRL-lpr mice, 941
MRL/MpJ-lpr/lpr Mice, 941
MR positive APCs, 334
MR regulates cell migration, 337
MS-FBP32, 442
MSFPB32 CRDs, 443
mSiglec-3/CD33, 395
mSiglec-E, 395
mSiglec-E (an ortholog of hSiglec-9), 395
mSiglec-E/MIS, 394, 395, 402
mSiglec-F, 395, 398, 402
mSiglec-G, 394, 396
mSiglec-H, 394, 396

M. smegmatis, 967
^{99m}Tc-labeled asialoglycoprotein analog, TcGSA (galactosyl-human serum albumin-diethylenetriamine-pentaacetic acid), 1051
M. tuberculosis, 730, 777, 781, 787, 790, 792
M. tuberculosis H37Ra, 730
M. tuberculosis-infected monocytes, 682
MTX on psoriatic skin lesions, 1018
M-type lectins, 6, 7
M-type phospholipase A2 receptor, 331, 750
M-type PLA2 receptor, 339–341
MUC-1, 785
Mucin function, 561
Mucin-type polypeptides GlyCAM-1, CD34, and MAdCAM-1, 561
Mucolipidosis (ML), 1042
Mucolipidosis type II and III, 1042
Mucopolysaccharides, 802
Mucopolysaccharidosis (MPS), 1043
Mucopolysaccharidosis IVA (MPS IVA; Morquio A syndrome), 1044
Mucopolysaccharidosis type IIIA (MPS IIIA), 1044
Mucopolysaccharidosis type IIIB (MPS-IIIB, Sanfilippo, 1044
Mucopolysaccharidosis type IIID, 1044
Mucopolysaccharidosis type VI (MPS VI), 1044
Mucopolysaccharidosis type VII (MPSVII), 1043, 1044
Mucosal addressin cell adhesion molecule-1 (MAdCAM-1), 563
Mucosal surfaces, 955
Mult1, 621
Multi-CTLD endocytic receptors, 480
Multiple coagulation factor deficiency 2 (MCFD2), 154
Multiple C-type lectin-like domains, 331
Multiple myeloma (MM), 294, 644, 1008
Multiple sclerosis, 360, 660, 839, 1006
Multiple sclerosis plaques, 371
Murine brevican gene, 830
Murine chromosome 2, 902
Murine dectin-1, 727
Murine galectin-3 (Mac-2) gene, 266
Murine homolog of MICL (mMICL), 735
Murine homologue of human C1qRp (AA4), 905
Murine homologues of DC-SIGN, 789
Murine intestinal nematode, 283
Murine mahogany gene, 888
Murine model of asthma, 965
Murine NKG2A,-B,-C, 640
Murine NKG2-C, 640
Murine NKG2D splice variants, 675
Murine Nkrp1 genes, 624
    Nkr-p1a, Nkr-p1b, Nkr-p1c, Nkr-p1d Nkr-p1f, 624
Murine or rat neuronal pentraxin 1 (NP1), 178
Murine PLN homing receptor/mLHR3, 556
Murine Reg-IIIα, Reg-IIIβ, Reg-III γ, and Reg-IIIδ genes, 858
Murine siglec-F, 366
Murine tetranectin, 883
Murine Ym1 (CHI3L3/ECF-L), 421
Muscle (M)-type sPLA2 receptors, 340
Musculoskeletal systems, 802
Mutant SP-A1(δAVC,C6S), 504
Mutation in aggrecan gene, 818
Mutations in PKD1 and PKD2, 1034
Mutations of the mouse attractin gene, 892
Mut-HA22-liposomes, 1053
mv rats, 894
Mycobacterial carbohydrates as ligands of DC-SIGN, L-SIGN and SIGNR1, 786
Mycobacterial cell-wall lipoglycans, 967
Mycobacterial glycolipid, trehalose dimycolate, 765
Mycobacterial infection, 787

Mycobacterium avium, 934
Mycobacterium leprae, 934
Mycobacterium tuberculosis (Mtb), 16, 17, 271, 336, 782, 934
Mycolic acids (MA), 271
Mycoplasma pneumoniae, 271, 545
MyD88, 858
Myelin-associated glycoprotein (MAG) (Siglec-4a), 352, 354
Myelin-associated glycoprotein (MAG), 368, 838
Myelinated axons, 831
Myelin basic protein gene, 250
Myelinogenesis, 353
Myelodysplastic syndromes (MDS), 680
Myeloid, 381
Myeloid dendritic cells (MyDCs), 656, 1050
Myeloid differentiation factor 88 (MyD88), 730
Myeloid inhibitory C-type lectin-like receptor (MICL), 735
Myeloid inhibitory siglec in mice, 396
Myeloperoxidase (MPO), 169
Myeloperoxidase (MPO)-anti-neutrophil cytoplasmic antibodies (ANCA), 1005
Myoblast differentiation, 222
Myocardial damage, 996
Myocardial infarction, 200, 586, 996
Myocarditis, 858
Myogenic development, 93

N

N-acetyl-D-glucosamine, 440
N-acetylgalactosamine 4-sulfatase (4S), 1044
N-acetylgalactosamine-6-sulfatase (GALNS), 1044
N-acetylglucosamine-6-sulphatase (Glc6S), 1044
N-acetylglucosaminyl 6-phosphomethylmannoside, 88
N-acetyllactosamine, 269, 271
N-acetyl-L-cysteine (NAC), 600
N-acetylneuraminic acid (Neu5Ac), 351, 364
NADase activity of CD38, 398
NAG9 model, 427
Naive CD8+ T cells, 678
Naming of secondary structure elements, 475
Narp, 178
Nasal polyposis (NP), 258, 334, 999
Nasopharyngeal carcinomas (NPC), 259
Native LDL (n-LDL), 599
Natriuretic peptide (BNP), 996
Natural cytotoxicity receptors, 693, 701
Natural killer cells, 619
Natural killer gene complex (NKC), 14, 694, 725, 737, 757
Natural or synthetic glucans, 727
N-cadherin, 621
N-cadherin-binding N-acetyl-galactosamine-phosphoryl-transferase, 829
N-CAM/NCAM neural cell adhesion molecule, 827, 829
Neck domain, 777
NECK domain of LOX-1, 740
Neck region (NR) (coiled-coil), 489
Nectin-2 (CD112), 701
Nectin-2 B7, 621
Nectin family, 701
Negative regulator of LPS-mediated inflammation, 277
Neisseria, 493
Neisseria gonorrhoeae, 5, 272, 282, 934
Neisseria meningitidis, 359, 365, 396, 399, 934
Nematocyst outer wall antigen (NOWA), 921
Nematode parasite *Trichuris trichiura*, 336
Neonatal chronic lung disease, 964
Neonatal mice, 517
Neonatal non-obese diabetic (NOD) mice, 566
Neonatal rat microglia, 904
Neonatal RDS, 962
Neonatal sepsis, 415
Neoplastic goblet cells, 870
Nephritides, 167, 171
Nephropathy (DN), 1001
Nerve growth factor, 856
Nerve microvasculitis and ischemic, 1000
Neural cell adhesion molecule (NCAM), 361
Neural F-box 42-kDa protein (NFB42), 132
Neural progenitor cell proliferation, 228
Neural stem cells (NSCs), 228
Neurexin, 180
Neurinomas, 15
Neurite formation in differentiating cortical neural cells, 891
Neurite outgrowth, 827
Neuroblastoma, 225
Neurocan, 802, 806, 808, 825–828, 833–835, 837, 838
Neurocan-deficient mice, 835
Neurocan FGF2, 827
Neurocan in embryonic chick brain, 836
Neurocan-like and 6B4 proteoglycan, 829
Neurodegenerative diseases, 857
Neuro-endocrine tumors, 253
Neurofibromas, 15
Neurogenesis, 229
Neuroglycan C (NGC), 838
Neurological diseases, 430, 888
Neuro/muscular disorders, 1004
Neuronal ceroid lipofuscinosis, 1044
Neuronal N-type PLA2 receptor honey bee, 342
Neuronal or N-type PLA2 receptor, 340
Neuronal pentraxin I (NPI), 178
Neuronal pentraxin II (NPII), 178
Neuronal plasticity, 163
Neuronal thread proteins (NTP), 856
Neuronglia antigen 2 (NG2), 825, 837, 838
Neuropilins, neurexin IV, 324
Neutral glycosphingolipids, including, 322
Neutropenia, 391, 947
Neutrophil activation, 273, 274, 610
Neutrophil adhesion, 272
Neutrophilic asthma, 965
Neutrophil phagocytosis by macrophages, 225
Neutrophil serine proteases cathepsin G, 959
Neutrophil serine proteinases inactivate SP-D, 541
NF-ELAM1 complex, 602
NF-IL6, 764
NF-kB, 17, 224
NF-kB, AP-1, and Oct-2, 361
NF-kB-mediated transcription, 597
NF-kB/p65, 276
NF-kB transcription factor, 996
NG2, 837, 838
Ng-CAM/L1, 827
Ng-CAM neuron-glia cell adhesion molecule (a chick homolog of L1), 827
NG2 DSD-1, 838
N-glycanase (PNGase), 123, 131
N-glycans as major ligands, 252
N-glycans on pituitary glyco-hormones, 335
N-glycan status modifies virus interaction, 783
N-glycolylneuraminic acid (Neu5Gc), 351, 364
N-glycolylneuraminic acid in human evolution, 364

N-glycosylation, 66
NI250, 838
Nicastrin (NCT), 32
Nicotine, 957
Niemann-Pick disease (NPD) type-A, B and C, 1041
Niemann-Pick disease type C (NPC), 1042
Niemann-Pick type C (NPC) fibroblasts, 99, 100
Nitrated SP-A, 518
Nitration and oxidation of SP-A, 976
Nitration of protein tyrosine residues, 976
Nitric oxide, 247, 539
Nitric oxide synthase (iNOS), 980
Nitric oxide synthase 2 (NOS2) isoform, 956
Nitrite production by rat AMφ, 514
Nitrofen exposure, 977
NK1.1 (NKR-P1C), 625, 628
NK2.1 antigen, 625
NKC, 642
NK cell, 619, 620
NK cell-mediated cytotoxicity, 680
NK cell-mediated killing of myeloma cells, 686
NK cell receptor protein 1 (NKR-P1) or KLRB1, 623
NK cell receptors (NKR), 9, 477, 480, 619
NK cells in female reproductive tract and pregnancy, 650
NK cells in T cell lymphomas, 659
NK cells in umbilical cord blood (CB), 679
NKC gene locus, 620
NK complex, 667
NKG2A, 639–642, 644, 646, 654, 758
NKG2-A, 693
NKG2-A and NKG2-F genes, 639
NKG2A c.338-90*A/*A, NKG2C102*Ser/*Ser, 661
NKG2A/CD94, 621, 639, 653
NKG2Ac.338-90*G/*G, 661
NKG2A, NKG2D, CD95, 677
NKG2B, 642, 643, 646
NKG2C, 620, 640, 642–644, 646, 656, 657, 758
NKG2C-CD94, 621, 639
NKG2-C,-D,-E, and-F genes, 639
NKG2CE, 641
NKG2CE gene, 620
NKG2C102*Phe/*Phe, 661
NKG2D, 476, 621, 632, 640, 641, 646, 657, 667, 673, 676–682, 686, 693, 694, 701, 758
NKG2D (KLRK1), 667
NKG2D (NKG2D-L), 675, 683
NKG2D (NKG2D-S), 675
NKG2D(CD314), 683
NKG2D72*Ala/*Ala genotypes in RA, 661
NKG2D-bearing cytotoxic lymphocytes, 669
NKG2D-DAP10 complex, 674, 676
NKG2D-deficient mice, 679
NKG2D homodimer, 675
NKG2D homodimers binding to MICA and ULBP3, 655
NKG2D in cytokine production, 678
NKG2D in immune protection and inflammatory disorders, 681
NKG2D in progression of autoimmune diabetes, 682
NKG2D ligands, 668
NKG2D ligands RAE-1, MULT1, and H60, 670
NKG2D ligand transcription, 671
NKG2D-mediated NK cell cytotoxicity, 671
NKG2D-mediated tumor rejection, 679
NKG2D-MICA, 674
NKG2D on γδ T cells, 677
NKG2D-RAE-1b, 673
NKG2D receptor complex and signaling, 675, 676

NKG2D72*Thr/*Thr genotypes, 661
NKG2D/ULBP3 complex, 671, 672
NKG2E, 644, 758
NKG2E gene, 620, 644
NK gene complex, 640
NKG2F, 758
NKG2 gene cluster, 633
NKG2 genes, 620
NKG2H, 644
NKG2I, 661
NKG2 receptors in monkey, 641
NKG2 subfamily C (KLRC), 30, 639, 641
NKIS, 676
NKp30, 621, 646, 657, 679, 694, 701, 702
NKp30(CD337), 683, 685
NKp40, 674
NKp42, 702
NKp44(CD336), 621, 657, 674, 683, 694, 702
NKp46, 621, 646, 656, 657, 674, 679, 682, 686, 694, 701–703
NKp80, 674, 693, 694, 701
NKp80 or killer cell lectin-like receptorsubfamily F-member 1 (KLRF1) or NKG2F/KLRC4/CLEC5C), 693, 694
NK receptor clusters, 725
NKR-P1, 543, 619, 623, 624, 632, 633
NKR-P1 (KLRB), 619, 702
NKR-P2, 668
NKRP1A, 758
NKRP-1A, 626, 628
NKR-P1A in cancer cells, 432
NKR-P1A in clinical disorders, 630
NKR-P1A is an activating receptor, 624
NKR-P1A$^+$ T cells, 625
NKR-P1B, 621, 624, 627
NKR-P1^{bright} /T-cell receptor (TCR), 632
NKR-P1C, 524, 626, 630, 758
NKR-P1$^+$ cells, 630
NKR-P1C$^+$ NK, 624
NKR-P1D, 621, 624, 627
NKR-P1^{dim} /TCRγδ/CD3$^+$/CD5$^+$, 632
NKR-P1family, 623
NKR-P1 selectively binds Gz, Gs, Gq/11, and Gi3, 628
N-linked glycosylation site in ligand recognition, 362
NMR, 38, 40, 854, 867, 869, 870, 894, 912, 913
NMR analysis of Sn, 357
NMR monitoring, 215
NMR of the sulphated (Su) Le^a, 559
NMR relaxation, 321
NMR structure, 40
NMR structure of HIP/PAP, 868
NMR studies, 133
NO, 358, 514
NO and superoxide anion radicals (O$^-$2), 423
NOC-7 and NOC-8, 199
NOD Cd93, 906
Nodes of Ranvier, 831
Nod-like receptors (NLR), 396
Nomenclature of PGs, 801
Non-alcoholic fatty liver disease (NAFLD), 282
Nonciliated oviductal epithelial cells, 427
Non-classical MHC-I molecule Qa-1b, 653
Nonclassical secretion pathway, 191
Non-endocytic molecule on mature dendritic cells, 343
Nonglycosylated firefly luciferase, 33
Non-HLA class I-specific receptors, 619
Non-Hodgkin's, 297
Non-Hodgkin's lymphoma, 1052

Non-immune organs, 334
Non-insulin-dependent diabetes mellitus, NIDDM, 1000
Non-lectin inhibitory receptors, 693
Nonneuroendocrine large cell carcinomas (LCCs), 970
Non-obese diabetic (NOD) mice, 849, 906
Non-pulmonary SP-A, 518
Non-pulmonary SP-D, 537
Nonseminomatous, 302
Nonsialylated Lewis X (Le^x) and Le^Y glycans, 775
Non-small-cell lung cancer (NSCLC), 299, 851, 969, 1008
Nonsmokers, 957
Non-specific interstitial pneumonia (NSIP), 956
Nonsteroidal anti-inflammatory agents, 1017
NO production, 980
Norfloxacin antibiotic, 1047
Normal-sized octadecamer (APP-II), 961
N-syndecan, 834
NTB-A, 701
N-terminal cysteine-rich (CR) domain, 334, 340
N-terminal sialoadhesin glycopeptide domain, 355
N-terminal V-set Ig-like domain of siglec-7, 388
N-terminal V-set Ig-like domains 3, 83, 393
Nuclear antigens, 173
Nuclear factor of activated T cells (NFAT), 389, 529, 1035
Nuclear hormone receptors, 43, 47
Nuclear magnetic resonance, 355
Null mice, 371
NWGR anti-death motif of Bcl-2, 284

O

Obesity, 893
Obesity-binding protein-1, 397
Obesity binding protein-2 (OB-BP2), 384
OB-receptor (Ob-R), 385
OCA-B, 234
Occupational levels of diesel exhaust (DE), 957
Ocil/Clr-b, 628
Ocil/Clr-b as ligand, 627
Ocil/Clr family of genes, 627
OCRL1, 99
Olfactory glomeruli, 228
Oligo and polysaccharide recognition by SP-D, 532
Oligodendrocyte (OL), 826, 827, 837, 838
Oligodendroglial membranes of myelin sheaths, 368
Oligodendrogliomas, 15, 229
Oligodendrogliopathy in multiple sclerosis, 368
Oligonucleotide antagonists, 569
Oligosaccharide sequence Neu5Ac α2,3Gal β1,3GalNAc, 357
OL-rich hippocampal fimbria, 837
Omentin of adipose tissue, 449
3-O-methyl-D-galactose, 3-O-methyl-D-fucose, 441
ONOO$^-$ nitrated SP-A, 518
Ontogeny of surfactant apoprotein D, 529
Oosteoclast inhibitory lectin, 628
O-phosphorylethanolamine (O-PE), 164
O-3 polyunsaturated fatty acids, 93
Opossum (*Monodelphis domestica*), 439
Opossum CD-MPR, 84
Opsonic activity of multimerized chimeras of rSP-D, 978
Opsonin, 174, 492
Opsonization, 19, 167, 173, 539, 540
Opsonizing effects, 413
Oral squamous cell carcinomas (OSCCs), 259
Orexin neurons, 178
Organization of CD33-rSiglec genes, 384
Organ of corti protein 1 (OCP1), 132

Organ transplantation, 741
Ornithine decarboxylase, 47
Orthologs of OLR1, CD69, KLRE, CLEC12B, and CLEC16p genes, 725
Orthologues to human NKG2D, 668
Orthotopic liver transplantation (OLTx), 432, 632
Orthotopic Xenograft model of colorectal cancer, 919
OS-9, 110, 138
O-specific polysaccharides (O-antigens), 416
Osteoarthritis (OA), 182, 279, 428, 429, 803, 804, 818, 1002
Osteoarthritis SF, 281
Osteoblastic activity, 887
Osteoblasts, 64
Osteoclast inhibitory lectin, 628
Osteopontin, collagens, and matrix, 819
Osteosarcoma cells, 217
Osteosarcomas, 565
3-O-sulfated galactose, 335
3-O-sulfated Le^a, 335
4-O-sulfated N-acetylgalactosamine, 332
O-sulfate esters, 570
Ovarian and breast cancer, 300
Ovarian cystadenoma, 559
Ovarian steroids, 217
Ovarian stroma, 862
Oviductal glycoprotein, 426
Oviductin, 423, 427
Ovine pulmonary adenocarcinoma, 970
Ovocleidin, 17, 854
Ovulatory process, 862
Oxamido-Neu5Ac [methyl α-9-(amino-oxalyl-amino)-9-deoxy-Neu5Ac], 389
Oxidant stress-induced endothelial dysfunction, 995
Oxidation intermediates of nitrite, 975, 976
Oxidation of LDL, 996
Oxidative burst, 233
Oxidative damage, 545
Oxidative stress and hyperoxia, 544
Oxidative stresses, 35
Oxidized galectin-1, 226, 230
Oxidized galectin-1 as therapy of ALS, 230
Oxidized low-density lipoprotein (Ox-LDL), 170, 217, 736, 996
Oxidoreductase ERp57, 29
Oxidoreductases, 10
Ox-LDL, 726, 740, 996
Ox-LDL-binding activity, 739
Oyster *Crassostrea angulata*, 445
Ozone-exposed Balb/c mice, 545

P

p53, 280
p97 AAA ATPase, 138
Pachytene spermatocytes, 65
PACS-2, 33
P. aeruginosa, 182, 337, 971
PA-IIL from, 446
    crystal structure of PA-IIL in complex with fucose, 446, 447
    Pseudomonas aeruginosa, 446
Paired inhibitory and activating receptors, 362, 383
Palmitoylation, 74, 75
Palmitoylation of H1, 711
1-Palmitoyl-2-linoleoyl phosphatidylcholine (PLPC), 515
Palmitoyl-protein thioesterase, 1044
PAMP, 14, 16, 396, 493
Panceatitis associated protein(PAP)/hepatocarcinoma-intestine pancreas (HIP), 865

Index

Pancreas-kidney transplantation (SPKT), 946
Pancreas regeneration, 851855
Pancreatic acinar cells, 863
Pancreatic acinar protein, 871
Pancreatic adenocarcinoma, 868
Pancreatic ductal adenocarcinoma, 301
Pancreatic elastases, 863
Pancreatic ER kinase (PERK), 32
Pancreatic juice, 852, 862
Pancreaticobiliary maljunction/choledochal cysts, 855
Pancreatic stone protein, 473, 475, 479, 852–854, 857, 866
Pancreatic stone protein 2 (PSP-2), 850, 852, 854
Pancreatic stone protein/lithostathine (PSP/LIT), 852
Pancreatic thread protein, 850, 855, 856
Pancreatic-type msPLA2, 341
Pancreatitis, 852, 863, 864
Pancreatitis-associated diabetes, 851
Pancreatitis-associated thread protein (PATP), 856
P- and E-selectin in differentiation of hematopoietic cells, 610
Paneth cells, 869
Pantroglodytes NK cells, 694
PAP, 849, 867, 869
PAP2, 861
PAP: a multifunctional protein, 863
PAP/HIP$^{-/-}$ mice, 864
PAP-I, 848, 857, 858, 859, 862, 864, 870
PAP-I: an anti-inflammatory cytokine, 863
PAP-I and PAP-III mRNA in the injured cortex, 861
PAP-II, 848, 851, 859, 862, 864
PAP-III, 859, 860, 862
Papillary carcinomas (PTC) adenomas, 293
Papillary microcarcinomas, 292
Papillary thyroid microcarcinomas, 292
PAP-I/Reg-2 protein (or HIP, p23), 861
PAP ligands, 867
PAP, Reg Iα and Regβ, 851
Paracoccidioides, 182
Paracrine/endocrine functions, 462
Paracrine tumor suppressor, 849
Paragangliomas, 870
Parasite ppGalNAc-Ts, 319
Parasites and amoeba, 1014
Parathyroid carcinoma, 300
Parathyroid hormone related protein (PTHrP), 46
Parkinson disease (PD), 1004
Parotid gland, 157
Parturition, 519
Pathogen-associated molecular patterns (PAMPs), 415
Pathogenesis of arthritis, 421
Pathogenesis of asthma, 514
Pathogenesis of cerebral vasospasm, 1004
Pathogenesis of cerebrovascular disease, 1005
Pathogenesis of pulmonary alveolar proteinosis, 960
Pathogenesis of SLE, 904
Pathogenesis protein 1 (RTVP-1), 921
Pathogenesis-related 1 proteins (CAP), 921
Pathogenic spirochaetes, 1013
Pathogen recognition, 934
Pathogens and antifungal defense, 729
Pathologies associated with PGs, 815
Pathophysiological role of CD94/NKG2 Complex, 658
Pathophysiology of ficolins, 415
Pathways for Golgi-to-ER transport, 152
Patients of high-altitude pulmonary edema, 973
Patients with CCPA or ABPA of Caucasian origin, 975
Patients with the acute brain infarction, 997

Pattern recognition receptor (PRR), 13, 16, 17, 282
p33/41 binds to
    fetuin, 464
    glycopeptides, 464
    heparin, 464
    or phosphatidylinositol (PI)/PC, 464
    phosphatidylcholine (PC), 464
    phosphatidylethanolamine (PE)/PC, 464
    phosphatidylserine (PS), 464
    phospholipid vesicles, 464
P. brassicae, 322
P. carinii, 337
PC-1 as a mechanosensor, 1029
PC-1 complexes with talin (TAL), 1032
    focal adhesion kinase (FAK), 1032
    paxillin (PAX), 1032
    vinculin (VINC), 1032
pDC, 362
PDCD6, 456
PDGF receptor, 918
PDI 33
pDNA-coated nanoparticles, 1050
PDX1-Cre-mediated pancreas inactivation of IGF-I gene, 851
PDX-1/INGAP-positive cells, 872
PECAM-1, 994, 999, 1005
PECAM-1 gene, 996
Pentadecapeptide from INGAP (INGAP-PP), 872
Pentameric and decameric forms, 172
Pentameric discs, 164
Pentraxin, 341
Pentraxin 3 (PTX3), 163, 179
Pentraxin function by lactic acid, 174
Pentraxins (PTX), 6, 163
Pentraxins as receptors, 175
Pentylene tetrazol (PTZ), 896
Peptide loading complex (PLC), 51
Peptide:N-glycanase, 109
Peptide 23/PAP, 848, 869, 870
Percutaneous transluminal mitral valvuloplasty (PTMV), 993
Perforin, 33
Periaxonal Schwann cell, 368
Pericytes, 918
Perinatal lamb respiratory syncytial virus (RSV) model, 537
Perineuronal nets (PNN), 825, 826, 830, 835
Periodontal ligament cells (PDL), 258
Periodontal surfaces, 955
Periodontitis, 1013
Peripheral arterial disease, 992
Peripheral axotomy, 231
Peripheral benzodiazepine receptor (PBR)
Peripheral blood (PB), 679
Peripheral lymph nodes (PLNs), 554, 992
Peripheral nervous system (PNS), 369, 370, 872
Peripheral neuropathies, 36
Peritoneal dialysis (PD), 946
Peritoneal metastasis, 871
Peritoneal NK cells, 661
Peritonitis, 946
Peritonitis virus (FIPV), 784
Perivitelline space after egg activation, 45
Perlecan, 802, 806, 816, 817, 819
Perlecan, agrin, and syndecan-4, 560
Permeabilization of bacteria by SPs, 545
Peroxisome proliferator-activated receptor γ (PPARγ), 45, 260, 598, 728
Peroxynitrite (ONOO$^-$), 518, 976

Pervanadate treatment, 393
p53 expression in Crt-deficient cells 45
PGN, 415
PGs in developing brain, 825
PHA, 4
Phagocyte-bacteria interactions, 336
Phagocytes, 42, 413
Phagocytic interaction site on MBL, 491
Phagocytic receptor, 415
Phagocytosis, 91, 166, 167, 183, 410, 539
Phagocytosis of apoptotic bodies, 400
Phagocytosis of microorganisms, 340
Phagosomelysosome, 336
PHA-L, 21
PH domain of SOS1, 676
Phenotypes associated with leukemia, 658
Pheochromocytomas, 302, 870
Pheomelanin, 893
Phorbol ester (PMA), 607
Phosphacan, 825, 826, 829, 833, 837, 838
Phosphacan-KS, 838
Phosphatases SHP-1 and SHP-2, 384, 735
Phosphatidylcholine, 167, 506, 510, 908
Phosphatidylethanolamine, 510
Phosphatidylglycerol (PG), 510, 511
Phosphatidyl-inositol, 510, 511
Phosphatidyl inositol (PI) and sphingomyelin (SM), 511
Phosphatidyl inositol-3 kinase, 365, 667, 676, 858, 909
Phosphatidylinositol 3-kinase/Akt (PI-3-K/AKT), 178
Phosphatidylinositol-mannosides (PIMs) of Mtb, 786
Phosphatidylinositol-specific phospholipase C (PI-PLC), 90, 671, 831
Phosphatidyl-myo-inositol mannosides (PIMs), 336
Phosphatidylserine (PS), 43, 215, 225, 232, 325, 510, 908
Phosphocholine, 166–168, 170
Phosphodiesterase (PD), 147
Phosphoethanolamine, 167, 168
Phospholipase A2 ligands, 342
Phospholipase A2 neurotoxins in snake venoms, 339
Phospholipase A2 (PLA2) receptor, 315, 335, 339, 729
Phospholipase Cγ regulation, 219
Phospholipases A2 as ligand of sPLA2 receptors, 341
Phospholipid composition, 957
Phosphomannosyl receptor, 63
Phosphomonoester linkage, 67
Phosphorylated β-glucuronidase (P-GUS), 1043
Phosphorylethanolamine, 167
Phosphotransferase (PT), 147
Phosphotyrosine, 44
Photoreceptor outer segments (POS), 461
PHPT patients, 1008
Phylogenetic analysis, 207
Physiological fluid flow, 1033
Pieris brassicae, 315, 322
Pierisin-1, 315, 321–323
Pierisin-2, 315, 322, 324
Pierisin-3 and-4, 324
Pierisin-1 induced apoptosis, 323
Pig granulosa cell lysates, 235
Pigmentation, 888, 889
Pigment-type switching, 892
PI3K/Akt, Erk, 539
PI-3 K (phosphoinositide 3-kinase)-mediated pathway, 428
PI3K-related protein kinases, 671
Pinched conformation and stacked conformation, 425
Pioglitazone, 598
PIP3, 677

Pituitary, 830
PKB/Akt, and FKHR, 516
PKC activation, 219
PKCε, 197
PKC pathway, 595
PKD (Ig-like), 479
PKD1 and PKD2 genes, 1028
PKD (or Ig-like) domains, 480, 1029
PKD2 gene on chromosome 4, 1027
Placental anticoagulant protein II, 461
Placental functions, 916
Placental syncytiotrophoblast, 916
Placental trophoblast, 46, 386
Planctomycetes phylum, 446
P-, L-, and E-selectin, 12
P-, L, and E-selectins overlapping leukocyte ligand specificities, 585
Plants pathogenesis proteins (PR-1), 922
Plasma CRP, 430
Plasmacytoid CD11c_CD123^{bright}, 765
Plasmacytoid dendritic cell (pDC), 385, 396, 656, 749
Plasmacytoid monocytes, 749
Plasma levels of vWF and sP-selectin, 1001
Plasminogen, 84, 88, 90, 167, 460, 883
Plasminogen activator inhibitor, 167, 170
Plasminogen kringle-4 protein domain, 885
Plasminogen-tetranectin complex, 884
Plasmodium falciparum, 934
Platelet α(IIb)β3 integrin (GP IIb/IIIa receptor), 993
Platelet-activating factor (PAF), 582
Platelet activation, 575, 732, 733, 991, 992
Platelet adhesion and activation, 575
Platelet aggregation, 411
Platelet-dense granules, 1040
Platelet-derived growth factor (PDGF), 819
Platelet-endothelium interaction, 742
Platelet glycoprotein (GP) Ib-V-IX receptor, 993
Platelet GPIIb/IIIa complex, 563
Platelet membrane receptors (integrins), 993
Platelet receptors GP VI and GP Ia/IIb, 993
Platelets (P-selectin), 992
Platlet CRP (pCRP), 171
PLC, 51, 91
PLCγ, 677
PLC-γ2, 676, 729
PLC-γ2-[Ca^{2+}]i, Raf-1-Erk1/2, 696
    PKC-p38 coupling pathways, 696
Pleckstrin homology (PH) domain-containing binding partner, 676
Pleiotropic effects, 167
Plexiform layers, 834
PLN addressin (PNAd), 563
p53/56lyn kinase, p85, 365
p27-mediated activation, 250
Pneumococcal C, 166
Pneumococcal infection, 272
Pneumocystis, 728
Pneumocystis (Pc)-mediated, 334
Pneumonitis, 955
Podocalyxin-like protein (PCLP), 607, 1007
Podoplanin (aggrus), 726, 732
Podoplanin as ligand, 732
Polar *vs.* nonpolar interactions, 655
Poliovirus receptor (PVR, CD155), 701
Pollinosis from *Parietaria judaica*, 999
Poly(amino acids), 1051
Poly(DL-lactide-coglycolide), 1050
Polyadenylation signals, AATAAA, 910

Polyadenylation site, 172, 177
Polycations, 167
Polycystic kidney disease (PKD) gene 1 (PKD1), 1027
Polycystic: kidney disease-1 promoter, 1028
Polycystin 1, 479, 1028
Polycystin-1 (PC-1), 1027
Polycystin 2, 1028, 1030
Polycystin-2 (PC-2), 1027
Polycystin-1 and polycystin-2 function together, 1032
Polycystin-2 (or TRPP2) functions include, 1030
    asymmetric gene expression, 1030
    cell proliferation, 1030
    mating behaviour, 1030
    mechanosensation, 1030
    sperm fertilization, 1030
Polycystin-L, 1031
Polycystin 1 (PKD1)-like or TRPP1, 1028
Polycystin 2 (PKD2)-like or TRPP2, 1028, 1030
Polycystins, 480, 1027, 1028
Polycystins and sperm physiology, 1034
Polycystin-1 (TRPP1) structural domains:leucine-rich repeats (LRRs), C-type lectin domain, LDL-A region, multiple Ig-like domains (or PKD domains), REJ domain, GPS domain, 11 transmembrane domains PLAT domain, a G-protein-binding motif, coiled-coil region, 1028
Polycythemia, 323
Polygalactose, 255
Polyinosinic-polycytidylic acid (poly IC), 254
Poly-(L-lysine citramide imide), 1047
Polymeric glyco-conjugates, 1046
Polymeric nanospheres (NS), 1047
Polymorphism in L-SIGN, 778
Polymorphism in NKG2 genes, 658
Polymorphism in type 2 diabetes mellitus, 564
Polymorphism of CD94, NKG2A, NKG2CE, and NKG2D genes, 620
Polymorphisms at codons 19, 62, and 133 in SP-A1, and 223 in SP-A2, 974
Polymorphisms for P-selectin, L-selectin, 994
Polymorphisms in SP-A1, SP-A2, SP-B, and SP-D genes in ARDS, 974
Polymorphisms of CLEC4M, 792
Polymorphisms of E-selectin, 994, 1004
Polymorphisms of SP-A1 and SP-A2, 962
Polymorphisms of SPs, 543
Polymorphonuclear leucocytes (PMNs), 410, 544
Polymyositis/dermatomyositis (PM/DM), 960
Polyoma virus infection, 657
Polypeptide HBGp82, 222
Poly(ADP-ribose)polymerase, 275, 323
Poly (ADP-ribose) synthetase/polymerase (PARP), 851
Pompe disease (glycogen storage disease type ii), 1040
Pompe disease fibroblasts, 101
p13.3 on chromosome 19, 563
Popliteal LN (PLN) expansion, 919
Porcine alveolar macrophages (PAM), 360
Porcine chromosome 14, 531
Porcine DC-SIGN, 774
Porcine Eustachian tube (ET), 531
Porcine lung: SP-D (Sftpd) and SP-A (Sftpa), 531
Porcine pancreatitis, 1005
Porcine SP-D, 539
Porcine SP-D (pSP-D), 543
Porphylomonas gingivalis, 258
Positive regulatory domains (PDI to PDIV), 601
Posterior colliculus, 833
Postpneumonectomy (PNX) lung, 956
Potential hyaluronan binding sites, 423

Potent inhibitors of MAG, 399
PP-13, 205
PPAR, 743
PPARγ, 338
PPARγ-agonist thiazolidinediones (TZDs), 851
ppGalNAc-T, 317
ppGalNAc-Ts (ppGalNAc-T-T7 and-T9 [now designated T10]), 317
ppGalNAc-T-T1, 318
p50/p58/p70 family of Ig-like receptors, 621
PP4-X, 461
P. rapae, 322
Pravastatin (Pravacol) and atorvastatin (Lipitor), 170
Preadipocytes, 260
Pre-B cell receptor, 222
Predictive value of sE-selectin in CAD, 996
Preeclampsia (PE), 387
Preeclamptic decidual cells, 236
Preeclamptic EVT, 236
Preeclamptic placentas, 236
Pre-eclamptic women, 651
Pregnancyassociated plasma protein A, 895
Prelysosomal endosome compartment, 91
Previous assignment of annexin 5 to 4q28-q32, 464
PRG2, 894
Primary biliary cirrhosis (PBC), 937, 1004
Primary cilia, 1027
Primary cilia of inv/inv mouse renal epithelial cells, 1033
Primary effusion lymphoma (PEL), 294
Primary forms of Pompe disease, 1040
Primary human primitive neuroectodermal tumor (PNETs), 856
Primary hyperparathyroidism, 1008
Primary lung adenocarcinoma, 970
Primary olfactory neurons, 199, 247
Primary open angle glaucoma (POAG), 253
Primary sclerosing cholangitis, 681
Primates MBL, 495
Prion encephalopathies, 63
Prion protein, 857
Pro-apoptotic cytokines, 281
Proapoptotic lipocortin I, 227
Procollagen transcripts, 216
Production of cytokines, 514
Pro-inflammatory, 167
Proinflammatory cytokines, 342, 496, 543, 849
Proinflammatory signal, 276
Prolactinomas, 301
Proliferative glomerulonephritis, 364
Proliferin, 87
Proline rich regions, 134, 248, 750, 752
Promoter elements, 82
Promoter of CD69 gene, 633
Promoter of the PAP-I gene, 861
Promoter sequence, 338, 640
Promotion of axonal regeneration, 230
Promyelocytic protein (MR60), 148
Pro-opiomelano-cortin (POMC)-derived ligands, 890
Propeller structures, 415
Proprionibacterium acnes, 934
Prosaposin, 91, 1043
PROSITE-type signature sequences, 198
Prostate cancer, 252, 254, 300
Proteasomal degradation, 389
Proteasome subunits, 43, 51
Protection against bacteria, 682
Protection of type II pneumocytes from apoptosis, 516
Protein A, 909

Protein C, 915
Protein C (inactive)
   decreases WBC, 916
   inhibits factor V and Vlll, 916
   and NK-kB, 916
   promotes fibrinolysis, 916
Protein C activation, 913
Protein-carbohydrate interactions, 221, 223, 268, 510, 1029
Protein disulfide isomerase (PDI), 10, 29, 33, 38, 40, 42
Protein domains of layilin, 882
Protein kinase B (PKB), 253
Protein kinase C (PKC), 456, 908, 1032
Protein kinase C-α,-δ,-γ,-η, and ζ, 280
Protein kinases, 904
Protein oxidation by chronic pulmonary diseases, 975
Protein p63 (CKAP4/ERGIC-63/CLIMP-63) I, 509
Protein-protein interactions, 362, 509, 1029
Proteins of cytosolc sorting (PACS), 33, 72, 97
Proteoglycan binding link proteins, 807, 831
Proteoglycan NG2, 197
Proteoglycans, 3, 341, 477
Proteoglycans (PGs), 801, 804, 825, 827
   in carcinogenesis, 818
     facilitate lipid accumulation, 816
Proteoglycans in central nervous system, 825
Proteoglycans in CNS injury response, 836
Proteoglycans in sensory organs, 833
Proteolytic shedding of NKG2D, 669
Protostomes, 351
Proto-, tandem-repeat, and chimera types of galectins, 191
Protozoa, 934
Proximal kinases, 630
Proximal promoter of SP-D gene, 529
Proximal promoter regions, 200
PS (phosphatidylserine), 738
P-selectin, 992, 993, 1005
P-selectin (GMP-140, PADGEM, CD62), 575–577
P-selectin and its ligands, 575
P-selectin as a target of heparin, 1011
P-selectin deficiency attenuates tumor growth and metastasis, 1010
P-selectin-deficient (P-sele$^{-/-}$) mice, 606, 1010
P-selectin-dependent and PSGL-1-independent rolling, 583
P-selectin expression, 1012
P-selectin glycoprotein ligand-1 (PSGL-1/CD162), 581, 583
P-selectin-sulfatide interaction, 993
Pseudogene of mouse CD-MPR, 67
Pseudogene SIGLECP16, 384
Pseudomonas aeroginosae, 14, 175, 447, 682, 934, 966, 998
Pseudotype driven infection of permissive cells in vitro, 789
Pseudoxanthoma elasticum (PXE), 1006
PSGL-1, 575, 580, 589
   carbohydrate structures on, 586
   expressed by primitive hematopoietic cells, 586
   GalNAc-Lewis^x and Neu5Aca2-3Galβ1-3GalNAc
     sequences, 586
   ligand specificities of, 585
   PSGL-1 gene, 585
   PSGL-1 VNTR polymorphism in CHD, 586
   signaling by, 588
   structural polymorphism in, 585
   tandem repeats (VNTR), 583
   TATA-less promoters, 585
PSGL-1 (CD162 P), 605, 993
PSGL-1 and relating ligands, 605
PSGL-1 binds L-selectin, 564
PSGL-1-dependent and-independent E-selectin rolling, 605
   P-selectin-dependent rolling, 605
PSGL-1 in the CD34$^+$ cells, 562
PSGL-1$^{-/-}$mice, 564, 605
PSGL-1: the major ligand for P-selectin, 583
   BACE1 acts on, 587
   bovine PSGL-1, 584
   eosinophil *vs.* neutrophil PSGL-1, 584
   genomic organization, 585
   GlyCAM-1 and, 587
   human (PSGL-1, CD162), 583
   lysosomal membrane glycoprotein (lamp)-1 and-2, 589
   mouse homolog of, 584
   PSGL-1-deficient mice, 584
   PSGL-1 in equine, 584
   role:mediates rolling of neutrophils on P-selectin, 587
PSI (plexins, Semaphorins and Integrins), 479
PSI, EGF and CUB domains, 480
Psoriasis, 661
Psoriatic arthritis (PsA), 1002
Psoriatic epithelium, 281
PSP/LIT, 847, 854, 857
PSP/LIT, etiology of AD, 857
PSP/LIT in Alzheimer's disease, 855
PSP/LIT/RegIα (heterodimerization), 860
PSP/Reg, 853
PSP/Reg γ, 849
p72syk kinase, 365
p32/TAP, 908
PTKs, PTPs and PLCγ1 in signal, 366
P-type lectins, 6, 7
Pulmonary alveolar microlithiasis (PAM), 971
Pulmonary alveolar proteinosis (PAP), 960
   SP-A and SP-D in BAL fluids and sputum diagnostic
     for PAP, 960
Pulmonary and rectal tumors, 251
Pulmonary cytokine-network, 955
Pulmonary emphysema, 964
Pulmonary fibrosis (PF), 955, 957
Pulmonary hypoplasia, 976
Pulmonary lipoidosis, 544
Pulmonary sarcoidosis, 945, 960
   KL-6, SP-A and SP-D levels in BALF, 960
Pulmonary SP-A: forms and functions, 501
Pulmonary SP-D, 12, 480
Pulmonary surfactant, 501, 955
Pulmonary surfactant protein A, 501
Pulmonary surfactant protein-D, 527
Pulmonary surfactant proteins, 501
Pulmonary surfactant-system, 955
Pulmonary tuberculosis, 182, 966
Pupae of Pieris rapae, 323
PU.1 sites, 338
PUVA treatment, 297
Pyrrolidine dithiocarbamate (PDTC), 600

Q
Qa-1b, 640, 643, 649, 653
Qa-1b-dependent modulation of DC and NK cell cross-talk, 656
Qa-1b/Qdm tetramer, 653
Qa-1/Qdm complex, 623
QBRICK/Frem1, 480, 922–924
Qdm, 653
Qdm (Qa-1 determinant modifier), 653

Quadruple-helical fibrils (QHF-litho), 857
Quaternary structure of SP-D in BALFs, 857
Quinic and shikimic acids, 1047

R

Rab9, 72, 1042
Rab11A, 756
Rabbit bladder galectin-4, 248
Rabbit P-selectin, 577
Rab GTPase, 756
Rabs (Rab5, Rab4, Rab7, Rab15, Rab22), 100
Ra chemotype strain of *S. typhimurium* (TV119), 410
RACK1, 732
RACK1b, 726
rADCC, 624
Radiation pneumonitis (RP), 956
Radiographic abnormalities in patients with IPF, 957
RAE-1, 621, 668–671, 677
Raf-1 and CARD9 dependent pathways, 728
RAG-1-deficient mice, 360
RAGE, 15, 280
Rainbow trout plasma intelectin, 450
Ralstonia solanacearum, 447, 448
RANTES-28G, 789
RARβ receptors, 218
Ra-reactive factor (RaRF), 494
Ra-reactive factor: a complex of MBL, 494
Ras GTPase activating protein, 456
Ras in cell transformation, 227
Ras-MEK-ERK cascade, 220, 227
Ras/Raf-1/phospho-MEK macrocomplex, 611
RA synovial fibroblasts (SF), 281
Rat ASGP-R, 709, 717
    three subunits, rat hepatic lectin (RHL) 1, 2, and 3, 709, 712
Rat asialoglycoprotein receptor, 709
Rat CD209b, 790
Rat cystitis model, 862
Rat GlyCAM, 561
Rat hepatic fucose binding protein (FBP), 440
Rat hepatic lectins (RHL) 1, 2, and 3, 479
    HHL1 and HHL2, 479
Rat liver MBP, 494
Rat LOX-1 gene, 737
Rat lung SP-D, 529
Rat MBP-A, 475, 645
Rat model of monoarthritis, 858
Rat NKG2A, 640
Rat NKR-P1A, 624
Rat olfactory placode, 834
Rat pancreatic islets, 870
Rat PAP-I, PAP-II, and PAP-III, 860, 862
Rat 4q33-q34, 847
Rat RAE-1-like transcript (RRLT), 670, 672
Rat zitter (zi) mutation, 893
R-cadherin Ocil, 621
RCA-I and RCA-II, 313
RDS in premature infants, 973
Reactive nitrogen intermediates, 967
Reactive non-malignant bile ductules, 867
Reactive oxygen and nitrogen-induced lung injury, 518
Reactive oxygen species (ROS), 599, 891, 993, 994
    endothelial cell NO synthase (eNOS), 994
    fatty acid peroxyl radical (R-COO·), 994
    hydrogen peroxide (H2O2), 994
    hydroxyl radical (HO·), 994
    and lipid radicals, 994
    nitric oxide (NO·), 994
    peroxynitrite (ONOO⁻), and hypochlorous acid (HOCl), 994
    superoxide anion (O_2-), 994
Receptor for activated C-kinase-1(RACK1), 726, 732
Receptor for C1q, 901, 908
Receptor for tenascin-C, 460
Receptor-mediated endocytosis, 333, 362
    via the clathrin-coated pit pathway, 709
Receptor-mediated uptake of mannan-coated particles, 1046
Receptors for defense collagens, 496
Receptor type protein tyrosine phosphatase RPTP ζβ, 838
Recognition of pathogens, 492
Recognition of pathogens and endocytosis, 365
Recognition of tumor glycans, 785
Recognition sites of CI-MPR, 88
Recombinant α-(L)-iduronidase, 1042
Recombinant f4S (rf4S), 1044
Recombinant trimeric neck + CRDs of human SP-D (NCRD), 978
Recombinant truncated, 87
Recruitment of phosphatases, 651
Recruitment of SHP-1 via phosphorylated ITIMs, 366
Recurrent bronchitis and SP-D, 956
Recurrent miscarriage, 463, 934
Recurrent respiratory infections, 415
Recurrent spontaneous abortions (RSA), 683
Recycling of ERGIC-53, 150
Reef shark, 883
REG, 478
Reg 1, 848
REGα, 859
REG 1α/Reg 1A, 850. 851
REG1A, REGL, PAP, 852
Regenerating gene family, 847
Regenerating (Reg) gene products, 847
Regenerating genes (Reg), 477
Regenerating islet derived-2 (REG-2), 857
Regenerating islet-derived-like/pancreatic stone protein-like/
    pancreatic thread protein-like, 847
Regenerating islet derived protein 1-β, 852
Regenerating (Reg) islet-derived Reg-IIIβ, 848
Regenerating protein Iα, 850
Regenerating skeletal muscles, 232
Regeneration of islets, 855
Reg family: four subtypes (type I, II, III, and IV), 847
Reg group of lectins, 480
Reg I, 847, 849, 852
REGIA (*Reg Iα*) *gene in human pancreas*, 850
REG Iα promoter region, 851
Reg Iβ, 850
REGIB (*Reg Iβ*) *gene in human pancreas*, 852
Reg II, 847, 849
Reg III, 848, 858–860, 862
Reg-III-α, 848, 859
REG-IIIα/HIP/PAP/REG-III 1α (REG-1α) proteins, 859
Reg IIIβ and Reg IV, 847
Reg-IIIδ. Reg-IIIδ, 859
Reg-IIIγ, 859, 863
Reg-III/PAP, 852, 853
REG-II/REG-2, 857
Reg I-knockout mice, 850
Reg-I, Reg-II, and Reg-III gene, 859
Reg IV, 848, 871
Reg IV (RELP), 870
Reg IV-knockdown mice, 871
Reg/lithostathin genes, 852

Reg/PAP, 849
REG proteins, 479, 847, 850
REG-related gene (REGL), 847, 852
REG related sequence (REGL), 850
Regulation by androgens, 47
Regulation by cytokines, 647
Regulation by IL-15, 647
Regulation of bad protein, 300
Regulation of CD22, 364
Regulation of CD94/NKG2, 647
Regulation of dectin-1, 727
Regulation of Gal-1 gene by
    estrogen, 217
    progesterone, 217
    retinoic acid, 217
    TSH, 217
    TSH rat thyroid, 217
Regulation of MBP gene, 490
Regulation of myotube growth, 232
Regulation of NK cell receptors, 647
Regulation of NKG2D, 684
Regulation of sialoadhesin, 358
Regulation of SP-D, 536
Regulation of TM activity, 910
Regulator of inflammation, 864, 915
Regulatory CD4+CD25+T cells, 731
Regulatory domains, 459
Regulatory SNPs in CHI3L1, 429
Regulatory T cell activity, 235
Regulatory T cells in controlling onset of diabetes in mice, 566
REJ, 480
REJ domain, 479
Release of L-selectin, 558
Rel proteins (NF-kB), 164
Renal agenesis, 923
Renal apical brush-border membrane (BBM), 91
Renal cilia, 1033
Renal defects, 923
Renal failure, 1005
Renin and pro-renin, 1045
(Pro)renin receptors, 65
Reperfusion injury, 998
Replacement therapy, 851
Repression of Gal-3, 276
Repressor cis element, (T/C)TCCCCT(A/C)RRC, 812
Reproductive tissues, 235, 278
Respiratory disorders, 430
Respiratory distress syndrome (RDS), 543, 961
Respiratory syncytial virus (RSV), 13, 939
Respiratory syndrome (SARS) virus, 791
Respiratory syndrome virus (PRRSV), 360
RET, HBME-1, and Gal-3, 293
Reticulocalbin proteins, 341
Retinal degeneration, 359
Retinal ganglion cell (RGC) projections, 179, 231
Retinal pigment epithelial (RPE) cells, 334, 461
Retinitis pigmentosa, 36
Retinoblastoma protein (RB), 529
Retinoic acid, 45, 89, 90, 669, 1016
Retinoic acid early transcript 1G (RAET1G), 668
Retinoic acid early transcript-1 proteins (RAET-1), 669
Retinoic acid early (RAE) transcripts, 669
Retinoic acid receptor-β, 280
Retinopathy in type 2 diabetic patients, 1001
Retinoschisin or RS1, 11, 12
Retrieval signal in yeast, 151

rfhSP-D treatment, 980
RF/RHD, 944
Rhabdomyosarcomas, 565
RHAMM (receptor for HA-mediated motility), 881
Rheology, 992
Rheumatic disease-associated autoantigen, 49
Rheumatic fever (RF), 943
Rheumatic heart disease (RHD), 943
Rheumatic mitral stenosis (MS), 993
Rheumatoid arthritis (RA), 168, 281, 282, 361, 364, 660, 680, 682, 904, 934, 959, 991, 1001
Rheumatoid disease (RD), 630, 1001
rhGAA, 1040
Rhinitis and nasal polyposis, 999
RHL-1, 11, 714
RHL2, 11
RHL3, 11
Rhodocytin, 726
Rhodocytin/aggretin, 732
Rhodopirellula baltica SH 20, 446
Rhodopsin mutant rats, 833
Ribosome inactivating protein (RIP), 313
Ricin, 3, 313, 331
Ricin-A,-B,-C,-D, and-E, 313
Ricin A chain (RTA), 314
Ricin B chain (RTB), 314
Ricin domain, 479, 480
Ricin homolog from Abrus precatorius, 314
Ricinus communis (castor bean), 313
Ricinus communis lectins, 313
Ring fingers, 134
Risk factor for cardiovascular events, 738
Risk of heart disease, 916
RNAi, 97
r-Nogo-66/MAG, 373
Roc1/Rbx1, 125
Rodent homologs of MICA and MICB, 668
Rodents chromosome 2 F-H1, 355
Role in fertilization, 904
Role in pathology, 909
Role in renal disease, 359
Role of CD62L$^+$ cells in gastric cancer, 566
Role of hypoxia, 600
ROS and O2-., 742
Rosiglitazone (Avandia), 170
Ro/SS-A (Ro) antigen, 52
Rosuvastatin, 169
Rough ER (rER), 58
Round spermatids, 65
RSV and other microbial agents, 957
RTVP-1 protein, 922
R-type lectin domain, 313, 331
R-type lectin families, 313
R-type lectins, 5, 313, 332
R-type lectins in butterflies, 322
Russell bodies, 154
Ryncolin 1, 411
Ryncolin 2 (rynchops ficolin), 411

S

Saccharides, 202
Saccharomyces cerevisiae, 272, 934
Sacculotubular network, 59
Sailic acid-binding lectins, 8
Salivary MG2 saccharides, 560

Salmonella, 493
Salmonella enteric, 168
Salmonella infection, 277
Salmonella montevideo, 934
Salmonella typhimurium, 182, 410, 411, 413, 414
Sambucus nigra agglutinin (SNA), 314
Sandhoff disease, 1043
Sandhoff disease model mice (SD mice), 1043
Sanfilippo D syndrome, 1044
SAP130 acts as a Mincle ligand, 765
Saposin-like protein family, 1041
Sarcoidosis (sarcoid, Besnier-Boeck disease or Besnier-Boeck-Schaumann disease), 661, 957, 960, 995
Sarcoidosis and MBL variants, 945
(S)-armepavine, 941
SARS-coronavirus spike protein, 789
SARS-CoV, 792
Saturated phosphatidylcholine (Sat PC), 544
SauFBP32, 442
S. aureus, 182, 540, 909
SBD (sugar binding domain), 127, 133
SBD-chitobiose complex, 133
Sca-1^+ c-kit$^+$ cells, 611
Scanning electron microscopy, 857
Scavenger receptors, 13
s-CD40 ligand, 996
S. cerevisiae, 113, 115, 120, 126, 761
SCF(Fbs1 and-2), 131
SCF(Fbs1) complex, 128
SCF-Fbs1, 123
SCF-Skp2 ligase complex (Cull, Skp1 and Roc1), 135
SCF-type E3 ubiquitin ligase complex, 123
Schistosoma, 767
Schistosoma infection, 283
Schistosoma mansoni, 782
Schistosoma mansoni antigens, 782
Schistosoma mansoni eggs, 283
Schistosome, 1004
Schistosomiasis, 283, 780
Schizophrenia, 371, 896
Schizosaccharomyces pombe, 35
Schwann cell-derived peripheral myelin protein-22 (PMP-22), 36
Schwann cell myelin protein (Siglec 4b), 352, 354, 369
Schwann cells, 279, 335
Sciatic nerve-derived versican-and other CS-PGs, 834
Scleroderma spectrum disorders (SSD), 959
SCOP, 6
SCP, 922
SCP domain, 479
SCP, EGF, EGF and CTLD (SEEC), 920
Scrapie infection, 860
Scytovirin, 21
Sea cucumber, 478
Sea raven, 478
Sea raven antifreeze protein, 475
Sec13/31, 60
Sec23/24, 60
Sec63, 114
SEC-23B, 47
Secondary structure of human PTP, 856
Secondary structure organisation, 480
Sec23p, 150
Secreted and membrane attractins, 888
Secretory ameloblasts, 64
Secretory forms of PSP, 853
Secretory pancreatic stone protein 2, 852

Secretory pathway, 147
Secretory PLA2's (sPLA2's), 341
Secretory stress proteins (SSP), 849
Secretory type IIA phospholipase A2 (sPLA2-IIA), 515
Secretory vesicles or late endosomes, 59
SEEC, 479, 480
Selectin-deficient mice (L$^{-/-}$, PE$^{-/-}$, LPE$^{-/-}$), 556, 604
Selectin family includes, 991
    E-selectin (CD62E), 991
    Lselectin (CD62L), 991
    P-selectin (CD62P), 991
Selectin ligands in cancer cells, 1007
Selectin/MBP chimeras, 604
Selectins, 6, 7, 9, 351, 352, 456, 477, 479, 480, 553, 992, 996
Selenocosmia huwena lectin-I, 12
SELEX, 569
Self-sialome, 402
SEM and TEM, 856
Seminal receptacles and spermathecae, 1034
Seminomas, 302
Semisynthetic sulfated tri mannose C-C linked dimers, 1018
Sensory and motor neurons, 228
Sepsis, 1014
Sepsis-induced acute lung injury, 999
Sequence of human galectin-1, 215
Serglycin, 89, 802
Serine proteases (MASP-1 and MASP-2), 296, 484–486
Serotype III group B streptococci (GBS), 415
Serovar Typhimurium, 168
Sertoli cells, 65
Sertoli-Leydig cell tumors, 278
Serum amyloid A (SAA), 996
Serum amyloid P component (SAP), 163, 171
Serum Clara cell 16 (CC16), 960
Serum coagulation factor V (FV), 576
Serum PAP: an indicator of pancreatic function, 870
Serum Reg IV, 871
Serum SP-A, 959, 963
Serum SP-D, 959, 960, 963
Serum SP-D and KL-6 levels: useful markers for diagnosis and evaluation of ILD, 959
sE-selectin, 998, 1002, 1005
Severe acute respiratory syndrome (SARS), 784, 789
Severe acute respiratory syndrome coronavirus (SARS-CoV), 792
Sex hormone-binding globulin, 180
SEZ-12, 896
SF2-associated p32, 908
SFTPA1, 972
SFTPA2, 972
S4GGnM-specific receptor (S4GGnM-R), 334
SGP28 (specific granule protein of 28 kDa), 922
Shark protein, 883
Shear stress, 372
Sheep sITLN1, 450
Short pentraxins, 163
SHP-1, 362, 381, 382, 399, 641, 651
SHP-2, 362, 396, 399, 641, 651
Sialic acid-binding Ig-like lectins, 352
Sialic acid-binding property of siglecs, 353
Sialic acids (Sias), 351
Sialic acids deuterostome, 351
Sialidase, 1043
Sialidases or neuraminidases, 145
Sialoadhesin, 334, 354
Sialoadhesin binds NeuAc α 2→ 3Gal β1→ 3(4) GlcNAc or NeuAc α2→ 3Gal β1→ 3GalNAc, 354

Sialoadhesin/siglec-1, 336, 353, 357
Sialoadhesin (Sn)/siglec-1 (CD169), 357
Sialoadhesin structure, 357
Sialomucin, 1007
Sialomucin podocalyxin-like protein or PCLP, 563
Sialooligosaccharide ligands, 400
Sialoside, 389, 396
α2-6-Sialylated glycan, 364
α2-3-Sialylated glycans, 253
Sialylated glycans, 356
Sialylated glycan structures, 352
Sialylated ligands and transducing signals, 611
Sialylation, 222
2,3 Sialyllactose, 357
Sialyl Lewis, 351
Sialyl-Lewis A (sLe a), 1010
Sialyl-Lewis antigens, 1010
Sialyl Lewis X (s-Le X), 992, 1007, 1010
Sialyl-Lewis x sequence 6-O-sulfated at N-acetylglucosamine as ligand for L-selectin, 559
Sialyl-Le^x, 354
Sialyl-Tn epitope, 397
α3-Sialyltransferase, 603
Sialyltransferases, 402
Sia-recognizing Ig-superfamily lectins, 352
sICAM-1, 1001
SI-CLP, 423
Siglec-1, 354, 397, 401
Siglec-2 (CD22), 365, 397, 401
Siglec-3 (CD33), 390, 397
Siglec-4, 367, 401
Siglec-5, 353, 366, 385, 386, 392, 397–399, 402
Siglec-5 (CD170), 384
Siglec-6, 353, 386, 397
Siglec-6 (OB-BP1), 386
Siglec-7 (p75/AIRM1), 366, 387, 389, 390, 398
Siglec-8, 353, 355, 389–391, 393, 395–398
Siglec-8 (SAF-2), 389
Siglec-9, 366, 387, 389, 392, 398, 399
Siglec-10, 353, 396
Siglec-11, 392, 393
Siglec-14, 373, 385
Siglec-15, 352, 354
Siglec 16, 362
Siglec-16, 393
Siglec-10,-11,-12, and-16, 392
Siglec-5: an inhibitory receptor, 385
Siglec expression on T lymphocytes, 384
Siglec-F, 390, 391, 396, 399
Siglec-F in cell signaling, 366
Siglec-F ligands, 395
Siglec-F-null mice, 395
Siglec-G, 365, 390, 396
Siglec-10 homology to siglec-5 and siglec-3/CD33, 392
Siglec-8 in Alzheimer's disease, 391
Siglec-11 in brain microglia, 384
Siglec-like gene (SLG)/S2V, 393
Siglec-like molecule (siglec-L1), 401
Siglec-8 long form (siglec-8L), 389
Siglec-4/MAG, 353, 397
Siglec-8-mediated apoptosis, 390
Siglec-mediated cell adhesion to gangliosides, 398
Siglec-5-mediated sialoglycan recognition, 385
Siglec-15 recognizes the Neu5Aca2-6GalNAca, 373.
Siglec-3-related siglecs in mice, 395
Siglecs-7, 383

Siglecs-14,-15, and-16, 383
Siglecs-3 and-5-13 in primates, 401
Siglecs as inhibitory receptors, 353
Siglecs as targets for immunotherapy, 400
Siglecs of sialoadhesin family, 354
Siglec-9 structure, 388
Siglec-5 structure with V-domain, 386
Siglec-10Sv2, 392
Siglec-10Sv3, 392
Siglec-10Sv4, 392
Signaling by DC-SIGN through Raf-1, 781
Signaling lymphocyte activated molecule (SLAM), 384
Signaling pathways, 628
Signaling pathways by Dectin-1, 728, 729
Signalosome, 766
Signal peptide (SP), 489
Signal transduction by CD94/NKG2, 651
Signet-ring cell carcinoma (SRCC), 871
SIGNR1 (CD209b), 789
SIGNR6, 790
SIGNR7, 775, 790
SIGNR8, 790
SIGNR1-deficient mice, 791
SIGN related 1 (SIGNR1) or CD209b, 790
SIGNR1 through SIGNR4, 789
Silica-induced lipoproteinosis, 957
Silicosis, 957
Similarity between C1Q and collectins/defense collagens, 495
    functional similarities, 496
    structural similarities, 495
Similarity of PAP to peptide 23, 869
Single chimera-type galectin, 191
Single nucleotide polymorphism (C→T), 917
Single nucleotide polymorphisms (SNP), 486, 529, 564, 764, 935
Single-strand conformation polymorphism, 828
Sinonasal adenocarcinomas
Sinus node dysfunction, 49
Sirolimus (SRL), 1012
Site-directed mutagenesis, 355
Site for SH-2-containing protein-tyrosine phosphatase-1 in CD22, 366
sITLN3c, 450
Six stranded (S1-S6) β-sheets, 255, 256
Six triple helices, 507
Sjogren syndrome (Sjs), 683, 851, 959
Skeletogenesis, 808
Skin proteoglycans, 805
Skp1, 125
Skp1-Cullin1-F-box protein-Roc1 (SCF) complex, 125
Skp1-Cullin1-Fbx2-Roc1 (SCF-Fbx2), 123
S-Lac lectin, 265
SLAM associated protein (SAP), 384
SLAM-like proteins, 386
SLAM motifs, 393
SLE, 168
SLE (systemic lupus erythematosus), 364
SLG-S, 393
Slime mold *Dictyostelium discoideum*, 67
Slp65, 766
S-MAG, 369
Small cell lung cancer (SCLC), 179, 299, 970
Small MBL-associated protein, 415
Small nuclear ribonucleoproteins (snRNPs), 167, 173
sMAP, a truncated protein of MASP-2, 410
Smooth ER (sER), 58
Sn (sialoadhesin) (siglec-1), 15, 352
Snake venom factor IX0X binding protein, 475

Snake venom toxin rhodocytin, 732
Sn-deficient mice, 359
S-nitrosylation of SP-D, 980
SNO-SP-D, 542
SNP, 545
SNP (rs735240, G>A; rs2287886, C>T), 788
SNP in LGALS2, 200
SNPs in CHI3L1, 429
SNPs in collagen region of SP-A2, 968
  and pulmonary tuberculosis, 968
SNPs in pulmonary diseases, 974
SNPs in selectin genes and IgA nephropathy, 1005
SNPs in the L-selectin gene (SELL1, SELL4, SELL5, SELL6, SELL10, and SELL11), 1005
SNPs of human gene coding SFTPA1, SFTPA2, and SFTPD, 972
snRNPs, 268
SNX-1, 98
SNX-2, 98
SNX5, 99
SNX6, 99
SOCS3 (suppressor of cytokine signaling 3), 389
Sodium butyrate, 217
Sodium salicylate, 1017
Soft tick, Ornithodoros moubata, 416
4-SO4-GalNAcβ1–4GlcNAcβ1–2Manα1-R, 335
4-SO4-GalNAc β1-R, 332
Solid and pseudopapillary tumor of pancreas, 301
S. olivaceoviridis E-86, 321
S. olivaceoviridis E-86 Xylanase: sugar binding structure, 321
Soluble MAG or s-MAG, 368
Soluble platelet agonists, 993
Soluble thrombomodulin (sTM), 1005
Solution NMR, 444
Solution structure of odorranalectin, 444
Solution structure of the C-terminal domain of galactose-binding lectin, 247
Solution structures of galectins, 215
Solvation/desolvation energies, 20
Sorcin, 456
Sortilin, 1042
Sortilin and MPRs, 99
Sorting motifs, 83
Sorting of cargo at TGN, 71
Soybean agglutinin, 4
SP-A, 483, 903, 955
SP-A1, 502, 519, 528
SP-A2, 502, 519, 528, 537
SP-A against oxidative damage, 973
SP-A, a marker for lung adenocarcinomas, 971
SP-A and rheumatoid arthritis, 975
SP-A and SP-B as interactive genetic determinants, 973
SP-A and SP-C, 544
SP-A and SP-D, 501, 956
SP-A and SP-D alleles associated with various diseases, 969
SP-A and SP-D as indicator, 956
  idiopathic pulmonary fibrosis (IPF), 956
  interstitial pneumonia with collagen vascular diseases (IPCD), 956
  of pneumonia, 956
  pulmonary alveolar proteinosis (PAP), 956
SP-A and SP-D deficient mice, 979
SP-A and SP-D double deficient mice, 979
SP-A and SP-D in BAL fluids, 956, 957
SP-A and SP-D modulate
  innate immune cell function, 539
  T-cell dependent inflammatory events, 539
SP-A-as a component of complement system, 513

SP-A binding with lipids, 510
SP-A binds directly to C1q, 513
SP-A deficiency in primate model of BPD, 962
SP-A deficient mice, 542
SP-A effects on inflammation of mite-sensitized mice, 977
SP-A facilitates phagocytosis, 512
SP-A gene polymorphisms and RDS, 972
SP-A in epithelial cells, 518
SP-A in female genital tract and during pregnancy, 519
SP-A inhibits sPLA2, 515
SP-A is assembled as large oligomer, 502
SP-A$^{-/-}$ mice, 512, 514
SP1 and CREB binding sites, 275
SP-A pseudogene, 528
SP-A receptor 210 (SP-R210), 508
SP-A specific receptor (SPAR), 509
SP-B, 501, 537, 955
SP-C, 501, 955
SP-D, 527, 537, 955, 964
SP-D (rfh SP-D), 977
SP-D/Cong(neck +CRD), 978
SP-D deficient (SP-D-/-) mice, 542–544, 965, 979
SP-D gene-deficient (DKO) mice, 542
SP-D in non-pulmonary tissues, 974
SP-D is cruciform, 502
SP-D$^{-/-}$ mice, 544
Specificity for high mannose and fucose, 760
Spermatozoa, 891
Sperm-coating glycoprotein, 904
Sperm-coating glycoprotein (SCP) domain, 920
Sperm-egg interactions, 45
SP1-family transcription factors, 422
Sphingomyelin, 167, 510
Sphingomyelinase, 628
Spinal cord injury, 836, 837
Spinal deformity, 884
SPKT graft survival, 946
sPLA2s, 342
Splice variant of mSiglec-F, 396
Splice variant of siglec-8, 252, 389
Splice variant of siglec-10, 392
Splice variants of annexin A4, 462
Splice variants of human versican, 806, 813
S. pneumonia, 168, 272, 274, 489, 517, 540, 782
S. pneumonia lectin, 446
S. pneumonia lectin SpGH98, 446
S. pombe, 112
Spondyloepiphyseal dysplasia (SED), 818
Spongiform degeneration, 891
Spongiform encephalopathies, 174
Spongiform neurodegeneration, 889
Spontaneous BLRB mammary carcinomas, 294
Spontaneous recurrent abortion, 917
Spotted garden slug, 8
26S proteasome, 123
SPR studies, 426
SPs as a tool for diagnosis of lung tumors, 970
SPs in lung cancer, 969
SP1 site, 338
Squamous cell carcinoma (SCC), 970, 1008
Src-dependent Syk activation of CD69 mediated signaling, 633
Src-family, 611
Src family kinases, 399
Src tyrosine kinases, 219
SSc with and without pulmonary arterial hypertension (PAH), 1004
Stabilin-1 and-2 (also known as FEEL-1/-2), 477

Stabilin-1 interacting chitinase-like protein (SICLP, CHID1), 422
Stable fibrin clot, 916
Staphylococcal enterotoxin A (SEA), 999
Staphylococcal enterotoxin B (SEB), 999
Staphylococcal lipoteichoic acids, 416
Staphylococcus aureus, 489, 934, 1013
Staphylococcus epidermidis, 934, 1013
Staphylococcus pyrogenes, 934
STAT1, 272
STAT3, 164, 272, 851, 858, 864
STAT5, 272
STAT6, 1033
Statin therapy, 169, 1017
STAT-6 knockout mice, 728
Stem cell growth factor (SCGF), 882, 888
Steroid receptors, 45
Steroids, 1018
ST6Gal-I knockout mouse, 361
ST6Gal$^{-/-}$ mice, 365
Stimulation of galectin-9, 254
Stimulatory receptors, NKR-P1A, NKR-P1C, and NKR-P1F, 625
sTLR4, 509
Streptococci, 934
Streptococcus pneumonia, 934
Streptococcus pneumoniae bacteraemia, 430
Streptococcus sanguis, 1013
Streptomyces lividans, 321
Streptozotocin-induced (STZ) diabetes, 858, 872
Streptozotocin induced diabetic, 281
Streptozotocin (STZ)-treated rats, 849
Stress regulator XBP-1, 698
Stroke, 992
Stroke prevention in AF, 993
Stromal cell ligands (ADAM15/fibronectin, 234
Stromal lymphocytes in tongue cancer, 681
Stronglocentrotus purpuratus, 192
Structural changes in SPs in PAP, 961
Structural classification of lectins, 5
Structural classification of proteins, 6
Structural elements, 340
Structural families, 5
Structurally dissimilar NK cell receptor families, 620
Structural organization of mannose receptor and Endo 180, 344
Structural organization of SP-A and SP-D, 503
Structural properties of SP-A, 502
Structural requirement for MAdCAM-1, 564
Structure-function relations, 332, 491
Structure: function relations of lung SP-D, 533
    D4 (CRD) domain in phospholipid interaction, 534
    D3 (neck) plus D4 (CRD) domains, 534
    ligand binding amino acids, 535
    ligand binding and immune cell-recognition, 535
    NH2 domain and collagenous region, 533
    NH2-terminal cysteines in collagen helix formation, 534
    trimeric neck-carbohydrate recognition domains (NCRDs), 533
Structure of *Anguilla anguilla* agglutinin asymmetric unit, 441
Structure of CD94, 645
Structure of chondroitin sulfate, 804
Structure of DC-SIGN, 776
Structure of Fbs1, 128
Structure of F-type domain 1 from *S. pnemoniae*, 441
Structure of human galectin-7, 203
Structure of human galectin-3 CRD, 270
Structure of MSFPB32, 443
Structure of N-terminal domain of sialoadhesin, 358
Structure of pentameric human SAP, 172

Structure of polycystin-1 and polycystin-2 (LRR), 1030
Structure of ULBP3, 672
Structures of NKG2D-ligand complexes, 672
S-type lectins, 6, 7
S-type LPS, 416
S. typhymurium, 182, 411
Sub-human primate NK cells, 694
Subset of fibroblast-like cells, 918
Subsets of acute myeloid leukemia cells, 392
Subsets of CD8 T cells, 644
Subsets of sialylated, sulfated mucins, 561
Substrate reduction therapy, 1040
Subtribes Pierina, 322, 324
Subunit of integrin, 222
Sudden infant death syndrome (SIDS), 972
Sugar binding subsites in galectins, 265
Sugar code, 15
Sulfate-and 2,3-NeuAc-containing glycans, 703
3-Sulfated galactosyl ceramide (sulfatide), 575
Sulfated gangliosides, 398
Sulfated glycans, 335
Sulfated glycosaminoglycan (SGAG)-binding proteins, 3
Sulfated glycosphingolipids, 249
Sulfated monosaccharides, 335
4-Sulfated N-acetylgalactosamine, 334
Sulfated N-acetyl galactosamine moiety, 532
Sulfated polylactosamine, 804
Sulfated proteoglycan, 575
Sulfatides (glycolipids), 248, 827
3-Sulfo-galactoside, 570
Sulfoglucuronylglycolipids (SGGLs), 827
Sulphated glycolipids, 827
Superrafts, 249
Suppression of obesity, 891
Suppressor of cytokine signaling (SOCS) proteins, 1053
    SOCS targets siglec 7, 1053
Suppressor of cytokine signalling (SOCS), 864
Supramolecular complexes, 249, 506
Surface activity of bovine lipid extract surfactant (BLES), 516
Surface film formation, 511
Surface 9-O-acetylation and recognition processes, 363
Surface plasmon resonance, 87, 827, 885
Surface tension, 539, 955
Surfactant, 539
Surfactant protein A (SP-A), 12, 13, 480, 483
Surfactant protein (SP)-A, B, C and D, 501, 955
Surfactant protein D (SP-D), 13, 480, 483, 527
Surfactant protein D: beneficial in protecting lungs from COPD, 541
Surfactant proteins, 341, 956
Surfactant proteins in non-ILD pulmonary conditions, 971
Survival benefits of heparin, 1011
Susceptibility to RSV, CoV and HTLV, 939
Susceptibility to T-cell Lymphotropic Virus, 939
sVCAM-1 levels, 1001
S2V Siglec, 393
Syk and protein kinase C, 234
Syk-independent pathways, 728
SYK-SLP-76 signaling pathway, 734
Synaptic-like microvesicles, 1040
Syndecans, 802
Syndrome/sepsis, 941
Synergistic action of A2t and arachidonic acid, 461
Synoviocytes, 182
Syntenic human region 19q13.1-13.3., 251
Synthesis of Lewis antigens, 604
Synthetic and semisynthetic ligosaccharides, 569

Synthetic glycopolymers and glycoproteins, 1045
Synthetic lactulose amines: anticancer agents, 306
Synthetic oligosaccharide, 363
Synthetic peptide KLGFFKR, 46
Synthetic PPAR-δ ligands GW0742 and GW501516, 598
Syrian hamster female protein, 175, 176
Systemic amyloidosis, 174
Systemic inflammatory response syndrome (SIRS), 941
Systemic lupus erythematosus (SLE), 167, 253, 361, 415, 904, 940, 959, 1002
Systemic scleroderma (SSd), 959
Systemic sclerosis or systemic scleroderma, 959, 1003
Systemic Th-2 response, 965
System related disorders, 1003

T
Tachylectin-1, 416
Tachylectin-2, 416
Tachylectin-3, 416
Tachylectin-4, 416, 444
Tachylectin-5, 416
Tachylectin-5A, 8, 412, 413, 416
Tachylectin-5B (TL5B), 8, 416
Tachylectins 6, 415, 416
Tachypleus tridentatus, 416
Tachypleus tridentatus tachylectin-5A (TL5A), 413, 416
TAFI TAFI *
TAG-1/axonin-1, 827, 829
Takayasu's arteritis (TA), 1006
Talin, 881
Tandem-repeat-type structures, 251
T antigen, 222, 298
T-antigen-binding site expressions, 298
Targeting hepatocytes, 1051
Targeting of mannose-6-phosphate, 1040
Targets for immunointervention, 192
TATAA, 811
TATA box, 172, 214, 361, 633, 737, 814, 852
TATA box (TTTAAA), 338
TATA-box element, 1028
TATA, CCAAT, and Sp1 binding sites, 907
TATA promoter element, 172
Tat-associated protein, or TAP, 906
TATATATA, 907
Tat protein, 599
Tay-Sachs disease, 1043
T cell apoptosis, 199, 200
T cell death pathways, 257
T cell immunoglobulin and mucin domain (TIM)-3, 255
T-cell immunoglobulin mucin-3, 255
T-cell leukemia, 258
T. circumcincta, 449
TCRγ9/δ2, 671
TCRγδ+ NKG2C+, 657
T1D in diabetic mice, 858
Teladorsagia circumcincta, 449, 450
Telencephalon, 63
Telomere, 829
TEM, 506
TEMs-1, 2,-3,-4,-5,-7, and-8, 918
Tenascin-C, 216, 456, 825, 827, 829
Tenascin-R (Tn-R), 814, 825–827, 829, 833, 835, 838
Tenascin-Ras ligands, 827
Tennis-racket or rod shaped cytoplasmic organelles, 756
Ternary complex, 825

Tertiary structure of ricin, 314
Testicans, 838
Testicular germ cell tumors, 302
Testis and sperm, 64
Tetra-antennary oligosaccharides, 397
Tetranectin (TN), 475, 478, 480, 882, 883
Tetranectin-binding site, 885
Tetranectin: enriched in the cartilage, 884
Tetranectin group, 479
Tetranectin group of lectins, 882
Tetranectin homologous protein, 888
Tetranectin: homotrimeric protein (TN3), 885
    apoTN3, 885
    role in the survival of islets in the liver, 884
    wide tissue distribution, 88
Tetrasialo-tetra-antennary complex-type oligosaccharides, 462
Tetraspanin superfamily, 268
Tetratricopeptide (TPR) repeats, 134
TFBP, 442
TGF-β, 100, 513, 514, 686, 819, 849
TGliPR, 921
TGN, 38, 62, 1045
TGN Golgins, 95
TGT haplotype, 564
Thalamocortical axons, 835
THBD gene, 910
Th2 cytokines, 964
The mannose6-phosphate receptor (M6PR) system, 1040
Therapeutic applications, 569
Therapeutic applications of ATRN/mahogany gene products, 894
Therapeutic effects of SP-A, SP-D and their chimeras, 977
Thermolysin type, 863
Thiazolidinedione, 259
Thioesterase (PPT), 1044
Thoracic duct lymphocytes (TDL), 554
THP-1 cells, 400
Th1-polarized skin inflammation, 258
Three-dimensional models of annexins (1, 2, 3, 5 and 7), 464
Three-dimensional structure of gC1qR, 908.
Threefold symmetric ß-trefoil, 334
Three forms of PAP in rat, 860
Three structural domains of Crt, 38
    carboxyl-terminal C-domain, 38
    central P-domain, 38–40
    N-domain, 38
Thrombin, 906, 911, 912, 916
Thrombin-activatable fibrinolysis inhibitor (TAFI), 915
Thrombin-binding domain, 915
Thrombin cleavage sites, 252
Thrombogenesis in atrial fibrillation, 993
Thrombomodulin, 480, 1008
s-Thrombomodulin (TM), 1002
Thrombomodulin deficiency, 916
Thrombomodulin group, 479
Thrombomodulin (TM) (CD141) or BDCA-3, 901, 910
Thrombospondin (TSP), 8, 44
Thromboxane A2, 993
α-Thrombin, 575
Thrombus, 575
Thyroglobulin (Tg), 64
Thyroid cancer, 1008
Thyroid FNA, 292
Thyroid follicle cells, 64
Thyroid hormone receptor α1 (TRα), 47
Thyroid hormone regulates MBL, 491
Thyroid malignancy, 203

Thyroid nodular diseases, 292
Thyroid-stimulating hormone, 47
Thyroid transcription factor-1 (TTF-1), 47
Thyrotrophin-releasing hormone 157
Thyrotropin, 332
Tick hemocytes, 416
Tight junctions (TJ), 1027, 1032
TIM-1, 257
Tim-3, 254, 256
TIM barrel domains, 423
TIM-barrel structure, 421
TIM-3-galectin-9 pathway, 256, 257
TIMP-3, 817
TIM-3 polymorphisms, 257
TIP4, 772, 795
Tissue fibrosis, 428
Tissue plasminogen activator, 456, 460
Tissue remodelling, 428, 886
Tissue-type plasminogen activator (t-PA), 339
Tissue type plasminogen activator or urokinase, 994
TKDNNLLGRFELSG (TKD) of Hsp70, 659
TL5A, 416
TLR2, 517, 730
TLR4, 517 , 730
TLR9, 628
TLR-activated plasmacytoid dendritic, 628
TLR7 and TLR9 ligands, 766
TLR9 (Toll-like receptor 9) ligation, 396
TLR8-mediated cytokine production, 763
TLR2-mediated signaling, 516
TM gene (THBD), 917
TM structure, 912
TM-thrombin complex, 914
TM thrombin complex crystal structure, 913
TN-deficient mice, 887
TNF-α, 358, 392, 514, 516, 517, 599, 728, 762–764, 767, 849, 862
TNF-R55, 596
TNF-R75, 596
TNF-related apoptosis-inducing ligand (TRAIL), 766
TN-R in the retina and optic nerve, 833
Tolerogenic DCs, 968
Toll-like receptor (TLR)-9, 679
Toll-like receptor (TLR), 13, 15, 277, 396, 672, 757, 773, 781
Toll like receptor 2 (TLR2), 334
Tomato moth *(Lacanobia oleracea)*, 433
Tooth: biglycan, decorin, versican, and link protein, cementum, 806, 807
Topoisomerase II, 595
Toxic metals and transition elements, 963
Toxoplasma gondii infection, 272
t q24.2-q31.3, 669
Trabecular meshwork (TM), 253
Trachea, prostate, pancreas,intestine and thymus, 519
Tracheostomy, 971
Traffic and secretory pathway, 152
TRAIL-mediated cytotoxic activity of PDC, 766
Transcriptional activity of TTF-1, 47
Transcriptional regulation, 626
Transcriptional regulation of SP-D, 529
Transcription factors CCAAT/enhancer-binding protein-β and -α, 260
Transduction in B cell activation, 366
Trans fatty acid, 600
Transforming growth factor-β (TGF-β), 672
Trans-golgi network (TGN), 33, 57, 62, 81, 1040, 1041
Transient middle cerebral artery occlusion (MCAO), 829, 998
Transient receptor potential (TRP) family of ion channels, 1028, 1030
Transient retinal ischemia, 833
Transitional ER (tER), 58
Transition in myoblast adhesion, 222
Translocation channel, 147
Transmembrane conductance regulator (CFTR), 35, 124
Transmembrane domain, 340
Transmembrane signaling, 611
Transplantation, 192
Transplant rejection, 1005
Transport arrest of mitotic cells, 152
Transport complexes, 60
Treatment of allergic skin diseases, 569
Treatment of cancer, 367
Trehalose-6,6-dibehenate (TDB), 765
Trehalose-6,60-dimycolate (TDM; also called cord factor), 765
Trembler-J (Tr-J) sciatic nerves, 36
Tri-acidic (EEE) clusters, 777
Triantennary, 397
Trichinella spirali, 489
Trichinella spiralis, 449
Trichuris muris, 336, 449
Triggering receptors, 694
Trimeresurus stejnegeri lectin (TSL), 478
Trimeric CRD and neck domain of SP-A, 505
Trimeric structure of TN (TN3), 886
Triosephosphateisomerase (TIM) barrel, 424
Tripeptidyl peptidase I, 91
Triple α-helical coiled-coil, 487
Triple helix of MBL, 491
Triple-stranded coiled coil of α helices, 491
Trisialotriantennary, 462
Troglitazone, 259, 598
Trophoblast, 651
Tropical calcific pancreatitis (TCP), 850
Trout species (e.g. *Oncorhynchus mykiss*) 439
TRPP2 function, 1035
TRPV4, 1033
Truncated form of MASP-2 (MAp19), 413, 492
Truncated forms of CD-MPR and CI-MPR, 84
Trypanosoma, 934
Trypanosoma cruzi, 272, 360, 365
Trypanosoma cruzi infection, 234, 283
TSG-6
TSG-14, 163, 178
TSG-14/PTX3, 178
T. spiralis, 449
TTATTTAT, 910
TTF-1, 529
TTF-1 and SP-A in differential diagnosis, 970
T. tridentatus, 416
Tube formation on matrigel, 918
Tuberculosis, 776, 960
    SP-A values, 960
Tuberculosis disease, 789
Tubulovesicular, 59, 60
Tubulovesicular cis-face, 59
Tumor angiogenesis, 919
Tumor-bearing mice, 949
Tumor endothelial marker-1, 917
Tumor immune surveillance, 679
Tumor necrosis factor-α, 8, 167, 541
Tumor necrosis factor-inducible protein (TSG-6), 477
Tumor vaccine-draining lymph node cells, 566
Tumor vasculature, 304
"T" *vs.* "Y" shapes, 505
12-stranded antiparallel β-sheet, 255

Two-domain F-type lectins, 439
Two fold rotational symmetry axis, 333
Two-hit hypothesis as a mechanism of cystogenesis, 1033
Two subsets of siglecs, 352
    and the CD33-related siglecs, 352
        great apes: thirteen functional siglecs, 352
        humans: CD33 and siglecs-5,-6,-7 (7/p75/AIRM1)-8,-9, 10,-11,-14 and-16, 352
        mice:CD33 and siglecs-E,-F,-G and-H, 352
    subset of the sialoadhesin (Sn), 352
Type 1 angiotensin II (AT1) receptor, 47
Type B syndrome, 1044
Type 1 diabetes in NOD mice, 567
Type 1 diabetes mellitus (T1DM), 281, 631
Type 2 diabetes mellitus (T2DM), 169, 281, 429, 1000
Type 2 domain, 479
Type I and type II C-type lectins, 473
Type I autophagic vacuoles, 91
Type I collagen, 344
Type I collagen helix, 530
Type II autophagic vacuoles, 91
Type II (non-insulin dependent) diabetes (NIDDM), 849
Type I interferons, IL-2, IL-12, IL-15 and IL-18, 619
Type II repeat of fibronectin, 82
Type I, type III and type IV collagens, 316, 335
Type 2 predendritic cells (pDC2), 749
Types of cell surface C1q-binding proteins (C1qR), 902
Tyrosinase related protein (TRP)-2, 1049
Tyrosine kinases, 234
Tyrosine phosphatases SHP-1 and SHP-2, 353
Tyrosine phosphorylation, 365
Tyrosine phosphorylation sites, 353
Tyrosyl phosphorylation of CD22, 366
Tyr phosphorylation of FAK and paxillin, 253
TZDs, 851

U

U18666A, 99, 100
Ub, 362
Ubiquitin conjugating enzyme (E2), 124
Ubiquitin ligase 32 (E3), 109, 124
Ubiquitin ligase complex, 123
Ubiquitin-like gene, 274
Ubiquitin sorting machinery, 97
UDP-GalNAc:GM3/GD3 N-acetylgalactosaminyltransferase, 370
UDP-GalNAc:polypeptide α-N-acetylgalactosaminyltransferases (ppGalNAcTs), 316, 320
UDPglucose:glucosyltransferase (UGGT), 32
UDP-glucose:glycoprotein glucosyltransferase (GT/UGT), 33, 109, 147
UGGT, 33
UL16, 621, 668
UL16 binding proteins (ULBP), 668, 670, 677, 683
ULBP1, 670, 671, 682
ULBP2, 671, 672
ULBP3, 682
ULBP3, and RAE-1b, 669
ULBP4 or RAET1E, 670–672
ULBP5 or RAET1G, 670
ULBPs-encoded by RAET1 genes, 670
Ulcerative colitis (UC), 428, 791, 851, 870, 1003
Ulex europaeus lectin, 21
Umbilical cord blood T cells, 697
Umbilical vein endothelial cells (HUVECs), 598
Unconjugated humanized murine mAb, 400

Unfolded proteins response (UPR), 32, 35, 49, 154
Unilamellar phospholipid vesicles, 511
Unique long 16 [UL-16-binding proteins (ULBP)], 669
uPAR, 88, 343
uPARAP/Endo180, 344
uPA:uPAR, 344
Upregulation of endosialin gene, 918
UPR-specific protein kinases, 32
Uptake of capsular polysaccharides (caps-PS) by APCs, 790
Uptake of Gal-3 in breast carcinoma cells, 294
Uridine 5'-diphosphate (UDP)-Glc:glycoprotein glucosyltransferase, 110
Uridine diphosphate (UDP)-glucose, 32
Uridine 50-diphosphate-N-acetylglucosamine, 1042
Urokinase plasminogen activator receptor, 907
Urokinase plasminogen activator receptor associated protein, 331, 343
Urokinase-type plasminogen activator (scu-PA), 915
Urokinase-type plasminogen activator receptor, 62, 84
Urokinase-type plasminogen activator receptor associated protein, CD280, 750
Urothelial carcinoma of the bladder (Ta/T1 BC), 851
Usual interstitial pneumonia (UIP), 956
UTR polymorphism, 788

V

Valproic acid, 217, 219
Valvular, 943
Vascular cell adhesion molecule-1 (VCAM-1), 992
Vascular dementia (VD), 1004
Vascular endothelial growth factor receptor 2, 456
Vascular SMC, 232
Vascular smooth muscle cells (VSMC), 47, 48, 428, 736, 738
Vasculitic disorders, 1006
Vasculitides, 168
Vasculitis, 182
Vasculitis *vs.* other diagnoses, 1001
Vav1, 677
Vav1, phospholipase C-γ (PLCγ2) and Erk1/2, 766
VCAM-1 (CD106), 554, 564, 565, 567, 598, 994–996, 999, 1000, 1004, 1005
VCAM-1 pathways, 1002
Vδ$^+$ T lymphocytes, 651
Veficolins Cerberus rynchops (dog-faced water snake), 411
VEGF, 998
Vein grafts atherosclerosis, 742
Veltuzumab (v-mab), 367
Vena cava *vs.* aorta, 868
Venom of the snake bothrops, 478
Venom vPLA2s, 341
Venous thrombosis, 992, 1016
Ventricular dysfunction, 280
Versican, 825, 826, 833, 835, 837, 838
Versican (and avian homologue PG-M), 825
Versican (or chondroitin sulfate proteoglycan 2), 802, 807, 812
Versican-a multifunctional molecule, 815
Versican, decorin, biglycan and lumican, 807
Versican expression, 813
Versican gene, 814
Versican in aortic wall, 817
Versican : in osteoarthritic cartilage, rheumatoid arthritis, 807
Versican interactions, 814
Versican isoforms, 812, 813
Versican-structure, 814
Versican V0, 825, 827
Versican V1, 825, 827, 831, 834

Versicon in human aneurysmal abdominal aortas, 817
Very late antigen-4 (VLA-4), 1000
Vesicular coat complexes, 59
Vesicular integral membrane protein-36 (VIP36), 145
Vesicular tubular clusters (VTCs), 57, 153
Vespid, 921
Vγ3 T cells, 640
Vγ5Vδ1$^+$ T, 670
Vγ9Vδ2 T cells, 649, 650, 678
Vγ5Vδ1 TCR$\gamma\delta^+$ intraepithelial T cells, 670
VHS domain, 97
Vibrio cholerae Sialidase, 146
Vicia graminea, 21
Vimentin, 197
Vinculin, 44, 881
VIP36, 6, 30, 32
VIP36 like (VIPL), 145
Viral-and bacterial-specific CD8 T cells, 644
Viral pneumonitis, 956
Viridans streptococci, 449
Vitamin D binding protein (DBP), 491
Vitamin D3 receptors, 45
Vitronectin, 232, 251, 906, 912
Vitronectin binding protein, 907
VLA-4, 991, 998, 1000, 1002, 1004
von Willebrand factor (vWF), 478, 479, 575, 993
VP4 145 Vps10, 119, 120
V-set domains, 363, 393
V-set N-terminal domain of sialoadhesin, 357
V-type domain, 702
Vulvar squamous lesions, 300
Vulvovaginal candidiasis, 934, 943
vWF, 733

W

Water cluster, 427
Weibel-Palade bodies, 575
West nile virus, 784
Wisteria floribunda, 826
Wnt/β-catenin, 47, 278
Wnt/β catenin signaling, 859
Wnt signaling, 44, 278
Wortmannin, 1045
Wound repair, 44
WSC domain, 479, 480

X

X-box-binding protein-1 (XBP-1), 698
XBP-1-/-T cells, 698
XCL-1, 451
XCL-2, 451
Xenoantigen, 270
Xenopus-cortical granule lectin, 284
Xenopus embryonic epidermal lectin (XEEL), 451
Xenopus laevis egg cortical granule, 450
Xenopus laevis pentraxin 20, 445
Xenopus oocyte cortical granule lectin (XCGL), 451
Xenopus oocyte lectin, XL35, 448–450
X-epilectin, 444
XL35, 451
X-linked systemic vasculopathy, 1006
XL-PXN1, 178
X-ray crystallographic, 133, 355, 478
Xray crystallography for apo-Langerin, 753
X-ray crystal structure of H1, 711, 911
X-ray structure, 416
X-ray structure of chitolectin, 424
X-ray structure of human Gal-1, 214
X-ray structures of M, L-and H-ficolins, 412
XTP3-B/Erlectin, 110, 111, 116, 119
Xylan binding domain (XBD), 321

Y

Yeast orthologs, 57
Yeasts and fungi, 1014
Y271F mutant fibrinogen domain, 412
YKL-39, 426
YKL-40, 422, 429
YKL-40 [chitinase 3-like protein 1 (CHI3L1)], 422
YKL-40 diagnostic and prognostic marker for solid tumors, 429
YKL-40 levels, 429
Ym1 (or CH13L3), 6, 426, 429
Ym2 (or CH13L4) 6, 426
Ym1 and Ym2 (Ym1/2), 426
Ym1 and Ym2 mRNA, 426
Ym1 and Ym2 proteins, 426
Ym1 structure, 427
Yos9p, 138
YXXL motif, 667, 733, 734

Z

z-DNA conformation, 852
Zebrafish intelectins (zINTLs), 450
Zebrafish model system, 401
Zebrafish orthologues FRAS1, FREM1, or FREM2, 924
Zinc fingers, 134
zINTL1-3, 450
zINTL4-7, 450
Zipperlik striations, 755
Zoonotic disease, 782

Printed by Publishers' Graphics LLC